2015 IEEE Computer Society Annual Symposium on VLSI (ISVLSI 2015)

Montpellier, France
8-10 July 2015

IEEE Catalog Number: CFP15179-POD
ISBN: 978-1-4799-8720-7

Copyright © 2015 by the Institute of Electrical and Electronic Engineers, Inc
All Rights Reserved

Copyright and Reprint Permissions: Abstracting is permitted with credit to the source. Libraries are permitted to photocopy beyond the limit of U.S. copyright law for private use of patrons those articles in this volume that carry a code at the bottom of the first page, provided the per-copy fee indicated in the code is paid through Copyright Clearance Center, 222 Rosewood Drive, Danvers, MA 01923.

For other copying, reprint or republication permission, write to IEEE Copyrights Manager, IEEE Service Center, 445 Hoes Lane, Piscataway, NJ 08854. All rights reserved.

***This publication is a representation of what appears in the IEEE Digital Libraries. Some format issues inherent in the e-media version may also appear in this print version.

IEEE Catalog Number: CFP15179-POD
ISBN 13: 978-1-4799-8720-7
ISSN: 2159-3469

Additional Copies of This Publication Are Available From:

Curran Associates, Inc
57 Morehouse Lane
Red Hook, NY 12571 USA
Phone: (845) 758-0400
Fax: (845) 758-2633
E-mail: curran@proceedings.com
Web: www.proceedings.com

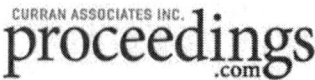

Proceedings

IEEE Computer Society
Annual Symposium on VLSI

ISVLSI 2015

Proceedings

IEEE Computer Society
Annual Symposium on VLSI

8–10 July 2015
Montpellier, France

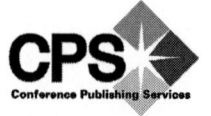

Los Alamitos, California

Washington • Tokyo

2015 IEEE Computer Society Annual Symposium on VLSI

ISVLSI 2015

Table of Contents

Message from the General and Program Chairs...xvii

Organizing Committee...xviii

Technical Program Committee..xx

Additional Reviewers...xxiii

Session 01: Computer Aided Design and Verification

Recurrence Relations Revisited: Scalable Verification of Bit Level Multiplier Circuits.....................................1
Amr Sayed-Ahmed, Ulrich Kühne, Daniel Große, and Rolf Drechsler

Hardware Verification Using Software Analyzers..7
Rajdeep Mukherjee, Daniel Kroening, and Tom Melham

Equivalence Checking Using Trace Partitioning..13
Rajdeep Mukherjee, Daniel Kroening, Tom Melham, and Mandayam Srivas

Session 02: Efficient Digital Designs

Joint Circuit-System Design Space Exploration of Multiplier Unit Structure
for Energy-Efficient Vector Processors...19
Ivan Ratković, Oscar Palomar, Milan Stanić, Milovan Duric, Djordje Pešić,
Osman Unsal, Adrian Cristal, and Mateo Valero

A Fine-Grained, Uniform, Energy-Efficient Delay Element for FD-SOI Technologies 27
Ajay Singhvi, Matheus T. Moreira, Ramy N. Tadros, Ney L. V. Calazans,
and Peter A. Beerel

Efficient Utilization of Imprecise Blocks for Hardware Implementation of a Gaussian
Filter..33
M. H. Seyed Javadi and H. R. Mahdiani

v

Session 03: Physical Design and Testing

A Detailed Routing-Aware Detailed Placement Technique 38
  Aysa Fakheri Tabrizi, Nima Karimpour Darav, Logan Rakai, Andrew Kennings,
  William Swartz, and Laleh Behjat

An Effective Chemical Mechanical Polishing Filling Approach 44
  Chuangwen Liu, Peishan Tu, Pangbo Wu, Haomo Tang, Yande Jiang, Jian Kuang,
  and Evangeline F. Y. Young

Conservatively Analyzing Transient Faults 50
  Niels Thole, Görschwin Fey, and Alberto Garcia-Ortiz

Session 04: FPGA and NoC Based Designs

Assessment of FPGA Implementations of One Sided Jacobi Algorithm for Singular
Value Decomposition 56
  Ali Ibrahim, Maurizio Valle, Luca Noli, and Hussein Chible

Index-Based Round-Robin Arbiter for NoC Routers 62
  Masoud Oveis-Gharan and Gul N. Khan

Session 05: Poster Session

VLSI Implementation of an Improved Multiplier for FFT Computation in Biomedical
Applications 68
  Arathi Ajay and R. Mary Lourde

Sub-Threshold SRAM Design in 14 Nm FinFET Technology with Improved Access
Time and Leakage Power 74
  Behzad Zeinali, Jens Kargaard Madsen, Praveen Raghavan, and Farshad Moradi

FPGA Based Novel High Speed DAQ System Design with Error Correction 80
  Swagata Mandal, Suman Sau, Amlan Chakrabarti, Jogendra Saini,
  Sushanta Kumar Pal, and Subhasish Chattopadhyay

High-Speed, Modified, Bulk stimulated, Ultra-Low-Voltage, Domino Inverter 86
  Ali Dadashi, Yngvar Berg, and Omid Mirmotahari

Modulo $2n \pm 1$ Fused Add-Multiply Units 91
  Constantinos Efstathiou, Kostas Tsoumanis, Kiamal Pekmestzi, and Ioannis Voyiatzis

High Throughput Floating Point Exponential Function Implemented in FPGA 97
  Peter Malik

Exploiting Circuit Duality to Speed up SAT 101
  Luca Amarù, Pierre-Emmanuel Gaillardon, Alan Mishchenko, Maciej Ciesielski,
  and Giovanni De Micheli

A New Method for Defining Monotone Staircases in VLSI Floorplans 107
  Bapi Kar, Susmita Sur-Kolay, and Chittarnjan Mandal

Logic Debugging of Arithmetic Circuits 113

    Samaneh Ghandali, Cunxi Yu, Duo Liu, Walter Brown, and Maciej Ciesielski

Mapping DAGs on Heterogeneous Platforms Using Logic-Based Benders
Decompostion 119

    A. Emeretlis, G. Theodoridis, P. Alefragis, and N. Voros

A Computational Primitive for Convolution Based on Coupled Oscillator Arrays 125

    Donald M. Chiarulli, Brandon Jennings, Yan Fang, Andrew Seel, and Steven P. Levitan

Homomorphic Data Isolation for Hardware Trojan Protection 131

    M. Tarek Ibn Ziad, Amr Alanwar, Yousra Alkabani, M. Watheq El-Kharashi,
    and Hassan Bedour

SecX: A Framework for Collecting Runtime Statistics for SoCs with Multiple
Accelerators 137

    Rajshekar Kalayappan and Smruti R. Sarangi

Low-Area Reed Decoding in a Generalized Concatenated Code Construction for PUFs 143

    Matthias Hiller, Ludwig Kürzinger, Georg Sigl, Sven Müelich, Sven Puchinger,
    and Martin Bossert

JSRAM: A Circuit-Level Technique for Trading-Off Robustness and Capacity
in Cache Memories 149

    Hamzeh Ahangari, Gulay Yalcin, Ozcan Ozturk, Osman Unsal, and Adrian Cristal

Reducing the Storage Requirements of a Set of Functional Test Sequences by Using
a Background Sequence 155

    Irith Pomeranz

A TMR Strategy with Enhanced Dependability Features Based on a Partial
Reconfiguration Flow 161

    Victor M. Gonçalves Martins, Paulo R. C. Villa, Horácio C. C. Neto,
    and Eduardo Augusto Bezerra

Low-Power and Low-Variability Programmable Delay Element and Its Application
to Post-Silicon Skew Tuning 167

    Daijiro Murooka, Yu Zhang, Qing Dong, and Shigetoshi Nakatake

A 10-Bit 500 MSPS Segmented DAC with Optimized Current Sources to Avoid
Mismatch Effect 172

    Santanu Sarkar and Swapna Banerjee

An Improved Dynamic Latch Based Comparator for 8-Bit Asynchronous SAR ADC 178

    Anush Bekal, Rohit Joshi, Manish Goswami, Babu R. Singh, and Ashok Srivatsava

Session 06: Ph.D. Forum

Translation Validation of Transformations of Embedded System Specifications Using
Equivalence Checking 183
    Kunal Banerjee, Chittaranjan Mandal, and Dipankar Sarkar

Design and Implementation of a Reversible Central Processing Unit 187
    Lafifa Jamal and Hafiz Md. Hasan Babu

An Algorithm Used in a Power Monitor to Mitigate Dark Silicon on VLSI Chip 191
    Zhou Zhao, Ashok Srivastava, Shaoming Chen, and Saraju P. Mohanty

Validating SPARK: High Level Synthesis Compiler 195
    Soumyadip Bandyopadhyay, Dipankar Sarkar, and Chittaranjan Mandal

Keynote 1

Enabling Scaling of Advanced CMOS Technologies: A Reliability Perspective 199
    Tanya Nigam and Andreas Kerber

Session 07: Special Session: IP Protection

Digital Right Management for IP Protection 200
    Jerome Rampon, Renaud Perillat, Lionel Torres, Pascal Benoit, Giorgio Di Natale,
    and Mario Barbareschi

Development of a Layout-Level Hardware Obfuscation Tool 204
    Shweta Malik, Georg T. Becker, Christof Paar, and Wayne P. Burleson

Reversible Denial-of-Service by Locking Gates Insertion for IP Cores Design
Protection 210
    Brice Colombier, Lilian Bossuet, and David Hély

Identification of IP Control Units by State Encoding 216
    Edward Jung and Seonho Choi

Session 08: Special Session: Biosignal Processing Embedded Systems

A Summary of Current and New Methods in Velocity Selective Recording (VSR)
of Electroneurogram (ENG) 221
    John Taylor, Benjamin Metcalfe, Chris Clarke, Daniel Chew, Thomas Nielsen,
    and Nick Donaldson

Resource Optimized Processor for Real-Time Neural Activity Monitoring 227
    Y. Bornat, A. Quotb, N. Lewis, and S. Renaud

In-silico Phantom Axon: Emulation of an Action Potential Propagating Along
Artificial Nerve Fiber 228
    Olivier Rossel, Fabien Soulier, Serge Bernard, David Guiraud, and Guy Cathébras

Session 09: Mixed-Signal and Optimization

A Simplified Phase Model for Oscillator Based Computing 231
Yan Fang, Victor V. Yashin, Donald M. Chiarulli, and Steven P. Levitan

A Statistical Approach to Probe Chaos from Noise in Analog and Mixed Signal
Designs 237
Ibtissem Seghaier, Mohamed H. Zaki, and Sofiène Tahar

Multi-objective Optimization of Floating Point Arithmetic Expressions Using Iterative
Factorization 243
Alireza Mahzoon and Bijan Alizadeh

Session 10: Digital Designs

Architecture for Dual-Mode Quadruple Precision Floating Point Adder 249
Manish Kumar Jaiswal, B. Sharat Chandra Varma, and Hayden K. H. So

VLSI Design of Edge-Preserving Coding Artifacts Reduction for Display Processing 255
*Zenghua Cheng, Xuchong Zhang, Huisheng Peng, Baolu Zhai, Hongbin Sun,
and Nanning Zheng*

A Custom Computing System for Finding Similarties in Complex Networks 262
*Christian Brugger, Valentin Grigorovici, Matthias Jung, Christian Weis,
Christian De Schryver, Katharina Anna Zweig, and Norbert When*

Session 11: Special Session: Minimizing Energy Consumption of Computing to the Limit

Heterogeneous Error-Resilient Scheme for Spectral Analysis in Ultra-Low Power
Wearable Electrocardiogram Devices 268
*Soumya Basu, P. Garcia del Valle, Georgios Karakonstantis, Giovanni Ansaloni,
and David Atienza*

Logic Switches Operating at the Minimum Energy of Computing 274
Francesco Orfei and Luca Gammaitoni

Synergistic Architecture and Programming Model Support for Approximate
Micropower Computing 280
Giuseppe Tagliavini, Davide Rossi, Luca Benini, and Andrea Marongiu

Session 12: Special Session: Unconventional Computing

Logic-in-Memory: A Nano Magnet Logic Implementation 286
*M. Cofano, G. Santoro, M. Vacca, D. Pala, G. Causapruno, F. Cairo, F. Riente,
G. Turvani, M. Ruo Roch, M. Graziano, and M. Zamboni*

Simscape Based Ultra-Fast Design Exploration of Graphene-Nanoelectronic Systems 292
Shital Joshi, Elias Kougianos, and Saraju P. Mohanty

Reversible Logic Based Mapping of Quaternary Sequential Circuits Using QGFSOP
Expression 297
Mozammel H. A. Khan and Himanshu Thapliyal

Session 13: Emerging Device Based Designs

Comparing Energy, Area, Delay Tradeoffs in Going Vertical with CMOS
and Asymmetric HTFETs 303
Moon Seok Kim, William Cane-Wissing, Jack Sampson, Suman Datta,
Vijaykrishnan Narayanan, and Sumeet K. Gupta

Novel UHF Passive Rectifier with Tunnel FET Devices 309
David Cavalheiro, Francesc Moll, and Stanimir Valtchev

Hybrid STT/CMOS Design of an Interrupt Based Instant On/Off Mechanism
for Low-Power SoC 315
Christophe Layer, Kotb Jabeur, Laurent Becker, Bernard Diény, Stéphane Gros,
Virgile Javerliac, Pierre Paoli, and Fabrice Bernard-Granger

Session 14: Special Session: Emerging Non-Volatile Memories

Radiative Effects on MRAM-Based Non-Volatile Elementary Structures 321
Jeremy Lopes, Gregory Di Pendina, Eldar Zianbetov, Edith Beigne, and Lionel Torres

RRAM Reliability/Performance Characterization through Array Architectures
Investigations 327
Cristian Zambelli, Alessandro Grossi, Piero Olivo, Christian Walczyk,
and Christian Wenger

Single-Ended and Differential MRAMs Based on Spin Hall Effect: A Layout-Aware
Design Perspective 333
Ahmedullah Aziz, William Cane-Wissing, Moon S. Kim, Suman Datta,
Vijaykrishnan Narayanan, and Sumeet K. Gupta

Session 15: Post-CMOS Computing Systems

Using Multiple-Input NEMS for Parallel A/D Conversion and Image Processing 339
Kaisheng Ma, Nandhini Chandramoorthy, Xueqing Li, Sumeet Kumar Gupta,
John Sampson, Yuan Xie, and Vijaykrishnan Narayanan

Implementing Data Structure Using DNA: An Alternative in Post CMOS Computing 345
Mayukh Sarkar and Prasun Ghosal

An Unbalanced Area Ratio Study for High Performance Monolithic 3D Integrated
Circuits 350
Hossam Sarhan, Sebastien Thuries, Olivier Billoint, and Fabien Clermidy

Session 16: Secure and Trusted Systems

Implementation of AES Using NVM Memories Based on Comparison Function 356
Jeremie Clement, Bruno Mussard, David Naccache, and Lionel Torres

Figure of Merits of 28nm Si Technologies for Implementing Laser Attack Resistant
Security Dedicated Circuits 362
S. de Castro, G. Di Natale, M.-L. Flottes, B. Rouzeyre, and J.-M. Dutertre

A Similarity Based Circuit Partitioning and Trimming Method to Defend
against Hardware Trojans 368
Yun Cheng, Ying Wang, Huawei Li, and Xiaowei Li

Session 17: Special session: Software Engineering for VLSI and Embedded Systems

On-Chip Instrumentation for Runtime Verification in Deeply Embedded Processors 374
Ciaran MacNamee and Donal Heffernan

Statistical Analysis of Resource Usage of Embedded Systems Modeled in EAST-ADL 380
Raluca Marinescu, Eduard Paul Enoiu, and Cristina Seceleanu

A Novel Architectural Pattern to Support the Development of Human-Robot
Interaction (HRI) Systems Integrating Haptic Interfaces and Gesture Recognition
Algorithms 386
Giuseppe Airò Farulla, Ludovico O. Russo, Vincenzo Gallifuoco, and Marco Indaco

Session 18: 3D and NoC Based Systems

TSV Placement and Core Mapping for 3D Mesh Based Network-on-Chip Design
Using Extended Kernighan-Lin Partitioning 392
*Kanchan Manna, Vadapalli Shanmukha Sri Teja, Santanu Chattopadhyay,
and Indranil Sengupta*

Achieving Memory Access Equalization Via Round-Trip Routing Latency Prediction
in 3D Many-Core NoCs 398
*Xiaowen Chen, Zhonghai Lu, Yang Li, Axel Jantsch, Xueqian Zhao, Shuming Chen,
Yang Guo, Zonglin Liu, Jianzhuang Lu, Jianghua Wan, Shuwei Sun, Shenggang Chen,
and Hu Chen*

Validating Delay Bounds in Networks on Chip: Tightness and Pitfalls 404
Alberto Saggio, Gaoming Du, Xueqian Zhao, and Zhonghai Lu

Session 19: Embeded System Design

Optimized Use of Parallel Programming Interfaces in Multithreaded Embedded
Architectures 410
  Arthur F. Lorenzon, Anderson L. Sartor, Márcia C. Cera,
  and Antonio Carlos Schneider Beck

The DRACON Embedded Many-Core: Hardware-Enhanced Run-Time Management
Using a Network of Dedicated Control Nodes 416
  Daniel Gregorek and Alberto Garcia-Ortiz

Backlog Bound Analysis for Virtual-Channel Routers 422
  Xueqian Zhao and Zhonghai Lu

Session 20: Digital System Design

A Timing Error Mitigation Technique for High Performance Designs 428
  Mehrnaz Ahmadi, Bijan Alizadeh, and Behjat Forouzandeh

RWT: Suppressing Write-Through Cost When Coherence is Not Needed 434
  Hao Liu, Clément Dévigne, Lucas Garcia, Quentin Meunier, Franck Wajsbürt,
  and Alain Greiner

Small FPGA Based Multiplication-Inversion Unit for Normal Basis Representation
in GF(2m) 440
  Jérémy Métairie, Arnaud Tisserand, and Emmanuel Casseau

Keynote 2

The Future of Nanoelectronics: New Materials, Architectures and Devices 446
  Heike Riel

Session 21: Special Session: Carbon-Based Materials for THz Nanoelectronics

Challenges and Perspectives of Nanoelectromagnetics in the THz Range 447
  S. A. Maksimenko, M. V. Shuba, P. P. Kuzhir, K. G. Batrakov, and G. Y. Slepyan

Semi-Classical Modelling of the Electron Transport in Carbon Nanotubes
and Graphene Nanoribbons for THz Range Applications 450
  Antonio Maffucci

Terahertz Applications of Carbon Nanotubes and Graphene Nanoribbons 456
  M. E. Portnoi, V. A. Saroka, R. R. Hartmann, and O. V. Kibis

Session 22: Special Session: Memory and Computing Units in Emerging Paradigm

Emerging Non-volatile Memory Technologies Exploration Flow for Processor
Architecture 460
    Sophiane Senni, Lionel Torres, Gilles Sassatelli, Abdoulaye Gamatie, and Bruno Mussard

Channel Modeling and Reliability Enhancement Design Techniques for STT-MRAM 461
    Liuyang Zhang, Wang Kang, Youguang Zhang, Yuanqing Cheng, Lang Zeng,
    Jacques-Olivier Klein, and Weisheng Zhao

STT-MRAM-Based Strong PUF Architecture 467
    Elena Ioana Vatajelu, Giorgio Di Natale, Lionel Torres, and Paolo Prinetto

Session 23: Special Session: Techniques and Trends for Energy Efficient and Ultra Low Power Digital

Approximate Computing: An Energy-Efficient Computing Technique for Error
Resilient Applications 473
    Kaushik Roy and Anand Raghunathan

Near/Sub-Threshold Circuits and Approximate Computing: The Perfect Combination
for Ultra-Low-Power Systems 476
    Jeremy Schlachter, Vincent Camus, and Christian Enz

Sub-Threshold Design and Architectural Choices 481
    Christian Piguet, Marc Pons, and Daniel Séverac

Session 24: Fault-Tolerant Design

A Novel Phase-Based Low Overhead Fault Tolerance Approach for VLIW Processors 485
    Anderson L. Sartor, Arthur F. Lorenzon, Luigi Carro, Fernanda Kastensmidt,
    Stephan Wong, and Antonio C. S. Beck

On the Design of a Fault Tolerant Ripple-Carry Adder with Controllable-Polarity
Transistors 491
    H. Ghasemzadeh, P. E. Gaillardon, J. Zhang, G. De Micheli, E. Sanchez,
    and M. Sonza Reorda

A Cellular Automata Based Fault Tolerant Approach in Designing Test Hardware
for L1 Cache Module 497
    Mousumi Saha and Biplab K. Sikdar

Session 25: Test for Digital Design

Diagnosis of Delay Faults Considering Hazards 503
    Yoshinobu Higami, Senling Wang, Hiroshi Takahashi, Shin-Ya Kobayashi,
    and Kewal K. Saluja

DONUT: A Double Node Upset Tolerant Latch 509

    Nikolaos Eftaxiopoulos, Nicholas Axelos, and Kiamal Pekmestzi

An ATPG Flow to Generate Crosstalk-Aware Path Delay Pattern 515

    Anu Asokan, Alberto Bosio, Arnaud Virazel, Luigi Dilillo, Patrick Girard,
    and Serge Pravossoudovitch

Session 26: Reliable Design Techniques

Analyzing the Impact of Frequency and Diverse Path Delays in the Time Vulnerability
Factor of Master-Slave D Flip-Flops 521

    Alexandra L. Zimpeck, Fernanda Lima Kastensmidt, and Ricardo Reis

Using Intra-Line Level Pairing for Graceful Degradation Support in PCMs 527

    Marjan Asadinia and Hamid Sarbazi-Azad

Using Configurable Bit-Width Voters to Mask Multiple Errors in Integrated Circuits 533

    Thiago Berticelli Ló, Fernanda Lima Kastensmidt, and Antonio Carlos Schneider Beck

Session 27: Special Session: Efficient Design of Manycore Embedded Systems

Communication-Aware Parallelization Strategies for High Performance Applications 539

    Imran Ashraf, Koen Bertels, Nader Khammassi, and Jean-Christophe Le Lann

Design of Fault-Tolerant and Reliable Networks-on-Chip 545

    Junshi Wang, Masoumeh Ebrahimi, Letian Huang, Axel Jantsch, and Guangjun Li

Design Exploration for next Generation High-Performance Manycore On-chip
Systems: Application to big.LITTLE Architectures 551

    Anastasiia Butko, Abdoulaye Gamatié, Gilles Sassatelli, Lionel Torres, and Michel Robert

Session 28: Special Session: Energy-Efficient Design Methods for Emerging Technologies

On Analysis of On-chip DC-DC Converters for Power Delivery Networks 557

    Ghizlane Mouslih, Aida Todri-Sanial, and Pascal Nouet

Multilevel Modeling Methodology for Reconfigurable Computing Systems Based
on Silicon Photonics 561

    Zhen Li, Sébastien Le Beux, Christelle Monat, Xavier Letartre, and Ian O'Connor

Multi-swarm Optimization of a Graphene FET Based Voltage Controlled Oscillator
Circuit 567

    Elias Kougianos, Shital Joshi, and Saraju P. Mohanty

Session 29: Reliable Circuits and Systems

Fast Stimuli Generation for Design Validation of RTL Circuits Using Binary Particle
Swarm Optimization 573
> Prateek Puri and Michael S. Hsiao

On the Performance Exploration of 3D NoCs with Resistive-Open TSVs 579
> Charles Effiong, Vianney Lapotre, Abdoulaye Gamatie, Gilles Sassatelli,
> Aida Todri-Sanial, and Khalid Latif

SymmTop: A Symmetric Circuit Topology for Ultra Low Power Wide
Temperature-Range Applications 585
> Elena K. Weinberg and Mircea R. Stan

Session 30: Power and Noise Aware Systems

Energy-Aware Computing via Adaptive Precision under Performance Constraints
in OFDM Wireless Receivers 591
> Fernando Cladera, Matthieu Gautier, and Olivier Sentieys

The Solar Cells and the Battery Charger System Using the Fast and Precise Analog
Maximum Power Point Tracking Circuits 597
> Yasuhiro Sugimoto

Session 31: Special Session: 3D Design Challenges and Perspectives

3D DFT Challenges and Solutions 603
> Yassine Fkih, Pascal Vivet, Marie-Lise Flottes, Bruno Rouzeyre, Giorgio Di Natale,
> and Juergen Schloeffel

Thermal Aspects and High-Level Explorations of 3D Stacked DRAMs 609
> Christian Weis, Matthias Jung, Omar Naji, Norbert When, Cristiano Santos,
> Pascal Vivet, and Andreas Hansson

Interconnect Challenges for 3D Multi-cores: From 3D Network-on-Chip to Cache
Interconnects 615
> P. Vivet, C. Bernard, E. Guthmuller, I. Miro-Panades, Y. Thonnart, and F. Clermidy

Session 32: Special Session: Test, Calibration and Tuning of Analog/RF Circuits

A Framework for Efficient Implementation of Analog/RF Alternate Test with Model
Redundancy 621
> S. Larguech, F. Azaïs, S. Bernard, M. Comte, V. Kerzérho, and M. Renovell

Test and Calibration of RF Circuits Using Built-in Non-intrusive Sensors 627
> Athanasios Dimakos, Martin Andraud, Louay Abdallah, Haralampos-S. Stratigopoulos,
> Emmanuel Simeu, and Salvador Mir

Silicon Demonstration of Statistical Post-Production Tuning 628
    Yichuan Lu, Kiruba Subramani, He Huang, Nathan Kupp, and Yiorgos Makris

Session 33: Signal Converter Circuits

Toward Adaptation of ADCs to Operating Conditions through On-chip Correction 634
    V. Kerzérho, L. Guillaume-Sage, F. Azaïs, M. Comte, M. Renovell, and S. Bernard

A Full-Swing CMOS Current Steering DAC with an Adaptive Cell and a Quaternary
Driver 640
    Yanghyeok Choi, Seonghyun Park, Jieun Yoo, Seol Namgung, and Minkyu Song

Flexible Ultra-Low-Voltage CMOS Circuit Design Applicable for Digital and Analog
Circuits Operating below 300mV 646
    Yngvar Berg and Omid Mirmotahari

Session 34: Analog Design and Test

A Linear Comparator-Based Fully Digital Delay Element 652
    Afshin Seraj, Mohammad Maymandi-Nejad, Parvin Bahmanyar, and Manoj Sachdev

Built-In Self Optimization for Variation Resilience of Analog Filters 656
    Jiafan Wang, Congyin Shi, Edgar Sanchez-Sinencio, and Jiang Hu

Author Index 662

Message from the General and Program Chairs

It is a distinct pleasure to welcome all the participants to Montpellier, France. The IEEE Computer Society Annual Symposium on VLSI (ISVLSI) continues its tradition as the premier forum for cross-cutting research in system architecture, digital, analog and mixed-signal (AMS) circuits, computer-aided design (CAD) and verification, testing, reliability, fault-tolerance and post-CMOS technologies. Over almost two decades the ISVLSI has been a unique forum promoting multidisciplinary research and new visionary approaches in the area of VLSI. The ISVLSI brings together leading scientists and researchers from academia and industry. ISVLSI has established a reputation in bringing well-known international scientists as speakers, and this trend will continue.

The primary goal of ISVLSI 2015 is to present the highest quality technical program to its attendees. A key element in attaining this goal has been the peer evaluation and selection process. To ensure quality review the papers were assigned to one of 6 tracks, namely the following: (1) Digital Circuits and FPGA based Designs (DCF) track, (2) Computer-Aided Design and Verification (CAD) track, (3) Emerging and Post-CMOS Technologies (EPT) track, (4) System Design and Security (SDS) track, (5) Testing, Reliability, and Fault-Tolerance (TRF) track, and (6) Analog and Mixed-Signal Circuits (AMS) track. The appropriate tracks were largely self-selected by the authors during the submission process. In a very few instances, the papers were re-assigned to a more appropriate track by the technical program chairs in consultation with the track chairs. A total of 136 submission were received for regular session consideration and 40 submissions were received for theme based special sessions to match the recent trend in VLSI circuit and system design. The submissions for ISVLSI 2015 were from 43 different countries from various parts of the globe with 3 highest number of submissions from India, France, and USA. The papers were mostly reviewed by program committee members in respective tracks. The technical program committee members sought external expert opinion as ad-hoc or sub-reviewers in many cases for the evaluation and final disposition of papers. The submissions were reviewed by 94 program committee members and 59 external or ad-hoc reviewers. On an average, each paper received 3 reviews. This resulted in the technical program featuring a total of 34 technical sessions which includes the following: 19 regular sessions, 13 special sessions, 1 poster session, and 1 Ph.D. forum session. The technical program also includes 2 keynote addresses, 1 per day, from eminent speakers from Industry. The 2 keynotes will be delivered by Heike Riel from IBM, Zurich, Switzerland and Tanya Nigam from GLOBALFOUNDRIES, USA.

We would like to thank the Technical Program Committee and all the reviewers for their dedication and hard work in preparing the final ISVLSI program. We gratefully thank all our authors, speakers, and session chairs for making ISVLSI 2015 a premier technical conference. Putting together ISVLSI 2015 has been a team effort. We thank the ISVLSI 2015 organizing Committee as well as ISVLSI Steering Committee for their diligent work and enduring support in putting together this program. We are immensely thankful to the technical committee on VLSI (TCVLSI) of IEEE-CS chaired by Saraju P. Mohanty who provided travel support from 4 student attendee as well as a best paper award. ISVLSI has been endorsed by TCVLSI for IEEE-CS sponsorship.

Finally, a special word of thanks goes to all of you, our conference attendees in Montpellier. We have worked hard to bring you a balanced, solid technical conference as well as provide a forum for informal exchanges during coffee breaks and the receptions. We wish you an enjoyable experience at the ISVLSI 2015 and look forward to future editions of ISVLSI.

Aida Todri-Sanial   Patrick Girard   Giorgio Di Natale   Saraju P. Mohanty   Marc Belleville
General Chairs   **Program Chairs**

Committees

Organizing Committee

General Chairs
Aida Todri-Sanial, *CNRS-LIRMM, France*
Giorgio Di Natale, *CNRS-LIRMM, France*
Patrick Girard, *CNRS-LIRMM, France*

Program Chairs
Saraju P. Mohanty, *University of North Texas, USA*
Marc Belleville, *CEA-LETI, France*

Publication Chairs
Mariane Comte, *University of Montpellier 2, France*
Lu Peng, *Louisiana State University, USA*

Financial Chair
Abdoulaye Gamatie, *CNRS-LIRMM, France*

Local Arrangement Chair
Nadine Azemard-Crestani, *CNRS LIRMM, France*

Special Session Chairs
Xin Li, *Carnegie Mellon University, USA*
Monica Pereira, *UFRN, Brazil*

Web Chairs
Theocharis Theocharides, *University of Cyprus, Cyprus*
Mike Borowczak, *Erebus Labs, USA*

Registration Chair
Vincent Kerzerho, *CNRS-LIRMM, France*

Industrial Liaison Chair
Lionel Torres, *University of Montpellier 2, France*

Ph.D. Forum Co-Chairs
Michael Hübner, *Ruhr University Bochum, Germany*
Patrick Haspel, *Strategic Academic Partnerships, Cadence Design Systems*

Publicity Chairs
Vasilis Pavlidis, *University of Manchester, UK*
Prasun Ghosal, *Indian Institute of Engineering Science and Technology, Shibpur, India*
Annajirao Garimella, *Intel Corporation, USA*
Nikolaos Voros, *Technological Educational Institute of Western Greece, Greece*

International Liaison Chairs
Brazil: Ricardo Reis, *Federal University of Rio Grande do Sul (UFRGS), Brazil*
China/South-East Asia: Jiang Xu, *Hong Kong University of Science and Technology, China*
Germany: Jürgen Becker, *Karlsruhe Institute of Technology, Germany*
Greece: Nicolas Sklavos, *Technological Educational Institute of Western Greece, Greece*
India: Susmita Sur-Kolay, *Indian Statistical Institute, Kolkata, India*

Steering Committee Chair
Nagarajan Ranganathan, *University of South Florida, USA*

Technical Program Committee

Analog and Mixed-Signal Circuits (AMS)
Chairs

Ashok Srivastava, *Louisiana State University*

Florence Azais, *LIRMM*

Members

Amir Zjajo, *Delft University of Technology*

Changzhi Li, *Texas Tech University*

Dhruva Ghai, *Oriental University*

Gildas Leger, *Instituto de Microelectronica de Sevilla (IMSE-CNM-CSIC)*

JosÃ© Machado Da Silva, *FEUP*

Manisha Goswami, *Indian Institute of Information Technology Allahabad*

Manisha Pattanaik, *IIIT Gwalior*

Maryam Baghini, *Indian Institute of Technology Bombay*

Steffen Paul, *University of Bremen*

Vincent Kerzerho, *LIRMM, CNRS/Univ. Montpellier 2*

Xin Li, *Carnegie Mellon University*

Computer Aided Design and Verification (CAD)
Chairs

Massimo Poncino, *Polytechnic University of Turin*

Mircea Stan, *University of Virginia*

Members

Aida Todri-Sanial, *CNRS-LIRMM*

Ke Wang, *University of Virginia*

Kunal Ganeshpure, *Mentor Graphics Corporation*

Michael Huebner, *Ruhr-University Bochum*

Nagi Naganathan, *Avago*

Shih-Hsu Huang, *Chung Yuan Christian University*

Theocharis Theocharides, *University of Cyprus*

Yann Thoma, *HEIG-VD*

Digital Circuits and FPGA Based Designs (DCF)

Chairs

Jia Di, *University of Arkansas*

Elaheh (Eli) Bozorgzadeh, *University of California Irvine*

Members

Bharat Joshi, *University of North Carolina at Charlotte*

Christophe Jego, *IMS Laboratory*

Christophe Bobda, *University of Arkansas*

David Bol, *Université catholique de Louvain*

Dimitrios Soudris, *NTUA*

Dr. Anirban Sengupta, *Indian Institute of Technology (IIT) Indore*

Hailong Jiao, *Eindhoven University of Technology*

Hao Zheng, *University of South Florida*

Ning Weng, *Southern Illinois University Carbondale*

Ron Demara, *University of Central Florida*

Shanq-Jang Ruan Ruan, *National Taiwan University of Science and Technology*

Srinidhi Kestur, *Broadcom Corporation*

Xuelian Liu, *Rensselaer Polytechnic Institute*

Emerging and post-CMOS Technologies (EPT)

Chairs

Sanjukta Bhanja, *University of South Florida*

Ian O'Connor, *University of Lyon*

Members

Arthur Nieuwoudt, *Synopsys*

Hai Li, *University of Pittsburgh*

Himanshu Thapliyal, *Qualcomm Technologies, Inc.*

Jiang Xu, *Hong Kong University of Science and Technology*

Kumar Yelamarthi, *Central Michigan University*

Lionel Torres, *LIRMM*

Prasun Ghosal, *Indian Institute of Engineering Science and Technology, Shibpur*

Rashmi Jha, *University of Toledo*

Saraju Mohanty, *University of North Texas*

Thomas Mikolajick, *TU Dresden*

System Design and Security (SDS)
Chairs
Garrett Rose, *University of Tennessee*
Ricardo Chaves, *IST TULisbon/INESC-ID*

Members
Apostolos Fournaris, *University of Patras*
Chengmo Yang, *University of Delaware*
David Hely, *Grenoble INP*
Guy Gogniat, *Université de Bretagne Sud - UEB*
Jude Ambrose, *University of New South Wales*
Luciano Ost, *University of Leicester*
Madhu Mutyam, *Indian Institute of Technology, Madras*
Mateus Rutzig, *Federal University of Santa Maria*
Mehran Kermani, *Rochester Institute of Technology*
Michail Maniatakos, *New York University Abu Dhabi*
Naghmeh Karimi, *New York University*
Nele Mentens, *KU Leuven*
Philippe Maurine, *CEA*
Rance Rodrigues, *University of Massachusetts at Amherst*
Tanguy Risset, *Citi, INSA-Lyon*
Yier Jin, *The University of Central Florida*

Testing, Reliability, Fault-Tolerance (TRF)
Chairs
Alberto Bosio, *LIRMM*
Lorena Anghel, *TIMA*

Members
Dong Xiang, *Tsinghua University*
Irith Pomeranz, *Purdue University*
Katherine-Shu-Min Li, *National Sun Yat-sen University*
Leticia-Maria Bolzani-Poehls, *PUCRS*
Matteo Sonza Reorda, *Politecnico di Torino*
Michel Renovell, *LIRMM*
Michele Portolan, *TIMA*
Mihalis Psarakis, *University of Piraeus*
Paolo Bernardi, *Politecnico di Torino*
Vishwani Agrawal, *Auburn University*
Xiaoqing Wen, *Kyushu Institute of Technology*
Zebo Peng, *Linkping University*

Additional Reviewers

Abhishek Srivastava
Alec Roelke
Álvaro Gómez-Pau
Andre Martins
Andrew Whetzel
Baohu Li
Bashar Hussein
Bei Zhang
Benedikt Janßen
Bertrand Le Gal
Bo Liu
Camille Leroux
Carolina Metzler
Chen Liu
Cheng-Liang Hsieh
Debashri Roy
Devarshi Das
Elena Ioana Vatajelu
Felipe Rosa
Festus Hategekimana
Guillaume Salagnac
Hailong Yao
Hoda Aghaei Khouzani
Jeremie Crenne
Jingwei Lu
Jones Yudi Mori
Josimar Sfreddo
Kaushik Mazumdar
Kele Shen
Kevin Marquet

Kimmo Järvinen
Laura Rozo Duque
Lionel Morel
Luca Bochi Saldanha
Mahanama Wickramasinghe
Mahima Arrawatia
Mayukh Sarkar
Mehdi Kabir
Miao He
Michael Guilherme Jordan
Muhammed Al Kadi
Naveen Kadayinti
Oscar Palomar
Osvaldo Navarro
Paulo Villa
Philip Gilley
Rizwan A. Ashraf
Rodrigo Possamai Bastos
Runjie Zhang
Shairfe Salahuddin
Tom Tracy
Tomas Grimm
Veeresh Dandur
Vijay Sheshadri
Xinfei Guo
Yuan Xue
Yuanqing Cheng
Zhenkun Yang
Zhou Jiang

2015 IEEE Computer Society Annual Symposium on VLSI

Recurrence Relations Revisited: Scalable Verification of Bit Level Multiplier Circuits

Amr Sayed-Ahmed*, Ulrich Kühne*, Daniel Große*, and Rolf Drechsler*†

* Faculty of Mathematics and Computer Science, University of Bremen, Germany
† Cyber-Physical Systems, DFKI GmbH, Bremen, Germany
Email: {asahmed, ulrichk, grosse, drechsle}@informatik.uni-bremen.de

Abstract—Although a lot of effort has been spent on verifying arithmetic designs, it is still a problem that has no general robust automated solution. One major challenge is verifying large scale multiplier circuits. For this purpose, we revisit the idea of using functional properties of the multiplication function, which can be expressed by recurrence equations. Then, instead of proving the equivalence of the implementation and a specification, the verification task is to show that the implementation satisfies the recurrence equation. We propose an approach which makes this verification task practically feasible for large scale multiplier circuits. Based on a combined add/multiply recurrence equation we can make efficient use of case splitting wrt. the partial products of the multiplier. As a result, the problem is split such that only a small part of the multiplier will be checked in every case, thereby avoiding redundant checks between the cases. Overall, our approach allows to verify a variety of multiplier designs in practical time. We present results for multipliers up to 128 bits.

I. INTRODUCTION

Verification of arithmetic circuits is still a big challenge. Since the famous floating point bug in Intel's Pentium processor [1], a lot of effort has been spent on the verification of arithmetic designs. The applied techniques can be divided into simulation-based verification and formal verification techniques. Simulation-based verification techniques can be used to discover bugs in arithmetic designs and they scale very well, but they cannot prove the absence of bugs. For the latter, formal methods are needed [2]. While large arithmetic circuits – in particular floating-point logic – have been verified using interactive theorem proving [3], this requires a huge amount of expert manual work. In this paper we concentrate on automated methods.

Automated equivalence checking – also used in our work – compares a gate level implementation against a reference implementation by combining them into one circuit, called a miter. Most equivalence checking tools exploit structural similarities between the two implementations and try to prove equivalence by decomposing the problem according to the internal equivalences (cut points) [4]. Therefore, equivalence checkers usually cannot deal with implementations that have few internal equivalences. This problem occurs especially for arithmetic circuits, where one function can be implemented in many different ways [5].

Formal verification of large multiplier designs is still a hard problem that has no general automated solution. Without any information on the high level structure of the netlists, property checking and many equivalence checking approaches fail to verify these circuits. Most successful techniques make use of knowledge about the architecture of the multiplier (see e.g. [6]). In this paper, we address the problem of verifying large multipliers on the gate level without any information

about their high level structure. For this problem, several approaches have been proposed in the past. However, most of these approaches have a scalability problem, cannot be applied on all types of multiplier architectures, or they fail if the circuit does not represent a correct multiplier function. The existing approaches can be divided mainly into three categories: 1) decision diagrams, 2) reverse engineering and structural methods, and 3) approaches based on special arithmetic properties of the realized function.

Decision diagrams provide a canonical representation for netlists of the *design under verification* (DUV) and the *reference implementation* (RI). The equivalence check then consists of checking that the resulting decision diagrams are identical. Unfortunately, many popular data structures like *Reduced Ordered Binary Decision Diagrams* (ROBDDs) [7] have exponential size for multiplication. *Multiplicative Binary Moment Diagrams* (*BMDs) [8] can represent the word level multiplier function in a compact way. However, exponential blowup can still occur for incorrect designs or during the construction of the *BMD from a bit-level circuit [9].

The second direction in the classification is based on reverse engineering. An approach in this direction has been proposed in [5]. It extracts adder structures from bit level netlists and builds half adder networks. This approach has mainly two drawbacks: First, it makes assumptions on the internal structure of the circuits that are not fulfilled by all multiplier architectures [10]. Second, the approach fails to build the half adder networks if the circuit does not represent a correct multiplier, leading to an inconclusive result. Nevertheless, this technique achieves promising results, e.g. it allows to verify a 48 bit multiplier in about 30 minutes. Other works in this direction use polynomial representations to verify arithmetic circuits [11], [12]. The verification problem is formulated as a reduction of polynomial equations using Gröbner basis. This technique has been used before in verification of Galois field multipliers [13]. The approach can verify large multipliers, however it suffers from the explosion of the number of polynomial terms, when verifying highly optimized circuits [11].

The third direction exploits arithmetic properties of the multiplier function to build a miter with many internal equivalences. The first approach in this context has been proposed by Fujita [14]. It is based on the fact that any function satisfying the recurrence relation $(X + 1) * Y = X * Y + Y$ is a multiplication. However, the original approach by Fujita does not scale, and it cannot verify multipliers larger than 16 bits (see also experiments later). A related approach [15] is based on case splitting by forcing a bit of one of multiplier operands to be zero and summing partial products that belong to this bit outside the multiplier. The problem of this idea is that the similarity of the nets inside the miter structure depends on the order of the partial products of the DUV, therefore this approach only works for certain implementations

978-1-4799-8720-7/15 $31.00 © 2015 IEEE

of multipliers.

The technique presented in this paper belongs to the third category. The proposed approach decomposes the verification of a multiplier such that in every case, the generation and the addition of one partial product of the DUV will be checked using a recurrence equation. In every case, the small difference between the compared implementations and the similar ways of carry propagation allow fast equivalence checking, regardless of the multiplier size. As the multiplier size increases, the number of cases will increase, while the complexity to check one case will remain almost the same. Our approach is able to verify a multiplier at the gate level without any information about its high level specification or the internal structure of the netlist. In addition, it is a general technique that can be applied on various different architectures of the DUV. Overall, it allows to verify large scale multipliers in practical time. The experiments show the ability of the proposed approach to verify 128 bits multiplier.

The remainder of the paper is structured as follows. In Section II, we explain the theoretical background of equivalence checking based on recurrence relations and review the original approach of Fujita as an example on this equivalence checking type. Section III discusses the details of our partial product approach. Experimental results are presented in Sections IV, and we conclude our work in Section V.

II. EQUIVALENCE CHECKING BASED ON RECURRENCE RELATIONS

In classical equivalence checking, the DUV and the *reference implementation* (RI) are combined in one circuit called miter by XOR-ing every output pin of the DUV with the respective output pin of the RI, and computing the OR of these functions. An efficient miter should have many similar nets, so the equivalence check can be partitioned into less complex sub-problems.

In this section, we review an equivalence checking that builds the RI from the DUV itself. Therefore it does not need a golden reference model and does not require any knowledge about the internal architecture of the DUV. This technique can be applied on any function which can be defined by a recurrence relation. These functions are also known as *primitive recursive (p.r.) functions* [16]. In the following, we briefly discuss p.r. functions, before showing how to exploit the recurrence relations for equivalence checking. Finally, we give an overview on Fujita's approach [14].

Consider the following theorem, as stated in [16]:

Theorem 1. Given $g \in \mathbb{N}^k \to \mathbb{N}$ and $h \in \mathbb{N}^{k+2} \to \mathbb{N}$, there is a **unique** $f \in \mathbb{N}^{k+1} \to \mathbb{N}$ satisfying:

$$f(0, \vec{B}) = g(\vec{B})$$

$$f(A + 1, \vec{B}) = h(f(A, \vec{B}), A, \vec{B})$$

for all $\vec{B} \in \mathbb{N}^k$ and $A \in \mathbb{N}$.

Based on this theorem, a unique p.r. function f is defined, if it can be constructed from other p.r. functions g and h by composition and primitive recursion. Let $\vec{B} = (B_0, B_1, \cdots, B_{k-1})$ and $A, B \in \mathbb{N}$. The basic p.r. functions are, a) the constant zero $C(\vec{B}) = 0$, b) the projection function $P_i(\vec{B}) = B_i$, and c) the successor function $S(A) = A + 1$. Known examples for p.r. functions are addition and multiplication. The addition function $Add(A, B)$ is defined according to Theorem 1 by the relations $Add(0, B) = B$ and $Add(A + 1, B) = S(Add(A, B))$. Note that the first relation is given in terms of a projection function, while the second one uses the successor function, which are both basic p.r. functions.

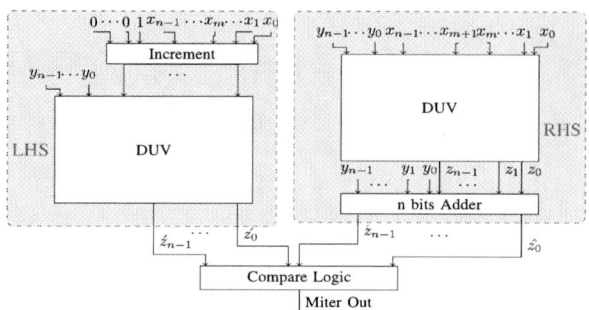

Fig. 1: The Equivalence Checking Miter of a Multiplier using Fujita's approach

This implies that Add is a p.r. function as well. Based on this, the multiplication function $Mult(A, B) = A * B$ is uniquely defined by $Mult(0, B) = 0$ and $Mult(A + 1, B) = Add(Mult(A, B), B)$, and is therefore also a p.r. function.

The uniqueness of a p.r. function thus defined gives rise to an equivalence checking method. Given some implementation F of a p.r. function f, we can check that F correctly realizes f by checking that it satisfies the recurrence relations of f. In case of the multiplication function, if we succeed to show that F satisfies the relations $F(0, B) = 0$ and $F(A + 1, B) = Add(F(A, B), B)$ for any inputs A and B, this implies that F is indeed a multiplication. For most p.r. functions, checking the first relation is relatively easy, while the second relation is verified by building a miter according to the recurrence relation and using standard equivalence checking techniques to show that both sides of the equation are equal for arbitrary inputs. In the following, we will refer to this approach as *equivalence checking based on recurrence relations* (ECRC).

An example of ECRC has been proposed by Fujita in [14]. The approach verifies a multiplier circuit by checking the equivalence of both sides of the recurrence relation:

$$Mult(X + 1, Y) = Mult(X, Y) + Y.$$

The resulting miter structure is shown in Figure 1. An implementation of the *left hand side* (LHS) of the equation is built by adding an increment circuit to the first operand X of the DUV, while the *right hand side* (RHS) is implemented by adding the second operand Y to to the output Z. The equivalence checking compares the two circuits in order to prove that the DUV is indeed a multiplier.

Although the two sides use the same circuit for multiplication, they have different inputs: $(X + 1, Y)$ on the LHS and (X, Y) on the RHS. Therefore the equivalence checker will not be able to find enough internal equivalence points. Fujita overcomes this problem by splitting the checking process into a series of sub-problems. The case split is performed wrt. to the position at which the last carry occurs on the output of the increment circuit $(X + 1)$. It can be observed that in the bit vector representing $X + 1$, the value of an input bit x_m will be inverted iff all lower bits $x_i, 0 \leq i \leq m - 1$ are equal to 1. Thus, for each case the increment circuit can be replaced by simple inverters for the $m + 1$ lower bits of X. The higher bits $x_i, i > m$ will not be modified. However, for larger multipliers beyond 16 bit, this simplification of the increment circuit does not lead to sufficient similarities to enable a scalable verification.

III. CHECKING PARTIAL PRODUCT APPROACH

In this section, we present our approach to verify an implementation of an integer multiplier. After some basic definitions

978-1-4799-8720-7/15 $31.00 © 2015 IEEE

of the multiplication function and the combined add/multiply function, we give an overview of the proposed approach. Then, we present the theoretical part of the approach. Finally, the implementation of the proposed approach will be introduced.

A. Basic Notions

We denote an integer multiplier as $Z = X * Y$, X and Y are the integer operands of the multiplier, Z is the integer result of the multiplier. These integer variables will be represented as vectors of Boolean variables, such that $X = \sum_{i=0}^{n-1} 2^i x_i$, $Y = \sum_{i=0}^{n-1} 2^i y_i$, and $Z = \sum_{i=0}^{2n-1} 2^i z_i$, where n is number of Boolean bits that represent each operand of the multiplier, 2^i is the weight of each Boolean variable. The multiplier $Z = X * Y$ can be expressed on the bit level as

$$\sum_{i=0}^{2n-1} 2^i z_i = \sum_{i=0}^{n-1} 2^i x_i * \sum_{i=0}^{n-1} 2^i y_i.$$

Typically, the multiplication of two operands is performed by generating *partial products*, which are then summed using an addition tree as shown in Figure 2.

Definition 1. A *partial product* is a bitwise multiplication $pp_{ci,ri} = y_{ci-ri} * x_{ri}$, where ri and ci are the row/column indices of the partial product in the addition tree. The partial product $pp_{ci,ri}$ has weight 2^{ci}, which is the product of the weights of x_{ri} and y_{ci-ri}. For input bit width n, we define that

$$pp_{ci,ri} = \begin{cases} y_{ci-ri} * x_{ri} & 0 \leq ci - ri \leq n - 1 \\ 0 & Otherwise \end{cases}$$

As an example, Figure 2 shows the partial product $pp_{4,4} = y_0 * x_4$.

The *partial products generator* generates partial products, such that every bit of the first operand is multiplied with all bits of the second operand. The generated partial products satisfy

$$\sum_{ci=0}^{2n-2} \left(2^{ci} * \sum_{ri=0}^{n-1} pp_{ci,ri} \right) = \sum_{i=0}^{n-1} 2^i x_i * \sum_{i=0}^{n-1} 2^i y_i.$$

The *addition tree* is the addition of partial products to generate the multiplier result Z. The addition is done according to the weights of the partial products, such that

$$\sum_{i=0}^{2n-1} 2^i z_i = \sum_{ci=0}^{2n-2} \left(2^{ci} * \sum_{ri=0}^{n-1} pp_{ci,ri} \right).$$

Note that $pp_{ci,ri}$ is equal to zero, if $ci - ri < 0$ or $ci - ri > n - 1$. The tree has two indices ri and ci, where ri is the index of the tree rows, while ci is the index of the tree columns. Partial products are ordered in the tree as shown in Figure 2, such that partial products have the same weight 2^{ci} if they belong to the same column ci. The output bit z_{ci} belongs to the column ci, having also weight 2^{ci}.

The *addition tree equation* is the equation that formulates the addition of the elements of every column in the tree and the cone that generates the output bit of a column. The addition of partial products in every column generates an output bit as a sum and carry bits. These carry bits are propagated to the next column. Therefore a column does not only take into account the partial products that belong to this column, but also carry bits propagated from the previous column. The addition of partial products in every column ci to the propagated carry bits from the previous column $ci - 1$, generates the output bit z_{ci} as sum and new carry bits to the next column $ci + 1$. We

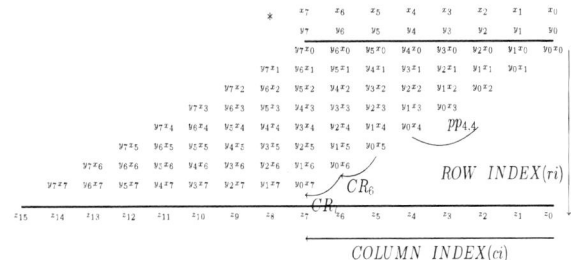

Fig. 2: Addition tree of a 8 bit multiplier

denote the summation of these carry bits as CR_{ci+1}, and the summation of the carry bits that propagate to the column ci as CR_{ci}. Based on that, we formulate the addition tree equation of column ci as

$$z_{ci} + 2 * CR_{ci+1} = CR_{ci} + \sum_{ri=0}^{n-1} pp_{ci,ri}. \tag{1}$$

As an example consider again Figure 2, where CR_6 is the summation of carry bits which propagate to column 6, and CR_7 is the summation of carry bits which are generated from column 6 and propagate to column 7.

In Section II we have explained how we can verify primitive recursive functions using the ECRC technique. In our approach, we use a p.r. function to verify the generation and addition of partial products. Verifying a multiplier then boils down to check the whole addition tree based on this method. For this purpose, we define the *combined multiply-add* (CMA) function, and show that CMA is a p.r. function, therefore its implementation can be verified using its recurrence relations.

Definition 2. The *combined multiply-add* (CMA) function $CMA(a_0, a_1, B) = a_0 * a_1 + B$, combines the bitwise multiplication function and the addition function, where B is an integer quantity, a_0 and a_1 are Boolean variables.

Lemma 1. The CMA function $CMA(a_0, a_1, B)$ is a p.r. function. **Proof:** According to Theorem 1, the CMA function can be defined by the relations $CMA(0, a_1, B) = B$ and $CMA(a_0 + 1, a_1, B) = Add(CMA(a_0, a_1, B), a_1)$. The first relation is a projection, while the addition function Add of the second relation is known to be a p.r. function. This implies that the CMA function is itself a p.r. function.

B. Overview of the Approach

As shown in the previous section a multiplier can be implemented as an addition tree of partial products, which are generated using bitwise multiplications. Based on that, the proposed approach applies case splitting by checking in each case the correct generation of only one partial product and the correct addition of this partial product to other partial products of the multiplier. We will refer to this partial product as *Partial Product Under Verification* (PPV). The combination of the generation and the addition of PPV can be defined using the CMA function from Definition 2. As the CMA function is a p.r. function, its implementation can be checked using ECRC.

The approach applies ECRC on the cone that is influenced by generating and adding PPV to other partial products. We refer to this cone as the PPV cone. This PPV cone can be described using the addition tree and the addition tree equation, as defined in Subsection III-A. As stated there, every partial product belongs to a column in the addition tree, and this column is formulated using the addition tree equation.

978-1-4799-8720-7/15 $31.00 © 2015 IEEE

The column containing the PPV receives carry bits from the previous column. Adding PPV to these carry bits and to other partial products of the column will generate a sum bit and carry bits, which propagate to the next column. Based on that the PPV cone of generating and adding PPV will be the gates that are represented by, a) the columns that produce carry bits to the column of PPV, b) the column that receives carry bits from the column of PPV, and c) the column of PPV itself.

The approach decomposes the verification into n^2 cases, which corresponds to the number of partial products in an n bit multiplier. In every case, it extracts the PPV cone and applies equivalence checking based on the CMA recurrence relation. This case split leads to small differences between the compared implementations of the constructed miter in each case and the approach avoids redundant checks between the cases, which allows for fast equivalence checking regardless of the multiplier size. In the following, we will refer to this approach as the *Checking Partial Product* (CPP) approach.

C. Mathematical Formulations

The goal of the approach in each case split is to verify the combination of generating and adding the PPV. The generation and the addition of PPV is a CMA function $a_0 * a_1 + B$. The approach verifies this function by extracting its implementation from the DUV (the PPV cone), and checking the consistency of the cone with the CMA relations. Consider Definition 1 and Eq. (1) to formulate a mathematical expression for the PPV cone. PPV is a partial product, therefore it can be expressed as $pp_{cv,rv} = x_{rv} * y_{cv-rv}$, where cv and rv are indices of PPV. cv and rv are ranging over the ranges of the column index ci and the row index ri. PPV belongs to a column in the addition tree which has been formulated using the addition tree equation. The addition tree equation of PPV after the extraction of PPV outside the summation of other partial products will be

$$z_{cv} + 2 * CR_{cv+1} = CR_{cv} + \sum_{\substack{ri=0 \\ ri \neq rv}}^{n-1} pp_{cv,ri} + pp_{cv,rv} \quad (2)$$

As can be seen from Eq. (2), adding PPV generates the carry CR_{cv+1} that propagates to column $cv + 1$. Therefore PPV addition has influence on column $cv + 1$. The addition tree equation of this column is

$$z_{cv+1} + 2 * CR_{cv+2} = CR_{cv+1} + \sum_{ri=0}^{n-1} pp_{cv+1,ri} \quad (3)$$

Because the structural relations between the addition tree equations are summations, the formulation of the PPV cone will be the summation of Eq. (2) that formulates the cone of z_{cv}, to Eq. (3) that formulates the cone of z_{cv}. Note that column $cv + 1$ has higher weight with 2^1 than column cv, therefore Eq. (3) will be multiplied by 2, then it is summed to Eq. (2), which leads to the PPV cone equation

$$z_{cv} + 2 * z_{cv+1} + 4 * CR_{cv+2} =$$
$$CR_{cv} + \sum_{\substack{ri=0 \\ ri \neq rv}}^{n-1} pp_{cv,ri} + \sum_{ri=0}^{n-1} 2 * pp_{cv+1,ri} + pp_{cv,rv} \quad (4)$$

Note that the term $2 * CR_{cv+1}$ is removed from the two sides of Eq. (4). We refer to the integer quantity that are added to PPV as

$$Q_{rv} = CR_{cv} + \sum_{\substack{ri=0 \\ ri \neq rv}}^{n-1} pp_{cv,ri} + \sum_{ri=0}^{n-1} 2 * pp_{cv+1,ri},$$

and we substitute the PPV term $pp_{cv,rv}$ by $x_{rv} * y_{cv-rv}$, which reformulates Eq. (4) to

$$z_{cv} + 2 * z_{cv+1} + 4 * CR_{cv+2} = x_{rv} * y_{cv-rv} + Q_{rv} \quad (5)$$

The right side of Eq. (5) formulates the mathematical expression of the PPV cone. Note that the PPV cone is a bitwise multiplication followed by addition, therefore it is an implementation of the CMA function. By replacing the term a_0 of the CMA recurrence relation with x_{rv}, the term a_1 with y_{cv-rv}, and the term B with Q_{rv}, we apply the CMA relations on the PPV cone, such that Eq. (5) becomes

$$z_{cv} + 2 * z_{cv+1} + 4 * CR_{cv+2} = CMA(x_{rv}, y_{cv-rv}, Q_{rv})$$

The initial relation of PPV will be $CMA(0, y_{cv-rv}, Q_{rv}) = Q_{rv}$, by assigning zero to x_{rv}. As the value of Q_{rv} depends on the value of other partial products of the multiplier, checking the initial relation for every partial product separately is not trivial. The approach overcomes this problem by checking the initial relations of all partial products together. Because in every initial relation $x_{rv} = 0$, checking all initial relations together, is done by assigning zeros to all x_i bits, which implies that $X = \sum_{i=0}^{n-1} 2^i x_i = 0$. At $X = 0$, the approach checks trivially that the result of the DUV is equal to zero, where $0 * Y = 0$. Therefore checking the initial CMA relations together is done, by checking the initial relation of the multiplier.

After checking the initial relation of PPV, the recurrence relation of PPV is checked which is

$$CMA(x_{rv} + 1, y_{cv-rv}, Q_{rv}) =$$
$$CMA(x_{rv}, y_{cv-rv}, Q_{rv}) + y_{cv-rv} \quad (6)$$

The approach checks this recurrence relation for every PPV cone, by applying equivalence checking. The approach builds a miter, such that the implementation of the miter represents the left side of the recurrence relation of PPV, while the other implementation represents the right side of this relation. The implementation details of the miter will be explained in the next subsection.

By checking the addition and the generation of all partial products in the multiplier, the approach announces the consistency of the DUV with the multiplication function.

D. Implementation

Implementing the recurrence relation of Eq. (6) as it is, in every splitting case, will suffer from redundant checks. The term $x_{rv} + 1$ propagates a carry bit of value one to other bits of X which has higher weights, like $x_{rv+1}, x_{rv+2}, \cdots, x_{n-1}$. As the value one will be added to these bits in other splitting cases, it is redundant to implement the term $x_{rv} + 1$ as an addition. To simplify the implementation of the equivalence checking and remove these redundant checks, the approach assigns zero to the bit x_{rv}, at this case $x_{rv} + 1$ does not generate a carry bit. Therefore the one bit adder $x_{rv} + 1$ in the left side of Eq. (6) is implemented by the XOR function $x_{rv} \oplus 1$, which is the inversion of x_{rv} ($\bar{x}_{rv}$). Because of this optimization the carry bit that results from $x_{rv} + 1$ to higher bits will be stopped, and building the equivalence checking miter will be easier. At $x_{rv} = 0$, still all patterns of the $pp_{cv,rv}$ are checked, where $pp_{cv,rv} = 1 * y_{cv-rv}$ in the implementation of of the left side of Eq. (6), and $pp_{cv,rv} = 0 * y_{cv-rv}$ in the implementation of the right side of Eq. (6), which implies that this assigning will not affect the coverage of the approach.

We implemented our approach using the tool ABC presented in [17]. The approach applies n^2 case splits. In every case, it builds a miter circuit based on Eq. (6), which checks

978-1-4799-8720-7/15 $31.00 © 2015 IEEE

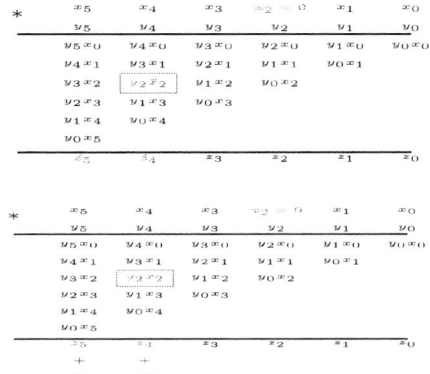

Fig. 3: The miter inputs for checking $pp_{4,2} = y_2 * x_2$

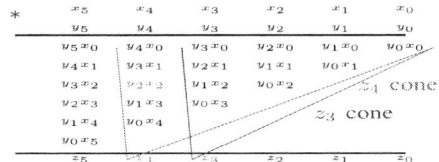

Fig. 4: The search of CPP approach for $pp_{4,2} = y_2 * x_2$

the generation and the addition of one partial product PPV, by applying the following steps:

1) It extracts two copies of the cone of $z_{cv} + 2 * z_{cv+1}$, which are the gates that are connected to the output bits z_{cv} and z_{cv+1}. These gates represent the PPV cone.
2) It assigns zero to the bit x_{rv} of the two copies, which is x_2 in the example of Figure 3 (upper and lower part).
3) In the first copy, it searches for the partial product $pp_{cv.rv}$ and inverts explicitly the bit x_{rv} of this partial product. This is shown in the upper part of Figure 3, where it inverts the bit x_2 of the partial product $x_2 * y_2$.
4) In the second copy, it adds to output bits z_{cv} and z_{cv+1}, the input bit y_{cv-rv}, using a 2 bit adder. This can be seen in the lower part of Figure 3, where the input bit y_2 is added to output bits $z_4 + 2 * z_5$.
5) The miter is completed with the usual comparison logic by XOR-ing and OR-ing the two output bits of the first copy with two output bits of the adder that is added to the second copy.

Checking the two implementations which are based on the recurrence relation of Eq. (6) is very fast because a) the carry CR_{cv} in both implementations propagates in the same way, b) the carry CR_{cv+1} which has different propagation ways in the compared implementations does not propagate through all the multiplier implementations, c) the difference between compared implementations is in the generation of PPV and the propagation way of CR_{cv+1}. Note that CR_{cv+1} propagates in the first copy through the multiplier implementation, while in the second copy it propagates such that parts from it propagate through the multiplier implementation and the other part propagates through the added Adder. Equivalence checking techniques like *rewriting* and *fraiging* [18] succeed to prove the equivalence (or non-equivalence) of many cases without resorting to the SAT solver. This explains the effectiveness of the CPP approach, even when applied to multipliers larger

than 32 bits.

By our case splitting scheme, a total of n^2 miters are constructed. These miters can be checked in parallel or serially, since the cases are independent of each other. We use parallelism for multipliers larger than 32 bits. In general, the CPP approach shows good results in the verification of multipliers of various architectures, as is discussed in the next section.

However, the approach suffers from some limitations. It cannot verify Booth recoding multipliers and some optimized multipliers. To clarify the approach limitations, consider Step 3 of the approach steps. The approach searches for the gate of the partial product $pp_{cv,rv}$. The search is done by comparing gates of cone z_{cv} and cone z_{cv-1}. The gate that belongs to the cone of z_{cv}, does not belong to the cone z_{cv-1}, and has the input bit x_{rv}, is the one that the approach searches for. Figure 4 shows an example of this search. After finding the gate of the PPV, the approach inverts the input bit x_{rv} of this gate. This inversion does not always affect on the partial product $pp_{cv,rv}$, but may affect other partial products, which is the case for multipliers based on Booth recoding and some optimized multipliers.

IV. EXPERIMENTAL RESULTS

Our approach is built on top of the ABC tool [18]. ABC compiles the miter circuit into an *And-Inverter Graph* (AIG) and applies structural reduction techniques like *fraiging* and *rewriting*. These techniques reduce effectively the size of the AIG if the miter has many similar internal nodes. At the backend of the equivalence checking procedure, the reduced AIG is converted to *Conjunctive Normal Form* (CNF) and the resulting instance is given to MiniSAT [19].

All experiments have been carried out on an Intel(R) Core(TM) i5-3320M CPU (2.6 GHz, 16 GByte) running Linux. For the experiments, we generated different multiplier architectures using the online tool *Arithmetic Module Generator* [20]. The multipliers are given as Verilog RTL code. The synthesis of the designs to a gate level netlists has been done using the *Yosys Open Synthesis Suite* [21] and ABC.

The multiplier architectures are categorized according to 1) the type of the partial products generator, 2) the partial products accumulator, and 3) the last stage adder. In our experiments, we use one type of partial products generator, namely *simple partial products* (denoted by SPP). The types of partial products accumulator are *array* (ARR), *wallace tree* (WLT), and *(4,2) compressor tree* (42CT). Finally, the chosen types of the last stage adder are *ripple carry adder* (RCA), *carry look-ahead adder* (CLA), and *brent-kung adder* (BKA). In the following the benchmarks are named according to their architecture features. For example, a circuit with simple partial products, a wallace tree as partial products accumulator, and a ripple carry adder as last stage adder will be labeled as SPP-WLT-RCA.

We conducted two types of experiments. The first one to verify the practical time of the approach in checking larger multipliers. The second one to show the capability of discovering bugs.

A. Equivalence Checking Results

We compare the run-times of our CPP approach against Fujita's approach. In the original approach of Fujita the equivalence is checked using ROBDDs. Here, – and for a fair comparison – we use Fujita's approach with ABC as backend for equivalence checking.

The first column of Table I shows the name of the circuit. The second column gives the number of inputs and output bits. The next two columns provide the run-times. Note that the run-times of Fujita's approach include only the verification

TABLE I: Run times for verification of multipliers

Benchmark	I/O bits	CPP [h:m:s]	Fujita [h:m:s]
SPP-ARR-RCA	16/32	00:01:23	00:00:31
SPP-WLT-CLA	16/32	00:00:46	00:09:08
SPP-WLT-BKA	16/32	00:00:52	00:10:05
SPP-42CT-CLA	16/32	00:00:44	00:49:37
SPP-42CT-BKA	16/32	00:00:43	00:17:56
SPP-ARR-RCA	32/64	02:34:40	11:09:18
SPP-WLT-CLA	32/64	00:15:12	TO
SPP-42CT-BKA	32/64	00:21:20	TO
SPP-ARR-RCA	48/96	20:32:12	TO
SPP-WLT-CLA	48/96	01:29:36	TO
SPP-42CT-BKA	48/96	01:20:00	TO
SPP-ARR-RCA	64/128	94:37:20	TO
SPP-WLT-CLA	64/128	05:46:40	TO
SPP-42CT-BKA	64/128	05:31:44	TO
SPP-42CT-BKA	128/255	78:11:12	TO

TABLE II: Results for models with injected faults

Benchmark	I/O bits	|AIG|	#Faults	$\varnothing$ runtime
SPP-ARR-RCA	16/32	2126	4252	2.56 s
SPP-WLT-CLA	16/32	2988	5976	1.80 s
SPP-42CT-BKA	16/32	2201	4402	0.45 s
SPP-WLT-CLA	32/64	14196	710	27.05 s
SPP-WLT-CLA	32/64	12741	319	59.72 s
SPP-42CT-BKA	32/64	9173	459	10.00 s

of the lower n bits (not the full $2n$ output bits). The time out (TO in the table) has been set to 100 hours. Please note that for a naive miter construction (one big miter) and then running the ABC command (CEC) *all* benchmarks timed out after 100 hours.

The experiments show that the verification time of the CPP approach depends not only on the size of the multiplier circuit, but also on the type of the partial products accumulator. The circuits with wallace tree or (4,2) compressor are verified in less time than those with array accumulator. As can be seen our approach verifies the correctness of the multipliers for up to 32 bits in practical time. Fujita's approach fails here already for the complex architectures. Furthermore, our approach allows the verification of multipliers up to 128 bits.

B. Fault Injection

In order to demonstrate the ability to discover bugs, we applied our approach to faulty designs that have been created by automatic fault injection. The faults have been injected in the netlist which is given as an AIG. We applied the CPP approach on different copies of each netlist where each copy contains one single fault. The approach has succeeded to discover all bugs. The results are summarized in Table II. The first two columns describes the type of the multiplier architecture (as explained in the previous section) and the bit width, respectively. The third column gives the size (number of nodes) of the AIG. The number of performed runs is given in the fourth column, and the average run-time needed to discover the bug is given in the last column.

For all 16 bit architectures, we systematically covered the whole AIG by injecting two faults for each node. For the larger designs, random gates were chosen with an even distribution over the netlist. For the 32 bit multipliers it can be observed that the run-times vary between fractions of a second and several minutes, where usually more than half of the bugs are discovered in less than a second.

To summarize, the results show that our approach works well for bug-hunting as well as for the verification of correct multiplier designs.

V. CONCLUSION

Verification of bit level multipliers still has no general automated solution. In this paper, we verify multipliers at the gate level without any information about their high level designs. We have developed an approach that allows to verify multiplier circuits up to 128 bits. The approach is based on functional properties of the multiplication function, which can be expressed as recurrence equation, as well as a new case splitting scheme. As a consequence enough similarities remain for the equivalence check of each case. Overall, the approach increases the scalability of equivalence checking to verify larger multipliers, however it cannot be applied for all types of multiplier architectures.

ACKNOWLEDGMENT

This work was supported in part by the German Research Foundation (DFG) (DR 287/23-1) within a Reinhart Koselleck project, by the University of Bremen's graduate school SyDe, funded by the German Excellence Initiative and by the German Federal Ministry of Education and Research (BMBF) within the project Effektiv under contract no. 01IS13022E.

REFERENCES

[1] H. P. Sharangpani and M. L. Barton, "Statistical analysis of floating point flaw in the pentium processor(1994)," Intel, Tech. Rep., Nov. 1994.
[2] A. Sayed-Ahmed, H. Fahmy, and U. Kühne, "Verification of the decimal floating-point square root operation," in *ETS*, 2014, pp. 1–2.
[3] J. Harrison, "Floating-point verification," *Journal of Universal Computer Science*, vol. 13, pp. 629–638, 2007.
[4] A. Kühlmann and F. Krohm, "Equivalence checking using cuts and heaps," in *DAC*, 1997, pp. 263–268.
[5] D. Stoffel and W. Kunz, "Equivalence checking of arithmetic circuits on the arithmetic bit level," *IEEE Trans. on CAD*, pp. 586–597, 2004.
[6] B. Xue, P. Chatterjee, and S. K. Shukla, "Simplification of c-rtl equivalent checking for fused multiply add unit using intermediate models," in *ASPDAC*, 2013, pp. 723–728.
[7] R. E. Bryant, "Symbolic boolean manipulation with ordered binary-decision diagrams," *ACM Computing Surveys 24(3)*, pp. 293–318, 1992.
[8] R. E. Bryant and Y. A. Chen, "Verification of arithmetic circuits with binary moment diagrams," in *DAC*, 1995.
[9] K. Hamaguchi, A. Morita, and S. Yajima, "Efficient construction of binary moment diagrams for verifying arithmetic circuits," in *ICCAD*, 1995, pp. 78–82.
[10] D. Stoffel, E. Karibaev, I. kufareva, and W. Kunz, *Advanced Formal Verification*, R. Drechsler, Ed. Kluwer Academic Publishers, 2004.
[11] M. Ciesielski, C. Yu, D. Liu, and W. Brown, "Verification of gate-level arithmetic circuits by function extraction," in *DAC*, 2015.
[12] F. Farahmandi and B. Alizadeh, "Gröbner basis based formal verification of large arithmetic circuits using gaussian elimination and cone-based polynomial extraction," *MICPRO Journal*, vol. 39, no. 2, pp. 83–96, 2015.
[13] J. Lv, P. Kalla, and F. Enescu, "Efficient gröbner basis reductions for formal verification of galois field multipliers," in *DATE*, 2012, pp. 899–904.
[14] M. Fujita, "Verification of arithmetic circuits by comparing two similar circuits," in *CAV*, ser. LNCS, vol. 1102. Springer Verlag, 1996, pp. 159–168.
[15] Y.-T. Chang and K.-T. Cheng, "Self-referential verification of gate-level implementations of arithmetic circuits," in *DAC*, 2002, pp. 311–316.
[16] N. Cutland, *Computability: An introduction to recursive function theory*. Cambridge university press, 1980.
[17] A. Mishchenko, S. Chatterjee, R. Brayton, and N. Een, "Improvements to combinational equivalence checking," in *ICCAD*, 2006, pp. 836–843.
[18] R. Brayton and A. Mishchenko, "Abc: An academic industrial-strength verification tool," in *CAV*, 2010, pp. 24–40.
[19] N. Een and N. Sörensson, "An extensible sat solver," in *SAT*, ser. LNCS, vol. 2919. Springer Verlag, 2004, pp. 502–518.
[20] "Arithmetic Module Generator Based on ACG," 2014, available at http://www.aoki.ecei.tohoku.ac.jp/arith/.
[21] C. Wolf, "Yosys Open Synthesis Suite," 2014, available at http://www.clifford.at/yosys/.

2015 IEEE Computer Society Annual Symposium on VLSI

Hardware Verification using Software Analyzers

Rajdeep Mukherjee
University of Oxford

Daniel Kroening
University of Oxford

Tom Melham
University of Oxford

Abstract—**Program analysis is a highly active area of research, and the capacity and precision of software analyzers is improving rapidly. We investigate the use of modern *software* verification tools for formal property checking of *hardware* given in Verilog at register-transfer level. To this end, we translate RTL Verilog into an equivalent word-level ANSI-C program, according to synthesis semantics. The property of interest is instrumented into the C program as an assertion. We subsequently apply three different software verification techniques—bounded model checking, path-based symbolic simulation and abstract interpretation—and compare their performance to conventional methods for property verification of hardware designs at netlist and register-transfer level. Our experimental results indicate that speedups of more than an order of magnitude are possible. To the best of our knowledge, this is the first attempt to perform property verification of hardware IPs given at register-transfer level using software verifiers.**

I. INTRODUCTION

Early tools for formal property checking of hardware converted the design into a netlist, typically represented using And-Inverter graphs (AIGs). This approach misses the opportunity to exploit the word-level structure of designs given at the register transfer level (RTL). Formal hardware verification tools have therefore now switched to a word-level representation of the transition relation. This change in the representation has enabled the use of modern solvers for Satisfiability Modulo Theories (SMT) [1] in the back-end of the tools [2]–[5].

Consider Bounded Model Checking (BMC) as an exemplar of the way contemporary formal verifiers for hardware work (Figure 1). The input Verilog design is first translated into a transition relation at register-transfer level, in which one transition corresponds to one clock cycle. The transition relation for a system and its specification are jointly unwound up to a user-defined depth to form a word-level formula. This formula is then given to a suitable SMT solver. If the formula is determined to be satisfiable, then there is a bug and the verifier extracts a trace of the circuit leading to the bug from the satisfying assignment.

The performance of word-level symbolic execution engines is determined by the level of abstraction of the symbolic expressions and the power of the rewrite engine used by the SMT solvers. Tools implementing this approach scale up to block level or small IP level [3], [4], for example, a FIFO controller or transceiver/receiver of a USB IP. They generally do not scale to large IPs, subsystems, or full SoCs.

In this paper, we argue that it is time for another, orthogonal, change in how designs are represented—this time to

---

Supported by EPSRC EP/J012564/1, ERC project 280053 and the Semiconductor Research Corporation (SRC) under task 2269.001.

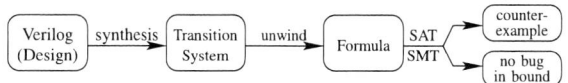

Fig. 1. Conventional flow for hardware property verification using BMC

enable and leverage leading-edge research in modern *software* analysis techniques in the *front-end* of our tools, as a complement better formal reasoning in the back-end. Specifically, we propose to translate circuit designs given initially in RTL into word-level ANSI-C programs that are equivalent according to synthesis semantics—and then undertake our formal analysis on this alternative software representation. The aim is a step-change in performance and scalability in the verification, since the new representation enables the direct application of a range of modern analysis techniques software, leveraging continued major advances in this technology and the sustained efforts of a large research community. In particular, effective analysis techniques such as abstract interpretation, software Bounded Model Checking, and symbolic execution [6]–[8] can be exploited.

Figure 2 gives an overview of the tool flow we propose. Note that the hardware design may be augmented with software, such as firmware or high-level models of surrounding IPs given in C. We first translate the Verilog design to ANSI-C and thus obtain a common representation in the early phase of the verification process. There is then a broad choice of software verification technologies that can be applied. In this paper, we evaluate three of these: software BMC, path-based forward symbolic simulation, and abstract interpretation. We observe substantial speed-ups compared to traditional approaches based on transition relations extracted from RTL.

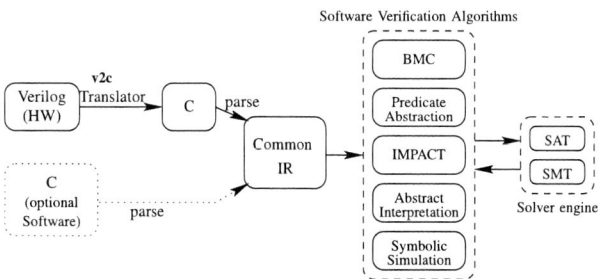

Fig. 2. Proposed new tool flow

The natural software language for our representation is C, so our approach bears some resemblance to methods for verifying system-level or transaction-level descriptions of hardware written in C/C++ or SystemC. Of course the idea of raising

978-1-4799-8720-7/15 $31.00 © 2015 IEEE

the level of abstraction above RTL by expressing designs in software has been widely advocated already—we will mention only [3], which highlights verification-related benefits in the context of SoC design. But we emphasise that these abstract models are usually written *manually* and are expensive to maintain—and often disconnected from a 'golden' RTL design model from which the chip is ultimately realised. Our method, by contrast, aims at a fully automatic verification technique for existing Verilog RTL, expressed at the register-transfer level of abstraction.

Contribution: This paper makes two main contributions:

1) We present a method for constructing a tool for checking properties of Verilog RTL. The proposed tool flow is significantly different from existing verification techniques, in that we translate RTL designs into software and build an analysis engine capable of reasoning with this representation. As we start with the same input files as conventional analysers, tools built this way can be integrated into current industrial formal verification flows. We do not require manually-written high-level models.

2) We evaluate three major exemplars of software analysis techniques within the proposed tool flow and offer a performance comparison to conventional RTL-based tools. To enable a full comparison, we implement the bounded model checking algorithm at different levels of design abstraction and in different representations—including netlists/AIGs, word-level RTL, and software given as C program. Our tool also permits a performance comparison between various SAT/SMT back-ends.

II. BACKGROUND

This section provides background on BMC and symbolic simulation for both hardware and software. We also briefly summarise options for back-end decision procedures.

A. Symbolic Hardware Simulation and Software Execution

Symbolic simulation of hardware designs aims to replace multiple simulation runs, each with different inputs, by a single symbolic simulation run that uses symbolic values as its inputs. The outcome is a set of symbolic expressions for the outputs of the design. And each assignment of values to the variables in these expressions produces a set of values observable on the outputs. The symbolic expressions are then passed to an appropriate symbolic reasoning engine together with the desired properties. Symbolic simulation can be performed on hardware at different abstraction levels—but also on software, where it is known as *symbolic execution*. We will discuss specific instances of symbolic simulation next.

B. Bounded Model Checking for Hardware

Bounded Model Checking was first introduced as a hardware verification technique [9]. The hardware model is typically at the register transfer level. The idea of BMC can be sketched as follows. Given a depth k and a set of error states F, BMC operates by unwinding the transition relation T up to depth k starting from initial state x_0, represented by an initial state predicate I. This results in the following formula:

$$I(x_0) \wedge T(x_0,x_1) \wedge \ldots \wedge T(x_{k-1},x_k) \wedge (F(x_1) \vee \ldots \vee F(x_k))$$

Verilog	Word-level Simulation
```module top(reset,a,b,x,y); output reg[3:0] x,y; input [3:0] a,b; input reset; always @(a or b) begin if(reset) begin x = 3'b0; y = 3'b0; end else begin if(a > b) x = a+b; else y = (a & 3) << b; end end endmodule```	$x'=ite(reset,0,$ $ite((a>b),a+b,x))$ $y'=ite(reset,0,$ $ite((a>b),y,(a\&3) << b))$

Fig. 3. Word-level symbolic simulation of Verilog

The formula is then checked for satisfiability using an efficient SAT or SMT procedure. If the formula is satisfiable, then it is possible for one or more of the error states to be reached. An error trace can be then extracted from the output of the SAT procedure. If the formula is not satisfiable, the system and its specification are further unwound. This process terminates when the length of the potential error trace exceeds a certain completeness threshold (it is sufficiently long to ensure that no trace exists) or when the SAT/SMT procedure exceeds its capacity limit.

The implementation details of BMC tools strongly depend on how the transition relation T, which constitutes a formal *model* of the design, is represented. The challenges related to the generation of formal models from designs given in hardware description languages are mostly shared among all analysis techniques for hardware. Most HDLs have, for example, both *simulation semantics* and *synthesis semantics*. Simulation semantics are typically based on an event queue. On the other hand, the synthesis semantics is closer to the actual hardware produced, and may uncover design flaws that go unnoticed during simulation. In addition, the semantics of the symbolic expressions generated by the tool differs, depending on the level of abstraction at which the analysis is carried out.

BMC on Netlists: One way to represent the transition relation is to use a netlist, which can be obtained from Verilog using standard behavioral synthesis. The netlist captures the effect of one clock period on the state-holding elements. The netlist consists of a network of and-gates, inverters, and memory elements referred to as registers. A typical way to represent a netlist is to use an *And-Inverter Graph* (AIG).

Word-level BMC: Word-level reasoning engines have motivated the use of word-level representations for the transition relation [3], [4]. The use of the term "word" refers to a *bit-vector* encoding of the registers wires, rather than representing them as individual bits. Consider the Verilog example in the left-hand column of Figure 3. The circuit has two state-holding registers, each of four bits. Thus, two next-state functions are generated, denoted by x' and y'. The branching in the input Verilog program yields expressions with the *ite* operator. As in the case of the netlist-based transition relation, the word-level transition relation encodes the effect of one clock period on the state-holding elements.

978-1-4799-8720-7/15 $31.00 © 2015 IEEE

Program	Path Constraint 1	Path Constraint 2	Path Constraint 3

`void top(){`
`  if(reset) {`
`    x=0;`
`    y=0;`
`  }`
`  else  {`
`    if(a > b)`
`      x=a+b;`
`    else`
`      y=(a & 3)<<b;`
`  }`
`}`
	$C_1 \equiv$
	$reset_1 \neq 0 \wedge$
	$x_2 = 0 \wedge$
	$y_2 = 0$
		$C_2 \equiv$
		$reset_1 = 0 \wedge$
		$b_1 \not\geq a_1 \wedge$
		$x_3 = a_1 + b_1$
			$C_3 \equiv$
			$reset_1 = 0 \wedge$
			$b_1 \geq a_1 \wedge$
			$y_3 = (a_1 \& 3) \ll b_1$

Fig. 4.   Single-path forward symbolic simulation

Program	Program Constraint

`void top(){`
`  if(reset){`
`    x=0;`
`    y=0;`
`  }`
`  else  {`
`    if(a > b)`
`      x=a+b;`
`    else`
`      y=(a & 3)<<b;`
`  }`
`}`
	$C \iff ((guard_1 = \neg(reset_1 = 0)) \wedge$
	$(x_2 = 0) \wedge (y_2 = 0) \wedge$
	$(x_3 = x_1) \wedge (y_3 = y_1) \wedge$
	$(guard_2 = \neg(b_1 >= a_1)) \wedge$
	$(x_4 = a_1 + b_1) \wedge (x_5 = x_3) \wedge$
	$(y_4 = (a_1 \& 3) \ll b_1) \wedge$
	$(x_6 = ite(guard_2, x_4, x_5)) \wedge$
	$(y_5 = ite(guard_2, y_3, y_4)) \wedge$
	$(x_7 = ite(guard_1, 0, x_6)) \wedge$
	$(y_6 = ite(guard_1, 0, y_5)))$

Fig. 5.   Translation of C program into bit-vector constraint

C. Path-based Symbolic Execution for Software

In the case of software, symbolic execution is a combination of traditional program testing together with symbolic methods [4], [10], [11]. Recent advances in SMT solvers and constraint solving have greatly promoted the use of symbolic execution for formal property verification and test generation in large software systems. Instead of unwinding the entire transition relation, path-based software analyzers perform forward symbolic execution along individual program paths up to a given depth. The resulting formula is then passed to the SAT/SMT solver. This basic approach has a range of applications. For example, it can be used to check arbitrary safety properties or to generate test vectors to achieve particular coverage goals.

We will use the software equivalent to the Verilog code in Figure 3 to illustrate symbolic execution on software. In their most basic form, symbolic simulators such as KLEE [11] use path-based encodings to explore precisely one path at a time. Figure 4 gives the three different path constraints corresponding to three paths in the program on the left. Note that all paths in this program are feasible. An advantage of this approach is that the generated formulas are often simple and can be solved effectively. But path-based exploration suffers from the *path explosion problem*, as the number of paths is in general exponential. A principal method to address this is *path merging*. The idea is to selectively merge the formulas that correspond to two (or more) paths at points of reconverging control-flow. As a result, the number of formulas is reduced.

D. BMC for Software

On the extreme end, the CBMC Bounded Model Checker *always* merges—generating only a *single* formula for a given unwinding bound k [12]. This formula is linear in the size of the program and linear in k even if there is an exponential number of paths in the program. The right-hand side of Figure 5 gives the outcome of the translation of our running example. The symbolic values of variables are computed as expressions over the initial values of variables x and y. The branching in the program yields expressions with the *ite* operator. The input program is translated into a bit-vector equation C that forms the set of constraints as given in Figure 5.

E. Bit-level Versus Word-level Proof Engines

The result of symbolic simulation or execution is a formula, which is given to a decision procedure. We briefly summarise the options for the decision procedure.

Bit-level Solvers: Current state-of-the-art propositional SAT solvers use a host of advanced techniques, such as Boolean constraint propagation, conflict-driven clause learning, non-chronological backtracking, pre-processing steps, and rapid restarts. These mechanisms have enable SAT solvers to reason very effectively about bit-level encodings of circuit designs. Our BMC engine uses MiniSat 2.2.0 as the default SAT solver. It is worth emphasising that bit-level reasoning engines can benefit from word-level input, as custom clause-level encodings into CNF can be used.

Word-level Solver Engines: Solver engines that perform word-level reasoning are now commonly referred to as solvers for *Satisfiability Modulo Theories* (SMT). Well-known solvers include Boolector, CVC4, MathSAT and Z3. SMT solvers offer a variety of *theories*, which determine the syntax and semantics of the input formulas. The theory typically used in verification of hardware or low-level software is the theory of bit-vectors combined with the theory of arrays. Solvers for this combination rely heavily on rewrite engines that exploit word-level equivalences between subformulas. When the word-level engine is unable to solve the problem using rewriting, the formula is incrementally translated to bit level and is then passed to the propositional SAT solver.

III. RTL Analysis via Translation into ANSI-C

A. Translating Verilog RTL to ANSI-C

The semiconductor industry is rapidly moving to building hardware designs above RTL to further increase design productivity [3], [13]. Although SoC designs are increasingly written at this higher level of abstraction, there is still a significant body of design IP blocks that are written in VHDL or Verilog. In this section, we briefly discuss the working of our tool V2C for translating Verilog to ANSI-C.

V2C translates synthesizable Verilog code to structurally equivalent word-level ANSI-C code based on a synthesis semantics interpretation of the Verilog code. We refrain from synthesis or simulation-based optimizations during the translation process in order to obtain a trustworthy translator. Figure 6 gives an example of translation performed by V2C. A very similar translation was proposed by Greaves [14], but we were unable to obtain a copy of his VTOC tool. Verilator is an open-source translator from Verilog to C++. But Verilator produces very large output code, which is acceptable for simulation but not suitable for formal analysis.

Verilog	C Program
| | ```
struct s_ff {
  _Bool q;
} sff;

struct s_en {
  _Bool ns;
  struct s_ff sff;
} sen;
``` |
| ```
module top(Din,En,CLK,Dout);
  wire cs; reg ns;
  input CLK, Din, En;
  output Dout;

  // Combinational Block
  assign Dout = cs;
  always @(Din or cs or En)
  begin
  if (En)
    ns = Din;
  else
    ns = cs;
  end
  ff ff1(ns,CLK,cs);
endmodule
``` | ```
// Combinational Block
void top(_Bool CLK, _Bool Din,
_Bool En, _Bool *Dout) {
  _Bool cs;
  if(En) sen.ns = Din;
  else sen.ns = cs;
  ff(CLK, sen.ns, &cs);
  *Dout = cs;
}
``` |
| ```
// Sequential Block
module ff(Din, CLK, Dout);
  input Din, CLK;
  output Dout;
  reg q;
  assign Dout = q;
  always @(posedge CLK)
    q <= Din;
endmodule
``` | ```
// Sequential Block
_Bool ff(_Bool CLK,
_Bool Din, _Bool *Dout) {
  _Bool tmp0;
  tmp0 = Din;
  sff.q = tmp0;
  *Dout = sff.q;
  return;
}
``` |

Fig. 6. Example for translation of Verilog RTL to ANSI-C with V2C

B. Path-explosion Problem in Hardware Simulation

The path explosion problem in RTL design is much more severe than in typical application software code. This is caused by structural properties of RTL, which models concurrent computations. Sequential behavior is obtained using clocked logic, which corresponds to one large outer loop. This leads to very complex paths, which are uncommon in software programs. Further, the number of paths usually grows exponentially with the number of clock cycles. All these factors make RTL simulation challenging and result in scalability issues for modern hardware verification tools [13].

C. Application of Software Verifiers

Once the software model for the hardware design is obtained, we can apply a range of formal software analysis tools. We now briefly discuss some of the back-end analysis engines given in Figure 2.

Application of BMC: SAT-based BMC techniques can be easily applied to both RTL and C-style software languages. We used a multi-path symbolic execution engine, CBMC, to verify the system-level models of hardware IPs. Here, the transition relation for a system and its specification are jointly unwound to obtain a formula, which is then checked for satisfiability using an efficient SAT or SMT procedure. This corresponds to replicating the basic blocks along the path k times, followed by a transformation of the concatenation of these blocks into static single assignment (SSA) form [12]. CBMC implements loop unrolling and uses bit-flattening to decide the resulting bit-vector formula. The tool also supports SMT solvers as proof engines in the backend.

Application of Path-Symex: We implemented a single-path forward symbolic execution engine, PATH-SYMEX, that prunes out infeasible paths on the fly and generates incremental path constraints along the feasible branch in a program. The tool uses MiniSat 2.2.0 to perform incremental SAT solving for the generated path constraints. PATH-SYMEX can be combined with property-based slicing techniques to further scale up property verification of system-level models.

Application of Abstract Interpretation: Static program analysis based on abstract interpretation [15] has been widely used to verify certain classes of properties for safety-critical systems. In abstract interpretation, a given program is analysed with respect to a set of given abstract domains. We are not aware of earlier attempts to apply analyzers that are based on abstract interpretation to the verification of properties of register-transfer level circuits.

IV. EXPERIMENTAL RESULTS

A. Experimental Setup

In this section, we report experimental results on the application of software verifiers to complex hardware IPs. All our experiments were performed on an Intel Xeon machine with 8 cores at 3.07 GHz with 48 GB RAM.

We use MiniSat 2.2.0 for purely propositional reasoning and the SMT solvers Z3 4.3.1, MathSAT 5.2.11, CVC4 1.4 and Yices 2.2.1 for word-level reasoning. We have observed only minor differences in the performance of the SMT solvers, and therefore omit the times for the runs with MathSAT, CVC4 and Yices.

We target two different classes of circuits: data-path intensive circuits, exemplified by DSP filters and arithmetic circuits, and control-intensive designs, represented by a USB PHY IP core, an Ethernet MAC IP, and an implementation of a cache coherence protocol from opencores.org. All benchmarks are available at www.cprover.org/hardware/v2c-isvlsi/ to enable other researchers to reproduce our results. We now provide a brief summary of the circuits and give examples of the properties we check.

B. Data-path Intensive Benchmarks

We verified properties of the implementation of a BCD-to-binary converter circuit, serial adder, parallel adder and an integer divider. Typical properties ranged from the absence of arithmetic overflow and underflow to the computation of the correct sum. The divider circuit implements a 16-bit restoring division algorithm. We also verified designs from the DSP domain, namely an FIR filter. The properties that were checked include the round-off error and several symmetry and periodicity properties.

C. Control-Intensive Benchmarks

USB PHY IP: The *USB PHY IP* core provides functions for interfacing to a USB 1.1 bus. This includes serial/parallel conversion, bit stuffing and unstuffing, NRZI encoding and decoding and a DPLL, and offers a UTMI interface. We verify the following properties for the USB PHY IP core:

- *Prop_1:* When rst=1, the UTMI interface of Transmitter and Receiver must be set *HIGH*. Thus the corresponding signals *TxReady_o* and *RxActive_o* of the Transmitter and Receiver must be both *HIGH*.
- *Prop_2:* In case of errors like like sync errors, bit-stuff error and byte-stuff error in transmitter and receiver module, the signals *TxError* and *RxError* must be flagged *HIGH*.

978-1-4799-8720-7/15 $31.00 © 2015 IEEE
```

- *Prop_3:* If the transmission is in progress, i.e. $tx\_ip = 1$, then the *data_done* signal must be asserted *FALSE*. If $tx\_ip = 0$, then *data_done* must be asserted *HIGH* in the next cycle.

Cache Coherence Protocol: The MESI protocol is a widely used cache coherence and memory coherence protocol in shared-memory MIMD multiprocessor systems. The protocol supports write-back caches and is used to maintain the consistency between the L1 and L2 caches in the microprocessors from the Intel Pentium family. The snoopy cache controller continuously monitors the activity of the bus and takes actions whenever a bus transaction involves data stored in their local caches. Each cache line in MESI protocol can be in one of the four states, *Modified, Exclusive, Shared* and *Invalid.* Each cache controller may take different action when there is a request for data read or data write from the CPU. These include *Read Hit, Read Miss, Write Hit* and *Write Miss.* We verified properties relating to the read miss and write miss actions.

- *Prop_1:* If a cache has a read miss and another cache has that line in the *exclusive* state, then the cache line is changed to be in a shared state in both caches after the read is complete.
- *Prop_2:* In case of write miss, the cache controller sends signal to other caches to see if they have a copy of the line. If another cache has that line in *exclusive* state, or if several other caches have the line in *shared* state, then the state of the line is changed to *invalid.*

Ethernet MAC IP: The tri-mode Ethernet MAC implements a MAC controller conforming to the IEEE 802.3 specification and supports serial PHY and parallel PHY interfaces. It also supports automated pause frame generation and termination as well as half-duplex for the 10 and 100 Mbps modes. We verified the encoding and the decoding features in the transceiver module and various operations relating to the block sync and word sync functionality.

D. Discussion

The verification times reported in Table I are in seconds. Best times for a particular benchmark are in bold font. We use two variants of each design; one is safe and one is unsafe.

Columns 1–4 in Table I give the name of benchmark and whether the *safe (y)* or *unsafe (n)* version of the design is used, the *bound* up to which the design is unwound and the number of properties verified for each design, respectively. For smaller benchmarks such as BCD_BINARY, the adders, the integer divider and the FSM design, the sequential depth can be computed easily and is used as the bound for BMC. However, for large designs, computing the sequential depth is hard and thus we chose a high value (50) as the bound. The pair x/y in column 4 gives the number of passing (x) and failing properties (y) for the unsafe designs. The timeout (TO) is set to one hour.

The columns with the heading "Conventional Analysis" give the runtimes for applying Bounded Model Checking as implemented in EBMC 4.2 and HW-CBMC 5.0 to the original Verilog code in three different modes: at the netlist level using MiniSat and at RTL using either MiniSat or Z3.

The columns with the heading "Analysis using Software Verifiers" give the runtimes for analysing the C code generated by V2C using three different software verification tools. The first two columns give the runtime of CBMC 5.0 on the C

program using the given bound. We report times using MiniSat and Z3. The second verifier is PATH-SYMEX 5.0, which is a path-based forward symbolic simulator. This technique heavily relies on incremental solving, which is not well supported by the SMT solvers that are currently available. We therefore only report the runtime when using the MiniSat backend, which has good support for incremental solving. Finally, the last two columns give the runtime taken by Astrée for unbounded and bounded analysis. Astrée implements abstract interpretation to verify safety properties of C source code [6].

We first discuss the results obtained with the conventional approaches. Our results reconfirm observations made in earlier work that the symbolic simulation of word-level RTL models outperforms netlist-based methods [3], [4]. However, our results also show that propositional SAT solvers still perform better than the SMT solvers for hardware property verification. SMT solvers strongly rely on built-in rewrite engines, which is known to be effective for certain equivalence checking tasks but ineffective for hardware property verification [4].

We now discuss the results obtained using our V2C tool and the software verifiers. Our experimental results highlight the potential of changing the design representation used in the verifiers. We observe that, for data-path intensive designs, BMC on the software models is on average 10.5 times faster than BMC on netlists. For control-intensive circuits, BMC on software models is on average 6.2 times faster than BMC on netlists. The symbolic simulators for software are consistently and significantly faster than their counterparts on word-level RTL or netlists. This holds true irrespective of the back-end used. This is quantified by the average speed-up of 10 times for the software model over the word-level RTL using BMC implementation with MiniSAT as back-end SAT solver. With Z3 as the backend SMT solver, we obtain an average speed-up of 6.2 times for the software model over the word-level RTL model. We furthermore observe that PATH-SYMEX performs better than CBMC for some of the data-path intensive benchmarks and a few control-intensive benchmarks like the FSM design and the USB PHY IP core. CBMC with MiniSat is able to prove all benchmarks in reasonable time.

Astrée is the fastest verifier for a broad variety of circuits. Unlike the other methods, however, Astrée may report "Verification Unknown" (denoted by ∗), as the abstract domains that are used by the tool have been explicitly designed for proving the absence of runtime errors and may not be sufficiently precise to prove our functional properties. We also perform bounded analysis using Atrée. This is done by annotating the loops in the program with an unwinding bound, which helps Astrée to compute more precise results. Comparing the netlist-based analysis with the best software verifier, we observe an average speed-up of 50.8. This is mostly because Atrée outperforms the netlist-based approaches very significantly on the control-intensive benchmarks. Atrée was never optimised for hardware analysis, and we thus believe that there is scope here for new tools that implement abstract interpretation using abstract domains developed specifically for this task, e.g., by applying abstract conflict driven learning [16].

V. RELATED WORK

There have been attempts to adapt software verification techniques to the analysis of hardware. For instance, Jain

Circuit	Safe	Bound	# Properties passing/failing	Conventional Analysis			Analysis using Software Verifiers				
				Netlist	RTL		CBMC		Path-Symex	Astrée	
				MiniSat	MiniSat	Z3	MiniSat	Z3	MiniSat	Unbounded	Bounded
Data-path Intensive Designs											
BCD_BINARY	y	8	4	0.47	0.46	0.51	0.14	0.12	0.13	**0.06**	0.08
	n		3/1	0.66	0.74	0.54	0.25	0.22	0.19	**0.03**	0.05
SERIAL_ADDER	y	10	20	66.12	68.61	56.19	6.47	5.86	**0.82**	*	3.93
	n		18/2	79.31	71.37	66.30	5.78	5.31	**0.97**	*	3.84
PARALLEL_ADDER	y	10	35	4.22	2.83	3.51	0.45	0.49	0.36	**0.22**	0.28
	n		30/5	3.89	2.91	2.78	0.54	0.45	0.37	**0.03**	0.05
FIR	y	50	30	7.91	4.67	6.89	0.37	2.97	0.27	**0.26**	0.28
	n		25/5	7.84	7.85	4.26	0.36	2.92	0.26	**0.13**	0.17
INTEGER_DIVISION	y	16	2	90.50	74.78	84.38	14.51	10.86	**1.66**	*	5.98
	n		1/1	15.86	10.56	68.74	15.23	7.15	0.27	*	**0.21**
Control-intensive Designs											
FSM	y	50	10	12.53	57.12	10.78	1.56	1.41	0.29	**0.18**	0.20
	n		7/3	13.75	58.17	10.23	1.57	1.65	0.41	**0.05**	0.07
CACHE COHERENCE	y	50	23	245.11	242.55	396.76	19.12	84.64	TO	*	10.49
	n		17/6	358.61	346.72	456.78	21.08	112.14		*	0.85
USB PHY IP CORE	y	50	25	342.78	337.17	TO	253.21	TO	33.68	**7.72**	17.58
	n		22/3	297.35	270.62		218.14		26.12	**0.13**	0.18
ETHERNET MAC IP	y	50	21	678.26	658.17	TO	571.84	TO	TO	*	**128.84**
	n		17/4	277.02	271.84		236.89			*	**18.24**

TABLE I.   HARDWARE PROPERTY VERIFICATION FOR DESIGNS DESCRIBED AT DIFFERENT LEVELS OF GRANULARITY
(TIMEOUT SET TO 1 HOUR, A * DENOTES "VERIFICATION UNKNOWN")

et al. [17] have applied predicate abstraction, a technique made popular by Microsoft's Driver Verifier, to Verilog RTL. But these techniques still operate on the conventional coarse-grained representation as a transition system; they are therefore limited in their scalability.

There is a large body of work in which software verification methods are applied to *high-level* models of circuits, e.g., given in SystemC or SpecC [18]–[20]. In [21], a high-level description is partially synthesised into RTL, followed by co-verification together with the remaining part of the SystemC model. The main problem of this approach is obtaining a suitable high-level model. The closest work to ours is [13], in which a hardware design at register-transfer level is analyzed using path-based symbolic simulation, similar to our experiments with PATH-SYMEX.

VI.   CONCLUSION

Demand for more scalable verification tools is ever growing. In this paper, we propose an alternative solution for verifying hardware, at the heart of which is a verifier for software. We suggest that verifiers should employ a software representation of circuits, which makes the advanced techniques in modern software analysis tools available. We evaluate this approach using three different program analysis techniques, which are BMC, path-based symbolic execution and abstract interpretation and observe performance gains between one and two orders of magnitude on our diverse set of benchmarks compared to conventional approaches based on netlists or RTL.

REFERENCES

[1] D. Kroening and O. Strichman, *Decision Procedures*. Springer, 2008.

[2] Z. S. Andraus, M. H. Liffiton, and K. A. Sakallah, "Reveal: A formal verification tool for Verilog designs," in *LPAR*, ser. LNCS, vol. 5330. Springer, 2008, pp. 343–352.

[3] M. Keating, *The Simple Art of SoC Design*. Springer, 2011.

[4] S. Sunkari, S. Chakraborty, V. M. Vedula, and K. Maneparambil, "A scalable symbolic simulator for Verilog RTL," in *MTV*, 2007, pp. 51–59.

[5] P. Bjesse, "A practical approach to word level model checking of industrial netlists," in *Computer Aided Verification (CAV)*, ser. LNCS, vol. 5123. Springer, 2008, pp. 446–458.

[6] B. Blanchet, P. Cousot, R. Cousot, J. Feret, L. Mauborgne, A. Miné, D. Monniaux, and X. Rival, "A static analyzer for large safety-critical software," in *Programming Language Design and Implementation (PLDI)*. ACM, 2003, pp. 196–207.

[7] T. Ball and S. K. Rajamani, "The SLAM project: debugging system software via static analysis," in *POPL*, 2002, pp. 1–3.

[8] D. Beyer, T. A. Henzinger, R. Jhala, and R. Majumdar, "The software model checker BLAST," *STTT*, vol. 9, no. 5-6, pp. 505–525, 2007.

[9] A. Biere, A. Cimatti, E. M. Clarke, O. Strichman, and Y. Zhu, "Bounded model checking," *Advances in Computers*, vol. 58, pp. 117–148, 2003.

[10] J. Yang and C. H. Seger, "Introduction to generalized symbolic trajectory evaluation," in *International Conference on Computer Design (ICCD)*. IEEE, 2001, pp. 360–367.

[11] C. Cadar, D. Dunbar, and D. R. Engler, "KLEE: unassisted and automatic generation of high-coverage tests for complex systems programs," in *OSDI*. USENIX, 2008, pp. 209–224.

[12] E. Clarke, D. Kroening, and F. Lerda, "A tool for checking ANSI-C programs," in *TACAS*, ser. LNCS. Springer, 2004, pp. 168–176.

[13] L. Liu and S. Vasudevan, "Scaling input stimulus generation through hybrid static and dynamic analysis of RTL," *ACM TODAES*, vol. 20, no. 1, pp. 4:1–4:33, 2014.

[14] D. J. Greaves, "A Verilog to C compiler," in *RSP*. IEEE Computer Society, 2000, pp. 122–127.

[15] P. Cousot, "Proving the absence of run-time errors in safety-critical avionics code," in *EMSOFT*, 2007, pp. 7–9.

[16] V. D'Silva, L. Haller, and D. Kroening, "Abstract conflict driven learning," in *Principles of Programming Languages (POPL)*. ACM, 2013, pp. 143–154.

[17] H. Jain, D. Kroening, N. Sharygina, and E. M. Clarke, "Word level predicate abstraction and refinement for verifying RTL Verilog," in *DAC*. ACM, 2005, pp. 445–450.

[18] E. Clarke, H. Jain, and D. Kroening, "Verification of SpecC using predicate abstraction," *Formal Methods in System Design (FMSD)*, vol. 30, no. 1, pp. 5–28, February 2007.

[19] N. Blanc and D. Kroening, "Race analysis for SystemC using Model Checking," *ACM Transactions on Design Automation of Electronic Systems (TODAES)*, vol. 15, no. 3, May 2010.

[20] A. Cimatti, I. Narasamdya, and M. Roveri, "Software model checking SystemC," *IEEE TCAD*, vol. 32, no. 5, pp. 774–787, 2013.

[21] D. Kroening and N. Sharygina, "Formal verification of SystemC by automatic hardware/software partitioning," in *MEMOCODE 2005*. IEEE, 2005, pp. 101–110.

978-1-4799-8720-7/15 $31.00 © 2015 IEEE

2015 IEEE Computer Society Annual Symposium on VLSI

Equivalence Checking using Trace Partitioning

Rajdeep Mukherjee
University of Oxford

Daniel Kroening
University of Oxford

Tom Melham
University of Oxford

Mandayam Srivas
Chennai Mathematical Institute

Abstract—One application of equivalence checking is to establish correspondence between a high-level, abstract design
and a low-level implementation. We propose a new partitioning
technique for the case in which the two designs are substantially
different and traditional equivalence-point insertion fails. The
partitioning is performed in tandem in both models, exploiting the
structure present in the high-level model. The approach generates
many but tractable SAT/SMT queries. We present experimental
data quantifying the benefit of our partitioning method for both
combinational and sequential equivalence checking of difficult
arithmetic circuits and control-intensive circuits.

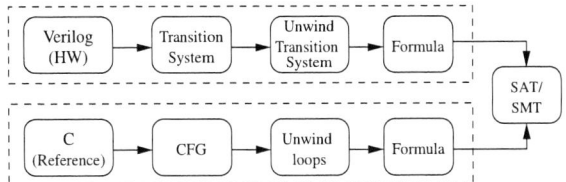

Fig. 1: Equivalence checking tool flow

I. INTRODUCTION

When a new device is designed, a "golden model" is often
written in a high-level programming language such as ANSI-C,
C++ or SystemC. This model is extensively simulated to ensure
both correct functionality and that performance targets are met.
Later, a Verilog or VHDL implementation is created. It is
essential to check that the C and the Verilog models are consistent. High-level models come in a broad variety of modeling
styles, which affects the choice of algorithm for performing
the consistency check. A combinational equivalence checker is
sufficient for a pair of models that are cycle accurate and have
the same set of state-holding elements. But when the high-level
model is not cycle-accurate or has a substantially different set
of state-holding elements, a sequential equivalence checker is
required [1], [2].

In this paper, we address the most general case and thus
most difficult variant of equivalence checking: we consider
the case in which the high-level and the low-level design are
substantially different. In this scenario, methods that rely on
equivalence points [3] are not very effective, and the equivalence checking problem becomes a general Hardware/Software
(HW/SW) co-verification problem.

A widely used method to do formal HW/SW co-verification
is to apply symbolic simulation to both models, and then
merge the resulting formulas. The low-level model is usually
provided in register-transfer level (RTL) Verilog and the high-
level models are written in ANSI-C or SystemC. Bounded
model checking (BMC) using propositional SAT or satisfiability modulo theories (SMT) can be easily applied to both RTL
and C-style languages [4], [5]. Figure 1 illustrates the flow of
a bounded co-verification tool, HW-CBMC. Here, the transition
relations for both models are unwound to obtain two formulas,
which are then conjoined and checked for satisfiability using
an efficient SAT or SMT procedure. The BMC approach relies
on a user-specified unwinding bound; if the bound is too small,
BMC is limited to bug-finding.

In practice, equivalence checking is limited to block-
level designs and does not scale up to IP-level or SoC-level
designs. State-of-the-art industrial equivalence checkers such
as HECTOR (Synopsys) [6] and SLEC (Calypto)[1], use combinations of bit-level and word-level solvers to show equivalence
between C and RTL models. Both HECTOR and SLEC exploit
structural similarities present in the input designs to simplify
the equivalence proofs. Thus, they strongly rely on word-level
proof engines and aggressive pre-processing/rewrite engines
to exploit the similarities present in the word-level formulas
derived from two input models. Furthermore, in order to deal
with difficult instances, they support manual case-splitting of
the design—for example input-based case-splitting or slicing.

In this paper, we experimentally evaluate the benefit of
search-space partitioning for both combinational equivalence
checking (CEC) and sequential equivalence checking (SEC).
Case-splitting followed by slicing is the most commonly
used simplification technique to decompose harder proofs into
simpler sub-proofs for equivalence checking for block level
circuits [7], [8], [6]. In the case of combinational equivalence
checking, partitioning is usually straight-forward: it is not
difficult for the engineer to split up the set of inputs into
suitably easy sub-cases.

Partitioning is, however, more subtle in the case of sequential designs. We use transaction or scenario-based partitions
for our sequential benchmarks. We do not rely on information
from a high-level synthesis stage to guide the equivalence verification and also do not make use of simplifying assumptions
about structural similarity of the two models.

To obtain broadly valid experimental data, we target two
different classes of circuits: data-path intensive designs, exemplified by circuits for floating-point arithmetic, and control-
path intensive designs, which are represented by a USB
physical IP core and an implementation of an Ethernet MAC
controller IP.

Contributions: We address the broad area of "symbolic co-
simulation" of two models, which are typically a high-level

Supported by EPSRC EP/J012564/1, ERC project 280053 and the Semiconductor Research Corporation (SRC) under task 2269.001.

[1]http://calypto.com/en/products/slec/

978-1-4799-8720-7/15 $31.00 © 2015 IEEE   13

reference design and a low-level implementation. We recognize that the existing symbolic methods do not scale to the full size of complex hardware IPs. This paper makes two main contributions that address this problem:

- We present a novel transaction/scenario-based trace partitioning technique for efficient symbolic co-simulation of hardware descriptions (in Verilog) together with a high-level reference model (in C). The technique is applicable to both combinational and sequential equivalence checking problems. The technique is aimed at the most difficult variant of equivalence checking in which the high-level reference design and the low-level implementation design are substantially different and traditional equivalence-point insertion technique fails.
- We experimentally evaluate the benefit of state-space partitioning for combinational and sequential equivalence checking of complex hardware IPs. To this end, we experimentally evaluate the performance of the word-level bounded equivalence checker, HW-CBMC, with and without partitioning. We benchmark the alternative approaches using a wide variety of circuits, ranging from difficult floating point arithmetic circuits to complex control-path intensive circuits exemplified by *USB PHY IP core* and *Ethernet MAC IP*.

II. BACKGROUND

This section introduces formalism for program traces, and reviews how to construct a miter for checking equivalence of a hardware and a software model, and briefly introduces Bounded Model Checking.

A. Traces

The semantics of program statements is defined relative to an environment function that maps program variables to their values, $Env = Var \rightarrow Val$. A statement s defines a function $\delta_s : \wp(Env) \rightarrow \wp(Env)$. A *state* is a location n with an environment ε_i. A *trace* is a sequence of states $(n_0, \varepsilon_0), \ldots, (n_k, \varepsilon_k)$ such that for all $0 \leq i < k$, (n_i, n_{i+1}) is a control-flow graph (CFG) edge and $\varepsilon_{i+1} \in \delta(n_i, n_{i+1})(\{\varepsilon_i\})$. A program is *safe* if there is no trace with $n_0 = init$ and $n_k = \frac{1}{7}$.

B. Miters for HW/SW Equivalence Checking

A miter circuit is built from two given circuits A and B as follows: identical inputs are fed into A and B, and the outputs of A and B are compared using a comparator. We consider the case in which one of the circuits is a software program. Figure 2 shows an example miter for checking combinational equivalence of a 32-bit floating-point adder/subtractor circuit. We provide the same floating-point numbers as inputs to the reference design (in C) and the hardware implementation (in RTL Verilog) using a function set_inputs(). Subsequently, we indicate that we want to perform a floating-point addition by setting isAdd=1. The results computed by the hardware design and the C reference model are compared using the compareFloat() function.

```
void miter(float f, float g) {
    // setting up the inputs to hardware FPU
    fp_add_sub.f = *(unsigned*)&f;
    fp_add_sub.g = *(unsigned*)&g;
    fp_add_sub.isAdd = 1;
    // propagates inputs of the hardware circuit
    set_inputs();
    // get result from hardware circuit
    float Verilog_result = *(float*)&fp_add_sub.result;
    // compute fp-add in Software with rounding mode RNE
    float C_result = add(RNE,f,g);
    // compare the outputs
    assert(compareFloat(C_result,Verilog_result));
}
```

Fig. 2: Miter for combinational equivalence checking for a 32-bit floating-point adder/subtractor for the case of addition

C. Bounded Model Checking

Bounded Model Checking (BMC) [5] can be briefly sketched as follows. Given a depth k and a set of error states F, BMC operates by unwinding the transition relation T up to depth k starting from initial state x_0, represented by an initial state predicate I. This forms the following formula:

$$I(x_0) \wedge T(x_0, x_1) \wedge \ldots \wedge T(x_{k-1}, x_k) \wedge (F(x_1) \vee \ldots \vee F(x_k))$$

A SAT solver is used to determine if the formula is satisfiable. If so, there exists an error trace of length at most k and the procedure terminates, reporting the error. Otherwise, the property holds true on the transition system up to the bound of k. The main challenge of this approach is scalability: the basic SAT/SMT approach works reasonably well up to the module level, but generates instances that are too large for bigger verification tasks. We present a technique for generating many but easy SAT/SMT instances to overcome this problem.

III. EQUIVALENCE CHECKING WITH TRACE PARTITIONING

Previous research has shown that case splitting and slicing make bit-level combinational equivalence checking easier [7], [8], [6]. The approach is straightforward to implement. But partitioning is more subtle in case of sequential designs. This section discusses the various trace partitioning techniques for application in the setting illustrated in Figure 3.

A. Trace Partitioning

Trace partitioning [9] is a method for independently analysing sets of program traces generated by a suitably chosen partitioning function, and was introduced in the context of abstract interpretation [10] as a way to increase the precision of a given abstract domain for a single program.

By contrast, we apply trace-driven partitioning in sequential equivalence checking with the goal to split the state space in a way that enables sufficiently tractable analysis of each of the partitions with the final SAT solver calls. The effort required for analysis increases linearly with the number of partitions generated by the partition function, and exponentially with the size of the partitions. The goal is thus a partitioning function that yields partitions of essentially constant difficulty.

The idea of trace partitioning is to distinguish traces using a function $\alpha : K \rightarrow \wp(S^*)$ that maps some set of tokens K to

978-1-4799-8720-7/15 $31.00 © 2015 IEEE   14

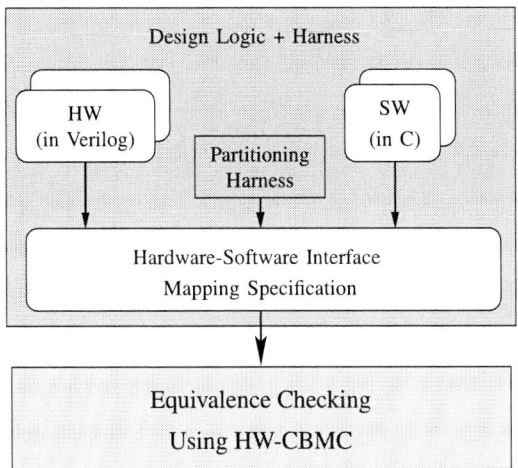

Fig. 3: Equivalence checking using trace partitioning

sets of traces S^*. We consider two extreme cases. If $|K| = 1$ and $\alpha(K) = S^*$, then there is no discrimination between traces. On the other hand, if $|K| = S^*$ and $\alpha(k) = id_{S^*}$ for $k \in K$ and id is the identity function, then each trace is considered in isolation and an analysis with this partition is basically an explicit exploration of all program traces. In practice, we aim at a partition that falls between the two extremes. We systematize the search for a suitable partition using *trace partitioning templates*, described next.

B. Trace Partitioning Templates

The choice of partition function helps to fine-tune the scalability of the analysis by performing coarser splitting where possible and only increasing the number of partitions where necessary. There exists several partitioning techniques in the literature [11], [12], [7] that can be used to partition the state-space of the design. In this paper, we present our experimental results based on input-based partition for CEC and transaction/scenario-based partition for SEC, which are described next.

Value, control-flow and length-based Partitioning: Value-based trace partitioning restricts the range of a variable at one or more program locations. For example, if x is a program variable, then the partition of the traces may depend on grouping the input values of the variable x into three separate cases $\{x < 0, x > 0, x = 0\}$. *Control-flow based trace partitioning* distinguishes traces according to control-flow history. Further, *length* or *parity* based trace partitioning is based on the length of a trace. In this case, the partition function is given by $\alpha : \{0, 1\} \to S^*$, where $\alpha(0)$ corresponds to those traces that execute the loop an even number of times and $\alpha(1)$ corresponds to those traces that execute the loop an odd number of times.

Input-based Partitioning: Input-based trace partitioning is a special case of value-based trace partitioning. To illustrate input-based partitioning [7], [8], [6], let us revisit the harness given in Fig. 2. The CPROVER_assume(c) statement instructs the HW-CBMC tool to restrict the analysis to only those paths

satisfying a given condition c. We can limit the analysis to those paths that are exercised by inputs where the rounding mode is nearest-even (RNE) and both input numbers are NaNs by adding the following statements:

```
CPROVER_assume(fp_add_sub.roundingMode==RNE);
CPROVER_assume(fp_add_sub.uf_nan);
CPROVER_assume(fp_add_sub.ug_nan);
```

Transaction/scenario-based Partitioning: We propose a novel partition technique, known as transaction/scenario-based trace partitioning, to systematize the search-space partitioning for scalable equivalence verification of complex IP designs. Let us consider a model M with input signals $\{I\}$ and output signals $\{O\}$. Here, a *transaction* can be defined by two events:

1) **Initiation event**, *init* – a condition on a subset of input signals, $\{I\}$.
2) **Completion event**, *comp* – a condition on a subset of output signals, $\{O\}$.

Both events, *init* and *comp*, can be possibly spread over time. But, assuming the duration of the transaction event is finite, there must exist some distinguishable last completion event. Now, recall the definition of a *trace* which is a sequence of states, $\{(n_0, \varepsilon_0), \ldots, (n_k, \varepsilon_k)\}$. For the purpose of illustration, let us assume that the $Var \in \varepsilon_0$ corresponds to I and $Var \in \varepsilon_k$ corresponds to O. A *transaction* can then be defined as a projection over a legal trace T of a model M. In other words, a *transaction* is an occurrence of a finite series of events of particular pattern satisfying *init* and *comp* and can be represented as a path constraint. A transaction-based partitioning is an enhancement of input-based case splitting. The reason being that a transaction may involve some input signals, output signals as well as some internal states of the model, as for example, device configuration registers.

A use-case scenario is defined as either a particular finite sequence of transactions or an infinite sequence of transactions satisfying a recurring pattern. This can be manifested in the form of a state machine or using a regular expression. Thus a *scenario* is characterized by the set of allowable control-flow paths in a model.

Figure 4 presents a fragment of a high-level model (in C) and two transactions. The high-level model consists of a single outer *while* loop that encodes complex control flow and interacts with the environment. This code fragment implements a high-level power management strategy in an IP, say a transmitter (*trans*). Assuming that only a small fragment of the logic is enabled inside *trans* module during power gating (i.e. *buf_out* is "don't care"). Thus, partitioning using transaction 1 (*voltage_level == 10*) simplifies the logic for the non-power gated components inside *trans* through the use of assumptions, which restricts the tool to those set of paths satisfying transaction 1 thereby making the final SAT solver query tractable. On the other hand, transaction 2 partitions the design to those set of paths which involves only the normal logic (*voltage_level == 20*) and prunes the logic involved for power gating and other modes of operation (e.g. TURN_OF, TURN_ON, STAND_BY). These transactions are fed to the tool as an assumption to partition the verification state-space.

978-1-4799-8720-7/15 $31.00 © 2015 IEEE

ANSI-C	Transactions
| ```c
#define threshold 15
if(reset) {
  mode=0,turn=0;
  feedback=0;
}
else { // code fragment for IP
// Trigger IP if env is set
if(env) {
  turn = 1; mode = 1;
  // check the voltage level
  if(voltage_level < threshold)
    power_gated = 1;
  else power_gated = 0;
  // check the low-power modes
  if(mode == STAND_BY ||
  mode == TURN_OFF) {
  // power gated logic.
  if(power_gated) {
    trans(reset,mode,power_gated
    ser_in,&buf_out);
    feedback = LOW; // no transmission
  }
  else { // normal logic
    trans(reset,mode,power_gated
    ser_in,&buf_out);
    feedback = buf_out;
}} }
``` | $Transaction1$ : <br> $(reset == 0) \wedge (env == 1)$ <br> $\wedge (mode == STAND\_BY)$ <br> $\wedge (voltage\_level == 10)$ <br><br> $Transaction2$ : <br> $(reset == 0) \wedge (env == 1)$ <br> $\wedge (mode == HIBERNATE)$ <br> $\wedge (voltage\_level == 20)$ |

Fig. 4: High-level model in C and some transactions

IV. EXPERIMENTAL RESULTS

We report experimental results for equivalence checking of two different classes of circuits. We perform trace-partitioning experiments on data-path intensive circuits, namely single-precision and double-precision floating point arithmetic circuits. As representatives for control-path intensive circuits we use an *USB_PHY IP* core and *Ethernet MAC IP* from opencores[2]. To enable other researchers to reproduce our results, all benchmarks are available online at www.cprover.org/hardware/partitioning-isvlsi/.

Data-path Intensive Benchmarks:

IEEE 754 Floating-point Arithmetic Circuits: We have developed both a C and a Verilog implementation of an IEEE-754 32-bit single-precision dual-path floating point adder/subtractor, which takes two 32-bit floating point numbers as input and returns a 32-bit floating point number as result. The floating-point design includes various modules like packing, unpacking, normalizing, rounding and handling of infinite, normal, subnormal, zero and NaN. We used Softfloat[3] as our reference implementation. Softfloat is a well-known software implementation of the IEC/IEEE standard for binary floating-point arithmetic.

Control-path Intensive Benchmarks:

USB_PHY IP: The USB 1.1 Physical Interface core IP provides all functions essential to interface to the USB 1.1 bus. This includes serial/parallel conversion, bit stuffing and unstuffing, NRZI encoding and decoding and a DPLL. The core USB PHY 1.1 supports the industry standard UTMI interface specification. The *phy_mode* signal in the core selects between single ended and differential *tx_phy* output. Currently, the PHY IP from opencores only operates in full speed mode. The required

[2]http://opencores.org/
[3]http://www.jhauser.us/arithmetic/SoftFloat.html

clock frequency is 48 MHz, from which the 12 MHz USB transmit and receive clocks are derived.

Ethernet MAC IP: The tri-mode Ethernet MAC implements a MAC controller conforming to IEEE 802.3 specification and supports serial PHY and parallel PHY interfaces. It also supports automated pause frame generation and termination as well as half-duplex for 10 and 100 Mbps mode. The default FIFO depth of the transmitter (*MAC_TX_FF_DEPTH*) and the receiver (*MAC_RX_FF_DEPTH*) is 9, which means that the FIFO can contain 512 words.

A. Experimental Setup

All our experiments were run on an Intel Xeon machine with 8 cores at 3.07 GHz with 48 GB RAM. All times in Table I and Table II are reported in seconds. CBMC[4] is used to perform bounded equivalence checking of C designs—the tool takes as input a C program with assertions. The tool HW-CBMC is built on top of CBMC. HW-CBMC is used for word-level bounded equivalence checking of designs at different levels of design abstraction—C-RTL models and RTL-RTL models. The tool takes as input a C program and a Verilog RTL implementation. The RTL hardware description is unrolled for each program step using a function call `next_timeframe()`. Other C-RTL and RTL-RTL SEC tools are HECTOR and SLEC, but these tools were not easily obtainable by us for experimentation.

Table I presents the timings for equivalence checking of IEEE-754 single-precision and double-precision floating point arithmetic circuits. Columns 1–4 give the name of benchmark, design sizes for reference and implementation models, and the total time for equivalence checking without partitioning, respectively. Columns 5–10 in Table I report the CEC time with input-based case splitting. Note that the total time in Table I indicates the time required to verify all possible combinations of the input numbers and roundingMode. The timeout is set to 16 hours for CEC. We note that the verification of the single-precision and double-precision multiplier and divider circuit did not terminate without partitioning.

Table II reports the effect of search-space partitioning for sequential equivalence checking of hardware IPs using HW-CBMC. Columns 1–3 give the name of benchmark and the size of the reference and implementation models, respectively. Columns 4–6 give the bound up to which the hardware transition system is unrolled, the unwinding limit for the reference design and the time taken by SEC without partitioning. Columns 7–8 give the run-times when using two different partition harnesses, TP1 and TP2, respectively. The timeout is set to 7 hours for SEC.

B. Discussion

HW-CBMC implements both bit-level and word-level equivalence checking engines. We observe that the bit-level equivalence checker performed badly when compared to the word-level equivalence checker, both for data-path and control-path intensive circuits. This holds true irrespectively of the back-end solver engines used. The default SAT solver engine used is MiniSAT 2.2.0[5]. HW-CBMC also supports different

[4]http://www.cprover.org/cbmc/
[5]http://minisat.se/

978-1-4799-8720-7/15 $31.00 © 2015 IEEE
```

backend SMT solvers, namely Z3, MathSAT, CVC4 and Yices. Our experience with word-level equivalence checkers suggests that SAT solvers still outperform SMT solvers, especially when the input models are substantially different. Thus, we report our results using word-level equivalence checker with MiniSAT 2.2.0. It is worth emphasising that bit-level reasoning engines can benefit from word-level input, as custom clause-level encodings into CNF can be used.

Combinational Equivalence Checking: In Table I, we report the results when using input-based case splitting based on rounding mode, *(RNE)*, and types of numbers *(subnormal, NaN, zero, infinity, normal)*. The top three rows reports C-RTL equivalence checking with three different reference implementations, *Dual-path, CBMC* and *Softfloat*. Note that the C and RTL designs are structurally different and thus techniques which rely on equivalence points fail [7], [8]. However, HW-CBMC is able to handle arbitrary designs and proves equivalence in a reasonable amount of time. The dominant times are required for the case of normal and subnormal numbers.

We highlight that trace partitioning has enabled us to verify the equivalence of single precision and double-precision floating-point multiplier and divider circuits, when, by contrast, the verification of these complex arithmetic circuits exceed the capacity of the state-of-the-art SAT solvers due to complex data-path logic. Our result in Table I reconfirms previous observations that input-based case splitting is an effective partitioning technique for block-level arithmetic circuits [7], [8], [6].

Sequential Equivalence Checking: Compared to the input-based case-splitting that is used to partition the search-space for combinational equivalence checking of block level arithmetic circuits, the partitioning for sequential hardware IPs is more subtle. Here, we perform partitioning based on transactions or what could be interpreted as a "use-case scenario". Consequently, the modular and hierarchical design structure of these IPs makes it easier to identify suitable transaction/scenario pairs for effective partitioning of the design. We now briefly discuss the various partitions used for our sequential benchmarks.

The modular structure of *USB_PHY IP* helped us to partition its operations based on different use-case scenarios in *transmitter* and the *receiver* module, which are identified as *TP1* and *TP2* respectively in Table II. *TP1* checks equivalence of the *DPLL* and the *NRZI Encoder* logic and equivalence of the output enable logic and output registers in the *transmitter* module. *TP2* performs equivalence checking of the *NRZI decoder* logic and the *serial to parallel converter* logic in *USB_PHY receiver* module. For the Ethernet MAC IP, *TP1* and *TP2* partition the design state-space based on different operation modes such as promiscuous (transparent) and non-promiscuous (filtered) operation, respectively.

It is important to note that the high-level use-case scenarios derived from the reference design (in software) are used to partition the verification space in hardware (RTL implementation). We emphasize that the discovery of a transaction/scenario pair is non-trivial. However, our experience with the application of different partitioning techniques (input-based and transaction/scenario based) with HW-CBMC tells us that the control paths in the reference model can be exploited to derive

suitable transaction/scenario pairs for effective partitioning of the design state-space in sequential equivalence checking. Our experimental results show the efficacy of this partition technique for equivalence checking on our benchmarks.

V. RELATED WORK

Over the past years, there have been several works that focus on equivalence checking between designs at different abstraction levels [13], [1], [2], [3], [6]

The work of [14], [6] proposed a theoretical framework for checking equivalence between system-level designs against an RTL design. All these techniques are tuned for the case of equivalence verification of structurally identical designs [7], [8]. Further, to scale up SEC, trace partitioning [9] has so far been usually performed in an ad-hoc fashion, the most simplest form of which is case-splitting followed by slicing [11], [7].

VI. CONCLUSION

In this paper, we experimentally evaluated the benefit of state-space partitioning for both combinational and sequential equivalence checking of two arbitrary input designs— the hardware descriptions (e.g., given in Verilog) together with a high-level reference model (in C). The key to scalability for SAT-based symbolic co-simulation techniques is to perform partitioning of both models, resulting in multiple but manageable queries for today's SAT/SMT solvers. Our experimental results show that equivalence checking with transaction/scenario-based partitioning can scale up existing SEC tools. We also experimentally evaluated the benefit of input-based case-splitting for CEC of complex arithmetic circuits. In future, we plan to developing a technique that effectively performs such partitioning in an automated way that is inspired by the decision heuristics and deduction methods in modern propositional SAT solvers.

REFERENCES

[1] C. A. J. van Eijk, "Sequential equivalence checking without state space traversal," in *Design, Automation and Test in Europe (DATE).* IEEE Computer Society, 1998, pp. 618–623.

[2] J. Baumgartner, H. Mony, V. Paruthi, R. Kanzelman, and G. Janssen, "Scalable sequential equivalence checking across arbitrary design transformations," in *International Conference on Computer Design (ICCD).* IEEE, 2006, pp. 259–266.

[3] W. Wu and M. S. Hsiao, "Mining global constraints for improving bounded sequential equivalence checking," in *Design Automation Conference (DAC).* ACM, 2006, pp. 743–748.

[4] A. Biere, A. Cimatti, E. M. Clarke, O. Strichman, and Y. Zhu, "Bounded model checking," *Advances in Computers*, vol. 58, pp. 117–148, 2003.

[5] E. M. Clarke, D. Kroening, and F. Lerda, "A tool for checking ANSI-C programs," in *Tools and Algorithms for the Construction and Analysis of Systems (TACAS)*, ser. LNCS, vol. 2988. Springer, 2004, pp. 168–176.

[6] A. Kölbl, R. Jacoby, H. Jain, and C. Pixley, "Solver technology for system-level to RTL equivalence checking," in *Design, Automation and Test in Europe (DATE).* IEEE, 2009, pp. 196–201.

[7] B. Xue, P. Chatterjee, and S. K. Shukla, "Simplification of C-RTL equivalent checking for fused multiply add unit using intermediate models," in *Asia and South Pacific Design Automation Conference (ASP-DAC).* IEEE, 2013, pp. 723–728.

[8] M. Fujita, "Verification of arithmetic circuits by comparing two similar circuits," in *Computer Aided Verification (CAV)*, ser. LNCS, vol. 1102. Springer, 1996, pp. 159–168.

978-1-4799-8720-7/15 $31.00 © 2015 IEEE

Circuit	# Lines of code		Combinational Equivalence Checking With Trace Partitioning						
			Without Partitioning	With Input-based Partitioning (Verification Time in Seconds)					
	Reference Model	Implementation Design	Total Time	Subnormal	Infinity	Zero	NaN	Normal	Total Time
	CBMC	RTL	C versus RTL Equivalence Checking (Single-Precision Dual-Path versus CBMC Adder)						
32-bit FP Adder	13	680	309.5	81.3	2.1	2.3	1.8	9.8	315.6
	Dual-path	RTL	C versus RTL Equivalence Checking (Single-Precision Dual-Path Adder)						
32-bit Adder	650	680	535.8	171	8.1	8.0	7.9	297.4	546.2
	Softfloat	RTL	C versus RTL Equivalence Checking (Single-Precision Dual-Path versus Softfloat Adder)						
32-bit FP Adder	1847	680	623.5	61.2	3.7	3.5	3.4	28.5	618.7
	RTL	Optimized RTL	RTL versus RTL Equivalence Checking (Dual-Path versus Optimized Dual-Path Adder)						
32-bit FP Adder	680	715	13.8	2.4	2.3	2.7	2.4	2.3	14.2
	C	C	C versus C Equivalence Checking (CBMC versus SoftFloat versus Dual-Path Adder)						
Dual-Path vs. CBMC	756	13	560.7	159.1	0.5	0.6	0.5	430.8	598.5
Softfloat vs. CBMC	1847	13	24.9	1.6	0.5	0.5	0.5	22.9	29.3
Softfloat vs.Dual-Path	1847	756	1145.8	98.7	0.5	0.5	0.5	1067.5	1174.8
	Softfloat	Slowfloat	C versus C Equivalence Checking (Single-Precision Softfloat versus Slowfloat)						
32-bit FP Multiplier	2136	1045	Did not Terminate	0.8	0.5	0.4	0.5	29866.3	29997.5
32-bit FP Divider	2369	1114	Did not Terminate	26789.4*	0.9	0.8	0.8	54341.7*	81733.6*
	Softfloat	Slowfloat	C versus C Equivalence Checking (Double-Precision Softfloat versus Slowfloat)						
64-bit FP Adder	2309	1174	4449.7	387.1	1.6	1.7	33.8	4056.2	4492.4
64-bit FP Multiplier	2383	1186	Did not Terminate	2.1	1.6	1.5	1.7	38764.5*	38824.4*

TABLE I: Run times for combinational equivalence checking for IEEE 754 floating-point arithmetic circuits (runs marked with * are limited to RNE+Type+Even+Odd, the timeout was set to 16 hours)

Circuit	# Lines of code		Sequential Equivalence Checking With Trace Partitioning				
	Reference Model	Implementation Design	Bound Depth	Unwind Limit	Without Partitioning	With Transaction/Scenario based Trace Partitioning	
					Total Time	TP1	TP2
	C	RTL	C versus RTL Equivalence Checking				
SERIAL ADDER	47	62	10	–	0.9	0.9*	0.9*
PIPELINED ADDER	52	68	10	–	0.8	0.8*	0.8*
ETHERNET MAC IP	1050	1200	100	15	Did not Terminate	21578.9	22451.3
USB_PHY IP	860	950	200	50	Did not Terminate	22681.4	21822.3
	RTL	Optimized RTL	RTL versus RTL Equivalence Checking				
SERIAL ADDER	62	54	10	–	0.6*	0.6*	0.6*
PIPELINED ADDER	68	58	10	–	0.5*	0.5*	0.5*
ETHERNET MAC IP	1200	1220	500	30	1384.5	956.2	948.7

TABLE II: Run times for sequential equivalence checking for control-intensive circuits (* – no partitioning possible, the timeout set was to 7 hours)

[9] X. Rival and L. Mauborgne, "The trace partitioning abstract domain," *ACM Trans. Program. Lang. Syst.*, vol. 29, no. 5, 2007.

[10] P. Cousot and R. Cousot, "Abstract interpretation: A unified lattice model for static analysis of programs by construction or approximation of fixpoints," in *Symposium on Principles of Programming Languages (POPL)*. ACM, 1977, pp. 238–252.

[11] C. Karfa, D. Sarkar, and C. Mandal, "Verification of datapath and controller generation phase in high-level synthesis of digital circuits," *IEEE Trans. on CAD of Integrated Circuits and Systems (TCAD)*, vol. 29, no. 3, pp. 479–492, 2010.

[12] L. Liu and S. Vasudevan, "Scaling RTL property checking using feasible path analysis and decomposition," in *Great Lakes Symposium on VLSI (GLSVLSI)*. ACM, 2013, pp. 173–178.

[13] E. M. Clarke, D. Kroening, and K. Yorav, "Behavioral consistency of C and verilog programs using bounded model checking," in *Design Automation Conference (DAC)*. ACM, 2003, pp. 368–371.

[14] Z. Khasidashvili, M. Skaba, D. Kaiss, and Z. Hanna, "Theoretical framework for compositional sequential hardware equivalence verification in presence of design constraints," in *International Conference on Computer-Aided Design (ICCAD)*. IEEE/ACM, 2004, pp. 58–65.

2015 IEEE Computer Society Annual Symposium on VLSI

Joint Circuit-System Design Space Exploration of Multiplier Unit Structure for Energy-Efficient Vector Processors

Ivan Ratković, *[‡] Oscar Palomar,* [‡] Milan Stanić,* [‡] Milovan Duric,* [‡] Djordje Pešić, [¶]
Osman Unsal, * Adrian Cristal, *[‡§] Mateo Valero *[‡]
* Barcelona Supercomputing Center [‡] BarcelonaTech (UPC) [¶] University of Belgrade [§] CSIC-IIIA

Abstract—Although touted as a power and energy-efficient solution for workloads that exhibit data-level parallelism, vector processors were not explored sufficiently from a low power perspective in the past. Therefore, there is a need for explorations of vector computational units from a low power angle. Multimedia workloads that are suitable for vector processing (such as image processing) typically have the multiplication as a fundamental operation. In this paper, we perform a joint circuit-architecture design space exploration of the vector multiplier unit (VMU). For this exploration, we use various circuit- and architecture-level parameters (e.g. multiplier family and maximum vector length), tools and simulators for a 40nm low power technology and the San Diego Vision Benchmark suite. We examine advantages and side effects of using multiple vector lanes and show how it performs across the frequency spectrum to achieve an energy- and thermal-efficient speed-up. As the final results of our exploration, we derive Pareto optimal VMU design points. Among other findings, our exploration reveals that *Wallace* VMU with 4 vector lanes and 2 pipeline stages is an optimal choice for fast and low power mobile vector processors, while single lane *Carry-Save Array* VMU is efficient for very low power and frequency requirements.

Keywords-Vector Processors; Low Power; Energy-Efficiency; Digital Arithmetic; Multipliers; Design Space Exploration; Joint Circuit-System Analysis

I. INTRODUCTION

Power dissipation has become the primary design constraint for almost all computer systems. For example, the main design goal of mobile, battery constrained, devices is energy efficiency, while data centers suffer from both power density and energy consumption issues. It is still expected that each new generation of microprocessors is better than the previous one. Since frequency does not scale with technology in an energy-efficient way anymore, we need changes at the architectural level that allow faster execution at the same envelope [1].

Vector processors are an inherently energy-efficient architecture [2, 3, 4] for applications that exhibit data-level parallelism (DLP), i.e. that operate on vectors of independent elements. They express DLP in a very compact form, thus removing redundant work (e.g. instruction fetch, decode and issue) for long vectors. It has been observed that the best vector-based machines are generally faster and/or more energy-efficient than scalar multicores [4].

There are server and mobile workloads with a lot of DLP, e.g. face, fingerprint and speech recognition. Taking this into account, we consider that low power vector processors are a promising fit for modern mobile devices and servers. Vector processors are particularly appealing for mobile processors because they support high degrees of parallelism at low cost and with low power dissipation [2, 5]. These computer systems have fairly different design goals than traditional vector supercomputers that prioritized performance [6]. Vector processors require new components that are energy and power efficient.

We explore the design space of the multiplier unit for vector processors (Vector Multiplier Unit - VMU). Multiplication is a fundamental operation in many algorithms that exhibits a lot of DLP, such as image processing. Therefore, its importance has increased in last decade due to ubiquitous media processing. Multipliers have large area, long latency and consume considerable power. High power dissipation in these structures is mainly due to the switching of a large number of gates during multiplication.

It has been shown that power dissipation of arithmetic units strongly depends on the input data' switching activity factor (α_I) and the usage of the units [7, 8, 9]. Therefore, α_I and VMU usage obviously affect arithmetic structure selection for a particular architecture and benchmark set. Activity factor is even more important in vector processing as data inside a vector is usually correlated [2]. There are architecture-related characteristics (such as the maximum vector length or number of lanes), specific only for vector processors, that significantly affect multiplier unit usage and input data activity factor, thus, power dissipation. Consequently, conclusions found for other architectures might not be valid for vector processors.

In this work, we perform comprehensive (power, delay, energy, area, power density) design space exploration of multipliers for vector processors. The exploration includes circuit-level parameters like the number of pipeline stages and clock-gating support, as well as architectural parameters such as the number of lanes. In order to join architectural-level information (e.g. microbenchmarks) with circuit-level outputs (e.g. VMU power measurements) we developed

978-1-4799-8720-7/15 $31.00 © 2015 IEEE   19

an automated and integrated architecture-circuit exploration framework (explained in detail in Section II-A) that consists of several simulators at different levels, and we perform this exploration with microbenchmarks generated with data obtained from vectorized applications of the San Diego Vision Benchmarks [10]. We observe how architectural-level parameters (e.g. vector length) affect the circuit-level metrics (e.g. multiplier's power dissipation) and how circuit-level parameters (e.g. multiplier's clock cycle) impact the execution time of a microbenchmark. The final goal of this design space exploration is to provide guidelines for designers of low power and energy efficient vector processors. Guidelines are based on Pareto-Optimal trade-off graphs of VMU design space that we get as result of our framework.

There is considerable prior circuit-level research on comparison of multipliers in the Energy-Delay space [11, 12, 13] However, our analysis differs as we include architecture-related parameters, additional parameters like clock-gating and more recent technology nodes, thus some of our conclusions are different. For example, we found that Carry-Save Array sometimes beat Wallace multipliers in terms of power when highly memory-bound workloads are considered (Section III). While there is very interesting research on the design of multiplier units for particular architectures such as DSPs [13, 14], their exploration demands are different, they do not target vector architectures neither utilize the same kind of architectural and circuit parameters as we do, thus, their analyses lead to distinct conclusions (Section V).

This paper presents the following contributions:

- A discussion of multipliers' power, timing and area characteristics, considering clock-gating, for various kinds of testbenchmarks and verified physical results for a recent technology (Section III).
- A study of vector multi-lane advantages and drawbacks for VMU. We found it as an effective solution when energy efficient speed-up is more important than low power (Section IV).
- Guidelines on VMU configuration design for different low power vector processors based on Pareto-optimal trade-off graphs that take into account both architectural- and circuit-level parameters. For instance, we find that Wallace VMU with 4 vector lanes and 2 pipeline stages is an adequate choice for performance oriented low power mobile vector processors.

II. Methodology

This section describes the exploration framework used to perform the design space exploration of VMU, the framework parameters and the test benchmarks that we use.

A. Exploration Framework

The framework is depicted on Figure 1. It includes architectural- (*VectorSim*) and circuit-level (*RC, Encounter, NCsim*) simulators and tools, as well as an interfacing tool

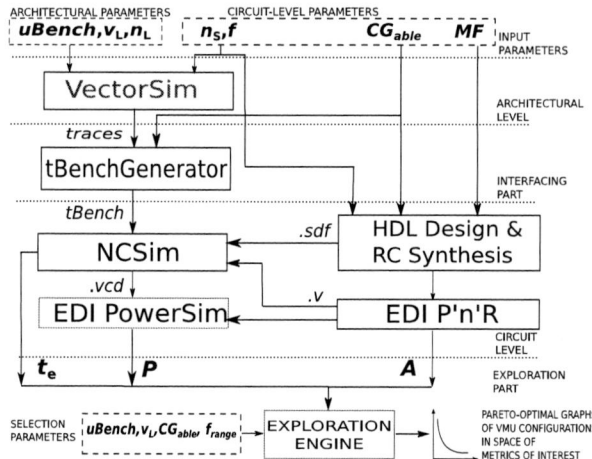

Figure 1. Block diagram of the framework's steps, parameters, and metrics.

Table I
METRICS OF INTEREST

Measured Metrics
P. The average power of VMU.
t_e. The execution time of a test benchmark *tBench*, also referred as Delay (D).
A. The area of multiplier(s).
Computed Metrics
P_d. Surface power density = P/A. It is proportional to the fourth power of temperature of the given surface by Stefan-Boltzmann law ($P_d \propto T^4$).
$E=PDP$. Power-Delay product. It is total energy spent in multiplier during a *tBench*.
P_dDP, EDP, E^2DP, and E^3DP are commonly used Power density and Energy-Delay products [15].

(*tBenchGenerator*) and an exploration engine. For various circuit- and architectural-level parameters (explained in detail in section II-B) we obtain the metrics of interest of our VMU (described in Table I).

At the architectural level we use a vector simulator *VectorSim* [16], that we modified, and we feed it with the architectural and some circuit parameters. This stage generates data and timing traces for the vector multiplications. We setup *VectorSim* to model an in-order 32-bit decoupled vector machine (separated vector and scalar execution units in the same core [17]). In-order execution is common in low power processors and performs more efficient in vector than in scalar processing as in vector architectures the drawbacks of in-order execution are diminished. The processor has one ALU per vector lane and implements chaining (vector equivalent of data forwarding) and dead time elimination (allowing to reuse the ALU immediately after the current instruction). Vector lanes are replicated ALUs that operate in parallel in a lockstepped fashion, where each vector lane is connected to a slice of the vector register file. Figure 2 shows a 2-lane VMU of a vector processor with vector length of 64 executing a vector multiplication instruction.

The interfacing part transforms this architectural-level information into verilog test benchmarks (*tBench*s) using our in-house our tool *tBenchGenerator*. Section II-C describes the different *tBench*s that we generate.

At the circuit level we incorporate all the circuit parameters into handcrafted HDL codes of multipliers which are supplied to Cadence RTL Compiler (*RC Synthesis*) to produce different multiplier's synthesized mapped netlists. We perform synthesis for all combinations of our circuit parameters in an automated way. For VMU configurations with multiple lanes, multipliers in all lanes are identical. We provide synthesized verilog netlists together with the physical layout information to Cadence Encounter Digital Implementation System in order to get placed and routed designs (*EDI P'n'R*). After this stage we obtain one metric of interest: area (*A*). The critical paths of designs with small slack are verified with Cadence Spectre in order to ensure there are no timing violations in the post-layout designs. This step is not shown in Figure 1 for the sake of simplicity.

The next step is to perform post-layout simulations of each multiplier in Cadence *NCSim* for each matching *tBench* with back-annotated delays using Standard Delay Format (sdf) files. This is done in order to obtain the execution time t_e, verify the synthesized designs and extract resulting switching activity information using Value Change Dump (vcd) files. The next step of the framework is precise estimation of power metrics (*P*) using *EDI Power Simulation*. The inputs are placed and routed designs and corresponding vcd files.

The final step is the exploration part, where the design points (VMU configurations) are supplied to the exploration engine (a combination of MATLAB and Perl scripts). The design points are first filtered with selection parameters: *uBench*, *CGable*, v_L, and frequency range (f_{range}). and then depicted using Pareto-optimal trade-off graphs in the space of desired metrics of interest (e.g. E-D space).

Since practically all existing vector processors are developed using standard cells approach [18], we implemented all the designs using low power TSMC40LP standard cell library. We set the tools to meet timing constraints while prioritizing power over area. The optimizations are applied using *high effort*. However, we prevent restructuring optimization techniques that could change the multiplier structure, in order to maintain the original topology of each multiplier family. The core density for layouts is initially set at 85% and decreased 5% each time if the routing does not succeed.

B. Framework Parameters

We first present the vector architecture specific parameters.

✦ *uBench* is a microbenchmark (kernel) vectorized by hand. We use three different integer *uBench*s from San Diego Vision Benchmark [10] run for *fullhd* inputs (described in Table II). These *uBench*s are used in mobile devices and

Table II
VECTORIZED MICROBENCHMARKS (*uBench*)

Disparity Map (computeSAD) computes depth information using dense stereo. It is used for robot vision for stereo vision.
Maximally Stable Regions (FitEllipses) does Blob detection in images. It is used for stereo matching and object recognition
Feature Tracking (ImageBlur) extract motion from a sequence of images. It is used for Robot vision for tracking.

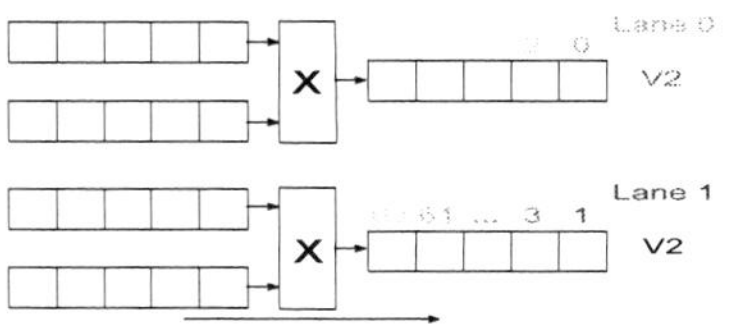

Figure 2. A 2-lane VMU executing VMUL V2<-V0,V1.

can also be found in server workloads. *uBench*s are multiplication intensive, so multiplier's impact on the *uBench*s' execution time is significant. We keep their size between 100k and 150k test vectors in order to keep circuit simulation time reasonable.

✦ v_L is the maximum vector length of the vector processor, and it indicates the number of elements stored in a vector register. Possible values are 16 and 128 to represent both extremes of short and long maximum vector lengths.

✦ n_L indicates the number of lanes. Possible values are 1, 2, and 4. We do not examine further doubling the number of lanes as it would not satisfy well a low power core budget.

We now present the circuit-level parameters which are primarily needed for the circuit-level part of the framework.

✦ *f* is the clock frequency of the multiplier and of the whole processor. We use (0.1-2.05) GHz as a practical frequency range for low power vector processors. Vector processors are usually able to afford high frequency due to two reasons: (1) control logic is simple so it is not a timing bottleneck and (2) they can afford deep pipelining because latency is hidden. f_{range} is a frequency range used as a selection parameter.

✦ *MF* is the multiplier family (algorithm). We consider only multiplier families that are suitable to be fully-pipelined. Structures that are not fully-pipelinable produce pipeline bubbles and this is inefficient for vector processors, as vector operands are streamlined to the vector ALU at the rate of one per cycle. For this reason, we do not include serial (with a feedback loop), but only parallel (fully-pipelined) structures. Therefore we consider array and tree *MF*s [19]. Carry-Save Array (*ar*) and Wallace (*wl*) [20] multipliers are chosen as representative examples for each of the mentioned *MF*s. Array multiplying algorithm is selected as it has a regular layout with simple interconnects (which became important in deep submicron design). Tree multipliers is chosen as they are supposed to be both timing- and area-efficient (delay proportional to the logarithm of the operand width). However, they have irregular layouts with complicated interconnects which can spoil their power, timing and area results in recent technology nodes. For partial product generation a simple AND network is chosen rather than Booth one, as we experimentally found that when it is attached to multiplier, it performs better in terms of area, timing, and power (a previous study confirmed these findings [21]). As the final stage adder, for all two considered multiplier families we choose Brent-Kung adder, as it has been shown to perform the best when clock-gating and pipelining are considered [22]. We implement our multipliers so they support 32-bit 2's complement integers. The output is 64-bit and it is left to the Instruction Set Architecture how this number is later handled.

✦ CG_{able} indicates whether an multiplier is implemented with or without support for clock-gating (CG or noCG, respectively). Clock-gating is quite common today, and makes pipelining more effective. We assume the clock-gating decision is done in the issue stage, similar to deterministic clock-gating approach [23]. Since we assume pipelined multipliers we implement fine-grained (stage-level) clock-gating. Each stage is activated with its Enable signals when the data on its inputs is valid. We use latch-based clock-gating cells from the cell library. We do not include power gating and voltage ramping techniques in this exploration as in our results the leakage is negligible (explained in Section III).

✦ n_S indicates the number of pipeline stages in the multiplier (pipeline depth). We only increase n_S for a particular MF if further pipelining provides noticeable f increase. In particular, we implement ar for 1-9 and wl for 1-5 stages. We consider a wide range of pipeline stages as the vector arithmetic units are typically fully and deeply pipelined, as in vector processing the throughput of an arithmetic units is more important than its latency. Additionally, pipelining is quite necessary for multipliers due to two reasons: (1) to achieve higher frequency, and (2) to reduce glitching. In parallel multipliers (especially tree ones), glitches or spurious transitions are typically generated because partial product bits arrive at the same time but are added serially and the input-to-output paths in adder cells have different delays. The input register (first pipeline stage) of an n-stage multiplier consists of 2x32 DFlipFlops for the input operands. Additionally, each n-stage multiplier has n-1 pipeline clocked registers which are used to save intermediate results. The size of each register depends on the MF structure and the place inside the multiplier where the register is inserted (i.e. on the way the multiplier is pipelined).

C. Test Benchmarks

We generate application-based verilog testbenchmarks (tBench) that are obtained from the uBenchs and are a function of all the parameters used in VectorSim. We consider four types of tBenchs:

Table III
WALLACE VS. ARRAY RATIO FOR AREA AND POWER

	Area (CGable)		Power (tBench)			
	noCG	CG	100%	noCG	CG	0%
Min	0.47	0.50	0.09	0.12	0.13	0.55
Max	1.08	1.08	0.98	0.99	1.03	1.52
GeomMean	0.64	0.66	0.21	0.24	0.29	0.80
GeomStdDev	1.28	1.22	1.75	1.65	1.70	1.30

✦ noCG is used to evaluate designs without clock-gating. The input values of the VMU are provided each cycle.

✦ CG enables evaluating designs with clock-gating support. Clock-gating is enabled when we have idle cycles. Here we need additional clock-gating signals, one per stage.

✦ 100% represents a case where the VMU is always busy, assuming that memory is fast enough to provide data on time and that consecutive multiplication instructions are independent. Therefore, execution time depends only on the characteristics of the VMU.

✦ 0% is used to evaluate the case when there is no vector multiplications in the uBench, i.e. clock-gating is always active.

100% has the highest input data activity factor, while 0% has the lowest among the tBenchs. CG and 0% are supplied to the multipliers with clock-gating, while the rest are supplied to multipliers without clock-gating logic integrated.

III. MULTIPLIERS' CHARACTERISTICS

We analyze the power, timing, and area characteristics of the examined multiplier families by observing the lowest power configurations for each pair of frequency and multiplier family. Table III provides statistics for area and power ratios of wl over ar multipliers for all four tBenchs of the lowest power configurations, with and without clock-gating mechanism integrated (CGable=CG and noCG). Power graphs of the lowest power configurations are shown on Figure 3. The goal here is to analyze characteristics of multipliers alone so we test configurations with one lane.

✦ Area. Except for slow designs ($f < 0.3\,\text{GHz}$), wl is more area efficient than ar, as it can be observed from Table III. For higher frequencies ar needs more pipeline stages (higher n_S) in order to achieve the same timing as wl, which increases area.

✦ Timing. Maximal achievable frequencies for CGable=noCG are 2.05 GHz and 1.85 GHz for wl with 5 stages and ar with 9 stages respectively. For CG both 5-stage wl and 9-stage ar achieve the same max frequency - 1.7 GHz. As wl has an irregular structure, its clock tree is also irregular, so the insertion of a clock gating mechanism affects overall timing results to a higher extent than in ar. Maximum frequency of multipliers would be higher if a high-performance technology is used instead of low power one.

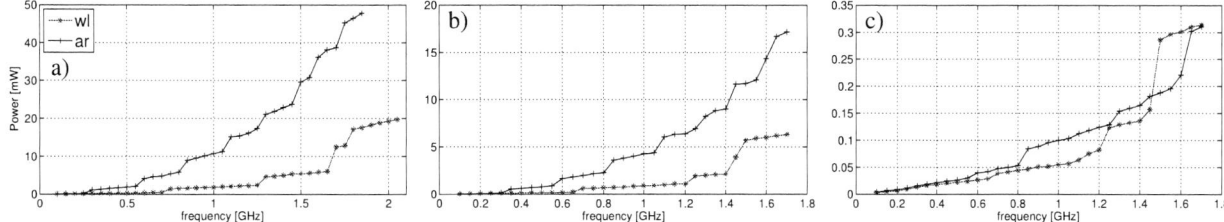

Figure 3. Power dissipation of the lowest power configurations of analyzed multiplier families for (a) *noCG*, (b) *CG*, and (c) *0%*, for v_L=128.

✦ *Power.* The total power of a VLSI multiplier is the sum of dynamic (P_{dyn}) and static (P_{stat}) power: $P = P_{dyn} + P_{stat}$, where P_{dyn} consists of switching, short-circuit and glitching power. Further, P is typically structurally divided into power of multiplier's clock tree and of the rest of the multiplier: $P = P_{ct} + P_{rest}$. We observe from the results that for all the *tBench*s, except for *0%* where it contributes with 20%, P_{stat} is almost negligible. It is an order of magnitude less than P_{dyn}, even for *CG*. The leakage is negligible due to the following reasons: (1) multipliers' topologies produce high switching (high P_{dyn}), (2) the technology that we use has low leakage, and (3) we optimize non-critical path logic for leakage using high-V_{TH} cells. Therefore, the power of a multiplier (P) is practically equal to its dynamic component ($P \approx P_{dyn}$).

Figure 3 shows the lowest power configurations for *noCG*, *CG*, and *0%* for v_L=128. Qualitatively, power graphs for v_L=16 and 128 have similar shapes for all *tBench*s. Absolute numbers indicate that power is slightly higher for v_L=16 in case of *noCG* and *100%* as in case of v_L=128 there is less correlation between data being executed (slightly higher α_I). However in case of *CG*, the difference is drastic, power dissipation for v_L=128 is almost 3 x higher than for 16. It is due to much better exploitation of the multiplier, so the clock-gating signal is active less often than for v_L=16. The graph for *100%* is not shown due to space constraints. It has the same shape as *noCG* one, but it is in average 5-10% higher.

The reduction of power achieved with clock-gating (*CG* vs. *noCG*) is in average around 5 x and 2 x for v_L=16 and 128 respectively. The reduction strongly depends on the multiplier usage so they vary across the *uBench*s. They are the highest for *Feature Tracking*, the lowest for *Disparity map*, while the savings for *Maximally Stable Regions* are in between and they are close to the average savings for all the three *uBench*s.

Power dissipation depends largely on n_S. Despite the glitch reduction achieved with pipelining, the overhead of adding registers is high. Therefore, for a given frequency, the lowest power configuration is almost always the one with the fewer stages. As a result of their topologies and critical paths, *ar* requires higher n_S than *wl* to achieve the same f. This is one of the reasons why *wl* in most of the cases has significantly lower power than *ar*.

The staircase shape of $P(f)$ graphs (Figure 3) is a result of pipelining. When we want to make a multiplier able to work on a higher frequency, we first try to achieve this with circuit optimization techniques (buffering, resizing, etc.). In this case, power increases approximately linearly with frequency. However, when it is not possible to further increase frequency with circuit optimization techniques, we apply pipelining (increase n_S). As a result we obtain these high gradient increases.

While for low frequencies *wl* and *ar* perform similarly, on the rest of the explored f scale, *wl* in most cases performs better. The only exception is *0%* where *ar* has lower power for $f > 1.45$. Thus, we can consider *ar* as a suitable choice for highly memory bound workloads when clock-gating is active most of the time, while for almost all other scenarios *wl* has lower power.

From Table III we can observe that geometrical mean (GM) of *wl*/*ar* power ratio differs among *tBench*s: GM(*100%*) < GM(*noCG*) < GM(*CG*) < GM(*0%*). There are two reasons for this: (1) Switching activity of *tBench*s. When switching activity of a *tBench* is high, P_{rest} is the dominant power component. P_{rest} is proportional to the area of the multiplier. (2) Clock-gating. The more clock-gating is active, the more significant P_{ct} is. When clock-gating is active most of the time, total P is practically defined by P_{ct}. In this case, designs with regular layout (thus less complicated clock-tree) perform better in terms of power. Thus, area inefficient and regular layout designs (*ar*) can compete in power efficiency with low area and irregular layout ones (*wl*) only for low activity *tBench*s, when clock-gating is active most of the time.

IV. MULTI-LANE EFFECTIVENESS

In order to fully understand the results expressed via Pareto-optimal trade-off graphs (Section V) we analyze the effectiveness of using multiple lanes, showing their behavior on the measured frequency range using averaged results of all *uBench*s. For a given combination of v_L, *tBench* and *f*, we pick the configuration with the lowest power for each n_L. The execution times for all *uBench*s are normalized before averaging in order to assure the correctness of averaged results. We comment on *uBench*s' results separately when there are interesting differences between them.

Table IV presents the ratios of power dissipated by *2-* and *4-lane* configurations over *1-lane* power. For the sake of simplicity, we choose two points in the frequency range explored, close to the extremes: 0.4 GHz and 1.6 GHz. Generally, there are three factors that define VMU power, and power ratios in case of vector multi-lane: (1) the amount of parallel multipliers (n_L), (2) switching activity factor of data at the VMU inputs (α_I), and (3) the usage of VMU. While the effect of the first one is obvious, the last two factors affect VMU power in less obvious ways, thus, require deeper analysis.

As data of consecutive elements inside a vector is typically correlated, α_I plays a more important role in vector than in scalar processing. The effect of α_I is clearly observable for *100%* where VMU usage is always the same (no VMU idleness). With multi-laning we actually slice a vector into n_L shorter vectors decreasing in that way correlation between data inside the vector (i.e. increasing α_I). Therefore, although we increase the amount of hardware 4 times, we actually increase power 5 times. Shorter vector lengths also produce some increase of α_I, therefore ratios are slightly higher for v_L=16 than for 128. We observe that power dependency on α_I decreases with frequency. The reason is reduced glitching. For high frequencies multiplier with more pipeline stages are used, and pipelining reduces glitching[24].

The notable increase of α_I with the increase of n_L is a consequence of the characteristics of image data. Neighboring pixels often have similar values. When processing them as a vector, this results in high correlation between consecutive values seen by the VMU, which translates into low α_I, even for short vectors. However, if data is processed on the multiple lanes, elements seen by one lane are interleaved, i.e. they are not neighbors anymore. For this reason, there is the aforementioned notable increase of α_I (and consequently power), in designs with multiple lanes.

The usage effects are most visible comparing the ratios of power dissipated for low and high frequencies in case of *CG*. The ratios are significantly lower for higher frequencies as for high frequencies the memory system becomes a bottleneck. Adding more lanes results in more idle cycles which causes clock-gating to be more active, thus average power is lower.

Power ratios for *CG* and noCG are notably higher for v_L=128 than for 16. As it has been already mentioned in Section III, in case of short v_L (16) we have lower average power as the VMU is more time idle. Adding more lanes means that the "effective" vector length is actually v_L/n_L which augments this effect. This is especially visible for *CG* where during idle cycles, the VMU is clock-gated.

Table V shows the speed-up obtained with 2 and 4 lanes over a *1-lane* VMU. We have chosen the same frequency

Table IV
POWER DISSIPATION RATIO OF 2- AND 4- OVER 1-LANE VMU

f[GHz]	0.4				1.6			
n_L	2		4		2		4	
v_L	16	128	16	128	16	128	16	128
noCG	1.93	2.19	3.73	4.5	2.19	2.27	4.36	4.5
CG	1.53	2.52	2.31	5.32	1.36	1.69	1.81	3.25
100%	2.24	2.24	5.02	5.01	2.40	2.31	4.96	4.77

Table V
SPEED-UP OF 2- AND 4- OVER 1-LANE VMU

f[GHz]	0.4				1.6			
n_L	2		4		2		4	
v_L	16	128	16	128	16	128	16	128
CG, noCG	1.23	1.74	1.43	2.59	1.16	1.48	1.31	1.80
100%	2	2	4	4	2	2	4	4

points as in the previous example: 0.4 GHz and 1.6 GHz. As can be expected from Amdahl's law, doubling the number of lanes provides exactly 2 x speedup for the *100%*, since there are multiplications every cycle. *CG* and *noCG* have practically the same results since the presence or absence of clock gating at a given frequency does not affect the execution time as long as they have the same n_S. In these two cases, the speed-up decreases pretty linearly with frequency. As a consequence, increasing the number of lanes is more effective for VMUs operating at lower frequencies (low power region). The reason is that in these *tBench*s memory access time is modeled (unlike in *100%*), and main memory access latency translates into more cycles when the processor operates at higher frequencies. Moreover, cache misses are more difficult to hide when the v_L is short, so speed-up degrades with frequency to a higher extent in this case.

EDP characteristics of multiple-lane VMUs are shown on Figure 4 for *CG* and both v_L=16 and 128. While for v_L=128 2- and *4-lane* VMUs are better solutions than *1-lane* in terms of EDP, for v_L=16 *1-* and *2-lane* VMUs are the most efficient in terms of EDP. Regarding the rest of PD^nP metrics, due to the lack of space, we do not present the figures but we highlight the most interesting observations. When considering energy (PDP), 1-lane outperforms 2- and 4-lane configurations for all *tBench*s. However, when we put the accent on the performance and consider ED^2P, the *4-lane* VMU performs the best for all v_L and *tBench*s. The $(f, ED^nP)(n>1)$ graphs have the minimum around 0.63 GHz and, as can be seen in Figure 4, this is actually the most energy-efficient operating region. Results for PD^nP are functions of P and t_e, thus they are directly explained with the discussion above. Regarding per *uBench* results, we have observed that the results for *Disparity map* are close to the average ones, while other two *uBench*s differ to some extent. *Feature Tracking* takes the most benefit of vector multi-lane being *4-lane* VMU more often the most efficient choice, while for *Maximally Stable Regions* the situation is

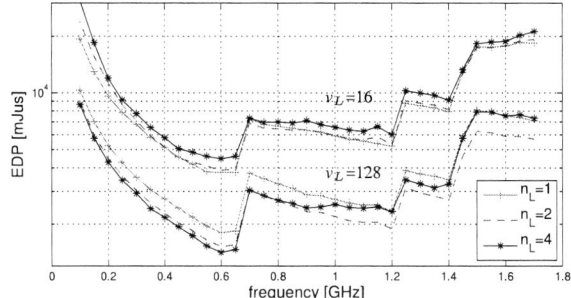

Figure 4. *EDP* of *1-,2-*, and *4-lane* VMU for *CG*.

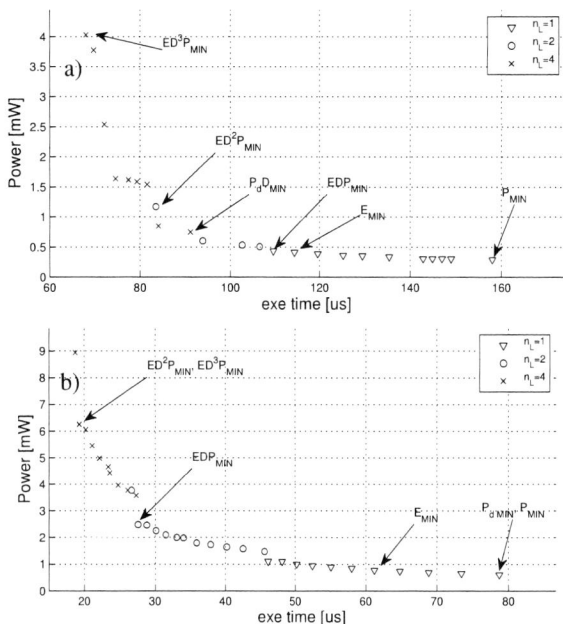

Figure 5. Pareto optimal trade-off graphs in $P - D$ space for (a) $v_L = 16$, $f_{range} = (0.1\,GHz - 1.0\,GHz)$ and (b) $v_L = 128$, $f_{range} = (0.7\,GHz - 1.7\,GHz)$.

the opposite and 1-lane performs better than in averaged results.

V. VMU Design Guidelines

This section presents guidelines on efficient VMU design for two groups of the targeted low power processors. The first group represent low power embedded processors with short v_L (16) that operate inside the lower frequency range (0.1 GHz-1.0 GHz). Embedded processors typically have restricted power budgets, therefore, they cannot afford long vector registers neither high frequencies. Examples of this market are ARM Cortex-Mx processors. The second group assumes low power mobile and low power throughput-oriented server processors with long v_L (128) that operate inside the higher frequency range (0.7 GHz-1.7 GHz). Mobile and server processors are expected to take full benefit of their workloads (that exhibit much DLP), thus they can afford long vector registers in order to achieve energy efficiency. Typical examples of this group are ARM Cortex-Ax and Hwacha [5] for mobile, and ARM-based servers [25] for server market. In this section we consider only VMUs with clock gating support as it is the most relevant from a low power implementation perspective. Additionally, we assume averaged results of all *uBenchs*.

Pareto optimal design points are shown on Figure 5, for both groups of processors in the Power−Execution time $(P - D)$ trade off design space. The most efficient designs for each metric of interest are listed in Table VI. Design points that perform the best for particular metrics of interest are highlighted in both Table VI and Figure 5 using bold font in the output metric column and arrows respectively. Pareto optimal design points for a given processor group are selected by filtering all design points with the selection parameters.

When we emphasize performance (ED^nP, $n>1$) vector multi-lane turns to be the most adequate choice. We observe it is an effective way to achieve energy-efficient speed-up, especially for $v_L=128$ due to its fair DLP exploitation. Vector multi-lane turns to be a more effective approach for fast and low power VMU design than frequency increase. Additionally, it also an efficient choice for achieving a

thermal-efficient speed-up (low P_dDP). Consequently, high n_S and high frequency design points do not appear as optimal design points for a given metric of interest. Therefore, *1-* and *2-stage* designs are always the best choices.

As it is expected from Figure 4, there are two efficient frequency points: 0.65 GHz and 1.2 GHz. After these break-even points, adding more lanes becomes a more effective approach than what has been increasing frequency. Regarding particular *MF*, as a consequence of discussed in Section III, *wl* is practically always more efficient than *ar* except when we look for ultra low power embedded processors market.

VI. Conclusions

We presented a joint circuit-system analysis, consisting of various simulators and tools, which tackles the problem of design of optimal VMU. We explored VMU design space in a comprehensive way including both circuit (e.g. clock-gating, multiplier family) and architectural parameters (e.g. number of vector lanes, maximum vector length) and showed how they direct optimal VMU configuration. All VMU design points were implemented and verified using physical EDA tools for a 40nm low power technology and simulated using the San Diego Vision Benchmark suite as input benchmark set. We found that Wallace multiplier topology perform the best for almost all the cases due to its area and timing efficiency. The exceptions are very low power design and memory-bound systems where Carry-Save Array outperform Wallace multipliers thanks to its

Table VI
SUMMARY OF OPTIMAL VMU CONFIGURATIONS

Selection Parameters			VMU Configuration				Output Metrics						
f_{range}[GHz]	v_L	Metric	n_L	AF	n_S	f[GHz]	t_e[μs]	P[mW]	E[mJ]	EDP[mJμs]	ED^2P[mJμs^2]	A[μm^2]	P_d[mW/m^2]
(0.1-1.0)	16	$P_{d\ min}$	1	wl	1	0.1	920.1	23.22x10^{-3}	21.36	19.66x10^3	18.09x10^6	6544	**3.55**
		P_dDP_{min}	4	wl	1	0.65	**125.6**	0.2924	36.73	4612	57.92x10^3	31.85x10^3	**9.18**
		P_{min}	1	ar	1	0.1	920.1	**22.83x10^{-3}**	21.01	19.33x10^3	17.78x10^6	6034	3.78
		E_{min}	1	wl	1	0.55	189.9	0.1058	**20.09**	3815	724.5x10^3	6707	15.77
		EDP_{min}	1	wl	1	0.6	181.1	0.1154	20.9	**3785**	685.5x10^3	7076	16.31
		ED^2P_{min}	2	wl	1	0.65	140.8	0.1993	28.05	3949	**555.9x10^3**	15.93x10^3	12.51
		ED^3P_{min}	4	wl	2	1	**98.83**	0.6736	66.57	6579	**650.2x10^3**	29.58x10^3	22.77
(0.7-1.7)	128	$P_{d\ min}$ P_{min}	1	wl	2	0.7	78.7	**0.6059**	47.69	3753	295.4x10^3	7224	**83.88**
		$P_dD_{min}P$	4	wl	2	0.7	**31.59**	3.055	96.5	3048	96.29x10^3	28.9x10^3	**105.7**
		E_{min}	1	wl	2	0.9	61.19	0.772	**47.24**	2891	176.9x10^3	7269	106.2
		EDP_{min}	2	wl	2	1.2	27.61	2.489	68.74	**1898**	52.42x10^3	17.29x10^3	144
		ED^2P_{min} ED^3P_{min}	4	wl	2	1.2	**19.32**	6.257	120.9	2336	**45.13x10^3**	34.58x10^3	180.9

regular interconnects. We showed that vector multi-lane approach beats increasing frequency as a measure to achieve energy- and thermal-efficient speed-up, especially for long vector lengths. Additionally, we analyzed the importance of considering the correlation between vector elements and VMU usage when using multiple vector lanes. Finally, we provided guidelines on optimal VMU design by highlighting the most efficient design points for all the metrics of interest for both low- and high-end low power vector processors. For instance, we observe that *2-lane, 2-stage wl* VMU running at 1.2 GHz performs the best in terms of *EDP* for the low power mobile market.

ACKNOWLEDGMENT

This work is supported in part by the Spanish Government under contract TIN2012-34557 and the European Unions Seventh Framework Programme under the ParaDIME project (GA no. 318693)

REFERENCES

[1] S. Borkar and A. A. Chien, "The future of microprocessors," *Commun. ACM*, vol. 54, no. 5, pp. 67–77, may 2011.
[2] K. Asanović, "Vector microprocessor," *PhD Thesis*, 1998.
[3] C. Lemuet *et al.*, "The potential energy efficiency of vector acceleration," in *SC '06*, nov., p. 1.
[4] Y. Lee *et al.*, "Exploring the tradeoffs between programmability and efficiency in data-parallel accelerators," in *ISCA'11*.
[5] ——, "A 45nm 1.3 ghz 16.7 double-precision gflops/w risc-v processor with vector accelerators," in *ESSCIRC 2014*, pp. 199–202.
[6] R. Espasa *et al.*, "Vector architectures: past, present and future," in *ISC1998*.
[7] Y. Liu and S. Furber, "The design of a low power asynchronous multiplier," in *ISLPED*, 2004, pp. 301–306.
[8] Z. Huang and M. D. Ercegovac, "High-performance low-power left-to-right array multiplier design," *IEEE TC*, vol. 54, no. 3, pp. 272–283, 2005.
[9] I. Ratković *et al.*, "On the selection of adder unit in energy efficient vector processing," in *ISQED*, 2013.

[10] S. K. Venkata *et al.*, "Sd-vbs: The san diego vision benchmark suite," in *IISWC*, 2009, pp. 55–64.
[11] T. K. Callaway and E. E. Swartzlander Jr, "Power-delay characteristics of cmos multipliers," in *ARITH'97*.
[12] W. J. Townsend *et al.*, "A comparison of dadda and wallace multiplier delays," in *SPIE*, 2003, pp. 552–560.
[13] A. Erdogan *et al.*, "Architectural trade-offs in the design of low power fir filtering cores," in *Circuits, Devices and Systems, IEEE*, vol. 151, no. 1, 2004, pp. 10–17.
[14] S. Gailhard *et al.*, "How to transform an architectural synthesis tool for low power vlsi designs," in *GLSVLSI'98*.
[15] M. Monchiero *et al.*, "Design space exploration for multicore architectures: a power/performance/thermal view," in *ICS '06*, pp. 177–186.
[16] M. Stanic *et al.*, "Valib and simplevector: tools for rapid initial research on vector architectures," in *CF*, 2014, p. 7.
[17] R. Espasa *et al.*, "Tarantula: a vector extension to the Alpha architecture," in *ISCA 29*, 2002, pp. 281–292.
[18] J. Hennessy and D. Patterson, *Computer Architecture, Appendix G*. MK, 2011.
[19] B. Parhami, *Computer Arithmetic*. Oxford UP, 2000.
[20] C. S. Wallace, "A suggestion for a fast multiplier," *Electronic Computers, IEEE Transactions on*, no. 1, pp. 14–17, 1964.
[21] D. Baran *et al.*, "Energy-efficient implementation of parallel cmos multipliers with improved compressors," in *ISLPED*, 2010, pp. 147–152.
[22] I. Ratković *et al.*, "Physically vs. physically aware estimation flow: Case study of design space exploration of adders," in *ISVLSI*, 2014.
[23] H. Li *et al.*, "Deterministic clock gating for microprocessor power reduction," ser. HPCA '03, 2003, pp. 113–.
[24] V. Srinivasan *et al.*, "Optimizing pipelines for power and performance," in *MICRO 35*, pp. 333–344.
[25] (2015) Euro server. http:www.euroserver-project.eu/.

2015 IEEE Computer Society Annual Symposium on VLSI

A Fine-Grained, Uniform, Energy-Efficient Delay Element for FD-SOI Technologies

Ajay Singhvi[*], Matheus T. Moreira[‡†‖], Ramy N. Tadros[†], Ney L. V. Calazans[‡‖], Peter A. Beerel[†§]

[*]Birla Institute of Technology and Science Pilani, Pilani Campus - Pilani, India.
[†]University of Southern California (USC) - Los Angeles, United States
[‡]Pontifícia Universidade Católica do Rio Grande do Sul (PUCRS) - Porto Alegre, Brazil
ajaysinghvi93@gmail.com {matheus.moreira, ney.calazans}@pucrs.br, {rtadros, pabeerel}@usc.edu

Abstract—**Contemporary digitally controlled delay elements trade off power overheads and delay quantization error. This paper proposes a new delay element that provides a balanced design that yields low power with low delay quantization error. The proposed element has a quasi linear delay characteristic, with uniform delay differences between adjacent codewords. The element employs and leverages the advantages offered by a 28nm FD-SOI technology, using its back body biasing feature to add an extra dimension to its programmability. To do so, a novel generic delay shift block is proposed, which enables incorporating both fine and coarse delays in a single delay element that can be easily integrated into digital systems, an advantage over hybrid delay elements that rely on analog design.**

I. INTRODUCTION

Delay Elements (DEs) are used in a variety of applications in VLSI systems and are typically employed to provide precise timing control and/or satisfy timing constraints. In synchronous systems, DEs support clock distribution and synchronization across different blocks, dealing with clock skew and jitter problems [1], [2]. Other uses include phase locked loops, digitally controlled oscillators [3], time-to-digital converters [4] and poly-phase clock generators [5]. DEs are also widely used in bundled data (BD) asynchronous systems, to control the timing of request and acknowledge signals between different blocks [6]. For some of these applications, like control circuits of 2-phase BD asynchronous designs, DEs require balanced rise and fall delays [6], [7]. Moreover, a typical concern in the design of DEs in modern technologies are the effects of process, voltage and temperature (PVT) variations. To account for those, DEs must be conservatively designed to have extra timing margins that can compromise performance. The alternative is to use programmable DEs.

Programmable DEs alleviate the detrimental effects of PVT variations in deep sub-micron technologies by providing a range of attainable delays to which the DE can be tuned post-silicon. The delay granularity provided by programmable DEs is an important concern. For instance, systems that require precise timing control, such as phase shift compensators [8], timing generators [9] and timing verniers [10], used for delay fault testing in automatic testing equipment, employ fine-grained DEs to ensure correct operation. In essence, the precision to which these DEs can be tuned affects the amount of timing margin they can effectively avoid. DEs can be controlled by either analog voltages (or currents), or digitally. Traditionally, analog-controlled DEs provide fine delay tuning, while digitally-controlled DEs provide coarse-grained delays, with their combination forming hybrid DEs. However, since this work deals primarily with low-power applications and high-performance digital VLSI circuits, the use of hybrid DEs is not considered, to avoid the high power consumption of the required analog circuitry, the switching noise at high

frequencies, and the challenges in the distribution of global analog signals in predominantly digital systems [6].

In contrast, this paper proposes a new circuit architecture and the use of FD-SOI technology to design fine-grained programmable DEs with balanced rise and fall delays. Section II discusses the state of the art in digitally-controlled DEs. Next, section III explains the design of the proposed DE to provide a quasi linear and monotonic delay characteristic, reducing the delay quantization error (DQE) to 12.57% from 269.92% presented by a state of the art DE. It also proposes the architecture of the delay shift inverter (DSI), which utilizes the FD-SOI back-body biasing feature [11], to provide fine-grained delays in a single DE structure that can be easily incorporated into digital systems without any of the problems posed by hybrid DEs. Section IV then discusses the methodology adopted for optimizing power consumption of the proposed design, resulting in significantly lower energy consumption when compared to existing DEs [12], [13]. Section V presents and discusses our experimental results, while Section VI draws a set of conclusions.

II. DIGITALLY-CONTROLLED DEs

Different digitally-controlled DE architectures exist in the literature, exploring trade-offs in terms of delay range, power consumption, and area. Some of the existing DE topologies are the thyristor-based [14], the transmission gate-based [15], the current starved [12] and the cascaded inverter-based [16] designs. Among these, thyristor-based designs provide delays in ranges from μs to ms. This is beyond the scope of this paper, which focuses on DEs that provide shorter delay ranges (from ps to a few ns). Moreover, it is difficult to control both rise and fall transitions in thyristor-based designs. As for the transmission gate-based DE, it suffers from poor signal integrity and modifications that alleviate the problem [15] add significant costs in terms of area and power.

Therefore, the focus here is on current starved and cascaded inverter-based designs. Attention is put on the directly-controlled current starved DE (DCCS-DE), analyzed in [12] and shown in Fig. 1. This DE falls under the category of current starved inverter (CSI)-based DEs, where current source transistors determine the current through an inverter. These transistors reduce the current through the inverter, or starve it, thereby increasing the delay of a signal propagating through it. This DE, as analyzed in [12] displays a few drawbacks, including a non-monotonic relationship between delay and the associated codewords which makes it difficult to predict the delay for a given codeword. Another problem is the non-uniform delay difference between successive codewords. Such non-linearity translates into large delay quantization error (DQE), which is problematic when one requires a delay not provided by any codeword.

In [6] the authors propose a modified DCCS-DE design to allow balanced rise and fall delays, which is beneficial

‖ PUCRS authors acknowledge the support of CNPq (grants 401839/2013-3 and 312556/2014-4) and CAPES (grant 2129/14-0).
§ Peter A. Beerel is also a Chief Scientist, Technology Development at Intel, Calabasas, CA 91302.

978-1-4799-8720-7/15 $31.00 © 2015 IEEE

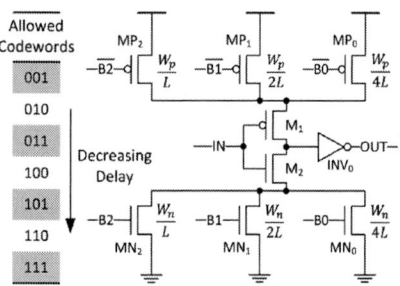

Fig. 1. The directly controlled current starved DE (DCCS-DE) [12].

for 2-phase BD asynchronous circuits. The modified design comprises two replicated DCCS-DEs in series, with signal conditioning inverters added to their inputs, to provide an acceptable slew rate, and inverters at their outputs, to provide the same load to each of the replicated DCCS-DEs. Unfortunately, the modified DCCS-DE still exhibits the other disadvantages of the original DE. Maymandi-Nejad and Sachdev proposed in [13] an alternative to cope with this problem: a programmable current mirror-based current starved DE (CMCS-DE) which has a linear delay characteristic. However, the CMCS-DE suffers from very large static power consumption, as discussed in [6], and cannot be employed in low power applications. In fact, a significant advantage of the DCCS-DE is its better energy efficiency when compared to other DE architectures, as [6] discusses, with its static power consumption being three orders of magnitude lower than that of the CMCS-DE.

The multiplexer-based DE (MUX-DE), depicted in Fig. 2 is often used in designs. Its popularity arises from a relatively simple design that can be implemented using standard cells. It also presents a linear delay characteristic. The codeword provided to the MUXes fixes the number of inverters in the signal path and hence its delay. This paper uses the MUX-DE as a comparison baseline design, to analyze the impact of the proposed modifications on the energy efficiency of the DCCS-DE over the MUX-DE.

III. PROPOSED DESIGN OF FINE-GRAINED DEs

A. Reducing Delay Quantization Error

Since the expected delay of a DE is uniformly distributed across the codewords, the delay quantization error (DQE), cited in Section II, is defined as the percentage deviation from the expected delay difference between any two adjacent codewords. In particular, the DQE is the maximum DQE across all adjacent codewords. The DQE is a handy metric for programmable DEs, since it encompasses the features of monotonicity, uniform delay distribution across codewords and the ability to predict the amount of delay provided by a particular codeword. Formally we have:

$$DD_{expected} = \frac{Delay\ Range}{N-1}, \qquad (1)$$

$$DQE = \frac{max(|DD_{measured,i} - DD_{expected}|)}{DD_{expected}} * 100\%, \qquad (2)$$

where Delay Range is the delay difference between the minimum and maximum delay settings and DD refers to delay difference. $DD_{measured,i}$ is the delay difference between the ith and $(i+1)^{th}$ adjacent codewords as observed in simulations, and $DD_{expected}$ is the ideal delay difference computed by (1). N is an integer representing the number of codewords that can be employed with a particular DE. A minimum DQE is required to enable the DE to be used efficiently across all codewords and possible delay values.

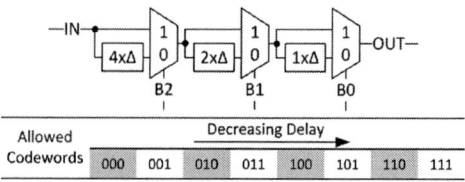

Fig. 2. The multiplexer based DE.

As previously mentioned, both the DCCS design of [12] and the modified one in [6] have unpredictable and non-monotonic delay behavior, which results in a large DQE. To minimize DQE while taking into consideration low power and high density results in state-of-the-art DEs, a new version of the DCCS-DE is proposed herein, based on the design presented on [6]. The new architecture, illustrated in Fig. 3, uses a one-hot code scheme instead of the binary codes used in [12] and [6]. This imposes the constraint that for a particular codeword, only one of the current source transistors is ON. In moving from a binary to one-hot code, the lengths of the current source transistors were altered to increase linearly (1L, 2L, 3L, 4L, ..., nL), instead of exponentially (1L, 2L, 4L, 8L, ..., $2^{(n-1)}$L), where n is the number of current source transistors, chosen on the basis of the amount of delay needed. The above changes ensure a constant delay difference between any two adjacent codewords, thereby minimizing DQE. This can be demonstrated mathematically as follows:

$$t_{pd} = C_L \frac{V_{ds}}{I_{ds}} \text{ and } I_{ds}\ \alpha\ \frac{1}{R_{ds}}, \qquad (3)$$

with

$$R_{ds}\ \alpha\ \text{L} \implies t_{pd}\ \alpha\ \text{L}. \qquad (4)$$

Thus, as L increases linearly for different codewords, the delay also increases linearly. This is different from the binary scheme used in previous works, where multiple parallel current source transistors could simultaneously be active. This implies summing currents together, which produces a non-linear relation between the total current and the codewords, and hence results in a non-linear delay behavior.

Regarding the MUX-DE, the design proposed in [6] uses a sum-of-product MUX implementation. This has an intrinsically linear delay behavior, because changes in the number of cascaded inverters from one codeword to the next are constant, ensuring a low DQE. Note that the MUX-DE still utilizes a binary codeword, as opposed to the one-hot scheme employed in the proposed DCCS-DE design.

B. Allowing Fine-Grain Tuning

The UTBB (Ultra Thin Body and Buried oxide) FD-SOI technology provides devices with better performance, lower leakage, and several power management design techniques. The Si film (the FDSOI ultra thin body) is very thin so that the depletion region continues to its end. This fully depleted (FD) body results in devices with low sub-threshold slope and low drain-induced barrier lowering (DIBL) figures. Transistors are normally controlled by the high-κ metal gate, which is called the *front gate*. Also, due to the very small width of the ultra thin body and buried oxide (or box), applying a potential from the back-body (or the back gate) has a large influence on the transistor's threshold voltage [11]. This is what is called back-body biasing, or just body biasing. There are two ways to employ body biasing: (i) forward body biasing (FBB), which decreases the threshold voltage for a faster mode of operation; and (ii) reverse body biasing (RBB), which increases the threshold voltage and, hence decreases the leakage current for power management purposes.

978-1-4799-8720-7/15 $31.00 © 2015 IEEE

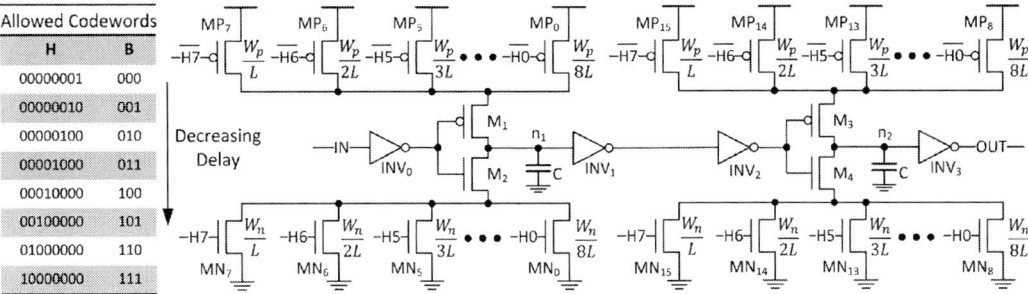

Allowed Codewords	
H	**B**
00000001	000
00000010	001
00000100	010
00001000	011
00010000	100
00100000	101
01000000	110
10000000	111

Fig. 3. The proposed DCCS-DE.

While body biasing is conventionally used either to reduce power consumption or to provide a performance boost, this paper employs it to provide fine-grained delay control. Therefore, we focus on RBB, as increasing the threshold voltage enables not only increasing the delay of transistors, but also reducing their leakage power, a side benefit to our techniques. But RBB is not applied to all transistors of the DE because each of them would get affected differently depending on its size. Moreover, this adds to the complexity of the design and, hence, the delay characteristic can change significantly. Also, it would increase the load that the bias voltage generating circuitry has to drive, resulting in more power consumption. Therefore, instead of employing RBB in each separate transistor, we propose the use of a Delay Shift Inverter (DSI) as shown in Fig. 4. The DSI is a conventional CMOS inverter with a programmable back body voltage that adjusts the threshold voltage of the inverter transistors, altering the current flowing through the inverter, changing its delay. Under normal operating conditions, the back-gate of the inverter pMOS transistor is connected to the core supply, while the back-gate of the nMOS is connected to ground. As illustrated in Fig. 4, depending on availability, additional body biasing voltages can be applied to (a) both pMOS and nMOS (b) only pMOS (c) only nMOS transistors. The delay shift provided by the DSI depends on two factors: (i) the change in the back body voltage, and (ii) transistors size. The number of delay shifts can be increased by additional body biasing voltages or by using differently sized DSIs. Section V explores this further.

DSIs can be easily incorporated into any existing DE architecture, as Fig. 5 illustrates. The intrinsic rise and fall delay characteristic of the original DE can also be maintained by cascading two DSIs in series as shown in Fig. 5, with buffers used to provide identical loads to both DSIs. The novelty of using the DSI is thus threefold: (i) it does not alter the original delay characteristics; (ii) it leads to less overhead in terms of area, as compared to replicating the DE architectures to increase the delay range; and (iii) it can be applied to any DE architecture. Hence, it serves as a good candidate to cope with the problems of using hybrid DE architectures to achieve precise and fine-grained delays. Moreover, applying body biasing to specific inverters is well-suited to the proposed DCCS-DE, because the biasing can be directly applied to the existing signal conditioning inverters (INV0 and INV2) of the DCCS-DE design (Fig. 3), instead of using additional area- and power-expensive DSI blocks.

Despite its advantages, a complication for the DSI is the generation and control of voltages from a domain other than the core supply and ground. The two most practical solutions are level shifters and voltage charge pumps. A level shifter is a simple circuit that shifts an input signal from its voltage domain to the provided reference domain and is commonly used to interface off-chip and on-chip voltage domains. On

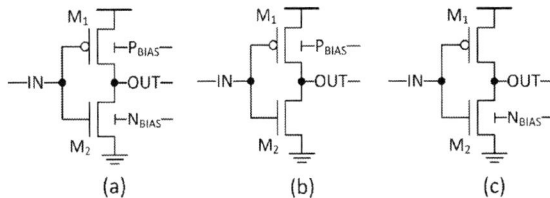

Fig. 4. Delay shift inverters (DSIs) with RBB applied to: (a) both pMOS and nMOS (b) only pMOS (c) only nMOS.

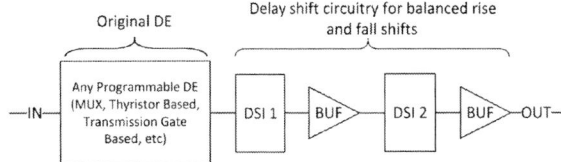

Fig. 5. The proposed architecture of fine-grained DEs.

the other hand, charge pumps use a complex arrangement of switching capacitors to pump charges up and generate higher voltage levels using only the input supply. Level shifters are small, simple, fast, and do not use any passive components. Charge pumps, on the other hand, consume significant amount of area due to the need of capacitors, power due to the need of switching clocks, and delay due to the time required to pump charges through several stages of capacitors. Also, its voltage output suffers from ripples [17]. The advantage of charge pumps is that they need no reference voltages.

Our target DSI, employs RBB only in the pMOS transistors (Fig. 4(b)), because the I/O voltage for the 28nm FD-SOI technology is 1.8V and it is easy to use this voltage internally to drive the pMOS RBB circuity. In particular, we propose to use level shifters to actively switch the pMOS back-body from the normal supply VDD=1V to the I/O voltage Vhigh=1.8V.

Fig. 6 shows our level shifter. Tran et al. [18] proposed the contention mitigated level shifter (CMLS), which was used in [19] for body biasing the LVT (flip-well) devices in FD-SOI. Since the devices used in the design of this paper's DEs are RVT (normal well), the low voltage connected to the CMOS inverter is VDD and not ground as in [19]. The circuit works as follows: when the input (IN) is low, M4 and M5 are on, while M3 and M6 are off. Then, the gate of M2 is discharged to ground, resulting in the charging of the input of INV1 to Vhigh and hence the output is only VDD. When the input (IN) is high, symmetrically the input of INV1 goes low, and hence the output is Vhigh. A conventional level shifter does not have transistors M3 and M4, but this entails a serious contention between the cross-coupled pMOS devices and the input coupling nMOS. Adding M3 and M4 reduces this contention and results in lower switching energy

978-1-4799-8720-7/15 $31.00 © 2015 IEEE   29

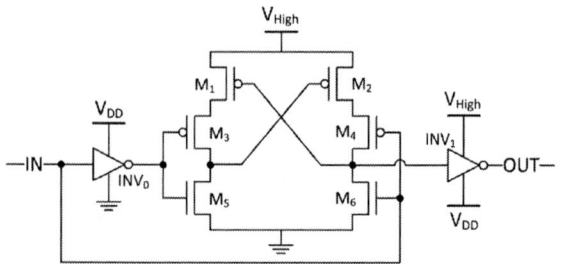

Fig. 6. The circuit diagram of the employed level shifter. Terminals shown across the CMOS inverters represent the connected high and low supply values.

and in faster switching. This is why this architecture is called CMLS [18]. It is worth mentioning that Vhigh=1.8V is the highest voltage value that can be used by such an architecture, because it results in 1.8V across the gates of the MOS devices, which is the maximum difference of potential allowed to avoid gate oxide breakdown. Regarding the overheads of the level shifter addition, leakage and area are relatively small, as is the switching delay, due to the use of contention mitigation.

IV. ENERGY AND LEAKAGE OPTIMIZATIONS

A. Energy Optimization

To minimize the DE energy consumption, an initial version was analyzed to determine the consumption in different parts of the circuit. Next, we redesigned the circuit in the most energy-efficient manner by optimizing each of these parts. As Equation (3) illustrates, for a given operating voltage the provided delay depends on two factors: the current through the CSI, and the output capacitive load, C_L. The former and the latter are controlled by the following parameters: *i)* the W and L of the current source transistors (MPx and MNx); *ii)* the W and L of the CSIs (Mx); *iii)* the external load capacitance, C; and *iv)* the input capacitance of the signal conditioning inverters. Parameters *i)-iv)* are to be tuned to get the required delay range, consuming the least energy per transition and having the lowest leakage.

To better understand the energy vs delay trade-offs for the above parameters, experiments were conducted in which each parameter was used to independently achieve a fixed delay range. The experiments show that increasing L of the current source transistors is the most energy-efficient manner to achieve the required delay range, as larger L results in lesser current and hence lesser energy. However, the maximum L that can be used is constrained by the layout rules of the technology and might not always be enough to get the desired delay range, especially for larger delays.

To increase the delay range, one may further decrease the current by increasing the L of the CSIs (parameter *ii)*, increase the output capacitance, by adding an external shunt capacitor (item *iii)*, or increase the size, and thus the input capacitance, of the signal conditioning inverters (parameter *iv)*. However, increasing L of the CSIs must be done conservatively, ensuring that the CSIs do not dominate the current source transistors by constraining the maximum current that can flow. This leaves two options, both implying the increase of the output capacitance. An added advantage of using parameters *iii)* and *iv)* is that these help mitigate the charge sharing problem present in the DCCS-DE from [12].

From experiments we conclude that increasing the input capacitance of the signal conditioning inverters yields the largest energy overhead. This is because increasing the input capacitance for these results in a slow slew rate, which in turn generates more short circuit current through the inverters,

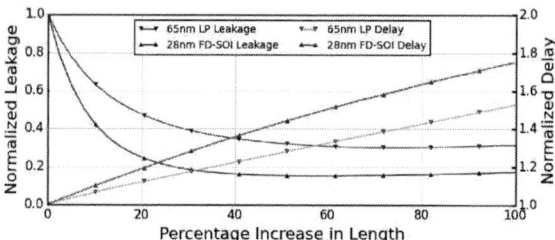

Fig. 7. Leakage vs delay trade-off in gate length biasing.

and hence leads to larger energy per transition (EPT). Thus, adding an external shunt capacitance at the output node is the preferred approach. In fact, an optimal combination of *i)* and *iii)* helps achieving the best energy-delay trade-off.

As for the MUX-DE, its delay range depends on two factors: *i)* the number of cascaded inverters; and *ii)* the W and L of these. The optimization of EPT for the MUX-DE is better done using approach *ii)*, i.e. increasing the lengths of the nMOS and pMOS transistors to meet the desired delay range, rather than adding more cascaded inverters. Approach *i)* is used only after reaching the maximum allowable L for a transistor, because a higher L would result in less current flowing through the inverters, increasing delay. Cascading smaller inverters results in additive current flowing through the DE, consuming more EPT.

B. Leakage Reduction

Gate length biasing [20] [21] is a promising technique for achieving substantial leakage reduction, and also requires no additional process steps. It involves increasing the length of the transistors to reduce leakage, at the cost of a delay increase. Gupta et al. [21] suggest a 10% upper bound on the increase in length to achieve the best trade-off for a bulk 130nm process. Experiments were run on an inverter in a bulk 65nm process as well as in the 28nm FD-SOI process to decide on a bound. The trade-off can be seen in Fig. 7, with greater reduction of leakage in the 28nm FD-SOI process as compared to the 65nm bulk CMOS technology, at the expense of an increase in delay.

This leakage vs delay trade-off is important when applying gate length biasing to transistors in the critical path, because of timing constraints that limit the increase in L to only 10%. However, the same limitation does not exist for DEs, because the delay range can always be tuned using other parameters like an external shunt capacitance or changing the number of DEs used to build the element. Thus, overlooking the 10% limitation advocated in [20], it can be observed from Fig. 7 that after roughly 40% increase in L, one obtains high leakage reduction, after which leakage reduction stagnates. Thus, while designing any of the DEs, the minimum L chosen is 40% greater than the technology's smallest L. Experiments on the DCCS-DE and the MUX-DE showed the same trend, with substantial leakage reduction.

V. EXPERIMENTS AND DISCUSSION

A 28nm FD-SOI CMOS technology with 1V supply was used for the DEs. All simulations employed the Cadence Spectre Simulator with a same environment across all designs, for fair comparisons. All simulations assumed an operating temperature of 27°C and typical process corners. DEs employed the techniques proposed in Section III as well as the power reduction optimizations of Section IV. Each DE was designed to have 8 different delay settings and provide an identical delay range.

Table I summarizes the trade-offs between DEs. The techniques from Section III-A significantly improve the delay

978-1-4799-8720-7/15 $31.00 © 2015 IEEE

TABLE I. TRADE-OFFS BETWEEN DEs FOR A 400PS RANGE.

DE	Original Binary DCCS-DE [6]	Proposed One-hot DCCS-DE	MUX-DE
DQE (%)	269.92	18.94	3.19
Avg. EPT (fJ)	0.73	1.01	2.71
Avg. Idle Power (nW)	0.30	0.12	0.28
Active Area (μm^2)	1.18	1.96	0.35

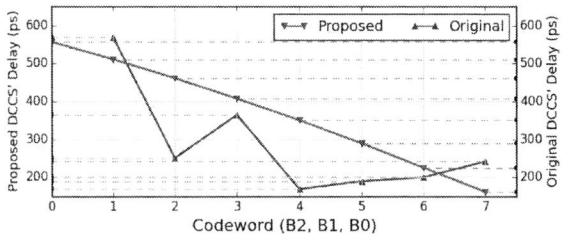

Fig. 8. Comparison of delay characteristics for the proposed DCCS-DE and the original DCCS-DE.

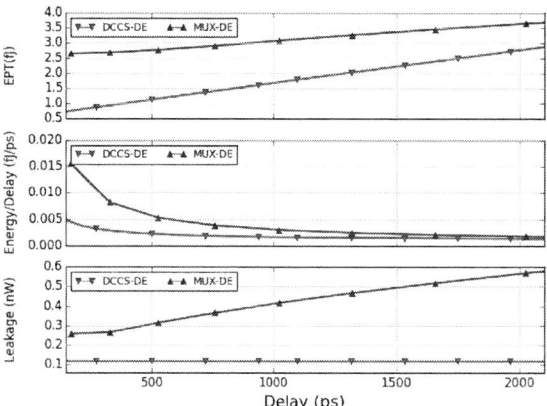

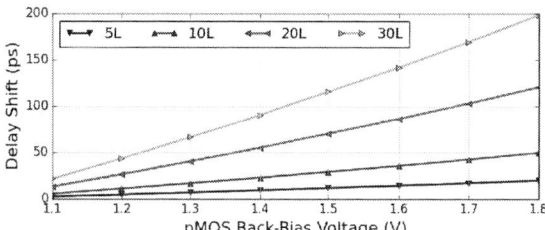

Fig. 9. Comparison of proposed DCCS-DE and MUX-DE: (a) EPT (b) $Energy/Delay$ (c) Leakage.

Fig. 10. Effect of L and body biasing voltages on delay shift.

characteristic of the DCCS-DE over the DCCS-DE proposed in [12] and [6]. Improvement is quantified using the definition of DQE in Equation 2. As Fig. 8 shows, the original DCCS-DE has a non-monotonic delay, which is problematic, as certain codewords might provide delays that are too close or too far from each other. This characteristic translates into a large DQE of 269.92%, making the original DCCS-DE unreliable for building a programmable DE. Note that the DQE for the original DCCS-DE was calculated after re-ordering the codewords, to provide a monotonically increasing delay characteristic; still, it presented high DQE. On the other hand, the proposed DCCS-DE has an almost linear delay characteristic, with nearly uniform delay difference between codewords, and does not require codeword reordering. This uniform delay difference enabled a much smaller DQE, 18.94%. Moreover, this DQE improvement comes without significant power overhead. The active area values in Table I are the sum of the W*L of all the transistors in the design. Furthermore, to make the comparison between the proposed and original DEs more pessimistic, any area and power overheads of the circuitry needed to re-order the original DE codewords are not considered when presenting results for the original binary DCCS-DE.

As Table I shows, the MUX-DE displays a better DQE of 3.19% as opposed to 18.94% of the proposed DCCS. This is due to the fact that the technique presented in Section III-A does not take non-idealities into account. For the 400ps delay range programmed using 8 codewords, this translates to an absolute error of 1.82ps, while the proposed DCCS-DE has a max deviation of 10.08ps from the ideal characteristic of having a uniform delay difference of $400/7 ps$ between adjacent codewords. However, with the aforementioned technique as a basis, the DQE achieved by the DCCS-DE can be improved by iteratively adjusting the L's of those current source transistors that contribute to the larger DQE. Moreover, as discussed later in this Section, the fine-graining technique proposed here further improves the DQE, and any issues arising due to minor deviations from the ideal characteristic can also be alleviated. On the other hand, the proposed DCCS-DE still consumes 2.68 times less energy than the MUX-DE.

The metric used for comparing energy efficiency is the average energy per transition for all codewords measured for a particular delay range. As Fig. 9(a) shows, the MUX-DE consumes nearly five times more energy than the proposed DCCS-DE for small delay ranges, due to more current being drawn by the cascaded inverters in the MUX-DE than the CSIs of the DCCS-DE. The disparity decreases as L of the cascaded

inverters increases to improve the delay range of the MUX-DE, with the energy advantage of the proposed DCCS-DE reducing by a factor of two for ranges larger than 2ns.

To better understand the relationship between delay range and EPT, Fig. 9(b) shows the $Energy/Delay$ relation between DEs. The results are consistent with the above discussion, as for delay ranges bigger than 2ns the energy spent per unit of delay becomes nearly equal for both. Next, the DEs idle power is compared. Leakage reduction is achieved for both DEs using the gate length biasing strategy from Section IV-B. As can be seen from Fig. 9(c), the DCCS-DE has a very low leakage power consumption of 0.12nW, which remains constant across delay ranges, due to the fact that the extended delay ranges were met using external shunt capacitors rather than more transistors. On the other hand, it was noticed that the MUX-DE has substantially higher leakage power consumption when compared to the DCCS-DE. This is attributable to the large transistor count of the MUX-DE, compared to the DCCS-DE.

The next set of experiments target enabling a fine-grained delay range of 400ps for the DCCS-DE and MUX-DE, In other words, the idea is to reduce the delay difference between two adjacent delay settings. As Section III-B mentioned, the amount of delay shift achieved by the DSI shown in Fig. 4 is controlled by the size as well as the magnitude of the additional body biasing voltage. Experiments were run to determine the optimal sizing and voltage. As Fig. 10 shows, the amount of delay shift increases as the body biasing voltage or the length of the transistor increases. Depending on the delay range and the application, the appropriate number and magnitude of body biasing voltages and transistor sizes can be chosen.

For the reasons elaborated in Section III-B, only an additional body biasing voltage of 1.8V is generated, using the contention mitigated level shifter shown in Fig. 6 to add an extra dimension of programmability. Moreover, for this application,

978-1-4799-8720-7/15 $31.00 © 2015 IEEE

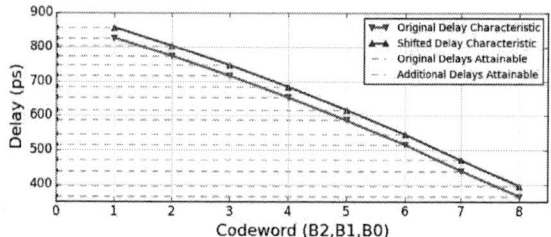

Fig. 11. Fine-grained delay characteristic.

TABLE II. TRADE-OFFS BETWEEN FINE-GRAINED DEs FOR A 400PS RANGE.

DE	One-hot DCCS-DE with 16 current sources	One-hot DCCS-DE with 8 current sources + DSI	MUX-DE with 15 buffers
DQE (%)	26.81	12.57	4.55
Avg. EPT (fJ)	1.03	1.57	5.08
Avg. Idle Power (nW)	0.22	0.16	0.40
Active Area (μm^2)	3.72	3.42	0.28

only the length of the DSI transistors was increased, as it would be the more energy-efficient solution. For a 400ps delay range, across eight codewords, the normal delay difference between each codeword would be $400/7\,ps \approx 57\,ps$. Thus, the length of the transistor chosen is one which corresponds to a delay shift of $400/14\,ps \approx 29\,ps$ for each codeword. The final fine-grained delay characteristic for the DCCS-DE can be seen in Fig. 11. Similar results were also observed for the MUX-DE. As the figure shows, the addition of a single body biasing voltage level doubles the resolution of the discrete delays offered by the DE. In the above experiment a delay step of $\approx 29ps$ is achieved as one moves from one setting to the next. Additional body biasing voltages can be used to further reduce the delay step size and make the achieved delay granularity finer.

In order to study the effect of using the DSI on the DQE, experiments were conducted on the DCCS-DE and MUX-DE. To ensure a fair comparison, the fine-grained structure was implemented in two flavors: one with the DSI and another without it. The two finer grained DCCS-DE designs were implemented by: (a) using 16 current source transistors sized as (1L, 1.5L, 2L, ... , 7.5L, 8L) instead of the original eight sized as (1L, 2L, ... , 7L, 8L); and (b) body biasing the signal conditioning inverters INV0 and INV2 of Fig 3 while still using only eight current source transistors. The MUX-DE design was re-implemented to have 16 codewords instead of the original eight (Fig. 2),so as to have a fair comparison and reduce the delay difference between adjacent codewords. The comparison of these designs appears in Table II.

As Table II shows, using the DSI with the DCCS-DE enables a DQE of 12.57%, which is less than half of that achieved when using additional current source transistors. Compared to the DCCS-DE with sixteen current source transistors, the one with the DSI has lower area and also consumes 3.23 times less energy than the fine-grained MUX-DE, further improving the energy efficiency of the DCCS-DE over the MUX-DE that was observed in Table I. Thus, adding the DSI to the DCCS not only presents a better DQE but also does not result in excessive overheads.

VI. CONCLUSION

This work presents and analyzes design modifications to the DCCS-DE. The proposed design has a linear monotonic delay behavior and low DQE, which is a considerable improvement over the previously discussed DCCS-DEs in [12] and

[6]. Additionally, the proposed DCCS-DE is significantly more energy-efficient than the current mirror based design proposed in [12] and [13]. And, it consumes less energy than the MUX-DE for delay ranges smaller than 2ns. The paper also proposes a generic DSI architecture, which utilizes the body biasing feature in 28nm UTBB FD-SOI technology to obtain fine-grained delays in a single DE structure, allowing the proposed architecture to be easily integrated into digital systems. This DSI architecture can be used to further improve the DQE of the delay element. Such advances enable leveraging the advantages of UTBB FD-SOI technologies for circuit design, and allow better design space exploration for applications that need low power DEs.

REFERENCES

[1] A. Chakraborty *et al.*, "Dynamic Thermal Clock Skew Compensation using Tunable Delay Buffers," *IEEE Trans. on VLSI Syst.*, vol. 16, no. 6, pp. 639–649, Jun. 2008.

[2] Y.-J. Jung *et al.*, "A dual-loop delay-locked loop using multiple voltage-controlled delay lines," *IEEE JSSC*, vol. 36, no. 5, pp. 784–791, May 2001.

[3] B.-M. Moon *et al.*, "Monotonic Wide-Range Digitally Controlled Oscillator Compensated for Supply Voltage Variation," *IEEE Trans. on Circ. and Syst. II: Express Briefs*, vol. 55, no. 10, pp. 1036–1040, Oct. 2008.

[4] G. Li *et al.*, "A high resolution time-to-digital converter using two-level vernier delay line technique," in *NSS/MIC*, 2007, pp. 276–280.

[5] H. Lin *et al.*, "New four-phase generation circuits for low-voltage charge pumps ," in *ISCAS*, 2001, pp. 504–507.

[6] G. Heck *et al.*, "Analysis and Optimization of Programmable Delay Elements for 2-Phase Bundled-Data Circuits," in *VLSID*, 2015, pp. 321–326.

[7] P. Beerel *et al.*, *A Designer's Guide to Asynchronous VLSI*. Cambridge University Press, 2010.

[8] T. Dogsa *et al.*, "Precision Delay Circuit for Analog Quadrature Signals in Sin/Cos Encoders," *IEEE Trans. on Instr. and Meas.*, vol. 63, no. 12, pp. 2795–2803, May 2014.

[9] K. Ryu *et al.*, "All-digital process-variation-calibrated timing generator for ATE with 1.95-ps resolution and a maximum 1.2-GHz test rate," in *ESSCIRC*, 2013, pp. 41–44.

[10] B. Arkin *et al.*, "Realizing a production ATE custom processor and timing IC containing 400 independent low-power and high-linearity timing verniers," in *ISSCC*, 2004, pp. 348–349.

[11] B. Pelloux-Prayer *et al.*, "Fine grain multi-VT co-integration methodology in UTBB FD-SOI technology ," in *VLSI-SoC*, 2013, pp. 168–173.

[12] M. Maymandi-Nejad and M. Sachdev, "A Digitally Programmable Delay Element: Design and Analysis," *IEEE Trans. on VLSI Syst.*, vol. 11, no. 5, pp. 871–878, Oct. 2003.

[13] M. Maymandi-Nejad and M. Sachdev, "A Monotonic Digitally Controlled Delay Element," *IEEE JSSC*, vol. 40, no. 11, pp. 2212–2219, Nov. 2005.

[14] G. Kim *et al.*, "A Low-voltage, Low-power CMOS Delay Element," *IEEE JSSC*, vol. 31, no. 7, pp. 966–971, Jul. 1996.

[15] N. Mahapatra *et al.*, "An Empirical and Analytical Comparison of Delay Elements and a New Delay Element Design," in *VLSI*, 2000, pp. 81–86.

[16] N. Mahapatra *et al.*, "Comparison and Analysis of Delay Elements," in *MWSCAS*, 2002, pp. 473–476.

[17] G. Palumbo and D. Pappalardo, "Charge Pump Circuits: An Overview on Design Strategies and Topologies," *IEEE Circ. and Syst. Mag.*, vol. 10, no. 1, pp. 31–45, Mar. 2010.

[18] C. Tran *et al.*, "Low-power High-speed Level Shifter Design for Block-level Dynamic Voltage Scaling Environment," in *ICICDT*, May 2005, pp. 229–232.

[19] J. Hamon and E. Beigne, "Automatic Leakage Control for Wide Range Performance QDI Asynchronous Circuits in FD-SOI Technology," in *ASYNC*, 2013, pp. 142–149.

[20] C. Lazzari *et al.*, "An automated design methodology for layout generation targeting power leakage minimization," in *ICECS*, 2009, pp. 81–84.

[21] P. Gupta *et al.*, "Gate-length biasing for runtime-leakage control," *IEEE Trans. on CAD of Int. Circ. and Syst.*, vol. 25, no. 8, pp. 1475–1485, Aug. 2006.

978-1-4799-8720-7/15 $31.00 © 2015 IEEE

2015 IEEE Computer Society Annual Symposium on VLSI

Efficient Utilization of Imprecise Blocks for Hardware Implementation of a Gaussian Filter

M. H. Seyed Javadi[a] and H. R. Mahdiani[b]

[a] EE Department, [b] CSE Department

Shahid Beheshti University

Tehran, IRAN

mahdiani@sbu.ac.ir, mahdiani@gmail.com

Abstract— Due to great advantages of the imprecise computational blocks in implementation of imprecision tolerant applications, a wide range of different imprecise structures as well as some of their new applications are introduced in recent years. However, due to inherent differences between imprecise and precise components, their utilization approaches should also be customized to gain the best performance. In this paper, a novel efficient approach for utilization of imprecise computational blocks is presented and applied to a Gaussian image filter hardware. This approach offers different imprecise implementations of the Gaussian filter which improve the filter area and delay up to 50% and 53% respectively while does not degrade the output quality below the excellent threshold.

Keywords— Imprecise Computational block, Gaussian Image Filter, Word-Length Optimization, LOA, VLSI.

I. INTRODUCTION

From about 2010 while the leading versions of the imprecise (or approximate) computational blocks were introduced [1] up to now, an important trend has been formed around this research area to introduce novel imprecise arithmetic components (with the main focus on addition and multiplication), as well as demonstrating their applicability in some limited applications [7][9][11]. The main reason which causes the popularity of this newly proposed type of arithmetic components is that these can be utilized to simultaneously enhance both the cost and also performance in hardware implementation of an important category of applications named as "imprecision tolerant" (IT) [2]. Based on definition, an IT application is inherently capable of tolerating some types of imprecision sources (which might be introduced due to environmental or input noise, faults, etc.) of course up to a threshold [2]. For example, a signal processing application which is devised for image, voice, or video coding and decoding does not rely on precise and deterministic values of pixels and voice samples [1]. The main contribution of the imprecise blocks is that they widely exploit the intrinsic tolerance of the IT applications to improve their implementation cost and performance [1]. In other words, they are designed to intentionally provide only "good enough" results of the computations instead of their tightly precise values. This relaxation lets applying significant simplifications on their architectures with respect to traditional precise components to improve their hardware cost, delay and power

consumption [1]. A normally neglected but very important issue when dealing with imprecise blocks is that goodness of the result totally depends on level of the imprecision tolerance of the application that varies from one application to another [2]. Therefore, the legal amount of imprecision which can be introduced inside an exploited imprecise block is a dictated factor from application side and this block should be able to obey this critical and variable limitation factor, to guarantee its fitness for different applications and preserve its generality [2].This important capability should be inserted in an imprecise block at its design level by embedding one or more structural parameters (beside Word-Length (WL)) named as "imprecision parameters". Increasing these parameters lead to higher relaxation in result approximation process and therefore higher imprecision, while also improves one or more physical properties such as its area, delay and power consumption. The main role of such parameters is to tune the amount of imprecision which is introduced inside the imprecise block based on recommendation of different IT applications while paying the least possible implementation costs. Therefore, tuning process of the imprecision parameters for a specific application is a mandatory issue which significantly improves imprecise block implementation cost and performance. It is a time consuming process which should be performed beside WL extraction in the design flow of imprecise block through extensive simulations. The incoming sections of the paper are arranged as the following. The next section provides a brief but complete survey of the recently proposed imprecise blocks and their utilization methods. Section III introduces a correct approach for development as well as exploitation and parameter tuning of the imprecise blocks by focusing on hardware implementation of the Gaussian image filtering as a crucial case study. The last section concludes the paper.

II. LITERATURE SURVEY AND PROBLEM DESCRIPTION

In this section, a detailed review is provided on recently proposed imprecise block architectures and their usage drawbacks in 15 published papers in this research area. Table I includes the summary of the results. While the second and third table columns indicate the type and number of imprecision parameters of the proposed architecture in each reference respectively, the fourth column includes the IT application which the proposed design is exploited to realize it. The last column indicates whether a fine tuning

978-1-4799-8720-7/15 $31.00 © 2015 IEEE

process for imprecision parameters is addresses in the paper or not. As it is clearly demonstrated in the table, more than 12% of the proposed imprecise adders and 44% of the proposed imprecise multipliers are designed without any imprecision parameters. This means that just similar to any precise component, the WL is their only single parameter and no further imprecision tuning is applicable, that limits their efficiency and applicability. To better illustrate this limitation, it is useful to refer to a similar situation in precise arithmetic blocks. A traditional precise component of the fixed WL cannot be exploited to implement different applications in an efficient manner. Due to its fixed WL, it cannot be tuned to provide higher precision for more accurate applications. On the other hand for some other applications which need less accuracy, it always provides higher unnecessary precision and pays extra costs in terms of area, delay and power.

TABLE I. SUMMARY OF THE RECENTLY PROPOSED IMPRECISE BLOCKS AND THEIR USAGE APPROACHES.

Ref. #	Introduced block type	# of Imp. Par.	Utilized IT App.	Imp. Par. Fine Tuning
[4]	Adder	1	FFT&IFFT	No
[5]	Adder	1	IDCT	No
[6]	Adder	0	-	-
[7]	Adder	1	IDCT, Filtering	No
[8]	Adder	1	IDCT, Filtering	No
[9]	Adder	1	Filtering	No
[10]	Adder	1	-	-
[1]	Adder	1	Neural, Fuzzy	Yes
[11]	Multiplier	0	Blending	-
[12]	Multiplier	1	DCT&IDCT	No
[13]	Multiplier	0	-	-
[14]	Multiplier	0	-	-
[8]	Multiplier	2	IDCT, Filtering	No
[15]	Multiplier	0	DCT, Filtering	-
[16]	Multiplier	1	-	-
[1]	Multiplier	2	Neural, Fuzzy	Yes
[17]	Multiplier	1	FIR Filter	Yes

The fourth table column indicates that more than 33% of the papers have not utilized the proposed architecture even in one application which is an important drawback. But the more important disadvantage of the listed papers is that (except for our works [1][17]), none of them, even those works who have both armed their blocks with imprecision parameters and also simulated the applicability of the proposed one in an application, have not addressed even a simple procedure for fine tuning of the imprecision parameters according to tolerance of the utilized application. In a technical viewpoint, this means that they have demonstrated only the feasibility and not the efficiency of the blocks in their design. The next section will focus on Gaussian image filtering application [3] as a crucial case study (due to number and also connection structure of the exploited blocks), to demonstrate a useful general approach for fine tuning of its many existing imprecision parameters to gain an acceptable performance with paying minimum possible cost. A wide range of results are also presented to demonstrate the importance of the fine tuning process on cost-performance of the implemented Gaussian filter.

III. IMPRECISION PARMETERS FINE TUNING APPROACH:HARDWARE IMPLEMNETATION OF A GAUSSIOAN FILTER

In this section, a brief representation of the Gaussian filter is presented first. A wide range of simulations is then included to show how the values of the block's imprecision parameters can be determined to gain the best result. The synthesis results are also presented to show the effects of the imprecision parameter fine tuning on Gaussian hardware cost and speed.

A. Gaussian Filter Architecture

The Gaussian filter is a convolution operator which is normally used for blurring images or removing their noise. A discrete version of this filter with the 3x3 window [3] shown in Fig. 1 is selected for implementation using imprecise adders. Fig. 2 also demonstrates the hardware structure of this filter. To form the imprecise model, all existing adders are instantiated by Lower-part-OR-Adder (LOA) adder [1]. Fig. 3 illustrates the structure of a LOA adder. It has a single imprecision parameter named as Lower-Part-Length (LPL) [1]. As it is shown in the figure, increasing the LPL causes higher imprecision while also leads to less hardware area and faster operation.

A1	A2	A3		1	2	1
A4	A5	A6	* 1/16	2	4	2
A7	A8	A9		1	2	1

Fig.1. Window coefficients of the 3x3 Gaussian filter.

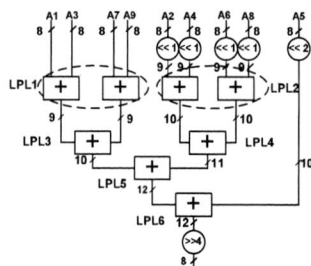

Fig.2. Hardware Architecture of the 3x3 Gaussian filter.

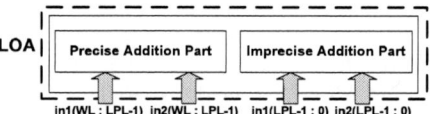

Fig.3. LOA imprecise adder structure.

As illustrated in Fig. 2, although the Gaussian filter only consists of adders, it should be treated as a challenging case for implementation. It relies on a 4-level addition depth and includes eight adders which results in eight imprecision parameters. All these together, increase interactions between different imprecision sources and also cause accumulation of the imprecision in final system. According to standards, the implemented filter is tuned to provide SNR of 32.04dB for excellent and 20 dB for acceptable image qualities [18].

978-1-4799-8720-7/15 $31.00 © 2015 IEEE

Although it is desirable to simultaneously optimize the system WL as well as all imprecision parameters, it is a very time consuming task which becomes infeasible when the number of parameters increases. Therefore, a greedy approach might be exploited to reduce the number of simultaneously under optimization parameters. As a good instance, it is desirable to perform WL optimization and determine its least acceptable value while all imprecision parameters are inactive (all LPLs are 0).Tuning of the imprecision parameters is then performed. In this example, choosing a suitable WL for the structure is straightforward. As it is shown in Fig. 2, the input WL is considered as 8 bits which is equal to the width of a gray-scale pixel value. To clearly focus only on imprecision, all imprecision sources which might be introduced due to finite WL effects are suppressed by increasing the WL by one after each addition to prevent overflow imprecision sources. By precisely looking to the filter structure (Fig. 2), it can also be seen that the four adders which reside in the first level can be categorized into two different groups, each one includes two identical adders whose parameters (WL and LPL) can be considered similar. This simplification is performed to only reduce the number of imprecision parameters in the design from 8 to 6 LPLs (as shown in Fig. 2) that ultimately significantly decreases the LPL extraction simulation time. In a generalized manner, a similar approach should be used in large circuits to reduce their simulation time

B. LPL Extraction Simulation results

This section performs two different types of exhaustive simulations with the purpose of fine tuning values of LPL1 to LPL6 for achieving excellent or acceptable output qualities. As the first step, some C simulations are run to discover the sensitivity of the imprecise model with respect to changes in a single LPL while all other LPLs are inactive and zero. This type of simulation helps the designer to have a better understanding about relative importance of a distinct imprecision parameter with respect to others. The next category of simulations on the other hand, determines the behavior of the model against simultaneous activation of all six imprecision parameters which helps to fine tune these parameters considering their complex interactions. All simulation results are performed on three well-known images, baboon, pepper, and Lena. Although each one of these three test cases has distinct features and details, their simulation results have a completely similar trend and differ only in some DC offset. Therefore, the average SNRs of the three images are computed as the final results.

1) Single LPL Change Analysis

In this section, the sensitivity of the Gaussian filter quality against changes in only one imprecision source is demonstrated (Fig. 4). In other words, it is assumed that the structure in Fig. 2, includes only one LOA (with its LPL as a none-zero value) at a time and all other adders are precise (their LPLs are all zero). Different regions for excellent, acceptable, and bad filter output qualities are also distinguished in the figure to provide a better insight about LPL values. As it is clear in Fig. 4, different LPLs have different effects on filter SNR. This simple observation

demonstrates the significance of fine tuning process of the imprecision parameters. It implies that blindly choosing constant and symmetric values for all LPLs without any further fine tuning efforts does not lead to maximum efficiency. Further analysis of the curves in Fig. 4 provides more useful results. The analysis can be justified by considering the position of the corresponding adders in filter structure (Fig. 2).As a useful analysis result, the curves show that (except for LPL2), as we move toward deeper adders in filter structure, their LPL has more destructive effects on whole system SNR. The reason is that for example increasing LPL1 degrades the result of a single introductory addition while increasing LPL6 destroys the final result of all additions which are performed in the system. The only exception is about LPL2 which although resides at the first level, has a higher significance because it operates on doubled input values due to its input shifts.

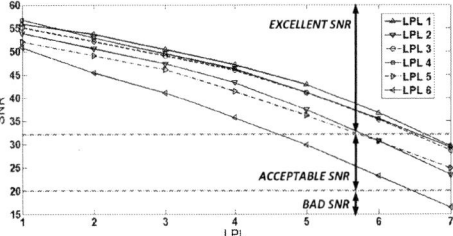

Fig. 4. SNR degradation with respect to an individual LPL.

2) Multiple LPL Change Analysis

It is shown in the previous section that increasing LPL in adders of the deeper layers causes higher damages on output quality of the Gaussian filter. In this section, exhaustive simulations are performed to simultaneously determine maximum values of the six LPL parameters to achieve excellent or acceptable output qualities with the minimum costs, while considering interactions between all active imprecision sources. Figure 5 includes the complete results of the exhaustive simulation.

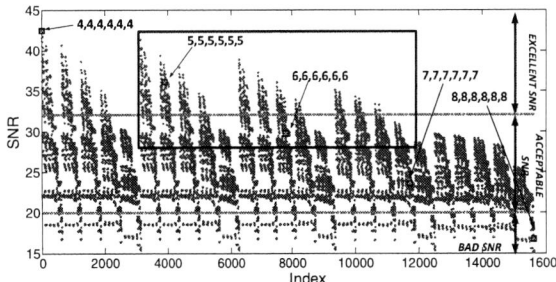

Fig. 5. Exhaustive test simulation results of the Gaussian filter for different LPLs.

To achieve Fig. 5, all LPLs are changed from 4 to 8 in nested loops in order of their importance (which is deduced from Fig. 4: LPL6, LPL5, LPL2, LPL3, LPL4, LPL1) and for each LPL set, the SNR of the resulting Gaussian filter is demonstrated in the figure with a dot. Fig. 6 also demonstrates the zoom-in version of the rectangle which is marked in Fig. 5 to provide a better understanding of different points. For more convenience, the LPL values of some important points are explicitly provided in format "LPL1, LPL2, ..., LPL6" on figures. Thus the first and the last index in Fig. 5 corresponds to LPL sets of (4,4,4,4,4,4) and (8,8,8,8,8,8) respectively. Different regions of excellent, acceptable, and bad output qualities are also distinguished in the figures. Despite its high number of imprecision parameters and also existing complexities in Gaussian filter structure; Fig. 5 demonstrates its regular behavior with respect to changes in multiple active LPLs. As a coarse tuning guideline, Fig. 5 shows that choosing all LPLs equal to 4 and 5 results in an excellent filter quality, choosing values 6 or 7 leads to an acceptable quality, and when all LPLs are equal to 8 the filter does not work properly. To better tune the structure imprecision however, Fig. 6. shows that there are many other fine tuned LPL sets whose SNRs fall above excellent quality threshold line and therefore can be exploited to achieve an excellent filter quality. However, by considering dependency of the results to the input image, choosing a point very near above the threshold lines (excellent or acceptable) increases the probability of the undesired filter quality when applying it on new images.

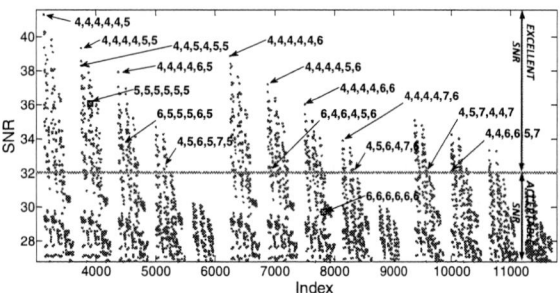

Fig. 6. Zoom-in version of the box shown in in Fig. 5.

C. LPL Extraction Synthesis results

Based on the unit-gate model and according to the area and delay of LOA adder, the gate-count and gate-delay of the Gaussian filter shown in Fig. 2 can be computed as follows. According to [1], while considering the gate-count of a Full-Adder as 6 gates, the gate-count of a WL-bit LOA adder with LPL imprecise bits is $6 \times (WL - LPL) + LPL + 1$. The total gate-count of the Gaussian filter can be computed according to below equation:

$$Gate\ Count_{GF} = 464 - 10(LPL1 + LPL2) \quad (1)$$
$$- 5(LPL3 + LPL4 + LPL5 + LPL6)$$

To determine the gate-delay of the Gaussian filter, there are two different longest paths in its circuit as shown in Fig. 2. While the first path passes through LPL1, LPL3, LPL5, and LPL6, the next path goes through LPL2, LPL4, LPL5, and LPL6. While according to [1], the gate-delay of a WL-bit LOA adder with LPL imprecise bits can be computed as $3.5 \times (WL - LPL) + 1$, total gate-delay of the Gaussian filter can be computed as:

$$Gate\ Delay_{GF} = Max\{\ 144 - 3.5(LPL1 + LPL3 + LPL5 + LPL6),\ 151 - 3.5(LPL2 + LPL4 + LPL5 + LPL6)\ \} \quad (2)$$

Based on (1) and (2), the third and fourth columns of Table II include area and delay of the precise (all LPLs equal to 0) and imprecise models of the Gaussian filter for a small selected number of LPL set values. The filter output SNR is also included in the last table column for more convenience. As it is clear, all imprecise models of #2 to #7 provide excellent qualities while their synthesis results are much better than precise implementation. However, they also offer a wide range of quality-area-delay trade-offs. While #2 LPL set results in 25% worse SNR with respect to the precise implementation, it also offers 34% better area and 37% better delay. On the other hand, #7 configuration trades 43% less SNR to gain 50% better area and 53% better delay with respect to precise implementation. From design viewpoint, a designer can choose anyone of these different LPL sets to achieve excellent image quality while saving different costs and achieving different speeds. For example, while both LPL sets #2 and #7 preserve the excellent output quality, the LPL set #7 provides 23% better area, 25% less delay and 42% improvement in area-delay-product with respect to LPL set #2.

Based on a brief analytical study, it is also simple to show the power efficiency of the imprecise Gaussian filter with respect to precise implementation. The dynamic power of a circuit as the dominant factor in its total power dissipation can be computed based on:

$$P_{Dynamic} = \propto \times\ C_L\ \times f\ \times VDD^2 \quad (3)$$

Where α is the activity factor and C_L is the load capacitance. As LOA structure has respectively lower number of gates, it also provides lower load capacitance and probably less activity factor with respect to precise adders [1], that results in power efficiency of the imprecise Gaussian implementation with respect to the precise implementation.

TABLE II. SYNTHESIS RESULTS AND OUTPUT QUALITIES OF THE PRECISE AND SOME IMPRECISE IMPLEMENTATIONS OF GAUSSIAN FILTER.

#	LPL Set	Area (Gate Count)	Delay (Gate Delay)	Area× Delay	SNR
1	0,0,0,0,0,0	464	151	70064	56.93
2	4,4,4,4,4,4	304	95	28880	42.44
3	5,5,4,4,5,4	279	88	24552	38.78
4	4,4,5,4,5,5	289	88	25432	38.33
5	5,5,5,5,5,5	264	81	21384	36.10
6	6,5,5,5,6,5	249	78	19298	33.85
7	6,6,5,5,6,6	234	71	16614	32.12
8	6,6,6,6,6,6	224	67	15008	29.73
9	7,7,7,7,7,7	184	53	9752	23.63
10	8,8,8,8,8,8	144	39	5616	16.72

IV. CONCLUSION

A new approach for efficient utilization of imprecise blocks is introduced in the paper and applied for hardware implementation of a Gaussian image filter. The synthesis and simulation results show that using this approach, it is possible to define the most suitable quality-physical properties trade-off when using imprecise blocks to implement an IT application.

V. REFERENCES

[1] Mahdiani, H.R., Ahmadi, A., Fakhraie, S.M., and Lucas, C.: 'Bio-inspired imprecise computational blocks for efficient VLSI implementation of soft-computing applications', IEEE Trans. Circuits Syst. I, vol. 57, no. 4, pp. 850-862, 2010.

[2] H. R. Mahdiani, S. M. Fakhraie, and C. Lucas, "Relaxed Fault-Tolerant Hardware Implementation of Neural Networks in the Presence of Multiple Transient Errors," *IEEE Trans. Neural Netw* vol. 23, pp. 1215-1228, 2012.

[3] R. C. Gonzales, and R. E. Woods, "Digital Image Processing, 2-nd Edition," Prentice Hall, 2002.

[4] N. Zhu, W. L. Goh, W. Zhang, K. S. Yeo, and Z. H. Kong, "Design of low-power high-speed truncation-error-tolerant adder and its application in digital signal processing," *IEEE Trans. VLSI Syst.* vol. 18, pp. 1225-1229, 2010.

[5] M. Weber, M. Putic, H. Zhang, J. Lach, and J. Huang, "Balancing adder for error tolerant applications," in *Circuits and Systems (ISCAS), 2013 IEEE Int. Symp. on*, 2013, pp. 3038-3041.

[6] Yang, Zhixi, et al. "Approximate XOR/XNOR-based adders for inexact computing." *IEEE Conf. on Nanotechnology.* 2013.

[7] J. Miao, K. He, A. Gerstlauer, and M. Orshansky, "Modeling and synthesis of quality-energy optimal approximate adders," *Int. Conf. on CAD*, 2012, pp. 728-735.

[8] P. Albicocco, G.-C. Cardarilli, A. Nannarelli, M. Petricca, and M. Re, "Imprecise arithmetic for low power image processing," in *Signals, Systems and Computers (ASILOMAR)*, 2012, pp. 983-987.

[9] A. B. Kahng and S. Kang, "Accuracy-configurable adder for approximate arithmetic designs," in *Proceedings of DAC*, 2012, pp. 820-825.

[10] K. Du, P. Varman, and K. Mohanram, "High performance reliable variable latency carry select addition," in *(DATE)*, 2012, pp. 1257-1262.

[11] A. Momeni, J. Han, P. Montuschi, and F. Lombardi, "Design and Analysis of Approximate Compressors for Multiplication.",accepted for publication in *IEEE Trans. Comput.* 2014

[12] K. Bhardwaj, P. S. Mane, and J. Henkel, "Power-and area-efficient Approximate Wallace Tree Multiplier for error-resilient systems," in *ISQED*, 2014, pp. 263-269.

[13] C. Liu, J. Han, and F. Lombardi, "A low-power, high-performance approximate multiplier with configurable partial error recovery," in *Proc. of conf. on Design, Automation & Test in Europe*, 2014, p. 95.

[14] C.-H. Lin and I.-C. Lin, "High accuracy approximate multiplier with error correction," in *ICCD*, 2013, pp. 33-38.

[15] P. Kulkarni, P. Gupta, and M. D. Ercegovac, "Trading accuracy for power in a multiplier architecture," *Journal of Low Power Electronics*, vol. 7, pp. 490-501, 2011.

[16] K. Y. Kyaw, W. L. Goh, and K. S. Yeo, "Low-power high-speed multiplier for error-tolerant application," in *EDSSC*, 2010, pp. 1-4.

[17] F. Farshchi, M. S. Abrishami, and S. M. Fakhraie, "New approximate multiplier for low power digital signal processing," in *CADS*, 2013, pp. 25-30.

[18] International Standard ISO12232, "Photography — Electronicstill picturecameras — Determination of ISOspeed," (1997).

A Detailed Routing-aware Detailed Placement Technique

Aysa Fakheri Tabrizi*, Nima Karimpour Darav*, Logan Rakai*, Andrew Kennings[†],
William Swartz[‡] and Laleh Behjat*
*Department of Electrical and Computer Engineering, University of Calgary, Canada
[†]Department of Electrical and Computer Engineering, University of Waterloo, Canada
[‡]Department of Electrical Engineering, University of Texas at Dallas, USA

Abstract—In this paper we propose a detailed placement algorithm targeting detailed routability for designs at or smaller than 22nm. The sheer number and complexity of routing design rules at these feature sizes preclude direct incorporation of detailed routing rules into a placement algorithm. However, using the detail routing information to guide the placement can significantly reduce the overall design time and improve the performance of the circuit. Our proposed detailed routing-aware detailed placement (DrDp) is developed as an add-on to the detailed placement process to improve detailed routability in a relatively short runtime. The proposed technique is added to the code obtained from one of the top three teams in the ISPD 2014 detailed routing-driven placement contest and tested on ISPD 2014 benchmark suite. Numerical results show that the proposed technique can improve the detailed routing quality with no significant change in detailed placement score, total wirelength or runtime.

I. INTRODUCTION

With the advances in VLSI technology, routability has become a significant issue and must be incorporated during all physical design stages. Hence, In recent years, the main goal of placement has shifted to routability rather than reducing wirelength [13], [7]. However, current routing-aware placers emphasize reducing the amount of resources used by global nets and the impact of local nets is ignored [16], [18]. Local nets are the nets that have all their pins inside a g-cell and are not routed during global routing. The local nets have a major effect on the designs' routability during the final detailed routing stage. According to [16], in ISPD 2011 placement contest on average 31.20% of the nets of the winners' solutions were local which will make the problem very hard to solve.

A typical placement flow consists of global placement, legalization, and detailed placement. Global placement performs an approximate distribution that allows overlapping. During legalization cells are aligned to sites in rows and overlaps are removed with minimizing perturbation to the global placement solution. Detailed placement performs local refinements while keeping the legal state.

In this paper, we propose a detailed placement technique that considers routability in detailed routing, *detailed routing-aware detailed placer* (DrDp). The main goal of the proposed technique is to reduce the number of conflicts that can occur in the detailed routing stage, by performing small changes during the placement stage. To achieve this goal, we chose to focus on only the detailed placement stage of the placement process

as it is the closest to detailed routing stage and the movements in the cells are generally local and happen in shorter distances.

First, we introduce a new net decomposition that considers route topologies versus routing estimation. Then we reduce the number of net segments and net bends by simply aligning the cells that are connected through such nets.

The proposed algorithm:

- Has a direct positive impact in detailed routing.

- Has low runtime.

The rest of the paper is organized as follows. Section II provides a literature review on detailed placement methods and explains the congestion modeling that is the inspiration of our algorithm. In Section III our detailed routing-aware detailed placer is described. Numerical results are presented in Section IV. Finally, in Section V, conclusion and future work are discussed.

II. BACKGROUND

Detailed placement aims to improve the quality of solution by moving and relocating cells within small regions while maintaining legality.

The optimization objective in detailed placement is frequently minimizing half-perimeter wirelength (HPWL). One of the popular time-proven methods for wirelength-driven detailed placement is the technique used in several placers [1], [3], [6] that exhaustively find the best ordering of cells within a window in a row. FastPlace-DP [14], included in FastPlace [17], attempts to further minimize wirelength by iteratively performing the three following steps: global cells swap using the optimal region; vertical moves towards optimal region; local window reordering. Finally, windows of cells are themselves reordered and merged to improve wirelength. FastPlace-DP has gained broad acceptance [5], [8], [9]. Branch-and-cut and branch-and-price placer in [11] aims to expand the size of windows used in optimal re-orderings. The authors of [4] formulate detailed placement as a minimization of the scaled HPWL used in the ICCAD 2013 Detailed Placement Contest [10] subject to a maximum cell displacement constraint. They propose to solve the formulation in two stages, a global move phase where cells are allowed to move into the best bin within the maximum displacement constraint, and a local move phase which performs moves between rows, local row reordering and compaction.

978-1-4799-8720-7/15 $31.00 © 2015 IEEE

More recent detailed placement algorithms have increased awareness of whitespace allocation or target density. These constraints are put in place to help with the subsequent stage of physical design, global routing. However, having a global routing solution does not guarantee a solution to detailed routing. A three step bin density-aware detailed placement algorithm is presented in [20]. Given a legalized placement, a greedy cell swapping step that reduces scaled HPWL while avoiding creating new overlaps is performed first. Then, a greedy, single row window re-ordering is performed, followed by a final cell swap step. The final cell swap begins by bloating cells in over-utilized bins and considers swaps only to neighboring bins with less utilization. Cell bloating has been used as a way to alleviate congestion since [2]. Finally, another swap-based detailed placer that constrains maximum cell displacement island presented in [15]. Swaps are made greedily based on a profit function that combines changes in HPWL, over-utilization, and overlap. Potential swaps are locally legalized before accepting to avoid problems legalizing after the completeness as may be conventionally done.

Linescan algorithms have been used to improve routability. A linescan algorithm presented in [12] assumes a scan line on the left vertical edge of the g-cell. The line is moved in small steps towards the right until it reaches to its right edge and the routing demand at every scan line is scanned. The routing demand within a g-cell at point p is $w_p = w_{gp} + w_{lp}$, where w_{gp} and w_{lp} are the routing demand caused by global and local routes at scan line p, respectively. When the scan line is very close to the edge, w_p is equal to w_{gp}. Since some global nets end at some point inside a g-cell and also at some points other global or local nets are added, the value of w_p is different for different scan lines inside a g-cell. The line scanning method is applied in a similar way for vertical tracks.

Optimizing the traditional HPWL objective in the discussed detailed placers is insufficient for success in routing and although the more recent density aware detailed placers aim global routing, having a global routing solution does not guarantee a solution to detailed routing. Indeed, as mentioned in the introduction, nearly one third of all nets are not even present in global routing. Therefore, in order to mitigate detailed routing failure, a scalable detailed placement algorithm described in the following section is designed. This algorithm specifically considers the local nets that can ultimately block the path to tapeout.

III. PROPOSED DETAILED ROUTING-AWARE DETAILED PLACEMENT

In this section first we propose our model for taking into account the congestion caused by local nets during the detailed placement. Afterwards, our alternative routing topology method based on a proposed net decomposition is introduced. And finally our proposed detailed routing-aware detailed placement heuristic is described and is followed by complexity analysis.

A. Local Congestion Modeling

In this work we aim to improve the detailed routability by reducing the congestion inside a g-cell. In order to reduce the congestion inside a g-cell, two objectives could be considered:

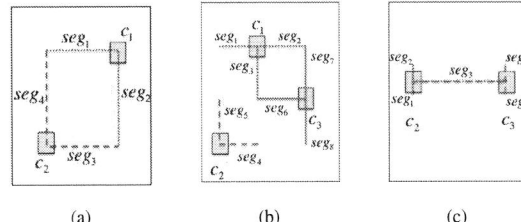

Fig. 1. Net decomposition for nets with (a) degree two: $n1$, (b) degree three: $n2$, and (c) direct net: $n3$

(1) To move the cell which is the end of a global net, as close as possible to g-cell boundary in order to minimize the occurrence of global nets intersecting the same scan line as local nets. (2) To place the cells of a local net as close as possible and shorten the length of the local net.

Since most of the cells have more than one pin and consequently are connected to more than one net, these moves can only happen if there is not any other net connected to the given cell in opposite direction. Therefore, in order to move a cell, all the nets connected to the cell should be considered and an appropriate model for the nets is needed. One approach is to use a routing estimation. This approach is not utilized in this work for following reasons: (1) Routing estimation during the placement is only a prediction for global routing and could be very different from exact routing at the end of placement. (2) Routing estimators ignore local nets which are an important part of detailed routing. (3) Avoiding routing estimation during the placement saves runtime. Instead, we suggest alternative routes modeling for nets that is based on a proposed net decomposition.

B. Net Decomposition and Alternative Routes Modeling

Our proposed algorithm considers different route topologies and makes the moves based on them. In this approach for each cell we consider all the nets connected to it and we decompose every net into vertical and horizontal segments as described in the following. Ideally, if there is no detour, a combination of those segments form the actual route. So during the detailed placement rather than selecting only one route that a routing estimator suggests, we keep track of the candidate routes. The net decomposition proposed in this work is explained by some examples in the following paragraphs.

An example of a degree two net is shown in Figure 1(a). In this figure two possible routing for n_1 are shown. Net n_1 is connected to c_1 either through seg_1 or seg_2, so in the proposed net decomposition, from the point of view of c_1, n_1 is decomposed to seg_1 and seg_2. This model prevents the move of c_1 upwards and towards the right later in the DrDp algorithm that is explained in Section III-C and consequently prevents congestion and wirelength. Likewise, from the point of view of c_2, n_1 is decomposed to seg_3 and seg_4.

Figure 1(b) shows the net decomposition for a net with degree three. The net segments from each cells' point of view are shown in this figure. Net n_2 is connected to c_1 through seg_1, seg_2 or seg_3. So $\{seg_1, seg_2, seg_3\}$ is the decomposition of n_2 from the point of view of c_1 . This model assures that

c_1 can not move horizontally during DrDp and can only move downwards and at most the length of seg_3. Net n_2 is connected to c_2 through seg_4 or seg_5. These segments start from c_2 and end when they reach to the level of the first pin connected to the net. So during DrDp, c_2 can only move in the direction of its segments and at most at the length of them. The net segments set of n_2 for c_3 consists of $\{seg_6, seg_7, seg_8\}$.

Algorithm 1 DrDp Algorithm

Require: Legalized placement solution
Ensure: Legalized final placement solution
1: Partition placement region into tiles $T = t_1, t_2, ..., t_n$
2: **for all** t **do**
3:     **for all** c in t **do**
4:         VerticalMove(c)
5:     **end for**
6:     **for all** c in t **do**
7:         HorizontalMove(c)
8:     **end for**
9: **end for**

Figure 1(c) shows a case where two pins are in the same vertical level and the ideal route for this net is a direct horizontal connection. In this case we assume imaginary vertical segments with very small lengths, in both upward and downward directions. These segments prevent the cell from moving vertically later during DrDp and preserve the direct horizontal connection. So, the net segments set of n_3 for c_1 consists of $\{seg_1, seg_2, seg_3\}$ and the net segments set of n_3 for c_2 consists of $\{seg_4, seg_5, seg_3\}$.

Likewise, for the pins that are connected to another pin at the same horizontal level, horizontal segments with very small lengths, in both left and right direction, are considered to prevent the cell from moving horizontally and preserve the direct vertical connection. Note that net decomposition considers ideal situations where there is no detour due to fixed cell or other design rule restrictions.

C. Algorithm Overview

The flow of our detailed routing-aware detailed placer is shown in Algorithm 1. In this algorithm the placement region is divided into tiles similar to the g-cells in global routing. All the moves will be performed inside the tiles. In each tile, first for all the cells in the tile starting from the left cell in the top row and moving to right and down, Vertical Moves are applied i.e. the potential cells for moving vertically are selected and are moved several rows above or below within the tile. Vertical Moves are explained in Section III-D. Next, Horizontal Moves are applied for all the cells in the tile in the same order. Horizontal Moves are explained in Section III-E. All the moves performed during this algorithm preserve legality i.e. during the DrDp the cells move to aligned placement sites and no overlap is allowed, and there is no need for an additional legalization step.

D. Vertical Move

The overview of Vertical Move of a cell is presented in Algorithm 2. For every cell in a tile, if all of the vertical segments of the given cell (line 1) are towards the same

---

Algorithm 2 Vertical Move Algorithm

Require: Cell c; boundaries (y_{up}, y_{down}) and $rows$ of tile t; location (x, y) , width (w), needed white spaces at left and right (wsl, wsr), and net information of all cells in t.
Ensure: New location of c if it is moved
1: Find vertical segments set of c, $SEG_v = \{seg_{v_1}, ..., seg_{v_p}\}$
2: **if** $\forall seg \in SEG_v$ are $upward$ **then**
3:     $seg_{min} \leftarrow min(SEG_v)$
4:     **if** $y_c + seg_{min} > y_{up}$ **then**
5:         $row_t \leftarrow$ highest row of t
6:     **else**
7:         $row_t \leftarrow row_{y_c + seg_{min}}$
8:     **end if**
9:     **repeat**
10:         $c_t \leftarrow$ the closest cell to (x_c, y_{row_t}) in row_t
11:         $c_l \leftarrow$ the first cell at the left of c_t
12:         $space_l \leftarrow x_{c_t} - x_{c_l}$
13:         $c_r \leftarrow$ the first cell at the right of c_t
14:         $space_r \leftarrow x_{c_r} - x_{c_t}$
15:         **if** $space_l > w_c + wsl_{c_t} + wsr_{c_l}$ **then**
16:             $x_c \leftarrow x_{c_t} - wsl_{c_t} - w_c/2$
17:             $y_c \leftarrow y_{c_t}$
18:         **else if** $space_r > w_c + wsr_{c_t} + wsl_{c_r}$ **then**
19:             $x_c \leftarrow x_{c_t} + wsr_{c_t} + w_c/2$
20:             $y_c \leftarrow y_{c_t}$
21:         **end if**
22:         $row_t \leftarrow row_t - 1$
23:     **until** $y_c = y_{c_t}$ **or** $row = row_c$
24: **else if** $\forall seg \in SEG$ are $downwards$ **then**
25:     Move $downwards$
26: **end if**

Algorithm 3 Horizontal Move Algorithm

Require: Cell c; boundaries (x_{right}, x_{left}) of tile t; location (x, y) , width (w), needed white spaces at left and right (wsl, wsr), and net information of all cells in t.
Ensure: New location of c if it is moved
1: Find horizontal segments set of c, $SEG_h = \{seg_{h_1}, ..., seg_{h_p}\}$
2: **if** $\forall seg \in SEG_h$ are $towards\ right$ **then**
3:     $seg_{min} \leftarrow min(SEG_h)$
4:     **if** $x_c + seg_{min} > x_{right}$ **then**
5:         $c_r \leftarrow$ the first cell at the right of c
6:         $x_c \leftarrow x_{c_r} - max(wsr_c, wsl_{c_r}) - w_c/2$
7:     **else**
8:         $x_c \leftarrow min(x_{c_r} - max(wsr_c, wsl_{c_r}) - w_c/2, x_c + l_h)$
9:     **end if**
10: **else if** $\forall seg \in SEG$ are $towards\ left$ **then**
11:     Move $towards\ left$
12: **end if**

---

direction (up/down) the cell is a candidate for Vertical Move. If all the segments are upwards (line 2) and none of them are local, i.e. the end of the shortest segment is beyond the tile upper boundary (line 3-4), the highest position available in the tile is found (line 5). Else if any of the vertical segments are local (line 6), the highest that the cell moves is the length of its shortest local vertical segment (line 7). Starting from the highest row, the closest cell to the given cell is found (line 10). If there is enough space beside that cell, considering the width of given cell and necessary whitespace between the cells, the given cell moves there (line 11-23). Likewise, If all the cells are downward (line 24), the cell moves downwards in a similar way.

E. Horizontal Move

After Vertical Move, Horizontal Move is applied on each cell in a tile. The algorithm of Horizontal Move of a cell is shown in Algorithm 3. In Horizontal Move, first all horizontal segments of the nets of the cell are found (line 1). If all of these segments are towards the same direction (line 2,10), the cell is a candidate cell for Horizontal Move. The cell moves in its segments direction until it reaches either the tile boundary, another cell considering whitespace requirements between cells or until the end of a local segment, whichever comes first (line 3-9).

F. Complexity Analysis

The time complexity of the proposed algorithm with fixed tile size is computed in this section. In Algorithm 1, each cell is considered as a candidate for Vertical Move and Horizontal Move. Suppose the time complexity of Vertical Move is equal to $T(p_c, d_e)$, where p_c and d_e denote the number of cell pins and the average degree of nets connected to cell pins, respectively. Since the algorithm of Horizontal Move is similar to the algorithm proposed for Vertical Move, the time complexity of Horizontal Move is equal to $T(p_c, d_e)$ as well. In Algorithm 1, for each tile t_i in the design, Vertical Move and Horizontal Move are applied on all cells $C(t_i)$ inside the tile. Since $\sum_{i=1}^{n} |C(t_i)|$ is equal to $|C|$ (total number of cells in the circuit), the time complexity of Algorithm 1 is in order of $|C|.T(p_c, d_e)$. In algorithms 2 (3), Finding vertical (horizontal) segments in line 1 has a time complexity equal to $O(p_c.d_e)$ and the rest of algorithms has the same time complexity. Therefore, we can write:

$$T(|P|, d_e) = O(|C|.p_c.d_e) = O(|P|.d_e) \tag{1}$$

Where $|P|$ is the total pins in the circuit. Since $1 < d_e < |P|$, we can say the proposed algorithm has a time complexity between $O(|P|)$ and $O(|P|^2)$. However, in most cases, d_e has a maximum value and is independent of $|P|$, meaning that the order of the proposed algorithm is almost linear. The empirical results demonstrate that the proposed algorithm is highly scalable and this can be observed in Table IV.

IV. RESULTS

Our placer has been developed in C++, and the experiments were conducted on a 64-bit CentOS (release 6.5) server with two Intel Xeon E5-2620 processors @ 2.00 GHz and 64 GB of RAM. The algorithm is evaluated on the ISPD 2014 detailed routing-driven placement contest benchmark suite A [19].

In Table I, the statistics of the benchmark circuits i.e. number of cells, number of nets, and density as well as the results of top teams of the contest, Team01 (T01), Team09 (T09), and Team10 (T10), obtained from the contest organizers, are presented. Evaluation in the contest is based on

four metrics: Detailed Placement Score (DP Score), Detail-Routed Wirelength Score (DR WL), Detailed Routing Score (DR Score), and runtime Score. In all categories lower numbers indicate better results. DP Score is a measure of the displacement in legalization that the evaluator tool makes to fix issues such as cell overlaps, cells not aligned on standards cell rows, some Design Rule Checking (DRC) violations, etc. DR WL refers to wirelength of final detail-routed circuit reported by the router. DR Score represents the cost of final detailed routing violations. DR Score is the main criteria of comparison and is a weighted sum of different types of routing violations. Types of routing violations and their weights in computing DR Score are given in table II. The objective of DrDp is improving the detailed routability (decreasing DR Score) without sacrificing the wirelength or adding DRC issues to be fixed during routing i.e. preserving DR WL and DP Score.

To evaluate the quality of the proposed method, we obtained the most recent placer code from Team10, one of the top three teams in the contest, and added the proposed technique as an extension to it. This placer uses FastPlace-DP [14] in its detailed placement stage and our proposed DrDp is added as the next step after FastPlace-DP. So, our starting point is a placement by a placer obtained from one of top three teams in the contest that uses FastPlace-DP in its detailed placement stage. The placed circuits are submitted to the contest website and evaluated by Mentor Graphics' Olympus commercial tool. Since other stages of placement, such as global placement, have significant impact on the result of placements, the effectiveness of the proposed extension is evaluated by focusing on the improvements made by the proposed algorithm. Therefore, in the following, the results of the proposed technique are compared to those of the starting point. DrDp could be added to other placers; however since the final placements files or placer codes of other teams from the contest were not available to us, we could not evaluate the improvement made by DrDp on different placements.

An important step before applying the DrDp is specifying the proper tile size. A large tile size results in more number of moves and also longer moves that can benefit the placement in terms of wirelength and via reduction. However, it increases the cell displacement and can result in placement congestion and pin density that can further lead to pin accessibility and pin short problems. On the other hand, by choosing a small tile size the number and length of moves might not be sufficient to reduce the detailed routing congestion and number of vias. Therefore, selecting the right tile size can significantly affect the performance of the algorithm. In this work, we have tried different tile sizes for the circuits and based on that, in order to have a unique strategy for all circuits, a tiling method is proposed and adopted in the following experiments. In the proposed method, DrDp is applied by assigning an initial tile size to the tiles. A set of stopping criteria is checked at the end of each iteration and if any of the criteria is met, the algorithm stops. Otherwise, the algorithm is repeated with a larger tile size. Increment in tile size continues until a stopping criterion is met. The recommended initial tile size (ITS) is $3H \times 3H$, where H is equal to the height of a single row. The tile size in the next iteration is $\alpha \times ITS$, where α is the iteration number. The stopping criterion are: 1) At least 95% of the eligible horizontal moves were accepted, and 2) The size of the tiles reached to its maximum of $6 \times ITS$. Eligible horizontal moves

TABLE I.   Circuit statistics and detailed routing results for benchmarks placed by top teams from ISPD14 contest

| Circuit | Number | | Dens | Placement | | | | | | | | |
| | Cells | Nets | % | DP Score | | | DR WL | | | DR Score | | |
				T01	T09	T10	T01	T09	T10	T01	T09	T10
des_perf_1	113018	112878	92	2.62	2.57	3.84	1.83	1.93	1.85	7610.6	1243.2	8964.0
des_perf_2	113018	112878	87	1.93	1.40	2.63	1.84	1.88	1.87	498.0	672.2	297.0
edit_dist_1	133235	133223	46	1.04	1.10	0.99	4.85	5.33	4.75	0.0	3366.8	99.4
edit_dist_2	133235	133223	49	1.10	1.16	0.98	4.71	5.09	4.75	2.2	1380.8	1064.4
fft	35291	33307	85	2.26	2.19	2.33	0.64	0.67	0.66	950.4	1032.0	956.4
matrix_mult	160127	158527	82	1.43	1.20	1.25	3.08	3.07	3.04	237.4	8955.0	811.4
pci_bridge32_1	31036	30835	86	1.52	1.45	1.40	0.36	0.35	0.37	0.0	0.0	2.0
pci_bridge32_2	31036	30835	87	2.08	1.51	1.45	0.37	0.37	0.37	2.0	1.0	3.0

TABLE II.   The weights of different types of routing violations in computing DR Score

Violation type	weight
Routing opens	1.0
Routing shorts	1.0
Routing blocked pin	1.0
DRC violation	0.2

are the cells that do not have any net segments in opposite horizontal directions. The constant number of iterations do not change the overall complexity order of the algorithm.

In the following, the performance of the proposed DrDp is first described by showing the moves inside a sample tile from one of the benchmark circuits by DrDp and then the numerical results for all the benchmarks are presented.

In Figure 2, the reduction in the local routing congestion is shown with a real example. In this figure a tile from mgc_des_perf_1 benchmark circuit, is shown before and after applying DrDp. Two Horizontal and one Vertical Move occurred within this tile. The cells that are moved are colored. The dashed lines show the bounding box of the seven nets that are affected by the moves. It can be seen that as the result of the moves, three of these nets are converted to the nets with no bends and could significantly impact the congestion. The bounding box of the three other nets are diminished and the length of the net with no bend is shortened.

In Table III the results of the placer with the proposed DrDp algorithm are compared to those of the starting point. In this table, the columns under DP Score, DR WL, and DR Score represent the results of starting point and DrDp, respectively. It can be observed from the table that DrDp improved the DR Score for all the designs with non-zero DR Scores and preserves the zero DR Score of the pci_bridge32_1 design, while making no significant changes in wirelength or DP Score.

By comparing the DR Score results of our starting point to those of top three teams from Table I, it can be seen that DrDp improved the placements that are already better than state-of-the-art results in terms of detailed routability, except for two benchmarks.

In Table IV, the statistics of the moves performed by DrDp are presented. The first columns of Vertical Moves and Horizontal Moves present the number of cells that are eligible to move in DrDp. The next two columns show the number of cells that actually moved and the percentage of the eligible cell that moved, respectively. It can be seen that among the possible moves in vertical direction only a small fraction is actually moved; however the majority of the cells that are eligible to move horizontally are moved. This is due to the

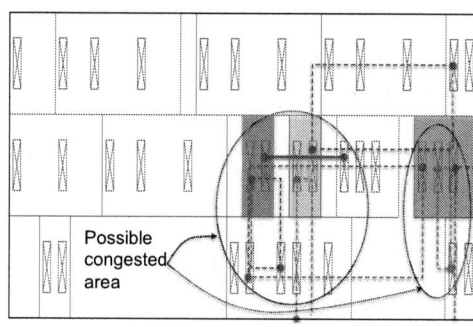

(a) Before applying DrDp

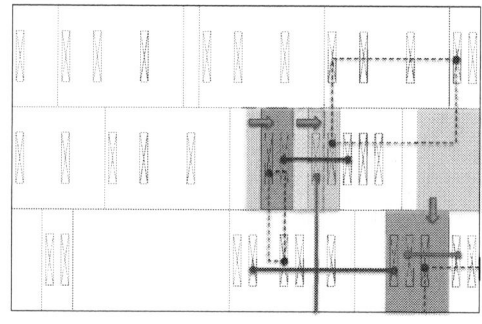

(b) After applying DrDp

Fig. 2. A tile from mgc_des_perf1: Location of the cells and bounding boxes of the nets affected by DrDp before and after applying DrDp algorithm

fact that Vertical Moves are restricted to align with rows of much coarser resolution and in order to move a cell vertically there must be enough whitespace in the target row. However, a cell eligible to move horizontally can move up to the next cell considering necessary whitespace between the cells; a much more easily satisfied requirement.

Finally, in the last block of the table, the information regarding the runtime of the proposed technique is shown. The first column of the block presents the total runtime of the placer without DrDp. Final two columns present the runtime that DrDp adds to the total runtime in seconds, and as a percentage of total runtime, respectively. It can be observed DrDp is performed in a very short runtime and added at most 1% of the runtime.

V. Conclusion

In this paper we proposed DrDp, a detailed placer targeting optimizing the detailed routability. Numerical results show

TABLE III. DETAILED ROUTING RESULTS BY ADDING PROPOSED DRDP TO THE PLACER OF ONE OF THE TOP TEAMS FROM THE CONTEST THAT USES FASTPLACE DP IN DETAILED ROUTING COMPARED TO THOSE OF STARTING POINT

	DP Score		DR WL		DR Score		
Circuit	Initial	DrDp	Initial	DrDp	Initial	DrDp	Imp %
mgc_des_perf_1	3.47	**3.06**	1.88	**1.86**	1157.2	**995.0**	14
mgc_des_perf_2	**2.71**	2.84	**1.90**	1.91	32.8	**22.8**	30
mgc_edit_dist_1	0.92	0.92	**4.93**	4.94	174.0	**166.0**	5
mgc_edit_dist_2	0.86	0.86	4.78	**4.77**	85.9	**79.0**	8
mgc_fft	2.42	**2.34**	0.67	0.67	385.3	**362.2**	6
mgc_matrix_mult	**1.22**	1.23	3.06	3.06	147.2	**144.5**	2
mgc_pci_bridge32_1	1.40	**1.39**	0.37	0.37	0.0	0.0	NA
mgc_pci_bridge32_2	**1.45**	1.47	0.38	**0.37**	1.5	**1.3**	17
Average	1.81	**1.76**	2.25	**2.24**	248.0	**221.3**	12

TABLE IV. DRDP MOVE STATISTICS AND RUNTIME

	Vertical Moves			Horizontal Moves			Runtime		
Circuit	Elig	Move		Elig	Move		Initial	Added by DrDp	
		#	%		#	%	(s)	(s)	%
mgc_des_perf_1	1527	65	4	382	364	95	325.8	1.1	0.3
mgc_des_perf_2	1036	76	7	545	518	95	341.1	1.0	0.3
mgc_edit_dist_1	591	160	27	934	884	95	757.2	6.4	0.8
mgc_edit_dist_2	615	213	35	911	867	95	758.4	2.4	0.3
mgc_fft	383	25	7	273	260	95	76.0	0.1	0.2
mgc_matrix_mult	2276	180	8	1074	1025	95	494.9	5.1	1.0
mgc_pci_bridge32_1	149	13	9	92	87	95	72.1	0.7	1.0
mgc_pci_bridge32_2	190	12	6	70	67	96	72.0	0.0	0.0

that the proposed technique can effectively improve the actual detailed routing score in a very short runtime with no significant change in detailed placement score and actual wirelength. Future work includes using machine learning techniques to find the best tile size for each design.

REFERENCES

[1] A. Agnihotri, S. Ono, C. Li, M. Yildiz, A. Khatkhate, C.-K. Koh, and P. Madden. Mixed block placement via fractional cut recursive bisection. *IEEE Transactions on Computer-Aided Design*, 24(5):748–761, 2005.

[2] U. Brenner and A. Rohe. An effective congestion-driven placement framework. *IEEE Transactions on Computer-Aided Design*, 22(4):387–394, 2003.

[3] T. Chan, J. Cong, J. R. Shinnerl, K. Sze, and M. Xie. mpl6: Enhanced multilevel mixed-size placement. In *In Proceedings of the International Symposium on Physical Design*, pages 212–214, 2006.

[4] W.-K. Chow, J. Kuang, X. He, W. Cai, and E. F. Young. Cell density-driven detailed placement with displacement constraint. In *Proceedings of the 2014 on International Symposium on Physical Design*, pages 3–10, 2014.

[5] X. He, T. Huang, W.-K. Chow, J. Kuang, K.-C. Lam, W. Cai, and E. F. Y. Young. Ripple 2.0: High quality routability-driven placement via global router integration. In *Proceedings of the Design Automation Conference*, DAC '13, pages 152:1–152:6, 2013.

[6] M.-K. Hsu, Y.-F. Chen, C.-C. Huang, T.-C. Chen, and Y.-W. Chang. Routability-driven placement for hierarchical mixed-size circuit designs. In *Proceedings of the Design Automation Conference*, pages 1–6, 2013.

[7] A. Kennings, N. Darav, and L. Behjat. Detailed placement accounting for technology constraints. In *VLSI-SoC*, pages 1–6, 2014.

[8] M. Kim, D. Lee, and I. Markov. Simpl: An effective placement algorithm. *IEEE Transactions on Computer-Aided Design*, 31(1):50–60, 2012.

[9] M.-C. Kim, N. Viswanathan, C. J. Alpert, I. L. Markov, and S. Ramji. Maple: Multilevel adaptive placement for mixed-size designs. In *Proceedings of the International Symposium on International Symposium on Physical Design*, pages 193–200, 2012.

[10] M.-C. Kim, N. Viswanathan, Z. Li, and C. Alpert. ICcad-2013 Cad contest in placement finishing and benchmark suite. In *International Conference on Computer-Aided Design*, pages 268–270, 2013.

[11] S. Li and C.-K. Koh. Mixed integer programming models for detailed placement. In *Proceedings of the International Symposium on Physical Design*, pages 87–94, 2012.

[12] Y. Li, L. M. Rakai, L. Behjat, and B. Swartz. Wirelength and congestion estimation for routability-driven placement. In *2011 International Workshop on System Level Interconnect Prediction,*, page 1, 2011.

[13] I. L. Markov, J. Hu, and M.-C. Kim. Progress and challenges in VLSI placement research. In *Proceedings of the International Conference on Computer-Aided Design*, pages 275–282, 2012.

[14] M. Pan, N. Viswanathan, and C. Chu. An efficient and effective detailed placement algorithm. In *Proceedings of the International Conference on Computer-Aided Design*, pages 48–55, 2005.

[15] S. Popovych, H.-H. Lai, C.-M. Wang, Y.-L. Li, W.-H. Liu, and T.-C. Wang. Density-aware detailed placement with instant legalization. In *Proceedings of the Design Automation Conference*, pages 1–6, 2014.

[16] H. Shojaei, A. Davoodi, and J. Linderoth. Planning for local net congestion in global routing. In *Proceedings of the International Symposium on International Symposium on Physical Design*, pages 85–92, 2013.

[17] N. Viswanathan, M. Pan, and C. Chu. Fastplace 3.0: A fast multilevel quadratic placement algorithm with placement congestion control. In *Proceedings of the Asia South Pacific Design Automation Conference*, 2007.

[18] Y. Wei, C. Sze, N. Viswanathan, Z. Li, C. Alpert, L. Reddy, A. Huber, G. Tellez, D. Keller, and S. Sapatnekar. Glare: Global and local wiring aware routability evaluation. In *Proceedings of the Design Automation Conference*, pages 768–773, 2012.

[19] V. Yutsis, I. S. Bustany, D. Chinnery, J. R. Shinnerl, and W.-H. Liu. Ispd 2014 benchmarks with sub-45nm technology rules for detailed-routing-driven placement. In *Proceedings of the 2014 on International Symposium on Physical Design*, ISPD '14, pages 161–168, 2014.

[20] Q. Zhou, J. Hu, and Q. Zhou. An effective iterative density aware detailed placement algorithm. In *Proceedings of the International Symposium on Circuits and Systems*, pages 1444–1447, 2014.

2015 IEEE Computer Society Annual Symposium on VLSI

An Effective Chemical Mechanical Polishing Filling Approach

Chuangwen Liu[1], Peishan Tu[1], Pangbo Wu[1], Haomo Tang[1], Yande Jiang[2], Jian Kuang[1], and Evangeline F.Y. Young[1]

[1]Department of Computer Science and Engineering, The Chinese University of Hong Kong
[1]{cwliu, pstu, pbwu1, hmtang2, jkuang, fyyoung}@cse.cuhk.edu.hk
[2]{yandejiang_nudt}@163.com

Abstract—**To reduce chip-scale topography variation, dummy fill is commonly used to improve the layout density uniformity. Previous work either sought the most uniform density distribution or sought to minimize the inserted dummy fills while satisfying certain density uniformity constraint. However, due to more stringent manufacturing challenges, more criteria, like line deviation and outlier, emerge at newer technology nodes. This paper presents a joint optimization scheme to consider variation, total fill, line deviation and outlier simultaneously. More specifically, first we decompose the rectilinear polygons and partition fillable regions into rectangles for easier processing. After decomposition, we insert dummy fills into the fillable rectangular regions optimizing the fill metrics simultaneously. We propose three approaches–*Fast Median approach*, *LP* approach and *Iterative* approach, which are much faster with better quality, compared with the results of the top three contestants in the ICCAD Contest 2014.**

Keywords—*Polygon Decomposition; Filling Optimization; Fill Insertion*

I. INTRODUCTION

To reduce manufacturing variation and also improve performance predictability and yield, layout should be made uniform with respect to certain density criteria in the *Chemical Mechanical Polishing (CMP)* process. Inserting dummy fills into sparse areas in a layout is a common way to make layout uniform. However, too many dummy fills inserted are not preferred because of the bad impact fills have on performance. Two common objectives are *Min–Var* and *Min–Fill*. For the *Min–Var* objective, variation is minimized. For the *Min–Fill* objective, the amount of inserted dummy fills is minimized while satisfying given variation bounds. Later on, some researchers also addressed the performance degradation issues caused by dummy fills, such as increase of coupling capacitance, delay and slack. However it is still believed that the first-order concern remained to improve uniformity of layout [8].

A number of previous works have been done on this *CMP* dummy fill problem. We can briefly classify them into three categories according to their optimization objectives. **(1)** Some previous works target at minimizing the variation. The work [6] has proposed a window-based solution to this dummy fill problem such that a density bound is preserved and variability across windows is minimized during insertion. The work [7] has proposed a *Monte-Carlo* based filling method using some fast dynamic data structures. The same group has later proposed another approach called the *hybrid hierarchical filling* to solve the *Min-Var* problem, which is proved to be very efficient, scalable and accurate in [9]. **(2)** Some previous works seek to minimize the total fill under some variation constraints. The paper [9] has worked on this *Min-Fill* objective with a *hybrid hierarchical filling* approach. A recent work [13] has proposed an innovative algorithm based on a polynomial time approximation scheme to solve the *Min-Fill* problem. **(3)** Other works address the impact of CMP dummy fill on performance, like coupling capacitance, delay or slack, crosstalk, dielectric properties. The work [5] has targeted fill optimization to reduce dielectric thickness variation. The work [10] has proposed a performance-driven fill insertion method, either to maximize the smallest slack or to minimize delay. Another work [12] has introduced a coupling-aware fill insertion method.

However the previous works only took the overall variation into consideration when trying to maximize the uniformity. Besides, there

The work described in this paper was partially supported by a grant from the Research Grants Council of the Hong Kong Special Administrative Region, China (Project No. CUHK418611).

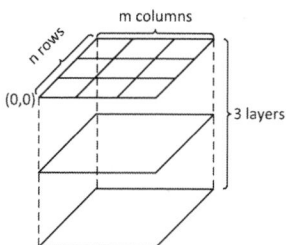

Fig. 1: An Illustration of the Layout

are no existing fast linear filling algorithms. As a result, the previous works are too slow to apply to real industrial benchmarks. To motivate the development of more effective and practical filling algorithms, ICCAD 2014 held a dummy fill contest [19] and released a suite of industrial benchmarks. Many conventional issues and emerging concerns were holistically modeled with new objectives–line variation and outlier variation, which leads to a brand new challenge.

In this paper, we propose two efficient methods to preprocess complex rectilinear fillable regions and then insert fills to optimize multiple variation related objectives simultaneously. Our contributions can be summarized as follows.

1. We propose and implement two fast polygon-decomposition algorithms applicable to complex rectilinear polygons. These algorithms are inspired by a previous work [1]. However, the algorithm *PTR (Polygon-To-Rectangle)* in [1] will fail in some cases and cannot handle the situation when two points overlap. We propose *I-PTR (Improved PTR)* and also our innovative algorithm – *Edge-based Decomposition* to resolve all these problems.

2. We propose three algorithms to solve this multi-objective dummy fill problem – *Fast Median, Linear Programming (LP)* and an iterative heuristic approach. *Fast Median* is an innovative algorithm in filling optimization. It is a very efficient $O(n)$ algorithm and we can show theoretically that its overall variation is close to the optimal solution. In the *Linear Programming* approach, we formulate the problem as a standard linear program and solve it with existing tools. In the iterative heuristic algorithm, we iteratively insert fills to minimize variations. The three algorithms are much faster with better quality, compared with the results of the top three contestants in the ICCAD Contest 2014.

The remainder of this paper is organized as follows. The problem formulation is in Section II. We give an overview of our methodology in Section III. Section IV, V and VI describe in details the decomposition, optimization and dummy fills insertion steps. Finally, experimental results will be shown in Section VII.

II. NOTATION AND PROBLEM FORMULATION

To measure variation, the whole layout is divided into non-overlapping windows starting from the lower left corner [6]. In this work, the windows are of size $20\mu m \times 20\mu m$, which is the value having been used in *CMP* modeling since several years ago. Besides, $K = 3$ metal layers are considered simultaneously. An illustration is shown in Fig. 1.

978-1-4799-8720-7/15 $31.00 © 2015 IEEE

In this problem, we are given a set of non-overlapping rectilinear regions in each layer such that fills can be inserted into these regions (*fillable region*) to optimize uniformity. Each layer is divided into $m \times n$ windows and each window i will have a *no-fill density d_i* given which is the density due to the layout features. The objective is to insert fills into those fillable regions to minimize:

- **Overall Variation** V, i.e., standard deviation of the chip density. Deviations are added up across all layers, i.e., $V = \sum_{k=1}^{K} \sigma_k$ where σ_k is the density standard deviation on layer k.

- **Line Variation** L, i.e., sum of density difference of each window from the corresponding column-based average . These variations are also added up across layers.

- **Outlier Variation** O, i.e., sum of density difference of each *outlier window* from the chip density mean where an *outlier window* is one with density more than $\pm 3\sigma_k$ from the chip density mean. These variations are also added up across layers.

- **Total Fills** F, i.e., sum of the fills inserted into the whole layout.

We formulate the problem as follows.

Multi-Objective Fill Insertion Given a set of non-overlapping rectilinear fillable regions on each of the given K layers, each layer is divided into $m \times n$ windows and each window i has a no-fill density d_i, insert fills into those fillable regions to minimize the cost:

$$cost = F + w_v \times V + w_l \times L + w_o \times O$$

where w_v, w_l and w_o are weights of the objectives. Aligning with the requirement of the ICCAD Contest 2014, the inserted fills must have a minimum width and spacing of $32nm$ and a minimum area of $4800nm^2$ [19]. The major parameters are listed in Table I and for simplicity, we define the parameters on one layer only.

III. An Overview of our Method

An overview of our approach is shown in Fig. 2. First we load a file of GDS format that defines the fillable regions as a set of complicated non-overlapping rectilinear polygons, and another text file that provides no-fill density information d_i of each window i. There are two major steps in our approach, *decomposition* and *fill optimization*.

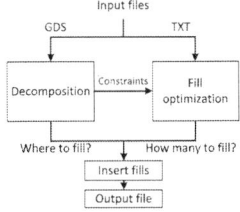

Fig. 2: An Overview of Our Method

TABLE I: Major Parameters

Name	Description
n	row number of windows on each layer
m	column number of windows on each layer
d_i	no-fill density of window i where $i = 1 \dots m \times n$
a_i	fillable area in window i divided by area of window, where $i = 1 \dots m \times n$
x_i	density of fill inserted in window i where $i = 1 \dots m \times n$
l_i	lower bound density of window i, which is equal to d_i
u_i	upper bound density of window i, which is equal to $d_i + a_i$
d_{min}	minimum of $x_i + d_i$ among all windows i
d_{max}	maximum of $x_i + d_i$ among all windows i
cl_j	minimum of $x_i + d_i$ among all windows i in column j
cu_j	maximum of $x_i + d_i$ among all windows i in column j
c_{dis}	the maximum $cu_j - cl_j$ among all columns $j = 1 \dots m$

This metric is an approximation of measuring deviations after lithographic tools implement an automated fix as described in [14].

In the decomposition step, we will decompose the fillable regions into rectangles for later processing. Inspired by the work in [1], we implement both *I-PTR* and a new decomposition algorithm–*Edge-based Decomposition* to divide the fillable regions into rectangles according to the window boundaries on one hand and to get more large rectangles formed on the other hand. We will see later why more large rectangles is beneficial. After this decomposition step, we can compute the total fillable area in each window and provide an upper bound on the dummy fill to be inserted in that window.

In the fill optimization step, given the no-fill density d_i and the upper bound constraints obtained from the decomposition step, we compute the fill inserted into each window to optimize the cost function that is a weighted sum of the four objective values. We will propose and show the results of three different methods. The *Fast Median* algorithm is the most efficient one and we will show theoretically that its solution is very close to the optimum in terms of *Overall Variation*. Another two approaches are based on *LP* and an iterative heuristic.

Based on the results obtained by the fill optimization step, we will insert fills in the rectangles decomposed from the fillable regions. To control the size of the output file, large rectangles will be filled first. This is why larger rectangles are preferred in the decomposition step. Details of our algorithm will be given in the following sections.

IV. Decomposition

In the decomposition process, a GDS file that defines the fillable regions in the form of rectilinear polygon is input. Some of these polygons are very complicated with thousands of vertices and possibly with holes inside. Therefore, we need a fast and effective method to decompose these polygons for later processing. There are two major steps as showed in Fig. 3. First, all the fillable regions are divided into rectangles. Then, we further divide these rectangular fillable regions according to the windows.

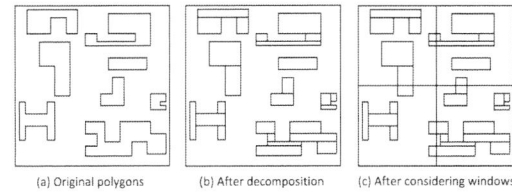

(a) Original polygons   (b) After decomposition   (c) After considering windows

Fig. 3: Steps in Decomposition

A lot of related work has been done in decomposing rectilinear polygon into rectangles [1][2][3][4][11][15]. A recent survey has shown that there is no "generally optimal" decomposition method [16]. While among all the work, the approach in [1] is both easy to implement and able to handle polygons with holes. As a result, it is widely adopted as a decomposition tool in recent works [17][18]. However, it is not found by any previous work that the approach *PTR* in [1] will actually fail in some cases. In the following, we will briefly describe *PTR*, followed by its drawback that some cases cannot be handled properly and finally, we will present *I-PTR* and also a new decomposition algorithm called *Edge-based Decomposition*.

A. PTR and I-PTR

PTR works as follows. Given a set A of coordinates defining a closed rectilinear region, it will first identify 3 points as follows.

- $P_k(x_k, y_k)$ where P_k is the leftmost point among all the lowest points in A.

- $P_l(x_l, y_l)$ where P_l is the second leftmost point among all the lowest points in A.

- $P_m(x_m, y_m)$ where P_m is the leftmost point among all the lowest points with $x_k \le x_m < x_l$ and $y_m > y_k$.

978-1-4799-8720-7/15 $31.00 © 2015 IEEE   45

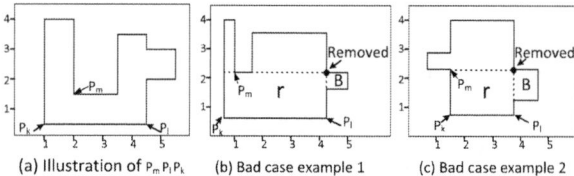

Fig. 4: Illustration of *PTR*

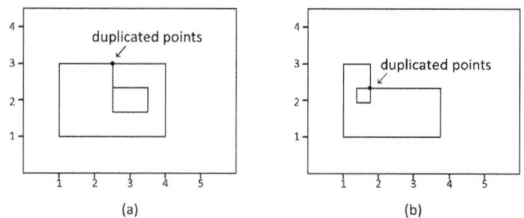

Fig. 5: Examples of Duplicated Points

The three points are illustrated in Fig. 4(a). These three points P_k, P_l and P_m define a rectangle r that will be taken as one of the rectangles obtained. Then, A is updated as follows. For each corner point P_i where $i = 1, 2, 3, 4$ of r, $A = A - \{P_i\}$ if $P_i \in A$; else, $A = A + \{P_i\}$. After the update, the process repeats until $A = \emptyset$. The proof of linear complexity can be found in [1]. In most cases, *PTR* is correct and fast. However, there are two drawbacks about this algorithm. First, the given coordinates must be unique, i.e., *PTR* does not support duplicated points, while duplicated points are common in our benchmarks. Some typical examples of duplicated points are shown in Fig. 5. Second, there are cases in which *PTR* will fail. Actually *PTR* cannot handle the suite of industry benchmarks released by the contest organizer for these two reasons.

The following is an analysis of this algorithm, through which we will see where and why it will fail.

From the algorithm, we can see that a very important step for *PTR* is extracting a rectangle r from the original polygon and then deleting or inserting points in set A. In fact, when we extract a rectangle from the polygon, we are changing the shape of the polygon. If the edges of the rectangle r overlap with some edges of the polygon, the overlapped parts of the edges should be removed from the polygon, and the remaining parts (not exist in the original polygon) should be added into the polygon. The step is pretty much like a *complement* operation on the edges. Fig. 6 shows an example of this *complement* operation. Segments BP_m, P_mP_k, P_kP_l, and P_lD are edges on the original polygon which overlap with the edges of the rectangle to be extracted. As a result, they are deleted from the polygon. On the contrary, segments BC and CD are inserted.

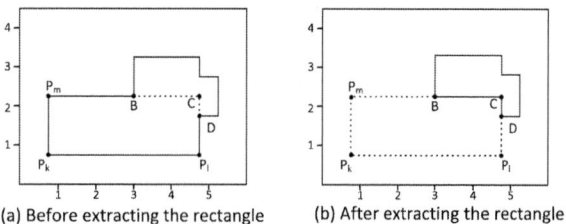

Fig. 6: Rectangle Extraction Process

This *complement* operation on edges causes the corresponding deletion or insertion of vertices. We have to insert the vertices of the rectangle

---

The work [1] does not show a proof of correctness.

into A if they are not in set A, e.g. point C in Fig. 6. On the other hand, we may need to delete or insert the points from A if they are already in A. In this case, whether to insert or delete depends on the shape of the resultant polygon surrounding the point after the extraction. Let us take the upper right vertex X of the extracted rectangle as an example for discussion. Note that analyzing the upper right vertex X will not lose generality since the four corners of the extracted rectangle r are symmetrical.

In total, there are seven cases for the upper right vertex X, which are illustrated in Fig. 7. The shadowed regions indicate the exterior of the polygon. The upper half and lower half are illustrations before rectangle extraction and after rectangle extraction respectively. In Fig. 7(a), it is obvious that the space has nothing after rectangle extraction. Hence, X should be deleted. In Fig. 7(b)(c), the point X lies on a segment after rectangle extraction. As a result, they should also be removed. The situation for Fig. 7(d) is more complicated. Not only that we should not delete the original point, instead we should insert one more point to describe the two parts remaining after rectangle extraction. In Fig. 7(e)(f)(g), we have originally two points to describe the two parts. After rectangle extraction, only one point is needed, so we delete one of the two points.

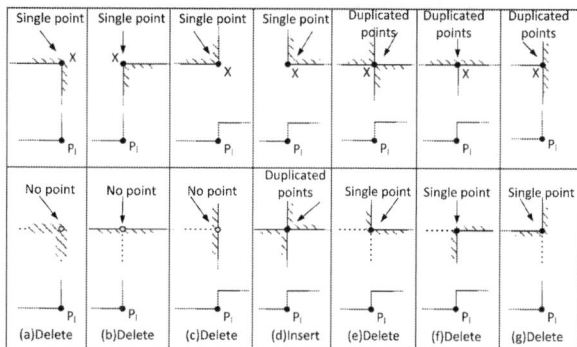

Fig. 7: Seven cases in which the upper-right vertex X of the extracted rectangle r already exists in A

We can see that *PTR* will fail in the case of Fig. 7(d). *PTR* always deletes the vertices of the extracted rectangle r when the vertices are in the set A, while we actually need to insert one more point in the case of Fig. 7(d). Two examples of Fig. 7(d) are shown in Fig. 4(b)(c). In these examples, the upper right corner of the rectangle r found will be removed by *PTR* because the same point already exists in A. However, this is incorrect since we still need this point to define other rectangles, e.g., rectangle B.

Our new algorithm–*I-PTR* will solve this problem by handling the seven cases and duplicated points appropriately. Since we have exhausted all the cases and deal with them correspondingly, our algorithm is thus correct. The pseudocode of *I-PTR* is described in Algorithm 1.

B. Edge-based Decomposition

Rather than point-based, our new edge-based algorithm can also handle duplicated points and all the cases correctly. Besides, it can produce more larger rectangles, which is good for both the filling optimization and controlling the output file size.

In *I-PTR*, we extract a rectangle by finding a point P_m defined as the leftmost point among all the lowest points with $x_k \leq x_m < x_l$ and $y_m > y_k$. This constraint will limit the height of the rectangle to the lowest point right above P_k. We will introduce our *Edge-based Decomposition* algorithm in the following. In *Edge-based Decomposition*, the height of the rectangle r to be extracted is not determined by a point. Instead, we will find all the lowest edges of the polygon higher than P_k and overlap with the bottom edge P_kP_l when projecting on the x axis. More formally, let H be the set of all horizontal edges of the polygon, and we are finding all the lowest edges $E_{m_1}, E_{m_2}, \cdots, E_{m_p}$ in H satisfying:

Algorithm 1 *I-PTR*

1: Initialize A //Let A be a linked list storing all the corner points of a polygon
2: Sort A w.r.t. y coordinates first and then w.r.t. x coordinates
3: **while** $A \neq NULL$ **do**
4:    **repeat**
5:       $P_k, P_l \leftarrow$ the first 2 points in A
6:       Delete P_k, P_l from A if $P_k = P_l$
7:       //If one of the duplicated points is left, i.e., the other duplicated point has been deleted, A will not be changed.
8:    **until** $P_k! = P_l$
9:    Determine P_m
10:   Extract the rectangle r based on P_k, P_l, P_m
11:   **for** each corner point $P_i, (i = 1, 2, 3, 4)$ of r **do**
12:      **if** $P_i \notin A$ **then**
13:        $A = A + \{P_i\}$
14:      **else if** P_i is case(d) in Fig. 7 **then**
15:        $A = A + \{P_i\}$
16:      **else**
17:        $A = A - \{P_i\}$
18:      **end if**
19:   **end for**
20: **end while**

- $y_{m_i} > y_k$, where y_{m_i} is the y coordinate of the edge E_{m_i} (note that $y_{m_1} = y_{m_2} = ... = y_{m_p}$).

- Interval $(l_{m_i}.r_{m_i})$ intersects with interval (x_k, x_l), where l_{m_i} and r_{m_i} are the x coordinates of the left and right end points of E_{m_i} respectively.

The y coordinate of the top edge of the rectangle r will be equal to y_{m_i}. Note that we do not regard "touching", i.e. sharing only one point, as overlapping.

After extracting the rectangle r, the shape of the polygon will change according to the edge *complement* operation explained in Section IV-A. Therefore we have to update H correspondingly after the extraction where H contains all the horizontal edges of the current polygon. $P_k P_l$ will always be removed as a whole, since it is an edge of the original polygon. For the top edge, we first need to remove every E_{m_i} totally or partly depending on how much it overlaps with the top edge of r, and then the segments along the top edge of r that are not in the polygon originally will be inserted.

The pseudocode of *Edge-based Decomposition* is described in Algorithm 2.

Algorithm 2 Edge-based Decomposition

1: $A \leftarrow$ all the corner points of a polygon
2: $H \leftarrow$ all horizontal edges of the polygon
3: Sort A w.r.t. y coordinates first and then w.r.t. x coordinates
4: **while** $A \neq NULL$ **do**
5:    **repeat**
6:       $P_k, P_l \leftarrow$ the first 2 points in A
7:       Delete P_k, P_l from A if $P_k = P_l$
8:    **until** $P_k! = P_l$
9:    Determine all E_{m_i} for $i = 1...p$
10:   Extract the rectangle r
11:   Calculate the new edges E_{n_i} by complementing E_{m_i} with the top edge of r
12:   Delete all E_{m_i} from H
13:   Insert all E_{n_i} into H
14:   Insert corner points of rectangle r into or delete them from A in the same way as in Algorithm 1
15: **end while**

C. Dividing by Windows

After the above efficient decomposition step, all the fillable regions are divided into rectangles. Next, we need to further divide these rectangular fillable regions according to the windows. That is, we need to figure out to which window a rectangle belongs and further divide the rectangle if it lies across more than one windows. We perform this step efficiently in $O(n)$ time where n is the total number of rectangles obtained after this window based cutting. For each rectangle r obtained from *I-PTR* or *Edge-based Decomposition*, we can quickly find out the

windows in which r's lower left corner and upper right corner lie within. From this information, we can compute the windows with which this rectangle overlaps and update to them the corresponding fillable regions.

V. FILL OPTIMIZATION

We propose three different approaches to solve this multi-objective fill insertion problem. Among all the objectives, a common goal is to minimize the total fill while at the same time to have the density of each window as close to (1) the chip density mean and to (2) the individual column-based density mean as much as possible. We explore three approaches, namely *Fast Median*, *LP based* and *Iterative*. They will be discussed in detail in the following sections.

A. Fast Median Approach

Fast Median is a very quick approach to minimize the overall variation V. It is also an innovative algorithm for filling optimization with $O(n)$ time complexity. We can show that its result is close to the optimum, and the experimental results in Section VII also confirm its good performance.

For each window i, we are given its no-fill density d_i, which will serve as the lower bound density l_i of window i. We can also obtain the total fillable area a_i in this window after the decomposition step in Section IV, which will provide us the upper bound density $u_i = d_i + a_i$ of window i. Suppose we can find an optimal chip density mean μ that minimizes V. We can just determine the inserted fill density x_i in each window i directly as follows. If $\mu < l_i$, $x_i = 0$, and the difference of the window density from the mean will be $l_i - \mu$. If $l_i \leq \mu \leq u_i$, $x_i = \mu - l_i$, and the difference will be 0. If $\mu > u_i$, $x_i = u_i - l_i$ and the difference will be $\mu - u_i$. Here, we can use a function $v(\mu)$ to represent the overall variance cost:

$$v(\mu) = \sum_{i=1}^{m \times n} \frac{|\mu - l_i| + |\mu - u_i| - (u_i - l_i)}{2}$$

Theorem 1: Consider the median M of the set of numbers, $\{l_i, u_i | i = 1 \ldots m \times n\}$. The function $v(\mu)$ achieves its minimum value when $\mu = M$.

Proof: First, the constant inside the summation sign can be removed when seeking the best μ. Therefore, the problem is the same as finding the μ to minimize $\sum_{i=1}^{m \times n} |\mu - l_i| + |\mu - u_i|$. This optimization problem has become a simple one as follows. Given $2 \times m \times n$ points, find a position to minimize the total distance from every point to that position. The optimal position is the median of all these points, which can be shown easily using induction.

- When there are only two points, the median of the two points (choose either one) is an optimal position.

- Suppose the statement holds for the case of $2 \times k$ points, we need to prove that it also holds for the case of $2 \times (k + 1)$. We divide the $2 \times (k + 1)$ points into two groups. One group only contains the leftmost and the rightmost points while the middle $2 \times k$ points form the other group. For the middle $2 \times k$ points, we know that its optimal position is the median point using the inductive hypothesis. This median point is also the optimal position of the leftmost and the rightmost points. So it is the optimal point for the whole $2 \times (k + 1)$ points.

By mathematical induction, the statement holds for all natural k. ∎

Analysis: In our implementation, we find the median M first and then adjust the density of each window considering M as the chip density mean. However after the adjustment, it is not guaranteed that the chip density mean is exactly equal to M. (Otherwise, optimality will be guaranteed according to the Theorem 1.) This is why the variation obtained from this method is a little bit worse than the optimum in the experiments. But we can see from the experiments that the mean is very close to M and the difference is smaller than 5%. During the process of *Fast Median*, we do the density adjustment of each window only once. So the time complexity is $O(n)$. As far as we know, this is the most efficient algorithm on filling optimization, and its quality is also very close to the optimum. As a result, this *Fast Median* algorithm performs well in terms of both running time and quality, which will also be confirmed in Section VII.

B. Iterative Approach

In the *Iterative* approach, we consider *Overall Variation V*, *Line Variation L*, *Outlier Variation O* and *Total Fills F* at the same time. For V and L, we adjust x_i to make $x_i + d_i$ of window i as close to the value of $\mu_{weighted}$ computed below as possible.

$$\mu_{weighted} = \alpha \times \mu_{chip} + (1 - \alpha) \times \mu_{column}$$

where μ_{chip} represents the chip density mean and μ_{column} represents the column-based density mean of window i, and α is a parameter to weight between these two variations. More specifically, when $\alpha = 0$, we only take column-based density mean into account. When $\alpha = 0.5$, we consider chip density mean and column-based density mean equally. When $\alpha = 0.9$, chip density mean plays a more important part.

To optimize O, we will check the density of each window in every iteration and adjust the density of the outlier windows back within $\pm 3\sigma$ from the chip density mean. In order to control F, we will stop the process if too many fills are inserted.

The detailed algorithm is described in Algorithm 3.

Algorithm 3 Iterative

---

while $iter \leq max\_iter$ and $\Delta var \geq min\_var$ and $num\_fill \leq max\_fill$ **do**
  for i from 1 to $m \times n$ **do**
    Determine $\mu_{weighted} = \alpha \times \mu_{chip} + (1 - \alpha) \times \mu_{column}$
    if $\mu_{weighted} \leq d_i + x_i$ **then**
      decrease x_i but x_i must be non-negative
    else
      increase x_i but x_i cannot exceed the maximum density of the window
    end if
    update μ_{chip} and μ_{column}
  end for
  $iter = iter + 1$, Determine Δvar and num_fill
end while

C. Linear Programming Approach

In the *LP* approach, we will take *Total Fills F*, *Overall Variation O* and *Line Variation L* into consideration. In order to reduce running time, we formulate the objectives of *Overall Variation O* and *Line Variation L* as differences between the corresponding maximum density and minimum density, as shown in the *LP* formulation below. For each window, the density includes the fill density x_i we will insert plus the no-fill density d_i. d_{max} and d_{min} are respectively the maximum and minimum density among all the windows. One of the objectives in the *LP* is to minimize the difference between d_{max} and d_{min}. For *Line Variation L*, we also compute the upper bound cu_j and the lower bound cl_j for each column j and try to minimize the difference between them. However, the order of magnitude of the total fills is larger than that of the *Overall Variation O* and the *Line Variation L*. Thus, we will multiply the later terms in the objective function by a value of $N = m \times n$. The weights s and r represent the relative importance of the terms in the objective function. In the experiments, we use $s = 2$ and $r = 2$.

Minimize $\sum_{i=1}^{m \times n} x_i + s \times N \times (d_{max} - d_{min}) + r \times N \times c_{dis}$
Subject to:
  $0 \leq x_i \leq a_i, \; i = 1, ..., N$
  $c_{dis} \geq cu_i - cl_i, \; i = 1, ..., m$
  $d_{min} \leq x_i + d_i \leq d_{max}, \; i = 1, ..., N$
  $cl_{i+1} \leq x_{i \times n+j} + d_{i \times n+j} \leq cu_{i+1},$
  $i = 0, ..., m-1, \; j = 1, ..., n$

Besides, we will have a post-processing step (for all the three approaches) to locate the outlier windows and adjust their fills back to within the $\pm 3\sigma$ limits if possible.

VI. INSERTION

In this step, we will insert fills based on the result of the optimization and decomposition steps. For each window i, we obtain x_i, the area we need to insert. We will then insert the fills rectangle by rectangle. For each rectangular fillable area, we will insert a dummy fill of size as large as possible, as long as it does not violate the design rules. In order to decrease the output file size, we will adopt the principle of large rectangle first. In this way, we can use less rectangles, which means smaller output file.

VII. EXPERIMENTAL RESULTS

The results of the top three teams in the ICCAD Contest 2014 were reported by the organizers. The configuration of the computer used can be found in [19]. Our programs were performed on a 64-bit Linux machine with Intel Xeon 3.4GHz CPU and 32GB memory. In our experiments, we used the suite of industrial benchmarks s, m and b released by the contest organizer. They have 3 layers with 8×20, 117×117, 325×53 windows on each layer respectively.

A. Fast Median, Iterative and LP

In this section, we will compare the results of *Fast Median*, *Iterative* and *LP*. The results are shown in Table II. The best result (minimum value) for each objective and benchmark is highlighted. Column 6 shows the normalized outlier variation and is computed as $V \times O$. Note that the outlier variation will decrease when the overall variation increase. In order to avoid minimizing the outlier variation by worsening the overall variation, the outlier variation is evaluated according to $V \times O$, instead of just based on the value of O.

From Table II, we can see that *Fast Median* is the fastest, especially in terms of the optimization runtime, and it uses a small amount of dummy fills. Although *Iterative* outperforms it in terms of V, L and $V \times O$ on average, the difference is very small. Furthermore, *Fast Median* is much better when applied to small benchmarks with $V = L = 0$. In conclusion, *Fast Median* is the best for its efficiency and good quality.

B. Fast Median and Top Three Contestants

In this section, we will compare *Fast Median* with the top three teams in the ICCAD Contest 2014, which are shown in Table III.

From Table III, we can see that the V and L results of *Fast Median* are all much smaller (at least 50%) than those of the top three teams. In terms of F and $V \times O$, *Fast Median* is better than the top 2 and top 3 teams. Top two teams insert only a very small amount of dummy fills. The top 1 team's chip density standard deviations are big and thus the $\pm 3\sigma$ limits are loose. However, a big chip density variation is not preferred in practice. On the other hand, *Fast Median* has very good variation control in all aspects and can run very fast. It is 25× faster than the fastest one among the top three teams. Besides, the organizer of the ICCAD Contest 2014 has helped us to evaluate the results of *Fast Median*, and our total score is 0.70, while the total scores of the top three teams are 0.63, 0.61 and 0.61 respectively. (The larger the score the better.) The contest evaluation considers even more factors which can be found in [19].

C. I-PTR and Edge-based Decomposition

In this section, we compare the new algorithm *Edge-based Decomposition* with *I-PTR*. Table IV shows that *Edge-based Decomposition* outperforms *I-PTR* in all of the four objectives–F, V, L and O. In this comparison, *Fast Median* was used for the fill optimization.

VIII. CONCLUSION

In this paper, we improved the layout density uniformity considering four objectives simultaneously: *Overall Variation*, *Line Variation*, *Outlier Variation* and *Total Fills*. For decomposition, we proposed two efficient algorithms *I-PTRS* based on [1] and *Edge-based Decomposition*. For optimization, we studied three approaches: *Fast Median*, *LP* and *Iterative*. After optimization, we inserted fills into the fillable regions in an order of larger rectangles first. The experimental results show that our

---

We have compared the running times of the same program on our computer and on the organizer's computers. The ratios are 1:1.42, 1:1.53 and 1:1.42 respectively for data sets s, m and b.

Total area of fills inserted divided by the whole area of the layout.

TABLE II: Comparison Between *Fast Median*, *Iterative* and *LP*

Benchmark	Approaches	F	V	L	$V \times O$	Optimization Runtime	Total Runtime(s)
s	*Fast Median*	5.69%	**0**	**0**	**0**	**0.00004**	**0.45**
	Iterative ($\alpha = 0.5$)	**5.23%**	0.001	0.044	0.00001	0.0006	0.48
	LP	5.69%	**0**	0.0001	**0**	0.0236	**0.45**
m	*Fast Median*	21.15%	0.2205	1157.87	50.31	**0.0012**	**12.43**
	Iterative ($\alpha = 0.5$)	23.26%	**0.2173**	**1120.66**	**46.48**	0.0139	12.94
	LP	**0.78%**	0.4377	5220.14	80.49	102.0300	109.70
b	*Fast Median*	35.01%	0.1072	421.86	9.74	**0.0018**	**9.90**
	Iterative ($\alpha = 0.5$)	37.67%	**0.097**	446.52	**5.83**	0.0146	11.79
	LP	**31.18%**	0.1322	543.88	107.68	28.9800	38.93
average	*Fast Median*	20.62%	0.1092	526.58	20.02	**0.0010**	**7.59**
	Iterative ($\alpha = 0.5$)	22.05%	**0.1051**	**522.41**	**17.44**	0.0148	8.40
	LP	**12.55%**	0.1900	1921.34	62.72	43.6779	49.69

TABLE III: Comparison Between *Fast Median* with Top Three Contestants

Benchmark	Approaches	F	V	L	$V \times O$	Total Runtime(s)
s	*Fast Median*	5.69%	**0**	**0**	**0**	**0.45**
	Top 1 Team	**3.74%**	0.028	3.145	**0**	7.38
	Top 2 Team	4.24%	0.007	0.389	0.00036	9.22
	Top 3 Team	14.29%	0.001	0.122	**0**	9.48
m	*Fast Median*	21.15%	**0.2205**	**1157.87**	50.31	**12.43**
	Top 1 Team	17.65%	0.2850	3108	**21.95**	532.41
	Top 2 Team	**13.38%**	0.2860	2310	66.64	263.65
	Top 3 Team	22.47%	0.2600	1884	56.16	302.22
b	*Fast Median*	35.01%	**0.1072**	**421.86**	9.74	**9.90**
	Top 1 Team	22.58%	0.3270	2277	**2.94**	290.81
	Top 2 Team	**12.38%**	0.3200	1667	48.32	292.48
	Top 3 Team	28.75%	0.2660	1426	29.53	267.77
average	*Fast Median*	20.62%(1)	**0.1092(1.0)**	**526.58(1.0)**	20.02(1.0)	**7.59(1.0)**
	Top 1 Team	14.66%(0.7)	0.2133(2.0)	1796.05(3.4)	**8.30(0.4)**	276.87(36)
	Top 2 Team	**10.00% (0.5)**	0.2043(1.9)	1325.80(2.5)	48.32(2.4)	188.45(25)
	Top 3 Team	21.83%(1.1)	0.1757(1.6)	1103.37(2.1)	38.32(1.9)	193.16(26)

The ratios of the running times for the same program on our computer and on organizer's computer for benchmark s, m, b are 1:1.42, 1:1.53, 1:1.42 respectively.

TABLE IV: Comparison Between I-PTR and Edge-based Decomposition

Benchmark	Approaches	F	V	L	$V \times O$
s	*I-PTR*	**5.69%**	**0**	**0**	**0**
	Edge-based	**5.69%**	**0**	**0**	**0**
m	*I-PTR*	21.31%	**0.2205**	1157.87	50.31
	Edge-based	**21.15%**	**0.2205**	**1151.81**	**50.04**
b	*I-PTR*	**34.86%**	0.1072	421.86	9.74
	Edge-based	35.01%	**0.1043**	**408.26**	**9.12**
average	*I-PTR*	20.62%	0.1092	526.58	20.02
	Edge-based	**20.61%**	**0.1083**	**520.02**	**19.72**

approach outperforms the top three contestants in the ICCAD Contest 2014. Our *Fast Median* is much faster with better quality. Besides, *Edge-based Decomposition* has better quality than *I-PTR*.

REFERENCES

[1] Gourley, Kevin D., and Douglas M. Green, "Polygon-to-rectangle conversion algorithm." in *IEEE COMP. GRAPHICS* &: *APPLIC.* 3, no. 1 : 31-32, 1983.

[2] Ferrari, L., P. V. Sankar, and Jack Sklansky. "Minimal rectangular partitions of digitized blobs." in *Computer vision, graphics, and image processing* 28.1: 58-71, 1984.

[3] Nahar, Surendra, and Sartaj Sahni. "Fast algorithm for polygon decomposition." in *Computer-Aided Design of Integrated Circuits and Systems*, IEEE Transactions on 7.4: 473-483, 1988.

[4] Liou, W. T., J. J. Tan, and R. C. Lee. "Minimum partitioning simple rectilinear polygons in O (n log log n)-time." in *Proceedings of the fifth annual symposium on Computational geometry*. ACM, 1989.

[5] B.E. Stine et al., "The Physical and Electrical Effects of Metal-Fill Patterning Practices for Oxide Chemical-Mechanical Polishing Processes." in *IEEE Trans. Electron Devices*, 45(3), : 665-679, 1998.

[6] Kahng, Andrew B., Gabriel Robins, Anish Singh, and Alexander Zelikovsky, "Filling algorithms and analyses for layout density control." in *Computer-Aided Design of Integrated Circuits and Systems, IEEE Transactions* on 18, no. 4 : 445-462, 1999.

[7] Chen, Yu, Andrew B. Kahng, Gabriel Robins, and Alex Zelikovsky. "Monte-Carlo algorithms for layout density control." in *Proceedings of the 2000 Asia and South Pacific Design Automation Conference*, pp. 523-528. ACM, 2000.

[8] Grobman, Warren, M. Thompson, R. Wang, C. Yuan, Ruiqi Tian, and E. Demircan."Reticle enhancement technology: Implications and challenges for physical design." in *Proceedings of the 38th annual Design Automation Conference*, pp. 73-78. ACM, 2001.

[9] Chen, Yu, Andrew B. Kahng, Gabriel Robins, and Alexander Zelikovsky. "Area fill synthesis for uniform layout density." in *Computer-Aided Design of Integrated Circuits and Systems, IEEE Transactions* on 21, no. 10 : 1132-1147, 2002.

[10] Chen, Yu, Puneet Gupta, and Andrew B. Kahng, "Performance-impact limited-area fill synthesis." in *Advanced Microelectronic Manufacturing, International Society for Optics and Photonics* pp. 75-86, 2003.

[11] Tseng, I-Lun, and Adam Postula. "An efficient algorithm for partitioning parameterized polygons into rectangles." in *Proceedings of the 16th ACM Great Lakes symposium on VLSI*. ACM, 2006.

[12] Kahng, Andrew B., and Rasit Onur Topaloglu, "Performance-aware CMP fill pattern optimization." Invited Paper, in *Proc. VMIC*, 2007.

[13] Feng, Chunyang, Hai Zhou, Changhao Yan, Jun Tao, and Xuan Zeng. "Efficient Approximation Algorithms for Chemical Mechanical Polishing Dummy Fill." in *Computer-Aided Design of Integrated Circuits and Systems, IEEE Transactions* on 30, no. 3 : 402-415, 2011.

[14] B. Liegl, B. Sapp, S. Greco, T.A. Brunner, N. Felix et al. "Predicting substrate-induced focus error," int *Journal of Micro/Nanolith. MEMS MOEMS* 9(4), 041311, 2011.

[15] Durocher, Stephane, and Saeed Mehrabi. "Computing Partitions of Rectilinear Polygons with Minimum Stabbing Number." in *Computing and Combinatorics. Springer Berlin Heidelberg.* 228-239, 2012

[16] Suk, Tomáš, Cyril Höschl IV, and Jan Flusser. "Decomposition of binary images—A survey and comparison." in *Pattern Recognition* 45.12: 4279-4291, 2012.

[17] Zhang, Ye, et al. "Layout decomposition with pairwise coloring for multiple patterning lithography." in *Proceedings of the International Conference on Computer-Aided Design*. IEEE Press, 2013.

[18] Zhao, Jinming, et al. "A new splitting graph construction algorithm for SIAR router." in *ASIC (ASICON), 2013 IEEE 10th International Conference on*. IEEE, 2013.

[19] Rasit O. Topaloglu. "ICCAD-2014 CAD Contest in Design for Manufacturability Flow for Advanced Semiconductor Nodes and Benchmark Suite" in *Proceedings of the 51th annual Design Automation Conference* ACM, 2014.

Conservatively Analyzing Transient Faults

Niels Thole*[†], Görschwin Fey*[†], Alberto Garcia-Ortiz[‡]
*Institute of Computer Science, University Bremen, Germany
{nthole, fey}@informatik.uni-bremen.de
[†]Institute of Space Systems, German Aerospace Center, Germany
[‡]Institute of Electrodynamics and Microelectronics, University Bremen, Germany
agarcia@item.uni-bremen.de

Abstract—Due to the decreasing size of transistors, the probability of transient errors and the variability of the transistor's characteristics in electrical circuits are continuously increasing. These issues demand for techniques to check the robustness of circuits and their behavior under transient faults and variability. We present a conservative algorithm to decide if a transient fault leads to erroneous output of a circuit. Our approach considers logical, timing, and electrical masking as well as variability in the gates. In experiments, we show the runtime of our implementation on the ISCAS-85 benchmarks and compare our approach to precise transistor-level simulations as well as fast logic level analysis.

I. INTRODUCTION

While new technologies facilitate the creation of more advanced systems, the systems become more susceptible to transient faults. External factors like cosmic radiation may induce glitches in the system, which can lead to erroneous behavior. A circuit needs to be analyzed to ensure that no erroneous output is produced under transient faults, i.e., that the circuit is robust. Otherwise vulnerable gates have to be determined. The effects of transient faults may be masked due to logic, timing or electrical effects. During the analysis, variation in the gates' parameters must be taken into account.

Numerous approaches have been proposed to analyze SETs. Some of them only consider logic masking [2, 3, 4, 10, 11] rather aiming at register or gate level verification. For analyzing delay faults timing information [9, 1] is added. Finally, [5] adds electrical masking while [8] focuses on electrical masking only. Moreover, some approaches do not aim at verification but provide error probabilities [5]. Simulation using Spice or the approach of [6] can only consider selected input stimuli. Beside all these previous efforts, variability effects have not been thoroughly discussed until now.

Our approach is the first to analyze a *Single Event Transient* (SET) under logic, electrical and timing masking including variation while considering all possible input assignments. Moreover, the analysis is conservative, i.e., if our approach decides that an SET may not cause an error this decision is safe under the given constraints for variation. Technically, we model the behavior using three-valued logic (0,1,X) where unknown values (X) conservatively approximate variation effects. The decision engine is based on *Boolean Satisfiability* (SAT). For brevity, we only consider combinational circuits. Along the lines of [2] the extension to sequential circuits is straightforward. Empirical results on ISCAS89 benchmarks show

This work has been supported by the University of Bremen's Graduate School SyDe, funded by the German Excellence Initiative.

the effectiveness of our approach. Comparisons against Spice simulations validate the conservativeness.

Application-wise our approach can characterize robustness of a design under variation effects and transient faults, exclude known safe corners from a more detailed analysis, or restrict the diagnosis of transient faults to certain gates.

Section II describes preliminaries. Section III explains our approach in detail. Section IV shows experimental results. Finally, Section V concludes this work.

II. PRELIMINARIES

Three-valued logic extends Boolean logic by a value X, meaning that it is not known if that variable is 1 or 0. Operations on variables are extended accordingly, e.g., $1 \wedge X = X$ and $0 \wedge X = 0$.

A *circuit* is given as a graph $C = (V, E)$ with inputs $I \subseteq V$ and outputs $O \subseteq V$. The function $op : V \rightarrow Op$ matches gates to their operations. The set Op contains possible functions of gates, i.e., $Op = \{f : \{0, 1, X\}^i \rightarrow \{0, 1, X\} | i \in \mathbb{N}\}$.

Discrete values represent the delays of the gates. This is no restriction since the length of a timestep can be normalized to accurately represent real delays. The functions $delay_{min} : V \rightarrow \mathbb{N}$ and $delay_{max} : V \rightarrow \mathbb{N}$ return the minimal and maximal delay of each gate, respectively.

During one clock cycle, the gates change their output values according to the new input values. At the end of the clock cycle, the new values of the flip flops and the primary outputs are defined by their inputs during the sampling window. If the effect of an SET changes the input value of a flip flop during that time, the flip flop saves a wrong value and the system can reach a faulty state. When a primary output is changed, the output of the system is erroneous. The first and last timesteps of the sampling window are given by the functions $sample_s : O \rightarrow \mathbb{N}$ and $sample_e : O \rightarrow \mathbb{N}$, respectively.

For symbolic modeling of the circuit, we use variables from a set *Var*. For the nominal circuit the output is given by the function $output_{nom} : O \times Var^{|I|} \rightarrow Var$.

For the description of the behavior of the circuit, we use *waveforms* similar to [9]. An example is given in Figure 1. Each gate is associated with a waveform and a *timeshift* value. The waveform is a vector $(v_1, v_2, \ldots, v_l) \in Var^l$ that describes the changes of the output of the gate. The timeshift describes an offset from timestep 0 for the first variable in the waveform. The second variable models the following timestep and so on. The logical value before and after the waveform remain identical to the first and last variable, respectively. For example, a constant primary input i is described by timeshift 0 and a waveform that contains only one element.

978-1-4799-8720-7/15 $31.00 © 2015 IEEE

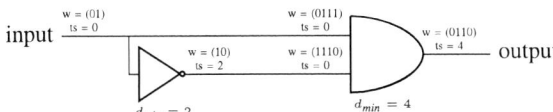

Fig. 1: An example for WaveSAT

To calculate the waveform of a gate g, the waveforms of the predecessor gates are needed. All these waveforms need to be aligned to the same timeshift and have the same length. To change the input's waveforms to fulfill this condition, *padding* is used. The minimal timeshift t_{min} among the inputs is determined. For an input with timeshift t, $t - t_{min}$ copies of the first variable are added in front of the waveform. This is feasible since the value does not change before t. The function $pad_{front} : (\mathbb{N} \times Var^*)^n \to (\mathbb{N} \times Var^*)^n$ performs this padding. The natural number describes the timeshift and the variables describe the waveforms. The value n is the number of predecessors of g and each waveform corresponds to one predecessor.

Afterwards, all waveforms are extended to the same length. Given the maximal length of the waveforms l_{max}, a waveform of length l is extended by adding $l_{max} - l$ copies of the last variable at the back to model the static value. This padding is executed by the function $pad_{back} : (\mathbb{N} \times Var^*)^n \to (\mathbb{N} \times Var^*)^n$.

After modifying the inputs and ensuring the same timeshift and length, the waveform of g is determined. The waveform of g is as long as the padded waveforms of the predecessors. The i-th variable is defined by using the operation of the gate $op(g)$ with the i-th variable of each waveform of the predecessors. The timeshift of g is obtained by adding the delay of the gate to t_{min}[1]. The function $apply_{op} : V \times (\mathbb{N} \times Var^*)^n \to V \times \mathbb{N} \times Var^l$ applies the gate's operation to the waveform with

$$apply_{op}(g, (t, (v_1^1, v_2^1, \ldots, v_l^1)), \ldots, (t, (v_1^n, v_2^n, \ldots, v_l^n))) =$$
$$(g, t + delay_{min}(g), (op(g)(v_1^1, \ldots v_1^n), \ldots, op(g)(v_l^1, \ldots, v_l^n)))$$

The complete process of padding and computing the new waveform is summarized in the function $wave : V \times (\mathbb{N} \times Var^*)^n \to V \times \mathbb{N} \times Var^*$ where

$$wave(g, (t^1, (v_1^1, v_2^1, \ldots, v_{l_1}^1)), \ldots, (t^n, (v_1^n, v_2^n, \ldots, v_{l_n}^n))) =$$
$$apply_{op}(g, pad_{back}(pad_{front}((t^1, (v_1^1, v_2^1, \ldots, v_{l_1}^1)), \ldots,$$
$$(t^n, (v_1^n, v_2^n, \ldots, v_{l_n}^n)))))$$

In our implementation of [9], we reuse the same variable for two timesteps t and $t + 1$ in a waveform if the input variables of the considered gate are equal at both timesteps.

III. CHECKING FOR ROBUSTNESS

Our approach checks the possibility that any output value changes during its sampling window due to a given SET. If counterexamples exist, our algorithm returns one. Otherwise, the circuit is guaranteed to be robust under the given SET, i.e., the given SET cannot affect the output values for any assignment of input values.

To ensure conservativeness in our analysis we use X-values to model uncertainties. For example, transitions caused by the

---

[1]Here, we use the model of [9] with a fixed delay. The maximal delay is taken into account in our approach in Section III.

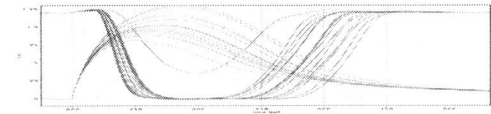

(a) Original Error Signals

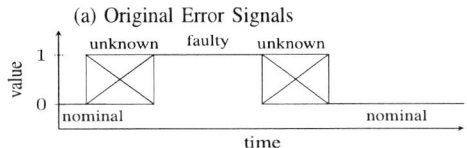

(b) Model of an SET using a three-valued logic

Fig. 2: Modeling an SET

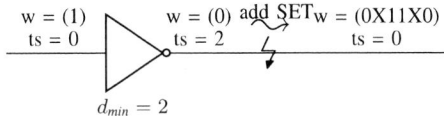

Fig. 3: Generating an SET in a gate

original SET in Figure 2a are approximated by X-values in the model of Figure 2b.

Our algorithm decides the robustness in three steps:

1) Define the waveform for every gate iteratively by using the function *propagate*
2) Compare the waveform of each output during its sampling window to the nominal value
3) Return a counterexample or "circuit is robust" if no counterexamples exist

The function *propagate* defines the waveform of a gate under the given inputs. The execution of *propagate* corresponds to the call of multiple functions: The function *wave* as introduced in the preliminaries describes the initial waveform of the gate. The function *varDelay* describes the variable delay of the gate due to variability and other factors. The SET is given by the function *addSET* and electrical masking is considered in the function *elecMask*. Thus, we define $propagate : V \times (\mathbb{N} \times Var^*)^* \to V \times \mathbb{N} \times Var^*$ as $propagate = elecMask \circ addSET \circ varDelay \circ wave$. These functions are introduced in the following sections.

A. Generation

An SET is given as tuple $(g_{SET}, s_{SET}, e_{SET}, o_{SET}) \in V \times \mathbb{N} \times \mathbb{N} \times \mathbb{N}$, which contains the information about the affected gate, the starting time, the end time of the SET, and the offset where the value is unknown in the beginning and unknown at the end of the time window. We assume the offset to be equal in the beginning and the end of the time window. This model of an SET allows us to consider the unknown rising and falling times of a faulty signal while still remaining on an abstract level.

The function $addSET : V \times \mathbb{N} \times Var^* \to V \times \mathbb{N} \times Var^*$ induces the effects of the SET in the affected gate g_{SET}. The inputs of $addSET$ contain the current gate, the waveform of the gate and its timeshift. If the gate equals g_{SET}, the SET is inserted into the waveform. Otherwise the inputs of this function are identical to its outputs.

In a first step, the algorithm ensures that the nominal behavior of the circuit before and after the SET is modeled

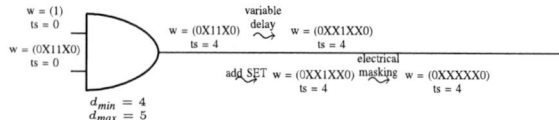

Fig. 4: Propagating a signal considering logical, timing, and electrical masking

within the waveform. Padding extends the waveform to start at timestep $s_{SET} - 1$ and to end at $e_{SET} + 1$ by applying the function $pad_{SET} : V \times \mathbb{N} \times Var^* \to V \times \mathbb{N} \times Var^*$ with

$$pad_{SET}(g, t, (v_1, \ldots, v_l)) =$$
$$(g, min(t, s_{SET} - 1), (\underbrace{v_1, \ldots, v_1}_{s_{SET} - t + 1 \text{ times}}, v_1, \ldots, v_l, \underbrace{v_l, \ldots, v_l}_{e_{SET} - t - l + 2 \text{ times}}))$$

where the function min returns the minimum value among the inputs. If $s_{SET} - t + 1$ or $e_{SET} - t - l + 2$ are less than 0, no variables are added at the corresponding location.

In a next step, the SET as seen in Figure 2b is inserted. The values in the offset are replaced with c_X which is set to X and the variables in between are replaced with the negation of the variable at that location. The outputs of *addSET* are the gate, the new waveform, and timeshift. This is done by the function $apply_{SET} : V \times \mathbb{N} \times Var^l \to V \times \mathbb{N} \times Var^l$ with

$$apply_{SET}(g, t, (v_1, \ldots, v_l)) =$$
$$(g, t, (v_1, \ldots, v_{s_{SET} - t}, \underbrace{c_X, \ldots, c_X}_{o_{SET} \text{ times}},$$
$$\neg v_{s_{SET} + o_{SET} - t + 1}, \ldots, \neg v_{e_{SET} - o_{SET} - t + 1},$$
$$\underbrace{c_X, \ldots, c_X}_{o_{SET} \text{ times}}, v_{e_{SET} - t + 2}, \ldots, v_l))$$

With these functions, *addSET* is defined as:
$$addSET(g, t, (v_1, \ldots, v_l)) =$$
$$\begin{cases} apply_{SET}(pad_{SET}(g, t, (v_1, \ldots, v_l))) & \text{if } g = g_{SET} \\ (g, t, (v_1, \ldots, v_l)) & \text{otherwise} \end{cases}$$

Example 1. Let us consider the circuit in Figure 1. Note that for our examples we use explicit values for better understandability, even though our algorithm considers the variables symbolically. Let the primary input be a constant 1, which corresponds to the waveform (1) and the timeshift 0. Let the SET be $(g_{not}, 1, 4, 1)$, where g_{not} is the not-gate in the circuit. This means the SET strikes in g_{not} at timestep 1 and lasts until timestep 4. The offset where the value of the signal becomes unknown is 1. The insertion of the SET is shown in Figure 3. Before the SET is inserted, the regular waveform of g_{not} is computed. The resulting waveform is (0) with timeshift 2. Afterwards, the waveform needs to be padded to include the timestep before and the timestep after the SET. The SET starts at timestep 1 and ends at timestep 4. Therefore the waveform needs to include the timesteps 1 and 5. The padded waveform is (000000) with timeshift 0. Inserting the SET sets the variables at the beginning and the end of the SET to X. The variables in between are negated. When the SET is inserted, the waveform changes to (0X11X0).

B. Propagation

This section describes the propagation of values in the circuit by introducing the waveforms for each gate and defining

the relation between the variables. The propagation considers the variable delays of the gates as well as electrical masking.

The real delay of a gate at a certain time depends on different factors like hardware variability, the input signals, or external influences. To approximate this behavior without modeling all details that affect the delay of a gate, we define a minimal and a maximal delay for each gate. If the output value of a gate at a certain timestep differs for different possible delays, the output value becomes unknown.

Let $diff = delay_{max}(g) - delay_{min}(g)$ be the difference between the minimal and the maximal delay of g. After generating the initial waveform, it is padded in the back for $diff$ variables to consider the latest possible output as well. This is realized by the function $pad_{delay} : V \times \mathbb{N} \times Var^* \to V \times \mathbb{N} \times Var^*$.

Example 2. Let us return to our example from Figure 1. In this example, we consider the and-gate pictured in Figure 4. The and-gate g has two predecessors with the waveforms (1) and (0X11X0). Each waveform of a predecessor has the timeshift 0. The resulting initial waveform of g is (0X11X0) with timeshift 4. Next, the variable delay is considered. The difference $diff$ between the maximal and the minimal delay is $diff = d_{max} - d_{min} = 5 - 4 = 1$. Therefore, we use padding to add one variable at the back, which results in the waveform (0X11X00).

For the variable delay, each variable in the waveform is compared to the *diff* previous variables. If the values are equal, the output at that time is identical for all applicable delays and remains the same. If the compared values are different, the resulting output is unknown and is set to X. The function $apply_{delay} : V \times \mathbb{N} \times Var^l \to V \times \mathbb{N} \times Var^l$ with $apply_{delay}(g, t, (v_1, \ldots, v_l)) = (g, t, (v'_1, \ldots, v'_l))$ describes this step. In this function, the new variables v'_i are defined as
$$v'_i = \begin{cases} v_i & \text{if } v_{i - diff} = \cdots = v_i \\ X & \text{otherwise} \end{cases}$$
The complete process of adding the variable delay is summarized in the function $varDelay : V \times \mathbb{N} \times Var^* \to V \times \mathbb{N} \times Var^*$ with $varDelay = apply_{delay} \circ pad_{delay}$.

Example 3. In our example from Figure 4, we compare the value of each variable to the value of the previous variable because $diff = 1$. If the values are equal, the variable is not changed, otherwise it is replaced with X. The first variable has no previous variable and remains unchanged. The third variable needs to be replaced with X because its value is 1 and the value of the previous variable is X. The fourth variable remains unchanged, because the value of the third and fourth variable are both 1. The resulting waveform is (0XX1XX0).

In a next step, the earlier described function *addSET* is used to induce the SET, if $g = g_{SET}$. Afterwards, electrical masking is applied by using the function $elecMask : V \times \mathbb{N} \times Var^l \to V \times \mathbb{N} \times Var^l$ presented in Section III-C.

Low level optimizations improve our algorithm by efficiently reducing the amount of used variables.

When a waveform consists of a single variable, the according signal is constant. If inputs of a gate g are constant, the output of g is also constant. We use a waveform with a single variable for g. The waveform with a single variable can be padded towards any timestep if needed. Furthermore, variable delays and electrical masking do not need to be considered for constant signals.

978-1-4799-8720-7/15 $31.00 © 2015 IEEE

Within the fanout of the SET, waveforms usually consist of five blocks of variables as long as there is no overlap. Either all of these variables will have the same value or the values will still correspond to the SET and have the form $vX\neg vXv$. When computing the variable delay, whenever different variables are compared, we use a variable from the second or fourth block instead of introducing additional variables. If the values of all variables are equal, it does not matter which variable is picked, and if the second and fourth block's value is X, the compared variables have different values and the output of X in the modified waveform is correct.

C. Electrical Masking

Electrical properties of the gates mask short glitches. A glitch is a change of a signal that lasts for a finite time and switches back to its original value afterwards.

Let the threshold t be the maximal duration of glitches masked by gate g. Every glitch shorter than or equal to t is removed by electrical masking. As simplification, we set t to half the minimal delay of g. Let us assume, there are two variables v_1 and v_2 with the same value val on the waveform and there are t or less variables between them. In this case, the variables between v_1 and v_2 need to be set to the value val to remove the glitch. If multiple glitches exist in the waveform, our process starts at the front and the processing of an earlier glitch can remove a later glitch.

To prepare the decision of electrical masking on the waveform $(v_1, \ldots, v_l)$, we introduce three vectors $\vec{v}^{is0} = (v_1^{is0}, \ldots v_l^{is0})$, $\vec{v}^{is1} = (v_1^{is1}, \ldots v_l^{is1})$, and $\vec{v}^{isX} = (v_1^{isX}, \ldots v_l^{isX})$. A variable v_i^{is0} is 1 iff v_i is equal to 0. The variables v_i^{is1} and v_i^{isX} are defined likewise for 1 and X, respectively.

Example 4. Let us apply this step to our example from Figure 4. The current waveform is (0XX1XX0). Since exactly the first and the last variable are equal to 0, the vector $\vec{v}^{is0} =$ (1000001). The other vectors are $\vec{v}^{is1} =$ (0001000) and $\vec{v}^{isX} =$ (0110110).

The following explanation describes how electrical masking towards 0 is handled. These operations are executed equivalently for 1 and X.

After calculating the vectors $\vec{v}^{is0}$, $\vec{v}^{is1}$, and $\vec{v}^{isX}$, we check for each variable v_i on the waveform, if it could be changed to 0 due to electrical masking. If v_j and v_k are the closest variables to v_i that are equal to 0 and have a distance of $k - j \leq t$ variables, v_i could be changed to 0.

For every variable $v_{i-t}, \ldots, v_{i-1}$, it is checked if that variable is the last variable before v_i that is equal to 0. A variable v_j is such a variable iff it is equal to 0 and all variables between v_j and v_i, i.e., $v_{j+1}, \ldots, v_{i-1}$, are not equal to 0. For this comparison, we use the prepared variables $\vec{v}^{is0}$:
$$v_j^{last\text{-}0} = v_j^{is0} \wedge \neg v_{j+1}^{is0} \wedge \cdots \wedge \neg v_{i-1}^{is0}$$
We check the t variables behind v_i similarly for the first variable after v_i that is equal to 0. For v_j after v_i the variable $v_j^{first\text{-}0}$ is defined:
$$v_j^{first\text{-}0} = v_j^{is0} \wedge \neg v_{j-1}^{is0} \wedge \cdots \wedge \neg v_{i+1}^{is0}$$

Example 5. The threshold for glitches t is half the minimal delay of g, i.e., in our example $t = delay_{min}(g)/2 = 2$. For our example, we consider the fourth variable v_4. The single 1 is a glitch that will be removed and replaced by X. Since

$t = 2$, we need to consider the two variables before and after v_4. Let us only consider electrical masking towards X. We need to compute the variables $v_2^{last\text{-}X}$, $v_3^{last\text{-}X}$, $v_5^{last\text{-}X}$, and $v_6^{last\text{-}X}$. The variable $v_2^{last\text{-}X}$ is 0 because there is another X between v_2 and v_4, i.e., v_3. Since v_3 is equal to X and there are no further variables between v_3 and v_4, $v_3^{last\text{-}X} = 1$. Similarly, $v_5^{last\text{-}X} = 1$ and $v_6^{last\text{-}X} = 0$ hold.

After determining the location of the closest variables to v_i that are 0, we can decide if it is possible, that v_i is masked towards 0. If any two variables $v_j^{last\text{-}0}$ and $v_k^{last\text{-}0}$ are equal to 1 and the difference between j and k is t or less, v_i could be changed to 0, which is presented by the variable $v_i^{potential\text{-}0}$:
$$v_i^{potential\text{-}0} = \bigvee_{j \in \{i-t,\ldots,i-1\}} \bigvee_{k \in \{i+1,\ldots,j+t+1\}} v_j^{last\text{-}0} \wedge v_k^{first\text{-}0}$$

Example 6. In our example, v_4 will be masked towards X, so we will compute $v_4^{potential\text{-}X}$:
$$v_4^{potential\text{-}X} = \bigvee_{j \in \{4-2,\ldots,4-1\}} \left(\bigvee_{k \in \{4+1,\ldots,j+2+1\}} (v_j^{last\text{-}X} \wedge v_k^{first\text{-}X}) \right)$$
$$= (v_2^{last\text{-}X} \wedge v_5^{first\text{-}X}) \vee (v_3^{last\text{-}X} \wedge v_5^{first\text{-}X}) \vee (v_3^{last\text{-}X} \wedge v_6^{first\text{-}X}) = 1$$

In case v_i could be changed into more than one value, we change v_i according to the earlier variable before v_i. This is realized by checking all possible combinations of variables:
$$v_i^{change\text{-}0} = v_i^{potential\text{-}0} \wedge$$
$$(\neg v_i^{potential\text{-}1} \vee \bigvee_{j \in \{i-t,\ldots,i-1\}} (v_j^{last\text{-}0} \wedge \neg \bigvee_{k \in \{i-t,\ldots,j-1\}} v_k^{first\text{-}1})) \wedge$$
$$(\neg v_i^{potential\text{-}X} \vee \bigvee_{j \in \{i-t,\ldots,i-1\}} (v_j^{last\text{-}0} \wedge \neg \bigvee_{k \in \{i-t,\ldots,j-1\}} v_k^{first\text{-}X}))$$

Example 7. In the example, it can be shown that $v_4^{potential\text{-}0} = v_4^{potential\text{-}1} = 0$. This leads to the conclusion, that $v_4^{change\text{-}X} = 1$. For every other variable than v_4, electrical masking will not change the value. The resulting waveform for g is (0XXXXX0).

The process of electrical masking is summarized in the function $elecMask : V \times \mathbb{N} \times Var^l \rightarrow V \times \mathbb{N} \times Var^l$ with $elecMask(g, t, (v_1, \ldots, v_l)) = (g, t, (v_1', \ldots, v_l'))$ where
$$v_i' = \begin{cases} val & \text{if } v_i^{change\text{-}val} = 1, \; val \in \{0, 1, X\} \\ v_i & \text{otherwise} \end{cases}$$
When two variables on the waveform next to each other are equal, the resulting variables from electrical masking will be equal as well. In those cases, we can reuse the variable that describes electrical masking in the previous timestep.

Additionally, we check the length of equal variables in a row before considering electrical masking. If this variable block is longer than t, electrical masking within that block is impossible and is not checked.

D. Observation of Erroneous Behavior

By executing the described steps for each gate, it is possible to represent the whole circuit in form of a SAT formula using three-valued logic. This formula is used to check if erroneous output in the sampling window is possible.

For the observation of an error, the nominal value of each gate is computed. If any output differs from the nominal output

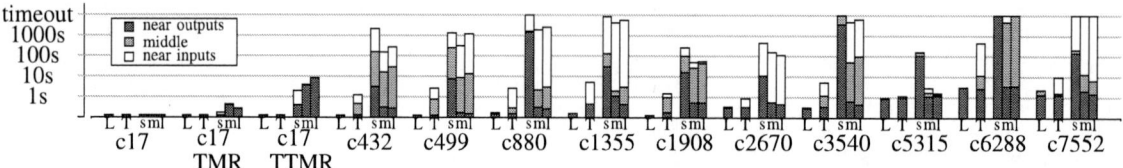

Fig. 5: Results of our experiments to determine the runtime of our algorithm with logical masking(L), timing masking(T) and electrical masking with a short(s), middle(m), and long(l) SET

in the given sampling window, an error occurs. This check is realized by the function $gate\text{-}error : O \times \mathbb{N} \times Var^* \to Var$ with

$$gate\text{-}error(g, t, (v_1, \ldots, v_l)) =$$

$$\bigvee_{start \leq i \leq end} (v_i \oplus output_{nom}(g, in_1, \ldots, in_{|I|})$$

where the variable in_i corresponds to the i-th input, $start = max(1, sample_s(g) - t + 1)$, and $end = min(l, sample_e(g) - t + 1)$. The variables $start$ and end correspond to the beginning and the end of the sampling window in relation to the timeshift.

The given SET can possibly lead to erroneous behavior if at least one of these checks returns 1. This is checked by or-operations over all these checks. The final variable *overall-error* describes, if an error occurs under the given SET:

$$overall\text{-}error = \bigvee_{g \in O} gate\text{-}error(g, t_g, \vec{v}_g)$$

The variable t_g describes the timeshift of an output gate g and $\vec{v}_g$ is its waveform.

To find a counterexample, the variable *overall-error* is set to 1. If the resulting SAT formula is satisfiable, the solution corresponds to a counterexample. Otherwise, the circuit is robust against the given SET since no assignment of variables exists that can lead to erroneous output in a sampling window.

Example 8. Let the sampling window for our check from the previous examples range from timesteps 8 to 12. The timeshift of the last and-gate g is 4. Therefore, the fifth variable corresponds to timestep 8 and so on. The last variable in the waveform corresponds to timestep 11. Therefore timestep 12 is not checked since the effect of the SET has passed by then. The nominal value of g is 0. Checking the waveform of g returns the following result:

$$overall\text{-}error = (v_5 \oplus 0) \vee (v_6 \oplus 0) \vee (v_7 \oplus 0) = 1$$

Thus, our circuit can possibly be affected by the SET.

IV. EXPERIMENTS

We use the ISCAS-85 benchmarks. To further evaluate our algorithm in comparison to transistor-level simulation, circuit c17 has been modified into two robust versions. One version uses *Triple Modular Redundancy* (TMR) to handle SETs. The original circuit is triplicated and a voter decides which output value is returned by using the value of the majority. The other version uses *Timed TMR* (TTMR) similar to [7]. The outputs of the original circuit are delayed by buffers. A voter decides similarly to TMR by using the current values, the value delayed by 3 buffers and the value delayed by 6 buffers. This method requires less overhead than TMR but still provides a certain level of robustness against timing errors.

We run the experiments on a Dual-Core AMD Opteron Processor 2222 SE with 3 GHz and 64 GB main memory.

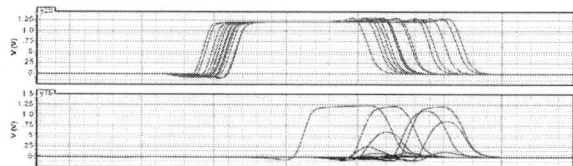

Fig. 6: Output of c17 under a glitch in primary input G3

Assignments	Spectre	Our algorithm
G1 = 1, G3 = 1	82ps	30ps
G1 = 1, G3 = 0	111ps	60ps
G1 = 0, G3 = 1	107ps	60ps
G1 = 0, G3 = 0	76ps	60ps

TABLE I: Minimal SET causing an error per input and method

The transistor-level simulation is done with the tool Spectre from Cadence using a commercial 65nm technology.

A. Validation

To validate the accuracy and functionality of the approach we test it against Spectre. We use c17 as a test case and perform a detailed Monte-Carlo simulation including on-chip variability for each transistor. The effect of the alpha particles is modeled as a double exponential current pulse with a parameterizable energy, as done in [6]. When all the inputs are set to 1 and an SET is induced into input G3 the effect is clearly visible in one output as seen in the top graph in Figure 6. However, the effects on the other output vary depending on the variability of the gates as seen on the lower graph. Unlike previous symbolic tools that do not consider variability, our algorithm will discover the possible error on the second output.

In a second experiment we validate that the results of our tool and Spectre are consistent for any input combination. Because of the long simulation time required by Spectre, we disabled the variability analysis. In this second experiment we used c17 with TTMR, the SET is induced into G10, a NAND-gate directly behind the inputs. Different strengths of particle strikes are simulated with Spectre for all possible input valuations. Due to the physical behavior the minimal strength of the particle strike and therefore the length of the SET that leads to an error differs depending on the inputs G1 and G3 of G10. The results of our algorithm depend on the expected output value of G10. Since there are two possible output values, we consider two different cases while Spectre considers four different cases, one for each possible input of G10. We adjusted our algorithm to return all counterexamples instead of one to check which assignments of variables are

978-1-4799-8720-7/15 $31.00 © 2015 IEEE   54

counterexamples for a given SET. We used timesteps of 5ps for our algorithm. The results of this experiment are shown in Table I. Since the Spectre simulations did not consider variability in this experiment, we set $d_{min} = d_{max}$ for all gates. The result of our algorithm is usually off by 50ps due to abstractions from the transistor level and conservative analysis of electrical masking. Besides the different minimal lengths, our algorithm returned exactly all counterexamples that were confirmed by Spectre to lead to an error. The transistor-level simulation for all possible inputs and particle strikes on c17 took a few hours while our algorithm took less than a minute, which is a significant speedup in comparison to Spectre.

B. Runtime

To evaluate the performance penalty for the accurate model, Figure 5 compares the runtime for different configurations of our algorithm: only logic masking (L), only logic and timing masking (T), and all effects. The logical check assigns a single value to each gate and flips the value of the gate affected by the SET. For each circuit, we randomly picked a gate close to the inputs, a gate close to the outputs, and a gate in between as position for the SET. The shade of a bar indicates the position. We inserted a short, a medium, and a long SET ranging from 2 to 10 timesteps. The logical check does not consider the length of the SET. For timing masking, the runtimes were similar. We show the average runtime. For the algorithm including electrical masking, we present one bar for each length (s,m,l). We tested every combination of length and position of an SET. Experiments that took more than 150 minutes were aborted.

In the unmodified circuits, a counterexample is found in most cases since the circuits provide no robustness. An exception are short SETs when electrical masking is considered. In this case, the SET is short enough to be masked soon after triggering and therefore does not change the behavior of the circuit. This is only discovered when we analyze electrical masking. The TMR version of c17 is not affected by an SET in our tests, which is expected, since TMR can handle SETs inside the circuit. This is discovered by all used algorithms. In the TTMR version, the existence of a counterexample depends on the length of the SET. The buffers can handle the short and middle SET, but not the long SET, since the long SET affects a delayed and a non-delayed signal at the same time. Since TTMR depends on timing, the logical check that does not consider timing always returns a counterexample despite the circuit being robust.

Checks including electrical masking take significantly more time due to the additional variables and clauses required to model the electrical masking. Without electrical masking, our algorithm handles most circuits of the benchmark in less than 10 seconds. The only exception is the multiplier c6288 which lasts up to 509 seconds. Most checks with electrical masking finish within the time limit of 150 minutes. For some of the larger circuits, our algorithm timed out. Additionally, when the effects of the SET split and overlap again, the runtime increases since most optimizations aim for a simplification of signals that do not include this behavior. This increase can be seen in c6288 where signals often split and overlap later on. The check for a short SET usually takes more time than the other lengths because the resulting SAT formula is not satisfiable. Finding a counterexample in the other cases is faster than verifying the unsatisfiability.

Further experiments were done to check the significance of our optimizations. In our experiments, the runtime decreased significantly with each optimization. Especially the optimization to simplify the waveforms of gates that are not in the fanout of the SET provided an average speedup of 40.

V. CONCLUSION

We presented a verification approach considering SETs under logic, timing, and electrical masking, including variation, and considering all possible input assignments. Validation against transistor-level simulations shows the conservativeness. Runtimes cannot directly be compared with previous techniques as the approach has unique characteristics. However, the approach is significantly faster compared to simulating individual input assignments.

Our approach can be exploited for characterizing a design under variation effects and can be complemented by expensive Spice simulations. These aspects remain future work.

REFERENCES

[1] Mehdi Dehbashi and Görschwin Fey. SAT-based speed-path debugging using waveforms. In *IEEE European Test Symposium*, pages 1–6, 2014.

[2] Stefan Frehse, Görschwin Fey, Eli Arbel, Karen Yorav, and Rolf Drechsler. Complete and effective robustness checking by means of interpolation. In *Formal Methods in Computer-Aided Design*, pages 82–90, 2012.

[3] Jie Han, Hao Chen, Jinghang Liang, Peican Zhu, Zhixi Yang, and Fabrizio Lombardi. A stochastic computational approach for accurate and efficient reliability evaluation. *Computers, IEEE Transactions on*, 63:1336–1350, 2014.

[4] Regis Leveugle. A new approach for early dependability evaluation based on formal property checking and controlled mutations. In *International On-Line Testing Symposium*, pages 260–265, 2005.

[5] Natasa Miskov-Zivanov and Diana Marculescu. Multiple transient faults in combinational and sequential circuits: A systematic approach. *Computer-Aided Design of Integrated Circuits and Systems, IEEE Transactions on*, pages 1614–1627, 2010.

[6] Kartik Mohanram. Simulation of transients caused by single-event upsets in combinational logic. In *IEEE International Test Conference*, pages 973–981, 2005.

[7] Michael Nicolaidis. Time redundancy based soft-error tolerance to rescue nanometer technologies. In *IEEE VLSI Test Symposium*, pages 86–94, 1999.

[8] Martin Omaña, Giacinto Papasso, Daniele Rossi, and Cecilia Metra. A model for transient fault propagation in combinatorial logic. In *IEEE International On-Line Testing Symposium*, pages 111–115, 2003.

[9] Matthias Sauer, Alexander Czutro, Ilia Polian, and Bernd Becker. Small-delay-fault ATPG with waveform accuracy. In *International Conference on Computer-Aided Design*, pages 30–36, 2012.

[10] Ashwin Seshia, Wenchao Li, and S Mitra. Verification-guided soft error resilience. In *Design, Automation Test in Europe Conference Exhibition*, pages 1–6, 2007.

[11] Michael Yoeli and Shlomo Rinon. Application of ternary algebra to the study of static hazards. *J. ACM*, pages 84–97, 1964.

2015 IEEE Computer Society Annual Symposium on VLSI

Assessment of FPGA Implementations of One Sided Jacobi Algorithm for Singular Value Decomposition

Ali Ibrahim[a,b], Maurizio Valle[a], Luca Noli[a]
DITEN, COSMIC Lab
[a]University of Genova
Genova, Italy
ibrahim.3ali@hotmail.com

Hussein Chible[b]
Microelectronics research Lab
[b]Lebanese University
Beirut, Lebanon

Abstract—**Nowadays many application domains require an embedded electronic system for tactile data processing. Our research aims to implement a real time embedded electronic system based on tensorial kernel approach for tactile data processing. Singular value decomposition represents the more computational expensive algorithm for the tensorial kernel approach. This paper presents an assessment of FPGA implementations of one sided Jacobi algorithm for singular value decomposition. Designs are implemented to handle an arbitrary m×n matrix with fixed point arithmetic. The results figure out an efficient implementation suitable for real time embedded applications.**

Keywords—FPGA implementations; Singular value decomposition; one sided Jacobi algorithm; Cordic method; Tactile data processing;

I. INTRODUCTION

For tactile data processing, many application domains require an embedded electronic system to be designed and fabricated together with the tactile sensors array (e.g. robotics, biomedical, and industrial automation) [1], [2]. The embedded electronic system has to provide a real time response when extracting and transmitting the structured information of the sensed applied stimuli.

Machine learning (ML) based on tensorial kernel approach has proven its effectiveness in processing tactile data [3]. The importance of this approach that it preserves the inherent tensorial structure of the signals provided by the sensing device. Computing tensorial kernel corresponds to computing the singular value decomposition of the unfolded matrices [4]. The singular value decomposition represents the bottleneck of the computation of tensorial kernel. In this perspective, our main goal is to implement a real time embedded electronic system based on tensorial kernel approach for tactile data processing. However, regarding to the complexity of the approach, and in order to comply with the constraints imposed by the applications, implementation methods have to be well assessed in order to achieve satisfied results in terms of area and time. Hence, as a result on the way to achieve our goal, this paper presents the assessment of the FPGA

implementations of the one sided Jacobi algorithm for singular value decomposition.

II. RELATED WORK

Because of the crucial role it plays in a wide range of signal processing applications, efficient computation of the singular value decomposition (SVD) is the subject of many publications. A VLSI Design of Singular Value Decomposition Processor for Portable Continuous-wave Diffusion Optical Tomography Systems is presented in [5], the design was implemented using 90nm CMOS process technology and simulation results verify the functionality of the JSVD design within the developed CW-DOT system. An implementation of SVD on a reconfigurable system made upon a Pentium processor and an FPGA-board plugged on a PCI slot of the PC for SVD computation of a 2 × 2 matrix is presented in [6]. In [7] a custom hardware design for computing SVD was presented. The hardware developed was suitable for computing SVD of a 2 × 2 complex-value matrix used in MIMO-OFDM standards. However, by using some of these techniques it is possible to obtain the SVD for small matrices only. Highly parallel accelerators such as Graphic Processing Unit (GPUs) and multi-core platforms have been employed to explore parallel SVD implementations, although these works only achieved speedups when the input matrices have dimensions greater than 1000 [8], [9]. In [10], a hardware efficient VLSI architecture for steering matrix computation using a hardware optimized Givens rotation SVD algorithm is presented. It achieves high processing throughput at low area by using a high speed given rotation unit based on CORDIC (Coordinate Rotation Digital Computer) algorithm.

An FPGA-based Singular Value Decomposition Processor is introduced in [11]. A mesh-connected array is proposed to be used with the CORDIC algorithm to compute the SVD. Generic FPGA-based hardware architecture for SVD computation is presented in [12]. It is dealing with large m×n matrices with fixed point arithmetic utilizing Hestenes approach and one-side Jacobi rotations. An FPGA implementation based on Jacobi method for singular value

978-1-4799-8720-7/15 $31.00 © 2015 IEEE

decomposition is presented in [13]; it introduced a floating-point Hestenes Jacobi architecture for SVD, which is capable of analyzing arbitrary sized matrices. The techniques in [12] and [13] provided as the more accurate and the fastest recent architectures respectively, assure the potential of the one sided Jacobi algorithm in the hardware implementation of the SVD. However, the one sided Jacobi algorithm could be implemented by using different methods, and in literature each architecture was based on a different method. Consequently, a tradeoff between the implementation methods in terms of area, time, and accuracy of computation for one sided Jacobi algorithm is still missing.

In this paper, we present three FPGA implementations of the one sided Jacobi algorithm for SVD based on different computation methods. The designs are implemented to handle an arbitrary m×n matrix with fixed point arithmetic. Implementation results figure out an efficient implementation suitable for real time embedded applications.

III. BACKGROUND

A. Singular Value Decomposition (SVD)

The singular value decomposition is a factorization of a real matrix into a product of three matrices, that is for a given m × n matrix A, the SVD is defined by:

$$A = USV^T \qquad (1)$$

Where U is an orthogonal m × m matrix, that is $U^TU = I_m$, it consists of n left singular vectors, and V is an orthogonal n × n matrix, that is $V^TV = I_n$, it contains n right singular vectors. The superscript T designates the transpose of the matrix. The m × n matrix S is a diagonal matrix diag(σ_0,.......,σ_{n-1}), where the σ_i are the singular values of A, they are the square roots of the Eigen values, and arranged in descending order.

B. One Sided Jacobi Algorithm

One sided Jacobi algorithm applies a sequence of rotations to the original matrix A, in order to reach the diagonal matrix S. For a given symmetric n×n matrix A = A_0, the Jacobi algorithm produces a sequence A_1, A_2... which eventually converge to a diagonal matrix with the Eigenvalues on the diagonal [14]. A_{i+1} is obtained each time from A_i by the transformation given by the formula

$$A_{i+1} = J(i, j, \theta)^T A_i J(i, j, \theta) \qquad (2)$$

where $J(i, j, \theta)$ is called a Jacobi rotation.

The Jacobi rotation $J(i, j, \theta)$ is introduced, for an index pair (i, j) and a rotation angle θ, as a square matrix that is equal to the identity matrix I plus four additional entries at the intersections of rows and columns i and j. The Jacobi rotations are calculated on every 2 × 2 matrix to zero out all non-zero off-diagonal elements of the original matrix. Considering a 2 × 2 example given by eq. (3), the rotation matrix J must be constructed such that it annihilates the off-diagonal elements w:

$$\begin{pmatrix} \hat{x} & 0 \\ 0 & \hat{y} \end{pmatrix} = \begin{pmatrix} cos\theta & -sin\theta \\ sin\theta & cos\theta \end{pmatrix}^T \begin{pmatrix} x & w \\ w & y \end{pmatrix} \begin{pmatrix} cos\theta & -sin\theta \\ sin\theta & cos\theta \end{pmatrix} \quad (3)$$

where $\hat{x}$ and $\hat{y}$ are the diagonal elements of the 2×2 matrix corresponding to x and y elements after rotation by the corresponding angle.

The one sided Jacobi algorithm provides high accuracy and fast convergence. It takes about 5-10 sweeps to converge which reduces the time and the number of resources to be used.

C. Cordic Algorithm

The Coordinate Rotation Digital Computer (CORDIC) is an iterative algorithm and extremely hardware friendly. It involves a set of Shift-Add operations for computing a very rich set of functions from the one basic set of equations [15]. CORDIC can be either operated in vectoring mode or in rotation mode. In the vectoring mode, CORDIC rotates the input vector through whatever angle is necessary to align the resultant vector with the horizontal axis. The result of the vectoring operation is a rotation angle and the scaled magnitude of the original vector. In rotation mode, the angle accumulator is initialized with the desired rotation angle (Z). The rotation decision (d_i) at each iteration is made to diminish the magnitude of the residual angle in the angle accumulator. The decision at each iteration is based on the sign of the residual angle after each step. The iteration equations are given by:

$$X_{i+1} = X_i - Y_i \times d_i \times 2^{-i}$$
$$Y_{i+1} = Y_i - X_i \times d_i \times 2^{-i}$$
$$Z_{i+1} = Z_i - d_i \times tan^{-1}(2^{-i}) \qquad (4)$$

Where:

$$\begin{cases} d_i = +1 \ if \ Y_i < 0, -1 \ otherwise \ for \ vectoring \ mode \\ and \\ d_i = -1 \ if \ Z_i < 0, +1 \ otherwise \ for \ rotation \ mode. \end{cases}$$

IV. HARDWARE IMPLEMENTATION

Cyclic Jacobi method [16] provides an inexpensive computational approach to compute the transformations, it consists in organizing the computations in sweeps within which each matrix element is annihilated once, and each sweep consists of n(n-1)/2 transformations. The one sided Jacobi computes the SVD through a pre- and post-multiplication by the Jacobi rotation matrix. For that, the complexity of this algorithm lies in the computation of cosine and sine functions and the management of the rotations. Fig. 1 shows a general flow diagram for the hardware implementation of the singular value decomposition algorithm. Different methods can be used to implement the phase solver block and the rotation blocks presented in the flow diagram. Three architectures have been implemented based on different methods: Architecture 1, 2, and 3 are shown in Fig. 2. The data of the input matrix $A_{m×n}$ is represented with N bit data resolution. A has to be multiplied by its transpose to achieve the square symmetric matrix $U_{n×n}$ represented as 2N+P bits, where P is a correction parameter preventing the overflow during the row by column multiplications. For small matrices the correction parameter P tends to 1 and it increases gradually when matrix dimension increases. Thus, for an input data resolution N =8 bits and for matrices with columns size

978-1-4799-8720-7/15 $31.00 © 2015 IEEE

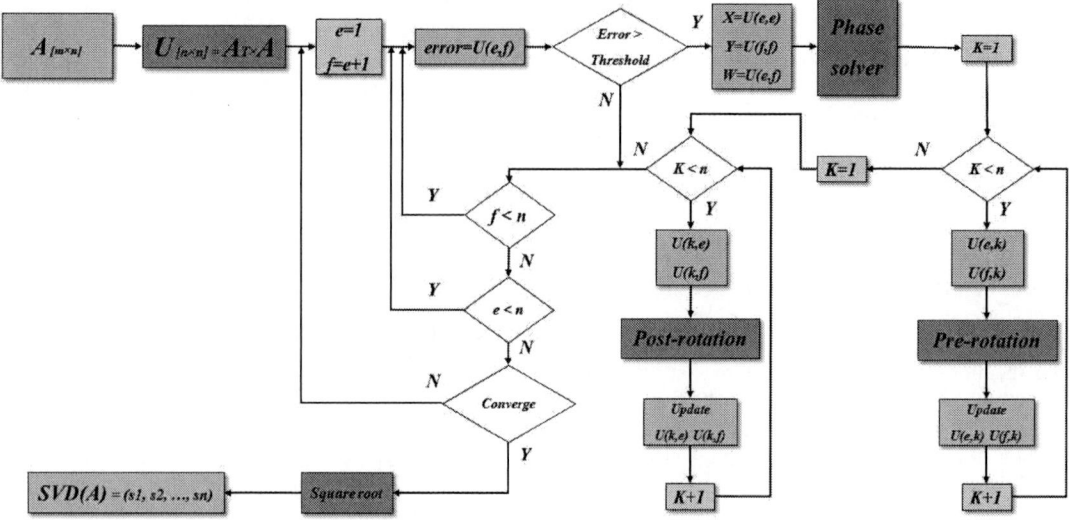

Fig. 1. General flow Diagram for the hardware implementation of the one sided Jacobi algorithm for SVD computation

up to 30, P could be set to 5 and the data of the matrix U is represented as 21 bits. The important question when dealing with the one sided Jacobi algorithm is when to stop the Jacobi iterations. If stopped too early, we could not have the convergence up to given accuracy. When stopped too late, unnecessary sweeps would be completed spending perhaps much more time and power than needed. The definition of a reliable stopping criterion is a difficult task and adds more complexity to the hardware implementation. That is because the stopping criterion needs a number of complex arithmetic operations (multiplication, division, and square root) in order to be implemented [17]. These arithmetic operations are

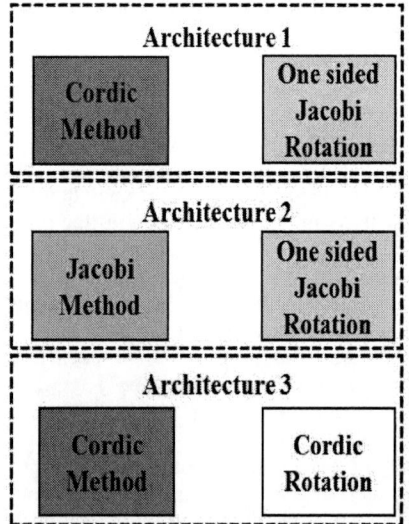

Fig. 2. Architectures definition based on the different implementation methods.

applied to the off-diagonal elements in order to estimate an error value. In order to decrease the complexity in our architectures, instead of implementing the stopping criterion, only the upper off-diagonal element U(e, f) is tested at each iteration and the threshold value has been set to zero. Using this technique had ensured the convergence of the Jacobi algorithm, for all tested resolutions and dimensions, after 8 iterations in the worst case and 4 iteration in the best case.

A. Phase Solver Block

Phase solver is the block which computes the sine and cosine functions or the angle of rotation. Two different methods can be used to implement the phase solver block: Jacobi method and CORDIC method.

1) Jacobi Method
Jacobi method is defined by the following equations:

$$\alpha = \frac{y-x}{2\,w} \qquad\qquad tg\theta = \frac{sign(\alpha)}{|\alpha|+\sqrt{1+\alpha^2}}$$

$$cos\theta = \frac{1}{\sqrt{1+tg^2}} \qquad\qquad sin\theta = tg\theta \times cos\theta \qquad (5)$$

where x, y, w represent respectively U(e, e), U(f, f) and U(e, f) of the flow diagram of figure 1. These equations are used to compute the sine and cosine functions without calculating the angle. They are introduced as the more accurate since they diminish the accumulation of the rounding errors. The implementation uses four adders and one subtracter, three dividers, three multipliers, and two square roots. So, the sine and cosine functions are computed each iteration starting from the elements of the matrix U and data is represented as 21 bits. Then in order to achieve accurate singular values, the sine and cosine functions are represented as fixed point 2.20 that is 2 bits for integer part and 20 bits for fractional part. In fact, this representation is the more adequate since it ensures high

978-1-4799-8720-7/15 $31.00 © 2015 IEEE

accuracy in the results. It has been selected after studying the variation effects of the fractional part of the trigonometric functions on the singular values.

2) Cordic Method

For this method, CORDIC processor has been made using LogiCORE generator provided by XILINX tool. Two blocks has been implemented to find the trigonometric functions: *Cordic_Phase* and *Cordic_Trigo*. The inputs to the *Cordic_Phase* block are x, y, and w. it utilizes CORDIC algorithm in vectoring mode to compute the angle of rotation. Then, *Cordic_Trigo* block uses the provided angle of rotation and applies CORDIC algorithm in its rotational mode to compute the sine and cosine functions. CORDIC method figure out the sine and cosine functions as fixed point 2.20 bits also.

B. *Rotation Blocks*

The rotation blocks consist on the pre- and post-rotation of in the flow diagram of the Fig. 1. Two different methods are used to implement the rotation blocks: one sided Jacobi rotation and CORDIC rotation.

1) One Sided Jacobi rotation

This method is based on the traditional one sided Jacobi rotation. It updates the row/column of the matrix after each pre/post-rotation by applying a row column multiplication to achieve the rotation. Fig. 3 shows the circuit of a single pre-rotation update. The inputs of this circuit are the elements of the two rows e and f of the matrix U (see Fig. 1) with the corresponding sine and cosine functions provided by the phase solver block. It uses four multipliers, one adder and one subtracter. Supposing that the element of the matrix U are all integers i.e. fixed point 21.0 bits. So, the pre/post-rotation is done by the multiplication of 21.0 bits by 2.20 bits (the sine and cosine functions). After each iteration data is rounded to reach again the same data representation (21.0 bits).

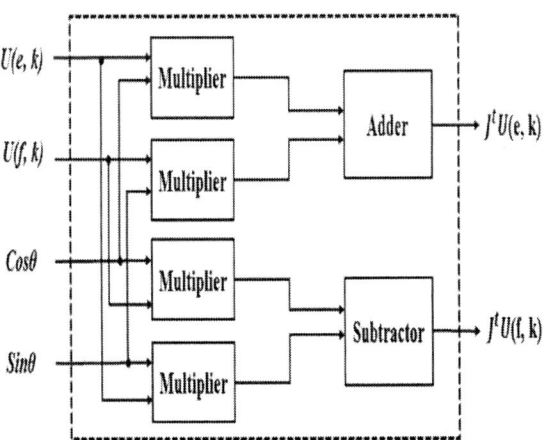

Fig. 3. The circuit of a single pre-rotation operation update. U(e, k) and U(f, k) are the kth elements of the rows e and f respectively, JᵗU(e, k) and JᵗU(f, k) represent the kth elements of the rows of the matrix U after the pre-rotation.

2) CORDIC Rotation

After computing the corresponding angle using the CORDIC method in vectoring mode (*Cordic_phase* block), CORDIC rotation rotates the rows/columns of the input matrix using CORDIC algorithm in its rotational mode. It maintains the same data representation in the output that is 21 bits. CORDIC rotation block has been also made using LogiCORE generator provided by XILINX tool.

C. *Square root Block*

Square root is fundamental to compute the singular values resulted from the Eigen values matrix after converging to a diagonal matrix. The implementation of a fixed point square root is done using a modified non-restoring algorithm for fixed point square root proposed in [18].

V. EXPERIMENTAL RESULTS

This section highlights the tests carried out to study the performance of the FPGA implementations for the described architectures. The study aims to compare the area, time and the accuracy of computation for the three architectures in order to figure out an efficient implementation.

A. *Accuracy of Computation*

The singular values resulted from the presented implementations have been compared with the singular values of the MATLAB SVD function in order to assure the correctness of the results. For each architecture, three matrices 8 × 7 with 6 bit input data resolution have been created randomly using MATLAB and the singular value decomposition has been computed. The percent error has been calculated using the following formula:

$$e' = \frac{|S_i - S_{Mi}|}{S_{Mi}} \times 100\% \qquad (6)$$

where S_i are the singular values corresponding to the pair row column i of the matrices resulted from the presented implementations, and S_{Mi} correspond to the same elements of the matrix yields from SVD MATLAB simulation. Then, the mean of the percent errors from the three matrices has been computed. Moreover, in order to assure the correctness of our margin of error, Fig. 4 shows a comparison of the percent error between our implementations and the implementation presented in the literature [12].

B. *Hardware Implementation results*

Six square matrices with 8 bit data resolution have been generated randomly using MATLAB: that is A(4,4) to A(9,9). Singular values for the generated matrices have been computed using the three architectures. The design has been implemented for a Xilinx SPARTAN-6 XC6slx16 FPGA device. Then, in order to evaluate the speed of implementations, elapsed time for each matrix size has been calculated using the following formula:

$$Elapsed\ time = N \times \frac{1}{f_{max}} \qquad (7)$$

where N is the number of clock cycles needed to compute the SVD, and f_{max} is the maximum frequency. Fig. 5 shows the

978-1-4799-8720-7/15 $31.00 © 2015 IEEE

graph of the elapsed time in terms of the matrix size for the implemented architectures. Moreover, Fig. 6 shows a bar plot for the percentage occupied area in terms of matrix size. In Table 1, the resources utilization of the singular value decomposition implementations for a matrix size of 160×8 are shown. Design have been implemented for a Virtex-5 XC5VLX330T FPGA device. It is clear that the area relation between the three architectures still valid for large matrices size.

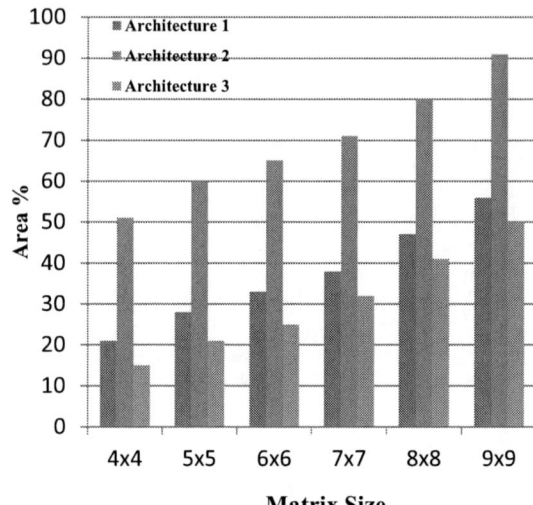

Fig. 6. Percentage area vs. matrix size for the three implemented architectures for FPGA Spartan 6 XC6SLX45.

TABLE 1. RESOURCE UTILIZATION OF THE IMPLEMENTED ARCHITECTURES FOR A MATRIX SIZE OF 160×8 USING VIRTEX-5 XC5VLX330T FPGA DEVICE.

Resource utilization	Slice registers	Slice LUTs	Maximum frequency (MHz)	Occupied area (%)
Architecture 1	25113	19542	146.022	20
Architecture 2	40305	26875	146.022	25
Architecture 3	25037	18103	147.230	18

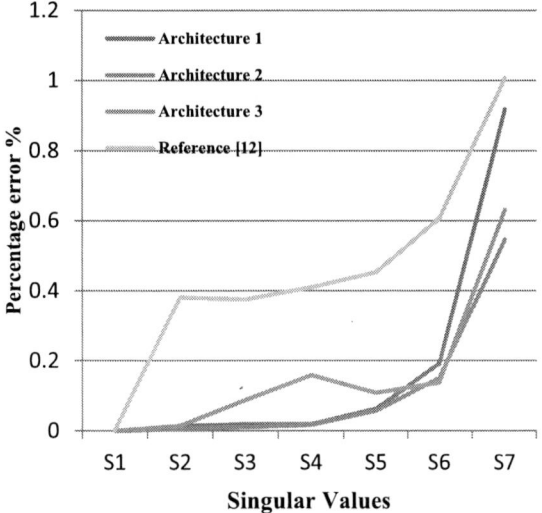

Fig. 4. Percent error of the computation of singular value decomposition for an 8 × 7 matrix. Si represent the singular values of the matrix A.

C. Discussion

The experimental results of the studied cases show that the presented implementations provide a high accuracy of computation of the singular value decomposition. The percent error of the computation is close in the three implementations also when compared with [12]. The hardware implementation of our architectures uses registers to memorize the matrices for the SVD computation. The fact which increases the percentage occupied area and maybe it is more appropriate to use a RAM (for limited matrix sizes) in order to decrease resource utilization. Nonetheless, as this same method is used for all three architectures, improving it will affect all implementations keeping the same result of comparison figured out by this study. Fig. 5 and Fig. 6 show that the time needed to compute the SVD using Architecture 2 is acceptable but the occupied area is very high. Although the method utilized to implement this architecture is introduced as the more accurate, other architectures show a very close margin of error. For that, using this architecture will be very expensive and will limit the implementation to the small matrices only. It is clear that Architecture 3 provides the best results in terms of area, however it needs the higher time to be executed. So, using Architecture 3 is suitable for applications which intend to diminish the area with acceptable time response. Moreover, Architecture 1 provides the best results in terms of time with a

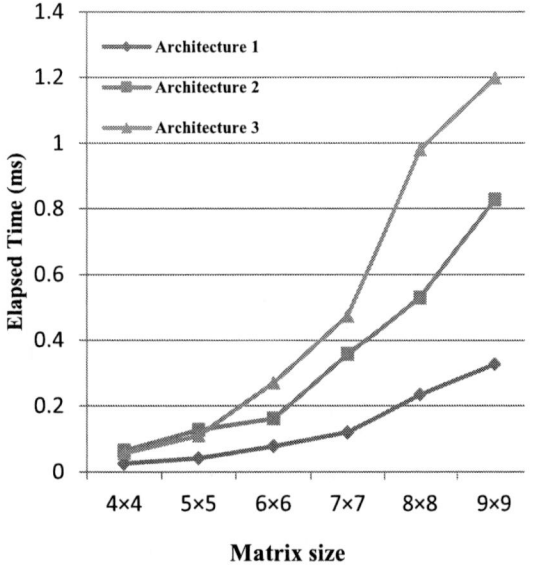

Fig. 5. Elapsed time in terms of the matrix size for the three implemented architectures.

satisfied percentage of the area occupied. For our goal represented by the implementation of machine learning based tensorial kernel for tactile data processing, a tradeoff between the resource utilization and timing is needed. Hence, Architecture 1 represents the tradeoff for the hardware implementation of the singular value decomposition based on one sided Jacobi algorithm.

VI. CONCLUSION AND FUTURE WORK

This paper presented three FPGA implementations of singular value decomposition dealing with an arbitrary m×n matrix with fixed point arithmetic. The work aimed to find an efficient implementation to compute the singular value decomposition using one sided Jacobi algorithm. It presented a comparison between three architectures in terms of area and time. Moreover, correctness of the accuracy of computation have been assured by comparing the percent errors with an implementation presented in literature [12]. The study figured out an efficient implementation suitable for real time embedded applications.

The selected implementation of the SVD is intended to be a part of the hardware implementation of a real time embedded electronic system for tactile sensors data processing. Next steps will consist on the implementation of the overall machine learning based on tensorial kernel approach [3].

ACKNOWLEDGMENT

This work is supported by the University of Genova – Progetti di Ricerca di Ateneo 2014: Electronic skin for prosthetic systems, resp. M. Valle.

REFERENCES

[1] Dahiya R, Mittendorfer P, Valle M, Cheng G, Lumelsky VJ. irections Toward Effective Utilization of Tactile Skin: A Review.IEEE SENSORS JOURNAL, VOL. 13, NO. 11; NOVEMBER 2013.

[2] L. Seminara, L. Pinna, A. Ibrahim, L. Noli, M. Capurro, S. Caviglia, P. Gastaldo, M. Valle. "Electronic Skin: achievements, issues and trends," 2ndInternational Conference on System-integrated Intelligence SysInt2014.

[3] P. Gastaldo, L. Pinna, L. Seminara, M. Valle, R. Zunino." A Tensor-Based Pattern-Recognition Framework for the Interpretation of Touch Modality in Artificial Skin Systems". IEEE SENSORS JOURNAL, VOL. 14, NO. 7, JULY 2014.

[4] Signoretto M, De Lathauwerb L, Suykens JAK. A kernel-based framework to tensorial Networks. vol. 24; 2011. pp. 861-874.

[5] S. Kang, S.Y. Wu, Y. H. Hsu, C. C. Fu, and W. C. Fang. "A VLSI Design of Singular Value Decomposition Processor for Portable Continuous-wave Diffusion Optical Tomography Systems". IEEE Life Science Systems and Applications Workshop (LiSSA), 2011.

[6] C Bodda, K. Danne, A. Linarth, "Efficient Implementation of the Singular Value Decomposition on a Reconfigurable System," Lect. Notes Comp. Sci., vol. 2778/2003, pp. 1123–1126.

[7] Y. Wang, K. Cunningham, P. Nagvajara, J. Johnson, "Singular Value Decomposition Hardware for MIMO: State of the Art and Custom Design," in Proc. 2010 International Conference on Reconfigurable Computing and FPGAs (ReConFIg), Dec. 2010, pp. 400–405.

[8] S. Lahabar and P. Narayanan, "Singular Value Decomposition on GPU using CUDA," in Proceedings of IEEE International Symposium on Parallel Distributed Processing, May 2009, pp. 1–10.

[9] A. Haidar, J. Kurzak, and P. Luszczek, "An improved parallel singular value algorithm and its implementation for multicore hardware," in Proceedings of SC13: International Conference for High Performance Computing, Networking, Storage and Analysis (SC '13), New York, NY, USA, Nov. 2013, pp. 90:1–90:12.

[10] C. Senning, C. Studer, P. Luethi, and W. Fichtner, "Hardware-efficient steering matrix computation architecture for MIMO communication systems," in Proc. IEEE Int. Symp. Circuits Syst., May 2008, pp. 304–307.

[11] W. Ma, D. M. Luke, M. E. Kaye, R. Doraiswami, "AN FPGA-BASED SINGULAR VALUE DECOMPOSITION PROCESSOR," IEEE CCECE/CCGEI, Ottawa, May 2006.

[12] L. M. Ledesma-Carrillo, E. Cabal-Yepez, R. de J. Romero-Troncoso, A. Garcia-Perez, R. A. Osornio-Rios, T. D. Carozzi. "Reconfigurable FPGA Based Unit for Singular Value Decomposition of Large m × n Matrices". International Conference on Reconfigurable Computing and FPGAs, 2011.

[13] X. Wang, J. Zambreno. "An FPGA Implementation of the Hestenes-Jacobi Algorithm for Singular Value Decomposition". IEEE 28th International Parallel & Distributed Processing Symposium Workshops, 2014.

[14] M. Hestenes, Inversion of matrices by biorthogonalisation and related results, J.SIAM6(1958)51–90.

[15] R. Andraka, "A Survey of CORDIC algorithms for FPGA based computers," in Proc. 1998 ACM/SIGDA 6th International Symposium on FPGA, Feb. 1998, pp. 191–200.

[16] J. Corr, K. Thompson, S. Weiss, John G. M. Whirter, I. K. Proudler, "Cyclic-by-Row Approximation of Iterative Polynomial EVD Algorithms," Sensor Signal Processing for Defence (SSPD), 2014.

[17] Zhou, B.B. ; Brent, R.P. ; Kahn, M. "Efficient one-sided Jacobi algorithms for singular value decomposition and the symmetric eigenproblem," Algorithms and Architectures for Parallel Processing, 1995. ICAPP 95. DOI: 10.1109/ICAPP.1995.472193.

[18] A. Nanhe, G. Gawali, S. Ahire, K. Sivasankaran. "Implementation of Fixed and Floating Point Square Root Using Nonrestoring Algorithm on FPGA". International Journal of Computer and Electrical Engineering, Vol. 5, No. 5, October 2013.

Index-based Round-Robin Arbiter for NoC Routers

Masoud Oveis-Gharan and Gul N. Khan
Department of Electrical and Computer Engineering
Ryerson University
Toronto, Ontario M5B 2K3 Canada
gnkhan@ee.ryerson.ca

Abstract—Scalable on-chip communication system such as Network-on-Chip (NoC) is needed to meet the communication demand of large number of SoC (System on Chip) cores. In the NoC router micro-architecture design, arbiter has become increasingly important due to its significant impact on the performance and efficiency of NoC systems. In this paper, we propose an Index-based Round Robin (IRR) arbiter that functions on the index format of input ports of the router. The microarchitecture of IRR arbiter scales logarithmically ($\log_2$) with the number of input ports as compared to a conventional round robin arbiter that scales with its input ports. The behavior and architecture of our arbiter leads to lower power consumption and chip area as well as higher performance characteristics.

Keywords—arbiter micro-architecture; critical path delay; network on chip; router arbiter design

I. INTRODUCTION

In digital system design, arbiters are used to allocate and access shared resources. Whenever a resource, such as a buffer, channel or a switch-port is shared, an arbiter is required to assign the access to the resource at a particular time. The most common usage of arbiters is the shared-bus arbitration of a bus-based system where multiple master modules can initiate their transactions. The modules must be arbitrated for access to the bus before initiating a transaction. In this paper, we investigate the arbiters used in NoC systems. An NoC system facilitates communication among IP cores of an SoC. It includes a network of switches (routers) that are interconnected by communication links as illustrated in Figure 1.a. Figure 1.b shows a typical NoC router that consists of some input and output ports, an arbiter and a crossbar switch [1].

The datapath of a router is made of port buffers, crossbar switch and interconnection structure, while the control unit of a router is mainly consists of arbiters. The structure of arbiter becomes even more complex with the utilization of Virtual Channel (VC) mechanism. Figure 2 shows two 4-input arbiter that can arbitrate the use of resources. The arbiter accepts n requests ($r_0, r_1, \ldots, r_{n-1}$), arbitrates among the asserted request lines, and selects an r_i for service, and then asserts the corresponding grant line, g_i. For example, assume the arbitration for the output port of a crossbar switch among a set of requests from the VCs of some input ports. The input-port VCs that have flits will issue request signals for having access to one of the desired output-port. Assume, there are 5 VCs and VCs 0, 2, and 4 assert their request lines, r_0, r_2, and r_4 respectively. The arbiter will then arbitrate and select one of these VCs for assigning the desired output-port. Assume the grant of VC2 (i.e. g_2) is asserted. VCs 0 and 4 lose the arbitration and must hold their requests active until they receive the grant signal for their output-ports.

Arbiters can be categorized in terms of fairness (weak, strong or FIFO) arbiters [1]. In a weak fairness arbiter, every request is eventually granted. The requests of a strong fairness arbiter will be granted equally often. The requests of FIFO fairness are granted in a first come first served basis. Moreover, arbiters in terms of priority can be grouped in two fixed and variable architectures. For a fixed priority arbiter, the priority of requests is established in a linear order. Figure 3a illustrates a 4-input fixed-priority arbiter where r_0 has the highest and r_3 has the least priority [1]. The architecture can be expanded to n-input arbiter where for each middle request, there is an arbiter cell consisting of two ANDs and an Inverter. The first and last arbiter cells can be simplified (see Figure 2a). For each request input ri, there is a carry input c_i, a grant output g_i, and a carry output, c_{i+1} where $i \in \{0,1, .. n-1\}$. Therefore, a low level c_i indicates that at least one of requests from r_0 to r_{i-1} was has been asserted. Moreover, in case that the request ri and carry c_i are high, the grant, g_i is set, and all the following grants i.e. g_{i+1} to g_{n-1} will become reset. It is obvious that the critical path of the circuit is from the first request, r_0 to the last grant, g_{n-1} due to propagation of carry from head to the tail of the arbiter. Fixed priority arbiters provide weak fairness arbitration because when a request is continuously asserted, none of its following requests will ever be served.

In order to have a fair iterative arbiter, we can use a variable priority arbiter as illustrated in Figure 2b. An OR gate and a priority input signal, p_i is added to each cell of the fixed priority arbiter shown in Figure 2a. When p_1 is set, its corresponding request, r_1 has high priority and the priority decreases from that point cyclically around the circular carry chain. Now we can create a fair iterative arbiter by changing its priority from cycle to cycle. In an n-input arbiter, if the grant, g_i (where $i \in \{0, 1, \ldots n-1\}$) is connected to the next priority vector p_{i+1}, a Round Robin (RoR) arbiter is created. Figure 3 illustrates a 4-input RoR arbiter. If a grant, g_1 becomes high at the current cycle, it causes p_1 to be set high on the next clock cycle. This leads the request, r_2 to become the highest priority at the next cycle, where the request, r_1 becomes the lowest priority. For the sake of simplicity, we assume that the arbitration cycle takes one clock cycle in all the architectures describe in this paper.

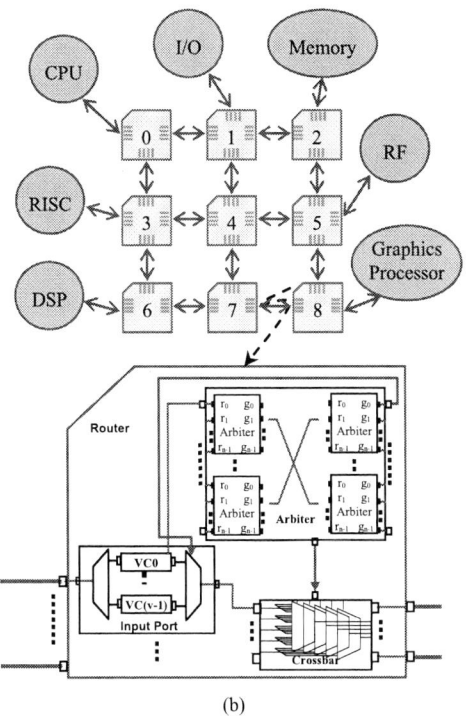

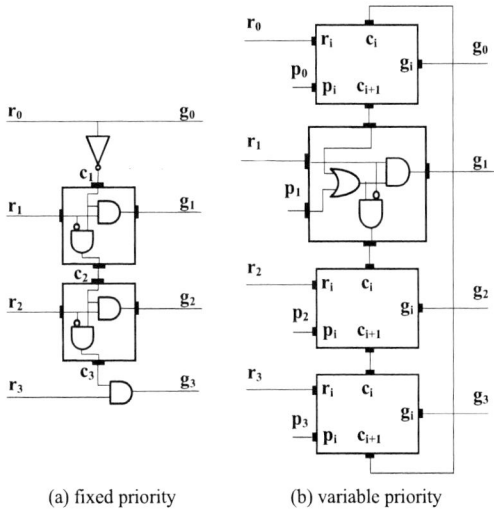

(b)

Figure 1. (a) 2D SoC mesh. (b) Wormhole NoC router.

(a) fixed priority   (b) variable priority

Figure 2. 4-input arbiter architectures.

The functionality of a round-robin arbiter can be explained as a request that is just granted will have the lowest priority on the next arbitration cycle [1]. The round robin arbiters are simple, easy to implement, and starvation free. When the input requests are large in numbers, the structure of round robin arbiter grows that leads to large chip area, higher power consumption, and critical path delay. In an NoC design, the critical path delay of arbiter usually dominates among the critical path delays of input-port and

crossbar switch due to the architectural complexity of arbiter as compared to those of port and crossbar switch. Therefore, the arbiter circuit determines the maximum frequency (or the speed), F_{max} of an NoC router. The critical impact of arbiter on the performance of the NoC system and the characteristic behaviour of round robin architectures have created a lot of interest of NoC researchers.

II. RELATED WORK

The architecture of a popular Matrix round robin arbiter is presented by Dally and Towles [1]. A 4-input Matrix arbiter architecture is shown in Figure 4. It implements a least recently served priority scheme where a request, r_i wins an arbitration. It resets the bits of row i and sets the bits of column i to make itself the lowest priority where $i \epsilon \{0,..3\}$. The Matrix arbiter is claimed to be useful for small number of inputs as it is fast, economical, and performs strong fairness arbitration. However, no evaluation is presented. Fu and Ling evaluated and compared the RoR and Matrix arbiters in terms of resource, performance and power consumption for an FPGA platform [2]. They concluded that the Matrix arbiter consumes more resource, same power but can process data more quickly than the RoR arbiter.

Zheng and Yang proposed a Parallel Round Robin Arbiter (PRRA) based on a simple binary search algorithm as illustrated for a 4-input PRRA in Figure 5 [3]. They further proposed an Improved PRRA (IPRRA) design where the output signals, gL and gR of PRRA are disconnected and directly ANDed with grant signals to generate new grant signals as shown in Figure 6. The IPRRA reduces the timing of PRRA significantly. A High speed and Decentralized Round robin Arbiter (HDRA) has been presented by Lee et al., which is illustrated in Figure 7 [4]. Each circuit enclosed with dash circle represents a filter circuit whose main components are a D flip-flop and a multiplexer. The filter circuit filters out the input without request or the one with request that has already been granted at that arbitration cycle. The un-filtered inputs with their requests participate in the arbitration again next cycle by setting its corresponding D-type flip-flops to 0 that are done by enabling the *ack* signals from higher lower level. The HDRA arbiter will reset itself asynchronously by the input *self_rst* from the root. The *sys_rst* indicates the system reset signal and is used initially before each arbitration cycle for all requests. A 4-input HDRA arbiter has a simpler circuit than a higher input HDRA architecture because the *act, rnext* and *self-rst* are connected together.

III. INDEX-BASED ROUND ROBIN ARBITER

In this paper, we present a new arbiter design called Index-based Round Robin (IRR) arbiter that employs a least recently served priority scheme and achieve strong fairness arbitration. The proposed arbiter has smaller arbitration delay, lower chip area and it also consumes less power as compared to the aforementioned arbiters. Before describing the IRR arbiter architecture, we introduce an inseparable and critical output in the arbiter design.

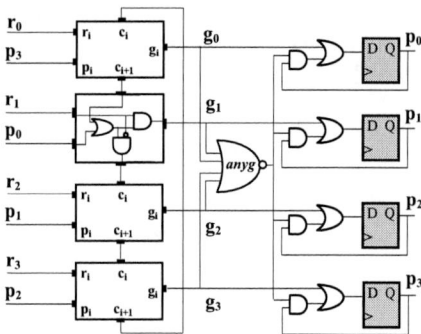

Figure 3. 4-input RoR arbiter architecture.

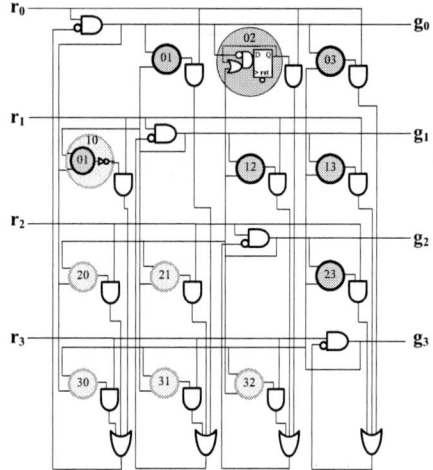

Figure 4. A 4-input Matrix arbiter.

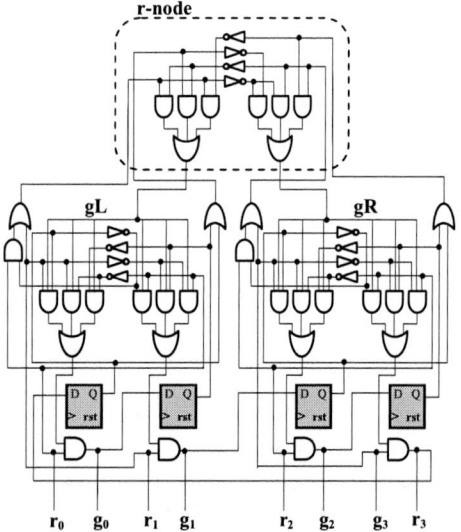

Figure 5. 4-input PRRA architecture.

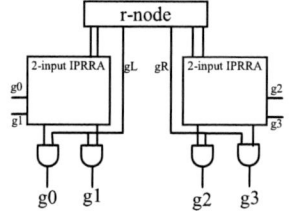

Figure 6. 4-input IPRRA structure.

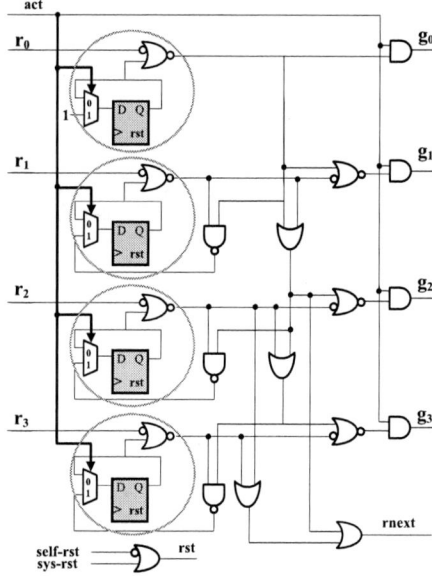

Figure 7. 4-input HDRA architecture.

A. Grant Index

All the arbiters have an output array, *grant* whose width is the same as that of input width. However, in practical designs, the index of *grant* signals, *g_id* is also generated that is used to address the granted request in some other components, such as control tables, multiplexers and memories used in NoC routers. When a router crossbar is made of multiplexers, the *g_id* can be connected to selection port of multiplexer to switch the granted input to the requested output port (see Figure 1.b). The width of *g_id* is the $\log_2$ of the width of *grant*. We used the *g_id* as the first output of our proposed design and due to lower width of *g_id*, our arbiter design is smaller and faster as compared to other arbiters. Due to the critical use of *g_id* in NoC design, we consider all the arbiters covered in this paper to generate both *grant* and *g_id* as outputs.

B. Fixed Priority Arbiter

Our fixed priority arbiter is simpler and economical as illustrated in Figure 8. The priority of requests is linear and in the ascending order where r_0 has the highest priority. The index of first asserted request is switched to the output as the index of *grant*, *g_id*. Then the *g_id* is decoded to create the *grant* signals. The last request, r_{n-1} has a simplified circuit where instead of being multiplexed like other requests, it is ANDed by g_{n-1}. Therefore, in case that only r_{n-1} is high, then only g_{n-1} becomes high.

C. Variable Priority Arbiter

If the *g_id* output of the fixed priority arbiter of Figure 8 is connected to the last multiplexer, each request behaves as it has the highest priority through ascending order of the loop. For example, for four requests (r_0, r_1, r_2 and r_3) the output of multiplexer, *M1* generates an index where r1 has the highest priority then r_2, r_3, and r_0. Therefore, by further multiplexing of these outputs, we can choose an input as the highest priority request as shown in Figure 9. For example, when *P*=1, the output of multiplexer *M1* is selected and the request, r_1 has the highest priority. In the case of no request asserted, or r_0 is asserted, the *g_id* issue the same value (i.e. zero). In order to separate these two conditions, ORing of requests, *any_r* is ANDed with the g_0 so that when all the requests are zero, all the *grant*s will become zero.

D. IRR Arbiter

If the next index of granted request is chosen for the next priority selection, the current granted request receives the least priority, and its next request receives the highest priority among all the requests. To accomplish it, the *g_id* array is stored in a register, and the output of the register is incremented and connected to the selection port of multiplexer, *MP* as shown in Figure 10. In this way, the arbiter follows the least recently served priority scheme or a round robin scheme. Figure 12 illustrates our proposed IRR arbiter that takes one clock cycle for arbitration. To keep the priority unchanged, the output of priority register, *next_g* is fed back through the SF multiplexer to the register to cater for no request. It guarantees a strong fairness arbitration in our design we discuss later.

E. IRR Arbiter Functional Behaviour

We present the functionality and behaviour of our round robin arbiter illustrated by its timing diagram in Figure 11. During time 1-6, a fixed input request, "1111" is applied and granted bit by bit per clock cycle. At time 6, the request is changed to "0000", i.e. no request is asserted. For no request situation, the priority of last granted request is recorded and applied when a new request is asserted. For example, at time 6, the priority of second bit of request is recorded and applied at time 8. Consequently, the forth bit of request is granted at time 8. We tested our arbiter along with some past arbiters (RoR, Matrix, PRRA, IPRRA, and HDRA) with the same testbench and same request scenario as illustrated in Figures 11 and 12. When no request is asserted, the RoR, Matrix and our IRR arbiters record the current priority shown in Figure 11. However, the PRRA, IPRRA and HDRA couldn't record the priority and show different waveforms shown in Figure 12. For no request condition for PRRA, IPRRA and HDRA waveforms, the highest priority is given to the least significant bit of the request. This arbitration behavior of PRRA, IPRRA and HDRA is due to no circuit to handle the no request

condition. Keeping the last priority during no request condition has direct impact on the fairness of an arbiter. The first advantage of our IRR arbiter is aits stronger fairness arbitration.

F. Simpler Incrementing Hardware

The adder in IRR architecture performs an incrementing operation and it can be built in a simple form. For incrementing an *n*-bit operand, one can connect *n* full adder cells in serial form, and feed 1 to carry input of the first cell and zero value as the second operand. Zero value for the 2nd operand enables us to use half adders instead of full adders.

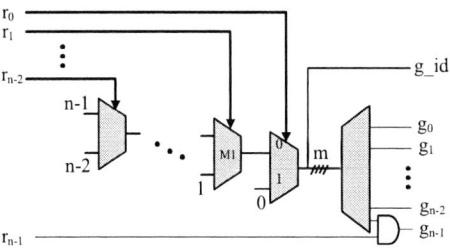

Figure 8. *n*-input fixed priority arbiter, where $m = \log_2(n)$

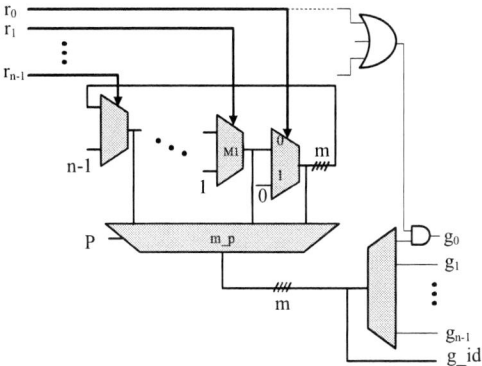

Figure 9. *n*-input variable priority arbiter.

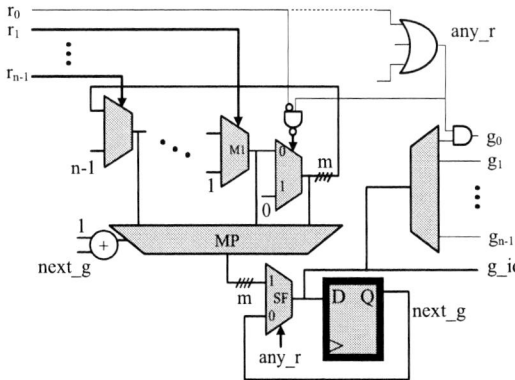

Figure 10. n-input IRR arbiter, *where $m = \log_2(n)$*.

978-1-4799-8720-7/15 $31.00 © 2015 IEEE

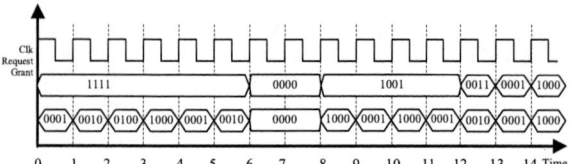

Figure 11. Input request scenarios for strong-fairness RoR arbiters.

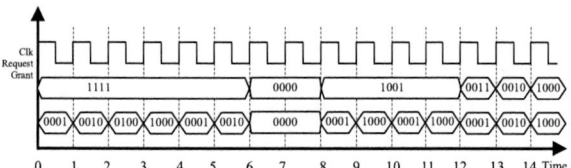

Figure 12. Input request scenarios for weak fairness RoR arbiters.

G. Analytical Comparison

We perform a hardware analysis to compare the expected performance and hardware overhead of the aforementioned round robin arbiters with our proposed IRR arbiter. We don't apply any algorithm to optimize the circuits as an Electronic Design Automation software does. The main figures of merit of an arbiter circuit are speed, area and power consumption. The usual measure for speed of an arbiter circuit is the delay time or maximum clock frequency (F_{max}). The clock frequency of an arbiter depends on the longest delay (critical path) between two registers clocked at the same time. The circuits of 4-input arbiters, which are shown in Figures 3, 4, 5, 6, 7 and 10, are decomposed at the gate level. The electrical parameters of the logic gates are derived from Synopsys 90nm Digital Standard Cell Library as listed in Table I. We calculated the summation of the areas and powers of all the cells of each arbiter to estimate their power and area. The power includes both static and dynamic powers. For speed estimation, the critical path delay between two registers of each circuit is calculated. We have further illustrated the critical path of each circuit through the number in parentheses for the last column of Table I. The increment is done by a half adder (XOR2x1) propagation delay.

As discussed earlier, RoR and Matrix arbiters have strong fairness, and the HDRA, PRRA and IPRRA has weak fairness. Therefore, we introduce a weak fairness version of IRR named IRR_WF for comparison with the HDRA, PRRA and IPRRA arbiters. The only difference between IRR and IRR_WF is the SF multiplexer shown in Figure 10 and it affects the critical path delay. We expect a faster IRR_WF arbiter than IRR. Table II list the characteristics of all the arbiters. IRR has the optimum performance and hardware overhead among all the listed arbiters. In terms of speed, the IRR can run 15% to 50% faster than the other arbiters. Moreover, the IRR saves the area from 10% to 47% and power from 1% to 44%. We also evaluate our IRR arbiter with the other arbiters in terms of area, power and timing by using the EDA tool.

TABLE I. ELECTRICAL PARAMETERS GATES FOR SYNOPSYS 90nm

Gate name	Propagation Delay (ps)	St. Power (nW)	Dy. Power (nW/MHz)	Area (μm²)
INVX1	38	88	12	6.5
AND2X1	85	298	19	7.4
AND3X1	119	297	34	8.3
NAND2X1	51	336	15	5.5
OR2X1	85	226	23	7.4
OR3X1	114	250	39	9.2
OR4X1	137	261	56	10.1
NOR2X1	64	170	15	6.5
MUX21X1	107	815	43	11.1
MUX41X1	168	827	58	23.0
DEC24X1	119	1238	66	29.5
XOR2X1	133	454	26	13.8
DFFARX1	217	620	100	32.2

TABLE II. CHARACTERISTICS OF 4-INPUT ARBITERS BASED ON TABLE I

Type of 4-input arbiters	Area (μm²)	Power (μW)	Critical Path Delay (ps)
IRR	294	296(282[d])	625 (217+133+168+107)
RoR	328	298(289[d])	1242(217+5*(85+85)+137+38)
Matrix	556	479(465[d])	747 (217 +2*38 +3*85+114+85)
IRR_WF	280	274(262[d])	518 (217+133+168)
HDRA	431	360(348[d])	609 (217 +64+51+85+85+107)
PRRA	510	493(479[d])	861(217+2*38+3*85+85+2*114)
IPRRA	528	488(473[d])	747 (217+2*38 +3*85+85+114)

Arbiters	Saving	Saving	Faster
IRR/RoR	10%	1%	50%
IRR/Matrix	47%	38%	16%
IRR_WF/HDRA	35%	24%	15%
IRR_WF/PRRA	45%	44%	40%
IRR_WF/IPRRA	47%	44%	31%

[d] dynamic power

IV. EXPERIMENTAL RESULTS

IRR (IRR_WF) arbiter is evaluated and compared with other arbiters. We consider IRR_WF for comparison with the HDRA, PRRA and IPRRA arbiters. To analyze the speed, area and power overhead, all the arbiters are implemented in structural RTL Verilog and synthesized using the Synopsys Design Compiler (DC) for 90nm Synopsys Generic Library and Altera FPGAs (Stratix V). The resulting designs operate at 400 MHz and 1.2 Volt. Different structures of each arbiter are synthesized, and their results for total cell area, critical path delay, and power (dynamic and static) consumption are listed in Tables III and IV. The Synopsys tool runs different algorithms iteratively to find an optimum architecture for a design. There is always a trade off among the three characteristics i.e. power, area and critical path of a design. This trade off behaviour dictates us to the efficient arbiter design that has smaller power, area and critical path delay. The critical path is the limiting factor preventing us from decreasing the clock period.

For a fair comparison, we group the arbiters into strong and weak fairness arbiters and present their results in Tables III and IV. Table III lists the IRR, RoR and Matrix arbiter characteristics. We listed two synthesised versions of IRR at 16 and 32-input configurations indicating the IRR having efficient area, power and timing characteristics than RoR and Matrix arbiters. It consumes on average 19% and 70%

less chip area, 44% and 74% less power (static and dynamic), and 43% and 7% shorter critical path delay as compared to RoR and Matrix arbiters respectively. The results of 4-input arbiters follow the analytical results of Table II.

Table IV data indicates that IRR_WF is more efficient as compared to PRRA and IPRRA arbiters for all the input configurations. It saves in average 16% and 22% less chip area, 46% and 45% less power, and 34% and 24% shorter critical path delay as compared to PRRA and IPRRA arbiters. The IRR_WF is also more efficient as compared to HDRA for input structures. It saves 13% less chip area, 44% less power, and 15% shorter critical path delay on average. Overall, IRR and IRR_WF arbiters consume the least power among all the arbiters due to its usage of fewer registers. The fewer number of registers of IRR also leads to simpler design and chip layout due to simple clock tree organization. To illustrate the clock tree advantage, we measured the F_{max} of IRR_WF and HDRA for 32-input configuration in FPGA. We have chosen the IRR_WF and HDRA because they are very competitive in ASIC design. The IRR_WF arbiter had 15% higher F_{max} as compared to HDRA in FPGA design.

V. CONCLUSION

We have presented a strong fairness round robin arbiter design, IRR. We proved that our design achieve a strong fairness arbitration for all input patterns, which is not guaranteed by some other previous designs such as HDRA [3], PRRA and IPRRA [4]. The key difference of our approach is its operation on the basis of index format of the input ports. This index based arbitration is simple, fast and with small hardware overhead. We did analytical comparison of the arbiters illustrated in Figures 3, 4, 5, 6, 7 and 10 and concluded that the IRR can execute 15% to 50% faster, requires 10% to 47% less area and consumes 1% to 44% less power as compared to the other arbiters. In ASIC and FPGA designs, the IRR arbiter is more economical and faster than other arbiters. Simulation results with the 90nm Synopsys Generic Library show that the IRR saves up to 70% area improvement, up to 74% power improvement, and up to 43% timing improvement over RoR and Matrix arbiters. The distinctive feature of our IRR arbiter design is its lower power consumption as compared to other arbiters due to its usage of fewer registers.

REFERENCES

[1] W. J. Dally and B. Towles. "Arbitration," In: Principles and Practices of Interconnection Networks. Morgan Kaufmann Publishers, 2004, pp. 349-362.

[2] Z. Fu and Xiang Ling, "The design and implementation of arbiters for Network-on-chips," Proc. 2nd Int. Conf. Industrial and Information Systems, Dalian, 2010 pp. 292-295.

[3] S. Q. Zheng and Mei Yang, "Algorithm-Hardware Codesign of Fast Parallel Round-Robin Arbiters", IEEE Trans Parallel and Distributed Systems, vol. 18, January 2007, pp. 84-95

[4] Yun-Lung Lee, Jer Min Jou, and Yen-Yu Chen, "A High-Speed and Decentralized Arbiter Design for NoC," Proc. IEEE/ACS Int. Conf. on Computer Systems and Applications, Rabat, 2009 pp. 350-353.

TABLE III. HARDWARE CHARACTERISTICS OF STRONG FAIRNESS ROUND ROBIN ARBITERS

Input	Design	ASIC design (90 nm Generic Library) (2.5ns)			FPGA design (Stratix V)	
		Total Cell Area (μm^2)	Power (μW)	Critical path(ns)	Comb. Element	Reg. (bits)
4	RoR	299	76	0.99	13	4
	IRR	295	53	0.54	9	2
	Matrix	431	80	0.56	16	6
8	RoR	1066	204	0.90	25	8
	IRR	705	105	0.66	37	3
	Matrix	1838	313	0.73	58	28
16	RoR	1846	251	1.46	56	16
	IRR	1390	155	1.24	95	4
		2140[a]	182[a]	0.71[a]		
	Matrix	7817	1242	0.73	238	120
32	RoR	3175	384	2.39	109	32
	IRR	2859	219	1.23	205	5
		4224[a]	351[a]	0.89[a]		
	Matrix	31498	4663	0.92	958	496
IRR/RoR		19% saving	44% saving	43% shorter	28% extra	73% saving
IRR/Matrix		70% saving	75% saving	7% shorter	48% saving	90% saving

[a] the second version of IRR for comparison with the Matrix arbiter

TABLE IV. HARDWARE CHARACTERISTICS OF WEAK FAIRNESS ROUND ROBIN ARBITERS

Input	Design	ASIC design (90 nm Generic Library) (2.5ns)			FPGA design (Stratix V)	
		Total Cell Area (μm^2)	Power (μW)	Critical path delay (ns)	Comb. element	Register (bits)
4	IRR_WF	278	45	0.48	9	2
	HDRA	352	117	0.6	11	4
	PRRA	344	85	0.79	9	4
	IPRRA	373	74	0.69	9	4
8	IRR_WF	667	100	0.61	37	3
	HDRA	754	172	0.72	28	8
	PRRA	777	158	1.02	24	8
	IPRRA	848	153	0.87	28	8
16	IRR_WF	1478	179	0.77	95	4
	HDRA	1588	249	0.85	62	16
	PRRA	1683	262	1.25	56	16
	IPRRA	1809	267	1.05	73	16
32	IRR_WF	2801	211	1.11	205	5
		3831[b]	315[b]	0.93[b]		
	HDRA	3204	398	0.94	125	32
	PRRA	3466	442	1.48	126	32
	IPRRA	3719	454	1.27	159	32
IRR_WF/HDRA at 4, 8 and 16-inp.		Saving 13%	Saving 44%	Shorter 15%	Extra 12%	Saving 63%
IRR_WF/PRRA		16%	46%	34%	33%	73%
IRR_WF/IPRRA		22%	45%	24%	23%	73%

[b] IRR_WR for comparison with the HDRA arbiter

2015 IEEE Computer Society Annual Symposium on VLSI

VLSI Implementation of an improved multiplier for FFT Computation in Biomedical Applications

Arathi Ajay[1], Dr. R. Mary Lourde[2]

Department of Electrical and Electronics Engineering
BITS Pilani, Dubai Campus
Dubai, UAE.
ajay.arathi@gmail.com[1], marylr@dubai.bits-pilani.ac.in[2]

Abstract—Discrete Fourier Transform (DFT) is a fundamental Digital Signal Processing domain transformation technique used in many applications for frequency analysis and frequency domain processing. Fast Fourier Transform (FFT) is used for signal processing applications. It consists of addition and multiplication operations, whose speed improvement will enhance the accuracy and performance of FFT computation for any application. It is an algorithm to compute Discrete Fourier Transform (DFT) and its inverse. DFT is obtained by decomposing a sequence of values into components of different frequencies. FFT can compute DFT in O(N log N) operations unlike DFT computation that takes O(N2) arithmetic operations . This reduces computation time by several orders of magnitude and the improvement is roughly proportional to N / log N. Present day Research focus is on performance improvement in computation of FFT specific to field of application. Many performance improvement studies are in progress to implement efficient FFT computation through better performing multipliers and adders.

Electroencephalographic (EEG) signals are invariably used for clinical diagnosis and conventional cognitive neuroscience. This work intends to contribute to a faster method of computation of FFT for analysis of EEG signals to classify Autistic data.

Keywords—FFT, Multiplier, EEG

I. INTRODUCTION

Modern applications are demanding high speed computations. Technology is coming close to theoretical limits on how fast computations can be done on a single chip. Multiple processors operating in parallel, performing different functions of a process and combining them at the end is the solution to this. Fourier Transform is the basis of many signal processing and communication applications. It is the tool for analysis of the signal in its frequency domain. Fourier transform has many applications, in fact any field of physical science that uses varying signals, such as engineering, physics, applied mathematics, and chemistry, will make use of Fourier series and Fourier transforms. Most of these fields nowadays make use of digital and discrete data. Thus the determination of Fourier Transform of discrete signals is of prime importance and such a transform is called Discrete Fourier Transform

(DFT). Fast Fourier Transform (FFT) is an efficient algorithm to evaluate DFT.

Discrete Fourier Transform has a wide range of applications. It is mainly used for converting discrete time domain signals to its frequency domain. However, the process of conversion is expensive and takes lot of time. Thus we go for Fast Fourier Transform which uses a divide-and-conquer approach to reduce computational complexity of DFT. DFT is defined as

$$X(k) = \sum_{n=0}^{N-1} x(n).W_N^{nK} \qquad (1)$$

where n,k $\in$ [0,N-1] and $W_N{}^K = e^{-j2\pi k/N}$ is the twiddle factor, x(n) is the nth sample of discrete time signal and X(k) is its frequency sample at kth instant.

Complexity of computation of DFT is O (N^2) unlike the computation of FFT which is O (N log N) butterfly operations. Hence DFT computation takes more time and is a costlier process. FFT algorithm deals with these complexities by exploiting regularities in the DFT algorithm. Radix-2 Algorithm is a common factor algorithm for N-point DFTs, where N is a power of 2. FFT computation in Radix-2 system takes place in $\log_2 N$ different steps, hence it enables pipelining in hardware design.

FFT computation involves addition and multiplication operations. As multipliers are slow performing hardware units, their performance directly affects the performance of the FFT hardware. Existing hardware multipliers are Serial Multiplier, Array Multiplier, Booth Multiplier, Wallace Tree Multiplier, Booth encoded Wallace tree multiplier, etc. Studies on performance optimizations of Booth Multiplier, Wallace Tree Multiplier and Booth encoded Wallace Tree Multipliers are in progress. Improvement in terms of Hardware Description Language (HDL), Floor planning, Routing, etc. are of main interest to Researchers. This work intends to compare the performance of main hardware multipliers and study and implement the most efficient multiplier for FFT computations in biomedical field.

FFTs are extensively used in data compressions, filtering signals, Signal spectral analysis, Image Processing, etc. In Biomedical field, Medical Imaging plays an important role for

978-1-4799-8720-7/15 $31.00 © 2015 IEEE   68

diagnostics of various health conditions. Electroencephalography (EEG) and electrocardiography (ECG) are various techniques to study the patterns of signals generated by brain and heart, respectively. These are further researched to clinically study human behaviour and heart functions. This work analyses spectral components of EEG signals and proposes an effective method to classify EEG levels of Autistic children.

II. FAST FOURIER TRANSFORM ALGORITHM

Fast Fourier Transform is an algorithm to compute Discrete Fourier Transform which diminishes the number of computations for n-point radix from N^2 to N logN arithmetic operations. There are two methods to compute DFT through FFT algorithm, namely, Decimation-in-time (DIT) and Decimation-in-frequency (DIF) FFT algorithms. In Radix-2 DIT-FFT, input signal is decimated into even-indexed and odd-indexed values such that the series x(n) where, n=0, 1,2,…N changes to x(2r) and x(2r+1) where r=0,1,2,….N/2-1. In Radix-2 DIF-FFT, the x(n) series is broken into x(n) for n=0,1,2….N/2 -1 and x(n) for n=N/2, N/2+1,….N-1.

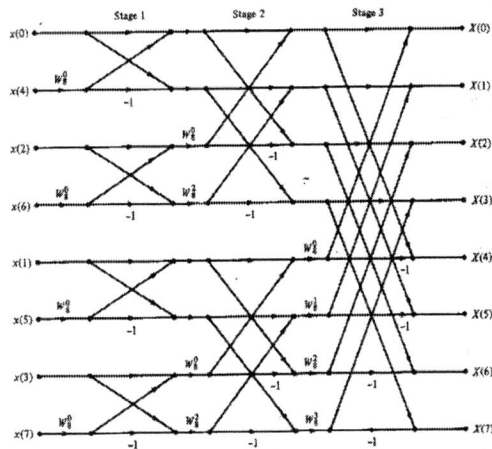

Fig. 1. Radix-2 8-point DIT-FFT algorithm

Fig. 1 shows the butterfly structure of an 8-point Radix-2 DIT-FFT algorithm. Eight time-domain inputs, namely, x0, x1, x2, x3, x4, x5, x6, x7 are transformed to their frequency components, namely, X0, X1, X2, X3, X4, X5, X6 and X7 in three stages. This paper implements Radix-2 8-point DIT-FFT algorithm.

III. MULTIPLIERS

Multipliers have large area, long propagation delays and consume power. Therefore, low-power multiplier design has an important role in design of low-power VLSI systems. Power refers to number of Joules dissipated over a certain amount of time whereas energy is the measure of the total number of Joules dissipated by a circuit [9]. In digital CMOS design, the well-known power-area/power-delay product is commonly used to assess the merits of designs [9].

Any multiplier design involves 3 steps. They are i) partial product generation ii) partial product reduction and addition iii)

final addition. The partial products are formed first either by using an algorithm or using AND gate for each bit of the multiplier with each bit of the multiplicand. The next step is reduction of these partial products. The third step is addition of the remaining partial products to yield the final product [9].

Hardware multipliers widely used are Booth Multipliers, Wallace Tree Multiplier and Booth Encoded Wallace Tree Multiplier. Area and Power parameters of Booth Multiplier and Wallace Tree Multiplier are studied and their advantages have been incorporated to develop the concept of Booth-encoded Wallace Tree Multiplication.

A. Booth Multiplier

Booth Multiplier implements Booth Algorithm, named after its originator, A. D. Booth. This algorithm is implemented for signed multiplication of integers and can be extended to real numbers. Algorithm is based on recording the multiplier to a recorded value leaving the multiplicand unchanged [6] i.e.

• Each '0' digit is retained in the recorded number until 1 is encountered on evaluating from LSB to MSB.

• Complement of 1 is inserted at every '1' digit in recorded number and all other succeeding 1's are complemented until a '0' is encountered.

• Then, replace the '0' with '1' and continue the process.

Two main drawbacks of Booth Algorithm are the inefficiency of the circuit when isolated 1's are encountered and difficulty in designing parallel multipliers as number of shift-and-add operations may vary. Hence Modified Booth Algorithm was developed by O. L. Macsorley. Modified Booth Algorithm is twice as fast as normal Booth Algorithm [4]. This modified Booth Encoding algorithm reduces the number of partial product rows to (N +2)/2 where N is the number of bits of Multiplier or Multiplicand [4].

B. Wallace Tree Multiplier

Wallace Tree Multiplier is one of the hardware multipliers used to accomplish high speed and low power multiplication to condense the number of partial products generated. There are two main techniques followed in designing Wallace Tree Multiplier. First technique is to consider all bits in each column at a time and compress them into two bits, namely, Sum and Carry. Second technique is to consider all bits in four rows at a time and compress them. Wallace Tree Multipliers use half adders, full adders, 4:2 and 3:2 compressors and a high speed adder [10].

Partial product generation, partial product addition and final addition are the three stages in a multiplier. In Wallace Tree Multiplier, the multiplicand is multiplied by the multiplier, bit-by-bit, to generate partial products. They are, then, added based on Wallace Tree structure to produce two rows of partial products which are finally added using any high speed adder. The critical path delay of Wallace Tree multiplier is proportional to the logarithm of the number of bits in the multiplier [9].

Algorithm for Multiplication of two signed integers is as follows:

978-1-4799-8720-7/15 $31.00 © 2015 IEEE

• Multiply (AND) each bit of one of the arguments,

• Reduce the number of partial products to two by layers of full adders and half adders (Compressors).

• Group the wires in two numbers, and add them with a conventional adder [9].

Wallace tree multiplication can be implemented only for signed integers and are avoided for low power applications as excess wiring consumes more power.

C. Booth-Encoded Wallace Tree Multiplier

Based on the comparative study of Area and Power parameters of Booth Multiplier and Wallace Tree Multiplier, Booth-encoded Wallace Tree Multiplier is chosen as an efficient multiplier for FFT computation. Table I. shows Area and Power performance parameters of the two multipliers which are coded in Verilog HDL and performance parameters are evaluated using IC Compiler tool from Synopsys Inc. Though Wallace Tree multiplier shows better performance in terms of Area and Power than Booth multiplier, its operation is limited to signed integers alone. As FFT computation in biomedical applications involve signed real numbers, Booth Algorithm is to be implemented for the multiplier.

TABLE I.   PERFORMANCE PARAMETERS OF 4X4 BOOTH MULTIPLIER
AND WALLACE TREE MULTIPLIER

Multiplier	Area(nm^2)	Power(µW)
Booth Multiplier	974.872	235.184
Wallace Tree Multiplier	600.868	180.504

Multiplication of the two operands, Multiplicand (MD) and Multiplier (MR) results in 2N bits for conventional multiplication. However, Booth encoded multiplier reduces number of partial products to (MD/2 -1) partial products [5]. Booth Algorithm is based on recording the multiplier to a recorded value leaving the multiplicand unchanged. It scans the multiplier operand and skips chains. It reduces the number of additions required to produce the result [4]. Booth-encoded Wallace Tree Multiplier has advantages of both Booth Multiplier and Wallace Tree Multiplier. This paper implements Booth encoding to increase speed of algebra by reducing number of partial products and Wallace Tree module for decreasing number of levels of addition.

1) Architecture of the Multiplier

In Booth encoded multiplier, number of partial products are reduced by grouping multiplier bits into pairs and selecting partial products from the set {0, M, 2 Hardware Description Language (HDL) implementation of Booth-encoded Wallace Tree Multiplier has been done in Verilog HDL. The outputs are correctly obtained for both signed and unsigned multiplication. Simulation has been done using Xilinx ISE tool. Generic Gate-level schematic has been obtained in IC

Compiler tool from Synopsys Inc. Fig. 3 show the simulation results of the Booth-encoded Wallace Tree Multiplier.

M, 3M} where M is the multiplicand. Modified Booth encoded multiplier, avoids the use of Carry Propagate Adder to calculate 3M, rather it utilizes Carry-Save-Adder. Thus, in Modified Booth Algorithm, number of partial products is reduced by a factor of two without a pre-adder to produce partial products [6]. Multiplier decoding is done such that multiples needed are in {0, M, 2M, 4M + -M} data set. These multiples can be generated using shift-and-complement methods.

The architecture of Booth-encoded Wallace Tree Multiplier consists of five blocks, namely, 2's Complement Generator, Booth Encoder, Partial Product Generator, Wallace Tree module and Carry Look-ahead Adder [7]. Booth encoder inspects each bit of the multiplicand and records the multiplier in terms of 0, 1 and complement of 1. As complement of 1 cannot be represented on hardware, operational equivalent of recorded multiplier is implemented based on Table II. Here, outputs of the encoder. x and z are defined as:

$$x = MR[i] \bullet MR[i-1] \qquad (2)$$

$$z = MR[i] \oplus MR[i-1] \qquad (3)$$

Where MR[i] and MR [i-1] corresponds to ith and i-1th bit of Multiplier, respectively. 2's complement generator takes the multiplicand MD as input and produces –MD as output .i.e. inverts all bits of multiplicand and uses a Ripple Carry Adder to generate 2's complement. Partial product generator generates appropriate partial products to be added in Wallace tree structure. Wallace Tree module adds all partial products. Addition is implemented using Carry Look Ahead Adder [7].

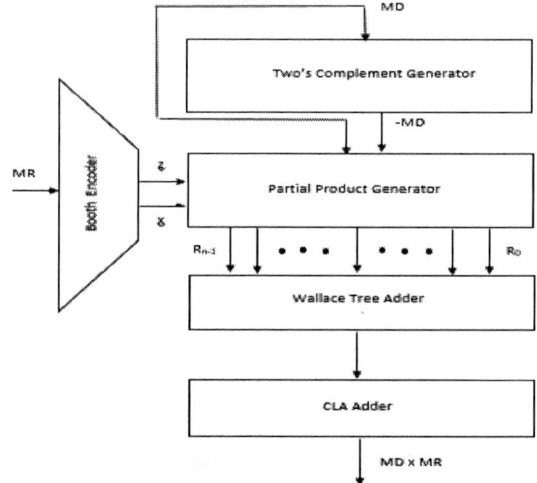

Fig. 2. Booth-encoded Wallace Tree Multiplier

978-1-4799-8720-7/15 $31.00 © 2015 IEEE

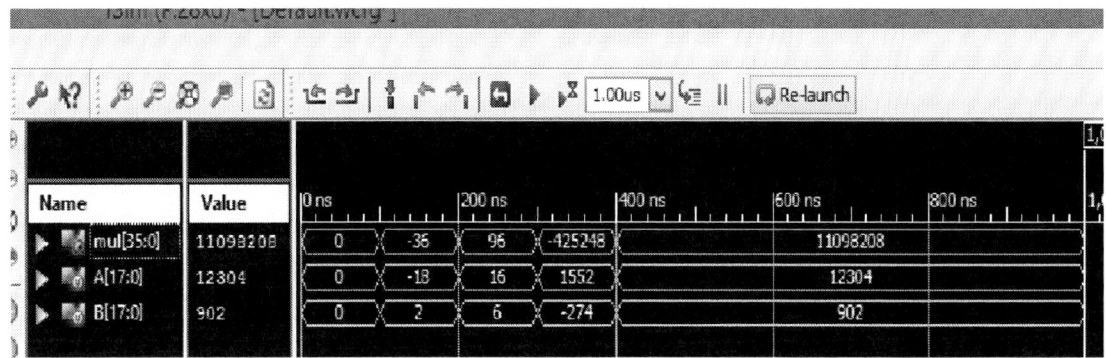

Fig. 3. Simulation Results of Booth-encoded Wallace Tree Multiplier

IV. FFT WITH BOOTH-ENCODED WALLACE TREE MULTIPLIER

FFT implementation using Booth-encoded Wallace Tree Multiplier is an efficient circuit with advantages of both Booth Multiplier and Wallace Tree Multiplier. As explained in II, Modified Booth Algorithm is used to calculate Partial Products as per Table II. Partial Product Generator provides minimum number of partial products which are added using Wallace Tree structure. Hence, efficient algorithm of multiplication is implemented.

TABLE II.   BOOTH ENCODING VALUES

MR[i]	MR[i-1]	Recorded y[i]	x	z	Assigned to partial product
0	0	0	0	0	0
0	1	1	0	1	MD. sign extended
1	0	1̄.	1	1	-MD, sign extended
1	1	0	0	0	0

Radix-2 8-point DIT-FFT has been coded in Verilog HDL for input values scaled up by a factor of 10000. There are ten inputs to the circuit, namely, eight time-domain samples,

Twiddle factor 'w_r' and the scaling factor 'k'. Outputs are separately obtained for real and imaginary parts of corresponding frequency components.

As 8-point corresponds to 3 stages ($8 = 2^3$), twiddle factor varies for each stage. Twiddle factor is calculated as in (1). For N=1, Twiddle factor value is {1}, for N=2, Twiddle factor values are {1, -j} and for N=3, Twiddle factor values are {1, -j, $(1/\sqrt{2})$ − $j(1/\sqrt{2})$, $(-1/\sqrt{2})$ - $j(1/\sqrt{2})$}. Depending on the accuracy of outputs required, scaling can be varied from 10 to any power of 10 and by increasing the number of significant digits of twiddle factor value of $(1/\sqrt{2})$. Simulation results shown in Fig.5 has been scaled up for a factor of k = 10000. Outputs and inputs have been correctly verified. Accuracy of ±0.1% is obtained for scaling improvement from 1000 to 10000.

Design Vision tool from Synopsys Inc. is a logic synthesis tool which inputs HDL design and synthesize out gate-level HDL net lists. The Generic Gate-level schematic diagram implemented in 90 nanometer technology using saed90nm_typ_ht.db library in IC Compiler tool from Synopsys Inc.

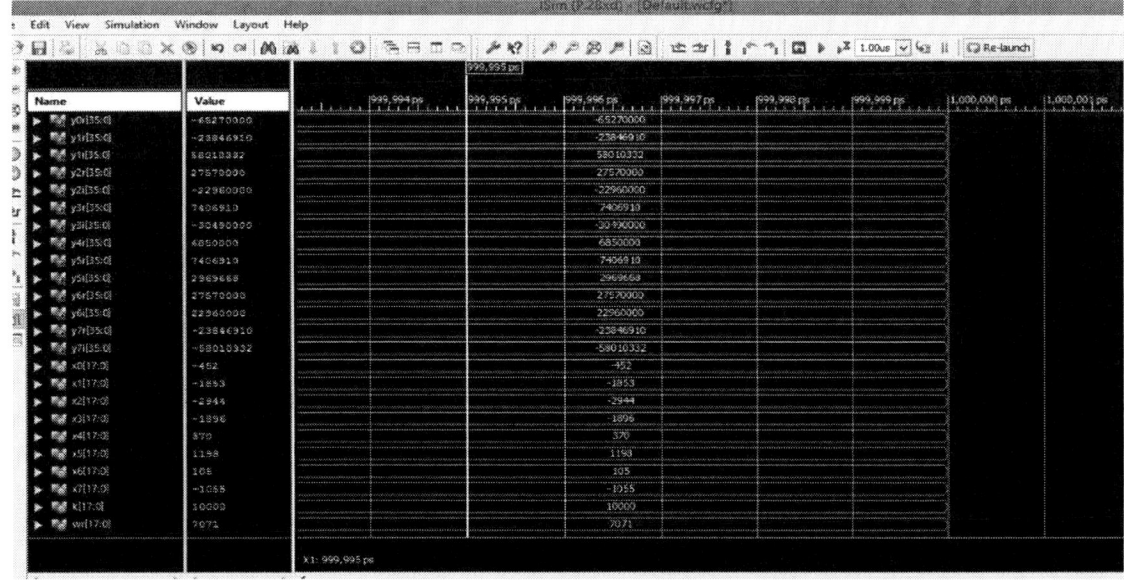

Fig. 4.   Simulation Results of FFT with Booth-encoded Wallace Tree Multiplier shown in Signed Decimal Radix

V. FFT AND EEG

Virtually all sciences in the world contribute to the maintenance of human health and the practice of medicine. Medical physicists and biomedical engineers support the effective utilization of this medical science and technology as their responsibilities to enhance human health care with the new development of the medical tools such as Electroencephalogram. Electroencephalography (EEG) is a mechanism of measuring electrical activity of the brain. Upon studying EEG signals, various health conditions can be monitored and diagnosed e.g. Brain diseases like Alzheimer, Tumors, Head injuries, Epilepsy, Dementia, Human Behaviour, etc. These signals are recorded from various positions on scalp through electrodes and conductive media. Diagnostic results are made from the spectral content of EEG signals. Here comes the importance of FFTs in EEG signal analysis. They are extensively used in neuroscience, cognitive science and cognitive psychology due to its capability to reflect both normal and abnormal electrical activity of brain [1].

EEG signals are typically represented either through rhythmic activity or transients. Rhythmic Activity of brain signals is divided into different bands of frequency, namely, Delta, Theta, Alpha, Beta, Gamma. Table III. shows the frequency range of each band and the different states of the person [1]. Amongst these bands, Alpha band is widely used for various diagnostic studies [8]. Real-time applications like alerting drivers about their drowsiness through EEG variations in Alpha band is one of its kind. EEG signal analysis are crucial for evaluating Epilepsy. In marketing field, EEG signals have been used to study customer response to various products in market as a feedback method and product improvement study, referred to as neuro-marketing.

TABLE III.   FREQUENCY BANDS IN EEG SIGNAL

EEG bands	Frequency band(in Hz)	Prominent
Delta	Less than 4	Deep sleep
Theta	4-7	Drowsiness
Alpha	8-13	Relaxed and awake
Beta	14-30	Sleep stages
Gamma	>30	Finger movements

In this work, EEG signal analysis is used to classify EEG levels of Autistic persons in Alpha band. Autism is a brain development disorder whose symptoms are visible since childhood. Autistic children due to seizures, show high delta and theta waveforms and low alpha waveform due to less metabolism. EEG signal pattern of normal persons are similar for similar age group, sex, race, food habits, environmental conditions, etc. However, frequency bands will remain the same for all people. Different types of special children show variations in EEG signals in Alpha band depending on their neurological function. Hence they can classified by comparing values of mid-frequency Alpha band of normal and Autistic persons.

EEG samples have been collected for a normal male aged at 21 years during closed eye condition based on 10 – 20 International Standards Electrode System with 19 channel electrode placement and 1 electrode as reference. FFT is computed for the two Occipital electrode position (O1 and O2) samples at a sampling frequency of 80 Hz in MATLAB. Power Spectral response is obtained to study and analyze Alpha band. Similarly, Radix-2 8-point DIT-FFT computation is performed using FFT implemented through Booth-encoded Wallace Tree Multiplier for the same sample.

Mid-frequency FFT value of Alpha band, at 10 Hz, are compared for the normal person. Values obtained through MATLAB and the FFT implemented circuit using Xilinx ISE is given in Table IV. Real-part values of the computed FFT at electrode O2 are plotted in Fig.7. Y-axis corresponds to real value of computed FFT and X-axis shows the corresponding frequency. Value at $X0 = -0.0006527$ volts is also shown in the figure.

TABLE IV.   VALUES OF 8-POINT FFT FROM MATLAB AND XILINX ISE

Out-put	FFT using MATLAB (O2 electrode position) x 10^-3 volts	FFT computed using VLSI circuit (O2 electrode position) x 10^-11 volts
X0	-0.6526 + j0.0000	-65270000 +j0
X1	-0.2385 + j0.5801	-23846910 + j58010332
X2	+0.2757 - j0.2296	+27570000 - j22960000
X3	+0.0741 - j0.0297	+7406910 - j2969668
X4	+0.0685 + j0.0000	+6850000+ j0
X5	+0.0741 + j0.0297	+7406910 + j2969668
X6	+0.2757 + j0.2296	+27570000 + j22960000
X7	-0.2385 - j0.5801	-23846910 – j58010332

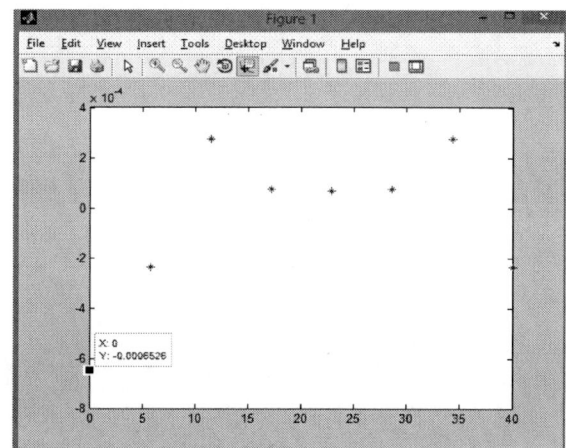

Fig. 5.   Real-part values of the FFT output from MATLAB

VI. RESULTS AND DISCUSSION

EEG data collected for a normal person has been sampled at 80 Hz in MATLAB. The mid-frequency alpha band value of the same at 10 Hz (x0) has been obtained as -0.0006526 volts. FFT has been implemented using Booth-encoder Wallace Tree multiplier to compute Radix-2 DIT-FFT algorithm. The same sample has been given as input to the FFT circuit in Test bench and the mid-frequency alpha band value at 10 Hz is obtained as $(-65270000 \times 10^{-11})$ volts. An accuracy of $\pm 0.1\%$ has been obtained through the implemented FFT circuit for all eight outputs obtained.

As per [1], Mid-frequency value of Autistic child is less than that of a normal child. The implemented FFT algorithm can be used to verify this result. Accuracy of computed Fast Fourier Transform value can be increased by increasing the significant digits of the twiddle factor ($1/\sqrt{2}$). The results obtained show an improvement in accuracy of +0.1% for the value of twiddle factor = 0.7071. Complexity of the circuit increases with increase in width of data.

VII. CONCLUSION AND FUTURE SCOPE

Generic-gate level implementation of Fast Fourier Transform using Booth-encoder Wallace Tree Multiplier has been done using VLSI 90nm technology. The circuit has been studied and analyzed for data accuracy and efficient performance. An accuracy of $\pm 0.1\%$ has been obtained for the twiddle factor = 0.7071. On increasing the number of significant digits of the twiddle factor, accuracy of output can be improved but it will share a trade-off with the area of the circuit. Physical level chip design and further optimization of the circuit in terms of area and power and increasing the number of points for FFT computation form the future scope of work.

Autistic EEG samples can be evaluated for the mid-frequency alpha band. This value is expected to be lower than normal person [1]. This circuit can thus be used to classify Autistic person based on EEG level and this work can be extended for its further analysis and verification.

ACKNOWLEDGMENT

The authors thank the Management and Department of Neurophysiology, Welcare Hospital, Dubai for providing medical data.

REFERENCES

[1] Sudirman (IEEE member), S. Saidin, N. Mat Safr, "Study of Electroencephalography signal of autism and Down syndrome children using FFT", 2010 IEEE Symposium on Industrial Electronics & Applications (ISIEA).

[2] Mokhtar Aboelaze, Member, IEEE, "An FPGA Based Low Power Multiplier for FFT in OFDM Systems using Precomputations", 2013 International Conference on ICT Convergence (s).

[3] Leif Sornmo, Pablo Laguna, "Bioelectrical Signal Processing in Cardiac and Neurological Applications, Copyright (c) 2005, Elsevier Inc.

[4] Sukhmeet Kaur1, Suman2 and Manpreet Signh Manna3, "Implementation of Modified Booth Algorithm (Radix 4) and its Comparison with Booth Algorithm (Radix-2)", Advance in Electronic and Electric Engineering, Volume 3, Number 6 (2013), pp. 683-690.

[5] Rahul D Kshirsagar, Aishwarya.E.V., Ahire Shashank Vishwanath, P Jayakrishnan, "Implementation of Pipelined Booth Encoded Wallace Tree Multiplier Architecture, International Conference on Green Computing, Communication and Conservation of Energy (ICGCE), 2013.

[6] Deepali Chandel, Gagan Kumawat, Pranay Lahoty, Vidhi Vart Chandrodaya, Shailendra Sharma, "Booth Multiplier: Ease of Multiplication", International Journal of Emerging Technology and Advanced Engineering, Volume 3, Issue 3, March 2013.

[7] B. Dinesh, V. Venketeshwaran, P. Kavinmalar, Dr. M Kathirvelu, "Comparison of Regular and Tree based Multiplier Architectures with Booth Encoding for 4-bits on Layout level using 45 nm technology", International Conference on Green Computing Communication and Electrical Engineering (ICGCCEE), March 2014.

[8] Samaneh Valipour1, A.D. Shaligram 2, G.R.Kulkarni3, "Spectral analysis of EEG signal for detection of alpha rhythm with open and closed eyes", International Journal of Engineering and Innovative Technology (IJEIT), Volume 3, Issue 6, December 2013.

[9] N V Vineela Maunika, M Vasuja Devi, *"A Dwindled Power and Delay of Wallace Tree Multiplier"*, International Journal of Engineering and Innovative Technology (IJEIT), Volume 2, Issue 4, October 2012.

[10] M.Ravindra Kumar, G Parameswara Rao, "Design and Implementation of 32*32 Bit High Level Wallace Tree Multiplier", International Journal of Innovative Research & Studies (IJIRS), Volume2, Issue 8, August 2013.

Sub-threshold SRAM Design in 14 nm FinFET Technology with Improved Access Time and Leakage Power

Behzad Zeinali[1], Jens Kargaard Madsen[1], Praveen Raghavan[2], Farshad Moradi[1]

[1] ICE-LAB, Department of Engineering, Aarhus University
Finlandsgade 22, 8200, Aarhus N, Denmark
[2] IMEC, Kapeldreef 75, B-3001, Leuven, Belgium
Email: {beze, jkm, moradi}@eng.au.dk, praveen.raghavan@imec.be

Abstract—In this paper, we propose a new sub-threshold 9T-Static Random Access Memory (SRAM) cell in 14 nm FinFET technology by which the access time of the memory cell is improved by at least 30 percent compared to the standard 8T-SRAM cell. Furthermore, the leakage current of the proposed SRAM cell is reduced by an assisting circuit. Simulation results show that the leakage of the proposed cell is reduced by 20% after using the technique. Furthermore, the read access time of the proposed cell is reduced by 30% compared to the 8T-SRAM cell while write margin and read noise margin of the proposed cell is not degraded. The maximum operating frequency for the proposed SRAM cell is 2.7 MHz at V_{DD}=200 mV.

Keywords—FinFET, Static Random Access Memories (SRAM), near/sub-threshold operating.

I. INTRODUCTION

Scaling the CMOS technology to sub-22nm has faced a great challenge due to the increasing process variations and leakage currents (e.g. sub-threshold and gate leakage). On the other hand, lowering the supply voltage to achieve a lower power consumption has aggravated the performance of integrated circuits such as SRAMs. SRAM arrays play an important role in future wearable and implantable electronics due to the requirement of larger on-chip SRAM with low leakage and better robustness. To this end, several candidates have been introduced to tackle the CMOS scaling issues [1]-[6]. One of the most promising devices is FinFET technology that has enabled a significant improvement in performance compared to other counterparts especially in sub-22nm scales. Due to the key features such as suppressed sub-threshold leakage, improved controllability of the channel due to the 3D gate structure, and better scalability, FinFETs have emerged as one of the leading candidate for next generation electronic devices. On the other hand, one of the key feature to achieve lower power consumption is supply voltage scaling down to near/sub-threshold region. The robust operation of FinFETs at low supply voltages has enabled designers to implement low-medium frequency applications in a low power mode. However, all functions of an SRAM such as read, write and hold margin are affected by increased process variation at low supply voltages that results in a lower yield. To improve the performance of the SRAM cell at low supply voltages, several FinFET SRAM solutions from device level [6] to circuit [7]-[9] and architecture level [10]-[12] have been proposed. Techniques such as negative bit-line [10], boosted word-line [13], and transient voltage collapse write assist [14] can be applied to the standard 6T-SRAM cell to improve the reliability of the cell which is defined by write margin (WM). On the other hand, to achieve a higher read static noise margin (RSNM), techniques such as level-programmable word-line driver (LPWD) [15], replica access transistor (RAT) [16] and SRAM cells such as 8T-SRAM have been proposed. However, most of these architectures suffer from degraded access time, especially at near/sub-threshold operation region. To alleviate these obstacles of scaling supply voltage to sub-threshold voltages, in this paper, a 9T-SRAM cell is proposed by which access time is reduced significantly while the WM and RSNM are not degraded.

The rest of the paper is organized as follows: In section II, the proposed SRAM cell is presented. The simulation results of the proposed 9T-SRAM cell are presented in section III. Different SRAM structures are compared in terms of area, leakage, read, and write performance in this section, as well. Section IV explores a technique to reduce the leakage current of the 9T-SRAM cell. In section V, the behavior of the proposed cell at different technology corners is explored compared to the 6T- and 8T-SRAM cell. Conclusions are drawn in section VI.

II. PROPOSED 9T-SRAM CELL

A SRAM cell must be designed to provide a stable read operation and a reliable write operation while these along with random variations confine the minimum operational power supply. The RSNM and WM determine cell stability and write ability, respectively, which need different requirements to become satisfactory.

The standard 6T-SRAM cell containing a cross coupled inverter latch and two access transistors controlled by word-line (WL) signals for write and read operations, achieves

proper read and write specification by using transistor sizing in super-threshold supply voltages. A stable read is achieved by keeping the pull-down devices stronger than the access transistors. On the other hand, for a reliable write operation, the access transistors are implemented using stronger devices than pull-up transistors. However, the sizing is obsolete for 6T-SRAM cell operating at very low supply voltages, i.e. near/sub-threshold, due to the exponential dependence of the sub-threshold current to the threshold voltage of the transistors. This dependence causes large changes in current due to even a small variation of threshold voltage. Therefore, other solutions must be considered to solve the trade-off between read and write. This is achieved by separating the read and write paths with the use of SRAM cells, such 8T-SRAM [10], [11].

In the 8T-SRAM cell, the read and write operations are decoupled and are separately performed by using the exclusive bit-lines. Furthermore, the read bit-line (RBL) becomes decoupled from the cell storage nodes, which prevents any disturbance in the storage node during read operation, hence, the 8T-SRAM has a high read and hold margin. However, at low voltages, due to the stacked n-FinFET transistors in read path and as a result the low driving current, the access time is significantly degraded that results in read failure at deep sub-threshold voltages. Furthermore, the operating supply voltage of the 8T-SRAM cell is also limited by access time, which is limited by the drivability of the stacked transistors.

To this end, the 9T-SRAM cell illustrated in Fig. 1 is proposed in which the access time is improved by removing the stacked configuration in the read path. In the proposed cell, the RWL is applied to the gate of a p-FinFET transistor (M_7) which functions as a switch to connect the storage node to the gate of transistor M_9. When the stored data is "1", M_9 turns on and the bit-line capacitance is discharged through the transistor M_9. On the other hand, when the data is "0", the transistor M_9 turns off and the bit-line voltage change depends on the OFF current of the transistor M_9. M_8 is utilized to ensure that M_9 remains OFF when the RWL is "0" (i.e. hold and write modes). During write, the proposed 9T-SRAM cell works similar to the 8T-SRAM cell. During hold, RWL and WWL are "0" where transistors M_5, M_6, M_7, and M_9 are OFF. This results in a similar hold static noise margin (HSNM) as 6T- and 8T-SRAM cells.

III. SIMULATION RESULTS AND DISCUSSION

To show the efficacy of the proposed 9T-SRAM cell, different features of the design are explored. The proposed cell is realized using 14 nm FinFET technology. The parameters of the FinFET technology, including the fin width (W_{fin}), fin height (H_{fin}) and equivalent gate dielectric thickness (EOT) are fixed. On the other hand, the gate length (L_g) could be adjusted from 20 nm up to 40 nm. The parameters of the n-FinFET and p-FinFET devices are tabulated in Table 1. To lower the leakage current and lower process variations, high-V_{th} transistors are used. In the following simulations, 6T-SRAM cell is sized as follows: the access and pull-up transistors use one fin while the pull-down transistors utilize 2-fin FinFET transistors. On the other hand, in the 8T-SRAM cell, only the pull-down transistor in the buffer uses 3-fin FinFET transistor while the rest of the transistors in the cell use 1-fin transistors. The proposed 9T-SRAM cell use only 1-fin FinFET transistors.

A. Read/Write Static Operation

The HSNM and RSNM are defined as the length of the side of the largest square that can fit into the lobes of the butterfly curve during hold and read operations, respectively. Figures 2 and 3 show the butterfly curves for the 6T-, 8T-, and the proposed 9T-SRAM cells during hold and read modes, respectively operating at power supply of 0.2 V and 0.4 V. As it is shown, 8T- and 9T-SRAM cells give similar RSNM and HSNM due to the isolated storage nodes from read bit-lines. On the other hand, the RSNM of the 6T-SRAM cell is degraded significantly, especially at low voltages. Figure 4 illustrates the hold and read SNMs at different supply voltages for the proposed SRAM cell in comparison with 6T- and 8T-SRAM cells. As it is shown in the figure, the RSNM and HSNM of the 9T- and 8T-SRAM cells are significantly higher in comparison to the 6T-SRAM cell for near/sub-threshold regions. Therefore, both 8T and 9T can operate at very low voltages in 14 nm FinFET technology.

Another metric to compare different SRAM structures is write margin. Different methods are used to measure the WM of a cell [17]. The method used in this paper is presented in [18]. In this method, the write word-line (WWL) is swept from 0 to V_{DD} while the storage nodes and their corresponding write bit-lines are biased at reciprocal voltages. In this test bench, the WM is defined as the difference between V_{DD} and WWL when the storage nodes flip as illustrated in Fig. 5. The simulation results for WM of the proposed SRAM cell at different supply voltages in comparison with 6T- and 8T-SRAM cells is depicted in Fig. 5. According to this figure, the WM of the 9T- and 8T-SRAM cells are higher than 6T-SRAM cell for near/sub-threshold regions. As shown in the figure, the WM of these cells is degraded significantly at near/sub-threshold region. Therefore, a write assist should be used to improve the WM.

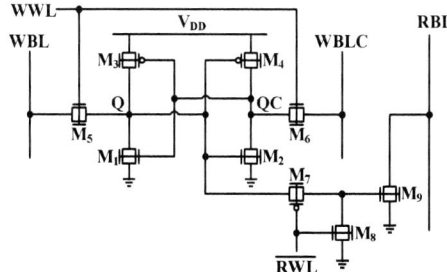

Fig. 1. The proposed 9T-SRAM cell.

TABLE I

14 NM FINFET PROCESS PARAMETERS

FinFET Device	1-fin FinFET Device @ V_{DD}=0.4 V L_g=24 nm, W_{fin}=10 nm, H_{fin}=30 nm, EOT=1.2 nm			
	I_{on} ($\mu A/\mu m$)	I_{off} ($pA/\mu m$)	SS ($mV/decade$)	DIBL (mV/V)
n-FinFET	38.7	260	69.6	29.9
p-FinFET	35.4	288	69.6	30.1

B. Read/Write Dynamic Operation

As described earlier, using one transistor instead of the conventional stacked configuration to discharge the RBL improves read access time, which is defined as the time required for discharging the bit-line to a specific voltage after asserting the RWL. This voltage is determined according to the sensitivity of the sense amplifier (SA). In this paper, a 70 mV bit-line voltage difference is considered to be sensed by the SA. Equivalently, the time between asserting the word-line signal and when the storage node voltages cross 90% of their final values is measured as write time. According to these definitions, in Figs. 6 and 7, read access time and write time of the 6T-, 8T- and the proposed 9T-SRAM cells are compared for supply voltage of 0.2 V, respectively. According to Fig. 6, the read access time of the proposed SRAM is 30% lower than its corresponding 8T-SRAM cell at V_{DD}=0.2 V. Furthermore, Fig. 7 illustrates that the WM of the proposed cell improves in comparison to the 8T-SRAM cell at V_{DD}=0.2 V which is due to using a 3-fin pull-down transistor in the read path of 8T cell. This 3-fin n-FinFET transistor increases the parasitic capacitance of storage node and hence increases the WM. However, size reduction of this transistor can degrade the read

access time of the 8T-SRAM cell even further. Besides, Fig. 8 illustrates the impact of using the proposed 9T-SRAM cell in lowering the read access time and write time in comparison to the 6T- and 8T-SRAM cells in different operating voltages. This figure proves the efficiency of the proposed SRAM cell to improve the read access time in all over the ranges of near/sub-threshold supply voltages.

C. Area

The layouts of the 6T-, 8T- and the proposed 9T-SRAM cells are shown in Fig. 9. The area of the 6T-, 8T- and the proposed 9T-SRAM are $0.252 \times 0.524~\mu m^2$, $0.252 \times 0.654~\mu m^2$, and $0.252 \times 0.756~\mu m^2$, respectively. As it is shown the proposed cell has a 44% and 15% area overhead compared to the 6T- and 8T-SRAM cells, respectively. It is worth to mention that the dynamic performance of the proposed cell is better than its 8T counterpart even at higher area for the 8T-SRAM. It is due to this fact that using even further fins in the pull-down transistor in the buffer, cannot improve the access time. On the other hand, using more fins for the other transistor in the buffer increases the bit-line capacitance.

D. Leakage Power

The main contribution of power consumption in SRAM arrays is due to the standby power. Figure 10 illustrates the leakage current of the 6T-, 8T- and the proposed 9T-SRAM cells at different power supply voltages. The figure shows the leakage currents during hold for data equals to "1" and "0".

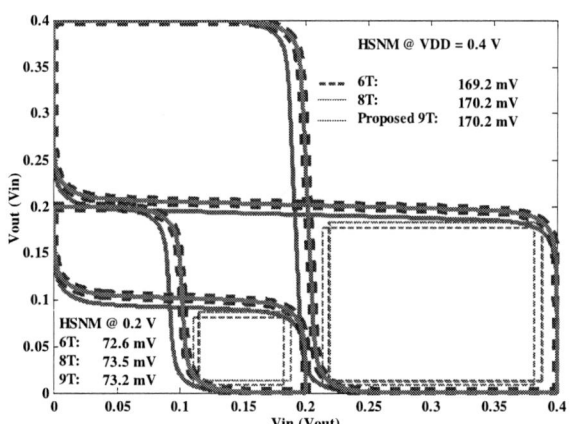

Fig. 2.   HSNM of the 6T-, 8T- and 9T-SRAM cells at two operating voltages of 0.2 V and 0.4 V.

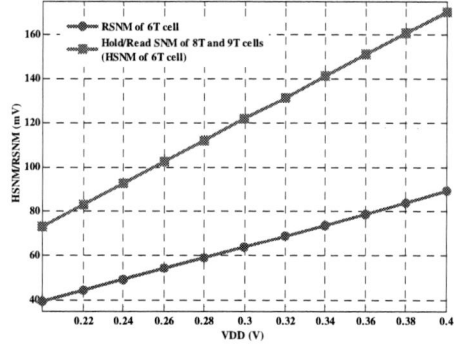

Fig. 4.   The Hold/Read SNM at different supply voltages for the cells.

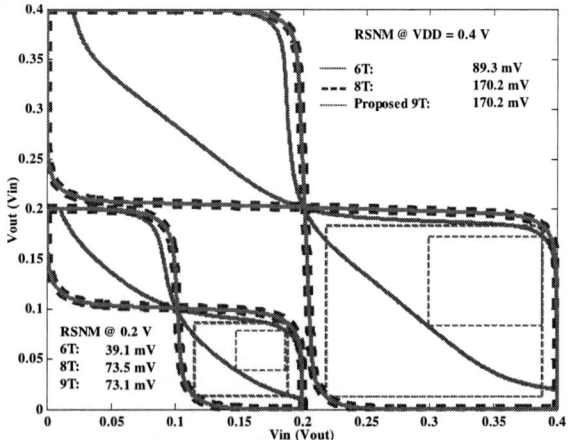

Fig. 3.   RSNM of the 6T-, 8T- and 9T-SRAM cells at two operating voltages of 0.2 V and 0.4 V.

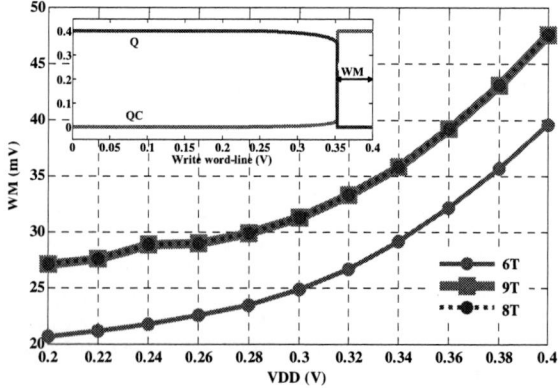

Fig. 5.   The WM at different supply voltages for the 6T, 8T and 9T cells.

According to this figure, the 9T-SRAM has a higher leakage power among all three configurations. Lower leakage provided by 8T-SRAM cell compared to the proposed 9T-SRAM cell is due to the stacked configuration of the read path. The higher leakage power of the proposed SRAM cell is mainly attributed to the sub-threshold current through M_9 where the V_{DS} is equal to V_{DD} during standby mode. To this end, we propose a read assist technique in which a short pulse is generated during read time while the bit-line is grounded for the rest of the cycle and hold mode. This technique significantly improves leakage power saving which is described in the next section.

IV. LEAKAGE REDUCTION ASSIST TECHNIQUE

According to the simulation results discussed in section III, the proposed 9T-SRAM cell suffers from a high leakage power compared to the 8T-SRAM cell due to the use of a single n-FinFET transistor in the read path. In 8T-SRAM cell, the leakage through access transistors depends on the stored value on the storage nodes. When the data on node Q is "0", the leakage current through the read path is minimum while Q="1" results in maximum leakage through the read path. On the other hand, for the 9T-SRAM cell, the gate voltage of the transistor M_9 is always connected to ground independent of the storage node values. However, the use of single transistor M_9 in the read path with the drain voltage of V_{DD} results in a high leakage current. To lower the leakage current through M_9, the RBL voltage is discharged to the ground during hold

mode, which results in a drain-to-source voltage of zero which leads to a significant reduction in leakage current. However, the voltage on RBL needs to be charged at the beginning of the read cycle. To this end, we propose a technique shown in Fig. 11(a) by which the leakage current through access transistor M_9 is reduced without any penalty on access time. The appropriate timing of the technique is sketched in Fig. 11(b) as well. By the use of this technique, the RBL is kept discharged through an n-FinFET transistor (M_P) during hold mode while, the capacitor (C_R) which can be shared between all the bit-lines, is charged through p-FinFET transistor (M_C).

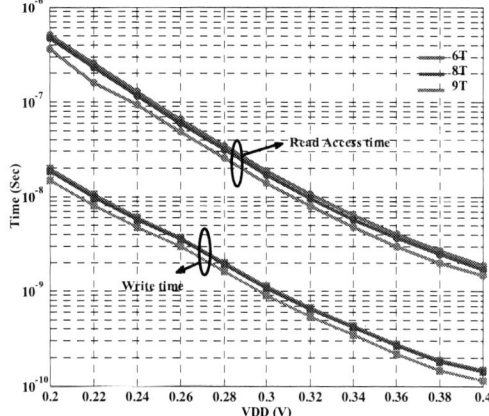

Fig. 8. The write/read access time at different supply voltages for the cells.

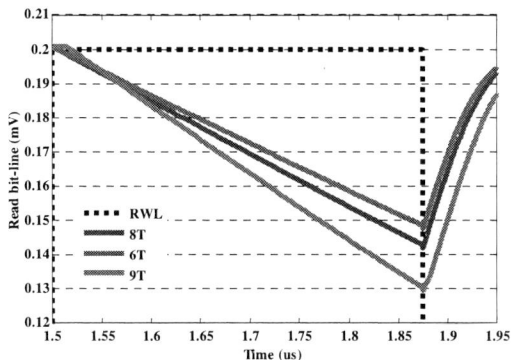

Fig. 6. The read access time of the 6T, 8T and 9T cells @ V_{DD}=0.2 V.

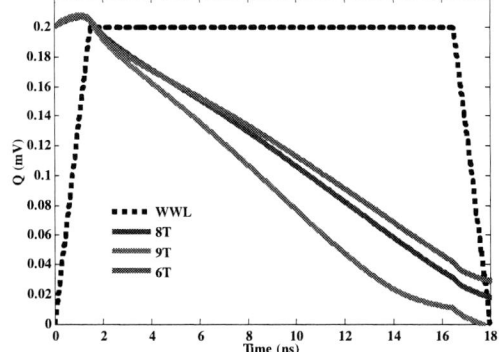

Fig. 7. The write time of the 6T-, 8T- and 9T-SRAM cells @ V_{DD}=0.2 V.

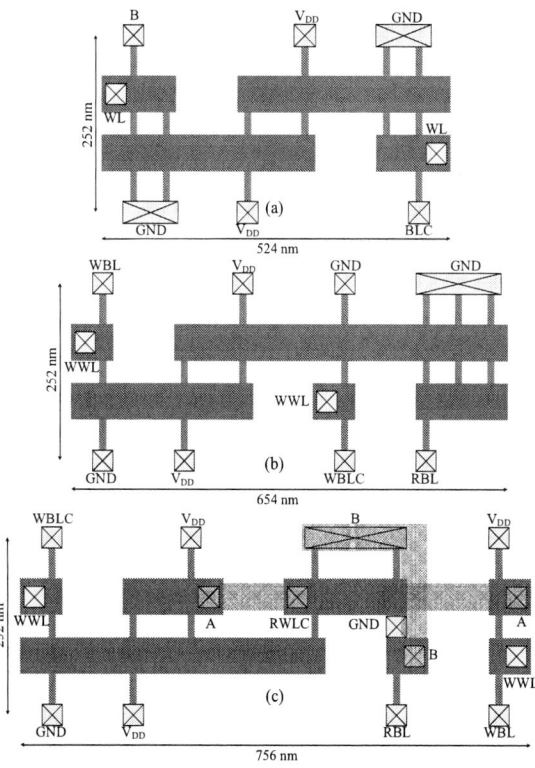

Fig. 9. The Layout of (a) 6T-, (b) 8T-, and (c) 9T-SRAM cells in 14 nm FinFET technology

978-1-4799-8720-7/15 $31.00 © 2015 IEEE

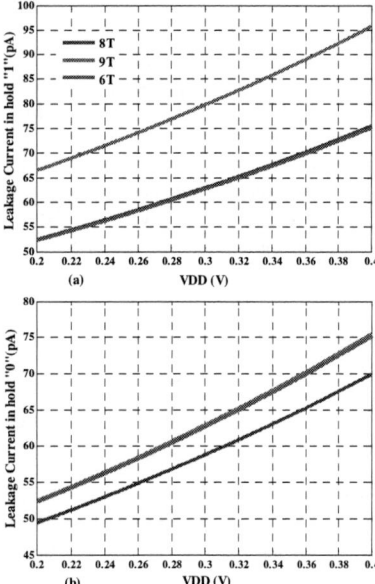

Fig. 10. Leakage current of the SRAM cells in the hold (a) "1", (b) "0".

After asserting column select (CS) signal, the enable signal (EN) is generated which connects the capacitor to the bit-line through a p-FinFET transistor (M_S). It is worth to mention that because of large transistor resistance in subthreshold region, it may take long to dump the charge from C_R to the bit-line, hence, M_S has to be a low-Vth transistor. After asserting the EN signal, C_R shares its charge with bit-line capacitance immediately and then becomes disconnected from the bit-line after a delay (t_{de}), i.e. 3 gate delay. Afterwards, the RWL is asserted and the RBL either is discharged or keeps its charge after required delay for generating the RWLC signal (t_{dr}). This delay is equal to 1 gate delay. It is worth mentioning that C_R should be enough larger than the bit-line capacitance, i.e. 20 times, in order to that RBL is approximately charged to V_{DD}. The simulation results for leakage reduction techniques at different corners are discussed in the next section.

V. PROCESS VARIATIONS

Here, the performance of the proposed cell at different corners (TT, SS, and FF) is studied. The value of C_R is chosen as 1 PF and all the results are presented for $V_{DD}=0.2$ V. Figure 12 illustrates the RSNM for the 6T-, 8T- and proposed 9T-SRAM cell for different corners. As it can be seen, the worst case scenario of RSNM is at FF corner that is attributed to the increased leakage through the pull-up transistor. However, for the SS corner, due to the further decrease in leakage current for the pull-up and access transistors compared to the pull-down transistors, a higher RSNM is obtained. The maximum change of the RSNM for the 6T, 8T, and the proposed 9T is 18%, 9%, and 7%, respectively. Figure 13 shows the read access time for the proposed SRAM cell in comparison to the 6T- and 8T-SRAM cells at the corners. As it is shown in Fig. 13, the proposed 9T-SRAM indicates the minimum access time for FF corner. The proposed SRAM cell shows higher variations due to the use of single transistor in the read path

compared to the stacked configuration of the 6T- and 8T-SRAM cells. The high access time variation is attributed to the exponential increase of the leakage current versus threshold voltage variations. As illustrated in Fig. 5, the WM of all three SRAM configurations is low due to the low driving current in near/sub-threshold region. To improve the WM of the SRAM cells, a write assist technique is mandatory. In this paper, transient voltage collapse write assist technique has been utilized [14]. Figure 14 shows the WM of the 6T-, 8T- and the proposed 9T-SRAM cells after applying the write improvement technique at the corners. The best value of WM can be achieved for FF corner due to the higher current of the access transistors which discharge the charge on the storage node. Furthermore, increased leakage current through the pull-down transistors increases the speed of discharging the storage node, as well.

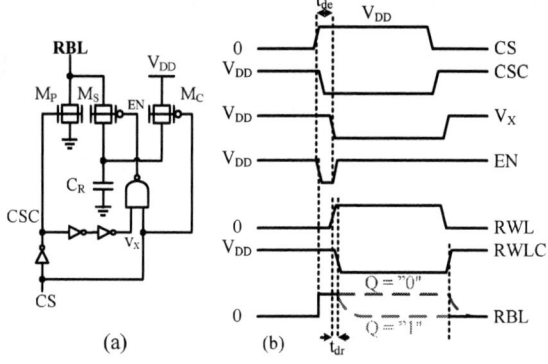

Fig. 11. (a) The proposed assist circuit for leakage reduction, (b) Timing diagram.

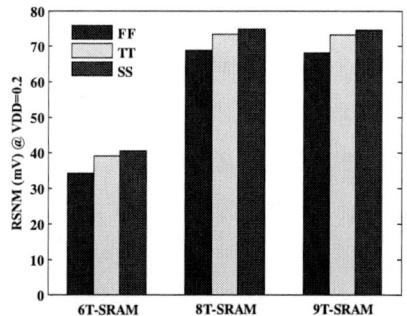

Fig. 12. The RSNM for TT, SS, and FF corners @ $V_{DD}=0.2$.

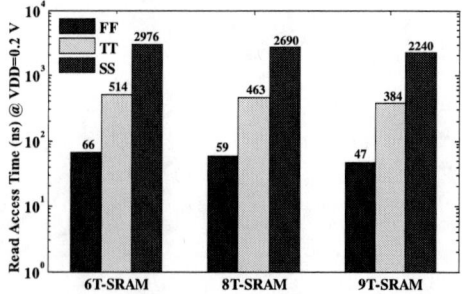

Fig. 13. The read access time for TT, SS, and FF corners @ $V_{DD}=0.2$.

978-1-4799-8720-7/15 $31.00 © 2015 IEEE

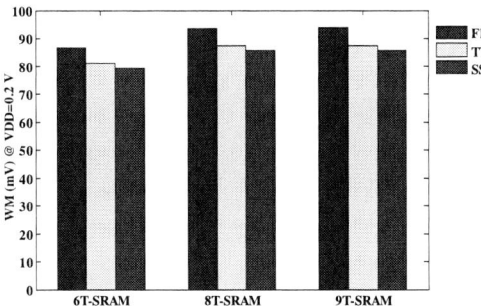

Fig. 14. The WM for TT, SS, and FF corners @ V_{DD}=0.2.

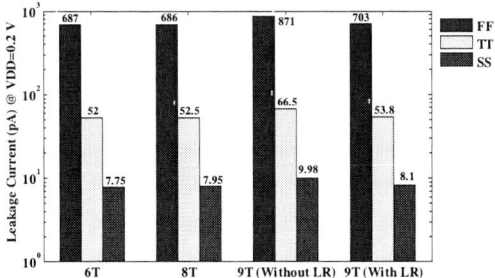

Fig. 15. The Leakage current for TT, SS, and FF corners @ V_{DD}=0.2.

Another metric to compare different SRAM cells is leakage current. Figure 15 shows the leakage current drawn from a supply voltage for the proposed 9T cell before and after applying the leakage reduction assist technique in comparison to the 6T- and 8T-SRAM cells. According to this figure, after applying the proposed technique, the leakage current decreases by 20%, which is comparable to the leakage current of the 8T-SRAM cell.

VI. CONCLUSION

In this paper, we proposed a new sub-threshold 9T-SRAM cell that reduces the access time by at least 30% compared to the standard 8T-SRAM cell. Furthermore, a leakage reduction assist technique was proposed by which the penalty of increased leakage due to the use of single transistor in the read path was resolved. In the proposed 9T-SRAM cell, RSNM and WM are similar to 8T-SRAM cell while the area is increased by 15%.

REFERENCES

[1] N. Collaert et al., "Multi-gate devices for the 32 nm technology node and beyond," in Proc. 37th ESSDERC, Sept. 2007, pp.143–146.

[2] K. Noda et al., "0.1 μm delta-doped MOSFET using postenergy implanting selective epitaxy," in VLSI Symp. Tech. Dig., 1994, pp. 19–20.

[3] R. Yan, A. Ourmazd, and K. Lee, "Scaling the Si MOSFET: From bulk to SOI to bulk," IEEE Trans. Electron Devices, vol. 39, no. 7, pp. 1704–1710, Jul. 1992.

[4] A. B. Sachid, and C. Hu, "Denser and more stable SRAM using FinFETs with multiple fin heights," IEEE Trans. Electron Devices, vol. 59, no. 8, pp. 2037-2041, Aug. 2012.

[5] S. A. Tawfik, and V. Kursun, "Multi-threshold voltage FinFET sequential circuits," IEEE Trans. Very Large Scale Integr. (VLSI) Syst.. vol. 19, no. 1, pp. 151-156, Jan. 2011.

[6] F. Moradi, S. K. Gupta, G. Panagopoulos, D. T. Wisland, H. Mahmoodi, and K. Roy, "Asymmetrically Doped FinFETs for Low-Power Robust SRAMs," IEEE Trans. Electron Devices, vol. 58, no. 12, pp. 4241–4249, Dec. 2011.

[7] C. -Y. Hsieh, M. -L. Fan, V. P. -H. Hu, P. Su, and C. -T. Chuang, "Independently-Controlled-Gate FinFET Schmitt Trigger Sub-Threshold SRAMs," IEEE Trans. Very Large Scale Integr. (VLSI) Syst., vol. 20, no. 7, pp. 1201–1210, Jul. 2012.

[8] Z. Liu, S. A. Tawfik, and V. Kursun, "An Independent-Gate FinFET SRAM Cell for High Data Stability and Enhanced Integration Density," in IEEE Int. SOC Conf., Sept. 2007, pp. 63–66.

[9] Y. B. Kim, Y. -B. Kim, and F. Lombardi, "Low Power 8T SRAM Using 32nm Independent Gate FinFET Technology," in IEEE Int. SOC Conf., Sept. 2008, pp. 247–250,.

[10] Y. -H. Chen et al., "A 16 nm 128 Mb SRAM in High- Metal-Gate FinFET Technology With Write-Assist Circuitry for Low-VMIN Applications," IEEE J. Solid-State Circuits, vol. 50, no. 1, pp. 1–8, Jan. 2015.

[11] E. Karl et al., "A 4.6 GHz 162 Mb SRAM design in 22 nm Tri-Gate CMOS technology with integrated read and write," IEEE J. Solid-State Circuits, vol. 48, no. 1, pp. 150–158, Jan. 2013.

[12] N. -C. Lien et al., "A 40 nm 512 kb Cross-Point 8 T Pipeline SRAM With Binary Word-Line Boosting Control, Ripple Bit-Line and Adaptive Data-Aware Write-Assist," IEEE Trans. Circuits Syst. I: Reg. Papers, vol. 61, no. 12, pp. 3416-3425, Dec. 2014.

[13] A. Raychowdhury et al., "PVT-and-aging adaptive wordline boosting for 8T SRAM power reduction," in Proc. IEEE Int. Solid-State Circuits Conf., Feb. 2010, pp. 352–353.

[14] N. Verma, and A. P. Chandrakasan, "A 256 kb 65 nm 8T subthreshold SRAM employing sense-amplifier redundancy," IEEE J. Solid-State Circuits, vol. 43, no. 1, pp. 141-149, Jan. 2008.

[15] O. Hirabayashi et al., "A Process Variation Tolerant Dual-PowerSupply SRAM Design with 0.179 μm² Cell in 40nm CMOS Using Level Programmable Wordline Driver," in IEEE Int. Solid-State Circuits Conf. Dig. Tech. Papers, Feb. 2009, pp. 458-459.

[16] K. Nii et al., "A 45-nm single-port and dual-port SRAM family with robust read/write stabilizing circuitry under DVFS environment," in IEEE Symp. VLSI Circuits, Jun. 2008, pp. 212-213.

[17] J. Wang, S. Nalam, and B. H. Calhoun, "Analyzing static and dynamic write margin for nanometer SRAMs," in Proc. Int. Symp. Low Power Electron. Design, Aug. 2008, pp. 129-134.

[18] N. Gierczynski, B. Borot, N. Planes, and H. Brut. "A new combined methodology for write-margin extraction of advanced SRAM," in IEEE Int. Conf. on Microelectronic Test Structures (ICMTS), Mar. 2007, pp. 97-100.

2015 IEEE Computer Society Annual Symposium on VLSI

FPGA based Novel High Speed DAQ System Design with Error Correction

Swagata Mandal*, Suman Sau [†], Amlan Chakrabarti [†], Jogendra Saini *,
Sushanta Kumar Pal * and Subhasish Chattopadhyay *
*Variable Energy Cyclotron Centre, Kolkata, India
*(swagata.mandal, sushant, sub)@vecc.gov.in
[†]A.K.Choudhury School of Information Technology, University of Calcutta, Kolkata, India
[†](ssakc_s, acakcs)@caluniv.ac.in

Abstract—Present state of the art applications in the area of high energy physics experiments (HEP), radar communication, satellite communication and bio medical instrumentation require fault resilient data acquisition (DAQ) system with the data rate in the order of Gbps. In order to keep the high speed DAQ system functional in such radiation environment where direct intervention of human is not possible, a robust and error free communication system is necessary. In this work we present an efficient DAQ design and its implementation on field programmable gate array (FPGA). The proposed DAQ system supports high speed data communication (∼4.8 Gbps) and achieves multi-bit error correction capabilities. BCH code (named after Raj Bose and D. K. RayChaudhuri) has been used for multi-bit error correction. The design has been implemented on Xilinx Kintex-7 board and is tested for board to board communication as well as for board to PC using PCIe (Peripheral Component Interconnect express) interface. To the best of our knowledge, the proposed FPGA based high speed DAQ system utilizing optical link and multi-bit error resiliency can be considered first of its kind. Performance estimation of the implemented DAQ system is done based on resource utilization, critical path delay, efficiency and bit error rate (BER).

Keywords—DAQ, FPGA, SEU, Error correction, PCIe, SGDMA

I. Introduction

High speed and fault resilient DAQ system is an integral part of the signal processing unit in some crucial real time applications like radar communication, HEP, satellite communication etc. In a traditional DAQ system Frontend Electronics (FEE) board captures data from the sensors through high speed LVDS link, processes it and sends it to storage device using high speed link like Ethernet, PCIe, fiber optic etc. for further analysis. Commonly faced problems in the traditional DAQ systems are low data rate [1] and prone to SEU in highly radiated area. Presently optical fiber and PCIe are the most suitable options in high speed data transmission over normal copper based LVDS line as they are susceptible to noise and interference.

In [2] the authors developed a new Gigabit Optical Serial Interface Protocol (GOSIP) for communication over optical fiber and implemented PCIe to optical link interface on FPGA for their DAQ system. This system gives a stable data rate of 1.6 Gbps. In [3]development of an optical link between two PCIe buses of computing nodes with data rate of 8.5

Gbit/s has been proposed. Authors used PCIe hard IP available in ALTERA Stratix IV FPGA board. Liansheng Liu *et.al* presented the development of a fibre channel node with PCIe interface for avionics environments in [4]. Each node consists of two modules: FPGA module and PowerPC module. Two nodes are connected by optical fiber through Small Form-factor Pluggable (SFP) interface and maximum data transfer rate achieved in this case was 2.125 Gbps. A high-speed data transmission protocol over optical fiber for real time data acquisition in Beijing Spectrometer III (BESIII) trigger system had been developed by Hao Xu *et.al* in [5]. Here they have used Multi Gigabit Transceiver (MGT) of Virtex-II Pro series FPGA for data transmission over optical fiber and achieved data rate of 1.75 Gbps. In [6], a high speed data transfer protocol named Serial Front Panel Data Port (SFPDP) over optical fiber is implemented on FPGA to capture the data from Digital Signal processor (DSP) through Extended Attachment Unit Interface (XAUI) where the optical link can work on three distinct speeds: 1.0625 Gbaud, 2.125 Gbaud, and 2.5 Gbaud. In [7], authors have used a bus master DMA along with a 4-lane generation 2 PCIe link to transfer the stream data from FPGA to PC. The highest data transfer rate achieved in this case is 784 Mbps.

Normally, SEU occurs when a charged particle hits and transfers sufficient energy to the silicon area of a circuit. The SEU mitigation techniques can be classified into two types: prevention and recovery. Prevention methods are mainly considered during the ASIC design. The recovery methods include online recovery mechanisms *e.g.* fault tolerant computing, error detecting/correcting code and online testing, which make the system more robust. Several error detection/correction techniques have been tried by many researchers. The concurrent error detection (CED) [8] is one of such technique where an extra error detection circuit is used along with the main circuit. When an error is detected, the main circuit recomputes or rolls back the whole operation from the beginning. Triple Modular Redundancy (TMR) [9] is another scheme where the same functional replica is used thrice and the final result is based on majority voting system. Above mentioned schemes are not applicable in real time because they are based on either time or space redundancy or combination of both. The error detection and correction (EDAC) codes play an important role in many successful SEU mitigation schemes. The SEU can be protected by using single error correcting hamming code, which may not

978-1-4799-8720-7/15 $31.00 © 2015 IEEE
80

be sufficient for reliable communication in high speed DAQ system. So multiple error correction codes can be applied in this kind of high speed DAQ. Several multiple forward error correction codes (FEC) have been presented in different papers *e.g.* BCH code [10], Reed Solomon code and Reed Muller code [11].

In all of the above mentioned works the authors did not discuss anything about SEU mitigation in high speed data acquisition system to work in an adverse environmental condition like Deep Space Experiment or in HEP experiment. The scope of our work takes into consideration an efficient design of the various stages in the DAQ chain to meet the high data rate requirements. Our high speed data acquisition system is protected from SEU by multi-bit error correcting code and interleaver. Scrambling is used here as line coding technique to maintain the DC balance and to obtain 20% extra throughput unlike 8b/10b coding in [2], [3], [4]. We have achieved a maximum data rate of 4.8 Gbps compared to 1.6 Gbps,2.125 Gbps,1.75 Gbps in [2], [4], [5] respectively. In this paper, our key contributions are:

- Efficient implementation of FPGA based high speed DAQ with optical link having multi-bit error correction capability.

- Performance analysis of the DAQ has been done using real time test set up having PCIe gen2 and scatter gather direct memory access (SGDMA).

The rest of the paper is organized as follows. Section II describes the full system design topology for the high speed DAQ. Experimental setup with performance evaluation are described in Section III followed by concluding remarks in Section IV.

II. SYSTEM DESIGN FOR HIGH SPEED DAQ

The main aim of the DAQ used in critical applications are to handle high data rate, error correction capabilities and efficient storage mechanism for future analysis. In high speed data transmission optical fiber is generally used as the communication media. Several multibit error correction methods for efficient communication had been discussed in Section I, where BCH coding is most suitable for random error correction. The interleaver block has been introduced after encoder block judiciously to enhance the error correction efficiency. In the receiver side data is directly transfered to PC through PCIe from the FPGA board. Functional blocks of the proposed system is shown in Figure 1. The details of each block have been discussed in the following subsections.

A. Scrambler/Descrambler

Scrambler is used to reduce the occurrence of long sequences of '1' (or '0') that maintains a good DC balance in input signal coming from the detector/sensor. This helps in accurate timing recovery on receiver equipment. It has a latency of one clock cycle but does not add any overhead in the system like the 8b/10b or 7b/8b line coding. Here 52 bit incoming data is divided into four blocks of 13 bit data and then each block is scrambled simultaneously using 13 bit polynomial. These scrambled data from all four blocks are combined together to produce the 52 bit scrambled output data.

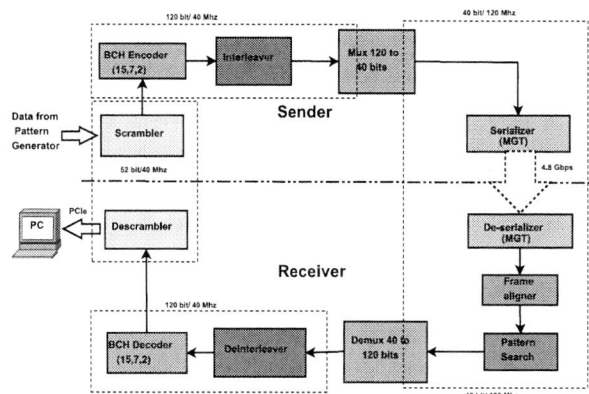

Fig. 1. Internal blocks of the proposed system

B. BCH Encoder/Decoder

BCH is a binary error correcting code. Here, BCH (15,7,2) code is used to correct the error due to SEU or Multiple bit upset (MBU). In this coding scheme 7 bit data is appended with 8 redundant bits for error correction. So the code rate (ratio of input data to coded data) is 0.467. 56 bit data (52 bit data with 4 bit header) is broken down into eight 7 bit data, which are encoded with BCH encoder in parallel. After encoding the 15 bit data from the eight blocks, they are assembled to generate a total 120 bit data. Though this block introduces one clock cycle latency in the system but increases the reliability in data transfer. The encoded data has been decoded in the following three steps: determination of the error locater polynomial, detection of error location using Chien Search Algorithm [10] and location of the data at the error position. In our present work we have designed the BCH encoding/decoding block as a custom hardware design. Instead of selecting BCH code with larger block size like (31, 26, 1) or (63, 57,1), we used eight BCH (15,7,2) in parallel for faster error correction without compromising the time complexity. Hence, each BCH decoder block can correct up to 2 bits of error within 7 bits of input. So the total $8 \times 2 = 16$ error can be corrected simultaneously using this technique without any extra resources. Similarly, we can use triple error correcting BCH code [10] but that will reduce the code rate.

C. Interleaver/De-interleaver

Interleaving is the reordering of the data that is to be transmitted, so that the consecutive bytes of data are distributed over a larger sequence of data to reduce the effect of burst error. Generally two types of interleaving strategies (Block interleaver and convolutional interleaver) are used in any communication system. Here we have used block interleaver. The first 120 bit data from the encoder is divided into two block of 60 bits of data and then interleaving operation is done on each 60 bit data using block interleaver. The whole process increases the code correction capabilities without any clock latency and overhead. De-interleaver process is used to reorder the data again in the receiver side.

D. MUX/DEMUX and Clock Domain Crossing

This block consists of dual port RAM and read-write controller. It breaks down 120 bit frame into three words of 40 bits width. It reduces bandwidth consumption keeping data rate same. Here, we have used 120 MHz clock to drive the multi-gigabit transceiver (MGT) available from Xilinx IP core to keep the data rate same with the internal blocks those are running with 40 MHz frequency. The data rate and clock frequency can be changed to any value according to the requirement. This block is used to synchronize the data rate between MGT and the other parts of the design. Figure 2 shows the architectural block diagram of the MUX-DEMUX and clock domain crossing.

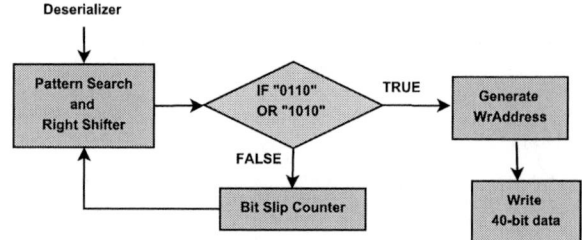

Fig. 3. Algorithm for Frame Aligner and Pattern Search

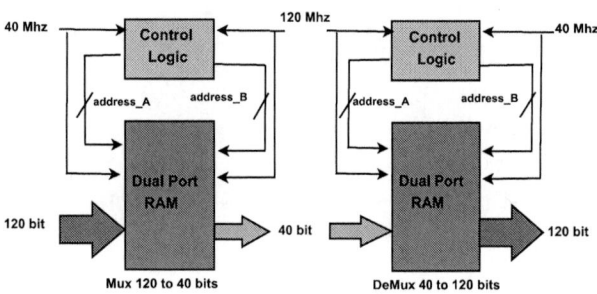

Fig. 2. Mux DeMux for clock domain crossing

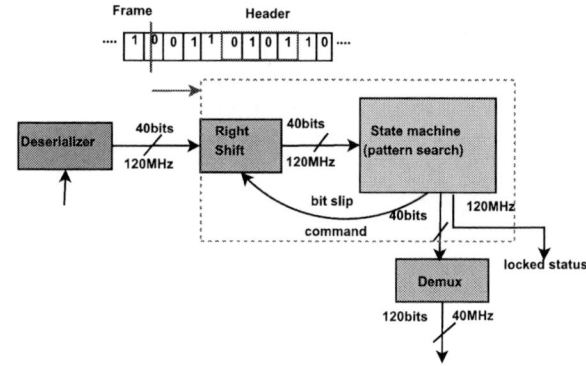

Fig. 4. Data flow diagrams of the Frame Aligner and Pattern Search block

E. Serializer/De-serializer

This block simply converts the parallel data to serial data, which is transmitted over the communication channel. It is inbuilt within the MGT. De-serializer simply converts the serial data to parallel data in the receiver side.

F. Frame Aligner and Pattern Search

In the receiver side the frame aligner block aligns the frames in a proper order by using frame header as an index. This frame header is detected by an efficient pattern search algorithm whose flow chart is given in Figure 3. There are two types of frame: Standard and Frame without FEC. The standard frame consists of four fields: Header field (4 bit), Slow Control (4 bit), Data field of width 48 bit, FEC field of width 64 bit. Slow Control (here, we merged with data field) field is reserved for controlling the DAQ chain in future. Whereas frame format without FEC consists of three field: Header field (4 bit), Slow control field (4 bit) and data field (112 bit). In standard frame format 1010 is used as header and 0101 is used as header in frame format without FEC. Frame format without FEC may be used for those applications where probability of error is low and high throughput is needed. The frame aligner and pattern search block consists of two sub blocks (Pattern search block, Right shifter block) as shown in the Figure 4. Right shifter block shifts the received data by one bit to the right from MSB side and sends it to the pattern search block. The pattern search block checks whether the header is received or not. Once the header is properly detected pattern search block will continuously search for another 32 subsequent headers of other frames to ensure the stability of the link and then the process gets terminated.

G. Data Transfer to Host PC through PCIe

The asynchronous Fast In Fast Out (FIFO) and SGDMA is used to transfer data from FPGA board to PC through PCIe. We have used PCIe gen2 Intellectual Property (IP) core available from Xilinx. Interconnection of FPGA to PC through PCIe is shown in Figure 5. Data is written into FIFO at a frequency of 120 MHz by which MGT is running and data will be read from FIFO at a frequency of 125 MHz by which PCIe core is running. In the PC side we capture the data by a program developed using windows software development kit (SDK) written in C language.

H. Data Flow Overview

The complete chain of the functional blocks as shown in Figure 1 for the high speed DAQ with multi-bit error correction (Considering two bits error correction) has been implemented on the FPGA board. Figure 6 shows the complete mechanism of standard frame generation and the error correction flow. At first, only 52 bit user data is scrambled by the scrambler block. These 52 bits of data is divided into four blocks of 13 bits data and scrambles each block parallely. The scrambled data with the 4 bit header is mapped in the input lines of the eight BCH (15,7,2) encoders, which are running parallel. Output of all the encoders are combined to get a frame of 120 bit data. This 120 bits of data are interleaved first and then goes to the next functional block that is the MUX. Interleaving is used to reduce the effect of burst error. But the header position is never changed in the frame format (red color in Figure 6) even after interleaving process, helps to synchronize the frame in the

978-1-4799-8720-7/15 $31.00 © 2015 IEEE   82

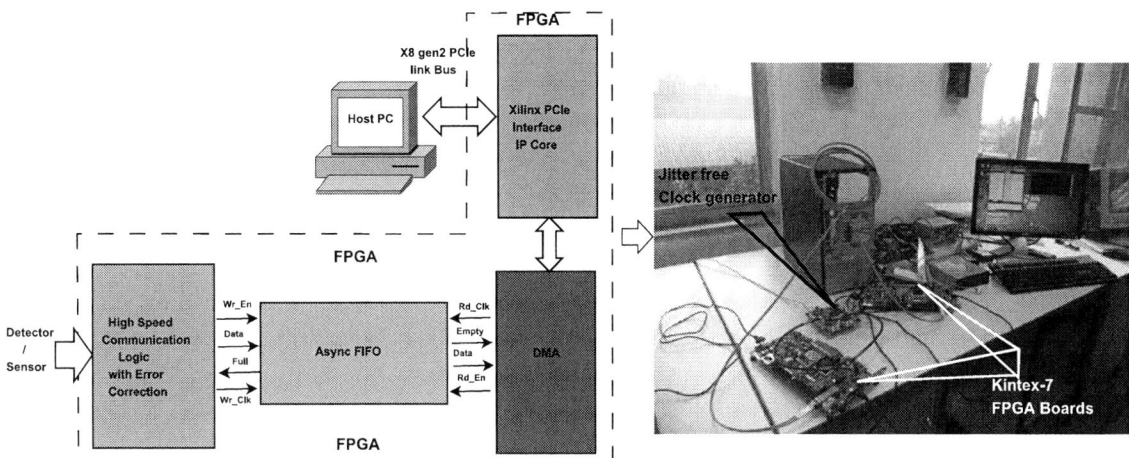

Fig. 5. PCIe interfacing with blocks and Experimental setup of proposed DAQ

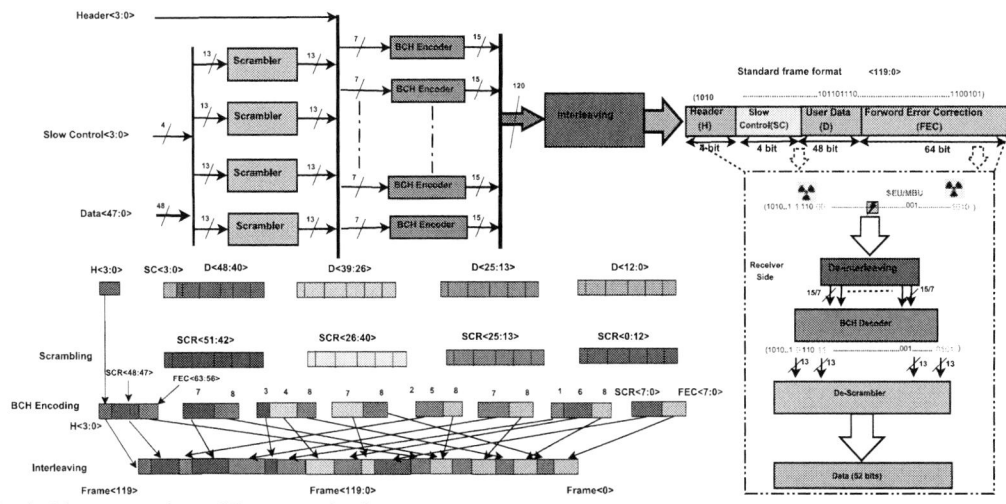

Fig. 6. Standard frame generation and Error correction flow

receiver side. In Mux-DeMux and clock domain crossing block a dual port RAM is used to write this 120 bits data using 40 MHz clock and read the same data at 120 MHz clock rate with 40 bit word size. So the data rate for writing ($40 \times 120 = 4.8$ Gbps) and reading ($120 \times 40 = 4.8$ Gbps) are same. The 40 bit data is serialized first and goes to the transmitter (TX) for transmitting over the optical fiber cable. In the receiver (RX) side functional blocks are Deserializer, DEMUX, De-interleaver, BCH Decoder (15, 7, 2) and Descrambler. They perform reverse function with respect to Serializer, MUX, Interleaver, BCH Encoder (15, 7, 2) and Scrambler respectively. The extra block frame aligner and pattern search in the receiver side is added in this chain whose functional description has been described in section II-F.

III. EXPERIMENTAL SETUP AND PERFORMANCE ANALYSIS

The full prototype of the DAQ chain is implemented in the Xilinx Kintex-7 boards (KC705 from Avnet) using the Xilinx ISE 14.5 platform and VHDL for design entry. We have used an external jitter cleaned clock source (CDCE62005EVM of TI) to drive MGT of two Kintex-7 boards. One Agilent

power supply has been used to drive the whole system. Two Kintex-7 boards are connected through single mode optical fiber using SFP from *Finisar* (FTLX8571D3BCL). For board to PC communication we have used eight lane PCIe gen2. Names of the various signals and their functions are given in Table I. The timing diagram of the transmitter and the receiver side are given in Figure 7. The block diagram and experimental setup of the system are shown Figure 5. We achieved maximum bit rate of 4.8 Gbps in our system. In standard mode, a frame contains only 52 bits of data, 64 bits for error correction (FEC) and 4 bits of header. 64 bits FEC field can correct upto 16 bits of error, as it is applied on 8 decoder blocks in parallel (2 bit error correction for each block). So the data rate achieved considering only the data field (D) in this mode is:

$$40MHz \times 52bits = 2.08Gbps$$

In frame format without FEC, where error correction code is not used, the frame can carry ($52 + 64 = 116$) bits of data out of 120 bits frame. So in this mode data rate is measured:

$$40MHz \times 116bits = 4.64Gbps$$

Hence, the data transfer efficiency for the above mention two modes are ($2.08/4.80$) × 100 = 43.33% and ($4.64/4.80 =$ 96.6% respectively.

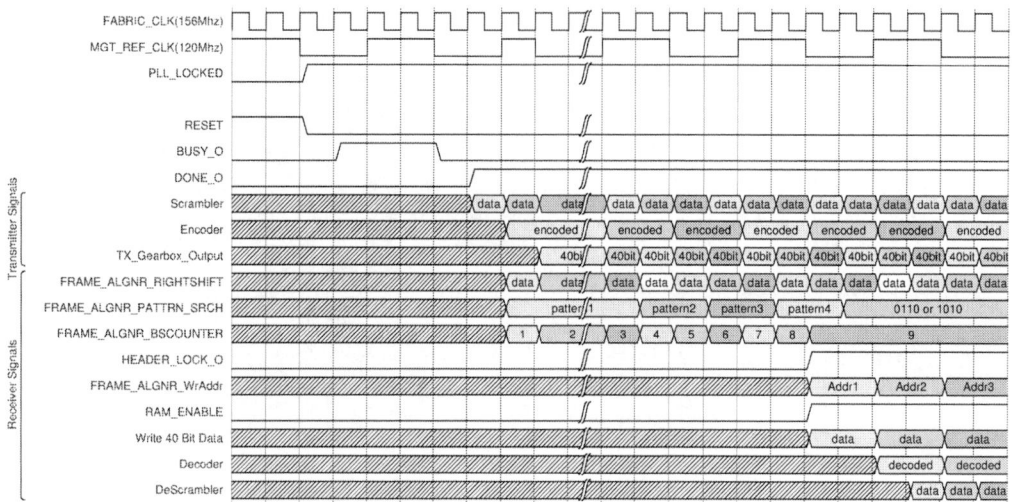

Fig. 7. Timing diagram of the transmitter and receiver signals

TABLE I.   DESCRIPTION OF THE SIGNALS USED IN TIMING DIAGRAM

Signal	Width	Function	Use
Fabric_Clk	1	Use to drive different blocks in DAQ	Both in Transmission and Receiver side
MGTREF_Clk	1	Use to drive MGT	Both in Transmission and Receiver side
PLL_Clk	1	Use to drive MGT	Both in Transmission and Receiver side
PLL_Locked	1	Output of PLL. It indicates PLL generate stable clock	Both in Transmission and Receiver side
RESET	1	Use to reset the whole system	Both in transmission and Receiver side
BUSY_O	1	Becomes high when System enters a process before ready	Both in transmission and Receiver side
Scrambler	52	This signal contains the data of output of scrambler block	In Transmission side only
Encoder	120	This signal contains the data after BCH encoding	In Transmission side only
MUX_Output	40	This signal contains output of MUX block which is 40 bit width	In Transmission side only
FRAME_ALIGNR _RIGHTSHIFT	40	Store the receive data after shifting right by one bit	In Receiver side only
FRAME_ALIGNR _PATTERN_SEARCH	4	Check whether header is matched or not	In Receiver side only
FRAME_ALIGNR _BSCOUNTER	5	Store the output of counter until header is not matched	In Receiver side only
Header_LOCK_O	1	Becomes high when the frame is locked	In Receiver side only
FRAME_ALIGNR _WrAddr	5	store the address of RAM where receive data will be written	In Receiver side only
RAM_ENABLE	1	Becomes high when RAM is Ready to perform	In Receiver side only
Write 40 bit Data	40	Store 40 bit data which is to be written in RAM	In Receiver side only
DECODER	52	Contains the decoded data	In Receiver side only
DESCRAMBLER	52	Contains output data of descrambler block	In Receiver side only

TABLE II.   COMPARISON WITH EXISTING WORKS

Device Used	Speed (Gbps)	Error correcting capability	Line coding used
Lattice SCM40[2]	1.60	Not mentioned	8b/10b
Altera Stratix IV[3]	8.50	Not mentioned	8b/10b
Altera EP2SGX90EF1152C5[4]	2.125	Not mentioned	8b/10b
Virtex-II Pro series FPGA[5]	1.75	CRC used for error detection only	8b/10b
Kintex-7 (our design)	**4.80**	**Multi bit error correction BCH code used**	**Scrambler**

Table II shows the comparison between our work and that of the other existing works. Existing system[3] gives better speed with respect to our implementation but it does not incorporate any error correction mechanism as that of ours.

Resource utilization for each functional block of the proposed DAQ system is given in Table III. In Figure 8 we show the critical time, which is the maximum delay time to get the output of a circuit for each of the circuit blocks. Power consumption is estimated using Xilinx Xpower tool and we

TABLE III.   RESOURCE UTILIZATION

Board	Module Name	Slice Register	Slice LUTs	LUT Flip Flop	BRAM
Kintex 7- 325t	BCH Encoder (15,7,2)	7//407600	951/203800	0	7/951
	BCH Decoder (15,7,2)	135/407600	446/203800	0	119/462 (25%)
	Scrambler	52	53	5	0
	Descrambler	104	56	5	0
	Interleaver	44	40	40	0
	DeInterleaver	201	82	80	0
	Frame Aligner	115	308	72	0
	Encryption (AES)	1311	4300	864	0
	Encryption (RSA)	116	31612	75	0
	PCIe	5882	5287	2694	10
	Top Module Without PCIe	3665	9003	1998	5
	Top Module With PCIe	8360	8555	3779	26

show the estimated average logic and signal power for the various model of the proposed design in Table IV. To the best of our knowledge, we are reporting the critical time and power consumption of this type of system for the first time.

978-1-4799-8720-7/15 $31.00 © 2015 IEEE

SEU error is random in nature. We have emulated the SEU error by generating random error on the input data using random error generator [12]. The simulation results of BER is shown in the Figure 9 with respect to the noise (Eb/N), which varies from 0 dB to 10 dB. We have assumed the power spectral density of noise as a Poison distribution. Figure 9 shows the efficiency of our system comprising of BCH code with interleaver and scrambler gives the best performance in presence of noise. The throughput of the DAQ system is measured as 4.8 Gbps in the Xilinx platform installed in Fedora OS.

TABLE IV.   MODULE WISE POWER CONSUMPTION

Board	Module Name	Logic Power(mW)	Signal Power(mW)
Kintex 7-325t	BCH Encoder(15,7,2)	0.02	0.01
	BCH Decoder(15,7,2)	0.05	0.07
	Scrambler	0.04	0.00
	Descrambler	0.01	0.00
	Interleaver	0.01	0.01
	DeInterleaver	0.01	0.02
	Frame Aligner	1.34	1.07
	PCIe	253.24	45.55
	Top Module Without PCIe	474.18	2.91
	Top Module With PCIe	304.24	56.31

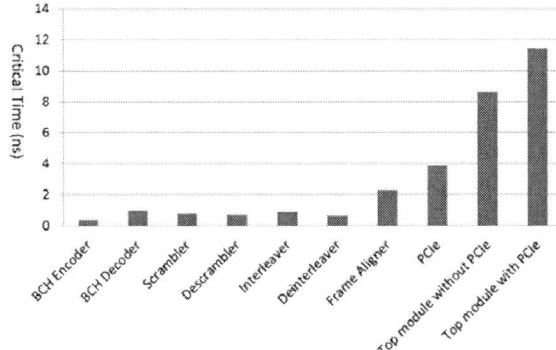

Fig. 8.   Critical time of different blocks

IV. CONCLUSION

In this work we have proposed a novel DAQ design for HEP experiments. The proposed DAQ supports high speed (in terms of Gbps) optical data communication and also corrects multi-bit error. The DAQ design has been implemented on Xilinx Kintex-7 board and real test setup has been developed involving board to board communication and PCIe interfacing with a host PC. A detailed performance analysis of the DAQ implementation is presented in terms of timing diagram, resource utilization and critical timing for of each of the blocks (FPGA), power consumption and BER. The proposed DAQ design and its implementation involving optical data communication and multi-bit error correction capability can be considered as first of its kind and can serve as a benchmark design in HEP DAQ. In future, we plan to use the concept of multi-pocessor into this system for more efficiency in data processing and test the setup in the radiation zone.

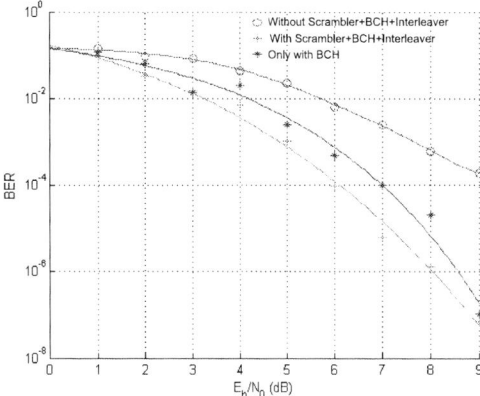

Fig. 9.   Study of of BER with noise

ACKNOWLEDGMENT

The authors like to acknowledge DAE, DST, GSI, CERN, TEQIP-II (CU) to provide necessary support for carrying out the research.

REFERENCES

[1] Wang Lixin, Song Wei, and Lv Chao. Implementation of high speed real time data acquisition and transfer system. In *Industrial Electronics and Applications, 2009. ICIEA 2009. 4th IEEE Conference on.* pages 382–386, May 2009.

[2] S. Minami, J. Hoffmann, N. Kurz, and W. Ott. Design and implementation of a data transfer protocol via optical fiber. *Nuclear Science, IEEE Transactions on,* 58(4):1816–1819, Aug 2011.

[3] E. Kadric, N. Manjikian, and Z. Zilic. An fpga implementation for a high-speed optical link with a pcie interface. In *SOC Conference (SOCC), 2012 IEEE International.* pages 83–87, Sept 2012.

[4] Liansheng Liu, Chuan Liu, Yu Peng, and Datong Liu. A design of fibre channel node with pci interface. In *Instrumentation and Measurement Technology Conference (I2MTC), 2013 IEEE International,* pages 1817–1822, May 2013.

[5] Hao Xu, Zhan'an Liu, Yunpeng Lu, Lu Li, Dixin Zhao, and Ya'nan Guo. Fpga based high speed data transmission with optical fiber in trigger system of besiii. In *Nuclear Science Symposium Conference Record, 2007. NSS '07. IEEE.* volume 1, pages 818–821, Oct 2007.

[6] C. Mattihalli. Design and realization of serial front panel data port (sf-pdp) protocol. In *Consumer Electronics, Communications and Networks (CECNet), 2012 2nd International Conference on,* pages 2505–2509, April 2012.

[7] H. Kavianipour and C. Bohm. High performance fpga-based scatter/gather dma interface for pcie. In *Nuclear Science Symposium and Medical Imaging Conference (NSS/MIC), 2012 IEEE,* pages 1517–1520, Oct 2012.

[8] Daniel P. Siewiorek and Robert S. Swarz. *Reliable Computer Systems (3rd Ed.): Design and Evaluation.* A. K. Peters, Ltd., Natick, MA, USA, 1998.

[9] A Gabrielli, G. De Robertis, D. Fiore, F. Loddo, and A Ranieri. Architecture of a slow-control asic for future high-energy physics experiments at slhc. *Nuclear Science, IEEE Transactions on,* 56(3):1163–1167, June 2009.

[10] R.T. Chien. Cyclic decoding procedures for bose- chaudhuri-hocquenghem codes. *Information Theory, IEEE Transactions on,* 10(4):357–363, 1964.

[11] B. Varghese, S. Sreelal, P. Vinod, and AR. Krishnan. Multiple bit error correction for high data rate aerospace applications. In *Information Communication Technologies (ICT), 2013 IEEE Conference on,* pages 1086–1090, April 2013.

[12] L. Antoni, R. Leveugle, and B. Feher. Using run-time reconfiguration for fault injection applications. *Instrumentation and Measurement, IEEE Transactions on,* 52(5):1468–1473, Oct 2003.

2015 IEEE Computer Society Annual Symposium on VLSI

High-Speed, Modified, Bulk stimulated, Ultra-Low-Voltage, Domino Inverter

Ali Dadashi, Yngvar Berg and Omid Mirmotahari

Nanoelectronics Research Group, Department of Informatics, University of Oslo
Blindern, Oslo, Norway
alidada@ifi.uio.no

Abstract— In this paper, a new Ultra low voltage (ULV) logic circuit based on the floating gate structure is presented. In this technique we utilized the bulks of the transistors to speed up the circuit. Using the proposed method, the speed of the circuit enhances by connecting the bulks of the evaluating and precharge devices to the clock, power supply (VDD) and input signals. The simulation results for the designed ULV logic in a typical *90nm* CMOS technology show more than 40% delay reduction. Higher speed in the lower supply voltages and robustness against process variations are the main advantages of the proposed approach in comparison to the previously reported FGULV and other ULV methods.

Index Terms— Ultra Low Voltage (ULV), Floating Gate; Speed, Bulk

I. INTRODUCTION

As the semiconductor industry grows, the demand for Ultra Low Voltage (ULV) circuits is increasing. Power reduction techniques are proposed to improve the battery life of the applications such as implantable biomedical systems laptop computers, personal digital assistants, and portable communication devices. One of the most important techniques to lower power consumption is supply voltage scaling. By scaling the supply voltage, the dynamic power is reduced significantly. For wireless sensor network applications, due to lower frequency rate, the supply voltage may be reduced below the threshold voltage. This operation is referred as sub-threshold design that uses the sub-threshold current as drive current to evaluate the inputs [1, 2]. Lowering the supply voltage reduces the sub-threshold current, exponentially. The optimal supply voltage for CMOS logic in terms of EDP (Energy delay product) is close to the threshold voltage of the nMOS transistor Vtn for a specific process, assuming that the threshold voltage of the pMOS transistor Vtp is approximately equal to Vtn [3]. Low voltage digital CMOS becomes more and more attractive, due to the general advances in process technology and due to the low power applications. The aggressive scaling of device dimensions and supply voltage in order to achieve greater transistor density and low power consumption results in degradation in the speed of the logic circuits due to reduced effective input voltage on gate source of the transistors. On one hand, the ever increasing market segment of portable electronic devices demands the availability of low-power building blocks that enable the implementation of long-lasting battery-operated systems. On the other hand, the general trend of increasing operating frequencies and circuit

complexity, in order to cope with the throughput needed in the modern high-performance processing applications, requires the design of very high-speed circuits. Several techniques for high speed and low voltage digital CMOS circuits have been presented in [4]. Floating-Gate (FG) gates have been proposed for ULV domino logic [3]. FG logic implemented in a modern CMOS process requires frequent initialization to avoid significant leakage. By using input floating capacitances to the transistor gate terminals, the semi-floating-gate (SFG) nodes can have a different DC level than provided by the supply voltage headroom [5]. There are several approaches for both analog and digital applications using FG CMOS logic proposed in [6-9]. The gates proposed in this paper are influenced by the ULV non-volatile FG circuits [10]. In this paper a new bulk stimulation technique is utilized to speed up the domino logic structures, by reducing the threshold voltage of transistors. Different topologies are studied and simulation results are reported. The proposed technique is applicable to other floating gate ULV (FGULV) domino logic structures like carry generators [11], NAND and NOR gates [12], and FGULV Flip-Flops [13].

This paper is organized as follows: in Sect.2, the simple domino FGULV inverter and also the proposed domino FGULV inverter circuit are presented and the circuit specifications are compared for the different topologies; in Sect.3, the simulation results for the different FGULV inverter

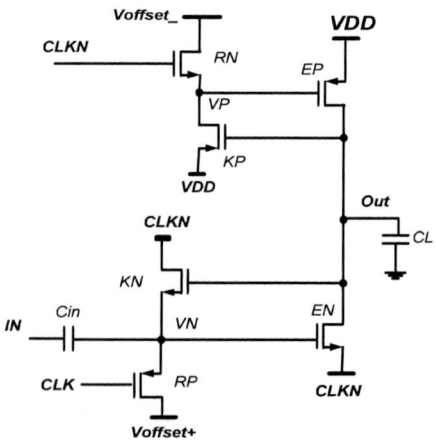

Fig. 1 Simple FGULV Domino inverter. (Precharg to 1).

978-1-4799-8720-7/15 $31.00 © 2015 IEEE   86

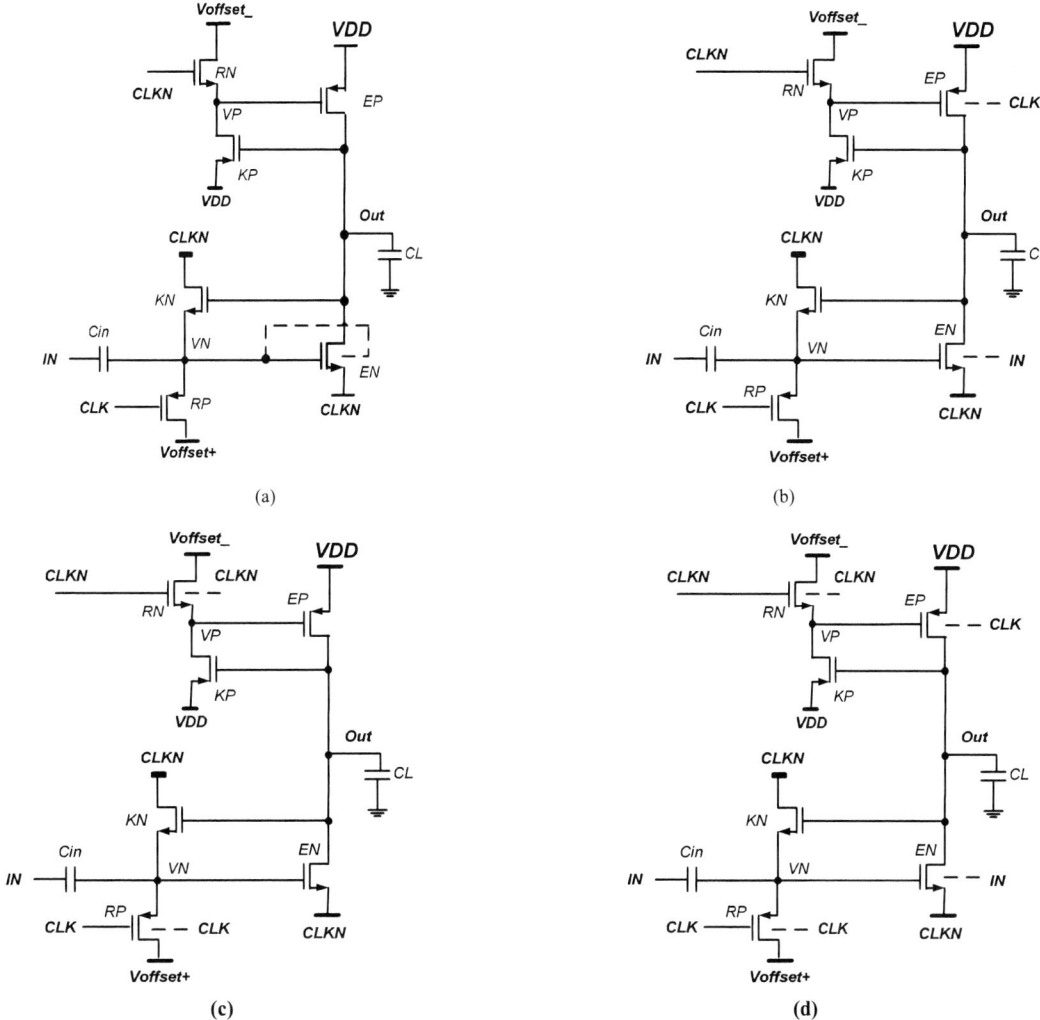

Fig. 2. Different domino FGULV inverters with bulk stimulating technique (a) Bulk of EN is connected to floating gate node (VN) (b) Bulk of EN is connected to the Input and the bulk of the EP devices is connected to the CLK signal. (c) Bulk of the RN is connected to the CLKN and the bulk of the RP is connected to CLK (d) The bulk of the precharge and evaluate devices are manipulated.

circuits are given, and compared with the simple FGULV inverter circuit; finally Sect.4 concludes the paper.

II. DOMINO FGULV LOGIC

The ULV domino gate presented in this paper is related to the FGULV domino logic style presented in [11-13]. The main purpose of the FGULV domino logic style is to increase the current level for the transistors at low supply voltages without increasing the transistor widths. We may increase the current level compared to complementary static CMOS using different initialization voltages to the gates and applying capacitive inputs. The extra load represented by the input capacitors (Cin) is less than the extra load given by increased transistor widths. The capacitive inputs lower the delay through increased transconductance while increased transistor widths

only reduce the parasitic delay. The proposed logic style may be used in the critical high speed and low voltage systems together with the conventional CMOS logic. In these topologies Voffset+ pins are connected to the VDD and Voffset- pins are connected to the GND and CLKN signal is the inversion of the CLK signal.

A. Simple FGULV inverter

The High-speed N-type FGULV domino inverter (precharge to 1) presented in [11], is shown in Figure 1. The clock signals CLK and CLKN are used both as control signals for the recharge transistors RP and RN, and as reference signals for NMOS evaluation transistor EN.

When CLK switches from 1 to 0, the circuit becomes in the precharge/recharge phase. During this phase, RP turns on and recharges the gate of EN transistor to 1. Meanwhile CKN switches from 0 to 1 which turns on RN and recharges the gate of EP to 0. Thus EP turns on and precharge the output node Vout to VDD. The keeper transistors, KN and KP, are inactive during this phase as the output node is precharged to1.

In the evaluation phase, clock signals CLK and CKN switch from 0 to 1 and 1 to 0 respectively. Both recharge transistors RP and RN switch off which make the charge on the gate terminals Vp and Vn floating. The output node Vout remains high (VDD) until a rising transition occurs at the input signal. The input signal Vin must be monotonically rising to ensure the correct operation for the N type domino inverter. This can only be satisfied if the input signal Vin is low at the beginning of the evaluation phase, and if Vin only makes a single transition from 0 to 1 in the evaluation phase. When this transition happens (Vin goes from 0 to 1), in the ideal case, the voltage of semi-floating gate (VN) increases up to 2VDD and this increases the current of EN in the evaluating phase and speed up the evaluation process. KN turns on when the output node gets a negative transition in the evaluation phase. This partially turns off the evaluation transistor (EN) and let the output node swings fully to GND. This helps to reduce the static current which directly impacts on the noise margin and the power consumption of the proposed logic. As it mentioned in the recent research reports, the FGULV logic demonstrates significant speed improvements in comparison to conventional static CMOS logic [8-15].

B. *Proposed domino FGULV inverters*

The proposed High-speed N-type FGULV domino inverters (precharge to 1)], are shown in Figure 2. In these topologies Voffset+ pins are connected to VDD and Voffset- pins are connected to GND. Body biasing technique is utilized in the different logic structures to reduce the threshold voltages of the devices. This threshold reduction helps the logic circuits to operate in the higher speed, especially in the ultra low voltage circuits (eg. [14]. In the Figure 2 (a), the bulk pin of evaluating device (EN) is connected to semi-floating gate node (VN). In this topology the threshold voltage of evaluating devices are reduced in the evaluating phase and this increase the ON current of the evaluating devices in the evaluating phase. However in this topology parasitic capacitance of VN node is increases and this causes to have larger parasitic capacitance at VN node. With larger parasitic capacitance at VN node, small portion of the input signal (IN) drops in VN node and this can reduce the speed of logic in the evaluating mode. So an optimum size should be chosen for devices and input capacitor (Cin) in order to maximize the speed.

Another way to manipulate the bulks of the devices in the FGULV is shown in Figure 2 (b). In this topology the bulk pin of evaluating device (EN) is connected to input signal and the bulk of EP device is connected to CLK signal. As mentioned before connecting the bulk of an NMOS device to VDD, reduces the threshold voltage of that device and increases the current of that device and finally reduces the delay in evaluating phase. The bulk of EP is connected to CLK signal. In the precharge phase, when EP is charging the output node, connecting the bulk of this device to the minimum possible

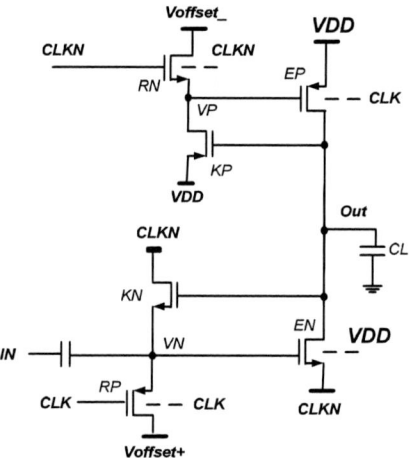

Fig 3. the Bulk of the Precharge devices are stimulated with the clock and the bulk of EN is connected to VDD.

voltage in the circuit (GND), reduces the threshold voltage of that device and this speed up the precharge process as well. In the evaluating phase, since CLK signal is high (VDD), the bulk of EP is connects to VDD and this device has bigger threshold voltage and less leakage current as well. In the topology shown in Figure 2(c), the bulks of RN and RP devices are manipulated to speed up the precharge process. Bulk of RP is connected to CLK signal. This connection reduces the threshold voltage of this device in the precharge phase and speeds up the precharge process for VN node. In the evaluating phase (CLK=1), in order to minimize the leakage current, the voltage of the bulk of this device (RP) goes high and it doesn't have small threshold voltage anymore. The bulk of RN is connected to CLKN signal. This connection reduces the threshold voltage of this device in the precharge phase and speeds up the precharge process for VP node. In the evaluating phase (CLK=1), in order to minimize the leakage current, the voltage of the bulk of this device (RN) goes low and so it will have the normal threshold voltage.

In the topology shown in Figure 2(d), the bulks of RN and RP devices are manipulated by clock signal, to speed up the precharge process. Also in this topology the bulk of the EN is connected to input signal to increase the current and speed of the evaluating process. The effectiveness of this structure should be evaluated in a chain of different inverters, since the parasitic bulk capacitance of the evaluating devices are added to the overall parasitic capacitance of the output node of the previous inverter, and this can reduce the overall speed of the chain. In the topology shown in Fig.3, the bulk of EN device is connected to VDD, and also the bulks of RN, EP and RP devices are manipulated by the clock signals.

III. SIMULATION RESULTS

In this section, the simulation results for the different FGULVs are presented and compared. The simulations for the designed FGULV logic inverters are done using cadence software in a typical *90nm* CMOS technology. Low threshold voltage devices are chosen to speed up the circuit. To verify the effect of the bulk stimulation method on the performance of the

978-1-4799-8720-7/15 $31.00 © 2015 IEEE

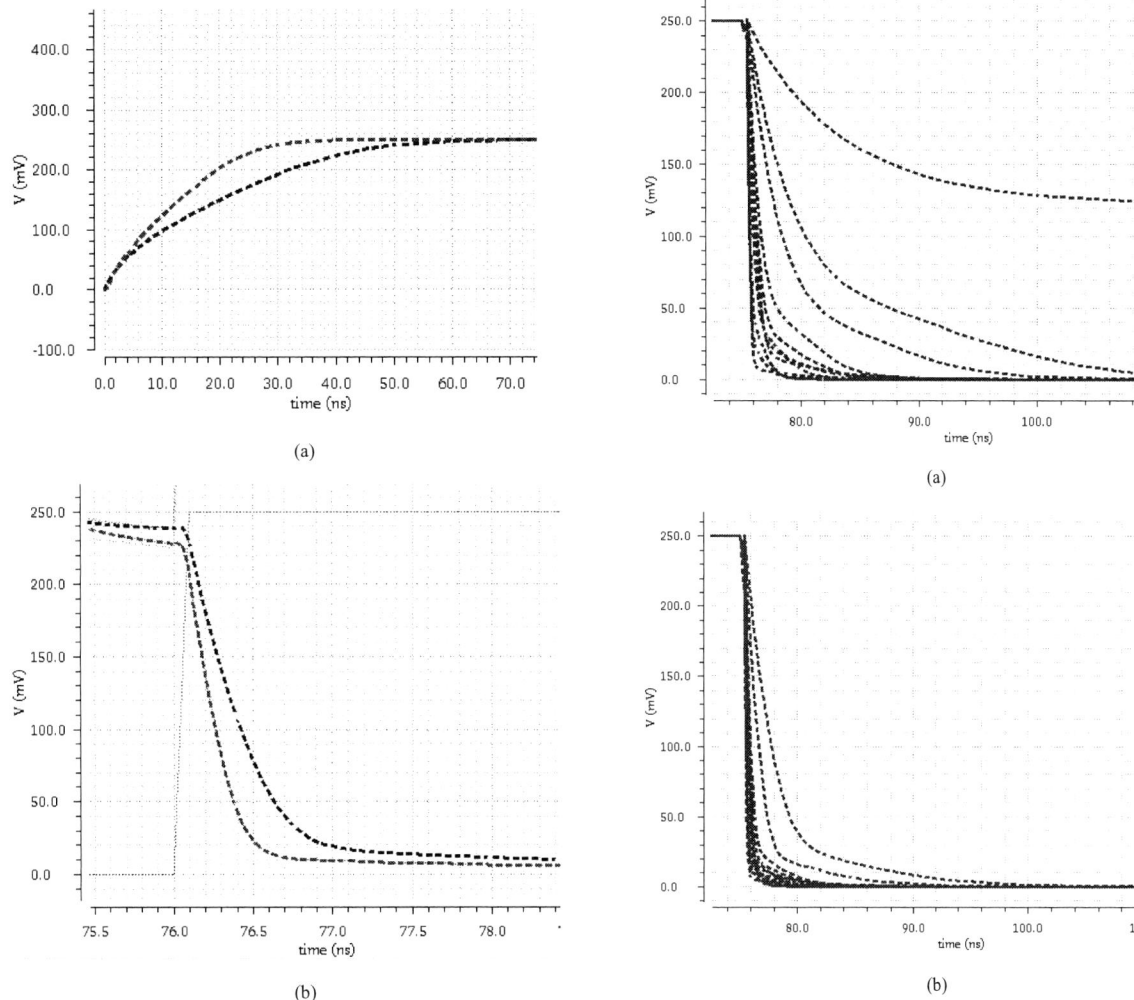

(a)

(b)

Fig 4. (a) The voltage of VN node in Precharge phase. (Blue is for proposed bulk stimulated precharge devices and black is for simple FGULV). (b) The voltage of output node in the evaluating phase. (Blue is for proposed bulk stimulated evaluating devices (EN) and black is for simple FGULV)

(a)

(b)

Fig 5. Monto- Carlo simulation results when VDD is 250mV. (a) The Output voltage in the evaluation phase for simple FGULV inverter shown in Fig.1. (b) The Output voltage in the evaluation phase for FGULV inverter shown in Fig.3.

FGULV, different FGULVs, shown in Fig.1, Fig.2 and Fig.3 are designed in the same size and power supply, and finally the characteristics are compared. In the all designed circuits, a 2fF capacitor is chosen for input capacitor (Cin). Simulation result shows that this size is optimum capacitor size for maximum speed, when the minimum size devices are used for inverter. Smaller capacitor (eg. 1fF) causes a voltage swing reduction from input signal to the semi-floating gate (VN) and this reduces the speed. Larger input capacitors also reduce the overall speed of the domino logic, since this capacitors adds to the output capacitance of the previous (precharge to 0) inverter. All the topologies shown in the Fig.2 and Fig.3 have been designed in a typical 90nm CMOS process with the same capacitor load. Simulation results show that for the evaluating phase, structure presented in the Fig. 3 has maximum speed. Fig.4 shows the transient simulations results of the designed

FGULV domino logic inverters in this paper (shown in Fig. 3) with 250mV power supply. As shown in Fig.4, the FGULV domino logic presented in this paper, achieve higher speed both in the precharge and evaluating phase. The proposed circuits are simulated in the different power supplies. Simulation results show that the proposed circuits are operating properly with power supplies down to 100 mV. In those low power supplies, the speed reduces significantly, structures become more sensitive to process variations and overall performance of the structure reduces. However, as mentioned in the previously reported papers (e.g. [8-13]), FGULV inverter is much faster and more robust than conventional static CMOS logic inverter. For supply voltages in the region from 200mV to 400mV, the delay of the simple FGULV inverter, reduces more than 96% comparing to the delay of the standard static CMOS inverter in the same device size [9]. Fig. 5 shows the Monte-Carlo simulation results for both simple and proposed FGULV.

Monte-Carlo simulations show that the proposed FGULVs are more robust than simple FGULV against process variations. Also simulation results shows better noise margin for the proposed FGULV. The simulation results for the proposed FGULV logics in the different power supply voltages shows significant speed enhancement for the both precharge and evaluating phases.

IV. CONCLUSION

In this paper, new inverters based on the FGULV domino logic structure are presented which uses bulk pins of the different devices in the original FGULV domino logic inverter structure to speed up the circuits. By manipulating the bulk voltages of the transistors, in the different topologies, the threshold voltages of these devices are reduced to speed up the circuits. Using the presented method, delay of the FGULV Domino logic inverter is reduced more than 40% in the both precharge and evaluating phases. Different topologies for the proposed technique are presented and advantages and disadvantages of the designs are studied.

REFERENCES

[1] B. A. Wang and A. Chandrakasan, "A 180mV FFT processor using subthreshold circuit techniques", IEEE ISSCC, pp.292-529, 2004.

[2] H. Soeleman and K. Roy, "Ultra-low power digital subthreshold logic circuits", IEEE ISLPED, pp.94-96, Aug. 1999.

[3] Chandrakasan A.P. Sheng S. Brodersen R.W.: "Low-power CMOS digital design" , IEEE Journal of Solid-State Circuits, Volume 27, Issue 4, April 1992 Page(s):473 – 484

[4] Verma N. Kwong J. Chandrakasan A.P.: "Nanometer MOSFET Variation in Minimum Energy Subthreshold Circuits" , IEEE Transactions on Electron Devices, Vol. 55, NO. 1, January 2008 Page(s):163 – 174

[5] Y. Berg, D. T. Wisland and T. S. Lande: "Ultra Low-Voltage/Low-Power Digital Floating-Gate Circuits", IEEE Transactions on Circuits and Systems, vol. 46, No. 7, pp. 930–936,july 1999

[6] K. Kotani, T. Shibata, M. Imai and T. Ohmi. "Clocked- Neuron-MOS Logic Circuits Employing Auto-Threshold-Adjustment", In IEEE

International Solid-State Circuits Conference (ISSCC), pp. 320-321,388, 1995.

[7] T. Shibata and T. Ohmi. " A Functional MOS Transistor Featuring Gate-Level Weighted Sum and Threshold Operations", In IEEE Transactions on Electron Devices, vol 39, 1992.

[8] Y. Berg and O.Mirmotahari "Ultra Low-Voltage and High Speed Dynamic and Static Precharge Logic", In proc. of the 11th Edition of IEEEFaible Tension Faible Consommation, June 6-8, 2012, Paris, France.

[9] Y. Berg, Tor S. Lande and Ø. Næss. "Programming Floating-Gate Circuits with UV-Activated Conductances", IEEE Transactions on Circuits and Systems -II: Analog and Digital Signal Processing, vol 48, no. 1,pp 12-19, 2001

[10] O. Mirmotahari, and Y. Berg "ROBUST LOW-POWER CMOS PRECHARGE LOGIC", In proc. of the 11th Edition of IEEE Faible Tension Faible Consommation, June, 2013, Paris, France.

[11] Y. Berg and O. Mirmotahari: "Static Ultra Low-Voltage and High Perfor-mance CMOS NAND and NOR Gates", Proceedings of the 10th WSEAS International Conference on CIRCUITS, SYSTEMS, ELECTRONICS, CONTROL & SIGNAL PROCESSING (CSECS '11). Montreux, Decem- ber 29-31, 2011.ISBN: 978-1-61804-062-6. s. 143-146.

[12] Y. Berg "Novel high speed differential CMOS flip-flop for ultra low-voltage appications ", In proc. of the 9th Edition of IEEE New Circuits and Systems Conference (NEWCAS), June 26-29, 2011, Bordeaux France.

[13] S. Narendra, J. Tschanz, J. Hofsheier, B. Bloechel, S. Vangal, Y. Hoskote, S. Tang, D. Somasekhar, A. Keshavarzi, V. Erraguntla, G. Dermer, N. Borkar, S. Borkar, and V. De, "Ultra-low voltage circuits and processor in 180 nm to 90 nm technologies with a swapped-body biasing technique," in IEEE Int. Solid-State Circuits Conf. Dig. Tech. Papers, Feb. 2004, pp. 156–157.

[14] Y. Berg and O.Mirmotahari. "Novel High-Speed and Ultra-Low-Voltage CMOS NAND and NOR Domino Gates", In proc. of the 5th internationalConference on Advances in Circuits, Electronics and Micro-electronics,August 19-24, 2012, Rome, Italy.

[15] Y. Berg and O.Mirmotahari. "Novel Static Ultra Low-Voltage and High Speed CMOS Boolean Gates", North atlantic university union: International Journal of Circuits, Systems and Signal Processing. ISSN 1998-4464. 6(4), s 249- 254.

2015 IEEE Computer Society Annual Symposium on VLSI

Modulo $2^n \pm 1$ Fused Add-Multiply Units

Constantinos Efstathiou
Department of Electrical
Engineering
Technological Institute of Athens
Athens, Greece
Email: cefsta@teiath.gr

Kostas Tsoumanis, Kiamal Pekmestzi
School of Electrical and
Computer Engineering
National technical University of Athens
Athens, Greece
Email: {kostastsoumanis, pekmes}@microlab.ntua.gr

Ioannis Voyiatzis
Department of Electrical
Engineering
Technological Institute of Athens
Athens, Greece
Email: voyageri@teiath.gr

Abstract—Complex arithmetic operations are widely used in Digital Signal Processing (DSP) applications. Targeting to increase performance, in this work, we focus on optimizing the design of the modulo $2^n \pm 1$ Add-Multiply (AM) operation. The proposed modulo $2^n \pm 1$ fused AM units incorporate an initial row of Half Adders which carry out the requested addition of two specified operands resulting to an intermediate Delayed Carry representation of their sum. The Delayed Carry represented vectors are then multiplied by the specified multiplicand and the partial products are driven to OR logic gates in pairs. Using the appropriate Carry-Save (CS) Adder trees, the resulting n-bit operands are reduced to a pair of CS vectors, which are finally added by a modulo $2^n - 1$ or a modulo $2^n + 1$ adder. Compared to the conventional designs of first instantiating a modulo $2^n \pm 1$ adder and then, driving its output to a modulo $2^n \pm 1$ multiplier, the proposed fused AM designs yield considerable reductions in terms of critical delay, area complexity and power consumption.

Keywords-Modulo $2^n - 1$; modulo $2^n + 1$; residue number system; RNS, addition; multiplication; fused add-multiply units.

I. INTRODUCTION

Residue number systems (RNS) [1], [2] eliminate the delay of carry propagation, thus, offering significant speedup and area reduction over the conventional 2's complement binary system. Such a feature becomes advantageous when repetitive arithmetic operations on long operands have to be performed. The RNS has been adopted in the design of Digital Signal Processors (DSP) [3], Finite Impulse Response (FIR) filters [4]–[6], Discrete Cosine Transform (DCT) [7], [8], Fast Fourier Transform (FFT) [9], communication components [10], [11] and cryptography [12], [13]. A recent work [6] has shown that the RNS-based implementation of FIR filters reduces power consumption compared to the one based on the conventional 2's complement number system. The RNS can be also used to design variation tolerant arithmetic circuits [14] and, therefore, they may be a promising candidate for building processing circuits in deep sub-micron technologies.

The moduli set $\langle 2^n - 1, 2^n, 2^n + 1 \rangle$ and its extensions have received significant attention because they offer simple and efficient implementations [15]. Extensive research has been conducted on both the modulo $2^n - 1$ and modulo

$2^n + 1$ addition [16]–[21] and multiplication [17], [22], [23]. Many DSP applications (e.g. Fast-Fourier Transform (FFT), FIR filters) perform a large number of Add-Multiply (AM) operations. The straightforward design of the Add-Multiply (AM) unit indicates that an adder is allocated first and then, its output is driven to the input of a multiplier. However, both the area and the critical delay of the AM unit increase significantly by such a design decision.

In this work, we propose the design of modulo $2^n \pm 1$ Fused Add-Multiply (FAM) units, which process the requested addition of two specified operands through an initial row of Half Adders (HA) producing an intermediate Delayed Carry representation of their sum [24]. The Delayed Carry represented vectors are then multiplied by the specified multiplicand and the partial products are driven to OR logic gates in pairs. The appropriate Carry-Save (CS) Adder trees reduce the resulting n-bit operands to a pair of CS vectors, which are finally added by a modulo $2^n - 1$ or a modulo $2^n + 1$ adder. Compared to the conventional modulo $2^n \pm 1$ AM designs, the proposed FAM units yield considerable reductions in terms of critical delay, area complexity and power consumption.

The rest of the paper is organized as follows: In Section II, we describe the design methodology for the proposed modulo $2^n \pm 1$ FAM units. Experimental results are given in Section III where a discussion on the comparison of the proposed modulo $2^n \pm 1$ FAM units with the conventional modulo $2^n \pm 1$ AM ones is also included. Section IV concludes our work.

II. DESIGN METHODS

A. Design of Modulo $2^n - 1$ Fused Add-Multiply Unit

Let us consider $A = a_{n-1} \ldots a_0$, $B = b_{n-1} \ldots b_0$ and $D = d_{n-1} \ldots d_0$ as the n-bit modulo $2^n - 1$ representations of three numbers in the range $[0, 2^n - 1)$. The proposed modulo $2^n - 1$ Fused Add-Multiply (FAM) unit computes the expression $|(A + B) \times D|_{2^n-1}$. The pair of the vectors (A, B) is converted to a Delayed-Carry Representation denoted as (U, V) by driving them into a row of n Half-Adders (HA) [24]. Thus, $|(A + B) \times D|_{2^n-1} = |(2U + V) \times D|_{2^n-1}$, where $U =$

978-1-4799-8720-7/15 $31.00 © 2015 IEEE   91

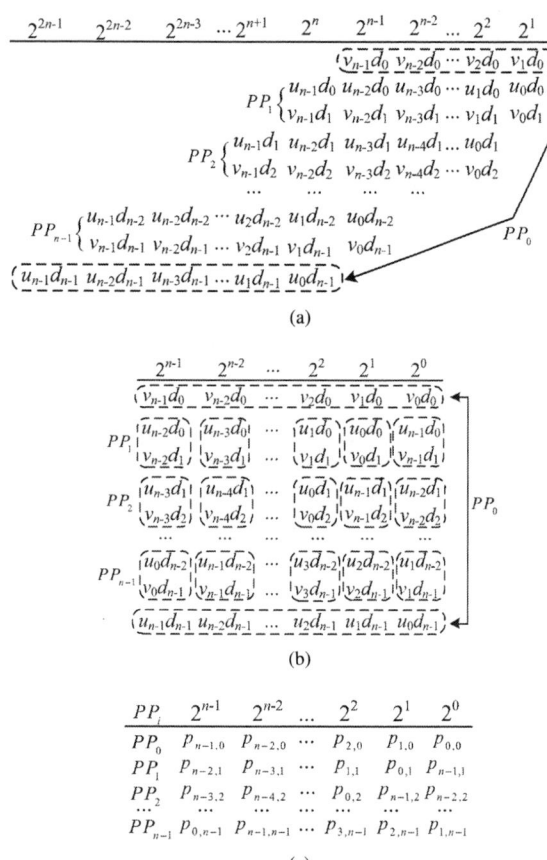

(a)

(b)

(c)

Figure 1. The a) initial, b) reconstructed consisting of n columns and c) final partial products matrix for the proposed modulo $2^n - 1$ FAM unit.

$u_{n-1} \ldots u_0$ and $V = v_{n-1} \ldots v_0$. For the computation of the expression $\left| (2U + V) \times D \right|_{2^n-1}$, the partial products of Fig. 1a are derived.

When $n \leq k \leq 2n - 1$, it holds that

$$
\begin{aligned}
\left| 2^k \right|_{2^n-1} &= \left| 2^n 2^{k-n} + 2^{k-n} - 2^{k-n} \right|_{2^n-1} \\
&= \left| (2^n - 1) 2^n + 2^{k-n} \right|_{2^n-1} \\
&= 2^{k-n}.
\end{aligned}
$$

Thus, considering the matrix of partial products exposed in Fig. 1a, each term with weight $2^n \leq 2^k \leq 2^{2n-1}$ can be repositioned to a column with weight 2^{k-n}. Then, the $n \times 2n$ partial products matrix of Fig. 1b is obtained.

Since the bits u_i, v_i are produced by half adders, they cannot have concurrently the value 1. Thus, the terms $u_i d_{|n+j-1|_n}$ and $v_i d_j$ can be driven to an OR logic gate instead of being added. The terms in the first row of the matrix of Fig. 1b are driven to an OR logic gate along with the corresponding terms of the last row. Thus, the partial products shown in Fig. 1c are derived, where $p_{i,j} = u_i d_{|n+j-1|_n} \vee v_i d_j$.

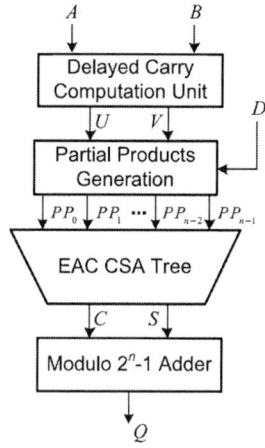

Figure 2. Block diagram of the proposed modulo $2^n - 1$ FAM unit.

The PP_i partial products of Fig. 1c are summed using an End Around Carry (EAC) Carry-Save Adder (CSA) tree [17] which outputs the n-bit vectors C, S. It holds that

$$
\left| (A + B) \times D \right|_{2^n-1} = \left| \sum_{i=0}^{n-1} PP_i \right|_{2^n-1} = \left| C + S \right|_{2^n-1}.
$$

(1)

The sum $\left| C + S \right|_{2^n-1}$ is computed by a final stage modulo $2^n - 1$ adder [16], [17], [25]. The block diagram of the proposed modulo $2^n - 1$ FAM unit is shown in Fig. 2.

B. Design of Weighted Modulo $2^n + 1$ Fused Add-Multiply Unit

Let us consider $A = a_n a_{n-1} \ldots a_0$, $B = b_n b_{n-1} \ldots b_0$ and $D = d_n d_{n-1} \ldots d_0$ consisting of $(n+1)$ bits, as the weighted modulo $2^n + 1$ representations of three numbers in the range $[0, 2^n + 1)$. The proposed weighted modulo $2^n + 1$ FAM unit computes the expression $\left| (A + B) \times D \right|_{2^n+1}$. The pair of the vectors (A, B) is converted to a Delayed-Carry Representation denoted as (U, V) by driving them into a row of $(n+1)$ HAs [24]. Thus, $\left| (A + B) \times D \right|_{2^n+1} = \left| (2U + V) \times D \right|_{2^n+1}$, where $U = u_n u_{n-1} \ldots u_0$ and $V = v_n v_{n-1} \ldots v_0$ are the weighted modulo $2^n + 1$ representations of the (U, V) vectors respectively. For the computation of the expression $\left| (2U + V) \times D \right|_{2^n+1}$, the partial products of Fig. 3a are derived.

Since the bits u_i, v_i are generated by half adders, they cannot have concurrently the value 1. Therefore, for $0 \leq i \leq n - 1$ and $1 \leq j \leq n$, the terms $u_i d_{j-1}$ and $v_i d_j$ can be driven to an OR logic gate instead of being added. The same observation holds for the terms $u_n d_i, v_n d_{i+1}, u_i d_n$. Considering that $p_{i,j} = u_i d_{j-1} \vee v_i d_j$, for $0 \leq i \leq n - 1$ and $1 \leq j \leq n$, $q_i = u_n d_{i-1} \vee v_n d_i \vee u_{i-1} d_n$, for $1 \leq i \leq n$, and $q_0 = v_n d_0$, the partial products of Fig. 3b are derived.

978-1-4799-8720-7/15 $31.00 © 2015 IEEE

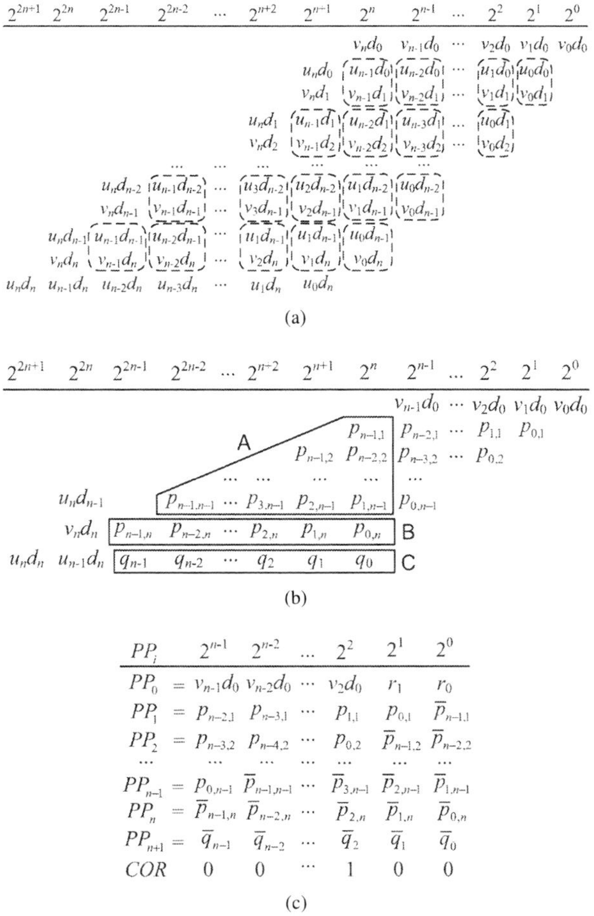

$$2^{2n+1} \quad 2^{2n} \quad 2^{2n-1} \quad 2^{2n-2} \quad \cdots \quad 2^{n+2} \quad 2^{n+1} \quad 2^n \quad 2^{n-1} \quad \cdots \quad 2^2 \quad 2^1 \quad 2^0$$

$$
\begin{array}{l}
v_n d_0 \quad v_{n-1}d_0 \cdots v_2 d_0 \; v_1 d_0 \; v_0 d_0 \\
u_n d_0 \; u_{n-1}d_0 \, u_{n-2}d_0 \cdots u_1 d_0 \, u_0 d_0 \\
v_n d_1 \; v_{n-1}d_1 \, v_{n-2}d_1 \cdots v_1 d_1 \, v_0 d_1 \\
u_n d_1 \; u_{n-1}d_1 \, u_{n-2}d_1 \, u_{n-3}d_1 \ldots u_0 d_1 \\
v_n d_2 \; v_{n-1}d_2 \, v_{n-2}d_2 \, v_{n-3}d_2 \ldots v_0 d_2 \\
\cdots \quad \cdots \\
u_n d_{n-2} \; u_{n-1}d_{n-2} \cdots u_3 d_{n-2} \, u_2 d_{n-2} \, u_1 d_{n-2} \, u_0 d_{n-2} \\
v_n d_{n-1} \; v_{n-1}d_{n-1} \cdots v_3 d_{n-1} \, v_2 d_{n-1} \, v_1 d_{n-1} \, v_0 d_{n-1} \\
u_n d_{n-1} \, u_{n-1}d_{n-1} \, u_{n-2}d_{n-1} \cdots u_1 d_{n-1} \, u_1 d_{n-1} \, u_0 d_{n-1} \\
v_n d_n \; v_{n-1}d_n \, v_{n-2}d_n \cdots v_2 d_n \; v_1 d_n \; v_0 d_n \\
u_n d_n \; u_{n-1}d_n \; u_{n-2}d_n \; u_{n-3}d_n \cdots u_1 d_n \; u_0 d_n
\end{array}
$$

(a)

$$2^{2n+1} \quad 2^{2n} \quad 2^{2n-1} \quad 2^{2n-2} \quad \cdots \quad 2^{n-2} \quad 2^{n+1} \quad 2^n \quad 2^{n-1} \quad \cdots \quad 2^2 \quad 2^1 \quad 2^0$$

$$
\begin{array}{l}
\qquad\qquad\qquad\qquad v_{n-1}d_0 \cdots v_2 d_0 \; v_1 d_0 \; v_0 d_0 \\
A \qquad\qquad p_{n-1,1} \; p_{n-2,1} \cdots p_{1,1} \; p_{0,1} \\
\qquad\qquad p_{n-1,2} \; p_{n-2,2} \cdots p_{3,2} \cdots p_{0,2} \\
\qquad\qquad \cdots \quad \cdots \quad \cdots \quad \cdots \\
u_n d_{n-1} \quad p_{n-1,n-1} \cdots p_{3,n-1} \, p_{2,n-1} \, p_{1,n-1} \qquad p_{0,n-1} \\
v_n d_n \quad p_{n-1,n} \; p_{n-2,n} \cdots p_{2,n} \; p_{1,n} \; p_{0,n} \qquad B \\
u_n d_n \; u_{n-1}d_n \quad q_{n-1} \; q_{n-2} \cdots q_2 \; q_1 \qquad q_0 \qquad C
\end{array}
$$

(b)

PP_i	2^{n-1}	2^{n-2}	$\cdots$	2^2	2^1	2^0
$PP_0 =$	$v_{n-1}d_0$	$v_{n-2}d_0$	$\cdots$	$v_2 d_0$	r_1	r_0
$PP_1 =$	$p_{n-2,1}$	$p_{n-3,1}$	$\cdots$	$p_{1,1}$	$p_{0,1}$	$\bar{p}_{n-1,1}$
$PP_2 =$	$p_{n-3,2}$	$p_{n-4,2}$	$\cdots$	$p_{0,2}$	$\bar{p}_{n-1,2}$	$\bar{p}_{n-2,2}$
$\cdots$						
$PP_{n-1} =$	$p_{0,n-1}$	$\bar{p}_{n-1,n-1}$	$\cdots$	$\bar{p}_{3,n-1}$	$\bar{p}_{2,n-1}$	$\bar{p}_{1,n-1}$
$PP_n =$	$\bar{p}_{n-1,n}$	$\bar{p}_{n-2,n}$	$\cdots$	$\bar{p}_{2,n}$	$\bar{p}_{1,n}$	$\bar{p}_{0,n}$
$PP_{n+1} =$	$\bar{q}_{n-1}$	$\bar{q}_{n-2}$	$\cdots$	$\bar{q}_2$	$\bar{q}_1$	$\bar{q}_0$
COR	0	0	$\cdots$	1	0	0

(c)

Figure 3. The a) initial, b) reduced and c) final partial products matrix for the proposed modulo $2^n + 1$ FAM unit.

When $n \le k \le 2n-2$, it holds $\left|2^k\right|_{2^n+1} = \left|-2^{k-n}\right|_{2^n+1}$. Thus, for $n \le i+j \le 2n-2$, we have that

$$
\begin{aligned}
\left| p_{i,j} 2^{i+j} \right|_{2^n+1} &= \left| -p_{i,j} 2^{i+j-n} \right|_{2^n+1} \\
&= \left| 2^{i+j-n} - p_{i,j} 2^{i+j-n} - 2^{i+j-n} \right|_{2^n+1} \\
&= \left| \bar{p}_{i,j} 2^{i+j-n} - 2^{i+j-n} \right|_{2^n+1}
\end{aligned}
$$

or $\left| \bar{p}_{i,j} 2^{i+j-n} \right|_{2^n+1} = \left| p_{i,j} 2^{i+j} + 2^{i+j-n} \right|_{2^n+1}$. Therefore, each term $p_{i,j}$ with weight $2^{i+j}, 1 \le i \le n-1, 1 \le j \le n-1$, in group A (Fig. 3b) is inverted and repositioned to a column with weight 2^{i+j-n} (Fig. 3c) requiring a correction of -2^{i+j-n}. The total correction that is required due to inverting and repositioning the terms of group A, is computed as $-(2^n - n - 1)$ [23].

Concerning the terms $p_{i,n}$ with weight $2^{i+n}, 0 \le i \le$ $n-1$, of group B (Fig. 3b), it holds that

$$
\begin{aligned}
\left| p_{n-1,n} 2^{2n-1} + \cdots + p_{0,n} 2^n \right|_{2^n+1} &= \\
&= \left| (p_{n-1,n} 2^{n-1} + \cdots + p_{0,n}) 2^n \right|_{2^n+1} = \\
&= \left| -(p_{n-1,n} 2^{n-1} + \cdots + p_{0,n}) \right|_{2^n+1} = \\
&= \left| 2^n + 1 - (p_{n-1,n} 2^{n-1} + \cdots + p_{0,n}) \right|_{2^n+1} = \\
&= \left| 2^n - 1 - (p_{n-1,n} 2^{n-1} + \cdots + p_{0,n}) + 2 \right|_{2^n+1} = \\
&= \left| \bar{p}_{n-1,n} 2^{n-1} + \cdots + \bar{p}_{0,n} + 2 \right|_{2^n+1}.
\end{aligned}
$$

Thus, each term of group B (Fig. 3b) is inverted and repositioned to a column with weight 2^i (Fig. 3c) requiring a total correction of $+2$. The same observation holds for the terms of group C.

Since $\left| 2^{2n} \right|_{2^n+1} = 1$, the terms with weight 2^{2n} in the reduced partial products matrix of Fig. 3b can be repositioned to the column with weight 1. The terms $v_0 d_0$, $u_n d_{n-1}$, $v_n d_n$, $u_{n-1}d_n$ cannot have concurrently the value 1 and therefore, are driven to an OR logic gate instead of being added. In the same way, the term $u_n d_n$ with weight 2^{2n+1} is repositioned to the column with weight 2 and is processed along with the term $v_1 d_0$ by an OR logic gate. Then, $n+2$ n-bit partial products PP_i are derived (Fig. 3c). In Fig. 3c, based on the aforementioned observations, $r_0 = v_0 d_0 \vee u_n d_{n-1} \vee v_n d_n \vee u_{n-1}d_n$ and $r_1 = u_n v_n \vee v_1 d_0$.

Since $\left| -(2^n - n - 1) \right|_{2^n+1} = n + 2$, it holds that

$$
\begin{aligned}
\left| (A + B) \times D \right|_{2^n+1} &= \left| PP_i - (2^n - n - 1) + 4 \right|_{2^n+1} \\
&= \left| PP_i + n + 6 \right|_{2^n+1} \qquad (2) \\
&= \left| PP_i + 4 + n + 1 + 1 \right|_{2^n+1}.
\end{aligned}
$$

The operands PP_i and the correction term $00\ldots0100 = 4$ consisting of n bits (Fig. 3c) are added by an Inverted End Around Carry (IEAC) CSA tree with output vectors C, S. The relation $\left| c_{n,i} \right|_{2^n+1} = \left| -c_{n,i} + 1 \right|_{2^n+1} = \left| c_{n,i} 2^n + 1 \right|_{2^n+1}$ implies that each $c_{n,i}$ output carry of the IEAC CSA tree, which is inverted and repositioned to the least significant bit position, adds an 1 to the result. Since $n+1$ carries are inverted and repositioned, the IEAC CSA tree computes the sum $\left| PP_i + 4 + n + 1 \right|_{2^n+1}$. Thus,

$$
\left| (A + B) \times D \right|_{2^n+1} = \left| C + S + 1 \right|_{2^n+1}. \qquad (3)
$$

The sum $\left| C + S + 1 \right|_{2^n+1}$ is computed by a modified diminished-1 modulo $2^n + 1$ adder [23], [26]. The block diagram of the proposed modulo $2^n + 1$ FAM units for weighted operands is shown in Fig. 4.

III. IMPLEMENTATION AND COMPARISONS

In this section, we present experimental evaluations of the modulo $2^n \pm 1$ AM operations (i.e. $\left| (A + B) \times D \right|_{2^n-1}$ and $\left| (A + B) \times D \right|_{2^n+1}$), which are implemented using conventional modulo $2^n \pm 1$ adders followed by modulo $2^n \pm 1$ multipliers and utilizing the proposed modulo $2^n \pm 1$ FAM units. The considered designs have been described in

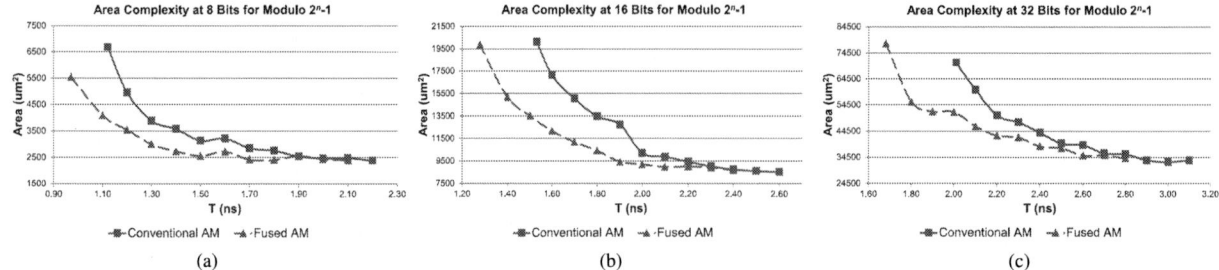

(a)   (b)   (c)

Figure 5.   The area complexity of both the conventional and the proposed fused modulo $2^n - 1$ AM designs for a) 8, b) 16 and c) 32 bits.

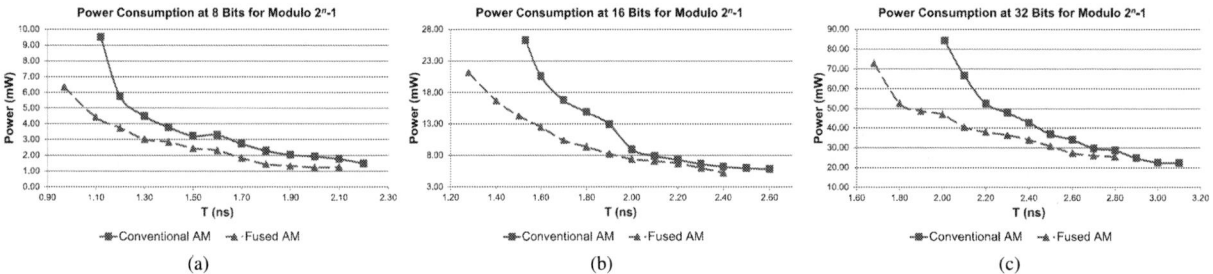

(a)   (b)   (c)

Figure 6.   The power consumption of both the conventional and the proposed fused modulo $2^n - 1$ AM designs for a) 8, b) 16 and c) 32 bits.

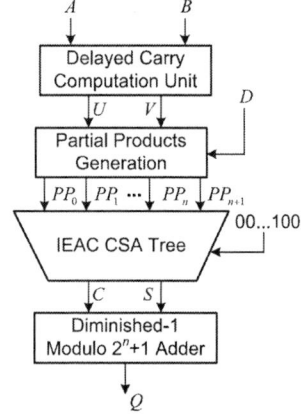

Figure 4.   Block diagram of the proposed modulo $2^n + 1$ FAM unit.

structural Verilog HDL with registered inputs and outputs for n=8, 16 and 32 bits, verified for their correctness and synthesized with the Synopsys Design Compiler tool using a Faraday 90nm standard cell library in typical conditions. The synthesis was conducted considering the highest degree of optimization in Synopsys Design Compiler. With respect to n, we first synthesized each design at the lowest achievable clock period and then, at 10 higher ones with increasing step of 0.10 ns targeting to explore how the designs behave considering different timing constraints in terms of area and power consumption. For each clock period, we simulated

the designs using ModelSim for the same set of 2^{16} random numbers in the range $[0, 2^n - 1)$ or $[0, 2^n + 1)$ for the modulo $2^n - 1$ or modulo $2^n + 1$ units respectively. The inputs were generated randomly with equal possibility of a bit to be 0 or 1. Finally, we used Synopsys PrimeTime-PX to calculate power consumption triggering the average mode of calculation.

Fig. 5 and 6 depict the area complexity and power consumption respectively for both the conventional and the proposed fused modulo $2^n - 1$ designs. The conventional implementation of the expression $|(A + B) \times D|_{2^n-1}$ includes a parallel-prefix modulo $2^n - 1$ adder [17] followed by a modulo $2^n - 1$ multiplier [22]. The critical delay of the proposed modulo $2^n - 1$ FAM approach is decreased compared to the conventional AM design (0.15 ns at 8 bits, 0.25 ns at 16 bits and 0.33 ns at 32 bits). Note that the larger the input bit-width is, the higher the values of decrease are. Such an observation is justified by the fact that the critical delays of the units that produce the Delayed-Carry representation and generate the partial products (Fig. 2) are constant and do not depend on the input bit-width while the critical delay of the modulo $2^n - 1$ adder of the conventional AM design increases when the input bit-width becomes larger. Also, with respect to the clock periods at which both the conventional and the proposed fused AM designs are synthesized, the proposed fused modulo $2^n - 1$ AM scheme delivers significant area and power gains which reach their highest values of 29.00% and 39.32% respectively when T=1.60ns at 16 bits.

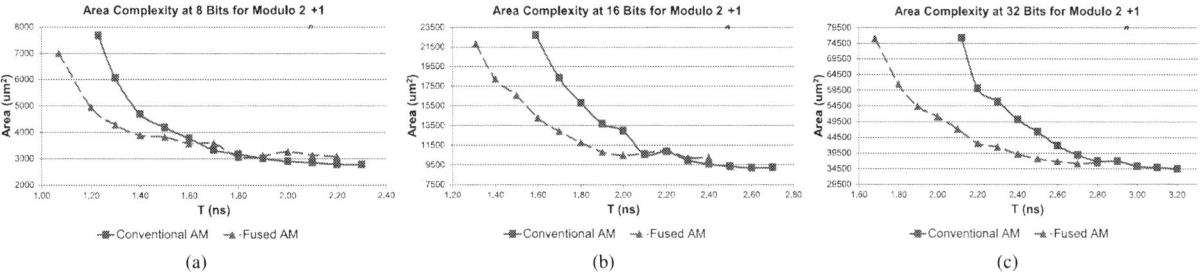

Figure 7. The area complexity of both the conventional and the proposed fused modulo $2^n + 1$ AM designs for a) 8, b) 16 and c) 32 bits.

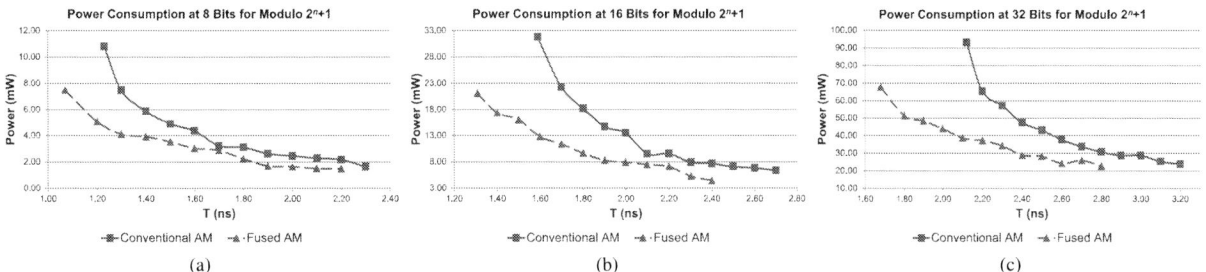

Figure 8. The power consumption of both the conventional and the proposed fused modulo $2^n + 1$ AM designs for a) 8, b) 16 and c) 32 bits.

In Fig. 7 and 8, the area occupation and power dissipation of the conventional and the proposed fused modulo $2^n + 1$ designs are illustrated, respectively. A parallel-prefix modulo $2^n + 1$ adder [17] and a subsequent modulo $2^n + 1$ multiplier [23] form the conventional implementation of the expression $|(A + B) \times D|_{2^n + 1}$. Compared to the conventional modulo $2^n + 1$ AM design, the critical delay of the proposed modulo $2^n + 1$ FAM approach is decreased (0.16 ns at 8 bits, 0.28 ns at 16 bits and 0.44 ns at 32 bits). As also noted in the previous paragraph for the modulo $2^n - 1$ units, the larger the input bit-width is, the higher the values of decrease are. The justification remains the same as for the modulo $2^n - 1$ designs. Moreover, with respect to the clock periods at which both the conventional and the proposed fused AM designs are synthesized, the proposed fused modulo $2^n + 1$ AM scheme delivers significant area and power gains which reach their highest values of 29.52% and 48.88% respectively when T=1.70ns at 16 bits.

IV. CONCLUSION

In this work, we proposed an optimized design for the modulo $2^n \pm 1$ AM operations. The requested addition of two specified operands is carried out using a row of HAs which produce an intermediate Delayed Carry representation of their sum. The Delayed Carry represented vectors are then multiplied by the specified multiplicand and the partial products are driven to OR logic gates in pairs. Compared to the straightforward designs of first allocating a modulo $2^n \pm 1$ adder and then, driving its output to a modulo $2^n \pm 1$ multiplier, the proposed fused AM designs yield significant reductions in terms of critical delay, area occupation and power dissipation.

ACKNOWLEDGMENT

This research has been co-funded by the European Union (European Social Fund) and Greek national resources under the framework of the "Archimedes III: Funding of Research Groups in TEI of Athens" project of the "Education & Lifelong Learning" Operational Programme.

REFERENCES

[1] P. V. Mohan, *Residue Number Systems: Algorithms and Architectures.* Norwell, MA, USA: Kluwer Academic Publishers, 2002.

[2] A. Omondi and B. Premkumar, *Residue Number Systems: Theory and Implementation.* Imperial College Press, 2007.

[3] R. Chaves and L. Sousa, "Rdsp: a risc dsp based on residue number system," in *Proc. Euromicro Symp. Digital System Design*, Sep. 2003, pp. 128–135.

[4] J. J. Ramrez and U. Meyer-Baese, "High performance, reduced complexity programmable rns-fpl merged fir filters," *Electronics Letters*, vol. 38, pp. 199–200, Feb. 2002.

[5] Y. Liu and E.-K. Lai, "Moduli set selection and cost estimation for rns-based fir filter and filter bank design," *Design Automation for Embedded Systems*, vol. 9, no. 2, pp. 123–139, 2004.

978-1-4799-8720-7/15 $31.00 © 2015 IEEE

[6] G. Bernocchi, G. Cardarilli, A. Del Re, A. Nannarelli, and M. Re, "Low-power adaptive filter based on rns components," in *IEEE Int. Symp. Circuits and Systems (ISCAS)*, May 2007, pp. 3211–3214.

[7] P. Fernandez, A. Garcia, J. Ramirez, and A. Lloris, "Fast rns-based 2d-dct computation on field-programmable devices," in *IEEE Workshop Signal Processing Systems (SiPS)*, 2000, pp. 365–373.

[8] P. Fernandez and A. Lloris, "Rns-based implementation of 8×8 point 2d-dct over field-programmable devices [image compression]," *Electronics Letters*, vol. 39, no. 1, pp. 21–23, Jan. 2003.

[9] G. Alia and E. Martinelli, "Optimal {VLSI} complexity design for high speed pipeline {FFT} using {RNS}," *Computers and Electrical Engineering*, vol. 24, no. 34, pp. 167–182, 1998.

[10] U. Meyer-Baese, A. Garcia, and F. Taylor, "Implementation of a communications channelizer using fpgas and rns arithmetic," *J. VLSI Signal Process. Syst.*, vol. 28, no. 1/2, pp. 115–128, May 2001.

[11] J. Ramirez, A. Garca, U. Meyer-Baese, and A. Lloris, "Fast rns fpl-based communications receiver design and implementation," in *Proc. 12th Int. Conf. Field-Programmable Logic and Applications*, 2002, vol. 2438, pp. 472–481.

[12] J. Bajard and L. Imbert, "A full rns implementation of rsa," *IEEE Trans. Comput.*, vol. 53, no. 6, pp. 769–774, Jun. 2004.

[13] D. Schinianakis, A. Kakarountas, and T. Stouraitis, "A new approach to elliptic curve cryptography: an rns architecture," in *IEEE Mediterranean Electrotechnical Conference (MELECON)*, May 2006, pp. 1241–1245.

[14] I. Kouretas and V. Paliouras, "Residue arithmetic for variation-tolerant design of multiply-add units," in *Proc. 19th Int. Conf. Integrated Circuit and System Design: Power and Timing Modeling, Optimization and Simulation (PATMOS)*, 2010, pp. 26–35.

[15] R. Chaves and L. Sousa, "Improving residue number system multiplication with more balanced moduli sets and enhanced modular arithmetic structures," *Computers Digital Techniques, IET*, vol. 1, no. 5, pp. 472–480, Sep. 2007.

[16] R. Zimmermann, "Binary adder architectures for cell-based vlsi and their synthesis," Ph.D. dissertation, Swiss Federal Institute of Technology (ETH) Zurich, Hartung-Gorre Verlag, 1998.

[17] ——, "Efficient vlsi implementation of modulo $2^n \pm 1$ addition and multiplication," in *Proc. 14th IEEE Symp. Computer Arithmetic*, 1999, pp. 158–167.

[18] G. Dimitrakopoulos, D. Nikolos, H. Vergos, D. Nikolos, and C. Efstathiou, "New architectures for modulo $2^n - 1$ adders," in *12th IEEE Int. Conf. Electronics, Circuits and Systems (ICECS)*, Dec. 2005, pp. 1–4.

[19] J. Chen and J. Stine, "Parallel prefix ling structures for modulo $2^n - 1$ addition," in *20th IEEE Int. Conf. Application-Specific Systems, Architectures and Processors (ASAP)*, Jul. 2009, pp. 16–23.

[20] H. Vergos, C. Efstathiou, and D. Nikolos, "Diminished-one modulo $2^n + 1$ adder design," *IEEE Trans. Comput.*, vol. 51, no. 12, pp. 1389–1399, Dec. 2002.

[21] H. Vergos and C. Efstathiou, "A unifying approach for weighted and diminished-1 modulo $2^n + 1$ addition," *IEEE Trans. Circuits Syst. II, Exp. Briefs*, vol. 55, no. 10, pp. 1041–1045, Oct. 2008.

[22] Z. Wang, G. Jullien, and W. Miller, "An algorithm for multiplication modulo $2^n - 1$," in *IEEE 39th Midwest Symp. Circuits and Systems*, vol. 3, Aug. 1996, pp. 1301–1304.

[23] H. Vergos and C. Efstathiou, "Design of efficient modulo $2^n + 1$ multipliers," *Computers Digital Techniques, IET*, vol. 1, no. 1, pp. 49–57, Jan. 2007.

[24] R. Zimmermann and D. Tran, "Optimized synthesis of sum-of-products," in *Rec. 37th Asilomar Conf. Signals, Systems and Computers*, vol. 1, Nov. 2003, pp. 867–872.

[25] C. Efstathiou, D. Nikolos, and J. Kalamatianos, "Area-time efficient modulo $2^n - 1$ adder design," *IEEE Trans. Circuits Syst. II, Analog Digit. Signal Process.*, vol. 41, no. 7, pp. 463–467, Jul. 1994.

[26] C. Efstathiou, N. Moschopoulos, I. Voyiatzis, and K. Pekmestzi, "On the design of modulo $2^n + 1$ dot product and generalized multiplyadd units," *Computers and Electrical Engineering*, vol. 39, no. 2, pp. 410–419, 2013.

2015 IEEE Computer Society Annual Symposium on VLSI

High throughput floating-point dividers implemented in FPGA

Peter Malík

Institute of Informatics, Slovak Academy of Sciences
Dubravska cesta 9, 845 07 Bratislava, Slovak Republic
Email: p.malik@savba.sk
http://ui.sav.sk/diag/index.php

Abstract—New high throughput floating-point dividers implemented in FPGA based on different fast computation division algorithms are proposed. The hardware implementations uses 32-bit floating-point single precision. The implementations include both multiplicative inverse and division. The proposed hardware implementations are designed with high computation speed and throughput. They are oriented for high computation demanding applications with multiple division computations in short sequences.

I. Introduction

The division is a mathematical operation that is commonly used in signal processing algorithms. The hardware implementation of division is not a trivial process. Several different algorithms are described in literature [1-5]. Slow computation algorithms require many iteration cycles and fast computation algorithms require complex computations within each iteration cycle. This makes the division much more complex operation in comparison to its inverse operation the multiplication. Many signal processing algorithms use divisions in complex iterative processes where the requirement is not only the precision but also maintaining the precision with very large data intervals. This can be achieved by using floating-point form in division calculations. The additional advantage is the higher versatility that allows the use of different data with no rescaling requirement.

This paper presents new implementations of three different fast computation division algorithms and theirs comparisons. The algorithms are implemented in software environment as division models. Then, they are implemented in hardware to FPGA. The presented implementation results include computation of multiplicative inverse and division. The software models use 64-bit floating-point double precision, and the hardware implementations utilize 32-bit floating-point single precision. The presented implementations are designed with high computation speed and throughput. They are suitable for high computation demanding applications with multiple division computations in short sequences.

The paper is organized as follows. The floating-point standard is given in Section 2. Section 3 characterizes division algorithms. Section 4 depicts software implementation of division models. Section 5 describes hardware implementations in FPGA. The paper is concluded in Section 6.

II. Floating-point standard

The most common floating-point format in use is 32-bit binary and 64-bit binary also known as single and double defined in IEEE 754-2008 standard. These formats were originally defined in IEEE 754-1985 standard and remained unchanged in the new standard extension IEEE 754-2008. These formats are composed of binary strings that are divided to three sections. The first section composed of the first bit represents the sign of coded number. The second section composed of the next eight bits in 32-bit binary and the next eleven bits in 64-bit binary represent the exponents of coded number. The third section composed of the remaining 23 bits in 32-bit binary and the next 52 bits in 64-bit binary represents the mantissa of coded number. The exponent uses base 2 so the minimal value is -127 and the maximal value 128 in 32-bit binary, and the minimal value is -1023 and the maximal value 1024 in 64-bit binary; however, these maximal values are reserved for special meanings. The minimal value represents zero and denormal numbers also known as subnumbers, and the maximal value represents infinity. The exponent coding mostly uses bias so zero exponent is represented by 127 in 32-bit binary and by 1023 in 64-bit binary.

III. Division algorithms

The division operation can be implemented as an inverse process of the multiplication operation. Denominator (divisor) is continually subtracted from numerator (divident); however, this requires very high number of iterations. If combined with shift operation, the number of iterations is reduced to the number of result valid bits. The more precise results require more iterations and therefore fast algorithms were developed e.g. Newton-Raphson division, Goldschmidt division and its variations. These algorithms converge much faster within one iteration; however, more mathematical operations are needed (usually several additions or subtractions and multiplications).

A. Newton-Raphson division

The Newton-Raphson division is based on calculating of multiplicative inverse (reciprocal) of denominator which is multiplied by numerator to produce the result [2]. The formula to calculate multiplicative inverse of the denominator is shown in (1) where X_i is multiplicative inverse at iteration i and D is denominator.

$$X_{i+1} = X_i + X_i(1 - DX_i) = X_i(2 - DX_i). \qquad (1)$$

978-1-4799-8720-7/15 $31.00 © 2015 IEEE

The Newton-Raphson division can be optimized by scaling the denominator to interval [0.5, 1]. The minimization of maximal relative error on this interval can be achieved by initialization shown in (2).

$$X_0 = \frac{48}{17} - \frac{32}{17}D. \tag{2}$$

B. Goldschmidt division

The Goldschmidt division is based on the continual multiplication of both the numerator and denominator by the same factor F_i with goal to converge the denominator to 1. At the same time, the numerator converges directly to the quotient [3]. The formulae is shown in (3) where Q is quotient, N is numerator, D is denominator and F_i is factor at iteration i.

$$Q = \frac{N}{D} \frac{\prod_i^n F_i}{\prod_i^n F_i}. \tag{3}$$

The factor for the next iteration is calculated by (4). The numerator and denominator for next iteration are calculated by (5).

$$F_{i+1} = 2 - D_i. \tag{4}$$

$$\frac{N_{i+1}}{D_{i+1}} = \frac{N_i}{D_i} \frac{F_{i+1}}{F_{i+1}}. \tag{5}$$

C. Goldschmidt division with binomial simplification

The Goldschmidt division can be simplified by the binomial theorem. When scaling N and D such D is from interval (0.5, 1] the substitution (6) and (7) can be used where D is the denominator, x is substituted variable and F_i is factor at iteration i.

$$D = 1 - x. \tag{6}$$

$$F_{i+1} = 1 + x^{2^i}. \tag{7}$$

This leads to (8) where N is numerator.

$$\frac{N}{1-x} = \frac{N(1+x)}{1-x^2} = \frac{N \prod_{i=0}^{n-1} 1 + x^{2^i}}{1 - x^{2^n}}. \tag{8}$$

IV. DIVISION MODELS

The fast division algorithms have been implemented in Matlab. These software models were used to analyse computation precision, convergence rate and dependence of computation precision on input values. All computation were processed with double precision.

A. Newton-Raphson division model

The Newton-Raphson division model uses complex initialization which requires several mathematical operations and therefore this initialization has been classified as the first iteration cycle. Table I shows the relative error of the Newton-Raphson division model. The convergence rate varies minimally in the interval [0.5, 1] except the boundary values 0.5 and 1. It can be seen that four iterations are needed to compute multiplicative inverse with single precision. The division is acquired by additional multiplication with the numerator in the next step. The computation of multiplicative inverse with double precision requires only one additional iteration cycle. The non zero values in the last column represent limits of the models and simulation process with double precision.

TABLE I.   RELATIVE ERROR OF NEWTON-RAPHSON DIVISION MODEL

Input value	Relative error at iteration cycle:				
	1	2	3	4	5
0.50	-0.058824	-0.0034602	-1.1973e-05	-1.4335e-10	0
0.55	-0.016471	-0.00027128	-7.3593e-08	-5.4956e-15	-1.2212e-16
0.60	0.016471	-0.00027128	-7.3593e-08	-5.4623e-15	-1.3323e-16
0.65	0.04	-0.0016	-2.56e-06	-6.5535e-12	1.4433e-16
0.70	0.054118	-0.0029287	-8.5774e-06	-7.3572e-11	0
0.75	0.058824	-0.0034602	-1.1973e-05	-1.4335e-10	1.6653e-16
0.80	0.054118	-0.0029287	-8.5774e-06	-7.3572e-11	0
0.85	0.04	-0.0016	-2.56e-06	-6.5535e-12	-1.8874e-16
0.90	0.016471	-0.00027128	-7.3593e-08	-5.3957e-15	0
0.95	-0.016471	-0.00027128	-7.3593e-08	-5.4845e-15	0
1.00	-0.058824	-0.0034602	-1.1973e-05	-1.4335e-10	0

TABLE II.   RELATIVE ERROR OF GOLDSCHMIDT DIVISION MODEL

Input value	Relative error at iteration cycle:				
	1	2	3	4	5
0.50	-0.25	-0.0625	-0.0039063	-1.5259e-05	-2.3283e-10
0.55	-0.2025	-0.041006	-0.0016815	-2.8275e-06	-7.9948e-12
0.60	-0.16	-0.0256	-0.00065536	-4.295e-07	-1.8439e-13
0.65	-0.1225	-0.015006	-0.00022519	-5.0709e-08	-2.5979e-15
0.70	-0.09	-0.0081	-6.561e-05	-4.3047e-09	1.5543e-16
0.75	-0.0625	-0.0039062	-1.5259e-05	-2.3283e-10	0
0.80	-0.04	-0.0016	-2.56e-06	-6.5535e-12	1.7764e-16
0.85	-0.0225	-0.00050625	-2.5629e-07	-6.587e-14	-1.8874e-16
0.90	-0.01	-0.0001	-1e-08	-1.9984e-16	-1.9984e-16
0.95	-0.0025	-6.25e-06	-3.9062e-11	2.1094e-16	2.1094e-16
1.00	0	0	0	0	0

B. Goldschmidt division model

The Goldschmidt division model uses very simple initialization and therefore is classified as zero iteration cycle. Table II shows the relative error of the Goldschmidt division model. The convergence rate differs in the interval [0.5, 1]. The slowest convergence has the boundary values 0.5, and the opposite boundary value 1 is calculated instantly. This decreasing dependence can be seen in the whole interval [0.5, 1]. The computation speed is dependant on the slowest convergence rate and therefore to compute multiplicative inverse with single precision five iterations are needed. The advantage is that the division can be acquired at the same time by changing initial conditions. The computation of multiplicative inverse with double precision requires only one additional iteration cycle. The sixth iteration is not shown in Table II but the limits of the models and simulation process with double precision can already be seen in the last column with larger input values.

C. Division model with binomial simplification

The Goldschmidt division model with binomial simplification uses simple initialization that requires a subtraction; however, due to its simplicity it is classified as zero iteration cycle. Table III shows the relative error of the Goldschmidt with binomial simplification division model. The convergence rate differs in the interval [0.5, 1]. The slowest convergence has the boundary values 0.5, and the opposite boundary value 1 is calculated instantly. This decreasing dependence is similar as for the standard Goldschmidt model. The computed values are also very similar. The first three columns with the relative error values are the same, and only in the last two columns small differences can be seen in comparison with the standard Goldschmidt model shown in Table II. Five iterations are needed to compute multiplicative inverse with single precision which is the same as for the standard Goldschmidt model.

TABLE III.   RELATIVE ERROR OF GOLDSCHMIDT DIVISION MODEL WITH BINOMIAL SIMPLIFICATION

Input value	Relative error at iteration cycle:				
	1	2	3	4	5
0.50	-0.25	-0.0625	-0.0039063	-1.5259e-05	-2.3283e-10
0.55	-0.2025	-0.041006	-0.0016815	-2.8275e-06	-7.9949e-12
0.60	-0.16	-0.0256	-0.00065536	-4.295e-07	-1.8452e-13
0.65	-0.1225	-0.015006	-0.00022519	-5.0709e-08	-2.5979e-15
0.70	-0.09	-0.0081	-6.561e-05	-4.3047e-09	1.5543e-16
0.75	-0.0625	-0.0039062	-1.5259e-05	-2.3283e-10	0
0.80	-0.04	-0.0016	-2.56e-06	-6.5535e-12	1.7764e-16
0.85	-0.0225	-0.00050625	-2.5629e-07	-6.6058e-14	-3.7748e-16
0.90	-0.01	-0.0001	-1e-08	0	0
0.95	-0.0025	6.25e-06	-3.9063e-11	0	0
1.00	0	0	0	0	0

The option to calculate the division at the same time by changing initial conditions is similarly valid. The computation of multiplicative inverse with double precision requires only one additional iteration cycle. The sixth iteration is not shown in Table III but the limits of the models and simulation process with double precision can already be seen in the last column with larger input values similarly as in Table II.

V.   HARDWARE IMPLEMENTATION IN FPGA

The hardware implementation in FPGA has been selected only as 32-bit implementation with single floating-point precision. All three division models described in the previous section have been designed and implemented. The internal hardware architectures are designed based on the software division models. These models require one addition/subtraction and two multiplications to calculate one iteration cycle. The reduced 32-bit floating-point multiplier and adder have been designed and implemented. These computation blocks have been simplified by removing support for denormal numbers. Removing the support for denormal numbers reduces the area of the computation blocks which allows to use more computation blocks in the same design.

All designs has been implemented in Xilinx Virtex4 SX35 FPGA with FF668 package. This FPGA contains 4-input look up tables (LUTs) and embedded DPS48 blocks. The presented results are generated with no use of DSP48 blocks. Table IV shows implementation results of all three division models designed to compute a multiplicative inverse only. The area consumed by computational logic (LUTs) is very similar in all three designs. This is caused by utilizing one 32-bit floating-point adder and two 32-bit floating-point multipliers in each of them. All three designs have been designed to high computation speed and throughput. This has been achieved by the parallel computation of several inputs at the same time. The Newton-Raphson implementation calculates three multiplicative inverses at the same time. The Goldschmidt and Goldschmidt with binomial simplification implementations calculate two multiplicative inverses at the same time. This is the reason why the Newton-Raphson implementation has one and half time higher computation time than the Goldschmidt implementation while they both have the same throughput. The similar reason is responsible for the one and half time increased number of registers (Flip Flops) in the Newton-Raphson implementation in comparison to Goldschmidt implementation. The Goldschmidt with binomial simplification implementation requires one clock cycle to calculate initial condition which

TABLE IV.   MULTIPLICATIVE INVERSE IMPLEMENTATION IN FPGA

	Newton-Raphson reciprocal	Goldschmidt reciprocal	Goldschmidt + binomial reciprocal
LUTs as logic	2352	2343	2350
LUTs as route-thru	96	98	103
Total LUTs	2448	2441	2453
Occupied Slices	1334	1273	1282
Flip Flops	291	161	195
Delay	13 * clk	11 * clk	13 * clk
Computation time	15 * clk	10 * clk	12 * clk
Throughput	0.2 / clk	0.2 / clk	0.1667 / clk
Delay (ns)	14.957	14.990	14.902
Maximum frequency (MHz)	66.858	66.711	67.105

TABLE V.   OUTPUT VALUES OF NEWTON-RAPHSON RECIPROCAL 32-BIT IMPLEMENTATION IN FPGA

Input value	Output values at iteration cycle:			
	1	2	3	4
3F000000	3FF0F0F0	3FFF1D3B	3FFFFF36	3FFFFFFD
3F200000	3FD2D2D2	3FCC9F71	3FCCCCC0	3FCCCCCC
3F400000	3FB4B4B4	3FAA137B	3FAAAA25	3FAAAAAA
3F600000	3F969696	3F9228BE	3F92491D	3F924923

utilizes the adder. This results in increased computation time to 12 clock cycles and reduced throughput to 0.1667 per clock cycle. The Table IV shows that the Goldschmidt implementation is the best from all three presented implementations.

Table V shows output values of Newton-Raphson reciprocal 32-bit implementation in FPGA. The first column represents input data shown in hexadecimal form. The output values are shown in hexadecimal form in the last four columns. Each column represents output data at specific iteration cycle. It can be seen that the number of valid bits is increasing with each next iteration cycle. The Table VI shows the relative errors of Newton-Raphson reciprocal 32-bit implementation in FPGA. The relative errors are calculated directly from data shown in Table V. Input values from the first column correspond directly to the input values shown in Table V. The relative error decreases with each next iteration cycle. The decreasing rate is slower in comparison with data shown in Table I due to 32-bit computational blocks used in the implementation.

Table VII shows output values of Goldschmidt reciprocal 32-bit implementation in FPGA. The first column represents input data shown in hexadecimal form which are the same as in Table V. Only the first four hexadecimal digits are shown to reduce the table width. The output values are shown in hexadecimal form in the last five columns. Each column represents output data at the specific iteration cycle. It can be seen that the number of valid bits is increasing with each next iteration cycle. The increasing rate is higher for input values closer to 1. The Table VIII shows the relative errors of Goldschmidt reciprocal 32-bit implementation in FPGA. The relative errors are calculated directly from data shown in Table VII. Input values from the first column correspond directly to the input values shown in Table VII. The relative

TABLE VI.   RELATIVE ERRORS OF NEWTON-RAPHSON RECIPROCAL 32-BIT IMPLEMENTATION IN FPGA

Input value	Relative error at iteration cycle:			
	1	2	3	4
0.500	-0.0588	-0.0035	-1.2040e-05	-1.7881e-07
0.625	0.0294	-8.6515e-04	-9.5367e-07	-5.9605e-08
0.750	0.0588	-0.0035	-1.1951e-05	-5.9605e-08
0.825	0.0294	-8.6519e-04	-7.8976e-07	-1.6391e-07

978-1-4799-8720-7/15 $31.00 © 2015 IEEE

TABLE VII. OUTPUT VALUES OF GOLDSCHMIDT RECIPROCAL 32-BIT IMPLEMENTATION IN FPGA

Input value	Output values at iteration cycle:				
	1	2	3	4	5
3F00	3FC00000	3FF00000	3FFF0000	3FFFFF00	3FFFFFFF
3F20	3FB00000	3FC8C000	3FCCB84C	3FCCCCC9	3FCCCCCC
3F40	3FA00000	3FAA0000	3FAAAA00	3FAAAAAA	3FAAAAAA
3F60	3F900000	3F924000	3F924924	3F924924	3F924924

TABLE VIII. RELATIVE ERRORS OF GOLDSCHMIDT RECIPROCAL 32-BIT IMPLEMENTATION IN FPGA

Input value	Relative error at iteration cycle:				
	1	2	3	4	5
0.500	-0.2500	-0.0625	-0.0039	-1.5259e-05	-5.9605e-08
0.625	-0.1406	-0.0198	-3.9107e-04	-2.8312e-07	-5.9605e-08
0.750	-0.0625	-0.0039	-1.5259e-05	-5.9605e-08	-5.9605e-08
0.825	-0.0156	-2.4414e-04	-5.9605e-08	-5.9605e-08	-5.9605e-08

error decreases with each next iteration cycle. The decreasing rate is slower in comparison with data shown in Table II due to 32-bit computational blocks used in the implementation.

Table IX shows output values of Goldschmidt with binomial simplification reciprocal 32-bit implementation in FPGA. The first column represents input data shown in hexadecimal form which are the same as in Table V. The output values are shown in hexadecimal form in the last five columns. These data are identical to data from Table VII. The Table X shows the relative errors of Goldschmidt with binomial simplification reciprocal 32-bit implementation in FPGA. The relative errors are calculated directly from data shown in Table IX and therefore are also identical to the data from Table VIII. The relative error decreases with each next iteration cycle; however, the decreasing rate is slower in comparison with data shown in Table III due to 32-bit computational blocks used in the implementation.

Table XI shows implementation results of two division implementations in FPGA. The first implementation representing 32-bit floating-point divider based on the Goldschmidt model is shown in the second column. The second implementation representing 32-bit floating-point divider based on the Goldschmidt with binomial simplification model is shown in the last column. It can be seen that the area consumed by computational logic (LUTs) is only slightly bigger in comparison to multiplicative inverse implementations from Table IV. The

TABLE IX. OUTPUT VALUES OF GOLDSCHMIDT WITH BINOMIAL SIMPLIFICATION RECIPROCAL 32-BIT IMPLEMENTATION IN FPGA

Input value	Output values at iteration cycle:				
	1	2	3	4	5
3F00	3FC00000	3FF00000	3FFF0000	3FFFFF00	3FFFFFFF
3F20	3FB00000	3FC8C000	3FCCB84C	3FCCCCC9	3FCCCCCC
3F40	3FA00000	3FAA0000	3FAAAA00	3FAAAAAA	3FAAAAAA
3F60	3F900000	3F924000	3F924924	3F924924	3F924924

TABLE X. RELATIVE ERRORS OF GOLDSCHMIDT WITH BINOMIAL SIMPLIFICATION RECIPROCAL 32-BIT IMPLEMENTATION IN FPGA

Input value	Relative error at iteration cycle:				
	1	2	3	4	5
0.500	-0.2500	-0.0625	-0.0039	-1.5259e-05	-5.9605e-08
0.625	-0.1406	-0.0198	-3.9107e-04	-2.8312e-07	-5.9605e-08
0.750	-0.0625	-0.0039	-1.5259e-05	-5.9605e-08	-5.9605e-08
0.825	-0.0156	-2.4414e-04	-5.9605e-08	-5.9605e-08	-5.9605e-08

TABLE XI. DIVISION IMPLEMENTATION IN FPGA

	Goldschmidt divider	Goldschmidt + binomial divider
LUTs as logic	2374	2381
LUTs as route-thru	98	103
Total LUTs	2472	2484
Occupied Slices	1285	1326
Flip Flops	193	291
Delay	11 * clk	13 * clk
Computation time	10 * clk	12 * clk
Throughput	0.2 / clk	0.1667 / clk
Delay (ns)	14.892	14.985
Maximum frequency (MHz)	67.150	66.733

transformation of calculating of the multiplicative inverse to divider based on the Goldschmidt model requires only light modifications of initial values and therefore the increased area is minimal. The number of used registers (Flip Flops) is slightly increased. This is caused by necessity to store additional input data that represent denominator values. The other parameters as delay, computation speed and throughput are the same in comparison to data from Table IV.

VI. CONCLUSION

New hardware implementations of three different fast computation division algorithms have been presented. The algorithms have been implemented as division models with 64-bit floating-point double precision. They also have been implemented in FPGA with 32-bit floating-point single precision. The hardware implementations include both multiplicative inverse and division. The result shows that the Goldschmidt implementation is the best from all presented implementations. The presented implementations are designed with high computation speed and throughput. They are suitable for high computation demanding applications with multiple division computations in short sequences.

ACKNOWLEDGMENT

This work has been supported by Slovak national project VEGA 2/0192/15.

REFERENCES

[1] A. R. Garca, L. P. Escalante, R. P. Michel, O. L. Gandara and J. Cortez, "Fast Fixed-Point Divider Based on Newton-Raphson Method and Piecewise Polynomial Approximation," in Proceedings International Conference on Reconfigurable Computing and FPGAs (ReConFig), Cancun, 2013, pp. 1-6.

[2] J. M. Muller, "Avoiding Double Roundings in Scaled Newton-Raphson Division," in Proceedings of Asilomar Conference on Signals, Systems and Computers, Pacific Grove, CA, 2013, pp. 396-399.

[3] I. Kong and E. E. Swartzlander, "A Rounding Method to Reduce the Required Multiplier Precision for Goldschmidt Division," IEEE Transactions on Computers, Vol. 59, No. 12, 2010, pp. 1703-1708.

[4] I. Kong and E. E. Swartzlander, "A Goldschmidt Division Method With Faster Than Quadratic Convergence," IEEE Transactions on Very Large Scale Integration (VLSI) System, Vol. 19, No. 4, 2011, pp. 696-700.

[5] T. Viitanen, P. Jaaskelainen and J. Takala, "Inexpensive Correctly Rounded Floating-point Division and Square Root with Input Scaling," in Proceedings of the IEEE Workshop on Signal Processing Systems (SiPS), Taipei City, 2013, pp. 159-164.

978-1-4799-8720-7/15 $31.00 © 2015 IEEE

2015 IEEE Computer Society Annual Symposium on VLSI

Exploiting Circuit Duality to Speed Up SAT

Luca Amarù[1], Pierre-Emmanuel Gaillardon[1], Alan Mishchenko[2], Maciej Ciesielski[3], Giovanni De Micheli[1]

Integrated Systems Laboratory, EPFL, Switzerland[1]

Department of EECS, University of California, Berkeley, USA[2]

Department of ECE, University of Massachusetts, Amherst, USA[3]

Abstract— In this paper, we establish a *non-trivial* duality between tautology and contradiction check to speed up circuit SAT. Tautology check determines if a logic circuit is *true* in every possible interpretation. Analogously, contradiction check determines if a logic circuit is *false* in every possible interpretation. A *trivial* transformation of a (tautology, contradiction) check problem into a (contradiction, tautology) check problem is the *inversion* of all outputs in a logic circuit. In this work, we show that *exact logic inversion* is not necessary. We give operator switching rules that selectively exchange tautologies with contradictions, and *viceversa*. Our approach collapses into logic *inversion* just for tautology and contradiction extreme points but generates *non-complementary* logic circuits in the other cases. This property enables computing benefits when an alternative, but equisolvable, instance of a problem is easier to solve than the original one. As a case study, we investigate the impact on SAT. There, our methodology generates a dual SAT instance solvable in parallel with the original one. This concept can be used on top of any other SAT approach and does not impose much overhead, except having to run two solvers instead of one, which is typically not a problem because multi-cores are widespread and computing resources are inexpensive. Experimental results show a 25% speed-up of SAT in a concurrent execution scenario. Also, statistical experiments confirmed that our runtime reduction is not of the random variation type.

I. INTRODUCTION

Inspecting the properties of logic circuits is pivotal to logic applications for computers and especially to *Electronic Design Automation* (EDA) [1]. There exists a large variety of properties to be checked in logic circuits, e.g., *unateness, linearity, symmetry, balancedness, monotonicity, thresholdness* and many others [2]. Basic characteristics are usually verified first to provide grounds for more involved tests. Tautology and contradiction are the most fundamental properties in logic circuits. A check for tautology determines if a logic circuit is *true* for all possible input patterns. Analogously, a check for contradiction determines if a logic circuit is *false* for all possible input patterns. While investigating elementary properties, tautology and contradiction check are difficult problems, i.e., co-NP-complete and NP-complete, respectively [3]. Indeed, both tautology and contradiction check are equivalent formulation of the Boolean *SATisfiability* (SAT) problem [3]. In this scenario, new efficient algorithms for tautology/contradiction check are key to push further the edge of computational limits, enabling larger logic circuits to be examined.

Tautology and contradiction check are dual problems. One can interchangeably check for tautology in place of contradiction by *inverting* all outputs in a logic circuit. In this *trivial* approach, the two obtained problems are fully complementary and there is no explicit computational advantage in solving one problem instead of the other.

In this paper, we show that *exact logic inversion* is not necessary for transforming tautology into contradiction, and *viceversa*. We give a set of operator switching rules that

selectively exchange tautologies with contradictions. A logic circuit modified by our rules is *inverted* just if identically *true* or *false* for all input combinations. In the other cases, it is not necessarily the complement of the original one. In a simple logic circuit made of AND, OR and INV logic operators, our switching rules swap AND/OR operator types. We give a set of rules for general logic circuits in the rest of this paper. Note that in this paper we mostly deal with single output circuits. For multi-output circuits, the same approach can be extended by ORing (contradiction) or ANDing (tautology) the outputs that need to be checked into a single one.

Our approach generates two different, but equisolvable, instances of the same problem. In this scenario, solving both of them in parallel enables a positive computation speed-up. Indeed, the instance solved first stops the other reducing the runtime. This concept can be used on top of any other checking approach and does not impose much overhead, except having to run two solvers instead of one, which is typically not a problem because multi-cores are wide-spread and computing resources are inexpensive. Note that other pallel checking techniques exist. For example, one can launch in parallel many randomized check runs on the same problem instance with the aim to hit the instance-intrinsic minimum runtime [4]. Instead, in our methodology, we create a different but equi-checkable instance that has a potentially lower minimum runtime. As a case study, we investigate the impact of our approach on SAT. There, by using *non-trivial* and *trivial* dualities in sequence, we create a dual SAT instance solvable in parallel with the original one. Experimental results show 25% speed-up of SAT, on average, in a concurrent execution scenario. Also, statistical experiments confirmed that our runtime reduction is not of the random variation type.

The remainder of this paper is organized as follows. Section II describes some background and discusses the motivation for this study. Section III presents theoretical results useful for the scope of this paper. Section IV proves our main result on the duality between tautology and contradiction check. Section V shows the benefits enabled by this duality in SAT solving. Section VI concludes the paper.

II. BACKGROUND AND MOTIVATION

This section first provides notation on logic circuits. Then, it gives a brief background on tautology checking from an EDA perspective. Finally, it discusses the motivation for this study.

A. Notation

A logic circuit is a *Directed Acyclic Graph* (DAG) representing a Boolean function, with nodes corresponding to logic gates and directed edges corresponding to wires connecting the gates. The *on-set* of a logic circuit is the set of input patterns evaluating to *true*. Analogously, the *off-set* of a logic circuit is the set of input patterns evaluating to *false*. Each

978-1-4799-8720-7/15 $31.00 © 2015 IEEE   101

logic gate is associated with a primitive Boolean function taken from a predefined set of *basis* logic operators, e.g., AND, OR, XOR, XNOR, INV, MAJ, MIN etc. Logic operators such as MAJ and MIN represent *self dual* Boolean functions, i.e., functions whose output complementation is equivalent to inputs complementation. A set of *basis* logic operators is said to be *universal*[1] if any Boolean function can be represented by a logic circuit equipped with those logic gates. For example, the *basis* set {OR, INV} is *universal* while the *basis* set {AND, MAJ} is not. Fig.1 shows a logic circuit for the

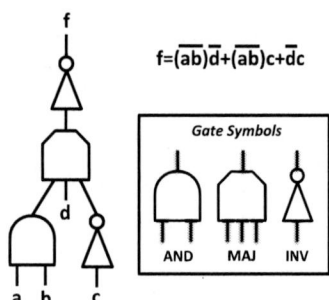

Fig. 1: Logic circuit example representing the function $f = \overline{(ab)\overline{d}} + \overline{(ab)}c + \overline{d}c$. The *basis* set is {AND, MAJ, INV}. The gates symbolic representation is shown in the box.

function $f = \overline{(ab)\overline{d}} + \overline{(ab)}c + \overline{d}c$ over the *universal basis* set {AND, MAJ, INV}.

B. Tautology Checking

Tautology checking, i.e., verifying whether a logic circuit is *true* in every possible interpretation, is an important task in computer science and at the core of EDA [5], [7]. Traditionally, tautology checking supports digital design verification through combinational equivalence checking [7]. Indeed, the equivalence between two logic circuits can be detected by XNOR-ing and checking for tautology. Logic synthesis also uses tautology checking to (i) highlight logic simplifications during optimization [5], [6] and to (ii) identify matching during technology mapping [8]. On a general basis, many EDA tasks requiring automated deduction are solved by tautology check routines.

Unfortunately, solving a tautology check problem can be a difficult task. In its most general formulation, the tautology check problem is co-NP-complete. A straightforward method to detect a tautology is the exhausitve exploration of a function truth table. This *naive* approach can declare a tautology only in exponential runtime. More intelligent methods have been developed in the past. Techniques based on cofactoring trees and binary recursion have been presented in [9]. Together with rules for pruning/simplifying the recursion step, these techniques reduced the checking runtime on several benchmarks. Another method, originally targeting propositional formulas, is Stalmarck's method [10] that rewrites a formula with a possibly smaller number of connectives. The derived equivalent formula is represented by triplets that are propagated to check for tautology. Unate recursive cofactoring trees and Stalmarck's method are as bad as any other tautology check method in

[1]In this work, the term *basis* does not share the same properties as in linear algebra. In particular, here not all the *basis* are *universal*.

the worst case but very efficient in real-life applications. With the rise of *Binary Decision Diagrams* (BDDs) [11], tautology check algorithms found an efficient canonical data structure explicitly showing the logic feature under investigation [12]. The BDD for a tautology is always a single node standing for the logic constant *true*. Hence, it is sufficient to build a BDD for a logic circuit and verify the resulting graph size (plus the output polarity) to solve a tautology check problem. Unfortunately, BDDs can be exponential in size for some functions (multipliers, hidden-weight bit, etc.). In the recent years, the advancements in SAT solving tools [13], [14] enabled more scalable approaches for tautology checking. Using the *trivial* duality between tautology and contradiction, SAT solvers can be used to determine if an *inverted* logic circuit is unsatisfiable (contradiction) and consequently if the original circuit is a tautology. Still, SAT solving is an NP-complete problem so checking for tautology with SAT is difficult in general.

C. Motivation

Tautology checking is a task surfing the edge of today's computing capabilities. Due to its co-NP-completeness, tautology checking aggressively consumes computational power when the size of the problem increases. To push further the boundary of examinable logic circuits, it is important to study new efficient checking methodologies. Indeed, even a narrow theoretical improvement can generate a speed-up equivalent to several years of technology evolution.

In this paper, we present a *non-trivial* duality between contradiction and tautology check problems that opens up new efficient solving opportunities.

III. PROPERTIES OF LOGIC CIRCUITS

In this section, we show properties of logic circuits with regard to their *on-set/off-set* balance and distribution. These theoretical results will serve as grounds for proving our main claim in the next section.

We initially focus on two *universal basis* sets: {AND, OR, INV} and {MAJ, INV}. We deal with richer *basis* sets later on. We first recall a known fact about majority operators.

Property A MAJ operator of n-variables, with n odd, can be configured as an $\lceil n/2 \rceil$-variables AND operator by biasing $\lfloor n/2 \rfloor$ inputs to logic *false* and can be configured as an $\lceil n/2 \rceil$-variables OR operator by biasing $\lfloor n/2 \rfloor$ inputs to logic *true*.

For the sake of clarity, an example of a three-input MAJ configuration in AND/OR is depicted by Fig. 2. Extended

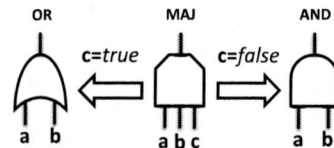

Fig. 2: AND/OR configuration of a three-input MAJ.

at the circuit level, such property enables the emulation of any {AND, OR, INV} logic circuit by a structurally identical {MAJ, INV} logic circuit. This result was previosuly shown in [15] where logic circuit over the *basis* set {AND, OR, INV} are called AND/OR-INV graphs and logic circuits over

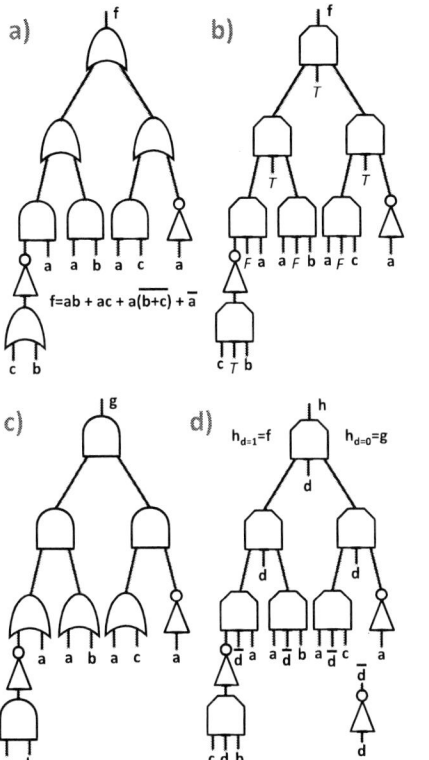

Fig. 3: Logic circuits examples. {AND, OR, INV} logic circuit representing $f = ab + ac + a\overline{(b+c)} + \overline{a}$ (a). {MAJ, INV} logic circuit emulating the circuit in (a) using constants (b). {AND, OR, INV} logic circuits derived from (a) by switching AND/OR operators (c). {MAJ, INV} logic circuit emulating the circuit in (a) using an fictitious input variable d (d).

the *basis* set {MAJ, INV} are called MAJ-INV graphs. An example of two structurally, and functionally, identical logic circuits over the *basis* sets {AND, OR, INV} and {MAJ, INV} is depicted by Fig. 3(a-b). The Boolean function represented in this example is $f = ab+ac+a\overline{(b+c)}+\overline{a}$. MAJ are configured to behave as AND/OR by fixing one input to *false*(F)/*true*(T), respectively. In place of biasing one input of the MAJ with a logic constant, it is also possible to introduce a fictitious input variable connected in regular/inverted polarity to substitute *true*(T)/*false*(F) constants, respectively. In this way, the function represented is changed but still including the original one when the fictitious input variable is assigned to *true*. Fig. 3(d) shows a logic circuit with a fictious input variable d replacing the logic constants in Fig. 3(b). The Boolean function represented there is h with property $h_{d=true} = f$.

Up to this point, we shown that {AND, OR, INV} logic circuits can be emulated by {MAJ, INV} logic circuits configured either by (i) logic constants or by (ii) a fictitious input variable. In the latter case, {MAJ, INV} logic circuits have all inputs assignable. With no logic constants appearing and all operators being *self-dual*, this particular class of logic circuits have a perfectly balanced *on-set/off-set* size. The following theorem formalizes this property.

Theorem 3.1: Logic circuits over the *universal basis* set

{MAJ, INV}, with all inputs assignable (no logic constants), have $|on\text{-}set|=2^{n-1}$ and $|off\text{-}set|=2^{n-1}$, with n being the number of input variables.

Proof MAJ and INV logic operators, with no constants, represent *self-dual* Boolean functions. In [2], it is shown that *self-dual* Boolean functions have an $|on\text{-}set|=|off\text{-}set|=2^{n-1}$, with n being the number of input variables. Also, it is shown in [2] that Boolean functions composed by *self-dual* Boolean functions are *self-dual* as well. This is indeed the case for {MAJ, INV} logic circuits with no constants in input. As these circuits represent *self-dual* Boolean functions, we can assert $|on\text{-}set|=|off\text{-}set|=2^{n-1}$. *Q.E.D.*

{MAJ, INV} logic circuits with no constants have a perfectly balanced partition between *on-set* size and *off-set* size. This is the case for the example in Fig. 3(d). Eventually, we know that by assigning d to *true* in such example circuit the *on-set/off-set* balance can be lost. Indeed, with *d=true* the {MAJ, INV} logic circuit then emulates the original {AND, OR, INV} logic circuit in Fig. 3(a), that could have different *on-set* size and *off-set* size. Still, it is possible to reclaim the perfect *on-set/off-set* balance by superposing the cases *d=true* and *d=false* in the {MAJ, INV} logic circuit. While we know precisely what the {MAJ, INV} logic circuit does when *d=true*, the case *d=false* is not as evident. We can intepret the case *d=false* as an inversion in the MAJ configuration polarity. This means that where a MAJ is configured as an AND (OR) node in *d=true*, it is instead configured as an OR (AND) node in *d=false*. In other words, *d=false* in the {MAJ, INV} logic circuit of Fig. 3(d) corresponds to switch AND/OR operator types in the original {AND, OR, INV} logic circuit of Fig. 3(a). The resulting AND/OR switched circuit is depicted by Fig. 3(c).

United by a common {MAJ, INV} generalization, {AND, OR, INV} logic circuits and their AND/OR switched versions share strong properties about *on-set/off-set* repartition. The following theorem states their relation.

Theorem 3.2: Let A be a logic circuit over the *universal basis* set {AND, OR, INV}. Let A' be a modified version of A, with AND/OR operators switched. The following identities hold $|on\text{-}set(A)|=|off\text{-}set(A')|$ and $|off\text{-}set(A)|=|on\text{-}set(A')|$.

Proof Say M a {MAJ, INV} logic circuit emulating A using an extra fictitious input variable, say d. $M_{d=1}$ is structurally and functionally equivalent to A, while $M_{d=0}$ is structurally and functionally equivalent to A'. From Theorem 3.1 we know that $|on\text{-}set(M)|=|off\text{-}set(M)|=2^{n-1}=2^m$, where m is the number of input variables in A and n the number of input variables in M, with $n = m + 1$ to take into account the extra fictitious input variable in M. We know by construction that $|on\text{-}set(M_{d=1})|+|on\text{-}set(M_{d=0})|=2^{n-1}=2^m$ and $|off\text{-}set(M_{d=1})|+|off\text{-}set(M_{d=0})|=2^{n-1}=2^m$. Again by construction we know that $M_{d=1}$ and $M_{d=0}$ can be substituted by A and A', respectively, in all equations. Owing to the basic definition of A and A' we have that $|on\text{-}set(A)|+|off\text{-}set(A)|=2^m$ and $|on\text{-}set(A')|+|off\text{-}set(A')|=2^m$. Expressing $|on\text{-}set(A)|$ as $2^m-|on\text{-}set(A')|$ and substituting this term in $|on\text{-}set(A)|+|off\text{-}set(A)|=2^m$ we get $2^m-|on\text{-}set(A')|+|off\text{-}set(A)|=2^m$ that can be simplified as $|off\text{-}set(A)|=|on\text{-}set(A')|$. This proves the first identity of the Theorem. The second identity can be proved analogously.

Q.E.D.

Informally, the previous theorem says that by switching AND/OR operators in an {AND, OR, INV} logic circuit we swap the *on-set* and *off-set* sizes. From a statistical perspective, this is equivalent to invert $Pr(A=true)$ with $Pr(A=false)$, under uniformly random input string of bits. While this also happens with *exact* logic *inversion*, here the actual distribution of the *on-set/off-set* elements is not necessarily complementary. In the next section, we show the implications of the theoretical results seen so far in tautology and contradiction check problems.

IV. FROM TAUTOLOGY TO CONTRADICTION AND BACK

Verifying whether a logic circuit is a tautology, a contradiction or a contingency[2] is an important task in logic applications for computers.

In this section, we show that tautology and contradiction check in logic circuits are dual and interchangeable problems that do not require *exact* logic *inversion per se*. We start by considering logic circuit over the *universal basis* set {AND, OR, INV} and we consider richer *basis* sets later on. The following theorem describes the *non-trivial* duality between tautology and contradiction in {AND, OR, INV} logic circuits.

Theorem 4.1: Let A be a logic circuit over the *universal basis* set {AND, OR, INV} representing a tautology (contradiction). The logic circuit A', obtained by switching AND/OR operations in A, represents a contradiction (tautology).

Proof If A represents a tautology then $|on\text{-}set(A)|=2^m$ and $|on\text{-}set(A)|=0$, with m being the number of inputs. Owing to Theorem 3.2 $|on\text{-}set(A')|=|off\text{-}set(A)|=0$ and $|off\text{-}set(A')|=|on\text{-}set(A)|=2^m$. It follows that A' is a contradiction. Analogous reasoning holds for contradiction to tautology transformation. *Q.E.D.*

Switching AND/ORs in an {AND, OR, INV} logic circuit is strictly equivalent to logic inversion only for tautology and contradiction. In the other cases, A and A' are not necessarily complementary. We give empirical evidences about this fact hereafter. Fig. 4 depicts the obtained results in a graph chart. We examined 17 random Boolean functions of four input variables, with *on-set size* ranging from 0 (contradiction) to 16 (tautology). We first compared the *on-set* size of the real inverted logic circuits with the *on-set size* of the AND/OR switched circuits. As expected, Theorem 3.2 holds and switching AND/OR operators results in exchanging the *on-set* and *off-set* sizes. This also happens with the real inverted circuits, but in that case also the actual *on-set/off-set* elements distribution is complementary. To verify what is the *on-set/off-set* elements distribution in general, we define a distance metric between the real inverted and AND/OR switched circuits. The distance metric is computed in two steps. First, the truth tables of the circuits are unrolled, using the same input order, and represented as binary strings. Second, the distance metric is measured as the Hamming distance[3] between those binary strings. For tautology and contradiction extremes the distance metric between AND/OR switched circuits and real inverted

[2]A logic circuit is a contigency when it is neither a tautology nor a contradiction [2]

[3]The Hamming distance between two binary strings, of equal size, is the number of positions at which the corresponding bits are different.

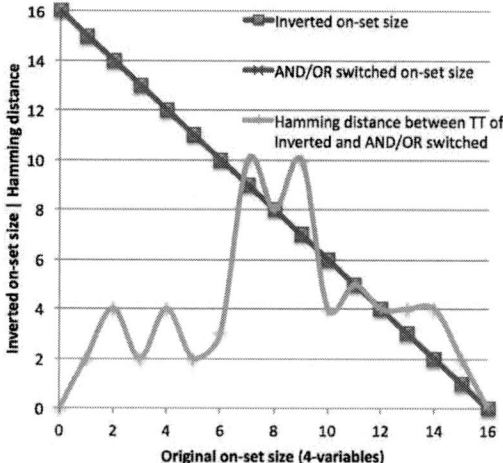

Fig. 4: Comparison between *real inverted* and *AND/OR switched* logic circuits representing 4-variable Boolean functions. The *on-set* size ranges from 0 to 2^4.

circuits is 0, as obvious consequence of Theorem 4.1. For other circuits, real inverted and AND/OR switched circuits are different, with distance metric ranging between 2 and 10.

As a practical intepretation of the matter discussed so far, we can get an answer for a tautology (contradiction) check problem by working on a functionally different and *non-complementary* structure than the original one under test. We explain hereafter why this fact is interesting. Suppose that the logic circuit we want to check is a contigency but algorithms for tautology (contradiction) are not efficient on it. If we just invert the outputs of this logic circuit and we run algorithms for contradiction (tautology) then we would likely face the same difficulty. However, if we switch AND/ORs in the logic circuit we get a functionally different and *non-complementary* structure. In this case, algorithms for contradiction (tautology) do not face by construction the same complexity. Exploiting this property, it is possible to speed-up a traditional tautology (contradiction) check problem. Still, Theorem 4.1 gurantees that if the original circuit is a tautology (contradiction) then the AND/OR switched version is a contradiction (tautology) preserving the checking correctness.

Recalling the example in Fig. 3(a), the original logic circuit represents a tautology. Consequently, the logic circuit in Fig. 3(c) represents a contradiction. These properties are verifiable by hand as the circuits considered are small. For an example which is a *contingency*, consider the {AND, OR, INV} circuit realization for $f = ab' + c'$ (contingency). By switching AND/ORs, we get $g = (a + b')c'$ which is different from both f or f', as preticted.

We now consider logic circuits with richer *basis* set functions than just {AND, OR, INV}. Our enlarged *basis* set includes {AND, OR, INV, MAJ, XOR, XNOR} logic operators. Other operators can always be decomposed into this *universal basis* set, or new switching rules can be derived. In the following, we extend the applicability of Theorem 4.1.

Theorem 4.2: Let A be a logic circuit over the *universal basis* set {AND, OR, INV, MAJ, XOR, XNOR} representing a tautology (contradiction). The logic circuit A', obtained by

TABLE I: Switching Rules for Tautology/Contradiction Check

Original Logic Operator	Switched Logic Operator
INV	INV
AND	OR
OR	AND
MAJ	MAJ
XOR	XNOR
XNOR	XOR

switching logic operators in A as per Table I, represents a contradiction (tautology).

Proof In order to prove the theorem, we need to show the switching rules just for XOR, XNOR and MAJ operators. AND/OR switching is already proved by Theorem 4.1. Consider the XOR operator decomposed in terms of {AND, OR, INV}: $f = a \oplus b = ab' + a'b$. Applying the duality in Theorem 4.1 we get $g = (a + b')(a' + b)$ that is indeed equivalent to a XNOR operator. This proves the XOR to XNOR switching and *viceversa*. Analogously, consider the MAJ operator decomposed in terms of {AND, OR, INV}: $f = ab + ac + bc$. Applying the duality in Theorem 4.1 we get $g = (a + b)(a + c)(b + c)$ that is still equivalent to a MAJ operator. Hence, MAJ operators do not need to be modified. *Q.E.D.*

Note that in a data structure for a computer program, the operator switching task does not require actual pre-processing of the logic circuit. Indeed, each time that a node in the DAG is evaluated an external flag word determines if the regular or switched operator type has to be retrieved from memory.

In the current section, we shown a *non-trivial* duality between contradiction and tautology check. In the next section, we study its application on Boolean satisfiability.

V. EXPERIMENTAL RESULTS

In this section, we exercise our *non-trivial* duality in Boolean *SATisfiability* (SAT) problems. First, we describe how to use the tautology/contradiction duality to generate a second (dual) equisatisfiable SAT instance. Second, we demonstrate that the dual instance can be solved faster than the regular one and the corresponding runtime reduction is not of the random variation type. Third, and last, we show experimental results for a concurrent regular/dual SAT execution scenario.

A. Boolean SAT and Tautology/Contradiction Duality

The Boolean SAT problem consists of determining whether there exists or not an interpretation evaluating to *true* a Boolean formula or circuit. The Boolean SAT problem is reciprocal to a check for contradiction. When contradiction check fails then Boolean SAT succeeds while when contradiction check succeeds then Boolean SAT fails. Instead of checking for Boolean SAT or for contradiction, one can use a dual transformation in the circuit and check for tautology. Such transformation can be either (i) *non-trivial*, i.e., switching logic operators in the circuit as per Table I or (ii) *trivial*, i.e., output complementation. If we use twice any dual transformation, we go back to the original problem domain (contradiction, SAT). Note that if we use twice the same dual transformation (*trivial-trival* or *non-trivial-non-trivial*) we obtain back exactly the original circuit. Instead, if we apply two different dual transformations in sequence (*trivial-non-trivial* or *non-trivial-trivial*) we obtain an equisatisfiable but not necessarily equivalent circuit. We use the latter approach

to generate a second equisatisfiable circuit, which we call the dual circuit. The dual circuit SAT can be solved in parallel with the regular one in a *"first finishing wins"* speculative strategy. Fig. 5 depicts the corresponding flow. We generate the

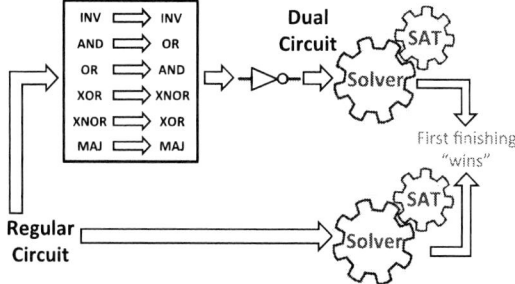

Fig. 5: Speculative parallel regular/dual circuit SAT flow.

dual circuit by first applying our *non-trivial duality* (switching rules in Table I) and finally complementing the outputs (*trivial duality*). Note that these operations ideally require no (or very little) computational overhead, as explained previously.

B. Verification of SAT Solving Advantage on the Dual Circuit

In our first set of experiments we focused on verifying whether the dual circuit can be easier to satisfy than the regular circuit. For this purpose, we modified MiniSat-C v1.14.1 [16] to read circuits in AIGER format [18] and to encode them in CNF internally via Tseitin transformation. The dual circuit is generated online during reading if a switch "-p" is given. We considered a large circuit (0.7 M nodes) over 1000 randomized (pseudo-random number generator seed) runs. Fig. 6 shows the runtime distributions for dual and regular SAT. The

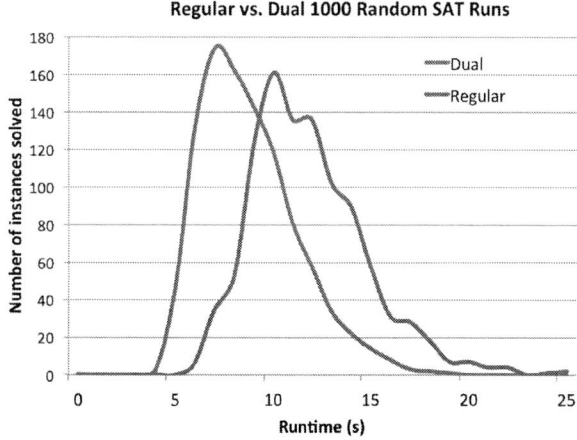

Fig. 6: 1000 randomized SAT runs for regular and dual circuit.

dual runtime distribution is clearly left-shifted (but partially overlapping) with respect to the regular runtime distribution. This confirms that (i) the dual circuit can be solved faster than the regular one and (ii) the runtime reduction is not of the random variation type.

C. Results for Concurrent Regular/Dual SAT Execution

In our second set of experiments (downloadable at [19]) we used ABC tool [17] to test our dual approach together

978-1-4799-8720-7/15 $31.00 © 2015 IEEE

TABLE II: Experimental Results for Regular *vs.* Dual SAT Solving
All runtimes are in seconds

Benchmark	I/O	Logic Size	Logic Depth	Runtime Regular	Runtime Dual	|Δ Runtime|	Best Runtime
hardsat1	4580/1	283539	392	186.35	58.9	127.35	58.9
hardsat2	4580/1	287635	392	51.1	191.87	140.77	51.1
hardsat3	198540/1	920927	267	0.94	1.1	0.16	0.94
hardsat4	2452/1	43962	436	68.82	20.53	48.29	20.53
hardsat5	5725/1	562027	464	40.91	22.72	18.19	22.72
hardsat6	3065/1	86085	437	37.51	64.24	26.73	37.51
hardsat7	372240/1	85596	151	4.8	3.68	1.12	3.68
Total *sat*	591182/7	2269771	2539	390.43	363.04	27.39	195.38
hardunsat1	61/1	448884	2181	26.72	27.22	0.50	26.72
hardunsat2	61/1	264263	2951	3.70	1.32	2.38	1.32
hardunsat3	61/1	451350	2181	27.8	20.33	7.47	20.33
hardunsat4	540/1	244660	1158	234.88	326.84	91.96	234.88
hardunsat5	2352/1	208221	439	7.61	7.65	0.04	7.65
hardunsat6	550/1	117820	423	142.28	137.94	4.34	137.94
Total *unsat*	3625/6	1735198	9333	442.99	521.30	78.31	428.80
Total	594807/13	4004969	11872	833.42	884.34	50.84	624.18
Norm. to Regular	–	–	–	1.00	1.06	–	0.75

with advanced techniques to speed-up SAT. Our custom set of benchmarks is derived by (i) unfolding SAT sequential problems (ii) encoding combinational equivalence check problems. All benchmarks are initially described in Verilog as a netlist of logic gates over the *basis* {AND, OR, INV, XOR, XNOR, MAJ}. The dual circuits are obtained by applying switching rules in Table I and inverting the output. The ABC script to read and run SAT on these benchmarks is: *read library.genlib; r -m input.v; st; write out.aig; &r out.aig; &ps; &write_cnf -K 4 out.cnf; dsat -p out.cnf*. Apart from standard I/O commands, note that *&write_cnf -K 4 out.cnf* generates a CNF using a technology mapping procedure and *dsat -p* calls MiniSat with variable polarity alignment.

Table II shows results for regular *vs.* dual SAT solving with our setup. For about half of the benchmarks (7/13) the dual instance concluded first while for the remaning ones (6/13) the regular instance was faster. The total regular runtime is quite close to the total dual runtime (just 6% of deviation). However, considering here the speculative parallel SAT flow in Fig. 5, we can ideally reduce the total runtime by about 25%. Note that this is an ideal projection into a parallel execution environment, with no overhead. We experimentally verified that the average overhead can be small (few percentage points) thanks to the intrinsic independence of the two tasks.

VI. CONCLUSIONS

In this paper, we have shown a *non-trivial* duality between tautology and contradiction check to speed up circuit SAT. On the one hand, tautology check determines if a logic circuit is *true* for all input combinations. On the other hand, contradiction check determines if a logic circuit is *false* for all input combinations. A *trivial* transformation of a (tautology, contradiction) check problem into a (contradiction, tautology) check problem is the *inversion* of all the outputs in a logic circuit. In this work, we proved that *exact* logic *inversion* is not necessary. By switching logic operator types in a logic circuit, following the rules presented in this paper, we can selectively exchange tautologies with contradictions. Our approach is equivalent to logic *inversion* just for tautology and contradiction extreme points. It generates *non-complementary* logic circuits in the other cases. Such property enables computing benefits when an alternative but equisolvable instance is easier to solve than the original one. As a case study, we

studied the impact on SAT. There, our methodology generated a dual SAT instance solvable in parallel with the original one. This concept can be used on top of any other SAT approach and does not impose much overhead, except having to run two solvers instead of one, which is typically not a problem because multi-cores are wide-spread and computing resources are inexpensive. Experimental results shown 25% speed-up of SAT in a concurrent execution scenario.

ACKNOWLEDGEMENTS

This work was supported by grant ERC-2009-AdG-246810.

REFERENCES

[1] G. De Micheli, *Synthesis and Optimization of Digital Circuits*, McGraw-Hill, New York, 1994.
[2] T. Sasao, *Switching Theory for Logic Synthesis*, Springer, 1999.
[3] M. R. Garey, D. S. Johnson, *Computers and Intractability– A Guide to the Theory of NP-Completeness*. W. H. Freeman and Company, 1979.
[4] A. E. Hyvarinen, *et al.*, *Incorporating clause learning in grid-based randomized SAT solving*, Journal on SAT (JSAT) 6, 223-244, 2009.
[5] R.K. Brayton, *Logic minimization algorithms for VLSI synthesis*, Vol. 2. Springer, 1984.
[6] R. Rudell, A. Sangiovanni-Vincentelli *Multiple-valued Minimization far PLA Optimization*, IEEE Trans. on CAD of ICs and Syst. 6.5: 727-750, 1987
[7] G. Hachtel, F. Somenzi, *Logic synthesis and verification algorithms*. Springer, 2006.
[8] L. Benini, G. De Micheli, *A survey of Boolean matching techniques for library binding*, ACM Transaction on DAES (TODAES), 2(3), 193-226, 1997.
[9] G. D. Hachtel, M. J. Reily, *Verification algorithms for VLSI synthesis*, IEEE Trans. on CAD of ICs and Syst. 7.5: 616-640, 1980
[10] G. Stalmarck, *A system for determining propositional logic theorems by applying values and rules to triplets that are generated from a formula*, Swedish Patent No. 467,076 (approved 1992); U.S. Patent No. 5,276,897 (approved 1994); European Patent No. 403,454 (approved 1995).
[11] R.E. Bryant, *Graph-based algorithms for Boolean function manipulation*, IEEE Trans. on Comp., C-35(8): 677-691, 1986.
[12] S. Malik, A. R. Wang, R. K. Brayton, A. Sangiovanni-Vincentelli, *Logic verification using binary decision diagrams in a logic synthesis environment*, Proc. ICCAD, 1988.
[13] C. P. Gomes, H. Kautz, A. Sabharwal, B. Selman, *Satisfiability solvers*, Handbook of Knowledge Representation 3 (2008): 89-134.
[14] http://www.satcompetition.org
[15] L. Amaru, P.-E. Gaillardon, G. De Micheli, *Majority-Inverter Graph: A Novel Data-Structure and Algorithms for Efficient Logic Optimization*, Proc. DAC, 2014.
[16] MiniSat SAT solver available online at http://minisat.se/MiniSat.html
[17] Berkeley Logic Synthesis and Verification Group, ABC: A System for Sequential Synthesis and Verification, http://www.eecs.berkeley.edu/ alanmi/abc/
[18] AIGER benchmarks available online at *http://fmv.jku.at/aiger/*.
[19] http://lsi.epfl.ch/DUALSAT

2015 IEEE Computer Society Annual Symposium on VLSI

A New Method for Defining Monotone Staircases in VLSI Floorplans

Bapi Kar*[†], Susmita Sur-Kolay*
*Indian Statistical Institute Kolkata
Kolkata, INDIA
Email: bapi.kar@gmail.com, ssk@isical.ac.in

Chittaranjan Mandal[†]
[†]Indian Institute of Technology Kharagpur
Kharagpur, INDIA
Email: chitta@iitkgp.ac.in

Abstract—Physical design of a chip typically entails the global routing (GR) step after detailed placement. In this step, the grid graph (GG) model is used widely for the one or two-bend pattern routing stage. While several iterations may be required for GR, these may be reduced with appropriate route planning at the floorplanning stage. In this case, the routing regions can be identified as recursive top-down bipartitioning by either straight cutlines or monotone staircases. The capacity of each region is estimated as the number nets that can fit in it without design rule violation. Given a floorplan, identifying a set of monotone staircases such that (a) the balance in area of each bipartition is maximized, (b) the number of nets cut, and (c) the number of bends in a monotone staircase are minimized, is a multi-objective optimization problem and is NP-hard. The existing heuristic methods are based on either maxflow or directed search on floorplan adjacency graph.

In this paper, we propose a new monotone staircase bipartitioning method using a randomized neighbor search technique. Compared to the earlier methods, our experimental results on a set of floorplanning benchmark circuits demonstrate on an average 3% improvement in the cost function, as well as 9% and 7% in the number of vias and congestion after global routing.

Keywords-**VLSI physical design, balanced bipartitioning of floorplans, monotone staircase, bend minimization, global routing.**

I. INTRODUCTION

In the existing physical design flow, global routing is an important stage and relies on the preceding placement solution. The routing regions are identified by dividing the layout into an set of m-by-m equal sized tiles and the corresponding capacity of each routing region is computed by considering the blockages in it. Many of the recent global routers employed pattern based routing in order to identify the routing paths through those tiles [16], [21]. In case of no feasible global routing is attained or the subsequent detailed routing fails to converge, a new placement solution or even a new floorplan has to be regenerated, either incrementally [7], [8], [20] or as a completely new solution. This leads to increased number of iterations and thus inferior design time-to-market. Although, few works have been proposed that addressed routability and congestion [3], [12] during the placement optimization stage. Some [13], [21] also address via minimization, but during the global routing stage.

On the other hand, recent efforts have been made to predict feasibility of an early global routing solution at the

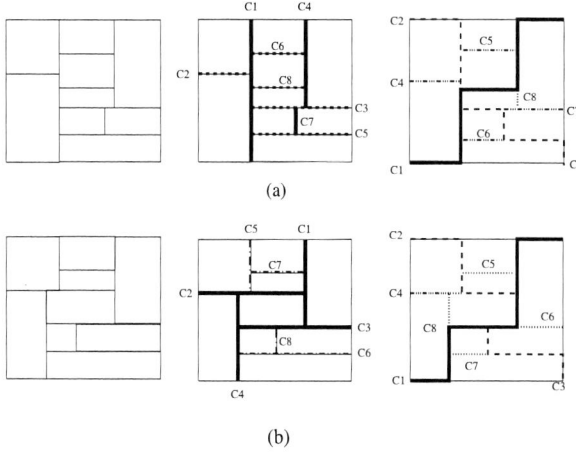

Figure 1: Illustrating hierarchical floorplan bipartitioning: by slicing cutlines and monotone staircases for (a) sliceable floorplan (b) non-sliceable floorplan

floorplanning stage [9], [10], [11]. This can be done by identifying the routing regions between boundary regions of the blocks in the form of monotone staircases, more suitable in the lower metal layers. This model can also be extended to upper metal layers as well, but does not utilize the free regions above the blocks.

Routing completion is of utmost importance in physical design. While a sliceable floorplan having straight cuts only has acyclic routing order assuring 100% routing, it was shown in [5], [19] that monotone staircase cuts can provide similar acyclic routing order in non-sliceable floorplans. Such staircase cuts can handle the switch-box routing problem [18] efficiently and also provide flexibility in re-sizing the routing regions during congestion reduction. In Figure 1, we illustrate the effectiveness in floorplan bipartitioning using monotone staircase cuts, irrespective of its *sliceability*, over slicing tree method. Interestingly, slicing tree method can be seen as a special case of monotone staircase bipartition.

The objectives considered in [4], [9], [14], [15] were (a) balancing the area (number) of blocks on either side of each partition, and (b) the number of cut nets due to this partition;

978-1-4799-8720-7/15 $31.00 © 2015 IEEE   107

while an extension was proposed in [11] to address via minimization problem by minimizing the number of bends in a monotone staircase. A suitable trade-off between these objectives is highly required in order to obtain an optimal monotone staircase. However, those existing bipartitioners are either maxflow [2] based [14], [15], [22] or directed search [9], [11] based greedy heuristics, and therefore do not necessarily guarantee an optimal solution. The net cut reflects the number of nets that should pass through the corresponding monotone staircase channel during global routing, also known as routing capacity. Maximizing the balance in area in turn implies the number of terminals and hence the routing area along the boundary of each block [15]. Clearly, minimizing the number of cut nets at any given hierarchy would push the nets to the subsequent level, but with an intent to distribute the nets homogeneously across the layout, for better routing completion and lesser congestion. Moreover, minimizing the number of cut nets at the higher level of hierarchy may lead to fewer longer nets passing through the corresponding staircases, leading to less cross talk among the nets.

Therefore, for a given floorplan, we propose a novel monotone staircase based floorplan bipartitioning method by introducing a novel randomized neighbor search technique, with an intent to find a better (optimal) monotone staircase. The main contributions in this work are:

(i) a randomized neighbor search technique is used that generates an wave-front, instead of earlier directed search or maxflow based techniques

(ii) increasing the scope of exploring more number of distinct monotone staircases as compared to earlier methods; identify the best among them as an optimal solution

Please note that this work considers the aforementioned objectives and the trade-off parameters [11] while solving for an optimal solution.

The organization of this paper is as follows: we start with preliminaries on monotone staircase bipartitioning framework based on a flooplan topology graph model in Section II, followed by Section III which discusses the proposed monotone staircase bipartitioning method. The experimental results and their analysis are covered in Section IV, followed by conclusive remarks in Section V.

II. Preliminaries on Monotone Staircase Cuts in a Floorplan

Before discussing the proposed monotone staircase based floorplan bipartitioning framework, we revisit the formulation of an unweighted directed graph $G_b(V_b, E_b)$, namely *block adjacency graph* (BAG) [14], for a set of n blocks $B = \{b_i\}$ and a given floorplan topological information F. This graph yields two different variations of monotone staircases: (a) *monotonically increasing staircase* (MIS) and

(b) *monotonically decreasing staircase* (MDS), as illustrated in Figure 2.

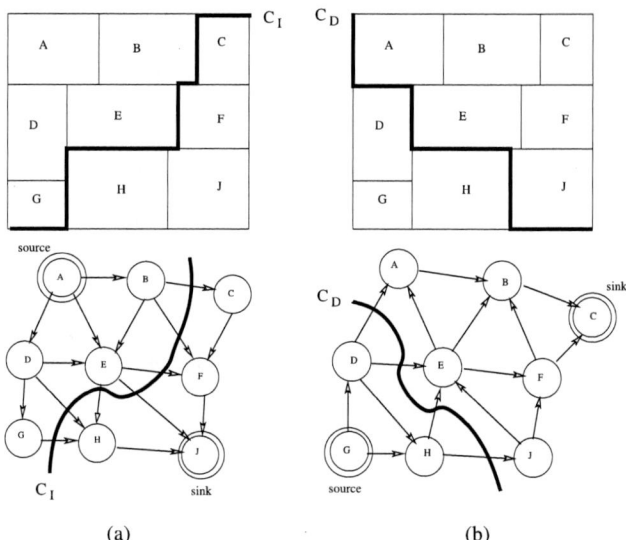

(a)   (b)

Figure 2: A floorplan and the corresponding block adjacency graph (BAG) for a (a) MIS cut, and (b) MDS cut [9]

A monotone staircase channel (MIS or MDS) in the given floorplan is obtained by identifying a cut on G_b (vide Figure 2). However, each cut must obey the following lemma [14], called *monotone staircase property* in order to identify a distinct monotone staircase channel in the floorplan. These channels are then used as the routing resources for global routing [10], [11].

Lemma 1: If $e_{ij} \in E_b$ is an arc in G_b, then there exists at least one monotone staircase in the floorplan such that the blocks b_i and b_j appear in the left and right partitions respectively, and there exists no staircase with b_i in the right partition and b_j in the left partition.

Proof: In [14]. ∎

III. Monotone Staircase Bipartition using a Randomized Neighbor Search Technique

It is known that the number of all possible monotone staircases in a given floorplan topology F for a set of n blocks is exponential in n, and the problem of finding an area balanced monotone staircase cut is NP-Hard [4], [15]. The problem of monotone staircase based balanced bipartitioning addresses a suitable trade-off between the stated objectives [11] as: (a) maximizing area of each bipartition, (b) minimizing the number of bends in the corresponding monotone staircase, and (c) the number of nets being cut by that bipartition. Until now, several greedy heuristic approaches have been proposed including the recent work [11] that emphasized on bend minimization. We reiterate the objectives of this optimization problem [11] as follows:

978-1-4799-8720-7/15 $31.00 © 2015 IEEE

1) balance ratio $\delta = \min(A_l, A_r)/\max(A_l, A_r)$ be maximized
2) the number of cut nets (k_c) be minimized, *and*
3) the number of bends (z) in the monotone staircase be minimized

where $A_{l(r)}$ denotes the area of the left (right) partition and $B_{l(r)}$ denotes the corresponding set of blocks in the left (right) partition. The number balanced version can be seen as a special case when the area of each block is almost equal i.e. having negligible variance such that they can be normalized to unity. In that case, δ is defined as $\min(n_l, n_r)/\max(n_l, n_r)$, where n_l (n_r) denotes the number of blocks in the left (right) partition.

In this work, we reuse the cost function defined in [11] with a pair of trade-off parameters (γ, β) as below:

$$Cost = \gamma.\delta + (1 - \gamma - \beta)(1 - k_c/k) + \beta(1 - z/z_{max}) \quad (1)$$

Here, z_{max} is the maximum number of bends possible in a monotone staircase between the corresponding end points. Maximizing $Cost$ as defined in Equation 1, with a judiciously chosen pair of (γ, β), may lead to an optimal monotone staircase.

Given a design with a set of n blocks B with a set of k nets N and a given floorplan topology F for it, each monotone staircase represents a subset of B. Therefore, the set of all possible monotone staircases S_m is a subset of the power set of B ($S_m \subseteq power(B)$). In other words, each element in $power(B)$ represents a distinct monotone staircase and must obey the *monotone staircase property* (Lemma 1). Now, we will study how (random) selection of a neighbor of a given vertex v_i (in G_b) that may lead to different sequences of monotone staircases. These sequences can be represented by different paths, not necessarily disjoint, from *START* to *STOP* in the *Hasse Diagram* [6] illustrated in Figure 3 (b) for a typical floorplan topology with $n=12$ and $|S_m|=56$. In this diagram, each node represents a distinct monotone staircase and edges represent a possible transition from one monotone staircase to another, either by an earlier greedy method or by the proposed randomized method. It is also evident that the length of such a path (*START* $\rightsquigarrow$ *STOP*) is always $n - 1$ [9]. However, the number of such paths grow exponentially with n and the sequences also differ due to different floorplan topology for the same B.

In Figure 3 (b), we study two different sequences of monotone staircases marked by *bold* and *dotted* lines. If the bold sequence does not contain an optimal monotone staircase, an alternative path may be explored in a hope to come across an optimal one. However, we can not apply brute force method due to exponentially large number of such sequences. Careful study of Figure 3 (b) shows that randomization at a suitable node, say $\{1, 2\}$, by choosing the block 3 randomly instead of 5 may guide to a different sequence containing a potentially optimal solution, say $S_{opt} = \{1, 2, 3, 4, 5, 9, 10\}$. In summary, several such

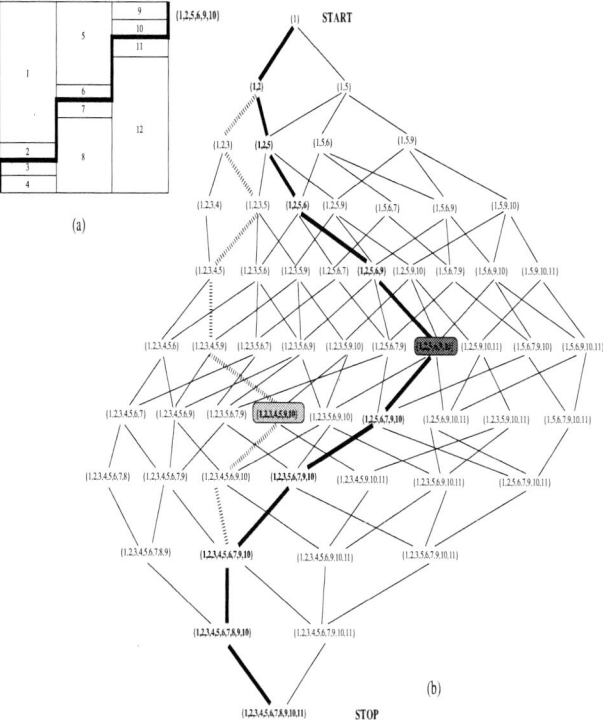

Figure 3: (a) A floorplan having 12 blocks with a near optimal staircase $\{1, 2, 5, 6, 9, 10\}$, and (b) a Hasse Diagram containing exponentially large number of sequences (paths) of monotone staircases: one path (dotted) containing the optimal monotone staircase $\{1, 2, 3, 4, 5, 9, 10\}$ and the other (bold) containing a near optimal monotone staircase $\{1, 2, 5, 6, 9, 10\}$

randomized selections may lead to an optimal solution; or at least gives us a scope of exploring a bigger solution space for identifying a better solution along a specific path (*START* $\rightsquigarrow$ *STOP*). In this work, a number trials are exercised for exploring different sequences. We consider a solution as an optimal (or the best) among the monotone staircases thus obtained, in terms of $Cost$ for given a (γ, β) pair (vide Equation 1). In order to contain the runtime within the same bound as in [11], we can not afford to have large number of trials. Instead, we restrict it to a reasonably small number and use random seeds during each trial.

The main contribution in this work is random indexing of the neighbors of a vertex v_i with out-degree p. Unlike greedy indexing from left to right, as depicted in Figure 4 (a) [9], [11], a neighbor v_j of v_i is indexed with a randomly chosen number $j \in [1, p]$ (vide Figure 4 (b)). It is clear that identifying one or more monotone staircases greedily while exploring the adjacent vertices of v_i takes p time. But, the following lemma shows that the average runtime improves in our case.

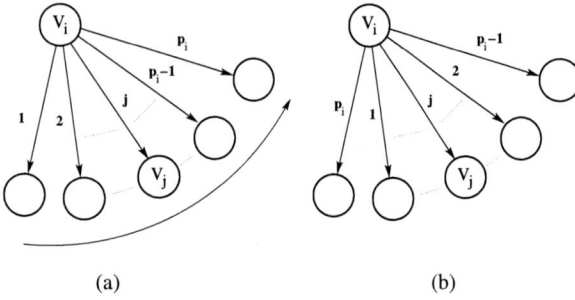

(a)   (b)

Figure 4: Neighbors of v_i (in G_b) with: (a) greedy indexing (left to right), and (b) random indexing

Lemma 2: For a given vertex v_i having out-degree p in G_b, the expected time $E[t_{adj}]$ to search its neighbors to identify one or more distinct monotone staircases is $(p+1)/2$.

Proof: Since all the neighbors of v_j are equally probable to be picked, with a probability of $1/p$, the expected runtime to search a particular neighbor v_j with random indexing $j \in [1, p]$ is:

$$E[t_{adj}] = \sum_{j=1}^{p} (1/p).j = (p+1)/2$$

∎

The best case occurs when all the p edges emanating from v_i obey the monotone staircase property (Lemma 1), resulting in p distinct monotone staircases. On the other hand, the worst case scenario is encountered when the number of edges that obey Lemma 1 is only 1. In general, there can be $[1, p]$ monotone staircases contributed by the neighbors of v_i.

Lemma 3: For a given vertex v_i in G_b with out-degree p, $O(p)$ distinct monotone staircases can be identified while obeying Lemma 1.

Proof: Since, all the edges have probability of $1/2$ while obeying Lemma 1, the expected number of monotone staircase
$$= 1/p(1 + 2 + \dots + (p-1) + p) = (p+1)/2$$
Hence, average $O(p)$ distinct monotone staircases can be identified. ∎

The pseudo-code for the proposed monotone staircase bipartitioning method *GenMSCut-Rand* is presented in Algorithm 1. Please note that this algorithm aims to identify an optimal monotone staircase only for a given hierarchy of the bipartition tree, namely MSC (*monotone staircase cut*) tree [11].

Theorem 1: The proposed randomized monotone staircase bipartitioning method GenMSCut-Rand takes $O(n^2 + nk)$ time.

Proof: Since G_b is a planar graph, the number of edges $|E_b|$ is $O(n)$ and $E_b = \sum_{i=1}^{n} p_i$, where p_i is the out-degree of v_i, it takes $O(n)$ time for searching distinct monotone

---

Algorithm 1 GenMSCut-Rand

Inputs: G_b, N, γ, β, *baltype*
Outputs: A monotone staircase with maximal area balance and minimal net cut and bend count in the resulting monotone staircase, with maximum $Cost$ (vide Eqn 1)

Define a Queue Q of size $2n$; $S = source(G_b)$
$\lambda = \varnothing$; $r = 0$
while ($r < 3$) **do**
  Initialize left partition, $L = \varnothing$, $Wt(L) = 0$
  $level\_no = 0$, Enqueue(S), S.level = $level\_no$
  Enqueue($\varnothing$)
  while (NOT_EMPTY(Q)) **do**
    Let v_i be the dequeued vertex and a vertex front $V_{list} = \varnothing$
    if ($v_i \neq \varnothing$) **then**
      $V_{list} \leftarrow \{v_i\}$
      /* build V_{list} */
    else
      while at least one cut edge in E_b exists for V_{list} **do**
        Generate a random seed based on trial number r
        Randomly choose a cut edge (v_i, v_j), such that $v_i \in V_{list}$ and $v_j \notin V_{list}$
        if ($IsValid\_Monotone\_Staircase$ = TRUE) **then**
          /* vide Lemma 1 */
          $L = L \cup \{v_i\}$; v_i.level = $level\_no$
          $Wt(L) \leftarrow Wt(L) + Wt(v_i)$; Enqueue($v_j$)
          findNetPartition(B_l, B_r, N)
          {it returns $N_l(N_r)$, N_c, $Z(Z_{max})$: set of left(right) nets, cut nets and bends (maximum possible bends) respectively}
          Calculate $Cost$ using Eqn. 1 for B_l, B_r, $k_c = |N_c|$, $k = |N|$, $z = |Z|$ and $z_{max} = |Z_{max}|$ and given (γ, β)
          $C_p = \langle Cost, B_l, B_r, N_l, N_r, N_c, Z, Z_{max} \rangle$
          $\lambda \leftarrow \lambda \bigcup \{C_p\}$
        end if
      end while
      Increment $level\_no$
      Enqueue($\varnothing$)
    end if
  end while
  Increment r
end while
Return optimal monotone staircase as the one with $C_{max} \in \lambda$ with maximum $Cost$ value.

---

staircases. We use $r=3$ trials intended to obtain 3 different sequences of monotone staircases. Also the net partitioning procedure takes $O(k)$ [9], while finding the number of bends account for $O(n)$ time [11]. Thus, the overall time taken by GenMSCut-Rand is $O(n(n + k))$, i.e., $O(n^2 + nk)$. ∎

In Theorem 1, it is shown that Algorithm 1 has the same time complexity as [11], but only a constant times higher due to multiple trials. We use the same recursive bipartitioning framework of [11] in order to identify a set of optimal monotone staircases for the entire floorplan, using MIS (MDS) cuts at the alternative levels of the hierarchy of MSC tree [9], [11] with $O(\log n)$ height. Therefore, it takes $O((n^2 + nk) \log n)$ time to generate the entire hierarchy of monotone staircases for a given floorplan.

IV. Experimental Results

Our algorithm was validated on MCNC and GSRC Hard floorplanning benchmark circuits presented in Table I. The floorplans for each of the circuits were generated using *Parquet* tool [17] with random seeds. The proposed algorithm was implemented in C and run on a Linux platform (2.8GHz, 4GB RAM).

We compare the performance of the proposed work and

Table I: MCNC and GSRC (Hard) Benchmark Circuits.

| Suite | Circuit | #Blocks | #Nets | | Avg. Net |
			original	modified	degree
MCNC	apte	9	97	44	3.500
	hp	11	83	44	3.545
	xerox	10	203	183	2.508
	ami33	33	123	84	4.154
	ami49	49	408	377	2.337
GSRC	n10	10	118	54	2.129
	n30	30	349	147	2.102
	n50	50	485	320	2.112
	n100	100	885	576	2.135
	n200	200	1585	1274	2.138
	n300	300	1893	1632	2.161

that presented in [11] for $\gamma \in [0.1, 0.7]$ and $\beta \in [0.0, 0.3]$, both in steps of 0.1, such that $\gamma + \beta <= 1$. To study different scenarios, we consider 4 different floorplan instances for each circuit, generated by Parquet Floorplanning tool [1], [17] with random seeds. The corresponding bipartitioning results are presented in Figure 5. The results show that our work outperforms [11] in almost all the objectives whose mean values are computed for all 4 instances of a circuit and the specified combinations of (γ, β) pair. This is due to the fact that our proposed randomization technique is effective with large number of blocks and may potentially yield a sequence of monotone staircases that presumably contains an optimal solution.

Our results of bipartitioning shows that the height of the corresponding MSC tree is contained within the tight bounds of $\log n$ and $2 \log n$ and hence is $O(\log n)$. We also observed that the runtime for bipartitioning in our case is no more than twice of that obtained for [11]. This is due to the fact that our runtime complexity as stated earlier is only a constant times higher than that of [11].

In addition to the values of the objectives and $Cost$ (vide Equation 1) as the measures for identifying an optimal monotone staircase, we conduct a comparative study on the impact of this work on the global routing (GR) metrics against that by [11]. For our purpose, we employ the only known early global router STAIRoute [10] that is known to utilize monotone staircases as the routing resources. In this experiment, STAIRoute is configured to run for 8 metal layers using reserved layer model for assigning the nets at different layers. Routability for each circuit has remained 100% for any specified (γ, β). Due to limited space, we compare the global routing metrics for $n300$ only, as depicted in Figures 6 for $\gamma \in [0.1, 0.7]$ and $\beta = 0.1$. The most important observation is that this work has improved via count, and average congestion (wACE4) [23] for most of the cases, while net length is slightly higher as compared to [11]. It is also to be noted that the worst congestion (wACE4) in some of the routing resources is much less than 100%.

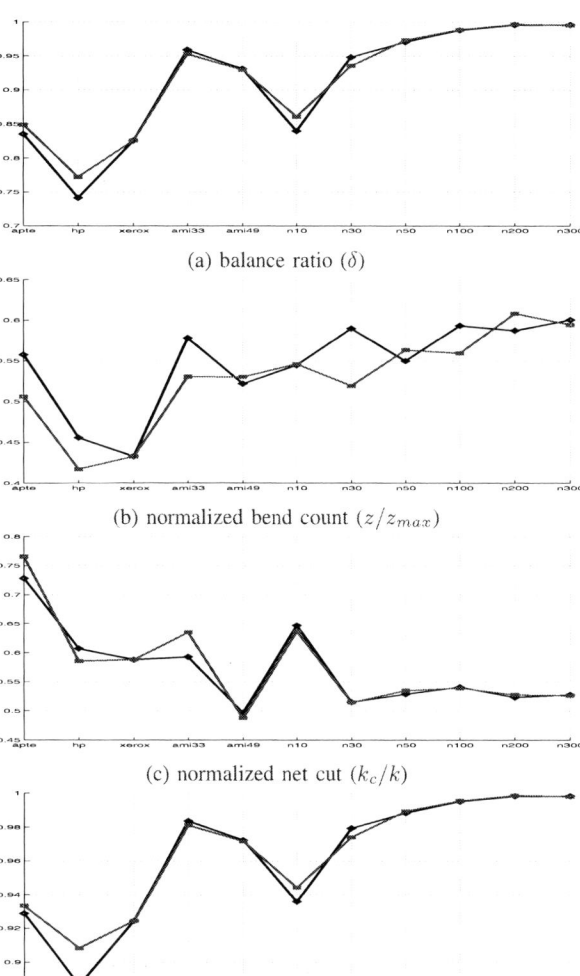

(a) balance ratio (δ)

(b) normalized bend count (z/z_{max})

(c) normalized net cut (k_c/k)

(d) $Cost$

Figure 5: Average (over all (γ, β) pairs and all 4 instances of a circuit) obtained by this work (■) vs that by [11] (♦)

V. CONCLUSION

In this paper, we propose a new method for identifying monotone staircases in floorplans, using randomized neighbor search technique. Our intention is to increase the search space containing more number of distinct monotone staircases than the earlier methods; so that an optimal monotone staircase can be found with better quality in terms of improved trade-off between the objectives: (a) area balance, (b) bend count, and (c) net cut. The impact of the trade-off parameters (γ, β) were studied on the said objectives and the results showed improvement over the latest monotone staircase bipartitioning method, but with increased overhead in runtime. Further experiments were conducted to study the impact on the global routing metrics such as routability, routed net length, via count and routing

978-1-4799-8720-7/15 $31.00 © 2015 IEEE

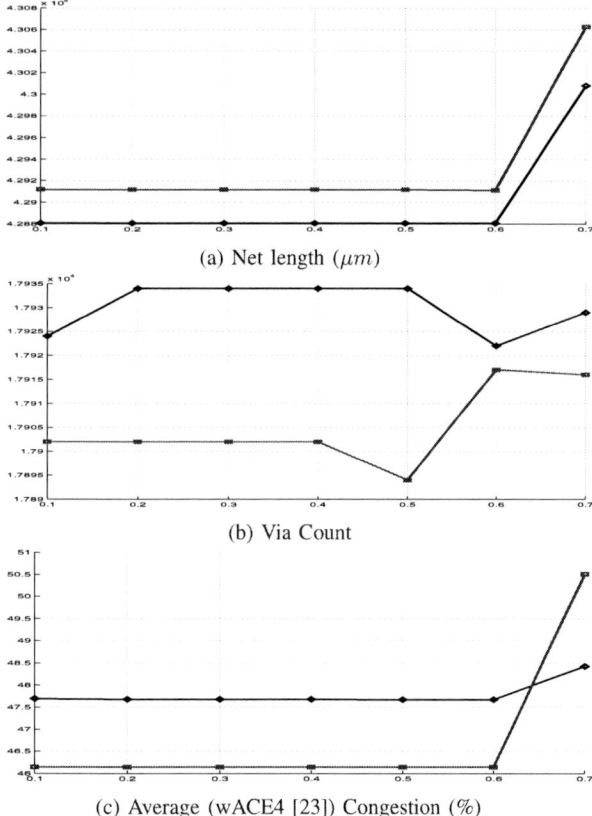

(a) Net length (μm)

(b) Via Count

(c) Average (wACE4 [23]) Congestion (%)

Figure 6: Comparing global routing metrics for $n300$ (for $\beta = 0.1$) vs γ: between this work (■), and [11] (♦)

congestion for a set of (γ, β) pairs. The results depicted improvement in via count and congestion, with a very small degradation in net length. Moreover, minimizing the number of cut nets at the higher level of hierarchy of monotone staircases may lead to fewer longer nets passing through the corresponding staircases, thereby leading to lesser cross talk among the nets. Therefore, it pertains to the fact that a judiciously chosen (γ, β) pair may achieve an optimal floorplan bipartitioning solution using monotone staircases.

ACKNOWLEDGMENT

The authors would like to thank the reviewers for their valuable feedback.

REFERENCES

[1] S. N. Adya and I. L. Markov, "Fixed-outline Floorplanning: Enabling Hierarchical Design", *IEEE Transactions on VLSI Systems, Vol. 11, No. 6, pp. 1120-1135, December 2003.*

[2] T. H. Cormen, C. E. Leiserson, R. L. Rivest and C. Stein, "Introduction to Algorithms", *3rd Edition, MIT Press, 2009.*

[3] C.-C. Chang et. al., "Multilevel Global Placement With Congestion Control',' *IEEE Transactions on Computer Aided Design of Integrated Circuits and Systems, Vol. 22, No. 4, pp. 395-409, April 2003.*

[4] P. Dasgupta, P. Pan, S.C. Nandy and B.B. Bhattacharya, "Monotone Bipartitioning Problem in a Planar Point Set with Applications to VLSI", *ACM Transactions on Design Automation of Electronic Systems, Vol. 7, No. 2, pp. 231-248, 2002.*

[5] M. Guruswamy and D. F. Wong, "Channel Routing Order for Building-Block Layout with Rectilinear Modules", *Proc. of IEEE International Conference on Computer-Aided Design (ICCAD), pp. 184-187, 1988.*

[6] *http://en.wikipedia.org/wiki/Hasse_diagram*

[7] X. He et. al., "Ripple 2.0: High Quality Routability-Driven Placement via Global Router Integration", *50th Design Automation Conference (DAC), 2013.*

[8] J. Hu, M.-C. Kim, and I.L. Markov, "Taming the Complexity of Coordinated Place and Route" *50th Design Automation Conference (DAC), 2013.*

[9] B. Kar, S. Sur-Kolay, S. H. Rangarajan and C. Mandal, "A Faster Hierarchical Balanced Bipartitioner for VLSI Floorplans using Monotone Staircase Cuts", *Proceedings of VDAT 2012, Lecture Notes in Computer Science (LNCS), Vol. 7373, pp. 327-336, 2012.*

[10] B. Kar, S. Sur-Kolay and C. Mandal, "STAIRoute: Global Routing using Monotone Staircase Channels", *Proceedings of International Symposium on VLSI (ISVLSI), pp. 90-95, 2013.*

[11] B. Kar, S. Sur-Kolay and C. Mandal, "Global Routing using Monotone Staircases with Minimal Bends", *Proceedings of International Conference on VLSI Design (VLSID), pp. 369-374, 2014.*

[12] W.-H. Liu, C.-K. Koh, and Y.-L Li, "Optimization of Placement Solutions for Routability", *50th Design Automation Conference (DAC), 2013.*

[13] J. Lu and C.-W. Sham, "LMgr: A Low-Memory Global Router with Dynamic Topology Update and Bending-Aware Optimum Path Search", *Proceedings of International Symposium on Quality Electronic Design (ISQED), pp. 231-238, 2013.*

[14] S. Majumder, S.C. Nandy and B.B. Bhattacharya, "On Finding a Staircase Channel with Minimum Crossing Nets in a VLSI Floorplan", *Journal of Circuits, Systems and Computers, Vol. 13, No. 5, pp.1019-1038, 2004.*

[15] S. Majumder, S. Sur-Kolay, B.B. Bhattacharya and S. Das, "Hierarchical Partitioning of VLSI Floorplans by Staircases", *ACM Transactions on Design Automation of Electronic Systems, Vol. 12, No. 1, Article 7, 2007.*

[16] M. Pan and C. Chu, "FastRoute: A Step to Integrate Global Routing into Placement", *Proceedings of IEEE International Conference on Computer Aided Design (ICCAD), November 2006.*

[17] "Parquet FloorPlanner, Rev-4.5", *http://vlsicad.eecs.umich.edu/BK/parquet,* University of Michigan, 2006.

[18] N. Sherwani, "Algorithms for VLSI Physical Design Automation", *Kluwer Academic Publishers, 1993.*

[19] S. Sur-Kolay and B. Bhattacharya, "The cycle structure of channel graphs in non-sliceable floorplans and a unified algorithm for feasible routing order," *Proceedings. of IEEE International Conference on Computer Design (ICCD), pp. 524-529, 1991.*

[20] N. Viswanathan and C. Chu, "FastPlace: Efficient Analytical Placement using Cell Shifting, Iterative Local Refinement and a Hybrid Net Model", *IEEE Transactions on Computer-Aided Design, Vol. 24, No. 5, pp. 722-733, 2005.*

[21] Y. Xu, Y. Zhang and C. Chu, FastRoute 4.0, "FastRoute 4.0: Global Router with Efficient Via Minimization", *Proceedings of Asia and South Pacific Design Automation Conference, pp. 576 - 581, 2009.*

[22] H. H. Yang and D. F. Wong, "Efficient network flow based min-cut balanced partitioning", *IEEE Transactions on Computer Aided Design and Integrated Circuits and Systems, vol. 15, no. 12, pp. 1533-1540, December 1996.*

[23] Y. Wei et. al, "GLARE: Global and local wiring aware routability evaluation", *49th Design Automation Conference (DAC), pp. 768-773, 2012.*

Logic Debugging of Arithmetic Circuits

Samaneh Ghandali, Cunxi Yu, Duo Liu, Walter Brown, Maciej Ciesielski
University of Massachusetts, Amherst, USA
{samaneh, ycunxi, duo, webrown, ciesiel}@umass.edu

Abstract— This paper presents a novel diagnosis and logic debugging method for gate-level arithmetic circuits. It detects logic bugs in a synthesized circuit caused by using a wrong gate ("gate replacement" error), which change the functionality of the circuit. The method is based on modeling the circuit in an algebraic domain and computing its algebraic "signature". The location and type of the bug is determined by comparing signatures computed in both directions, using forward (PI to PO) and backward (PO to PI) rewriting. It will also perform automatic correction for the detected bugs. The approach is demonstrated and tested on a set of integer combinational arithmetic circuits.

Keywords— *Formal verification; Logic debugging; Arithmetic circuits.*

I. INTRODUCTION

As today's VLSI designs grow in complexity and size, design errors become more frequent and difficult to track [1]. The process of verifying the functional correctness of a design, determining the source of potential errors and correcting those errors, can take up to 70% of the overall design time [2]. Recent developments have automated most of the verification tasks, but debugging, i.e., error localization and correction, still remains a resource intensive, manually conducted process [4]. Efficient automated debugging techniques are necessary to complement and enhance the verification techniques.

Traditional automated debugging solutions for hardware designs are based on simulation, critical path tracing [5], BDDs and *BMDs [6]. Recent automated debugging methods tend to rely on SAT solvers. In [7] error detection is facilitated by adding corrector models to the circuit implementation and mapping it into a Boolean formula in CNF. By solving the resulting SAT problem, a set of suspect error locations are obtained. This approach, however, is restricted by the performance and capacity of the available SAT solvers. Other techniques, such as those based on Quantified Boolean Formula (QBF) [8], abstraction and refinement in error localization [9], and maximum satisfiability [10], [11], are used to improve SAT-based debugging method. However, the performance of these methods and their capability to handle large designs remain limited by SAT. In order to reduce the number of SAT solver calls, the concept of reverse dominators was introduced in [12] to allow for early pruning of non-solution areas of the problem search space. FPGA-based debugging methodology, proposed in [13], [14], [15], locally modifies the circuit structure.

The diagnosis problem for sequential circuits is typically structured as bounded model checking (BMC) [16], [17] and formulated as a SAT problem. In [17], resynthesis method guided by counterexamples performs gate-level circuit repair, based on error traces composed of input vectors and output responses. The method described in [18],[19] introduces an abstraction and refinement algorithm for design debugging built upon a time-windowing framework to manage excessive error trace lengths. Non-modeled portions of the trace are approximated using a path directed abstraction that represents structural circuit paths. Due to the inherent iterative nature of the algorithm, performance remains the crucial issue in this work.

A debugging technique applicable to divider circuits is proposed in [20]. It is based on a "reverse-engineering" mechanism of extracting a high level arithmetic model, called Functional Bit Level Adder (FBLA), which may be difficult to obtain in synthesized circuits. Furthermore, these methods can only reason about the correctness of the quotient part of the result using iterative subtraction model, but not about the entire divider circuit. In [21] verification of RTL code to optimize assertion coverage is proposed. Their algorithm can be used to isolate only those statements that are covered by an assertion as the most likely location of the bug. However, the problem of generating assertions remains open.

In this paper, we introduce a novel diagnosis and logic debugging method for gate-level arithmetic circuits. The proposed method is part of the functional verification approach proposed in [3]. It detects the logical bugs, caused by using a wrong gate ("gate replacement" bug) or inversion of an internal signal, that change the functionality of the circuit. It will also perform automatic correction for the detected bugs. The approach assumes a "single-gate" replacement error, caused by using a wrong gate, but it can correct multiple independent bugs. It consists of three phases: 1) the circuit is scanned forward from the primary inputs (PI) to primary outputs (PO) and an algebraic expression (signature) is derived for each cut (a set of signals that separate PI from PO); 2) the circuit is scanned backward from PO to PI and an algebraic signature is generated for each cut; 3) the difference between the two expressions, Δ_i, at each cut is then computed. A non-zero Δ_i for a given cut indicates inconsistency between the two expressions, showing that there is a bug located at this cut. The value of Δ_i is then analyzed to determine the source of the bug and to correct it. Under certain conditions several bugs in the same cut can be corrected simultaneously.

The rest of this paper is organized as follows. Section II describes preliminaries. Section III explains in detail the proposed diagnosis and debugging method. Section IV presents

experimental results and Section V provides summary and conclusions.

II. PRELIMINARIES

We follow the arithmetic verification approach proposed in [3], with the circuit modeled as a network of basic logic gates (AND, OR, XOR, INV, etc.). Each gate is represented as a pseudo-Boolean polynomial $poly[X]$, with Boolean variables $X = \{x_1, ..., x_n\}$ and integer coefficients from Z_2^n. The following equations summarize algebraic representation of basic Boolean operators:

$$\neg a = 1 - a$$
$$a \wedge b = a\,b$$
$$a \vee b = a + b - a\,b \qquad (1)$$
$$a \oplus b = a + b - 2\,a\,b$$

Definition 1 *(Input Signature)*: The input signature, Sig_{in}, is a polynomial in primary input variables that uniquely represents an integer function computed by the circuit, i.e., its specification. For example, an n-bit binary adder with inputs $\{a_0,...,a_{n-1},b_0,...,b_{n-1}\}$, is described by $Sig_{in} = \sum_{i=0}^{n-1} 2^i a_i + \sum_{i=0}^{n-1} 2^i b_i$. The input signature of a 2-bit signed multiplier is $Sig_{in} = (-2a_1+a_0)(-2b_1+b_0) = 4a_1b_1 - 2a_0b_1 - 2a_1b_0 + a_0b_0$, etc. The integer *coefficients* (weights) associated with the circuit signals are uniquely determined by the intended circuit function (specification). For example, in an adder, the coefficients of the primary inputs at bit position i are $c(a_i) = c(b_i) = 2^i$.

Definition 2 *(Output Signature)*: the output signature, Sig_{out}, of the circuit is defined as a polynomial in the primary output signals. Such a polynomial is uniquely determined by an n-bit encoding of the output provided by the designer. For example, the output signature of the 2-bit signed multiplier is $-8z_3+4z_2+2z_1+z_0$. In general, an output signature of an unsigned arithmetic circuit with n output bits is represented as a linear polynomial, $Sig_{out} = \sum_{i=0}^{n-1} 2^i z_i$. The coefficients of the primary outputs are also unique, defined by the known output encoding.

Definition 3 *(Cut Signature)*: The *cut* is a set of signals that separates PI from PO. The *signature* of a cut is a polynomial expression in signal variables of the cut that represents the integer number computed by the circuit.

The selection of the cuts is an important issue in this approach, as it affects the efficiency of finding the bugs. In the worst case, two cuts may only differ by a single gate. For illustration purpose we assume that the cuts are determined by the topological ordering of signals w.r.t. PI, but other choices exist (determining the best set is part of the future work).

Example 1: Figure 1 shows a two-bit adder, with $Sig_{in} = 2a_1+2b_1+a_0+b_0$, $Sig_{out} = 4r_2+2r_1+r_0$, and "topological" cuts, labeled $f_0,...,f_3$. The meaning of Δ_i in the figure will be explained in Section III.C. This circuit will be used as a running example in the paper.

As shown in [3], in a bug-free arithmetic circuit, the expressions for any two cuts, although expressed by different polynomials, always evaluate to the same value, i.e., $f(cut_i) = f(cut_j)$, for any $\{i, j\}$. This fundamental property of the arithmetic circuit serves as basis of the proposed diagnostics and debugging approach.

III. BUG IDENTIFICATION

Our debugging method consists of: computing cut signatures by forward rewriting, backward rewriting, and comparing the pairs of signatures for each cut to identify and fix the bugs.

A. Forward (PI-PO) rewriting

The forward (PI-PO) rewriting starts by dividing the initial polynomial, Sig_{in}, by the polynomials describing the logic gates connected to the PI signals. The goal is to replace the input variables associated with the PI gates with an expression involving the corresponding gate outputs. This produces an expression in the new set of variables, moving away from PI. While in principle this can be done one gate at a time, one can eliminate several gates at once to speed up the process. To do this division efficiently, knowledge of the signal *coefficients* (weights) is needed. We explain how to calculate the required coefficients of the newly introduced variables using the structures shown in Figure 2.

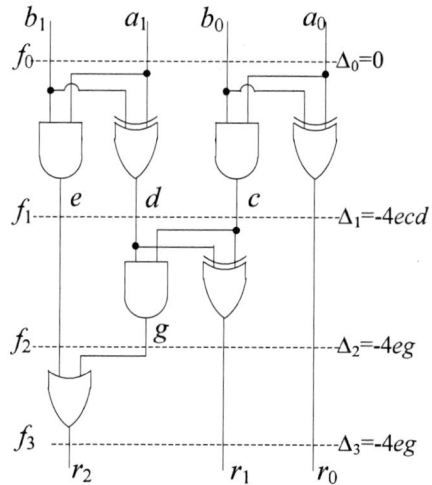

Fig. 1. Two-bit adder circuit with topological cuts.

Figure 2(a) shows a half-adder (HA) circuit, consisting of a pair (XOR, AND) with common variables. Using the notation in the figure and the algebraic representation of the logic gates given in Eq.(1), the computation of the output coefficients of the half-adder circuit with inputs (a,b) and outputs (m,n) is performed as follows: $Sig_{in}(HA) = c_1a + c_1b$

$Sig_{out}(HA) = c_2m + c_3n = c_2(a+b-2ab) + c_3ab$

Since $Sig_{in}(HA) = Sig_{out}(HA)$, we have:

$c_1a + c_1b - c_2(a+b-2ab) - c_3ab = 0$.

By regrouping the variables as follows

$a(c_1 - c_2) + b(c_1 - c_2) + ab(2c_2 - c_3) = 0$

and solving the above equation for c_2 and c_3, we obtain:

$c_2 = c_1$ and $c_3 = 2c_1$.

The second structure shown in Fig. 2(b) consists of an XOR gate and an OR gate. Using similar approach, we obtain: $c_2 = -$

c_1 and $c_3 = 2c_1$. Similarly, the structure in Fig. 2(c), consisting of an OR and an AND gate, produces the coefficients: $c_2 = c_1$ and $c_3 = c_1$.

Coefficients of the individual gates can be derived similarly. It can be shown that the inputs to an OR or XOR gate must have the same coefficients, $c_1 = c_2$, otherwise the algebraic equation for this gate will not be satisfied; the coefficient c_3 of the output is equal to those of the input. In contrast, the input coefficients c_1, c_2 of an AND gate can be different, and the output coefficient $c_3 = c_1 \cdot c_2$.

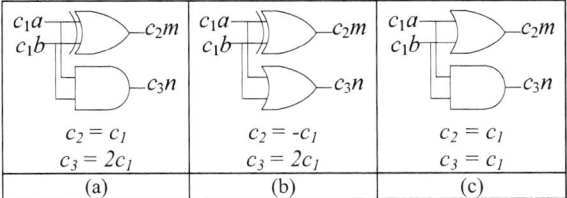

Fig. 2. Calculation of signal coefficients.

Computing cut signatures: The pseudo code of the algorithm for forward rewriting equation for each cut is shown in **Algorithm 1**. The input to this algorithm is the circuit with its input signature, and its output is a set of cut equations. Equation of the first cut is the same as the specification, i.e., $f(cut_0) = Sig_{in}$. Obtaining cut_i from cut_{i-1} works as follows. Let x_j be an output signal of gate g_j in cut_i and $poly(g_j)$ be the polynomial expression describing its logic function (*c.f.* Eq.(1)), so that $x_j = poly(g_j)$. For each gate g_j in cut_i we add variable x_j to the current cut and subtract the algebraic expression $poly(g_j)$ representing this gate, without changing the arithmetic function of the cut. Equation (2) shows the computation of $f(cut_i)$ from $f(cut_{i-1})$.

$$f\left(cut_i\right) = f\left(cut_{i-1}\right) + \sum_{j=1}^{gates \in cut_i} c_j \left(x_j - poly\left(g_j\right)\right) \quad (2)$$

Here, i is the index of the cut, and c_j is the coefficient of gate j of cut_i. Lines 3-7 of Algorithm 1 describe the computation of each cut. For simplicity, we write f_i instead of $f(cut_i)$. Each f_i is initialized with f_{i-1}; using the structure in Fig. 2, for each gate j of cut_i, the gate coefficient is calculated so as to eliminate the gate inputs (line 6); finally, equation of cut_i is computed (line 7), based on Eq. (2).

Algorithm Forward (PI-PO) rewriting (Sig_{in}, $Circuit$)
1 Compute all cuts of the Circuit;
2 $f_0 = Sig_{in}$;
3 **for** $i = 1$ to # of cuts
4    $f_i = f_{i-1}$;
5   **for** $j = 1$ to # of gates of cut_i
6   c_j = compute coefficient of g_j;
7       $f_i = f_i + c_j(x_j - poly(g_j))$;

Algorithm 1. Forward (PI-PO) rewriting

Example 2: Consider again the circuit in Fig. 1. By applying the forward (PI-PO) rewriting algorithm, we obtain the following equations for each cut:

$$f_0 = 2b_1 + 2a_1 + b_0 + a_0$$

$$\begin{aligned} f_1 &= 2b_1 + 2a_1 + b_0 + a_0 + 4(e - a_1 b_1) + 2(d - (a_1 + b_1 \\ &\quad - 2a_1 b)) + 2(c - a_0 b_0) + (r_0 - (a_0 + b_0 - 2a_0 b_0)) \\ &= 4e + 2d + 2c + r_0 \end{aligned}$$

$$\begin{aligned} f_2 &= 4e + 2d + 2c + r_0 + 4(g - dc) + 2(r_1 - (d + c - 2dc)) \\ &= 4e + 4g + 2r_1 + r_0 \end{aligned}$$

$$\begin{aligned} f_3 &= 4e + 4g + 2r_1 + r_0 + 4(r_2 - (e + g - eg)) \\ &= 4r_2 + 2r_1 + r_0 + 4eg \end{aligned}$$

Note that such computed f_3 is different than the expected output signature, $4r_2 + 2r_1 + r_0$. Specifically, it contains the term $4eg$, associated with variables of cut_2. We call such a term a *Residual Expression (RE)*. In a correct circuit, RE should be zero. This can be proved by a straightforward rewriting of $4eg$ up to the PI variables:

$$4eg = 4(a_1 b_1)(dc) = 4(a_1 b_1)(a_1 + b_1 - 2a_1 b_1)(a_0 b_0) = 0$$

Then, $f_3 = 4r_2 + 2r_1 + r_0$, indicating that the circuit is correct. The reason for the existence of a residual expression in a correct circuit is that the polynomial division used by forward rewriting does not take into account the Boolean nature of the circuit signals. To avoid RE one would need to divide the polynomials by a set of polynomials $\langle x^2 - x \rangle$, called *ideals*, for each signal x in the circuit, to guarantee that $x=0,1$. This method, often employed by symbolic algebra approach, is too costly and inefficient for this work.

B. Backward (PO-PI) rewriting

The backward (PO-PI) rewriting is conceptually simpler, basically a reversed symbolic simulation. Starting at the PO with Sig_{out}, it creates a new cut signature by replacing an output signal x_j of gate g_j with its corresponding algebraic expression: $x_j \rightarrow poly(g_j)$. Here the signal coefficients are known, provided by the binary encoding of the PO signals.

Example 3: The cut expressions for the two-bit adder in Fig. 1 computed in PO-PI fashion are as follows:

$$f_3 = 4r_2 + 2r_1 + r_0$$

$$f_2 = 4(g + e - eg) + 2r_1 + r_0 = 4g + 4e - 4eg + 2r_1 + r_0$$

$$\begin{aligned} f_1 &= 4e + 4(cd) - 4e(cd) + 2(c + d - 2cd) + r_0 \\ &= 4e + 2d + 2c + r_0 - 4ecd \end{aligned}$$

$$\begin{aligned} f_0 &= 4e + 2d + 2c + r_0 - 4ecd \\ &= 4(a_1 b_1) + 2(a_1 + b_1 - 2a_1 b_1) + 2(a_0 b_0) + (a_0 + b_0 \\ &\quad - 2a_0 b_0) - 4(a_1 b_1)(a_0 b_0)(a_1 + b_1 - 2a_1 b_1) \\ &= 2a_1 + 2b_1 + a_0 + b_0 \end{aligned}$$

The computed signature at the PI matches the expected specification, Sig_{in}, so the circuit is correct. Note that the backward rewriting will never produce a residual expression. This is because the algebraic model (1) of Boolean gates correctly represents the binary value of the gate signal. This convenience comes at a cost of a potentially exponential explosion of the signature size during backward rewriting.

C. Computing the signature difference (Δ_i)

At this point, a pair of expressions is generated for each cut of the circuit: one computed by the forward and the other by

978-1-4799-8720-7/15 $31.00 © 2015 IEEE

the backward rewriting. The difference Δ_i between the two expressions is defined as follows:

$$\Delta_i = f_{i,BACK} - f_{i,FOR}; \quad 0 \leq i \leq cuts \qquad (3)$$

Here, i is the index of cut_i, $f_{i,BACK}$ is the signature of cut_i in the PO-PI direction, and $f_{i,FOR}$ is the signature of cut_i computed by the PI-PO rewriting. If the circuit contains no bug, the value of Δ_i at each cut is equal to zero; otherwise, the circuit contains a bug. The expression of Δ_i will be used to identify and to correct the bug.

Example 4: Consider the adder circuit in Examples 2 and 3 again. The values of parameter Δ_i for each cut of the circuit are shown in Fig. 1. As can be seen, Δ_3, Δ_2 and Δ_1 are non-zero polynomials. However, as explained in Example 3, *4eg* and *4ecd* are zero functions (expressions that evaluate to zero), so Δ_3, Δ_2 and Δ_1 also reduce to zero.

D. The Debugging Algorithm

In this phase, the circuit is analyzed and verified against the given specification to either confirm its correctness or to find and locate the bug. In principle, the circuit is correct (satisfies its specification) if the signature obtained by backward rewriting matches the given input signature (specification). Alternatively, the signature obtained by the forward rewriting should match the given output signature, provided that the residual expression *RE* generated during this rewriting is proven to be zero (as explained earlier, this can be done by a local backward rewriting of the RE expression up to PI). In this paper we consider a particular type of a bug, namely *gate replacement*, i.e., using a wrong gate in the circuit. In practice, in the presence of a bug, the size of the computed signature may become prohibitively large, and the goal is to locate the cut at which the bug (faulty gate) resides. If the bug is located at some cut_i, then the value of Δ_i for this cut will be nonzero. This is illustrated by the following example.

Example 5: Let us intentionally insert a bug into the two-bit adder circuit in Fig. 1, by replacing the AND gate with inputs (c,d) with an OR gate. The resulting buggy circuit is shown in Fig. 3. The cut equations for this circuit are as follows.

Forward (PI-PO) rewriting:

$$f_0 = 2b_1 + 2a_1 + b_0 + a_0$$
$$f_1 = 4e + 2d + 2c + r_0$$
$$f_2 = 4e + 4g - 2r_1 + r_0$$
$$f_3 = 4r_2 - 2r_1 + r_0 + 4eg$$

Backward (PO-PI) rewriting:

$$f_3 = 4r_2 + 2r_1 + r_0$$
$$f_2 = 4e + 4g - 4eg + 2r_1 + r_0$$
$$f_1 = 4e + 6d + 6c - 8cd - 4ec - 4ed + 4ecd + r_0$$
$$f_0 = 6a_1 + 6b_1 - 8a_1b_1 + a_0 + b_0 + 4a_0b_0 - 8a_0b_0a_{11}$$
$$\quad - 8a_0b_0b + 12a_0b_0a_1b_1$$

The values of parameter Δ_i for each cut of the buggy circuit are calculated using Eq. (2) and shown in Fig. 3. The type and the location of the bug can be obtained by assessing the value of Δ_i for each cut. Assume initially that each cut has only a

single bug (the constraint to be removed later). **Table I** shows the difference in the signatures between the cut with the correct gate and the cut with the wrong gate. As an example, consider the entry $(a+b-2ab)$ in the 1st row (AND) and 2nd column (OR) of the table. It reflects the difference between the correct AND gate (ab) and the wrong OR gate $(a+b-ab)$. That is, if in a given cut, an AND gate is replaced with an OR gate, then $\Delta_i = (a+b-ab)-ab =a+b-2ab$, where a, b are the gate inputs. Conversely, if for some cut computed by the algorithm, $\Delta_i =a+b-2ab$, this means that it contains an OR gate while it should contain an AND.

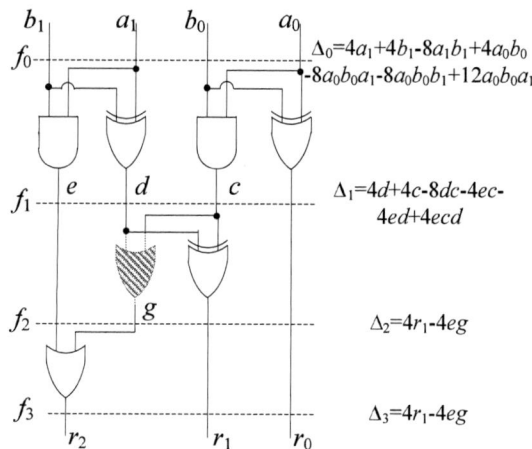

$$\Delta_0=4a_1+4b_1-8a_1b_1+4a_0b_0$$
$$-8a_0b_0a_1-8a_0b_0b_1+12a_0b_0a_1$$

$$\Delta_1=4d+4c-8dc-4ec-4ed+4ecd$$

$$\Delta_2=4r_1-4eg$$

$$\Delta_3=4r_1-4eg$$

Fig. 3. Two-bit adder with a bug: OR instead of AND

The remaining entries in the table give expressions for different bugs for a single-gate replacement. To detect the location of the bug in a given cut we need to check if Δ_i contains any of the expressions in Table I (replaced by the appropriate variable names). If this is the case, the location of the bug is detected and can be corrected as specified in the table. Otherwise, either the bug originates at a different cut, or it has a different nature, not considered in this model.

TABLE I.   EXPRESSIONS CAUSED BY GATE REPLACEMENT ERROR

Correct \ Buggy	AND	OR	XOR
AND		$a+b-2ab$	$a+b-3ab$
OR	$-a-b+2ab$		$-ab$
XOR	$-a-b+3ab$	ab	

Table I can be readily extended to other types of logic gates, such as the complex And-Or-Invert gates used in standard cell implementations.

The pseudo code of the Debugging Algorithm is shown in **Algorithm 2**. The equation of each cut is first computed in a PI-PO direction (line 1) and then by PO-PI rewriting (line 2). Then, Δ_i is calculated for each cut. If $\Delta_0 =0$, the circuit is correct, otherwise Δ_i of other cuts are assessed and their expressions are checked against those in Table I (lines 3-8). The detected bugs at each cut are stored in the set *Bug-list,* which consists of the potential bug locations.

978-1-4799-8720-7/15 $31.00 © 2015 IEEE   116

The expression of each Δ_i is then modified by adding to it expression $(f_{i,FOR} - f_{i+1,FOR})$, where $f_{i,FOR}$ and $f_{i+1,FOR}$ are the expressions obtained by *forward* rewriting of cut_i and cut_{i+1}, respectively (lines 9-11). This is done to account for the residual expression generated during the forward rewriting between cut_i and cut_{i+1}, since during the forward rewriting, the residual expression that is based on the variables of cut_i actually appears in cut_{i+1}. Note that only those terms of expression $(f_{i,FOR} - f_{i+1,FOR})$ that belong to cut_i are added to Δ_i. The new Δ_i is examined again to check for the buggy expressions (lines 12-13).

Algorithm *Debugging*
1 Scan forward (PI-PO) and write equation for each cut_i: $f_{i,FOR}$;
2 Scan backward (PO-PI) and write equation for each cut_i: $f_{i,BACK}$;
3 **For** each cut_i
4   compute $\Delta_i = f_{i,BACK} - f_{i,FOR}$;
5   **If** ($\Delta_0 == 0$)
6   **Return** "No Bug";
7   **If** Δ_i contains expression in Table I
8   add detected bug to Bug_list;
9 **If** (bug_list is empty)
10   **For** each cut_i
11                 $\Delta_i = \Delta_i + (f_{i,FOR} - f_{i+1,FOR})$;
12   **If** Δ_i contains an expression in Table I
13   add detected bug to Bug_list;
14 **If** (Bug_list is empty)
15   **Return** "Bug cannot be detected by our method";
16 **For** each member of Bug_list
17   Correct the circuit;
18   Compute Δ_0;
19   **If** ($\Delta_0 == 0$)
20   **Return** "Bug is detected and corrected";

Algorithm 2. Pseudo code of the Debugging Algorithm

The *Bug-list* shows all potential bug locations. If after modifying the Δ_i expressions the *Bug-list* remains empty, this means that our debugging algorithm cannot detect the bug (lines 14-15). To detect the exact location of the bug and correct the circuit, the first bug of the *Bug-list* is replaced with the corresponding correct gate from Table I. Then the circuit is verified by re-computing Δ_0. If $\Delta_0 \neq 0$, we consider the next bug from the list and to correct it in the buggy circuit (lines 19-20). This process is repeated until the circuit becomes correct, i.e., until $\Delta_0 = 0$.

Example 6: Continuing with Example 5, we compute Δ_i for all the cuts, trying to match their variables with one of the expressions in Table I. Recall that only the signals from the given cut must be used in the matching. At cut_3, with signals $\{r_2, r_1, r_0\}$, $\Delta_3 = 4r_1 - 4eg$; but $4r_1$ does not match any of the expressions in the table. At cut_2, with signals $\{e, g, r_1, r_0\}$, we have $\Delta_2 = 4r_1 - 4eg$. Here we find that $-4eg$ matches an expression in the table ($-ab$), at the OR/XOR entry of the table. However, cut_2 does not have any XOR gate, so this cannot be the source of the bug. At cut_1, with signals $\{e, d, c, r_0\}$, we have $\Delta_1 = 4d + 4c - 8dc - 4ec - 4ed + 4ecd$. Here $4d + 4c - 8dc$ matches the buggy expression $(a + b - 2ab)$ with coefficient 4, indicating that an OR gate was used instead of an AND gate. As shown in Fig. 4, there is an OR gate at cut_1 with inputs $\{c, d\}$ so a bug is detected. This bug is then corrected by replacing the OR with

an AND. The other terms ($-4ec$, $-4ed$) match the expression in the table (OR-XOR gate replacement). But at cut_1 there are no gates with inputs $\{e, c\}$ or $\{e, d\}$; hence no additional bug is reported. In a similar fashion, we can verify that there are no bugs in cut_0. Therefore, the only bug detected is at cut_1, caused by the OR gate in place of an AND. To correct the circuit, we just need to replace the OR gate with inputs (c, d) with an AND gate with the same inputs.

IV. EXPERIMENTAL RESULTS

The algorithm has been implemented in C#. The experiments were conducted on a PC with Intel 1.80-GHz Core i7 processor and 6 GB of memory under Windows 8. We tested gate-level circuits of arithmetic functions: $F_1 = A + B$ and $F_2 = A \times B$, with bit-widths ranging from 32 to 128 bits. Several bugs (erroneous gates) were inserted in the middle of each circuit. Note that the bugs located near PI are easiest to detect (there is no residual expression) and the bugs inserted near POs are most difficult to detect (the signature with backward rewriting may explode in size). Table II and III show the results for the two circuits containing multiple bugs.

TABLE II.   DEBUGGING OF $F_1 = A + B$ WITH MULTIPLE BUGS

Bit-width	# Gates	# Bugs	Memory	CPU time (sec)
32	414	1	4.2 MB	1.46
		3	4.2 MB	1.56
		5	4.2MB	1.61
64	810	1	4.6 MB	3.40
		3	4.6 MB	3.51
		5	4.6 MB	3.58
128	1,662	1	6.3 MB	6.70
		3	6.3 MB	6.92
		5	6.3 MB	6.98

Our method can detect and correct every inserted bug in all the instances of the tested circuits in a reasonable time. The table demonstrates a linear CPU time dependence in the number of bugs (for a small number of bugs performed in this experiment).

TABLE III.   DEBUGGING OF $F_2 = A \times B$ WITH MULTIPLE BUGS

Bit-width	# Gates	# Bugs	Memory	CPU Time (sec)
32	8,062	1	8.9 MB	18.32
		3	9.5 MB	23.50
		5	11.2 MB	30.78
64	32,512	1	85 MB	184.50
		3	91 MB	189.43
		5	95 MB	194.45
128	131,072	1	122 MB	1927.40
		3	131 MB	2027.56
		4	143 MB	2136.86

V. CONCLUSIONS

The goal of this work was to provide a proof of concept for identifying and correcting bugs caused by gate replacement in gate-level arithmetic circuits. Despite its preliminary nature, the initial results demonstrate the validity and potential of the proposed approach for solving practical problems. The limitation of the method is generation of cuts as a means to locate the bugs. Determining the best set of cuts to improve the efficiency of the method is the major goal of our future work. One possibility is to adopt a "binary search" approach, by sampling the circuit with selected cuts and checking if their signature is correct. This may need to be supported by random backward simulation to help qualify the cut as correct or incorrect. The next cut in the sequence will then be placed halfway between the faulty one and the PO and the search for bugs will continue in this area in a similar fashion. The method is applicable to locating multiple bugs associated as long as they correspond to disjoint sets of variables.

ACKNOWLEDGMENT

This work was supported by a grant from the National Science Foundation, award No. CCF-1319496.

REFERENCES

[1] M.F. Ali, S. Safarpour, A. Veneris, M.S. Abadir, "Post-Verification Debugging of Hierarchical Designs," IEEE/ACM International Conference on Computer-Aided Design (ICCAD), pp. 871-876, 2005.

[2] Y. Chen, S. Safarpour, A. Veneris, J.M. Silva, "Spatial and temporal design debug using partial MaxSAT," Great Lakes symposium on VLSI (GLVLSI), pp. 345-350, 2009.

[3] M. Ciesielski, C. Yu, W. Brown, D. Liu, "Verification of Gate-level Arithmetic Circuits by Function Extraction," ACM Design Automation Conference (DAC-2015), 2015.

[4] Y. Yang, S. Sinha, A. Veneris, R. Brayton, "Automating Logic Rectification by Approximate SPFDs," Asia and South Pacific Design Automation Conference (ASP-DAC), pp. 402-407, 2007.

[5] M. Abramovici, P.R. Menon, D.T. Miller, "Critical path tracing-an alternative to fault simulation," Design Automation Conference (DAC), 1983, pp. 214–220.

[6] R.E. Bryant, Y-A. Chen, "Verification of Arithmetic Functions with Binary Moment Diagrams," Design Automation Conference (DAC), pp. 535–541, 1995.

[7] A. Smith, A. Veneris, M.F. Ali, A. Viglas, "Fault diagnosis and logic debugging using boolean satisfiability," IEEE Trans. on Computer-Aided Design of Integrated Circuits and Systems, vol. 24, no. 10, pp. 1606–1621, 2005.

[8] H. Mangassarian, A. Veneris, S. Safarpour, M. Benedetti, D. Smith, "A performance-driven QBF-based iterative logic array representation with applications to verification, debug and test," IEEE/ACM International Conference on Computer-Aided Design (ICCAD), pp. 240–245, 2007.

[9] S. Safarpour, A. Veneris, "Abstraction and refinement techniques in automated design debugging," Seventh International Workshop on Microprocessor Test and Verification (MTV), , pp. 88–93, 2006.

[10] S. Safarpour, H. Mangassarian, A. Veneris, M.H. Liffiton, K.A. Sakallah, "Improved design debugging using maximum satisfiability," Formal Methods in Computer Aided Design (FMCAD), pp. 13–19, 2007.

[11] Y. Chen, S. Safarpour, J. Marques-Silva, A. Veneris, "Automated design debugging with maximum satisfiability," IEEE Transactions on Computer-Aided Design of Integrated Circuits and Systems, vol. 29, no. 11, pp. 1804–1817, 2010.

[12] B. Le, H. Mangassarian, B. Keng, A. Veneris, "Non-solution implications using reverse domination in a modern sat-based debugging environment," Design, Automation & Test in Europe Conference (DATE), pp. 629–634, 2012.

[13] M. Kubo, M. Fujita, "Debug methodology for arithmetic circuits on FPGAs," IEEE International Conference on Field-Programmable Technology (FPT), pp. 236–242, 2002.

[14] S. Yang, H. Shim, W. Yang, C-M Kyung, "A new RTL debugging methodology in FPGA-based verification platform," Asia-Pacific Conference on Advanced System Integrated Circuits, pp. 180–183, 2004.

[15] W. Li, Z.J. Song, A.W. Ruan, C.Q. Li, D.S. Yu, "A two-mode debugging system for vlsi designs using xilinx FPGA," International Conference on Computational Problem-Solving (ICCP), pp. 84–88, 2012.

[16] M.K. Ganai, A. Gupta, "Efficient BMC for multi-clock systems with clocked specifications," Asia and South Pacific Design Automation Conf. (ASP-DAC), pp. 310–315, 2007.

[17] B. Keng, S. Safarpour, A. Veneris, "Bounded model debugging," IEEE Transactions on Computer-Aided Design of Integrated Circuits and Systems, vol. 29, no. 11, pp. 1790–1803, 2010.

[18] B. Keng and A. Veneris, "Path directed abstraction and refinement in SAT-based design debugging," Design Automation Conference (DAC), 2012, pp. 947–954.

[19] B. Keng, A.G. Veneris, "Path-Directed Abstraction and Refinement for SAT-Based Design Debugging," IEEE Trans. on Computer-Aided Design of Integrated Circuits and Systems, vol. 32, n. 1o, pp. 1609-1622, 2013.

[20] M.H. Haghbayan, B. Alizadeh, A.M. Rahmani, P. Liljeberg, H. Tenhunen, "Automated formal approach for debugging dividers using dynamic specification," IEEE International Symposium on Defect and Fault Tolerance in VLSI and Nanotechnology Systems (DFT), pp. 264–269, 2014.

[21] V. Athavale, S. Ma, S. Hertz, S. Vasudevan, "Code coverage of assertions using rtl source code analysis," Design Automation Conference (DAC), pp. 1–6, 2014.

Mapping DAGs on Heterogeneous Platforms Using Logic-Based Benders Decomposition

A. Emeretlis, G. Theodoridis
Dept. of Electrical & Computer Engineering
University of Patras, Patras, Greece

P. Alefragis, N. Voros
Dept. of Computer & Informatics Engineering
TEI of Western Greece, Antirio, Greece

Abstract— Efficient mapping of DAGs on heterogeneous multicore platforms is a key component for modern embedded applications. An approach based on the Benders decomposition principle that uses a heuristic pre-solver and Integer Linear and Constraint Programming methods to find proven-optimal solutions is introduced. We present multiple cuts generation schemes, that improve the performance of the solution process, and extensive experimental results, that show significant speedups compared to the pure ILP-based method.

Keywords — Multicore architectures, DAGs mapping, Benders decomposition, ILP – CP optimization

I. INTRODUCTION

To satisfy the strict time requirements of the applications, modern computing platforms include a number of heterogeneous cores, which are used for the parallel execution of the tasks, while their heterogeneity is utilized to efficiently address the task's diverse functionality. However, to exploit these features, proper methods to map the applications on these platforms are needed. By the term mapping, it is meant the finding of an assignment of the tasks to the cores and an execution scheduling of them to optimize one or more metrics (e.g. execution time, power dissipation).

When the application is static, its characteristics (e.g. tasks, tasks' dependencies, execution time of each task per core) are known and fixed and can be described by static dataflow graphs. Using these graphs and taking into account the features of the target platform, a static mapping can be produced. Since the mapping is developed once and does not change during the product's lifetime, reasonable effort and time can be spent to get an optimal or a near-optimal mapping. This case includes many real-life applications, for instance, typical DSP and multimedia ones.

This problem has been studied in the past [1] and the presented methods can be classified in two main classes. The first one includes algorithms that are based on: i) heuristics, such as list scheduling [2], [3] or node duplication [4], and ii) stochastic search algorithms, such as genetic algorithms [5], [6]. The main advantage of these methods is their low complexity, which allows them to find a mapping in reduced time even for large and complex dataflow graphs. However, they have the following drawbacks. Since they are based on heuristics, they cannot guarantee that the produced mapping is the optimal one. If the assumptions on which they are based are not satisfied then, the solution may be far from the optimal

one. The second category includes methods that always produce an optimal solution. In this case, the mapping problem is modeled as an optimization problem using Integer Linear Programming (ILP) [7], [8] or Constraint Programming (CP) models [9]. Their main advantage is that they always produce an optimal mapping. However, they suffer from large complexity, which prohibits their use in complex applications.

In the last decade, a lot of research effort has been spent in the fields of Operations Research and Artificial Intelligence to develop methods to solve complex optimization problems in reduced time. These methods are based on hybrid ILP-CP models and sophisticated algorithms and have been used in complex optimization problems achieving significant speedups [10], [11]. Yet, they have not been applied extensively in the field of multicore platforms. To the best of our knowledge, only one group has followed this approach achieving remarkable speedups compared to methods that are based on pure ILP or CP models [12]. However, in these work the target multicore platform was homogeneous, which facilitates the solution process of the introduced method. In a homogeneous platform, it is easier to find and exclude some permutations of each solution that produce the same or worse objective values, reducing significantly the solution space.

In this paper, a hybrid model, based on the Benders decomposition approach [13] is introduced to map static DAGs on heterogeneous multicore platforms. The whole optimization problem is decomposed in two sub-problems where the first sub-one is responsible for the assignment of the tasks to the cores, while the second one undertakes the scheduling of the tasks. The process is iterative and terminates when no better assignment can be found or a time limit is met. The model employs two complementary optimization techniques namely the ILP and CP ones and a set of novel constraints that are dynamically generated during the iterations. In addition, a heuristic is applied as pre-processing step to provide a good initial solution. Based on extensive experimental results, it is proved that the proposed model outperforms the corresponding ILP one in terms of the solution time and quality of the returned solution.

The paper is organized as follows. In Section II the problem is defined and the corresponding ILP model is presented, while in Section III the Benders decomposition approach is discussed. In Sections IV and V, the proposed method and the experimental results are presented and discussed. Finally, Section VI concludes the paper.

978-1-4799-8720-7/15 $31.00 © 2015 IEEE

II. PROBLEM DEFINITION AND ILP MODEL

A heterogeneous multicore platform equipped with a rich and low-latency interconnection network is assumed. It is also assumed that the cores $P = \{1, 2, ..., m\}$ communicate asynchronously without the intervention of the involved cores. Hence, compared to the tasks' execution time, the communication overhead is small and is ignored. However, the method can be extended to address this issue. An embedded application is usually described by a Hierarchical Task Graph [24] that introduces a hierarchy of interconnected kernels. At each hierarchy level, multiple Directed Acyclic Graphs (DAGs) are used to describe the application's kernels. In this work, we focus on solving optimally the problem of mapping kernels on heterogeneous multicore platforms.

Let $G = (V, E)$ be a DAG of where $V = \{v_1, v_2, ..., v_n\}$ is the set of nodes (tasks) and $E = \{e_1, e_2, ..., e_{nE}\}$ the set of edges, which describe tasks' dependencies. For each edge, $e = (v_i, v_j)$, v_i and v_j are its head and tail, respectively. It is assumed that: i) a task starts its execution when all its predecessor tasks have finished their execution, ii) each task is executed by one core with no preemption, and iii) many tasks can be executed by the same core in non-overlapped time intervals. The goal is to find a proven-optimal mapping that minimizes the total execution time (makespan).

For this problem, formulated by (1)-(5), ILP solvers can find an optimal solution given plenty of time.

$$\min t_{sink}^s \tag{1}$$

$$\forall v_i \in V \sum_{p \in P} x_{i,p} = 1 \tag{2}$$

$$\forall e = (v_i, v_j) \in E \ \ t_i^s + \sum_{p \in P} (D_{i,p} x_{i,p}) \leq t_j^s \tag{3}$$

$$\forall v_j \in ON(v_i) \ \ t_i^s + \sum_{p \in P}(D_{i,p} x_{i,p}) \leq t_j^s + (3 - x_{ip} - x_{jp} - a_{ij})M \tag{4}$$

$$\forall v_j \in ON(v_i) \ \ t_j^s + \sum_{p \in P}(D_{j,p} x_{j,p}) \leq t_i^s + (2 - x_{ip} - x_{jp} + a_{ij})M \tag{5}$$

In this formulation, t_{sink}^s is the start time of the sink node. The sink node is a zero execution time virtual task to which all tasks with zero out-degree connect. Also, $x_{i,p}$ is a variable that equals to 1 when task v_i is assigned to core p, otherwise $x_{i,p}=0$. Also, t_i^s denotes the start time and $D_{i,p}$ is the execution time of task v_i in core p. Hence, (2) ensures that each task is assigned to one core. Equation (3) enforces that for an edge $e = (v_i, v_j)$, the task v_j starts its execution after the execution of task v_i. Also, $ON(v_i)$ is the set of tasks that are independent of task v_i. A non-overlapped in time execution sequence between two independent tasks v_i, v_j is imposed by (4) and (5), when they are assigned to the same core. Specifically, a_{ij} is a binary decision variable that equals to 1 when v_j is executed after v_i, and 0 otherwise. Finally, M is a large integer constant.

In the general case, this ILP model cannot be solved in short time because in the worst case m^n combinations of $x_{i,p}$ variables must be examined, while the big-Ms in (4), (5) lead to large integrality gaps, which increase the solution time. The proposed method reduces significantly the solution time following the Logic-Based Benders decomposition approach [13], which is briefly presented in the following.

III. LOGIC-BASED BENDERS DECOMPOSITION

The Logic-Based Benders decomposition approach is an iterative process that aims at reducing the solution time of complex optimization problems. The main idea is to create a sequence of two sub-problems, where the second sub-problem uses the solution of the first one. The process terminates if no better solution can be found for the first sub-problem or the whole solution space has been examined. Compared to the solution of the whole problem, this approach is more efficient in terms of complexity and solution time since smaller problems are solved iteratively.

Assume an optimization problem (Primal Problem–PP) in the form of (6)

$$\begin{aligned} \min \ z &= f(x, y) \\ s.t. \ \ &C(x, y) \end{aligned} \tag{6}$$

where x, y are the variables, $z = f(x, y)$ is the cost function, and $C(x,y)$ is the constraints set. The principle is to decompose the PP into two sub-problems that are solved separately, whereas the process is iterative (see Fig. 1). Specifically, two sub-problems namely, the Master Problem (MP) and the Sub-Problem (SP) are developed. The MP contains only a subset of the total variables (x variables) and a sub-set, $C_1(x)$, of the constraints, while the cost function, $g(x)$, is a relaxation of the global cost function, $f(x,y)$. On the other hand, the SP is almost *identical* to the PP.

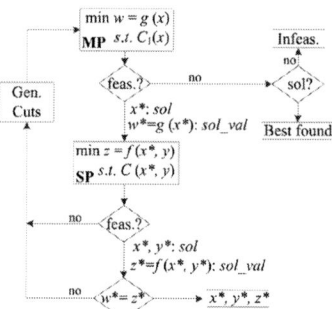

Fig. 1. Benders Decomposition

Initially, the MP is solved and assume that the returned solution and cost value are $x = x^*$ and $w^* = g(x^*)$, respectively. Then, the SP is solved with $x = x^*$. The SP may be feasible, producing a solution (x^*, y^*) with cost value $z^* = f(x^*, y^*)$, or it may be infeasible. In either case, the SP's solution/ infeasibility is studied and extra constraints, called *Benders Cuts*, are generated by inference. These constraints are inserted to the constraints set of the MP. Their purpose is: i) to prohibit the MP to generate in a next iteration the same or another solution that makes the SP infeasible and ii) to guide the MP to produce a solution, which will result in a better global solution when used by the SP. In addition, when the SP is solved and the current solution z^* is better than any previous-found, which means that $z^* = z_{best}$, two additional cuts $w < z^*$ and $z < z^*$ are inserted to the constraints sets of MP and SP, respectively, to impose them to search for a better solution in the next iterations. The process is iterative and terminates if the current solution of the MP is satisfied by the SP, which means that the values of their cost values are the same ($w^* = z^*$), or the MP becomes infeasible. In the second case, the global solution is the best-found one if it exists; otherwise, the PP is infeasible.

IV. PROPOSED APPROACH

Although the Benders decomposition approach is promising in terms of complexity and solution time, many questions are arisen. How should the PP be decomposed so that the MP and SP are characterized by reduced complexity? How much must the MP be relaxed to quickly provide the SP with efficient solutions? Which model/solver must be used for each sub-problem? How are the Benders cuts generated? These questions are discussed in the following.

A. Problem's Decomposition – Hybrid Model

As our problem consists of two sub-problems namely, the assignment of the tasks to the cores and their scheduling per core, a direct decomposition could be the above assignment and scheduling sub-problems to be the MP and SP, respectively. However, in this case, the MP solves the assignment problem ignoring its impact to the scheduling one. This results in a loose interaction between the MP and SP, increasing the iterations and the solution time. Hence, the MP must include a part of the scheduling problem.

Given a solution of the assignment sub-problem, the scheduling one has to: i) respect the tasks' dependencies, and ii) find an optimum execution sequence when independent tasks are assigned to the same core (*sequencing problem*). The first constraint can be easily handled by the MP as the tasks' dependencies are described by the edges of the DAG and their satisfaction is simple. Hence, it is proposed that the MP undertakes to solve totally the assignment sub-problem and partially the scheduling one taking into account only the tasks' dependencies. Hence, it has to optimize (1) subject to (2), (3). The solution of the MP is the assignment of the tasks to cores, an initial start time for each task and an initial makespan value, w, which both may be modified after the SP's solution.

To find the makespan, z, the SP has to solve the scheduling problem (i.e. to satisfy the tasks' dependencies and solve the sequencing problem) respecting the MP's solution (assignment of the tasks to the cores). To achieve this, (1) must be optimized with respect to (3)-(5). Notice that (2) is excluded from the SP, because it refers to the assignment sub-problem, which is solved by the MP. The solution of the SP is the start time, of each task and the final makespan value, z.

The next issue is the choice of the model for the MP and SP. It is known that in mathematical models (e.g. ILP) a major difficulty arises when big-Ms are used, since they cause large integrality gaps and poor LP relaxations. These drawbacks have as result the increase of time required to find a solution and prove its optimality. On the other hand, the ILP solvers are very strong in combining inequality constraints such as these of (3) and produce stronger ones (cutting planes), which prune significantly the search space. As our MP has the above features, we model it by an ILP model described by (1)-(3).

As mentioned, the SP is responsible to solve the sequencing problem and to satisfy the tasks' dependencies. It

has also to check if the scheduling provided by the MP is satisfied. Since this includes constraints' satisfaction checking, we adopt a Constraint Programming (CP) model [14] because these models have been developed to solve the Constraint Satisfaction Problem (CSP). They include strong constraints propagation algorithms and efficient algorithms suitable for the scheduling problem. However, as they target to the CSP and due to the lack of LP relaxations, they are not so efficient when they are used alone to solve an optimization problem. Nonetheless, the combination of the ILP and CP is a good choice as it combines the strengths of these two approaches [10], [11]. In our case, the CP model of the SP is:

$$\min t^s_{sink} \tag{7}$$

$$\forall \ e = (v_i, v_j) \in E \ \ t^s_i + D_i \le t^s_j \tag{8}$$

$$\forall \ v_j \in ON(v_i) \,\big|\, X_i = X_j \ \ t^s_i + D_i \le t^s_j \lor t^s_j + D_j \le t^s_i \tag{9}$$

where t^s_{sink}, t^s_i are in agreement with the initial ILP model. X_i and D_i are positive integers that denote the assignment and execution time of task v_i, respectively, whose values are derived by the solution of MP. The objective function and the precedence constraints are the same as in the MP. In more details, (8) corresponds to (3), while (9) replaces (4), (5) and states a disjunctive relation on independent tasks that are assigned on the same processor.

B. Benders Cuts Generation

In any method based on the Benders decomposition principle, the generation of the Benders cuts is the key point since they strongly affect the solution time. These cuts must be: i) *safe* to not exclude an optimal solution and ii) *strong* to prune drastically the search space. In the following, the generation of these cuts is discussed in details.

1) Critical Path Cuts

As mentioned, the MP produces a solution for the assignment sub-problem in every iteration. To search the whole solution space, the previous-found MP's solutions must be excluded in the next iterations. This is achieved by inserting in the MP in each iteration the constraint $x_{1p} + x_{2p} + \ldots + x_{np} \le n-1$ ($x_{ip} = 1, p \in P$). These constraints are called *assignment cuts*. However, this method is inefficient as only one assignment solution is excluded each time; thus stronger cuts are needed. The goal is to generate a set of cuts called *cover cuts* that each of them will dominate a number of assignment cuts. In the general case, if there are n tasks and m cores, if l assignment variables ($l < n$) are included in the cover cut, then $(n - l)^m$ possible MP's solution are excluded at once.

Assume that there are three tasks ($i = 1, 2, 3$) and two cores ($p = 1, 2$), the current solution is $x_{11} = x_{21} = x_{31} = 1$, and the assignment $x_{11} = x_{21} = 1$ must be forbidden. In this case, the assignment cut is $x_{11} + x_{21} + x_{31} \le 2$, which is not strong as it is satisfied for the cases $x_{11} = x_{21} = x_{32} = 1$ and $x_{11} = x_{21} = x_{33} = 1$. Thus, using this cut two iterations with assignments $x_{11} = x_{21} = x_{32} = 1$ and $x_{11} = x_{21} = x_{33} = 1$ may be performed until achieving our goal. However, using the constraint $x_{11} + x_{21} \le 1$ the goal is achieved at once.

The above idea is captured by the proposed method as follows. In every iteration and after the solution of the SP, the initial DAG is augmented with extra edges, producing the *Scheduled-DAG* (S-DAG). Specifically, extra directed edges are introduced between two independent nodes that are assigned to the same core depicting their execution sequence. Then, the S-DAG is traversed and all the critical paths are found. For each critical path, the corresponding cover cut is generated and inserted in the Benders cuts pool of the MP. We call these cuts as *Critical Path Cuts*. In the general case, the number of the tasks, l, in each critical path is smaller than the total number of tasks, n, which means that $(n\text{-}l)^m$ assignment cuts are covered by each critical path cut.

2) Knapsack Cuts

Due to the adopted decomposition, the MP ignores the sequencing problem and assigns many independent tasks on the same core in overlapped time intervals, which after the SP's solution will be executed sequentially in time increasing the makespan. To address this issue, additional cuts are used.

Assume that the MP has produced a solution where three tasks ($i = 1, 2, 3$) are assigned to core 1 ($p = 1$), while their execution times are $D_{11} = D_{21} = D_{31} = 2$ and their release times, RL, are $RL_1 = RL_2 = RL_3 = 0$. If all tasks must be executed with a Deadline $DL = 3$, at most one task should be assigned to this core. In this case, the cut that prevents the above assignment is the knapsack constraint $2x_{11} + 2x_{21} + 2x_{31} \leq 3$. In addition, this cut implies the cover cuts $x_{11}+x_{21}<2$, $x_{11}+x_{31}<2$, and $x_{21}+x_{31}<2$, which as discussed are strong cuts. Thus, in the general case, a set of strong cover cuts are implied by a knapsack cut. Consequently, if the knapsack cuts are produced efficiently, the solution time will strongly benefit.

To generate such type of cuts, a novel multiple knapsack problem is introduced, where each core is a knapsack the capacity of which equals to the duration of a time interval called *knapsack time interval*, while the items' weights are the execution times of the tasks assigned to the core. A simple approach to solve this knapsack problem is to define a time interval in which the lower and upper bounds are the minimum RL and the maximum DL time, respectively, of the tasks assigned to this core. Next, the total execution time of these tasks is calculated and if it exceeds the duration of the knapsack time interval, the corresponding cut is produced. However, this approach has a major limitation because it ignores the interaction of the RL and DL times of the tasks.

Assume that three tasks ($i = 1, 2, 3$) with execution times $D_{11} = D_{21} = D_{31} = 2$ are assigned to core $p=1$ and their RL and DL times are $RL_1 = 0$, $DL_1 = 6$, $RL_2 = 2$, $DL_2 = 4$, and $RL_3 = 3$, $DL_3 = 5$. In this case, the knapsack interval is $[0, 6]$ and the constraint $2x_{11} + 2x_{21} + 2x_{31} \leq 6$ is the knapsack cut, which is satisfied for $x_{11} = x_{21} = x_{31} = 1$. This means that the above tasks can be executed by core 1 in the time interval $[0, 6]$. However, studying their RL and DL times, it is derived that this is infeasible since tasks 2 and 3 cannot be executed in the above time interval satisfying their RL and DL times. Hence, another constraint is needed. The constraint that achieves this is the $2x_{21} + 2x_{31} \leq 3$, which implies the constraint $x_{21} + x_{31} \leq 1$ that prevents the assignment $x_{21} = x_{31} = 1$. In [15] a three-phase algorithm was proposed to address such cases and generate the corresponding knapsack. However, these algorithms assume that the RL and DL times of the tasks are known in advance,

which does not hold in our case. To overcome this limitation, the procedure shown in Fig. 2 is proposed.

```
Knapsack_cuts  (in:G=(V,E),   D_{i,p}  out:knapsack
    cuts)
1. RL_{source} = 0,  DL_{sink} = z-1  (z>0)
2. traverse the DAG downwards
3. for each  v_i ∈ V ,      solve RL_i
4. traverse the DAG upwards
5. for each  v_i ∈ V ,      solve DL_i
6. Gen_Phase1_equations  (in:G=(V,E),   D_{i,p},   RL_i,
       DL_i  out:Phase1_eq)
7. Gen_Phase2_equations  (in:G=(V,E),   D_{i,p},   RL_i,
       DL_i  out:Phase2_eq)
8. Gen_Phase3_equations  (in:G=(V,E),   D_{i,p},   RL_i,
       DL_i  out:Phase3_eq)
9. knapsack_eq=Phase1_eq ∪ Phase2_eq ∪ Phase3_eq
   // Start Benders iterations //
10. z_{best}=□,,  knapsach_cuts = { }
11. for each iteration
12.   z' = current_solution
13.   if z'<z_{best} then z'= z_{best}
14.     DL_{sink}=z_{best}-1
15.     check_violated_knapsack_equations
16.     knapsack_cuts=knapsack_cuts ∪
                {violated_knapsack_equations}
17.   end if
18. end for
```

Fig. 2. Knapsack cuts generation and application

In a pre-processing step (Lines 2-5), the DAG is traversed downwards and upwards and the RL_i and DL_i times for each task v_i are computed by (10), (11). In the cases that the number of incoming or outgoing edges exceeds the number of the available processing cores, a full mapping problem including the sequencing sub-problem is solved per node. The solution improves the estimated RL_i and DL_i times and subsequently we avoid the generation of loose knapsack intervals.

$$RL_i = \max_{(v_j, v_i) \in E} (RL_j + \min_{p \in P} D_i, p) \tag{10}$$

$$DL_i = \max_{(v_i, v_j) \in E} (DL_j - \min_{p \in P} D_j, p) \tag{11}$$

Using the computed RL_i and DL_i times, the algorithms of [15] are applied and the candidate knapsack cuts are generated (lines 6-8) and stored (line 9). As the sink node is the first node which is accessed when the DAG is traversed upwards, each DL_i time of the remaining tasks is expressed as a function to DL_{sink}. Thus, in contrast to [15], the candidate knapsack cuts are parameterized in terms of DL_{sink}. Then, z_{best} is updated during the iterations (Line 13) and in each iteration, DL_{sink} is reduced by one, $DL_{sink} = z_{best}$ -1, (line 14). Next, based on the new DL_{sink} value the activated knapsack cuts (those that violate the candidate knapsack equations) are found and stored (Lines 15, 16).

C. Generation of the initial solution

An additional point of the Benders decomposition approach is the generation of the initial solution. If strong Benders cuts are generated, the method is very efficient for optimizing a good initial solution. However, if the initial solution is far from the optimal one, then a large number of iterations is required due to the relaxation of the MP. To overcome this limitation, an additional pre-processing step is introduced that uses a heuristic to provide a good initial solution for the first iteration. In this work, the HEFT heuristic [2] is used for this purpose. Specifically, the solution provided by the HEFT heuristic is used as an initial solution and the knapsacks cuts (lines 10-18, Fig. 2) are generated. Then, the original iterations of the Benders approach are applied.

V. EXPERIMENTAL RESULTS

The pure ILP model ((1)-(5)) and the proposed one with its routines for the generation of the Benders cuts were developed in FICO Xpress Optimization suite [16]. The introduced model was applied in synthetic DAGs and in a DAG of real-life application and it was compared with the ILP model in terms of solution time. Since the HEFT heuristic is applied as pre-processing step in the proposed model providing an initial upper bound for the makespan value, to have a fairer comparison, the same pre-processing step was applied to the pure ILP model, which is called HEFT-ILP (HILP) model, hereafter. Also, in all the experiments, a time limit (TL) equal to two hours was set. All the experiments were run on an i7 6-core PC operating at 3.2 GHz with 16 GB installed memory.

For the case of synthetic DAGs, four experiments were performed. In the first one, 10 DAGs with 20 nodes each mapped on a platform with two cores, while in the second one, 10 different DAGs of 20 nodes each mapped on a platform with five cores. Similarly, the third and fourth experiments refer to the mapping of DAGs with 30 nodes on platforms with four and six cores, respectively. All the DAGs are generated by an in-house tool and are available in [17]. The experimental results are shown in Tables I and II. Specifically, the solution times of the HILP and the proposed (Prop.) model, which includes the time for the generation of the Benders cuts, are provided. Also, the makespan value returned by each model within the time limit along with the one delivered by the HEFT heuristic are included. However, when each of the models fails to solve the problem within the time limit the returned makespan value in not proven-optimal.

Based on the results of Tables I, II the following conclusions are derived. For the cases where DAGs with 20 nodes each are mapped to platforms with two and five cores, the proposed model finds an optimal solution for all the cases within the specified time limit. In contrast, the HILP model returns a proven-optimal solution for a significantly smaller

subset. Specifically, when the platform includes two and five cores, the HILP model finds the optimal solution only for the DAGs 1-4 and 11-14, respectively. In addition, when DAGs with 30 nodes are mapped in platforms with four and six cores, the proposed model finds the optimal solution for the DAGs 1-5 and 11-18, respectively. On the other hand, the HILP returns an optimal solution only for the DAGs 11-14. Thus, the first outcome is that the proposed model outperforms the HILP one since it is able to find an optimal solution for significantly more cases.

Studying the solution time of the proposed model, it is concluded that for all the DAGs of Table I it does not exceed 2 minutes, except for the DAG 20, which requires 5 minutes. In addition, for the DAGs of Table II, the solution time is also below 5 minutes in most of the cases. Nevertheless, there are cases (e.g. DAGs 17-18) that the time reaches or exceeds one hour. Comparing the solution times of the two models, it can be seen that the proposed model outperforms the HILP one in most of the cases. For the DAGs 1, 2, and 11 of Table I, the HILP model is faster. However, these times are very small, while the solution time of the proposed model includes the time for the generation of the cuts and the synchronization of the ILP and CP solvers. The HILP model in faster only in DAGs 3 and 14 in Tables I and II, respectively.

Studying the cases where both models fail to find an optimal solution, the proposed model systematically returns a better solution (DAGs 6-10 and 19-20 in Table II). For the cases where the proposed model is solved and the HILP one is not solved, in many cases the HILP has found a solution value that is equal to that of the proposed model, but it has not proved its optimality in the time limit. This is a well-known limitation of the ILP models. However, there are also cases (DAGs 2-5 in Table II) where the HILP solution is worse.

The proposed method highly benefits from the quality of the solution provided by the HEFT heuristic, which is improved by 0.3% up to 26%. Yet, in many cases most of the time is spent to prove the optimality of the solution even though the improvement is small compared to the initial solution. For example, for the case of DAG 18 in Table II, the solution is improved by only 0.3% while the time limit is almost reached. This is caused by the inability in a heterogeneous system to detect and exclude the permutations

TABLE I.   EXPERIMENTAL RESULTS FOR DAGS WITH 20 NODES

	Platform with 2 cores						Platform with 5 cores					
DAG	**Time (sec.)**		**Solution**			**Iterations**	**DAG**	**Time (sec.)**		**Solution**			**Iterations**
	HILP	Prop.	HILP	Prop.	HEFT	Prop.		HILP	Prop.	HILP	Prop.	HEFT	Prop.
1	2.8	13.5	295	295	350	75	11	0.2	1.3	224	224	239	22
2	7.5	16.5	304	304	378	83	12	26.2	20.6	120	120	131	44
3	63.7	125.8	348	348	377	220	13	1945	51.4	118	118	126	39
4	257.4	3.2	288	288	295	25	14	2576.8	21.5	95	95	118	28
5	TL	2.9	313	311	330	22	15	TL	11.4	103	102	123	61
6	TL	15.7	277	277	294	19	16	TL	27.1	122	122	139	62
7	TL	46	289	289	364	14	17	TL	63	121	120	126	29
8	TL	55.9	277	277	293	12	18	TL	98.8	121	121	123	11
9	TL	59.3	285	285	295	36	19	TL	111.8	130	130	136	87
10	TL	74.1	273	273	289	8	20	TL	301.5	106	103	117	144

TL: Time Limit (7200 sec.)

TABLE II.   EXPERIMENTAL RESULTS FOR DAGS WITH 30 NODES

| DAG | Platform with 4 cores | | | | | | DAG | Platform with 6 cores | | | | | |
| | Time (sec) | | Solution | | | Iterations | | Time (sec) | | Solution | | | Iterations |
	HILP	Prop.	HILP	Prop.	HEFT	Prop.		HILP	Prop.	HILP	Prop.	HEFT	Prop.
1	TL	15.1	1994	1993	2066	79	11	87.5	36.9	1829	1829	1900	48
2	TL	180.1	3518	3475	3749	478	12	342	66.3	1638	1638	1712	225
3	TL	253.9	2178	2160	2328	220	13	1934.4	245.5	1563	1563	1631	359
4	TL	328.3	3972	3932	4072	1175	14	379.5	1148.5	1688	1688	1767	1321
5	TL	3472.8	1835	1815	2040	2921	15	TL	258.3	1801	1799	1861	379
6	TL	TL	4181	4142	4398	2972	16	TL	1095.3	1754	1754	1780	549
7	TL	TL	3614	3575	3803	1492	17	TL	6004.9	3016	3016	3055	5912
8	TL	TL	3295	3231	3367	1937	18	TL	6710.3	2731	2726	2733	1291
9	TL	TL	1973	1973	2075	4559	19	TL	TL	1726	1709	1764	1502
10	TL	TL	2148	2174	2268	1174	20	TL	TL	2656	2639	2706	852

TL: Time Limit (7200 sec.)

of the current solution that lead to the same or worse objective values. However, the proposed method is more efficient in improving the solution or finding the optimal one.

To further evaluate our approach, we used the top level task graph of a real application from Fraunhofer IOSB using the ALMA tool chain [18]. The application employs an image processing algorithm for detecting keypoints using scale-invariant features for multi-object tracking based on the Scale-Invariant Feature Transform (SIFT) [19]. The application was run through the ALMA tool chain and execution times in cycles for the KAHRISMA platform [20] were obtained. As it is shown in Table III, compared to the HILP, a speed up of 2.2x in the solution time is achieved by the proposed method.

TABLE III.   EXPERIMENTAL RESULTS FOR REAL-LIFE APPLICATION

| Platform with 4 cores | | | | |
| DAGs Nodes | Time (sec.) | | Makespan (10^3cycles) | |
	HILP	Prop.	HILP	Prop.
118	45.7	20.6	1006.22	1006.22

VI. CONCLUSIONS

A hybrid ILP-CP model augmented with a heuristic as a pre-processing step has been presented for mapping static applications onto heterogeneous multicore platforms aiming at minimizing the makespan. It is shown that the proposed model outperforms the corresponding ILP-based one in terms of solution time. In a future work, the model will be extended to take into account the data transfer delay between the tasks and it will be applied in larger and more complex DAGs.

ACKNOWLEDGMENT

This work was funded by the European Commission through the Horizon 2020 Program (H2020/PHC-19-2014) under the Research and Innovation Action "RADIO – Robots in assisted living environments: Unobtrusive, efficient, reliable and modular solutions for independent aging".

REFERENCES

[1] L. C. Canon, E. Jannot, R. Sakellariou, and KW. Zheng, "Comparative Evaluation of the Robustness of DAG Scheduling", in Grid Computing: Achievements and Prospects, Springer, pp. 73-84, 2008.

[2] H. Topcuoglu, S. Harri, and M-Y Wu, "Performance-Effective and Low-Complexity Task Scheduling for Heterogeneous Computing", in IEEE Trans. on Parallel and Distributed Systems,. 13 (3), pp. 260-274, 2002.

[3] L. F. Bittencourt, R. Sakellariou, and E. Madeira, "DAG Scheduling using a Lookahead Variant of the Heterogeneous Earliest Finish Time Algorithm", in Euromicro Int. Conf. on Parallel, Distributed and Network-Based Processing, pp. 27-34, 2010.

[4] S. Ranaweera and D. P. Agrawal, "A Task Duplication Based Scheduling Algorithm for Heterogeneous Systems", in Parallel and Distributed Processing Symposium, pp. 445-450, 2000.

[5] S Goupta, G.Agarwal, and V. Kumar "Task Scheduling in Multiprocessor System using Genetic Algorithm", in Int. Conf. on Machine Learning and Computing, pp. 267-271, 2010.

[6] O.L Sathappan, P. Chitra and M. Prabhu, "Modified Genetic Algorithm for Multi-objective Task Scheduling on Heterogeneous Computing System", in Int. Journal of Information technology, Communications and Convergence, 1 (2), pp. 146-158, 2011.

[7] G. Theodoridis, N. Vassiliadis, and S. Nikolaidis, "An Integer Linear Programming Model for Mapping Applications on Hybrid Systems", in IET Computers & Digital Techniques, 3 (1), pp. 33-42, Jan. 2009.

[8] G. Raravi, V. Nelis, "Task Assignment for Heterogeneous Multiprocessors", in ACM Trans. on Embedded Computing Systems, Vol. 13, No5, Article 159, 2014.

[9] K. Martin, C. Wolinski, K. Kuchcinski, A. Floch, and F. Charot, "Constraint Programming Approach to Reconfigurable Processor Extension Generation and Application Compilation", in ACM Trans. on Reconfigurable Technology, (TRETS), 5 (2), June 2012.

[10] J. N. Hooker, "Integrated Methods for Optimization", Springer, 2011.

[11] P. van Hentenryck and M. Milano, "Hybrid Optimization: The Ten Years of CPAIOR", Springer, 2011.

[12] L.Benini, M. Lombardi, M. Milano, Mar. Ruggiero, "Optimal Resource Allocation and Scheduling for the CELL BE Platform", in Annals of Operations Research, Springer, 184 (1), pp. 51-77, 2011.

[13] J. Hooker and G. Ottosson, "Logic-Based Benders Decomposition", in Mathematical Programming, 96, pp. 33-60, 2003.

[14] F.Rossi, P. van Beek, and T. Walsh, "Handbook of Constraint Programming", Elsevier Science, 2006.

[15] C Maravelias, "A Decomposition Framework for the Scheduling of Single- and Multi-Stage Process", in Computers and Chemical Engineering, (30), pp. 407-420, 2006.

[16] http://www.fico.com/en/products/fico-xpress-optimization-suite/

[17] https://dl.dropboxusercontent.com/u/3067415/ISVLSI_2015_DAGs.rar

[18] T. Stripf et al., "Compiling Scilab to high performance embedded multicore systems", in Microprocessors and Microsystems, Vol.37, No.8, pp. 1033–1049, 2013.

[19] David G. Lowe, "Distinctive Image Features from Scale-invariant Keypoints", Int. J. Comput. Vis. 60, 2 (2004), pp. 91–110, 2004.

[20] T. Stripf, R. Koenig, and J. Becker. "A cycle-approximate, mixed-ISA simulator for the KAHRISMA architecture", in Design, Automation & Test in Europe Conference & Exhibition (DATE). pp. 21–26, 2012.

2015 IEEE Computer Society Annual Symposium on VLSI

A Computational Primitive for Convolution based on Coupled Oscillator Arrays

Donald M. Chiarulli[1], Brandon Jennings[2], Yan Fang[2], Andrew Seel[3], and Steve P. Levitan[3]

[1]Department of Computer Science
[2]Computer Engineering Graduate Program
[3]Department of Electrical and Computer Engineering
University of Pittsburgh
Pittsburgh, PA 15260

Abstract— In this paper we present a new computational primitive for convolution using coupled oscillator arrays. It is based on a Degree of Match (DOM) operation for pairs of vectors of analog voltage encoded data. The convolution operation is synthesized from these DOM circuits and computes the precise mathematical convolution of the two vectors. We present an example circuit design for the DOM operator and SPICE simulations of the circuit behavior. A MATLAB model is curve fit to an inverted copy of this output and shown to be roughly equivalent to the square of the Euclidian distance between the vectors. Using a parameterized version of this model we conducted a MATLAB study to analyze the accuracy of the DOM and convolution operations under variations in oscillator array symmetry, locking range, and additive noise. We also show a small example of convolution with an edge enhancement filter.

Keywords—coupled oscillators; convolution; analog computing; vector matching;

I. Introduction

This paper describes a new of computational primitive for convolution based on the phyical properties of coupled oscillators systems. Coupled oscillator systems have been studied for centuries, having been first described by Christian Huygens in 1673 [1] with his observations about the spontanious synchronization of coupled pendulum clocks. Since then, numerous other models of interacting oscillatory systems have been reported, spanning the neural, mechanical, magnetic, and electronic oscillator domains [2] [3] [4] [5] [6]. Of more recent interest for Post-CMOS systems are emerging nano-devices including Spin-Torque Oscillators (STOs) [7] and Resonant Body Oscillators (RBOs) [8], and vanadium dioxide devices (VO$_2$) [9] with coupling based on magnetic, substrate and direct electrical interactions. These are enabling technologies for nano-scale low-power oscillator arrays. Other research has also demonstrated computational primitives that use coupled oscillator arrays for pattern matching computations [10] and in a variety of associative memories [11][12][13].

Mathematical convolution is key to many image and signal processing algorithms. An oscillator based implementation will enable new computational architectures that mimic the human visual [14] or auditory cortex [15]. To the authors' knowledge there has been only one prior published work that approximates a convolution operator using coupled oscillator arrays [16]. Our solution provides an exact result.

II. A Coupled Oscillator Degree of Match Circuit

In this paper, our focus is on a specific type of electrically coupled, voltage controlled, oscillator array with an associated detector that provides a metric for the **Degree of Match** (DOM) between two vectors of values represented by analog voltages, two segments of image pixels for example. Our example implementation is in CMOS, but the design should be readily transferable to emerging device technology when available.

Figure 1 shows a block diagram of the DOM circuit. Each oscillator in the array is a 2-port voltage controlled oscillator with an input control port and bidirectional output/coupling port. The input to the DOM circuit is two vectors of analog voltages $(v_1...v_n)$ and $(v_1'..v_n')$. The control input of each oscillator is driven by the pairwise difference of the individual voltages $(v_i'-v_i)$. The oscillator outputs are directly coupled through a resistor network and the voltage at the common node is amplified, rectified, and integrated as a measure of the relative synchronization of the oscillators and hence the degree of match of the input vectors. This circuit is designed assuming relativity small oscillator arrays with sizes on the order of 16 to 64 oscillators. This is consistent with both current technology limits and support for a direct electrical coupling structure.

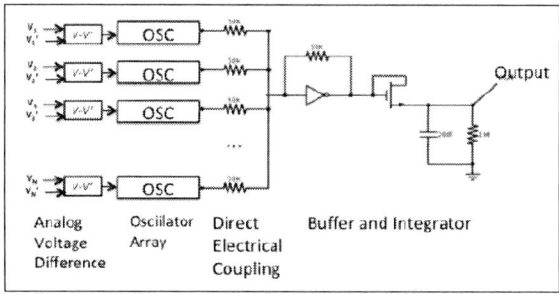

Figure 1: Block Diagram of Coupled Oscillator Degree of Match (DOM) circuit

To understand the behavior of the DOM circuit in a nano-scale CMOS technology, we designed the circuits shown in Figure 2 and Figure 4, using the Arizona State University 7nm finfet technology models [17][18]. Figure 2 is the voltage-controlled oscillator. In this design, the input control voltage is scaled and used to control the charging current on the timing capacitor. There is a positive feedback loop between the PMOS and NMOS transistors in a thyristor configuration which switches when the capacitor charges above the *CoupleIn* node.

978-1-4799-8720-7/15 $31.00 © 2015 IEEE

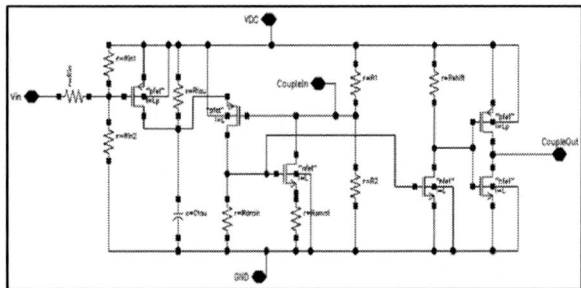

Figure 2: Voltage Controlled Oscillator Circuit

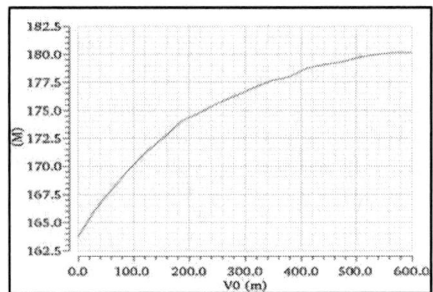

Figure 3: Frequency response of oscillator circuit versus control voltage

This dumps the charge and resets the circuit for another charging cycle. The output is buffered and level shifted to feed the degree of match network. The frequency range in his configuration is about 160-180MHz for an input range of 0-600mV.

Figure 3 shows the frequency response of the oscillator for a sweep of the input control voltage. In the tests we use 300mV as a "nominal" input value and use an input range of 100mV to 500mV.

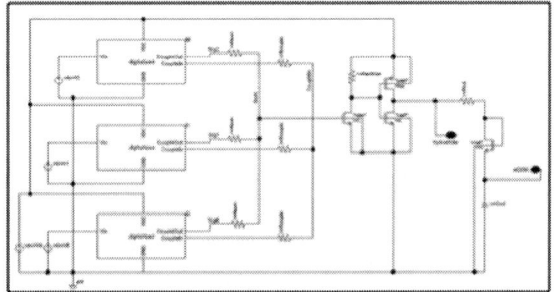

Figure 4: DOM circuit including three oscillators, coupling network and detector

Figure 4 shows a coupled network of three oscillators and the detector circuit. The oscillators are resistively coupled at their *CoupleIn* nodes. This node is the threshold-setting node for each oscillator and, therefore, a large resistance of 10 MOhm will provide mutual feedback between oscillators to produce the weak coupling mechanism essential for the frequency pulling and locking needed. The sensing of the oscillators is performed on another node, *CoupleOut,* to isolate the sensitive coupling network from the degree of match circuit itself. The DOM circuit simply sums the outputs of the

oscillators, level shifts and amplifies the result, and then rectifies and integrates the output. The integrated value is sampled (not shown) to produce a stable degree of match value corresponding to the relative frequencies of the oscillators, and thus the relative values of the elements of the input vectors, Vx and Vy.

TABLE I.   RELAXATION OSCILLATOR CIRCUIT PARAMETERS

Oscillator		*Level Shifter*	
Rtau	1.1MOhm	Rshift	400k
Ctau	1fF		
Ltrans	20nm	*Coupling*	
Vdc	.6V	Rsum	10MOhm
Rdrain	4.8MOhm	Rcouple	10MOhm
Rsmall	0 Ohm		
R1	147kOhm	*DoM*	
R2	30kOhm	RshiftOut	400kOhm
		Rout	10MOhm
VCOBias		Cout	5fF
Rin	900k		
Rin1	79.4k		
Rin2	117.4k		

Table 1 shows the parameter values for the oscillators and the degree of match circuit. Each oscillator consumes about 4.45uW with a 0.6V power supply. Convergence is detected after about 200ns, yielding 890fJ per oscillator for each match

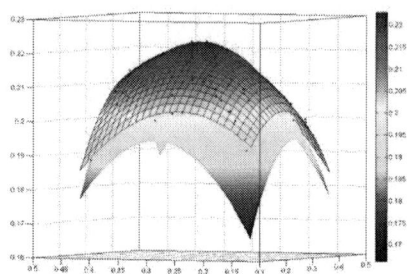

Figure 5: Plot of DOM from SPECTRE simulation data (scatter points), fitting surfaces, $L_2{}^2$ (smooth) and polynomial (grid) for input voltage range $V_{(x1-y1)}, V_{(x2-y2)}$ = 100mV to 500mV and $V_{(x3-y3)}$ fixed at 300mV

operation.

Figure 5 is a scatter plot of the results of a set of SPECTRE simulations with control voltage values between 0.1V and 0.5V for two of the VCOs ($V_{(x1-y1)}$. and $V_{(x2-u2)}$) and a fixed value of 0.3V for the third (V_{x3-y3}). For each data point, the DOM value was sampled after 200ns. This is compared to two fitting functions. The first is a curve fit 2nd degree polynomial: (subtraction subscripts $v_{(xi-yi)}$ abbreviated to v_i) $F(v_1, v_2) = .137 + .2682v_1 + .2739v_2 - .6064v_1^2 + .3555v_1v_2 - .6216v_2^2$ which has an RMSE of 0.0011 to the sampled data. This function is plotted as the solid surface in Figure 5.

The first fit equation is in the general form of a bivariate quadratic polynomial. Since the circuit is intended as a distance metric, we applied a second quadratic polynomial fitting function derived from the equation of the well-known Euclidian (L_2) distance metric, squared and inverted, namely

$$F(Vx, Vy) = max - [(Vx_1 - Vy_1)^2 + (Vx_2 - Vy_2)^2]$$

where max is the maximum value of the function and accounts for the inversion of the function over the range and Vx_i, Vy_i = 0.3V. This function is plotted in three dimensions as the hashed surface in Figure 5 and in a two dimensional cross section in Figure 6. The RMS error between the two fitting functions is 0.0069.

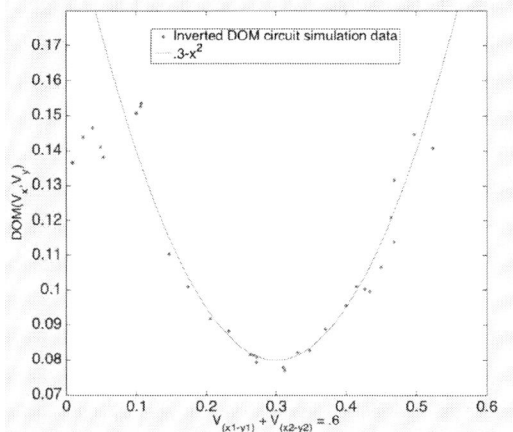

Figure 6: Inverted cross sectional plot of $L_2{}^2$ fitting surface shown in Figure 5 with corresponding simulation data points

Given this behavior and the fact that the inputs are the element-by-element difference between two vectors, the DOM circuit behaves as a distance metric that can be modeled as the Euclidean distance squared ($L_2{}^2$) between the two vectors.

$$DOM(A, B) = L_2{}^2(A, B)$$

$$= \sum_{i=1}^{n} (a_i - b_i)^2 \qquad (1)$$

For the balance of this paper we will use this model based on the assumption that the inversion can be built into the output of the DOM circuit. We will simply refer to it as the $L_2{}^2$ model.

III. OSCILLATOR BASED CONVOLUTION

The $L_2{}^2$ model by itself is a key behavior that can be exploited as a computational primitive for **template matching** and for **distance metrics**. It enables a large variety of image processing algorithms and vector distance based classifiers to be accelerated or in some cases directly computed using oscillator DOM primitives.

However, it is also possible to directly implement the more powerful vector **convolution** primitive with oscillators by making simple algebraic transformation of equation (1). By expanding and rearranging this equation, we can derive an expression for the convolution of A and B in terms of three oscillator based DOM circuits.

$$DOM(A, B) = (A - B)^2$$

$$= A^2 - 2AB - B^2$$

$$\frac{(A - B)^2 - A^2 - B^2}{-2} = AB$$

$$\frac{DOM(A, B) - DOM(A, 0) - DOM(B, 0)}{-2} = AB \qquad (2)$$

In this implementation the first DOM circuit computes $L_2{}^2$ of A and B, a second computes of A and 0, and a third of B and 0. Subtraction, division by two, and inversion are implemented in analog support circuitry.

An oscillator implementation of a convolution kernel enables a wide variety of other signal processing and image processing primitives to be implemented or accelerated. This includes key kernels for filtering, spectral transforms, and convolutional neural networks.

We will validate this analysis in the results section. However, we first describe our effort to improve the fidelity of the $L_2{}^2$ model for DOM circuits in real implementations.

IV. PARAMETERIZED MATLAB MODEL

Although the curve fit in section III captures the $L_2{}^2$ behavior of a simulated DOM circuit, particular oscillator circuits and implementations in specific technology nodes are likely to show a high degree of variability in the actual circuit behavior. To capture this variability we have added several parameters to the $L_2{}^2$ model. Specifically, beginning with equation (1) as a precise $L_2{}^2$ behavioral model for the DOM of two vectors, we add three parameters that model, coupling asymmetry (CA), locking range (LR), and Noise (N).

The CA parameter is a vector of the same length as the input vectors with each element corresponding to the relative coupling strength of each oscillator in the cluster. These variations between the oscillators can be intentionally designed into the DOM circuit, but are more likely to be due to the processing tolerances of a particular technology node. In the model it is simply a coefficient on each term in the summation.

$$DOM(A, B) = \sum_{i=1}^{n} ca_i(a_i - b_i)^2$$

The Locking region (LR) parameter models that fact that oscillator clusters couple over a small range of frequencies rather than a single frequency. This means that contrary to the frequency response curve in Figure 3, a coupled oscillator array will synchonize and lock to common frequency and stay at that frequency over a small range of voltage input variations. By definition this happens when individual vector differences are small and thus the relative difference between individual oscillator frequencies are within the locking range of the oscillator array. Under these circumstances, a pair of vectors with a small relative $L_2{}^2$ distance cannot be distinguished from any other pair of similarly separated vectors since both pairs induce synchronization and have the same DOM output. This behavior is modeled as a scalar (low) threshold value at the DOM output. Values below the LR threshold are forced to the locking value of zero. Values above the threshold output are unaffected. The modified model equations are as follows.

978-1-4799-8720-7/15 $31.00 © 2015 IEEE

$$DOM(A,B) = \sum_{i=1}^{n} ca_i(a_i - b_i)^2$$
$$when \left(L_2{}^2 > LR\right) else \ 0$$

Finally, the noise model parameter controls additive white Gaussian noise included at the model output.

$$DOM(A,B) = \begin{bmatrix} \sum_{i=1}^{n} ca_i(a_i - b_i)^2 \\ when \left(L_2{}^2 > LR\right) else \ 0 \end{bmatrix}$$
$$+ [N * randn()]$$

Figure 7 shows example plots of the DOM model output for selected values of each of the model parameters. To facilitate 3D plotting, the example consists of two oscillators with difference inputs swept from -1 to 1 across the x and y axis. The z axis is the DOM model output. The subplot on the top left is the base $L_2{}^2$ model with no distortions, specifically, CA = [1 1], LR=0, and N=0. In the top left subplot, the CA vector is set to [.75 1] to model asymmetry between the two oscillators. This value is perhaps unrealistically large, but it was chosen for visual clarity. The bottom left subplot shows the impact of the locking range (LR) parameter, in this case set to .15 of the full range, again somewhat large for visual clarity. Note the flattening of the bottom of the parabola as small values of the inputs near zero (i.e. less than LR) and become indistinguishable. Finally, the bottom right subplot shows the model with white Gaussian noise with SNR = 12db.

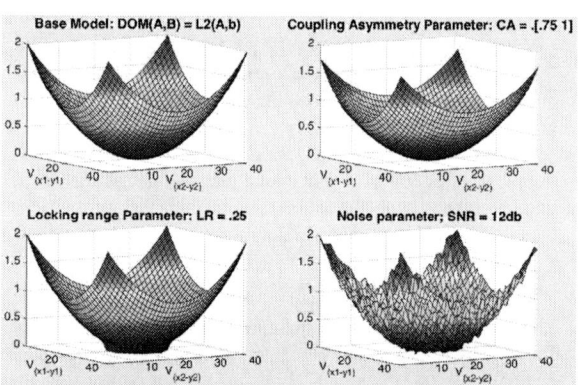

Figure 7: DOM behavioral model and examples of three modeling parameters

Together these three parameters capture the most likely modes in which the actual DOM circuit will vary from the basic $L_2{}^2$ behavior. In the next section we present the results of a study that measures the sensitivity of the computational primitives to each of the model parameters.

V. MODELS OF OSCILLATOR BASED COMPUTATIONAL PRIMITIVES

In this section we present the results of a MATLAB study in which the parameterized $L_2{}^2$ model was analyzed for the accuracy of the DOM and convolution functions. These functions were compared to corresponding MATLAB functions and once equivalence between the oscillator and conventional versions was established, we studied the sensitivity of the results to each of the model parameters discussed in the previous section.

In each study, the A and B input vectors consisted of 64 randomly generated elements each in the range 0 to 1. Each study evaluated the functions across the complete range of possible $L_2{}^2$ values (0 to 64) with a step size of one. Random A and B vectors corresponding to each value of $L_2{}^2$ were generated using the following algorithm. Working backwards from equation (1) we started with a vector of 64 random numbers between 0 and 1 constrained to sum to the target $L_2{}^2$ value. The element wise square root of this vector was then computed to generate the vector (A-B). Since the allowable range of (A-B) elements is -1 to 1, individual elements were randomly selected for sign inversion. Next, a second constrained random vector is assigned to A with constraints imposed such that each element in A and the corresponding element in B (computed algebraically) must be within the range 0 to 1 and (A-B) must be equal to the corresponding element in the first random array.

Using this algorithm, random A and B vectors corresponding to integer values of $L_2{}^2$ from 1 to 64 were generated and applied to the parameterized DOM and convolution functions. The results for the DOM function are shown in Figures 8 and 9. The results for convolution are shown in Figures 10 and 11.

The plots in Figure 8 are the output of the DOM function plotted versus a linear metric for the composition of the A and B vectors, $sum(abs(A + B))$. Figure 9 shows the normalized squared error between the oscillator based function and the equivalent MATLAB direct computation, plotted versus the corresponding value of $L_2{}^2$. Normalized squared error is the square of the difference between the two values divided by the square of the MATLAB computed value. Figures 10 and 11 show the same information for the convolution function plotted versus $L_2{}^2$ throughout.

The four subplots in each figure correspond to individual studies that each sweep a different model parameter. The upper left subplot shows the base case $L_2{}^2$ model (CA={1} LR=0 N=0). These results establish the equivalence between the $L_2{}^2$ model for oscillator implementations and the MATLAB as the normalized error is zero for all values of $L_2{}^2$ in Figure 9.

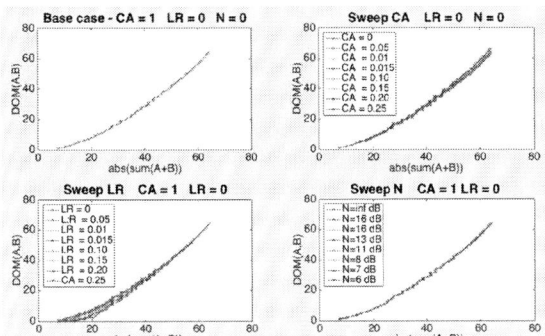

Figure 8: DOM versus sum(abs(A+B)): base case (upper left) and tests of three model parameters, CA (upper right) LR(lower left) and N (lower right)

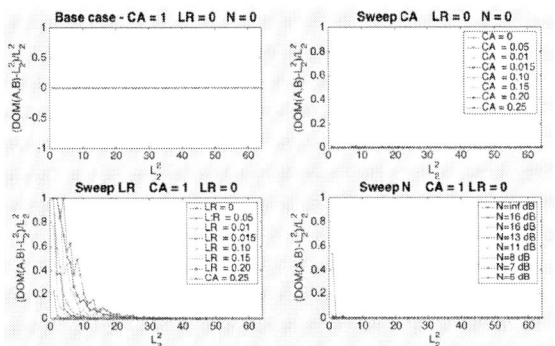

Figure 9: Normalized Squared Error DOM L_2^2 model compared to MATLAB plotted vs L_2^2

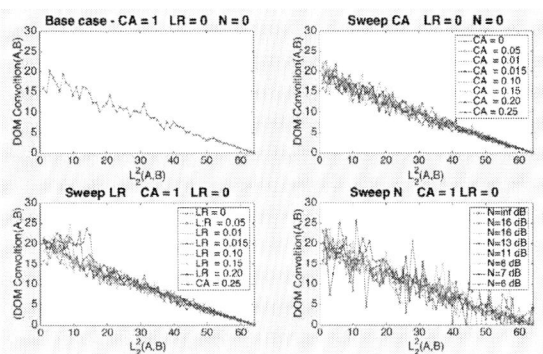

Figure 10: Convolution function versus L_2^2: base case (upper left) and tests of three model parameters, CA (upper right) LR(lower left) and N (lower right)

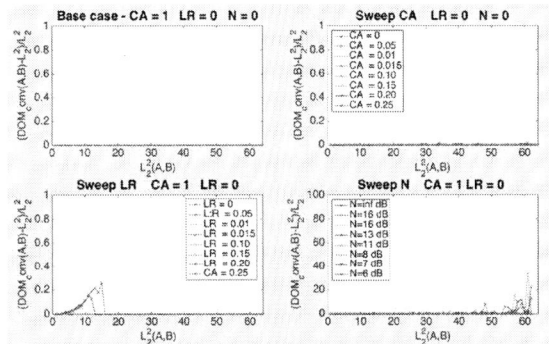

Figure 11: Normalized Squared Error for DOM Convolation^2 relative to MATLAB plotted vs L_2^2

The upper right subplot in each figure examines the impact of the CA model parameter. In this study, each coefficient in the L_2^2 summation is varied randomly with a Gaussian distribution centered at 1 and with a range of (1-CA) to (1+CA). Values are plotted for CA values ranging from .05 to .25. These results show the DOM and convolution functions are relatively robust for small to moderate variations in CA. Much of this error tolerance comes from the fact that a relatively large, 64 oscillator array that was modeled. Obviously, individual oscillator errors will be more significant in smaller oscillator clusters, however we can typically expect that process variations would be less pronounced in the smaller area encompassed by a smaller array.

One side effect of the algorithm for generating the A and B vectors is a boundary condition such that the vectors at the extremes, ($L_2^2 = 0$ and $L_2^2 = 64$) tend to be populated primarily with zeros and ones in patterns such that A-B is also 0 or 1. The side effect of this is that the MATLAB computed expected outputs for DOM and convolution fiunction outputs are zero or very small thus leading to an artificially large normalized error even for small actual error. As a result the y-axis of the normalized error plots have been artificially set below these anomolous values to enhance the display of the more meaningful non-boundary condition results.

The lower left subplot is an analysis of the locking range parameter The LR parameter is expressed as a fraction of the output dynamic range. Thus it shows the flattened shelves in the DOM function corresponding to the fixed low-threshold output for each LR value. Similarly, the output of the convolution function shows greater variation for lower L_2^2 as larger numbers of A,B vector pairs evaluate within the locking range. These variations are reflected in the normalized error plots. However, these plots also verify that computations outside of the locking range are largely unaffected. For example, convolution shows significant errors for values below the locking range, with convolution somewhat less affected than DOM because of the contributions of the individual vector computations. In any case, these results suggest overall that for oscillator computation it is best to design oscillator arrays with as small a locking region as possible. This is a somewhat counterintuitive result.

Finally, the lower right subplot shows the impact of increasing the signal to noise model parameter N. As expected additive white Gaussian noise at the output of the L_2^2 model propagates through the DOM and convolution functions in proportion to the noise model. The noise parameter, N, is also expressed as a fraction of the output dynamic range, thus, we see more significant

978-1-4799-8720-7/15 $31.00 © 2015 IEEE

errors values at the extremes, small DOM at L_2^2 near 0 and small convoltion results at L_2^2 near 64.

As a final examination of the effectives of the convolution primitives under stresses induced by sweeping the L_2^2 model parameters, Figures 12 and 13 show the results of a Gabor filtering operation on a small subimage "chip" used in an image processing algorithm. Gabor filtering is used to implement rotational invariance as it convolves a series of edge enhancement filters, each with edges oriented at a different

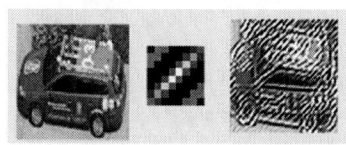

Figure 12: Image chip, Gabor filter kernel, and MATLAB convn() output for example below

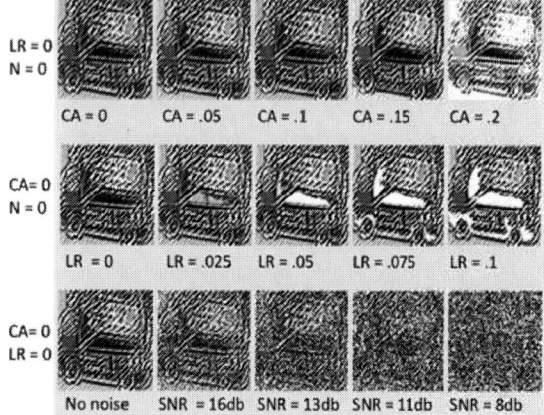

Figure 13: Three slide tray strips of oscillator convolved image with Gabor filter kernel, each with successful more stressed model parameters

angle, with the original image. Figure 12 shows the original 120x120 pixel inage chip to the left, an 8x8 Gabor filter kernel oriented to 45 degrees in the center, and the MATLAB convolved resulting image generated with the *convn()* function on the right. Figure 13 is a set of three slide tray strips that show the same output patch generated using oscillator based convolution for increasingly stressed L_2^2 models that sweep the CA, LR, and N parameters respectively. The leftmost chip in each strip is the base case convolution model and the rightmost is the most stressed version. With the exception of the noise parameter, the results are relatively robust, exposing most of the edges even in the most stressed models. Noise at the output of the DOM function is more problematic, as it would be for any analog computation, oscillator based or otherwise.

VI. CONCLUSIONS AND FUTURE RESEARCH

Our ultimate goal is to adopt oscillator based acceleration into an image-processing pipeline of a vision system. As such, we are currently working on similar model based implementations of the spectral transformations such as the FFT and DCT. In addition we are looking at oscillator accelerated K-nearest neighbor classifiers. On a separate track we are currently

designing test chips with hardware implemenions of the DOM and convolution functions to verify our L_2^2 model.

ACKNOWLEDGMENTS

This research is supported in part by grants from the Defense Advanced Research Projects Agency under the UPSIDE program and the National Science Foundation under grant CCF-1317373.

VII. REFERENCES

[1] Horologium oscillatorium (1673)

[2] Han, S. K., Kurrer, C., and Kuramoto, Y., Dephasing and bursting in coupled neural oscillators. *Physical Review Letters, 75* (17), 3190, 1995.

[3] Horvath, A., Synchronization in cellular spin torque oscillator arrays. *Cellular Nanoscale Networks and their applications (CNNA), 13th International Workshop on. IEEE*, 2012.

[4] Shibata, T., *et al.*, (2012). CMOS supporting circuitries for nano-oscillator-based associative memories. *13th International Workshop on Cellular Nanoscale Networks and their Applications (CNNA)* , 1-5. Turin, Italy, August 29-31, 2012.

[5] Hoppensteadt F.C. and Izhikevich E.M. (2001) Synchronization of MEMS Resonators and Mechanical Neurocomputing. *IEEE Transactions On Circuits and Systems I,* 48:133-138

[6] Van Der Pol, B. (1927). Forced oscillations in a circuit with non-linear resistance. *The London, Edinburgh, and Dublin Philosophical Magazine and Journal of Science, 3* (13), 65-80.

[7] Kaka, S., Mutual phase-locking of microwave spin torque nano-oscillators. *Nature , 437* (15), 389-392, 2005.

[8] Weinstein, D., and Bhave, S. A. A resonant body transistor, Nano Lett. 2010, 10, 1234–1237 DOI: 10.1021/nl9037517.

[9] Nikhil Shukla, et al., , Synchronized charge oscillations in correlated electron systems, Scientific Reports 4, Article number: 4964 doi:10.1038/srep04964 14 May 2014.

[10] Levitan S. P., et al., Non-Boolean associative architectures based on nano-oscillators," 13th IEEE Int'l Workshop on Cellular Nanoscale Networks & Their Applications (CNNA 2012), pp. 1-6, Turin, Italy, August 29-31, 2012.

[11] Csaba, G., Spin torque oscillator (STO) models for applications in associative memories. *Cellular Nanoscale Networks and Their Applications (CNNA)*, 2011.

[12] Nikonov, D. D., Coupled-oscillator associative memory array operation. *arXiv preprint arXiv , 1304* (6125), 2013.

[13] Csaba, G., and Porod, W., Computational study of spin-torque oscillator interactions for non-Boolean computing applications. Magnetics, IEEE Transactions on , 49 (7), 4447-4451, 2013.

[14] Ronny Meir and Pierre Baldi; Computing with Arrays of Coupled Oscillators: An Application to Preattentive Text ure Discrimination (doi:10.1162/neco.1990.2.4.458)

[15] Giraud Mamessier, Anne-Lise, Poeppel, David. Cortical oscillations and speech processing: emerging computational principles and operations. Nature Neuroscience, 2012, vol. 15, no. 4, p. 511-7

[16] Nikonov, Dmitri E., Ian A. Young, and George I. Bourianoff. "Convolutional Networks for Image Processing by Coupled Oscillator Arrays." *arXiv preprint arXiv:1409.4469* (2014).

[17] ASU Predictive Techology Models URL: http://ptm.asu.edu/

[18] S. Sinha, G. Yeric, V. Chandra, B. Cline, Y. Cao, "Exploring sub-20nm FinFET design with predictive technology models," Proceedings of the 49th Annual Design Automation Conference, July 2012, p. 283-288

Homomorphic Data Isolation for Hardware Trojan Protection

M. Tarek Ibn Ziad*, Amr Alanwar†, Yousra Alkabani*, M. Watheq El-Kharashi* and Hassan Bedour*

*Department of Computer and Systems Engineering, Ain Shams University, Cairo, Egypt
†Department of Electrical Engineering, UCLA, Los Angeles, CA, USA
Email: mohamed.tarek@eng.asu.edu.eg, alanwar@ucla.edu,
{yousra.alkabani, watheq.elkharashi, hassan.bedour}@eng.asu.edu.eg

Abstract—The interest in homomorphic encryption/decryption is increasing due to its excellent security properties and operating facilities. It allows operating on data without revealing its content. In this work, we suggest using homomorphism for Hardware Trojan protection. We implement two partial homomorphic designs based on ElGamal encryption/decryption scheme. The first design is a multiplicative homomorphic, whereas the second one is an additive homomorphic. We implement the proposed designs on a low-cost Xilinx Spartan-6 FPGA. Area utilization, delay, and power consumption are reported for both designs. Furthermore, we introduce a dual-circuit design that combines the two earlier designs using resource sharing in order to have minimum area cost. Experimental results show that our dual-circuit design saves 35% of the logic resources compared to a regular design without resource sharing. The saving in power consumption is 20%, whereas the number of cycles needed remains almost the same.

Keywords- ElGamal Encryption, Hardware Trojan, Homomorphism, Security

I. INTRODUCTION

Increasing the complexity of systems proclaims the out-sourced manufacturing concept nowadays. This raises a lot of trust issues in the design industry in many directions. Anyone with access to any step of the manufacturing process could alter the final product to inject a Hardware Trojan. Malicious circuitry can be injected by fabrication facilities or third party IP owners. This threatens design community as the fabrication process is obscured from designers and the details of third party IPs are hidden to protect IP owners' rights [1].

Hardware Trojan appears to be one of the most important topics as the use of silicon chips in different applications becomes very popular, varying from cell phones, cars, to strategically important military devices. It is important to provide methods that resolve the trust issues between fabrication facilities, designers, and end-users. End-users need to make sure that products are not controlled by unknown entities, are stable enough, and will not leak critical information. Maintaining technology secrets of the fabrication facilities and design royalties of third party IP owners raises the difficulty of Hardware Trojan detection and protection. Homomorphic encryption may be used to solve this issue and defeat Hardware Trojans.

In general, homomorphic encryption is a type of encryption, which allows specific types of operations to be carried out

on ciphertext and generates an encrypted result which, when decrypted, matches the result of operations performed on the plaintext. This is a desirable feature that has been utilized in many modern systems [2], [3]. In this work, we introduce the idea of using homomorphism to defeat Hardware Trojan injected in third party IPs. Consider the case where a third party IP is needed to carry out some operation on data A and will produce output data B. Fig. 1 shows the ideal world, where the third party IP does not have any access to the real data as it is homomorphically encrypted. This will give us the capability to carry out the required operation by the third party IP without revealing the original data. Thus, we can retrieve the result B after the decryption process. Full homomorphism (FH) evaluates any arbitrary depth circuit on ciphertexts, whereas partial homomorphism (PH) supports one type of operations only, addition or multiplication.

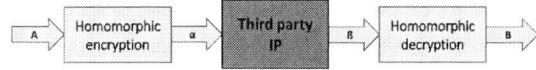

Fig. 1: Homomorphic encryption to protect from hardware Trojan.

The key contributions of this paper include:
1) Discussing new ideas to have a blind data processing by the third party IP with a minimum cost.
2) Implementing ElGamal encryption scheme, which is multiplicative homomorphic and the CRT-based Elgamal (CEG) encryption scheme, which is additive homomorphic, on a low-cost FPGA and showing the resource utilization, performance, and power analysis of both schemes.
3) Introducing a dual-circuit design that supports both, multiplicative and additive homomorphic properties and providing the obtained savings on area and power over a regular design that has no resource sharing.

The rest of the paper is organized as follows. Section II summarizes the related work. Overview about homomorphism and the utilized schemes are given in Section III. Hardware Trojan protection using PH is introduced in Section IV. Experimental evaluation and estimation of the overhead of the proposed methods are shown in Section V. Section VI concludes the work.

II. RELATED WORK

Recently, some architectural methodologies try to increase the chances of the activation of a Hardware Trojan during testing. Salmani *et. al* increased the Trojan activity by inserting dummy flip-flops in the design [4]. They chose the locations of the inserted flip-flops based on a transition probability threshold. Rajendran *et al.* introduced a methodology for securing all the gates of the design using ring oscillators [5]. They added extra logic that converts paths of the circuit into ring oscillators. Changes in the frequency of the ring oscillators were used to detect the presence of Trojans. Al-Anwar *et al.* in [6] developed a novel method for the protection against a hardware Spyware that depends basically on decreasing the probability of seeking sensitive information. They introduced multiplexing between multiple variants implementation. Then, they used cyclic redundancy check (CRC) to detect the infected IP. In [7], Al-Anwar *et al.* suggested obfuscating the output of the suspected IP before sending out data, then undoing that obfuscation at the input of the receiver in order to protect data from leaking and avoid injected triggering. They introduced using either RC4 or a simple obfuscating function. Ibn Ziad *et al.* injects a Hardware Trojan in a voting machine to tamper voting results [8]. The attack depends mainly on the unused bits. They provided a protection technique against the proposed attack and showed its overhead. Al-Anwar *et al.* in [9] introduced multiplexing reconfigurable IPs' outputs and CRC Trojan detection scheme (MCRC) method in order to decrease the probability of leaking critical information by a hacked IP. They also suggested using partial reconfiguration technology to remove an infected IP.

Side-channel dependent methodologies for Hardware Trojan detection aim to localize the impact of the Trojan on the circuit without activating it. Their main idea is to try to detect the presence of a Trojan with high probability via detecting the overload of the Trojan circuit on different circuit parameters; such as the delay or the power as compared to a non-infected circuit. Rad *et al.* studied the impact of a Trojan on the power supply transient current of an IC using statistical methods [10]. Jin and Makris used path delay analysis to detect Trojans [11]. Moreover, gate-level characterization techniques accompanied by statistical methods were used to detect Hardware Trojans [12].

Unfortunately, all Hardware Trojan detection methods require the presence of a non-infected (golden) chip. That requirement represents a real problem as it is feasible only if the design does not contain third party IPs [13]. But, if the system designer integrates third party IPs in the design, these methods become less practical. Zhang and Tehranipoor tried to provide an alternative to using a golden design by using code coverage analysis, formal verification, and ATPG methods to achieve high confidence in whether the circuit is Trojan-free or Trojan-inserted [14]. Baumgarten *et al.* suggested using reconfigurable logic barriers within a design to prevent the activation and operation of Hardware Trojans inserted during the manufacturing stage of an IC [15].

III. BACKGROUND

The aim of this section is to give a brief description about the idea of homomorphism, survey existing partial homomorphic encryption schemes, and discuss ElGamal security scheme.

A. Partial Homomorphism (PH)

PH has been known for many years. It offers the ability to perform a certain type of operations, addition or multiplication, on ciphertexts without revealing data. For example, let us consider the two messages, m_1 and m_2, where both messages are encrypted and their ciphertexts are given by $E(m_1)$ and $E(m_2)$, respectively. If the multiplication of the two ciphertexts is equivalent to the ciphertext of the multiplication of the two messages as shown in (1), we call this a multiplicative homomorphic scheme. On the other hand, if the multiplication of the two ciphertexts equals the ciphertext of the addition of the two messages as shown in (2), we call this an additive homomorphic scheme.

$$E(m_1) \times E(m_2) = E(m_1 \times m_2) \qquad (1)$$

$$E(m_1) \times E(m_2) = E(m_1 + m_2) \qquad (2)$$

It is worth mentioning that PH is different from FH, which allows the efficient evaluation of an arbitrary depth circuit (composed of additions and multiplications) to be evaluated directly on ciphertexts. The first full homomorphic encryption (FHE) scheme was introduced by Gentry [16] in 2009. Since then, there has been some work done toward obtaining efficient hardware implementations of FHE schemes. Hardware building blocks for the lattice-based cryptosystem were considered by Göttert *et al.* [17]. Also, Pöppelmann and Güneysu introduced an efficient hardware implementation of ring-learning-with-errors (RLWE) based encryption [18]. However, these schemes are not practical for this application due to its very large ciphertext and public key sizes. Thus, we focus on PH in this paper.

One of the earliest discoveries in the context of PH is the Goldwasser-Micali cryptosystem [19], whose security is based on the quadratic residuosity problem. It allows homomorphic evaluation of a bitwise exclusive-or. Other additive homomorphic encryption schemes that provide semantic security are Benaloh [20] and Paillier [21]. On the other hand, there exist two well-known schemes, which are multiplicative homomorphic schemes. The first one is the Rivest-Shamir-Adleman (RSA) [22], which is one of the most widely used public-key cryptosystems. The second is ElGamal encryption scheme [23], which is the selected cryptosystem to be used in our work.

B. ElGamal Scheme

ElGamal public-key cryptography algorithm is considered to be one of the efficient and popular algorithms that provides a high level of security. To illustrate its functionality, let us consider that a user called *Alice* wants to send a private message m to another user *Bob*. ElGamal process works as follows. *Bob* generates his keys. He chooses a secret random exponent k and a generator g. So, his public key is (g, h) where $h = g^k(mod\ n)$ and n is a large prime. *Alice* has to encrypt the message m before sending it to *Bob*. She generates a random exponent l and sends the ordered pair (C_1, C_2) to *Bob*, where C_1 and C_2 are defined as (3).

$$C_1 = g^l(mod\ n), C_2 = h^l \times m(mod\ n) \qquad (3)$$

Bob can easily decrypt the ciphertext using (4).

$$m = C_1^{-k} \times C_2(mod\ n) \qquad (4)$$

This encryption scheme is homomorphic with respect to multiplication as if (x_1, y_1) and (x_2, y_2) are valid encryptions for messages m_1 and m_2, with the same key, then $(x_1 x_2, y_1 y_2)$ is a valid encryption of $m_1 m_2$. Hu *et al.* proposed a simple modification to make ElGamal additively homomorphic by placing the message m in the exponent [24]. So, if we encrypt two messages m_1 and m_2 using (3) but multiply h^l with g^m instead of m, the multiplication of the two ciphertexts results in a valid encryption of $g^{m_1+m_2}$. The problem here is that recovering the message involves solving a discrete logarithm problem (DLP) and this is precisely the problem whose difficulty ensures security. To solve this problem, they introduced a new scheme, called CRT-based ElGamal Scheme (CEG), which uses the Chinese Remainder Theorem (CRT) to replace one DLP in a large space by several similar problems in a more tractable search space. This allows for easily obtaining $m_1 + m_2$, while retaining the full security of the scheme, as shown later.

C. CRT-based ElGamal (CEG) Scheme

To illustrate how CEG works, let us reuse the previous example of *Alice* and *Bob*. In the first step, *Bob* also chooses a secret random exponent k and a generator g. He also chooses d_i for $i = 1, \ldots, t$ such that $gcd(d_i, d_j) = 1$ for $i \neq j$. So, his public key is $(g, h, (d_1, \ldots, d_t))$, where $h = g^k(mod\ n)$ and n is a large prime. For encryption, *Alice* sends the encryption of message m as a t-tuple of pairs (C_1, C_2) by using (5).

$$C_1 = g^{l_i}(mod\ n), C_2 = h^{l_i} \times g^{m_i}(mod\ n) \qquad (5)$$

where $m_i = m\ (mod\ d_i)$ and l_i is a generated random exponent for $i = 1, \ldots, t$. *Bob* can decrypt the ciphertext using (6) and (7).

$$m = CRT^{-1}[(log_g(C_{2_i} \times C_{1_i}^{-k}(mod\ n)), i = 1, \ldots, t)] \quad (6)$$

$$CRT^{-1}[C_i] = \sum_{i=1}^{t} C_i \frac{d}{d_i}(\frac{d}{d_i}^{-1} mod\ d_i)\ mod\ d \qquad (7)$$

Correctness and efficiency of the illustrated scheme is discussed in details in [24]. As a part of this work, we implement the CEG scheme in hardware and show its resource utilization and power consumption.

IV. HARDWARE TROJAN PROTECTION USING PH

Here, we introduce our suggested methods for defeating Hardware Trojan in third party IPs. First, we propose two schemes that support PH for the third party IP, which performs one type of operation (multiplication only or addition only). Then, we combine the two methods in a dual-circuit design that supports both multiplication and addition to satisfy applications that utilize the two operations.

A. Sufficient PH Support

Upon classifying the IPs based on processing type, one concludes that there is no need to afford the high cost of FH if the third party IP does only one type of operation. It is totally sufficient to have PH encryption/decryption before/after the suspected IP. In other words, if the suspected IP is used in an electronic voting system and only does addition operation to count votes on the server side [8], it is enough to support one of the additively homomorphic schemes mentioned before in Subsection III-A. For non computational suspected IPs, it is adequate to do simple obfuscation functions before the suspected IP and do the inverse of that function afterwords. Here, we discuss two partial homomorphic hardware implementations based on ElGamal encryption scheme described in Subsection III-B. The first implementation is the main ElGamal encryption/decryption scheme [23], which is a multiplicative homomorphic scheme. The second one is the CEG scheme [24], which is an additive homomorphic scheme.

1) Elgamal Scheme Implementation: Fig. 2 shows the block diagram for our implementation of ElGamal encryption/decryption scheme. The encryption module consists of two Montgomery modular multipliers, two Montgomery modular exponentiators, and a finite state machine (FSM) controller that is responsible for synchronizing other components' inputs and outputs to perform the encryption operations defined in (3). The decryption module consists of one Montgomery modular exponentiator, one modular divider, and a FSM controller that is also responsible for synchronizing other components' inputs and outputs to perform the decryption operations defined in (4). Both modules use a clock and reset signals as inputs. Reset and done signals are utilized to indicate the start and the end of module operations. The message m, ciphertexts $C_1\ and\ C_2$, and the public key h are all k bits vectors, where k is a user-defined integer.

Montgomery multipliers were used in the design as the Montgomery's algorithm [25] is the most widely used algorithm for efficient modular multiplication. Other multiplication methods like the *multiply and reduce* and *double, add, and reduce* are computationally more complex [26]. The binary Montgomery multiplier employs only simple addition, subtraction, and shift operation to avoid trial division, which is a critical and time-consuming operation in conventional modular multiplication. In fact, this multiplier computes $Z = X \times Y \times R^{-1} mod\ M$ instead of $Z = X \times Y mod\ M$, where R is a chosen integer that should be a power of two and relatively prime to M. So, in this case, the operands need to

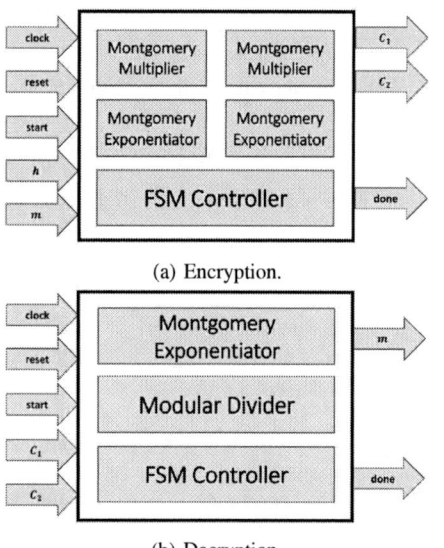

(a) Encryption.

(b) Decryption.

Fig. 2: Block diagram for ElGamal encryption/decryption scheme.

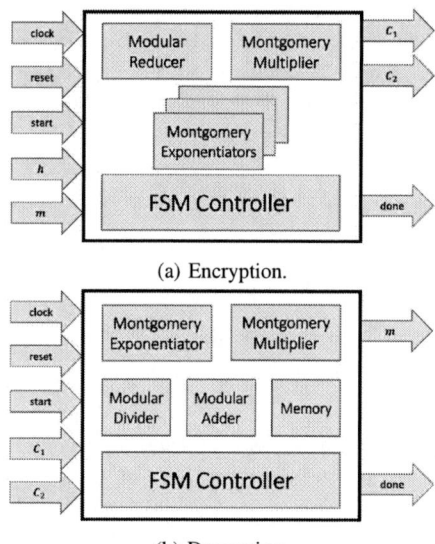

(a) Encryption.

(b) Decryption.

Fig. 3: Block diagram for the CRT-based ElGamal (CEG) encryption/decryption scheme.

be converted into and out of Montgomery's domain each time this multiplier is used.

In general, modular exponentiation is usually accomplished by performing repeated modular multiplications. For our modular exponentiators, the LSB-first algorithm using Montgomery multiplication is used. This algorithm computes $Z = Y^X mod M$ in k executions of a loop that, in turn, includes at most two Montgomery multiplication operations, which are executed concurrently. That improves the performance of the module [26].

The decryption part of the scheme includes the usage of a modular divider module. We implemented the plus-minus algorithm as it gives the shortest computation time with a cost-effective area [27]. The key generation module consists mainly of a Montgomery exponentiation circuit and a true random number generator (TRNG) module, which is not in the scope of this paper. Finally, it is worth noting that the usage of only one multiplier and one exponentiator is enough to achieve the desired encryption results, but that results in a high critical path delay.

2) CEG Scheme Implementation: Fig. 3 shows the block diagram for our implementation of the CEG encryption/decryption scheme. This design is quietly different from ElGamal design discussed before as the encryption operations defined in (5) requires the usage of multiple Montgomery exponentiators. As the timing delay needed by one exponentiator is more than the delay of a single multiplier, the FSM controller is modified to utilize only one Montgomery multiplier. A modular reducer circuit is used to handle the operation of reducing m into several m_i based on the relation of $m_i = m(mod\ d_i)$ for $i = 1, \dots, t$.

For the decryption module, it consists of one Montgomery

modular exponentiator, one Montgomery modular multiplier, one modular divider, one modular adder, FSM controller, and a single block of memory used to facilitate the implementation of the inverse CRT needed in (6) [24]. Input and output vectors are now $k \times t$ *bits* instead of k *bits*, where t is the number of ciphertext pairs.

B. Dual-Circuit Design

The main motivation for this design is that some third party IPs require the usage of more than one single type of operation. For instance, an IP may need to perform both addition and multiplication but not at the same time. One can imagine the functionality of that IP as a simple ALU that uses a selection line to switch its mode between two different operations. In this case, using one type of partial homomorphic schemes would not be sufficient. We have to implement two different schemes, such as implementing the two schemes described above, in order to prevent the attacker from revealing the ALU input and output data. We suggest a solution for this issue by combining the two previously proposed schemes, ElGamal and the CEG, in a single dual-circuit design. Thus, the proposed design supports both additive and multiplicative homomorphism.

Furthermore, we try to share resources as much as we can between the two schemes in order to have minimal design cost. For example, computing C_1 in (3) and (5) needs an exponentiation operation. The same situation occurs when computing C_2 as we need an exponentiation operation followed by a multiplication operation. The only difference is that the modified versions in (5) reuse their modules many times based on the value of t. Thus, we use the duality concept that enables us of sharing as much resources between the two circuits in order to reduce the design area. As the CEG scheme uses the same basic blocks of ElGamal scheme with some additional blocks, we depend on the same architecture shown in Fig. 3 and add a *select* signal that chooses between the multiplicative homomorphic one and the additive homomorphic algorithms. The FSM controller is modified to be able of handling the two cases with the same building modules. The case is the same for the key generation and decryption modules.

By using this simple idea, we manage to decrease the area cost a lot and allow for the two homomorphic properties to be available on a single module. That completely solves the issue of the third party IP, which needs to perform both addition and multiplication operation. Moreover, another possible example for an application that needs the availability of both homomorphic operations is when we have two unique IPs in a design and the first IP performs addition while the second IP performs multiplication. Assuming that both IPs will not work on the same time, one can instantiate only one instance of our dual-circuit module and control its functionality to perform the needed operation of any of the two IPs, when needed, with the minimal cost in area and power consumption.

V. Experimental Results

This section evaluates the performance of our proposed methods, described in Section IV, in terms of resource utilization, delay, and power consumption. The proposed methods are implemented on Xilinx Spartan-6 XC6SLX75 with FGG484 package and -2 speed grade. The area and performance results are obtained from the Xilinx ISE 14.6 tool after place and route analysis. The power is calculated using Xilinx Xpower Analyzer with 100 MHz clock.

A. PH Schemes Results

Table I shows the top-level module resource utilization of our two partial homomorphic encryption/decryption schemes, ElGamal and CEG, using vectors of size equals 8 bits.

Table II shows the maximum operating frequency of the two proposed partial homomorphic schemes along with the needed number of cycles to finish their work.

TABLE I: Resource utilization of ElGamal and CRT-based ElGamal (CEG) encryption/decryption schemes for k = 8 bits.

	Encryption		Decryption	
	ElGamal	CEG	ElGamal	CEG
Number of Registers	295	614	207	364
Number of LUTs	420	715	259	442
Number of BRAMs	0	0	0	1

TABLE II: Timing performance of ElGamal and CRT-based ElGamal (CEG) encryption/decryption schemes for k = 8 bits.

	Encryption		Decryption	
	ElGamal	CEG	ElGamal	CEG
Frequency (MHz)	161.277	164.352	123.870	121.862
No. of Cycles	171	480	153	512

TABLE III: Power consumption (mW) of ElGamal and CEG encryption/decryption schemes for k = 8 bits.

	Encryption		Decryption	
	ElGamal	CEG	ElGamal	CEG
Clocks	5.65	7.87	4.21	5.87
Logic	3.84	5.47	2.70	3.69
Signals	2.82	4.69	2.01	3.23
BRAMs	0.00	0.00	0.00	0.74
IOs	16.51	8.99	5.23	2.74
Leakage	65.00	65.00	64.00	64.00
Total	93.82	92.02	78.15	80.27

From power prospective, Table III shows the power analysis for ElGamal encryption/decryption scheme and the CRT-based one. It was found that the dynamic power slightly decreased in case of encryption and increased in case of decryption due to the usage of the memory component and its logic controller in decryption. The leakage power remains constant in the both cases.

B. Dual-Circuit Design Results

Here, we compare the results of our proposed dual-circuit design to using regular two IPs, one for ElGamal and another for CEG design without any resource sharing between them. We want to address the effect of our resource sharing. In order to differentiate between the two designs, we call the first design, *Dual ElGamal*, while the second design is called *Regular ElGamal*.

Firstly, Table IV shows the area reduction that results from using our *Dual ElGamal* design over *Regular ElGamal* design. The area reduction column is calculated using (8). It is clear that the idea of dual-circuit design has greatly improved the usage of hardware resources.

$$Reduction(\%) = \frac{Regular\ area - Dual\ area}{Regular\ area} \times 100. \quad (8)$$

Table V gives the maximum operating frequency of our *Dual ElGamal* design and the *Regular ElGamal* design using vectors of size $k = 8\ bits$. The number of cycles here represents the clock cycles needed to perform one multiplicative homomorphic operation followed by one additive homomorphic operation. The needed number of cycles to get the final output is the same in both designs, except that the encryption part of our dual designs utilizes more clock cycles.

TABLE IV: Area reduction of our Dual ElGamal design over the Regular ElGamal design for k = 8 bits.

	Encryption			Decryption		
	Regular ElGamal	Dual ElGamal	Area reduction (%)	Regular ElGamal	Dual ElGamal	Area reduction (%)
Registers	909	635	30.14	536	364	32.09
LUTs	1137	735	35.36	626	457	26.99
BRAMs	0	0	00.00	1	1	00.00

TABLE V: Timing comparisons between our Dual ElGamal design and the Regular ElGamal design for k = 8 bits.

	Encryption		Decryption	
	Regular	Dual	Regular	Dual
Frequency (MHz)	161.277	158.51	117.099	121.344
No. of Cycles	651	662	665	665

TABLE VI: Power consumption (mW) of our Dual ElGamal design and the Regular ElGamal design for k = 8 bits.

	Encryption		Decryption	
	Regular	Dual	Regular	Dual
Clocks	11.78	6.89	8.78	4.86
Logic	9.25	6.29	5.91	3.82
Signals	8.14	6.02	5.67	3.49
BRAMs	0.00	0.00	0.74	0.74
IOs	25.27	10.83	5.67	3.61
Leakage	65.00	65.00	65.00	64.00
Total		119.44	95.03	91.77	80.52

That is due to the usage of only one Montgomery multiplier instead of two, as illustrated in Section IV.

From power prospective, Table VI shows the power analysis for our *Dual ElGamal* design and the *Regular ElGamal* design. The usage of the duality idea results in an obvious improvement in total power consumption as it eliminates the power consumed by the duplicated modules. The savings in power consumption are 20.44% for encryption and 12.26% for decryption.

VI. CONCLUSION

In this work, we highlighted the importance of homomorphic encryption in defeating Hardware Trojans in third party IPs. As PH is sufficient enough with some third party IPs, we implemented two designs that supports PH (multiplicative only and additive only) based on ElGamal encryption/decryption scheme.

Furthermore, we integrated the two designs together and introduced a dual-circuit design that achieved a great improvement in area and power over a regular design that combines two IPs, one for ElGamal and another for CEG, without any resource sharing between them. Our architectures were implemented on a low-cost Xilinx Spartan-6 FPGA and area, delay, and power results were reported.

REFERENCES

[1] M. Tehranipoor and F. Koushanfar. A Survey of Hardware Trojan Taxonomy and Detection. *IEEE Design & Test of Computers*, 27(1):10–25, January 2010.

[2] Y. Gahi, M. Guennoun, Z. Guennoun, and K. El-Khatib. An Encrypted Trust-based Routing Protocol. In *IEEE Conference on Open Systems (ICOS)*, 2012, October 2012.

[3] D. Hrestak and S. Picek. Homomorphic Encryption in the Cloud. In *37th International Convention on Information and Communication Technology, Electronics and Microelectronics (MIPRO)*, 2014, pages 1400–1404, May 2014.

[4] H. Salmani, M. Tehranipoor, and J. Plusquellic. New design strategy for improving hardware Trojan detection and reducing Trojan activation time. In *Proceedings of the IEEE International Workshop on Hardware-Oriented Security and Trust, HOST '09*, pages 66–73, July 2009.

[5] J. Rajendran, V. Jyothi, O. Sinanoglu, and R. Karri. Design and analysis of ring oscillator based Design-for-Trust technique. In *VLSI Test Symposium (VTS), 2011 IEEE 29th*, pages 105–110, May 2011.

[6] A. Al-Anwar, Y. Alkabani, M.W. El-Kharashi, and H. Bedour. Defeating hardware spyware in third party IPs. In *Saudi International Electronics, Communications and Photonics Conference (SIECPC)*, April 2013.

[7] A. Al-Anwar, Y. Alkabani, M.W. El-Kharashi, and H. Bedour. Hardware Trojan Protection for Third Party IPs on FPGA. In *16th EUROMICRO Conference on Digital System Design*, pages 662–665, Santander, Spain, September 2013.

[8] M. Tarek Ibn Ziad, A Al-Anwar, Y. Alkabani, M. W. El-Kharashi, and H. Bedour. E-Voting Attacks and Countermeasures. In *28th International Conference on Advanced Information Networking and Applications Workshops (WAINA)A-2014*, pages 269–274, Victoria, BC, Canada, May 2014.

[9] A Al-Anwar, Y. Alkabani, M.W. El-Kharashi, and H. Bedour. Hardware Trojan detection methodology for FPGA. In *IEEE Pacific Rim Conference on Communications, Computers and Signal Processing (PACRIM)*, pages 177–182, Victoria, BC, Canada, August 2013.

[10] R. Rad, J. Plusquellic, and M. Tehranipoor. A sensitivity analysis of power signal methods for detecting hardware Trojans under real process and environmental conditions. *IEEE Trans. Very Large Scale Integr. Syst.*, 18:1735–1744, December 2010.

[11] Y. Jin and Y. Makris. Hardware Trojan detection using path delay fingerprint. In *Proceedings of the IEEE International Workshop on Hardware-Oriented Security and Trust*, pages 51–57, June 2008.

[12] S. Wei and M. Potkonjak. Scalable Hardware Trojan Diagnosis. *IEEE Transactions on Very Large Scale Integration (VLSI) Systems*, 20(6):1049–1057, June 2012.

[13] M. Tehranipoor, H. Salmani, X. Zhang, X. Wang, R. Karri, J. Rajendran, and K. Rosenfeld. Trustworthy Hardware: Trojan Detection and Design-for-Trust Challenges. *Computer*, 44(7):66–74, July 2011.

[14] X. Zhang and M. Tehranipoor. Case study: Detecting Hardware Trojans in Third-party Digital IP Cores. In *2011 IEEE International Symposium on Hardware-Oriented Security and Trust (HOST)*, pages 67–70, San Diego, CA, USA, June 2011.

[15] A. Baumgarten, A. Tyagi, and J. Zambreno. Preventing IC Piracy Using Reconfigurable Logic Barriers. *IEEE Design & Test of Computers*, 27(1):66–75, January 2010.

[16] C. Gentry. *A Fully Homomorphic Encryption Scheme*. PhD thesis, Stanford University, 2009.

[17] N. Göttert, T. Feller, M. Schneider, J. Buchmann, and S. Huss. On the Design of Hardware Building Blocks for Modern Lattice-based Encryption Schemes. In *Proceedings of the 14th International Conference on Cryptographic Hardware and Embedded Systems, CHES'12*, pages 512–529, Leuven, Belgium, 2012.

[18] T. Pöppelmann and T. Güneysu. Towards Efficient Arithmetic for Lattice-Based Cryptography on Reconfigurable Hardware. In *Progress in Cryptology - LATINCRYPT 2012*, volume 7533 of *Lecture Notes in Computer Science*, pages 139–158. 2012.

[19] S. Goldwasser and S. Micali. Probabilistic encryption. *Journal of Computer and System Sciences*, 28(2):270–299, 1984.

[20] J. B. Clarkson. Dense Probabilistic Encryption. In *Proceedings of the Workshop on Selected Areas of Cryptography*, pages 120–128, 1994.

[21] P. Paillier. Public-key Cryptosystems Based on Composite Degree Residuosity Classes. In *Proceedings of the 17th International Conference on Theory and Application of Cryptographic Techniques, EUROCRYPT'99*, pages 223–238, New York, USA, 1999.

[22] R. L. Rivest, A. Shamir, and L. Adleman. A Method for Obtaining Digital Signatures and Public-key Cryptosystems. *Commun. ACM*, 21(2):120–126, February 1978.

[23] T. El Gamal. A Public Key Cryptosystem and a Signature Scheme Based on Discrete Logarithms. In *Advances in Cryptology*, volume 196 of *Lecture Notes in Computer Science*, pages 10–18. 1985.

[24] Y. Hu, W. J. Martin, and B. Sunar. Enhanced Flexibility for Homomorphic Encryption Schemes via CRT. In *International Conference on Applied Cryptography and Network Security, ACNS*, pages 93–110, Singapore, June 2012.

[25] P. L. Montgomery. Modular Multiplication Without Trial Division. *Mathematics of Computation*, 44(170):519–521, 1985.

[26] J. P. Deschamps. *Hardware Implementation of Finite-Field Arithmetic*. McGraw-Hill, Inc., New York, USA, 2009.

[27] J. P. Deschamps and G. Sutter. Hardware Implementation of Finite-Field Division. *Acta Applicandae Mathematicae*, 93(1-3):119–147, 2006.

978-1-4799-8720-7/15 $31.00 © 2015 IEEE

2015 IEEE Computer Society Annual Symposium on VLSI

SecX: A Framework for Collecting Runtime Statistics for SoCs with Multiple Accelerators

Rajshekar Kalayappan and Smruti R. Sarangi

Department of Computer Science & Engineering, Indian Institute of Technology, New Delhi, India
{rajshekark,srsarangi}@cse.iitd.ac.in

Abstract—We are moving into an era where large SoCs will have a portfolio of different kinds of cores and accelerators. Many of these computational elements might be designed by third parties. In this setting, it is beneficial to collect accurate runtime information such that we can diagnose performance problems, verify and report correctness issues, and collect usage scenarios of third party hardware. This problem is non-trivial if we consider the possibility of defective or malicious elements in the chip. We design an architecture, *SecX*, which helps us collect various metrics of potential interest in a fair, reliable and secure fashion. These logs can subsequently be made available to users, IP vendors, and the system integrator through a trusted third party called the *auditor*. The performance (0.03%), power (1.04W) and area (0.32%) overheads of our scheme are minimal.

I. INTRODUCTION

Designers are increasingly shifting towards incorporating more and more specialized accelerators on chip, instead of having many general purpose cores [1]. These accelerators can perform a variety of tasks such as encryption, compression, XML parsing, network packet processing, and regular expression matching. Such multi-core processors with a host of accelerators are already being manufactured by major processor vendors. Examples of such processors include Intel Ivybridge, IBM PowerEn [2], and IBM Power7.

In the future, we expect to see this trend continuing, and a section of the research community believes [2] that a lot of these accelerators will be designed by third parties. In this context, we note that there will be multiple parties involved in the process of designing an SOC – a *user* defines the system requirements, a *system integrator* designs a system that can meet the requirements by using its own designs and sourcing third party IPs from multiple *accelerator vendors*. It is expected that a heterogeneous system comprising of a myriad of such components will have a host of timing, correctness, security, and performance issues (as reported in [3, 4]). As a result, there is a need for a reporting (auditing) framework to collect information from processors in the field, and make it available to the various parties for offline analyses with user consent. Such a framework finds diverse applications in debugging, malware analysis, and reporting of usage scenarios as explained in Table I. Current approaches are either at a very high level such as performance counters, or are at a very low level such as scan chains and similar trace collection buffers. We desire a hardware solution that operates at the interface of the cores and the accelerators, and logs the interaction between these entities.

An added requirement of such a framework is that it has to be trusted by all the parties (users, system integrators, vendors), and should be tamper proof. This is important as all parties stand to gain from manipulation of logs. For instance, consider a scenario where the system integrator is in charge of the auditing. Now suppose a user's task fails to meet the deadline because of a poorly designed interconnect. The system integrator may instead fabricate logs that implicate some of the accelerators for the delay. This can have an adverse effect on the accelerator vendors from a business

standpoint. Note that we assume a trusted foundry in the rest of the paper. There are works that focus on fabrication at untrusted foundries [5]; however, this is beyond the scope of this paper.

Fair, Tamper-Proof, Secure and Reliable Information Gathering: We design such a reporting framework called *SecX*, that efficiently collects and disseminates useful runtime information to the concerned parties, in a manner that is fair to all concerned, that cannot be illegally influenced by anyone, and that does not disclose anything sensitive.

We define four metrics that capture the framework's requirements: QoS (quality of service), QoE (quality of environment), HoI (hash of input), and HoO (hash of output). The QoS metric refers to accelerator throughput or the duration of jobs, depending upon the context. The QoE metric describes the runtime environment perceived by an accelerator. This includes the memory latency, and available memory bandwidth for the accelerator. HoI is the hash of the input to an accelerator, and the HoO metric is the hash of the output. These four metrics are collected by a third party called the *auditor*, one who is trusted by all the parties (vendors, users, and system integrators). The *auditor* makes the logs available to the various parties on-demand, who benefit from them as shown in Table I. Note that the concept of a trusted third party is not unrealistic. Intel's Trusted Execution Technology (TXT) uses a secure cryptographic processor [7] provided by a trusted third party that both hardware and software developers trust.

The system integrator and the IP vendor incorporate *auditor*-provided macros in their designs. During runtime, the auditing hardware elements co-ordinate to fairly, securely and reliably log the four metrics for each task run on an accelerator. The logs are then transferred to the servers of the *auditor*. The different parties then request the *auditor* for the logs. Authentication and data integrity is ensured by *SecX*. At the moment, we collect general architectural parameters; however our framework can capture very specific parameters also. In this paper, we look at the fundamental issue of guaranteeing the correctness and consistency of the collected information, and utilizing it gainfully (Table I) to find and report performance and correctness problems.

II. RELATED WORK

The field of DFD (design for debug) architectures is very extensive. Over the last ten years researchers have proposed a lot of techniques to collect debugging information (in the form of logs), and analyze them for defects(representative papers [8, 9]). Chen et al. [10] create accelerator specific debug logs (in hardware) for both correctness and performance problems. Our work is different from this line of research as for us the collection of logs is difficult because of potentially malicious components in the system.

Another related line of research is the efforts made to detect and/or protect against malicious hardware. Hicks et al. [11] highlighted the problem of malicious hardware accelerators. Subse-

978-1-4799-8720-7/15 $31.00 © 2015 IEEE   137

TABLE I
NEED FOR RUNTIME INFORMATION REGARDING DIFFERENT ON-CHIP COMPONENTS

Application	Description	Realization Through SecX
Improving Accelerator Designs	Accelerator designers will find usage scenarios useful, e.g., *typical* latencies and throughputs of the accelerator's accesses to the memory. They can use this to improve their pipeline designs – how best to overlap compute/communicate phases to reduce run time and/or power.	QoS and QoE help the *vendor* to study performance under different environment conditions
Analyzing System Throughput	System integrators will find extremely useful the breakdown of delays spent in different parts of the chip. For each task, we can know the time consumed at the cores, accelerators, memory system and the interconnect. This will aid in the identification of bottlenecks and improvement of future designs.	QoS and QoE help the *system integrator* to study performance of different components, both her own and externally sourced
Handling Correctness Issues	The accelerators may give incorrect results. Defects going undetected through post-silicon validation is well documented. The user will like a reporting framework to raise a concern with the vendor.	*User* detects functional bug (possibly offline); reports to *vendor*; uses HoI and HoO to corroborate her claim
Handling Timing Issues	An accelerator may take too long to complete a task. This, again, is something that cannot be caught during the validation phase for many accelerator classes. E.g., the run time of a linear programming accelerator is closely tied to the actual input. In a soft real time scenario, deterministic accelerator behavior is needed to meet deadlines.	*User* detects timing bug (possibly offline); reports to the *vendor*; uses QoS to corroborate her claim
Handling Security Issues	The accelerators can also possibly be deliberately malicious [6] – they can deliberately compute wrongly, compute slowly, disrupt the rest of the chip (DoS attacks) or perform data theft.	Computing wrongly/slowly is proven using QoS/QoE/HoI/HoO, thus discouraging accelerators; DoS and data theft are prevented by gateways (Section III-B)
Enforcing Service Level Agreements	What if the user falsely reports correctness or timing issues? What if the system integrator/accelerator vendor falsely blames the other for a missed deadline? This could damage the reputation of an accelerator vendor / system integrator and hamper her future in the industry.	*auditor* signed logs prevent the parties from making false claims

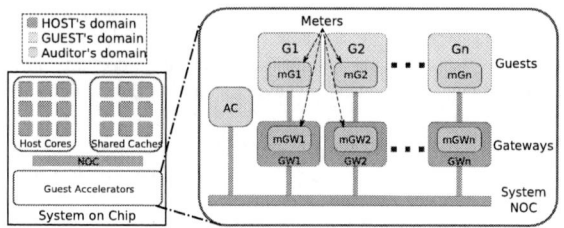

Figure 1.   Hardware architecture

quently, there have been several proposals to detect and remove such hardware using a host of techniques such as security assertions, and pattern mining [12]. HybridOS [13] seeks to isolate defective or malicious accelerators at the OS level. Our contribution can enrich these techniques with high quality runtime information.

III. ARCHITECTURE

A. Hardware Overview

A SoC can be viewed as being composed of circuitry belonging to different domains – *host*, *guest* and *auditor*, as shown in Figure 1. The *host* domain refers to all components designed by the system integrator. The system integrator procures third-party accelerators from IP vendors in the form of hard macros. These third party accelerators constitute the *guest* domain. Each guest is allowed access to the rest of the chip only through host-maintained gateways, as shown in the figure. The auditor provides meters (in the form of hard macros) to both the accelerator vendor and the system integrator. The latter two integrate the meters into their designs as shown in Figure 1. The meters and the auditor-comptroller(AC) together form *SecX*, and constitute the *auditor* domain.

As can be seen in Figure 1, for every accelerator in the system, there are two meters – one embedded in the guest design (mG) and the other in the host gateway (mGW). The need for dual-metering is explained in Section IV-B2. The two meters work together to collect the runtime statistics in a fair, secure and reliable manner. The collected logs are maintained by the central Auditor-Comptroller (AC).

B. Hardware Components

Auditor-Comptroller (AC): The Auditor-Comptroller (AC) is a small structure that handles the functionalities of receiving, validating, processing, storing and serving logs. Logs are maintained for each task run on each guest. The contents of each log are as

TABLE II
DESCRIPTION OF THE LOGS

Timestamp	16 B	job-id	8 B
guest-id	1 B	[up to 256 guests]	
job-type	1 B	[up to 256 job types]	
QoS.Latency	4 B	QoS.Throughput (B/s)	4 B
QoE	256 B	[$R \times nbins \times 4$ (R=4, nbins=16)]	
HoI	128 B	HoO	128 B
Checksum	32 B	[SHA-256]	
Total	**578 Bytes**	[R = No. of resources]	

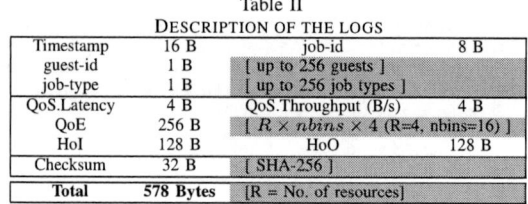

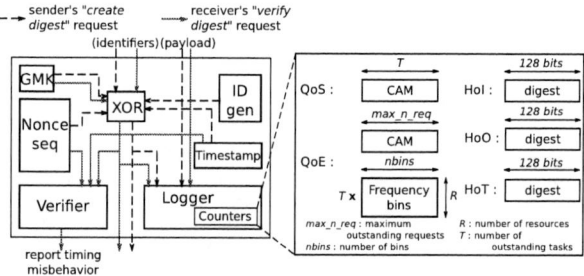

Figure 2.   Structure of a meter

shown in Table II. Each log is tagged with a timestamp and a job-id. A hash-based message authentication code (HMAC) (based on 256-bit SHA) of the log is made to ensure integrity and authenticity. The AC has its own limited non-volatile storage to store the logs. Periodically, it encrypts and stores detailed auditor logs in the host's storage device. The ACs in all chips in the field are expected to periodically upload the logs to the auditor's servers. The various parties – system integrator, IP vendor and user – may contact the auditor to obtain the logs.

Meters: The schematic of a meter is shown in Figure 2. The meters are responsible for the collection of the QoS, QoE, HoI, and HoO metrics. We use a CAM array to store the starting time of each active task on the accelerator. When a task finishes, we access the CAM to find its starting time, and subsequently compute its duration. For QoE, we compute a distribution of resource access latencies using binning. Each bin is essentially a frequency count (4 bytes) of the number of accesses whose latencies fall in the particular bin. The number of bins, and the range of values that a particular bin corresponds to is accelerator/resource specific. *SecX* imposes no restriction in this regard. Multiple outstanding resource accesses are possible – here again a CAM is employed to store the starting time of each resource access request. Additionally, the meter is capable of performing tabulation hashing to compute

978-1-4799-8720-7/15 $31.00 © 2015 IEEE   138

HoI, HoO and HoT (hash of all traffic – see Section IV-B3). QoS, QoE, HoI, HoO are maintained only at mGW; HoT is computed at both the meters. The meters and the AC are capable of generating time-stamps, and have dedicated circuitry to perform different cryptographic operations (see Section IV-B5). The parameters and sizes of all the structures are shown in Table VI (in Section V).

Gateways: Gateways (GWs) are host-maintained circuits that form the interface for a guest to the rest of the chip. This allows the host to prevent any unauthorized access or signaling by the guest. Gateways essentially implement resource access lists. They also maintain a small TLB to translate the guest's memory requests (see Section IV-A).

IV. SYSTEM OPERATION

A. Functional Layer

1) Job Creation: The accelerators expose a software API that allows users to access them. A function to access a guest has the following form:
*int **issue**(int taskcode, void * input, void * output, int ip-length, int op-length)*

When the user executes this function, a host core sends a "job create" message to the guest containing the *taskcode* (task type, as the guest may be capable of performing more than one task), memory addresses of the input and the output, and the lengths of the input and the output segments.

After receiving this message, the gateway configures the access lists that it maintains, and it forwards the message to its meter *mGW*. mGW generates a job-id and records the job start time (required for the QoS metric), and initializes the frequency bins for the QoE metric. It also initializes the input and output hashes. The computation of the input/output hashes requires the mGW to be able to distinguish between guest accesses to the input data, output data and the temporary working area. We assume that the input and output memory regions cannot be used as temporary work-areas, and cannot be modified by other cores while the job is running (due to memory consistency issues). The mGW maintains a table of memory ranges for the input and output data of each active job. It accesses this table to find out if a memory request is reading an input, or writing to an output. It can then add the data read or written by the memory request to the HoI or HoO. Note that these are permutation independent hashes (order of accesses does not matter). Subsequently, the request (timestamped by the mGW) is forwarded to the guest, where its meter verifies that its current time is within τ cycles of the recorded job start time. If the time difference exceeds a threshold, then most likely there is a malicious delay, and the job is dropped.

2) Job Execution: During the execution of a job, the guest reads input data, performs computations on it. We assume the Rigel memory model [1] that enforces software based coherence at task boundaries. For security reasons, the gateway maintains a 32 entry 4-way associative TLB that maintains a portion of the host's TLB. This TLB is used to translate virtual addresses into physical addresses. If there is a TLB miss, then a request is sent to the host core. Now, when a guest wishes to read or write data, it sends a request to its meter (mG). It timestamps the message, and sends it to mGW. The mGW checks for malicious delays by comparing the timestamps, and then updates the HoO if it is a write request. The gateway then proceeds to construct a memory request by accessing its TLB. After the request returns (in the case of a read), mGW

notes the time, computes the duration of the request, and updates HoI . Subsequently, the gateway forwards the response to the guest. mG follows the same protocol (same as while sending the message) to verify the timestamps, and check for malicious delays. Along with accessing memory, accelerators can access other resources as well (such as I/O devices). For all communication between the guest and the gateway, mG and mGW both individually maintain a permutation-independent hash of all traffic (*HoT*). These hashes are used to identify any data corruption in the communication.

3) Job Completion: The guest sends a job completion message, containing its computed HoT, to the GW once it is done with its computation. The GW forwards the message to the mGW, who compares its HoT against the one received to detect any data corruption. The mGW then computes the latency of the job, and the average throughput (output bytes/second). It compiles the various components of the QoS, QoE, and the input/output hashes to form the job log. It then creates a digest (SHA) of the log. The log, along with the digest is then sent to the AC via the GW. The AC, upon receipt of the log stores it in its local storage.

B. Security Layer

Owing to the fact that multiple parties are involved, a variety of misbehavior patterns and fairness issues are possible. The different parties can potentially project each other in bad light, which can have significant repercussions. The *SecX* framework, which all parties trust, must prevent this. In this section, we will extend the functionality explained in the previous section with security and cryptographic aspects [14] to form a complete solution.

1) SecX Setup: We begin with explaining the setup of *SecX*. When the chip is powered on for the first time (or maybe after burn-in tests), the AC needs to verify that all the meters in the system are genuine. We suggest the employment of Physically Unclonable Functions (PUFs) in the meters. A PUF typically uses some random phenomena such as lithographic variations to compute a unique key for a circuit. We assume that the fabrication facility is trusted, and it can record the PUF based keys for each meter after fabrication (see [15] for more details). To verify that the meters are genuine, the AC can aggregate the random keys, and send them to a remote server, which can validate the keys. After verification, the AC generates a Global Meter Key GMK (shared between all the *SecX* hardware). It employs the technique of public PUFs suggested in [16] to securely distribute the GMK to all meters. Additionally, the GMK may be periodically updated to improve security. After receiving the GMK, each mGW sets up a nonce sequence (sequence of random values) and communicates this (securely, using the shared GMK) to its companion mG.

2) Handling Malicious Delays: We begin by explaining the need for a dual-metering scheme at the host-guest interface. Guest resource request/response messages cross the guest-host domain. If the latency is measured on the host side, the host can delay the sending of the response to the guest after the latency measurement, in a bid to affect the QoS. If the latency is measured on the guest side, the guest can delay the sending of the request after getting it timestamped (or delay the sending of the response to its meter), in a bid to affect the QoE. To guard against these, meters on both sides are required. The meter at the sending side timestamps the message, and the meter at the receiving side verifies that no undue delay was introduced. It is to the advantage of the sender to send the message to its meter as soon as possible for timestamping.

Similarly, it is to the advantage of the receiver to use the message as soon as possible after timestamp verification at its meter. Any delay induced in between timestamping at the sender, and verification at the receiver, by either party, is detected by the dual-meter system.

Only the meters can encrypt the timestamps because they have a unique key to perform the XOR based encryption (see Section IV-B5). This prevents parties from generating their own timestamps. It must be noted that this scheme only allows the detection of misbehavior. Identifying the guilty party is a harder problem, and is something we are pursuing. Table III shows the complete guest resource access protocol (functional and security aspects). This detection of malicious delays has a minimal effect on performance.

3) Handling Malicious Data Corruption: The host/guest can potentially provide the other with incorrect data, and accuse the latter of incorrect functionality. This is countered by having *SecX* maintain a hash of all traffic between the host and the guest (HoT). Just like in the case of malicious delays, computing the hash at only one of the meters proves insufficient. The solution is to maintain the HoT at both the host and guest meters, and compare the two HoTs at the time of job completion. A mismatch indicates a data corruption at some point of the host-guest communication. The HoT is not stored as part of the logs.

4) Authentication and Integrity: In our example system, the NoC is in the host's domain, allowing it access to all messages sent/received. This could compromise fairness, allowing for false accusations of a guest being defective. Thus, the logs, before transfer from the mGW to the AC, are encrypted. These encrypted logs are stored in the host's storage. SHA-based signatures, with a secret key known to all the auditor hardware is used in this case.

5) Cryptography Support: We use multiple types of encryption depending upon the security threats, and the time required for encryption and decryption. We use fast XOR based encryption for encrypting the timestamps of memory requests and responses, which are frequent messages. We use SHA encryption for a more infrequent class of messages related to the job completion protocol. Tabulation hashing is used to create digests of the input, output and all traffic, which is off the critical path. Lastly, for the AC's communication with the auditor's servers, we need to use one of the slowest (100+ cycles) yet strongest methods: public key encryption.

- **Simple XOR** : When messages move from one domain (host/guest) to the other, the two parties may induce delays to affect the QoS/QoE in their favor. To check this, a dual metering system, where their attached timestamps are checked (explained in Section IV-B2), is used. Now, these timestamps may be spoofed by the malicious party to escape detection. To protect against this, the timestamps have to be encrypted using a key private to the pair of meters. Also, these cross-domain messages are frequent. Therefore, the encryption mechanism must be light-weight. For this reason, a simple XOR-based encryption, along with nonce sequences is used.
- **SHA** : We employ SHA based encryption when integrity, authenticity and non-repudiation of messages is important. For each sender-receiver pair, a secret key is set up that is shared between the two parties. If party A sends a message to B, the message is attached with a digest of the message using the shared key. This digest allows B to verify that the sender was indeed A, and that the message was not tampered with. *SecX* employs SHA to sign logs.

Table IV
SIMULATION PARAMETERS

Parameter	Value	Parameter	Value
System Configuration			
Cores	24		
Accelerators	24	Accelerator Types	6
Technology	14 nm	Frequency	3.4 GHz
Shared Elements			
L3 cache	64MB (32 banks)	Main Memory Latency	200 cycles
NoC and Traffic			
Topology	2-D Torus	Routing Alg.	dyn XY routing
Flit size	16 bytes	Hop-latency	1 cycle
Router-Latency	3 cycles		
General Purpose Core Configuration			
Retire Width	4	Issue Width	6
Private L1 i-cache, d-cache			
Size	32 kB	Latency	4 cycles
Private L2 Unified Cache			
Size	256 kB	Latency	12 cycles
Cryptographic Circuitry			
XOR Encryption	4 cycles	SHA Encryption	41.04ns [17]

- **Tabulation Hashing** : Simple tabulation hashing is used by the meters to maintain hashes of the input/output/traffic of the accelerator. Apart from being simple to compute, this hashing technique has the desirable property of permutation-independence. The payload is broken down into fixed size tokens (8 bits). This token is used to index a precomputed secret table (4kB; known to the pair of meters only) to get a code (128 bits). This code is XORed with the current contents of the digest (128 bits) to give the new digest value.
- **RSA** : The communication between the AC and the auditor's servers is encrypted using RSA encryption.
- **Nonce Sequence** : The host/guest may perform a replay attack to induce a malicious delay. We avoid replay attacks by tagging messages with a nonce. The meter pairs share a secret nonce sequence.

V. EVALUATION : OVERHEAD OF AUDITING

In this section, we show that the performance, area and power overheads of *SecX* are minimal. We consider a system of 24 host cores and 24 guest accelerators with a 32-bank 64 MB LLC. The simulation parameters were derived from the designs of Intel Sandybridge, and IBM Power7 (see Table IV). A portfolio of 6 popular types of accelerators is considered – Table V compares their hardware and software implementations (scaled to $14nm$ using [24]). For the software timing study (Table V), single-threaded applications of the corresponding library were simulated using the Tejas simulator [25, 26]. To make the simulation more accurate, the relevant data was assumed to be in the LLC before-hand (no cold start misses to main memory). For the hardware timing study, we incorporated software implementations of the hardware accelerators in Tejas to give a heterogeneous chip comprising of general purpose cores and accelerators. We used Cadence tools to synthesize the control logic using the UMC $90nm$ standard cell library. Due scaling to $14nm$ was performed [24], and we used Cacti 5.3 [27] to estimate the area and delay of memory structures.

A. Performance Overhead

In this section, we present the performance overhead of performing secure, tamper-proof auditing. The utilization of the logs, be it to detect functional bugs or timing issues, is done offline.

We analyze the different stages of the auditing process and measure overheads. During the job issue phase, the initialization of the various counters and hashes is on the critical path. We estimate

Table III
GUEST RESOURCE REQUEST PROTOCOL

Resource Access				
Request				
(1)	guest → mG	$job\_id \| resource\_id \| payload$	$\|$ is the concatenation operator	
(2a)	mG → guest	$digest = XOR(GMK, job\_id \| req\_id$ $\| resource\_id \| cur\_time \| nonce)$	GMK – key known to all third-party hardware A unique req_id is generated by mG nonce → next number in a secret sequence of numbers shared between mG and mGW	
(2b)	mG		updates HoT (off critical path)	
(3)	guest → GW	$digest \| payload$		
(4a)	GW		verifies if resource access is legal	
(4b)	GW → mGW	$digest \| payload$		
(5)	mGW	$XOR(GMK, digest)$	decrypts and checks for timing misbehavior	
(6a)	mGW → GW	req_id		
(6b)	mGW		records req_id, $send\_time$, $resource\_id$ (off critical path)	
(6c)	mGW		updates HoT (off critical path)	
(7)	GW → resource	$req\_id \| payload$	with a TLB access	
Response				
(1)	resource → GW	$req\_id \| payload$		
(2)	GW → mGW	$req\_id \| payload$		
(3)	mGW → GW	$digest = XOR(GMK, req\_id \| cur\_time \| nonce)$		
(4a)	GW → guest	$digest \| payload$		
(4b)	mGW		uses $recv\_time$ to update QoE latency distribution (off critical path)	
(4c)	mGW		updates HoI and HoO (off critical path)	
(4d)	mGW		updates HoT (off critical path)	
(5)	guest → mG	$digest \| payload$		
(6)	mG	$XOR(GMK, digest)$	decrypts and checks for timing misbehavior	
(7a)	mG → guest	req_id		
(7a)	mG		updates HoT (off critical path)	
(8)	guest		guest uses payload	

Table V
ASIC V/S SOFTWARE IMPLEMENTATIONS OF ACCELERATORS

Task	Details	Citation	Technology	Original		Scaled to 14nm		Software	
				Latency	Area	Latency	Area	Details	Latency
FFT	complex 1024-point FFT	[18]	0.18 μm	3.2 μs	0.686 mm^2	2.3 μs	0.024mm^2	FFTW library	32.72 μs
Sort	sorting 3969 longs	[19]	0.13 μm	32 μs	2 mm^2	17.01 μs	0.07mm^2	libstdc++6	442.49 μs
JPEG	compressing 640 × 480 pixel bmp	[20]	0.09 μm	1.11 ms	0.062 mm^2	0.65 ms	0.004mm^2	Independent JPEG Group's library	2.16 ms
MD5	512 byte input	[21]	0.13 μm	1.96 μs	0.081 mm^2	1.04 μs	0.003mm^2	OpenSSL/Crypto	9.76 μs
AES	512 byte input	[22]	0.13 μm	1.25 μs	0.015 mm^2	0.753 μs	0.0005mm^2	library	4.749 μs
RSA	128 byte input	[23]	0.5 μm	325 ms	3 mm^2	133.7 ms	0.011mm^2		258 ms

this to be less than 15 cycles. During job execution, each resource access has to go through the dual-metering system. The critical path is the nonce retrieval (163 ps) and the XOR encryption (31 ps) at the sender's side, the transfer to the receiver (30 ps), and the XOR decryption at the receiver (31 ps). This would add a delay of 2 cycles (1 each for request/response) to an LLC access. When the job has completed, the logs are aggregated and signed using SHA. The logs are sent from the mGW to the AC, which stores it appropriately. These tasks can overlap with the host's utilization of the guest's output, and hence, are off the critical path.

Table V shows that software is roughly 2-20X slower on average. The additional delay of 2 cycles represents a worst-case overhead of 4% for a LLC read access (48 cycle, mean access time). However, full system simulations done using Tejas, revealed an average overhead of just 0.03%. The reason for this is as follows: the increase in resource access latency translates to a decrease in accelerator performance only if the parameters of the access (e.g., memory address for an LLC) depend upon the result of some computation. If there is no such dependency, then the access can be overlapped with another access or some computation, thus masking the auditing overhead. The nature of tasks typically accelerated are ones with statically determinable access patterns, allowing for significant overlap.

The NoC bandwidth overhead is minimal because the only usage is the transfer of logs from the meters to the AC. That is, per accelerated task, a mere 578 bytes (see Table II) is transferred.

B. Area Overhead

The overhead of auditing is dependent on the number of accelerators incorporated. The prime contributors to the area are the memory elements (volatile and non-volatile) that are required for the cryptographic operations and for storing/processing logs (in the case of the AC). The other significant contributors are the cryptographic circuits: SHA and RSA. Area estimates are obtained from ASIC implementations of published designs [17, 23]. Table VI gives the area breakup of the SecX hardware. The total additional area is $1.267mm^2$, for a system with 24 guests. Assuming a reference die size of $400mm^2$, the area overhead is 0.32%.

C. Power Overhead

The peak power consumption of SecX is estimated at 1.04W (TDP of 8-core Intel E5-2687W chip is 150W). Cacti [27] was used to estimate power for memory elements. References [29, 30, 31] were used for the major logic components. This estimate is assuming all 24 guests are operational, and the associated meters are working with a 100% duty cycle. In practice, the power consumed would be much lower as firstly, not all guests are simultaneously active, and secondly, the meters are inactive during the guest's compute phase.

VI. CONCLUSION AND FUTURE WORK

In this paper, we presented a mechanism for measuring useful metrics (QoS, QoE, HoI, HoO), an architecture for validating this information, and ultimately saving it in tamper proof logs. The

Table VI
AREA ESTIMATION OF AUXILIARY STRUCTURES

Component	Formula	Area Estimate	Details
Major Components of the Gateway Meter			
SHA		$0.003mm^2$	[17]
Memory for Tabulation Hashing Table	$2^8 \times 128bits$		
Memory for QoS	$T \times 8\ bytes$		T:number of tasks guest can run at the
Memory for QoE	$R \times T \times nbins \times 4\ bytes$		same time; R:number of different
Memory for I/O/T hashes	$T \times 3 \times 1024\ bits$		resources guest makes use of; $nbins$:
Memory for Nonce Sequence	$1024 \times 1\ bytes$		number of bins in frequency binning
Memory Total	Taking $R = 4, T = 4, nbins = 16$	$0.010mm^2$	8kB cache (Cacti [27])
Logic		$946.92\mu m^2$	
Sum		$0.014mm^2$	
Major Components of the Guest Meter			
Memory for Tabulation Hashing Table	$2^8 \times 128bits$		
Memory for HoT	$T \times 1024\ bits$		
Memory for Nonce Sequence	$1024 \times 1\ bytes$		
Memory Total	Taking $T = 4$	$0.010mm^2$	8kB cache (Cacti [27])
Logic		$296.98\mu m^2$	
Sum		$0.010mm^2$	
Major Components of the Auditor-Comptroller			
SHA		$0.003mm^2$	[17]
RSA		$0.014mm^2$	[23]
Memory for log processing		$0.010mm^2$	8kB cache (Cacti [27])
Non-volatile Memory for Audit Logs	$NJA \times 586\ bytes$; Taking $NJA = 7000$		NJA: number of jobs audited; refer Table II for log size
Non-volatile Memory Total		$0.664mm^2$	4MB 32nm RERAM [28]
Sum		$0.691mm^2$	
Total Area of Additional Hardware	Taking $num\_guests = 24$	$1.267mm^2$	
Area Overhead	Assuming a base chip area of $400mm^2$	0.32%	

performance (0.03%), power (1.04W) and area (0.32%) overheads are minimal. We wish to extend this work to consider an even larger set of possible attacks, and make minimal assumptions about trusted components.

REFERENCES

[1] J. H. Kelm, D. R. Johnson, W. Tuohy, S. S. Lumetta, and S. J. Patel, "Cohesion: An adaptive hybrid memory model for accelerators," *Micro*, 2011.

[2] R. Hou, L. Zhang, M. C. Huang, K. Wang, H. Franke, Y. Ge, and X. Chang, "Efficient data streaming with on-chip accelerators: Opportunities and challenges," in *HPCA*, 2011.

[3] M. Birnbaum and H. Sachs, "How vsia answers the soc dilemma," *Computer*, vol. 32, 1999.

[4] E. Haritan, H. Yagi, W. Wolf, T. Hattori, P. Paulin, A. Nohl, D. Wingard, and M. Muller, "Multicore design is the challenge! what is the solution?" in *DAC*, 2008.

[5] "Trusted foundry program." [Online]. Available: http://www.dmea. osd.mil/trustedic.html

[6] M. Hicks, M. Finnicum, S. T. King, M. Martin, and J. M. Smith, "Overcoming an untrusted computing base: Detecting and removing malicious hardware automatically," in *SP*, 2010.

[7] S. L. Kinney, *Trusted platform module basics: using TPM in embedded systems*. Newnes, 2006.

[8] B. Vermeulen and S. K. Goel, "Design for debug: catching design errors in digital chips," *IEEE Design & Test of Computers*, 2002.

[9] D. Josephson, "The good, the bad, and the ugly of silicon debug," in *DAC*, 2006.

[10] Y.-T. Chen, J. Cong, M. A. Ghodrat, M. Huang, C. Liu, B. Xiao, and Y. Zou, "Accelerator-rich cmps: From concept to real hardware," in *ICCD*, 2013.

[11] M. Hicks, M. Finnicum, S. King, M. Martin, and J. Smith, "Overcoming an untrusted computing base: Detecting and removing malicious hardware automatically," in *SP*, May 2010.

[12] M. Bilzor, T. Huffmire, C. Irvine, and T. Levin, "Security checkers: Detecting processor malicious inclusions at runtime," in *HOST*, 2011.

[13] J. H. Kelm and S. S. Lumetta, "Hybridos: Runtime support for reconfigurable accelerators," in *FPGA*, 2008.

[14] S. William and W. Stallings, *Cryptography and Network Security, 5/E*. Pearson Education India, 2006.

[15] Y. Alkabani and F. Koushanfar, "Active hardware metering for intellectual property protection and security." in *USENIX Security*, 2007.

[16] N. Beckmann and M. Potkonjak, "Hardware-based public-key cryptography with public physically unclonable functions," in *Information Hiding*, 2009.

[17] L. Dadda, M. Macchetti, and J. Owen, "The design of a high speed asic unit for the hash function sha-256 (384, 512)," in *DATE*, 2004.

[18] Y. Peng, "A parallel architecture for vlsi implementation of fft processor," in *ASIC*, 2003.

[19] N. Tabrizi and N. Bagherzadeh, "An asic design of a novel pipelined and parallel sorting accelerator for a multiprocessor-on-a-chip," in *ASIC*, 2005.

[20] M. Papadonikolakis, V. Pantazis, and A. Kakarountas, "Efficient high-performance asic implementation of jpeg-ls encoder," in *DATE*, 2007.

[21] A. Satoh and T. Inoue, "Asic hardware focused comparison for hash functions md5, ripemd-160, and shs," in *ITCC*, 2005.

[22] P. Hamalainen, T. Alho, M. Hannikainen, and T. D. Hamalainen, "Design and implementation of low-area and low-power aes encryption hardware core," in *DSD*, 2006.

[23] Z. Keija, X. ke, W. Yang, and M. Hao, "A novel asic implementation of rsa algorithm," in *ASIC*, 2003.

[24] W. Huang, K. Rajamani, M. Stan, and K. Skadron, "Scaling with design constraints: Predicting the future of big chips," *Micro*, 2011.

[25] G. Malhotra, P. Aggarwal, A. Sagar, and S. R. Sarangi, "ParTejas: A parallel simulator for multicore processors," in *ISPASS*, 2014.

[26] S. Sarangi, R. Kalayappan, P. Kallurkar, and S. Goel, "Tejas simulator : Validation against hardware," 2015. [Online]. Available: http://arxiv.org/abs/1501.07420

[27] S. Thoziyoor, N. Muralimanohar, J. H. Ahn, and N. P. Jouppi, "Cacti 5.1," HP Laboratories, Tech. Rep., 2008.

[28] X. Dong, C. Xu, Y. Xie, and N. Jouppi, "Nvsim: A circuit-level performance, energy, and area model for emerging nonvolatile memory," *TCAD*, 2012.

[29] A. T. Tran and B. M. Baas, "Design of an energy-efficient 32-bit adder operating at subthreshold voltages in 45-nm cmos," in *ICCE*, 2010.

[30] A. Sajid, A. Nafees, and S. Rahman, "Design and implementation of low power 8-bit carry-look ahead adder using static cmos logic and adiabatic logic," in *IJITCS*, 2013.

[31] A. K. Pandey, R. A. Mishra, and R. K. Nagaria, "Leakage power analysis of domino xor gate," in *ISRN Electronics*, 2013.

2015 IEEE Computer Society Annual Symposium on VLSI

Low-Area Reed Decoding in a Generalized Concatenated Code Construction for PUFs

Matthias Hiller*, Ludwig Kürzinger†, Georg Sigl*, Sven Müelich‡, Sven Puchinger‡ and Martin Bossert‡

* Institute for Security in Information Technology, Technische Universität München, Munich, Germany

† Fraunhofer Institute for Applied and Integrated Security (AISEC), Garching near Munich, Germany

‡ Institute of Communications Engineering, Ulm University, Ulm, Germany

{matthias.hiller, sigl}@tum.de, ludwig.kuerzinger@aisec.fraunhofer.de,

{sven.mueelich, sven.puchinger, martin.bossert}@uni-ulm.de

Abstract—Physical Unclonable Functions (PUFs) enable secure key storage for integrated circuits and FPGAs. PUF responses are noisy such that error correction is required to generate stable cryptographic keys. One popular approach is to use error-correcting codes.

We present an area-optimized VLSI implementation of a recent Generalized Concatenated (GC) code construction using Reed–Muller codes. Reed–Muller codes have the advantage that there exist very efficient decoders. Our new Reed decoding implementation makes extensive use of a circular shift register. The functionality is extended so that it can also handle erasure symbols to improve the error correction capability.

The overall GC code decoder occupies less than 110 slices and two block RAMs on an entry-level FPGA, and has a key error probability of 1.5×10^{-9}. The slice count is reduced by 50% compared to the reference implementation.

Index Terms—Physical Unclonable Functions (PUFs), Error Correction, Generalized Concatenated Codes, Reed–Muller Code, Reed Decoding, VLSI, FPGA.

I. INTRODUCTION

With the increasing usage of FPGAs in real-world commercial products, the protection of private data, as well as intellectual property on FPGAs becomes a more and more critical issue [1]. Physical Unclonable Functions (PUFs) [2] evaluate manufacturing variations to generate unique secrets that can be used for secure key generation. This permits to protect FPGAs with security primitives generated in their own, native FPGA fabric. The costs for deriving a secure key from a PUF are determined by the area that is occupied for the required building block.

Several PUF types are available for FPGAs, such as the Ring Oscillator (RO) PUF [3], the Butterfly PUF [4] or the SRAM PUF [5] that evaluate different physical effects and implement different measurement methods to generate the secrets. They all have in common that they require error correction, so helper data is generated to map the random PUF response into a codeword of an error correcting code (ECC) [6], [7], [8], [9].

The code-offset fuzzy extractor [6] is a popular construction that selects a random codeword of an ECC and XORs it with the PUF response. The result is stored as external helper data and does not reveal which of the possible codewords was chosen. Later when the device is operated in the field, the key that was derived from the initial PUF response is reproduced again. A new PUF response is read out and XORed with the

helper data to map the PUF response in the proximity of a codeword of the ECC. This enables to correct the errors in the PUF response. XORing the helper data again creates the original PUF response that is hashed to the cryptographic key.

In this work, we assume the PUF type, specification and fuzzy extractor as given and focus on efficient ECC design that is optimized for the strict area constraints in the PUF context.

There already exist several FPGA implementations using different code classes [5], [10], [11], [7], [8], [12], [13]. In PUFKY [12], Maes *et al.* presented a stand-alone module for FPGAs containing an RO PUF and the necessary error correction. We will use the scenario as reference because it is especially demanding since the target key error probability is set to the very low value of 10^{-9}. Reading out the ROs takes several milliseconds such that the implementation cannot be used for highly time sensitive applications. Therefore, we will optimize our ECC module fully for area and not for speed as well.

The constructions of Müelich *et al.* [14] and Puchinger *et al.* [15] improve the efficiency of the error correction for the scenario in PUFKY. The area-optimized hardware implementation in this work is based on their Generalized Concatenated (GC) code construction with Reed–Muller codes [16], [17].

Reed decoding is a very memory efficient approach to decode Reed–Muller codes [18]. Our extended and optimized Reed decoder that is also able to handle unreliable codeword symbols as erasures is the core component of the discussed design. The goal is to break down the decoding algorithm into simple operations and perform them sequentially. As a result, we are able to reduce the slice count of the decoder by up to 50% compared to the reference implementation in [12].

Our contributions:

- Design of the first PUF error correction module using generalized code concatenation over multiple codes
- Serialized Reed decoder architecture for efficient decoding of Reed–Muller codes
- Area-optimized hardware implementation and synthesis for an entry-level Xilinx Spartan-6 FPGA
- Comparison to the state-of-the-art implementation

978-1-4799-8720-7/15 $31.00 © 2015 IEEE

Outline:

Related ECC implementations for PUFs are discussed in Section II. Section III presents the GC code construction and Section IV gives an introduction to Reed–Muller codes and the Reed decoding algorithm. Our design decisions and the implementation are presented in Section V. In Section VI, our synthesis results are compared to the reference implementation.

II. RELATED WORK

In previous work, area-optimized decoders for different code classes were already developed for error correction on FPGAs in the PUF context.

Maes *et al.* [11] presented a Reed–Muller decoder for PUFs. The core of the decoder is a coprocessor with a custom instruction set that carries out the recursive decoding algorithm, such that a large number of intermediate results has to be stored. The standard Reed–Muller code is decoded as generalized multiple concatenated code [19], [17]. However, in the referenced work the actual code concatenation is still carried out in the conventional way where the codes are decoded separately. In contrast, we will apply the more powerful and generic concept of generalized code concatenation over different codes.

The BCH decoder by van Herrewege and Verbauwhede [20] used in the PUFKY reference [12] is also based on a custom instruction set. However, BCH decoding has a relatively high arithmetic complexity so that it is hard to further reduce the area consumption.

Seesaw by Hiller *et al.* [21] is a Viterbi decoder for convolutional codes that uses two mostly redundant data sets stored in block RAM. A small data path reads data from one data set and writes it to the other. After the update is complete, the data flow toggles and new results are written back into the other region.

In contrast to related work, our Reed decoder combines both, simple arithmetics and no need to store large intermediate results.

III. GENERALIZED CONCATENATED CODE CONSTRUCTION

This section discusses the GC code construction based on Reed–Muller codes that was presented by Müelich *et al.* in [14] and explained in more detail by Puchinger *et al.* in [15].

A. Generalized Concatenated (GC) Codes

Long codes have better error correction properties than short codes, but come with an increased decoding complexity. This results in high implementation area and cost. Code concatenation is an approach to construct more powerful, larger codes from smaller codes that can still be easily decoded.

Typically, an information sequence is first encoded with an outer code. In the next step, the codeword of the outer code is encoded with an inner code. In reverse, the inner code is decoded before the outer code.

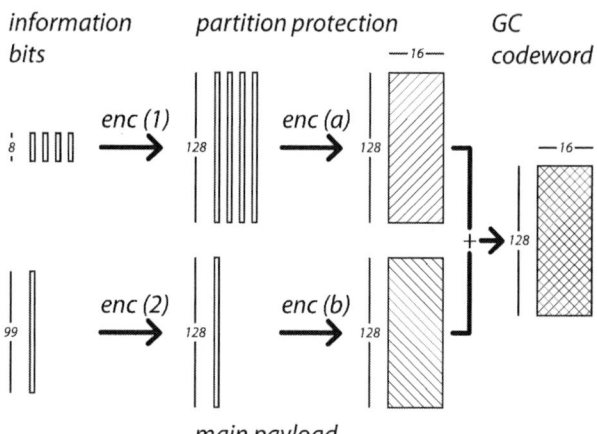

Figure 1: Encoding steps of the GC code construction

Generalized code concatenation [17] interweaves the previously strictly separated decoding steps to increase the code dimension or code distance without increasing the code length[1]. The inner code is split into several equally sized subsets of codewords, or partitions, with an increased minimum distance within each partition. The first outer code protects the information which partition was selected. In the next step, the decoding within each partition can be carried out with the increased distance which results in a significantly reduced error probability.

Figure 1 shows the GC construction that is implemented in this work. The inner code with code length 16 operates row-wise. It is separated into two parts a and b, which are combined with two different outer codes of length 128, represented in columns. The first outer code (1) with a large minimum distance protects the partitioning information encoded with a. The second outer code (2) carries the major payload of the construction and is encoded with part b of the inner code.

The discussed construction exploits the advantages using a GC code construction to reduce the complexity of the decoder of an error correction for the scenario used in PUFKY [12]. The setting with 14% PUF bit error probability and 99% entropy was chosen, so that at least 130 corrected PUF bits are required to generate a 128 bit key with an error probability around 10^{-9}. More advanced GC code examples for this scenario can be found in [15].

B. Decoding Procedure

Fig. 2 illustrates the steps how to decode the 2048-bit codeword with the errors caused by the variation in the PUF response with generalized code concatenation to recover the final 131 information bits that are later hashed to the cryptographic key.

---

[1]Typically codes are characterized by the parameters (n, k, d). The code length n is the length of the codewords. The code dimension k characterizes the number of information bits per codeword and the minimum distance between codewords d quantifies the error correction capability.

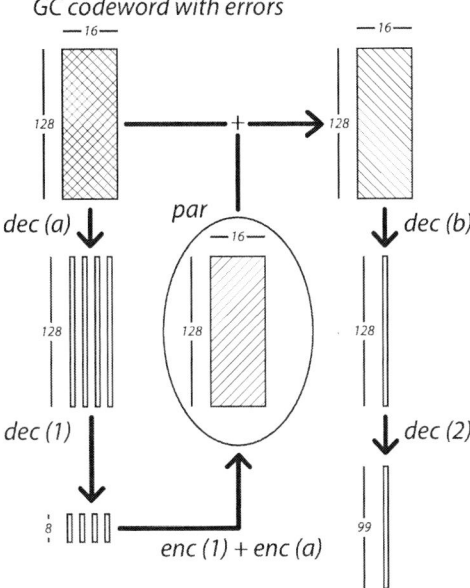

GC codeword with errors

recovered information bits

Figure 2: Decoding of the GC code construction

1) First inner decoding (dec (a)): First, the 2048 code bits are decoded row-wise as 128 codewords of the inner $(16, 5, 8)$ code with maximum-likelihood decoding. If a unique result is found, the first four bits containing the partitioning information are written into one line of a 4×128 matrix while the fifth bit is discarded. If the result is ambiguous, i.e. two codewords are equally likely, the 4-bit line is filled with erasure symbols. For the next decoder stage filling a blank is much easier than detecting and correcting an error at an unknown position. So, using erasures reduces decoding errors in subsequent steps.

2) First outer decoding (dec (1)), partition protection: In the second step, the columns of the 4×128 matrix are decoded with the $(128, 8, 64)$ code. Each column yields 8 bits of the final PUF key.

This code has a very high minimum distance of 64 and thus a powerful error correction capability. In addition to recovering the information bits, the corrected information bits are forwarded as partitioning information to the following, most crucial decoding step.

3) Partition Selection (par): The corrected 4×8 information bits are first re-encoded column-wise with enc (1) to 4×128 bits and then with enc (a) row-wise to a 16×128 matrix. Adding this matrix to the GC codeword with errors selects the partition, so that the resulting matrix only contains the result of enc (b), which has a higher code distance, and still the errors. This removes the hatching in one direction in Fig. 2.

4) Second inner decoding (dec (b)): Removing the effects of the other 4 information bits per row leaves either only ones or only zeros in each row. Instead of decoding as $(16, 5, 8)$

code, we can now decode as $(16, 1, 16)$ repetition code with twice the Hamming distance between the codewords. Therefore, this step reduces the error probability significantly such that a dec (2) only requires a smaller distance than dec (1) to achieve the same bit error probability.

Toy example: A $(4, 2, 2)$ code has the codewords $0000_b, 0011_b, 1100_b, 1111_b$ and distance two. If we know that a correct codeword has to be either 0000_b or 1111_b, we can decode with distance 4 instead of 2. The difference to our code is that we partition 32 codewords into 16 pairs.

5) Second outer decoding (dec (2)), payload: In the last step, the remaining 128-bit column is decoded with the $(128, 99, 8)$ code which contributes the largest amount of information bits to the final key, namely 99.

Note that the bit error and erasure probability in Figure 2 always decreases from the top to the bottom. A more detailed and illustrated description can be found in [15].

In comparison to [12], this construction has the main advantage that the RM decoding only operates on $GF(2)$ instead of $GF(2^8)$, which promises a lower implementation complexity. In addition, the code size is slightly decreased.

C. Block Error Probability

An upper bound on the key error probability was deduced in [15] based on the channel models of the intermediate decoding steps. Assuming a PUF response with bit error rate of 14% as input, the resulting upper bound on the key error probability is

$$p_{\text{key error}} = 1.48 \cdot 10^{-9}. \tag{1}$$

Using other algorithms such as generalized minimum distance decoding or recursive algorithms can further decrease the key error probability [15]. However, these algorithms have the drawback of increased memory requirements and more complex hardware designs.

IV. REED–MULLER CODES AND THEIR DECODING

A. Reed–Muller Codes

Reed–Muller (RM) codes are characterized by two integer parameters r and m with $m > 0$ and $0 \leq r \leq m$. In the following, we will use a variable r' with $r' \in \{0, ..., r\}$.

For $\mathcal{RM}(0, m)$ codes, the generator matrix[2] $G_{\mathcal{RM}(0,m)}$ contains one row filled with 2^m ones. Incrementing r' to one adds m rows with m orthogonal vectors $\mathbf{v}_1, ... \mathbf{v}_m$ to the generator matrix. For each subsequent r', $\binom{m}{r'}$ vectors are created by multiplying r' of the m basic vectors component-wise in all possible combinations and appended as rows to the generator matrix.

For example, the first row of the $\mathcal{RM}(2, 3)$ code generator matrix in Eq. 2 contains the all-one row vector from $r' = 0$. For $r' = 1$, the next lines contain the three orthogonal row vectors. The part of the matrix with $r' = 2$ contains all

---

[2]The generator matrix G of an ECC is an $n \times k$ matrix and characterizes the mapping between information word and codeword of a code. $c^n = i^k \cdot G$

possible products of two basic row vectors. Note that the logical AND of $\mathbf{v}_2$ and $\mathbf{v}_3$ is written as $\mathbf{v}_{32}$ and equals 1 when the corresponding components of $\mathbf{v}_2$ and $\mathbf{v}_3$ both also are equal to 1.

$$G_{\mathcal{RM}(2,3)} = \begin{pmatrix} 11111111 \\ 00001111 \\ 00110011 \\ 01010101 \\ 00000011 \\ 00000101 \\ 00010001 \end{pmatrix} = \begin{pmatrix} \mathbf{v}_0 \\ \mathbf{v}_3 \\ \mathbf{v}_2 \\ \mathbf{v}_1 \\ \mathbf{v}_{32} \\ \mathbf{v}_{31} \\ \mathbf{v}_{21} \end{pmatrix} \quad (2)$$

Following the nomenclature for the rows of the generator matrix, the corresponding information bits to the lines are denoted as $i_0, i_3, i_2, i_1, i_{32}, i_{31}, i_{21}$. These bits are encoded to a vector of 2^m codeword symbols, namely $\mathbf{c} = c_0 \dots c_7$.

In terms of conventional code parameters (n, k, d), the code length is given by $n = 2^m$ and its minimum distance is $d = 2^{m-r}$. The code dimension calculates to $k = \sum_{r'=0}^{r} \binom{m}{r'}$.

The three relevant codes used in this work are the $\mathcal{RM}(1,4)$, the $\mathcal{RM}(1,7)$ and the $\mathcal{RM}(4,7)$ codes. Their code parameters are $(16, 5, 8)$, $(128, 8, 64)$ and $(128, 99, 8)$ respectively.

B. Reed Decoding Algorithm

Similar to the step-wise construction of the generator matrix, the Reed decoding algorithm [18], [16] successively subtracts the parts with higher r' from the codeword. It is highly suited for a low-area decoder implementation because instead of decoding the entire codeword at once or in large blocks, each information bit is decoded separately using the corresponding parity-check equations. This approach is also known as Hadamard transformation.

1) Parity-check equations: Mapping the codeword symbols to parity-check equations is based on the geometrical description of RM codes [16]. This formal approach places the codeword symbols as points in $GF(2^m)$ and assigns the information bits to disjunct subspaces.

In more practical terms, in the previous example the information bit i_{21} can be extracted from the received word with the following two parity-check equations:

$$i_{21} = c_0 + c_1 + c_2 + c_3 = c_4 + c_5 + c_6 + c_7 \quad (3)$$

For each information bit, $2^{m-r'}$ independent parity-check equations can be generated. This equals the Hamming weight of the corresponding row in G. A hardware-efficient generation procedure to derive the corresponding parity check equations for each information bit is presented in the next section.

2) Parity-check calculation and majority logic decision: Larger RM codes typically have several parity-check equations for each information bit. After recovering the information bit candidates for all equations, a majority voting is performed to identify the most likely information bit. Note that each error can only affect one parity-check equation.

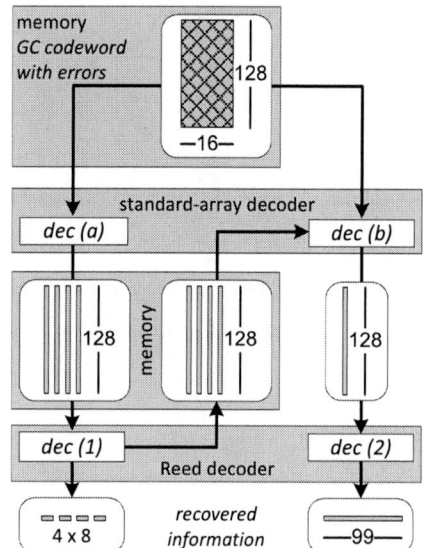

Figure 3: Hardware architecture overview of the GC RM decoder

3) Row reduction: Looking again at the generator matrix $G_{\mathcal{RM}(2,3)}$, we can see that the ones in the last row are also part of the fourth row, for example. If $i_{21} = 1$, we would add 2 errors to the decoding of the i_1. Let $\mathbf{c}'$ be the received word. To remove these errors caused by previous information bits, we add $i_{21} \cdot \mathbf{v}_{21}$ and create a new intermediate received word $\mathbf{c}'$ according to

$$\mathbf{c}' := \mathbf{c}' + i_{21} \cdot \mathbf{v}_{21} \quad (4)$$

This procedure is repeated for all information bits until the top row is reached.

V. DECODER DESIGN

The GC RM code decoder in Figure 3 contains two main components that perform the decoding steps discussed in Section III and implements directly the decoding steps described in Figure 2.

From a high-level perspective, the 2048-bit input is processed by the two decoders of the inner code, designated as the two standard-array decoders $dec\ (a)$ and $dec\ (b)$. The Reed decoder decodes both outer codes $\mathcal{RM}(1,7)$ and $\mathcal{RM}(4,7)$, denoted as $dec\ (1)$ and $dec\ (2)$.

The 4×128 input symbols of $dec\ (1)$, output from $dec\ (a)$, are stored in memory as intermediate values, as well as the corrected codewords from $dec\ (1)$. Then, $dec\ (b)$ reads both the input GC word and the corrected codewords from $dec\ (1)$ from the memory modules. Its output is directly fed into the Reed decoder, now configured as $dec\ (2)$.

A. $\mathcal{RM}(1,4)$ Standard-Array Decoder

The $\mathcal{RM}(1,4)$ decoder is a maximum-likelihood standard-array decoder [17] which generates all 32 possible codewords and compares them to the received vector. The output consists of five information bits and one erasure bit.

Counting from 0 to 31, a counter value is encoded to a codeword and then the Hamming distance between the generated codeword and the received word is calculated. If the distance is less than $d/2$, the counter value determines the decoded 5-bit information vector. If exactly $d/2$ errors occurred, so that two codewords are equally likely, the decoder sets the erasure bit to mark all 5 information bits as erasures. Note that only four of the information bits and the erasure bit are forwarded to $dec\,(1)$.

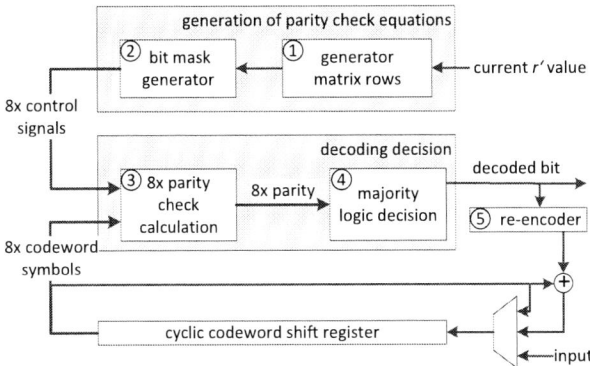

Figure 4: Hardware architecture of the Reed decoder

B. $\mathcal{RM}(r, 7)$ Reed Decoder

By going through the generator matrix line by line from the bottom to the top, the Reed decoder recovers the information bits one by one based on parity-check equations. For each decoded information bit, the corresponding generator matrix line is subtracted from the received word in the shift register. The efficient generation of parity check equations is the most complex part of the design. Figure 4 shows the data flow through the submodules of our Reed decoder.

Codeword shift register: During initialization, the received word is fed through the multiplexer into the central shift register of length 128.

In the basic case, the codeword symbols are stored as binary values. Our design contains an additional shift register containing the corresponding erasure bits.

1) Generator matrix row generation: Each row decoding operation starts with the generation of a compact representation of the corresponding generator matrix row. For every r' value with $r' \in \{r, ..., 0\}$, the module generates each of the $\binom{7}{r'}$ rows using a 7-bit counter and a comparator that checks if the Hamming weight of the counter value equals r'. This corresponds to a 7-bit vector with one-hot encoding and permits to derive all other rows from the basic rows v_1 to v_7 and combinations thereof.

2) Bit mask generation: For each information bit, all 128 codeword symbols have to be assigned to exactly one parity check equation. The regular structure of the RM code enables a very compact design using a counter that enumerates the codeword symbols. Bit masks, that are generated from the previous 7-bit vector, trigger different parity-check calculating (PCC) modules depending on the current counter value. Our decoder calculates 8 equations in parallel.

3) Parity-check calculation: When a PCC module is triggered, the current symbol at the output of the shift register is XORed onto the state of the PCC module. However, as soon as an erasure is detected, the outcome is always set to erasure.

4) Majority logic decision: The results of all parity-checks are evaluated by using a majority logic decision. It is convenient to sum up the parities in two's complement representation. The value -1 corresponds to a one, $+1$ to a zero and an erasure is represented as $+0$. The sign-bit of the sum determines the recovered information bit.

For $N_\otimes$ erasures and N_{err} errors in the received word, the correct codeword is recovered as long as $N_\otimes + 2 \cdot N_{err} < d$ holds.

5) Re-encoding: After recovering the information bit, it is encoded again and subtracted from the codeword shift register.

Example: The following example demonstrates the decoding process of our Reed decoder based on the previous $\mathcal{RM}(2, 3)$ example. For decoding information bit i_3, the corresponding 3-bit vector representation of its generator matrix row is 100_b.

As a representation for the 4 parity-check equations, the decoder then generates 4 bit masks for the two rightmost bits, which are the zero bits in the 3-bit vector. The corresponding bit masks are: $x00_b, x01_b, x10_b, x11_b$.

To calculate the parity checks, the codeword symbols are enumerated from 0 to 7. The decoder combines the four bit masks with the binary value of the codeword symbol indices, resulting in the following 4 parity-check equations:

$$i_3 = c_0 + c_4 = c_1 + c_5 = c_2 + c_6 = c_3 + c_7 \qquad (5)$$

C. GC Error Correction

The previous decoding of the columns with the $\mathcal{RM}(1, 7)$ code uncovered incorrect decoding decisions of the first inner decoding that relate to part a of the $\mathcal{RM}(1, 4)$ code.

Equivalent to choosing the correct partition, the corrected values are re-encoded by a hard-coded generator matrix inside the $\mathcal{RM}(1, 4)$ decoder and added to the original PUF response. The $(16, 1, 16)$ repetition decoder evaluates the input based on its Hamming weight. A Hamming weight of 8 is decoded as erasure, and a Hamming weight below or above 8 decodes to a 0 resp. 1.

VI. SYNTHESIS RESULTS

The discussed modules were implemented for low-end Xilinx Spartan-6 FPGAs to generate results consistent to [12]. The goal is to consume as few permanent resources as possible on an FPGA without dynamic reconfiguration. Slices can only be allocated once during initialization. Therefore, we make less restrictive use of block RAM, that could be shared with other modules whenever the GC code module is not active.

	Reed Dec standard	Reed Dec erasure
Slices Total	62	73
Registers	94	106
LUTs	183	184

Table I: Synthesis results of the $\mathcal{RM}(4,7)$ Reed Decoder in default and erasure configuration for Xilinx Spartan-6 FPGAs

Table I shows a comparison between our Reed decoder in the basic design with binary codeword symbols and the extension that can also handle erasures. Only 11 additional slices are necessary. Note that the circular buffers are synthesized as shift registers that are counted as LUTs and not as registers. The Reed decoder takes $10,500$ clock cycles for the $\mathcal{RM}(1,7)$ code configuration and $50,000$ cycles for the $\mathcal{RM}(4,7)$ code. Therefore, decoding the five 128-bit codewords of the GC code construction in the Reed decoder takes $92,000$ clock cycles in total.

	Std. Arr. Dec	Reed Dec	Entire Module
Slices Total	17	73	107
Registers	15	106	156
LUTs	37	184	317
8B Block Rams	0	0	2
Clock Cycles	100	$\leq 50,000$	$108,000$

Table II: Synthesis results of the GC code decoder implementation for Xilinx Spartan-6 FPGAs

The synthesis results for the entire decoder in Table II show that the Reed decoder is the most complex module using roughly 70% of the overall resources.

	PUFKY	GC RM Block RAM	GC RM Distr. RAM
Slices	221	107	179
8B Block Rams	0	2	0
Clock Cycles	50,831	108,000	108,000
Key Error Prob.	$1 \cdot 10^{-9}$	$1.48 \cdot 10^{-9}$	$1.48 \cdot 10^{-9}$

Table III: Implementation Comparison for Xilinx XC6SLX45 Spartan-6 FPGAs

Table III compares our GC RM code decoder, synthesized using block RAM or distributed RAM, to the original PUFKY [12] publication. Our approach pays off in a 50% reduction in slice count with a minor increase in error probability. Using distributed RAM, we are still able to achieve a 20% decrease in slice count. The maximum frequencies of our designs are $78.7\ MHz$, and $86.3\ MHz$ respectively. Note that reading out the RO PUF in the reference implementation takes roughly $250,000$ clock cycles such that the increase by $50,000$ cycles only has a minor influence on the total key generation time.

VII. CONCLUSION

Generalized code concatenation enables more area-efficient error correction for PUFs through a reduced code complexity. In this work, we have shown with an FPGA implementation that the size of the error correction can be reduced by 50% compared to the reference implementation. This compact implementation size was mainly achieved by developing an optimized sequential Reed decoder architecture for Reed–Muller codes that is constructed around a circular buffer.

ACKNOWLEDGEMENTS

This work was partly funded by the German Federal Ministry of Education and Research (BMBF) in the project SIBASE through grant numbers 01IS13020A and 01IS13020C.

REFERENCES

[1] S. Trimberger and J. J. Moore, "FPGA security: From features to capabilities to trusted systems," in *DAC*, 2014.
[2] C. Herder, M. Yu, F. Koushanfar, and S. Devadas, "Physical unclonable functions and applications: A tutorial," *Proc. IEEE*, vol. 102, no. 8, pp. 1126–1141, 2014.
[3] G. E. Suh and S. Devadas, "Physical unclonable functions for device authentication and secret key generation," in *DAC*, 2007, pp. 9–14.
[4] S. S. Kumar, J. Guajardo, R. Maes, G. J. Schrijen, and P. Tuyls, "Extended abstract: The butterfly PUF protecting IP on every FPGA," in *HOST*, 2008, pp. 67–70.
[5] J. Guajardo, S. S. Kumar, G. J. Schrijen, and P. Tuyls, "FPGA intrinsic PUFs and their use for IP protection," in *CHES*, 2007, pp. 63–80.
[6] Y. Dodis, L. Reyzin, and A. Smith, "Fuzzy extractors: How to generate strong keys from biometrics and other noisy data," in *EUROCRYPT*, 2004, pp. 523–540.
[7] M. Yu and S. Devadas, "Secure and robust error correction for physical unclonable functions," *IEEE D&T*, vol. 27, no. 1, pp. 48–65, 2010.
[8] M. Hiller, D. Merli, F. Stumpf, and G. Sigl, "Complementary IBS: Application specific error correction for PUFs," in *HOST*, 2012, pp. 1–6.
[9] M. Hiller, M. Weiner, L. Rodrigues Lima, M. Birkner, and G. Sigl, "Breaking through fixed PUF block limitations with differential sequence coding and convolutional codes," in *TrustED*, 2013, pp. 43–54.
[10] C. Bösch, J. Guajardo, A.-R. Sadeghi, J. Shokrollahi, and P. Tuyls, "Efficient helper data key extractor on FPGAs," in *CHES*, 2008, pp. 181–197.
[11] R. Maes, P. Tuyls, and I. Verbauwhede, "Low-overhead implementation of a soft decision helper data algorithm for SRAM PUFs," in *CHES*, 2009, pp. 332–347.
[12] R. Maes, A. Van Herrewege, and I. Verbauwhede, "PUFKY: A fully functional PUF-based cryptographic key generator," in *CHES*, 2012, pp. 302–319.
[13] M. Hiller and G. Sigl, "Increasing the efficiency of syndrome coding for PUFs with helper data compression," in *DATE*, 2014.
[14] S. Müelich, S. Puchinger, M. Bossert, M. Hiller, and G. Sigl, "Error correction for physical unclonable functions using generalized concatenated codes," in *ACCT*, 2014.
[15] S. Puchinger, S. Müelich, M. Bossert, M. Hiller, and G. Sigl, "On error correction for physical unclonable functions," in *SCC*, 2015.
[16] F. J. MacWilliams and N. J. A. Sloane, *The theory of error-correcting codes*. North-Holland, 1977.
[17] M. Bossert, *Channel Coding for Telecommunications*. New York: Wiley, 1999.
[18] I. S. Reed, "A class of multiple-error-correcting codes and the decoding scheme," *IRE Trans. on Inf. Th.*, vol. 4, no. 4, pp. 38–49, 1954.
[19] G. Schnabl and M. Bossert, "Soft-decision decoding of Reed-Muller codes as generalized multiple concatenated codes," *IEEE Trans. on Inf. Th.*, vol. 41, no. 1, pp. 304–308, 1995.
[20] A. Van Herrewege and I. Verbauwhede, "Tiny application-specific programmable processor for BCH decoding," in *SoC*, 2012, pp. 1–4.
[21] M. Hiller, L. Rodrigues Lima, and G. Sigl, "Seesaw: An area-optimized FPGA viterbi decoder for PUFs," in *DSD*, 2014, pp. 387–393.

2015 IEEE Computer Society Annual Symposium on VLSI

JSRAM: A Circuit-level Technique for Trading-off Robustness and Capacity in Cache Memories

Hamzeh Ahangari [#1], Gulay Yalcin [#2], Ozcan Ozturk [#1], Osman Unsal [#2], Adrian Cristal [#2]

Department of Computer Engineering, Bilkent University, Ankara, Turkey

Barcelona Supercomputing Center, Barcelona, Spain

Abstract— **In this paper, we propose Joint SRAM (JSRAM) cell, a novel circuit-level hardening solution for having a flexible trade-off between cache capacity and reliability. Our solution aims at addressing both permanent and transient faults in SRAM cells. JSRAM technique enhances read-stability by increasing critical Read Static Noise Margin (RSNM) to decrease faults when circuit operates at lower VDD voltages, while improving yield. It also increases hold-stability to mitigate against soft-errors, due to noise or radiation when circuit works in harsh environmental conditions. Moreover, our approach is resistant to Multi Bit Upsets (MBUs) while it is fully immune against single faults due to a novel self-correcting technique. Specifically in fault prone condition, some user pre-selected cache lines and ways are combined together vertically and horizontally at circuit level to form more robust cache cells. Four cells are joined by joiner circuits to form one robust cell. Joining cells can be selected far enough to cope with MBUs. In the case of soft-errors, this technique can be interpreted as *virtually doubling transistor size*. In the case of permanent faults, it can be seen as *increasing β ratio* to improve readability. This approach provides versatile and flexible solution with negligible performance penalty and low area overhead. This technique requires one extra transistor per cell for joiner circuit. However, extra transistor is only involved in the reliable mode. Therefore it can be described as 6T+1T SRAM cell. While our implementation is on 6T SRAM with 22nm technology, it is applicable to other SRAM structures and technologies.**

Keywords- SRAM Stability; Cache Reliability; Low Power Cache

I. INTRODUCTION

Today, Static-RAM (SRAM) based memories are the building blocks for the most critical processor structures such as caches and registers [11]. Moreover, SRAM cells are prone to myriad types of faults such as those caused by voltage scaling, yield lost or soft errors [1]. Thus, improving the stability as well as the reliability of SRAM cells, which conventionally is measured by static noise margin (SNM) gauge, is always under the spotlight [22].

Reliability concerns in SRAM cells originate from multiple sources. First, due to power consumption concerns in today's electronic devices, a variety of power saving techniques like dynamic voltage scaling (DVS) and dynamic frequency scaling are applied to processors to avoid excessive heating and to improve the battery lifetime especially in ubiquitous mobile devices. Quadratic relation between power and voltage makes supply voltage scaling decisively effective in power reduction by trading off performance when low-power is preferred over high-performance. However, voltage reduction by silicon geometry scaling or DVS causes reduction in static noise margin (SNM). Reducing SNM is equivalent to less stability, which means increased susceptibility to noise as presented in Figure 1(a) [1]. Therefore, SRAM cells become less stable and more vulnerable to adverse disturbing factors in lower voltages. Number of faults increases severely when the voltage is scaled down. This effect prevents further opportunities for voltage reduction and consequently hinders additional power saving. Hence, providing reliability solutions while operating at lower voltages in cache memories has attracted a lot of attention in recent years in order to provide substantial power saving [1, 21, 22].

The second significant source of reliability concerns in SRAM cells is random manufacturing process variation which raises instability in smaller dimensions. In the ideal case, transistor parameters should be nominal values and SRAM cells should have a symmetric SNM butterfly diagram. However, random manufacturing variations lead to fluctuations in parameters around nominal value, like Vth [22]. This process variation contributes to asymmetric cell SNM distribution, and consequently less stable cells [1, 20] as shown in Figure 1(b) [1]. Moreover, due to the severity of variation, some of the cells come out of manufacturing process permanently faulty, and thereby causing production yield loss [22].

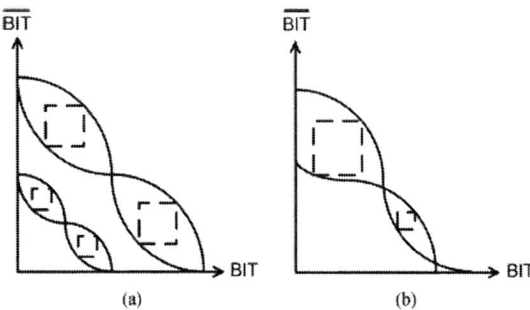

Figure 1. SNM degradation because of (a) VDD scaling (b) process variation [1].

Even in the absence of process variation or voltage scaling, reliability may decrease because of extraneous adverse factors

including noise and high energy particles [3, 5]. High energy particles may originate from impurities inside the chip package or from cosmic radiations; thus, it is not possible to avoid these, especially in high altitudes [4]. However, those particle strikes can corrupt data stored in memory and can cause so-called soft errors. Moreover, the likelihood of a single particle strike causing Multiple-Bit Upsets (MBU) also increases for recent technology nodes [2, 4]. This blunts the effectiveness of previous solutions such as DICE [6, 9] and reinforce the call for new approaches in Single Event Upset (SEU) mitigation [2, 3, 5]. Due to increasing importance of soft errors, decreasing the sensitivity against radiation is desirable both in terrestrial or space applications [3].

In this paper, we propose Joint SRAM (JSRAM) cell, a novel solution for having a flexible trade-off between cache capacity and robustness at the cache line and way granularity. JSRAM is a simple low-overhead circuit idea which joins together four selected SRAM cells to make a single but more robust cell when reliability becomes a concern (i.e. when applications require high reliability or when the voltage is decreased). The main contributions of this study are followings:

- We introduce Joint SRAM (JSRAM), a novel design which tolerates both transient and permanent faults.

- We design JSRAM in a way that it provides flexible trade-off between cache capacity and robustness. In fault-tolerant mode it provides a full immunity against single faults. Moreover it copes with Multiple Bit Upsets (MBUs).

- We evaluated JSRAM by using HSPICE simulator with 22nm. Our results show that Read and Hold SNM are improved up to 217% and 40% respectively at Vdd=0.3 using NMOS transistor as joiner circuit.

II. BACKGROUND

In this section, we present the conventional technique of stability measurement and nomenclature of SRAM faults.

A. Static Noise Margin

Static Noise Margin (SNM) is the traditional gauge for measuring stability of an SRAM cell. By definition, SNM for SRAM cell is defined as maximum tolerable simultaneous DC noise on both CMOS invertor's inputs [20]. Any stronger noise will toggle the value stored in the memory cell. The graphical maximum square technique and its formulation is the prevalent way of calculating SNM in circuit simulators [20]. Main adverse factors which degrade SNM are process variations and voltage scaling [1, 22]. SNM at the time of reading cell's value (Read SNM or RSNM) is considered as worst case SNM and it is the main parameter considered for improvement [21, 22].

B. SEU, MBU and SER

A single event upset (SEU) is an unwanted upset in electronic circuits caused by high energy particle strikes and leads to change of data. These errors are also called soft errors [3, 5]. These faults are considered as non-destructive non-persistent faults. Not only circuits work above ground level, like satellites are exposed to, but also in terrestrial devices, they have become a real concern [2]. Cosmic radiations and impurities inside chip package are two main sources of SEU [4].

By technology feature size shrinking, the charges stored inside transistors also shrink. This tends to increase sensitivity against single events. On the other hand, smaller dimensions mean smaller area being exposed to radiation and, thereby less upset. It has been shown that these opposing factors almost cancel each other and soft error rate (SER) per bit does not change significantly in newer technologies [4]. However, dense integration made by silicon geometry scaling means more cells will fit in unit area and so does more SER per device [2, 4]. Two other arguments for SER increase per device in smaller technologies are: 1) devices are more susceptible to multiple bit upsets (MBU) and 2) compared to past generations of technology, particles of lower energy are able to cause an SEU [2].

III. RELATED WORKS

There have been many efforts devoted to improve reliability and stability of storage cells including SRAM cells in cache, targeting persistent and transient faults. Solutions can be classified into three main levels, device level, circuit level, and system level [3]. Device level techniques are fundamental ones applied in manufacturing process such as error-aware transistor positioning. Circuit level techniques usually rely on using extra transistors or gates (i.e. comes with area penalty) for hardening against soft errors [6] or increasing SNM boundaries [21]. Among the studies on hardening, DICE [6] is one of the most important techniques. Although many of the efforts were designed based on (or compared to) DICE in recent years [7, 8], today it is known that this idea is no longer as effective [3, 9]. Many of the previous techniques, like DICE, aimed to tolerate just one upset. However, in today's deep sub-micron technology, the probability of having multiple bit upsets (MBU) is increasingly intensified [2]. Although we have also been inspired by DICE in joining SRAM cells together, our idea is completely different in multiple ways. First, DICE is a fixed SRAM cell structure while JSRAM is flexible in joining four SRAM cells or not. Then in normal high performance and less reliable mode, area overhead of DICE is almost double. Second, DICE uses double transistors which provide single point of failure immunity, however it is still vulnerable to MBU [9]. On the other hand, JSRAM is resistant to MBU while it is immune against single faults too. Additionally, our JSRAM addresses permanent faults too by decreasing noise margins. One structural difference between DICE and JSRAM is functionality of DICE is based on feed-back loop [6], whereas our approach is based on traditional transistor size ratios inside SRAM cells.

A variety of system level solutions were also proposed. While area and performance penalty in these techniques can be excessive and non-negligible, sometimes they are the only solutions available. We survey these solutions in the next three paragraphs.

Dense and regular pattern of cache memory provides possibility of using Error Correcting Code (ECC). But generally the reliability gain achieved by devoting more bits to ECC grows slowly, thereby limiting the effectiveness of ECC against multiple errors. Particularly, this is true for correcting multiple bit upsets (MBU), which are prevalent in today's extremely dense circuits. Simple ECC, like parity or Hamming code, is usually combined with other techniques to be more effective. 2D ECC [13] and methods adaptively sacrificing cache capacity for storing ECC codes [15, 17] are examples of applying ECC in caches. However, most of these techniques suffer from high-level overheads such as encoding/decoding time and energy.

Replication techniques like triple modular redundancy (TMR) are another important class in which some solutions are proposed. While effective and simple, area overhead is

978-1-4799-8720-7/15 $31.00 © 2015 IEEE

significant in such approaches. Flexicache [14] proposes a flexible in cache replication to achieve more reliability in low operating voltage level, while keeping cache capacity intact in high voltage. The drawback in this solution is abrupt transition of cache size to half and then one third. In Flexicache, all the cache lines have to be replicated in reliable mode; while in JSRAM each cache line can be individually selected for replication.

Bit-fix [19] combines two consecutive cache's physical lines in architectural level to make one logical line. If faulty bits in two lines do not fall into same index, this technique creates one healthy line from two partially faulty lines. Positions of faulty bits are stored in a quarter of cache ways. Bit-fix only targets permanent faults aimed to reach lower supply voltage. Since our solution combines lines too, it is in a sense similar to bit-fix idea. However, our approach is in lower circuit layer with less performance overhead. JSRAM makes unstable cells more robust, rather than filtering them out. It also mitigates soft-errors and imposes no limitation to have higher layer solutions like bit-fix or ECC concurrently.

A series of studies proposed cross-layer hybrid-cache memories [16, 18]. In these techniques, cache is implemented by different type of SRAM cells in a spectrum from big and robust cells to small and weaker cells. For example, in [16] 6T cells with six different sizes are used, whereas in [18], 6T and 10T cells are used to implement cache. Considering the fact that, in low power modes, less performance and less cache capacity is needed, weak part of cache memory is turned off gradually in lower voltages to preserve reliability. However, above architectures unnecessarily use big and power hungry cells in high-performance mode, which makes them only appropriate for applications spending most of their time in low power mode [18].

Quality of Bit (QoB) [10] proposed combining two SRAM cells to achieve more reliability or performance. While JSRAM provides similar static noise margin as QoB, it introduces two additional novelties: 1) JSRAM provides full immunity against single faults by a novel technique 2) by joining far enough cells, JSRAM is capable to cope with MBU. However QoB focused on performance improvement by joining adjacent cells.

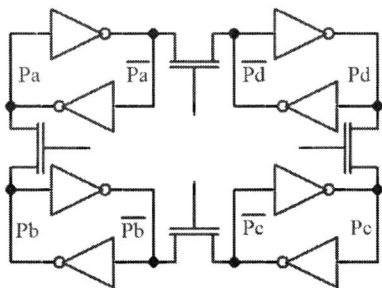

Figure 2. JSRAM cell with NMOS joiner circuit. Each node is driven by two cells. JSRAM forms a symmetric structure.

IV. DESIGN OF JSRAM

Normal SRAM cell has two internal nodes that hold strong opposite digital values by a feed-back loop. JSRAM (Joint SRAM) is constructed by combining four cells in circuit level by means of four joiner circuits as depicted in Figure 2. Each joiner circuit connects two equivalent internal nodes of two distinct cells to each other, here PA to PB, $\overline{PA}$ to $\overline{PD}$ and so on. The logic behind this idea is that the coupled nodes which hold same value,

electrically support each other if any instability threatens one of them. Each of four cells receives an electrical support from two directly connected cells.

In normal mode, switches (joiner circuits) are off. Therefore, cells A, B, C and D are four separate cells, where each belongs to different cache lines or ways, and work independently without interfering with each other. When, by any of previously mentioned reasons, stability degrades, the joiner circuitries can be turned on to electrically join those four cells to each other. Because of special circuit behavior, this technique constructs one but more robust SRAM cell. By applying this technique, some part of the cache can become fully reliable with fine granularity. For instance, JSRAM can be utilized in order to make tag areas highly reliable since any fault in these areas are more likely to impact the execution significantly compared to data areas. Encryption keys are another data types suitable to be stored in JSRAM cell. Joiner circuit is placed between cache lines vertically and between ways inside a line horizontally to join them as needed.

A. Joiner Circuit

In choosing the proper joiner circuit, the electrical resistance is a critical issue. If the resistance is too low like a zero-ohm ideal switch, then the four cells can support each other more strongly. However the negative side is the error propagation happens more easily too. On the other hand, if the resistance is chosen too high, the error propagates hardly but cells receive less support from each other as well. In practice, joiner circuit can be chosen as simple as a pass transistor. In simulation we investigated different types of switches including NMOS and PMOS, all of them with various ranges of (W/L) sizes, to find the proper choice. In both cases, area overhead is 4T for four cells, or equivalently 1T for one cell. We call this 6T+1T cell.

B. Static Noise Margin Improvement

Shrinking of SRAM cell's noise margin during read access (Read-SNM) puts most critical limitation on stability of cell than other factors [22]. During read operation both bitlines are pre-recharged to Vdd and the word line becomes '1' to connect bitlines to internal nodes. Drop of voltage on bitlines is detected by sense amplifiers. Read operation is performed on the half of the cell that keeps zero value. The path formed by access and pull down transistor from Vdd to ground and voltage division on elements on this path pulls the voltage of internal node upward from ideal zero. This decreases read static noise margin (RSNM) which means decrease maximum tolerable noise on internal node. One general way for improving RSNM is to increase β ratio (or cell ratio), the ratio of pull down transistor over access transistor. However, this also increases cell area.

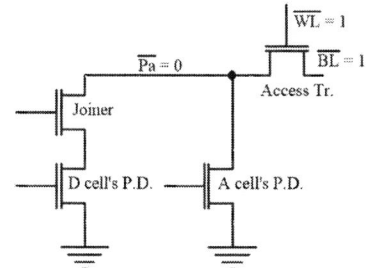

Figure 3. Virtually increasing cell ratio by parallelizing pull down transistors of two cells.

JSRAM technique provides a smart way to virtually increasing β ratio to improve RSNM. Let's assume internal node $\overline{PA}$ stores '0' and a read operation is ongoing on this node. As Figure 3 illustrates, by turning on the joiners, other than pull down NMOS on $\overline{PA}$ node, another path from $\overline{PA}$ to ground is formed via pull down NMOS of $\overline{PD}$ node. This parallelization is equivalent to decrease the impedance between node $\overline{PA}$ and ground or equivalently *'virtually increasing size of pull down transistor of node $\overline{PA}$'*.

In addition to increased cell ratio, the intrinsic tendency of those cells directly surrounded cell A, i.e. here cell B and D, to hold their own value, creates a strong resistance against changing value of cell A. JSRAM shows a special behavior similar to Schmitt Trigger circuits in increasing SNM greatly.

Effect of radiation strikes is usually modeled as short current passes through OFF transistors [12]. The voltage division over two ON transistors in same inverter may pass the critical point and toggle cell's value. In tackling against errors caused by adverse environmental condition, JSRAM shows more robustness by same mechanism and arguments similar to RSNM. If one OFF transistor turns on momentarily, instead of only competing against its own already ON counterpart transistor, it faces two parallel ones. This is translated to less probability of corrupting the value and more robustness of cell. In simulation it is shown that less critical Hold-SNM (HSNM) is also improved by JSRAM to mitigate transient faults. However, since HSNM value is usually higher than RSNM, the amount of improvement is not as much as the RSNM case.

C. Single Fault Immunity and MBU Mitigation

Although JSRAM structure shows enhanced robustness against faults to assist reaching lower operating voltages or working in harsh conditions, but generally occurrence of faults are unavoidable. In this regard, JSRAM proposes two techniques.

First, to avoid or decrease the corruption probability of more than one bit inside JSRAM by single particle strike (MBU effect), four cells can be chosen far enough to guarantee a safe distance. For instance cache line 0 can be combined with line 3, 1 with 4, and so on. However, more distance implies more wiring overhead.

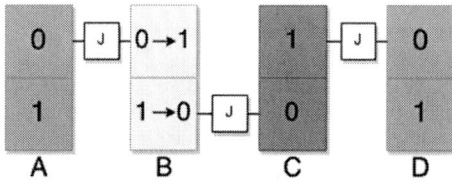

Figure 4. Faulty cell C corrupts cell B by toggling weaker good '1' with stronger bad '0'. Strong intact '0' values in A and D form two barriers against fault propagation. Later those strong '0' values will recover genuine values of B and C.

Second, based on traditional characteristics of normal SRAM cell, JSRAM shows full immunity against single faults. This novel technique is based on this fact that in normal SRAM cell, pull down transistors usually are stronger than pull up transistors. Equivalently logical '0' is more potent than logical '1'. When SRAM cells are connected to each other in a chain, like what is depicted in Figure 4, strong intact '0' values form two barriers around the faulty cell to prevent fault propagation. Hence the fault

can propagate at most one cell. Here fault inside cell C propagates to cell B. After disappearance of the noise source, the two potent good '0' values, here in cells A and D, will recover those (at most) two faulty cells. As mentioned previously, proper choosing of joiner circuit is necessary and are investigated. Circular connection of four cells as illustrated before in Figure 2, constitutes a symmetric and single fault immune structure.

This point should be mentioned that circuit level JSRAM has no conflict with other conventional higher level techniques. Benefits of high level solutions like simple ECC and in cache replication can effectively be exploited alongside with JSRAM.

D. Benefits of JSRAM

JSRAM solution brings all below features together:

- Versatility and Flexibility: In reliable mode, JSRAM addresses both permanent and transient faults simultaneously. In normal mode it works like 6T SRAM cell benefiting from high cache capacity and no power penalty. While whole of cache works in normal mode, some more error-prone lines or ways can be configured to work in reliable mode.

- High Performance: Since SRAM is at the circuit level, unlike fully high-level techniques like ECC based ones, it has negligible performance overhead. Moreover it imposes no limitation against having other upper layer solutions like ECC concurrently. JSRAM is a data redundancy based technique but without explicit majority voter.

- Low power and area: In reliable mode, each joiner circuit connects two nodes with same logic value (same voltage level). Thus, in joiner circuit, sensible power is only dissipated when there is an ongoing effort to resist against a fault. This is unlike other hardening or yield improvement techniques, which employ extra transistors permanently and which in turn waste more power [6, 7, 21]. Area overhead is as small as 1 extra transistor per cell, i.e. 6T+1T. However in normal mode extra transistor is not utilized.

- Immune against single faults: Any single fault inside JSRAM cell immediately and automatically is corrected.

- Resistant against Multiple Bit Upset (MBU): Nominated cells can be selected far enough to mitigate against MBU by more routing overhead.

V. EVALUATION

Simulations were performed in HSPICE simulator with 22nm predictive technology model library [24]. Transistor sizes for typical 22nm SRAM cell were chosen from [23]. Ratio values are: Cell ratio $= W_{PD}/W_{PG} = 2.02$, and pull up ratio $= W_{PU}/W_{PG} = 1.18$. In this section, we evaluate employing various types of elements as joiner circuit and show the improvement obtained in read and hold SNM. After that a time domain simulation for automatic fault correction process is illustrated.

For calculating noise margins, we applied conventional method of inserting two sources of identical DC noise between two inverters of victim SRAM cell and performed a DC simulation [20]. In our circuit, two sources of noise are inserted between inverters of cell A, i.e. between node A and $\overline{A}$ as illustrated in Figure 5.

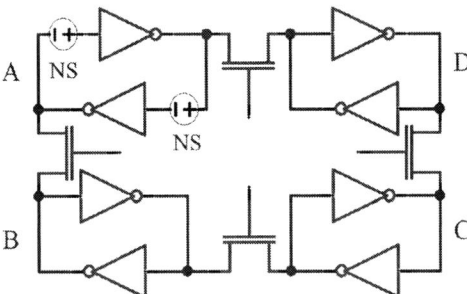

Figure 5. JSRAM with identical static noise source.

Unlike other implementations on SNM [21, 22], JSRAM shows a special behavior similar to Schmitt Trigger circuits. Resistance imposed by adjacent cells, here cells B and D, against changing the value of cell A describes this behavior. Therefore, calculating SNM has some subtitles as follow. Voltage transfer characteristic (VTC) diagram of each inverter is dependent on the value stored in JSRAM. Because of this, we ran two DC-simulations, one sweeps DC voltage value from high to low and the other sweeps in reverse order (Figure 6(b)). Moreover, we carefully confirmed each obtained noise margin simulation result with time domain simulation. Self-error correction simulation is depicted in Figure 7.

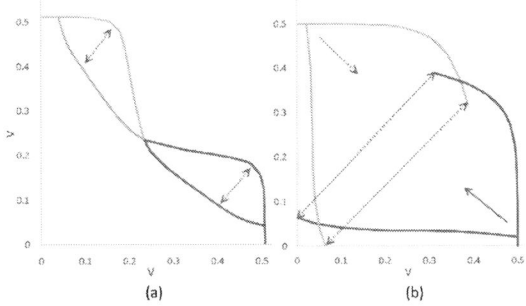

Figure 6. Read-SNM at Vdd=0.5. Maximum limit of improvement is %310 and is obtained by 0-ohm switch joiner. (a): Normal cell with RSNM=80mv (b): JSRAM with 0-ohm switch joiner and with RSNM=325mv. Schmitt Trigger behavior is observable in (b).

NMOS and PMOS switch types were investigated as joiner circuit. For obtaining maximum feasible SNM improvement by JSRAM technique, we also performed simulation with an ideal 0-ohm switch only for comparison purpose. Two different sizes for each transistor were chosen: NMOS1, NMOS2, PMOS1 and PMOS2, where NMOSX means X times of access transistor size and PMOSX means X times of pull-up transistor size. For each circuit configuration, Vdd voltage interval is swept from 0.3V to 0.8V.

In the case of Read-SNM, as depicted in Figure 8, there is considerable improvement in RSNM value with all joiner circuits. As expected, PMOS is not a good choice. Because in read operation the node storing logical '0' should be supported and PMOS does not pass strong '0'. Obviously, in all cases, bigger transistors provide better connectivity and better results, but require more area. Less critical hold stability is also investigated. As shown in Figure 9, Hold-SNM generally has higher values than Read-SNM, whereas improvements over HSNM are less in comparison to critical RSNM. Moreover, with

the same size transistors, PMOS shows slightly better results than NMOS in lower voltages. Simulation showed that resistance value of PMOS1 is too high and in some voltage values, self-correction is not done by PMOS1. As explained before, excessive electrical isolation makes PMOS1 an inappropriate choice as joiner circuit.

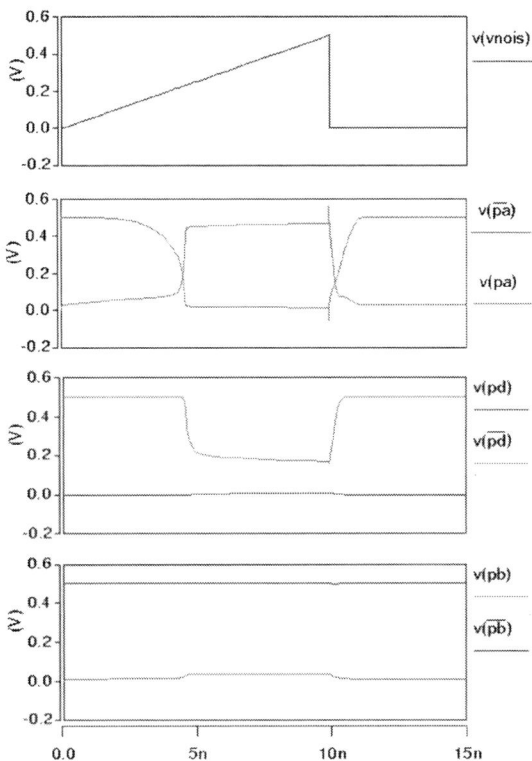

Figure 7. Self-error correction in JSRAM with NMOS1 joiner circuit: First, value of node 'a' is toggled when noise passes RSNM. Later value is automatically corrected by forces imposed from adjacent cells, when noise disappears.

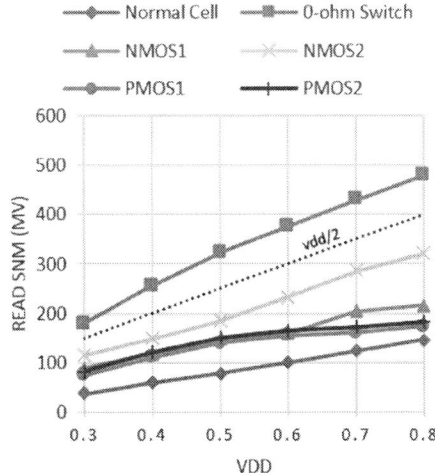

Figure 8. Read Static Noise Margin (RSNM) with different joiner options.

978-1-4799-8720-7/15 $31.00 © 2015 IEEE

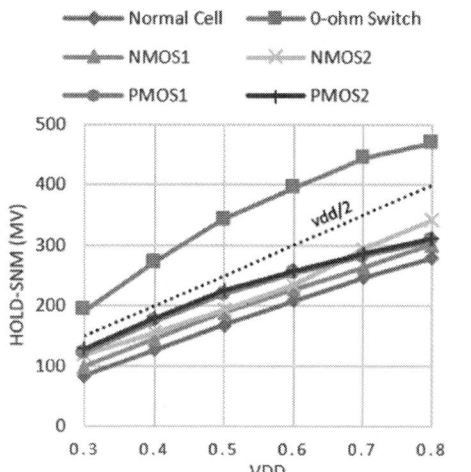

Figure 9. Hold Static Noise Margin (HSNM) with different joiner options.

VI. CONCLUSION

In this work we propose JSRAM, a novel and simple low power circuit technique for having a flexible trade-off between capacity and robustness inside SRAM based memories. With special circuit design, we achieve promising improvements in Static Noise Margin values, especially in near threshold Vdd and in critical Read-SNM. This assists to reach lower operating voltages. Moreover JSRAM is armed with a novel self-correcting technique to immediately correct any occurrence of single faults inside its joining cells. Area overhead for having JSRAM structure is four transistors per four cells or equivalently one transistor per each SRAM cell, where this extra transistor is not in use in normal execution. Experimental evaluation shows that our approach improves reliability with minimal overhead.

VII. ACKNOWLEDGMENT

This research was supported in part by European Cooperation in Science and Technology (COST). Support was done via Median group and grant donated under IC1103 action in summer of 2014. Additionally it was supported by the FP7 ParaDIME Project, grant agreement no. 318693.

VIII. REFERENCES

[1] Chen, Gregory, et al. "Yield-driven near-threshold SRAM design." Very Large Scale Integration (VLSI) Systems, IEEE Transactions on 18.11 (2010): 1590-1598.

[2] Ibe, Eishi, et al. "Impact of scaling on neutron-induced soft error in SRAMs from a 250 nm to a 22 nm design rule." Electron Devices, IEEE Transactions on 57.7 (2010): 1527-1538.

[3] Dodd, P. E., et al. "Current and future challenges in radiation effects on CMOS electronics." Nuclear Science, IEEE Transactions on 57.4 (2010): 1747-1763.

[4] Baumann, Robert. "Soft errors in advanced computer systems." Design & Test of Computers, IEEE 22.3 (2005): 258-266.

[5] Ferlet-Cavrois, Veronique, Lloyd W. Massengill, and Pascale Gouker. "Single event transients in digital CMOS—A review." Nuclear Science, IEEE Transactions on 60.3 (2013): 1767-1790.

[6] Calin, Th, Michael Nicolaidis, and Raoul Velazco. "Upset hardened memory design for submicron CMOS technology." IEEE Trans. Nucl. Sci., vol. 43, pp.2874 -2878 1996.

[7] Jahinuzzaman, Shah M., David J. Rennie, and Manoj Sachdev. "A soft error tolerant 10T SRAM bit-cell with differential read capability." Nuclear Science, IEEE Transactions on 56.6 (2009): 3768-3773.

[8] Krueger, Dan, Erin Francom, and Jack Langsdorf. "Circuit Design for Voltage Scaling and SER Immunity on a Quad-Core ItaniumÂ® Processor." Solid-State Circuits Conference, 2008. ISSCC 2008. Digest of Technical Papers. IEEE International. IEEE, 2008.

[9] Loveless, T. D., et al. "Neutron-and proton-induced single event upsets for D-and DICE-flip/flop designs at a 40 nm technology node." Nuclear Science, IEEE Transactions on 58.3 (2011): 1008-1014.

[10] Fujiwara, Hidehiro, et al. "Quality of a bit (QoB): A new concept in dependable SRAM." Quality Electronic Design, 2008. ISQED 2008. 9th International Symposium on. IEEE, 2008.

[11] Maric, Bojan, Jaume Abella, and Mateo Valero. "Efficient cache architectures for reliable hybrid voltage operation using EDC codes." Proceedings of the Conference on Design, Automation and Test in Europe. EDA Consortium, 2013.

[12] Nan, Haiqing, and Ken Choi. "High performance, low cost, and robust soft error tolerant latch designs for nanoscale CMOS technology." Circuits and Systems I: Regular Papers, IEEE Transactions on 59.7 (2012): 1445-1457.

[13] Kim, Jangwoo, et al. "Multi-bit error tolerant caches using two-dimensional error coding." Proceedings of the 40th Annual IEEE/ACM International Symposium on Microarchitecture. IEEE Computer Society, 2007.

[14] Yalcin, Gulay, et al. "Flexicache: Highly Reliable and Low Power Cache under Supply Voltage Scaling." High Performance Computing. Springer Berlin Heidelberg, 2014.

[15] Miller, Timothy N., et al. "Parichute: Generalized turbocode-based error correction for near-threshold caches." Proceedings of the 2010 43rd Annual IEEE/ACM International Symposium on Microarchitecture. IEEE Computer Society, 2010.

[16] Ghasemi, Hamid Reza, Stark C. Draper, and Nam Sung Kim. "Low-voltage on-chip cache architecture using heterogeneous cell sizes for high-performance processors." High Performance Computer Architecture (HPCA), 2011 IEEE 17th International Symposium on. IEEE, 2011.

[17] Chishti, Zeshan, et al. "Improving cache lifetime reliability at ultra-low voltages." Proceedings of the 42nd Annual IEEE/ACM International Symposium on Microarchitecture. ACM, 2009.

[18] Maric, Bojan, Jaume Abella, and Mateo Valero. "APPLE: adaptive performance-predictable low-energy caches for reliable hybrid voltage operation." Design Automation Conference (DAC), 2013 50th ACM/EDAC/IEEE. IEEE, 2013.

[19] Wilkerson, Chris, et al. "Trading off cache capacity for reliability to enable low voltage operation." Computer Architecture, 2008. ISCA'08. 35th International Symposium on. IEEE, 2008.

[20] Seevinck, Evert, Frans J. List, and Jan Lohstroh. "Static-noise margin analysis of MOS SRAM cells." Solid-State Circuits, IEEE Journal of 22.5 (1987): 748-754.

[21] Chang, Leland, et al. "Stable SRAM cell design for the 32 nm node and beyond." VLSI Technology, 2005. Digest of Technical Papers. 2005 Symposium on. IEEE, 2005.

[22] Calhoun, Benton H., and Anantha P. Chandrakasan. "Static noise margin variation for sub-threshold SRAM in 65-nm CMOS." Solid-State Circuits, IEEE Journal of 41.7 (2006): 1673-1679.

[23] Shin, Changhwan. Advanced MOSFET designs and implications for SRAM scaling. Diss. University of California, Berkeley, 2011.

[24] Predictive technology model, http://ptm.asu.edu

2015 IEEE Computer Society Annual Symposium on VLSI

Reducing the Storage Requirements of a Set of Functional Test Sequences by Using a Background Sequence

Irith Pomeranz
School of Electrical & Computer Eng.
Purdue University
W. Lafayette, IN 47907, U.S.A.

Abstract - **An existing storage scheme for a functional test sequence T uses a vector B, called a background vector, to store only the differences between the test vectors of T and B. After selecting B, modifying T based on B reduces the storage requirements of T. This paper describes a storage scheme for a *set* of test sequences T that uses a test *sequence B*, called a background *sequence*, to store only the differences between the sequences in T and B. The use of a background sequence allows more flexibility in matching B to T. By selecting B from the sequences in T, storage of B does not increase the storage requirements of T, and B itself contributes to the fault coverage. The paper describes a procedure for selecting a background sequence B from a set of sequences T, and for modifying T (including B) in order to reduce the storage requirements of T.**

I. INTRODUCTION

Functional test sequences are applied to a circuit in its functional mode of operation. A large set of functional test sequences may be generated for the purpose of simulation-based design verification [1]. They can also be generated by sequential test generation procedures targeting gate-level faults. Functional test sequences can be used as manufacturing tests or for speed binning. As manufacturing tests they may detect defects that are not detected by scan-based tests [2]. They also avoid overtesting of delay faults by creating functional operation conditions during test application [3]-[5].

Given a set of functional test sequences, it is possible to concatenate them into a single sequence [6]-[7]. During the concatenation process, it is possible to exclude subsequences that become unnecessary for detecting target faults. As a result, the length of a single sequence can be significantly smaller than the total length of the sequences in the set. However, a set of test sequences provides the flexibility to apply different sequences at different times in order to detect or diagnose different faults. This paper considers the case where a set of test sequences is used.

To facilitate at-speed test application, functional test sequences may be generated on-chip using a software-based self-test procedure [8]-[11]. For a higher fault coverage, functional test sequences can be generated off-line and stored on-chip in a compressed form. A software procedure [8]-[11] can be used for decompressing the sequences before they are applied to the circuit. In this context, it is important to use a compact set of functional test sequences.

The procedures from [12]-[16] accept a set of functional test sequences T. They produce a compact set of sequences by selecting a subset of T such that the subset detects the same target faults as the complete set T. The procedure from [12] also removes the last vectors of a sequence if these vectors are not necessary for detecting any faults. These procedures reduce the test data volume and the test application time.

Compression of functional test sequences for on-chip storage can be accomplished by encoding of their test vectors [17]. However, general encoding procedures do not utilize the flexibility to modify a functional test sequence such that it would detect the same target faults, but would be more amenable to compression. Such an approach was described in [18]. Considering a single functional test sequence T, the approach described in [18] associates with T a primary input vector B called a background vector. The test sequence T is stored by storing only the differences between its test vectors and B. This requires storage of the clock cycles and primary inputs where T is different from B. A procedure for computing B based on a given functional test sequence T was described in [18]. In addition, a procedure for modifying T so as to reduce its storage requirements based on B was described. Modification of T consists of three steps. The first step adds at the end of T test vectors that are similar to B. This increases the storage requirements of T slightly, but it also increases the flexibility of changing the entries of T such that they are equal to the corresponding entries of B. The second step complements entries of T that are different from the corresponding entries of B in order to reduce the storage requirements of T. The third step applies static test compaction to remove unnecessary test vectors from T, and thus reduce its storage requirements and the test application time.

The effectiveness of this approach in reducing the storage requirements of a functional test sequence is demonstrated in [18] for benchmark circuits. Since the modification of T is performed without reducing its fault coverage with respect to a target fault model, it is expected to preserve its advantages with respect to the detection of unmodeled faults.

A single functional test sequence is compressed in [18] by using a single background vector (the procedure is also extended to use two background vectors). Because of the advantages of using a set of functional test sequences, this paper describes a storage scheme for a *set* of test sequences $T = \{T_0, T_1, \cdots, T_{N-1}\}$. The approach described in this paper uses a background *sequence B* to store only

978-1-4799-8720-7/15 $31.00 © 2015 IEEE   155

the differences between every sequence in T and B. Moreover, it uses one of the sequences in T as a background sequence. The use of a background sequence from T, instead of a background vector, has the following advantages.

(1) In a single background vector, let $B(j)$ be the value of primary input j. The value of $B(j)$ is based on the value of primary input j in all the clock cycles of all the functional test sequences in T. In a background sequence, let $B(u, j)$ be the value of primary input j at clock cycle u. The value of $B(u, j)$ is based on the value of primary input j only in clock cycle u of all the functional test sequences in T. Suppose, for simplicity of discussion, that every sequence in T has length L. With a background vector, $B(j)$ is based on NL entries of N sequences in L clock cycles. With a background sequence, $B(u, j)$ is based on N entries of N sequences. Therefore, a background sequence can be matched to T better than a single background vector. This implies smaller numbers of differences between T and B when B is a background sequence than when it is a background vector, and more flexibility to modify T so as to reduce its storage requirements.

(2) When a background vector B is used, the vector does not achieve any fault coverage. In addition, a test sequence where all the test vectors are equal to B typically has a low fault coverage. In contrast, a background sequence can detect a significant percentage of faults. With a background sequence that is included in T, storage of the background sequence does not add to the storage requirements of T. Moreover, all the sequences in T (including B) can be modified together so as to reduce the storage requirements.

The paper describes procedures for selecting a background sequence B, and for modifying T based on B in order to reduce the storage requirements of T. The background sequence B is modified as part of the modification of T. In addition, the procedure replaces B with a different sequence from T when the modification of T allows B to be removed without losing fault coverage. There is no analogous procedure in [18] for modifying or replacing the background vector.

The paper focuses on the compression of a set of functional test sequences. The compressed test data is assumed to be stored on-chip, and decompressed using a software procedure [8]-[11]. Output compaction is assumed for the output sequences that the circuit produces. The output storage requirements are not affected by the modification of T. They may be reduced when the number of sequences in T is reduced.

The paper is organized as follows. Section II describes the selection of a background sequence B for a set of functional test sequences T. Section III describes the procedure for modifying T based on B. Section IV presents experimental results.

II. BACKGROUND SEQUENCE

This section describes the selection of a background sequence B for a set of functional test sequences $T = \{T_0, T_1, \cdots, T_{N-1}\}$. It first introduces notation and describes a

test compaction procedure that is applied to T in order to ensure that it does not contain unnecessary sequences. It then discusses the storage requirements of T. Based on this discussion it describes the selection of B.

A. Notation and Compaction Procedure

For $0 \leq i < N$, let the length of T_i be L_i. For $0 \leq u < L_i$, let $T_i(u)$ be the test vector at clock cycle u of T_i. For a circuit with n primary inputs, and for $0 \leq j < n$, let $T_i(u, j)$ be the value of primary input j at clock cycle u of T_i. With this notation, $T_i = T_i(0) \, T_i(1) \cdots T_i(L_i - 1)$, and $T_i(u) = T_i(u, 0) \, T_i(u, 1) \cdots T_i(u, n - 1)$.

Let $L = \max \{L_i : 0 \leq i < N\}$ be the highest length of a functional test sequence in T. The length of the background sequence B is L. In this way, every sequence in T can be described based on its differences from B. With the notation above, $B = B(0) \, B(1) \cdots B(L-1)$, and $B(u) = B(u, 0) \, B(u, 1) \cdots B(u, n-1)$ for $0 \leq u < L$.

To ensure that the initial set of test sequences T is as compact as possible, the procedure described in this paper checks whether any of the sequences can be removed from T without reducing the fault coverage. Such a sequence is removed in order to reduce the storage requirements of T. The same compaction procedure is applied during the process of modifying T. The test compaction procedure proceeds as follows.

Let F be the set of target faults. During every fault simulation run (fault simulation is always carried out with fault dropping), the procedure associates with every fault $f \in F$ the index of the sequence in T that detects it. This index is denoted by $det(f)$. The compaction procedure considers the sequences in T one at a time. When T_i is considered, F_i initially consists of every fault $f \in F$ such that $det(f) = i$. The procedure simulates F_i under every test sequence $T_k \in T - \{T_i\}$. If a fault $f \in F_i$ is detected by T_k, the procedure removes f from F_i and assigns $det(f) = k$. This leaves in F_i the faults that are only detected by T_i. If $F_i = \phi$, the procedure removes T_i from T.

Considering the initial set of test sequences T, the run time for simulating and compacting T is denoted by RT_{init}. The run time for modifying B and T will be normalized to this run time in order to obtain an indication of the computational effort of the modification procedure.

B. Storage Requirements

Without using a background sequence B, storage of a functional test sequence T_i with length L_i for a circuit with n primary inputs requires nL_i bits. Storage of the set T requires $n \sum_{i=0}^{N-1} L_i$ bits. This is referred to as conventional storage of T. The number of bits for conventional storage is denoted by CST.

With a background sequence B, a sequence T_i is stored by storing a list of entries where T_i is different from B. The list is denoted by $DIFF_i$. It includes an entry (u, j) for every clock cycle u and primary input j such that $T_i(u, j) \neq B(u, j)$. The number of entries in $DIFF_i$ is denoted by δ_i.

Storage of a clock cycle u requires $\lceil \log_2 L \rceil$ bits. Storage of a primary input j requires $\lceil \log_2 n \rceil$ bits. Thus, storage of $DIFF_i$ requires $(\lceil \log_2 L \rceil + \lceil \log_2 n \rceil)\delta_i$ bits. In addition, storage of B requires nL bits. The number of bits required for storing T is thus

$$(\lceil \log_2 L \rceil + \lceil \log_2 n \rceil) \sum_{i=0}^{N-1} \delta_i + nL.$$

This is referred to as background storage, and the number of bits is denoted by BST.

The preferred storage scheme for T depends on the relationship between CST and BST. The conventional storage scheme is preferred when $CST \leq BST$. The background storage scheme is preferred when $CST > BST$.

C. Selection of B

The procedure described in this subsection selects one of the test sequences from T to be the background sequence B. The selected sequence is one of the longest sequences in T that minimizes the background storage requirements of T. Using the notation introduced in Section II.A, the length of the background sequence is L.

For every sequence $T_k \in T$ such that $L_k = L$, the procedure computes the value of $\Delta_k = \sum_{i=0}^{N-1} \delta_i$ with $B = T_k$. This is the number of entries where the sequences in T differ from $B = T_k$, and it determines the storage requirements of T when $B = T_k$. The procedure selects the sequence $T_k \in T$ for which Δ_k is the minimum as the background sequence for T.

For illustration, the first two columns of Table I show a set of test sequences $T = \{T_0, T_1, \cdots, T_9\}$ for ISCAS-89 benchmark $s27$. The circuit has $n = 4$ primary inputs, and the maximum length of a functional test sequence is $L = 3$. The total length of all the sequences is 26. Storage of the test sequences using the conventional storage scheme requires $CST = 4 \cdot 26 = 104$ bits.

TABLE I. Functional Test Sequences

i	T_i			δ_i
0	0010	1101		2
1	0110	1001		2
2	1010	0001		1
3	1000	0000	0000	4
4	0100	1001		3
5	0010	0000	1001	2
6	1000	0010	0001	6
7	0010	0001	1000	0
8	0010	0001	0100	2
9	0010	1001	0000	2

The sequences that can be used as background sequences in this example are T_3, T_5, T_6, T_7, T_8 and T_9, with corresponding numbers of differences $\Delta_3 = 34$, $\Delta_5 = 30$, $\Delta_6 = 44$, $\Delta_7 = 24$, $\Delta_8 = 26$ and $\Delta_9 = 24$. The value $\Delta_7 = 24$ is obtained with the values of δ_i shown in the last column of Table I. For example, $\delta_0 = 2$ with $DIFF_0 = \{(1,0),$ $(1,1)\}$ since T_0 differs from T_7 in the first two bits of the second vector. The procedure selects T_7 as the background sequence. With $\Delta_7 = 24$, four bits to describe an entry (u, j), and 12 bits to store T_7 in full, the background storage requirements are $BST = 24 \cdot 4 + 12 = 108$. Modifying T (and B) will reduce these storage requirements.

III. MODIFICATION OF THE SET OF SEQUENCES

This section describes the procedure for modifying a set of test sequences $T = \{T_0, T_1, \cdots, T_{N-1}\}$ so as to reduce the storage requirements based on a background sequence B. The set of target faults that T detects is denoted by F. Throughout the modification procedure, T continues to detect all the faults in F. As part of the modification of T, the background sequence B is also modified.

A. Overall Flow

The modification of T consists of complementing entries of B, complementing entries of sequences other than B, and padding the sequences in T with vectors from B. For simplicity, $T - B$ refers to the set $T - \{B\}$.

The modification steps of T are applied in the order illustrated by Figure 1. Padding is applied only once. The complementation steps are applied iteratively before padding, and after padding. The index of an iteration is denoted by I. The number of complemented bits in an iteration is denoted by c_B for the background sequence B, and by c_{T-B} for sequences in $T - B$. The iterative application of the complementation steps ends when they are unable to complement any additional values, and $c_B + c_{T-B} = 0$.

Entries of B are complemented before entries of $T - B$ in every iteration since complementing an entry of B affects the sets of differences for all the sequences in T, while complementing an entry of a sequence $T_i \in T - B$ affects only the set of differences for T_i. Thus, complementing an entry of B has a more significant effect on the storage requirements of T.

At the end of every iteration, the procedure applies the test compaction procedure described in Section II.A in order to remove from T sequences that are not necessary for maintaining the fault coverage. The test compaction procedure may remove the background sequence B from T. This may happen even before padding because the complementation of values in $T - B$ results in values from B appearing in all the sequences of T. This may cause the sequences in $T - B$ to detect all the faults that are detected by B, allowing B to be removed. In addition, padding copies vectors from B into the sequences in $T - B$. This increases the possibility that the sequences in $T - B$ will detect all the faults that B detects, and B will be removed from T. After padding, the sequences in $T - B$ contain unnecessary vectors, which can be modified to be equal to B. This, again, allows them to detect all the faults that are detected by B, making B unnecessary.

To accommodate the case where B is removed from T, the background sequence is selected again after test compaction. If the previous background sequence was not removed by test compaction, the same sequence is likely to be selected again as a background sequence. Otherwise, the procedure will select one of the longest sequences in T that minimizes the background storage requirements of T. Changing the background sequence may cause a temporary increase in the storage requirements of T because the sequences in T are not matched to B. This may occur for several iterations. However, eventually the procedure will

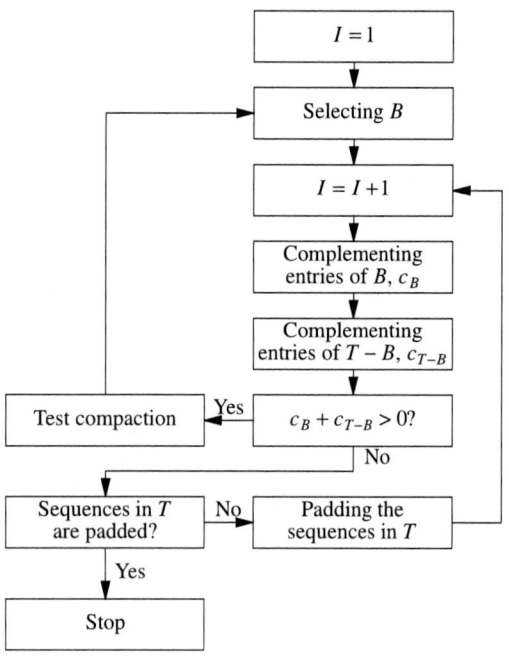

Figure 1. Procedure for Modifying T

not be able to remove the background sequence. It will then perform iterations where it modifies T based on the same background sequence B without changing it, and the background storage requirements will be reduced.

The index I of an iteration is initialized to one only once, and padding is assigned a unique iteration number by incrementing I after padding. An iteration ends when I is incremented, and storage requirements are computed.

The padding and complementation steps are described in the following subsections.

B. Padding the Sequences in T

In the step described in this subsection, the procedure pads the sequences in T by using test vectors from B. Padding proceeds as follows.

For every sequence $T_i \in T$, and for $u = L_i, \cdots, L$, the procedure assigns $T_i(u) = B(u)$.

Padding does not increase the background storage requirements of T since it does not add entries where a functional test sequence T_i is different from B. It does not reduce the fault coverage. It may increase the number of faults that some sequences detect (without increasing the overall number of detected faults), thus allowing sequences to be modified or even removed from T. This reduces the background storage requirements. Sequences are modified and removed in the steps that complement entries of T and in the test compaction step that follows.

Padding the functional test sequences in T increases the test application time as follows. Before padding, the

total number of clock cycles required for applying the sequences in T is $\sum_{i=0}^{N-1} L_i$. After padding, all the sequences are of length L, and the total number of clock cycles required for applying the sequences in T is NL. The increase in test application time is not expected to be an issue when the sequences are applied at-speed from an on-chip memory. The fact that all the sequences are of the same length is an advantage in making the test application process uniform across all the sequences.

C. Complementing Entries of $T - B$

The step described in this subsection complements entries of the sequences in $T - B$ that are different from the corresponding entries of B in order to reduce the sets $DIFF_i$, and thus reduce the background storage requirements of T. For every test sequence $T_i \in T$, the procedure applies the following process.

It first finds the set of faults $F_i \in F$ that are detected only by T_i. These faults must continue to be detected by T_i in order to ensure that T continues to detect all the faults in F. The procedure finds F_i as described in Section II.A.

Next, the procedure considers every entry (u, j) such that $T_i(u, j) \neq B(u, j)$. For every such entry, the procedure complements the value of $T_i(u, j)$. It then simulates F_i under T_i. If all the faults in F_i are detected, the procedure accepts the complemented value of $T_i(u, j)$, and $DIFF_i$ is reduced by one entry. Otherwise, the procedure stops after finding the first fault in F_i that is not detected by T_i. In this case, the procedure complements $T_i(u, j)$ again in order to restore its previous value.

In the worst case, to modify a test sequence T_i, the procedure performs fault simulation of F under T_i a number of times that is equal to nL_i. In effect, F_i is significantly smaller than F, and only a subset of the entries of T_i are considered for modification.

D. Complementing Entries of B

The step described in this subsection complements entries of B in order to reduce the differences between the sequences in $T - B$ and B.

Let $B = T_k \in T$. The procedure first finds the set of faults $F_k \in F$ that are detected only by T_k as described in Section II.A.

Next, for every entry (u, j) such that $0 \leq u < L$ and $0 \leq i < n$, the procedure finds the number of sequences $T_i \in T$ such that $T_i(u, j) = 0$, and the number of sequences $T_i \in T$ such that $T_i(u, j) = 1$. The numbers are denoted by $m_0(u, j)$ and $m_1(u, j)$, respectively. With $T_k(u, j) = 0$, the number of entries where the sequences in T differ from B is $m_1(u, j)$. With $T_k(u, j) = 1$, the number of entries where the sequences in T differ from B is $m_0(u, j)$. If $m_0(u, j) > m_1(u, j)$, the storage requirements will be lower with $T_k(u, j) = 0$. If $T_k(u, j) = 1$, it is possible to reduce the storage requirements of T by assigning $T_k(u, j) = 0$. Similarly, if $m_0(u, j) < m_1(u, j)$ and $T_k(u, j) = 0$, it is possible to reduce the storage requirements of T by assigning $T_k(u, j) = 1$. In both cases, the procedure includes (u, j) in a set denoted by UJ. If

978-1-4799-8720-7/15 $31.00 © 2015 IEEE   158

$T_k(u, j) \in UJ$ and $T_k(u, j)$ is complemented, the reduction in $\sum_{i=0}^{N-1} \delta_i$ will be $|m_0(u, j) - m_1(u, j)|$.

The procedure arranges the entries in UJ by order of decreasing $|m_0(u, j) - m_1(u, j)|$. This gives a preference to complementing an entry with a larger contribution to the background storage requirements. Considering the entries in this order, the procedure applies the following steps for every $(u, j) \in UJ$.

It complements the value of $T_k(u, j)$. It then simulates F_k under T_k. If all the faults in F_k are detected, the procedure accepts the complemented value of $T_k(u, j)$. Otherwise, the procedure stops after finding the first fault in F_k that is not detected by T_k. In this case, it complements $T_k(u, j)$ again in order to restore its previous value.

IV. EXPERIMENTAL RESULTS

The procedure was applied to sets of functional test sequences that were computed for single stuck-at faults in ISCAS-89, ITC-99 and IWLS-05 benchmark circuits. The circuits are ones that are testable by functional test sequences. The sets were compacted as described in Section II. The sets detect all the single stuck-at faults that are known to be detectable in all the benchmark circuits. Single stuck-at faults are also considered during the modification procedure.

A comparison with [18] is not possible since [18] addresses the compression of a single sequence, while this paper considers a set of sequences. A comparison is performed with the conventional storage scheme.

The results are shown in Table II as follows. There are three rows for every circuit. The first row corresponds to the initial set of test sequences. The second row corresponds to the final set of sequences that is produced by modifying T and B iteratively without padding. The third row corresponds to the final set of sequences that is produced by modifying T and B iteratively after padding.

Each row contains the following information. Column pi shows the number of primary inputs. Column I shows the index of the iteration of the modification procedure for which results are reported in the row. The value $I = 1$ refers to the initial set of test sequences. Column pad has a 0 when the sequences are not padded, and a 1 when the sequences are padded.

Column len shows the length of the longest sequence in T. Column seq shows the number of sequences in T.

Column $storage$ subcolumn ST shows the best storage scheme for T, where C stands for conventional storage, and B stands for storage based on a background sequence. Subcolumn $bits$ shows the number of bits required for storing the set. Subcolumn $ratio$ shows the number of bits as a fraction of the number of bits required for conventional storage of the initial set of test sequences.

Column $f.c.$ shows the stuck-at fault coverage of T in order to stress that it does not change. Column $cycles$ shows the number of clock cycles required for applying the sequences in T. This is the total length of all the sequences. It is increased by padding, and decreased when sequences are removed from T by the modification

procedure. Column $ntime$ shows the normalized run time.

The following points can be seen from Table II. The conventional storage scheme is preferred for the initial set of test sequences. After modifying T and B, the background storage scheme is preferred, and the storage requirements are reduced significantly.

Padding allows the storage requirements to be reduced further for most of the circuits considered. The cost of padding is an increased number of clock cycles for test application. However, padding also allows the number of sequences to be reduced, compensating for the increased test sequence lengths.

There are many cases where B is removed from T, and needs to be replaced by another sequence. In the case of *simple_spi*, the first iteration of modifying T removes three sequences from T, including the background sequence. The longest sequence in T is shorter, and therefore, the second background sequence is shorter than the first. The storage requirements are reduced even though a new background sequence is used. They are reduced further by modifying T based on the new background sequence.

V. CONCLUDING REMARKS

This paper described a storage scheme for a set of test sequences T that uses a background sequence B to store only the differences between T and B. The paper described procedures for selecting a background sequence B from T, and for modifying T in order to reduce its storage requirements based on B. The modification of T included complementing entries of B, complementing entries of the other sequences in T, padding of the test sequences in T, and test compaction. The use of a background sequence B, the selection of B from T, and the modification of B to reduce the storage requirements, were unique to the procedure described in this paper. Experimental results for benchmark circuits demonstrated significant reductions in storage requirements.

REFERENCES

[1] W. K. Lam, *Hardware Design Verification: Simulation* and *Formal Method – Based Approaches*, Prentice Hall, 2008.

[2] P. C. Maxwell, R. C. Aitken, K. R. Kollitz and A. C. Brown, "IDDQ and AC Scan: The War Against Unmodelled Defects", in Proc. Intl. Test Conf., 1996, pp. 250-258.

[3] J. Rearick, "Too Much Delay Fault Coverage is a Bad Thing", in Proc. Intl. Test Conf., 2001, pp. 624-633.

[4] J. Saxena, K. M. Butler, V. B. Jayaram, S. Kundu, N. V. Arvind, P. Sreeprakash and M. Hachinger, "A Case Study of IR-Drop in Structured At-Speed Testing", in Proc. Intl. Test Conf., 2003, pp. 1098-1104.

[5] S. Sde-Paz and E. Salomon, "Frequency and Power Correlation between At-Speed Scan and Functional Tests", in Proc. Intl. Test Conf., 2008, Paper 13.3, pp. 1-9.

[6] R. K. Roy, T. M. Niermann, J. H. Patel, J. A. Abraham and R. A. Saleh, "Compaction of ATPG-Generated Test Sequences for Sequential Circuits", in Proc. Intl. Conf. on Computer-Aided Design, 1988, pp. 382-385.

[7] I. Pomeranz, "Concatenation of Functional Test Subsequences for Improved Fault Coverage and Reduced Test Length", IEEE Trans. on Computers, June 2012, pp. 899-904.

[8] L. Chen and S. Dey, "Software-Based Self-Testing Methodology for Processor Cores", IEEE Trans. on Computer-Aided Design, Mar. 2001, pp. 369-380.

[9] P. Parvathala, K. Maneparambil and W. Lindsay, "FRITS - A Microprocessor Functional BIST Method" in Proc. Intl. Test Conf., 2002, pp. 590-598.

[10] N. Kranitis, A. Paschalis, D. Gizopoulos and G. Xenoulis, "Software-Based Self-Testing of Embedded Processors" IEEE Trans. on Computers, Apr. 2005, pp. 461-475.

[11] M. Nakazato, S. Ohtake, M. Inoue and H. Fujiwara, "Design for Testability of Software-Based Self-Test for Processors", in Proc. Asian Test Symp., 2006, pp. 375-380.

[12] F. Corno, P. Prinetto, M. Rebaudengo and M. Sonza Reorda, "New Static Compaction Techniques of Test Sequences for Sequential Circuits", in Proc. European Design and Test Conf., 1997, pp. 37-43.

[13] M. Dimopoulos and P. Linardis, "Efficient Static compaction of Test Sequence Sets through the Application of Set Covering Techniques", in Proc. Design, Autom. and Test in Europe Conf., 2004, pp. 194-199.

[14] S. Park, L. Chen, P. Parvathala, S. Patil and I. Pomeranz, "A Functional Coverage Metric for Estimating the Gate-Level Fault Coverage of Functional Tests", in Proc. Intl. Test Conf., Paper 27.1, 2006, pp. 1-10.

[15] I. Pomeranz, P. K. Parvathala and S. Patil, "Estimating the Fault Coverage of Functional Test Sequences Without Fault Simulation", in Proc. Asian Test Symp., 2007, pp. 25-32.

[16] H. Fang, K. Chakrabarty, A. Jas, S. Patil and C. Tirumurti, "RT-Level Deviation-Based Grading of Functional Test Sequences", in Proc. VLSI Test Symp., 2009, pp. 264-269.

[17] I. Koren and C. M. Krishna, *Fault − Tolerant Systems*, Morgan Kaufman Publishers, 2007.

[18] I. Pomeranz and S. M. Reddy, "Reducing the Storage Requirements of a Test Sequence by Using One or Two Background Vectors", IEEE Trans. on VLSI Systems, Oct. 2011, pp. 1755-1764.

TABLE II. Experimental Results

circuit	pi	I	pad	len	seq	ST	storage bits	ratio	f.c.	cycles	ntime
s820	18	1	0	239	21	C	10602	1.00	95.76	589	1.00
s820	18	5	0	239	19	B	5446	0.51	95.76	580	19.25
s820	18	10	1	239	15	B	5290	0.50	95.76	3585	23.98
s1196	14	1	0	24	93	C	11116	1.00	99.76	794	1.00
s1196	14	6	0	24	85	B	6465	0.58	99.76	761	8.74
s1196	14	65	1	24	28	B	5313	0.48	99.76	672	151.99
s1423	17	1	0	485	18	C	27812	1.00	93.33	1636	1.00
s1423	17	5	0	485	15	B	12095	0.43	93.33	1497	9.74
s1423	17	11	1	485	12	B	11129	0.40	93.33	5820	15.73
s5378	35	1	0	330	11	C	49490	1.00	79.10	1414	1.00
s5378	35	5	0	330	9	B	15450	0.31	79.10	1327	6.16
s5378	35	14	1	330	4	B	14295	0.29	79.10	1320	8.34
s35932	35	1	0	50	10	C	14630	1.00	89.78	418	1.00
s35932	35	6	0	50	9	B	2614	0.18	89.78	368	2.89
s35932	35	8	1	50	9	B	2614	0.18	89.78	450	2.99
b14	33	1	0	213	116	C	212586	1.00	89.42	6442	1.00
b14	33	7	0	213	100	B	56015	0.26	89.42	5875	6.22
b14	33	33	1	213	73	B	52291	0.25	89.42	15549	57.49
b15	36	1	0	769	79	C	575892	1.00	92.57	15997	1.00
b15	36	7	0	769	77	B	98004	0.17	92.57	15897	7.11
b15	36	14	1	769	69	B	82516	0.14	92.57	53061	14.10
b20	33	1	0	884	194	C	692505	1.00	90.78	20985	1.00
b20	33	9	0	884	162	B	140548	0.20	90.78	17873	7.63
b20	33	41	1	884	108	B	102708	0.15	90.78	95472	102.60
aes_core	258	1	0	96	45	C	477816	1.00	99.16	1852	1.00
aes_core	258	6	0	95	39	B	125614	0.26	99.16	1683	17.32
aes_core	258	20	1	95	31	B	92078	0.19	99.16	2945	48.42
des_area	239	1	0	24	97	C	87235	1.00	99.88	365	1.00
des_area	239	8	0	24	81	B	20660	0.24	99.88	318	9.64
des_area	239	18	1	24	67	B	12587	0.14	99.88	1608	35.32
i2c	17	1	0	747	12	C	46478	1.00	78.78	2734	1.00
i2c	17	5	0	747	12	B	16014	0.34	78.78	2734	9.47
i2c	17	15	1	747	6	B	15384	0.33	78.78	4482	16.55
simple_spi	15	1	0	857	12	C	96525	1.00	92.62	6435	1.00
simple_spi	15	7	0	840	9	B	22008	0.23	92.62	5448	38.91
simple_spi	15	11	1	840	8	B	20188	0.21	92.62	6720	41.52
usb_phy	14	1	0	422	12	C	27636	1.00	70.85	1974	1.00
usb_phy	14	4	0	422	11	B	9314	0.34	70.85	1778	10.82
usb_phy	14	12	1	422	7	B	9041	0.33	70.85	2954	14.77
am2910	20	1	0	491	21	C	40040	1.00	91.93	2002	1.00
am2910	20	7	0	491	18	B	18458	0.46	91.93	1794	9.81
am2910	20	15	1	491	12	B	16610	0.41	91.93	5892	20.44
div16	33	1	0	665	42	C	69894	1.00	84.30	2118	1.00
div16	33	5	0	665	40	B	27113	0.39	84.30	2067	23.47
div16	33	9	1	665	36	B	26217	0.38	84.30	23940	56.97

2015 IEEE Computer Society Annual Symposium on VLSI

A TMR Strategy with Enhanced Dependability Features Based on a Partial Reconfiguration Flow

Victor M. Gonçalves Martins*[†], Paulo R. C. Villa*, Horácio C. C. Neto[†], Eduardo Augusto Bezerra*

*Department of Electrical Engineering, Federal University of Santa Catarina (UFSC), Florianopolis, Brazil
{victor.martins, paulo.villa, eduardo.bezerra}@eel.ufsc.br
[†]Electronic System Design and Automation, INESC-ID, Lisbon, Portugal
{marte, hcn}@inesc-id.pt

Abstract—Reliability is an important concern in many *Field Programmable Gate Array* (FPGA) based designs, mainly in applications which must be free of faults, and need to be available for long periods of time. *Triple Modular Redundancy* (TMR) is a well known approach for a system to tolerate possible faults, but a permanent fault in the TMR hardware may diminish the redundancy advantage. In this paper, we present a low cost solution to improve some dependability figures of a TMR system. The methodology comprises a design flow, which is used in the implementation of the proposed *Partial Bitstream for Multiple Partitions* (PB4MP) mechanism. The strategy allows modules's relocation in different *Reconfigurable Partition*s (RPs) using only one bitstream per module, enabling fault prevention and fault removal (recovery) actions. The results show that with a minor increase in the digital design configuration memory, it is possible to improve considerably the TMR availability, reliability and maintainability.

Keywords—*Partial Reconfiguration; Fault Prevention and Removal; FPGA Reliability*

I. INTRODUCTION

The usage of *Field Programmable Gate Array*s (FPGAs) in critical applications, where eventual faults may result in a system crash, is increasing. Even when all the tests in the manufacturing process are performed successfully, there is no assurance that the design is perpetually immune to permanent faults. In fact, these devices also suffer from ageing, which results in a natural physical degradation process [1]. The ageing speed depends on several factors such as the maturity of the technology used, manufacturing quality control, temperature and thermal shock (related to the environment where it operates), operating voltage, dimensions, and even design complexity [1]. As the FPGA resources density keeps increasing and with the recent *Fin Field-Effect-Transistor* (FinFET) technology generation, reliability concerns, mainly due to the *Negative Bias Temperature Instability* (NBTI), become a relevant factor [2]. Faults created by NBTI do not radically damage the FPGA resources, but they increase its response time [3].

Availability, reliability and maintainability are important dependability attributes in any system [4]. The goal of this work is to add permanent faults prevention and to strengthen the recover capability of an FPGA based system, in order to improve these dependability features, regarding the ageing degradation process.

In this paper, a new dual strategy applied to a *Triple Modular Redundancy* (TMR) [1] system is proposed. The TMR

architecture has autonomous capabilities for *Fault Detection, Isolation and Recovery* (FDIR) [5]. To make this possible a design flow has been developed, in order to implement the proposed *Partial Bitstream for Multiple Partitions* (PB4MP) mechanism. Our approach is mainly applied to embedded FPGA designs using hardcore or softcore processors [6]. Although, it has been developed targeting the Virtex-6 Xilinx XC6VLX240T device, the proposed methodology can be used in the latest Xilinx families with *Partial Reconfiguration* (PR) support. The results show that the proposed TMR mechanism can detect and tolerate permanent faults (hard errors), with only a slight increase in the amount of memory required by a processor. The amount of memory may become negligible when the system itself runs on an *Operating System* (OS), such as the *Embedded Parallel Operating System* (EPOS) [7]. Fig. 1 shows the block diagram of an FPGA based system implemented using the proposed strategy. The TMR approach is applied to the *Advanced Encryption Standard* (AES) application at the bottom of Fig. 1. The voter is implemented in software, running on the MIPS CPU. In case of a permanent fault, the AES modules can be reallocated to any FPGA RP, as the the proposed strategy ensures that the modules will have identical physical interfaces.

The remaining of the paper is organized as follows. Section II describes the state-of-the-art. Section III presents the work's contributions, and also the implemented case study. Section IV introduces the proposed partial reconfiguration design flow (PB4MP). Section V presents the proposed architecture, the flow adaptation process, and the correspondent processing and memory requirements. Section VI summarizes the results obtained from the case study implementation. Section VII concludes the paper and discusses the future work.

II. RELATED WORK

This section discusses the related work regarding FPGA designs based on FDIR [5] approaches. Available techniques and technology that allow the system implementation are also discussed.

A. Fault Detection, Isolation and Recovery (FDIR)

The use of TMR [1] approaches is not new. The work in [8] presents a system with TMR which uses the FPGA PR capacities. In their work the fault detection is performed automatically. If a faulty module is detected, the corresponding partition is excluded, the module is allocated to an alternate

978-1-4799-8720-7/15 $31.00 © 2015 IEEE   161

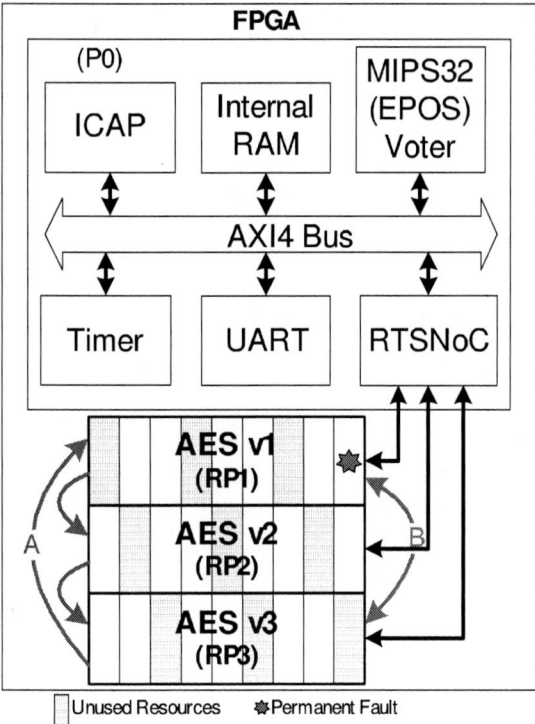

Fig. 1. System Diagram

partition and the system returns to its normal operation. However, this approach sacrifices a considerable amount of resources for module relocation. Furthermore, the relocation process requires a partial bitstream library, which requires external memory.

In another project [9], although without TMR, a framework called *Dynamic Partially Reconfigurable* (DPR) was developed. This platform is made of several *Reconfigurable Partitions* (RPs) called slots. These slots are allocated hardware modules, which are identified by the author as *Collaborative Macro-Functional Units* (CMFUs). Each CMFU has a *Built-In Self-Test* (BIST) module allowing periodic testing. When a fault is detected by the test, the CMFU is relocated to another free slot, and the previous slot is marked as faulty.

B. FPGA Configuration Manipulation

The RP advent brought tremendous flexibility on how we can use the FPGA resources. The work on [10] shows that an embedded microprocessor running an OS can observe and control the Flip-Flops of an FPGA using its *Internal Configuration Access Port* (ICAP).

The *Partial Reconfiguration* (PR) flow on PlanAhead tool from Xilinx [11] allows to create RPs individually and to generate a partial bitstream for each module on each RP. This individuality is necessary, mainly, because:

- the routing between external resources of an RP crosses the RP itself. Therefore, the partial bitstream used in an RP must accommodate not only the implementation of the desired module (the module's logic

design itself), but also part of the external resources interconnection routing;

- there is no control over the routing between the module implemented in the RP, and the remaining parts of the FPGA system. As a result, two modules (in two different RP) with the same interface, may have different routings.

Regarding the first issue, Xilinx provides the *Isolation Design Flow* (IDF) [12], which can be used to ensure that the implemented module in an RP stays completely isolated from the rest of the FPGA implementation.

For the second issue, the work in [13] presents a flow that tries to solve the compatibility of RP interfaces. It performs a normal PR flow through the end of the "Place and Route" process. Next, using the `pr2ucf` command, it extracts the positions of Proxy Logics from all RPs. Then, it chooses a set of Proxy Logics of one partition and builds a constraints file forcing the Proxy Logics locations of the remaining partitions to have the same relative location. Following the same methodology, the Directed Routing Constraints tool from Xilinx FPGA Editor extracts the routing of signals that are part of the interface of the reference partition. Then, with this information, it makes the routing of the signals that compose the interface of the remaining PRs. This is performed by adding rules in the same constraints file. Finally, the whole flow is rerun with the updated constraints file and the bitstream with the desired implementation is generated.

Although very interesting, this process is manual and does not cover the following:

- Situations where one output signal from one partition has a fanout greater than one. In this scenario the net has multiple destinations;

- Situations where the source of the input signal of a partition has a fanout greater than one. In this situation, the module input is just one of the several destinations of the net. Often the same net may be the input for multiple partitions;

- The clock tree must be relatively the same. As each clock region of an FPGA has multiple clock trees, during the Placement and Routing step, different clock trees may be selected.

III. CONTRIBUTIONS AND CASE STUDY

The main contribution of this work is to propose a way to implement a TMR system targeting permanent faults (i.e. hard errors). The TMR has to have fault prevention and fault recovery capabilities. Another important contribution is the increased reliability using the proposed partial reconfiguration design flow (PB4MP) which allows swapping modules between RPs and to manage the used resources, with extra precision constraints. This new mechanism results in two important benefits: it provides the TMR system with the capability to recover from permanent faults without wasting FPGA resources (by using extra partitions as proposed by Montminy [8] and Dumitriu [9]); and it reduces the NBTI degradation in the new FinFET based FPGAs. The main issue considered is the system capability *to recover* from permanent

978-1-4799-8720-7/15 $31.00 © 2015 IEEE

faults, which may have been originated from a strong damage in the device, or from delay faults that do not destroy the hardware resource but they may increment its response time. The capability *to prevent* permanent faults is also our target, but it is not scrutinized in this work. A feasible solution to deal with soft errors as *Single Event Upsets* (SEUs) is to add to the design a *Soft Error Mitigation* (SEM) technique, such as in [14].

The block diagram in Fig. 1 shows the System-on-Chip Platform [6] implemented as a case study, which is used to explore and to evaluate the proposed approach. The platform supports dynamic reconfiguration and it was implemented in a Xilinx XC6VLX240T FPGA. It is based on a MIPS32 ISA, the Plasma softcore, which is freely available at Opencores [15]. It employs the *Advanced eXtensible Interface 4* (AXI4) family of protocols, which is becoming the industry's standard for bus-based interconnections. The hardware reconfiguration is done by the *Internal Configuration Access Port* (ICAP) interface, that is interconnected through the AXI4 and managed by the *Embedded Parallel Operating System* (EPOS) [7] running on Plasma, which provides the necessary run-time support to implement the system features and the communication with the remaining modules. RP1, RP2 and RP3 represent the RPs that contain in this case, one AES component each one (three copies, in a TMR configuration). The three AESs are implemented following the resources distribution discussed in Section V-B. We have chosen a *Real-Time Star Network-on-Chip* (RTSNoC) [16] based interconnection scheme for connecting the reconfigurable partitions to the *Central Processor Unit* (CPU) where the voter is running (implemented in software). Each AES is connected to a port of an RTSNoC router and one of the RTSNoC router ports is connected to an AXI4 bridge.

IV. PB4MP DESIGN FLOW

The proposed Partial Reconfiguration mechanism is complex and requires a deep knowledge of the target Virtex-6 Family [17]. Before presenting the proposed flow, some important details regarding the Partial Reconfiguration and Isolation process are reviewed in this section.

A. FPGA Configuration Generation Process

A full understanding of the PR and Isolation process is essential in order to implement PB4MP. There are four main features present in the PR flow that are essential for our purpose:

- Xilinx synthesis tools for PR add a primitive *LUT1* to every input and output of an RP. They are designed by Proxy Logic and their position on the FPGA can be controlled by a constraint [11];

- Adding the flag `Isolated="true"` in the xpartition.pxml file, the tools force the module implemented in this RP to be fully isolated from the rest of the system [12];

- Using the constraint `SCC_BUFFER` in the signals that connect to the Proxy Logics in the system side, ensures that the net sources that connect the modules

implemented in RPs to the rest of the system have a fanout=1 [12];

- The Directed Routing Constraints allow us to compel the routing of a net in the FPGA [18].

B. Previous Flow Requirements

Solutions as the ones proposed by Montminy [8] and Dumitriu [9] that use the normal PR flow, imply an unavoidable limitation: it is necessary to generate a partial bitstream for each module in each RP. It means that, if there are N modules, M RPs and we wish to allocate any module in any RP, it will be necessary to create a library with $N \times M$ partial bitstreams.

The PB4MP allows to overcome this major limitation. Using a compilation of knowledge, from IDF [12], previous third party work [13], and Xilinx documentation [11] [18], we have developed a design flow.

The proposed flow has the following requirements:

- All RPs need to include exactly the same FPGA resources with the same physical distribution;

- The vertical RP limits (up and down) need to be coincident with the FPGA clock regions;

- The RPs resource areas can include I/O resources only if they are not used by the system;

Following these constraints, the flow permits implementing an FPGA system where all RPs have the same physical interface, which allows to use only one partial bitstream for each module that can be allocated in any RP. Therefore only N partial bitstreams are required, instead of the $N \times M$ required in the aforementioned works.

C. Design Flow Sequence

Fig. 2 shows the full design flow of a PB4MP-based FPGA system. It is based on the standard Xilinx tools flow (ISE and PlanAhead) and requires only a small user intervention at the initial phase of the hardware design, as in the IDF process [12]. The remaining steps are automated.

The base of the flow is the sequence Synthesis → Floor Plan → Translation → Mapping → Placement and Routing → Configuration Bitstream Generation. It is important that the Hardware Description Language specification is structured to correctly assign each module to the appropriate RP. This includes adding primitives `BUFHCE` and deciding between adding the normal `SCC_BUFFER`, or a primitive LUT2 that allows to disable all interface signals. If the second option is chosen, each physical LUT6 can implement two LUT2, which helps to reduce the FPGA resource overhead. Another step that needs the designer attention is the Floor Plan, when deciding the RPs location. It is imperative to follow the requirements listed in Section IV-B.

After this stage the PB4MP Tool generates the first constraints group. These constraints will be responsible for the buffers inclusion (`SCC_BUFFER` or LUT2). This time the tool does not block the resources for these buffers. It only defines a window of possible locations for them, as shown in Fig. 3.

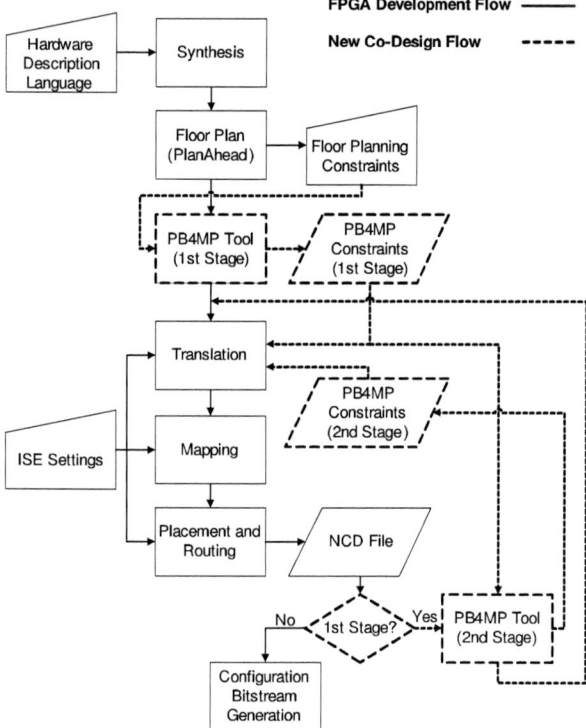

Fig. 2. FPGA Development + Design Flow

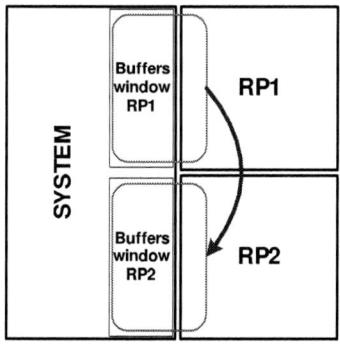

Fig. 3. Partition interface process

This is a way to give some freedom for the next steps to obtain a good placement and routing.

Initially, the Placement and Routing operation generates an NCD file. The PB4MP tool is then executed again and the nets that interconnect the modules in the RPs to the remaining parts of the system are extracted using the Directed Routing Constraints tool from `fpga_edline.exe` (which is the command line version of FPGA editor, used for making changes in the NCD file). With this information, as show in Fig. 3, the tool chooses one group of interconnect nets to replicate (automatically or defined by the designer), and with this information, a second constraints group is created. The sequence, Translation → Mapping → Placement and Routing is executed again. At the end of these two stages the final

bitstream is generated and can be loaded into the FPGA.

V. Proposed TMR based on PB4MP

In this paper, a system using TMR with capabilities to prevent and recover from permanent faults is proposed. As it is developed using the PB4MP tool, its reliability increase and maintainability added are obtained with no additional FPGA resources.

A. TMR Architecture

Basically, the TMR mechanism implies implementing the same module/function three times. The redundant modules run in parallel, performing the same computation. The results from the three modules are analyzed by another module, responsible for voting the correct result. This way, even if a module produces wrong results, the other two will guarantee the correct operation. The weak point in this strategy is the voter, as if it fails, the whole TMR strategy will also fail. The major advantage of TMR is its error detection and correction feature. An important drawback is the need to employ three times more hardware [1].

B. The PB4MP Adaptation for TMR Purpose

To take advantage of the PR design strategy in the implementation of the TMR system, an extra constraints list is added to the flow process. The list is built using the `CONFIG PROHIBIT` constraint, which provides the option to select the FPGA resources that are not to be used. In the selected policy implemented for this first version, a full resource column is removed from every three columns, delimited by the FPGA rows organization [17]. To assure the desired reliability and maintainability levels, this process never excludes two columns with the same relative position in different RPs, as shown at the bottom in Fig. 1.

This distribution allows two management resources processes. Process **A** that allows to swap the resources in use, helping to mitigate the NBTI degradation [2]. Process **B** where, if a module presents a permanent fault, it is able to replace the RP by another module that does not use the same resource where the permanent fault has been detected. The `CONFIG PROHIBIT` constraint excludes the resource (*Configurable Logic Block* (CLB), *Digital Signal Processing* (DSP), BRAMs, etc) but does not exclude the associated switching box. However, if the resource is not used, there is no routing for that resource, which means the switch box is not used, or it is used in a different way.

This distribution may look as a waste of 33% of the RP resources, however, in an FPGA system as a result of the routing congestion this is not a real overhead. When the routing is limited to the RP size, this congestion is bigger. Therefore and in practice, the proposed approach does not imply additional hardware consumption.

C. Memory Usage and Recovery Execution Time

The memory usage and the execution time of the software that implements the recovery procedure depend on several details of the adopted strategy. The basis of the strategy is the same, i.e. swapping modules between RPs, but it can be done in two different ways:

- The modules are available externally to the FPGA in a memory holding the partial bitstreams;

- The swap is implemented by moving the correspondent part of the FPGA configuration memory.

The adopted aternative is the second one, as it does not need extra hardware (the external memory). Nevertheless, this solution can be more or less demanding on the memory and runtime requirements. If the OS has a considerable amount of available free memory, it is possible to read and write all the corresponding module configuration memory. But if the free memory is very limited the option is to read and write the configuration, frame by frame. In this work we adopt the latter solution.

In order to evaluate the memory consumption and the run time execution, a set of equations was defined based on TABLE I and TABLE II. TABLE I describes all software functions and its memory usage (compiled using gcc for MIPS32 architecture with -O3 optimization flag) and the run time requirements for the Virtex-6 family. The *Bytes* column shows the memory usage required for the code part (M_{CODE}). The *Run time* column details all the parameters that influence the execution time indicating the number of clock cycles. TABLE II details all hardware parameters that influence the run time and the memory usage (M_{DATA}).

TABLE I.   SOFTWARE ROUTINES WITH INDIVIDUAL MEMORY AND EXECUTION TIMES CHARACTERISTICS

Software Routine Description	Bytes	Run time (Cycles)
RP_swap(rp1, rp2): Function that uses all routines below to implement the algorithm to swap two RPs.	418	– – –
XHwICAP Base Driver: Base Driver to communicate with XHwICAP (include Init(), SelfTest() and GetConfigReg() functions).	7012	– – –
XHwICAP DeviceReadFrame(): Routine to read one frame from the FPGA configuration memory.	554	NC_{RF} (3720)
XHwICAP DeviceWriteFrame(): Routine to write one frame to the FPGA configuration memory.	784	NC_{WF} (4010)
Total M_{CODE}	8768	– – –

TABLE II.   PARTITION CHARACTERISTICS

RP Parameters	Parameters description for each hardware module
$Size_{Frame}$	Number of bytes in one frame.
N_{RPs}	Number of RPs in the system.
$N_{RPFrames}$	Number of frames that have one full RP configuration.
$Frequency$	Frequency value of the system clock (used by the processing unit).

The memory required is divided in two parts: code (M_{CODE} detailed in TABLE I) and temporary data (M_{DATA}). The Data section stores the initial addresses of the RPs, four bytes for each one, and temporarily backs-up the FPGA configuration memory content from the RPs being swapped. In the low cost option, where there is one read/write operation for each frame, it is required twice the frame size $Size_{Frame}$ (equation 1a). For the worst case, when the swap is done with only one whole read/write operation for each RP, it is required the double of the size of the configuration memory of an RP. This means twice the frame size $Size_{Frame}$ for each of the $N_{RPFrames}$ frames (equation 1b).

$$M_{DATA} = N_{RPs} \times 4 + Size_{Frame} \times 2 \qquad (1a)$$
$$M_{DATA} = N_{RPs} \times 4 + Size_{Frame} \times N_{RPFrames} \times 2 \qquad (1b)$$

Equation 2 gives the amount of memory required to implement the recovery ability. The total amount depends on the FPGA family, as the frame size ($Size_{Frame}$), and on the number of frames to be moved in each read/write operation.

$$MEM_{Total} = M_{CODE} + M_{DATA} \qquad (2)$$

The total number of clock cycles required to execute a complete swap between two RPs (NC_{Total}) is determined by two factors (equation 3): FPGA family, because of the frame size $Size_{Frame}$, which influences the number of clock cycles spent by the functions *XHwICAP DeviceReadFrame()* (NC_{RF}) and *XHwICAP DeviceWriteFrame()* (NC_{WF}); and the RP size $N_{RPFrames}$, that defines how many frames need to be read and written.

$$NC_{Total} = (NC_{RF} + NC_{WF}) \times N_{RPFrames} \times 2 \qquad (3)$$

After obtaining the total number of cycles, equation 4 gives the required time necessary to swap two RPs in seconds (T_{Swap}).

$$T_{Swap} = \frac{NC_{Total}}{Frequency} \qquad (4)$$

VI.   EXPERIMENTAL RESULTS AND DISCUSSION

In the System-on-Chip case study shown in Fig. 1, the input data is sent to the three AESs modules (TMR). After the processing time, the results are retrieved from the modules. The three results are analysed in one vote function, implemented in software. If a mismatch is detected, the faulty module is identified and another RP is selected to swap with it. Before performing the swaping, the input clock signals are disabled on both RPs. The voting function is running on the Plasma CPU, making it a single point of failure. To overtake this drawback, a solution, based on [19], is under development as a related work aiming to improve the CPU dependability figures.

Each AES module occupies 4,172 *Look Up Tables* (LUTs), which in the Virtex-6 family [17] means that a minimum of 522 CLBs is required. Because a section of a CLB column has 40 CLBs, at least 14 of such CLB columns are needed. In practice, and due to placement and routing requirements (without any CONFIG PROHIBIT constraint), the module implementation required 23 CLB columns. This fact results in the opportunity to implement the strategy presented at the bottom part of Fig. 1, without wasting extra resources in the second stage and with no routing congestion. The characteristics of each RP are presented in TABLE III.

Using equation 1a we obtained $M_{DATA} = 660$ bytes, and from equation 2 we concluded that our TMR approach needs only 9,498 bytes of extra memory from the system. Regarding the runtime, a swapping of two RP requires 23,561,040

TABLE III.   AES RP CHARACTERISTICS

RP Parameters	Parameters Values
$Size_{Frame}$	324
N_{RPs}	3
$N_{RPFrames}$	1524 (1012 for CLBs + 512 for BRAMs)
$Frequency$	50MHz

clock cycles (equation 3), which correspond to 471 ms (using equation 4).

Regarding the RP in Fig. 1, this means 1.4 seconds for each process **A** actuation, and 471 ms for a process **B**. Although this is still a considerable time, the probability of the process **B** swap be necessary is relatively low. The process **A** period will be very large (days or weeks) and can be proceeded when this system function is idle.

Assuming that the AESs modules consume 66% of the resources in each RP, we can define $Fail2$ ={The FPGA resources area where the three RPs have two permanent faults that make the TMR fail irremediably} and calculate the probability $\mathbb{P}(Fail2)$ for each TMR presented in TABLE IV.

TABLE IV.   TMRS COMPARISON

TMR Version	$\mathbb{P}(Fail2)$	Extra Hardware
Standard TMR [1]	$\frac{13}{15} = 86,7\%$	0%
Montminy [8]	$\frac{9}{13} = 69,2\%$	PR Overhead + 25%
With PB4MP	$\frac{2}{5} = 40\%$	PR Overhead

The first system compared on TABLE IV is a standard TMR version, without any recovery mechanism. The second system is the TMR proposed by Montminy [8] with only one extra RP for the reallocation process. The last one is our PB4MP-based proposal. As indicated, our PB4MP-based system is the solution that provides better availability, by far, while using only the normal extra hardware associated to the PR implementation and requiring less than 10Kbytes of extra memory.

VII.   CONCLUSIONS AND FUTURE WORK

This work's main objective is to devise a TMR system using the proposed design flow (PB4MP), which is based on the Xilinx ISE flow. The proposed approach increases considerably some dependability attributes of the TMR system with minimal memory and hardware increment.

The PB4MP approach can be used in other FPGA systems where extended life time periods are important. With no TMR strategy the modules must have test mechanisms, such as [10], in order to implement the FDIR policy.

As a future work, PB4MP will be adapted to the Vivado flow. Also, it is planned an investigation on to what extent the tool can enhance the FPGA life cycle by reducing stress effects as, for instance, NBTI.

The CPU shown in Fig. 1, as discussed before, runs the TMR voter in software, and it is a single point of failure for the system as a whole. In an on-going work at UFSC (Brazil) in cooperation with UNSW (Australia), a rollback/recovery strategy based on checkpoints is under development, in order to provide error recovery functionality to the processor [19].

ACKNOWLEDGEMENTS

This work has been partly funded by the Brazilian National Council for Scientific and Technological Development (CNPq).

REFERENCES

[1] Dhiraj K. Pradhan. *Fault-tolerant computer system design*. Prentice-Hall, 1996.

[2] Ying Zhang et al. Mitigating NBTI degradation on FinFET GPUs through exploiting device heterogeneity. In *Proc. of the IEEE Computer Society Annual Symposium on VLSI (ISVLSI)*, pages 577–582, Tampa, Finland, Jul. 2014.

[3] Rajeev Kumar Mishra et al. Analysis and impacts of negative bias temperature instability (NBTI). In *Proc. of the IEEE Students Conference on Electrical, Electronics and Computer Science (SCEECS)*, pages 1–4, Bhopal, India, Mar. 2012.

[4] Algirdas Avizienis et al. Basic concepts and taxonomy of dependable and secure computing. In *Proc. of the IEEE Transactions on Dependable and Secure Computing*, pages 11–33, Jan. 2004.

[5] Andrea Guiotto et al. SMART-FDIR: use of artificial intelligence in the implementation of a satellite fdir. ftp://ftp.estec.esa.nl/pub/wme/anonymous/wme/Web/SmartFDIR2003.pdf, Jun. 2003. Project coordened by Alenia Spazio (ALS) with participation of Politecnico di Milano (POLIMI).

[6] Tiago Rogerio Muck and Antonio Augusto Frohlich. Seamless integration of HW/SW components in a HLS-based SoC design environment. In *Proc. of the International Symposium on Rapid System Prototyping (RSP)*, pages 109–115, Montreal, Canada, Oct. 2013.

[7] The EPOS Project. Embedded parallel operating system. http://epos.lisha.ufsc.br, 2014.

[8] David P. Montminy et al. Using relocatable bitstreams for fault tolerance. In *Proc. of the Adaptive Hardware and Systems (AHS 2007). NASA/ESA Conference*, pages 701–708, Edinburgh, Scotland, Aug. 2007.

[9] Victor Dumitriu and Lev Kirischian. SoC self-integration mechanism for dynamic reconfigurable systems based on collaborative macrofunction units. In *Proc. of the International Conference on Reconfigurable Computing and FPGAs (ReConFig '13)*, pages 1–7, Cancun, Mexico, Dec. 2013.

[10] Victor M. G. Martins et al. Low cost fault detector guided by permanent faults at the end of FPGAs life cycle. In *Proc. of the IEEE Latin America Test Workshop (LATW)*, pages 1–6, Fortaleza, Brasil, Mar. 2014.

[11] Xilinx Inc. Partial reconfiguration tutorial: PlanAhead design tool v14.5, Apr. 2013. www.xilinx.com/support/documentation/sw_manuals/xilinx14_7/PlanAhead_Tutorial_Partial_Reconfiguration.pdf.

[12] Xilinx Inc. Isolation design flow. www.xilinx.com/applications/isolation-design-flow.html, 2014.

[13] Yoshihiro Ichinomiya et al. Designing flexible reconfigurable regions to relocate partial bitstreams. In *Proc. of the IEEE Annual International Symposium on Field-Programmable Custom Computing Machines (FCCM)*, page 241, Toronto, Canada, May 2012.

[14] Xilinx Inc. LogiCORE IP soft error mitigation controller user guide v3.1, Oct. 2011. www.xilinx.com/support/documentation/ip_documentation/sem/v3_1/ug764_sem.pdf.

[15] OpenCores. Opencores. http://opencores.org, Nov. 2014.

[16] Marcelo Daniel Berejuck. Worst-case Latency Network-on-chip for Real-time Systems. Technical report, Federal University of Santa Catarina, Florianopolis, Brazil, 2015. PhD thesis.

[17] Xilinx Inc. Virtex-6 FPGA configuration user guide v3.8 (ug360). www.xilinx.com/support/documentation/user_guides/ug360.pdf, Aug. 2014.

[18] Xilinx Inc. Constraints guide v14.5 (ug625). www.xilinx.com/support/documentation/sw_manuals/xilinx14_7/cgd.pdf, Apr. 2013.

[19] Tuo Li et al. DHASER: Dynamic heterogeneous adaptation for soft-error resiliency in ASIP-based multi-core systems. In *Proc. of the IEEE/ACM International Conference on Computer-Aided Design (ICCAD)*, pages 646–653, San Jose, USA, Nov. 2013.

2015 IEEE Computer Society Annual Symposium on VLSI

Low-power and low-variability programmable delay element and its application to post-silicon skew tuning

Daijiro Murooka, Yu Zhang, Qing Dong, Shigetoshi Nakatake

Department of Information and Media Engineering, The University of Kitakyushu, Japan

Abstract—The post-silicon tuning introducing programmable delay elements (PDEs) to mitigate the manufacturing variability on the delay is promising. This work presents a novel PDE based on the channel-length decomposition, and reveals that it contributes to the low-power and low-variability comparing with a conventional inverter-chain-type. In addition, in a model of a clock tree along with the PDEs, we propose a mechanism for post-silicon tuning of a skew between a pair of flip-flops by a multilevel DLL employing our PDEs of multiple delay steps. In experiments, our proposed mechanism provides a high tunability even under the variability of PDE itself. Furthermore, we demonstrate our mechanism can be used for various clock systems by easy extensions.

I. INTRODUCTION

A timing error caused by the manufacturing variability is becoming critical for digital circuits. A programmable delay element (PDE) can generate various delays by controlling the number of active delay elements from the outside of the chip. Hence, it is used to resolve the timing error by the post-silicon tuning technology [1][6][7]. If we place PDEs on a path from a clock source driver to each flip-flop (FF), we can change the arrival time at the FF after manufacturing. It is well known that the linearity and the relative accuracy are significant characteristics for PDEs.

For the post-silicon delay tuning, however, a PDE should be paid attention to its variability and should be quite a low-power when it is used in a clock network. In addition, because it is an important issue how to capture skews to be tuned after manufacturing, we need to address a mechanism for tuning the skews by PDEs including how to set the delay to PDE.

In this paper, we present a low-power and low-variability programmable delay elements as well as a mechanism for clock skew tuning consistent with our PDEs. The contributions of this paper are summarized as follows;

(A) - We propose a novel PDE employing channel length decomposed transistors, like a clocked inverter, which is composed of a series of pMOS subtransistors and that of nMOS. Compared to a conventional inverter-chain-type PDE which iterates charging and discharging, our PDE always only one charging or discharging when generating any delay, so that it contributes to the low-power. Besides, because a series of channel-length decomposed subtransistors is regarded as a large transistor, the size-dependent variability is suppressed to be low [2]. We verify the variability by the Monte Carlo simulation. Plus, we fabricate the PDEs with various sizes and demonstrate the measurement results related to the linearity and the relative accuracy.

(B) - We provide a post-silicon skew tuning mechanism introducing a multilevel delay locked loop (DLL) to synchronize the timing at a clock source and a specified FF. PDEs are placed

from the source to the FF of a clock tree. The key point is that our multilevel DLL configures a loop by employing the series of PDEs placed on the clock tree. Our mechanism consists of two modes, skew-calibration and deskew-clocking. In the skew-calibration mode, a loop is configured such that a clock signal from the source passes the DLL to synchronize the skew between the source and the FF. In the deskew-clocking mode, the loop is broken and the DLL becomes a sleep. The stored code of the PDEs in the previous mode still corresponds to an appropriate delay for a zero-skew.

(C) - We illustrate the tunability by our multilevel DLL even when the PDE has a large variability. To our best knowledge, this is the first work to consider the PDE itself variability. In experiments, the Monte Carlo simulations on 0.6um/3.3V and 0.18um/3.3V processes show that our tuning mechanism can reduce the clock skews even though the PDE in the clock tree is affected by the process variability. Furthermore, we demonstrate applicability of our proposed mechanism for various clock systems.

The rest of this paper is organized as follows; Section II describes the variability and power consumption of the proposed PDE and the simulation and the measurement results. Section III is devoted into describing a mechanism to employ the multilevel DLL for reducing the clock skew and demonstrating the simulation results under the PDE variability. Section IV presents extensions of our mechanism based on multilevel DLL to various clock systems. Section V concludes this work.

II. PROGRAMMABLE DELAY ELEMENT

A timing design is essential for designs of digital circuits. A programmable delay element (PDE) is a technology to change the intrinsic delay after manufacturing, and it is used for the post-silicon tuning to resolve a timing issue due to the manufacturing variability.

Several types of PDEs have been proposed so far, but the most popular type is an inverter-chain-type. It has a good linearity and a useful model for designs, but suffers from the power saving and the suppression of the variability. This paper proposes a simple but new PDE with better characteristics with respect to the power and the variability.

A. Channel-length decomposition PDE

A schematic of a PDE proposed in this paper is illustrated in Fig. 1. It is constituted, like a clocked inverter, such that pMOS transistors of the same size are connected in serial, as well as nMOS transistors. By the control signals (D1-D8), the number of active transistors can be changed, and it results in changing of the intrinsic delay. Focusing on the series of pMOS transistors, we can regard this series such that a large transistor is decomposed into a set of subtransistors along the

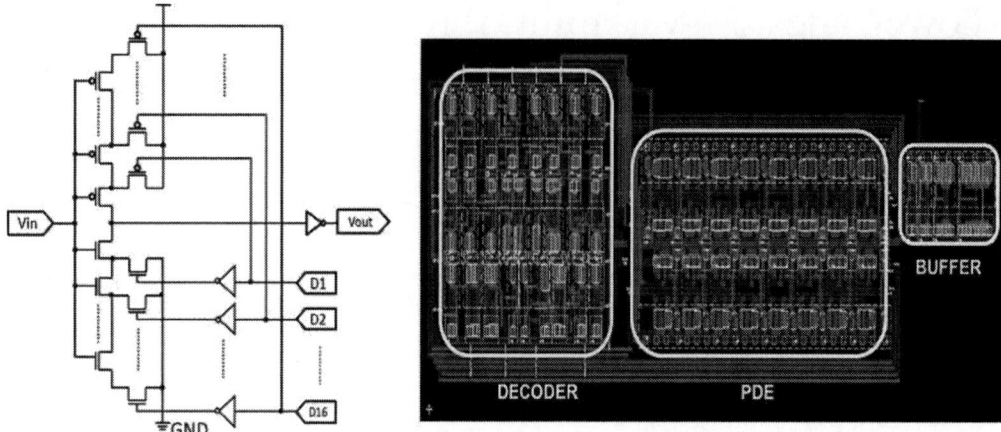

Fig. 1. (a) Schematic of PDE, (b) Layout of PDE with decoder and buffer

channel-length. The nMOS transistors can be also regarded analogously. This is why we call our PDE channel-length decomposition type.

Besides, since each subtransistor has the same channel-length, the delay is proportional to the number of active sub-transistors passed by the current when charging or discharging. Note that, in a DC model, a subtransistor is regarded as a resistor, and the delay is proportional to the RC value in the model when considering the first moment of the circuit, like the Elmore delay model.

B. Variability analysis

We analize the variability for our proposed PDE and an inverter-chain-type as the existing PDE by the Monte Carlo simulation. In the simulation, both PDEs have the same size for each transistor, we observe the difference with respect to the deviation of the variability distribution. The transistor sizes are shown in Table I. In addition, the variability to the control switch blocks for both PDEs are not given.

The simulation is set by 5σ normal distribution where ($\mu = 12\mu m, \sigma = 3\mu$) of pMOS and ($\mu = 6\mu m, \sigma = 3\mu$) of nMOS with respect to the channel width of each transistor. μ and σ are the average and the standard deviation, respectively. The resultant distribution of the delay of each PDE is shown in Fig. 2. It can be observed that σ of the distribution for our proposed PDE is smaller than that for the inverter-chain-type. This is because the transistor numbers of the the inverter-chain-type PDE and our proposed PDE corresponding to the same delay are 64 and 34, respectively.

TABLE I. CHANNEL-SIZE OF EACH TRANSISTOR OF PDEs

	pMOS	nMOS
our PDE	L=0.6[um], W=12[um]	L=0.6[um], W=6[um]
Inverter-chain-type	L=0.6[um], W=12[um]	L=0.6[um], W=6[um]

C. Power consumption analysis

The inverter-chain-type, current-stave-type, DCDE-type and thyristor-type are known as the conventional PDEs

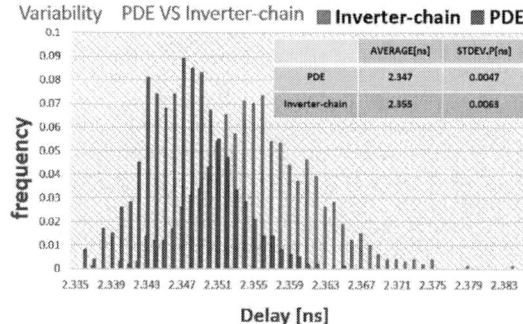

Fig. 2. Variability of PDE and Inverter-chain

[3][4][5]. Each circuit mechanism can achieve a good linearity as shown in Fig. 3, although it is hard for each to generate a specified absolute delay value.

The inverter-chain is the most frequent type, and a charging and a discharging are repeated at every time when the signal passes through each inverter of multi-stage. This means the power consumption increases according to the amount of delay to be generated. On the other hand, in our PDE, a charging and a discharging occur only once even when increasing the delay. As as result, our PDE achieves the further low power consumption as shown in Fig. 4.

D. Measurement result

We demonstrate the measurement results of various size PDEs in Fig. 5. Each PDE provides a good linearity along the code. These results imply the skews should be tune in terms of not absolute delay values but relative values.

III. APPLICATION TO POST-SILICON SKEW TUNING

In this section, we present a mechanism how to capture a clock skew and how to tune the skew after manufacturing by using our proposed PDE of channel-length decomposed type.

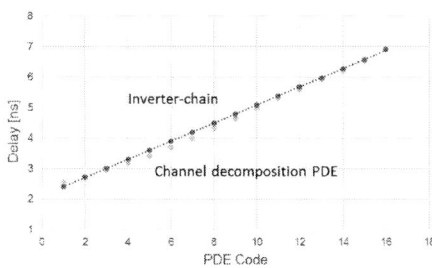

Fig. 3. The linearity of the delay values generated by the inverter-chain-type and our PDE

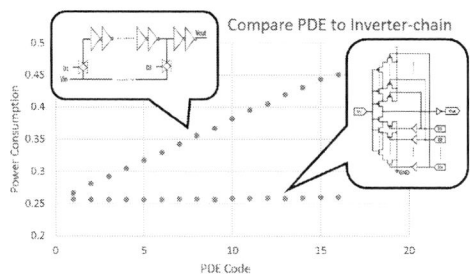

Fig. 4. The power consumption to Inverter-chain-type and our PDE

A. Multilevel delay locked loop

A key component of our mechanism for skew tuning is DLL (Delay Locked Loop). Also, we propose a DLL circuit incorporating channel-length decomposed PDEs, which is called multilevel DLL. A diagram of the proposed multilevel DLL is shown Fig. 6. The DLL detects the delay difference between two input signals by PD (Phase Detector), and control the number of PDEs and the number of active channels of each PDE according to the delay difference. The output the PDE is fed back as one of the input signals, and the DLL gradually decreases the delay difference. Because a total delay of PDEs should cover the maximum delay difference, we introduce a series of eight PDEs for delay tuning. We call it 8-level DLL shown Fig. 6.

B. Mechanism for skew tuning

In post-silicon tuning, we place the PDE on a clock tree, and align skews from a clock source to different FFs. Since it is hard to measure the arrival time from the clock source to the FF after chip fabrication, we need a circuit mechanism to capture the skew.

In a conventional tuning mechanism shown in [6][7], a connection with a phase detector (PD) to synchronize the timing at FFs placed adjacently by using adjustable delay buffer (ADB) on the clock tree. This idea contributes to reduce the wire-length of the connection for the PD. However, it could occur that a slight skew between adjacent FFs after tuning is transmitted and accumulated, and it results in a large skew between FFs apart from each other.

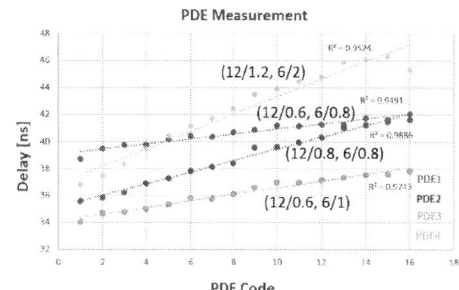

Fig. 5. The measurement result of the delay of our PDEs

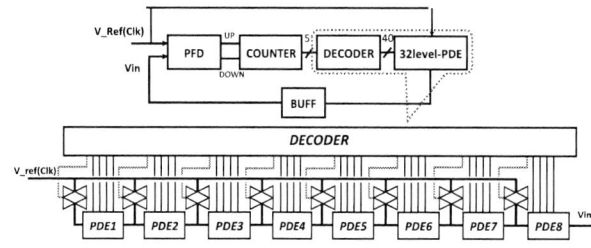

Fig. 6. The proposed multilevel DLL: 8-level DLL

In this paper, we introduce a mechanism employing the multilevel DLL with our PDE. In our mechanism, a series of PDEs are shared by the clock tree and the DLL. The PDEs are placed on the clock tree, but they are low-power as described above. Besides, unlike [6][7], we make a connection with a PD of a multilevel DLL between each FF and the clock source. This means each FF is synchronized the timing at the clock source, and it does not occur a large skew even for FFs apart from each other.

C. Simulation for skew tuning

In this section, we illustrate the tunability by our multilevel DLL under the variability of the PDE itself. In our simulation, we compare two cases, Case1 and Case2 as shown in Fig. 7 and Fig. 8, respectively.

Case1: This case assumes a clock tree has no tuning mechanism. In a path from the clock source to the FF (FF-A), buffers are placed Buf-A1 and Buf-A2. As well, Buf-B1 and Buf-B2 are placed on a path to FF-B.

Case2: This case is for the tunability by our mechanism. A signal from the clock source is input to the multilevel DLL, and the output of the DLL is connected to FF-A via the PDE-A. In addition, a signal from a point just after PDE-A is also input to the DLL for the synchronization. An analogous connection is configured for PDE-B and FF-B. The mechanism provides a real-time tunability for clock skews.

As a preliminary, we give the size of each buffer is shown in Table II and observe the waveforms at the input of FF-A and FF-B in both cases. As shown in Table III, the skew between FF-A and FF-B in Case2 is much smaller than that in Case1.

978-1-4799-8720-7/15 $31.00 © 2015 IEEE

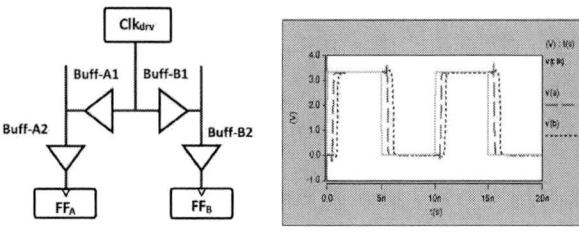

Fig. 7.   Case1:Skew when I gave a buffer variability

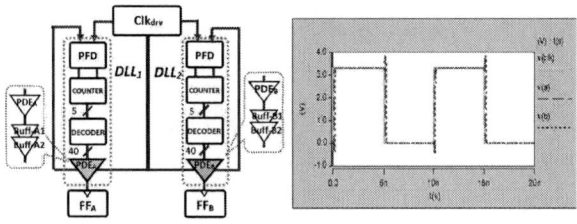

Fig. 8.   Case2:Skew adjustment between the FF by conventional DLL

D. Simulation under PDE variability

One of the most important motivations of the post-silicon tuning is to resolve the variability issue. However, Most of studies related to the post-silicon tuning do not pay attention to the PDE itself variability. This is the first work to consider the PDE itself variability.

As for two cases shown in Fig. 7 and Fig. 8, we apply the Monte Carlo simulation under the assumption that Buf-A1, Buf-A2, Buf-B1 and Buf-B2 have the variation, and observe the skew distribution between FF-A and FF-B.

We use the channel-width as the variability parameters for the buffers(Case1) and PDEs(Case2) on the clock tree. The numbers of transistors given the variability is 112 and 128 for Case1 and Case2, respectively. The transistor size of the buffers (Case1) and PDEs (Case2) are the same as pMOS is W/L = 12u/1.5u, and nMOS is W/L = 6u/2u. Besides, for both cases, the fluctuation of channel-width is described as PMOS(W) = AGAUSS(12u 4u 5), NMOS(W) = AGAUSS(6u 2u 5)) in the HSPICE simulation. Changing the parameters, 1000 times simulations are executed.

The resultant distributions of the skews by the simulation on 0.6um/3.3V and on 0.18um/3.3V processes are shown in Fig. 10 and Fig. 11, respectively. Our mechanism can obviously reduce the deviation of the skew for both processes, but the effect for the tunability (i.e. the difference between Case1 and Case2) on the finer process (0.18um) is much larger that on the 0.6um process. On the other hand, even for every case, we cannot achieve a complete zero skew. This means a conventional tuning mechanism in [6][7] has a risk of increasing a skew between FFs apart from each other.

IV.   APPLICABILITY TO VARIOUS CLOCK SYSTEMS

Our proposed mechanism for post-silicon tuning can be easily extended to various clock systems. In the previous

TABLE II.   BUFFER SIZE

buffer	nMOS W/L(μm)	pMOS W/L (μm)
Buf-A1	2.8×4/0.7	5.6×4/0.6
Buf-A2	2.8×2/0.7	5.6×2/0.6
Buf-B1	2.8×8/0.7	5.6×8/0.6
Buf-B2	2.8×2/0.7	5.6×2/0.6

TABLE III.   SKEW OF BETWEEN FF AND FF

	FF-A	FF-B	skew
Case1	0.48ns	1.17ns	0.69ns
Case2	0.14ns	0.25ns	0.11ns

section, the mechanism provides a real-time tuning for zero-skew. Sometimes, however, such a real-time tuning cannot be accepted from the viewpoint of the power consumption. Accordingly, our mechanism can have a flexibility for the operating mode as follows; Introducing switch S_A and S_B as shown in Fig. 12(a), our mechanism can consist of two modes, skew-calibration and deskew-clocking. In the skew-calibration mode, a loop is configured for synchronization of the skew by the DLL. In the deskew-clocking mode, the loop is broken and the DLL becomes a sleep for power saving. In this mode, we need a set of FFs to to store the code of the PDEs corresponding to an appropriate delay for a zero-skew. The FFs are denoted by $FF_{A1} \cdots FF_{An}$ for the loop of FF_A in the figure.

Furthermore, it is known that a clock speed can be accelerated by controlling of skews of FFs[8][9]. As described in this paper, however, the variability of PDE itself makes it difficult to distribute appropriate skews to FFs. Then, we introduce PDEs nearby the clock source to generate the specified skews. At the point X_R in Fig. 12(a), PDE_{A0} and PDE_{B0} are added as shown in the figure (b). It means that the delay of PDE_{A0} and PDE_{B0} correspond to the skews of FF_A and FF_B, respectively. These PDEs also have the variability, but they are placed closely each other as well as a large size to reduce the variability.

In addition, compared to conventional mechanism as in [6][7], our mechanism may increase the overhead for tuning of a clock system for a large number of FFs. However, as shown in Fig. 12 (c), we can make groups of FFs according to FFs' placement, and provide a DLL based mechanism to each group.

V.   CONCLUSION

We propose a low-power and low-variability programmable delay elements and clarify the performance by simulation and measurement. Introducing the PDE, we present a post-silicon tuning mechanism along with the multilevel DLL. The DLL configures a loop by employing the series of PDEs placed on the clock tree, and attains the low variability even though the PDE itself in the clock tree is affected by the variability. In the comparison with a conventional clock skew tuning mechanism, the simulation result demonstrates the higher tunability of our mechanism. In addition, we demonstrate our proposed mechanism can apply to various clock systems by easy extensions.

978-1-4799-8720-7/15 $31.00 © 2015 IEEE

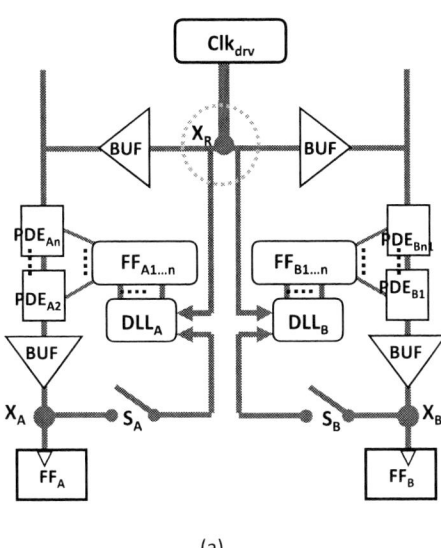

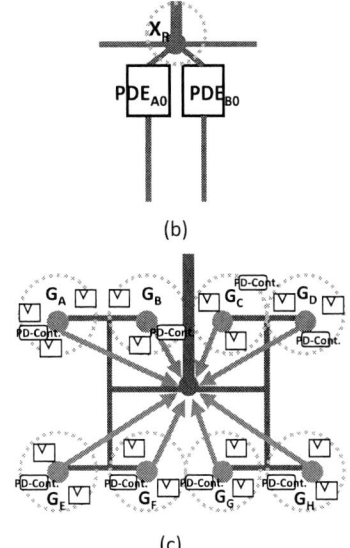

(b)

(c)

Fig. 12.   Our mechanism and various clock systems

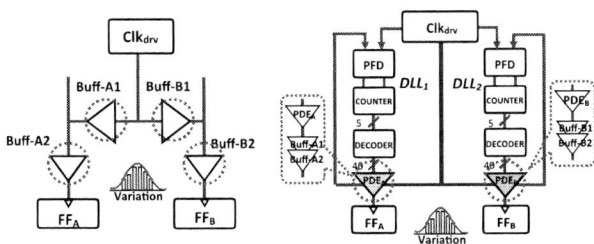

Fig. 9.   Variability of Case1 VS Case2

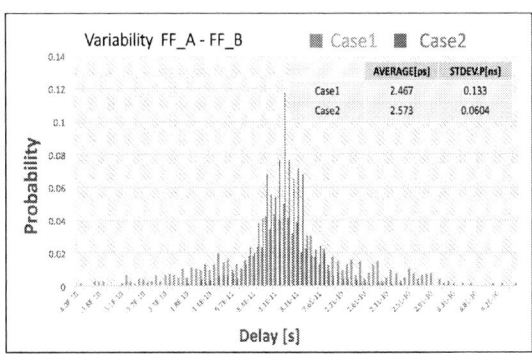

Fig. 11.   Variability of Case1 VS Case2 (0.18um)

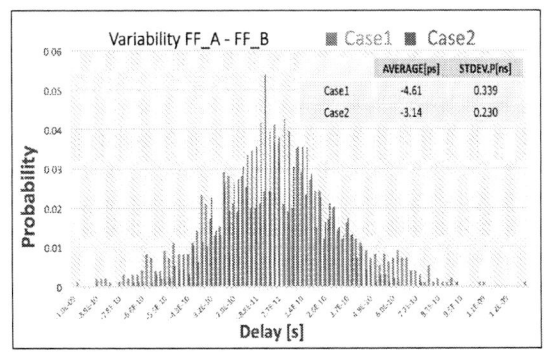

Fig. 10.   Variability of Case1 VS Case2 (0.6um)

REFERENCES

[1] Yuko Hashizume, Yasuhiro Takashima, Yuichi Nakamura, *Post-Silicon Clock-Timing Tuning Based on Statistical Estimation* IEICE Transactions on Fundamentals, Vol.E91-A, No.9, pp.2322-2327, 2008.

[2] Bo Yang, Qing Dong, Jing Li, Shigetoshi Nakatake, *Structured Analog Circuit Design and MOS Transistor Decomposition for High Accuracy Applications*, ICCAD, pp.721-728, 2010.

[3] M. Maymandi-Nejad and M. Sachdev, *A digitally programmable delay element: Design and analysis* IEEE Transaction on Very Large Scale Integration (VLSI) Systems, Vol.11, No.5, pp. 871-878, Oct. 2003.

[4] Junmou Zhang, Simon R. Cooper, Andrew R.LaPietra, Michael W. Mattern, Robert M.Guidash, Eby G.Friedman *A Low Power Thyristor-Based CMOS Programmable Delay Element*, ISCAS, pp.769-772, 2004.

[5] Jung-Lin Yang, Chih-Wei Chao, Sung-Min Lin, *Tunable Delay Element for Low Power VLSI Circuit Design* IEEE TENCON 2006.

[6] Yu-Chien Kao, Hsuan-Ming Chou, Kun-Ting Tsai, Shih-Chieh Chang, *A Efficient Phase Detector Connection Structure for the Skew Synchronization System*, DAC'10, pp.729-734, 2010.

[7] Mac Y.C.Kao, Kun-Ting Tsai, Shih-Chieh Chang, *A Robust Architecture for Post-Silicon Skew Tuning*, ICCAD, pp.774-778, 2011.

[8] Yukihide Kohira, Atsushi Takahashi, *Clock Period Minimization Method of Semi-Synchronous Circuits by Delay Insertion*, IEICE Transaction on Fundamentals, Vol.E88-A, No.4, pp.892-898, 2005.

[9] Yukihide Kohira, Shuhei Tani, Atsushi Takahashi, *Minimization of Delay Insertion in Clock Period Improvement in General-Synchronous Framework* IEICE Transaction on Fundamentals, Vol.E92-A, No.4, pp.1106-1114, 2009.

978-1-4799-8720-7/15 $31.00 © 2015 IEEE

2015 IEEE Computer Society Annual Symposium on VLSI

A 10-bit 500 MSPS Segmented DAC with Optimized Current Sources to Avoid Mismatch Effect

Santanu Sarkar
IEEE Member, Department of ECE
National Institute of Technology, Rourkela
Rourkela, India
Email: sarkars@nitrkl.ac.in

Swapna Banerjee
IEEE Member, Department of E & ECE
Indian Institute of Technology, Kharagpur
Kharagpur, India
Email: swapna@ece.iitkgp.ernet.in

Abstract—This paper describes the design techniques of a segmented current steering (CS) digital-to-analog converter (DAC) with optimum sizing of the current sources. The DAC has been designed in 0.18 μm CMOS n-well technology provided by National Semiconductor. The 10-bit DAC is segmented as 5+5, where the 5-LSB bits are implemented in binary and the 5-MSB bits are implemented in unary architecture. The matching of the unit current sources plays an important role in determining the overall linearity of the DAC. Static linearity of the DAC can be improved by using larger area of the current source transistors, sacrificing the dynamic performances. At high frequency the spectral performance of the DAC degrades due to the increased parasitic. In this work the current sources are designed with optimum sizes to achieve improved static as well as dynamic performances. In simulation, the DAC achieves a maximum DNL of 0.248 LSB and a maximum INL of 0.440 LSB. The DAC achieves a maximum spurious free dynamic range (SFDR) of 59.79 dB for 5.37 MHz signal in mismatch environment at 500 MSPS sampling rate. The DAC shows a Nyquist SFDR of 57 dB at 500 MSPS sampling rate with mismatch. The DAC consumes only 17.85 mW of power for Nyquist signal at 500 MSPS sampling rate with 1.8 V supply.

Index Terms—Current steering DAC; Segmented DAC; Matching; INL; DNL; Low Power; SFDR; Nyquist Signal Frequency;

I. Introduction

Due to the recent growth in 3G and 4G communication systems the demand of low power, single chip wireless transceivers for portable terminals is growing rapidly. To realize the concept of always connectivity, the wireless transceiver uses reconfigurable baseband and RF architectures to operate at different standards (GSM, UMTS, WLANs, Bluetooth) [1]. With the advancement of digital CMOS technology, it is now possible to implement the reconfigurable and multimode digital transmitters for wireless communications. Present transceiver systems, use several complex digital modulation techniques. Inclusion of more features into the system demands more spectral purity from the output signal. As DAC is the first analog block in the signal chain, it determines the performance of the overall system [1]. Unless design with high care, the DAC could become the bottleneck for the signal path [2]. For that reason the demand of high performance and high

speed DAC has increased over the years [3]. Typical block diagram of a direct transmitter system is shown in Fig. 1, where the digital I and Q baseband signals are converted to analog form by using two baseband DACs [4].

For high speed applications current steering architecture is most suitable as it is inherently fast and can drive a typical resistive load of 50 Ω without using any output buffer [5]. Conventional CS-DACs are classified in three sub-categories, namely, unary, binary and segmented architectures [6]. For DACs with larger resolution, segmented CS-architectures are used. Segmented DAC, which is composed of both unary and binary DACs [7], [8] retains the advantages of both architectures. Generally segmented DAC is divided in two parts, the MSB section and the LSB section [6]. To achieve better linearity the MSB section is implemented in unary DAC, and to save the total area, LSB section is implemented in binary DAC.

The performance of the current steering DAC depends on

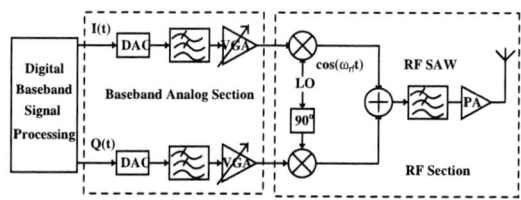

Fig. 1. Block diagram of a typical direct transmitter system

the matching of the current sources [9]. Use of larger transistors to implement the current sources helps to improve the matching [10], [11] among the current cells. Use of larger transistors increase the overall area and implementation cost of the DAC. Also due to increased parasitic the performance of the DAC degrades at high frequency. For that reason, it is always better to implement the current sources with optimum size to improve the static as well as the dynamic performances. In this work the optimum sizes of the current source transistors are determined to achieve improved spectral performances. To

978-1-4799-8720-7/15 $31.00 © 2015 IEEE   172

reduce the correlated noise of the current sources due to the common bias cells, distributed bias sources are used for the 10-bit DAC. The remaining sections of the paper are organized as follows. The architecture and the design of the different sub-blocks of the DAC are described in Section II. The simulation results of the proposed DAC are described in Section III. Finally, Section IV, concludes the paper.

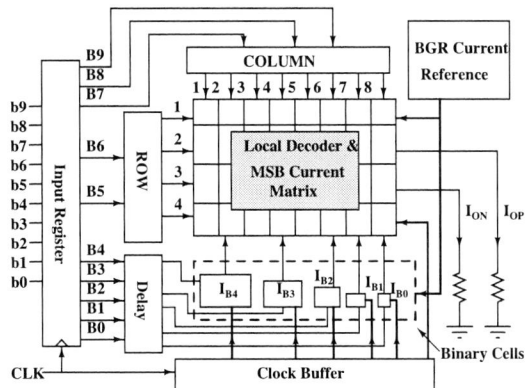

Fig. 2. Block diagram of the 10-bit segmented DAC

II. ARCHITECTURE

The block diagram of the proposed DAC is shown in Fig. 2. In the figure, (B0-B4) the five-least significant bits (LSB) are used to control the 5-bit sub-binary DAC. The binary sub-DAC is implemented with five binary weighted current sources of amplitudes I, 2I, 4I, 8I and 16I respectively. Here, I is the weight of the LSB current cell and the value of I is chosen as 5 μA. The (B5-B9) five-most significant bits (MSB) are used to generate the binary-to-thermometric decoder outputs to control the MSB unit weighted cells. The 5-bit MSB-unary DAC is composed of 31 unit cells of weight 32I. A 4×8 current matrix is used to realize the current cells in a rectangular geometrical structure. The 3-MSB bits (B7-B9) are sent to the 3-to-8 binary-to-thermometric column decoder and the next two-bits (B5-B6) are sent to the 2-to-4 binary-to-thermometric row decoder. The row and column decoder outputs are again combined in the local decoder matrix to generate the final control signals for the MSB unit cells. As the binary sub-DAC does not require any decoder, a delay equalizer block is designed for the binary section to synchronize the control signals of the two blocks. Cascoded PMOS transistors are used to implement the current sources and differential current switches are used to generate the current outputs. An on chip bandgap current reference (BGR) circuit is used to generate the bias voltages for the current sources.

There are several digital blocks in the digital-to-analog data converter. Important digital blocks include input registers, row and column decoders, local decoders and final re-timing latch. The analog part of the DAC includes, current sources, current switches and current reference circuit. The design technique

of the important building blocks are described in the next sub-sections.

A. Design of the Input Register

In the proposed DAC, 10-bit input registers are used to synchronise the digital data with respect to the internal clock. A master-slave positive edge triggered D-flip flop is used as the input register [12]. A multiplexer-based latch is used for that purpose and the schematic is shown in Fig. 3. The multiplexer is implemented by using transmission gates. When the clock is low, the master stage is transparent and the D-input is passed to the master stage output. During this phase the slave is in hold mode and keeps the value of the previous clock by using feedback. On the rising edge of the clock the slave stage samples the data of the master stage output and the master stage stops sampling the input. The value of the output is the value of the D-input right before the rising edge of the clock and so it acts as a positive edge-triggered register. Static CMOS design style is used to achieve a robust design.

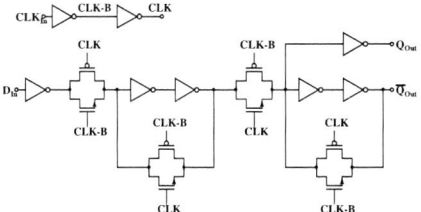

Fig. 3. Schematic of the D-flip flop

B. Design of the Row and Column Decoder

The proposed DAC uses five MSB bits for binary-to-thermometric decoder design. For the 10-bit inputs the first three MSB bits (B7-B9) are used for column decoding and next two bits (B5-B6) are used in row decoding. For the 32 decoder outputs an 4×8 array is used for a rectangular layout area. The 3-to-8 column decoder generates eight decoded outputs which control corresponding eight columns. The 2-to-4 row decoder generates control signals corresponding to the four rows. The schematic of the row and column decoders are shown in Fig. 4(a) and Fig. 4(b) respectively.

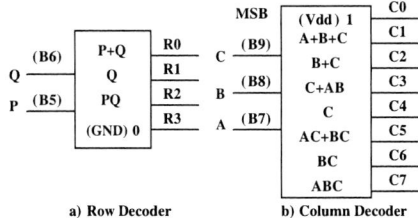

Fig. 4. Schematic of the row and column decoders

C. Design of the Local Decoder

As shown in Fig. 2, the row and column decoder outputs are again combined in the local decoder matrix to generate the final control signals for the current switches. In the proposed 10-bit DAC there are 31 number of MSB unit cells. The local decoding logic of the MSB unit cells are equivalent to an AND-OR gate function [8] as shown in Fig. 5.

D. Design of the Final Re-timing Latch

The circuit of the final re-timing latch is shown in Fig. 6 [11], [12]. As PMOS transistors are used for the design of the differential current switches, the latch is designed with PMOS drivers. Care should be taken to design the slightly overlapped switch control signals to ensure make-before-brake operation of the differential current switches. Otherwise, for any transition if both switches are off-for a small moment, the common source node of the switches will be charged up to V_{DD}. For the next transition, the node capacitance first need to be discharged to presume normal operation and thus creates large glitches and transition delay degrading the performance of the DAC. Extra NMOS transistors are used in the design of the latch to generate the overlapped differential switch control signals by modifying the rise and fall times [11].

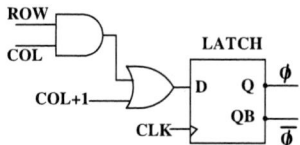

Fig. 5. Local decoder for the 31 MSB unit current cells

E. Design of the Current Sources

In CS-DAC, the INL is mainly determined by the matching of the current sources. Increasing the sizes of the current source transistors help to improve the matching of the current cells. Although, over-sizing may degrade the performance of the DAC at high signal frequencies. It is always better to design the current sources with optimum size to maintain high performances at all signal frequencies. To predict the INL within a certain boundary the INL yield is used as the percentage of the working DACs with an INL smaller than 0.5 LSB [13]. The relative standard deviation $\sigma(I)/I$, of unit current sources should satisfy the relationship given in Eq. 1, to achieve an upper bound of INL ($INL_{Upper-Bound}$) [14]. In the equation, *inv-normal* is the inverse cumulative normal distribution and N is the resolution of the DAC. The minimum required area of the current sources is obtained following Eq. 2 [15]:

$$\frac{\sigma^2(I)}{I^2} \leq \frac{INL_{Upper-Bound}}{inv-normal(0.5 + \frac{INL_{yield}}{2}).\sqrt{2^{(N-1)}}} \quad (1)$$

$$(W \times L)_{min} = \frac{\frac{4A_{Vt}^2}{(V_{GS}-V_t)^2} + A_\beta^2}{2 \times \frac{\sigma^2(I)}{I^2}} \quad (2)$$

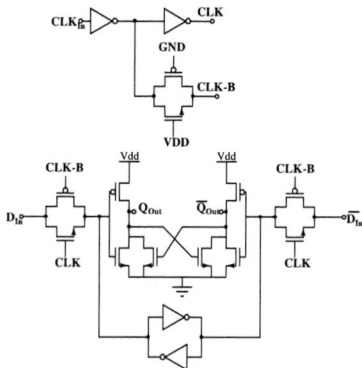

Fig. 6. Final latch circuit

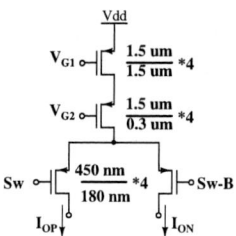

Fig. 7. 20 μA current source with differential current switches

where A_{Vt} and A_β are process mismatch parameters. V_{GS} is the MOS transistor's gate-to-source voltage and V_t is the threshold voltage. To reduce the minimum required area of the current cells, large overdrive voltage of 457 mV is used . For the LSB current of 5 μA the size the LSB cell is chosen as 1.5 μm$\times$1.5 μm, satisfying Eq. 2. The size of the cascode transistor is chosen as 1.5 μm$\times$0.3 μm. For 10 μA current the number of finger of the PMOS transistors are increased to two and for 20 μA current the number of finger is four. The finger number increases accordingly with the weight of the current sources. The schematic of the current source for 20 μA is shown in Fig. 7.

F. Design of the Current Switches

For the proposed DAC, PMOS transistors are used to design the differential current switches. The differential switch pairs are designed with minimum length. The size of the switch transistor for LSB current source is 450 nm$\times$180 nm. For current switches of 10 μA current cells the number of finger of the PMOS switch transistors are increased to two. For 20 μA current the number of finger is four as shown in Fig. 7. For the rest of the weighted current sources, the size of the switch transistors increases accordingly.

G. Design of the Current Reference Circuit

To bias the current sources a bandgap current reference circuit is designed. The schematic of the BGR is shown in Fig. 8. A two stage operational trans-conductance amplifier (OTA) has been designed for the band gap current reference.

978-1-4799-8720-7/15 $31.00 © 2015 IEEE

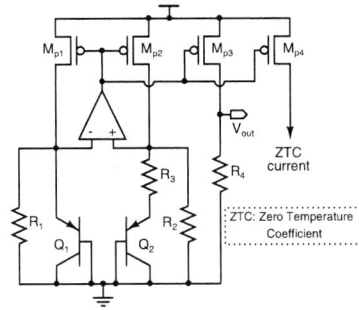

Fig. 8. Band-gap current reference circuit

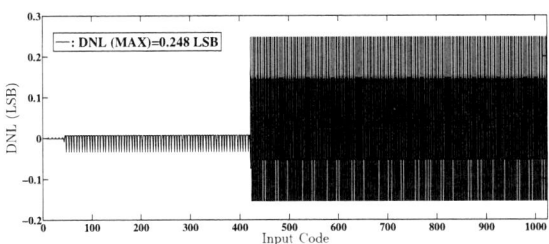

Fig. 9. Simulated DNL plot of the DAC

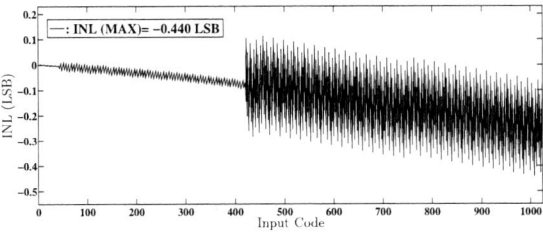

Fig. 10. Simulated INL plot of the DAC

III. SIMULATION RESULTS

In Spice simulation at normal condition the segmented DAC achieves a maximum DNL of 0.248 LSB and an INL 0.440 LSB, as shown in Fig. 9 and Fig. 10 respectively. Cadence Virtuoso 6.1.4 ADE XL tool is used for the Monte Carlo-mismatch analysis of the proposed DAC. Mismatch effect is considered for the whole DAC and same mismatch seed is used for all simulations result in this paper. The DAC achieves a SFDR of 59.79 dB in mismatch environment for 5.37 MHz signal at 500 MSPS sampling rate, as shown in Fig. 11. The DAC achieves a mismatch simulated SFDR of 57.98 dB, for 15.136 MHz signal at 500 MSPS sampling rate and the output spectrum is shown in Fig. 12. The DAC shows a Nyquist SFDR of 57 dB for 249.02 MHz at 500 MSPS sampling rate and the corresponding frequency spectrum is shown in Fig. 13.

In present wireless transmitters, several multi-carrier digital modulation techniques are used and the modulated signals

are directly fed into the DAC. In such case, a two-tone test gives better information about the DAC's performances for real modulated signals [11]. The DAC produces an IM3 of 50.66 dB for dual tone test with 245.60 and 247.56 MHz signals at 500 MSPS sampling rate as shown in Fig. 14. The 5+5 segmented DAC consumes 17.85 mW of power at Nyquist frequency for 500 MSPS sampling rate with 1.8 V supply.

The performance metrics of the proposed 10-bit DAC are

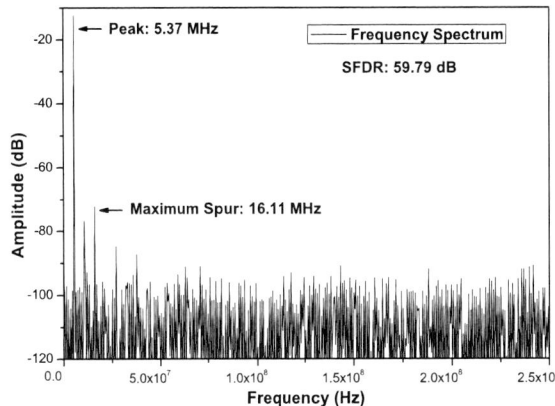

Fig. 11. Mismatch simulated output spectrum showing 59.79 dB SFDR for 5.37 MHz signal sampled at 500 MSPS

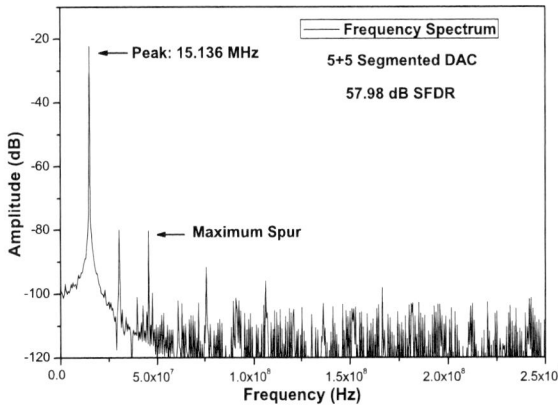

Fig. 12. Frequency spectrum showing 57.98 dB SFDR for 15.136 MHz signal, considering the effect of mismatch

summarized in Table I at the end of this section. To evaluate the overall performance of the proposed DAC, the figure of merit (FOM) described in Eq. 3 [16] is used. The proposed 10-bit DAC achieves 1.35×10^4 GHz/mW of FOM in Table I. The FOM of the DAC can be improved further by increasing the percentage of segmentation, where more number of bits will be implemented in unary architecture.

$$FOM = \frac{2^{\frac{SFDR_{dc}-1.76}{6.02}} \times 2^{\frac{SFDR_{Nyquist}-1.76}{6.02}} \times f_{clk}}{P_{total} - \frac{1}{2}I_{load}^2 \times R_{load}} \quad (3)$$

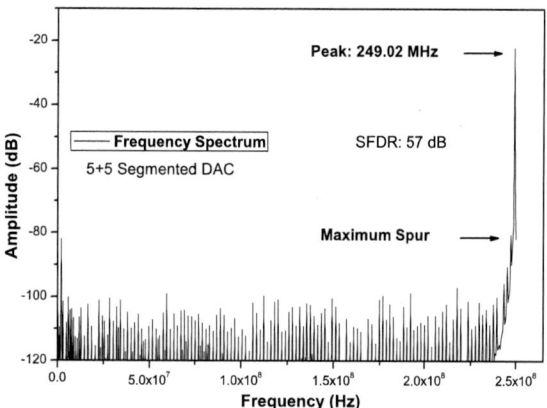

Fig. 13. Frequency spectrum of the DAC showing 57 dB of Nyquist SFDR for 500 MSPS clock rate with mismatch

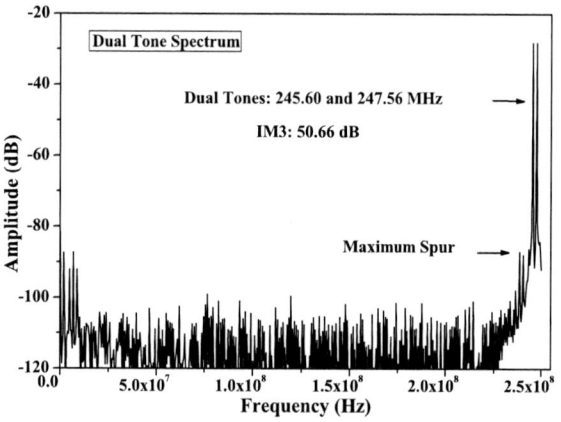

Fig. 14. Dual tone spectrum showing 50.66 dB of IM3 for 245.60 and 247.56 MHz signals, considering the mismatch effect

TABLE I
SUMMARY OF THE SIMULATION RESULTS OF THE PROPOSED 10-BIT DAC

Sampling Rate	500 MSPS
DNL (LSB)	0.248 LSB
INL (LSB)	0.440 LSB
Power Supply	1.8 V
Power Dissipation (at Nyquist Signal)	17.85 mW@500 MSPS
SFDR	59.79 dB@5.37 MHz
	57 dB@249.02 MHz
IM3	50.66 dB@245.60 and 247.56 MHz
FOM	1.35×10^4 GHz/mW
Process	0.18 μm CMOS

IV. CONCLUSIONS

In this paper a 10 bit 500 MS/s segmented current steering DAC is presented. The DAC has been designed for wireless transmitters for reconfigurable multi-standard terminals. To improve matching between the unit current sources optimum transistor size is selected and an on chip bandgap current reference circuit is used for biasing the current sources. The DAC has been designed using 0.18 μm five-metal CMOS process. The DAC exhibits 10-bit intrinsic linearity. Simulation results show that the DAC achieves a maximum DNL of 0.248 LSB and an INL of 0.440 LSB. The DAC achieves an IM3 of 50.66 dB for dual tone test with 245.60 and 247.56 MHz signals at 500 MSPS sampling rate considering the effect of mismatch. The proposed DAC achieves a maximum mismatch simulated SFDR of 59.79 dB for 5.37 MHz signal at 500 MSPS sampling rate. The DAC achieves a Nyquist SFDR of 57 dB for 249.02 MHz signal at 500 MSPS sampling rate even at mismatch environment. The driving load for the DAC is considered as 50 Ω resistance. The simulated maximum power dissipation of the DAC is only 17.85 mW at Nyquist signal frequency sampled with 500 MSPS for 1.8 V DC supply. The DAC achieves moderate FOM in Table I without using any additional complex design techniques.

ACKNOWLEDGMENT

The authors would like to thank National Semiconductor Corporation for providing the CMOS technology. The authors would also like to thank Advanced VLSI Design Laboratory (AVLSI) at IIT Kharagpur.

REFERENCES

[1] MAX5887, *3.3V, 14-Bit, 500MSPS High Dynamic Performance DAC with Differential LVDS Inputs, 19-2777; Rev 2; 12/03* . Maxim Integrated Products, 2003.

[2] Y. Li and Z. Li, "A low-power 6-bit D/A converter design for WSN transceivers," in *IEEE 13th International Conference on Communication Technology (ICCT)*, Sept 2011, pp. 307–310.

[3] G. I. Radulov, P. J. Quinn, and A. H. M. van Roermund , "A 28-nm CMOS 1 V 3.5 GS/s 6-bit DAC With Signal-Independent Delta-I Noise DfT Scheme," *IEEE Transactions on Very Large Scale Integration (VLSI) Systems*, vol. -Early Edition, no. 99, pp. 1–10, 2014.

[4] B. Razavi, *RF Microelectronics*. Prentice Hall, Inc., 1998, no. ISBN 0138875715.

[5] M. Khafaji, C. Scheytt, U. Jorges, C. Carta, D. Micusik, and F. Ellinger, "SFDR considerations for current steering high-speed digital to analog converters," in *IEEE Bipolar/BiCMOS Circuits and Technology Meeting (BCTM)*, Sept 2012, pp. 1–4.

[6] C. H. Lin and K. Bult, "A 10-b, 500-MSample/s CMOS DAC in $0.6mm^2$," *IEEE Journal of Solid-State Circuits*, vol. 33, pp. 1948–1958, Dec 1998.

[7] S. Sarkar, R. S. Prasad, V. Belde, S. K. Dey and S. Banerjee, "An 8-bit, 1.8V, 500MS/s CMOS DAC with a Novel Four Stage Current Steering Architecture," in *IEEE Proc. on International Symposium on Circuits and Systems (ISCAS)*, May 2008, pp. 149–152.

[8] S. Sarkar and S. Banerjee, "An 8-bit 1.8V 500MSPS CMOS Segmented Current Steering DAC ," in *IEEE Computer Society Annual Symposium on VLSI (ISVLSI)*, May 2009, pp. 268–273.

[9] W. Lin and T. Kuo, "A Compact Dynamic-Performance-Improved Current-Steering DAC Random Rotation-based Binary-Weighted Selection," *IEEE Journal of Solid-State Circuits*, vol. 47, pp. 443–453, Feb 2012.

[10] J. Deveugele and M. S. J. Steyaert, "A 10-b 250-MS/s Binary-Weighted Current-Steering DAC ," *IEEE Journal of Solid-State Circuits*, vol. 41, pp. 320–329, Feb 2006.

[11] X. Wu, P. Palmers and M. S. J. Steyaert , "A 130 nm CMOS 6-Bit Full Nyquist 3-GS/s DAC ," *IEEE Journal of Solid-State Circuits*, vol. 43, pp. 2396–2403, Nov 2008.

[12] J. M. Rabaey, A. Chnadrakasan and B. Nikolic, *Digital Integrated Circuits*, 2nd ed. Prentice Hall of India, New Delhi, 2006.

[13] A. Van den Bosch, M. Steyarert and W. Sansen , "An Accurate Statistical Yield Model for CMOS Current-Steering D/A Converters ," *Analog Integrated Circuits and Signal Processing*, vol. 29, pp. 173–180, Mar 2001.

[14] A. Van den Bosch, M. A. F. Borremans, M. S. J. Steyarert and W. Sansen , "A 10-b 1-GSample/s Nquist Current-Steering CMOS D/A Converter ," *IEEE Journal of Solid-State Circuits*, vol. 36, pp. 315–324, Mar 2001.

[15] M. Pelgrom, A. Duinmaijer and A. Welbers , "Matching Properties of MOS Transistors ," *IEEE Journal of Solid-State Circuits*, vol. 24, pp. 1443–1440, Oct 1989.

[16] M. Clara, W. Klatzer, B. Seger, A. D. Giandomenico, and L. Gori, "A 1.5V 200MS/s 13b 25mW DAC with Randomized Nested Background Calibration in 0.13 μm CMOS," in *Digest of Technical Papers, IEEE International Solid-State Circuits Conference (ISSCC)*, Feb 2007, pp. 250–252.

2015 IEEE Computer Society Annual Symposium on VLSI

An Improved Dynamic Latch Based Comparator for 8-bit Asynchronous SAR ADC

Anush Bekal[1], Rohit Joshi[1], Manish Goswami[1*], Babu.R.Singh[1] and Ashok Srivatsava[2]

[1]Microelectronics Division, Indian Institute of Information Technology (IIIT-A), Allahabad, INDIA

[2]Division of Electrical and Computer Engineering, Louisiana State University, Baton Rouge, LA 70803, U.S.A

Email:{rs132}{imi2013004}{manishgoswami}{brs}@iiita.ac.in[1], eesriv@lsu.edu[2]

Abstract—High speed analog to digital converters (ADC), memory sense amplifiers, RFID applications, data receivers with low power and area efficient designs has attracted a wide variety of dynamic comparators. This paper presents an improved design for a dynamic latch based comparator in achieving higher speed of conversion targeting 8-bit asynchronous successive approximation register (ASAR) ADC. The comparator has two different stages comprising of a dynamic differential input gain stage and an output latch. The objective of improving the speed of conversion is done by removing the dead time required for reset in the differential input stage. In the proposed work the output node in the differential gain stage requires lesser time to regain higher charge potential. The proposed methodology has been designed and simulated using 180nm CMOS technology operated on a single 1V power supply and achieves complete 8-bit conversion in 75nsec.

Keywords—*Dynamic clocked comparator, Latched comparator, analog to digital converter (ADC), Asynchronous SAR ADC (ASAR), digital-to-analog converter (DAC).*

I. INTRODUCTION

Comparators are considered to be one of the rudimentary building blocks in most ADCs. Major criterion for an ADC design is high speed, low power and lesser real estate over the chip [1]. In the conventional dynamic latched comparators a small input-voltage difference at the input terminals is pulled up to a full scale digital level in a short span of time, by a positive feedback mechanism (regenerative latch). Due to the random offset errors and internal parasitic/external load capacitances mismatches they suffer a lot of accuracy issues [2] [3] [4]. To overcome this inaccuracy issue, the conventional architecture used a separate preamplifier stage anteceding the positive feedback stage due to which it could amplify a small difference in the input voltage to a full scale digital output, negating the kickback noise [5]. But for an ADC with a feature of high speed and low power application, a comparator without the preamplifier is preferred since it suffers from high static power dissipation [6].

The present work deals with increasing the speed of conversion for an 8-bit ASAR ADC which is done by incubating a modified dynamic latch based comparator as proposed. In the reset phase the output nodes has to be charged up to the initial supply voltage level. This charging of the output nodes induces latency in the process of comparison. We propose a modified approach to tackle this dead time required to reset and improve the speed of comparison. This will in turn make the ASAR ADC to fasten the process of conversion.

In the paper the sections are categorized as follows: Section II describes the working of comparator, proposed dynamic latch based comparator and proposed comparator performance, Section III discusses the overall architecture of low power 8-bit ASAR ADC, discusses the simulation results of ASAR ADC and comparison with other proposed architectures while Section IV concludes the paper.

II. PROPOSED DYNAMIC LATCH BASED COMPARATOR

The comparator [7] in Figure 1 has a clk that is used to operate the circuit in two different modes i.e. reset phase and the evaluation phase. During the reset phase, when clk=0 the output nodes charges to supply voltage while in the evaluation phase, when clk=1 the output nodes generate an input dependent output. This information is then transmitted to the regenerative latch stage which will then convert to the full-scale digital values.

However the major pitfall of comparator design is the dead time required to reset the output nodes of the differential gain stage back to supply voltage. Since the circuit has to wait for the next reset phase it significantly requires a larger time to charge the output node capacitances. This will slower the process of comparison which limits the speed of the comparator.

The proposed design is shown in Figure 2. It retains all important features of a comparator as can be obtained in [7] besides offering high speed and removing dead time issue. It differs from Figure 1 with respect to the switching transistors SM1,SM2 which avoids dynamic power consumption and the control transistor CM1,CM2 which provides a extra mechanism to keep either of the outputs in the differential pair charged up to VDD. The two cross coupled control transistors CM1 and CM2 are used which at the beginning of the reset phase (clk=0) are turned OFF since pMOS transistors M4 and M5 pre-charge the output node to VDD. During the evaluation phase (clk=1) the tail transistor nMOS M1 turns ON and the output nodes start to drop at different rates based on the input voltages at transistors M2 and M3. Figure 3 shows a pictorial representation of the comparison process. The higher input voltage transistor will draw more current which drops the voltage at its drain terminal, turning ON either of the control transistors high keeping one control transistor completely OFF. This arrangement will make one of the output terminal discharge completely to the ground potential and keeps the other output node at a higher potential.

978-1-4799-8720-7/15 $31.00 © 2015 IEEE

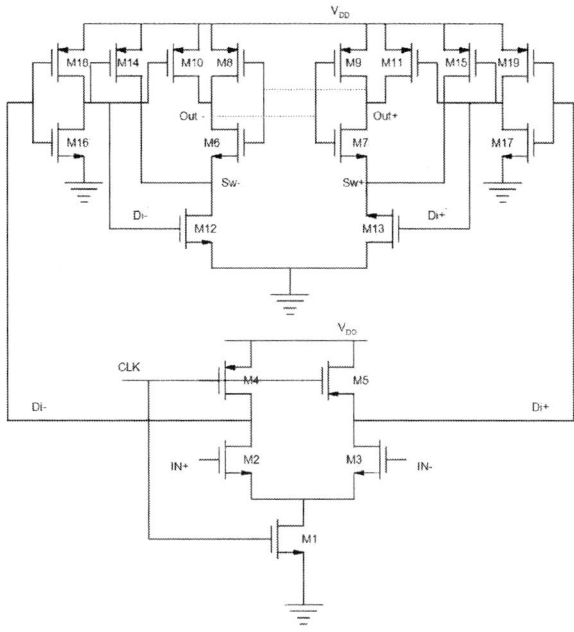

Fig. 1. Original latch based comparator[7].

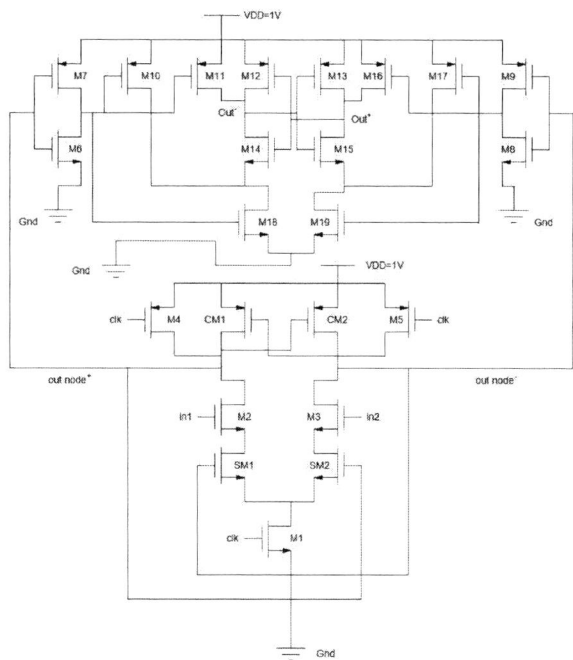

Fig. 2. Proposed latch based comparator.

In the conventional approach, the output nodes was just the function of input transistor transconductance and input voltage difference. This created a timing constraint during the next reset phase due to fact that the output node capacitances was again needed to be charged back to VDD. Extra two nMOS transistors SM1 and SM2 acting as switches are also used in the proposed work to avoid the direct consumption of current from VDD to ground through the tail transistor M1 resulting in saving static power consumption. The design does not use any specific low voltage transistors.

The current flowing through the input transistors M2 and M3 of the differential gain stage can be related from Eq. 1 and Eq. 2

$$V_{Di-}(t) = V_{DD} - \frac{I_{D2}}{C_{Di-}} t \qquad (1)$$

$$V_{Di+}(t) = V_{DD} - \frac{I_{D3}}{C_{Di+}} t \qquad (2)$$

Therefore the total differential gain can be expressed as a function of total charge on the output capacitive load given by Eq. 3

$$A_V(t) = \frac{\Delta V_{Di+}(t)}{\Delta V_{in}} \qquad (3)$$

where V_D are the voltages across the output capacitive loads, I_D are the currents through the transistors M2 and M3, V_{DD} is the supply voltage, C_D is the output capacitive load, A_V is total differential gain.

A. Delay Analysis

The delay analysis for the proposed architecture is given as:

$$t_{\text{delay}} = t_O + t_{\text{latch}} \qquad (4)$$

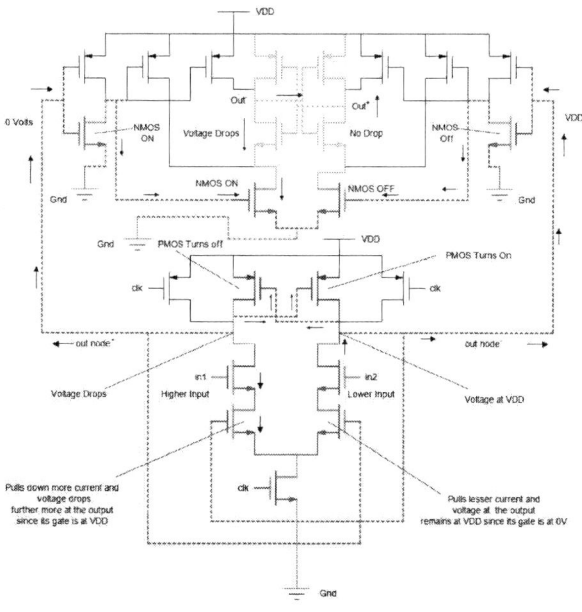

Fig. 3. Pictorial representation of the comparator during evaluation process.

The discharge delay t_O is defined as the time taken by the capacitive load to discharge C_{load} till either of pMOS transistors M12/M13 turns ON and the t_{latch} is defined as latching delay of the regenerative circuit. The discharge delay is given as

$$t_O = \frac{C_{L_{out}} V_{thn}}{I_{D1}/2} \qquad (5)$$

978-1-4799-8720-7/15 $31.00 © 2015 IEEE   179

here C_{Lout} is the output capacitive load, I_{D1} is the current through the transistor M1. It must be noted that $\frac{I_{D1}}{2}$ current flows through the transistor M1. This equation is derived from [8] given by

$$\frac{I}{C} = \frac{V}{t} \qquad (6)$$

The other delay t_{latch} can be formulated as

$$t_{latch} = C_{L_{out}} \left[\frac{1}{g_{m,latch}} + \frac{1}{g_{m18}, g_{m19}} \right] ln \frac{\Delta V_{out}}{\Delta V_O} \qquad (7)$$

$\Delta V_{out} = V_{DD}/2$, ΔV_O is the output voltage difference, $g_{m,latch}$ is the effective transconductance of the latch, g_{m18}/g_{m19} are the transconductance of transistors M18/M19.

The proposed architecture has a merit of enhanced latch transconductance i.e in the proposed work the output node from the differential gain stage will charge up to VDD at the evaluation phase and turn ON one of the intermediate transistor, thus enhancing the regenerative process.

It is the latching delay of two cross coupled inverters which is assumed to have a voltage swing of VDD/2. The transistors M18 and M19 create imbalance in the voltage levels during the comparison phase which adds up along with effective transconductance of the latch.

Figure 4 shows the transient response of the modified comparator (for two inputs in1= 500mV and in2= 525mV). Due to the control transistors and the extra switches either of the outputs pull up to VDD or discharge completely to the ground. This information is then passed on to the latch to hold this value for a certain instant of time and due to the cross coupled configuration (positive feedback) it will furthermore regenerate the signal into the correct logic levels. During the reset phase both the outputs (out1 and out2) are charged up to the supply voltage while during evaluation phase output out1 remains at a higher potential and out2 is completely discharged.

Figure 5 shows the transient analysis of a comparator subjected to a single-ended sinusoidal input of 100MHz frequency having an amplitude of 650mV. It can be seen that when out2 node remains high, out1 node drops down to zero potential and regains to high potential when clk=0. The major advantage of proposed work is that only one output node has to be charged up in the reset phase since during the evaluation phase due to the previous stature of the control transistors one of the output node has already been pulled up to VDD. This speeds up the process of comparison which adds a merit of fast conversion process.

III. LOW POWER 8-BIT ASAR ADC

Figure 6 shows the block representation of the complete low power 8-bit ASAR ADC design using proposed comparator to trigger and complete the process of conversion in a faster mode. The comprehensive working can be summarized as the comparator outputs being connected to a XOR circuitry which will be used as a ready pulse (clock) to be used by the SAR and in turn claims the design to be asynchronous. By this approach we avoid an external clock required by the SAR. As a preset condition the SAR register is set to a mid-value

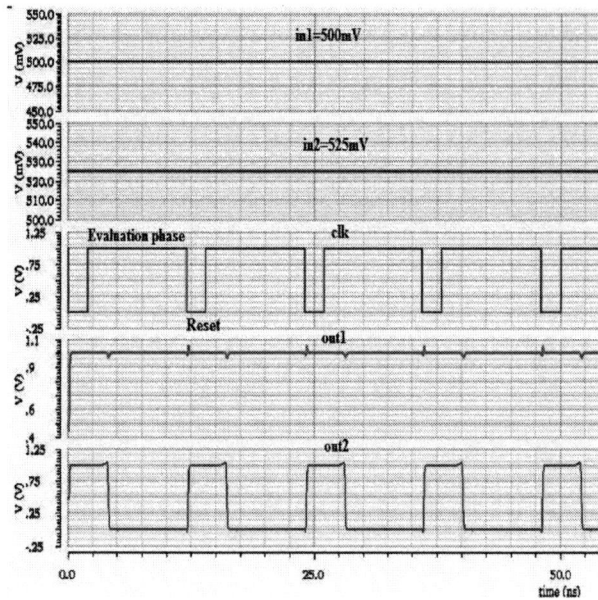

Fig. 4. Transient response of the proposed comparator.

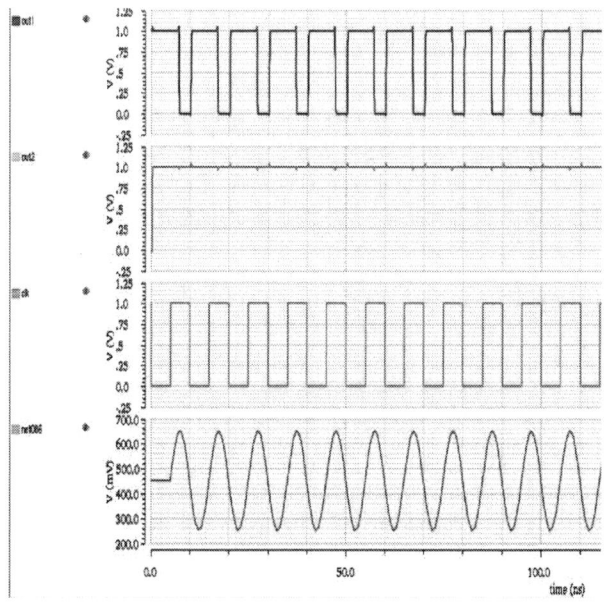

Fig. 5. Transient response of the proposed comparator.

range based on the resolution (N=no.of bits) of the ADC. The SAR logic will now trigger and shift to the correct logic levels based on the clock signal provided by the XOR circuitry. One of the comparator output is connected to the data input of the sequencer logic of SAR register. A 8-bit ADC will require 8 ready signals to complete one conversion. The complete working and description of this can be found from [9] [10]. Figure 7 shows the block diagram of 8-bit charge scaling DAC which is a combination of two 4-bit charge scaling subDACs. It has two sections namely the MSB array and the LSB array connected with a scaling capacitor.

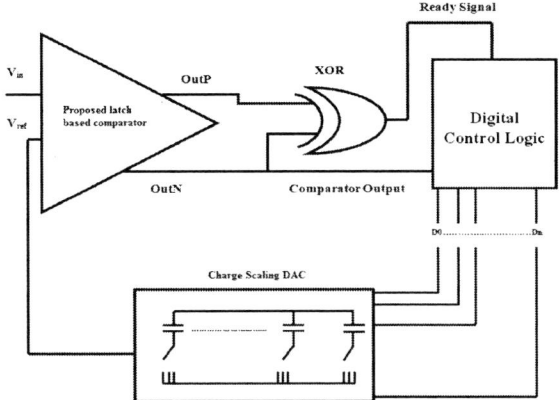

Fig. 6. Low power 8-bit Asynchronous SAR ADC design using Charge Scaling DAC.

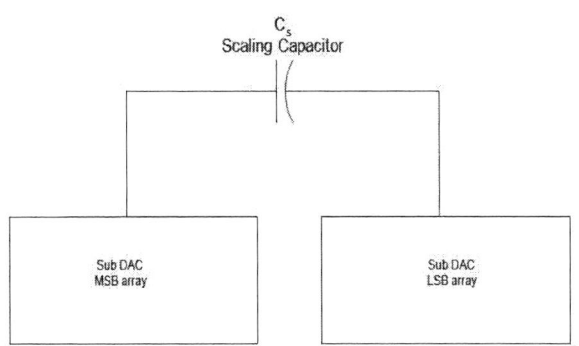

Fig. 7. Charge Scaling DAC.

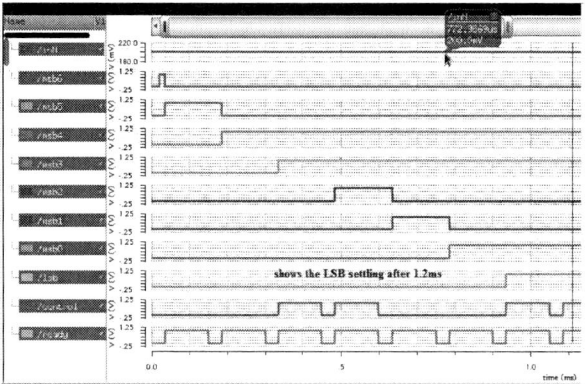

Fig. 8. Simulation results showing the complete conversion of ASAR ADC with static latch based comparator.

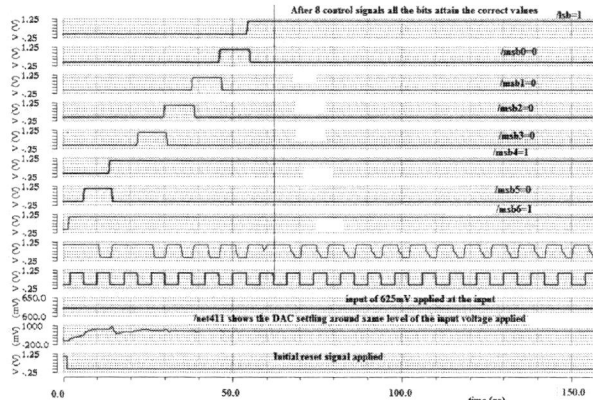

Fig. 9. Simulation results showing the complete conversion of ASAR ADC with improved dynamic latched comparator.

A. Simulation results.

Figure 8 presents the simulation result of 8-bit ASAR ADC with static latch based comparator [9]. The design consumes 1.2ms for one full cycle conversion. Figure 9 shows the transient analysis of the entire ASAR ADC using proposed dynamic latch based comparator for an input of 625mV. As shown in the Figure 9, the proposed design achieves a faster level of conversion process in comparison with Figure 8. Other inputs are not shown for brevity. After 8 ready signals generated by the comparator output, all the individual bits in the SAR register attain the correct logic levels.

Figure 10 shows the schematic simulation result of DAC for an input of 600mV. The voltage settling around the same range of the DAC is shown in the y-axis with respect to the time on the x-axis. Initially the circuit starts up from the reset condition for which the SAR has been fixed to the mid-value range of the resolution (N=8), corresponds to the value of $128=10000000_2$. In this case the supply voltage is 1V which implies the corresponding mid-range to be 500mV. The DAC settles around the same range of the input applied which implies the error is lesser than 0.5 LSB.

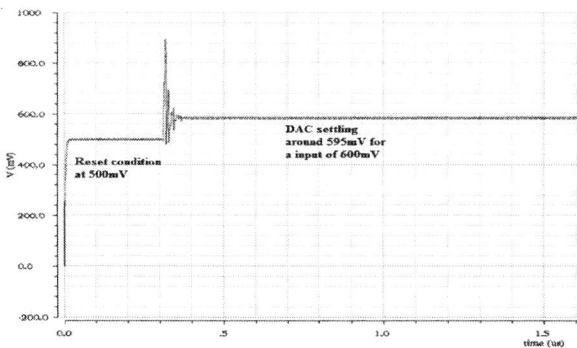

Fig. 10. Simulation results showing the DAC settling for a input of 600mV.

B. Comparison with other works

Below table summarizes comparison of the present design with other existing architectures.

978-1-4799-8720-7/15 $31.00 © 2015 IEEE

Table of comparison				
	Ref [11]	Ref [12]	Ref [13]	**This work**
Technology(nm)	180	180	180	**180**
Supply Voltage (V)	1.8	1.8	1.8	**1**
Resolution (bits)	12	-	8	**8**
Power Dissipation(uW)	72	2100	225	**47.18**
Sampling Rate(MSPS)	2	1.1	3	**13**
Conversion Time (ns)	500	909	333	**75**

IV. CONCLUSION

The proposed work presents an improved dynamic latch based comparator for 8-bit ASAR ADC. The dead time required to reset the output nodes was removed by making use of the proposed architecture in the input differential gain stage. This results in improved process of comparison and it also accelerated conversion process of the ADC. The static power dissipation was saved by using extra switches. The simulation results confirm the entire 8-bit conversion process to be completed in 75ns with a total power dissipation of $47.18\mu W$.

ACKNOWLEDGEMENT

The authors would like to thank Prof. Somenath Biswas, Director IIIT-A and Dr. Manjunath Bhandary, President Sahyadri College of Engineering & Management for lab facilities, moral support and encouragement respectively.

REFERENCES

[1] B. Goll and H. Zimmermann, "A comparator with reduced delay time in 65-nm CMOS for supply voltages down to 0.65," *IEEE Trans. Circuits Syst. II, Exp. Briefs,* vol. 56, No. 11, pp. 810-814, Nov. 2009.

[2] Pelgrom et.al., "Matching properties of MOS transistors," *IEEE Journal of Solid-State Circuits,* vol. 24, No. 5, pp. 1433-1439, Oct. 1989.

[3] Jun He et.al., "Analyses of Static and Dynamic Random Offset Voltages in Dynamic Comparators," *IEEE Transactions on Circuits and Systems I: Regular Papers,* vol. 56, No. 5, pp. 911-919, May. 2009.

[4] A. Nikoozadeh and B. Murmann, "An Analysis of Latch Comparator Offset Due to Load Capacitor Mismatch," *IEEE Transactions on Circuits and Systems II: Express Briefs,* vol. 53, No. 12, pp. 1398-1402, Dec. 2006.

[5] P. Figueiredo and J. Vital, "Kickback noise reduction techniques for CMOS latched comparators," *IEEE Transactions on Circuits and Systems II: Express Briefs,* vol. 53, No. 7, pp. 541-545, July. 2006.

[6] B. Murmann, P. Nikaeen, J. Connelly, W. Dutton, "Impact of Scaling on Analog Performance and Associated Modeling Needs," *IEEE Transactions on Electron Devices,* vol. 53, No. 9, pp. 2160-2167, Sept. 2006.

[7] H. Jeon and Y. Kim, "A novel low-power, low-offset and high-speed CMOS dynamic latched comparator," *Analog Integr Circ Sig Process,* vol. 70, pp. 337-346, July. 2011.

[8] R. Jacob Baker, "Clocked Circuits,"*CMOS Circuit Design, Layout and Simulation,* 2nd edition, IEEE press, pp. 375-389, 2007.

[9] A. Bekal, M. Goswami, B. Singh, D. Pal, "A low power 8-bit Asynchronous SAR ADC design using Charge Scaling DAC," *International Symposium on Electronic System Design(ISED),* Dec. 2014.

[10] S. Tabassum, A. Bekal, M. Goswami, "A Low Power Preamplifier Latch based Comparator Using 180nm CMOS Technology," *IEEE Asia Pacific Conference on Postgraduate Research in Microelectronics and Electronics (PrimeAsia),*pp. 208-212, Dec. 2013.

[11] Y. Siyu, Z. Hui, F. Wenhui, Y. Ting, H. Zhiliang, "A low power 12-bit 200-kS/s SAR ADC with a differential time domain comparator," *Journal of Semiconductors,* vol. 32, No. 3, Mar. 2011.

[12] C. Yingying and L.Dongmei, "A 1.8V 1.1MS/s 96.1 dB-SFDR successive approximation register analog-to-digital converter with calibration," *Journal of Semiconductors,* vol. 34, No. 4, Apr. 2013.

[13] P. Otfinowski, "A 2.5MS/s 225 W 8-bit Charge Redistribution SAR ADC for Multichannel Applications,"*International Conference, Mixed Design of Integrated Circuits and Systems(MIXDES),* June. 2010.

2015 IEEE Computer Society Annual Symposium on VLSI

Translation Validation of Transformations of Embedded System Specifications using Equivalence Checking

Kunal Banerjee, Chittaranjan Mandal, Dipankar Sarkar
Department of Computer Science and Engineering,
Indian Institute of Technology Kharagpur,
Kharagpur, West Bengal, India.
Email: {kunalb, chitta, ds}@cse.iitkgp.ernet.in

Abstract—In the last two decades extensive research has been conducted addressing the design methodology of embedded systems and their verification. The initial behavioural specification of an embedded system goes through significant optimizing transformations, automated and also human-guided, before being mapped to an architecture. Establishing the validity of these transformations is crucial to ensure that the intended behaviour of a system has not been faultily altered during synthesis. State-of-the-art verification methods fail to cope with the complexity of the problem. So, we have devised some efficient translation validation methodologies to handle diverse code transformations; in the initial part of our work, we have worked with the Finite State Machine with Datapath (FSMD) model and its extension to validate various code motion techniques; in the latter part, we have designed an equivalence checking method around the Array Data-Dependence Graph (ADDG) model, which provides a more suitable framework to reason about index spaces of array variables, to verify loop transformations and arithmetic transformations in the presence of recurrences; we have also shown how to relate our path based equivalence checking mechanisms with that of bisimulation based verification techniques by deriving bisimulation relations from the outputs of our equivalence checkers.

Keywords—*Code Transformations, Translation Validation, Finite State Machine with Datapath (FSMD), Bisimulation, Array Data-Dependence Graph (ADDG), Recurrence.*

I. INTRODUCTION

In the last two decades extensive research has been conducted addressing the design methodology of embedded systems. Application areas of such systems include, but are not limited to, cars, telecommunication equipment, medical systems, consumer electronics, robotics, the authentic systems, etc. Due to increasing design complexity of such systems, designers advocate construction of an initial meta-model (specification) of the overall system which is transformed through a sequence of refinements eventually leading to the final implementation. So, it is necessary to devise some efficient verification methodology that demonstrates behavioural equivalence between the specification and the implementation. Also, it is important to develop proper abstractions for modeling such complex systems so that the relevant aspects of the behaviours are captured for verification. Our work mainly focuses on these two issues that arise during embedded system design verification.

Sections II–V briefly explain the contributions of our work. Section VI concludes the paper by outlining some possible future extensions of our work.

II. VERIFICATION OF CODE MOTION TECHNIQUES USING SYMBOLIC VALUE PROPAGATION (SVP)

Scheduling optimization is a crucial step in design of embedded systems. Code motion based transformations, whereby operations are moved across basic block boundaries, constitute important components of such optimizations in order to save hardware resources or meet hard or soft deadlines [1], [2], [3]. Since code motions change the data flow of a program considerably, verifying correctness of the optimization process poses serious challenge. The bisimulation based method presented in [4], [5] has been applied successfully to verify structure preserving code motions. However, this method cannot be applied when the control structure of a program gets modified by path based schedulers [6], [7]. To alleviate this limitation, path extension based equivalence checkers for programs represented using the FSMD model were proposed [8], [9], [10], [11]. A path, however, cannot be extended across a loop by definition of path cover [12]. Therefore, all these methods fail in the case of code motions across loops, whereupon some code segment before a loop body is placed after the loop body, or vice-versa. Literature survey reveals that almost no work exists to tackle code motions across loops; one that exists [13] assumes that there exists an injective function from the nodes of the original code to the nodes of the transformed code. Such a mapping may not hold for practical synthesis tools; even if it holds, it is hard to obtain such a mapping from the synthesis tool.

Our initial objective was to develop a unified verification approach for code motion techniques, including code motions across loops, and control structure modifications without requiring any information from the transformation engine. This combination of features had not been achieved by any single verification technique earlier. A preliminary version of our work appears in [14] which has later been modified considerably to handle speculative code motions and dynamic loop scheduling [7] – the updated technique can be found in [15]. In addition to uniform and non-uniform code motion techniques, this work aims at verifying code motions across loops by propagating the (symbolic) variable values through all the subsequent path segments if mismatch in the values of some live variables is detected. Repeated propagation of

978-1-4799-8720-7/15 $31.00 © 2015 IEEE

symbolic values is possible until an equivalent path or a final path segment ending in the reset state is reached. In the latter case, any prevailing discrepancy in values indicates that the original and the transformed behaviours are not equivalent; otherwise they are. The variables whose values are propagated beyond a loop must be invariant to that loop for valid code motions across loops. The loop invariance of such values can be ascertained by comparing the propagated values that are obtained while entering the loop and after one traversal of the loop. The method has been implemented and satisfactorily tested on the outputs of a basic block based scheduler [16], a path based scheduler [6] and the high-level synthesis tool SPARK [17] for some benchmark examples.

Along with non-structure preserving transformations that involve path merging/splitting (as introduced by the schedulers[6], [7]), the uniform code motion techniques that also can be verified by our technique include boosting up, boosting down, duplicating up, duplicating down and useful move – a comprehensive study on the classification of these transformations can be found in [18]; the supported non-uniform code motion techniques include speculation, reverse speculation, safe speculation, code motions across loops, etc. To determine the equivalence of a pair of paths, our equivalence checker employs the normalization process described in [19] to represent the conditions of execution and the data transformations of the paths. This normalization technique further aids in verifying the following arithmetic code transformations: associative, commutative, distributive transformations, copy and constant propagation, common subexpression elimination, arithmetic expression simplification, partial evaluation, constant folding/unfolding, redundant computation elimination, etc. It is important to note that the computational complexity of the method presented in [15] has been analyzed and found to be no worse than that for [11], i.e., the SVP based method of [15] is capable of handling more sophisticated transformations than [11] without incurring any extra overhead of time complexity; in fact, as demonstrated in [15], the implementation of the SVP based equivalence checker has been found to take less execution time in establishing equivalence than those of the path extension based equivalence checkers, [9] and [11].

III. Extending the FSMD framework for validating code motions of array-handling programs

A significant deficiency of the above mentioned equivalence checkers for the FSMD model is their inability to handle an important class of programs, namely those involving arrays. This is so because the underlying FSMD model does not provide formalism to capture array variables in its datapath. The data flow analysis for array-handling programs is notably more complex than those involving only scalars. For example, consider two sequential statements $a[i] \Leftarrow 10$ and $a[j] \Leftarrow 20$, now if $i = j$ then the second statement qualifies as an *overwrite*, otherwise it does not; unavailability of relevant information to resolve such relationships between index variables may result in exponential number of case analyses. Moreover, obtaining the condition of execution and the data transformation of a path by applying simple substitution as outlined by Dijkstra's weakest precondition computation may become more expensive in the presence of arrays; conditional clauses need to be associated depicting equality/inequality of

the index expressions of the array references in the predicate as it gets transformed through the array assignment statements in the path.

In [20], we have introduced a new model, namely Finite State Machine with Datapath *having Arrays* (FSMDA), which is an extension of the FSMD model equipped to handle arrays. To alleviate the problem of determining overwrite/non-overwrite, the SVP based method described in [14], [15] is enhanced to propagate the values assumed by index variables[1] in some path to its subsequent paths (in spite of a match); to resolve the problem of computing the path characteristics, the well-known McCarthy's *read* and *write* functions [21] (originally known as *access* and *change*, respectively) have been borrowed to represent assignment and conditional statements involving arrays that easily capture the sequence of transformations carried out on the elements of an array and also allow uniform substitution policy for both scalars and array variables. An improvisation of the normalization process [19] is also suggested in [20] to represent arithmetic expressions involving arrays in normalized forms.

Experimental results are found to be encouraging and attest the effectiveness of the method. Our tool can be downloaded from http://cse.iitkgp.ac.in/~kunban/ EquivalenceChecker.tar.gz along with the benchmarks, installation and usage guidelines. It is pertinent to note that the formalism of our model allows operations in non-single assignment form and data-dependent control flow which have been serious limitations for previous methods that have attempted equivalence checking of array-intensive programs [22], [23], [24]. It is also to be noted that our tool detected a bug in the implementation of copy propagation for array variables in the SPARK [17] tool as reported in [20].

IV. Deriving bisimulation relations from path based equivalence checkers

Constructing bisimulation relations between programs as a means of translation validation has been an active field of study. Translation validation for an optimizing compiler by obtaining simulation relations between programs and their translated versions was first demonstrated by Necula in [25]. The procedure broadly consists of two algorithms – an inference algorithm and a checking algorithm. The inference algorithm collects a set of constraints (representing the simulation relation) in a forward scan of the two programs and then the checking algorithm checks the validity of these constraints. This work is enhanced by Kundu et al. [4], [5] to validate the high-level synthesis process. Unlike Necula's approach, Kundu et al.'s procedure uses a general theorem prover, rather than specialized solvers and simplifiers, and is thus more modular. As already mentioned in Section II, a major limitation of these methods [25], [4], [5] is that they cannot handle non-structure preserving transformations such as those introduced by path based schedulers [6], [7]. Our research group has developed two path based equivalence checkers – the path extension based equivalence checker [8], [9], [11] and the (more general) SVP based equivalence checker [14], [15] – as alternative translation validation techniques to alleviate this drawback.

---

[1]Index variables are basically the "scalar" variables which occur in some index expression of some array variable.

However, transformations such as loop shifting [26] that can be handled by bisimulation based methods of [4], [5] by repeated strengthening of the simulation relation over the related loops of the source and the target codes still elude the path based equivalence checking methods. This process of strengthening the simulation relation iteratively until a fixed point is reached (i.e., the relation becomes strong enough to imply the equivalence), however, may not terminate. On the contrary, a single pass is used to determine the equivalence/non-equivalence for a pair of paths in the path based approaches and hence, the number of paths in a program being finite, these methods are guaranteed to terminate. Thus, we find that both bisimulation and path based approaches have their own merits and demerits, and therefore, both have found application in the field of translation validation of untrusted compilers. However, the conventionality of bisimulation as the approach for equivalence checking raises the natural question of examining whether path based equivalence checking yields a bisimulation relation or not.

In [27], we have shown how a bisimulation relation can be derived from an output of the path extension based equivalence checker; we plan to communicate the entire work containing the procedure for deriving a bisimulation relation from an output of the SVP based equivalence checker as well to a journal in near future. It is to be noted that none of the earlier methods that establish equivalence through construction of bisimulation relations has been shown to tackle code motion across loops; our work demonstrates, for the first time, the existence of a bisimulation relation under such a situation.

V. AN EQUIVALENCE CHECKING MECHANISM FOR HANDLING RECURRENCES IN ARRAY-INTENSIVE PROGRAMS

Loop transformations together with arithmetic transformations are applied extensively in the domain of multimedia and signal processing applications to obtain better performance in terms of energy, area and/or execution time. The work reported in [28], for example, applies loop fusion and loop tiling to several nested loops and parallelizes the resulting code across different processors for multimedia applications. Minimization of the total energy while satisfying the performance requirements for applications with multi-dimensional nested loops is targeted in [29]. Application of arithmetic transformations can improve the performance of computationally intensive applications as suggested in [30], [31]. Often loop transformation and arithmetic transformation techniques are applied dynamically since application of one may create scope of application of several other techniques. In all these cases, it is crucial to ensure that the intended behaviour of the program has not been altered wrongly during transformation.

Verification of loop transformations in array-intensive programs has drawn a significant amount of investigation. A translation validation approach based on transformation specific rules is proposed in [32] for verification of loop interchange, skewing, tiling and reversal transformations. The main drawback of this approach is that it requires information such as the list of transformations applied and their order of application; however, such information need not be readily available from the synthesis tools. A symbolic simulation based approach is proposed in [33] for verification of loop transformations

for programs with no recurrence and with affine indices and bounds. However, this method does not handle any arithmetic transformation that may be applied along with loop transformations. Basically, for multiple occurrences of an array in an expression, this method loses track between each occurrence of that array and its corresponding indices in the presence of arithmetic transformations. Another approach for verifying array-intensive programs can be to use off-the-shelf SMT solvers or theorem provers since the equivalence between two programs can be modeled with a formula such that the validity of the formula implies the equivalence [34]. Although SMT solvers and theorem provers can efficiently handle linear arithmetic, they are not equally suitable for handling non-linear arithmetic. Since array-intensive programs often contain non-linear arithmetic, these tools are found to be inadequate for establishing equivalence of such programs [34]. The works reported in [35], [22], [36] consider a restricted class of programs which must have static control-flows, valid schedules, affine indices and bounds and single assignment forms. In [22], [36], the original and the transformed behaviours are modeled as ADDGs and the correctness of the loop transformations is established by showing the equivalence between the two ADDGs. These works are capable of handling a wide variety of loop transformation techniques without taking any information from the synthesis tools. The method proposed in [37], [23] extends the ADDG model to a dependence graph model to handle recurrences along with associative and commutative operations. All the above methods, however, fail if the transformed behaviour is obtained from the original behaviour by application of arithmetic transformations such as, distributive transformations, arithmetic expression simplification, common sub-expression elimination, constant unfolding, etc., along with loop transformations. The work reported in [38], [24] furnishes an ADDG based method which compares ADDGs at slice-level rather than path-level as performed in [36] and employs a normalization technique [19] for the arithmetic expressions to verify a wide variety of loop transformations and a wide range of arithmetic transformations applied together in array-intensive programs. However, it cannot verify programs involving recurrences because recurrences lead to cycles in the ADDGs which is otherwise a directed acyclic graph (DAG). The presence of cycles makes the existing data-dependence analysis and simplification (through closed-form representations) of the data transformations in ADDGs inapplicable.

Our work reported in [39] provides a unified equivalence checking framework based on ADDGs to handle loop and arithmetic transformations along with recurrences. The validation scheme proposed here isolates the suitable subgraphs (arising from recurrences) in the ADDGs from the acyclic portions and treats them separately; each cyclic subgraph in the original ADDG is compared with its corresponding subgraph in the transformed ADDG in isolation and if all such pairs of subgraphs are found equivalent, then the entire ADDGs (with the subgraphs replaced by equivalent uninterpreted functions of proper arities) are compared using the conventional technique of [24]. Limitations of currently available schemes are thus overcome to handle a broader spectrum of array-intensive programs. The experimental results demonstrate the efficacy of the method. We intend to submit an extended version of this work to a journal containing further enhancement in the algorithm to additionally handle mutual recurrences along with formal treatments of the algorithm's correctness and complexity issues.

VI. CONCLUSION

Here we have given the overview of some translation validation methods involving the FSMD model, its extension and the ADDG model to cover verification of programs ranging over a diverse domain; we have also proposed techniques for relating (apparently different) translation validation approaches by deriving bisimulation relations from the outputs of the path extension based and the SVP based equivalence checkers. A limitation of our FSMDA based verification method is that it does not allow arrays containing other arrays in their subscripts, whereas the ADDG based method does not support recurrences that employ reductions [40]. Extending our techniques to alleviate the current limitations seems to be promising future endeavour.

ACKNOWLEDGMENT

The work of K. Banerjee was supported by TCS Research Fellowship.

REFERENCES

[1] S. Gupta, N. Dutt, R. Gupta, and A. Nicolau, "Using global code motions to improve the quality of results for high-level synthesis," *IEEE Trans on CAD of ICS*, vol. 23, no. 2, pp. 302–312, Feb 2004.

[2] L. C. V. Dos. Santos and J. Jress, "A reordering technique for efficient code motion," in *DAC*, 1999, pp. 296–299.

[3] G. Lakshminarayana, A. Raghunathan, and N. Jha, "Incorporating speculative execution into scheduling of control-flow-intensive design," *IEEE Trans on CAD of ICS*, vol. 19, no. 3, pp. 308–324, March 2000.

[4] S. Kundu, S. Lerner, and R. Gupta, "Validating high-level synthesis," in *CAV*, 2008, pp. 459–472.

[5] ——, "Translation validation of high-level synthesis," *IEEE Trans on CAD of ICS*, vol. 29, no. 4, pp. 566–579, 2010.

[6] R. Camposano, "Path-based scheduling for synthesis," *IEEE Trans on CAD of ICS*, vol. 10, no. 1, pp. 85–93, Jan. 1991.

[7] M. Rahmouni and A. A. Jerraya, "Formulation and evaluation of scheduling techniques for control flow graphs," in *EURO-DAC*, 1995, pp. 386–391.

[8] C. Karfa, C. Mandal, D. Sarkar, S. R. Pentakota, and C. Reade, "A formal verification method of scheduling in high-level synthesis," in *ISQED*, 2006, pp. 71–78.

[9] C. Karfa, D. Sarkar, C. Mandal, and P. Kumar, "An equivalence-checking method for scheduling verification in high-level synthesis," *IEEE Trans on CAD of ICS*, vol. 27, pp. 556–569, 2008.

[10] C.-H. Lee, C.-H. Shih, J.-D. Huang, and J.-Y. Jou, "Equivalence checking of scheduling with speculative code transformations in high-level synthesis," in *ASP-DAC*, 2011, pp. 497–502.

[11] C. Karfa, C. Mandal, and D. Sarkar, "Formal verification of code motion techniques using data-flow-driven equivalence checking," *ACM Trans. Design Autom. Electr. Syst.*, vol. 17, no. 3, p. 30, 2012.

[12] R. W. Floyd, "Assigning meaning to programs," in *Proceedings the 19th Symposium on Applied Mathematics*, 1967, pp. 19–32.

[13] J.-B. Tristan and X. Leroy, "Verified validation of lazy code motion," in *PLDI*, 2009, pp. 316–326.

[14] K. Banerjee, C. Karfa, D. Sarkar, and C. Mandal, "A value propagation based equivalence checking method for verification of code motion techniques," in *ISED*, 2012, pp. 67–71.

[15] ——, "Verification of code motion techniques using value propagation," *IEEE Trans. on CAD of ICS*, vol. 33, no. 8, pp. 1180–1193, 2014.

[16] C. A. Mandal and R. M. Zimmer, "A genetic algorithm for the synthesis of structured data paths," in *VLSI Design*, 2000, pp. 206–211.

[17] S. Gupta, N. Dutt, R. Gupta, and A. Nicolau, "SPARK: A high-level synthesis framework for applying parallelizing compiler transformations," in *VLSI Design*, 2003, pp. 461–466.

[18] M. Rim, Y. Fann, and R. Jain, "Global scheduling with code-motions for high-level synthesis applications," *IEEE Trans. VLSI Syst.*, vol. 3, no. 3, pp. 379–392, 1995.

[19] D. Sarkar and S. De Sarkar, "A theorem prover for verifying iterative programs over integers," *IEEE Trans Software. Engg.*, vol. 15, no. 12, pp. 1550–1566, 1989.

[20] K. Banerjee, D. Sarkar, and C. Mandal, "Extending the FSMD framework for validating code motions of array-handling programs," *IEEE Trans. on CAD of ICS*, vol. 33, no. 12, pp. 2015–2019, 2014.

[21] J. McCarthy, "Towards a mathematical science of computation," in *IFIP Congress*, 1962, pp. 21–28.

[22] K. C. Shashidhar, M. Bruynooghe, F. Catthoor, and G. Janssens, "Functional equivalence checking for verification of algebraic transformations on array-intensive source code," in *DATE*, 2005, pp. 1310–1315.

[23] S. Verdoolaege, G. Janssens, and M. Bruynooghe, "Equivalence checking of static affine programs using widening to handle recurrences," *ACM Trans. Program. Lang. Syst.*, vol. 34, no. 3, 2012.

[24] C. Karfa, K. Banerjee, D. Sarkar, and C. Mandal, "Verification of loop and arithmetic transformations of array-intensive behaviours," *IEEE Trans. on CAD of ICS*, vol. 32, no. 11, pp. 1787–1800, 2013.

[25] G. C. Necula, "Translation validation for an optimizing compiler," in *PLDI*, 2000, pp. 83–94.

[26] A. Darte and G. Huard, "Loop shifting for loop compaction," *J. Parallel Programming*, vol. 28, no. 5, pp. 499–534, 2000.

[27] K. Banerjee, C. Mandal, and D. Sarkar, "Deriving bisimulation relations from path extension based equivalence checkers," in *WEPL*, 2015, pp. 1–2.

[28] Y. Bouchebaba, B. Girodias, G. Nicolescu, E. M. Aboulhamid, B. Lavigueur, and P. G. Paulin, "MPSoC memory optimization using program transformation," *ACM Trans. Design Autom. Electr. Syst.*, vol. 12, no. 4, 2007.

[29] I. Kadayif, M. T. Kandemir, G. Chen, O. Ozturk, M. Karaköy, and U. Sezer, "Optimizing array-intensive applications for on-chip multiprocessors," *IEEE Trans. Parallel Distrib. Syst.*, vol. 16, no. 5, pp. 396–411, 2005.

[30] M. Potkonjak, S. Dey, Z. Iqbal, and A. C. Parker, "High performance embedded system optimization using algebraic and generalized retiming techniques," in *ICCD*, 1993, pp. 498–504.

[31] J. Zory and F. Coelho, "Using algebraic transformations to optimize expression evaluation in scientific codes," in *IEEE PACT*, 1998, pp. 376–384.

[32] L. D. Zuck, A. Pnueli, B. Goldberg, C. W. Barrett, Y. Fang, and Y. Hu, "Translation and run-time validation of loop transformations," *Formal Methods in System Design*, vol. 27, no. 3, pp. 335–360, 2005.

[33] T. Matsumoto, K. Seto, and M. Fujita, "Formal equivalence checking for loop optimization in C programs without unrolling," in *ACST*, 2007, pp. 43–48.

[34] C. Karfa, K. Banerjee, D. Sarkar, and C. Mandal, "Experimentation with SMT solvers and theorem provers for verification of loop and arithmetic transformations," in *I-CARE*, 2013, pp. 3:1–3:4.

[35] D. Barthou, P. Feautrier, and X. Redon, "On the equivalence of two systems of affine recurrence equations (research note)," in *Euro-Par Conference on Parallel Processing*, 2002, pp. 309–313.

[36] K. C. Shashidhar, "Efficient automatic verification of loop and data-flow transformations by functional equivalence checking," Ph.D. dissertation, Katholieke Universiteit Leuven, 2008.

[37] S. Verdoolaege, G. Janssens, and M. Bruynooghe, "Equivalence checking of static affine programs using widening to handle recurrences," in *CAV*, 2009, pp. 599–613.

[38] C. Karfa, K. Banerjee, D. Sarkar, and C. Mandal, "Equivalence checking of array-intensive programs," in *ISVLSI*, 2011, pp. 156–161.

[39] K. Banerjee, "An equivalence checking mechanism for handling recurrences in array-intensive programs," in *POPL: Student Research Competition*, 2015, pp. 1–2.

[40] G. Iooss, C. Alias, and S. V. Rajopadhye, "On program equivalence with reductions," in *Static Analysis*, 2014, pp. 168–183.

2015 IEEE Computer Society Annual Symposium on VLSI

Design and Implementation of a Reversible Central Processing Unit

Lafifa Jamal and Hafiz Md. Hasan Babu

Department of Computer Science and Engineering, University of Dhaka, Dhaka-1000, Bangladesh

Email: lafifa@yahoo.com and hafizbabu@hotmail.com

Abstract—This work addresses the reversible circuit design using novel modularization approach by presenting architecture of a logically reversible processor based on the Von Neumann architecture that can operate with very low power consumption, protection of power analysis attack and long span of life due to less heat dissipation. The organization and architecture of the proposed processor is designed from scratch. Sequential algorithms are proposed to produce the components of the reversible processor. The capabilities of the new processor is determined, the datapath layout is designed to handle the necessary capabilities, the instruction format is defined and the necessary logic is also constructed to control the datapath. To estimate the execution time of the algorithm, we consider the computational complexity, memory access patterns and the complexity of the instructions. Existing component designs are compared with the proposed components and theorems and lemmas are presented to prove the superiority of the proposed architecture. The proposed design is simulated and the simulation result verifies the correctness of the proposed design.

Index Terms—Reversible Logic, Reversible CPU, Quantum Cost, Garbage Output.

I. INTRODUCTION

Power crisis is a vital problem in today's world. In recent years, the growing market of electronic systems suffers from power dissipation and heat removal problem. If more and more power is dissipated, system becomes overheated which reduces the life time of the electronic system. The need of micro-electronic circuits with low power dissipation leads to the implementation of reversible logic circuit. Bennett [1] proved that the one-to-one mapping between the inputs and outputs of reversible circuit drastically reduces the power consumption and heat dissipation of a circuit. Today security in digital computing and communications is of prime importance and therefore cryptographic protocols play a major role.

David [2] proved that reversibility plays a vital role in quantum computation. Quantum gates and reversible logic gates are closely related to each other. Like classical reversible circuits, the number of input quantum bits must be equal to the number of output quantum bits. The quantum gates and circuits must be reversible. So, the quantum circuits can directly be designed from reversible circuits.

Processor design is indeed a difficult task and thinking the organization and architecture of the design in reversible way requires a lot of works. The irreversible processors dissipate a significant amount of heat and they require more power than the corresponding reversible one. A reversible processor can overcome these problems. As reversible computing is a new

era, very few works [3], [4] regarding reversible processor have been done. As well as, there is no established patent work in this area. The existing reversible processor architectures [3], [4] lack rigor and completeness. These parameters motivate this work of reversible processor design. In this work, we present a logical reversible processor design architecture and algorithms that try to fill up the gap between realization of the reversible architecture and completeness of the design.

II. BASIC DEFINITIONS

In this section, we present the basic definitions regarding reversible logic.

A. Reversible Gate

Reversible gate is an n-input and n-output (denoted by $n \times n$) circuit that produces a unique output pattern for each possible input pattern. In other words, reversible gates are circuits in which the number of outputs is equal to the number of inputs and there is a one-to-one correspondence between the vector of inputs and outputs.

B. Garbage Output

Unwanted or unused output of a reversible gate (or circuit) is known as **garbage output**, i.e., the output(s) which is(are) needed only to maintain the reversibility is (are) known as garbage output(s).

C. Quantum Cost

The **quantum cost** can be derived by substituting the reversible gates of a circuit by a cascade of elementary quantum gates [5]. Elementary quantum gates realize quantum circuits that are inherently reversible and manipulate qubits rather than pure logic values. The state of a qubit for two pure logic states can be expressed as $|\psi> = \alpha| 0 > + \beta| 1 >$, where $| 0 >$ and $| 1 >$ denote 0 and 1, respectively, and and α and β are complex numbers such that $| \alpha |^2 + | \beta |^2 = 1$. The most used elementary quantum gates are the NOT gate, the controlled-NOT (CNOT) gate, the controlled-V gate and the controlled-V+ gate.

III. PROPOSED WORK

The internal arrangement of a microprocessor varies depending on the design and the intended purposes of the microprocessor. A microprocessor includes an arithmetic logic unit (ALU) and a control unit (CU) section. These two sections are connected to memory and I/O by buses which

978-1-4799-8720-7/15 $31.00 © 2015 IEEE   187

carry information and signal between the units. They have a defined datapath. The ALU performs the arithmetic and logical operations. The control logic section retrieves instruction operation codes from memory, and initiates whatever sequence of operations of the ALU requires to carry out the instruction. A single operation code affects many individual datapaths, registers, and other elements of the processor. Following steps are considered to design the desired reversible central processing unit (CPU):

- Design the overall structure of the reversible CPU.
- Layout the datapath to handle all of the operations of the reversible CPU.
- Design the reversible realizations of the flip-flops.
- Design the reversible memory circuits (such as buffer registers and counter circuits) using the proposed reversible flip-flops of the previous step.
- Design the arithmetic circuits such as adder, multiplier, divider, comparator etc.
- Design the reversible realization of ALU.
- Design the reversible control unit of the processor by designing an efficient instruction decoder.
- Construct the necessary logic to control the datapath.
- Realize the overall architecture and organization of the proposed reversible processor.
- Analysis of the proposed reversible central processing unit in terms of cost and performance efficiency.
- Simulate the design using Microwind DSCH 3.5 [6] and CMOS 45 nm Open Cell Library [7] software.

The architecture of the proposed reversible processor is shown in Fig. 1.

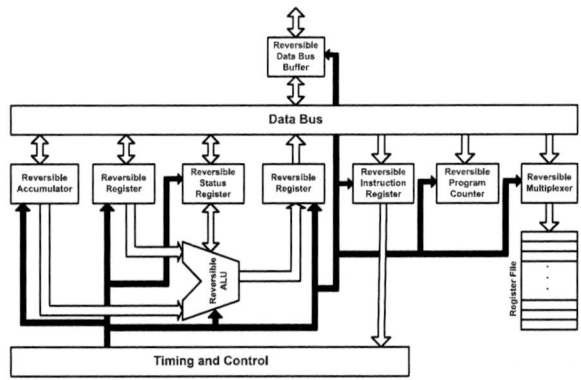

Fig. 1. Outline of the Design of the Proposed Reversible Processor

The datapath of the proposed reversible processor is shown in Fig. 2. The control signals are not shown in the diagram for simplicity.

According to the block diagram and the datapath design the proposed reversible processor has been divided into small components. The reversible realization of the proposed components are discussed in the following subsections.

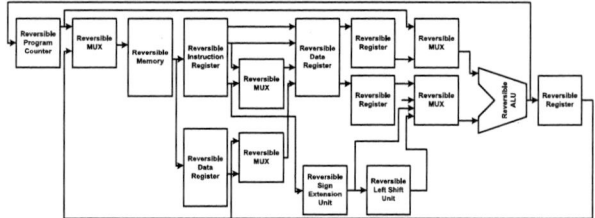

Fig. 2. Datapath Design of the Proposed Reversible Processor

A. Proposed Reversible Memory Components

Flip-flops (FF) are the basic elements of a register. We use template matching algorithm to propose a reversible J-K FF with minimum quantum cost. This J-K FF is used to design the instruction register. We propose a new reversible gate, namely BJ gate [8], to design a reversible J-K FF. The proposed reversible J-K FF requires only one gate, it produces only one garbage output and it has 12 quantum cost. The design of the proposed J-K FF is shown in Fig. 3.

The proposed reversible J-K FF achieves the improvement of 66.6% in terms of number of gates, 66.6% in terms of garbage outputs, 66.6% in terms of delay and 4.28% in terms of quantum cost over the existing best one [9].

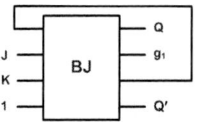

Fig. 3. Proposed Reversible J-K Flip-Flop

This proposed J-K FF is used to design the registers. A reversible 16-bit register can be designed using sixteen HNFG gates and sixteen proposed J-K FFs. The J-K FFs take the inputs through HNFG gates and with the change of the clock pulse they produce the normal and complemented outputs.

The sequence counter is also designed using four proposed J-K FFs and four Feynman gates. The J-K FFs change states with the positive clock edge and the counter counts from 0 to 15. This optimized design of sequence counter produces 4 garbage outputs.

B. Proposed Reversible Multiplexer

Reversible Multiplexer (Mux) is required to select a certain input from various input sources. A reversible 4-to-1 mux is designed with Fredkin gates. The proposed design of 4-to-1 mux is shown in Fig. 4. The proposed reversible 4-to-1 mux achieves the improvement of 57.14% in terms of number of gates, 54.54% in terms of garbage outputs, 73.68% in terms of quantum cost and 57.14% in terms of delay over the existing best one [10].

C. Proposed Reversible Decoder

The control unit requires decoders to decode the instructions. We propose the design of an n-to-2^n decoder. To design

978-1-4799-8720-7/15 $31.00 © 2015 IEEE   188

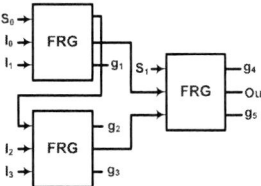

Fig. 4. Proposed Reversible 4-to-1 Multiplexer

this decoder at first we propose a reversible 2-to-4 decoder which was used to construct a 3-to-8 decoder and so on. The design of an improved 2-to-4 decoder was very challenging as an existing 2-to-4 decoder [11] requires only one gate and it does not produce any garbage outputs. So, there was no scope to improve gate count, garbage outputs and delay. Exhaustive search was used to find the reversible gate with optimal solution which produces the minimum quantum cost. We propose a new reversible gate, namely HL gate [8] to design a 2-to-4 decoder. The proposed 2-to-4 decoder requires only one gate without any garbage output which has 7 quantum cost. Our proposed design has the same gate count, garbage outputs and delay. But it has improved quantum cost. A reversible 3-to-8 decoder can be designed using one 2-to-4 reversible decoder and four Fredkin gates. The achievement of the design of 3-to-8 design was greater as the existing best 3-to-8 decoder [12] is not much optimized. The designs of the proposed reversible 2-to-4 decoder and 3-to-8 decoder are shown in Fig. 5(a) and Fig. 5(b) respectively.

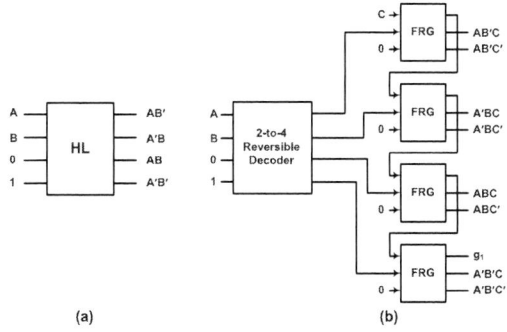

Fig. 5. Proposed Reversible (a) 2-to-4 Decoder (b) 3-to-8 Decoder

D. Proposed Reversible Control Unit

A control unit is a circuit that directs operations within the computer's processor by directing the input and output of a computer system. The control unit consists of two decoders, a sequence counter, and a number of control logic gates. It fetches the instruction from instruction register. The inputs to the control logic gates come from two decoders, flip-flop and instruction register. The outputs of the control logic circuit are: signals to control the inputs of the registers, signals to control the read and write inputs of memory and signals to set, clear or complement the flip-flops.

A step by step design technique has been proposed to design the control unit which connects all the proposed components with some other extra circuitry. The block diagram of the proposed control unit is shown in Fig. 6.

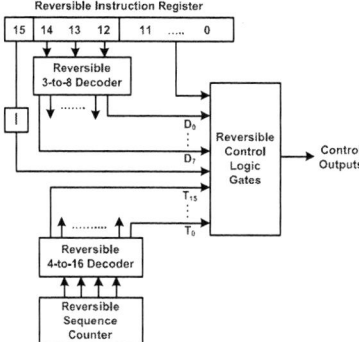

Fig. 6. Proposed Reversible Control Unit

E. Proposed Reversible Comparator

A magnitude comparator takes two numbers as input in binary form and determines whether one number is greater than, less than or equal to the other number. Comparators are used in central processing unit. We propose a compact and improved algorithm for constructing a compact reversible n-bit binary comparator circuit. In order to improve the design, we use quantum cost minimization algorithm to design an n-bit comparator. Reduction rules and algorithms are applied by template matching to minimize the quantum cost. The whole comparator is divided into three modules. They are as follows: We propose an MSB comparator circuit for comparing the nth bit (MSB) of two n-bit numbers. Then, a single-bit GE (greater or equal) comparator cell has been designed to generate greater and equal signal for the remaining $(n-1)$ bits of two numbers with the previous comparison result of MSB. A single-bit LT (less than) comparator cell is designed to determine the less than signal. These three components with minimum quantum cost are then cascaded with each other using efficient techniques to design 2-bit and n-bit reversible binary comparators more compactly. A reversible 2-bit binary comparator consists of a proposed reversible MSB comparator, a single-bit GE comparator and a single-bit LT comparator circuit. The reversible design of 2-bit binary comparator is shown in Fig. 7. The detail design of the individual blocks are not shown here due to space constraints.

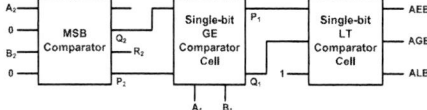

Fig. 7. Proposed Reversible 2-bit comparator

The proposed reversible 64-bit comparator achieves the improvement of 24.4% in terms of number of gates, 19.9%

978-1-4799-8720-7/15 $31.00 © 2015 IEEE

in terms of garbage outputs, 7.7% in terms of quantum cost, 25.77% in terms of area and 3.43% in terms of power over the existing best one [13].

F. Proposed Reversible Multiplier

Multiplier circuit is used in a processor. Multiplier circuit mainly has two components: Partial Product Generation (PPG) circuit and the Multi-Operand Addition (MOA) circuit. A multiplier circuit can be optimized in two ways. The algorithm of the multiplication can be optimized, as well as, the construction procedure of the PPG circuit and MOA circuit can be optimized. Firstly, we construct efficient template using reversible gates to reduce the quantum cost and delay. Secondly, we use an efficient searching technique to find the appropriate positions of different components of the proposed circuit. Finally, we propose two efficient algorithms which produce high speed minimum cost PPG circuit and MOA circuit of the proposed multiplier.

The comparative study shows that the proposed reversible 4×4 multiplier achieves the improvement of 26.32% in terms of number of gates, 12.5% in terms of garbage outputs, 17% in terms of quantum cost and 20.97% in terms of constant inputs over the existing best one [14]. The proposed reversible $n \times n$ multiplier requires $n(2n - 1)$ gates, $4n(n - 1) + 1$ garbage outputs and $17n(n-1)+1$ quantum cost; where as the best existing fault tolerant reversible multiplier [14] requires $n(2n - 1) + \lfloor 5n/2 \rfloor$ gates, $2n(n - 1)$ garbage outputs and $n(19n - 17) + 7 + 2\lfloor n/2 \rfloor$ quantum cost.

G. Proposed Reversible Divider

Divider circuit is used in a processor. We propose two design techniques to construct reversible n-bit floating point nonrestoring division circuit. In the first approach of the reversible divider circuit, we propose carry propagate 2's complement adder and shifters. To speed up the division technique, we propose a major modification in the second approach of the reversible divider. We omit the carry propagation to avoid the propagation delay and introduce two new vectors and a single carry look-ahead adder circuit. Subtraction is implemented using 2's complement adder. Both of the approaches can handle floating point numbers. Exhaustive search method is used to find the appropriate positions of reversible gates to obtain the minimum cost parameters for a specific function that are used to design the reversible divider circuit more compactly. The comparative study shows that the proposed reversible conventional 4-bit divider achieves the improvement of 62.92% in terms of number of gates, 70.97% in terms of garbage outputs and 74.83% in terms of quantum cost over the existing best one [15]. The proposed reversible high speed 4-bit divider achieves the improvement of 22.47% in terms of number of gates, 17.74% in terms of garbage outputs and 43.87% in terms of quantum cost over the existing best one [15].

IV. CONCLUSIONS

In this work, reversible logic syntheses with the minimum cost factors are carried out for the components of the reversible processor. Many important contributions have been made in the literature towards the reversible implementations of arithmetic and logical structures; however, there have not been many efforts directed towards efficient approaches for designing reversible ALU (Arithmetic Logic Unit). Currently we are working on the design of the reversible ALU. We will propose an efficient approach to design the reversible ALU. Finally all the components of the reversible CPU (Central Processing Unit) will be interfaced together with necessary connecting circuits to get the complete reversible CPU. The proposed reversible CPU can make a significant contribution in the field of low power reversible computing and quantum computing.

ACKNOWLEDGMENT

This work was done under the assistance of Prime Minister's Research and Higher Studies Fund given by the Prime Minister's Office and ICT Innovation Fund given by the Ministry of ICT, Government of the People's Republic of Bangladesh.

REFERENCES

[1] C. H. Bennett, "Logical reversibility of computation," *IBM J. Research and Development*, vol. 17, pp. 525–532, 1973.

[2] D. P. DiVincenzo, "Quantum gates and circuits," *Proceedings of the Royal Society of London A: Mathematical, Physical and Engineering Sciences*, vol. 454, no. 1969, pp. 261–276, 1998.

[3] H. Axelsen, R. Glck, and T. Yokoyama, "Reversible machine code and its abstract processor architecture," in *Computer Science Theory and Applications*, ser. Lecture Notes in Computer Science. Springer Berlin Heidelberg, 2007, vol. 4649, pp. 56–69.

[4] M. Thomsen, H. Axelsen, and R. Glck, "A reversible processor architecture and its reversible logic design," in *Reversible Computation*, ser. Lecture Notes in Computer Science. Springer Berlin Heidelberg, 2012, vol. 7165, pp. 30–42.

[5] M. Nielsen and I. Chuang, *Quantum Computation and Quantum Information*. Cambridge Univ. Press, 2000.

[6] Microwind dsch - schematic editor and digital simulator. [Online]. Available: http://www.microwind.net/dsch.ph

[7] Cmos 45 nm open cell library. [Online]. Available: http://www.si2.org/openeda.si2.org/projects/nangatelib

[8] L. Jamal, M. M. Alam, and H. M. H. Babu, "An efficient approach to design a reversible control unit of a processor," *Sustainable Computing: Informatics and Systems*, vol. 3, no. 4, pp. 286 – 294, 2013.

[9] A. S. M. Sayem and M. Ueda, "Optimization of reversible sequential circuits," *Journal of Computing*, vol. 2, no. 6, pp. 208 – 214, 2010.

[10] M. Polash and S. Sultana, "Design of a lut-based reversible field programmable gate array," *Journal of computing*, vol. 2, no. 10, pp. 103–108, 2010.

[11] M. Nachtigal and N. Ranganathan, "Design and analysis of a novel reversible encoder/decoder," in *Nanotechnology (IEEE-NANO), 2011 11th IEEE Conference on*, Aug 2011, pp. 1543–1546.

[12] M. Shamsujjoha and H. Babu, "A low power fault tolerant reversible decoder using mos transistors," in *VLSI Design and 2013 12th International Conference on Embedded Systems (VLSID), 2013 26th International Conference on*, Jan 2013, pp. 368–373.

[13] C. Vudadha, P. Phaneendra, V. Sreehari, S. Ahmed, N. Muthukrishnan, and M. Srinivas, "Design of prefix-based optimal reversible comparator," in *VLSI (ISVLSI), 2012 IEEE Computer Society Annual Symposium on*, Aug 2012, pp. 201–206.

[14] X. Qi, F. Chen, K. Zuo, L. Guo, and Y. L. M. Hu, "Design of fast fault tolerant reversible signed multiplier," *International Journal of Physical Sciences*, vol. 7(17), pp. 2506–2514, 2012.

[15] N. Nayeem, A. Hossain, M. Haque, L. Jamal, and H. Babu, "Novel reversible division hardware," in *Circuits and Systems, 2009. MWSCAS '09. 52nd IEEE International Midwest Symposium on*, 2009, pp. 1134 –1138.

978-1-4799-8720-7/15 $31.00 © 2015 IEEE

2015 IEEE Computer Society Annual Symposium on VLSI

An Algorithm Used in a Power Monitor to Mitigate Dark Silicon on VLSI Chip

Zhou Zhao, Ashok Srivastava, and Shaoming Chen
Division of Electrical & Computer Engineering
Louisiana State University, Baton Rouge, LA 70803, USA
{zzhao13, eesriv, schen26}@lsu.edu

Saraju Mohanty
Department of Computer Science and Engineering,
University of North Texas, Denton, TX 76207, USA
{saraju.mohanty@unt.edu}

Abstract—**Data with increasing bandwidth requires future general-purpose as well as application specific microprocessors to improve performance endlessly. Transistor scaling, novel transistor structures, novel state-of-art VLSI design techniques and new computer architectures are the key drivers for boosting power and performance of microprocessors. Unfortunately, the processor cooling technique is unable to keep pace with higher density of transistors and high performance. For appropriate trade-offs between performance and limitation of power dissipation, dark silicon has appeared in the current processors. With the number of transistors increasing in future chips, we could envision that next generation processors might be getting darker and darker. This compromise could reduce multiple-core processors' efficiency. In this paper, power dissipation and circuit optimization are discussed in an attempt to mitigate dark silicon for future processors. A power monitor and its algorithm are proposed mainly to explain how to efficiently regulate voltage and power in the future processors with multiple cores.**

Keywords—*Dark silicon; Transistor scaling; Power dissipation; Power monitor; Multicore processors*

I. INTRODUCTION

Down scaling of transistors promotes high density of transistors in a microprocessor design. Meanwhile in last 20 years, clock frequency as well as the required LAN bandwidth LAN increased significantly due to the development of advanced communication technology [1]. Dark silicon, which refers to a part of transistors in a chip which drops work frequency and compromises with the limitation of cooling technique, increases in a chip [2]. Dark silicon largely occupies the entire chip and seriously influences work performance of processors, especially in advanced multiple-core designs [3]. It can be anticipated that dark silicon will become larger in chip if there is no implementation of novel processor topology as well as invention of state-of-art circuit. Even worse, dark silicon might result in failure of MOS scaling [4].

To solve this issue, some architecture-based solutions have been proposed. Miller et al. [5] proposed a power switch to deliver more current for single core with the purpose of improving performance at the cost of increasing low power consumption. Chen et al. [6] set power wire to work in two modes including traditional power mode and data mode. This novel topology obviously increases bandwidth of data but does not increase power consumption. For portable devices, Goulding-Hotta, et al. [7] proposed a novel "greendroid" concept to solve dark silicon issue for Android smartphones.

These current solutions are implemented in architecture level based on abstract computer architecture analysis. This work also gives some mathematical analysis from circuit point of view and proposes a feasible solution at the circuit level. Section II shows analysis regarding power dissipation, which is the key factor resulting in dark silicon. Section III proposes an on-chip power monitor and its relative algorithm to regulate work mode of a multiple-core processor facing different power situation. Results are summarized in Section IV.

II. MODELING OF POWER DISSIPATION COMPONENTS

The size of transistor still shrinks due to advanced fabrication technology. Nanometer process makes transistor to switch at higher frequencies. Unfortunately, transistor scaling has negative impact on power dissipation of chip and challenges current cooling techniques. It is to be mentioned that the power dissipation per unit area but not the total power dissipation, restricts the development of processors which brings dark silicon into a chip. Power dissipation per unit area has increased in recent years in that the speed of reducing power consumption is in a way slower than that of reducing chip area. Meanwhile, current cooling technique has met bottleneck and will not be suitable for future processors [8]. One exception is that liquid cooling technique performs well in the condition of high temperature but the cost is too high to be used widely. Hence, further reducing power dissipation is needed to mitigate dark silicon.

The problem of static power dissipation is very serious in nanometer technology with low power supply due to leakage current. The quantum tunneling effect mainly leads to a large leakage current [9, 10]. With transistor scaling, quantum tunneling effect would make carriers to lose control under gate voltage thus become a serious issue which is a main source of leakage current when transistor is off. The leakage current can be expressed as follows [10]:

$$I_{leak} = I_{sub} + I_{ox} = K_1 W e^{-V_{th}/nV_0}\left(1 - e^{-V/V_0}\right) + K_2 W\left(\frac{V}{T_{ox}}\right)e^{-\alpha T_{ox}/V} \quad (1)$$

The leakage current includes subthreshold leakage and gate oxide leakage. It can be noted that if supply voltage and threshold voltage increase with transistor scaling, leakage current will contribute more to total power dissipation. The thermal voltage, V_0 increases with temperature thus increasing leakage. Thus, if transistor scaling continues and there is no

978-1-4799-8720-7/15 $31.00 © 2015 IEEE

191

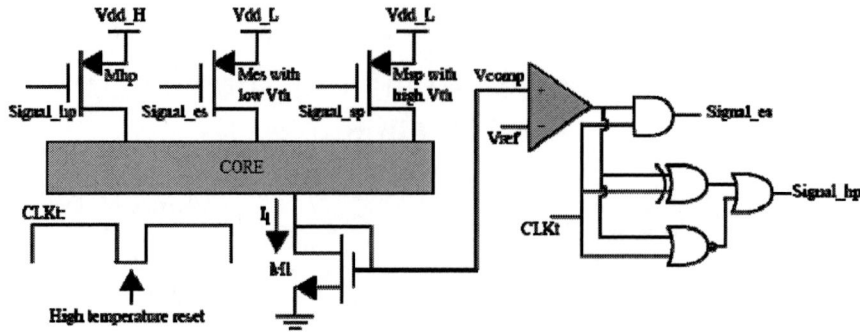

Figure 1: The proposed power dissipation monitor

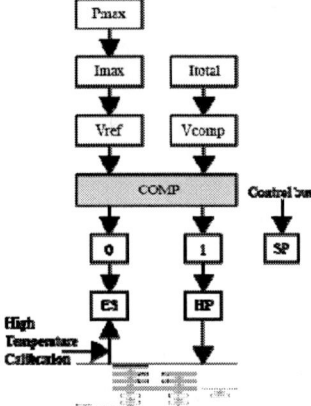

Fig. 2: The proposed algorithm for dark silicon reduction.

breakthrough in chip cooling techniques, leakage current in VLSI chip will be more serious and even force more areas of a processor to change to dark silicon.

Besides above static power dissipation, dynamic power dissipation is the other issue. The equation for calculation of dynamic power is shown as follows [11]:

$$P_{dynamic} = \frac{N_{work}}{N_{work} + N_{sleep}} C_{total} V_{dd}^2 f \qquad (2)$$

where N_{work} and Ns_{leep} represent the number of on transistors and transistors in sleep-mode, respectively. Future CPUs will be operating at higher clock frequencies. But V_{dd} won't decrease much because of design difficulty at extremely low power. Hence, dynamic power dissipation would not reduce. Equation (2) can be modified as follows:

$$P_{dynamic} = \frac{N_{work}}{N_{work} + N_{sleep}} C_{total} V_{dd} R_{total} f I_{total} \qquad (3)$$

Based on Eqn. (3) it is clearly indicated that reducing current seems to be a deserved method. To reduce current,

increasing total resistance might be a feasible method. However, in VLSI design, W/L, which could change resistance directly, has been set to a small value. Therefore, this method is not feasible. Another consideration is the propagation delay. For a CMOS inverter, the delay time, t_{pd} is the average of t_{pLH} and t_{pHL} which is expressed as follows:

$$t_{pLH} = t_{pHL} = 2 \frac{C_L}{\beta(V_{DD} - V_{th})} [\frac{V_{th} - 0.1V_{DD}}{V_{DD} - V_{th}} + \frac{1}{2} \ln\{\frac{19V_{DD} - 20V_{th}}{V_{DD}}\}] \qquad (4)$$

where V_{th} is threshold voltage and it is assumed to be same for both n- and p-MOSFETs. Symmetrical transconductance parameter, β is also assumed same for both devices for design. It can be seen that decreasing current will seriously restrict bandwidth for processor. However, simultaneously decreasing supply voltage and threshold voltage might be a potential method to reduce power dissipation and avoid reducing bandwidth.

The third contribution to total power dissipation results from interconnects wires, which can't be ignored. To sum up, total power consumption can be shown as follows:

$$P_{total} = \frac{N_{work}}{N_{work} + N_{sleep}} C_{total} V_{dd}^2 f + \sum I_{leak} * V_{dd} + \sum R_{wire} I_{wire}^2 \qquad (5)$$

Equation (5) tells us, that besides inventing some energy-saving underlying circuits such as the variety of gate circuits, to reduce power dissipation and thus solve dark silicon issue, many parameters can be optimized and looked as potential breakthrough facing many tradeoffs between performance and power dissipation.

III. PROPOSED POWER REGULATOR FOR THE PROCESSORS

In [12], the importance of voltage regulator brings to the next generation CPU. The work in [13] gives a multiple threshold voltage regulator to reduce static power dissipation. A novel power monitor, including a feedback system, to monitor and control work mode of a CPU for different total current and to predict temperature, is described as follows.

978-1-4799-8720-7/15 $31.00 © 2015 IEEE

The essence of the proposed design is using a feedback system to monitor and modify work mode of a CPU. Figure 1 shows the schematic. It can be seen that this feedback system is not very complex compared to a single core. This simple feedback system could make single core to have three working states, which are high performance mode, energy saving mode, and sleep mode.

The first mode is the high performance mode in which the core works with the high clock frequency as well as full supply voltage. The second mode is used when the single core either meets the limitation of power dissipation or achieves high temperature which could seriously damage the chip. The last mode is for the multiple core system, in which both work core and sleep core might exist at the same time. M_{hp}, M_{es} and M_{sp} represent three switches which connect to high performance mode, energy saving mode and sleep mode, respectively. The proposed power monitor can be used to let chip correctly switch between the first two modes, and control signal out of single core can determine whether the core works in the sleep mode.

The proposed design is used for letting a chip avoid bottleneck power dissipation which leads to and work in extremely high temperatures. Therefore, first we introduce an equation which could estimate maximum power dissipation of a chip for a given chip area, fabrication process, and cooling technique as follows:

$$P_{\max} \approx P_{unit-area} S \alpha_1 \qquad (6)$$

$P_{unit-area}$ is determined by the fabrication process and cooling technique. S is the total area of a chip. The correction factor is α_1 which has a value of less than 1 to prevent the chip from crossing maximum power dissipation. Then using Eqn. (5) and (6), maximum current is estimated as follows:

$$I_{\max} \approx \frac{N_{work}}{N_{work}+N_{sleep}} C_{total} V_{dd} f + \sum I_{leak} + \frac{V_{dd}}{R_{wire}} \qquad (7)$$

where, C_{total} and I_{leak} are described by the following:

$$C_{total} \approx \alpha_2 WLC_{ox} N_{total} \qquad (8)$$

$$\sum I_{leak} \approx I_s e^{\frac{qV_{gs}}{nkT}} \left(1 - e^{-\frac{qV_{ds}}{kT}}\right) N_{sleep} \alpha_3 \qquad (9)$$

In Eqn. (8), α_2 is larger than 1 and represents that capacitor includes gate capacitor and bulk capacitor. α_3 is a correction parameter to account for several off-transistors in a loop with only one value of current. So α_3 should be lower than 1. The maximum current can be calculated from Eqn. (7), (8), (9) with some approximations. From the proposed structure, M1 should work in saturation region due to its diode configuration which gives gate-source voltage to compare with the reference voltage, V_{ref}. Using this maximum value of current, V_{ref} can be calculated as follows:

$$V_{ref} = \sqrt{\frac{2I_{\max}}{\mu C_{ox} \frac{W_1}{L_1}}} + V_{th} \qquad (10)$$

For the actual current, the voltage to be compared can be obtained from the following expression:

$$V_{comp} = \sqrt{\frac{2I_{total}}{\mu C_{ox} \frac{W_1}{L_1}}} + V_{th} \qquad (11)$$

If output of the comparator is low, which means total current is larger than the maximum current which is restricted by the fabrication process and cooling technique. The feedback connects to energy saving mode in which the core works in relative low frequency with a small supply voltage. In this case, due to a small supply voltage, threshold voltage of transistor in the core should be reduced in order to avoid serious frequency drop which is the essence of dark silicon. Making tradeoff between frequency, design complexity and static power consumption, only M_{es} is designed with low threshold voltage.

If output of comparator is high, which means the core works in a safe state with an acceptable power dissipation. In this case, the core connects to the high performance mode. However, even if the core always keep distance from the limitation of power consumption, heat the core generates will be increasing all the time and the temperature of chip will go high. This high temperature will be negative for the performance of the chip. Therefore, to avoid this phenomenon, a temperature calibration part is added. CLK_t is the key part to force the core to work in energy saving mode if the core has been working in high performance mode for a long time. If CLK_t is high, the core is normally controlled by the feedback system as previously explained. When CLK_t is low, and the core still works in high performance mode, that implies the chip would be very hot, the additional logic circuit makes the core switch to energy saving mode to reduce temperature. Note that the three-state buffer connected to energy saving mode can't generate wrong control for the core. Determining time of CLK_t being high or low is very difficult since temperature changing in a VLSI chip relies on many complicated factors. Monitoring temperature of a real CPU is a feasible method to determine T_{high} and T_{low}.

The last mode is sleep mode which is used for future CPUs with multiple core which could work asynchronously. To reduce leakage current, threshold voltage of Msp should be set to a high value. This mode is controlled by the control line out of one core. To sum up, Fig. 2 shows the total process of the proposed algorithm. The first step is to calculate I_{max} for a given power consumption. Then compare I_{max} and actual current Itotal. If output of comparator is 0, the core will work in energy saving mode and if output of comparator is 1, the core will safely stay in high performance mode. Further, if the core works in high performance mode for a very long time,

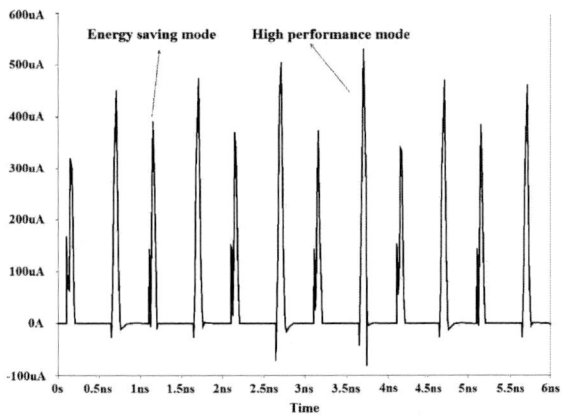

Figure 3: Simulation of mode switching.

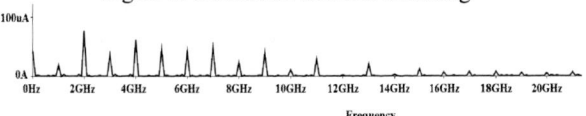

Figure 4: Frequency response of switching process.

temperature calibration, which is CLK_t will change to 0. In this case, the core will be forced to work in energy saving mode to reduce temperature of the chip. In next cycle of CLK_t, because CLK_t returns to 1, judging what mode the core works, will depend on the comparator again. Therefore, this feedback system uses few transistors, compared to the core, to achieve switching work mode of the core according to different power dissipation and temperature.

To verify the function of the proposed power monitor, an RC model of a microprocessor with 4 cores is needed to make. Using the parameters of various Intel processors [14], 16 nm predictive transistor model [15] and the trend of processors' performance, the virtual processor in this verification is determined as an ultra-low power chip used in portable devices with 0.9V voltage supply, 0.6 mA maximum current, 3GHz clock frequency, 90 million transistors. Then use equation (8) and the equation used for calculating the resistance of a transistor working in linear region, the entire capacitance and resistance of the virtual processor are estimated as 1020nF and 1500Ω, respectively. Figure 3 shows the results of mode switching, which is ideally defined as 1ns, high performance mode, energy saving mode and sleep mode last around 0.2 ns, 0.2ns, and 0.5ns, respectively. It can be seen from the figure that the entire switching process is correct. The problem with this power monitor is that switch delay is large and current loss is over 0.1mA. Figure 4 shows the Fourier transform. It can be seen that in the proposed design, the maximum frequency of switching work mode is 13GHz, which is way larger than the clock frequency regulated before.

IV. CONCLUSIONS AND FUTURE RESEARCH

In this paper, dark silicon issue, power dissipation and its impact on chip are analyzed. From the analysis, potential method to reduce power dissipation and mitigate dark silicon on a chip is proposed. At the circuit level, a power monitor and its algorithm are proposed at very low cost compared to a complex core with numerous transistors. Due to time limitation, we only substituted microprocessor by a RC circuit to demonstrate usefulness of our proposed method. The simulation at an abstract level which lacks simulation of a real core cannot safely demonstrate that our idea is correct and feasible. Our future work would include real core also include more circuits to a microprocessor in an attempt to reduce power dissipation without large frequency drop.

ACKNOWLEDGMENT

Part of the work is supported under NSF grant 1422408. Authors thankfully acknowledge help from Dr. Lu Peng on the problem of addressing dark silicon and finding solution.

REFERENCES

[1] Danowitz, Andrew, et al. "CPU DB: recording microprocessor history", *Communications of the ACM*, 55.4 (2012): 55-63.

[2] Goulding-Hotta, Nathan, et al. "The greendroid mobile application processor: An architecture for silicon's dark future", *IEEE Micro*, 31.2 (2011): 86-95.

[3] Esmaeilzadeh, Hadi, et al. "Dark silicon and the end of multicore scaling", in *Proceedings of the 38th IEEE Annual International Symposium on. Computer Architecture (ISCA)*, 2011, 2011.

[4] Taylor, Michael B. "A landscape of the new dark silicon design regime", *IEEE Micro*, 33.5 (2013): 8-19.

[5] Miller, Timothy N., et al. "Booster: Reactive core acceleration for mitigating the effects of process variation and application imbalance in low-voltage chips", in *Proceedings of the 18th IEEE International Symposium on High Performance Computer Architecture*, 2012.

[6] Chen, Shaoming, et al. "Increasing off-chip bandwidth in multi-core processors with switchable pins", in *Proceeding of the 41st Annual International Symposium on Computer architecuture*, 2014.

[7] Goulding-Hotta, Nathan, et al. "Greendroid: An architecture for the dark silicon age", in *Proceedings of the 17th Asia and South Pacific Design Automation Conference (ASP-DAC)*, 2012.

[8] Hardavellas, Nikos, et al. "Toward dark silicon in servers", *IEEE Micro*, 31.EPFL-ARTICLE-168285 (2011): 6-15.

[9] Abbas, Zia, and Mauro Olivieri. "Impact of technology scaling on leakage power in nano-scale bulk CMOS digital standard cells", *Microelectronics Journal*, 45.2 (2014): 179-195.

[10] S. P. Mohanty, Nanoelectronic Mixed-Signal System Design, McGraw-Hill, 2015, ISBN: 978-0071825719 and 0071825711.

[11] S. P. Mohanty, N. Ranganathan, and S. K. Chappidi, "Peak Power Minimization through Datapath Scheduling", in *Proceedings of the IEEE CS Annual Symposium on VLSI (ISVLSI)*, pp.121-126, 2003.

[12] López, Toni, Reinhold Elferich, and Eduard Alarcón, "Voltage regulators for next generation microprocessors", Springer, 2010.

[13] Kao, James, Siva Narendra, and Anantha Chandrakasan. "Subthreshold leakage modeling and reduction techniques" in *Proceedings of the IEEE/ACM International Conference on Computer-Aided Design*, 2002.

[14] http://ark.intel.com/#@Processors

[15] http://ptm.asu.edu/modelcard/LP/16nm_LP.pm

2015 IEEE Computer Society Annual Symposium on VLSI

Validating SPARK: High Level Synthesis Compiler

Soumyadip Bandyopadhyay, Dipankar Sarkar and Chittaranjan Mandal
Indian Institute of Technology, Kharagpur
Email: (soumyadip, ds, chitta)@cse.iitkgp.ernet.in

Abstract—**Embedded systems have found applications in diverse domains. Due to the criticality of their operations, verification of embedded systems is a necessity. With the advancement of multi-core and multiprocessor systems, there has been a paradigm shift to incorporate these features in embedded systems as well. The initial behavioural specification of a system goes through significant optimizing transformations using automated high level synthesis (HLS) compiler like SPARK, before being mapped to an architecture. Establishing the validity of these transformations is crucial to ensure that correct optimizations are applied during synthesis. To model parallel behaviours, especially in embedded systems, the use of PRES+ models is advocated for. In this paper, two path based equivalence checking methods for untimed PRES+ models are given. The experimental results demonstrate the efficiency the of the method.**

Keywords—**Equivalence checking; PRES+ model; FSMD model; SPARK; HLS**

I. INTRODUCTION

Embedded systems have become almost omnipresent in diverse application domains, from hand-held devices to automobiles. In general, they are reactive, application-specific, and have real-time constraints. Hence, they must be efficient and dependable in their performance. In recent times, with the advancement of multi-core and multiprocessor systems, there has been a paradigm shift to incorporate these features in embedded systems as well. Considerable effort has been made to adapt concurrent applications to embedded systems using their programming [1]. However, such programs have to be much sophisticated to exploit the potential concurrency of the system in which they are implemented, leading to intricacies that are difficult to analyze. Hence, there is a growing concern to enhance the current methods for designing and analyzing complex embedded systems.

Modeling and formal verification of embedded systems has received much attention in the last two decades. A comprehensive list of models proposed to represent embedded systems and their validation can be found in [2]. These models encompass a broad range of styles, characteristics, and application domains and include extensions of finite state machines, data flow graphs, communication processes and Petri nets. Petri nets are especially suited for modeling concurrent behaviours. The PRES+ model (Petri net based Representation for Embedded Systems) enhances the classical Petri net model to capture computation over integers, reals and general data structures; it captures concurrency and timing behaviour of embedded systems by allowing the tokens to carry information [3]. This modeling formalism has a well defined semantics for precise representation of systems.

A typical synthesis flow of complex systems like VLSI circuits or embedded systems comprises several phases. In each phase, the input behavioural specification (of the system to be designed) gets transformed/refined to optimize time and physical resources. To ensure the correctness of a translation phase, the intended behaviour needs to be preserved in the process. Behavioural verification involves demonstrating the equivalence between the original behaviour and the transformed behaviour. In computational terms, it is required to show that all the computations represented by the original behavioural description, and exactly those, are captured by the transformed behaviour. Our objective is to verify the transformation steps.

Several code transformation techniques such as, code motions, common sub-expression elimination, dead code elimination, etc., may be applied at the preprocessing stage of embedded system synthesis. The effectiveness of speculation in improving schedule lengths for designs with control flow has been demonstrated in [4]. List scheduling based on condition vectors (CVLS) [5] improves resource sharing among mutually exclusive operations. Recent work [6] supports generalized code motions during scheduling of designs with control flow. Literature [7] reports the effects of several global code motion techniques in the design.

Application of code motion techniques increases verification challenges significantly. Some verification techniques [8] are able to verify code transformations that occur within the basic blocks. However, they are inadequate for transformations where a code moves beyond the basic block boundary. Some recent works [9], [10] target verification of such code motions. For example, a path recomposition based finite state machines with data-path (FSMD) equivalence checking method has been proposed in [10] to verify speculative code motions. The equivalence checking method reported in [11] is strong enough to handle scenarios where the control structure of the input behaviour is modified by the scheduler.

However, the above mentioned verification techniques are applicable for sequential behaviours. These techniques cannot be readily adapted to model and verify parallel behaviours. To model parallel behaviours, especially in embedded systems, the use of PRES+ models is advocated for. The major *contributions of the present work is to devise an efficient equivalence checker for verifying several code transformations which are applied in high level synthesis compiler like SPARK.*

The paper is organized as follows. Section II demonstrates the framework of the method. Section III describes the basic issues of equivalence checking problem with an example. Section IV describes the solution of the given equivalence

978-1-4799-8720-7/15 $31.00 © 2015 IEEE

checking problem. Section V demonstrates the experimental results. The paper is finally concluded in section VI.

II. FRAMEWORK

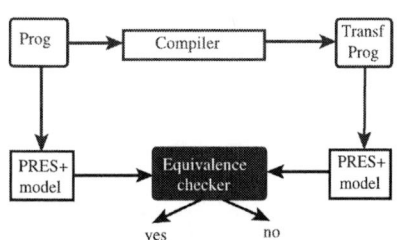

Fig. 1: Basic framework

Figure 1 depicts the actual framework of the present work. A high level program is compiled using some compiler transformation techniques and generates an optimized intermediate code. For program analysis, it is necessary to translate a program to its equivalent formal model representation. As the main target of the work is to validate code optimizing and several parallelizing transformations, hence, parallel model of computation (MoC) is necessary. In this work, PRES+ model whose underlying structure is Petri net model is chosen as parallel MoC. Therefore, PRES+ models are constructed from both original and transformed programs. The automated PRES+ construction method is reported in [12]. Here, our main task is to devise a *PRES+ equivalence checker* which takes two PRES+ models as inputs and returned either "yes" or "no" response. If the equivalence checker gives "yes" response, two programs are equivalent, i.e., particular transformation which is carried out by the compiler is correct. If it gives "no" response, the two programs *may not be equivalent*. Therefore, our equivalence checking method is sound but not complete.

III. A MOTIVATING EXAMPLE

Before describing a path based equivalence checking mechanism between two PRES+ models, in this section, we underline through an example some of the issues.

```
int i = 1, j = 1;        int i = 1, j = 1;        int i = 1, j ;
int k;                   int k;                   int k;
while (i*7 <=100)        while ((j+1)*11 <=100)   j = i;
    i++ ;                    j++;                 #parbegin
while ((j+1)*11 <=100)   while (i*7 <=100)        while (i*7 <=100)
    j++;                     i++ ;                    i++ ;
k = i+j;                     j++ ;              || while ((j+1)*11 <=100)
                         k = i+j;                    j++;
                                                 #parend
                                                 k = i+j;

      (a)                      (b)                      (c)
```

Fig. 2: Initial and Transformed Behaviour

Example 1. Petri net based models of programs: *Figure 2 (a) represents an initial program which computes $\lceil \frac{100}{7} \rceil + \lfloor \frac{100}{11} \rfloor$. Figure 2(b) and 2(c) pertain to programs transformed using loop swapping transformation and thread level parallelizing transformations, respectively. Figures 3 (a), 3(b) and 3(c) depict the PRES+ models corresponding to the programs given in Figures 2(a), 2(b) and 2(c), respectively. Automated model construction from high level program to its equivalent PRES+ model is reported in [3]. In Figure 3(a), the in-ports p_1 and p_2 contain respectively the initial values*

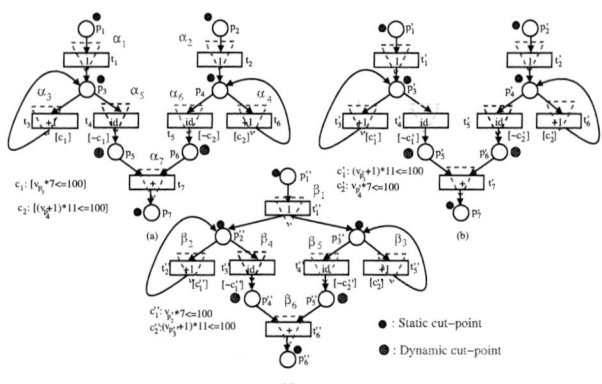

Fig. 3: Initial and transformed PRES+ models

of the input variables i and j of Figure 2(a). A computation of the model proceeds identically as an execution of a Petri net model, i.e., through firing of the enabled transitions resulting in progress of tokens with proper values from the input places of the enabled transitions to their output places at each step. A marking comprises places having tokens at any point in a computation. Note that there are two loops in Figure 2(a) where the loop variables are independent of each other and are accordingly depicted as parallel loops in the PRES+ model. Thus a PRES+ model captures the data dependence to depict independent operations in parallel while retaining the control dependence whenever needed (e.g., among the iterations of the loops). As we compare the models of Figures 3(a) and 3(b), we notice that they are structurally almost identical. Hence, less effort is expected to be required in course of equivalence checking, which is the ultimate goal of this work.

Let us now address the issues in finding equivalence between the PRES+ models of Figures 3(a) and 3(c). In Figure 3(a), the fragments $p_1.(p_3)^n.p_5$ computes the first term $\lceil \frac{100}{7} \rceil$ and the fragment $p_2.(p_4)^m.p_6$ computes the second term $\lfloor \frac{100}{11} \rfloor$; correspondingly, in Figure. 3(c), $p_1''.(p_2'')^n.p_4''$ computes the first term and $p_1''.(p_3'')^m.p_5''$ computes the second term ($n \neq m$). In general, in a computation, the number of traversals of a loop in a PRES+ model depends upon the input token values. Since, equivalence has to be established for all computations, the notion of finite computation paths is used so that any computation can be captured in terms of these paths. To do so, we cut the loops by introducing cut-points so that every loop is cut in at least one cut-point; each path will originate from a set of cut-points and extend up to a cut-point without having any intermediating cut-points.

In Figure. 3(a), suppose the cut-points are p_1, p_2 (as in-ports), p_3, p_4 (for cutting the loops) and p_7 (for the out-port). A path is represented as a sequence of maximally parallalizable transitions; hence, the corresponding paths are as follows: $\alpha_1 = \langle\{t_1\}\rangle, \alpha_2 = \langle\{t_2\}\rangle, \alpha_3 = \langle\{t_3\}\rangle, \alpha_4 = \langle\{t_6\}\rangle$ and $\alpha_5 = \langle\{t_4, t_5\}, \{t_7\}\rangle$.

Capturing computation in terms of paths: *In two ways, any computation can be represented as in terms of path. The strategies are as follows:*

978-1-4799-8720-7/15 $31.00 © 2015 IEEE   196

In Fig. 3(a), suppose the cut-points are p_1, p_2 (as in-ports), p_3, p_4 (for cutting the loops) and p_7 (for the out-port); a path is represented as a sequence of maximally parallalizable transitions. Therefore, the corresponding paths are as follows: $\alpha_1 = \langle\{t_1\}\rangle, \alpha_2 = \langle\{t_2\}\rangle, \alpha_3 = \langle\{t_3\}\rangle, \alpha_4 = \langle\{t_6\}\rangle$ and $\alpha_5 = \langle\{t_4, t_5\}, \{t_7\}\rangle$. Let a computation μ_{p_7} of the out-port p_7, depicted as a sequence of markings, be $\langle\{t_1, t_2\}, (\{t_3, t_6\})^m, \{t_3, t_5\}, \{t_3\}^n, \{t_4\}, \{t_7\}\rangle$; the computation, however, cannot be obtained as a sequence of concatenation of paths from the set $\{\alpha_1, \alpha_2, \alpha_3, \alpha_4, \alpha_5\}$. Instead, suppose the cut-points are $p_1, p_2, p_3, p_4, p_5, p_6$ and p_7; the corresponding paths are also depicted in Figure 3(a) using inverted dotted triangular boxes, such as $\alpha_7 = \langle\{t_7\}\rangle$. Now, μ_{p_7} can be represented as $(\alpha_1 \parallel \alpha_2).(\alpha_3 \parallel \alpha_4)^m.(\alpha_3 \parallel \alpha_6).(\alpha_3)^n.\alpha_5.\alpha_7$. Thus, introducing extra cut-points judiciously is an important step. Similarly, in Figure. 3(c), the paths are as shown by inverted dotted triangular boxes with cut-points $\{p_1'', p_2'', p_3'', p_4'', p_5'', p_6''\}$. Two entities are used to characterize a path α: (1) the condition R_α of execution along the path and (2) the data transformation r_α for the path α; these are computed by the forward substitution method. For the paths α_1 and α_3 in Figure 3(a), $R_{\alpha_1}(v_{p_1})$: "true" and $R_{\alpha_3}(v_{p_3})$: $v_{p_3} * 7 \leq 100$; and the data transformations are $r_{\alpha_1}(v_{p_1}) = 1$ and $r_{\alpha_3}(v_{p_3}) = v_{p_3} + 1$.

On the other way, in Figure 3(a), if p_1, p_2, p_3, p_4 and p_7 are cut-points, then the paths are $\alpha_1 = \langle\{t_1\}\rangle, \alpha_2 = \langle\{t_2\}\rangle, \alpha_3 = \langle\{t_3\}\rangle, \alpha_4 = \langle\{t_6\}\rangle$ and $\alpha_5 = \langle\{t_4, t_5\}, \{t_7\}\rangle$. Let a computation μ_{p_7} of the out-port p_7, depicted as a sequence of maximally parallelizable transitions $\langle\{t_1, t_2\}, (\{t_3, t_6\})^m, (\{t_3, t_5\}), (\{t_3\})^n, \{t_4\}, \{t_7\}\rangle$; let us examine how the computation μ_{p_7} can be captured in terms of the paths α_1 through α_5 of Figure 3(a). To do so the sequence μ_{p_7} can be rewritten as $\langle\{t_1, t_2\}, \{t_3, t_6\}^m, \{t_3\}^{n+1}, \{t_5\}, \{t_4\}, \{t_7\}\rangle = \langle\{t_1, t_2\}, \{t_3, t_6\}^m, \{t_3\}^{n+1}, \{t_4, t_5\}, \{t_7\}\rangle = \langle(\alpha_1 \parallel \alpha_2).(\alpha_3 \parallel \alpha_4)^m.(\alpha_3)^{n+1}.\alpha_5\rangle$. Similarly, in Figure 3(c), the the set of cut-points are $\{p_1'', p_2'', p_3'', p_6''\}$ and the corresponding paths are $\beta_1 = \langle\{t_1''\}\rangle, \beta_2 = \langle\{t_2''\}\rangle, \beta_3 = \langle\{t_5''\}\rangle$ and $\beta_4 = \langle\{t_3'', t_4''\}, \{t_6''\}\rangle$.

These two strategies are collectively verified in almost all of transformations which are applied in SPARK compiler. Although those transformations are also applicable for normal GCC compiler.

Equivalence checking procedure: *If cut-points $\{p_1, p_2, p_3, p_4, p_5, p_6\}$. The equivalence checking procedure is as follows:*

Let $f_{in} \subseteq inP_0 \times inP_1$ be $\{\langle p_1, p_1''\rangle, \langle p_2, p_1''\rangle\}$; let $f_{out} : outP_0 \leftrightarrow outP_1$ be $p_7 \mapsto p_6''$. For each path of Figure 3(a), the equivalent path of Figure 3(c) is obtained by the following steps:

For the path α_1: We first identify candidate paths whose pre-places have some correspondence with those of α_1. The path β_1 is chosen as the only candidate path for α_1 because $\langle {}^\circ\alpha_1, {}^\circ\beta_1\rangle \in f_{in}$, As $R_{\alpha_1} \equiv R_{\beta_1}$ and $r_{\alpha_1} = r_{\beta_1}$, it is inferred that $\alpha_1 \simeq \beta_1$ (α_1 is equivalent to β_1). Also, for finding the candidate for the subsequent paths, the post place $p_3 (= \alpha_1^\circ)$

is made to have correspondence with β_1°, denoted as $\alpha_1^\circ \to \beta_1^\circ$, say. Similarly, it is also inferred that $\alpha_2 \simeq \beta_1$.

For the path α_3: Since the pre-places of the path $\alpha_3 = \{p_3\}$ is not an in-port, a different method is used to select the candidate paths of α_3. Now, $\alpha_1 \simeq \beta_1$ and ${}^\circ\alpha_3 = \alpha_1^\circ$ has correspondence with $\beta_1^\circ = {}^\circ\beta_2 = {}^\circ\beta_3 = {}^\circ\beta_4 = {}^\circ\beta_5$; hence the candidate paths for α_3 are the paths $\beta_2, \beta_3, \beta_4$ and β_5. However, only for the path β_2, $R_{\beta_2} \equiv R_{\alpha_3}$ and $r_{\beta_2} = r_{\alpha_3}$; hence, it is inferred that $\alpha_3 \simeq \beta_2$. Similarly, it is found that $\alpha_4 \simeq \beta_3$, $\alpha_5 \simeq \beta_4$, $\alpha_6 \simeq \beta_5$ and $\alpha_7 \simeq \beta_6$. Since all the paths of the original behaviour have some equivalent paths in the transformed behaviour, and vice versa, the models are asserted to be equivalent. We can also establish the equivalence between the models of Figures 3(a) and 3(b) identically. Path extension: *Code motion transformations move code segments beyond the basic block boundaries; consequently, some paths of one model may be found to have no equivalent paths in the other model. Such paths will have to be extended through its subsequent path(s) till paths equivalent to the resulting concatenated path(s) are obtained. The idea of path extension is similar to that of path based FSMD equivalence checking mechanism [11]. Intricacies, however, arise due to the presence of paths parallel to the path being extended. In this method, paths are generated by preserving both executional trace and strict parallelism, hence scheduling type transformations are handled by this method.*

If the cut-points are P_1, p_2, p_3, p_4 and p_7, the equivalence checking The equivalence checking procedure is as follows:

Let $f_{in} \subseteq inP_0 \times inP_1$ be $\{\langle p_1, p_1''\rangle, \langle p_2, p_1''\rangle\}$; let $f_{out} : outP_0 \leftrightarrow outP_1$ be $p_7 \mapsto p_6''$. For each path of Figure 3(a), the equivalent path of Figure 3(c) is obtained by the following steps:

For the path α_1: The first task is to identify paths , called candidate paths, whose pre-places have some correspondence with the path α_1. The path β_1 is chosen as the only candidate path for checking equivalence with α_1 because $\langle {}^\circ\alpha_1, {}^\circ\beta_1\rangle \in f_{in}$, As $R_{\alpha_1} \equiv R_{\beta_1}$ and $r_{\alpha_1} = r_{\beta_1}$, it is inferred that $\alpha_1 \simeq \beta_1$. Also, for finding the candidate for the subsequent paths, α_1° is made to have correspondence with β_1°, denoted as $\alpha_1^\circ \to \beta_1^\circ$, say. Similarly, it is also inferred that $\alpha_2 \simeq \beta_1$.

For the path α_3: Since the pre-places of the path $\alpha_3 = \{p_3\}$ is not an in-port, i.e., ${}^\circ\alpha_3 \notin inP_0$, a different method is used to select the candidate paths of α_3. Now, ${}^\circ\alpha_3 \subseteq \alpha_1^\circ$ has correspondence with $\beta_1^\circ = {}^\circ\beta_4 \supseteq {}^\circ\beta_2, {}^\circ\beta_3$; hence the candidate paths for α_3 are β_2, β_3 and β_4. However, only for the path β_2, $R_{\beta_2} \equiv R_{\alpha_3}$ and $r_{\beta_2} = r_{\alpha_3}$, hence, it is inferred that $\alpha_3 \simeq \beta_2$. Similarly, it is found that $\alpha_4 \simeq \beta_3$ and $\alpha_5 \simeq \beta_4$. Since all the paths of the original behaviour have some equivalent paths in the transformed behaviour, and vice versa, the models are ascertained to be equivalent. We can also establish the equivalence between the models of Figures 3(a) and 3(b), identically. In this method, the costly path extension is not needed. Therefore, the equivalence checking method is faster. In this method, some scheduling type transformations which are carried out by SPARK cannot be handled.

IV. PROPOSED SOLUTION

We have already mentioned the two strategies for validation the transformations occurring in SPARK compiler. There are two challenges in each strategy; the path construction and equivalence checking. The solution of the strategies are as follows:

A. Strategy I

When the equivalence checking is carried out using maximum number of cut-points where executional trace and parallelism is strictly captured, the corresponding theory for validating path based equivalence checking is *any computation can be represented as concatenation of parallel paths*. In this method, during path construction, the procedure checks the marking contains at least one cut-point, if the condition is true, mark all places in the particular marking are marked as dynamic cut-points and paths are constructed using backward cone of foci method. It is to be noted that path construction and designation of extra cut-points goes in hand in hand. For equivalence checking, there are two major steps (1) checking correspondence and (2) path equivalence. As in this method small size paths are generated, hence during equivalence checking phase path extension is needed.

B. Strategy II

When the equivalence checking is carried out using minimum number of cut-points where executional trace is captured strictly but parallelism is not captured vividly, the corresponding theory for validating path based equivalence checking is *for any computation μ_p of an out-port p, there exists a reorganized sequence μ_p^r such that $\mu_p \simeq \mu_p^r$*. In this method, during path construction, the procedure checks the marking contains any one cut-point, if the condition is true, paths are constructed using backward cone of foci method. For equivalence checking, there are two major steps (1) checking correspondence and (2) path equivalence. In this method, as less number of paths are generate, the costly path extension is not needed.

V. RESULTS

The equivalence checking procedures have been implemented in C and satisfactorily tested on several sequential [6] and parallel benchmarks. The translation is carried out by one HLS (high level synthesis) compiler, i.e., SPARK [6] and two thread level parallel compilers PLuTO and Par4All . For checking equivalence between two paths SMT solver ($Z3$) has been used. For sequential benchmarks, we have compared our both methods with the method reported in [11], in all cases one of our method is found to be two times faster than the reported method and our other method is somewhat faster than the reported method [11]. During experimentation with parallel benchmarks, our equivalence checker has identified a bug of the PLuTo compiler (possibly due to faulty usage of a variable name in the source program).

VI. CONCLUSION AND FUTURE WORK

Two path based equivalence checking methods for PRES+ models are described through one example for loop swapping techniques as well as thread level parallelizing transformations for scalar programs. Our both proposed method can handle both uniform and non-uniform code transformations as well as thread level parallelizing transformations. When our equivalence checking is carried out with maximum number of cut-points, various scheduling type of transformations are handled. On the contrary, when our equivalence checking is carried out with minimum number of cut-points, it fails to handle scheduling type of transformations.

The *limitations* of both the two method are that it cannot handle loop-shifting, software pipelining based transformations as well as several loop transformations for array handling programs. It can be enhanced in future to encompass these transformations.

ACKNOWLEDGMENT

This work is supported by TCS research fellowship of Soumyadip Bandyopadhyay and DST Project No: $SB/EMEQ - 281/2013$.

REFERENCES

[1] David Gay, Philip Levis, Robert von Behren, Matt Welsh, Eric Brewer, and David Culler. The nesc language: A holistic approach to networked embedded systems. In *PLDI*, pages 1–11, 2003.

[2] Stephen Edwards, Luciano Lavagno, Edward A. Lee, and Alberto Sangiovanni-Vincentelli. Design of embedded systems: Formal models, validation, and synthesis. In *Proceedings of the IEEE*, pages 366–390, 1997.

[3] Luis Alejandro Cortés, Petru Eles, and Zebo Peng. Verification of embedded systems using a petri net based representation. In *ISSS '00: Proceedings of the 13th international symposium on System synthesis*, pages 149–155, Washington, DC, USA, 2000. IEEE Computer Society.

[4] Luiz C. V. Dos. Santos and J.A.G Jress. A reordering technique for efficient code motion. In *Procs. of the 36th ACM/IEEE Design Automation Conference*, DAC '99, pages 296–299, New York, NY, USA, 1999. ACM.

[5] K. Wakabayashi and H. Tanaka. Global scheduling independent of control dependencies based on condition vectors. In *DAC '92: Proceedings of the 29th ACM/IEEE conference on Design automation*, pages 112–115, 1992.

[6] S. Gupta, N. Dutt, R Gupta, and A. Nicolau. Spark: a high-level synthesis framework for applying parallelizing compiler transformations. In *Proc. of Int. Conf. on VLSI Design*, pages 461–466, Washington, DC, USA, Jan 2003. IEEE Computer Society.

[7] S. Gupta, N. Dutt, R Gupta, and A. Nicolau. Using global code motions to improve the quality of results for high-level synthesis. *IEEE Transactions on CAD of ICS*, 23(2):302–312, Feb 2004.

[8] Rajesh Radhakrishnan, Elena Teica, and Ranga Vemuri. Verification of basic block schedules using rtl transformations. In *Proceedings of the 11th IFIP WG 10.5 Advanced Research Working Conference on Correct Hardware Design and Verification Methods*, CHARME '01, pages 173–178, London, UK, 2001. Springer-Verlag.

[9] S. Kundu, S. Lerner, and R. Gupta. Validating high-level synthesis. In *Proceedings of the 20th international conference on Computer Aided Verification*, CAV '08, pages 459–472, Berlin, Heidelberg, 2008. Springer-Verlag.

[10] Youngsik Kim and Nazanin Mansouri. Automated formal verification of scheduling with speculative code motions. In *Proceedings of the 18th ACM Great Lakes symposium on VLSI*, GLSVLSI '08, pages 95–100, New York, NY, USA, 2008. ACM.

[11] K. Banerjee, C. Karfa, D. Sarkar, and C. Mandal. Verification of code motion techniques using value propagation. *Computer-Aided Design of Integrated Circuits and Systems, IEEE Transactions on*, 33(8), Aug 2014.

[12] L.A Cortes, P. Eles, and Zebo Peng. Verification of embedded systems using a petri net based representation. In *System Synthesis, 2000. Proceedings. The 13th International Symposium on*, pages 149–155, 2000.

Enabling scaling of advanced CMOS technologies: A reliability perspective

Tanya Nigam* and Andreas Kerber^

GLOBALFOUNDRIES, *Santa Clara, CA 95054 and ^Malta NY, 12020.

Email:tanya.nigam@globalfoundries.com

ABSTRACT — As CMOS technologies continue to scale and devices become more interconnected, new reliability challenges are emerging. With Internet-of-Things, semiconductor devices will be ubiquitous and used under diverse environmental conditions. In this paper we will review the device level scaling challenges from the reliability perspective, which include new materials, variability in ever decreasing dimensions, and methodology enhancements needed to provide reliable solutions across different product segments. The critical need for product level reliability assessment will be highlighted to provide additional margin for the consumer market.

INDEX TERMS — HFO_2, HK MG, BTI, SILC, TDDB, RO, SRAM.

SYNOPSIS

As we move towards the sub 20 nm regime, a convergence of application is occurring in the product segment. As shown in Fig. 1, a typical product like a cell phone currently provides not only enhanced computing power (CPU and GPU) but also encompass many sensors such as accelerometers, gyroscopes, and microphone. As we move towards IoT, connecting devices and ensuring a secure and reliable data collection/transfer will be critical. All of this must come at a reduced cost and shorter development times. The trade-off between power and performance becomes more demanding as well. Reliability of individual components may need a re-evaluation and assessing over all product reliability where failure of any one of the diverse components leads to product failure needs to be addressed.

The power performance trade off in device scaling, has been addressed via material changes and device architecture change. Material changes include introduction of Hf-based HK dielectric stacks to replace SiON gate dielectric. Ultra Low K dielectrics for Back end of Line dielectrics and material/interface optimization for EM. For sub 10nm technologies new channel materials such III-V and SiGe are being explored along with FINFET and Gate all around Nanowire. Each of these approaches brings new challenges and solutions to technology scaling. A third component of scaling is managing variability both at Time zero and those induced due to defect generation under operating conditions. Understanding and modeling these variability components is also critical for future nodes. Finally, it is critical to correlate the device level learning to product operation. Starting block for such studies can be

ring oscillators for logic and small array of SRAM for memory. Implication of device degradation over time on Logic and SRAM Vmin needs to be understood and appropriate guard band is needed during technology definition.

In this presentation, we summarize reliability challenges due to material change in current and future technology nodes. We will provide an overview of key physical mechanisms which are impacted due to change in gate dielectric stack from SiON to HK MG. A short overview of the changes in BEOL material and its impact on electro-migration and ILD TDDB will be discussed. Move to FINFETs with III-V channel material will be briefly touched. Impact of variability both time zero and post stress will be elaborated. Current and future approaches to bridge the gap between Wafer Level Reliability and product level reliability challenges will be highlighted.

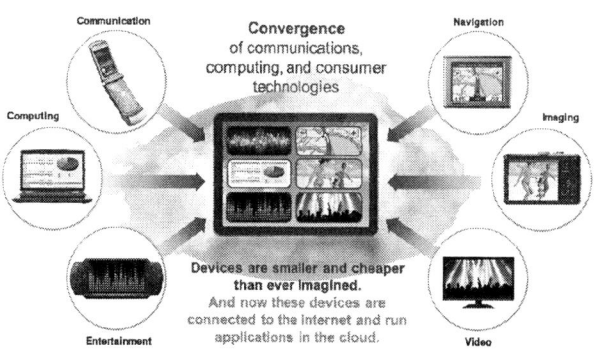

Figure 1 Convergence of multiple computing elements and sensors in a single product requires more robust assessment of reliability and leads to new mechanism for product fail.

2015 IEEE Computer Society Annual Symposium on VLSI

Digital Right Management for IP Protection

Jerome Rampon,
Renaud Perillat
Algodone, Montpellier,
France
jerome@algodone.com
renaud@algodone.com

Lionel Torres,
Pascal Benoit,
Giorgio Di Natale
LIRMM, Univ. Montpellier/CNRS,
UMR 5506, France
Email: {name.surname}@lirmm.fr

Mario Barbareschi
University of Naples "Federico II,
Italy"
mario.barbareschi@unina.it

Abstract—**The Digital Rights Management (DRM) principle is generally well known for the exchange of files (music, video), or software management. Specialized solutions concerning vprofessional software are behind a business called "Software License Managemen". The concept of DRM can be transposed to other areas, in particular to the design of hardware devices. Although the concept of DRM is allowed and widely used in the field of software, this is not the case for Integrated Circuit design (no industrial solutions to date are actually proposed). We propose into this paper a new approach based on a original way to extract Physically Unclonable Function which is at the heart of the DRM proposal.**

I. INTRODUCTION

The Digital Rights Management (DRM) is mostly known for the exchange of media files and proprietary software. Specific solutions (e.g., flexlm [1], safenet [2], rlm [3]) are available for professional software, and are at the source of a sector called Software License Management (SLM). The concept of DRM can also be applied to other fields, and in the particular to the one of hardware design. While the DRM concept is accepted and widely used in the software area, this is not the case for the hardware. Indeed there are no industrial solutions proposed so far, and this becomes an international concern to protect these digital rights, like the recent DARPA projects to fight against counterfeiting and improve the traceability of integrated circuits [4].

Chip counterfeiting is less mediatized than other problems; nevertheless its consequences can be fatal when we look at the number of electronic devices in a connected hospital or objects related to human. For instance, between 2006 and 2010, the US retailer VisionTech circuits delivered almost 60,000 counterfeit circuits to its clients, including the US Navy, Raytheon Missile System [5]. Many cases of counterfeit circuits used in sensitive applications have been reported in the US (military equipment) [6] and are increasingly relayed in the press [7]. The problem of counterfeit integrated circuits has increased significantly in recent years. The number of electronic circuits counterfeits seized by US Customs from 2001 to 2011 has been approximately multiplied by 700 [7]. Between 2007 and 2010, US Customs sized 5.6 million counterfeit electronic products [8]. The counterfeiting IC market is a evaluated at of 7% of the semiconductor market [9], representing a loss of about $ 22 billion in 2014 for the legal industry. It is therefore crucial and strategic to implement

research projects to protect the intellectual property (IP) of IC designers.

In 2013, about 2,000 new ASIC circuits of projects were performed, with high design costs, and nearly 100,000 projects with FPGA circuits to lower design costs, were initiated These circuits are found in many consumer electronics (smartphones, tablets, laptops, game consoles, connected objects, etc.) and even safety-critical systems (cars, avionics, military, space, nuclear). For 15 years, nearly all of these new circuits use predefined blocks called Intellectual Property (IPs) because of their numerous benefits (proven functionality, compatibility, performance, time-to-market, cost and patent enforcement). The design-re-use trend is growing steadily and now it is possible to find more than 100 IP-based elements in a new circuit (e.g., memory, processors, peripherals, communication protocols). The share of IP components represents an average percentage around 70% of the circuit, up to 90

to developers that supply IPs, it is essential to provide an automated control system of IP rights. However, for the IP market it is not possible to observe directly the inner parts of the hardware, since IPs are virtual components when looking at their physical implementations. However, it is clear that IP vendors are threatened in the same way as the circuit provider and undoubtedly they are even more at risk because the copy of an IP is much simpler than a circuit.

One solution proposed herein is to develop a solution to provide a DRM system for integrated circuits. The solution allows an IP vendor to take under control the number of sold IPs. Indeed, since no commercial DRM solutions for hardware exist yet, IP Vendors do not have any feedback on how much the IP is used by a customer. The latter can then produce an unlimited number of chips without acknowledging the IP Vendors on the actual number of integrated circuits that are deployed in the market.

The Figure 1 explains the selling process of an IP. It includes two parts, the IP vendor, proposing the intellectual property. The IP customer, using this one. Once the IP customer has the IP, the IP vendor has no real view on how the customer uses the IP.

Unlike other proposals in the literature, such as [10], [11], to protect its IPs, IP vendors simply need to insert a Algodone smart lock IP, which is fully compatible with traditional digital design and verification methodologies. Once the chip is

978-1-4799-8720-7/15 $31.00 © 2015 IEEE

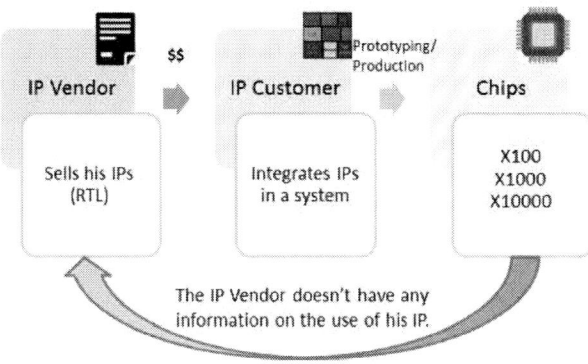

Fig. 1. IP Vendor customer chain.

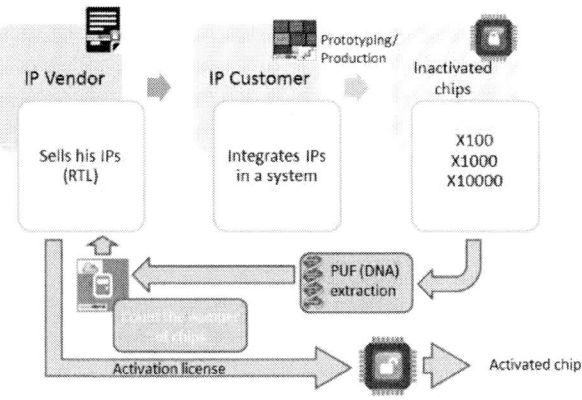

Fig. 2. Main principle of the DRM process.

manufactured or the FPGA programmed, each distinct physical instance requires a unique runtime license key to activate the functions protected with smart locks. Even though an IP customer may produce thousands or millions of identical devices, each one of them actually requires a unique runtime license key for its activation. The main advantage of our DRM solution is its simplicity: design once, and activate every single IP use.

The Figure 2 shows the selling process of an IP using our smart lock IP. It includes a third party, proposed by Algodone Company, which provides the smart lock IP to the IP vendor. The IP vendor inserts Algodone's IP in it own IP and then he can offer it to his customer. Now the IP is protected, and once the chip is manufactured, the IP customer extracts the IC DNA of the chip in order to buy the license and activate it.

This principle of DRM is well known for managing audio or video file types for access to music or movies in particular around the issues of illegal downloading (excessive infringe-

ment to the early 2000s). The methods that apply (secure platform type *iTunes*, package deals of type *deezer*) do not counteract the primary need that is addressed in the IP authentication, circuits and fight against counterfeiting. Similarly, in the software field, the business model of software vendors imposes to control their income. Said SLM protection systems have established for 30 years in this sector. Many people are familiar to write software license codes obtained from the provider to switch modes from demonstration on 30 days to a permanent mode. The most common and accessible example is probably the Microsoft Office suite. What is proposed herein is close to SLM concept, in taking the principle of master-slave type used by SLM solutions. It should be noted that marking solutions (identification, bar code, RFID) are available on the market, however the advantage of the approach is to be much more integrated (and indeed secure). The key point under the IP activation will be tied to the ability to ensure the uniqueness of the considered component. The main primitive that is used in the smart lock IP is a Physically Unclonable Function (PUF). The rest of the paper proposes the current strategy developed in order to generate PUFs in a simple, generic and effective way. Section II presents the main characteristics of PUFs, while Section III proposes a novel approach based on the use of several ring oscillators, perfectly suitable to FPGAs. Section IV concludes this paper.

II. PHYSICAL UNCLONABLE FUNCTION

The Physical Unclonable Function (PUF) is a hardware security primitive that exploits some physical randomness, introduced explicitly or indirectly into a device, to generate device unique information used to address security related problems. Basically, a PUF is a function that one-way maps a set of inputs, namely challenges, to a set of outputs, defined as responses, forming a set of challenge-response pairs (CRPs) that are unique for each device where the PUF has been implemented. Moreover, PUFs should be unclonable and tamperevident, meaning that it is unfeasible for an attacker to build another PUF that provides the same original CRPs set, and that invasive attacks destroy the PUF, as they can be easily detected. Furthermore, PUF responses should be persistent and unpredictable, and it should not be possible to discover the applied challenge given its output response. Among all types of PUFs, this paper focuses on silicon PUFs, introduced by Gassend et al. [12], which exploit variations in integrated circuit manufacturing, inherently random across different dies and wafers, to generate robust, unclonable, unpredictable, and chip-specific outputs. Due to their underlying mechanism, silicon PUFs do not require any alteration of manufacturing processes, and several designs are available for both Field Programmable Gate Arrays (FPGAs) and Application Specific Integrated Circuits (ASICs). Since exploited properties are electrical, the responses are inherently affected by noise. The environmental and working conditions, such as the temperature and the supplied voltage, can dramatically alter PUFs responses, making them non-suitable secure primitives because of lack of reliability. According to different physical sources of

978-1-4799-8720-7/15 $31.00 © 2015 IEEE

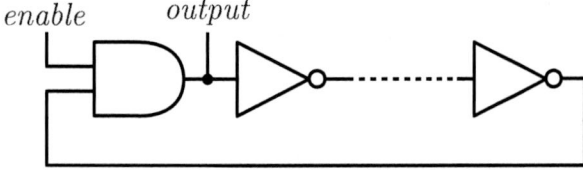

Fig. 3. Ring Oscillator loop controlled by a AND control gate.

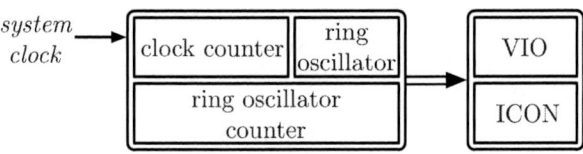

Fig. 4. Schematic view of the design adopted to extract the frequency from a target RO.

imperfections, silicon PUFs can be categorized in delay-based PUFs and memory-based PUFs. While the former requires additional hardware resources, because it involves a time measurement, the latter is based on the random power-up values of storage elements, inherently present in practically any Integrated Circuit (IC) device. The most discussed memory-based PUF there are the SRAM PUF [13] and the D flip-flop PUF [14]. As for the delay-based category, there are the Arbiter PUF [LIM05], the Ring Oscillator (RO) PUF [15], [16], [17], the Butterfly PUF [18], and the Anderson PUF [19] for the delay-based family. SRAM PUFs are not suitable for all FPGA families, hence the introduction of D flip-flop PUFs. Arbiter PUFs and Butterfly PUFs require layout symmetry that is hard to satisfy on FPGAs due to the lack of routing control. Conversely, the RO PUF is suitable for every silicon technology, and hence can be easily adopted as PUF technique for both the FPGA and ASIC technology. Indeed, the RO is a primitive easily implementable in terms of design, because does not require symmetry for the place and route. In order to integrate the PUF mechanism in the DRM scheme defined above, we established as PUF architecture the one based on ROs, in order to enable the Algodones approach on every silicon technology. But, as anticipated before, external conditions might change PUF responses during time and, consequently, cause a fault in licensing protocol.

In the next section, we illustrate, through experimental result, how the temperature impacts the RO frequency.

III. RING OSCILLATOR WITH TEMPERATURE VARIATIONS

A ring composes the RO structure, that we adopt in this paper (Figure 3), with an odd number of inverting stages and by a control gate, which interrupts the ring in order to enable or disable the oscillations. The output of the control gate is exploited to obtain the oscillating signal. If the inverting stages are odd, the control gate must be an and-gate, otherwise a nand-gate.

In order to measure the RO frequency value, we exploit a simple architecture, similar to the ROPUF case, which involves only two counters. One counter establishes the time window in which the frequency measurement has to be accomplished, hence it is timed by the system clock, which frequency is C, and it is configured to count up to a maximum fixed value T. The other is fed with the RO output, so it counts the edges (rising or falling) of the oscillations. When the first clock reaches the established maximum counting value, the RO is disabled and the RO frequency can be obtained as $R = C/T$, with R the number of counted oscillations edges. Due to the uncertainty on the system clock phases, this measurement introduces a measurement error $\varepsilon = \pm C/(2 \times T)$.

The proposed architecture has been implemented on Xilinx Spartan-6 XC6LX16 (45nm) devices within Nexys3 boards. Each configurable logic block (CLB) of Spartan-6 technology contains a pair of slices, Slice L and Slice X or Slice M and Slice X, and all of them are equipped with four 6-input look-up tables (LUTs). The two different slices couples are alternated among CLBs columns. The targeted FPGA device has a CLB array composed of 18 columns and 60 rows. The clock counter width is 18 bits and the RO counter width is fixed to 24 bits. This choice enables to keep the RO and the two counters bounded in a block of 12 CLB, such that it can be easily placed over allowed FPGA positions. Furthermore, each implemented entity is fixed in position within the cell and in used basic elements. This implies that each cell translation does not alter the design at the netlist level description. In order to read the bits contained in the RO counter, the design architecture is instrumented with Xilinx Chipscope, in particular with one Virtual IO port and ICON, and, by adopting the Xilinx software libraries (Figure 4). A PC in which the board is plugged-in can elaborate the value. Like the RO measurement cell, all involved Chipscope cores are bounded in a fixed shape and constrained in location and basic elements. This allowed us to exclude from the measuring the variability introduced by differences which can be generated by the synthesis tool whenever, generating the architecture from the description language, some place and routing solutions are

picked each time a RO happens.

The working temperature for an integrated circuit is directly related to the signal propagation delay. Indeed, high working temperatures cause speed degradation. In order to analyze the effect of temperature changing on ROs frequencies, we measured frequencies of 5 stages ROs under 5 different external fixed temperature: 0°C, 13.3°C, 26.6° C, 40°C, 53.3°C and 66.6°C. To this aim, the FPGA is placed in a thermal chamber which keeps the temperature with a precision of 0.1°C. When the thermal chamber has reached the right temperature, we have waited 30 minutes in order to be sure that the die has uniformly reached the same external temperature before starting each test campaign. Figure 5 illustrates all measured frequencies varying the temperature within the range [0°C, 66.6°C]. They are inversely proportional to temperature values and the relationship between them is quite close to be a linear function. The only observable exception is after 40°C because the curve start to be more descendant. To better analyze the relationship between the temperature value and the

978-1-4799-8720-7/15 $31.00 © 2015 IEEE

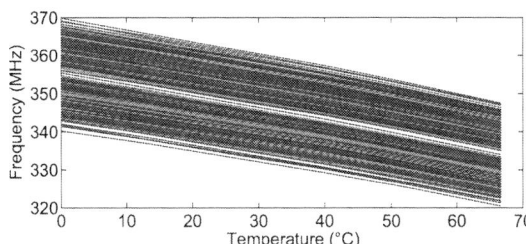

Fig. 5. Values of all ROs frequencies for different temperature.

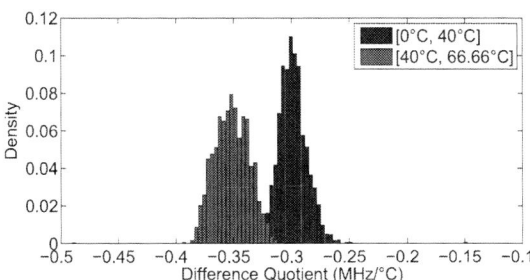

Fig. 6. Difference quotient distributions evaluated before and after 40C.

ROs frequencies, we can consider the difference quotient for each temperature range, namely how the frequency decreases increasing the temperature of $13.33°C$. Figure 6 shows two distribution difference quotients calculated for all ROs. The blue histogram is related to the temperatures less than $40°C$, while the red to the temperature greater than $40°C$. They indicate that the average values of difference quotients are respectively -0.29 MHz/$°C$ and -0.36 MHz/$°C$. Both are distributed with a standard deviation of -0.013 MHz/$°C$.

IV. CONCLUSION

In this paper we have shown that ROs frequencies are tight coupled not only with design parameters, but also with their working conditions. In particular, we have illustrated the role played by temperature. This study has demonstrated that temperature effect is quite linear and could be considered as an important result for PUF extraction. Of course other parameters should be considered (power supply, noise, and so on). But for DRM application for IP protection, certainly a weak PUF will be sufficient and a simple model could be considered to obtain IP authentication process.

REFERENCES

[1] FlexLM, "Solution for applications producers," www.flexerasoftware.com/.

[2] SafeNet, "World leading data protection and software monetization," www.safenet-inc.com/.

[3] RLM, "Reprise license manager," www.reprisesoftware.com.

[4] DARPA, "Tiny, cheap, foolproof: Seeking new component to counter counterfeit electronics," www.darpa.mil/NewsEvents/Releases/2014/02/24.aspx.

[5] A. Z. P.E. Chaudhry, *Protecting Your Intellectual Property Rights*, 2013, the Global Growth of Counterfeit Trade.

[6] S. Maynard, "Trusted foundry be safe. be sure. be trusted," *Trusted Manufacturing of Integrated Circuits for the Department of Defenses*, October 2010, www.trustedfoundryprogram.org.

[7] C. Gorman, "Counterfeit chips on the rise," June 2012.

[8] AGMA, "Alliance for gray markets and counterfeit adatement," www.agmaglobal.org.

[9] S. T. M. Pecht, "Bogus! electronic manufacturing and consumers confront a rising tide of counterfeit electronics," May 2006.

[10] A. Cilardo, M. Barbareschi, and A. Mazzeo, "Secure distribution infrastructure for hardware digital contents," *IET Computers & Digital Techniques*, vol. 8, no. 6, pp. 300–310, 2014.

[11] R. Maes, D. Schellekens, and I. Verbauwhede, "A pay-per-use licensing scheme for hardware ip cores in recent sram-based fpgas," *Information Forensics and Security, IEEE Transactions on*, vol. 7, no. 1, pp. 98–108, 2012.

[12] B. Gassend, D. Clarke, M. Van Dijk, and S. Devadas, "Silicon physical random functions," in *Proceedings of the 9th ACM conference on Computer and communications security*. ACM, 2002, pp. 148–160.

[13] D. E. Holcomb, W. P. Burleson, and K. Fu, "Power-up sram state as an identifying fingerprint and source of true random numbers," *Computers, IEEE Transactions on*, vol. 58, no. 9, pp. 1198–1210, 2009.

[14] V. van der Leest, G.-J. Schrijen, H. Handschuh, and P. Tuyls, "Hardware intrinsic security from d flip-flops," in *Proceedings of the fifth ACM workshop on Scalable trusted computing*. ACM, 2010, pp. 53–62.

[15] G. E. Suh and S. Devadas, "Physical unclonable functions for device authentication and secret key generation," in *Proceedings of the 44th annual Design Automation Conference*. ACM, 2007, pp. 9–14.

[16] A. Maiti, J. Casarona, L. McHale, and P. Schaumont, "A large scale characterization of ro-puf," in *Hardware-Oriented Security and Trust (HOST), 2010 IEEE International Symposium on*. IEEE, 2010, pp. 94–99.

[17] A. Maiti and P. Schaumont, "Improved ring oscillator puf: An fpga-friendly secure primitive," *Journal of cryptology*, vol. 24, no. 2, pp. 375–397, 2011.

[18] S. S. Kumar, J. Guajardo, R. Maes, G.-J. Schrijen, and P. Tuyls, "The butterfly puf protecting ip on every fpga," in *Hardware-Oriented Security and Trust, 2008. HOST 2008. IEEE International Workshop on*. IEEE, 2008, pp. 67–70.

[19] J. H. Anderson, "A puf design for secure fpga-based embedded systems," in *Proceedings of the 2010 Asia and South Pacific Design Automation Conference*. IEEE Press, 2010, pp. 1–6.

2015 IEEE Computer Society Annual Symposium on VLSI

Development of a Layout-Level Hardware Obfuscation Tool

Shweta Malik[1], Georg T. Becker[2], Christof Paar[1,2], Wayne P. Burleson[1]
[1] *University of Massachusetts Amherst*
[2] *Horst Görtz Institute for IT-Security, Ruhr University Bochum*

Abstract—While hardware obfuscation has been used in industry for many years, very few scientific papers discuss layout-level obfuscation. The main aim of this paper is to start a discussion about hardware obfuscation in the academic community and point out open research problems. In particular, we introduce a very flexible layout-level obfuscation tool that we use as a case study for hardware obfuscation. In this obfuscation tool, a small custom-made obfuscell is used in conjunction with a standard cell to build a new obfuscated standard cell library called *Obfusgates*. This standard cell library can be used to synthesize any HDL code with standard synthesis tools, e.g. Synopsis Design Compiler. However, only obfuscating the functionality of individual gates is not enough. Not only the functionality of individual gates, but also their connectivity, leaks important important information about the design. In our tool we therefore designed the obfuscation gates to include a large number of "dummy wires". Due to these dummy wires, the connectivity of the gates in addition to their logic functionality is obfuscated. We argue that this aspect of obfuscation is of great importance in practice and that there are many interesting open research questions related to this.

Keywords-Layout-level hardware obfuscation, reverse-engineering

I. Introduction

There are many reasons why hardware companies want to prevent their designs from being reverse-engineered. One of the main reasons is to prevent intellectual property (IP) theft and counterfeiting. IP theft and counterfeits products are huge problem in the IT industry. According to a SEMI report[1] approximately $4 billion is lost due to IP infringement, which includes counterfeiting through reverse engineering, theft of trade secrets and trade marks. Furthermore, for security critical devices, attackers can use hardware reverse-engineering to reveal design details that can compromise the security of the system. For example, an attacker might be able to identify the deployed security mechanisms using reverse-engineering and how to circumvent them.

One approach to combat reverse-engineering is the use of layout-level hardware obfuscation. In layout-level hardware obfuscation the design is implemented using gates that are designed to be especially hard to reverse-engineer. While layout-level hardware obfuscation has

been of interest to the industry till the 90s, there has been very little work on the topic in academia. Most scientific papers dealing with layout-level obfuscation have only been published in the last two years. The goal of this paper is to close this gap and and to initiate more scientific debates on layout-level obfuscation. The tool we introduce in this paper is based on an obfuscation gate that uses dopant modifications to combat optical reverse-engineering. Using dopant modifications to prevent reverse-engineering is not new and the industry and is already been used for years in the industry [2], [3].

The idea behind dopant obfuscation is to build gates which are identical on all routing layers (metal and polysilicon) and only differ in the dopant polarity of the active area. Depending on the dopant configurations, the gates can have many different logic functions. Since it is harder to determine the dopant polarity than the metal or polysilicon layer, an attacker can not differentiate between the different gates, and hence cannot reverse-engineer the design. At least this is the goal. However, while it is more difficult to determine the dopant polarity of a gate, it is not impossible. For example, recently Sugawara *et al.* showed how such a dopant obfuscation cell can be reverse-engineered using a SEM by looking at the contact layer [4].

II. Related Work

Layout-level hardware Obfuscation has been of interest to industry since the 90s [3] and has been used in many chips in the past. One of the companies providing layout-level hardware obfuscation techniques is for example SypherMedia International (SMI). They use two fundamentally different approaches for layout-level obfuscation [2], [5]. In the first approach, they use custom made obfuscation gates that are used instead of a standard cell library to obfuscate the designs. The second approach is to modify existing cells of a standard cell library to create obfuscated cells that look identical to the standard cells. In the first approach a single obfuscation gate can have many different functionality and hence offers more obfuscation per gate. The advantage of the second approach is that the attacker does not know which parts of the design is being obfuscated or that obfuscation was used at all. Hence, if only a few obfuscation gates are used in the design, the second

978-1-4799-8720-7/15 $31.00 © 2015 IEEE

approach is favorable, while the first approach should be used if an entire part of a design is being obfuscated. The individual gates are being obfuscated based on the dopant polarity, similar as it is being proposed in this paper. A technique suggested by Rajendran *et al.* uses a mixture of real and dummy contacts to camouflage the functionality of the obfuscated gates [6]. The main idea behind this technique is that extra effort is needed to reverse-engineer the contact layer. In [7] an obfuscation gate that is also based on the dopant polarity has been proposed. In this approach a single 2-input look-alike gate is used. This gate is basically a 2-input look-up-table. The value of the look up table can be configured based on the dopant polarity and hence their obfuscation gate can function as any 2-input function without changing any other layer than the dopant layer.

However, the authors also showed at CHES 2014 that detecting such a dopant-based obfuscation gate can be reverse engineered using a technique called passive voltage contrast using a SEM [4]. In their proposed reverse-engineering method, they use a technique from failure analysis called passive voltage contrast to successfully reverse-engineer such dopant based obfuscation gates. Hence, while currently proposed circuit level obfuscation techniques increase the costs of reverse-engineering, they cannot completely prevent it. At a higher level, the paper by Rajendran *et al.* looks at how to efficiently reverse-engineer designs efficiently in which only a few gates are being obfuscated using camouflage gates [6]. But all in all, there are relatively few scientific papers about layout-level hardware obfuscation which is quite surprising considering the attention it has gained in the industry.

In the following we will first introduce our obfuscation tool and will then present some overhead results for different designs. In Section V the importance of not only obfuscating the individual gates but also the connectivity between gates is highlighted at the example of the PRESENT block cipher. The paper is concluded with a discussion of open research problems.

III. Overview of Obfusgate

The basic tool-flow of the obfuscation tool is depicted in Figure 1. A Obfuscation Library is created in which each gate only differs in the dopant layer but is identical in all other layers (i.e. polysilicon, metal, contacts, wells and active area). This library can be used like any other standard cell library to synthesis any HDL code using standard tools such as Synopsys Design Compiler. After synthesis, an additional step called the "Wiring Step" is added to the standard design flow. The Wiring Step adds randomness to the design process, so that calling the obfuscation tool twice with the same HDL code does not result in the same obfuscated design. Besides adding

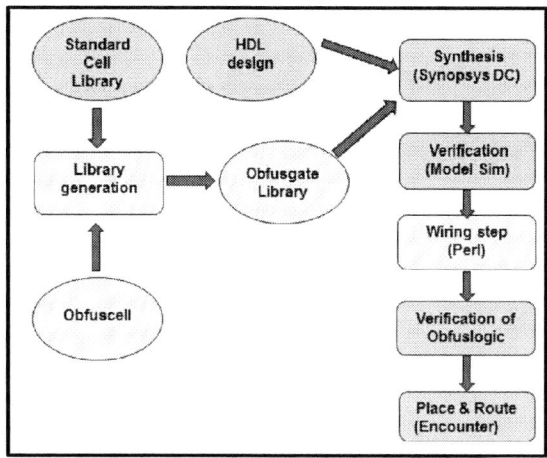

Figure 1. Overview of the obfuscation tool-flow. In a one-time process a Obfuscation Library is created based on the custom Obfuscell and the standard cell library. This is used as an input to the standard synthesize step. After synthesize, an additional step called the "wiring" step is used to introduce to make non-deterministic changes to the design and obfuscate the connectivity using dummy wires and inputs.

randomness to the design process, the main purpose of the Wiring Step is to obfuscate the connectivity of the design.

The obfuscation library is based on an obfuscation gate called "Obfusgate", which is depicted in Figure 2. It consists of five so-called "Obfuscells" and a 4-input NAND gate. Depending on the dopant polarity within the active area of the Obfuscell, the Obfuscell can have four different logic functions. The Obfuscell is either an inverter, a buffer, or it outputs a constant 1 or 0. The Obfuscell with different configurations is depicted in Figure 3 and 4. The design is similar to the dopant-level hardware Trojans from [8] and is based on the same idea as the obfuscation gates used in [5], [7]. Depending on how the Obfuscells are configured, the Obfusgate can have many different logic functions. In total, 162 different configurations, each with a unique logic behavior are possible with the Obfusgate as depicted in Figure 2. In particular, by setting one of the Obfuscells that are connected to an input of the Obfusgate to a constant '1', this input has effectively been turned into a "Dummy Input". That is, the value of the corresponding input does not influence the output of the Obfusgate. Therefore, any signal can be connected to such an input, creating "Dummy Wires". In practice, many Dummy Inputs are available in such a design, since 2-input gates are the most commonly used gates in a typical design. For example, in our case study of an AES SBox, more than half of the gates were two-input gates.

Our obfuscation tool uses these dummy inputs and

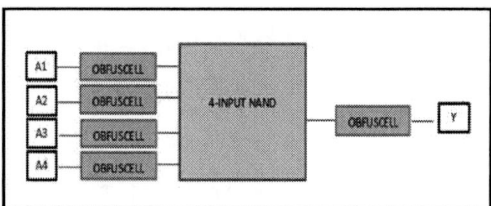

Figure 2. Schematic of a single Obfusgate that consists of 5 Obfuscells together with a 4-input NAND gate. Depending on the configuration of the Obfuscells the Obfusgate can have 162 different logic functions.

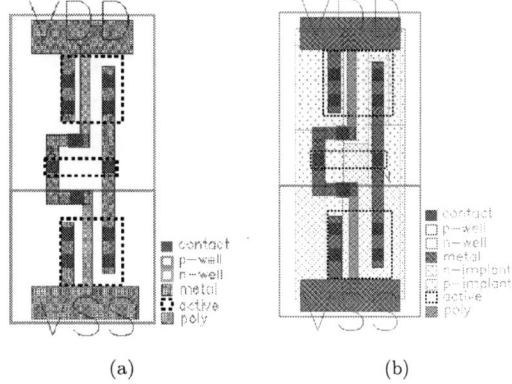

(a)   (b)

Figure 3. Layout view of the obfuscell. On the left one can see the basic structure of the obfuscell. The obfuscell has three active regions who's dopant polarity defines the logic function of the gate. On the right the dopant configuration is shown that will result in an "Always 1 gate".

dummy wires to connect otherwise unconnected gates and blocks. Due to the large amount of dummy wires, an attacker will not be able to determine the connectivity of the gates. This makes it much harder for an attacker to determine individual blocks in the design and therefore to identify the used architecture or structure. This aspect of the obfuscation tool will be discussed in more detail in Section V. One advantage of our approach is that it is very flexible and different gates can be used in the center of the Obfusgate. For example, using a 3-input NAND gate instead of the 4-input NAND gate can decrease the introduced area overhead of the Obfusgates at the cost of less obfuscation. The main advantage is that to do such trade-offs, no additional custom cell design is needed. Once the Obfuscell has been designed, it can be combined easily and automatically with any logic gate of a standard cell library. We used only one Obfusgate design based on a 4-input NAND gate in our implementation. However, if the designer accepts more than one look-alike gates (i.e. different obfusgate designs), additional obfuscation gates can be integrated into the library easily. For example, the designer could add an Obfusgate based on a 6-input And-Or-Inverter

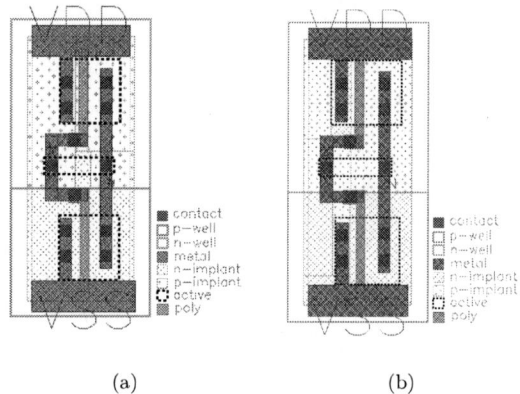

(a)   (b)

Figure 4. Layout view of the obfuscell when configure (a) as an inverter or (b) as a buffer in which the output is the same as the input.

gate (AOI222). In this case, the number of gates after the synthesis step can be reduced since additional gates are available to the synthesize tool. But this comes with the penalty that some information about the used gates is revealed. The fact that only a single custom made obfuscell is needed also makes porting the obfuscation tool to different technologies very simple and cheap. Furthermore, we have automated the generation of the library so that when a new library is created, only the middle gate needs to be defined. Everything else can be done by the tool.

IV. IMPLEMENTATION RESULTS

We implemented our design in 45nm technology using the Nangate Open Cell Library [9]. Synthesis was done using Synopsys Design Compiler and place and route was performed using Cadence Encounter and the custom cell design using Cadence Virtuoso. The wiring step and the library generation was performed using a Perl script and the design has been verified using Modelsim. In the first experiment we obfuscated an SBox of the Advanced Encryption Standard (AES) using our tool. The AES SBoxes is a function with 8-input bits that produces 8 output bits. We used a look-up table implementation for the AES SBox in HDL and used a design synthesized using the entire Nangate Open Cell Library as a reference implementation. Two different Obfusgates were tested, one based on a 4-input NAND and one based on a 2-input AND gate. For an Obfusgate based on a 4-input NAND gate, in total $(3)^4 \cdot (2) = 162$ different configurations for one gate are possible. For the 2-input AND gate there are only $3^2 * 2 = 18$ possible combinations. In both case this also includes the gates that output a constant '1' and constant '0' which we actually did not need in our design. We used Cadence Encounter for the place and route step and also used

978-1-4799-8720-7/15 $31.00 © 2015 IEEE   206

Encounter for an estimation of the power and delay overhead introduced by the obfuscation tool. Table II summarizes the overhead for the AES SBox. We also tested the tool for the round function of the lightweight block cipher PRESENT[10][1]. The results of this analysis is are also included in Table II. The number of combinations is simply the number of combinations per gate to the power of the used number of gates in the design. The area overhead for the 4-input NAND gate based design is an increase by a factor of 4.25 compared to the unobfuscated design for the PRESENT implementation while it is 7.09 for the AES S-Box design. For both the designs the overhead is smaller if Obfusgate based on a 2-input AND gate is used. However, please note that that in the 2-input AND gate design no dummy wires are used and hence the obfuscation is considerably weaker. It is also interesting to note that due to the large number of dummy wires, the area bottleneck for the AES SBox was actually not the gate area but the the routing overhead. That is, to be able to route the design, the gate density had to be reduced during the place and routing step from 0.7 to 0.5. Please note that in our current design any net can be chosen as a dummy input. While this is good for obfuscation reasons, this obviously adds considerable overhead when designs become larger. Hence, for larger designs, the selection of dummy inputs need to be adjusted to reduce the routing overhead.

Table I
OVERHEAD AND PERFORMANCE FOR DIFFERENT OBFUSCATION LIBRARIES FOR AES 8-BIT SBOX (421 GATES) AND THE ROUND FUNCTION OF PRESENT (383 GATES)

Design	AES SBox		Present Round	
Library	NAND4	AND2	NAND4	AND2
No. of gates	865	650	626	511
Number of combinations	2^{3096}	2^{4758}	2^{2241}	2^{3741}
Area increase	7.09x	6.03x	4.25x	3.2x
Power increase	6.45x	4.59x	5.65x	2.59x
Delay increase	3.12x	2.5x	2.51x	2.12x

Besides these two case studies based on cryptographic building blocks we also tested our tool using the ITC'99 benchmark circuits. The results of this analysis are presented in Table II. For the used benchmark circuits, the area increase was between a factor of 3.2x to 7.5x with an average of 5.87x. The results from the benchmark circuits suggest that the overhead increases for larger designs. In general, the introduced overhead is quite large. While the overhead can be reduced e.g. by using Obfusgates with less inputs or not obfuscating every gate in the design, this also comes with a decreased level of obfuscation. Unfortunately, obfuscation is very hard to express as a metric as the next example will show.

[1]To be more precise, we only implemented the Substitution and Permutation Layer without the key addition.

Table II
OVERHEAD FOR DIFFERENT ITC'99 BENCHMARK CIRCUITS WHEN USING OBNAND4 OBFUSCATION GATES IN COMPARISON TO USING AN UNOBFUSCATED STANDARD CELL LIBRARY (NANGATE OPENCELL LIBRARY)/

	Area in um^2 (OBNAND4)	Area in um^2 (standard cells)	Area increase
b01	324.74	77.532	4.18x
b02	177.66	48.272	3.6x
b03	1103.81	324.912	3.39x
b04	5639.94	1035.496	5.44x
b05	5319.13	895.67	5.93x
b06	316.42	101.47	3.11x
b07	3961.80	653.68	6.06x
b08	1035.34	273.27	3.7x
b09	1261.05	318.44	3.96x
b11	7056.84	890.76	7.7x

V. OBFUSCATING THE CONNECTIVITY

Compared to other approaches, e.g. in [7], [5] the NAND4 based Obfusgate will likely result in a larger area overhead. However, compared to these approaches, our approach adds considerable amount of obfuscation in terms of Dummy Wires. More than half of the obfuscation gates in the AES SBox and PRESENT round functions were 2-input gates. Each Obfusgate that is configured as a 2-input gate has two Dummy Inputs and hence also two Dummy Wires. In the case of our implementation of the Substitution and Permutation Layer of PRESENT, 941 dummy wires and 1103 normal wires are used. This very large amount of dummy wires effectively hide the connectivity and hence the structure of the design. To illustrate how important dummy wires are, we would like to use the round function of the block cipher PRESENT as an example [10]. The present round function is depicted in Figure 5. It consists of a key addition, a Substitution Layer and a Permutation Layer. In the key addition the 64 bit state is XORed with the 64 bit round key. The Substitution Layer consists of 16 identical 4-bit SBoxes. The Permutation Layer is a permutation that can be realized simply by wiring. If an obfuscation approach is used in which each gate is replaced by an obfuscated gate without dummy wires, the logic functionality of the design would be obfuscated. That is, an attacker would not know the individual gate functions. However, the connectivity of the gates are not hidden and hence the attacker can still see the inputs and outputs of each gate. He can use this information to build a graph, in which each node is a obfuscated gate with the input and outputs being the edges of the graph. From this graph, the attacker can derive the basic structure of the design. In particular, an attacker can derive the structure of the round function as depicted in Figure 6 by looking at this connectivity graph. The attacker could not derive the functionality of one of the SBoxes. However, from the connectivity the attacker would see each SBox as a set of gates that have 4 inputs and produce 4 outputs. Since a 4-bit SBox is a 4-input function with 4 outputs the attacker would know that the design uses 16 4-bit SBoxes. Similarly, the attacker would not know that the key addition is an XOR. But he could see that each bit of the state registers goes into one or more gates that only get a single input from the round key as an additional input. Hence, the attacker would know that the key addition is a 2-input function with one input being the state register and the other the round key. The Permutation Layer is not obfuscated at all, since it does not consist of combination logic but only wiring. Hence, it would be rather simple for an attacker to derive the basic structure of the used encryption function. From this information the attacker can make educated guesses to recover more functionality and derive the used algorithm. Furthermore, the basic structure is very helpful as a starting point for other reverse-engineering techniques such as side-channel based reverse engineering or reverse-engineering based on scan chain outputs.

The heavy used of Dummy Wires in our Obfusgate prevents the attacker from deriving such a structure. In the resulting obfuscated design, each output bit of each Sbox will depend on nearly all of the 64 state register. Therefore the attacker would not know that the design is based on 4-bit SBoxes, as he would just see one large graph with lots of edges.

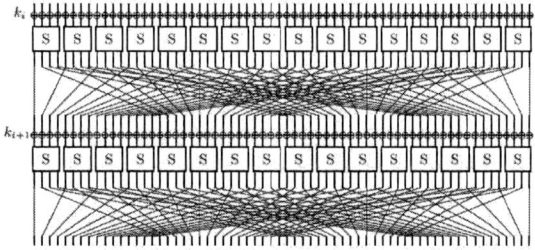

Figure 5. Figure of the PRESENT round function taken from [10]. It consists of a key addition layer, a Substitution Layer consisting of 16 identical 4-bit SBoxes and a Permutation Layer consisting of simple wiring.

VI. DISCUSSIONS AND OPEN RESEARCH PROBLEMS

In this paper we have introduced a new obfuscation tool flow that not only tries to hide the functionality of individual gates, but also their connectivity. One aim of this paper is to give a rough estimation of the area penalty of using obfuscation as a baseline for future reference. Unfortunately it is very hard to compare different designs in terms of overhead and obfuscation level. For example, one could reduce the overhead by using a obfuscation gate with less inputs and hence less

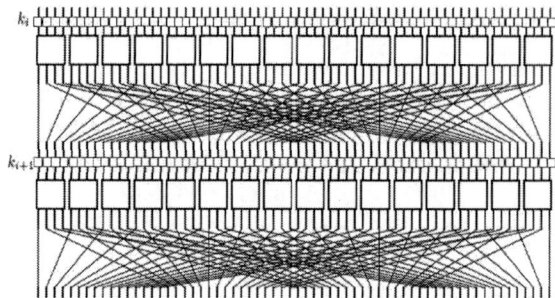

Figure 6. Illustration what an attacker could deduct from an obfuscated design that only obfuscate individual gates without introducing dummy wires. The main structure can still be derived from the connectivity graph, while only the individual functions are obfuscated.

dummy wires. However, this would reduce the amount of obfuscation of the connectivity and hence it would be easier to derive the structure of the obfuscated design. However, currently no good metric for this obfuscation is available. Simply counting the number of possible gate combinations to determine the brute-force complexity is not a suitable metric. For example, Table I we counted for each obfuscated design the possible configurations. If one wants to try to find the correct configuration using brute-force, even for the 2-input Obfusgate design this results in a ridiculous high number of 2^{3741} for the PRESENT round function. But as pointed out in the previous section, it would be fairly easy to derive the internal structure of this design if the 2-input obfusgate would be used, due to the missing dummy wires. Hence, such numbers are only misleading and do not tell much about the actual level of obfuscation. A meaningful metric of obfuscation is therefore a very difficult but also important open research problem.

Furthermore, in the current design the Dummy Wires are simply connected to any of the available nets. It seems that especially for larger designs different strategies for the connection of dummy wires need to be examined. Otherwise the area overhead would increase drastically. The area penalty of the dummy wires can be reduced greatly by taking the placing of the gates into account when choosing Dummy Inputs. But only taking Dummy Inputs that are close by would not efficiently hide the structure of the design since gates belonging to one part of the design are usually placed close to each other. Therefore one should also try to optimize the level of obfuscation by adopting a wiring mechanism that tries to optimize the connectivity graph. However, the first open research question is already how such an optimized connectivity graph should looks like. This is also of great importance when methods based on small 2-input obfuscation gates are used. In these 2-

978-1-4799-8720-7/15 $31.00 © 2015 IEEE

input obfuscation libraries dummy wires can be realized by introducing dummy gates into the design. The big question is then how many dummy gates are needed and how these dummy gates should be connected to the design to optimize the obfuscation of the connectivity. It seems that this is an entire research area on its own that heavily relies on graph theory.

How to construct obfuscation gates at the circuit level that are impossible or at least extremely hard to reverse-engineer using optical methods is another interesting line of research. The current state of the art in the academic literature, dopant polarity changes and dummy contacts, add to the reverse-engineering costs but are far from impossible for a well equipped attacker. Hence, new and alternative approaches are needed. Furthermore, relatively little is known of the costs of reverse-engineering such obfuscated designs. One reason for this is that very few public information is available of what hardware reverse-engineering companies are currently capable of. In that regard, the fault-testing community might be able to provide a lot of insight, since many techniques to optically detect faults can be used to reverse-engineer such obfuscated gates. This has also been pointed out by Sugawara *et al.* in the only academic paper discussing such a reverse-engineering attack we are aware of [4]. We would like to note that these different research directions can be explored independently from each other.

To summarize, while layout-level hardware obfuscation has been studied by the industry for 20 years, the academic community has only recently started to look into this research area. Therefore, there are still many open research problems, in both circuit level obfuscation as well as how to efficiently use these gates to maximize the level of obfuscation. Ideally, these open research questions do not only attract the attention of circuit designers but also of computer scientist and reverse-engineers.

REFERENCES

[1] "Innovation at risk: Intellectual property challenges and opportunities," white paper, Semiconductor Equipment and Materials International, June 2008.

[2] S. International, "Circuit camouflage technology - smi ip protection and anti-tamper technologies," White Paper Version 1.9.8j, March 2012.

[3] J. P. Baukus, L. W. Chow, and W. M. Clark, "Digital circuit with transistor geometry and channel stops providing camouflage against reverse engineering," Patent US 5 783 846, 06 21, 1998.

[4] T. Sugawara, D. Suzuki, R. Fujii, S. Tawa, R. Hori, M. Shiozaki, and T. Fujino, "Reversing stealthy dopant-level circuits," in *Cryptographic Hardware and Embedded Systems (CHES 2014)*, ser. LNCS. Springer, 2014, vol. 8731, pp. 112–126.

[5] R. P. Cocchi, J. P. Baukus, L. W. Chow, and B. J. Wang, "Circuit camouflage integration for hardware ip protection," in *Proceedings of the 51st Annual Design Automation Conference (DAC 14)*. New York, NY, USA: ACM, 2014, pp. 153:1–153:5.

[6] J. Rajendran, M. Sam, O. Sinanoglu, and R. Karri, "Security analysis of integrated circuit camouflaging," in *Proceedings of the 2013 ACM SIGSAC conference on Computer & communications security*. ACM, 2013, pp. 709–720.

[7] M. Shiozaki, R. Hori, and T. Fujino, "Diffusion programmable device : The device to prevent reverse engineering," *IACR Cryptology ePrint Archive*, vol. 2014, p. 109, 2014.

[8] G. T. Becker, F. Regazzoni, C. Paar, and W. P. Burleson, "Stealthy dopant-level hardware trojans," in *Cryptographic Hardware and Embedded Systems (CHES 2013)*, ser. LNCS. Springer, 2013.

[9] N. Inc., "Nangate open cell library, version pdkv1_3_v2010_12," http://www.si2.org/openeda.si2. org/projects/nangatelib, August 2011.

[10] A. Bogdanov, L. R. Knudsen, G. Leander, C. Paar, A. Poschmann, M. J. Robshaw, Y. Seurin, and C. Vikkelsoe, "Present: An ultra-lightweight block cipher," in *Proceedings of the 9th International Workshop on Cryptographic Hardware and Embedded Systems (CHES 07)*. Berlin, Heidelberg: Springer-Verlag, 2007, pp. 450–466.

2015 IEEE Computer Society Annual Symposium on VLSI

Reversible Denial-of-Service by Locking Gates Insertion for IP Cores Design Protection

Brice Colombier, Lilian Bossuet
Hubert Curien Laboratory, UMR CNRS 5516, University of Lyon
42000 Saint-Étienne - France
{b.colombier, lilian.bossuet}@univ-st-etienne.fr

David Hély
LCIS, Grenoble Institute of Technology
26000 Valence - France
david.hely@lcis.grenoble-inp.fr

Abstract—Nowadays, electronics systems design is a complex process. A design-and-reuse model has been adopted, and the vast majority of designers integrates third party intellectual property (IP) cores in their design in order to reduce time to market. Due to their immaterial form and high market value, IP cores are exposed to threats such as cloning and illegal copying. In order to fight these threats, we propose to achieve functional locking, equivalent to a triggerable and reversible denial-of-service. This is done by inserting locking gates at specific locations in the netlist, allowing to force outputs at a fixed value. We developed a new method based on graph exploration techniques for locking gates insertion. It selects candidate nodes ten thousand times faster than state-of-the-art fault analysis-based logic masking techniques. Methods are then compared on ISCAS'85 combinational benchmarks.

Keywords—Intellectual property protection, logic masking, functional locking, graph analysis.

I. INTRODUCTION

Electronics system design has become a complex and demanding task. Due to a shorter time to market, a design-and-reuse paradigm has been widely adopted. It allows system integrators to design electronics systems faster, using functional building blocks. Companies have specialized in providing these pieces of design, known as intellectual property (IP) cores. The question of a fair IP cores distribution system, however, remains open. Indeed, an IP core designer must disclose the design entirely to the system integrator. Since IP cores can be extremely valuable and are provided as data file, they are a prime target for people with malicious intent. Multiple cases of counterfeiting and non-contracted overbuilding have been reported in recent years, and the trend is growing [1], [2], [3]. Facing such threats, the development of an efficient design data protection scheme is necessary.

An increasingly studied solution is to make the system unusable unless it has been unlocked beforehand. The activation procedure is initiated by the designer. In case the design has been obtained illegally, it remains locked and unusable, until the activation procedure is carried on. To ensure appropriate security, the activation requires a key. This secret key is delivered by the designer, who can thus precisely control how many instances of the design have been instantiated. In order to lock the design, a solution is to insert extra logic gates on specific nodes, that will modify them if the wrong values are applied to the gates' inputs. For instance, XOR/XNOR gates can be used to invert the value of the nodes if the wrong key is applied [4]. In this first article, however, gates are placed

randomly. A better placement algorithm, based on fault analysis [5], allows the designer to select more suitable nodes to lock, in order to have a stronger locking power with lower area overhead.

Instead of using XOR or XNOR gates to cause disturbances to the circuit, we propose to achieve a reversible denial-of-service by forcing the outputs to a fixed value if the wrong key is applied. For this purpose, (N)AND and (N)OR gates are used instead, and inserted as deep as possible in the netlist, yet achieving total functional locking. These logic gates are used to force specific nodes to logical 0 or 1. We developed an algorithm, based on graph exploration, to select the best nodes in a netlist for locking gates insertion. Our method enables us to select optimal candidate nodes ten thousand times faster than the fault analysis-based technique. Moreover, since it conducts an exact analysis and is not based on any simulation, the optimal subset of nodes to modify is reached, and the analysis does not rely on any external software.

The remainder of this paper is organized as follows. Section II provides background of active protection schemes based on extra gates insertion. It gives formal definitions for logic masking and functional locking, that are the two protection methods investigated here. Section III presents the new insertion method used to select the nodes of a netlist for functional locking. A comparison is presented in Section IV. It aims at showing the differences between our insertion method and the state-of-the-art. Section V discusses design challenges that must be addressed during protection schemes development. Finally, the paper is concluded in Section VI.

II. RELATED WORK

In 2007, the first paper describing an active protection scheme [6] proposes to add extra dummy states before the original start state of the IP core's finite state machine (FSM). A key is required to allow transition through the extra states. Therefore, access to the original FSM was controlled. This way, access to the normal behaviour of the system depends on a secret key, that is only available from the designer. This solution, however, is cryptographically weak, and requires the original design to contain an FSM. Another option is to act on the combinational part of the circuit, and add XOR/XNOR gates at specific locations. Similarly, applying the wrong key on these gates inverts the associated value, and therefore the behaviour of the circuit is altered. This is the principle of logic masking.

978-1-4799-8720-7/15 $31.00 © 2015 IEEE

Logic masking: XOR or XNOR gates are inserted on the data path in order to change the logical behaviour of the circuit if the wrong key is applied to these additional gates. In this case, some internal nodes have their value inverted. Ideally, the output obtained when a wrong key is applied is not correlated with the original output.

This was first proposed in [4], in which the authors place logic masking gates randomly. Using the random placement method, the correlation between the normal and masked output reduces slowly with respect to the number of masking gates inserted, and the area overhead to obtain low correlation is not negligible. A better placement technique was then necessary. In [5], the authors elaborate on this idea, and propose another method to place the masking gates. The point is to place these gates more efficiently, in order to reduce correlation faster and thus limit the area overhead for an identical level of disturbance. They develop a placement method based on fault analysis, called logic encryption. In order to determine on which nodes of a design XOR or XNOR gates should be placed, they compute a metric, named Fault Impact, for each node. This metric takes into account the number of patterns that detect a s-a-0 and s-a-1 fault for this node, and the total number of output bits that get affected by these faults. Then, the node for which the Fault Impact is maximal is selected. An XOR or XNOR gate is added on it, according to a user defined key, and the process is carried out again until the number of output bits that are affected by a fault reaches 50% on average. This criterion is fulfilled when the Hamming distance between correct and masked output vectors is close to 50%. Once this value is obtained, the logic is considered as "encrypted", according to [5] (Section V-C will discuss it in more details). This method can not be referred to as logic encryption though, due to the lack of security proofs and the absence of ciphering and deciphering steps. Instead, this protection technique should be classified as logic masking. What we propose here is to use another method to alter the outputs of the system, called functional locking:

Functional locking: a triggerable and reversible denial-of-service. The circuit can be rendered useless, and then come back to its normal behaviour. This is achieved by forcing the outputs of the netlist to logical 0 or 1 after inserting AND, NAND, OR or NOR locking gates in the netlist.

In both cases, a method is required to select the appropriate nodes to act on, i.e. the ones on which an XOR or XNOR will be placed for masking, or a locking gate for functional locking. In the next section, we present the placement method we developed. It detects sequence of gates in the netlist that can propagate a locking value to one or more outputs.

III. PROPOSED GATES-INSERTION METHOD FOR FUNCTIONAL LOCKING: GRAPH EXPLORATION FOR GATES SEQUENCES

A. Principle

In order to force the output of a logic gate to logical 0 or 1, a specific value must be present on one of its inputs. For instance, the output of a NAND gate is forced to 1 if one input is set to 0. Similarly, the output of an OR gate is forced to 1 if one input is set to 1. Therefore, it follows that all logic gates are able to propagate a locking value if one of their input is set to the right value.

AND: Set one input to **0** forces the output to **0**
NAND: Set one input to **0** forces the output to **1**
OR: Set one input to **1** forces the output to **1**
NOR: Set one input to **1** forces the output to **0**

The AND and NAND gates can propagate a locking value if one of their inputs is set to 0. The OR and NOR gates can propagate a locking value if one of their inputs is set to 1. Hence, it is the value on one of the inputs that determines if a gate propagates the locking value. This value only depends on the preceding gate. An example showing how a sequence of gates (AND → NAND → OR) can propagate a locking value is shown in Fig. 1:

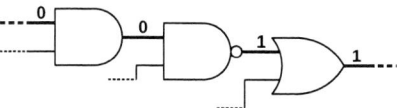

Fig. 1. Propagation of a locking value in a sequence of gates

In order to lock the outputs of a netlist, we must identify such sequences of gates that lead to an output. For that, we compute the two following features for all nodes in the netlist:

- V_{forced}: this is the value at which the node will be forced at. It depends on which type of logic gate precedes the node. For example, if the node is the output of an OR gate, then $V_{forced} = 1$.

- V_{locks}: this is the value at which the node should be forced to propagate the locking It depends on which type of logic gate succeeds to the node. For instance, if the node is the input of a NAND gate, then $V_{locks} = 0$. It should be noted that if a node has a fan-out higher than 1 and spans NAND and OR gates for example, then $V_{locks} = \{0, 1\}$.

A summary of the V_{forced} and V_{locks} values for a node G depending on the type of the preceding and following gates is given in Table I.

Preceding gate	$V_{forced}(G)$		Following gate	$V_{locks}(G)$
AND	0		AND	0
NAND	1		NAND	0
OR	1		OR	1
NOR	0		NOR	1

TABLE I. V_{forced} AND V_{locks} VALUES FOR A NODE G, DEPENDING ON THE PRECEDING AND FOLLOWING LOGIC GATE

It follows that, for sequences that propagate a locking value, Criterion 1 is fulfilled for all the nodes in the sequence.

$$\textbf{Criterion: } V_{forced} \in V_{locks} \qquad (1)$$

To represent the netlist more easily and identify interesting sequences of gates, we use graph exploration techniques. They are presented in the following subsection.

978-1-4799-8720-7/15 $31.00 © 2015 IEEE

B. Graph building

First, a directed acyclic graph is generated from the netlist. Logic gates are represented as follows: inputs and outputs are vertices connected by edges labelled after the logic gate type. This is illustrated in Fig. 2.

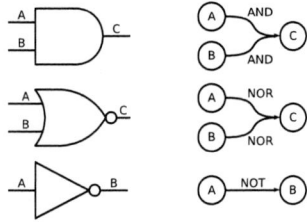

Fig. 2.   Logic gates conversion to vertices and edges

This process is carried out on the original netlist. An example of this conversion is shown in Fig. 3.

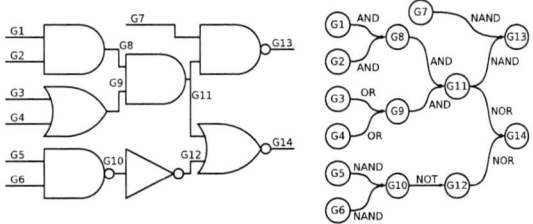

Fig. 3.   Sample from a netlist and its equivalent graph

In order to identify interesting sequences in the netlist, we use the previously built graph and compute V_{forced} and V_{locks} for each vertex. Here are the values of V_{forced} and V_{locks} for all the nodes of the netlist in Fig. 3:

Node	V_{forced}	V_{locks}	Node	V_{forced}	V_{locks}
G1	—	0	G8	0	0
G2	—	0	G9	1	0
G3	—	1	G10	1	0
G4	—	1	G11	0	{0, 1}
G5	—	0	G12	0	1
G6	—	0	G13	1	—
G7	—	0	G14	0	—

TABLE II.   EXAMPLES OF V_{forced} AND V_{locks}

According to these values, nodes that do not belong to locking sequences are deleted. Inputs and output nodes are retained. Then nodes that do not fulfil Criterion 1, i.e. $V_{forced} \notin V_{locks}$ are removed from the graph, except if one of their direct successors fulfils Criterion 1 or is an output. Incoming edges should be removed in that case though, because the node cannot propagate the locking value. The outcome is a graph in which all connected vertices can propagate a locking value. This graph, however, is usually disconnected. Connected components that do not include an output of the netlist are removed, since they are not useful for functional locking.

In the previous example, G1 to G6, G13 and G14 are retained since they are inputs or outputs. G9, G10 and G12 do not satisfy Criterion 1, but G9 is retained since its direct successor,

G11, fulfils Criterion 11. G9 incoming edges are deleted though. G12 is retained too since its successor, G14, is an output. G3, G4, G5 and G6 are then in a connected component with no outputs, and are therefore deleted. Eventually we obtain the graph shown in Fig. 4. A sequence of gates that can propagate a locking value is also highlighted.

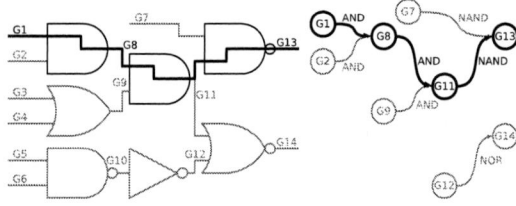

Fig. 4.   Sample from a netlist and its equivalent graph after analysis. One locking sequence is highlighted

C. Graph analysis for optimal locking nodes selection

The point is now to select the nodes to lock. This selection process is independent of the Boolean function. It will be taken into account later, when locking gates are inserted (Section III-D). The three different types of connected components that compose the final graph are presented in Fig. 5.

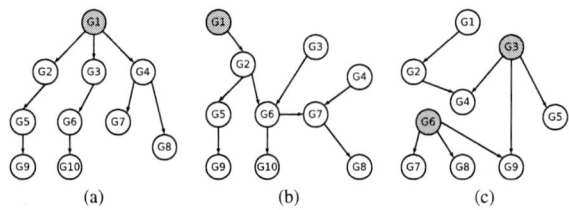

Fig. 5.   The three different types of connected components. The optimal vertex to select for functional locking is shown in grey

The case shown in Fig. 5(a), where there is only one source vertex, is rare. A source vertex has an empty in-neighbourhood. Selecting the locking node here is trivial: since the graph is directed and acyclic, only one vertex has all the others as successors. Forcing this node will lock all the outputs of the connected component. In Fig. 5(a), forcing G1 will lock the outputs G7, G8, G9 and G10. The second situation, shown in Fig. 5(b) is more common: multiple source vertices span multiple outputs. Only one of them, however, spans all the outputs of the connected component. Therefore, the corresponding node is selected. If several vertices span all the outputs, then the farthest vertex from all the outputs is the best candidate. It is the case in Fig. 5(b) where G1 and G2 span all the outputs. G1, however, is the farthest vertex from the outputs (G8, G9 and G10) and is preferred to G2. The last case, depicted in Fig. 5(c) requires more computation. Multiple vertices span multiple outputs, but none of them spans them all. The best nodes to select are the ones that span the greatest number of outputs. The number of outputs among each source vertex successors is computed, and source vertices are sorted accordingly. Then they are greedily selected until all the outputs are locked. In the graph shown in Fig. 4, the nodes to lock are G1 and G12. The complete procedure for graph building and analysis is summarized in Algorithm 1.

978-1-4799-8720-7/15 $31.00 © 2015 IEEE   212

Algorithm 1: Optimal nodes selection by graph analysis

Input: Netlist *file*
Output: List of nodes to lock *list_lock*

```
// Build graph
for line in file do
    Add vertex to graph for every new node in line
    Add edges to graph for Boolean function in line
    if "OUTPUT" in line then
        Add vertex to list_outputs

for vertex in graph do
    Compute V_forced(vertex) and V_locks(vertex)

// Select vertices
for vertex in graph do
    delete ← False
    if vertex is neither an input nor an output then
        if V_forced(vertex) ∉ V_locks(vertex) then
            delete ← True
            for succ in successors(vertex) do
                if V_forced(succ) ∈ V_locks(succ) then
                    delete ← False
                    Remove vertex incoming edges

    if delete = True then
        Remove vertex from graph

// Identify connected components
list_cc ← Clustering(graph)

// Analyse connected components
for cc in list_cc do
    if cc contains no output then
        Remove cc from graph
    else
        Identify source_vertices in cc
        if #source_vertices = 1 then
            add source to list_lock
        else
            Identify outputs in cc
            for source in source_vertices do
                Compute distance from source to outputs
            if one source spans all outputs then
                add source to list_lock
            else if multiple source spans all outputs then
                add farthest source to list_lock
            else
                while some outputs are unlocked do
                    add the source that spans the greatest
                    ↪ number of outputs to list_lock
```

Return: *list_lock*

D. Locking gates insertion

We now have a list of nodes to lock, and their respective V_{locks} value. If $V_{locks} = 0$, an AND or a NOR gate is inserted. If $V_{locks} = 1$, an OR or a NAND gate is inserted. In the previous example, we insert a NOR gate on G1 to force it at 0, and an OR gate on G12 to force it at 1. Two locking inputs are added: K1 and K2. The modified netlist is shown in Fig. 6.

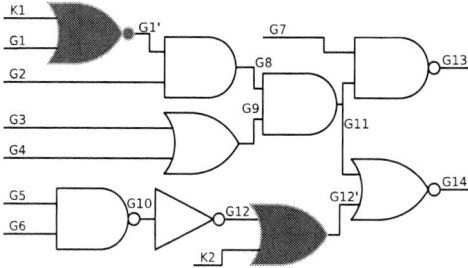

Fig. 6.   Modified sample from a netlist with locking gates inserted

In case $V_{locks} = \{0, 1\}$, then the node has a fan-out superior to 1, and two locking gates must be inserted. One AND or NOR gate is inserted to lock part of the fan-out to 0. One NAND or OR gate is inserted to lock part of the fan-out to 1.

IV. EXPERIMENTAL RESULTS

A. Real case

The implementation of Algorithm 1 has been done in Python. The *igraph* extension is used for graph analysis. We validated our method on ISCAS'85 benchmarks, and typically obtained the type of connected component shown in Fig. 7. Forcing G27 (light grey vertex) to 0 locks the seven outputs (white vertices) contained in this connected component: G809, G656, G820, G636, G845, G704 and G717. For instance, forcing G27 to 0 will force G3176 to 0, that will force G635 to 0, that will force G636 to 1.

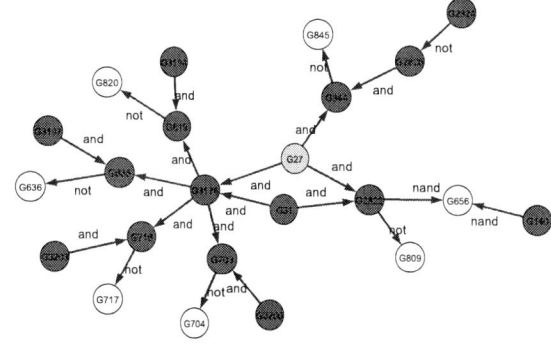

Fig. 7.   A connected component from the graph obtained after c5315 netlist analysis

B. Security analysis

Correlation is used to evaluate the security of the protection schemes. Let S be the output of the system. $S_{protected}$ is the output when the protection scheme is active. Correlation is computed using Pearson's correlation coefficient:

$$\rho_{S,S_{protected}} = \frac{cov(S, S_{protected})}{\sigma_S \sigma_{S_{protected}}}$$

When considering functional locking though, since the outputs are stuck at a fixed value, $\sigma_{S_{protected}} = 0$. Thus Pearson's correlation coefficient cannot be computed. Nevertheless, since

978-1-4799-8720-7/15 $31.00 © 2015 IEEE   213

the output is fixed, it provides no information about the underlying locking scheme. This is why functional locking can be analysed as if the correlation had a zero value.

	Key size	Random [4]	Fault analysis [5]	Graph analysis
c432	32 bits	0.272	0.012	0
7 outputs	64 bits	0.153	0.019	0
189 nodes	128 bits	0.026	0.014	0
c5315	32 bits	0.902	0.554	0
123 outputs	64 bits	0.873	0.357	0
2362 nodes	128 bits	0.820	0.277	0
c7552	32 bits	0.952	0.254	0
108 outputs	64 bits	0.920	0.235	0
3612 nodes	128 bits	0.761	0.217	0

TABLE III.   CORRELATION OBTAINED FOR THREE BENCHMARKS AND DIFFERENT PLACEMENT TECHNIQUES WITH RESPECT TO THE KEY SIZE

As expected, increasing the key size reduces correlation and makes the protection more effective (see Table III). The main drawback that is revealed here is that the efficiency of logic masking drops drastically for large circuits. For instance, using a 128-bit key instead of a 32-bit one only reduces correlation from 0.254 to 0.217 using fault analysis on c7552, that comprises 3612 nodes. It shows that for large netlists, correlation cannot reach optimal values, at least not with keys of reasonable length. It follows that the security can not be proven for such circuits. On the contrary, by identifying gates sequences in the netlist and applying functional locking, it is proven that outputs are locked by forcing the candidate nodes to a specific value. Consequently, since security requirements are so hard to fulfil with logic masking, one should use functional locking instead, and rely on a separate cryptographic function to establish provable security, as it has been proposed in [5].

C. Area overhead

The second criterion used to evaluate the protection schemes is the area overhead, since we want it to be as low as possible. For a fair comparison, ISCAS'85 combinational benchmarks have been used as references. We give the amount of extra gates. Results are depicted in Fig. 8. Graph exploration-based functional locking requires slightly more than 3% extra gates on average to achieve total functional locking of the netlist. For small circuits such as c432, comprising 189 nodes, functional locking requires 3.27% extra gates, when logic masking requires three times more gates : +10,46%. For large circuits, the overheads come closer and are similar for logic masking (+5% for c5315, +1.62% for c7552) or functional locking (+3.57% for c5315, +2.70% for c7552). Unlike logic masking, however, that is only partially efficient for large circuits, functional locking ensures total locking.

D. Computation time

Computation times are given here for the placement method described in [5] and for graph analysis. The workstation we used embeds an Intel Core i5-4570 operating at 3.20GHz and a 16Gb RAM. A log plot of the computation times for all the benchmarks is given in Fig. 9. As we can see, the proposed graph exploration method is ten thousand times faster than fault analysis on average. This is a valuable result. Indeed, those protection schemes could then be integrated EDA tools, in order

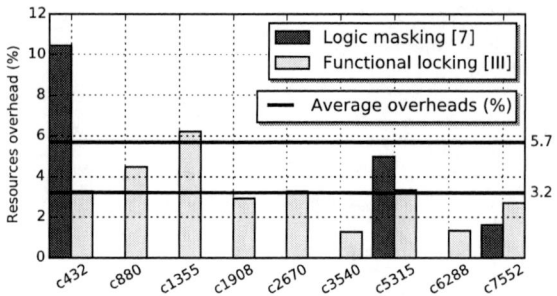

Fig. 8.   Area overhead for the two protection schemes

for the designer to be able to add protection to its design on the fly. Even for the bigger netlist, c7552, that contains more than 3600 nodes, the graph exploration method takes only 2.5s to identify gates sequences and return a list of candidate nodes for functional locking. Conversely, fault analysis takes several hours on the same netlist.

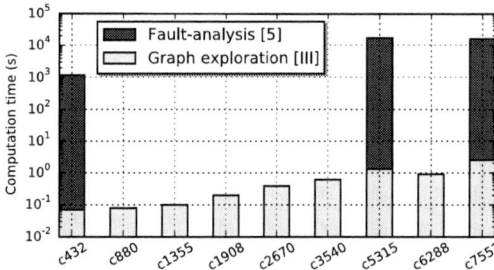

Fig. 9.   Computation time required to analyse the original netlist for the two techniques

The main advantage of the proposed methods is to not depend on any external software to conduct the analysis. Only Python scripts were used to analyse the netlists, providing optimal locking gates placement and generating the modified, lockable netlist. It considerably speeds up the analysis in comparison with fault analysis-based method, that employs fault analysis and simulation tools.

V.   DISCUSSION

A. Exact or simulation-based placement methods

The placement technique proposed in [5] selects the best candidates with the help of a fault simulator. It evaluates the Fault Impact after applying one thousand random input patterns to the netlists. It requires circuit simulation, however, to derive the nodes to modify for logic masking. Therefore, this method is sub-optimal, since it is based on a partial exploration of the design space. In comparison, our technique based on graph exploration for gates sequences is optimal. Consequently it always provides the optimal subset of nodes that should be locked.

B. Security margin for functional locking

In III, we only gave details about how to select as few nodes as possible in order to lock all the outputs of a design. In the

978-1-4799-8720-7/15 $31.00 © 2015 IEEE

context of real-life designs protection, however, the designer may want to add extra locking gates on the locking propagation path. This ensures that the locking value is set by multiple gates, and is harder to cancel. Actually, all the vertices that are present in the final subgraphs can be modified to lock outputs. Therefore, the designer can choose precisely the number of gates to be involved in the locking process. For example, as shown in Fig. 7, G27 can lock all the outputs. All the other nodes (dark grey vertices) are candidates too, although they do not lock as many outputs as G27. They can be used to add redundancy to the locking scheme, and provide better security through stronger locking. As it is the case for all the proposed protection schemes, the final implantation is a trade-off between security and area, power and timing overhead. The minimum and maximum number of locking gates that can be selected for each netlist is shown in Fig. 10.

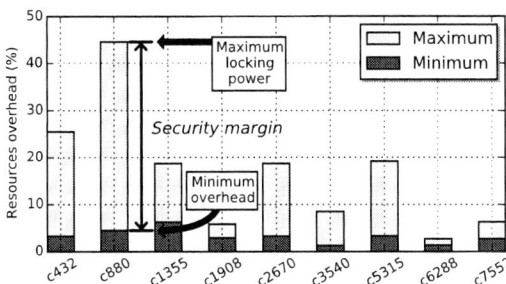

Fig. 10. Resources used to protect the netlist by functional locking using the graph analysis method

C. On the necessary separation between the security primitive and the locking/masking module

The design flow of most of the protection schemes found in the literature is the following: after designing a strong and efficient locking scheme, the authors do their best to highlight cryptographic properties of their system. For instance in [5], the authors look for a 50% Hamming distance between the original and masked output vectors. There are some circuits, however, for which this criterion can be satisfied, that bring absolutely no security, cryptographically speaking. Therefore, we believe that instead of trying to develop cryptographic and locking functions out of the same module, those two essential parts of the protection scheme must be explicitly separated. This separation is only suggested in [5]. On the one hand, a provably secure cryptographic primitive should be selected to guarantee safe access to the protection scheme. Examples of lightweight ciphers can be easily found, and some of them have been extensively studied, such as PRESENT [7]. Moreover, the advantage of using a dedicated cipher is that the key can be different for all instances of a given design, and can be derived from a PUF for example [8]. This is an important security requirement for design data protection, since unlocking one device should not be of any help in unlocking others. On the other hand, a locking scheme with a small footprint is used. Its only purpose is to make the system useless when the wrong key is applied to the cryptographic function, nothing more. This configuration allows the designer to choose the security and locking functions independently. A crucial point here is how these two primitives will be combined to form the complete protection scheme. A one-bit signal triggering the activation of the locking scheme is not suitable. It could compromise the security of the whole scheme, since bit flipping is achievable using a laser for example [9]. A multiple-bit signal should be used instead, to avoid this issue. The description of a complete protection scheme, however, is beyond the scope of this paper.

VI. CONCLUSION

This paper discusses a new approach to select candidate nodes of a netlist for functional locking. The proposed method has proven to be much more computationally effective than the state-of-the-art, yet achieving similar area overhead. In addition, it conducts an exact analysis using clearly defined parameters, instead of using incomplete simulation. Next, the necessity of a strict distinction between security and locking functions was highlighted, and should be taken into account in future protection schemes design. Thus paired with a secure yet lightweight cryptographic function, functional locking techniques form a strong design data protection scheme, that allows the designer to keep control on its design after it has been sold.

ACKNOWLEDGEMENT

The work presented in this paper was realized in the frame of the SALWARE project number ANR-13-JS03-0003 supported by the French "Agence Nationale de la Recherche" and by the French "Fondation de Recherche pour l'Aéronautique et l'Espace", funding for this project was also provided by a grant from "La Région Rhône-Alpes".

The authors would also like to thank Jeyavijayan Rajendran [5] for providing his results and netlists for comparison.

REFERENCES

[1] Frontier-Economics, "Estimating the global economic and social impacts of counterfeiting and piracy," Business Action to Stop Counterfeiting and Piracy (BASCAP), Tech. Rep., 2011.

[2] IHS Technology. (2012, April) Top 5 most counterfeited parts represent a $169 billion potential challenge for global semiconductor market. IHS Technology.

[3] U. Guin, K. Huang, D. DiMase, J. M. Carulli, M. Tehranipoor, and Y. Makris, "Counterfeit integrated circuits: A rising threat in the global semiconductor supply chain," Proceedings of the IEEE, vol. 102, no. 8, pp. 1207–1228, 2014.

[4] J. A. Roy, F. Koushanfar, and I. Markov, "EPIC: Ending piracy of integrated circuits," in Design, Automation and Test in Europe, 2008, pp. 1069–1074.

[5] J. Rajendran, H. Zhang, G. S. Rose, Y. Pino, O. Sinanoglu, and R. Karri, "Fault analysis-based logic encryption," IEEE Transactions on Computers, October 2013.

[6] Y. Alkabani and F. Koushanfar, "Active hardware metering for intellectual property protection and security," in USENIX Security, Boston MA, USA, August 2007, pp. 291–306.

[7] A. Bogdanov, L. R. Knudsen, G. Leander, C. Paar, A. Poschmann, M. J. B. Robshaw, Y. Seurin, and C. Vikkelsoe, "PRESENT: an ultra-lightweight block cipher," in International Workshop on Cryptographic Hardware and Embedded Systems, Vienna, Austria, September 2007, pp. 450–466.

[8] L. Bossuet, X. Ngo, Z. Cherif, and V. Fischer, "A PUF based on transient effect ring oscillator and insensitive to locking phenomenon," IEEE Transaction on Emerging Topics in Computing, vol. 2, no. 1, pp. 30–36, 2013.

[9] S. P. Skorobogatov and R. J. Anderson, "Optical fault induction attacks," in International Workshop on Cryptographic Hardware and Embedded Systems, San Fransisco CA, USA, August 2002.

978-1-4799-8720-7/15 $31.00 © 2015 IEEE

2015 IEEE Computer Society Annual Symposium on VLSI

Identification of IP Control Units by State Encoding

Edward Jung
Department of Computer Science
Kennesaw State University
Marietta, GA, USA
ejung4@kennesaw.edu

Seonho Choi
Department of Computer Science
Bowie State University
Bowie, MD, USA
schoi@bowiestate.edu

Abstract—**Synchronous finite state machines (FSMs) are the backbone of an embedded controller design. Partially due to the reuse of design in System on Chip (SOC), there has been the growing need for protecting hardware design intellectual property (IP). We propose non-destructive embedding method for protecting the design IPs of an embedded controller modeled as a synchronous FSM. Unlike many other embedding schemes which require adding non-functional states and/or state transitions, the proposed method does not require such additions. The central idea is based on embedding a designer's watermark at a FSM-level, as opposed to a state level. The embedding process is done by a hierarchical state encoding method. The proposed method is favorably evaluated in most of the criteria used in the field of hardware IP protection.**

Keywords—IP protection; fsm watermarks; state encoding

I. INTRODUCTION

Hardware design reuse has been the viable solution to deal with the ever-increasing logic density in the semiconductor industry. Hardware design intellectual property (IP) is a design unit that can be viewed as an independent subcomponent of a complete design (e.g., SOC design.) Examples of design unit include abstract algorithm, technique, or methodology that can make the design better as well as physical design blocks such as embedded control units. This can result in new products to be on the market in time and at a cost-effective way. The newly developed subcomponents can also be tested and deposited as new design IPs in the IP library for future reuse. Despite the attractiveness of reuse-based design, IP owners and vendors have encountered IP piracy and infringement. In this paper, we address the problem of protecting embedded hardware designs modeled as a synchronous finite state machine (FSM) and present a new type of FSM IP watermarking solution.

Most of the embedded hardware IP protection solutions [4] [7][8][10][11][12] have been developed by hiding secret information that is explicitly added to the circuit in order to prevent illegal use of the IC. Another characteristic of the previous work is to use secret information at a state level. For instance, a designer's secret information (i.e., a watermark) can be defined in terms of a set of visited states (usually non-functional states), which are traveled upon by applying specific input signals. In an analogy, this type of approach would be similar to adding tattoos (either big or small) on the external surface of the human skin to hide the designer's secret information. Then, the tattoos are verified at a later stage in order to prove the ownership of the design IP. For the survey of FSM

IP watermarking, refer to [1]. For the recent work on IP protection, refer to [2].

We challenge this traditional design philosophy by raising the question: *Would it be feasible to embed a watermark at a FSM level without adding non-functional entities (e.g., states, state transitions)?* We show that, under a certain condition, this is feasible. We analyze the necessary and sufficient condition based on the theory of machine decomposition [5][6].

In general, the task of watermarking consists of two processes: (1) embedding and (2) verification. The uniqueness of the proposed method lies in non-destructiveness. Here, *non-destructiveness* is meant to satisfy the following two conditions: (1) neither new (i.e., redundant) states nor additional state transitions are added to construct any form of watermark in the embedding process, and (2) no extra circuitry within the FSM system is required to check internal states in the verification process. For the description and proposed solutions for the verification process, refer to [9]. In this paper, we shall focus on the embedding method. The underlying idea is based on a FSM-level watermark using the state encoding scheme.

This paper is organized as follows. Preliminary information is given in Section II. Section III describes the embedding method. Analysis is done in Section IV. Finally, we provide the conclusions and future work in Section V.

II. PRELIMINARY

A. Partitions on States

For convenience, the definitions are summarized below [5].

Definition 1: A *synchronous finite state machine* (or *sequential machine*) is a quintuple $M = (S, I, O, \delta, \lambda)$ where (i) S is finite nonempty set of states; (ii) I is a finite nonempty set of inputs; (iii) O is a finite nonempty set of outputs; (iv) $\delta: S \times I \to S$ is the transition function; (v) $\lambda: S \times I \to O$ (*Mealy type*), $\lambda: S \to O$ (*Moore type*).

A Mealy-type machine can be converted into a Moore-type machine, and vice versa. Unless it is needed to distinguish one type from the other, Mealy-type machine is assumed in the paper.

Definition 2: A *state machine* is a triplet $M = (S, I, \delta)$ where (i) S is a finite nonempty set of states; (ii) I is a finite nonempty set of inputs; (iii) $\delta: S \times I \to S$ is a transition function.

Definition 3: A *partition* φ on S of the machine $M = (S, I, O, \delta, \lambda)$ is a collection of disjoint subsets of S whose set

This material is based upon work supported by, or in part by, the U. S. Army Research Laboratory and the U. S. Army Research Office under contract/grant number W911NF1210060 and W911NF1310143.

978-1-4799-8720-7/15 $31.00 © 2015 IEEE   216

union is S. That is, $\varphi = \{B_\alpha\}$ such that $B_\alpha \cap B_\beta = \emptyset$ for $\alpha \neq \beta$ and $\cup\{B_\alpha\} = S$.

We refer to the sets of φ as *blocks* of φ and designate the block containing s by $B_\varphi(s)$. In writing out a partition, we distinguish blocks with bars and semicolons.

Example 1: If $S = \{1,2,3,4,5,6,7,8\}$ and partition φ on S has blocks $\{1,3,4,5\}$, $\{2,6\}$, and $\{7,8\}$, then we write $\varphi = \{\overline{1,3,4,5}; \overline{2,6}; \overline{7,8}\}$.

Note that φ is a set and has elements like any other set. For φ_1 and φ_2 on S, we say that φ_2 is *larger than or equal to* φ_1, and write $\varphi_1 \leq \varphi_2$, if and only if every block of φ_1 is contained in a block of φ_2.

Example 2: If $\varphi_1 = \{\overline{1,2}; \overline{3,4}; \overline{5,6}; \overline{7,8}\}$ and $\varphi_2 = \{\overline{1,2,3,4}; \overline{5,6,7,8}\}$, then $\varphi_1 \leq \varphi_2$.

Definition 4: Two partitions φ_1 and φ_2 are *equal*, $\varphi_1 = \varphi_2$, if and only if $\varphi_1 \leq \varphi_2$ and $\varphi_2 \leq \varphi_1$.

We write $s \equiv t(\varphi)$ if and only if s and t are contained in the same block of φ. That is, $s \equiv t(\varphi)$ if and only if $B_\varphi(s) = B_\varphi(t)$.

Definition 5: A partition φ on the set of states of $M = (S, I, O, \delta, \lambda)$ is a *closed* partition if and only if $s \equiv t(\varphi)$ implies that $\delta(s, a) \equiv \delta(t, a)(\varphi)$ for all a in I.

Now, we can define a "multiplication" operation (denoted by $\cdot$) on partitions on a set.

Definition 6: If φ_1 and φ_2 are partitions on S, then $\varphi_1 \cdot \varphi_2$ is the partition on S such that $s \equiv t(\varphi_1 \cdot \varphi_2)$ if and only if $s \equiv t(\varphi_1)$ and $s \equiv t(\varphi_2)$.

Example 3: If $\varphi_1 = \{\overline{1,2}; \overline{3,4}; \overline{5,6}; \overline{7,8,9}\}$ and $\varphi_2 = \{\overline{1,6}; \overline{2,3}; \overline{4,5}; \overline{7,8}; \overline{9}\}$, then $\varphi_1 \cdot \varphi_2 = \{\overline{1}; \overline{2}; \overline{3}; \overline{4}; \overline{5}; \overline{6}; \overline{7,8}; \overline{9}\}$.

Definition 7: If partition φ_1 on S has the same singleton element as S, then the partition φ_1 is called a "*zero*" partition (denoted by $\emptyset$). If partition φ_2 on S has one block containing all elements in S, then the partition φ_2 is called an "*identity*" partition.

Definition 8: If partition φ on S is either "zero" or "identity" partition, then the partition φ is called a "*trivial*" partition. Otherwise, it is called a "*non-trivial*" partition.

Example 4: If $S = \{1,2,3,4,5,6,7,8\}$ and partition $\varphi_1 = \{\overline{1}; \overline{2}; \overline{3}; \overline{4}; \overline{5}; \overline{6}; \overline{7}; \overline{8}\}$, then $\varphi_1 = \emptyset$. If $S = \{1,2,3,4,5,6,7,8\}$ and partition $\varphi_2 = \{\overline{1,2,3,4,5,6,7,8}\}$, then φ_2 is an identity partition. Both φ_1 and φ_2 are "trivial." Both φ_1 and φ_2 in Example 3 are "non-trivial."

Definition 9: A partition $\varphi = \{\overline{B_1}; \overline{B_2}; ...; \overline{B_t}\}$ on S is an *input-consistent partition* if $\delta(\overline{B_i}, a) = \delta(\overline{B_i}, b)$ for a, b in I, and $i = 1,2,....t$.

That is, the state behavior of a component with an input-consistent partition is independent of input I.

B. Motivational Example

To illustrate the basic idea of the proposed method, we use the FSM M as shown in Fig. 1 [6]. Note that (x/z) indicates an input x and an output z, and " − " denotes a *don't-care* condition.

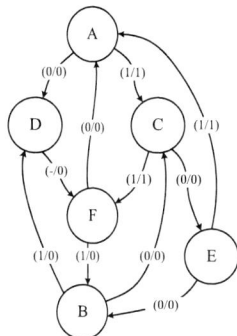

Fig. 1. An example FSM M [6].

The M has the six states $\{A, B, C, D, E, F\}$, input x, and output z. The state transition and the output function are described in the diagram. Fig. 2 shows the extraction of M_p from M where M_p shows a cycle of three states: α, β, γ. The successor M_s is to preserve the original functionality of M. This is similar to a series decomposition [6]. Note M_p possesses a cyclic state behavior. This can be viewed as extracting M_p in a given M. We propose to use the extracted M_p as the watermarked FSM.

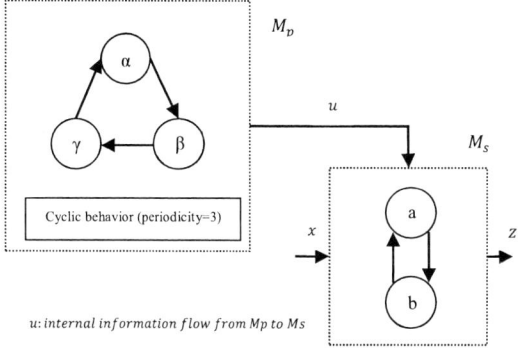

u: internal information flow from Mp to Ms

Fig. 2. Extraction of M_p from M.

Using the machine decomposition and state encoding methods (i.e., state assignment problem) [5][6], the following state encoding can extract M_p from M: $e_1 = \{(A, 000), (B, 001), (C, 010), (D, 011), (E, 100), (F, 101)\}$ with $\{(\alpha, 00), (\beta, 01), (\gamma, 10)\}$ and $\{(a, 0), (b,1)\}$. Note that the original states are realized by two internal states: $A = (\alpha, a)$, $B = (\alpha, b)$, $C = (\beta, a)$, $D = (\beta, b)$, $E = (\gamma, a)$, $F = (\gamma, b)$. For instance, the state "A" can be realized by two internal states "α" and "a" using three flip flops.

978-1-4799-8720-7/15 $31.00 © 2015 IEEE   217

III. METHOD FOR EMBEDDING WATERMARKS

A. State Encoding

The basic ideas lie in (1) performing a state encoding (e_i) and (2) extracting a watermarked FSM (FSM_w) possessing a property (p_j). The conceptual diagram is shown in Fig. 3.

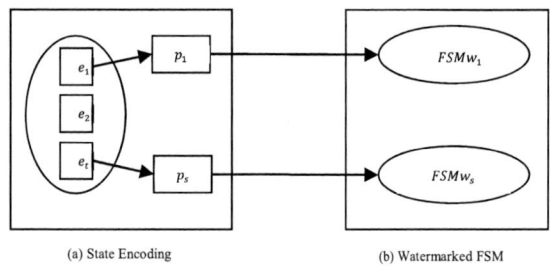

(a) State Encoding   (b) Watermarked FSM

Fig. 3. Conceptual diagram of showing two main phases

For a given n-state FSM and m, the number of variables (or flip flops), there exists potentially many different state encodings. Let Ω be the set of all possible encodings: $\Omega = \{e_1, e_2, ..., e_t\}$. The encoding space Ω can be quite large. The number of flip flops, which is a variable (i.e., a watermark design parameter), makes a direct impact on the size of the encoding space (i.e., $|\Omega| = t$). One possible case is "$n = m$" and this is known as "hot encoding," where each state is implemented by a distinct flip flop. Another possibility is to use a "minimum" number of flip flops, which will guarantee the usage of minimum storage elements ($m = \lceil \log_2 n \rceil$). In general, $t = |\Omega| \neq s$, indicating that there is not always possible to produce a property p_i by every state encoding e_i.

In this paper, we are fundamentally interested in exploring "non-destructive" solutions. Thus, we shall focus on using the minimum number of flip flops. *However, the basic idea of the proposed method is general and it can be extended for any value of "m", including "$m \geq n$".* Also, the particular property of interest, denoted by P^*, is one that has the cyclic state behavior with *maximal* periodicity.

Example 5 ($n = 6$, $m = 3$): For M in Fig. 1, using the minimum number of flip flops, $|\Omega| = t = 8 \times 7 \times 6 \times 5 \times 4 \times 3 = 20{,}160$, there are six states (A, B, C, D, E, F), three flip flops, and eight binary codes {000, 001, 010, 011, 100, 101, 110, 111} are available for the state encodings.

Hierarchical State Encoding: In a hierarchical encoding (h = level of hierarchy), multiple state encoding steps are applied. A two-tier state encoding ($h = 2$) is considered in this paper. The basic idea can be expanded to a multi-tier state encoding, if needed. This is similar to the traditional state assignment [6].

$$e_i = e_i^1 + e_i^2 + \cdots + e_i^h = \bigcup_{j=1}^{h} e_i^j \qquad (1)$$

In a two-tier state encoding, $e_i = e_i^1 + e_i^2$. In the initial state encoding (e_i^1), the goal is to extract the property P^*. Generally, in this step, a set of blocks of state is encoded. During the next step of state encoding (e_i^2), the number of states in each

block is encoded. The final state encoding e_i is then the concatenation of e_i^1 and e_i^2.

Example 6 [Two-tier State Encoding]: For M in Fig. 1, $e_i^1 = \{(\alpha, 00), (\beta, 01), (\gamma, 10)\}$ and $e_i^2 = \{(a, 0), (b, 1)\}$. Then, $e_i = \{(A, 000), (B, 001), (C, 010), (D, 011), (E, 100), (F, 101)\}$.

B. Watermarked FSM

A watermarked FSM (FSM_w) can be constructed as a state machine using e_i^1. For a given e_i^1, we can construct the corresponding state machine. The process for embedding (i.e., constructing) a watermarked FSM is described in Fig. 4. The input to the algorithm is an "n"-state FSM. The output is a watermarked FSM, FSM_w, which possesses the property P^*. Note that FSM_w is a state machine with a triplet $FSM_w = (S_w, I_w, \delta_w)$.

Input: a "n"-state FSM = $(S, I, O, \delta, \lambda), |S| = n$
Requirement: $m = \lceil \log_2 n \rceil$ /* minimum no. of binary codes */
Output: a watermarked FSM$_w$ = (S_w, I_w, δ_w) /* state machine */

Procedure:
1. Find p_{max}, the *maximal* input-consistent periodicity
2. Construct a state machine FSM$_w$ = (S_w, I_w, δ_w) as follow:
3.   The set of states: $S_w = \{s_1^*, s_2^*, ..., s_{p_{max}}^*\}$
4.   The input: $I_w = \Phi$
5.   The state transition: $\delta_w(s_1^*, a) = s_2^*$, $\delta_w(s_2^*, a) = s_3^*$, $\delta_w(s_3^*, a) = s_4^*, ..., \delta_w(s_{p_{max}}^*, a) = s_1^*$ for $\forall a \in I$

Fig. 4. Extracting a watermarked FSM (FSM_w).

The critical step is to find the maximal input-consistent periodicity (Step 1). The algorithm of finding p_{max} is shown in Fig. 5. The underlying idea of finding p_{max} is based on [6].

Input: a "n"-state FSM = $(S, I, O, \delta, \lambda), |S| = n$
Output: p_{max} /* Maximal periodicity */

Procedure:
1.   Find a set of a closed partition $\pi = \{\pi_1, \pi_2, ... \pi_\varphi\}$ and a *nontrivial* input-consistent partition τ on S, where $\pi_i \geq \tau$, for $i = 1, 2, ... \phi$.
2.   Determine the *smallest* closed partition π_{min} from $\pi = \{\pi_1, \pi_2, ... \pi_\varphi\}$.
3.   Let $\pi_{min} = \{B_1; B_2; ..... ; B_q\} = \tau$.
4.   The number of blocks in π_{min}, q, is the *maximal* periodicity p_{max}.

Fig. 5. Finding the maximal periodicity (p_{max})

The complexity of the algorithm for finding p_{max} is $O(n^2)$, a polynomial time, since a pair-wise state operation is required for the n states ($i.e, |S| = n$) in Step 1. Step 3 shows that the number of blocks in the smallest closed partition π_{min} is q ($< n$) and the nontrivial input-consistent partition τ can be set to π_{min} since this equality satisfies the condition $\pi_i \geq \tau$ in Step 1. Both determining the smallest closed partition in Step 2 and finding the number of blocks in π_{min} in Step 4 are straight forward and can be done in $O(n)$.

Example 7 [Maximal Periodicity]: For M in Fig. 1, the maximal periodicity is $p_{max} = 3$ since $\pi_{min} = \tau = \{\overline{A,B}; \overline{C,D}; \overline{E,F}\} = \{\alpha; \beta; \gamma\} = \pi_1$ and the number of blocks in π_{min} is 3. Note that there exists other candidates of input-consistent partitions, namely $\pi_2 = \{\overline{A,B,C,D}; \overline{E,F}\} = \{\overline{\alpha,\beta}; \overline{\gamma}\}$, $\pi_3 = \{\overline{A,B,E,F}; \overline{C,D}\} = \{\overline{\alpha,\gamma}; \overline{\beta}\}$, $\pi_4 = \{\overline{C,D,E,F}; \overline{A,B}\} = \{\overline{\beta,\gamma}; \overline{\alpha}\}$, and $\pi_5 = \{\overline{A,B,C,D,E,F}\} = \{\overline{\alpha,\beta,\gamma}\}$. However, π_5 is a *trivial* input-consistent partition since it combines all states into a single block, which implies that no useful information is processed with this state partition. All other partitions above are nontrivial input-consistent partitions. However, the *smallest* closed partition is π_1.

Hierarchical (Two-tier) State Encoding: Upon determining the partition π_{min} and the maximal periodicity p_{max}, a state encoding e_i can be performed by assigning the minimum number of binary codes to the blocks of the smallest partition. In a tier-1 state encoding, the assignment with a minimum number of binary bits can be made on the blocks of state $\{B_1; B_2; \ldots; B_q\}$ in π_{min}. This is described in Step 1 and Step 2 in Fig. 6. Pseudo-code of tier-1 state encoding (step 1 and 2) for the blocks in the smallest closed partition is associated with FSM_w. Pseudo-code of tier-2 state encoding (step 3-5) for the blocks of partition μ is associated with the residual FSM_r. Note that FSM_r is to preserve the original functionatility of M. Once the smallest closed partition (π_{min}) and nontrivial input-consistent partition (τ) are determined, it is simple to perform the initial state encoding (e_i^1) using the minimum number of binary bits (in Step 1 and Step 2). The complexity of tier-1 state encoding procedure is $O(q)$. Note that Step 2 might not produce a unique state encoding.

Input: State partition $\pi_{min} = \tau = \{B_1; B_2; \ldots; B_q\}$, $q < n$.
Requirement: Use the minimum number of binary codes
Output: State encoding of (e_i^1) and (e_i^2)

Procedure:
1. Let $m_w = \lceil \log_2 q \rceil$
2. Perform one-to-one mapping using m_w *binary bits* on $\{B_1; B_2; \ldots; B_q\}$.
3. Find another state partition μ such that $\pi_{min} \cdot \mu = \emptyset$; Let $\mu = \{b_1; b_2; \ldots; b_r\}$.
4. Let $m_r = \lceil \log_2 r \rceil$
5. Perform one-to-one mapping using m_r *binary bits* on $\{b_1; b_2; \ldots; b_r\}$.

Fig. 6. Pseudo-code of tier-1 and tier-2 state encoding

Example 8 [State Encoding; Tier-1]: From Example 7, $q = 3$, $m_w = 2$, $e_i^1 = \{(\alpha, 00), (\beta, 01), (\gamma, 10)\}$. Note that there exists other tier-1 state encodings possible such as $\{(\alpha, 01), (\beta, 10), (\gamma, 11)\}$.

To complete the state encoding e_i, a tier-2 state encoding e_i^2 needs to be done. Informally, e_i^2 should distinguish each state in every block B_i in $\{B_1; B_2; \ldots; B_q\}$. This can be done with another partition $\mu = \{\overline{A,C,E}; \overline{B,D,F}\}$, as an example.

A general procedure for a tier-2 state encoding e_i^2 is described in Step (3)-(5) in Fig. 6. Given π_{min}, finding another partition μ satisfying the condition (i.e., $\pi_{min} \cdot \mu = \emptyset$) is straight forward since it can split the states in each block of π_{min} into an individual state. Step 3 can be done in polynomial time, $O(q \cdot |B_i|)$, where q is the number of blocks in π_{min} and $|B_i|$ is the maximum number of states in the block. Step 4 and 5 are similar to the tier-1 state encoding procedure.

Example 9 [State Encoding; Tier-2]: From Example 7 and Example 8, $\mu = \{\overline{A,C,E}; \overline{B,D,F}\} = \{b_1; b_2\}$ since $\pi_{min} \cdot \mu = \emptyset$. Also, $r = 2$, $m_r = 1$, $e_i^2 = \{(b_1, 0), (b_2, 1)\}$.

The final state encoding e_i is made using the *concatenation* of two tiers of state encodings e_i^1 and e_i^2.

Example 10 [Final State Encoding]: $e_i = e_i^1 + e_i^2 = \{(A, 000), (B, 001), (C, 010), (D, 011), (E, 100), (F, 101)\}$.

Note that the watermarked FSM (FSM_w) is a state machine that behaves as an input-independent $\lceil \log_2 q \rceil$ − bit counter.

IV. ANALYSIS

In this section, we provide the analysis of the proposed approach. The limitation of the proposed approach and the potential solution are also discussed.

A. Existence of Watermarked FSM

We provide the analysis for the existence of the watermarked FSM possessing the property P* based on [6].

Theorem 1: If a closed partition π and a nontrivial input-consistent partition τ_i on S in a *FSM = (I, O, S, δ, λ)* which satisfy the condition of $\pi \geq \tau_i$, then the watermarked $FSM_w = (S_w, I_w, \delta_w)$ possessing the property P* exists.

Proof: If the FSM possesses a closed partition π such that $\pi \geq \tau_i$, then, for a given state S_i and every input in I, the next states must be in the same block of τ_i, and therefore in the same block of π. Consequently, for a given initial state, the block of π in which the state of FSM_w is contained after any finite input sequence depends only on the initial block and on the length of the sequence. Furthermore, there must exist a cycle of states such that $\delta_w(s_1, x) = \delta_w(s_2, x) = \ldots = \delta_w(s_p, x)$ for any input x in I_w. Then, the maximal periodicity ($p_{max} = p$) of a cycle can be chosen. □

B. Attack Analysis

IP watermarking attacks can be further categorized in the main classes below [3].

1) Removal Attacks: Removal attacks are divided into either elimination attacks or masking attacks.

a) Elimination attacks: the watermark can be eliminated completely by an attacker. For instance, the attacker tries to estimate the watermark and subtract it from the watermarked design. In the proposed scheme, it will be technically infeasible to eliminate the watermarked FSM without affecting the original functionality since the watermarked FSM itself performs the sub-computation.

b) Masking attacks: masking attacks aim at distorting the watermark detector to disable its ability to sense the presence of the watermark. This attack is more related to the watermark verification process.

2) False Positive (Probability of Coincidence or Watermark Collision): In [12], the probability of coincidence was defined as "the odds that an unintended watermark is detected in a design." Often, this is used as a measure of watermark validity. In [9], we demonstrated using a set of simple FSMs that the collisions practically may not occur.

3) Embedding Attacks (Forging): Watermark designers need to find techniques to protect their designs from intruders to embed another watermark in the design. In the proposed scheme, it is likely that the functionality of a FSM may change and thus detectable, if another watermark is added.

C. Limitation

The existence of P^* is not guaranteed for *all* FSMs (see Theorem 1). The nature of the limitation is rooted from using a strong property (i.e., maximal periodicity and input-consistency). In this case, we may consider relaxing these requirements at the cost of lowing security level (i.e., to embed *less* useful data for IP protection.)

V. CONCLUSIONS AND FUTURE WORK

For a class of deterministic finite-state machines (D-FSMs), we have proposed the new watermarking embedding method at a FSM-level. We showed that a watermarked FSM, which possesses a property of cyclic state behavior with maximal periodicity, can be extracted using a hierarchical state encoding. The proposed work was carried out based on the observation that this property may be used as an IP owner's unique property for the FSM IP. However, the proposed method does not provide a universal solution (Section IV.C). Using state-splitting [6] may resolve this issue, but at the cost of relaxing the non-destructiveness.

Another interesting problem that should be investigated further is to formally analyze a false positive collision that may arise during the embedding process. This problem can be described as follows: *Given a FSM M which inherently possesses the property of a cyclic state behavior with periodicity p, does another FSM M' exist such that M' has the same property and $M' \neq M$?* We plan to work on this important problem.

ACKNOWLEDGMENT

The authors would like to thank Prof. Lilian Bossuet for his insightful, constructive comments, and many thought-provoking discussions.

REFERENCES

[1] Abdel-Hamid, A. T., Tahar, S., and Aboulhamid, E. M. A survey on IP watermarking techniques. In *Design Automation for Embedded Systems*, Vol. 9, (2004), Springer Science+Business Media, Berlin, 211-227.

[2] Bossuet, L. and Hely, D. Salware: Salutary hardware to design trusted IC. In *Proc. of the Trustworthy Manufacturing and Utilization of Secure Devices Workshop* (TRUDEVICE '13). Avignon, France, (2013), 30-31.

[3] Cox, I. J., Miller, M. L., Bloom, J. A., and Honsinger, C. *Digital watermarking.* (1998), Morgan Kaufmann Publishers.

[4] Cui, A., Chang, C.H, Tahar, S., and Abdel-Hamid, A.T.. A Robust FSM Watermarking Scheme for IP Protection of Sequential Circuit Design. *IEEE Trans. Comp.-Aided Des. Integ. Cir. Sys.* 30, 5 (May 2011), 678-690.

[5] Hartmanis, J. and Stearns, R.E. *Algebraic Structure Theory of Sequential Machines*, Prentice-Hall, Inc., (1966), Upper Saddle River, NJ, USA.

[6] Kohavi, Z. *Switching and Finite Automata Theory* (2nd ed.). (1978), McGraw-Hill.

[7] Koushanfar, F. and Alkabani, Y. Provably secure obfuscation of diverse watermarks for sequential circuits. In *Proc. of the Inter. Symp. on Hardware-Oriented Security and Trust* (HOST '10), (2010), 42-47.

[8] Lewandowski, M., Meana, R., Morrison, M., and Katkoori, S. 2012. A Novel Method for Watermarking Sequential Circuits. In *Proc. of the IEEE Int. Symp. on Hardware-Oriented Sec. and Trust*, San Francisco, CA. (2012), 21-24.

[9] Marchand, C., Bossuet, L., and Jung, E., "IP Watermarking Verification Based On Power Consumption Analysis," *Proc. of the 27th IEEE Int. Sys.-on-Chip Conf. (IEEE SOCC '14)*, Las Vegas, September 2014.

[10] Oliveira, A. L. Robust techniques for watermarking sequential circuit designs. In *Proc. of the 36th annual ACM/IEEE Design Automation Conf.* (DAC '99), Mary Jane Irwin (Ed.). ACM, New York, NY, USA, (1999), 837-842.

[11] Oliveira, A. L. Techniques for the creation of digital watermarks in sequential circuit design. *IEEE Trans. Comp.-Aided Des. Integ. Cir. Sys.* Vol. 20, Issue 9, (September 2001), 1101-1117.

[12] Torunoglu, I. and Charbon, E. Watermarking-based copyright protection of sequential functions. *IEEE Journal of Solid-State Circuits.* Vol. 35, Issue 3 (2000), 434-440.

2015 IEEE Computer Society Annual Symposium on VLSI

A Summary of Current and New Methods in Velocity Selective Recording (VSR) of Electroneurogram (ENG)

John Taylor, Benjamin Metcalfe & Chris Clarke
Department of Electronics and Electrical Engineering
University of Bath
Bath, BA2 7AY, UK
E-mail: j.t.taylor@bath.ac.uk

Thomas Nielsen
Department of Health Science and Technology
Aalborg University
Aalborg, Denmark

Daniel Chew
Cambridge Centre for Brain Repair
University of Cambridge
Cambridge, CB2 0PY, UK

Nick Donaldson
Department of Medical Physics and Bioengineering
University College London
London, WC1, UK

Abstract — **This paper describes the theory of *velocity selective recording* (VSR) of neural signals including some new developments. In particular new limits on available selectivity using bandpass filters are introduced and discussed. Existing work has focussed primarily on electrically evoked *compound action potentials* (CAPs) where only a single evoked response per velocity is recorded. This paper extends the theory of VSR to naturally occurring neural signals recorded from *rat* and describes a practical method to estimate the level of activity (firing rates) within particular velocity ranges.**

Keywords – Electroneurogram; Velocity Selective Recording; Multielectrode Cuff

I. INTRODUCTION

Velocity selective recording (VSR) is a technique that allows discrimination of action potentials (APs) based both on the direction of propagation (*afferent* or *efferent*) and *conduction velocity* (CV), without the need to submit the nerve to invasive and potentially damaging procedures. Since, for myelinated fibres, there is a well-known relationship between CV and axonal diameter (Erlanger and Gasser 1937 [1]), the VSR method can be used to extract activity related to physiological function from the whole neurogram [2]-[5]. While CV can be calculated using only a single pair of electrodes (a dipole) it has been shown that the *velocity selectivity* of a system, i.e. its ability to discriminate between populations with adjacent CVs propagating simultaneously along the same nerve, can be increased by using multiple electrodes. These can be practically implemented using a *multi-electrode cuff* (MEC), which is now available as a component for use in implantable neuroprostheses. The method provides a viable interface for neural recording systems that have potential use in a range of prosthetic devices, for example the *'Bioelectronic Medicines'* currently being advocated by GlaxoSmithKline [6]. In addition, information about conduction velocity may be useful for neuroscientists wishing to study nerve conduction disorders within the peripheral nervous system.

To date, recordings from cuff-based systems have been made on the assumption that individual APs (spikes) are generally not visible in the time records of individual channels due to poor signal-to-noise ratio. To counter this problem a technique based on bandpass filters was introduced that enables the velocity spectrum of the data to be calculated. This method works well for *compound action potentials* (CAPs), typically evoked by electrical stimulation, but cannot be easily extended to the case of naturally-occurring ENG mostly due to the very small amplitudes of the APs, at least when recorded using nerve cuffs [2]. In addition, in natural ENG, information is encoded as neural firing rates and so it is necessary to determine these rates in a particular velocity band, rather than just the relative amplitudes of activity between bands, which is generally the case for CAPs. These are very demanding requirements on the recording system. However, some recent experimental data obtained from the *vagus* nerve of a *pig* has indicated that, contrary to previous assumptions, simple linear processing of the outputs of an MEC does in fact reveal spike-like patterns of signals in the time record. This observation suggests the possibility of new methods to determine both the activated velocity bands and the level of activity within those bands.

In this paper we begin by surveying existing VSR techniques and propose some fundamental limits on available performance that extends current theoretical knowledge. We then discuss the extension of VSR to the recording of natural ENG and describe our new method of *velocity spectral density* (VSD). This is validated using both simulated data and some preliminary experimental results in *rat*.

II. THEORY AND FUNDAMENTAL LIMITS OF VSR

A. Basic Principles

The basic principle of VSR is to transform time domain recordings of neural signals into the *velocity domain*. The velocity spectrum, which is computed from the time record of the neural data, shows both the direction of AP propagation (either *efferent* or *afferent*) as well as the level of excitation of each fibre population within the nerve. A simple process termed *delay-and-add* is commonly used to carry out this

978-1-4799-8720-7/15 $31.00 © 2015 IEEE

transformation [2]. Recordings are made at equal distances along the nerve using a MEC or other linear electrode array such as a microchannel. Each channel is delayed relative to the first channel by an interval that depends on both the electrode spacing and the propagation velocity of the signal. So if the delay between the first two channels is dt the delay between the first and third channels is $2 \cdot dt$ and so on. Delay-and-add operates by inserting variable delays τ (where τ is an integer multiple of dt) into the channels to effectively cancel the naturally occurring delays after which the channels are summed resulting in a single signal. As the delay is swept through a range values the output goes through a peak when $\tau = d/v$ where d is the inter-electrode spacing and v is the CV of the AP, called the *matched velocity*. So when matching occurs, a pulse (such as an AP) of a particular duration and amplitude on one channel becomes a pulse of the same width and N-times the amplitude when the outputs of N channels are summed.

The velocity selectivity of the system is therefore increased by a factor N and the signal to noise ratio by $\sqrt{N}$, assuming the noise sources in each channel are uncorrelated [3]. This leads naturally to the concept of the *velocity quality factor Q_v* described in the next section. The resulting velocity profile is called the *intrinsic velocity spectrum* (IVS).

Figure 1 shows the schematic of a typical interface between an MEC and the delay-and-add signal processing. The electrodes are grouped in two ranks: initially in *dipole* pairs ('single differential') and these in turn are connected as *tripoles* ('double differential') although only the dipole circuits are shown in the figure since that was the system used in the experiments described in Section III. These differential arrangements are used to suppress common mode interference signals, most notably the *electromyogram* (EMG). For analysis purposes the MEC is considered to be a linear time invariant (LTI) system with transfer function $H(f, v)$ transforming neural signals (input) into electrical ones (output). The input to the MEC is a *trans-membrane action potential* function (TMAP), $V_m(t)$, with the corresponding spectrum $V_m(f)$. The resulting *single fibre action potential*

(SFAP) is a propagating wave with the time dependence of the underlying TMAP function, the relationship between the two being explained in [2]. For the purpose of simulation, we represent the TMAP function and its spectrum by the Fourier transform pair [2]:

$$V_m(t) = At^n e^{-Bt}$$

$$V_m(f) = \frac{n!\,A}{\left(B + j2\pi f\right)^{n+1}} \qquad (1)$$

where A, B and n are constants and f is frequency. The output $Y(f, v)$, which is a function of both frequency and velocity, is given by eqn (2):

$$Y(f,v) = \left| \frac{\sin\left(N\pi f\left(\frac{d}{v} - \tau\right)\right)}{\sin\left(\pi f\left(\frac{d}{v} - \tau\right)\right)} \right| \cdot 4\frac{R_e}{R_a}\sin^2\left(\frac{\pi f d}{v}\right) \cdot \left| \frac{An!}{\left(B + j2\pi f\right)^{n+1}} \right| \qquad (2)$$

This equation describes the output of a cuff with N tripoles, electrode spacing d and propagation velocity v. R_a, the intra-axonal resistance per unit length, has been assumed to be large compared to R_e, the extra-axonal resistances per unit length inside the cuff. Equation 2 is the product of the spectrum of the TMAP ($V_m(f)$), the transfer function of one tripole ($H_o(f,v)$), and the transfer function of the delay-and-add block ($G(f,v)$). At *matched velocities* (i.e. where $\tau = d/v$ and $v = v_o$ in eqn (2)), eqn (2) reduces to:

$$Y(f,v_0) = 4N\frac{R_e}{R_a}\sin^2\left(\frac{\pi f d}{v_0}\right) \cdot \left| \frac{An!}{\left(B + j2\pi f\right)^{n+1}} \right| \qquad (3)$$

The placement of a *bandpass filter* (BPF) in each channel fixes f to the centre frequency of the BPF reducing Y to a function of velocity *only* (this issue was also examined in [7]). Finally, in order to quantify the velocity selectivity, we define a *velocity quality factor, Q_v*, by analogy with linear systems in the frequency domain:

$$Q_v = \frac{v_0}{v_{3+} - v_{3-}} \cong \frac{N\pi d}{2.64}\left(\frac{f_0}{v_0}\right) \qquad (4)$$

where v_0 is the matched velocity and v_{3+} and v_{3-} are the upper and lower 3 dB points respectively. Note that although the proposed arrangement requires a BPF at the output of each channel, in practice, due to the linearity of the processes involved, the summation and filtering operations can be reversed. This leads to a much simpler and more practical arrangement where only a single BPF is required for each velocity band of interest (Fig 1).

B. Limits on available velocity selectivity; selectivity bandwidth

In order to determine upper and lower bounds on the velocity selectivity generated by an MEC-based recording system, we make the fundamental assumption that BPFs are always used and hence eqn (2) reduces to $Y(v)$, f appearing as a constant (f_0).

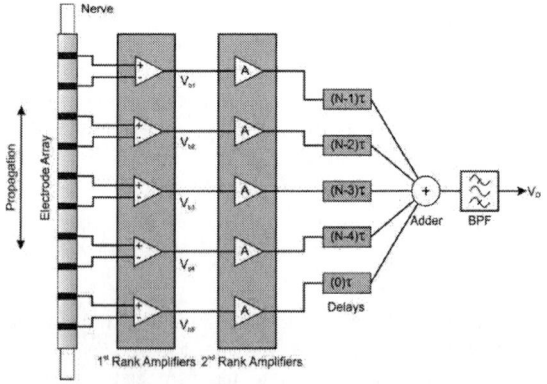

Figure 1: Simplified schematic of the amplifier configuration used to capture neural signals from an MEC. In this arrangement $2N$ electrodes results in N dipole signals. N is typically about 5. In an alternative configuration the electrode connections to the first rank amplifiers are interleaved resulting in $2N - 1$ dipoles.

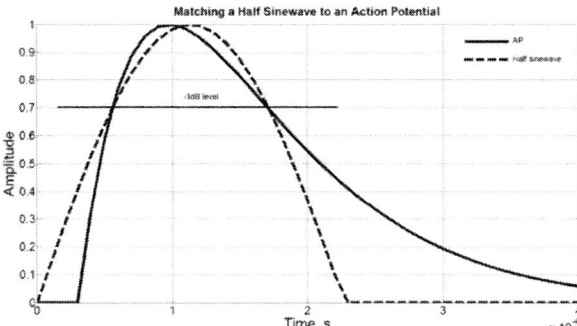

Figure 2. Fitting half a sinewave (dashed line) to an AP (solid line) to find the lower bound on velocity selectivity. The matching is in the time domain at the -3 dB points of the curves.

Using eqn (4) calculation of the range of available velocity selectivity (expressed as Q_v) at a velocity v_0 reduces to finding the range of possible values of f_0, which might appropriately be called the *selectivity bandwidth*.

(a) Lower bound: this is taken to be the frequency of a sinewave whose width most nearly matches the positive phase of an AP of the same amplitude, as illustrated in Fig 2. This is an approximation to the *intrinsic velocity selectivity* (IVS). The matching is carried out at the -3 dB points of both waveforms and it can be shown that, to a good level of approximation, the equivalent frequency f_L is given by:

$$f_L \cong \frac{B}{8} \qquad (5)$$

Where the parameter B relates to $V_m(t)$ in eqn (1). So, for example, if $B = 15$ kHz, $f_L = 1.875$ kHz (approximated to 2 kHz for simulation) and hence substituting into eqn (4) with $d = 3$ mm and $N = 10$, the lower bound on Q_v at $v_0 = 30$ m/s is 2.2.

(b) Upper bound: In [4] we noted that the upper bound on velocity selectivity is set by noise considerations because the spectrum of the signal (see eqn (3)) decreases monotonically with frequency eventually merging with the noise floor of the system. Clearly the signal has no energy left beyond this frequency to drive a BPF and so it seems reasonable to choose this 'noise corner frequency' as the upper frequency limit that also determines the maximum available velocity selectivity. In the example in Fig 3, the spectrum of a single monopolar AP is plotted with and without additive white noise. The form of the AP is given by eqn (1) with $A = 40,774$ V, $B = 15,000$ Hz and $n = 1$. These constants are chosen to be representative of a mammalian AP normalised to peak amplitude of unity. The noise generator employed produces zero mean Gaussian white noise with instantaneous power σ^2 and so for a sequence length L the mean power is σ^2/L and the *rms* power is $\sigma/\sqrt{L}$. For example if $\sigma = 0.1$ and $L = 1024$, the *rms* power is 3.125×10^{-3} W.

Note that the spectrum of the noise is actually flat and the oscillations shown are an artefact of the FFT process for a finite length sequence.

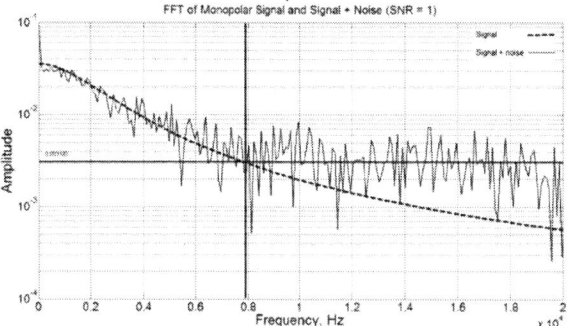

Figure 3. Spectrum (1024 point FFT) of a single monopolar AP (dashed curve) and the same with additive white Gaussian noise (solid curve). In this example $\sigma = 0.1$ (SNR ≈ 1) and so the noise floor is 3.125×10^{-3} V. The spectra intersect at a frequency of approximately 7.7 kHz. This 'noise corner' frequency is taken to be the maximum frequency at which a BPF can operate and therefore defines the maximum available velocity selectivity.

The spectrum of the TMAP function is described by the continuous time Fourier Transform (CTFT) in eqn (3) but it should be noted that the spectrum shown in Fig 3 was calculated using a 1024 point FFT with a sampling frequency of 100 kHz. The two representations are very similar for frequencies well below the Nyquist limit (50 kHz in this case) but closer to this limit the two plots diverge somewhat. However, the single sided CTFT produces a particularly simple analytical expression for the signal and was adopted in this paper:

$$V_m(\omega) = \frac{2F_s}{L} \frac{A}{(B + j\omega)^2}$$

and so:

$$|V_m(\omega)| = \frac{2F_s}{L} \frac{A}{B^2 + \omega^2} \qquad (6)$$

In order to calculate the noise corner frequency ω_{lim} it is necessary to set $|V_m(\omega)|$ equal to $\sigma/\sqrt{L}$ and solve for ω_{lim}:

$$\omega_{\lim} = 2\pi f_{\lim} = B\sqrt{\frac{2AF_s}{\sigma\sqrt{L}B^2} - 1}$$

we can write the complete expression for the *selectivity bandwidth*:

$$\frac{B}{8} \le f_0 \le \frac{B}{2\pi}\sqrt{\frac{2AF_s}{\sigma\sqrt{L}B^2} - 1} \qquad (7)$$

It is useful to express this in terms of the *signal to noise ratio* (SNR). The average (*rms*) signal power is calculated from the TMAP function as follows:

$$P_s = A\sqrt{\frac{1}{\tau}\int_0^\tau t^2 e^{-2Bt}dt} \cong \frac{A}{2\sqrt{\tau}B^{3/2}}$$

where τ, the length of the sequence is $LT_s = L/F_s$. Combining this with the expression for the *rms* noise:

$$SNR = \frac{A\sqrt{F_s}}{2\sigma\sqrt{L}B^{3/2}} \qquad (8)$$

978-1-4799-8720-7/15 $31.00 © 2015 IEEE   223

Using this expression, eqn (7) can be rewritten in a more compact form:

$$\frac{B}{8} \le f_0 \le \frac{B}{2\pi} \sqrt{4\left(\frac{F_s}{B}\right)^{0.5}(SNR)-1}$$

and, finally, recalling that SNR increases as a function of $\sqrt{N}$, we can write:

$$\frac{B}{8} \le f_0 \le \frac{B}{2\pi} \sqrt{4\left(\frac{NF_s}{B}\right)^{0.5}(SNR)-1} \qquad (9)$$

Since the ratio of velocity selectivity enhancement, R_s, is proportional to this *selectivity bandwidth* (see eqn (4)), we can write:

$$R_s \cong \frac{8}{\pi}\left(\frac{NF_s}{B}\right)^{0.25}(SNR)^{0.5} \qquad (10)$$

So, e.g., with the parameter values given above and $SNR = 1$, $R_s = 11.7$, a significant level of velocity selectivity enhancement. Note that R_s changes very slowly as the parameters N, F_s and B are varied but more rapidly with SNR.

C. Simulation results

The 10-channel system discussed above was simulated using MATLAB for the three values of SNR tabulated plus the noiseless case (i.e. SNR → ∞). The resulting selectivity parameters are given in Table I. Fig 4 shows the IVS for a single AP propagating at 30 m/s for these four values of SNR. For the case where $SNR = 10$ the profile is indistinguishable from the noiseless case. For $SNR = 1$ there is some degradation of performance but for $SNR = 0.1$ the method fails completely. This is entirely consistent with the values given in Table I since the delay-and-add process fails when f_{lim} falls below the lower limit given by (5).

Table I
Relationship between SNR and the Selectivity Bandwidth

calculated from eqn (8)		f_{lim} (kHz) calculated from eqn (9)		Selectivity BW (kHz)	
σ	SNR	N = 1	N = 10	N = 1	N = 10
0.01	11	25.3	45.16	23.3	43.16
0.1	1.1	7.7	14.12	5.7	12.12
1.0	0.11	0.88	3.85	-	1.85

D. Experimental results

Some preliminary experiments were carried out to assess the applicability of VSR to recording naturally evoked (physiological) ENG using an MEC. The right *vagus* nerve of a Danish Landrace *pig* was fitted with an MEC of length 4 cm containing ten annular electrodes with a pitch of 3.5 mm. Bipolar measurements of ENG were made with the animal at rest. A short segment of each of the resulting nine channels of data (original duration 2 minutes and sampled at 100 kS/s with 16-bits resolution) is shown in Fig 5.

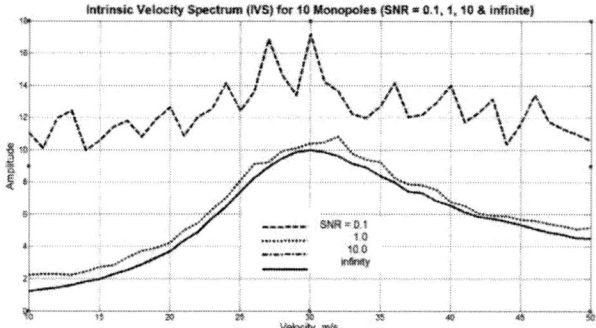

Figure 4. IVS for a 10 channel system with added white noise. Three values of *SNR* are indicated (10, 1 and 0.1) in addition to the noiseless case. Note that for *SNR* = 10 the profile is indistinguishable from the noiseless case. For *SNR* = 1 there is some degradation of performance but for *SNR* = 0.1 the method fails completely.

Examination of the traces in Fig 5 shows no discernable features that could be attributable to ENG activity. However when delay-and-add is applied, as shown in Fig 6 for matched velocities in the range 31 – 45 m/s with an interval of 2 m/s, correlated peaks are clearly visible revealing several excited populations including the one illustrated in the figure at 37 m/s.

The ability to recover correlated data from nerve cuff recordings using an MEC results directly from the improvement in SNR provided by the delay-and-add process as discussed above. This result is significant as it suggests not only that practical VSR systems can be used to record physiological data using MECs but also suggests a technique to extract information from the nervous system based on axonal firing rates. Such a method is discussed in the next section.

III. VELOCITY SPECTRAL DENSITY

A. Basic method

VSR is well suited to the analysis of single event APs such as those typically found in electrically evoked CAPs. Information within the nervous system is transmitted in an encoded fashion; the number of APs propagating through a single axon per second is representative of the sensory (analogue) input signal to that axon [8]. This feature is a direct result of the *all-or-nothing* nature of the neuron. As an example, the *afferent* fibres that contain information about the fullness of the bladder have been measured in man to propagate at a velocity of 41 m/s with a baseline firing rate of about 15 APs per 200 ms and a rate representing a full bladder of about 40 APs per 200 ms [9]. The contribution of this section of the paper is to propose an extension to VSR to include information about the number of APs occurring in a given time period. The result is called the method of *velocity spectral density* (VSD) and provides a measure of the time varying activity within a band of conduction velocities [10]. One method for extracting both conduction velocity and neuronal firing rates from a nerve recording is to use a sliding time window of sufficient length to enclose only a single AP. Delay-and-add can then be applied to extract the IVS of the

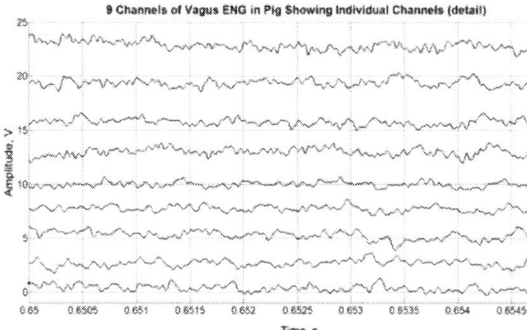

Figure 5. A short segment of nine bipolar channels of recorded ENG from the right *vagus* nerve of a *pig* (the animal was at rest). An MEC of length 4 cm fitted with ten annular electrodes with a pitch of 3.5 mm. Note that there are no discernable features attributable to ENG activity.

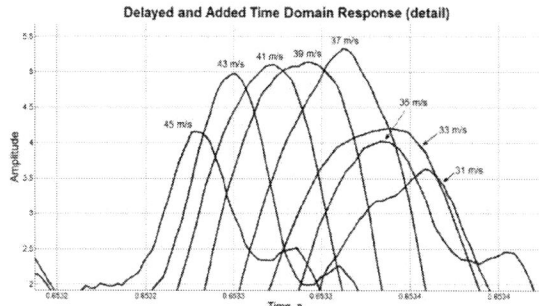

Figure 6. Detail of the data in Fig 5 after delay-and-add has been applied for a range of matched velocities from 31 to 45 m/s in steps of 2 m/s. Note that correlated peaks are clearly visible.

window contents and thus identify the most likely conduction velocity for the AP based on the velocity of the peak value, V_{peak}. This process could be repeated as the window is moved along the time record and the firing rates extracted by simply counting the number of occurrences of each velocity but this has two significant drawbacks. Firstly, the window must only contain a single AP, otherwise only the AP with the largest amplitude will be identified as the largest peak in the IVS. Secondly the windowing function must be carefully selected to avoid *velocity spectral leakage* (VSL), an effect that is similar to spectral leakage in the frequency domain, resulting from the time domain window failing to encompass the AP fully. A more robust method has been developed that does not require the use of a sliding time window and so avoids these issues. The new method by which both conduction velocity and neuronal firing rates can be extracted is described in the following steps.

1. A set of time records of arbitrary length is processed using the delay-and-add method as described above. The values of *dt* used can be selected, based on the required velocity range and resolution. For example a velocity range of 10 - 50 m/s with an electrode spacing of 1 mm requires *dt* values in the range 20 - 100µs. If the resolution is 1 m/s then this will result in a set of 41 V_D waveforms, one for each velocity and formed by simple summation of the delayed signals. The form of V_D for each value of *dt* over 5 channels of raw data is given by:

$$V_D(t, dt) = \sum_{i=1}^{5} V_{Bi}(t(i-1) \cdot dt)$$

2. A simple hard noise threshold is applied that removes any samples below the system noise floor. At present the noise floor for experimental results is computed from the input-referred noise as measured during the experiments.

3. In order to identify an AP the relationship between V_D for neighbouring values of *dt* must be examined. Each V_D waveform is passed through a filter that detects the centroid of each AP [10]. This filter is

implemented as a linear finite impulse response (FIR) filter with impulse response h[n] given in eqn (11):

$$h[n] = -(2/N)n + 1 \qquad (11)$$

This is a linear function of gradient -2/N where N is the width of the filter and *n* is the current index of the discrete-time samples. The function *h[n]* varies in amplitude from +1 to -1 where N is chosen to be at least as wide as a single AP in the time domain. Since in practice the APs are neither regular nor symmetric the centroid represents a more robust method for locating the midpoint of the AP than taking the maximum value as has been done previously. The centroid can be considered as the geometric centre of any two dimensional region, in this case the area under the AP as bounded by the *x* axis. It is necessary to separate the positive and negative phases of the AP before locating the centroid, and this was achieved via half wave rectification of the signal. Computing the centroid considers the contribution from every sample as opposed to the single samples used in peak detection and so it is more robust against noise and interference.

4. Finally a detection algorithm is applied that examines each velocity response for the criterion $V_{D-1} < V_D > V_{D+1}$. If this criterion is met the histogram for the current window can be incremented at the velocity V_D.

The signal processing requirements for VSD are conceptually minimal. In a practical system designed for implementation within a VLSI architecture the largest components are the delay lines associated with the delay-and-add process and the centroid FIR filter. Generally speaking the band-pass filters are realised using 4th order Butterworth structures for which both area and power efficient versions are readily available. The driving force behind implantable systems is power consumption and the methods described within this paper are well suited to low power VLSI implementations using pre-existing and scalable technologies.

978-1-4799-8720-7/15 $31.00 © 2015 IEEE   225

B. Preliminary measured results

In order to provide some preliminary validation of the VSD process, acute *in-vivo* recordings were made from a *rat* [11]. These experiments were part of a larger study that is not included here. The recording setup consisted of five bipolar recordings taken from a set of six hook electrodes placed on a fascicle of the L5 *dorsal* root. For consistency with the simulated data, the electrode spacing was 1 mm and the sample rate was 500 kS/s. The data were captured and processed using MATLAB in the same manner as the simulated data reported above. The recordings were of length 250 ms and were made both with and without cutaneous stimulation of the L5 dermatome. In order to identify the effect of cutaneous stimulation of the dermatome, direct electrical stimulation was used to identify the conduction velocities of the relevant *afferents*.

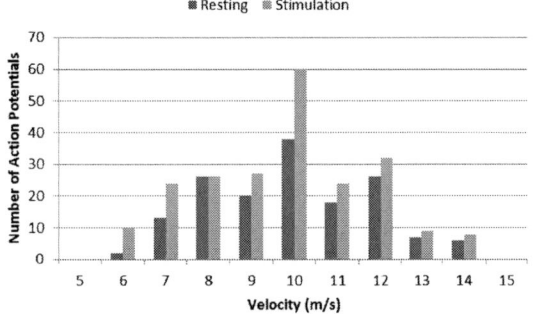

Figure 7. VSD histogram recorded from 250ms of physiological recordings from *rat* during both a resting state and during cutaneous stimulation.

VSD analysis applied to the data showed consistent levels of activity in both sets of natural recordings with a significant increase in the number of APs propagating at 10 m/s during cutaneous stimulation (Fig. 7). This was in agreement with direct electrical stimulation of the dermatome where CAPs were observed with a conduction velocity of 10 m/s. The velocity band 10 m/s – 17 m/s is within the accepted range of conduction velocities for the *Aδ afferent* fibres in rat, which are responsible for light touch sensation.

IV. CONCLUSION

This paper has reviewed two related topics of current interest in neural recording. Firstly, the method of *velocity selective*

recording (VSR) was reviewed and new performance limits were introduced. Secondly, a new method that extends significantly the capabilities of VSR using an automated detection system and a histogram-based analysis of neuron firing rates, was described. The method generates a detailed overview of the firing rates of neurons based on conduction velocity and direction of propagation. This was demonstrated for both simulated data and *in-vivo* physiological recordings from the L5 dorsal root in *rat*.

V. ACKNOWLEDGEMENTS

This work was generously supported by the Brian Nicholson PhD scholarship.

VI. REFERENCES

[1] H. Gasser, "The classification of nerve fibers.," *Ohio Journal of Science*, vol. 41, no. 3, pp. 145–159, 1941.
[2] Taylor J., Donaldson N., & Winter J. (2004) "The use of multiple-electrode nerve cuffs for low velocity and velocity-selective neural recording." *Med. & Biol. Eng. & Comput.*, 42 (5), 634-43.
[3] N. Donaldson, R. Rieger, M. Schuettler, and J. Taylor, "Noise and selectivity of velocity-selective multi-electrode nerve cuffs.," *Med. Biol. Eng. Comput.*, vol. 46, no. 10, pp. 1005–18, Oct. 2008.
[4] J. Taylor, M. Schuettler, C. Clarke, and N. Donaldson, "The theory of velocity selective neural recording: a study based on simulation.," *Med. Biol. Eng. Comput.*, vol. 50, no. 3, pp. 309–18, Mar. 2012.
[5] Yoshida, K., Kurstjens, G., & Hennings, K. (2009) "Experimental validation of the nerve conduction velocity selective recording technique using a multi-contact cuff electrode." *Medical Engineering & Physics* 31 1261-1270.
[6] K. Famm, B. Litt, K. J. Tracey, E. S. Boyden, and M. Slaoui, "Drug discovery: a jump-start for electroceuticals.," *Nature*, vol. 496, no. 7444, pp. 159–61, Apr. 2013.
[7] Karimi, F, Seydnejad, S, "Velocity selective neural signal recording using a space-time electrode array", *IEEE Transactions on Neural Systems and Rehabilitation Engineering,* issue 99, 2014.
[8] H. Milner-Brown, R. Stein, and R. Yemm, "Changes in firing rate of human motor units during linearly changing voluntary contractions," *The Journal of physiology*, pp. 371–390, 1973.
[9] G. Schalow, G. Zäch, and R. Warzok, "Classification of human peripheral nerve fibre groups by conduction velocity and nerve fibre diameter is preserved following spinal cord lesion," *Journal of the autonomic nervous system*, vol. 1838, no. 6, 1995.
[10] B. Metcalfe, D. Chew, C. Clarke, N. Donaldson, and J. Taylor, "An enhancement to velocity selective discrimination of neural recordings: Extraction of neuronal firing rates.," *Annu. Int. Conf. IEEE Eng. Med. Biol. Soc.*, vol. 2014, pp. 4111–4, Aug. 2014.
[11] B. Metcalfe, D. Chew, C. Clarke, N. Donaldson, and J. Taylor, "Fibre-selective discrimination of physiological ENG using velocity selective recording: Report on pilot rat experiments," *Annu. Int. Conf. IEEE Eng. Med. Biol. Soc.*, vol. 2014, pp. 2645–8, Aug. 2014.

Resource optimized processor for Real-time neural activity monitoring

Y. Bornat, A. Quotb, N. Lewis, S. Renaud
Bordeaux INP, IMS, UMR 5218,
Univ. Bordeaux, IMS, UMR 5218,
CNRS, IMS, UMR 5218,
F-33405 Talence, France
Yannick.bornat@ims-bordeaux.fr
Noelle.lewis@ims-bordeaux.fr
Sylvie.renaud@ims-bordeaux.fr

Keywords-Real-time computing; neural activity detection, closed-loop systems

ABSTRACT

Closing the loop between living tissues and electronics is at the basis of the treatment of numerous pathologies or disabilities with prosthetic devices: artificial pancreas, neuroprostheses, BCI, etc. Biosignals are acquired and processed to detect a signature that finally controls actuators (insulin delivery pump, electrical stimulator, etc.) This signature is mostly based on action potentials from excitable cells or other spike shaped waveforms recognition.

The in vivo application of such a paradigm implies severe constraints on speed and consumption at all loop processing stages. In the research described in this paper, we optimized computation in the first stage of an acquisition system dedicated to closed-loop living/artificial experiments. The described functions are wavelet-based spike detection and slow signal filtering. Most of these functions are well known from the digital processing point of view like in [1]. We focus here on the optimization of these algorithms to reduce hardware resources while keeping a strong constraint on the computation time, and easy scalability for massive multichannel recordings. Our main constraint is to keep real-time computing. Depending on the final application, this means that each computation step must be completed within the sampling period of the signal, or that all computation steps must be completed.

The real-time implementation of algorithms in hardware requires at least two steps: a deep change in the computation technique and the replacement of floating point operations by fixed point operations whenever it is possible.

The first step implies the exploitation of model properties that provide the expected results with a better fit of the constraints inherited from circuit design. It implies a complete pre-implantation simulation to determine the eventual loss of the computation change. Unlike generic signal processing problems, computing on biosignals allow the assumption that signals are continuous. Then, the chosen techniques generally come from the field of analog design [2]. The main issues are stability constraints that may affect regulation loops.

The second step requires a study of the signal dynamics to check the usefulness of eventual floating point operations. Using fixed point operations reduces system complexity dramatically and therefore improves the performance figures. An extreme optimization may be achieved replacing some constant multipliers by bit shift operations. Is such a case, the implementation effort is much higher and the gain is limited to the hardware resources implied in the saved multiplier.

The architecture is implemented on FPGA for prototyping and evaluation with living cells recorded on a 60 multi-electrode array. We describe the minimal requirements of the algorithms, present the computation architecture, and the required hardware resources, as well as the evolution of these resources depending on the number of recording channels.

[1] Hwan, K. K., and June, K. S. (2003). *A wavelet-based method for action potential detection from extracellular neural signal recording with low signal-to-noise ratio.* IEEE Trans. Biomed. Eng. 50, 999–1011.

[2] R. R. Harrison, *A low-power integrated circuit for adaptive detection of action potentials in noisy signals*, Engineering in Medicine and Biology Society. Proceedings of the 25th Annual International Conference of the IEEE in Seattle, 2003-2130, doi:10.1126/science.1065467.

2015 IEEE Computer Society Annual Symposium on VLSI

In-silico Phantom Axon: Emulation of an Action Potential Propagating Along Artificial Nerve Fiber

Olivier Rossel*, Fabien Soulier†, Serge Bernard‡, David Guiraud*, Guy Cathébras†

*INRIA/DEMAR, LIRMM, Montpellier, France
†Université Montpellier, LIRMM, Montpellier, France
‡CNRS, LIRMM, Montpellier, France
{1- 4}@lirmm.fr

Abstract—This paper presents an original method to emulate a single fiber action potential in a quasi-infinite conductive volume, suitable for reproducible testing of bio-potential recording systems. The model is accurate for reproducing the bio-potential even for a small electrode to fiber radial distances. Established current activities of axon is used and programmed in the developed FPGA-based instrument, the model takes into consideration action potential propagation properties, electrode to fiber radial distances, medium conductivity. This paper investigates differences in the action potential amplitude for two longitudinal probe positions one in front of a node of Ranvier (NOR) and one between two NOR, for a large range of radial distances. Results are reported and compared with simulation with a correlation level of 97.6 %. The model is realistic enough to help the design of new recording systems.

I. INTRODUCTION

ElectroNeuroGram (ENG) recordings were investigated for years in order to provide useful data to control neuroprosthetic devices [1] [2] [3] For chronic applications, signals can be acquired through cuff electrodes with various geometries.The number and location of contacts become critical to get richer signals. The sizes of contacts are becoming smaller, distances between them could be reduced, increasing the spatial discretization.

About cuff electrode design, researches focus on increasing the efficiency of ENG measurements, such as optimizing Signal to Noise Ratio (SNR) [4] [5] [6], increasing selectivity linked to axon location (spatial selective recording SSR) and types of fiber (velocities selective recording VSR). SSR and VSR are mainly achieved through a wide variety of designs dealing with the number and position of contacts [7], [8] and [9] [10] (respectively). Generally the theoretical development are based on single fiber action potential (SFAP) properties and are then generalized to compound action potential (CAP). Each design has to be validated. First, electrode prototypes are designed and tested through numerical models. Then, theoretical benefit of new electrode prototypes are confronted to *in-vivo* experimental validation. But it exists an intermediate step: the validation through emulation,being increasingly used.

The emulation of nerve property through *in-silico* experiments presents an intermediate validation step. It avoids the constraints of *in-vivo* experiments but is closer to the real-case compared to numerical model. It often consists in emulating the recording of one SFAP or CAP [11] [12], or a single

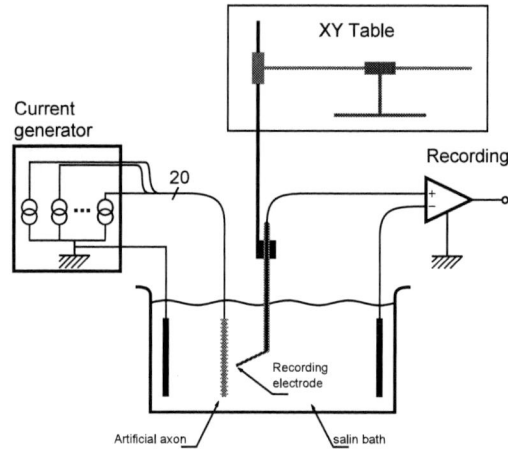

Figure 1. Experiment principle.

NOR [13]. L. Andreassen [13] proposed a method based on this principle to study the spatial transfer function of a cuff electrode. This paper demonstrates that SFAP can be built knowing this transfer function and the current AP on each NOR of an axon. In [11] [12] electrical model of nerve to AP and ENG recording is presented, the model is suitable for conduction velocity, and test different VSR.

The main focus of this work, is to develop a prototype able to emulate an axon fiber at the node of Ranvier (NOR) level allowing SFAP representation.

II. METHOD

We developed a 20 channels programmable and controlled current sources, to mimic ionic currents flowing through 20 NOR of a natural axon. To do so, a simulation of electrical activity of a nerve fiber is computed, the resulting ionic current are programmed in the device. The programmed current flow through 20 contacts disposed in line, placed in a saline bath, it is the phantom axon. Then, measurement of extracellular potential in the saline is carried out using a monopolar probe. A XY table controlled by the computer is used for setting the position of the measuring probe.

978-1-4799-8720-7/15 $31.00 © 2015 IEEE

A. Current sources and measurement setup description

To generate the current on each artificial NOR (Fig. 1), we designed a programmable current generator with 20 independent channels. The desired current is set and programmed on a FPGA (Field Effect Programmable Array, XILINX *Nexys 2*) that controls 20 12-bit digital-to-analog converters (*Analog Device AD7564*). These converters produce a differential current. A current mirror based on bipolar transistors is used to provide a bidirectional current (positive or negative) from the differential output of the DAC. These currents are delivered on the contacts of the electrode. It is well known that DC component in the generated currents could lead to rapid deterioration of electrode contacts. We thus added a serial capacitor on each output channel to ensure a null charge balance. To cancel voltage drift, a discharge circuit is added between the output of the current generator and the electrode. At the end of a current generation phase, the discharge circuit switches the capacitor from the current generator to a resistor pulled to the ground voltage. Finally, no charge is accumulated and no voltage drift occurs.

The linearity of the multichannel current generator is limited by the characteristics of the DAC (accuracy of ± 1 LSB and ± 4 LSB of gain error) and the current mirror. To increase the accuracy, a digital correction is added to minimize errors and allow full consistency between the 20 channels. The correction relies on a look-up table (LUT) filled up from a calibration phase: first, real output from each code are measured, and then a corrected code is determined. Finally, we designed a highly accurate current generator, with a full scale of $800 \, \mu A$ (-400 to $+400 \, \mu A$), and an absolute error less than $0.4 \, \mu A$ (regardless of the digital code and the channel). Thus the relative error lies below 0.05%.

The contact of the monopolar probe is composed of a section of a stainless steel wire with a diameter of $75 \, \mu m$, surrounded by an insulating sheath. We perform a differential measurement with a reference electrode formed by a large contact located far enough to consiedr point like measurement.

A *Multichannel MPA32I* amplifier with a gain of 10, is used for amplifying the signal. It is then acquired via a Data Acquisition (DAQ) card from *National Instruments* (NI6259) connected to a computer. The measurement protocol follows these steps: first, a signal generated by the FPGA is used to switch on the current generator and the signal acquisition device. Then, data from the amplifier are recorded by the DAQ. Finally the DAQ generates a signal ($Trig - Acquisition$) to allow displacement of the probe to a new position.

B. Current configuration

We used a model of axon to generate simulated values of the transmembrane current as a function of time. We used the study [14] to set realistic parameters for the modeled axon, simulated with the *Neuron* software (http://www.neuron.yale.edu/neuron/).

We modeled an axon with 100 NOR, set the sample frequency to 400 kHz then we computed 1.28 ms of axon activity (corresponding to 512 time samples. To ensure a favorable

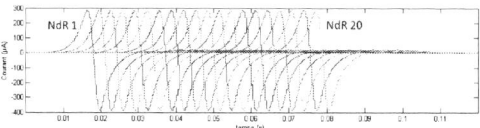

Figure 2. Current delivered by the 20 sources on the 20 contacts of the phantom axon. Only the first 300 time samples are represented, corresponding to the first 0.75 ms over 1.28 ms of the simulation.

SNR, the amplitude of nodal currents is set at $400 \, \mu A$ (it is around tens of nanoampere in a natural axon). The 20 central computed transmenbrane currents are saved in a database to be used as inputs to the 20 channels of the current generator. This database corresponds to a $20 * 512$ matrix Fig. 2). The current configuration of the 20 contacts is maintained during the measuring time.

C. Phantom axon

To represent an axon of $8.7 \, \mu m$ of diameter, with distance between NOR around $1.2 \, mm$, we used an electrode with 20 contacts, disposed in line, the distance between contact being $1.2 \, mm$. Practically we use a cochlear electrode (courtesy from MXM-Neurelec, Vallauris, France). The phantom axon is placed in a saline bath (sodium chloride 0.9%), which conductivity is close to *in-vivo* environment. The monopolar probe is used to measure the electric potential close to the fiber. This probe stays near the middle of the phantom, to avoid end effects. To get an accurate image of the potential around the phantom axon, the position of the measuring probe is controlled and adjusted via a micrometer screw system.

The XY table controlled by the computer (Fig.1) is used for setting the position of the measuring probe. Two longitudinal positions of the probe are set, the first one in front of a NOR and the second one equidistant to two adjacent NOR. 41 radial positions of the probe are set from $250 \mu m$ to $1000 \mu m$ (step of $250 \mu m$).

III. RESULTS

The measured signals are substantially identical to those expected. For example the correlation coefficient between measured and simulated potentials for the radial distances of $250 \, \mu m$ for two longitudinal positions; in front of a NOR and equidistant to two adjacent NOR (not shown) are respectively 99.8% and 99.6%. Comparison of the signals is then extended to all the measurement sites the total correlation index amounted to 97.6%.

For this example, it can be noticed that SFAP amplitude recorded in front of the NOR is twice that one measured between two NOR, ($0.6 \, \mu V$ Vs $0.3 \, \mu V$ reported in Fig. 3).

Quantitative results are presented in the Fig. 3, it shows the relationship between SFAP amplitude and electrode to fiber distance. A logarithmic scale representation is used for the radial distances and for the peak-to-peak value of AP .

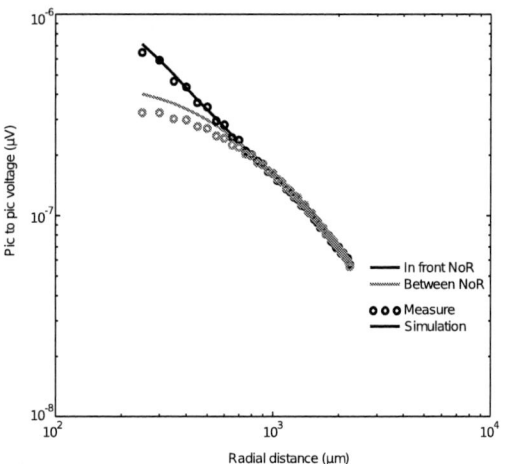

Figure 3. Comparisons between the measured attenuation in front of a NOR and between two NOR.

IV. DISCUSSION

Concerning the experimental setup and the phantom axon validity, the main difference between the phantom and a real axon is the size of the NOR. But using the duality of the transfer equation, modeling point-small NOR and a probe electrode of length L gives the same results as modeling NOR with a length of L and a point-small probe electrode.

The first contribution is that the measurements performed with the phantom are very close to the simulated potentials. The index of correlation between the two of 97.6 % is very high, proving the relevance of the simulation model and the technique used to emulate an artificial axon. Based on this validation, we demonstrate through both simulation and emulation that: i) the amplitude of the SFAP measured between two NOR is lower than in front of a NOR, for the same radial distance. ii) this effect is only observable at small radial distances, whereas for higher radial distances the difference decreases down to zero. This difference in amplitude of the AP could even be used to give an indirect estimation of the electrode-to-fiber distance.

So, selective and efficient recordings, sensitive to longitudinal positioning should be designed with small contacts and inter distances lower than 1 mm.

V. CONCLUSION

The simulation model together with the emulation were validated and gave consistent results. These results are also consistent with the literature. This setup can be used to evaluate and validate a large span of applications, such as recording device for SSR as well as VSR.

More complex configurations, such the cuff behavior are hard to simulate numerically but can easily be investigated with the phantom setup. As well, further experimental recording designs that can not be validated in true *in-vivo* environments can be evaluated with the phantom axon.

VI. ACKNOWLEDGEMENT

This work was supported by Axa research foundation.

REFERENCES

[1] X. Navarro, T. B. Krueger, N. Lago, S. Micera, T. Stieglitz, and P. Dario, "A critical review of interfaces with the peripheral nervous system for the control of neuroprostheses and hybrid bionic systems," *Journal of the Peripheral Nervous System*, vol. 10, no. 3, pp. 229–258, Sep. 2005. [Online]. Available: http://doi.org/10.1111/j.1085-9489.2005.10303.x

[2] G. S. Dhillon and K. W. Horch, "Direct neural sensory feedback and control of a prosthetic arm," *Neural Systems and Rehabilitation Engineering, IEEE Transactions on*, vol. 13, no. 4, pp. 468–472, Dec. 2005. [Online]. Available: http://doi.org/10.1109/TNSRE.2005.856072

[3] P. M. Rossini, S. Micera, A. Benvenuto, J. Carpaneto, G. Cavallo, L. Citi, C. Cipriani, L. Denaro, V. Denaro, G. Di Pino, F. Ferreri, E. Guglielmelli, K.-P. Hoffmann, S. Raspopovic, J. Rigosa, L. Rossini, M. Tombini, and P. Dario, "Double nerve intraneural interface implant on a human amputee for robotic hand control." *Clinical neurophysiology : official journal of the International Federation of Clinical Neurophysiology*, vol. 121, no. 5, pp. 777–783, May 2010. [Online]. Available: http://doi.org/10.1016/j.clinph.2010.01.001

[4] Z. Yang, Q. Zhao, E. Keefer, and W. Liu, "Noise characterization, modeling, and reduction for in vivo neural recording," in *Advances in Neural Information Processing Systems 22*, Y. Bengio, D. Schuurmans, J. Lafferty, C. K. I. Williams, and A. Culotta, Eds., 2009, pp. 2160–2168. [Online]. Available: http://books.nips.cc/papers/files/nips22/NIPS2009_0025.pdf

[5] E. Huigen, A. Peper, and C. a. Grimbergen, "Investigation into the origin of the noise of surface electrodes." *Medical and Biological Engineering and Computing*, vol. 40, no. 3, pp. 332–8, May 2002.

[6] X. Liu, A. Demosthenous, and N. Donaldson, "Platinum electrode noise in the ENG spectrum." *Medical and Biological Engineering and Computing*, vol. 46, no. 10, pp. 997–1003, Oct. 2008. [Online]. Available: http://www.ncbi.nlm.nih.gov/pubmed/18777185

[7] J. Zariffa, M. K. Nagai, Z. J. Daskalakis, and M. R. Popovic, "Influence of the number and location of recording contacts on the selectivity of a nerve cuff electrode," *Neural Systems and Rehabilitation Engineering, IEEE Transactions on*, vol. 17, no. 5, pp. 420–427, Oct. 2009. [Online]. Available: http://doi.org/10.1109/TNSRE.2009.2023301

[8] O. Rossel, F. Soulier, S. Bernard, and G. Cathébras, "New electrode layout for internal selectivity of nerves," in *Engineering in Medicine and Biology Society, 2009. EMBC 2009. Annual International Conference of the IEEE*, 2009, pp. 3798–3801. [Online]. Available: http://doi.org/10.1109/IEMBS.2009.5334437

[9] J. Taylor, N. Donaldson, and J. Winter, "Multiple-electrode nerve cuffs for low-velocity and velocity-selective neural recording," *Medical and Biological Engineering and Computing*, vol. 42, no. 5, pp. 634–643, 2004. [Online]. Available: http://doi.org/10.1007/BF02347545

[10] J. Taylor, M. Schuettler, C. Clarke, and N. Donaldson, "The theory of velocity selective neural recording: a study based on simulation," *Medical and Biological Engineering and Computing*, vol. 50, no. 3, pp. 309–318, Feb. 2012. [Online]. Available: http://doi.org/10.1007/s11517-012-0874-z

[11] R. Rieger, M. Schuettler, and S.-c. Chuang, "A device for emulating cuff recordings of action potentials propagating along peripheral nerves," vol. 22, no. 5, pp. 937–945, 2014.

[12] A. Al-Shueli, C. Clarke, N. Donaldson, and J. Taylor, "Improved Signal Processing Methods for Velocity Selective Neural Recording Using Multi-Electrode Cuffs," ... *Circuits and Systems*, ..., vol. 8, no. 3, pp. 401–410, 2014. [Online]. Available: http://ieeexplore.ieee.org/xpls/abs_all.jsp?arnumber=6605584

[13] L. N. S. Andreasen, J. J. Struijk, and M. K. Haugland, "An artificial nerve fiber for evaluation of nerve cuff electrodes," in *Engineering in Medicine and Biology Society, 1997. Proceedings of the 19th Annual International Conference of the IEEE*, vol. 5, 1997, pp. 1997–1999 vol.5.

[14] C. C. McIntyre, A. G. Richardson, and W. M. Grill, "Modeling the excitability of mammalian nerve fibers: Influence of afterpotentials on the recovery cycle," *Journal of Neurophysiology*, vol. 87, no. 2, pp. 995–1006, 2002. [Online]. Available: http://jn.physiology.org/content/87/2/995.abstract

2015 IEEE Computer Society Annual Symposium on VLSI

A Simplified Phase Model for Oscillator Based Computing

[1]Yan Fang, [2]Victor. V. Yashin, [3]Donald M. Chiarulli, [1]Steven P. Levitan

[1]Department of Electrical and Computer Engineering, [2]Department of Chemical Engineering, [3]Department of Computer Science

University of Pittsburgh

Pittsburgh, PA 15260

Abstract—Building oscillator based computing systems with emerging nano-device technologies has become a promising solution for unconventional computing tasks like computer vision and pattern recognition. However, simulation and analysis of these systems is both time and compute intensive due to the nonlinearity of new devices and the complex behavior of coupled oscillators. In order to speed up the simulation of coupled oscillator systems, we propose a simplified phase model to perform phase and frequency synchronization prediction based on a synthesis of earlier models. Our model can predict the frequency locking behavior with several orders of magnitude speedup compared to direct evaluation, enabling the effective and efficient simulation of the large numbers of oscillators required for practical computing systems.

Keywords—Non-Boolean Computing, Oscillator Based Computing, Coupled Oscillators, Phase Model, Synchronization.

1. INTRODUCTION

Pursuing high-density, low-power, high-speed computing systems for the post-CMOS era drives researchers to exploit the potential of emerging nano-device technologies. Based on recent advances in emerging nano-devices such as spin torque oscillators (STO) [1][2], resonate body transistor oscillators (RBO) [3], and vanadium oxide oscillators (VO2) [4], systems built from coupled nano-oscillators have become promising candidates for next generation computing structures used in intelligent information processing [5]. Inspired by the interaction between neural oscillations that occurs at different scales in biological systems, Hoppensteadt and Izhikevich developed an associative memory model of coupled oscillators by using phase locked loops and provide examples about how this dynamic system performs recognition by forming attractor basins at the minima of a Lyapunov energy function [6].

The essential idea of utilizing coupled oscillator systems to perform computation lays in the energy transfer in a dynamical system. Initialized with input information, a number of oscillators interact and exchange energy with each other, making the whole dynamical system converge from a perturbed state to a stable state. This process brings several advantages: First, it provides a high level multi-dimension norm, like the Euclidean distance, between sets of input vectors. Compared to the Hamming distance computed with exclusive-OR Boolean operations, the oscillator clusters are capable of processing matching operations in parallel and giving a robust pattern match computation. The high frequency of oscillations of new devices like STOs means that the systems can converge extremely fast. More importantly, this computing structure is very suitable for large high dimension data sets due to its high scalability and the degree of match that spans all of the dimensions of a input vector without any arithmetic calculations. Recently, research on

these systems has been conducted not only from the perspective of devices and circuits [7][2], but also at the level of algorithms and architectures and present structures that address the scalability of oscillator computing models [8].

In order to design and build these systems, we need a good understanding of the behavior of coupled oscillators, including the synchronization and de-synchronization between oscillators, the prediction of the coupled oscillators' frequencies, and their relation to a "degree of match" function in pattern matching. Unfortunately, accurate simulation and analysis from the device level to the architecture level is difficult and computationally expensive for current EDA/CAD tools. This is because of both the complexity of the nano-oscillators' device models and the short time scales necessary to capture the coupling dynamics. Thus, circuit and system designers cannot model systems at very large scales, while algorithm designers and architects can only use approximate distance norms instead of using the oscillators' true behaviors. To address this problem, we focus on modeling the ensemble behavior of coupled oscillators.

Winfree [9] and Kuramoto [10] made fundamental contributions to the understanding of these systems. Hoppensteadt and Izhikevich studied weakly coupled oscillator networks and unified previous work by using the idea of the phase resetting curve (PRC) [11]. Recently, Roychowdhury proposed a nonlinear phase model based on a perturbation projection vector (PPV) to study the perturbations due to noise of electronic oscillators [12] and used this to model coupled oscillators. These models were widely used in modeling the injection locking of oscillators to external signals [13]. Mafizzoni developed this model one step further and analyzed weakly coupled oscillators from the perspective of multi-frequency analysis [14]. The contribution of this work is the synthesis of oscillator phase model with the PPV model to provide a new model that simplifies the analysis of large systems.

The rest of this paper is organized as follows. First, we study different phase models for coupled oscillators and show the equivalency between them. Then, we combine ideas from these models to abstract a simplified model for oscillator based pattern matching operations. Finally, we provide comparative simulation results and conclude with observations about the effectiveness of our model and future work.

2. PHASE MODELS

In this section, we review phase models proposed in previous work, including the PPV model [12], Winfree and Malkin's approach summarized by Izhikevich [11]. By discussing the

978-1-4799-8720-7/15 $31.00 © 2015 IEEE   231

relation between these models and combining their advantages, we develop our new model.

2.1 PPV Model

We start the introduction of the model with the PPV model, since it is derived from the demand for a theoretical analysis of oscillator circuits perturbed by noise and therefore is easier to understand from an electrical engineering perspective compared to the other models.

The PPV model [12] starts with the general differential equation of an oscillator,

$$\dot{x}(t) = f\big(x(t)\big) + P(t) \qquad (1)$$

where x(t) is the vector of oscillation states and $\dot{x}$(t) is their derivatives. In real circuits, these states are usually the voltages or currents of nodes. P(t) represents an external perturbation on the oscillation, which can be noise, signal injection, or a coupling term from other oscillators, as in our case. f is the nonlinear function that describes the oscillation and t is time. Then, the response solution from this equation can be written as,

$$x_c(t) = x_s(t + \alpha(t)) \qquad (2)$$

where $x_s(t)$ is the oscillator's natural response without any perturbation, namely $P(t) = 0$, while $x_c(t)$ is the response with perturbation. $\alpha(t)$ represents the time shift of phase that is caused by the perturbation. Hence equation (2) reveals the phase relation between the natural response and perturbed response of the oscillator. According to the PPV model, $\alpha(t)$ can be obtained by solving the equation,

$$\dot{\alpha}(t) = \Gamma(t + \alpha(t))C(t) \qquad (3)$$

where $\Gamma(t)$ is the perturbation projection vector (PPV) that describes the perturbation response of the oscillators. The PPV can be thought of as the time-varying sensitivity of the induced time shift to the given injected perturbation [13]. The theoretical derivation and proof of the PPV method can be found in [12]. The PPV $\Gamma(t)$ is usually a periodic function that can be obtained either numerically or analytically [12][13]. After acquiring $\Gamma(t)$, we know the time shift of phase $\alpha(t)$ from the solution of equation (3) and then solve equation (2).

This model actually does not predict the "phase change" but a time shift function of oscillation response. Thus, it is difficult for us to use this model to directly predict the frequency shift in a coupled oscillator system. The next models we introduce provide us further insight.

2.2 Izhikevich's Model

In [11], the phase model of weakly coupled nonlinear oscillators is explored from an abstract view, by unifying some earlier models. Differing from the PPV model, this model assumes all the oscillators share the same free-running frequency, so the "phase" of nonlinear coupled oscillators can be normalized and defined as,

$$\theta(t) = t + \varphi(t) \qquad (4)$$

Taking the derivative from both sides, we have

$$\dot{\theta}(t) = 1 + \dot{\varphi}(t) \qquad (5)$$

In equations (4)(5), $\theta(t)$ is the defined phase, a periodic function with period T = 1. $\varphi(t)$ is called the phase deviation, caused by the coupling from other oscillators. We can notice that when there is no coupling term, the phase term is simply time, t, and the free-running frequency is normalized to 1. The derivative $\dot{\varphi}(t)$ represents the change of phase deviation, namely the frequency shift, due to the coupling effect.

In order to map this model to various nonlinear oscillators, the key point lies in the phase deviation $\varphi(t)$, described by:

$$\dot{\varphi}(t) = Q(\theta(t))P(t) \qquad (6)$$

which has a similar form to equation (3). $P(t)$ is the same external injection signal to the oscillators (i.e., the coupling term). The function $Q(\theta)$ is called the phase response curve or phase resetting curve (PRC), representing how sensitive the phase deviation is in response to P(t) at a specific phase $\theta(t)$. Thus (5) can also be written as,

$$\dot{\theta}(t) = 1 + Q(\theta(t))P(t) \qquad (7)$$

The mechanism of phase and frequency of coupled oscillators revealed by this equation was discovered multiple times in the early studies of oscillator phase models. But the researchers named it PRC in different ways and exploited different methods to derive and utilize it.

A theorem first proposed by Malkin in 1949 [17] and later abstracted by Hoppenstead in 1997 indicates that $Q(\theta)$ can be computed by solving the linear adjoint equation,

$$\dot{Q}(t) = -\mathcal{J}[f\big(x(t)\big)]^T Q \qquad (8)$$

where $\mathcal{J}[f\big(x(t)\big)]^T$ is the transposed Jacobian matrix of the oscillation function f in (1). This theorem is identical to Kuramoto's approach in 1984, where the gradient of phase plays the role of PRC.

In Winfree's work (1967), PRC was experimentally approached by applying a pulse stimuli with amplitude A. Then a function called the linear response or sensitivity function $Z(\theta)$ was measured by observing the phase shift caused by the stimuli, and PRC is $Z(\theta)$ divided by amplitude A.

$$PRC(\theta) \approx \frac{Z(\theta)}{A} \qquad (9)$$

2.3 Our Simplified Model

From the previous models discussed above, if we pair equations (1)(4), (2)(5), (3)(6) and compare them, we can notice that they have the same pattern because the intrinsic method behind these models is the same, which is to quantify how the oscillation is affected by the external perturbation. However, these models use different methods to calculate this term, either an analytical derivation or a numerical measurement. In addition, the PPV model studies the oscillation variables while Izhikevich's model focuses on the "phase" of nonlinear oscillators.

From the view of solving the practical problem we are addressing, the pattern matching operation is performed by frequency shifting or frequency locking of oscillators caused by coupling. The elements of the pattern vectors are represented by frequencies and the degree of match is evaluated by how well the oscillators synchronize. Thus, for the purpose of predicting the frequencies of coupled oscillators, we introduce the phase definition idea from Izhikevich's model into the PPV model.

We assume we have n oscillators with different frequencies: $\omega_0, \omega_1, \omega_2 \dots \omega_{n-1}$. Since the equation (4) and (5) require that oscillators run at the natural frequencies normalized to 1, we scale these frequencies to $1, \lambda_1, \lambda_2 \dots \lambda_{n-1}$, where $\lambda_i = \frac{\omega_i}{\omega_0} = \frac{T_0}{T_i}$. So for an arbitrary oscillator i, in the PPV model, equation (2) can be changed into the phase form similar to (4):

$$\theta_i(t) = \lambda_i\big(t + \alpha(t)\big) = \lambda_i t + \varphi_i(t), \qquad (10)$$

with

$$\varphi_i(t) = \lambda_i\big(\alpha(t)\big), \qquad (11)$$

(10) indicates the relation of phase deviation in Izhikevich's model and time shift of phase in the PPV model. Taking the derivative of (9) we get,

$$\dot{\theta}_i(t) = \lambda_i + \lambda_i\,\dot{\alpha}(t), \qquad (12)$$

substituting with (3) we have,

$$\dot{\theta}_i(t) = \lambda_i + \lambda_i\,\Gamma(\theta_i(t))P_i(t) \qquad (13)$$

where $\Gamma(\theta_i(t))$ is still the PPV in (3) but determined by the phase, instead of time. This equation transfers the PPV from a time domain to the phase domain and replaces the simulation time span into the number of oscillation cycles. $C_i(t)$ is the coupling term in this model, defined as:

$$C_i(t) = \sum_{j=0}^{n} g_{ij}x_s(\theta_j(t)) \qquad (14)$$

where g_{ij} is the coupling coefficient and j is the index of other oscillators in the system. Therefore, solving (12) can provide us the frequency and phase response of a coupled oscillator system.

In our simplified model, the PPV function is also equivalent to the PRC function obtained from other methods. In the next section we give examples of these methods. We note that it is useful to have several methods available because some methods might prove inaccurate or fail to converge for specific nonlinear oscillator systems.

3. OSCILLATOR EXAMPLES

In this section we use the nonlinear ring oscillators as an example and demonstrate three different methods to obtain their PPV/PRC function.

A simple ring oscillator consists of three inverters and RC circuits, shown in Figure 1(a) with its analytical model given in [13]. The voltage derivatives of the three nodes are,

$$\dot{v}_i(t) = \frac{f(v_{i^-}(t)) - v_i(t)}{RC}, i = 1,2,3 \qquad (15)$$

where i^- is the previous node of i and $f(v(t))$ is the simplified response of an inverter:

$$f(v) = \begin{cases} +1, & if\ v > 0 \\ -1, otherwise \end{cases} \qquad (16)$$

By normalizing the standard frequency and period to 1, we can write the voltage state response of three nodes from [18] into,

$$v_1(t) = \begin{cases} 1 - \psi e^{-\gamma t},\ 0 \leq t \leq 1/2 \\ -1 + \psi e^{-\gamma\left(t-\frac{1}{2}\right)}, 1/2 \leq t \leq 1 \end{cases}$$

$$v_2(t) = v_1\left(t - \frac{2}{3}\right), v_3(t) = v_1\left(t - \frac{1}{3}\right) \qquad (17)$$

where $\psi = \frac{1+\sqrt{5}}{2}, \gamma = 6\ln(\psi)$. Because the frequency here is 1, $RC = 1/\gamma$. Similarly, the PPV equation in [18] can be analytically solved as,

$$\Gamma_1(t) = \Gamma_3\left(t - \frac{2}{3}\right)$$

$$\Gamma_2(t) = \Gamma_3\left(t - \frac{1}{3}\right)$$

$$\Gamma_3(t) = \gamma^{-1}\frac{1+\psi^3}{4-2\psi^3}\left(\psi + 2\left[-u(t) + (-1 + 2\psi^{-1})u(t - \frac{1}{2})\right]e^{\gamma t}\right) \quad (18)$$

Figure 1(b) shows the waveform of oscillation response and PPV of node v3. From this plot we see that the PPV function for ring oscillators is periodic but not linear.

Sometimes, it is difficult to obtain the state response and PPV function directly from the ODE. For these cases we can obtain the corresponding PRC by applying Malkin's approach numerically. Solving equation (8) is actually very similar to the analytical derivation of PPV in [15]. However, when the analytical method does not work, we can use a technique called backward integration to obtain the Jacobian matrix [13]. Figure 2 shows the results of PRC from this method with a backwards integral of 4 cycles. The PRC in the first cycle is the one used.

In a few cases when a system's Jacobian matrix does not exist, Winfree's approach could be the only choice, especially for those nonlinear oscillators with complex mathematical

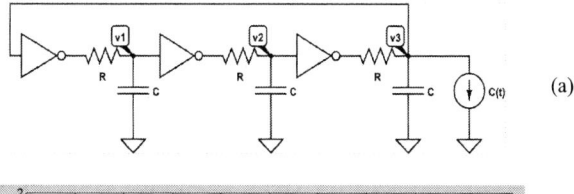

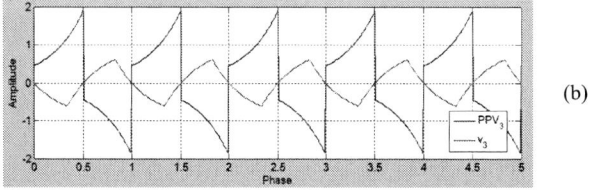

Figure 1. Ring oscillator model. (a) Simple schematic model; (b) Output waveform (red) and (PPV) (blue) at node v3.

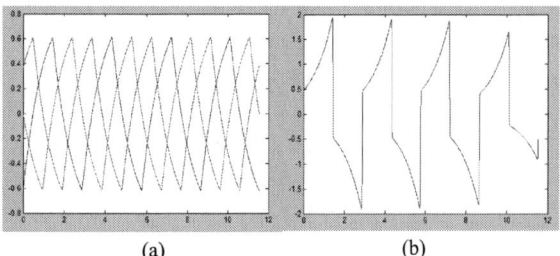

(a)   (b)

Figure 2. Malkin's approach for PRC. (x-axis: phase, y-axis: amplitude) (a) Oscillation waveform of voltage at three nodes; (b) PRC obtained from backward integral.

models, even though this method is "experimental" and tends to be inaccurate. As an example, we apply Winfree's method for the ring oscillator by adding a pulse stimulus with small amplitude and measuring the phase resetting curve step by step. Figure 3 illustrates the PRC generated by Winfree's method with different stimuli amplitudes. The glitches in the curve show the problems of this approach.

However, it is worth noting that the PRC amplitude here is proportional to the stimuli amplitude by factor of 2, which not only corresponds to equation (9) but also fits the PPV/PRC amplitude of the previous two methods in Figure 1(b) and Figure 2 (b). These three examples for phase deviation of the ring oscillator indicate that PPV and PRC functions are identical to each other and enhance the foundation of our model.

4. EXPERIMENTS AND SIMULATION

4.1 System Configuration

In this section, we apply our simplified phase model to coupled oscillator systems and analyze their synchronization behavior. We also compare the performance and efficiency of the models obtained by the different approaches as well as accuracy and speedup compared to the direct simulation of the oscillator systems.

Figure 4 gives a circuit example of our coupled oscillator system for pattern matching. For ring oscillators, the frequency of each oscillator is adjusted by two input control voltages and the coupling node is the output node of the third inverter. These oscillators are coupled to each other through a common node. The coupling component could be resistor or capacitor, chosen by the circuit designer, where larger

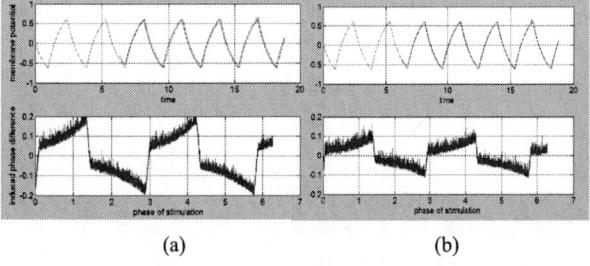

(a)   (b)

Figure 3. Winfree's approach for PRC, (a) Stimuli Amplitude=0.1, PRC Amplitude=0.2; (b) Stimuli Amplitude=0.05, PRC Amplitude=0.1.

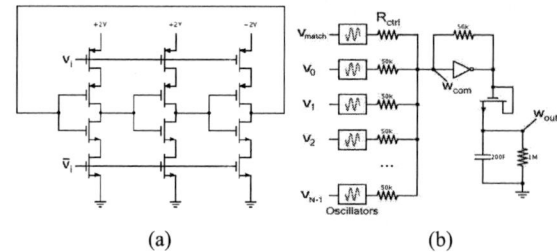

(a)   (b)

Figure 4. Coupled oscillator system. (a) Voltage controlled ring oscillator; (b) Oscillator coupled together through one common node with synchronization detection circuits.

resistors would give weaker coupling strength. The detection circuits that read out the degree of synchronization can be found in [8].

As we discussed in the previous section, such a weakly coupled system can be simply described by equations (13) and (14). In this structure the coupling strengths between oscillators are identical. Thus, for each oscillator (with coupling at node v3 in Figure 1) we have,

$$\dot{\theta}_t = \lambda_i + \varepsilon \lambda_i \, \Gamma_{node3}(\theta_i) \sum_{j=0}^{n} v_{node3}(\theta_j), i, j \in [1, n] \ (19)$$

where ε is the coupling coefficient.

4.2 Oscillator Behavior Analysis

We start with a three oscillator system (n = 3). It is convenient for us to predict the final frequency of each oscillator by solving (19) numerically. If $\dot{\theta}_1 = \dot{\theta}_2 = \dot{\theta}_3$, the system is synchronized and frequency locked. We use Matlab to run a simulation of equation (11). Figure 5 illustrates an example of locking and non-locking systems. For these examples, we set the initial frequencies to be: λ = [1, 0.95, 1.05] but use two different ε, 0.2 and 0.4. In the left plot, ε = 0.4 and the final frequencies are [0.8717, 0.8717, 0.8717]; in the right plot, ε = 0.2 and the final frequencies are [0.9644, 0.9298, 1.0263].

From (19) and these simulation results, it is worth noting that the conditions for frequency locking are determined by the coupling coefficient ε and the scaling ratio between each oscillators' free running frequency λ_i, not the absolute value of the frequencies. This interesting phenomenon is important for the design of future oscillator based computing systems. It implies that devices with high frequencies can provide wider bandwidth for information coding. Also, even when the

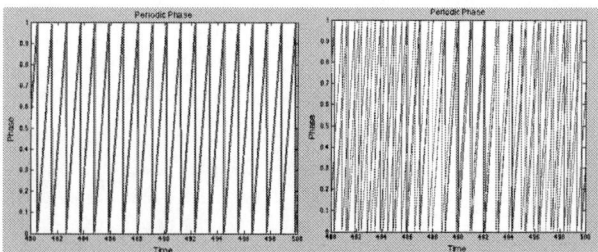

Figure 5. Simulations of three coupled oscillators with natural frequencies λ = [1.0, 0.95, 1.05]. Left: ε=0.4, Right: ε=0.2.

978-1-4799-8720-7/15 $31.00 © 2015 IEEE   234

oscillators fail to lock with each other, their frequencies are pulled closer to each other. It is based on this observation that we say that the degree of synchronization can provide a metric to measure the distance or similarity of each oscillator's initial frequency (and thus input) as discussed below.

To verify this point, we run the simulation multiple times by fixing the first oscillator's frequency and sweeping the frequency of the other two from 0.8 to 1.2. We use $1 - \{(f_2 - f_1)^2 + (f_3 - f_1)^2\}$ as the function to evaluate how well the oscillators synchronize. While there are many possible metrics, like standard deviation of frequencies, that can analyze the behavior of coupled oscillators, here we choose the inverted Euclidean distance. We use this function to capture the synchronization of oscillators in order to show the difference between the phase model and direct simulations and thus to verify the capability of our phase model in simulation of oscillator based computing. Note that f here is the coupled scaled frequency. In this test, we use three different methods to obtain the PPV/PRC for our model, as we did in section 3. In addition, we directly simulate the coupled oscillator without

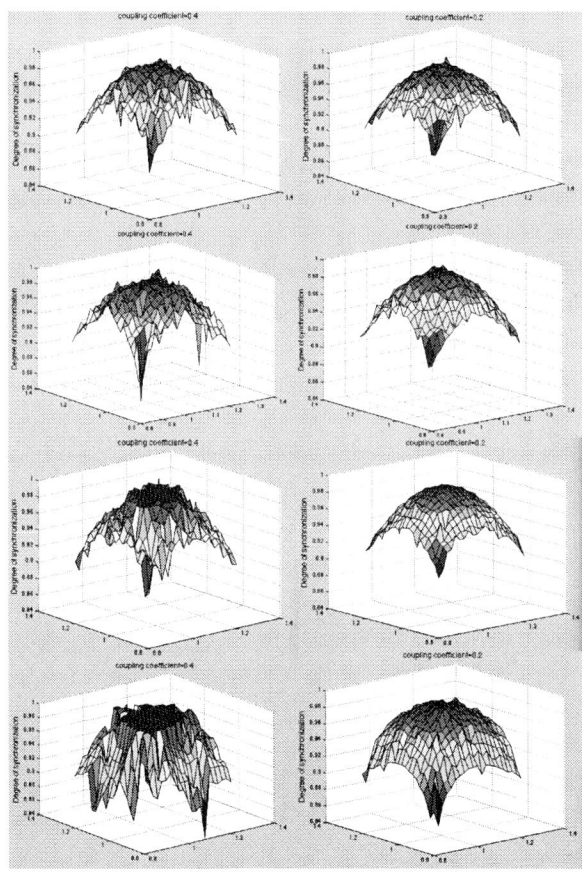

Figure 6. Degree of synchronization as inverted Euclidean distance, initial frequencies for a range $\lambda_{2,3} = [0.8 \text{ to } 1.2]$ and a fixed frequency of $\lambda_1 = 1.0$, Left column: ε=0.4, Right column: ε=0.2; Row 1: Winfree's approach, Row 2: Malkin's approach, Row 3: Analytical PPV, Row 4: Direct simulation.

using any phase model as a performance standard. Figure 6 demonstrates all the cases with 3D plots of the degree of synchronization for the two cases of ε = 0.4 and ε = 0.2. When the coupling strength is weak, the surface is smooth and the initial frequencies are easier to differentiate, while stronger coupling gives us more nonlinear features and a wider locking range. The flat area on the top of the surface indicates the highest degree of synchronization, giving us the frequency locking range for the simulation sweeps. Hence, a very strong coupling system may lack differentiation for pattern matching or nearest neighbor searching. But for clustering operations like image segmentation, stronger coupling strength can provide better resistance to noise.

Due to the fast simulation speed, our model is also very suitable for simulating systems with large numbers of coupled oscillators. To understand how the number of oscillators can influence the synchronization, we run the simulations with the same two dimensional frequency sweeping but different numbers of oscillators. Since we cannot show a plot of higher dimensional frequency sweeping, we keep the frequencies of all but two of the oscillators fixed to 1 and sweep the last two. Figure 7 shows the results for n=3, 4, 8, and 16. In these simulations, we use the analytical PPV as the PPV/PRC function.

From these results we can notice that larger numbers of oscillators with the same frequency gives a larger frequency locking range. Namely, the effective coupling strength to those oscillators with different frequencies becomes stronger because the system's stable state is close to the state of the majority of the oscillators. This could either be an advantage or disadvantage in the design of oscillator based systems, depending on the application and computation required.

4.3 Performance and Speedup

For performance comparisons of our phase model, we calculate the root mean square error between the direct simulation of the oscillators in each case of Figure 6 and the analytical PPV. As we can see in Table 1, the error is relatively small compared with the absolute value of degree of synchronization. This shows that our phase model is compatible with different methods for PPV/PRC and robust to the variations between these methods. Compared to the direct oscillator simulation, the analytical PPV generates the smallest error while Winfree's experimental method gives the largest error, and Malkin's numerical method lies in the middle.

Table 1. RMSE of simulations based on different methods

Comparing	Winfree's	Malkin's	Analytical PPV
Oscillator, $\varepsilon = 0.2$	127e-04	133e-04	121e-04
Oscillator, $\varepsilon = 0.4$	276e-04	297e-04	271e-04
Analytical PPV, $\varepsilon = 0.2$	58e-04	63e-04	/
Analytical PPV, $\varepsilon = 0.4$	128e-04	142e-04	/

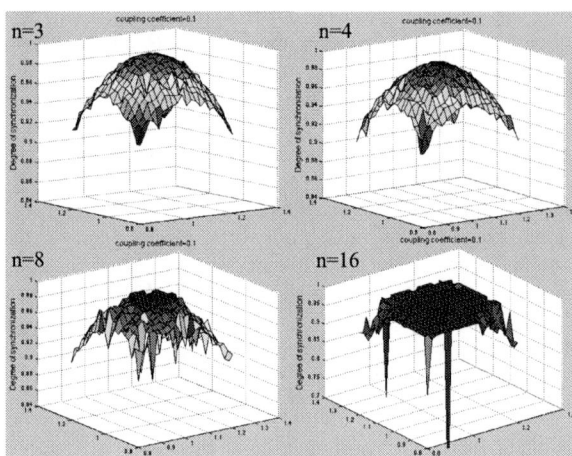

Figure 7. Simulations of different numbers of coupled oscillators. All oscillators but two are kept at frequency 1, while the last two are swept.

We next evaluate the efficiency of our model by comparing simulation speed to the direct simulation of the oscillator network. The speedup here is defined by the ratio of real time for simulation of two methods (both in Matlab) for the same length of simulation time. Since we always initialize the systems with random phases, we run the test of each configuration 100 times and average the speedup for evaluation. The results in Figure 8 only serve as an approximation of the efficiency of our model because the simulation of coupled oscillator systems is affected by multiple factors, such as the oscillator model, initial states, and the convergence process. Nevertheless, we still observe very promising speedups from our simplified phase model, similar to [14]. The speedup comes from the fact that the differential equations of the original oscillators are nonlinear but the simplified phase models are linear equations with much simpler periodic functions. A second advantage is that in the phase domain, simulation is done by fractions of cycles, rather than time steps so that run time is frequency invariant. This is an advantage in simulation when compared to the regular PPV model where the user must optimize the time step for accurate simulation vs. performance.

5. CONCLUSIONS

In this paper, we introduce the problems of oscillator simulation encountered in the design of oscillator based computing systems. To address this problem, we review previous phase models and propose a reduced simple phase model. We apply our model in the analysis of ring oscillators. The results show that our model is capable of predicting the frequency and phase of coupled oscillator systems with small errors compared to direct simulation of the oscillator model. The main contribution of our model is simplifying the nonlinear equations and transferring simulation from traditional time domain into phase domain. This provides very promising simulation speedup as the size of the system increases. Furthermore, we demonstrate that this model is particularly suitable for simulation and analysis for oscillator

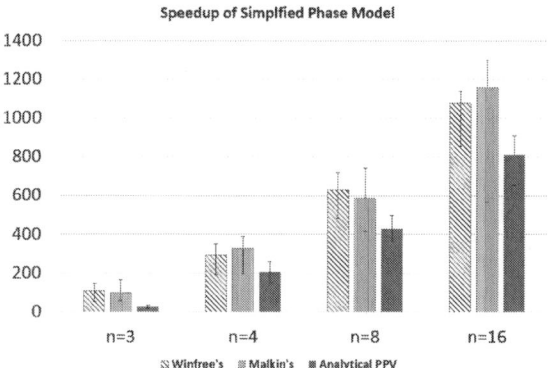

Figure 8. Speed-up (factor) of simplified phase model over direct simulation for different PPV/PRC methods.

based computing operations like pattern matching and nearest neighbor search.

6. ACKNOWLEDGEMENTS

This work was supported in part by the National Science Foundation under grants DMR-1344178 and CCF-1317373.

7. REFERENCES

[1] M.R. Pufall, W. H. Rippard, S. Kaka, T. J. Silva, S. E. Russek, "Frequency modulation of spin-transfer oscillators", *Applied Physics Letters*,86(8), 082506-082506, 2005.

[2] Locatelli, N., Cros, V., & Grollier, J. (2014). Spin-torque building blocks. *Nature materials*, *13*(1), 11-20.

[3] Weinstein, D., & Bhave, S. A. (2010). The resonant body transistor. *Nano letters*, *10*(4), 1234-1237.

[4] Shukla, N., et al. (2014). Synchronized charge oscillations in correlated electron systems. *Scientific Reports, 4*.

[5] Cotter, M. J., Fang, Y., Levitan, S. P., Chiarulli, D. M., & Narayanan, V. (2014, July). Computational architectures based on coupled oscillators. In *VLSI (ISVLSI), 2014 IEEE Computer Society Annual Symposium on* (pp. 130-135).

[6] Hoppensteadt, F. C., & Izhikevich, E. M. (2000). Pattern recognition via synchronization in phase-locked loop neural networks. *IEEE Transactions on Neural Networks*, *11*(3), 734-738.

[7] Shibata, T., Zhang, R., Levitan, S. P., Nikonov, D. E., & Bourianoff, G. I. (2012). CMOS supporting circuitries for nano-oscillator-based associative memories. In *Cellular Nanoscale Networks and Their Applications (CNNA), 13th IEEE International Workshop on* (pp. 1-5).

[8] Levitan, S. P., Fang, Y., Dash, D. H., Shibata, T., Nikonov, D. E., & Bourianoff, G. I. (2012). Non-Boolean associative architectures based on nano-oscillators. In *Cellular Nanoscale Networks and Their Applications (CNNA), 13th IEEE International Workshop on* (pp. 1-6).

[9] Winfree, A. T. (1967). Biological rhythms and the behavior of populations of coupled oscillators. *Journal of theoretical biology*, *16*(1), 15-42.

[10] Kuramoto, Y. (1984). *Chemical oscillations, waves, and turbulence*. Springer.

[11] Izhikevich, E. M. (2007).Synchronization, Chapter 10 in *Dynamical systems in neuroscience*. MIT press.

[12] Demir, A., Mehrotra, A., & Roychowdhury, J. (2000). Phase noise in oscillators: a unifying theory and numerical methods for characterization. *Circuits and Systems I: Fundamental Theory and Applications, IEEE Transactions on*,47(5), 655-674.

[13] Lai, X., & Roychowdhury, J. (2005, September). Analytical equations for predicting injection locking in LC and ring oscillators. In *Custom Integrated Circuits Conference, 2005. Proceedings of the IEEE 2005* (pp. 461-464).

[14] Maffezzoni, Paolo. "Synchronization analysis of two weakly coupled oscillators through a PPV macromodel." *Circuits and Systems I: Regular Papers, IEEE Transactions on* 57.3 (2010): 654-663.

A Statistical Approach to Probe Chaos from Noise in Analog and Mixed Signal Designs

Ibtissem Seghaier, Mohamed H. Zaki, and Sofiène Tahar
Department of Electrical and Computer Engineering
Concordia University, Montreal, Quebec, Canada
Email:{seghaier, mzaki, tahar}@ece.concordia.ca

Abstract—Chaotic circuits have gained increasing attention in many engineering applications. Qualitative measures such as Lyapunov Exponent (LE) are the most common methods for identifying chaotic behavior. However, the use of these measures is limited due to the short output signal length and its contamination by noise. In this paper, we propose a novel methodology for modeling and detecting chaotic vs stochastic behavior in AMS designs. First, the design is modeled using a system of recurrence equations for analog and digital parts. Second, a surrogate generation method is performed. The obtained surrogates are a typical realization of the circuit output under the hypothesis that the circuits exhibits noise. Next, hypothesis testing with Gaussian Kernel measure as test statistic is conducted over these surrogates and the original circuit output to statistically assess the circuit behavior. The effectiveness of the proposed methodology is illustrated on several AMS circuits such as PLL or Colpitts oscillator. The obtained results show sufficient improvements over the existing methods. For instance, comparing with the LE method, our approach is an order of magnitude faster and provides a more accurate detection of the chaotic circuit behavior.

I. INTRODUCTION

Chaotic circuits have potential applications in various fields such as communication [1], signal processing [2], and neural networks [3]. This is mainly because they present an unpredictable behavior that resembles noise with a broadband spectrum, so they can be deployed in encryption and random noise generation. Chaos can also be used to enhance Analog and Mixed Signal (AMS) circuits performance such as reduction of idle channel tones and pattern noise in Σ-Δ modulators [4] and stabilization of PLL by broadening its capture range [5]. Nevertheless, due to the nanometer-scale technologies, AMS designs manifest large scale process variations that cannot be reduced by foundries. This deviation in the geometrical and electrical device parameters can compromise the circuit's usefulness. Thereby, it might behave chaotically instead of periodically on a parameter change. This will affect the circuit performance (e.g., PLL locking, circuit stability, etc.) and consequently cause design failure. Experiments showed that in specific operation conditions (i.e., circuit parameter, initial conditions, and input signals) even remarkably simple nonlinear circuits can exhibit chaotic behavior [6].

When irregularity is observed in an AMS design output, designers assumes that the circuit exhibits stochastic noise which is stubbornly present in such designs. However, this could emerge from purely deterministic chaotic nonlinear

circuit model. Because noise detection is a contentious issue for designers and there is a lack of efficient noise verification tools in AMS designs, a chaotic circuit could be analyzed to be noisy erroneously. It is fundamental therefore to thoroughly investigate and append the real source of an aberrant circuit behavior. To this end, there is a pressing need for handy chaos detection tools in AMS designs. Such tools can be used to probe the circuit dynamics at early stage and so assess the observed behavior (chaotic or stochastic) of the design.

In this paper, we propose a novel statistical verification methodology of AMS designs behavior, mainly chaotic and noisy dynamics. To speed up the verification runtime, the analog and digital components of the AMS circuit are modeled in a unified environment. The behavior of the circuit is described as function of the preceding state variables terms using Extended-System of Recurrence Equations (E-SRE) [7]. We then elucidate the intended property to be verified within the ambit of a null hypothesis H_0. Thereafter, we generate several artificial output of the circuit called surrogates. The generated surrogates should comply with the property being verified H_0 while preserving some features of the real circuit output. A discriminating statistic, namely the Gaussian Kernel measure, is then conducted for the original circuit output and all the generated surrogates. If the computed original output and surrogates statistics are significantly different, hypothesis testing technique reports a rejection of the null hypothesis H_0. As a consequence, we conclude that the circuit dynamic does not comply with its property.

The remainder of this paper is organized as follows: Related work is discussed in Section II. Section III provides an overview of the proposed methodology. Experimental results for the analysis of chaotic features on a Colpitts oscillator circuit, a third order Σ-Δ modulators and, a Phase Locked Loop are reported in Section IV. Section V summarizes the contributions of this paper and provides future work hints.

II. RELATED WORK

The study of chaotic features in electronic circuits is the subject of an extensive research. Chaos detection is complicated by the lack of an accepted formal definition of chaos. For instance, despite the fact that chaotic features have been observed in Σ-Δ modulators, no literature provides satisfyingly rigorous proof of the presence of chaos in such circuits for orders higher than two [8].

To demonstrate the chaotic behavior of electronic circuits, qualitative metrics such as bifurcation diagrams have been proposed. This measure permits to visualize the different behaviors of the system as parameters change. Visualizing such diagram for high dimensional systems is very difficult and resources hungry. The use of qualitative methods, namely Lyapunov Exponent measure [9], is another paradigm that has been adopted to quantify chaos. It indicates the average rates of convergence or divergence of nearby trajectories. A positive exponent implies divergence and is indicative of chaotic dynamics while a negative one implies convergence and is said to be periodic.

Lyapunov Exponent is defined as a limit when time t approaches infinity (Equation (1)), one encounters fundamental difficulties using it for a circuit simulated for a limited time.

$$\lambda_i = \lim_{t \to \infty} \frac{1}{t} ln|\sigma_i(t)|, \forall i \in [1,..,n] \qquad (1)$$

Where $\{\sigma_i\}_{i=1}^n$ are the eigenvalues of the Jacobian matrix of the circuit. This technique is hampered by technical issues related to the signal length and its contamination by noise (known as Perron effects); Hence, direct application of this measure on the circuit output might be inappropriate. To circumvent these shortcomings, we present, to the best of our knowledge, the first methodology to statistically investigate determinism in apparently stochastic AMS design behavior. Unlike the Lyapunov Exponent measure [9], the proposed methodology can also be adopted for circuit level AMS circuits outputs.

III. Proposed Methodology

An overview of our proposed methodology to statistically probe the dynamics (deterministic vs stochastic) of AMS designs is shown in Fig. 1. The AMS circuit behavior is modeled as a System of Recurrence Equations (SREs) that describes its behavior with/without noise. Recurrence equations are the discrete version of an analog differential equations. For the discrete components, the SREs are extended to E-SRE [7] by expressing them with *if-else* logical formulas as follows:

$$X_i(n) = f_i(X_j(n - \gamma)), \forall i, j, n \in \mathbb{Z} \qquad (2)$$

where f_i is a generalized *If-formula*. In this work, we are considering only thermal noise excitation that adheres to a Gaussian distribution with mean m, and standard deviation σ. The obtained model is evaluated for specific environment constraints, namely the initial values of the voltage and current state variables and simulation parameters (such as the total simulation time, and the simulation step size). Thereafter, we elucidate the property of interest ($\mathcal{P}$) that the circuit should comply with. The property to be verified in this paper can be phrased as follows: "*Is the observed random like behavior of the AMS design due to noisy or chaotic behavior?*". Hence, we define a null hypothesis, denoted by H_0, which assumes that the circuit exhibits stochastic noise and an alternative hypothesis H_1 that assumes the circuits to be purely deterministic. To verify the above mentioned hypotheses, the idea is to generate artificial circuit outputs called surrogates that are realizations

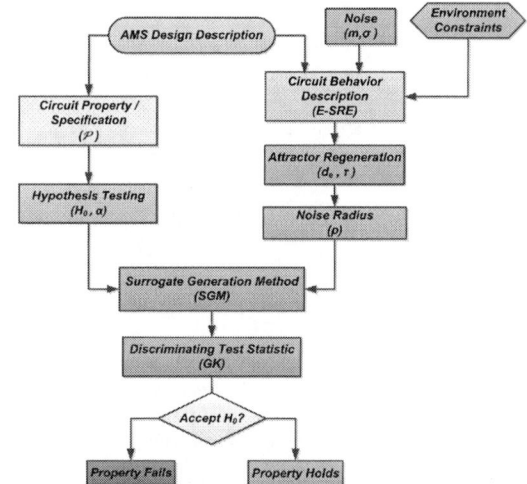

Fig. 1. Proposed statistical behavioral verification methodology

of what would the circuit output be if it was consistent with the property $\mathcal{P}$. Hence, these surrogates serve as a useful model against which to verify the real circuit output. They are carefully constructed from the circuit output so as to be free from any chaotic process while preserving some features of the original output.

To do so, we appeal to a different mathematical representation of the circuit behavior known as phase diagram in the nonlinear system theory. A phase diagram displays the circuit variables against each other and leaves time as an implicit dimension not explicitly graphed. The subset of this phase space toward which the circuit tends to evolve regardless of the initial conditions is called an attractor. This attractor will be used to predict the chaotic dynamics of the AMS circuit in order to consider them in the surrogates generation later. We applied the non uniform embedded window [10] and the false nearest neighbor method [11] for establishing optimal embedding parameters (d_e, τ) for the attractor reconstruction from the circuit output as depicted in Fig. 1. Thereafter, we determine the noise radius ρ that is the amount of noise that will obliterate the attractor of the surrogates. The best selection of this parameter is very important for the accuracy of the results. ρ is computed according to the suggestions in [12]. These parameters (d_e, τ, ρ) together with the hypothesis H_0 will be passed to the Surrogate Generation Method (SGM). A number of surrogates N_S will be generated using this method (more details will be given later in Section III-A). Those surrogates must preserve the coarse deterministic features of the real output (such as periodicity) while satisfying the null hypothesis H_0. Therefore, chaotic structure by fine scale dynamics will be altered by random noise with level ρ. To verify the AMS circuit behaviors, we perform hypothesis testing technique for a given confidence level α in order to derive the acceptance region for the noisy behavior expressed with the null hypothesis H_0. The acceptance region concludes that if the original circuit output test statistic is located outside this region, H_0 is rejected. In other words, hypothesis testing approach reports a significant difference between the original output and its surrogates in term of Gaussian Kernel (GK) measure (see Section III-B for more details). The rejection of the null hypothesis implies the acceptance of the alternative hypothesis H_1 that the circuit behavior exhibits chaos.

A. Surrogate Generation Method

We extend the Pseudo Periodic Surrogate (PPS) method, developed first in [12] to study dynamics of human electrocardiogram (ECG), to verify AMS circuits behavior. Algorithm 1 illustrates the surrogates generation procedures in order to verify aberrant behaviors of AMS designs described as Extended-System of Recurrence Equations (E-SREs).

Algorithm 1 Surrogate Generation Algorithm

Require: E-SRE(X), ρ, d_e, τ, N_S

1: $N \leftarrow length\ (X)$;
2: $\tilde{N} \leftarrow N - (d_e - 1)\tau$;
3: $d_w \leftarrow d_e\tau - 1$;
4: $\{Z_t\}_{t=1}^{\tilde{N}} \leftarrow embed\ (\text{E-SRE}(X), d_e, \tau)$;
5: $\mathcal{A} = \{z_t/\ t = 1, 2, ..., \tilde{N}\}$;
6: **for** $k = 1 \rightarrow N_S$ **do**
7:     **for** $j = 1 \rightarrow N - d_w$ **do**
8:         $i \leftarrow 1$;
9:         $s_1 \in \mathcal{A}$;
10:         **while** $i < n$ **do**
11:             $d_j = \|s_i - z_j\|$;
12:             $\omega_j = e^{-\frac{d_j}{\rho}}$;
13:             $p_j \leftarrow \omega_j / \sum_k \omega_k$;
14:             $P(s_{i+1} = z_t) \propto p_j$;
15:             $s_{i+1} = z_j$;
16:             $i \leftarrow i + 1$;
17:         **end while**
18:     **end for**
19:     $\{(s_t)_k\} \equiv \{(s_1)_k, (s_2)_k, ..., (s_N)_k\}$;
20: **end for**

The Algorithm requires: an E-SRE model of the circuit for the state variables X with/without thermal noise in some or all circuit components denoted by *E-SRE(X)*, the noise radius ρ, the embedding dimension d_e, the embedding lag τ, and the number of surrogates to be generated N_S. The algorithm begins with state space reconstruction of the circuit dynamics (line 4). It consists of representing the dynamical features of the circuit output E-SRE(X) in an alternative domain namely an Euclidian space $\mathbb{R}^{d_e}$ where d_e is the embedding dimension. By doing so, the points in $\mathbb{R}^{d_e}$ form an attractor $\mathcal{A}$ (line 5) that gives intuition about the circuit dynamics. Thereafter, embedding points of neighboring trajectories in the obtained attractor are used to create a new attractor with noisy trajectories (lines 10-17); The algorithm chooses an initial condition s_1 randomly from the reconstructed attractor $\mathcal{A}$ (line 9). For the following noisy attractor point, a near neighbor $z_j \in \mathcal{A}$ is then chosen with a probability commensurate to the noise radius ρ (line 14). The introduction of this dynamical

noise by the surrogate generation algorithm will obliterate any deterministic dynamics of the circuit while preserving periodicity. Hence, chaotic circuit dynamics lead to distinct trends of their surrogates produced by this method.

B. Gaussian Kernel Test Statistic

The Gaussian Kernel (GK) test is a measure of correlation dimension d_c which is the dimensionality of the circuit attractor $\mathcal{A}$ [13]. It is mathematically defined by Equation (3). It uses the Gaussian kernel function (Equation (4)) that is more convenient for calculating the effect of Gaussian noise.

$$d_c = \lim_{h \to 0} \lim_{N \to 0+\infty} \frac{log\ \hat{T}_m(h)}{log\ h} \qquad (3)$$

$$where\ \ \hat{T}_m = \frac{1}{N} \sum_i \sum_{j \neq i} \left(\frac{1}{N-1} e^{\frac{\|x_i - x_j\|^2}{4h^2}}\right) \qquad (4)$$

where h denotes the bandwidth, and N denotes the number of estimation points. Our choice for this test is explained by the fact that it has been proven to provide a rigorous estimation of correlation dimension even for a noise level 50% higher than the ideal signal.

IV. APPLICATIONS

In this section, we report the results of the application of our methodology on three AMS circuits. All computation and circuit models were performed in a MATLAB environment and were run on a 64-bit Windows 7 machine with 2.8 GHz processor and 24 GB memory. The type of hypothesis testing used is the one tailed test with the level of significance α=5%.

A. Colpitts Oscillator

A Colpitts oscillator is a combination of a transistor amplifier and an LC circuit as shown in Fig. 2.The Colpitts circuit behavior has been reported to exhibit chaotic behavior [14]. We model its behavior by the following E-SREs:

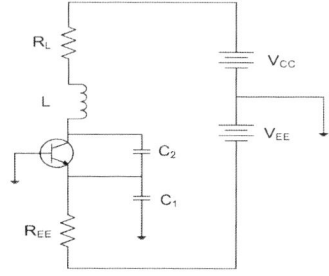

Fig. 2. Colpitts oscillator

$$
\begin{aligned}
i_B(n) &= if(V_{BE} > V_{Th}, \frac{V_{BE}(n) - V_{Th}}{R_{ON}}, 0) \\
i_c(n) &= if(true, \beta i_B(n), 0) \\
V_{CE}(n+1) &= if(true, V_{CE}(n) + \delta_t \frac{i_L(n) - i_c}{C_1}, 0) \\
V_{BE}(n+1) &= if(true, V_{BE}(n) - \frac{\delta_t}{C_2} \\
&\quad (\frac{V_{EE} + V_{BE}(n)}{R_{EE}} + i_L + i_B)), 1) \\
i_L(n+1) &= if(true, i_L(n) + \delta_t(V_{CC} - V_{CE}(n) + \\
&\quad V_{BE}(n) - i_L(n)R_L), 0)
\end{aligned}
\qquad (5)
$$

978-1-4799-8720-7/15 $31.00 © 2015 IEEE   239

Table I summarizes the simulation and surrogate generation parameters for the Colpitts circuit. Fig. 3 illustrates both the original and reconstructed attractor of the Colpitts oscillator behavior using the embedding dimension (d_e, τ) given in Table I. The similarity of both attractors demonstrates the appropriate choice of embedding parameters.

TABLE I
SIMULATION PARAMETERS OF THE COLPITTS CIRCUIT

Parameter	Value	Parameter	Value
R_{EE}	0.904	R_L	35
C_1, C_2	54e-9	R_{EE}	400
R_{ON}	100	V_{CC}	5
V_{EE}	-5	V_{Th}	0.75
β	94	d_e	5
τ	3	ρ	0.003
N_S	100	N	3600

The importance of an adequate selection of the noise radius ρ is shown in Fig. 4. For instance, if ρ is too large ($\rho = 0.01$), the surrogate generation algorithm will introduce too much randomization and the surrogates will no longer resemble the circuit output z (see Fig. 4(c)). Conversely if ρ is too small ($\rho = 0.001$), the algorithm will introduce insufficient randomization, and surrogates will be identical to the output as shown in Fig. 4(b).

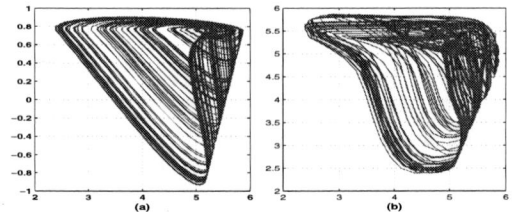

Fig. 3. Original attractor of Colpitts output (a), reconstructed attractor (b)

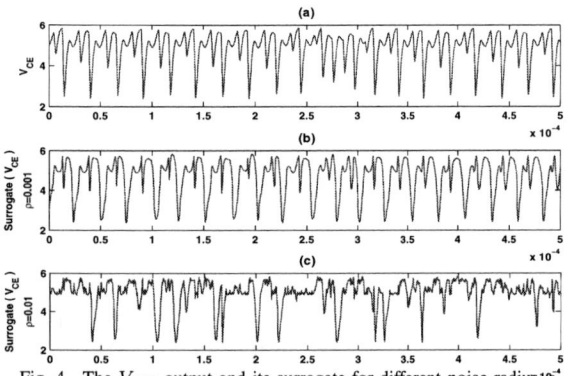

Fig. 4. The V_{CE} output and its surrogate for different noise radius ρ.

Figure 5 depicts the GK correlation dimension d_c of the V_{CE} output (dashed line) and its corresponding 100 surrogates (dotted line). It can be observed that our approach successfully probes the chaotic behavior of the Colpitts circuit. For instance, the $d_c(V_{CE})$ is significantly different from those of the surrogates and so it falls in the rejection region (see Fig. 5). This leads to rejection of the noisy dynamics hypothesis and consequently proves the chaotic circuit dynamics.

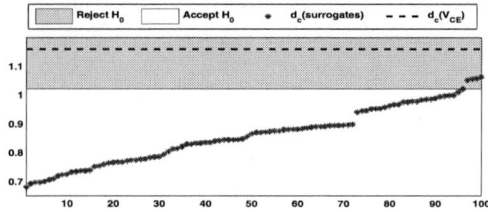

Fig. 5. Analysis results for chaotic Colpitts circuit

B. Third Order Σ-Δ Modulator

We consider the third order Σ-Δ modulator depicted in Fig. 6 and modeled as a system of E-SREs given by Equations (6).

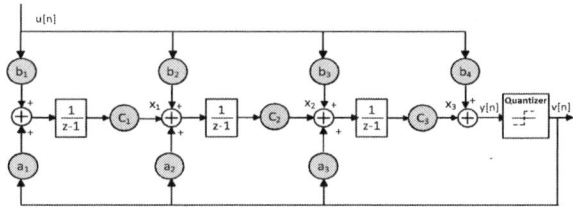

Fig. 6. Third Order Σ-Δ Modulator

$$
\begin{aligned}
v(k+1) &= if(P_{th}(k), -a, a) \\
x_1(k+1) &= if(P_{th}(k), x_1(k) + b_1 u(k) - a_1 a, \\
&\qquad x_1(k) + b_1 u(k) + a_1 a) \qquad (6) \\
x_2(k+1) &= if(P_{th}(k), c_1 x_1(k) + b_2 u(k) + x_2(k) - \\
&\qquad a_2 a, \; c_1 x_1(k) + x_2(k) + b_2 u(k) + a_2 a) \\
x_3(k+1) &= if(P_{th}(k), c_2 x_2(k) + x_3(k) + b_3 u(k) - \\
&\qquad a_3 a, \; c_2 x_2(k) + x_3(k) + b_3 u(k) + a_3 a) \\
\text{where } P_{th}(k) &= c_3 x_3(k) + u(k) \geq 0
\end{aligned}
$$

The following parameters of the circuit were computed using the Delta Sigma MATLAB Toolbox [15]:

$$
a = 2, \; A = B = \begin{pmatrix} 0.044 \\ 0.2881 \\ 0.7997 \end{pmatrix}, \; C = \begin{pmatrix} 1 \\ 1 \\ 1 \end{pmatrix}
$$

In [8], the author proved that Σ-Δ modulators can reproduce chaos if it is fed by a chaotic input. This is a very important feature of Σ-Δ modulators in communication applications such as encryption and cryptography. Therefore, we will use our methodology to verify the Σ-Δ modulator given in Fig. 6, with the chaotic input fed from the Colpitts Oscillator studied in Section IV-A. Fig. 7 shows the time variation of the quantized output V for sinusoidal input (Fig. 7 (a)) and a chaotic input signal (Fig. 7(b)).

By using GK correlation dimension, we verify the circuit using the proposed methodology in the presence of thermal noise and chaos. Our results shown in Fig. 8(b) are in good agreement with the results in [8]. Indeed, the correlation dimension of the output V (dashed line) is very different from its corresponding surrogates (dotted line). This violates the hypothesis that the apparently random output is generated from a noisy modulator. In contrast, in the presence of thermal noise Fig. 8(a), there was no apparent distinction between the two. Hence, H_0 holds and consequently the circuit exhibits noise.

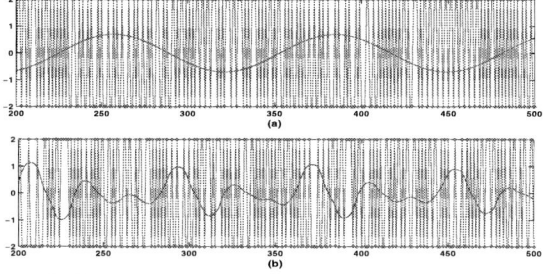

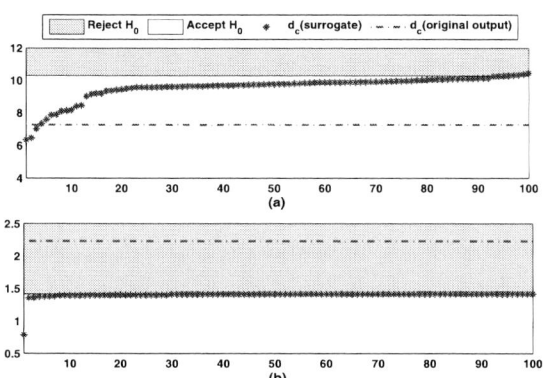

Fig. 7. Quantized sinusoidale wave (a) and chaotic (b) inputs

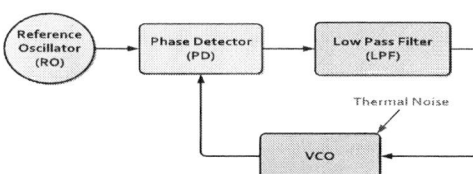

Fig. 8. Chaos verification for noisy modulator (a), and modulator fed with chaotic input (b)

C. Phase Locked Loop

PLLs are widely used circuits as modulators and demodulators in communication systems. In this section, we verify a third order PLL that serves as FM demodulator [16]. In this PLL, a multiplier Phase Detector (PD) and a resonant Low Pass Filter (LPF) are deployed as shown in Fig. 9.

![Conventional PLL block diagram with Reference Oscillator (RO), Phase Detector (PD), Low Pass Filter (LPF), VCO and Thermal Noise]

Fig. 9. Conventional PLL block diagram

The PLL dynamics are governed by the following E-SREs:

$$
\begin{aligned}
\varphi(n+1) &= if(true,\ \varphi(n) + \delta_t\ f_n,\ 0) \qquad (7)\\
\Psi(n+1) &= if(true,\ \Psi(n) - \delta_t\ m\ f_n\ sin(\varphi(n)),\ \pi)\\
x(n+1) &= if(true,\ x(n) + \delta_t\ (\Omega_n - k_n\ z(n)),\ 0)\\
y(n+1) &= if(true,\ y(n) + \delta_t\ (sin(x(n) - \Psi(n)) +\\
&\quad (g-2)\ y(n) - \frac{g-1}{g}\ z(n)),\ 0)\\
z(n+1) &= if(true,\ z(n) + \delta_t\ (g\ y(n) - z(n)), 1)
\end{aligned}
$$

where the state variables φ, Ψ, x, y, and z stand for modulating signal, frequency of modulation, phase difference

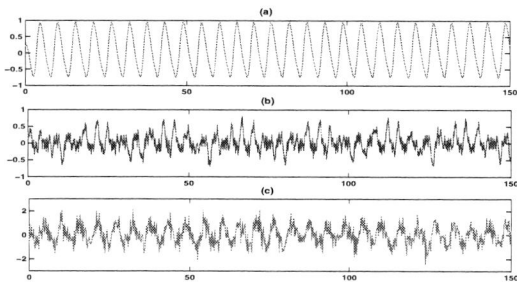

Fig. 10. Time variation of Y for periodic (a), chaotic (b), and noisy (c) regimes

between PLL input and VCO output, PD output, and LPF output, respectively. As the control parameter m changes, the dynamics of the PLL circuit change till culminating to chaotic regime. For instance, the circuit operates in periodic regime for $m = 0$ while chaotic dynamics occur for $m = 10$. Figure 10 depicts the time domain behavior of the PD output y for periodic (panel(a)), chaotic (panel(b)), and noisy behavior (panel(c)). It can be remarked that the chaotic output reveals a similar behavior to the noisy output simulated with thermal noise in the VCO. This demonstrates the need to assess the real source of random-like behavior observed in nonlinear AMS circuits during the design process. A phase diagram of the PLL attractor is depicted in Fig. 11.

TABLE II
SIMULATION PARAMETERS OF THE PLL CIRCUIT

Parameter	Value	Description
f_n	0.904	normalized frequency of the modulating signal
m	10	modulating index
Ω_n	1.2	normalized detuning
d_e	5	embedding dimension
k_n	0.6511	normalized loop gain
g	1.728	filter gain
τ	10	embedding lag
ρ	0.0170	noise radius
N_s	100	number of surrogates

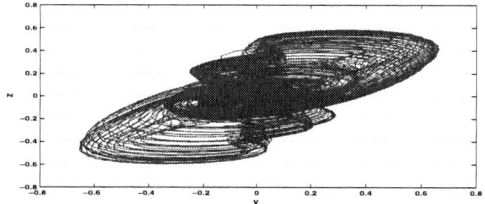

Fig. 11. Attractor of the PLL circuit during chaotic regime

The GK correlation dimension acquired from the the PLL circuit output y (dashed line) and 100 surrogates (dotted line) is shown in Fig. 12 for chaotic behavior ($m=10$) and in Fig. 13 for noisy behavior ($m=0$). A good qualitative agreement between the results of our methodology and the theory [16] is demonstrated; Indeed, the correlation dimension of the original output (dashed line) is very different from its corresponding surrogates (dotted line) in the chaotic case. This violates the hypothesis that the apparently random output is generated from a noisy circuit and hence indicates the deterministic dynamics

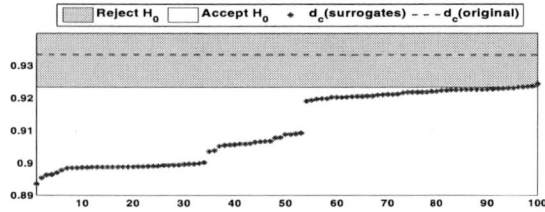

Fig. 12. Verification of PLL in chaotic regime

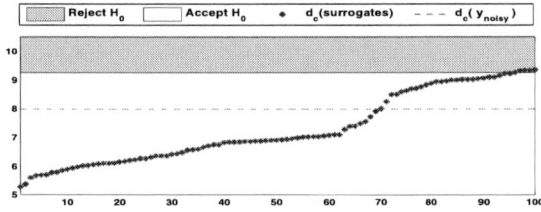

Fig. 13. Verification of PLL in Noisy regime

of the circuit. Consequently, the proposed methodology was able to successfully distinguish the noise like behavior exhibited by deterministic chaotic circuit from the stochastic noisy PLL.

D. Comparison with Lyapunov Exponent Method

In order to demonstrate the efficiency of the proposed methodology, the Lyapunov Exponent (LE) measure was carried out for the previously analyzed circuits under the same simulation conditions and for the same circuits outputs. The results of chaos verification and simulation time are recapitulated in Table III. The obtained results using our approach are in good agreement with those obtained with the LE technique for the Coplitts circuit and Σ-Δ Modulator. However, a failure to discriminate the noisy behavior of PLL has been detected (see Table III); Thermal noise in the VCO creates sensitivity to initial conditions of the PLL design that triggered the finding of a positive Lyapunov exponent which is a signature of chaos. In fact, a maximum exponent $\lambda = +0.0154$ has been obtained while the circuit exhibits thermal noise in the VCO and not chaotic behavior. Moreover, a simulation time acceleration is obtained using our methodology. Indeed, when adopting our approach, the simulation time was minimized from thousands of seconds to hundreds of seconds as shown in Table III.

TABLE III
ACCURACY AND SIMULATION TIME COMPARISON

	Colpitts Oscillator		3rd order Σ-Δ Modulator		PLL	
	Our Method	LE	Our Method	LE	Our Method	LE
Simulation Time [Sec.]	164.7	709.7	196.3	985.4	247.3	1213.4
Chaos\Noise Detection	√	√	√	√	√	×

×: Failure in detecting circuit dynamics.
√: Successfully detecting circuit dynamics.

V. CONCLUSIONS

In this paper, a novel methodology to statistically assess chaos from noise in AMS circuits is proposed. The circuit is modeled using Extended System of Recurrence Equations. The verification approach is based on hypothesis testing and surrogate generation method to decide whether to reject or accept the hypothesis that the unpredictable circuit behavior emerges from noisy design. Our methodology has been successfully employed on a Colpitts oscillator, a third order Σ-Δ modulator and a third order PLL circuit. The main advantages of the proposed methodology are: (1) It improves robustness to thermal noise. For instance, our approach successfully discriminates noise in PLL while traditional techniques such as LE method fails to do so; (2) It sufficiently reduces the simulation time to around 5 times compared to the LE method; and (3) It is applicable to simulation traces when no mathematical model of the design is available. We believe that the proposed methodology will be a handy tool to give insight to designers about the onset of chaos in AMS circuits.

As future work, we plan to verify larger AMS circuits with other types of noise, such as $1/f$ noise and jitter. By doing so, we will be able to assess the limitations of the proposed methodology. Additionally, we aim to extend the proposed approach to discriminate simple chaos from hyperchaos.

REFERENCES

[1] J. Yang and F. Zhu, "Synchronization for uncertain chaotic systems with channel noise and chaos-based secure communications," in *Unifying Electrical and Electronics Engineering*, 2014, pp. 1483–1491.

[2] Y. Zhang, W. Hu, L. Wang, and L. Zhang, "A novel random stepped frequency radar using chaos," in *IEEE Radar Conference*, 2014, pp. 0662–0665.

[3] M. Stern, J. Aljadeff, and T. O. Sharpee, "Chaos in heterogeneous neural networks: II. multiple activity modes," *BMC Neuroscience*, vol. 15, no. 1, p. O21, 2014.

[4] M. J. Borkowski, T. A. Riley, J. Hakkinen, and J. Kostamovaara, "A practical delta-sigma modulator design method based on periodical behavior analysis," *IEEE Transactions on Circuits and Systems II*, vol. 52, no. 10, pp. 626–630, 2005.

[5] B. De and B. Sarkar, "Nonlinear dynamics of a nonlinear amplifier-based delayed PLL incorporating additional phase modulator," *International Journal of Electronics*, vol. 95, no. 9, pp. 939–949, 2008.

[6] M. Šalamon, "Chaotic electronic circuits in cryptography," in *Applied cryptography and network security*, 2012, pp. 295–320.

[7] G. Al Sammane, M. H. Zaki, Z. J. Dong, and S. Tahar, "Towards assertion based verification of analog and mixed signal designs using PSL," *Forum on specification & Design Languages*, pp. 293–298, 2007.

[8] D. Campbell, "The sigma-delta modulator as a chaotic nonlinear dynamical system," Ph.D. dissertation, University of Waterloo, 2007.

[9] B. Muthuswamy and P. Kokate et al., "Memristor-based chaotic circuits," *IETE Technical Review*, vol. 26, no. 6, p. 417, 2009.

[10] M. Small and C. Tse, "Optimal embedding parameters: a modelling paradigm," *Physica D: Nonlinear Phenomena*, vol. 194, no. 3, pp. 283–296, 2004.

[11] M. B. Kennel, R. Brown, and H. D. Abarbanel, "Determining embedding dimension for phase-space reconstruction using a geometrical construction," *Physical Review A*, vol. 45, no. 6, p. 3403, 1992.

[12] M. Small et al., "Surrogate test for pseudoperiodic time series data," *Physical Review Letters*, vol. 87, no. 18, p. 188101, 2001.

[13] D. Yu et al., "Efficient implementation of the gaussian kernel algorithm in estimating invariants and noise level from noisy time series data," *Physical Review*, vol. 61, no. 4, p. 3750, 2000.

[14] O. Tsakiridis, D. Syvridis, E. Zervas, and J. Stonham, "Chaotic operation of a colpitts oscillator in the presence of parasitic capacitances." *WSEAS Transactions on Electronics*, vol. 1, no. 2, pp. 416–421, 2004.

[15] R. Schreier, "Delta sigma toolbox," *http://www. mathworks. com/matlabcentral/fileexchange/19*, 2000.

[16] B. C. Sarkar and S. Chakraborty, "Chaotic dynamics of a third order PLL with resonant low pass filter in face of CW and FM input signals," *International Journal on Communication*, vol. 3, no. 1, pp. 62–66, 2012.

2015 IEEE Computer Society Annual Symposium on VLSI

Multi-objective Optimization of Floating Point Arithmetic Expressions Using Iterative Factorization

Alireza Mahzoon* and Bijan Alizadeh*[†]

*School of Electrical and Computer Engineering, College of Engineering, University of Tehran, Iran
[†]School of Computer Science, Institute for Research in Fundamental Sciences (IPM), P.o.Box 19395-5746, Tehran, Iran
{a_mahzoon, b.alizadeh}@ut.ac.ir

Abstract— **Nowadays some High Level Synthesis (HLS) tools are introduced which are able to generate Hardware Description Language (HDL) codes from high level floating point arithmetic expressions for implementation on FPGAs. Before this conversion, changing the form of high level expressions usually leads to significant improvements in the final implementation in terms of accuracy, resource usage and latency. In this paper, we introduce a method to find better forms of a floating point arithmetic expression in terms of accuracy, resource usage and delay. We first come up with a solution to estimate delay and resource usage in an FPGA for a specific expression. Then we extract equivalent expressions by iterative factorization and to reduce the complexity, we select only those expressions generated by associativity rule which are better in terms of accuracy. Finally, we propose an algorithm to choose expressions from the set of equivalent expressions in terms of accuracy, resource usage and delay while let designers apply their accuracy, area and delay constraints. The results show that our proposed method in comparison with existing methods such as SOAP [10], has up to 35.2% overall improvement in terms of runtime for generating expressions and choosing expressions from the set of final expressions.**

Keywords—High level synthesis; Floating point; Optimization; Accuracy; Area; Delay

I. INTRODUCTION

Nowadays floating point representation is the most common approach for storing and manipulating real numbers in computers. Floating point representation has a number of advantages over fixed point representation, dynamic ranges for representing real numbers is one of these advantages. This representation however has some problems; less accuracy especially for large numbers and hardware complexity in operations like addition and multiplication are two important disadvantages of floating point representation.

Computations in floating point representation are based on IEEE 754 standard [1]. Exact computations are approximated by the floating point arithmetic which relies on a finite representation of the numbers. Although this approximation is often accurate enough, in some cases it may lead to irrelevant or too inaccurate results. Furthermore, when arithmetic expressions are implemented on Field Programmable Gate Arrays (FPGAs), other quality factors like resource usage and delay should be taken into account as well. Floating point computations are not intuitive and it is not possible to obtain more accurate expressions analytically. For example, if $x \in [0, 0.01]$ and $y \in [10, 20]$ then expression $(x+y)^2$ is more accurate when implemented in the form of x^2+y^2+2xy because its error is approximately $[-2.7 \times 10^{-4}, 2.7 \times 10^{-4}]$ in single precision. Although in the form of $(x+y) \times (x+y)$, error is approximately $[-3.7 \times 10^{-4}, 3.7 \times 10^{-4}]$, it has the minimum delay and uses the fewest resources.

High level synthesis (HLS) is the process of converting a high level representation of an application (usually in C, C++ or MATLAB) into a Register Transfer Level (RTL) implementation [2]. HLS has recently gained more attention especially in floating point implementations and a variety of HLS tools are introduced which are able to generate HDL codes for floating point expressions. Hence, improvement of the accuracy, area and delay of expressions before synthesis, leads to better performance of final implementation. Authors of [3] and [4] proposed a method to formally verify and optimize polynomials to reduce the complexity of polynomial datapaths in terms of the number of multipliers and adders. Authors of [5] and [6] proposed a system-level approach to improve the area and delay of datapath designs that perform polynomial computations. Authors of [7] have introduced a method to extract all equivalent expressions and compare them in terms of accuracy in a specific depth. Authors of [8] and [9] have introduced the abstract interpretation of equivalent expressions and proposed a suitable structure to represent an exponential number of equivalent expressions and a method has been used to search the structure and select a high accurate expression. Authors of [10] have used a Pareto frontier algorithm to choose a non-dominated set of expressions which covers whole design space in terms of accuracy and area. Authors of [11] added delay to Pareto algorithm. The method in [10], however, is not able to deal with the delay. Furthermore, this method consumes a considerable amount of times to process complex expressions with a large number of equivalent expressions.

In this paper, we optimize the accuracy, area and delay of a floating point expression simultaneously. For this purpose, we should first find equivalent expressions for a given expression. Equivalent expressions are those expressions which do not differ in real numbers but in floating point numbers, they may produce different results for the same inputs. Then we propose a method which can choose the most suitable expressions in terms of accuracy, resource usage, delay or a balance of them. Unlike [10], we aim only one design point and extract only those expressions which are desirable for designer in terms of accuracy, area or delay. A kind of semantics introduced in [12] is used to evaluate the accuracy of expressions. Using this semantics we calculate floating point, real and error values for a specific expression. To calculate delay and area we should have an estimation of the delay and also resources used in FPGAs for implementing an expression. To do so, we propose algorithms to estimate delay and the number of resources with respect to common subexpressions.

Fig. 1 shows our proposed optimization methodology. In the first phase (*Phase 1* in Fig. 1) to generate equivalent expressions a fast factorization method is proposed. The given expression is expanded to its widest form and then with

978-1-4799-8720-7/15 $31.00 © 2015 IEEE   243

iterative factorization, equivalent expressions are generated. To reduce the number of equivalent expressions and the complexity of generating expressions, we propose a solution to reduce the number of expressions generated by associativity rule. The results in Section IV (runtime for *Phase 1*) show that our method is faster than those of [7] and [10].

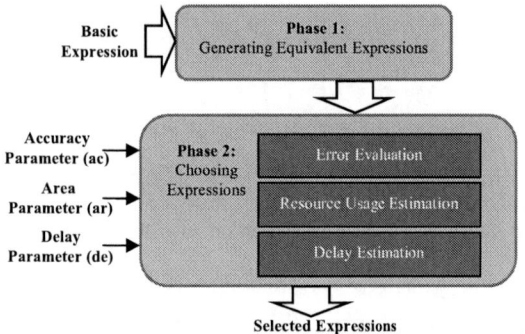

Fig. 1. Proposed optimization flow

In the second phase (*Phase 2* in Fig. 1) to select expressions, an elective algorithm is proposed where we define three parameters for accuracy, area and delay which can be set by the designer. Based on these parameters and after error evaluation, resource usage estimation and delay estimation of each expression, some expressions are selected and the others are eliminated. Please note that in contrast to [10], our proposed elective algorithm can deal with the delay constraint. Furthermore, its time complexity is less than that of [10]. Hence, the main contributions of this paper are as follows:

- Generating equivalent expressions with iterative factorization and introducing a solution to reduce the number of expressions generated by associativity rule.

- Proposing an algorithm to estimate delay and the number of resources used in FPGAs.

- Introducing an elective algorithm to choose expressions in terms of accuracy, resource usage and delay.

The rest of the paper is organized as follows. Section II introduces preliminaries needed to evaluate error values. Our proposed optimization flow including resource and delay estimations, equivalent expressions generation and selecting the best expressions is presented in Section III. Experimental results are reported and discussed in Section IV and Section V concludes the paper.

II. PRELIMINARIES

In this section basic concepts of intervals are introduced and useful methods to compute floating point error are explained.

A. Interval Arithmetic

In error value computations, all inputs are assumed to be intervals and error values which are consequences of computations should be intervals too. Interval arithmetic is the best way to verify the safety of floating point computations. A wide interval for error indicates risk of inaccuracy in computations. Addition, subtraction and multiplication operations can be defined in interval arithmetic as follows:

$$
\begin{aligned}
[a,b]+[c,d] &= [a+c,b+d] \\
[a,b]-[c,d] &= [a-d,b-c] \\
[a,b]\times[c,d] &= [\min(s),\max(s)]
\end{aligned}
\tag{1}
$$

where $s = \{a\times c,\ a\times d,\ b\times c,\ b\times d\}$.

B. Error Analysis

A radix-β floating-point number x is a number of the following form:

$$
(-1)^s m\beta^e
\tag{2}
$$

where s is the sign and can be 0 or 1, β is the radix of the floating point system, m such that $|m|<\beta$ is the significand and e is the exponent [13]. Note that in IEEE-754-1985, β is 2.

Floating point computations usually have errors which can be propagated along the computations. A part of error that can appear both in inputs and further computations is called round-off error. Real values, used in inputs or generated in computations, usually do not have a floating-point representation and should be rounded to one of the nearest floating-point numbers and therefore this will generate round-off errors. IEEE-754 has four rounding modes: round to nearest, round toward $+\infty$, round toward $-\infty$ and round toward zero [1]. The last three rounding modes are called directed rounding modes. If $[a, b]$ is an interval in real values, the round-off error will be:

$$
\downarrow_\circ^\# [a,b] = [-z,z]
\quad
\begin{aligned}
z &= \max(ulp(a), ulp(b))/2 \\
&\text{for round to nearest mode} \\
z &= \max(ulp(a), ulp(b)) \\
&\text{for directed modes}
\end{aligned}
\tag{3}
$$

In (3), the function $\downarrow_\circ^\#$: *Interval* $\rightarrow$ *Interval* determines the range of round-off error due to the floating-point computation under one of the rounding modes. Note that, function $ulp(x)$ characterizes the distance between two adjacent floating point values f_1 and f_2 satisfying $f_1 \le x \le f_2$ [14]. According to Goldberg's definition in [15] if $x \in [\beta^e, \beta^{e+1})$, then:

$$
ulp(x) = \beta^{\max(e,e_{\min})-p+1}
\tag{4}
$$

where p is the number of significand bits and e_{min} is specified by the floating point format.

In this paper a triplet ($f^\#$, $r^\#$, $e^\#$) is used for all inputs and values generated at each step of computations, where $f^\#$ and $r^\#$ are intervals used to present floating-point and real values for a specific computation or for inputs and $e^\#$ is a real interval for error value generated and propagated at each step. Addition, subtraction and multiplication can be defined for this triplet as follows:

$$
\begin{aligned}
&\left(f_1^\#,r_1^\#,e_1^\#\right)+\left(f_2^\#,r_2^\#,e_2^\#\right)= \\
&\qquad \left(\uparrow_\circ^\#\left(f_1^\#+f_2^\#\right),r_1^\#+r_2^\#,e_1^\#+e_2^\#+\downarrow_\circ^\#\left(f_1^\#+f_2^\#\right)\right) \\
&\left(f_1^\#,r_1^\#,e_1^\#\right)-\left(f_2^\#,r_2^\#,e_2^\#\right)= \\
&\qquad \left(\uparrow_\circ^\#\left(f_1^\#-f_2^\#\right),r_1^\#-r_2^\#,e_1^\#-e_2^\#+\downarrow_\circ^\#\left(f_1^\#-f_2^\#\right)\right) \\
&\left(f_1^\#,r_1^\#,e_1^\#\right)\times\left(f_2^\#,r_2^\#,e_2^\#\right)= \\
&\qquad \left(\uparrow_\circ^\#\left(f_1^\#\times f_2^\#\right),r_1^\#\times r_2^\#,f_1^\#\times e_2^\#+f_2^\#\times e_1^\#+e_1^\#\times e_2^\#+\downarrow_\circ^\#\left(f_1^\#\times f_2^\#\right)\right)
\end{aligned}
\tag{5}
$$

978-1-4799-8720-7/15 $31.00 © 2015 IEEE

Please note that, in (5), the function $\uparrow_\circ^\#$: *Interval* $\rightarrow Interval_F$ computes the floating point bound from a real bound, by rounding *a* and *b* of the input interval [*a*, *b*] under one of the four rounding modes explained before. We use function *ErrorEvaluation* in the rest of the paper to get error upper-bound of an expression.

III. PROPOSED OPTIMIZATION FLOW

In this section, we first propose an algorithm to estimate delay and the number of resources used in FPGAs, then we explain how to generate equivalent expressions and finally an algorithm is proposed to choose better expressions for hardware implementation on FPGA in terms of accuracy, area, delay or a balance of them.

A. Resource Usage and Delay Estimation

Before explaining two main phases of our optimization methodology, we need to explain how to estimate the delay and the number of resources used in FPGAs for a given expression. For doing so, based on given expression a graph called *Operation Tree* is generated.

Definition 1: Operation Tree (OT): Graph *T* is called an *Operation Tree* with *N* nodes where each node n_l contains: 1) a label *l* which is a unique number, 2) a floating point expression *e* and 3) an operator $o \in \{+, -, *, NO\}$ which is the top level operator in the expression. *NO* indicates no operator. Note that $n_l.e$ and $n_l.o$ return the expression and the operator of n_l, respectively. Each parent node n_l has two children left and right (denoted by $n_l.left$ and $n_l.right$, respectively) which are sub-expressions generated by decomposing the parent expression.

Definition 2: Equivalent Nodes (EN): Nodes with the same expressions and operators are called equivalent nodes.

In order to clarify *Definition 1 and Definition 2*, let us consider $ab+(ab+c)^2$. Fig. 2 shows its operation tree where leaf nodes with no operator (*NO*) have been shown with a different color. In this tree, nodes 2, 8 and 10 are *EN*.

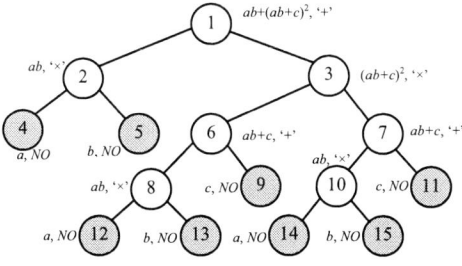

Fig. 2. Operation tree for $ab+(ab+c)^2$ (*NO*: No Operator)

Fig. 3 shows our proposed algorithm for resource usage estimation which gets operation tree *T*, *Adder_area* and *Mult_area* (estimated resources for an adder and a multiplier, respectively) as inputs and generates the number of estimated resources in FPGA as output. Although the main goal is to count the number of adders and multipliers, we need to avoid counting *EN* which are seen as common sub-expressions. For example in $ab+(ab+c)^2$, *ab* is a common sub-expression and

should be implemented only one time. To do so, we first define a flag for each node and unmark it (line 2). Then for each unmarked node *n*, if its operator is an adder or subtractor, the number of adders, i.e., *Add_num*, increases (lines 4-5) and if its operator is a multiplier, the number of multipliers, i.e., *Mult_num*, increases (lines 8-9). Note that, during traversing the tree, equivalent nodes (*EN*) are detected and marked as visited (lines 6-7, 10-11 and 12).

Fig. 4 shows our proposed algorithm for delay estimation which gets *T* (operation tree of a given expression), *Adder_delay* and *Mult_delay* (estimated delay for an adder and a multiplier, respectively) as inputs and returns the delay as output. The main goal is to estimate the delay of the critical path in *Operation Tree*. To do so, *DelayEst* is recursively called to compute the accumulated delay for each node which is the maximum delay from leaf nodes to the current node.

Please note that *Adder_area*, *Mult_area*, *Adder_delay* and *Mult_delay* used in Fig. 3 and 4 are calculated for different mantissa and exponent bits using FloPoCo [16] which is an open-source C++ framework for generating arithmetic datapaths and its output is a synthesizable VHDL code. These results are obtained after synthesizing VHDL codes generated by FloPoCo targeting Stratix III FPGA. We assume that datapaths are not pipelined and DSP blocks are not used.

It should be noticed that many factors such as optimizations done by FloPoCo, variety of floating point implementations and using DSP blocks, may change the number of resources and delay. Taking all of these factors into account, however, increases the run time significantly. Because our goal in this work is to compare different expressions in terms of resource usage and delay, this estimation would be enough.

```
ResourceEst(Tree T, int Adder_area, int Mult_area)
  Inputs:   T: operation tree of a given expression
            Adder_area: estimated resources in FPGA for an adder
            Mult_area: estimated resources in FPGA for a multiplier
  Output:   estimated resources (area)
  1   Add_num = 0;   Mult_num = 0;
  2   Unmark all nodes in T; // set a flag for all nodes
  3   for each ( n ∈ T ) do
  4       if (n is not marked and n.o = '+' or '-') then
  5           ++Add_num;
  6           for each (m∈T & m ≠ n) do
  7               if (m and n are ENs) then      Mark m;
  8       if (n is not marked and n.o = '×') then
  9           ++Mult_num;
  10          for each (m∈T & m ≠ n) do
  11              if (m and n are ENs) then      Mark m;
  12      Mark n;
  13  return  Adder_area×(Add_num)+Mult_area×(Mult_num);
```

Fig. 3. Resource usage estimation in FPGAs for a specific expression

```
DelayEst(Tree T, int Adder_delay, int Mult_delay)
  Inputs:   T: operation tree of a given expression
            Adder_delay: estimated delay in FPGA for an adder
            Mult_delay: estimated delay in FPGA for a multiplier
  Output:   estimated delay
  1   pn = Get parent node of T;
  2   if (pn.o = NO) then return 0;
  3   left_delay = DelayEst(pn.left, Adder_delay, Mult_delay);
  4   right_delay = DelayEst(pn.right, Adder_delay, Mult_delay);
  5   max_delay = MAX(left_delay, right_delay);
  6   if (pn.o = '+' / '-') then  return Adder_delay + max_delay;
  7   if (pn.o = '×') then   return Mult_delay + max_delay;
```

Fig. 4. Proposed algorithm to estimate delay in FPGAs for a given expression

978-1-4799-8720-7/15 $31.00 © 2015 IEEE   245

B. Phase 1: Equivelnce Expressions Generation

As mentioned before, the first phase of our proposed optimization methodology is to generate equivalent expressions. Authors of [7] and [10] make use of a series of arithmetic rules for generating equivalent expressions. These rules are as follows:

$$\text{Associativity } (\circ): \ \left(e_1 \circ e_2\right) \circ e_3 \rhd e_1 \circ \left(e_2 \circ e_3\right)$$
$$\text{Commutativity } (\circ): \ e_1 \circ e_2 \rhd e_2 \circ e_1$$
$$\text{Distributivity}: \ e_1 \times \left(e_2 + e_3\right) \rhd e_1 \times e_2 + e_1 \times e_3 \qquad (6)$$
$$\text{Factorization}: \ e_1 \times e_2 + e_1 \times e_3 \rhd e_1 \times \left(e_2 + e_3\right)$$

where $\rhd$ shows equivalency and $\circ \in \{+, -, \times\}$. To obtain all equivalent expressions, these arithmetic rules are applied iteratively to each expression generated at each step and this process continues until no new expression is generated. This solution has two main problems: 1) applying these rules on expressions at each step will generate a large number of redundant expressions that need to be eliminated (a time consuming process), and 2) when associativity rule is applied to an expression, it will generate a large number of equivalent expressions. For example the parsing of a simple summation of n variables could result in $(2n-3)!!=1 \times 3 \times 5 \times \ldots \times (2n-3)$ distinct expressions [8]. For example, $a+b+c+d$ has 15 distinct forms which are generated by applying associativity rule. This method is called *Exhaustive Generation of Floating point Expressions* (*EGFE*) in the rest of the paper.

To solve the first problem we have proposed an iterative factorization technique to generate equivalent expressions. This method is called *Factorization based Generation of Floating point Expressions* (*FGFE*). Initially the main expression is expanded to its widest form, and then function *Factor:* $\{expr\} \rightarrow \{expr\}$ is applied to expressions iteratively to extract common sub-expressions until no new expression is generated. For example, the process of iterative factorization for expression $a(a+b)+b(a+b)$ is as follows:

$$a(a+b)+b(a+b) \xrightarrow{\text{expand}} a^2 + b^2 + 2ab \xrightarrow{\text{Factor}} \begin{array}{l} a(a+b) + b^2 + ab \\ b(a+b) + a^2 + ab \end{array}$$

$$\begin{array}{l} a(a+b)+b(a+b) \\ \xrightarrow{\text{Factor}} \quad a(2a+b)+b^2 \quad \xrightarrow{\text{Factor}} (a+b)^2 \\ b(2b+a)+a^2 \end{array} \qquad (7)$$

Note that in contrast to existing methods such as *EGFE* that apply factorization and distributivity functions on expressions at each step, in our proposed method (*FGFE*) only factorization function is applied. Hence, *FGFE* is better than *EGFE* in terms of complexity by an order of 2^n where n is the number of steps needed to generate equivalent expressions which has direct relationship with the complexity of the expression.

To solve the second problem, we do not take associativity rule into account in the process of generating equivalent expressions. At the end of the process, however, we have expressions in which order of additions and multiplications for variables are not fixed. The order of additions and multiplications for each expression is determined in such a way that an accurate expression can be obtained. To do so, we

should find a solution to specify the order of additions and multiplications in a sequence of numbers in order to have a high accurate result. For example in expression $x+y+z$ where $x \in [0, 0.001]$, $y \in [0, 0.01]$ and $z \in [0, 1]$, the form $(x+y)+z$ is more accurate than $(x+z)+y$ and $(y+z)+x$ and therefore is selected.

Theorem 1: Suppose f_1 and f_2 are two floating point numbers and $Er=\downarrow_\circ \left(f_1 + f_2\right)$ is the error of addition. The greater value for $|f_1 \text{-} f_2|$ (distance between two numbers) results in greater Er.

Proof: Let us consider two floating point numbers $f_1 = m_1 2^{e_1}$ and $f_2 = m_2 2^{e_2}$ so that $e_1 \geq e_2$. The addition of these two numbers will be as follows.

$$S = \left(m_1 + m_2 2^{e_2 - e_1}\right) 2^{e_1} \qquad (8)$$

In fact the mantissa of smaller number (m_2) should be shifted $e_2\text{-}e_1$ bits and this means the lost of mantissa bits. So greater distance between two numbers in addition causes more bits lost in shifting and this results in greater error. ∎

To determine the order of additions, initially all values are sorted increasingly and then add together because adding numbers with same magnitude will generate less error according to *Theorem 1*. To determine the order of multiplications, we use the heuristic discussed in [9]. This heuristic is a greedy algorithm which searches at each step two values such that the error for multiplication of these values is minimum, it replaces both terms by the result of multiplication. This process continues until the result of multiplication is obtained.

C. Phase 2: Choosing Better Expressions

After generating equivalent expressions, those expressions with better accuracy, area and delay should be chosen. To do so we can follow two different strategies. In the first strategy, the goal is to choose expressions which cover whole design space in terms of accuracy, area and delay; and designers find their desirable expressions in this set. Authors of [10] and [11] have used this strategy to choose expressions. In the second strategy, designers emphasize on only one-design point with some defined constraints and only those expressions with specific amount of accuracy, area and delay is chosen. We use this strategy to choose expressions and show this strategy is faster.

Note that although the Pareto Frontier algorithm introduced in [10] gives a set of equivalent expressions which covers whole design space in terms of accuracy and area, it is not able to deal with the delay. Furthermore, [10] has a sorting procedure with time complexity of $O(nlog(n))$ where n is the number of equivalent expressions. To alleviate this issue, we have proposed an elective algorithm that supports accuracy, area and delay optimizations with $O(n)$ complexity. That is because our proposed algorithm uses only minimum and maximum finding procedure. So our proposed algorithm is faster especially for a large number of equivalent expressions.

Fig. 5 shows our proposed elective algorithm which gets equivalent expressions and three parameters ac (accuracy), ar

(area) and *de* (delay) such that $0 \le ac, ar, de \le 1$. Based on these parameters some expressions are selected and the others are removed. Please note that if *ac* (*ar* or *de*) is closer to 1, it means higher importance of the accuracy (area or delay) for the designer. The proposed algorithm initially finds maximum and minimum values of error, area and delay from equivalent expressions (lines 1-3 of Fig. 5). Then we define three constraints *E*, *S* and *D* for the accuracy, area and delay based on *ac*, *ar* and *de* and the minimum and maximum values obtained before (lines 4-6). Then the error, resource usage and delay are calculated for each expression (line 8) and those expressions with the error, area and delay less than *E*, *S* and *D* are selected and other expressions are eliminated (lines 7-11).

ElectiveSelectionofExpressions (float ac, float ar, float de, expr set $\mathcal{E}$)

Inputs: *ac: accuracy parameter between 0 and 1*
  ar: area parameter between 0 and 1
  de = delay parameter between 0 and 1
  $\mathcal{E}$ *= generated equivalent expressions*

Output: *result = selected expressions for FPGA implementation*

1 $e_{min} = min\_error(\mathcal{E})$; $e_{max} = max\_error(\mathcal{E})$;
2 $s_{min} = min\_area(\mathcal{E})$; $s_{max} = max\_area(\mathcal{E})$;
3 $d_{min} = min\_delay(\mathcal{E})$; $d_{max} = max\_delay(\mathcal{E})$;
4 $E = (1-ac) \times (e_{max}-e_{min}) + e_{min} = (1-ac) \times e_{max} + ac \times e_{min}$;
5 $S = (1-ar) \times (s_{max}-s_{min}) + s_{min} = (1-ar) \times s_{max} + ar \times s_{min}$;
6 $D = (1-de) \times (d_{max}-d_{min}) + d_{min} = (1-de) \times d_{max} + de \times d_{min}$;
7 **for** $i \in \{1,2,...,n\}$ **do**
8   **if** $(ErrorEvaluation(t_i) \le E)$ &
      $(ResourceEst(t_i, Adder\_area, Mult\_area) \le S)$ &
      $(DelayEst(t_i, Adder\_delay, Mult\_delay) \le D)$ **then**
9      $result = result \cup \{t_i\}$;
10 **end for**
11 **return** *result*;

Fig. 5. Elective algorithm to select expressions in terms of accuracy, area and delay

IV. EXPERIMENTAL RESULTS

Our proposed methods have been implemented in C++ under Linux. For computing error values we make use of GNU Multiple Precision Arithmetic Library (GMP) [17] and its extension multi-precision floating-point library MPFR [18] to access floating-point rounding modes and arbitrary precision floating-point computations. GiNaC library [19] is also used to manipulate expressions and subexpressions. GiNaC has useful functions to extract subexpressions and factorize common subexpressions. In all cases we have used single precision and double precision formats and rounding to nearest mode while other formats and rounding modes can be used as well. All experiments were carried out on an Intel 2.67GHz core i5 with 1GB main memory running Linux operating system. In order to evaluate the effectiveness of the proposed methods, we have employed different polynomials extracted from real embedded systems such as digital signal processing, image processing and communication applications.

In Our experiment, we have considered the following expressions:

$$Exp1 = z^2 + xz + (2z+x)x + (x+z)y + (x+y)y,$$
$$Exp2 = (x+y)(yz+yt) + x(y(z+t) + xz + xt),$$
$$Exp3 = xz + (y+z)t + (x+t)x + (2x+y+z)y,$$

Quadratic filters (Quad) for DSP applications, degree-5 filter (D5flt) that is used in communication applications, anti-aliasing function (Anti) that is used in MP3 decoder and digital image rejection unit (Image) benchmark that is used in image processing applications. Table I shows the results in single precision and double precision formats, where $x \in [0, 0.01]$, $y \in [10, 20]$, $z \in [0, 1]$ and $t \in [1, 2]$ and the target FPGA is Stratix III. In single precision format, we emphasized on accuracy and area ($ac=0.6$, $ar=0.6$, $de=0.1$), so the average error and area of the final expressions are improved significantly (bolded results for single precision in Table I). In double precision format, we emphasized on accuracy and delay ($ac=0.6$, $ar=0.1$, $de=0.6$), so the average error and delay of the final expressions are improved (bolded results for double precision). The number of generated equivalent expressions (TNE) and the number of selected expressions (RNE) after applying proposed method are shown in Table I.

Table I. The reults of applying our proposed method with different *er*, *ar* and *de* parameters on expressions (EXP: expressions; TNE: total number of expressions; RNE: reduced number of expressions)

EXP	Format	Error	Area (ALUT)	Delay (ns)	TNE	Our proposed method						
						ac	ar	de	RNE	Avg. Error	Avg. Area (ALUT)	Avg. Delay (ns)
Exp1	Single precision	3.931×10^{-4}	8357	368	698	0.6	0.6	0.1	2	$\mathbf{3.217 \times 10^{-4}}$	**4413**	258
	Double precision	7.322×10^{-13}	29042	484		0.6	0.1	0.6	10	$\mathbf{6.049 \times 10^{-13}}$	18991	**330**
Exp2	Single precision	4.055×10^{-4}	7436	368	1377	0.6	0.6	0.1	7	$\mathbf{3.541 \times 10^{-4}}$	**5907**	297
	Double precision	7.554×10^{-13}	24800	486		0.6	0.1	0.6	9	$\mathbf{6.396 \times 10^{-13}}$	22008	**338**
Exp3	Single precision	1.844×10^{-3}	9261	296	5884	0.6	0.6	0.1	52	$\mathbf{1.588 \times 10^{-3}}$	**7436**	287
	Double precision	3.434×10^{-12}	35568	380		0.6	0.1	0.6	31	$\mathbf{2.997 \times 10^{-12}}$	30841	**303**
Quad	Single precision	1.525×10^{-3}	7402	296	15	0.6	0.6	0.1	1	$\mathbf{1.427 \times 10^{-3}}$	**6481**	241
	Double precision	2.841×10^{-12}	29368	380		0.6	0.1	0.6	2	$\mathbf{2.659 \times 10^{-12}}$	27247	**306**
D5flt	Single precision	4.524×10^{-3}	13397	465	320	0.6	0.6	0.1	2	$\mathbf{3.807 \times 10^{-3}}$	**9713**	418
	Double precision	8.418×10^{-12}	60041	580		0.6	0.1	0.6	1	$\mathbf{7.133 \times 10^{-12}}$	55799	**496**
Anti	Single precision	9.498×10^{-3}	12024	520	3685	0.6	0.6	0.1	22	$\mathbf{8.647 \times 10^{-3}}$	**10455**	488
	Double precision	1.757×10^{-11}	52536	622		0.6	0.1	0.6	18	$\mathbf{1.648 \times 10^{-11}}$	50187	**532**
Image	Single precision	7.372×10^{-4}	9244	372	4306	0.6	0.6	0.1	32	$\mathbf{5.386 \times 10^{-4}}$	**7890**	317
	Double precision	1.528×10^{-12}	41115	465		0.6	0.1	0.6	25	$\mathbf{1.218 \times 10^{-12}}$	36873	**360**

978-1-4799-8720-7/15 $31.00 © 2015 IEEE

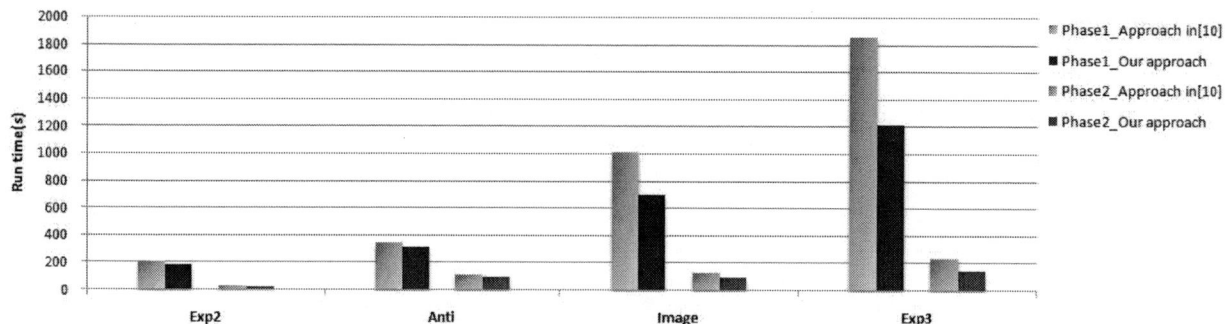

Fig. 6. Comparison between our proposed method and [10] in terms of runtimes (*Phase 1* and *Phase 2*).

Fig. 6 shows the runtimes of *Phase 1* (generating equivalent expressions) and *Phase 2* (choosing better expressions) for method in [10] and our proposed method. Fig. 6 shows that although for simple expressions with small number of equivalent expressions like Exp2 and Anti, the improvements are not desirable, in more complex expressions like Image and Exp3; we have up to 34% and 40% improvements for *Phase 1* and *Phase 2*, respectively.

V. CONCLUSION

In this paper, we have introduced a novel optimization method to generate equivalent floating point expressions and select better expressions in terms of accuracy, resource usage and delay where designer's constraints have been taken into account. This method unlike [10] and [11] aims only one design point and by changing accuracy, area and delay parameters the accuracy, area and delay of the final results change as well. The way we generate equivalent expressions and also our elective algorithm used to select expressions are faster than those of [10]. Our proposed optimization methods have up to 35.2% run time improvement in comparison with the existing methods such as [10].

In the future, we plan to extend resource usage and delay computations by considering the pipelining added to the implementation in order to improve the latency in floating point units. Moreover, we plan to extend our work in order to verify the floating point circuits. We hope to make use of verification techniques introduced in [20]. Note that our tool would be available to the HLS community soon.

REFERENCES

[1] ANSI/IEEE, "IEEE Standard for Floating-Point Arithmetic," Microprocessor Standards Committee of the IEEE Computer Society, Tech. Rep., Aug. 2008.

[2] P. Coussy and A. Morawiec, *High-Level Synthesis: from Algorithm to Digital Circuit*, 1st ed. Springer-Verlag, 2008.

[3] B. Alizadeh and M. Fujita, "Modular Datapath Optimization and Verification based on Modular-HED," *IEEE Trans. CAD*, vol. 29, no. 9, pp. 1422-1435, 2010.

[4] S. Ghandali, B. Alizadeh, Z. Navabi, M. Fujita, " Polynomial datapath synthesis and optimization based on vanishing polynomial over Z_2^m and algebraic techniques," in *Proceedings of the International Conference on Formal Methods and Models for Codesign, MEMOCODE '12*. IEEE, pp. 65-73, 2012.

[5] S. Ghandali, B. Alizadeh, M. Fujita, Z. Navabi, " RTL datapath optimization using system-level transformations," in *Proceedings of the*

International Conference on Quality Electronic Design, ISQED '14. IEEE, pp. 309-316, 2014.

[6] S. Ghandali, B. Alizadeh, M. Fujita and Z. Navabi, "Automatic high-level data-flow synthesis and optimization of polynomial datapaths using functional decomposition," *IEEE Trans. on Computers*, 2015.

[7] M. Martel, "Semantics-based transformation of arithmetic expressions," in *Static Analysis, Lecture Notes in Computer Science*, vol. 4634. Springer-Verlag, pp. 298-314, 2007.

[8] A. Ioualalen and M. Martel, "A new abstract domain for the representation of mathematically equivalent expressions," in *Proceedings of the 19th International Conference on Static Analysis, SAS '12*. Springer-Verlag, pp. 75-93, 2012.

[9] A. Ioualalen and M. Martel, "Synthesizing accurate floating- point formulas," in *Proceedings of the 24th International Conference on Application-Specific Systems, Architectures and Processors, ASAP '13*. IEEE, pp. 113-116, 2013.

[10] X. Gao, S. Balys, G. Constantinides, "Soap: structural optimization of arithmetic expressions for high-level synthesis," in *Proceedings of the International Conference on Field-Programmable Technology, FPT '13*. IEEE, pp. 112-119, 2013.

[11] A. Mahzoon, B. Alizadeh and M. Fujita, "HOFEX: high level optimization of floating point expressions for implementation on FPGAs," in *Proceedings of the International Workshop on Logic and Synthesis, IWLS '15*. 2015

[12] E. Goubault, S. Putot, "Static Analysis of Finite Precision Computations," in *proceeding of the 12th International Conference on Verification, Model Checking, and Abstract Interpretation, VMCAI '11*. Springer-Verlag, pp. 232-247, 2011.

[13] J.-M. Muller, N. Brisebarre, F. de Dinechin, C.-P. Jeannerod, V. Lef evre, G. Melquiond, N. Revol, D. Stehl´e, and S. Torres. *Handbook of Floating-Point Arithmetic*. Birkhauser Boston, 2010.

[14] J.-M. Muller, "On the definition of ulp(x)," in *Research report, Laboratoire de l'Informatique du Paralli elisme, RR2005-09*, February 2005.

[15] D. Goldberg, "What every computer scientist should know about floating-point arithmetic," *ACM Computing Surveys (CSUR)*, vol. 23, no. 1, pp. 5-48, 1991.

[16] F. de Dinechin and B. Pasca, "Designing custom arithmetic data paths with FloPoCo," *IEEE Design and Test of Computers*, vol. 28, no. 4, pp. 18-27, 2011.

[17] T. Granlund et al., "GMP, the GNU multiple precision arithmetic library," 1991. [Online]. Available: http://gmplib.org/

[18] L. Fousse, G. Hanrot, V. Lefei vre, P. P'elissier, and P. Zimmermann, "MPFR: A multiple-precision binary floating-point library with correct rounding," *ACM Transactions on Mathematical Software (TOMS)*, vol. 33, no. 2, pp. 13, 2007.

[19] C. Bauer *et al.*, "Ginac is not a CAS," 1999. [online]. Available: http://ginac.de/

[20] F. Farahmandi and B. Alizadeh, "Groebner basis based formal verification of large arithmetic circuits using Gaussian elemination and cone-based polynomial extraction," *Microprocessors and Microsystems*, Elsevier, vol. 39, No. 2, pp. 83-96, 2015.

978-1-4799-8720-7/15 $31.00 © 2015 IEEE

2015 IEEE Computer Society Annual Symposium on VLSI

Architecture for Dual-Mode Quadruple Precision Floating Point Adder

Manish Kumar Jaiswal, B. Sharat Chandra Varma, and Hayden K.-H So
Department of EEE, The University of Hong Kong, Hong Kong
Email: {manishkj, varma, hso}@eee.hku.hk

Abstract—This paper presents a configurable dual-mode architecture for floating point (F.P.) adder. The architecture (named as QPdDP) works in dual-mode which can operates either for quadruple precision or dual (two-parallel) double precision. The architecture follows the standard state-of-the-art flow for floating point adder. It is aimed for the computation of normal as well as sub-normal operands, along with the support for the exceptional case handling. The key sub-components in the architecture are re-designed & optimized for on-the-fly dual-mode processing, which enables efficient resource sharing for dual precision operands. The data-path is optimized for minimal multiplexing circuitry overhead. The presented dual-mode architecture provide SIMD support for double precision operands, along with high (quadruple) precision support. The proposed architecture is synthesized using UMC $90nm$ technology ASIC implementation. It is compared with the best available literature works, and have shown better design metrics in terms of area, period and *area × period*, along with more computational support.

Keywords-Floating Point Addition, Configurable Architecture, Dual-Mode Arithmetic, ASIC, Digital Arithmetic.

I. INTRODUCTION

Floating point (FP) number system [1], due to its wide dynamic range, is a common choice for a large set of scientific, engineering and numerical processing computations. Generally, the performance of these computations greatly depends on the underlying floating point arithmetic processing unit. Furthermore, the availability for double precision (DP) computation is not enough and the demand for high precision arithmetic is increasing in many application areas [2], [3].

The contemporary processing units achieve high performance requirement by using multiple units of single precision and double precision arithmetic hardware. Like, the Synergistic Processing Element (SPE) in Cell-BE processor [4] contains a vector array of 4 single precision and an array of 2 double precision. The ARM VFU co-processor (VFU9-S) [5] provides a vector array of 16 single precision FP units and 8 double precision vector array. Similarly, it can be seen in recent Intel Xeon Phi^{TM} and Nvidia Kepler^{TM} GK110 [6]. In general, these computing systems contain separate units/arrays for single precision and double precision computations. However, if an unified and configurable unit can support a double precision with dual/two-parallel single precision (DPdSP) arithmetic, or quadruple precision (QP) with dual/two-parallel double precision (QPdDP) arithmetic, it can save a large silicon area in the above computing

machines. In view of above, this paper is aimed towards the design of a configurable dual-mode floating adder/subtractor architecture, with high precision support.

Some literature [7], [8], [9] have proposed dual-mode architectures for adder. These works have tried to improve the resource utilization for the hardware with multi-precision computational support. However, the overhead of extra hardware, and un-optimized data-path and resource sharing lead to large overhead of area and period metrics. Furthermore, they have limited support only for normal operands. The dual-mode adder architectures of [7], [8] used a large number of multiplexers (to support dual mode) at various level of architecture, and have less tuned data path for dual mode operation. Further the extra use of resources (like more adders/subtractors for exponent & mantissa, relatively larger dual shifters, extra mantissa normalizing shifters for dual mode support) made their area & period overhead larger. Some recent literature [10], [11] have also worked on the dual-mode architectures, but with low precision support.

This paper proposes an architecture for dual-mode QPdDP (quadruple precision with dual/two-parallel double precision) adder/subtractor arithmetic. The computational sub-components are designed for configurable dual-mode support. The data-path is tuned for better resource sharing and to minimize the multiplexing circuitry. The proposed architecture provides full support for normal as well as sub-normal operands computation, exceptional case handling, and with round-to-nearest rounding method. Other rounding methods can also be easily included. A pipelined architecture is designed and synthesized using $90nm$ standard cell based ASIC implementation. The proposed architecture is compared with the best available literature.

II. PROPOSED ARCHITECTURE OF QUADRUPLE PRECISION / DUAL (TWO-PARALLEL) DOUBLE PRECISION (QPdDP) ADDER/SUBTRACTOR

The present work on the dual-mode floating point adder architecture follows the basic single-path algorithm for this computation. A floating point arithmetic computation involves computing separately the sign, exponent and mantissa part of the operands, and later combine them after rounding and normalization [1]. The standard format for floating point numbers are as follows:

$$SP : \overbrace{1-bit}^{\text{Sign}}\ \overbrace{8-bit}^{\text{Exponent}}\ \overbrace{23-bit}^{\text{Mantissa}}$$

978-1-4799-8720-7/15 $31.00 © 2015 IEEE   249

$$DP : \overbrace{1-bit}^{\text{Sign}} \overbrace{11-bit}^{\text{Exponent}} \overbrace{52-bit}^{\text{Mantissa}}$$

$$QP : \overbrace{1-bit}^{\text{Sign}} \overbrace{15-bit}^{\text{Exponent}} \overbrace{112-bit}^{\text{Mantissa}}$$

A basic state-of-the-art computational flow of the floating point adder is shown in the Algorithm 1. Here, steps 6-7 and step-22 are require for sub-normal processing. In this work, each steps of the flow are constructed for the support of the dual-mode operation with resource sharing and tuned data-path with minimum multiplexing circuitry.

Algorithm 1 F.P. Adder Computational Flow [1]

1: (*IN*1, *IN*2) Input Operands;
2: **Data Extraction & Exceptional Check-up:**
3:      {S1(Sign1), E1(Exponent1), M1(Mantissa1)} ← *IN*1
4:      {S2, E2, M2} ← *IN*2
5:      Check for INFINITY, NAN
6:      Check for SUB-NORMALs
7:      Update Exponents & Mantissa's MSB for SUB-NORMALs
8: **COMPARE, SWAP & Dynamic Right SHIFT:**
9:      $IN1\_gt\_IN2 \leftarrow \{E1, M1\} \geq \{E2, M2\}$
10:     Large_E,M ← IN1_gt_IN2 ? E1,M1 : E2,M2
11:     Small_E,M ← IN1_gt_IN2 ? E2,M2 : E1,M1
12:     Right_Shift ← Large_E - Small_E
13:     Small_M ← Small_M >> Right_Shift
14: **Mantissa Computation:**
15:     $OP \leftarrow \overline{S1 \oplus S2}$
16:     **if** OP == 1 **then**
17:         Add_M ← Large_M + Small_M
18:     **else**
19:         Add_M ← Large_M - Small_M
20: **Leading-One-Detection & Dynamic Left SHIFT:**
21:     Left_Shift ← LOD(Add_M)
22:     Left_Shift ← Adjustment for SUB-NORMAL or Underflow
23:         Add_M ← Add_M << Left_Shift
24: **Normalization & Rounding:**
25:     Mantissa Normalization & Compute Rounding ULP based on Guard, Round & Sticky Bit
26:     Add_M ← Add_M + ULP
27:     Large_E ← Large_E + Add_M[MSB] - Left_Shift
28: **Finalizing Output:**
29:     Update Exponent & Mantissa for Exceptional Cases
30:     Determine Final Output

The architecture for proposed dual-mode QPdDP adder is shown in Fig. 1. The input/output register for this architecture is assumed as shown in Fig. 2. The two 128-bit input operands, contain either 1 set of quadruple precision or 2 sets of double precision operands. Based on the mode deciding control signal (qp_dp), the dual-mode architecture switched to either quadruple precision or dual (two-parallel) double precision computation mode (qp_dp: 1 → QP Mode, qp_dp: 0 → Dual DP Mode). All the computational steps in QPdDP dual mode adder are discussed below in detail.

The data-extraction, sub-normal and exceptional handling are shown in the Fig. 3. Based on the precision format, the sign, exponent and mantissa parts of the operands are extracted for both, the quadruple precision and double precision.

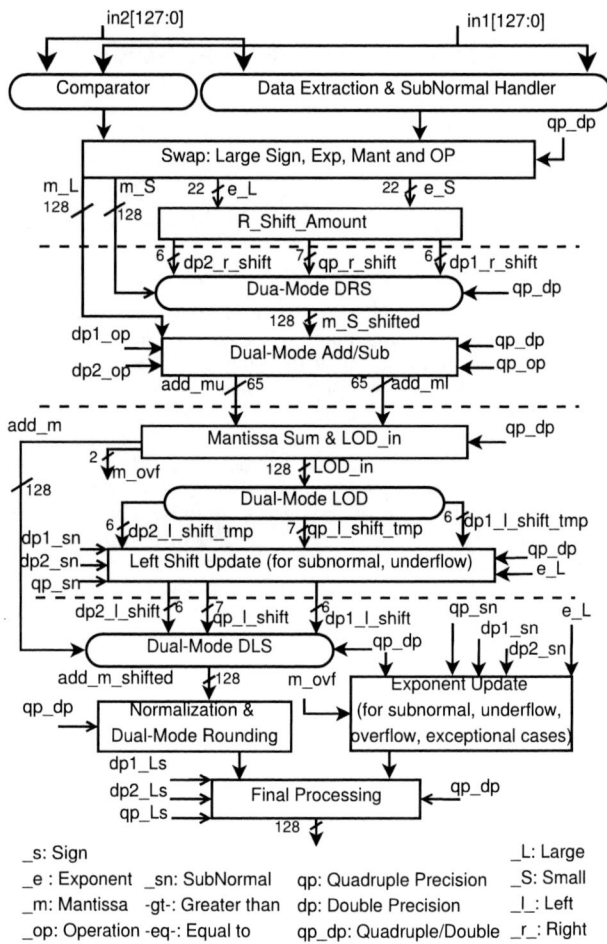

_s: Sign   **_L:** Large
_e: Exponent   **_sn:** SubNormal   **qp:** Quadruple Precision   **_S:** Small
_m: Mantissa   **-gt-:** Greater than   **dp:** Double Precision   **_l:** Left
_op: Operation **-eq-:** Equal to   **qp_dp:** Quadruple/Double   **_r:** Right

Figure 1: QPdDP Adder Architecture.

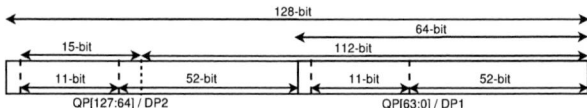

Figure 2: QPdDP Adder: Input / Output Register Format.

dp1_sn1=~|in1[62:52]	dp2_sn1=~|in1[126:116]	qp_sn1=~|in1[115:112] & dp2_sn1
dp1_sn2=~|in2[62:52]	dp2_sn2=~|in2[126:116]	qp_sn2=~|in2[115:112] & dp2_sn2
dp1_sn=dp1_sn1 & dp1_sn2	dp2_sn=dp1_sn1 & dp2_sn2	qp_sn = qp_sn1 & qp_sn2
dp1_e1={in1[62:53],in1[52]	dp1_sn1}		dp1_m1={~dp1_sn1,in1[51:0]}
dp1_e2={in1[62:53],in2[52]	dp1_sn2}		dp1_m2={~dp1_sn2,in2[51:0]}
dp2_e1={in1[126:117],in1[116]	dp2_sn1}		dp2_m1={~dp2_sn1,in1[115:64]}
dp2_e2={in2[126:117],in2[116]	dp2_sn2}		dp2_m2={~dp2_sn2,in2[115:64]}
qp_e1={in1[126:113],in1[112]	qp_sn1}		qp_m1={~qp_sn1,in1[111:0]}
qp_e2={in2[126:113],in2[112]	qp_sn2}		qp_m2={~qp_sn2,in2[111:0]}

Figure 3: QPdDP Adder: Data Extraction and Subnormal Handler.

As shown in Fig. 2 that the exponent portion of QP and

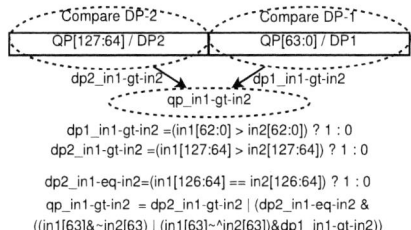

$$dp1\_in1\text{-}gt\text{-}in2 = (in1[62:0] > in2[62:0]) ? 1 : 0$$
$$dp2\_in1\text{-}gt\text{-}in2 = (in1[127:64] > in2[127:64]) ? 1 : 0$$

$$dp2\_in1\text{-}eq\text{-}in2 = (in1[126:64] == in2[126:64]) ? 1 : 0$$
$$qp\_in1\text{-}gt\text{-}in2 = dp2\_in1\text{-}gt\text{-}in2 \mid (dp2\_in1\text{-}eq\text{-}in2 \,\&$$
$$((in1[63]\&\sim in2[63]) \mid (in1[63]\sim\char`\^ in2[63])\&dp1\_in1\text{-}gt\text{-}in2))$$

Figure 4: QPdDP Adder: Comparator.

second DP (DP-2) operand are overlapped.

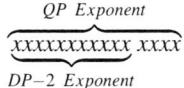

This scenario is used to share the resources related to sub-normal, `infinity`, and `NaN` checks computations of QP and second DP operands (the checks of sub-normal is shown in the Fig. 3, similarly the checks for `infinity` and `NaN` are handled). After these exceptional checks the exponent and mantissa are updated accordingly. In comparison to only QP computation, this unit requires extra related resources for first DP (DP-1) operands.

The dual-mode comparator unit for dual-mode QPdDP adder is shown in Fig. 4. The comparator unit determines which operand is large and which one is small. This unit is shared among the QP and both DP operands. It comprises of two comparator units for both DPs operands, which generates their corresponding comparison results. These DP results are further combined to form QP comparison. In terms of resources, this comparator unit requires similar resources as needed in only QP comparator, and there is no area overhead in this unit.

The next computational unit in this architecture is the Dual-Mode SWAP, which generates large sign (effectively output sign-bit), small & large exponents, small & large mantissas and effective operations (to be performed between large and small mantissas). This computational unit is shown in Fig. 5. For SWAP, in general to handle both DPs and QP, it needs four 11-bit (for both DP exponents), two 15-bit (for QP exponents), four 53-bit (for both DP mantissas) and two 113-bit (for QP mantissa) SWAP components for all the computations of this section. However, to minimize the swapping overhead, the unified exponents, mantissas and greater-than control signals are generated, by multiplexing either of the quadruple precision or both double precision operands (as shown in Fig. 5). This is an important step included in the dual-mode QPdDP architectural flow, which helps to design a tuned data-path computation in later stages, with reduced multiplexing circuitry. Using these unified exponents, mantissas and greater-than control signals, it requires only four 11-bit (for exponents) and four 64-bit (for mantissas) SWAP circuitry for entire processing. Effectively,

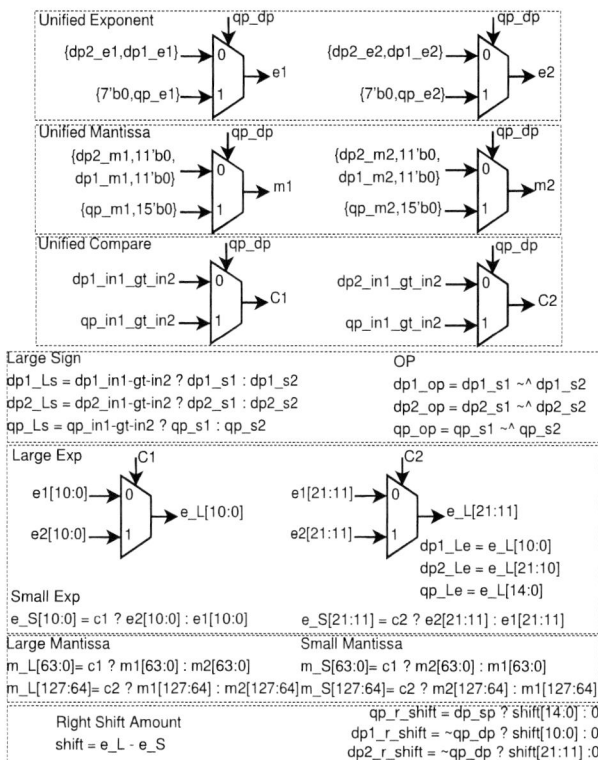

Figure 5: QPdDP Adder: SWAP - Large Sign, Exponent, Mantissa and OPERATION; Right Shift Amount.

it needs SWAP components slightly more than it requires for only QP (only QP requires two 15-bit SWAP for exponents and two 113-bit SWAP for mantissas), along with extra multiplexing circuitry needed to generate unified signals, however, facilitates the tuned data-path processing. Further, among extra appended LSB ZEROs in mantissa multiplexing (for m1 and m2), 3-bit are for Guard, Round and Sticky bit computations in rounding phase, and remaining can provide extended precision support to the operands.

The m_L contains mantissa of either large QP operand or both of large DP operands. Similarly, m_S contains small mantissas. Likewise, e_L contains large exponent, and e_S contains small exponents, either of QP or both DP operands.

Now, the small mantissa needs right shifting by the difference of large and small exponents. The right shift amount for small mantissas are determined using the component shown in Fig. 5. In general, it requires two 11-bit subtractors for both double precision and one 15-bit subtractor for quadruple precision. However, because of effective multiplexing of operands in SWAP section, it needs only one a 22-bit subtractor. A subtraction of unified large exponent (e_L) and unified small exponent (e_S) will produce right shift amount either for quadruple precision or for both double precision. For right shift amount, compared

978-1-4799-8720-7/15 $31.00 © 2015 IEEE   251

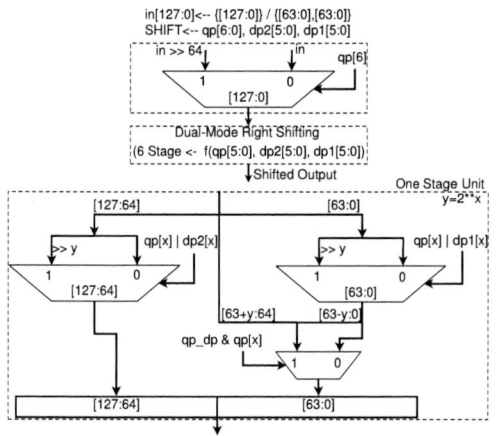

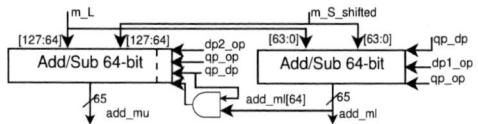

Figure 6: QPdDP Dual Mode Dynamic Right Shifter (DRS).

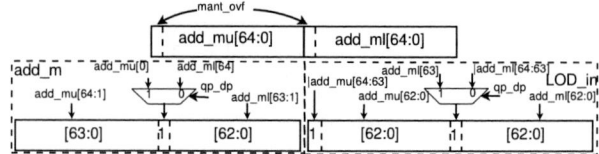

Figure 8: QPdDP Dual Mode Mantissa SUM and LOD_in.

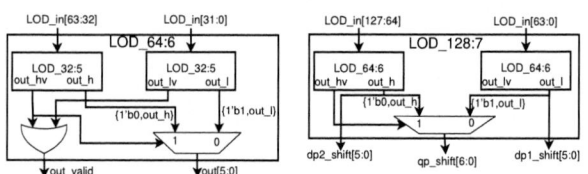

Figure 9: QPdDP Dual Mode Leading-One-Detector.

Figure 7: QPdDP Adder: Dual Mode Mantissa Addition/Subtraction.

to only quadruple precision, it requires extra resources for 7-bit subtraction. Other processing in this section are bit-wise operations, and are done separately for all operands.

For right shifting of small mantissas of quadruple and both double precision operands, a dual-mode dynamic right shifter (DRS) is designed. The QPdDP dual-mode dynamic right shifter is shown in Fig. 6, which is used to right-shift the small mantissas of either QP or both DPs. The initial step in it right-shifts the operand by 64-bit in case of QP mode with its true shift bit. The later 6-stages in it works in dual mode, either for QP or for both DPs operands. Each dual-mode stage contains two shifters for each of 64-bit blocks, which right-shifts their inputs corresponding to their shifting bit (either for quadruple or double precision). Each of these stages also include one multiplexer which selects between lower shifting output or their combination with primary input to the stage, based on the mode of the operation.

Further to the right shifting of small mantissas, the core operation of mantissa addition/subtraction fall in the computational flow. The large mantissas and right-shifted small mantissas undergo addition/subtraction based on their effective operation. This computation is performed in dual-mode using two 64-bit integer adder-subtraction unit, which individually works for each double precision, and collectively works for quadruple precision computation (as shown in Fig. 7). This unit generates the lower and upper parts of addition/subtraction separately. This component requires effectively similar resources as present in only QP adder.

The lower and upper mantissa addition/subtraction results

generated in previous unit combined in "Mantissa SUM and LOD_in unit", to provide the actual sum (either for QP or both DPs), mantissa overflow, and the input for next level unit, leading-one-detector (LOD). This unit is shown in Fig. 8.

The mantissa sum now requires to check for any underflow, which requires a leading-one-detector (LOD), and further a dynamic left shifter for mantissa. This situation occurs when two very close mantissa undergoes subtraction operation. The LOD requires to compute the left-shift amount. In present context, the dual-mode leading-one-detector for QPdDP processing is shown in Fig. 9. The input of LOD is either a QP LOD_in or two DP LOD_in. The dual mode LOD is designed in a hierarchical manner, which leads to 64-bit LOD. It is comprised of two 64-bit LOD. The individual 64-bit LOD provides left shift information for both DP operands, and collectively for QP operand. It effectively requires resources equivalent to that of only QP LOD.

The left shift amount, thus generated from LOD, is then updated for sub-normal input cases (both sub-normal input operands) and underflow cases (if left shift amount exceeds or is equal to the corresponding large exponent). For both sub-normal input operand case, the corresponding left shift is forced to zero, and for the underflow case, the corresponding left shift will be equal to corresponding large exponent decremented by one. For the exponent decrements, one of the related subtractor is shared for the QP and first DP, as done in the case of computation of right shift amount. This becomes possible because the required LSBs of e_L are shared among the exponents of QP and first DP. This exponent decrements requires one 7-bit (shared for QP and a DP) and one 6-bit (for another DP) decrement. All other computations, related to left shift update need to be computed separately for QP and both DPs. With true qp_dp, both DPs' left shift are set to zero, and for false qp_dp the QP left shift is forced to zero.

978-1-4799-8720-7/15 $31.00 © 2015 IEEE   252

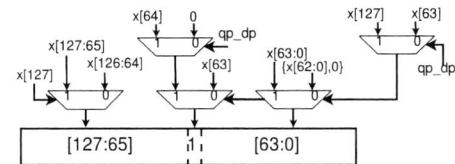

Figure 10: QPdDP Dual Mode 1-Bit Left Shifter.

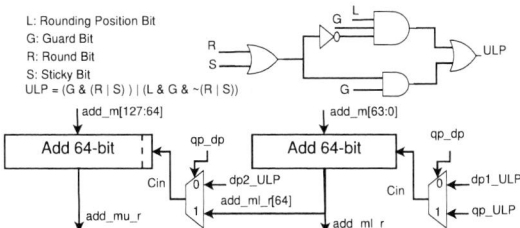

Figure 11: QPdDP Dual Mode ULP Addition.

The mantissa sum is then shifted left using a dual-mode dynamic left shifter (DLS). The basic design concept for dual-mode DLS architecture is similar to the dual-mode DRS, except that there is change in the shifting direction. (Architecture of DLS is not shown due to space limitation).

The output of dual-mode DLS then undergoes 1-bit left shifting (normalization), in-case of mantissa overflow in mantissa-addition. The dual-mode 1-bit left-shifter unit is shown in Fig. 10. It either performs a 1-bit left shifting for QP mode, or carries out 1-bit left-shifting for both DPs, with-respect-to their corresponding mantissa overflow. The resource requirement for this unit is similar to that of a only QP 1-bit shifter, except two 1-bit 2:1 MUX.

The output from 1-bit left shifter is further processed for rounding computation and ULP-addition (Fig. 11). In present work, the round-to-nearest method is included, however, other method can be included easily. The rounding ULP computations are done based on LSB precision bit, Guard bit, Round bit and Sticky bit. Here, the ULP computation is required for separately for each of QP and both DP. However, the ULP-addition is shared among both, as shown in Fig. 11.

Parallel to above mantissa processing, in Exponent-update unit, the exponents are updated for mantissa overflow and mantissa underflow. In this, the large exponents need to be incremented by one or decremented by left shift amount ($LargeExp + mant\_ovf - Left\_Shift$). Since large exponent (e_L) either contains large QP exponent or both DPs exponents, this update is shared for the QP and DP-1, by sharing a subtractor, similar to left shift update computation. In effect it requires a 15-bit shared subtractor and a 11-bit subtractor for DP-2. Thus it needs an extra 11-bit subtractor for DP-2 processing, and a 7-bit multiplexer for left shift amount multiplexing for the shared subtractor, as an overhead over only QP processing.

Finally, the exponents and mantissas are updated for underflow, overflow, sub-normal and exceptional cases to produce the final output, and each requires separate units for QP and both DPs. For overflow, the exponent will be set to infinity and mantissa will be set to zero, and for underflow case exponent will be set to zero and mantissa will take its related computed value. The computed signs, exponents and mantissas of quadruple precision and both double precision are finally multiplexed to produce the final 128-bit output, which either contains a QP output or two DP outputs.

III. IMPLEMENTATION RESULTS AND COMPARISONS

The proposed dual-mode QPdDP adder architecture is synthesized using UMC90 *nm* standard-cell based ASIC platform, using Synopsys Design Compiler. An architecture for QP only and DP only adder is also designed (using similar data path computational flow) and synthesized for area & period overhead measurements. These architectures are designed with four pipeline stages (as shown in Fig. 1). The implementation details are shown in Table-I. Architectures are synthesized for best possible period. The functionality of proposed architecture is verified using 5-millions random test cases in each mode, with all possible pairs of operands (normal, sub-normal, exceptional cases).

The proposed dual-mode QPdDP adder architecture requires approximately 17% more hardware resources and roughly 5.45% extra period than only DP adder. However, in comparison with a combination of 1-unit QP only and two-units of DP only adder, the proposed QPdDP adder requires approximately 35.86% smaller area ((QP+2*DP-QPdDP)/(QP+2*DP)).

A comparison of dual-mode QPdDP architecture with previous works is shown in Table-II. The comparisons are carried out in terms of % area-overhead and % period-overhead over corresponding only QP adder. Moreover, for a technological independent comparison, gate-equivalent or scaled area equivalent, and "Fan-Out-of-4' (FO4) delay are used. An unified comparison of *area × period* is also performed.

A dual-mode QPdDP adder architecture is presented by A. Akkas [7] with 3 & 6 pipelining stages, using 250 *nm* technology. It requires approximately 15% more area and roughly 8 − 14% more period than their only QP design. The proposed QPdDP architecture has similar area-overhead, with smaller period overhead. Moreover, the *area × period* of proposed architecture is much smaller than QPdDP adder of [7]. Furthermore, the architectures shown in [7] does not support sub-normal operands computation and exceptional case handling.

A 110 *nm* based dual-mode QPdDP adder is proposed by [8], with 3-stage and 5-stage pipelines. These architectures do not provide computational support for sub-normal operands and without any exceptional case handling. For their architectures, the area-overhead ranges between

Table I: ASIC Implementation Details

	DP	QP	QPdDP
Latency	4	4	4
Area(μm^2)	31863	76779	90116
Area(gates)	10621	25593	30038
Period(ns)	0.95	1.1	1.16
Period(FO4)	21.11	24.44	25.78
Power(mw)	7.26	12.87	16.93

Table II: Comparison of QPdDP Architecture with Related Work

	[7] 250nm		[8] 110nm		Proposed 90nm
Latency	3	6	3	5	4
Area OH[1]	15.3%	14.01%	35.80%	27.31%	17.37%
Period OH[1]	14.12%	8.71%	18.65%	10.11%	5.45%
Scaled Area[2]	-	-	239250	199723	90116
Gate Count[3]	26967	33702	-	-	30038
Period (FO4)[4]	65.28	35.92	28.9	17.81	25.78
Area $\times$ Period (10^6) [#1]	-	-	6.91	3.55	2.32
Area $\times$ Period (10^6) [#2]	1.76	1.21	-	-	0.77

[1] Area/Period OH = (QPdDP - QP) / QP
[2] in μm^2 @ 90nm = (Area @ 110nm) * $(90/110)^2$
[3] Based on minimum size inverter
[4] 1 FO4 (ns) $\approx$ (Tech. in μm) / 2
[#1] Scaled Area $\times$ Period (FO4),   [#2] Gate Count $\times$ Period (FO4)

$27-35\%$ and period overhead is approximately $10-18\%$. Compared to this work, the proposed dual-mode QPdDP architecture outperforms them in terms of design overheads, as well as in terms of design metrics: the area, period and *area $\times$ period*.

Thus, compared to previous works, the proposed dual-mode QPdDP adder architecture has smaller area-overhead and period-overhead when compared to only QP adder. The proposed QPdDP architecture shows an improvement of approximately 50% in terms of unified metrics *area $\times$ period* products.

IV. CONCLUSIONS

A configurable architecture for dual-mode floating point adder arithmetic is presented in this paper. The proposed dual-mode QPdDP adder architecture provides normal & sub-normal computational support and exceptional case handling. The data path and sub-components in the architecture are constructed/re-designed for on-the-fly dual-mode processing, with minimal required multiplexing. The presented dual-mode QPdDP adder architecture needs approximately 17% more resources and 5.45% more period than the QP only adder. When compared with the best literature work, the proposed dual-mode design has approximately 50% smaller *area $\times$ period* product, and has smaller area & period overhead over only QP adder. It also provides more computational support than previous literature work. Our future work is aiming towards a tri-mode adder architecture, which along with proposed computation can also be configured to handle four-parallel single precision computation.

V. ACKNOWLEDGMENTS

This work is party supported by the "The University of Hong Kong" grant (Project Code. 201409176200), the "Research Grants Council" of Hong Kong (Project ECS 720012E), and the "Croucher Innovation Award" 2013.

REFERENCES

[1] "IEEE Standard for Floating-Point Arithmetic," Tech. Rep., Aug. 2008.

[2] F. de Dinechin and G. Villard, "High precision numerical accuracy in physics research," *Nuclear Instruments and Methods in Physics Research Section A: Accelerators, Spectrometers, Detectors and Associated Equipment*, vol. 559, no. 1, pp. 207–210, 2006.

[3] D. H. Bailey, R. Barrio, and J. M. Borwein, "High-precision computation: Mathematical physics and dynamics," *Applied Mathematics and Computation*, vol. 218, no. 20, pp. 10106–10121, 2012.

[4] H.-J. Oh, S. Mueller, C. Jacobi, K. Tran, S. Cottier, B. Michael, H. Nishikawa, Y. Totsuka, T. Namatame, N. Yano, T. Machida, and S. H.Dhong, "A fully pipelined single-precision floating-point unit in the synergistic processor element of a cell processor," *Solid-State Circuits, IEEE Journal of*, vol. 41, no. 4, pp. 759–771, 2006.

[5] NXP Semiconductors, "AN10902 : Using the LPC32xx VFP," in *Application note*, Feb 2010. [Online]. Available: www.nxp.com/documents/application_note/AN10902.pdf

[6] Nvidia, "NVIDIA's Next Generation CUDA^{TM} Compute Architecture: Kepler^{TM} GK110," in *White Paper*, 2014. [Online]. Available: www.nvidia.com/content/PDF/kepler/ NVIDIA-Kepler-GK110-Architecture-Whitepaper.pdf

[7] A. Akkas, "Dual-Mode Quadruple Precision Floating-Point Adder," *Digital Systems Design, Euromicro Symposium on*, vol. 0, pp. 211–220, 2006.

[8] ——, "Dual-mode floating-point adder architectures," *Journal of Systems Architecture*, vol. 54, no. 12, pp. 1129–1142, Dec. 2008.

[9] M. Ozbilen and M. Gok, "A multi-precision floating-point adder," in *Research in Microelectronics and Electronics, 2008. PRIME 2008. Ph.D.*, 2008, pp. 117–120.

[10] M. Jaiswal, R. Cheung, M. Balakrishnan, and K. Paul, "Unified architecture for double/two-parallel single precision floating point adder," *Circuits and Systems II: Express Briefs, IEEE Transactions on*, vol. 61, no. 7, pp. 521–525, July 2014.

[11] ——, "Configurable architecture for double/two-parallel single precision floating point division," in *VLSI (ISVLSI), 2014 IEEE Computer Society Annual Symposium on*, July 2014, pp. 332–337.

VLSI Design of Edge-Preserving Coding Artifacts Reduction for Display Processing

Zenghua Cheng[†], Xuchong Zhang[‡], Huisheng Peng[††], Baolu Zhai[‡‡], Hongbin Sun[††‡] and Nanning Zheng[‡]

[†]Department of Microelectronics, [‡] Institute of AIR, Xi'an Jiaotong University, Xi'an, Shaanxi, China
[††]HiSilicon Technologies Co., Ltd., Shanghai, China   [‡‡]Spreadtrum Communications Inc., Tianjin, China
*Corresponding author, email: hsun@mail.xjtu.edu.cn

Abstract—**Although coding artifacts reduction is one of indispensable modules in video display processing system, its VLSI design is still rarely studied. Existing coding artifacts reduction approaches mainly focus on in-loop techniques, the other post-processing methods have not addressed the challenges posed by display system and are actually inapplicable to real-time display processing. This paper proposes an efficient algorithm which can reduce both blocking and ringing artifacts while effectively preserving edge and texture regions. The proposed algorithm is inherently hardware-friendly, and is implemented with a fully pipelined VLSI architecture which is synchronized to pixel clock and can be easily integrated into display processing system. Evaluation results demonstrate that, the image quality of the proposed algorithm outperforms four previously proposed post-processing approaches, and its VLSI circuit can achieve the real-time performance of 1080P@60Hz at the cost of less than 63K logic gates and 15.4KB memories.**

Keywords-deblocking; deringing; display processing; VLSI;

I. INTRODUCTION

With the rapid development and popularity of internet and smart TVs, internet video has become one of the major video source for TV content. Various block-based video coding standards, such as H.26X and MPEG-X, are exploited to compress and transfer video over internet. However, the compressed video inevitably suffers from the well-known blocking and ringing artifacts, particularly when the video is highly compressed. The blocking artifact is the grid noise along the block boundaries in a relatively homogeneous area, and is a major issue of block-based coding techniques [1]. The ringing artifact is an effect of Gibbs phenomenon due to the truncation of high-frequency coefficients by coarse quantization [2], and is often visible along sharp edges as rippling or ghost effects. As the size and resolution of display panel have been steadily increased, these artifacts tend to be more noticeable and should be carefully addressed.

Although the adaptive in-loop deblocking filter has been defined and adopted into advanced video coding standards, such as H.264 [3] and H.265 [4], coding artifacts reduction is still one of indispensable modules in video display processing [5]. This is mainly because post-processing methods do not require any modification to existing standards and can work independently after the image is decoded. Hence, post-processing methods have the relatively wider applicability and can support various video coding standards. In addition, post-processing methods can also efficiently reduce

annoying ringing artifacts as well. Therefore, coding artifacts reduction in the post-processing stage has also attracted much attention and been extensively exploited.

Nevertheless, the great majority of the previously proposed coding artifacts reduction methods are actually not applicable to real-time video display system. Many of these algorithms [6], [7] are extremely computational intensive and hence are infeasible for real-time implementation. Some less computational intensive algorithms, such as POCS-based algorithms [8], [9], usually invoke loop routines that require high delay and hence are only suitable for software implementation. However, video display processing is not only an application-specific VLSI system, but also poses extra challenges to coding artifacts reduction algorithm design. In video display system, coding information including block boundaries are completely unknown, and only those algorithms that can reduce both blocking and ringing artifacts while not destroying edge and texture are practically desirable. Therefore, although some spatial adaptive filtering algorithms [10]–[12] seems to be efficient for hardware implementation, the aforementioned issues have not been well addressed. In particular, how to implement the desired coding artifacts reduction with efficient hardware design is rarely studied [13], [14] and needs further exploration, especially for high-definition display system.

This paper proposes an edge-preserving coding artifacts reduction algorithm that can efficiently compensate both blocking and ringing artifacts. In addition, the proposed algorithm can adaptively detect the location and strength of blocking and ringing artifacts from the decoded images, hence is able to well address the challenges posed by display system. We further implement the proposed algorithm with a fully pipelined VLSI design, which employs the line-buffer-based design and can seamlessly conform to VESA display standard. Therefore, the proposed VLSI design can be easily integrated into video display processing system. Experimental results demonstrate that, the proposed algorithm outperforms four previously proposed post-processing approaches in terms of both objective and subjective tests, and its VLSI circuit can achieve the real-time performance of 1080P@60Hz with less than 63K logic gates and 15.4KB memories.

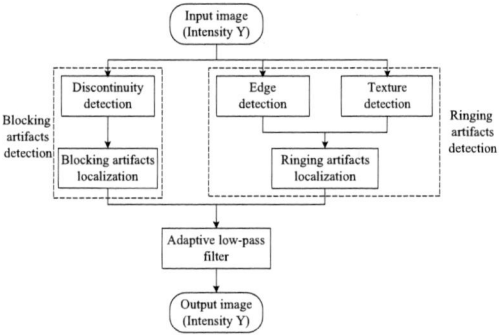

Figure 1. Overall flow of the proposed coding artifacts reduction algorithm.

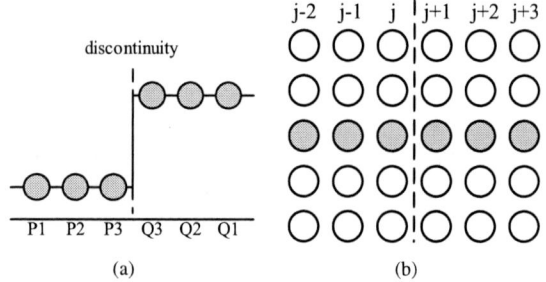

Figure 2. Horizontal discontinuity detection

II. PROPOSED ALGORITHM

A. Overall Flow

As the proposed coding artifacts reduction algorithm is designed for video display processing system, coding information, such as QP, prediction mode, block size and block boundary, are completely unknown. Moreover, we would like to well preserve the edge and texture region when applying low-pass filtering to the decoded images. Therefore, the proposed algorithm firstly detect and allocate blocking and ringing artifacts, and only applying low-pass filtering to pixels belonging to artifacts. Fig. 1 shows the overview of the proposed edge-preserving coding artifacts reduction algorithm, which includes three major modules, i.e. blocking artifact detection, ringing artifact detection, and adaptive low-pass filter.

For blocking artifact detection, most of the previously proposed post-processing deblocking algorithms [1], [2], [12], [14] assume the blocking boundaries are given and only apply low-pass filtering to boundary pixels. However, blocking artifacts may also occur on inner-block regions, due to motion-compensated prediction among consecutive video frames. Moreover, block boundaries may also be unknown as video may have been cropped or shifted before being delivering to display system. Therefore, in the proposed algorithm, the blocking artifact is allocated by employing discontinuity detection over the whole image. In addition, edge detection and texture detection are employed to allocate ringing artifact. Finally, the adaptive low-pass filter compensates both blocking and ringing artifacts simultaneously according to the location of artifacts.

B. Blocking Artifacts Detection

Since blocking artifacts are essentially the discontinuity on both the horizontal and vertical directions in the decoded images, we can allocate blocking artifacts by exploiting discontinuity detection. Let us take horizontal discontinuity detection for example to explain the detection procedure in detail, as illustrated in Fig. 2. P_1, P_2, P_3, Q_3, Q_2 and Q_1 are six adjacent pixels in the same image line. Then,

D_1, D_2, D_3, D_4 and D_5 are used to represent absolute luminance differences between each two neighboring pixels, i.e. $|P_1 - P_2|$, $|P_2 - P_3|$, $|P_3 - Q_3|$, $|Q_3 - Q_2|$ and $|Q_2 - Q_1|$, respectively. If all the three conditions in (1) are satisfied, horizontal discontinuity may exist between P_3 and Q_3.

$$\begin{cases} \text{Condition1} : D_3 < T_{blk\_1} \\ \text{Condition2} : max(D_1, D_2, D_4, D_5) < T_{blk\_2} \\ \text{Condition3} : D_3 > max(D_1, D_2, D_4, D_5) \end{cases} \quad (1)$$

We employ the same aforementioned horizontal discontinuity detection to 5 adjacent image lines to improve the detection robustness, as shown in Fig. 2(b). If 3 adjacent lines of 5 detected lines satisfy all the conditions in (1) simultaneously, we confirm horizontal discontinuity exist at current pixel (i,j). Then, we further compare D_3 with threshold T_{blk_3} to determine the strength of discontinuity. Finally, pixels are classified into three classes, i.e. non-discontinuity, weak-discontinuity, and strong-discontinuity, as illustrated in (2), where D_h represents the map of horizontal discontinuities.

$$D_h(i,j) = \begin{cases} 2'b11, & \text{strong-discontinuity}, D_3 \geq T_{blk\_3} \\ 2'b10, & \text{weak-discontinuity}, D_3 < T_{blk\_3} \\ 2'b00, & \text{non-discontinuity}, (1) \text{ is not satisfied} \end{cases} \quad (2)$$

Then, we locate pixels that belong to blocking artifacts and should be filtered in the next stage. If current pixel (i,j) belongs to the class of strong-discontinuity or weak-discontinuity according to $D_h(i,j)$, all its neighboring pixels in a 3×6 window are marked as blocking artifacts, and classified as (3).

$$B_{maph}(i,j) = \begin{cases} 2'b11, & \text{strong blocking artifact} \\ 2'b10, & \text{weak blocking artifact} \\ 2'b00, & \text{no blocking artifact} \end{cases} \quad (3)$$

Similarly, the map of vertical discontinuities D_v and blocking artifacts B_{mapv} can be detected in the same way, except that we only use four vertical adjacent pixels for D_v and 6×3 window for B_{mapv} to save the usage of line buffers. The overall blocking artifacts map B_{map} is generated according to B_{maph} and B_{mapv}. In our following

978-1-4799-8720-7/15 $31.00 © 2015 IEEE

experiments, T_{blk_1}, T_{blk_2} and T_{blk_3} are set as 16, 10 and 8, respectively.

$$B_{map}(i,j) = B_{maph}(i,j)||B_{mapv}(i,j) \qquad (4)$$

C. Ringing Artifacts Detection

It has been known that, ringing artifacts mainly occur along edges, especially sharp edges on the flat background. Moreover, we should also protect the edges and texture regions from being destroyed when removing ringing artifacts. Therefore, edge and texture detections are two necessary steps for ringing artifacts detection.

For edge detection, we use Sobel operators and calculate the average gradient as follows.

$$G_{avg} = (|Gx| + |Gy|)/2 \qquad (5)$$

Where G_x and G_y represent the horizontal and vertical outputs of Sobel operators in a neighboring 3x3 window, respectively. Then, we compare G_{avg} of each pixel with two thresholds TH_{edg1} and TH_{edg2} to construct the edge map E_{map}, which represents whether the corresponding pixel belongs to edge and how sharp it is as follows.

$$E_{map}(i,j) = \begin{cases} 2'b11, & \text{strong-edge, if } G_{avg} > TH_{edg1} \\ 2'b10, & \text{weak-edge, else if } G_{avg} > TH_{edg2} \\ 2'b00, & \text{non-edge, else} \end{cases}$$
$$(6)$$

In our following experiments, TH_{edg1} and TH_{edg2} are set as 168 and 78, respectively.

For texture detection, we calculate sum of absolute difference (SAD) in a neighboring 3x3 window. Then we compare SAD of each pixel with two thresholds TH_{tex1} and TH_{tex2} to construct the texture map T_{map}, which represents whether the corresponding pixel belongs to texture region and its strength as follows.

$$T_{map}(i,j) = \begin{cases} 2'b11, & \text{strong-texture, if } SAD > TH_{tex1} \\ 2'b10, & \text{weak-texture, else if } SAD > TH_{tex2} \\ 2'b00, & \text{non-texture, else} \end{cases}$$
$$(7)$$

In our following experiments, we set TH_{tex1} and TH_{tex2} as 96 and 24, respectively.

In general, ringing artifacts are observed as weak texture. Hence, pixels belonging to flat region or strong texture should not be ringing artifacts. While pixels belonging to weak texture may be ringing artifacts or normal texture. We should further use edge map E_{map} to allocate ringing artifacts, as ringing artifacts mainly appear along edges. Accordingly, we construct a neighboring 7×7 window for each pixel and determine the ringing artifact map R_{map} as

follows.

$$R_{map}(i,j) = \begin{cases} 2'b11, & \text{no ringring artifact (edge region),} \\ & \text{if } E_{map}(i,j)[1]||T_{map}(i,j)[0] \\ 2'b10, & \text{strong ringing artifact,} \\ & \text{else if } (NUM_{stedge} \geq TH_{num1}) \\ & \&\&(NUM_{flat} \geq TH_{num3}) \\ 2'b01, & \text{weak ringing artifact,} \\ & \text{else if } (NUM_{edge} \geq TH_{num2}) \\ & \&\&(NUM_{flat} \geq TH_{num3}) \\ 2'b00, & \text{no ringring artifact,} \\ & \text{else} \end{cases}$$
$$(8)$$

where NUM_{stedge}, NUM_{edge} and NUM_{flat} represent the number of pixels that belong to strong edges, edges and non-texture in the 7×7 window, respectively. The classification is also based on the observation that, ringing artifacts mainly occur along edges, especially sharp edges on the flat background. In our following experiments, TH_{num1}, TH_{num2}, and TH_{num3} are set as 3, 5, and 14, respectively.

D. Adaptive Low-Pass Filter

After obtaining the B_{map} and R_{map} for the overall image, we employ an adaptive low-pass filter to reduce both blocking and ringing artifacts. Moreover, the adaptive low-pass filter is only applied to the location of blocking and ringing artifact, hence the edge and texture region can be well preserved. The adaptive low-pass filter is essentially a simplified bilateral filter [15], which takes into consideration both the geometric closeness and intensity similarity, as illustrated in (9).

$$Y_{fir}(m,n) = \frac{\sum\limits_{(i,j) \in W} \omega(i,j) \cdot Y(i,j)}{\sum\limits_{(i,j) \in W} \omega(i,j)}, \qquad (9)$$
$$and \quad \omega(i,j) = \omega_1(i,j) \cdot \omega_2(i,j).$$

where $Y(i,j)$ represents the intensity value, (m,n) represents the central pixel and (i,j) represents all the pixels in the filtering window W. $\omega(i,j)$ is weighting coefficient, which includes two ingredients, i.e. geometric weight $\omega_1(i,j)$ and intensity weight $\omega_2(i,j)$. In the proposed adaptive low-pass filter, $\omega_1(i,j)$ and $\omega_2(i,j)$ are calculated with simple arithmetic operations to reduce the hardware resource consumption.

$$\omega_1(i,j) = DISTANCE - (|i - m| + |j - n|), \qquad (10)$$

$$\omega_2(i,j) = \begin{cases} 2, & \text{if } |Y(i,j) - Y(m,n)| < TH_{filter}/2 \\ 1, & \text{else if } |Y(i,j) - Y(m,n)| < TH_{filter} \\ 0, & \text{else.} \end{cases} \qquad (11)$$

Where DISTANCE is twice as large as the maximum Cityblock Distance between the central and corner pixels in the filtering window W. TH_{filter} is an adaptive threshold defined as follows.

$$TH_{filter} = max(TH_{fir}, \alpha \cdot Y(m,n)), \qquad (12)$$

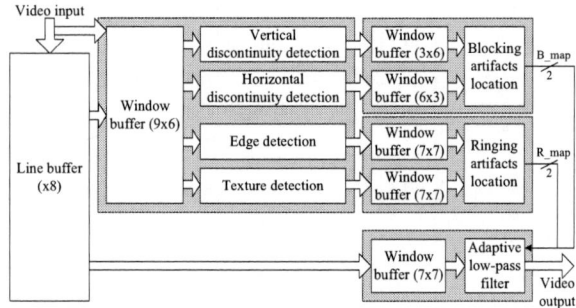

Figure 3. The overall VLSI architecture design.

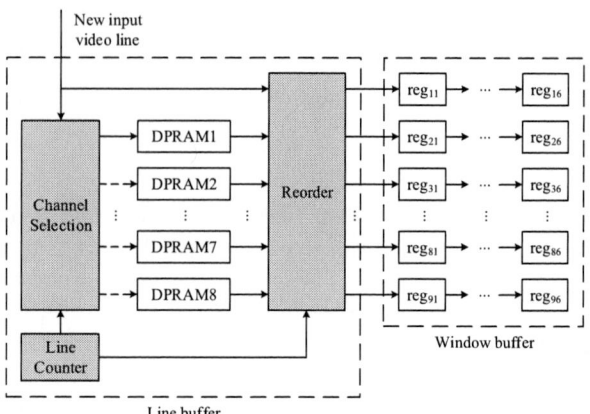

Figure 4. Memory hierarchy design for window-based processing.

Where α is the scale factor to $Y(m,n)$, and TH_{fir} is a threshold to guarantee TH_{filter} is above certain level in low-intensity region. In particular, the size of W, TH_{fir} and α are adaptively adjusted according to the strength of blocking or ringing artifacts. In our following experiments, for strong blocking or ringing artifacts, $W = 7 \times 7$, $TH_{fir} = 12$ and $\alpha = 3/32$; for weak blocking or ringing artifacts, $W = 5\times5$, $TH_{fir} = 6$ and $\alpha = 3/64$.

III. Proposed VLSI Architecture

The VLSI architecture of the proposed coding artifacts reduction is implemented for real-time video display processing application [16], which is usually integrated in TV or mobile SoC to enhance image quality. The output of display processing pipeline is usually directly connected to the following display panel, it thereby should meet certain timing standard, such as VESA standard. Therefore, the output video frames of display processing are transferred pixel by pixel in a horizontal raster-scanned manner at the defined frame rate and pixel clock rate. These timing requirements inherently favor the pipelined architecture. As a consequence, the proposed VLSI circuit is designed as a fully pipelined architecture and is synchronized to the pixel clock.

The overall architecture is illustrated in Fig. 3. As the proposed coding artifacts reduction algorithm is inherent-ly hardware-friendly, it provides a favorable base for the VLSI implementation. The proposed VLSI architecture is implemented as seven major modules, including horizon-tal&vertical discontinuity detections and blocking artifact localization for blocking artifact detection, edge&texture detections and ringing artifact localization for ringring arti-fact detection, and adaptive low-pass filter. As the proposed coding artifacts reduction algorithm is inherently based on local filtering, its VLSI architecture is mainly designed as window-based stream processing for both artifact detection and low-pass filter. The key to this window-based image processing is its memory hierarchy design, which caches the streaming data from external memory and provides all the neighboring pixels of the processing pixel simultaneously.

Fig. 4 illustrates the memory hierarchy design, which exploits the two-level buffering, i.e. line buffer and window buffer. Line buffer is implemented with dual-port SRAMs (DPRAMs), and each DPRAM can store one row of input image frame. A window buffer is a rectangle-shaped set of shift registers, and each register can store one pixel intensity value of the input image. As the window buffer is built by registers, it is able to guarantee instant access to all its elements at each pixel clock. This means that all the pixels in the window buffer can be accessed simultaneously to support the window-based processing. In addition, two extra modules, i.e. channel selection and reorder are also added to schedule the write and read of line buffers, thereby reducing write accesses to DPRAMs and the corresponding energy consumption. As the length of row buffer should be equal to the full width of the input image, its area cost is non-trivial for high definition video processing. In addition, as the proposed coding artifacts reduction algorithm employs several window-based processing modules that should be processed sequentially, the straightforward implementation tends to consume two much line buffers. Therefore, we construct a relatively larger window buffer (9×6) to not only enable the line buffer sharing among discontinuity, edge and texture detection, but also further eliminate the line buffer requirement of following modules, i.e. blocking and ringing artifact locations. Although these techniques may take extra logic gates, they can significantly save the usage of line buffers, thereby reducing the overall hardware overhead.

The proposed VLSI architecture is designed as fully pipelined implementation that is synchronized to pixel clock of video sequences. The maximum working frequency is set as 150 MHz to meet the throughput requirement of HDTV 1920×1080@60Hz. Inside the VLSI architecture, complex arithmetic units, such as multipliers and adders, are implemented as multi-clock pipelined circuits to shorten the delay path and match the pixel clock. The total clock

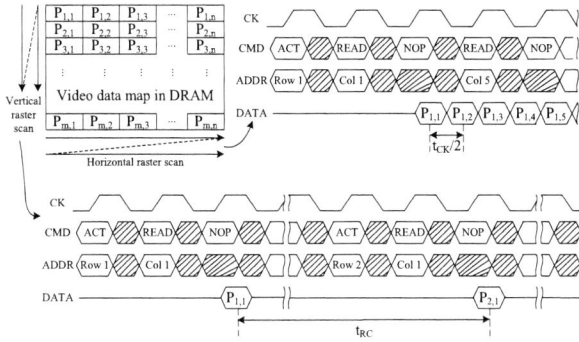

Figure 5. External DRAM access timing: horizontal vs. vertical scan.

cycle of the pipelined coding artifacts reduction design is 42, hence the delay between input and output is 42 clock cycles. More important, as the overall VLSI circuit processes video frames pixel-by-pixel in a horizontal raster-scanned manner, the proposed architecture design is seamlessly compatible with the VESA timing standard and hence can be easily integrated into video display processing system.

We note that the line-buffer-based design and horizontal raster scan format employed in the proposed VLSI architecture tend to consume more on-chip hardware resources, compared with the block-based design [14]. The block-based design proposed to use vertical raster scan and process all the pixels in current block before loading new block. Hence, it is able to eliminate the on-chip line buffers and save some logic gates, thereby significantly reducing the on-chip overhead, especially for memory usage. However, the timing of the block-based processing does not conform with VESA standard, and hence the block-based design actually can not be directly integrated into video display processing system. More important, the modification from horizontal raster scan to vertical raster scan will remarkably degrade the external memory access, thereby significantly increasing the external memory bandwidth requirement. Fig. 5 illustrates external DRAM access timing diagram for both horizontal and vertical raster scans. As the horizontal raster scan can effectively utilize the burst mode of DRAM, the minimum delay between two sequential accessed pixels is only half of the DRAM clock cycle t_{CK}. While the vertical raster scan has to continuously activate the row addresses in the same bank, and the minimum delay is activate-to-activate delay t_{RC}. For the main stream DDR3 DRAM chip [17], t_{CK} can reach up to 1.25ns, whereas t_{RC} is only as long as 48ns. According to our evaluation in Section IV, the vertically scanned block-based design actually can not support HD video processing due to the bottleneck of external memory access. Of course, we can also use tile-based design to balance between external memory bandwidth and on-chip memory usage. However, we choose the line-buffer-based

architecture to conform with VESA standard and simplify the integration of the proposed design in video display processing system.

IV. EXPERIMENTAL RESULTS

A. Algorithm Simulation

For the algorithm simulation, we use 13 test sequences downloaded from http://media.xiph.org/video/derf/, including foreman, mobile, hall, mother-daughter, news, et al., with format of CIF 4:2:0. All the test sequences contain 150 frames, and are compressed by H.264 reference software (JM18.4). For simplicity, the rate control is adopted, and only the first frame is encoded as an intra-coded frame (I-VOP) and the coding frame structure is with only one I frame initially followed by P frames. We compare the proposed coding artifacts reduction algorithm with four computation-efficient post-processing works, i.e. Zhai [10], Chebbo [11], Yeh [12] and Vinh [14]. These referred methods are implemented by us in MATLAB following the description given in the cited patent or papers.

The objective test compares video sequences obtained after compression and then processing them with different methods, with original reference video sequences. The differences between the processed and reference video sequences are evaluated by the metrics of peak signal noise ratio (PSNR) and mean structure similarity index (MSSIM) [18]. Table I illustrates the PSNR and MSSIM comparisons among different methods at various bit rates. We can tell that, the proposed coding artifacts reduction algorithm outperforms the other methods in terms of both PSNR and MSSIM. For the subjective test, we further compare the visual quality of video sequences produced by the proposed algorithm with that of other methods. Part of visual quality comparison results are shown in Fig. 6 and Fig. 7, where Fig. 6 mainly demonstrates for deblocking and Fig. 7 mainly demonstrates for deringing. We can see that, the proposed coding artifacts reduction algorithm shows better image visual quality compared with the other algorithms, that it can significantly reduce the blocking and ringing artifacts while well preserving the edge and texture region.

B. Hardware Implementation

The proposed VLSI architecture has been designed with Verilog and synthesized using Synopsys tools with TSMC 0.13um standard cell and memory library. The VLSI implementation results are summarized in Table II. The depth of input video is designed as 8 bits per channel. The maximum working frequency of the proposed VLSI circuit is set as 150MHz, which means that the maximum video processing throughput is 1920×1080@60Hz. According to the synthesis report, the overall VLSI circuit design only costs less than 63K logic gates and 15.4KB memories. In addition, as we explained in Section III, the proposed VLSI

978-1-4799-8720-7/15 $31.00 © 2015 IEEE

Table I
PSNR AND MSSIM COMPARISONS AMONG DIFFERENT METHODS AT VARIOUS BIT RATES.

	Sequences	Foreman			Mobile			Hall			Mother-daughter			News		
	Bit rate (kbps)	140	180	220	280	320	360	80	120	160	80	120	160	100	140	180
PSNR	Compressed	31.37	32.58	33.51	26.35	27.02	27.54	35.22	36.33	37.05	36.56	38.54	39.76	34.85	36.51	37.51
	Zhai [10]	31.46	32.67	33.59	26.44	27.11	27.64	35.30	36.40	37.12	36.64	38.63	39.84	34.93	36.59	37.58
	Chebbo [11]	31.47	32.68	33.60	26.43	27.09	27.63	35.31	36.41	37.13	36.64	38.64	39.85	34.94	36.60	37.59
	Yeh [12]	31.49	32.70	33.62	26.46	27.12	27.65	35.33	36.43	37.14	36.67	38.67	39.88	34.97	36.63	37.63
	Vinh [14]	31.48	32.70	33.61	26.46	27.12	27.65	35.33	36.42	37.14	36.66	38.65	39.86	34.96	36.63	37.61
	Proposed	31.50	32.72	33.64	26.48	27.14	27.67	35.35	36.44	37.16	36.67	38.67	39.88	34.98	36.65	37.64
MSSIM	Compressed	0.838	0.870	0.889	0.867	0.882	0.895	0.937	0.943	0.947	0.925	0.946	0.955	0.942	0.954	0.959
	Zhai [10]	0.842	0.873	0.890	0.868	0.884	0.896	0.939	0.944	0.948	0.925	0.946	0.956	0.943	0.954	0.959
	Chebbo [11]	0.843	0.873	0.890	0.868	0.883	0.896	0.939	0.944	0.948	0.925	0.946	0.956	0.944	0.955	0.960
	Yeh [12]	0.844	0.875	0.892	0.870	0.886	0.897	0.941	0.946	0.949	0.927	0.947	0.957	0.945	0.956	0.962
	Vinh [14]	0.843	0.874	0.891	0.868	0.884	0.896	0.940	0.945	0.948	0.925	0.945	0.956	0.944	0.955	0.960
	Proposed	0.845	0.876	0.893	0.871	0.887	0.898	0.942	0.947	0.950	0.928	0.947	0.957	0.946	0.957	0.962

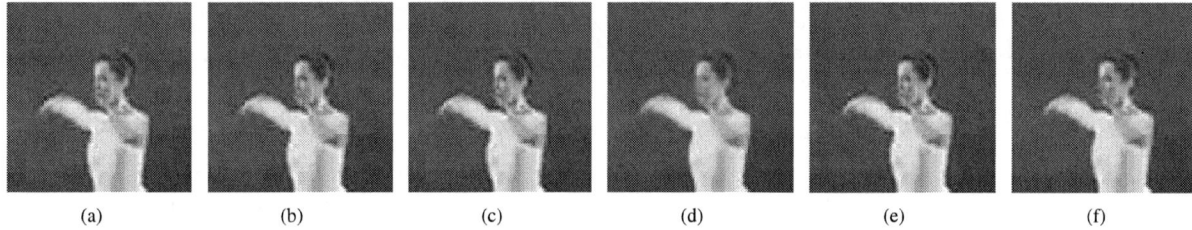

Figure 6. Subjective comparison among different methods for 'news' sequence. (a) Compressed (b) Zhai (c) Chebbo (d) Yeh (e) Vinh (f) Proposed.

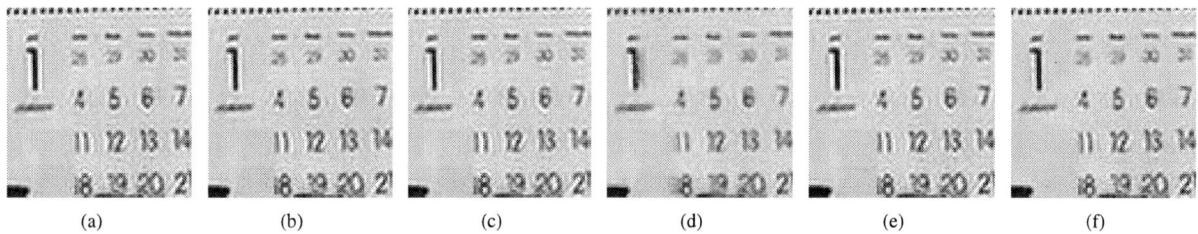

Figure 7. Subjective comparison among different methods for 'mobile' sequence. (a) Compressed (b) Zhai (c) Chebbo (d) Yeh (e) Vinh (f) Proposed.

circuit is fully pipelined and synchronized to the pixel clock of output video, and hence it can seamlessly support the VESA-compatible display system.

Fang [13] and Vinh [14] are the only two references that discuss VLSI design of post-processing blocking artifact reduction, according to our extensive literature research. Fang's design is implemented for SD video (NTSC@30Hz), and hence its hardware overhead is relatively very low. Although Vinh's design is implemented for HD video, it only integrates deblocking function and its maximum throughput is only 1080P@30Hz. With only 12.7K extra logic gates, the proposed design can reduce both blocking and ringring artifacts, and outperform Vinh's design in both objective, subjective tests and video processing throughput.

In addition, Vinh's design seems to consume much less on-chip memories as it employs block-based design and vertical raster scan. However, as we analyzed in Section III, the external memory bandwidth is the significant bottleneck for this block-based design. Targeting for a DDR3-1600 DRAM chip [17], the proposed VLSI design only takes 1.30ms to load one 1080P frame from external memory, while Vinh's design has to take 99.5ms (less than 11 frame/second). This further demonstrates that Vinh's design is actually infeasible for full HD video processing, and the proposed VLSI design is a viable and cost-efficient solution.

V. CONCLUSION

This paper proposes an efficient post-processing algorithm that can reduce both blocking and ringing artifacts while

Table II
VLSI IMPLEMENTATION RESULT COMPARISON.

Items	Fang [13]	Vinh [14]	Proposed
Deblocking	Yes	Yes	Yes
Deringing	No	No	Yes
Process	0.35um	0.25um	0.13um
Architecture	Unknown	Block-based	Line-buf-based
Frequency	100MHz	150MHz	150MHz
Video format	NTSC@30Hz	1080P@30Hz	1080P@60Hz
Pipeline	Yes	192 stages	42 stages
Gate count	5.2K	50.1K	62.7K
On-chip memory	Unknown	310B	15,360B
Memory bandwidth[1]	Unknown	99.5ms/frame	1.30ms/frame

[1] The memory bandwidth is evaluated as the delay required for loading a 1920×1080 video frame from external DDR3-1600 DRAM memory.

preserving the edge and texture. The proposed algorithm is implemented as a fully pipelined VLSI circuit employing the line-buffer-based architecture, which conforms to VESA display standard and can be easily integrated into display processing system. Experimental results show that, the proposed algorithm can provide better image quality than compared approaches, and the VLSI circuit can achieve real-time video processing up to 1080P@60Hz with reasonably low cost.

ACKNOWLEDGMENT

This research was supported by the National Science and Technology Major Project of China (No. 2013ZX01033001-001). H. Peng and B. Zhai participated in this work when they were graduate students at Xi'an Jiaotong University.

REFERENCES

[1] M.-Y. Shen and C.-C. J. Kuo, "Review of postprocessing techniques for compression artifact removal," *Journal of Visual Communication and Image Representation*, vol. 9, no. 1, pp. 2–14, Mar. 1998.

[2] H. Park and Y. L. Lee, "A postprocessing method for reducing quantization effects in low bit-rate moving picture coding," *IEEE Transactions on Circuits and Systems for Video Technology*, vol. 9, no. 1, pp. 161–171, Feb. 1999.

[3] P. List, A. Joch, J. Lainema, G. Bjontegaard, and M. Karczewicz, "Adaptive deblocking filter," *IEEE Transactions on Circuits and Systems for Video Technology*, vol. 13, no. 7, pp. 614–619, Jul. 2003.

[4] G. J. Sullivan, J. Ohm, W.-J. Han, and T. Wiegand, "Overview of the high efficiency video coding (HEVC) standard," *IEEE Transactions on Circuits and Systems for Video Technology*, vol. 22, no. 12, pp. 1649–1668, Dec. 2012.

[5] J. E. Caviedes, "Intelligent sharpness enhancement for video post-processing," in *the Proceedings of European Signal Processing Conference*, Sep. 2006.

[6] T. Meier, K. N. Ngan, and G. Crebbin, "Reduction of blocking artifacts in image and video coding," *IEEE Transactions on Circuits and Systems for Video Technology*, vol. 9, no. 3, pp. 490–500, Apr. 1999.

[7] Y. Zhao, G. Cheng, and S. Yu, "Postprocessing technique for blocking artifacts reduction in DCT domain," *Electronics Letters*, vol. 40, no. 19, pp. 1175–1176, Sep. 2004.

[8] T. Kartalov, Z. A. Ivanovski, L. Panovski, and L. J. Karam, "An adaptive POCS algorithm for compression artifacts removal," in *the Proceedings of International Symposium on Signal Processing and Its Applications*, Feb. 2007, pp. 1–4.

[9] Y. Kim, C.-S. Park, and S.-J. Ko, "Fast POCS based postprocessing technique for HDTV," *IEEE Transactions on Consumer Electronics*, vol. 49, no. 4, pp. 1438–1447, Nov. 2003.

[10] F. Zhai, W. P. Chang, and H. E. Lu, "Method and apparatus for reducing ringing artifacts," Feb. 2011, US Patent 7,894,685.

[11] S. Chebbo, P. Durieux, and B. Pesquet-Popescu, "Adaptive deringing and mosquito noise reducer," in *the Proceedings of International Workshop on Video Processing and Quality Metrics*, 2010.

[12] C.-H. Yeh, S.-J. Jiang, T.-F. Ku, M.-J. Chen, and J.-A. Jhu, "Post-processing deblocking filter algorithm for various video decoders," *IET Image Processing*, vol. 6, no. 5, pp. 534–547, Jul. 2012.

[13] H.-C. Fang, T.-C. Wang, and L.-G. Chen, "Real-time deblocking filter for MPEG-4 systems," in *the Proceedings of Asia-Pacific Conference on Circuits and Systems*, 2002, pp. 541–544.

[14] T. Q. Vinh and Y.-C. Kim, "Edge-preserving algorithm for block artifact reduction and its pipelined architecture," *ETRI Journal*, vol. 32, no. 31, pp. 380–389, Jun. 2010.

[15] C. Tomasi and R. Manduchi, "Bilateral filtering for gray and color images," in *the Proceedings of International Conference on Computer Vision*, Jan. 1998, pp. 839–846.

[16] J. E. Caviedes, "The evolution of video processing technology and its main drivers," *Proceedings of the IEEE*, vol. 100, no. 4, pp. 872–877, Feb. 2012.

[17] Mircon, *DDR3 SDRAM datasheet MT41J256M8HX-125*, 2013.

[18] Z. Wang, A. C. Bovik, H. R. Sheikh, and E. P. Simoncelli, "Image quality assessment: from error visibility to structural similarity," *IEEE Transactions on Image Processing*, vol. 13, no. 4, pp. 600–612, Apr. 2004.

A Custom Computing System for Finding Similarties in Complex Networks

Christian Brugger*, Valentin Grigorovici*, Matthias Jung*, Christian Weis*,
Christian De Schryver*, Katharina Anna Zweig‡, Norbert Wehn*
*Microelectronic System Design Research Group, University of Kaiserslautern, Germany
‡Graph Theory & Complex Network Analysis Group, University of Kaiserslautern, Germany
{brugger, jungma, weis, schryver, wehn}@eit.uni-kl.de,
valentin.grigorovici@gmail.com, zweig@cs.uni-kl.de

ABSTRACT

Complex graphs are at the heart of today's big data challenges like recommendation systems, customer behavior modeling, or incident detection systems. One reoccurring task in these fields is the extraction of network motifs, reoccurring and statistically significant subgraphs. In this work we propose a precisely tailored embedded architecture for computing similarities based on one special network motif, the co-occurrence. It is based on efficient and scalable building blocks that exploit well-tuned algorithmic refinements and an optimized graph data representation approach. On chip, our solution features a customized cache design and a lightweight data path that allows the system to perform over 10,000 graph operations per cycle on each chip. We provide detailed area, energy, and timing results for a 28 nm ASIC process and DDR3 memory devices. Compared to an Intel cluster, our proposed solution uses 44x less memory and is 224x more energy efficient.

1. INTRODUCTION

Many applications in the big data context are based on fast and reliable identification of so-called *network motifs* in large networks, i.e., those subgraphs whose occurrence is significantly higher than expected in a random graph model. This enables analyzing large-scale biological data in bioinformatics, the analysis of connections in social networks, incident detection, and general graph data cleaning procedures by link assessment [1].

In this work, we consider a special variant of motifs, the so-called *co-occurrence* (*coocc*) which is defined as the number of common neighbors of two nodes in a graph. It can, for example, be used to clean biological high-throughput data or to build e-commerce applications like recommendation systems [1]. However, computing the *coocc* on standard central processing unit (CPU)- and graphics processor unit (GPU)-based architectures is very time and energy consuming.

In this paper we present a dedicated architecture for network motif detection based on the *coocc*. It is problem-independent and universally applicable to a wide application range, for instance as a special accelerator device in bigger system contexts. Due to its modular approach, the proposed design can also be enhanced to other motifs in the future.

Our results show that the proposed architecture clearly outperforms standard CPU server nodes, both with respect to throughput and energy, but also total memory requirements. Compared to one 6-core Intel Xeon X5680 CPU, it is 226x faster with an equivalent power consumption. We demonstrate the performance of our design with the Netflix data set [2, 3] and show that one application specific integrated circuit (ASIC) instance computes the same results in roughly the same time as a 10-node Intel cluster but requires only 2.3% of total memory and less than 0.5% of energy. These superior characteristics allow in particular the use of our architecture in power- and space-limited data-centers and for constructing motif detection systems targeted to process very large graphs with reasonable power consumption and system costs.

In particular, the novel contributions in this paper are:

- We present the first hardware architecture for estimating similarities in graphs with *fixed degree sequence models* [1].
- We introduce algorithms and data paths for *coocc* calculations that massively exploit linear data access patterns.
- We show an efficient cache design that allows the architecture to be scaled without requiring a higher memory bandwidth.
- As results, we give detailed area, energy, and timing results for a 28 nm ASIC process and DDR3 memory devices, including a comparison to an Intel compute cluster.

2. BACKGROUND AND RELATED WORK

Everybody who has used social networks knows the famous "Do you also know ...?" function. One could argue that a large part of the success of modern social networks is based on this feature. But how does it work? Which persons should be for instance recommended specifically to you?

A powerful way is to look at the social network itself. There, one can calculate how similar you are to each person in the graph, only using the topology of the connections. Based on that, one can then suggest the top ten most similar people with respect to the direct neighborhood.

In this paper we consider the following method: In the example given in Fig. 1, the similarity between you and Liam is based on the number of common friends you have with her, the so-called *co-occurrence*: $coocc(you, Liam) = 3$. Now we have to ask whether the number three is significant. Assume

---

Permission to make digital or hard copies of all or part of this work for personal or classroom use is granted without fee provided that copies are not made or distributed for profit or commercial advantage and that copies bear this notice and the full citation on the first page. To copy otherwise, to republish, to post on servers or to redistribute to lists, requires prior specific permission and/or a fee.

978-1-4799-8720-7/15 $31.00 © 2015 IEEE

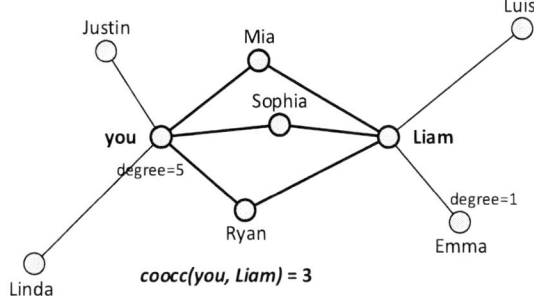

Figure 1: The *co-occurrence* (*coocc*) between you and Liam is defined as the number of shared friends you have, in this case three.

you and Liam have thousands of friends, versus you have only three. For this we create random graphs based on the same degree sequence and the premise that nodes have no similarities. In our example the sorted sequence of degrees, the number of edges a node has, is $\{1, 1, 1, 1, 2, 2, 2, 5, 5\}$.

To get the random graphs we swap a sufficient number of pairs of edges, drawn uniformly at random, if and only if no multiple edges would arise due to the swap. This generates independent graphs with the same degree sequence, see Fig. 2. Generating many of such graphs we can calculate the expected *coocc* of this so-called *fixed degree sequence model (FDSM)* [1, 4]. With that information we can judge how significant the *coocc* in our original graph is.

2.1 Similarity measures

In general, there exist many ways to quantify the similarity, most of them are various normalization of this *coocc*. In the Jaccard-index, this value is normalized by the cardinality of the joint neighborhood. Pearson's product moment correlation coefficient uses the co-variance of the two vectors, normalized by the product of their standard deviations. However, these classic similarities do not take into account the general distribution of the data. Complex network analysis has shown that in many cases, the number of neighbors is heavily skewed, some authors even suggest that this so-called *degree distribution* is a scale-free distribution [5]. This induces a heavily skewed *coocc* as well in most real-world data sets, which is why the field of complex network analysis suggests to compare the *observed coocc* (number of common neighbors) with the *expected coocc* in an ensemble of random graphs with the same degree distribution [6, 4]. This approach is called FDSM. In this setting, the observed *coocc* can be corrected by the expected *coocc*; the resulting difference is called the *leverage*. The leverage can then be normalized by the standard deviation of the expected distribution to yield the *z*-score [6]. An alternative approach is to use the empirical *p*-value, i.e., the probability to pick a random graph instance in which the *coocc* is at least as high as in the observed network. The usage of an elaborate random graph model to assess the statistical significance of an observed *coocc* by either the leverage, the z-score, or the p-values has shown to produce results with higher quality in many domains and is the basis of this work [7, 1].

2.2 Computing leverage, p-value, z-score

Given a bipartite graph $G((V_l, V_r), E)$ with vertices V_l

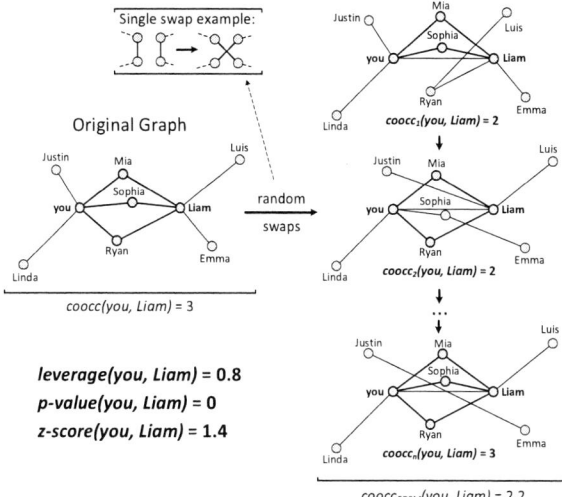

leverage(you, Liam) = 0.8
p-value(you, Liam) = 0
z-score(you, Liam) = 1.4

Figure 2: One hundred random graphs (right) with the same degree sequence as the initial graph (left) obtained by series of 40 swaps. The leverage is the difference of the original *coocc* and the average *coocc* of the random graphs samples (right).

and V_r and edges E, V_l being the side of interest, we define $coocc_i(u, v)$ as the *coocc* for the graph sample G_i. Algorithm 1 shows the full scheme. This way of generating graphs is also called *Markov chain Monte Carlo (MCMC)* [8]. The similarity measures for all $u, v \in V_l$ are defined as:

$$coocc(u, v) = \sum_{w \in V_r} \begin{cases} 1, & \text{if } (u, w) \in E \text{ and } (v, w) \in E \\ 0, & \text{otherwise} \end{cases},$$

$$coocc_{FDSM}(u, v) = \text{mean}\left(\{coocc_i(u, v)\}_{i=1,..,|samples|}\right),$$

$$leverage(u, v) = coocc(u, v) - coocc_{FDSM}(u, v), \quad (1)$$

$$p\text{-}value(u, v) = \sum_{i=1}^{|samples|} \begin{cases} 1, & \text{if } coocc_i(u, v) > coocc(u, v) \\ 0, & \text{otherwise} \end{cases},$$

$$z\text{-}score(u, v) = \frac{leverage(u, v)}{\text{stddev}\left(\{coocc_i(u, v)\}_{i=1,..,|samples|}\right)}.$$

The higher the leverage or the z-score, or the lower the p-value, the more similar the nodes are considered to be. Algorithm 1 shows all the steps, Fig. 2 illustrates the procedure with the friend-recommendation example. For our similarity measures the complexity of Algorithm 1 is $O(|V_l|^2 \cdot |V_r|)$ what makes it very challenging to come up with a scalable implementation.

2.3 Related work

Network motif detection is actively investigated in current research, mainly from the algorithmic point of view. From the implementation side, nearly all available work deals with mapping the motif detection problem on parallel CPU and GPU based clusters [9, 10]. To the best of our knowledge there are no publications for accelerating motif detection, in particular by a dedicated accelerator engine optimized for this task. We were only able to detect a few works about hardware implementations for general graph processing, but none of them considering the specific case of *coocc*s.

There are frameworks like GraphGen by Nurvitadhi et al.

978-1-4799-8720-7/15 $31.00 © 2015 IEEE   263

Data: Graph $G((V_l, V_r); E)$ with vertices V_l and V_r and edges E, V_l being the vertices of interest;

Result: Leverage, p-value, z-score for all pairs of vertices $(u, v) \in (V_l \times V_l)$;

Calculate $coocc(u, v) \, \forall \, (u, v) \in (V_l \times V_l)$; $G_0 := G$;

for $i := 1$ *to* $|samples|$ **do**

$\quad G_i := G_{i-1}$;

$\quad$ *Swap randomization:*

$\quad$ **for** $|swaps|$ **do**

$\qquad$ Choose two edges at random in G_i and swap them, if no duplicate edge arises from the swap;

$\quad$ **end**

$\quad$ *Coocc computation:*

$\quad$ Calculate $coocc_i(u, v) \, \forall \, (u, v) \in (V_l \times V_l)$;

end

Calculate leverage, p-value and z-score according (1);

Algorithm 1: The complete Link Assessment algorithm, calculating the similarity measures

---

presented in 2014 [11] or the Graphlet Counting Case Study from Betkaoui et al. in 2011 [12] that generate specific data processing engines for particular graph operations. Both approaches aim at optimizing the memory accessing schemes for the dynamic random-access memory (DRAM) in order to fully exploit the available memory bandwidths. However, they are not application tailored and cover a broad range of graph problems instead of optimizing the performance for motif detection.

3. EMBEDDED ARCHITECTURE

Nowadays it is a general trend in high-performance computing (HPC) to enhance current HPC systems with dedicated accelerators that are optimized for time and energy critical tasks what leads to more and more heterogeneous computing architectures. This methodology is adapted from embedded computing where the crucial design points are mainly energy, power, and throughput efficiency. In this paper we present an energy-optimized accelerator architecture for computing similarity measures based on a network-motif-approach in graphs. In detail, it consists of three main parts:

- Swap randomization: generate samples from the FDSM.
- *Coocc* calculation: calculate the *co-occurrence* of all pairs for each random graph sample.
- Similarity measures: generate leverage, p-value, and z-score based on the *coocc*.

Fig. 3 gives an overview of the architecture described next. The global aim of our design is to minimize the number of random DRAM accesses. For that reason, selecting appropriate data structures is the most crucial point. Instead of using adjacency lists, we use an adjacency matrix to store the graph. In that structure, accessing individually nodes directly corresponds to one access, as we exactly know where to look.

Although the graphs we are dealing with are sparse, we are storing them in adjacency matrices and not in adjacency lists. While this might sound surprising, matrices have so many advantages for the architecture that it outweights the disadvantages by far, as we will see in Section 3.2 and 3.3.

3.1 Data structures

We are using three data structures: an adjacency matrix holding the graph, an adjacency list holding the edges, and a result matrix.

Given a bipartite graph $G(V_l, V_r; E)$ consisting of the ver-

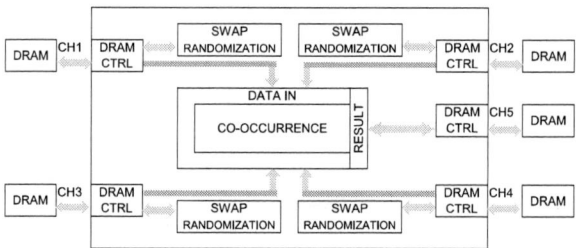

Figure 3: Overall architecture. The swap randomization blocks and the *co-occurrence* work on datasets residing in each of the DRAMs in a round-robin fashion, while multiple swap randomization blocks can work in parallel. The result is stored separate.

Table 1: Partial result matrix entry

Variable	Required bits
$coocc(u, v)$	$\log_2(|V_r|)$
$\sum_i coocc_i(u, v)$	$\log_2(|V_r| \cdot |samples|)$
$\sum_i coocc_i(u, v)^2$	$\log_2(|V_r|^2 \cdot |samples|)$
p-value count	$\log_2(|samples|)$

tices V_l and V_r and the edges E, an adjacency matrix $A = (V_l \times V_r)$ is stored. An entry in the matrix is $A_{u,v} = 1$ if $(u, v) \in E$. It is sufficient to store A with one bit per entry and total storage requirement of $|V_l| \cdot |V_r|$ bit. Furthermore, the edges are stored as a list in L. An *edge* is defined as the pair of vertex indices which are connected by it. The edge list requires $|E|(\log_2 |V_l| + \log_2 |V_r|)$ bits.

The result matrix contains partial results for calculating the p-value and z-score for all the pairs in V_l. For each pair $(u, v) \in (V_l \times V_l)$ we store an entry in an upper triangular matrix, see Table 1. In a post-processing step, all similarity measures can be efficiently calculated with these partial results on a CPU.

3.2 Swap randomization

for $|swaps|$ **do**

$\quad$ Generate two random numbers (RNs): $a, b \in [1, |E|]$;

$\quad$ Read the two edges: $u, x := L_a$ and $v, y := L_b$;

$\quad$ Check existence of swapped edges by reading:

$\quad$ $k := A_{u,y}$ and $l := A_{v,x}$;

$\quad$ **if** *both edges do not exist:* $k = l = 0$ **then**

$\qquad$ Swap the edges by writing: $A_{u,x} := 0$; $A_{v,y} := 0$;

$\qquad$ $A_{u,y} := 1$; $A_{v,x} := 1$;

$\qquad$ Update edge list by writing: $L_a := u, y$;

$\qquad$ $L_b := v, x$;

$\quad$ **end**

end

Algorithm 2: Implementation of the swap randomization step in Algorithm 1

We sample graphs from the FDSM based on MCMC. A new sample G_i can be generated from the previous one G_{i-1} by randomly swapping $|swaps|$ edges, starting from the original graph. In our architecture, the swapping is realized in a finite-state machine (FSM) implementing Algorithm 2.

For generating the RNs, we use a Mersenne twister (MT)

19937 algorithm [13]. In this formulation, four random reads and six random writes are necessary per swap.

3.3 Coocc calculation

Calculating the *coocc* between all pairs of vertices in V_l in a naive way requires to load the same data many times, making memory a bottleneck. For example, calculating the *coocc* between $u, v \in V_l$ requires edges connected to u and v, or in other words the two rows u and v of the matrix A. When the *coocc* is later calculated between u and w, the same row A_u needs to be loaded. This leaves huge potential for an optimized memory hierarchy and algorithms to miminizing data transfer. Solving this is our first major contribution.

First we add a row-cache to the *coocc* module. Second we structure the accesses in such a way that, once a row is loaded into cache, it is not necessary to load it a second time. Furthermore we use multiple modules to parallelize the computations.

Having k parallel *coocc* units, we use their caches to store a consecutive block of k rows $A_u, .., A_{u+k-1}$. Then we stream one by one all following rows through the *coocc* modules, starting with A_{u+k}. With each new row A_v the modules can calculate the *coocc* of all pairs of the cached rows $(u, v), .., (u+k-1, v)$. Algorithm 3 formalizes the scheme.

Data: Graph $G((V_l, V_r); E)$ stored as adjacency matrix
$\quad A = (V_l \times V_r)$, V_l being of the vertices of interest;
Result: coocc for all pairs of vertices $(u, v) \in (V_l \times V_l)$;
for $u := 1$ **to** $|V_l|$ **step** K **do**
$\quad$ k := 0;
$\quad$ **for** $v := u$ **to** $|V_l|$ **do**
$\quad\quad$ Stream row A_v from external memory;
$\quad\quad$ **if** $k \geq 1$ **then**
$\quad\quad\quad$ Compare the streamed row with all
$\quad\quad\quad$ previously cached rows 1 to k and calculate
$\quad\quad\quad$ the *coocc* for the pairs:
$\quad\quad\quad$ $(u, v), .., (u+k-1, v)$;
$\quad\quad$ **end**
$\quad\quad$ **if** $k < K$ **then**
$\quad\quad\quad$ k := k + 1;
$\quad\quad\quad$ Store the streamed row in cache k;
$\quad\quad$ **end**
$\quad$ **end**
end

Algorithm 3: Implementation of the *coocc* computation step in Algorithm 1 for K *coocc* modules

---

Looking closely, you will notice that this scheme also solves the scaling problem. While adding m times more modules reduces the runtime by a factor of m, it does not increase the requirements for external bandwidth, still only one row has to be streamed through all the blocks at each given time. This allows us to place hundreds of *coocc* units next to each other, providing massive speedups.

To calculate the actual *coocc*s, we propose efficient data paths, our second major contribution. While standard CPU and GPU architectures only standard data types can be used, on the ASIC we are much more flexible. Assuming that our rows are large, we receive them in blocks of data, l bits per cycle. Having stored A_u in the caches (LMEM) and streaming A_v, the *coocc*(u, v) can be obtained by counting all vertices that are connected to both u and v. The row A_u at position w is "1", if and only if there is an edge between u and w, for A_v equivalently. Counting common edges is equivalent to computing the cardinality of $A_u \& A_v$. This requires a lot of additions.

Fig. 4(a) shows our efficient data path tailored to this task, consisting of an adder tree and accumulator. It is able to process l edges per cycle. In the first stage, it uses only 1 bit adders, 2 bit adders in the second stage, and so on.

3.4 Similarity measures

Once all blocks of the row A_w have been streamed, the final *coocc* value is computed. Based on that, all partial results from Table 1 can be updated. This involves reading the entry from memory, updating, and writing it back, see Fig. 4(c). It is worth noticing that this block does not have to operate in each clock cycle, only once per complete row, so that most of the operators can be shared among multiple *coocc* computation blocks. After all samples have been computed, the final similarity measures can be calculated on a CPU by going once over the result data, which only takes a few seconds.

Calculating the initial *coocc* is also performed on chip at the beginning and stored in the partial results.

3.5 Parallelization

Parallelization is easily possible by using multiple instances with each instance working on independent samples. For k instances this reduces the total time by a factor of k. They start all with the same initial graph and the results can be easily combined at the very end by summation.

4. ASIC IMPLEMENTATION

We have implemented our architecture on register-transfer level (RTL) with SystemVerilog and synthesized it with the Synopsys Design Compiler for an advanced 28 nm low-power technology node. Place&Route was performed with the Cadence SoC Encounter. Based on the Netflix data we have extracted realistic activities that we fed into Synopsys PrimeTime to obtain accurate power estimates.

We have integrated three 64 bit DDR3 memory controllers, one for the result and two for storing the graph. At each given time, the random swapping block is operating on one controller and the *coocc* modules on the others. When both are finished they switch over.

In the *coocc* module, the edge caches are 64 kB each, targeting a frequency of 400 MHz. For a 64 bit double data rate (DDR) channel at 800 MHz we get 256 edges per cycle when running the *coocc* units at 400 MHz. That means the adder tree has a width of 128 adders at the top and a depth of seven stages. Pipelining the tree was not required for our application. The operations to calculate the partial results for the similarity are designed with 64 bit for the squares and 32 bit for the rest, being shared over the cell.

We have synthesized four *coocc* modules in a single cell and then combined them in a grid of 5 times 12. In total, our ASIC architecture has 240 *coocc* modules. To distribute the data to the caches or to stream further rows of the matrix a tree-like replication network is used, while for the results a shift register over the whole chip is used. That makes the architecture perfectly scalable. In total, our architecture performs $240 \cdot 256 = 61,440$ graph operations per cycle.

The rest of the design is occupied by memory controllers and IO, see Fig. 5. Table 2 shows the resources of each module. For the memory controllers we have estimated the numbers based on the corresponding publications [14, 15]. The whole chip has a size of $51.2 \, \text{mm}^2$ and an average power consumption of 11.7 W.

We assume that the final system is equipped with two

978-1-4799-8720-7/15 $31.00 © 2015 IEEE

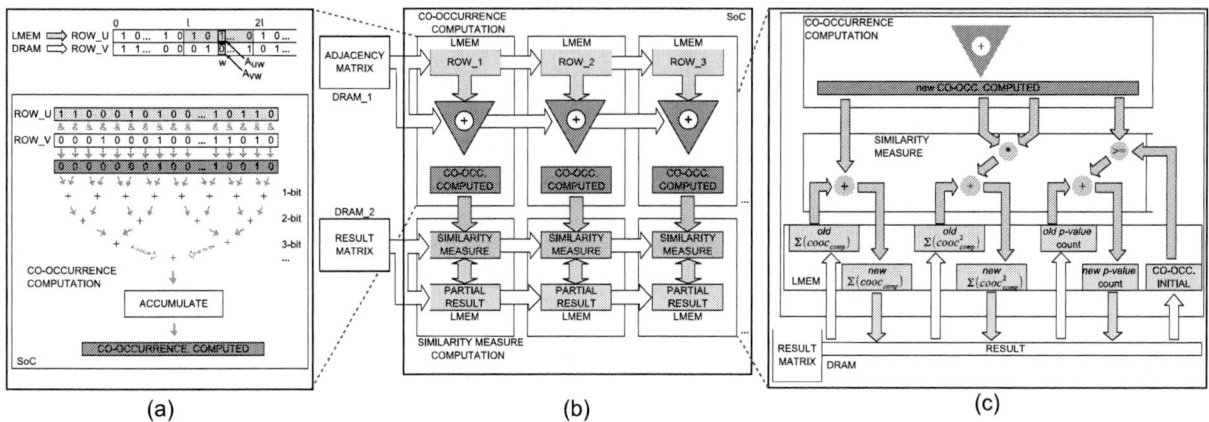

Figure 4: The *coocc* and result module (b) works on one dataset after another, always updating the same result. It loads one row of the graph into the caches (LMEM) and first calculates the *coocc* before calculating the similarity measures. The *coocc* module (a) consists of an efficient adder tree operating on blocks of l edges per cycle. While the similarity measures (c) consists of several arithmetic blocks it is only called once per row, making it possible to share most of the resources.

Table 2: ASIC Ressources

Component	Size [mm^2]	Frequency [MHz]	Power[1] [W]
Swap randomization	0.01	400	0.002
4 *coocc* cell	0.572	400	0.123
DDR3 controller	4.8	800	0.8
IO and interconnect	2.4	400	0.56
One DDR3 DIMM during:			
swapping	–	–	1.8
coocc computation	–	–	2.2
result processing	–	–	1.5
Our chip[2]	51.2	400	11.7[3]

[1]fully utilized for single components
[2]240 *coocc* modules, 1 swap module, 3 DDR3 controllers
[3]based on activities running the Netflix dataset

2 GB DDR3-1600 DIMMs for the graphs and a 4 GB module for the result. We have modeled the timing of the DIMMs with DRAMSys tool [16] and estimated power with DRAM-Power tool [17]. As inputs to the tools we created DRAM access patterns for all the modules, specifying in which order, when, and at which address the DRAM is accessed. Based on those trace files the tools provided us with exact timings and power estimations given in Table 2.

Note that neither the architecture nor the algorithm makes any assumption about the properties of the graph: Every graph can be processed by the ASIC. Of course as in every computing architecture the graph needs to fit into system memory. The number of *coocc* units was chosen such that the chip area is around 50 mm^2, a reasonable target for today's IP-cores. At runtime, the input graph is folded in space and time onto the available *coocc* units and buffers.

5. CLUSTER COMPARISON

To demonstrate the performance of our design we have calculated the similarity measures for the Netflix dataset [2, 3]. Netflix, a commercial video streaming service, has released 100,480,507 user ratings for all of their 17,700 movies

from 480,189 users. While users give ratings in the range of 1-5, we have extracted an input graph with edges between users and movies whenever the rating is 4 or 5 only. As a result, the input graph has 17,769 movies, 478,615 users, and 56,919,190 edges. In this case V_l will be the movies, V_r the users, and E the 4, 5 ratings.

The swap randomization module of our design takes 2.14 s to generate a new random graph by swapping the dataset with $|swaps| = |user| \log |user| = 6,259,639$. The *coocc* module, working in parallel, takes 3.25 s to calculate the *coocc* partial results from one graph. During this time the result memory controller is active in 20 % of the time. We have used a total number of samples $|samples| = 10,000$ that ensure sufficient convergence. In total, our design takes 9.0 hours to process the Netflix data.

For a system level comparison, we have included the power of our design and DRAM as well as a 20 % overhead of 2.63 W to account for necessary board components (ethernet, clocks) and the power supply. Post-processing and calculating the final similarity measures for the Netflix data takes on an Intel node 0.325 seconds or 120 J, during the rest of the time the CPU is free to use for other purposes.

To make a fair comparison, we have spent a lot of effort on optimizing the parallel cluster implementation. Two skilled persons spent three months full time for this reference work. Optimization involved selection of algorithm that minimize computing time for the given memory resources, removing locks by data partitioning and data access linearization. Since swapping is hard to parallelize, each core works on its own sample during swapping, generating 12 samples on one server node. Afterwards, the partial results are updated one sample after another, while all the 12 cores work in parallel on one sample to reduce memory requirements. Among nodes, the parallelization is the same as for the chips with each node working on independent samples. Communication is performed over InfiniBand only required to distribute the data at the beginning and aggregate the results at the very end of the algorithm.

For the cluster implementation, the swapping works on the same adjacency matrix A as in the proposed architecture to minimize random accesses. At the same time an

Table 3: Cluster ASIC comparison

Implementation	Memory[1] [GB]	Runtime [hour]	Power [W]	Energy [MJ]
10 node Intel cluster[2]	202/480	8.5	3700	114
This work[3]	4.6/8	9.0	15.8	0.51
Improvement	44x	0.95x		224x

[1] used/available memory

[2] each node: 2×Intel Xeon X5680 @ 12×3.33 GHz, 32 nm; 48 GB DDR3 memory

[3] node inclduing: ASIC with 240 modules, 28 nm; 8 GB DDR3 memory; board (ethernet, clocks), power supply.

adjacency list is kept in memory that contains the user ids of the people who have rated the film for each movie. This adjacency list is used to calculate the *coocc*, since using the matrix for *coocc* calculations is very inefficient on the CPU.

For the cluster we have measured the average system power with a power meter during operation. Note that the power measurements for both the cluster and proposed architecture consider the entire link assessment algorithm, assuming the input and result are in memory. The results are listed in Table 3.

6. CONCLUSION

Discovering similarities in graphs is an important task in many big data applications. However, for large graphs this is in general a very time and memory consuming job on standard computing clusters. In this work we introduce a dedicated hardware architecture that implements the three basic steps for this tasks: swap randomization, *coocc* computation, and calculating the similarity measure. To the best of our knowledge, it is the first tailored hardware architecture for computing similarity measures based on a network-motif-approach in graphs. Our design is universally applicable to a large range of big data applications like bioinformatics, recommendation systems, friend suggestions in social networks, or general graph cleaning as it is. Due to its modular structure, it can be easily enhanced to other network motifs and similarity measures.

Since the main limitation in this application is in general the memory bandwidth utilization to the DRAM, we have optimized our architecture for minimal DRAM accesses. For that purpose, our proposed design features well-matched data structures: Instead of using adjacency lists as normally

done in CPU clusters, we implement the complete adjacency matrix, but with 1 bit entries only. This is not practical in generic computing platforms in this way. With that approach we minimize the data transmission in general and in particular the DRAM bandwidth.

We have synthesized our design for a 28 nm process and could achieve a clock rate of 400 MHz for a die size of 51.2 mm^2 and a power consumption of 11.7 W.

As a result, the proposed architecture is both energy and memory efficient. In comparison to a 10-node standard dual-socket Intel cluster, our design achieves the same throughput with less than 0.5% of power and therefore energy per task. At the same time, it requires only 2.3% of DRAM.

7. REFERENCES

[1] Katharina Anna Zweig et al. *A systematic approach to the one-mode projection of bipartite graphs.* Social Network Analysis and Mining, 1(3):187–218, 2011.

[2] http://www.netflixprize.com/, last access: 2014-12-01.

[3] James Bennett et al. *The netflix prize.* In Proceedings of KDD cup and workshop, volume 2007, page 35, 2007.

[4] Aristides Gionis, et al. *Assessing data mining results via swap randomization.* ACM Transactions on Knowledge Discovery from Data, 1(3):article no. 14, 2007.

[5] Albert-László Barabási et al. *Emergence of Scaling in Random Networks.* Science, 286(5439):509–512, 1999.

[6] Ron Milo, et al. *Network Motifs: Simple Building Blocks of Complex Networks.* Science, 298:824–827, 2002.

[7] Emőke-Ágnes Horvát, et al. *A network-based method to assess the statistical significance of mild co-regulation effects.* PLOS ONE, 8(9):e73413, 2013.

[8] George W. Cobb et al. *An Application of Markov Chain Monte Carlo to Community Ecology.* The American Mathematical Monthly, 110:265–288, 2003.

[9] Pawan Harish et al. *Accelerating Large Graph Algorithms on the GPU Using CUDA.* In Srinivas Aluru, et al., editors, High Performance Computing (HiPC), volume 4873 of Lecture Notes in Computer Science, pages 197–208. Springer Berlin Heidelberg, 2007.

[10] B.A. Miller, et al. *A scalable signal processing architecture for massive graph analysis.* In Proceedings of the 2012 IEEE International Conference on Acoustics, Speech and Signal Processing (ICASSP), pages 5329–5332, March 2012.

[11] Eriko Nurvitadhi, et al. *GraphGen: An FPGA Framework for Vertex-Centric Graph Computation.* In Proceedings of the 2014 IEEE 22nd Annual International Symposium on Field-Programmable Custom Computing Machines (FCCM), pages 25–28, May 2014.

[12] B. Betkaoui, et al. *A framework for FPGA acceleration of large graph problems: Graphlet counting case study.* In Proceedings of the 2011 International Conference on Field-Programmable Technology (FPT), pages 1–8, Dec 2011.

[13] Makoto Matsumoto et al. *Mersenne Twister: A 623-Dimensionally Equidistributed Uniform Pseudo-Random Number Generator.* ACM Trans. Model. Comput. Simul., 8(1):3–30, January 1998.

[14] Jason Howard, et al. *A 48-core IA-32 message-passing processor with DVFS in 45nm CMOS.* In Solid-State Circuits Conference Digest of Technical Papers (ISSCC), 2010 IEEE International, pages 108–109. IEEE, 2010.

[15] Denis Dutoit, et al. *A 0.9 pJ/bit, 12.8 GByte/s WideIO memory interface in a 3D-IC NoC-based MPSoC.* In VLSI Technology (VLSIT), 2013 Symposium on, pages C22–C23. IEEE, 2013.

[16] Matthias Jung, et al. *TLM modelling of 3D stacked wide I/O DRAM subsystems: a virtual platform for memory controller design space exploration.* In Proceedings of the 2013 Workshop on Rapid Simulation and Performance Evaluation: Methods and Tools, RAPIDO '13, pages 5:1–5:6, New York, NY, USA, 2013. ACM.

[17] Karthik Chandrasekar, et al. *Improved Power Modeling of DDR SDRAMs.* In proc. DSD'11, 2011.

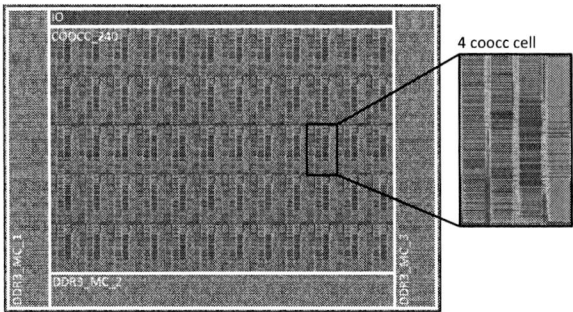

Figure 5: ASIC floorplan of our 28 nm chip. The chip consists of 240 *coocc* modules, three DRAM controllers and IO logic. The swap randomization block is not visible here due to its small size.

2015 IEEE Computer Society Annual Symposium on VLSI

Heterogeneous Error-Resilient Scheme for Spectral Analysis in Ultra-Low Power Wearable Electrocardiogram Devices

Soumya Basu[1], P. Garcia del Valle[1], Georgios Karakonstantis[2], Giovanni Ansaloni[3] *and* David Atienza[1]

[1]Embedded Systems Laboratory, École Polytechnique Fédérale de Lausanne, Switzerland

[2]School of Electronics, Electrical Engineering and Computer Science, Queen's University Belfast, United Kingdom

[3]Università della Svizzera Italiana, Faculty of Informatics, Lugano, Switzerland

Abstract— **Wearable devices performing advanced bio-signal analysis algorithms are aimed to foster a revolution in healthcare provision of chronic cardiac diseases. In this context, energy efficiency is of paramount importance, as long-term monitoring must be ensured while relying on a tiny power source. Operating at a scaled supply voltage, just above the threshold voltage, effectively helps in saving substantial energy, but it makes circuits, and especially memories, more prone to errors, threatening the correct execution of algorithms. The use of error detection and correction codes may help to protect the entire memory content, however it incurs in large area and energy overheads which may not be compatible with the tight energy budgets of wearable systems.**

To cope with this challenge, in this paper we propose to limit the overhead of traditional schemes by selectively detecting and correcting errors only in data highly impacting the end-to-end quality of service of ultra-low power wearable electrocardiogram (ECG) devices. This partition adopts the protection of either significant words or significant bits of each data element, according to the application characteristics (statistical properties of the data in the application buffers), and its impact in determining the output.

The proposed heterogeneous error protection scheme in real ECG signals allows substantial energy savings (11% in wearable devices) compared to state-of-the-art approaches, like ECC, in which the whole memory is protected against errors. At the same time, it also results in negligible output quality degradation in the evaluated power spectrum analysis application of ECG signals.

Keywords— *ultra-low power, embedded systems, wearable health monitors, error tolerance, power spectrum analysis.*

I. INTRODUCTION AND MOTIVATION

The emergence of wearable devices for long-term acquisition of cardiac signals (electrocardiogram or ECG) promises a paradigm shift in the monitoring of chronic heart-related conditions. Functionalities of state-of-the-art wearable cardiac sensors are not limited to sensing and (wirelessly) transmitting the acquired data, but they also provide advanced Digital Signal Processing (DSP) capabilities to analyze bio-signals on-node and extract clinically-relevant features [1, 2, 3]. Diverse applications have been proposed, ranging from the automated detection of epileptic seizures [12] to the predictive risk assessment of atrial fibrillations [13].

In this context, power spectral analysis (PSA) of the heart rate variability (HRV) is among the most widely employed

strategies, as it allows the monitoring of various health conditions associated with the heart as well as other organs [4, 5], providing valuable frequency-domain medical indicators. The implementation of PSA on ultra-low power embedded devices requires a carefully tailored digital architecture.

A key element in these devices are the memory components, where a significant amount of energy is consumed [1]. To maximize the energy efficiency, an effective approach is to scale down the supply voltage (Vdd). Aggressive supply voltage scaling, leads to quadratic energy savings, but makes circuits (and especially SRAM cells) prone to errors compounding the reliability issues present in nanometer technologies. Larger memory bit-cells [15] and error detection and correction errors (ECC) [17] can help in dealing with errors induced by a scaled Vdd. However, such mechanisms impose large energy and area overheads.

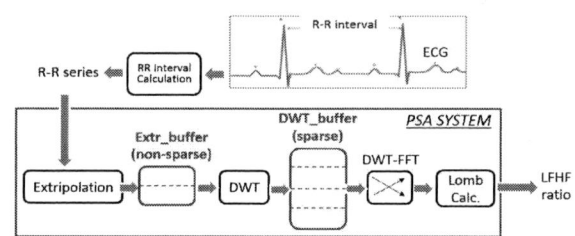

Figure 1: Block Scheme of the PSA Application.

Ensuring the correctness of run-time execution in digital systems is a major challenge, due to the increase in variability derived from technology scaling and near-threshold voltage supplies. A striking alternative, often referred to as approximate computing, is to take advantage of the resilience nature and statistical properties of bio-signal DSP applications such as filtering, features extraction etc. As an energy-saving strategy, the approximate computing paradigm relaxes reliability constraints when errors have a negligible impact from an application perspective. Algorithms in the embedded health monitoring domain operate on noisy acquisitions, while often presenting statistical or qualitative outputs [18]. In such scenarios, in this paper we extend the observations from our previous work [18], and advocate for not needing to provide 100% exactness in all cases, which is also extremely expensive from an energy efficiency viewpoint.

978-1-4799-8720-7/15 $31.00 © 2015 IEEE

In this paper, we propose to investigate the application of the approximate paradigm in bio-signal analysis and feature extraction applications and utilize their statistical properties for limiting the overhead of classical ECC schemes. Our contributions can be briefly described as:

1) We study the statistical properties of a PSA of heart rate variability system and classify data elements in the intermediate steps of the algorithm into significant and less-significant based on their contribution to output quality.

2) We apply a significance-based memory protection scheme in PSA systems and evaluate the energy gains and quality compared to traditional full ECC scheme.

3) We exploit the statistical properties of the intermediate elements produced by the PSA system and propose an alternative mitigation scheme. The scheme, rather than trying to correct the errors, only limits their impact on the output quality by replacing the erroneous data with the expected value that is the best available statistical information for the incorrect data.

4) We compare the proposed scheme with existing techniques such as ECC and analyze the energy savings and the quality loss on benchmarks of ECG records in wearable embedded devices.

The rest of this paper is structured as follows: Section II describes the target PSA application in the context of ECG signals. Section III describes the impact of scaled voltages in memories and presents state-of-the-art techniques used to address them. Section III describes the proposed approach for smart wearable sensors. It also analyzes the sparsity of the content of the buffers employed by an optimized PSA implementation, justifying our choice of a hybrid protection scheme. Section IV presents experimental evidence. It also presents experimental evidence showcasing how our proposed scheme can achieve a high quality-of-service in wearable embedded systems, in the presence of errors, by protecting a small fraction of the memory buffers.

II. POWER SPECTRAL ANALYSIS (PSA)

The power spectral analysis (PSA) of the HRV, has been proposed as a powerful technique to evaluate the autonomic control of the heart rate [10]. Such a system, shown in Figure 1, is composed of 4 essential steps: a) in the first step the time difference between consecutive heartbeats (known as RR intervals) is extracted by processing the recorded ECG within a fixed size window. Due to the non-periodic nature of the RR intervals, Fast Lomb method is considered as the most suitable method for estimating the power spectrum of cardiac signals [6]. According to Fast Lomb, in the next step b) the estimated RR intervals are extrapolated to a fixed size window (i.e. 512 samples), which are then in step c) processed for estimating specific trigonometric functions [6]. A fast method needs to be used at this step for estimating these complex functions, such as the Fast-Fourier-Transform (FFT). In our implementation, we used a Wavelet-based FFT, which was shown recently to

be able to take advantage of the sparsity (most elements close to zero) of bio-signals and lead to substantial energy gains [6].

In this modified algorithm initially a Discrete Wavelet Transform (DWT) of size N is applied on the input signal, followed by butterfly operations similar to the conventional FFT but with modified, simpler coefficients [6] as depicted also in Figure 1. Finally, the Lomb calculator combines the data, estimating the real-time power spectrum information. In the clinical practice, the most used metric derived from PSA is the ratio between the power in low-frequencies (LFP, defined as 0.04 – 0.15 Hz) and high-frequencies (HFP, 0.15 – 0.4 Hz), with *LFHF Ratio=LFP/HFP*. A deviation of the *LFHF Ratio* above or below normal values is indicative of various health issues [10].

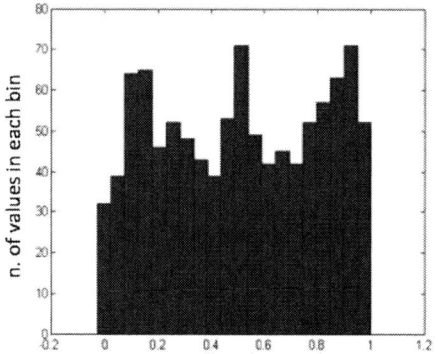

Figure 2: Histogram of `Extr_buffer` Data (normalized), Distributed in 20 bins. `Extr_buffer` Presents a Non-Sparse Distribution.

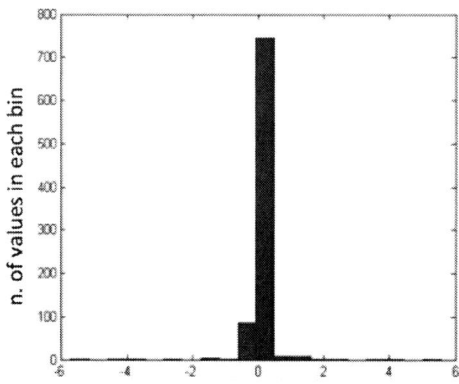

Figure 3: Histogram of `DWT_buffer` Data (normalized), Distributed in 20 bins. `DWT_buffer` Presents a Sparse Distribution.

The buffers required by this implementation are the output of the extrapolation step (composed by two sub-arrays) and the output of the DWT transform (composed of four sub-arrays).

978-1-4799-8720-7/15 $31.00 © 2015 IEEE

III. Unreliable Memories and Proposed Approach

As discussed in Section I, aggressive voltage scaling can induce a non-zero probability of erroneous reads and writes to the memory subsystem. In [7], a bit-flip probability of 0.22% and 0.07% are reported for 6-transistors SRAMs supplied at 0.6V and 0.65V respectively, implemented on a 40nm technology. More resilient memory topologies (such as 8-transistors and SCMEMs) do allow reliable operations at these voltage levels, however they do incur high area and energy overhead [7]. Alternatively, error detection and correction techniques can be employed to recover from bit-flip events, but they also present non-negligible added complexity, from an area as well as energy perspective.

Our proposed method minimizes such overhead by judiciously employing detection and correction of errors depending on the criticality of the stored data, providing high correctness guarantees only to its most critical part, as dictated by the application characteristics.

Figures 2 and 3 show the typical data distribution across the two buffers used in the PSA application. Figure 3 highlights how the elements of the `DWT_buffer` are mostly centered on zero, and therefore are sparse, while elements of `Extr_buffer` (c.f.: Figure 2) have a non-sparse distribution.

Different data distribution patterns require different protection approaches. Intuitively, the most suited protection scheme for the non-sparse `Extr_buffer` is to protect the Most Significant Bits (MSBs) of every word with an ECC code, as they will have a larger influence on the output. Conversely, a small, but non-zero, probability of a bit-flip in the Least Significant Bits (LSBs) can be allowed (Figure 4).

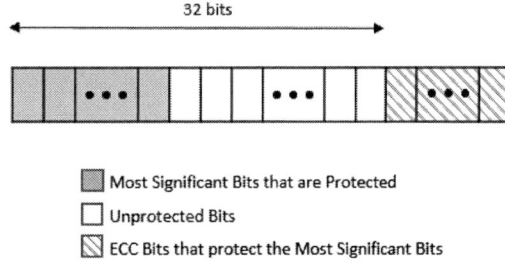

Most Significant Bits that are Protected

Unprotected Bits

ECC Bits that protect the Most Significant Bits

Figure 4: A Memory Word in the `Extr_buffer`

In the case of `DWT_buffer`, we discriminate between significant and non-significant *words*, instead of significant bits. In fact, as most of the elements of this buffer are close to zero, it is possible to replace them with their expected value (zero) if an error occurs, which can be detected by employing a simple parity check. For the rest (which, in the PSA application, reside in the low-frequency range) a more expensive error correction capability must be provided; in our case, ECC (Figure 5).

Such partition between significant and non-significant words can be performed statically, i.e. independently from the particular window of inputs being processed. It is in fact derived from the inherent properties of the DWT transform, and the resulting separation of the processed data into high and low frequencies.

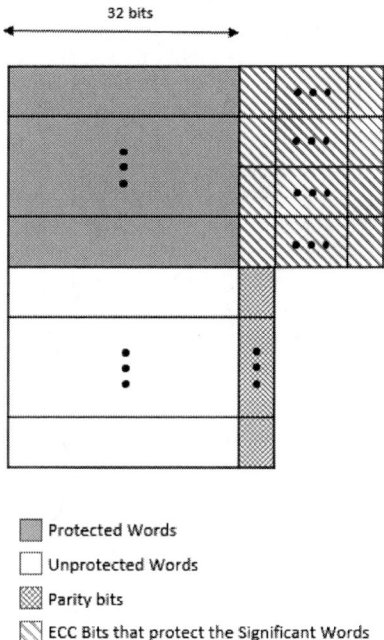

Protected Words

Unprotected Words

Parity bits

ECC Bits that protect the Significant Words

Figure 5: Block Diagram of the `DWT_buffer`

It must be noted that a difference exists between the strategies adopted for *non-significant bits* in non-sparse buffers (e.g.: LSBs in `Extr_buffer`) and *non-significant words* in sparse buffers (e.g.: near-zero values in `DWT_buffer`). In the first case, errors will be completely undetected while, in the second, a parity check will detect the error and invalidate the corresponding word, but no correction will be required. In both cases, the aim is to have a negligible deviation in the end results of the application, while greatly diminishing the protection overhead. In the case of *significant bits and words*, each bit-flip is detected and corrected, because it highly affects the quality of the output.

IV. Experimental Setup and Results

To evaluate the proposed heterogeneous protection scheme, we developed a high-level fault simulation environment, executing the entire target application. In this way, we evaluated the impact of errors in the intermediate buffers on the quality of the PSA output, which was compared, under different protection schemes, to a fault-free execution. Single bit-flip errors in the buffers are considered with probabilities of 0.07% and 0.22%, corresponding to the behavior of a 6-transistor SRAM at 0.65V and 0.6V, respectively [6].

Input ECG data was retrieved from the PAF prediction challenge database, available on the Physionet portal [8]. The database includes 100 recordings, of 30 minutes each. We have considered input data windows of 2, 4 and 6 minutes, with an overlap of 1, 3 and 5 minutes, respectively. Results from each recording and each window size are averaged in the presented results.

978-1-4799-8720-7/15 $31.00 © 2015 IEEE

To obtain fair results, we employed error masks, forcing bit-flips in random locations of the buffers if they reference to un-protected regions. Different error masks are employed for each processed input window and for each buffer, but the same set is used across all protection configurations. For all buffers, data is represented with 32 bits words. For the `Extr_buffer`, we explored a protection of the 8, 16, or 32 (all) most significant bits, while for `DWT_buffer`, we assumed a protection of the 5%, 10% or 15% of the most significant memory words.

In the following sections, we explored the output degradation induced by bit-flips in the buffers (IV-A), the energy overhead of different protection schemes (IV-B) and the trade-off between energy efficiency and quality of service (IV-C).

A. Error Analysis

Figures 6 and 7 compare the percentage error in the computation of the LFHF ratio at supply voltages of 0.65V (bit-flip probability = 0.07%) and 0.6V (bit-flip probability = 0.22%), respectively.

Results highlight that the selective protection of significant words in sparse buffers can guarantee high-quality results with little overhead. In fact, in the case of a voltage supply of 0.65V (Figure 6), less than 1% error in the LFHF ratio can be achieved by protecting only the most significant 15% of the significant words in the `DWT_buffer`, when the non-sparse `Extr_buffer` is error free. As expected, by reducing the ratio between significant and non-significant words in `DWT_buffer`, the PSA error increases. Nevertheless, it still remains rather low (3%) even when only the 5% most significant words are protected.

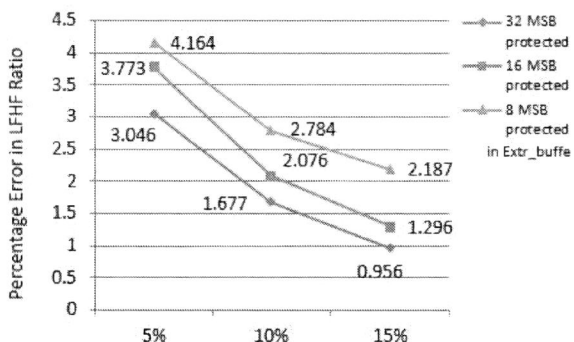

Figure 6: Percentage of Error in the Calculation of the LFP/HFP Ratio under Different Protection Schemes at 0.65V Supply

Conversely, the protection of the non-sparse `Extr_buffer` presents more challenges; even when protection against bit-flips is provided for the 16 MSBs of each word (which corresponds to protecting half of the buffer content), still a noticeable decrease in quality of service can be noted (square-dotted orange Line in Figure 6).

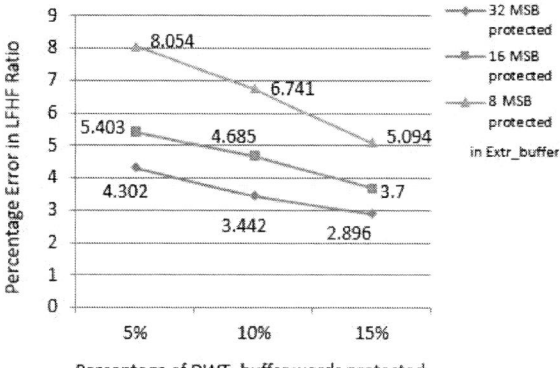

Figure 7: Percentage Error in the Calculation of the LFP/HFP Ratio under Different Protection Schemes at 0.60V Supply

The same trends can be noticed in Figure 7 for a lower voltage supply of 0.6V, and corresponding higher bit-flip probability. Interestingly, even in this case the deviation in the LFHF ratio, with respect to a fault-free execution, can be bounded to 5% by allowing errors in the 16 LSBs of `Extr_buffer` and only checking (but not correcting) errors in 90% of `DWT_buffer`.

B. Energy Analysis

We comparatively evaluated the energy overhead induced by the memory protection configurations by modeling different schemes using CACTI [11] in the McPAT framework [17]. The `Extr_buffer` requires two buffers, each of 8K Bytes, while the `DWT_buffer` is composed of four buffers of the same size. The operating temperature was assumed to be 300K. The technology node employed in the simulation of the memories was 40nm. All wirings were considered to be global, and the interconnect projection was taken as conservative. The memories have a single port used for both reading and writing. Also, for our purpose, we assumed that each of the memories consists of a single bank connected using a bus.

Relevant metrics, output of the CACTI model, consist of the dynamic read and write energies per access and the leakage power of the target memory configurations. To derive the corresponding dynamic energy, we retrieved the number of accesses to the different buffers from the high-level model of the application. In addition, the leakage energy was estimated by considering the execution time of an optimized version of PSA running on an ARM Cortex M4 processor running at 180MHz.

Additional storage is required to support data protection. In the case of `Extr_buffer` (protection of most significant bits), considering one error detection and correction per memory word, 6 extra ECC bits are required for each word when all 32 data bits are protected. Fewer ECC bits are used when only the most significant part of each word is protected: 5 and 4 bits in the case of 16- and 8- MSB protection, respectively. A maximum of one error occurring in a memory word has been assumed and the number of ECC bits are chosen accordingly as described before.

In any case, larger number of errors in each word can be also assumed which will require higher order ECC with more redundant bits to detect and correct all the errors. Note that this will lead to higher overheads for the conventional case as opposed to the proposed scheme which acts (for the less-significant words) irrespective of the number of errors replacing erroneous words based on the detection of a single error. Therefore, focusing on one failure per word proves the efficacy of the proposed scheme for the popular case of single bit flips while being fair to the conventional schemes.

In the case of DWT_buffer (protection of most significant words), a single parity bit is employed for non-significant words, allowing error detection. Single error correction is instead supported for significant words by employing a 6-bit error correction code. To simulate this heterogeneous structure in CACTI, we considered the two corner cases where all data is either employing 1-bit parity or 6-bit ECC. To derive the dynamic energy per read access of intermediate configurations, we employed the formula as described in (1).

$$E_t = p * E_p + (1 - p) * E_u \qquad (1)$$

Where E_p is the read energy per access in the protected memory, and E_u is the read energy per access in the unprotected memory. Also, p is the percentage of considered significant words. E_t is the net read energy per access in the heterogeneous memory. The write energy per access and the leakage power of the hybrid memory were also calculated in the same manner.

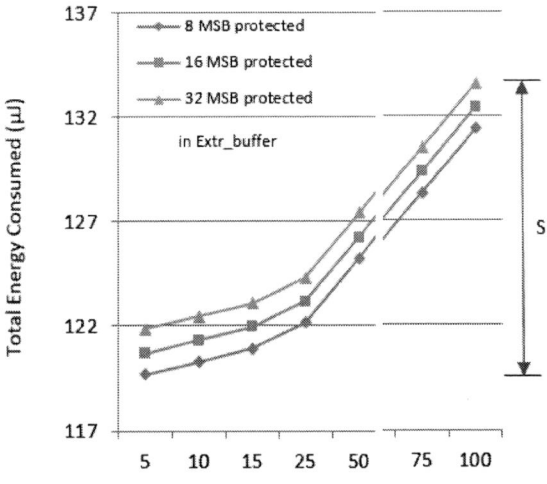

Figure 8: Total Energy Consumption at 0.65V Supply for Different Memory Protection Schemes for an execution time ≈ 2.67s

Figure 8 represents a scenario with the maximum possible number of accesses to the hybrid memories for running the application using the RR intervals from a patient recording of 5 minutes. This figure shows the energy consumption of the memory subsystem in an optimized embedded system where power management makes the memory being active only 2.67s for the processing of a five minutes recording for a patient. The variable S represents the energy saved with respect to the baseline configuration where 8 MS bits protected in the Extr_buffer and 5% significant words protected in the DWT_buffer. It shows that full protection of all data entails a significant energy overhead of at least 12% with respect to baseline configuration. On the other hand, by having a 16-bit MSB protection in Extr_buffer and 10% of significant words protection in DWT buffer, only a 1.4% energy overhead is acquired. As reported in Figure 6, this scheme induces a negligible quality-of-service degradation of just 2% at 0.65V.

C. Energy / Quality-of-Service Trade-off

The proposed methodology can be employed to devise the proper memory configuration when an error constraint is given for the target signal processing application. Alternatively, it can be used to devise the most accurate solution for a fixed energy budget.

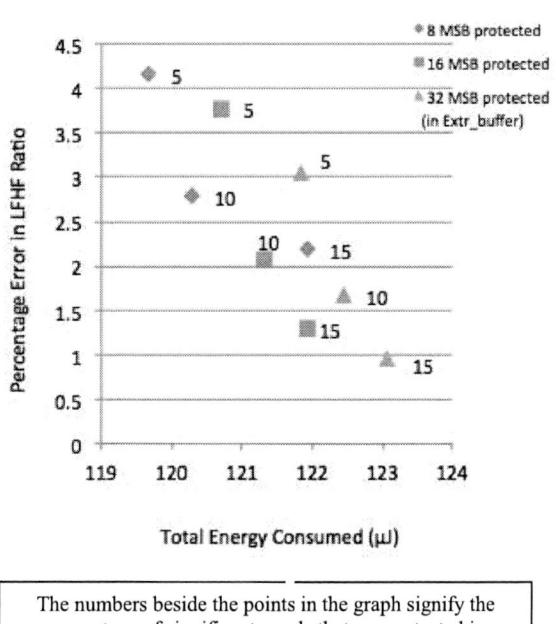

The numbers beside the points in the graph signify the percentage of significant words that are protected in the DWT_buffer

Figure 9: Percentage Error in LFHF Ratio and the corresponding energy consumptions under different memory protection schemes at 0.65V

Figure 9 illustrates this trade-off by plotting the total energy consumed in the buffers for the PSA application against the percentage of error in the computed HFLF error. A percentage of bit-flips of 0.07% is considered (Vdd = 0.65V). For a maximum tolerable error of 2.5%, a solution with 16 MSBs protected in the Extr_buffer and the 10% most significant data protected in the DWT_buffer is the most energy efficient. Conversely, for an energy budget of 120 μJ,

978-1-4799-8720-7/15 $31.00 © 2015 IEEE   272

the smallest error is achieved by protecting 10% of the significant data and 8 MSBs in the two buffers.

V. CONCLUSION

In this paper, we have explored the energy benefits that can be obtained by applying hybrid data protection schemes to embedded memories for ultra-low power wearable monitoring systems. Working within a well-known, specific application domain allows designers to implement algorithms that exploit significance-based computing to effectively reduce the energy consumption, while keeping the output error under a certain threshold and providing outputs of high enough quality. Our experiments show that by adopting different correctness guaranties in words and bits of varying *significance* from a digital signal processing viewpoint, the proposed approach can effectively reduce the energy overhead implicit in data protection, while minimally impacting the end-to-end quality of service of the target Power Spectral Analysis application.

The illustrated methodology is applicable in many real-world applications in the embedded health monitoring domain beyond PSA, because they share the same characteristics of processing noisy inputs, providing statistical or qualitative outputs and adopting a sparse representation in intermediate buffers.

Experimental evidence highlights that heterogeneous protection reduces approximately by 11% the energy budget of the data memory used to store the buffers in real-life wearable ECG analysis systems. Moreover, this new proposed approach tolerates high error rates, potentially allowing more aggressive voltage/frequency scaling at system level. Hence, this observation opens promising venues to be explored in the development of ultra-low power multi-modal wearable embedded systems.

ACKNOWLEDGEMENTS

This work has been partially supported by the EC FP7 FET SCoRPiO project (grant no. 323872), and ONR-G (grant no. N62909-14-1-N072).

REFERENCES

[1] Braojos, R., Mamaghanian, H., Junior, A. D., Ansaloni, G., Atienza, D., Rincón, F. J. and Murali, S. "Ultra-Low Power Design of Wearable Cardiac Monitoring Systems". In Proceedings of DAC, June 2014.

[2] Hao, Y. and Foster, R. "Wireless body sensor networks for health-monitoring applications." Physiological measurement, 29(11), R27, 2008.

[3] Massé, F., Bussel, M. V., Serteyn, A., Arends, J. and Penders, J. "Miniaturized wireless ECG monitor for real-time detection of epileptic seizures." ACM Transactions on Embedded Computing Systems (TECS), Vol. 12, no. 4, 102, 2013.

[4] Sörnmo, L., and Laguna, "P. Bioelectrical signal processing in cardiac and neurological applications." Academic Press, 2005.

[5] Chou, C. C., Tseng, S. Y., Chua, E., Lee, Y. C., Fang, W. C. and Huang, H. C., "Advanced ECG processor with HRV analysis for real-time portable health monitoring." In Consumer Electronics-Berlin (ICCE-Berlin), pp. 172-175. September 2011.

[6] Karakonstantis, G., Sankaranarayanan, A., Sabry, M. M., Atienza, D., and Burg, A., "A quality-scalable and energy-efficient approach for spectral analysis of heart rate variability." In Proceedings of Design, Automation and Test in Europe Conference and Exhibition (DATE), pp. 1-6, March 2014.

[7] Bortolotti, D., Mamaghanian, H., Bartolini, A., Ashouei, M., Stuijt, J., Atienza Alonso, D., and Benini, L., "Approximate compressed sensing: ultra-low power biosignal processing via aggressive voltage scaling on a hybrid memory multi-core processor." In Proceedings of 2014 IEEE International Symposium on Low Power Electronics and Design (ISLPED 2014) .Vol. 1, pp. 40-45.

[8] PhysioBank. Available online: www.physionet.org/physiobank/.

[9] Niskanen, J. P., Tarvainen, M. P., Ranta-Aho, P. O., and Karjalainen, P. A, "Software for advanced HRV analysis." Computer methods and programs in biomedicine, 76(1), pp 73-81. 2004.

[10] Akselrod, S., Gordon, D., Ubel, F. A., Shannon, D. C., Berger, A. C., and Cohen, R. J., "Power spectrum analysis of heart rate fluctuation: a quantitative probe of beat-to-beat cardiovascular control." Science, 213(4504), pp 220-222. 1981.

[11] Muralimanohar, Naveen, Rajeev Balasubramanian, and Norman P. Jouppi. "CACTI 6.0: A tool to model large caches." HP Laboratories. 2009.

[12] Massé, Fabien, Martien Van Bussel, Aline Serteyn, Johan Arends, and Julien Penders. "Miniaturized wireless ECG monitor for real-time detection of epileptic seizures." ACM Transactions on Embedded Computing Systems (TECS) 12, no. 4, Article 102 (July 2013), 21 pages.

[13] J. Milosevic, A. Dittrich, A. Ferrante, M. Malek and D. C. Rojas Quirós et al., "Risk Assessment of Atrial Fibrillation: a Failure Prediction Approach", In Proceedings of Computers in Cardiology (CinC), September 2014, Cambridge, MA, pp. 801-804.

[14] Zidong Du; Lingamneni, A.; Yunji Chen; Palem, K.; Temam, O.; Chengyong Wu, "Leveraging the error resilience of machine-learning applications for designing highly energy efficient accelerators," In Proceedings of Asia and South Pacific Design Automation Conference (ASP-DAC), January 2014, pp. 201-206,

[15] Sharma, V.; Cosemans, S.; Ashouei, M.; Huisken, J.; Catthoor, F.; Dehaene, W., "8T SRAM with Mimicked Negative Bit-lines and Charge Limited Sequential sense amplifier for wireless sensor nodes," In Proceedings of ESSCIRC, September 2011, pp. 531-534,

[16] Mohamed M. Sabry, et al.; "OCEAN: An Optimized HW/SW Reliability Mitigation Approach for Scratchpad Memories in Real-Time SoCs". ACM Trans. Embed. Comput. Syst. 13, 4s, Article 138, April 2014, 26 pages.

[17] Sheng Li et al.; "McPAT: An Integrated Power, Area, and Timing Modeling Framework for Multicore and Manycore Architectures". In Proceedings of the 42nd Annual IEEE/ACM International Symposium on Microarchitecture (MICRO 42). ACM, New York, NY, USA, pp. 469-480.

[18] Mohamed Sabry et al.; "Design of energy efficient and dependable health monitoring systems under unreliable nanometer technologies", In Proceedings of the 7th International Conference on Body Area Networks(BodyNets '12). ICST (Institute for Computer Sciences, Social-Informatics and Telecommunications Engineering), ICST, Brussels, Belgium, Belgium, pp. 52-58.

Logic Switches Operating at the Minimum Energy of Computing

Francesco Orfei, Luca Gammaitoni

NiPS Lab, Dept. of Physics, University of Perugia
Via A. Pascoli 1, 06123 Perugia, Italy
francesco.orfei@nipslab.org

Abstract— In modern computers information is processed through binary switches, usually realized with transistors, i.e. microelectronic devices. Thus binary switches represent a paradigmatic example of "small scale physical systems" employed in the processing of information. In the last forty years the semiconductor industry has been driven by its ability to scale down the size of the CMOS devices, to increase computing capability and reducing the power dissipated in heat. Thus we assisted to a continuous reduction in the supply voltage and electrical noise induced errors became an important issue to take into account for the performance of the device. Here we discuss the relationship between the energy required by an advanced ultra low power logic gate and the associated error rate at different operating frequencies.

Keywords—binary switches; CMOS; logic gate; scaling; error rate; energy; computing; noise

I. INTRODUCTION

Computation is energy demanding: this statement is clear to whoever uses an electronic device, but has deeper implications than expected. As a matter of fact, the energy issue is at the moment one of the major limitation in the development of both small autonomous wireless sensor nodes and large scale supercomputers. In the former case, computing devices are intended to be small (micro-nano scales) and self powered through some energy harvesting techniques [1-2]. The main problem is that micro and nano energy harvesters are not yet able to efficiently power off-the-shelf sensors, processors and wireless transmitters: they generally require power at mW scale, but these harvester are able to convert µWs of power, not enough. In the latter case, each processor of a supercomputer requires enormous amount of energy, because power, the product of current and voltage, has to be fed into the system to properly operate it. Consequently the heat produced while operating must be removed from the chip to preserve functionality. From these two different examples it's clear that lowering energy consumption in computing devices is then one of the main goals of ICT industry [3].

On the other hand, this may be limited or forbidden by physics: from the early work of Landauer [4], it is well understood that resetting one bit of information with no error requires a minimum heat production of $K_B T\ ln(2)$, where T is the temperature of the bit-encoding device and K_B is Boltzmann constant. This seems to apply to nowadays logic gates too. In an extensive interpretation of the Landauer principle, some authors [5-6] claim that a NAND logic gate

must dissipate $K_B T\ ln(2)$ because it reduces the information between the input and the output. However only the theoretical prediction on the heat produced by resetting one bit of information (Landauer principle) has been confirmed experimentally [7], while there is still no proof that a NAND gates generates $K_B T\ ln(2)$ of heat per bit of information lost during computation. Additionally, it has been recently suggested that the energy efficiency of the bit resetting operation can be improved allowing a finite error probability during operation [3, 8]. This has never been tested neither as a fundamental physical limit, nor as a technological shortcut available with the off-the-shelf logic gates.

Logic switches are the fundamental component of each logic gate. Logic gates can be used in any digital circuit, computer, portable equipment or telecommunication device. Portable equipment requires a very small package to fit in its compact slim casing, and a low supply voltage to meet the requirements of CPUs and MCUs, as well as low power consumption to extend battery life. For example, smart phones, as well as many other portable devices, become more complex, more features are offered to the user, but more power is consumed by systems in both active and standby mode. Consequently, selecting low-power devices for portable electronics imposes new challenges in the areas of core voltage, energy management, and battery lifetime. Increasing microprocessor complexity generally increases power consumption and migration to low-power CMOS logic achieves the lowest power consumption in both dynamic and static mode situations.

In this work a NAND gate has been tested to evaluate its performances at variable bit rate when its supply voltage is decreased below the minimum voltage specified by the producer. In this way it has been possible to identify the minimum supply voltage before error rate rises up to a non negligible limit. Reducing the supply voltage has an impact on the total power dissipation, but it is important to understand the quantitative correlation between the power saving and the error rate when the device is under-powered. It has been also investigated the behavior of the error generation trying to find if it can be predicted or not.

II. ENERGY DISSIPATION IN LOGIC SWITCHES

Energy consumption during computation has become a matter of strategic importance for modern ICT and its impact on future society. In the last forty years the progress of the semiconductor industry has been driven by its ability to cost

effectively scale down the size of the CMOS-FET [9] switches while increasing their density. This has been accompanied by a continuing increase in energy consumption and heat generation up to a point where the power dissipated in heat during computation has become a serious limitation [10, 11]. An information processor can be seen as a computing engine that transforms incoming energy flow into useful work and also produces some heat [3, 12]. It has been shown that information processing is intimately related to energy management ("information is physical")[4]. The ultimate limit on the minimum energy per switch resetting is set at $K_B T$ ln2 (approximately 10^{-21} J at room temperature) [4, 13] while there is no fundamental limit associated with the switching operation per sè [3]. This limit is known to be the Shannon-von Neumann-Landauer (SNL) limit. As Landauer argued, this minimum amount cannot be reduced to zero if some information is discarded (erased) during the computation process. The reason is directly linked to thermodynamics: if erasing information decreases the overall entropy of the system, this cannot be done without dissipating heat of at least $K_B T$ ln2 J per bit erased [4, 13]. It is important to review the sources of energy consumption in state of the art ICT systems, since this may offer insights for possible directions to improve their energy efficiency.

A. Fundamental limits

An ICT device can be seen as a machine [14, 15] that receives in input information and energy (in the form of work), and after processing them gives information and energy (in the form of heat) as output. A traditional thermal machine is a device that processes energy. More precisely it transforms energy in the form of heat into work for industrial applications. An ICT device is a slightly more complex machine because it processes energy and information at the same time. More specifically, it inputs a certain amount of data and some energy in the form of work and outputs a reduced amount of information and the same quantity of energy, although in the form of heat. In modern computers the information is processed via networks of logic gates. In a real computer, the logic gate function is realized by some physical components. The bit value is represented by some physical entity (signal) such as an electric current or voltage, light intensity, magnetic field etc. Such signal inputs to the logic gate device and go through a transformation to represent at the output the desired bit value. The minimum energy to operate a physical switch can be reduced to zero provided that the quantity of information associated to the physical entropy in the switch is not decreased during the switch transformation. This condition has been pointed out initially by John von Neumann in a lecture in 1949 [6] and subsequently focused by R. Landauer [7] and extended, sometimes loosing the connection between physical entropy and quantity of information, by C. H. Bennet [13].

The reason is the following: the switch is a macroscopic apparatus composed by many elementary parts (atoms) and thus can be considered a thermodynamic system approximately at equilibrium with the environment. This implies that its transformations are subjected to the laws of thermodynamics

[3]. A switch event is a change from an initial condition to a final condition. During this change the exchanges of energy and entropy need to be accounted for. If the switch is thermally isolated from the external environment, then there can be no transfer of heat. If it is possible to suppose for a moment that a switching event can be performed without any work from outside, then the only balance that needs to be taken into account is the change in entropy. This change is measured by the change in the macroscopic configuration of the switch. If the change is from state open to state close, then there is only one initial configuration and one final configuration. There is therefore no net change in the number of configurations and thus no change in entropy according to Boltzmann. If now a status change is imposed to put the switch into a *close* (or *open*, same reasoning) condition, the number of configurations is halved and thus there is a change in entropy according to Boltzmann given by the (1):

$$S_f - S_i = k_B(\ln 1 - \ln 2) = -k_B \ln 2 \qquad (1)$$

where k_B is Boltzmann's constant (1.38×10^{-23} J/K).
The change in entropy is associated with a change in heat via the relation (2):

$$T dS \geq dQ \qquad (2)$$

B. Practical limits

Two components determine the power consumption in a CMOS circuit: static power consumption and dynamic power consumption [16]. The first is proportional to the current flowing into the device as a consequence of leakages; the second one is related to the current required during switching. CMOS devices have very low static power consumption because parasitic diodes are reverse biased and only their leakage currents contribute to static power consumption. The leakage current (I_{lkg}) of the diode is described by the following equation (3):

$$I_{lkg} = I_S \left(e^{\frac{qV}{kT}} - 1 \right) \qquad (3)$$

Where:

I_S = reverse saturation current (A)
V = diode voltage (V)
k = Boltzmann's constant (1.38×10^{-23} J/K)
q = electronic charge (1.602×10^{-19} C)
T = temperature (K)

Static power consumption is the product of the device leakage current and the supply voltage. Total static power consumption P_S can be obtained as:

$$P_S = \sum (\text{leakage current}) \cdot (\text{supply voltage}) \qquad (4)$$

Dynamic power consumption is due to the current that flows only when the transistors are switching from one logic state to another. This is a result of the current required to charge the internal nodes (switching current) plus the through current (current that flows from the supply voltage V_{CC} to GND when the p-channel transistor and n-channel transistor turn on briefly at the same time during the logic transition). The frequency at which the device is switching, plus the rise

978-1-4799-8720-7/15 $31.00 © 2015 IEEE

and fall times of the input signal, as well as the internal nodes of the device, have a direct effect on the duration of the current spike. For fast input transition rates, the through current of the gate is negligible compared to the switching current. For this reason, the dynamic supply current is governed by the internal capacitance of the IC and the charge and discharge current of the load capacitance. If we consider the energy required to drive an inverter's output high and low, this is the same of loading and discharging a capacitor C_L through a resistor R_L. It is possible to evaluate the energy stored and dissipated in the resistor and capacitor as:

$$E_{R_L} = \int_0^\infty \frac{V_{R_L}^2(t)}{R_L} dt = \frac{V}{R_L} \int_0^\infty \left[e^{\left(-\frac{t}{R_L C_L}\right)} \right]^2 dt = \frac{C_L V^2}{2} \quad (5)$$

$$E_{C_L} = \frac{C_L V^2}{2} \quad (6)$$

The energy dissipated for one switch operation is given by the equation:

$$E_{SW} = E_{R_L} + E_{C_L} = C_L V^2 = C_{pd} V^2 \quad (7)$$

Considering NSW working at the frequency f_l, transient power consumption can be calculated using equation (8).

$$P_T = \alpha \cdot C_{pd} \cdot V_{CC}^2 \cdot f_I \cdot N_{SW} \quad (8)$$

where:

P_T = transient power consumption (W)

α = activity factor, $\alpha=1$ for a square wave switching and $\alpha<0.5$ in typical digital circuits

C_{pd} = dynamic power-dissipation capacitance (F)

V_{CC} = supply voltage (V)

f_I = input signal frequency (Hz)

N_{SW} = number of bits switching

III. THE RELATION ENERGY VERSUS ERROR RATE

In the datasheet of logic gates nothing is written about the relation between the energy dissipated and the error rate. It is important, in fact, that each device can work without producing unexpected or unpredictable results. It is also important that the device works demanding the minimum possible energy. It has been observed, discussed and demonstrated that there is a relation between energy and information, that the practical amount of energy is technology dependant and that the amount of energy required for information processing is directly proportional to the velocity of the processing itself. Great advancements have been made on electronic devices, and today's technology requires millions times less energy that in the past. But we are still quite far from the theoretical Landauer's limit. What to do?

It has been suggested that a significant advantage could be reached by lowering the gate voltage. In this case it is expected that keeping the same technology and reducing the energy per bit can lead to an increase of the error rate because of the thermal noise and the barrier crossing effect. Unfortunately not much is reported in a typical product datasheet, so it can be

interesting to make a test to experimentally obtain the relation between the error rate and the energy consumption due to voltage lowering.

IV. EXPERIMENTAL SETUP

Aim of the experiment is to obtain a relationship between the bit error-rate and the power consumed by a logic device, a SN74AUP1G00 [17] two inputs single NAND gate from Texas Instruments, under several test conditions. The experimental setup is composed by a stable voltage supply, a variable voltage supply, a signal generator, a digital comparator, a temperature sensor, some current to voltage converters and an asynchronous counter. The block diagram of the experimental setup is depicted in Fig. 1. Two of GUT boards (GUT Gate Under Test board) are used in parallel. In the first one, the NAND gate is always powered at the nominal voltage of 3.3 V: in this way no errors are expected during its working time. In the other board, instead, the NAND gate is powered from a variable voltage power supply. In this way it is possible to control the supply voltage of the logic device to the minimum value of 0.4 V. An EX-OR is used as bit comparator to detect errors. At its input the output signals of the GUT boards are presented. When the two inputs are not equal, the output of the comparator goes to the logic level "1". This is the error detection condition required during this experiment. Unfortunately the output of the EX-OR gate is asynchronous, or in other words, it changes state as soon as the inputs change state and a synchronization circuitry is required to correctly measure the number of errors.

A Flip-Flop type D has been used to sample and hold the output of the comparator for the entire time interval of one bit. It is clocked by the main system's clock, an arbitrary programmable signal generator. To minimize the propagation delay difference between the two signal paths to the comparator, two GUT boards are used in parallel and the same signal generator produces the bit sequence to evaluate the performances of the NAND gates. At the output from these boards the signals are time aligned except for a small time interval due to the different propagation delay of the two NAND. At lower voltage supply the propagation delay rises. A SN74AUP1G00 is the NAND gate used for the experiment.

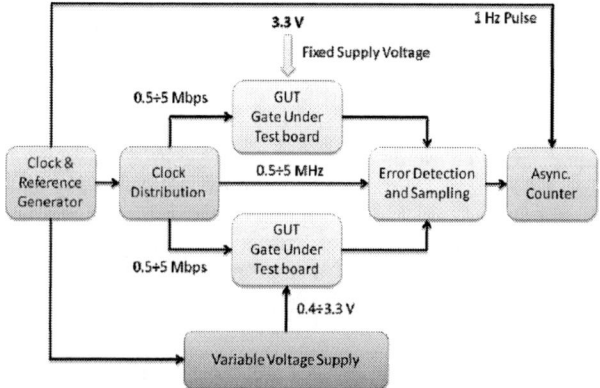

Fig. 1. Block diagram of the system designed for the evaluation of the error rate.

AUP (advanced ultra-low-power CMOS) is one of the lowest power families now on the market. Its operating power supply range is extending from 0.8 V to 3.6 V, with a 3.5-ns typical propagation delay, along with a reduced current drive capability of 4 mA. Typical C_{pd} for one NAND gate (SN74AUP1G00) is 4 pF at 3.3 V. The typical quiescent current ICC is around 0.5 µA (0.9 µA max) when the input voltage is equal to the supply voltage, and it increases of up to 40 µA per each gate when the input voltage is lower than the supply voltage. The highest operating frequency is 190 MHz.

The dynamic current of the NAND gate is measured through the voltage across a precision shunt resistor. Its value is 10 Ω / 0.125 W and the precision is ±0.1%. To have an easier reading, the small voltage is amplified by a differential amplifier: the gain of the amplifier is set to 100.

V. TEST PROCEDURES

The bit rate for the tests has been chosen low enough to not have issues with the propagation delay. This means that delays on the two different signal paths can be neglected with respect the bit duration. Moreover the expected delay introduced by the under-powered gate will be much bigger than the small difference in time between the digital signals through the two NAND gates.

It is well known that reducing the power supply voltage saves energy but the propagation delay generally increases. The first measurement set is devoted to obtain the dynamic current required by the NAND gate during its operation. A second one, instead, is devoted to count the number of errors of the under powered gate. The counting of the errors is performed by a Racal-Dana 1992 Nanosecond Universal Counter. A third measurement is required to investigate the voltage gaps between the high (V_H) and low (V_L) voltage levels of the output signal from the NAND gate. This is an important measurement to try to understand the reason why errors rise up.

The procedure to perform the measurement of the dynamic currents, the errors and the voltage gaps is the following:
1. set the initial reference voltage to obtain a stable power supply of the 0.4 V;
2. set the initial bit rate to have 0.5 Mbps and apply the bit sequence to the input pin of the NAND gate;
3. check the voltage supply to be the desired one;
4. read the current to voltage converter output, the number of errors and the voltage gap;
5. increase the bit rate;
6. repeat the point 3 to 5 till the bitrate arrives to 5 Mbps;
7. increase the voltage reference of a small amount (10 to 25 mV);
8. repeat the steps from the point 2 to the point 7 till to arrive to the maximum level of the supply voltage of 3.3 V.

The bit rates used for the measurements are: 0.5 Mbps, 1 Mbps, 1.5 Mbps, 2 Mbps, 3 Mbps, 4 Mbps and 5 Mbps. All the tests have been performed at 25°C.

VI. MEASUREMENTS AND RESULTS

Several measurements have been performed following the procedure previously described. Anyway a preliminary test of the system has been conducted to check that each part was properly working. In particular, Fig. 2 depicts the propagation delay difference between the two channels of the GUT boards.

As it can be seen the time difference is only about 200 ps, meaning that this will not be the source of errors, at least at low baud rates, and the system can be correctly used to perform all the tests on the gate. In this preliminary test all the two gate were normally powered at 3.3 V. Fig. 3 depicts the correlation among the dynamic current, the error rate and the voltage gap between the two logic states at different bit rate and variable supply voltage (0.4 V to 3.3 V). From these measurements is possible to obtain an estimation of the overall (internal and external load capacitance) power dissipation capacitance C_e.

Considering a dynamic current of 200 µA at 2.5 V and 5 Mbps, the value of C_e is given by the equation (9) and it is in good agreement with the predictions.

$$C_e = \frac{I_{CC}}{V_{CC} \cdot f_o} \approx 16\,pF \qquad (9)$$

As it can be seen, when the supply voltage is greater than 0.8 V no errors are measurable, as expected. Moreover, the curves of the error rates and of the voltage gap converge on the right part of the graphs, meaning that the system is working properly.

More interesting are the left parts of the graphs, as depicted in Fig. 4, Fig. 5 and Fig. 7: these figures show the dynamic currents, the error rates and the voltage gaps respectively at different bit rates and when the supply voltage ranges between 0.4 V and 0.8 V: we were interested to investigate the error probability in extreme low power operating condition. Fig. 4 depicts the dynamic current of the NAND gate at different bit rates as function of the supply voltage. As it can be noted, there are points where the slope of the curves changes.

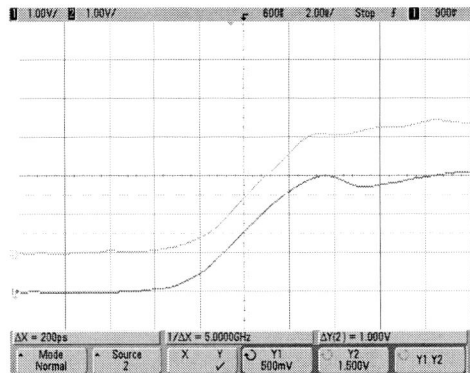

Fig. 2. Propagation delay difference between the two channels at 3.3 V: red line represents the signal coming from the variable voltage power supply, green line represents the constant 3.3 V powered signal. The time difference is 200 ps (it has been measured in the point where the signals reach 1.5 V).

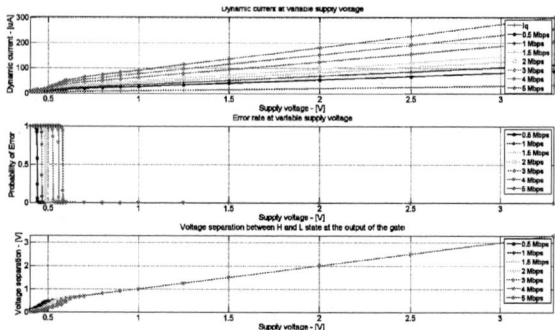

Fig. 3. Results of the measurements of the dynamic current, of the error rate and of the voltage gap between the high and low level of the output of the NAND gate.

These points are few tens of mV higher than the point where the error rates rise (see Fig. 5). These points can be interpolated by and exponential curve as function of the supply voltage: this means that higher bit rates require higher supply voltages.

$$I_c = 0.08774 e^{(10.64 \cdot V_{CC})} \quad (10)$$

Fig. 5 depicts the error rates curves as function of the bit rate and the supply voltage. As it can be easily seen, all the curves can be approximated by an *erfc* function defined as in the (11):

$$erfc(V_{CC}) = \frac{2}{\sqrt{\pi}} \int_{V_{CC}}^{\infty} e^{-t^2} dt \quad (11)$$

All these error functions are parallel to each other and they appear to be only translated: it was possible to obtain the translation coefficients values V_{Tr} and they are depicted in Fig. 6 as function of the bit rate: they are linearly dependant by the bit rate B_R:

$$V_{Tr} = 0.0315 \cdot B_R + 0.428 \quad (12)$$

Fig. 7 depicts the voltage differences between the high and low level of the output of the NAND gate as function of the supply voltage and at different bitrates. As it can be seen, all the curves converge to the supply voltage when the device is working correctly, it means for supply voltage over the corner voltage V_C discussed before. As it is trivial to imagine, they converge again when the supply voltage goes to zero: no more energy. In the middle of these two convergences, they exhibit a "s" shape trend and some considerations can be done correlating these curves with the probability of error. First consider the voltage separation at 5 Mbps and its value in the point where the error probability rises rapidly, it means approximately at 0.58 V of voltage supply. From the graphs of Fig. 7 this value can be found to be approximately 0.35 V. Anyway at this level of voltage the errors are no more negligible. Instead, at a supply voltage of 0.59 V the error probability is close to zero and the voltage separation is close to 0.4 V. Compared to the supply voltage, it is approximately 32% lower. In terms of power of the output signal from the NAND gate, this means a power saving of approximately 54%, and the device is still working correctly.

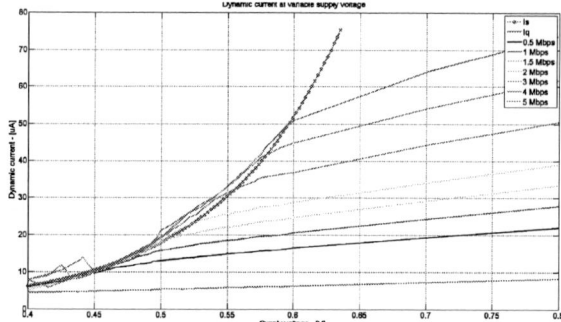

Fig. 4. Dynamic currents of the NAND gate. The plot shows the static current Iq and the dynamic currents in the 0.4 V to 0.8 V voltage range at different bit rates. The "o" marked line is an exponential curve passing for the point where the dynamic current slope changes significantly.

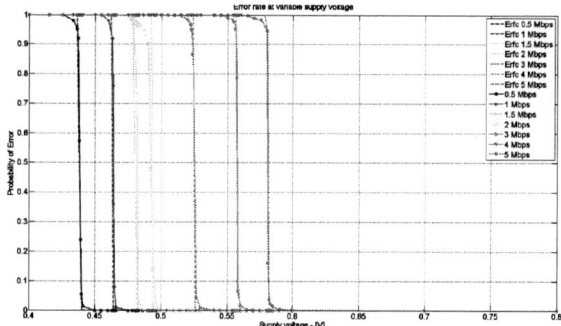

Fig. 5. Plot of the error rate measurement as function of the supply voltage, from 0.4 V to 0.8 V, at different bit rates. The dashed lines are the corresponding complementary *erfc* functions.

Same considerations can be done for the 3 Mbps curve. At 0.525 V its probability of error is raising rapidly and its voltage separation is approximately 0.28 V. Let's choose a point where the errors are close to zero, for example 0.54 V. At this supply voltage the voltage separation is approximately 35 % lower, it is close to 0.35 V. Considering that at this voltage level and at this bit rate the logic gate is still working, the power saving on the output signal is approximately 58%.

VII. CONCLUSIONS

In this work a NAND logic gate has been studied to investigate its error-rate behavior in its nominal voltage supply range and in under-powering conditions. The dynamic current, the error probability and the voltage separation between the logic states of the output have been studied and some remarkable features can be noted:

- The dynamic current $I_{CC}(V_{CC})$ as a function of the supply voltage V_{CC}, grows linearly for $V_{CC} > V_{TI}$, a threshold value that assumes different values for different bit rates BR, $V_{TI} = V_{TI}(BR)$.
- For $V_{CC} < V_{TI}(BR)$ the $I_{CC}(V_{CC})$ decreases almost exponentially with an exponent that is approximately the same for different bit rates.

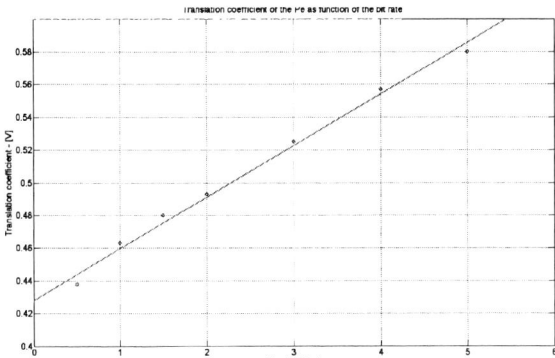

Fig. 6. Plot of the translation coefficients used to approximate the probability of error measurements with an *erfc* distribution as function of the bit rate. In blue the coefficient used, in red their linear interpolation.

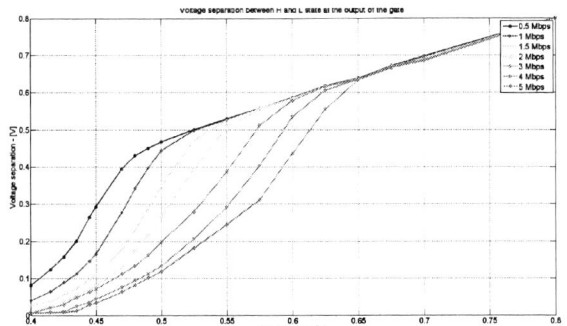

Fig. 7. Voltage gaps of the output voltage of the NAND gate between the high and low state as function of the supply voltage at different bit rates, in the supply voltage range 0.4 V to 0.8 V.

- The error probability $P_e(V_{CC})$ as a function of the supply voltage V_{CC}, decreases rapidly for $V_{CC} > V_{T2}$. V_{T2} assumes different values for different bit rates BR, $V_{T2} = V_{T2}(BR)$
- For $V_{CC} < V_{T2}(BR)$ the error probability $P_e(V_{CC}) = 1$, it stays maximum regardless the value of the supply voltage, i.e. the response is totally random.
- $V_{T1}(BR)$ is not necessarily equal to $V_{T2}(BR)$.
- $V_{T1}(BR)$ and $V_{T2}(BR)$ grows approximated proportional with BR.

Based on these considerations, it is possible to propose a phenomenological model for $I_{CC}(V_{CC})$ and $P_e(V_{CC})$. First it can be noted the absence of a step-like behavior for $V_{CC} = V_{T2}$, i.e.: $P_e(V_{CC}) = 0$ for $V_{CC} > V_{T2}$ and $P_e(V_{CC}) = 1$ for $V_{CC} < V_{T2}$. This is not the case here, and $P_e(V_{CC})$ looks like an exponential, far enough from a step-like behavior. This seems to indicate the role played by the voltage/threshold fluctuations and the device failure appears to be a stochastic phenomenon other than a deterministic event. A more realistic model of the two states of equilibrium can thus be proposed by introducing a bistable potential, where the distance between the two wells and the height of the barrier is supply voltage dependent, as it can be seen in Fig. 8.

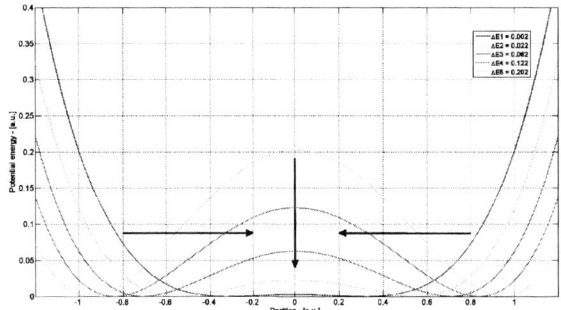

Fig. 8. Double well potential energy of a hypothetical particle in a particular bistable system. The height of the barrier decreases when the distance between the equilibrium positions decreases: at the lower limit of zero height the equilibrium positions are no more distinguishable.

REFERENCES

[1] Sustainable Energy Harvesting Technologies - Past, Present and Future, Edited by Yen Kheng Tan, ISBN 978-953-307-438-2, 268 pages, Publisher: InTech, Chapters published December 22, 2011 under CC BY 3.0 license, http://www.intechopen.com/books/sustainable-energy-harvesting-technologies-past-present-and-future

[2] Energy Harvesting Technologies, Shashank Priya, Daniel J. Inman, Spinger ISBN: 978-0-387-76463-4 (Print) 978-0-387-76464-1 (Online)

[3] Towards zero-power ICT, L. Gammaitoni, D. Chiuchiù, M. Madami, G. Carlotti, Nanotechnology, 2015 (in print)

[4] Irreversibility and heat generation in the computing process, R. Landauer, IBM Journal of Research and Development, Volume 5 Issue 3, July 1961 Pages 183-191

[5] The thermodynamics of computation—a review, Charles Bennett, 1982, International Journal of Theoretical Physics

[6] Fredkin, Edward; Toffoli, Tommaso (1982), "Conservative logic", International Journal of Theoretical Physics 21 (3-4): 219–253

[7] Experimental verification of Landauer's principle linking information and thermodynamics Antoine Bérut, Artak Arakelyan, Artyom Petrosyan, Sergio Ciliberto, Raoul Dillenschneider, Eric Lutz, Nature 483, 187–189

[8] Beating the Landauer's limit by trading energy with uncertainty, Luca Gammaitoni, arXiv:1111.2937 [cond-mat.mtrl-sci]

[9] Y. Taur and T. H. Ning, Fundamentals of Modern VLSI Devices (1998 Cambridge University Press)

[10] K. Bernstein, R.K. Cavin, W. Porod, A. Seabaugh, J. Welser, "Device and architecture outlook for beyond CMOS switches, Proc. IEEE 98 (2010) 2169-2184

[11] J. J. Welser, G. I. Bourianoff, V. V. Zhirnov, R. K. Cavin, "The quest for the next information processing technology", J. Nanoparticle Res. 10 (2008) 1-10

[12] S. Shankar, R. K. Cavin, and V. V. Zhirnov, "Computation from Devices to System-Level Thermodynamics," Electrochem. Soc. Transactions 25 (2009) 421-431

[13] C. H. Bennett, "The thermodynamics of computation - a review", Int. J. Theoretical Physics 21 (1982) 905-940

[14] Luca Gammaitoni (2012): There's plenty of energy at the bottom (micro and nano scale nonlinear noise harvesting), Contemporary Physics, 53:2, 119-135

[15] L. Gammaitoni, 2011, arXiv:1111.2937; L. Gammaitoni, Nanoenergy Letters, 5, 10, 2013.

[16] V. Zhirnov, R. Cavin and L. Gammaitoni (2014). Minimum Energy of Computing, Fundamental Considerations, ICT - Energy - Concepts Towards Zero - Power Information and Communication Technology, InTech, DOI: 10.5772/57346

[17] SN74AUP1G00 datasheet, from http://www.ti.com/product/sn74aup1g00?qgpn=sn74aup1g00

Synergistic Architecture and Programming Model Support for Approximate Micropower Computing

Giuseppe Tagliavini
and Davide Rossi
DEI, University of Bologna, Italy
Email: {giuseppe.tagliavini, davide.rossi}@unibo.it

Andrea Marongiu
and Luca Benini
DEI, University of Bologna, Italy
ETH Zurich, Switzerland
Email: {a.marongiu, luca.benini}@iis.ee.ethz.ch

Abstract—Energy consumption is a major constraining factor for embedded multi-core systems. Using aggressive voltage scaling can reduce power consumption, but memory operations become unreliable. Several embedded applications exhibit inherent tolerance to computation approximation, for which indicating parts that can tolerate errors has proven a viable way to reduce energy consumption. In this work we propose an extension to OpenMP to specify what regions of code and data are tolerant to approximation. A compiler pass places data into memory regions with different reliability guarantees according to their tolerance to errors. The voltage supply level is dynamically adjusted according to tolerance policies, with the overall goal of minimizing energy in full compliance with precision constraints.

Keywords—OpenMP, approximate computing, micropower computing, many-core accelerators.

I. INTRODUCTION

Traditionally, ultra-low power (ULP) systems are mostly based on single-core microcontrollers, where low-power operation correlates with low computational requirements. This assumption is less and less true for today's deeply embedded sensing systems [12], which execute computation-demanding tasks that are inherently parallel. To match conflicting requests for energy consumption and performance, several (ultra) low-power multi-core systems have been recently proposed [17] [14] [23] in application fields such as industrial automation, mission critical systems, pervasive video infrastructures and bio-medical computing.

For such systems, the memory system typically consumes more than 50% of the total chip power, and are thus the key component to be optimized. Using aggressive voltage scaling can reduce energy, but memory operations become unreliable due to the lack of sufficient static noise margin [7]. Approximate computing has recently emerged as a promising approach to design energy-efficient digital systems working at unreliable voltage levels. The notion of approximate computing [13] refers to a set of techniques ranging from programming language- to transistor-level, with a common aim to allow computing systems to save energy to the detriment of the *quality* of the computed results. Approximate computing is a promising approach in the ULP domain when applications exhibit inherent tolerance to errors, i.e. the property to produce *acceptable* outputs despite some computations execute in an inaccurate manner within controlled thresholds [22], [19] [16]. The error tolerance may derive from several factors [8]:

- some algorithms process large amounts of redundant input data, that in turn may contain inherent noise;
- errors can easily get averaged down or averaged out in statistical and aggregating algorithms;

- some algorithms use iterative refinement techniques, and subsequent iterations may correct errors introduced in previous iterations;
- some algorithms do not have a single "golden" result, they may return an element of a set of multiple solutions that are equally acceptable;
- less-than-perfect results are perceived as correct by the users.

In this work we propose a set of HW/SW approximation techniques for PULP, a ultra-low-power multi-core platform implemented in 28 nm UTBB FD-SOI technology [23]. PULP is a scalable, clustered many-core computing platform that features a parametric number of Processing Elements (PE) per cluster, sharing a multi-banked tightly coupled L1 data memory (TCDM, a scratchpad).

At the HW level, our proposal focuses on energy saving techniques for the TCDM, while the PEs work at the most energy efficient operating point (0.5V @ 20MHz). We consider two different memory technologies, Standard Cell Memory (SCM) and six-transistor Static RAM (6T-SRAM). Although 6T-SRAM provides a much better storage density than SCM, its minimum operating voltage is much higher (0.8V as opposed to 0.5V). Accessing 6T-SRAM below 0.8V results in a flip-bit error with a certain probability. In our solution, the TCDM is physically partitioned into two different regions, implemented with the two technologies. With this layout, the operating voltage of 6T-SRAM can be safely lowered whenever data located in this region is not accessed. In addition, to leverage approximation tolerance in applications in a controlled manner we provide a transparent mechanism to split multi-byte data into multiple banks at the architecture level. We call the associated address range *tolerant memory*. The most significant bits (MSB) of a word are stored in the reliable, low-voltage SCM, while the least significant bits (LSB) are stored in the 6T-SRAM. This allows to access 6T-SRAM cells at low voltage for error-tolerant computations, bounding the error to the LSB. Our results demonstrate that this approach provides much better precision than just dropping the LSB.

At the SW (programming model) level, we provide constructs for specifying regions of code and data that are tolerant to approximation in a C program. A compiler optimization pass (based on Clang+LLVM [1] [18]) places data into different memory types according to its tolerance to errors, activating the tolerant memory region and inserting voltage switch points for 6T-SRAM when this does not violate precision constraints. A greedy allocation algorithm uses a simple heuristic to minimize the power consumption due to data accesses when other policies are not possible.

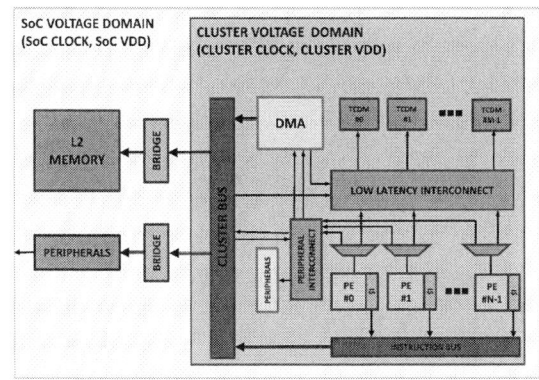

(a) The baseline PULP architecture

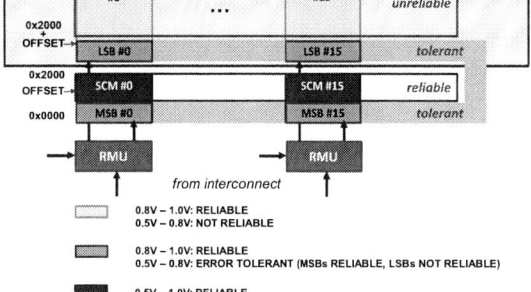

(b) Hybrid TCDM organization

Fig. 1: The PULP architecture and the extended L1 memory system

Using cycle accurate simulation models of the target platform annotated with power numbers extracted from a silicon implementation, we demonstrate that our hybrid memory architecture can reduce by 25% on average the energy consumption, very close to the "ideal" savings achievable with SCM cells only.

II. HARDWARE ARCHITECTURE

ULP systems are largely based on microcontrollers featuring simple, cache-less cores (e.g., Cortex M0 or M4), coupled to simple support for power management and a standard set of peripherals. The Parallel processing Ultra-Low-Power platform (PULP) [23] aims at providing a significant boost to the peak performance that ULP systems can achieve by coupling the multi-core paradigm to the most advanced FDSOI design technology and associated techniques for energy efficiency (near-threshold computing, body biasing, etc.). We describe here the PULP platform (Sec. II-A) and the extensions to the memory system for computation approximation (Sec. II-B).

A. The PULP Architecture

PULP is a scalable, many-core computing fabric, organized as a set of *clusters*. Figure 1a shows the main building blocks of single-cluster PULP SoC. Multiple clusters can be interconnected at the top level to share L2 memory and peripherals for off-chip communication. Within a cluster, a parametric number of Processing Elements (PEs) based on an optimized OpenRISC microarchitecture [2] share a L1 multi-banked tightly coupled data memory (TCDM). The TCDM is configured as a shared scratchpad memory, featuring as many R/W ports as the number of memory banks. This allows concurrent access to memory locations mapped on different banks, via a one-cycle-latency logarithmic interconnect implementing word-level interleaving to reduce contention. In addition, each PE is equipped with a private instruction cache. A lightweight, multi-channel DMA enables fast and flexible communication with other clusters, the L2 memory and external peripherals [24] (the latter leverages a dedicated peripheral interconnect).

The specific platform instance considered in this work consists of 8 PEs and 16 TCDM banks of 2.5KB each (40KB total), plus 64KB of L2 memory. Each core features 1 Kbyte of I$. Two voltage domains are considered: i) the SoC domain includes the L2 memory and peripherals; ii) the

cluster domain includes the PEs the TCDM, the DMA and the cluster interconnect. A 28nm STMicroelectronics UTBB FD-SOI implementation of the platform in this configuration can operate at 20MHz @ 0.5V, and at 100MHz @ 0.6V.

For improved energy efficiency, we extend this baseline platform with a hybrid L1 memory system that supports computation approximation in strict cooperation with the programming model.

B. L1 Memory System Extensions

On-chip memory is traditionally implemented with 6T-SRAM in super-threshold designs. When operating close to the threshold voltage, standard cell memory (SCM) has demonstrated a better tradeoff between reliability, energy efficiency, area and portability among technology nodes [26]. In particular, although 6T-SRAM cells provide a much better storage density than SCM ($\sim$ 3x), their minimum operating voltage is much higher (0.8V-1.0V) [9], while SCMs are reliable over all the operating voltage range of the architecture (0.5V – 1.0V). Accessing 6T-SRAM at a low voltage may results in a flip-bit error, as shown in Table I (values estimated for our chip).

Voltage [V]	0.50	0.55	0.6	0.65	0.7	0.75
P(bit-flip)	0.0037	0.0012	0.0003	5.24e-5	4.35e-6	6.16e-8

TABLE I: Probability of bit-flip errors in 6T-SRAM.

Both memory types have similar retention voltages (0.25V and 0.29V), above which data persistence is guaranteed. SCM can thus operate at the same low voltage of the logic in a reliable way, with the key benefit of providing much smaller energy/access ($\sim$ 4x) [21]. Based on these observations and on the evidence that we cannot afford to build the entire TCDM with SCM, we propose a hybrid L1 memory design.

We organize the TCDM in two different regions: 16 128x32 SCM cuts plus 16 standard 512x32 6T SRAM cuts. Reliability Management Units (RMU) are introduced in the path between the low-latency interconnect and the TCDM. RMUs are simple combinational logic blocks that allow to remap the address space of the TCDM into three distinct logical memory regions:

- *reliable* – mapped in the SCM, reliable at any operating point.

```c
int main()
{
    #pragma omp tolerant var
    int sparse_A[];
    int i = 0, index;

    while ( func(i) )
    {
        ...
        index = compute_index();
        #pragma omp tolerant computation
        sparse_A[index] = compute_element();
        ...
        update(i);
    }
}
```

Fig. 2: A sparse matrix computation with tolerant directives.

- *unreliable* – mapped into the 6T-SRAM, reliable above 0.8V.
- *tolerant* – mapped in the SCM (most significant bits - MSB) and in the 6T-SRAM (least significant bits - LSB), errors can show up on the LSB below 0.8V.

The *tolerant* region guarantees a correctness threshold for data when executing approximate code regions.

Figure 1b shows the new TCDM design and the logical regions. A third voltage domain is considered for the *unreliable* region (6T-SRAM cuts). Level shifters are introduced at the boundaries between voltage domains, i.e., when the SRAM is operating at 0.8V and the logic at a lower voltage[1]. The RMUs provide access to the *tolerant* region only at word or half-word level. In both cases, MSB (upper 16 or 8 bits) are placed into SCM cells, while LSB (lower 16 or 8 bits) are placed into 6T-SRAM cells.

The offset that defines the boundaries between the three logical regions can be configured by writing into a memory-mapped peripheral, accessible by every PE. Thus, the memory map can be re-configured on-the-fly by the software, enabling application-specific optimizations.

III. PROGRAMMING MODEL AND COMPILER

In the target platform the program code is stored in L2 and accessed via private I$. The latter is implemented with SCM cuts, and thus is always reliable. The focus of our techniques in thus only on data, which resides in the TCDM[2]. To conveniently control energy-efficient data mapping, we propose an extension to an OpenMP implementation for embedded multicore systems [20], in the form of two new directives:

- `#pragma tolerant var` – coupled to a variable declaration to specify it tolerates approximation.
- `#pragma tolerant computation` – coupled to a program statement to specify the memory addressed therein can be accessed in tolerant (low-voltage) mode.

Figure 2 shows an example of use of these directives. In sparse matrix computation, matrix indexes cannot be approximated (a single error implies wrong element choice), while matrix elements may tolerate approximation. Thus, we declare the `sparse_A` array and the element computation as tolerant, while `index` and its computation are not tolerant. Declaring pointers with the `tolerant var` directives is not supported,

---

[1]The power overhead of the level shifters is < 1% of a memory bank.
[2]The DMA can be used to move data between the L2 and the TCDM

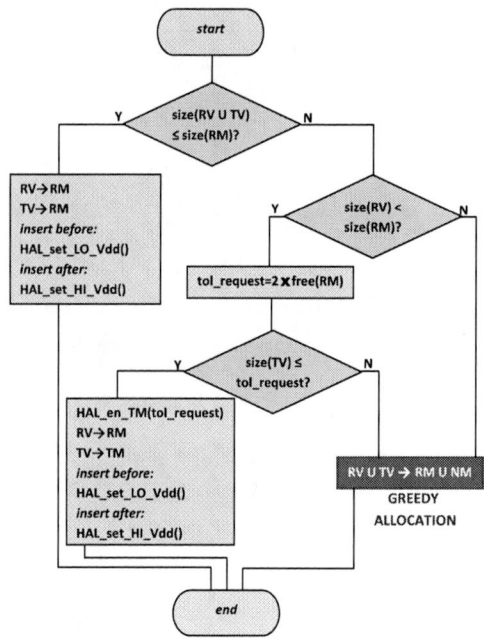

Fig. 3: Flowchart of the reliable allocation algorithm.

as an error on pointer dereferencing would lead to a random memory area access.

Our compiler (based on *Clang + LLVM*) transforms the directives into annotated tokens in the intermediate representation (IR). These annotations are used by an allocation optimization pass which operates as follows. Ideally, if all the data could be allocated in the reliable SCM we could keep the voltage low for the entire program life. However this is typically not possible, as the SCM is very small, so we need to identify what can go in SRAM and in which conditions it is safe to lower the voltage. The focus of the optimization is on the `tolerant computation` regions, since out of these regions the voltage must be kept high anyhow (again, unless everything fits in the SCM). Within `tolerant computation` regions, under low voltage operating mode:

1) tolerant variables can safely have their LSB in SRAM (we call these variables *TV*)
2) non-tolerant variables can only go in SCM (we call these variables *RV*)

The remaining variables are those whose live ranges do not overlap with the `tolerant computation` regions, so it is safe to lower the voltage within such regions even if they reside in SRAM (we call these variables *SV*).

First, standard *use-defs* analysis is applied to each node of the call graph, to extract usage estimates of global and local variables[3]. For each variable the following attributes are determined:

- *tolerant* – a flag that specifies whether the variable is tagged as tolerant or not;
- *scope* – a flag that specifies whether the variable is local or global;

---

[3]Static use-defs analysis limits the applicability of the approach to programs written as a single translation unit. Inter-Procedural Analysis and Link-Time Optimization approaches can be used to avoid this limitation.

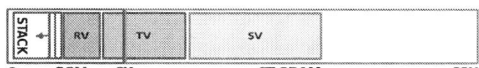

Fig. 4: Example of greedy allocation.

- *class* – the type of the variable (scalar/array/structure);
- *use intervals* – live range within the current scope;
- *tolerant use interval* – live range within tolerant code regions;
- *alloca* – for local variables, a reference to the declaration statement;
- *usedefs* – use-defs chains for the variable within the current scope.

Live ranges are computed with respect to call-graph nodes in case of global variables, and to basic blocks in case of local variables.

The flowchart in Figure 3 describes our **reliable allocation algorithm**. *RM* is the reliable memory (SCM), *NM* is the unreliable memory (6T-SRAM), and finally *TM* is the tolerant memory that can be optionally activated. For each `tolerant computation` region, the algorithm computes the memory size of RV∪TV. If this value is less than or equal to the size of the reliable memory region, (i) all data are allocated in RM, (ii) TM is not activated, and (iii) the voltage of NM is switched down without introducing any level of approximation. If the previous check fails, it considers the memory size of RV. If this value is less than the size of RM, it verifies whether the memory size of TV is less than or equal to the maximum size that we can set for TM to avoid segment overlaps (i.e., the amount of free reliable memory multiplied by two). If this last check is successful, the compiler forces the allocation of RV variables in RM, and TM is properly set to allocate TV variables. In this case, (i) the voltage of NM can be switched down just before executing the tolerant computation, and switched on to previous value after finishing, using proper API calls provided by the hardware abstraction layer (HAL); at the same time, (ii) TM is activated by another HAL call, and its offset is then propagated to the linker.

In the fall-through case of reliable allocation algorithm, when RV does not fit RM or TV does not fit TM, we execute a **greedy allocation algorithm**. In this case we do not take into account the tolerant flag, first of all we allocate the stack in RM and then we start allocating all the other structures, again prioritizing RM allocation, because accesses to this memory area imply a minor power consumption. Figure 4 shows an example of a typical allocation obtained by this algorithm. In any case, the core stacks are always allocated in the reliable memory, and also local variables that are tagged as tolerant are allocated onto the stack when their size is less than a limit value (the default value in current implementation is 16 bytes). This reduces the payload of the allocation algorithm with a minimum impact on the final memory layout. At link time, tolerant data are mapped at the proper memory address using a specific ELF segment. This implies that local variables allocated in tolerant memory are no longer allocated on the stack, and the stack reference in the function code is replaced with a unique global instance.

IV. EXPERIMENTAL EVALUATION

A. Simulation Setup and Methodology

The proposed architecture has been modeled in Virtual SoC [6], a SystemC-based cycle-accurate virtual platform for the

acceleration of massively parallel heterogeneous System-On-Chips. The energy consumptions of all the key components of the platform (cores, I$, SRAM banks, SCM banks, interconnect) are extracted through back-annotated simulations performed on the post place&route netlist, and fed into the power models of the VSOC SystemC simulator (see Table III). These numbers are derived by the first silicon implementation of PULP, realized using STMicroelectronics 28nm UTBB FD-SOI technology. This approach couples the advantages of very accurate power models with the simulation speed of the SystemC model, that allows to perform a wide exploration utilizing real-life benchmarks. In average, the virtual platform shows a maximum error in timing accuracy below 6% with respect to a complete RTL simulation of the same benchmark.

	SCM@0.5V	6T@0.5V	6T@0.8V
Read	0.609	2.766	5.347
Write	0.813	2.719	5.316
Leakage	0.002	0.009	0.018

TABLE III: Comparison of dynamic and leakage energy for SCM and 6T-SRAM. All values are reported in pJ.

B. Benchmark Suite

To assess the results of our solution, we consider the benchmarks of Table II, which have been implemented in C using OpenMP directives to split the workload over the 8 available cores. Considering the intrinsic data parallel nature of the selected benchmarks, in all cases we use a `omp parallel for` directive with a static chunking pattern. The algorithm execution is always marked as tolerant, while the tagging of data structures is detailed for each benchmark. The first two columns of table IV report a detailed analysis of the memory footprint of the benchmarks and the related number of read/write operations on the TCDM. In the experiments we take into account four reference platforms:

- *Base* – A PULP architecture with uniform memory cells (6T-RAM)
- *Ideal* – A PULP architecture with uniform memory cells (SCM)
- *Hybrid* – Our architecture with hybrid memory, with support to greedy allocation.
- *Hybrid tolerant* – Like the previous one, with support for voltage switching and tolerant region activation.

	Memory (KB)	R/W (words)	Zero-ing error	Flip-bit error
MM	RM: 4.0 (50%) TM: 8.0 (33%)	67728	8.37e-3	5.50e-7
LU	RM: 2.9 (36%) TM: 9.8 (41%)	428	2.39e-1	2.26e-5
SM	RM: 2.0 (25%) TM: 7.1 (30%)	2752	2.61e-2	1.15e-4
CV	RM: 4.0 (50%) TM: 4.0 (16%)	43088	5.35e-6	4.54e-10
DCT	RM: 1.0 (13%) TM: 8.0 (33%)	65536	3.55	1.29e-2
SOR	RM: 1.5 (19%) TM: 12.5 (52%)	99404	6.83e-3	3.65e-5
DIJ	RM: 1.0 (13%) TM: 14.0 (58%)	24207	8.92e-5	4.85e-11

TABLE IV: Benchmark results: memory footprint, number of TCDM accesses, MSE with zero-ing, MSE for flip-bit

	Description	Annotations
MM	Matrix multiplication	Matrix operands are tolerant
LU	LU decomposition	Matrix is tolerant, row pointers/pivots are reliable
SM	Sparse matrix multiplication with a vector	Matrix is tolerant, indexes/vectors are reliable
CV	Convolution of a matrix with a 5x5 kernel	Matrix operands are tolerant
DCT	DCT transform	Coefficients are tolerant, intermediate/output values are reliable
SOR	Successive over-relaxation	Matrix is tolerant, row pointers/pivots are reliable
DIJ	Minimum path using Dijkstra algorithm	Connectivity data are reliable, weights are tolerant

TABLE II: Benchmarks.

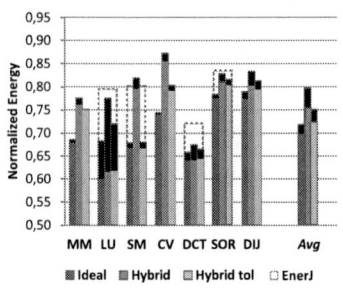

(a) Normalized energy consumption.

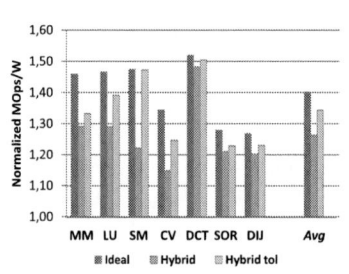

(b) Normalized operation efficiency.

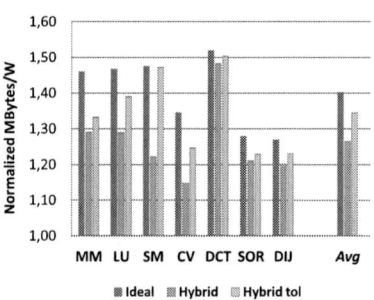

(c) Normalized data transfer efficiency.

Fig. 5: Experimental results

C. Accuracy

Approximate algorithms can "absorb" errors which occur with a probability lower than a given threshold throughout the entire computation. To assess the accuracy of our approach, we compared the use of unreliable LSB with a more drastic alternative, that is not computing them at all. To make the two solutions fully comparable, we used the same algorithms and the same data type, but in the second case we forced the LSBs to zero.

The last two columns in Table IV reports mean squared errors (MSE) for both approaches (referred to an implementation where data is kept on a reliable memory area). For all applications our approach achieves at least two orders of magnitude smaller MSE.

D. Energy Consumption

Figure 5a shows the energy consumption of the benchmarks on each platform. All values are normalized to the result of the *Base* architecture, and the rightmost set of bars shows the average. Bars are broken down in leakage (upper part) and dynamic energy (lower part).

The *Ideal* platform allows on average a 28% energy reduction w.r.t. the *Base* platform. This platform is not a real alternative due to the area cost of SCMs, but the related value is an upper bound to the energy reduction that we can obtain with any other method, whereas the *Base* platform is a lower bound.

The energy reduction of the *Hybrid* architecture is around 20% in the average case, to which the *Hybrid tolerant* platform sums an additional 5% saving. Overall our approach gets very close to the *Ideal*. Figure 5a also reports the energy consumption estimated by *EnerJ* [25] for a subset of common benchmarks (LU, SM, DCT, SOR)[4]. EnerJ consists mainly

[4]normalized energy values as reported in [25]

of a software approach that considers different methods of approximation (DRAM, SRAM, ALU, FPU) gleaned from the literature, while in our platform we do not have a DRAM and we do not support any form of approximation of hardware computation units. However, since SRAM is the main contributor to energy consumption for ULP systems, in general our solution performs better in terms of energy efficiency.

Figure 5b shows the computation efficiency, that is the number of operations executed per power unit (MOps/W) (numbers normalized to the *Base* platform). On average, the *Ideal* platform achieves a 40% improvement for this metric. The average improvement for the *Hybrid* platform is 27%, and 35% for the *Hybrid tolerant*. Again, our approach gets very close to the *Ideal*.

Figure 5c shows the number of read/write accesses performed in TCDM memory per power unit (MBytes/W), normalized to the *Base* platform. The values of this chart are really close to the ones reported for computation efficiency. This is an important outcome for our work, because it confirms that memory operations are the main limiting factor to reduce power consumption, and our approach leads to an equivalent increase of the computation power efficiency.

V. RELATED WORK

Many studies have shown that approximate computing is an amenable solution for applications in the ULP domain [22] [19] [16], and this facilitates the translation of such algorithms into energy-efficient hardware implementations. Architecture-level approaches include inserting control knobs in the design [10] and dual-voltage supply for SRAM and logic [13]. Focusing on the characteristics of ULP systems (where most of the energy is spent on on-chip memory), our approach is limited to 6T-SRAM memories, while other voltage domains are designed to work safety at low voltage. In the context of ULP multi-core systems this assumption simplifies the overall

978-1-4799-8720-7/15 $31.00 © 2015 IEEE

model with a minimum loss of generality.

Other HW approaches include approximate arithmetic blocks [15], hybrid memory architecture for MPEG-4 video processors [9] and dynamically reconfigurable SRAM array for low-power mobile multimedia application [11]. Such approaches share some commonalities with our techniques, but are clearly optimized for specific applications and lack the flexibility of our techniques.

Raising power concerns to the software enables a range of energy savings opportunities. Some approaches explore a set of abstractions for managing system resources under energy constraints [27] [5], targeting complete frameworks that put together architecture-, OS- and application-design. Approximate code transformations at the compiler level have been addressed in several works, targeting automatic identification/auto-tuning of tolerant code [3] [4] or code annotations [25]. The latter approach, EnerJ, is the closest to ours and probably the most known approach for approximate programming. For this reason we compare with this approach in our results (see Figure 5a). The comparison, limited to the results reported in [25], shows that our solution performs ≈6% better in terms of energy efficiency.

VI. CONCLUSION

In this work we propose a HW/SW approach to enable approximate computing on a multi-core ULP architecture. On the HW side we propose a hybrid L1 data memory design, composed of SCM and 6T-SRAM cells. We provide an additional mechanism to split multi-byte data into multiple banks, introducing a memory region to host variables that can tolerate computation approximation.

At the software level, we support OpenMP extensions for specifying what regions of code and data are tolerant to approximation. A compiler pass allocates data in various memory regions to allow for voltage reduction and, ultimately, energy savings. The experimental results show that our hybrid memory architecture can reduce by 25% on average the energy consumption, which is quite close to the 28% "ideal" reduction that we could achieve if the TCDM was entirely implemented with SCM cells.

VII. ACKNOWLEDGEMENTS

This work has been supported by the EU-funded research projects PHIDIAS (g.a. 318013) and P-SOCRATES (g.a. 611016).

REFERENCES

[1] "clang: a C language family frontend for LLVM ," http://clang.llvm.org/, accessed: 2015-04-15.

[2] "OpenRISC project," http://www.opencores.org/or1k/, accessed: 2015-04-15.

[3] A. Agarwal, M. Rinard, S. Sidiroglou, S. Misailovic, and H. Hoffmann, "Using code perforation to improve performance, reduce energy consumption, and respond to failures," in *MIT CSAIL Technical Reports*, 2009.

[4] J. Ansel *et al.*, "Language and compiler support for auto-tuning variable-accuracy algorithms," in *Proceedings of the 9th Annual IEEE/ACM International Symposium on Code Generation and Optimization*, 2011.

[5] W. Baek and T. Chilimbi, "Green: A system for supporting energy-conscious programming using principled approximation," TR-2009-089, Microsoft Research, Tech. Rep., 2009.

[6] D. Bortolotti, C. Pinto, A. Marongiu, M. Ruggiero, and L. Benini, "VirtualSoC: A Full-System Simulation Environment for Massively Parallel Heterogeneous System-on-Chip." in *IPDPS Workshops*, 2013.

[7] B. H. Calhoun and A. Chandrakasan, "Analyzing static noise margin for sub-threshold SRAM in 65nm CMOS," in *Proceedings of the 31st European Conference on Solid-State Circuits*, 2005.

[8] S. T. Chakradhar and A. Raghunathan, "Best-effort computing: rethinking parallel software and hardware," in *Proceedings of the 47th Design Automation Conference*, 2010.

[9] I. J. Chang, D. Mohapatra, and K. Roy, "A priority-based 6T/8T hybrid SRAM architecture for aggressive voltage scaling in video applications," *IEEE Transactions on Circuits and Systems for Video Technology*, 2011.

[10] V. K. Chippa, D. Mohapatra, A. Raghunathan, K. Roy, and S. T. Chakradhar, "Scalable effort hardware design: exploiting algorithmic resilience for energy efficiency," in *Proceedings of the 47th Design Automation Conference*, 2010.

[11] M. Cho, J. Schlessman, W. Wolf, and S. Mukhopadhyay, "Reconfigurable SRAM architecture with spatial voltage scaling for low power mobile multimedia applications," *IEEE Transactions on Very Large Scale Integration (VLSI) Systems*, 2011.

[12] A. Y. Dogan, J. Constantin, D. Atienza, A. Burg, and L. Benini, "Low-power processor architecture exploration for online biomedical signal analysis," *Circuits, Devices & Systems, IET*, 2012.

[13] H. Esmaeilzadeh, A. Sampson, L. Ceze, and D. Burger, "Architecture support for disciplined approximate programming," in *Proceedings of the 17th International Conference on Architectural Support for Programming Languages and Operating Systems*, 2012.

[14] D. Fick *et al.*, "Centip3De: A 3930DMIPS/W configurable near-threshold 3D stacked system with 64 ARM Cortex-M3 cores," in *IEEE International Conference on Solid-State Circuits*, 2012.

[15] J. Han and M. Orshansky, "Approximate computing: An emerging paradigm for energy-efficient design," in *18th IEEE European Test Symposium (ETS)*, 2013.

[16] R. Hegde and N. R. Shanbhag, "Energy-efficient signal processing via algorithmic noise-tolerance," in *Proceedings of the 1999 International Symposium on Low power Electronics and Design*, 1999.

[17] E. Krimer, R. Pawlowski, M. Erez, and P. Chiang, "Synctium: a near-threshold stream processor for energy-constrained parallel applications," *IEEE Computer Architecture Letters*, 2010.

[18] C. Lattner and V. Adve, "LLVM: A Compilation Framework for Lifelong Program Analysis and Transformation," in *International Symposium on Code Generation and Optimization*, 2004.

[19] J. T. Ludwig, S. H. Nawab, and A. P. Chandrakasan, "Low-power digital filtering using approximate processing," *IEEE Journal of Solid-State Circuits*, 1996.

[20] A. Marongiu and L. Benini, "An OpenMP compiler for efficient use of distributed scratchpad memory in MPSoCs," *IEEE Transactions on Computers*, 2012.

[21] P. Meinerzhagen, S. Y. Sherazi, A. Burg, and J. N. Rodrigues, "Benchmarking of Standard-Cell Based Memories in the Sub-Domain in 65-nm CMOS Technology," *IEEE Journal on Emerging and Selected Topics in Circuits and Systems*, 2011.

[22] S. H. Nawab, A. V. Oppenheim, A. P. Chandrakasan, J. M. Winograd, and J. T. Ludwig, "Approximate signal processing," *Journal of VLSI Signal Processing Systems for Signal, Image and Video Technology*, 1997.

[23] D. Rossi *et al.*, "Energy efficient parallel computing on the PULP platform with support for OpenMP," in *Electrical & Electronics Engineers in Israel (IEEEI), 2014 IEEE 28th Convention of*, 2014.

[24] D. Rossi, I. Loi, G. Haugou, and L. Benini, "Ultra low latency lightweight DMA for tightly coupled multi core clusters," *Proceedings of the 11th ACM Conference on Computing Frontiers*, 2014.

[25] A. Sampson *et al.*, "EnerJ: Approximate data types for safe and general low-power computation," in *Proceedings of the 32nd ACM SIGPLAN Conference on Programming Language Design and Implementation*, 2011.

[26] A. Teman, D. Rossi, P. Meinerzhagen, L. Benini, and A. Burg, "Controlled placement of standard cell memory arrays for high density and low power in 28nm FD-SOI," in *Design Automation Conference (ASP-DAC), 2015 20th Asia and South Pacific*, 2015.

[27] H. Zeng, C. S. Ellis, A. R. Lebeck, and A. Vahdat, "ECOSystem: Managing energy as a first class operating system resource," in *Proceedings of the 10th International Conference on Architectural Support for Programming Languages and Operating Systems*, 2002.

978-1-4799-8720-7/15 $31.00 © 2015 IEEE

2015 IEEE Computer Society Annual Symposium on VLSI

Logic-In-Memory: A NanoMagnet Logic Implementation

M. Cofano, G. Santoro, M. Vacca, D. Pala, G. Causapruno,
F. Cairo, F. Riente, G. Turvani, M. Ruo Roch and M. Zamboni
Politecnico di Torino, Dep. of Electronics and Telecommunications

M. Graziano
London Centre for Nanotechnology,
University College London

Abstract—In most computational systems memory access represents a relevant bottleneck for circuits performance. The execution speed of algorithms is severely limited by memory access time. An emerging technology like NanoMagnet Logic (NML), where its magnetic nature leads to an intrinsic memory ability, represents therefore a very promising opportunity to solve this issue. NanoMagnet Logic is the ideal candidate to implement the so called Logic-In-Memory (LIM) architecture. But how is it possible to organize an architecture where logic and memory are mixed and not separated entities?

In this paper we try to address this issue presenting our recent developments on LIM architectures. We originally conceived a LIM architecture without considering any technological constraints. Here we present the first adaptation of that architecture to NanoMagnet Logic technology. The architecture is based on an array of identical cells developed on three virtual layers, one for logic, one for memory and one for information routing. These three virtual layers are mapped on two physical layers exploiting all our recent improvements on NanoMagnet Logic technology, which are validated with the help of low level simulations. The structure has been tested implementing two different algorithms, a sort algorithm and an image manipulation algorithm. A complete characterization in terms of area and power is reported. The structure here presented is therefore the first step of an ongoing effort directed toward the development of truly innovative architectures.

Index Terms—NanoMagnet Logic, Logic-In-Memory, Parallel Computing, Multi-Layer Cirtcuits.

I. INTRODUCTION

In the last decades CMOS technology has undergone a considerable evolution. Circuits have become incredibly small and fast, allowing to fit an increasing number of functionalities in a single chip. Most of this growth was due to the scaling process, with transistors constantly shrinking in size. The advantages provided by transistors scaling has shadowed other problems that affect VLSI circuits and that are not improved by the scaling process itself. A significant problem of modern computational systems is in fact the "Memory Wall" problem [1], the huge performance loss due to memory access times. Currently a processing unit can handle an enormous amount of operations, but memories are not fast enough to provide them the required data. As a consequence system performance are limited by memories. The scaling process makes things worse, because the performance gap between logic and memory further increases reducing transistor size. To attenuate this problem, current computational systems use caching systems and hierarchical memories [2]. However, this is not enough,

and a change in computational paradigm to overcome Von Neumann limitations is required. This is a particularly favorable moment of history to work on this topic, because transistors scaling is ending and new emerging technologies are being developed. The scaling process end means that the focal point of future developments is shifting from device to architecture. The emerging of new technologies, particularly magnetic-based technologies like NanoMagnet Logic (NML) [3], means that new features, impossible to achieve with CMOS transistors, can now be integrated inside VLSI circuits.

This work represents the sum of our recent efforts in trying to reach the goal of a "true" logic-in-memory (LIM) circuit. We have developed an architecture made by a big array of computational units. Each unit is made by three "virtual" layers, a logic core, a memory core and a routing plane. Since memory is locally embedded in each processing unit, this architecture provides several benefits in case of parallel algorithms with a strong local interaction among neighbor elements. The technology of choice is NanoMagnet Logic. The LIM architecture here presented does not fully exploit the potential of a magnetic technology, however we describe several enhancement that we have made toward this application. The background on NML logic and our improvements are described in Section II. The general structure of LIM architecture is instead described in Section III. To test the effectiveness of our idea we have adapted and implemented two algorithms on the LIM structure, the odd-even sort and the binomial filter, both highlighted in Section III. Finally Section IV describes the NML implementation. The circuit is designed and simulated using an RTL model. At the same time our advancement on NML technology allow us to design circuits using two physical layers, instead of planar circuits as normally happens in NML technology. The architecture is analyzed in terms of power and area. We believe that the solution here proposed represents a promising new beginning for architectures development, leading the research on circuits for computation on unexplored paths.

II. BACKGROUND ON NML

NanoMagnet Logic is an emerging technology that uses single domain nanomagnets with only two stable states to represent logic values (Figure 1.A) [4]. The innovative key point, compared to transistors based technologies, is based on a signal propagation that does not rely on voltages or currents.

978-1-4799-8720-7/15 $31.00 © 2015 IEEE   286

Signals propagate through the circuit thanks to magnetody-namic coupling among neighbor magnets [3]. A NML wire is therefore composed simply by chaining the desired number of magnets (Figure 1.A) [5]. The advantages of this technology are related to its magnetic nature. A magnet maintains its state also without an applied power supply. The consequence is that in NML logic a magnet can act as a logic or memory element. Moreover it has no static power consumption and a potential very low dynamic power absorption [6]. The main characteristics of NML technology is the need of an external mean to switch magnets from one stable state to the other [7]. Magnets are forced in an intermediate unstable state, allowing them to overcome the energy barrier and switch to a different state depending on the value of neighbor magnets [8]. This system is called clock, because it gives to the technology a synchronous behavior similar to clocked CMOS circuits. Several techniques can be used to force magnets in the RESET state, like an external magnetic field [7] or the Spin Hall Effect [9], a spin-torque coupling with a current flowing through the magnet itself [10] or a mechanical stress applied through a voltage and piezoresistive substrate [6]. Regardless of the mechanism used a multiphase clock system must be used [11]. Circuits are divided in areas composed by a limited number of magnets called clock zones. At each clock zone a different clock signal is applied. The multiphase clock system allows to safely propagate signals in NML circuits in presence of thermal noise [12].

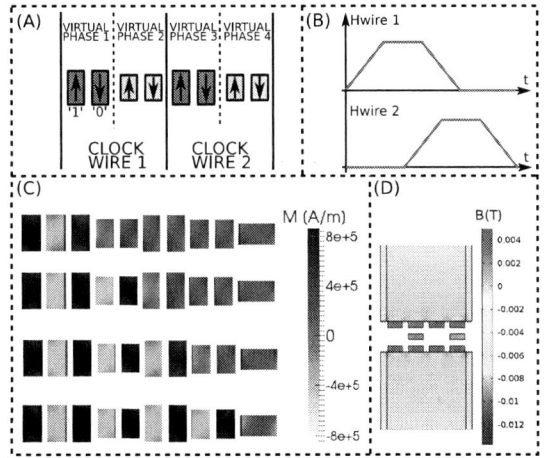

Fig. 1. NML fundamentals. A) Single domain nanomagnets encode the information. Virtual clock zones can be created with magnets of different size. B) Clock signals applied to the circuit. C) Micromagnetic simulation of a NML wire with virtual clock phases. D) Comsol simulation of 3D NML structure, with copper wires placed above and below magnets.

Among the clock solutions only the magnetic field [3] and the Spin Hall Effect [9] clock where experimentally demonstrated. Both shares the same technological frame, where the clock is generated by a current flowing through a wire placed under the magnets plane. The difference among these two clocks lies in the current value required and the material used for the wires. More details will be provided in Section IV.

We base our LIM design on these two clock solutions. In this paper we present a major improvement to NML clock, the virtual clock system. The working principle is depicted in Figure 1.A. Magnets with a bigger aspect ratio require a higher magnetic field to be reset. As a consequence we have built a wire chaining magnets with different aspect ratios ($50x100nm^2$ and $50x80^2$ in Figure 1.A). Two physical wires are used to generate the clock field, the waveforms applied are depicted in Figure 1.B). This particular solution allows us to use only two wires to create four "virtual" clock phases. Each virtual phase correspond to magnets with a different aspect ratio (Figure 1.A). In Figure 1.C the low level simulation obtained with NMAG [13] software is reported. Magnets start to switch from left to right, the two bigger magnets in the first virtual clock phase and then the two smaller magnets in the second virtual clock phase. Following the waveforms of Figure 1.B, the next magnets in the chain that switch are the bigger magnets corresponding to the third virtual clock phase and then the last two small magnets of the fourth virtual clock phase.

The complete explanation and validation of this clock system is out of the scope of this paper. However we commonly employ it in our designs because it provides several benefits to NML circuits. It simplifies the clock generation network, only two clock wires are required to generate the four clock phases necessary to assure a correct signals propagation. It enables to safely use the majority voter, the basic logic gate of this technology, also considering the constraints related to the fabrication of clock wire (more details are in Section IV). It allows us to build magnetic interconnections in NML circuits using domain walls, as we demonstrated in [14], greatly reducing interconnections overhead. Finally, placing clock wires above and under the magnets, as suggested in [3], it is possible to build multilayer NML circuits. In this case we limit our design to two logic layers, with a third separation layer between them. Figure I.D shows a Comsol Multiphysics [15] simulation of such structure. The section view depicts two clock wires of 400nm thickness and three layer of magnets. The current value used is 3mA and the magnetic flux density varies in the range of 2-3mT, depending on the layer distance. This value of magnetic flux density is sufficient to reset magnets. As will be described in Section IV, the possibility of using two logic layers allow us to greatly reduce area thanks to lower interconnection overhead.

III. THE LOGIC-IN-MEMORY STRUCTURE

The *Logic-In-Memory* can be seen as a smart memory where data are not only stored but also elaborated. The basic element (*cell*, Figure 2.A) embeds memory and logic with an information routing component. The architecture has a matrix-like structure (*grid*, Figure 2.A) where each element works autonomously in parallel.

The cell is composed by three "virtual" layers, memory, logic and routing. With the definition "virtual" we mean that these three layers can be mapped on different physical layers, depending on the possibility offered by the target technology.

978-1-4799-8720-7/15 $31.00 © 2015 IEEE   287

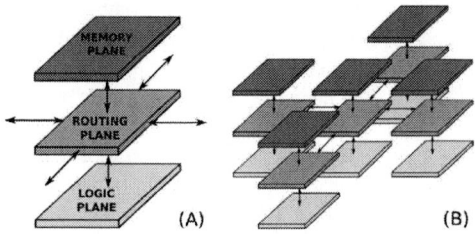

Fig. 2. (A) Logic In Memory (LIM) basic cell, three virtual layers are used, a memory plane, a logic plane and the routing plane. (B) Example of cells connection. Every LIM cell is connected with neighbor cells on 4 sides through the routing plane.

The *Logic Plane* hosts all the computational blocks required for the algorithm execution. It is a unit that must be designed ad-hoc for each implemented algorithm. The *Memory Plane* stores memory cells that can be read and written with simple operations. The number of memory locations depends on the algorithm implemented. In the case of the algorithms described in this paper, few memory locations are needed, making the memory plane size negligible compared to the other planes. The *Routing Plane* handles the communication and data exchange with neighbor cells. The routing plane is the same for all applications, while the memory plane size and the logic plane structures depends on the algorithms implemented.

Information are exchanged between cells through messages, called *words*. Each *word* is a complex bit string with many fields used to identify important information, like the operation type and cells address. Each *word* is composed by the following fields.

- TAG: it identifies the operation type.
- TCA (Target Cell Address): it is the address of the message origin cell inside the grid.
- WA (Word Address): it specifies the memory location to access;
- DATA: it is the data to be computed or stored.
- DEST: it represents the message destination cell address inside the grid

The instruction set of each is composed by several operations.

- Local write/read: local operations to allow the logic to write or read data from the memory.
- Toc-toc write/read: they allow the communication among two different cells; each cell can write or read a data of adjacent cells.
- Remote write/read: remote operations are the slowest ones and are used to allow the communication between non adjacent cells.
- Logic-logic: these operations allows to the logic planes of two adjacent cells to directly exchange data without accessing their memory planes.

This kind of structure is particularly suited to execute algorithms where a high number of memory accesses have to be performed, since the logic is embedded in the memory and therefore the access is less expensive. Moreover, since each cell is connected to the adjacent ones by means of local interconnections avoiding global wires, the LIM architecture is particularly adapted to NML technology where interconnections overhead wastes a lot of area. In the following the routing plane and the logic planes for the two algorithms are described. The description of the memory plane is not reported because it is made simply by an array of memory cells.

A. The routing plane

The block diagram of the routing plane is depicted in Figure 3. To simplify the architecture, the routing plane has been designed with only one FSM. It manages the input requests without solving conflicts in case of simultaneous request coming from different cells. To solve this issue a priority management block has been created. The highest priority has been given to the logic and then, in descending order, to the requests coming from the north, west, south and east cell respectively. If one or more requests cannot be satisfied, an acknowledge bit is sent to the request-sender. The key role of the routing plane is to discriminate the operation type and to identify the sender and the receiver. Neighbor cells are identified by two bits (from 00 of the east cell to 11 of the north cell in clockwise order). This two bits represent a sort of relative address, that each cell uses to identify its neighbors. In conjunction to the relative address, an absolute address system is used to identify each cell. A number indicates the cell position in terms of column and row. The routing plane can be divided in four main regions. 1) The input interface is responsible for sinking all requests and managing their priorities. 2) The priority manager checks if the FSM is available to execute an operation. In the affirmative case it selects the request with the highest priority and sends it to the selection unit. If a request cannot be satisfied, acknowledge bits are sent to the senders. 3) The selection unit identifies the operation to be executed. 4) The last region is the output interface. It is responsible of sending the data to the correct destination.

B. The odd-even sort logic plane

As a first case of study the logic plane has been designed to execute the odd-even sort algorithm. It is a comparison sorting algorithm where every even and odd couple of numbers are compared. If numbers are not in the exact order, they are swapped. The main advantage of this algorithm is its ability to simultaneously compare all even or odd couples of numbers exploiting the parallel computation. The logic plane for this application (depicted in Figure 4) is composed by registers, a FSM, an adder to compare the value of the two data and two counters: one to switch from the operating mode to the stand-by mode and one to stop the execution.

Only even cells start the computation while the odd ones are in stand-by. After the algorithm execution, the even cells go in stand-by mode while the odd ones run the sorting. The process is repeated until all data are sorted. The pseudo-code of the algorithm, in order to implement it inside the LIM architecture, is reported in the following.

978-1-4799-8720-7/15 $31.00 © 2015 IEEE

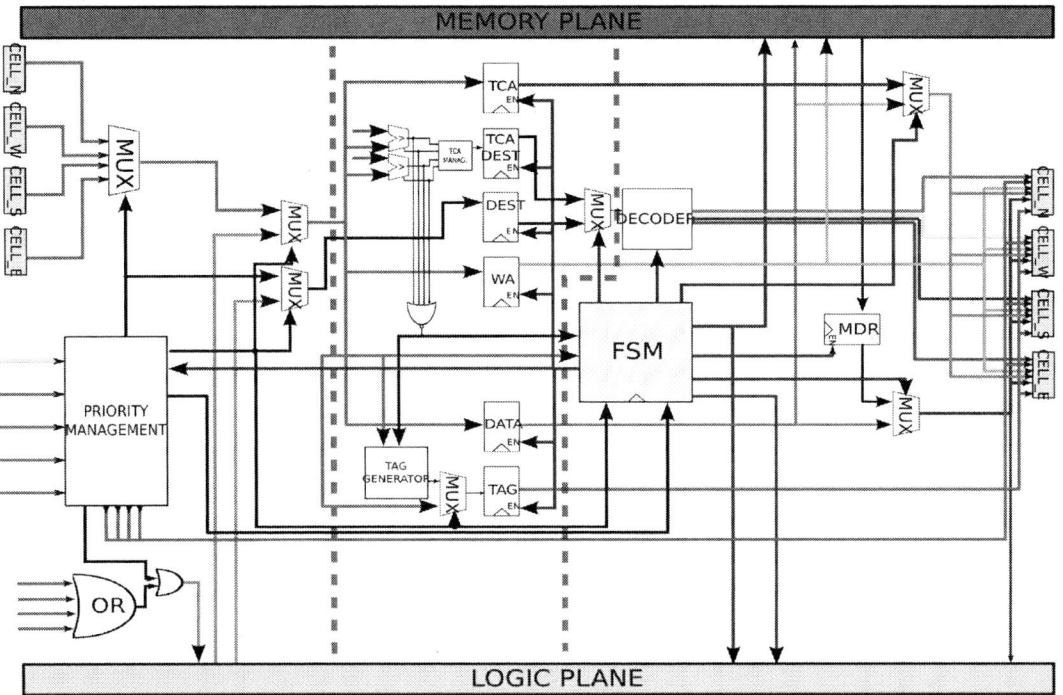

Fig. 3. Routing plane block diagram. Multiplexers, a priority management component and destination selection blocks are used to route signals to logic and memory planes and to neighbor cells. The plane behavior is controlled by a dedicated finite state machine (FSM).

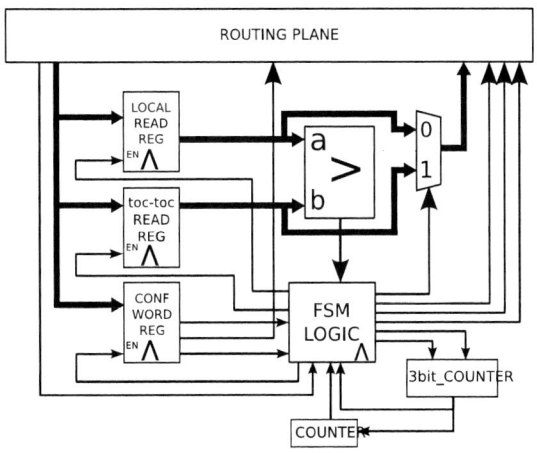

Fig. 4. Logic plane block diagram for the odd-even sort algorithm. The logic plane is composed by a comparator, a multiplexer, counters, registers and an FSM control unit.

1) Read the configuration word;
2) Read the local data;
3) Read the adjacent cell data;
4) Compare the two data;
5) Exchange data if necessary;
6) Stand-by until the execution restarts.

C. The binomial filter logic plane

As a second case of study a binomial filter, mostly used for signal processing, has been implemented. The logic plane (depicted in Figure 5) is composed of registers, a FSM and an adder to sum all read data.

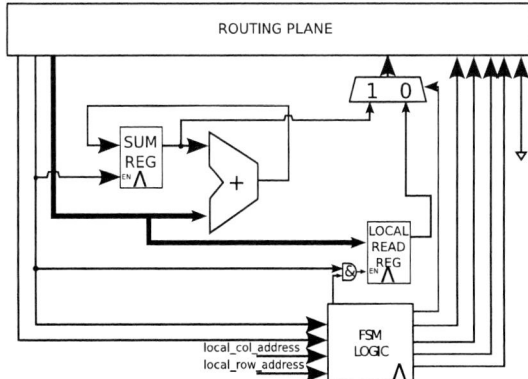

Fig. 5. Logic plane block diagram for binomial filter algorithm. The logic plane is composed by an adder, two registers, a multiplexer and an FSM control unit. The datapath is simpler compared to the odd-even sort algorithm.

The pseudo-code of the algorithm is reported in the following.

1) Read the local data;
2) Read the eight neighbouring cells data;

978-1-4799-8720-7/15 $31.00 © 2015 IEEE

3) Sum all data and divide the result by 16;
4) Write the result in the local memory.

IV. THE LOGIC-IN-MEMORY ARCHITECTURE

The Logic-In-Memory architecture has been implemented using NML technology applying the enhancements described in Section I. To design and simulate the NML circuit, an RTL model has been developed [16]: as depicted in Figure 6, registers simulate the propagation delay of signals through consecutive clock zones, while ideal logic gates model the logic function. Majority voters and inverters are sufficient to represent every logic function. An inverter is obtained linking consecutively an odd number of magnets. AND/OR gates can be obtained from a majority voter fixing the polarization of one input to a constant logic '0' or logic '1', respectively. Figure 6

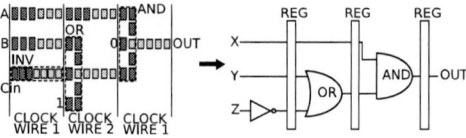

Fig. 6. RTL model of an NML circuit. Registers are used to model the propagation delay introduced by each clock zone whereas ideal gates are used to represent the logic function.

highlights the majority voter implementation using the virtual clock. The three input wires belong to the first virtual phase (bigger magnets). The central and output magnets belong to the second virtual phase (smaller magnets). As a consequence the central magnet switch only after all input signals have successfully propagate. This solution greatly improves the reliability of majority voters considering constraints related to clock wires fabrication [5].

The NML implementation of the Logic-In-Memory architecture is based on the use of two physical layers and virtual clocking. In Figure 7 is depicted, as an example, a 3D full adder. The adder develops horizontally on two physical layers (layers 1 and 3), thus, it can be thought of as a three-dimensional structure. Layer 1 and layer 3 are connected through an intermediate level on which only few magnets (those necessary to the communication between layers) are allowed. The role played by magnets on Layer 2 is the same as the one played VIAs in CMOS circuits. Thanks to virtual clocking and multilayer structures the circuit area greatly decreases. The full adder here presented is more than 4 times smaller than the full adder described in [14].

Figure 8 depicts instead the 3D layout of a ripple carry adder. The layout has been designed according to the following constraints: 1) clock wire length equal to 6 magnets and 2) maximum number of consecutive magnets belonging to the same virtual phase equal to 5 [12].

The LIM architecture is based on three virtual layers, but the technology allow us to use only two physical layers. As a consequence we found that the optimal solution was to map every plane independently on two physical layers. The cell and the array were then assembled by aligning the three virtual

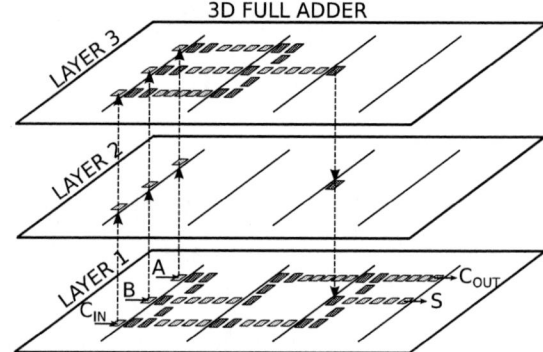

Fig. 7. Layout of a 3D full adder.

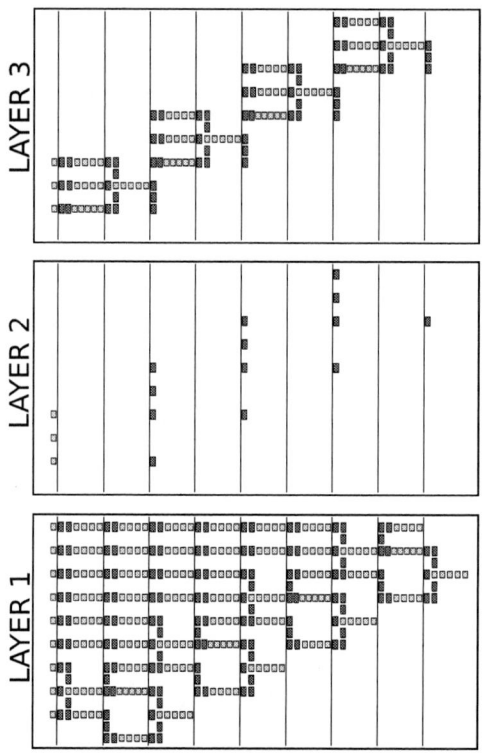

Fig. 8. NML layout implementation of a 4 bit ripple carry adder. Two physical layers and virtual clocking have been used.

planes. Figure 9 depicts the generic organization of a 4×4 grid of LIM cells with the block scheme of a cell outlined. The block scheme and the dimensions refer to a PE in which the logic plane implements the odd-even sort algorithm. It can be noticed that the routing plane occupies most of the area of the cell. This is mainly due to two reasons. First there is an high number of interconnections, that waste a considerable portion of area even if solutions (such as domain walls and two physical layers) to compact the layout are exploited. Secondly, an high parallelism is adopted, implying that a large number of NML wires travels in parallel and the area that they take is not at all negligible. The area occupied by the "virtual" logic plane is comparable to the routing plane, while the memory is much smaller. Each processing element (PE) has an estimated height of $33.1\,\mu m$ and a length of $66.4\,\mu m$.

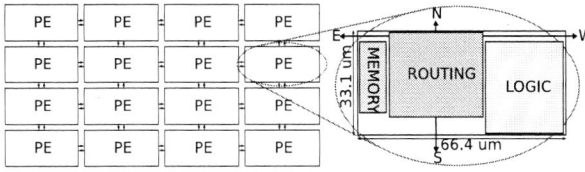

Fig. 9. 4×4 grid of LIM cells. The block scheme of each PE is outlined. Each "virtual" plane is mapped independently on two physical layers. The whole structure is then assembled by aligning the "virtual" planes together.

The area occupation and the power consumption have been estimated for both implementations. Results are summarized in TABLE I. The area was evaluated by knowing the total height and width in terms of number of magnets and considering the physical height and width of magnets as reported in Section II. Power dissipation was evaluated estimating the clock wires length starting from circuit area. More details on how to evaluate power consumption can be found in [16]. Power was evaluated considering the magnetic field clock [3], using copper wires of 400nm thickness and 420nm width and a current of 6 mA for each wire [3]. Power was also estimated in case of Spin Hall Effect clock, considering wires made by a complex heterostructure where the resistance value is dominated by the 10nm layer of Tantalum. In [9] authors use a current of 2mA for a clock wire with a width of $1\mu m$ and an height of 10nm. In this case we use wires with a width of 420nm and an height of 10nm, thus the current value that we use is 0.8mA.

TABLE I
AREA OCCUPATION AND POWER CONSUMPTION ESTIMATION OF A SINGLE CELL THAT EXECUTES, IN ONE CASE, THE ODD-EVEN SORT ALGORITHM AND, IN THE OTHER, THE BINOMIAL FILTER ONE.

Circuit	Area (μm^2)	Power (mW)	
		Magnetic field	Spin Hall Effect
Odd-even sort	2196.2	18.4	101.24
Binomial filter	2129.8	18.4	101.24

Considering the two algorithms the cell area occupation and power consumption is roughly the same. The two clock systems instead lead to the same circuits organization, therefore the area does not change considering one clock system instead of the other. Power consumption is however a different case, being 5 times lower in the magnetic field clock. If further development of this solution will allow to use the Spin Hall Effect clock with low resistance materials instead of Tantalum, power consumption will be greatly reduced. Regarding time, circuits works with a frequency of 100MHz, the instructions execution time varies from 79 to 131 clock cycles. This is

mainly due the intrinsic pipelined nature of this technology which reduces performance in case of feedback signals [11].

V. CONCLUSIONS

We have developed a logic-in-memory architecture to exploit the potential of emerging technologies. We have implemented the LIM architecture using NML technology as target, bringing, at the same time, several enhancements to the technology itself. Performance obtained are very good, especially in terms of area and power consumption. We are now working on technological solution that allow us to implement NML circuits with many layers. At the same time we are redesigning the architecture keeping in mind its pipelined behavior in order to improve circuit performance.

VI. ACKNOLEDGEMENTS

We thank Alessandra Fioriti, Michele Giacobbe, Giuseppe Casuccio and Massimo Caligaris for their valuable contribution on this project.

REFERENCES

[1] A. Nowatzyk, P. Fong, and A. Saulsbury, "Missing the memory wall: The case for processor/memory integration," *1996 23rd Annual International Symposium on Computer Architecture*, p. 90, May 1996.

[2] P. Jacob, A. Zia, O. Erdogan, P. Belemjian, J. Kim, M. Chu, R. Kraft, J. McDonald, and K. Bernstein. "Mitigating memory wall effects in high-clock-rate and multicore cmos 3-d processor memory stacks," *Proceedings of the IEEE*, vol. 97. no. 1, pp. 108–122, Jan 2009.

[3] M. Niemier and al., "Nanomagnet logic: progress toward system-level integration," *J. Phys.: Condens. Matter*, vol. 23, p. 34, Nov. 2011.

[4] R. Cowburn and M. Welland, "Room temperature magnetic quantum cellular automata," *Science*, vol. 287, pp. 1466–1468, 2000.

[5] M. Vacca, D. Vighetti, M. Mascarino, L. Amaru, M. Graziano, and M. Zamboni, "Magnetic QCA Majority Voter Feasibility Analysis ," *2011 7th Conference on Ph.D. Research in Microelectronics and Electronics (PRIME)*, pp. 229–232, 2011.

[6] M. Vacca, M. Graziano, L. D. Crescenzo, A. Chiolerio, A. Lamberti. D. Balma, G. Canavese, F. Celegato, E. Enrico, P. Tiberto, L. Boarino, and M. Zamboni, "Magnetoelastic clock system for nanomagnet logic," *Ieee Transaction On Nanotechnology*, vol. 13, no. 5, September 2014.

[7] M. Niemier, X. Hu, M. Alam, G. Bernstein, M. P. W. Porod, and J. DeAngelis, "Clocking Structures and Power Analysis for nanomagnet-Based Logic Devices," in *Int.Symp. on Low Power Electronics and Design.* Portland-Oregon, USA: IEEE, 2007, pp. 26–31.

[8] M. Awais, M. Vacca, M. Graziano, and G. Masera, "FFT Implementation using QCA," *2012 19th IEEE International Conference on Electronics, Circuits and Systems (ICECS)*, pp. 741–744, 2012.

[9] D. Bhowmik, L. You, and S. Salahuddin, "Spin hall effect clocking of nanomagnetic logic without a magnetic field," *Nature Nanotechnology Letter*, vol. 9, pp. 59–63, 2014.

[10] J. Das, S. Alam, and S. Bhanja, "Ultra-low power hybrid cmos-magnetic logic architecture," *Trans. on Computer And Systems*, 2011.

[11] J. Wang, M. Vacca, M. Graziano, and M. Zamboni, "Biosequences analysis on NanoMagnet Logic," *International Conference on IC Design and Technology*, pp. 131–134, May 2013.

[12] G. Csaba and W. Porod, "Behavior of Nanomagnet Logic in the Presence of Thermal Noise," in *International Workshop on Computational Electronics.* Pisa, Italy: IEEE, 2010, pp. 1–4.

[13] T. Fischbacher, M. Franchin, G. Bordignon, and H. Fangohr, "A Systematic Approach to Multiphysics Extensions of Finite-Element-Based Micromagnetic Simulations: Nmag," *IEEE Transactions on Magnetics*, vol. 43, no. 6, pp. Available on–line, 2007.

[14] F. Cairo, M. Vacca, M. Graziano, and M. Zamboni, "Domain magnet logic (dml): A new approach to magnetic circuits," in *IEEE International Conference on Nanotechnology*, 2014.

[15] "Comsol Multiphysics," http://www.comsol.com/.

[16] M. Vacca, M. Graziano, and M. Zamboni, "Nanomagnetic Logic Microprocessor: Hierarchical Power Model," *IEEE Transactions on VLSI Systems*, vol. 21, no. 8, pp. 1410–1420, Aug. 2012.

978-1-4799-8720-7/15 $31.00 © 2015 IEEE

2015 IEEE Computer Society Annual Symposium on VLSI

Simscape based Ultra-Fast Design Exploration of Graphene-Nanoelectronic Systems

Shital Joshi
Computer Science and Engineering
University of North Texas, USA.
Email: ShitalJoshi@my.unt.edu

Elias Kougianos
Electrical Engineering Technology
University of North Texas. USA.
Email: eliask@unt.edu

Saraju P. Mohanty
Computer Science and Engineering
University of North Texas, USA.
Email: saraju.mohanty@unt.edu

Abstract—This paper presents non-EDA based design simulation using Simscape® for graphene based nanoelectronic systems. The objective of this paper is to explore ultra-fast design for analog devices using substitutes to conventional but time intensive EDA simulations such as SPICE. A GFET behavioral model is presented where the model is based on the drift-diffusion conduction mechanism of the dual-gate device. The kink region of the $I - V$ characteristic is modeled via a displacement current. A case study design circuit, an all graphene based low noise amplifier (LNA) is presented. The results show this to be a viable alternative approach to simulate GFET based circuits and systems in addition to existing SPICE, VHDL-AMS or Verilog-A based flows. To the best of the authors' knowledge, this is the first ever paper presenting a Simscape® model of a GFET device and also performing design exploration of GFET based RF circuits using Simscape® .

Index Terms—Nanoelectronic System, Design Simulation, Simscape® Modeling, Graphene Field Effect Transistor, All-Graphene Low-Noise Amplifier

I. INTRODUCTION AND MOTIVATION

Nanoelectronic devices are some of the most complicated structures produced by modern manufacturing technology. With transistor count already reaching billion in a die, it is becoming extremely challenging to shrink the transistor further down, particularly at 22 nm and below, in order to follow Moore's law. However relentless demand for smaller, cheaper and lower power consumption electronic products has forced designers to explore in depth two possible options: (1) combine analogue, radio frequency (RF) and digital components into a single chip, leading to analog mixed signal systems on chip (AMS-SoC) [1], [2] and (2) search for alternative technologies which can replace silicon to meet future demand. The former option necessitates a fusion of two design domains, namely analog and digital while the latter demands alternative material to silicon. AMS-SoC designs undergo a series of simulationa with SPICE in order to accurately simulate their behavior at layout level (with parasitics) [3]. However the major drawbacks associated with the SPICE simulation are the following: (1) heavy computational needs which prolongs the design time and increases the non-recurrent design cost of the chip, (2) the need of fab data or TCAD simulations which may not always be available for new or emerging technologies, and (3) design optimization support is very limited. In order to overcome these difficulties associated with SPICE based flows,

this paper proposes a non-EDA design flow. This Simscape® based design flow offers two distinct advantages over conventional SPICE: (1) it can model emerging technologies without the need for fab data, (2) it provides fast and easy optimization at system level.

Motivated by the requirements to design smarter devices, researchers are actively opting for an alternative technology to conventional CMOS. Some of these technologies that are promising as an alternative to silicon FET are graphene FET, tunneling FET, BisFET, and spinFET [4]. This paper considers the graphene FET for design of nanoelectronic systems. A Simscape® [5] bahavioral model of a GFET based low-noise amplifier (LNA) is modeled. The LNA has significant applications in real-life circuits and systems. The model accuracy is verified with data available from MATLAB® [6], SPICE [7], VHDL-AMS [8] or Verilog-A [9] models presented in the existing literature.

The rest of this paper is organized in the following manner: Section II discusses the novel contributions of the current paper. Section III discusses background information on GFETs and GFET based circuits. Section IV presents the Simscape model that was developed as part of this work. Section V presents a GFET based RF circuit where an LNA design used as case study. Section VI presents our conclusions and directions for future research.

II. NOVEL CONTRIBUTIONS OF THIS PAPER

The **novel contributions of this paper** to the state-of-art are the following:

1) A Simscape® based ultra-fast design exploration which is a non-EDA flow and a paradigm shift.
2) Modeling of graphene FET devices using Simscape® .
3) Modeling of a GFET based LNA using the Simscape® graphical environment.
4) Experimental validation of the Simscape® device level models with existing VHDL-AMS or Verilog-A models.
5) Characterization of GFET based case study circuits and comparison with Verilog-A based designs.

Fig. 1 shows a comparative perspective of the proposed non-EDA based design flow with the well-accepted conventional EDA based flow. In the conventional flow, fab data or process information for a specific technology is needed,

978-1-4799-8720-7/15 $31.00 © 2015 IEEE 292

which introduces delay in its implementation for emerging technologies. Fig. 1 shows the conventional EDA based design flow where the compact model is first derived from the fab data or technology computer-aided design (TCAD) simulation. The compact model is then converted into a compiled or Verilog-A model, which is used in SPICE. These are quite effort intensive tasks and may increase the design cost as well as slow the simulation process. For large circuits, the SPICE based EDA design flow has heavy computational requirements and does not have comprehensive design optimization options. However, the proposed Simscape® based non-EDA design flow does not require any fab data. Instead it relies on the first principle models published in the physics/semiconductor literatures. As shown in Fig. 1(b), the device level models presented in the MATLAB®, Simulink®, or Simscape® languages are simulated in the MATLAB® engine. The target circuit or system is designed using the Simscape® graphical based tool interface.

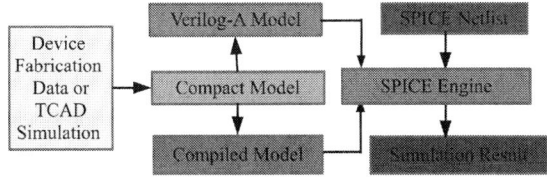

(a) SPICE based Design Simulation Flow

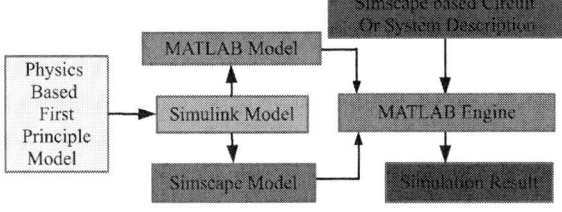

(b) Simscape based Design Simulation Flow

Fig. 1. SPICE versus Simscape® based Design Simulation Flow.

III. GRAPHENE BASED NANOELECTRONICS

Due to thermal fluctuation caused by the heat generated in nanoelectronic circuits, silicon below 10 nm can not exist as a crystalline solid and it becomes amorphous. Thus due to continuous downsizing of the channel size in modern circuitry, silicon will soon reach its fundamental limitations and a suitable substitute needs to be investigated. Carbon falls in the same group with silicon in the periodic table (VI-A column) and due to its impressive allotrope, it is being treated as a suitable substitute for silicon.

In recent years, various forms of carbon structures have been studied such as three dimensional (diamond, graphite), two dimensional (graphene), one dimensional (nanotube) and zero dimensional (fullerones) [10]–[12]. Among those structures, nanotubes and graphene have been the dominant contenders to substitute silicon. The charge carrier mobility of of graphene

is found to be above 200,000 cm²/Vs at room temperature. Apart from this, graphene shares the same set of processing techniques as used for silicon. Due to these advantages graphene is considered in this paper.

A. Graphene FET: Structure

Several graphene transistor structures such as back-gated graphene transistor, dual-gate graphene transistor and epitaxial graphene from SiC and transistor structures have been studied in recent years. However in this paper, the dual-gate graphene transistor is used and Fig. 2 shows a cross-sectional view of such a GFET structure. A single layer of graphene is placed on top of a SiO_2 substrate, which is separated from the top gate by a thin layer of oxide. A back gate lies below the substrate which controls the resistance of a symmetric source-drain arrangement. By applying an electric field perpendicular to the graphene channel, a tunable bandgap can be opened which can be then modulated by the gate voltage. There have been three regions of operation identified for a bilayer GFET: triode region, unipolar saturation region and the ambipolar saturation region. The drain-source $I - V$ characteristics for the triode and the uniploar saturation region are equivalent to that of the MOSFET. However in the ambipolar saturation region, with increase in the drain voltage, there is an increase in the drain current. Such a behavior in the ambipolar region is referred as a second linear region. Ambipolar condition occurs when the electrons and holes have the same contribution to the total current whereas in other regions, either electron or hole dominates the total current.

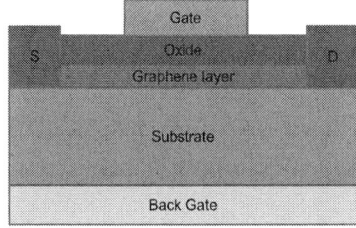

Fig. 2. Dual-gate graphene field effect transistor (GFET) cross-section. The graphene layer is indicated in green, below the oxide layer.

B. Graphene FET Based Circuits

Due to unique properties of graphene such as high carrier mobility and saturation velocity, high stability, low noise, graphene makes a good candidate for high frequency electronic applications. However there are some challenges associated with bulk graphene namely it is difficult to turn off the transistors, and the I_{on}/I_{off} ratio is low. These challenges makes them like unrealistic for digital circuits, since these applications need to save power during the off state. However this is not the case for analog and radio frequency (RF) applications. These applications always operate in the ON-state and also high carrier mobility (hence frequency) is the main requirement for these application over power saving. So, GFETs are well suited for these applications. Currently,

978-1-4799-8720-7/15 $31.00 © 2015 IEEE

expensive group II-IV elements are used for high electron mobility transistors (HEMTs) for high carrier mobility and saturation velocities. So GFET can be a cost effective solution to these applications.

Furthermore, reports have suggested that using sophisticated technology, these GFETs can also be used for digital circuits. For example, dual-gate and bi-layer GFETs in [13] were measured to have an I_{on}/I_{off} ratio of 100. Thus reports have shown GFETs being used in all analog, digital as well as RF circuits such as inverters [14], [15] and [16], frequency multipliers [17], [18], and [19], an RF mixer [20], amplifiers [21], and [22], a photo detector [23], and low-noise amplifiers (LNAs) [24] and [25]. The concept of wafer scalable analog circuits was verified with an RF mixer example in [26]. In the current work, an LNA is chosen as case study circuit for GFET based circuit design utilizing the Simscape® graphene model.

IV. SIMSCAPE® MODELING OF GRAPHENE FET

Simscape® is part of MATLAB®/Simulink® which is capable of multi-domain, multi-discipline modeling and simulation of physical systems. The Simscape® framework includes a variety physical component libraries. Essentially, there are two ways to build custom Simscape® models for emerging devices like the GFET: (i) a graphical method using fundamental Simulink®/Simscape® blocks, or (ii) textually with the Simscape® physical modeling language. Since the later approach offers better portability and is easier to maintain, the current paper adopts this approach. In addition, this approach makes hierarchical modeling and simulation of complex system easier as well. In order to compare the proposed Simscape® model with well accepted EDA model, the paper considers the GFET model based on the VHDL-AMS model from [8] and the Verilog-A model from [9]. The Simscape® simulation results show good agreement with the prior results presented in [6], [8], [27]–[29].

Fig. 3 and Fig. 4 show the GFET drain-to-source current-voltage characteristics. For a negative V_{bs}, the GFET source/drain region is P-type [6], [8], [27] and the mobility $\mu = 700$ cm^2/V/s, $R_s = 800\Omega$, and $E_c = 4.5$ kV/cm. In Fig. 3, top-gate voltages of 0 V, -1.5 V, -1.9 V and -3 V were used and V_{ds} is varied from 0 to -3 V. For positive V_{bs}, the source/drain region is n-type with $\mu = 1200$ cm^2/V/s, $R_s = 1500$ Ω, and $E_c = 15$ kV/cm. In Fig. 4, top-gate voltages of -0.8 V, -1.3 V, -1.8 V, -2.3 V, and -2.8 V were used and V_{ds} is varied from 0 to -3 V. The device parameters selected are based on published results [13]. The resultant $I-V$ characteristic curves are similar to the results obtained in VHDL-AMS [6], [8] and in Verilog-A models discussed in [9].

V. GRAPHENE FET BASED AMPLIFIER CIRCUIT DESIGN

As a specific demonstration of Simscape® based design simulation flow, an amplifier circuit simulation is presented in this Section. Design simulation of widely used circuits such as LC-VCO has been similarly performed, but has not been presented in this paper for brevity.

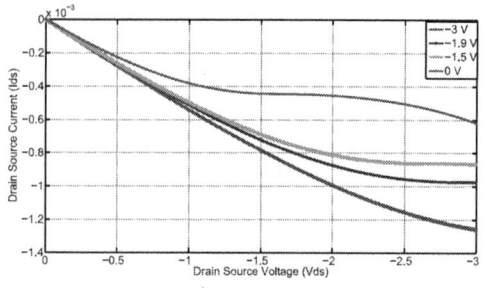

(a) I-V Characteristics

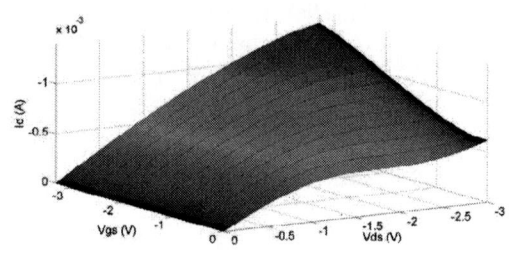

(b) Surface Plot

Fig. 3. $I-V$ characteristics for discrete and continuous values of V_{gs} for P-type GFET.

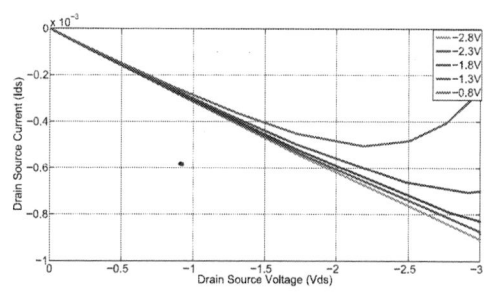

(a) I-V Characteristics

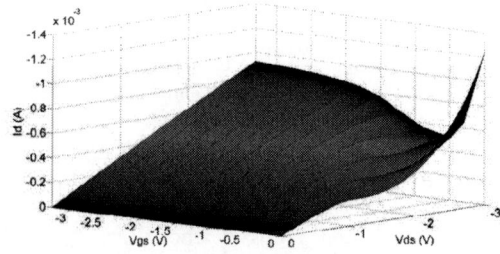

(b) Surface Plot

Fig. 4. $I-V$ characteristics for discrete and continuous values of V_{gs} for N-type GFET.

A. Theoretical Perspective

The Low-Noise Amplifier (LNA) is an electronic component which is used to amplify very weak signals in RF circuits. It is usually located very close to the detection device so as to reduce losses and is highly susceptible to electromagnetic, thermal and transmission noise. Since it can detect even a small change at the receiver, the LNA must have intrinsically low noise. The four important design parameters in LNA design are the following: (1) gain, (2) noise figure (NF), (3) non-linearity, and (4) impedance matching. Low NF results in better signal reception and in order to process signals effectively, high gain is necessary. If the LNA does not have high gain then the signal will be affected by the noise in the LNA circuit itself. However the general design trade-off is to minimize the noise of the circuit by accepting fairly low gain. The low gain is alleviated in subsequent stages of the receiver.

Fig. 5 shows the schematic diagram of a simple possible all-graphene LNA consisting of a common source amplifier and a load [25]. In the circuit, G_1 acts as a load and G_2 acts as a common source amplifier transistor. The circuit is amenable to exhaustive design exploration. This simple circuit is used as a test case to validate the Simscape® model developed for the current paper.

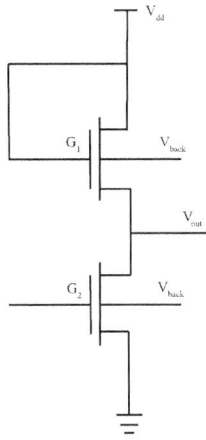

Fig. 5. Schematic of a GFET based LNA circuit. G_1 is the load transistor and G_2 is the amplifier.

The graphene nanoribbon widths W_1 and W_2 for devices G_1 and G_2 are chosen as the design variables. The gain (G), bandwidth (F_T) and power consumption (P_{LNA}) are considered as the figures-of-merit (FoMs) of the LNA.

B. Simscape® Modeling of the LNA

Fig. 6 shows the Simscape® experimental setup for LNA circuit characterization. In the LNA simulation, GFETs with $T_{ox} = 1$ nm, $H_{sub} = 2.85$ nm, and $L = 50$ nm were used. W was varied for both transistors for all simulations.

The result in Fig. 7 shows the inverse relationship between the band-gap and gain as demonstrated in [25]. Similarly, Fig. 8 shows how the bandwidth and the load resistance are

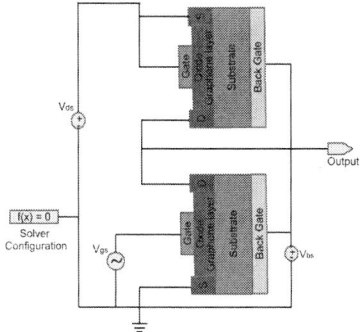

Fig. 6. Simscape® Experimental Setup for LNA Simulation.

related under constant gain G=15.75 dB. Finally, as shown in Fig. 9, the GFET based LNA is shown to have a small-signal bandwidth of 3.119 GHz. Table I summarizes the basic characteristics of the LNA for two different transistor sizes.

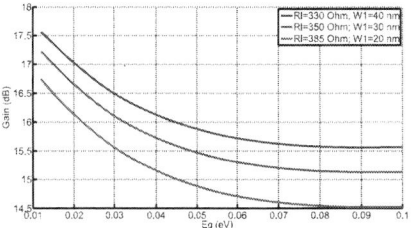

Fig. 7. Gain (G) vs. E_g at different R_L values.

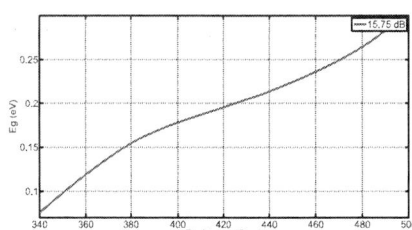

Fig. 8. E_g vs. R_L at constant G = 15.75 dB.

TABLE I
GFET BASED AMPLIFIER FIGURES-OF-MERIT

Parameters	Values	Values
W_1	20 nm	30 nm
W_2	10 nm	15 nm
Gain (G)	14.54 dB	15.41 dB
Bandwidth (f_T)	3.12 GHz	3.12 GHz
Power (P_{LNA})	23.8 mW	27.2 mW

VI. CONCLUSIONS AND FUTURE RESEARCH

A Simscape® based behavioral model of graphene FETs, suitable for design exploration at high levels of abstraction,

978-1-4799-8720-7/15 $31.00 © 2015 IEEE

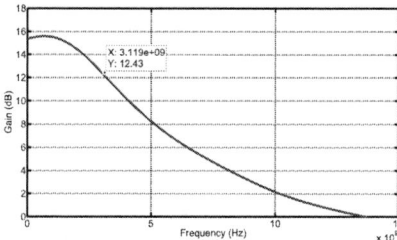

Fig. 9. The simulated frequency characteristics of the GFET based LNA.

has been presented in this paper. The model has been verified by extensive $I - V$ characterization of the GFET. A study circuit (LNA) is considered and the results are compared with the well-accepted EDA models. The results obtained in this paper show that the Simscape® based model can be used as a substitute for more detailed but time consuming EDA simulations such as SPICE with Verilog-A and VHDL-AMS models. Thus the ability to perform mixed high-level (behavioral) and transistor-level simulations for RF systems with an integrated design environment provides RF designers with unique design exploration and verification tools.

As a future research, additional functionalities for noise, transfer function and non-linear RF analyses such as periodic and quasi-periodic steady state can be incorporated within the Simscape® model. Optimization techniques using particle swarm-based optimization (PSO) algorithms such as artificial bee colony and ant colony optimization for GFET based circuits will be explored within MATLAB®/Simscape®.

Acknowledgments

The authors would like to thank UNT graduates and students Dr. G. Zheng, Mr. M. Gautam, and Mr. A. Khan for their help and inputs on this paper.

References

[1] S. P. Mohanty, *Nanoelectronic Mixed-Signal System Design*. McGraw-Hill Education. 2015.

[2] W. Fu and A. Fayed, "Power Conversion Schemes in Nanometer CMOS Mixed-Signal SoCs," in *Proceedings of the IEEE International Symposium on Circuits and Systems*, 2014, pp. 606–609.

[3] K. Gulati, J. F. Croix, S. Khatri, and R. Shastry, "Fast Circuit Simulation on Graphics Processing Units," in *Proceedings of the Asia and South Pacific Design Automation Conference*, 2009, pp. 403–408.

[4] D. E. Nikonov and I. A. Young, "Overview of Beyond-CMOS Devices and a Uniform Methodology for Their Benchmarking," *Proceedings of the IEEE*, vol. 101. no. 12, pp. 2498–2533, Dec 2013.

[5] "Simscape™ Multi-domain Physical System Simulation," http://www.mathworks.com/products/simscape/, last Accessed on 09/26/2014.

[6] I. Meric, M. Y. Han, A. F. Young, B. Ozyilmaz, P. Kim, and K. L. Shepard, "Current saturation in zero-bandgap, top-gated graphene field effect transistors," *Nature Nanotechnology*, vol. 3, pp. 654–659, November 2008.

[7] R. Doe, "Graphene Field Effect Transistor Modelling," http://www.cnt.ecs.soton.ac.uk/gfet_web/gfet_web.html, Jan 2013.

[8] I. J. Umoh and T. J. Kazmierski, "VHDL-AMS Model of A Dual Gate Graphene FET," in *Proceedings of the Forum on Specification and Design Languages*, 2011, pp. 1–5.

[9] M. A. Khan, S. P. Mohanty, and E. Kougianos, "Statistical Process Variation Analysis of a Graphene FET based LC-VCO for WLAN Applications," in *Proceedings of the 15th IEEE International Symposium on Quality Electronic Design (ISQED)*, 2014, pp. 569–574.

[10] A. K. Geim and K. S. Novoselov, "The Rise of Graphene," *Nature Materials*, vol. 6, no. 3, pp. 183–191, 2007.

[11] K. S. Novoselov, A. K. Geim, S. V. Morozov, D. Jiang, M. I. K. I. V. Grigorieva. S. V. Dubonos, and A. A. Firsov, "Two-dimensional gas of massless dirac fermions in graphene," *Nature*, vol. 438, no. 7065, pp. 197–200, 2005.

[12] K. S. Novoselov, A. K. Geim, S. V. Morozov, D. Jiang, Y. Zhang, S. V. Dubonos, I. V. Grigorieva, and A. A. Firsov, "Electric Field Effect in Atomically Thin Carbon Films," *Science*, vol. 306, pp. 666–669, 2004.

[13] F. Xia, D. B. Farmer, Y.-m. Lin, and P. Avouris, "Graphene Field-Effect Transistors with High On/Off Current Ratio and Large Transport Band Gap at Room Temperature," *Nano Letters*, vol. 10, no. 2, pp. 715–718, 2010.

[14] N. Harada, K. Yagi, S. Sato, and N. Yokoyama, "A Polarity-Controllable Graphene Inverter," *Applied Physics Letters*, vol. 96, no. 1, pp. 012 102–012 102, 2010.

[15] L. Rizzi, M. Bianchi, A. Behnam, E. Carrion, E. Guerriero, L. Polloni, E. Pop, and R. Sordan, "Cascading Wafer-Scale Integrated Graphene Complementary Inverters under Ambient Conditions," *Nano Letters*, vol. 12, p. 3948, 2012.

[16] F. Traversi, V. Russo, and R. Sordan, "Integrated Complementary Graphene Inverter," *Applied Physics Letters*, vol. 94, no. 22, pp. 223 312–223 312–3, June 2009.

[17] H. Wang, D. Nezich, J. Kong, and T. Palacios, "Graphene Frequency Multipliers," *IEEE Electron Device Letters*, vol. 30, no. 5, pp. 547–549, May 2009.

[18] H. Wang, A. Hsu, K. K. Kim, J. Kong, and T. Palacios, "Gigahertz Ambipolar Frequency Multiplier Based on CVD Graphene," in *Proceedings of the IEEE International Electron Devices Meeting*, 2010, pp. 23.6.1–23.6.4.

[19] M. E. Ramon, K. N. Parrish, S. F. Chowdhury, C. W. Magnuson, H. C. P. Movva. R. S. Ruoff, S. K. Banerjee, and D. Akinwande, "Three-Gigahertz Graphene Frequency Doubler on Quartz Operating Beyond the Transit Frequency," *IEEE Transactions on Nanotechnology*, vol. 11, no. 5, pp. 877 –883, September 2012.

[20] H. Wang, A. Hsu, J. Wu, J. Kong, and T. Palacios, "Graphene-Based Ambipolar RF Mixers," *IEEE Electron Device Letters*, vol. 31, no. 9, pp. 906–908, September 2010.

[21] X. Yang, G. Liu, A. A. Balandin, and K. Mohanram, "Triple-Mode Single-Transistor Graphene Amplifier and Its Applications," *ACS Nano*, vol. 4, no. 10, pp. 5532–5538, 2010.

[22] S.-J. Han, K. A. Jenkins, A. Valdes Garcia, A. D. Franklin, A. A. Bol, and W. Haensch, "High-Frequency Graphene Voltage Amplifier," *Nano Letters*, vol. 11, no. 9, pp. 3690–3693, 2011.

[23] T. Mueller, F. Xia, and P. Avouris, "Graphene photodetectors for high-speed optical communications," *Nature Photonics*. vol. 4, no. 5, pp. 297–301, 2010.

[24] S. Das and J. Appenzeller, "On the Importance of Bandgap Formation in Graphene for Analog Device Applications," *IEEE Transactions on Nanotechnology*, vol. 10, no. 5, pp. 1093–1098, September 2011.

[25] ——, "An All-graphene Radio Frequency Low Noise Amplifier," in *Proceedings of the IEEE Radio Frequency Integrated Circuits Symposium (RFIC)*, 2011, pp. 1–4.

[26] Y. M. Lin, A. Valdes-Garcia, S. J. Han, D. B. Farmer, I. Meric, Y. Sun, Y. Wu, C. Dimitrakopoulos, A. Grill, P. Avouris *et al.*, "Wafer-Scale Graphene Integrated Circuit," *Science*, vol. 332, no. 6035, pp. 1294–1297, 2011.

[27] B. W. Scott and J. Leburton, "Modeling of the Output and Transfer Characteristics of Graphene Field-Effect Transistors," *IEEE Transactions on Nanotechnology*, vol. 10, no. 5, pp. 1113–1119, September 2011.

[28] F. Schwierz, "Graphene transistors," *Nature Nanotechnology*, vol. 5, May 2010.

[29] J. Kedzierski, P.-L. Hsu, A. Reina, J. Kong, P. Healey, P. Wyatt, and C. Keast, "Graphene-on-Insulator Transistors Made Using C on Ni Chemical-Vapor Deposition," *IEEE Electron Device Letters*, vol. 30, no. 7, pp. 745–747, July 2009.

978-1-4799-8720-7/15 $31.00 © 2015 IEEE

2015 IEEE Computer Society Annual Symposium on VLSI

Reversible Logic Based Mapping of Quaternary Sequential Circuits Using QGFSOP Expression

Mozammel H A Khan
Department of Computer Science and Engineering
East West University
Aftabnagar, Dhaka, Bangladesh
mhakhan@ewubd.edu

Himanshu Thapliyal
Department of Electrical and Computer Engineering
University of Kentucky
Lexington, KY, USA
hthapliyal@uky.edu

Abstract—Quaternary encoded binary circuits are more compact than their binary counterpart. Although several methods for designing binary reversible sequential circuits are presented, to the best of our knowledge, no design method for quaternary reversible sequential circuits has yet been reported in the literature. In this paper, we propose a design method for quaternary sequential circuits where the present state outputs are directly fedback to the next state determination circuit and that circuit is realized using QGFSOP expression as a cascade of one-digit, M-S, Feynman, and Toffoli gates. We also develop methods for making the sequential circuit falling-edge triggered and presettable using a quaternary Fredkin gate. As design examples, we present designs for up/down counters and universal registers. As there are no previous designs of quaternary sequential circuits, the closest comparison is made with designs of two-digit quaternary counter and universal register with designs of equivalent four-bit binary counter and universal register, respectively. In comparison to the equivalent binary designs, the proposed method requires less ancilla inputs with an increase in quantum cost.

Keywords - Quaternary reversible sequential circuit, Toffoli gate realization, Fredkin gate realization, QGFSOP expression

I. INTRODUCTION

For a given binary function, a quaternary encoded realization requires half the number of primary input lines compared to a binary realization by encoding two bits into a quaternary digit. In quantum technology, interaction of more than two particles is nearly impossible to control. Therefore, the number of coupled quantum bits is a major limitation. Thus, in quantum technology, quaternary encoded realization of binary functions may be a favorable choice. Quaternary quantum circuits are realizable using ion-trap quantum technology [1]. The multiple-valued controlled gates proposed in [1] are commonly known as Muthukrishnan-Stroud (M-S) gates.

A number of customized designs of medium scale quaternary reversible combinational circuits are reported in [2-6]. A generalized method of designing arbitrary quaternary reversible combinational circuits using quaternary Galios field sum of products (QGFSOP) expression is reported in [7, 8]. All these designs are based on elementary quantum one-digit and M-S gates.

Though several design methods of binary reversible sequential circuits are reported in literature [9-13], as far as we know, no design method for quaternary reversible sequential circuits has yet been reported in the literature. In

this paper, we propose a design method for quaternary reversible sequential circuits, where the present state outputs are directly fedback to the next state determination circuit and that circuit is realized using QGFSOP expression as a cascade of one-digit, M-S, Feynman, and Toffoli gates. We also develop methods for making the sequential circuit falling-edge triggered and presettable using a quaternary Fredkin gate. For this purpose, we propose a new realization method for multiple-input Toffoli gates using one-digit and M-S gates, which requires up to 95.22% less quantum costs (number of elementary quantum gates) than that in [14] for up to $n = 10$ inputs and requires only $(n-3)$ zero-initialized ancilla inputs (working inputs other than the function inputs). For the first time in the literature, we propose realization of Fredkin gate using one-digit and M-S gates, which requires quantum cost of 18.

As design examples, we present designs for up/down counters and universal registers. As there are no previous designs of quaternary sequential circuits, the closest comparison is made with designs of two-digit quaternary counter and universal register with designs of equivalent four-bit binary counter and universal register in [13], respectively. The comparison shows that for both cases, quaternary designs require more quantum cost. Although the quaternary counter design requires slightly more ancilla inputs than the binary counter design, the quaternary register design requires nearly half the ancilla inputs than the binary register design.

The rest of the paper is organized as follows: In section II, we introduce the concept of QGFSOP expressions. We propose realizations for multiple-controlled Toffoli gates and Fredkin gate using one-digit and M-S gates in section III. In section IV, we discuss the reversible logic based mapping of quaternary sequential circuit using QGFSOP expression. Finally, we conclude the paper in section V.

II. BACKGROUND ON QGFSOP EXPRESSIONS

In the proposed mapping of quaternary sequential circuit, the next state is generated as a combinational function of the directly fedback present state output and the external input using QGFSOP expressions. Therefore, understanding of QGFSOP expression is very important. In this section, we briefly introduce the background of QGFSOP expression. Readers are referred to [7, 8] for more details.

The basis of a QGFSOP expression is the GF(4) arithmetic, which has a set of values $Q = \{0, 1, 2, 3\}$ and two binary operations: addition (denoted by $\oplus$) and

978-1-4799-8720-7/15 $31.00 © 2015 IEEE   297

multiplication (denoted by · or juxtaposition). The GF(4) operations are shown in Table I. These operations are both commutative and associative. The multiplication operation is distributive over the addition operation.

TABLE I.   GF(4) Addition (⊕) and Multiplication (·) Tables

⊕	0	1	2	3		·	0	1	2	3
0	0	1	2	3		0	0	0	0	0
1	1	0	3	2		1	0	1	2	3
2	2	3	0	1		2	0	2	3	1
3	3	2	1	0		3	0	3	1	2

There are 4! = 24 possible permutations of 0, 1, 2, and 3. Therefore, there are 24 possible permutative unitary transforms of a quaternary logic variable. These transforms are shown using truth tables in Table II. Each of the 24 transforms of Table II changes the value of a variable in reversible manner. Thus, we can define 24 reversible literals of a variable corresponding to 24 transforms. These literals are normally represented as x^z or $[z]x$, where x is the variable and $[z]$ is the transform. For example, x^{+1} or $[+1]x$ is the literal of the variable x corresponding to the transform $[+1]$. In this paper, when the context explicitly identifies the variable, we will represent a literal by its output vector. For example, the literal x^{+1} is represented by its output vector $[1032]$.

TABLE II.   Quaternary Permutative Unitary Transforms

Transform [z]	Input 0 1 2 3	Transform [z]	Input 0 1 2 3	Transform [z]	Input 0 1 2 3
[+0]	0 1 2 3	[13]	0 3 2 1	[123]	0 2 3 1
[+1]	1 0 3 2	[23]	0 1 3 2	[132]	0 3 1 2
[+2]	2 3 0 1	[012]	1 2 0 3	[0123]	1 2 3 0
[+3]	3 2 1 0	[013]	1 3 2 0	[0132]	1 3 0 2
[01]	1 0 2 3	[021]	2 0 1 3	[0213]	2 3 1 0
[02]	2 1 0 3	[023]	2 1 3 0	[0231]	2 0 3 1
[03]	3 1 2 0	[031]	3 0 2 1	[0312]	3 2 0 1
[12]	0 2 1 3	[032]	3 1 0 2	[0321]	3 0 1 2

A quaternary Galios field (QGF) product is defined to be a GF(4) product of two or more literals of one or more variables. For example, $x^{+3}x^{0213}y^{23}$ is a QGF product of three literals x^{+3}, x^{0213}, and y^{23} of two variables x and y. We define QGF product of two or three literals of the same variable as composite literals. For example, $x^{+1}x^{+2}x^{+3}$ is a composite literal. We also call the sum of the form $(1 \oplus$ composite-literal) as composite literal. For example, $(1 \oplus x^{+2}x^{23})$ is a composite literal. All the composite literals are irreversible in nature. We also represent a composite literal by its output vector. For example, the composite literal $x^{+1}x^{+2}x^{+3}$ is represented by its output vector $[1000]$ and the composite literal $(1 \oplus x^{+2}x^{23})$ is represented by its output vector $[1213]$. A missing variable is represented as $[----]$. We represent a QGF product using output vectors of the associated literals. For example, if $x^{+3}z^{23}$ is a QGF product of three variables x, y, and z, where the variable y is missing, then it is represented as $[3210][----][0132]$.

A QGFSOP expression is defined to be a GF(4) sum of two or more QGF products. For example, $F(x, y) = x^{+3}x^{0213}y^{23}$

$\oplus$ $xx^{+2}x^{+3}y^{+3}y^{0213}$ $\oplus$ y^{13} is a QGFSOP expression. This QGFSOP expression is represented using output vectors of the associated literals as follows:

$[1110][0132]$

$[0100][1110]$

$[----][0321]$

III.   Quaternary Reversible Gates

In the proposed mapping technique of quaternary reversible sequential circuit, the QGFSOP expression representing the next state function will be realized as a cascade of one-digit, M-S, Feynman, and Toffoli gates. Moreover, the sequential circuit will be made falling-edge triggered and presettable using Fredkin gate. Therefore, understanding of these gates is very important. In this section, we will introduce these gates.

Each of the transforms of Table II can be realized as a one-digit gate [1]. The symbol of this gate is shown in Fig. 1(a), where z is any of the transforms. The quaternary M-S gate proposed in [1] is a controlled gate, whose symbol is shown in Fig. 1(b). The input x is the control input and appears at the output as $P = x$. The input y is the target input. When $x = 3$, the target output is $Q = [z]y$, otherwise $Q = y$. The one-digit and M-S gates are the elementary quantum gates and there quantum cost is assumed to be one.

The symbol of the quaternary Feynman gate is shown in Fig. 2(a). The input x is the control input and appears at the output as $P = x$. The input y is the target input. The target output $Q = x \oplus y$. The Feynman gate is a macro-level gate and is realized using one-digit and M-S gates. Realization of the Feynman gate is reproduced in Fig. 2(b) from [14], whose quantum cost is six.

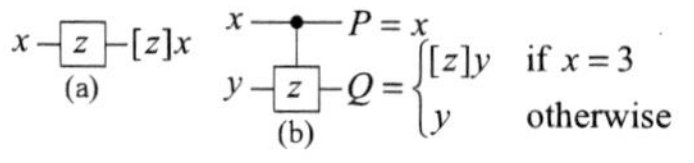

Figure 1.   Symbol of (a) one-digit gate and (b) M-S gate.

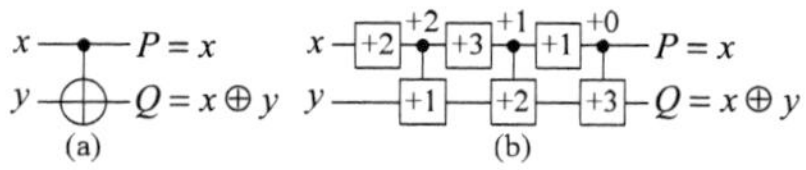

Figure 2.   (a) Symbol and (b) realization of the Feynman gate.

For making the realizations of Toffoli and Fredkin gates modular, we introduce the concept of quaternary controlled Feynman gate, whose symbol is shown in Fig. 3(a). To avoid conflict with the symbol of Toffoli gate, the controlling point of the input x is represented by a ×. Outputs $P = x$ and $Q = y$. When $x \neq 3$ (that is, $xx^{+1}x^{+2} = 0$), then the target output $R = z$ making the gate an identity gate. When $x = 3$ (that is, $xx^{+1}x^{+2} = 1$), then the target output $R = y \oplus z$ making the gate Feynman gate controlled by the controlling input x. If the three one-digit gates of Fig. 2(b) are replaced by their corresponding M-S gates, then the realization will become controlled Feynman gate as shown in Fig. 3(b). If $x \neq 3$, then the three upper M-S gates are inactive and $Q = y$. In this case, if $y \neq 3$, then the three bottom M-S gates are inactive and $R = z$. If $y = 3$, then all three bottom M-S gates are active and $R = 1 \oplus 2 \oplus 3 \oplus z = 0 \oplus z = z$. Thus, if $x \neq 3$, then $P = x$, $Q = y$, and $R = z$ making the gate an identity gate. If $x = 3$, then all the three upper M-S gates are active and the circuit becomes equivalent to that of Fig. 2(b) making it a controlled Feynman gate. The quantum cost of the controlled Feynman gate is six.

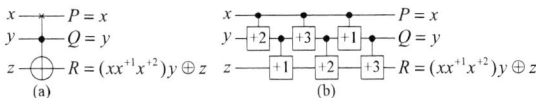

Figure 3. (a) Symbol and (b) realization of the controlled Feynman gate.

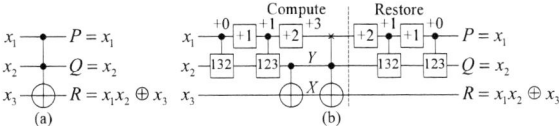

Figure 4. (a) Symbol and (b) realization of the three-input Toffoli gate.

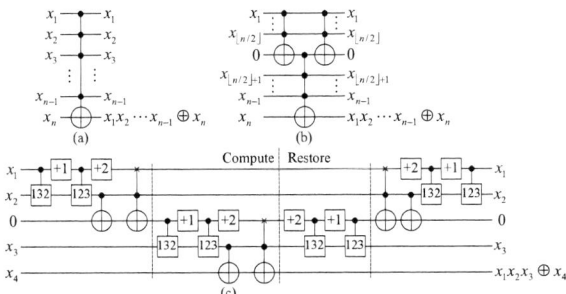

Figure 5. (a) Symbol and (b) architecture for realization of an n-input Toffoli gate. (c) Realization of a four-input Toffoli gate.

The symbol of quaternary three-input Toffoli gate is shown in Fig. 4(a). The inputs x_1 and x_2 are control inputs and appear at the output as $P = x_1$ and $Q = x_2$, respectively. The input x_3 is the target input. The target output $R = x_1x_2 \oplus x_3$. The Toffoli gate is a macro-level gate and is realized using one-digit and M-S gates. We propose a new realization of the three-input Toffoli gate in Fig. 4(b) that requires quantum cost of 20, which is less than that of [14]. When $x_1 = 0$, both the M-S gates of the compute part are inactive and $Y = x_2$, $X = Y \oplus x_3 = x_2 \oplus x_3$. The controlled Feynman gate is active and $R = Y \oplus X = x_2 \oplus x_2 \oplus x_3 = 0 \oplus x_3 = 0x_2 \oplus x_3 = x_1x_2 \oplus x_3$. When $x_1 = 1$, both the M-S gates of the compute part are inactive and $Y = x_2$, $X = Y \oplus x_3 = x_2 \oplus x_3$. The controlled Feynman gate is inactive and $R = X = x_2 \oplus x_3 = 1x_2 \oplus x_3 = x_1x_2 \oplus x_3$. When $x_1 = 2$, the second M-S gate of the compute part is active and $Y = [123]x_2 = 2x_2$, $X = Y \oplus x_3 = 2x_2 \oplus x_3$. The controlled Feynman gate is inactive and $R = X = 2x_2 \oplus x_3 = x_1x_2 \oplus x_3$. When $x_1 = 3$, the first M-S gate of the compute part is active and $Y = [132]x_2 = 3x_2$, $X = Y \oplus x_3 = 3x_2 \oplus x_3$. The controlled Feynman gate is inactive and $R = X = 3x_2 \oplus x_3 = x_1x_2 \oplus x_3$. Thus, for all values of x_1, the compute part computes $R = x_1x_2 \oplus x_3$. The restore part restores the values of x_1 and x_2 at the output. The last transform along x_1 is +0, so $P = [+0]x_1 = x_1$. When $x_1 = 0$ or

$x_1 = 1$, then none of the M-S gates of the compute part and the restore part is active, so $Q = x_2$. When $x_1 = 2$, the second M-S gate of the compute part with transform [123] and the first M-S gate of the restore part with transform [132] are active. The transforms [123] and [132] are inverse of each other and $Q = [132][123]x_2 = [132]2x_2 = 3 \cdot 2x_2 = x_2$. When $x_1 = 3$, the first M-S gate of the compute part with transform [132] and the second M-S gate of the restore part with transform [123] are active and $Q = [123][132]x_2 = [123]3x_2 = 2 \cdot 3x_2 = x_2$. Thus, for all values of x_1, $Q = x_2$. Therefore, the circuit in Fig. 4(b) realizes a three-input Toffoli gate. The quantum cost of the compute part is 16 and that of the restore part is 4 making a total of quantum cost as 20.

A Toffoli gate may have more than three inputs. The symbol of an n-input ($n > 3$) Toffoli gate is shown in Fig. 5(a). Realization of the n-input Toffoli gate using two $(\lfloor n/2 \rfloor + 1)$-input Toffoli gates, one $(\lceil n/2 \rceil + 1)$-input Toffoli gate, and one zero-initialized ancilla input is shown in Fig. 5(b). The first $(\lfloor n/2 \rfloor + 1)$-input Toffoli gate produces $x_1x_2...x_{\lfloor n/2 \rfloor}$ along the ancilla input. The second $(\lfloor n/2 \rfloor + 1)$-input Toffoli gate restores the ancilla input. The $(\lceil n/2 \rceil + 1)$-input Toffoli gate produces $x_1x_2...x_{n-1} \oplus x_n$ at the target output. The first $(\lfloor n/2 \rfloor + 1)$-input Toffoli gate is realized using the compute part of the Toffoli gate and the second $(\lfloor n/2 \rfloor + 1)$-input Toffoli gate is realized using the compute part of the Toffoli gate by reversing the gate sequence and replacing each gate by its inverse gate, so that the inputs are restored at the outputs. The one-digit gates with transforms [+1], [+2], and [+3] are self-inverse gates. The Feynman and the controlled Feynman gates are also self-inverse gates. The M-S gates with transforms [123] and [132] are inverse of each other. The $(\lceil n/2 \rceil + 1)$-input Toffoli gate is realized using both compute and restore parts of the Toffoli gate. The quantum cost of an n-input Toffoli gate is $16(2n - 5) + 4$ and it requires $n - 3$ ancilla inputs. As an example of realizing an n-input Toffoli gate using the method of Fig. 5(b), the realization of a four-input Toffoli gate is shown in Fig. 5(c). The quantum cost and the number of ancilla inputs for up to $n = 10$ are computed and tabulated in Table III. The quantum cost is compared with that in [14]. We find that the present realization technique reduces the quantum cost in the range of 21.21% to 95.22% with the use of only $n - 3$ ancilla inputs.

The symbol of the quaternary Fredkin gate is shown in Fig. 6(a). The input x is the control input. The inputs y and z are the target inputs. The outputs are $P = x$, $Q = x^{+3}x^{0213}y \oplus xx^{+1}x^{+2}z$, and $R = xx^{+1}x^{+2}y \oplus x^{+3}x^{0213}z$. If the control input $x \neq 3$, then $xx^{+1}x^{+2} = 0$ and $x^{+3}x^{0213} = 1$. In this case the target outputs are $Q = 1y \oplus 0z = y$ and $R = 0y \oplus 1z = z$. Thus the target inputs y and z are passed directly to their outputs. If $x = 3$, then $xx^{+1}x^{+2} = 1$ and $x^{+3}x^{0213} = 0$. In this case the target outputs are $Q = 0y \oplus 1z = z$ and $R = 1y \oplus 0z = y$. Thus target inputs y and z are swapped at the outputs. Realization of the Fredkin gate is shown in Fig. 6(b). If $x \neq 3$, then all three controlled Feynman gates will act as identity gates and the outputs will be $P = x$, $Q = y$, and $R = z$. If $x = 3$, then all three controlled Feynman gates will be active and the outputs

978-1-4799-8720-7/15 $31.00 © 2015 IEEE 299

will be $P = x$, $Q = y \oplus (y \oplus z) = z$, and $R = z \oplus (y \oplus z) = y$. The quantum cost of this realization is 18.

TABLE III. QUANTUM COST (QC) AND NUMBER OF ANCILLA INPUTS (AI) FOR REALIZATION OF MULTIPLE-INPUT TOFFOLI GATES

n	QC [14]	QC (Present)	AI (Present)	% QC Improvement Over [14]
3	26	20	0	23.08
4	66	52	1	21.21
5	146	84	2	42.47
6	306	116	3	62.09
7	626	148	4	76.36
8	1266	180	5	85.78
9	2546	212	6	91.67
10	5106	244	7	95.22

If a composite literal has two or more a, where $a \in \{0, 1, 2, 3\}$, in the output vector, then the composite literal is realized along a a-initialized ancilla input using one-digit and M-S gates. For example, the realization of composite literal $(+0,23) = [0111] = 1 \oplus [1000]$ is shown in Fig. 7(a), which requires three quantum cost. Similarly, the realization of $(1\oplus+0,01) = [1123] = 1 \oplus [0032]$ is shown in Fig. 7(b), which requires four quantum cost.

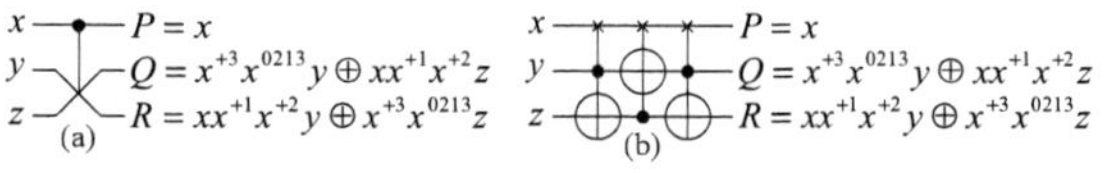

Figure 6. (a) Symbol and (b) realization of the Fredkin gate.

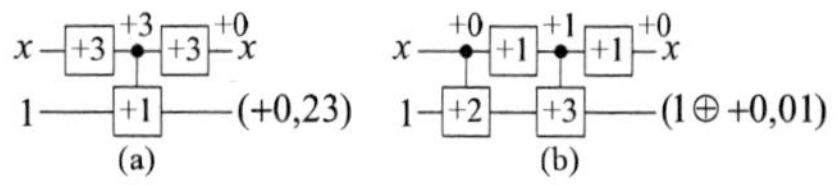

Figure 7. Realization of composite literals (a) $(+0,23)$ and (b) $(1\oplus+0,01)$.

IV. REVERSIBLE MAPPING OF SEQUENTIAL CIRCUITS

Sequential circuit has memory called state of the circuit. The next state and the output of the sequential circuit are determined as combinational functions of the present state and the external input. The state is changed when the clock is applied to the circuit.

No method has yet been reported in the literature for designing quaternary reversible sequential circuits. Here we propose a special technique of designing quaternary reversible sequential circuits, where the present state is directly fedback to generate the next state. Let Q and Q^+ be the present state and the next state, respectively. Then Q^+ can be expressed as $Q^+ = Q \oplus Q \oplus Q^+ = Q \oplus Q^*$, where $Q^* = Q \oplus Q^+$ is the modified next state. Now, Q^* is determined as the combinational function of the present state and the external input. Let C be the clock. Then the next state is realized in the next state determination circuit as $Q^+ = Q \oplus CC^{+1}C^{+2}Q^*$. When $C \neq 3$ (i. e., $CC^{+1}C^{+2} = 0$), then $Q^+ = Q$, which provides the feedback of the present state through the next state determination circuit to maintain the present state. When $C = 3$ (i. e. $CC^{+1}C^{+2} = 1$), then $Q^+ = Q \oplus Q^*$, which generates the next state in the next state determination circuit. This generated next state is then passed to the present

state output through a Fredkin gate when C goes from $C = 3$ to $C \neq 3$ (referred to as falling-edge), which then becomes the present state in the next clock. Thus, the circuit becomes a falling-edge triggered sequential circuit. The present state can also be asynchronously loaded using another Fredkin gate.

The proposed method of reversible mapping of sequential circuit is illustrated using a two-digit up/down counter. The state outputs are $Q1$ and $Q0$. The mode selector input M determines the direction of count. When $M \neq 3$, then it counts up. When $M = 3$, then it counts down. The output vector for the next state $Q0^+(M, Q1, Q0)$ is

[1230 1230 1230 1230 1230 1230 1230 1230 1230 1230 1230 1230 3012 3012 3012 3012]

Now, the output vector for the modified next state $Q0^*(M, Q1, Q0) = Q0(M, Q1, Q0) \oplus Q0^+(M, Q1, Q0)$ is

[1313 1313 1313 1313 1313 1313 1313 1313 1313 1313 1313 1313 3131 3131 3131 3131]

Now, the modified next state $Q0^*(M, Q1, Q0)$ is minimized as QGFSOP expression using the technique of [8], but the QGFSOP expression is represented using the notation adopted in this paper. The resulting minimized QGFSOP expressions is

$Q0^*(M, Q1, Q0) =$
[- - - -][- - - -][3131]
[2220][- - - -][- - - -]

Similarly, the resulting QGFSOP expression for the modified next state $Q1^*(M, Q1, Q0)$ is

$Q1^*(M, Q1, Q0) =$
[- - - -][3131][1000]
[1110][1010][3001]
[1110][0101][1003]

Now, the QGFSOP expressions for the next state equations are determined using the formula $Q^+ = Q \oplus CC^{+1}C^{+2}Q^*$. The minimized QGFSOP expressions for the next state equations of the two-digit up/down counter are

$Q0^+(C, M, Q1, Q0) =$
[- - - -][- - - -][- - - -][0123]
[0001][- - - -][- - - -][3131]
[0001][2220][- - - -][- - - -]

$Q1^+(C, M, Q1, Q0) =$
[- - - -][- - - -][0123][- - - -]
[0001][- - - -][3131][1000]
[0001][1110][1010][3001]
[0001][1110][0101][1003]

A reversible mapping of the two-digit up/down counter from the above QGFSOP expressions is shown in Fig. 8. Using the QGFSOP expressions of $Q1^+$ and $Q0^+$, we have

978-1-4799-8720-7/15 $31.00 © 2015 IEEE

realized the "Next state determination" part in Fig. 8. In the "Feedback" part in Fig. 8, the two present states are copied using two Feynman gates and fedback to the "Next state determination" part of the circuit. In the "Load" part in Fig. 8, a mechanism has been provided for asynchronous parallel loading of the output states. When $L = 3$, the $D1$ and $D0$ data inputs will be loaded to the state output $Q1$ and $Q0$, respectively, through the Fredkin gates and they will become the present state. Then these newly loaded output states will be fed-back to the "Next state determination" part by the "Feedback" part of the circuit. As discussed latter, the "Next state determination" part determines the next state when the clock $C = 3$. Thus the output state loading should be made asynchronously only when $C \neq 3$ so that the loaded states are fed-back before the next state determination starts. When $L \neq 3$ then the Fredkin gates will pass the two state values provided by the "Edge trigger" part to maintain the output states unchanged. The QGFSOP expressions of next states $Q1^+$ and $Q0^+$ are realized in the "Next state determination" part in Fig. 8 as function of the fed-back present states $Q1$ and $Q0$, the mode selector input M, and the clock C. As discussed earlier, when $C \neq 3$ (i. e., $CC^{+1}C^{+2} = 0$), then $Q^+ = Q \oplus CC^{+1}C^{+2}Q^* = Q$. In this case, the two outputs of the "Next state determination" part will simply be equal to the two fed-back present state values. When $C = 3$ (i. e. $CC^{+1}C^{+2} = 1$), then $Q^+ = Q \oplus CC^{+1}C^{+2}Q^* = Q \oplus Q^*$. In this case, the "Next state determination" part will determine the two next states. In the "Edge trigger" part of Fig. 8, a mechanism is provided for falling edge triggering of the circuit. When $C = 3$ then the Fredkin gate will provide the feedback to maintain the state output unchanged. At the same time, the "Next state determination" part will determine the next state. After that when the clock goes from $C = 3$ to $C \neq 3$ (falling-edge of the clock) then the Fredkin gates will pass the outputs of the "Next state determination" part to the state outputs through the "Load" part provided $L \neq 3$ is maintained. Thus, the sequential circuit becomes a falling-edge triggered circuit. The quantum cost of the circuit in Fig. 8 is 379.

A post-synthesis reduction of quantum cost is possible. As shown in the dashed box in Fig. 8, the top two control lines of two five-input Toffoli gates are the same. If these two Toffoli gates are decomposed, then we see from Fig. 5(c) that the last six gates of the restore part of the first Toffoli gate are in the reverse order of the first six gates of the compute part of the second Toffoli gate with each gate replaced by its inverse gate. So, these 12 gates will produce an identity effect and they can be eliminated. The quantum cost of these 12 gates is 32. Thus, the final quantum cost of Fig. 8 is 379 - 32 = 347. The circuit in Fig. 8 needs 14 ancilla inputs. In addition, the five-input Toffoli gate will require two more ancilla inputs. Thus, the total ancilla inputs are 14 + 2 = 16.

We have mapped a two-digit universal register as shown in Fig. 9. The state outputs are $Q1$ and $Q0$. The "Load" part of the circuit provides asynchronous load facility for parallel input. When $L = 3$ then the parallel input data $D1$ and $D0$ are loaded to the state outputs $Q1$ and $Q0$, respectively. Parallel outputs are taken from state outputs $Q1$ and $Q0$. The mode selector input M determines the direction of shift. When $M \neq$

3 then the circuit shifts the data right. In this case, the serial input is D_R and the serial output is $Q0$. When $M = 3$ then the circuit shifts the data left. In this case, the serial input is D_L and the serial output is $Q1$. Thus, the circuit is a two-digit universal register. The QGFSOP expressions for the next state equations are

$$Q0^+(C, M, D_R, D_L, Q1, Q0)=$$
$$[- - - -][- - - -][- - - -][- - - -][- - - -][0123]$$
$$[0001][- - - -][- - - -][- - - -][- - - -][0123]$$
$$[0001][0001][- - - -][0123][- - - -][- - - -]$$
$$[0001][1110][- - - -][- - - -][0123][- - - -]$$

$$Q1^+(C, M, D_R, D_L, Q1, Q0)=$$
$$[- - - -][- - - -][- - - -][- - - -][0123][- - - -]$$
$$[0001][- - - -][- - - -][- - - -][0123][- - - -]$$
$$[0001][0001][- - - -][- - - -][- - - -][0123]$$
$$[0001][1110][0123][- - - -][- - - -][- - - -]$$

Based on the above next state equations, a reversible mapping of a two-digit universal register is shown in Fig. 9. The quantum cost of the circuit in Fig. 9 is 359. After post-synthesis cost reduction the quantum cost becomes 295. The circuit has nine ancilla inputs. The four-input Toffoli gate realization requires one ancilla input. Thus, the total ancilla inputs are 9 + 1 = 10.

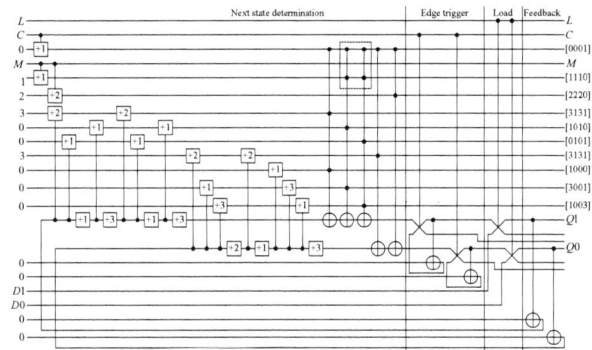

Figure 8. Reversible realization of a two-digit up/down counter with falling-edge triggering and parallel load.

The quantum costs and ancilla inputs for up/down counters and universal registers are calculated and summarized in Table IV. The designs of two-digit up/down counter and universal register are compared with their equivalent four-bit up/down counter and universal register from [13] in Table V. The quaternary designs require up to four-fold larger quantum cost than equivalent binary designs. However, quaternary universal register design requires almost half the ancilla inputs than equivalent binary design.

V. CONCLUSION

A quaternary encoded binary logic circuit requires 50% fewer primary inputs than its binary counterpart. Quaternary reversible circuits are realizable using ion-trap quantum

978-1-4799-8720-7/15 $31.00 © 2015 IEEE

technology [1]. Therefore, efficient design methods of quaternary circuits are of great interest to researchers.

Though several designs of customized reversible quaternary combinational circuits [2-6] and a general technique of synthesizing reversible arbitrary quaternary combinational circuits [7, 8] are reported in the literature, to the best of our knowledge, no general design method of reversible quaternary sequential circuits has been reported yet. For the first time, we have proposed a QGFSOP-based design technique for a quaternary reversible falling-edge triggered sequential circuit. The QGFSOP expressions are minimized using the method of [8]. The QGFSOP expressions are realized as a cascade of one-digit, M-S, Feynman, and Toffoli gates. For this purpose, we have proposed a new realization technique of multiple-input Toffoli gates using one-digit and M-S gates. The proposed Toffoli gate realizations are by far the best and requires up to 95.22% less quantum costs than that in [14] for up to $n = 10$ inputs and requires only $(n - 3)$ zero-initialized ancilla inputs. A post-synthesis technique is used to reduce the quantum costs of the circuits by decomposing Feynman and Toffoli gates into one-digit and M-S gates. The design method also requires use of quaternary Fredkin gate. For this purpose, for the first time in the literature, we have proposed realizations of Fredkin gates using one-digit and M-S gates.

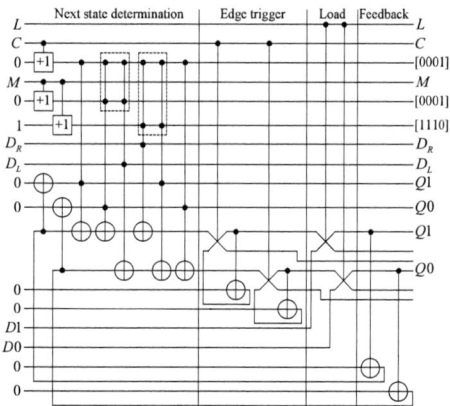

Figure 9.   Reversible realization of two-digit universal register with falling-edge triggering.

TABLE IV.   EXPERIMENTAL RESULTS OF REVERSIBLE SEQUENTIAL CIRCUIT DESIGN

Circuit	Initial Cost	Final Cost	Ancilla
Two-digit up/down counter	379	347	16
Three-digit up/down counter	752	751	24
Four-digit up/down counter	1217	1216	30
Two-digit universal register	359	295	10
Three-digit universal register	537	535	13
Four-digit universal register	715	711	16

The quantum costs and ancilla inputs of several sequential circuits are reported. In comparison to the equivalent binary designs in [13], this method requires more quantum costs but less ancilla inputs. Therefore, the proposed reversible mapping technique of quaternary sequential circuits is suitable for the quantum technology where the number of quantum digits is a major limitation.

TABLE V.   COMPARISON OF QUATERNARY SEQUENTIAL CIRCUIT DESIGNS WITH EQUIVALENT BINARY DESIGNS IN [13]

Quaternary Designs			Binary Designs in [13]		
Circuit	Cost	Ancilla	Circuit	Cost	Ancilla
Two-digit up/down counter	347	16	Four-bit up/down counter	85	14
Two-digit universal register	295	10	Four-bit universal register	112	17

REFERENCES

[1] A. Muthukrishnan and C. R. Stroud Jr., "Multivalued logic gates for quantum computation," *Phy. Rev. A*, vol. 62, no. 5, pp. 052309/1-8, 2000.

[2] M. H. A. Khan, "Reversible realization of quaternary decoder, multiplexer, and demultiplexer circuits", *Engineering Letters*, vol. 15, no. 2, pp. 203-207, 2007.

[3] M. H. A. Khan, "A recursive method for synthesizing quantum/reversible quaternary parallel adder/subtractor with look-ahead carry", *J. Systems Architecture*, vol. 54, no. 12, pp. 1113-1121, Dec. 2008.

[4] M. H. A. Khan, "Synthesis of quaternary reversible/quantum comparators", *J. Systems Architecture*, vol. 54, no. 10, pp. 977-982, Oct. 2008.

[5] M. M. Khan, A. K. Biswas, S. Chowdhury, M. Tanzid, K. M. Mohsin, M. Hasan, and A. I. Kahn, "Quantum realization of some quaternary circuits," Proc. *2008 IEEE Region 10 Conference*, 2008, pp. 1-5.

[6] S. B. Mandal, A. Chakrabarti, and S. Sur-Kolay, "A synthesis method for quaternary quantum logic circuits," Proc. *Progress in VLSI Design and Test*, LNCS 7373, 2012, pp. 270-280.

[7] M. H. A. Khan and M. A. Perkowski, "GF(4) based synthesis of quaternary reversible/quantum logic circuits", *J. Multiple-Valued Logic Soft Computing*, vol. 13, pp. 583-603, 2007.

[8] M. H. A. Khan, N. K. Siddika, and M. A. Perkowski, "Minimization of quaternary Galois field sum of products expression for multi-output quaternary logic function using quaternary Galois field decision diagram", Proc. *IEEE Intl. Symp. Multiple-Valued Logic*, 2008, pp. 125-130.

[9] M.-L. Chuang and C.-Y. Wang, "Synthesis of reversible sequential elements," *ACM J. Emerging Tech.*, vol. 3, no. 4, pp. 1-19, 2008.

[10] H. Thapliyal and N. Ranganathan, "Design of reversible sequential circuits optimizing quantum cost, delay and garbage outputs," *ACM J. Emerg. Technol. Comput. Syst.*, vol. 6, no. 4, pp. 14:1–14:35, Dec. 2010.

[11] H. Thapliyal and N. Ranganathan, "Reversible logic-based concurrently testable latches for molecular QCA," *IEEE Trans. Nanotechnol.*, vol. 9, no. 1, pp. 62–69, Jan. 2010.

[12] H. Thapliyal, N. Ranganathan, and S. Kotiyal, "Design of testable reversible sequential circuits," *IEEE Trans. VLSI*, vol. 21, no. 7, pp. 1201 - 1209, June 2013.

[13] M. H. A. Khan, "Design of reversible synchronous sequential circuits using pseudo Reed-Muller expressions," IEEE Trans. VLSI, vol. 22, no. 11, pp. 2278-2286, 2014.

[14] M. H. A. Khan, "Quantum realization of multiple-valued Feynman and Toffoli gates without ancilla input," Proc. *IEEE Intl. Symp. Multiple-Valued Logic*, 2009, pp. 103-108.

978-1-4799-8720-7/15 $31.00 © 2015 IEEE

2015 IEEE Computer Society Annual Symposium on VLSI

Comparing Energy, Area, Delay Tradeoffs in Going Vertical with CMOS and Asymmetric HTFETs

Moon Seok Kim[†], William Cane-Wissing[‡], Jack Sampson[†], Suman Datta[‡], Vijaykrishnan Narayanan[†], Sumeet K. Gupta[‡]

Dept. Computer Science & Engineering[†] and Dept. Electrical Engineering[‡]

The Pennsylvania State University, University Park, PA U.S.A. 16802

mqk5211@cse.psu.edu, wgc5032@psu.edu, sampson@cse.psu.edu, sdatta@engr.psu.edu, vijay@cse.psu.edu, skg157@engr.psu.edu

Abstract—**Vertical transistors are one of the promising alternatives to standard lateral device structures in future technologies due to benefits in terms of reduced footprint and feasibility of fabrication of heterojunction structures. While such device-level benefits have been widely explored, the circuit and layout-level implications of vertical transistors require further analysis. In this work, we carry out a systematic layout and circuit analysis for 20nm vertical transistors, namely symmetrical vertical MOSFET and asymmetrical heterojunction tunnel FET (HTFET), and present a detailed comparison with 20nm FinFETs.**

Our analysis clearly outlines the differences from the perspective of layouts and the performance/power of standard cells. The absence of width quantization in vertical FETs and steep switching characteristics in HTFETs result in larger drive strengths compared to FinFETs. However, for high fan-in cells, vertical transistors show area overheads due to infeasibility of contact sharing in parallel and series transistors. For each type of device, we synthesized a 32-bit carry look ahead adder and compared energy, delay and area, taking into account layout differences due to the device structures. Our analysis shows that in spite of area overhead for some cells, high drive-strength in HTFET cells brings advantages in both area and energy over both FinFETs and vertical MOSFETs at $V_{DD} < 0.6$V.

Keywords—*Vertical FETs; Standard cell library; cell design; symmetric vertical MOSFETs; tunnel FET; asymmetric vertical TFET; FinFET; Layout; Area; Energy; Delay; trade-off.*

I. INTRODUCTION

As the transistor scaling continues in sub-20nm technology nodes, standard cell library design is encountering increasing challenges due to issues such as physical fabrication and layout restrictions, increased leakage current, rise in parasitic resistance and capacitance, and increased susceptibility of circuits to process variations [1-4]. In order to counter such issues, alternate transistor architectures are being explored, amongst which vertical FETs show an immense promise. A vertical transistor structure has the advantages of (a) a reduced device footprint and (b) the feasibility of the implementation of heterojunction tunnel FETs (HTFETs), which have emerged as promising candidates for low power applications. Moreover, unlike FinFETs, vertically fabricated FETs do not suffer from width quantization, leading to an enhanced design flexibility [5, 6]. While vertical FETs show a lot of benefits at the device level, their advantages at the circuit level are not entirely clear. The reason is that with a stacked source and drain regions in vertical FETs, several layout restrictions such as infeasibility of sharing of some contacts [7] may lead to larger layout area. In order to understand the benefits of the vertical FETs at the circuit level, there is a need to perform a systematic analysis of vertical FET based layouts and investigate the impact on circuit power and performance.

Additionally, it is well known that the HTFETs show superior performance compared to FinFETs at a low voltage (V_{DD}) due to the steep switching characteristics [8, 9, 10]. While there have been several studies defining the range of $V_{DD}s$ for which TFETs are suitable in comparison to FinFETs, no study has explored the comparison of vertical FETs with TFETs. Since vertical FETs show an immense promise for the future technologies, it is important to understand benefits and trade-offs for TFETs in comparison with vertical MOSFETs.

In this paper, we perform a detailed analysis of the power, performance and area of circuits based on FinFETs, vertical MOSFETs and HTFETs, all in 20nm technology node [10, 11]. Since a standard cell library is widely used for high-level designs including functional circuit synthesis, static timing and power analysis, and area optimization in computer-aided-designs (CAD) [1, 4], we develop a library of cells for the three technologies. A detailed layout analysis is performed by quantifying the area for vertical MOSFETs and HTFETs in comparison with FinFETs. We build a liberty-formatted standard cell library [12] by selecting the appropriate number of fins or transistor width for the pull-up and pull-down networks of the logic cells. After that, we use the lambda (λ)-based layout design rules [13] to characterize the lateral FinFET, symmetric vertical MOSFET and asymmetric HTFET logic cell layouts. All the cell layouts are designed using the same height to help with floorplanning flexibility and eventually area reduction. The standard cell library is used to synthesize a carry lookahead adder (CLA) with FinFETs, vertical symmetrical MOSFETs and asymmetrical HTFETs. We characterize and compare the power, performance and area at different V_{DD} to comprehensively evaluate benefits and trade-offs of the three devices.

In this paper, we make the following contributions: (a) We perform a detailed layout analysis for vertical symmetric, asymmetric FETs and FinFETs to investigate implications of the symmetric and asymmetric vertical device structures on the layouts of standard cells in comparison to FinFETs. Our analysis shows that vertical FETs exhibit smaller area than FinFETs in some logic cells due to a smaller device footprint. However, for cell with fan-in > 3 and cells with drive > 3, vertical FETs show area overhead. (b) We evaluate advantages and overheads of asymmetric HTFET based cells in terms of area, energy and delay in comparison with symmetrical vertical FETs and FinFETs. This evaluation reveals that HTFETs can bring significant improvements in area, energy and delay at low V_{DD} (< 0.6V). (c) We extend our study for area, energy and delay comparisons of a CLA, synthesized with our in-house standard cell libraries at various V_{DD} to scrutinize the optimal design space for each technology.

The rest of this paper is organized as follows. Section II introduces device structures of FinFETs, symmetric vertical MOSFETs and asymmetric HTFETs. Section III presents the

978-1-4799-8720-7/15 $31.00 © 2015 IEEE   303

layout analysis for a standard cells based on the three different technologies in a 7.5-track (7.5T) architecture. In Section IV, synthesis results for inverter chains and CLA comparing the energy, delay and area tradeoffs are elaborated. We conclude the paper in Section V.

II. DEVICE STRUCTURE

In this section, we introduce the structures of each of the three devices viz. FinFET, symmetric vertical MOSFET, and asymmetric HTFET and lay the groundwork for the rest of the paper.

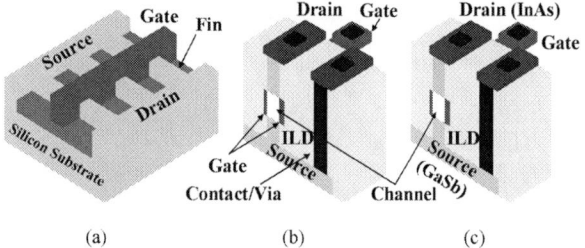

(a)   (b)   (c)

Fig. 1. Device structures of (a) FinFETs [13], (b) symmetric vertical MOSFET [6, 7, 14] and (c) asymmetric HTFETs [8, 9, 10].

A FinFET is composed of fins that form a path for current flow between the source and drain of a transistor. The gate controls the three sides of the fin along its height and fin thickness [2, 4, 13] as shown in Fig. 1(a). Because of the nature of the structure, the electrical width is quantized by the number of fins. Due to the ultra-thin body, FinFETs exhibit better electrostatic control of the channel and electrical characteristics such as low leakage and higher ON current due to reduced short-channel effects over conventional MOSFETs [2, 14].

Recently, vertically fabricated MOSFETs have been widely investigated to further overcome scaling issues [5, 6, 14]. In this type of device structure, the current flow is perpendicular to the wafer [5-7] (unlike the lateral current flow in FinFETs). Further, the gate and channel lengths are independent of lithographic patterning methods [15]. Fig. 1(b) shows the 3-dimensional (3-D) device structure of the vertical MOSFETs. The high-k/metal double-gate (DG) is between the vertically aligned source and drain regions and the gate metal is extended to form gate contacts. The source and drain contacts are formed on the top or bottom active regions. In a symmetrical vertical MOSFET, the source and drain are inter-changeable. As a result, the symmetric vertical MOSFET allows current to flow in both directions. This feature enables layouts using the symmetric vertical MOSFETs to have reduced area compared to HTFETs because transistor connections are more flexible in series (as we will show later).

Fig. 1(c) depicts the vertical device structure of asymmetric HTFETs. It is very similar to symmetric vertical MOSFETs. Like vertical MOSFETs, the gate deposition and patterning is not limited by conventional lithography [9]. The main physical difference is that HTFETs are essentially comprised of gated p-i-n tunnel diode with asymmetrical GaSb-InAs (III-V) source/drain doping [8, 9, 10]. The on and off switching is driven by the gate voltage induced band-to-band tunneling (BTBT) at the source-channel junction [8, 9]. The asymmetrical feature results in current flow in one direction, also known as "uni-directional conduction behavior". This is unlike the other two devices mentioned above, which exhibit bi-directional current. This difference has important implications in layout design.

III. LAYOUT ANALYSIS

In this section, we present the layout implications for symmetric vertical MOSFETs and asymmetric vertical TFETs, and perform a detailed comparison with lateral FinFETs. We consider several combinational and sequential logic gates in our analysis and develop a standard cell library based on the three devices. The appropriate cell-height is a critical design consideration for a standard cell library and is dependent on the design requirements with regard to the optimization of density, power and performance. For the subsequent analysis, we select a 7.5-track (7.5T) architecture, which is typically used for low-power designs [16]. The layouts are based on λ-based layout design rules, which are derived from the scalable CMOS design rules [13]. Here, λ is defined by half of the minimum feature size associated with the physical dimensions of a technology. The design rules are supplemented by the information on structural dimensions of Intel 14nm and 22nm technologies [17]. The device dimensions and parameters used in this analysis are presented in Fig. 2.

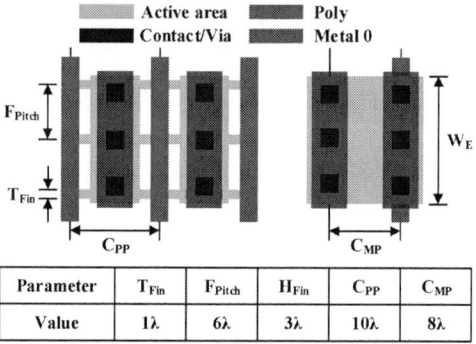

Parameter	T_{Fin}	F_{Pitch}	H_{Fin}	C_{PP}	C_{MP}
Value	1λ	6λ	3λ	10λ	8λ

T_{FIN}: The thickness of Fin, F_{Pitch}: The Fin pitch
H_{FIN}: The Height of the Fin, C_{PP}: Constant poly pitch
C_{MP}: Constant metal pitch, W_E: Electrical width across active

Fig. 2. Device dimensions and parameters for the layout representation.

A. Electrical Width Comparison

A key parameter in the design of a standard cell library is the maximum electrical width of n- and p-devices that can fit in a given cell height in a single-finger layout. Subsequently, this width will be referred to as W_{EMAX}. The significance of W_{EMAX} is that it determines the drive strength of the logic gates for a fixed area or the layout area of the logic gates for a fixed drive strength. In the 7.5T architecture, the cell height is equal to 60λ. For FinFETs, 10 fins can be formed in this cell height (e.g., $60\lambda = 10$ fins $\times$ 6λ fin pitch) as is evident from the ratio between the metal pitch and fin pitch (Fig. 1). Out of 10 fins, two fins are etched for gate contact placement in between the n- and p-devices and for internal routing. Two more fins (one each on the top and bottom cell boundary) are etched to enable abutment of cells. Hence, the total number of active fins is 6. Note that the p- and n-type FinFETs have equal strength due to the hole mobility improvement from the adoption of SiGe for the p-type transistor [2]. Hence, W_{EMAX} for FinFET-based cells in a 7.5T architecture is equivalent to N = 3 fins (Fig. 2) and is given by [18]:

$$W_{EMAX} = N \times (2 \times H_{Fin} + T_{Fin}) \qquad (1)$$

where N is the number of fins for either n-type or p-type FETs, H_{Fin} is the height of a fin, T_{Fin} is the thickness of a fin, and F_{Pitch}

978-1-4799-8720-7/15 $31.00 © 2015 IEEE   304

is the fin pitch. Therefore, in a 7.5T architecture, the maximum W_{EMAX} is 21λ for a FinFET based inverter.

For vertical FETs, W_{EMAX} is $2 \times W_E$ (see Fig. 2 for the definition of W_E) due to the double gate structure. Based on the design rules, we obtain the total width of the active region or the sum of the width of n- and p- devices to be 38λ. For bi-directional vertical FETs, we assume the same strength of n- and p- devices, as in FinFETs. Hence, $W_E = 19λ$ and $W_{EMAX} = 38λ$ for symmetrical vertical FETs. It is important to note that W_{EMAX} is significant larger compared to that of a FinFET.

In case of asymmetric HTFETs, p- and n-type transistors have different current per transistor width (unlike symmetrical vertical FETs and FinFETs) due to the difference of a tunneling probability as well as other material characteristics (e.g., electron and hole mobility) [2]. The size of each transistor is obtained by balancing rise and fall delays with the assumption of the ratio of 1.8:1 for the width of p- and n-type transistors. W_{EMAX} of a HTFET inverter (defined as the maximum width for an n-type device), is obtained as 27λ. Hence, a HTFET inverter shows larger W_{EMAX} than FinFET inverter but less W_{EMAX} than symmetric vertical MOSFETs. However, due to high current driving ability of HTFETs at low voltages [7], the smaller W_{EMAX} compared to the symmetric vertical MOSFET may not be a bottleneck for low voltage circuit operation. Also, the symmetrical vertical FET based logic cells have the larger width at a given height, leading to higher driving ability (I_{DS}) in comparison with FinFET logic cells so that the vertical FET logic cells can reduce the size of the buffer chain to meet the performance requirements while driving a large load.

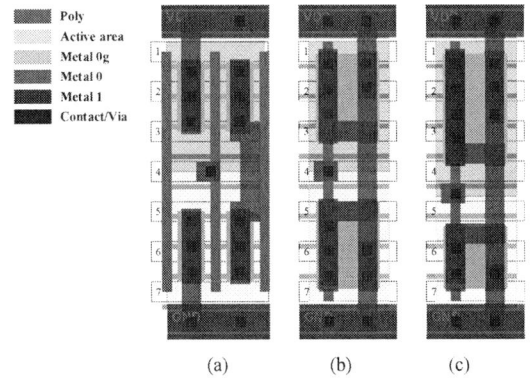

Fig. 3. The maximum electrical width (W_{EMAX}) in the single finger inverter layouts: (a) lateral FinFETs, (b) symmetric vertical FETs, (c) HTFETs.

B. Area Comparison of a single-finger inverter (INVX1)

Fig. 3 shows the layout of a single-finger inverter. The layouts exhibit difference not only in W_{EMAX} (as discussed in the previous sub-section) but also in the cell width or horizontal dimension of the cell. For FinFET-based inverters, the cell width is determined by the poly pitch or C_{PP}. However, in vertical FETs, the gate metal/poly deposition is performed around the vertical pillar which forms the source/channel/drain and does not require conventional lithography and patterning. Therefore, unlike FinFETs, the horizontal dimension of the vertical FET based inverter is determined by the metal pitch (C_{MP}). Since cell-widths of single-finger inverters based on FinFETs and vertical FETs are $2C_{PP}$ and $2C_{MP}$, respectively,

and typically $C_{MP} < C_{PP}$ [17], the layout area for vertical FETs is lower compared to FinFETs. The cell-area of a single-finger inverter is obtained as $1.2 \times 10^3 \lambda^2$ and $1.0 \times 10^3 \lambda^2$ for the FinFET inverter and vertical FET inverters, respectively. For more complex cells with larger drive or higher fan-in, the cell width is a multiple of poly pitch in case FinFETs and metal pitch in case of the vertical transistors, including HTFETs, as we will discuss later.

C. Parallel Connection

In a standard cell library, a number of transistors are connected either in series or in parallel to (a) perform NAND, NOR and other complex Boolean functions and (b) to construct the multiple fingers in order to increase the drive strength. For parallel transistor connections, the FinFETs share the same active/diffusion regions between adjacent transistors, resulting in fewer diffusion breaks and smaller area [7]. As an example, (n-1) diffusion regions will be shared by adjacent transistors if n transistors connected in parallel.

For symmetric vertical MOSFETs in parallel, the bottom diffusion region can be shared. The shared bottom source is connected through the via contact between two top diffusion regions with retaining the constant 8λ metal pitch. In case of the asymmetric vertical TFETs, it has same layout footprint with symmetric vertical MOSFETs. In [7], the author introduces the compact layout strategy for vertical FETs, which removes the contacts between top drains. However, we keep the shared source with contacts while maintaining the C_{MP} between two top drains to avoid the significant large parasitic resistance.

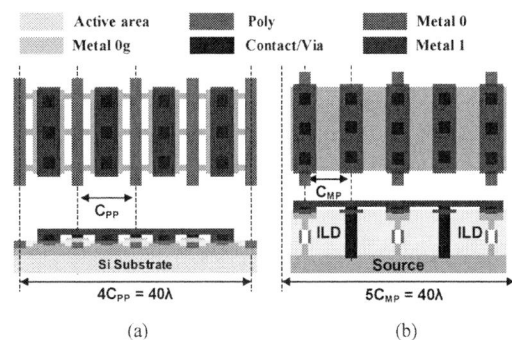

Fig. 4. Layout view of parallel connections for three transistors: The (a) lateral FinFETs and (b) Symmetric/Asymmetric FETs.

The example of three transistors connected in parallel for lateral FinFETs and both the vertical FETs is shown in Fig. 4(a). For the lateral FinFET, the cell width is equal to $(n+1) \cdot C_{PP}$ when n transistors are connected in parallel. The cell width of n parallel chains in symmetric/asymmetric vertical FETs is $(2n-1) \cdot C_{MP}$. Fig. 4 shows the cell width versus a number of parallel chains. As expected, two parallel transistor chains in the symmetric/asymmetric FETs have less width than lateral FinFETs. However, the cell width of both vertical FETs exceed that of lateral FETs as the number of parallel transistors exceed three. It is evident that the area of vertical FET based cells will be larger than lateral FET based cells when either the drive-strength or the number of inputs increases.

D. Series Connection

Unlike the transistor connections of lateral FinFETs in series, the isolation between diffusion regions is necessarily

978-1-4799-8720-7/15 $31.00 © 2015 IEEE

required for both symmetrical/asymmetrical vertical FETs. Three stacked transistors are shown, for example, in Fig. 5. In symmetric vertical MOSFETs, the bottom diffusion region is shared where forming two stacked transistors, and third stacked transistor is isolated from two stacked transistors to form the source/drain connection. Hence, it needs five metal pitches to implement the series connection while the FinFETs need four poly pitches. For asymmetric HTFETs, the isolation penalty because of the uni-directional conduction behavior is aggravated when transistors are connected in series as shown in Fig. 5. Since the C_{MP} and C_{PP} are equal to 8λ and 10λ, respectively, the cell widths in three stacked transistors are 40λ, 40λ, and 48λ for the FinFETs, symmetric vertical MOSFETs and asymmetric HTFETs. The cell widths can be simply calculated with the equations in Fig. 5.

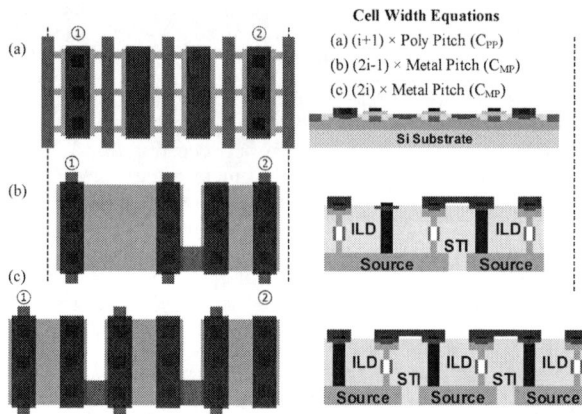

Cell Width Equations
(a) $(i+1) \times$ Poly Pitch (C_{PP})
(b) $(2i-1) \times$ Metal Pitch (C_{MP})
(c) $(2i) \times$ Metal Pitch (C_{MP})

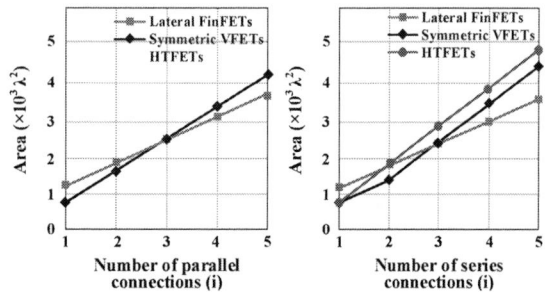

Fig. 5. Layout top- and cross-sectional views of series connections for three transistors: The (a) lateral FinFETs, (b) Symmetric vertical MOSFETs and (c) Asymmetric HTFETs.

Fig. 6. Area vs. number of parallel and series chains.

Fig. 6 shows area difference with respect to a number of parallel and series connections. For both the cases, symmetric vertical FETs have less area than FinFETs when the number of parallel/series connections is less than three. However, the area of vertical FETs exceeds the area of FinFETs as the number of connections increases due to (a) the isolation penalty for series connections and (b) contact placement for sharing sources (or drains) for parallel transistors. For HTFETs, the trends for parallel connections is the same as vertical FETs. However, series connection have an additional isolation penalty as discussed earlier and therefore, HTFETs exhibit larger area compared to FinFETs when the number of series connections exceed two.

TABLE I. AREA CHANGES IN COMPARISON WITH FinFET BASED CELLS

Cell	Symmetric Vertical CMOS	Asymmetric HTFETs
INVX1	-20 %	-20 %
INVX2	-20 %	-20 %
INVX4	12 %	12 %
NAND2_X1	-20 %	6 %
NAND2_X2	-6.7 %	-6.7 %
AOI21_X1	-20 %	-14 %
AOI22_X1	-7 %	-2 %
DFF w Reset	-26 %	0 %
DFF w/o Reset	-27 %	-6 %

*negative: area decreases, *positive: area increases.

Considering parallel and series connections, the layout analysis in logic cells for the standard cell library show that an area overhead is observed in some cells with the asymmetric/ symmetric vertical FETs. Table I compares the area changes for cells based on the vertical FETs against the FinFET based cells. Most single finger cells (e.g., X1) based on vertical FETs have area benefit compared to cells based on FinFETs. However, like parallel and series connections, the area advantage for asymmetric and symmetric vertical FETs degrades as the number of inputs and drive-strength increases. For the area comparison between vertical FETs, and asymmetric HTFETs, the latter exhibits larger area due to the asymmetric device structure. Fig. 7 shows, for instance area overheads for the and-or-inverter (AOI) and NAND logic gates according to a number of inputs and drive-strength in vertical FETs.

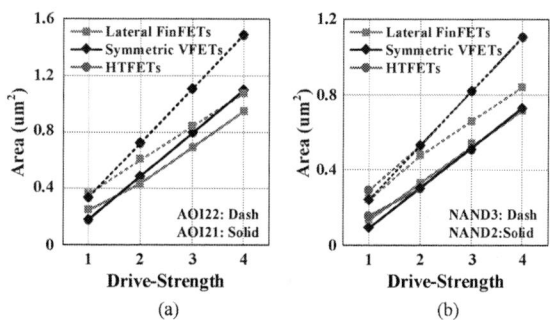

Fig. 7. The area overhead of AOI and NAND induced from a nature of vertical devices compared to FinFETs.

IV. AREA, DELAY AND ENERGY COMPARISON

In Section III, the area comparison of standard cells based on the three devices is described with a detailed analysis of impacts of the device structure on layouts. While vertical FETs show benefits over FinFETs in terms of larger W_{EMAX}, issues like larger area for high fan-in gates make it unclear whether vertical FETs still offer an explicit advantage over FinFETs in large digital designs. In order to evaluate benefits and overheads beyond the layout area, we compared the delay, area and energy of a three-stage inverter (INVX1) chain and a two-stage 32-bit CLA at various V_{DD} using standard cell libraries based on three devices. All benchmark inverter chains and CLAs are synthesized using Synopsys Design Compiler [16]. Note that standard cell libraries can support the full range of voltages used in this paper. Also, all the standard cell libraries are designed with the same fixed-height (7.5T). Fig. 8 presents the simple diagram for the library characterization process and logic synthesis flow for the 32-bit CLAs. From cell layouts, the information with regard to electrical parameters (width, length and capacitance) is extracted for generating SPICE netlists. We perform SPICE simulations to characterize each logic cell for the three devices, which are modeled using Verilog-A with 20nm gate-length [8, 10] employing lookup-tables (LUT). The liberty format of each standard cell library is constructed with 7x7 LUTs for a propagation delay and energy.

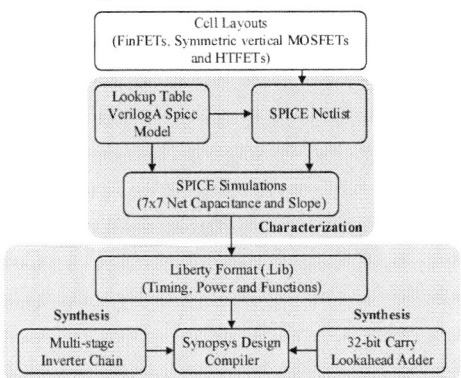

Fig. 8. A simple diagram of the library characterization process and logic synthesis flow for 32-bit CLA.

A. Energy, delay and area analysis for an inverter

Fig. 9(a) shows the drive current as a function of V_{DD} in an INVX1 based on FinFETs, symmetric vertical MOSFETs and HTFETs. Note that the n-type and p-type transistors in symmetric vertical MOSFETs and FinFETs have same current per width and each INVX1 drives a FO4 load. Due to larger W_{EMAX}, vertical FETs, including symmetric vertical MOSFETs and HTFETs, show a larger drive strength over FinFETs. Steep switching characteristics of HTFETs leads to a moderate decrease in the current as V_{DD} decreases. However, FinFET and symmetric vertical MOSFET INVX1, operating at $V_{DD} < 0.5$V are severely limited by their inherent low drive-ability in the near/sub-threshold region, leading to lower current compared to HTFETs. Consequently FinFETs and symmetric vertical FETs show delay that is over 12x and 7x higher compared to the HTFET inverter at $V_{DD} = 0.4$V, respectively as presented in Fig. 9(b). However, a cross-over at 0.6V is observed for the symmetric vertical MOSFET and HTFET inverters. For FinFETs, low drive-ability results in significant delay degradation in comparison with HTFETs at $V_{DD} < 0.7$V, as shown in Fig. 9(b). However, the delay difference between FinFETs and HTFETs reduces as V_{DD} approaches 0.7V, even though the FinFET inverter has less W_{EMAX}. It can also be observed that FinFET based inverters exhibit larger delay in comparison to a symmetric vertical FET based inverter for the entire range of V_{DD}.

Fig. 10(a) shows the required area of the three inverters at iso-delay. A single finger HTFET inverter is used as a baseline which corresponds to delay of 50 ps at $V_{DD} = 0.3$V. To achieve the same performance at $V_{DD} = 0.3$V, lower drive strength of FinFETs compared to HTFETs necessitates the use of INVX30 resulting in 21 times (21x) increase in area. Similarly, for symmetric vertical FETs, INVX12 meets the performance

requirements, which corresponds to an area increase of 12 times (12x) with respect to HTFETs. Fig. 10(b) depicts the energy vs. delay comparisons. To meet the target performance, either V_{DD} or area should be increased for both symmetric vertical MOSFET and FinFET inverters compared to the HTFET inverter. Like inverters, this behavior is also conspicuous in other cells, including both combinational and sequential logic gates. In order to match the performance of HTFET cells, the FinFET and symmetric vertical MOSFET cells have to operate at a higher V_{DD}, resulting in significant energy overhead compared to HTFETs. With the insights developed in this sub-section, we discuss the results for a CLA, next.

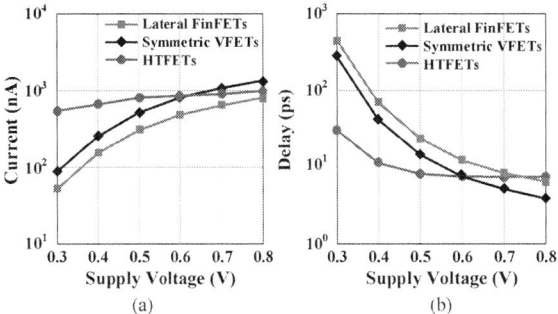

Fig. 9. (a) Current and (b) propagation delay comparisons of a single figure inverter (INVX1) vs. V_{DD} for FinFETs, symmetric vertical FETs and HTFETs when INVX1 has an iso-area.

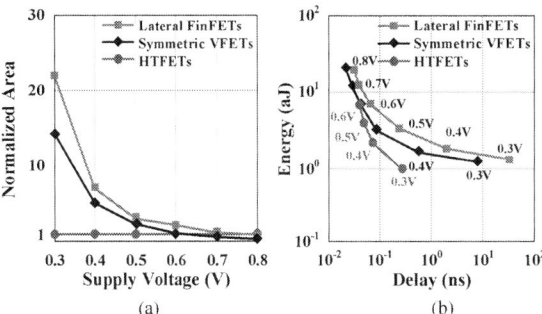

Fig. 10. (a) Normalized area vs. V_{DD} with an iso-delay and (b) energy-delay comparisons INVX1 for FinFETs, symmetric vertical FETs and HTFETs.

B. Implementation of a two-stage 32-bit CLA and Evaluation

To evaluate and compare the trends in energy, area and delay induced from the nature of vertical device structure and steep-slope characteristics in HTFETs at a low V_{DD}, we synthesized the 32-bit CLA circuit and benchmarked designs based on FinFETs, symmetric vertical MOSFETs and HTFETs.

A single-stage CLA commonly requires large fan-in gates, and a multi-stage CLA adder is widely used to avoid such large fan-in requirements. In this paper, a two-stage 32-bit CLA is implemented using eight 4-bit CLAs. Fig. 11(a) presents the delay vs. energy for different V_{DD} in the 32-bit CLA while maintaining the same area. In order to meet the requirements of a small delay at iso-area, higher V_{DD} is desired for both symmetric vertical MOSFETs and FinFETs; however it leads to higher power dissipation, resulting in significant energy overheads. The aforementioned differences in energy reduce as the delay increases. Fig. 11(b) presents the normalized area vs. delay at $V_{DD} = 0.5$V. As the delay increases, the area reduces due to the reduction in the usage of cells with large drive. A cross-over after a certain delay for symmetrical vertical MOSFET and

978-1-4799-8720-7/15 $31.00 © 2015 IEEE   307

HTFET CLAs is due to the usage of small drive cells in the symmetric vertical FETs. However, FinFETs, compared to both symmetrical vertical MOSFET and HTFET CLAs, still need larger area to meet the same performance. Fig. 11(c) shows the energy comparison with respect to the area at iso-delay. Both FinFET and symmetric vertical MOSFET based CLAs require higher V_{DD} (e.g., 0.6V for FinFET and symmetric vertical MOSFET designs when HTFETs operate at V_{DD} = 0.5V) to preserve the small area, resulting in higher energy overhead compared to the HTFET CLA.

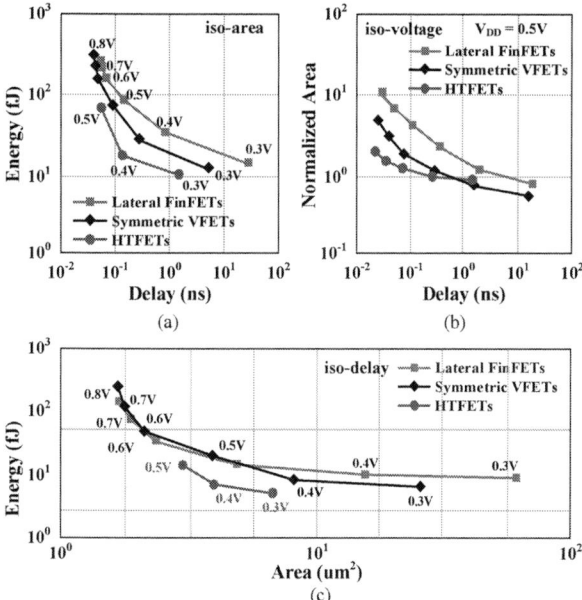

Fig. 11. The critical path delay vs. (a) energy at an iso-area, (b) normalized area at an iso-voltage, and the (c) area vs. energy at an iso-delay.

For an iso-delay comparison at a high V_{DD} such as 0.8V (when the target performance is large), the CLA circuits synthesized with FinFETs and symmetric vertical MOSFETs have a smaller area compared to HTFETs. Further, comparing symmetric vertical MOSFET and FinFET CLAs, the vertical MOSFET CLA has smaller area than FinFET CLA but slightly higher energy at V_{DD} > 0.5V. However, larger area to overcome lower drive strength of the FinFET CLA at low V_{DD} (< 0.5V) make both area and energy larger than the vertical MOSFET CLA. Hence, at high V_{DD}, symmetric vertical MOSFETs show better performance than both FinFETs and HTFETs while maintaining smaller area.

To sum up, the advantages of HTFETs for low voltage operation can be clearly observed in Fig. 11. At iso-delay and iso-area, FinFET and symmetric vertical CMOS designs require higher V_{DD} to meet performance requirements, resulting in significant energy overheads compared to HTFET designs, as presented in Fig. 11. For V_{DD} > 0.8V, symmetrical vertical FETs show the best performance and area efficiency.

V. CONCLUSION

In this paper, we examined the implications of vertical device structures such as the symmetric vertical MOSFET and HTFET on layouts compared to the lateral FinFETs in logic cells for a standard cell library with a fixed height of 7.5T. We also compared the area benefits and tradeoffs by considering parallel

and series connections. Further, we synthesized a two-stage 32-bit carry lookahead adder (CLA) with our in-house standard cell libraries based on all three 20 nm devices to carry out benchmarking and comparison in terms of area, energy, and performance. Our evaluation demonstrates significant area, energy and performance benefits of adopting HTFETs for future energy-constrained system designs while retaining small area at a low V_{DD} (< 0.6V). However, above 0.6V, symmetric vertical MOSFETs exhibit better performance and smaller area but their energy-efficiency is worse than FinFETs and HTFETs.

ACKNOWLEDGMENT

This work was supported in part by the Center for Low Energy Systems Technology (LEAST), one of the six SRC STARnet Centers, sponsored by MARCO and DARPA. The authors also acknowledge support from the College of Engineering.

REFERENCES

[1] T. Cui, Q. Xie, Y. Wang, S. Nazarian, M. Pedram, "7nm FinFET standard cell layout characterization and power density prediction in near- and super-threshold voltage regimes," *International on Green Computing Conference (IGCC)*, Nov. 2014.

[2] B. Yu et. al, "FinFET scaling to 10 nm gate length," International Electron Devices Meeting, *IEDM '02.*, Dec. 2002.

[3] S. K. Gupta, W. Cho, A. A. Goud, K. Yogendra, K. Roy, "Design space exploration of FinFETs in sub-10nm technologies for energy-efficient near-threshold circuits," *Device Research Conference (DRC)*, 2013.

[4] Q. Xie, X. Lin, Y. Wang, M.J. Dousti, A. Shafaei, M. Ghasemi-Gol, M. Pedram, "5nm FinFET Standard Cell Library Optimization and Circuit Synthesis in Near-and Super-Threshold Voltage Regimes," *2014 IEEE Computer Society Annual Symposium on VLSI (ISVLSI)*, July 2014.

[5] K. Tomioka, M. Yoshimura, T. Fukui, "A III–V nanowire channel on silicon for high-performance vertical transistors," *Nature*, Aug. 2012.

[6] A Moers, J. Turning the world vertical: MOSFETs with current flow perpendicular to the wafer surface, *Springer*, pp. 531,537, 2013.

[7] W. Wang, P. Gupta, "Efficient layout generation and evaluation of vertical channel devices," *IEEE/ACM International Conference on Computer-Aided Design (ICCAD)*, Nov. 2014.

[8] H. Liu, S. Datta, V. Narayanan, "Steep switching tunnel FET: A promise to extend the energy efficient roadmap for post-CMOS digital and analog/RF applications," *IEEE International Symposium on Low Power Electronics and Design (ISLPED)*, Sept. 2013.

[9] D.K. Mohata, et. al, "Demonstration of improved heteroepitaxy, scaled gate stack and reduced interface states enabling heterojunction tunnel FETs with high drive current and high on-off ratio," *Symposium on VLSI Technology (VLSIT)*, June 2012.

[10] H. Liu, V. Saripalli, V. Narayanan, and S. Datta. (May 2014). III-V Tunnel FET Model 1.0.1. [Online]. Available: https://nanohub.org/publications/12

[11] Predictive Technology Model. [Online]. Available: http://ptm.asu.edu/

[12] Synopsys Inc., Liberty Library Modeling, http://www.synopsys.com/community/interoperability/pages/libertylibmodel.aspx.

[13] Vendor-independent, MOSIS Scalable CMOS Design Rules (revision 8), May, 2009, https://www.mosis.com/files/scmos/scmos.pdf.

[14] A. Biddle, Jason S.T. Chen, "FinFET Technology – Understanding and Productizing a New Transistor," A joint whitepaper from TSMC and Synopsys, April 2013.

[15] J.M.Hergenrother et al, "The Vertical Replacement-Gate (VRG) MOSFET: a 50-nm vertical MOSFET with lithography-independent gate length," International Electron Devices Meeting, *IEDM*, Dec. 1999.

[16] Design Compiler, Synopsys, http://www.synopsys.com/Tools/implementation/RTLSynthesis/DCGraphical/Pages/default.aspx.

[17] Natarajan, S. et. al, "A 14nm logic technology featuring 2nd-generation FinFET, air-gapped interconnects, self-aligned double patterning and a 0.0588 µm2 SRAM cell size," *IEEE International Electron Devices Meeting, IEDM*, Dec. 2014.

[18] K. G. Anil, K. Henson, S. Biesemans, N. Collaert, "Layout density analysis of FinFETs," *33rd Conference on European Solid-State Device Research, 2003. ESSDERC '03.*, pp.139,142, 16-18 Sept. 2003.

978-1-4799-8720-7/15 $31.00 © 2015 IEEE

Novel UHF passive rectifier with Tunnel FET devices

David Cavalheiro, Francesc Moll
Department of Electronic Engineering
Universitat Politècnica de Catalunya,
Barcelona, Spain
david.manuel.nunes@estudiant.upc.edu
francesc.moll@upc.edu

Stanimir Valtchev
Department of Electric Engineering
Universidade Nova de Lisboa-FCT
Lisbon, Portugal
ssv@fct.unl.pt

Abstract—**The increase of the losses in UHF passive rectifiers with Tunnel FET devices at large RF AC amplitudes are mainly due to the high reverse current inherent of this technology when subjected to high reverse bias conditions. In this work, a new UHF passive rectifier circuit is proposed, with the purpose of reducing the reverse current suffered by Tunnel FET devices at large RF AC amplitudes. Compared to the differential-drive rectifier, the proposed topology is shown to improve the output voltage and power conversion efficiency at similar RF voltage/power conditions as well as the transmission distance for RFID applications.**

Keywords—Energy Harvesting; Passive Rectifier; RFID; Tunnel FET; UHF; Ultra-low Power.

I. INTRODUCTION

Several low-power applications benefit from the surrounding radiated energy in order to power their circuits. RFID tags and biomedical implants are examples of radio-frequency (RF) powered circuits that can be placed in areas of difficult access, and therefore, the constant replacement of their batteries is undesired. However, the low power conversion efficiency (PCE) demonstrated by UHF passive rectifiers at low RF input power levels (below -30 dBm) along with the limited available RF power from the surrounding environment constrains the operation range of these circuits in both distance of operation and computational capability [1-7].

At such low power levels, the losses in the front-end rectifiers are mainly due to the high forward losses at each diode/transistor present in the rectification process. Schottky diodes present low threshold voltage values (0.2-0.3 V) and for this reason they are often found in Dickson multiplier structures [1]. However, the incompatibility with CMOS processes has resulted in the use of diode-connected MOSFET configurations for IC applications [2].

In order to increase the range of available power operation in conventional passive rectifiers, higher power efficiencies are required, especially at sub-mW levels of the available RF power, where the use of current MOSFET technology is shown to be inefficient [3-6]. Conventional transistors applied in the front-end rectifiers of RF powered circuits are characterized by a minimum subthreshold-slope swing (SS) of 60 mV/ dec (at room temperature). This characteristic limits the current at low voltage values in the front-end rectifier.

The steep-slope Tunnel FET (TFET) device has been shown to increase the efficiency of passive rectifiers at lower

RF power levels compared to the use of FinFET technologies [6-7]. For example, an RF passive rectifier circuit with TFETs can improve the PCE up to 70 % at -39 dBm, compared to the 7 % achieved with the FinFET technology at similar load conditions [6]. However, when the *p-i-n* structure of the TFET device (Fig. 1) is largely reverse biased (both V_{GS} and V_{DS} negative for n-type TFET, and V_{GS}, V_{DS} positive for p-type TFET), the reverse current can be increased by several orders of magnitude, thus limiting the PCE of the rectifier circuit due to the resultant reverse losses. In order to alleviate these losses, a different passive rectifier topology is proposed in this work.

The structure of the work is as follows. Section II introduces the TFET carrier injection mechanism. Section III discusses the problems of applying TFET devices in passive rectifiers. Section IV proposes a novel UHF passive rectifier topology for the specific characteristics of TFET devices. Section V compares the performance between a TFET differential-drive rectifier and the proposed TFET rectifier. The last section presents the conclusions.

II. THE TUNNEL FET DEVICE

A. Physical Characteristics

Unlike the conventional MOSFET, the TFET device is designed as a reverse-biased gated *p-i-n* diode. For an n-type TFET (n-TFET) the source (drain) region presents a highly doped p-type (n-type) semiconductor as shown in Fig. 1. For this configuration, the tunneling current is generated at the source-channel interface (Fig. 2 b). In the p-TFET the drain (source) presents a p-type (n-type) doped semiconductor.

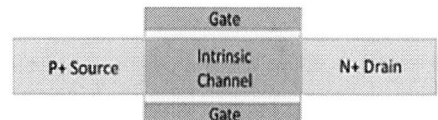

Fig. 1 Double-gate n-TFET structure

B. Band-to-Band Tunneling and Drift Diffusion

In TFET devices, the carrier injection mechanism does not follow the laws of thermionic injection as in conventional MOSFET devices. In Fig. 2, the band-to-band tunneling (BTBT) injection mechanism and drift diffusion of a 20 nm Si n-TFET device with the same source/drain doping concentration ($N_{A,D} = 1 \cdot 10^{20} cm^{-3}$) is presented:

978-1-4799-8720-7/15 $31.00 © 2015 IEEE

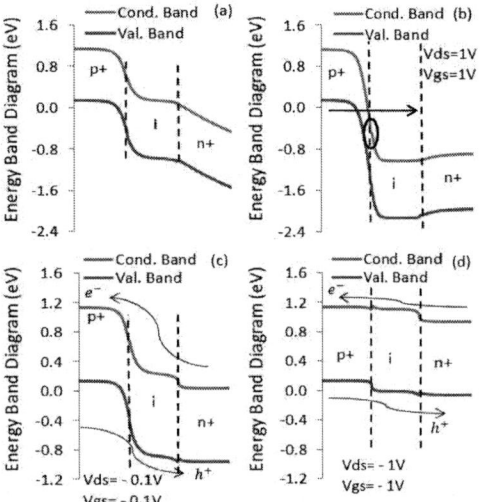

Fig. 2 Energy band diagram of a Si n-TFET, a) Equilibrium State; b) BTBT injection with forward biasing; c) Low drift diffusion injection with low reverse biasing; d) High drift diffusion injection with high reverse biasing

In the equilibrium state ($V_{GS}=V_{DS}=0V$), both regions in the n-TFET are doped such that the valence band in the p^+ type region is located above the Fermi level and the conduction band in the n^+ type region is located below the Fermi level. When no voltage is applied to the gate, the tunneling barrier between the source and the channel region is high (Fig 2 a). This will result in a low BTBT probability as expressed in (1), and a consequent low tunneling generation rate (TGR) between the regions (2). According to (3), the tunneling current is directly dependent on the TGR [8].

$$T_{b2b} = \exp\left(-\frac{4}{3} \cdot \frac{\sqrt{2 \cdot m \cdot m_0} \cdot E_g^{3/2}}{q \cdot h_{eff} \cdot E_{(x,y)}}\right) \quad (1)$$

$$TGR = T_{b2b} \cdot (f_c - f_v) \cdot E_{(x,y)} \quad (2)$$

$$I_{on} = \frac{\mu_{tun} \cdot N_C \cdot q^2}{k \cdot T} \iint TGR \cdot E_{(x,y)} dxdy \quad (3)$$

In passive rectifiers, it is assumed that the n-type transistors present a similar voltage polarity on both the gate and drain regions while the p-type transistors present a similar voltage polarity on both the gate and source regions. For the case of forward biased n-TFET devices, the increase of both gate and drain voltage values decrease the tunneling barrier in the source-channel interface as shown in Fig 2 b). Both the conduction and valence band in the channel region bend down, thus increasing the tunneling probability of carriers under the valence band of the p^+ region to tunnel through the channel to the empty states of the conduction band in the n^+ region. For p-type devices, the decrease of both gate and source voltage values increases the BTBT probability of carriers to tunnel through the channel from the n^+ to the p^+ region (not shown).

In TFET devices, the reverse bias characteristic leads to two different carrier injection mechanisms. At low reverse bias, the increase of both energy band diagrams on both channel and drain regions can produce both reverse BTBT and drift injection. The latter is shown in Fig. 2 c). At large reverse bias, only the drift diffusion mechanism is produced, as shown in

Fig. 2 d). The reverse current produced either by the reverse BTBT and drift diffusion mechanism under reverse bias conditions degrade the TFET transistor performance when applied in rectifiers. The following sections discuss the limitations of using TFET devices in passive rectifiers and how these limitations can be attenuated by the rectifier proposed in this work.

III. TFET IN ENERGY HARVESTING PASSIVE RECTIFIERS

This section discusses the limitation of using Tunnel FET devices in UHF passive rectifiers. For reference, the differential-drive passive rectifier (DDPR) circuit presented in Fig. 3 a) has been presented as a viable solution for RF energy harvesting at low power levels (~mW range) with conventional transistors [3-5]. The application of TFET devices in the DDPR topology was also investigated in [6-7]. The authors have shown by simulations higher power conversion efficiency at low RF voltage amplitudes (sub-0.35 V_{AC}) compared to the use of the FinFET technology. However, and as expected, at higher RF amplitudes (and consequent high reverse bias) the power conversion efficiency of the circuit is degraded. This degradation, although not explicitly stated by the cited references, is not only due to the TFET forward losses but also due to the increase of the reverse current suffered by the TFET transistors during the "reverse" state on both regions of operation (Fig. 3 b).

Considering the state-of-the-art DDPR of Fig. 3 a) and during the first region of operation of Fig. 3 b), the node RF^+ always presents a voltage higher than that of the node RF^-. This behavior results in the transistor conditions presented in Fig. 4 a): both transistors M2 and M3 are in the "on state" while M1 and M4 are in the "off state". According to Table I and II at steady-state conditions, the transistors in the "on state" are characterized by a $V_{GS}=2 V_{DS}$. The same biasing conditions are presented in the transistors during the "off state".

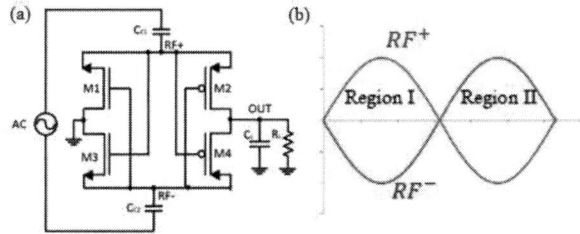

Fig. 3 a) Differential-drive passive rectifier (DDPR); b) Regions of operation

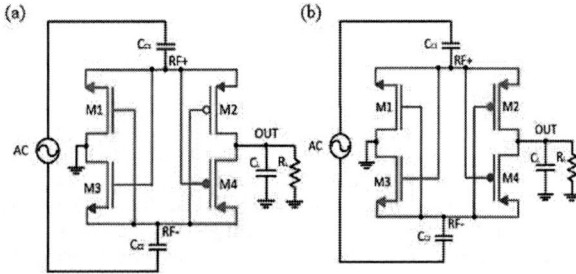

Fig. 4 Different regions of operation: a) Region I, b) Region II. Transistors in green are in the "on" state and red in the "off" state

978-1-4799-8720-7/15 $31.00 © 2015 IEEE

Table 1 Steady-state bias conditions of the DDPR in region I

Region I	State	V_{GS}	V_{DS}
M1 (n)	Off	$RF^- - RF^+ < 0$	$RF^+ < 0$
M2 (p)	On	$RF^- - RF^+ < 0$	$Vout - RF^+ < 0$
M3 (n)	On	$RF^+ - RF^- > 0$	$RF^- > 0$
M4 (p)	Off	$RF^+ - RF^- > 0$	$Vout - RF^- > 0$

In the first region of operation and according to Table I, the increase of the RF magnitude (RF^+-RF^-) will result in the increase of the reverse bias V_{DS} of the transistors in the "off state" M1 and M4. With a similar polarity than that of V_{GS} these two transistors are conducting reverse current as explained in section II B. Higher RF AC magnitudes will result in higher reverse losses of the rectifier and PCE degradation.

In the second region of operation and according to Table II, the reverse losses of the rectifier at high RF AC magnitudes will result due to the increase of the reverse bias V_{DS} of transistors M2 and M3.

Table 2. Steady-state bias conditions of the DDPR in region II

Region II	State	V_{GS}	V_{DS}
M1 (n)	On	$RF^- - RF^+ > 0$	$- RF^+ > 0$
M2 (p)	Off	$RF^- - RF^+ > 0$	$Vout - RF^+ > 0$
M3 (n)	Off	$RF^+ - RF^- < 0$	$- RF^- > 0$
M4 (p)	On	$RF^+ - RF^- < 0$	$Vout - RF^- < 0$

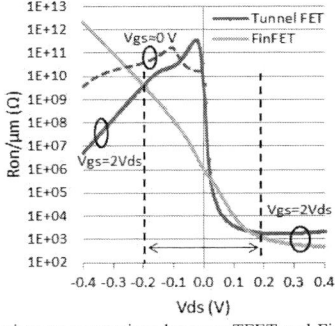

Fig. 5 Internal resistance comparison between TFET and FinFET. The TFET presents the following characteristics: Lg=40nm, P$^+$ GaSb, N_A=4x10^{19} cm^{-3}, N$^+$ InAs, N_D=2x10^{17} cm^{-3}, T_{CH}=5 nm, T_{OX} =2.5 nm (HfO2), EOT=1 nm, Φ_M=4 eV. FinFET device is presented as a triple-gate configuration, bulk material, EOT=0.84 nm, fin height=28 nm, fin width=15 nm and gate length of 22 nm.

In order to understand the degree of reverse losses in the DDPR due to the reverse current conducted by the transistors under reverse bias conditions, a comparison between the internal resistance of a Tunnel FET and FinFET device is presented in Fig. 5 for the biasing condition V_{GS}=2V_{DS}. The simulation results of the TFET device are based on heterojunction structure with III-V materials [9-10]. As expressed in (1) the use of materials with low energy band gap materials can increase the BTBT probability of the TFET device, thus increasing the tunneling current at similar voltage values compared to the use of higher energy band gap materials as silicon.

According to Fig. 5, there is a V_{DS} bias region (in between the dotted lines) where the TFET device is expected to improve the power conversion efficiency of UHF passive rectifiers due to its lower internal resistance in the forward region and higher internal resistance in the reverse region compared to the FinFET device. However, when largely reverse biased (V_{DS} below -0.2 V), the drift diffusion current explained in section II B dominates the reverse current, and degrades the PCE of the rectifier. In Fig. 5, it is shown that under reverse V_{DS} bias, a V_{GS} forced to 0 V can increase the internal resistance, thus attenuating the reverse current (dotted curve). This way, a viable solution to apply TFET devices in rectifiers passes through forcing the V_{GS} of the devices in the "off state" to values close to 0 V.

IV. PROPOSED UHF PASSIVE RECTIFIER CIRCUIT

In Fig. 6, an UHF passive rectifier circuit designed for the special characteristics of Tunnel FET devices is proposed. The rectifier maintains the forward characteristics of the previous differential-drive circuit, attenuating the reverse losses suffered from the transistors during the "off state". Compared to the previous rectifier, it maintains four main transistors M1-M4 and requires two auxiliary transistors M1$_{aux}$, M2$_{aux}$ (p and n-type) and two auxiliary capacitors.

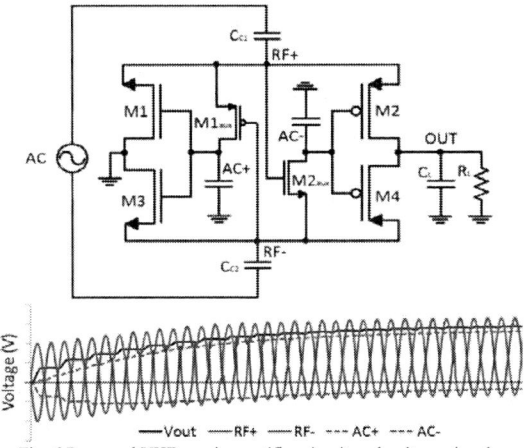

Fig. 6 Proposed UHF passive rectifier circuit and voltage signals

A. Fist Region of Operation ($RF^+ > RF^-$)

Considering steady-state conditions, during the first region of operation the two auxiliary transistors are in the "on state" allowing the auxiliary capacitors to charge the nodes AC$^+$ and AC$^-$ to the maximum and minimum voltage values of the nodes RF$^+$ and RF$^-$ respectively. This behavior is observed in Fig. 6, for an ideal case (no output load, no losses in the transistors). In this region, and similarly to the previous DDPR, the main transistors M2 and M3 are forward biased "on state" while transistors M3 and M4 are reverse biased "off state". With the proposed rectifier, the DC voltage at the gate of the four main transistors will allow a lower average value of V_{GS} (in magnitude) for the reverse biased transistors (less reverse losses) and a higher average value of V_{GS} for the forward biased transistors (less forward losses).

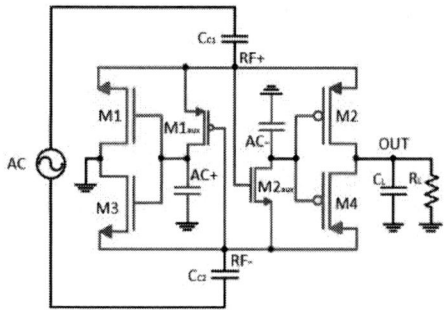

Fig. 7 First region of operation in the proposed rectifier. Transistors in green are in the "on" state and red in the "off" state

With this configuration, the V_{DS} values of the main four transistors remain the same as the counterpart DDPR. As seen in Table 3, there are possible conditions for reverse current to occur in the reverse biased transistors M1 and M4. However, as AC^+ and AC^- present DC values, the period of time that the reverse current is verified is lower compared to the previous rectifier. In this region of operation, there is also a specific region where the auxiliary transistors enter into a reverse bias condition, conducting reverse current.

Table 3 Steady-state bias conditions of the proposed rectifier in region I

Region I	State	V_{GS}	V_{DS}	Rev. Current
M1 (n)	Off	$AC^+ - RF^+$	$-RF^+ < 0$	$AC^+ > RF^+$
M2 (p)	On	$AC^- - RF^+ < 0$	$Vout - RF^+ < 0$	-
M3 (n)	On	$AC^+ - RF^- > 0$	$-RF^- > 0$	-
M4 (p)	Off	$AC^- - RF^-$	$Vout - RF^- > 0$	$AC^- < RF^-$
M1$_{aux}$ (p)	On	$RF^- - RF^+ < 0$	$AC^+ - RF^+$	$RF^+ < AC^+$
M2$_{aux}$ (n)	On	$RF^+ - RF^- > 0$	$AC^- - RF^-$	$RF^- < RF^-$

B. Second Region of Operation ($RF^- > RF^+$)

During the second region of operation, the two auxiliary transistors are reverse biased "off state" and therefore, with a sufficient auxiliary capacitance value, the previous voltage at nodes AC^+ and AC^- is retained during the entire region. As seen in Table 4, the V_{DS} of the four main transistors remains the same than those of the previous rectifier. In this region, the average value of V_{GS} in the reverse biased transistors M2 and M3 is lower (in magnitude) than the values presented by the previous rectifier.

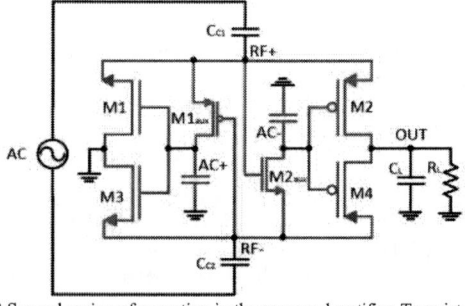

Fig. 8 Second region of operation in the proposed rectifier. Transistors in green are in the "on" state and red in the "off" state

Table 4 Steady-state bias conditions of the proposed rectifier in region II

Region II	State	V_{GS}	V_{DS}	Rev. Current
M1 (n)	On	$AC^- - RF^+ > 0$	$-RF^+ > 0$	-
M2 (p)	Off	$AC^- - RF^+$	$Vout - RF^+ > 0$	$RF^+ > AC^-$
M3 (n)	Off	$AC^+ - RF^-$	$-RF^- < 0$	$AC^+ > RF^-$
M4 (p)	On	$AC^- - RF^- < 0$	$Vout - RF^- < 0$	-
M1$_{aux}$ (p)	Off	$RF^- - RF^+ > 0$	$AC^+ - RF^+ > 0$	At large RF^+
M2$_{aux}$ (n)	Off	$RF^+ - RF^- < 0$	$AC^- - RF^- < 0$	At large RF^-

V. SIMULATION RESULTS

In this section, the performance comparison between the state-of-the-art differential-drive passive rectifier DDPR of Fig. 3 and the proposed rectifier of Fig. 6 is presented. Both rectifiers are simulated with GaSb-InAs Tunnel FET devices [9-10] (characteristics presented in Fig. 5). For both rectifiers, transistors M1-M4 are simulated with channel widths of 1 μm. The auxiliary transistor widths of the proposed rectifier are 400 nm. In both rectifiers, the coupling and load capacitors present a capacitance of 1 pF. The auxiliary capacitors of the proposed circuit are simulated with capacitances of 0.1 pF. The frequency of the AC source is 900 MHz.

Fig. 9 and Fig. 10 present respectively the output voltage and power conversion efficiency of the two rectifiers in study, in function of the RF input voltage. For the simulations, two loads were chosen at the output: 100 kΩ and 10 kΩ

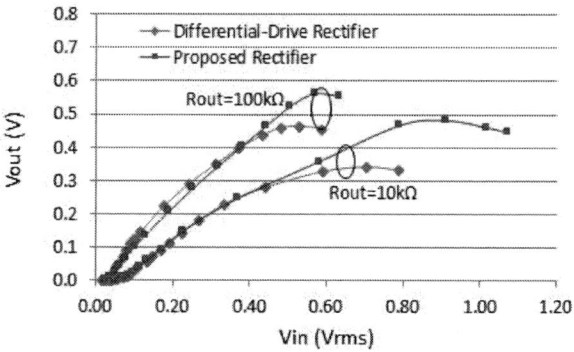

Fig. 9 Output voltage of the rectifiers in function of the RF input voltage

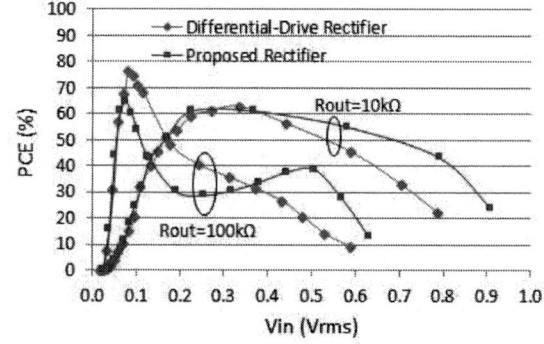

Fig. 10 PCE of the rectifiers in function of the RF input voltage

978-1-4799-8720-7/15 $31.00 © 2015 IEEE

In Fig. 9, it is observed that the proposed rectifier allows for higher DC output voltage at large RF input magnitudes, considering both loads. The reduction of both BTBT and drift diffusion current of the reverse biased transistors allows for higher current in the output load of the circuit, and therefore a higher voltage between the load terminals. At low RF voltage magnitudes, the output voltage of the rectifiers at both load conditions is similar.

As presented in Fig. 10, the reduction of the reverse losses of the proposed rectifier allows for a higher PCE at larger RF input voltage values compared to the DDPR. However, when considering low RF voltage values, the PCE performance of both rectifiers differs according to the load. Under the highest output impedance the losses of the auxiliary circuitry (charging losses and transistor losses) degrade the PCE of the proposed rectifier. As indicated in Table 3 and 4, when the auxiliary transistors are reversed biased, their V_{GS} values are directly dependent on the RF^+ and RF^- voltage values. For the considered load of 100 kΩ, the PCE of the proposed converter is shown lower at the input voltage range of 0.1 V and 0.45 V. At higher input voltage, the increase of the output current outperforms the current losses resultant by the auxiliary circuitry, thus increasing the PCE.

The "M" shape of the PCE for a load of 100 kΩ is explained due to the specific characteristics of TFET devices. When reverse biased, the reverse BTBT current is highly dependent on V_{GS}, increasing first at low V_{GS} and then decreasing at high V_{GS}. Consequently, at low input voltage values the reverse current of the auxiliary transistors is higher than that observed at higher voltage values. When considering the load of 10 kΩ, the higher output current required at the output surpasses the reverse current during all the RF input voltage range considered, and therefore no "M" shape is observed.

In Fig. 11, the PCE of both rectifiers in function of the output current is presented. For both cases, the increase of the output current degrades the efficiency of the rectifiers due to the transistor forward losses. Considering the load of 10 kΩ, the proposed rectifier presents a wider range of output current in terms of efficiency compared to the DDPR.

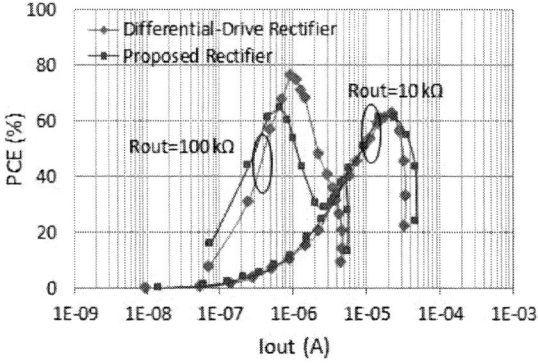

Fig. 11 Power conversion efficiency of the rectifiers in function of the output current

Under the load condition of 100 kΩ, the "M" shape presented in Fig. 10 is reflected in the PCE of the proposed rectifier. In between the range of 700nA-3µA, the proposed converter is shown less efficient.

In Fig. 12, it is shown that at similar load conditions, and at higher RF input power values, the proposed rectifier allows for higher output voltage values compared to the DDPR For an output load of 100 kΩ and 10 kΩ, this behavior is respectively observed at -22 dBm and -18 dBm.

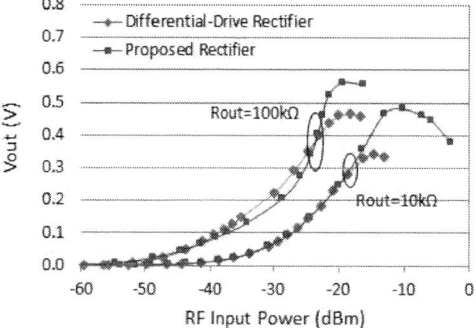

Fig. 12 Rectifier output voltage in function of the RF input power (dBm)

Considering the proposed rectifier topology, the consequent increase of the output voltage and output current allows for a higher PCE at large RF input power levels as shown in Fig. 13. As explained in Fig. 10, when considering a load of 100 kΩ the auxiliary circuitry degrades the PCE at low RF input voltage. This degradation is observed for the RF input power range of -42 to -25 dBm.

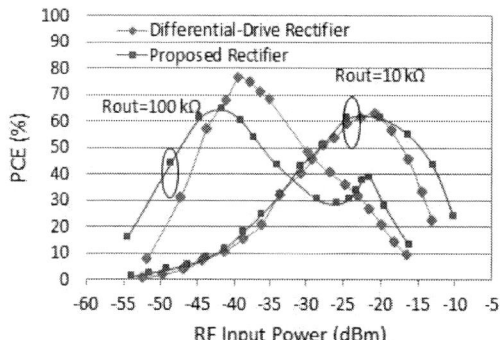

Fig. 13 Power conversion efficiency of the rectifiers in function of the RF input power (dBm)

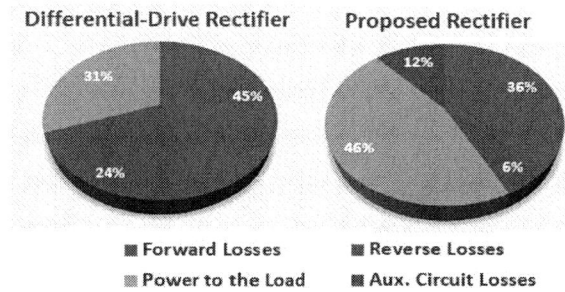

Fig. 14 RF power distribution in the rectifiers considering an output load of 10 kΩ and an RF input power of -13 dBm

In Fig. 14, the RF input power distribution of both rectifiers is presented, considering an output load of 10 kΩ and an input RF power level of -13 dBm. At this RF power level, it is observed that the differential-drive rectifier suffers from higher reverse losses (24 %) compared to the proposed rectifier (6 %), It is also observed that the auxiliary circuit of the proposed rectifier is consuming 12 % of the total RF input power. Despite these losses, an increase from 31 % to 46 % in power efficiency is observed with the proposed rectifier.

For reference, the performance of a differential-drive rectifier with FinFET devices (characteristics presented in Fig. 5) is compared with the above mentioned two rectifiers with Tunnel FET devices. The results are presented in Fig. 15.

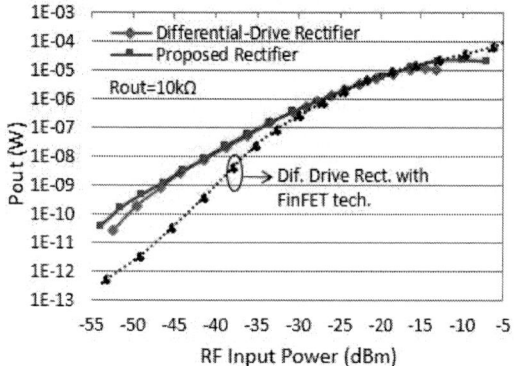

Fig. 15 Rectifier output power in function of the RF input power (dBm)

At low RF input power levels (below -30 dBm), the application of FinFET devices in the differential-drive rectifier is shown to be less efficient than the application of TFET devices due to the difficulty of conventional transistors in conducting the same levels of current as TFETs at low voltage values (sub-0.25 V).

For RFID applications, the use of Tunnel FET devices in passive rectifiers can improve the transmission distances as shown in Fig. 16. According to the Friis transmission equation [3] and compared to the differential-drive rectifier, the proposed solution can increase the range of transmission distances at RF input power levels below -45 dBm.

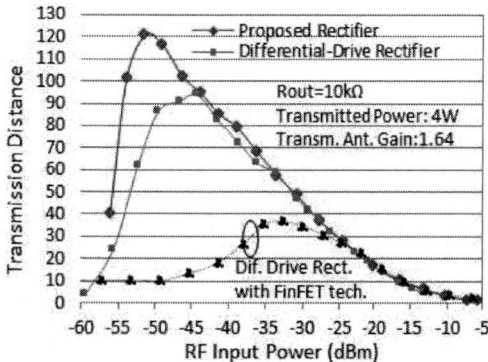

Fig. 16 Transmission distance in function of RF input power

VI. CONCLUSIONS

In this work, a novel UHF passive rectifier topology for the application of Tunnel FET devices is proposed. It was shown by simulations that compared to the differential-drive rectifier topology, the proposed rectifier allows for higher output voltage and power conversion efficiency at higher RF input power levels. This is possible due to the reduction of the drift diffusion and BTBT current of the TFET devices during their reverse biasing state and a consequent decrease of the reverse losses of the rectifier.

Considering low RF voltage amplitudes, the increase of the output load in the proposed rectifier is expected to degrade the PCE compared to the differential-drive rectifier topology due to the losses suffered by the auxiliary circuitry.

In summary, the proposed rectifier for Tunnel FET devices can improve the RF energy harvesting field due to the possibility of delivering more power to a load for a wider RF input power range compared to the differential-drive rectifier. For RFID applications, the proposed rectifier can increase the transmission distance at a wider range of RF power level.

ACKNOWLEDGMENT

The authors wish to acknowledge the support for this research, coming from the Portuguese funding institution FCT (Fundação para a Ciência e a Tecnologia) and Spanish Ministry of Economy (MINECO) and ERDF funds through project TEC2013-45638-C3-2-R (Maragda).

REFERENCES

[1] Jinpeng, S. et al. "Design and implementation of an ultra-low power passive UHF RFID tag", *J. Semiconductors*, vol. 33, no. 11, pp.115011, 2012.

[2] D.-S. Liu et al. "New Analysis and Design of a RF Rectifier for RFID and Implantable Devices," *Sensors*, vol.11, no.7, pp.6494-6508, 2011.

[3] S. Jinpeng et al. "A passive RFID tag with a dynamic-VTH-cancellation rectifier," *J. Semiconductors*, vol.34, no.9, p.95005, Sep. 2013.

[4] P. Burasa, N. G. Constantin and K. Wu, "High-Efficiency Wideband Rectifier for Single-Chip Batteryless Active Millimeter-Wave Identification (MMID) Tag in 65-nm Bulk CMOS Technology," *Microwave Theory and Techniques, IEEE Trans. on* , vol.62, no.4, pp.1005,1011, April 2014.

[5] Mandal, S. & Sarpeshkar, R "Low-Power CMOS Rectifier Design for RFID Applications", *Circuits Syst., IEEE Trans.* vol.54, no.6, pp.1177-1188, 2007.

[6] X. Li, U. Hei, K. Ma, V. Narayanan, H. Liu and S. Datta, "RF-powered systems using steep-slope devices," *New Circuits and Systems Conference (NEWCAS), 2014 IEEE 12th International*, pp.73-76, 2014.

[7] H. Liu, R. Vaddi, S. Datta and V. Narayanan, "Tunnel FET-based ultra-low power, high-sensitivity UHF RFID rectifier," *Low Power Electronics and Design (ISLPED), 2013 IEEE Int. Symposium on*, pp.157-162, 2013.

[8] M. Graef, T. Holtij, F. Hain, A. Kloes and B. Iñiguez, "A 2D closed form model for the electrostatics in hetro-junction double-gate tunnel-FETs for calculation of band-to-band tunneling current," *Microelec. J.*, vol. 45, no. 9, pp.1144-1153, 2014.

[9] V. Saripalli, A. Mishra, S. Datta and V. Narayanan, "An energy-efficient heterogeneous CMP based on hybrid TFET-CMOS cores," *Proceedings of the 48th Design Automation Conference (DAC)*, pp.729-734, 2011.

[10] V. Saripalli, S. Datta, V. Narayanan and J.P. Kulkarni, "Variation tolerant ultra low-power heterojunction tunnel FET SRAM design", *IEEE/ACM International Symposium on Nanoscale Architectures*, pp.45-52, 2011.

Hybrid STT/CMOS Design of an Interrupt based Instant On/Off Mechanism for Low-Power SoC

Christophe Layer, Kotb Jabeur
Laurent Becker, Bernard Diény
SPINTEC (CEA/CNRS/Univ. Grenoble Alpes)
17 rue des Martyrs, 38054 Grenoble, France
{name.surname}@cea.fr

Stéphane Gros, Virgile Javerliac,
Pierre Paoli, Fabrice Bernard-Granger
EVADERIS – MINATEC Entreprises
7 Parvis Louis Néel, 38040 Grenoble, France
{name.surname}@evaderis.com

Abstract—**This paper describes the design and the evaluation of a low-power System-on-Chip (SoC) in an advanced hybrid 40nm magnetic/CMOS technology node. Without external memory interface, the processor of the SoC benefits from a privileged access to the embedded NVM (Non-Volatile Memory), providing means for internal data storage and integrity thanks to its inherent non-volatility. Furthermore, a method based on an IRQ (Interrupt Request) controls the instant-on/off features of the SoC at assembler level through the use of NVM elements and improves the whole system in terms of power consumption and functionality enhancements, compared to an equivalent system relying on standard volatile memory blocks only. We discuss our simulation results on the basis of still image compression benchmarks at various data throughputs and show the benefits of NVM even for rather computation intensive algorithms.**

Keywords—STT-MRAM; non-volatile memory; hybrid magnetic-CMOS; low-power system-on-chip; check-pointing interrupt;

I. INTRODUCTION

Low-power embedded systems have become the ubiquitous body of the Internet of Things (IoT), a paradigm virtually consisting of connected sensor nodes processing and exchanging data. However, these components have to meet constraints related to a rigorous energetic environment in which increasing data amounts have to be handled almost autonomously using a severely limited power budget. Several studies [1]–[3] revealed that the two major consumers of power are the processor and the memory. In modern processors, memories are usually implemented using Static Random Access Memories (SRAM) or embedded Dynamic RAM (eDRAM) as in lower level caches [4]. For on-chip memory design however, the significant standby power due to high cell-leakage is still a major bottleneck and memory power has become a large concern. While some work has explored the use of Spin-Transfer Torque Magnetic RAM (STT-MRAM) in the cache hierarchy, through architectural techniques such as hybrid caches [5], other studies were carried out to analyze the complete replacement of the main DRAM memory by the STT-MRAM [6]. Embedded memory technologies such as STT-MRAM are hence considered to be a very promising candidate thanks to their intrinsic non-volatility, high density and high endurance [7]–[9]. For embedded systems in the IoT era, semiconductor fabs offer low-power variants and invest into research for new technologies. Encouraging results were then obtained especially in terms of energy consumption thanks to the non-volatility of the STT-MRAM technology which reduces the standby

power as the device can potentially be usefully turned off even for a very short amount of time (instant-on/off). Following these considerations, we describe in this paper the use a fully characterized library of hybrid CMOS/magnetic cells to redesign the basic memory blocks of a typical SoC which includes a 32-bit processor, different communication interfaces and a system bus that connects all the elements together. The methodology relying on the elementary Magnetic Tunnel Junction (MTJ) device is illustrated in section II, as well as the design of the non-volatile register and memory banks. Section III depicts the architecture of the system and describes the check-pointing mechanism in details. Section IV finally presents benchmarking results using different algorithms with varying data throughput for still image compression.

II. CUSTOM DESIGN FLOW OF HYBRID CIRCUITS

A. Magnetic Tunnel Junction devices

Physical Intellectual Properties (IPs) including standard cells, logic and many variants of memories represent the puzzle pieces gathered by system designers to implement a full SoC. The choice of the IP largely depends on the targeted applications and the trade-off between speed, power consumption, area and yield. From a system engineering point of view, the achievement of a SoC that takes advantage from the non-volatility of spintronic devices requires the conception of hybrid CMOS/Magnetic IP-blocks to be a part of the digital design flow. As we target to design IPs based on STT-MTJs, this section illustrates the structure of the device as well as its combination with CMOS based circuit architectures in a full custom style in order to be exploited by digital designers. Cornerstone of MRAMs and magnetic logic devices such as

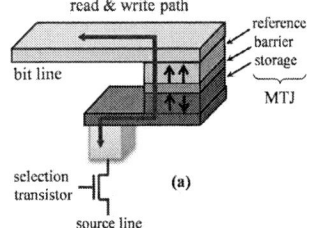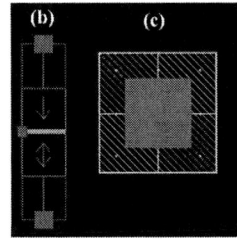

Fig. 1. STT-MTJ base structure: (a) physical view showing three stacked layers placed on top of a selection transistor, (b) symbol and (c) layout views as they appear in the PDK.

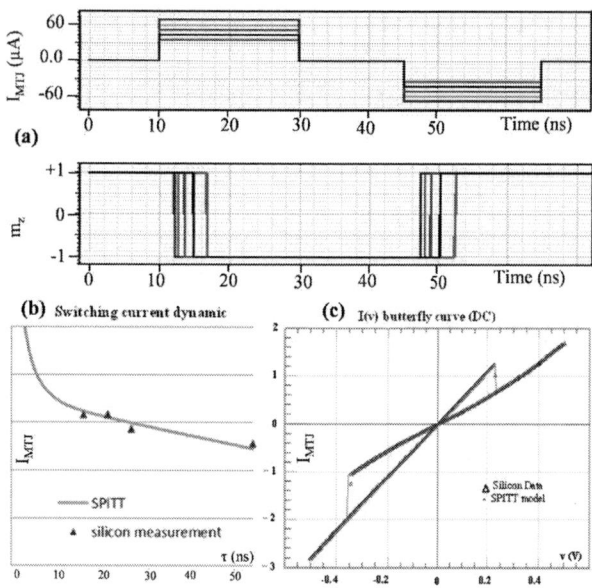

(b) Switching current dynamic   **(c)** I(v) butterfly curve (DC)

Fig. 2. STT-MTJ in the analog design flow: (a) validation of the model with simulation of the magnetization state $m_z(t)$ according to the amplitude and the direction of the applied current pulse, (b) the switching current and (c) calibration of the SPITT [11] model using raw foundry data.

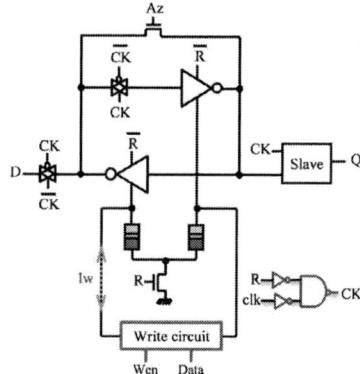

Fig. 3. Schematic of a NVFF including two MTJs with opposite configuration states written in a differential way by the current I_w.

Fig. 4. Layout of a single bit NVFF including an STT-MTJ differential pair in 40nm LP hybrid CMOS technology according to the schematic in Fig. 3.

latches and flip-flops, an MTJ is a nano-pillar which consists of two Ferro-Magnetic (FM) layers separated by a thin layer of insulator which represents the tunnel barrier, as depicted in Fig. 1(a). The first FM layer is a pinned layer with a fixed magnetic state which acts as a reference while the second FM layer has a free magnetic state which acts as a storage layer. The storage layer is programmable according to the direction of the spin-polarized current flowing through the MTJ. Thus, its magnetization state can be switched between two states with respect to the reference layer, either parallel or anti-parallel. When the reference and the storage layers have parallel magnetization states, the resistance of the MTJ is $R_{\min}$ and it can be assimilated to a logic zero. When the reference and the storage layers have anti-parallel magnetization states, the resistance of the MTJ is $R_{\max}$ and it can be assimilated to a logic one. At a technological level, a single mask process is required to integrate the MTJ material layers. Fig. 1(b) and Fig. 1(c) represent the symbol and the layout views used to run hybrid CMOS/magnetic circuit simulations.

B. Compact modeling of the STT-MTJ device

Standard microelectronics design suites enable chip designers to conceive and analyze entire semiconductor chips. Thus, in order to realize hybrid CMOS/MTJ circuits, it is necessary to integrate the STT-MTJ devices into standard microelectronics design suites and ensure their compatibility and stability with the existing environment. Therefore an accurate and fast SPICE compact model of the STT-MTJ must be used for analog electrical simulations within the Process Design Kit (PDK) for the hybrid CMOS/magnetic technology, as presented in [10]. A typical foundry-specific CMOS PDK contains technology files for layout and physical verification, as well as standard cells for the design of complex logic cir-

cuits. In addition, our magnetic PDK contains a compact model of the MTJ named SPITT (Spintec's spIn Transfer Torque model) [11]. It has been developed by CEA-Spintec laboratory jointly with the company eVaderis which is currently adding a methodology to write a complete model card including mismatch and corners cases based on real silicon data. Hence SPITT is a physics-based, accurate, scalable, robust and predictive model which takes into account the temperature, the bias voltage and the impedance load of the MTJ. Furthermore, we have described in [12] the two possible strategies to efficiently model spintronic devices while highlighting pros and cons of the different modeling strategies. Fig. 2 summarizes the integration of the STT-MTJ device in standard design suites based on an accurate physical model, a symbol view and a layout aspect to allow the evaluation of the static and transient behaviors of the device at circuit level. Fig. 2(a) shows the magnetization switching from parallel ($m_z = 1$) to anti-parallel ($m_z = -1$) and back, for various amplitudes of the applied current I_{MTJ}. The model is able to predict the delay of the switching which depends on the current intensity and direction. Fig. 2(b) represents the switching current dynamics obtained from the simulation of the model for a case study of a 40nm STT-MTJ diameter. Fig. 2(c) illustrates SPITT calibration with raw foundry-data with the same vertical axis.

C. Non-volatile hybrid CMOS/magnetic latches

As one of the basic element of digital systems, Flip-Flops (FF) are widely used in synchronous designs and have a direct impact on speed and power consumption. Fig. 3 shows the schematic of a Non-Volatile FF (NVFF) architecture, whereas many variants are thoroughly described in [13]–[15]. This cell offers the usual CMOS-FF functionality with its typical internal dual inverters structure, plus the storing and restoring the magnetic data by exploiting the non-volatility asset of the

978-1-4799-8720-7/15 $31.00 © 2015 IEEE

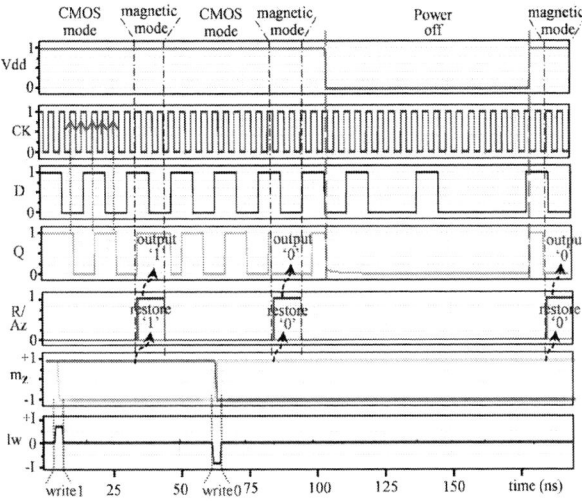

Fig. 5. Transient simulations of CMOS/magnetic NVFF for which two operating modes are possible. In CMOS mode, the output Q of the NVFF follows the input data D at every rising clock edge (CK). In the magnetic mode, an MTJ-read operation is performed by enabling Az and R signals. An MTJ-write is driven by the write circuit.

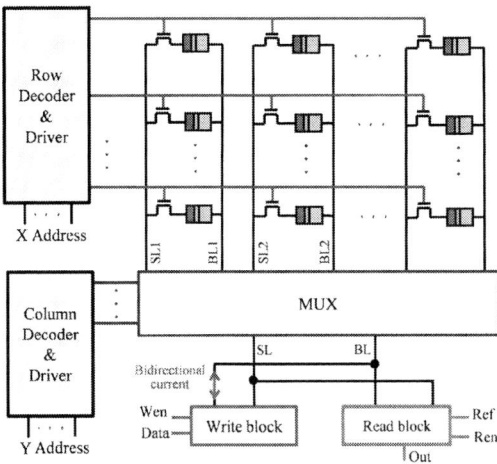

Fig. 6. Architecture of the STT-MRAM memory array based on 1T1MTJ bit cell, according to Fig. 1. Writing an MTJ is performed using a bidirectional current flowing through an n-type selection transistor, while reading the magnetic state of the MTJ is achieved using reference comparison.

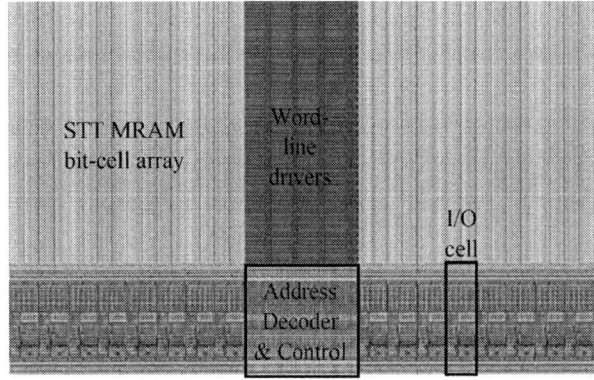

Fig. 7. Partial layout of a memory block based on an 1T1MTJ matrix structure of STT-MTJ in 40nm LP hybrid CMOS technology. Note that an I/O cell includes the write drivers, the column decoders and the sense amplifiers.

STT-MTJ when the circuit is powered off. Fig. 5 illustrates the two possible functional modes of the NVFF validated through electrical simulations using the STT-MTJ model described in the previous section. In the standard CMOS mode, the appliance of a positive current pulse I_w writes the logic one data by setting the magnetization m_z of the two MTJs up ($+1$) and down (-1), respectively. The output Q of the NVFF takes the value of the input data D at each rising edge of the clock signal CK. When the signal R (restore) is high, the data already written in the MTJs is restored using a potential equalizer signal that resets the cross-coupled inverters. Fig. 5 shows the state inversion phase during the first restore operation, where Q outputs the logic one stored as magnetic state m_z. Then, during the second restore, the logic zero data is loaded back. Fig. 5 confirms the non-volatility criteria of the NVFF, where the last data stored in the MTJs is loaded back after a power-off of the cell. This cell was designed and fully characterized at device level, using the SPITT compact model of the MTJ.

D. Magnetic Random Access Memory (MRAM)

Since its announcement, a lot of work on MRAM design optimization has been achieved and multiple circuit architectures have been proposed for the writing and the reading circuitry, the reference circuit and the bit array organization. The work in [16] overviews MRAM design considerations and involves several associated design studies and scaling challenges. Fig. 6 shows the 1T1MTJ array architecture of the MRAM. Each MTJ is connected in series with an n-type FET. A row decoder is required for the selection of the word line according to the memory address (X-address). A column decoder selects the column designated by the memory address (Y-address). The write block is used to configure the logic state of the bit cell by passing a positive or negative current through the selected MTJ. The read block determines the logic value of the bit cell by comparing the current on the bit line (BL) to a reference current. It is usually composed of

a sense amplifier (SA) preceded by a differential amplifier. For a write operation, the write enable signal (Wen) is activated. Indeed, the write operation in STT-MRAM is bidirectional. To perform a read operation, Wen must be low so that the write block is disconnected from the source line (SL) and the bit line (BL) of the memory array. During a read operation, the current sensed on the BL is compared to a reference value to determine the value stored in the bit-cell. Based on this architecture, a library of MRAMs ranging from 16kB to 256kB has been designed and fully characterized at device level. To guarantee an industrially-high yield of the designed circuits, 6σ Monte-Carlo simulations had to be carried out. For each cell, a behavioral model with the corresponding library describing the timing has been furnished, as well as optimized layout views to be fully integrated into the standard digital design flow used in section III. Fig. 7 shows the partial layout of our own STT-MRAM memory instance as presented schematically in Fig. 6. It has been designed in 40nm low-power CMOS technology node and is fully compatible with our standard design flow.

978-1-4799-8720-7/15 $31.00 © 2015 IEEE   317

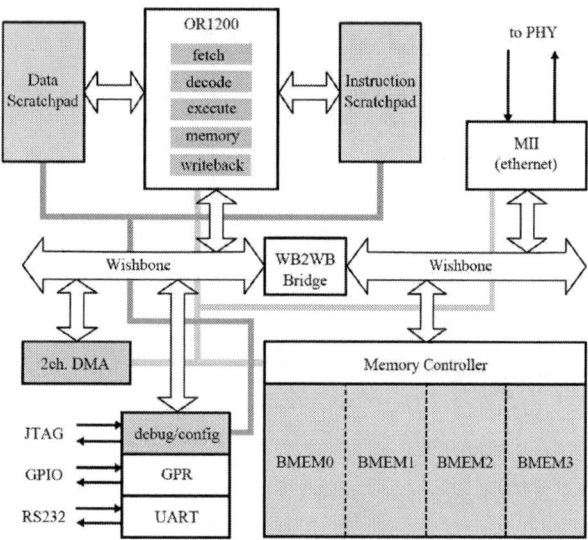

Fig. 8. Architecture of the complete system including the OpenRISC with its two 2Mb fast scratchpad memories and the peripherals connected over a Wishbone bus, including four 2Mb dense block memories (BMEM0-3).

III. SYSTEM-ON-CHIP DESIGN

A. Embedded hybrid CPU and memory integration

MRAM is a promising non-volatile technology that combines lower power consumption and higher performances in terms of speed, compared to other non-volatile technologies [6]. In order for a SoC to get the full benefit of embedding MRAM, it needs to be associated with a processor that has an efficient memory interface and computing means of processing data accordingly. Suitable candidates are found in the lower end of the 32-bit CPU class, with enough computing resources to enable any application, and a modular architecture offering a high potential for power saving. Therefore we used in this work our own modified version of the well-known OpenRISC processor [17], which is a 32-bit scalar RISC (Reduced Instruction Set Computer) with Harvard micro-architecture, a 5-stage integer pipeline (fetch, decode, execute, memory, write back) and a virtual memory support, as seen in Fig. 8.

Cache memory is used within processors as a fast buffer in front of a slower memory in the aim of increasing the throughput and reducing the latency of the data and instructions transfers. In the memory hierarchy of a processor system, caches are located between the register-file running at the frequency of the core, and the slower but denser main-memory, which is usually found off-chip due to its technological incompatibility with the CMOS process of the core. As an alternative to caches, scratchpad memories provide a very fast data and code buffer directly mapped onto the address space of the processor. They allow programmers to use efficient and quick data load/store operations, even though their usage is not hardware controlled, contrary to caches. Furthermore, code can also be allocated to this memory to permit direct instruction fetches. Our modified OpenRISC processor provides a fast interface to a scratchpad memory located just behind the MMU (Memory Management Unit), that makes instruction fetches and data load/stores very fast and independent from the peripherals connected to the

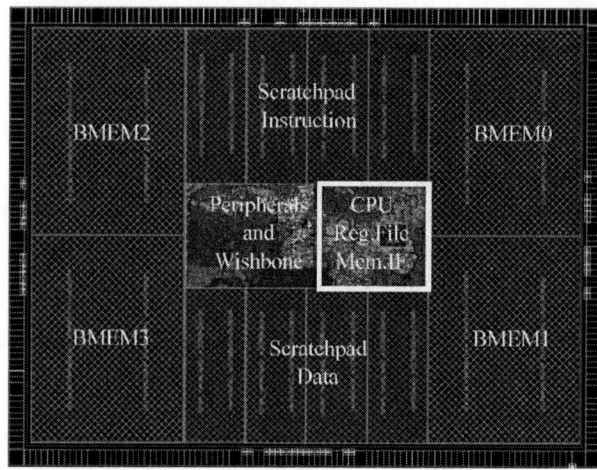

Fig. 9. Layout of the NV-SoC including the OpenRISC, the peripherals and the embedded STT-MRAM blocks in 40nm LP hybrid CMOS technology. The size of the layout is $2840\mu m \times 2280\mu m$ for an area of about $6.47mm^2$

Wishbone interface, thus remaining a protected private area for the processor.

B. Instant-ON/OFF mechanism

A regular way for a CPU to save its context is by writing it in the stack which is in SRAM usually. The context is lost when the chip is powered down, but if the SRAM is replaced by an STT-MRAM, the context is saved even without current supply and it can be recovered at power up. Checkpointing mechanisms in traditional systems save the state of process running on volatile memory to a file on a non-volatile storage. Beyond recent results showing that check-pointing performance based on non-volatile memories can be order of magnitudes faster than traditional flash-based approaches [9], [15], [18], we have developed a ground structure in assembly code on the basis of a regular IRQ (Interrupt Request) routine of context saving and restoring. In this section, we present the interesting parts of the code related to such a routine. First, we show the boot section of every standard program (section _reset located at address 0x100 for the OpenRISC). Note that the register r0 is architecturally meant to store zero and might not be writable in some implementations of the OpenRISC. This section is extended with the check of an internal non-volatile register BOOT_STAT (lines 6 to 9) which contains the value INSTANT_ON in case the programmable instant-off IRQ was triggered, as explained below. This value is then checked and if so, the program branches to the label _irq_inst_on. If not, the program continues its execution, i.e., the regular boot sequence, at address _start.

```
1  .global _reset
2  .section .vectors, "ax"
3  .org 0x100
4  _reset :
5    l.andi  r0, r0, 0        // Ensure r0 is zero
6    l.lwz   r10, BOOT_STAT(r0) // Read boot status reg
7    l.sfeqi r10, INSTANT_ON   // Check if flag is set
8    l.bf    _irq_inst_on     //    then branch to
9    l.nop                    //    restore part else
10   l.j     _start           //    continue w/ boot
```

Secondly, the _start section continues with the regular reset procedure, hence preparing the configuration of the register file and the BSS segment (see [17] for more details).

```
20   .section .text
21   _start :
22      // Reset stack and frame pointers
23      l.andi   r1, r0, 0
24      l.andi   r2, r0, 0
25      [···]
26      // Clear BSS segments
27      l.movhi  r4, hi(_bss_start)
28      l.ori    r4, r4, lo(_bss_start)
29      l.movhi  r5, hi(_bss_end)
30      l.ori    r5, r5, lo(_bss_end)
31      [···]
```

When the IRQ is triggered, the program branches at address _irq_inst_off as seen below. The exception registers EPCR, ESR and EEAR are saved with the last values of the Program Counter (PC), the Status Register (SR) and the last Effective Address (EA) accessed as seen from lines 50 to 55. The GPRs, SPRs and exception registers are saved into the stack, which is in QMEM, and hence non-volatile. Note that at this point, the only difference with a regular IRQ routine is the saving of CPU configuration contained in SPRs.

Before powering down the chip, we need to make sure that at power up the CPU can restore the last context. Therefore a small hardware-defined location (BOOT_STAT) is used to store a boot status value (INSTANT_ON) and the new stack pointer (SP_INSTANT_ON), lines 63 to 66. The endless loop permits the system to be properly shutdown, preventing any other instruction to be executed by the CPU, and any data to be modified in memory.

```
40   _irq_inst_off :
41      // Clear trigger bit
42      l.mtspr  r0, r0, SPR_PICSR(0)
43      // Create a new stack frame in QMEM
44      l.sw     GPR1 - PWOFF_STACK_FRAME(r1), r1
45      l.addi   r1, r1, -PWOFF_STACK_FRAME
46      // Save 30 GPRs into QMEM
47      l.sw     GPR2(r1), r2
48      [···]
50      // Save EPCR , ESR , EEAR into QMEM
50      l.mfspr  r13, r0, SPR_EPCR_BASE
51      l.mfspr  r14, r0, SPR_ESR_BASE
52      l.mfspr  r15, r0, SPR_EEAR_BASE
53      l.sw     EPCR(r1), r13
54      l.sw     ESR (r1), r14
55      l.sw     EEAR(r1), r15
56      // Save group 9 of SPR (PIC) to QMEM
57      l.mfspr  r13, r0, SPR_PICMR0
58      l.mfspr  r14, r0, SPR_PICSR0
59      l.sw     PICMR0(r1), r13
60      l.sw     PICSR0(r1), r14
61      [···]
62      // Save stack pointer into QMEM
63      l.sw     SP_INSTANT_ON(r0), r1
64      // Write INSTANT_ON status at BOOT_STAT
65      l.ori    r10,r0, INSTANT_ON
66      l.sw     BOOT_STAT(r0), r10
67      // Write CPU_POWER_OFF command to PMU reg
68      l.ori    r10,r0, CPU_POWER_OFF
69      l.sw     PMU_CTRL(r0), r10
70      // Loop here now waiting for shutdown
71      l.j      .
```

The code for the restoring part of the IRQ is listed below starting from the reference line _irq_inst_on. It corresponds to the end of a regular IRQ routine, which performs

the write-back of the GPRs, SPRs and exception registers (EPC, ESR and EEAR) with their values previously saved in the stack, from lines 83 to 93. Finally, at the end of the routine, the instruction l.rfe (i.e., return from exception) writes the Program Counter (PC) and the Status Register (SR) with their respective exception registers (EPCR and ESR), so that the program execution continues from where it was before triggering the instant-off interruption.

```
72   _irq_inst_on :
73      // Reset boot status bit
74      l.sw     BOOT_STAT (r0),r0
75      // Get stack pointer from before power down
76      l.lwz    r1, SP_INSTANT_ON( r0)
77      // Restore SPRs from QMEM stack
78      l.lw     r13, PICMR0 (r1)
79      l.lw     r14, PICSR0 (r1)
80      l.mtspr  r0, r13, SPR_PICMR0
81      l.mtspr  r0, r14, SPR_PICSR0
82      // Restore EPCR, ESR, EEAR from QMEM stack
83      l.lw     r13, EPCR(r1)
84      l.lw     r14, ESR(r1)
85      l.lw     r15, EEAR(r1)
86      l.mtspr  r0, r13, SPR_EPCR_BASE
87      l.mtspr  r0, r14, SPR_ESR_BASE
88      l.mtspr  r0, r15, SPR_EEAR_BASE
89      // Restore 32 GPRs from QMEM stack
90      l.lwz    r1,  GPR1(r1)
91      [···]
92      l.lwz    r31, GPR31(r1)
93      // Return from exception to normal execution
94      l.rfe
```

IV. IMPLEMENTATION AND BENCHMARKING RESULTS

The SoC depicted in Fig. 8 has been compiled and synthesized with our own designs for STT-MRAM (Fig. 6/Fig. 7) and NVFF (Fig. 3/Fig. 4) on the one hand, and with standard SRAM blocks of the same size in the 40nm low-power CMOS technology from ST-Microelectronics on the other hand. We did not consider a non-volatile SoC embedding NOR-flash instead of RAM since their energy consumption is way too high and their speed too low, especially when writing data. Reaching the target frequency of 400MHz in both cases, the design is DRC and LVS clean for all memory and logic blocks, and the final layout is given in Fig. 9. The layout measures $2840\mu m \times 2280\mu m$ for an area of approximately $6.47mm^2$. The main part of the logic includes the bus and the peripherals (on the left) as well as the OpenRISC processor and the register file (on the right). The logic area is surrounded by the fields of STT-MRAM blocks as seen in Fig. 7.

In the benchmarks, we estimated the energy consumption of the CPU, the memory blocks and all the logic of the chip, but made abstraction of the analog modules such as the DC converters, the clock generator and the pads. We have simulated the performance and the behavior of the system using different data compression scenarios consisting in four kinds of algorithms: JPEG [19] encoding with high compression ratio, PNG [20] encoding with lossless data compression, Huffman [21] encoding which is quite fast, and LZW [22] which provides a dictionary for a faster later decoding. For this we let the input data throughput vary from 128kB/s up to realistic 512kB/s. Given as an example, a 256×256 pixel color bitmap of 192kB, as seen in Fig. 10 left, captured

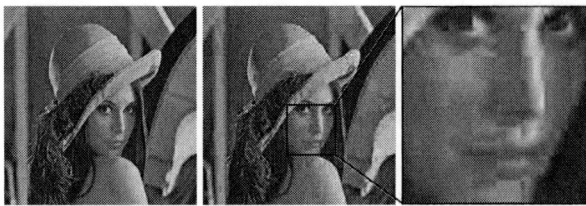

Fig. 10. Test picture "lena" in 256×256 px in original (left) and after JPEG compression using integer arithmetics (right) with details magnified 4×.

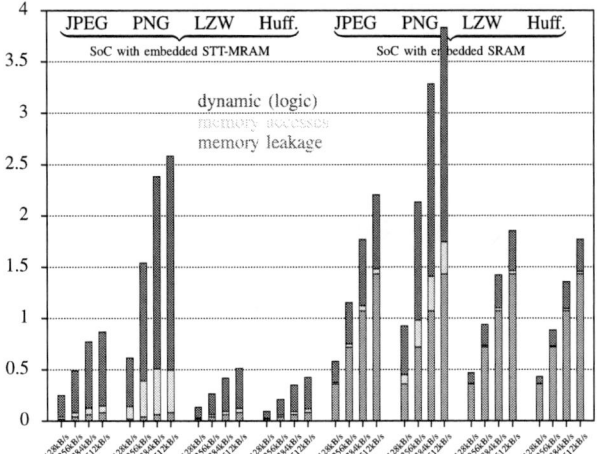

Fig. 11. Energy consumption benchmark in arbitrary units for data compression with varying input bandwidth ranging from 128kB/s up to 512kB/s, using JPEG, PNG, LZW and Huffman encoding, comparing the same SoC embedding STT-MRAM (left) and SRAM (right) memory blocks.

twice a second by a camera sensor and sent to our CPU for compression, gives a benchmark throughput of 384kB/s. As a functional verification step, Fig. 10 shows the details of the JPEG still picture compression that operates on 8×8 pixel blocks which passe through an integer based Discrete Cosine Transform (DCT), a quantization and Huffman coding processes. For the tests, input and output data is stored in the block memories (BMEM) and the data-scratchpad is used as working memory during execution, according to Fig. 8. We rely on power and clock gating when possible, hence lowering the power consumption. No data is stored on external memories. Using standards IC development and RTL (Register Transfer Level) simulation tools and the corresponding low power 40nm libraries, we can estimate the power consumption of our pre-silicon model for standard SRAM IPs and for our own STT-MRAM blocks. Fig. 11 reports the energy numbers in arbitrary units for the four algorithms executed on the SoC for one second. It appears that memory leakage is the main difference between the two hardware models, but also that computation intensive applications with a large amount of memory accesses such as PNG compression benefit from STT-MRAM technology even though reading and writing an MTJ requires more energy than an SRAM bit-cell.

V. CONCLUSION

For a system embedding MRAM blocks as data storage components, the gain in terms of power consumption is huge

compared to embedded SRAM blocks, especially when the system can be turned off most of the time, typically 99%. This is due to the leakage current of the SRAM. For data intensive applications in which data must be stored before being later processed, STT-MRAM undeniably provides a huge energy-saving advantage against all other types of memory devices. However for computation intensive programs, even though the active power dominates, the use of NVM still contributes to reduce the overall consumption. Our benchmark showed that our hybrid CMOS/STT NV-SoC running 32-bit applications for still image compression achieves a power reduction factor of up to 2.5, thanks to optimized instant-on/off procedures.

ACKNOWLEDGMENT

The authors would like to thank the European Union for the financial support through the ERC Advanced Grant HYMAGINE nr. 246942.

REFERENCES

[1] L. Minas, B. Ellison, "The Problem of Power Consumption in Servers", Hillsboro, Intel Press, 2009

[2] T. Vogelsang, "Understanding the Energy Consumption of Dynamic Random Access Memories", IEEE MICRO, pp. 363-374, 2010

[3] J. Carter, K. Rajamani, "Designing Energy-Efficient Servers and Data Centers", IEEE Computer, 43(7), 2010

[4] M. Bohr, "The New Era of Scaling in an SoC World", Intl. Solid-State Circuits Conference (ISSCC), Digest Tech. Papers, pp. 23,28, 2009

[5] M. Rasquinha et al., "An Energy Efficient Cache Design Using Spin Torque Transfer (STT) RAM", Proc. ISLPED, 2010

[6] E. Kültürsay et al., "Evaluating STT-RAM as an Energy-Efficient Main Memory Alternative", IEEE, Proc. ISPASS, 2013

[7] A. Khvalkovskiy et al., "Basic Principles of STT-MRAM Cell Operation in Memory Arrays" J. Applied Physics, Vol. 46, Iss. 7, 074001, 2013

[8] N. Sakimura et al., "Nonvolatile Magnetic Flip-Flop for Standby power free SoCs", Proc. IEEE Custom IC Conf., pp. 355-358, 2008

[9] H. Yoda et al., "Progress of STT-MRAM Techno. and the Effect on Normally-off Comp. Systems", Proc. IEEE IEDM, pp. 1131-1134, 2012

[10] G. Di Pendina et al., "A Hybrid Magnetic/CMOS PDK for the Design of Low-Power Logic Circuits", J. Appl. Phys., Vol. 111, 2012

[11] F. Bernard-Granger et al., "SPITT: an MTJ SPICE compact model for STT-MRAM", MOS-AK Workshop, DATE Conf., 2015

[12] K. Jabeur et al., "Comparison of Verilog-A compact Modeling Strategies for Spintronics Devices", IEEE Electronics Letters, 2014

[13] K. Huang et al., "Low Power Computing using STT-MRAM", Proc. Non-Volatile Memory Technology Symp. (NVMTS), pp. 1-6, 2014

[14] J.-M. Portal, M. Bocquet, M. Moreau et al., "An Overview of Non-Volatile Flip-Flops Based on Emerging Memory Technologies", J. Electronic Science and Technology, Vol. 12, No. 2, 2014

[15] W. Wei et al.,"Design of a Nonvolatile 7T1R SRAM Cell for Instant-on Operation", IEEE Trans. Nanotech., Vol. 13, No.5, pp. 905-916, 2014

[16] T. M. Maffitt et al., "Design Considerations for MRAM", IBM Journal Res. Dev., Vol. 50, No. 1, 2006

[17] D. Lampret et al., "OpenRISC 1000 Architecture Manual", rev. 2012. Available from the OpenCores website: http://opencores.org/or1k

[18] J. Sun et al., "A Checkpointing and Instant-On Mechanism for an Embedded System based on Non-Volatile Memories", Proc. IEEE Comp. Comm. and IT Applications Conf., pp. 173-178, 2014

[19] T. Noda et al., "A JPEG Still Picture Compression LSI", Digest IEEE Symp. VLSI Circuits, pp. 33-34, 1991

[20] G. Roelofs et al., PNG website, IEC std. http://www.libpng.org/pub/png

[21] D. A. Huffman, "A Method for the Construction of Minimum-Redundancy Codes", Proc. IRE, pp. 1098-1102, 1952

[22] J. Ziv, A. Lempel, "A Universal Algorithm for Sequential Data Compression", IEEE Trans. Information Theory, 23(3):337-343, 1977

978-1-4799-8720-7/15 $31.00 © 2015 IEEE

2015 IEEE Computer Society Annual Symposium on VLSI

Radiative Effects on MRAM-Based Non-Volatile Elementary Structures

(Invited Paper)

Jeremy Lopes*[†§], Gregory Di Pendina*, Eldar Zianbetov*, Edith Beigne[‡] and Lionel Torres[§]

*Univ. Grenoble Alpes, INAC-SPINTEC,F-38000 Grenoble, France

CNRS, SPINTEC, F-38000 Grenoble, France

CEA, INAC-SPINTEC, F-38000 Grenoble, France

[†]CNES, Service Environnement et Composants nouveaux DCT/AQ/EC

[‡]CEA LETI, Minatec, Grenoble, France

[§]LIRMM, UMR CNRS 5506, University of Montpellier, France

Abstract—Radiation robust circuit design for harsh environments like space is a big challenge for today engineers and researchers. As circuits become more and more complex and CMOS processes get denser and smaller, their immunity towards particle strikes decreases drastically. This work has for objective to improve the System on Chip (SoC) robustness against particle attacks targeting very advanced processes. This should be possible combining three already proven robust design techniques: Asynchronous communication, Silicon on Insulator (SOI) technologies and Spintronics. The combination of these three techniques should give some fundamentally new architecture with higher performances than what is available today in terms of robustness but also in terms of speed, consumption and surface.

Index Terms—Asynchronous integrated circuit Design, Radiation hardening, Radiative particle injection, 28nm FD-SOI, Spintronics, STT-MRAM, SOT-MRAM.

I. INTRODUCTION

Today, there are several ways to develop microelectronic circuits adapted for space applications that meet the harsh constraints of temperature range, voltage and most importantly immunity to radiation, whether in terms of technical design or manufacturing process. The aim of this work is to improve circuit robustness targeting several very advanced techniques and approaches of microelectronics to design architectures adapted to this type of radiative environment. The techniques used here are: Asynchronous communication, Silicon on Insulator (SOI) technologies, magnetic memory (MRAM) technology and hardening by design. Such an assembly would be totally innovative and should benefit without precedent, in terms of silicon area, consumption and robustness which is the main target of our work.

This paper is divided as followed: In section II, we will present a general background on the techniques used. Section III gives transistor level simulation results on the performance of asynchronous logic in a harsh radiative environment in comparison with synchronous logic. The incorporation of Magnetic Tunnel Junctions (MTJs) in asynchronous and synchronous memorising elements is studied in section IV, as well as an in-depth comparison of Spin Transfer Torque (STT) and Spin

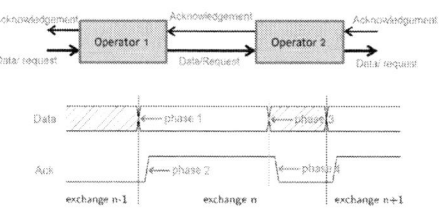

Fig. 1. Example of an asynchronous communication between two combinational blocks using a 4 phase protocol with a three state coding.

Orbit Torque (SOT) MTJs towards particles strikes. Finally Section V concludes this paper and offers perspectives for our future work.

II. BACKGROUND

A. Asynchronous Design

The asynchronous design is based on a specific communication protocol between purely combinational logic blocks of an integrated circuit. There are two main protocols used for the communication between operators: The 2 phase protocol and the 4 phase protocol. Each of these blocks has a request and an acknowledgement signal for communication between the previous or following block, i.e. when the calculation is complete (Fig. 1). Any change of data is acknowledged by the acknowledgement signal and any change to this signal is acknowledged by a data change and so on, this is called handshaking [1]. One question remains: how do we detect the presence of data and how do we generate a signal that indicates the end of the operation? The answer is in the adoption of a particular encoding for the data. It is impossible to use a single line per bit, because it makes it impossible to detect data that has the same value as the previous data state. The solution is the use of two lines per bit and the creation of a request signal associated with the data [2].

- Double rail coding: With two lines per data bit, four states are available to express the logic values '0' or '1'. The two most commonly used codes are the three state codes and the four state codes [3]. The three state coding is best

978-1-4799-8720-7/15 $31.00 © 2015 IEEE   321

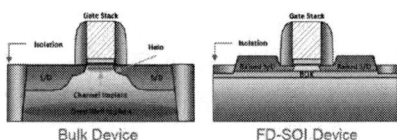

Bulk Device   FD-SOI Device

Source: https://www.semiwiki.com/forum/content/3599-soi-really-less-expensive.html

Fig. 2. Cut view of a transistor in both Bulk and FD-SOI technologies.

suited for the 4 phase protocol and the four state coding is best suited for the 2 phase protocol.

• Bundled data coding: Double rail coding is complicated and takes a lot of time to set-up as it doubles the number of wires. In order to simplify asynchronous design it is possible for multi-bit data to have a unique request line.

In addition to having a particularly low consumption compared to the synchronous circuits and being almost insensitive to delays and therefore to changes in manufacturing process, asynchronous circuits can be considered more robust towards SEE than traditional design [4].

Globally Asynchronous Locally Synchronous (GALS) architecture are on the other hand composed of large synchronous blocks which communicate with each other on an asynchronous basis [5]. This architecture allows designers to use common tools to design synchronous operators and then integrate them into a larger asynchronous circuit where each synchronous block communicates with the others using the asynchronous protocol [6]. Each synchronous operator is basicly surrounded by an asynchronous wrapper. This wrapper is the interface between synchronous and asynchronous worlds.

B. Silicon on Insulator

It is widely recognized in the field of microelectronics that integrated circuits manufactured on SOI substrates are more robust to radiation due to the smaller volume of the active areas of the transistors [7] and also thanks to the possibility of local insulation provided by the manufacturing process (Fig. 2). The radiation-induced errors are very localized and therefore cannot be spread from one block to another. SOI also opens the world of body biasing [8]. Body biasing is the possibility to locally polarise the bulk on a transistor in a way that influences its threshold voltage (Vt). By applying a positive or negative body biasing voltage (vbb), we can increase or decrease the transistor speed. This is a fundamental feature that helps to dissipate external energy provided by particle strikes.

C. Spintronics

Spintronics is the science of placing ferromagnetic materials on the route of electrons and using their spin to influence the mobility of the electrons in the materials [9]. The electron's spin has two states depending on the direction of the angular momentum. If the angular momentum is clockwise then it is called "spin up", if it is counter clockwise then it is called "spin down" [10]. When a current passes through

STT	SOT
+ Size / Density (2 terminals)	- Size / Density (3 terminals)
+ Low writing current	± Switching needs high current density (could decrease in the future)
- Common read and write paths	+ Separate read and write paths
- Stress on the MTJ barrier during writing	- Less mature technology
- Read operation can switch the magnetization state	- Small magnetic field is required to avoid stochastic switching
- Non-symmetrical current to switch from parallel to anti-parallel	+ Symmetrical current to switch from parallel to anti-parallel
	+ Writing speed
	+ Choice of R_p/R_{ap} values

TABLE I

COMPARATIVE TABLE OF STT AND SOT

a ferromagnetic material with a majority of "spin up" the electrons with a "spin down" are slowed, stopped or even reflected, allowing only the electrons with an "spin up" to pass through the material. This means that depending on the majority spin, a ferromagnetic material can be more or less resistive, this is called the Magnetoresistance effect.

The central device of all MRAMs today is the MTJ, which is based on the Tunnel Magnetoresistance effect (TMR) and which is intrinsically immune to radiations [11]. The MTJ is composed of two ferromagnetic layers, a reference layer (with a fixed magnetisation) and a storage layer (with a freely imposable magnetisation) with a thin insulator (barrier) between them. Depending on the relative magnetic orientation of the two layers (parallel or anti-parallel) the structure resistance changes being either high or low. This is caused by the electrons being able to pass more or less easily through the device and the thin insulator. Fig. 3 shows a simplified structure of the MTJ [12]. Changing the magnetic orientation of the storage layer makes it possible to have the MTJ in two different states: Parallel, or Anti-parallel. Depending in which state the MTJ is in, it has a different resistance Rp (parallel) or Rap (anti-parallel) [10].

Today the two main switching techniques that are either on the way of being integrated into industrial products or that have a large potential are respectively, Spin Transfer Torque switching (STT) and Spin Orbit Torque switching (SOT). The SOT process is still in the early stages of development. STT switching uses a polarized current injected through the 2-terminal MTJ to switch the magnetization of the storage layer [13]. Where as SOT uses a current passing under the storage

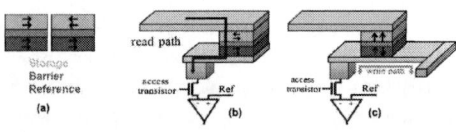

Fig. 3. (a) Simplified structure of a Magnetic Tunnel Junction: Right Parallel (R_p), Left Anti-parallel (R_{ap}); (b) STT-MTJ (Spin Transfer Torque); (c) SOT-MTJ (Spin Orbit Torque).

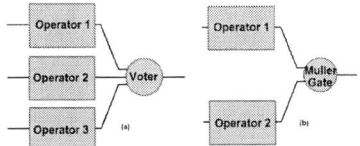

Fig. 4. Example of material redundancy: (a) TMR (Triple Modular Redundancy), (b) DMR (Dual Modular Redundancy).

layer of the 3-terminal MTJ, in a conducting line [14] (Fig. 3). Table 1 summarizes the main pros and cons of both STT-MTJ and SOT-MTJ. For instance SOT has separate reading and writing paths leading to high reliability and very high endurance, while STT share the same path. On the other hand STT with its 2-terminal structure is smaller than SOT that has 3-terminals.

D. Hardening by design

In addition to the means of protection against radiation described above, it is possible to adopt a system level approach through redundancy, error detection and correction, or data scrubbing.

This can be done either materially or by software. The material redundancy method is the replication of system resources. The most commonly used method is TMR (Triple Modular Redundancy) that enables the detection and correction of errors [15], but DMR (Dual Modular Redundancy) [16] has also attracted the attention of designers because it requires only one replication of the system (Fig. 4). However it permits only the detection and not the correction of errors. Software redundancy [17] is a sequential execution of a same task followed by a vote. This results in a loss of performance without the material part of the application being modified. Data scrubbing is the periodic inspection of memory cells looking for errors. If errors are detected, they are corrected using redundant data [18].

E. Radiations

Even thought space is an empty place, there is in reality hundreds of thousands of energetic particles of various kinds (Protons, Heavy ions, Electrons) [19] that are mainly produced by the sun. In our study we chose to simulate the effect of heavy ions since they are one of the most energetic and dangerous particles for electronic circuits orbiting the earth [20]. By studding the effect of heavy ions we cover the effect of the other particles which have lower energies. Heavy ions are produced by solar wind or flares and have energies from a few keV to $10MeV$ (Fig. 5).

We simulate the heavy ions effect by a current pulse that is determined by the following equation [21]:

$$I_{inj} = \frac{Q_{inj}}{\tau_1 - \tau_2}(e^{\frac{-t}{\tau_1}} - e^{\frac{-t}{\tau_2}})$$

Where Q_{inj} is the charge in Coulombs of a particle that vary from -2pC to 2pC, τ_1 and τ_2 are material dependant

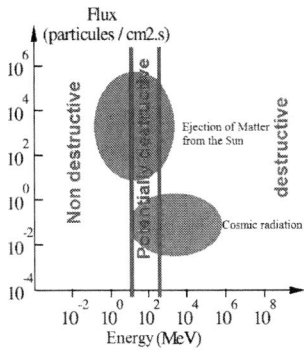

Fig. 5. Particle concentration and energies around the earth.

constants, here respectively 150ps and 50ps (which are the values of deposit and evacuation of electrical charges brought by a radiative particle). Fig. 6 gives a few plotted curves of the equation for several values of Q_{inj}.

The effect of a heavy ion hitting a transistor has been documented in the past and has shown that the most sensitive strike location of a transistor is the drain [22] (Fig. 8). In our simulations we consider the worst case: we inject particles that have energies at the limit of what is potentially destructive and consider that all the particles hit the drains of the transistors.

III. CMOS CHARACTERISTIC'S FOR RADIATION HARDENING

This study has for objective to determine the performance of asynchronous logic in a harsh radiative environment using very advanced processes compared to synchronous logic (FD-SOI). We designed two memory elements of synchronous and asynchronous design. A flip-flop and a half-buffer for respectively synchronous and asynchronous design techniques. These two structures conduct the same role in the two worlds and were designed to have the same timing characteristics so they could be compared coherently.

All simulation results presented in this paper were run with Spectre electrical simulator under Cadence Analog Design Environment platform using the 28nm FD-SOI technology from ST Microelectronics.

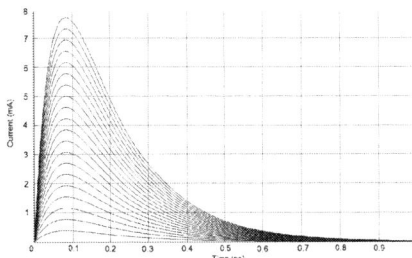

Fig. 6. Examples of the induced current pulses generated by a particle impact waveforms simulated.

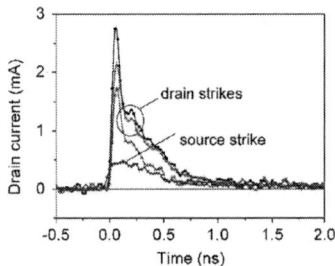

Fig. 7. Examples of transients measured for different ion strike locations along the source drain axis of a 0.25 μm bulk transistor. The gate length is 0.25 μm and the transistor width is 25 μm. [22]

A. Particle injection

To simulate the effect induced by a particle hitting the drain we inject a current pulse which has the same characteristics as a heavy ion strike to the drain of a transistor (Fig. 8). This current pulse will be sequentially injected in all the sensitive nodes of the evaluated circuit to determine the global sensitivity of the circuit. A sensitive node being any internal node except nodes connected to the supply voltage, ground, or any of the inputs or outputs.

B. Influence of the supply voltage

We conducted simulations of particle strikes at different supply voltages. From the simulations we have observed the inverse relationships between supply voltage and the amplitude on the induced pulse. In other words, as we increased the supply voltage for different particle strikes with identical energies, the effect of the particles decreased. This can be explained by the fact that, as the supply voltage increases the conductance of the transistor increases and also eases the dissipation of external energies.

C. Impact of the threshold voltage

In the 28nm FD-SOI design kit we have the choice between LVT (Low Vt) and RVT (Regular Vt) transistors. In our initial hypothesis RVT transistors would be more robust to pulses induced by particles strikes than LVT because of their superior Vt, thus less sensitive to commutations. The simulation results demonstrated that LVT are finally more robust. This can be explained by the fact that LVT transistors have a higher leakage current than RVT, so the induced current pulses are evacuated quicker due to the lower resistance.

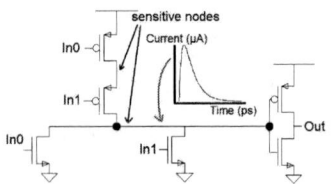

Fig. 8. Example of a particle injection (current pulse) in an OR gate.

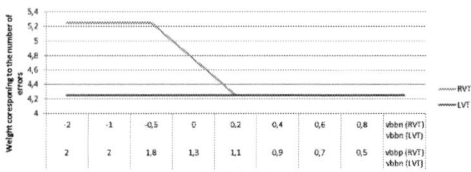

Fig. 9. Graphical representation of the number of particles induced errors depending on the Body Biasing.

D. Influence of body biasing

FD-SOI enables us to use body biasing to play around with the Vt of transistors. We then looked at the influence of vbb (body biasing voltage) in the case of particle induced errors. In RVT transistors, as vbb increased (boost configuration, Vt decreases) the global number of errors induced by heavy ion strikes on the circuit decreased. As the transistors response is boosted the accumulated charges caused by particle strikes can be evacuated quicker. Concerning LVT transistors, being already more robust to particles than RVT, increasing or decreasing vbb has no further effect on them (Fig. 9).

IV. THE EFFECT OF NON-VOLATILITY IN A SIMPLE STRUCTURE

The next step in our study was to see how the introduction of MTJs in the half-buffer and flip-flop would affect their robustness towards particle strikes.

A. Non-volatile Flip-Flop

To design a non-volatile (NV) flip-flop we used the classic flip-flop from the previous section and replaced one of the latches with a NV latch. We chose to use a dual MTJ Latch cell (Hass) for its robustness towards radiation [23]. Fig. 10 shows the architecture used for both STT and SOT NV latches.

During the writhing phase the MTJs are disconnected from the rest of the circuit with $WR = 1$ ($NWR = 0$), this allows the write command signal (WR_cmd) to transit through the junctions in opposite directions thus writhing the two MTJs in opposite magnetic states. Reading is achieved by the RD signal. When RD is high Q and nQ are forced to the same value (about $vdd/2$), and since the resistances of the MTJs are different, the latch becomes unstable and unbalanced, forcing Q to GND or VDD depending on the state of the MTJ, and NQ to VDD or GND.

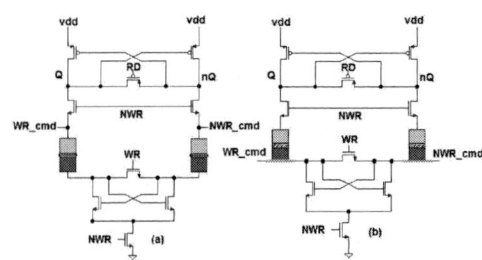

Fig. 10. Dual MTJ Latch cell(Hass): (a) STT version, (b) SOT version

978-1-4799-8720-7/15 $31.00 © 2015 IEEE   324

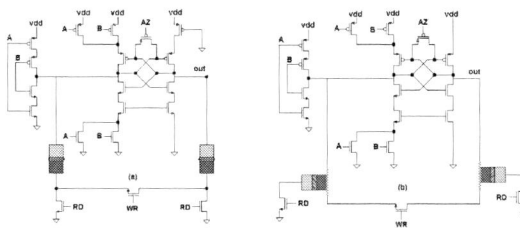

Circuit type	Number of Errors
Flip-Flop	0.41
NV STT Flip-Flop	0.47
NV SOT Flip-Flop	0.43
Half-Buffer	0.35
NV STT Half-Buffer	0.82
NV SOT Half-Buffer	0.66

TABLE II
COMPARATIVE TABLE OF SYNCHRONE/ASYNCHRONE,
VOLATILE/NON-VOLATILE AND STT/SOT

Fig. 11.   Non-volatile C-element: (a) STT version, (b) SOT version

B. Non-volatile Half-Buffer

The design of a NV half-buffer was achieved by replacing the C-elements by NV C-elements previously designed in our laboratory [24]. Fig. 11 show the architecture used for both STT and SOT NV C-elements.

The reading and writing functionality for the MTJs are the same as described for flip-flops in the previous subsection.

C. Simulation results

Current pulses were injected into sensitive nodes as describes earlier (Fig.8). Depending on the simulated circuit (volatile/non-volatile, synchronous/asynchronous) the number of sensitive nodes varies from 15 to 24. For each sensitive node we conducted successive simulations for all the possible input combinations, this lead to around 2000 individual simulations. The outputs of the flip-flop and half-buffer as well as the Magnetic states of the MTJs where compared to simulation results previously obtained without error injection using SOT-MTJ and STT-MTJ compact models described in Verilog A language [25].

Simulation results showed that due the high energies involved in some particles strikes, an error induced by a particle could write or even destroy an MTJ depending on where the particle hits. Some work have shown [11] that MTJs are robust against radiations, but like integrated circuit, it is the CMOS transistors that are sensitive towards particle strikes and indirectly causes bit flip on MTJ memory cells. Indeed, the current generated is high enough during a non-negligible time to induce such a behaviour. Moreover, an important thing to keep in mind when designing a circuit for radiative environments is the size of the transistors. The bigger the transistors are the less a particle hit will effect it and its surroundings. This can be explained by the fact that as transistors get bigger their resistance decreases, thus the amplitude of the voltage created by the current pulse will be lower.

To render MTJ more robust to radiative environment a solution is to increase there dimensions:

1) Influence of the MTJs dimensions for STT: As the dimensions of the MTJ increases more current is needed to write the junction. This is explained by the resistance of the MTJ decreasing as its size increases.

2) Influence of the strip dimensions for SOT: The switching mechanism for SOT-MTJs is dependant on the current density that passes under the junction in the conduction line. If the dimension of the conduction line are increased the current density that passes under the junction for a same given current will be lessened. Thus increasing the dimensions until a point where the current pulses induced by particles strikes will not create a sufficient current density under the junction to write the junction its a way of increasing is robustness.

D. Results and Interpretation

The inclusion of MTJ in the flip-flop or half-buffer has not rendered them more robust to particles strikes. This can be explained in the following manor: The addition of non-volatility in an elementary circuit increases the number of sensitive nodes as well as the global complexity. The more sensitive nodes the more a circuit is sensitive to errors induced by particles.

Table II gives a synchrone/asynchrone, volatile/non-volatile and STT/SOT comparison to particles strikes. The number of errors is determined by the actual number of errors visualised at the circuits outputs and/or Magnetic states. These numbers are normalised to the number of inputs and sensitive nodes so as to have a coherent comparison. The junctions used had a writing stripe of $200nm \times 100nm \times 5nm$ for SOT and diameter of $40nm$ for both STT and SOT. The energy needed to write such an STT-MTJ with a current pulse of $44\mu A/2.7ns$ is $0.15pJ$. This energy is twice as less than the energy needed for the flip-flop to change its output ($0.3pJ$). To render the junction more robust towards particle strikes, the MTJ switching energy should be increased. This can be done by increasing the dimensions of the MTJ or through customising the magnetic post-process.

We can see that volatile half-buffer is very slightly more robust than volatile flip-flop, and the addition of MTJs decreases there robustness. We can also see that SOT is also very slightly more robust than STT. This is due to the writing strip, more current is needed to write the junction so it is less sensitive than STT. The NV half-buffer is a lot less robust than the NV flip-flop. This can be explained because the Hass structure used in the flip-flop is the most robust architecture compared to a classical half buffer structure which is composed of two NV Muller cells each one having two differential MTJs.

978-1-4799-8720-7/15 $31.00 © 2015 IEEE

V. GENERAL CONCLUSION AND PERSPECTIVES

This work has allowed us to fully understand the influence of heavy ion strikes on transistors in elementary circuits. We have seen the influence of the supply voltage and body biasing on both LVT and RVT transistors when confronted with particle strikes. The simulation results have demonstrated that using the highest supply voltage permitted by the technology, in conjunction with LVT transistors is the best option to harden a system at transistor level. We also investigated the influence of MTJ in elementary circuits. The integration of MTJs in elementary circuits is not the solution to harden circuits even tough they are intrinsically immune to radiation. There integration at an architectural level with standard structures to limit the number of sensitive nodes could be the solution. Concerning the differences between synchronous and asynchronous design, it is difficult to determine which is more robust at transistor level. The differences are based on the communication protocols, the robustness of one design method over the other should become visible in a more complex architecture such as a pipeline structure.

Our next step will be to compare both asynchronous and synchronous architectures in a pipeline to see how the errors are propagated through the different stages; this should give us an in-depth comparison of both asynchronous and synchronous design towards radiative environments. This will lead to a study on hardening techniques by design such as DMR (Dual Modular Redundancy) and TMR (Triple Modular Redundancy). Indeed, the use of non-volatility thanks to MTJs should allow the restoration of secure data after a particle strike. This will allow us to compare both synchronous and asynchronous architectures in new conditions to determine which is more robust in harsh radiation environments. We will then evaluate a one MTJ one resistor rather than a two MTJ structure to see the difference in behaviour. Flowed by a study of the Mean Time To Failure (MTTF) of all the different combinations of volatile/non-volatile, synchronous/asynchronous. Finally we will investigate the hardening of MTJs by customisation, either at design or process level.

ACKNOWLEDGMENT

We would like to thank the CEA and CNES for their financial support for this work.

REFERENCES

[1] Renaudin, Marc and Rigaud, Jean Baptiste, *Etat de lart sur la conception des circuits asynchrones: perspectives pour lintégration des systèmes complexes*, rapport de recherche interne TIMA/STMicroelectronics, 2000.

[2] Vivet, Pascal, *Une méthodologie de conception de circuits intégrés quasi-insensibles aux délais: application à l'étude et à la réalisation d'un processeur RISC 16-bit asynchrone*, Ph.D. thesis, Institut National Polytechnique de Grenoble-INPG, 2001.

[3] Moreira, Matheus Trevisan and Calazans, Ney Laert Vilar, *Proposal of an Exploration of Asynchronous Circuits Templates and their Applications*, 2014.

[4] Jeitler, Marcus and Lechner, Jakob, *Comparing the robustness of synchronous and asynchronous circuits by fault injection*, 2009.

[5] Muttersbach, Jens and Villiger, Thomas and Fichtner, Wolfgang, *Practical design of globally-asynchronous locally-synchronous systems*, Advanced Research in Asynchronous Circuits and Systems, 2000.(ASYNC 2000) Proceedings. Sixth International Symposium on, pp. 52-95.

[6] Teehan, Paul and Greenstreet, Mark and Lemieux, Guy, *A survey and taxonomy of GALS design styles*, Design & Test of Computers, IEEE, 2007, vol. 24, no 5, pp. 418-428.

[7] Roche, Philippe and Autran, Jean-Luc and Gasiot, Gilles and Munteanu, Daniela, *Technology downscaling worsening radiation effects in bulk: SOI to the rescue*, Nuclear Science, IEEE international electron device meeting (IEDM2013). Washington DC, 2013, pp. 766-769.

[8] Flatresse, Philippe and Giraud, Bastien and Noel, J and Pelloux-Prayer, Bertrand and Giner, Fabien and Arora, D and Arnaud, Franck and Planes, Nicolas and others, *Ultra-wide body-bias range LDPC decoder in 28nm UTBB FDSOI technology*, Solid-State Circuits Conference Digest of Technical Papers (ISSCC), 2013 IEEE International, 2013, pp. 424-425.

[9] Kang, Wang and Zhao, Weisheng and Wang, Zhaohao and Klein, Jacques-Olivier and others, Dafiné and Chappert, Claude, *An overview of spin-based integrated circuits*, ASP-DAC, 2014, pp 676-683.

[10] Di Pendina, Gregory, *Conception innovante et développement d'outils de conception d'ASIC pour Technologie Hybride CMOS/Magnétique*, Université de Grenoble, 2012.

[11] Conraux, Yann, *Préparation et caractérisation d'un alliage amorphe ferrimagnétique de GdCo entrant dans la conception de jonctions tunnel magnétiques Résistance des jonctions tunnel magnétiques aux rayonnements ionisants*, Ph.D., Université Joseph-Fourier-Grenoble, 2005.

[12] Sousa, Ricardo C and Prejbeanu, I Lucian, *Non-volatile magnetic random access memories (MRAM)*, Comptes Rendus Physique, 2005, vol. 6, no 9, pp. 1013-1021.

[13] Dieny, Bernard and Sousa, RC and Herault, J and Papusoi, C and Prenat, G and Ebels, U and Houssameddine, D and others, *Spin-transfer effect and its use in spintronic components*, International Journal of Nanotechnology, 2010, vol. 7, no 4, pp. 591-614.

[14] Jabeur, K and Buda-Prejbeanu, LD and Prenat, G and Di Pendina, G, *Study of two writing schemes for a magnetic tunnel junction based on spin orbit torque*, International Journal of Electronics Science and Engineering, 2013, vol. 7, no 8, pp. 501-507.

[15] Di, Jia, *A framework on mitigating single event upset using delay-insensitive asynchronous circuits*, Region 5 Technical Conference, 2007 IEEE, 2007, pp. 354-357.

[16] Gong, Rui and Chen, Wei and Liu, Fang and Dai, Kui and Wang, Zhiying, *A new approach to single event effect tolerance based on asynchronous circuit technique*, Journal of Electronic Testing, 2008, vol. 24, n 1-3, pp. 57-65.

[17] Bolchini, Cristiana and Miele, Antonio and Rebaudengo, Maurizio and Salice, Fabio and Sciuto, Donatella and Sterpone, Luca and Violante, Massimo, *Software and hardware techniques for SEU detection in IP processors*, Journal of Electronic Testing, 2008, vol.24, no 1-3, pp.35-44.

[18] Goncalves, Olivier.*Conception sur mesure d'un FPGA durci aux radiations base de mmoires magntiques*. Ph.D. thesis, Université de Grenoble, 2013.

[19] Peronnard, Paul. *Méthodes et outils pour l'évaluation de la sensibilité de circuits intégrés avancés face aux radiations naturelles*, Ph.D. thesis, Université Joseph-Fourier-Grenoble I, 2009.

[20] Robert C. Baumann, *Soft Errors in Advanced Semiconductor Devices - Part I: The Three Radiation Sources*, Device and Materials reliability, IEEE Transactions on, 2001, vol.1, no 1, pp.17-22.

[21] Zhao, Weisheng and Deng, Erya and Klein, Jacques-Olivier and Cheng, Yuanqing and Ravelosona, Dafiné and others, *A radiation hardened hybrid spintronic/CMOS nonvolatile unit using magnetic tunnel junctions*, Journal of Physics D: Applied Physics, 2014, vol.47, no 40, pp. 405003.

[22] Ferlet-Cavrois, V and Paillet, P and Gaillardin, M and Lambert, D and Baggio, J and Schwank, JR and Vizkelethy, and others, *Statistical analysis of the charge collected in SOI and bulk devices under heavy Iion and proton irradiationImplications for digital SETs*, Nuclear Science, IEEE Transactions on, 2006, vol. 53, no 6, pp. 3242-3252.

[23] Raphael Martins Brum, *Une tude des mmoires magnetiques appliques aux processeurs et FPGAs*, Ph.D. thesis, Université Montpellier 2, 2014.

[24] Zianbetov, Eldar and Beigne, Edith and Di Pendina, Gregory, *Non-Volatility For Ultra-Low Power Asynchronous Circuits in Hybrid CMOS/Magnetic Technologys*, 2015.

[25] http://www.mos-ak.org/grenoble_2015/presentations/T5_Bernard-Granger_MOS-AK_Grenoble_2015.pdf.

RRAM Reliability/Performance Characterization Through Array Architectures Investigations

Cristian Zambelli, Alessandro Grossi, and Piero Olivo
Università di Ferrara, Dipartimento di Ingegneria, ENDIF
Via Saragat 1, Ferrara, 44122, Italy
E-mail: cristian.zambelli@unife.it

Christian Walczyk, and Christian Wenger
IHP Microelectronics
Im Technologiepark 25, 15236 Frankfurt (Oder), Germany
E-mail: christian.wenger@ihp-microelectronics.com

Abstract—The reliability and performance characterization of each non-volatile memory technology requires the thorough investigation of dedicated array test structures that mimic the real operations of a fully functional integrated product. This makes no exception also for emerging non-volatile memories like the Resistive Random Access Memory (RRAM) concept. An extensive electrical characterization activity performed on test vehicles manufactured in a CMOS backend-of-line process allowed the first glance estimation of operation modes and reliability threats typical of this technology. In this paper, it is provided a review of the most important issues like forming instabilities, optimal set/reset operation finding, and read disturb to provide a guideline either for a further technology optimization or an efficient algorithms co-design to handle these reliability/performance threats.

Index Terms—RRAM; test-structures; array; reliability; performance

I. INTRODUCTION

The Resistive Random Access Memories (RRAM) are a promising candidate to become a key memory technology in several applications. From a technological standpoint, embedded HfO_2-based RRAM devices are interesting because they offer compatibility with the standard CMOS backend-of-line (BEOL) process scheme and very fast operation times, mostly below the 100 ns limit. Extensive characterizations have been performed in the framework of a concept-validation for possible replacement of existing non-volatile memory technologies, trying to ease the evolution from single cell test structures to fully functional integrated array products [1]–[4].

In the last decade, most of the analysis, especially those devoted to provide a solid understanding of the physical mechanisms ruling the RRAM operations, have been performed on simple 1T-1R architectures where a select transistor or a diode is connected in series to the resistive element [3]. However, although this demonstrated competitive features with respect to the traditional floating gate-based Flash technology, single devices are not ideal to study the statistical distribution of the inter-cell variability of memory elements. Moreover, that solution does not allow a thorough characterization of the typical issues evidenced in a memory product such as disturbs, cells interaction, sub-optimal writing algorithms or cell faults due to process induced variability.

In this work we will present a review of the most important issues retrieved during the electrical characterization of 4kbits

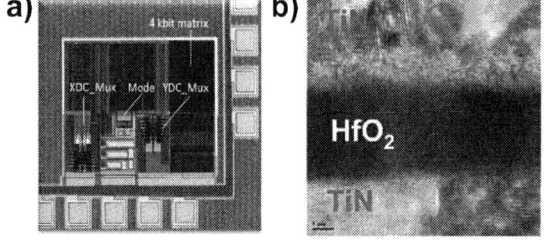

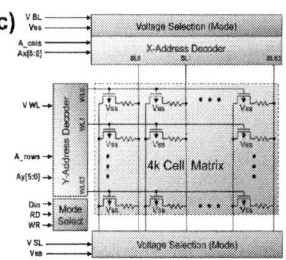

Fig. 1. Microphotograph of the 4kbits memory array with control circuits (a). Cross-Sectional STEM Image of the integrated MIM stack in the ReRAM Cell (b). Simplified block diagram of the memory array (c).

RRAM memory array test structures with the associated control circuitry designed in a 0.25 μm BiCMOS technology node [5]. The results will show that some important reliability threats that could severely limit the ramp-up of the RRAM arrays toward a technology mature level, can be addressed by simply acting on the operative conditions of the cells within the array and on the algorithms that handle the read/write operations. The common analysis performed in these test arrays are mainly related to the evaluation of the performance and of the reliability features of the technology, through the application of voltage/current waveforms at dedicated test pads for the following operations: forming, set, reset, and read. The applied waveforms shape, duration, and amplitude determine the behavior of the array either on a short time-scale (e.g., characterization of the read window, determination of the variability, etc.) or on a longer time-scale (e.g., endurance and retention evaluations).

978-1-4799-8720-7/15 $31.00 © 2015 IEEE

II. TEST VEHICLE MANUFACTURING

The structure of the 4kbits memory array (see Fig. 1) is described by four architectural blocks: the array of 4096 1T-1R RRAM cells; a wordline (WL) address decoder (XDC MUX); a bitline (BL) address decoder (YDC MUX); and an operation control circuitry (Mode) to handle read and write operation commands. The memory cells are constituted by a select NMOS transistor featuring W=1.14 μm and L=0.24 μm in series to a variable resistor connected to the bitlines.

The variable resistor is a Metal-Insulator-Metal (MIM) stack fabricated on 150 nm TiN bottom electrodes deposited by magnetron sputtering with sheet resistances in the order of 10-50 Ωsq^{-1} directly on the last metallization of the BiCMOS process. Next, HfO$_2$ films of 9nm thickness were grown in an AVD chamber at 320 °C using Hf[N(MeEt)]$_4$ precursor and O$_2$ as reactive gas. Finally, 10 nm Ti and 150 nm TiN were sputtered onto the HfO$_2$ layer [6], [7]. To investigate the impact of the MIM area on the memory performances and variability two different arrays have been integrated using 0.6 μm^2 and 1 μm^2 resistor area, respectively.

III. FORMING OPERATION VARIABILITY AND INSTABILITIES

RRAM behavior is based on the possibility of electrically modifying the conductance of a Metal-Insulator-Metal (MIM) stack: the Set operation switches the cell into a high conductive state, whereas Reset brings the cell back to a low conductive state. Some technologies like HfO$_2$-based RRAM require a preliminary forming operation to activate such a switching behavior by creating a conductive filament (CF) in the dielectric material [5], [8]–[10]. Even if such forming process is performed just once, it plays a fundamental role in determining the system performance [10]. A deep understanding of the forming process allows recognising faulty cells in the array from scratch and to get a first glance insight on the cells reliability and performances during lifetime. Forming usually consists in the application of a quasi-DC sweep on the BL up to $V_{BL} = 3.5$ V with step voltage equal to 0.025 V. To prevent hard breakdown, the saturation current of the select transistor is controlled by the WL voltage fixed at $V_{WL} = 1.4$ V, which translates into a compliance current almost equal to 300 μA. The forming process in the array could be accelerated by selecting multiple rows and/or columns simultaneously using the Mode circuitry. After the operation it is possible to read the array content by applying $V_{WL} = 1.4$ V and a read voltage V_{BL} considerably lower than the switching voltages requested for the set and reset operations.

Fig. 2 depicts the variability of the forming voltage distribution and the cumulative probability data of the read currents measured on the entire array before and after the forming operation. The large variability in forming voltages indicates the peculiar behavior for each of the cells in the array. Before forming the read currents are distributed around a mean value $\mu = 4.03$ μA with a standard deviation $\sigma = 0.48$ μA, whereas after forming the average current were distributed around $\mu = 30.31$ μA with a standard deviation $\sigma = 0.23$ μA. These

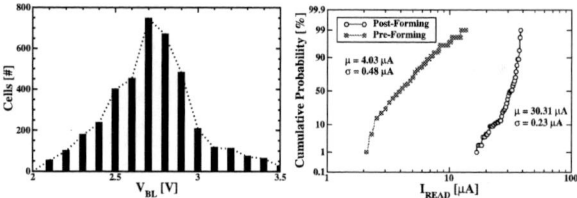

Fig. 2. Forming voltage distribution (left) and distributions of the read current in a ReRAM array with 1μm^2 MIM area before and after forming (right) [7].

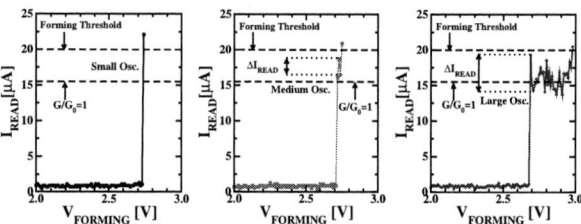

Fig. 3. Three different behaviors observed during forming process: small (left), medium (centre) and large (right) read-verify current oscillations [11].

results are obtained on arrays based on 1 μm^2 MIM area, however similar results are found in 0.6 μm^2 MIM area arrays. The analysis of the pre-forming distribution allows an indirect insight on the process induced variability in the MIM stack. In this technology, pre-forming currents larger than 10 μA usually indicate leaky cells due to fabrication issues such as the intrinsic variability of the HfO$_2$ deposition process [5]. This source of variability is also responsible for forming failures (i.e., cells that are unable to be formed), which is considered as a major contribution for the array yield loss (i.e., 40% of the cells in the array). A possible solution for this reliability threat has been proposed in [7] by using a forming-retry operation on the cells that are not able to create a CF, increasing the yield up to 99%.

Another issue retrieved during forming operation is the intrinsic instability of the created CF that could impact on the successive set/reset switching operations [11]. Indeed, by monitoring the cells forming behavior through an incremental pulse forming technique it is observed that the read current during forming could exhibit, in some cells, an oscillatory behavior (see Fig. 3). These oscillations interpreted either as the charging of a trap close to the surface of the conductive filament (CF) or the movement of an atom/defect in the filament [12], has been investigated in terms of reliability and cell-to-cell variability during 1k endurance cycles and 100k stress pulses in different cycling conditions.

Fig. 4 shows the cumulative distributions of the resistance ratio, set and reset switching voltages calculated after cycling. Resistance ratio is calculated as the ratio of the set state read current and reset state read current I_{LRS}/I_{HRS} at $V_{read} = 0.2V$. The cells formed with smaller oscillations are shown to require higher V_{SET} and V_{RES} after 1k cycles: that means small oscillations correspond to wider filaments. The Resistance Ratio, V_{SET}, V_{RES} average values and dispersion

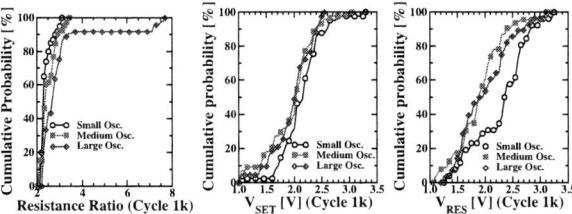

Fig. 4. Resistance ratio, V_{SET}, V_{RES} cumulative distributions for the different forming oscillations groups calculated on cycled devices [11].

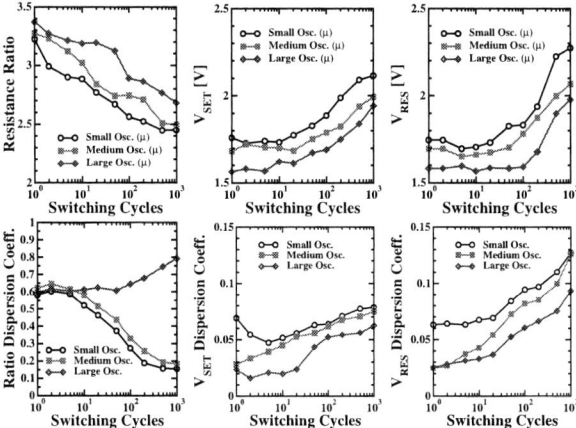

Fig. 5. Resistance Ratio, V_{SET} and V_{RES} average values and dispersion coefficients calculated during cycling [11].

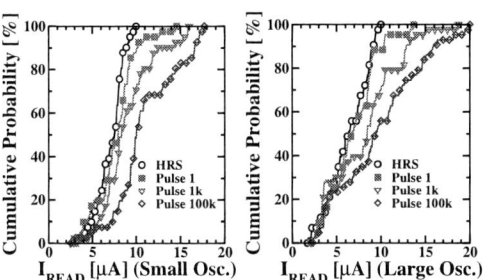

Fig. 6. Cumulative distributions of the read currents (with $V_{read} = 0.2V$) measured during set stress on HRS after different number of disturb pulses, at endurance cycle 1 [11].

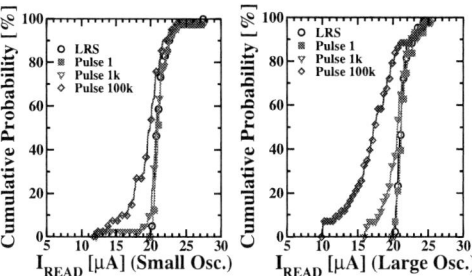

Fig. 7. Cumulative distributions of the read currents (with $V_{read} = 0.2V$) measured during reset stress on LRS after different number of disturb pulses, at endurance cycle 1 [11].

reason of the parameters dispersion could be the root mean square surface roughness of HfO_2 films due to the columnar structure of the TiN bottom metal electrode [13].

To evaluate the disturbs immunity of each cells group, 100k reset stress pulses have been applied after set with $V_{stress,res} = 0.8V$, $T_{stress,res} = 10\mu s$ and 100k set stress pulses after reset with $V_{stress,set} = 0.8V$, $T_{stress,set} = 10\mu s$ at different cycles. Set/reset stress voltage pulses with 0.8V have been used since it's almost half of the average set/reset switching voltage measured on fresh devices. Cumulative distributions of the read currents measured after reset (HRS), set (LRS) and during set and reset stress on fresh devices are reported in Fig. 6 and Fig. 7, respectively: in both cases cells formed with larger current oscillations show a lower disturb immunity. That reveals larger fluctuations indicate a not so well formed filament thus more prone to exhibit lower immunity.

All these findings summarize the importance of the forming operation in the lifetime of a RRAM array.

IV. OPTIMAL SET/RESET OPERATION

The average set and reset characteristics in a RRAM array feature the same variability observed in forming [6], [14]. Moreover, as usually evidenced in RRAM technology [1], the read current I_{HRS} shows a larger range of variability compared to I_{LRS}, as evidenced in Fig. 8 showing the cumulative distributions of the set/reset switching voltages calculated on the entire array. These results indicate that an optimization of the set/reset operations is mandatory to reduce the impact of the device variability, whereas minimizing the array yield loss due to non-switching cells. To this purpose, the analysis in [15] compared DC and pulsed set/reset operations featuring different durations and voltages. A set of 10000 set/reset cycles has been considered for the analysis.

SET operation in DC mode has been performed increasing the bitline voltage V_{BL} from 0 to 3.5V with $V_{step} = 0.1V$ ($T_{step,DC} = 50\mu s$) and the wordline voltage fixed to V_{WL} = 1.4V. RESET operation in DC mode has been performed increasing the source line voltage V_{SL} from 0 to 3.5V, with $V_{step} = 0.1V$ ($T_{step,DC} = 50\mu s$) and $V_{WL} = 2.5V$. In pulsed mode operation the wordline voltage has been fixed to V_{WL}

coefficients calculated during cycling are reported in Fig. 5. To evaluate the cell-to-cell variability the dispersion coefficient of I_{LRS} and I_{HRS} distributions, defined as (σ^2/μ), has been used. Resistance ratio of cells with large forming oscillations show both higher average value and dispersion coefficient in all cycling conditions: that means large fluctuations correspond to narrower filaments. V_{SET}, V_{RES} average values and dispersion coefficients are shown to increase during cycling: switching voltages on cells formed with large oscillations show lower average values and dispersion in all cycling conditions. This indicates cells with lower V_{SET}, V_{RES} have a not fully developed filament: this explains the large fluctuations. One

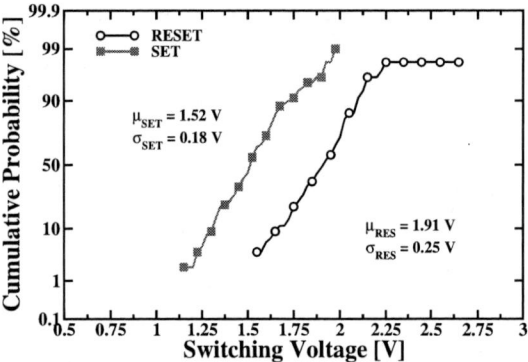

Fig. 8. Cumulative distribution of the set and reset switching voltages in the entire RRAM array [15].

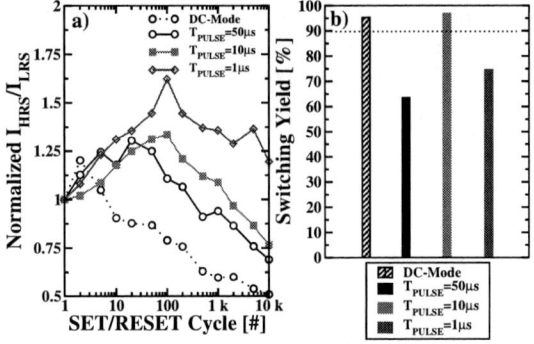

Fig. 9. Normalized read current Ratio (a) and Switching Yield (b) evaluation for different set/reset modes (DC and pulses with different durations) during cycling. V_{pulse} = 3V for pulsed modes [15].

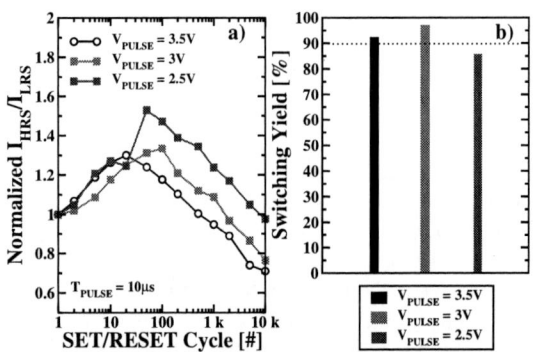

Fig. 10. Normalized read current Ratio (a) and Switching Yield (b) for different set/reset pulse amplitudes during cycling. T_{pulse} = 10μs [15].

= 1.4V during SET and V_{WL} = 2.8V during RESET, while different bitline/sourceline voltages and durations have been investigated.

In Fig. 9 a comparison between DC and pulsed mode with different durations at fixed V_{pulse} = 3V is depicted. Fig. 9a shows I_{HRS}/I_{LRS}, normalized with respect to that calculated at cycle 1, as a function of the set/reset cycle

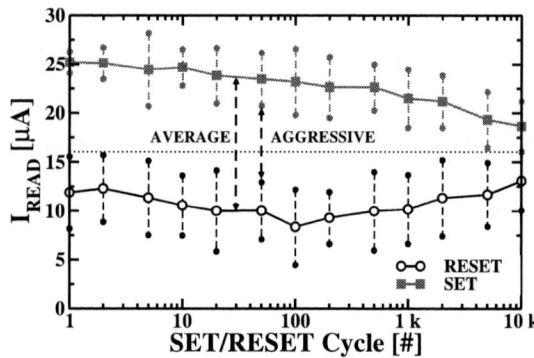

Fig. 11. Set and reset read current behavior during cycling with V_{pulse} = 3V, T_{pulse} = 10μs. Average and aggressive read window calculation points are indicated [15].

number for different pulse durations. In all cases a non-monotonic behavior is observed, eventually ending up with a significant I_{HRS}/I_{LRS} reduction with the exception of the shortest pulse duration (T_{pulse} = 1μs). Fig. 9b shows the switching yield (i.e., the percentage of cells in the array that actually toggles between set/reset states) of each set/reset mode providing an interesting trade-off: pulses with a too short or too long duration result in a lower yield compared to an average timing condition. Similar considerations can be derived by the analysis of Fig. 10, where the dependencies of the normalized I_{HRS}/I_{LRS} and that of the switching yield are evaluated in cycling for different pulse voltages considering the optimal pulse duration (T_{pulse} = 10μs).

From a physical point of view, this phenomenon can be explained as follows: while pulses with too low voltages or durations create too small filaments showing low current in set condition, too high voltages or durations create too big filaments hard to disrupt in the following reset operation. Both cases result in a lower yield compared to an average condition.

Starting from the best pulse conditions (T_{pulse} = 10μs, V_{pulse} = 3V) the read window closure has been analyzed as a function of set/reset cycling (see Fig. 11). Current reading has been performed at V_{WL} = 1.4V, $V_{BL,read}$ = 0.2V, T_{read} = 10μs. The average read current trend and the standard deviations are plotted for set and reset. It can be observed that the device variability of the cells in the array remains almost constant during cycling. Fig. 12 shows the read window ($I_{LRS} - I_{HRS}$) closure calculated using both the array average and aggressive (i.e., considering the worst-case condition) conditions. In this study, the endurance failure criterion is defined as the point where the aggressive read window case falls below 3μA [3], that is the limit for the sense amplifiers to discriminate between states. The read window show the same behavior for each pulse condition: an increase can be observed during the first cycles due to a variability reduction, followed by a closure after the degradation of the HfO$_2$ material stack. Short pulse durations and voltages result in a smaller read window due to a higher device variability caused

978-1-4799-8720-7/15 $31.00 © 2015 IEEE   330

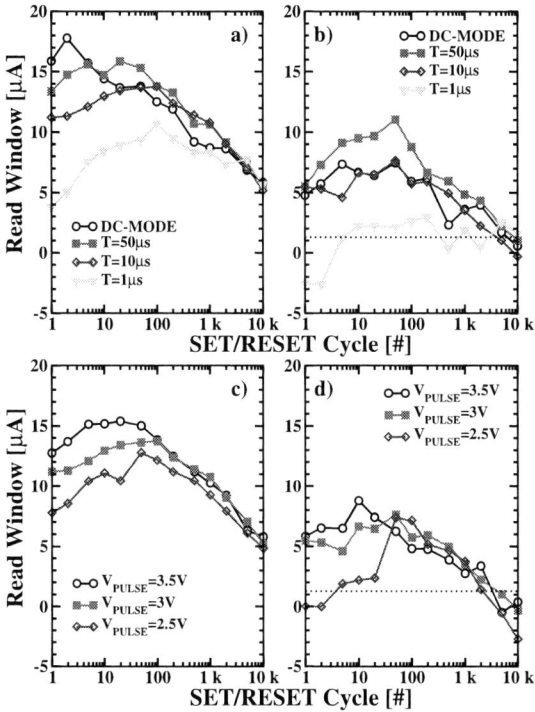

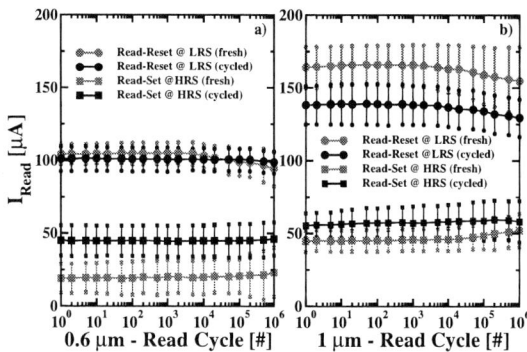

Fig. 13. $V_{dd}/2$ pulse stress effect measured on 0.6 μm^2 (a) and 1 μm^2 4kbit RRAM devices (b), in both set (LRS) and reset (HRS) condition.

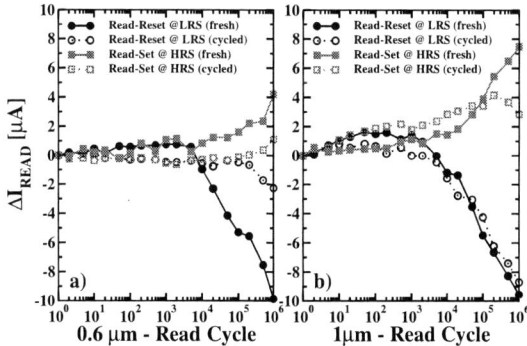

Fig. 14. Average read current variation measured during $V_{dd}/2$ pulse stress measured on 0.6 μm^2 (a) and 1 μm^2 4kbit RRAM devices (b).

Fig. 12. Read Window trends in cycling. Average (left column) and aggressive (right column) measurements: a) and b) same conditions as Fig. 9 c) and d) same conditions as in Fig. 10. The limit for the set/reset discrimination is depicted at read window = $3\mu A$ [15].

by incomplete set/reset switching.

V. READ DISTURBS AND INSTABILITIES

One major issue in RRAM technology is read instability [16]: consecutive reads on the same memory cell can yield widely fluctuating results, and/or cause permanent changes to the resistance itself. This behavior has been attributed to numerous physical mechanisms such as, random telegraph noise (RTN) due to capture and emission of trapped electrons [17], disturb due to the read electric field [18], diffusion of traps/vacancies [19], and retention/relaxation effects [20]. Conventionally, read instability has been measured on single-cell structures, with focus on specific physical mechanisms. An investigation of the disturb has been performed also using Costant Voltage Stress (CVS) on a large amount of samples, although the conditions used were not representative of a real array utilization [21].

On crossbar-based arrays, which is one of the potential array integration topologies offered by RRAM technology [3], [4], unselected WLs and BLs can be grounded or biased with a $V_{dd}/3$ or $V_{dd}/2$ scheme. In order to evaluate the impact on unselected WLs and BLs during set/reset operation in the worst-case condition, the $V_{dd}/2$ biasing effect on reset and set wordlines has been evaluated on 0.6 μm^2 and 1 μm^2 RRAM arrays. 10^6 $V_{RESET}/2$ pulse have been applied on set wordlines, while 10^6 $V_{SET}/2$ pulse have been applied on

reset wordlines where V_{RESET} and V_{SET} are the average reset and set switching voltages, respectively. The disturb effect has been evaluated on both fresh and cycled devices, after 10k set/reset pulse operations with $V_{pulse} = 3$ V, $T_{pulse} = 10$ μs. Fig. 13 shows the average set/reset read currents and their standard deviation measured during 10^6 stress pulse on 0.6 μm^2 (Fig. 13a) and 1 μm^2 (Fig. 13b) devices, for both fresh and cycled test chips. The dielectric material degradation in the MIM stack makes reset and set switching less effective, reducing the stress sensibility as well. The average current variation observed during stress is depicted in Fig. 14. The $V_{dd}/2$ stress caused a higher read current shift on fresh devices, for both 0.6 μm^2 and 1 μm^2 4kbits RRAM devices.

Ideally, considering device and circuit design margins, read resistance variation should be less than 10%. Error Correction Codes can also assist in recovery from less frequent, larger resistance fluctuations, but the occurrence of the resistance variation should be less than 1% for effective data integrity. The Read Error Rate, calculated as the fraction of cells showing a resistance variation higher than 10% during $V_{dd}/2$ stress is depicted in Fig. 15. Fresh devices show a higher error rate than cycled devices (after 10k set/reset cycles) for both 0.6 μm^2 and 1 μm^2 arrays. Although the average read current variation is higher in 1 μm^2 array, the error rate is lower with

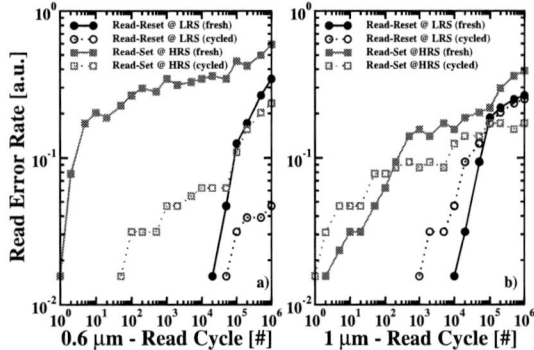

Fig. 15. Read Error Rate calculated on on 0.6 μm^2 (a) and 1 μm^2 4kbit RRAM devices (b). Full and dotted lines refers to fresh and cycled devices, respectively.

respect to that of 0.6 μm^2 devices because of a higher average set and reset currents that render the fluctuations less effective. Read disturb with set polarity stress on fresh devices in reset state is the operation that shows the highest read error rate for both 0.6 μm^2 (a) and 1 μm^2 (b) technologies, due to the conformation of the CF.

VI. CONCLUSIONS

In this review paper it was presented a detailed electrical characterization of different RRAM arrays manufactured in a compatible BiCMOS process. The analysis was entirely focused on the reliability and performance assessment of the integrated RRAM technology, through the depiction of the major issues retrieved during the characterization. Concerning the forming operation it was presented the relationship between this preliminary operation and the lifetime behavior of the array, by tackling also the typical cell-to-cell variability features. This activity lead to the search of the optimal set/reset parameters to improve the read window budget and the cycling features. Finally, the analysis of the read disturbs with their implications in advanced cross-bar architectures was evaluated showing that the integration of Error Correction Codes along with the RRAM memory is mandatory to guarantee the full data integrity.

ACKNOWLEDGEMENTS

This work was supported by the European Union's H2020 research and innovation programme under grant agreement N° 640073.

REFERENCES

[1] A. Chen and M. Lin, "Variability of resistive switching memories and its impact on crossbar array performance," in *IEEE Int. Reliability Physics Symposium (IRPS)*, 2011, pp. MY.7.1–MY.7.4.

[2] Y. Chen, B. Govoreanu, L. Goux, R. Degraeve, A. Fantini, G. Kar, D. Wouters, G. Groeseneken, J. Kittl, M. Jurczak, and L. Altimime, "Balancing set/reset pulse for $>10^{10}$ endurance in hfo2/hf 1t1r bipolar rram," *IEEE Trans. on Electron Devices*, vol. 59, no. 12, pp. 3243–3249, 2012.

[3] R. Waser, *Nanoelectronics and Information Technology: Advanced Electronic Materials and Novel Devices.* Wiley-VCH, 2012.

[4] E. Vatajelu, H. Aziza, and C. Zambelli, "Nonvolatile memories: Present and future challenges," in *Int. Design Test Symposium (IDT)*, 2014, pp. 61–66.

[5] D. Walczyk, T. Bertaud, M. Sowinska, M. Lukosius, M. Schubert, A. Fox, D. Wolansky, A. Scheit, M. Fraschke, G. Schoof, C. Wolf, R. Kraemer, B. Tillack, R. Korolevych, V. Stikanov, C. Wenger, T. Schroeder, and C. Walczyk, "Resistive switching behavior in tin/hfo2/ti/tin devices," in *Int. Semiconductor Conference Dresden-Grenoble (ISCDG)*, 2012, pp. 143–146.

[6] D. Walczyk, C. Walczyk, T. Schroeder, T. Bertaud, M. Sowiska, M. Lukosius, M. Fraschke, B. Tillack, and C. Wenger, "Resistive switching characteristics of cmos embedded hfo2-based 1t1r cells," *Microelectronic Engineering*, vol. 88, no. 7, pp. 1133–1135, 2011.

[7] C. Zambelli, A. Grossi, D. Walczyk, T. Bertaud, B. Tillack, T. Schroeder, V. Stikanov, P. Olivo, and C. Walczyk, "Statistical analysis of resistive switching characteristics in ReRAM test arrays," in *IEEE Int. Conf. on Microelectronics Test Structures (ICMTS)*, Mar 2014, pp. 27–31.

[8] N. Raghavan, A. Fantini, R. Degraeve, P. Roussel, L. Goux, B. Govoreanu, D. Wouters, G. Groeseneken, and M. Jurczak, "Statistical insight into controlled forming and forming free stacks for HfOx RRAM," *Microelectronic Engineering*, vol. 109, pp. 177 – 181, 2013.

[9] C. Chen, H. Shih, C. Wu, C. Lin, P. Chiu, S. Sheu, and F. Chen, "RRAM defect modeling and failure analysis based on march test and a novel squeeze-search scheme," *IEEE Trans. on Computers*, vol. 64, no. 99, pp. 1–1, 2014.

[10] P. Lorenzi, R. Rao, and F. Irrera, "Forming kinetics in HfO2-based RRAM cells," *IEEE Trans. on Electron Devices*, vol. 60, no. 1, pp. 438–443, 2013.

[11] A. Grossi, C. Zambelli, P. Olivo, E. Miranda, V. Stikanov, T. Schroeder, C. Walczyk, and C. Wenger, "Relationship among current fluctuations during forming, cell-to-cell variability and reliability in rram arrays," in *IEEE Int. Memory Workshop (IMW)*, May 2015, pp. 1–4.

[12] S. Balatti, S. Ambrogio, A. Cubeta, A. Calderoni, N. Ramaswamy, and D. Ielmini, "Voltage-dependent random telegraph noise (RTN) in HfOx resistive RAM," in *IEEE Int. Reliability Physics Symposium (IRPS)*, June 2014, pp. MY.4.1–MY.4.6.

[13] P.-S. Chen, Y.-S. Chen, H.-Y. Lee, T.-Y. Wu, P.-Y. Gu, F. Chen, and M.-J. Tsai, "Impact of flattened TiN electrode on the memory performance of HfO2 based resistive memory," *Electrochem. and Solid-State Letters*, vol. 15, no. 4, pp. H136–H139, 2012.

[14] E. Miranda, C. Walczyk, C. Wenger, and T. Schroeder, "Model for the resistive switching effect in hfo2 mim structures based on the transmission properties of narrow constrictions," *IEEE Electron Device Letters*, vol. 31, no. 6, pp. 609–611, 2010.

[15] C. Zambelli, A. Grossi, P. Olivo, D. Walczyk, J. Dabrowski, B. Tillack, T. Schroeder, R. Kraemer, V. Stikanov, and C. Walczyk, "Electrical characterization of read window in ReRAM arrays under different SET/RESET cycling conditions," in *IEEE Int. Memory Workshop (IMW)*, May 2014, pp. 1–4.

[16] T. Iwasaki, S. Ning, and K. Takeuchi, "Stability conditioning to enhance read stability 10x in 50nm alxoy reram," in *IEEE Int. Memory Workshop (IMW)*, May 2013, pp. 44–47.

[17] Y.-H. Tseng, W. C. Shen, and C.-J. Lin, "Modeling of electron conduction in contact resistive random access memory devices as random telegraph noise," *Journal of Applied Physics*, vol. 111, no. 7, pp. 073 701–073 701–5, Apr 2012.

[18] W. Liu, K. L. Pey, N. Raghavan, X. Wu, M. Bosman, and T. Kauerauf, "Random telegraph noise reduction in metal gate high-k stacks by bipolar switching and the performance boosting technique," in *IEEE Int. Reliability Physics Symposium (IRPS)*, April 2011, pp. 3A.1.1–3A.1.8.

[19] H.-Y. Lee, Y.-S. Chen, P.-S. Chen, P.-Y. Gu, Y.-Y. Hsu, W.-H. Liu, W.-S. Chen, C.-H. Tsai, F. Chen, C.-H. Lien, and M.-J. Tsai, "Comprehensively study of read disturb immunity and optimal read scheme for high speed hfox based rram with a ti layer," in *Int. Symp. on VLSI Technology Systems and Applications (VLSI-TSA)*, April 2010, pp. 132–133.

[20] D. Ielmini, "Modeling the universal set/reset characteristics of bipolar rram by field- and temperature-driven filament growth," *IEEE Trans. on Electron Devices*, vol. 58, no. 12, pp. 4309–4317, Dec 2011.

[21] T. Diokh, E. Le-Roux, S. Jeannot, M. Gros-Jean, P. Candelier, J. Nodin, V. Jousseaume, L. Perniola, H. Grampeix, T. Cabout, E. Jalaguier, M. Guillermet, and B. De Salvo, "Investigation of the impact of the oxide thickness and reset conditions on disturb in hfo2-rram integrated in a 65nm cmos technology," in *IEEE Int. Reliability Physics Symposium (IRPS)*, April 2013, pp. 5E.4.1–5E.4.4.

978-1-4799-8720-7/15 $31.00 © 2015 IEEE

2015 IEEE Computer Society Annual Symposium on VLSI

Single-Ended and Differential MRAMs based on Spin Hall Effect: A Layout-Aware Design Perspective

(Invited Paper)

Ahmedullah Aziz, William Cane-Wissing, Moon S. Kim, Suman Datta, Vijaykrishnan Narayanan and Sumeet K. Gupta
School of Electrical Engineering and Computer Science
The Pennsylvania State University, University Park PA 16802. Email: skgupta@psu.edu

Abstract—Spin-based memories have shown an immense promise for on-chip memory applications due to the possibilities of introducing non-volatility in caches employing a CMOS compatible process. Non-volatility leads to zero-standby leakage. However, at the same time, exploration of energy-efficient read and write mechanisms is important to lower the overall energy-consumption of an MRAM cache. Magnetization control employing spin-hall effect is one of the most promising approaches to enhance the write energy efficiency. A separate path for the read which comprises of a magnetic tunnel junction (MTJ) offers the benefits of simultaneous optimization of the read and write operations. In fact, spin-hall effect also leads to the possibility of designing cells with differential sensing. However, the advantages of a differential read over a single-ended read in terms of higher read speed and better noise immunity comes at the cost of larger number of access transistors, which *may* translate to lower integration density.

In this paper, we perform a comparative analysis of spin-hall effect (SHE) based MRAM cells with single-ended and differential read mechanisms in terms of the cell area, read performance and read stability. We perform a detailed layout analysis based on FinFET technology to evaluate the impact of introducing differential sensing on cell area. Our analysis shows that when the layout area is determined by the pitch of the bit-line and source-line metal tracks, the differential cell shows 1.5X increase in the cell area compared to the single-ended SHE MRAM. The area increase is 2X if the read access transistor determines the layout footprint. However, the differential sensing offers the advantages of ~48% increase in the read performance along with 6% to 9% boost in the read stability compared to the single-ended SHE MRAM. At iso-area, the differential cell shows ~24% lower read time and 12% higher read disturb margin with a similar write performance and power. Our analysis also presents other layout-driven perspectives on the design of differential and single-ended SHE MRAMs.

Keywords— *Differential sensing, layout, MRAMs, single-ended sensing, spin hall effect, spin memories.*

I. INTRODUCTION

Memories based on electron spin employing nano-magnets to store the binary data have received widespread attention in the recent past due to their attractive attributes like non-volatility, zero stand-by leakage, high integration density and compatibility with the CMOS process [1]. The nano-magnet that stores the information in terms of its magnetization (also known as the free layer or FL) is typically used in conjunction with another nano-magnet with a fixed magnetization (called pinned layer or PL). A thin tunneling oxide (typically MgO [2]) is sandwiched between the two nano-magnets (Fig. 1(a)) forming a magnetic tunnel junction (MTJ). The MTJ exhibits two stable states: (a) parallel (P) configuration, in which the magnetization

of FL and PL are in the same direction and (b) anti-parallel (AP) state, in which the FL and PL magnetization vectors point in the opposite directions. P state of the MTJ has a lower resistance compared to the AP state [1]. This resistance difference is typically expressed in terms of the tunneling magneto-resistance (TMR) and determines the distinguishability between the two states. TMR-based read is one of the most widely employed mechanisms for data sensing in spin-based memories because it offers much larger distinguishability in comparison to other effects like giant magneto-resistance (GMR) [3]. Moreover, MTJ structure has the benefit of achieving the switching of FL magnetization with current induced spin-transfer torque (STT) [4]. Hence, an MTJ can be used for effective data sensing as well as programming. For array multiplexing, standard STT MRAMs employ an access transistor in series with the MTJ [1], as shown in Fig. 1(a). The cell area is determined by the access transistor since it has a much larger footprint compared to the MTJ. With the cell requiring just one transistor, the integration density can be significantly higher in comparison to SRAMs, which typically are designed using six transistors [5]. However, STT MRAMs exhibit higher write energy consumption and lower performance compared to SRAMs [6]. Moreover, the read operation in standard STT MRAMs is single-ended as compared to a more noise-immune and faster differential sensing in SRAMs [7].

In response to the need for lowering the write power of MRAMs, energy efficient magnetization controlling phenomena other than STT are being explored [8], amongst which spin-hall effect (SHE) [9] shows an immense promise. Higher spin-injection efficiency compared to STT and lower write voltages due to a metallic write path leads to significant reduction in the write power of an SHE-based MRAM [10]. The read path retains the MTJ structure and the separation of the read-write path allows simultaneous optimization of the read and write operations. Moreover, SHE offers an opportunity to design cells with differential read, as proposed in [10], which offers the benefits of higher read performance. However, the advantages of differential sensing come at the cost of larger number of transistors. In order to properly evaluate the pros and cons of the differential and single-ended SHE-MRAMs, there is a need for an extensive comparison of the read operation coupled with the cell area analysis. Further, the impact of the cell layouts on the design of the cell and the memory devices need to be thoroughly investigated.

In this paper, we perform a comparative analysis of the single-ended and differential SHE MRAMs by quantifying the cell area through detailed layout analysis. We also compare the cell area of SHE MRAMs with standard 1T-1R STT MRAM and 6T SRAM. Based on the insights obtained from the layout

978-1-4799-8720-7/15 $31.00 © 2015 IEEE   333

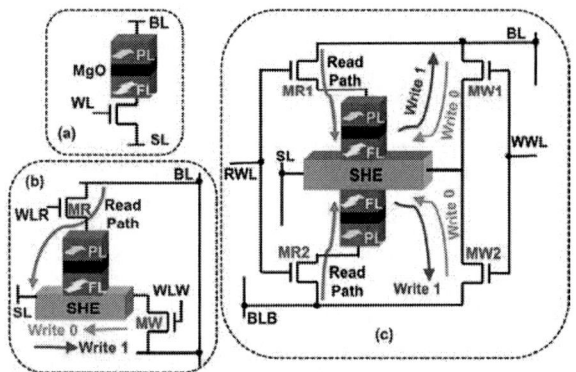

Fig. 1 (a) Standard 1T-1R STT MRAM (b) single ended Spin-Hall Effect (SHE) based MRAM and (c) differential Spin-Hall Effect (SHE) based MRAM [10]

comparison, we discuss the benefits and trade-offs of single-ended and differential SHE MRAMs by performing extensive read analysis. We show that at iso-area, the differential cell achieves better read performance and stability compared to its single-ended counterpart.

The rest of the paper is organized as follows. In Section II, we review the single-ended and differential SHE-MRAMs. Section III presents the layout analysis and discusses the area comparison of SHE MRAMs, standard STT MRAMs and SRAMs. In Section IV, we present the read analysis for SHE MRAMs. Section V discusses the iso-area comparison. Section VI presents design perspectives for the differential and single-ended SHE MRAMs and summarizes this work.

II. SPIN-HALL EFFECT (SHE) - BASED MRAMs

Figs. 1(b, c) show the single-ended and differential MRAMs based on spin-hall effect (SHE) [10]. In the single-ended cell, the free layer (FL) is formed on top of a heavy metal such as tungsten(W), platinum (Pt) or tantalum (Ta) which exhibits SHE. The write operation is performed by passing a current through the heavy metal, which generates spin current and leads to the accumulation of spin on its surfaces. The accumulated spins interact with the free layer to generate a torque, which leads to the switching of the FL magnetization. The direction of the current determines the FL magnetization [11] and by applying a bi-directional current, the magnetization of FL can be switched between the two stable states. The read path of the single-ended SHE MRAM comprises of an MTJ structure. The state of the cell (which is defined by the resistance of the P and AP configurations) can be sensed by the following mechanisms: (a) by applying a fixed read voltage across the cell, sensing the cell current and comparing it to a reference current [12] or (b) by applying a fixed read current and comparing the voltage developed on the bit-line (BL) to a reference voltage [13]. The separation of read-write paths allows simultaneous and independent optimization of the MTJ for the read operation and the heavy metal for the write operation. Moreover, high spin-injection efficiency of SHE and a metallic write path in the memory device leads to high write energy efficiency, while retaining the advantage of large TMR of the MTJ for the read operation. However, separate read-write paths mandate the use of two access transistors (MR and MW - Fig. 1(b)), in order to avoid sneak current in the unaccessed cells. This *may* lead to an increase in the cell area compared to

a standard STT MRAM, which employs a single access transistor. We will discuss the layout area implications, in detail, in the next section.

In the differential SHE MRAM [10], two free layers, one each on the top and bottom surfaces of the heavy metal, store complementary bits. The complementarity of the bits is maintained by SHE. The spin current induced by the electric current in the heavy metal leads to the accumulation of opposite spins on the top and bottom surfaces. As a result, the magnetization vectors of the two free layers always point in the opposite directions. The write operation of a differential SHE MRAM is identical to that of the single-ended SHE MRAM. For the read operation, two MTJs (Fig. 1(c)) are employed. The magnetization of the pinned layers of the MTJs are in the same direction. Due to the opposite magnetization of the complementary FL layers, one of the MTJs is in the P state, while the other in the AP state. By applying a read voltage across the two MTJs, the difference in their currents can be sensed. Alternatively, a fixed read current can be applied and the polarity of the voltage difference developed across the two bit-lines (BL and BLB- Fig. 1(c)) can be used to determine the state of the MTJs. Thus, the differential SHE MRAM is self-referenced, which leads to lower access time and better noise immunity compared to the single-ended SHE MRAM. However, differential sensing and separation of read-write paths require two read access transistors (MR1 and MR2) and a write access transistor. Note that in Fig. 1(c), the write access transistor is split into two transistors (MW1 and MW2) connected to BL and BLB in order to balance the load on the bit-lines for robust differential sensing. The total width of the write access transistor is the sum of the widths of MW1 and MW2.

The differential SHE MRAM offers distinct advantages over the single-ended SHE MRAM in terms of faster and more robust read operation. However, an increase in the number of access transistors is expected to be detrimental to its integration density compared to a single-ended cell. For proper evaluation of the two SHE MRAM cells, a detailed and comparative quantification of the dependence of area and read performance on the width of the access transistors is required. In the subsequent sections, we carry out an extensive layout analysis, based on which we perform an iso-area comparison of the read performance and stability of the single-ended and differential SHE-MRAMs.

III. LAYOUT ANALYSIS

In order to understand the layout implications of separating the read-write paths and introducing differential sensing in the SHE MRAM cells, we first present a discussion on the layouts of a standard 1T-1R STT MRAM to define the baseline. We then carry out the layout analysis of the single-ended and differential SHE MRAMs and derive general expressions for the cell area in terms of the layout parameters. We choose FinFET technology for the access transistor, considering its suitability for the present and future process generations [14]. The layouts employ λ-based design rules [15], where λ is defined as half of the minimum feature size associated with the physical dimensions of a technology. The λ-based design rules are supplemented by the information on poly, metal and fin pitches derived from Intel

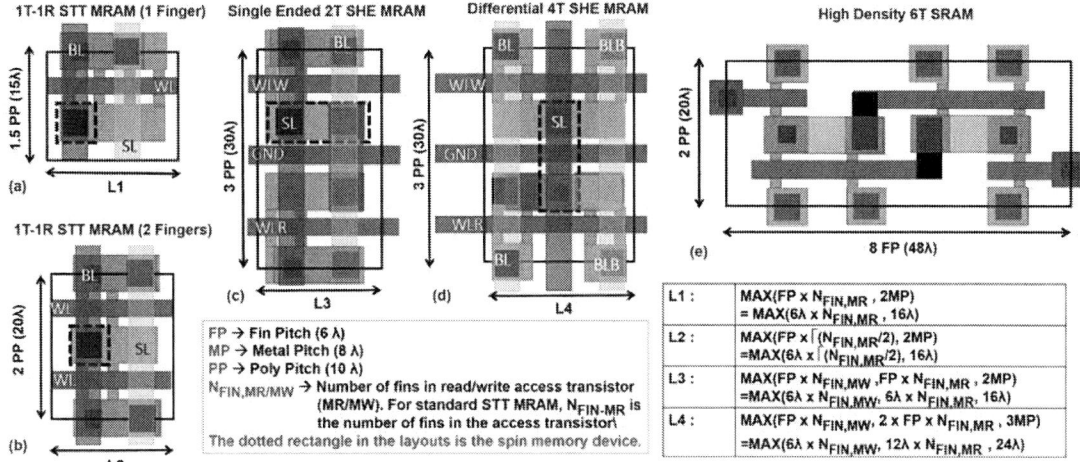

Fig. 2 Layouts of (a) 1T-1R STT MRAM with single finger access transistor (b) 1T-1R STT MRAM with two finger access transistor (c) single ended 2T SHE MRAM (d) differential 4T SHE MRAM and (e) high density 6T SRAM.

14nm technology [14]. The width of the access transistors is defined in terms of the number of fins. Subsequently, we denote the number of fins in the read and write access transistors of SHE-MRAMs by $N_{FIN,MR}$ and $N_{FIN,MW}$, respectively. For the standard STT MRAM, the number of fins in the access transistor is denoted by $N_{FIN,MR}$. Note that in the differential SHE MRAM, $N_{FIN,MW}$ is the total number of fins in the write access transistor, which is the sum of the number of fins in MW1 and MW2. It may also be mentioned that the number of fins in the two read access transistors of the differential SHE MRAM (MR1 and MR2) are equal and N_{FIN-MR} denotes the number of fins in *each* of the access transistors.

Fig. 2(a, b) shows the one-finger and two-finger layouts of the standard 1T-1R STT MRAM [16]. The vertical dimension of a one-finger layout is equal to1.5 poly pitches (PP) while that of a two-finger layout is 2 PP. The horizontal dimensions of the layout may be determined by the metal pitch (MP) or the fin pitch (FP). For a single-finger layout, $N_{FIN,MR} < 2 \times MP/FP$ leads to a metal-pitch limited layout, resulting in the independence of the cell area on the number of fins in the access transistor (Fig. 3(a)) [16]. For $N_{FIN,MR} > 2 \times MP/FP$, the cell area becomes a function of the number of fins (Fig. 3(a)). Note that a factor of 2 in the in-equalities is due to two vertically routed metals (BL and SL) in the layout. For a two-finger layout, the number of fins in the access transistor are divided between the two fingers. Therefore, for $N_{FIN,MR} < 4 \times MP/FP$ (Fig. 3(a)), the horizontal dimension is limited by the metal pitch, making the cell area independent of the number of fins. It is also noteworthy, that for $N_{FIN,MR} > 4 \times MP/FP$ (the region where cell area is a function of the number of fins), same cell area is obtained if the number of fins is rounded up to an even value. For instance, $N_{FIN,MR}=5$ and 6 yield the same area, as shown in Fig. 3. This is due to the fact that for odd number of fins, the finger, which has more fins determines the area. Hence, increasing the fin count by one in the other finger does not affect the layout footprint. Comparing the trends for a one- and two-finger layouts, the latter yields lower area for $N_{FIN,MR} >= 4$. With the understanding of the layouts of a standard 1T-1R STT MRAM, let us perform the layout and cell area analysis for the SHE MRAMs.

In a single ended SHE-MRAM, employment of separate read and write access transistors lead to an increase in the vertical dimension of the layout compared to 1T-1R STT MRAMs (Fig. 2(c)). With the layout height equal to 3 PP, each of the read and write access transistors need to be laid out with a single-finger. The layout is metal pitch limited for $N_{FIN,MR}$ and $N_{FIN,MW} < 2 \times MP/FP$. Otherwise, the access transistor, which has larger number of fins determines the cell area (Fig. 4).

A differential SHE-MRAM uses four access transistors: (a) two read access transistor (MR1 and MR2), each with number of fins equal to $N_{FIN,MR}$ and (b) two write access transistors (MW1 and MW2), each with number of fins equal to $N_{FIN,MW}/2$. (Note that the number of fins in MW1 and MW2 need to be same to balance the load on BL and BLB for robust differential sensing). MR1 and MW1 are abutted to MR2 and MW2 in the horizontal direction (Fig. 2(d)) to avoid a large increase in the layout height and the bit-line capacitance. Therefore, the vertical dimension of the layout of a differential SHE-MRAM is equal to that of a single-ended SHE-MRAM (=3 PP). However, the differential SHE-MRAM required three vertically routed metal tracks (SL, BL and BLB). Therefore, the layout is metal-pitch – limited for $N_{FIN,MW} < 3 \times MP/FP$ and $N_{FIN,MR} < 1.5 \times MP/FP$ (Fig. 4). The cell area is determined by the write access transistors if $N_{FIN,MW} > 2 \times N_{FIN,MR}$. Otherwise, the read access transistor determines the cell area. Due to two read access transistors, the rate of increase in the cell area with increasing $N_{FIN,MR}$ is double the rate for the single-ended SHE-MRAM.

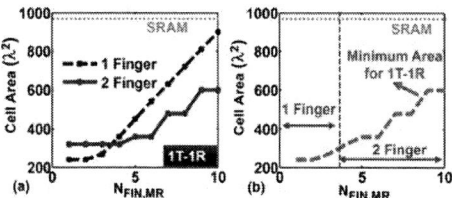

Fig. 3 (a) Cell area versus number of fins in the access transistor ($N_{FIN,MR}$) for a 1T-1R STT MRAM with one-finger and two-finger layouts (b) minimum area (considering one-finger and two-finger layouts) versus $N_{FIN,MR}$. Comparison with high density SRAMs is also shown.

978-1-4799-8720-7/15 $31.00 © 2015 IEEE

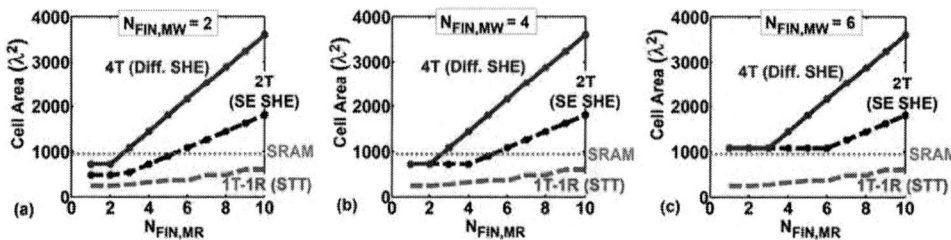

Fig. 4 Cell area versus number of fins in the read access transistor ($N_{FIN,MR}$) comparing single-ended spin-hall effect (SE SHE) MRAM and differential spin-hall effect (Diff. SHE) MRAM for different number of fins in the write access transistor ($N_{FIN,MW}$): (a) $N_{FIN,MW}$ = 2 (b) $N_{FIN,MW}$ = 4 (c) $N_{FIN,MW}$ = 6. Comparison with 1T-1R STT MRAM and high density SRAM is also shown.

From the discussion above and the plots in Fig. 4, it can be concluded that differential SHE MRAM has the same area as the single-ended SHE-MRAM only when the write access transistor determines the layout footprint. If the layout is metal-pitch limited, differential SHE-MRAM cell is 1.5X larger than its single-ended counterpart. If the layout footprint is determined by the read access transistor, the cell area of the differential cell is twice that of the single-ended SHE cell. Comparison with the standard 1T-1R STT MRAM shows a considerable increase in area of the SHE cells due to the following reasons. Firstly, separation of the read-write paths and consequently, increase in the number of access transistors results in increase in the vertical dimension of the layout. Secondly, for the standard STT MRAM, one has an option of switching to a two-finger layout (yielding lower area for large $N_{FIN,MR}$) while for SHE MRAMs, each of the access transistors can only be laid out with a single-finger. In spite of the area increase, SHE MRAMs have advantages over 1T-1R MRAMs due to the separation of the read-write paths and mitigation of the design conflicts, which has been extensively analyzed in [10]. In this paper, we will focus on the comparison of the single-ended and differential SHE MRAMs.

We also show the comparison of the cell area of 1T-1R and SHE MRAMs with a *high density* 6T SRAM cell (Fig. 2(e)) [17] in Fig. 4. It can be observed that 1T-1R MRAM has a significantly lower area compared to SRAM. Single-ended SHE MRAM shows area benefits for $N_{FIN,MW}$ <=5 and $N_{FIN,MR}$ <=5. For a differential SHE MRAM, cell area reduction with respect to SRAMs is achieved for $N_{FIN,MW}$ <=4 and $N_{FIN,MR}$ <=2. Note, that only even values of $N_{FIN,MW}$ are permissible in the differential SHE MRAM due to the constraint that number of fins in MW1 and MW2 should be equal for balancing the load on BL and BLB for differential sensing.

Single ended SHE MRAMs have a distinct advantage over differential SHE MRAMs in terms of lower cell area. However, read performance and robustness is expected to be more superior for the differential MRAMs. In order to properly evaluate the pros and cons of the two cells, we carry out a detailed read analysis in the next section. Based on the insights gained in this section and the results from the read analysis, we quantify the read performance and stability of the single-ended and differential cells at iso-area.

IV. READ ANALYSIS

The read operation of the SHE MRAM cells is analyzed by applying a constant read current and sensing the voltage developed on the bit-lines of the cell. For the single-ended cell, the bit-line voltage is compared against a reference voltage,

while for the differential cell, the voltage difference developed between BL and BLB is monitored. We define the read time (t_{READ}) as the time to develop a voltage differential of 70mV. Note, for the single ended cell, this voltage differential is between BL voltage and the reference. Therefore, a cell with MTJ in the AP state has 140mV higher BL voltage compared to a cell with the MTJ in the P state. In contrast, the differential cell requires 70mV voltage difference between the bit-lines corresponding to AP and P states of the MTJ, due to self-reference provided by the complementary bit-lines. We also analyze the read disturb margin (RDM) of the two cells. RDM is a measure of the read stability and is defined as follows: RDM = $1 - I_P/I_{C,P \to AP}$ [1]. Here I_P is the current through the MTJ in the P state and $I_{C,P \to AP}$ is the critical current required to flip the MTJ state from P to AP. Note that the read current direction in the SHE MRAM cells (Fig. 1) is such that only P$\to$AP state transition is likely [1]. In order to obtain I_P, we calculate the MTJ current averaged over a period of t_{READ}. The critical current used for the calculation of RDM corresponds to the switching time equal to t_{READ}. The simulations are carried out using physics based transport models (calibrated to experiments) for the MTJs and spin hall effect [1, 10, 18]. The transport models are coupled with Landau Lifshitz Gilbert Slonczewski (LLGS) equations. For the access transistors, 20nm predictive technology models [19] are used. The bit-line capacitance comprising of the interconnect capacitance and the capacitance of the read and write access transistors of the unaccessed cells in a column is considered in our analysis. The simulation parameters are summarized in Table 1.

Separation of the read-write paths in SHE-MRAMs allows the optimization of the oxide thickness of the MTJ ($T_{OX,MTJ}$), independent of the write operation. This is in contrast to the standard 1T-1R MRAM, in which $T_{OX,MTJ}$ may be determined by the write speed requirements. In this paper, we optimize $T_{OX,MTJ}$ for SHE MRAMs considering the read time (t_{READ}). Fig. 5 shows the dependence of t_{READ} on $T_{OX,MTJ}$. A non-monotonic trend can be observed for $N_{FIN,MR}$=2, which is explained as follows. As $T_{OX,MTJ}$ increases, the cell TMR (= $[R_{AP}-R_P]/[R_P+R_{MOS}]$) increases [20]. (Here R_P and R_{AP} are the

Table 1 Simulation Parameters

Simulation Parameters	Value
Dimensions of free layer	85nm x 30nm x 3nm
Dimensions of heavy metal (spin hall metal:Tungsten)	85nm x 60nm x 2.8nm
Sat. Magnetization	700 (kA/m)
Energy Barrier	56 $k_B T$
Access Transistor Tech.	20nm FinFET

Other parameters for spin hall effect in W taken from [10]

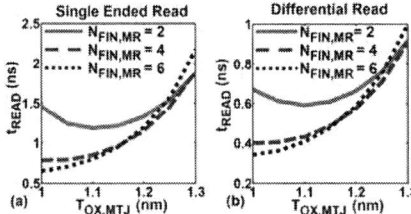

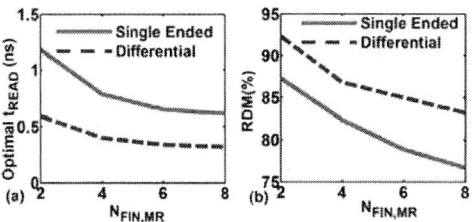

Fig. 5 Read time (t_{READ}) versus MTJ oxide thickness ($T_{OX,MTJ}$) for different number of fins in the read access transistor ($N_{FIN,MR}$) for (a) single-ended and (b) differential SHE MRAM cells.

MTJ resistance in the P and AP states and R_{MOS} is the resistance of the access transistor). Increase in cell TMR with $T_{OX,MTJ}$ tends to decrease t_{READ} due to the enhancement in the distinguishability of the states. However, at the same time, increase in $T_{OX,MTJ}$ leads to larger MTJ resistance, which reduces the current discharging the bit-line. This tends to increase t_{READ}. With these two factors changing t_{READ} in different directions, a non-monotonic trend is obtained. In other words, an optimal $T_{OX,MTJ}$ (~1.1nm for $N_{FIN,MR}$=2 in Fig. 5) minimizes the read time. This behavior is observed for both single-ended and differential cells.

As $N_{FIN,MR}$ increases, the optimal value of $T_{OX,MTJ}$ reduces. (In Fig. 5, the optimal values for $N_{FIN,MR}$ =4 and 6 are below 1nm). The reason is explained as follows. With an increase in $N_{FIN,MR}$, R_{MOS} reduces. Therefore, the sensitivity of the cell TMR with respect to $T_{OX,MTJ}$ reduces. This may be understood by considering the extreme values of R_{MOS}. If R_{MOS} >> MTJ resistance, cell TMR ~ $[R_{AP}-R_P]/R_{MOS}$. Increase in $T_{OX,MTJ}$ increases the numerator, leading to larger sensitivity of cell TMR. If R_{MOS} << MTJ resistance, cell TMR ~ $[R_{AP}-R_P]/R_P$. Thus, increase in $T_{OX,MTJ}$ increases both the numerator and the denominator and lowers the sensitivity of cell TMR. Thus, as $N_{FIN,MR}$ increases, cell TMR becomes less sensitive to $T_{OX,MTJ}$. In addition, reduction in R_{MOS} due to an increase in $N_{FIN,MR}$ makes the current discharging the bit-line more sensitive to the MTJ resistance (which is controlled by $T_{OX,MTJ}$). In other words, as $N_{FIN,MR}$ increases, the impact of $T_{OX,MTJ}$ on the discharging current becomes more dominant compared to the effect on cell TMR. This leads to reduction in the optimal value of $T_{OX,MTJ}$. It may be noted that in our analysis we only consider the optimal values of $T_{OX,MTJ}$ that are greater than or equal to 1nm. In case, a monotonic increase of t_{READ} with respect to $T_{OX,MTJ}$ is observed for $T_{OX,MTJ}$ > 1nm (as in Fig. 5 for $N_{FIN,MR}$ = 4 and 6), we set the optimal $T_{OX,MTJ}$ to be 1nm.

In Fig. 6 (a), we compare the optimal t_{READ} for the single-ended and differential cells. The results show ~48% reduction in the read time for the differential cell. We also compare the RDM of the two cells across a range of $N_{FIN,MR}$ in Fig. 6(b). The RDM values correspond to the optimal values of $T_{OX,MTJ}$ obtained for each $N_{FIN,MR}$. The differential cell shows 6% to 9% enhancement in the read disturb margin compared to the single-ended cell. This is due to reduction in t_{READ} achieved because of differential sensing, which increases the critical current for switching ($I_{CP \rightarrow AP}$), making it less likely for the accidental P→AP switching to occur.

To sum up, the differential cell exhibits faster read as well as superior read stability compared to the single-ended cell. However, as discussed in Section III, this comes at the cost of

Fig. 6 (a) Read time (t_{READ}) and (b) read disturb margin (RDM) corresponding to the optimal MTJ oxide thickness ($T_{OX,MTJ}$) versus the number of fins in the read access transistor ($N_{FIN,MR}$) comparing single-ended and differential SHE MRAM cells. The latter cell shows lower read time and larger disturb margin.

increase in area. In the next section, we perform an iso-area comparison of the two cells to gain insights into the benefits and trade-offs of the SHE MRAM cells.

V. ISO-AREA READ COMPARISON

The layout analysis in Section III (Figs. 2-4) clearly indicates that the number of fins in the read access transistors ($N_{FIN,MR}$) of the differential cell needs to be reduced in order to match the cell area with the single-ended cell. In order to obtain the values of $N_{FIN,MR}$ for iso-area analysis, we consider $N_{FIN,MW}$ = 2, so that the impact of the write access transistors on the bit-line capacitance is minimized. It is important to mention that unlike standard STT MRAM, lower width of the access transistor may be used in SHE MRAMs due to lower critical current and an all-metallic write path in the memory device [10]. From the equations in Fig. 2 and the plot in Fig. 4(a) (corresponding to $N_{FIN,MW}$=2), it can be observed that $N_{FIN,MR}$=2 for the differential cell and $N_{FIN,MR}$=4 for the single-ended cell yield the same area. This combination also leads to lower area compared to a high density SRAM, as discussed before. For each N_{FIN_MR}, we determine the optimal T_{OX_MTJ} (Fig. 5) and obtain the corresponding t_{READ} and RDM. The results are shown in Fig. 7(a) for different number of cells in a column ($N_{CELLS,COL}$). At iso-area, the differential cell shows consistent reduction (25% to 40%) in the read time. This is mainly due to the self-reference provided by the differential BL and BLB. Another reason for performance enhancement in the differential cell is the reduction in the bit-line capacitance due to lower N_{FIN_MR}. The performance benefits of the differential cell increases with the increase in the number of cells in a column. Comparison of RDM in Fig. 7(b) shows 7% to 28% increase for the differential cell compared to the single-ended cell. This increase is mainly attributed to lower $N_{FIN,MR}$ and larger $T_{OX,MTJ}$ (see Fig. 7) in the differential SHE MRAM. It may be noted that an increase in $N_{CELLS,COL}$ (and consequently, increase in the bit-line capacitance) leads to reduction in the RDM. This is due to higher read time and increase in the charge on the bit-line capacitance. The cell is disturbed for a longer duration, which increases the likelihood of accidental flipping of the bit. It is also noteworthy that the differential cell achieves larger improvement in RDM for higher $N_{CELLS,COL}$.

VI. DESIGN PERSPECTIVES AND SUMMARY

The analyses in the previous sections clearly show the benefits of the differential SHE MRAM in terms of 25% to 40% faster and 7% to 28% more stable read operation compared to

978-1-4799-8720-7/15 $31.00 © 2015 IEEE

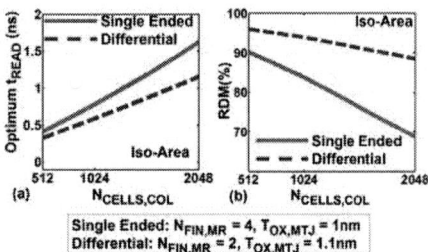

Fig. 7 Iso-area comparison of (a) read time (t_{READ}) and (b) read disturb margin (RDM) of a single-ended and differential SHE MRAM cells. Differential cell shows improvement in read speed and stability at iso-area across different number of cells in a column ($N_{CELLS,COL}$)

its single-ended counterpart at iso-area. With the same width of the write access transistor, the write performance and power is similar for the two cells [10]. The comparative study in this paper also points to some other design perspectives, which we will briefly highlight in this section. The layout analysis in Section III shows that the design options for the differential cell are constrained to a larger extent than the single ended cell. For instance, to obtain an area advantage over SRAMs, the read access transistor of the differential cell must not have more than 2 fins. On the other hand, the number of fins in the single-ended MRAM may be increased to 5 or even 6 (see Fig. 4) without incurring significant area penalty compared to SRAMs. Larger $N_{FIN,MR}$ can lead to higher tolerance of the read operation of the single-ended cells to *random* process variations [21]. Furthermore, for low area designs, the single-ended cells can achieve a maximum cell area reduction of 50% with respect to SRAMs, while the differential cell offers a maximum area reduction of only 25% (Fig. 4). Another design constraint is with respect to the number of fins in the write access transistor ($N_{FIN,MW}$). As discussed in Section III, $N_{FIN,MW}$ in the differential cell must have an even value so that the write access transistor can be split into two transistors to balance the load on BL and BLB. Therefore, the number of fins in the write access transistor of the differential cell can only be increased in steps of two. In other words, width quantization in FinFETs [22] is aggravated while optimizing the write operation of the differential cell.

In spite of such design constraints, the differential cell achieves improvement in the read speed and disturb margin over the single-ended cell at iso-area with a similar write performance and power. Furthermore, our analysis shows that in comparison to the standard STT MRAMs, SHE MRAMs achieve 85% lower write power at matched write performance. Moreover, the thickness of the MTJ oxide in the SHE MRAMs can be optimized independently to achieve an increase in the read performance along with enhancement in the stability. With the advantages of the differential SHE MRAM, it can be a promising alternative to the standard STT MRAM to alleviate the issues of high write power and mitigate the design conflicts at the cost of lower integration density. A single-ended SHE MRAM has its own advantages. It offers design options for trading-off the read performance to mitigate cell area penalty (observed in a differential cell), while enhancing the write efficiency compared to standard STT MRAMs.

ACKNOWLEDGMENT

Aziz, Cane-Wissing, Datta and Gupta acknowledge support from the Department of Electrical Engineering, Material

Research Institute and the College of Engineering at Penn State. Kim and Narayanan acknowledge support from the Center for Low Energy Systems Technology (LEAST), one of the six SRC STARnet Centers, sponsored by MARCO and DARPA. The authors are also thankful to Dr. Xuanyao Fong and Prof. Kaushik Roy (Purdue University) for helpful discussions.

REFERENCES

[1] X. Fong, S. H. Choday, K. Roy, "Bit-Cell Level Optimization for Non-volatile Memories Using Magnetic Tunnel Junctions and Spin-Transfer Torque Switching," *IEEE Trans. Nanotechnology*, vol.11, no.1,2012.

[2] D. Djayaprawira et. al, "230% room-temperature magnetoresistance in CoFeB/MgO/CoFeB magnetic tunnel junctions" *Appl. Phys. Lett.* 86, 092502 (2005).

[3] Z. H. Xiong, et al, "Giant magnetoresistance in organic spin-valves", *Nature* 427, 821-824, Feb. 2004.

[4] J.C. Slonczewski, "Current-driven excitation of magnetic multilayers", *Journal of Magnetism and Magnetic Materials*, 159 (1996) L 1 -L7.

[5] B. S. Haran, et al, "22 nm technology compatible fully functional 0.1 μm² 6T-SRAM cell," *Int. Electron Dev. Meeting*, pp.1,4, 15-17 Dec. 2008.

[6] S. P. Park, et al, "Future Cache Design using STT MRAMs for Improved Energy Efficiency: Devices, Circuits and Architecture", *Design Automation Conference*, pp: 492-497, 2012.

[7] S. K. Gupta, A. Raychowdhury and K. Roy, "Digital computation in sub-threshold regime for ultra-low power operation: A device-circuit-system co-design perspective", *Proceedings of the IEEE*, vol. 98, no. 2, 2010.

[8] W G Wang and C L Chien, "Voltage-induced switching in magnetic tunnel junctions with perpendicular magnetic anisotropy", *J. Phys. D: Appl. Phys.* 46 074004, 2013.

[9] L Liu, et al, "Spin-torque switching with the giant spin Hall effect of tantalum", *Science* 336 (6081), 555-558.

[10] Y. Kim S. H. Choday, and K. Roy, "DSH-MRAM: Differential Spin Hall MRAM for On-Chip Memories," *IEEE Electron Device Letters*, vol.34, no.10, pp.1259-1261, Oct. 2013.

[11] L. Liu, et al, "Current-Induced Switching of Perpendicularly Magnetized Magnetic Layers Using Spin Torque from the Spin Hall Effect", *Phys. Rev. Lett.* 109, 096602, 2012.

[12] Chia-Tsung Cheng; Yu-Chang Tsai and Kuo-Hsing Cheng, "A high-speed current mode sense amplifier for Spin-Torque Transfer Magnetic Random Access Memory," *IEEE International Midwest Symposium on Circuits and Systems (MWSCAS)*, pp.181,184, Aug. 2010.

[13] J. P. Kim, et al, "A 45nm 1Mb embedded STT-MRAM with design techniques to minimize read-disturbance," *Symposium on VLSI Circuits (VLSIC)*, pp.296,297, June 2011

[14] S. Natarajan et. al, "A 14nm logic technology featuring 2nd-generation FinFET, air-gapped interconnects, self-aligned double patterning and a 0.0588 μm2 SRAM cell size," *IEEE Int. Electron Dev. Meeting*, 2014.

[15] K. G. Anil, K. Henson, S. Biesemans, N. Collaert, "Layout density analysis of FinFETs," *33rd Conference on European Solid-State Device Research, 2003. ESSDERC '03.*, pp.139,142, 16-18 Sept. 2003.

[16] S. K. Gupta, et al, "Layout-Aware Optimization of STT MRAMs", *Design Automation and Test in Europe Conference*, pp: 1455-1458, 2012.

[17] E. Karl, et al, "A 4.6GHz 162Mb SRAM design in 22nm tri-gate CMOS technology with integrated active VMIN-enhancing assist circuitry," *International Solid-State Circuits Conference*, pp.230,232, Feb. 2012

[18] X. Fong et al, 2013, "SPICE Models for Magnetic Tunnel Junctions Based on Monodomain Approximation," https://nanohub.org/resources/19048.

[19] http://ptm.asu.edu/

[20] N. N. Mojumder; et al, "A Three-Terminal Dual-Pillar STT-MRAM for High-Performance Robust Memory Applications," *IEEE Trans. Electron Devices*, vol.58, no.5, pp.1508,1516, May 2011.

[21] D. Lekshmanan, et al, "FinFET SRAM: Optimizing Silicon Fin Thickness and Fin Ratio to Improve Stability at iso Area," *Custom Integrated Circuits Conference*, pp.623,626, 16-19 Sept. 2007.

[22] Chung-Hsun Lin, et al, "Modeling of width-quantization-induced variations in logic FinFETs for 22nm and beyond," *Symposium on VLSI Technology* pp.16,17, 14-16 June 2011.

2015 IEEE Computer Society Annual Symposium on VLSI

Using Multiple-Input NEMS for Parallel A/D Conversion and Image Processing

Kaisheng Ma[1], Nandhini Chandramoorthy[1], Xueqing Li[1], Sumeet Kumar Gupta [2], John Sampson[1], Yuan Xie[3],
Vijaykrishnan Narayanan[1]

[1]Dept. of Computer Science and Engineering, [2]Dept. of Electrical Engineering, The Pennsylvania State University
[3]Dept. of Electrical and Computer Engineering, University of California at Santa Barbara
Email: {kxm505,nic5090,lixueq,sampson,vijay}@cse.psu.edu, skgupta@psu.edu, yuanxie@ece.ucsb.edu

ABSTRACT

The technology advancements in semiconductor process have led to rapid progress in design and fabrication of multiple-input Nano-electro-mechanical relays/switches (NEM relays/NEMS). This work explores the design space of implementing image processing algorithms using a hybrid multiple-input NEMS–CMOS architecture. Different from the existing the approaches of building logic gates (e.g. INV, AND, OR, NAND, NOR, etc.) as conventional CMOS circuits, using NEMS, this work takes advantages of the electrical and mechanical physical features of the NEM relays to implement image processing algorithms. Simulation results show that the multiple-input NEM relays can be applied for decision-making and compare-select image processing algorithms. Moreover, we show that NEM relays can operate as parallel analog-to-digital converters (ADCs) with advantages of leakage reduction and power efficiency.

Keywords

ADC, computation in physics, image processing, NEMS.

1. INTRODUCTION

For nano-scaled device technology, NEMS relays demonstrate prominent advantages due to their non-leakage feature, aggressive supply voltage(VDD) scaling [1-5]. In traditional works, various digital integrated circuit building blocks including logic, memory, and clocking structures implemented purely with NEMS has been demonstrated [1-5]. But in this work, we open another path for using the NEMS as physical computing devices. By taking advantages of the mechanical and electrical features of the NEMS, the physical force can be used to make the computation. And this process can been done momentarily. All the other supporting circuits are used to read and transform the inputs and the results to digital signals.

In spite of the zero leakage advantage, the NEMS technology still faces several challenges [1-2]. One is the relatively lower speed (tens of ns to ms) compared to CMOS (less than 1 ns). This is because the switching between the off- and on-states is based on the mechanical movement of the electrode. Recently, a higher speed has been predicted with the device size scaling [3]. The other challenge is the higher on-state source/drain resistance which may further limit the applications. According to the Scaled Model in [2], the NEMS on-state resistance is in the range of 540~900 Ohm, while the most advanced NEMS has shown an on-state resistance of approximately 500 Ohm [3]. The third challenge is the comparatively larger cell area (about 50x50 um^2) [1][2]. As for the duration issue in the previous works, in the most advanced NEMS it is generally considered not a disadvantage with the duration up to 60 billion times [2].

Many works have focused on designing NEMS-based logic circuits to achieve similar functions to the conventional CMOS [1, 2, 5, 6]. Some basic logic cells have been designed including INV, AND, OR, NAND, NOR, XOR, latch, 32-bit full adder, etc. Multi-input NEMS design has been proposed to improve the original NEMS device [1]. In this case, more complex logic functions could be implemented with less delay [1].

In traditional digital image processing platforms, the pixel digitization and other higher-level digital signal processing are carried out separately by sensors and digital processors, which results in high transmission bandwidth with redundant information [8-14]. In the contrast, in this work, we explore the design space of implementing image processing algorithms using 3D integration of image sensors, CMOS and multiple-input NEMS. Instead of building logic gates (e.g. INV, AND, OR, NAND, NOR, etc.) in conventional CMOS circuits, this work takes advantages of the electrical, mechanical, and physical force features of the NEMS to implement certain image processing algorithms. With cooperative small-scale CMOS circuits, the multiple-input NEMS array demonstrates the potential to work as a parallel power-efficient analog-to-digital converter (ADC) array, and is capable of image processing such as Gaussian filtering, edge detection, saliency, as well as decision-making and compare-select image processing algorithms.

In the rest of this paper, Section 2 introduces the NEM relay physical layout and models. Section 3 describes how to use NEMS array to implement different image processing algorithms. Section 4 provides the simulation results and discussions. Section 5 benchmarks the speed and power performance, and integration with CMOS architecture. Section 6 concludes this paper.

2. NEMS MODEL INTRODUCTION

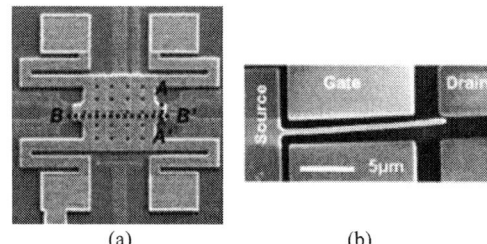

(a)   (b)

Figure 1. Two types of NEMS structure from [1] and [5].

The NEMS physical layout and electrical model have been described in detail in [2, 5, 6]. Figure 1 illustrates the physical layout of the NEMS [1][5]. Figure 1(a) is electrically a four terminal devices, including gate, drain, source and body, which is similar to a conventional bulk MOSFET. As in Figure 1, the gate

978-1-4799-8720-7/15 $31.00 © 2015 IEEE   339

is movable and suspended by spring-like flexures above the body, drain and source electrodes [1]. When a sufficiently large voltage is applied to the gate, the electrostatic force pulls the gate towards the bottom plate. As a result, the drain and source electrodes are connected through the channel, making the NEMS work in the "turned-on" state in Figure 2(a). Without the sufficiently large gate voltage, the source and drain electrodes are disconnected from the channel and the NEMS is working in the "turned-off" state. Figure 1(b) is another structure with two gates. In this paper, we propose a NEMS structure based on it, as shown in Figure 2(b). By splitting the two gates into smaller ones, this structure supports multiple inputs. Because the structure's gates are in two opposite directions, the force is equivalent to subtract operation.

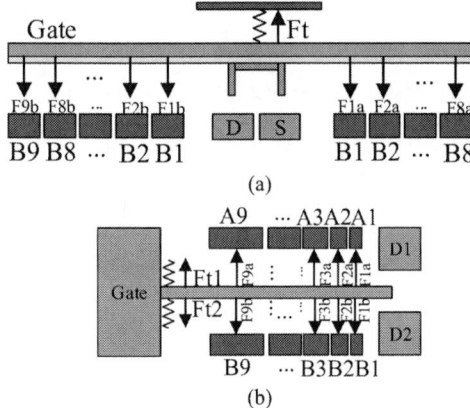

(a)

(b)

Figure 2. Illustration of the two different NEMS devices: (a) Structure based on Figure 1(a) for add operation. (b) Structure based on Figure 1(b) for add and subtract operation.

In the following part of this section, the physical features of the NEMS based on Figure 1(a) will be introduced as an example. Because the NEMS operation states are determined by the physical contact between the channel and the drain/source terminals, a high on-to-off current ratio is achievable. Different from the electrical switching mechanism in CMOS transistors, the switching delay of an equivalent NEMS is comparatively larger. Despite of such a larger delay, the NEMS has significantly lower leakage, and higher energy efficiency.

The behavior of the NEMS could be accurately modeled through a nonlinear second-order differential equation to describe and predict the mechanical and electrical actions.

$$m \cdot x'' = F(x) - b \cdot x' - k \cdot x, \tag{1}$$

where m is the gate mass, x is the displacement of the gate, b is the damping coefficient, k is the effective spring constant of the gate structure, and $Felec(x)$ is the nonlinear electrical force between the gate and the body:

$$F(x) = \frac{\varepsilon_0 A_{ov} V_{gb}^2}{2(g_0 - x)^2}. \tag{2}$$

In (2), ε_0 is the permittivity of free space, A_{ov} is the overlap area between the gate and body electrodes, g_0 is the normal gap between the electrodes without the electrical force, and V_{gb} is the gate-body voltage. The required voltage to snap the gate-body structure shut is called the pull-in voltage, V_{pi}:

$$V_{pi} = \sqrt{\frac{8}{27} \cdot \frac{k g_0^3}{\varepsilon_0 A_{ov}}}. \tag{3}$$

The turn-on delay is determined by (1). The gate mass m and the spring constant k affect the mechanical turn-on delay t_{mech} in a way that could be modeled as:

$$t_{mech} = \sqrt{\frac{m}{k}} \cdot \left(\frac{V_{pi}}{|V_{gb}|} \right). \tag{4}$$

To accurately model the switching behavior of the NEMS, electrical parameters are also extracted. The contact resistance between the channel and the source/drain R_{con} is modeled as

$$R_{con} = \frac{4\rho\lambda}{3A_r}, \tag{5}$$

where ρ is the contacting material resistivity, λ is the mean free path of electrons in the contact material, and A_r is the effective contact area given by

$$A_r \approx \frac{F_{elec}(g_d)}{\xi H}. \tag{6}$$

Some key parameter values are summarized in Table I, with the others the same as [2]. The gate-body capacitance (C_{gb} and C_{gc}) is modeled as

$$C_{gb}(x) = \frac{\varepsilon_0 A_{ov}}{g_0 - x}, \quad C_{gc} = \frac{\kappa_{gox} \varepsilon_0 A_{ch}}{t_{gox}}, \tag{7}$$

where κ_{gox} is the relative permittivity of the gate oxide, A_{ch} is the gate-channel overlap area, and t_{gox} is the gate oxide thickness.

Table 1. Parameter of the NEMS in this paper [2]

Parameter	Device before scaling	Scaled Device
A_{ov} [μm²]	450	0.77
R_{on} [Ω/contact]	~0.1	40-400
R_{pox} [Ω/contact]	500	500
$t_{mech\_turn-on}$ [μS]	34	0.02-0.08
$t_{mech\_turn-off}$ [μS]	3	0.002-0.008
t_{elec} [pS]	304.4	2.5-3.5
Duration [times]	60e10	unknown
Stability	2000g	unknown

3. HYBRID CMOS & NEMS IMAGE PROCESSING (NEMSIP)

In this section, NEMS IMAGE PROCESSING (NEMSIP) is proposed for various image processing algorithms. The key idea of NEMSIP is computing in physics. All the CMOS circuits are designed as supporting peripherals.

3.1 NEMSIP for Figure 1(a) Structure

In NEMS, the electrical force in (2) could be rewritten as:

$$F(x) = \frac{\varepsilon_0}{2(g_0 - x)^2} * \left[\sum_{i=1}^{n} A_{ovi} * (V_g - V_{bi})^2 \right], \tag{8}$$

where V_{bi} is output from the CCD or CMOS sensor array, V_g is the gate voltage, A_{ovi} is the area of inputs for each pixel in the back layer, and x is the distance between the gate layer and back layer. Once the voltage is applied to NEMS, the computation is carried out by physics. All the CMOS peripherals are functioning for data reading and A/D conversion.

Based on such NEMS computation mechanism, Figure 3 illustrates the readout circuits and ADCs. There is an 8-bit self-adder, which adds by 1 on each rising clock edge. This adder output is then converted to an analog voltage by the DAC. It is apparent that all NEMS gates are connected together and driven

978-1-4799-8720-7/15 $31.00 © 2015 IEEE   340

by the same DAC output. Therefore, only one adder and one DAC are required as the CMOS circuits. The adder output is then fed to an 8-bit flip-flop chain. With the voltage between the NEMS back and gate layers increasing, the force will enhance until the two layers are connected through the channel to turn on the NEMS and trigger the flip-flop to maintain the digital data. To ensure that the DAC and the NEMS stabilize within each clock cycle, the clock cycle should be carefully designed to meet the NEMS turn-on and turn-off timing requirement.

The speed of this NEMS-based parallel ADC array could be improved by replacing the self-adder by a self-subtracter. When the dimples pump up, the digital outputs would be ready. The improved speed originates from the fact that the mechanical turn-off delay is much smaller than the turn-on delay [14], in Table 1.

We propose an efficient DAC implementation for the NEMS-based ADC, as shown in Figure 4. Compared with Figure 3, the DAC is replaced with a constant current to charge a capacitor. Thus, the voltage of the capacitor increases linearly with time. A buffer is added between the charger and the NEMS gate for the purpose of isolation. The constant charging is designed in a way that the charging time matches the time required for the self-adder to add up to 255 from 0. Every time when the self-adder becomes 8'b0, the capacitor will be completely discharged to prepare for the next computation readout. Similarly, if a self-subtracter is used instead of the self-adder, the constant-current charger should be replaced by a constant-current discharger accordingly.

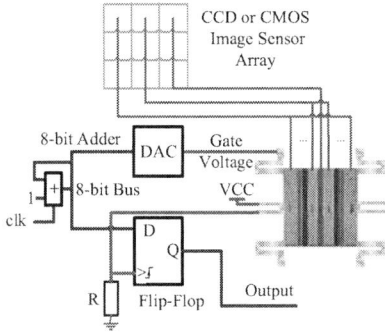

Figure 3. NEMSIP with a DAC.

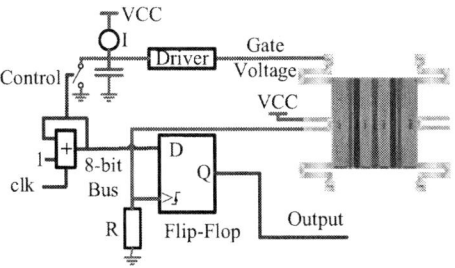

Figure 4. NEMS-based ADC structure with a constant current charger.

Another structure is based on the successive-approximation-register (SAR) structure, in which the NEMS comparison result is used to control the DAC so that the output of the DAC could approximate to the final state by at most N steps for an N-bit ADC. The advantage of this structure is high speed, as compared with

the structure in Figure 3 or Figure 4 which needs 2^N steps of DAC output changes in the worst case to finish the analog-to-digital conversion. The disadvantage is more power consumption due to the feed-back operations which results in a dynamic change in the output of the DAC and consequently the loss of the simple DAC implementation shown in Figure 4. For the image processing tasks in this paper, we'll use the structure in Figure 3 and Figure 4 which provides sufficient performance while enables a quick evaluation of the parallel ADC application using NEMS.

In such a CMOS-NEMS hybrid system, the NEMS are working as comparators, which turn the system into parallel ADCs. Straightforwardly, these comparators could be implemented in CMOS circuits as well. However, using CMOS circuits to implement such comparators will result in significant power consumption in terms of energy per comparison. Moreover, such NEMS-based comparators have zero standby leakage current, which outperforms the CMOS comparators.

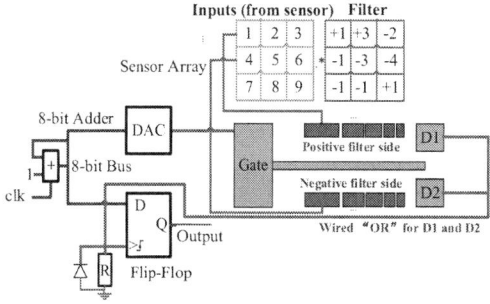

Figure 5. NEMSIP using the NEMS in Figure 1(b).

3.2 NEMSIP for Figure 1(b) Structure

Using the NEMS structure in Figure 1(b) instead of Figure 1(a), the subtraction operation could be efficiently implemented as shown in Figure 5. The NEMS in Figure 5 has two opposite force directions, achieving a subtraction operation. The temporal differential computation can also be implemented by this Figure 5. structure within 2 phrases. In phrase 1, the pixels are connected to positive filter side, while in phrase 2, connected to negative filter side. In this way, temporal differential can be computed. The computation function becomes

$$F(x) = \frac{\varepsilon_0}{2(g_0 - x)^2} \left[\sum_{pi=1}^{n} A_{ov,pi}(V_g - V_{b,pi})^2 - \sum_{nj=1}^{n} A_{ov,nj}(V_g - V_{b,nj})^2 \right], \quad (9)$$

where pi represents the positive gate controls in the NEMS and nj represents the negative gate controls.

The voltage of the components $D1$ and $D2$ in Figure 5 should be high enough to be capable of triggering the flip-flops. With such a voltage range requirement, the structure in Figure 5 is more suitable for compare-and-select algorithms, such as the motion estimation in JPEG and H.264.

3.3 PIPELINED NEMSIP

The challenge of the structure in Figure 5 is the routing and wiring for NEMS, especially in the scaled devices. To mitigate challenge, we propose the pipelined NEMSIP in Figure 6 to implement a 5*5 filter with three 3*3 NEMS. In the Pipeline Phase 1 in Figure 6, the 25-pixel input window is divided into three groups with 9, 9 and 7 pixels. The outputs are analog gate voltage instead of digital bits. Between Pipeline 1 and Pipeline2, a capacitor is used to store the analog charge. SW2 is

978-1-4799-8720-7/15 $31.00 © 2015 IEEE   341

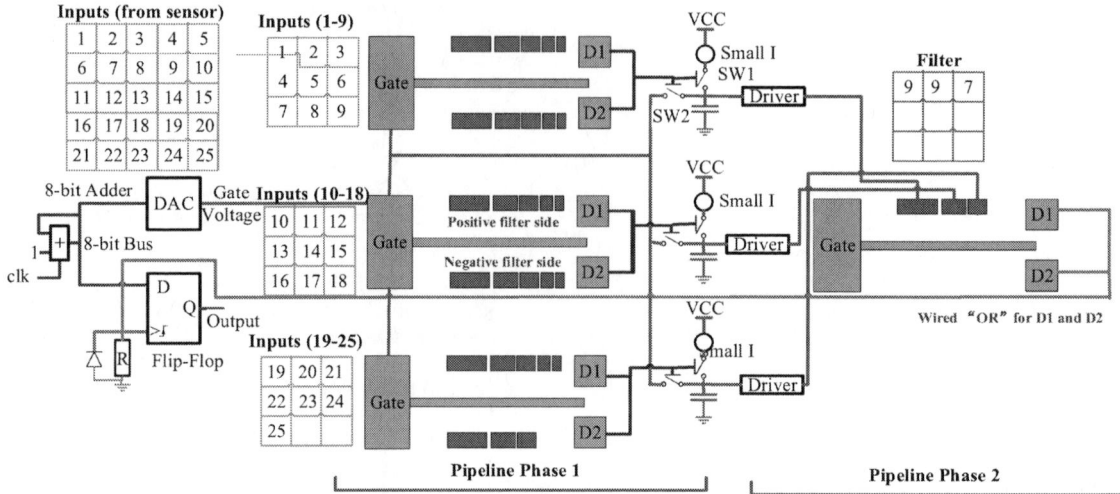

Figure 6. The pipelined NEMSIP structure to implement a 5*5 filter with 3*3 NEMS.

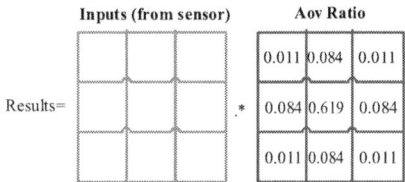

Figure 7. Gaussian filter implementation.

Figure 8. Gaussian filter results comparison. (a), Lenna grey scale 128*128, as inputs to NEMSIP. (b), Matlab results. (c), NEMSIP result. (d), Larger input scale. (e), Smaller input scale affects darkness. (f),Larger scale of gate voltage affects output to be darker. (g), Smaller scale of gate affects output quality. (h), high-speed results, with other setups same as (c). (i), high-speed mode with adjusted voltage scales.

turned-off, and SW1 is turned-on during Phase 1. In Pipeline Phase 2, the analog results from Phase 1 are inputs. During Phase 2, SW2 is turned-on, while SW1 is turned-off for the current source to compensate the capacitor leakage. The filter from Phase two is 9, 9, 7 to represent the weight.

4. MAPPING AND RESULTS

In this section, several image processing and video compression algorithms are mapped to NEMSIP. In this work, the mechanical and electrical parameters in [2][5][6], as well as the delay by the DAC and wires, are employed in the simulation platform. Although the model employed is a behavioral model and does not include second-order effects, the simulations based on this model proves the potential of the proposed NEMSIP in image applications.

4.1 POSITIVE FILTERS - GAUSSIAN FILTER

Gaussian filter is a commonly used basic method to reduce noise. When implemented with NEMSIP, the sensor output voltages are connected to the back layer of NEMS, and the A_{ov} ratio is another multiplier, as shown in Figure 7. The simulation results are shown in Figure 8. As shown in Figure 8(c), it is comparable to the Matlab result in Figure 8(b). Figure 8(i) is the result of the high-speed (on-to-off) mode. The quality in the high speed mode is similar to that of the low-speed (off-to-on) mode. The high-speed mode is very fast because the pump switching distance to break is less than 1nm distance. Just a little gap can turn off the connection.

4.2 POSITIVE AND NEGATIVE FILTERS - EDGE DETECTION

By changing the Ratio A_{ov}, an edge detection function could be implemented with NEMSIP. Various types of input are implemented and shown in Figure 9, including "Text" in Figure 9(a), "Cartoon" in Figure 9(d), "Dice" with different levels of distance and details in Figure 9(g), "Lenna" in Figure 9(j), "Car license" in Figure 9(m). Results show that the NEMSIP implementation could detect the edges in Figure 9(c, f, i, l, o), but the effect is weaker than the Matlab canny algorithm in Figure 9(b, e, h, k, n). These differences originate from the algorithm differences, not the implementation.

4.3 REDUCING MULTIPLE FILTERS - EDGE DETECTION

In edge detection, Gaussian filtering is useful to suppress the noise. The A_{vo} could also be tuned to include Gaussian filter into edge detection. Figure 10 shows the edge detection results using NEMSIP. Compared with Figure 9(o), Figure 10 gives preferred results. It is also observed that the NEMSIP could realize more complicated edge detection kernels through combining more than one path into one single NEMSIP processing step.

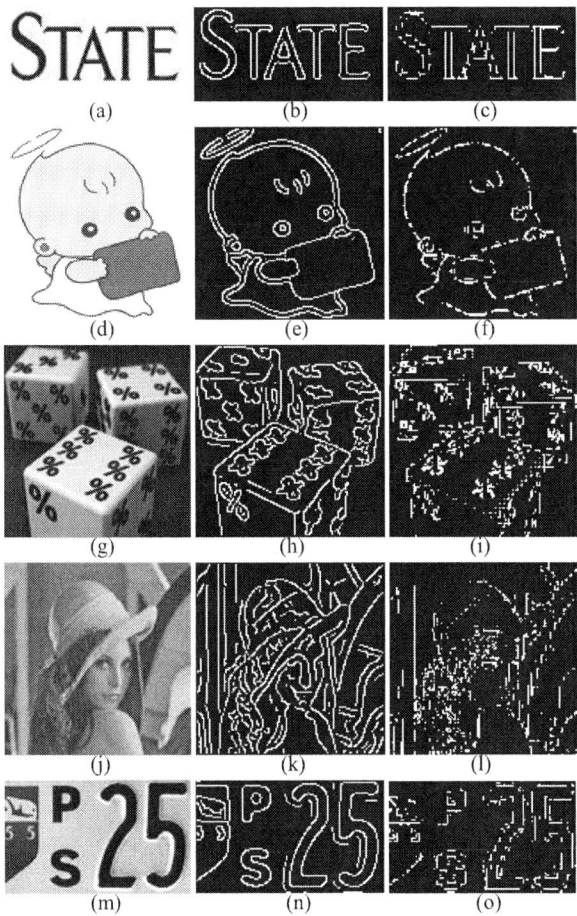

Figure 9. Edge detection results comparison. (a) Text image from sensor, as inputs to NEMSIP. (b) Matlab canny edge detection result. (c) NEMSIP edge detection result. (d) Cartoon inputs. (e) Matlab canny edge detection for cartoon in (d). (f) NEMSIP outputs for cartoon. (g) dice inputs. (h) Matlab canny edge detection for dice. (i) NEMSIP outputs for dice. (j) Lenna 128*128 inputs. (k) Matlab canny edge detection for Lenna. (l) NEMSIP outputs for Lenna. (m) Car license inputs. (n) Matlab canny edge detection result. (o) NEMSIP outputs for Car license.

4.4 MAPPING LARGE WINDOW IN NEMSIP FOR MOTION ESTIMATION

The widely-used motion estimation method calculates the sum of the absolute difference (SAD) block by block to search for the best matched block to reduce the difference between the source and the target [16]. NEMSIP is able to achieve the same quality with the full search algorithm with a delay in the order of nanoseconds. Figure 11(a, b, c) shows the JPEG inner frame

motion estimation with the Pipelined NESMIP. Figure 11(a) is the input to NEMSIP, and the results in Figure 11(b, c) show that NEMSIP is able to find the best matched block. Figure 11(d, e) shows two frames from the testbench "football", in which the human and ball are moving fast, and the NEMSIP can still find the best matched block as shown in Figure 11(f).

Figure 10. Edge detection with Gaussian filter.

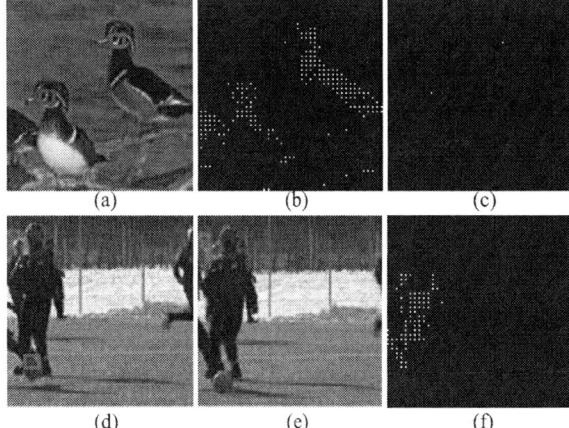

Figure 11. Motion estimation results for Figure 6. (a) Motion estimation input. (b) Results to identify the red block at a certain low gate voltage threshold. (c) By scanning the gate voltage from high to low, the best matched point is obtained (as well as the motion vector). (d, e) Inter frame motion estimation for H.264 video compression algorithm with the red block in (d) as the source block. (f) NEMSIP motion estimation results.

5. SPEED, POWER, and 3D INTEGRATION

In this section, NEMSIP is compared with traditional CMOS solutions in terms of speed and power. With the near-sensor processing by NEMSIP, the transmission bandwidth between the sensor chip and the digital signal processor could be reduced significantly after pre-processing by NEMSIP.

5.1 SPEED EVALUATION

The speed of NEMSIP is determined by the NEMS turn-on/off time t_{mesh} and t_{elec}. The time required to process one frame, T, could be obtained based on Figure 12(a):

$$T=(2^8+1)*(t_{adder} + t_{DAC} + t_{mesh} + t_{elec} + t_{flip-flop}), \quad (10)$$

where t_{adder}, t_{DAC}, and $t_{flip-flop}$ are the delay of the adder, the DAC, and the flip-flop, respectively. By applying the high-speed (on-to-off) mode, the t_{mesh} could be reduced significantly. The reset time could be optimized through the Ping-Pong loop shown in Figure 12(b). In the Ping loop, the adder increases from 0 to 255; while in the Pong loop the substractor decreases from 255 to 0.

Table 2 shows the speed comparison between NEMSIP and CMOS solutions. The NEMSIP solution is 7.6 times faster than the pipelined CMOS structure. For the scaled model, NEMSIP is 16 times faster than CMOS.

Furthermore, for NEMSIP, the speed is independent on the image scale, while the pipelined CMOS solution highly depends on the image scale. For NEMSIP, the image processing does not require additional CMOS-based ADC which, in contrast, is inevitable for the CMOS solution. In this test, the entire image is stored in the memory, and the read delay from the memory is not included.

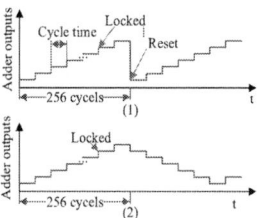

Figure 12. Timing strategies. (1) Reset solution. (2) Ping-Pong.

Table 2. Speed Comparison between NEMSIP and CMOS

Type	Speed (frame/second)
Current NEMSIP	233k
NEMSIP scaled to 90nm	488k
32nm CMOS [a]	30.5k

[a]: The CMOS is an optimized pipelined structure with one clock cycle per pixel output, 32nm process in tt corner, standard V_t, 1.05 V power supply, and 500MHz clock.

5.2 POWER EVALUATION

The power in NEMSIP is mainly consumed by the following parts: the adder, flip-flops, and charging and discharging of the NEMS capacitor. The NEMS could be treated as a panel capacitor. Each computation charges or discharges the capacitor once. For the current NEMS device, the V_{pi} is measured to be 8~10V, which is lowered to 40mV after scaling to 90nm. It is observed that the NEMSIP power estimation in this paper is significantly different from [2], in which NEMS was used to function similarly to traditional transistors switches.

Table 3 shows the power consumption comparison between NEMSIP and CMOS solutions in Table 2. The NEMSIP results are calculated, and 32nm CMOS results are from RTL level with Design Compiler. For the same frame rate, the 8-bit adder and the 128*128*8 flip-flops operates at 7.8MHz. The NEMSIP solution using the existing NEMS consumes only 5.9% of the power by the CMOS solution for the same frame rate.

Table 3. Power Comparison between NEMSIP and CMOS [a]

Type	Power
Current NEMSIP	0.2626mW for 128*128 NEMS + 0.286mW for 128*128*8 flip-flops + 2.9uW for adder
32nm CMOS	9.325mW total power

[a]: The NEMSIP is operating at the same frame frequency of 30.5k frame/second. The 32nm CMOS solution area is 0.45mm^2.

5.3 3D INTEGRATION

When integrated with the CCD or CMOS sensor layer on the top and the CMOS digital processor on the bottom, the sandwiched NEMSIP layer achieves near-sensor image pre-processing. Between every two adjacent layers, an insulator layer is used for isolation. Similar to the integration in [3], through-silicon via (TSV) is not necessary, and the top metal layer interconnection is sufficient for such integration with high interconnection density.

6. CONCLUSION

This paper uses NEMS as weighted multiple-input energy-efficient comparators to implement parallel ADCs for near-sensor

image processing. The proposed NEMSIP solution has almost no leakage by the NEMS, and no conventional ADCs are required. The current NEMS and scaled NEMS could operate at a speed of 16 times faster than the pipelined CMOS solution, while consuming only 5.9% of the CMOS power even if the ADC power of the CMOS solution is not included. Furthermore, the bandwidth of the near-sensor NEMSIP processing could also be significantly reduced with less transmitted data. In general, the NEMSIP is promising in bringing energy and performance benefits in future image processing.

7. ACKNOWLEDGEMENTS

This work was supported in part by the LEAST, one of the six SRC STARnet Centers, sponsored by MARCO and DARPA. This work was also supported in part by NSF Expeditions in Computing Award 1317560.

8. REFERENCES

[1] Jaeseok Jeon et al. Multiple-Input Relay Design for More Compact Implementation of Digital Logic Circuits. Electron Device Letters. 2012.

[2] Spencer, M. et al. Demonstration of Integrated Micro-Electro-Mechanical Relay Circuits for VLSI Applications. JSSC, vol.46, no.1, pp.308,320. 2011

[3] V. Pott et al. Mechanical Computing Redux: Relays for Integrated Circuit Applications. Proceedings IEEE, vol.98, no.12, pp.2076,2094.2010

[4] J. Daesung et al. Combinational Logic Design Using Six-Terminal NEM Relays. CADICS, vol.32, no.5, pp.653,666. 2013

[5] C. Soogine et al. Nanoelectromechanical (NEM) relays integrated with CMOS SRAM for improved stability and low leakage. ICCAD. 2009

[6] S. Yong-Ha et al. High-performance MEMS relay using a stacked-electrode structure and a levering and torsional spring for power applications. Micro Electro Mechanical Systems (MEMS). 2012

[7] B. Pruvost et al. 3-D Design and Analysis of Functional NEMS-gate MOSFETs and SETs. IEEE Transactions on Nanotechnology. 2007

[8] Alaghi, A. et al. Stochastic circuits for real-time image-processing applications. Design Automation Conference. pp.1,6. 2013

[9] Elouardi, A. et al. A Smart Sensor for Image Processing: Towards a System on Chip. ISIE, vol.4, no., pp.2857,2862.

[10] X. Zhang et al. A second generation 3D integrated feature-extracting image sensor. IEEE Sensors, pp.1933,1936.

[11] M. Sarkar, Biologically Inspired CMOS Image Sensor for Fast Motion and Polarization Detection. IEEESJ, vol.13, no.3, pp.1065,1073.

[12] Y. Chin et al. A 0.5V 34.4uW 14.28kfps 105dB smart image sensor with array-level analog signal processing," SSCC, pp.97,100, 11-13 Nov. 2013

[13] K. Kiyoyama et al 2013. A block-parallel ADC with digital noise cancelling for 3-D stacked CMOS image sensor. 3D Systems Integration Conference (3DIC), 2013 IEEE International, pp.1,4.

[14] Lie, D. et al. Analysis of the Performance, Power, and Noise Characteristics of a CMOS Image Sensor With 3-D Integrated Image Compression Unit. CPMT, vol.4, no.2, pp.198,208. 2014

[15] F.D. Oliveira et al. CMOS Imager With Focal-Plane Analog Image Compression Combining DPCM and VQ. Circuits and Systems I: Regular Papers, IEEE Transactions on, vol.60, no.5, pp.1331,1344. 2013

[16] K. Ma et al. A Novel Multi-direction Fast Parallel Search (MFPS) Method for Motion Estimation in Video Compression and Its Hardware Implementation. The IEEE 10th International Conference on ASIC, 2013.

[17] L. Areekath et al 2013. Sensor assisted Motion Estimation. Engineering and Systems (SCES), pp.1,6, 12-14 .

[18] Chen, X. et al 2011. Sensor-Assisted Video Encoding for Mobile Devices in Real-World Environments. IEEE Transactions on Circuits and Systems for Video Technology, vol.21, no.3, pp.335,349.

[19] L. Kang-Wook Lee et al 2013. Die-Level 3-D Integration Technology for Rapid Prototyping of High-Performance Multifunctionality Hetero-Integrated Systems. Electron Devices,vol.60, no.11, pp.3842-8.

[20] X. Cui et al, Research on power model of multi-mode FinFET standard cell, ICSICT, pp.1,3, Oct. 29 2012-Nov. 1 2012

[21] K. Ma et al, Key characterization factors of accurate power modeling for FinFET circuits. Science China Information Sciences, 2015, 58(2): 1-13.

[22] K. Ma et al, Independently-Controlled-Gate FinFET 6T SRAM Cell Design for Leakage Current Reduction and Enhanced Read Access Speed, ISVLSI, pp.296,301, 9-11 July 2014

978-1-4799-8720-7/15 $31.00 © 2015 IEEE

2015 IEEE Computer Society Annual Symposium on VLSI

Implementing Data Structure Using DNA: An Alternative in Post CMOS Computing

Mayukh Sarkar, Prasun Ghosal

Indian Institute of Engineering Science and Technology, Shibpur, Howrah 711103, WB, INDIA

Email: mayukh@it.iiests.ac.in, p_ghosal@it.iiests.ac.in

Abstract—DNA computing has attracted the eyes of many researchers in recent years to solve NP-complete problems. It over-performs conventional computers due to its inherent massively parallelism nature. But to make it generally applicable, problems those are very much implementable on conventional computers, should also be implementable on a DNA computer. To implement those problems, data structures are unavoidable. In this work, possible implementations of several data structures viz. stack, queue, list with insertions and deletions at random indexes have been presented. Additionally, maps and possible operations on them have been proposed. Proposed implementations are kept extremely simple, and can be practically implemented quite easily. This work is expected to serve as an important step towards the applicability of a DNA computer and to prove its efficacy as a promising alternative in tomorrows post CMOS computing paradigm.

Index Terms—**DNA Computing, Post CMOS Technology, Data Structure Implementation, Emerging Technologies**

I. INTRODUCTION

The DNA (Deoxyribonucleic acid), found in the living cells, is composed of four bases, namely Adenine(A), Guanine(G), Thiamine(T) and Cytosine(C). The order of these bases is unique in each individual, and determines the unique characteristics of that particular individual. Each base is attached to its neighbour base in the sequence via phosphate bonding. The base, sugar and the phosphate are together called a nucleotide. Two DNA sequences bond with each other via hydrogen bonding between each Watson-Crick complementary base pairs (A with T, and C with G), forming DNA double helix. Each DNA strand has two ends : 5'-end and 3'-end, that determine the polarity of the DNA strand. During the formation of DNA double strand, two complementary single strands bond with each other in anti-parallel fashion. Several molecular biological operations can be performed on DNA, discussed in Section IV.

The problems are solved on a DNA computer by encoding the problems using A, T, G and C, synthesizing corresponding DNA strands, and performing several operations on those strands by methods available in a molecular biology laboratory. The main power of DNA computing over conventional ones are as follows.

1) Massively parallel operation. A single test tube of DNA can contain trillions of DNA strands, and all strands respond to the biological operations in parallel.
2) The information density of DNA is huge over silicon. Estimated storage capacity of 2.2 petabytes per gram of DNA has been reported in [1].

A. Emergence of DNA Computing

In recent years, computing using DNA molecules, namely, DNA computing, has been proved to be a powerful emerging technology to solve hard computational problem. In 1994, Adleman [2] has shown the use of this powerful tool in solving Travelling Salesman problem, a difficult combinatorial search problem. In the very next year (1995), Lipton [3], in a very similar way, solved the SAT problem, by representing all possible combination of values in a graph. Lipton's DNA algorithm for solving the SAT problem was a linear one, which is impossible for a conventional computer at this stage. In the same year(1995), Boneh et.al. [4] broke DES, a famous encryption method, using DNA. In 1997, Ouyang et. al. [5] solved maximal clique problem given a six vertex graph. Along with these experiments, several other NP-complete problems, such as Graph Coloring [6], Bin Packing [7] etc. have been solved using DNA computing. All these experiments have shown that, inherently massive parallelism and high data storage density have made DNA computing a powerful technique to solve such problems over conventional computer.

Overall organization of the rest of the paper is as follows. Novel contributions have been summarized in section II. Section III briefly mentions the present status and importance of research works on realization of DNA computers and thereby the motivation behind the present work. Different Bio molecular operations and their mathematical modeling have been presented in section IV. Setion V, VI, and VII describes the implementations of Stacks and Queues, Lists, and Maps using DNA. Experimental feasibility of different Bio molecular operations have been presented in section VIII. Finally, Section IX concludes the paper with possible future directions.

II. NOVEL CONTRIBUTIONS OF THE WORK

In this paper, some of the data structures, viz., stack, queue, list and map have been implemented. Even though DNA computer is proven powerful over conventional computer for computationally hard problems with huge search-base, but to make it generally applicable, these problems which can be implemented on conventional computer, should also be implementable on DNA computer. But to solve these problems, several abstract data types and techniques associated with them are unavoidable. These data structures define the data storage in proper fashion, for better applicability. But true

978-1-4799-8720-7/15 $31.00 © 2015 IEEE   345

to the best of authors' knowledge, implementations of these data structures for DNA computer have not been addressed except in [8], and they have implemented stack and queue in the work, with a complete stack or queue being encoded on a double strand.

In this work, not only stack and queue, also a list with insertion and deletions at random indexes, and map have been implemented. The data structures are being stored as a group of DNA single strands in a test tube, rather than a single strand. Also, implementations are much simpler and easier.

III. REALIZATION OF DNA COMPUTER: RELATED RESEARCH AND MOTIVATION

In 2002, researchers from Weizmann Institute of Science, Israel, have developed a programmable molecular computing machine composed of enzymes and DNA molecules. A year later, the team advanced one step further, as in the new device, the single DNA molecule that provides the computer with the input data also provides all the necessary fuel [9]. In 2004, Benenson et. al. [10] described an autonomous bio-molecular computer that logically analyses the levels of messenger RNA species, and in response produces a molecule capable of affecting levels of gene expression. This computer, in theory, would be capable of diagnosing cancer, and producing anti-cancer drug. In 2013, Goldman et. al. [11] encoded computer files totaling 739 kB of hard disk storage and with an estimated Shannon information of 5.2×10^6 bits into a DNA code, synthesized, sequenced and then reconstructed the original data with 100% accuracy. In the same year, bio-engineers at Stanford University created the first biological transistors, named *transcriptor*, using DNA and RNA [12]. On 26th October, 2014, Israeli scientists, in collaboration with researchers from around the world, have developed DNA strands capable of carrying electrical charges for DNA-based electrical circuits [13] [14]. They have reported reproducible charge transport in guanine-quadruplex(G4) DNA molecules adsorbed on a mica substrate,and have measured currents of tens of picoamperes to more than 100pA in the G4-DNA over distances ranging from tens of nanometres to more than 100nm.

All these latest advancements towards the realization of a DNA computer expedites the development of a promising alternative of tomorrows computing hardware in post CMOS era. Successful implementation of such a computing device needs the support of more efficient algorithms that eventually induces the necessity of efficient implementations of different data structures.

IV. BIO-MOLECULAR OPERATIONS AND MATHEMATICAL MODELING

Each molecular biological operations on DNA strand filled test tube is considered to be run in O(1) time. In this section, the possible operations on DNA strands have been stated, and mathematical functional form of those operations have been given.

Single DNA strand G will be written as G, whereas double strand of G and $\overline{G}$ is represented as $\left(\dfrac{G}{\overline{G}} \right)$, where $\overline{G}$ is the Watson-Crick complement of G.

The following operations are possible on DNA strands.

1) *Merge(T1, T2)* - Pouring contents of test tube T2 in the test tube T1. T1 contains $T1 \cup T2$.
2) *Copy(T1, T2)* - Copying contents of test tube T1 in an empty test tube T2.
3) *Anneal(T1)* - All possible complementary strands $S_1 S_2$ and $\overline{S_1 S_2}$ in T1 gets joined together to form double strands $\left(\dfrac{S_1 S_2}{\overline{S_1 S_2}} \right)$. This is performed by cooling up the solution.
4) *Denature(T1)* - Denatures all double strands $\left(\dfrac{S_1 S_2}{\overline{S_1 S_2}} \right)$ present in T1 in single strands $S_1 S_2$ and $\overline{S_1 S_2}$. This is performed by heating up the solution.
5) *Cut(T1, $R_1 R_2$)* - All double stranded DNA in T1, containing a string $\left(\dfrac{S_1 R_1 R_2 S_2}{\overline{S_1 R_1 R_2 S_2}} \right)$, will be cut into two parts $\left(\dfrac{S_1 R_1}{\overline{S_1 R_1}} \right)$ and $\left(\dfrac{R_2 S_2}{\overline{R_2 S_2}} \right)$. As not all DNA string can act as restriction site, so some fixed strings $R_1 R_2$, $C_1 C_2$ or $K_1 K_2$ will only be used as restriction site.
6) *Append(T1, strand)* - This will append the strand *strand* at the end of each strand in T1.
7) *Separate_Length(T1, l, T2)* - This operation will separate the strands of length l from T1, and put them in T2.
8) *Separate_String(T1, G, T2)* - This operation will separate all strands containing G as substring, and put them in T2.

V. STACK AND QUEUE USING DNA

A stack can be considered as a Last-In-First-Out (LIFO) data structure, which has only a single entry and exit point, named *top*. All elements are pushed into the stack through the top, and also popped through the top. Hence, the elements which enters last, exits first.

On the other hand, a queue is a First-In-First-Out (FIFO) data structure, which has two points, one for entry of the elements, and the other one for exit. Elements which enters first , are also removed first.

For the implementations, it can be assumed, without any loss of generality, that the DNA encoded strands of the elements are of equal lengths. To assure this, a restriction site is appended at the end of all DNA strands via ligation enzyme, and the shorter DNA strands can be made of equal length with the longest strand by ligating additional dummy nucleotide bases after the restriction site.

As for example, there are m elements to be inserted into the data structure, and they are encoded as DNA strands $E_1, E_2, \cdots, E_m$ with E_s being the longest DNA strand. Let us assume, some fixed restriction site $R_1 R_2$ is ligated at the end of all DNA strands, making the length of ith strand as $(E_i + R)$ for $1 \leq i \leq m$, where R is the length of the restriction site $R_1 R_2$. To assure that, all DNA strands have the same length, dummy DNA strand of length $(E_s - E_j)$ is ligated at the end of jth DNA strand for $1 \leq j \leq m$ *and* $j \neq s$.

A. Insertion of element into Stack and Queue

For the implementation of the data structures, some unique DNA strand P is needed to be available in some test tube. The data structure is assumed to be built up in test tube T. Now, the insertion of element in the data structures can be performed using the following algorithm.

```
1: procedure INSERT(TestTube T, Element E_i)
2:     if T is not empty then
3:         Add P into T and ligate.
4:     end if
5:     Pour E_i into T.
6: end procedure
```

Fig. 1: Inserting element into data structure (*push* for Stack and *insert* for Queue)

As an example, if the elements $E_1, E_2, \cdots, E_m$ are inserted into the data structure one by one, according to the insertion algorithm, the elements will be inserted as follows.

- Prior to the addition of the first element E_1, T was empty. So simply pour E_1 into T. So, after this addition, $T = \{E_1\}$.
- When the next element E_2 will be added, T already contains E_1. Ligate P with the elements of T and add E_2. So, after this operation, $T = \{E_2, E_1 P\}$.
- Similarly, after adding the third element, the elements of T will be $T = \{E_3, E_2 P, E_1 P^{(2)}\}$.

In this manner, after adding all of the m elements, the elements of the test tube T will be, $T = \{E_m, E_{m-1}P, E_{m-2}P^{(2)}, \cdots, E_2 P^{(m-2)}, E_1 P^{(m-1)}\}$, where $P^{(i)}$ represents concatenation of i number of DNA strands P, i.e., $\underbrace{PPP \cdots P}_{i}$. So, the length of the ith element inserted into the data structure will be $[E + (i-1)P]$, where E is the length of the elements after adding the restriction site $R_1 R_2$ and equalized by adding dummy nucleotide bases, and P is the length of the DNA strand P.

B. Deletion of element from Stack and Queue

Removing an element from the top of the stack is a trivial operation. As the length of the DNA strands decreases towards the top of the stack, the top element can be removed by simply performing Gel Electrophoresis and separating the band corresponding to the DNA strand having the smallest length.

Similarly an element can be removed from the end of the queue by simply separating the band corresponding to the DNA strands having largest length.

After separating the desired band, the original element can be obtained using the following procedure.

1) The obtained DNA strand is annealed with the $\overline{R_1 R_2}$, the Watson-Crick complement of the restriction site $R_1 R_2$.
2) The annealed strand is cut at the proper location using the restriction enzyme, which detects the restriction site $R_1 R_2$.
3) The solution is heated to denature.
4) The DNA strand containing R_1 is separated. The output contains the original element with R_1 attached at the end.

VI. LIST USING DNA

List can be implemented in the similar manner as stack and queue, but to insert element at any location other than the front, algorithm 1 needs to be modified.

Assume, after inserting m elements, the contents of the test tube is as,
$T = \{E_1 P^{(m-1)}, E_2 P^{(m-2)}, E_3 P^{(m-3)}, \cdots, E_i P^{(m-i)}, E_{i+1}P^{(m-i-1)}, \cdots, E_{m-1}P, E_m\}$, with 1 being the beginning index having the element E_1. Now, if some element E_j needs to be inserted into the list at index i, E_j needs to appended with $P^{(m-i)}$, and all the elements before the ith index needs to have an additional P at the end. To obtain this, the following sequence of operations needs to be carried out.

1) Length of the DNA strand at index i is given by

$$[E + (m - i) * P] \tag{1}$$

All the elements at index $\leq i$ needs to be appended with an additional P. So, Gel Electrophoresis is performed and all the bands having length higher than or equal to the length $[E + (m - i) * P]$ are separated into another test tube T_h. Rest of the solution is kept as it is in T.

2) Pour P into T_h and ligate using ligation enzyme to append P at the end of all the DNA strands.

3) Append P at the end of E_j, $(m - i)$ number of times and keep in a test tube T_e.

4) Pour T_e and T_h back in T.

The final contents of the test tube T will become,
$T = \{E_1 P^{(m)}, E_2 P^{(m-1)}, E_3 P^{(m-2)}, \cdots, E_i P^{(m-i+1)}, E_j P^{(m-i)}, E_{i+1}P^{(m-i-1)}, \cdots, E_{m-1}P, E_m\}$, having E_j inserted at the desired position.

Removal of an element from the list is the same as in the removal of element from a stack or a queue. The only difference is that, during Gel Electrophoresis, rather than separating the band having maximum (*remove* from queue) or minimum length(*pop* from stack), the band of the DNA strands with length $[E + (m - i) * P]$ needs to be separated to remove the element at index i.

VII. MAP USING DNA

Map, an abstract data type, is a collection of $(key, value)$ pair, where each key is unique, and the retrieval of a value is performed using the key.

Let, $(K_1, V_1), (K_2, V_2), \cdots, (K_m, V_m)$ be the pairs to be inserted into the map. Now , let, $E_{K1}, E_{K2}, \cdots, E_{Km}$ be the DNA strands encoding the keys $K_1, K_2, \cdots, K_m$, and $E_{V1}, E_{V2}, \cdots, E_{Vm}$ be the DNA strands encoding the values $V_1, V_2, \cdots, V_m$ respectively. The $(key, value)$ pair (K_i, V_i) will be formed as the single DNA strand $E_{Ki} R_1 R_2 E_{Vi}$.

The possible operations on a map, viz., *reassign*, *remove* and *lookup* operations are implemented as specified in the next three subsections.

A. Lookup

In *lookup* operation, the value of a pair (K_s, V_s) is retrieved using the value of the key K_s. This retrieval is performed by simply separating the DNA strand containing the string E_{Ks}. This separation by string can be performed using Watson-Crick complement of E_{Ks} on a magnetic bead, and letting the required DNA strand annealed with the bead. The bead is then washed into another solution, and the strands are denatured by heating up. The solution thus contains the DNA strands $E_{Ks}R_1R_2E_{Vs}$. To retrieve the value V_s only, the resulting solution is annealed with $\overline{R_1R_2}$, cut at the specific location using the restriction enzyme, and separating the strands containing R_2.

B. Remove

In *remove* operation, some particular pair (K_s, V_s) is removed from the map. The implementation of this operation is trivial, viz., simply by performing the magnetic bead separation using the Watson-Crick complement of the given key E_{Ks}.

C. Reassign

To reassign the value of some particular pair (K_s, V_s) to the value of V_t, the required pair is first separated using the magnetic bead separation with the Watson-Crick complement of the key E_{Ks}, and kept in the test tube T_s. The separated pair is then annealed with $\overline{R_1R_2}$, and cut at the specific location using the restriction enzyme. The strand is cut into two parts, $E_{Ks}R_1$ and R_2E_{Vs}. The R_2E_{Vs} part is removed from the solution using magnetic bead containing $\overline{R_2}$.

In another test tube T_t, R_2 is kept, and the DNA strand E_{Vt} is appended at the end of R_2 using ligation enzyme. The solution in T_t is then poured in T_s and ligated using ligation enzyme again. The resulting solution in T_s will be formed as $E_{Ks}R_1R_2E_{Vt}$. This solution is then poured back in the main test tube containing the map.

VIII. Experimental Feasibility

The molecular biological operations being used in the work are already proven feasible operations on DNA solutions. A theoretical computational model based on the RDNA model [15] have been assumed here with some additional operations, such as *Append* and *Separate_DS*. Fujiwara et.al. [16] uses the RDNA theoretical model to perform arithmetic and logic operations with DNA strands. Xiao et.al. [17] has used this mathematical model to perform DNA representation of elements in $\{0, 1\}^n$. Tsai et.al. [18] has used this model to construct parallel adder.

The operations comprising the theoretical model used in this work, can be performed practically in a biomolecular laboratory as follows.

Merge : Two test tubes T_1 and T_2 containing DNA solutions can be merged by simply pouring the contents of T_2 in T_1. This leaves T_2 empty, so a backup of T_2 can be made

by copying, before merging and after merge, the backup can be poured back in T_2.

Copy : The copy of DNA solution residing in a test tube T_2 can be made in a previously empty test tube T_1 using *Polymerase Chain Reaction*.

Anneal : If a test tube containing DNA strands is cooled down, in the course of the normal, random molecular motion, the complementary strands come closer to each other and stick tightly to form DNA double strands.

Denature : A tube of DNA strands dissolved into water, when heated up at a suitably high temperature, the hydrogen bonds between the strands are broken to separate the double strands into two single strands.

Cut (Cleavage) : DNA double strands can be cut at or near specific nucleotide sequence using *restriction enzyme* found in bacteria. Such nucleotide sequence is called *recognition site*. Each restriction enzyme can identify different recognition site in DNA, and hence not all nucleotide sequence can be used as recognition site. As for example, cutting using *EcoRI*, *BamHI* etc. enzymes produce sticky ends, whereas, *SmaI*, *PvuII* etc. produce blunt ends.

Append : Appending a DNA strand at the end of another DNA strand is a trivial operation, that can be performed by a specific enzyme, *DNA ligase*, that joins two DNA strands by catalysing the formation of phosphodiester bond.

Separate_Length : DNA strands in a solution can be separated by length using Gel Electrophoresis. DNA strands of different lengths move through the gel in different speeds, and hence can be separated by separating the band of the desired length DNA strands.

Separate_String : DNA strands in a solution can be separated, based on the presence of a particular string, by magnetic bead separation procedure, where a magnetic bead containing the Watson-Crick complement of of the string, is hold into the solution and DNA strands containing the string are attracted towards the bead.

IX. Conclusion and Future Direction

In this paper, DNA implementations of all primary data structures viz. stack, queue, list and map have been proposed. For solving several problems solvable on conventional computer, data structures are unavoidable. Moreover, implementations proposed in this paper are based on simple to implement bio-molecular operations. So, with these data structures, DNA computing is expected to advance one step forward towards the general applicability. Future works may be extended towards the implementation of other data structures.

References

[1] N. Goldman, P. Bertone, S. Chen, C. Dessimoz, E. M. LeProust, B. Sipos, and E. Birney, "Towards practical, high-capacity, low-

978-1-4799-8720-7/15 $31.00 © 2015 IEEE

maintenance information storage in synthesized DNA," *Nature*, vol. 494, pp. 77–80, Feb. 2013.

[2] L.M.Adleman, "Molecular computation of solutions to combinatorial problems," *Science*, vol. 266, pp. 1021–1024, Nov. 1994.

[3] R.J.Lipton, "DNA solution of hard computational problems," *Science*, vol. 268, pp. 542–545, Apr. 1995.

[4] D. Boneh and C. Dunworth and R. J. Lipton, "Breaking DES Using a Molecular Computer," in *DIMACS workshop on DNA computing*, 1995.

[5] Q.Ouyang, P.D.Kaplan, S.Liu, and A.Libchaber, "DNA solution of the maximal clique problem," *Science*, vol. 278, pp. 446–449, Oct. 1997.

[6] Y.Liu, J.Xu, L.Pan, and S.Wang, "DNA solution of a graph coloring problem," *J Chem Inf Comput Sci.*, vol. 42, pp. 524–528, May-Jun 2002.

[7] C.A.A.Sanches and N.Y.Soma, "A polynomial-time DNA computing solution for the bin-packing problem," *Applied Mathematics and Computation*, vol. 215, pp. 2055–2062, 2009.

[8] W. Li, Y. Ding, Y. Huang, and X. Ye, "Linear data structure for DNA computer," in *Fuzzy Systems and Knowledge Discovery, 2009. FSKD '09. Sixth International Conference on*, vol. 7, Aug 2009, pp. 141–145.

[9] "Computer made from DNA and enzymes," http://news.nationalgeographic.com/news/2003/02/0224_030224_DNA computer.html, accessed: 2014-11-05.

[10] Y. Benenson, B. Gil, U. Ben-Dor, R. Adar, and E. Shapiro, "An autonomous molecular computer for logical control of gene expression," *Nature*, vol. 429, pp. 423–429, May 2004.

[11] N. Goldman, P. Bertone, S. Chen, C. Dessimoz, E. M. LeProust, B. Sipos, and E. Birney, "Towards practical, high-capacity, low-maintenance information storage in synthesized DNA," *Nature*, vol. 494, pp. 77–80, Feb. 2013.

[12] "Stanford creates biological transistors, the final step towards computers inside living cells," http://www.extremetech.com/extreme/152074-stanford-creates-biological-transistors-the-final-step-towards-computers-inside-living-cells, accessed: 2014-11-05.

[13] Livshits, Gideon I. and Stern, Avigail and Rotem, Dvir and Borovok, Natalia and Eidelshtein, Gennady and Migliore, Agostino and Penzo, Erika and Wind, Shalom J. and Di Felice, Rosa and Skourtis, Spiros S. and Cuevas, Juan Carlos and Gurevich, Leonid and Kotlyar, Alexander B. and Porath, Danny, "Long-range charge transport in single G-quadruplex DNA molecules," *Nature Nanotechnology*, October 2014.

[14] "Israeli scientists achieve breakthrough in dna computing," http://www.haaretz.com/life/science-medicine/1.623755, accessed: 2014-11-05.

[15] J. H. Reif, "Parallel biomolecular computation: Models and simulations," *Algorithmica*, vol. 25, pp. 21 322–3, 1995.

[16] A. Fujiwara, K. Matsumoto, and W. Chen, "Addressable procedures for logic and arithmetic operations with DNA strands," in *International Parallel and Distributed Processing Symposium, 2003*, April 2003.

[17] D. Xiao, W. Li, J. Yu, X. Zhang, Z. Zhang, and L. He, "Procedures for a dynamical system on $\{0,1\}^n$ with DNA molecules," *Biosystems*, vol. 84, no. 3, pp. 207 – 216, 2006.

[18] S. Tsai, W.-L. Chang, and S.-H. Ho, "Constructing bio-molecular parallel adder with basic logic operations in the Adleman-Liption model," in *International Conference on Convergence Information Technology, 2007*, Nov 2007, pp. 925–930.

2015 IEEE Computer Society Annual Symposium on VLSI

An Unbalanced Area Ratio Study for High Performance Monolithic 3D Integrated Circuits

Hossam Sarhan, Sebastien Thuries, Olivier Billoint and Fabien Clermidy

Univ. Grenoble Alpes, F-38000 Grenoble, France
CEA, LETI, MINATEC Campus, F-38054 Grenoble, France
{hossam.sarhan; sebastien.thuries; olivier.billoint; fabien.clermidy}@cea.fr

Abstract— **Monolithic 3D (M3D) integration technology offers fine grain gate level stacking capability compared to 3D Through Silicon Vias (3D-TSV) which is well adapted for coarse-grain applications. As a result, design partitioning, i.e. which cell on which tier, highly affects the 3D design performance. Previous partitioning methodologies focus on minimizing number of 3D interconnects for equal area ratio between the stacked partitions. This paper demonstrates that un-balancing the tier to tier area ratio of the M3D design brings better performance than classical balanced 3D design approach. Our study highlights that neither balanced area ratio nor the number of 3D interconnections remains mandatory criteria for M3D. We show that our technique can achieve up to 24% performance improvement compared to 2D and 15% better performance than the state-of-the-art technique without extra power penalty.**

Keywords—3D IC; Monolithic 3D; Gate-level; Area-Ratio

I. INTRODUCTION

The race of CMOS miniaturization has been following Moore's Law over the past four decades, boosting the semiconductor industry by reducing fabrication costs and increasing overall performance of the integrated circuits. Beyond 28nm, cost scaling will no more be possible due to lower Back End Of Line (BEOL) performance compared to previous nodes, leading to increase the power consumption and to reduce the maximum operating frequency [1].The growing impact of on-chip interconnect design shows a need for other approaches which could change the way cells are connected, in order to reduce the impact of wiring. Stacking in three dimensions by using Through Silicon Vias (TSV) is a possible solution. However, it is well adapted for coarse grain integrated applications such as 3D network-on-chip [2] and memory-on-logic [3] but not for fine-grain partitioning needed for real 3D design. Very high interconnect density will never be achieved by 3D TSV as the fabrication process needs an alignment between two wafers with tools not compatible with BEOL required accuracy.

Alternatively, Monolithic 3D (M3D) integration technology is a sequential process allowing the fabrication of a second active layer directly on top of the first one in a single flow featuring ultra-high alignment accuracy [4]. Table 1 shows a comparison between M3D, high-density TSV (HD-TV) [8] and high-density copper-to-copper bonding (Cu-Cu) [9, 10]. The pitch of the M3D inter-tier via (M3D-VIA) is 20 to 50 times smaller than TSV or Cu-Cu. Thus, fine grain 3D integration can be achieved, at standard cell or transistor level as presented in different studies [5-7].

As stated before, power consumption and delay of 2D-ICs are more and more dominated by interconnect parasitics [1, 2]. Consequently, splitting long wires using 3D technology is a good technique to shorten them for reducing power consumption and increasing performance of the design. However, an efficient 3D partitioning technique is needed to specify wires splitting and to assign standard cells on different tiers.

Previous 3D partitioning techniques are constrained by minimizing the number of 3D connections with balanced area ratio between top and bottom tiers [11-13]. These techniques are achieved without taking into consideration the performance of the obtained partitioning.

That is why our main contributions in this study are following:

- Definition of the Physical Aware Partitioning (PAP) methodology to early validate the allocation of a cell to the appropriate tier, in addition to introduce the concept of unbalanced area ratio between top and bottom tiers.

- Definition of the Bi-Directional Partitioning (BDP) algorithm as the core of the PAP methodology to automatically cut long wires of the 2D design leading to unbalance the area ratio of the resulting 3D design.

- Assessment of performance gains for M3D designs when using the proposed methodology.

Table 1: Comparison between different 3D technologies

3D Via	HD-TSV [8]	Cu-Cu [9, 10]	M3D [4]
Diameter	3 µm	1.7-3µm	0.1-0.25 µm
Pitch	10 µm	3.4-10 µm	0.2-0.5µm*
Density (per mm^2)	10,000	10,000	~25 x10^6

* Pitch is assumed to be double the 3D inter-tier via width

This work was funded thanks to the French national program 'programme d'Investissements d'Avenir, IRT Nanoelec' ANR-10-AIRT-05

978-1-4799-8720-7/15 $31.00 © 2015 IEEE   350

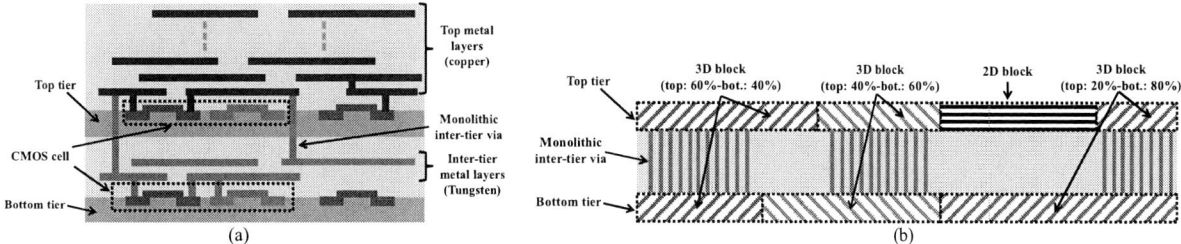

(a)   (b)

Figure 1: (a) Monolithic 3D integration technology, (b) hybrid 3D floor-planning with two 60-40% blocks, 80-20% block and 2D block.

II. PROBLEM DEFINITION

There are two main approaches for designing 3D blocks in the literature. The first one consists in partitioning by hand the micro-architecture of the 2D design into 3D, such as in [16, 17]. This approach requires the designer to have a preliminary knowledge and understanding of the 2D architecture, and consequently it takes long time that is not compatible with complex and large-scale designs. The second approach consists in partitioning the 2D netlist to get the 3D netlists using a partitioning algorithm. First, the netlist is converted to an intermediate representation form, such as hypergraph representation. In the hypergraph representation each node represents a standard cell and the edges represent the interconnections between cells. The hypergraph is then partitioned. Reference [15] uses hMetis [14] as the partitioning tool and targets balanced area ratio between 3D layers while minimizing the number of 3D connections (Min-Cut).

As introduced earlier, the cost of the BEOL becomes more and more a strong limitation to achieve high performance in 2D design. On the other hand, M3D integration breaks the limitation in terms of number of 3D connections. The need for equal area between layers is no longer a constraint for 3D floor-planning because different blocks can have different 3D area ratios while the whole architecture is well equilibrated to avoid any impact on die packaging. Figure 1 (a) shows a cross-section of M3D integration technology. Figure 1 (b) shows an example of a hybrid 3D floor-planning using blocks with different top-bottom area ratio blocks. Using such floor-planning avoids any extra area in the whole chip packaging due to the unbalanced blocks.

To illustrate the effect of 3D partitioning; we assume a partitioning case of a small circuit with just 5 nodes, i.e. 5 standard cells. Figure 2 (a) shows the hyper-graph representation of our example; where the weights on the nodes represent the area of each standard cell and the weights on the edges represent the wire length. The first way of partitioning tends to get balanced area with minimum number of cut. Figure 2(b) shows a balanced-area partitioning. In this technique, small weighted edges are cut (i.e. short wires) which may degrade the overall 3D design performance. A second way is to partition is to cut only the long wires without considering the resulting area ratio (Figure 2(c)).This example illustrates that cutting only the long wires, even if the area ratio between the top and bottom tiers is not equal, is better than targeting balanced area ratio which may lead to cut short wires as well.

In this work, we show that there is an area-performance trade-off for 3D block design techniques using a new 2D netlist partitioning approach. To achieve that, a physical aware partitioning methodology has been defined where only long critical paths are cut which improves the 3D design performance.

III. PHYSICAL-AWARE PARTITIONING METHODOLOGY

The automated Physical-Aware Partitioning (PAP) methodology consists of 4 main steps leading to partition a 2D netlist into two 3D netlists. Figure 3 shows the work flow of the PAP methodology. At the beginning the 2D netlist is obtained by a conventional synthesize process based on 28 nm FDSOI CMOS technology. Starting from the 2D netlist is

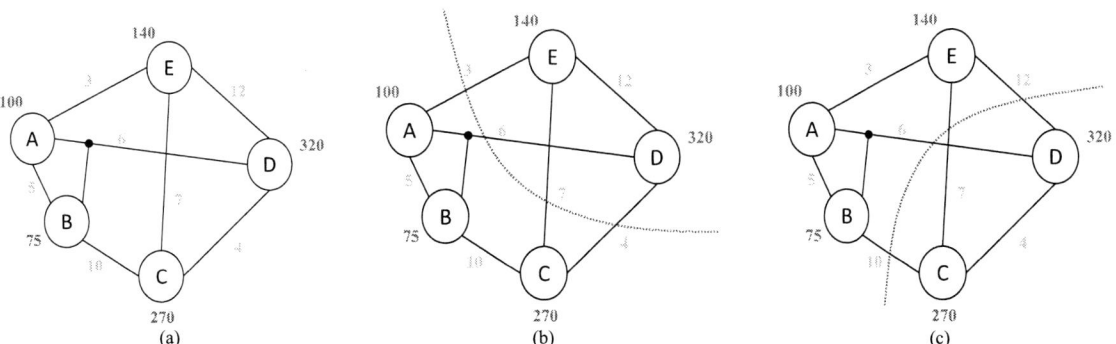

(a)   (b)   (c)

Figure 2: (a) An example of small circuit in a hypergraph representation; where (b) balanced area ratio partitioning, and (c) unbalanced area ratio partitioning (using critical length threshold equals 6)

978-1-4799-8720-7/15 $31.00 © 2015 IEEE   351

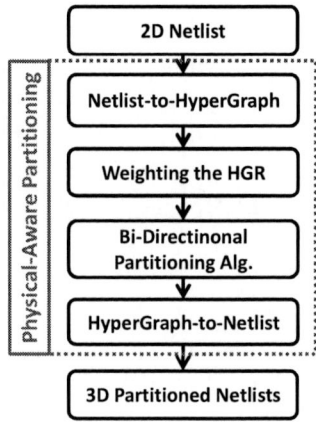

Figure 3: Physical-Aware Partitioning (PAP) methodology

helpful for the next steps because of the connectivity between each standard cell. The netlist is then passed through each step: netlist-to-hypergraph conversion, weighting the hypergraph, hypergraph partitioning, and finally hypergraph to netlist conversion.

A. Netlist-to-Hypergraph Conversion

The 3D partitioning flow starts with netlist to hypergraph representation. Figure 2 (a) shows an example of a hyper-graph representation. In the netlist to hypergraph conversion each standard cell is represented as a vertex node where these nodes are connected with each other with hyper-edges which represent wiring interconnections between the cells. A hyper-graph to netlist conversion is done at the end of the flow to generate 3D netlists from the partitioned hypergraph.

B. Weighting the HGR

Then the hypergraph (HGR) is passed by a weighting phase, at which every cell and wire takes a numerical value representing its weight in the partitioning algorithm. These numerical values represent physical parameters of the design. Each standard cell is represented by its physical cell area to control the area ratio between top and bottom die of the 3D design.

On the other hand, each interconnection is represented by its 2D wire length, so that the partitioning algorithm starts to cut the longest wires first. The wire length values can be obtained using a wire length model, such as the half-perimeter wire length (HPWL) model. These wire length values are then used in the partitioning flow to arrange wiring interconnects from the longest to the shortest one.

IV. BI-DIRECTIONAL PARTITIONING ALGORITHM

The main step of the PAP methodology is the partitioning phase of the 2D hypergraph. For the partitioning, a bi-directional partitioning algorithm has been developed which targets cutting long wires first. Algorithm 1 shows the detailed procedure of the partitioning algorithm. The first step in the hypergraph partitioning phase is to generate a "fixed_cell" list.

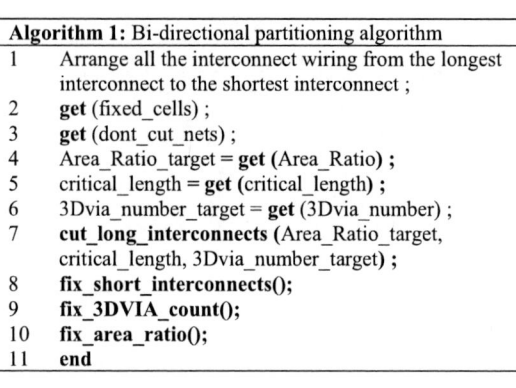

Algorithm 1: Bi-directional partitioning algorithm
1   Arrange all the interconnect wiring from the longest interconnect to the shortest interconnect ;
2   **get** (fixed_cells) ;
3   **get** (dont_cut_nets) ;
4   Area_Ratio_target = **get** (Area_Ratio) **;**
5   critical_length = **get** (critical_length) **;**
6   3Dvia_number_target = **get** (3Dvia_number) ;
7   **cut_long_interconnects** (Area_Ratio_target, critical_length, 3Dvia_number_target) **;**
8   **fix_short_interconnects();**
9   **fix_3DVIA_count();**
10   **fix_area_ratio();**
11   **end**

This fixed_cell list is generated according to the user specifications. For example, in case of having all the design ports in one die then a list of all the standard cells connected to the design ports is generated as a fixed_cell list.

Another example, in case of having synchronous clock tree in one die, then a list of all the clocked standard cells are generated as a fixed_cell list. The generated fixed_cell list is accompanied with an indication to the fixing die, either the top or the bottom die.

After generating the fixed cells, the main partitioning algorithm is applied. First, all wires are arranged from the longest to the shortest one. Then all the fixed cells are marked using the list generated in the previous step. After that, each wire is being cut starting from the longest to the shortest ones, by order. After each cut process, some of the standard cells are moved to the other die which creates additional M3D via. This cutting process is stopped either if it reaches a given critical length as a cut threshold, or reaching a given surface area of the bottom die. The critical length threshold is defined as a function of the resistive and capacitive values of the physical wiring compared to the resistance and capacitance of the M3D. The critical length threshold is a technology depending parameter.

Design interconnections are arranged from the highest weight (longest) to the less weight (shortest).

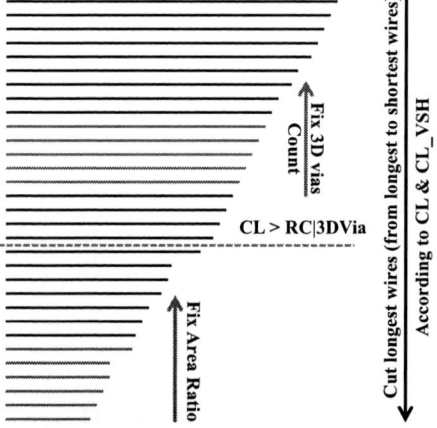

Figure 4: Bi-directional partitioning technique

After finishing cutting the long wires, three fixing steps are needed according to the user criteria given.

1. First, restoring any short wire (below than certain threshold) that has been unintentionally cut due to moving cells to the second tier.

2. Second, number of the M3D-VIAs is fixed. If resultant number of the M3D-VIAs is greater to a given value, then shortest cut wires are restored. Fixing number of M3D-VIAs is done from shortest cut wires to the longest.

3. Finally, area ratio between partitioned parts is fixed. If the moved cell area is less than the given value, then the standard cells connected to the shortest uncut wire will be moved to the bottom die. The direction of fixing the area ratio is from the shortest to the longest wire till the given area ratio is met.

Figure 4 shows the arranged wires of the design where the direction of the partitioning, i.e. cutting wires, is performed from the longest wires to the shortest ones, while the direction of fixing the resultant number of M3D-VIA and surface area ratio is performed from the shortest wires to the longest ones. By using this cutting and restoring procedure, each standard cell is placed either on the top tier or the bottom tier.

The order of complexity of the partitioning algorithm is $O(n^2)$ which depends on number of standard cells and consequently wiring interconnects of the block. We have tested the partitioning algorithm for designing different 3D blocks of complexity up to 100K gates.

V. RESULTS

A. Evaluation Methodology

To evaluate our approach, we have implemented a min-cut partitioning case with balanced (50-50) area ratio using hMetis [14] and set it as our reference point. Our proposed physical-aware partitioning (PAP) methodology has been implemented with unbalancing the area ratio and relaxing the number of M3D vias.

We have explored as well the cases of balanced area ratio using PAP methodology, however to be focus, we will discuss only the best performance results using unbalancing area ratio with PAP with a comparison to the reference points.

Our benchmark circuits are composed of growing complexity blocks: open-MSP microcontroller, Reconfigurable Fast Fourier Transform (Reconf-FFT), Low-Density Parity Check (LDPC) decoder, and triple Data Encryption Standard (DES-3). The Reconf-FFT block has several configurable parameters, such as the coefficient and input sizes, which makes it a control-oriented block. The openMSP, as well as, is a control-oriented block. On the other hand, LDPC and DES-3 are computing-oriented blocks.

To validate our different partitioning cases with timing and power results, we have decided to prototype the physical implementation of our test-bench blocks using a commercial 3D tool with FDSOI 28nm technology and M3D technology parameters extracted from the work demonstrated in [4].

The commercial prototyping tool used is fully aware of physical implementation floor-planning of both tiers. Figure 5 shows a snapshot for the physical implementation of one of our benchmarks, the LDPC block, in hMetis balanced case (50-50) and PAP unbalanced case (top: 75%, bottom: 25%). All the results presented in Table 2 are directly outputted by the tool including routing report, Total Wire Length (TWL), time slack (i.e. Performance) and Power.

In all test cases, we use M3D-VIA with a resistance of 2Ω and a capacitance of 0.1fF [5]. For the 3D design partitioning, we have converted the resistance and capacitance parasitics (RC) of the M3D-VIA to an equivalent wire length on the targeted FDSOI 28nm technology, and then apply this wire length as the cut threshold to our PAP methodology. By applying this critical length, no wire can be replaced by a M3D-VIA with a higher RC delay to guarantee gaining in performances. Partitioning parameters have been set to keep the clocked cells on the top tier to avoid clock issues on both tiers.

The density of 3DVIAs for each test case is calculated to ensure its value is within the technology 3DVIA density limitations as shown previously in the technology parameters in Table 1. Consequently increasing number of 3DVIAs for the PAP cases has no technology violation, thanks to the high density 3DVIA provided by the M3D technology.

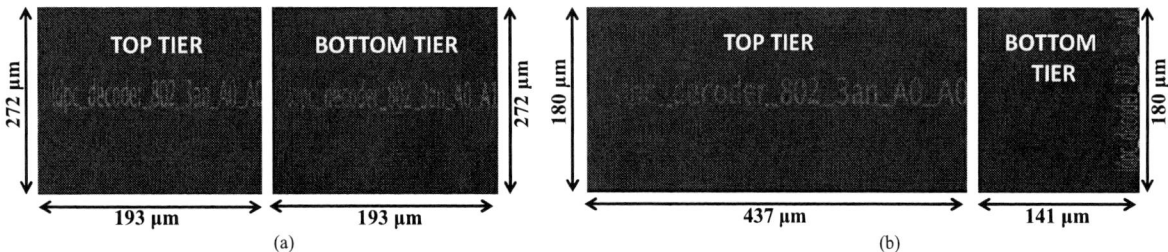

(a)   (b)

Figure 5: Physical prototype implementation snapshots of 3D LDPC block in two cases (a) 50-50 and (b) 75-25

978-1-4799-8720-7/15 $31.00 © 2015 IEEE

Table 2: Results comparison between 2D and 3D cases (hMetis and PAP partitioning) for openMSP, Reconf-FFT, LDPC and DES-3 blocks

		No. Standard Cells	Area (μm^2)		TWL (μm)	No. M3D-VIA	Max. Perf (GHz)	Perf. Gain (%)	Power @2D Max-Perf. (mW)	Power Gain (%)
			Top	Bottom						
openMSP	2D	6122	11390		91278	NA	1.21	NA	11.54	NA
	hMetis [14]	5456	6000	6000	89642	1110	1.24	2.5%	11.53	0.1%
	PAP (31-69)	5466	7875	3530	89630	3775	1.42	17.5%	11.74	-1.7%
Reconf-FFT	2D	29673	27600		420547	NA	1.33	NA	52.50	NA
	hMetis [14]	20093	13853	13853	316490	1158	1.45	9.0%	30.87	41%
	PAP (40-60)	20604	16656	10902	321269	13894	1.64	24.0%	28.75	45%
LDPC	2D	68179	88800		3277363	NA	1.58	NA	103.5	NA
	hMetis [14]	56896	52496	52496	1431432	12511	1.74	9.9%	98.3	5.2%
	PAP (25-75)	56896	78660	25380	965889	26917	1.76	11.5%	100.7	3.0%
DES-3	2D	106989	176400		1588509	NA	2.15	NA	185.4	NA
	hMetis [14]	61386	57000	57000	428599	5439	2.35	9.2%	160.9	13.2%
	PAP (50-50)	61386	57000	57000	357048	9518	2.41	12.0%	158.2	15.0%

B. Performance Results

PAP approach highlights that better performances can be obtained at unbalanced area ratio. The unbalancing area ratio varies depending on the block design, i.e. the ratio between number of long wires and short wires. In case of openMSP better performance is obtained at (35-65) area ratio, while the area ratio is better at (40-60) for the Recon-FFT, and (25-75) for the LDPC blocks. On the other hand, the DES-3 block is kept at (50-50) balanced area ratio to obtain its best performance. However PAP with balanced area ratio results better performance than the balanced area ratio using hMetis.

Table 2 summarizes the 2D reference results, the 3D results using hMetis with equal area ratio and 3D results using our PAP methodology with un-balanced area ratio at the max performance point obtained. The performance is calculated by removing the positive time slack to the frequency max of the clock used. The frequency max is the highest clock frequency we target in the trial before getting high congestion effect.

For control-oriented block like openMSP, we can observe that increasing the number of M3D-VIAS and un-balancing the area ratio by 31% can bring 17.4% better performance than the 2D design but also 15% more performance than classical min-cut balanced area ratio using hMetis. Similarly for the Reconfigurable-FFT block, unbalancing the area ratio by 40% can achieve 24% better performance than 2D where hMetis increases the performance only by 9%. It appears that the control-oriented blocks offer better performance gains by unbalancing area ratio due to the presence of long control wiring interconnects which is difficult to be optimized in 2D.

On the other hand, for the computing-oriented blocks like the LDPC, we show that unbalancing area ratio with our PAP methodology can increase performances by 11.5% compared to 2D which is slightly better than the hMetis. Similarly for the

DES-3 block, we demonstrate that our PAP methodology offers better performance, even at equal area ratio, compared to hMetis by 3% based on Table 2.

C. Power Results

For the power analysis, we have compared the iso-performance power results, i.e. at the same 2D max-performance point. The results shown in Table 2 highlight that going into M3D can reduce the power consumption up to 40% using hMetis methodology and up to 45% using PAP unbalanced methodology for a complex block like FFT. However, applying hMetis or PAP methodology for a low complex block like openMSP does not guarantee any reduction of power consumption. The block is too small to gain by cutting long wires as we can see with TWL results and number of standard cells.

Figure 6 shows the power-performance curves for openMSP, Reconf-FFT, LDPC and DES-3 blocks. The curves show that going to 3D always guarantee reaching better performances compared to 2D.

VI. SUMMARY AND CONCLUSION

In this work, we have demonstrated that partitioning the design with M3D integration technology involves a trade-off between number of M3D-VIAS, area ratio, power consumption and overall performance of the 3D design. Minimizing number of M3D-VIA and targeting balanced area ratio are no longer efficient as main constraints to partition designs for high density 3D technologies. We have proposed an unbalanced area ratio technique to improve the performance gain by partitioning long wires of the design without taking into account the area balance as a constraint. A hybrid floor-planning for different blocks is shown to avoid any extra space in the whole chip area. Additionally, we have defined a physical-aware partitioning (PAP) methodology based on a bi-

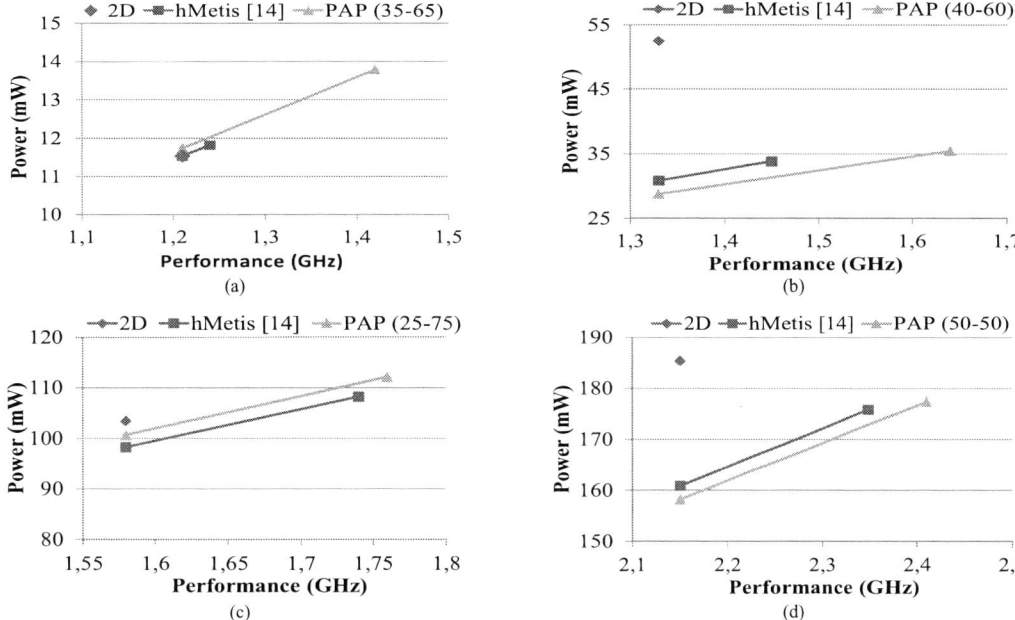

Figure 6: Power-Performance results for (a) openMSP, (b) Reconf-FFT, (c) LDPC and (d) DES-3 blocks.

directional partitioning algorithm to decide which wire has to be cut and which cell is assigned to which tier. We have applied PAP methodology on a set of benchmarks, where control-orient blocks show more gain compared to computing-oriented blocks. The physical implementation results show that we can gain in performance up to 24% compared to 2D design and up to 15% more compared to area balanced and min-cut 3D design. On the other hand for iso-performance cases, we show that 3D blocks with unbalanced area ratio reduce power consumption up to 45% compared to 2D and up to 5% more compared to area balanced and min-cut 3D designs.

VII. REFERENCES

[1] International Technology Roadmap for Semiconductors - Interconnect – 2011.

[2] A.G. Wassal, H.H. Sarhan, A. ElSherief. "Novel 3D memory-centric NoC architecture for transaction-based SoC applications." In Proc. SIECPC, 2011.

[3] Dutoit, Denis, et al. "A 0.9 pJ/bit, 12.8 GByte/s WideIO memory interface in a 3D-IC NoC-based MPSoC." VLSI Technology (VLSIT), 2013 Symposium on. IEEE, 2013.

[4] Batude, P., et al. "3D sequential integration opportunities and technology optimization." Interconnect Technology Conference/Advanced Metallization Conference (IITC/AMC), 2014 IEEE International. IEEE, 2014.

[5] S. Panth, K. Samadi, Y. Du and S. Lim, "Power-Performances Study of Block-Level Monolithic 3D ICs Considering Inter-Tier Variations", Design Automation Conference (DAC), 2014 51st ACM/EDAC/IEEE, 2014.

[6] Billoint, O., et al. "A comprehensive study of monolithic 3D cell on cell design using commercial 2D tool." Proceedings of the 2015 Design, Automation & Test in Europe Conference & Exhibition. EDA Consortium, 2015.

[7] Sarhan, Hossam, et al. "3DCoB: A new design approach for Monolithic 3D Integrated circuits." ASP-DAC. 2014.

[8] H. Chaabouni, et al., "Investigation on TSV impact on 65nm CMOS devices and circuits", Electron Devices Meeting (IEDM), IEEE International, 2010.

[9] R. Taibi, et al., "Full Characterization of Cu/Cu Direct Bonding for 3D Integration", Electronic Components and Technology Conference (ECTC), 2010. IEEE, 2010.

[10] Tezzaron Semiconductors, 2011.

[11] Ghosal, P.; Chatterjee, S., "Partitioning in 3D ICs: A TSV aware strategy with area balancing," Devices, Circuits and Systems (ICDCS), 2012 International Conference on, 2012.

[12] Sawicki, S.; Wilke, G.; Johann, M.; Reis, R., "A cells and I/O pins partitioning refinement algorithm for 3D VLSI circuits," Electronics, Circuits, and Systems, 2009. ICECS 2009. 16th IEEE International Conference on , vol., no., pp.852,855, 13-16 Dec. 2009.

[13] Yu Cheng Hu; Yin Lin Chung; Mely Chen Chi, "A multilevel multilayer partitioning algorithm for three dimensional integrated circuits," Quality Electronic Design (ISQED), 2010 11th International Symposium on , vol., no., pp.483,487, 22-24 March 2010.

[14] Karypis, G., Aggarwal, R., Kumar, V., and Shekhar, S. "Multilevel hypergraph partitioning: applications in VLSI domain", Very Large Scale Integration (VLSI) Systems, IEEE Transactions on, 1999, 69-79.

[15] Thorolfsson, T., Lipa, S., & Franzon, P. D. A 10.35 mW/GFlop stacked SAR DSP unit using fine-grain partitioned 3D integration. In Custom Integrated Circuits Conference (CICC), 2012 IEEE.

[16] Tada, Jubee, et al. "A middle-grain circuit partitioning strategy for 3-D integrated floating-point multipliers." 3D Systems Integration Conference (3DIC), IEEE, 2012.

[17] Egawa, Ryusuke, et al. "Evaluation of fine grain 3-D integrated arithmetic units." 3D System Integration, 2009. 3DIC 2009. IEEE International Conference on. IEEE, 2009.

2015 IEEE Computer Society Annual Symposium on VLSI

Implementation of AES Using NVM Memories Based on Comparison Function

Jeremie CLEMENT, Bruno MUSSARD
Crocus Technology
Rousset, France
Email: {jclement, bmussard}@crocus-technology.com

David NACCACHE
Ecole Normale Supérieure
Paris, France
Email: david.naccache@ens.fr

Lionel TORRES
LIRMM, UMR CNRS
Université de Montpellier, France
Email: lionel.torres@lirmm.fr

Abstract—Nowadays, cryptography is widely used in many different devices (smartphones, tablets, computers, etc.). One of the most used cryptographic algorithms is AES (Advanced Encryption Standard). Crocus Technology is developing a technology called Match-In-Place^{TM}(MIP). This is a memory based on comparison function. With this technology, data stored in the memory can be compared to an input data. The main expectation of this technology is that, during this comparison, no information leaks from the memory, which is interesting against side channel attacks. This assumption has been verified by simulation but not in practice yet. This paper proposes a new secure AES implementation based on the MIP technology. The use of this particular technology allows to protect the key in a secure environment to prevent attackers from retrieving it. MIP also allows a very low silicon area implementation of AES.

Keywords-Advanced Encryption Standard (AES), MRAM, cryptography, security, memory

I. INTRODUCTION

In 2001, the National Institute of Standards and Technology (NIST) specified the Advanced Encryption Standard (AES). AES is a symmetric block cipher designed by Daemen and Rijmen in [1]. Nowadays, AES is widely used in a variety of applications and devices. So, many works were performed to secure AES against side-channel attacks ([2], [3], [4]).

Crocus Technology is a semiconductor company developing Magnetoresistive Random-Access Memories (MRAM). They are currently developing a technology called Match-In-Place (MIP). MIP is a memory which has an embedded comparison function. It consists in comparing an input data, given to the memory to a secret data stored in the memory. The comparison operation itself takes place in the memory. The advantage of MIP is that the sensitive data never leaves the memory, so it is impossible for an attacker to intercept this data during the comparison operation. Each MIP cell is a non-volatile memory cell combined with a virtual XOR gate. Multiple cells can be connected in series to form a NAND chain acting as a linear MIP engine. This technology can be used to secure user authentication (for example by comparing passwords in a secure way). As all operations take place in the memory, there is no data leakage, so it is impossible to retrieve sensitive data (such as a key or a password) using power analysis. Although this security

features have been proved by simulation, these assumptions still have to be verified on actual MIP chips.

In this paper, a secure AES implementation is described, using MIP to secure the key. In this implementation, MIP cells are used in different ways :

- As a Content Addressable Memory (CAM) to replace AES S-Boxes;
- As a bitwise XOR by processing bit-to-bit matching operation;

By integrating MIP into AES operations, the key and the processed intermediate data during the different operations are secured. This algorithm was implemented in Verilog and simulated using the Icarus Verilog simulator.

The rest of this paper is organized as follows. Section 2 recalls AES operations. Section 3 describes the MIP technology. Section 4 explains how MIP is connected into AES algorithm. Section 5 shows the simulation results of the proposed implementation. Section 6 concludes.

II. AES OPERATION

AES encryption and decryption are based on four different transformations which are repeated N_r times. AES has a fixed 128 bits data block and a key size of 128, 192 or 256 bits. For the sake of simplicity, only the 128-bit key variant was implemented, so the rest of this paper only refers to AES-128. For this variant the number of rounds is $N_r = 10$. Fig. 1 describes the flow of operations for AES encryption. Encryption and decryption flows are slightly different as, compared to encryption, decryption executes the reverse transformations in the reverse order.

AES considers a 128-bit block data as a 4×4 byte state. In parallel to encryption or decryption, the 128-bit key is processed by a key schedule operation to generate N_r round subkeys. Each round subkey and the cipher key are considered as a 4×4 byte state. A full round is decomposed into four steps.

A. The SubBytes step

In the SubBytes step, each byte $a_{i,j}$ in the state matrix is replaced with another byte $S(a_{i,j})$ using an 8-bit substitution box (called S-box). This operation provides the non-linearity of the cipher. Many works on improving S-Boxes have been realized. Composite field S-Boxes are widely used, as they

978-1-4799-8720-7/15 $31.00 © 2015 IEEE   356

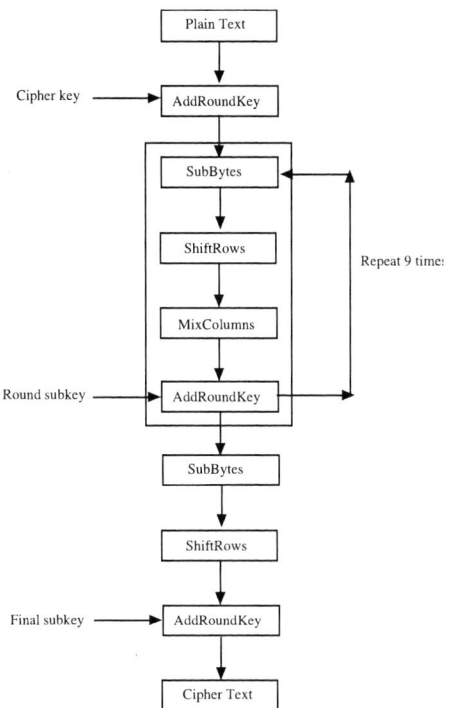

Figure 1. AES encryption flow

allow to minimize the gate count and improve performances on S-Box calculation. Many works on composite fields focus on the tower field $\mathbb{F}((2^2)^2)^2$ ([5], [6], [7]).

B. The ShiftRows step

The ShiftRows step cyclically shifts the bytes in each row of the state by a certain offset. The first row of the state is left unchanged. Then, the row n is shifted left circularly by $n - 1$ bytes. For decryption, the bytes are shifted circularly the same number of positions but to the right.

C. The MixColumns step

In this step, each column of the 4×4 byte state is treated as a polynomial over $GF(2^8)$. It is then multiplied with the polynomial $c(x) = 3x^3 + x^2 + x + 2$ modulo $x^4 + 1$. Usually, the MixColumns step is merged with the ShiftRows step, as these two steps provide diffusion of the cipher. This step is not present in the final round.

D. The AddRoundKey step

In this step, each byte of the state is simply XORed (exclusive OR operation) with the corresponding byte of the current round subkey.

E. The key schedule operation

Executed in parallel to the main AES operation, this operation derives the main key into N_r round subkeys. This operation creates a 176 byte array called the expanded key.

Several operations are performed to generate 4-byte words W_i where $i > N_k$ and N_k is the number of 4-byte words in the cipher key, $N_k = 4$ for a 128-bit key:

- $W_i = K_i$ for $0 \leq i < N_k$, where K_i is the ith 4-byte word from the cipher key,
- $W_i = W_{i-1} \oplus W_{i-N_k}$ for $i > N_k$ and i is not a multiple of N_k,
- If i is a multiple of N_k, the word W_{i-1} is first rotated circularly 8 bits to the left. Then it is transformed using the S-Box. Finally the leftmost byte of the result is XORed with a round-dependent constant $R_{con} = 2^{r-1}$ in $GF(2^8)$, where r is the current round number.

Once all the expanded key array is filled, it is divided into $N_r + 1$ 128-bit round subkeys. In the 128-bit key version of AES, 11 round subkeys are generated in this way, the first round subkey being the cipher key.

To improve the security of this algorithm, a technology called MIP was used.

III. THE MATCH-IN-PLACE TECHNOLOGY

A. Description of MIP technology

MIP [8] is a device developed by Crocus Technology, based on its Magnetic Logic Unit (MLU) [9], to authenticate users without exposing any confidential data to external attackers. The MLU technology is an evolution of the Thermally Assisted Switching MRAM described in [10]. In a TAS-MRAM, each Magnetic Tunnel Junction (MTJ) is composed of a storage layer and a fixed reference layer. In the MLU, the fixed reference layer of the MTJs is replaced with a ferromagnetic free layer, referred as the sense layer. By doing so, this self-reference MRAM has the particularity to perform logic comparisons. Considering the storage layer and the sense layer as two different inputs, it can perform a XOR operation by measuring the resistance of the memory cells. By connecting the self-referenced MRAM cells to form a NAND chain, it is possible to create a MIP engine, comparing the data stored into the memory, which is a sensitive data (passwords for example) and the data given as input. A mismatch between the two data results in a higher resistance than a match. Fig. 2 shows an example of a MIP engine composed of four MLU cells connected as a NAND chain. In this example, the stored data is "1010" and the input data is "1100". As the stored data and the input data

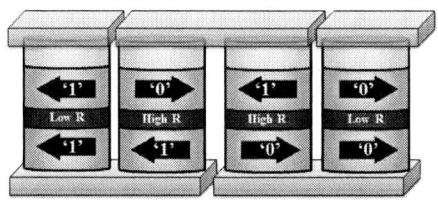

Figure 2. MIP composed of 4 chained MLU cells

978-1-4799-8720-7/15 $31.00 © 2015 IEEE

are different, the total resistance of the chain will be higher than the minimum resistance. If the input data match the stored data, all the cells would have a low resistance, then the total resistance of the chain would be minimal.

For AES implementation, instead of matching, two modified applications of the MIP were used. First it was used to perform a XOR operation between two data words. MIP XOR blocks contain 32 MTJs. Each MTJ performs a matching operation on one bit, and is not connected to the other MTJs of the block. In the end of the matching operation, 32 resistance values are obtained. Each value is then interpreted as bits, '0' for a low resistance (if the stored data and the input data are identical), '1' for a high resistance (if the stored data and the input data are different). So, a 32-bit value is obtained, result of the XOR operation between the stored data word and the input data word. Four of these blocks are used for the AddRoundKey operation (which performs a XOR operation between a 128-bit key and a 128-bit data). Four other blocks are used for the MixColumns operation, and another one for the key schedule operation.

The MIP can also be used as a fast associative memory like a Content Addressable Memory (CAM) used in routers and search engines for example. By placing many MIP engines in parallel, it makes the search faster. The input data is compared to all the entries of the CAM, handled by MIP. All the comparisons are made in parallel. When this is done, only one engine will return a perfect match notified by the chain which has the minimum resistance. With this the incoming packet can be directly routed to the recipient address. Fig.3 describes an example of CAM using MIP technology. In this example, there are 4 MIP engines. The input pattern "1010" is then compared to the memory. For AES implementation, two CAM blocks are used :

- The SBox block : It contains 2048 MTJs divided into 256 engines, each one composed of 8 MTJs. This block is used for the SubBytes step and the key schedule operation. The input of this block is the current byte processed from the state.
- The R_{con} block : It is used to determine the round constant used in the key schedule. It contains 10 MIP engines. Each engine is a NAND chain of 8 MTJs. In each engine is stored a possible round number (from 1 to 10). The input of each of these MIP engines is the current round number.

This technology has two advantages. First, the comparison operation is performed into the memory. The secret data stored into the memory never leaves it during this operation. The second advantage is that it provides a protection against side-channel attacks.

B. Side channel attack on MIP technology

In order to demonstrate the security features provided by MIP, a Differential Power Analysis (DPA) have been simulated based on the MIP netlists. This attack focused

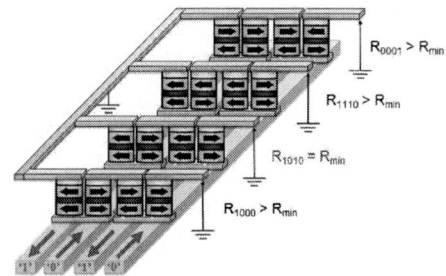

Figure 3.   Example of a CAM using MIP

on the matching of two 32-bit words using MIP. The MIP matching operation was simulated using Spectre in order to obtain simulated power traces. Using these traces, a DPA attack was conducted in order to try to retrieve the stored key. To perform this attack, a random 8-bit key has been stored in MIP memory. This key is unknown to the attacker. The size of the key (8-bits) has been chosen because it fits the size of the matching used for the S-Boxes. Then all the possibilities for 8-bits were proposed as input of the MIP. Only one of these inputs results as a match. The attack has then been realized using these 256 power traces. But, whatever the input of the MIP operation was, the power traces were all similar, which led the attack to fail. Thus, if a DPA attack fails on a 32-bit matching operation performed by MIP, the proposed AES implementation can be also considered secure against these attacks as it uses the same principles. The MIP technology is still in development. Differential Fault Analysis (DFA) attacks have not been tested on MIP as no functional test chips are available yet.

IV. INTEGRATING MIP IN AES IMPLEMENTATION

In order to make use of the protection offered by MIP against side channel attacks, AES algorithm was implemented using this technology. The proposed implementation was made using the Verilog Hardware Description Language. At every step of the algorithm, the integration of MIP provides additional security features, preventing an attacker from reading the cipher key or the internal state. Another advantage is that these operations can be executed directly into the MIP memory. In other words, the cipher key and data (plaintext or ciphertext) are stored into the memory, which triggers AES operation. Once it is finished, the memory outputs the result of the operation.

A. The AddRoundKey step : A secure Xor operation

The AddRoundKey step is just a Xor between two 128 bits values (the key and the data). To do this, 4 blocks of MIP XOR are used. Each block is composed of 32 bit cells. Each cell is not connected to the others and acts as a XOR gate. The round subkey is stored into the MIP memory as soon as it is generated, so it is kept safe during the entire operation. Then, the data is given to the blocks. For each block, the 32

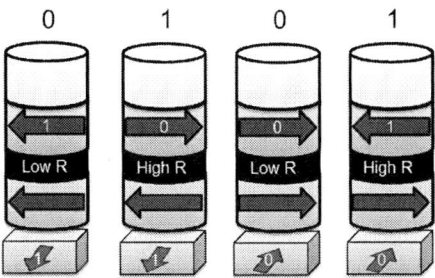

Figure 4. Example of 4-bit XOR operation managed by MIP

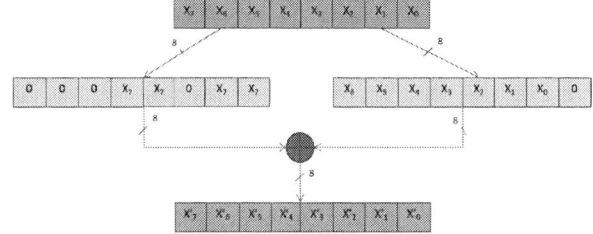

Figure 5. Multiplication by two using only XOR operations.

bits will check separately if the given data bit match with the corresponding key bit. These 32 operations are made in parallel. At the end of this operation, 32 resistance values were obtained, each value is then interpreted as bits, '0' for a low resistance (if the key and the data are identical), '1' for a high resistance (if the key and the data are different). So a 32-bit match value per block is obtained, each bit of this value corresponding to each 1-bit MIP operation. This match value gives the result of a Xor operation between the key and the data. Fig. 4 shows the XOR operation between two 4-bit data. The stored data ('1001') is Xored with an input data ('1100'). All the 1-bit matching operation are processed in parallel and in the end, the result of the XOR operation is obtained ('0101').

B. The SubBytes step : The S-Box

In classical AES implementations, there are two ways to calculate the output of the S-Box based on the inputs :

- Computation on-the-fly of the output value
- Use of a look-up table

For the proposed implementation, the look-up table approach is chosen. Here the MIP SBox block is used. At the beginning of AES operation, all the MIP NAND chains are programmed with a single value, corresponding to the different possible input values for the S-Box (from $0x01$ to $0xFF$). Then when the SubBytes operation starts, all the NAND chains receive the same input byte which is the input of the S-Box. Thus only one MIP engine will result as a match between the the input of the S-Box and their corresponding unique data. According to the number of the matching instance and if it is an encryption of a decryption process (due to the differences between the encryption and decryption look-up tables), the output of the S-Box will be different. In the proposed implementation the ShiftRows step is merged with the SubBytes step. Indeed all the output values of the S-Boxes, are written at their new location in the state, according to the ShiftRows step.

C. The MixColumns step : Make Multiplications with Xors

This step consists in a polynomial multiplication of each column of the internal state. But it can also be expressed

as the product between a fixed matrix and a column of the state. The positions of the bytes into the state are shifted according to the ShiftRows step described earlier. Hence, the new positions of the bytes in the state are taken into account to determine the inputs of the MixColumns step. Here is described the matrix multiplication used for the MixColumns step for the encryption process. A,B,C,D are the input bytes of the current column of the state and Y_1, Y_2, Y_3, Y_4 are the output bytes, constituting the new state column.

$$\begin{bmatrix} 2 & 3 & 1 & 1 \\ 1 & 2 & 3 & 1 \\ 1 & 1 & 2 & 3 \\ 3 & 1 & 1 & 2 \end{bmatrix} \begin{bmatrix} A \\ B \\ C \\ D \end{bmatrix} = \begin{bmatrix} Y_1 = 2A \oplus 3B \oplus C \oplus D \\ Y_2 = A \oplus 2B \oplus 3C \oplus D \\ Y_3 = A \oplus B \oplus 2C \oplus 3D \\ Y_4 = 3A \oplus B \oplus C \oplus 2D \end{bmatrix}$$

Here, the XOR operations are made with the MIP by using the same method than in the AddRoundKey step, in order to preserve the inputs corresponding to the state from attackers. To do the multiplication by 2, a technique that only involves XORs is used as described in Fig. 5. In this figure the input (X) and the output (X') are bytes with $X = X_7X_6X_5X_4X_3X_2X_1X_0$ and $X' = 2*X$. As all the operations are made in $GF(2^8)$, the multiplication by 3 is just a multiplication by two combined with a XOR. For the proposed implementation, four 32-bit XOR blocks of MIP are used. Each of these blocks performs the Mix-Columns operation on a different column of the state and are executed in parallel.

There are two interests of doing that. The first one is the security of the state, which will be protected against side-channel attacks during this step. The second interest is that this entire step can be implemented into a memory, which means that one just need to give the state data to the memory and it outputs the new state.

D. The key schedule operation: Different uses of MIP

For the key schedule operation, many different MIP blocks are used:

- The main S-Box is reused as for the SubBytes step described earlier.
- Another S-Box on the same principle than the SubBytes S-Box is used. This one is called the R_{con} S-Box and is used to determine the round constants. This S-Box takes

as input the number of the current round, and outputs the current round constant. There are 10 entries for this S-Box, so 10 blocks of 8 chained-bits of MIP are used to handle the inputs.

- A MIP block is used for the bitwise XOR operations by having 32 bit cells, each one unconnected with the others, as described earlier.

In this implementation, the key schedule operation executes in parallel of the main AES operation. At each round of AES operation, the next round subkey is generated. Integrating MIP into this operation provides an additional security feature, by preventing an attacker from extracting a key during this operation. Another advantage of using MIP for logical operations is that the entire key schedule operation can be made by the MIP memory. To do this one just need to store into the memory the current round subkey and the round number. Once the operation is finished, it outputs the new round subkey at a predefined address into the memory. This address can be reused in the AddRoundKey step, the round subkey being stored into a MIP memory, allowing to perform the XOR with the data without the subkey being read by the processor and extracted from the memory.

V. Results and Future Works

A. Simulation Results

This architecture has been implemented in Verilog. A Known Answer Test (KAT) vector has been used in order to verify that the proposed implementation gives the same result than a classical AES implementation.

Compared to classical Verilog AES implementations, the described method protects the cipher key and the S-Box from being collected by an attacker, due to the protection provided by MIP. Besides, the main operations of this implementation only uses the characteristics of the Magnetic Logic Unit technology used by the MIP and can be entirely executed into the memory, the processor is just interfacing with the MIP memory without executing these operations. In return, there is a lack of performances compared to other AES implementations.

Table I describes the area and the delay taken by the different steps of the proposed implementation. The delay is the execution time for one round of the algorithm. Due to the fact that all these operations are executed directly into the MIP memory, the area expressed in this table is an estimation based on the number of MTJs used for each step.

The proposed implementation only takes into account the AES-128 variant. In the cases of AES-192 and AES-256 variants, the needed area would be increased due to the increase on the key size, although performances decrease due to the fact that the key schedule has more steps than in the AES-128. However, the other steps of the algorithm are unchanged, the only difference would be in the key schedule part.

Table I
ESTIMATIONS ON AREA AND DELAY FOR EACH AES STEP

Step	Number of MTJ	Area (μm^2)	Delay per round (ns)
AddRoundKey	128	3.7	100
SubBytes	2048	78.7	2410
MixColumns	512	14.74	780
KeySchedule	112	3.9	1200
Total	**2416**	**101**	**4490**

Table II
COMPARISON OF AES IMPLEMENTATIONS (*: ESTIMATIONS)

Implementation	Throughput (Mbps)	Area (mm^2)	Energy ($\mu W/MHz$)
This paper	3.2	0.01*	261*
Verbauwhede et al. [11]	1600	3.96	432
Hwang et al. [12]	3840	2.45	4000
Hodjat et al. [13]	3840	0.79	1080
Sumanth et al. [14]	32820	0.74	534.30

Table II compares the proposed implementation with other AES implementations on throughput, area and energy. The values for the area and energy consumption for the implementation proposed in this paper are estimated values. The area presented in this table is the total area estimation of the proposed implementation. As shown in Table II, the proposed implementation suffers from a lack of performances compared to other implementations, but compensates it by having a very low area. The energy consumption estimation is realized using the technique described in [15]. This technique calculates the energy spent by a system during the execution of an algorithm using the following formula : $E = I * V_{DD} * N * \tau$, where E is the energy consumed by the implementation in Joules, I is the average current consumption per clock cycle, V_{DD} is the operating voltage of the MIP, N is the number of clock cycles taken for the execution of this algorithm and τ is the clock period. According to this method, the proposed implementation spends 1.440 μJ with a clock frequency at 40 MHz. When E is obtained, it is divided by the execution time in order to get the power consumption in W. In order to have a more accurate comparison with the other implementations, the power consumption in Table II is expressed in $\mu W/MHz$.

B. Future works

One of the possible future works will be making a compromise between the loss of performances and the gain of security. To do this, the use of MIP will be restrained to only sensitive operations, the AddRoundKey and the S-Box. The other AES operations will be executed by the processor. The last step will be to test this implementation on actual MIP chips.

978-1-4799-8720-7/15 $31.00 © 2015 IEEE

VI. Conclusion

In this paper a new technology called Match-In-Place (MIP), which will be used in secure functions, is introduced. This technology, which is an evolution of MRAM memories, allows secure matching operations such as password comparison, to be directly executed into a secure memory. This way, sensitive data are never exposed to an external attacker. A simulated DPA attack has been performed on MIP and failed, proving that MIP is secure. But this assumption was only observed on simulated power traces and is still to be proven on actual MIP chips, which are not available yet. MIP can also be used to secure look-up tables searching operations and to execute a secure XOR operation. Here, this technology has been integrated into a well-known and very used cryptographic algorithm, AES. This algorithm was implemented in Verilog and compared to a classical implementation. The described method has the advantage to perform the operations directly into the memory. Also once the secret key is stored into the memory, it is locked up and never leaves it, preserving it from eavesdropping. This method is also low power and low area, based on estimations. But the main drawback of this method is that it suffers from a lack of performances, compared to classical Verilog implementations of AES.

This work is not finished yet, and many improvements can be done on this implementation. Then it is possible to reduce the use of MIP to critical functions in order to make a compromise between the performance loss and the security. Critical functions would be the AddRoundKey step, because the key is directly used and combined to the message, and the main S-Box, which introduces the confusion in the algorithm. Finally, when actual MIP chips are functional, tests must be made in order to validate the results of the simulated DPA attack.

References

[1] J. Daemen and V. Rijmen, "Aes proposal: Rijndael," AES submission, 1999.

[2] M.-L. Akkar and C. Giraud, "An implementation of DES and AES, secure against some attacks," in *Cryptographic Hardware and Embedded Systems CHES 2001*, ser. Lecture Notes in Computer Science, vol. 2162. Springer Berlin / Heidelberg, 2001, pp. 309–318.

[3] R. K., K. W., P. M., and Y. K., "Fault-based side-channel cryptanalysis tolerant rijndael symmetric block cipher architecture," in *16th IEEE International Symposium on Defect and Fault-Tolerance in VLSI Systems (DFT 2001)*, 2001, pp. 427–435.

[4] G. Bertoni, L. Breveglieri, I. Koren, P. Maistri, and V. Piuri, "Error analysis and detection procedures for a hardware implementation of the advanced encryption standard," *IEEE Transactions on Computers*, vol. 52, no. 4, pp. 492–505, April 2003.

[5] D. Canright, "A very compact s-box for aes," in *Cryptographic Hardware and Embedded Systems CHES 2005*, ser. Lecture Notes in Computer Science. Springer Berlin Heidelberg, 2005, vol. 3659, pp. 441–455.

[6] X. Zhang and K. Parhi, "On the optimum constructions of composite field for the aes algorithm," *Circuits and Systems II: Express Briefs, IEEE Transactions on*, vol. 53, no. 10, pp. 1153–1157, Oct 2006.

[7] Y. Nogami, K. Nekado, T. Toyota, N. Hongo, and Y. Morikawa, "Mixed bases for efficient inversion in $f((2^2)^2)^2$ and conversion matrices of subbytes of aes," *IEICE Transactions*, vol. 94-A, no. 6, pp. 1318–1327, 2011.

[8] B. Cambou, "Match-in-place^{TM}, a novel way to perform secure and fast user authentication," White Paper, 2012.

[9] Q. Stainer, L. Lombard, K. Mackay, R. C. Sous, I. L. Prejbeanu, and B. Dieny, "Mram with soft reference layer: In-stack combination of memory and logic functions," IEEE International Memory Workshop (IMW), pp. 84–87, 2013.

[10] I. L. Prejbeanu, M. Kerekes, R. C. Sousa, H. Sibuet, O. Redon, B. Dieny, and J. P. Nozires, "Thermally assisted mram," *Journal of Physics: Condensed Matter*, vol. 19, no. 16, p. 165218, 2007.

[11] I. Verbauwhede, P. Schaumont, and H. Kuo, "Design and performance testing of a 2.29 gb/s rijndael processor," *IEEE Journal of Solid-State Circuits*, vol. 38, no. 3, pp. 569–572, 2003.

[12] D. Hwang, K. Tiri, A. Hodjat, B.-C. Lai, S. Yang, P. Schaumont, and I. Verbauwhede, "AES-based security coprocessor IC in 0.18-μm CMOS with resistance to differential power analysis side-channel attacks," *IEEE Journal of Solid-State Circuits*, vol. 41, no. 4, pp. 781–792, Apr 2006.

[13] A. Hodjat, D. Hwang, B. Lai, K. Tiri, and I. Verbauwhede, "A 3.84 gbits/s aes crypto coprocessor with modes of operation in a 0.18um cmos technology," in *Proceedings of the 15th ACM Great Lakes symposium on VLSI (GLSVLSI 2005)*, 2005, pp. 60–63.

[14] S. Sumanth Kumar Redy, R. Sakthivel, and P. Praneeth, "Vlsi implementation of aes crypto processor for high throughput," *International Journal of Advanced Engineering Sciences And Technologies (IJAEST)*, vol. 6, no. 1, pp. 22–26, 2011.

[15] K. Naik and D. S. L. Wei, "Software implementation strategies for power-conscious systems," *Mob. Netw. Appl.*, vol. 6, no. 3, pp. 291–305, Jun. 2001.

2015 IEEE Computer Society Annual Symposium on VLSI

Figure of merits of 28nm Si technologies for implementing laser attack resistant security dedicated circuits

S.de Castro, G.Di Natale, M.-L. Flottes, B.Rouzeyre
MIC
LIRMM CNRS
Montpellier, France

J.-M.Dutertre
LSAS
ENSMSE
Gardanne, France

Abstract—Among all means to attack a security dedicated circuit, fault injection by means of laser illumination is a very efficient one. The laser beam creates electrons/holes pairs along its way through silicon. Collection of these charges creates a transient current and thus may induce a fault in the circuit. Nevertheless the collection efficiency depends on various parameters including the technology used to implement the circuit. Here, Bulk and Fully Depleted Silicon on Insulator (FD-SOI) 28nm technologies are compared in terms of sensitivity against laser injection. Laser induced current measurements are performed on a ST 28nm technology bulk and FDSOI NMOS transistors. It comes out that FD-SOI structures show less sensitivity to laser injection and thus should be further explored for security dedicated circuits implementations.

Index Terms—hardware security; laser; fault injection; CMOS technology;

I. INTRODUCTION

The development of security-dedicated circuits, e.g. smart-cards or crypto-processors, goes together with hardware attacks development intending to retrieve secret information. Among all means to perform such attacks laser illumination is one of them, in particular the so-called fault attacks that rely on disrupting the target normal functional operation (e.g. [1]). Indeed, laser is particularly adequate since it offers a good accuracy in time and space to perform a precise disruption of the circuit [2].

If laser photon energy is higher than silicons bandgap, electron/hole pairs are generated as silicon is gone through by a laser beam. These charges are then put in movement and collected by the transistor.

The charge generation in silicon along the laser beam is illustrated on a PN junction in Fig. 1. Two different induced current type can be distinguished, the drift current and the diffusion current. In order to maintain the same charge carriers concentration over the substrate, charges newly released due to laser illumination diffuse then inducing diffusion current. PN junctions biasing creates an electromagnetic field. This field put charges in movement inducing a drift current. Diffusion current last longer than drift current but has a lower amplitude. The PN junction charge space zone amplifies this high current of short duration.

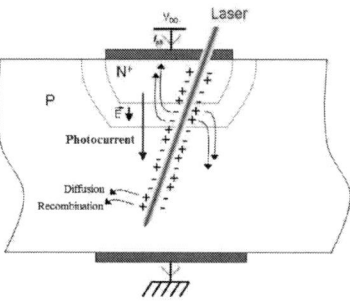

Fig. 1: Charge generation and induced current due to laser injection on a PN junction [3]

Assuming this PN junction corresponds to an OFF-state NMOS transistors drain in an inverter, the induced transient current may discharge the gates output capacitance and thus creates a voltage transient on the gate output, which temporarily switches from 1 to 0. If the transient propagates to memory element(s), the transient fault turns into a single or multiple bit errors from which the attacker can retrieve a secret information, e.g. key bits in a crypto-coprocessor [1][4].

As further discussed in section III, sensitivity to laser injection depends on the underlying CMOS technology. 90 nm bulk technology and 28nm FDSOI have been compared in [5], from which it comes out 90 nm technology is far more sensitive to laser attacks than 28nm FDSO technology.In this paper, we focus our study on the laser injection sensitivity on two up to date technologies, namely 28nm FD-SOI and 28nm bulk from STMicroelectronics.

FDSOI process differs from bulk by the insertion of insulated areas around the transistor channel. This allows reducing leakage and thus making these transistors more efficient. Moreover, the channel is thinner and made of intrinsic silicon. Thus the current flowing through logic gate is smaller than in bulk technology. As a side effect, the switching delay is faster in FDSOI. The FDSOI process and more details are given in [6].

978-1-4799-8720-7/15 $31.00 © 2015 IEEE

For our experiments, two transistors of the same size are illuminated, one for each technology, while keeping laser parameters identical. Transistors illumination is performed from the backside of the chip, in order to avoid shadowing effects of upper metal lines. Backside injection means that silicon is gone through by a laser beam from the substrate to the metal lines above. An infrared source (1064nm) has been used in order to loose as few energy as possible in the substrate. While this technique requires mechanical thinning of the substrate, it is generally preferred to "frontside injection" because metal lines on front side act like mirrors, reflecting the laser beam and preventing it to reach the silicon [7].

For the sake of conciseness, we report figure of merits of both 28nm technologies thanks to experiments on PMOS transistors only. Similar experiments on NMOS transistors have been performed and lead to the same conclusion.

This paper is organized as follows. FD-SOI and bulk PMOS transistors structures on which experiments are performed are given in section II. Section III deals with the laser induced current model. Section IV describes the experimental setup. Comparative results are given in section V. Section VI concludes the paper.

II. FD-SOI VS BULK STRUCTURE OF A 28NM ST PMOS TRANSISTOR

A. 28nm PMOS FD-SOI and bulk structure

PMOS transistors structures used here are developed and provided by STMicroelectronics.

A CMOS cross-sectional view of the two structures is given in Fig. 2.

In Fig. 2, red areas represent N doped region (Nwell), blue areas represent P doped region (drain, source and substrate). The STI and box used in FD-SOI structures are made of an insulator. FD-SOI structures channel is completely insulated from the substrate conversely to the bulk structure. Moreover, for the FD-SOI structure, the channel is made of intrinsic silicon.

B. Expected laser injection effects

In this section, the different charges contributions are discussed from a theoretical point of view.

In Fig. 3, the induced currents that result from laser illumination are highlighted (green arrows).

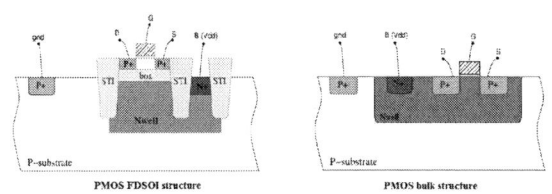

Fig. 2: Cross-sectional views of FD-SOI and bulk 28nm PMOS transistors

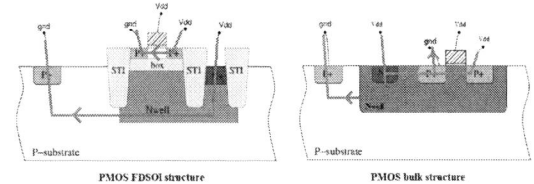

Fig. 3: Induced currents due to laser illumination for both structures

In a FDSOI transistor, only two currents can be induced by the laser, as illustrated on left side of Fig. 3. Those two currents are independent. The first one, which goes from source to drain, impacts the data path. The second current, which goes from Nwell to substrate, changes the electric potential of the Nwell. This change can alter the transistor function (e.g. threshold voltage). The threshold voltages modification modifies the transistors switching delay. This phenomenon may cause faults due to timing constraint violation in the circuits logical data path.

Conversely, in a bulk structure a laser generates many currents, which interact with each other. Thus there is a competition between all these currents. The result of this competition depends on experimental parameters such as the spot size and the lasers distance from the transistor.

These parameters and the model used to describe the laser-induced currents are detailed in the next section.

III. LASER INDUCED CURRENT MODELING

A laser-induced current model is defined in order to predict the current that flows through each PN junctions constitutive of the transistor. Such a model can be used in a simulator to predict laser injections effect on a circuit [7]. The laser-induced current can be modeled by a current source. One current source is used for each charge collection points (drain, source, Nwell, substrate). The equation of the current source Iph is given in (1) [8].

$$I_{ph}(t) = [a(P).V_r + b(P)].A.\alpha_{top}.\omega_{thick}.\Omega_{shape}(t) \quad (1)$$

All functions except $\Omega_{shape}(t)$, represent the impact of a parameter on the amplitude of the induced current. Functions a and b represent the laser beam power dependence, V_r the reverse bias voltage of the PN junction, A the surface of the PN junction, α_{top} the distance of the laser beam to the center of the PN junction, and ω_{thick} the wafer thickness. All the constant coefficients of these functions depend actually on the technology used to implement the transistor. For instance, those coefficients have been set for 90nm bulk technology in [8]. Finally the function $\Omega_{shape}(t)$ represents the shape of the induced current over time. This shape does not depend on the technology used (FDSOI/bulk) nor on the technology node. It is intrinsic to the PN junction physical behavior under illumination. More details of this behavior corresponding to a response of a photodiode (PN junction) to an optical pulse

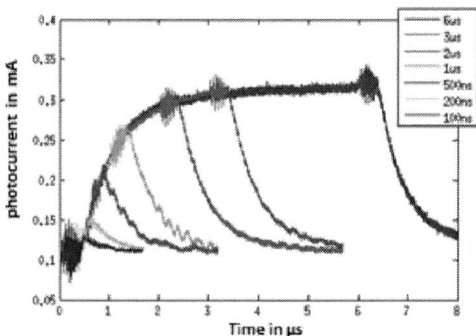

Fig. 4: Laser induced current over time for different illumination durations

input can be found in [9]. Fig. 4 shows the induced current measure over time for different illumination duration (from another transistor than the one studied here).

IV. MEASUREMENT EQUIPMENT

In this section, measurement methods and parameters are detailed. First, laser injections parameters and information about transistors are given. Section IV-B deals with measurement methods used to evaluate the induced current generated flowing through the transistor.

A. Laser and transistors description

The FD-SOI and bulk transistors used for experiments are built using a 28nm CMOS technology developed by STMicroelectronics for both technologies. Experiments results are reported for two PMOS transistors (FD-SOI and bulk). Both transistors have a channel width of $3\mu m$ and a channel length of $1\mu m$.

Our laser bench allows performing backside injection with a infrared laser source ($\lambda = 1064nm$). Laser beam used has a square spot size of $1\mu m * 1\mu m$. For the transistors used here, the induced current typically reaches its maximum for laser illumination longer than 500ns (see Fig. 4). In order to be sure to actually measure the maximum current in spite of the jitter between the laser and the oscilloscope and avoid measurements noise, transistors are illuminated during $50\mu s$. This long illuminations duration allows having a current amplitude that is large enough to distinguish signal from measurement noise.

B. Measurement circuit

The goal of the experiments was to measure the induced currents amplitude flowing through the transistor. The transistor is set on off mode (the gate is connected to Vdd) and a resistor ($1k\Omega$) is connected to the transistors drain. Other experiments have been performed with on mode (gate connected to gnd) but are not presented in this paper. The results of those experiments lead to the same observations as the one shown here. Fig. 5 depicts the measurement circuit used.

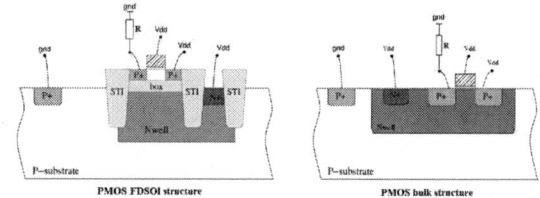

Fig. 5: Current measurement circuit

During transistors illumination, the voltage over the resistance R is measured. The transistors induced current flowing is deduced from the Ohm law.

The resistor used for experiments has an accuracy of $\pm1\%$. Moreover, the voltage accuracy of the oscilloscope is $\pm1\%$ of full scale. This measurement variation is smaller than the induced current measured.

Among all parameters that impact the induced current amplitude (laser energy, duration, supply voltage, etc.), here the distance between the laser beam and the transistor effect is focused (α_{top} in (1)). All the other parameters were set constant during the experiments, even the circuits temperature in order to maintain the same transistors performance during the whole experiments.

Fig. 6 depicts the performed experiments. The laser scans the transistor by $1\mu m$ step in x and y directions. All other laser parameters stay identical. Only the horizontal distances effect on the induced current amplitude is thus measured.

V. TECHNOLOGY SENSITIVITY TO LASER INDUCED CURRENT

The induced currents maximum amplitude as a function of the distance between laser beam and transistors center is compared for both transistor structures. For confidentiality reasons, all the following measures are scaled using an arbitrary unit noted a.u.

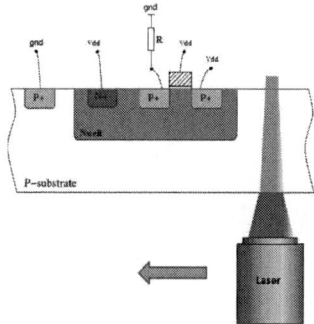

Fig. 6: Experiment to determine the effect of the horizontal distance on induced current (not scaled)

978-1-4799-8720-7/15 $31.00 © 2015 IEEE

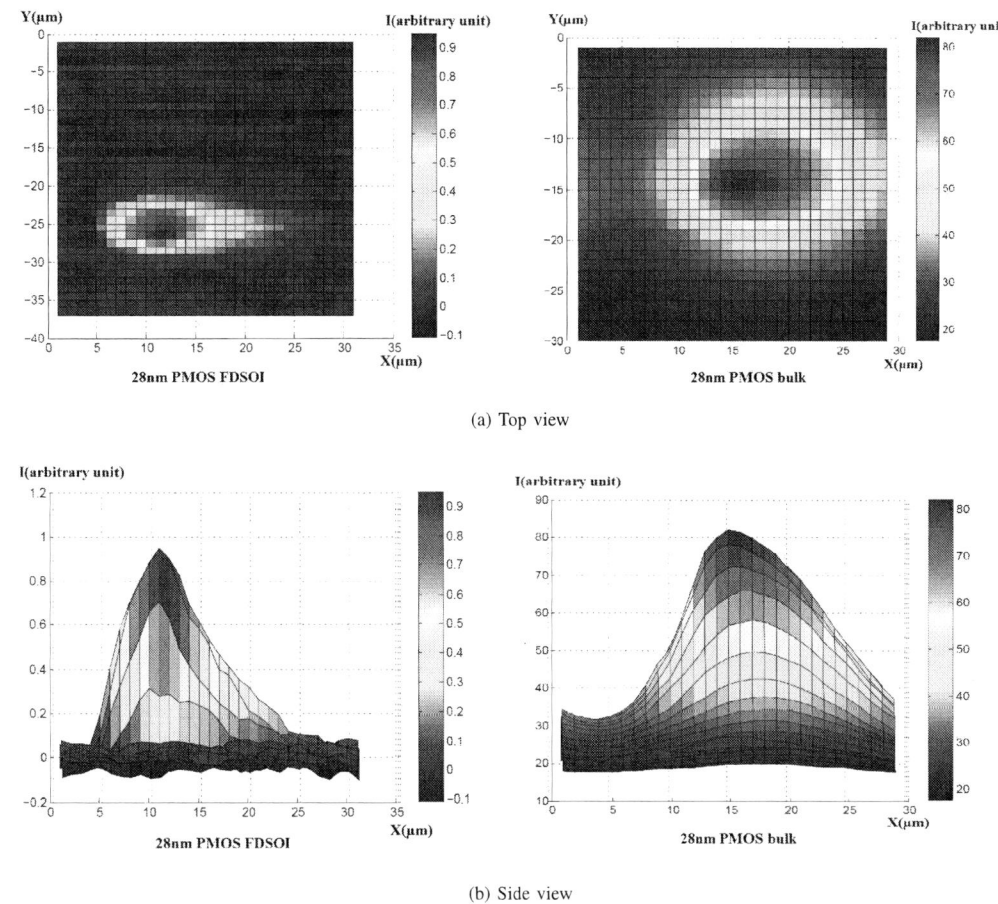

(a) Top view

(b) Side view

Fig. 7: Induced current amplitude for bulk and FD-SOI 28nm PMOS vs the distance

A. 28nm PMOS FDSOI vs 28nm PMOS bulk

Fig. 7a gives the maximum amplitude of the drain-collected induced current for each laser injection position for both PMOS structures.

The experiment performed here measures the induced current flowing through FD-SOI and bulk transistors drain. Red areas correspond to the more sensitive zones.

As it can be seen in Fig. 7a, that the red and yellow area is wider for the bulk structure than for the FDSOI structure ($10\mu m * 13\mu m$ for bulk against $5\mu m * 7\mu m$ for FDSOI).

These results have to be compared to the PMOS size. Its size is about $3\mu m * 1\mu m$. In FDSOI transistor, as the laser spot is not anymore above the transistor, induced currents maximum amplitude collapses.

The bulk transistors wide red area is due to the connection between the Nwell and the drain. Indeed the Nwell is wider than the transistor, thus it extends the area of laser injection effect. This shows that the distance sensitivity is less important for FDSOI transistors than for bulk ones.

Fig. 7b gives a side view of the previous experimentation. This allows comparing the maximum induced current for the same PMOS transistor in FDSOI and bulk. It comes out that the bulk transistors induced current is much higher (81a.u) than for the FDSOI transistor (0.9a.u).

The explanation of this difference is that in FDSOIs transistor, the charges that are collected at the drain only come from the charges generated in the channel as depict in Fig. 3. For the bulk, the current collected comes from the source and the Nwell of the transistor, which represents a wider volume of charges.

Nevertheless, in terms of fault effects, these induced currents have to be compared with the current needed to charge (or discharge) the output capacitance (i.e input capacitance of the downstream logic gate) without laser illumination.

Two inverters are connected together, with the PMOS transistor sized as the one used for experimental measurements. The required current necessary to change the logic output of the second inverter is computed thanks to electrical simulation, the result of this simulation is given in Table I. In FDSOI structure, the necessary current is about 0.2 a.u. For the Bulk structure, this current is about 9 a.u.

It comes out that, even if the FDSOI is less sensitive

978-1-4799-8720-7/15 $31.00 © 2015 IEEE

TABLE I: Necessary current and induced current comparison

	FDSOI	Bulk
Maximal laser induced current	0.9 a.u	81 a.u
Necessary current to charge the output capacitance	0.2 a.u	9 a.u

than the bulk, fault injection can be performed in a secure dedicated circuit using 28nm FDSOI technology. Nevertheless, the parameters used in the reported experiments do not reflect parameters used to perform a fault injection used for an attack, in particular the duration of the laser illumination.

Indeed, in the previous experiments, the illumination time ($50\mu s$) is long enough to obtain the highest induced current amplitude but certainly would overlap several circuits clock periods, leading to unusable errors (for an attacker perspective). To perform fault injection, the illumination time has to be of the clock period range (range of ns). If the illumination last shorter, the induced currents amplitude decreases too.

As the illumination time becomes shorter, the transistor becomes less sensitive to laser injection.

So, the amplitude of the induced current decreases by 3 when illumination time goes from 500ns to 250ns for the 28nm PMOS bulk transistors as depicted in Fig. 8. This result is the same for bulk and FDSOI transistors.

B. 28nm NMOS FDSOI vs 28nm PMOS FDSOI

In this subsection, a comparison is made between a NMOS and PMOS in 28nm technology. The size and the experimental parameters are set as the previous experiments.

Fig. 9 presents the induced currents maximum amplitude depending on the distance of the laser spot.

This experiment confirms that the 28nm FDSOI NMOS has the same sensitivity features as the 28nm FDSOI PMOS.

The difference of the induced currents maximum amplitude value between those transistors can be explained by the fact that they have the same size ($3\mu m * 1\mu m$). However, in order

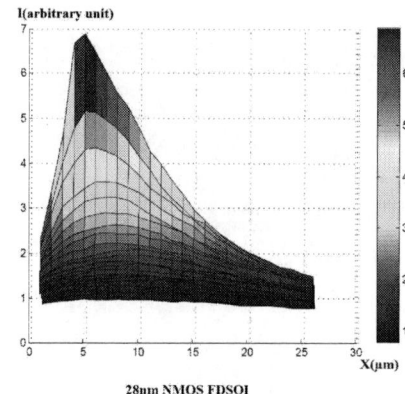

Fig. 9: 28nm FDSOI NMOS induced current amplitude

to drive the same current, the PMOS has to be around 3 times wider than the NMOS.

In spite of this difference, one can see that these sensitivity features comes from the FDSOI technology and are not bound to the 28nm FDSOI PMOS.

VI. CONCLUSION

In this paper, results about the laser injection sensitivity for FD-SOI and bulk 28nm transistors are given. These results tend to prove that the 28nm ST FD-SOI technology is less sensitive than 28nm bulk technology to laser injection.

This result in favor of FD-SOI technology is due on one hand to the presence of the insulator surrounding the channel. One of the insulators effects is to limit the volume of charges, thus reducing the induced current that flows through the drain. The other one is that when the laser spot is not above the FDSOI transistor the transistor is no more disturbed.

All these results tend to confirm that FD-SOI technology is a better option to implement security dedicated circuits than bulk technology.

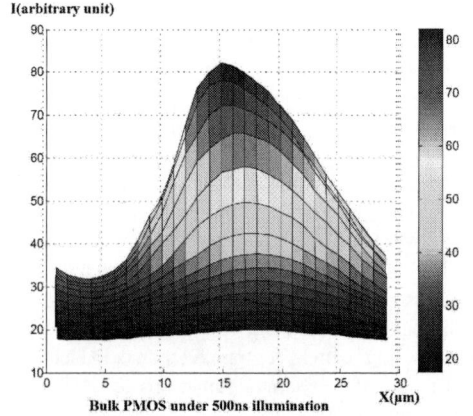 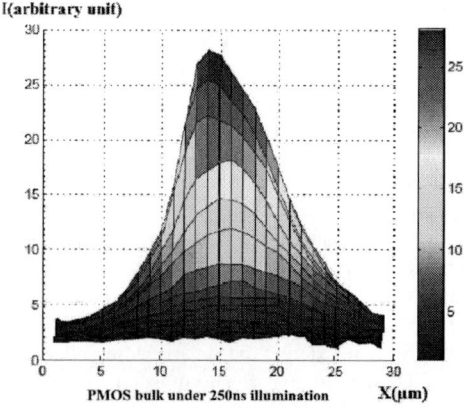

Fig. 8: Induced current of 28nm bulk PMOS under illumination of 500ns and 250ns

ACKNOWLEDGMENT

This work has been supported by the French Agence Nationale de la Recherche under contract ANR LIESSE (ANR-12-INSE-0008-01)

REFERENCES

[1] S. Skorobogatov and R. J. Anderson, "Optical fault induction attacks.in cryptographic hardware and embedded systems," in *CHES'02*, 2002.

[2] C. Roscian, J.-M. Dutertre, and A. Tria, "Frontside laser fault injection on cryptosystems application to the aes' last round," in *IEEE International Symposium, Hardware-Oriented Security and Trust (HOST))*, United States, Austin, Jun. 2013.

[3] A. Douin, "Contribution la modlisation et au dveloppement de techniques de test et d'analyse dynamiques de circuits intgrs par faisceau laser puls," Ph.D. dissertation, lectronique Bordeaux 1 2008, 2008. [Online]. Available: http://www.theses.fr/2008BOR13564

[4] C. Giraud, "Dfa on aes," in *Advanced Encryption Standard AES*, ser. Lecture Notes in Computer Science, H. Dobbertin, V. Rijmen, and A. Sowa, Eds. Springer Berlin Heidelberg, 2005, vol. 3373, pp. 27–41. [Online]. Available: http://dx.doi.org/10.1007/11506447_4

[5] J.-M. DUTERTRE, S. De Castro, A. Sarafianos, N. Boher, B. Rouzeyre, M. Lisart, J. Damiens, P. Candelier, M.-L. Flottes, and G. Di Natale, "Laser attacks on integrated circuits: from cmos to fd-soi," in *Design and Technology of Integrated Systems in Nanoscale Era (DTIS)*, Santorin, Greece, May 2014. [Online]. Available: http://hal-emse.ccsd.cnrs.fr/emse-01099042

[6] C. Fenouillet-Beranger, S. Denorme, P. Perreau, C. Buj, O. Faynot, F. Andrieu, L. Tosti, S. Barnola, T. Salvetat, X. Garros, M. Cass, F. Allain, N. Loubet. L. Pham-Nguyen, E. Deloffre, M. Gros-Jean, R. Beneyton, C. Laviron, M. Marin, C. Leyris, S. Haendler, F. Leverd, P. Gouraud, P. Scheiblin, L. Clement, R. Pantel, S. Deleonibus, and T. Skotnicki, "Fdsoi devices with thin box and ground plane integration for 32nm node and below," *Solid-State Electronics*, vol. 53, no. 7, pp. 730 – 734, 2009, papers Selected from the 38th European Solid-State Device Research Conference ESSDERC08. [Online]. Available: http://www.sciencedirect.com/science/article/pii/S003811010900077X

[7] F. Lu, G. Di Natale, M.-L. Flottes, B. Rouzeyre, and G. Hubert, "Layout-aware laser fault injection simulation and modeling: From physical level to gate level," in *Design Technology of Integrated Systems In Nanoscale Era (DTIS), 2014 9th IEEE International Conference On*, May 2014, pp. 1–6.

[8] S. Sarafianos, Alexandre, "Fault injections by laser impulsion in secured microcontrollers," Theses, Ecole Nationale Supérieure des Mines de Saint-Etienne, Sep. 2013. [Online]. Available: https://tel.archives-ouvertes.fr/tel-00944943

[9] J. Senior, *Optical Fiber Communications: Principles and Practice*. Pearson Education, 2009.

2015 IEEE Computer Society Annual Symposium on VLSI

A Similarity Based Circuit Partitioning and Trimming Method to Defend Against Hardware Trojans

Yun Cheng [1,2],   Ying Wang [1],   Huawei Li [1],   Xiaowei Li [1]

[1]State Key Laboratory of Computer Architecture

Institute of Computing Technology，Chinese Academy of Sciences, Beijing, China

[2]University of Chinese Academy of Sciences, Beijing, China

{chengyun, wangying2009, lihuawei, lxw}@ict.ac.cn

Abstract

Nowadays critical integrated circuits are facing the threat of being invaded, counterfeited and pirated by various malicious adversaries. The topic of Hardware Trojan has gained increasing attention in the past decade. Particularly, a plenty of countermeasure approaches have been investigated to obfuscate the design to avoid hostile circuit insertion during the manufacturing stage. Inspired by the observation that circuits have their own internal structure similarity which can be used to obfuscate the design and obstruct malicious insertions, we firstly adopt the metric called circuit similarity to measure the obfuscation degree of circuits, and then propose a novel circuit similarity based partitioning and trimming method to transform the original design into similar sub-circuits. After the procedure of partitioning and trimming, it will be much harder for the attackers to hack certain sub-circuits or gates. Experimental results show that our method can obfuscate the circuit to a desirable circuit similarity with about 30% wire overhead.

1. Introduction

Due to the separation of integrated circuits (ICs) design and manufacture, there exist security issues that chips are vulnerable to the attacks from adversary foundries. The foundries have the access to the chip's physical design and can adversely draw out the original gate level netlist if the cell libraries are known a priori. If the hostile foundries want, the actual functions of the circuit design may be revealed by various reverse engineering technologies. Consequently, malicious circuits may be carefully injected to weaken the performance, steal critical signals or even destroy the chip in certain working mode [1-2]. Moreover, Due to the lack of observability, it is almost impossible to validate the consistency between the silicon after tap-out and the physical design provided by the designer.

To address the threat of possibly inserted malicious hardware during the manufacturing stage, passive detection methods and active obfuscation methods have been explored to guarantee hardware trustability. The most typical and widely studied passive detection method is side channel analyzing (SCA) [3-5]. The critical paths' delay, power consuming or supply current can be characterized as the signature and used to distinguish the good dies and abnormal dies with Trojans. However, those methods need golden signatures which are obtained from the golden dies that are free of malicious hardware insertion. Obviously, golden dies are quite difficult to obtain in practice. Worse still, the existence of severe process variation (PV) greatly affects the detection accuracy and undermines its applicability if the

This work was supported in part by National Natural Science Foundation of China under Grant No. 61432017, 61176040, and 61221062, and in part by National Basic Research Program of China (973) under Grant 2011CB302501.

factors of PV are not well considered.

Instead of focusing on the passive detection of the malicious hardware after the manufacture stage, alternative methods have been proposed to intentionally obfuscate the original circuit before the manufacturing by making the design incomplete or ambiguous to the malicious adversary [6]. Some circuit locking and split manufacturing methods have been discussed in the literature [7-15]. However, these methods either work for special circuit designs or raise too much hardware overhead, and they fail to take advantage of the circuit's internal characteristics to enhance the effects of obfuscation.

With the observation that the circuits have their own internal structure similarity which can be used to increase the difficulty of malicious attacks, we propose a novel obfuscate methodology through similarity-based circuit partitioning and trimming. Based on the circuit topology, the proposed method splits the original design into several sub-circuits and obfuscates the cut-edges between those sub-circuits. In this way, it becomes hard to differentiate similar sub-circuits especially after minor circuit trimming to increase the obfuscation degree of such sub-circuits. If all of the similar sub-circuits are isolated or further enhanced by other obfuscation methods, the circuit becomes unrecognizable to the attackers even when they have access to netlist. A metric called circuit similarity is adopted to measure the obfuscation degree. Its value stands for the average amount of similar sub-circuits that each sub-circuit obtains, and it indicates the complexity and overhead of injecting stealthy Trojan circuits.

The contribution of our work is twofold.

1. Security-oriented circuit partitioning is firstly explored based on graph theory. The proposed method brings a new direction of design obfuscation by exploring the topology similarity, and tries to maximize obfuscation degree among all sub-circuits. Then it will be much harder to insert malicious circuits to the desired sites.

2. Post-partition circuit trimming method is investigated based on equivalence transformation. To optimize the obfuscation degree of the whole circuits after partition, we propose the sub-circuit trimming method to improve the obfuscation by circuits transforming. After transforming, the obfuscation degree can be improved by 20%.

The rest of the paper is organized as follows. Section 2 discusses previous work in circuit obfuscation, and indicates the salient features of this work. In Section 3, we describe the circuit similarity based hardware obfuscation scheme in details, and develop a novel iterative circuit partition flow to find similar sub-circuits. In Section 4 we present the post-partition circuit trimming method to optimize the obfuscation degree. Section 5 shows the results for a set of benchmark circuits. We conclude in Section 6.

978-1-4799-8720-7/15 $31.00 © 2015 IEEE

2. Related Work and Motivation

Preventing the hardware from malicious alterations is the major trust concern during the manufacturing stage, typically for security critical hardware. Therefore design obfuscations have been investigated to increase the hardware security from different perceptions. The key insights of obfuscations are either to make certain critical hardware components more complicated to be analyzed through a variety of locking technologies, or to separate the whole hardware into several half-baked parts through split fabrication technologies. The state-of-the-art circuit obfuscation methods can be broadly classified into the following two categories.

Key based circuit obfuscation: key based obfuscation method is initially used to avoid hardware pirate [7-9]. Only when the correct key or start up sequence is applied to the circuit, the circuit can work correctly. Without the right key or sequence, the circuit will run in the obfuscated mode and it is quite difficult for the adversary to know the real hidden function. Usually it uses XOR, NOR or MUX gates to lock the original circuit. However, the method still can't get rid of the danger of being invaded during the manufacturing stage. Malicious attackers may delve into the locking circuit, then crack or bypass the obfuscation circuit by modeling and simulation. Besides, it is also possible to circumvent the obfuscation structure, because attackers are able to identify some important sub-circuits with certain structure signatures and just inject malicious hardware into those identified sub-circuits.

Structural isolation based circuit obfuscation: split manufacturing and 3D manufacturing methods have been developed to divide the original whole design into two or more parts and only hand out one part to the untrustworthy foundry. Split manufacturing is first proposed by AMD to reduce the impacts of the manufacturing defects in the metal interconnect layers [10]. Since hardware security problem becomes serious and attracts increasing attentions, some researchers observed that split manufacturing is one possible solution to prevent malicious attack from adversaries. Typically split manufacturing will separate the original die into two parts: the Front End of Line (FEOL) and the Back End of Line (BEOL). The FEOL level contains all cell instances and several lowest levels of metal lines. The BEOL level contains the rest upper levels of metal lines [11-12]. FEOL has higher level of complexity and smaller feature size than BEOL. It can be delivered to the unreliable foundry with more advanced technology. Meanwhile, trustable foundry can complete the BEOL on existing FEOL. Because of lacking of the top level wires, even though the attackers know the whole cell instances and local wires on FEOL, it is still impossible to guess the exact function of the design and to insert malicious alterations to the right location. Then the security of the design is greatly improved by the split manufacturing method. However, the challenge of split manufacturing is the process compatibility. The BEOL foundry must take over the undone process left by the FEOL foundry and ensure process compatibility. Different foundries have distinct manufacturing processes and to make these processes compatible needs much work and significant cost. It is alternative to obfuscate the circuit with three-dimensional (3D) integration, which is also a promising technique to enhance the hardware security. A 3D IC consists of two or more independently manufactured tiers that are vertically stacked on top of each other. Through-silicon via (TSV) is used to connect those tiers. In a secure 3D chip of two tiers, the bottom tier is used for the all logic realizing and most wire connection, and the top tier can be trusted to utilize the confidential connections [14]. By lifting certain wires from un-trusted bottom tier to the trusted top tier, it can make every cell in the bottom tier indistinguishable with many other cells from both gate type and wire connections. Hence the attacker cannot insert malicious hardware to the desired location with little cost. [14] also proposes a notion of hardware security, called k-security. A k-security gate means there are at least k-1 other gates in the circuit indistinguishable from this gate and the k-security circuit means all gates across the circuit are k-security. However, to achieve a k-security circuit, the proposed greedy wire lifting algorithm greatly increases the total wire length. Lots of wires are needed to be lifted and connected in the top tier. According to the results given by [14], the total wire length is increased by several times, and wire delays are also increased. Hence, the unreasonable wiring will bring serious timing problems and performance reduction.

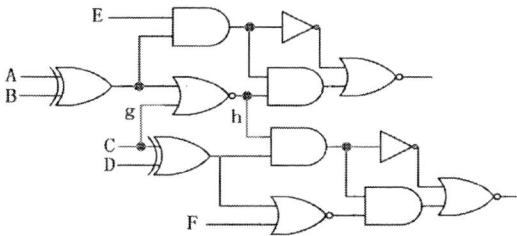

Figure 1: Example of Circuit Similarity

To improve the design security with a low hardware cost and low process complexity, we propose a novel circuit similarity based obfuscation method to defend against hardware Trojans. Through careful observation and analysis, some small sub-circuits among the design can be found similar in the structure. Supposing cut-edges between those sub-circuits can be made indistinguishable by certain obfuscation methods, those sub-circuits will become harder to be distinguished from each other. For example, as shown in Figure 1, if the wire *g* and wire *h* are taken away, then the original circuits will be parted into two sub-circuits and the two sub-circuits are exactly the same. Even though the adversary can make out the functionality of the sub-circuits, the adversary will still have two possible locations to insert malicious hardware, which will greatly increase the difficulty of attacking. To exploit and utilize the security brought by internal similarity, the proposed method splits the original design into several sub-designs and obfuscates the cut-edges between those sub-designs. The selected similar sub-designs are separated out and can be trimmed to increase the similarity degree. Previous work pays greater attention on the external manufacturing process and ignores the circuit's internal characteristic. Our method first considers the internal characteristic and explores it to obfuscate the design by circuit partition.

3. Circuit Similarity based Partitioning

3.1. Circuit Similarity

To ensure trustable hardware during manufacturing stage, our strategy is to extract more similar sub-circuits from the

978-1-4799-8720-7/15 $31.00 © 2015 IEEE

original design. To properly conduct the similarity based obfuscation method, choosing an appropriate abstraction level of the hardware design is of great importance and should be considered at the very beginning. Indeed, the hardware can be described in different abstraction formats. It can be described as algorithms and functions from high abstraction and usually shown as the control and data flow graphs (CDFGs). But, CDFGs cannot stand for the actual circuit' realization. The same sub-flows in the CDFGs do not indicate the same sub-circuits in the chip, and they are probably mapped to the same hardware component to save cost. Then the designs are transformed into RTL with more implementation details and usually expressed as codes in terms of hardware description language (HDL). It is possible to find some modules that are referenced multiple times in those HDL codes and they will be transformed to same components in the circuit's realization. But it is unusual to have many common modules in an ordinary design. Computing the similarity degree among different modules in HDL codes is also nearly impossible. The lowest design abstraction level is the physical descriptions, for example, GDSII file, which includes the layout and the wiring information. However, it is especially difficult to extract similar parts from the multi-layer layout which contains 3D information. Though similarity information is difficult to be extracted from CDFGs, HDL codes or GDSII files, it can be seen from Figure 1 that finding the similar circuits from the gate level is feasible and intuitive. This observation motivates us to employ the gate netlist as the target format to explore hardware design similarity.

The insight of the design similarity for the FPGA designs has been explored in the literature [16]. An algorithm to detect the global topological similarity between two FPGA designs was developed and an incremental physical design flow was applied to utilize the older place and route information to simplify optimization and save computing effort of newer physical design. Different to [16] which targets to find design similarity between two FPGA designs, we attempt to utilize the internal structure similarity of ASICs and employ it to obfuscate the circuits in the form of standard cells instead of LUTs.

To clearly and quantifiably describe the concept of the hardware security considering the structure similarity, we propose a metric called *circuit similarity*. Circuit similarity manifests the actual similarity degree of two circuits' structure, including both the topology and all containing logic gates. The bigger the circuit similarity is, the harder it is to differentiate similar sub-circuits and attack particular cells in one specific sub-circuit.

To calculate the circuit similarity of two circuits, we adopt graph similarity technology in graph theory. The gate level netlist can be viewed as a directed graph. The standard cells are converted into vertices and the wires are converted into directed edges between vertices. In graph theory, graph similarity measurements already have many wide-ranging applications like the analysis of social networks and the comparison of chemical structures. A vertex V_A in graph G_A and a vertex V_B in graph G_B are considered as similar vertices if their corresponding neighborhoods of G_A and G_B are similar, which results in iterative computing methods for the similarity degree of the vertex pairs between these graphs [17]. As shown in Figure 2, assuming that the two

sub-circuits are extracted from a large circuit, they are intuitively similar in the structure. Gate *a* in sub-circuit **I** and Gate *h* in sub-circuit **II** are identical, for their same gate type and their same fan-in and fan-out connections. The gate *c* and gate *j* have a little difference in different types of fan-outs, and if the OR gate *e* is separated from sub-circuit **I** and the INVERTER gate *l* is separated from sub-circuit **II**, the two sub-circuits will be exactly the same. They will be indistinguishable after the differences are separated away.

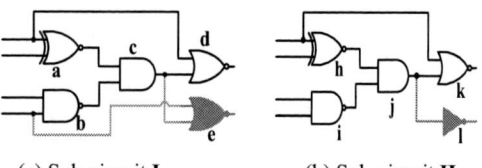

(a) Sub-circuit **I**   (b) Sub-circuit **II**

Figure 2: An example of circuit similarity

Algorithm 1 shows the iterative calculation method of the circuit similarity. The similarity of all gate pairs between two circuits should be computed. Like vertex similarity in graph theory, gate similarity is not only related to their gate types, but also closely related to their connecting gates, so their inputs and outputs are also considered during gate similarity calculation. Supposing gate v_i is selected from circuit **C** and gate v_j is from circuit C^*, the following table shows some notations which will be used in the algorithm.

Table 1: Some notations in Algorithm 1

T	the iteration number
$S_{i,j}^T$	the gate similarity of gate pair (v_i, v_j)
$M(S^T)$	the gate similarity matrix of all gate pairs in iteration **T**
CS^T	the circuit similarity of (C, C^*) in iteration **T**
$in(v_i), out(v_i)$	the fan-in and fan-out group of v_i
$in(v_j), out(v_j)$	the fan-in and fan-out group of v_j
π	one candidate gate mapping scheme for gates connected with v_i or v_j
$\pi(v)$	the matched gate of v in this mapping scheme π

To calculate $S_{i,j}^T$, the gates in the fan-in and fan-out group of (v_i, v_j) should be one-to-one matched. Inputs are mapped to inputs and outputs are mapped to outputs. Each candidate π can calculate one mapping similarity which is the sum of gate similarity of all matched gate pairs, and the best mapping scheme with the biggest mapping similarity is adopted to compute the final gate similarity of (v_i, v_j). After the gate similarity of all gate pairs have been computed, $M(S^T)$ can be updated. Hungarian algorithm is usually adopted to solve assignment problem in a weighted bipartite graph, which will be used to find the optimal matching scheme. The final result maps the gate in C to its similar gates in C^* and can maximize the total gate similarity of all matched gate pairs. Some gates that does not find similar gate will obtain zero gate similarity. Finally CS^T is calculated as the average gate similarity under the optimum scheme.

α is a real number between 0 and 1, which is used to adjust the weight of the older gate similarity and new connecting similarity. When the difference between two circuit similarity values calculated in two adjacent iterations is smaller than the predefined threshold ε or the total number of iterations is bigger than parameter P, the current result will be returned

978-1-4799-8720-7/15 $31.00 © 2015 IEEE   370

as the final circuit similarity. At the beginning of the calculation, the initial matrix of gate similarity $M(S^0)$ will be calculated by just considering the gate type. Different gate types will lead to zero initial gate similarity.

Algorithm 1: Circuit Similarity of Circuit C and C*

1   Initialize $M(S^0)$

2   do

3   for each gate pair (v_i, v_j) in all gate pairs:

$$S_{i,j}^T = (1-\alpha) * S_{i,j}^{T-1} + \alpha * \frac{\max_\pi \left\{ \sum_{v \in (in(v_i) \cup out(v_i))} S_{v,\pi(v)}^{T-1} \right\} * 2}{\left(|in(v_i)| + |out(v_i)| + |in(v_j)| + |out(v_j)| \right)}$$

      end for

4   T++ ;

5 $CS^{(T-1)} = CS^T$;

6 $CS^T <=$ the average gate similarity of the best mapping of the matrix $M(S^T)$;

      end do

7   while $(|CS^T - CS^{(T-1)}| > \varepsilon$ and $T < P)$;

8   return CS^T

3.2. Problem formulation

The structure similarity between individual circuits can be calculated through the above iterative methods. However, how to explore the internal similarity inside one particular circuit is much more challenging. There are no feasible solutions to analyze internal sub-graph similarity in graph theory so far. We propose a practical way based on circuit partition to meet this challenge. The challenge can be transformed as follows. Given a circuit C = {V, E}, where V is the whole set of gates and E is the whole set of directed wires between gates, optimally divide C into k disjoint sub-graphs $C_i = \{V_i, E_i\}$ { i ∈ (1,2 ... k) }. $V_i \cap V_j = \Phi$ for any i ≠ j, $V = \bigcup_{i=0}^{k} V_i$, $E = \bigcup_{i=0}^{k} E_i$. The total number of sub-circuits, k, is not prior known. Then the problem is changed to find one particular partition method which can reach a maximum average circuit similarity. In our method, the average circuit similarity is the sum of weighted circuit similarity of all sub-circuit pairs. The corresponding weight is the ratio between the number of the sub-circuit pair's vertices and the number of the original circuit's vertices.

After circuit partition, the edges between the sub-circuits, referred as cut-edges, should be obfuscated to conceal the logic connections. Otherwise the connection information may be exposed to the adversary and the similar partition will be ineffective. Hence, our partition goal is to find the particular partition scheme which does not only result in the best mean circuit similarity, but should also meet the pre-defined constraint of cut-edge cost (i.e., number of cut edges), which is referred as **Cons.**

The intra-similarity of sub-circuits can be calculated using Algorithm 1. However considering that if the similarity between two circuits is too small, the circuits are easy to be distinguished and the similarity value will be meaningless. Therefore only the circuit similarity that is larger than a certain threshold **Thres**, is taken as effective similarity and considered in the final similarity calculation. The effective similarity of (C_i, C_j) is calculated by equation (1).

$$CS_{eff}(C_i, C_j) = \begin{cases} CS(C_i, C_j); & \text{if } CS(C_i, C_j) \geq \textbf{Thres} \\ 0; & \text{if } CS(C_i, C_j) < \textbf{Thres} \end{cases} \quad (1)$$

As a consequence, the following equations show the formulation for circuit similarity based partition question.

Maximize **CS_{mean}**

Subject to

$$CS_{mean} = \sum_{i=0}^{N} \sum_{j=i+1}^{N} CS_{eff}(C_i, C_j) * \left(\frac{V(C_i) + V(C_j)}{V(C)} \right) \quad (2)$$

$$\sum cut\text{-}edges < \textbf{Cons} \quad (3)$$

3.3. Proposed Method

The obfuscation technologies that can be used to hide the cut-edges include 3D fabrication and split fabrication. Both attempt to lift the wires to a standalone trustable fabrication tier. Other obfuscation methods such as MUX based method are not powerful enough because they may bring large cost and fail to hide the logic connection between sub-circuits. The obfuscation cost is mainly spent in obfuscating the cut-edges. When the obfuscation budget has been determined, the number of the cut-edges that can be obfuscated is also constrained.

Circuit partitioning is a basic problem in VLSI design and has been studied by many researchers. Lots of algorithms have been proposed [18]. The Fiduccia–Mattheyses (FM) algorithm is the most widely used iterative approaches to part the circuit into several parts. It progresses by moving one cell at a time between the blocks of the partition. However, in our similarity based partition scenario, the partition problem becomes more difficult, because conventional partition methods only target the minimum cut-edges and don't need to target the maximum mean circuit similarity at the same time. Those partition solutions obtained under the priority of cut-edge cost may lead to a bad mean circuit similarity. So we propose a novel partition method based on iterative circuit similarity improvement.

First, we use the conventional partition tool to part the original netlist into several sub-circuits based on the minimum cut-edges. Then the circuit similarity scores between those sub-parts will be computed using the proposed computing method. Those sub-parts which are very similar and obtain high circuit similarity values that are bigger than a certain threshold will be separated out from original circuit and become individual sub-circuits. Then the leftover netlist is got to be a new circuit under partition. We will repeat the above loop until the final circuit is small enough or the cut-edge cost is run out. The proposed flow is shown in Figure 3.

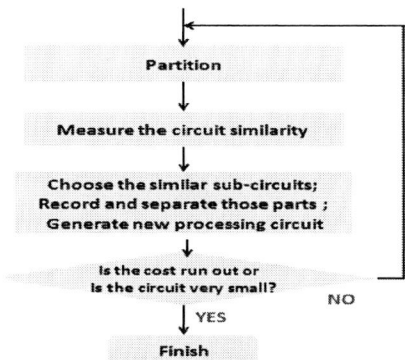

Figure 3: Proposed Iterative Partition Flow

978-1-4799-8720-7/15 $31.00 © 2015 IEEE

4. Post-partition Trimming

Exactly identical sub-circuits could achieve the optimal obfuscation effects but are rarely present in the actual circuit. To improve proposed partition scheme and obtain better obfuscation effects, we explore the post-partition trimming method to increase the circuit similarity among similar sub-circuits.

We classify the difference between two similar circuits into two broad types: (a) lacking of certain cells and ports; (b) different logic implementations. Both differences are shown in Figure 4. Sub-circuit B is different from sub-circuit A for lacking of the INVERTER gate 7; sub-circuit C is different from A for that gate 5 is an AND gate, not an INVERTER gate.

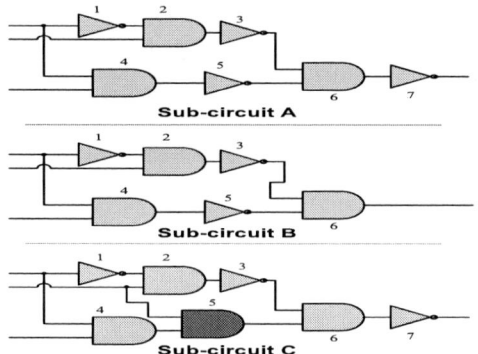

Figure 4: Similar Sub-circuits

The degree of difference between circuit pairs can be measured by the circuit similarity, and we use this new metric to evaluate our trimming scheme. During the trimming operation, it ensures that the circuit original functions will not be changed and their differences are eliminated at the same time. Hence, adding configurable cells is an intuitive method to migrate the differences. In our methods, we choose to add MUXs and corresponding control signals to original circuits. By configuring the control signals with different values in different circuits, it can be achieved that using the same circuit to realize different logic functions. If multiple MUXs have been added to the sub-circuits, then internal control signals with the same control value can share one control port to reduce trimming cost. For example, the circuit after trimming for sub-circuit A, B and C in Figure 4 is shown in Figure 5. Two MUXs and two control signal are added for trimming.

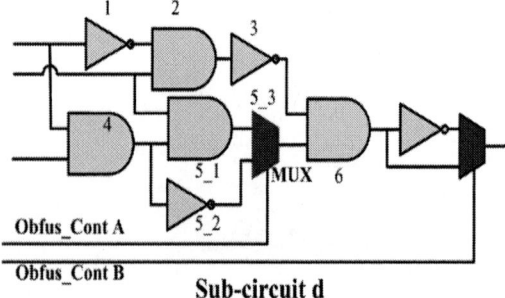

Sub-circuit d

Figure 5: Sub-circuits after Trimming

5. Experiments

To evaluate the effectiveness of the proposed method, we take the ISCAS85 and ISCAS89 circuits as our test benches to enhance the circuit security degree. For the ISCAS85 circuits, two synthesized formats are used on different cell libraries. In the first experiment, the ISCAS85 circuits are synthesized including only AND and INVERTER gates. We use the ***abc*** tool [19] written by the Berkeley logic synthesis and verification group to synthesize those test benches into netlists. In the second experiment, we directly use .bench format circuits which consist of AND, OR and INVERTER gates. After the netlists are obtained, we choose the ***metis*** tool [20] to part the netlists into certain sub-circuits for its high partition speed and outstanding results. The experiments are conducted on a computer with Intel Core i5-2400 CPU and 4G RAM. The runtime is not long for small circuits; however it takes much longer time to conduct on larger circuits. Most of time is spent on the calculation of the circuit similarity for the computational complexity is squares of the circuit size. To accelerate the processing speed, we use some heuristic information to speed up the similarity computing, such as the information about the gate type, the gate level and the sub-circuit scale. Especially, when we process very large circuits, we divide the circuits into several small circuits and manipulate each small circuit as an individual circuit. Consequently the runtime can be greatly reduced.

As Table 2 shown, the first two columns list the circuit name and the number of inputs/outputs. The similarity, the number of new generated cut-edges, the wire overhead (penc %) and the runtime for two format circuits are shown in the following columns. The wire overhead is the ratio between the number of new cut-edges and the number of original wires. It can be seen that different netlist can lead to different circuit security value. The circuit similarity is also related with the synthesizing results, which will be studied in future. Most of the circuits in .bench format can reach a circuit similarity value bigger than 1 with wire cost below 30%. We also show the partition process for the .bench circuits in Figure 6. Each line stands for one circuit. The x axis stands for the cost and the y axis stands for the similarity. It is clear that the similarity can be improved with the cost increasing. Different cost budget will lead to different obfuscation degree. If the wire cost is limited to 20% of the original wires, some circuits will still obtain an acceptable circuit similarity value. The ISCAS89 experiments can be conducted in the same way. The results are shown in Figure 7. For sequential circuits, D flip flop is added to the basic gate types. We also apply our method on the DES circuit from openrisc website [21]. The DES circuit has 33505 gates. The results show that it achieves a circuit similarity of 18.7 with a wire cost about 23%. From these results, we demonstrate our method is powerful to obfuscate the netlist with less than 30% wire cost.

We also evaluate the proposed trimming method. The differences between similar circuit pairs are first identified. The different functions in both circuits are combined using MUXs to construct new hybrid logic which is adopted to replace original implementations. After trimming the circuit similarity between similar circuits can be increases by about 20%. However, it needs much manual work to optimize the trimming procedure to obtain good similarity enhancement with low hardware cost.

978-1-4799-8720-7/15 $31.00 © 2015 IEEE   372

Table 2: Experimental Results

Circuit Name	PI/ PO	Format 1 (only contains AND and INVERTER)				Format 2 (.BENCH)			
		Similarity	Cut-Edges	penc(%)	Time /s	Similarity	Cut-Edges	penc(%)	Time /s
c432	36 / 7	1.02	211	36.3%	68	**1.22**	149	**32.8%**	**9**
c499	41 / 32	**1.48**	313	**25.5%**	68	0.94	101	41.4%	**4**
c880	60 / 26	1.03	316	**34.7%**	70	**1.27**	781	38.1%	**44**
c1355	41 / 32	1.68	382	26.1%	91	**2.39**	235	**15.7%**	**10**
c1908	33 / 25	1.00	352	30.0%	131	**1.77**	374	**19.7%**	**37**
c2670	233/140	1.01	680	32.7%	454	**1.68**	457	**19.4%**	**110**
c3540	50 / 22	**2.56**	817	**31.7%**	446	2.12	1233	36.3%	**342**
c5315	178/123	**2.37**	1396	28.8%	487	2.27	1216	**23.2%**	**237**
c6288	32 / 32	6.38	1506	21.6%	524	**19.7**	1212	**17.4%**	**330**
c7552	207/ 108	3.38	1558	26.0%	925	**4.48**	1565	**21.7%**	**439**

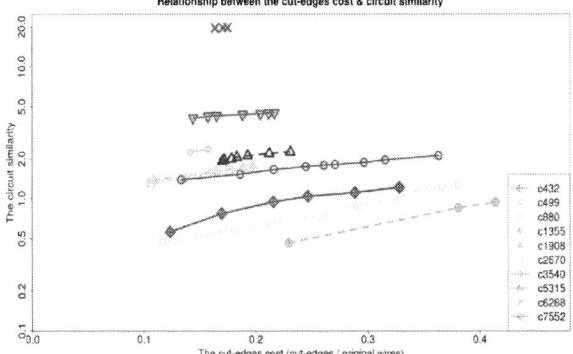

Figure 6: The partition process of ISCAS85 bench circuits

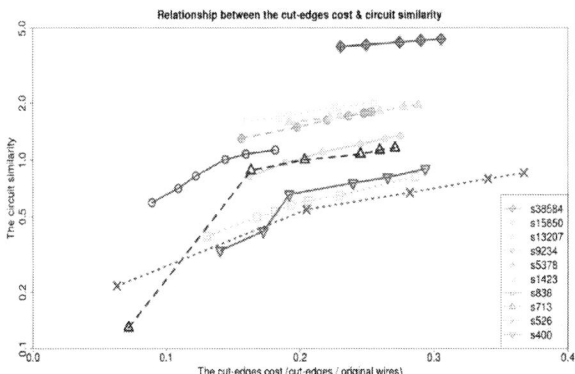

Figure 7: The partition process of ISCAS89 bench circuits

6. Conclusion

On the observation that designs has their own internal structure similarity which can be used to obfuscate the design and obstruct malicious insertions, we adopt the metric called circuit similarity to measure the obfuscation degree of similar circuit, and also propose a novel circuit similarity based partition and trimming method to transform the original design into several similar sub-designs, then it will be much harder to invade certain sub-circuits or gates. Experiments on the ISCAS circuits demonstrate that it can obfuscate the circuit to a desirable circuit similarity with about 30% cut-edge cost.

Related Work

[1] M. Tehranipoor and F. Koushanfar, "A Survey of Hardware Trojan Taxonomy and Detection," IEEE Design and Test of Computers, 2010, pp. 10-25.

[2] X. Wang, M. Tehranipoor, J. Plusquellic, "Detecting malicious inclusions in secure hardware: Challenges and solutions," HOST, 2008, pp. 15–19.

[3] D. Agrawal, S. Baktir, and D. Karakoyunlu, P. Rohatgi, and B. Sunar, "Trojan Detection using IC Fingerprinting," IEEE Symp. on Security and Privacy (SP), 2007, pp. 296-310.

[4] R. M. Rad, X. Wang, M. Tehranipoor, and J. Plusquellic, "Power supply signal calibration techniques for improving detection resolution to hardware trojans," ICCAD, 2008, pp. 632–639.

[5] Y. Jin and Y. Makris, "Hardware trojan detection using path delay fingerprint," HOST, 2008, pp. 51–57.

[6] R. Chakraborty and S. Bhunia, "Security against Hardware Trojan through a Novel Application of Design Obfuscation," ICCAD, 2009, pp. 113-116.

[7] J. Roy, F. Koushanfar, and I. Markov, "EPIC: Ending piracy of integrated circuits," DATE, 2008, pp. 1069-1074.

[8] J. Roy, F. Koushanfar, and I. Markov, "Protecting bus-based hardware IP by secret sharing," DAC, 2008, pp. 846–851.

[9]A. Baumgarten, A. Tyagi, and J. Zambreno, "Preventing IC Piracy Using Reconfigurable Logic Barriers," IEEE Design and Test of Computers, vol. 27, no. 1, 2010, pp. 66 – 75.

[10] R.W Jarvis and M. G. McIntyre, "Split manufacturing method for advanced semiconductor circuits", US Patent no. 7195931, 2004.

[11] J. Rajendrany, O. Sinanogluz, and R. Karriy, "Is Split Manufacturing Secure?, " DATE, 2013, pp. 1259–1264.

[12] K. Vaidyanathan, B. Das; E. Sumbul; R. Liu; L. Pileggi, "Building trusted ICs using split fabrication," HOST, 2014, pp. 1-6.

[13] M.Jagasivamani, P.Gadfort; M.Sika; M.Bajura;M. Fritze, "Split-fabrication obfuscation: Metrics and techniques, " HOST, 2014, pp. 7–12.

[14] F. Imeson, A. Emtenan, S. Garg, and M. Tripunitara, "Securing Computer Hardware Using 3D Integrated Circuit (IC) Technology and Split Manufacturing for Obfuscation," 22nd USENIX Security Symposium, 2013, pp. 495-510.

[15] J. Valamehr, T. Sherwood, R. Kastner, D. Marangoni-Simonsen, T. Huffmire, C. Irvine and T. Levin, "A 3-D Split Manufacturing Approach to Trust worthy System Development", IEEE Transactions on Computer-Aided Design of Integrated Circuits and Systems, vol. 32, no. 4, April 2013, pp. 611-615.

[16] X. Shi, D. Zeng, Y. Hu, G. Lin and O. Zaiane, "Enhancement of incremental design for FPGAs using circuit similarity," ISQED, 2011, pp. 1–8.

[17] S. Melnik, H. Garcia-Molina and E. Rahm, "Similarity flooding: a versatile graph matching algorithm and its application to schema matching", ICDE, 2002, pp. 117 – 128.

[18] C. J. Alpert and A. B. Kahng, "Recent directions in netlist partitioning: A survey," VLSI J. Integr., vol. 19, 1995, pp. 1–81.

[19] Berkeley Logic Synthesis and Verification Group, ABC: A System for Sequential Synthesis and Verification, Release 70930. http://www.eecs.berkeley.edu/~alanmi/abc/

[20] G. Karypis ang V. Kumar, "A Fast and Highly Quality Multilevel Scheme for Partitioning Irregular Graphs", SIAM Journal on Scientific Computing, 1999, Vol. 20, No. 1, pp. 359—392.

[21] http://opencores.org/

On-chip Instrumentation for Runtime Verification in Deeply Embedded Processors

Ciaran MacNamee [1] and Donal Heffernan [2]
Department of Electronic and Computer Engineering
University of Limerick
Limerick, Ireland
[1] ciaran.macnamee@ul.ie, [2] donal.heffernan@ul.ie

Abstract—**In the field of safety-critical applications, the industry has seen a dramatic increase in system integration and complexity for on-chip solutions, and this provides new challenges in gaining deep testability access within such products. IC developers have responded to this challenge by creating sophisticated on-chip trace/test/debug modules to facilitate real-time access to application programs' features in deeply embedded pipeline and cache based architectures for embedded-core-based chips. However, the emerging safety requirements for complex systems demand more stringent solutions for confirming confidence in systems and this leads to a requirement for the formal verification of some key system properties. In this paper the authors propose that the existing architectures for on-chip trace/test/debug modules can be extended to support runtime verification monitoring, working towards the formal verification of system properties of interest. A real benefit in the proposed solution is the provision of a lifecycle use for the on-chip debug logic, thus offering an improved return-on-investment prospect by justifying the resident logic with the benefit gained by lifelong runtime verification features.**

Keywords—monitors; runtime verification; on-chip debug; safety critical; embedded processor

I. INTRODUCTION

In recent years there has been a significantly increased demand for safe and reliable embedded electronic systems, with performance guarantees throughout a system's entire lifecycle [1]. Integrated circuits developers have seen a dramatic increase in system integration and complexity, which has compounded the requirements for comprehensive test and verification. Over a long period of time, many solutions and approaches have evolved to address such problems, including solutions to the problems of testing, debugging and verification of embedded-core-based chips at the manufacturing stage [2], the functional validation and debug schemes for multicore microprocessors and complex systems on-chip [3], [4], and the requirement to guarantee functional safety properties throughout a product's lifecycle, including deployment and end-of-use stages [5], [6]. To support such complex requirements, it is now commonplace to find on-chip test and debug support logic in large scale integrated circuits,

even though this logic may only be used for a tiny fraction of the useful lifetime of the chip. The fact that semiconductor manufacturers are willing to bear the cost of this extra logic in production devices attests to the essential role that they play [7], [8]. Recent developments relating to complex on-chip test and debug features are moving towards improved support for the access and control of multi-core systems at the chip level, and they can now include Performance Measurement enhancements [8].

Emerging stringent safety requirements for complex embedded processing systems demand new solutions that often extend to a requirement for the formal verification of some key properties. Unfortunately, the formal verification of embedded programs is unsolvable in general, as theorem prover solutions do not in practice scale to large embedded software systems [9]. A more realisable approach is to formally observe some selected proof obligations during the actual operating phase of a system. The concept of formally monitoring specific properties during a program's execution phase is referred to as runtime verification [10]; where a monitor is designed to detect or to react to property violations so as to provide a level of fault protection. In this context, a monitor is an entity that observes the behaviour of a system to confirm a predefined correct behaviour.

A number of monitor implementations have been reported in the literature over a long period [11]. Recent work on small embedded runtime verification monitors for microcontrollers is presented by Reinbacher et al [12]. Some of the emerging technical challenges relate to the development of economical on-chip monitor architectures to support data access that is deeply embedded within IC devices that have no external visibility. Thus, finding the required data that needs to be observed in a microcontroller's on-chip memory, analysing that data in real-time, non-intrusively, represent formidable problems in the design of a monitor circuit.

In this paper, the authors suggest that current industry standard on-chip debug support logic could be augmented to provide greater support for runtime verification schemes. For current on-chip debug solutions, manufacturers have already solved the problem of access to low level on-chip features and thus can already cope with complications such as processor pipelines and code or data caches. It is likely that the addition of architectural enhancements to such on-chip debug logic

systems could represent a promising pathway towards supporting purposeful on-chip monitoring schemes. Moreover, given that such debug logic blocks are likely to be present but dormant in deployed production silicon, providing a lifecycle use for such debug logic offers an improved return-on-investment prospect by justifying the logic with the benefit gained by lifelong runtime verification features.

The remainder of this paper is organised as follows: Section II summarises some previous work relating to the runtime monitoring of software. Section III presents some exemplar monitors based on simple processor systems. Section IV discusses issues relating to monitor implementations targeting embedded pipelined processors, with code and data caches. Section V shows how a simple closely coupled coprocessor and an on-chip debug module can be used to support monitor-related tasks. Finally Section VI presents conclusions and future work.

II. REPORTED MONITOR IMPLEMENTATIONS

A taxonomy of runtime software-fault monitoring tools is presented by Delgado et al [25] and Watterson et al [15]. Some of the fundamental concepts in monitoring are detailed by Nutt [26] and Plattner [27]. Examples of runtime verification systems are detailed by Gates et al [28] and Mok and Liu [29]. Schemes for runtime verification monitors, as a 'lightweight' formal verification solution, are proposed by Havelund and Roşu [30], Drusinsky [31] and Sammapun et al [32]. Much of the published work on runtime verification suggests that the monitored properties should be expressed formally, for example using a Linear-time Temporal Logic (LTL) [16].

The Monitoring and Checking (MaC) framework described by Lee et al [33] and by Kim et al [34] suggests that a means must also be provided to automatically relate low-level observations of program execution behaviour to the relevant monitored properties. Havelund and Roşu [35] present a runtime verification tool named PathExplorer (PAX) that uses LTL, and includes a synthesis algorithm to generate a verification algorithm from finite trace linear past-time temporal logic (ptLTL) formulae. Drusinsky [18] uses Metric Temporal Logic (MTL) in the Temporal Rover monitor.

However, on the specific topic of on-chip instrumentation for software monitors in deeply embedded systems there appears to be a somewhat limited set of literature. El Shobaki et al [36] propose a monitor scheme where the work is aimed at SoC design verification, rather than monitoring of application software. MacNamee et al [37] describe embedded monitoring and debug approaches using the Nexus and IEEE1149.1 interfaces and they propose an embedded requirements-based monitor in [38]. Fryer [39] examines minimally-intrusive monitoring in the context of complex embedded systems. Fogarty et al. [20] propose solutions for the on-chip support for software verification and debug in multi-core embedded systems. Some small embedded runtime monitor implementations have been reported by Heffernan et al [13] and Reinbacher et al [21]. A runtime verification monitor for functional safety properties in automotive embedded systems is presented in [17].

III. PREVIOUS WORK

The authors carried out previous projects on the development of runtime monitor solutions. Some of this work is presented to show the implementation features of such monitors.

A. A Monitor for Timing Properties

An on-chip runtime monitor was developed [13] where a digital circuit is synthesised to realise the monitor function that checks timing properties against predefined time bounds, on a state by state basis during a program's execution. The application program design is represented by timed automata. The digital circuit for the monitor is generated automatically and is instantiated within the fabric of an SoC. In the prototype development, the system is formally verified using the Uppaal [29] model checker tool. During execution the monitor continuously checks the program's timing behaviour and reports any violations via the monitor's interface.

The monitor is aware of state changes during runtime as the program's memory entry address, and other data, can be found for each state. On entry to a new state a *ReadMEMState* register is updated to report the actual state identifier number. Figure 1a) illustrates an abstract architecture for the monitor, which is built on the Xilinx Spartan IIIe device. The microcomputer is a simple 8-bit PicoBlaze soft core processor.

B. A More Expressive Monitor

The authors developed a more expressive monitor that implements a non-intrusive scheme to monitor runtime behaviour [17]. Runtime events, which model atomic propositions (AP), are assessed in real time. The trace of runtime events is considered to be a finite word over the AP, and is checked for inclusion in the language generated by the relevant ptLTL formulae.

Figure 1b) illustrates an abstract architecture for the on-chip monitor system. This application program is not instrumented and in this sense the monitor scheme is not intrusive. The hardware monitor is attached to the microcomputer's system bus. The microcomputer and the monitor co-reside on a single SoC device.

The Filter/AP-Eval block is a logic circuit for on-the-fly evaluation of each atomic proposition. The Event Recogniser block evaluates the observed variables associated with an event to define a unique state transition event that will represent a state change. The Runtime Checker block is a runtime formula checker to establish truth of a ptLTL formula in a trace t up to the current state. The full monitor was implemented with some 3500 gates in an FPGA.

The complexity of the Filter/AP-Eval block is particularly important. The on-the-fly range check detection of the variables that represent atomic propositions becomes complex as the number of atomic propositions is increased. Reinbacher et al [21] report a similar monitor architecture but with a special atomic proposition checker design, based on the use of constraints that are similar to Logahedra [41], and claims a requirement of 290 logic cells per individual atomic checker.

Based on the above experience with discovering the complexities of detecting on-the-fly address and data changes in the context of simple low-speed microcontrollers, the authors were encouraged to investigate the potential of applying such on-the-fly event detection in much more complex processors that operate at high speeds, but which offer additional logic support by way of debug logic blocks.

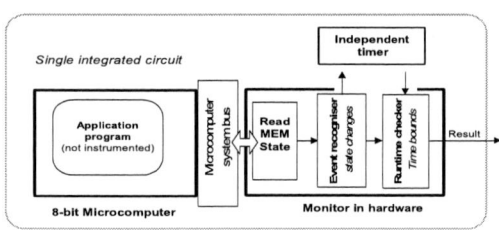

a) Timing properties monitor

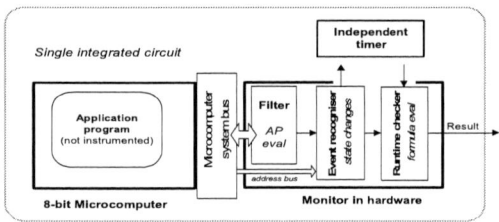

b) Formal logic based monitor

Figure 1 Single IC integrated runtime monitor schemes

IV. MONITORING EMBEDDED PROCESSORS

The monitors described in the previous section have in common the fact that the information to be captured and analysed is relatively easily available. For example in Figure 1 the microcomputer system bus can be used to capture code fetch or data access cycles for input to the monitor. Similarly, the processor used in [12] and [21] is the MCS-51 core. The authors show that its data access cycles are visible using on-chip buses. In the examples in [17] the required information is easily accessible from a CAN (Controller Area Network) bus interface.

More powerful 32-bit processors usually use pipelined instruction execution and often contain code and data caches to improve performance. Monitors for pipelined 32-bit processor cores, which may have on-core caches, must take these factors into account.

A. Debug and Trace

Monitor implementation challenges posed by 32-bit pipelined processors relate to the difficulty in capturing and analysing the application program's required information, whether this is code execution details or the values of relevant data variables. Similar but not identical issues confront designers of debug and test equipment, thus processor core providers have extensive experience of providing on-chip development hooks to enable software debugging tasks. The code and data trace features in on-chip debug logic are of particular interest to a monitor implementation design. The key

point about code and data trace support provided by core designers such as ARM [8], [22], [23] is that code execution and data access visibility problems have already been solved by their debug and trace support systems. Therefore monitors for processors such as these must either use the outputs of the trace modules, or better still, be incorporated into them.

B. Related Challenges and Applications

Because of the high bandwidth required for off-chip capture of trace data by instruments like logic analysers or trace port analysers (TPAs) [8], or size constraints if on-chip memory is used for trace capture, the volume of data captured and traced must be minimised. Two approaches are common, both used in the ETM [22]. Firstly, trace control logic defines a number of event recognisers that can enable or disable acquisition of instruction execution or data access cycles, and secondly trace information tends to be emitted in highly compacted formats, with the understanding that an offline tool can reconstruct code execution flow by reference to a static image of the code being executed.

Event recognisers in the latest ETM [22] include address comparators, which can be used for recognising instruction fetch or execution, or data reads or writes, and data value comparators. They can work on single addresses or single data values, or can be set up to be active when address or data values are within predefined ranges of values. The ability to trigger on a data read or write at a given address of data within a defined range is important for identifying atomic propositions for monitoring.

Figure 1 [13] shows a monitor implementation which was implemented in an FPGA. The monitor used blocks defined as ReadMemState or 'Filter AP Eval' to capture the AP set. Figure 2 shows how the ETM event recognisers could be used as 'building blocks' for the monitor AP events recognition. The obvious objection to this scheme is that the number of event recognisers in a given ETM implementation is limited. ARM suggests their Mediumplus ETM should have 4 pairs of address comparators and 2 data value comparators. However as long as the number of these comparators is not limited by the ETM architecture then their presence in a possible monitor design might convince an implementer to make a greater number available for that purpose. Monitor systems are not best suited to monitoring large numbers of properties; they work best when used to monitor a comparatively small number of important properties, which means that the number of required recogniser blocks can be limited. Nonetheless the limited number of address and data value comparators is an important factor in determining whether monitoring solutions can have practical implementations.

A solution to these limitations is based on the observation that the event recognisers do not all need to be active simultaneously. Unlike debugging and testing scenarios, the monitoring events of interest change as a function of the state of the system. A monitor generated for ptLTL analysis is only concerned with the previous state and the current state, so only the events relevant to the current state are of immediate interest. Thus if the event recognisers, address and data comparators can be reconfigured on-the-fly by the monitor system, then a limited number of comparators may suffice for a full monitor implementation without loss of event coverage. The conditions to be met are that the total number of event recognisers available should be greater than or equal to the number required to cover the state with the maximum number of relevant events. This is explored in Section V.

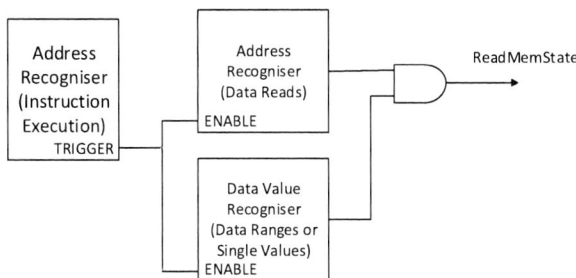

Figure 2 ReadMemState generated by Event Recognisers

Even if the necessary recogniser blocks are not available other ways of identifying the monitor atomic propositions are possible. If data trace is supported then online analysis of the information may be possible. In on-chip trace systems code execution is usually traced in the form of program discontinuities so that reference to the original object and source code is required to reconstruct the code execution history. It is unlikely, therefore, that online analysis of code trace would be feasible.

When data read or write cycles are tracedthe data addresses are less truncated [22] because more information is required even for offline analysis. As a result captured data trace (unlike instruction trace) information can be analysed by an online monitor with access to the trace. The Trace Data Analysis block of Figure 3 identifies state transition events from traced data. Alternatively, a monitor may use low overhead *printf* style tracing, as will be discussed next.

C. Intrusion and Instrumentation

On-chip trace units are not always implemented. In this case, there are still some options. For example, a Cortex-M3 based microcontroller may feature an Instrumentation Trace Macrocell (ITM), even if an ETM has not been implemented. In fact ETM itself also supports "Instrumentation Resources" [22]. These features support low overhead *printf* style tracing. An output is generated by a single CPU instruction and either triggers an event for controlling trace (ETM) or outputs a defined event for trace capture (ITM). The level of intrusion can be insignificant and instrumentation instructions can be retained in the released code. This low-overhead output messaging provides the basis for a minimally intrusive form of code instrumentation, where the trace messages from the ITM or the Instruction Resources provide the platform for a monitor. Figure 3 shows how a monitor fits into to overall ETM configuration in an ARM processor.

The proposed monitor scheme uses a modified trace system, similar to ETM. While adding extra gates to the design, they allow the on-chip trace logic to be used after product deployment in the field, allowing it to add value to the released system and to enable a range of monitor applications not previously considered practical. The extra gates are used

for two purposes: recognising events, when the events are more complex than single instruction fetches or single data reads or writes, and online analysis of the captured data. The online analysis is implemented in the monitor. The next section shows an example monitor that uses online analysis of results that are captured from a debugging module.

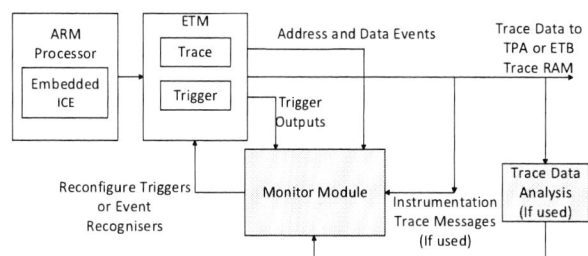

Figure 3 Software monitor used with ETM outputs

V. EXAMPLE MONITOR

Many of the components required to implement a monitor are available in the on-chip debug and trace modules of embedded microcontrollers. Event recognisers for instruction fetch or execution addresses, data access addresses and data values or ranges, can identify a number of events of interest for capture, triggering or analysis. Timer/counters and state sequencers may also be found within debug or trace modules.

Unlike conventional debugging, a monitor's events of interest change dynamically, and since the number of address and data comparators is limited, the monitor needs to identify the current state of the target system and define new event sets on the fly. This means reconfiguring the event recogniser logic for the events relevant to the current state. Experiments carried out by our research group, reported by Fogarty [20], [40], using the XGATE IO coprocessor of the MC9S12X microcontroller combined with the S12DBG debug module to monitor application code running on a CPU12X processor, demonstrate the technique.

The 16-bit Freescale MC9S12X microcontroller with its CPU12X processor and XGATE coprocessor architecture was developed by the manufacturer so that the XGATE would relieve the main CPU12X of many low level time critical tasks, such as buffering data or servicing on-chip peripheral devices [24]. The XGATE was not developed for monitoring purposes but with the S12XDBG debug module, the authors saw the opportunity to experimentally use it to monitor certain time-critical properties and also to carry out runtime verification tasks for application program code executing on the CPU12X.

The significance of this monitor design experiment is that it showed that even an I/O coprocessor such as XGATE is sufficiently programmable to support a range of instrumentation options, with relatively low implementation overhead [40]. Like the proposed on-chip trace based monitor of Section IV the coprocessor based monitor reduces the volume of data to be reported by analysing the data on-the-fly and reporting the analysis results rather than raw data. Figure 4 [20] illustrates the general concept.

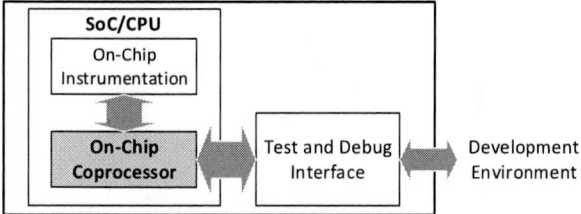

Figure 4 Development Platform using on-chip co-processor [20]

Event recognition logic typically uses comparators made of XOR gates or equivalents. This is a considerable overhead so chip designers are reluctant to include many of them. For a monitor scheme based on comparators provided by on-chip debugging systems this could lead to unacceptable limitations on the number of state transitions to be monitored. However, a software monitor does not need to monitor all states and all state transitions at all times. Provided that there are enough address and data comparators to cover the maximum number of events in the worst-case number of transitions, and provided that the comparator can be reprogrammed before the next possible state transition then a monitor can be implemented. Our group implemented a monitor with this configuration using S12XDBG and XGATE, as in Figure 5.

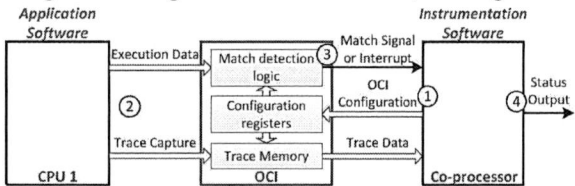

Figure 5. Combined use of coprocessor on-chip debug and trace data [40]

In Figure 5, the co-processor, event recogniser and trace system interact as follows: (1) the instrumentation software configures the event recogniser to trigger when the application software in CPU1 reaches an event of interest; (2) the on-chip trace hardware captures relevant trace data; (3) the event recogniser logic detects the event occurrence and notifies the co-processor; (4) the co-processor analyses the captured data against the monitor requirements to determine if execution is proceeding correctly, providing a status output if necessary. The process can be repeated, or the event recogniser can be reconfigured for other events. The re-use and reconfiguration of the event recognition logic overcomes issues associated with the limited number of comparator circuits.

VI. DISCUSSIONS, CONCLUSIONS AND FUTURE WORK

Many researchers, who have studied runtime verification, conclude that hardware based monitors are preferable because their intrusion effect on the temporal and overall performance of the target system is minimised. The use of a modal temporal logic, such as ptLTL, is a good framework for expressing a monitored system's proof obligations, in part because the volume of data to be analysed can be reduced based on the state-by-state trace scheme. In spite of this, the size of an effective monitor circuit may be high, considering the need to recognise instruction execution and variable data

values. Detecting such properties is difficult for embedded processors that have pipelined execution engines and instruction and data caches. Problems of identifying instruction execution and data read/write events have already been solved by incorporating on-chip debugging and trace acquisition logic into microcontroller cores. It is proposed that a monitor scheme for an embedded processor with these architectural features can use the features provided by on-chip debug and trace logic, and ideally the scheme would be built onto the structure of the debug module. Unlike conventional debugging and trace systems, the monitor analyses data on the fly. While custom logic could be used to implement specific monitor solutions, a better approach might involve a programmable processor, with the monitor implemented as software running on the processor. The limited number of event recognisers in on-chip debug modules means that the monitor needs to reconfigure the recognisers for the code and data events based on the state of the target system. This necessary feature requires extra logic in an on-chip trace module. It is noted, however, that support for a monitor would add value to on-chip debug and trace blocks. Results obtained using the CPU12X and S12XDBG/XGATE combination establish that a monitoring system is a viable proposition. These observations raise a number of questions:

- Intrusion. System monitors should not intrude on the performance of the target system. Non-intrusion is often understood as maintaining the temporal behaviour of the target by avoiding code instrumentation. An on-chip monitor may intrude on performance in other ways, however, notably in terms of power consumption. Power consumption figures for the XGATE based monitor are given in [40].

- Reconfiguration. A monitor unit needs to reconfigure the event recognisers as the target system traverses a set of states, but this raises a question of how quickly the reconfiguration needs to be done, which in turn raises the question of the minimum time in which a state transition may occur and of the time duration of any state. These properties are sometimes hard to define, being processor as well as application system dependent.

- End use of a monitor. Simply monitoring a system does not ensure its safety: it merely records the absence of a deviation from the expected behaviour. The presence of a monitor in a system, when deployed in its end use environment, raises the question of what should be done in the case of a violation.

REFERENCES

[1] SGS-TUV Saar: 'Functional Safety.' http://www.sgs-tuev-saar.com/en/functional-safety.html, accessed 27 Feb, 2015.

[2] Zorian, Y., Marinissen, E.J., Dey, S.: 'Testing embedded-core-based system chips', IEEE Comput., 1999, 32, (6), pp. 52–60.

[3] Foster, T.J., Lastor, D.L., Singh, P.: 'First silicon functional validation and debug of multicore microprocessors', IEEE Trans. VLSI Syst., 2007, 15, (5), pp. 495–504

978-1-4799-8720-7/15 $31.00 © 2015 IEEE

[4] Hopkins, A.B.T., McDonald-Maier, K.D.: 'Debug support for complex systems on-chip: a review', IEE Proc., Comput. Digit. Tech., 2006, 153, (4), pp. 197–207

[5] IEC 61508:2010 (all parts): 'Functional Safety of electrical/electronic/programmable electronic safety-related systems'. 2010

[6] ISO 26262:2011 (all parts): 'Road vehicles – functional safety. International standard'. 2011

[7] MIPS Technologies, Inc.: 'EJTAG Specification', 2010

[8] ARM Limited: 'Coresight Technical Introduction White Paper', 2013. Document Number: ARM-EPM-039795, http://infocenter.arm.com/help/topic/com.arm.doc.epm039795/coresight _technical_introduction_EPM_039795.pdf

[9] Havelund, K., Goldberg, A.: 'Verify your runs'. Verified Software: Theories, Tools, Experiments (VSTTE'05) Zurich, Switzerland, October 2005, pp. 10–13.

[10] Pike, L., Niller, S.,Wegmann, N.: 'Runtime verification for ultra-critical systems'. Runtime Verification (pp. 310–324). Springer, Berlin, Heidelberg. January 2012, pp. 310–324.

[11] Calingaert, P.: 'System performance evaluation: survey and appraisal', Commun. ACM, 1967, 10, (1), pp. 12–18.

[12] Reinbacher,T, Horauer, M, Steininger, A. 'A Runtime Verification Unit for Microcontrollers', System, Software, SoC and Silicon Debug Conference (S4D), 2012, pp. 1 – 6

[13] D. Heffernan, D., Shaheen, S., Watterson, C. 'Monitoring embedded software timing properties with an SoC-resident monitor', IET Software, 3(2): 140–153, 2009

[14] Uppaal home page: www.uppaal.com/

[15] Watterson, C and Heffernan, D. 'Runtime verification and monitoring of embedded systems', IET Software, 1(5):172–179, 2007.

[16] Pneuli, A.: 'The temporal logic of programs'. Proc. 18th IEEE Symp. Foundations of Computer Science (FOCS 1977), 1977, pp. 46–77

[17] Heffernan, D., MacNamee, C. , Fogarty, P. 'Runtime verification monitoring for automotive embedded systems using the ISO 26262 functional safety standard as a guide for the definition of the monitored properties', IET Software, 8(5): 193-203, 2014.

[18] Drusinsky D.: 'Monitoring temporal rules combined with time series'. Proc. 15th Int. Conf. on Computer-Aided Verification (CAV'03), Boulder, Colorado, USA, 8–12 July 2003 (Lecture Notes in Computer Science (LNCS)) vol. 2725, pp. 114–117

[19] Reinbacher, T, Rozier, K.Y., Schumann, J: 'Temporal-Logic Based Runtime Observer Pairs for System Health Management of Real-Time Systems'. In *Tools and Algorithms for the Construction and Analysis of Systems*, pp. 357-372. Springer Berlin Heidelberg, 2014.

[20] Fogarty, P., Heffernan, D., MacNamee, C. 'On-chip support for software verification and debug in multi-core embedded systems', IET Software, 7(1):56-64, 2013.

[21] Reinbacher, T., Brauer, J., Schachinger, D, Steininger, A and Kowalewski, S. 'Automated Test-Trace Inspection for Microcontroller Binary Code'. In *Runtime Verification*, pp. 239-244. Springer Berlin Heidelberg, 2012.

[22] ARM Limited: 'Embedded Trace Macrocell (ETMv1.0 to ETMv3.5) Architecture Specification', 2011. Document Number: ARM IHI 0014Q, http://infocenter.arm.com/help/topic/com.arm.doc.ihi0014q/IHI0014Q_e tm_architecture_spec.pdf , accessed 2 March 2015

[23] Yiu, J: 'The Definitive Guide to the ARM Cortex-M3', 2nd Edition. Newnes, 2009. Amsterdam.

[24] Freescale Inc, 'S12XE: High Performance 16-bit MCUs for Automotive and Industrial Applications, http://www.freescale.com/webapp/sps/site/prod_summary.jsp?code=S12 XE

[25] Delgado, N., Gates, A. Q. and Roach, S.: 'A Taxonomy and Catalog of Runtime Software-Fault Monitoring Tools', IEEE Transactions on Software Engineering, 2004, 30(12), pp. 859-872.

[26] Nutt, G.J.: 'Tutorial: computer system monitors', IEEE Comput., 1975, 8, (11), pp. 51–61

[27] Plattner, B. (1984) 'Real-Time Execution Monitoring', IEEE Transactions on Software Engineering, 1984, 10(6), pp. 756-764.

[28] Gates, A. Q., Roach, S., Mondragon, O. and Delgado, N. : 'DynaMICs: Comprehensive Support for Run-Time Monitoring', in Havelund, K. and Roşu, G., eds., Proceedings First Workshop on Runtime Verification (RV'2001) (held in conjunction with the 13th Conference on Computer Aided Verification, CAV '01), Paris, France, 23 July, 2001, Electronic Notes in Theoretical Computer Science (ENTCS), Amsterdam, The Netherlands: Elsevier, Vol. 55(2), pp. 164-180.

[29] Mok, A. K. and Liu, G.: 'Efficient Run-Time Monitoring of Timing Constraints', in Proceedings 3rd IEEE Real-Time Technology and Applications Symposium (RTAS '97), Montréal, Canada, 9-11 June, 1997, Washington, DC, USA: IEEE Computer Society Press, pp. 252-262.

[30] Havelund, K. and Roşu, G.: 'Synthesizing Monitors for Safety Properties', in Katoen, J.-P. and Stevens, P., eds., Proceedings 8th International Conference on Tools and Algorithms for the Construction and Analysis of Systems (TACAS 2002) (Held as Part of the Joint European Conference on Theory and Practice of Software, ETAPS 2002), Grenoble, France, 8-12 April, 2002, Lecture Notes in Computer Science (LNCS), London, UK: Springer-Verlag, Vol. 2280, 342-356.

[31] Drusinsky, D.: 'The Temporal Rover and the ATG Rover', in Havelund, K., Penix, J. and Visser, W., eds., Proceedings 7th International SPIN Workshop on SPIN Model Checking and Software Verification, Stanford, California, USA, 30 August - 1 September, 2000, Lecture Notes in Computer Science (LNCS), London, UK: Springer-Verlag, Vol. 1885, 323-330.

[32] Sammapun, U., Lee, I., Sokolsky, O: 'Checking Correctness At Runtime using Real-Time Java', Proceedings of the 3rd Workshop on Java Technologies for Real-time and Embedded Systems (JTRES'05), San Diego, CA, October, 2005.

[33] Lee, I., Kannan, S., Kim, M., Sokolsky, O. and Viswanathan, M.: 'Runtime Assurance Based on Formal Specifications', in Arabnia, H. R., ed., Proceedings International Conference on Parallel and Distributed Processing Techniques and Applications (PDPTA 1997), Las Vegas, Nevada, USA, 30 June - 3 July, 1997, Athens, Georgia, USA: CSREA Press.

[34] Kim, M., Viswanathan, M., Ben-Abdallah, H., Kannan, S., Lee, I. and Sokolsky, O. 'Formally Specified Monitoring of Temporal Properties', in Proceedings 11th Euromicro Conference on Real-Time Systems (Euromicro RTS'99), York, England, UK, 9-11 June, 1999, Los Alamitos, California, USA: IEEE Computer Society Press, pp. 114-122.

[35] Havelund, K. and Roşu, G.: 'An Overview of the Runtime Verification Tool Java PathExplorer', Formal Methods in System Design, 2004, 24(2), pp. 189-215.

[36] El Shobaki, M. and Lindh, L. : 'A Hardware and Software Monitor for High-Level System-on-Chip Verification', in Proceedings IEEE 2001 2nd International Symposium on Quality Electronic Design (ISQED 2001), San Jose, California, USA, 26-28 March, 2001, Washington, DC, USA: IEEE Computer Society Press, pp. 56-61.

[37] MacNamee, C. and Heffernan, D.: 'Emerging on-chip debugging techniques for real-time embedded systems', Computing & Control Engineering Journal, 2000, 11(6), pp. 295-303.

[38] MacNamee, C. and Heffernan, D.: 'Implementation Approaches for Requirements-Based Monitors for Embedded Systems', in Proceedings IEEE IC Test Workshop (ICTW 2004), Limerick, Ireland, 13-14 September, 2004, Limerick, Ireland: E&CE Dept., University of Limerick.

[39] Fryer, R. E. : 'FPGA Based CPU Instrumentation for Hard Real-Time Embedded System Testing', ACM SIGBED Review, 2005, 2(2), pp. 39-42

[40] Fogarty, Padraig. 'Utilizing Multicore Architectures to Enhance Software Verification in Real-Time Embedded Systems' PhD Dissertation, University of Limerick, Limerick, 2013.

[41] Howe, J., King, A.: Logahedra: A new weakly relational domain. In: ATVA, LNCS, vol. 5799, pp. 306-320. Springer (2009)

978-1-4799-8720-7/15 $31.00 © 2015 IEEE

2015 IEEE Computer Society Annual Symposium on VLSI

Statistical Analysis of Resource Usage of Embedded Systems Modeled in EAST-ADL

Raluca Marinescu*, Eduard Paul Enoiu*, Cristina Seceleanu*
*Mälardalen University, Västerås, Sweden, <firstname.lastname>@mdh.se

Abstract—**The growing complexity of modern automotive embedded systems requires new techniques for model-based design that take into consideration both software and hardware constraints, and enable verification at early stages of development. In this context, EAST-ADL has been developed as a domain-specific language dedicated to modeling functional-, software-, and hardware- architecture of automotive systems. This language offers convenient abstractions that support modeling of function, as well as relevant extra-functional properties, like timing and resource usage. These features make it a suitable framework for reasoning about the system's behavior. By providing formal semantics to the EAST-ADL language, as a network of priced timed automata, it becomes possible to reason about feasibility and worst-case resource consumption of the embedded components. In this paper, we show how to analyze such embedded systems modeled in EAST-ADL by using statistical model-checking. We report our experience from applying this approach to an industrial Brake-by-Wire system prototype.**

Index Terms—**embedded systems, EAST-ADL, statistical model-checking, resource usage.**

I. INTRODUCTION

As modern embedded systems contain an increasing number of software and hardware features, significant problems [1] arise in the integration of new sub-systems due to the emergent behavior of the whole system. It is not enough to have correct sub-systems, they must also be properly integrated in a system that meets both its functional and extra-functional requirements, such as real-time performance and resource consumption.

Integration and analysis of extra-functional requirements have been tackled in several domains [2], [3], [4]. Consider, for instance, the automotive domain where a change to a software sub-system or the substitution of a hardware component can affect the system's latency, energy and memory utilization. The modus operandi in industrial practice [1] shows that analyzing resource usage attributes is often done ad-hoc by industrial engineers, by using disjoint specifications. This trend could result in a costly and risky integration phase during the development of embedded systems.

In this context, architectural models that can be introduced earlier in the development process provide a holistic system description that captures the structure of the system, as well as related extra-functional information, e.g., timing properties, triggering annotations, and resource annotations. EAST-ADL [5] is a domain-specific language for modeling software and hardware specifications in the automotive domain. The approach supports modeling of resource usage, which enables the

early analysis of extra-functional requirements like feasibility and worst-case resource consumption of components.

In this paper, we show how efficient verification techniques, like statistical model-checking, can be applied on high-level design artifacts, such as EAST-ADL models, to provide early information on the resource consumption of an automotive embedded system. To achieve this, we propose a methodology that uses our existing tool VITAL [6], in which it is possible to obtain formal models in form of networks of timed automata [7] from EAST-ADL architectural models. Here, we extend such networks with resource annotations based on the information provided in the architectural model, thus creating networks of priced timed automata, which are extensions of timed automata with costs. Consequently, we can employ UPPAAL SMC [8], the UPPAAL extension for statistical model-checking, to analyze the resource usage of automotive systems described in EAST-ADL. The results of the analysis, which we apply on a Brake-by-Wire industrial prototype, provide valuable information on the system's resource-usage prior to the actual implementation.

This paper is structured as follows. We start by briefly presenting the EAST-ADL architectural language and the tools involved in the verification process (i.e., the VITAL tool and the UPPAAL SMC model-checker) in Section II. Our resource usage analysis methodology is described in Section III. In Section IV we show how EAST-ADL functions can be analyzed in terms of resource-usage. Next, in Section V we introduce our case-study, the Brake-by-Wire (BBW) system, and we show how the proposed methodology is applied on this industrial system. We end the paper by discussing related works in Section VI, and by presenting our conclusions in Section VII.

II. PRELIMINARIES

In this section, we present the tools and frameworks used in our analysis methodology.

A. EAST-ADL Language

EAST-ADL [5] is an AUTOSAR[1][9] compatible architectural description language dedicated to the development of automotive embedded systems. The functionality of the system is defined at four levels of abstraction, as follows: (i) the *Vehicle Level*, the highest level of abstraction, describes the electronic features as they are perceived externally, (ii) the

[1]AUTOSAR stands for AUTomotive Open System ARchitecture developed by manufactures as a standard in the automotive domain.

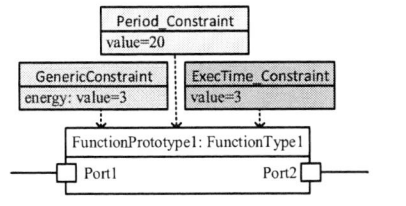
(a) A generic EAST-ADL *FunctionPrototype*.

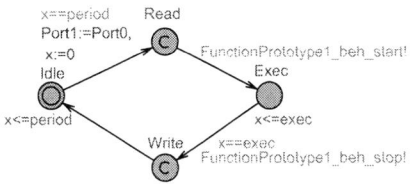
(b) The interface timed automaton.

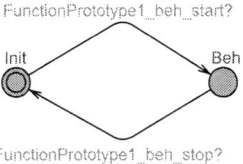
(c) Behavior template.

Fig. 1: The ViTAL transformation.

Analysis Level provides an abstract functional representation of the architecture, (iii) the *Design Level* provides a detailed functional representation of the architecture, together with the allocation of these elements onto the hardware platform, and (iv) the *Implementation Level* provides the implementation of the system using AUTOSAR elements.

At each abstraction level, the system model uses components, each a *FunctionType*, which describe the functional elements of the system. The *FunctionType* has: (i) ports that receive and provide data, respectively, (ii) a trigger, either time-based or event-based, and (iii) an associated behavior. Each of these components is instantiated as one or more of type *FunctionPrototype*, which are connected to provide the system model. The execution of each *FunctionPrototype* is based on the "*read-execute-write*" semantics, and the associated behavior can be defined using different notations and tools (e.g., Simulink or UPPAAL PORT timed automata [10]). The model can be extended with a *GenericConstraint* annotation, which allows the system designer to specify various extra-functional properties, such as energy consumption or memory utilization.

B. The ViTAL tool

VITAL [6] is an analysis tool that enables formal verification of EAST-ADL models based on model-checking techniques. To achieve this, the tool provides a transformation from an architectural model to: (i) UPPAAL PORT timed automata (for component-based verification with the UPPAAL PORT model-checker) [10], and (ii) UPPAAL timed automata (for verification with the UPPAAL model-checker) [6]. In this paper, we will use the latter formalism. Further information on timed automata theory can be found elsewhere [7].

To formally verify that the EAST-ADL model meets its requirements, we represent the architectural elements as UPPAAL timed automata, by an automatic transformation within VITAL. Each *FunctionPrototype* (see Figure 1a) is transformed into a network of two synchronized automata: (i) an interface timed automaton (Figure 1b), dedicated to the elements of the EAST-ADL component interface, and (ii) a behavior timed automaton (Figure 1c), dedicated to the behavior of the EAST-ADL component. For more details we refer the reader to our previous work [6].

The result of this transformation is a network of timed automata that can be analyzed with the UPPAAL model-checker. In this paper, we extend this transformation to also include resource annotations.

C. UPPAAL SMC

UPPAAL SMC [8] is an extension of UPPAAL that enables the analysis of different performance properties of networks of priced timed automata with stochastic semantics. Statistical model-checking generates stochastic simulations to estimate probabilities and probability distributions over time with given confidence levels, so the technique scales better than exact symbolic model-checking.

Priced timed automata (PTA) are extensions of timed automata with cost variables that can evolve at integer rates (also $\neq 1$) and are used in this paper to capture the resource usage. The resource usage is modeled via a function $P : (L \cup E) \to \mathbb{N}$, where L is a finite set of locations and E is the set of edges of the PTA model, which assigns costs to both locations and edges. A network of PTA (NPTA) can be expressed as a composition of n PTA over clocks and actions; the PTA synchronize on send-receive actions (i.e., *send b!* is complementary to *receive b?*) and can use shared variables in guards (boolean conditions that enable the execution of edges).

UPPAAL SMC uses an extension of WMTL [11] to verify properties like:

- *Hypothesis testing*: check if the probability to reach state ϕ within cost $x \leq C$ is greater or equal to a certain threshold p ($Pr(\Diamond_{x \leq C}\phi) \geq p$),
- *Probability evaluation*: calculate the probability $Pr(\Diamond_{x \leq C}\phi)$ for a given NPTA,
- *Probability comparison*: is $P(\Diamond_{x \leq C}\phi_1) > P(\Diamond_{y \leq D}\phi_2)$?

III. METHODOLOGY OVERVIEW OF RESOURCE ANALYSIS

In this section, we propose an approach to resource usage analysis for EAST-ADL with UPPAAL SMC. The methodology presented in this paper is tailored to EAST-ADL, and it contains the following steps (also mirrored in Figure 2):

1) *Model Transformation.* To analyze the EAST-ADL model, we map it to a finite-state formal model suitable for model-checking. For this, we use VITAL to automatically obtain a network of timed automata (see Section II-B).

2) *Resource Annotation.* We annotate the resulting model such that the resource usage information in the EAST-ADL model (as *GenericConstraint*) is expressed in the PTA formalism.

3) *Requirement Formalization.* We formalize the extra-functional properties such that they can be verified with the UPPAAL SMC model-checker.

978-1-4799-8720-7/15 $31.00 © 2015 IEEE
381

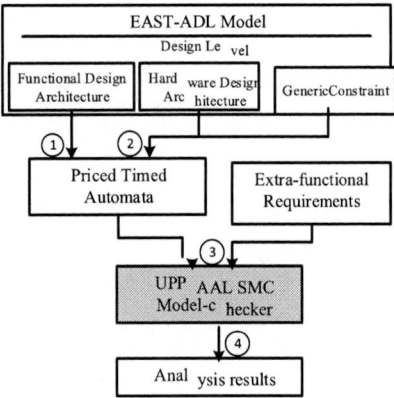

Fig. 2: Methodology overview.

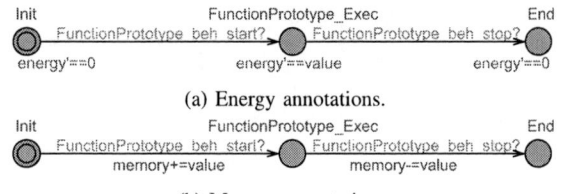

(a) Energy annotations.

(b) Memory annotations.

Fig. 3: The resource annotations.

4) *Analysis Results Generation.* We use UPPAAL SMC to generate analysis results. This model-checker requires as input the resulting model of Step 1 and 2, and the properties expressed in Step 3. Currently, U PPAAL SMC supports queries related to the stochastic interpretation of PTA including the visualization of expressions along the simulated runs.

While UPPAAL SMC is a viable tool for statistical model-checking, it is not directly tailored to analyzing resource usage in EAST-ADL. We demonstrate our solution to this, by discussing the above steps in the following section.

IV. ANALYZING RESOURCE -USAGE OF EAST-ADL FUNCTIONS

A *GenericConstraint* allows the E AST-ADL model to be annotated with different types of resources, like *memoryConsumption, powerConsumption* , *weight, developmentCost*, etc. Resources have different properties that impact the model and the resource analysis. The consumption of a resource c is the accumulated resource usage up to some point in time, calculated based on the rate of consumption over time c'. Based on this, resources can be classified [3] as: (i) continuous ($c' = n$, with $n \in \mathbb{Z} - \{\infty\}$), and (ii) discrete ($c' = 0$ or $c' = \infty$).

In this paper we focus on two types of EAST-ADL resource usage: *energyConsumption*, of a continuous resource, and *memoryConsumption*, of a discrete resource.

A. The Priced Timed Automata Model

To be able to analyze the resource usage of continuous and discrete resources of systems modeled in EAST-ADL, one needs to create the corresponding formal model as a network of PTA. For this, we use the VITAL tool, which provides automatic transformation from the EAST-ADL model to a timed automata model. We extend the resulting formal model with the corresponding resource annotations, to obtain the PTA model that can be used by UPPAAL SMC. Concretely, we extend the timed automata network with a monitor automaton that contains all the resource annotations of the EAST-ADL model, including energy consumption and memory usage.

The monitor is a loop-free PTA that follows the execution of the system, which is achieved through the already existing synchronization channels *FunctionPrototype_beh_start* and *FunctionPrototype_beh_stop*. Assuming an architectural model consisting of only one *FunctionPrototype*, we annotate the monitor as follows:

- With a continuous resource, that is, energy, whose consumption is increasing with time and, in our case, is consumed in the *FunctionPrototype_Exec* location. The rate of consumption is $energy' == value$ (see Figure 3a), where $value$ is provided in the EAST-ADL model;
- With a discrete resource, that is, static memory, which is used instantaneously, so its usage does not depend on time. We encode the memory allocation via the edge assignment *memory+=value*, and memory deallocation via the edge assignment *memory-=value* (see Figure 3b).

B. Resource Analysis

In the following section, we present four types of resource analysis performed on EAST-ADL, using our methodology.

1) Simulation: This technique provides graphical visualization of behavior over a predefined number of runs of the system model. Simulation can be formulated as the property:

$$simulate\ n[bound]\{E_1, .., E_k\}$$

where n is the number of simulations to be performed, $bound$ is the time bound on the simulations, and $E_1, .., E_k$ are the expressions to be monitored.

2) Resource Feasibility Analysis: Feasibility analysis is used to verify if a certain resource usage stays within the available resource amounts provided by the platform. The verification is achieved by examining the probability distribution of a particular resource, which is formulated as a probability evaluation, as follows:

$$Pr[bound](\psi)$$

where $bound$ defines the time bound for the runs, and ψ is either of the form <> q (eventually q), or [] q (always q), where q is a state predicate.

3) Worst-case Resource Consumption Analysis: The worst-case resource consumption analysis returns a path that will eventually reach a certain behavior at a maximal cost. This problem is reduced to maximizing the resource cost function such that the following property is satisfied:

$$E[bound; n](max : expr)$$

978-1-4799-8720-7/15 $31.00 © 2015 IEEE

where *bound* is the time bound, n gives the number of runs, and *expr* is the expression to be evaluated.

4) Resource Leakage: A resource leak, in our case a memory leak, occurs when the system uses a resource that is unable to be released back for later usage (even though it should do so). To this end, we perform a probability evaluation using UPPAAL SMC to detect possible memory leakage.

V. APPLYING SMC ON THE BRAKE-BY-WIRE SYSTEM

In this section, we apply our methodology to provide resource analysis for an industrial prototype system used previously in our research [6], [10], namely the Brake-by-Wire system.

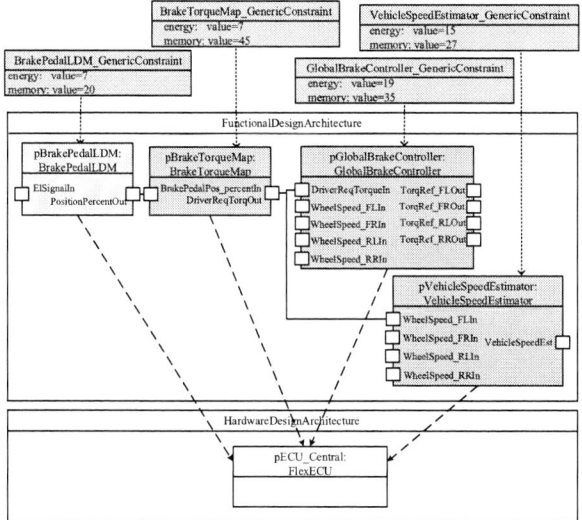

Fig. 4: The EAST-ADL model of the BBW system.

A. Brake-by-Wire System

The Brake-by-Wire (BBW) system is a braking system equipped with an Anti-Lock Braking (ABS) function, and without any mechanical connectors between the brake pedal and the brake actuators. A sensor attached to the brake pedal reads its position, which is used to compute the desired global brake torque. At each wheel, a sensor measures the speed of the wheel, which is used by the ABS algorithm together with the brake torque and the estimated vehicle speed to compute the actual brake torque that will be sent to the actuator. The ABS algorithm computes the slip rate s based on the following equation:

$$s = (v - w \times R)/v$$

where v is the speed of the vehicle, w is the speed of the wheel, and R is the radius of the wheel. The friction coefficient has a nonlinear relationship with the slip rate: when s starts increasing, the friction coefficient also increases, and its value reaches the peak when s is around 0.2. After that, further increase in s reduces the friction coefficient of the wheel. For this reason, if s is greater than 0.2 the brake actuator is released

and no brake is applied, otherwise the requested brake torque is used. Figure 4 presents a part of the BBW system modeled in EAST-ADL, at Design Level, which is allocated to the pedal ECU. This model is extended with annotations for energy and memory consumptions, respectively, as a *GenericConstraint* each.

B. Applying ViTAL on the Brake-By-Wire System

In this section, we apply the proposed methodology and its tool support on the BBW system. We use the VITAL tool to automatically transform the EAST-ADL model of the BBW system into a network of timed automata. The timed automaton modeling the interface behavior of one of the BBW components, the pBrakePedalLDM *FunctionPrototype*, is depicted in Figure 5.

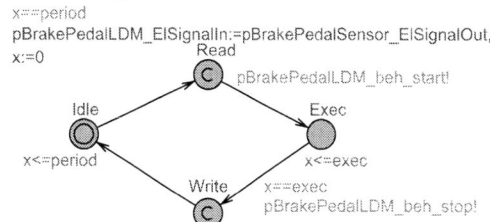

Fig. 5: The interface automaton of the pBrakePedalLDM.

C. The Monitor

Being equipped with the formal model of the BBW system, we can focus on the hardware elements of the platform, and their available resources, as they are represented in EAST-ADL. For the BBW system, we are interested in the *pECU_Central* component shown in the *Hardware Design Architecture*. This ECU is dedicated to the following pedal computational elements (depicted in Figure 4): *pBrakePedalLDM*, *pBrake-TorqueMap*, *pGlobalBrakeController*, and *pVehicleSpeedEstimator*, which are allocated to this ECU.

To analyze the resource consumption of the *pECU_Central* component (actually of the four allocated components), we have implemented a resource monitor depicted in Figure 6. The energy is consumed during the execution of the ECU. For the *pBrakePedalLDM* component, the model is annotated with energy'==6.3+random(1.4), which represents the consumption rate inherited from EAST-ADL $\pm$ 10% tolerance. In a similar manner, the monitor is annotated with its memory usage. Since memory is a static, discrete resource, it is allocated before the component is executed, and it is deallocated at the end of the execution.

D. Analysis of Energy Consumption

Figure 7 depicts the simulation of the energy consumption of the four components allocated on the *pECU_Central* hardware component. Concretely, the plot shows one stochastic simulation carried out for 50 time units, obtained by using the following query on the model: simulate 1[<=50]{energy, memory, t}.

We are also interested in the mean energy consumption and its distribution over runs bounded by a certain value. To obtain

978-1-4799-8720-7/15 $31.00 © 2015 IEEE   383

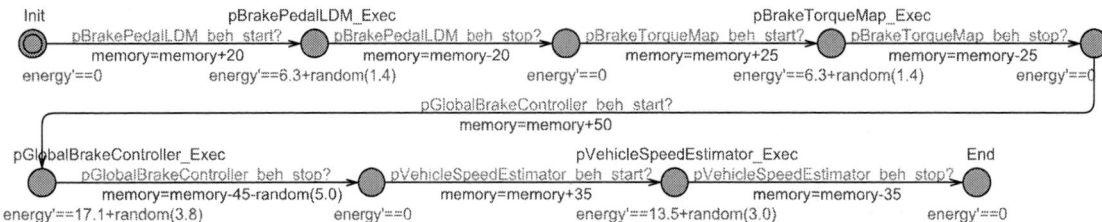

Fig. 6: Allocation of the software elements on the hardware platform.

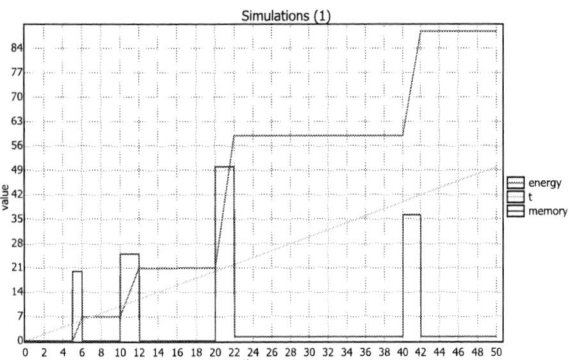

Fig. 7: Simulation of the energy consumption and memory utilization on pECU_Central.

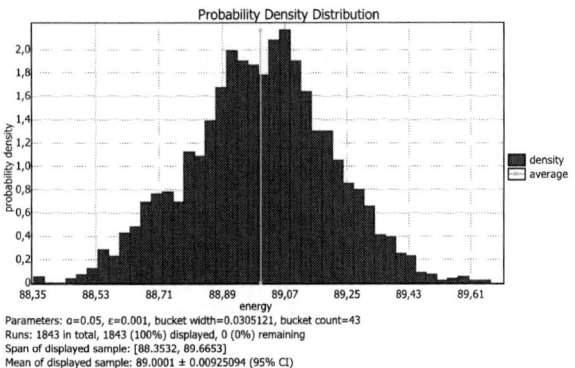

Fig. 8: Estimated energy probability distribution.

this, we check the query Pr[energy<=100](<> Monitor.End), where we assume that the bound of the actual energy is a realistic value that covers the reachable range for all runs. This value is based on the previous simulation of the model. Using UPPAAL SMC, we record the distribution of the energy consumption over 1843 runs, as shown in Figure 8. For increased accuracy, the energy consumption is checked with $\alpha = 0.05$ (the default value) and $\epsilon = 0.001$, which are parameters that improve the precision of the assessed runs. The mean value of the energy consumption is estimated at about 89 energy units, with the minimum value of the energy being approximated at 88.3532 and the maximum value of the energy being 89.6653 for the *pECU_Central* component.

In addition, we evaluate the maximum expected value for the energy. For this, we simulate the system over 2000 runs, trying to maximize the energy consumption, with the query E[t<=50, 2000](**max** : energy). The mean value provided by UPPAAL SMC for the maximum consumption is 89.0019. We note here that 89.0001 is the mean value of the energy distribution.

E. Analysis of Memory Usage

In addition to the energy analysis, in Figure 7 we depict the simulation of the memory usage of the four components allocated onto *pECU_Central*. As observed on the plot, not all the memory is deallocated by the *pGlobalBrakeController*, meaning that we encounter a memory leak. In order to check this with UPPAAL SMC, we use the following query against the model: Pr[t<=50] (<>Monitor.End **and** memory==0.0). The verifier returns the probability that the system will not suffer from a memory leak, in our case this being between 0 and 0.0019.

F. Discussion

In Table I, we present the overall results of our resource analysis methodology applied to the BBW system. This table lists–for each resource, analysis type and property to be checked–the number of states explored during model-checking, together with the time and memory used, as well as the number of simulation runs. Regarding energy analysis, UPPAAL SMC is able to find a solution by exploring 5068 states in 31 ms, using 27164 KB, within a single simulation of the model. We can observe that energy feasibility analysis and worst-case energy consumption analysis are computationally very expensive compared to the other properties. Nevertheless, this shows how capable UPPAAL SMC is in analyzing a realistic industrial system. Our observations could allow an industrial designer to gain deeper understanding in the system's resource behavior, and consequently adjust and optimize both software and hardware designs accordingly.

VI. RELATED WORK

Recently, there has been a growing interest in developing analysis and testing techniques to enhance the adoption of architectural description languages into industrial practice. Related work has also been carried out with respect to the analysis of embedded resources [12], using UML notations intended to complement architecture description languages. For instance, the work of Mallet et al. [13] can been seen

978-1-4799-8720-7/15 $31.00 © 2015 IEEE   384

Resource	Analysis Type	Property	States Explored	Time (ms)	Memory (KiB)	Runs
Energy	*Simulation*	simulate 1[<=50]{energy}	5068	31	27164	1
	Feasibility Analysis	Pr[energy<=100](<> Monitor.End)	7841965	51995	27372	1843
	Worst-Case Energy Consumption	E[t<=50,2000](**max** : energy)	10136000	71277	27364	2000
Memory	*Simulation*	simulate n[<=50]{memory}	5068	47	27064	1
	Memory Leak	Pr[t<=50](<>Monitor.End **and** memory==0.0)	182448	1170	27356	36

TABLE I: Overall Results for Resource Analysis

as one of the major efforts in modeling of embedded systems and their resource usage. Targeting a similar goal, we have already implemented a methodology for formal analysis and verification of EAST-ADL [10]. In addition, in a recent paper we have proposed an extension of EAST-ADL for modeling and analysis of a system's resource-usage [14], which we have analyzed exhaustively with UPPAAL. In this paper, we have improved our previous work by proposing a method that employs statistical model checking for resource usage analysis of EAST-ADL models, in an attempt to increase the verification's scalability. To illustrate our approach, we have applied the proposed methodology on an industrial Brake-By-Wire system.

Other researchers have tried to address the problem of establishing a generic formal foundation for modeling and analysis of resources in embedded systems [15], from low-level code resource estimates [16] to higher level UML and formal approaches [17]. To continue this trend, we have focused on using abstract resource usage information at architectural levels of EAST-ADL, in order to provide to industrial systems designers analysis means for simulating and optimizing the system's overall resource usage.

VII. CONCLUSIONS AND FUTURE WORK

In this paper, we have proposed a methodology for analyzing resource usage for EAST-ADL, using the UPPAAL SMC model-checker. We presented a case-study on which our methodology is applied to transform and analyze an industrial Brake-By-Wire system prototype. We have shown how the initial system can be transformed and annotated using the priced-timed automata formalism, and how the resource-wise behavior is semantically translated and statistically model-checked using UPPAAL SMC.

Regarding the analysis, we have shown how to analyze the usage of resources expressed in EAST-ADL within the priced-timed automata framework. In this setting, we have simulated the energy and memory usage of various BBW components, and have also performed resource feasibility analysis, as well as derive analysis results for worst-case energy consumption and memory leakage. Such results can help a system designer to gain a deeper understanding of the resource behavior of embedded systems modeled in EAST-ADL. As future work, we plan to implement this methodology in the VITAL tool and apply it to other real-world embedded systems.

REFERENCES

[1] B. Graaf, M. Lormans, and H. Toetenel, "Embedded Software Engineering: The State Of The Practice," *IEEE Software*, vol. 20, no. 6, 2003.

[2] J. Mylopoulos, L. Chung, and B. Nixon, "Representing and Using Nonfunctional Requirements: A Process-Oriented Approach," *IEEE Transactions on Software Engineering*, vol. 18, no. 6, 1992.

[3] C. Seceleanu, A. Vulgarakis, and P. Pettersson, "Remes: A Resource Model for Embedded Systems," in *International Conference on Engineering of Complex Computer Systems*. IEEE, 2009.

[4] T. Šimunić, L. Benini, and G. De Micheli, "Cycle-accurate Simulation of Energy Consumption in Embedded Systems," in *Proceedings of the Design Automation Conference*. ACM, 1999.

[5] H. Blom, H. Lönn, F. Hagl, Y. Papadopoulos, M.-O. Reiser, C.-J. Sjöstedt, D.-J. Chen, F. Tagliabò, S. Torchiaro, and S. Tucci, "EAST-ADL: An Architecture Description Language for Automotive Software-Intensive Systems," *EAST-ADL White Paper*, vol. 1, 2013.

[6] R. Marinescu, H. Kaijser, M. Mikučionis, C. Seceleanu, H. Lönn, and A. David, "Analyzing Industrial Architectural Models by Simulation and Model-Checking," in *Third International Workshop on Formal Techniques for Safety-Critical Systems*. Springer, 2014.

[7] R. Alur, "Timed Automata," in *International Conference on Computer Aided Verification*. Springer, 1999.

[8] P. Bulychev, A. David, K. G. Larsen, M. Mikučionis, D. B. Poulsen, A. Legay, and Z. Wang, "UPPAAL-SMC: Statistical Model Checking for Priced Timed Automata," *Workshop on Quantitative Aspects of Programming Languages and Systems*, 2012.

[9] (2014) The AUTomotive Open System ARchitecture (AUTOSAR) Standard. Available from http://www.autosar.org/.

[10] E.-Y. Kang, E. P. Enoiu, R. Marinescu, C. Seceleanu, P.-Y. Schobbens, and P. Pettersson, "A Methodology for Formal Analysis and Verification of EAST-ADL Models," *Reliability Engineering and System Safety*, vol. 120, 2013.

[11] P. Bulychev, A. David, K. G. Larsen, A. Legay, G. Li, and D. B. Poulsen, "Rewrite-Based Statistical Model Checking of WMTL," in *Runtime Verification*. Springer, 2013.

[12] H. H. Ammar, V. Cortellessa, and A. Ibrahim, "Modeling Resources in a UML-Based Simulative Environment," in *International Conference on Computer Systems and Applications*. IEEE, 2001.

[13] F. Mallet, M.-A. Peraldi-Frati, and C. André, "Marte CCSL to execute East-ADL Timing Requirements," in *International Symposium on Object/Component/Service-Oriented Real-Time Distributed Computing*. IEEE, 2009, pp. 249–253.

[14] R. Marinescu and E. P. Enoiu, "Extending EAST-ADL for Modeling and Analysis of System's Resource-Usage," in *Computer Software and Applications Conference Workshops*. IEEE, 2012.

[15] A. Vulgarakis and C. Seceleanu, "Embedded Systems Resources: Views on Modeling and Analysis," in *International Workshop On Component-Based Design Of Resource-Constrained Systems*, July 2008.

[16] J. Muskens and M. Chaudron, "Prediction of Run-Time Resource Consumption in Multi-task Component-Based Software Systems," in *Component-Based Software Engineering*, 2004.

[17] S. Becker, H. Koziolek, and R. Reussnerl, "The Palladio Component Model for Model-Driven Performance Prediction," *Journal of Systems and Software*, 2009.

978-1-4799-8720-7/15 $31.00 © 2015 IEEE

2015 IEEE Computer Society Annual Symposium on VLSI

A Novel Architectural Pattern To Support The Development Of Human-Robot Interaction (HRI) Systems Integrating Haptic Interfaces And Gesture Recognition Algorithms

Giuseppe Airò Farulla, Ludovico O. Russo, Vincenzo Gallifuoco
Politecnico di Torino
Corso Duca degli Abruzzi 24
I-10129 Torino TO, Italy
E-mail: {name.surname}@polito.it

Marco Indaco
Politecnico di Torino
Lero - The Irish Software Engineering Research Centre
Limerick, Ireland
E-mail: marco.indaco@polito.it

Abstract—Haptic and robotic interfaces are recently gaining momentum to be pervasively integrated in modern everyday life. In fact, they can be employed in several different fields, ranging from manipulation of small and dangerous objects to rehabilitation, assistive and service technologies, and are also integrated in mission critical systems. Modern research is rapidly shifting to investigate novel and more intuitive ways of controlling these interfaces. In particular, gesture-based control is one of the most interesting scenario for Human-Robot Interaction (HRI), since we human perceive gestures as a natural way of interaction with the external world. In this work we present a novel architectural pattern, entirely based on the Robotic Operating System (ROS), to support the development of applications and systems where computer vision techniques are applied to control robotic interfaces. As case study, the presented pattern is used to develop and assess the overall PARLOMA system. PARLOMA project aims at developing a system to enable remote communication between deaf-blind subjects. The system is designed to send, remotely and in real-time, messages in tactile Sign Language from a sender to a deaf-blind recipient (or many recipients) by integrating hand tracking and gesture recognition algorithms coupled with bio-mimetic haptic interfaces.

Keywords—Human-Robot Interaction, Gesture Recognition, Haptic Interfaces, Service Robotics

I. INTRODUCTION

Human-Robot Interaction (HRI) is the interdisciplinary study of dynamic interactions between humans and robots. It deals with designing and implementing systems that can be easily and intuitively controlled by humans. In the last years, technological advances in modern low-cost sensing technologies have pushed the research on HRI toward new scientific challenges devoted to develop more natural and intuitive interaction techniques. Interaction requires some sort of communication between humans and robots [1]. Communication between a human and a robot may take several forms, as it typically depends on the desired application, ranging from mobility assistance to physical manipulation and social interaction. The term Socially Interactive Robotics (SIR) was used in [2] to describe robots whose main task was some form of interaction with human beings; [3] defines social robots as agents embedded in a society of robots or humans, capable

of engaging social interactions through communication and learning.

Scientific research in HRI comes from a variety of fields including electronics, mechanics, robotics, natural language understanding, computer vision and assistance. Assistive Robotics (AR) has largely referred to the development of (often bio-inspired) robots to give aid or support to a human user, especially in the case of users with physical disabilities [4]. Research into AR includes rehabilitation robots [5], [6], [7], mobility aides [8], [9], [10] and companion robots and platforms to be used in hospitals and homes [11], [12], [13]. Recently, research is also investigating psychological aspects behind HRI patterns [14], to assess the importance of robots involvement (with particular regards to humanoid and bio-mimetic ones) in bringing assistance to children, elderly people and disabled [15].

Moreover, the usage of humanoid robotic systems is envisaged also in many other fields, ranging from tele-presence to tele-rehabilitation and tele-manipulation. In this context, one of the most interesting research area is investigating gesture recognition systems intended as part of the control architecture for an humanoid robot [16]. Research on computer vision applied to robotics investigates the development of algorithms and techniques to control robots through simple gestures [17]. Computer vision holds great promise for helping both disabled persons and individuals at unease with technology to interact with machines and more in particular with robotic interfaces [18].

Usually, human beings interact with machines using cumbersome interfaces (e.g., a joystick, a mouse) that may limit the naturalness of interaction with robot. This is especially true in the field of assistive technologies. Elderly and disabled people may experiment serious difficulties in interacting with robots by using a joystick. On the other hand, human beings use gestures to interact with surroundings since the earliest stages of their development. From there the interest in researching on Hand Gestures Recognition (HGR). The appeal of HGR-based interfaces derives from their flexibility and customizability. The recent availability of innovative, low-cost, off-the-shelf input devices (e.g., structured light cameras) have enabled

978-1-4799-8720-7/15 $31.00 © 2015 IEEE

access to a finer-grained set of input data that can be exploited to enhance HGR algorithms. Nevertheless, hand detection is an extremely complex problem due to variable hand shape, lighting condition, skin color and hand size. Major tracking challenges come from occlusions, cluttered environments and rapid motions causing motion blur. The problem is even more complex when investigating marker-less unaided hand tracking. On the other hand, for any robot that is designed to interact in a natural manner with humans an unconstrained hand-tracker is necessary, and it should not require any special initialization procedure, since they may represent an hindrance for categories of users with special needs. Addressing these challenges is of relevance when developing tools and softwares that are supposed to be reliable and dependable [17].

Therefore the development of advanced applications combining robotic interfaces controlled by hand gestures requires the integration of several heterogeneous devices such as cameras, multiple sensors and robotic interfaces. As a conclusion, an architectural pattern to simplify HRI application development which involves the integration of different technologies is of utmost importance.

In this paper, we propose a novel architectural pattern to support application development combining heterogeneous nodes (intended as input cameras, computational resources, robotic interfaces) that interoperate and easily exchange real-time information, commands and status updates. The full system is based on the Robot Operating System (ROS). ROS is nowadays becoming the de-facto standard for robotic software development. In fact, as better explained in SectionII, the ROS framework makes it easy to write modular and distributed applications.

As case study, the proposed pattern has been adopted to develop the PARLOMA system, which aims to develop a remote communication system for deaf-blind people based on a single depth camera and a bio-mimetic robotic hand.

The remaining of the paper is organized as follows: Section II presents ROS, Section III describes the architectural pattern, Section IV presents the PARLOMA project as our case study, Section V presents experimental results and finally Section VI concludes the paper and presents planned future work.

II. ROS

ROS [19] is an open-source, meta-operating system for robot software development, providing a collection of packages, software building tools and an architecture for distributed inter-process and inter-machine communication. It can be thought as a collection of software frameworks specifically designed to help developers in writing software for robotic interfaces [20]. For the functionalities it offers (e.g., hardware abstraction, low-level device control, implementation of commonly used functionality, message-passing between processes, package management) it can be assimilated to a a regular operating system [21].

ROS-based software is composed of nodes, messages, topics, and services. *Nodes* are processes, i.e., resources that perform computation and tasks. Any real system is typically comprised of many nodes, which communicate in a peer-to-peer way. Communication is ensured even when nodes

are physically executed in different and distributed machines. Nodes communicate with each other by passing messages.

Messages are modeled as data structures, containing data typed as primitive types (e.g., integer, boolean) as well as arrays of primitive types and constants. *Topics* are named buses, identified by a string, over which nodes exchange messages. Topics have anonymous publish/subscribe semantics: a node sends a message by publishing it to a given topic; another node that is interested in a certain data has just to subscribe to the appropriate topic. There may be multiple concurrent publishers and subscribers for a single topic, and a single node may publish and/or subscribe to multiple topics. In general, publishers and subscribers are not aware of each others existence.

Finally, *services* are thought to deliver broadcast synchronous transactions, which can simplify the design of some nodes. Services are represented by a name and a pair of messages: one for the request and one for the response.

The whole ROS framework is composed of three main macro components:

1) ROS client library implementations;
2) language- and platform-independent tools used for building and distributing ROS-based software;
3) packages containing application-related code which uses one or more ROS client libraries. These packages implement commonly used functionality and applications such as hardware drivers, robot models, datatypes, planning, perception, simultaneous localization and mapping, simulation tools.

Both the language-independent tools and the main client libraries (developed to allow compatibility with C++, Python, and LISP code) are released as open source software.

III. THE PROPOSED ARCHITECTURAL PATTERN

We propose an architectural pattern to support the development of HGR-based HRI systems leveraging on all the functionalities offered by ROS. Our proposal is organized in three blocks: (i) *Input Module*, responsible for acquiring frames from the input camera, tracking hands and recognizing gestures; (ii) gesture conversion and transmission; and finally the (iii) *Output Module*, responsible for gesture synthesis through an haptic interface. This architectural pattern is depicted in Figure 1. The *Input Module* aims at gathering data from a depth camera and at recognizing gestures performed by the user. Fast and reliable remote transmission is ensured by the ROS framework. The *Output Module* aims at controlling the output robotic and haptic interface(s). The *Gesture Converter* node offers the desired abstraction, since it generates specific commands to proper control various and several robotic interfaces.

In addition, since ROS provides hardware abstraction for machines executing nodes and it implements a distributed paradigm, remote communication is achieved in a simple manner, and our proposal is ready to control different robotic interfaces and input cameras even in the case where multiple I/O devices are connected. In the following, our proposal will be detailed.

978-1-4799-8720-7/15 $31.00 © 2015 IEEE

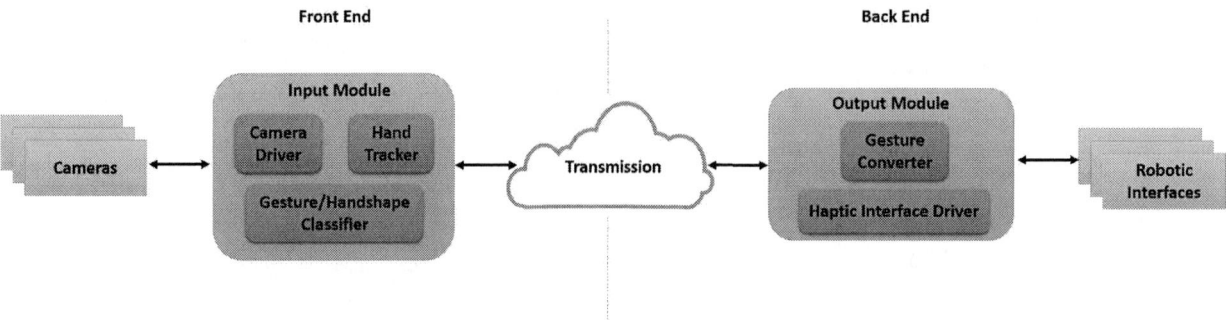

Fig. 1: The Pipeline. The *Input Module* is in charge of performing hand-shape recognition using the depth image of the signer. It is composed of three ROS nodes and the depth camera itself. The recognized gesture is sent over the network to the *Output Module*, which is composed of a robotic interface and two ROS nodes.

A. Input Module

The *Input Module* is in charge of extrapolating and processing in real-time 3D information captured by the input camera to recognize the gesture performed by the user. It consists of three ROS nodes, namely the *Camera Driver* node, the *Hand Tracker* node and the *Gesture/Handshape Classifier* node.

The *Camera Driver* node reads the image stream from the input camera and publishes it within a ROS topic. In our proposed solution, this node is thought to be as generic as possible, to support video streams coming from a wide range of sensors (e.g, a single RGB camera, a depth camera, a multi-camera system, etc.) and of different types (e.g., 3D information, RGB, grayscale, etc.), even if successive nodes use only depth information for reliability.

The *Hand Tracker* node computes the hand joints' positions from depth images and can be composed of different sub modules depending on the implemented computer vision technique.

The *Handshape/Gesture Classifier* node is devoted to classify and recognize gestures the user performs in front of the input camera, and is fed on inputs from the previous nodes.

B. Output Module

The *Output Module* is in charge of reproducing the recognized hand-shape by using the output interface(s). This module is composed by two main ROS nodes, namely the *Gesture Converter* and the *Haptic Interface Driver* one. Our architecture allows the system to have different output devices connected, enabling one user to simultaneously control many remote interfaces. In fact, several interfaces, if available, can receive the same inputs and perform the same gesture. Nodes are hereby detailed:

- the *Gesture Converter* node is in charge of converting recognized gestures in generic poses (i.e., intermediate representations that do not target a specific robotic interface). This is achieved through a pre-compiled dictionary of possible poses that associates gestures as received from the network with a corresponding set of commands;

- the *Haptic Interface Driver* node is devoted to control the robotic hand with specific commands depending upon the employed robotic interface. It receives as input a set of commands from the *Gesture Converter* node and performs specific algorithms of inverse kinematics and collision avoidance. Collision avoidance is of extreme importance for mission critical tasks, i.e., when there is the necessity that the robotic interface keeps working for a long time without errors and without getting entangled. Behavior of the robotic interface (e.g., allowed movements and working area) is described in a XML file that the driver takes as input and parses to produce proper commands for the given interface. Leveraging on such information, this node discards input stimuli related to tasks and movements that the robotic interface cannot accomplish due to mechanical constraints.

The *Haptic Interface Driver* hides physical details of the attached robotic interface to the other nodes. As already stated, the proposed pattern also enables one-to-many communication. In this case, multiple *Haptic Interface Driver* nodes need to be instantiated according to the different simultaneously-operated actuators.

IV. CASE STUDY

This Section illustrates how the proposed pattern has been successfully employed to develop the PARLOMA system.

A. The PARLOMA project

PARLOMA [22] aims at developing a remote communication system (i.e., a telephone) for deaf-blind people. It is specifically thought for people who use any tactile Sign Language (t-SL, intended as tactile-based transpositions of a Sign Language) as their main communication system. t-SL requires the two interlocutors to be in the same location (otherwise tactile exploration cannot happen) and forces pairwise communication (i.e., one-to-many communication is not possible), causing severe limitations to the communication among deaf-blind people.

The development of a remote communication system specifically thought for the needs of deaf-blind individuals (and

978-1-4799-8720-7/15 $31.00 © 2015 IEEE

their relatives) is envisaged to dramatically improve their social inclusion and participation in active society, guaranteeing a massive access to information and to interaction with the community.

PARLOMA integrates haptic and robotic interfaces together with marker-less hand-tracking algorithms and allows reception and remote reproduction of t-SL [23]. To transmit signs remotely, reliably and in real-time, PARLOMA was developed to track a single signer's hand and to reproduce the performed hand-shape by using one 3-D printed bio-mimetic robotic hand. The only input system is composed of a RGB-D camera; as stated in Section III, only depth information is used (RGB stream is discarded, as too influenced by varying lighting and environmental conditions). When the system boots up, the user is required to wave his right hand in front of the camera to initialize the hand tracking algorithm; then he can freely sign, as no calibration task is needed. Depth stream from the camera is processed to reconstruct the handshape and recognize the gesture, which is later sent to the robotic hand. Finally, the robotic hand performs the received gesture, mimicking the behavior of the user hand. In this way, a deaf-blind recipient can place his hands over the robotic one, explore it and understand the meaning of the performed gesture. By collecting all the gestures, the whole message is received. In the following, the development of the entire system based on our novel architectural pattern is presented.

B. The Overall System Implementation

We have developed an implementation of the proposed pattern using C++ and Python programming languages. Nodes, in particular, are distributed in two machines:

1) a laptop (a Macbook PRO late 2011, mounting an Intel Core i7@2.7GHz CPU, 4GB of RAM and an Intel HD Graphics 3000 512MB GPU), which is connected to an Asus XtionPRO RGB-D camera and is in charge of managing the *Input Module* and of transmitting the recognized gestures to the Internet;

2) a Raspberry Pi, which is connected to the Internet through an USB-WiFi dongle and linked to the robotic hand; it is in charge of managing the *Output Module*.

We customized the *Hand Tracker*, *Handshape/Gesture Classifier* and the *Haptic Interface Driver* nodes for the needs of the case study.

The *Hand Tracker* node is a modified implementation of the algorithm proposed in [24], where the authors propose a full-DoF appearance-based hand tracking approach [25] by employing uses a Random Forest (RF) classifier [26].

This node extracts the hand skeleton and publishes the 3D position of each joint of the reconstructed skeleton with respect to the camera reference frame. It accomplishes three main tasks performed sequentially on each incoming frame, as shown in Figure 2. In detail, the first task is accomplished by the *Hand Segmenter* block, which aims to separate the hand (i.e., foreground) from the background. The second task is executed by the *Hand Labeller* block, which runs an appearance-based algorithm to recognize different parts of the hand; this is done to isolate hand joints. The *Joints Position Estimator* block

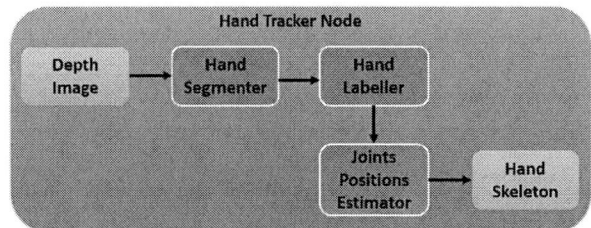

Fig. 2: The *Hand Tracker* node. This node computes the hand joints' positions from depth images. It consists of three sequential tasks. The first one segments the hand (foreground) from the background. The second one classifies each pixel of the foreground to the hand region that it should belong to. The last one extrapolates the regions' centroids corresponding to the joints' positions.

approximates the joints' 3D positions, starting from the input depth measurements and from the hand skeleton as estimated from the first block. In fashion of what done in [25], in our approach a RF classifier is used to label pixels of the depth image according to the area of the hand they belong to. Then each area is processed to find the position of its corresponding joint. At the end of the clustering process, the algorithm outputs the 3D position of each joint of the hand (except when some joint is not visible, for instance in the case of self-occlusions). Our labeling algorithm can recognize 22 different parts of the hand, namely the palm, the wrist and four joints for each of the fingers. The joints' positions are approximated applying the Mean Shift (MS) [27] clustering algorithm on the hand sub-parts. This approach shows promising results: experiments with real-world depth map images demonstrate that it can properly label most parts of the hand in real-time with the already stated hardware configuration.

When hands regions have been processed and labeled, there is a successive task of the gesture recognition. This is necessary because outcomes from the labeling procedure tend to be noisy, thus they cannot be used to directly control any robotic interface, especially when reliability is of utmost importance for the intended task.

The *Handshape/Gesture Classifier* node is devoted to classify and recognize the hand gesture the user is making. This second classification is not performed directly on the array of joints' positions, but rather on a pattern containing the joint-to-joint Euclidean distances for all pairs of joints of the hand, in fashion of [22]. Then, a second RF classifier is used to evaluate the probability for a pose to actually reproduce one of the accepted gestures. The gesture is set as recognized and thus sent over the network only if the posterior probability associated with that hand-shape is above a p_{th} threshold and if it is recognized over N consecutive frames. In our experiments, we set $p_{th} = 0.3$ and $N = 5$.

Finally, the hand-shape is encapsulated in a network message sent to the *Output Module*. Secure and lossless remote communication is guaranteed by the ROS communication libraries (secure socket layer SSL).

978-1-4799-8720-7/15 $31.00 © 2015 IEEE

The *Haptic Interface Driver* is in charge of controlling the robotic hand on the basis of the input it receives. This node produces a serial message for each input it receives. Each message is composed of an integer activation code, followed by the list of fingers positions (i.e, valid positions get an integer value between 0 and 180).

V. EXPERIMENTAL RESULTS

This Section presents the results of the performed experiments. We have tested the transmission efficiency of the whole system (Section V-A) by measuring the loss of information during the recognition and reproduction tasks in a test-case scenario.

These experiments focused on hand-shapes, i.e., static configurations of the hand. We choose finger-spelling as the hand-shapes source. Finger-spelling consists of spelling complex language words using the manual alphabet. This choice because the usage of finger-spelling in the t-SL lexicon is typically wide. In addition, our main concern was to validate the transmission pipeline in its entirety to prove the efficacy of our proposed pattern.

A. Transmission Evaluation

Seven subjects with no previous knowledge about the project and few expertise in LIS (four men, three women, mean age 30 years, range ± four years) volunteered to test our application together with a LIS signer (female, 24 years old). Since the status of the project is still in its early development stages, the test was not assessed by deaf-blind users.

Each experiment was performed as hereby described. A list of 125 hand-shapes (LoS) to be reproduced was randomly generated at run-time. The LoS appeared on a monitor nearby the LIS expert signer. She reproduced in front of the input camera each hand-shape with her right hand. To simulate a real-use case scenario, the LIS expert was placed in unfavorable lighting conditions, with moving people behind. Recognized hand-shapes were saved for successive processing in a dedicated list (SRI).

To simulate remote communication via the Internet, t Raspberry Pi receiving the ROS messages was placed in a different room in the same building. It continuously decoded received messages and generated the commands needed to control the hand. The task for the subject who was in this room (volunteers were introduced one by one) was to visually recognize the hand-shape actually performed by the robotic hand. The hand reproduced each hand-shape for five seconds. After this period of time, it returned to the rest position (with all the fingers opened) before the next sign was reproduced. Answers given by the volunteers were collected by an experimenter and recorded and saved in a separate list (SRS).

The experiment lasted for approximately 10 minutes per person. At the end, the experimenters asked the volunteers for their comments, especially asking for particular difficulties that they experienced and their opinions on the usability of the hand for their purposes. No major problem has been highlighted in this phase. On the other hand, as a consequence of the constrained number of degrees of freedom actually employed in the robotic hand, few volunteers reported minor difficulties in discriminating similar signs when more precise subtle movements (e.g., little bending) would have been useful for signs' correct identification.

During the experiments, all possible feedback was collected and stored. Pipeline latency (e.g., delay from gesture acquisition to its reproduction) was never higher than one second, thus demonstrating the feasibility of a real-time remote communication.

Overall collected results (summarized in Figure 3) demonstrate the feasibility of our system. In total, 875 hand-shapes were performed by the hand (125 hand-shapes per seven volunteers). The average accuracy of the acquisition module was 88.14%; however, few errors reported for the *Input Module* were actually caused by the expert in LIS performing the wrong hand-shape. No errors where registered for the transmission and conversion systems. On 82.78% of occasions, the signs were correctly recognized by the volunteers and the total accuracy of the system was thus 73.32%.

Achieved results along the whole pipeline are illustrated in Fig. 3, which highlights:

- *Recognition Efficiency*, referring to the percentage of hand-shapes correctly recognized by the *Input Module*. This comparison was to estimate the effectiveness of the recognition module. Here, we have noticed that errors were mainly due to classification errors, finger occlusions occasionally deceiving the recognition algorithm and noisy data from the input camera;

- *Reproduction Efficiency*, referring to the percentage of hand-shapes correctly recognized by the subjects. Here, we have noticed that errors were mainly due to the fact that volunteers tend to confuse an hand-shape for a similar one, and we are confident that this will not influence communication for proficients in t-SL.

- *Transmission Efficiency*, referring to the efficiency of the whole experimental apparatus. The accuracy of the entire pipeline from the hand-shape produced by the signer to the visual recognition by the subjects is significantly high.

VI. DISCUSSION AND CONCLUSIONS

In this paper we have presented a novel architectural pattern supporting the development of HRI applications based on haptic and robotic interface controlled by human gestures. Basic idea is to support developers of robotic applications by giving them open source resources that they can freely use and easily integrate, while only taking care of implementing custom hand tracking algorithms (related to the application needs) and specific drivers for I/O interfaces. The pattern has been used to implement the PARLOMA system. Future works will deal with the extension of the pattern to assist the development of hand tracking algorithms.

ACKNOWLEDGEMENT

This research was partially supported by the "Smart Cities and Social Innovation Under 30" programme of the Italian Ministry of Research and Universities through the PARLOMA Project (SIN_00132). This work has been partially supported by CINI AsTech Lab.

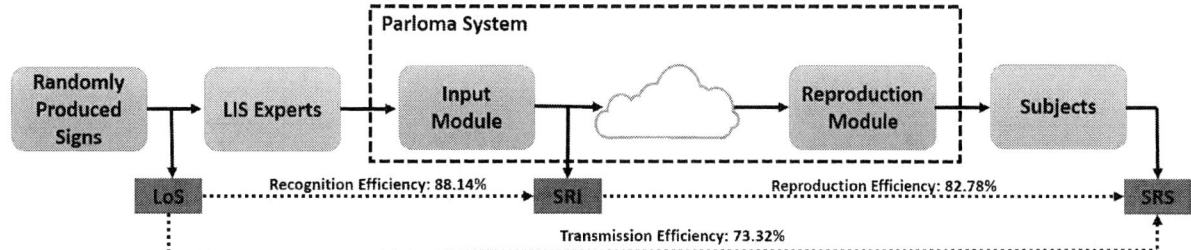

Fig. 3: Sign recognition and reproduction efficiency along the pipeline of the experimental apparatus.

REFERENCES

[1] M. A. Goodrich and A. C. Schultz, "Human-robot interaction: a survey," *Foundations and trends in human-computer interaction*, vol. 1, no. 3, pp. 203–275, 2007.

[2] T. Fong, I. Nourbakhsh, and K. Dautenhahn, "A survey of socially interactive robots," *Robotics and autonomous systems*, vol. 42, no. 3, pp. 143–166, 2003.

[3] K. Dautenhahn and A. Billard, "Bringing up robots orthe psychology of socially intelligent robots: From theory to implementation," in *Proceedings of the third annual conference on Autonomous Agents.* ACM, 1999, pp. 366–367.

[4] D. Miller, "Assistive robotics: An overview," in *Assistive Technology and Artificial Intelligence*, ser. Lecture Notes in Computer Science, V. Mittal, H. Yanco, J. Aronis, and R. Simpson, Eds. Springer Berlin Heidelberg, 1998, vol. 1458, pp. 126–136. [Online]. Available: http://dx.doi.org/10.1007/BFb0055975

[5] L. W. Forrester, A. Roy, R. N. Goodman, J. Rietschel, J. E. Barton, H. I. Krebs, and R. F. Macko, "Clinical application of a modular ankle robot for stroke rehabilitation," *NeuroRehabilitation*, vol. 33, no. 1, pp. 85–97, 2013.

[6] A. Pennycott, D. Wyss, H. Vallery, V. Klamroth-Marganska, R. Riener *et al.*, "Towards more effective robotic gait training for stroke rehabilitation: a review," *J Neuroeng Rehabil*, vol. 9, p. 65, 2012.

[7] R. C. Loureiro, W. S. Harwin, K. Nagai, and M. Johnson, "Advances in upper limb stroke rehabilitation: a technology push," *Medical & biological engineering & computing*, vol. 49, no. 10, pp. 1103–1118, 2011.

[8] J. Schneider, W. Stork, S. Irgenfried, and H. Worn, "A multimodal human machine interface for a robotic mobility aid," in *Automation, Robotics and Applications (ICARA), 2015 6th International Conference on.* IEEE, 2015, pp. 289–294.

[9] S. Takahara and S. Jeong, "Prototype design of robotic mobility aid to assist elderly's standing-sitting, walking, and wheelchair driving in daily life," in *Control, Automation and Systems (ICCAS), 2014 14th International Conference on.* IEEE, 2014, pp. 470–473.

[10] M. M. Martins, C. P. Santos, A. Frizera-Neto, and R. Ceres, "Assistive mobility devices focusing on smart walkers: classification and review," *Robotics and Autonomous Systems*, vol. 60, no. 4, pp. 548–562, 2012.

[11] C. A. A. Calderon, E. R. MOHAN, and B. S. NG, "Development of a hospital mobile platform for logistics tasks," *Digital Communications and Networks*, 2015.

[12] I. Baroni, M. Nalin, P. Baxter, C. Pozzi, E. Oleari, A. Sanna, and T. Belpaeme, "What a robotic companion could do for a diabetic child," in *Robot and Human Interactive Communication, 2014 RO-MAN: The 23rd IEEE International Symposium on.* IEEE, 2014, pp. 936–941.

[13] M. Nalin, I. Baroni, A. Sanna, and C. Pozzi, "Robotic companion for diabetic children: emotional and educational support to diabetic children, through an interactive robot," in *Proceedings of the 11th International Conference on Interaction Design and Children.* ACM, 2012, pp. 260–263.

[14] P. H. Kahn Jr, T. Kanda, H. Ishiguro, B. T. Gill, S. Shen, H. E. Gary, and J. H. Ruckert, "Will people keep the secret of a humanoid robot?: Psychological intimacy in hri," in *Proceedings of the Tenth Annual ACM/IEEE International Conference on Human-Robot Interaction.* ACM, 2015, pp. 173–180.

[15] P. H. Kahn Jr, T. Kanda, H. Ishiguro, N. G. Freier, R. L. Severson, B. T. Gill, J. H. Ruckert, and S. Shen, "robovie, you'll have to go into the closet now: Children's social and moral relationships with a humanoid robot," *Developmental psychology*, vol. 48, no. 2, p. 303, 2012.

[16] A. Aly and A. Tapus, "Gestures imitation with a mobile robot in the context of human-robot interaction (hri) for children with autism," in *3rd Workshop for Young Researchers on Human-Friendly Robotics, Tübingen, Germany*, 2010.

[17] J. P. Wachs, M. Kölsch, H. Stern, and Y. Edan, "Vision-based hand-gesture applications," *Communications of the ACM*, vol. 54, no. 2, pp. 60–71, 2011.

[18] R. Manduchi and J. Coughlan, "(computer) vision without sight," *Communications of the ACM*, vol. 55, no. 1, pp. 96–104, 2012.

[19] "Ros (robot operating system)," http://www.ros.org, accessed on 03/23/2015.

[20] M. Quigley, K. Conley, B. Gerkey, J. Faust, T. Foote, J. Leibs, R. Wheeler, and A. Y. Ng, "Ros: an open-source robot operating system," in *ICRA workshop on open source software*, vol. 3, no. 3.2, 2009, p. 5.

[21] "Ros introduction," http://wiki.ros.org/ROS/Introduction, accessed on 03/23/2015.

[22] G. Airò Farulla *et al.*, "Real-time single camera hand gesture recognition system for remote deaf-blind communication," in *Augmented and Virtual Reality*, ser. Lecture Notes in Computer Science, L. T. De Paolis and A. Mongelli, Eds. Springer International Publishing, 2014, pp. 35–52. [Online]. Available: http://dx.doi.org/10.1007/978-3-319-13969-2_3

[23] L. O. Russo *et al.*, "PARLOMA - A Novel Human-Robot Interaction System for Deaf-blind Remote Communication," *International Journal of Advanced Robotic Systems*, (In Press).

[24] C. Keskin, F. Kıraç, Y. E. Kara, and L. Akarun, "Real time hand pose estimation using depth sensors," in *Consumer Depth Cameras for Computer Vision.* Springer, 2013, pp. 119–137.

[25] V. Rodriguez-Galiano, B. Ghimire, J. Rogan, M. Chica-Olmo, and J. Rigol-Sanchez, "An assessment of the effectiveness of a random forest classifier for land-cover classification," *ISPRS Journal of Photogrammetry and Remote Sensing*, vol. 67, pp. 93–104, 2012.

[26] L. Breiman, "Random forests," *Machine learning*, vol. 45, no. 1, pp. 5–32, 2001.

[27] D. Comaniciu and P. Meer, "Mean shift: A robust approach toward feature space analysis," *Pattern Analysis and Machine Intelligence, IEEE Transactions on*, vol. 24, no. 5, pp. 603–619, 2002.

978-1-4799-8720-7/15 $31.00 © 2015 IEEE

2015 IEEE Computer Society Annual Symposium on VLSI

TSV Placement and Core Mapping for 3D Mesh Based Network-on-Chip Design Using Extended Kernighan-Lin Partitioning*

Kanchan Manna[1], Vadapalli Shanmukha Sri Teja[2], Santanu Chattopadhyay[3] and Indranil Sengupta[4]

[1]School of Information Technology, [2,3]Dept. of Electronics & Elect. Comm. Engg. and [4]Dept. of Computer Science

Indian Institute of Technology Kharagpur, India - 721 302.

e-mail: {[1]kanchanm@sit, [3]santanu@ece, [4]isg@cse}.iitkgp.ernet.in, [2]vvssteja@gmail.com

Abstract—Three-dimensional (3D) Network-on-Chip (NoC) based designs can utilize communication in vertical dimension to reduce distance between cores. Vertical connections are best implemented using Through-Silicon-Via (TSV). However, TSV geometry restricts the number of 3D routers in any layer of the die. This work proposes a strategy to select the TSV positions. This has been augmented by developing a core mapping procedure based on the Kernighan-Lin graph bi-partitioning algorithm, improved via an iterative improvement phase. The overall approach shows promising results compared to the existing mapping and TSV placement algorithms.

I. INTRODUCTION

Emerging 3D integration of VLSI chips can build complex and highly connected systems. Several silicon-dies are stacked and connected via vertical interconnections. Length, delay and power consumption between vertical and horizontal interconnects are typically asymmetric in nature [1], vertical interconnects outperforming horizontal ones.

To design a 3D-NoC based system, the routers should have the ability to support 3D layout. The easiest way to extend a 2D router to 3D is to add two more ports to connect two additional neighbours in the vertical direction. The communication channel in the vertical direction of an extended router can be implemented by several techniques, such as, wire bonding, micro-bump, contact-less interconnection and Through-Silicon-Via (TSV) [2]. Among them, TSV is the most viable solution due to its low latency and low power consumption [1]. However, the TSV overhead such as area, manufacturing cost, routing congestion and yield loss, can increase significantly, with increase in the number of TSVs [1], [3]. Moreover, as per ITRS, the maximum number of TSVs in a high-performance 3D chip is about 1,000 in 2012 and expected to increase further by 1,000, every year [4]. Therefore, the TSVs are expected to be limited in number. The stringent constraints on the number of TSVs as well as their position can make 3D-NoC-routers heterogenous. Some routers in a layer can have TSVs whereas, others do not. A minimum distance has to be maintained between two adjacent 3D-routers to take care of the TSV geometry. This can increase the average distance among the cores and reduce the bisection width, compared to a fully connected 3D-NoC.

Degradation of these parameters have negative impact on the overall system performance. A smart *application mapping* strategy determining the association of cores with routers, together with proper placement of TSVs can play a major role in mitigating the problem.

The mapping problem is NP-hard [5]. Several heuristic methods have been developed to solve it. A detailed survey of these approaches can be found in [6].

This work explores the Kernighan-Lin (KL) graph partitioning approach [7] to solve the combined problem of TSV placement and mapping for 3D-NoC design. The classical KL algorithm has been augmented with an iterative improvement step. It may be noted that the KL-partitioning strategy has been used in [8] for two-dimensional NoC design. The present work extends it to three-dimensional NoC design, integrating the TSV placement problem with it. TSVs are placed by detailed consultation with the application. As a result, it evolves as a complete solution for the 3D-NoC design problem and also incorporates several improvement operations on the initial mapping solution. To the best of our knowledge, this is the first attempt to solve the TSV placement problem together with mapping strategy using Kernighan-Lin partitioning algorithm. The overall problem can be stated as follows.

To design a 3D-NoC based system for an application with constraints on the number of TSVs in each layer and distance between two adjacent TSVs. The association of cores and TSVs with routers have to be determined minimizing the overall communication cost.

The paper is organized as follows. Section II surveys the related works. The architecture and its corresponding routing algorithm have been described in Section III. The proposed strategy for application mapping and TSV placement has been described in Section IV. Experimental results have been presented in Section V. Section VI concludes the work.

II. RELATED WORKS

In [1], authors have presented a study on system performance, number of TSV usage between layers and their configuration. System performance has been measured based on three types of TSV configurations, such as full, quarter and one-eight. A trade-off has been inferred between system performance and manufacturing cost. Another study is based on sharing the TSVs among the neighbouring routers in each

---

*This work is partially supported by the Department of Science and Technology, Govt. of India (SB/S3/EECE/058/2013; dt. 26-08-13).

978-1-4799-8720-7/15 $31.00 © 2015 IEEE

392

layer of 3D-NoC [9]. Neighbouring routers share the TSVs in time division multiplexed fashion. Here, a set of four routers are considered as neighbours of each other. In [10], four routers are grouped into a single virtual group. For each group, a single TSV is allocated. To transfer data in vertical direction, a router can use TSV either in its group or the neighbouring group, based on the current state.

In [11], authors have presented a study that uses serialization of TSVs to reduce their number. TSV-virtualization can also reduce the number of TSVs used in 3D-NoC, as has been discussed in [12]. Here, the authors have proposed TSV-virtualization scheme for multi-protocol-interconnect in 3D-ICs. In this approach, TSVs are clocked at a much higher rate than conventional intra-layer links.

The present work has chosen partially-connected-3D-mesh [13] as the target architecture for mapping an application. The Elevator-first routing algorithm for this architecture has been proposed in [13] which is deadlock free. Section III describes the target architecture together with the routing algorithm. All the reported works in the literature put TSVs without detailed consultation with the application. This has motivated us to propose an application mapping strategy integrated with the TSV placement policy.

III. ARCHITECTURE AND ROUTING ALGORITHM

Fig 1 shows an example of partially connected 3D-mesh architecture, used in this work. Here, r and c represent the router and core in the topology respectively.

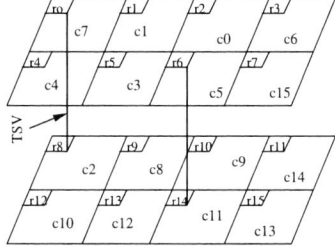

Fig. 1: A 2-layers Partially Connected 3D-mesh-NoC
This work assumes that data can flow in both the directions through a TSV. Individual layers maintain a mesh topology. It has been assumed that upto 25% routers in each layer of a 3D-NoC can support TSVs (similar to [9]). TSVs are spread out including a constraint that no two TSVs can be within one hop distance of each other. That is, to place a TSV (making the router 3D), at least two hop distance should be maintained between the adjacent 3D routers, within a layer. It may be noted that the number, data flow direction and position of TSVs can be varied from one layer to another, though not explored in the current work. Further, individual layers may maintain their own NoC topologies.

IV. PROPOSED STRATEGY FOR TSV PLACEMENT AND APPLICATION MAPPING

Application mapping refers to determining the association of cores to routers. It is a special case of Quadratic Assignment Problem (QAP) [14]. Moreover, proper placement of TSVs into die is intractable which can be transformed into Uncapacited Facility Location Problem [15]. Therefore, the

overall problem is intractable. Our proposed solution strategy works in two phases. In the first phase, it is assumed that all routers in the NoC are 3D with TSVs. The mapping algorithm proposed in this paper is run on this topology to produce a core-to-router attachment. After this, individual TSVs are inspected for traffic flow through them. If the design constraint specifies that only $t\%$ routers in each layer can be 3D, we need to modify the remaining routers to 2D. For this, top $t\%$ 3D routers of a layer are preserved which carry the maximum traffic. Moreover, these TSV locations are replicated in successive layers. Care is also taken to ensure that no two neighbouring routers are 3D in nature. The mapping problem is now solved again on this modified 3D topology to arrive at the final solution of the integrated mapping and TSV placement problem. Thus, the mapping algorithm described next works with predefined TSV positions and map the cores into the architecture. The application is represented in the form of a core graph with nodes corresponding to the participating cores and edges identifying the communication requirements between them (in terms of bits per seconds). In the topology graph, nodes correspond to the routers and edges to the links in the topology.

Algorithm 1: KL_Map_and_TSV_place

Input: Core graph C and Topology graph T
Output: Mapping of C to T along with position of TSVs in T satisfying restrictions on TSV numbers and places.

1: TSV_position← All routers in T
2: Call KL_Partition($G, Core\_List$) //Basic KL partitioning
3: Map clusters onto topology T //Initial mapping
4: Call Improvement_Phase(G,T) //An iterative phase
5: Sort TSVs on decreasing order of usage
6: Update TSV_positions to hold only top 25% TSV positions giving topology $\acute{T}$
7: Map clusters onto modified topology $\acute{T}$
8: Call Improvement_Phase($C, \acute{T}$)
9: **return** Mapping

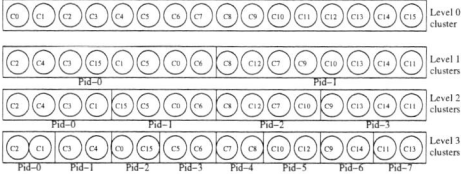

Fig. 2: Partition of all levels of a core graph contains 16 cores

At first, KL_Map_and_TSV_place algorithm partitions the core graph using KL-partitioning procedure and produces a number of clusters each having only two cores. Now, it maps each such cluster to the topology T. To arrive at a good solution, it tries local changes by applying *swapping* and *flipping* operations in procedure Improvement_Phase. Next, it analyzes the traffic flow through each TSV and sorts the TSVs on descending order of traffic flow. Now, it preserves the top 25% TSVs in the individual layers of the topology. Care is also taken so that no two TSVs are placed within one-hop distance of each other. The modified topology is

978-1-4799-8720-7/15 $31.00 © 2015 IEEE

noted as $\acute{T}$ and the clusters that are coming from KL-partition are mapped onto this modified topology. To achieve better mapping solution on this modified topology, we once again run iterative improvement phase and get the final solution.

A. KL-partitioning

Kerninghan and Lin [7] has proposed a balanced graph bi-partitioning algorithm. Starting with a random initial partition, it performs pairwise swappings of vertices between the partitions to minimize the cut size, till no further improvement is possible. A pair of vertices are exchanged if it improves the cut size.

This work considers the number of nodes in a core graph to be an exact power of 2. Otherwise, it incorporates a number of dummy-cores into the core graph. Dummy-cores are connected with all cores including themselves. An edge with cost zero connects a core to a dummy core. Edges between dummy cores are assigned cost infinity. The dummy cores are removed at the end of the mapping phase.

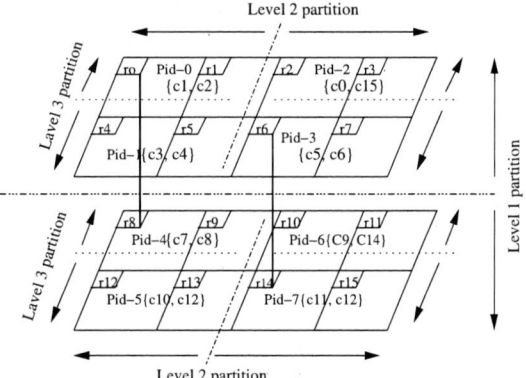

Fig. 3: Partition within mesh topology

The algorithm takes a core graph (G), level (l) and the set of cores to be partitioned in $Partitioned\_list$ as inputs. It is invoked with $l = 0$. It starts partitioning and computes sets of clusters at different levels. At the begining, it forms a single cluster, cluster[0], a single set consisting of all cores in C. In the next level, it is partitioned into two sets of equal sizes. Cost of the partition is taken to be equal to the sum of edge costs (in the core graph) of cores belonging to two sides of the partition. Thus, cluster[1] is a collection of two sets. The cores belonging to a set should be mapped on NoC in the closest proximity of each other, compared to two cores belonging to two different sets. In general, a set in cluster[k] gives rise to two ($k + 1$) level disjoint sets in cluster[$k + 1$]. This process continues till individual sets in a cluster hold only 2 cores. Fig 2 shows an example of the partitioning process.

Algorithm 2: KL-partitioning

1: **procedure** KL-PARTITIONING(G, l, C)
Input: Core graph $G(C, E)$, l : level of partitioning, $partitioned\_list$: the subset of cores in C to be partitioned
Output: A list of cluster; $cluster_i$ is the ith level collection of disjoint sets of cores, initialized to ϕ

2:    $Cluster_l \leftarrow Cluster_l \bigcup Partitioned\_list$
3:    **if** $Partitioned\_list$ contains only 2 number of cores **then**
4:       **return**
5:    **end if**
6:    $(p_1, p_2) \leftarrow$ Random partitioning of cores in $Partitioned\_list$
7:    Call $KL(G, p_1, p_2)$
8:    Call $KL - partitioning(G, l + 1, p_1)$
9:    Call $KL - partitioning(G, l + 1, p_2)$
10: **end procedure**

1: **procedure** KL(G, p_1, p_2)
2:    $present\_partition \leftarrow best\_partition \leftarrow (p_1, p_2)$
3:    Unlock all cores
4:    **while** unlock_core_exist($present\_partition$) **do**
5:       swap$\leftarrow$ select_next_move($present\_partition$)
6:       $best\_partition \leftarrow$ get_better_partition(best_partition, present_partition)
7:       **if** not(cost_fct($best\_partition$) < cost_fct(p_1, p_2)) **then**
8:          **return** (p_1, p_2) // Terminate, no improvement
9:       **else**// do another iteration
10:          $(p_1, p_2) \leftarrow best\_partition$
11:          Unlock all cores
12:       **end if**
13:    **end while**
14: **end procedure**
15: **procedure** SELECT_NEXT_MOVE(P)
16:    **for** each unlocked ($c_i \epsilon p_1, c_j \epsilon p_2$) **do**
17:       Append costlog.Cost_Fct(swap(P, c_i, c_j))
18:    **end for**
19:    **return** (c_i, c_j swap in cost log with lowest cost)
20: **end procedure**

B. Initial mapping

The final partitions generated, are mapped to the respective partition IDs of the mesh-based NoC topology, as shown in Fig 3. For example, core pair (c_1, c_2) is mapped to the router pair (r_0, r_1). To achieve better solution, iterative improvement is applied on this initial mapping.

C. Iterative improvement

At the end of the partitioning and initial mapping phase, a number of clusters at different levels have been created on the mesh topology. Let, l be the total number of levels that have been created. In the following, Improvement_Phase algorithm has been presented that chooses two partitions (say, M and N) and attempts various local changes. Partition M and N are chosen in such a way that they have common parent and level. At first, the cores of partition N are flipped, keeping M unchanged. The cores are flipped first along horizontal axis, then along vertical axis and finally again along horizontal axis, creating three new mapping candidates (a total of 4 mapping candidates). The cores in partition M are now swapped along the horizontal axis. The flippings of partition N are repeated generating four more mappings. In M-partition, swapping is done two more times, first along vertical and then along

978-1-4799-8720-7/15 $31.00 © 2015 IEEE   394

horizontal axes. Each such case generates 4 number of new mapping candidates via flipping of cores in partition N. Therefore, the entire process creates 16 number of mapping candidates among cores in partitions M and N. The mapping with the least local communication cost (computed by considering cores in M and N only) is taken. The strategy is then repeated on partition M and N but considering improvement in the global cost for the entire network. The process is continued with next pair of partitions. If all the partitions in the current level have been completed, the algorithm considers next lower level clusters, till it reaches the level-0 cluster.

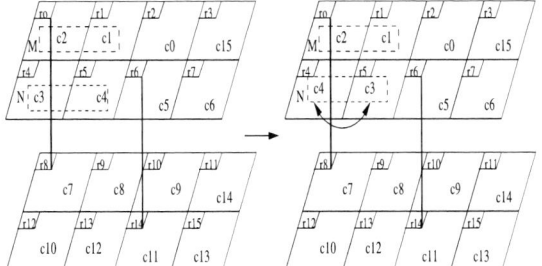

Fig. 4: Improvement operation at level-3 partition

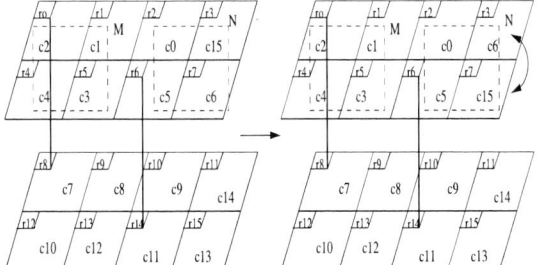

Fig. 5: Improvement operation at level-2 partition

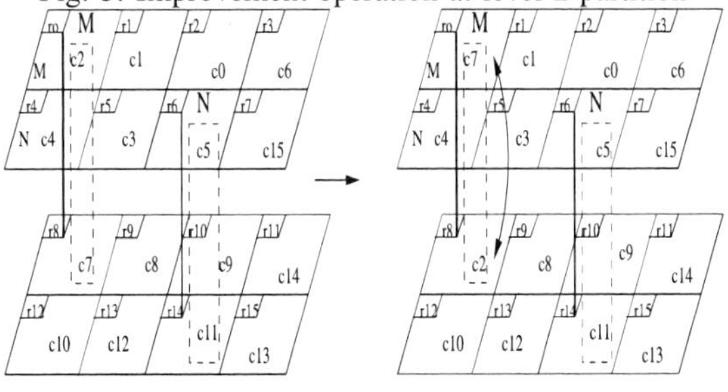

Fig. 6: Improvement operation at level-1 partition

1: **procedure** IMPROVEMENT_PHASE(G, T)
2:     **for** each level starting from higher level **do**
3:        **for** each unlocked partition at this level **do**
4:           Choose two partitions (say M and N) which have common parent
5:           Now, perform flipping operation on cores of partition N in the order such as horizontal, vertical and horizontal with respect to cores in partition M and notes the local communication cost
6:           Swap the cores in partition M horizontally and repeat step 5
7:           Swap cores modified M vertically and repeat step 5
8:           Swap cores in modified M horizontal and repeat step 5
9:           The modified mapping of M and n is identified as $\acute{M}$ and $\acute{N}$ which gives minimum communication cost
10:          Repeat step 5 through 9 for $M = \acute{M}$ and $N = \acute{N}$ considering global cost minimization
11:          Lock the partition M and N
12:       **end for**
13:    **end for**
14: **end procedure**

Let us consider an application mapping in which Improvement_Phase starts with level-3 followed by level-2 and 1. Fig 4 shows level-3 operations on partitions M and N. The flipped operation is performed on partition N. As there is only a single row in each partition, further flips within them will not result in any better cost. The final mapping after all iterations in level-3 has been shown in Fig. 4. Similar flippings are performed at other levels as well. The final mapping solution has been shown in Fig 1.

V. RESULTS AND DISCUSSION

TABLE I: Benchmark applications and their mesh sizes

Benchmarks	No. of Cores	3D Mesh
PIP	8	$2 \times 2 \times 2$
263ENC-MP3DEC	12	$2 \times 3 \times 2$
VOPD	16	$2 \times 4 \times 2$
DVOPD	32	$4 \times 4 \times 2$
G1	64	$4 \times 8 \times 2, 4 \times 4 \times 4$
G2	64	$4 \times 8 \times 2, 4 \times 4 \times 4$
G3	64	$4 \times 8 \times 2, 4 \times 4 \times 4$
G4	128	$8 \times 8 \times 2, 4 \times 8 \times 4$
G5	128	$8 \times 8 \times 2, 4 \times 8 \times 4$

This section describes the experimental results obtained for a set of benchmark NoC systems and compares them with some existing approaches. It has used some of standard benchmarks, such as, `VOPD`, `263ENC-MP3DEC`, `PIP`, and `DVOPD` [5], [6] and generated some large 64-core and 128-core applications using the task graph generation tool TGFF [6], [8]. These benchmarks are noted as `G1-G5` [8] in Table I. The TGFF parameters used for their generation are as follows. The bandwidths are varied from 50 MB/s to 150 MB/s for some graphs and 10 MB/s to 1500 MB/s for others. For generating both high and low communication graphs, in-out degrees of nodes are varied from 1 to 8. Number of start nodes are also varied to generate different graphs and to see the effect of mapping and TSV placement solution upon them. The bandwidth values for the edges are also generated randomly to get heterogeneous communication behaviour of cores. The graphs are mapped onto 3D mesh networks of 2 and 4 layers, as noted in Table I. While generating the mapping solutions, static performance of mapping has been evaluated via the communication cost metric. Communication cost is measured as the product of bandwidth requirement of a pair of cores and hop count between corresponding routers, summed over the edges of the application core graph. Dynamic performance (in terms of throughput, latency and energy consumption) of mapping solutions have been obtained via the 3D-NoC simulator, Noxim [16].

978-1-4799-8720-7/15 $31.00 © 2015 IEEE

TABLE II: Communication cost for different applications with various configurations TSV-locations

Benchmarks	PSMAP [17]	Squeezing [9]	NMAP [5]				Proposed KL_Map_and_TSV_place			
	100%	25%	100%	25%			100%	25%		
	Fully	Symm.	Fully	Symm.	Rand.	Intl.	Fully	Symm.	Rand.	Intl.
PIP	640	768	640	896	896	896	640	768	768	768
263ENC-MP3DEC	230.21	230.21	230.94	230.21	268	230.21	230.47	230.47	230.4	230.47
VOPD	4135	4157	4199	4281	4642	4279	4167	4189	4189	4189
DVOPD	10032	10307	9914	10506	10740	10404	9618	9784	9838	9726
G1	42465.1	50626	36800.49	39387.91	45362.3	39387.91	36567.65	39651.87	42876.54	38576.63
G2	8035.2	9814.02	7684.29	7713.99	8361.63	7579.85	6700.5	6898.52	7117.9	6847.48
G3	111673.1	129985	105054.88	111384.42	124357.06	111384.42	102501.54	108654.56	113456.65	105423.65
G4	160396.69	166271.2	191295.61	201561.16	216127.15	145123.3	125219.16	171259.12	180156.17	134161.3
G5	74391.23	75128.27	76929.12	78251.15	80152.51	53489.71	49110.13	76210.15	77516.12	52374.61
Avg.	1.14	1.25	1.15	1.23	1.32	1.12	1	1.15	1.18	1.05

TABLE III: Comparison of CPU time between DPSO and Our Map_and_TSV_place

Benchmarks	CPU time in s	
	PSMAP [17]	Proposed KL_Map_and_TSV_place
G1	139.25	40.45
G3	180.88	58.67
G4	257.74	90.47

TABLE IV: Comparison of communication cost between before and after re-mapping with of 25% highly communicated TSVs in 3D-NoC

Benchmarks	Fully	25% TSVs	Re-map with 25% TSVs
PIP	640	1024	768
263ENC-MP3DEC	230.47	232.49	230.47
VOPD	4167	4231	4189
DVOPD	9618	10480	9726
G1	36567.65	40929.49	38576.63
G2	6700.5	6946.72	6847.48
G3	102501.54	113947.39	105423.65
G4	125219.16	182713.12	134161.3
G5	49110.13	79261.50	52374.61
Imp. over Fully KL_Map_and_TSV_place	—	-22.83	-5.11

A. Results on different TSV distributions and mapping strategies

This section presents a few results to highlight the efficiency of our mapping strategy KL_Map_and_TSV_place, noted in Section IV. As noted earlier, the algorithm takes as input the TSV positions and generates a mapping optimizing the communication cost. Table II notes the corresponding results. This work has experimented with four different types of TSV distributions. The results marked "Fully" assume all routers to be 3D in nature, each having a TSV. The columns marked "Symm." assume an uniform distribution of 25% TSVs, whereas, "Rand." corresponds to random distribution of 25% TSVs with restriction that no two adjacent routers are having TSVs. The TSV can be placed in consultation with application, as mentioned in Section IV. The values presented in the columns marked "Intl.", are the communication cost of application, remapped onto the partially connected 3D-mesh network. The work has been compared with the PSMAP [17], Squeezing [9] and NMAP [5]. Out of these three, NMAP is a constructive heuristic algorithm, whereas, PSMAP is a particle swarm optimization based approach. We have extended NMAP and PSMAP, originally proposed for 2D-NoC to 3D. PSMAP assumes a fully connected 3D architecture. From Table II, it can be noted that compared to the KL_Map_and_TSV_place fully connected version, PSMAP and NMAP show 14% and 15%, on an average, increase in communication cost. However, compared to KL_Map_and_TSV_place fully connected version, works Squeezing [9], NMAP ("Intl.") and KL_Map_and_TSV_place ("Intl.") are about 25%, 12% and 5% inferior, on an average. This shows the merit of the proposed mapping together with TSV placement approach. Table III compares the CPU time requirements of PSMAP and the proposed approach for a number of benchmarks. On an average our approach requires one-third the time needed for PSMAP.

B. Impact of TSV position selection

Now, we enumerate the results of integrated TSV position selection and mapping. Table IV notes the corresponding results. The column marked "Fully" corresponds to the situation in which all routers are 3D in nature. As suggested in Section IV, we next keep only 25% highly utilized TSVs. Communication cost values are recomputed such that interlayer message flow uses elevator-first algorithm [13]. Naturally, communication cost degrades. Next, it performs a remapping of cores with the current TSV distribution. Compared to the fully connected configuration, the 25% TSV case without remapping increases communication cost by (-)22.83%, on an average. A remapping of cores, now, creates a final solution with communication cost degradation restricted to (-)5.11% with respect to the fully connected situation.

C. Dynamic performance of different mapping and TSV configurations

To measure the proficiency of individual mapping techniques together with different TSV configuration in partially-connected 3D-mesh-NoC, this work has simulated each of the NoC system using Noxim-3D simulator. Synthetic self-similar traffic has been generated, by obeying the communication requirement of cores in the application. Self-similar traffic has been observed in typical video and networking applications [18]. Table V shows the results of throughput, latency and average packet energy (μJ) for the benchmarks. As it can be noted from the table, fully connected 3D NoC gives the best performance in terms of all the three factors. Using 25% TSVs results in degradation of all the three parameters.

978-1-4799-8720-7/15 $31.00 © 2015 IEEE

TABLE V: Comparison of throughput, latency and energy (μJ) of different mapping and TSV placement strategies TSV distribution

Parameters	Proposed KL_Map_and_TSV_place with											
	Benchmarks: PIP				Benchmarks: 263ENC-MP3DEC				Benchmarks: VOPD			
	Fully	Symm.	Rand.	Intl.	Fully	Symm.	Rand.	Intl.	Fully	Symm.	Rand.	Intl.
Throughput	0.77	0.68	0.54	0.76	0.85	0.72	0.64	0.73	0.77	0.76	0.72	0.77
Latency	77508.4	82036.8	84357.1	79135.2	61963	67982.5	68197	65870.8	74252	75448.3	77564	74339.4
PKT. Energy	10.2	11.5	11.5	11.3	10.7	10.8	11.3	10.7	11.0	11.7	11.8	11.6
	Benchmarks: DVOPD				Benchmarks: G1				Benchmarks: G2			
Throughput	0.74	0.69	0.66	0.74	0.69	0.62	0.56	0.69	0.83	0.79	0.77	0.80
Latency	76387.8	77468.6	77388.2	76994	78435.8	81577.6	82020.5	80862.8	82639.3	82834.8	84461.6	82437
PKT. Energy	11.3	11.9	12.6	11.8	13.2	14.0	14.0	13.5	11.6	12.2	12.4	11.9
	Benchmarks: G3				Benchmarks: G4				Benchmarks: G5			
Throughput	0.61	0.51	0.49	0.61	0.79	0.75	0.72	0.78	0.84	0.83	0.79	0.83
Latency	78352.6	81766.8	83049.9	79719.9	82624.3	82979	83183.5	82722.3	82686	83372	83976	83273.2
PKT. Energy	13.8	15.2	15.2	14.4	13.1	13.6	13.7	13.1	11.6	11.9	12.0	11.7

Intelligent TSV placement with core remapping improves the solution quality. Compared to the fully connected version, in "25% Symm." case throughput, latency and packet energy deteriorate by 8.86%, 3.45% and 6.22% respectively. If the 25% TSVs are placed randomly, the corresponding degradations are 16.54%, 4.54% and 8.12%. Intelligent TSV placement improves the solution quality, with degradation factors improving to 3.62%, 6.08% and 3.08% for throughput, latency and average packet energy respectively.

D. Result for higher number of layers

Table VI shows the communication cost comparison of mapping technique together with TSV placement strategy for four-layer NoC realization. It can be observed that the proposed strategy makes a significant improvement in communication cost over strategies, such as, PSMAP (14%), Squeezing (22%) and NMAP (10%). Dynamic performance comparison (in terms of throughput, latency and power) shows similar trends as in Table V. The results are not included due to shortage of space.

TABLE VI: Communication cost comparison of different integrated mapping and TSV placement techniques with 4 number of layers

Bench marks	100%	25%		
	Fully	Symm.	Intl.	
	PSMAP [17]	Squeezing [9]	NMAP [5]	KL_Map_and_ TSV_place
G1	44305.92	54251.15	52507.96	45622.87
G2	9804.02	9912.07	8459.07	6847.07
G3	128547	131721	159312.9	157012
G4	148357.69	173242.6	152142.6	149812.1
G5	79732.18	77481.6	57361.02	53241.18
Avg.	1.14	1.22	1.10	1

VI. CONCLUSION

In this paper, we have presented a strategy to design 3D mesh based NoC with restricted vertical interconnects via TSVs. An application mapping policy developed around Kernighan-Lin bi-partitioning approach has been designed and integrated with the TSV position selection. The approach shows promising results. The future work includes handling other 3D design issues, such as, thermal-aware design.

REFERENCES

[1] T. C. Xu, P. Liljeberg, and H. Tenhunen, "Optimal Number and Placement of Through Silicon Vias in 3D Network-on-Chip," in *IEEE 14th International Symposium on Design and Diagnostics of Electronic Circuits & Systems (DDECS)*, 2011, pp. 105 –110.

[2] W. R. Davis, J. Wilson, S. Mick, J. Xu, H. Hua, C. Mineo, A. M. Sule, M. Steer, and P. D. Franzon, "Demystifying 3D ICs: the pros and cons of going vertical," *IEEE Design & Test of Computers*, vol. 22, pp. 498–510, 2005.

[3] S. Pasricha, "A Framework for TSV Serialization-aware Synthesis of Application Specific 3D Networks-on-Chip," in *VLSI Design (VLSID), 2012 25th International Conference on*, 2012, pp. 268–273.

[4] Semiconductor Industry Association et al., "The International Technology Roadmap for Semiconductors (ITRS)," 2007.

[5] S. Murali and G. D. Micheli, "Bandwidth Constrained Mapping of Cores onto NoC Architectures," in *Proceedings of Design, Automation and Test in Europe Conference and Exhibition (DATE)*, 2004, pp. 896–901.

[6] P. K. Sahu and S. Chattopadyay, "A Survey on Application Mapping Strategies for Network-on-Chip Design," *Journal of System Architecture, Elsevier*, vol. 59, issue 1, pp. 60–76, 2013.

[7] B. Kernighan and S. Lin, "An Efficient Heuristic Procedure for Partitioning Graphs," *Bell System Technical Journal*, vol. 49, no. 2, pp. 291–307, 1970.

[8] P. K. Sahu, K. Manna, N. Shah, and S. Chattopadhyay, "Extending kernighanlin partitioning heuristic for application mapping onto network-on-chip," *Journal of Systems Architecture*, vol. 60, pp. 562–578, 2014.

[9] C. Liu, L. Zhang, Y. Han, and X. Li, "Vertical interconnects squeezing in symmetric 3D mesh Network-on-Chip," in *Procedings of Design Automation Conference (ASP-DAC)*, 2011, pp. 357–362.

[10] Y. J. Hwang, J. H. Lee, and T. H. Han, "3D Network-on-Chip system communication using minimum number of TSVs," in *Procedings of ICT Convergence (ICTC)*, 2011, pp. 517–522.

[11] S. Pasricha, "Exploring Serial Vertical Interconnects for 3D ICs," in *Proceding of IEEE/ACM Design Automation Conference (DAC 2009)*, 2009, pp. 581–586.

[12] F. Miller, T. Wild, and A.Herkersdorf, "TSV-virtualization for Multi-protocol-Interconnect in 3D-ICs," in *Proceding of Euromicro Conference on Digital System Design (DSD)*, 2012, pp. 374–381.

[13] F. Dubois, A. Sheibanyrad, F. P. trot, and M. Bahmani, "Elevator-First: A Deadlock-Free Distributed Routing Algorithm for Vertically Partially Connected 3D-NoCs," *IEEE Transaction On Computer*, vol. 62, no. 3, pp. 609–615, 2013.

[14] L. Congying, Z. Huanping, and Y. Xinfeng, "Particle swarm optimization algorithm for quadratic assignment problem," in *Proceedings of IEEE Int. Conference Computer Science Networking Technolgy*, 2011, pp. 1728–1731.

[15] A. R. Guner and M. Sevkli, "A discrete particle swarm optimization algorithm for uncapacited facility location problem," *Journal of Artificial Evolution and Application*, pp. 1–9, 2008.

[16] K. Jheng, C. Chao, H. Wang, and A. Wu, "http://access.ee.ntu.edu.tw/noxim/index.html." 2013.

[17] P. K. Sahu, P. Venkatesh, S. Gollapalli, and S. Chattopadhyay, "Application mapping onto mesh structured network-on-chip using particle swarm optimization." in *IEEE ISVLSI*, 2011, pp. 335–336.

[18] G. V. Varatkar and R. Marculescu, "On-chip traffic modelling and synthesis for MPEG-2 video applications," in *IEEE Trans. Very Large Scale Integr. (VLSI) Syst.*, vol. 12(1), 2004, pp. 108–119.

Achieving Memory Access Equalization via Round-trip Routing Latency Prediction in 3D Many-core NoCs

Xiaowen Chen[†,‡], Zhonghai Lu[‡], Yang Li[¶], Axel Jantsch[§], Xueqian Zhao[‡], Shuming Chen[†], Yang Guo[†],
Zonglin Liu[†], Jianzhuang Lu[†], Jianghua Wan[†], Shuwei Sun[†], Shenggang Chen[†], Hu Chen[†]

[†]College of Computer, National University of Defense Technology, 410073, Changsha, China
[‡]Department of Electronic Systems, KTH Royal Institute of Technology, 16440 Kista, Stockholm, Sweden
[§]Institute of Computer Technology, Vienna University of Technology, 1040 Vienna, Austria
[¶]College of Computer, National University of Defense Technology, 410073, Jilin, China

Abstract—3D many-core NoCs are emerging architectures for future high-performance single chips due to its integration of many processor cores and memories by stacking multiple layers. In such architecture, because processor cores and memories reside in different locations (center, corner, edge, etc.), memory accesses behave differently due to their different communication distances, and the performance (latency) gap of different memory accesses becomes larger as the network size is scaled up. This phenomenon may lead to very high latencies suffered from by some memory accesses, thus degrading the system performance. To achieve high performance, it is crucial to reduce the number of memory accesses with very high latencies. However, this should be done with care since shortening the latency of one memory access can worsen the latency of another as a result of shared network resources. Therefore, the goal should focus on narrowing the latency difference of memory accesses. In the paper, we address the goal by proposing to prioritize the memory access packets based on predicting the round-trip routing latencies of memory accesses. The communication distance and the number of the occupied items in the buffers in the remaining routing path are used to predict the round-trip latency of a memory access. The predicted round-trip routing latency is used as the base to arbitrate the memory access packets so that the memory access with potential high latency can be transferred as early and fast as possible, thus equalizing the memory access latencies as much as possible. Experiments with varied network sizes and packet injection rates prove that our approach can achieve the goal of memory access equalization and outperforms the classic round-robin arbitration in terms of maximum latency, average latency, and LSD[1]. In the experiments, the maximum improvement of the maximum latency, the average latency and the LSD are 80%, 14%, and 45% respectively.

I. INTRODUCTION

3D die stacking technology is an emerging solution for future high-performance single chip design, because it continuously increases the number of processor cores and memories by stacking multiple processor layers and memory layers, and reduces the long wire latency and improves the memory bandwidth by using massive TSVs (Through Silicon Vias). Such many processor cores and memories are connected by pipelined communication networks (called Network-on-Chip (NoC)[1]) rather than buses, so hundreds or thousands of communications can go on concurrently at any time. Such architecture is referred to as 3D many-core NoC, which has attracted great attentions. For instance, Kim, et. al. proposed a 3D multi-core system with one 64-core layer and one 256KB SRAM layer using Tezzaron TSV/bonding technology[2]. Fick, et. al. designed a low-power 64-core system, named Centip3De, which has two stacked dies with a layer of 64 ARM Cortex-M3 cores and a cache layer[3]. Furthermore, they extended Centip3De

[1]LSD: Latency Standard Deviation measures the amount of variation or dispersion from the average latency. A low standard deviation indicates that the latencies tend to be very close to the mean. Therefore, LSD is suitable to evaluate the memory access equalization.

to be a 7-layer 3D many-core system including 2 processor layers with 128 cores in total, 2 Cache/SRAM layers, and 3 DRAM layers[4]. Wordeman, et. al. proposed a prototype of a 3D system with a memory layer and a processor-like logic layer[5]. Yuang, et. al. proposed a 3D mesh NoC, which tightly mixes memories and processor cores so as to improve the memory access performance[6]. All of these researches use NoC as the communication infrastructure and TSV to connect multiple layers.

In 3D many-core NoCs, since processor cores and memories reside in different locations (center, corner, edge, etc.), the memories are asymmetric so as to be a kind of NUMA (Non Uniform Memory Access)[7][8] architecture and memory accesses behave differently due to their different communication distances. As the network size is scaled up and the number of processor cores and memories increases, the communication distance difference of memory accesses becomes larger and a large increased number of memory accesses worsen the network contention and congestion, thus the performance (latency) gap of different memory accesses becomes larger. This phenomenon may lead to very high latencies suffered from by some memory accesses that are the bottleneck of the system and even extremely degrade the system performance. To achieve high performance in emerging 3D many-core NoCs, it is crucial to reduce the number of memory accesses with very high latencies. However, this should be done with care as shortening one memory access's latency can worsen the latency of another because the network resources such as router buffers and communication links are shared. Therefore, the goal should focus on balancing the latencies of memory accesses (i.e. narrowing the latency difference of memory accesses) as well as ensuring a low average latency value, which is referred to as the "memory access equalization" problem in 3D many-core NoCs.

In the paper, we address the goal by proposing to prioritize the memory access packets on the basis of predicting the round-trip routing latencies of memory accesses in the network. A round-trip routing latency of a memory access is its elapsed time in the network, which is equal to the time of the memory request (read or write) routing in the outward trip from the source to the destination plus the time of the corresponding memory response (read data or write acknowledgement) routing back to the source from the destination in the return trip. The basic idea of our approach is that the memory access with longer round-trip routing latency should go first when it competes with other memory accesses. The round-trip routing latency includes the elapsed routing time and the future routing time, which is unknown when a memory access is traversing in the network and hence requires prediction. We use the communication distance and the number of the occupied items in the buffers in the remaining routing path to predict the

future routing time of a memory access on-the-fly. Experiments demonstrate that our approach can achieve the goal of memory access equalization and has performance improvement in terms of maximum latency, average latency, and LSD in comparison to the classic round-robin arbitration.

The rest of the paper is organized as follows. Section II discusses the related work. Section III details our approach containing the target architecture, motivation, and router design supporting the round-trip routing latency prediction of memory accesses. Section IV reports the experimental results and the performance analysis. Finally, we conclude in Section V.

II. RELATED WORK

For memory access equalization, prior work first studied it in the context of computer systems[9], then some literatures turned their eyes to the chip and paid attentions to studying the equalization of memory accesses to the off-chip SDRAM, because off-chip SDRAM has high capacity and serves a large number of memory accesses so that it is a hotspot/high congestion region with possible heavy contention and the latencies of memory accesses may be so different. Usually, the equalization of memory accesses to the off-chip SDRAM is studied in two aspects: balancing the memory access performance in memory interface to the off-chip SDRAM[10][11][12] or on-chip routers[13]. For the first aspect, In [10], targeting Chip Multiprocessors, Mutlu, et al. presented a fair memory access scheduling mechanism in SDRAM interface, which balances the ratio of memory access latency between shared and alone cases in memory access latency among cores. According to NoC-based multicores, some literatures utilized the on-chip network information to perform fair scheduling of the memory requests in SDRAM interface so as to achieve memory access equalization. For instance, Both [11] and [12] proposed to prioritize requests in SDRAM memory interfaces according to the congestion information such that requests from less-congested regions prioritize over the other requests. The definition and the concrete implementation of their congestion information are different. For the second aspect, memory requests are considered to be scheduled during their transmission in on-chip networks in order to avoid large congestion near the SDRAM interface so that the memory access latency would be more balanced [13]. As the memory size in a single chip increases, on-chip memory access performance gradually attracts researchers. In [14], Pimpalkhute et. al. proposed a holistic solution for intelligently scheduling network packets (on-chip cache requests) and memory packets (off-chip requests) to optimize overall system performance. They balance the latency performance between the on-chip requests and off-chip requests. Different from them, we focus on balancing the latencies amongst the on-chip memory accesses. In [15], Sharifi et. al. addressed balancing latencies of memory accesses issued by an application in an execution phase. They prioritized memory response messages such that, in a given period of time, messages of an application that experience higher latencies than the average message latency of that application are expedited. However, they divided a whole memory access into two parts: outward trip for memory request (read or write) and return trip for memory response (read data or write acknowledgement) and treated them separately. Their scheduling scheme only prioritizes the memory response messages and balances the return trip latencies. Different from them, our approach considers the two parts of an memory access as a whole and the round-trip routing latency is used to prioritize memory accesses so as to achieve the goal of memory access equalization,

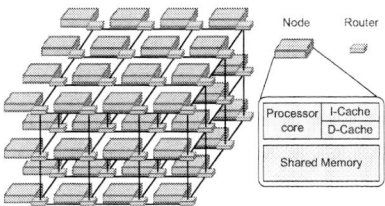

Fig. 1. A homogenous 4×4×3 3D many-core NoC

since considering the entire round trip of a memory access is more reasonable. Besides, to the best of our knowledge, there is little work studying the memory access equalization in the context of 3D many-core NoCs.

III. OUR APPROACH

A. Target Architecture: 3D many-core NoCs

Fig. 1 shows an example of our target architecture: homogenous 3D many-core NoCs. The example system is composed of 4×4×3 nodes interconnected via a packet-switched 3D mesh network. Each node is connected to a router. Routers are interconnected with bidirectional links. In homogenous 3D many-core NoCs, all nodes are identical and so are the routers. A node contains a processor core and a shared memory that is visible to all nodes. Memories are distributed and shared so that the centralized memory organization and hence the hotspot area are avoided.

B. Motivation

In such architecture shown in Fig. 1, a large number of memory accesses are generated. Memory accesses behave differently due to their different communication distances. For instance, the up left corner node in the top layer accesses the shared memory in its neighboring node in 2 hops for a round-trip memory read, but it takes 16 hops over the network if it reads a data in the shared memory of the bottom right corner node in the bottom layer. The latency difference results in unequal memory access and some memory accesses with very high latencies, thus negatively affecting the system performance. The impact becomes worse when the network size is scaled up, because the latency gap of different memory accesses becomes bigger. Because memory accesses traverse and contend in the network, improving one memory access's latency can worsen the latency of another. Therefore, we are motivated to focus on memory access equalization in 3D many-core NoCs through balancing the latencies of memory accesses (i.e. narrowing the latency difference of memory accesses) while ensuring a low average latency value.

An entire memory access is a round-trip one containing two parts: memory request (read or write) in outward trip and memory response (read data or write acknowledgement) in return trip, so the performance of a memory access includes that of the two parts. We envision that, to achieve the goal of memory access equalization, it is better to consider the total routing latency of the round-trip of a memory access as the base to prioritize the memory accesses when they contend in the routers (i.e. the memory access with longer round-trip routing latency gains the link successfully and go first), because the round-trip routing latency represents the full performance of a memory access and the two parts should not be considered separately.

978-1-4799-8720-7/15 $31.00 © 2015 IEEE

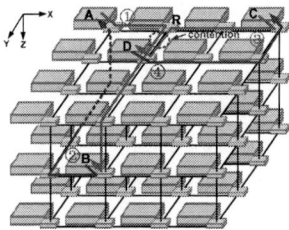

Fig. 2. A short motivation example: two memory accesses traverse and contend.

Let us take Fig. 2 as an example. Fig. 2 shows a $4 \times 4 \times 3$ 3D mesh NoC, which is packet-switched, performs deterministic DOR[2] X-Y-Z routing and takes one cycle for one hop. In the figure, Node A starts a memory access to the shared memory in Node B, which contains two parts: the memory request from Node A to Node B (the red line ①) and the memory response from Node B to Node A (the blue line②). Meanwhile, a memory access is started from Node C to the shared memory in Node D, which contains two parts: the memory request from Node D to Node C (the red line ③) and the memory response from Node C to Node D (the blue line ④). Assume that memory request ① and ③ contend at router R for the downstream link and memory request ③ occupies the link successfully, therefore memory request ① has to wait one cycle in router R due to its failure. Finally, memory access (A-B) takes 6 cycles[3] for its outward trip and 5 cycles for its return trip, thus 11 cycles in total, while memory access (C-D) takes 6 cycles for its round trip. The average latency of the two memory accesses is 8.5 cycles, and both the two memory accesses deviate from the average latency by 2.5 cycles. Actually, when at router R, memory access (A-B) has taken 1 cycles and memory access (C-D) has used 2 cycles. Although the elapsed time of memory access (A-B) is less than that of memory access (C-D), the former's total time is bigger than the latter's. if the total time (round-trip routing latency) is used as the base for arbitration, memory request ① obtains the link of router R successfully. Therefore, memory access (A-B) takes 10 cycles, while memory access (C-D) takes 7 cycles with 1-cycle waiting time at router R. Although the average latency is still 8.5 cycles, the deviation of both memory accesses from the average latency becomes smaller to be 1.5 cycle. Thus, the difference of the two memory accesses is narrowed and memory access equalization is achieved. Fig. 2 is an intrinsic case to describing our motivation and idea. As we know, when a memory access is being routed in the network, the past routing time is known but the future routing time is unknown, so the round-trip latency (equaling the past routing time plus the future routing time) is needed to be predicted. The next subsection describes our router design supporting round-trip routing latency prediction of memory accesses in detail.

C. Router Design Supporting Memory Access Equalization

The communication infrastructure in our target architecture is a packet-switched 3D mesh network with deterministic DOR X-Y-Z routing, thus preventing cyclic dependencies and avoiding network deadlock. Memory access equalization is supported in routers during the transmission of memory access packets in the network.

(I) Router Microarchitecture

[2] Dimension-Ordered Routing
[3] For calculation simplicity, only the time elapsed in the network rather than in the node is considered in the example.

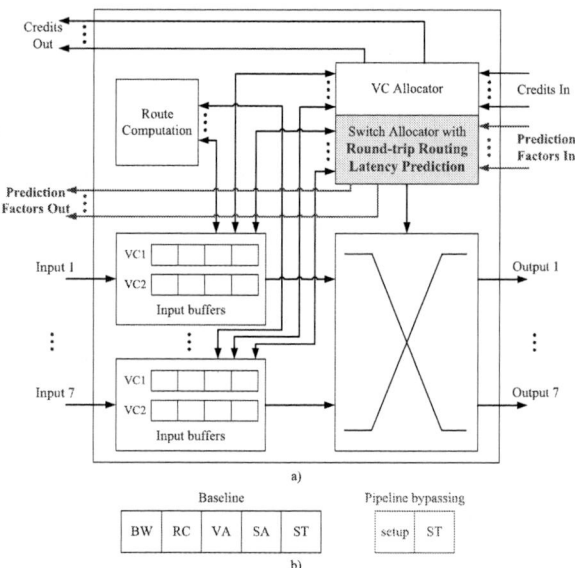

Fig. 3. a) A packet-switched credit-based virtual channel router supporting round-trip routing latency prediction; b) Pipeline stages in baseline and pipeline bypassing

Fig. 3 a) shows the microarchitecture of our router, which is a state-of-the-art packet-switched credit-based Virtual Channel (VC) [16] router. It is enhanced by supporting round-trip routing latency prediction implemented in the Switch Allocator (SA) module. Two virtual channels (VC1 for memory request packets and VC2 for memory response packets) are used to break deadlocks induced by the dependency of memory requests and memory responses. The router is designed as a typical microarchitecture with 5 logical stage pipeline[17], as shown in Fig. 3 b). A packet (its type can be "memory request" or "memory response"), upon arriving at an input port, is first written into the input buffer according to its input VC in the Buffer Write (BW) pipeline stage. In the next stage, the routing logic performs Route Computation (RC) to determine the output port for the packet. The packet then arbitrates for a VC corresponding to its output port in the VC Allocation (VA) stage. Upon successful allocation of a VC, the packet proceeds to the Switch Allocation (SA) stage where it arbitrates for the switch input and output ports. On winning the output port, the packet is then read from the input buffer and proceeds to the Switch Traversal (ST) stage, where it traverses the crossbar to be sent over the physical link finally. To accelerate the packet transmission speed over the router, a mechanism called "Pipeline bypassing"[17] is adopted. The BW, RC, VA and SA stages are combined and performed in the first stage which is named the "setup stage" where the crossbar is set up for packet traversal in the next cycle while simultaneously allocating a free VC corresponding to the desired output port. Therefore, moving one hop takes 1 clock cycle in our NoC. If there is a port conflict in the switch allocation between the packets in the two VCs, the packet with the higher priority is prioritized over the other. To support memory access equalization, we use the round-trip routing latencies of memory accesses as the prioritization base, since considering the outward trip and the return trip of an memory as a whole is more reasonable. The packet with longer round-trip routing latency has higher priority to go through the router. When the packet is traversing in the network, how much time will be taken in its future routing path is not known exactly but can be estimated. The next describes how to predict the round-

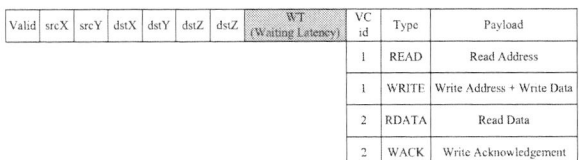

Valid	srcX	srcY	dstX	dstY	dstZ	dstZ	WT (Waiting Latency)	VC id	Type	Payload
								1	READ	Read Address
								1	WRITE	Write Address + Write Data
								2	RDATA	Read Data
								2	WACK	Write Acknowledgement

Fig. 4. Packet format

trip routing latency of a memory access.

(II) Predicting Round-trip Routing Latency

An entire memory access is a round-trip one containing two parts: memory request (read or write) in outward trip and memory response (read data or write acknowledgement) in return trip, so its appearance is a memory request packet in the first "memory request" phase of its transmission and a memory response packet in the second "memory response" phase of its transmission. The round-trip routing latency (notated as L) of a memory access can calculated by Formula (1).

$$L = (DL_p + WL_p) + (DL_f + WL_f) \qquad (1)$$

The part in the first parentheses is the time that a memory access has consumed during its past transmission, which contains DL_p and WL_p representing the distance latency and the waiting latency[4] in the past transmission respectively. The part in the second parentheses is the time of the remaining transmission of a memory access, which includes DL_f and WL_f represents the distance latency and the waiting latency in the future transmission respectively. Because our network adopts the deterministic DOR X-Y-Z routing strategy, DL_f is deterministic. Therefore, Formula (1) is refined as:

$$L = DL_t + WL_p + WL_f \qquad (2)$$

where DL_t is the total round-trip distance latency that is equal to DL_p plus DL_f.

To predict the round-trip routing latency, we need to calculate DL_t, WL_p, and WL_f.

(i) Calculating DL_t

DL_t is known in the deterministic routing network and can be calculated by Formula (3) according to the coordinates of the source and the destination.

$$DL_t = 2 \cdot (|X_{src} - X_{dst}| + |Y_{src} - Y_{dst}| + |Z_{src} - Z_{dst}|) \quad (3)$$

where X_{src}, Y_{src}, and Z_{src} are the X, Y, and Z coordinates of the source, and X_{dst}, Y_{dst}, and Z_{dst} are the X, Y, and Z coordinates of the destination. They can be extracted from the packet (see Fig. 4).

(ii) Obtaining WL_p

WL_p is obtained from the "waiting latency (WL)" field in the packet. Fig. 4 illustrates the packet format. When a memory access starts, "WL" is initialized as zero in its memory request packet. The initial value of "WL" in its memory response packet is equal to the "WL" value of its memory request packet at the time when it reaches the destination shared memory. "WL" is incremented by 1 per clock cycle when the memory request packet or the memory response packet is blocked in the buffer due to the arbitration failure.

(iii) Estimating WL_f

---

[4]Distance latency is the transmission time of a packet without any contention, which is determined by the hop count and the clock cycles per hop. Waiting latency is the time consumed by a packet when it has to wait in the buffer due to its failure of winning the arbitration.

WL_f represents the possible waiting latency of a memory access in its future transmission. To estimate WL_f, we propose to use the number of the occupied items of the related input buffers in the downstream routers along the remaining routing path of a memory access. For instance, when a packet (notated as A) is going to its 1st downstream router (notated as R) through the inport whose input buffer has 2 packets (the first and the second are notated as $B1$ and $B2$ respectively), it will have to wait 1 hop (1 cycle in our design) for the departure of packet $B2$ until it passes through router R, since packet $B1$ leaves router R at the same time when packet A enters router R. Therefore, the future waiting time of packet A in its 1st downstream router R is considered to be 2. WL_f is estimated to be the sum of the count of the occupied items of the related input buffers in all downstream routers along the remaining routing path. Two steps are used to estimate WL_f as follows:

1) The list (notated as $\mathbb{R}$) of the downstream routers in the remaining routing path is obtained according to the packet's source and destination coordinates.
2) WL_f is calculated by Formula (4).

$$WL_f = \sum_{R \in \mathbb{R}} FWL(R) \qquad (4)$$

where $FWL(R)$ is the function of calculating the future waiting latency in the downstream router R and shown in Formula (5).

$$FWL(R) = \begin{cases} \eta_R - \delta_R, & when \ \eta_R > \delta_R \\ 0, & when \ \eta_R \leqslant \delta_R \end{cases} \qquad (5)$$

where δ_R is the hop count from the current router to the downstream router R and η_R is the number of the occupied items of the related input buffer in the downstream router R.

The two-step prediction method above considers the potential waiting latency due to the Head-of-Line (HoL) blocking induced by the packets that have existed in the input buffers in the remaining routing path. It does not predict the possible waiting latency due to the resource contention with other packets because the arrival time of other packets is undetermined and contention is hard to be described properly and estimated accurately.

(III) Implementing the prediction method

The two-step prediction method is generic and can be simplified when in concrete implementation. From Formula (5), we can see that the packet will not wait in the downstream router R if the number of packets in the related input buffer in the downstream router R is less than or equal to the hop count between the current router and the downstream router R. Therefore, the first step can be simplified to get the nearest 3 downstream routers in the remaining routing path since the depth of the input buffer in our router is only 4. Fig. 5 shows the pseudocode of estimating WL_f in our router. According to the X, Y, and Z coordinates of the current router and the destination of the packet, $DownstreamRouterExist()$ judges whether the downstream router exists, while $GetDownstreamRouter()$ gets the downstream router. In the router design, all of η_R in the nearest 3 routers come together to be the "Prediction Factors In" input and all of η_R in the current router form the "Prediction Factors Out" output, as shown in Fig. 3. η_R keeps track of the number of the occupied items of the related input buffer in the downstream router R. Its implementation is similar with the "Credit" signal. Each input buffer has a corresponding η_R counter, which increments when a new packet goes into the buffer and decrements when a packet departs the input buffer. η_R is the counter value.

```
WL_f = 0;
if DownstreamRouterExist(1) == true then
    R1 = GetDownstreamRouter(1);
    WL_f = WL_f + (η_R1 − 1);
end if //the 1st downstream router
if DownstreamRouterExist(2) == true then
    R2 = GetDownstreamRouter(2);
    WL_f = WL_f + (η_R2 − 2);
end if //the 2nd downstream router
if DownstreamRouterExist(3) == true then
    R3 = GetDownstreamRouter(3);
    WL_f = WL_f + (η_R3 − 3);
end if //the 3rd downstream router
```

Fig. 5.   Pseudocode of estimating WL_f

D. Hardware Cost

The router design is synthesized in Synopsys® Design Compiler with TSMC® 45nm process. Table I lists the logic synthesis results excluding the wire cost. As stated in the previous paragraph, the extra hardware for implementing the prediction method is some η_R counters, so the extra hardware overhead is marginal. The extra area values are shown in the parentheses in Table I. The extra hardware does not degrade the frequency.

TABLE I
LOGIC SYNTHESIS RESULTS OF THE ROUTER

	Area	Frequency
Combinational Logic	$34029.30\mu m^2$ ($798.35\mu m^2$)	1.96 GHz
(RC+VA+SA+ST)	(36.17k NAND gates)	
Sequential Logic	$24545.19\mu m^2$ ($593.21\mu m^2$)	(0.51 ns)
(Input Buffers)	(26.09k NAND gates)	

Note: The area of a NAND gate with two inputs is 0.9408 μm^2.

IV. EXPERIMENTS AND RESULTS

A. Experimental Setup

We implement a cycle-accurate homogenous many-core NoC simulator with Verilog, as shown in Fig. 1. The simulator models the processor cores, the shared memories and the NoC. The NoC has a 3D mesh topology and its size is configurable. The router is designed as described in Subsection III-C. To evaluate the performance of our proposal, uniform synthetic traffic patterns are considered. The random traffic represents the most generic case, where each processor core sends in-order memory accesses to the shared memories distributed in all nodes with a uniform probability. The target memories are selected randomly. In experiments, each processor core begins to generate 10,000 memory requests (read or write) to randomly selected destination shared memories when an experiment starts, and an experiment finishes after all processor cores receive their related 10,000 memory responses (read data or write acknowledgement). All of the experiments are performed with a variety of network sizes and packet injection rates. For performance comparison, we take the classic widely used round-robin arbitration as the counterpart and evaluate maximum latency (ML), average latency (AL), and latency standard deviation (LSD) that are defined by Formula (6), (7), and (8) respectively:

$$ML = \max(\{L_1, L_2, \cdots, L_N\}) \quad (6)$$

$$AL = \frac{1}{N}\sum_{i=1}^{N} L_i \quad (7)$$

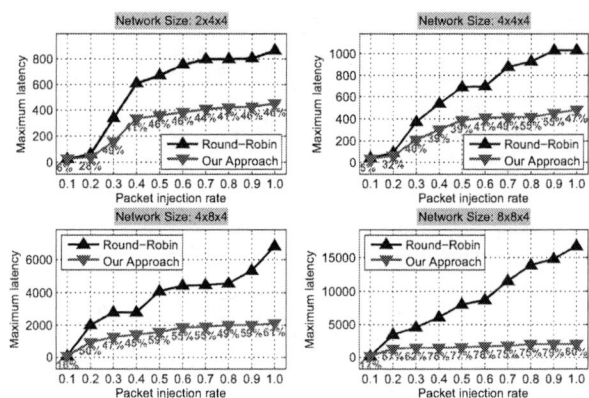

Fig. 6.   Comparison of maximum latency between round-robin arbitration and our approach under varied network sizes and packet injection rates

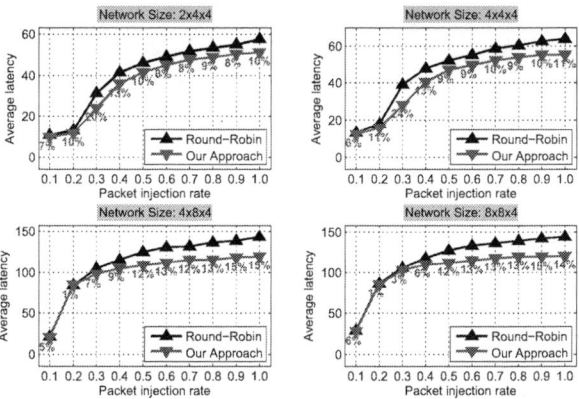

Fig. 7.   Comparison of average latency between round-robin arbitration and our approach under varied network sizes and packet injection rates

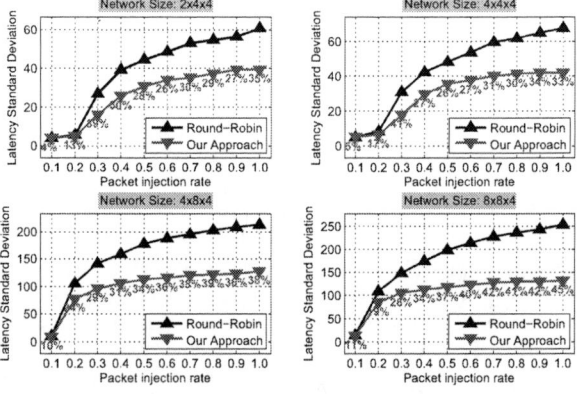

Fig. 8.   Comparison of latency standard deviation between round-robin arbitration and our approach under varied network sizes and packet injection rates

$$LSD = \sqrt{\frac{1}{N}\sum_{i=1}^{N}(L_i - AL)^2} \quad (8)$$

where N is the total number of memory accesses and L_i is the round-trip routing latency of the memory access with the id of i.

B. Performance Evaluation

Fig. 6, 7, and 8 respectively plot maximum latency, average latency, and LSD under different network sizes and packet injection rates. The percentages in the figures indicate the performance improvement of our approach in comparison to the round-robin arbitration. From these figures, we can see that:

- Compared with the classic round-robin arbitration, our approach can has lower maximum latency, average latency, and LSD, thus making the latencies of memory accesses more balanced and achieving the goal of memory access equalization.

- As the network size is scaled up and the packet injection rates increases, our approach can basically gain more performance improvement in terms of maximum latency, average latency and LSD, meaning that, under large-scale network size with a large number of memory accesses, the classic round-robin arbitration has large latency gap of different memory accesses and our approach can balance the latencies of memory accesses well. For instance, under the network size of $8 \times 8 \times 4$ and the packet injection rate of 1.0, in comparison to the round-robin arbitration, the maximum latency, the average latency and the LSD are improved by 80%, 14%, and 45% respectively, which are the maximum performance improvement we obtain in the experiments.

To further evaluate the memory access equalization, we collect the number of memory accesses with different round-trip routing latencies. Fig. 9 exhibits the latency dispersion of memory accesses. From the figure, we can see that:

- Our approach reduces the number of memory accesses with high latencies in comparison to the classic round-robin arbitration.

- Because the total number of memory accesses are the same in both our approach and the classic round-robin arbitration, our approach increases the number of memory accesses with low latencies.

Therefore, our approach makes the latency dispersion curve be narrower and higher so that the memory access equalization is achieved.

V. Concluding Remark

In 3D many-core NoCs, as the network size is scaled up and the number of memory accesses increases largely, the performance (latency) gap of different memory accesses becomes bigger. Some memory accesses with very high latencies exist, thus negatively affecting the system performance. To achieve the goal of memory access equalization (i.e. balancing the latencies of memory accesses and reducing the number of memory accesses with high latencies), the paper proposes to prioritize the memory access packets through predicting their round-trip routing latencies. The communication distance and the number of the occupied items in the input buffers in the remaining routing path are used to predict the future possible waiting time of a memory access on-the-fly, which is part of the round-trip routing latency. Experiments with varied network sizes and packet injection rates demonstrate that our approach outperforms the classic round-robin arbitration in terms of maximum latency, average latency, and LSD and can achieve the goal of memory access equalization. In the future, we plan to apply real application workloads to evaluate the performance of our proposal, and link and optimize our approach on other irregular or heterogeneous 3D many-core NoCs.

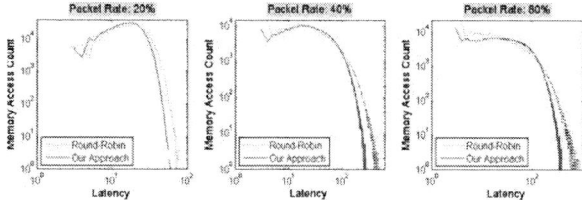

Fig. 9. Comparison of memory access latency dispersion between round-robin arbitration and our approach under varied packet injection rates in $4 \times 4 \times 4$ network.

Acknowledgment

The research is partially supported by the Hunan Natural Science Foundation of China (No. 2015JJ3017), and the Doctoral Program of the Ministry of Education in China (No. 20134307120034), and the National Natural Science Foundation of China (No. 61402500).

References

[1] J. D. Owens, W. J. Dally *et al.*, "Research challenges for on-chip interconnection networks," *IEEE MICRO*, vol. 27, no. 5, pp. 96–108, Oct. 2007.

[2] D. H. Kim, K.Athikulwongse, H. M. *et al.*, "3d-maps: 3d massively parallel processor with stacked memory," in *IEEE International Solid-State Circuits Conference, Digest of Technical Papers*, Feb. 2012, pp. 188–189.

[3] D. Fick, R. Dreslinski, B. Giridhar *et al.*, "Centip3de: A 3930 dmips/w configurable near-threshold 3d stacked system with 64 arm cortex-m3 cores," in *IEEE International Solid-State Circuits Conference, Digest of Technical Papers*, Feb. 2012, pp. 190–191.

[4] R. Dreslinski, D. Fick, B. Giridhar *et al.*, "Centip3de: A 64-core, 3d stacked near-threshold system," *IEEE Micro*, vol. 33. no. 2, pp. 8–16. Feb. 2013.

[5] M. Wordeman, J. Silberman, G. Maier *et al.*, "A 3d system prototype of an edram cache stacked over processor-like logic using through-silicon vias," in *IEEE International Solid-State Circuits Conference, Digest of Technical Papers*, Feb. 2012, pp. 186–187.

[6] Y. Zhang, L. Li, Z. Lu *et al.*, "Performance and network power evaluation of tightly mixed sram nuca for 3d multi-core network on chips," in *Proceedings of IEEE International Symposium on Circuits and Systems (ISCAS)*, Jun. 2014, pp. 1961–1964.

[7] D. Genius, "Measuring memory access latency for software objects in a numa system-on-chip architecture," in *Proceedings of the 8th International Workshop on Reconfigurable and Communication-Centric Systems-on-Chip (ReCoSoC)*, Jul. 2013, pp. 1–8.

[8] Z. Majo and T. R. Gross, "Memory system performance in a numa multicore multiprocessor," in *Proceedings of the 4th Annual International Conference on Systems and Storage*, May 2011, pp. 1–10.

[9] S. Rixner, W. Dally. U. Kapasi. P. Mattson, and J. Owens. "Memory access scheduling," in *Proceedings of the 27th International Symposium on Computer Architecture (ISCA)*, Feb. 2000, pp. 128–138.

[10] O. Mutlu and T. Moscibroda, "Stall-time fair memory access scheduling for chip multiprocessor," in *Proc. of the 40th Annual IEEE/ACM International Symposium on Microarchitecture (MICRO)*, Dec. 2007, pp. 146–160.

[11] M. Daneshtalab, M. Ebrahimi, J. Plosila, and H. Tenhunen, "Cars: Congestion-aware request scheduler for network interfaces in noc-based manycore systems," in *Proc. of the Design, Automation and Test in Europe Conf. (DATE'13)*, Mar. 2013, pp. 1048–1051.

[12] D. Kim, S. Yoo, and S. Lee, "A network congestion-aware memory controller," in *Proc. of the 4th ACM/IEEE International Symposium on Networks-on-Chip*, May 2010, pp. 257–264.

[13] W. Jang and D. Pan, "An sdram-aware router for networks-on-chip," *IEEE Trans. on Computer-Aided Design of Integrated Circuits and Systems*, vol. 29, no. 10, pp. 1572–1585, Oct. 2010.

[14] T. Pimpalkhute and S. Pasricha, "Noc scheduling for improved application-aware and memory-aware transfers in multi-core systems," in *Proc. of the 27th International Conference on VLSI Design and the 13th International Conference on Embedded Systems*, Jan. 2014, pp. 234–239.

[15] A. Sharifi, E. Kultursay, M. Kandemir, and C. Das, "Addressing end-to-end memory access latency in noc-based multicores," in *Proc. of the 45th Annual IEEE/ACM International Symposium on Microarchitecture (MICRO)*, Dec. 2012, pp. 294–304.

[16] W. J. Dally, "Virtual-channel flow control," *IEEE Transactions on Parallel and Distributed Systems*, vol. 3, no. 2. pp. 194–205, Feb. 1992.

[17] A. Kumar. L. Peh, P. Kundu, and N. Jha, "Express virtual channels: Towards the ideal interconnection fabric," in *Proc. of the 34th Annual Int'l Symp. on Computer Architecture (ISCA)*, Dec. 2007, pp. 150–161.

Validating Delay Bounds in Networks on Chip: Tightness and Pitfalls

Alberto Saggio*, Gaoming Du[†], Xueqian Zhao* and Zhonghai Lu*

*Department of Electronics and Embedded Systems, KTH Royal Institute of Technology, Stockholm, Sweden.
[†]Institute of VLSI design, Hefei University of Technology, Hefei, China, 230009.
E-mail: *{saggio, xueq, zhonghai}@kth.se; [†]dugaoming@hfut.edu.cn

Abstract—**Analytical methods for estimating on-chip network performance can be very useful to accelerate and simplify the design process of Network on Chips. However, in order to increase the confidence in these approaches it is fundamental to perform systematic studies that assess their potential.**

We present a methodical investigation on the tightness between analytical end-to-end delay bounds and worst-case simulation latencies in various scenarios. We first introduce our network calculus based analytical technique to derive per-flow communication delay bounds. Then, we examine the worst-case performance analysis process in NoCs outlining the major aspects that affect the tightness. Finally, experimental results confirm our deductions and allow us to provide general guidelines to avoid pitfalls in the validation process of analytical delay bounds.

I. INTRODUCTION

Today, NoCs are in the limelight to provide the parallelism, re-usability and flexibility needed to decrease the actual productivity gap [1]. Despite their huge potential, NoCs intrinsically define a wide design space that may make the design process hard, long and, thus, expensive both in time and costs. To deal with real-time applications the situation becomes even worse; the on-chip network designs must consider strict worst-case requirements for data transfers and provide guarantees on performance. To include such information in the design process two different tasks need to be performed, i.e., (i) to specify the communication requirements between cores and (ii) to evaluate the on-chip network performance to yield a feedback mechanism. The first point is generally achieved by using high level abstract models able to, e.g., capture the average rate and the burstiness of the traffic streams, while the second one generally requires tedious and time consuming simulations. This could be a fierce bottleneck because the design process requires several iterations and, moreover, many solutions must be explored each time. Although simulations can be performed at different abstraction levels (obviously accepting different accuracies), even the fastest simulations can lead to unacceptable loss of time. Furthermore, due to the fact that it is impossible for timing constraints to simulate every possible scenario, simulations cannot provide 100% guarantee that performance requirements are met. These considerations emphasize the necessity to research into analytical methods capable of rapidly estimating on-chip network performance and, thus, of going "*beyond simulation*" [1].

Network Calculus [4] has been proposed as a formal approach to evaluate performance bounds of NoCs. Previous works have demonstrated both the correctness of the bounds obtained and the Network Calculus' capability to provide worst-case performance guarantees for application specific NoCs [2] [3]. However, one important issue has been insufficiently investigated on previous research and needs to be emphasized in order to generate confidence in applying Network Calculus theory with NoCs: a tightness study of the analytical bounds. This motivates us to evaluate the gap between analytical delay bounds and worst-case simulation values in various contexts and to draw conclusions on the analytical bounds validation process. Nevertheless, this can be a very hard task, because it is laborious to discover the worst-case scenarios and, thus, compare the analytical bounds with the real unknown worst-case values.

This problem has led us to carrying out a second parallel investigation all over this work: are simulations truly mimicking the worst-case scenario? In fact, experiments cannot be conducted unilaterally without losing their consistency.

In this work, we first introduce a network calculus based analytical technique to derive per-flow communication delay bounds. Then, we discuss the complexity of finding the worst-case through simulations and we elucidate our preferred simulation approach. Finally, we analyze the main factors that affect the tightness and we compare analytical results against time consuming worst-case simulations. Our experimental setups aim to unravel the relationship between the tightness and the basic network and traffic's parameters. In addition, our results allow us to underline pitfalls in the validation process and to provide guidelines for successful and trustworthy validations.

The paper's contributions can be summarized as follows:

(i) We identify and reason about the main aspects that affect the tightness between analytical bounds and worst-case simulations.

(ii) We use a network calculus based analytical technique to conduct experiments in specific setups that help us both to grasp the tightness' behavior and to expose pitfalls in the validation process.

(iii) By reasoning on the results we provide basic guidelines for successful and trustworthy validations.

The remainder of the paper is organized as follows. Section II reviews related work. In Section III, we describe the basic Network Calculus concepts and results, and we present the general approach for calculating the analytical bounds for a tagged traffic flow. In Section IV, we analyze in detail the

tightness and all the aspects involved. Experiments and results are presented and discussed in Section V. Finally, we draw conclusions in Section VI.

II. RELATED WORK

Network calculus was pioneered by Cruz [4] in 1991. Since then, it has been developed into a comprehensive theory and laid a solid foundation for QoS analysis and provisioning in ATM and Internet (both differentiated and integrated services) [5] [6]. A stochastic version of network calculus was developed in [7]. Recently, it has been applied to real-time embedded systems [8], to SoCs [10] and to NoCs [2] [3] [12].

Realtime calculus [8] combines network calculus with real-time scheduling theory to reason about the worst-case delay and backlog bounds under various scheduling policies (fixed priority, TDMA, EDF, FIFO, etc.) in a compositional way. The potential of saving resources when relaxing hard QoS requirements was studied in [10]. Through a probabilistic study on multimedia playout buffer, Raman et al. showed that using reduced delay requirement can enormously save on-chip memory requirement.

The application of network calculus to on-chip network performance analysis is a very recent effort. In [2] [3], analysis techniques have been developed to analyze deterministic worst-case performance in NoCs. As the network problem is essentially a contention problem, the work in [3] categories basic network contention patterns for buffered traffic flows and derives their individual analytical models. Then, based on the basic analytical models, they use a contention-tree model to capture all possible interferences (combined contention patterns) for a flow and derive per-flow end-to-end delay bound by iteratively utilizing the basic analytical models. In [2], besides buffer-resultant contention, they further consider link sharing and credit-based feedback control in the analysis. They propose an analysis method which transforms a feedback analysis model into an open-ended analysis model, for which the delay-bound derivation technique developed in [3] can be applied. In [12], based on xMAS (eXecutable Micro-Architectural Specification) [9] which is a formal framework modeling communication fabrics, a per-flow delay bound analysis procedure using network calculus is presented.

Though the application of network calculus in on-chip networks covers various aspects, few works systematically study the tightness of delay bounds. In [2], the per-flow delay bounds are reported within 86% to 96% of simulated results. However, there lacks of systematic study and analysis of the gap between analytical results and simulated results. Needless to say, tightness is a crucial issue since loose bounds mean inferior performance, which may require large resources to compensate, likely leading to the over-dimensioning of NoCs. Zhao and Lu studied the tightness of NoC delay bound analyzed by network calculus in [13]. They propose a heuristic approach to guide the search of flow and service parameter configurations for worst-case scenarios in simulations.

III. END-TO-END DELAY BOUND ANALYSIS

A. Basic Results of Network Calculus

In network calculus, traffic and network elements are modelled following the abstraction of the *arrival curve* and the *service curve*, respectively. The arrival curve sets the upper bound on the injected traffic within any time interval whereas the service curve defines the lower bound on the available service capability. Throughout the paper, we assume that all flows can be characterized with a linear arrival curve [4] $\alpha(t) = rt + b$, where b bounds the burstiness and r the average (sustainable) rate, and that all network elements (i.e., the routers) provide a service curve equal to the latency-rate function [11] $\beta_{R,T} = R(t - T)^+$, where R is the minimum service rate, T the maximum processing latency and x^+ equals 0 if $x \leq 0$, else x. Under the stability condition ($R \geq r$), when a flow f with a linear arrival curve is served by a node guaranteeing a latency-rate server $\beta_{R,T}$ [5], the maximum delay for the flow is bounded by the equation: $\bar{D} = \frac{b}{R} + T$.

B. Delay Bound Analysis

We call *tag flow* and *contention flows* respectively the flow for which we shall derive its delay bound and the other flows that share resources with it. Our aim is to obtain a simple closed-form formula to derive the upper delay bound for the tag flow and then use it in the experiments to judge the tightness' behavior in different conditions.

Based on [2] [3], we summarize the main procedure for calculating the per-flow delay bound as follows:

1. Specify tag flow and contention flows.
2. Calculate the *equivalent service curve* (ESC) of the tag flow. In order to get the ESC the basic principle is to remove the effect of contention flows from the tag flow path and to calculate the leftover service. In [2], three basic contention patterns in NoC, namely nested, parallel and crossed, are identified and analyzed. Following [2] [3], the system ESC of the tag flow can be calculated by resolving the contention patterns.
3. Finally, the worst-case delay for the tag flow can be calculated using the delay bound equation [5].

C. An Example

Figure 1 shows an example with four flows in a 4×4 mesh NoC. Its delay bound derivation is presented without providing all analytical details.

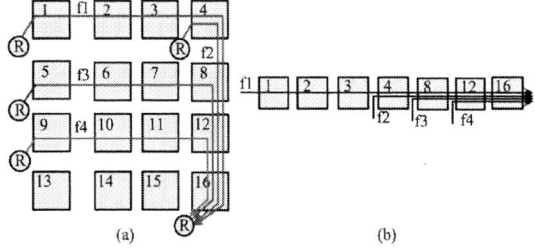

Fig. 1. (a) Four flows in a 4×4 Mesh NoC; (b) Contention map.

978-1-4799-8720-7/15 $31.00 © 2015 IEEE

Step 1. Specify the tag flow: We select f_1 as the tag flow and f_2, f_3 and f_4 as contention flows. The injection of flow f_i is constrained in node n by a linear arrival curve $\alpha_n^{fi} = r_i t + b_i$.

Step 2. Calculate the ESC of the tag flow: The goal is to obtain the ESC of f_1 in the format $\beta_{eq}^{f1} = R_{eq}^{f1}(t - T_{eq}^{f1})^+$ so that the equivalent minimum service rate R_{eq}^{f1} and the equivalent maximum processing latency T_{eq}^{f1} can be used in the delay bound equation above.

The contention relationship is shown in figure 1(b). Tag flow f_1 traverses through node 1, 2, 3, 4, 8, 12 and 16. Contention flows f_2, f_3 and f_4 are injected at nodes 4, 8 and 12, respectively. We begin by expressing the service that nodes 12 and 16 can provide for all flows as $\beta_{12} \otimes \beta_{16}$, where $\otimes$ is the min-plus convolution [5]. To include node 8 we first need to remove f_4 out of the router chain and get the leftover service for the remaining flows. This can be expressed as $\in \left(\beta_{12} \otimes \beta_{16}, \alpha_{12}^{f4} \right)$, where $\in (\beta, \alpha)$ is the function to find the ESC. Hence, the service that nodes 8,12 and 16 provide for f_1, f_2 and f_3 is $\in \left(\beta_{12} \otimes \beta_{16}, \alpha_{12}^{f4} \right) \otimes \beta_8$. We continue this procedure until we obtain the service that nodes 1, 2, 3, 4, 8, 12 and 16 provide for f_1, i.e., the desired β_{eq}^{f1}. However, this result is function of the arrival curves α_4^{f2}, α_8^{f3} and α_{12}^{f4} which first need to be calculated as $\alpha_4^{f2} = r_2 t + b_2$, $\alpha_8^{f3} = r_3 t + b_3 + r_3 (T_5 + T_6 + T_7)$ and $\alpha_{12}^{f4} = r_4 t + b_4 + r_4 (T_9 + T_{10} + T_{11})$, where T_i is the maximum processing latency at node i.

Finally, with the equations above we can express the ESC of the tag flow through:

$$R_{eq}^{f1} = R - r_2 - r_3 - r_4;$$

$$T_{eq}^{f1} = 7T + \frac{b_2}{R - r_3 - r_4} + \frac{b_3 + 3r_3 T}{R - r_4} + \frac{b_4 + 3r_4 T}{R}.$$

Step 3. Calculate the tag flow's delay bound: The delay bound of f_1 can be obtained as: $D^{f1} = T_{eq}^{f1} + \frac{b_1}{R_{eq}^{f1}}$.

IV. General discussion

Latency can be decomposed into the sum of two independent terms: (i) one directly dependent from fixed network and traffic physical factors and (ii) one that depends on the actual traffic pattern. The first one, known as *zero-load latency*, is of easy derivation and represents the delay without any contention (lower bound); the second one introduces uncertainty and the need for the *worst-case* analysis of latency.

A. Worst-case

Finding the worst-case traffic pattern inside a network is a non-trivial task due to the high variety of different contention scenarios that can occur. In particular, there are mainly two traffic aspects to consider hereof.

- *Spatial communication pattern*: Two flows can experience a contention only if they share a link in their source-destination paths; ergo, the routing strategy is binding.
- *Temporal traffic behavior*: Other than a spatial condition, one contention also needs that the two flows attempt to use a link in the same cycle. Obviously, this depends on

when the flows inject flits in the network. In this sense the permutation of the flows' temporal injection patterns is a powerful instrument to search the worst-case scenario.

B. Worst-case Simulations

Unfortunately, the real worst-case is not known a priori, hence we rely on long worst-case simulations to validate analytical delay bounds. This emphasizes the need to conduct trustworthy simulations able to accurately approximate the real worst-case. In particular, three aspects need to be considered.

1) Traffic generation: Two approaches are mainly used, i.e., (i) application workload and (ii) synthetic generation. Although it is a common debate to decide which one should be used in specifics contexts, these two techniques have very different characteristics and lead to very different simulation behavior. As the names suggest, while the second one directly controls the injection process using a "fake" traffic, the first one uses traffic injection produced by real applications. This means that without inserting an injection control mechanism, the only way to increase its probability of randomly obtaining the worst-case is to drastically raise simulation time leading to day-long simulations. For our methodical study its utilization is even less advisable; the experiments must be carried out respecting the specifications with a very high accuracy and repeatability of the results, otherwise the tightness study will lose consistency.

2) Resource allocation: Flow contentions are safely resolved by using arbitration policies that, however, introduce nondeterministic outcomes in the network. This exponentially increases the number of dynamics that needs to be explored and therefore requires to drastically raise the simulation time. In particular, to fully guarantee the simulated worst-case result one should simulate all the traffic injection patterns with all the possible arbitration outcomes (de facto impossible in most cases). This explains why often simulations seem mimicking the worst-case but are still not tight with the analytical bounds.

3) Exploration strategy: Given the traffic specification of the flows, finding the worst-case scenario relies on a detailed exploration both of the network and of the temporal traffic behavior. The exploration strategy is, therefore, very important and must be chosen properly.

Working on the permutation of the injection traffic pattern is usually the best choice and basically consists in: (i) *generation* of one specific traffic pattern; (ii) *simulation* for a fixed time; (iii) *mutation* of the pattern; (iv) *loop* of the last two operations. Distinct algorithms, then, relies on different ideas exploited to mutate the pattern towards the worst-case. Basically, it is possible to (i) induce random variation of a starting traffic pattern and simulate it for a long time, (ii) try to explore all the possible traffic pattern (*exhaustive permutation*) and (iii) realize an experience-based algorithm (heuristics, genetic, etc.). While the first method is very inaccurate and the third one is very difficult to realize in this field due to plenty of local optimal solutions, the second one guarantees the best exploration and a great repeatability of the results (all key aspects in validating analytical bounds). This, however,

978-1-4799-8720-7/15 $31.00 © 2015 IEEE   406

does not come without any drawback. In fact, the timing consumption of this method greatly increases with the complexity of the NoC and, moreover, no 100% guarantees of finding the real worst-case can be provided due to nondeterministic events (e.g., arbitration outcomes) that require many simulations of the same traffic pattern.

C. Tightness

In this work, we consider the tightness as *the difference between the derived analytical delay bound and the simulated worst-case latency*. We mainly identify the following factors that affect the tightness.

- *Resource allocation*: Its already discussed unpredictability in simulations reduces the tightness at every increase of the number of contentions experienced in the network.
- *Bottleneck links, link throughput and link sharing*: The more one link is used, the easier it is to experience contentions and waiting times due to its allocation. Therefore, bottleneck links, link sharing and in general the link throughput can affect the tightness.
- *Flow burstiness*: A tag flow with a very long burst injection does not experience high delays if the contention flows have short burst injections and, thus, few flits to content with. This means that the distribution of the flits in the traffic injection pattern greatly affects the number of contentions and, due to the allocation unpredictability, the tightness.
- *Routing strategy*: To successfully apply the closed-form formula of our analytical method we consider only deterministic routing algorithms. Moreover, non-deterministic routings exponentially increase the number of possible scenarios "growing" from a single injection pattern making it even harder to simulate the real worst-case.
- *Simulation strategy*: We use the exhaustive approach to provide the highest guarantees on the accuracy and repeatability of the simulations. Indeed, with application workload approaches can be difficult to converge the tightness due to inflexible traffic injection.

V. Experiments

We perform experiments on three different size mesh NoCs (Fig. 2). The tightness is expressed as a *gap* (i.e., the difference between analytical and simulation values), thus a lower (higher) gap implies a tighter (less tight) analytical bound. The

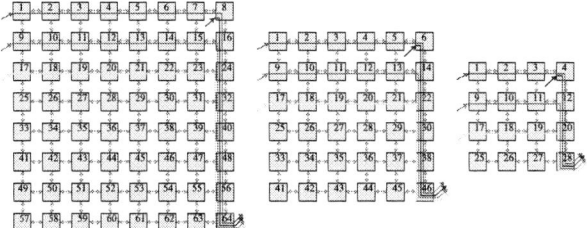

Fig. 2. Synthetic communication platforms used in the experiments.

designed experimental platforms use 4 flows each. We set the flow injected in node 1 as tag flow, thus leaving the others as contention flows. All the router nodes have a service curve of $\beta(t) = \delta \otimes \gamma_{0,1}(t) = 1 \cdot (t-2)^+$, i.e., every router can forward one flit per cycle with a delay of two cycles.

We use SoCLib [14] to realize cycle-accurate simulation platforms for the NoCs. The network performs wormhole switching with the STALL/GO flow control. Each router guarantees the service curve of $\beta(t)$, uses the packet-by-packet round-robin scheme [2] for link scheduling and adopts the deterministic X-Y routing algorithm. We build the router with big enough buffer size and assume zero link delay in all the experiments.

According to the reflections presented in Section IV, we carry out the simulation experiments using synthetic traffic and an exhaustive permutation algorithm that:

1. computes all possible traffic injection patterns;
2. selects one pattern and pre-loads the network abiding by the traffic specifications;
3. runs the selected traffic pattern;
4. re-iterates until all patterns have been run multiple times.

The traffic specifications for each flow (i.e., the arrival curves) can be expressed as a periodic injection of the maximum amount of flits allowed. As an example, an arrival curve of $\alpha(t) = 0.05t + 1$ can be translated in the injection of 1 flit every 20 cycles. To simplify the reasoning over the worst-case experiments (in particular concerning the link throughput), all the following traffic specifications of each flow are expressed as a maximum number of flits over a fixed period of 20 cycles. Moreover, the all-to-one communication fashion helps further to make conclusions about saturation and bottleneck links in the on-chip network. Indeed, in this kind of communication, all the flows must share at least the output link, meaning that this must be one bottleneck link (or one of the many bottleneck links, depending on the flows path).

A. Link Throughput, Bottleneck link and Saturation Point

1) Setup: In order to draw conclusions about the influence of different link throughput on the tightness, we start identifying the bottleneck link (intended as in communication network, i.e., a link that for a given data flow is fully utilized) in the experimental NoCs, then we derive the respective traffic specification that saturates that link and finally we obtain and plot the analytical and simulation results with the variation of its throughput.

Thanks to the fixed flit injection period of 20 cycles per flow, when a link is fully utilized any rise in the injection rate of the flits passing through that link would cause an unbounded delay, meaning that we would work beyond the saturation point of the network. Therefore, we increase the throughput of the bottleneck link from a very low one (i.e., 4 flit every 20 cycles across the bottleneck link) until we reach the saturation point (i.e., 20 flits every 20 cycles across the bottleneck link). To increase the throughput we raise equally the flits injection among the flows, thus varying from 1 flit every 20 cycles per flow to 5 flits every 20 cycles per flow.

2) Results: Fig. 3 depicts both the analytical bounds and the worst-case simulation delays for the wider NoC used. As expected, the analytical bounds totally constrain the simulated values. The two curves are very tight when the throughput is low and then they gradually separate while reaching the saturation point where the distance is the 36.17% of the analytical bound. In Fig. 4 the tightness is explicitly shown and confirms this trend.

We explain this behavior considering that when we inject few flits per flow in the network (low link throughput) we witness very few contentions and thus a very predictable and easy to "mimic with simulations" scenario. However, if we start to congest the network and we reach the saturation point, we saturate links that need to be allocated by the routers in a round-robin fashion; this increases the unpredictability in the network leading to both higher analytical delay bounds and less trustworthy simulations in terms of worst-cases. In fact, to obtain an acceptable probability to mimic the worst-case, every single traffic pattern needs to be simulated many times leading to either longer or less accurate simulations.

Therefore, this elucidates why the two curves tend to diverge towards the saturation condition. When coping with increasing unpredictability in the network, the analytical bounds (which are by definition upper-limits that must not be exceeded) tend to over-bound, whereas the worst case simulations (which rely on the observed worst-case value) tend to under-estimate.

B. Flow burstiness

1) Setup: With this experiment we want to evaluate the variation of the tightness when the traffic specifications of the tag flow are modified. To estimate properly this aspect it is important to keep a constant network congestion level while changing the injection parameters of the tag flow, otherwise we would shuffle also the effect discussed in the previous experiment and, thus, lose objectiveness. Therefore, we conduct the experiment at the saturation point, that in our case means maintaining the bottleneck link throughput of 1 flit per cycle. In order to achieve this, every time the number of flits of the tag flow is increased (or decreased) the number of flits of the contention flows are accordingly decreased (or increased). Hence, we first vary the tag flow injection from 1 flit to 10 flits every 20 cycles, then we derive the respective traffic specification for the analytical method and, finally, we obtain and plot the analytical and simulation results in the same conditions.

2) Results: Fig. 5 shows the results for the 8x8 mesh NoC. Once again, we can see that the simulated delays are totally constrained by the calculated delay bounds. From the figure it is evident that when we increase the number of flits injected by the tag flow both the analytical bounds and the simulation results decrease. This is because we hold the same congestion level by decreasing the number of flit injected by the contention flows, or to be more precise we accordingly decrease the contention flows' burstiness. As discussed in Section IV, this makes the tag flow always experience fewer contentions.

The tightness behavior is shown in Fig. 6. As depicted, the difference between analytical and simulation values gets smaller with the increment of the number of flits injected per period by the tag flow. This is exactly what we expect and exposes a pitfall: *validating the analytical bounds at the saturation point is a necessary but not sufficient condition to generate confidence on the results.* The flows' burstiness has a direct impact on the tightness and, in particular, it is important to alter the ratio between the tag flow burstiness and the contention flows' burstiness. Indeed, in this setup we experience and depict the widest gap between the curves.

C. Mesh size

We conduct all the previous experiments on three different size meshes in order to estimate the variation of the tightness with the size of the on-chip network.

Results are shown in Fig. 7 and Fig. 8 respectively for the first and the second case study. As depicted, the three meshes exhibit an almost identical gap; this let us imply that *the tightness is not affected by the size of the mesh.* The truly relevant aspect is the contention relationship between the flows and, thus, if we keep it similar while increasing the size of the NoC, then we shall not appreciate differences in the tightness.

We believe that this is an important result that verifies that our analytical method maintains the same accuracy regardless of the on-chip network dimension. Furthermore, this setup enhances the time convenience of analytical bounds; whereas a formal approach is not affected by the size of the NoC, the simulation speed can drastically decrease.

D. Links sharing

1) Setup: In the 4x4 mesh NoC we move to node 12 the ejection point of the two flows injected at node 4 and node 9, and we perform the same throughput analysis conducted in the first experiment. This drastically reduces both the traffic

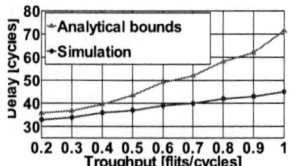

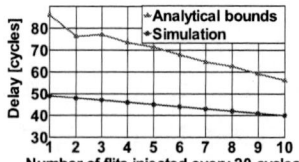

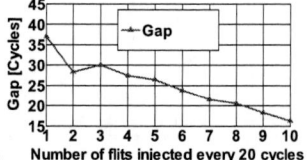

Fig. 3. Analytical bounds and simulation results with the variation of the bottleneck link throughput in the 8x8 mesh NoC.

Fig. 4. Tightness behavior with the variation of the bottleneck link throughput in the 8x8 mesh NoC.

Fig. 5. Analytical bounds and simulation results with the variation of the number of flits injected by the tag flow every 20 cycles (8x8 mesh).

Fig. 6. Tightness behavior with the variation of the number of flits injected by the tag flow every 20 cycles in the 8x8 mesh NoC.

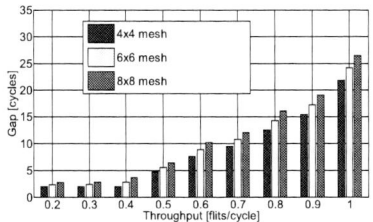
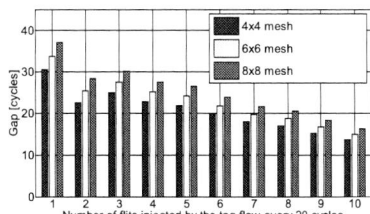
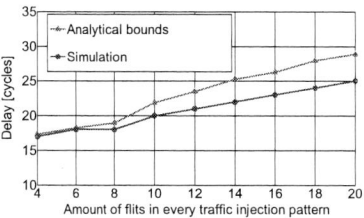

Fig. 7. Tightness of different size mesh NoCs considering the variation of the bottleneck link throughput.

Fig. 8. Tightness of different size mesh NoCs considering the variation of the number of flits injected by the tag flow every 20 cycles.

Fig. 9. Analytical bounds and simulation results with the variation of the bottleneck link throughput.

through the previous bottleneck links (i.e., links from node 12 to node 28) and the number of links shared by the flows.

2) Results: Results are shown in Fig. 9. As depicted, the tightness does not exhibit significant variation and, moreover, the comparison with the values in the first experiment reveals much more tight results. Therefore, this setup had a positive effect; we still guide the network towards saturation but the reduced link sharing between tag flow and contention flows directly affects the contentions pattern improving the reproducibility of the outcome and thus the tightness. Driving a network towards saturation is an important scenario for analytical bounds validations. Nevertheless, it is fundamental to combine it with different links sharing's configurations.

VI. CONCLUSION

Validating analytical end-to-end delay bounds is a non-trivial task that should be conducted in a methodical way to avoid pitfalls which could generate misleading results. In this paper we have presented a systematic investigation of the tightness between analytical delay bounds and worst-case simulation latencies with a particular attention in outlining the dominant aspects that affect its behavior. This is an imperative step that has allowed us to define a set of scenarios in which the analytical bounds can be validated to provide confidence in the results.

Experiments have suggested that the tightness is mainly influenced by stochastic events that both lead to higher analytical bounds and less trustworthy simulations in terms of worst-case scenarios. Furthermore, misleading results have exposed pitfalls in the validation process: with the exclusion of the flow burstiness experiment we have gathered wrongly promising findings that have validated the accuracy of our analytical method; nevertheless, the flow burstiness experimental setup has clearly demonstrated how the tightness can drastically change and affect the outcome of the validation process.

Working at the saturation point can be crucial but only if wisely combined with other factors: (i) the ratio between the tag flow and the contention flows' burstiness can have a big impact on the tightness at saturation conditions and (ii) the network saturation must be combined with appropriate link sharing patterns between the flows. In particular, our experiments have shown that the more contentions are possible with a specific nework configuration, the wider becomes the gap between simulation results and analytical delay bounds.

As a result of the considerations hereof, we believe that succsessful validations should:

- show results with a gradual variation of link throughput in the network;
- run tests at the saturation point and with a variation of the flow burstiness (with particular attention to the ratio between tag flow and contention flows);
- run tests at the saturation point and with a variation of the link sharing to judge the respective tightness variation;
- show the tightness' independence from the mesh size.

We consider this work as a crucial step towards comprehending what really affects the tightness and in which setups the analytical bounds should be validated to assess their real quality and re-usability. In the future we plan to extend this investigation with the effects of queue sizes, buffer sharing and other aspects that we assumed herein constant (e.g., arbitration algorithms, routing strategies, NoC topologies).

REFERENCES

[1] J. Owens et al., "Research challenges for on-chip interconnection networks," in *IEEE Micro*, vol.27, no.5, pp.96-108, 2007.
[2] Y. Qian, Z. Lu, and W. Dou, "Analysis of worst-case delay bounds for best-effort communication in wormhole networks on chip," in *Proc. 3rd ACM/IEEE Int. Symp. on NOCs*, San Diego, CA, May 2009.
[3] Y. Qian, Z. Lu, and W. Dou, "Analysis of worst-case delay bounds for on-chip packet-switching networks," in *IEEE Trans. Comput.-aided Des. Integr. Circuits Syst. (TCAD)*, vol.29, no.5, pp.802-815, May 2010.
[4] R.L. Cruz. "A calculus for network delay, part I: Network elements in isolation; part II: Network analysis," in *IEEE Trans. Inf. Theory*, 1991
[5] J.-Y. Le Boudec and P. Thiran, "Network Calculus: A Theory of Deterministic Queuing Systems for the Internet," in *Lect. notes Comput. Science 2050*, Berlin, Germany: Springer-Verlag, 2004.
[6] C. Chang, "Performance Guarantees in Communication Networks," *Springer-Verlag*, 2000.
[7] Y. Jiang and Y. Liu, "Stochastic Network Calculus," *Springer*, 2008.
[8] E. Wandeler, L. Thiele, M. Verhoef, and P. Lieverse, "System architecture evaluation using modular performance analysis: a case study," in *Soft. Tools for Techn. Transfer (STTT)*, vol.8, iss.6, pp.649-667, 2006.
[9] S. Chatterjee, M. Kishinevsky, and U. Ogras, "xMAS: Quick formal modeling of communication fabrics to enable verification," in *IEEE Design & Test of Computers*, vol.29, no.3, pp.80-88, June 2012.
[10] B. Raman et al., "On buffering with stochastic guarantees in resource constrained media players," in *Proc. CODES+ISSS*, pp.169-178, 2011.
[11] D. Stiliadis and A. Varma, "Latency-rate servers: a general model for analysis of traffic scheduling algorithms," in *IEEE/ACM Trans. Netw.*, vol.6, no.5, pp.611-624, Oct. 1998.
[12] X. Zhao and Z. Lu, "Per-Flow Delay Bound Analysis Based on a Formalized Micro-architectural Model," in *Proc. 7th ACM/IEEE Int. Symp. on NOCs*, Tempe Arizona, USA, April 2013.
[13] X. Zhao and Z. Lu, "Empowering Study of Delay Bound Tightness with Simulated Annealing," in *Proc. Des. Autom. Test in Eur. Conf.*, 2014
[14] Soclib simulation environment. On-line, available at http://soclib.lip6.fr/.

Optimized Use of Parallel Programming Interfaces in Multithreaded Embedded Architectures

Arthur F. Lorenzon[1], Anderson L. Sartor[1], Márcia C. Cera[2] and Antonio Carlos Schneider Beck[1]

[1]Federal University of Rio Grande do Sul – Institute of Informatic, Porto Alegre – Brazil

[2]Federal University of Pampa – Campus Alegrete, Alegrete – Brazil

marciacera@unipampa.edu.br, {aflorenzon, alsartor, caco}@inf.ufrgs.br

Abstract— Thread-level parallelism (TLP) exploitation for embedded systems has been a challenge for software developers: while it is necessary to take advantage of the availability of multiple cores, it is also mandatory to consume less energy. To speed up the development process and make it as transparent as possible, software designers use parallel programming interfaces (PPIs). However, as will be shown in this paper, each one implements different ways to exchange data, influencing performance, energy consumption and energy-delay product (EDP), which varies across different embedded processors. By evaluating four PPIs and three multicore processors, we demonstrate that it is possible to save up to 62% in energy consumption and achieve up to 88% of EDP improvements by just switching the PPI; and that the efficiency (i.e., the best possible use of the available resources) decreases as the number of threads increases in almost all cases, but at distinct rates.

Keywords—Parallel programming interfaces, Performance and energy efficiency evaluation, Multithreaded embedded architectures

I. INTRODUCTION

High-end embedded systems have been built on top of multicore processors to meet applications demands for more computing power [22]. Moreover, there is the concern of reducing energy consumption, since most embedded devices are mobile and heavily dependent on battery power. The increase in performance is a result of TLP exploitation, which involves multiple processors simultaneously executing different parts of the same program. To speed up the development process and make it as transparent as possible to the programmer, different PPIs are used (e.g., OpenMP – Open Multi-Processing [2], Pthreads - POSIX Threads [3], or MPI – Message Passing Interface [4]). However, each one of these has different characteristics with respect to the management (i.e., creation and finalization of threads/processes), distribution of workload, and synchronization.

In any of the previously mentioned PPIs, threads/processes communicate during execution by the use of shared memory: through shared variables (OpenMP and Pthreads) or send/receive primitives (MPI is implemented using a shared memory context [19]). These memory regions are usually more distant from the processor (e.g., L3 cache and main memory) and have higher delay and power consumption when compared to memories that are closer to the processor (e.g., registers, L1 and L2 caches) [1].

Considering the aforementioned scenario, one can infer that the more communication a parallel application has, the more energy it will consume. On the other hand, performance will increase as a result of the parallelization, which may lead to energy reductions. However, this performance increase is not linear and sometimes does not scale with the number of threads, due to synchronization and bandwidth of the communication bus [5].

In this work, we assess the influence of the PPIs on the energy efficiency of parallel applications with different demands of communication for embedded processors. Additionally, we show that the number of executed instructions and amount of memory accesses impact performance, energy consumption, and EDP. We demonstrate that, in some cases, the resource efficiency (i.e., the best possible use of the available resources) decreases as more TLP is exploited, which may be not acceptable for embedded systems. Most importantly, we show that by only switching PPI, huge savings in EDP can be achieved in a significant number of cases.

We consider 9 multi-threading interfaces: Pthreads; MPI-1; MPI-2; and OpenMP with different Iteration Distributions Policies (IDPs: static, dynamic and guided, and two granularities (fine and coarse); We use 11 benchmarks, classified according to communication rate (low and high), each one split in 2, 3, 4 and 8 threads/processes; resulting in a total of 36 possible combinations per application, executed on three embedded processors (ARM A8, A9, and Atom), with distinct number of cores.

The remainder of the paper is organized as follows. Related work is discussed in Section 2. The four PPIs are briefly described in Section 3. The benchmarks, their parallel implementation, and the execution environment are presented in Section 4. Section 5 discusses the results. Finally, Section 6 draws the final considerations.

II. RELATED WORK

Few studies investigated the potential of using PPIs to speedup embedded applications. The authors in [7] obtained speedups of up to 111% with 2 threads over the sequential version using OpenMP in the parallelization of two applications running on a Dual-core ARM Cortex-A9. The work presented in [8] evaluated the performance of three Symmetric Multiprocessing cores for embedded systems using OpenMP. The results indicated that non memory-intensive applications exhibit large speedups by parallelization. None of them took into account energy consumption.

Considering only sequential applications running on a single processor, the authors in [9] compared the impact of different microarchitectures and ISAs (Instruction Set

Architectures) on performance, power, energy and EDP for both general purpose and embedded systems. A great number of experiments were performed, evaluating the number of executed instructions and processor cycles, average instruction length, number of memory accesses (cache misses and hits), execution time and so on. The authors demonstrated that there is a significant difference between processors implementing distinct ISAs: while the Cortex-A9 presents the best trade-off considering EDP, the Intel Core i7 is the most effective when it comes to execution time. These results, however, are mainly dependent of the microarchitecture (e.g.: issue width, accuracy of the branch predictor) than the ISA itself.

The authors in [10] compared OpenMP with MPI on High Performance Computing (HPC) systems using different CPU clock frequencies. They demonstrate that the PPI has a major impact on the execution time and energy consumption of the applications. The authors in [11] evaluated OpenMP applications on HPC systems, showing that energy consumption can vary because of several factors such as the algorithm, compiler, optimization level, and number of threads. Even though performance was improved thanks to the TLP exploitation, energy consumption was substantially increased.

Although the authors have shown in [6] that embedded processors consume less energy to execute HPC applications compared to classical server processors, the use of PPIs has been little explored in this field. Thus, this article extends the aforementioned works by investigating which PPI (OpenMP, Pthreads, MPI-1 and MPI-2) optimizes the use of the available resources considering different embedded processors, number of cores and applications with distinct behaviors. We consider the number of memory accesses and executed instructions, performance, energy consumption, EDP, scalability and efficiency; and how each one influences the others.

III. PARALLEL PROGRAMMING INTERFACES

OpenMP is a PPI for shared-memory in C/C++ and Fortran, which consists of a set of compiler directives, library functions, and environment variables [2]. Parallelism is exploited through the insertion of directives in the sequential code that inform the compiler how and which parts of the code should be executed in parallel. *Pthreads* is a standard PPI for C/C++, which functions allow fine adjustment in the grain size of the workload. Thus, the creation/termination of the threads, the distribution of workload and the control of execution are fully defined by the programmer [3]. *MPI* is a standard message passing library standard for C/C++ and Fortran [4] that implements optimizations to provide communication in shared memory environments [19][20]. MPI is similar to Pthreads regarding the explicit exploitation of parallelism. In *MPI-1*, all processes are created at the beginning of the execution and the number of processes does not change throughout program execution; in *MPI-2*, processes are created at runtime and the number of processes may change during the execution [15].

The communication between threads in OpenMP and Pthreads occurs through shared variables in memory that can be accessed by all threads. To ensure data integrity, synchronization operations are used to control access to these variables by multiple threads. In OpenMP, synchronization may be implicit (implied barrier at the end of a parallel region),

or explicit (synchronization constructs) to the programmer. Whenever there is a synchronization point, OpenMP threads enter in a busy-waiting state, i.e., they access the shared memory repeatedly until the end of synchronization [13]. Pthreads synchronization is done by blocking threads with mutexes, which are inserted in the code by the programmer. In this case, threads lose the processor and wait on standby until the end of the synchronization, when they are rescheduled for execution [13]. Communication between MPI processes, on the other hand, occurs through send/receive operations (point-to-point or collective ones), which are likewise explicitly handled by the programmers. When MPI programs are executed on shared memory architectures, message transmissions are done as shared memory accesses, in which messages are broken into fragments that are pushed and popped in FIFO queues of each MPI process [19].

IV. METHODOLOGY

A. Benchmarks

Eleven parallel benchmarks were designed, with different demands of communication. They were classified into two classes according to the amount of operations to exchange data (barriers, locks/unlocks and operations to create/finalize threads): *Low Communication* (LC) and *High Communication* (HC). Table I quantifies the communication rate for each benchmark, considering 2, 3, 4 and 8 threads/processes. It also shows their inputs sizes. LC applications have little communication among threads/processes. In this case, communication occurs only to distribute the workload and to join the final result (as it is shown in Figure 1.a). On the other hand, HC applications require a larger amount of operations to ensure proper data exchange among threads/processes, as it is shown in Figure 1.b.

The applications were implemented using the C language. Since the way the parallel applications is written may influence its behavior during execution, we have followed the guidelines indicated by [2], [3], [4], [14], and [15]. Thus, the OpenMP implementations were parallelized using parallel loops, splitting the number of loops iterations (*for*) between the threads. As the authors discuss in [2], this approach is ideal for applications that computes on uni and bi-dimensional structures, which is the case. On the other hand, in Pthreads and MPI implementations, the loops iterations (*for*) were distributed based on the best workload balancing between threads/processes.

B. Execution Environment

The experiments were performed on three different embedded processors as shown in Table II: Intel Atom, which is becoming popular on mobile phones and tables; ARM Cortex-A8 and A9, which dominate the industry of embedded processors. To obtain the data directly from the hardware counters of these processors, without the influence of the operating system (i.e., function calls, interruptions, etc.), we used the Performance Application Programming Interface (PAPI [18]). With the counters, we can gather the *number of executed instructions, data/instructions memory accesses* and *number of executed cycles* to calculate performance and energy, as it will be further explained.

TABLE I. Main Characteristics of Benchmarks

Benchmarks		Operations exchange data (total per nº of threads/processes)*				Input Size
		2	3	4	8	
LC	Calc. of the Number PI	4	6	8	16	4 billions
	Harmonic Series	8	12	16	32	100000 elem.
	Dijkstra	4	6	8	16	2048 x 2048
	Similarity of Histograms	4	6	8	16	1920 x 1080
	Matrix Multiplication	4	6	8	16	2048 x 2048
HC	Jacobi	4004	6006	8008	16016	2048 x 2048
	LU-Decomposition	8192	12288	16398	32776	2048 x 2048
	Game of Life	414	621	1079	1625	4096 x 4096
	Gram-Schmidt	3009277	4604284	6385952	12472634	2048 x 2048
	Odd-Even Sort	300004	450006	600008	1200016	150000 elem.
	Turing Ring	16000	24000	32000	64000	2048 x 2048

*Considered operations: *mutex, barriers, and operations to create/finalize threads/processes.*

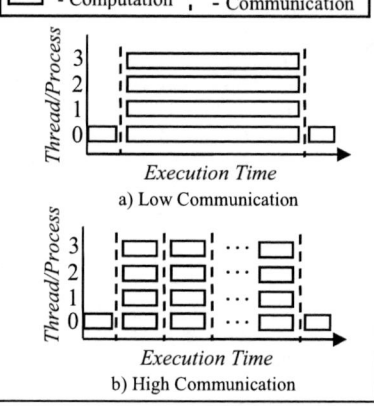

Fig 1. Behavior of Benchmarks

Due to restrictions of the hardware counters present in the ARM processors, it is not possible to extract data of the L2 caches accesses, so they are not considered for the energy consumption (but they are for performance, since we have the total number of executed cycles). Therefore, we are considering that every miss in L1 cache spends the energy for one access to the main memory, even for the Atom processor, so as not to bias the results. According to [21], the number of misses in the L1 cache is extremely low (less than 2%) for sequential applications, when one considers the cache sizes used in this work, which would reflect it even lower rates for the L2. Hence, as most of time the main memory is used for the communication among threads/processes, we strongly believe that the impact of not considering the L2 cache for energy consumption is negligible.

The energy consumption was calculated using the data provided by the authors in [9] (for the processor) and the Cacti Tool [16] (for the memory system), as shown in Table II. To estimate the total energy consumption (Et), we have taken into account the energy consumption of executed instructions (E_{inst}), cache and main memory (E_{mem}), and static energy (E_{static}), as given by (1).

$$Et = E_{inst} + E_{mem} + E_{static} \quad (1)$$

To find the energy consumed by the instructions, (2) was used, where I_{exe} is the number of executed instructions multiplied by the average energy spent by each one of them (E_{inst}).

$$E_{inst} = I_{exe} * E_{inst} \quad (2)$$

The total energy consumption for the memory system was obtained with (3), where ($DC_{acc} * E_{DC}$) is the energy spent by accessing the data cache memory; ($IC_{acc} * E_{IC}$) is the same, but for the instruction cache; and ($TC_{miss} * E_{main}$) is the energy spent by each main memory access.

$$E_{mem} = (DC_{acc} * E_{DC}) + (IC_{acc} * E_{IC}) + (TC_{miss} * E_{main}) \quad (3)$$

The static consumption of each component is given by (4). As static power is consumed while there is circuit activity, it must be considered during all execution time: (#Cycles) of the application divided by the operating frequency (Freq). We have considered the static consumption of the processor (S_{CPU}),

data (S_{DC}) and instruction (S_{IC}) caches, and main memory (S_{MAIN}).

$$E_{static} = \left(\frac{\#Cycles}{Freq}\right) * (S_{CPU} + S_{DC} + S_{IC} + S_{MAIN}) \quad (4)$$

The results presented in Section 5 consider the average of 10 executions, with a standard deviation of less than 1% for each benchmark. Their input sizes are described in Table I. The programs were split into 2, 3, 4, and 8 threads/processes. Although the processors support only four threads (Table II) and are not commercially available in an 8-core configuration, it is possible to approximate the results by assuming that at each synchronization point, the runtime for that period (i.e., in between two synchronization points) is the number of the slowest thread/process.

The compiler used was GCC-4.7.3 without optimization flags, in order to minimize the influence of the compiler. The following distributions were used: OpenMPI 1.6, OpenMP 3.0 and Pthreads/POSIX.1-2008, running on the Linux Debian operating system. OpenMP was executed with three different IDPs (*static, guided* and *dynamic*) and two granularities: *fine* and *coarse*. As observed in [23], the IDP *static* with *coarse* granularity presented the best results among all OpenMP configurations. Therefore, we consider only this setup when comparing OpenMP with other PPIs.

TABLE II. Main Characteristics of the Processors

	Intel Atom	Cortex-A8	Cortex-A9
#Cores	2	4	4
#Threads	4	4	4
Frequency	1.6 GHz	1.0 GHz	1.2 GHz
Cache L1D/I	24/32 KB	32 KB	32 KB
Cache L2	1 MB	1 MB	1 MB
RAM	2 GB	1 GB	1 GB
Processor – Static Power	0.484 W	0.17 W	0.25 W
RAM – Static Power	0.149 W	0.120 W	0.120 W
L1-D – Static Power	0.008 W	0.011 W	0.011 W
L1-I – Static Power	0.011 W	0.011 W	0.011 W
L1-D (Energy per access)	0.014 nJ	0.017 nJ	0.017 nJ
L1-I (Energy per access)	0.017 nJ	0.017 nJ	0.017 nJ
RAM (Energy per access)	3.944 nJ	2.772 nJ	2.772 nJ
Energy per Instruction	0.413 nJ	0.304 nJ	0.290 nJ

V. RESULTS

A. Performance and Energy Consumption

Figures 2 and 3 show the execution time and energy consumption of each PPI, using the geometric mean. For the LC programs, the performance (Figure 2a) of the PPIs are similar and very close to ideal, since the applications are CPU-intensive, amortizing the impact of communication, regardless the PPI used. Moreover, as this class presents low communication rate, the energy consumed by each PPI on each processor as the number of threads/processes increases is also similar, as show in Figure 3a. However, when the number of threads/processes increases, the overhead of managing the parallelization (e.g.: creation/termination of threads or process) starts to be an important factor [13], which impacts the most for MPI-1 and MPI-2. For HC programs (Figure 2b and 3b), the way how each PPI performs communication between threads/processes greatly influences the results.

Let us discuss the results of the HC programs running on the Atom processor. Pthreads showed to be the best alternative with a larger advantage over the other PPIs. Comparing with OpenMP, the difference comes from the workload distribution, and mainly from the amount of shared memory accesses performed by threads in the busy-waiting state. MPI implementations are the worst alternative: the more communication the application has, the more performance degrades and energy consumption increases, reinforcing that the impact of communication is larger with this PPI.

On the other hand, when one considers the execution of HC applications on ARM processors, OpenMP achieved the best results in performance and energy savings. Comparing with Pthreads, the difference is related to the impact of context switching imposed by the use of mutex to ensure synchronization, since the performance of this mechanism depends of the architecture and operating system used [17]. The behavior of the MPI applications follows the same one observed in the Atom processor, but with a larger difference between MPI-1 and MPI-2.

In order to better understand why MPI is always worse

TABLE III. PERCENTAGE OF TIME SPENT ON SYNCHRONIZATION POINTS

PPI	Number of Threads/Processes			
	2	3	4	8
OpenMP	0.03%	0.20%	0.60%	1.41%
Pthreads	0.02%	0.18%	0.56%	1.23%
MPI-1	0.47%	0.81%	1.58%	3.11%
MPI-2	8.81%	16.21%	19.15%	33.26%

than the other PPIs for HC programs, we used the Intel® Parallel Studio XE 2015 – Cluster Edition to collect runtime data on the Intel Atom Processor (this tool is only available for Intel processors). Table III shows the percentage of time spent with synchronization points by each number of threads/processes to ensure correct communication and synchronization. For OpenMP and Pthreads, which use shared variables for the communication, the increase in the number of threads does not have a significant impact on the application performance. This is because the threads do not need to send data to each other, but only guarantee exclusive access in the memory region that is involved in the data exchange. On the other hand, in MPI implementations, when the number of processes increase, the number of executed instructions for communication (which reflects the send/receive operations) also increases (as for the synchronization time). Moreover, the time spent by the dynamic processes creation offered by MPI-2 adds up to the already increased overhead, leading to huge performance and energy losses.

Finally, analyzing the whole scenario, for most cases when the memory system is not stressed much, the parallel versions consumed less energy than its sequential version. It occurs because the static energy consumption of the memory system decreases as the performance increases. Let us consider an example where the sequential version would take 100 seconds to execute. The static energy consumption of the CPU and memory system would be both multiplied by 100 seconds. On the other hand, if its parallel version with four threads/processes would take 25 seconds to execute, while the static consumption of the CPU is 25 seconds multiplied by 4 cores (which would be similar to the sequential version), the static energy consumption of the memory system would be

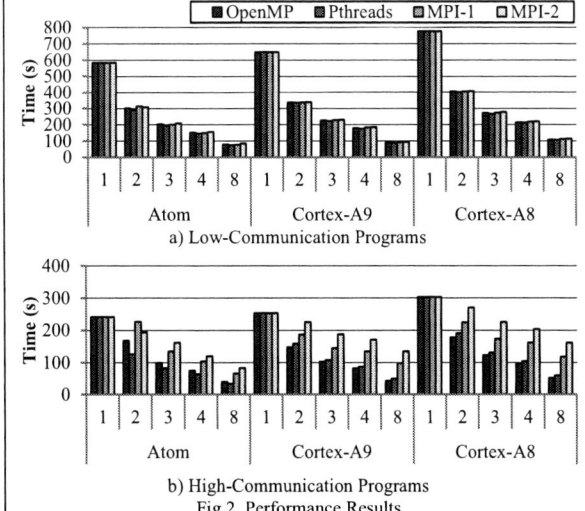

a) Low-Communication Programs

b) High-Communication Programs
Fig 2. Performance Results

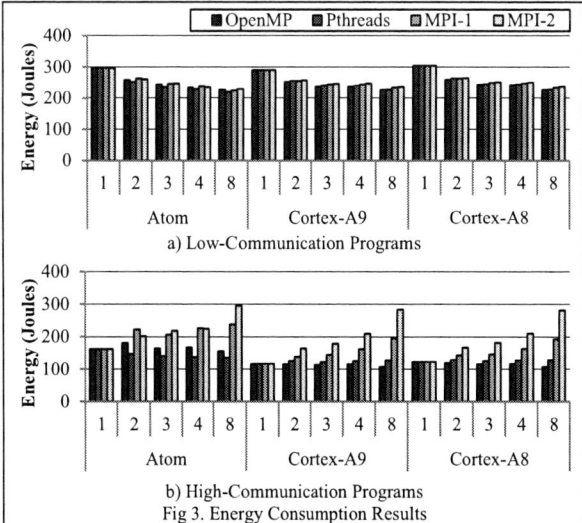

a) Low-Communication Programs

b) High-Communication Programs
Fig 3. Energy Consumption Results

only multiplied by 25 seconds, because this would be the total time than the memory was active.

B. Resource Efficiency in Energy-Delay Product

Even though for very few cases the energy consumption decreases and in most cases there are performance improvements as result of parallelization, in embedded system another factor is important: the best possible use of the available resources. Therefore, *resource efficiency* is here defined as the best possible use of the processors, considering the sequential execution of the application on a single core as baseline. For example, a program that is split in four threads/processes may be faster and less efficient than one split in only two threads/processes. In this case, each of the four processors was used less (e.g.: they spent more time waiting for synchronization) during program execution than each of the two processors. This can be extrapolated to energy or Energy-Delay Product, for instance. In this work, we use the EDP as metric, which is obtained by multiplying the execution time with the energy consumption, so it can correlate energy and performance within one value only.

The resource efficiency is given by (5), where *Rseq* corresponds to the EDP of the processor when executing the sequential version of the application. *Rpar* is the same as the previous, but for each thread of the parallel version, and *NT* is the number of threads/processes.

$$R_{ef} = \frac{R_{Seq}}{R_{Par} \times NT} \qquad (5)$$

It is possible to note in Fig. 4a that for LC programs, all PPIs were able to achieve an EDP efficiency over than 100%, which means that the available resources were better utilized in the parallel versions than in their sequential counterparts. This behavior occurs because their performances gains are close to ideal, while there are energy savings because of the reductions in the static energy consumption of the memory system. However, this scenario totally changes when it comes to HC applications. With the exception of Pthreads applications executing on the Atom processor, all the other PPIs have EDP efficiency below 80%, regardless the number of threads/processes used. This means that, even though there are performance improvements when executing these parallel applications in embedded systems, the best choice if one considers both performance and energy is almost always the sequential version. Therefore, depending on the performance requirements of the application, it is better to use one core only even if there are more available.

Figure 5 presents the resource efficiency considering the EDP average of each benchmark class for the Atom and ARM Cortex-A9 (from now, we will consider this processor only, given that the results of both ARM processors are very similar). However, only the best PPI for each benchmark class/processor was selected. It highlights the fact that choosing the right PPI may lead to better utilization of computational resources when compared to the sequential version. It also shows that the processor scales differently according to the benchmark class as the number of threads/processes increase. Most importantly, the figure demonstrates how different the use of resources can be when comparing the two processors. For instance, the ARM processor is 27% less efficient on average than the Atom for HC applications, when one considers the best PPI for each one.

C. Optimized use of PPIs

Table IV indicates which PPI is the best choice in terms of performance, energy consumption, and EDP for each processor, number of threads/processes and benchmark class. Here, we are not considering efficiency anymore, but raw numbers. In addition, it also shows how much better the best choice is compared to the second place and worst PPI for each class. In some cases, it is possible to note that more than one PPI is the best choice, which means that the results between them are similar and the difference is within the standard deviation of 1%. In cases where two columns are merged, the results are the same.

It may be noted that for LC applications, the difference between the best and the worst choice in the most cases is less than 7% on both processors. This shows that the particular characteristics of each PPI are not significant to influence results. Therefore, for this class of applications, the programmer may take into account secondary factors during the development process, such as ease of programmability, code reuse, and so on. However, the same is not true for HC applications, where there are two different scenarios. Pthreads is the best choice for the Atom processor. In the most

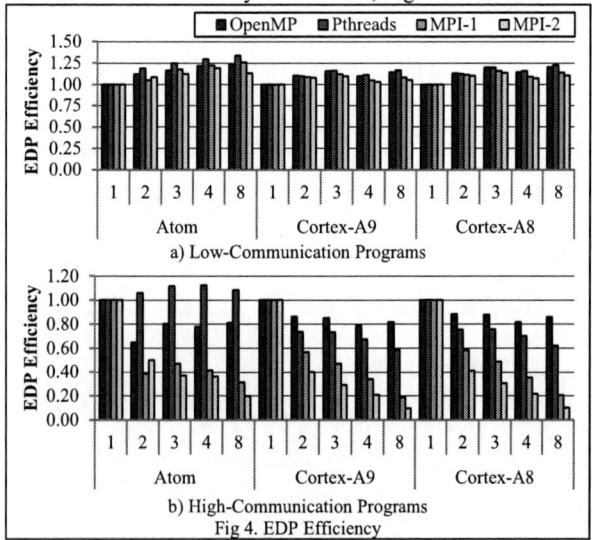

a) Low-Communication Programs

b) High-Communication Programs

Fig 4. EDP Efficiency

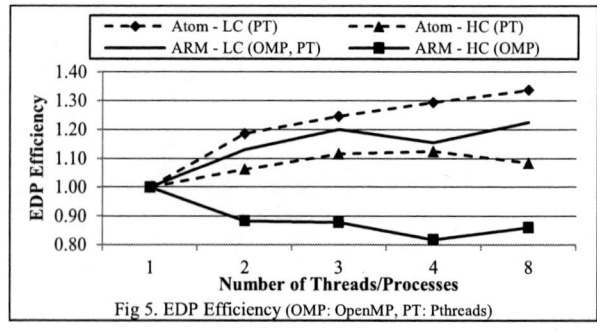

Fig 5. EDP Efficiency (OMP: OpenMP, PT: Pthreads)

TABLE IV. OPTIMIZED USE OF PARALLEL PROGRAMMING INTERFACES FOR PERFORMANCE, ENERGY, AND ENERGY-DELAY PRODUCT

Benchmarks	T/P	Performance						Energy						Energy Delay Product					
		Intel Atom			ARM			Intel Atom			ARM			Intel Atom			ARM		
		Best	2nd	Worst	Best	2nd	Worst	Best	2nd	Worst	Best	2nd	Worst	Best	2nd	Worst	Best	2nd	Worst
LC	2	PT	3%	7%	OMP, PT, MPI-1, MPI-2			PT	3%	4%	OMP PT	2%	2%	PT	6%	11%	OMP PT	2%	2%
	3		2%	6%	OMP, PT	2%	3%		3%	4%		2%	3%		3%	6%		3%	5%
	4		2%	5%	PT	2%	5%		3%	4%		3%	4%		6%	6%		6%	7%
	8		4%	12%		3%	6%		2%	4%		3%	4%		2%	7%	PT	2%	10%
HC	2	PT	25%	45%	OMP	7%	34%	PT	18%	34%	OMP	8%	29%	PT	39%	64%	OMP	15%	54%
	3		17%	49%		6%	45%		14%	36%		8%	37%		28%	67%		14%	36%
	4		16%	48%		6%	52%		18%	39%		9%	45%		31%	68%		14%	74%
	8		15%	60%		14%	69%		13%	54%		16%	62%		25%	82%		28%	88%

* OMP: OpenMP; PT: Pthreads

significant case, it is 60% faster, saves 54% of energy and improves EDP in 82%. On the other hand, for the ARM processor, OpenMP is the best alternative, being up to 69% faster, bringing energy savings of up to 62% and EDP improvements of up to 88%. This means that by just switching the PPI it is possible to obtain huge gains in all metrics.

VI. CONCLUSIONS

This work analyzed the influence of four widely used PPIs (Pthreads, OpenMP, MPI-1 and MPI-2) on three embedded processors, taking into account different metrics: performance, energy consumption and EDP. It showed that, depending on the behavior of the program, the use of a specific PPI results in huge gains. For instance, by using OpenMP on the ARM processor, it is possible to achieve improvements of 69% on performance, energy savings of up to 62% and EDP improvements of 88% over the worst PPI, when one considers the most significant case for each metric. We also showed that the influence of the PPIs increases as the memory system is more stressed (i.e., when there are more communication and synchronization).

As future work, we will expand our benchmark set to cover a larger range of application behaviors and consider other factors, such as Instruction-Level Parallelism and applications that are more control or data flow oriented. We will also consider heterogeneous architectures (e.g., System on Chips) and dynamic hardware techniques used to lower energy consumption. We will also compare the impact of using different compilers (e.g., GCC and LLVM – Low Level Virtual Machine) with different levels of optimization.

REFERENCES

[1] V. A. Korthikanti and G. Agha, "Towards optimizing energy costs of algorithms for shared memory architectures". *In 22nd Annu. ACM Symp. On Parallelism in Algorithms and Architectures*, pp. 157-165, 2010.

[2] B. Chapman, G. Jost and R. Van Der Pas, *Using OpenMP: Portable Shared Memory Parallel Programming*. The MIT Press, 2008.

[3] D. R. Butenhof, *Programming with POSIX Threads*. Addison-Wesley Longman Publishing Co., 1997.

[4] W. Gropp, et. al, *MPI - The Complete Reference*. The MIT Press, 1998

[5] M. A. Suleman, M. K. Qureshi and Y. N. Patt, "Feedback-driven threading: power-efficient and high-performance execution of multi-threaded workloads on CMPs," *In 13th Int. Conf.on Architectural Support for Programming Lang. and Oper. Systems*, pp. 277-286, 2008.

[6] L. Stanisic, B. Videau, J. Cronsioe, A. Degomme, V. Marangozova-Martin, A. Legrand, J-F. Méhaut. "Performance analysis of HPC applications on low-power embedded platforms," *In Design, Automation & Test in Europe Conference & Exhibition (DATE)*, pp. 475-480, 2013.

[7] K. M. Lee, T. H. Song, S-H. Yoon, K. H. Know, J-W. Jeon, "OpenMP parallel programming using dual-core embedded system," *In 11th Int. Conf. on Control, Automation and Systems (ICCAS)*, pp. 762-766, 2011.

[8] T. Hanawa, M. Sato, J. Lee, T. Imada, H. Kimura and T. Boku, "Evaluation of multicore processors for embedded systems by parallel benchmark program using openmp," *In Lecture Notes on Computer Science*, v. 5568, pp. 15-27, 2009.

[9] E. Blem, J. Menon, and K. Sankaralingam, "Power struggles: revisiting the RISC vs. CISC debate on contemporary ARM and x86 architectures," *In IEEE 19th HPCA*, pp. 1-12. 2013.

[10] J. Balladini, R. Suppi, D. Rexachs and E. Luque, "Impact of parallel programming models and CPUs clock frequency on energy consumption of HPC systems," *In IEEE/ACS AICCSA*, pp. 16-21, 2011.

[11] A. K. Porterfield,. S. L. Oliviear, S. Bhalachandra, J. Prins. "Power measurement and concurrency throttling for energy reduction in OpenMP programs," *In IEEE 27th IPDPSW & PhD Forum*. pp 884 – 891, 2013.

[12] C. Gao, A. Gutierrez, R. G. Dreslinski, T. Mudge, K. FLautner and G. Blake. "A study of thread level parallelism on mobile devices," *In IEEE ISPASS*, pp. 126-127, 2014.

[13] A. S. Tanenbaum, and A. S. Woodhul. *Operating Systems: Design and Implementation*. Prentice-Hall, 2009

[14] I. T. Foster. *Designing and Building Parallel Programs – Concepts and Tools for Parallel Software Engineering*. Addison-Wesley Press 1995

[15] W. Gropp, E. Lusk, R. Thakur. *Using MPI-2: Advanced Features of the Message Passing Interface*. MIT Press, Cambridge, MA. 1999

[16] CACTI. Retrieved from: http://www.cs.utah.edu/~rajeev/cacti6/

[17] H. Akkan, M. Lang, L. Ionkov, "HPC runtime support for fast and power efficient locking and synchronization," *In IEEE Int. Conf. on Cluster Computing (CLUSTER)*. pp. 1-7 2013

[18] S. Browne. J. Dongarra, N. Garner, G. Ho, P. Mucci. "A portable programming interface for performance evaluation on modern processors," *In Int. Journal of High Performance Computing Applications*. Vol 14, number 3, pp. 189-204, Fall 2000.

[19] A. Chandramowlishwaran., K. Knobe., R. Vuduc: "Performance evaluation of concurrent collections on high-performance multicore computing systems," IEEE PDP, vol., no., pp.1,12, 19-23 April 2010

[20] D. Buono; T. De Matteis,; G. Mencagli; M. Vanneschi, "Optimizing message-passing on multicore architectures using hardware multi-threading," 22nd Euromicro Int. Conf. on Parallel, Distributed and Network-Based Processing, vol., no., pp.262,270, 12-14 Feb. 2014

[21] M.R. Guthaus, J.S. Ringenberg, D. Ernst, T.M. Austin, T. Mudge, R.B. Brown,, "MiBench: A Free, Commercially Representative Embedded Benchmark Suite. 4th Workshop on Workload Characterization", Austin, TX, Dec. 2001

[22] A. C. S. Beck, C. A. L. Lisboa, L. Carro. Adaptable Embedded Systems. Springer Publishing Company, Incorporated. 2012.

[23] A. F. Lorenzon, M. C. Cera, A. C. S. Beck, "Performance and Energy Evaluation of Different Multi-Threading Interfaces in Embedded and General Purpose Systems". *In Journal of Signal Processing Systems*. pages 1-13. Springer US. 2014.

2015 IEEE Computer Society Annual Symposium on VLSI

The DRACON Embedded Many-Core: Hardware-enhanced run-time Management using a Network of Dedicated Control Nodes

Daniel Gregorek, Alberto Garcia-Ortiz
Integrated Digital Systems Group
ITEM, University of Bremen, Germany
{gregorek,agarcia}@item.uni-bremen.de

Abstract—**Many-core systems provide abundant computing power for parallel applications. The run-time manager of an embedded system has to efficiently exploit the available resources while guaranteeing a high responsiveness. We propose a dedicated hardware infrastructure to improve the scalability and responsiveness of a run-time task manager. The hardware enhancements constitute a hierarchy of global and local control nodes which communicate by means of message passing. The global nodes facilitate a distributed task manager which performs the task scheduling and a flexible task synchronization scheme at run-time. A low-latency interface between the run-time system and the processing cores is provided by the local nodes.**

Based on simulations using a SystemC model, we demonstrate the advantages of our approach in terms of application performance. The design feasibility is substantiated by means of gate-level analysis. We compare our results against state-of-the-art software and hardware-based run-time management systems.

Keywords-**run-time task manager; embedded many core; hardware enhancements;**

I. INTRODUCTION

Power-density and reliability are becoming primary concerns in ultra-deep-submicron chip design. Many-core architectures are a promising candidate to address the upcoming challenges [2]. Dynamic hardware faults or changes of the user requirements make run-time task scheduling a necessary feature of the many-core system. But finding an optimal schedule for parallel applications usually requires an unaffordable amount of resources (NP complete) leading to the employment of heuristics for the run-time manager.

The demands for high performance, low power and deterministic response time advice for a hardware implemented solution of the run-time task manager [14][8]. Goal of the dedicated hardware is to reduce the management overhead and improve the overall system performance. It is therefore predicted [16], that hardware support for run-time system management will attain into mainstream for many-core platforms. Especially for embedded systems, where low-power is a crucial design constraint, the application specific hardware implementation of a task manager can become necessary.

Different architectural approaches for the dedicated system management have been reported in the literature. Centralized approaches have been shown to lack scalability for the on-line computation with an increasingly large number of cores [13]. Also, traffic hot-spots may become a bottleneck for such architectures. On the other hand, the overhead introduced by a fully-distributed approach can supersede the potential benefits. As a consequence, choosing the right granularity for an on-chip task manager remains an insisting optimization problem.

The idea of having a separate network for task synchronization has been described before by Herkersdorf [9]. State-of-the-Art many-core designs contain such dedicated hardware for synchronization (e.g. Tilera [1], Intel SCC [18], Kalray [6] and STHORM [23]). But to our best knowledge, we are the first to present and to analyze a *full-fledged* task management infrastructure using a dedicated hardware network. Our contribution, the DRACON many-core (**D**edicated **R**un-Time **A**rchitecture **Co**ntrol **N**etwork) focuses on low latency and predictability for run-time task management. The control nodes implement a message passing protocol for communication. The task management is charged with scheduling and synchronization of user tasks. Key feature of the dedicated network is the capability to perform parallel computation and to minimize the overall management latency.

The remainder of this paper is organized as follows: Sec. II discusses related work from the domain of run-time task management, Sec. III and Sec. IV present our system architecture and give details about the task manager. In Sec. V we show and analyze our experimental results and finally Sec. VI concludes the paper.

II. RELATED WORK

The design space for run-time system management can be categorized into distributed versus centralized architectures, as well as into hardware versus software implementations [16]. Further, we localize along the hardware-software axis programmable approaches with optimized instruction sets.

A representative for a symmetric software-based run-time system is the Linux OS kernel. Linux has been shown to provide good scalability and high performance, even in the many-core domain [3]. Due to the shared-memory design, Linux depends on high-performance cache coherence and fine grain lock access.

A centralized OS-specific circuit resolving time critical task dependencies at run-time has been implemented by Nexus++ [5]. It has been designed to overcome a performance bottleneck involved by a software-based Master/Slave run-time system. An optimized processor architecture has been proposed by TMU [22] which employs an application specific instruction set optimized for task management. The TMU contains the capability to perform look-ahead computation. However, both solutions lack scalability due to the drawbacks of centralism in computation and communication.

A globally distributed and dedicated hardware approach has been implemented by Isonet [13]. Isonet applies a fully-distributed network of dedicated management nodes for hard-

978-1-4799-8720-7/15 $31.00 © 2015 IEEE   416

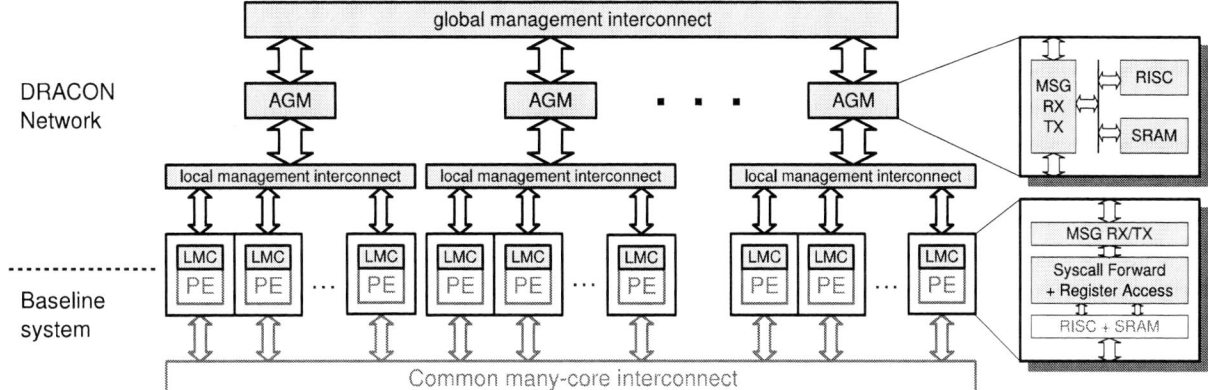

Fig. 1: Architecture of the dedicated control network for task management (drawn gray) on top of a baseline many-core system (drawn dimmed). The DRACON network consists of global nodes (AGM) and local nodes (LMC). The nodes communicate by means of message passing and perform the task scheduling and synchronization at run-time. The processing elements are connected by a common many-core interconnect.

ware supported load balancing. Using independent tasks, the approach claims to be scalable for more than 1024 cores. Yet, due to a limited synchronization scheme, the Isonet nodes may not find a globally optimal load balance. A lightweight HW/SW run-time framework for task management is presented by ARTM [17]. But, while ARTM uses the hardware semaphores included in the STHORM many-core architecture their evaluation is limited to intra-cluster task synchronization. As an advancement (see Tab. I) we propose the DRACON architecture as a full-fledged and global-view HW/SW run-time task manager.

TABLE I: DRACON vs. related work.

Arch.	Performance	Scalability	Area Overhead
Software	o	+	++
Nexus++	+	o	o
Isonet	+	+	-
DRACON	++	+	-

III. SYSTEM ARCHITECTURE

We propose a hierarchical and distributed network for run-time task management. The network enhances a common many-core system and is hierarchical by means of a clustering of the resources to be managed. It is also distributed, since each cluster can be managed autonomously. Fig. 1 gives an outline of our proposed system architecture. The hardware enhancements establish an infrastructure of dedicated management resources (drawn gray). The baseline many-core system (drawn dimmed) consists of numerous processing elements (PE), connected by a common multiprocessor-interconnect. As a low-latency interface, we closely couple a local management controller (LMC) to each PE. The LMCs are connected to a local management interconnect, which constitutes a cluster of PE + LMC pairs. Each cluster is controlled by an autonomous global manager (AGM). Communication between the AGMs is done via a global management interconnect. For the sake of readability we explain the particular hardware and software components side-by-side to our SystemC model.

A. Baseline System

As an initial baseline, we consider a *homogeneous* many-core system for our analysis. The many-core consists of RISC-like processing elements which are connected by a common many-core interconnect. For the SystemC model we consider the common interconnect to be a simplified Network-on-Chip without virtual channels having a mesh topology and XY-routing. The baseline system is not necessarily clustered by itself, the hierarchy is only constituted by the additional hardware enhancements. We employ a functional model for the processing elements, which interprets a task-based programming language (see also Sec. V). The employed network model considers effects like hop-distance, communication-volume and link utilization. The baseline NoC is only responsible for data transfer between the PEs (task communication). The NoC is not connected to any external memory or I/O, which is out of the scope of this paper.

B. Autonomous Global Manager

Each instance of the AGMs realizes one instance of the task manager on a dedicated processor and further contains an interface to the dedicated messaging protocol. The AGMs run a OS based on Micro-C/OS-II [11] which has been extended to a light-weight and distributed multicore OS. The AGMs implement the run-time task scheduling, the task synchronization and a cluster status communication mechanism (see Sec. IV). They monitor the activity (number of running tasks), and the number of mapped tasks inside their cluster. Each AGM has its private address-space, the communication between the AGMs and between one AGM and LMC is determined by the message protocol presented in Sec. III-D.

C. Local Management Controller

The local management controller (LMC) is constituted by a messaging interface, a system-call dispatcher and a tightly-coupled interface to the PE. The dedicated hardware LMC can be implemented with low area overhead (see Tab. V) and operates in parallel to the PE. Any system-call from a user task is fetched by the LMC and dispatched to its global node by means of a dedicated message. Due to the dedicated

TABLE II: System calls with inputs, outputs, and type of required context-wwitch.

Name	in	out	Context switch	Description
os-task-spawn	imem, dmem, cnt		SW only	Recursively spawn number of child tasks (cnt) with given instruction- (imem) and data-memory (dmem) address.
os-task-exit	addr		SW only	Send a signal to the synchronization barrier given by addr and terminate recursive child task. The parent task is restarted, if all child tasks have finished.
os-info-send	key, value		SW only	Send 32-Bit value to key.
os-info-recv	key, cnt	value(s)	SW and LMC	Receive 32-Bit value(s) by key. Blocks the task until all values are available.
os-get-pe		pe-id	none	Get numeric identifier of PE

infrastructure for the task management the PE is discharged from the execution of the OS service.

D. Messaging Protocol

We send messages via the dedicated interconnects to implement the communication between the hardware nodes. Each message has a header and one or more 32-Bit data fields. The header contains the message type, at least the source address, the message priority and a broadcast flag. The size of the message header depends on the actual hardware configuration (i.e. number of nodes/address-width) and the direction of the message. In the following we give a short overview about the implemented messages ordered by the direction of the conversation. A comprehensive reference can be found at the end of the paper (Tab. VI).

LMC to AGM: Most of the messages from the LMCs to the AGMs directly correspond to the system calls given in Tab. II and are send from a local controller to its global node. Messages from an LMC have a unique destination and therefore have a reduced header.

AGM to LMC: We send the message msg-task-start from an AGM to a LMC. The message transports a task identifier (tskd), and the task's stack-pointer as message data. It invokes the start of a task at the LMC. If the task has previously raised a system call, the message also transports the return value(s) of the call.

AGM to AGM: An AGM may also request to start a task at a remote cluster, therefore the message msg-task-give can be send from one AGM to another AGM. Since communicating task may be located inside different clusters, the task synchronization presented in Sec. IV-C also requires inter-AGM communication. To broadcast the current cluster workload status (see Sec. IV-D) the AGM uses the message msg-beacon.

IV. TASK MANAGEMENT

Principal duty of the task manager is the scheduling and synchronization of the user tasks. Each task is considered to be sequential while the whole application is defined by a task graph [21] and consists of numerous depending tasks. The task graphs may contain task-level parallelism which must be exploited by the run-time task manager.

The run-time task management requires computation time to service the system-calls and to schedule (map) user tasks to processing elements [20]. Fig. 2 illustrate the inherent advantages of our hardware-enhanced run-time manager versus a software-based approach (for a more general HW-SW comparison see also [16]). A software approach will, at least initially, service a system call at the local processing element and requires more context switches (Fig. 2a). The hardware

enhanced DRACON task manager forwards a system call (Fig. 2b) by means of a message to the AGM. The AGM services the system calls and may pre-compute the scheduling before going idle. If a PE becomes idle, the next tasks can be dispatched immediately. The responsiveness of the management system is therefore improved and the PEs are given more time for processing the user task. The response-time of the user applications is increased, while the standard deviation of the response-time is decreased.

Throughout this paper the user tasks can access the system calls given in Tab. II. Most of the system-calls require a context switch in a software-based implementation. Using the hardware enhancements only the call to os-info-recv triggers a context switch while the calling task is blocked and transported to its AGM.

A. Recursive Task Fork

Since our targeted task scheduling problems consist of task sets having a large number of tasks, we use a recursive task spawning/fork strategy [12]. Every recursive task spawns two additional system tasks and then blocks until its child's have terminated. The recursive start-up follows a dynamic cluster mapping procedure which tries to equally distribute the recursive system tasks onto the clusters. After the binary fork-tree has stopped to expand, the actual child tasks of the application are spawned. The final number of working child tasks is fixed and determined by the application profile.

B. Task Scheduling

The spatial and temporal task scheduling is performed on a cluster-basis by the AGMs. Each AGM contains one ready queue for waiting tasks. Our scheduling is quasi-preemptive: a running user task may be preempted by a higher priority system task but not by another user task. We use different scheduling algorithms depending on the application type. The algorithms have different time complexities depending on the number of ready tasks n or the number of PEs m. Usually, there is a trade-off between the time complexity and the quality of the scheduling algorithm [10].

(1) Round-Robin: We apply round-robin scheduling for independent tasks. The tasks are mapped and dispatched to the first idle PE. Scheduling time complexity is $\mathcal{O}(1)$.

(2) Max-Bottom-Level: For applications containing task dependencies we use a Max-Bottom-Level-First algorithm [21]. The ready tasks are stored inside a priority queue which is implemented as a Red-Black Tree and has a logarithmic time-complexity for inserting and removing [4]. The tasks are mapped and dispatched to the first idle PE. Time complexity is $\mathcal{O}(\log n)$.

978-1-4799-8720-7/15 $31.00 © 2015 IEEE

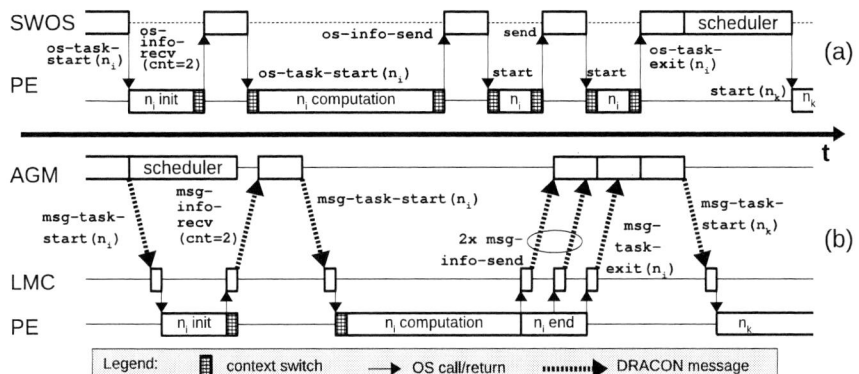

Fig. 2: Comparison of task management for one task n_i having two input and two output edges. The input data has arrived before the start of the given sequence. The software approach (a) runs at the same PE and interrupts the user task for OS service. The software approach schedules as late as possible.
The hardware-enhanced solution (b) implements the same OS interface but sends messages between the LMC and the AGM. The hardware approach requires less interuption of the user task and schedules as soon as possible in parallel to the PE. The outcome is an earlier dispatching of the next waiting task n_k.

(3) Nearest-Idle: To address the Manhattan distance between communicating tasks we apply the Max-Bottom-Level task selection. The selected task is mapped and dispatched to an idle PE nearest to the location of the application's parent task. Scheduling time complexity is $\mathcal{O}(\log n + \log m)$.

C. Task Synchronization

To have a flexible task synchronization sub-system which is able to work in a distributed environment, we developed a communication infrastructure based on key-value hash tables. Each global management node maintains one hash table autonomously at run-time. A task synchronization event is assigned to a unique key. User tasks can request to send a (non-empty) value to an arbitrary key. To allow tasks to wait until all of their input values have arrived, the system-call os-info-recv has been implemented. A task may therefore wait for a specific amount of values and then becomes ready for scheduling.

The synchronization sub-system does not contain critical races and utilizes a dynamic *localize-at-receive* mechanism: Every time a read request occurs, and the key position is not yet known or has changed, the position of the new key is broadcasted by the AGM via the global interconnect. The other AGMs read that broadcast and update their key positions accordingly. The broadcast is fully transparent to the user tasks, and allows their synchronization, even when tasks change their position from one cluster to another. Any value which is send to a key, who's position is not yet discovered, is pending until the required broadcast occurs. As soon as the position of a given key is known, a transport of the waiting values to the discovered position is triggered. In our implementation, we use the task synchronization sub-system to transport memory addresses between communicating tasks.

D. Cluster Status Communication

To perform the run-time task scheduling in an efficient manner it depends on reliable information about the global and local workload status. We use a broadcast message to inform all global managers about the local workload and apply a threshold-based mechanism to decide, whether to send that broadcasting message. The mechanism triggers the transmission, when the change in the number of locally active tasks reaches a certain threshold Δn_{th}.

TABLE III: Benchmark characteristics: (a) number of workers, (b) total amount of work, (c) maximum task workload, (d) number of communication edges, (e) total sum of communication volume, (f) maximum edge communication volume.

Name	(a)	(b) Ticks	(c) Ticks	(d)	(e) 32b	(f) 32b
Indep.	1e4	2e7	2e3	0	0	0
Horiz.	8160	1.632e8	2e4	8092	0	0
Sparse	96	309760	5440	67	13668	204

V. EXPERIMENTAL RESULTS

In our analysis we consider three different types of parallel applications. Their characteristic values are given in Tab. III. The first one assumes totally independent child tasks, the second has horizontal dependencies between child tasks, and the third one (Sparse) is a real-world task graph and is taken from the MCSL benchmark suite [15]. We use a transaction-level framework [7] implemented in SystemC to evaluate our hardware/software architecture. It uses a blocking transaction level methodology and TLM 2.0 sockets. The framework employs a task-based programming model to describe the behaviour of the user applications. The tasks are defined by trace instructions which are interpreted by the PE model. The traces consider the execution time, input edges and output edges of the tasks as well as their communication volume. Task communication is modeled by means of reading via the common NoC from the remote memory of a processing element.

For comparison with current state-of-the-art, we employ a symmetric software-based operating system and a centralized hardware implementation [5]. Since Isonet is only targeting on applications having independent tasks we do not compare to them. Both the reference measurements and DRACON provide the same syscall interface as given in Tab. II and make use of the same scheduling algorithms. In contrast, the software OS is only running at the baseline hardware of the many-core system. Due to the symmetry, one instance of the software OS is running at each PE. The software OS uses shared resources for scheduling, which are protected by fine-grained locks. The OS must acquire a lock before reading from the resource but we allow an asynchronous write to the shared resource. The locking ensures sequential read but allows concurrent computation and write to the shared resources. We do not consider any further memory contention or communication overhead for the software OS but assume fully coherent caches

978-1-4799-8720-7/15 $31.00 © 2015 IEEE

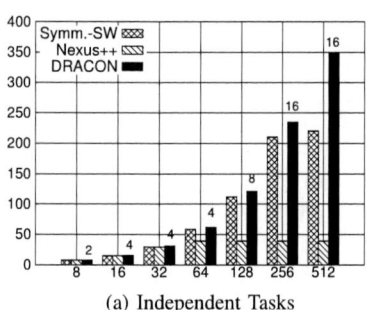
(a) Independent Tasks

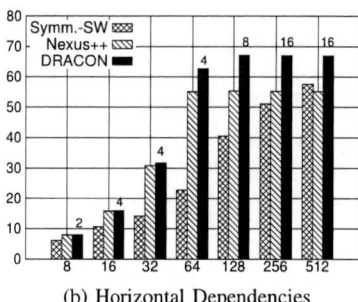
(b) Horizontal Dependencies

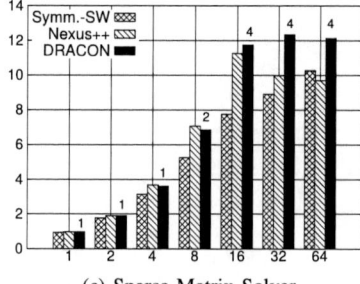
(c) Sparse-Matrix Solver

Fig. 3: Speedup versus number of PEs. The number of AGMs is given on the top of each DRACON column

TABLE IV: Timing parameters of the transaction-level model

Description	Value	Stakeholders
Scheduling delay coefficient	24 Ticks	SW-OS, AGM
System call	8 Ticks	SW-OS, LMC
Context switch	16 Ticks	PE
Task-Info Synchronization	16 Ticks	SW-OS, AGM
Lock access	8 Ticks	SW-OS
Software FIFO access	4 Ticks	SW-OS, AGM
TX/RX of message	4 Ticks	AGM, LMC
Local man. interconnect	64 Bit	AGM, LMC
Global man. interconnect	64 Bit	AGM
Common interconnect	128 Bit	Baseline NoC

in the background. For the second reference measurement we compare to the centralized hardware implementation Nexus++. We use a single AGM and calibrate our model according to the speedup values available for Nexus++ [5] and the synthetic benchmark containing horizontal task dependencies.

Tab. IV shows the parameters, which have been set for our transaction-level model. Both the DRACON architecture as well as the software reference have to cope with the same timing parameters for the task scheduling and task synchronization and the same baseline hardware. In our current work the management interconnects are implemented as shared buses, but our architecture allows to use other topologies as well. As a metric for the performance we measure the application speedup $S = t_{seq}/t_{par}$. In a first instance, each benchmark is run individually while the number of processing elements is increased. The resulting values for the speedup S are given in the Fig. 3. DRACON provides good scalability for fine-grain **independent tasks (Fig. 3a)** and the analyzed number of up to 512 PEs. The software approach reveals a scalability bottleneck due to a locking in the task mapping part of the scheduler. The speedup for the centralized approach Nexus++ does not scale beyond 64 PEs. The **horizontal benchmark (Fig. 3b)** contains more coarse-grained tasks and has a maximum parallelism of 68. The parallelism is almost fully exploited by DRACON. Nexus++ benefits from the larger task size and exhibits good performance. The software approach must compete for computing power at the PEs and requires a much higher number of PEs to achieve the same speedup. The **sparse-matrix solver (Fig. 3c)** contains moderate inter-task communication and has a theoretical speedup limit of $S = 15.78$ due to it's critical path in the task graph. For 16 PEs both DRACON as well as Nexus++ reveal their strengths due to its parallel scheduling capacities. For a larger number of PEs the distance between communicating tasks increases and the speedup slightly decreases.

In Tab. V we provide values for the gate-level area overhead of one AGM, LMC and PE. The values have been obtained using an industrial 65nm low-power process. Each AGM and each PE contain one dedicated RISC processor which is implemented as mLite/PLASMA CPU [19]. We consider 32-Bit RX buffers for message communication inside each AGM. Assuming a system of 256 PEs and 16 AGMs the RX buffers have a size of 16 entries which gives one RX entry per connected node at the local and the global dedicated interconnect. To address the impact of on-chip memory we also consider 4kB of SRAM per PE and AGM.

TABLE V: Area $[\mu m^2]$ for 65nm technology

Unit		Total	Comb.	Non-comb.	f_{max}
AGM		81003.8	20072.8	60931.0	
	core	30397.3	17313.1	13084.2	502.5 MHz
	mem	50606.5	2759.7	47846.8	
PE		63367.9	14246.2	49121.7	
	core	27627.8	14241.9	13385.9	529.1 MHz
	mem	35740.1	4.3	35735.8	
LMC		858.6	375.5	483.1	543.4 MHz

As a final evaluation we run all of the three benchmarks simultaneously for 10 iterations on a many-core system containing 256 processing elements. The DRACON network is configured to use 16 AGMs and the threshold for the cluster status communication has been set to $\Delta n_{th} = 4$. We measure the linear average for the application speedup $\overline{S}$ and the standard deviation σ of the normalized application response-time t_{par}/t_{seq}. The results of the simulations and the approximated total area are given in Fig. 4. The DRACON architecture has an average speedup $\overline{S}$ which is 20.56% higher compared to the software approach while at the same time the standard deviation is reduced by a factor of 2.96. The centralized approach Nexus++ is simply overburdened due to the large number of PEs. The area overhead of DRACON is approximated to be 9.4% compared to a common software approach.

VI. CONCLUSION

This paper presents DRACON, a scalable network of control nodes for run-time task management on embedded many-cores. The network of control nodes performs the communication and computation for task management in parallel to the processing cores. This is exploited to prepare task management decisions without disturbing the user application and to improve the responsiveness of the system.

978-1-4799-8720-7/15 $31.00 © 2015 IEEE

TABLE VI: DRACON messages. Size is given in 32-bit words

Name	Size	Parameters	Dir.	Description
msg-task-start	2+	tskd, stack, size	AGM to LMC	Start task at destination LMC by means of the task descriptor (tskd), stack pointer (stack). The message may contain additional data words
msg-task-spawn	4	parent, imem, dmem, cnt	LMC to AGM	Spawn new tasks of count cnt with given parent, instruction- and data memory address
msg-task-give	4	parent, imem, dmem, rcsv	AGM to AGM	Request start of a tasks at another AGM. The parameter rcsv is required to implement the recursive task fork mechanism
msg-task-exit	2	tskd, join	LMC to AGM, AGM to AGM	Terminate task and/or signalize to join barrier join
msg-beacon	1	status	AGM to AGM	Broadcast load status
msg-info-recv	2	tskd, key, cnt	LMC to AGM	Let task receive cnt synchronization values for key.
msg-info-send	2	key, value	LMC to AGM, AGM to AGM	Send synchronization value to key
msg-info-cast	1	key	AGM to AGM	Broadcast key position

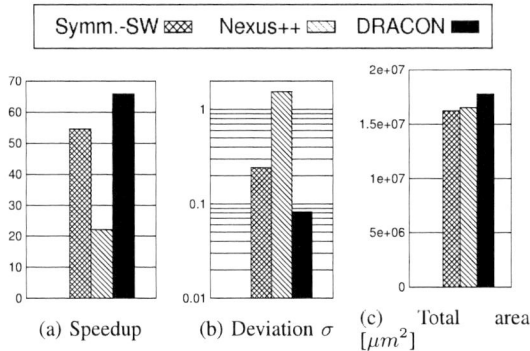

Fig. 4: Comparison of speedup, deviation of response-time and total area for a symmetric software-approach, a centralized hardware task manager (Nexus++) and DRACON on 256 PEs.

Our performance evaluation employs a transaction-level SystemC model and a task-based programming language. We implemented fundamental task scheduling and task synchronization algorithms to realize the distributed and hardware-enhanced task management. The results fortify the DRACON perspective to efficiently handle the complexity of run-time management for ultra-deep-submicron chip designs.

For 64 processing cores and horizontal task dependencies we achieve a 12% higher application speedup compared to a centralized hardware implementation and 70% higher speedup compared to a software approach. The 10% area investment for DRACON, although not negligible, is justified by the much higher performance outcome and better predictability. For 256 processing cores, DRACON outperforms a software approach by 20% in terms of speedup and has around three times smaller standard deviation of application response time when running a mixture of benchmarks.

REFERENCES

[1] S. Bell, B. Edwards, J. Amann, R. Conlin, K. Joyce, V. Leung, J. MacKay, M. Reif, L. Bao, J. Brown, et al. Tile64-processor: A 64-core soc with mesh interconnect. In *Solid-State Circuits Conference, ISSCC*, pages 88–598. IEEE, 2008.
[2] S. Borkar. Thousand core chips - a technology perspective. In *DAC*, 2007.
[3] S. Boyd-Wickizer, A. T. Clements, Y. Mao, A. Pesterev, M. F. Kaashoek, R. Morris, and N. Zeldovich. An analysis of linux scalability to many cores. 2010.
[4] T. H. Cormen, C. E. Leiserson, R. L. Rivest, C. Stein, et al. *Introduction to algorithms*, volume 2. MIT press Cambridge, 2001.

[5] T. Dallou and B. Juurlink. Hardware-based task dependency resolution for the starss programming model. In *Parallel Processing Workshops (ICPPW), 41st International Conference on*. IEEE, 2012.
[6] B. D. d. Dinechin, P. G. d. Massas, G. Lager, C. Léger, B. Orgogozo, J. Reybert, and T. Strudel. A distributed run-time environment for the kalray mppa-256 integrated manycore processor. *Procedia Computer Science*, 18:1654–1663, 2013.
[7] D. Gregorek and A. Garcia-Ortiz. A transaction-level framework for design-space exploration of hardware-enhanced operating systems. In *International Symposium on System-on-Chip (SOC 2014)*. IEEE, 2014.
[8] N. Gupta, S. Mandal, J. Malave, A. Mandal, and R. Mahapatra. A hardware scheduler for real time multiprocessor system on chip. In *VLSI Design, 2010. VLSID'10. 23rd International Conference on*, pages 264–269. IEEE, 2010.
[9] A. Herkersdorf, A. Lankes, M. Meitinger, R. Ohlendorf, S. Wallentowitz, T. Wild, and J. Zeppenfeld. Hardware support for efficient resource utilization in manycore processor systems. In *Multiprocessor System-on-Chip*, pages 57–87. Springer, 2011.
[10] Y.-K. Kwok and I. Ahmad. Benchmarking and comparison of the task graph scheduling algorithms. *Journal of Parallel and Distributed Computing*, 59(3):381–422, 1999.
[11] J. J. Labrosse. *MicroC/OS-II*. R & D Books, 1998.
[12] D. Lea. A java fork/join framework. In *Proceedings of the ACM 2000 conference on Java Grande*, pages 36–43. ACM, 2000.
[13] J. Lee, C. Nicopoulos, H. G. Lee, S. Panth, S. K. Lim, and J. Kim. Isonet: Hardware-based job queue management for many-core architectures. *Very Large Scale Integration (VLSI) Systems, IEEE Transactions on*, 21(6):1080–1093, 2013.
[14] L. Lindh. Fastchart-a fast time deterministic cpu and hardware based real-time-kernel. In *Real Time Systems, 1991. Proceedings., Euromicro'91 Workshop on*, pages 36–40. IEEE, 1991.
[15] W. Liu. J. Xu, X. Wu, Y. Ye. X. Wang, W. Zhang. M. Nikdast, and Z. Wang. A noc traffic suite based on real applications. In *VLSI (ISVLSI), IEEE Computer Society Annual Symposium on*, pages 66–71. IEEE. 2011.
[16] V. Nollet, D. Verkest, and H. Corporaal. A safari through the mpsoc run-time management jungle. *Journal of Signal Processing Systems*, 60(2):251–268, 2010.
[17] M. Ojail, R. David, Y. Lhuillier, and A. Guerre. Artm: a lightweight fork-join framework for many-core embedded systems. In *Conference on Design, Automation and Test in Europe, DATE*, pages 1510–1515. EDA Consortium, 2013.
[18] P. Reble, S. Lankes, F. Zeitz, and T. Bemmerl. Evaluation of hardware synchronization support of the scc many-core processor. In *4th USENIX Workshop on Hot Topics in Parallelism (HotPar 12), Berkeley, CA, USA*, 2012.
[19] S. Rhoads. Plasma-most mips i (tm) opcodes: overview. *Internet: http://opencores. org/project, plasma [May 2, 2012]*, 2006.
[20] A. K. Singh, M. Shafique, A. Kumar, and J. Henkel. Mapping on multi/many-core systems: survey of current and emerging trends. In *50th Annual Design Automation Conference (DAC)*. ACM, 2013.
[21] O. Sinnen. *Task scheduling for parallel systems*, volume 60. John Wiley & Sons, 2007.
[22] M. Sjalander, A. Terechko, and M. Duranton. A look-ahead task management unit for embedded multi-core architectures. In *Digital System Design Architectures, Methods and Tools, 2008. DSD'08. 11th EUROMICRO Conference on*, pages 149–157. IEEE, 2008.
[23] F. Thabet, Y. Lhuillier, C. Andriamisaina, J.-M. Philippe, and R. David. An efficient and flexible hardware support for accelerating synchronization operations on the sthorm many-core architecture. In *Design, Automation & Test in Europe Conference & Exhibition (DATE), 2013*, pages 531–534. IEEE, 2013.

978-1-4799-8720-7/15 $31.00 © 2015 IEEE

2015 IEEE Computer Society Annual Symposium on VLSI

Backlog Bound Analysis for Virtual-Channel Routers

Xueqian Zhao and Zhonghai Lu
Department of Electronics and Embedded Systems, School for ICT
KTH Royal Institute of Technology, Stockholm, Sweden
Email: {xueq, zhonghai}@kth.se

Abstract—Backlog bound analysis is crucial for predicting buffer sizing boundary in on-chip virtual-channel routers. However, the complicated resource contention among traffic flows makes the analysis difficult. Because conventional simulation-based approaches are generally incapable of investigating the worst-case scenarios for the backlog bounds, we propose a formal analysis technique. We identify basic buffer use scenarios and propose corresponding analysis models to formally deduce per-buffer backlog bound using network calculus. A topology independent analysis technique is developed to convey the per-buffer backlog bound analysis step by step. We further develop an algorithm to automate the analysis procedure with polynomial complexity. A case study shows how to apply the technique and the analytical bounds are tight.

I. Introduction

Network-on-Chip (NoC) has been a promising solution for the global interconnect in CMPs and MPSoC designs. For NoCs, buffer sizing is one of the major concerns due to the design tradeoffs among performance, power dissipation and cost. Buffers consume a considerable portion of the router's power consumption as well as the area [1]. In Intel's Terascale 80-core chip [2], the input buffers consume up to 22% of the router power. Thus buffer analysis and sizing have been a critical problem for NoC designs.

Theoretically, any non-zero size can be used for buffer design in on-chip routers by applying local handshaking or feedback flow control to prevent buffer overflow. However, to enable fully utilizing the channel capacity, the buffer size has to cover at least the credit round trip time [3]. Therefore the credit round-trip time determines the lower bound of virtual channels (VCs) in the routers. The question we are after in the paper is the upper bound of VCs. This is important because we want to achieve maximal network performance with minimum buffering cost. In general, with buffered flow control schemes, NoCs achieve more efficient flow control by adding buffers, thus improving delay and throughput [3]. However the improvement saturates upon reaching a threshold value called backlog bound in the paper. Above the threshold, adding buffers becomes in vain.

Simulation based approach is widely used to analyse the buffer boundary because it can model router details and accurate behavior of a network. However, simulations are usually time consuming and difficult to cover all the system states. By contrast, formal analysis is more competent in investigating how network parameters and traffic scenarios affect the buffer requirements. By closed-form formulas, it provides a once-for-all solution with less efforts rather than simulating various configurations and scenarios.

Based on *network calculus* [4][5], we use a formal analysis approach to calculate the backlog bounds of the ingress buffers for on-chip VC routers. Our contributions can be summarized as follows.

(1) In order to decompose complex communication and contention scenarios, we identify the basic use cases of buffers and give formal models of these cases.

(2) We propose an analysis procedure to analyse the backlog bounds given traffic flows on a VC network and develop an algorithm to automate the analysis procedure.

(3) We show the application and tightness of the proposed analysis technique through a case study of an on-chip Video Object Plane (VOP) decoder.

The remainder of the paper is organized as follows. Section II reviews related work. We give notations and network calculus basics in Section III. The backlog bound analysis problem is described in Section IV. In Section V, we present the elementary buffer use cases and models, followed by the analysis procedure and algorithm in Section VI. A realistic SoC application is studied in Section VII. Finally, we conclude in Section VIII.

II. Related Work

Several methods have been used to dimension the buffer size either with iterative simulations or by formal analysis. The simulation based method in [6] allocates the buffer sizes by iterative trials. Although it tries to find the minimized buffer size, the upper bound of the required buffer size is not explored, and it becomes time exhaustive for a design with complicated contentions or a large system size.

Given traffic characteristics or conforming the traffic to specific constraints (for example, applying regulators [7]), the buffer backlogs in an NoC can be formally analysed. Probabilistic methods, such as Poisson process [8] and queueing theory [1], can be used to analyse the buffer utilization. However, they assume independent traffic arrivals, which is not so reasonable in on-chip networks because the contentions among the flows will result in interdependencies between the packets from different flows. Data flow analysis models such as synchronous data flow (SDF) [9] and cyclo-static dataflow (CSDF) [10] provide another formal approach to calculate the buffer requirement in a system with the flow dependencies well captured. However, SDF and CSDF are generally restrictive in static firing rules.

Network Calculus (NC) is another formalism to reason about delay bound and backlog bounds [4][5]. A complex non-linear queueing system can be approximated as a linear one in min-plus algebra [4]. Besides, NC does not require

978-1-4799-8720-7/15 $31.00 © 2015 IEEE   422

accurate characterization of traffic arrivals as long as the traffic conforms to an upper-bound called *arrival curve*. In recent years NC has been applied for per-flow delay bound analysis in on-chip networks [11][15]. NC is also used to help customize the buffer size for application specific NoCs [12]. Recently a worst-case backlog analysis technique is presented in [13]. This technique first generates a *directed contention graph* (DCG) and then traverses the graph to derive backlog bound. However it is restricted to one buffer per router inport. The DCG can directly reflect the flow route and contention, but it has no explicit buffer and server information, making it difficult to handle a typical router with multiple VCs. In contrast, our technique considers a typical router with multiple VCs. It is based on *NC analysis graph* which captures not only flow route and contention but also explicit buffer and server information.

III. NETWORK CALCULUS BASICS

A. Notations

To facilitate the presentation, we first define the symbols and notations in Table I.

TABLE I
SYMBOLS AND DEFINITIONS

Symbol	Explanation
f_i	The ith flow
$f_{i,j}$	Aggregate flow composed of f_i and f_j
f_{ij}	Flow from node i to node j
α_i	Arrival curve of f_i with burstiness b_i and rate r_i
$\alpha_{i,j}^*$	Output arrival curve of aggregate flow $f_{i,j}$
β_i	ith service curve
B_{iX_j}	jth virtual channel of direction X in router i
B_i	The ith buffer
$\bar{B}_i$	Backlog bound of buffer B_i
ϕ_i	Weight of flow f_i
β_i^j	Equivalent service curve (ESC) of β_i for flow f_j
$\wedge$	The *minimum* operation, e.g. $a \wedge b = \min\{a, b\}$
$\otimes$	*Min-plus* convolution, defined as $(f \otimes g)(t) = \inf_{0 \le u \le t}\{f(t-u) + g(u)\}$
$\oslash$	*Min-plus* de-convolution, defined as $(f \oslash g)(t) = \sup_{u \ge 0}\{f(t+u) - g(u)\}$
$[x]^+$	The maximum of x and 0, namely $[x]^+ = \max\{x, 0\}$
$\delta_T(t)$	Burst-delay function, $\delta_T(t) = +\infty$ if $t > T$, else 0
$v(g_1, g_2)$	Function to calculate the maximum vertical distance between curves g_1 and g_2

B. Network Calculus Concepts

In network calculus, *arrival curve* and *service curve* are two of the most fundamental concepts. Arrival curve defines the upper bound of the injected amount of traffic within any time interval. We use linear arrival curve [7] $\alpha_i = \gamma_{b_i, r_i} = r_i t + b_i$ when $t > 0$ otherwise 0 as the arrival curve of flow f_i, where r_i is the sustainable arrival rate and b_i the burstiness. Here, a *flow* is a unicast stream of data from a source to a destination. A flow can be shaped to conform to a linear arrival curve by applying a leaky-bucket regulator [7]. As proved in [4], such a regulator does not add delay to the flow. Figure 1(a) shows an example of linear arrival curve $\alpha(t) = \gamma_{b,r}$ which constrains a periodic on-off traffic $R(t)$.

Service curve models a network service element (server), e.g. a router with various channel arbitration policies, expressing the minimum service capability. The latency-rate function $\beta_{C,T} = C[t-T]^+$ can model a variety of network elements

where C is the minimum service rate and T the maximum processing latency [14]. Consider a flow f with linear arrival curve $\gamma_{b,r}$, the backlog bound $\bar{B}$ is calculated as $v(\alpha(t), \beta(t))$ which is graphically the largest vertical distance between $\alpha(t)$ and $\beta(t)$ as shown in Figure 1(b). The output arrival curve after f is served by a server with service curve $\beta_{C,T}$ is calculated as $\gamma_{b+rT,r}$ [4].

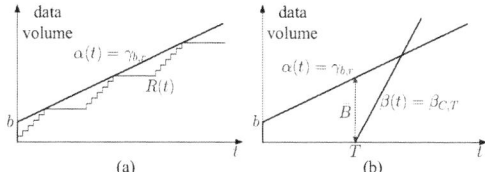

Fig. 1. (a) Linear arrival curve; (b) Latency-rate service curve and backlog bound.

Buffer sharing is a common resource contention among the flows in a network. We use the left-over service curve (*Theorem 6.2.1* in [4]) to resolve buffer sharing. Consider that two flows f_1 and f_2 aggregate into a shared FIFO (First-In First-Out) buffer and then are served by $\beta_1 = \beta_{C_1, T_1}$. we can calculate the *equivalent service curve* (ESC) of f_1 as $\beta_1^1 = [\beta_1 - \alpha_2]^+ = \beta_{C_1 - r_2, T_1 + b_2/C_1}$, and similarly for f_2.

Next, we exemplify the backlog bound analysis problem.

IV. THE BACKLOG BOUND ANALYSIS PROBLEM

A. Problem Illustration

Figure 2 shows a 3×3 mesh network together with the router micro-architecture showing only the relevant components. The router is a typical input-queuing virtual channel (VC) router with each input physical channel (link) equipped with two VCs. Three flows f_1, f_2 and f_3 are routed by deterministic routing in the network. The VCs are statically allocated to each flow. We denote R_i ($i = 1 \dots 9$) as the ith router in the mesh, B_{iX_j} as the jth VC of direction X in router R_i. For example, B_{4N_1} refers to the *first* VC of the *northern* input in the *fourth* router R_4.

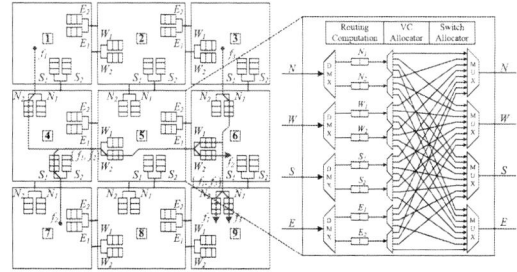

Fig. 2. An on-chip network example with flows annotated.

Both link sharing and buffer sharing exist in Figure 2. Flows f_1 and f_2 aggregate in R_4 and have link sharing at the eastern output channel. Then they aggregate into the same VC B_{5W_2}, thus have buffer sharing. f_1 and f_3 also have link sharing at the southern output channel of R_6, but not share any buffer because they eventually go to different VCs in R_9.

Our goal is to calculate the backlog bound of each VC involved in the communication. The initial step is to model the network-flow information as illustrated in Figure 2 to produce an NC analysis graph or model. As presented in

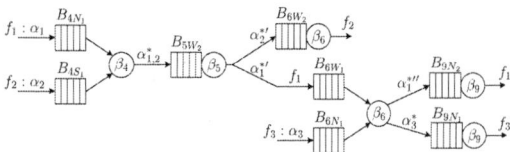

Fig. 3. The NC analysis graph of the illustrative example.

[15], this is fairly straightforward. We map a VC to a queue and multiplexer or link to a server β. Then we obtain the NC analysis graph for the example as illustrated in Figure 3. Looking into the graph closely, we can make two observations: (1) The backlog bound analysis is a *local* problem, which means, to get per-buffer backlog bound, we only need to know the input arrival to a buffer and the service provided to the buffer. Indeed this is in contrast to per-flow delay bound analysis [11][15], which is an *end-to-end* problem since it has to consider the servicing behavior from source to destination. (2) Because of the local nature, the backlog analysis problem is rather *regular* in the sense that we can decompose any complex buffer and link sharing scenarios into primitive cases, and further develop a regular analysis procedure which can be automated via an algorithm.

In the following, we discuss the basic resource-sharing cases and their NC analysis models.

B. Assumptions

We make the following basic assumptions:

- Flow f_i conforms to a linear arrival curve $\alpha_i = \gamma_{b_i, r_i}$, where r_i and b_i are the rate and the burstiness, respectively; otherwise this condition can be maintained by employing a leaky-bucket regulator [7].
- The network uses deterministic routing and static VC allocation such that each flow has a deterministic route. Buffers are served in the FIFO order.
- Network service elements such as link, multiplexer and packet sink are modeled by the latency-rate service curve $\beta_{C,T}$. The link is work-conserving with capacity C.

V. ELEMENTARY ANALYSIS CASES AND MODELS

A. Single Flow Cases

- **Case 1: A single flow passes a single node**

This is the simplest case, with just one node. The NC model is shown in Figure 4.

Fig. 4. A single flow through a single node.

We can calculate the backlog bound of B_1 as

$$\bar{B}_1 = v(\alpha, \beta_1) = b + rT_1. \tag{1}$$

Though this case is very simple, it tells two important principles for studying the backlog bounds of buffers: 1) As a local problem, the backlog bound model needs node by node analysis; and 2) We can deduce the backlog bound as long as both the arrival curve and the service curve of the server node are given or obtained. In the successive cases, these two principles are applied to deduce the backlog bound for each buffer.

- **Case 2: A single flow passes two nodes in a tandem**

The second case is shown in Figure 5. The input of the second node is from the output of the first node. In order to calculate $\bar{B}_2$, we need to calculate the output arrival curve of β_1 rather than directly use the arrival curve parameters.

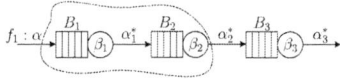

Fig. 5. Single flow through two nodes.

The calculation of $\bar{B}_1$ is the same as in case 1. We focus on the calculation of $\bar{B}_2$. First, we calculate the output arrival curve of the first node. Due to the finite capacity C of a real channel, the output cannot exceed $\gamma_{1,C}$ according to Lemma 1 in [16]. Thus output arrival curve α_1^* is calculated as

$$\alpha_1^* = (\alpha \oslash \beta_1) \wedge \gamma_{1,C} = \gamma_{b+rT_1,r} \wedge \gamma_{1,C}, \tag{2}$$

where b is the burstiness of the flow, r the arrival rate, and T_1 the latency of service curve β_1. Thus we calculate the backlog bound of B_2 as

$$\bar{B}_2 = v(\alpha_1^*, \beta_2) = v(\gamma_{b+rT_1,r} \wedge \gamma_{1,C}, \beta_{C_2,T_2})$$
$$= \begin{cases} (C-r)t_0 + rT_2 + 1, & T_2 \geq t_0 \\ (C-C_2)t_0 + C_2T_2 + 1, & T_2 < t_0 \end{cases}, \tag{3}$$

where $t_0 = (b + rT_1 - 1)/(C - r)$.

- **Case 3: A single flow passes three nodes in a tandem**

In case 2, we deduce the formula of the backlog bound considering the finite channel capacity. In this case, we continue to add the third node after the first two nodes. The calculation of $\bar{B}_1$ and $\bar{B}_2$ are the same as in case 2.

However, for $\bar{B}_3$, according to the principles proposed in case 1, we first calculate the output arrival curve α_2^* and we should consider two nodes before β_3. We treat all the nodes before the third node as a whole to tighten the output arrival curve to $\bar{B}_3$ as highlighted by the dashed line in Figure 6.

Fig. 6. Single flow through three nodes.

The channel capacity is greater than all the service rates, thus it does not affect the burstiness of the flow. The ESC of β_1 and β_2 is $\hat{\beta}_{12} = \beta_1 \otimes \beta_2 = \beta_{C_{12},T_{12}}$, where $C_{12} = C_1 \wedge C_2$, $T_{12} = T_1 + T_2$. Then the output arrival curve α_2^* can be calculated as $\alpha_2^* = \min\{\alpha \oslash \hat{\beta}_{12}, \gamma_{1,C}\} = \gamma_{b+rT_{12},r} \wedge \gamma_{1,C}$. Thus we calculate the backlog bound of B_3 as

$$\bar{B}_3 = v(\gamma_{b+rT_{12},r} \wedge \gamma_{1,C}, \beta_{C_3,T_3})$$
$$= \begin{cases} (C-r)t_0 + rT_3 + 1, & T_3 \geq t_0 \\ (C-C_3)t_0 + C_3T_3 + 1, & T_3 < t_0 \end{cases}, \tag{4}$$

where $t_0 = (b + rT_{12} - 1)/(C - r)$.

This case suggests a way to extend the backlog bound analysis model for complex cases. We always focus on the target node, and treat the nodes before it as a subsystem to calculate the ESC.

B. Aggregate Flow Cases

- **Case 4: Two flows merge into one aggregate flow**

Figure 7 shows the NC model where the aggregate flow $f_{1,2}$ is composed of flow f_1 with arrival curve α_1 and flow f_2 with arrival curve α_2.

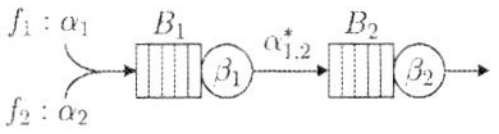

Fig. 7. Two flows aggregate through nodes.

We use the aggregate flow parameters to calculate the backlog bound. The calculation method for $\bar{B}_1$ and $\bar{B}_2$ is the same as in case 2 in Section V-A. According to the property of superposition, the arrival curve of the aggregate flow is $\alpha_{12} = \alpha_1 + \alpha_2 = \gamma_{b_{12}, r_{12}}$, where $b_{12} = b_1 + b_2$ and $r_{12} = r_1 + r_2$. Then $\bar{B}_1$ is calculated as

$$\bar{B}_1 = v(\alpha_{12}, \beta_1) = b_{12} + r_{12}T_1, \qquad (5)$$

and $\bar{B}_2$ is

$$\bar{B}_2 = \begin{cases} (C - r_{12})t_0 + r_{12}T_2 + 1, & T_2 \geq t_0 \\ (C - C_2)t_0 + C_2T_2 + 1, & T_2 < t_0 \end{cases}, \qquad (6)$$

where $t_0 = (b_{12} + r_{12}T_1 - 1)/(C - r_{12})$.

- Case 5: Two flows merge and two separate flows out

We consider the scenario that the aggregate flow is split after the buffer sharing as shown in Figure 8. $f_{1,2}$ is split into two parts: the packets of f_1 and f_2 feed into B_2 and B_3, respectively. The calculation of $\bar{B}_1$ is the same as in case 4. We focus on $\bar{B}_2$ and B_3.

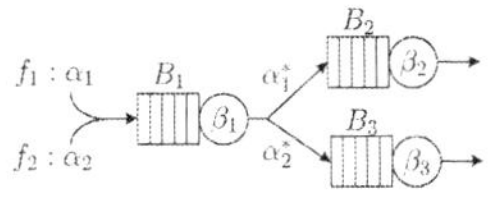

Fig. 8. Aggregate input with split output flows.

To calculate the output arrival curve α_1^*, we first calculate the ESC of β_1 for f_1 as $\hat{\beta}_1^1 = [\beta_1 - \alpha_2]^+ = \beta_{C_1^1, T_1^1}$, where $C_1^1 = C_1 - r_2$ and $T_1^1 = T_1 + \frac{b_2}{C_1}$. Then we calculate α_1^* as $\alpha_1^* = (\alpha_1 \oslash \hat{\beta}_1^1) \wedge \gamma_{1,C}$. Thus we obtain

$$\bar{B}_2 = v(\alpha_1^*, \beta_2) = \begin{cases} (C - r_1)t_0 + r_1T_2 + 1, & T_2 \geq t_0 \\ (C - C_2)t_0 + C_2T_2 + 1, & T_2 < t_0 \end{cases}, \qquad (7)$$

where $t_0 = (b_1 + r_1T_1^1 - 1)/(C - r_1)$. $\bar{B}_3$ is calculated in the same way and obtains the similar formula.

- Case 6: Two separate flows in and one aggregate flow out

As shown in Figure 9(a), packets are injected into B_1 and B_2 from two individual flows f_1 and f_2. Then these two flows merge at the server β_1 and share the output channel. β_1 is the service provided by the multiplexing channel with service rate of C and latency of 0.

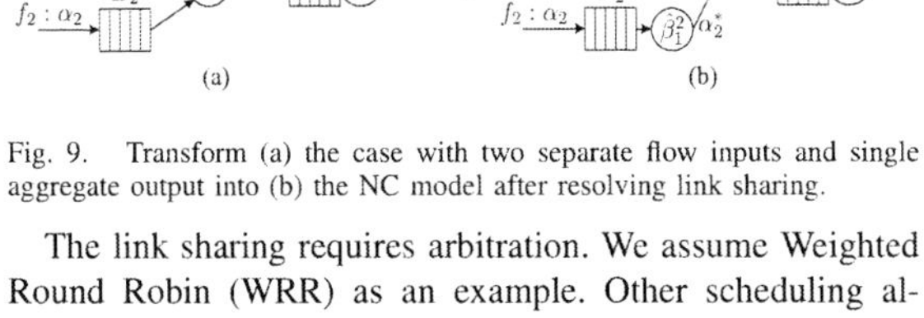

Fig. 9. Transform (a) the case with two separate flow inputs and single aggregate output into (b) the NC model after resolving link sharing.

The link sharing requires arbitration. We assume Weighted Round Robin (WRR) as an example. Other scheduling algorithms such as FIFO, priority-based polices can be used as well [4]. Let ϕ_1 and ϕ_2 be the weights of f_1 and f_2 respectively, we calculate the ESC of β_1 for f_1 as $\hat{\beta}_1^1 = \frac{\phi_1}{\phi_1 + \phi_2}\beta_{C,0} \otimes \delta_{\phi_2} = \beta_{C_1^1, T_1^1}$, where $C_1^1 = \frac{\phi_1}{\phi_1 + \phi_2}C$, $T_1^1 = \phi_2$

and for f_2 as $\hat{\beta}_1^2 = \frac{\phi_2}{\phi_1 + \phi_2}\beta_{C,0} \otimes \delta_{\phi_1} = \beta_{C_1^2, T_1^2}$, where $C_1^2 = \frac{\phi_2}{\phi_1 + \phi_2}C$ and $T_1^2 = \phi_1$. The NC model is transformed into the one shown in Figure 9(b).

Then we obtain the backlog bound of B_1 and B_2 as $\bar{B}_1 = v(\alpha_1, \hat{\beta}_1^1) = b_1 + r_1T_1^1$ and $\bar{B}_2 = v(\alpha_2, \hat{\beta}_1^2) = b_2 + r_2T_1^2$. The output arrival curves after β_1^1 and β_1^2 are $\alpha_1^* = (\alpha_1 \oslash \hat{\beta}_1^1) \wedge \gamma_{1,C} = \gamma_{b_1 + r_1T_1^1, r_1} \wedge \gamma_{1,C}$ and $\alpha_2^* = (\alpha_2 \oslash \hat{\beta}_1^2) \wedge \gamma_{1,C} = \gamma_{b_2 + r_2T_1^2, r_2} \wedge \gamma_{1,C}$. Thus the output arrival curve before B_3 is $\alpha_{1,2}^* = (\alpha_1^* + \alpha_2^*) \wedge \gamma_{1,C} = \gamma_{b_{12}, r_{12}} \wedge \gamma_{1,C}$, where $b_{12} = b_1 + b_2 + r_1T_1^1 + r_2T_1^2$ and $r_{12} = r_1 + r_2$. Then we obtain the backlog bound $\bar{B}_3$ as

$$\bar{B}_3 = \begin{cases} (C - r_{12})t_0 + r_{12}T_2 + 1, & T_2 \geq t_0 \\ (C - C_2)t_0 + C_2T_2 + 1, & T_2 < t_0 \end{cases}, \qquad (8)$$

where $t_0 = (b_{12} - 1)/(C - r_{12})$.

C. Summary and Discussion

We propose six basic scenarios and the corresponding network calculus models to calculate the backlog bounds. From the perspective of calculation, *Case 2*, *Case 5* and *Case 6* are able to cover all the contention scenarios and calculation methods. *Case 2* provides the concatenation model for buffers and servers in a tandem. The split model is proposed in *Case 5* to resolve buffer sharing. Then *Case 6* gives the merge model to resolve link sharing.

However, *Case 1*, *Case 3* and *Case 4* are still meaningful in implying important principles to complete the calculation models. *Case 1* shows the basic calculation model and implies that the backlog bound analysis requires node by node analysis and can be calculated once both of the arrival curve and the service curve are obtained. *Case 3* tells that we can treat all the nodes before the analyzed buffer as a subsystem, thus the calculation of the output arrival curve before the buffer can be significantly simplified. *Case 4* uses the superposition of arrival curves when the analyzed buffer has aggregate input and aggregate output. In this case the aggregate flow is treated as a whole.

For all the basic models, an implicit stability condition should be met especially for flow aggregations. The flow arrival rate should not surpass the ESC rate provided by the system. Otherwise the backlog goes to infinity.

VI. Analysis Procedure & Algorithm

A. Analysis Procedure

We propose iterative analysis steps to calculate the backlog bounds for the buffers in the network. The calculation status is updated after each iteration. Thus intermediate results, e.g. ESCs and output arrival curves, are reused for the uncalculated buffers in the next iteration. The analysis procedure is composed of three main steps, described as follows:

Step 1: Resolve all the link sharing contentions and calculate the corresponding ESCs with the method in *Case 6*.

Step 2: Go through all the buffers whose backlog bounds are not calculated yet.

Step 2.1: If both the arrival curve and the service curve are available (or given), calculate the backlog bound with the model in *Case 2*.

Step 2.2: Calculate the output arrival curve after this buffer. In case of buffer sharing, if the buffer still outputs an aggregate flow, calculate the output arrival curve as in *Case*

4. If the aggregate input is split into separate flows, calculate the left-over service curve for each flow, then calculate the output arrival curves for each flow as in *Case 5.*

Step 3: If all the buffer backlog bounds are calculated, stop; otherwise, return to **Step 2**.

B. Analysis Example

Using the procedure, we have calculated the bounds for the buffers in the example (see Figure 2 and 3). The calculation consists of five iterations as summarized in Table II. We omit the calculation details due to the space limitation.

TABLE II
OVERVIEW OF THE CALCULATION PROCEDURE

Iteration	Output (backlog bounds)	Output arrival curves
Initial	Resolve link sharing	$\alpha_1, \alpha_2, \alpha_3$
1	$\bar{B}_{4N_1}, \bar{B}_{4S_1}, \bar{B}_{6N_1}$	$\alpha^*_{1,2}, \alpha^*_3$
2	$\bar{B}_{5W_2}, \bar{B}_{9N_1}$	$\alpha^{*\prime}_1, \alpha^{*\prime}_2$
3	$\bar{B}_{6W_1}, \bar{B}_{6W_2}$	$\alpha^{*\prime\prime}_1$
4	$\bar{B}_{9N_2}$	–

As an example, we obtain the closed-form backlog bound formula of B_{5W_2} in iteration 2 as

$$\bar{B}_{5W_2} = \begin{cases} b^*_{12} + (r_1 + r_2)T_5, & T_5 \geq t_{12} \\ (C - C_5)t_{12} + C_5 T_5 + 1, & T_5 < t_{12} \end{cases}, \quad (9)$$

where $b^*_{12} = b_1 + b_2 + r_1(T_4 + \phi_2) + r_2(T_4 + \phi_1)$, $t_{12} = (b^*_{12} - 1)/(C - r_1 - r_2)$.

C. Analysis Algorithm

We present Algorithm 1 to automate the backlog bound analysis procedure. Denote $\mathcal{P} = \{p_i\}$ the set of flow paths, $\mathcal{B} = \{B_i\}$ the set of buffers and $\bar{\mathcal{B}} = \{\bar{B}_i\}$ the set of backlog bounds, $\mathcal{B}_{nc}$ the set of buffers whose bounds are not calculated yet, F_{B_i} the set of flows feeding B_i, β_{B_i} the set of service curves at B_i and $\alpha^*_{B_i}$ the set of (output) arrival curves before B_i. We define $v_i = < \alpha^*_{B_i}, B_i, \beta_{B_i}, F_{B_i} >$ as the vertex corresponding to B_i in the backlog analysis graph $G = < V, E >$ where $V = \{v_i\}$ and $E = \{e_i\}$, and $e_i = < v_i, v_j >$ as the edge between v_i and v_j.

Algorithm 1 has two parts. It first constructs the NC analysis graph G according to the set of buffers, service curves, and the path information of the flows in the network (Line 1 to 13). Let n be the number of buffers to be analysed in the network, m be the number of flows, k be the largest length of all the paths of the flows. In Algorithm 1, the *for-loop* initializes V with the complexity of $\mathcal{O}(n)$. Function $GenerateEdges(p_i)$ generates the edges of G according to the path information with complexity of $\mathcal{O}(k)$. Thus the *for-loop* which generates G has complexity of $\mathcal{O}(m(n + k))$. The second part of Algorithm 1 calculates the backlog bounds iteratively until no buffer is left, namely $\mathcal{B}_{nc} = \Phi$ (Line 14 to 29). The set of the backlog bounds $\bar{\mathcal{B}}$ is returned, where $\bar{\mathcal{B}} = \{\bar{B}_i | B_i \in \mathcal{B}\}$.

The link sharing is resolved with the complexity of $\mathcal{O}(n^2)$. Function $ResolveLinkSharing(v_i, v_j)$ calculates ESCs with the method in *Case 5* and updates the service curve information in vertices v_i and v_j. The *while-loop* iteratively calculates the backlog bounds as in **Step 2** and **Step 3** in Section VI-A. In the worst case, only one buffer is calculated each time, and the outer loop executes n times while the internal *for-loop* runs $\sum_{i=1}^{n} = n(n + 1)/2$ times. Thus the complexity

---

Algorithm 1: Calculate the backlog bounds of buffers

Input: $\mathcal{B}, \beta_B, \mathcal{P}$
Output: The set of all the backlog bounds $\bar{\mathcal{B}}$
Initialize: $V \leftarrow \Phi, E \leftarrow \Phi, \bar{\mathcal{B}} \leftarrow \Phi, \mathcal{B}_{nc} \leftarrow \mathcal{B}$

```
 1  foreach Bᵢ ∈ ℬ do
 2  │   vᵢ ← < Φ, Bᵢ, Φ, Φ > ;  // Initialize V
 3  end
 4  foreach pᵢ ∈ 𝒫 do
 5  │   if Bⱼ is in pᵢ then
 6  │   │   vⱼ.F_Bⱼ ← vⱼ.F_Bⱼ ∪ {fᵢ};
 7  │   │   vⱼ.β_Bⱼ ← vⱼ.β_Bⱼ ∪ {βᵢ} ;
 8  │   │   if Bⱼ is the first node then
 9  │   │   │   vⱼ.α*_Bⱼ ← vⱼ.α*_Bⱼ ∪ {αᵢ} ;
10  │   │   end
11  │   end
12  │   E ← E ∪ GenerateEdges(pᵢ) ;
13  end
14  foreach vᵢ ∈ V do
15  │   foreach vⱼ ∈ V and j ≠ i do
16  │   │   if vᵢ.β_Bᵢ equals vⱼ.β_Bⱼ then
17  │   │   │   ResolveLinkSharing(vᵢ, vⱼ);
18  │   │   end
19  │   end
20  end
21  while ℬ_nc ≠ Φ do
22  │   foreach vᵢ.Bᵢ ∈ ℬ_nc do
23  │   │   if Ready(vᵢ.α*_Bᵢ, vᵢ.F_Bᵢ) and vᵢ.β_Bᵢ ≠ Φ then
24  │   │   │   B̄ ← B̄ ∪ CalcBound(vᵢ) ;
25  │   │   │   ℬ_nc ← ℬ_nc \ {Bᵢ} ;  // Update ℬ_nc
26  │   │   │   UpdateArrivalCurves(vᵢ, Next(vᵢ)) ;
27  │   │   end
28  │   end
29  end
30  return B̄ ;
```

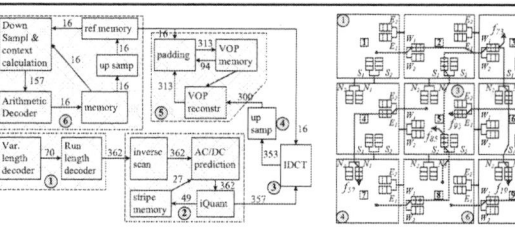

Fig. 10. Block diagram of the VOP decoder with bandwidth requirements annotated.

Fig. 11. Task-to-node mapping of the VOP decoder.

of the iterative calculation is $\mathcal{O}(n^2)$. We obtain the overall complexity of Algorithm 1 as $\mathcal{O}(n^2 + mn + mk)$. Considering an $N \times N$ mesh network, we have $n \propto N^2$, $m \propto N^2$, and $k \propto N$. Thus the complexity is $\mathcal{O}(N^4)$ with respect to the network dimension.

VII. A CASE STUDY

A. Experiment Description and Setup

We use a realistic SoC application, the Video Object Plane (VOP) decoder [17], as an example to validate and show the potential of the backlog analysis method. Figure 10 depicts the block diagram of the VOP decoder with the required bandwidths (in MB/s) annotated on the communication channels. The blocks are grouped into six task clusters which are enclosed by the dashed polygons.

As shown in Figure 11, the tasks are then mapped to a mesh NoC that is implemented in Verilog HDL for cycle-accurate simulation. For simplicity, the processing and memory elements are not shown in Figure 11. The network adopts source routing algorithm. The router has the same structure as shown in Figure 2. In the case of channel contention, round-robin arbitration is used. For node i in the NoC, we denote

$B_{iN}/B_{iE}/B_{iS}/B_{iW}$ as the northern/eastern/southern/western ingress buffer, i.e. B_{5S} represents the southern ingress buffer of node 5. Figure 11 also illustrates the inter-task flows and how the buffers are utilized by these flows. We analyse the backlog bounds for B_{2W}, B_{2S}, B_{3W}, B_{3S}, B_{4E}, B_{5E}, B_{5S}, B_{6N}, B_{6S}, B_{7N}, B_{8W}, B_{9W} and B_{9N} in this case.

We first convert the bandwidth requirement of each flow into corresponding traffic injection rate in packets/cycle. We assume that the NoC works in a medium power saving mode with frequency of 300 MHz. All packets have the same length and each packet comprises one flit with a uniform size of 32 bits. The channel capacity is 1.0 flit/cycle. Each flow injection follows the linear arrival model. Thus we obtain the arrival curves of all flows as listed in Table III, where b_i is the burstiness of arrival curve α_i for $i = 0 \ldots 5$.

TABLE III
FLOW CONFIGURATIONS

Flow	Arrival Curve	Flow	Arrival Curve
f_{19}	$\alpha_0 = 0.302t + b_0$	f_{73}	$\alpha_1 = 0.250t + b_1$
f_{85}	$\alpha_2 = 0.013t + b_2$	f_{83}	$\alpha_3 = 0.013t + b_3$
f_{95}	$\alpha_4 = 0.298t + b_4$	f_{57}	$\alpha_5 = 0.294t + b_5$

B. Results and Discussions

We report two groups of representative results to validate the effectiveness of our proposed method. In the first simulation, the flow burstiness are set as $b_i = 16$ for $i = 0 \ldots 5$. The packet sinking service is configured as $C_s = 0.8$ and $T_s = 15$. In the other simulation, we set $b_i = 32$ for $i = 0 \ldots 5$, $C_s = 0.7$ and $T_s = 15$. Each simulation runs for $500,000$ cycles under a specific configuration to collect the maximum backlogs. The calculated backlog bounds and the simulated maximum backlogs of the two simulations are listed in Table IV. We denote the calculated backlog bound as B, and the simulated maximum backlog as B_{max}. The results in Table IV indicate that the calculated backlogs can bound the maximum simulated backlogs tightly. The difference between the simulated and theoretical bounds is due to the limitation of simulation, as the occurrence of worst case requires exhaustive simulations of traffic and service configurations, which is hard to fulfill in experiments. We observe that not all the simulated backlogs closely approach the calculated bounds simultaneously. This is not surprising because the formal analysis covers all cases but the worst case backlogs may not occur for all the buffers simultaneously under the same simulation configuration. In fact, we can observe that analytic backlog bounds which are not tight in one simulation configuration may become tight in another simulation configuration. For example, B_{3W} in result I has a difference of 9 (18%) but in result II only 2 (1.9%).

TABLE IV
TWO GROUPS OF RESULTS: CALCULATED BOUNDS AND SIMULATED MAXIMUM BACKLOGS

	B_i	B_{2W}	B_{2S}	B_{3W}	B_{3S}	B_{4E}	B_{5E}	B_{5S}	B_{6N}	B_{6S}	B_{7N}	B_{8W}	B_{9W}	B_{9N}
I	$\bar{B}_i$	13	26	49	17	1	19	37	1	1	18	12	10	20
	B_{max}^i	12	23	40	16	1	17	26	1	1	17	11	9	17
II	$\bar{B}_i$	24	48	102	22	1	29	79	1	1	25	22	18	34
	B_{max}^i	23	44	100	17	1	17	39	1	1	17	21	17	17

VIII. CONCLUSION

Formal analysis of the backlog bound of virtual channels in on-chip routers provides quick insight into the buffer sizing boundary and the non-uniformity of buffer requirements. In this work, we identify basic buffer use scenarios and develop corresponding analysis models with network calculus. A complex communication case can be decomposed into the basic cases, and further be analysed to calculate the backlog bounds step by step following the proposed analysis procedure. We exemplify the analysis procedure by an illustrative example. An algorithm is then developed to automate the per-buffer backlog bound analysis with polynomial complexity. The case study of an on-chip VOP decoder shows the application and fidelity of our analysis method. Moreover, the proposed method is topology independent, thus can be applied to various NoC designs with regular or irregular topology.

REFERENCES

[1] Jingcao Hu, Umit Y. Ogras, and Radu Marculescu. System-level buffer allocation for application-specific networks-on-chip router design. *IEEE Trans. Computer-Aided Design*, 25(12):2919 –2933, Dec. 2006.

[2] Y. Hoskote, S. Vangal, A. Singh, N. Borkar, and S. Borkar. A 5-GHz mesh interconnect for a teraflops processor. *IEEE Micro*, 27(5):51–61, Sep.-Oct. 2007.

[3] William James Dally and Brian Towles. *Principles and Practices of Interconnection Networks*. Morgan Kaufman Publishers, 2004.

[4] Jean-Yves Le Boudec and Patrick Thiran. *Network Calculus: A Theory of Deterministic Queuing Systems for the Internet*. Number 2050 in LNCS, 2004.

[5] Chengshang Chang. *Performance Guarantees in Communication Networks*. Springer-Verlag, 2000.

[6] A. S. Kumar, M. P. Kumar, S. Murali, V. Kamakoti, L. Benini, and G. De Micheli. A simulation based buffer sizing algorithm for network on chips. In *Proc. IEEE Computer Society Annual Symp. VLSI (ISVLSI)*, pages 206–211, 2011.

[7] Rene L. Cruz. A calculus for network delay, part I: Network elements in isolation; part II: Network analysis. *IEEE Trans. Inf. Theory*, 37(1), Jan. 1991.

[8] M. A. Al Faruque and J. Henkel. Minimizing virtual channel buffer for routers in on-chip communication architectures. In *Proc. Design, Automation and Test in Europe*, pages 1238–1243, 2008.

[9] A. Hansson, M. Wiggers, A. Moonen, K. Goossens, and M. Bekooij. Applying dataflow analysis to dimension buffers for guaranteed performance in networks on chip. In *Proc. 2nd ACM/IEEE Int. Symp. Networks-on-Chip*, pages 211 –212, Apr. 2008.

[10] Andreas Hansson, Maarten Wiggers, Arno Moonen, Kees Goossens, and Marco Bekooij. Enabling application-level performance guarantees in network-based systems on chip by applying dataflow analysis. *IET Computers & Digital Techniques*, 3(5):398–412, 2009.

[11] Yue Qian, Zhonghai Lu, and Wenhua Dou. Analysis of worst-case delay bounds for on-chip packet-switching networks. *IEEE Trans. Computer-Aided Design*, 29(5):802–815, 2010.

[12] A. Chariete, M. Bakhouya, J. Gaber, and M. Wack. A buffer size customization approach for application-specific NoC design. In *Int. Conf. High Performance Computing and Simulation*, pages 224–228, 2013.

[13] Junhui Wang, Yue Qian, Yi Wang, Zili Shao, Wenhua Dou, and Qiang Dou. Analysis of worst-case backlog bounds for networks-on-chip. *Journal of Systems Architecture*, 60(6):494–508, 2014.

[14] D. Stiliadis and A. Varma. Latency-rate servers: a general model for analysis of traffic scheduling algorithms. *IEEE/ACM Trans. Networking*, 6(5):611–624, 1998.

[15] Xueqian Zhao and Zhonghai Lu. Per-flow delay bound analysis based on a formalized microarchitectural model. In *Proc. 7th ACM/IEEE Int. Symp. Networks-on-Chip*, pages 1–8, April 2013.

[16] Yuming Jiang. Delay bounds for a network of guaranteed rate servers with FIFO aggregation. *Computer Networks*, 40(6):683 – 694, 2002.

[17] S. Murali and G. De Micheli. Bandwidth-constrained mapping of cores onto NoC architectures. In *Proc. Conf. Design, Automation and Test in Europe*, pages 896 – 901, Feb. 2004.

978-1-4799-8720-7/15 $31.00 © 2015 IEEE

2015 IEEE Computer Society Annual Symposium on VLSI

A Timing Error Mitigation Technique for High Performance Designs

Mehrnaz Ahmadi, Bijan Alizadeh, Behjat Forouzandeh

School of Electrical and Computer Engineering, College of Engineering
University of Tehran
Tehran, Iran
m.ahmady@ut.ac.ir, b.alizadeh@ut.ac.ir, bforooz@ut.ac.ir

Abstract – Dynamic flip-flop conversion (DFFC) is a time borrowing method which converts the critical flip-flops into transparent latches to allow timing slacks pass between pipeline stages of given circuits. However, our previous DFFC methods [13] [14] suffer from false error prediction. It means even when there is no setup time violation, our previous method incorrectly issues timing error. In this paper we present an improved DFFC method which consumes less power and unlike previous DFFC methods does not suffer from false errors. Also we investigate the effectiveness of our proposed DFFC method on different benchmarks by considering all existing critical paths, instead of applying our method only to one of critical paths as done in [13][14]. The results show that our proposed method improves maximum allowable frequency 7.54% on average while the previous method in [13] [14] increases maximum allowable frequency 4.24% on average. Furthermore, it consumes on average 19.59 microwatt less power per critical path compared to previous DFFC methods [13][14].

Keywords—Dynamic flip-flop conversion; time borrowing, transparency window; hold time violation (HTV); setup time violation (STV)

I. INTRODUCTION

The demand for high performance design with low power consumption has been significantly increasing over the past few years. Traditionally, in worst case design methodology the maximum allowable frequency (MAF), in which a circuit works, is computed based on the delay of the longest paths (critical paths) in the circuit. Such critical paths, which are rarely activated, make us keep the frequency low and that is why we lose the performance. It is obvious that by increasing the clock frequency we achieve better performance. However, the problem would be the fact that critical paths are not able to finish their job and therefore setup/hold time violation would occur. In order to address these timing violation issues three main approaches including error detection and correction, retiming and time borrowing have been proposed.

Various timing error detection and correction techniques at different level of abstractions have been proposed to address timing violation issues [1] [5] [18]. In such techniques a timing error is detected at each flip-flop using a transition detector circuit and then a recovery mechanism corrects the data after some clock cycles [1] [2]. Such techniques, however, do not prevent timing errors. Retiming is a technique to prevent timing violations by moving flip-

flops backwards or forwards in the paths. Such a technique is not efficient because it may increase the number of flip-flops and its verification is a big challenging [3] [4]. A well-known and commonly used technique to avoid timing violation is time borrowing technique [5]. The main idea is the fact that a critical stage in which the setup time is violated can borrow some time from the following stage if it has enough timing slack.

The level sensitive behavior and small overhead of latches make them a good choice to tolerate timing violations [6]. Since latches are level sensitive, the late data from critical paths can cross the latches when they are transparent. In [7] the timing yield has been improved by replacing some flip-flops with latches. Although the overhead of latches is smaller than that of flip-flops, the number of required resources increases which results in an extra area overhead. Moreover, latch based designs are very prone to hold time violations (HTV) due to large transparency window. Pulsed latches have been used in [6] to avoid HTV on Field Programmable Gate Array (FPGA). By lowering the duty cycle of the clock in these latches, hold time violations have been reduced. In [8] a special flip-flop and a clock shifter together help to prevent timing violations. When timing error happens this flip-flop allows time borrowing during a time-borrowing window (TBW) and generates a flag that shows time borrowing has happened. The clock shifter elastically stretches the clock period in the next clock cycle to pay back for the borrowed time in the next clock cycle and prevents the next stage in pipeline to become critical.

In [9] [10], soft-edge flip-flops (SEFF) have been used which have a small window of transparency instead of a hard edge, allowing limited cycle stealing on critical paths, and thus compensating for delay variations. This, however, results in more power and area overhead. An optimization method has been proposed in [10] to find the optimal window size of SEFFs in order to get the minimum power-delay product (PDP). Clock skewing or clock stretching is another technique that has been used in [11] [12]. In [11] a circuit that dynamically detects timing violations and feeds a delayed clock signal to the destination flip-flop has been proposed. It also inserts donor flip-flops which are empty stages at the end of each stage. Whenever time borrowing is needed the clock pulse of the flip-flop at the end of critical path called "critical flip-flop" is skewed and donor flip-flop

978-1-4799-8720-7/15 $31.00 © 2015 IEEE

next to critical flip-flop is activated to donate its time slack to slow path. In [12] an optimization algorithm has been proposed to find the optimal delay values of each flip-flop. It uses variable delay chains and assigns a particular clock signal to each flip-flop so that time slacks can be transferred between different stages [13]. DFFC is another time borrowing technique which utilizes the advantages of latches in flip-flop based design in a dynamic manner. The first and second types of DFFC (Type A and Type B) have been proposed in [13] and [14], respectively. However, these two methods suffer from false errors which means that DFFC Types A and B incorrectly issue a timing error while actually there is no timing violation. Note that false errors create a large transparency window which may result in HTV. That is because if a short path ends to the critical flip-flop, it overwrites the correct value that was received before transparency window opens and therefore HTV happens. In this paper we present a modified version of DFFC (Type C) to alleviate the false error problem. Hence, our main contributions in this paper in comparison with [13] and [14] are as follows:

- Improving DFFCs in [13][14] by modifying Data Arrive Detector (DAD) block in such a way that it remains the transparency window close if the data arrives before the rising edge of the clock. This way, we not only solve the false error problem but also reduce the power consumption.

- Evaluating improved DFFC (Type C) on benchmark circuits considering all path structures inside them, while Types A and B have been applied to one of critical paths of a given circuit.

The rest of this paper is organized as follows. In Section II, we present our proposed DFFC method (Type C) in more details. Section III discusses challenges for architecture level implementation of time borrowing methods on a complete system. The experimental results are reported in Section IV and finally, Section V concludes the paper.

II. DYNAMIC FLIP-FLOP CONVERSION (DFFC)

DFFC is a time borrowing method which takes the advantage of transparency window of latches in a flip-flop based design. Unlike latch based designs which are very prone to HTV, the circuit is designed based on flip-flop timing character. Whenever setup time violation is predicted the critical flip-flops dynamically convert into transparent latches and the late data is stored in destination flip-flop.

A. DFFC Types A and B

Suppose all flip-flops sample the data at the rising edge of the clock. Fig. 1 shows two Types of DFFC, i.e. Type A [13] and Type B [14]. If the critical path is slow, then it cannot finalize the value at the destination flip-flop's input by the end of clock period. So, after half of the clock period the value in the middle of the slow path is not finalized yet. By checking the transition on a node in the middle of the

path after falling edge of clock the late data arrival is predicted.

The node which is located in the middle of the critical path ("mid" point) is connected to timing violation predictor (TVP) block. This block stores the value at the falling edge of a clock via prediction flip-flop and detects any transition on "mid" point during the second half of the clock period. As shown in Fig. 1(a), if the value of "mid" point changes, "Err" signal goes high and it stays high until the next falling edge of the clock. Once the error is predicted in DFFC Type A, master latch clock is inverted using an XOR gate. So, the flip-flop is turned into two latches that act like one transparent latch when the clock is high and late data from critical path can cross the transparency window. When the output of TVP block becomes zero the critical flip-flop goes back to normal operation. In this type of DFFC, HTV may happen because the transparency window stays open for half of the clock period. To resolve this problem, in DFFC Type B shown in Fig. 1(b), a DAD block is added to the circuit. This block detects any transition on destination flip-flop input and closes the transparency widow before the data from short path overwrites the previous value. After predicting error by TVP block "Conv" signal becomes one and master latch clock is inverted and the critical flip-flop is turned into a latch that is transparent when the clock is high. When the late data arrives at the input of critical flip-flop the transition of signal is detected by DAD so "Conv" signal becomes zero and the critical flip-flop goes back to normal operation. Note that, in Type B, HTV is prevented because the transparency window is not fixed and closes when data arrives at destination flip-flop.

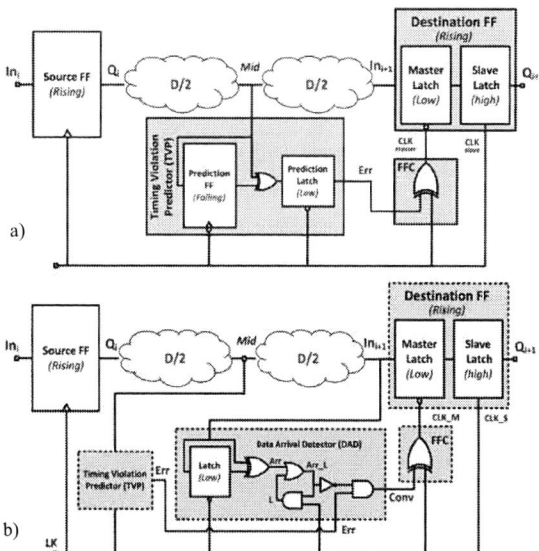

Fig. 1. DFFC circuit, a) Type A [13], b) Type B [14].

In [13] and [14], it is proven that DFFC Types A and B can result in a higher yield and better performance compared to existing methods like SEFF [10] and dynamic clock stretching [12]. However these two types of DFFC both suffer from false error prediction of TVP block [14]. Because in these methods the "mid" point is selected a little after the half delay in asymmetrical critical path [13], false errors may happen. This may result in a wrong prediction of error while the path may not violate any timing constraint and data arrives before the rising edge of the clock. In the case of a false error the transparency window falsely opens until either the data from short path overwrites the value in destination flip-flop or the next falling edge of clock closes the transparency window. In order to solve this problem, we have proposed DFFC Type C which will be explained in more details in the following subsection.

B. DFFC Type C

Fig. 2 shows DFFC Type C which is an improved version of DFFC Type B to solve the false error without any additional hardware. The TVP block is used to predict the timing violation. The main difference between Types B and C is in their DAD block. When false error happens the DAD block in Type C is able to detect data arrival on destination flip-flop's input before rising edge of clock and prevents the transparency window at destination flip-flop from opening while in Type B the DAD block is not able to detect data arrival on destination flip-flop before the rising edge of the clock.

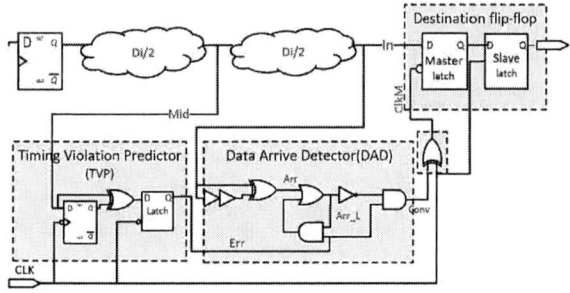

Fig. 2. Circuit diagram of improved DFFC (Type C).

In normal operation mode "Arr", "Arr_L" and "Err" signals are low. When error is predicted by TVP block ("Err"=1) one of two cases happen: 1) Correct error detection: With "Err"=1, "Conv" signal goes high and with rising edge of clock the transparency window opens. When the data arrives at input of destination flip-flop, an Impulse is created on "Arr". Then "Arr_L" goes high and "Conv" becomes zero which closes the transparency window and destination flip-flop goes back to normal work. 2) False error prediction: With "Err"=1, "Conv" signal goes high. Then the data arrives at input of destination flip-flop. An Impulse is created on "Arr". "Arr_L" goes high and "Conv" becomes zero so after rising edge of clock the transparency window is not created. Fig. 3 shows the internal signals for both DFFC Types B and C in the presence of false errors. The "In" signal is the destination flip-flop's input. False errors occur when the data arrives at the destination flip-flop before the rising edge of the clock. Since DAD in Type B

does not detect the signal arrival, a large transparency window is created. In Fig. 3 it is colored in red.

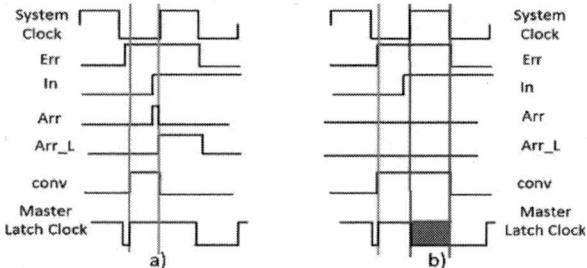

Fig. 3. Internal signals of DAD block in the presence of false errors: (a) Type C, (b) Type B.

III. IMPLEMENTATION ON A COMPLETE CIRCUIT

Although time borrowing methods may be effective in single path structures, it is important to make sure that they can improve the performance of the sequential circuits and they are able to increase the clock frequency of a digital circuit. In [8], a clock stretching method was validated using a prototype test-chip designed in 180-nm CMOS technology. In [3], retiming and time borrowing method using pulsed latches were performed on several benchmark circuits. In [15], selectively clocked skewed logic was simulated on 16-bit Carry Look ahead Adder (CLA) using TSMC 0.25μm technology. In previous works [13] [14], DFFC was evaluated only on single critical paths and its effectiveness for applying to a sequential circuit with various forms of critical paths is not evaluated.

In this work we have implemented DFFC Type C on complete sequential circuits including real pipeline stages and feedback paths and then compared the results with those of DFFC Type B. Fig. 4 shows some challenging path structures between flip-flops in two ITC'99 benchmarks [18], b05 and b12. As shown in this figure, challenging paths are categorized into five categories that are explained in the following subsections.

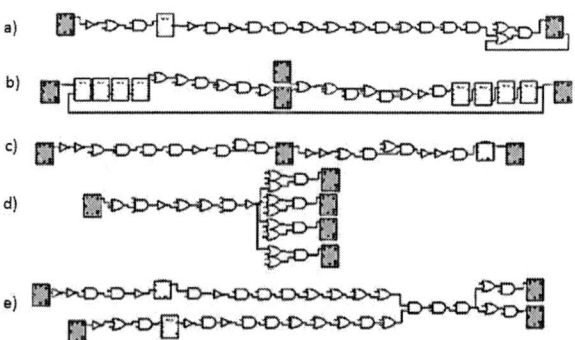

Fig. 4. Some challenging paths in b05 and b12 circuits, a) short feedback, b) critical feedback, c) consequence critical, and d) critical fan-out, and e) critical fan-in.

A. Short Feedback Path

Fig. 4(a) shows a short feedback path. Some methods like latch based designs [6] [11] and DFFC Type A [13] are

978-1-4799-8720-7/15 $31.00 © 2015 IEEE

not useful in this case because the wide and fixed transparency window creates a combinational feedback loop that may make the data unstable. Since in DFFC Types B and C the transparency window closes with the first data arrival on the input of the flip-flop, the data would not be unstable. As a result, our method can be used for feedback paths that are not followed by another critical path. Note that short feedback can cause HTV in DFFC Type A. We need to insert large delay elements in short feedback paths so that the data from short path arrives at destination flip-flop no earlier than falling edge of clock. In DFFC Types B and C, however, we close the transparency window once the data arrives at the input of the critical flip-flop. If the data from the short path arrives before the transparency window closes by DAD block then the data stored in critical flip-flop will be overwritten by the data from short path. So, the delay of the short path should be larger than the delay of DAD block. This will reduce the number of delay elements needed to increase the path delay of such short paths to avoid HTV.

B. Critical Feedback Path

Fig. 4(b) shows a critical feedback. It is a feedback path which is also a critical path. Most of the time borrowing methods including DFFC Types A, B and C are not effective in this case. Because these methods we borrow time from the next stage in the pipeline and the time slack from the next stage is transferred to the critical stage. In cases where the next stage does not have any time slack we cannot borrow time from next stage. In this structure the next stage after a critical feedback is a critical path (again critical feedback path) and therefore it is a problematic path structure for time borrowing.

C. Consequence Critical Path

Fig. 4(c) shows a consequence critical path. It is a critical path followed by another critical path in a pipeline stage. In this structure time borrowing methods like all Types of DFFC is not applicable exactly for the same reason that was described in critical feedback structure. In brief the next stage in pipeline cannot lend any time to the previous stage and therefore this structure is also a problematic path structure for the former critical path but still can be used for the latter one.

D. Critical Fan-out Path

Fig. 4(d) shows critical fan-outs. In these structures a fan-out exists at the end of a critical path so that all destination flip-flops would be critical. DFFC can be very efficient when it comes to re-convergence fan-outs. In these structures only one TVP blocks is used and the "Err" signal generated by TVP can be shared between DAD blocks. In these cases area and power overhead is reduced.

E. Critical Fan-in Path

Fig. 4(e) shows critical fan-in structure. In such structures that convergence fan-in exist in critical path and all branches that converge are critical and the "mid" point of each branch is before converging node, we have two options to implement DFFC method: 1) using one TVP block for each path which increases the power and area

overhead, and 2) choosing converging node as "mid" point which may cause false error prediction of TVP block. In these cases DFFC Type C is very efficient since it can remove false errors. In Section IV we will explain how such challenging path structures affect applying time borrowing methods.

IV. EXPERIMENTAL RESULTS

In order to investigate the effectiveness of the proposed structure (DFFC Type C), it has been implemented on several ITC'99 benchmark circuits. The behavioral description of each benchmark was synthesized using Nangate open cell library [16]. By using a synthesis tool, the gate level netlist is created and the delay of each path is calculated. We have used such information given by synthesis tool in order to find the short and critical paths of the circuit. The gate level description was then simulated. Note that if any short feedback exists, enough buffers are inserted so that the delay of the short paths becomes more than that of DAD block. To evaluate our proposed micro-architecture, we have taken two experiments into account: 1) Calculating maximum allowable frequency (MAF) for all benchmark circuits we have considered in this work, and 2) estimating power consumption.

A. MAF Calculation

In the first experiment, in order to calculate MAF, after identifying all critical paths in the circuit we choose the longest path which limits the maximum frequency of the circuit. Then we check if this critical path is a problematic path discussed in Section III, i.e. check to see whether it is a critical feedback path or a consequence critical path. After that, if there are any critical fan-in or fan-out we choose the "mid" point based on given power and area constraints. Finally, we are ready to add TVP and DAD blocks to the circuit.

In the next step, we reduce the clock period until the next longest path becomes critical and needs time borrowing. Then we perform all the steps described above for this critical path. This process continues until we find a critical path with problematic path structure that does not allow time borrowing. Some other problems may occur that do not allow us to use DFFC in a complete circuit. For example, let us consider the critical path shown in Fig. 5(a). This critical path is asymmetrical which means there is a large sub circuit, e.g. full adder, in the middle of the path that makes us place "mid" point very close to the destination flip-flop. In such case the delay of "mid" point plus the delays of DAD and TVP blocks would be larger than 1.5 times of the period of the clock. This causes "Conv" signal to go high after the falling edge of the next clock cycle. At the falling edge of the clock "Err" signal goes down which causes "Conv" signal to fall immediately (see Fig. 5(b)). The master latch is not transparent in the first half of the next clock period and therefore time borrowing is not possible. We refer to this issue as *Late Conversion* (LC) in the rest of this section. Note that, if in a critical fan-in path we choose the converging point as "mid" point which is near to the destination flip-flop, the delay of "mid" point

978-1-4799-8720-7/15 $31.00 © 2015 IEEE   431

would be large compared to the clock period and therefore LC happens. Also when sequential depth of the circuit is small and the delay of DAD and TVP blocks is large in comparison with the clock period LC situation arises.

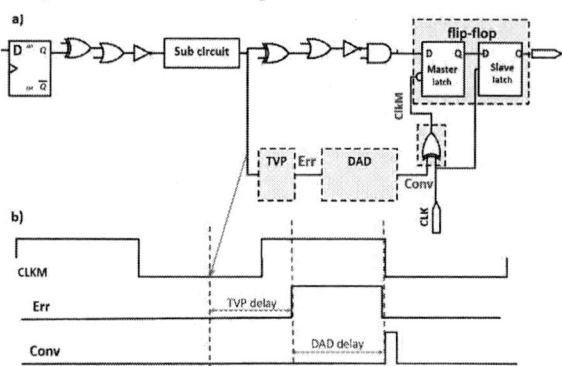

Fig. 5. Late Conversion (LC) situation a) Asymmetrical circuit b) waveforms.

We have performed the above mentioned steps on five ITC'99 benchmark circuits. In the following we explain the challenges we faced in each benchmark circuit to increase MAF:

- **b03:** This circuit has short feedback path structure, so a buffer was inserted to avoid HTV. Because it has symmetrical critical path structures no false errors occurs using DFFC Type B. Hence, MAF improvement for both DFFC Types B and C are the same. We were able to increase MAF until the small sequential depth of the circuit causes LC.
- **b05:** In this circuit we can see critical fan-in structure. By choosing converging point in critical fan-in structures as "mid" point to decrease power and area overhead, false error occurs using DFFC Type B. These false errors were removed using DFFC Type C. This way, we were able to increase MAF until the converging point becomes close to the destination flip flop and LC happens.
- **b11:** This circuit has symmetrical critical path structures so no false errors occurs using DFFC Type B and therefore MAF improvement for both DFFC Types B and C are the same. We were able to increase MAF until the small sequential depth of the circuit causes LC.
- **b12:** In this circuit, we have faced critical fan-in structures. We chose converging point as "mid" point to decrease power and area overhead, but false error occurs using DFFC Type B. False errors were removed using DFFC Type C and we could increase MAF until the next critical path became a consequence critical path.
- **b14:** In this circuit, signal glitches cause the DAD block to close transparency window. The results show that when signal glitches happen DFFC method is not very effective.

In Table I the total number of flip-flops (column #flip-flops), gates (#gates), critical fan-in paths (#critical fan-in), and critical fan-out paths (#critical fan-out) in each

benchmark circuit is reported. Column #RemovedFalseErrors indicates how many false errors may happen if we increase the clock frequency of the circuit from MAF calculated using DFFC Type B to MAF calculated using Type C. The MAF column presents the MAF values before using any time borrowing method (sub column No DFFC), after using DFFC Type C (sub column TypeC), and after using DFFC Type B (sub column TypeB). Our results show that DFFC Type C can improve MAF of the circuit 7.54% on average which is on average 3.3% higher compared to that of DFFC type B.

Our simulations show that if asymmetrical paths and converging fan-in paths, which may cause false errors, exist in the circuit DFFC Type C can increase MAF more than DFFC Type B. In other cases the MAF using DFFC Type B and C is almost the same.

B. Power Estimation

In the second experiment, we have used power compiler to estimate power consumption. The power of overall circuit was calculated for the same number of clock cycles and the same clock frequency. Table II reports the dynamic power and leakage power of benchmark circuits for three cases: using DFFC Type C, using DFFC Type B and without using any time borrowing method. The first and second columns show the frequency of system and the number of critical paths (CP) that use time borrowing, respectively. Dynamic power consists of cell internal power (CIP) and net switching power (NSP). CIP is the dynamic power consumed by charging and discharging of capacitance and loads inside logic blocks and NSP is the dynamic power consumption on interconnects. The results show that CIP and NSP have decreased by 11.56% and 18.31% on average, respectively, using DFFC Type C. Also, the leakage power is slightly higher in DFFC Type B compared to DFFC Type C. The reason for lowering dynamic power in DFFC Type C is that in its DAD blocks the internal gates are evaluated only when error is detected and "Err" signal is high while in Type B we have more clock gating and during each clock cycle the gates in DAD block are evaluated. This will reduce the switching activity of DAD in Type C and total power over head decreases compared to that of Type B. Another reason for that is whenever false errors happen in DFFC Type B a large transparency window in critical flip-flop is created which increases power consumption while in DFFC Type C this transparency window is avoided.

V. CONCLUSION

We have presented an improved DFFC structure for time borrowing purposes which consumes less power than previous DFFC structures and does not suffer from false errors. Also we have investigated the application of our proposed micro-architecture on different benchmark circuits considering all their path structures. Our results show that our micro-architecture can improve the performance by 7.4% on average while using DFFC Type B [14] only 4.2% improvement is achieved.

978-1-4799-8720-7/15 $31.00 © 2015 IEEE   432

Table I. Comparison between DFFC Type C and Type B in terms of maximum allowable frequency (MAF).

Circuit	# flip-flops	# gates	#critical fan-in	#critical fan-out	#RemovedFalse Errors	MAF (GHz)			Improvement%	
						No DFFC	TypeC	TypeB	TypeC	TypeB
b03	30	150	0	0	0	1.66	1.724	1.724	3.3	3.3
b05	34	608	2	6	8	1.28	1.388	1.282	7.7	0.0
b11	30	366	0	0	0	1.19	1.310	1.310	10.6	10.6
b12	121	1000	6	0	27	1.22	1.388	1.250	13.8	2.4
b14	245	10,098	0	0	0	0.39	0.396	0.406	2.3	4.9
Average MAF improvement of DFFC Types B and C in comparison with No DFFC									7.54	4.24

Table II. Comparison between DFFC Type C and Type B in terms of power consumption.

Circuit	F(GHz)	#CP	Dynamic Power (µw)						Leakage Power (µw)			power improvement per critical path(µw)
			CIP (µw)			NSP (µw)						
			C	B	No DFFC	C	B	No DFFC	C	B	No DFFC	
b03	1.66	3	419.06	560.96	341.44	151.63	190.90	83.74	4.61	5.39	5.61	60.39
b05	1.21	8	376.32	458.93	279.34	27.56	57.32	18.53	12.86	13.19	10.88	14.12
b11	1.19	3	439.51	511.12	343.82	276.21	319.93	129.26	11.74	11.63	11.79	38.44
b12	1.21	4	1120.7	1156.1	870.03	230.49	244.46	158.85	24.90	24.90	23.22	12.32
b14	0.39	3	2887.7	2814.9	2795.5	3134.2	3125.3	3129	146.60	145.66	144.86	-27.3
Average power improvement of Type C compared to Type B			11.56%			18.31%			3.08%			19.59

REFERENCES

[1] S. Das, "RAZOR: Variability-Tolerant design methodology for low-power and robust computing," Ph.D dissertation, Dep. Computer Science and Engineering, University of Michigan, 2009.

[2] D. Ernst, N.S. Kim, S. Das, S.Pant, R. Rao, T. Pham, "Razor: A Low-Power pipeline based on circuit-level timing speculation," in IEEE International Symposium on Microarchitecture (MICRO), pp. 7-18, 2003.

[3] L. Seonggwan, S. Paik, and Y. Shin, "Retiming and time borrowing: Optimizing high-performance pulsed-latch-based circuits" in IEEE/ACM International Conference on Computer-Aided Design (ICCAD), pp. 375-380, 2009.

[4] D. P. Singh , S. D. Brown, "Integrated retiming and placement for field programmable gate arrays," in ACM/SIGDA International Symposium on Field-Programmable Gate Arrays (FPGA), pp. 67-76, 2002.

[5] M. Choudhury, V. Chandra, R. Aitken, and K. Mohanram, "Time-Borrowing circuit designs and hardware prototyping for timing error resilience," in IEEE Transactions on Computers (TC), vol. 63, no. 2 , pp. 497-509, 2014.

[6] J. H. Anderson. B. Teng, "Latch-based performance optimization for field-programmable gate arrays," in IEEE Transactions on Computer-Aided Design of Integrated Circuits and Systems (TCAD), vol. 32, no. 5 , pp. 667-680, 2013.

[7] Y. Chen and Y. Xie, "Tolerating process variations in high-level synthesis using transparent latches " in Proc. of Asia and South Pacific Design Automation Conference (ASP-DAC), pp. 73-78, 2009.

[8] K. Chae, CH. Lee, and S. Mukhopadhyay, "Timing error prevention using elastic clocking," in IEEE International Conference on IC Design & Technology (ICICDT), pp. 1-4, 2011.

[9] M.Wieckowski, M.P.Young, C. Tokunaga, D.W Kim, Z. Foo, D. Sylvester, and D.Blaauw, "Timing yield enhancement through soft edge flip-flop based design," in Custom Integrated Circuit Conference (CICC), pp.543-546, 2008.

[10] V. Joshi, "Soft-edge flip-flops for improved timing yield: design and optimization," in IEEE/ACM International Conference on Computer-Aided Design (ICCAD), pp. 667-673, 2007

[11] A. Tiwari, S. R. Sarangi, and J. Torrellas, "ReCycle: Pipeline adaptation to tolerate process variation," in ACM SIGARCH Computer Architecture News, vol. 35, no. 2, pp. 323-334, 2007

[12] V. Mahalingam, "Dynamic clock stretching for variation compensation in vlsi circuit design," in ACM Journal on Emerging Technologies in Computing Systems (JETC), vol. 8, no. 16, pp. 1-13, 2012.

[13] M. Nejat, B. Alizadeh, A. Afzali-Kusha, "Dynamic Flip-Flop Conversion to Tolerate Process Variation in Low Power Circuits," in Proc. of Design, Automation & Test in Europe (DATE), pp. 1-4, 2014.

[14] M. Nejat, B. Alizadeh, and A. Afzali-Kusha, "Dynamic Flip-Flop Conversion: A Time-Borrowing Method for Performance Improvement of Low-Power Digital Circuits Prone to Variations," in IEEE Transactions on Very Large Scale Integration (VLSI) Systems, 2015 (to appear).

[15] N. Sirisantana , K. Roy, "A Time Borrowing Selectively Clocked Skewed Logic for High-Performance Circuits in Scaled Technologies," in Proc. of European Solid-State Circuits Conference (ESSCIRC), pp.181-184, 2003.

[16] http://www.si2.org/openeda.si2.org/projects/nangatelib/, [Accessed 2015].

[17] M. Fojtik, D. Fick, Y. Kim, N. Pinckney, D. M. Harris, D. Blaauw, and D. Sylvester, "Bubble Razor: eliminating timing margins in an ARM Cortex-M3 processor in 45 nm CMOS using architecturally independent error detection and correction," in IEEE Journal of Solid-State Circuits, vol. 48, no. 1, pp.66-81, 2013.

[18] http://www.cad.polito.it/downloads/tools/itc99.html. [Accessed 2015].

2015 IEEE Computer Society Annual Symposium on VLSI

RWT: Suppressing Write-Through Cost when Coherence is not Needed

Hao Liu, Clément Dévigne, Lucas Garcia, Quentin Meunier, Franck Wajsbürt, Alain Greiner
Laboratoire d'Informatique de Paris 6 UMR 7606
Université Pierre et Marie Curie UPMC – Sorbonne Universités,
4 Place Jussieu, 75252 Paris Cedex 05 France
Email: lucas.garcia@polytechnique.edu, *firstname.lastname*@lip6.fr

Abstract—In shared-memory multicore architectures, handling a write cache operation is more complicated than in single-processor systems. A cache line may be present in more than one private L1 cache. Any cache willing to write this line must inform all the other sharers. Therefore, it is necessary to implement a cache coherence protocol for multicore architectures.

At present, directory based protocols are popular cache coherence protocols in both industry and academic domains because of their reduced coherence traffic compared to snooping protocols, at the expense of an indirection. The write policy – write through or write back – is crucial in the protocol design.

The write-through policy reduces the bandwidth because it augments the write traffic in the interconnection network, and also augments the energy consumption. However, it can efficiently solve the false sharing problem via write updates. In this paper, we introduce a new way to reduce the write traffic of a write-through coherence protocol by combining write-through coherence with a write-back policy for non coherent lines. The baseline write-through used as reference is a scalable hybrid invalidate/update protocol.

Simulation results show that with our enhanced protocol, we can reduce at least by 50% the write traffic in the interconnection network, and gain up to 20% performance compared with the baseline write-through protocol.

Index Terms—System-on-Chip; Many-core Architecture; Shared Memory Programming; Hardware Cache Coherence; Network-on-Chip; Write-Through; Write-Back; Released Write-Through

I. INTRODUCTION

The problem of cache coherence protocols in shared-memory multicore architectures is still an active domain of research. Thanks to technology advances, we can put more and more cores in one single chip [1], making it impossible to use bus as an interconnect in the architecture. Instead, most multicore or manycore architectures now use networks-on-chip (NoC) [2] as interconnect [3], [4]. With a NoC, snooping coherence can only come at the price of many broadcasts; and if this approach has been experimented (e.g. in [5]), directory-based coherence protocols are the most commonly used in multicore architectures.

One of the most popular cache coherence protocol, the write-back MESI [6], supports directory based implementations, like the MOESI protocol in AMD architecture Opteron [7], and the GOLS protocol in Intel architecture Xeon Phi [8]. The MESI protocol uses the write-back policy to avoid propagating writes to the lower levels of cache hierarchy. However, it can lead to a performance diminution when a write targets a line not present in the L1 cache: this is because the

write-back policy implements a *write allocate* policy which needs to fetch the missing line before performing the write. During this time, the processor is blocked. On the contrary, a write-through based protocol always sends the write command to the lower level of cache hierarchy, and thus is not blocked in case of miss. However, this increases the write traffic and consumes more energy than a write-back policy.

This work extends currently submitted work on the TSAR architecture [9]. The coherence protocol used in this architecture, called DHCCP (Dynamic Hybrid Cache Coherence Protocol), is a directory-based write-through protocol whose coherence traffic scales with the number of cores. However, DHCCP has two main disadvantages:

- First, the high number of writes from L1 caches to L2 caches limits the number of processors per cluster
- Second, these writes induce a high power consumption

Our objective is to reduce this write traffic by modifying DHCCP, while keeping the advantage of write-through coherence.

To achieve this, we propose a modified version of this coherence protocol, called RWT for *Released Write-Through*. The idea of RWT is to enable coherency only when it is necessary. This necessity will be determined fully in hardware, with a per line granularity in the L2 cache. Thus, a coherent line will use the same version of DHCCP as presented in [10]. However, for lines which have only one copy and thus do not need coherence, a write-back approach is used. This technique allows to reduce the traffic related to writes for the majority of the lines which are present in a single cache, while keeping a scalable coherence protocol for shared lines. A line can then be in one of two states:

- Non-coherent write-back
- Coherent write-through

We can notice that the state change can only occur from non-coherent state to coherent state. The state of a line is reset when it is evicted from the L2 cache.

The rest of the article is organized as follows: section II discusses related works; section III presents the mechanism of RWT; section IV presents the simulation environment and the experimental results; finally, section V concludes.

II. RELATED WORK

Cache coherence protocols have been studied for a long time. In traditional multiprocessor systems in which processors

978-1-4799-8720-7/15 $31.00 © 2015 IEEE   434

shared a bus, the MESI protocol and its variants have proven to be the most efficient [11]. However, this problem recently regained in interest with the arrival of manycore systems, which replaced buses interconnects with NoCs of varying topology, offering a higher bandwidth, but a higher latency. A possible answer to the problem is to use software coherence and let programmers invalidate shared data when needed [12], or use different programming paradigms like message passing. However, it is now more and more accepted that hardware supported on-chip coherence will be part of future multi- and manycore chips [13]. Yet, the question of the coherence protocol remains open, and the write-through strategy could be an actual answer for manycore systems because of NoCs characteristics: [14] shows that a write-through invalidate coherence scheme can perform as well as a write-back MESI one over a NoC, while being simpler to design; besides, the commercially available Tilera architecture [4] uses a write-through protocol between tiles.

Other approaches try to reduce network latency in case of a write miss. Token Coherence [15], [16] is a technique which allows to avoid indirection at the L2 level by broadcasting requests on all the mesh, while detecting race conditions and guaranteeing correctness by a token mechanism. This is an optimistic approach using potentially failing requests, enabling a fast common case. The DicoCMP protocol and its variants [17], [18] store the sharers in both the L1 and the L2 caches. The approach used consists in that a requesting writer L1 cache must send invalidations directly to the other L1 cache owning a copy instead of the L2 cache (home tile), at the price of tracking the list of sharers for each line in the L1 caches.

While these approaches are interesting, they put efforts to minimize write-miss latency, which is almost non-existing with write-through protocols, since the latter just need to propagate writes to the lower level of the memory hierarchy without blocking the requesting L1 or processor.

[19] presents a mechanism including a policy manager able to detect when a cache powers up/down, allowing a write-back policy when a single cache is activated, and a write-through policy otherwise. While this work uses a hybrid policy like RWT, it has a cache granularity which does not fit well with manycore systems. At the opposite, RWT allows several active caches to access memory in write-back mode, provided the caches access different pieces of data.

Finally, [20] describes a hybrid technique for cache co-herence, combining a coherent strategy with a non coherent one. The main difference with RWT is that this approach is based on the operating system for detecting that data is shared; this in turn involves to handle TLB misses by software – while TSAR allows a hardware handling – and restricts the strategy granularity to pages, whereas RWT allows a per line granularity.

III. PRINCIPLES OF RWT

A. Baseline Write-Through

The write-through coherence protocol DHCCP used as baseline is based on the following idea: when the number of

copies in L1 caches is below a certain threshold, an explicit list of the sharers is kept in the L2, while above this threshold, only the number of copies is known. The explicit copies are stored in a L2 data structure shared among all lines. Thus, when the number of copies is low, a multicast update strategy is used upon receiving a write, whereas when the number of copies is high, a broadcast invalidation is sent to all L1 caches. In the following, this protocol is considered as a write-through coherent protocol and is not fully described, as this article focuses on the write-back strategy for non coherent lines, and the coupling with the write-through coherent lines. Finally, despite the fact that we only considered DHCCP as a baseline, the principles of RWT could be applied to any directory-based write-through coherent protocol.

B. Modifying Cleanups to Add a Write-Back Mechanism

In DHCCP, the distributed shared L2 cache directory keeps coherence information for every cache line, and this requires the inclusive property of L1 caches in L2 caches. A *cleanup* request sent by a L1 cache can happen in two scenarios:

- the line has been evicted spontaneously by the L1 cache
- the L1 cache has received an invalidation request from L2 cache and responds with a *cleanup*

The *cleanup* request does not include any data in DHCCP, because the L2 cache line is always up to date. In RWT, we need to keep this inclusive property for non-coherent lines: if a line is locally modified in a L1 cache and not present in a L2 cache, a miss in another L1 on this line will result in an incoherence when the line is then retrieved by the L2. It implies that when a non-coherent line is evicted from a L2, the L1 owning the copy must both invalidate its copy and send the up-to-date values to the L2.

When a L1 cache spontaneously evicts a dirty line or when it receives an invalidation request targeting a dirty line, the *cleanup* needs to be done in parallel with propagating up-to-date values to the L2. This can *a priori* be done *via* three different ways:

- the L1 can send a write with the new values in parallel with the cleanup, thus using two different networks (one for the cleanup and one for the write);
- the L2 can interpret the write with the new values as a response to its coherence request;
- the L1 can send the new values directly in the response to the coherence request.

The first solution has the drawback that since the two transactions use different networks, their order of arrival is not defined, adding complexity in the protocol. The second solution can actually not be achieved easily because it actually presents a risk of deadlock. For these reasons, we decided to modify the existing *cleanup* request to include a potential write-back of dirty lines; this request is called *cleanup-data*. For coherent lines and the non-coherent clean lines, we use the classic *cleanup* requests as in DHCCP – i.e. without data.

C. Non-Coherent and Coherent States in L1 Caches

The RWT protocol defines two states for a valid cache line: the non-coherent (NC) state and the coherent (C) state. This

978-1-4799-8720-7/15 $31.00 © 2015 IEEE

state is determined by the L2 cache, and is sent along with a miss response to a requesting L1 cache. When a line is fetched from memory by the L2 cache, the requesting L1 cache obtains the first copy of this line, and thus is granted NC state.

When a write hit happens in a L1 cache, the strategy used depend on the line state.

- the write-through policy is chosen for a line in C state. The L1 cache line is updated and the write is also sent to the L2 cache.
- the write-back policy is chosen for a line in NC state. The L1 cache line is locally modified and the lines become *dirty*: there is no write sent to the L2 cache.

Because the write-allocate strategy blocks the requesting processor, write misses in RWT are directly sent to the corresponding L2 cache. Figure 1 models the decision diagram associated with the actions taken by a L1 cache upon receiving a processor request.

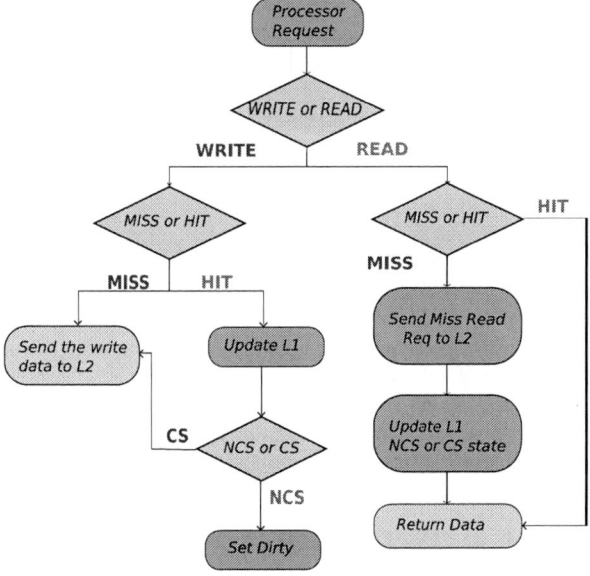

Figure 1: Decision Diagram of a L1 Cache upon Receiving a Processor Request

D. Switching From Non-Coherent to Coherent State

A line in a L2 cache is initially in the NC state, thus there can be only one copy in a given L1 cache. A key point in RWT is the detection by the L2 that a line currently in NC state ought to be in C state. This happens when a L1 cache requests a copy of (*resp.* issues a write on) a line which is in NC state and already has a copy in another L1 (it is possible that a shared line only has one copy). This triggers the L2 cache to send an invalidation request to the actual owner of the copy, which responds with a *cleanup-data* if the line is dirty, and *cleanup* otherwise (see Figure 2). When the L2 cache receives the *cleanup*, it changes the state for this line from NC state to C state, thus the L2 cache sends the copy with the C state information to the L1 cache requesting the line. Figure 2 illustrates this line state change mechanism in RWT.

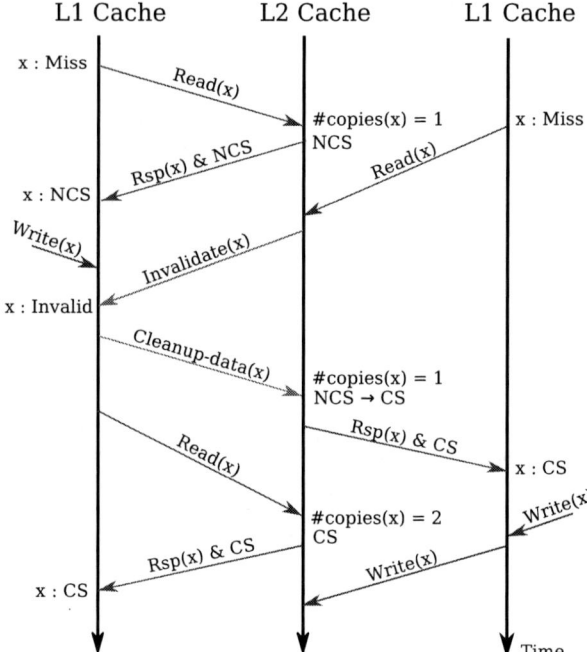

Figure 2: Line State Change Mechanism: From Non-Coherent State (NCS) to Coherent State (CS), for a Line Named x

RWT does not allow a line in C state to be switched back in NC state: the rationale behind this is that even if the number of copies goes back down to one, there is a high probability that another cache will access it in the near future since this line contains shared data. Of course, when a line is evicted from a L2 and then fetched back from memory, its status is reset to NC state. Thus, this line lives in C state until it is evicted from the L2 cache.

To avoid an unnecessary switch overhead for shared cache lines containing instructions – which are almost never modified – RWT has been implemented only for the data part of the L1 cache.

IV. EVALUATION

A. Architecture

We implemented our proposed RWT cache coherence protocol at the cycle-accurate level, by modifying the models of L1 and L2 caches of the original TSAR components. The TSAR model is written using a library of cycle-accurate components called SoCLib [21], which uses the SystemC standard [22].

Figure 3 shows an overview of the TSAR architecture. It is a clustered architecture with a 2D mesh topology using a network-on-chip. Each cluster contains 4 Mips cores, a shared L2 cache, and a local crossbar connected to the router. There are separate networks for direct requests (reads and writes) and coherence requests in order to avoid deadlocks. Table I shows the configuration parameters for the TSAR architecture.

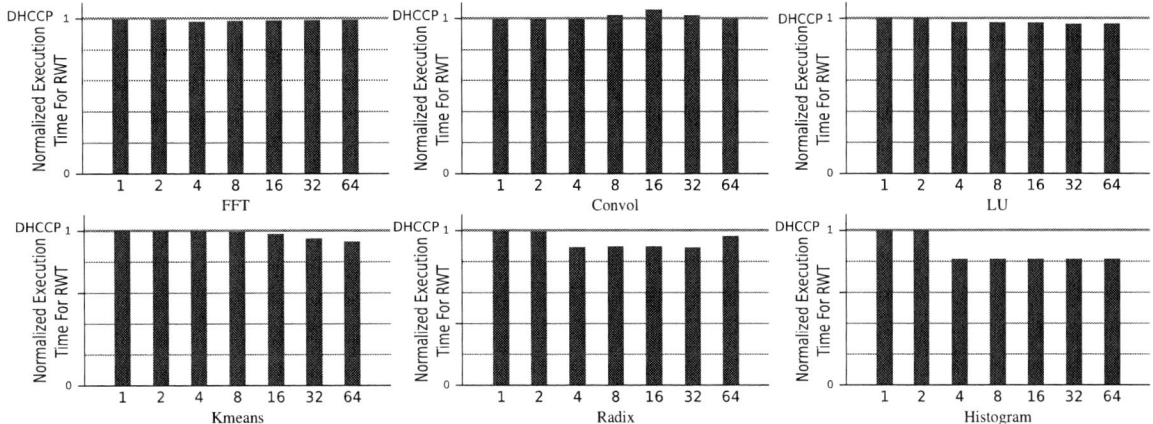

Figure 4: Execution Times (in cycles) for RWT Normalized w.r.t. Times with DHCCP

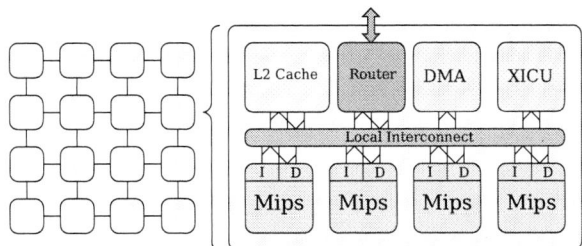

Figure 3: Overview of the Clustered TSAR Architecture used for Implementing RWT

Table I: Hardware Main Characteristics

Mesh Size	up to 4 × 4 clusters
L1 Cache Sets (I & D)	64
L1 Cache Words (I & D)	16
L1 Cache Ways (I & D)	4
L2 Cache Sets	256
L2 Cache Words	16
L2 Cache Ways	16
TLB Sets (I & D)	8
TLB Ways (I & D)	8

B. Applications

Applications used for evaluations are FFT and LU from the Splash-2 suite [23], Histogram and Kmeans from the Phoenix-2 benchmark suite [24], and Convol, which is an image filtering program performing a convolution filter in X and Y directions. Table II shows the configuration for each application.

All these benchmarks have been run over an operating system called GietVM, which is developed in our laboratory and supports the physical placement of software objects *via*

virtual memory. We used this feature to spread data physically in clusters, trying to improve the locality of data accesses when possible.

Table II: Applications Parameters

Application	Input Data
Histogram	25 MB image (3408 x 2556)
Convol	1024 x 1024 image
Radix	262,144 keys (default)
FFT	2^{18} Complex Points
LU	512 × 512 elements
Kmeans	10,000 points

C. Measurements

Other experiments currently in review have shown the scalability of DHCCP up to 64 cores, both in performance and coherence traffic. Therefore, these experiments focus more on the comparison between RWT and DHCCP.

We compared the performance of RWT relatively to DHCCP, as well as the traffic in the interconnection networks, for reads, writes, and coherence requests. Precisely, the measures used to evaluate our two protocols are:

- **Execution time**: the execution times correspond to the parallel phase of each application.
  Execution time = timestamp of the last thread finishing the parallel phase - timestamp of first thread starting the parallel phase
- **Traffic**: in L2 caches, we implemented several counters to measure the *cost* of each type of request. This cost is a product between a number of flits and a distance, and is defined as follows:
  - For requests going from a L1 cache to its local L2 cache, we define that the cost is equal to the number of flits, because there is only one access in the local

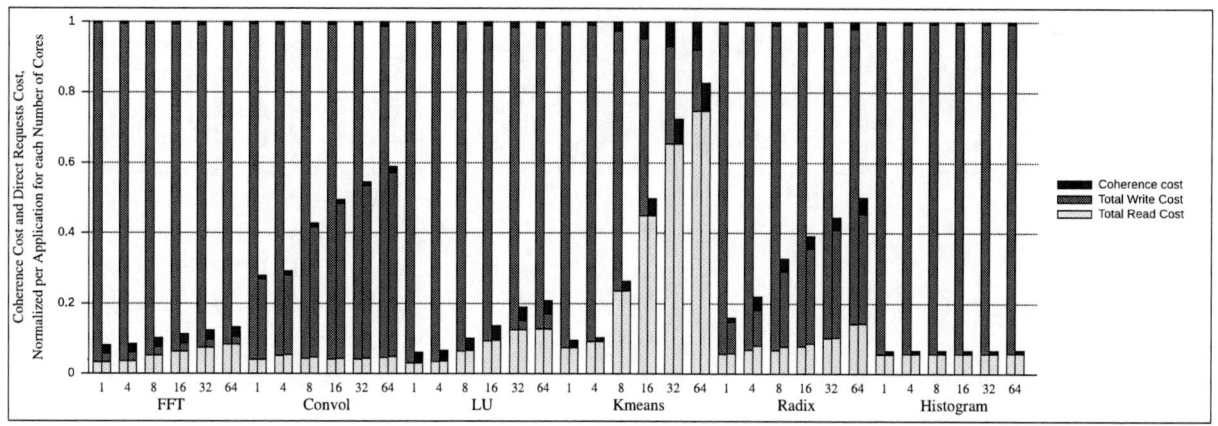

Figure 5: Comparison of Read, Write and Coherence Cost between DHCCP (Left Bar) and RWT (Right Bar), Normalized w.r.t. DHCCP total cost

crossbar.

Local cost = Nflits x 1

– For requests going from a L1 cache to a remote L2 cache, the cost includes 2 accesses to the local crossbar (an access from the local crossbar to the router and another access from the router to the local crossbar) and the hops for propagating the request from a router to another.

Remote cost = Nflits x (Nhops + 2)

The rationale behind the cost is that it should be closely related to the energy consumption, and therefore be as low as possible.

D. Results

Figure 4 shows the execution time with RWT for the 6 considered benchmarks. These times are normalized per number of cores and per application w.r.t. times on DHCCP.

For 5 out of the 6 benchmarks, RWT either has identical performances of improves them, up to 20%. Results on Convol can be explained by the nature of the application, in the sense that all the internal image buffers – 5 in total – are constituted by lines shared by different writers only (then later read by a single reader). Thus, when DHCCP only propagates writes to memory, RWT adds a coherence overhead since each line is first sent in non-coherent mode, then has its status switched to coherent, inducing a non-negligible overhead. Yet, for most applications, this case remains rare enough so that this overhead is largely balanced by RWT's overall gain.

Figure 5 shows the cost of Reads, Writes, and coherence for both RWT and DHCCP. Results are displayed normalized w.r.t. the sum of the costs for DHCCP. We can observe that all applications benefit from RWT's decrease in writes, and that the write cost is entirely removed in 4 of the 6 applications. This results in low or very low overall costs for all applications compared to DHCCP. Also, if the deletion of writes in RWT translates into a small increase in the coherence traffic (since for non-coherent lines, data is propagated in memory *via cleanup-data* requests), we can notice that the overhead in

RWT's coherence cost is negligible compared to the write cost it allowed to suppress.

The rationale behind RWT is that lines do not change their state often: either they are private to a processor, which is the most common case, or they are shared once and then stay in the L2 cache long enough so that the switch cost is amortized. Private lines typically contain stack variables, but also many shared variables which are actually distributed per thread. Writes on private lines are either buffered in case of hit – which is the best case –, or propagated in case of miss – like for the write-through. To confirm that line state switches are rare events, we measured the percentage of reads and writes resulting in a line state change. Figure 6 shows the results for the same configurations as before: we can observe that this percentages for *reads* barely exceeds 3% for LU, while remaining very low for other applications: less than 1% in most cases. *Write* misses triggering a line state change are negligible. This results confirm our hypothesis that line state changes are rare and explain the good results of RWT.

These encouraging results lead us to consider integrating RWT as the main coherence protocol in the TSAR architecture.

E. Hardware Overhead

RWT adds little hardware overhead compared to DHCCP. In fact, the number of per-line bits of metadata is unchanged in the L1 cache, while it is only increased by 1 in the L2 cache for storing the NC or C state. The major overhead comes from the fact that RWT requires a cleanup FIFO in the L1 cache which contains a full cache line (64 bytes), whereas it is not necessary in DHCCP as cleanup cannot contain data. This overhead still remains negligible, meaning that RWT manages to reuse most of DHCCP hardware efficiently.

V. CONCLUSION

In this paper, we proposed a new hybrid hardware cache coherence protocol called RWT, which allows to reduce the write traffic while not degrading the performances of a scalable write-through coherence protocol. RWT mixes a non-coherent

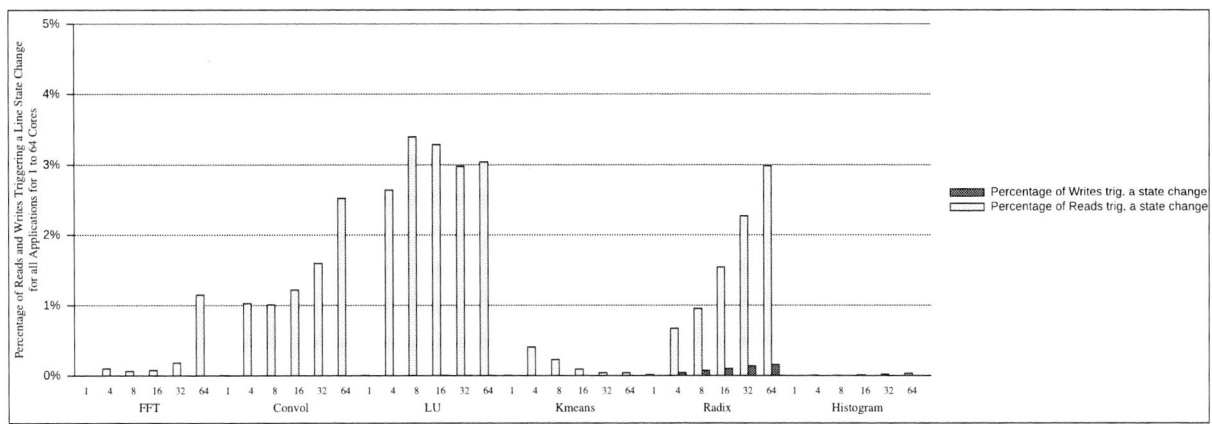

Figure 6: Percentage of Read Requests (Left) and Write Requests (Right) Triggering a Line State Change from NC State to C State

write-back strategy for non-shared lines, and a write-through coherent strategy for shared lines, the choice being made entirely in hardware.

Our results show that it reduces more than 50% of write traffic in average, while reducing execution time by 5% in average compared to the baseline protocol DHCCP. Further experiments need to be done on a larger set of applications, and for platforms containing more than a hundred cores. Indeed, RWT targets manycore platforms like the TSAR architecture. Yet, we are confident the significant reduction in cost brought by RWT will continue as the number of cores grows. Future work also include the implementation of some optimizations in RWT, especially the line replacement policy in the L2 cache, as we can take advantage of the small eviction overhead of non-coherent lines to favor eviction of such lines over coherent ones.

REFERENCES

[1] J. M. Rabaey, "Scaling the power wall: Revisiting the low-power design rules," *Keynote speech at SoC*, vol. 7, 2007.

[2] G. De Micheli and L. Benini, *Networks on chips: technology and tools.* Academic Press, 2006.

[3] G. Kurian, J. E. Miller, J. Psota, J. Eastep, J. Liu, J. Michel, L. C. Kimerling, and A. Agarwal, "Atac: a 1000-core cache-coherent processor with on-chip optical network," in *Proceedings of the 19th international conference on Parallel architectures and compilation techniques.* ACM, 2010, pp. 477–488.

[4] C. Ramey, "Tile-gx100 manycore processor: Acceleration interfaces and architecture," *Tilera Corporation*, 2011.

[5] J. Howard, S. Dighe, Y. Hoskote, S. Vangal, D. Finan, G. Ruhl, D. Jenkins, H. Wilson, N. Borkar, G. Schrom *et al.*, "A 48-core ia-32 message-passing processor with dvfs in 45nm cmos," in *Solid-State Circuits Conference Digest of Technical Papers (ISSCC), 2010 IEEE International.* IEEE, 2010, pp. 108–109.

[6] M. S. Papamarcos and J. H. Patel, "A low-overhead coherence solution for multiprocessors with private cache memories," *ACM SIGARCH Computer Architecture News*, vol. 12, no. 3, pp. 348–354, 1984.

[7] P. Conway, N. Kalyanasundharam, G. Donley, K. Lepak, and B. Hughes, "Cache hierarchy and memory subsystem of the amd opteron processor," *IEEE micro*, vol. 30, no. 2, pp. 16–29, 2010.

[8] S. R. Garea and T. Hoefler, "Modelling communications in cache coherent systems," *Technical Report*, 2013.

[9] *TSAR: Tera-Scale Multiprocessor ARchitecture*, Available: https://www-soc.lip6.fr/trac/tsar, 2009.

[10] Y. Gao, "Generic cache controller for a massively parallel manycore architecture using coherent shared memory," Ph.D. dissertation, Université Pierre et Marie Curie (UPMC), 2011.

[11] M. Loghi, M. Poncino, and L. Benini, "Cache coherence tradeoffs in shared-memory mpsocs," *ACM Transactions on Embedded Computing Systems (TECS)*, vol. 5, no. 2, pp. 383–407, 2006.

[12] X. Zhou, H. Chen, S. Luo, Y. Gao, S. Yan, W. Liu, B. Lewis, and B. Saha, "A case for software managed coherence in manycore processors," in *Poster on 2nd USENIX Workshop on Hot Topics in Parallelism HotPar10*, 2010.

[13] M. M. Martin, M. D. Hill, and D. J. Sorin, "Why on-chip cache coherence is here to stay," *Communications of the ACM*, vol. 55, no. 7, pp. 78–89, 2012.

[14] P. Guironnet de Massas and F. Pétrot, "Comparison of memory write policies for noc based multicore cache coherent systems," in *Design, Automation and Test in Europe (DATE).* IEEE, 2008, pp. 997–1002.

[15] M. M. Martin, M. D. Hill, and D. A. Wood, "Token coherence: decoupling performance and correctness," in *Computer Architecture, 2003. Proceedings. 30th Annual International Symposium on.* IEEE, 2003, pp. 182–193.

[16] M. R. Marty, J. D. Bingham, M. D. Hill, A. J. Hu, M. M. Martin, and D. A. Wood, "Improving multiple-cmp systems using token coherence," in *High-Performance Computer Architecture, 2005. HPCA-11. 11th International Symposium on.* IEEE, 2005, pp. 328–339.

[17] A. Ros, M. E. Acacio, and J. M. García, "A direct coherence protocol for many-core chip multiprocessors," *Parallel and Distributed Systems, IEEE Transactions on*, vol. 21, no. 12, pp. 1779–1792, 2010.

[18] A. Ros, M. E. Acacio, and J. M. García, "Cache coherence protocols for many-core cmps," *Parallel and Distributed Computing*, 2010.

[19] P. Sassone, C. Koob, D. Vantrease, S. Venkumahanti, and L. Codrescu, "Hybrid write-through/write-back cache policy managers, and related systems and methods," Jul. 18 2013, uS Patent App. 13/470,643. [Online]. Available: http://www.google.com/patents/US20130185511

[20] B. A. Cuesta, A. Ros, M. E. Gómez, A. Robles, and J. F. Duato, "Increasing the effectiveness of directory caches by deactivating coherence for private memory blocks," in *ACM SIGARCH Computer Architecture News*, vol. 39, no. 3. ACM, 2011, pp. 93–104.

[21] The Soclib Consortium, *SoCLib: an open platform for virtual prototyping of multi-processors system on chip.* Available: http://www.soclib.fr, 2008.

[22] *SystemC Reference Manual*, Synopsys Inc., http://www.systemc.org.

[23] S. C. Woo, M. Ohara, E. Torrie, J. P. Singh, and A. Gupta, "The SPLASH-2 programs: Characterization and methodological considerations," in *Proceedings of the 22nd Annual International Symposium on Computer Architecture.* New York: ACM Press, 1995, pp. 24–37.

[24] C. Ranger, R. Raghuraman, A. Penmetsa, G. Bradski, and C. Kozyrakis, "Evaluating mapreduce for multi-core and multiprocessor systems," in *High Performance Computer Architecture, 2007. HPCA 2007. IEEE 13th International Symposium on.* IEEE, 2007, pp. 13–24.

978-1-4799-8720-7/15 $31.00 © 2015 IEEE

2015 IEEE Computer Society Annual Symposium on VLSI

Small FPGA based Multiplication-Inversion Unit for Normal Basis Representation in GF(2^m)

Jérémy Métairie, Arnaud Tisserand, *Senior Member, IEEE,* and Emmanuel Casseau

Abstract—Halving methods have been proposed for parallel implementation of ECC primitives on multicore processors. In hardware, they can also provide protection against some side channel attacks (thanks to parallel independent operations). But they require affine coordinates for curve points and costly inversions. We propose a new combined multiplication-inversion unit for binary field extensions and halving based ECC methods optimized for FPGAs. We target small area solutions compared to very fast but costly ones from state-of-art. Our solution is based on permuted normal basis, Massey-Omura multiplication and Itoh-Tsujii inversion algorithms. Our FPGA implementations show better efficiency for large fields.

Index Terms—finite-field arithmetic, binary fields, normal basis, multiplication, inversion, ECC, halving scalar multiplication

I. Introduction

FINITE fields [1] are widely used in cryptography. Efficient arithmetic over finite fields is a key element for implementing public-key cryptosystems. Binary field extensions GF(2^m) are widely used in hardware implementations of elliptic curve cryptography (ECC [2]) due to their higher speed and smaller silicon area compared to GF(p) based solutions.

Scalar multiplication using *point halving* has been proposed in [3] to provide more parallelism. There, operations can be divided into two independent sequences and performed in parallel [4]. See [5] for recent work in this domain. This can also provide higher resilience against side channel attacks (see [6] for instance) since two shorter and independent scalar multiplications are performed in parallel. This requires the curve points to be represented using affine coordinates. Then, more inversion operations have to be computed than in projective coordinates [2, Sec. 3.2]. Representing field elements using normal basis will also provide very efficient square and square-root operations (*i.e.* circular shifts) which are also important for halving methods.

In this paper, we present a combined multiplication-inversion unit (MIU). State of art GF(2^m) inversion and related operations are recalled in Sec. II. Sec. III presents our GF(2^m) inversion algorithm. It uses a new representation for field elements called *permuted normal basis*, a modified Massey-Omura multiplication [7] which allows efficient inversion through Itoh-Tsujii algorithm [8]. Our proposed hardware unit presents a higher internal parallelism without the large area penalty used in state-of-art solutions. We target small solutions instead of faster but very costly operators since ECC primitives

J. Métairie and A. Tisserand are with CNRS, IRISA, University Rennes 1, INRIA in Lannion France.
E. Casseau is with University Rennes 1, IRISA, INRIA in Lannion France.

for high security levels (*e.g.* $m = 571$) may not be used frequently. It offers a new speed-area trade-off in arithmetic units with higher throughput for halving based ECC accelerators. We evaluated our solution using GF(2^m) fields recommended by NIST [9]. Our MIU architecture and its implementation results are presented in Sec. IV. Our implementations have been performed on a Spartan-6 LX75T and Virtex-4 LX100 FPGAs for comparison purpose. Finally, Sec. V concludes the paper.

II. State of the Art

A. Representations of GF(2^m) Elements

There exist two popular representations: *normal basis* and *polynomial basis* [1]. In normal basis, $A \in$ GF(2^m) is encoded as $\sum_{i=0}^{m-1} a_i \beta^{2^i}$ where coefficients a_i belong to GF(2) and β is a special element. $A = [a_0, a_1, \ldots, a_{m-1}]$ denotes the vector of coefficients. Square and square-root operations can be easily computed using circular shifts on the vector of coefficients. In this work, we use normal basis.

In polynomial basis, A is encoded as $\sum_{i=0}^{m-1} a_i' x^i$ where coefficients $a_i' \in$ GF(2). Additions and multiplications are polynomial operations modulo the irreducible polynomial f in GF(2)[x]. There exist other representations with particular properties such as Dickson basis [10] or dual basis [11].

B. GF(2^m) Addition and Multiplication

GF(2^m) addition can be efficiently performed using parallel XOR gates on all coefficients without any carry propagation. In normal basis, most of multiplication algorithms are based on matrix-vector product formulations of the basic multiplication. This leads to larger multipliers compared to polynomial basis ones. Massey and Omura (MO) proposed a multiplication method (see Algo. 1 from [7]), which computes $P = A \times B$, based on matrix-vector product where the constant matrix M_0 is only composed of GF(2) elements. Notation ROL(x, n) (ROR) stands for n-bit left (right) circular shift of x. Notation $P[i]$ is the i-th bit of P (starting with LSBs). At line 3 in Algo. 1, M_0 is a $m \times m$ binary, constant, symmetric and very sparse matrix. Then multiplying by M_0 just uses XOR trees. A serial-output MO multiplier requires a block of multiplication by M_0 (noted $\times M_0$), two ROL registers and a dot product operator. A last ROL register can be used to provide a parallel output if needed.

We use Gaussian normal basis (GNB, a specific type of normal basis) due to its very low hardware complexity for the implementation of multiplication by M_0 [12]. The number of "1" in M_0 is given by $C_m \leq t \times m - t + 1$ where t is the

978-1-4799-8720-7/15 $31.00 © 2015 IEEE   440

Algorithm 1: Original Massey-Omura multiplication [7].

Operands: A, B in $GF(2^m)$ represented in normal basis
Result: $P = A \times B$

1 $P \leftarrow 0$
2 **for** i **from** 0 **to** $m - 1$ **do**
3    $P[0] \leftarrow A \times M_0 \times B^T$
4    $A \leftarrow \text{ROL}(A, 1)$; $B \leftarrow \text{ROL}(B, 1)$; $P \leftarrow \text{ROL}(P, 1)$
5 **return** P

TABLE I
FPGA AREA EVALUATION OF THE BLOCK "MULTIPLICATION BY M_0"
(SPARTAN-6 LX75T).

m / t	163 / 4	233 / 2	283 / 6	409 / 4	571 /10
C_m /#LUT	645 / 159	465 / 232	1677 / 282	1629 / 408	5637 / 1128

field type (the smallest integer such that $p = m \times t + 1$ is prime and $\gcd(\frac{tm}{k}, m) = 1$ where k is the multiplicative order of 2 mod p). The number of XOR gates in the $\times M_0$ block is $C_m - m$ (see Tab. I for a FPGA area evaluation).

The original Massey-Omura multiplier produces one result bit every clock cycle. Using two parallel $\times M_0$ blocks produces two result bits per clock cycle with an important area overhead. For instance, multiplication in $GF(2^{233})$ only requires $117 = \lceil 233/2 \rceil$ clock cycles with an area overhead close to 2.

In [13] some redundancies in d parallel copies of the $\times M_0$ block are used to provide multiple bits of the result at each clock cycle without the full area penalty. The number of XOR gates is then $d(t(m - (d + 1)/2) + (d - 1)/2)$.

Parallel-output multipliers have been proposed in normal basis [14], [15]. P is the final accumulation of the partial products which is available after m clock cycles. In [16], a serial-input/parallel-output multiplier using w-bit sub-words computes the result in $\lceil m/w \rceil$ clock cycles.

C. GF(2^m) Inversion Algorithms

Two main methods are used for $GF(2^m)$ inversion: *Euclidean algorithm* and *Fermat's little theorem* (FLT). Up to now, Euclide based solutions are rarely used in hardware [17]. FLT states that $A^{2^m-1} = 1$, then $A^{-1} = A^{2^m-2}$. Thus, an exponentiation can compute the inverse of $A \in GF(2^m)^*$ using the standard *square and multiply* algorithm.

Itoh and Tsujii (IT) proposed in [8] an efficient way to perform this exponentiation noting that $2^m - 2 = (111 \cdots 110)_2$ and using addition chains. An addition chain is a sequence of additions where all operands are selected among the previously

---

TABLE II
INVERSION EXAMPLE OF A IN $GF(2^7)$ USING IT ALGORITHM.

$T_0 = A^{(1)_2}$	$u_0 = 1$
$T_1 = T_0^2 \times T_0 = A^{(10)_2} \times A^{(1)_2} = A^{(11)_2}$	$u_1 = u_0 + u_0$
$T_2 = T_1^2 \times T_0 = A^{(110)_2} \times A^{(1)_2} = A^{(111)_2}$	$u_2 = u_1 + u_0$
$T_3 = T_2^{2^3} \times T_2 = A^{(111000)_2} \times A^{(111)_2} = A^{(111111)_2}$	$u_3 = u_2 + u_2$
$A^{-1} = T_3^2 = A^{(1111110)_2}$	

computed terms [18, Sec. 4.6.3]. Each term is the Hamming weight of the exponent written in the binary representation. Tab. II provides an example in $GF(2^7)$ with both the addition chain and the corresponding sequence of multiplication and square operations (where T_i is the i-th intermediate product). In practice, the efficiency of IT based inversion mostly relies on the multiplier efficiency.

Recently in [19], a new IT inversion operator has been proposed using the hybrid multiplier from [16] which performs $A \times B \times C$ in $\lceil m/w \rceil + 1$ clock cycles with w-bit sub-words.

In [20], the authors generalize the concept of addition chains introducing the k-addition chains. They use such a tool to improve the work from [19]. Very recently in [21], a parallel-IT algorithm has been proposed to speed-up the IT-sequence using a second multiplier.

III. PROPOSED SOLUTION

In ECC halving based scalar multiplication, there is an inversion for each scalar bit equal to "1": on average every 0.5 key bit without key recoding or every 0.33 key bit using non-adjacent form (NAF) recoding, see [2, Sec. 3.3]. Then a standalone inversion unit would lead to underused silicon. We designed a new *combined multiplication-inversion unit* (MIU) to support both multiplication and inversion $GF(2^m)$ operations in GNB. It uses the IT method, then only multiplications and squares are required for field inversion. As squaring is very cheap in GNB (*i.e.* just a ROR), implementing a MIU requires a field multiplier, a ROR, a local register and a small controller. Our MIU has two operation modes: i) standard $GF(2^m)$ multiplication where operands A and B are multiplied to produce $P = A \times B$; ii) IT based $GF(2^m)$ inversion where multiplier and ROR blocks are used to compute the multiplications and squares (with intermediate values in the local register). We improved the inversion speed by using optimized multiplications for specific terms of the IT exponentiation. Our MIU produces two result bits per clock cycle with only one MO multiplication block. The obtained speed-up is about 20% and not 100% for two reasons: first, not all terms in the IT exponentiation sequence have this specific form; second, some of result bits are produced several times.

The $GF(2^m)$ operand(s) and result of our MIU, for $P = A \times B$ or $R = A^{-1}$ operations, are represented using a variation of GNB with permuted coefficients proposed in Sec. III-C. Some values are stored using w-bit sub-words.

A. Optimizing Large Shifter in IT Exponentiation on FPGA

During the IT exponentiation [8], the intermediate products in $GF(2^m)$ must be shifted by several shift amounts (*e.g.* for $m = 571$ there are 13 different shift amounts). This requires a very large m-bit barrel shifter ($m \in [163, 571]$ bits in ECC). As an example, for $GF(2^{571})$ on a Spartan 6 FPGA, a 571-bit shifter (optimized for the 13 shift amounts) requires 3425 LUTs while a Massey-Omura multiplier requires 2246 LUTs. Furthermore, the shift amounts are different for each field size m, then one must use a specific shifter for each m.

We replace all these large shifters by one hardwired block RAM (BRAM) of the FPGA. Instead of storing each bit of

Algorithm 2: w-bit words Massey-Omura multiplication.

Operands: $A, B \in \mathrm{GF}(2^m)$
Result: $P = A \times B$
1 $G \leftarrow 0$
2 **for** i **from** 0 **to** $m + w - 2$ **do**
3    $G \leftarrow \mathtt{SHL}(G, 1);$ $G[w-1] \leftarrow A \times M_0 \times B^T$
4   **if** $i \geq (w-1)$ **then**
5      $P[i - w + 1] \leftarrow G$ (*write in BRAM*)
6    $A \leftarrow \mathtt{ROL}(A, 1);$ $B \leftarrow \mathtt{ROL}(B, 1)$
7 **return** P

Algorithm 3: Massey-Omura Multiplication in PNB.

Operands: $A, B \in \mathrm{GF}(2^m)$ in PNB, and k
Result: $P = A \times B$
1 $C \leftarrow \mathtt{ROL}(A, -k);$ $G \leftarrow 0;$ $H \leftarrow 0;$ $j \leftarrow k \cdot \theta^{-1} \bmod m$
2 **for** i **from** 0 **to** $(w + N(k) - 2)$ **do**
3    $G \leftarrow \mathtt{SHL}(G, 1);$ $H \leftarrow \mathtt{SHL}(H, 1)$
4    $V \leftarrow A \times M_0'$
5    $G[w-1] \leftarrow V \times B^T;$ $H[w-1] \leftarrow V \times C^T$
6   **if** $i \geq (w-1)$ **then**
7      $P[i - w + 1] \leftarrow G$ (*write in DP-BRAM*)
8      $P[j + i - w + 1] \leftarrow H$ (*write in DP-BRAM*)
9    $A \leftarrow \mathtt{ROL}(A, 1);$ $B \leftarrow \mathtt{ROL}(B, 1);$ $C \leftarrow \mathtt{ROL}(C, 1)$
10 **return** P

---

the intermediate m-bit product P in a register, we duplicate them w times in the BRAM using the following patterns: $[p_0, p_1, \ldots, p_{w-1}]$ at address @ $= 0$; $[p_1, p_2, \ldots, p_w]$ at @ $= 1$; $[p_2, p_3, \ldots, p_{w+1}]$; $\ldots$; $[p_{m-1}, p_0, \ldots, p_{w-2}]$ at @ $= m - 1$. BRAMs in recent FPGAs are large enough to support the $m \cdot w$ bits. For instance, with $w = 32$ (a typical width for BRAMs), 7456 bits are required for $m = 233$ and 18272 bits for $m = 571$. In a low-cost Spartan-6 FPGA, 18Kb BRAMs are available, then for the largest field $m = 571$, one needs two BRAMs. The MO multiplier is a sequential operator with m clock cycles for producing each intermediate P. We use a w-bit temporary register to store $[p_i, p_{i+1}, \ldots, p_{i+w-1}]$ to feed the BRAM, and this register is shifted by 1 bit for the next cycle. The total cycle count for each product in the IT sequence is $m + w$. Using our BRAM based solution, shifting is performed by reading the words at addresses $(i + \alpha w) \bmod m$ for $\alpha \in \{0, 1, 2, \ldots, \lfloor m/w \rfloor\}$, where i is the shift amount. In Algo. 2, we adapt the MO multiplication algorithm to our BRAM based shifting solution (where SHL is a left shift).

This method seems to be equivalent to a simple shift register iteratively used over m clock cycles. But it will allow us to provide, for all m parameters, two shifted values for two different shift amounts at each clock cycle using a dual-port BRAM (DP-BRAM).

B. Support of Specific Patterns in the Exponentiation Chain

In the Itoh-Tsujii exponentiation sequence, the specific multiplication pattern (SMP) $A^{2^k} \times A$ frequently appears. It corresponds to the term $U_i = U_j + U_j$ in the addition chain. We modify the MO multiplication algorithm, and operator, to support both standard multiplications $A \times B$ as well as optimized SMPs. During the SMP computation, we notice that $\mathtt{ROL}(A, -i) \times M_0$ and $M_0 \times \mathtt{ROL}(A, -i)^T$ are computed. Since these two values are equal (one is the transpose of the other), we save one matrix-vector multiplication for each iteration of the algorithm. At every clock cycle, we compute $V = M_0 \times \mathtt{ROL}(A, -i)^T$ and return $(p_i = \mathtt{ROL}(A, k - i) \times V, p_{i+k} = \mathtt{ROL}(A, -i - k) \times V)$. For instance, if $k = 1$ then the outputs will be at CC(0): (p_0, p_1); CC(1): (p_1, p_2); $\ldots$; CC($m - 2$): (p_{m-2}, p_{m-1}) where CC(r) is clock cycle number r. We produce the output bits serially, like the original MO, but 2 bits at each clock cycle. This allows to efficiently overlap successive computations as in [19]. Producing most

of the result bits twice will not be a problem using the trick presented in the next subsection.

C. Multiplication and Inversion using Permuted Normal Basis

During the computation of a SMP in the IT sequence, using the modified MO from Sec. III-B leads to some redundancies in the produced bits. Using the example from last paragraph of Sec. III-B, a SMP with $k = 1$ produces $2m - 2$ bits instead of m for the result of $A^{2^k} \times A$ (redundancy ≈ 2). This is due to the constant shift amount equal to 1 used in the MO algorithm.

We propose a modified MO algorithm with a new constant shift amount $\theta > 1$ leading to a lower overall redundancy level during a complete IT exponentiation. We wrote a Python program to find θ which minimizes the inversion time for a given field $\mathrm{GF}(2^m)$. We compute an addition chain for m using the basic binary method: $U_i = U_{i-1} + U_{i-1}$ or $U_i = U_{i-1} + U_0$ with $U_0 = 1$. Then our program exhaustively tests the m possible shift amounts. It returns θ leading to the smallest number of clock cycles for the whole inversion based on all SMPs $A^{2^k} \times A$ in the IT sequence.

Addition chains produced by the basic binary method are not always the shortest ones but they lead to a smaller number of intermediate m-bit registers. In practice, all our chains are the shortest or at most 1-term longer than the shortest chains from [22].

We introduce the *permuted normal basis* (PNB) representation where element $A = [a_0, a_1, a_2, \ldots, a_{m-1}]$ is represented by $A' = [a_0, a_\theta, a_{2\theta \bmod m}, \ldots, a_{(m-1)\theta \bmod m}]$. Our adaptation of the MO multiplication to PNB is detailed at Algo. 3 where $N(k)$ denotes the clock cycle count in SMP $A^{2^k} \times A$.

In our modified Algo. 3, matrix M_0' is the adapted matrix M_0 for PNB representation. M_0' is also symmetric and is a permutation of the rows and columns of M_0. Row i in M_0 is at row $i\theta^{-1} \bmod m$ in M_0' and column j in M_0 is at column $j\theta^{-1} \bmod m$ in M_0' (where θ^{-1} is the modular inverse of $\theta \bmod m$). An example is given in Fig. 1 for $\mathrm{GF}(2^7)$.

Our algorithm produces the two result bits $(p_{i\theta}, p_{i\theta+k})$ at every clock cycle. Since m is prime, the indexes $i\theta \bmod m$ generate $\mathrm{GF}(m)$ seen as an additive group. Consequently, there exists for each $i < m$ an integer $j < m$ such that $(i\theta + k) \bmod m = j\theta \bmod m$. We sequentially write the result bits into two w-bit words in a DP-

BRAM with at @ $= i$, $[p_{i\theta}, p_{(i+1)\theta}, \ldots, p_{(i+w-1)\theta}]$ and @ $= j$, $[p_{j\theta}, p_{(j+1)\theta}, \ldots, p_{(j+w-1)\theta}]$ (all the indexes are computed modulo m).

Let us present a complete example for the toy field $GF(2^7)$. $m = 7$ leads to the chain $(U) = (1, 2, 3, 6)$ (see Tab. II). In this chain, there are two SMPs: $k = 1$ (for $2 = 1 + 1$) and $k = 3$ (for $6 = 3 + 3$). Our program returns $\theta = 2$. For SMP $k = 1$, the cycles are: CC(0): (p_0, p_1); CC(1): (p_2, p_3); CC(2): (p_4, p_5); CC(3): $(p_6, \underline{p_0})$ and $\overline{N}(1) = 4$ (redundancies are underlined). For SMP $k = 3$, the cycles are: CC(0): (p_0, p_3); CC(1): (p_2, p_5); CC(2): $(p_4, \underline{p_0})$; CC(3): $(p_6, \underline{p_2})$; CC(4): $(\underline{p_1}, \underline{p_4})$; and $\overline{N}(3) = 5$. Using our PNB MO multiplication, a $GF(2^7)$ inversion requires $4 + 7 + 5 = 16$ clock cycles instead of $7 + 7 + 7 = 21$ for the classical MO algorithm.

For the cryptographic field $GF(2^{163})$, the addition chain is $(U) = (1, 2, 4, 5, 10, 20, 40, 80, 81, 162)$ and our program returns $\theta = 72$. For this field, the SMPs lead to $N(1) = 120$, $N(2) = 86$, $N(5) = 111$, $N(10) = 104$, $N(20) = 118$, $N(40) = 90$ and $N(81) = 103$.

D. Cost Estimation

We denote N_{SMP} the number of SMPs in the IT sequence. We compute the number of clock cycles in an inversion as $C_{\text{inv}} = C_{\text{IO}} + C_{\text{SMP}} + C_{\text{nonSMP}}$ where:

- $C_{\text{IO}} = 2\lceil \frac{m}{w} \rceil$ is time for reading the operand A and writing the result $R = A^{-1}$;
- $C_{\text{SMP}} = D + 2N_{\text{SMP}}\lceil \frac{m}{w} \rceil$, with D the computation time in the SMPs evaluated by our program and reported in Tab. III, and the second term the duration of internal memory transfers;
- $C_{\text{nonSMP}} = (|(U)| - N_{\text{SMP}}) \cdot (m + \lceil \frac{m}{w} \rceil)$ for computations and internal memory transfers of non-SMP terms in the IT sequence ($|(U)|$ is the length of chain (U)).

In Tab. III, we report θ parameters, the estimated inversion time (in clock cycles) of both original and PNB algorithms for finite fields recommended by NIST [9] and parameter $w = 32$ (selected for our target FPGAs, see Sec. IV). For the largest fields, our Python program generates θ in less than 5 minutes on a mid-range computer (3 GHz Xeon processor with 8 GB).

Using PNB, IT-based inversion and modified MO multiplication from Algo. 3, we obtain 20 % faster $GF(2^m)$

$$M_0 = \begin{pmatrix} 0&1&0&0&0&0&0 \\ 1&0&1&0&0&1&1 \\ 0&1&0&1&1&1&0 \\ 0&0&1&0&0&1&0 \\ 0&0&1&0&0&0&1 \\ 0&1&1&1&0&0&1 \\ 0&1&0&0&1&1&1 \end{pmatrix} \Rightarrow M_0' = \begin{pmatrix} 0&0&0&0&0&1&0 \\ 0&0&0&1&1&0&0 \\ 0&0&1&0&1&1&1 \\ 0&1&0&0&1&1&1 \\ 0&1&1&1&0&1&0 \\ 1&0&1&1&1&0&0 \\ 0&1&0&1&1&0&0 \end{pmatrix}$$

Fig. 1. M_0 and the corresponding M_0' in PNB for $GF(2^7)$ and $\theta = 3$.

TABLE III
INVERSION DETAILS AND COMPARISON FOR NIST FIELDS.

| m | θ | D | inversion time | | speed-up |
			original	PNB	
163	72	732	1702	1335	$\approx 21\%$
233	36	1046	2667	2082	$\approx 21\%$
283	28	1431	3522	2761	$\approx 21\%$
409	35	2263	5090	4185	$\approx 17\%$
571	171	3221	8282	5973	$\approx 27\%$

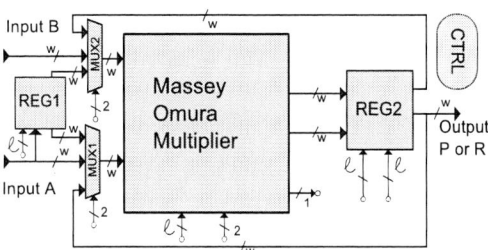

Fig. 2. Architecture of our multiplication-inversion unit (MIU).

inversions compared to the original IT [8]. For PNB $GF(2^m)$ multiplication, our MIU is as fast as the original MO with a small area overhead. Addition in $GF(2^m)$ has exactly the same time and area costs in both representations (GNB and PNB). Squaring in PNB $GF(2^m)$ is also a constant shift but where the shift amount is θ instead of 1. Conversion between PNB and GNB (for both directions) is a permutation of the coefficients. In parallel architectures this can be done using routing. In sequential implementations with several chunks per field element, this can be done using several registers accesses and routing. In practice, no conversion is required during the scalar multiplication.

IV. IMPLEMENTATION DETAILS AND RESULTS ON FPGAS

The architecture of our combined multiplication-inversion unit (MIU) is presented at Fig. 2 where $\ell = \lceil \log_2 m \rceil$ is the address size for field elements at bit level. Our MIU supports both SMP $A^{2^k} \times A$ and standard multiplications $A \times B$. In the MIU, the internal address computations (addition and reduction modulo m) are pipelined to reach higher frequencies. Its main computation block is our modified MO multiplier for PNB representation from Algo. 3 and illustrated at Fig. 3. This block requires two dot product operators (the central rectangles in Fig. 3) instead of one (in the original MO multiplier) as well as three m-bit shift registers instead of two. It also requires two small w-bit registers for temporary transfers to the BRAM. In Fig. 2, the m-bit register REG1 stores A during the complete IT sequence when computing A^{-1}. REG2, in BRAM, stores all intermediate/final products of the IT sequence. We used $w = 32$ to fit actual BRAM sizes in our target FPGAs. Our MIU operator is scalable for larger w (for other targets). Our solution is optimized for FPGAs with BRAMs. In case of ASIC implementation, we need to study an appropriate architecture.

All internal operations and transfers in our MIU are scheduled by the high-level finite states machine (FSM) presented at Fig. 4. It schedules the complete IT algorithm with:

- Initial state LOAD_OP loads the operand(s) A in PNB $GF(2^m)$ for inversion (and possibly B for multiplication) sequentially using w-bit sub-words;
- MULT starts a standard MO multiplication $A \times B$;
- DONE is reached when the computation is finished (and the user can serially read the result);
- REG×REG starts the computation of the first product $A \times A$ in the IT sequence using the modified MO multiplier

978-1-4799-8720-7/15 $31.00 © 2015 IEEE

from Algo. 3. The result A^2 is stored in the BRAM using the redundant representation from Sec. III-A.

- BRAM×BRAM starts the computation of a SMP $A^{2^k} \times A$;
- REG×BRAM starts the computation of a non-SMP multiplication $A \times B$;
- WAIT is reached at the end of one step in the IT sequence;
- NEXT determines what is the next step in the IT sequence (or the end of the computation).

Tab. IV and Tab. V report implementation results of $GF(2^m)$ inversions using our MIU on Spartan-6 LX75T and Virtex-4 LX100 FPGAs respectively. Three inversion algorithms have been fully implemented and validated: MO1 uses a single $\times M_0$ block with the original MO multiplier from [7]; RM2 uses two $\times M_0$ blocks with the modified multiplier proposed in [13] (which eliminates t redundant XOR gates and produces two output w-bit words at addresses $(i, i + \lceil m/2 \rceil)$); In Tab. IV, V and VI, lines with a star (*) correspond to state-of-art algorithms for multiplication and inversion from [7] and [13] we implemented and optimized on the same FPGA than our PNB solution. PNB corresponds to our MIU architecture at Fig. 2 with one $\times M_0$ block from Algo. 3 (Fig. 3) in PNB representation. The three versions have been implemented using the architecture of Fig. 2 with the corresponding internal multiplier. The inversion duration is reported in column named "Time". Bottom of Tab. V presents comparisons with state-of-art algorithms from [19] and [21]. Our solution is about 10 times slower but it is more than 10 times smaller. On a Virtex-4 LX100 (a mid-range device), there are 98304 LUTs. Then state-of-art solutions from [19] and [21] require 86 and 57% of the device just for multiplication and inversion respectively.

The RM2 solution (with a large internal multiplier) leads to important frequency reduction compared to the original MO1 solution. Our PNB solution has a smaller impact on the circuit frequency. This can be an advantage when designing complete halving based accelerators. Our MIU with PNB representation seems to be interesting for the larger fields. For instance, in $GF(2^{571})$, we obtain a better area–speed trade-off than the RM2 for a cost similar to the simple MO1. Comparing various computation units is not an easy task when multiple area–speed trade-offs are possible. The efficiency of our MIU also

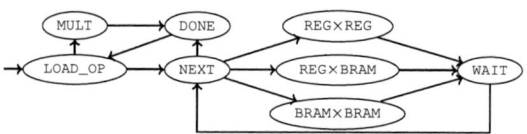

Fig. 4. Finite-state machine of the proposed multiplication-inversion unit.

TABLE IV
FPGA IMPLEMENTATION RESULTS FOR OUR MULTIPLICATION-INVERSION UNIT WITH VARIOUS INTERNAL MULTIPLIERS (SPARTAN 6 LX75T).

m	Algo.	Area Slices (LUT, FF)	Freq. MHz	Time μs	#block RAM
163	MO1 [7]*	337 (1004, 763)	217	7.84	1
	RM2 [13]*	423 (1348, 862)	140	6.04	1
	our PNB	469 (1411, 1034)	196	5.56	1
233	MO1 [7]*	420 (1231, 966)	229	11.6	1
	RM2 [13]*	569 (1765, 1078)	138	9.66	1
	our PNB	526 (1703, 1296)	181	10.0	1
283	MO1 [7]*	484 (1659, 1101)	160	22.0	1
	RM2 [13]*	668 (2164, 1209)	131	13.4	1
	our PNB	719 (2230, 1498)	159	15.3	1
409	MO1 [7]*	647 (2178, 1501)	163	31.2	1
	RM2 [13]*	917 (2768, 1610)	139	18.3	1
	our PNB	980 (3167, 1993)	159	22.3	1
571	MO1 [7]*	941 (3336, 1968)	109	75.0	2
	RM2 [13]*	1656 (4911, 2727)	80	53.8	2
	our PNB	1190 (4422, 2634)	94	63.5	2

depends on the targeted FPGA: the best results are obtained for the most recent one (this is probably due to a better use of 6-input LUTs and more important routing resources).

Our solution actually factorizes some matrix-vector products in the SMPs. For large values of t, our solution leads to more factorizations of sub-expressions. We believe that the PNB solution could be used for fields whose type t is greater or equal to 10.

We estimated the cost and performance of various halving based ECC scalar multiplication algorithms. Paper [3] presents a rough estimation using r-NAF key recoding: $mI/(r+1) + mM(1 + 3/(r+1)) \mu s$ where I is the inversion time and M the multiplication time (given in microseconds). We estimated the area cost and the computation time of all operators used in halving based methods (field addition, field multiplication-inversion, field square and square-root, trace, and other small specific operators). We compared several algorithms, the corresponding results are reported in Tab. VI. To compare the efficiency of various algorithms and area–time trade-offs, we also report the *area–time product* (ATP): the number of LUTs multiplied by the scalar multiplication time. The area results are very similar for both NAF and 3-NAF key recoding units (the area difference is about a few LUTs). Tab. VI shows that our PNB solution is more efficient when considering both computation time and silicon area. We think our approach can be used in applications where high security level is required on small circuits.

All algorithms have been validated using intensive functional simulation (at bit level) in Maple. All architectures have been validated using VHDL simulations using Modelsim.

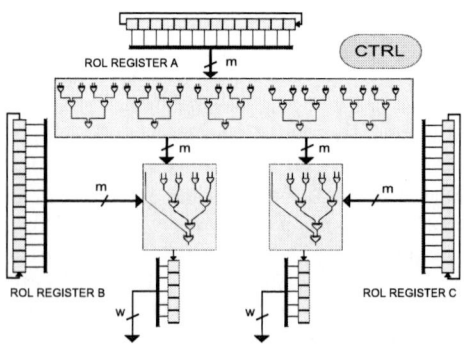

Fig. 3. High-level architecture of our modified Massey-Omura multiplier for PNB representation.

978-1-4799-8720-7/15 $31.00 © 2015 IEEE

TABLE V

FPGA IMPLEMENTATION RESULTS FOR OUR MULTIPLICATION-INVERSION UNIT WITH VARIOUS INTERNAL MULTIPLIERS (VIRTEX-4 LX100).

m	Algo.	Area Slices (LUT, FF)	Freq. MHz	Time μs	#block RAM
163	MO1 [7]*	906 (1636, 743)	250	6.80	1
	RM2 [13]*	1220 (2068, 808)	210	3.99	1
	our PNB	1227 (2274, 1058)	247	4.12	1
233	MO1 [7]*	1256 (2233, 1009)	202	13.2	1
	RM2 [13]*	1430 (2435, 1016)	204	6.53	1
	our PNB	1654 (2792, 1329)	212	8.22	1
283	MO1 [7]*	1577 (2839, 1149)	191	18.4	1
	RM2 [13]*	1924 (2435, 1147)	165	9.80	1
	our PNB	2073 (3741, 1525)	186	12.8	1
409	MO1 [7]*	2283 (2839, 1149)	169	31.1	1
	RM2 [13]*	2729 (4833, 1532)	150	15.9	1
	our PNB	2482 (4627, 2024)	155	21.4	1
571	MO1 [7]*	3378 (5615, 2016)	125	64.4	2
	RM2 [13]*	4976 (9445, 2090)	107	38.7	2
	our PNB	4308 (5928, 2650)	125	47.7	2
571	Hybrid ($d = 13$) [19]	#LUTs = 85268	74	4.98	–
	Parallel ($d = 13$) [21]	#LUTs = 56657	82	5.00	–

TABLE VI

COST/PERFORMANCE ESTIMATIONS FOR VARIOUS HALVING BASED ECC SCALAR MULTIPLICATIONS AND $m = 571$ ON A VIRTEX-4.

	Algorithm	halving ms	area #LUTs	ATP $\times 10^{-3}$
NAF	MO1 [7]*	17.3	5742	95
	RM2 [13]*	13.0	9572	122
	our PNB	**14.3**	6055	**82**
	Parallel IT (d=13) [21]	1.59	56784	90
	Hybrid IT (d=13) [19]	1.60	85395	136
3-NAF	MO1 [7]*	14.6		79
	RM2 [13]*	8.95		76
	our PNB	**11.3**	similar	**65**
	Parallel IT (d=13) [21]	1.34		74
	Hybrid IT (d=13) [19]	1.40		119

V. CONCLUSION

We proposed a new hardware unit, called MIU, combining $GF(2^m)$ multiplication and inversion operations for halving based ECC cryptosystems. We proposed a new representation of the field elements called *permuted normal basis* (PNB), modifications of the Massey-Omura algorithm for multiplication, and the Itoh-Tsujii algorithm for inversion. Our solution leads to about 20 % theoretical speed-up over previous works. It has been implemented on FPGAs and seems to be interesting for the larger fields. It also leads to a higher frequency compared to parallel solutions from state-of-art. On halving based ECC scalar multiplication, our PNB solution leads to a better area-time efficiency for large fields.

In the future, we plan to study extensions with multiple internal multiplication blocks, mixing our PNB and parallel-IT [21] solutions and also try to adapt our algorithm (especially the optimization of very large shifters) to ASIC targets. We believe that PNB representation and associated algorithms could be efficiently used for finite fields with a type $t \geq 10$.

ACKNOWLEDGMENT

This work has been partially supported by the PAVOIS project (ANR 12 BS02 002 01).

REFERENCES

[1] R. Lidl and H. Niederreiter, *Introduction to Finite Fields and Their Applications*, 2nd ed. Cambridge University Press, 1994.

[2] D. Hankerson, A. Menezes, and S. Vanstone, *Guide to Elliptic Curve Cryptography*. Springer, 2004.

[3] E. W. Knudsen, "Elliptic scalar multiplication using point halving," in *Proc. Int. Conf. Theory and Application of Cryptology and Information Security (ASIACRYPT)*, Nov. 1999, pp. 135–149.

[4] T. Oliveira, D. F. Aranha, J. Lopez, and F. Rodriguez-Henriquez, "Fast point multiplication algorithms for binary elliptic curves with and without precomputation," IACR ePrint, Tech. Rep. 427, Jun. 2014.

[5] C. Negre and J.-M. Robert, "New parallel approaches for scalar multiplication in elliptic curve over fields of small characteristic," LIRMM, University of Perpignan UPVD, Tech. Rep., 2013.

[6] S. Mangard, E. Oswald, and T. Popp, *Power Analysis Attacks: Revealing the Secrets of Smart Cards*. Springer, 2007.

[7] J. K. Omura and J. L. Massey, "Computational method and apparatus for finite field arithmetic," US Patent US4587627 A, May 1986.

[8] T. Itoh and S. Tsujii, "A fast algorithm for computing multiplicative inverses in $GF(2^m)$ using normal bases," *Information and Computation*, vol. 78, no. 3, pp. 171–177, Sep. 1988.

[9] NIST, "FIPS 186-2, digital signature standard (DSS)," 2000.

[10] M. A. Hasan and C. Negre, "Low space complexity multiplication over binary fields with dickson polynomial representation," *IEEE Transactions on Computers*, vol. 60, no. 4, pp. 602–607, Apr. 2011.

[11] H. W. Chang, W.-Y. Liang, and C. W. Chiou, "Low cost dual-basis multiplier over $GF(2^m)$ using multiplexer approach," in *Knowledge Discovery and Data Mining*. Springer, 2012, pp. 185–192.

[12] Q. Liao, "The Gaussian normal basis and its trace basis over finite fields," *J. Number Theory*, vol. 132, no. 7, pp. 1507–1518, Jul. 2012.

[13] A. Reyhani-Masoleh, "Efficient algorithms and architectures for field multiplication using Gaussian normal bases," *IEEE Transactions on Computers*, vol. 55, no. 1, pp. 34–47, 2006.

[14] G.-L. Feng, "A VLSI architecture for fast inversion in $GF(2^m)$," *IEEE Transactions on Computers*, vol. 38, no. 10, pp. 1383–1386, Oct. 1989.

[15] G. B. Agnew, R. C. Mullin, I. M. Onyszchuk, and S. A. Vanstone, "An implementation for a fast public-key cryptosystem," *Journal of Cryptology*, vol. 3, no. 2, pp. 63–79, Jan. 1991.

[16] R. Azarderakhsh and A. Reyhani-Masoleh, "Low-complexity multiplier architectures for single and hybrid-double multiplications in Gaussian normal bases," *IEEE Transactions on Computers*, vol. 62, no. 4, pp. 744–757, Apr. 2013.

[17] W. Drescher, K. Bachmann, and G. Fettweis, "VLSI architecture for non-sequential inversion over $GF(2^m)$ using the euclidean algorithm," in *Proc. Conf. Signal Processing Applications and Technology*, 1997.

[18] D. E. Knuth, *Seminumerical Algorithms*, 3rd ed., ser. The Art of Computer Programming. Addison-Wesley, 1997, vol. 2.

[19] R. Azarderakhsh, K. Jarvinen, and V. Dimitrov, "Fast inversion in $GF(2^m)$ with normal basis using hybrid-double multipliers," *IEEE Transactions on Computers*, vol. 63, no. 4, pp. 1041–1047, Apr. 2014.

[20] P. Jarvinen, S. Dimitrov, and R. Azarderakhsh, "A generalization of addition chains and fast inversions in binary fields," *IEEE Transactions on Computers*, 2015, accepted paper.

[21] J. Hu, W. G. J. Wei, and R. Cheung, "Fast and generic inversion architectures over $GF(2^m)$ using modified Itoh-Tsujii algorithms," *IEEE Transactions on Circuits and Systems II: Express Briefs*, 2015, accepted paper.

[22] M. D. Bielefeld University, "Shortest addition chains," Website, 2011, http://wwwhomes.uni-bielefeld.de/achim/addition_chain.html.

[23] D. Hankerson, J. Lopez, and A. Menezes, "Field inversion and point halving revisited," *IEEE Transactions on Computers*, vol. 53, no. 8, pp. 1047–1059, 2004.

978-1-4799-8720-7/15 $31.00 © 2015 IEEE

The Future of Nanoelectronics:
New Materials, Architectures and Devices

Heike Riel

IBM Research – Zurich
Säumerstrasse 4, 8803 Rüschlikon, Switzerland
hei@zurich.ibm.com

Technological breakthroughs have led to enormous improvements in performance, power, functionality and cost of computing devices and have thus enabled 50 years of Moore's law. Cost per function has decreased several thousand fold, while system performance and reliability have been improved dramatically. Today conventional silicon transistor scaling is approaching fundamental physical limits. For example, the increasing power dissipation on the chip level is one of the key challenges. Rising leakage currents and the increasing difficulty to further reduce the supply voltage have impacted the passive and active power dissipation, limiting the overall performance. Therefore a key attribute of any new device that may be considered for replacing the conventional field-effect transistor (FET) is reduced power dissipation. In that respect new strategies, including the use of novel materials, innovative device architectures and device concepts need to be explored and assessed.

Attention has turned to III–V compound semiconductors that are well positioned to replace silicon as the base material in logic switching devices. Their outstanding electron transport properties and the possibility to tune heterostructures provide tremendous opportunities to engineer novel nanometerscale logic transistors. The scaling constraints require an evolution from planar III–V metal oxide semiconductor field-effect transistors (MOSFETs) toward transistor channels with a three-dimensional structure, such as nanowire FETs, to achieve future performance needs for complementary metal oxide semiconductor (CMOS) nodes beyond 10 nm.

Carbon nanotubes (CNT) represent another class of semiconductor materials possessing transport properties more attractive than silicon to lower operation voltage and thus power consumption of MOSFETs. The superior low-voltage performance of sub-10 nm CNT transistors proves the viability of nanotubes for consideration in future aggressively scaled transistor technologies. Still challenges such as precise positioning and low contact resistance remain for large-scale integration.

Further device innovations are required to increase energy efficiency. This could be addressed by devices with a steeper subthreshold slope compared to MOSFETs to enable scaling of voltage supply and future low-power designs. In that regard, tunnel FETs (TFETs) are very promising as they allow to achieve a subthreshold swing of below 60 mV/dec. at room temperature by utilizing band-to-band-tunneling (BTBT) for charge injection. To achieve the required TFET performance boosters such as heterostructures are needed to lower the effective tunnel barrier and enable steep slope and high on-currents.

This presentation describes the challenges and recent progress toward the most prominent candidates for becoming the next nanoelectronic switch where new materials, architectures and devices are crucial.

Challenges and Perspectives of Nanoelectromagnetics in the THz range

S.A. Maksimenko, M.V. Shuba, P. P. Kuzhir,
K.G. Batrakov

Institute for Nuclear Problems,
Belarusian State University
Minsk, Belarus
sergey.maksimenko@gmail.com

G.Y. Slepyan

School of Electrical Engineering,
Tel-Aviv University
Tel-Aviv, Israel
gregory_slepyan@yahoo.com

Abstract—Current trends in nanoelectromagnetics - a new field of research synthesizing classical radio physics and electrodynamics with quantum theory of solids, physical kinetics, quantum chemistry, computational mathematics - are analyzed with the focus on applications in the terahertz frequency range. The analysis is exemplified by the discussion of the THz response peculiarities of carbon nanostructures and composite materials with nanocarbon fillers. Some electromagnetic applications of nancaerbons are discussed, such as nanoantennas, amplifiers and generators of terahertz radiation, THz range functional materials for radar absorbing and reflective coatings.

Keywords—nanoelectromagnetics; terahertz frequency range; nanocarbon; carbon nanotubes; graphene

I. INTRODUCTION

Since the very beginning, one of the main trend in the development of radio physics is the increase of the operating frequency: from microwaves - to mm-and submm-waves. In the later case, due to technological limitations, realization of passive components with characteristic dimensions comparable to the wavelength, becomes problematic. This has led to the refusal to use electrical circuits with lumped constants and transmission lines with TEM-waves in this frequency range. The problem is not only small size of elements. For example, the realization of hollow single-mode waveguides and resonators becomes unrealistic because surface treatment with the required purity class is impossible, that leads to unacceptably high level of heat losses during the scattering of working modes on the randomly irregular surfaces. Actually, the dimensions of circuit elements are accepted to be significantly higher than the wavelength, which led to a multimode nature of the electromagnetic field. It causes difficulties in frequency tuning, limits range of operating frequencies, provokes high sensitivity of the elements to errors of their setting. Therefore, physics of mm- and submm-range borrowed the principles of the field formation in optics of visible light. Accordingly, lens and aperture lines were used as the transmission lines, and Fabry-Perot interferometers were used as resonators.

A great progress has been achieved during last two decade in the synthesis and fabrication of different nanostructured materials with fascinating mechanical, electronic and optical properties irreducible to that of bulk media. Accompanied by the impressive parallel development of the characterization techniques and instrumentation, this process necessitates the revision of traditional concepts of physics and chemistry of condensed matter, adapting them to peculiarities of nanoworld and significantly extending our knowledge of the nature of solids and our capabilities to control their properties.

The appearance of nano-sized elements and nanostructures has fundamentally changed the main trend bringing lumped elements and circuits into frequency ranges, which have never assumed to be suitable for them: terahertz, IR and even optical range. Indeed, the circuit elements were designed whose dimensions are much smaller than the wavelength, or comparable to it. In particular, single-mode components (photonic crystals, transmission lines, microcavities, antennas) and lumped elements (capacitors, inductors, resistors, interconnects) became possible for high-frequency ranges. Thus, the reverse flow becomes evident: the principles of the field formation typical for low-frequency microwave and RF engineering are extended to the terahertz-to-optical range.

II. ANTENNA RESONANCES IN CARBON NANOTUBES

There are different physical mechanisms responsible for the advanced electromagnetic properties of nanostructures in terahertz range. In general, they are related to spatial confinement of the charge carriers motion to the sizes comparable with the de Broglie wavelenbgth, what thereby produces a discrete spectrum of energy states in one or several directions. As a result, specific dispersion laws manifest themselves in nanostructures providing their unusual electromagnetic response. In particular, quantum mechanical calculations of the conductivity of a single-walled carbon nanotube (SWCNT) followed by the solution of the impedance boundary-value problem predicts [1] a strong, up to 100 times, slowing-down of surface waves (plasmons) in SWCNTs. Figure 1 shows the slow-wave coefficient $\beta=k/h$ for the axially symmetric ($l=0$) surface wave (plasmon) in the metallic CNT. As one can see, in the THz and IR regimes the nanotube permits propagation of slowly decaying surface waves with almost frequency-independent β and the phase velocity v_{ph} much smaller than the speed of light in vacuum: $v_{ph}/c = 0.02$.

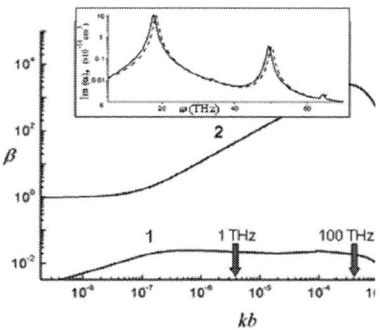

Fig. 1 The slow-wave coefficient β in a (9,0) metallic zigzag CNT for relaxation time τ=3 ps and T=295 K [1]. 1: Re(β); 2: Re(β)/Im(β). Inset: antenna resonances in polarizability of a 0.6 μm length (18,0) zigzag CNT [2].

These properties make CNTs prospective for the design of IR and THz nanoantennas [2]-[4] (see inset in Fig. 1), allow us to propose finite-length hypothesis [5] for the interpretation of experimental results on THz absorption in CNT-based composites [6]-[10], which demonstrated non-Drude behavior of the conductivity with pronounced terahertz peak. Experimental verification of the hypothesis has been reported for the first time in [11]. In our experiments [11] the frequency shift of the terahertz conductivity peak is shown to be dependent of the average length of CNTs in samples and thus is unambiguously attributable to the excitation of localized plasmon resonance, what is agrees well with calculations for a relevant physical model. This conclusion has been supported independently in alternative experiment [12], which is based on the idea that only metallic nanotubes contribute into terahertz peak if its nature is related to antenna (plasmon) resonance, while substitutional doping of CNTs increases "metallicity" of nanotubes. Theoretical model and experimental demonstration of the increase were reported in [13] and [14], respectively. Figure 2 demonstrate the role of doping, comparing calculations of THz-range response of a 1/2 mixture of metallic and semiconducting CNTs [13] with the optical density measurements in CNT thin films of different CNT bundle diameters and lengths [14].

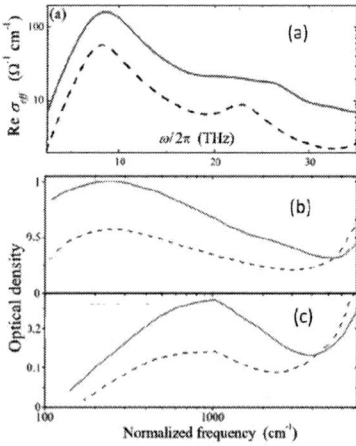

Fig. 2 The increases of "metallicity" in nanotube collections under effect of doping. Theoretical model (a) [13] and experimental verification (b)-(c) [14].

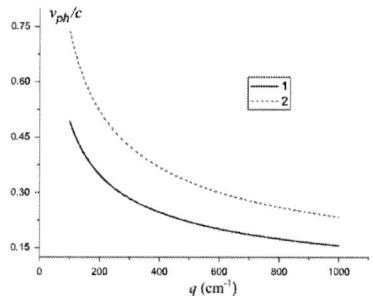

Fig. 3 The surface wave phase speed v_{ph}/c versus the wave number q (1/cm) in a single-layer graphene for the densities (1) 10^{12} cm^{-2} and (2) 5×10^{12} cm^{-2} of the doping electrons [16].

The plasmon nature of terahertz conductivity peak was also reported in [15] on the basis of experiments analogous to that carried out in [11]. Thus, one can conclude that antenna (plasmon) nature of the terahertz conductivity peak is established. Moreover, experimental results reported in [11] are actually direct demonstration of CNT as THz antenna. Also, the correlation between CNT length and THz peak position allows us to propose the peak position measurements for express determination of the average length of CNTs in a collection.

Using the many-body formalism and the tight-binding approach, we also investigated surface plasmon-polariton waves in graphene [16]. In single-layer graphene, the phase speed is about three to five times smaller than the speed of light in vacuum, see Fig. 3. Reduction of the phase speed by as much as 300 times is possible in a graphene structure with two spatially expanded monolayers, because inter-layer tunneling is suppressed, and the interlayer distance can be used to tune the plasmon frequency and the phase speed.

III. CARBON NANOTUBE AS A TRAVELING WAVE TUBE

The proposition to use a CNT as lasing basic element [17] opens a new hopeful field in the nanosized laser device engineering. In [18], the electric-field-induced heating of the electron gas is proposed to achieve coherent light emission in CNT in the infrared and THz frequency ranges. The basic concept to use the kinetic energy of the directed electron beam for stimulated emission producing in CNT has been presented in [19]-[20] for the first time.

There is a wide class of macroscopic devices utilizing an electron beam as a source of laser radiation. These are klystrons, traveling wave tubes (TWT) and backward wave oscillators, free electron lasers (FEL). All they are based on a few basic phenomena: the synchronous interaction of moving electrons with electromagnetic wave, the electron beam modulation and coherent radiation from the produced electron bunches. For the coherent generation, a high vacuum must be maintained in the region of the electron-beam–electromagnetic wave interaction. Otherwise, collisions of electrons with atoms move electrons out of the synchronism and, consequently, lasing is not reached. From this point of view CNTs are unique objects since they exhibit ballistic electrical conduction at room temperature with mean-free paths on the order of microns and even tens of microns [21].

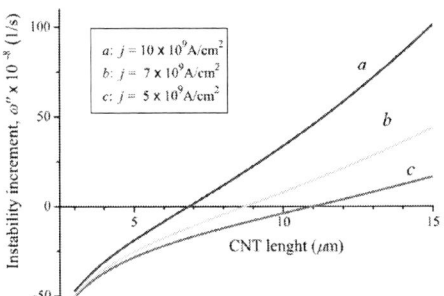

Fig. 4 Instability increment *vs* nanotube length at different electron current densities [20].

Therefore, electrons can emit coherently from the whole CNT length which is typically 1–10 μm. Besides, single-walled and multiwalled carbon nanotubes can carry a high current density of the order of 10^9–10^{10} A/cm^2, see, e.g., [22]. Lastly, metallic CNTs exhibit a strong, as large as 50–100 times, slowing down of surface electromagnetic waves [1]. Thus, a combination in CNTs of three key properties: (i) ballisticity of the electron flow over typical CNT length, (ii) extremely high current-carrying capacity, and (iii) strong slowing down of surface electromagnetic waves, allows proposing them as candidates for the development of nanosized Cherenkov-type emitters, i.e., nano-TWT. It follows from Fig. 4 [20] that the gain for CNT is extremely large as compared with macroscopic electronic devices. For chosen parameters, the generation development starts when the CNT length is about 6 μm or larger, which is technologically routine range. Therefore, we demonstrate that the development of CNT-based nano-TWT or nano-FEL is already possible at the current stage of nanotechnology.

IV. CONCLUSION

Here we made a brief overview of results obtained on studying and theoretical modeling of electromagnetic response properties of nanocarbon structures. Referring to original publications for details of the analysis, in the overview we endeavored to present several basic physical effects which emerge from the interaction of light with different nano objects and which are due to characteristic sizes of objects. However, the main goal of the work is to demonstrate as theoretical models lead to ideas basic for the development of elemental basis of the future nano-optics, nano-photonics, quantum informatics, etc. It should be stresses, that the applicability of the methodology and approaches developed in our original works and presented here is not restricted to carbon nanostructures considered here. Quite the contrary, from the very beginning the methodology was developed as a general platform adapting methods of classical electrodynamics to nanoscale systems and oriented on the solution of a wide class of electromagnetic problems of nanoworld.

ACKNOWLEDGMENT

This research was partially supported by the BRFFR under Project No. F14R-026, and EU FP7 under Projects FP7-318617 FAEMCAR and FP7-612285 CANTOR.

REFERENCES

[1] G.Ya. Slepyan, S.A. Maksimenko, A. Lakhtakia, O. Yevtushenko, and A.V. Gusakov, "Electrodynamics of carbon nanotubes: Dynamic conductivity, impedance boundary conditions, and surface wave propagation," Phys. Rev. B, vol. 60, pp. 17136–17149, Dec. 1999.

[2] G.Ya. Slepyan, M.V. Shuba, S.A. Maksimenko, and A. Lakhtakia, "Theory of optical scattering by achiral carbon nanotubes and their potential as optical nanoantennas," Phys. Rev. B, vol. 73, art. no 195416, May 2006.

[3] G.W. Hanson, "Fundamental transmitting properties of carbon nanotube antennas," IEEE Trans. Antennas Propag., vol. **53**, pp. 3426–3435, Nov. 2005.

[4] P.J. Burke, S. Li, and Z. Yu, "Quantitative theory of nanowire and nanotube antenna performance," IEEE Trans. Nanotechnol., vol. **5**, pp. 314–334, Oct. 2006.

[5] G.Ya. Slepyan, M.V. Shuba, S.A. Maksimenko, C. Thomsen, and A. Lakhtakia, "Terahertz conductivity peak in composite materials containing carbon nanotubes: Theory and interpretation of experiment," Phys. Rev. B, vol. 81, art. no 205423, 2010.

[6] F. Bommeli, et al., "Evidence of anisotropic metallic behaviour in the optical properties of carbon nanotubes", Solid State Comm. vol. 99, pp. 513-517, Aug 1996.

[7] A. Ugawa, A.G. Rinzler, "Far-infrared gaps in single-wall carbon nanotubes", Phys. Rev. B, vol. 60, art. no R11605, 1999.

[8] F. Borondics, et al., "Charge dynamics in transparent single-walled carbon nanotube films, from optical transmission measurements," Phys. Rev. B, vol. 74, art. no 045431, 2006.

[9] T. Kampfrath et al., "Mechanism of the far-infrared absorption of carbon-nanotube films," Phys. Rev. Lett., vol. 101, art. no 267403, Dec. 2008.

[10] N. Akima, et al., "Strong anisotropy in the far-infrared absorption spectra of stretch-aligned single-walled carbon nanotubes," Adv. Mater., vol. 18, pp. 1166–1169, 2006.

[11] M. V. Shuba, et al., "Experimental evidence of localized plasmon resonance in composite materials containing single-wall carbon nanotubes," Phys. Rev. B, vol. 85, 165435, 2012.

[12] Q. Zhang, et al., "Plasmonic nature of the terahertz conductivity peak in single-wall carbon nanotubes," Nano Lett., vol. 13, pp. 5991-5996, 2013.

[13] A. M. Nemilentsau, et al., "Substitutional doping of carbon nanotubes to control their electromagnetic characteristics," Phys. Rev. B, vol. 82, art. no 235424, 2010.

[14] M.V. Shuba, et al., "Effects of inclusion dimensions and p-type doping in the terahertz spectra of composite materials containing bundles of single-wall carbon nanotubes," J. Nanophoton., vol. 6, art. no 061707, 2012.

[15] T. Morimoto, et al., "Length-dependent plasmon resonance in single-walled carbon nanotubes," ACS Nano, vol. 8, pp. 9897-9904, 2014.

[16] K.G. Batrakov, V.A. Saroka, S.A. Maksimenko and C. Thomsen, "Plasmon polariton slowing down in graphene structures," J. Nanophoton., vol. 6, art. no 061719, 2012

[17] J. Chen, et al., "Bright Infrared Emission from Electrically Induced Excitons in Carbon Nanotubes", Science vol. 310, pp.1171- 1173, 2006.

[18] O. Kibis, M. Portnoi, "Carbon nanotubes: A new type of emitter in the terahertz range," Techn. Phys. Lett. vol. 31. pp. 671-673, 2005.

[19] K.G. Batrakov, P.P. Kuzhir, and S.A. Maksimenko, "Radiative instability of electron beam in carbon nanotubes," Proc. SPIE vol. 6328, art. no 632801, 2006.

[20] K.G. Batrakov, S.A. Maksimenko, P.P. Kuzhir and C. Thomsen, "Carbon nanotube as a Cherenkov-type light emitter and free electron laser," Phys. Rev. B, vol. 79, art. no 125408, 2009.

[21] S. Frank, P. Poncharal, Z.L. Wang, and W.A. de Heer, "Carbon nanotube quantum resistors," Science vol. 280, pp. 1744-1746, June 1998.

[22] R. Vajtai, B.Q. Wei, Z.J. Zhang, et al., "Building carbon nanotubes and their smart architectures," Smart Mater. Struct. vol. 11, pp. 691- 698, Sept. 2002.

2015 IEEE Computer Society Annual Symposium on VLSI

Semi-Classical Modelling of the Electron Transport in Carbon Nanotubes and Graphene Nanoribbons for THz Range Applications

Antonio Maffucci

Department of Electrical and Information Engineering
University of Cassino and Southern Lazio
Cassino, Italy
maffucci@unicas.it

Abstract—**This paper deals with the possible use of carbon-based nano-interconnects for terahertz applications. The electrodynamics of nano-interconnects, either made of carbon nanotubes or graphene nanoribbons, may be conveniently described by means of a semi-classical transport model, based on the modified Boltzmann transport equation. In this paper, this model is revised in order to include the effects of tunneling between adjacent carbon nanotubes or graphene layers. The final transmission line model is modified taking into account new terms for the line matrix parameters, directly related to the tunneling. Such terms introduce resonances and dispersion which potentially reduce the applicability of such interconnects. The effects of tunneling in the terahertz range is investigated, with reference to two examples of carbon nano-interconnects.**

Keywords - Carbon nanotubes; Graphene Nanoribbons, Nanointerconnects; Terahertz interconnects; Tunneling effect.

I. INTRODUCTION

The Terahertz range (0.1 – 10 THz) or submillimeter wave range (1-0.03 mm) is one of the last unexplored frontiers in electronics, which has gained more and more interest in the recent literature [1]-[2]. There are many interesting features associated to the THz electromagnetic waves that make them excellent candidates, for instance, for sensing or imaging applications. For examples, the THz waves are proposed for safe biological tissue screening (being non-ionizing radiations), for characterization of semiconductor materials, and for failure analysis of VLSI circuits [1].

From a technological point of view, accessing the THz region has been a challenge for decades, due to its nature of "boundary" region between electronics and photonics: classical electronics devices exhibit a dramatic drop of performance for frequencies above 100 GHz, due to the degradation of the transport properties. Conversely, classical photonics devices fail for frequencies below 10 THz, due to the steep increase of the conversion of the photon energy to the thermal energy. The term "THz gap" has been used in the past to indicate the lack of efficient sources of submillimeter-wave power sources, detectors, and low-loss interconnects in such a range, compared to the well-assessed RF/microwave

technology (electronics) or infrared or far infrared technology (photonics). This difficulty has meant that the THz range has only begun to be explored thoroughly over the last decade.

Nowadays, the rapid progress in nanoscience and nanotechnology is giving an unprecedented push to the THz technology. A promising technological solution is related to the use of new nanomaterials like those based on carbon, such as *carbon nanotubes* (CNTs) or *graphene nanoribbons* (GNRs), whose outstanding electrical, thermal and mechanical properties for THz optoelectronics [3]-[4]. Carbon-based materials are proposed for realizing devices like ballistic transistors [5], sources like plasmon oscillators [6], waveguides and nanoantennas [7]-[8].

A bottle-neck for the THz technology has been the lack of low-loss interconnects, even for chip to chip links. This limit imposes that the circuit elements like sources, detectors, antennas, etc., must be either placed in direct proximity to one another or coupled by free space optical links. Attempts have been made to transmit THz signals by using metallic waveguides, like the parallel-plate waveguide in [9], or the low-loss dielectric ribbons in [10]. The packaging realized with conventional materials, although the losses can be lowered to reasonable values, results in too big dimensions of the realized module, hard suitable for compact low-cost THz systems.

The use of GNRs and CNTs as VLSI interconnects has been widely investigated in these years for the microwave range. Theoretical results predict outstanding electrical and thermal performance of such interconnects [11]-[15]. In the meanwhile, the first real world applications of CNT and GNR interconnects have been demonstrated [16]-[19]. Recently, these carbon interconnects have been proposed for the THz packaging. In [20] two innovative solutions have been proposed: a flip-chip technology with contacts between carbon nanotubes realized via Van der Waals forces, and a wireless interconnect technology, based on monopole carbon nanotube antennas.

A typical arrangement is depicted in Fig.1: an on-chip trace is realized by means of a bundle of N carbon nanotubes or an array of N graphene nanoribbons. Due to the huge resistance introduced by the contacts, an efficient

978-1-4799-8720-7/15 $31.00 © 2015 IEEE   450

realization of carbon interconnects requires the use of N conductors in parallel, i.e. a CNT bundle or a GNR array. Similarly, vias or TSVs are realized by bundles of CNTs [15].

In view of using the carbon-based interconnects in the THz range, there is the need to reconsider the circuit models widely adopted in the microwave range. The propagation of electrical signals along carbon-based interconnects may be effectively modeled in the frame of the transmission line (TL) theory [21]-[25]. The quantum effects are taken into account by novel electrical parameters (like the *kinetic inductance* and the *quantum capacitance*) which add to the classical electromagnetic ones. These models may be derived from a semi-classical electrodynamic model of the electron transport along such nanostructures, as shown in [24]-[26]. In all the above TL models the tunneling between adjacent CNT shells or GNR layers may be neglected, since it does not have influence for frequencies up to THz.

In this paper we propose a reformulation of the TL model, which can take into account the tunneling effect. In Section II we briefly recall the electrodynamic model, shown in details in [24]-[26]. Then, we formulate the inclusion of the tunneling effect. Section III is devoted to the modified TL model, whereas in Section IV the CNT or GNR interconnects are studied in the THz range.

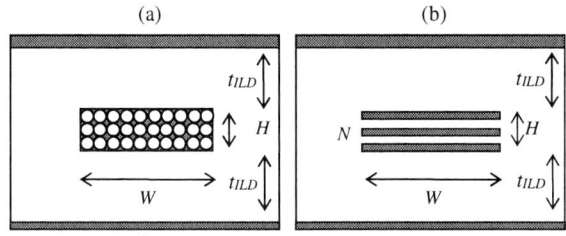

Figure 1. Typical horizontal interconnect made by:
(a) a CNT bundle; (b) a GNR array.

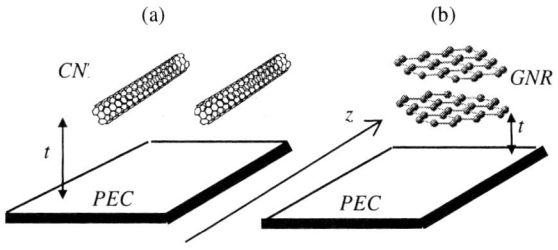

Figure 2. A nano-interconnect with two signal lines above a perfect ground. The lines are made by: (a) CNTs or (b) GNRs.

II. TRANSPORT EQUATION

In the following, for the sake of simplicity, we consider the coupled lines in Fig.2, with two signal lines and a common ground. The signal lines are made by a pair of CNTs of diameter D, or by a pair of GNRs of width W. As in [24]-[26], to derive a TL model, we should couple Maxwell equations to a suitable transport equation (a constitutive relation). Such a relation should model the transport of the free electrons (the π – electrons) along the nanostructures.

The cross-section sizes of the interconnects of our interest are typically large enough (at least 1 nm in the quantum confined directions) to have local crystal structures and to allow using *semi-classical transport model*. The electrons are regarded as classical particles but the movement takes place in a spatially periodic potential, where they move between two collisions according to the Boltzmann transport equation. The electrons are quantum confined laterally and thus occupy quantized energy levels, instead of the traditional continuum of energy levels in bulk materials. In the following, the N quantized energy subbands are labeled by a band index μ and a wave vector $\mathbf{k}=k\mathbf{u}$, where $\mathbf{u}$ is the unit vector oriented along the carbon lattice.

Let us consider time-harmonic longitudinal electric field, $E_z(z,t)$, and current density, $J_z(z,t)$, namely in the form $X_z(z,t) = \mathrm{Re}\left\{\hat{X}_z \exp[i(\omega t - \beta z)]\right\}$ being β the wavenumber, and let us assume that:

(i) the CNT diameter D or the GNR width W are electrically small;
(ii) the transverse currents may be neglected;
(iii) only intraband transitions are considered, whereas interband ones are not allowed.

Conditions (i) and (iii) are fully satisfied in THz range, since the transverse dimensions of the CNT or GNR are far below the minimum wavelength (0.03 mm at 10 THz), and the interband transitions occur at optical frequencies. Condition (ii) is fulfilled for large aspect-ratio interconnects.

Let us consider the n-th signal line (n=1,2). If the tunneling effect is neglected, it is possible to introduce the following generalized Ohm's law in the wavenumber domain [24]-[26]:

$$\hat{J}_{z,n}(\beta,\omega) = \hat{\sigma}_n(\beta,\omega)\hat{E}_{z,n}(\beta,\omega) \quad n=1,2, \quad (1)$$

where $\hat{\sigma}_n$ is the generalized longitudinal conductivity:

$$\hat{\sigma}_n(\beta,\omega) \cong -i\frac{2e^2}{\pi X}\frac{v_F}{\hbar}\frac{1}{\omega-iv_n}M_n\left[1+\alpha_n\left(\frac{v_F\beta}{\omega-iv}\right)^2\right], \quad (2)$$

where e is the electron charge, $\hbar$ is the Planck constant, v_F the Fermi velocity, usually of the order of 10^6 m/s, and v_n is the collision frequency. In addition, the quantity X is given by $X=D$ for CNTs and $X = \pi W$ for GNRs, whereas the quantity M_n represents the number of conducting channels,

978-1-4799-8720-7/15 $31.00 © 2015 IEEE 451

evaluated by the following integral over all the subbands:

$$M_n = \frac{\hbar}{v_F} \sum_{\mu=1}^{N} \int_0^{\pi/l} \left[v_{\mu,n}^{(+)2} \left(-\frac{dF}{dE_{\mu,n}^{(+)}} \right) + v_{\mu,n}^{(-)2} \left(-\frac{dF}{dE_{\mu,n}^{(-)}} \right) \right] dk. \quad (3)$$

Here $E_{\mu,n}^{(\pm)}$ and $v_{\mu,n}^{(\pm)}$ represent the energy level and the electron velocity associated to the μ-th subband, respectively. In (3), $F[E]$ is the Dirac-Fermi function:

$$F[E] = \frac{1}{e^{E/k_B T} + 1} \quad (4)$$

being k_B the Boltzmann constant and T the absolute temperature. Finally, the function α_n in (2) is equal to 1 for CNTs, whereas for GNR it is [26]:

$$\alpha_n = \frac{\hbar}{M_n v_F^3} \sum_{\mu=0}^{N-1} \int_0^{\pi/l} \left[v_{\mu,n}^{(+)4} \left(-\frac{dF}{dE_{\mu,n}^{(+)}} \right) + v_{\mu,n}^{(-)4} \left(-\frac{dF}{dE_{\mu,n}^{(-)}} \right) \right] dk \quad (5)$$

In the following, we assume that the two CNTs (GNRs) have the same chirality and the same diameter (width), therefore all the parameters in (2)-(4) are the same, in particular: $M_1 = M_2$, $\alpha_1 = \alpha_2$, and $v_1 = v_2$.

Let us introduce the tunneling effect between the two CNTs or the two GNRs. The generalized Ohm's law (1) modifies as follows [7],[25]:

$$\hat{J}_{z1} = \hat{\sigma}_{11}(\beta,\omega)\hat{E}_{z1} + \hat{\sigma}_{12}(\beta,\omega)\hat{E}_{z2} \quad (6)$$

$$\hat{J}_{z2} = \hat{\sigma}_{21}(\beta,\omega))\hat{E}_{z1} + \hat{\sigma}_{22}(\beta,\omega)\hat{E}_{z2} \quad (7)$$

where $\hat{\sigma}_{11} = \hat{\sigma}_{22} = \hat{\sigma}_s$ are the self conductivities given as in (2), whereas the mutual conductivities $\hat{\sigma}_{12} = \hat{\sigma}_{21} = \hat{\sigma}_m$ may be expressed as follows [25]:

$$\hat{\sigma}_m(\beta,\omega) \cong -i\frac{2e^2}{\pi\hbar v_F X}\frac{\omega_t \omega'}{\beta^2}\frac{1}{1-(\beta v_F/\omega')^2}\hat{\gamma}(\beta,\omega), \quad (8)$$

where $\omega' = \omega + i\nu$, and

$$\hat{\gamma}(\beta,\omega) = \sum_{k=1,2} \frac{\omega'}{\beta^2} \frac{2\omega_t + (-1)^k \left[1 + (\beta v_F/\omega')^2 \right]}{(2\omega_t + (-1)^k \omega') - \beta^2 v_F^2}. \quad (9)$$

In (8)-(9), a fundamental role is played by the tunneling frequency ω_t, related to the binding energy $\hbar\omega_t$ due to delocalization of the $\pi-$electrons for the tunnel effect. For CNT shells or graphene layers separated by the Van der Waals distance $\delta = 0.34$ nm, a measured energy of $\hbar\omega_t \approx 35$ meV is reported, corresponding to a tunneling frequency of $\omega_t \approx 10^{13}$ rad/s [25].

III. TRANSMISSION LINE MODEL

As shown in [24]-[26], in case of negligible tunneling effect, by coupling the constitutive relation (2) to Maxwell equations it is possible to model the interconnects in Fig. 2 as multiconductor lossy TLs, whose per-unit-length (p.u.l.) matrix parameters are given by:

$$\mathbf{R} = \nu L_k (\mathbf{U} + \mathbf{K})^{-1}, \quad \mathbf{L} = (L_k \mathbf{U} + \mathbf{L}_M)(\mathbf{U} + \mathbf{K})^{-1}, \quad \mathbf{C} = \mathbf{C}_E. \quad (10)$$

Here $\mathbf{U}$ is the unitary matrix, $\mathbf{R}$, $\mathbf{L}_M$ and $\mathbf{C}_E$ are the electrical resistance, magnetic inductance and electrostatic capacitance matrices, respectively, whereas the matrix $\mathbf{K}$ is:

$$\mathbf{K} = \mathbf{C}_E \frac{1}{C_q} \frac{1}{1 - i\nu/\omega}. \quad (11)$$

The p.u.l. *kinetic inductance* L_k and *quantum capacitance* C_q are given by ($R_0 = 12.9$ kΩ is the *quantum resistance*):

$$L_k = \frac{R_0}{2v_F M}, \quad C_q = \frac{v_F^2}{L_k}, \quad (12)$$

If the tunneling effect is taken into account, we should add to the inductance and capacitance matrices the operators $L_{tun}(\mathbf{I})$ and $C_{tun}(\mathbf{V})$, acting on the vector distributions of currents and voltages $\mathbf{I}(z,\omega)$ and $\mathbf{V}(z,\omega)$ as follows [25]:

$$\mathbf{L}_{tun}\{\mathbf{I}\} = \frac{1}{4\pi} \frac{\omega_t \omega'}{\omega v_F} \left(\frac{c}{v_F} \right)^2 \mathbf{\Theta} \mathbf{L}_M \int_{-\infty}^{+\infty} K_1(z - z') \mathbf{I}(z',\omega) dz', \quad (13)$$

$$\mathbf{C}_{tun}^{-1}\{\mathbf{V}\} = -\frac{\omega_t \omega'}{2v_F} \mathbf{\Theta} \mathbf{C}_E^{-1} \int_{-\infty}^{+\infty} K_2(z - z') \mathbf{V}(z',\omega) dz', \quad (14)$$

$$\mathbf{\Theta} = \begin{bmatrix} 1 & -1 \\ -1 & 1 \end{bmatrix}. \quad (15)$$

The kernels in (14) are defined as

$$K_1(z;\omega') = \sin\left(\frac{2\omega_t}{v_F}|z| \right) \exp\left(-i\frac{\omega'}{v_F}|z| \right) \quad (16)$$

$$K_2(z;\omega') = \frac{1}{2i} \exp\left(-i\frac{\omega'}{v_F}|z| \right)$$
$$\times \left[\frac{1}{2\omega_t - \omega'} \exp\left(i\frac{2\omega_t}{v_F}|z| \right) + \frac{1}{2\omega_t + \omega'} \exp\left(-i\frac{2\omega_t}{v_F}|z| \right) \right]. \quad (17)$$

Note that the operator $\mathbf{L}_{tun}$ also adds to the kinetic inductance $L_k\mathbf{U}$ in the modified expression of the resistance matrix $\mathbf{R}$. The operators (13) and (14) introduce a spatial and frequency dispersion.

IV. CARBON INTERCONNECTS IN THE THZ RANGE

Let us briefly review a well-known result, namely the possibility to have resonances in the THz range related to the line length. To this end, let us refer to the CNT line in Fig.2a, considering a metallic isolated CNT of diameter $D = 2$ nm (single-walled CNT), at a distance of $t = 8$ nm from the ground, embedded in a dielectric with permittivity $\varepsilon_r = 2.2$. The electrical and quantum parameters for such a line are reported in Table I.

Usually, the kinetic inductance dominates the magnetic one, being 3-4 order of magnitude higher, which is also our case. As for the capacitances, the quantum capacitance is approximately one order of magnitude smaller than the electrostatic one. In addition, the collision frequency is usually of the order of $\nu \approx 10^{12}$ Hz, therefore in the THz range $(0.1 - 10$ THz$)$ we can assume that the parameters in (10) reduce to:

$$R \approx \nu L_k, \quad L \approx L_k + L_M, \quad C = C_E. \quad (18)$$

A consequence of the huge value of the kinetic inductance is a low propagation velocity and a high characteristic impedance: the values of these two parameters normalized to an ideally scaled PEC interconnect of the same dimensions would be

$$\frac{c_{CNT}}{c} \approx \sqrt{\frac{L_m}{L_k}} = 0.012, \quad \frac{Z_{CNT}}{Z_C} \approx \sqrt{\frac{L_k}{L_m}} = 81.91. \quad (19)$$

TABLE I.   P.U.L. PARAMETERS FOR THE CNT LINE

Lm	Lk	Ce	Cq
[nH/mm]	*[μH/mm]*	*[pF/mm]*	*[pF/mm]*
0.55	3.69	0.04	0.35

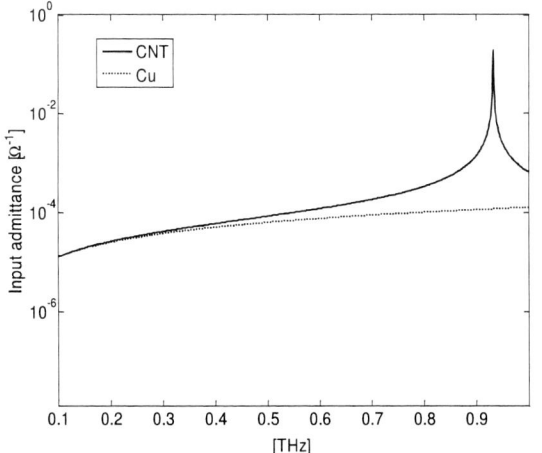

Figure 3. Input admittance for a single line made by a CNT above a PEC ground, vs a single line made by a copper wire above a PEC ground.

The low propagation velocity suggests the possibility to excite antenna resonances in the THz range. Figure 3 shows the resonance peak exhibited by the input admittance of a 1μm-long line, short-circuited at the far end.

The possibility of using isolated CNTs as nano-antennas in the THz range has been theoretically predicted [8], experimentally observed [27], and recently applied, for instance, in an on-chip link [20]. Considering the multiconductor case, this qualitative behavior does not change as far as we neglect the tunneling effect, and the only interaction is of electromagnetic nature (imposed by the mutual terms of inductance and capacitance matrices).

The above considerations impose a limit in the line length of a carbon interconnect to be used in the THz range: therefore, in the following we assume a length of up to 100nm, for which there is no resonance related to the length.

Let us consider the tunneling effect, focusing on the effect of the additional term $\mathbf{L}_{tun}$ (13) in the resistance and inductance matrices, assuming again that the quantum capacitance is such that in (10) it is $\mathbf{U} + \mathbf{K} \approx \mathbf{U}$. Having assumed lengths up to 100 nm, in the THz range the line is electrically short and it is possible to disregard the spatial variation of the distribution of the currents along it. In other words, we can neglect the spatial dispersion introduced by (13) and focus on the frequency dispersion. In addition, since $\nu \approx 10^{12}$ Hz, in the THz range it is also $\omega' \approx \omega$. Thus, the tunneling operator may be approximated as:

$$\mathbf{L}_{tun}(\mathbf{I}(\omega)) \approx \frac{1}{\pi} \left(\frac{c}{v_F} \right)^2 \frac{\omega_t^2}{4\omega_t^2 - \omega^2} \begin{bmatrix} 1 & -1 \\ -1 & 1 \end{bmatrix} \mathbf{L}_M \mathbf{I}(\omega). \quad (20)$$

Let us consider a pair of metallic single-walled CNTs of length 50 nm, diameter $D = 2$nm, intertube distance of 0.34nm, distance from ground $t = 4D$. The line is embedded in a dielectric with permittivity $\varepsilon_r = 2.2$. Figure 4 shows the self and mutual impedances of such a coupled line, computed assuming a tunneling frequency $\omega_t = 10^{13}$ rad/s.

The figure shows that a new resonance arises in the THz range as a consequence of the tunneling effect. Note that the tubes are separated by the minimum possible distance (the Van der Waals distance, $\delta = 0.34$ nm), which is the condition for the maximum coupling via tunneling. However, a minimum distance is also the most desirable conditions for realizing good quality CNT interconnects, since it provide the maximum bundle density (e.g., [12]-[15]). Therefore, there is a trade-off between a good quality interconnect and of low resonance peaks due to tunneling.

Let us now consider a pair of metallic GNRs, as in Fig.2b, assuming again a length of 50 nm, a width $W = 2$nm, an interlayer distance of 0.34 nm, a distance from the ground $t = 4D$, and $\varepsilon_r = 2.2$. Figure 5 shows the self and mutual impedances of such a line, assuming $\omega_t = 10^{13}$ Hz. The behavior is the same as that of the CNT lines, in Fig.4.

978-1-4799-8720-7/15 $31.00 © 2015 IEEE   453

The tunneling effect has a different impact if the line is fed by a differential mode or a common mode current. Given the matrix (15), if the signal lines carry a common mode current, the self and mutual terms related to tunneling cancel themselves, whereas in case of differential feeding they sum up. This result is shown in Fig.6, where the input voltage is calculated for the CNT line, assuming a feeding current of 0.01mA and assuming that the two far ends are terminated on open circuits. This suggests that the tunneling resonances may be canceled out in multiconductor carbon lines, provided that they are ideally put in parallel and that the current flowing through them is exactly the same.

In Fig.7 the input voltages computed assuming a perfect common mode pair of currents is compared to that obtained by considering slight misbalances, expressed in terms of the relative difference $\Delta I = \left| (I_1 - I_2) \right| / \left| I_2 \right|$.

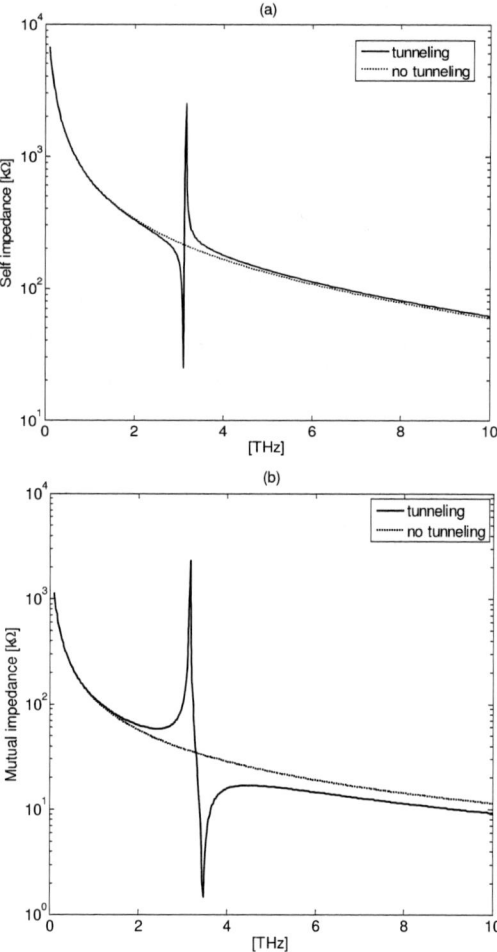

Figure 5. Absolute value of the self (a) and mutual (b) impedance for a coupled line made by two GNRs, with or without the tunneling effect.

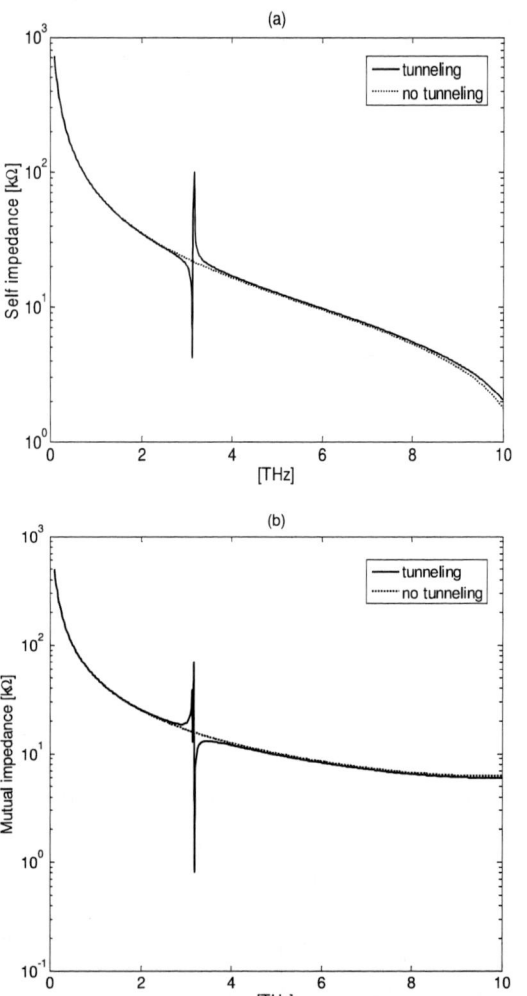

Figure 4. Absolute value of the self (a) and mutual (b) impedance for a coupled line made by two CNTs, with or without the tunneling effect.

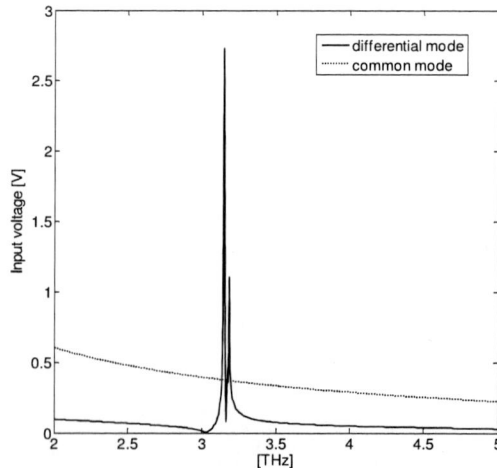

Figure 6. Input voltage of the CNT line fed by either a differential mode and a common mode current of 0.01 mA, with the far ends left open.

978-1-4799-8720-7/15 $31.00 © 2015 IEEE   454

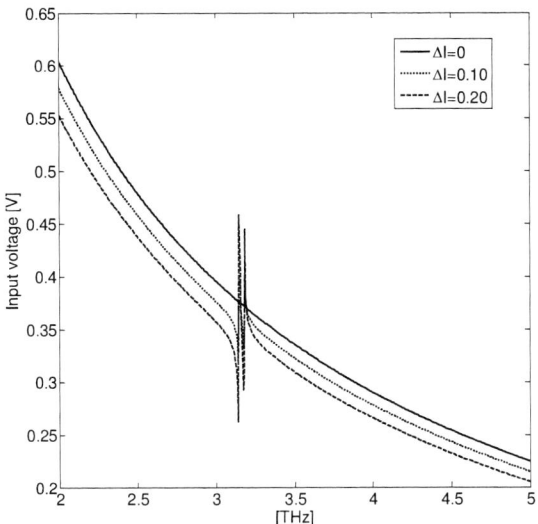

Figure 7. Input voltage of the CNT line fed by a common mode current of 0.01mA, with the far ends left open, in balanced and unbalanced cases.

V. CONCLUSIONS

The THz response of nano-interconnects made by carbon nanotubes and graphene nanoribbons have been investigated, by means of an equivalent lossy transmission line model. In particular, the TL model has been reformulated to include the coupling between the lines due to tunneling effect. In this frame, this phenomenon is described by additional terms for the resistance, inductance and capacitance matrices which introduce a spatial and frequency dispersion.

The analysis in the THz range highlights the presence of resonances and dispersion even for line lengths smaller than those associated to the well-known THz resonances. Therefore, the tunneling effect strongly reduces the frequency range of utilization of carbon interconnects (for a fixed length), or equivalently reduces their lengths (for fixed operating frequencies). This unwanted coupling may be perfectly canceled out if the signal lines carry exactly the same currents, but should be taken into account in case of misbalance between such currents.

REFERENCES

[1] P. H. Siegel, "Terahertz technology," IEEE Trans. Microwave Theory Tech., Vol. 50, pp.910–928, 2002.

[2] S. Kiyomi, Teraherts optoelectronics, Berlin: Springer, 2005.

[3] J. A. Berres and G. W. Hanson, "Multiwall Carbon Nanotubes at RF-THz Frequencies: Scattering, Shielding, Effective Conductivity, and Power Dissipation," IEEE Trans. Antennas and Prop., Vol. 59, pp.3098-3103, 2011.

[4] K. G. Batrakov, O. V. Kibis, P. Kuzhir, M. Rosenau da Costa, and M. E. Portnoi, "Terahertz processes in carbon nanotubes," Journal of Nanophotonics, Vol.4, p.041665, 2010.

[5] P.J. Burke, "AC performance of nanoelectronics: towards a ballistic THz nanotube transistor," Solid-state El., Vol.48, pp,1981-1986, 2004

[6] F. Rana, "Graphene Terahertz Plasmon Oscillators," IEEE Trans. on Nanotechnology, Vol. 7, pp. 91-99, 2008.

[7] M. V. Shuba, G. Y. Slepyan, S. A. Maksimenko, C. Thomsen, and A. Lakhtakia, "Theory of multiwall carbon nanotubes as waveguides and antennas in the infrared and the visible regimes," Phys. Rev. B, vol. 79, no. 15, pp. 155403–155403, 2009.

[8] G. W. Hanson, "Fundamental transmitting properties of carbon nanotube antennas," IEEE Trans. Ant. Prop. Vol.53, pp.3426–3435, 2005

[9] R. Mendis, and D. Grischkowsky, "THz interconnect with low loss and low group velocity dispersion," IEEE Microwave & Wireless Components Lett., Vol. 11, pp. 444-446, 2001.

[10] C. Yeh, F. Shimabukuro, and P. H. Siegel, "Low-loss terahertz ribbon waveguides," Applied Optics, Vol. 44, pp.5937-5946, 2005.

[11] A. Maffucci, G. Miano and F. Villone, "Performance Comparison Between Metallic Carbon Nanotube and Copper Nano-interconnects," IEEE Trans. on Advanced Packag., Vol.31, pp. 692-699, Nov. 2008.

[12] A. Naeemi and J. D. Meindl, "Performance Modeling for Single- and Multiwall Carbon Nanotubes as Signal and Power Interconnects in Gigascale Systems," IEEE Trans. on Electron Devices, Vol. 55, N.10, pp. 2574-2582, Oct. 2008.

[13] H. Li, N. Srivastava, J.-F. Mao, W.-Y. Yin, K. Banerjee, "Carbon Nanotube Vias: Does Ballistic Electron–Phonon Transport Imply Improved Performance and Reliability?," IEEE Trans. on Electron Devices, vol.58, n.8, pp.2689 – 2701, 2011.

[14] H. Li, C. Xu, N. Srivastava, and K. Banerjee, "Carbon Nanomaterials for Next-Generation Interconnects and Passives: Physics, Status, and Prospects," IEEE Trans. on Electr. Dev., Vol.56, pp.1799-1821, 2009.

[15] A. G. Chiariello, A. Maffucci, G. Miano, "Circuit Models of Carbon-based Interconnects for Nanopackaging, IEEE Trans. on Components, Packaging and Manufacturing, Vol.3, pp.1926-1937, Nov.2013

[16] I. Soga, D. Kondo, Y. Yamaguchi, T. Iwai, M. Mizukoshi, Y. Awano, K. Yube, and T. Fujii "Carbon nanotube bumps for LSI interconnect," in Proc. of Electronic Comp. and Techn. Conf., pp.1390-1394, 2008.

[17] X. Chen, et al., "Fully Integrated Graphene and Carbon Nanotube Interconnects for Gigahertz High-Speed CMOS Electronics," IEEE Trans. Electr. Devices, Vol.57, pp.3137-3143, 2010.

[18] L. Ding, S. Liang, T. Pei, Z. Zhang, S. Wang, W. Zhou, J. Liu, L-M. Peng, "Carbon nanotube based ultra-low voltage integrated circuits: Scaling down to 0.4 V," Applied Physics Letters, vol. 100, n.26, pp.263116 - 263116-5, 2012.

[19] M.M. Shulaker, et al., "Carbon nanotube computer," Nature, Vol.501, 526-530, 2013.

[20] C. Brun, et al., "Carbon Nanostructures Dedicated to Millimeter-Wave to THz Interconnects," IEEE Trans. on Terahertz Science and Technology, in press, available in on-line preprint, 2015.

[21] P.J. Burke, "An RF circuit model for carbon nanotubes," IEEE Trans. on Nanotechnology, Vol.2, N. 1, pp.55-58, 2003.

[22] A. Naeemi and J. D. Meindl, "Compact physical models for multiwall carbon-nanotube interconnects," IEEE Electron. Devices Lett., vol. 27, pp. 338–340, May 2006.

[23] C. Xu, H. Li, and K. Banerjee, "Modeling, analysis, and design of graphene nano-ribbon interconnects," IEEE Trans. Electron Devices, vol.56, pp.1567–1578, 2009.

[24] G.Miano, C.Forestiere, A. Maffucci, S.A. Maksimenko, G. Slepyan, "Signal propagation in single wall carbon nanotubes of arbitrary chirality," IEEE Trans. on Nanotech., vol.10, pp. 135-149, Jan. 2011.

[25] C. Forestiere, A. Maffucci, S. A. Maksimenko, G. Miano, G.Y. Slepyan, "Transmission Line Model for Multiwall Carbon Nanotubes with Intershell Tunneling," IEEE Trans. on Nanotechnology, vol.11, no.3, pp. 554-564, May 2012.

[26] A. Maffucci, G. Miano, "A general frame for modeling the electrical propagation along graphene nanoribbons, carbon nanotubes and metal nanowires," Computer Model. & New Techn., Vol.19, p-8-14, 2015.

[27] G. Y. Slepyan, M. V. Shuba, and S. A. Maksimenko, "Terahertz conductivity peak in composite materials containing carbon nanotubes: Theory and interpretation of experiment," Phys. Rev. B, Vol. 81, pp.035020, 2010.

2015 IEEE Computer Society Annual Symposium on VLSI

Terahertz Applications of Carbon Nanotubes and Graphene Nanoribbons

M. E. Portnoi, V. A. Saroka

School of Physics
University of Exeter
Exeter, United Kingdom
m.e.portnoi@exeter.ac.uk

R. R. Hartmann

Physics Department
De La Salle University
Manila, Philippines
richard.hartmann@dlsu.edu.ph

O. V. Kibis

Applied and Theoretical Physics
Novosibirsk State Tech. University
Novosibirsk, Russia
oleg.kibis@nstu.ru

Abstract—**The results of the theoretical study of interband THz transitions in narrow-gap carbon nanotubes and graphene nanoribbons are reported. We consider dipole transitions across magnetically-induced gaps in armchair nanotubes, curvature-induced gaps in quasi-metallic nanotubes and edge-effect induced gaps in armchair nanoribbons. A giant enhancement of the transition matrix elements is discovered for all three types of nanostructures.**

Keywords—carbon nanotubes; graphene nanoribbons; THz transitions

I. Introduction

Creating reliable and portable coherent sources and sensitive detectors of terahertz (THz) radiation is one of the most formidable tasks of modern device physics [1]. Potential applications of THz spectroscopy range from medical imaging and security to astrophysics and cosmology. The unique position of the THz range in the gap between the parts of electromagnetic spectrum which are accessible by either electronic or optical devices, leads to an unprecedented diversity in approaches to bridging this gap [2-5]. One of the latest trends in THz technology [6] is to use carbon nanomaterials as building blocks of high-frequency devices. In particular, there are a growing number of proposals using carbon nanotubes for THz applications including several schemes [7-11] put forward by some of the authors of the present work.

In this paper we focus on interband THz transitions in graphene-based nanostructures. In Section II we show that a magnetic field applied along the axis of a metallic (armchair) single-wall carbon nanotube (SWNT) not only opens a band gap, which typically corresponds to the THz frequency range, but also allows dipole optical transitions across this gap and provide a simple analytic expression for the probability of this transition. In Section III we discuss similar transitions across the curvature-induced gap in very narrow-gap (quasi-metallic) SWNTs. In Section IV we discuss hitherto overlooked dipole transitions across a band gap opened due to edge effects in armchair graphene nanoribbons (GNRs). We also discuss the practical implications of these effects as well as conditions for their experimental observation.

This work was supported by EU FP7 ITN NOTEDEV (Grant No. FP7-607521) and EU FP7 IRSES projects QOCaN (Grant No FP7-316432), CANTOR (Grant No FP7-612285), and InterNoM (Grant No FP7-612624).

II. Magnetically-Induced Gaps and Transitions

The Aharonov-Bohm effect [12] (also known as Ehrenberg-Siday-Aharonov-Bohm effect [13]) was one of the first phenomena discussed in relation to SWNTs [14]. However, a field of the order of 10^4 Tesla applied along the nanotube axis is needed to observe a single period in magnetically induced oscillations of a bandgap for a typical SWNT. In experimentally-attainable fields, more subtle effects, such as the magnetic brightening of "dark" excitons in semiconducting SWNTs [15] and the modulation of the magnetoresistance in field-effect transistors based on quasi-metallic SWNTs [16] have been observed. The most dramatic effect occurs when a realistic magnetic field is applied to a truly gapless highly-symmetric armchair [11] nanotube.

In armchair SWNTs a longitudinal magnetic field (field applied along the nanotube axis) opens an energy gap near the Fermi level. This gap can be easily calculated in the nearest-neighbour tight binding approximation [17], in which the influence of a magnetic field is accounted for by adding the number $f = \Phi / \Phi_0 = eBR^2 / (2\hbar)$ (here Φ is the magnetic flux through the nanotube, $\Phi_0 = h/e$ is the flux quantum and R is the nanotube radius) to the angular momentum quantum number l, characterizing the lowest conduction and highest valence band branches of the SWNT energy spectrum [5]. A similar procedure was justified for an arbitrary periodic potential by Luttinger [12] almost six decades ago. For an armchair (n,n) nanotube (see Refs. [17] for classification of SWNTs) the gap opened by the longitudinal magnetic field is given by [11,18]

$$\varepsilon_g = 2\gamma_0 \left| \sin\left(\frac{f}{n} \pi \right) \right|, \qquad (1)$$

where $\gamma_0 = 3\,\text{eV}$ is the commonly used tight-binding matrix element for graphene [17]. For a $(10,10)$ SWNT this gap corresponds to 1 THz when the nanotube is subjected to a field of approximately 6 T. For attainable magnetic fields, the gap

978-1-4799-8720-7/15 $31.00 © 2015 IEEE   456

grows with increasing both the magnetic field and the nanotube radius. Whereas the field dependence of the band gap is well-understood [14], the drastic change in the probability of optical transitions caused by the magnetic field has been completely overlooked. In the absence of a magnetic field, optical transitions between the top valence subband and the lowest conduction subband in armchair SWNTs are strictly forbidden by symmetry. However, the same longitudinal magnetic field, which opens a gap in the armchair SWNT spectrum, allows the dipole optical transitions. In the frame of the nearest-neighbor tight binding model, one can show that for an (n,n) armchair nanotube the squared matrix element of the velocity operator between the states at the edge of the gap opened by the magnetic field is given by a simple analytic expression [11,18]:

$$\left|\left\langle \Psi_n^v \left| \hat{v}_z \right| \Psi_i^c \right\rangle\right| = \frac{4}{3}\left[1 - \frac{1}{4}\cos^2\left(\frac{f}{n}\pi\right)\right]v_F^2, \quad (2)$$

where $v_F = 3\gamma_0 a_{C-C}/(2\hbar) \approx 9.8 \times 10^5$ m/s is the Fermi velocity of graphene (here $a_{C-C} = 1.42\,A$ is the nearest-neighbour distance between carbon atoms in graphene). For experimentally attainable magnetic fields, when the magnetic flux through the SWNT is much smaller than the flux quantum, the absolute value of the velocity operator is close to v_F. Equation (2) is relevant to the transitions between the highest valence subband and the lowest conduction subband only for $f \leq 1/2$, since for higher values of f the order of the nanotube subbands is changed. Notably, the same equation can be used to obtain the maximum value of the velocity operator in any armchair SWNT for the transitions polarized along its axis: this value cannot exceed $2v_F/\sqrt{3}$. The electron (hole) energy spectrum near the bottom (top) of the gap produced by the magnetic field is parabolic as a function of carrier momentum along the nanotube axis. This dispersion results in the van-Hove singularity in the reduced density of states, which in turn leads to a very sharp absorption maximum near the band edge and, correspondingly, to a very high sensitivity of the photocurrent to photon frequency.

The same effect can be used for the generation of a very narrow emission line having the peak frequency tuneable by the applied magnetic field. A population inversion can be achieved, for example, by optical pumping, with the exciting light polarized normally to the nanotube axis. of the current designations.

III. Curvature-Induced Gaps and Transitions

In the absence of curvature, in the Dirac point, optical transitions between the top valence subband and the bottom conduction subband in metallic SWNTs are strictly forbidden by symmetry within the simple zone-folding model of the π-electron graphene spectrum. However, for zigzag quasi-metallic SWNTs (which would be metallic in a zone-folding scheme), dipole optical transitions are indeed allowed due to the gap opened in their energy spectrum by intrinsic curvature,

which is of the order of THz [17, 18] for a typical quasi-metallic nanotube. These transitions rates are several orders of magnitude larger than the previously considered interband THz transitions calculated in a model that neglects curvature [9]. As with the armchair nanotube, a magnetic field applied along the nanotube can be used to modify the optical selection rules. However, in quasi-metallic tubes, the magnetic field creates two different size bandgaps, and therefore, two peaks in the absorption spectra. Arrays of armchair and quasi- metallic SWNTs could be used as the building blocks of THz radiation detectors, which would have a high sensitivity in the photocurrent to photon frequency. Furthermore, since the bandgap of such SWNTs can be controlled by the size of the applied magnetic field, they are tuneable. Again the application for THz detection requires a very low level of doping and low temperatures. The same set-up can be used for THz emitters if a quasi-metallic nanotube is excited optically with the light polarized along its axis. Photoexcited carriers will quickly relax in energy to the states near the curvature-induced gap maintaining inversion of population. A suitably arranged SWNT array in a THz cavity can be then used as a source of the coherent THz radiation under the carefully designed conditions when THz gain exceeds free-carrier (plasmonic) losses, which are known to be quite strong in metallic SWNTs [19].

IV. Gaps and Transitions in Graphene Nanoribbons

Graphene nanoribbons can be viewed as unrolled SWNTs. The main consequence of such unrolling is the formation of the edges. The presence of edges in GNRs results in electrons being confined in the direction perpendicular to the ribbons axis instead of being effectively free in SWNTs due to the possibility of circular motion in the equivalent direction perpendicular to the tube axis. This fundamental difference between two quasi-one-dimensional structures leads to different types of boundary conditions imposed on their electron wave functions. Thus, electrons of a SWNT are described by angularly-periodic wave functions, whereas electrons of a GNR are described by finite wave functions vanishing at the region outside the ribbon. Nevertheless, as was noticed by White [20] the electronic properties a GNR are equivalent to those of the SWNT obtained by rolling up of the twice wider ribbon. The only difference is a double degeneracy of the SWNT bands compared to the GNR bands and the absence of some bands above 1 eV in the GNR band structure compared to the SWNT band structure. In particular, the low energy spectra of zigzag SWNTs $(n,0)$ are identical to electronic spectra of armchair GNR containing $n+1$ pairs of carbon atoms in the unit cell. Moreover, for each quasi-metallic zigzag SWNT, there is a quasi-metallic armchair GNR. In spite of the absence of the curvature, metallic GNRs are indeed quasi-metallic. They have a small gap of about several tenth of meV due to the relaxation of C-C bonds at the edge of the ribbon [21]. Such a change of edge bonds can be incorporated into the tight-binding model of pi-electrons as a hopping integral correction [22, 23].

With the help of the analytical model developed in Ref. [23], we calculated velocity matrix elements characterizing dipole transitions of quasi-metallic armchair GNRs with and

978-1-4799-8720-7/15 $31.00 © 2015 IEEE

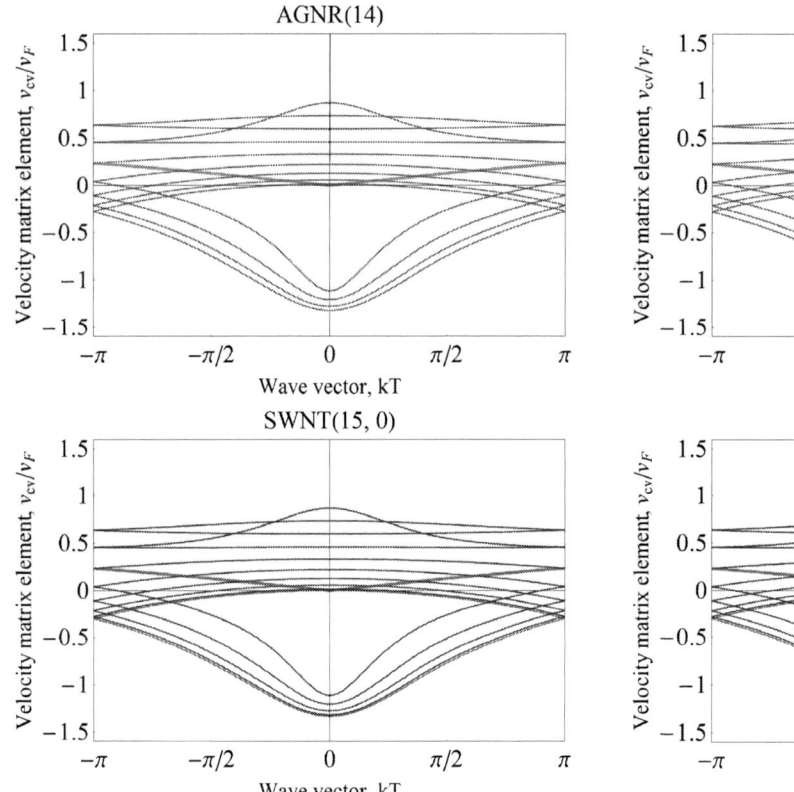

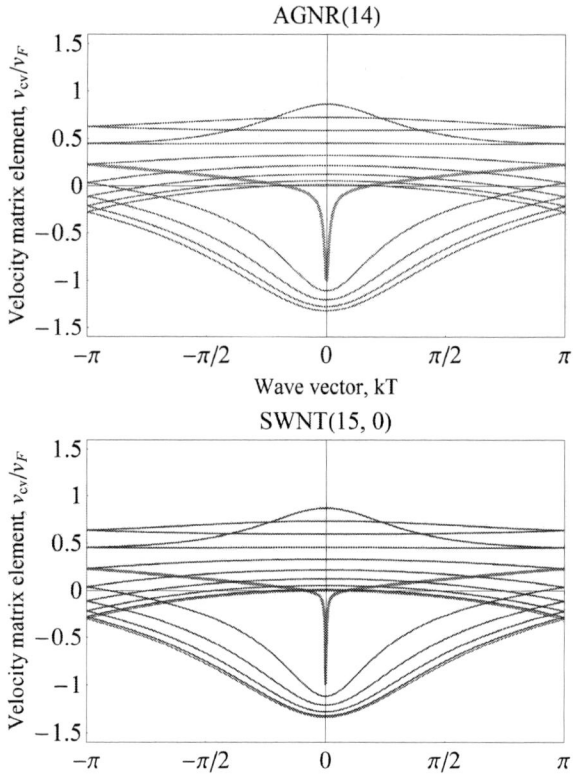

Fig. 1. Velocity matrix element as a function of wave vector in the first Brillouin zone for armchair GNR and zigzag SWNT without taking into account edge and curvature effects. Velocity matrix element between highest valence and lowest conduction subband is highlighted by red color.

Fig. 2. Velocity matrix element as a function of electron wave vector in the first Brillouin zone for armchair GNR and coresponding zigzag SWNT with the edge and curvature effects taken into account. Velocity matrix element between highest valence and lowest conduction subband is highlighted by red color.

without taking into account the edge effect and compare the results to calculations for quasi-metallic zigzag SWNTs with and without taking into account the curvature effect of the tube surface.

As can be seen in Fig.1, the above mentioned equivalence of GNRs and SWNTs with respect to electronic properties preserves also in optical properties. All curves for the GNR have equivalent ones in the plot for the SWNT. However, there are few curves in the bottom part of the plot for the SWNT that do not correspond to any curves for the GNR. Moreover, although it is unnoticeable in the Figure, all curves for SWNT corresponding to curves for the GNR are double degenerate. It is also seen from Fig.1 that if one does not allow for perturbations like curvature and edge effects the optical transitions between highest valence and lowest conduction sub bands for quasi-metallic SWNT and GNR are forbidden at the centre of the Brillouin zone and increase monotonically towards the edge of the Brillouin zone.

As shown in Fig. 2, both above mentioned effects do not affect the majority of curves for velocity matrix elements between valence and conduction subbands, except those closest to the Fermi level. However, even in this case the effect is

confined to the narrow region around the centre of the Brillouin zone. Nevertheless, for this particular region the observed changes are extremely pronounced. The previously forbidden optical transition across the narrow band gap in the SWNT and the GNR is now allowed in the narrow region near the centre of the Brillouin zone, and a pronounce peak in the velocity matrix element as a function of electron wave vector is observed. It is worth noting that in both cases the heights of the peaks are comparable to the greatest values of velocity matrix elements for other allowed optical transitions. Speaking of the differences, one must point out that the peak for GNR caused by the edge effects is slightly wider, than the peak for SWNT caused by the curvature effects.

Taking into account that the band gap of GNRs are dependent on the ribbon width, which can be easily controlled, for instance by using STM lithography [24], the frequency of the transition can be finely engineered. Up to recently GNRs had a problem in the control of the quality of their edges. Fortunately, this problem has been resolved and nowadays narrow armchair GNRs can be synthesized with an atomic precision [25]. Moreover, the recent experimental observation of room-temperature magnetic order in narrow zigzag GNRs of

7 nm width gives us evidence of high quality of ribbons in the latest samples and provides an indirect proof of the quality of their edges produced by STM lithography [26].

In addition to careful engineering of the GNR parameters, the frequency of the interband transition can be tuned by external fields. Whereas, for the case of SWNTs a strong magnetic field along the nanotube axis is needed, a more accessible electric field tuning [27] can be used for GNRs.

ACKNOWLEDGMENT

The authors are grateful to C. A. Downing for fruitful comments.

REFERENCES

[1] G. M. Lee and M. C. Wanke, "Searching for a solid-state terahertz technology," Science, vol. 316, pp. 64-65, April 2007.

[2] B. Ferguson and X. C. Zhang, "Materials for terahertz science and technology," Nat. Mater., vol. 1, pp. 26-33, September 2002.

[3] E. A. Avrutin and M. E. Portnoi, "Estimate of the lifetimes of nonequilibrium carriers in a semiconductor irradiated with heavy ions," Sov. Phys. Semicond., vol.22, pp. 968-970, August 1988.

[4] V. V. Kruglyak and M. E. Portnoi, "Generation of femtosecond current pulses using the inverse magneto-optical Faraday effect," Tech. Phys. Lett., vol. 31, 1047-1048, December 2005.

[5] S. A. Mikhailov, "Non-linear electromagnetic response of graphene," Europhys. Lett., vol. 79, 27002, July 2007.

[6] R. R. Hartmann, J. Kono, and M. E. Portnoi,"Terahertz science and technology of carbon nanomaterials," Nanotechnology, vol. 25, 322001, July 2014.

[7] O. V. Kibis, D. G .W. Parfitt, and M. E. Portnoi, "Superlattice properties of carbon nanotubes in a transverse electric field," Phys. Rev. B, vol. 71, 035411, January 2005.

[8] O. V. Kibis and M. E. Portnoi, "Carbon nanotubes: A new type of emitter in the terahertz range," Tech. Phys. Lett., vol. 31, pp. 671-672, August 2005.

[9] O. V. Kibis, M. Rosenau da Costa, and M. E. Portnoi, "Generation of terahertz radation by hot electrons in carbon nanotubes," Nano Lett., vol. 7, pp. 3414-3417, November 2007.

[10] M. E. Portnoi, O. V. Kibis, M. Rosenau da Costa, "Terahertz applications of carbon nanotubes," Superlattices Microstruct., vol.43, pp. 399-407, May 2008.

[11] M. E. Portnoi, M. Rosenau da Costa, O. V. Kibis, I. A. Shelykh, "Magnetically controlled terahertz absorption and emission in carbon nanotubes," Intern. J. Modern Phys. B, vol. 23, pp. 2846-2850, May 2009.

[12] Y. Aharonov and D. Bohm, "Significance of electromagnetic potentials in the quantum theory," Phys. Rev., vol. 115, pp. 485-491, August 1959.

[13] W. Ehrenberg, R. E. Siday, "The refractive index in electron optics and the principles of dynamics," Proc. Phys. Soc. B, vol. 62, pp. 8-21 January 1949.

[14] H. Ajiki and T. Ando, "Electronic states of carbon nanotubes," J. Phys. Soc. Jpn., vol. 62, pp. 1255-1266, April 1993.

[15] J. Shaver and J. Kono, "Temperature-dependent magneto-photoluminescence spectroscopy of carbon nanotubes: evidence for dark excitons," Laser & Photon. Rev., vol. 1, pp. 260-274, November 2007.

[16] G. Fedorov, A. Tselev, D. Jiménez, S. Latil, N. G. Kalugin, P. Barbara, D. Smirnov, and S. Roche, "Magnetically induced field effect in carbon nanotube devices," Nano Lett., vol. 7, pp. 960-964, March 2007.

[17] R. Saito, G. Dresselhaus, and M. S. Dresselhaus, Physical Properties of Carbon Nanotubes. London: Imperial College Press, 1998.

[18] R. R. Hartmann and M. E. Portnoi, Optoelectronic Properties of Carbon-based Nanostructures: Steering Electrons in Graphene by Electromagnetic Fields. Saarbrücken: LAP Lambert Academic Publishing, 2011.

[19] M. V. Shuba, A. G. Paddubskaya, A. O. Plyushch, P. P. Kuzhir, G. Ya. Slepyan, S. A. Maksimenko, V. K. Ksenevich, P. Buka, D. Seliuta, I. Kasalynas, J. Macutkevic, G. Valusis, C. Thomsen, and A. Lakhtakia, "Experimental evidence of localized plasmon resonance in composite materials containing single-wall carbon nanotubes," Phys. Rev. B, vol. 85, 165435, April 2012.

[20] C. T. White, J. Li, D. Gunlycke, and J. W. Mintmire, "Hidden one-electron interactions in carbon nanotubes revealed in graphene nanostrips," Nano Lett., vol. 7, pp. 825-830, February 2007.

[21] Y.-W. Son, M. L. Cohen, and S. G. Louie, "Energy gaps in graphene nanoribbons," Phys. Rev. Lett., vol. 97, 216803, November 2006.

[22] D. Gunlycke and C. White, "Tight-binding energy dispersions of armchair-edge graphene nanostrips," Phys. Rev. B, vol. 77, 115116, March 2008.

[23] H. Zheng, Z. Wang, T. Luo, Q. Shi, and J. Chen, "Analytical study of electronic structure in armchair graphene nanoribbons," Phys. Rev. B, vol. 75, 165414, April 2007.

[24] L. Tapasztó, G. Dobrik, P. Lambin, and L. P. Biró, "Tailoring the atomic structure of graphene nanoribbons by scanning tunnelling microscope lithography," Nat. Nanotechnol., vol. 3, pp. 397-401, June 2008.

[25] J. Cai, P. Ruffieux, R. Jaafar, M. Bieri, T. Braun, S. Blankenburg, M. Muoth, A. P. Seitsonen, M. Saleh, X. Feng, K. Müllen, and R. Fasel, "Atomically precise bottom-up fabrication of graphene nanoribbons," Nature, vol. 466, pp. 470-473, July 2010.

[26] G. Z. Magda, X. Jin, I. Hagymási, P. Vancsó, Z. Osváth, P. Nemes-Incze, C. Hwang, L. P. Biró, and L. Tapasztó, "Room-temperature magnetic order on zigzag edges of narrow graphene nanoribbons," Nature, vol. 514, pp. 608-611, October 2014.

[27] V. A. Saroka, K. G. Batrakov, V. A. Demin, and L. A. Chernozatonskii, "Band gaps in jagged and straight graphene nanoribbons tunable by an external electric field," J. Phys. Condens. Matter, vol. 27, 145305, March 2015.

978-1-4799-8720-7/15 $31.00 © 2015 IEEE

2015 IEEE Computer Society Annual Symposium on VLSI

Emerging Non-Volatile Memory Technologies Exploration Flow For Processor Architecture

Sophiane Senni[1,2], Lionel Torres[2], Gilles Sassatelli[2]
and Abdoulaye Gamatie[2]
LIRMM – UMR CNRS 5506 – University of Montpellier
Montpellier, France
{lastname[2]}@lirmm.fr

Bruno Mussard
Crocus technology
Rousset, France
{ssenni[1], bmussard}@crocus-technology.com

Abstract— Most die area of today's systems-on-chips is occupied by memories. Hence, a significant proportion of total power is spent on memory systems. Moreover, since processing elements have to be fed with instructions and data from memories, memory plays a key role for system's performance. As a result, memories are a critical part of future embedded systems. Continuing CMOS scaling leads to manufacturing constraints and power consumption issues for the current three main memory technologies, i.e. SRAM, DRAM and FLASH, which compromises further evolution in upcoming technology node. To face these challenges, new non-volatile memory technologies emerged in recent years. Among these technologies, magnetic RAM (MRAM) is a promising candidate to replace existing memories since it combines non-volatility, high scalability, high density, low latency, and low leakage. This paper describes an evaluation flow to explore next generation of the memory hierarchy of processor-based systems using new non-volatile memory technologies.

Keywords—MRAM, NVM, Memory hierarhy, VLSI, SoC, Embedded Systems

I. Introduction

Because of its low access time, SRAM is currently the most suitable memory technology to design the upper level of cache memories to reach the best performance, particularly for multiprocessor architecture. Current issue of SRAM decreasing the technology node is the high leakage current leading to high power dissipation. For decades, DRAM has been used for main memory since it is slower but has higher density than SRAM. This technology is also power consuming due to its mandatory refresh policy. In addition, DRAM faces to manufacturing constraints for the most advanced node. At a lower level of the memory hierarchy, FLASH memory is used for its high storage and non-volatility capabilities. To overcome performance and power challenges of this multi-core era, new non-volatile memory technologies (NVMs) emerged over the past few years. While being non-volatile, MRAM is suitable to become a universal memory as it combines good scalability, low leakage, low access time and high density. However, despite the many attractive features of MRAM, two main issues are still under intensive investigation: write latency and write energy. Compared to SRAM, MRAM write latency is around three to ten times

higher, as well as MRAM write energy due to its high current requirement to switch the bit cell.

This paper presents an exploration flow to evaluate integration of MRAM into the memory hierarchy of processor architectures. Both performance and energy are analyzed using both architecture-level and circuit-level tools. Useful information about the memory activity is extracted to better understand the results.

II. MRAM Basics

MRAM bit is a magnetic tunnel junction (MTJ) which consists of two ferromagnetic layers separated by a thin insulating barrier. The information is stored as the magnetic orientation of one of the two layers, called the free layer (FL). The other layer, called the reference layer (RF), provides the fixed reference magnetic orientation required for reading and writing. Four methods have been proposed to switch the orientation of the FL: toggle [1], thermally assisted switching (TAS) [2], spin transfer torque (STT) [3] and spin orbit torque (SOT) [4].

The toggle write scheme consists of a specific timing sequence of the write-current pulses through the conductive lines to switch the magnetic orientation of the FL to its opposite direction.

TAS-MRAM uses an anti-ferromagnetic layer to block the magnetic orientation of the FL under a threshold temperature. To switch the bit cell, a select transistor provides a flow of current to heat the MTJ above the blocking temperature thereby enabling storage of new information thanks to application of a magnetic field.

STT-MRAM uses the spin transfer torque effect to switch the magnetic orientation of the FL. A highly spin polarized current flowing through the MTJ causes a "torque" applied by the injected electron spins on the magnetization of the FL.

SOT is the most recent MRAM technology. Contrary to STT-MRAM, this new technique uses a three-terminal structure to separate the read and write paths. The physical effect responsible for the reversal of magnetization of the FL is not yet fully understood. According to some authors, the Rashba effect [5] or the spin hall effect [6] could explain the switch in magnetization of the storage layer.

978-1-4799-8720-7/15 $31.00 © 2015 IEEE   460a

III. NVM EXPLORATION FLOW

To evaluate the impact of integrating MRAM into the memory hierarchy of processor architecture, a framework based on both circuit-level and architecture-level tools is needed (See Fig. 1). A circuit-level tool needs to provide characteristics of a complete memory circuit (i.e. including data array and peripheral circuits). An architecture-level tool simulates a complete processor-based system with its memory hierarchy.

For area, performance, and energy evaluations, the minimum information required is:

- Circuit-level requirements
  - Access latency (read/write)
  - Access energy (read/write)
  - Static power
  - Area
- Architecture-level requirements:
  - Execution time of the simulated applications
  - Amount of memory transactions for each level of the memory hierarchy (reads/writes)

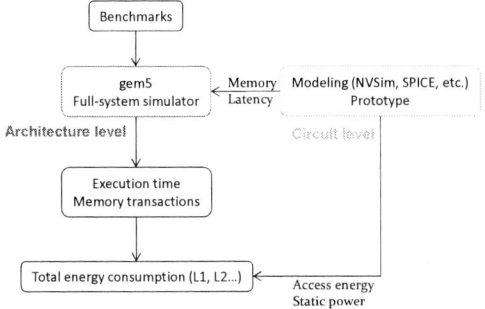

Fig. 1. NVM exploration framework

This section presents a framework based on gem5 [7], a processor architecture simulator widely used by the research community. gem5 currently supports most commercial Instruction Set Architectures (ISA) including ARM, ALPHA, MIPS, Power, SPARC, and x86. gem5 is able to simulate a complete processor-based system with devices and operating system in full system mode (i.e. nothing is emulated). The use of gem5 make it possible to define the total processor system architecture, including memory hierarchy specifications: cache size, cache and main memory latencies, etc. Execution time and memory transactions can be extracted for a given application, i.e. cache read/write accesses including cache hits and misses. In addition, the cache miss rate, the cache miss latency and the memory bandwidth can be monitored over time to better understand the activity of the memory. Hence, a fine-grain analysis of performance and energy results for each simulated workload is possible.

To calibrate the memory hierarchy in terms of access latency, NVSim [9] was used, a circuit-level model for NVM performance, energy, and area estimation, which supports different NVM technologies including STT-MRAM (planar), resistive RAM (RRAM), phase-change RAM (PCRAM). It also models the volatile SRAM memory. NVSim needs two input files to estimate the performance, energy and area of a complete memory circuit:

- An input file specifying the memory cell properties (memory technology, cell area, etc.)
- An input file specifying the memory module parameter (cache or RAM, memory size, etc.)

Using NVSim, a rapid estimation of electrical features of a complete memory chip is possible including read/write access time, read/write access energy and static power. The estimation error is ≤24% [9]. If more precise values are desired, the results of the SPICE simulation of a design or the electrical features of a real prototype can be easily used.

IV. EXPERIMENTAL SETUP

As a case study, some applications of SPLASH-2 benchmark suite [10] were used to explore STT-MRAM and TAS-MRAM based caches for quad-core processor ARM architecture. SPLASH-2 workloads are mostly in the area of high performance computing (HPC). Table I shows the architecture configuration and Table II provides details on the simulated workloads.

TABLE I.   ARCHITECTURE CONFIGURATION

Hierarchy Level	Configuration
Processor	4-core, 1 GHz, 32-bit RISC ARMv7 (Linux OS)
L1 I/D cache	Private, 32kB, 4-way associative, 64B cache line
L2 cache	Shared, 512kB, 8-way associative, 64B cache line
Main memory	DRAM, DDR3, 100-cycle latency

TABLE II.   SPLASH-2 WORKLOADS

Workloads	Input set
barnes	16K Particles, Timestep = 0.25, Tolerance 1.0
fmm	16K Particles, Timestep = 5
fft	2^{20} total complex data points
lu1	Contiguous blocks, 512x512 Matrix, Block = 16
lu2	Non-contiguous blocks, 512x512 Matrix, Block = 16
ocean1	Contiguous partitions, 514x514 Grid
ocean2	Non-contiguous partitions, 258x258 Grid
radix	4M Keys, Radix = 4K

Note that the cache latency parameters in gem5 were calibrated using simulation results of NVSim for both SRAM and STT-MRAM, while for TAS-MRAM, outcomes from a real prototype were used thanks to support provided by Crocus Technology. To take into account the state of the art of MRAM technology and to evaluate performance and energy fairly, we compare 45 nm STT-MRAM results with a baseline 45 nm SRAM cache, and 130 nm TAS-MRAM results with a baseline 120 nm SRAM cache.

V. PERFORMANCE EVALUATION

Table III shows the latencies of a 512 kB L2 cache for the three memory technologies concerned. As expected, both MRAM technologies have higher write latency than SRAM. Regarding hit latency, STT-MRAM is faster than SRAM (45 nm). This is not surprising since STT-MRAM is denser than SRAM. As a result, for the same capacity, the total L2 cache area for STT-MRAM is smaller than for SRAM (see Table III) resulting in a shorter bit line delay. This difference in hit latency in favor of STT-MRAM is only noticeable in the case of large cache capacity. On the other hand, TAS-MRAM write and hit latencies are respectively 8.5 and 6 times higher than those of SRAM (120 nm).

TABLE III. 512 KB L2 CACHE FEATURES

Technology	Latency		Energy			Cache area	
	Hit (ns)	Write (ns)	Hit (nJ)	Write (nJ)	Leakage (mW)	Total (mm²)	Cell (F²)
45 nm SRAM	4.28	2.87	0.27	0.02	320	1.36	146
45 nm STT	2.61	6.25	0.28	0.05	23	0.82	57
120nm SRAM	5.95	4.14	1.05	0.08	82	9.7	146
130 nm TAS	35	35	1.96	4.62	10	11.7	35

TABLE IV. 32 KB L1 CACHE FEATURES

Technology	Latency		Energy			Cache area	
	Hit (ns)	Write (ns)	Hit (nJ)	Write (nJ)	Leakage (mW)	Total (mm²)	Cell (F²)
45 nm SRAM	1.25	1.05	0.024	0.006	22	0.091	146
45 nm STT	1.94	5.94	0.095	0.04	3.3	0.117	57

The difference between the latency parameter in SRAM and MRAM will of course depend on the frequency used by the processor. In this study, the frequency used for the processor was 1GHz. Table V shows the access latencies in terms of CPU cycles for L1 and L2. Since TAS-MRAM was evaluated only for L2 cache, L1 latencies for TAS-MRAM and the baseline 120 nm SRAM are not shown.

TABLE V. CACHE LATENCY FOR A 1GHZ PROCESSOR

Technology	Latency (CPU cycle)			
	32kB L1		512kB L2	
	Hit	Write	Hit	Write
45nm SRAM	2	2	5	3
45nm STT-MRAM	2	6	3	7
120nm SRAM	-	-	6	5
130nm TAS-MRAM	-	-	35	35

Fig. 2 shows the execution time of SPLASH-2 workloads for both STT-MRAM and TAS-MRAM based L2 caches. Fig. 2 shows that the performance of STT-MRAM-based L2 scenario is similar and sometimes better (*ocean1, ocean2*) than the baseline. This is because STT-MRAM has a smaller hit latency than its SRAM equivalent.

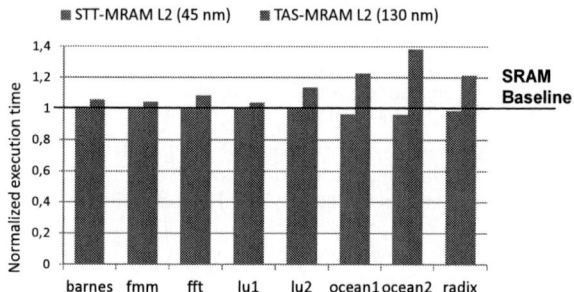

Fig. 2. Execution time with the MRAM-based L2 cache

TAS-MRAM-based L2 performance penalties from 3% (for *lu1*) to 38% (for *ocean2*) were observed. This is understandable since TAS-MRAM has higher access latency than STT-MRAM for both read and write operations. To better understand these results, the L2 cache miss rate was monitored over time (see Fig. 3). For *ocean2*, a high L2 cache miss rate is observed explaining the high penalty on the execution time using TAS-MRAM. For other workloads, such as *barnes*, the small penalty on the execution time is justified by a lower L2 cache miss rate compared to the *ocean2* workload.

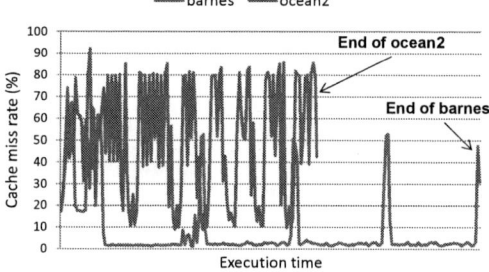

Fig. 3. L2 cache miss rate for barnes and ocean2 workloads

Fig. 4 shows the execution time with different configurations of STT-MRAM-based L1 cache: a first scenario in which both the I-Cache and D-Cache are based on STT-MRAM, a second scenario with STT-MRAM-based I-Cache only and, a third scenario with STT-MRAM-based D-Cache only.

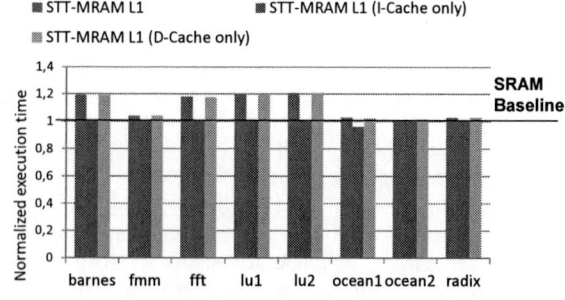

Fig. 4. Execution time of MRAM-based L1 cache

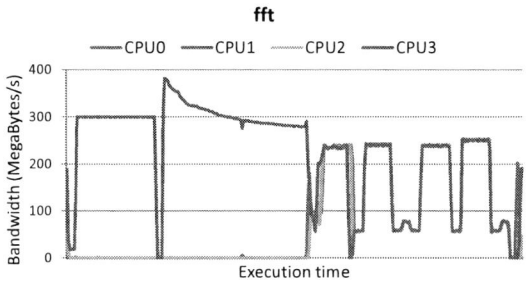
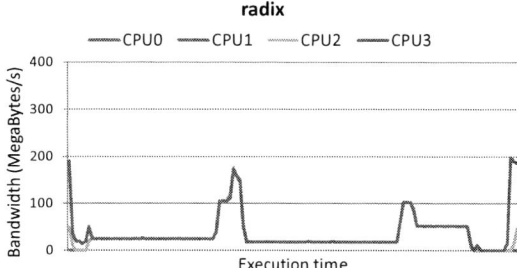

Fig. 5. D-Cache write bandwidth for fft and radix workloads

As already observed in Table V, the L1 hit latency is the same (in terms of CPU cycles) with both technologies. Therefore, STT-MRAM is slower than SRAM only in write operations. Since I-Cache is read only, replacing SRAM by STT-MRAM only in the I-Cache does not affect performance. Penalties are only noticed in the other scenarios in which the D-Cache is based on STT-MRAM. For some workloads (*barnes*, *fft*, *lu1*, *lu2*), using STT-MRAM in the D-cache reduces overall performance by around 20% due to its high write latency. For others, such as *radix*, the execution time penalty is very small, even with a STT-MRAM-based D-Cache. This can be explained by analyzing the cache write bandwidth of the D-Cache (see Fig. 5), which clearly shows that D-Cache is more frequently accessed in write for *fft* than for *radix* (for example). Hence, the impact of the long write latency of STT-MRAM on the execution time is more visible for the *fft* workload. Overall, the simulation results showed that the execution time penalty of STT-MRAM-based D-Cache does not exceed 21%, because the L1 cache is much more accessed in read for the simulated workloads.

VI. ENERGY EVALUATION

Table III provides energy consumption of SRAM, STT-MRAM and TAS-MRAM based L2 caches, while Table IV shows energy consumption of SRAM and STT-MRAM based L1. Regarding L2 energy consumption, using MRAM instead of SRAM results in higher write energy for both STT-MRAM and TAS-MRAM. STT-MRAM hit energy is very similar to that of SRAM, whereas a TAS-MRAM hit consumes around 2 times more energy than SRAM. Concerning L1 energy consumption, STT-MRAM consumes around 4 times and 7 times more energy than SRAM for hit and write operations, respectively.

However, in terms of static power, MRAM has a considerable advantage over SRAM (see Table III and IV): a 45 nm STT-MRAM-based L2 consumes over one order of magnitude less power than a 45 nm SRAM-based L2, while a TAS-MRAM-based L2 consumes around 8 times less power than a 120 nm SRAM-based L2. For L1 cache, a significant gain in static power is also obtained by replacing SRAM with STT-MRAM due to the zero leakage of the MTJ. This is because most of the static power of memories comes from cell arrays. Since MRAM cell has zero standby power and the CMOS access transistor does not require power supply, all the

static power in MRAM-based memory is due to peripheral circuitry such as address decoding, drivers and sense amplifiers.

Fig. 6 and Fig. 7 display total L2 energy consumption and total L1 energy consumption, respectively. Concerning L2, simulation results showed using STT-MRAM is 92% more energy efficient, on average, than SRAM, and using TAS-MRAM, 63% energy gain on average were observed. For total L1 energy consumption (i.e. including for the L1 cache of each core), replacing SRAM with STT-MRAM does not gain as much energy as with the L2 cache. The reason is the L1 cache is much more accessed than L2 cache. As a result, the dynamic energy impact of STT-MRAM is more visible. It will be recalled that previous analysis showed that STT-MRAM consumes respectively around 4 times and 7 times more energy than SRAM for hit and write operations in L1 (See Table IV). Fig. 7 shows average energy gains of 38% for STT-MRAM in both I-Cache and D-Cache, of 9% for STT-MRAM in I-Cache only, and of 24% for STT-MRAM in D-Cache only. This notable difference in energy consumption between the two technologies is explained by the low leakage power of MRAM compared to SRAM. Simulations results showed that replacing SRAM with MRAM can dramatically reduce the total energy consumption in the cache. This make MRAM-based cache memory an attractive alternative for energy efficient systems since the performance remains reasonable.

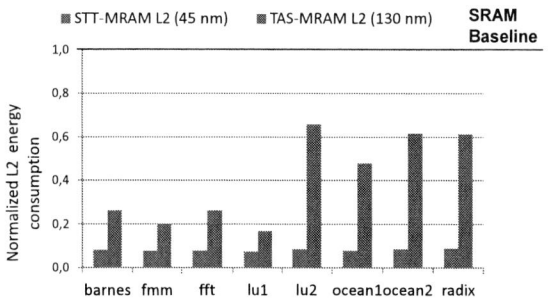

Fig. 6. MRAM-based L2 energy consumption

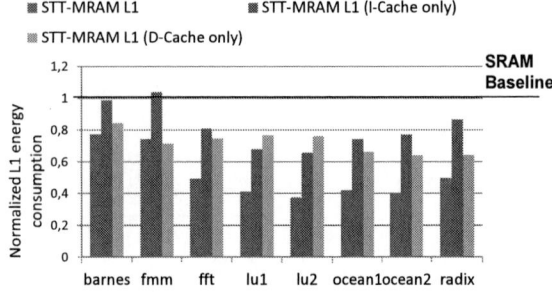

Fig. 7. MRAM-based L1 energy consumption

To better understand the large variations observed in L2 energy consumption between the simulated workloads using TAS-MRAM, the cache bandwidth is monitored over time (see Fig. 8) to see how often the L2 is accessed in read and write (in Bytes per second). Fig. 6 shows for instance a notable difference in energy consumption between *lu1* and *lu2* using TAS-MRAM. This is because the L2 read and write bandwidths for *lu1* are significantly lower than those for *lu2*. As a result, total TAS-MRAM-based L2 energy consumption is lower for *lu1* than for *lu2*. This kind of analysis (i.e. the cache bandwidth analysis) can also be done for the L1 cache to better understand the results on the energy consumption between the simulated workloads. It is not shown in this study for the sake of brevity.

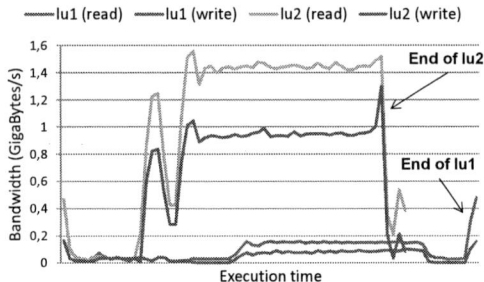

Fig. 8. L2 bandwidth for lu1 and lu2 worloads

VII. EXPLORATION FOR DIFFERENT NUMBER OF CORES

This section aims at analyzing the energy impact of MRAM-based cache when the number of cores is changed: quad-core, dual-core, and single-core. The results shown in this section are the average L1/L2 energy consumption over all the simulated workloads. Performance analysis is not detailed because simulation results reveal that the execution time penalty of MRAM-based cache (compared to the baseline) does not change significantly when increasing the number of cores from one to four. It will be recalled that the L2 is shared for multi-core architectures.

Fig. 9 and Fig. 10 depict respectively total L2 energy consumption and total L1 energy consumption for 4-core, 2-core and 1-core processor architectures. Substantial variations are noticed when the number of cores is changed. Regarding L2 cache in Fig. 9, changing from 4-core to 1-core architecture increases the energy consumption gain by 14% using TAS-MRAM instead of SRAM. On the other hand, no significant change is noticed with STT-MRAM-based L2 when the number of cores is changed.

Fig. 10 shows that when SRAM is replaced by STT-MRAM in L1, changing from 4-core to 1-core architecture increases the energy consumption gain by 18% for STT-MRAM-based I-Cache only, by 4% for STT-MRAM-based D-Cache only, and by 24% for STT-MRAM in both I-Cache and D-Cache.

To better understand this trend (i.e. the energy consumption gain over SRAM-based cache increases when the number of cores is reduced), the cache bandwidth is monitored over time (see Fig. 11) for 4-core, 2-core and 1-core architecture. The analysis is done only for the L2 and only for one workload, since analysis for L1 and for other workloads result in the same conclusions.

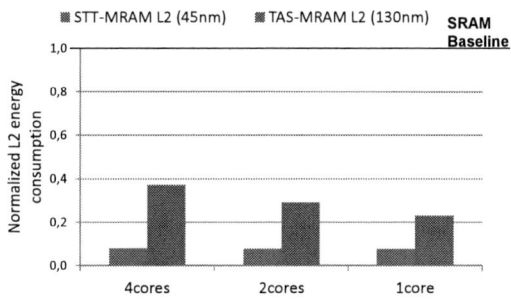

Fig. 9. MRAM-based L2 energy consumption for different number of cores

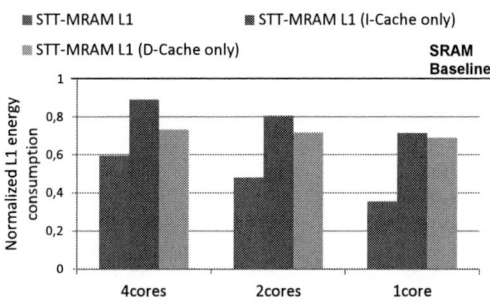

Fig. 10. MRAM-based L1 energy consumption for different number of cores

978-1-4799-8720-7/15 $31.00 © 2015 IEEE   460e

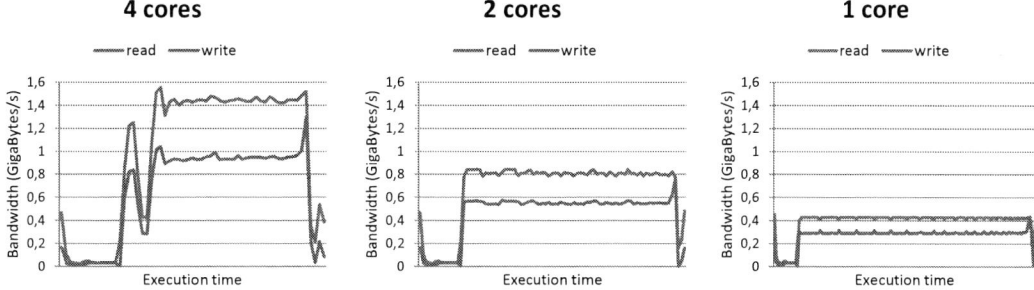

Fig. 11. L2 bandwidth for different number of cores (lu2 workload)

Fig.11 shows that the L2 bandwidth is reduced by around 2 when the number of cores is decreased from 4 to 2, and from 2 to 1. The dynamic part of total L2 energy consumption then is lower for 1-core than for 4-core architecture. Consequently, the loss due to the high dynamic energy of MRAM is reduced. This explains the higher energy gain using MRAM instead of SRAM in L2 for 1-core than for 2-core or 4-core architecture (see Fig. 9). This is particularly visible for TAS-MRAM because it has higher dynamic energy than SRAM for both read and write, whereas for STT-MRAM, this is the case only for writes.

VIII. RELATED WORK

A few studies on integrating NVMs into the memory hierarchy of processor architectures were made in [11], [12], and [13] also using the gem5 simulator. Contrary to these investigations, we do not limit the analysis of the performance and energy of the MRAM-based cache to a direct comparison with that of SRAM-based cache, but we rather observe and analyze memory activity over time to better understand the performance and energy issues.

IX. CONCLUSIONS

This paper presented a NVM exploration flow using the gem5 processor architecture simulator to evaluate integration of NVM into a memory hierarchy. Both performance and energy are analyzed using useful information about the memory traffic. Simulations show it is possible to significantly reduce the total energy consumption of caches thanks to the low leakage power of MRAM.

Concerning the future work, evaluation of MRAM at register level is envisaged to not only analyze the performance, energy, and area metrics, but also to explore new possible applications using non-volatile registers inside a processor.

ACKNOWLEDGMENT

The Authors wish to acknowledge all people from ADAC team at LIRMM and people from Crocus technology for their support in this work.

REFERENCES

[1] B.N. Engel et al., "A 4-Mb Toggle MRAM Based on a Novel Bit and Switching Method," in IEEE Transactions on Magnetics, vol. 41, no. 1, January 2005.

[2] I.L. Prejbeanu et al., "Thermally assisted MRAM," in Journal of Physics: Condensed Matter, vol. 19, no. 16, 2007.

[3] A.V. Khvalkovskiy et al., « Basic principles of STT-MRAM cell operation in memory arrays," in Journal of Physics D: Applied Physics, vol. 46, no. 7, 2013.

[4] D. Gambardella and I. M. Miron, "Current-induced spin-orbit-torques," in Philosophical transactions A of the Royal Society : Mathematical, Physical and Engineering Sciences, vol. 369, no. 1948, pp. 3175-3197, Aug. 2011.

[5] I. M. Miron et al., "Current-driven spin torque induced by the Rashba effect in aferromagnetic metal layer," in Nature Materials, vol. 9, pp. 230-234, Jan. 2010.

[6] L. Liu et al., "Spin-Torque Switching with the Giant Spin Hall Effect of Tantalum," in Science, vol. 336, no. 6081, May 2012.

[7] N. Binkert, S. Sardashti, R. Sen, K. Sewell, M. Shoaib, N. Vaish,M. D. Hill, D. A. Wood, B. Beckmann, G. Black, S. K. Reinhardt, A. Saidi, A. Basu, J. Hestness, D. R. Hower, and T. Krishna, "The gem5 simulator," ACM SIGARCH Computer Architecture News, vol. 39, no. 2, pp. 1–7, Aug. 2011.

[8] A. Butko, R. Garibotti, L. Ost, and G. Sassatelli, "Accuracy Evaluation of GEM5 Simulator System," in the proceedings of the 7th International Workshop on Reconfigurable Communication-Centric Systems-on-Chip (ReCoSoC), pp. 1–7, 2012.

[9] X. Dong, C. Xu, Y. Xie, and N. P. Jouppi, "NVSim: A Circuit-Level Performance, Energy, and Area Model for Emerging Nonvolatile Memory," IEEE Transactions on Computer-Aided Design of Integrated Circuits and Systems, vol. 31, no. 7, pp. 994-1007, Jul. 2012

[10] S.C. Woo, M. Ohara, E. Torrie, J.P. Singh and A. Gupta, "The SPLASH-2 Programs: Characterization and Methodological Considerations," in Proceedings of the 22nd Annual International Symposium on Computer Architecture, pp. 24–36, June 1995.

[11] J. Wang, X. Dong and Y. Xie, "OAP: an obstruction-aware cache management policy for STT-RAM last-level caches," in Design, Automation & Test in Europe Conference & Exhibition (DATE), pp. 847-852, March 2013.

[12] R. Bishnoi, M. Ebrahimi, F. Oboril and M. Tahoori, "Architectural Aspects in Design and Analysis of SOT-Based Memories," in the 19th Asia and South Pacific Design Automation Conference (ASP-DAC), pp. 700-707, January 2014.

[13] E. Arima et al., "Fine-Grain Power-Gating on STT-MRAM Peripheral Circuits with Locality-aware Access Control," in The Memory Forum (in conjounction with the 41st International Symposium on Computer Architecture), June 2014, unpublished.

978-1-4799-8720-7/15 $31.00 © 2015 IEEE

2015 IEEE Computer Society Annual Symposium on VLSI

Channel Modeling and Reliability Enhancement Design Techniques for STT-MRAM

Liuyang Zhang[1], Wang Kang[1], Youguang Zhang[1], Yuanqing Cheng[1], Lang Zeng[1],
Jacques-Olivier Klein[2], Weisheng Zhao[*1,2]

[1] Spintronics Interdisciplinary Center (SIC)
Beihang University
Beijing, 100191, China
liuyang.zhang@buaa.edu.cn

[2] Institut d'Electronique Fondamentale (IEF)
Univ. Paris-Sud, CNRS
Orsay, 91405, France
weisheng.zhao@buaa.edu.cn

Abstract—Spin transfer torque magnetic random access memory (STT-MRAM) has emerged as a potential candidate for the next-generation universal memory technology. However, many challenges still exist that block its commercialization and application. One of the main challenges is the reliability concerns, especially as technology scales down to nanometer nodes. For example, the intrinsic stochastic switching characteristics and asymmetry of the STT driven magnetic tunnel junction (MTJ), as well as the process, voltage and temperature (PVT) variations of the manufacturing process. Generally, there exists a technical gap between the device-level designers and system-level circuit or system designers for the STT-MRAM optimizations. To our knowledge, no efficient evaluation tools have been proposed to provide an effective link between them until now. In this paper, we aim to firstly give quantitative analyses of the reliability issues of STT-MRAM, taking into consideration the physical properties and manufacturing process. Secondly, based on these analysis results, a channel model from informative perspective would then be proposed. This model provides an effective simulation tool to evaluate the reliability performance of STT-MRAM. Finally, reliability enhancement design considerations and strategies are presented for the STT-MRAM circuit and system designers.

Keywords—Channel modeling; Nonvolatile memory; Reliability design ; STT-MRAM

I. INTRODUCTION

Spin transfer torque magnetic random access memory (STT-MRAM) has many merits, such as nonvolatility, lower power, high endurance and high integration density etc., and has been considered as one of the most potential candidates to replace conventional SRAM or (and) DRAM for the next-generation memory technologies[1-5]. However, our investigation shows that certain issues and challenges still exist and remain to be efficiently addressed before its wide commercialization and application. One of the main challenges of STT-MRAM is the reliability concerns, especially when under the nanometer technology nodes (e.g., below 65 nm). Not only the intrinsic physical properties of the magnetic tunnel junction (MTJ, basic element of STT-MRAM), e.g., stochastic switching characteristics and asymmetry of the spin transfer torque (STT) driven magnetic tunnel junction (MTJ), but also the manufacturing process and environment fluctuation, e.g., process, voltage and temperature (PVT) variations, all may result in STT-MRAM reliability issues.

Generally, a good memory design optimization should take multi-level (from device, circuit to system) co-design strategies into consideration. Especially for the upper system-level design, a good knowledge of the device-level or (and) system-level information helps to achieve better performance in a more efficient way. Unfortunately, to our knowledge, no such tools has been proposed until now to provide an effective link among different-level designers. The main work of this paper is to develop a channel model of STT-RMAM to bridge the technical gap among them. In this case, the system-level designers, e.g., error correction coding (ECC) designers, are able to obtain the device-level error information, thus designing effective reliability enhancement techniques with optimal overhead. On the other hand, the system-level reliability requirements and resource limits can also help the device-level designers, e.g., device designers, to optimize the manufacturing process and design corners.

The rest of this paper is organized as follows. Basics of STT-MRAM are introduced, and failure sources are analyzed in section II. The STT-MRAM operation channel model is proposed, and then simulation results are given in section III. In section IV, some design considerations and strategies for STT-MRAM reliability enhancement are suggested. Finally section V concludes this paper.

II. BASICS AND FAILURES ANALYSIS OF STT-MRAM

A. Basics of STT-MRAM

The STT-MRAM cell consists of magnetic tunnel junctions (MTJ) and transistors, as shown in Fig.1 (a), i.e., one MTJ connecting with one transistor in series (1T1MTJ cell structure).

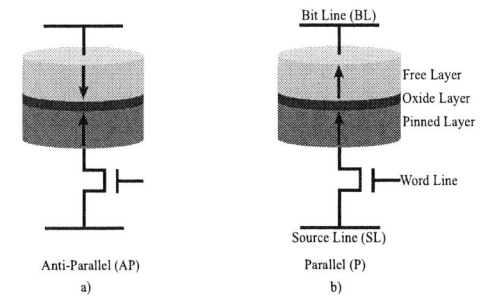

Fig. 1. STT-MRAM cell structures.

978-1-4799-8720-7/15 $31.00 © 2015 IEEE

A MTJ has the sandwich structure with one oxide barrier layer between two ferromagnetic layers as shown in Fig.1 (b). If the magnetization directions of the two layers are parallel, the MTJ is in low resistance state; whereas if their directions are anti-parallel, the MTJ is then in high resistance state [6][7]. The transistor is an access device, acting as a switch in the cell. If the memory cell is accessed, the transistor provides either a bi-directional driving current for writing operations or directional current for reading operations. The driving ability of the transistor has a deep relationship with the reliability of STT-MRAM, especially when the technology node shrinks down to the deep nanometer nodes.

B. Failures of STT-MRAM

The failure sources of STT-MRAM mainly include: (a) the intrinsic stochastic switching characteristics; (b) asymmetry of the STT driven MTJ; (c) the process, voltage and temperature (PVT) variations of the manufacturing process [8-11]. Due to these failures, the desired data may not be stored into or read from the STT-MRAM correctly [12]. These failures can be categorized into writing errors, reading errors and flipping errors, as shown in TABLE I. Next we will give an explanation of the errors occurring in STT-MRAM.

TABLE I.   OPERATION FAILURES IN STT-MRAM

Operation	Bit In Cell	Target Bit	Read-Out Bit	Error Type
Writing	1	0	1	Writing failure
Writing	0	1	0	Writing failure
Reading	1	1	0	Decision error or Flipping error
Reading	0	0	1	Decision error

Writing errors occur when the writing current cannot meet the switching threshold current or the writing pulse duration is not long enough. Generally speaking, according to the magnitude of writing current, the switching of MTJs could be classified into three types, including: (a) if the writing current amplitude is larger than the switching threshold current, the MTJ switching is mainly decided by spin activation; (b) if the writing current is lower than the switching threshold current but higher than 0.8 times of it, the MTJ switching would result from both spin activation and thermal activation; (c) if the writing current is lower than 0.8 times of the switching threshold current, the MTJ switching would be decided only by thermal activation [13-15]. All writing errors could be tackled by increasing the driving current amplitude or (and) pulse duration [16][17].

Reading errors include two types: decision error and flipping error. Decision errors occur when the sensing margin is not large enough to overcome the input-offset of the sensing circuit. Increasing reading current is one way to improve sensing margin, but which may bring about flipping errors, i.e., erroneous switching of the MTJ during read operations. Fortunately, as the reading current is generally directional, there is only one direction of flipping error, e.g., occurring in reading bit '1' with flipping error '1'->'0'.

III. STT-MRAM OPERATION CHANNEL MODELING

In order to evaluate and optimize the reliability of STT-MRAM, This section presents a STT-MRAM operation channel model, which mimics the writing and reading operations of STT-MRAM. This operation channel model includes two parts, i.e., write operation channel model and read operation channel model. Then based on this operation channel model, we will give some reliability analysis results.

A. STT-MRAM Operation Channel Model

In the operation channel model, we take the 1T1MTJ cell structure as example and take into consideration the failure sources discussed above.

1) *Writing operation channel model*: The writing operation channel model is illustrated in Fig.2 and is described by the following steps.

a) According to the target bit w, generate the writing current value I_w while considering the PVT variations of the driving transistor, $w=\{0, 1\}$;

b) Calculate the critical switching current I_{c0} with concerning the PVT variations of the MTJ;

c) Compare the writing current value I_w with the critical switching current I_{c0}. If I_w is larger than I_{c0}, the switching would be mainly decided by spin activation, then go to step *d)*; otherwise, go to step *e)*;

d) Calculate the time τ_{w1} for writing the bit into MTJ correctly. If τ_{w1} is shorter than the writing pulse width τ_w, then the bit is thought to be written into the MTJ successfully, otherwise it fails to write the bit ;

e) Calculate the time τ_{w2} for writing the bit into MTJ correctly. If τ_{w2} is shorter than the writing pulse width τ_w, then the bit is thought to be written into the MTJ successfully, otherwise it fails to write the bit.

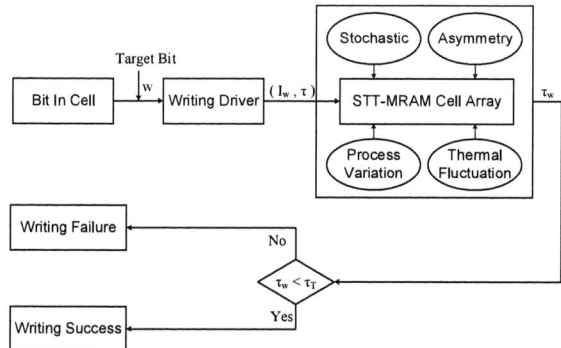

Fig. 2.   Framework of the writing operation channel model.

For different bits written into the MTJ, the writing current would also be different. The critical switching current I_{c0} and the switching duration τ_{w1} and τ_{w2} have strong relationships with the intrinsic properties of the MTJ, which can be calculated by (1), (2) and (3) according to [18-21].

$$I_{C0} = \alpha \frac{\gamma e}{\mu_B g}\left(\mu_0 M_s\right)H_K V = 2\alpha \frac{\gamma e}{\mu_B g}E \tag{1}$$

$$\frac{1}{\tau_1} = \left[\frac{2}{C + \ln(\pi^2 \Delta)}\right]\frac{\mu_B P}{em_f(1+P^2)}(I_W - I_{C0}) \qquad (2)$$

$$\tau_2 = \tau_0 \exp\left(\frac{E}{K_B T}\left(1 - \frac{I_W}{I_{C0}}\right)\right) \qquad (3)$$

where α is the damping constant, γ is the gyromagnetic ratio, e is the elementary charge, μ_B is the Bohr magneton constant, $\mu_0 M_S$ is the magnetization ratio of the free layer, H_K is the magnetic anisotropy field, V is the bulk of free layer, P is the spin polarization percentage of the tunnel current. The factor g is the spin polarization efficiency, which decides the asymmetry of the current. C is the Euler constant, m_f is the magnetic torque of free layer. τ_0 is the attempt period, which relates with materials and spin relax time.

In this model, the PVT variations of the MTJ and transistor are regarded as following the normal distribution for simplicity and they can be modified based on the practical manufacturing conditions.

2) Reading operation channel model: The reading operation channel model is illustrated in Fig.3 and is described by the following steps.

a) Generate the reading current value I_r for the cell being read, and I_{ref} for the reference cell while considering the process variations of MTJ and transistor;

b) According to the bit being read out from the MTJ, calculate the critical switching current I_{c0} with concerning the process variations of the MTJ;

c) Compare the reading current value I_r with the critical switching current I_{c0}. If I_r is larger than I_{c0}, go to step *d)*; otherwise, go to step *e)*;

d) Calculate the time τ_{r1} for switching current state of the MTJ. If τ_{r1} is shorter than the reading pulse width τ_r, which is assumed that the bit stored in current MTJ has been disturbed, or there is no reading disturbance occurring, and go to step *f)*;

e) Calculate the time τ_{r2} for switching current state of the MTJ. If τ_{r2} is shorter than the reading pulse width τ_r, which is assumed that the bit stored in current MTJ has been disturbed, or there is no reading disturbance occurring, and go to step *f)*;

f) According to the bit being reading, calculate the resistance R_{cell} with considering the PVT variations of MTJ and transistor, and the voltage $V_{cell} = I_r * R_{cell}$ of the cell;

g) Calculate the reference voltage $V_{ref} = I_{ref} * R_{ref}$ of the reference cell;

h) Calculate the sensing margin $V_{SM} = |V_{cell} - V_{ref}|$.

i) Compare the sensing margin with the sensing margin threshold (offset). If V_{SM} is larger than the sensing margin threshold, then we consider it a read success; otherwise a read failure.

In the reading operation model, reading current is bi-directional for simulation experiments; the low resistance R_P of MTJ with the CoFeB/MgO/CoFeB structure could be described by (4) with no bias voltage [18][20][25]; for sake of reducing the computing complexity, the bias voltage is not considered when calculating the high resistance R_{AP}, which is given in (5).

$$R_P = \frac{t_{ox}}{F \cdot \overline{\varphi}^{-1/2} \cdot A}\exp\left(1.025 \cdot t_{ox} \cdot \overline{\varphi}^{-1/2}\right) \qquad (4)$$

$$R_{AP} = R_P \cdot (1 + TMR) \qquad (5)$$

Where F is a factor obtained from the resistance area product value depending on the materials of MTJ. In this model, F is given the value 332.2. $\overline{\varphi}$ is the oxide layer barrier height.

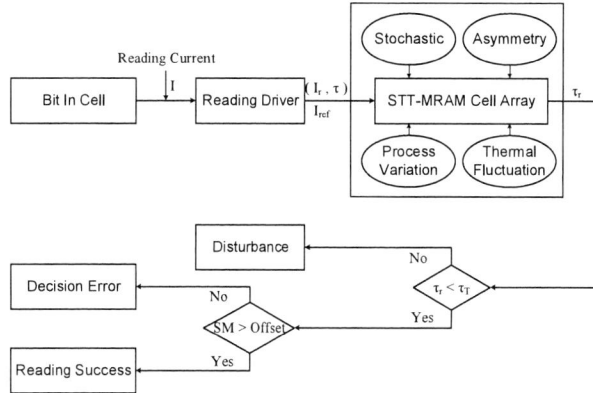

Fig. 3. Framework of the reading operation channel model.

B. Simulation Results of Operation Channel Model

Based on the proposed operation channel model, simulations are carried out to validate its functionality. The simulation results also reveal the failure sources on the STT-MRAM reliability performance. The typical parameters used in these simulations are listed in TABLE II.

TABLE II. KEY PARAMETERS OF PROCESS OF MTJ[a] AND TRANSISTOR

Parameter	Description	Default Value
a	Diameter of MTJ cross-section	40 nm
t_{ox}	Thickness of oxide layer	0.85 nm
t_f	Thickness of free layer	1.3nm
TMR	Tunneling Magneto-Resistance	70%
R.A	Product of resistance and area	10 $\Omega.\mu m^2$
I_{C0}	Critical switching current of MTJ	55 μA
Δ	Thermal stability factor	60
L	Length of transistor channel	40 nm
W	Width of transistor of channel	200 nm
V_{th}	Threshold voltage of transistor	0.6 V
V	Bulk of MTJ	-
T	Temperature	300 K

a. The MTJ has the structure of CoFeB/MgO/CoFeB.

1) Asymmetry of Writing Operation: Writing current shows asymmetric characteristics when writing bits '0' and '1'. The asymmetry is caused by the different spin polarization efficiency in parallel and anti-parallel state.

From Fig.4, the average current of writing '1' is higher than writing '0'. Through model computing, the critical current is about 26 μA for writing '0', and 75 μA for writing '1'. Providing a 10 ns writing pulse duration, the current should be about two times the critical current if no error occurs. In designing STT-MRAM, this asymmetry property should be considered for performance optimization.

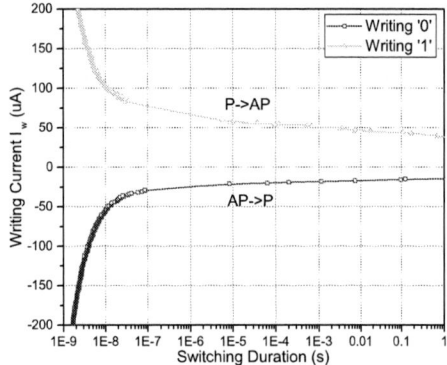

Fig. 4. The asymmtery of writing operation.

Writing operation delay from '0' to '1' or from '1' to '0' is decided by spin activation and thermal activation of MTJs. Therefore, the writing error rate has a different distribution for the different writing pulse width T_{wp}. As shown in Fig.5, large writing current amplitude is needed for finishing writing '1' under the same writing pulse width, because the energy barrier is higher than writing '0'. Furthermore, large current would be favorable for fast writing in the cost of power consumption.

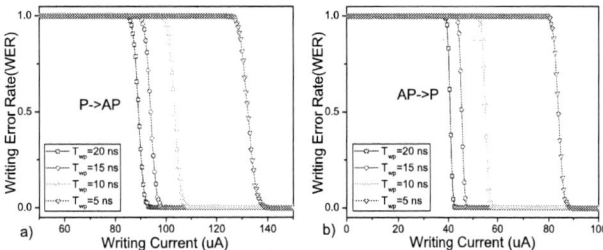

Fig. 5. Writing error rate under different T_{wp}.

2) Reading Error Rate: As shown in Fig.6, when read current smaller than a threshold (e.g., 80 uA for reading bit '0'), the read error rate is decreasing with the increase of the read current amplitude, this is because the sensing margian grows as discuseed above. In addition, a large TMR ratio results in lower read error rate, as a large TMR ratio lead to also a large sensing margin. However, as the read current continuously increases, the read error rate grows rapidly, indicating the occurance of flipping errors. It can also be noted that we can obtain a large sensing margin with relatively large read current and low flipping error rate when the direction of reading current is parallel with the current dirction of writing '1', as the write threshold current for bit "1" is larger than that of bit "0".

3) Process variation of MTJ size: As shown in Fig.7, the curves of writing error rate would have a sharp descending with

small diameter deviation of MTJ σ_{MTJ}; as for the reading operation, σ_{MTJ} has nearly no impact on reading '0' and '1' with restricting reading current to a proper range. However, keeping on increasing reading current will result in reading disturbance.

For the writing operation, increasing writing current could eliminate the influence caused by the process variation of MTJ size. However, increasing large reading current is not a way to eliminate the error of reading disturbance, for tackling it some joint means are needed.

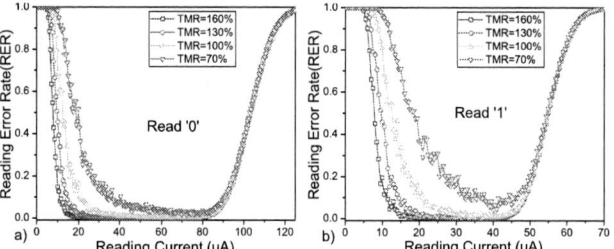

Fig. 6. Reading error rate under different TMR.

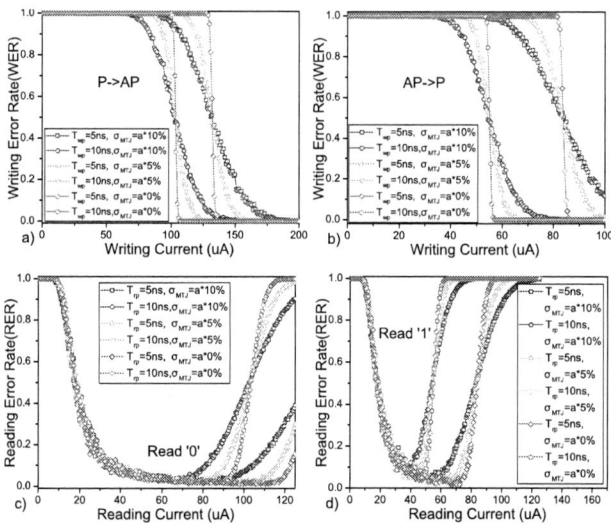

Fig. 7. Writing and reading error rate under diameter deviation of MTJ.

4) Process Variation of Free Layer Height: Writing and reading error rates are also related with the height of MTJ free layer, but their impacts can be ignored when compared with diameter deviation of MTJ. As shown in Fig.8, the main influence of the height deviation of MTJ free layer σ_{tf} focuses on the writing operation and the disturbance in reading operation. Low height deviation of MTJ free layer would enlarge the sensing margin for reading decision and result in a sharper descending of the writing error rate.

Like the diameter of MTJ, these process variations should be shortened as possible for getting a sharp threshold during device fabricating.

5) Process Variation of Oxide Layer Thickness: According to [22-25], the resistance of MTJ relates to the thickness of the oxide layer, as described in (4) and (5).

978-1-4799-8720-7/15 $31.00 © 2015 IEEE   464

In Fig.9, the thickness deviation of oxide layer σt_{ox} has nearly no impact on the reading error rate. The STT effect can only be detected when the thickness of the oxide layer is small enough. In 40 nm technology node of MTJ, the oxide layers are usually set to 1 nm.

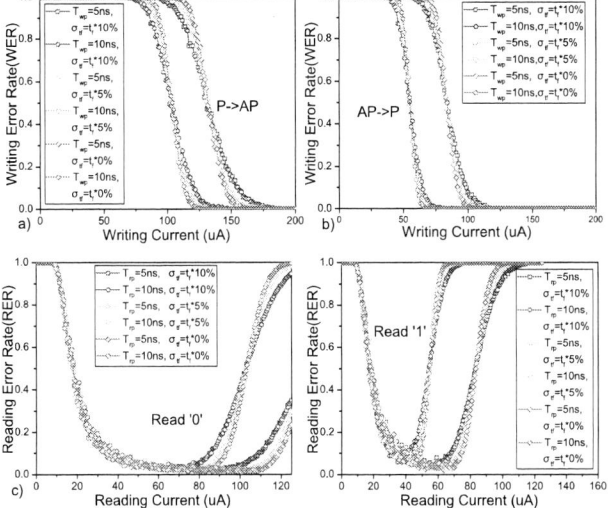

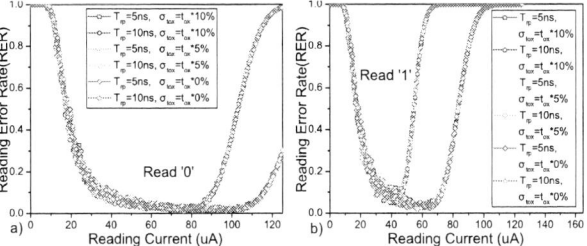

Fig. 8. Writing and reading error rate under σ_{tf}.

Fig. 9. Reading error rate under σt_{ox}.

6) *Process Variation of Transistor Channel Width:* Writing or reading currents are both provided by the transistor connected in series with the MTJ, which contributes the main responsibility to the writing and reading error.

Generally, the variation of driving ability is caused by the variations of the channel width. The relationship between transistor channel width and writing/reading error rates is described in Fig.10. From Fig.10, 280 nm may be the most suitable size for transistor channel length at 40 nm technology node, writing and reading error rate are both at the lowest level. Setting the transistor channel length longer than 280 nm will result in reading disturbance when reading '1'. Therefore, the transistor channel size must be considered properly according to the driving ability needed by the MTJ to switch its state with error free.

The operation channel model provides the information about intrinsic properties and process variation impacting on writing and reading error rates. Through this operation channel, the means of controlling the key parameters to enhance reliability of writing and reading operation can be acquired. For given conditions, the data of writing and reading error rates can also be obtained through this model; at the same time, the channel model provides knowledge for circuit and system level designers to find suitable reliability enhancement solutions.

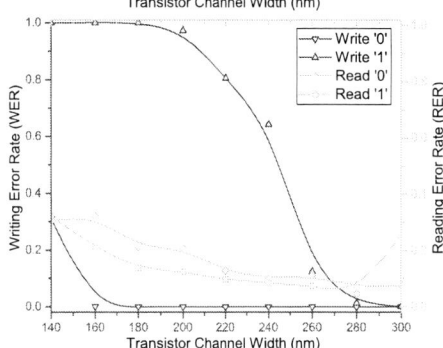

Fig. 10. Writing and reading error rate under transistor channel length.

IV. RELIABILITY ENHANCEMENT DESIGN STRATEGIES

This section presents some implications and considerations by employing the proposed operation channel model for STT-MRAM reliability enhancement design.

Error correction code (ECC) is one of the effective system-level techniques used to improve reliability and yield of the integrated circuits and has been widely employed for memory chips[26][27]. The principle of ECC is to add redundant bits (parity check bits) to detect and correct possible errors[28]. Given a raw bit error rate (BER), say p_b, the final BER after ECC, say P_{ECC}, can be calculated using the binomial distribution as (7),

$$P_{ECC} = \sum_{j=t+1}^{n} \binom{n}{j} (p_b)^j (1 - p_b)^{n-j} \qquad (7)$$

Where t and n are respectively the error correction capability and codeword size of the given ECC. Fig. 11 shows the final BER by applying three different ECCs with $n = 256$ and $t = 1,2,3$ respectively. As can be seen, the ECC can achieve a significant BER reduction. Given the target final BER and the raw BER, we can design the corresponding ECC strategy. Nevertheless, this reliability improvement is at the cost of hardware overhead (storage space for the redundant parity check bit). In addition, the raw BER is very important for an efficient ECC design and it is usually unavailable for system-level designers. For example, for $n = 256$, the ECC with $t = 3$ costs 8 more parity check bits than the ECC with $t = 2$. With the help of our proposed channel model, we could obtain effectively this raw BER by considering the practical device and circuit conditions.

Another important performance of STT-MRAM is the chip area. As discussed above, given the final BER requirement, device-level, circuit-level and system-level techniques can be employed to improve the STT-MRAM reliability performance, e.g., increasing the transistor size from device-level design, write/read circuit optimization from the circuit-level design, ECC from system-level design, but all these techniques are with the cost of area overhead. Therefore, an area-efficient design strategy either using single-level design technique or multi-level co-design scheme can be obtained by using our proposed channel model.

Based on the knowledge of the access transistor size and the read/write circuit, the memory cell area and raw BER can be estimated. Then given the target final BER requirement, the required ECC can be design and the corresponding area can also be estimated. Finally, the optimal multi-level co-design strategy for the required reliability performance can be achieved.

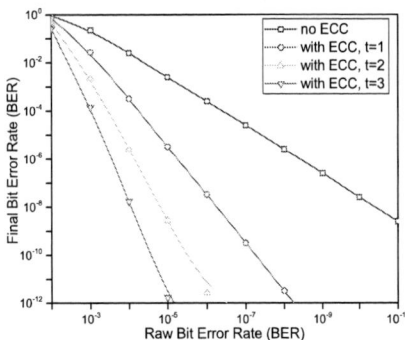

Fig. 11. Final BER by applying ECC.

V. CONCLUSION

This paper proposed an operation channel model of the 1T1MTJ structure cell, which could be used to evaluate the reliability of writing and reading operation for STT-MRAM. Based on the BER data obtained from this channel model, the reliability optimization in single level or multi-level could be carried out. According to the optimized reliability solution, this channel model would provide implications on how to tune the key parameters to decrease writing and reading error rates. Furthermore, this operation channel model bridges the gap between device-level designers and circuit or (and) system-level designers to achieve optimal reliability performance.

REFERENCES

[1] N.S. Kim, T. Austin, D. Baauw, T. Mudge, K. Flautner and et al., "Leakage current: Moore's law meets static power," Computer, vol. 36, pp. 68-75, 2003.

[2] Z. Diao, Z. Li, S. Wang, Y. Ding, A. Panchula and et al., "Spin-transfer torque switching in magnetic tunnel junctions and spin-transfer torque random access memory," Journal of Physics: Condensed Matter, vol. 19, p. 165209, 2007.

[3] Y. Huai, "Spin-transfer torque MRAM (STT-MRAM): challenges and prospects," AAPPS Bulletin, vol. 18, pp. 33-40, 2008.

[4] S.A. Wolf, D.D. Awschalom, R.A. Buhrman, J.M. Daughton, S. von Molnár and et al., "Spintronics: a spin-based electronics vision for the future," Science, vol. 294, pp. 1488-1495, 2001.

[5] C. Chappert, A. Fert and F.N. Van Dau, "The emergence of spin electronics in data storage," Nature materials, vol. 6, pp. 813-823, 2007.

[6] J.C. Slonczewski, "Current-driven excitation of magnetic multilayers," Journal of Magnetism and Magnetic Materials, vol. 159, pp. L1-L7, 1996.

[7] S. Ikeda, J. Hayakawa, Y.M. Lee, F. Matsukura, Y. Ohno and at al., "Magnetic tunnel junctions for spintronic memories and beyond," IEEE Transaction on Electron Devices, vol. 54, pp. 991-1002, 2007.

[8] W.S. Zhao, Y. Zhang, T. Devolder, J.-O. Klein, D. Ravelosona and at al., "Failure and reliability analysis of STT-MRAM," Microelectronics Reliability, vol. 52, pp. 1848-1852, 2012.

[9] Y.J. Zhang, X.B. Wang, H. Li, and Y.R. Chen, "STT-RAM cell optimization considering MTJ and CMOS variations," IEEE Transaction on Magnetics, vol. 47, pp. 2962-2965, 2011.

[10] W. Kang, L.Y. Zhang, W.S. Zhao, J.-O. Klein, Y.G. Zhang and at al., "Yield and reliability improvement techniques for emerging nonvolatile

STT-MRAM," IEEE Journal on Emerging and Selected Topics in Circuits and Systems, vol. 5, pp. 28-39, 2015.

[11] X. Liu, D. Mazumdar, W. Shen, B.D. Schrag, and G. Xiao, "Thermal stability of magnetic tunneling junctions with MgO barriers for high temperature spintronics," Applied physics letters, vol. 89, p. 023504, 2006.

[12] Y. Ye, T. Liu, M. Chen, S. Nassif and Y. Cao, "Statistical modeling and simulation of threshold variation under random dopant fluctuations and line-edge roughness," IEEE Transaction on Very Large Scale Integration (VLSI) Systems, vol. 19, pp. 987-996, 2011.

[13] J.Z. Sun, S.L. Brown, W. Chen, E.A. Delenia, M.C. Gaidis and et al., "Spin-torque switching efficiency in CoFeB-MgO based tunnel junctions," Physical Review B, vol. 88, p. 104426, 2013.

[14] Y. Zhang, W.S. Zhao, D. Ravelosona, J.-O. Klein, J. Kim and at al., "Compact modeling of perpendicular-anisotropy CoFeB/MgO magnetic tunnel junctions," IEEE Transaction on Electron Device, vol. 59, pp. 819-826, 2012.

[15] X. Wang, Y. Zheng, H. Xi, and D. Dimitrov, "Thermal fluctuation effects on spin torque induced switching: mean and variations," Journal of Applied Physics, vol. 103, p. 034507, 2008.

[16] X. Fong, Y. Kim, S. H. Choday, and K. Roy, "Failure mitigation techniques for 1T-1MTJ spin-transfer torque MRAM bit-cells," IEEE Transactions on Very Large Scale Integration (VLSI) Systems, vol. 22, pp. 384-395, 2014.

[17] W. Guo, G. Prenat, V. Javerliac, M. El Baraji, N. de Mestier and et al., "SPICE modelling of magnetic tunnel junctions written by spin-transfer torque," Journal of Physics D: Applied Physics, vol. 43, p. 215001, 2010.

[18] S. Ikeda, K. Miura, H. Yamamoto, K. Mizunuma, H.D. Gan and et al. "A perpendicular-anisotropy CoFeB-MgO magnetic tunnel junction," Nature materials, vol.9, pp. 721-724, 2010.

[19] Y. Wang, Y. Zhang, E.Y. Deng, J.-O. Klein, L.A.B. Naviner and et al., "Compact model of magnetic tunnel junction with stochastic spin transfer torque switching for reliability analyses," Microelectronics Reliability, vol. 54, pp. 1774-1778, 2014.

[20] M. Gajek, J.J. Nowak, J.Z. Sun, P.L. Trouilloud, E.J.O Sullivan and et al., "Spin torque switching of 20 nm magnetic tunnel junctions with perpendicular anisotropy," Applied Physics Letters, vol. 100, p. 132408, 2012.

[21] Y. Zhang, W.S. Zhao, G. Prenat, J.-O. Klein, C. Chappert and et al., "Electrical modeling of stochastic spin transfer torque writing in magnetic tunnel junctions for memory and logic applications," IEEE Transaction on Magnetics, vol. 49, pp. 4375-4378, 2013.

[22] W.S. Zhao, T. Devolder, Y. Lakys, J.-O. Klein, C. Chappert and et al., "Design considerations and strategies for high-reliable STT-MRAM," Microelectronics Reliability , vol. 51, pp. 1454-1458, 2011.

[23] A.F. Vincent, N. Locatelli, J.-O. Klein, W.S. Zhao, S. Galdin-Retailleau and et al., "Analytical Macrospin Modeling of the Stochastic Switching Time of Spin-Transfer Torque Devices," IEEE Transaction on Electron Devices, vol. 62, pp. 164-170, 2015.

[24] G.D. Fuchs, I.N. Krivorotov, P.M. Braganca, N.C. Emley, A.G.F. Garcia and et al, "Adjustable spin torque in magnetic tunnel junctions with two fixed layers," Applied Physics Letters, vol. 86, p.152509, 2005.

[25] W.S. Zhao, J. Duval, J.-O. Klein, C. Chappert, "A compact model of magnetic tunnel junction (MTJ) switched by thermally assisted spin transfer torque (TAS+STT)," Nanoscale Research Letters, vol. 6, p. 368, 2011.

[26] W. Kang, Z. Li, Y.G. Zhang, J.-O Klein, C. Chappert and at al., "Variant-tolerant and disturbance-free sensing circuit for deep nanometer STT-MRAM," IEEE Transaction on Nanotechnology, vol. 13, pp. 1088-1092, 2014.

[27] N.S. Kim, S.C. Draper, S.T. Zhou, S. Katariya, H.R. Ghasemi and et al., "Analyzing the impact of joint optimization of cell size, redundancy, and ECC on low-voltage SRAM array total area," IEEE Transactions on Very Large Scale Integration (VLSI) Systems, vol. 20, pp. 2333-2337, 2012.

[28] W. Kang, W.S. Zhao, Z. Wang, Y. Zhang, J.-O. Klein and at al., "A low-cost built-in error correction circuit design for STT-MRAM reliability improvement," Microelectronics Reliability, vol. 53, pp. 1224-1229, 2013

2015 IEEE Computer Society Annual Symposium on VLSI

STT-MRAM-Based Strong PUF Architecture

Elena Ioana Vatajelu[1], Giorgio Di Natale[2], Lionel Torres[2], Paolo Prinetto[1]

[1]Politecnico di Torino,
Dip. di Automatica e Informatica, Turin, Italy

[2]LIRMM, University of Montpellier/CNRS
Montpellier, France

Abstract— **Physically Unclonable Functions (PUFs) are emerging cryptographic primitives used to implement low-cost device authentication and secure secret key generation. Weak PUFs (i.e., devices able to generate a single signature or to deal with a limited number of challenges) and Strong PUFs (i.e., devices able to deal with multiple challenges) are widely discussed in literature. The most investigated solutions today are based on CMOS devices. However, with the rapid development of low power, high density, high performance SoCs new technologies have emerged. One of the most promising choices to replace charge-based computing is the spin-based computing, due to its reduced read/write latency and high CMOS integration capability. In this paper, we propose an innovative PUF design based on Spin-Transfer-Torque-MRAM memory. We exploit the high variability affecting the electrical resistance of the Magnetic Tunnel Junction (MTJ) device and special configurations of the reference cells for the read operation. The reference cell array is configured in such a way to allow for normal memory operation, and, when required, allow for signature generation (PUF solution). Switching between memory and PUF is done by controlling the state of the referencing array. In this way, we can generate multiple challenges for the implementation of a Strong PUF solution. We demonstrate that the proposed solution is robust, unclonable and unpredictable.**

Keywords — Emerging Memory Technology, MRAMs, Phisical Unclonable Functions PUFs, Security

I. INTRODUCTION

Today, Physical Unclonable Functions (PUFs) exploit intrinsic manufacturing variability of the CMOS fabrication process to generate a signature unique to each single device. When PUFs are used to generate secret keys, they have to generate a single signature. In order to guarantee the security of the generated secret key, the signature must be unique from device to device. Moreover, for a same device, it must be robust with respect to aging and environmental variations (temperature, voltage noise, electromagnetic interferences) in order to generate always the same value. PUFs used for secret key generation are called *weak PUFs*. One of the most investigated solutions exploits the randomness of the state of SRAMs after power-on, since they provide high security (i.e., high inter-chip variation) and high stability (i.e., low intra-chip variation).

When used for device authentication, the function implemented by the PUF depends on some additional input values (*challenge*) that allow generating different responses for each challenge. The set of Challenge/Response Pairs

(CRPs) must be unique for a single device, being this set the overall signature of the device. The authentication of the device is done by storing in a secure server some randomly chosen challenge/response pairs and by interrogating the device with that set of challenges. If the responses generated by the device are equal to those stored in the server, the device is authenticated. PUFs with a very large number of possible CRPs are defined as *strong PUFs*. In order to guarantee the unclonability of the PUF and to avoid exhaustive attacks, the device must be able to generate at least 2^{80} CRPs.

The implementation of strong PUFs usually exploits the variability in the delay of some nets within the circuit. More in particular, the most investigated solutions are those based on Arbiter PUFs (where the delay of two identical paths is measured) or those based on ring oscillators (where two supposedly identical frequencies are compared). These PUFs are strong because they insert in the architecture a high number of configurable switches or ring oscillators. No solutions based on memories are nowadays available for strong PUF, since even a big memory would not satisfy the requirements of having more than 2^{80} CRPs. Indeed, such a PUF would require a memory array of 2^{80} cells.

Very few studies exist for PUFs based on emerging memory technologies. While there are some proposals exploiting resistive memories [1][2][3], only two solution have been so far proposed for magnetic memories [4][5], however, both this solutions describe weak PUF implementations.

In this paper, building on our previous work [5], we propose a novel strong PUF implementation, which takes advantage of the intrinsic variability of the STT-MRAM cell electrical characteristics, with added challenges to provide strength to the resulted signature. The proposed PUF solution is reliable (has the ability to be the same under different environmental conditions for one device), unclonable (has the ability to be different between devices) and unpredictable. Our PUF solution is unclonable and unpredictable since it is based on fabrication-induced variability (which cannot be controlled, not even by the manufacturer, hence resulting in unclonable devices) and we show that the PUFs generated for different devices exhibit high mismatch probability. To the best of our knowledge, this is the first attempt of designing a strong PUF based on the use of memories.

This paper is organized as follows. In Section II, we

978-1-4799-8720-7/15 $31.00 © 2015 IEEE
467

present the basic characteristics and operation principles of the STT-MRAM memory and a short description of the STT-MRAM-based weak PUF implementation. In Sections III and IV, we present the principle underlying the strong PUF architecture and some simulation results, respectively. Section V concludes the paper.

II. STT-MRAM BASED WEAK PUF IMPLEMENTATION

This section includes a brief description of the STT-MRAM cell operation principle (since it is the main element of our proposed PUF implementation) and a summary of the previously proposed weak PUF implementation, on which the present work is built.

A. STT-MRAM Technology

In magnetic type memories, information is stored into the Magnetic Tunneling Junction (MTJ), usually composed of two ferromagnetic layers separated by a tunneling oxide barrier. One ferromagnetic layer has a pinned magnetization direction (*fixed ferromagnetic* layer). The magnetization of the second ferromagnetic layer is unpinned (*free ferromagnetic* layer), thus it can be programmed. The relative Magnetization Directions (MDs) of the two ferromagnetic layers determine the electrical resistance of the MTJ device. When the magnetization directions of the two layers are parallel, the MTJ device exhibits low electrical resistance (R_L), while when they are anti-parallel, the MTJ device exhibits high electrical resistance (R_H) [6]. In a Spin Transfer Torque device, the magnetization direction of the free layer is controlled by a current spin-polarized by the magnetization of the fixed ferromagnetic layer.

In a magnetic-type random access memory cell, an NMOS transistor (1T) is used to access the MTJ element (1MTJ). In this topology (1T1MTJ), the cell is accessed and controlled by the corresponding control lines, i.e., Bit Line (BL), Source Line (SL) and Word Line (WL) as depicted in Fig. 1. The state of the cell is changed by forcing a sufficiently large current (I_{MTJ}) through the MTJ device, able to switch the magnetic orientation of the free-layer. During read operation, small voltage bias is applied on the memory cell; therefore, a current passes through the device. This current (I_R) is inversely proportional to the electrical resistance of the MTJ device. By using a sense amplifier (SA), the reading current (I_R) is compared against a reference value (I_{ref}) and a decision is made on the memorized state. For a reliable read operation, the reference current is set to be equally apart from the nominal values of I_{R0} (current during read '0' operation) and I_{R1} (current during read '1' operation), i.e., $I_{ref}=(I_{R1}+I_{R0})/2$, to assure maximum sensing margins for both logic states.

The reference current (I_{ref}) can be generated in different ways, from using a current source, to using pinned or unpinned MTJ devices. In our approach, we generate the reference current using a reference array with unpinned MTJ devices, identical to the ones of the cells to be read [7]. The reference MTJ elements are connected in a serial/parallel configuration in which half of the MTJs are set to anti-

parallel, while the other half are set to parallel magnetization. The serial/parallel configuration allows for the averaging of the low and the high resistances of the MTJ devices. The equivalent resistance of the reference array is, in this case, equal to the value mean value between the high and low MTJ resistance. The reference array is connected to the SA through an NMOS transistor identical to the one providing access to the bit cell. The minimum reference array has 4 MTJ devices, connected 2 by 2 in series and the resulting branches connected in parallel. This solution for reference current generation has high reliability due to the fact that the reference current is obtained from averaging in this way, partially compensating the effect of variability on I_{ref}. To further mitigate this effect, a larger number of reference cells can be used. These cells are spatially distributed and interleaved with the active cells to capture the variability profile of the memory array. In this way, we maximize the probability that the reference current (I_{ref}) is equally apart from the read currents I_{R0} and I_{R1}, respectively.

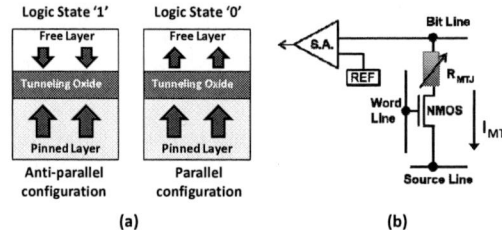

Fig. 1. The STT-MRAM Memory cell: a) MTJ configurations; b) Electric circuit of 1T1MTJ structure.

In the next subsection, we briefly describe the STT MRAM-based weak PUF, introduced in [5], based on which we build the proposed strong PUF implementation.

B. Weak PUF Implementation

This solution takes advantage of the high variability and uncontrollability of the MTJ resistance in anti-parallel magnetization and that of the transistor's strength. For this, we resort to a memory array consisting of *N active cells* (to generate the PUF value), and *M reference cells* (to generate the reference current for read operations). Since the active and the reference cells have the same design characteristics and are fabricated in the same way, they are similarly affected by variability.

Our PUF solution exploits the differential sensing during read operation, based on read current comparison against a reference value. While in a classic MRAM the reference current is set to be half distance between high and low resistances, in this solution we force all reference cells to be in the anti-parallel state. We then write all active cells to '1' (anti-parallel state). In this scenario, the differential sense amplifier will compare two currents with the same nominal values. The result of the differential sensing will thus only depend on the variability affecting both reference and active cells, and it will be randomly distributed among different devices. The overall procedure is described in Fig. 2.

978-1-4799-8720-7/15 $31.00 © 2015 IEEE

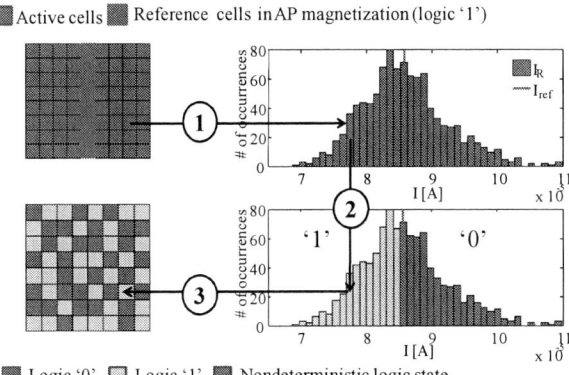

Active cells ▨ Reference cells in AP magnetization (logic '1')

■ Logic '0' □ Logic '1' ▨ Nondeterministic logic state

Fig. 2. The implementation strategy of the proposed PUF solution: 1) Write all cells to '1'; 2) Read each cell; 3) Use the read value

The viability of this PUF solution has been proven by simulation. Details on the simulation environment, procedure and the obtained results are completely described in [5]. The schematic representation of our design is shown in Fig. 3. Here, we emphasize the active memory array and the reference cells set. The number of active cells in our memory array is $\#A=Ma \times N$, where Ma represents the number of cells in a row, while N represents the number of cells in a column in our array and Mr represents the number of MTJ devices in the reference set. The read operation is performed cell by cell; the resulting signature is a binary vector of size equal to the total number of cells in the active memory array ($\#A$).

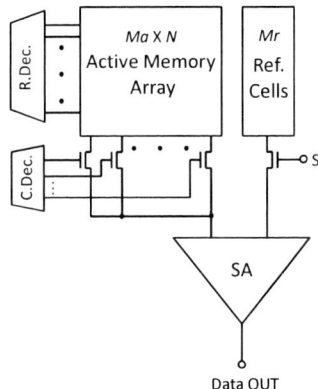

Fig 3. Schematic representation of the weak PUF implementation [5]

The high fabrication-induced variability affecting the electric characteristics of the MTJ device guarantees the unpredictability, the unclonability and the uniqueness of the PUF solution. Concerning its reliability, since the MTJ and access transistor parameter values are set by fabrication, the read current distribution and the reference current drift in the same way for one device operated under the same environmental conditions (temperature and voltage). This characteristic guarantees high reproducibility of the results. However, there is an uncertain sensing zone for the sense amplifier. Indeed, the bits falling in this zone, i.e. the bits for which $|I_R-I_{ref}|$ is smaller than the sensing margin of the SA, can induce a meta-stable state, which will be randomly

stabilized to '0' or '1' depending on the noise in the circuit. These cells can be read randomly as '1' or '0' at different runs of the algorithm, and are denoted as nondeterministic active cells in Fig 2. This is the main reliability issue of the described PUF implementation and its effect can be reduced by careful design of the SA.

With this implementation, one signature is generated per device. In the next section, we describe the proposed PUF architecture, which allows for generating multiple challenge/response pairs per device, i.e., a strong PUF implementation.

III. PROPOSED STT-MRAM-BASED STRONG PUF IMPLEMENTATION

In this paper, we propose a new architecture for STT-MRAM based PUF implementation, in which we take advantage of the specificity of the read operation and previously described signature generation, and of a dedicated column decoder to generate challenge/response pairs. The signature generation is based on comparing two currents, which nominally have the same value, but due to process variability they are un-equal. The sense amplifier has the role to sense and amplify this difference, which will eventually be interpreted as 0 or 1.

The proposed architecture consists of an active memory array designed with Ma rows and N columns and a reference array designed with N groups of reference cells. All reference cell groups are identical and consist of Mr MTJ devices. These Mr MTJ devices are connected in the serial/parallel configuration described in the previous section, and are used to generate the reference current. The schematic representation of our proposed architecture is depicted in Fig. 4.

The novelty of the proposed approach consists in using multiple reference cell groups (the number of reference cell groups is equal to the number of columns in the active memory array). In this situation, different signatures can be generated by activating, at the same time, one or more columns in both the active and reference arrays. The underlying principle follows the same previously described idea of comparing two currents, nominally identical but in actuality different due to process variability. More precisely, for each active row Ma, the signatures can be generated by:

- Reading one column at a time using one reference group (this is in fact how the weak PUF is generated),

- Reading one column at a time using each of the reference groups (in this case, for each combination column/reference group we obtain a set of signatures),

- Reading two columns at a time using a pair of reference groups. In this case, the current fed to the sense amplifier is double compared to the previous cases, since the two columns (both in the active and reference arrays) are in parallel. Therefore, the condition of having nominally identical currents compared by the SA is met (in this case, we obtain new signature from any combination of 2 columns/2 reference array).

978-1-4799-8720-7/15 $31.00 © 2015 IEEE   469

- Combining even more columns together. The previous case (i.e., two active columns and two reference columns) can be extended to larger number of columns to be electrically mixed in parallel. The maximum number of columns we can combine for signature generation is N (the total number of columns in the memory array), in which case we will also use N reference groups.

A strong PUF requires a high number of challenges. In our solution, a challenge is defined as a combination of columns (for active and reference cell groups), and a row of the active memory array. The Column Decoder Controller (*C.Dec Control* in Fig. 4) has the role of generating the combinations and selecting the corresponding active columns and reference cell groups used for the current comparison, and, therefore, for generating the response. In the described implementation, one Column Decoder Controller was used (its repetition in Fig. 4 is just for illustration purposes, in fact, the two decoders are the same device). The total number of column combinations that can be obtained in this way is expressed in equation (1).

$$Comb = \sum_{1 \le k \le n} \binom{n}{k} = 2^n - 1 \qquad (1)$$

This number of combination grows exponentially with the number of columns, as required for a strong PUF. For instance, a memory array composed of 2^{18} rows and 64 columns would allow a total number of CRP equal to 2^{82}.

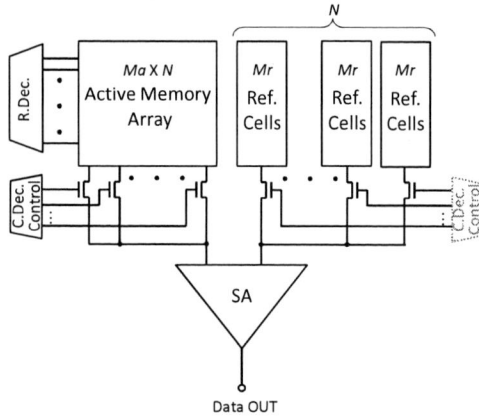

Fig 4. Schematic representation of the proposed strong PUF implementation

To implement the proposed PUF solution, we first program all cells (both active and reference) to the anti-parallel state. Then, for the selected row, the combination of columns is generated and the resulting column(s) and reference cell group(s) are selected. The comparison in the sense amplifier is performed between currents coming from one column and one reference group or currents coming from sets of columns and reference groups. In the first case, the resulting read operations are identical to the ones obtained when the weak PUF implementation is used. In the second situation, the currents compared in the sense amplifier are the sum of the currents coming from the combined columns against the sum of currents coming from the combined reference groups. This implementation works if and only if the number of selected

columns is equal to the number of reference groups. This constraint is assured in our solution since we use the same Column Decoder Controller for both the active and reference arrays.

IV. SIMULATION BASED PUF EVALUATION

A PUF device has to be easy to produce, practically impossible to duplicate (unique), and when a physical stimulus is applied, the device has to have an unpredictable but repeatable response.

Traditionally, the reliability (repeatability) of a PUF is evaluated by analyzing the differences between the signatures obtained after different runs of the *same* PUF. Ideally, the resulting signatures are identical; however, small numbers (*less than 10%* of the bits are different at different runs in the case of SRAM-based PUFs) are accepted since error-correcting codes and protocols using helper data can be implemented at low cost.

The unpredictability of a PUF is evaluated by computing the *Hamming Weight* (*HW*) *of the response* and the *Hamming Distance* between the signatures obtained from *different* PUFs. Ideally the PUF should exhibit bias neither towards 0 nor 1, i.e., the *fractional Hamming Weight* of one response should be *0.5*. The *fractional Hamming Distance* (HD_U) between two responses should be *0.5*, since this value allows for the maximum number of different responses. In the case of strong PUFs, it is also important to evaluate the unpredictability of the challenge/response pairs. We use the previously described Hamming metrics to evaluate the unpredictability of CRPs.

We performed statistical analyses to estimate the values of the three Hamming metrics as follows: for the robustness estimation we estimate the number of unreliable cells in the array (i.e. the number of cells which, during read operation generate a current very similar to the reference current, the difference $|I_R - I_{ref}|$ falls within the sensing margin of the sense amplifier); for the unpredictability estimation we evaluated 500 samples of our proposed STT-MRAM-based PUF. This statistical analysis yields statistically distributed Hamming metrics. Therefore, our unpredictable strong STT-MRAM-based PUF has to meet the following specifications:

- Fractional Hamming weight:
  $\mu(fHW) \simeq 0.5$, $\sigma(fHW)$ very small
- Fractional Hamming distance for unpredictability:
  $\mu(fHD_U) \simeq 0.5$, $\sigma(fHD_U)$ very small
  $min(fHD_U) > max(fHD_R)$

where $\mu(.)$ represents the mean value of the distribution and $\sigma(.)$ its standard deviation.

A physical STT-MRAM-based PUF is currently infeasible due to the novelty of the STT-MRAM technology and the advanced manufacturing techniques required to produce such design. Therefore our results cannot be validated on silicon and we will resort to electrical simulations.

The relevant geometrical and electrical parameters of the STT-MRAM bit-cell are extracted from literature [8][9][10] and summarized in Table I. All parameters are assumed to follow Gaussian distributions.

TABLE I.   PARAMETERS OF THE STT-MRAM CELL

Parameters	Mean(μ)	σ/μ
NMOS transistor width (W)	210nm	10%
NMOS transistor length (L)	45nm	10%
NMOS threshold voltage (W)	275mV	10%
Cross section area of MTJ free layer (A)	90x180$\Omega\cdot\mu m^2$	5%
Low MTJ resistance (R_L)	1600 Ω	9.3%
High MTJ resistance (R_H)	3200 Ω	10.3%

In order to apply a challenge to the device, we first program all MTJ devices in the reference cell groups to anti-parallel state (logic '1'). Next, we write all cells in the active memory array to logic '1'. The write operation is performed in 40ns to allow enough energy to be stored by the MTJ device, and in this way the cell to be correctly written. Next, the read operations are performed.

When more than one column is connected to the sense amplifier, the read current increases linearly with the number of columns connected in parallel. The same holds for the reference cell groups connected in parallel to generate the reference current. The reliability issue introduced in the response by the sensing margin of the sense amplifier is still present, but the number of cells in the nondeterministic region (marked in red in Fig. 2) decreases. This is due to the fact that the current of the signal fed to the sense amplifier is larger; hence the probability that the differential read current is smaller than the sense margin of the SA decreases. The percentage of cells exhibiting nondeterministic behaviors from the total numbers of cells generating the signature is illustrated in Fig. 5 as a function of number of columns connected in parallel during the read operation. The simulations have been performed assuming an active memory array with $Ma = 1024$ rows and $N = 16$ columns. One reference cell group is composed of $Mr = 64$ MTJ devices and we use a total of $N = 16$ reference groups.

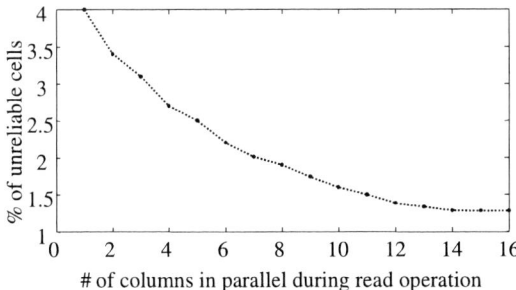

Fig. 5 - Reliability dependence on the number of columns connected in parallel during the read operation

The results show great reliability improvement when the responses are obtained by combining several columns and reference cell groups. For instance, when the response is obtained by combining (in parallel) any 8 columns from the available 16 columns and comparing the resulting current against the current generated by combining (in parallel) any 8 groups from the available 16 reference cell groups, the number of unreliable cells is half of the number of unreliable cells found when the response is obtained by reading only one column.

As stated in the previous section, the possible challenges are obtained by combining columns and reference cell groups. Theoretically, one combination of columns compared against all permutations of combinations of reference cell groups provides us with another challenge. However, out of these challenges only one is relevant, the rest will provide very similar results. To prove this statement, we generated the responses for one column (all cells in a row) compared against each of the reference cell groups. The results are depicted in Fig. 6. We have observed that for large number of cells in one group of the reference array (i.e. $Mr > 64$) the signatures generated comparing the same active cells against each of the reference groups are practically the same (fractional hamming distance is under 10%). It is worth noting that when one group of the reference array has a small number of MTJ devices ($Mr = 4$) the signatures the fractional hamming distance between the obtained signatures is around 20%. This is due to the fact that the reference currents over all reference groups have a larger distribution around the nominal value due to the fact that the high variability affecting the MTJ device is not completely compensated by the series/parallel reference design. This effect is reduced when large number of MTJ devices are used to generate the reference current.

We have established, so far, that our PUF implementation is reliable, since only a small fraction of the active cells (<5%) are unreliable under all challenges. We have also shown that the number of possible challenges is limited by the inefficiency of using permutations.

The next and last step in our analysis is to evaluate the unpredictability of the PUF solution. We perform this analysis in tow steps: first, we evaluate the unpredictability of CRPs of a single device (by evaluating the fractional hamming distance between responses corresponding to different challenges - Fig. 7) and second, we evaluate the unpredictability of the entire set of CRPs over different devices (by evaluating the fractional hamming distance between all responses obtained for different devices - Fig. 8).

Our statistical analysis results demonstrate the high degree of unpredictability of our proposed PUF solution. When we compare different CRPs of a single device (Fig. 7) we observe that the mean fractional Hamming distance between responses is very close to 50% (which is the ideal condition for unpredictability). However, some outlier cases are observed for which the responses differ at a lower extent. The probability of occurrence of these outliers has been observed to be significantly reduced when the array is designed with a larger number of columns (and, therefore of reference cell groups) or when a lower number of MTJ devices in one reference group is used (i.e. larger N or smaller Mr). However, both these cases come with tradeoffs: increasing N leads to a bigger device, while decreasing Mr can lead to reduced reliability. These issues have to be accounted for at design time.

When the unpredictability of the entire CRPs set is evaluated from device to device, the results look even more promising (Fig. 8). Large differences are observed in all cases, with fractional Hamming Distance > 40% with small variance. When the number of MTJ devices in one reference group (Mr) is small we observe that the mean value of the fractional Hamming Distance is close to 50% (the ideal value), however there is a large variance of its values. For larger values of Mr, on the other hand, we observe a good value of the fractional Hamming distance, with extremely low variance and reduced number of outliers. These results show that our proposed PUF solution is unpredictable and unique between devices.

All results show that special attention has to be played at design time, when deciding upon the size of the active memory array and the number of MTJ devices in one reference group in such a way to minimize the unreliability and increase the unpredictability and uniqueness of the responses.

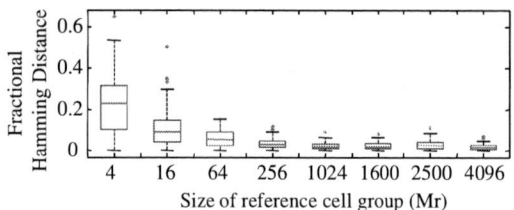

Fig. 6 - Reproducibility of the response assuming that one column is accessed at a time, with permutations of reference cell groups

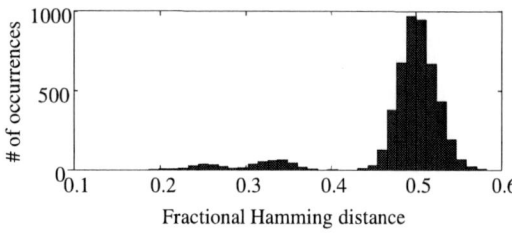

Fig.7 - The distribution of the fractional hamming distance metric obtained for CRPs (simulations obtained from a PUF implemented using Ma = 512 active cell rows, N = 32 active cell columns and reference cell groups, respectively and Mr = 64 MTJ devices in a reference cell group)

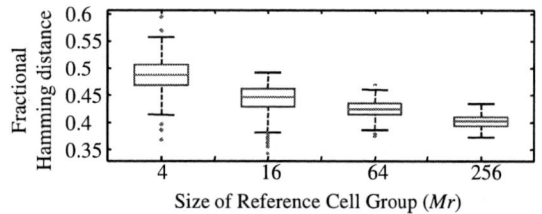

Fig.8 - Box-plots of the fractional hamming metric distributions for PUF unpredictability evaluation for different circuit sizing.

V. CONCLUSIONS

In this work, we proposed an innovative strong PUF solution based on the emerging STT MRAM technology. The basic PUF properties are guarantee by the high variability affecting the electrical resistance of the MTJ device. In order to implement a very high number of challenges, we propose a scheme where columns of the memory array can be electrically connected in parallel, by considering all possible combinations of columns.

We show that the STT-MARM-based architecture is a potential strong PUF solution since statistical simulations demonstrate high robustness, and limited sensitiveness to environmental variations. While the unpredictability is demonstrated by resorting to Hamming Distance calculations, this solution still requires to be validated under know attacks based on Machine Learning.

REFERENCES

[1] P. Koeberl, Ü. Kocabaş, A.-R. Sadeghi, "Memristor PUFs: a new generation of memory-based physically unclonable functions", In Proceedings of the Conference on Design, Automation and Test in Europe (DATE '13). Dresden, Germany, pp. 428-431.

[2] G. S. Rose, et al., "Foundations of Memristor Based PUF Architectures," NANOARCH, July 2013.

[3] W. Che, J. Plusquellic, S. Bhunia, "A Non-Volatile Memory Based Physically Unclonable Function without Helper Data", accepted to ICCAD'14.

[4] Z. Le, F. Xuanyao, C. Chip-Hong, K. Zhi Hui, and K. Roy. 2014. Highly reliable memory-based Physical Unclonable Function using Spin-Transfer Torque MRAM. *IEEE International Symposium on Circuits and Systems (ISCAS).* pp. 2169-2172. (June 2014)

[5] E. I. Vatajelu, G. Di Natale, M. Indaco, and P. Prinetto. 2015. STT-MRAM-Based PUF. To be published in the *Proceedings of the Conference on Design, Automation and Test in Europe (DATE).* (Mar. 2015)

[6] M. Hosomi, et al., "A novel nonvolatile memory with spin torque transfer magnetization switching: spin-ram," in IEEE International Electron Devices Meeting IEDM Technical Digest 2005, pp. 459–462.

[7] M. Durlam, et al., "A 1-Mbit MRAM based on 1T1MTJ bit cell integrated with copper interconnects," IEEE Journal of Solid-State Circuits, 2003, vol.38, no.5, pp.769-773.

[8] R. Ubal, J. Sahuquillo, S. Petit, H. Hassan, P. Lopez, "Leakage Current Reduction in Data Caches on Embedded Systems," Intelligent Pervasive Computing Conference (IPC), 2007, pp.45-50

[9] L. Jing, L. Haixin, S. Salahuddin, K. Roy, "Variation-tolerant Spin-Torque Transfer (STT) MRAM array for yield enhancement," IEEE Custom Integrated Circuits Conference (CICC), 2008, pp.193-196.

[10] S. Zhenyu, L. Hai, C. Yiran, W. Xiaobin, "Variation tolerant sensing scheme of Spin-Transfer Torque Memory for yield improvement," IEEE/ACM International Conference on Computer-Aided Design (ICCAD), 2010, pp.432-437.

Approximate Computing: An Energy-Efficient Computing Technique for Error Resilient Applications

Kaushik Roy and Anand Raghunathan

School of Electrical and Computer Engineering, Purdue University

{kaushik, raghunathan}@purdue.edu

Approximate computing has become a well-known computing technique in recent years. It relies on the ability of many systems and applications to self-heal or to tolerate some loss of quality or optimality in the computed result. The main idea is to exploit the inherent error resiliency or error tolerance of the system to achieve energy efficiency, or in other words, trading accuracy with energy consumption. Such trade-off, in most cases, is also associated with performance improvements like faster operations, area reduction etc. Fortunately, most of the heavy workloads nowadays exhibit intrinsic application resilience [1],[2]. In most multimedia applications, the final output is interpreted by human senses, which are not perfect. This fact averts the need to produce exact outputs. Recent research efforts have quantitatively ascertained the high degree of inherent resilience in many applications. For example, our analysis of a benchmark suite of 12 recognition, vision and multimedia applications shows that on average, 83% of the runtime is spent in computations that can tolerate at least some degree of approximation [3]. Therefore, there is a potential to leverage inherent resilience in a broad context. Let us consider digital signal processing (DSP). In multi-media applications, DSP blocks implement image, sound and video processing algorithms, where the final output is either an image or sound or a video for human senses. When interpreting an image or a sound or a video, human beings have limited perceptual capacities. This allows the system to be flexible in producing quality outputs. As an example, the relaxation on numerical precision provides some freedom to carry out imprecise or approximate computation. Another example of approximate signal processing is the utilization of incremental refinement. In [4], number of ideas and approaches regarding the use of incremental refinement to develop energy efficient structures for the basic application-specific DSP systems such as FFT, DFT, DCT, FIR, IIR etc. have been discussed. It is evident that the tradeoffs may allow approximate computing to handle tasks beyond what we can do with traditional computing.

There are different levels of design abstraction where approximate computing methods can be implemented [4]-[10]. In this paper we will briefly describe different approaches that we have developed to implement approximate hardware for error resilient applications.

Circuits: There are various ways to implement approximation at the circuit level. One commonly practiced method is designing circuits that operate gracefully under overscaled conditions (timing/voltage). In [11] voltage overscaling (VOS) differs from traditional voltage scaling, in that we do not scale the clock frequency, thereby intentionally causing the critical paths to violate the clock period [12]. Note, however, we do not have to take precise corrective measures to preserve the accuracy of computation as the applications that we consider are inherently error resilient [13]. Mitigating the impact of overscaling to the most significant bits, which results in unacceptable loss in quality, is the primary goal. We have implemented an energy-efficient recognition and mining (RM) processor based on the proposed approach [11].

Another way of hardware approximation is designing functionally approximate circuits that differ from the golden specification but contain fewer transistors or gates [14]-[17]. Various imprecise or approximate Full Adder (FA) cells with reduced complexity and internal node capacitances at the transistor level have been designed with minimal errors in its truth table [14]. In SALSA [15] we automatically synthesize approximate circuits ranging from arithmetic building blocks (adders, multipliers, MAC) to entire datapaths (DCT, FIR, IIR, SAD, FFT Butterfly, Euclidean distance), demonstrating scalability and significant improvements in area and power. Hybrid architecture of accurate and approximate building blocks can be utilized to construct highly energy efficient hardware implementations of various applications [14]-[17].

Approximate computation can be carried out at several levels of design hierarchy. The notion behind most efforts is to identify computations that need to be protected and those that can tolerate some errors. We presented a low power reconfigurable DCT architecture which is based on the computation-energy/image-quality tradeoff [18]. Results show that the reconfigurable DCT leads to 47% power savings without seriously compromising the image quality [18]. In [19], the idea of identifying hardware building blocks that exhibit more graceful degradation under voltage overs-scaling is put into practice. The work analyses several commonly encountered algorithms used in DSP, recognition, and mining to identify their computational kernels as meta-functions. For each such application, we found that there exists a computational kernel where the algorithm spends up to 95% of its computation time, and therefore consumes most of the circuit energy. Importantly, all identified meta-functions use the accumulator which becomes the first block to experience timing violation under VOS. By making the accumulator more VOS-friendly,

using dynamic segmentation and delay budgeting of chained units, the quality-energy trade-offs are improved for the meta-functions.

Different strategies are being used to deal with errors that result from overscaling. In some designs, the results produced by the blocks which are subject to timing errors are not directly accepted. Rather, computation is terminated early and intermediate results impacted by timing errors are discarded entirely [20],[21]. From the aspect of gate-level design, such techniques ensure timing correctness of all digital operations.

Approximate memory is another source of energy efficiency. In [22] we proposed a system level design of approximate memory. We applied unequal error protection scheme to memory design that allows significant power reduction by disabling adaptively, less crucial banks of the memory. Furthermore, the utilization of robust 8T cells in crucial banks allowed voltage over-scaling resulting in further power reduction in the system memory [31].

Architecture: At the architecture level, identifying computations for approximation that have lower impact on the quality of results is the main objective. Since all computations do not have equal significance on the output, run-time complexity of programs can be reduced by skipping insignificant computations [4], or by reducing communication between threads to allow better scalability [23]. Neuromorphic system architectures are appropriate candidates for approximate computations. In this context, a systematic transformation, AxNN [24], exploits some of the key properties of artificial neural network such as backpropagation and training to mitigate the impact of approximations on output quality.

Significance-driven computation (SDC) [25] (where important computation elements are better protected than the not so important elements) and scalable effort hardware design [11],[2],[26]-[28] can lead to highly energy efficient implementations. An important feature of such system can be quality configurability, i.e., the gravity of approximation can be controlled based on desired output quality. In the domain of programmable processors, approximate computing may be realized through quality programmable processors, wherein the notion of quality is codified into the natural hardware/software interface. As a first embodiment, we developed a quality- programmable vector processor 'QUORA' that demonstrated significant energy improvements for applications from the recognition, mining, and search domains [29].

Design Approach: The efficacy of approximate computing depends on the simplicity of the techniques applied, i.e., the overall method should be realized without substantial increase in the design effort so that the overhead is minimal. To ensure this, supporting design methodologies and tools must be developed. Tools which can ascertain the parts of an application that are compliant to approximate computing, and identify specific approximate computing techniques that are most apposite to these parts have been developed and has in-built models of approximate computing techniques [3]. In [30], verification tools are presented that can verify "quality" and/or approximate equivalence of hardware and software implementations.

Also based on arbitrary RTL specifications and quality constraints, synthesis tools that can produce "correct by construction" approximate circuits have been developed [15]-[17].

In summary, paradigms like approximate computing have enabled a vast scope of implementation strategies for error resilient circuits. Ubiquitous presence of error resilient applications provides designers with a unique dimension for optimizing power consumption and area of a circuit. The techniques we have proposed at various levels of design abstraction need to be synthesized into a unified, cross-layer framework. We believe that the proposed concepts can be applied to realize highly energy-efficient implementations of a wide range of Multimedia, Recognition and Data Mining algorithms.

Acknowledgement: This research was funded in part by National Science Foundation.

REFERENCES

[1] S. Chakradhar et. al. "Best-effort computing: Re-thinking parallel software and hardware." In *Proc. DAC*, 2010.

[2] V. K. Chippa et. al. "Approximate computing: An integrated hardware approach." In *Proc. ASILOMAR*, pages 111–117, 2013.

[3] V. K. Chippa et. al. "Analysis and characterization of inherent application resilience for approximate computing." In *Proc. DAC*, pages 1–9, 2013.

[4] S. H. Nawab et. al. "Approximate signal processing." *Journal of VLSI signal processing systems for signal, image and video technology*, pages 177–200, 1997.

[5] R. Hegde et. al. "Energy-efficient signal processing via algorithmic noise- tolerance." In *Proc. ISLPED*, pages 30–35, 1999.

[6] K. Palem et. al. "Ten years of building broken chips: The physics and engineering of inexact computing." *ACM Trans. on Embedded Computing Systems*, pages 1–23, 2013.

[7] H. Esmaeilzadeh et. al. "Architecture support for disciplined approximate programming." In *Proc. ASPLOS*, pages 301–312, 2012.

[8] S. Sidiroglou-Douskos et. al. "Managing performance vs. accuracy trade-offs with loop perforation." In *Proc. ACM SIGSOFT symposium*, pages 124–134, 2011.

[9] S. Narayanan et. al. "Scalable stochastic processors." In *Proc. DATE*, pages 335–338, 2010.

[10] J. Han et. al. "Approximate computing: An emerging paradigm for energy-efficient design." In *Proc. ETS*, pages 1–6, 2013.

[11] V. K. Chippa et. al. "Scalable effort hardware design." *IEEE Trans. On VLSI Systems*, pages 2004–2016, Sept 2014.

[12] N. Shanbhag, "Reliable and energy-efficient digital signal processing," In *Proc. DAC*, pp. 830–835, 2002.

[13] L. Leem et al., "Error resilient system architecture (ERSA) for probabilistic applications," In *Proc. DATE*, 2010.

[14] V. Gupta et. al. "IMPACT: imprecise adders for low-power approximate computing." In *Proc. ISLPED*, pages 409–414, 2011.

[15] S. Venkataramani et. al. "SALSA: systematic logic synthesis of approximate circuits." In *Proc. DAC*, 2012.

[16] S. Venkataramani et. al. "Substitute-and-simplify: A unified design paradigm for approximate and quality configurable circuits." In *Proc. DATE*, pages 1367–1372, 2013.

[17] A. Ranjan et. al. "ASLAN: synthesis of approximate sequential circuits." In *Proc. DATE*, pages 1–6, 2014.

[18] J. Park et. al. "A low power reconfigurable DCT architecture to trade off image quality for computational complexity." In *Proc. ICASSP*, vol. 5, pp. V-17, 2004.

[19] D. Mohapatra et al. "Design of voltage-scalable meta-functions for approximate computing." In *Proc. DATE*, pp. 1-6, 2011.

[20] G. Karakonstantis et. al. "System level DSP synthesis using voltage overscaling, unequal error protection & adaptive quality tuning." In *Proc. SiPS, IEEE Workshop on*, pp. 133-138. IEEE, 2009.

[21] N. Banerjee et. al. "Process variation tolerant low power DCT architecture." In *Proc. DATE, EDA Consortium*, 2007.

[22] G. Karakonstantis et. al. "Logic and memory design based on unequal error protection for voltage-scalable, robust and adaptive DSP systems." *Journal of Signal Processing Systems 68*, no.3 pp: 415-431. 2012

[23] J. Meng et. al. "Best-effort parallel execution framework for recognition and mining applications." In *Proc. IPDPS*, 2009.

[24] S. Venkataramani et. al. "AxNN: energy-efficient neuromorphic systems using approximate computing." In *Proc. ISLPED*, pages 27–32, 2014.

[25] D. Mohapatra et. al. "Significance driven computation: A voltage-scalable, variation-aware, quality-tuning motion estimator." In *Proc. ISLPED*, pages 195–200, 2009.

[26] V. K. Chippa et. al. "Scalable effort hardware design: Exploiting algorithmic resilience for energy efficiency." In *Proc. DAC*, pages 555–560, 2010.

[27] V. K. Chippa et. al. "Dynamic effort scaling: Managing the quality-efficiency tradeoff." In *Proc. DAC*, 2011.

[28] V. K. Chippa, S. Venkataramani, K. Roy, and A. Raghunathan. "StoRM: A stochastic recognition and mining processor." In *Proc. ISLPED*, pages 39–44, 2014.

[29] S. Venkataramani et. al. "Quality programmable vector processors for approximate computing." In *Proc. MICRO*, pages 1–12, 2013.

[30] R. Venkatesan et. al. "MACACO: modeling and analysis of circuits for approximate computing." In *Proc. ICCAD*, pages 667–673, 2011.

[31] I. Chang, D. Mohapatra, and K. Roy, "Priority-Based 6T/8T Hybrid SRAM Architecture for Aggressive Voltage Scaling in Video Application," *IEEE Transactions on Circuits and Systems for Video Technology*, pp. 101-112, February 2011.

Near/Sub-Threshold Circuits and Approximate Computing: the Perfect Combination for Ultra-Low-Power Systems

Jeremy Schlachter[†], Vincent Camus[†], Christian Enz

Integrated Circuits Laboratory (ICLAB)

Ecole Polytechnique Fédérale de Lausanne (EPFL), Switzerland

jeremy.schlachter@epfl.ch, vincent.camus@epfl.ch

[†]these authors contributed equally to this work

Abstract—**While sub/near-threshold design offers the minimal power and energy consumption, such approach strongly deteriorates circuit performances and robustness against PVT (process/voltage/temperature) variations, leading to gigantic speed penalties and large silicon areas. Inexact and approximate circuit design can address these issues by trading calculation accuracy for better silicon area, circuit speed and even better power consumption. This paper reviews and proposes improvements for two approximate computing techniques applicable to arithmetic circuits: gate-level pruning and carry speculation. A critical study is then carried out considering several error metrics, and for the first time, those techniques are combined to produce approximate adders showing even higher gains at similar error levels. It is then shown that those techniques can be applied to a sub-threshold library to mitigate the large variability.**

Keywords—*Approximate computing, inexact circuits, speculative adder, pruning, sub-threshold.*

I. INTRODUCTION

Sub/near-threshold circuit are capable of ultra-low power and the best possible energy-efficiency compared to conventional super-threshold circuits. However, with this approach, the Process Voltage and Temperature (PVT) variations become huge and thus, significantly complexify the design process to meet timing and area constraints. Moreover, the circuit speed is drastically reduced and memories require a large area to be functional in the sub-threshold domain. It would therefore only be possible to execute low-complexity applications on a sub/near-threshold hardware. A potential solution to overcome these limitations is the use of inexact circuits. Indeed, approximate circuits are capable of significant power, area and delay reductions compared to their conventional counterparts. Applying inexact design techniques on sub/near-threshold hardware, can significantly relax the design constraints, speed up the nominal frequency of operation and eventually further increase the energy efficiency at the cost of some occasional errors.

This article reviews two inexact design techniques that can easily be implemented in a standard near/sub-threshold digital flow: probabilistic pruning and carry speculation. A critical comparison is carried out and finally, the demonstration is made that the two techniques can be combined to further reduce power

consumption, critical path delay and silicon area. The remainder of this paper is organised as follow: section II reminds the state-of-the-art for the two design techniques, section III re-defines the error metrics and provides a functional description of the pruning and the speculation technique, section IV carries out a comparative study considering several error metrics and depicts how both design techniques can be combined to save even more silicon area, power and critical path delay. Finally, section V demonstrates that inexact design can help to overcome the limitations of sub-threshold circuits.

II. STATE-OF-THE-ART

A. Probabilistic Pruning

Probabilistic pruning is a design technique consisting in removing circuit blocks or elements such as full adder cells, gate clusters or single gates in order to trade exactness of computation against power, area and delay savings without any overhead. The decision of pruning one of these elements is based on two parameters: the significance, which is a structural parameter, and the activity determined by hardware simulations. The amount of error is simply proportional to the number of pruned elements. This technique was first introduced in [1], where full adder cells were removed from various adder architectures, resulting in gains of up to 7.5 X in Energy-Delay-Area Product (EDAP), for a 10 % relative error magnitude.

Later on, this technique has been improved by integrating it in the standard digital flow using existing tools, and by applying pruning at the gate level [2]. This finer granularity enables an order of magnitude area and power savings for a 64-bit adder with 10 % relative error magnitude. It has also been shown that 25 % power and area reduction can be achieved for 16-bit multipliers.

B. Speculative Adders

Speculative adders [3] exploit the fact that carry propagate sequences in additions are typically short, making it possible to estimate intermediate carries using a limited number of previous stages. They split the binary addition into several subpaths executed concurrently for higher execution speed and energy efficiency, but at the risk of generating occasionally incorrect results. Thus, the critical path of the adder can be divided in two or more shorter paths, relaxing constraints over the entire design and improving the speed, area and power beyond the theoretical bounds of exact adders.

This work was supported by the NanoTera "IcySoC" project and the Swiss National Science Foundation grant No 200021-144418.

A number of speculative adders have been proposed in literature with different approaches in order to reduce the error frequency or magnitude. The ETAII adder [4] consists of regular sub-adder blocks with input carries speculated from Carry Look Ahead (CLA) blocks of the same length. In the ETAIIM version, several of the most significant CLA blocks are chained to increase accuracy. The ETBA adder [5], direct descendent of the ETAIIM, adds variable speculation signs and sub-adder sum balancing multiplexed blocks to mitigate relative errors. The ETAIV [6] and CSA [7] adders have enhanced accuracy by considering two prior carry speculation blocks instead of one, coupled respectively with a carry select or a carry skip technique, with the latter also using sum balancing over several sub-adder blocks. On the other side, ISA [8] and CSC [9] adders have recently improved circuit performance and efficiency by introducing off-critical path error reduction techniques. The ISA adder concept [8] has also proposed an optimal and generalized approach of speculative compensated adders, encompassing aforementioned adders, and has introduced a simple methodology to allow designers to generate efficient architectures from a delay-accuracy specification.

III. INEXACT ARITHMETIC CIRCUIT DESIGN

A. Gate-Level Pruning

Gate-Level Pruning is a CAD technique to automatically generate inexact circuits starting from a conventional design by adding only one small step in the digital design flow. The CAD framework is presented in Fig. 1.

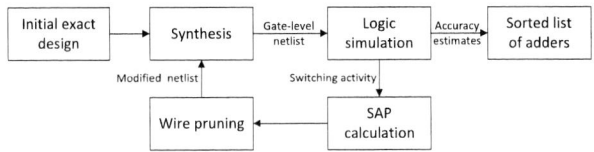

Fig. 1: CAD framework for Gate-Level Pruning.

Any exact circuit can be represented by a directed acyclic graph as depicted in Fig. 2, where the nodes are components such as gates, and whose edges are wires. The decision to prune a node is based on two criteria: the significance, which is a structural parameter, and the activity or toggle count. The nodes with the lowest Significance-Activity Product (SAP) are pruned first. By doing so, the error magnitude grows with the amount of pruning. Alternatively, depending on the application's requirement, the designer may choose to prune nodes according to the activity only, in order to minimize the error rate.

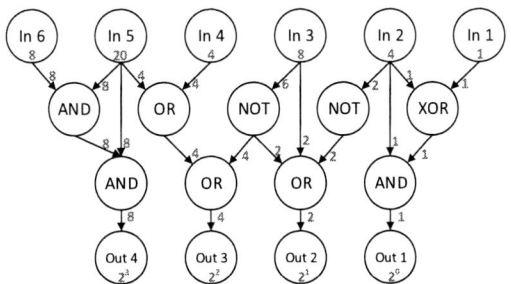

Fig. 2: Directed acyclic graph representation of a gate level netlist and the associated significance attribution

The activity of each wire is extracted from the SAIF file (Switching Activity Interchange Format) obtained through gate-level hardware simulations. This file contains the toggle count of each wire, as well as the time spent at the logic levels 0 and 1 respectively. In order to get an accurate activity estimation, the system should be simulated with an input stimulus representative of the *real operation* of the circuit. The more the simulation is realistic, the more the toggle count is accurate and leads to an efficient pruning.

The significance of each primary output is set by the designer depending on the application's requirement. In this paper, pruning is applied on several arithmetic circuits where each primary output is weighted by a power of two. It is therefore worth applying a weighted significance attribution, where each bit position has a significance two times higher than the previous when moving from the Least Significant Bit (LSB) to the Most Significant Bit (MSB). Reverse topological graph traversal is then performed to compute each nodes' significances as follows:

$$\sigma_i = \sum \sigma_{desc(i)} \tag{1}$$

where σ_i is the significance of the node i and $\sigma_{desc(i)}$ is the significance of the direct descendants of node i. An example of weighted significance attribution is shown in Fig. 2.

Once the significance and activity is determined, the nodes, i.e. gates and their corresponding wires, are ranked according to their SAP. The ones with the lowest SAP are disconnected from the verilog netlist, and an incremental re-synthesis is performed in order to remove or replace the unconnected gates.

B. Inexact Speculative Adders

1) General Concept

The general block diagram of an Inexact Speculative Adder (ISA) adder is depicted in Fig. 3. An ISA splits the carry propagation chain in multiple paths executed concurrently. Each path consists of a carry speculator block (SPEC), a sub-adder block (ADD) and an error compensation block (COMP). For each of these SPEC-ADD-COMP paths, the different blocks have the following functions:

- **SPEC** – The speculator block generates a partial carry signal from a limited number of operand bits in a carry look-ahead approach and sourced by either a static or a dynamic input. When a propagate chain covers the full SPEC block, the exact carry cannot be speculated from the partial product and the output carry is guessed at the

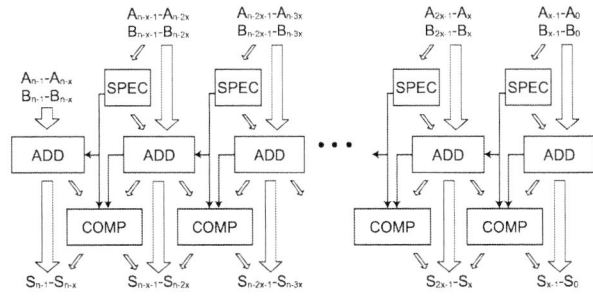

Fig. 3: General block diagram of an Inexact Speculative Adder (ISA). Each speculative segment consists of a carry speculator (SPEC), a regular adder (ADD) and an error compensation block (COMP).

978-1-4799-8720-7/15 $31.00 © 2015 IEEE

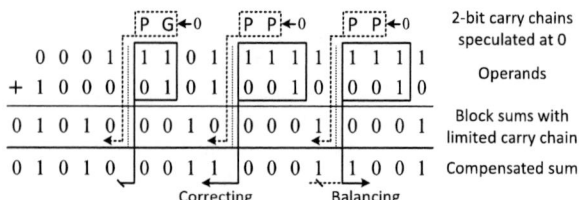

Fig. 4: Example of ISA addition arithmetic with 2-bit speculation, 1-bit correction and 1-bit error reduction. Faults only occur in the two right-hand paths. The 1st LSB of the central path can be corrected. The 1st LSB of the right path cannot be corrected, so the 1st MSB of the preceding sum is flipped.

input value. As long propagate sequences are uncommon in uniform input distribution [4], the probability of fault decreases when increasing the size of this block.

- *ADD* – The sub-adder block calculates local sums from the speculated carry of the SPEC block.

- *COMP* – Without compensation, an internal overflow caused by an inconsistent carry could lead to a massive error. Therefore, the COMP block detects those speculation faults by comparing the carry generated from the SPEC with the carry-out coming from the prior ADD block. It then compensates faulty sums either by attempting to correct a few bits of the local sum or by reducing relative error over a few bits of the preceding sum.

The first speculative path, operating on the LSBs of the adder, does not have SPEC nor COMP blocks since it uses directly the adder carry-in. The achieved addition arithmetic is illustrated in Fig. 4.

2) Error Compensation

The COMP's error correction technique, introduced in [8], consists in incrementing or decrementing only a small group of LSBs of the local sum in order to compensate the erroneous speculated carry. In most cases, this can fully resolve carry errors. In the case where those stages are all in propagate modes, correction is impossible as it would lead to an internal overflow. In that case, the uncorrected bits, having a higher significance than the error bit, ensure a low relative error of the result. Using the COMP's error correction technique thus reduces both error rate and relative error. The correction hardware is executed concurrently to the local addition, thus this technique impacts minimally the critical path of the adder.

The COMP's error reduction technique consists in balancing a group of MSBs of the preceding sub-adders in opposite direction than the error. This technique, similarly as in [5], has been intensively employed in literature. But to avoid high relative errors and better control the worst-case error (RE_{MAX}), it relies on large SPEC block directly lying in the critical path of the adder.

3) Design Strategy

The ISA offers a general topology of speculative compensated addition inclusive of the state-of-the-art and that allows an optimal balance between circuit and accuracy specifications.

A design methodology through a delay-accuracy approach is presented in Fig. 5. The adequate delay tradeoff is mainly obtained by sizing SPEC and ADD blocks, principal slack elements of the ISA. Then, the COMP's error correction and error reduction techniques enable to tune and fit the accuracy

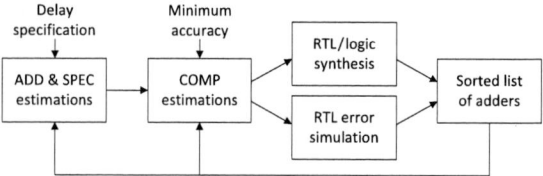

Fig. 5: CAD framework for ISA design.

requirements at the cost of hardware overhead and with a minimum delay penalty for multiplexing the result on a few compensated bits.

Adders in literature describe particular cases of implementation excessively considering either performances or errors. In the ISA architecture, the speculation overhead can be traded for longer sub-adders while fitting the same delay requirement. It is then possible to use fewer speculative paths and limit the in-critical path speculation-compensation overhead to a few bits of each path while fitting the accuracy requirement. This approach allows notable improvements in circuit performances [8].

IV. RESULTS AND COMPARATIVE STUDY

A. Accuracy Metrics

The metrics used to characterize approximate adders in this work are based on the relative error (RE), defined as:

$$RE = \left| \frac{S_{approx} - S_{correct}}{S_{correct}} \right| \qquad (2)$$

where S_{approx} and $S_{correct}$ are respectively the approximate and correct sums of an addition. In [1], three interesting metrics are defined:

- *Error Rate* – The error rate corresponds the ratio of erroneous computations over the entire set of computations and is defined as follow:

$$Error\ Rate = \frac{Number\ of\ erroneous\ computations}{Total\ number\ of\ computations} \qquad (3)$$

- *Relative Error RMS (RE_{RMS})* – The root mean square (RMS) of RE is a good estimator of accuracy and is interesting for many applications, particularly in image and video processing. It is defined as:

$$RE_{RMS} = \sqrt{\frac{1}{N} \sum_{n=1}^{N} RE_n{}^2} \qquad (4)$$

- *Maximum Relative Error (RE_{MAX})* – RE_{MAX} represents the largest relative error of an adder and defines its worst case accuracy. It is here obtained over a set of computations.

B. Pruning or Speculation Applied Individually

In order to perform a comparative study, both techniques have been applied on 32-bit adders synthesized in a 65 nm UMC technology library and all the designs have been simulated with a set of five million uniformly distributed random inputs. Comparisons of error characteristics and normalized costs in terms of energy and Power-Delay-Area Product (PDAP) are shown for a selection of pruned and speculative adders synthesized at 3.3 GHz in Fig. 6a and 6c and synthesized at 1.3 GHz in Fig. 6d and 6f.

Only speculative adders with regular structures have been synthesized (i.e. 2x16, 4x8, 8x4 and 16x2 bit concurrent

978-1-4799-8720-7/15 $31.00 © 2015 IEEE

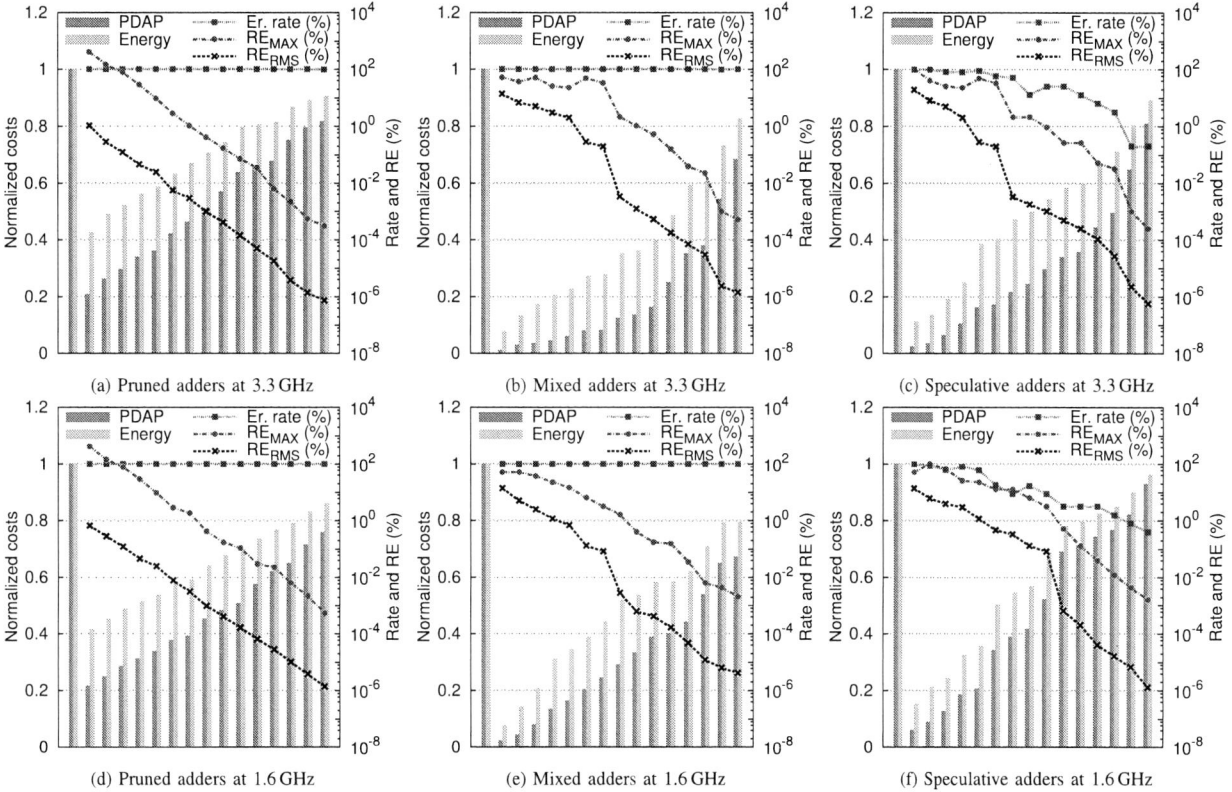

Fig. 6: Error characteristics and normalized cost of 32-bit pruned, mixed and speculative adders synthesized at 3.3 GHz and 1.6 GHz.

paths) with diverse error characteristics. For this reason, the displayed characteristics present steps corresponding to changes of structure sizes. Generally, ISA adders built out of small sub-adders show high errors and high savings (on the left of the figures), whereas ISA with large sub-adders are preferred for low errors and lower savings (on the right of figures).

Fig. 6 clearly show that the two techniques have a different impact on the output quality. The error rate of pruned adders rapidly reaches 100 %, the reason being that in the first steps of the pruning process, some of the least significant outputs are removed. On the other hand, in speculative adders, a small speculation-correction overhead leads to a decrease of the error rate despite lower circuit efficiency.

For both techniques and frequencies, the RE_{RMS} and the RE_{MAX} grow with an exponentially trend versus circuit savings. The ISA adders on the left of figures have a low RE_{MAX}, but this one does not follow the same exponential trend as it is expensive to control. Thus, a gap appears between RE_{MAX} and RE_{RMS} when the constraints on the circuit become too high (at 1 % for 3.3 GHz and 10^{-3} % at 1.6 GHz).

Timing constraint has a significant influence on the result obtained with the two techniques. Fig. 6a and Fig. 6c show that at high frequency of 3.3 GHz, and for a relative PDAP cost of 0.42, the RE_{MAX} and the RE_{RMS} of the pruned adder are equal to 4 % and 0.008 % respectively. In comparison, the speculative adder having a similar PDAP has a RE_{MAX} of 10^{-1} % and a RE_{RMS} of 10^{-4} %. This could lead to the conclusion that the speculation technique can achieve similar energy savings than the pruning technique, at a much higher accuracy level.

However, Fig. 6d and Fig. 6f actually depict the opposite trend when using a slightly lower frequency of 1.6 GHz. Hence, a more extensive comparative study might show that the two depicted design techniques might produce uncorrelated errors, and therefore could be combined to get additive savings.

C. Mixing of Pruning and Speculation

All the previous paragraphs were discussing the individual use of the pruning and the speculation techniques on 32 bit adders, once at a frequency of 3.3 GHz, and once at 1.6 GHz. The results clearly show that the pruned and the speculative adders produce different types of errors. It is therefore worth combining the two techniques. Since the pruning methodology is only one additional step in the standard digital flow, the general methodology to combine speculative circuits and pruning is:

- Synthesize speculative circuits at the desired speed constraint starting from the HDL description.

- Apply the pruning methodology to the gate-level netlists of the speculative circuits.

Fig. 6b and Fig. 6e show the normalized costs as well as the error characteristics of *pruned speculative* (mixed) 32-bit adders synthesized at 3.3 GHz and 1.6 GHz respectively. It is very interesting to see that the RE_{RMS} and the RE_{MAX} follow almost the same trend for the mixed adders than for the speculative adders. This is due to the fact that the two techniques produce different types of errors: in speculative adders which are cutted into sub-blocks, errors are rare but can occur as often on

TABLE I: Comparison of the three inexact design techniques.

Specifications		Normalised PDAP		
Frequency	RE_{RMS}	Pruning	Mixed	Speculative
	$3 . 10^{-6}$ %	0.76	**0.54**	0.64
3.3 GHz	10^{-3} %	0.52	**0.14**	0.3
	2 %	0.2	**0.05**	0.1
	10^{-5} %	0.64	**0.54**	0.78
1.6 GHz	10^{-1} %	0.3	**0.2**	0.43
	1 %	0.2	**0.14**	0.2

LSBs than on MSBs, the latter having a significant impact on the error magnitude. In opposition, the weighted significance attribution of the pruning methodology leads to a large number of errors — and thus an error rate 100 % for most of the pruned adders — but those are limited to the LSBs and have a small impact on the error magnitude. Hence, the assumption can be made that the errors resulting from pruning and speculation are uncorrelated, and thus, equalizing the error levels resulting from each technique enables additive area, power and delay savings. This is verified by simulation, all the *pruned speculative* adders depicted Fig. 6b and 6e have a lower PDAP cost and consume less energy than the pruned or the speculative adders.

Table I summarises the PDAP reduction obtained through the use of the three techniques, namely pruning, speculation and the mixing of both. It is shown that a *pruned speculative* adder with only 2 % relative error magnitude has a normalised PDAP of 0.05, which is a factor 20 improvement. Moreover, energy consumption is reduced by a factor 4 compared to the exact 32 bit adder synthesized at the same speed.

V. INEXACT DESIGN IN THE SUB-THRESHOLD DOMAIN

On the one hand, the demonstration has been made that inexact design techniques applied individually, or even combined, in a UMC 65 nm technology can lead to drastic gains in energy consumption, silicon area and critical path delay. On the other hand, sub-threshold suffer from large variability, low speed of operation and require a large silicon area. In order to verify that sub-threshold circuits can benefit from approximate computing, gate-level pruning has been applied on an 32 bit adder synthesized with a 180 nm sub-threshold library, at a frequency of 22.7 MHz. Sub-threshold pruned adders Fig. 7 show similar savings and error characteristics than the ones synthesized with the UMC 65 nm technology. In this regards, speculative and *pruned speculative* might show similar gains in the sub-threshold domain than in the standard UMC 65 nm library.

VI. CONCLUSION

This paper reviewed and compared two well established techniques for generating approximate hardware: carry speculation and gate-level pruning. It has been shown that both can achieve up to 85 % PDAP reduction for a RMS relative error of 1 %. Moreover, since the two techniques clearly have

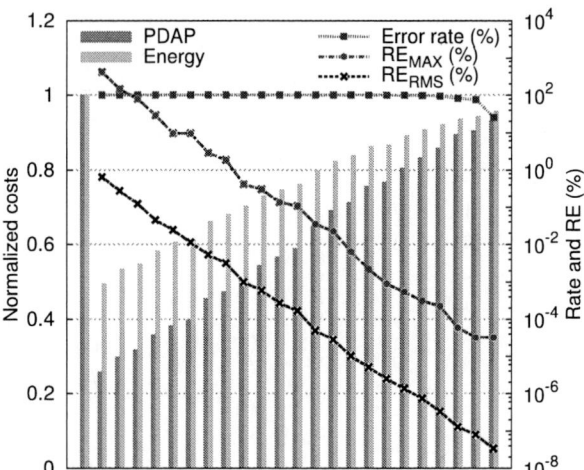

Fig. 7: Pruned 32-bit adder synthesized with a 180 nm sub-threshold library

a different impact on the accuracy of the generated adders, they can be combined to achieve up to a factor 20 saving in EDAP for only 2 % RE_{RMS}. These improvements can clearly be exploited to mitigate the large area and PVT variability as well as the low speed of sub-threshold circuits: pruned sub-threshold 32-bit adders can reduce the PDAP by a factor 4 compared to exact ones and a future work might demonstrate that the use of *pruned speculative* adders in the sub-threshold domain would lead to even higher gains.

REFERENCES

[1] A. Lingamneni, C. Enz, J. L. Nagel, K. Palem, and C. Piguet, "Energy parsimonious circuit design through probabilistic pruning," in *Design, Automation Test in Europe Conference Exhibition (DATE), 2011*, March 2011, pp. 1–6.

[2] J. Schlachter, V. Camus, C. Enz, and K. Palem, "Automatic Generation of Inexact Digital Circuits by Gate-level Pruning," in *Circuits and Systems (ISCAS), 2015 IEEE International Symposium on*, May 2015.

[3] T. Liu and S.-L. Lu, "Performance Improvement with Circuit-level Speculation," in *Microarchitecture, 2000. MICRO-33. Proceedings. 33rd Annual IEEE/ACM International Symposium on*, 2000, pp. 348–355.

[4] N. Zhu, W.-L. Goh, and K.-S. Yeo, "An Enhanced Low-power High-speed Adder For Error-tolerant Application," in *Integrated Circuits (ISIC), Proc. of the 2009 12th International Symposium on*, Dec 2009, pp. 69–72.

[5] M. Weber, M. Putic, H. Zhang, J. Lach, and J. Huang, "Balancing Adder for Error Tolerant Applications," in *Circuits and Systems (ISCAS), 2013 IEEE International Symposium on*, May 2013, pp. 3038–3041.

[6] N. Zhu, W.-L. Goh, G. Wang, and K.-S. Yeo, "Enhanced Low-power High-speed Adder for Error-tolerant Application," in *SoC Design Conference (ISOCC), 2010 International*, Nov 2010, pp. 323–327.

[7] Y. Kim, Y. Zhang, and P. Li, "An Energy Efficient Approximate Adder with Carry Skip for Error Resilient Neuromorphic VLSI Systems," in *Computer-Aided Design (ICCAD), 2013 IEEE/ACM International Conference on*, Nov 2013, pp. 130–137.

[8] V. Camus, J. Schlachter, and C. Enz, "Energy-Efficient Inexact Speculative Adder with High Performance and Accuracy Control," in *Circuits and Systems (ISCAS), 2015 IEEE International Symposium on*, May 2015.

[9] J. Hu and W. Qian, "A New Approximate Adder with Low Relative Error and Correct Sign Calculation," in *Design, Automation and Test in Europe (DATE), 2015 IEEE Conference and Exhibition on*, March 2015.

978-1-4799-8720-7/15 $31.00 © 2015 IEEE

Sub-Threshold Design and Architectural Choices

Christian Piguet, Marc Pons, Daniel Séverac
CSEM Centre Suisse d'Électronique et Microtechnique SA
Neuchâtel, Switzerland

Abstract— Standard cell design and memory design need to be optimized for sub-threshold operation. It is interesting to revisit digital block architectures when implemented using these sub-threshold basic bricks. Out of many possible architectures for the same logic function (i.e. multiplier), it turns out that there are optimal sub-threshold architectures.

Keywords—CMOS, IC, sub-near-threshold, architectures, parallel, serial, energy reduction

I. INTRODUCTION

There is a clear trend today towards ultra-low-power devices for IoT (Internet of Things) for wireless sensor nodes, for wearable and for implanted medical devices. These devices will be supplied either by miniature batteries, for which energy reduction is key, or by energy harvesting mechanisms, for which the amount of available energy is critical.

All these applications generally require very low performances that sub- or near-threshold designs can reach, while maximizing energy efficiency by power supply voltage tuning (sub- or near-threshold operation can be defined respectively as supplying the circuits at a voltage *Vdd* lower than the threshold voltage *Vth* of transistors or just above). Another advantage of sub-threshold design is to extend the functionality down to ultra-low-voltage (0.4 V to 0.3 V in 180nm technology). For example, RFID can work at higher distance as voltage level required for operation is reduced.

CMOS logic running at so low supply voltages do requires special standard cell libraries and memories. Section II will present how these sub-near-threshold libraries and memories have been designed. A significant issue is the variability, as these sub-near-threshold circuits are very sensitive to the *Vth* variations of the transistors. Delays are roughly inversely proportional to *Vdd- Vth*. With *Vth* close to *Vdd*, any very small *Vth* variation will have a significant impact on the delay.

Another very important question is to determine if there is an optimal sub-near-threshold architecture for a given logic function. It is the goal of this paper to explore solutions to this question. Section III will present the variability issue and its impact on digital architectures like processor cores or arithmetic units.

Section IV will raise the following question related to architectures: is it better in sub-near-threshold to design serial architectures or parallel architectures for the same task? Several experiments will be described to answer to this difficult question.

II. SUB-NEAR-THRESHOLD DESIGN

A. Transitor sizing for sub-threshold operation

The design of sub-near-threshold basic cells is different compared to conventional cells at nominal voltage. It is due to the well-known RSCE (Reverse Short Channel Effect) and INWE (Inverse Narrow Width Effect). Notice that DIBL (Drain Induced Barrier Lowering) effect is negligible in sub-threshold.

RSCE is a *Vth* reduction if the channel length L of transistors is increased (about 20%), which imply a speed increase (counter intuitive). When L is small, the MOS halos are touching each other, so *Vth* is higher. When L is larger, halos do not touch and *Vth* is reduced. In addition, a larger L is beneficial for reducing sensitivity to *Vth* variability.

Fig. 1 shows this RSCE for a 180 nm technology. The upper and red curve for N-MOS starts at 180 nm, the delay is maximum for L=300 nm, but decreases significantly for L=400 or 500 nm. There is no RSCE effect for P-MOS (the other curve).

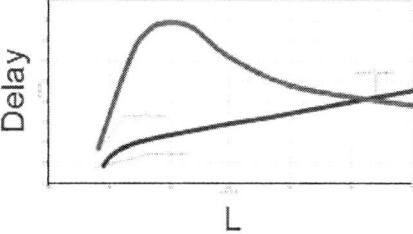

Fig. 1. Delay versus L in a 180 nm, RSCE for N-MOS (upper and red curve), no RSCE for P-MOS (lower and blue curve).

The choice of L for N-MOS and P-MOS is a tradeoff. L should not be smaller than 350 nm for avoiding too much variability, it can be 600 nm for the same N-MOS and P-MOS delay, or smaller for layout area, for instance selecting 500 nm for N-MOS and 350 nm for P-MOS.

Regarding INWE, it is a *Vth* reduction if the channel width W of transistors is reduced, so a speed increase, due to STI (Shallow Trench Isolation). Increasing W to have more current is in fact not so beneficial. It is better to have two smaller W transistors than a single big one. For instance, for a N-MOS of W=420 nm, doubling the W to 840 nm reduces only slightly the delay (0.9X) while implementing two parallel N-MOS (called transistor fingers) of 420 nm each nearly reduces the delay by 2X.

B. Sub-threshold Standard Cell Library Design

CSEM has designed several standard cell libraries (180 nm, 65 nm) based on previous observations for sizing in sub-threshold operation [1]. Silicon area of resulting circuits is increased roughly by a factor 2X (compared to standard voltage library) and speed is significantly degraded. Signal slopes are degraded, and not in the same strength for rising and falling. Hence the clock tree constraints are difficult to achieve for flip-flop designs, resulting in increasing number of buffers and, of course, silicon area and power consumption.

To deal with these issues, two techniques have been used for the design of sub-threshold circuits: a reduced number of standard cells [2] and a latch-based design [3].

Reducing the number of cells to 30 or 40 has two interesting advantages. The first one is obviously a reduction of the design effort. The second one is a speed increase, due to the fact that the library does not contain complex and very slow cells. It is known that complex cell decomposition in several simple cells increases the speed [4], which is even more obvious for sub-threshold design.

The latch-based design consists in using latches and not flip-flops, requiring two clock trees for master and slave latches with significant relaxed timing constraints (no hold time constraints) due to less parasitic capacitances and non-overlapping clock scheme [3]. This is also key for sub-threshold design.

III. SUB-THRESHOLD ARCHITECTURES AND PVT VARIATIONS

It is well-known that sub-threshold logic is much more sensitive to PVT variations (process, voltage and temperature). Mitigating these effects is generally performed at transistor and circuit level. But it is quite interesting to determine if one can increase the resistance to PVT by choosing better architectures for a given logic function.

A first example is provided by multicore architectures that are more resistant to technology variations than single core architectures. The reason is that memory latency and memory bandwidth are much more important for determining throughput in multicore architectures than in single core architectures. In single core architectures, for achieving the same throughput, the core frequency is the main parameter, and it has to be significantly increased compared to multicore architectures. As the core frequency is much more sensitive to technology variations, single core architecture are much more sensitive to technology variations.

Another example is provided by a full adder for which one has to choose the best architecture (ripple carry, carry look ahead) and Vdd for mitigating process variations [5]. As shown in Fig. 2, a ripple carry architecture is better (2X less sensitivity to variations) at 500 mV compared to carry look ahead architecture at 400 mV exhibiting the same performances in speed, dynamic and leakage power.

So it is an example for which the choice of the architecture (ripple carry instead of carry look ahead) for different conditions (for instance a different Vdd) is an interesting design parameter.

	CLA @ 400 mV	RCA @ 400 mV	RCA @ 500 mV
Performance	Fast	Slow	Fast
Dynamic Power	Normal	Low	Normal
Leakage Power	Normal	Low	Normal
Sensitivity to variations	Sensitive	Sensitive	~2X less Sensitive

Fig. 2. Comparison of full adder architecture (CLA: Cary Look-ahead Adder, RCA: Ripple Cary Adder) depending on *Vdd* (400 mV or 500 mV).

IV. SERIAL OR PARALLEL ARCHITECTURE DESIGN

The question is which is the best architecture (i.e. serial or parallel) for a given logic function in sub-threshold design regarding total power consumption and total energy. We have to distinguish "total power consumption" and "total energy". The first one is for applications continuously running while the second one is for devices which are idle most of the time and running for short duty cycles.

A. Total Power Consumption

Parallelization is good for reducing dynamic power while reducing *Vdd*. It is the case for instance for devices operating continuously with energy harvesting devices. As sub-near-threshold is mainly used for reducing dynamic power, it could be good to increase the parallelism. Serial design with less logic gates is good for reducing parasitic capacitance, leakage and silicon area, but in principle not so well adapted for sub-near-threshold (due to speed issues, more cycles to perform the job). Two scenarios can be applied [6]: the best architecture for free selection of *Vdd* and *Vth*, or the best architecture when *Vdd* and *Vth* are fixed by the application.

The first scenario consists in choosing the architecture and the couple *Vdd* and *Vth*. These working conditions will be called optimal working point. The location of this optimal working point and its associated total power consumption are tightly related to architectural and technology parameters. As can be observed in Fig. 3 the ratio of dynamic (*Pdyn*) over static (*Pstat*) power at the minimum of total power consumption (*Ptot*) is not constant and depends on some parameters (e.g. the activity *a* but also on the logical depth *LD*, working frequency, etc.).

Without loss of generality we can write:

$$Pdyn = k_1 * Pstat$$

where the constant k_1 is expected to depend in a complex fashion on architectural parameters.

A first approximation [7, 8] was obtained by considering that the optimum of the total power consumption is obtained with the same amount of static and dynamic power (k_1=1). From the results shown in Fig. 3, this approximation is clearly too rough. The parameter k_1 can be obtained by searching the minimum of *Ptot* depending on *Vdd*. The value of k_1 is roughly located between 1 and 5. Fig. 3 illustrates the fact that reducing the activity *a* allows reducing *Ptot*, whereas it tends to increase the optimal *Vdd* and *Vth*. As architectural modifications will change simultaneously several factors (not just the activity *a*), it is necessary to develop a methodology to evaluate the influence of such transformations on *Ptot*. The originality of this methodology is that at this optimal point there is a given ratio between dynamic and static power, which is strongly correlated to the ratio between active and static currents *Ion/Ioff* of the technology. Moreover, due to the increase of the leakage current in the modern technologies, this ratio *Ion/Ioff* is becoming smaller and smaller. On the other hand, the optimal ratio *Ion/Ioff* is dependent on architectural parameters, which are the activity *a* and the logical depth *LD* (number of gates in series). An approximation for *Ion/Ioff* is given by:

$$Ion/Ioff = k_1 * LD/a$$

Note that k_1 is itself depending on architectural parameters such as *LD* and *a*.

If the *Ion/Ioff* ratio is small (100 to 500), one can see that *LD* has to be small and activity *a* quite large! This can lead to a paradigm shift, as activity reduction has until now been a primary goal, due to the fact that only dynamic power was considered. By optimizing both static and dynamic power, the goal of a low-power design methodology is obviously not always to reduce as much as possible the activity. In fact, very inactive gates or transistors doing nearly nothing are still leaky devices. So it is not optimal to have too many gates or transistors with an extremely low activity. On the other hand, if the reduction of activity is not due to an increase in the number cells but rather to a reduction of the number of transitions, both nominal and optimal total power will benefit from it.

Eleven 16-bit multipliers architectures have been described and synthesized. Their power consumption was analyzed using the same stimuli for all circuits and the optimal *Vdd* and *Vth* were computed. All circuits work at 31.25 MHz, and were synthesized with Synopsys in the UMC 0.18μm technology.

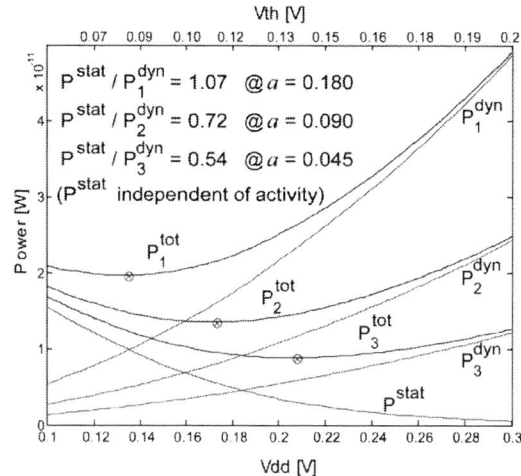

Fig. 3. Total power consumption of a square transistor for the UMC 0.18 um technology. The optimal working point is marked by the cross mark.

The various architectures are the following: Ripple Carry Array or RCA (basic, pipelined 2, pipelined 4, parallel 2, parallel 4), Wallace Tree (basic, parallel 2, parallel 4) and Sequential (basic, 4x16 Wallace, parallel) [6].

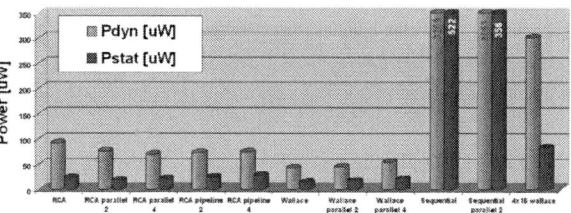

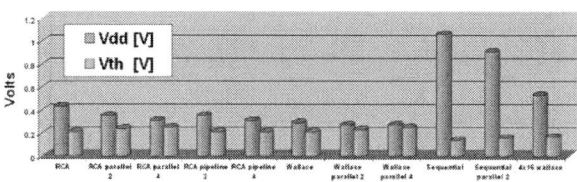

Fig. 4. Optimal total power and corresponding Vdd/Vth couple for eleven 16-bit multiplier architectures at their optimal working point (UMC 0.18 μm, 31.25 MHz).

The architecture providing the lowest total power consumption is the structure based on the Wallace tree, operating at very low *Vdd* and *Vth* (Wallace, *Vdd*=290 mV, *Vth*=210 mV). As could be expected at these low *Vdd* and *Vth*, the *Ion/Ioff* ratio for optimal total power consumption is rather small, approximately between 100 and 1'000 (compared to several million at nominal *Vdd*). Moreover, the percentage of static power over the total power is between 15% and 30%, i.e. *k1* between 3 and 6, far from *k1*=1. The fact that *Ion/Ioff* is very close to *k1*LD/a* shows that the previous formula is a good approximation.

Fig. 4 shows the results in total power consumption for these eleven multiplier architectures as well as the corresponding optimal Vdd and Vth. Parallelizing the RCA (i.e. dividing its LD and a) shows improvements to total power. Similarly, using a two-stage pipelined RCA (i.e. dividing its LD, but keeping activity a almost constant) reduces the total power compared to the basic RCA, as could be expected. However, the register overhead becomes too important with a four-stage pipeline leading to a poorer solution. Although increasing the parallelism of the RCA was beneficial, increasing the parallelism of the Wallace structure does not lead to a better solution. In fact, the total power consumption becomes a little bit higher due to the increased number of cells, which are inactive and leaky. Despite a quite small number of cells, sequential architectures implemented using the add-shift algorithm, present a very large total power due to their lack of speed, requiring a very high Vdd and a very small Vth to fulfil the speed requirements.

A second scenario considers Vdd and Vth as fixed by the application [6]. Even in this case, one can show that an optimal architecture exists. For instance, for multipliers, if $Vdd= 1.0$ V and $Vth = 0.1$ V, it is a sequential architecture that is the winner. It is due to the very small Vth and significant leakage, so it is mandatory to reduce the number of logic gates.

In this experiment, one can see that near-threshold parallel architectures are the best choice for reducing the total power consumption of circuits running at the same speed.

B. Energy: To shoot for a serial architecture?

Generally, sub-near-threshold circuits are designed for energy reduction, as speed performances are very low, so applications like IoT or medical devices are the ones for which sub-near-threshold are quite attractive. Reducing Vdd but keeping the same Vth, delays are significantly increased and also leakage is increased due to larger operation times. So there is also an optimal point for total energy (dynamic and leakage) which depends on Vdd and the chosen architecture.

Reference [9] claims that in this context serial architectures are better than parallel architectures. The paper compares 1-bit, 16-bit and 32-bit ripple carry adders at 300 mV in 22 nm CMOS. The 1-bit adder is obviously much simpler than the 32-bit adder, so silicon area and leakage are significantly reduced. However, for energy, the comparison is not performed at the same speed, but one accepts that the 1-bit adder is running slowly. A priori, it would be 32X slower, requiring 32 cycles instead of one cycle for the 32-bit adder. But as the 1-bit adder is much simpler, a cycle could be faster, and finally, the speed reduction is only 11X, with a reduction of 32X in leakage (due to the reduction of the number of logic gates), with also a reduction of the active energy (40% due to less parasitic capacitances, shorter critical path, less glitches, short interconnect). Even if leakage in sleep mode is 10% of leakage in active mode, if the duty cycle is very low, the sleep mode leakage is so large compared to dynamic and static energy in active mode that the important number is this 32X leakage reduction in sleep mode.

For applications without sleep modes, the paper [9] claims an energy reduction of 40% for 1-bit design compared to 32-bit design. It is not what we have seen with the multiplier designs in the previous experiment (but we perform the comparison for the same speed, which is completely different!!). But anyway, serial architectures will be attractive if it is possible to accept a quite longer delay for achieving some operations (like 11X speed reduction in the adder). If one has tough timing constraints, parallelization is good for achieving the required speed while reducing total power. One has also to look at Fig. 2 in which a more serial adder design was also a quite good architecture. This is also mentioned in [9] where a slight increase in Vdd drastically increases the speed of sub-threshold serial architectures. It shows that even serial architectures (1-bit adder) can be as fast as parallel ones (32-bit adder) with lower energy. This is true in sub-threshold, but not in super-threshold for which an increase of Vdd does not increase so much the speed.

V. CONCLUSION

It is interesting to see that architecture designers can also contribute to sub-near-threshold design. Near-threshold parallel architectures are suitable for reduction of power if the comparison is performed at the same speed. Sub-threshold serial architectures are suitable for energy reduction if one accepts to run slower (IoT) in case of very large duty cycle.

ACKNOWLEDGMENT

This work has been conducted through a Swiss Nanotera project called IcySoC, with the following partners: CSEM, EPFL, ETHZ, EM Microelectronics, Switzerland.

REFERENCES

[1] Marc Pons, Jean-Luc Nagel, Daniel Séverac, Marc Morgan, Daniel Sigg, Pierre-François Ruedi and Christian Piguet, " Ultra Low-Power Standard Cell Design using Planar Bulk CMOS operating in Subthreshold", PATMOS 2013, Kaiserslautern, September 9-11, 2013

[2] J.-M. Masgonty, S. Cserveny, C. Arm, P.-D. Pfister, and C. Piguet, "Low-Power Low-Voltage Standard Cell Libraries with a Limited Number of Cells," in PATMOS, 2001.

[3] C. Arm, J-M. Masgonty, C. Piguet, "Double-Latch Clocking Scheme for Low-Power I.P. Cores", PATMOS 2000, Goettingen, Germany, September 13-15, 2000.

[4] Christian Piguet, Chapter 7, "Logic Circuits and Standard Cells", in Low-Power Electronics Design, CRC Press, November 2004

[5] Bahman Kheradmand Boroujeni, "Process Variation and Leakage Power", Ph.D. work, EPFL, Lausanne Switzerland, 2011

[6] C. Schuster, "Leakage Aware Digital Design Optimization for Minimal Total Power Consumption in Nanometer CMOS Technologies," Ph. D. work, University of Neuchâtel, Swotzerland,

[7] K. Nose, T. Sakurai, "Optimization of Vdd and Vth for Low-Power and High-Speed Applications", Proceedings of the Asia and South Pacific Design Automation Conference, (2000) January 25-28, pp. 469-474, Yokohama, Japan

[8] R.W. Brodersen, et al.,"Methods for True Power Minimization", Proceedings of the Int'l Conf. on Computer Aided Design, (2002) November, pp. 35-42. San Jose, California

[9] Khanna and Calhoun, "Serial Sub-threshold Circuits for Ultra-Low-Power Systems", ISPLED 2009

2015 IEEE Computer Society Annual Symposium on VLSI

A Novel Phase-based Low Overhead Fault Tolerance Approach for VLIW Processors

Anderson L. Sartor[1], Arthur F. Lorenzon[1], Luigi Carro[1], Fernanda Kastensmidt[1], Stephan Wong[2] and Antonio C. S. Beck[1]

[1] Institute of Informatics, Federal University of Rio Grande do Sul (UFRGS), Brazil
{alsartor, aflorenzon, carro, fglima, caco}@inf.ufrgs.br
[2] Computer Engineering Laboratory, Faculty of EEMCS, Delft University of Technology, The Netherlands
j.s.s.m.wong@tudelft.nl

Abstract— **Because of technology scaling, the soft error rate has been increasing in digital circuits, which in turn affects system reliability. Therefore, modern processors, including VLIW architectures, must have means to mitigate such effects to guarantee reliable computation. In this scenario, our work proposes two new low overhead fault tolerance approaches, with zero latency detection, that correct soft errors in the pipelines of a configurable VLIW processor. Each approach has a distinct way to detect errors, but they both utilize the same rollback mechanism. The first utilizes redundant hardware by having specialized duplicated pipelines. The second uses idle issue slots to execute duplicated instructions and does this by first identifying phases within an application. Our implementation does not require changes to the binary code and has negligible performance losses. It has 50% of area overhead with 35% power dissipation for the full pipeline duplication, and only 7% of extra area when using idle resources. We compared our approach to related work and demonstrate that we are more efficient when one considers the area, performance, power dissipation and error coverage altogether.**

Keywords—fault tolerance; VLIW; soft errors; configurable processor

I. INTRODUCTION

Technology scaling has been allowing high logic integration and performance improvements of integrated circuits, as higher frequencies can be achieved. However, their reliability is compromised as they get more susceptible to soft errors [1]. Soft errors affect processors by modifying values stored in memory elements (such as pipeline registers, register files, and control registers) and are caused by numerous energetic particles such as protons and heavy ions from space or neutron and alpha particles at ground-level. In order to harden the processor against soft errors, fault-tolerant techniques implemented in hardware and software are mandatory to detect those errors and correct their effects before a failure in the system is observed [2].

Very Long Instruction Word (VLIW) processors are representative examples of current architectures that may suffer from the aforementioned issues. VLIW processors exploit Instruction-Level Parallelism (ILP) by means of a compiler, executing several operations per cycle depending on the processor's issue width and the intrinsic ILP available in the application. They occupy less area and dissipate less power when compared to superscalar processors, because their

hardware is much simpler, e.g., the instruction queue, reorder buffer, and dependency-checking hardware are not needed.

In this paper, we propose two approaches for detecting and correcting soft errors in VLIW pipelines. The first approach is the full duplication based on a 4-issue configuration, with four main pipelines and four duplicated ones. The other is a phase-based configurable mechanism implemented in hardware that uses idle issue slots to execute duplicated instructions based on the application profiling. It is implemented on an 8-issue processor that can have a variable number of duplicated instructions depending on the application phase: between zero (no fault tolerance) and four (full duplication). These techniques are based on a modified dual modular redundancy (DMR) with instruction rollback mechanism, called DMRr.

In order to validate our proposed techniques, a fault injection campaign was performed in several benchmarks on four configurations of the VLIW processor: unprotected 4-issue, 4-issue with DMRr, unprotected 8-issue, and configurable DMRr 8-issue. We evaluated the results considering several axes, such as error coverage, area, power dissipation, code size, and performance. We also compared our approach to others, showing that we are more efficient when one considers these axes altogether.

The remainder of this paper is organized as follows. Section II presents the fault tolerance techniques for VLIW processors proposed in this work. Section III evaluates the implementation and shows the results. Section IV discusses related works, while Section V concludes this work.

II. PROPOSED PIPELINE DMR WITH INSTRUCTION ROLLBACK

The focus of this work is to protect the pipelines of a VLIW processor against soft errors. The pipelines occupy about 45% of the total area and the register file the other 55% (on the tested 4- and 8-issue configurations); however, the register file can be protected with parity or ECC. The pipelines are duplicated in hardware, and a checker compares the results (i.e., all output signals) of the main and the duplicated pipeline. Therefore, the checker will detect if there is any error in the execution (e.g., error in an arithmetic operation, jump address of a branch, or in the values of a memory operation). The destination register, the register file's and memory's write enable signal are also compared by a checker.

978-1-4799-8720-7/15 $31.00 © 2015 IEEE

In order to not only detect an error, but also correct it, a rollback mechanism is used. When a mismatch is found in any of the compared signals, the rollback executes the last instruction again, and it flushes the pipeline in order to prevent data corruption. As the checker has zero latency error detection, the memory, and the register file will not be corrupted in case of an error, because the writing to these components will be disabled. The rollback overhead has a fixed cost of 5 cycles (because of the pipeline length of five stages), which is negligible when compared to the total number of cycles of an application. In addition, this cost is only paid in case of an error, having no overhead otherwise. Both the checkers and the rollback mechanism do not affect the critical path of the processor, as they operate in parallel to the pipelines, and the area overhead for each checker is of about 1%.

The VLIW processor used in this work is the ρ-VEX softcore VLIW processor [3], implemented in VHDL. The ρ-VEX core has a five-stage pipeline, and it can be configured to have different number of issue slots (e.g., 2, 4, or 8). Each issue may contain different functional units from the following set: ALU (always present), multiplier, memory, and branch units. In this work, the 4-issue version has 4 ALUs, 2 multipliers, 1 memory unit and 1 branch unit; while the 8-issue has 8 ALUs, 4 multipliers, 1 memory unit, and 1 branch unit, which are similar to other VLIW processors (e.g., Intel Itanium [4]).

As already mentioned, two configurations using the DMR with instruction rollback (DMRr) are evaluated:

A. 4+4-issue

In this configuration, the 4-issue VLIW processor has its pipelines duplicated, and checkers compare the results from the main pipelines to the ones from the duplicated pipelines. Therefore, there is full duplication with dedicated pipelines to execute replicated instructions, as depicted in Fig. 1. This process is totally done in hardware, and the compiler is not aware of it.

B. Phase-configurable

Based on the 8-issue VLIW, idle issue slots during a whole given program phase (i.e., a sequence of instructions words that always have NOPs in specific issue slots) are used to execute duplicated instructions from other pipelines, and their results are checked. Consequently, it is a configurable fault tolerance mechanism, since it may have none (eight issue slots without duplication) to full duplication (four main issue slots and four duplicated ones), depending on the available resources.

The 8-issue configuration for this approach is depicted in Fig. 2. In order to keep the low overhead (area and delay), the duplication pairs are statically placed, i.e., pipeline 0 with pipeline 4, pipeline 1 with pipeline 5, and so on. Therefore, the issue slots are combined in a way that the first four pipelines are compared with the four last ones. For example, if the pipeline 6 is going to execute NOPs in a given phase, then it will execute the duplicated instructions from the pipeline 2 instead. Every functional unit is capable of executing both main program instructions and, in the case of pipelines P4, P5, P6, and P7, duplicated instructions from another pipeline. The exceptions are the memory unit in pipeline 4 and the branch unit in the pipeline 7: these units execute only duplicated instructions, since

the ρ-VEX does not support more than one memory or branch operations per cycle. The compiler used in this work (HP VEX compiler) schedules the instructions starting from the lower issue slots (from 0 to 7), considering the availability of the functional units. Therefore, our approach in combining the pairs of the duplicated pipelines will efficiently exploit this scheduling mechanism.

For this approach to work, first, the application must be profiled in order to detect the phases. After that, a table indexed by the program counter and containing the configuration of each application's phase is statically created (a mechanism to dynamically detect phases will be developed in the future). The phase configuration represents the function of each pipeline in a given phase, informing whether each issue slot will execute regular instructions of the application or execute duplicated instructions of another pipeline. Based on this table, the processor will dynamically change the function of the pipelines and will enable or disable the checkers in each phase.

The simulation was conducted in the Mentor Graphic's ModelSim. In this way, it was possible to save which instruction was executed in each cycle to generate the application profile. The benchmark set chosen is composed of the following 10 applications, which comprises a subset of the WCET benchmark suite [5]: ADPCM, CJPEG, CRC, DFT, Expint, FIR, Matrix Multiplication, NDES, Sums (recursively executes multiple additions on an array) and x264. The profiling was performed for all applications from our benchmark set. The results for five benchmarks are depicted in Fig. 3. The dots demonstrate when a given pipeline, identified by its ID (Y-axis), is being used in a

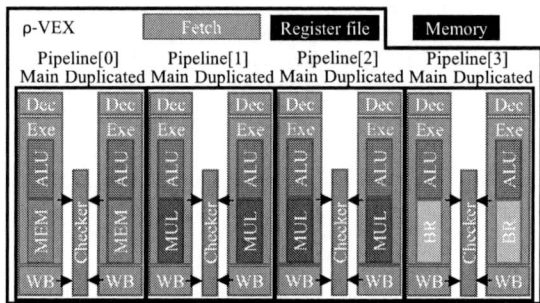

Fig. 1. DMRr 4+4-issue configuration

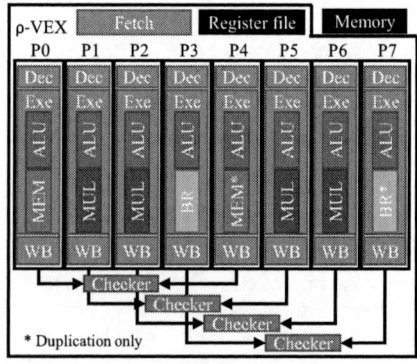

Fig. 2. DMRr phase-configurable duplication

978-1-4799-8720-7/15 $31.00 © 2015 IEEE

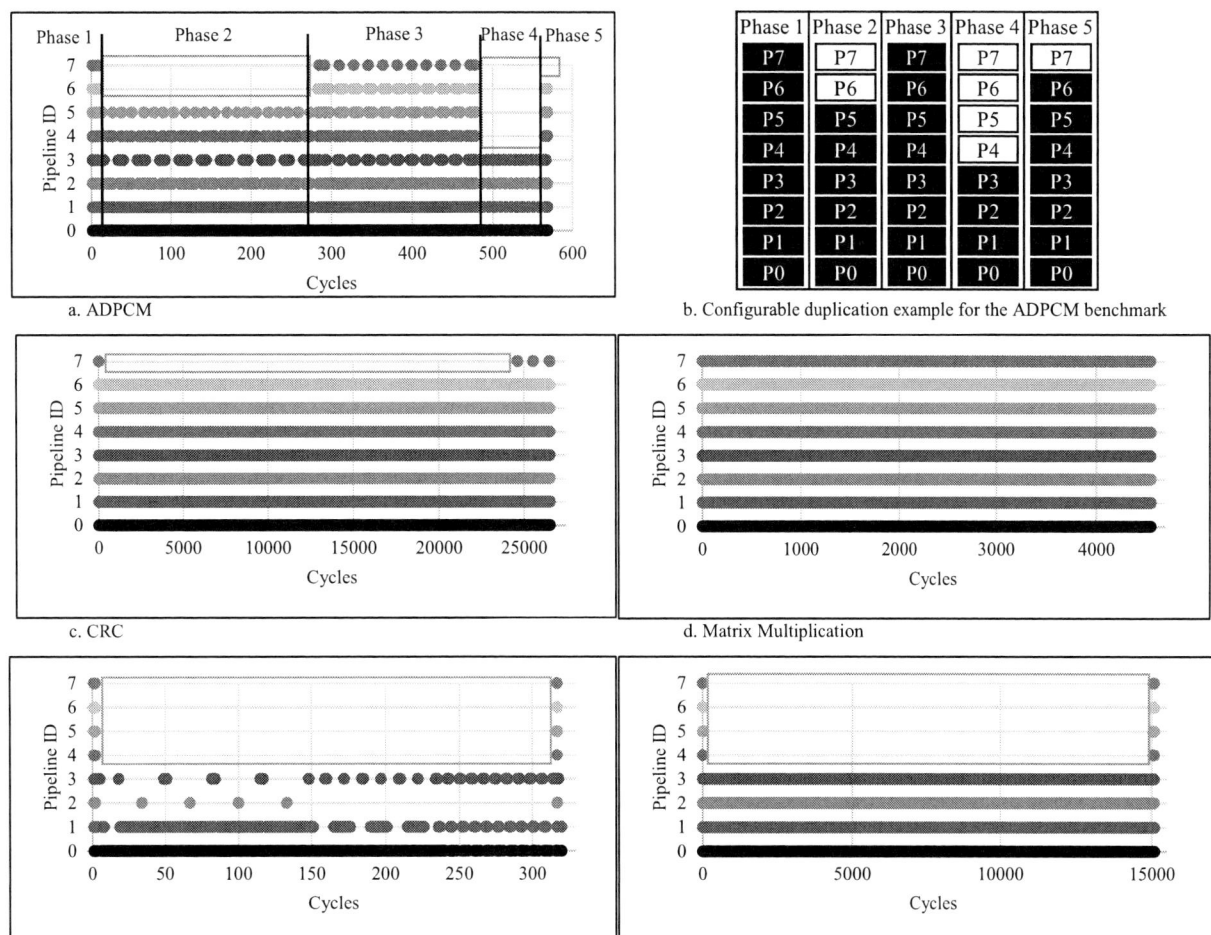

Fig. 3. Issue utilization and configurable duplication

given moment of the application's execution (X-axis). The profiling for the other five benchmarks has a similar behavior to the one from the Matrix Multiplication benchmark.

The idle phases that were used to execute duplicated instructions are highlighted in Fig. 3. An example of each phase for the ADPCM benchmark is depicted in Fig. 3b: the P0-P7 represent the pipelines 0 to 7. The white background on the pipelines indicate that they are executing duplicated instructions from the other pipeline in their pairs (as discussed in Fig. 2), and the black background represents that the pipeline is executing main program instructions (no duplication). On this example, there are phases with full duplication (phase 4), partial duplication (phases 2 and 5) and no duplication (phases 1 and 3). As it can be noticed, the ADPCM, CRC, Sums, and x264 benchmarks have phases when some issue slots are not utilized. On the other hand, as the Matrix Multiplication, CJPEG, DFT, Expint, FIR, and NDES benchmarks do not have such phases, they cannot take advantage of the phase-configurable approach, because it would have the same behavior as the unprotected 8-issue, once no issue slot would be duplicated and compared. Therefore, this second approach, even though it has no costs in

terms of performance and negligible power overhead, can be only used when the application has phases with lower levels of ILP than the processor supports.

III. RESULTS

In this section, the results will be discussed regarding the fault tolerance, area, performance, and power dissipation of the proposed mechanism and compared to those from the unprotected processor.

A. Error Coverage Analysis

An extended fault injection campaign was performed to inject soft errors in the pipeline, making it possible to evaluate the failure rate of the processor. Scripts in TCL (Tool Command Language) were created to inject transient faults in all pipelines and checkers of the processor. The script chooses a random bit from an arbitrary signal to be flipped at a random time during the execution of the application. The fault duration is one clock cycle, and one fault is injected per application's execution. The total number of injected faults was 2.882.376 (so there was the same number of application executions), varying from 2.750 to 322.495 injected faults per benchmark for each configuration.

The variation in the number of injected faults (executions) per benchmark is due to the difference between their simulation times. Also, the faults were injected randomly in both protected and unprotected versions because their hardware are different. Therefore, it would not be correct to randomly inject a fault in the unprotected version and inject the same fault in the protected version.

There are three possible reasons for failures in the system:

- **Data failure:** happens when there is a mismatch between the memory dump from the application and the golden memory dump. The dumps are compared once the application ends its execution.

- **Data flow failure:** happens when the application does not stop within the number of cycles that it should (the one from the application without any failures).

- **Simulation failure:** happens when some specific signals at specific times are flipped and crashes the simulation, (i.e.: ModelSim's simulation is aborted without finishing the execution of the application).

These failures are checked when the application finishes its execution. Not all injected errors cause failures: some of the errors are application masked, and some of the errors are detected by our proposed technique. Table I presents the failure rate of the chosen applications in all four VLIW configurations. The unprotected 8-issue has a lower failure rate than the unprotected 4-issue due to the elevated number of NOPs in the VLIW instruction; therefore, the probability of a flipping bit affecting the result of an instruction is lower than on the 4-issue configuration. Also, the 4+4-issue decreases the failure rate by 121 times when compared to the 4-issue; and the phase-configurable by 8 times when compared to the 8-issue configuration.

On average, the 4+4-issue has a 99.95% error coverage, while the phase-configurable has 99.55%. The latter has a lower coverage because of the ADPCM and CRC benchmarks. In the former, there are three significant phases in which no duplication can be done because there are no idle issue slots (i.e.: they achieve the maximum ILP possible). In the latter, in most of the times only one issue is idle and available to execute duplicated instructions. For the benchmarks that present a significant phase with four idle issue slots, as the example of Sums and x264 benchmarks, the coverage is close to the one achieved by the full duplication: 99.89% and 99.93%, respectively. Hence, when the application has idle phases to execute duplicated instructions, it is not only possible to provide the same level of fault tolerance as in the case of dedicated duplication; but also, keep the 8-issue's performance, which is better than the one from 4-issue configuration as more functional units are available to exploit the application's ILP.

The distribution of failures for each application running on each configuration is depicted in Fig. 4. Both protected and unprotected versions have a similar probability of simulation failures, because the simulation failure interrupts the execution of the application. Therefore, as the protected version is able to correct most of the errors that may lead to a data or control flow

TABLE I. FAILURE RATE WITH DIFFERENT CORE CONFIGURATIONS

			# Injected faults	# Errors	Failure rate (%)
Unprotected	4-issue	ADPCM	287,398	19,911	6.93
		CJPEG	250,441	23,911	9.55
		CRC	79,346	4,128	5.20
		DFT	52,531	2,432	4.63
		Expint	69,221	2,911	4.21
		FIR	5,649	618	10.94
		Matrix	59,906	5,934	9.91
		NDES	35,651	1,424	3.99
		Sums	322,495	17,814	5.52
		x264	55,502	2,891	5.21
	8-issue	ADPCM	143,077	5,234	3.66
		CJPEG	154,789	9,399	6.07
		CRC	43,444	1,280	2.95
		DFT	32,383	869	2.68
		Expint	53,949	1,281	2.37
		FIR	2,797	166	5.93
		Matrix	60,287	3,423	5.68
		NDES	20,574	430	2.09
		Sums	200,363	5,937	2.96
		x264	28,807	847	2.94
Protected	4+4-issue	ADPCM	183,168	103	0.06
		CJPEG	160,527	28	0.02
		CRC	39,157	22	0.06
		DFT	21,054	15	0.07
		Expint	70,396	37	0.05
		FIR	2,750	1	0.04
		Matrix	33,760	28	0.08
		NDES	17,755	7	0.04
		Sums	239,239	94	0.04
		x264	11,961	11	0.09
	Configurable	ADPCM	36,958	366	0.99
		CRC	16,720	107	0.64
		Sums	65,221	75	0.11
		x264	14,448	10	0.07
Average	Unprotected 4-issue				6.61
	Unprotected 8-issue				3.73
	4+4-issue				0.05
	Configurable				0.45

failure, the proportion of simulation failures in the protected configurations is bigger than the one from the unprotected ones.

B. Area and power dissipation evaluation

The synthesis tools used to evaluate area, performance and power were: the Xilinx ISE synthesis tool to obtain the FPGA area and frequency to the Virtex 6 - XC6VLX240T FPGA; and the Cadence Encounter RTL compiler to obtain power dissipation and ASIC (Application Specific Integrated Circuit) area, using a 65nm CMOS cell library from STMicroeletronics.

Table II presents the processor area for both FPGA and ASIC and power dissipation of the processor for all VLIW configurations. The operation frequency for these configurations was of 65MHz, and the speed up from the 8-issue configuration when compared to the 4-issue, for the aforementioned benchmark set, varies from 0.1% to 23.6%, with an average of 5.4%.

As it can be observed, the overhead of the full duplication (4+4-issue) is small in terms of area and power dissipation when compared to the 4-issue. The area overhead is of 30% for the FPGA (given in LUTs) and 50% for the ASIC cells, while the power dissipation overhead is of 35%. The overhead is almost

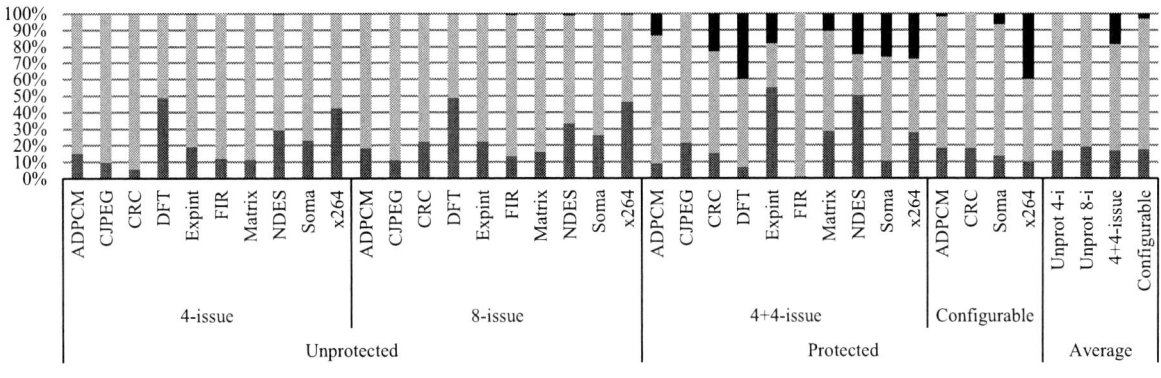

Fig. 4. Failure distribution

TABLE II. AREA AND POWER DISSIPATION COMPARISON

		FPGA		ASIC	
		Registers	LUTs	Cells	Power dissip. (nW)
Unprotected	4-issue	3,058	16,006	28,041	2,298,962.51
	8-issue	3,974	35,075	68,977	7,556,643.89
Protected	4+4-issue	4,102	20,819	42,121	3,109,613.33
	Configurable	4,133	35,973	73,988	7,484,929.91

negligible when one compares the phase-configurable approach with the base 8-issue configuration: 3% for the FPGA and 7% for the ASIC; with almost no overhead in power dissipation. This overhead comes from the checkers only, since no extra circuitry, such as duplicated functional units, is necessary for the latter approach.

IV. RELATED WORK

Several works have been proposed for the detection and correction of soft errors in VLIW and superscalar processors. These works aim to improve the fault tolerance of the target system, typically based on redundancy, which may be implemented either in software, in hardware, or both.

Dual modular redundancy (DMR) with rollback was used by [6] and [7] to detect and correct errors. These works used checkpoints in order to rollback, whenever an error is detected, to a state in which the execution was correct. Therefore, the latency to detect the error on these approaches will vary according to the periodicity of the checkpoints. On the other hand, the DMRr has zero latency detection as it compares the results at all times and executes again only the instruction that presented the error. Therefore, in addition to the zero latency detection, the control structure of the rollback is much simpler than the ones that use checkpoints.

Another common approach is to triplicate a processor and use a majority voter to vote the correct answer (triple modular redundancy - TMR), as implemented in [8] and [9]. In these cases, they only triplicate the functional units of a VLIW processor rather than the entire processor; therefore, reducing the area and power dissipation costs. In [8], both hardware and software needed to be changed, and if the two operations (main and duplicated) compute different results, the operation is

executed a third time. This approach is called Reduced TMR. Both approaches only cover errors that happen in the computation of a given operation, therefore, not begin able to detect an error if the operands were wrong in a stage prior to the execution or after the operation was computed by the functional unit. DMRr has higher coverage than [9], occupying less area and dissipating less power than [8] and [9], and the proposed approach does not change the binary code of the application, as it is done in [8]. In [10], the authors propose a similar approach to [8], with the instruction replication done in software. In the same way, the replication is done partially to some instructions in order to avoid significant performance losses, but it also affects the capacity of providing fault tolerance. Therefore, the code is changed, performance is affected as well as the error coverage, even though there is no area or power overheads. In [11] the authors also propose a TMR approach on the synchronous flip-flops, with an area and power dissipation overhead higher than the one created by the DMRr, even when comparing to the full duplication configuration.

In [12] and [13], the authors propose a software-based redundancy based on duplication with comparison (DWC) for VLIW data paths aiming to reduce the performance overhead by using the idle functional units. However, these techniques still present performance degradation and increase code size, as they are in software. Authors in [14] propose an optimization to the DWC's generated code by reducing the impact of the basic block fragmentation caused by the check instructions, having lower, but still not negligible, performance degradation than the previous two techniques. Authors in [15] propose a compiler assisted multiple word retry scheme for VLIW architectures, and the authors evaluate the performance and code growth of having a rollback mechanism with different number of instructions.

The main limitations of software-based redundancy are the increase in the code size, energy consumption and performance overheads that come with it. On the other hand, hardware-based redundancy approaches increase area, power dissipation with little or no performance overhead. The approach proposed in this paper, even though implemented in hardware, has low overhead in area and power dissipation.

978-1-4799-8720-7/15 $31.00 © 2015 IEEE

Table III presents the comparison, in several axes, between the results from the DMRr and the other works previously discussed in this section. These axes comprise error coverage, area, performance, power dissipation and code size increase. As it can be noticed, the DMRr has the lowest area and power dissipation overhead when compared to other hardware-based techniques (**in bold**). Software-based techniques (*in italic*) naturally do not affect the area nor the power dissipation, but they create a performance overhead and increase the code size, both affecting total energy consumption of the system, as the application will take longer to execute, and the memory will be more stressed when executing more instructions. In addition, several hardware-based techniques also have performance and/or code size overheads, besides the power and area overheads, impacting even more on the energy consumption of the system.

By using DMRr, on the other hand, the application's code is not changed, the performance overhead is negligible, and the power and area overheads are much lower than the other techniques. Therefore, being able to provide good coverage at a low cost, considering all axes: performance, power dissipation, energy consumption, and area. In addition, for benchmarks that have idle phases, the configurable duplication can be used in order to execute duplicated instructions in the idle pipelines, having only 7% of area overhead and negligible overhead on the other axes.

V. CONCLUSIONS AND FUTURE WORK

The pipelines of a VLIW processor occupy a significant area of the processor core; therefore, it is important to protect them against soft errors with minimum cost while providing good coverage. In this work, a fault tolerance mechanism based on duplication and instruction rollback is proposed, which is able to not only detect a fault, as a conventional DMR approach, but also correct the error by executing the faulty instruction again via rollback. The performance overhead that a rollback causes is negligible (5 cycles) compared to the application's total number of cycles. In addition, this mechanism is able to provide fault tolerance with a minimum cost, by using idle resources of the VLIW processor or by duplicating the pipelines with low area and power overhead.

As future work, we will consider two more processor configurations: a phase-based approach with *dynamic* phase detection and an instruction-adaptive configuration that will duplicate instructions during runtime whenever a pipeline is executing a NOP instruction. In addition, a mechanism to detect permanent faults will be developed.

REFERENCES

[1] P. Shivakumar, M. Kistler, S. Keckler, D. Burger and L. Alvisi, "Modeling the effect of technology trends on the soft error rate of combinational logic," *International Conference on Dependable Systems and Networks*, pp. 389-398, 2002.

[2] A. C. S. Beck, C. A. L. Lisbôa and L. Carro, Adaptable Embedded Systems, Springer Publishing Company, 2012.

[3] S. Wong, T. v. As and G. Brown, "ρ-VEX: A reconfigurable and extensible softcore VLIW processor," *International Conference on ICECE Technology*, pp. 369-372, 2008.

[4] H. Sharangpani and K. Arora, "Itanium processor microarchitecture," *IEEE Micro*, vol. 20, no. 5, pp. 24-43, 2000.

[5] J. Gustafsson, A. Betts, A. Ermedahl and B. Lisper, "The Mälardalen WCET benchmarks – past, present and future," *International Workshop on Worst-Case Execution Time Analysis (WCET)*, p. 137–147, 2010.

[6] R. Xiaoguang, X. Xinhai, W. Qian, C. Juan, W. Miao and Y. Xuejun, "GS-DMR: Low-overhead soft error detection scheme for stencil-based computation," *Parallel Computing*, vol. 41, pp. 50-65, 2015.

[7] J.-M. Yang and S. Kwak, "A checkpoint scheme with task duplication considering transient and permanent faults," *Industrial Engineering and Engineering Management (IEEM), IEEE International Conference on*, pp. 606-610, 2010.

[8] M. Scholzel, "Reduced Triple Modular Redundancy for built-in self-repair in VLIW-processors," *Signal Processing Algorithms, Architectures, Arrangements and Applications*, pp. 21-26, 2007.

[9] Y.-Y. Chen and K.-L. Leu, "Reliable data path design of VLIW processor cores with comprehensive error-coverage assessment," *Microprocessors and Microsystems*, vol. 34, no. 1, pp. 49-64, 2010.

[10] Y. Li, J. Lee, Y. Ko, K. Lee and Y. Paek, "Compiler-directed instruction duplication for soft error detection," *Workshop on Synthesis And System Integration of Mixed Information technologies*, pp. 54-59, 2012.

[11] F. Anjam and S. Wong, "Configurable fault-tolerance for a configurable VLIW processor," *Reconfigurable Computing: Architectures, Tools and Applications*, vol. 7806, pp. 167-178, 2013.

[12] C. Bolchini, "A software methodology for detecting hardware faults in VLIW data paths," *IEEE Transactions on Reliability*, vol. 52, no. 4, pp. 458-468, 2003.

[13] J. Hu, F. Li, V. Degalahal, M. Kandemir, N. Vijaykrishnan and M. J. Irwin, "Compiler-assisted soft error detection under performance and energy constraints in embedded systems," *ACM Transactions on Embedded Computing Systems (TECS)*, vol. 8, no. 4, p. 27, 2009.

[14] K. Mitropoulou, V. Porpodas and M. Cintra, "DRIFT: Decoupled CompileR-Based Instruction-Level Fault-Tolerance," *Languages and Compilers for Parallel Computing*, pp. 217-233, 2014.

[15] S.-K. Chen and W. Fuchs, "Compiler-assisted multiple instruction word retry for VLIW architectures," *IEEE Transactions on Parallel and Distributed Systems*, vol. 12, no. 12, pp. 1293-1304, 2001.

TABLE III. VLIW FAULT TOLERANCE TECHNIQUES COMPARISON

Technique	Error Coverage	Area overhead	Performance overhead	Power dissipation overhead	Code size increase
DMRr - 4+4 configuration	~100%	50%	~0%	35%	0%
DMRr - Configurable	~100%	7%	~0%	~0%	0%
DMR with rollback [5] and [6]	~100%	0%	51%-100%	0%	100%
TMR	~100%	200%	~0%	~200%	0%
Partial TMR [8]	95%-99%	100%	0.6%-34.3%	~100%	0%
Reduced TMR [7]	~100%	100%	0%-100%	~100%	> 0%
Reduced TMR - SW [9]	~100%	0%	30%-60%	0%	100%
Flip-flops TMR [10]	~100%	200%	~0%	~200%	0%
DWC - SW [11] and [12]	~100%	0%	28%-106%	0%	109%-217%
DWC opt. - SW [13]	~100%	0%	29%	0%	100%-150%

On the Design of a Fault Tolerant Ripple-Carry Adder with Controllable-Polarity Transistors

H. Ghasemzadeh, P.-E. Gaillardon, J. Zhang, G. De Micheli
Integrated Systems Laboratory (LSI)
Ecole Polytechnique Fédérale de Lausanne (EPFL)
1015 Lausanne, Switzerland

E. Sanchez, M. Sonza Reorda
Dip. Automatica e Informatica
Politecnico di Torino
Torino, Italy

Abstract— **This paper first explores the effects of faults on circuits implemented with controllable-polarity transistors. We propose a new fault model that suits the characteristics of these devices, and report the results of a SPICE-based analysis of the effects of faults on the behavior of some basic gates implemented with them. Hence, we show that the considered devices are able to intrinsically tolerate a rather high number of faults. We finally exploit this property to build a robust and scalable adder whose area, performance and leakage power characteristics are improved by 15%, 18% and 12%, respectively, when compared to an equivalent FinFET solution at 22-nm technology node.**

Keywords—Controllable-polarity Transistors, Fault model, Fault-tolerant adder

I. INTRODUCTION

In the recent years, many novel *Field-Effect Transistor* (FET) technologies have been proposed and evaluated in order to overcome the ultimate limits of conventional silicon-based *Integrated Circuits* (ICs). While most of them improve the structure and materials of FETs to boost their intrinsic performances, an alternative approach increases the functionality of the individual device for a constant area [1].

One of the most promising devices with enhanced functionalities is the controllable-polarity transistor. Exploiting a *dual-gate* structure, controllable-polarity transistors can be electrostatically configured to be either *n*- or *p*-type [2][3]. The functionality of such a device is logically biconditional on both gate values and enables a compact realization of XOR/MAJ-based logic functions, which are not implementable in CMOS in a compact form [4][5]. Controllable-polarity devices can be fabricated in many different technologies, from pure silicon [2][3][6] to carbon electronics [7][8]. In particular, a top-down fabrication process showing full compatibility with industrial fabrication techniques has been employed in [3] to demonstrate the feasibility of the approach. Basic logic gates exploiting the enhanced expressiveness of the technology has been demonstrated in [9], making the practical usage of this technology even closer. In addition to showing interests in the realization of compact logic elements, emerging technologies with enhanced functionalities can also introduce novel opportunities in terms of fault tolerance.

In this paper, we first perform an analysis on the effects of possible permanent faults affecting a generic controllable-polarity device and studied in details in [10]. In order to take into account the specific characteristics of the controllable-polarity devices, this analysis has to be performed at the transistor level. We propose a new fault model that takes into account the specific characteristics of these devices, extending the popular stuck-open/stuck-short fault model traditionally used at that level. Then, we analyze the behavior of circuits based on controllable-polarity devices when permanent device faults are present, and identify the conditions for their detection/masking. Results show that a high number of faults are masked, thus making this new technology particularly interesting from a reliability point of view.

Performing this analysis at the transistor level allowed us to express the behavior of each gate when any of the possible faults affecting each of its transistor arise. We use this information to forecast the behavior of more complex circuits composed out of the above gates, thereby achieving the same precision than a transistor level analysis but with a much lower computational complexity.

Based on the results of the previous analysis, we also propose in this paper a fault-tolerant ripple-carry adder architecture exploiting XOR/MAJ logic gates built entirely with controllable-polarity transistors. In order to guarantee a high degree of resiliency with respect to single and double permanent faults in every single stage, we combine the intrinsic resiliency of the controllable-polarity-based circuits with the usage of the *Triple Modular Redundancy* (TMR) architecture. Although faster solutions are often adopted to implement adders, e.g., based on the Kogge-Stone architecture [11] and its fault tolerant version [12], the TMR version of the ripple-carry adder still represents the reference to compare with, especially when the parallelism is limited and power is not a major issue (in the latter case, solutions based on reversible logic are often adopted [13]).

Experimental validation shows that the full-adder architecture we propose is able to tolerate all possible single faults and a very high percentage of the double ones. In addition, it proves that the proposed solution provides a 15%, 18% and 12% gain in area, performance and leakage power with respect to similar architectures implemented in FinFET technology at 22-nm technology node. Finally, the proposed architecture is significantly cheaper with respect to solutions based on hardening the circuit at the transistor level, such as those proposed in [14], whose area overhead 4× the unhardened circuit.

978-1-4799-8720-7/15 $31.00 © 2015 IEEE

This paper is organized as follows. Section II gives some background related to controllable-polarity transistors. Section III introduces a new fault model suited to controllable-polarity devices and analyzes the conditions for its detection/masking. Section IV proposes a fault tolerant architecture for a 1-bit adder that can be exploited to build cost-effective fault-tolerant adders. Section V reports the results of a quantitative analysis of the characteristics of the proposed architecture. Finally, Section VI draws some conclusions.

II. BACKGROUND

Transistors with controllable polarity are *Double-Independent Gate* (DIG) FETs having one gate controlling on-line the device polarity. Transistors with controllable polarity have been experimentally fabricated in several novel technologies, such as carbon nanotubes [7], graphene [8] and *Silicon NanoWires* (SiNWs) [2][3][6].

In DIG devices, one gate electrode, denoted the *Control Gate* (CG), acts conventionally by turning *on* and *off* the device. The other electrode, denoted the *Polarity Gate* (PG) acts on the side regions of the device, dynamically switching the device polarity between *n*-type (PG=1) and *p*-type (PG=0). The behavior of this device is illustrated in Fig. 1.

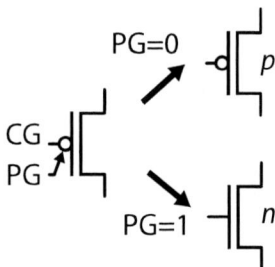

Fig. 1. Controllable-polarity transistor behavior.

Using controllable-polarity devices, it is possible to build very compact arithmetic logic gates, such as *eXclusive OR* (XOR) [4] and *MAJority* (MAJ) [15]. For instance, a 2-input XOR gate requires only 4 transistors [4] instead of the 8 required by the traditional full-swing static CMOS implementation [16]. This compactness can be leveraged in adder implementations, as reported in Fig. 2, where we show a full adder composed of only 8 controllable polarity transistors. This circuit exploits 3-input XOR and MAJ gates to implement the sum and the carry, respectively. Note that the proposed cells exploit a transmission-gate design. We will see, in the following, that this introduces a degree of redundancy at the gate level, that is beneficial from a robustness perspective. A self-checking ripple-carry adder architecture, exploiting this adder structure, is proposed in [15]. This architecture is far less expensive in terms of area than comparable CMOS architectures.

In this paper, we make one step forward with respect to [15]. In addition to exploit the reduced area cost offered by

controllable-polarity devices, we take into account their intrinsic capabilities in masking, i.e., tolerating, faults and use them to build a fault-tolerant adder.

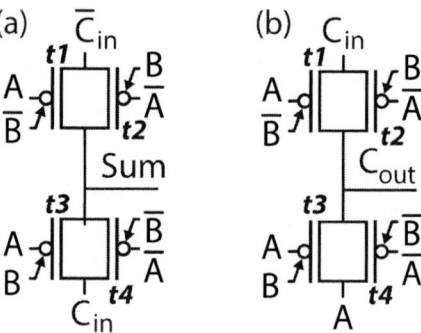

Fig. 2. Realization of 1-bit full adder using controllable-polarity transistors. (a) 3-input XOR – Sum=$A \oplus B \oplus C_{in}$, (b) 3-input MAJ – C_{out}=MAJ(A,B,C_{in}).

III. EVALUATING CONTROLLABLE-POLARITY CIRCUIT ROBUSTNESS

In this section, we introduce a new fault-model suited to controllable-polarity transistors, study the robustness of operations for XOR and MAJ logic gates exploiting controllable-polarity transistors, and extend the results of this analysis to circuits composed of different gates.

A. Fault Model

The robustness evaluation of circuits based on controllable-polarity devices cannot be performed by relying on usual fault models and tools, e.g., working at the gate level [17]. Indeed, when new technologies are introduced, it is common to envisage a lower-level approach, e.g., resorting to transistor-level fault models [18]. In such a case, the most common solution lies in inductive fault analysis of the device as well as layout-based defect map extraction for feasible fabrication shortcomings [10].

In this work, we only consider the defects that completely change the functionality [10], e.g., change the polarity of a transistor from *p*-type to *n*-type. Defects affecting the performances but keeping the functionality untouched are out of the scope of this paper. These defects can be modeled by generalizing bridge defects to the two gates composing our transistors [10]. Therefore, we introduce a new fault model that generalizes the *stuck-at* model for the mentioned bridge defects:

- *stuck-at-0* on CG (CG/0), *stuck-at-1* on CG (CG/1)

This defect is similar to what happens in the current technology. Depending on the polarity of the transistor, such defect will lead to a *Stuck-Open* (SO) or *Stuck-Short* (SS) behavior of the device.

- *stuck-at-0* on PG (PG/0), *stuck-at-1* on PG (PG/1):

This defect affects the polarity of the device. The device will be either *stuck-at-n* or *stuck-at-p*, affecting the logic operation.

The new fault model straightforwardly extends the traditional transistor-level fault model, where the gate can be either stuck-at-0 or stuck-at-1, and takes into account the specific characteristics of controllable polarity transistors. Each of these faults corresponds to forcing to 0 or 1 the value of the corresponding controllable-polarity transistor input signal. We denote this fault model as *CG/PG fault model*.

B. XOR/MAJ Gates Robustness

In order to evaluate the robustness of a circuit implemented with controllable-polarity devices, we need to evaluate the behavior of the basic logic primitives, when a CG/PG fault occurs in any of the transistors of the gate.

1) Methodology

The 3-input XOR and MAJ logic gates, shown in Fig. 2-a and Fig. 2-b, respectively, have been characterized using electrical simulations. Fig. 3 summarizes the simulation setup that we used for fault analysis.

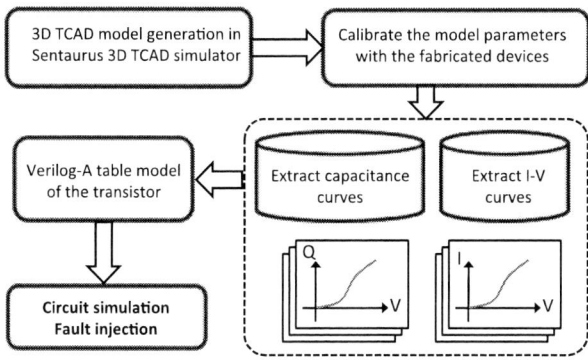

Fig. 3. Fault Analysis Methodology.

The logic gates are realized using SiNW-based controllable-polarity transistors [3]. A simple table-based compact model of the device is used with HSPICE simulator. The model is extracted using TCAD simulations of a 22-nm device, as shown in [3]. In the simulation experiments, the V_{dd} value was fixed to 1.2V, which is in line with the technological results. The output of the gates is loaded with a fixed 1fF capacitance.

First, we identify the input voltage ranges associated to Boolean input *0* (V_{IL}) and *1* (V_{IH}), according to the definitions given in [16]. Defining the input voltage boundaries will help us to identify a faulty gate behavior in presence of a transistor-level fault. We report the obtained points for the logic gates:

- $V_{IH} = 0.600V$
- $V_{IL} = 0.540V$.

Therefore, the two logic gates will correctly behave when the output voltages for Boolean output *0* (V_{OL}) and *1* (V_{OH}) are in the following ranges:

- $0.600V < V_{OH} < 1.2V$
- $0V < V_{OL} < 0.540V$.

The identified ranges are used to classify the output values of the different gates. Then, the behavior of the logic gates under all possible CG/PG faults is computed by using DC operating points analyses for all possible input conditions.

2) XOR/MAJ Gates Behavior under CG/PG Faults

Tables I and II report the simulated DC operating points of the 3-input XOR and MAJ gates, respectively, when the gates are fault-free and when each of the CG/PG faults are injected in the different transistors (*t1* to *t4*). The CG/PG fault injection induces different behaviors classified under three categories:

- correct behavior (highlighted in green) when a CG/PG fault is not excited by the applied input vector;
- masked-fault behavior (highlighted in blue) when a CG/PG fault is excited and induces a reduction of the noise margin at the output of the gate, but does not induce a faulty gate behavior as the output voltage is still in the correct V_{OH} and V_{OL} range;
- faulty behavior (highlighted in red) when a CG/PG fault induces an incorrect value at the output of the gate.

Considering the 3-input XOR (Table I), the results indicate that 8 CG/PG faults out of 16 lead to a faulty gate behavior that is observable at the gate output for at least one input combination. The remaining 8 CG/PG faults are always masked. Moreover, the 8 detectable faults produce a faulty output when 4 out of 8 possible input values are applied (*001, 010, 100,* and *111*). With the other 4 input combinations (*000, 011, 101,* and *110*), the circuit always produces the correct output no matter the presence of a fault.

Controllable-polarity transistors have 4 different modes of operations: *on n*-type, *off n*-type, *on p*-type and *off p*-type. A CG/PG fault restricts the number of operations of the device but does not fully lock it in a unique mode. This property is unique to the class of controllable-polarity transistors and unachievable with standard transistors. This has a positive impact on the fault tolerance of the overall gate circuit. As an example, we can consider the PG/0 fault on *t4* in the 3-input XOR under input values *000*. Under fault-free conditions, the bottom transmission-gate is *on*, with *t3* configured as *p*-type and *t4* as *n*-type. *t4* propagates properly the logic *0*. However, when a PG/0 fault affects *t4*, *t4* polarity switches to *p*-type. In this condition, the logic *0* cannot be fully propagated, but is still transmitted with limited voltage degradation. Such degradation reduces the noise margin of the gate, but does not induce a faulty behavior.

Similarly, for the 3-input MAJ (Table II), the results indicate that the number of faulty behaviors is very small: 12 faults out of 16 are always masked. Moreover, the 4 faulty conditions do produce a difference in the output voltage only when 2 of the 8 possible input combinations are applied (*011* and *110*). For the 6 remaining input combinations, the CG/PG faults never produce any output misbehavior.

TABLE I. Output voltage values of the 3-input XOR gate.

Input	Fault-free	t1				t2				t3				t4			
		CG/0	CG/1	PG/0	PG/1	CG/0	CG/1	PG/0	PG/1	CG/0	CG/1	PG/0	PG/1	CG/0	CG/1	PG/0	PG/1
000	0	0.32	0	0	0.27	0	0.27	0.32	0	0	0	0	0	0.14	0	0.13	0
001	1.2	0.83	1.2	1.2	0.36	1.2	0.36	0.83	1.2	1.2	1.1	1.2	1.1	1.2	1.2	1.2	1.2
010	1.2	1.2	1.1	1.2	1.1	1.2	1.2	1.2	1.2	0.83	1.2	1.2	0.36	1.2	0.36	0.83	1.2
011	0	0	0	0	0	0.14	0	0.13	0	0.32	0	0	0.27	0	0.27	0.32	0
100	1.2	1.2	1.2	1.2	1.2	1.2	1.1	1.2	1.1	1.2	0.36	0.83	1.2	0.83	1.2	1.2	0.36
101	0	0.14	0	0.13	0	0	0	0	0	0	0.27	0.32	0	0.32	0	0	0.27
110	0	0	0.27	0.32	0	0.32	0	0	0.27	0.14	0	0.13	0	0	0	0	0
111	1.2	1.2	0.36	0.83	1.2	0.83	1.2	1.2	0.36	1.2	1.2	1.2	1.2	1.2	1.1	1.2	1.1

TABLE II. Output voltage values of the 3-input MAJ gate.

Input	Fault-free	t1				t2				t3				t4			
		CG/0	CG/1	PG/0	PG/1	CG/0	CG/1	PG/0	PG/1	CG/0	CG/1	PG/0	PG/1	CG/0	CG/1	PG/0	PG/1
000	0	0	0	0	0	0	0	0	0	0	0	0	0	0	0	0	0
001	0	0.32	0	0	0.27	0	0.27	0.32	0	0	0	0	0	0.14	0	0.13	0
010	0	0	0	0	0	0	0	0	0	0	0	0	0	0	0	0	0
011	1.2	1.2	1.1	1.2	1.1	1.2	1.2	1.2	1.2	0.83	1.2	1.2	0.36	1.2	0.36	0.83	1.2
100	0	0.14	0	0.13	0	0	0	0	0	0	0.27	0.32	0	0.32	0	0	0.27
101	1.2	1.2	1.2	1.2	1.2	1.2	1.2	1.2	1.2	1.2	1.2	1.2	1.2	1.2	1.2	1.2	1.2
110	1.2	1.2	0.36	0.83	1.2	0.83	1.2	1.2	0.36	1.2	1.2	1.2	1.2	1.2	1.1	1.2	1.1
111	1.2	1.2	1.2	1.2	1.2	1.2	1.2	1.2	1.2	1.2	1.2	1.2	1.2	1.2	1.2	1.2	1.2

C. Circuit-level Analysis

Based on the results of the detailed transistor-level analysis presented so far, we can describe the behavior of each possible logic gate for each possible input combination and for each possible fault affecting each of the internal transistors. Therefore, we build a detailed model of the fault-free and faulty behavior of each gate. Using these models, we then determine the fault-free and faulty behavior of a larger circuit composed of different gates working at the logic level. This allows us to ignore the details of the underlying transistor-level structure, without loosing in accuracy. As a matter of facts, the approach we use in the remaining parts of this paper is based on developing VHDL models for each gate (with suitable control signals to inject each possible fault), and combining them to extensively analyze the behavior of larger circuits.

IV. Fault-Tolerant ripple-carry Adder Architecture

Knowing the behavior of the 3-input XOR and MAJ gates exploiting controllable-polarity transistors, and using the approach we just described, we now investigate the possibility to implement a fault-tolerant ripple-carry adder architecture based on these primitives.

We consider the 1-bit adder circuit represented in Fig. 4 and consisting of two 3-input XOR gates for the sum generation and two 3-input MAJ gates for the carry generation. As compared to the simpler adder of Fig. 2, this circuit generates both the sum and carry signals and their inverted versions in a unique logic level. This allows us to create ripple-carry adder structures without adding any inverters to drive the next stage. Note that the inverted sum is used for realizing a fault-tolerant version of the adder as described later and that, due to the transmission gates, buffers will be required every 4 stages.

Using the approach described in Section III.C, we first created a logic model of the proposed adder in VHDL combining the models of each composing gate, and used it to gather simulation results for every one of the proposed faults during an exhaustive simulation. Results (not reported here in details for sake of conciseness) show that most of the possible CG/PG faults (40 out of 64) are masked in this structure.

In order to make the adder fault-tolerant with respect to all the possible faults and under all the input conditions, we propose a *Triple Module Redundancy* (TMR)-based 1-bit adder architecture, shown in Fig. 5. Note that, in this figure, the inverted input and internal signals are not represented for the sake of visibility. The key characteristics of the proposed fault tolerant adder are:

- Each 1-bit adder of Fig. 4 is triplicated and voted. In this way, any single fault affecting a single adder can be tolerated and does not propagate to the following stages of the adder.
- The inputs to each replica, labeled from 1 to 3, are permuted. In this way, even if the same fault affects more than a single replica, this does not evolve into a common mode fault, and the circuit behaves correctly.
- Each of the 3 output signals, i.e., the sum and the carry signals (regular and complemented to cascade further stages)

978-1-4799-8720-7/15 $31.00 © 2015 IEEE

is voted. Since the majority voter has been shown in the previous section to never fail with the *000* and *111* input combinations, the voter never fails when the three 1-bit adders are fault-free. This guarantees that the 1-bit adder never produces a faulty output in the presence of a single fault affecting an adder or a voter.

- Given the above properties, one can easily build a ripple-carry adder out of the proposed 1-bit adder, knowing that fault effects cannot propagate from one stage to another.

The proposed architecture provides significant benefits in terms of fault tolerance. Thanks to triplication, it can mask any single fault in the three 1-bit adders and in the voter, which is never stimulated with an input value able to excite a fault inside it, unless another fault exists in an adder. Clearly, faults on the inputs A and B of the single cell of the full adder cannot be tolerated (unless they are in turn triplicated).
Thanks to the input permutation and to the robustness of the 1-bit adder, less than 1% of the possible 4,032 (64×63) double faults affecting a couple of replicas produce a faulty behavior.

Moreover, based on the previous analysis we can state that only 192 faults can produce a failure out of the set of 6,912 (144×48) double faults composed of one fault in an adder, and another in a voter.

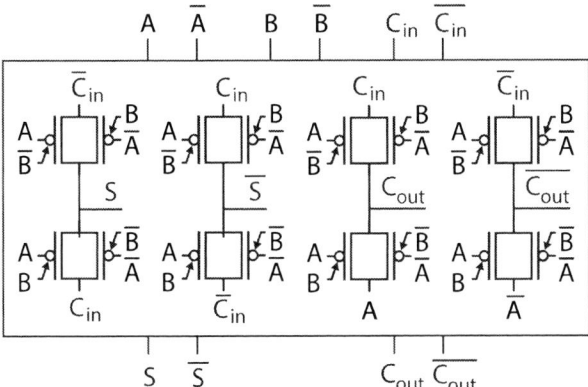

Fig. 4. 1-bit adder with generation of inverted sum and carry.

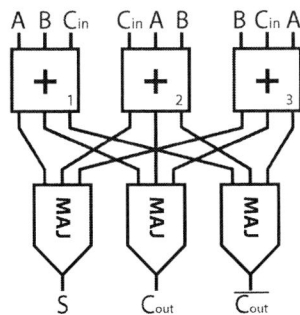

Fig. 5. Fault tolerant 1-bit adder exploiting TMR and MAJ voters. Inverted signals are not represented for the sake of visibility.

V. QUANTITATIVE ANALYSIS

In order to provide the reader with some more details about the performance and characteristics of the proposed architecture, we first performed some experimental analysis, aimed at checking its behavior in the presence of single and double faults.

Results of this analysis (performed by combining at the gate level the results reported in the previous Sections) confirmed that all single faults are masked, either by the characteristics of the controllable-polarity gate implementation, or by the TMR architecture.

From the circuit-level performance perspective, we compared the proposed circuit implemented using SiNWFETs with its equivalent CMOS FinFET 20-nm LSTP counterpart using electrical simulations. The load capacitance for the two circuits is set to 1fF. We consider the area, the worst-case delay and the leakage power. Note that dynamic power is not considered due to a lack of precision in the considered compact model. The circuit-level results are summarized in Table III.

TABLE III. FAULT-TOLERANT 1-BIT ADDER PERFORMANCES

20-nm node	# Transistors	Area (μm²)	Delay (ps)	Leakage Power (nW)
FinFET LSTP	108	5.89	371	23.84
SiNWFET	60	4.98	304	21.06
Gain	44%	15.5%	18.1%	11.6%

The proposed implementation requires 16 controllable-polarity transistors for each 1-bit adder, plus 12 transistors for the 3 majority voters. Hence, 60 transistors are required for the proposed fault-tolerant 1-bit adder. By applying the same design principles with transmission-gate CMOS, we obtained 24 transistors for a 1-bit full adder. Note that the reference structure also generates all the inverted signals required to cascade the different adder stages. Smallest implementations can be identified for both CMOS and controllable-polarity transistors if dedicated inverters are used to generate the inverted signals. Then, a TMR-based implementation in CMOS technology would require 3×24 transistors, plus the cost for the majority voter on the data output, accounting for 3×12 transistors. In total, 108 transistors would thus be required. Hence, the proposed solution requires 44% less transistors. When considering the area of the two adders, the proposed solution requires 4.98μm² as compared to 5.89μm² for its equivalent FinFET implementation. This leads to a gain of 15% in area. The gain is reduced compared to the simple transistor count, as controllable-polarity transistors are bigger than FinFETs, due to the additional polarity terminals. The proposed solution is also significantly less expensive than the one proposed in [19], which proposes a fault-tolerant architecture for the voter consisting of an XOR and a multiplexer.

Finally, the proposed solution can be easily used to build up an adder with whichever data parallelism n, whose total cost scales linearly with n. Since we demonstrated that single faults affecting one stage do not propagate to the following ones, the level of fault tolerance of the final adder is not affected by its parallelism.

From a performance perspective, the proposed implementation is shown to be faster with a 18% reduction of the worst case delay. This is accounted to the reduced number of stacked transistors coming from the use of controllable polarity transistors. Finally, the leakage power is reduced by 12%, thanks to the good electrostatic control offered by the SiNWFETs.

VI. CONCLUSIONS

Controllable-polarity transistors offer many advantages to implement arithmetic logic gates at a reduced implementation cost. Besides the implementation compactness, an important parameter to consider is the robustness with respect to possible faults. In this paper, we performed such an analysis and showed that circuits based on controllable-polarity transistors can tolerate a large number of faults. Thanks to this property, they can be used to build effective structures demonstrating large fault tolerance, in addition to area, power and speed improvements. In particular, we showed that the SiNWFET implementation of a fault tolerant 1-bit adder (that can be easily used to build an adder of any size) is 15% smaller, 18% faster, and 12% less power consuming than the corresponding CMOS solution. This module can be used to build a ripple-carry adder of any length able to tolerate any single permanent fault and most of the possible double faults in any of its stages.

ACKNOWLEDGMENTS

This work has been partially supported by the ERC senior grant NanoSys ERC-2009-AdG-246810.

REFERENCES

[1] K. Bernstein, R. K. Cavin, W. Porod, A. Seabaugh, J. Welser, "Device and Architecture Outlook for Beyond CMOS Switches," *Proc. of the IEEE*, 98(12):2169-2184, 2010.

[2] A. Heinzig *et al.*, "Reconfigurable Silicon Nanowire Transistors," *Nano Letters*, 12:119-124, 2011.

[3] M. De Marchi *et al.*, "Polarity Control in Double-Gate Gate-All-Around Vertically Stacked Silicon Nanowire FETs", *IEDM Tech. Dig.*, 2012.

[4] M.H. Ben-Jamaa *et al.*, "An Efficient Gate Library for Ambipolar CNTFET Logic", *IEEE Trans. on CAD*, 30(2):242-245, 2011.

[5] P.-E. Gaillardon *et al.*, "Vertically-stacked double-gate nanowire FETs with controllable polarity: from devices to regular ASICs", *DATE Tech. Dig.*, 2013.

[6] J. Appenzeller *et al.*, "Dual-gate silicon nanowire transistors with nickel silicide contacts," *IEDM Tech. Dig.*, 2006.

[7] Y.-M. Lin *et al.*, "High-Performance Carbon Nanotube Field-Effect Transistor with Tunable Polarities," *IEEE Trans. on NANO*, 4:481-489, 2005.

[8] N. Harada, K. Yagi, S. Sato and N. Yokoyama, "A polarity-controllable graphene inverter," *Applied Physics Letters*, 96:12102, 2010.

[9] M. De Marchi *et al.*, "Configurable Logic Gates Using Polarity Controlled Silicon Nanowire Gate-All-Around FETs," *IEEE EDL*, 35(8):880-882, 2014.

[10] H. Ghasemzadeh Mohammadi, P.-E. Gaillardon, G. De Micheli, "Fault Modeling in Controllable Polarity Silicon Nanowire Circuits," *DATE Tech. Dig.*, 2015.

[11] P. M. Kogge, H. S. Stone, "A Parallel Algorithm for the Efficient Solution of a General Class of Recurrence Equations," *IEEE Trans. on Computers*, C-22(8):786-793, 1973.

[12] S. Ghosh, P. Ndai, K. Roy, "A Novel Low Overhead Fault Tolerant Kogge-Stone Adder using Adaptive Clocking," *DATE Tech. Dig.*, 2008.

[13] S. K. Mitra, A. R. Chowdhury, "Minimum Cost Fault Tolerant Adder Circuits in Reversible Logic Synthesis," *VLSI Design Tech. Dig.*, 2012.

[14] L. Anghel, M. Nicolaidis, "Defects Tolerant Logic Gates for Unreliable Future Nanotechnologies," *IWANN Tech. Dig.*, 2007.

[15] O. Turkyilmaz *et al.*, "Self-checking ripple-carry adder with Ambipolar Silicon NanoWire FET", *ISCAS Tech. Dig.*, 2013.

[16] J. Rabaey, A. Chandrakasan, B. Nikolic, "Digital Integrated Circuits: A design perspective," *Prentice-Hall Publication*, 2003.

[17] M. L. Bushnell, V. D. Agrawal, "Essentials of Electronic Testing for Digital, Memory and Mixed-Signal VLSI Circuits", *Springer*, 2000.

[18] F. Hapke *et al.*, "Cell-Aware Test", *IEEE Trans. on CAD*, 33(9):1396-1409, 2014.

[19] T. Ban, L. A. de Barros Naviner, "A Simple Fault-tolerant Digital Voter Circuit in TMR Nanoarchitectures," *NEWCAS Tech. Dig.*, 2010.

A Cellular Automata Based Fault Tolerant Approach in Designing Test Hardware for L1 Cache Module

Mousumi Saha
Department of Computer Applications
National Institute of Technology
Durgapur, India
msaha.nitd@gmail.com

Biplab K Sikdar
Department of Computer Science and Technology
Indian Institute of Engineering Science and Technology
Shibpur, India
biplab@cs.becs.ac.in

Abstract—Although fault detection can be successfully made using 2-state 3-neighborhood null boundary cellular automata (as dealt with in our previous work), self correcting property, however, can not be achieved this way. The present work deals with 2-state 5-neighborhood CA instead, which has been found to be self correcting with a much higher efficiency. This fault tolerant approach in designing the test hardware for testing the cache module in chip multiprocessors (CMPs), has been found to achieve 100% accuracy in single stuck-at one fault. It can also challenge multiple stuck-at one fault and can give 22% accuracy.

Keywords-CMPs; cache testing; cellular automata; March test.

I. INTRODUCTION

The developed von Neumann computing model gives the maximum performance limit [1] of computing speed. The effort to achieve higher performance has led to various modifications in the processor design technique. This results into an inevitable transition toward multicore architecture, the Chip Multiprocessors (CMPs).Chip Multiprocessors (CMPs) have been widely adopted [2] [3] as the building block for future computer systems. CMPs designers integrate multiple, potentially simpler, processor cores on a single chip to improve the overall throughput while reducing power consumption and design complexity.

A tiled CMPs architecture consists of a number of replicated tiles connected over a direct switched network [4]. Each tile contains a processing core with primary caches (both I- and D-caches), a slice of the L2 cache, and a connection to the on-chip network. Besides being physically distributed, the L2 cache is also shared among the different processing cores. Therefore, some accesses to the L2 cache will be sent to the local slice while the rest will be serviced by remote slices (L2 NUCA architecture). In addition, the L2 cache tags store the directory information needed to ensure coherence between the L1 caches.

However, the increasing number of cores in CMPs add to the reliability and dependability of design. In multithreaded multicore processors, caches are in multiple levels and multi-bank architectures occupying almost 90% of the relative chip area. This points to the fact that the on-line testing is essential for CMPs [5]. So, an innovation is necessary to find more accurate solution.

In 80s, Wolfram [6] studied the behavior of simple 1-dimensional CA. A special class of Wolfram's 3-neighborhood CA [7] is found to be effective for efficient fault detection in VLSI circuits.

March algorithm gives us [8] the basic approach for fault detection in memory. This basic design has already been used by the present author [4] in 3-neighborhood CA for fault detection of CMPs L1 cache. The test design enables identification of the region of defective cache module in the CMPs and memorizes the fault. A further exploration on the basis of a 5-neighborhood CA has been done in the present work to implement the characteristics of fault tolerance or self correction in test structure considering the single stuck-at fault and multiple stuck-at fault. The next section provides the brief on cellular automata(CA) relevant for the current work.

II. CELLULAR AUTOMATA

A Cellular Automaton (CA) evolves in discrete space and time, and can be viewed as an autonomous finite state machine (FSM). Each cell stores a discrete variable at time t that refers to the present state (PS). The next state (NS) of the cell at $(t+1)$ is affected by its state and the states of its $neighbors$ at t. The 1-dimensional CA, where a cell is having two states - 0 or 1 and the next state of ith CA cell in 5-neighborhood is

$$S_i^{t+1} = f_i(S_{i-2}^t, S_{i-1}^t, S_i^t, S_{i+1}^t, S_{i+2}^t)$$ S_{i-2}^t, S_{i-1}^t, S_i^t, S_{i+1}^t and S_{i+2}^t are the present states of the left most neighbour situated after the left neighbor, the left neighbor, self, the right neighbor and the right most neighbour situated after the right neighbor of the ith cell at time t respectively and f_i is the next state function. Present state of CA at time t with all the states of the cells taken together, can be represented as $\mathcal{S}^t = (S_1^t, S_2^t, \cdots, S_n^t)$ at time t. Therefore, the next state of an $n-$cell CA is

$$S^{t+1} = (f_1(S_0^t, S_1^t, S_2^t, S_3^t, S_4^t), f_2(S_1^t, S_2^t, S_3^t, S_4^t, S_5^t), \cdots, f_n(S_{n-2}^t, S_{n-1}^t, S_n^t, S_{n+1}^t, S_{n+2}^t))$$

Table I
RMTs OF RULE 51, 4294967294 AND 4294967295 IN 5-NEIGHBORHOOD

PS	11111	11110	..	01110	01101	...	00111	00110	00101	00100	00011	00010	00001	00000	Rule
RMT	(31)	(30)	...	(14)	(13)	...	(7)	(6)	(5)	(4)	(3)	(2)	(1)	(0)	
NS	0	0	0	0	0	0	0	0	1	1	0	0	1	1	51
NS	1	1	1	1	1	1	1	1	1	1	1	1	1	0	4294967294
NS	1	1	1	1	1	1	1	1	1	1	1	1	1	1	4294967295

The next state function of the ith CA cell can be expressed in the form of a truth table (Table I). The decimal equivalent of the $2^5 = 32$ outputs (NSs) is called 'rule' R_i. In a 2-state 5-neighborhood CA, there can be 2^{32} (4294967296) rules. Three such rules 51, 4294967294 and 4294967295 are illustrated in Table I. The first row lists the combinations of PSs of $(i-2)^{th}$, $(i-1)^{th}$, i^{th}, $(i+1)^{th}$ and $(i+2)^{th}$ cells at t. The last three rows indicate the NSs of ith cell at $(t+1)$. A combination $S_{i-2}^t, S_{i-1}^t, S_i^t, S_{i+1}^t, S_{i+2}^t$ of PS_i is reformed as the rule minterm (RMT). The column 00011 of Table I is the 3rd RMT. The next states corresponding to this RMT are 0 for rule 51, and 1 for 4294967294 & 4294967295.

In the current design, we employ the null boundary CA. As the leftmost cell is considered without any neighbour to its left, two neighbours to its left are to be considered as zeroes. The same condition will apply to the right most neighbour, where its two right neighbours are considered to be zeroes. A CA is synthesized with a set of rules R. The set of rules R = $< R_1, R_2, \cdots, R_i, \cdots, R_n >$, that configures the cells of a CA, and is called the rule vector of the CA. If all the rules, R_is, are same, then the CA is uniform otherwise it is a non-uniform/hybrid.

III. DESIGN DETAILS

In the current design, we consider effective realization of March algorithm [5] for determining correctness of the cache function. The March algorithm first writes 0 (1) to each memory cell. Then, a read operation r0 (r1) is performed on each cell. If the read bits are all 0s (1s), the memory is fault free. Otherwise, the memory is faulty.

The hardware realization of March algorithm as reported in [8] demands n 2-input XOR gates and n 2-input ORs. At each read '0' (1) from a cache word, it compares the content of n cache cells with n 0s (1s), pre-stored in n FFs, to decide on the faults in a cell of the word.

In the proposed CA based test hardware (Fig.1), we realize the March C^- developed over the cellular automata(CA) structure. A read operation 'r0' or 'r1' of March algorithm stores the n-bit word, read from the cache, to a register RG. These n bits are used to set the cell rules of an n-cell CA. The ith-bit RG_i sets rule for the ith CA cell. If the CA is then run for some definite time steps, it settles to an attractor state. For a fault in the cache, the least significant cell of the CA (Sig_i) becomes '1'. On the other hand, Sig_i is '0' if

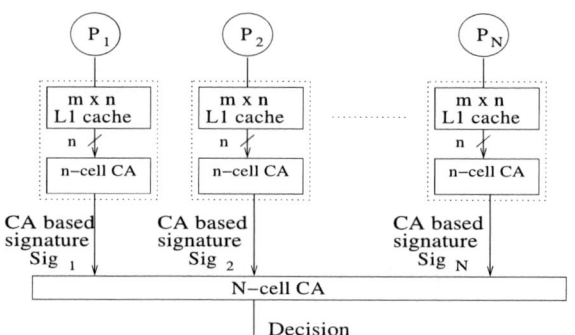

Figure 1. Architecture of CA based test hardware for NUCA in CMPs

the cache module is fault free. That is, by sensing only the Sig_i, cache module can be declared as faulty or fault free.

The signature generated from each test module is then used to set the ith cell CA rule of the N-cell CA(Fig. 1). That is, the signature response is tested with an initial n-cell CA. Having done this it is applied on an N-cell CA (since N processors have N number of L1 cache modules), different from initial n-cell CA, to see the applicability of the signatures(Fig. 1). The new N-cell CA structure is similar to the initial one. If signatures are found to be correct, it can be concluded that the cache module are working correctly. The next section reports the detailed design of the proposed test structure.

Design Requirements: In [4], it has already been shown that the design requirement for fault detection was for n-cell CA that can settle to a single length cycle attractor X. Further, for fault free and faulty cache word, we need to form different CAs so that the effect of fault can induce lsb of X as '1' for faulty and '0' for fault free conditions.

The fault tolerant design needs the consideration of a 5-neighbourhood CA instead of 3-neighbourhood CA [4] to implement the fault tolerance in addition to the basic fault detection capacity. So, the design requirement for the present work is to figure out a CA which will stop stuck-at zero fault in test hardware , masking the cache module fault and will also stop propagation of stuck-at one fault in test hardware to the lsb of CA. In this way, it can produce a correct information regarding the cache module.

The CA chosen for the fault free cache word settles to an attractor 0 (all 0s state) if the CA is initialized with all 0s state (Fig. 2). In all other faulty cases, the CA selected settles

978-1-4799-8720-7/15 $31.00 © 2015 IEEE   498

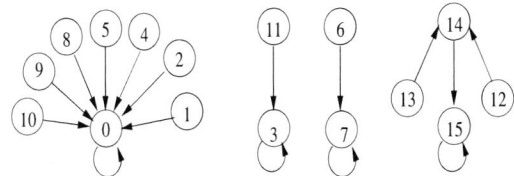

Figure 2. uniform CA configured with rule 4290834624

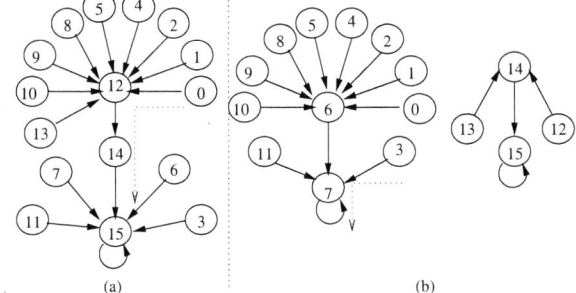

(a)   (b)

Figure 3. Hybrid CA configured with rule 4290834624, 4294967295

to attractor 1 with *lsb* 1(Fig. 3(a)(b)). That is, an incidence of fault in cache module is translated as the switch from 0-basin to another basin with *lsb* 1. The next section reports the identification of CA rules that can be selected for the test hardware.

A. Single length cycle attractor CA rules

Nature of a CA is directly related to the nature of its RMTs. The RMTs of a rule R are of two types - passive and active. An RMT r of a rule is passive if at time $(t+1)$ a CA cell remains in the same state as in time t (0 to 0 or 1 to 1) . An RMT r of a rule is active if a CA cell flips its state (1 to 0 or 0 to 1) . From Table I, it can be seen that for rule 51, the RMTs 2, 3 and 5 are passive where the RMts 0, 1, 4 are active. The active and passive RMTs play a role to form single length cycle attractor(i.e. fixed point attractor).

The following theorems are introduced that help to reduce the search space to identify the rules that only form single length cycle attractor CA.

Relationship between RMTs of two consecutive cells: A CA state can be viewed as a sequence of RMTs. For example, the state 1100 for a null-boundary CA can be viewed as the sequence <6,12,24,16>. A 5-bit imaginary window is considered that slides over the state. To get the ith RMT, the window is loaded with $(i-2)^{th}$, $(i-1)^{th}$, i^{th}, $(i+1)^{th}$ and $(i+2)^{th}$ bits of the state. Again to get the $(i+1)^{th}$ RMT, the window is shifted one bit to the right. Thus RMTs of two consecutive cells differ only in the LSB. For example if the ith RMT is 3(00011) then the $(i+1)^{th}$ RMT can either be 6(00110) or 7(00111), thus giving rise to the relationship between RMTs of ith and $(i+1)^{th}$ cells of a given state. The

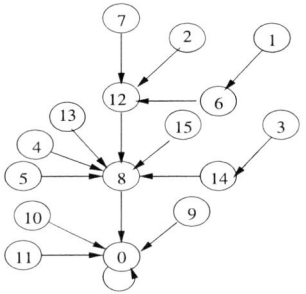

Figure 5. State transition of even numbered rule 206

relationship between two consecutive RMTs in a sequence of RMTs is given in Table II.

Table II
RELATIONSHIP BETWEEN TWO CONSECUTIVE RMTS

i^{th}	$(i+1)^{th}$
0,4, 16, 20	0,1, 8, 9
1,5, 17, 21	2,3, 10, 11
2,6, 18, 22	4,5, 12, 13
3,7, 19, 23	6,7, 14,15
8, 24	16, 17
9, 25	18, 19
10, 26	20, 21
11,27	22, 23
12,28	24, 25
13,29	26, 27
14,30	28, 29
15, 31	30, 31

Theorem 1: If the RMTs 7, 15, 28, 30, and 31 are passive in a CA rule, it can allow formation of at least one fixed point attractor -that is, all in 1s state.

Proof: The CA having one state of all 1s at time t, corresponds to RMT sequence sq $< 7, 15, 31, 30, 28 >$. If all the RMTs in the RMT sequence sq are simultaneously passive then CA would have next state of all 1s at time $(t+1)$ and obviously it can form all 1s fixed point attractor, otherwise any one RMT in the RMT sequence Sq is active it can not form all 1s attractor. Then CA can transit to another state at time $(t+1)$. ■

Theorem 2: An even numbered CA rule can allow formation of at least one fixed point attractor, i.e. the state 0.

Proof: Any even rule has a 0 in the *lsb* (11001110 for rule 206). Thus the RMT 0 for an even 0 will always have the next state value 0 (i.e., it is apassive RMT). The state 0 corresponds to $< 0000 >$ RMT sequence. Since the state 0 of an even rule contains all passive RMTs, hence state 0 is a fixed point attractor for an even rule of a null boundary CA. ■

Lemma 3: If the RMTs 7, 15, 28, 30, and 31 are passive in an even numbered CA rule, it can allow formation of at least two fixed point attractors -that is, all 0s state and all 1s state.

978-1-4799-8720-7/15 $31.00 © 2015 IEEE   499

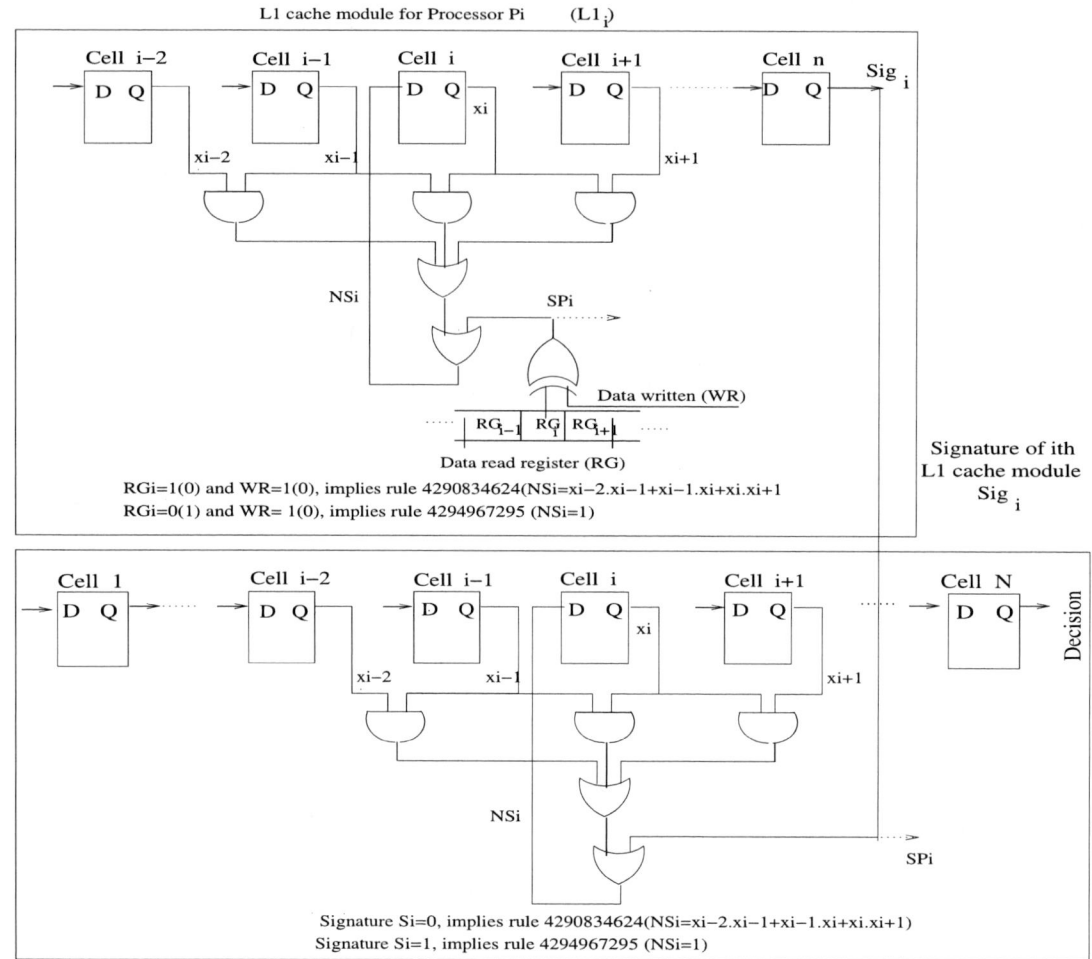

Figure 4. CA hardware test logic

Proof: The CA sinthesized with even numbered rule R always forms all 0s fixed point attractor (Theorem 2) and if sumultaneously that rule R consists of passive RMTs 7, 15, 28, 30, and 31 then it can also form another fixed point attractor of all 1s (Theorem 1). Obviously the even numbered rule with passive RMTs 7, 15, 28, 30, and 31 can allow formation of at least two fixed point attractors -that is, all 0s state and all 1s state. ∎

B. The fault tolerant test hardware design

In order to design the fault tolerant test hardware for CMPs cache system, two stages are envisaged. First stage activity is to test the performance of the cache and second stage activity is to locate the faulty region (Fig. 4). The analysis of first stage test results (signature) indicates the final decision in the second stage in assessing the overall performance as well as locating the fault.

The architecture of the test hardware realizing March C$^-$ is shown in Fig. 1. For an n-bit cache module, we employ

n-cell CA. The design of n-cell CA hardware is shown in Fig.4. Once 0 is written (WR=0) to each cache cell M_i, the data bit read (r0) from cell M_i (RG$_i$ of Fig.4) is used to set the rule of ith CA cell as well as $(i+1)^{th}$ CA cell. When WR=0 and RG$_i$=0 implies there is no fault then it is encoded as rule 4290834624 (NS$_i$ = $x_{i-2} . x_{i-1} + x_{i-1} . x_i + x_i . x_{i+1}$). When there is a fault in M_i, RG$_i$=1 -that is, NS$_i$=1, then rule 4294967295 is set for the ith as well as $(i+1)^{th}$ CA cell. On the other hand, when 1 is written (WR=1) to each cache cell M_i, the read bit RG$_i$=1 is encoded as the rule 4290834624. For a fault in M_i, RG$_i$=0 -that is, NS$_i$=1, the rule is set as 4294967295 (Fig.4).

The activity of second stage is also similar to the first stage with the difference that in the second stage N (no of processors in CMPs) is used instead of n (no of bits in cache word). For N no of L1 cache modules, we employ an N-cell CA. The signature generated from the ith L1 cache module (Sig$_i$) is used to set the rule of the ith and $(i+1)^{th}$

CA cell simultaneously. The $Sig_i=0$ is encoded as the rule 4290834624. When there is a fault in cache module $Sig_i=1$, that is, $NS_i = 1$, the rule is set as 4294967295 (Fig. 4) for the ith and $(i+1)^{th}$ CA cell.

The CA, thus constructed, is run for t-steps ($t <= N$), after initializing with all 0s seed. The CA for a fault free cache module is a uniform CA constructed with rule 4290834624 and so, it reaches the attractor state 0 (Fig. 2). On the other hand, for fault at one or more cache modules, the CA is a hybrid CA (with rule 4290834624 and 4294967295) and it reaches a non-zero attractor with lsb=1 (Fig. 3(a), (b)) within t-steps. Therefore, by sensing the lsb of the CA, we can detect faults in the cache module.

C. Performance analysis:

The fault tolerant test hardware design (Fig. 4) targets the detection of fault and faulty region of cache module in CMPs including the achievement of fault tolerance. The classical method [8] and the 3-neighborhood method [4] can not give the correct result when the fault is in test hardware itself. In both the methods, the stuck-at one (single or multiple) faults in test hardware in any position always indicates fault in cache module even if the cache module is fault free.

However, in the fault tolerant approach, the single stuck-at one fault in test hardware, can not propagate it to the error line because, the applied rule configures the design of test hardware in such a way that it has the inherent property of constructing single length cycle attractor with lsb 0 (Fig. 2 and Fig 6). If it is assumed that, the cache module is fault free and the test hardware is having a single stuck-at one fault, then 4-bit cell CA can have states 1, 2, 4 and 8. So from the state transition diagram of rule 4290834624 (Fig. 2 and Fig. 6), we can observe that the CA configured by that rule can form single length cycle attractor with lsb 0. It simultaneously has a property to propagate the error of cache module by configuring the error bit consequtively two 1s through SP_i of Fig. 4 (Fig. 3(b) shows that 0 basin shifts to the attractor 7 with lsb 1). Thus ultimately, we can get the correct behaviour of cache module even when the test hardware has single stuck-at one fault itself. The X-OR logic and 3-neighborhood logic do not work in this case at all when we find 100% fault tolerance achievement for 5-neighborhood based fault tolerant logic. This, however, does not include faults at the error line(lsb). Fault at error line is an issue to be discussed separately.

This new design approach has also been found to be quite receptive in some cases of multiple stuck-at one fault. The design can tolerate the multiple scattered stuck-at-one fault but it can not tolerate consequitive stuck-at one fault. The calculation of test hardware tolerance can be done with the following theorem and formulae.

Theorem 4: For a string of length n, the number of binary strings having scattered 1s, denoted $f(n)$ may be obtained as

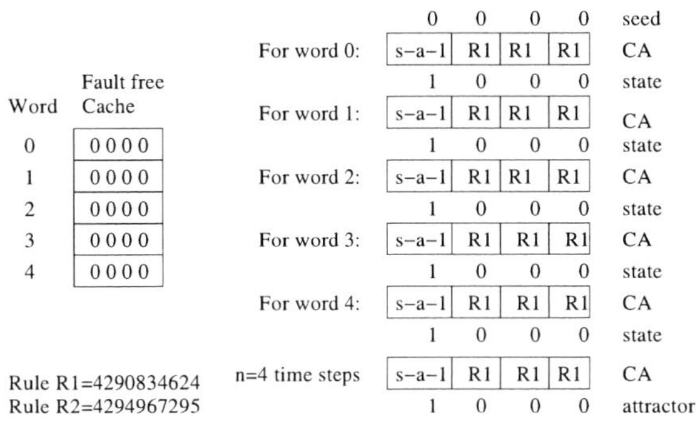

Figure 6. Function of test hardware when it has stuck-at 1 fault

$$f(1) = 2,$$
$$f(2) = 3,$$
$$f(n) = f(n-1) + f(n-2), n >= 3$$

Where $f(n-1)$ and $f(n-2)$ are no of posssible scattered 1s strings for $(n-1)$ bit string and $(n-2)$ bit string consequitively.

Proof: Let us have n bit string. The end bit (lsb) will be either 0 or 1. For $n = 1$, the possible strings $S = \{0, 1\}$ and the number of strings $|S| = 2$. For $n = 2$, $S = \{00, 01, 10\}$ and $|S| = 3$ etc.
For arbitrary $n(>= 3)$,
$Case 1$: If the lsb is 0 then $(n-1)^{th}$ bit will be 0 or 1 i.e. there are no restriction on $(n-1)^{th}$ bit. So, $f(n-1)$ no of n-bit string ends with 0 and contains no two consecutive 1s.
$Case 2$: if the lsb bit is 1 then $(n-1)^{th}$ bit must be 0. Hence no restriction on $(n-2)^{th}$ bit. Hence $f(n-2)$ no of n-bit string ends with 1 and contain no consecutive 1s. Since these two cases are mutually exclusive, hence we can apply addition principle. Therefore , $f(n) = f(n-1) + f(n-2), n >= 3$ and $f(1) = 2, f(2) = 3$. ∎

We need to keep the last digit fixed at 0. Hence all the strings that end with 1 will be discarded. Therefore except last digit, there are no restrictions for all the other $(n-1)$ digits , hence for n-bit string patterns, we need to calculate value of $f(n-1)$, i.e., if we have a string of length 4, we need to calculate the value of $f(3)$ since the last bit is fixed at 0. We store the result in a variable k_n.
i.e., $k_n = f(n-1)$
The number(Kn) gives all such strings of length n having scattered 1s and the last bit fixed at 0. Even it consists of strings having all 0s and all single ones. We have seen that the test hardware can tolerate single stuck-at one fault. Later we will explain the achievement of fault tolerance considering stuck-at zero fault. Hence the no of strings having all 0s and single 1 (except lsb1) is equal to n and so the number of scattered ones strings without the all 0s and single 1s strings are
$$k'_n = f(n) - n$$
Hence, the fault toleranance has been calculated on the

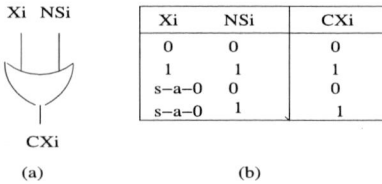

Xi	NSi	CXi
0	0	0
1	1	1
s−a−0	0	0
s−a−0	1	1

(a)　　　　　　　(b)

Figure 7.　(a) Checker and mask unit, (b) Function table

basis of scattered ones excepting the case ending in one with respect to the total no of multiple ones since we want to find out the fault tolerance considering the multiple stuck at fault of test hardware. The approach can achieve 22% multiple stuck-at one fault tolerance considering the test hardware for testing 8 bit cache word. The 16 bit and 32 bit cache word size can achieve the same tolerance i.e. 22% with the help of CA segmentation technique.

When cache module is faulty, the stuck-at zero fault in test hardware can mask the fault of cache module and gives an incorrect decision. In the classical [8] and 3-neighborhood method [4], if Stuck-at zero fault in test hardware is situated to the right of the faulty bit of cache word then the fault can propagate to the error line i.e. correct decision can be made. We can not take correct decision if Stuck-at zero fault in test hardware is coinsided with the faulty bit of cache word and Stuck-at zero fault in test hardware is situated to the left of the faulty bit of cache word. So, here out of these three situation the classical method and 3-neighborhood method can tolerate only one (i.e., 33% tolerance can achive). The five neighborhood method can successfully tolerate in all three situations (including multiple stuck-at 0) with the help of an additional checker and mask unit.

Checker and mask unit: The output X_i of any cell C_i can be checked whether it has stuck-at 0 fault with the help of NS_i. The cheker and mask unit Fig. 7(a) is explained with the help of function Table 7(b). If NS_i is 1 at time t then X_i has to be the state of 1 at time $(t + 1)$. So, the checker and mask unit(CMU) can be added to the ouput line of each cell and it can check whether the that output line has stuck-at 0. If it has stuck-at 0 then it can mask and generate the correct output value of cell (CX_i). The single stuck-at 0 as well as multiple stuck-at 0 of test hardware can not disturb to give the information about cache module. Fig. 8 shows that the one of the cache word has a fault and test hardware has a stuck at 0 fault that is situated to the left of cache faulty bit and here the test hardware can detect the fault of cache module efficiently. We can resolve the stuck-at 0 fault at error line using checker and mask unit. The Table III shows that the method proposed in this work can properly address and solve all the possible conditions.

IV. CONCLUSION

This work proposes an efficient test hardware for March algorithm. The solution is developed around the regular structure of

Rule R1=4290834624
Rule R2=4294967295

Word	Faulty Cache
0	0000
1	0000
2	0100
3	0000
4	0000

For word 0

X	0	0	s−a−0	0
SP	0	0	0	0
NS	0	0	0	0
CX	0	0	0	0
CA	R1	R1	s−a−0	R1

For word 1

X	0	0	s−a−0	0
SP	0	0	0	0
NS	0	0	0	0
CX	0	0	0	0
CA	R1	R1	s−a−0	R1

For word 2

X	0	0	s−a−0	0
SP	0	1	0	0
NS	0	1	1	0
CX	0	1	1	0
CA	R1	R2	s−a−0	R1

For word 3

X	0	1	s−a−0	0
SP	0	0	0	0
NS	0	1	1	1
CX	0	1	1	1
CA	R1	R1	s−a−0	R1

For word 4

X	0	1	s−a−0	1
SP	0	0	0	0
NS	0	1	1	1
CX	0	1	1	1
CA	R1	R1	s−a−0	R1

Figure 8.　Function of Cheker and Mask unit

Table III
LEVEL OF ACCURACY

Fault type	Classical logic [8]	3-neighborhood logic [4]	Fault tolerant logic
Single Stuck-at one	0%	0%	100%
Multiple Stuck-at one	0%	0%	22%
Single Stuck-at zero	33%	33%	100%
Multiple Stuck-at zero	33%	33%	100%

5-neighborhood CA. The special class of CA rules, forming single length cycle attractor, are chosen for the design. This enables the high speed testing of cache modules in tile CMPs architecture at low cost and at a decision efficiency of 100% in single stuck-at one fault and 22% in multiple stuck-at fault.

REFERENCES

[1] K. Olukotun and L. Hammond, *The Future of Microprocessor*, Magazine Queue - Multiprocessors, Volume 3 Issue 7, Pages 26-29, September, 2005.

[2] AMD. *AMD Multi-core*, http://multicore.amd.com/en/, 2006.

[3] Intel. *Multi-core from Intel*, http://www.intel.com/multi-core/, 2006.

[4] M. Saha and B. K. Sikdar, *An Efficient Method for Testing of L1 Cache Module in Tiled CMPs Architecture at Low Cost*, VLSI-SATA, 2015

[5] G. Theodorou, N. Kranitis, A. Paschalis and D. Gizopolos, *Software-based Self Test Methodology for On-Line Testing of L1 Caches in Multithreaded Multicore Architectures*, IEEE Transactions on VLSI Systems, Vol. 21, No.4, 2013.

[6] S. Wolfram, *Cellular Automata and Complexity – Collected Papers*. Addison Wesley, 1994.

[7] P. P. Chaudhuri, D. R. Chowdhury, S. Nandi and S. Chatterjee, *Additive Cellular Automata - Theory and Applications*, volume 1. IEEE Computer Society Press, USA, ISBN 0-8186-7717-1, 1997.

[8] P. Baanen, *Testing word oriented embedded RAMs using built-in self test*, IEEE 1988.

2015 IEEE Computer Society Annual Symposium on VLSI

Diagnosis of Delay Faults Considering Hazards

Yoshinobu Higami, Senling Wang, Hiroshi Takahashi, Shin-ya Kobayashi
Graduate School of Science and Engineering
Ehime University
Matsuyama, Ehime 790-8577, Japan
Email: higami@cs.ehime-u.ac.jp

Kewal K. Saluja
Dept. of Electrical and Computer Engineering
University of Wisconsin - Madison
Madison, WI 535706-1691
Email: saluja@engr.wisc.edu

Abstract—It is very difficult, if not impossible, to design hazard free circuits in view of substantial delay uncertainties of gates and interconnects implemented in submicron technologies. In this paper, we propose diagnosis methods for gate delay faults for such circuits. The fault simulation method employed by us uses eight values and calculates logic values as well as earliest transition times and latest transition times. It can deal with hazard signals more accurately than conventional methods. The proposed method uses a fault dictionary to deduce candidate faults which sufficiently explain the output responses of a circuit under diagnosis.

I. INTRODUCTION

In delay fault simulation, logic values for two consecutive test patterns need to be calculated. Conventionally, four logic values, 0, 1, R (0→ 1: Rise) and F (1 → 0: Fall), have been used in the fault simulation for two pattern tests. In real circuits, hazard signals happen depending on timing of signal transitions. Fault simulation without considering hazards may bring incorrect fault coverage or incorrect fault diagnosis results. For the purpose of more accurate calculation, extended values have been introduced in some researches. In [1] and [2], six values have been introduced, which consider the possibility of hazards. In [3], seven values have been used, and also the latest transition times have been calculated. However, hazard signals have not been explicitly represented. Hazard-based detection conditions have been discussed in [4] and [5]. When fault simulation is performed by taking hazard signals into account, resultant fault coverage may be improved. For example, if the faulty line produces a hazard signal 1 → 0 → 1, and if a certain path to a flip-flop is sensitized, then a slow-to rise delay fault will be detected depending on the amount of the delay. However, if hazard signals were not considered, the fault might be regarded as undetected.

In order to diagnose delay defects accurately, timing of signals needs to be considered. In [6], diagnostic fault simulation has been performed using timing information. Signal arrival time, signal propagation delay and possible range of the delay induced by a defect have been considered. However, effects of hazards have not been considered.

In this paper, we propose two diagnosis methods for gate delay faults by a fault dictionary based approach. The fault dictionary is created by the fault simulation that takes hazard signals into account using eight values. The simulation calculates output responses as well as the earliest transition times and the latest transition times. In the proposed diagnosis methods, certain rules are introduced to deduce candidate faults. The rules find contradiction between the output responses of a

circuit under diagnosis (CUD) and the information about candidate faults in the dictionary. The methods target single gate delay faults and double gate delay faults. In the experiments for benchmark circuits, the proposed diagnosis methods are applied for CUDs in which a randomly selected delay fault with varying amounts of additional delay is injected.

The rest of this paper is organized as follows. Section II describes eight value simulation that can deal with hazard signals. Section III gives some definitions. Section IV explains how to deduce candidate faults using examples. Section V and Section VI explain the proposed diagnosis methods for single delay faults and for double delay faults, respectively. Section VII gives experimental results for benchmark circuits. Section VIII concludes this paper.

II. EIGHT VALUE FAULT SIMULATION CONSIDERING HAZARD SIGNALS

First we describe the fault simulation proposed in [3], referred to as **seven value simulation** hereafter. In that simulation, seven logic values, 0, 1, X, R, F, B and $\overline{B}$ are used. Values 0 and 1 denote stable 0 and 1, X denotes an unknown value. R and F denote a rising transition and a falling transition, respectively. B and $\overline{B}$ also denote a rising transition and a falling transition, respectively, but they indicate that the values are propagated from a faulty site. The seven value simulation calculates not only logic values but also latest transition times. Now consider the calculation at a 2 input AND gate. Fig. 1 shows signal waveforms on output c and inputs a and b. Suppose that one unit time is added when the signal is propagated through the gate. The seven value simulation has no value that represents hazard signal(s) explicitly, and signal values are represented only by focusing on the latest transitions. The signals on a and b are represented as R and F with the latest transition times 4 and 3, respectively. As a result of AND gate calculation, output c produces 0, which means that information about hazard signal is lost.

The proposed simulation uses eight values: 0, 1, R, F, RF FR, UU, DD. Values 0, 1, R and F denote stable 0, stable 1, a rising transition and a falling transition, respectively. These values do not include any hazards. Value RF, FR, UU and DD denote hazard signals, as shown in Fig. 2, where et and lt denote earliest transition time and latest transition time, respectively. When an initial value and the final value are both 0 (1), value RF (FR) is assigned. Whereas, when an initial value is 0 (1) and the final value is 1 (0), value UU (DD) is assigned. In the proposed simulation, latest transition times as well as earliest transition times are calculated.

978-1-4799-8720-7/15 $31.00 © 2015 IEEE   503

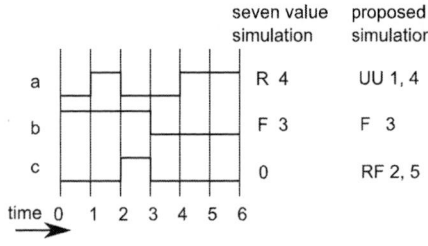

Fig. 1. Waveforms and values

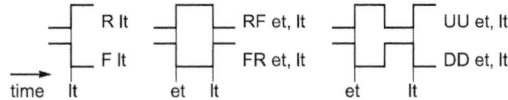

Fig. 2. Signal values

AND operation using the eight values is shown in Table I (a), (b) and Table II. When one input is either one of RF, FR, UU and DD, the output value depends on the earliest transition time and the latest transition time. The operations of the values including RF, FR, UU or DD are shown for each case of transition times in Table I (b). Earliest transition time and latest transition time are calculated by referring to Table II. Variable $et_{1(2)}$ and $lt_{1(2)}$ denote earliest transition time and latest transition time, respectively. We use another table for the case of $et_1 = et_2$ in AND operation, and use other tables in OR, and NOT operation, but due to the space limitation we do not show such tables.

An example of the AND output c with two inputs a and b, which produce RF and UU, respectively, is illustrated in Fig. 3. When a's latest transition time lt_1 is earlier than b's earliest transition time et_2, the output c produces 0, as shown in Fig. 3(a). When $et_1 < et_2 < lt_1 < lt_2$, the output c produces RF with earliest transition time $et_2 + d$ and latest transition time $lt_1 + d$, as shown in Fig. 3(b), where d denotes a propagation delay of the gate.

In the example of Fig. 1, value at a is represented by UU with earliest transition time 1 and latest transition time 4, and value at b is represented by F with latest transition time 3. Value at output c is represented by RF with earliest transition time 2 and latest transition time 5. The proposed simulation can represent the signals more accurately than the seven value simulation. We must make it clear that the proposed simulation does not always represent the signal waveform exactly. While the latest transition time in the real signal at c is 3, in the proposed simulation it is 5. The limited set of values can not always represent real signals exactly. However, the proposed simulation can represent signal values somewhat more accurately than the conventional simulation. When a falling transition delay fault is assumed on c, it is activated in the proposed simulation, while it is not activated in the seven value simulation.

III. DEFINITIONS

In the following definitions, z denotes a primary output (PO) or a flip-flop (FF), f denotes a delay fault, and t denotes a test pattern.

TABLE I. AND OPERATION

(a) Part - I

	0	1	R with lt_1	F with lt_1	RF	FR	UU	DD
0	0	0	0	0	0	0	0	0
1	0	1	R	F	RF	FR	UU	DD
R with lt_2	0	R	R	0 (if $lt_1 \leq lt_2$) RF (otherwise)	(shown in Table I (b))			
F with lt_2	0	F	RF (if $lt_1 < lt_2$) 0 (otherwise)	F				

(b) Part - II

		RF with et_1,lt_1	FR with et_1,lt_1	UU with et_1,lt_1	DD with et_1,lt_1
R with lt_2	cond1	0	R	R	0
	cond2	0	R	R	0
	cond3	RF	R	R	RF
F with lt_2	cond1	RF	DD	RF	DD
	cond2	RF	F	RF	DD
	cond3	RF	F	RF	DD
RF with et_2,lt_2	cond1	0	RF	RF	0
	cond2	0	RF	RF	0
	cond3	RF	RF	RF	RF
	cond4	RF	0	RF	RF
	cond5	RF	0	RF	RF
FR with et_2,lt_2	cond1	RF	FR	UU	DD
	cond2	RF	FR	UU	DD
	cond3	RF	FR	UU	DD
	cond4	RF	FR	UU	F
	cond5	RF	FR	UU	DD
UU with et_2,lt_2	cond1	0	UU	UU	0
	cond2	0	UU	UU	0
	cond3	RF	UU	UU	RF
	cond4	RF	R	UU	RF
	cond5	RF	R	UU	RF
DD with et_2,lt_2	cond1	RF	DD	RF	DD
	cond2	RF	DD	RF	DD
	cond3	RF	DD	RF	DD
	cond4	RF	F	RF	DD
	cond5	RF	F	RF	DD

cond1: $et_1 < lt_1 < et_2$
cond2: $et_1 < lt_1 = et_2$
cond3: $et_1 < et_2 < lt_1 < lt_2$
cond4: $et_1 < et_2 < lt_1$ & $lt_1 = lt_2$
cond5: $et_1 < et_2$ & $lt_2 < lt_1$

[**Definition 1**] An output z is called a **CP_output** when a *pass* response (i.e., no erroneous value) is observed at z for t in the responses of a CUD.

[**Definition 2**] An output z is called a **CF_output** when a *fail* response (i.e., an erroneous value) is observed at z for t in the response of a CUD.

[**Definition 3**] An output z is called a **DP_output** when z produces a *pass* response (no erroneous value) for f and t in the dictionary.

[**Definition 4**] An output z is called a **RF_output** when z produces value R or F with latest transition time larger than the fault free one for f and t in the dictionary.

[**Definition 5**] An output z is called a **DF_output** when z is a RF_output or z produces one of RF, FR, UU or DD for f and t in the dictionary.

[**Definition 6**] $Lt(t, z, f)$ is the **latest transition time** that is denoted with respect to t, z and f in the dictionary.

[**Definition 7 (Minimum latest transition time for f)**] Variable $Min\_Lt(f)$ is defined as follows.

$$Min\_Lt(f) = \min\{Lt(t, z, f)\} \text{ for } ^\forall t \text{ and } ^\forall z$$

which is a CF_output and a RF_output.

TABLE II. EARLIEST TRANSITION TIME AND LATEST TRANSITION TIME AT AND OUTPUT

		RF with et_1, lt_1		FR with et_1, lt_1		UU with et_1, lt_1		DD with et_1, lt_1	
R with lt_2	cond1	-	-	-	et_2	-	et_2	-	-
	cond2	-	-	-	lt_1	-	lt_1	-	-
	cond3	et_1	lt_1	-	lt_1	-	lt_1	et_1	lt_1
F with lt_2	cond1	et_1	lt_1	et_1	et_2	et_1	et_2	et_1	lt_1
	cond2	et_1	lt_1	-	et_1	et_1	et_2	et_1	lt_1
	cond3	et_1	et_2	-	et_1	et_1	et_2	et_1	et_2
RF with et_2, lt_2	cond1	-	-	et_2	lt_2	et_2	lt_2	-	-
	cond2	-	-	lt_1	lt_2	lt_1	lt_2	-	-
	cond3	et_2	lt_1	lt_1	lt_2	lt_1	lt_2	et_2	lt_1
	cond4	et_2	lt_1	-	-	et_2	lt_2	et_2	lt_1
	cond5	et_2	lt_2	-	-	et_2	lt_2	et_2	lt_2
FR with et_2, lt_2	cond1	et_1	lt_1	et_1	lt_2	et_1	lt_2	et_1	lt_1
	cond2	et_1	lt_1	et_1	lt_2	et_1	lt_2	et_1	lt_1
	cond3	et_1	et_2	et_1	lt_2	et_1	lt_2	et_1	et_2
	cond4	et_1	et_2	et_1	lt_1	et_1	lt_1	-	et_1
	cond5	et_1	lt_1	et_1	lt_1	et_1	lt_1	et_1	lt_1
UU with et_2, lt_2	cond1	-	-	et_2	lt_2	et_2	lt_2	-	-
	cond2	-	-	lt_1	lt_2	lt_1	lt_2	-	-
	cond3	et_2	lt_1	lt_1	lt_2	lt_1	lt_2	-	lt_1
	cond4	et_2	lt_1	-	lt_1	et_2	lt_1	et_2	lt_1
	cond5	et_2	lt_1	-	lt_1	et_2	lt_1	lt_2	lt_1
DD with et_2, lt_2	cond1	et_1	lt_1	et_1	lt_2	et_1	lt_2	et_1	lt_1
	cond2	et_1	lt_1	et_1	lt_2	et_1	lt_2	et_1	lt_1
	cond3	et_1	lt_1	et_1	lt_2	et_1	lt_2	et_1	lt_1
	cond4	et_1	lt_1	-	et_2	et_1	et_2	et_1	lt_1
	cond5	et_1	lt_2	-	et_2	et_1	lt_2	et_1	lt_2

cond1: $et_1 < lt_1 < et_2$
cond2: $et_1 < lt_1 = et_2$
cond3: $et_1 < et_2 < lt_1 < lt_2$
cond4: $et_1 < et_2 < lt_1$ & $lt_1 = lt_2$
cond5: $et_1 < et_2$ & $lt_2 < lt_1$

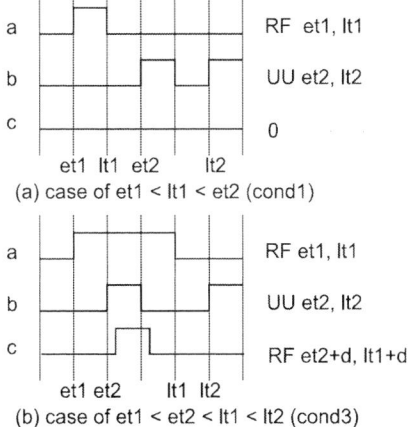

Fig. 3. Example of AND operation

[**Definition 8 (No. of CUD-pass and dictionary-RF outputs)**] $Num\_CPRF(t, f)$ is the number of outputs for test pattern t and fault f that satisfy the followings.

- The outputs are CP_outputs as well as RF_outputs.

- $Min\_Lt(f) \le Lt(t, z, f)$.

[**Definition 9 (Set of CUD-fail outputs)**] Z_{CF} is a set of 2-tuples (t, z) such that z is a CF_output for $^\forall t$.

[**Definition 10 (Set of dictionary-fail outputs)**] $Z_{DF}(f)$ is

a set of 2-tuples (t, z) such that z is a DF_output for $^\forall t$ and f in the dictionary.

IV. EXAMPLE OF DIAGNOSIS

The proposed method uses a fault dictionary to deduce candidate faults by investigating whether the information in the dictionary explains the output responses of a CUD without contradictions. In this subsection, we explain how to deduce candidate faults by using an example and various scenarios. Table III shows an example of a fault dictionary, where f_1, f_2 and f_3 are candidate faults and simulation results of values on FFs for a certain test pattern are shown. In the table, values, earliest transition times and latest transition times are shown for each candidate fault. For example, when fault f_2 is injected, FF_1 produces value R with latest transition time 20. When fault f_3 is injected, FF_1 produces value R with latest transition time 30, and FF_2 produces value UU with earliest transition time 15 and latest transition time 35. The table also shows that a fail response and a pass response is observed at FF_1 and FF_2 in a CUD, respectively. The fault dictionary is created by inserting each fault with a certain amount of delay. Obviously, the amount of delay may be different between the fault existing in the CUD and each candidate fault in the fault dictionary.

Now let's consider the case of fault f_1. The dictionary shows that no erroneous output is observed at FF_1 for f_1. If a larger amount of delay is injected at the faulty site of f_1 in the simulation, then an erroneous output might be propagated. However, we have confirmed through a number of simulations that such a case rarely happens. As a result, the possibility that fault f_1 exists in the CUD is extremely low, and thus f_1 should be removed from the candidate fault set.

Next consider the case of fault f_2. The dictionary shows FF_1 produces a transition value, which can be a fail response, when f_2 is injected, and this information explains the output response on FF_1 in the CUD. However, the information of FF_2 in the dictionary conflicts with the output response in the CUD. If fault f_2 existed in the CUD and if FF_1 produced a fail response, then FF_2 would also produce a fail response. This is because a transition value is propagated to FF_2 later than FF_1. Therefore, f_2 cannot be a candidate.

From this example, the following rule is derived. Now let z_1 and z_2 be FFs. Suppose that a CUD produces a fail response at z_1 and a pass response at z_2, respectively, when a certain test pattern is applied. Also suppose that z_1 and z_2 produces either R or F with latest transition time lt_1 and lt_2, respectively, for fault f in the dictionary. If $lt_1 \le lt_2$, then fault f cannot be a candidate.

Next consider the case of fault f_3. In the dictionary FF_1 and FF_2 produce R and UU with the latest transition time 30 and 35, respectively. If we apply a similar rule as the above one, the fault f_3 will be removed from the candidate fault set. However, when the situation as shown in Fig. 4 occurs, fault f_3 can be a candidate. The figure shows FF_1 produces a fail output, and FF_2 produces a pass output when the output responses are captured at time 20. (Fault free values at FF_1 and FF_2 are supposed to be 1.) If the output responses are captured at time 30, FF_2 produces a fail output, which contradicts the fact that the CUD produces a pass response at FF_2. It is noted that

TABLE III.   An Example of a Dictionary

	Faults			
	f_1	f_2	f_3	CUD
FF_1	fault free	R 20	R 30	Fail
FF_2	fault free	R 30	UU 15 35	Pass

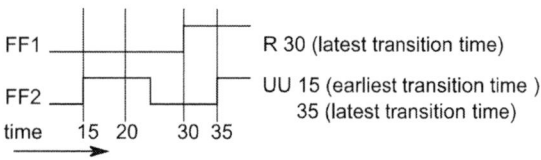

Fig. 4.   Example of a hazard signal

hazard signals RF, FR, UU and DD imply uncertain values at the timing before the latest transition time.

V.   Diagnosis method for single delay faults

In this section, we explain a diagnosis method for single delay faults. The diagnosis method applies a dictionary based approach, where output values and latest transition times are investigated to check whether contradictions are derived or not. We introduce some rules, by which candidate faults are deduced.

[**Rule 1**] If there is at least one output which is a CF_output as well as DP_output for fault f, then f is removed from the candidate fault set.

This rule is explained by the example of fault f_1 in Section IV. A fault dictionary is created with setting an amount of injected delay to a specific one. For example, suppose that a delay fault f with an amount of delay d is injected to create a dictionary, and that a pass response is obtained at a certain output z. If fault f is injected with the amount of delay larger or smaller than d, the output response at z may be turned to be fail one. However, the previous research [7] has indicated that such a case rarely happens.

[**Rule 2**] If there is at least one output z which satisfies all the followings, then f is removed from the candidate fault set.

- Output z is a CP_output as well as RF_output.

- $Min\_Lt(f) \le Lt(t, z, f)$ for test pattern t, output z and fault f.

This rule is explained by the example of fault f_2 in Section IV. In general cases, there are more than one CP_outputs for each fault, and thus the minimum latest transition time $Min\_Lt(f)$ is compared. Even when output z produces a hazard signal and the above second condition is satisfied, fault f is not removed from the candidate fault set. This is because a hazard signal, RF, FR, UU or DD implies that the corresponding line produces either 0 or 1 before the latest transition time. The overall flow of the diagnosis method is shown in Fig. V.

VI.   Diagnosis method for double delay faults

In this section, we explain a diagnosis method for double delay faults. The method uses a fault dictionary which includes

Diagnosis flow for single delay faults
```
/* F_cand: A set of candidate faults */
 1 : for ( every fault f in F_cand )
 2 :     Set a large number to Min_Lt(f);
 3 :     for ( every pattern t )
 4 :         for ( every output z )
 5 :             if ( z is a CF_output and a DP_output )
 6 :                 Remove f from F_cand;  /* Rule 1 */
 7 :             else if ( z is a CF_output and a RF_output )
 8 :                 Update Min_Lt(f);
 9 :         end for
10 :     end for
11 : end for
12 : for ( every fault f in F_cand )
13 :     for ( every pattern t )
14 :         for ( every output z )
15 :             if ( z is a CP_output and a RF_output, and
                     Min_Lt(f) ≤ Lt(t, z, f) )
16 :                 Remove f from F_cand;  /* Rule 2 */
17 :         end for
18 :     end for
19 : end for
```

Fig. 5.   Diagnosis flow for single delay faults

the information on each single delay fault. Rule 1, explained in Sec. V, is also applied for deducing candidate faults. According to the rule, when the output which is a CF_output produces a pass response in the dictionary, a corresponding fault is removed from the candidate set. When it comes to double fault assumption, the output which has a pass response in the dictionary may produce a fail response in a CUD with very low probability. However, experimental results show such a case rarely happens.

Due to the fault masking, a method using Rule 2 does not deduce candidate faults successfully in the diagnosis for double delay faults. Fault masking means termination of fault effect propagation, which occurs when the effects of the two single faults are propagated to a same gate. Even though a fault produces a fail output under the single fault assumption, such fail output may disappear when another single fault is activated simultaneously. However, empirically it is known that the fault masking does not occur so often. The proposed method counts the number of times of contradictions on Min_Lt and Lt, in an extended manner of Rule 2. When the number of times of the contradictions exceeds a threshold value, the corresponding fault is removed from the candidate fault set. Thus, we formulate Rule 3.

[**Rule 3**] If the following equation is satisfied, then f is removed from the candidate set. Here N_{TH} is a threshold value.

$$\sum_t Num\_CPRF(t, f) > N_{TH} \qquad (1)$$

In order to reduce the number of candidate faults, another rule is introduced. The rule is based on the observation that all the CF_outputs are covered by the DF_outputs of two candidate faults.

[**Rule 4**] If the following equation is satisfied, then fault f is removed from the candidate set, where F is a candidate

TABLE IV. AN EXAMPLE OF A GATE DELAY FAULT AND A CLOCK DELAY FAULT DICTIONARY

pattern	output	dictionary			CUD
		f_1	f_2	f_3	
t_1	FF_1	R	-	-	Fail
	FF_2	-	R	R	Fail
t_2	FF_1	F	-	-	Fail
	FF_2	-	F	-	Fail

Symbol '-' means fault free value is propagated.

fault set, and f and g are single delay faults.

$$Z_{CF} \not\subseteq Z_{DF}(f) \cup Z_{DF}(g) \text{ for } \forall g \in F \ (f \neq g) \qquad (2)$$

Example of application of Rule 4:
Application of Rule 4 is explained using an example of Table IV. In the CUD, FF_1 and FF_2 produces a fail response for both test patterns t_1 and t_2. According to this fact, $Z_{CF} = \{(t_1, FF_1), (t_1, FF_2), (t_2, FF_1), (t_2, FF_2)\}$ is derived. In the dictionary, FF_1 produces fail responses for fault f_1 and test pattern t_1, t_2, and FF_2 produces a fail response for fault f_2 and test pattern t_1, t_2. From these facts, $Z_{DF}(f_1) = \{(t_1, FF_1), (t_2, FF_1)\}$ and $Z_{DF}(f_2) = \{(t_1, FF_2), (t_2, FF_2)\}$ are derived. Since $Z_{CF} \subseteq Z_{DF}(f_1) \cup Z_{DF}(f_2)$, fault f_1 and f_2 can be candidates. This result comes from the observation that combination of fault f_1 and f_2 explain sufficiently the output responses in the CUD.

Next, consider the case of fault f_3.

$$Z_{DF}(f_1) \cup Z_{DF}(f_3) = \{(t_1, FF_1), (t_1, FF_2), (t_2, FF_1)\}$$
$$Z_{DF}(f_2) \cup Z_{DF}(f_3) = \{(t_1, FF_2), (t_2, FF_2)\}$$

are derived. Since Z_{CF} is a subset of neither $Z_{DF}(f_1) \cup Z_{DF}(f_3)$ nor $Z_{DF}(f_2) \cup Z_{DF}(f_3)$, fault f_3 is removed from the candidate set. This comes from the observation that any combination of f_3 does not sufficiently explain the output responses in the CUD. The flow of the diagnosis method for double delay faults is shown in Fig. 6.

VII. EXPERIMENTAL RESULTS

A. Output responses considering effects of hazards

We carried out experiments for ISCAS'89 benchmark circuits. The output responses of CUDs were created using the simulator that deals with hazard signals. When an output produces a hazard signal, RF, FR, UU or DD, and test observation timing is earlier than the latest transition time of the hazard signal, we do not know whether the value captured at the test observation timing is 0 or 1, as Fig. 7 shows. We only know that UU produces 0 before the earliest transition time, and 1 after the latest transition time. Therefore, we decide the value of a hazard signal to be 0 or 1 with probability p_h, when test observation time is later than the earliest transition time and earlier than the latest transition time of the hazard signal.

B. Results of diagnosis for single delay faults

In this subsection, we show experimental results for diagnosis for single delay faults. We selected single delay faults randomly to create CUDs. For each benchmark circuit, 50 different CUDs were created. As diagnostic test patterns,

Diagnosis flow for double delay faults
```
/* F_cand: A set of candidate single delay faults */
 1 : for ( every fault f in F_cand )
 2 :     Set a large number to Min_Lt(f);
 3 :     for ( every pattern t )
 4 :         for ( every output z )
 5 :             if ( z is a CF_output and a DP_output )
 6 :                 Remove f from F_cand;  /* Rule 1 */
 7 :             else if ( z is a CF_output and a RF_output )
 8 :                 Update Min_Lt(f);
 9 :         end for
10 :     end for
11 : end for
12 : for ( every fault f in F_cand )
13 :     for ( every pattern t )
14 :         if ( ∑_t Num_CPRF(t, f) > N_TH )
15 :             Remove f from F_cand;  /* Rule 3 */
16 :     end for
17 : end for
18 : for ( every fault f_i in F_cand )
19 :     for ( every fault f_j ( f_i ≠ f_j ) in F_cand )
20 :         if ( Z_CF ⊆ Z_DF(f_i) ∪ Z_DF(f_j) )
21 :             Go to Line 18
22 :     end for
23 :     Remove f_i from F_cand;  /* Rule 4 */
24 : end for
```

Fig. 6. Diagnosis flow for double delay faults

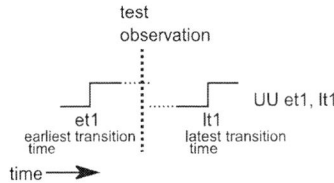

Fig. 7. Uncertainty of logic value in a hazard signal

transition test patterns generated by an in-house tool were used. Hazard signal probability p_h was set to 50%. We do not show other cases of p_h due to the space limitation.

Table V shows the diagnosis results for single gate delay faults. In the table, from left to right, circuit name, average number of fail test patterns, average number of candidate faults, maximum number of candidate faults, the number of CUDs for which one candidate fault was deduced, the number of CUDs for which the candidate fault set did not include the fault existing in the CUD, and the amount of delay injected in the CUD were shown. The amount of delay was shown by percentage for the test cycle. The test cycle was one unit plus the units of time which is equivalent to the number of gates existing along the longest sensitized path. Since the effects of small delay faults are propagated to limited outputs, the average number and the maximum number of candidate faults were relatively large with 30% delay. However, the results show that the average number of candidate faults was less than 20 for every circuit.

Table VI shows the results by the diagnosis method without considering effects of hazards [7]. Each column has same meanings in Table V, except for column "no", which shows the

978-1-4799-8720-7/15 $31.00 © 2015 IEEE

TABLE V.   Diagnosis results for single gate delay faults

circuit	pat	cand	max	one	not	delay (%)
s9234	3.9	12.7	28	2	0	30
s9234	6.5	8.9	28	6	0	50
s9234	9.9	5.0	28	13	1	80
s13207	6.4	24	121	6	0	30
s13207	8.1	2.8	10	16	0	50
s13207	9.4	2.7	10	21	0	80
s15850	2.5	19.1	40	3	0	30
s15850	4.2	9.2	50	9	0	50
s15850	12.0	3.3	10	18	1	80
s35932	4.4	4.4	15	7	0	30
s35932	5.0	4.1	13	9	0	50
s35932	6.3	2.7	16	13	0	80
s38417	6.6	9.0	51	13	3	30
s38417	11.4	4.3	26	14	0	50
s38417	12.9	1.9	15	31	0	80
s38584	8.9	11.3	79	14	1	30
s38584	11.9	2.9	15	20	0	50
s38584	10.6	3.9	12	10	0	80

TABLE VI.   Results by the diagnosis method without considering hazard signals

circuit	pat	cand	max	no	one	not	delay(%)
s9234	9.9	2.6	9	21	10	3	80
s13207	9.4	2.4	8	28	12	0	80
s15850	12.0	2.6	10	25	13	2	80
s35932	6.3	2.5	12	20	8	1	80
s38417	12.9	1.2	4	29	19	0	80
s38584	10.6	2.9	6	22	9	1	80

TABLE VII.   Diagnosis results for double delay faults with various N_{TH}

circuit	pat	ave	dbl	one	not	delay(%)	N_{TH}
s9234	21.4	16.6	24	22	4	80	1
s9234	21.4	19.3	24	23	3	80	2
s9234	21.4	22.1	25	22	3	80	3
s9234	21.4	25.1	25	22	3	80	5
s9234	21.4	31.2	25	22	3	80	10
s13207	17.6	18.9	33	17	0	80	1
s13207	17.6	22.5	33	17	0	80	2
s13207	17.6	25.2	34	16	0	80	3
s13207	17.6	29.3	34	16	0	80	5
s13207	17.6	31.5	34	16	0	80	10

TABLE VIII.   Results for double delay faults

circuit	pat	cand	dbl	one	not	delay(%)
s9234	7.8	27.1	8	24	18	30
s9234	12.3	24.5	21	20	9	50
s9234	24.5	22.7	25	22	3	80
s13207	10.8	59.7	8	36	6	30
s13207	16.4	26.4	34	16	0	50
s13207	19.8	24.8	34	16	0	80
s15850	4.0	76.2	17	24	9	30
s15850	10.7	24.6	39	9	2	50
s15850	15.5	16.7	45	4	1	80
s35932	7.4	13.4	44	5	1	30
s35932	8.6	13.5	45	5	0	50
s35932	10.9	10.7	45	5	0	80
s38417	12.5	40.9	14	29	7	30
s38417	19.5	21.8	32	13	5	50
s38417	28.4	16.4	38	9	3	80
s38584	16.3	21.9	5	31	14	30
s38584	25.0	12.2	35	15	0	50
s38584	27.3	11.7	39	11	0	80

number of CUDs for which no candidate faults were deduced. The amount of delay was 80% of the test cycle. It is found that the number of CUDs for which no candidate faults were deduced was large.

C. Results of diagnosis for double delay faults

In this subsection, we show experimental results of diagnosis for double delay faults. First we carried out the experiments with various N_{TH}, which is used in Rule 3, for s9234 and s13207. Table VII shows the results. Column "dbl" shows the number of CUDs for which two single delay faults existing in the CUD were included in the final candidate fault set. Column "one" shows the number of CUDs for which either one single delay fault among the faults existing in the CUD were included in the final candidate fault set. Column "not" shows the number of CUDs for which neither single delay fault existing in the CUD was included in the final candidate fault set. The smaller N_{TH} was, the smaller the average number of candidate faults. However, when $N_{TH} = 1$, the number of CUDs for which candidate fault set included neither single delay fault existing in the CUD was a little large. Therefore, in the other experiments, we set 2 to N_{TH}.

Table VIII shows the diagnosis results for double delay faults. Each column has same meanings as in Table VII. It is found that less than 30 candidate faults were deduced for every circuit when the amount of delay was 50 and 80 %. When the amount of delay was 30 %, relatively large number of candidate fualts were deduced. This is because the effects of small amount of delays were propagated to limited outputs. As a result, it was difficult to distinguish between the faults in CUDs and other faults.

VIII.   Conclusion

In this paper, we proposed diagnosis methods for delay faults considering hazard signals to improve the diagnostic accuracy. Our methods use a fault dictionary which is created by the simulator considering hazard signals, and they can deduce candidate faults successfully for CUDs which are not diagnosed successfully by the conventional method without considering the effects of hazards. In our future work, we will develop a diagnostic test generation method considering the effects of hazard signals.

References

[1] G. L. Smith, "Model for Delay Faults Based upon Path," in *Proc. Int. Test Conf.*, 1985, pp. 342–349.

[2] D. Dumas, P. Girard, C. Landrault, and S. Pravossoudovitch, "An Implicit Delay Fault Simulation Method with Approximate Detection Threshold Calculation," in *Proc. Int. Test Conf.*, 1993, pp. 705–713.

[3] H. Takahashi, K. O. Boateng, and Y. Takamatsu, "A Method of Generating Tests with Linearity Property for Gate Delay Faults in Combinational Circuits," *IEICE Trans. on Information and Systems*, vol. E82-D, pp. 1466–1473, 1999.

[4] I. Pomeranz and S. M. Reddy, "Hazard-based Detection Conditions for Improved Transition Path Delay Fault Coverage," *IEEE Trans. on Computer-Aided Design*, vol. 29, pp. 1449–1453, 2010.

[5] ——, "Hazard-based Detection Conditions for Improved Transition Fault Coverage of Scan-based Tests," *IEEE Trans. on VLSI Systems*, vol. 18, pp. 333–337, 2010.

[6] Z. Wang, M. M. Marek-Sadowska, K.-H. Tsai, and J. Rajski, "Delay-Fault Diagnosis Using Timing Information," *IEEE Trans. on Computer-Aided Design of Integrated Circuits and Systems*, vol. 24, pp. 1315–1325, 2005.

[7] Y. Higami, H. Takahashi, S. Kobayashi, and K. K. Saluja, "Diagnosis of Gate Delay Faults in the Presence of Clock Delay Faults," in *IEEE Computer Society Annual Symp. on VLSI*, 2014, pp. 320–325.

2015 IEEE Computer Society Annual Symposium on VLSI

DONUT: A Double Node Upset Tolerant Latch

Nikolaos Eftaxiopoulos*, Nicholas Axelos† and Kiamal Pekmestzi‡

Department of Computer Science
School of Electrical and Computer Engineering
National Technical University of Athens, Greece
<eftaxiopoulos*,njaxel†,pekmes‡>@microlab.ntua.gr

Abstract—**In this paper we propose the novel DONUT (DOuble Node Upset Tolerant) latch, a soft error tolerant latch for SNUs (Single Node Upsets) and DNUs (Double Node Upsets). The latch comprises four DICE cells in a multi-interlocked scheme where each DICE cell is always guaranteed two stable nodes, rendering the DONUT latch resilient to SNUs and DNUs. The proposed latch design merges efficiently the four DICE cells saving 25% of the required transistors. Regarding the power dissipation and propagation delay, simulation results showed that the DONUT latch performs better compared to other BISER-based latches that are SNU or DNU tolerant.**

Index Terms—**soft error; charge sharing; radiation tolerant latch; double node upset**

I. INTRODUCTION

Technology miniaturization has had adverse effects on the susceptibility of integrated circuits to radiation induced errors. The shrinking dimensions, aggressive voltage scaling and reduced parasitics, all otherwise desirable properties from a performance, power and area perspective, contribute to the susceptibility of storage elements (memory cells, latches and flip-flops) to energetic particle hits. Until recently, the problem was prominent on SRAM memories with their densely packed arrays and minimum feature size transistors. Recently however, soft errors in latches and flip-flops have become a concern. Affected latches (as well as flip-flop comprising latches), being in abundance in modern processing units, propagate erroneous values to be consumed by later logic stages [1]–[5].

Soft errors are caused by energetic particles hitting the semiconducting materials, thereby generating electron-hole pairs along their trajectory. The generated charge is then collected by nearby junctions via drift and diffusion mechanisms, resulting in voltage transients referred as SEUs (Single Event Upsets) [6], [7]. Storage elements are particularly susceptible to SEUs, since the inherent positive feedback mechanisms amplify the transients. If the induced charge is over a threshold, a.k.a. critical charge Q_{crit}, the transient at the node of impact flips the storage element, resulting in a soft error. The critical charge Q_{crit} is a technology dependent parameter that deteriorates as V_{dd} drops. Soft error resilience is therefore important in microprocessors, where dynamic voltage frequency scaling techniques are common [8], [9].

Conventional soft error resilient designs assume single node charge collection and mitigate the radiation induced transient by utilizing redundancy, dual interlocking, guard gates or by introducing delays in the feedback mechanism [10]–[16].

However, recent studies indicate that with technology scaling and the proximity of devices, the generated charge by a single particle hit can be collected by adjacent junction fields, rendering the above assumption invalid [9]. Layout oriented techniques have also been proposed that relocate adjacent susceptible transistors in order to avoid multiple charge collection [17], [18]. Nevertheless, such solutions may be effective in SRAM arrays but not in the confined area of a latch or a flip-flop cell [16]. Furthermore, storage elements featuring soft error resilience already come at the price of increased power, area and delay penalties. Any additional resilience features that provide protection from dual node upsets must be assessed with respect to the imposed overheads.

In this paper, we present the DONUT (DOuble Node Upset Tolerant) latch, a novel topology that tolerates SNUs (Single Node Upsets) and DNUs (Double Node Upsets) caused by charge sharing. The proposed latch, although based on the principle of four interconnected dual interlocked cells, allows 25% less transistors than the equivalent of four DICE cells and provides the SNU and DNU tolerance with moderate overheads compared to similar solutions. In Section II we discuss existing solutions that provide SNU and partial or full DNU tolerance. In Section III we introduce the DONUT latch and perform simulations to validate normal operation and SNU and DNU tolerance. Section IV compares the proposed solutions with the latches discussed in Section II in terms of propagation delay, dynamic energy, energy×delay product and static energy. Section V concludes our work.

II. BACKGROUND

Upon a particle strike, the generated charge along the particle trajectory is collected by junction fields in the proximity of the impact, causing voltage fluctuations to the interconnected nodes. In bulk CMOS circuits, transistors are only susceptible to charge collection when being reverse biased in the OFF state (in the ON state any excess charge is shunted to the power rails). Moreover, with technology scaling and shrinking geometries, several studies have shown that apart from the struck node, charge collection by neighboring passive nodes is possible [6], [19], [20]. Passive collection efficiency is a strong function of the distance between the active and the passive nodes as well as the device types [20]. Devices within the same well and in the proximity of the incidence particle, i.e. NMOS to NMOS / PMOS to PMOS, have been found more

978-1-4799-8720-7/15 $31.00 © 2015 IEEE

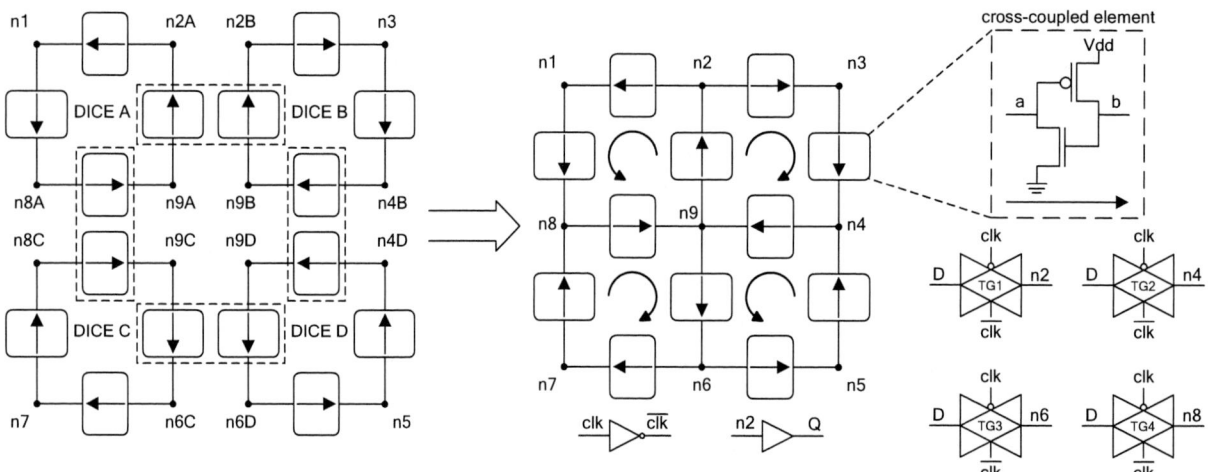

Fig. 1. The DONUT (Double Node Upset Tolerant) latch.

susceptible to charge sharing than across well boundaries, i.e. any NMOS-PMOS pair [21].

Considering the above, we revise several conventional latches (see Table I) based on their resilience properties to SNUs (Single Node Upsets) and DNUs (Double Node Upsets).

A. DICE

The DICE cell [10] is a robust soft error resilient design, utilizing two interlocked latches that provide the necessary redundancy to mitigate single node SEU transients. By keeping the complementary values in two latch pairs, in case of a single node upset or neighboring nodes upset due to radiation, the cell always retains a source of uncorrupted state that restores the disturbed node through its positive feedback [17]. However, a strike affecting non-adjacent nodes will flip the cell. Hence, the DICE cell is considered practically immune to SNUs and partially resilient to DNUs [16].

B. BISER

The BISER latch [11] is a design utilizing a pair of D-latches that drive a C-element, which in turn drives a weak keeper. Upon a single upset, the C-element is driven by two complementary values and therefore does not propagate the erroneous value, while the weak keeper retains the correct state. During normal operation, both latches capture the same input and drive the C-element accordingly, which is designed with enough driving strength to override the weak keeper. Finally, an upset on the weak keeper quickly recovers from the voltage fluctuation due to the driving strength of the C-element. BISER latches effectively block errors from SNUs [22], but cannot do so when an upset affects both latches, or a latch and the weak keeper due to charge sharing.

C. FERST

The FERST latch [14] utilizes two parallel latches comprising two C-elements and clocked transmission gates, driving

a third C-element and a weak keeper at the output. The multiplexing of the two parallel latches with each other and the inputs does not allow the propagation of errors to the output, which is retained by the weak keeper. Similarly to the BISER latch, the FERST latch is very efficient at blocking the propagation of an SNU erroneous state, however is susceptible to DNUs.

D. HIPER

The HIPER latch [15] utilizes two independent feedback loops and triplicates the internal nodes that drive a C-element, which determines the correct output based on the value of the majority of the internal nodes. Since the design is inherently based on the principle of triplication, it is very effective towards SNU resilience, but susceptible to a DNU.

E. DNCS

The DNCS latch [16] uses two DICE latches and a C-element with a weak keeper at the output. With that approach, the DNCS latch combines the soft error resilience properties of the DICE and the BISER latch for SNUs as well as providing full DNU resilience.

TABLE I
EXAMPLES OF SNU AND DNU TOLERANT LATCHES.

	DICE	BISER	FERST	HIPER	DNCS
SNU tolerant	✓	✓	✓	✓	✓
DNU tolerant	✗	✗	✗	✗	✓
Reference	[10]	[11]	[14]	[15]	[16]

III. THE DONUT LATCH

A. Concept

The proposed latch (demonstrated in Fig.1) comprises four DICE latches (DICE A, B, C and D) where nodes are driven by cross-coupled elements as follows:

978-1-4799-8720-7/15 $31.00 © 2015 IEEE

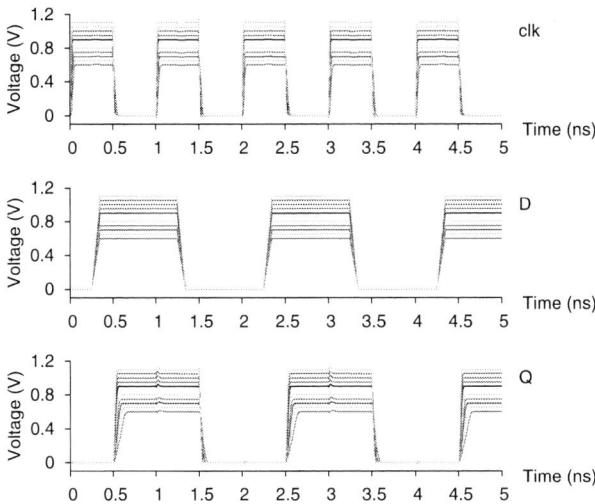

Fig. 2. DONUT latch capturing input D and feeding output Q at 1GHz across a voltage range of V_{dd}=0.6 to 1.1V. The latch is transparent when CLK='0'.

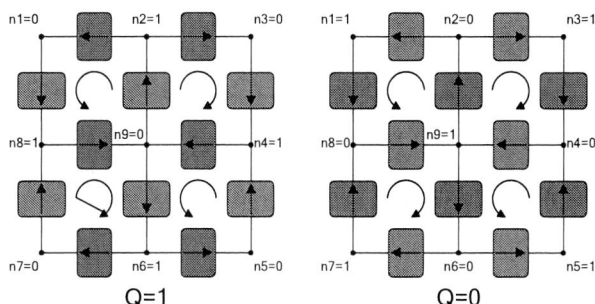

Fig. 3. DONUT latch cross-coupled element states for Q=1 and Q=0. Green color indicates cross-coupled element is ON, red color is OFF.

- n1, n3, n5, n7 are driven by two cross-coupled elements
- n2A & n2B joined to n2, n4B & n4D joined to n4, n6C & n6D joined to n6, and n8A & n8C joined to n8, are driven by three cross-coupled elements
- n9A & n9B & n9C & n9D joined to n9 are driven by four cross-coupled elements

Joining the aforementioned nodes as described above, allows us to merge the four pairs of cross-coupled elements, as denoted by the dashed lines in the left part of Fig.1. Thus, we produce the circuit topology shown in the right part of Fig.1. The latch inputs n2, n4, n6 and n8 are driven through the clocked transmission gates TG1, TG2, TG3 and TG4 respectively and the output Q is taken from node n2.

In normal operation and when the latch is in transparent mode (CLK=0), the transmission gates TG1-TG4 propagate the input signal D to all four DICE latches via nodes n2, n4, n6 and n8 and update the output Q. When the latch is in opaque mode (CLK=1), the four DICE latches store the same data and retain the state of output Q.

B. Normal Operation

The proposed latch has been simulated in SPICE with models derived from a 1.1V UMC 65nm technology. All transistors gate lengths have been set at L=60 nm, with PMOS widths set to W=180 nm and NMOS widths set to W=120 nm. Simulation results for normal operation with a switching input at 1GHz and variable V_{dd} are shown in Fig.2, with the latch operating successfully down to V_{min}=0.6V.

C. Soft Error Modeling

The critical charge Q_{crit} is the minimum induced charge required to flip the contents of the storage element. A particle strike incurs a flip when the induced charge overcomes the intrinsic charge stored in the parasitic capacitance of the affected node (static component) and the driving strength of

the pull-up or pull-down keeper transistor that is conducting prior to the flip (dynamic component). Calculating the critical charge Q_{crit} at the circuit level is possible by carrying out SPICE simulations with a stimulus providing the generated transient due to the particle strike. At the netlist level, that stimulus is usually an exponential current source connected at the sensitive node [23]–[26]. The transient has a small rising time in the range of 0.1-10 ps and a longer falling time in the range of 10-200 ps. The shape of the pulse (magnitude and rising/falling times) strongly depend on the localization of the incident, i.e. a strike directly at the drain area produces a high-peak narrow pulse, whereas one occurring further away results in a low-peak wider pulse [27]. However, the exponential current source parameters are process and design dependent and using approximate values can present a great variability in results [26], [28]. In the present work and in order to extract preliminary results, we use a simple rectangular current pulse approximation. Indicative pulse widths and charges have been defined from relevant research at the 90nm and the 65nm technology nodes [29]–[31].

D. SNU & DNU Mitigation

The DONUT cell, being based on the DICE design, inherits the excellent resilience properties of dual interlocking against SNUs. Regarding DNUs and considering the two possible states shown in Fig.3, either nodes (n8,n9,n4) when Q=1 or nodes (n2,n9,n6) when Q=0 are driven by two ON (a='0' b='1') cross-coupled elements (green colored elements in Fig.3). Therefore, all DICE cells (A, B, C, D) contain at least two nodes with each driven by two ON cross-coupled elements, rendering the DONUT latch resilient to DNUs.

There are seven possible types of DNUs to consider, with each type specifying the number of nodes affected on each DICE cell:

(a) Two DICE cells have 1 affected node and two DICE cells have 0 affected nodes (e.g. strike in nodes n1,n3). There are 6 node pair combinations of this type.

(b) Three DICE cells have 1 affected node and one DICE cell has 0 affected nodes (e.g. strike in nodes n3,n8). There are 8 node pair combinations of this type.

978-1-4799-8720-7/15 $31.00 © 2015 IEEE

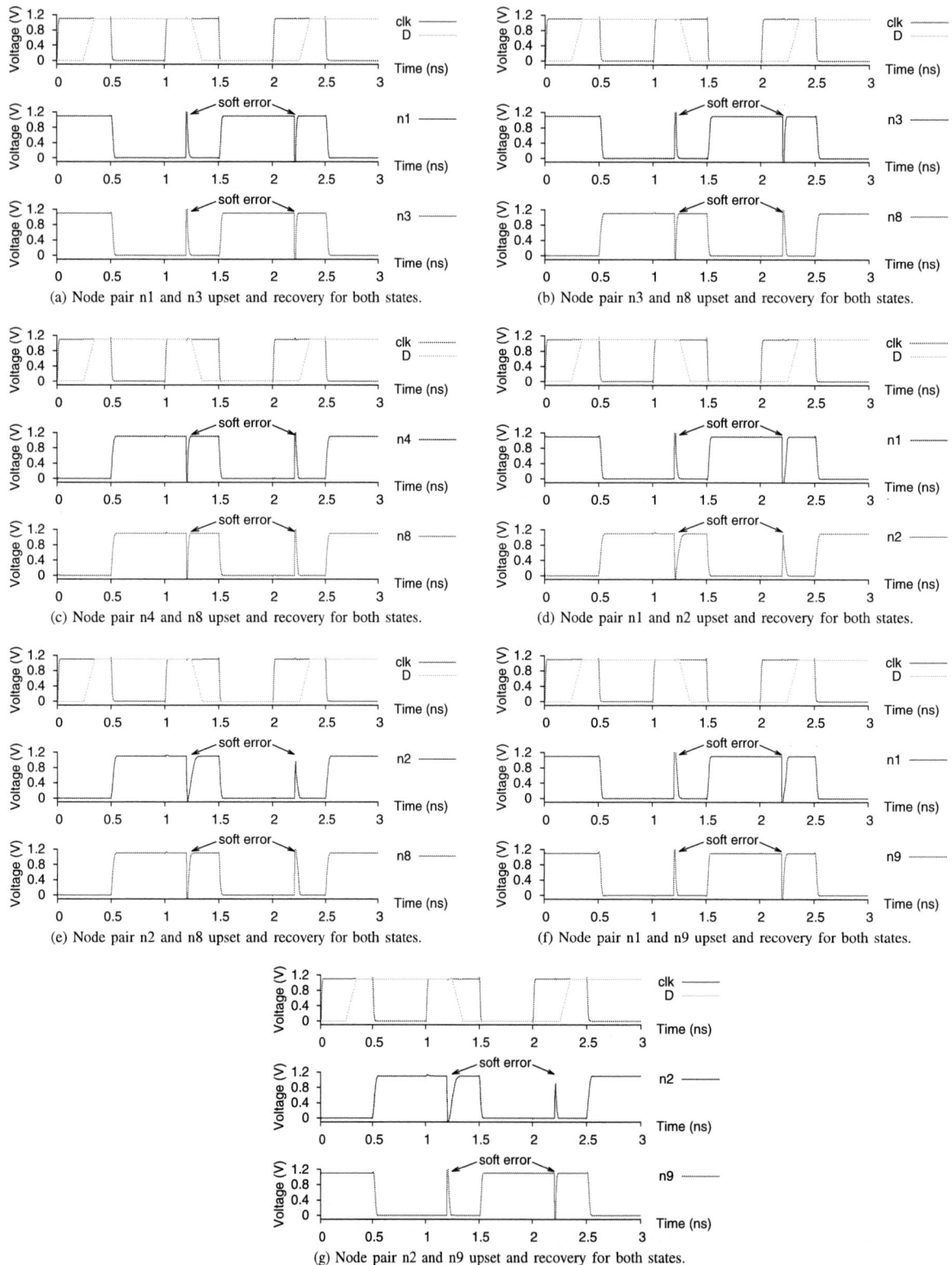

Fig. 4. Strikes on node pairs that represent all the possible DNU types for both states (Q='0' and Q='1').

978-1-4799-8720-7/15 $31.00 © 2015 IEEE

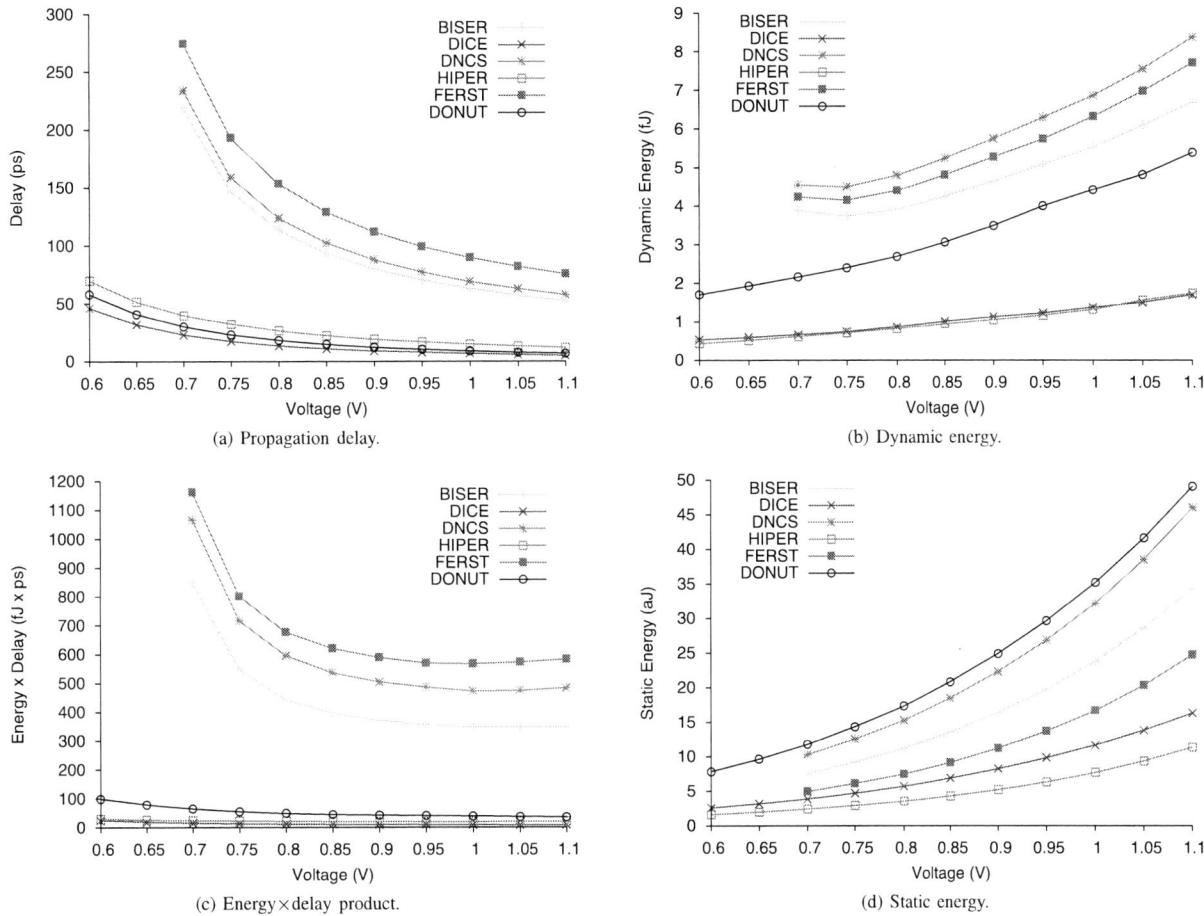

Fig. 5. Energy and delay characterization of the proposed DONUT latch compared to the latches mentioned in Section II.

(c) Four DICE cells have 1 affected node (e.g. strike in nodes n4,n8). There are 2 node pair combinations of this type.

(d) One DICE cell has 2 affected nodes, one DICE cell has 1 affected node and two DICE cells have 0 affected nodes (e.g. strike in nodes n1,n2). There are 8 node pair combinations of this type.

(e) One DICE cell has 2 affected nodes, two DICE cells have 1 affected node and one DICE cell has 0 affected nodes (e.g. strike in nodes n2,n8). There are 4 node pair combinations of this type.

(f) One DICE cell has 2 affected nodes and three DICE cells have 1 affected node (e.g. strike in nodes n1,n9). There are 4 node pair combinations of this type.

(g) Two DICE cells have 2 affected nodes and two DICE cells have 1 affected node (e.g. strike in nodes n2,n9). There are 4 node pair combinations of this type.

The latch has been simulated in SPICE for particle strikes when opaque (CLK='1') for all possible node pair combinations and for both states (nodes n1,n3,n5,n7,n9 holding a '0' with nodes n2,n4,n6,n8 holding a '1' and vice versa).

Simulation results, as shown in Fig.4, correspond to the aforementioned listed DNU types and demonstrate that the proposed latch provides full soft error tolerance to DNUs.

IV. OVERHEADS

The proposed latch has been evaluated against the soft error resilient latches analyzed in Section II in terms of propagation delay, dynamic energy, energy$\times$delay product and static energy across a dynamic voltage range of 0.5V to 1.1V. All latches have been designed with transistor models from the same UMC 65nm technology and compared using the unified approach suggested in [32]. The propagation delay has been evaluated as the average $t_Q - t_{CLK}$ of both $0{\rightarrow}1$ and $1{\rightarrow}0$ transitions (both measured at 50% of the transition). As shown in Fig.5a, compared to latches deriving features from the BISER principle (i.e. the BISER itself, FERST and DNCS designs with a V_{min}=0.7V), the DONUT latch, as well as having lower V_{min}=0.6V, is at least $\times$6.12 faster, slightly faster than HIPER and only slower than DICE. In terms of dynamic energy, as shown in Fig.5b, the DONUT latch is at least $\times$1.24 more energy efficient than BISER, DNCS and

FERST and worse than HIPER and DICE. Considering the energy×delay product shown in Fig.5c, the proposed latch is slightly worse than the top performers HIPER and DICE across the examined voltage range, featuring however soft error resilience to both SNUs and DNUs in contrast to the compared counterparts except DNCS. With respect to static energy, static dissipation is a function of the number of transistors in the design, which reflects in the results shown in Fig.5d.

V. CONCLUSION

Due to charge sharing, double node upset resilient latches are essential to retain circuit reliability in nanometer technologies. In this paper, we presented the DONUT (DOuble Node Upset Tolerant) latch, a design based on a four DICE interlocking scheme that provides single node upset and double node upset tolerance. Simulation results showed that the proposed design exhibits its soft error tolerance properties at little energy×delay penalty compared to other SNU tolerant latches (HIPER & DICE) and performs better than BISER-based latches that are SNU or DNU tolerant.

ACKNOWLEDGMENT

This research has been co-financed by the European Union (European Social Fund ESF) and Greek national funds through the Operational Program "Education and Lifelong Learning" of the National Strategic Reference Framework (NSRF) - Research Funding Program: Thales -UPATRAS-REIN. Investing in knowledge society through the European Social Fund.

REFERENCES

[1] P. E. Dodd and L. W. Massengill, "Basic mechanisms and modeling of single-event upset in digital microelectronics," *IEEE Trans. Nucl. Sci.*, vol. 50, no. 3, pp. 583–602, 2003.

[2] M. Nicolaidis, "Design for soft error mitigation," *IEEE Trans. Dev. Mat. Rel.*, vol. 5, no. 3, pp. 405–418, 2005.

[3] B. S. Gill, C. Papachristou, F. G. Wolff, and N. Seifert, "Node sensitivity analysis for soft errors in CMOS logic," in *Int. Test Conf.*, 2005, pp. 9 pp. ——972.

[4] T. Heijmen, P. Roche, G. Gasiot, K. R. Forbes, and D. Giot, "A Comprehensive Study on the Soft-Error Rate of Flip-Flops From 90-nm Production Libraries," *IEEE Trans. Dev. Mat. Rel.*, vol. 7, no. 1, pp. 84–96, 2007.

[5] P. E. Dodd, M. R. Shaneyfelt, J. R. Schwank, and J. A. Felix, "Current and Future Challenges in Radiation Effects on CMOS Electronics," *IEEE Trans. Nucl. Sci.*, vol. 57, no. 4, pp. 1747–1763, 2010.

[6] R. C. Baumann, "Radiation-induced soft errors in advanced semiconductor technologies," *IEEE Trans. Dev. Mat. Rel.*, vol. 5, no. 3, pp. 305–316, 2005.

[7] V. Chandra and R. Aitken, "Impact of voltage scaling on nanoscale SRAM reliability," in *Design Automation and Test Europe Conf.*, 2009, pp. 387–392.

[8] P. E. Dodd, "Device simulation of charge collection and single-event upset," *IEEE Trans. Nucl. Sci.*, vol. 43, no. 2, pp. 561–575, 1996.

[9] A. Dixit and A. Wood, "The impact of new technology on soft error rates," in *Int. Rel. Physics Symp.*, 2011, pp. 5B.4.1–5B.4.7.

[10] T. Calin, M. Nicolaidis, and R. Velazco, "Upset hardened memory design for submicron CMOS technology," *IEEE Trans. Nucl. Sci.*, vol. 43, no. 6, pp. 2874–2878, 1996.

[11] M. Zhang, S. Mitra, T. M. Mak, N. Seifert, N. J. Wang, Q. Shi, K. S. Kim, N. R. Shanbhag, and S. J. Patel, "Sequential Element Design With Built-In Soft Error Resilience," *IEEE Trans. Very Large Scale Integr. (VLSI) Syst.*, vol. 14, no. 12, pp. 1368–1378, 2006.

[12] R. Naseer and J. Draper, "DF-DICE: a scalable solution for soft error tolerant circuit design," in *Int. Symp. Circuits and Syst.*, 2006, p. 4 pp.

[13] M. Nicolaidis, R. Perez, and D. Alexandrescu, "Low-Cost Highly-Robust Hardened Cells Using Blocking Feedback Transistors," in *VLSI Test Symp.*, Apr. 2008, pp. 371–376.

[14] M. Fazeli, S. G. Miremadi, A. Ejlali, and A. Patooghy, "Low energy single event upset/single event transient-tolerant latch for deep subMicron technologies," *IET Computers Digital Techniques*, vol. 3, no. 3, pp. 289–303, May 2009.

[15] M. Omana, D. Rossi, and C. Metra, "High-Performance Robust Latches," *Computers, IEEE Transactions on*, vol. 59, no. 11, pp. 1455–1465, 2010.

[16] K. Katsarou and Y. Tsiatouhas, "Double node charge sharing SEU tolerant latch design," in *Int. On-Line Testing Symp.*, 2014, pp. 122–127.

[17] M. Haghi and J. Draper, "The 90 nm Double-DICE storage element to reduce Single-Event upsets," in *Int. Midwest Symp. Circuits and Syst.*, 2009, pp. 463–466.

[18] H.-H. K. Lee, K. Lilja, M. Bounasser, P. Relangi, I. R. Linscott, U. S. Inan, and S. Mitra, "LEAP: Layout Design through Error-Aware Transistor Positioning for soft-error resilient sequential cell design," in *Int. Rel. Physics Symp.*, 2010, pp. 203–212.

[19] B. D. Olson, D. R. Ball, K. M. Warren, L. W. Massengill, N. F. Haddad, S. E. Doyle, and D. McMorrow, "Simultaneous single event charge sharing and parasitic bipolar conduction in a highly-scaled SRAM design," *IEEE Trans. Nucl. Sci.*, vol. 52, no. 6, pp. 2132–2136, 2005.

[20] O. A. Amusan, A. F. Witulski, L. W. Massengill, B. L. Bhuva, P. R. Fleming, M. L. Alles, A. L. Sternberg, J. D. Black, and R. D. Schrimpf, "Charge Collection and Charge Sharing in a 130 nm CMOS Technology," *IEEE Trans. Nucl. Sci.*, vol. 53, no. 6, pp. 3253–3258, 2006.

[21] O. A. Amusan, L. W. Massengill, M. P. Baze, A. L. Sternberg, A. F. Witulski, B. L. Bhuva, and J. D. Black, "Single Event Upsets in Deep-Submicrometer Technologies Due to Charge Sharing," *IEEE Trans. Dev. Mat. Rel.*, vol. 8, no. 3, pp. 582–589, 2008.

[22] N. Seifert, V. Ambrose, B. Gill, Q. Shi, R. Allmon, C. Recchia, S. Mukherjee, N. Nassif, J. Krause, J. Pickholtz, and A. Balasubramanian, "On the radiation-induced soft error performance of hardened sequential elements in advanced bulk CMOS technologies," in *Int. Rel. Physics Symp.*, May 2010, pp. 188–197.

[23] G. R. Srinivasan, P. C. Murley, and H. K. Tang, "Accurate, predictive modeling of soft error rate due to cosmic rays and chip alpha radiation," in *Annu. Int. Rel. Physics Symp.*, 1994, pp. 12–16.

[24] L. B. Freeman, "Critical charge calculations for a bipolar SRAM array," *IBM J. Research and Develop.*, vol. 40, no. 1, pp. 119–129, 1996.

[25] T. Merelle, H. Chabane, J.-M. Palau, K. Castellani-Coulie, F. Wrobel, F. Saigne, B. Sagnes, J. Boch, J. R. Vaille, G. Gasiot, P. Roche, M.-C. Palau, and T. Carriere, "Criterion for SEU Occurrence in SRAM Deduced From Circuit and Device Simulations in Case of Neutron-Induced SER," *IEEE Trans. Nucl. Sci.*, vol. 52, no. 4, pp. 1148–1155, 2005.

[26] T. Heijmen, D. Giot, and P. Roche, "Factors that impact the critical charge of memory elements," in *Int. On-Line Testing Symp.*, 2006, pp. 57–62.

[27] L. Zeng and P. Beckett, "Soft Error Rate Estimation in Deep Sub-micron CMOS," in *Pacific Rim Int. Symp. Dependable Computing*, 2007, pp. 210–216.

[28] P. Jain, "Judicious Choice of Waveform Parameters and Accurate Estimation of Critical Charge for Logic SER," in *Workshop Dependable Secure Nanocomputing*, 2007.

[29] R. Naseer, Y. Boulghassoul, J. Draper, S. DasGupta, and A. Witulski, "Critical Charge Characterization for Soft Error Rate Modeling in 90nm SRAM," in *Int. Symp. Circuits and Syst.*, May 2007, pp. 1879–1882.

[30] Y. Shiyanovskii, F. Wolff, and C. Papachristou, "SRAM Cell Design Protected from SEU Upsets," in *Int. On-Line Testing Symp.*, 2008, pp. 169–170.

[31] R. Garg, C. Nagpal, and S. P. Khatri, "A fast, analytical estimator for the SEU-induced pulse width in combinational designs," in *Design Automation and Test Europe Conf.*, 2008, pp. 918–923.

[32] V. Stojanovic, V. G. Oklobdzija, and R. Bajwa, "A unified approach in the analysis of latches and flip-flops for low-power systems," in *Int. Symp. Low Power Electron. and Design*, 1998, pp. 227–232.

978-1-4799-8720-7/15 $31.00 © 2015 IEEE

2015 IEEE Computer Society Annual Symposium on VLSI

An ATPG Flow to Generate Crosstalk-Aware Path Delay Pattern*

Anu Asokan, Alberto Bosio, Arnaud Virazel, Luigi Dilillo, Patrick Girard, Serge Pravossoudovitch
LIRMM - University of Montpellier/CNRS - Montpellier, France
Email: <lastname>@lirmm.fr

Abstract—With technology scaling, the growing impact of signal integrity issues imposes significant challenges in integrated circuit testing. They cause delay in the supply and ground networks, crosstalk noise between multiple interconnects, variations in substrate and thermal noise parameters of the circuit, etc. Therefore, the consideration of signal integrity issues during pattern generation ensures a better path delay fault coverage. In this paper, we propose a constrained Automatic Test Pattern Generation(ATPG) flow to test path delay fault by simultaneously coupling multiple aggressors surrounding a delay sensitive victim path. This flow uses the most effective aggressors for pattern generation based on crosstalk information predetermined from the layout of the circuit. We developed the constrained ATPG flow by customizing the existing scan-based techniques to generate functionally testable path delay faults. The proposed ATPG flow is applied to ITC'99 benchmark circuits. The comparison results with SPICE-based simulations shows the effectiveness of our pattern in determining worst-case path delay and also in terms of computation time.

I. INTRODUCTION

With technology scaling, the growing impact of signal integrity issues imposes significant challenges in integrated circuit (IC) testing. These issues increase the number of delay-related defects, consequently affecting the circuit's performance, functionality and reliability. Therefore, a serious examination at the chip design and test stage is required to avoid field failures of IC products. Signal integrity issues are due to the increase in the frequency and circuit density (miniature devices and larger die area) at lower technology, which adversely introduces noise effects in circuit paths. Among the different noise sources, crosstalk noise, supply noise, substrate and thermal noise; crosstalk noise, highly contributes to an additional delay to a circuit path [1].

Always, there is a concern in the semiconductor industry for crosstalk noise and delay-aware ATPGs. With the objective of reducing computational time in pattern generation procedure, delay fault models are based on zero interconnect delay in ATPGs. They target to capture as many faults as possible with lesser number of patterns in reduced time. In other words, interconnect delays which are due to crosstalk noise are neglected during ATPG test. Therefore, testing with patterns generated by ATPG tools may create low quality products. Different delay fault models that are currently employed in capturing delay defects are path delay and transition delay

faults. Among these, the path delay fault model can be utilized for testing failures due to crosstalk noise.

Crosstalk noise is due to the coupling between the victim net (i.e., signal net considered) and aggressor nets (i.e., near neighboring nets). This expands to the entire victim path (consists of many victim nets) with distributed delay variations. A single victim net may have many numbers of aggressor nets with the same or different input signal transitions and varying arrival time (at gate inputs) with respect to the victim net. Multiple aggressors can speed-up or slow down the worst-case path delay in the victim net [2]. Opposite signal transition maximizes the delay slow-down [3] compared to the same signal transitions and stable inputs (i.e., stable 0 and stable 1) in the aggressor nets. We determine aggressor nets from the layout of the circuit as it is a realistic circuit structure that will be manufactured and IC tested. In our work, the existing path delay fault at launch-off-capture in ATPG is customized to test crosstalk noise. This will be an effective alternative to SPICE-based simulation in delay testing [4] in terms of computational time. In this paper, we focus on pattern generation by constraining the aggressor nets to replicate the crosstalk noise.

In this paper, an effective ATPG flow is presented to generate crosstalk-aware delay patterns during at-speed test. We also demonstrate that a timing-aware ATPG is incapable of considering the patterns causing worst-case path delay. This flow can consider any number of aggressor nets surrounding the victim path. The proposed flow utilizes standard commercial ATPG tools for pattern generation and evaluation. Results from our ATPG flow shows that an effective pattern can be generated and tested to obtain worst-case path delay with lesser computational time.

The rest of the paper is organised as follows. Section II, demonstrates a motivational experiment to indicate the importance of adding crosstalk effects during path delay test. In Section III, the detailed flow of our constrained ATPG method of generating patterns in the presence of multiple-aggressors are presented. Experimental results are shown in Section IV. Finally, we conclude our work in Section V with future directions.

II. MOTIVATIONAL EXPERIMENT

The main objective behind our work is shown in this section. Essentially, our goal is to show that a standard ATPG tool may

---

*This work has been funded by the French government under the framework of the ENIAC "ELESIS" european program.

978-1-4799-8720-7/15 $31.00 © 2015 IEEE

not be able to generate good patterns that can capture worst-case path delay in the presence of crosstalk. Our experimental results will show that such situation occurs simply because any commercial ATPG tool does not take into account any physical design parameters as they work only on gate level simulation for pattern generation.

We perform SPICE simulations on s27, ISCAS89 benchmark circuit as shown in Fig.1(a). The SPICE level circuit includes parasitics in interconnects which is usually ignored at gate level simulation. Measured path delays for all input patterns obtained from SPICE simulations helps us to understand which patterns and under which conditions will provide the worst-case path delay. Such input patterns are then compared with a pattern generated by ATPG and their mismatches are highlighted. Hence, we provide evidence that a circuit may escape the test due to lower quality in the applied patterns.

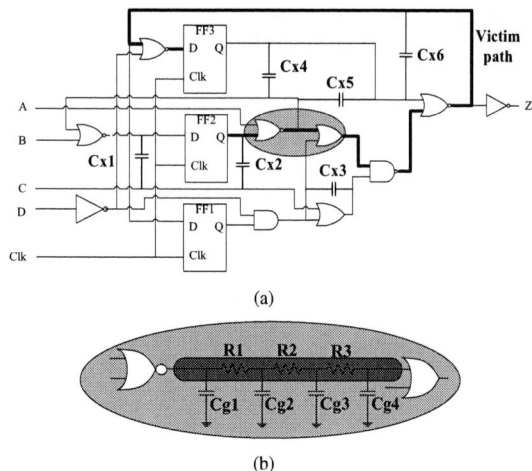

(a)

(b)

Fig. 1. (a) s27 benchmark circuit-under-test (b) 3π network model

We utilize a sample s27 circuit for analyzing the impact of crosstalk noise, shown in Fig.1(a). This circuit consists of 10 gates and 3 D-type flipflops with 4 inputs and a single output. CMOS models in the circuit are taken from the Predictive Technology Model (PTM) for 90nm technology [5]. Interconnecting nets between all the gates and flipflops are modeled as 3π network [6] as given in Fig.1(b). Interconnect parasitics are determined after the placement and routing of the circuit netlist and they consist of resistances (R1, R2, R3) and ground coupling capacitances (C_{g1}, C_{g2}, C_{g3}, C_{g4}). Cross-coupling capacitances (C_{X1}, C_{X2}, C_{X3}, C_{X4}, C_{X5}, C_{X6}) between victim and aggressor nets are ignored. We identify the all aggressor nets for the entire victim path.

A sketch of a circuit layout is illustrated in Fig.2 with interconnecting nets routed in different metal layers. From this, we can see that a single victim net can get affected by many aggressor nets in different metal layers and each aggressor nets may have varying signal transition with respect to the victim net. We estimated and included crosstalk capacitances, performed SPICE simulation and measured path delay.

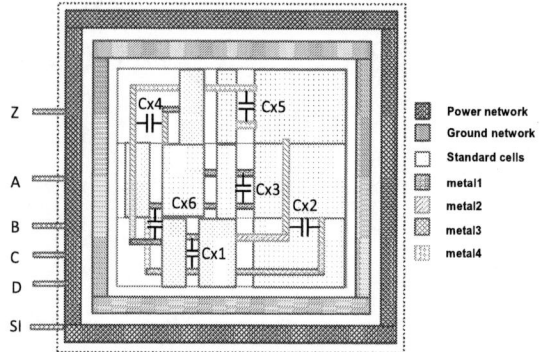

Fig. 2. Layout of a circuit

Victim path we examine, is highlighted in Fig. 1(a). Worst-case path delay is observed at FF3/D (for a falling transition) for the combinational victim path from FF2/Q (for a rising transition) by giving all possible vector pair transitions at the circuit inputs (A,B,C,D). Vector pair transitions include transitions in the same direction, opposite direction, stable 0 or stable 1 condition at the victim path. The vector pair (V1,V2) to the flipflops (FF1,FF2,FF3) and the worst-case path delay (δ_1) obtained from the SPICE simulation are listed in column 2-5 of Table I.

TABLE I
PATTERN COMPARISON AND DELAY VARIATION

| | SPICE | | | | ATPG | | | | δ_v |
	FF 1	FF 2	FF 3	δ_1 (ns)	FF 1	FF 2	FF 3	δ_2 (ns)	(%)
V1	1	0	1	0.289	1	0	1	0.286	1.05
V2	1	1	0		0	1	0		

$$\delta_v = \frac{(\delta_1 - \delta_2)}{\delta_1} \times 100 \qquad (1)$$

We developed a sequential circuit module in verilog similar to the one shown in Fig.1(a). The circuit is tested in an ATPG tool with the existing path delay fault model. Gate and flipflop models are obtained from the 90nm standard cell library. As, it is not possible to include interconnect models during pattern generation, so is impossible to analyze crosstalk-induced delay fault. The vector pair (V1,V2) to the flipflops (FF1,FF2,FF3) and the worst-case path delay (δ_2) obtained from the ATPG simulation of the pattern generated by the tool are listed in column 6-9 of Table I.

The path delay variation (δ_v) is calculated as the mean delay difference between the worst-case delays obtained from SPICE (δ_1) and ATPG (δ_2) patterns, as givin in equation 1. The comparison of patterns obtained after SPICE simulation and patterns generated by the ATPG tool, clearly shows a mismatch. Also, the delay variation (δ_v) of 1.05% (slow-down) as listed in column 10 of Table I implies that a good pattern can escape path delay fault testing due to the exclusion of crosstalk

978-1-4799-8720-7/15 $31.00 © 2015 IEEE

parameters. This indicates a serious notice on customizing the existing path delay fault testing in ATPG.

In the next section, we propose a flow to generate patterns that can capture worst-case path delay under the impact of crosstalk noise.

III. CONSTRAINED ATPG FLOW FOR PATTERN GENERATION

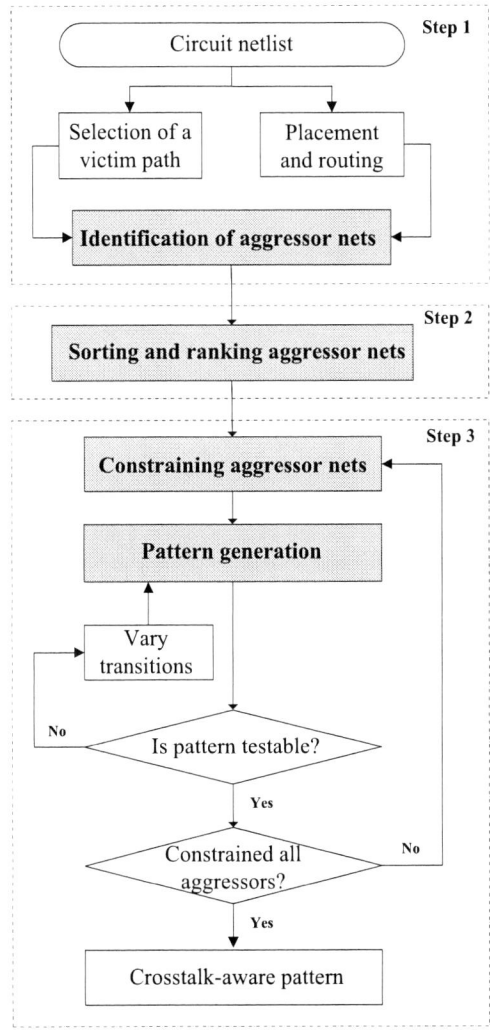

Fig. 3. Constrained ATPG flow

In this section, we present our flow for generating a crosstalk-aware pattern that can be effective in path delay fault testing. The flow in Fig. 3 consists of three major steps: (1) Identification of aggressor nets to a victim path, (2) Sorting and ranking aggressor nets based on their impact, and (3) Constraining aggressor nets and pattern generation for a victim path. We describe this method on b08, ITC'99 benchmark circuit. The relevance of the proposed flow shows

a great promise to reduce the gap between the path delay patterns generated by ATPG tools and crosstalk-aware patterns identified from exhaustive SPICE simulations.

A. Identification of aggressor nets

We utilize industrial tools such as Cadence Encounter RC Compiler [7], Synopsys PrimeTime Static Timing Analysis (STA) tool [8] and a Cadence SOC Encounter tool [7] to generate a circuit netlist, select a longest victim path and to place and route the circuit netlist, respectively. From the circuit layout, all the aggressor nets are determined for a victim net based on minimal distance from the victim net and area of the aggressor nets. Each aggressor net may reside in different metal layers with varying length. Therefore, determining aggressor nets from the layout of the circuit gives a proper estimation of the crosstalk impact. For a selected victim path in b08, the 22 aggressor nets identified for the 5 victim nets are represented as in Table II.

TABLE II
AGGRESSOR NET RANKING

Victim net	Aggressor net	Rank
Vn1	An1	12
	An2	20
	An3	6
	An4	13
Vn2	NA*	
Vn3	An5	2
	An6	1
	An7	10
	An8	16
	An9	18
	An10	15
	An11	21
	An12	9
	An13	7
	An14	17
	An15	5
Vn4	An16	19
	An17	11
	An18	4
	An19	3
	An20	14
	An21	8
Vn5	An22	22

NA* - No aggressor net for this victim net.

B. Sorting and ranking aggressor nets

Firstly, aggressor nets identified are sorted based on their impact on the victim net, i.e., the higher cross-coupling capacitance will have higher impact. Capacitance values measured are between 0.02fF - 2fF. Secondly, we rank aggressor nets, i.e., nets with lower capacitances are given a lower rank as

TABLE III
AGGRESSOR NET TRANSITION AND DELAY MEASUREMENT

Rank	Vnet[a]	C_v^b	Anet[c]	C_1^d	Fault class	Δ_v^e	C_2^f	Fault class	Δ_v	C_3^g	Fault class	Δ_v
1	Vn3	R[h]	An6	F[i]	UD/AU[j]		0	DT[k]	0%			
2	Vn3	R	An5	F	UD/AU		0	DT	0%			
3	Vn4	F	An19	R	UD/AU		0	UD/AU		1	DT	0%
4	Vn4	F	An18	R	UD/AU		0	UD/AU		1	DT	-0.04%
5	Vn3	R	An15	F	UD/AU		0	DT	-0.04%			
6	Vn1	R	An3	F	UD/AU		0	DT	-0.04%			
7	Vn3	R	An13	F	UD/AU		0	UD/AU		1	DT	-0.04%
8	Vn4	F	An21	R	DT	-0.68%						
9	Vn3	R	An12	F	UD/AU		0	DT	-0.70%			
10	Vn3	R	An7	F	UD/AU		0	DT	-0.72%			
11	Vn4	F	An17	R	UD/AU		0	UD/AU		1	DT	-0.76%
12	Vn1	R	An1	F	UD/AU		0	DT	-0.79%			
13	Vn1	R	An4	F	UD/AU		0	DT	-0.79%			
14	Vn4	F	An20	R	UD/AU		0	UD/AU		1	DT	-0.80%
15	Vn3	R	An10	F	UD/AU		0	UD/AU		1	DT	-0.80%
16	Vn3	R	An8	F	UD/AU		0	UD/AU		1	DT	-0.94%
17	Vn3	R	An14	F	UD/AU		0	DT	-0.97%			
18	Vn3	R	An9	F	DT	-1.59%	0	DT				
19	Vn4	F	An16	R	UD/AU		0	DT	-1.71%			
20	Vn1	R	An2	F	UD/AU		0	UD/AU		1	DT	-1.73%
21	Vn3	R	An11	F	UD/AU		0	UD/AU		1	DT	-1.73%
22	Vn5	F	An22	R	UD/AU		0	UD/AU		1	DT	-1.73%

[a] victim nets, [b] transition in the victim net, [c] aggressor nets, [d] contraining aggressor net with opposite trasition, [e] path delay variation for a constrained ATPG pattern, [f] contraining aggressor net with stable 0, [g] contraining aggressor net with stable 1, [h] indicates a rising transition, [i] indicates a falling transition, [j] ATPG fault that is either undetectable or untestable, [k] ATPG detected fault class that is testable.

they have minimal influence on the victim path delay [9]. Their ranking order is shown in column 3 of Table II. Some of the victim nets that act as an aggressor net is ignored to ensure a transition at launch-off-capture, so as to capture a path delay fault.

C. Constraining aggressor nets and pattern generation

Our major contribution is in customizing the standard ATPG method of generating a path delay fault pattern. The existing ATPG is unaware of crosstalk impact on delay of a victim path. Therefore, we aim at generating a crosstalk-aware pattern by forcing constraints during pattern generation to aggressor nets that impact the victim path. In general, crosstalk issue can be resolved by adding their impact during SPICE simulations, but it is computationally exhaustive. We compare our method with the X-filling method [6] in terms of (1) computational time of generating a path delay fault pattern, and (2) validating the effectiveness of the pattern by verifying worst-case path delay of pattern. The X-filling method is based on selectively filling don't care bits in the pattern generated by the ATPG.

In this step, we constrain all the aggressor nets based on the signal transition corresponding to each victim net in the entire path. Also, aggressor net with the higher rank (based on their impact) is constrained first. For a victim net with a Rising

(R) transition pattern, the opposite transition pattern on the aggressor net, i.e., a Falling (F) constrains can maximize the crosstalk-induced delay slow-down [10]. We check whether it is a testable pattern or not i.e., so that it can assure a signal transition propagation at the launch cycle in at-speed test [11]. Sometimes, this doesn't give a successful ATPG pattern generation. So, we then constrain with a stable (0 and 1) condition on the aggressor net. We ignore the same direction transition pattern as it may only induce a minimal delay slow-down or speed-up in the victim net. All aggressor nets are constrained and a final pattern is generated as shown in Step 3 of Fig. 3, which can be path delay fault tested in the presence of multiple aggressors.

Table III details the steps of the proposed ATPG flow on b08 benchmark circuit. Aggressor net ranking on column 1, victim net, their transition pattern (i.e., rising signal, R or falling signal, F) on column 2-3 and aggressor nets, their transition pattern (i.e., opposite transition, stable 0, or stable 1) in subsequent columns. From the different fault class, such as detected (DT), possibly detected (PT), undetectable (UD), not detected (ND) and ATPG untestable (AU), we select the detected faults. The corresponding path delay values (δ_v) are measured in SPICE, in order to show the effectiveness

TABLE IV
PATTERN COMPARISON AND PATH DELAY VARIATION

	Vp^a	FF^b 1	FF 2	FF 3	FF 4	FF 5	FF 6	FF 7	FF 8	FF 9	FF 10	FF 11	FF 12	FF 13	FF 14	FF 15	FF 16	FF 17	FF 18	FF 19	FF 20	FF 21	δ_v^c (%)
C_{atpg}^d	V1	1	X	X	1	0	X	X	1	0	X	0	X	X	X	X	0	1	X	X	1	0	-1.73
	V2	X	1	X	0	1	X	X	X	0	0	0	0	0	0	0	0	0	X	X	0	1	
bt^e	V1	X	1	0	1	X	X	X	X	0	0	0	0	0	1	1	1	0	0	1	1	0	-2.12
	V2	X	0	X	0	0	X	X	X	0	0	0	0	0	0	1	0	1	0	0	1	1	

^{a}pattern (V1,V2) to scan-flipflops, ^{b}scan-flipflops from 1 to 21 from the synthesized DFT scan chain structure, ^{c}path delay variation, ^{d}by constrained ATPG method, ^{e}by X-filling method.

TABLE V
PATTERN TRANSITION IN AGGRESSOR NETS

	Vp^a	An 6	An 5	An 19	An 18	An 15	An 3	An 13	An 21	An 12	An 7	An 17	An 1	An 4	An 20	An 10	An 8	An 14	An 9	An 16	An 2	An 11	An 22
C_{atpg}	V1	0	0	1	1	0	0	1	0	0	0	1	0	0	1	1	1	0	1	0	0	1	1
	V2	0	0	1	1	0	0	1	1	0	0	1	0	0	1	1	1	0	0	0	1	1	1
bt	V1	X	0	1	X	X	0	1	X	X	X	X	X	X	X	X	1	0	X	1	0	1	1
	V2	X	0	1	X	X	0	1	X	X	X	X	X	X	X	X	1	0	0	1	1	1	1

of our pattern. By simulating this pattern, we obtain the maximum possible worst-case delay of -1.73% on the victim path. Negative value shows the slow-down impact.

Patterns to scan-flipflops (from FF1 to FF21) sensitize the victim path and generates a pattern. FF21/Q to FF4/D is the longest path under test. Given vectors to victim nets (from Vn1 to Vn5) in the victim path, $< V1, V2 >$ are $< 01011, 10100 >$. In Table IV, we compare the patterns generated by our flow (C_{atpg}) with patterns identified by the X-filling (bt) method, utilizing exhaustive SPICE. The path delay variation (δ_v) obtained by each method in shown in the last column of Table IV. ATPG generates a random pattern to test path delay fault. Both method pattern delays are compared with this pattern delay to obtain the path delay variation. By this, we show that our flow is able to achieve a delay value (-1.73%) closer to the exhaustive X-filling method (-2.12%), hence an effective pattern generation flow.

In Table V, we show a pattern comparison in aggressor nets between the two methods. As aggressor nets are either primary input nets, scan input nets or other interconnecting nets between standard cells, a constrained transition can force to change the generated pattern. Therefore, constraining these nets with opposite or stable 0, 1 transition, modifies the pattern, generated by the tool finally. From the vector pair colored in Table V, we can see that 13 aggressor nets are undergoing changes in transition of patterns.

In Fig. 4, we show the delay variation in the victim path due to the impact of the aggressor net. Aggressors are constrained one after the other based on their rank, mentioned in our flow. From the plot, we can see that there are two successful opposite pattern transitions (first at An21 and then at An9) that are highly maximizing the slow-down of victim path delay. Path delay variation is comparatively low after constraining other aggressor nets.

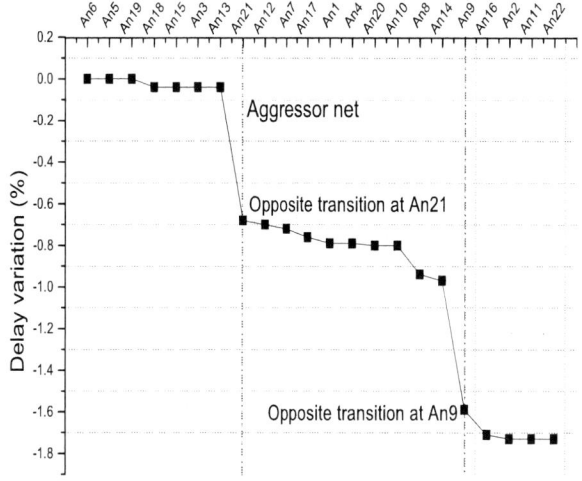

Fig. 4. Path delay variation plot of a victim path

IV. EXPERIMENTAL RESULTS

In this section, we describe the experimental results after applying our constrained ATPG flow on 9 full-scan version of ITC'99 benchmark circuits [12]. Our aim is to generate a pattern that can be effectively path delay fault tested. The pattern should be capable of capturing worst-case delay in the victim path due to the multiple aggressors and multiple aggressor switching in victim path. This flow can be applied to any size of the circuit. Our method of pattern generation is fully based on ATPG tool. We could save a lot of computational time by this flow, as this is focused on avoiding the exhaustive SPICE simulation in any sized circuits.

TABLE VI

CIRCUIT DESCRIPTION AND PATH DELAY VARIATION RESULTS FOR ITC'99 BENCHMARK CIRCUITS

Ckt	#Gates	#FF's	$\#Vp^a$	$\#X^b$	$\#bt_X^c$	$\#An^d$	bt		C_{atpg}	
							δ_v	t_r	δ_v	t_r
b01	49	5	5	2	1	8	-0.01%	$40s^e$	-0.01%	58s
b02	28	4	4	2	1	10	-0.11%	25s	-0.24%	46s
b03	160	31	30	31	10	11	-2.06%	170342s	-2.87%	53s
b05	998	33	34	29	16	36	-3.34%	9133840s	-1.88%	278s
b06	56	8	8	6	1	6	-0.17%	353s	-0.17%	30s
b07	441	41	41	35	20	41	-4.92%	818469s	-3.81%	201s
b08	183	21	21	26	14	22	-2.12%	291094s	-1.73%	104s
b09	170	27	28	10	3	13	-1.63%	1298s	-1.86%	68s
b10	206	17	17	22	10	27	-2.59%	185661s	-2.23%	188s

[a] number of victim paths in a circuit, [b] number of X-bits (don't care bits) as inputs to scan-flipflops, [c] number of X-bits that is filled with X-filling method, [d] number of aggressor nets, [e] computational time in seconds.

Details of experimental results are summarized in Table VI. Number of gates, scan flip-flops, victim paths, X-bits in the scan pattern and primary input pattern in the circuit netlist are given in column 2-5, respectively. In column 6, the minimum number of relevant X-bits that need to filled in X-filling method are shown. Number of all possible aggressor nets to the victim path are given in column 7. Column 8-9 depict the path delay variation and computational time for producing the patterns by X-filling method (by exhaustive SPICE simulation). We haven't shown some results on the X-filling method, as the simulation in a large sized circuit takes days to complete. Similarly, in column 10-11, we show the results obtained after applying our ATPG flow on pattern generation.

Our results show that the proposed ATPG flow is able to generate an effective pattern that can provide a delay value nearest to the expected (achieved by using a SPICE simulation pattern). Also, we have shown that our flow can be executed with extremely low computational time. For instance, pattern identification on a comparatively bigger circuit like b05 circuit require many days to estimate a single worst-case delay pattern, whereas this ATPG flow takes only 278 seconds to generate a pattern that can be path delay fault tested in the presence of multi-aggressor crosstalk noise. We acknowledge that the proposed flow is simple to implement, and it offers many benefits in terms of computational time and worst-case path delay patterns.

V. CONCLUSIONS

In this paper, we present an ATPG flow of pattern generation targeting path delay fault under the impact of crosstalk noise. The flow is based on constraining transition signals on all aggressor nets nearer to the tested path. This method eliminates the exhaustive SPICE simulation and it can produce patterns which can be utilized to test the worst-case path delay in a circuit. Our flow is implemented on ITC'99 benchmark circuits and the results were shown by comparison with SPICE simulation. By this flow, we also show the effectiveness in computational time and pattern quality (can be utilized for crosstalk-aware path delay fault testing). As future work, we plan to extend this flow to add the impacts of other signal integrity issues such as supply noise, substrate and thermal noise on pattern generation.

REFERENCES

[1] A. Rubio, N. Itazaki, X. Xu, and K. Kinoshita, "An approach to the analysis and detection of crosstalk faults in digital VLSI circuits," in *Computer-Aided Design of Integrated Circuits and Systems, IEEE Transactions on*, 1994, pp. 387–395.

[2] M. Zhang, H. Li, and X. Li, "Path delay test generation toward activation of worst case coupling effects," in *IEEE Transactions on VLSI Systems*, 2011, pp. 1969–1982.

[3] K. Peng, M. Yilmaz, K. Chakrabarty, and M. Tehranipoor, "Crosstalk-and Process Variations-Aware High-Quality Tests for Small-Delay Defects," in *IEEE Transactions on VLSI Systems*, 2013, pp. 1129–1142.

[4] M. Chen and A. Orailoglu, "Examining Timing Path Robustness Under Wide-Bandwidth Power Supply Noise Through Multi-Functional-Cycle Delay Test," in *Very Large Scale Integration (VLSI) Systems, IEEE Transactions on*, 2014, pp. 734–746.

[5] Predictive Technology Model (PTM). [Online]. Available: http://ptm.asu.edu/

[6] A. Asokan, A. Todri-Sanial, A. Bosio, L. Dilillo, P. Girard, S. Pravossoudovitch, and A. Virazel, "Path Delay Test in the Presence of Multi-Aggressor Crosstalk, Power Supply Noise and Ground Bounce," in *Design and Diagnostics of Electronic Circuits and Systems,. 17th*, 2015.

[7] [Online]. Available: http://www.cadence.com/products/pages/default.aspx

[8] PrimeTime User Guide, Synopsys, Inc., Mountain View, CA,May 1999.

[9] A. Asokan, A. Todri-Sanial, A. Bosio, L. Dilillo, P. Girard, S. Pravossoudovitch, and A. Virazel, "A Delay Probability Metric for Input Pattern Ranking Under Process Variation and Supply Noise," in *VLSI (ISVLSI), 2014 IEEE Computer Society Annual Symposium on*, 2014, pp. 226–231.

[10] D. Gope and D. Walker, "Maximizing crosstalk-induced slowdown during path delay test," in *Computer Design (ICCD), 2012 IEEE 30th International Conference on*, 2012, pp. 159–166.

[11] H. Li, P. Shen, and X. Li, "Robust test generation for precise crosstalk-induced path delay faults," in *VLSI Test Symposium, 2006. Proceedings. 24th IEEE*, 2006, pp. 6 pp. – 305.

[12] [Online]. Available: http://www.cad.polito.it/downloads/tools/itc99.html

978-1-4799-8720-7/15 $31.00 © 2015 IEEE

Analyzing the Impact of Frequency and Diverse Path Delays in the Time Vulnerability Factor of Master-Slave D Flip-Flops

Alexandra L. Zimpeck, Fernanda Lima Kastensmidt, Ricardo Reis

Instituto de Informática – PPGC - Universidade Federal do Rio Grande do Sul (UFRGS)

{alzimpeck, fglima, reis}@inf.ufrgs.br

Abstract— **Soft errors are becoming a major concern in integrated circuits fabricated in nanometer technology working in dependable applications. The goal of this paper is to determine the dependency of soft errors in integrated circuits with its operating frequency and variety of delays in the combinational logic paths. Each circuit flip-flop has a different Time Vulnerability Factor (TVF) that can be measured by electrical simulations based on the delay of the combinational logic path that is connected to that flip-flop. The TVF values of the master and slave latches can vary from 50% to 0% of the clock period according to the logic propagation delay and slack presented in the circuit, the operating frequency and technology process. In this work, the analysis was performed by electrical simulation using sequential designs described in 16nm, 22nm and 32nm nanometer technologies. Results show that the probability of SEU occurrence decreases with the increase of frequency and flips-flops connected to the critical paths present the lowest TVFs. This information can be easily integrated in design tools to help identifying the most vulnerable flip-flops in circuits before mitigate or replace the flip-flops by radiation hardened ones.**

Keywords- Window of Vulnerability; Timing Vulnerability Factor; Soft Erros; Radiation Effects

I. INTRODUCTION

Technology scaling has leaded to drastically voltage supply reduction, transistor size shrinking and high transistor density, which have been contributing to increase the vulnerability to soft errors [1,2]. One of the main challenges of the semiconductor industry in recent years is to cope with the interaction of ionized particles in the silicon [3]. According to the energy of the ionized particles hitting the silicon, transient pulses may occur provoking transient upsets that may be interpreted as functional errors [4]. A transient pulse arises from the interaction of energetic particles that deposits enough charge near a sensitive reverse biased drain of a transistor. If the collected charge exceeds the minimum charge needed (Q_{crit}) a transient voltage pulse appears in the stroke node. When this pulse occurs inside a memory element, such as a latch, it may invert the original stored value, producing in this case a Single Event Upset (SEU) or bit-flip [5]. Therefore, it is important to predict the behavior of the flip-flops to determine amount of time that these devices are sensitive to upsets.

The time that sequential elements are vulnerable to upset is defined as Window of Vulnerability (WOV) [6] and the Timing Vulnerability Factor (TVF) corresponds the fraction of the WOV, which a device is vulnerable to upset according to the clock period [7]. Consequently, the TVF depends on the process technology, the logic propagation delay between two flip-flops and the chosen operating clock frequency. Lower is the TVF of a flip-flop, lower is the probability of SEU occurs in that flip-flop and consequently, less vulnerable to SEUs is it. When taking into account an integrated circuit, each flip-flop has a different TVF that can be measured by electrical simulations based on the delay of the combinational logic path that is connected to that flip-flop.

This work analyzes in details the Window of Vulnerability and the Timing Vulnerability Factor of standard Master-Slave (MS) D flip-flop under bit-flip in some case-study circuits. The goals are: 1) determine the dependency between the SEU occurrences in a flip-flop with its operating frequency and 2) analyze the fault propagation to the next stage going through a combinational logic with a certain propagation delay. The proposed study is very important to show in details how to analyze the vulnerability of a flip-flop under SEU and the real impact of the technology and the chosen operating frequency in the sensitivity of the design under radiation.

In this approach, electrical simulations in HSPICE tool [8] are performed to analyze the circuit response to a soft error mechanism. For the experiments, we adopt 32nm, 22nm and 16nm CMOS process technologies from Berkeley Predictive Transistor Model (PTM) [9]. Results show that the probability of SEU decreases with the increase of frequency. Furthermore, electrical simulations have demonstrated the impact of the combinational logic delay has on TVF of the master and slave latches. Based on the results, designers can take in consideration different propagation delays and operating frequencies according to the process technology used in order to reduce the TVF of the flip-flops. The information of TVF of each flip-flop can be easily integrated in design tools to help identifying the most vulnerable flip-flops in circuits before mitigate or replace the flip-flops by radiation hardened ones.

The paper is organized as follows. Section II analysis the TVF in MS D Flip-Flops. Case-study pipeline designs are presented in Section III. Section IV describes the simulation results and analysis of experiments. Finally, conclusions are presented in Section VI.

II. ANALYZING THE TIMING VULNERABILITY FACTOR (TVF) IN MS FLIP-FLOPS

When energetic particle impacts a sensitive node of a transistor affecting the NP junction, the electron-hole pair is generated and the charge collated mechanism produces a transient pulse that lasts until the collected charge is eliminated through VDD or GND, returning the logic node to its correct state. When a latch holds a value, it has two transistors in "on" state and two transistors in "off" state. Therefore, there are always two sensitive nodes in the memory cell. When the particle strikes in these nodes, the energy by the particle can provoke a transistor to switch "on" if the collected charge exceeds the critical charge (Q_{crit}). This effect is known as Single Event Upset (SEU) and generates a bit-flip, i.e., an inversion in the stored value in memory elements.

In real circuits, the flip-flops are usually designed as master-slave D flip-flops that are composed of two latches. The energetic particle can hit the master or the slave latch of the D flip-flop provoking a bit-flip in one of the latches. But not always the bit-flip will be seen in the output of the flip-flop or at the input of the next stage flip-flop due to Window of Vulnerability (WOV) effect and Timing Vulnerability Factor (TVF).

Fig. 1 depicts two possible induced faults cases that involve a strike occurring directly in sequential circuit. In the first case (a), the particle hits the master latch and in the second case (b), the particle hits the slave latch. In both cases, this is characterized as SEU and provokes a bit-flip in the latch. SEU in the D flip-flop is only important if this SEU can propagate through the combinational logic path and reach the output of next stage of the flip-flops. The node for the TVF analysis is the output node of the last flip-flop.

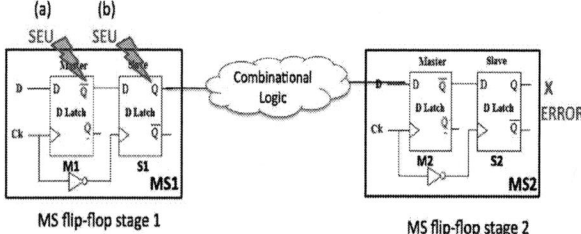

Fig. 1 Two possible induced faults cases with strike in sequential circuit

Here in this work, we are investigating only the WOV and not other masking effects such as electrical and logic masking that may occur also to reduce the effect of SEU.

The Window of Vulnerability (WOV) is defined as the time where a transient fault can generate a bit-flip in the master or slave latch and this bit-flip can be seen at the last flip-flop output. The Master-Slave Flip-Flop is not sensitive to soft errors during the entire clock period. When considering a rising edge MS D flip-flop, the total WOV is the contributions of both the master and slave parts. The master latch is susceptible to upsets when the clock is high, i.e., when the master latch is in hold mode. For the slave latch, the opposite happens, it is susceptible to upsets when the clock is low, that corresponds a hold mode.

As seen in Fig. 2, the master latch is vulnerable in a portion of the first clock phase (T_{phase1}), while the slave latch is vulnerable in a portion of time in the second clock phase (T_{phase2}) [10]. It is important to emphasize that upsets that occur outside the WOV limits of the master or slave latches do not propagate in time to be seen in the next stage flip-flop and should not contribute to soft error rate.

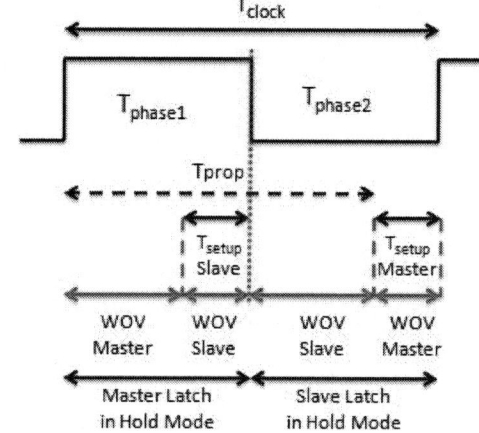

Fig. 2 Window of Vulnerability for Master and Slave latches working in a rising edge MS D flip-flop

When there is no combinational logic between flip-flops from stage 1 to stage 2, for example as in a shift register of pipelines, the propagation delay (T_{prop}) is zero. In this case, T_{phase} is divided only in two parts composed for WOV and setup time (i.e., time that input must fall or rise before the clock edge [11]). In this case, the WOV of the master and slave latches is equal to Tclock/2. This means that this is the worst-case scenario when the total WOV of the flip-flop is T_{clock}.

When there is combinational logic between flip-flops from stage 1 and stage 2, for example as in pipelines with arithmetic operations in between, the propagation delay T_{prop} is different to zero and it contributes to reduce the WOV of the slave and master latches. If there is a very large combinational delay, it will take a greater amount of time for the error to propagate to the next flip-flop.

As first discussed in [6], the WOV strongly depends on the number of combinational gates in the logic path between two sequential elements. There are cases that the propagation delay is just a small percentage of the clock period and then the WOV of the slave latch is reduced, but the WOV of the master latch is entirely preserved. In some cases, the propagation delay represents a large percentage of the clock period, which causes the WOV of the slave latch to be equal to zero, and the WOV of master latch can be reduced drastically.

The Timing Vulnerability Factor (TVF) corresponds to the fraction of time according to the clock period, which a device is vulnerable to upset [7]. In case of MS D Flip-Flop design, even master and slave TVF values are different for the same propagation delay $(T_{prop} > 0)$ [5]. The TVF of the master and slave latches vary from 0% to 50%, depending on the size of the combinational logic block. The following equations determine the impact caused for the combinational logic for the TVF of master and slave latches.

$$TVF_{Slave} = (T_{clock}/2) \text{ and } TVF_{Master} = (T_{clock}/2),$$
$$\text{if } T_{prop} = 0 \tag{1}$$

$$TVF_{Slave} = ((T_{clock}/2 - T_{setup\_Master} - T_{prop}) + T_{setup\_Slave})$$
$$\text{and } TVF_{Master} = (T_{clock}/2),$$
$$\text{if } T_{prop} \leq ((T_{clock}/2 - T_{setup\_Master}) + T_{setup\_Slave}) \tag{2}$$

$$TVF_{Slave} \approx 0 \text{ or } TVF_{Slave} = 0 \text{ and } TVF_{Master} \downarrow,$$
$$\text{if } T_{prop} > ((T_{clock}/2 - T_{setup\_Master}) + T_{setup\_Slave}) \tag{3}$$

As discussed, there are two main dependencies. The first is the combinational path delay T_{prop}. Fig. 3 (a) shows the effect of T_{prop} in the TVF in the three cases presented. Note that higher is the T_{prop}, lower is the WOV of the slave. In some cases, the TVF of the slave latch can be zero due to the high combinational path delay. The second dependency is the operating frequency that the circuit actually runs. Sometimes, the circuit can operate at 2 GHz but due to power limitations, the circuits operate at a lower frequency. This brings some slack in the combinational paths. The total TVF of flip-flop has a strong dependency with its operating frequency.

This brings some slack in the combinational paths. The total TVF of flip-flop has a strong dependency with its operating frequency. Fig. 3 (b) shows the effect of the TVF when T_{clock} increases for the same propagation delay used in case (a). According to the equations, applications that run with lower frequencies and consequently larger clock period, introduce higher TVF.

Some analyses of TVF according to the frequency have been shown in related papers [5,6,12], but here in this paper we show in details how to measure and to analyze this impact and also the tendency of the TVF according not only to the operating frequency but to different combinational path delays and slacks with very clear graphs that can be used in design tools to select the most sensitive flip-flops in the circuit.

III. CASE-STUDY PIPELINE DESIGNS

In order to analyze the TVF of master slave flip-flops, it is necessary to implement the flip-flop in a pipeline design without and with logic between the stages working in different frequencies. Different pipeline stages were designed at Spice level. Each one has two standard Master-Slave Flip-Flops separated by a chain of inverters varying from 0, 3, 6, 9 and 12 inverters. The case-study flip-flops are master slave flip-flops working in the rising edge composed of logic gates as described in Fig. 4, totalizing 20 transistors NMOS and 20 transistors PMOS.

For the experiments, we use the technological High Performance (HP) models for 32nm, 22nm and 16nm from Berkeley Predictive Transistor Model – PTM [9]. All simulations were performed in HSPICE tool [8] and were inserted only one fault at a time. We analyze each pipeline circuit operating at 4GHz, 3GHz, 2GHz, 1GHz, 500MHz and 250 MHz clock frequencies.

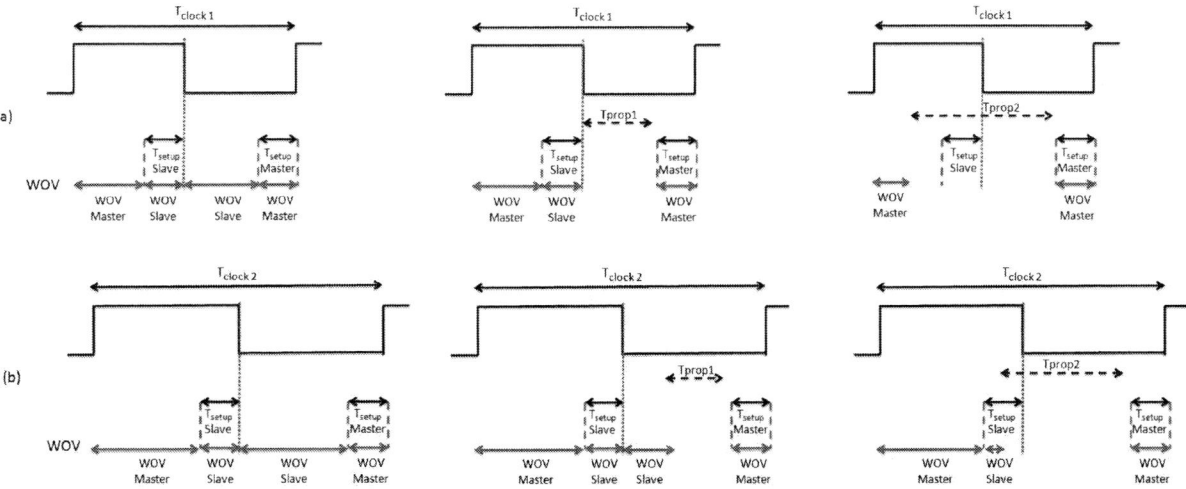

Fig. 3 (a) Effect of the TVF for different propagation delays Tprop (Tprop = 0, Tprop = Tprop1 and Tprop = Tprop2, where Tprop2 >Tprop1) at clock period (Tclock1) and (b) Effect of the TVF when Tclock increases (Tclock2 > Tclock1) for the same propagation delays Tprop (Tprop = 0, Tprop = Tprop1 and Tprop = Tprop2)

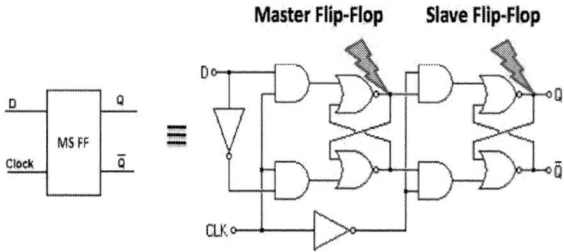

Fig. 4 Standard Master-Slave D Flip-Flop

Fig 5 shows a circuit with different amount of combinational logic between the MS flip-flops. Even though it operates in the same operating frequency, cases (a) and (b) have different TVF values. It happens because flip-flops in critical path have lower TVF (case b) than flip-flops placed in logic paths with large slack (case a). The transient fault is modeled using a double exponential current and injected in the sensitive nodes of the circuit [13]. In this work, we injected the transient current in nodes QM1 of master and QS1 of slave.

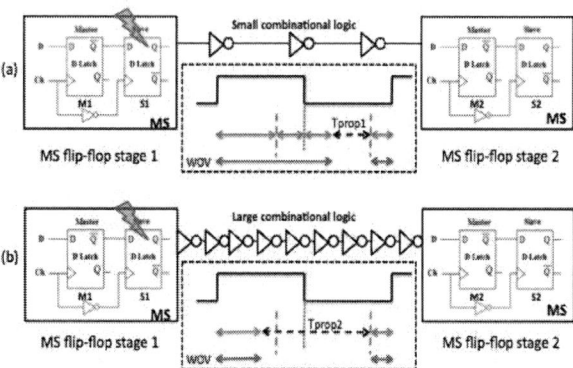

Fig. 5 MS flip-flop in the pipeline design described in Spice level with different amount of combinational logic between the sequential stages.

To compute the TVF of a standard Master-Slave D Flip-Flop, a charge is injected during one clock cycle, which is divided up into N time-steps [6]. According with [14], there are many values for pulse width SET that may invert the original stored value, even when exposed to the same amount of charge. In this work, the average pulse widths are 100ps.

In order to determine the minimum collected charge that corrupts the state of each node by means of electrical simulations, the magnitude of the current pulse is increased until an upset condition can be observed at the output of sequential [15]. Fig 6 shows simulation results when the magnitude of the current pulse reaches the critical charge and a bit-flip occurs in the (a) master latch and in the (b) slave latch.

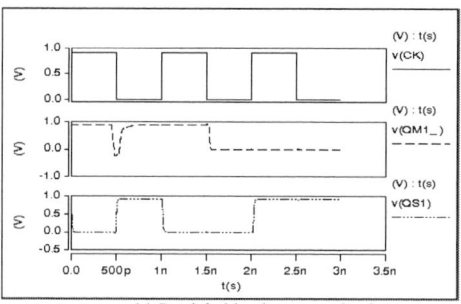

(a) Particle hits the master latch

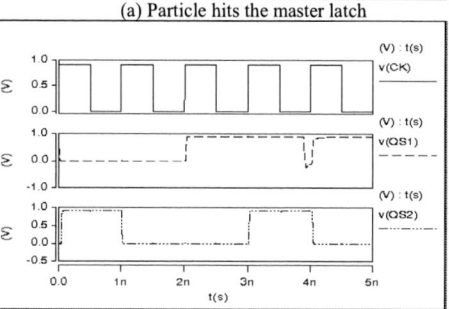

(b) Particle hits the slave latch
Fig. 6 Latch waveforms as a response to SEU

IV. SIMULATION RESULTS

The measured TVF for the slave latches of the case-study circuits are presented in Fig. 7, 8 and 9 for 32nm, 22nm and 16nm nanometer technologies, respectively. The graphs show the TVF curves for six different operating frequencies. For example, if the circuit operates at 0.25GHz, all the slave latches connected to combinational path delays of 2052ps or more have 0% of TVF. If the circuit operates at 4GHz, for instance, all the slave latches connected to combinational path delays of 340ps or more have 0% of TVF. In this way, it is possible to see that each slave latch in the circuit will present different TVF according to the delay of the path that it is connected to and the operating frequency of the circuit.

We know that if a circuit has a critical path delay of 1ns, this means that the circuit can operate in any frequency up to 1GHz. By analyzing the measured graphs for 32nm technology, one can see that in order to achieve the lowest TVF for the slave latches, it is better to operate in 1GHz where all latches connected to paths longer than 690ps have 0% TVF, while if the same circuit operates at 0.25GHz, only latches connected to path delays over 1ns will have 0% TVF. In summary, this graphs shows that reducing frequency to reduce power for instance is a bad choice for reliability as it may increase the TVF for the slave latches. One also can notice that the impact of frequency, path delays and slack gets slightly more intense with reduced technology.

The TVF of a master latch also may decreases with the increase of the combinational logic delay, but it is not so affected as the slave latch is. The TVF of master latch is usually constant and equal to 50%. But it can be reduced when the flip-flop is connected to large combinational logic path delays that are usually larger then at least half of the clock period.

978-1-4799-8720-7/15 $31.00 © 2015 IEEE   524

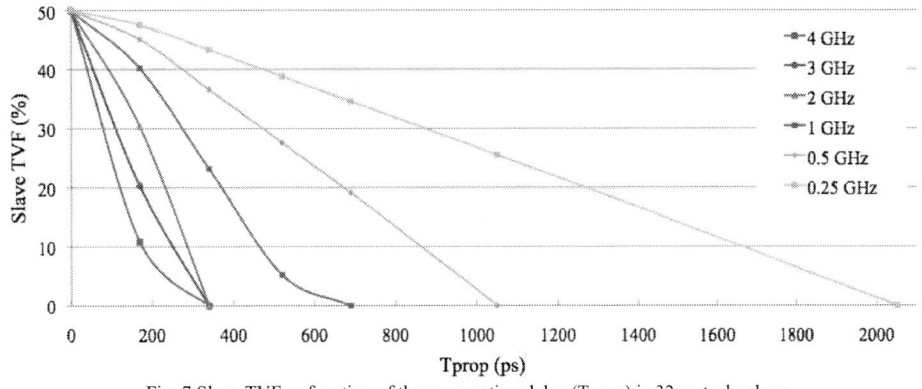

Fig. 7 Slave TVF as function of the propagation delay (Tprop) in 32nm technology

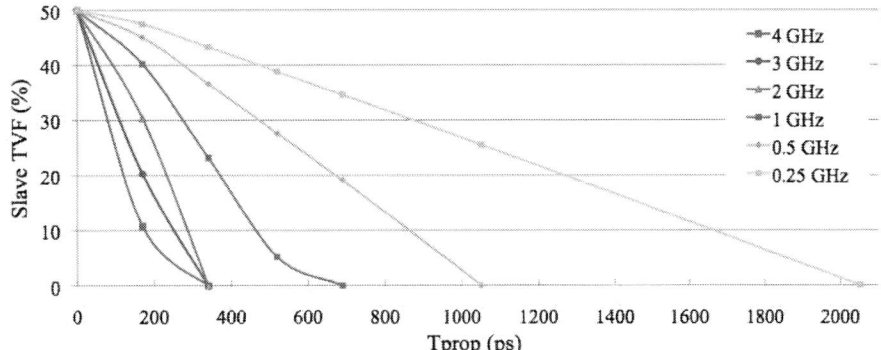

Fig. 8 Slave TVF as function of the propagation delay (Tprop) in 22nm technology

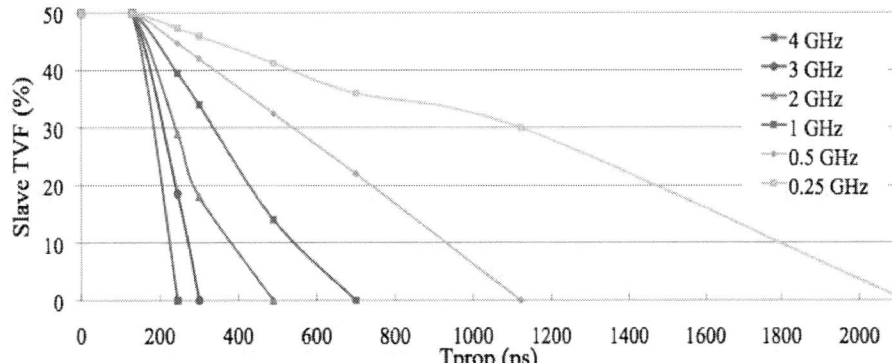

Fig. 9 Slave TVF as function of the propagation delay (Tprop) in 16nm technology

In addition, the TVF of master latches does not reach zero, but a minimum value around 25% for critical combinational logic paths with very reduced slacks. But this value changes with the process technology and flip-flop topology.

When analyzing the results of this work, three important conclusions can be highlighted. First, results indicate that master latches are more vulnerable to soft errors than slave latches. The TVF of master latch is usually higher than the TVF of the slave latch. Thus, is more important to designers develop master latches robust to bit-flips using DICE cell, for example [16].

Secondly, the dependency between the SEU occurrences with the circuit operating frequency is significant. According with the analyzed technologies, we conclude that the circuit must operate at higher frequency possible according to the critical path delays in order to obtain the lowest TVF values for the slave latches.

By changing the technology but not the frequency, the TVF curves practically do not change. Fig 10 shows the plot of the slave latch TVF curves for different node technology running at the same frequency. Note that by changing the technology from 32nm to 16nm but maintaining the frequency at 1GHz, there is no significant reduction in the

TVF. Only when the circuit can operate at 2GHz in 16nm, the TVF drastically reduces and even reaches 0% for path delays higher than 132ps.

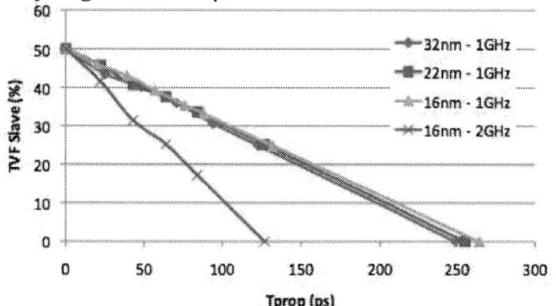

Fig. 10. Slave TVF as function of the different technology nodes at 1GHz and 2GHz

Fig. 11 shows the plot of the slave latch TVF curves for 32nm nanometer technology in different frequencies. When a circuit operates at 0.5GHz, in order to reach 0% TVF is necessary long combinational path delay such as 1052ps. Nevertheless, when the circuit operates at higher frequencies, as 2GHz, the TVF drastically reduces and even reaches 0% for path delays higher than 340ps.

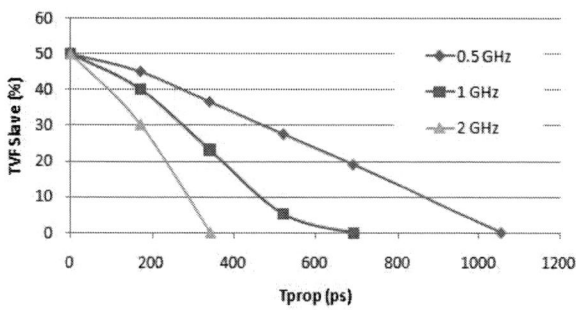

Fig. 11 Slave TVF as function of the different clock frequencies at 32nm technology

Thirdly, results have shown that each slave latch presents a different TVF according to the combination path delay connected to. The shorter combinational logic paths are more sensitive than larger combinational logic paths. The main reason for this is because a bit-flip generated in the slave of MS1 flip-flop still has to arrive at least setup time before the clock edge at the MS2 flip-flop. For long delays, the bit-flip does not arrives in time to be latched at the next stage and consequently it does not become a soft error at the next flip-flop stage of the pipeline. By plotting a histogram of the integrated circuits path delays, it is possible to analyze the amount of flip-flops with high and low TVFs.

V. CONCLUSIONS

Master slave flip-flops present a different TVF values for the master and slave latches. Master latches present a higher TVF compared to slave latches. The slave latches present TVF that strongly depend to the operating frequency of the circuit and the propagation delay between the flip-flops stages. The slave latch TVF can be easily reduced due to the combinational path delay. If a circuit must be redesigned to improve reliability, the master latches must definitely incorporate techniques for radiation hardened as redundant transistors, and the logical paths must have minimal slack as possible.

REFERENCES

[1] L. Anghel and M. Nicolaidis, Defects Tolerant Logic Gates for Unreliable Future Nanotechnologies, Lecture Notes in Computer Science, v. 4507, Berlin, 2007, pp. 422-429.

[2] M. Orshansky, S. Nassif and D. Boning, Design for Manufacturability and Statistical Design, Springer, 2008.

[3] R. C. Bauman, Radiation-induced soft errors in advanced semiconductor technologies. Published in: IEEE Transaction on Device and Materials Reliability, vol. 5, no. 3, pp. 305-316, 2005.

[4] M. V. O'BRYAN, Single Event Effects – NASA Radition Effects & Analysis. Acess in: http://radhome.gsfc.nasa.gov/radhome/see.htm. Acesso em: outubro, 2014.

[5] A. Bramnik, A. Sherban and N. Seifert, Timing Vulnerabiblity Factors of Sequential Elements in Moderns Microprocessors. IEEE 19th International On-line Test Symposium (IOLTS), pp. 55-60, 2013.

[6] N. Seifert and N. Tam. Timing Vulnerability Factors of Sequentials. IEEE Transactions on Device and Materials Reliability, Vol. 4, no. 3, 2004.

[7] S. Kim and A. K. Somani, "Soft Error Sensitivity Characterization for Microprocessor Dependability Enhancement Strategy," Proceedings of the International Conference on Dependable Systems and Networks (DSN), 2002.

[8] SYNOPSYS. Avalaible in: http://www.synopsys.com.

[9] PTM. Predictive Technology Model. Avalaible in: http://ptm.asu.edu/.

[10] S. Mukherjee. Architecture Design for Soft Errors. Ed. Morgan Kaufmann, 1st edition, 2008.

[11] M. M. Ghahroodi, M. Zeolinski, R. Wong and S. Wen, Timing Vulnerability Factors of Ultra Deep-Sub-Micron CMOS. 6TH IEEE European Test Symposium, 2011.

[12] T. Heijmen, Soft-Error Vulnerability of Sub-100-nm Flip-Flops. 14th IEEE International On-line Test Symposium, 2008.

[13] G. C. Messenger, Collection of charge on junction nodes from ion tracks, IEEE Transactions on Nuclear Science, 1982, pp. 2024-2031.

[14] G. Hubert et al. SET and SEU Analyses Based on Experiments and Multi-Physics Modeling Applied to the ATMEL CMOS Library in 180 and 90-nm Technological Nodes, IEEE Transactions on Nuclear Science, V. 6, 2014.

[15] H. T. Nguyen, Y. Yagil, N. Seifert and M. Reitsma. Chip-Level Soft Error Estimation Method. IEEE Transactions on Device and Materials Reliability, vol. 5, no. 3, 2005.

[16] M. D'Alessio, M. Ottavi and F. Lombardi, Design of Nanometric CMOS Memory Cell for Hardening to a Single Event with a Multiple-Node Upset, IEEE Transactions on Device and Materials Reliability, vol. 14, no. 1, 2014.

[17] J. Fang and S. Sapatnekar, The Impact of BTI Variations on Timing in Digital Logic Circuis, IEEE Transactions on Device and Materials Reliability, V. 13, 2012.

Using Intra-line Level Pairing for Graceful Degradation Support in PCMs

Marjan Asadinia* and Hamid Sarbazi-Azad[†‡]

*Science and Engineering Department,
Sharif University of Technology, International Branch, Kish, Iran
Email: asadinia@kish.sharif.edu

[†]Computer Engineering Department,
Sharif University of Technology, Tehran, Iran
Email: azad@sharif.edu

[‡]School of Computer Science,
Institute of Research in Fundamental Sciences (IPM), Tehran, Iran
Email: azad@ipm.ir

Abstract—In Phase-Change Memory (PCM), the number of writes a cell can take before wearing-out is limited and highly varied due to unbalanced write traffic and process variation. After the failure of weak cells and in presence of large number of failed lines, some techniques have been proposed to further prolong the lifetime of a PCM device by remapping failed lines to spares and salvage a PCM device with graceful degradation. Others rely on handling failures through inter-line pairing. Observations reveal that most of cells in a line are healthy when the line is marked as faulty by any of these proposals. To overcome this deficiency, we propose Intra-line Level Pairing (ILP), a technique that mitigates the problem of fast failure of lines by coupling faulty parts of a line onto other healthy parts of the same line. The target part of the line is programmed in the Multi-Level Cell (MLC) mode to keep data of both the faulty and target parts. Evaluation results for multi-threaded and multi-program workloads reveal noticeable improvement in time-to-failure and performance over existing techniques. Note that ILP is also orthogonal to all known line-level and page-level techniques.

Keywords-Phase Change Memory; Multi-Level Cell; Intra-line Level Pairing.

I. INTRODUCTION

Compared to competitive non-volatile memories, Phase Change Memory (PCM) technology offers a highly-scalable alternative to DRAM [10], [16], [21]. A PCM cell uses chalcogenide material which is composed of Ge, Sb, and Te. The material has two different states (low resistive crystalline state and high resistive amorphous state). Although PCM has negligible leakage, better scalability, and comparable read speed to DRAM, it suffers from lifetime problem: a PCM cell can wear-out after a limited number of writes (10^7-10^8 per cell in 32nm prototypes [1]). The cell is stuck at 1 or 0 state after a cell reaches its lifetime limit. Process variation is another challenging issue on PCM cells' lifetime and results in early failures of many cells compare to a cell built with less process variation. Past researches present a differential write scheme [5], [21], or wear-leveling mechanism in order to evenly spread the writes [15]. Such

approaches do not guarantee better memory lifetime, but write traffic to individual cells is reduced.

To recover from stuck-at faults in PCM, some prior works use operating system page remapping [7] and error correcting schemes [17], [18], [20], [13]. Among all of these methods, only DRM [7] try to reuse pages or lines that have already been disabled to correct others that are still enabled. This solution can simply recycle two faulty lines to a healthy line. In this scheme, recycling depends on the cell failure pattern which relies on the fact that the two lines must not have faulty bits in the same position. This dependency leads to DRM can not always find a compatible line to recover a faulty line. For this reason, it is more important to use an error protection mechanisms like ECP [17] and SAFER [18] for individual data lines within the PCM chip. A bit failure is corrected by recording its position and value in ECP. However, large storage overhead (12% for tolerating up to 6 bit failures in a 64B line) and the reserved pointers provide main deficiencies for ECP. SAFER also partitions a block into some groups. If each group consists of one faulty cell, the cell can store data; when the value being written into the cell is opposite to the cell's stuck-at value, a flag bit is set to indicate the happening of inversion and data is stored in an inverted form.

Our observations reveal that all previous researches has costly allocation of large number of pointers to a faulty PCM line. They also rely on OS support and has the necessity of multiple accesses to different PCM lines or error correcting unit, so they can cause to degrade memory performance and PCM lifetime. Moreover, in presence of highly coefficient variation of process technology, current wear-leveling algorithms cannot do much, they just reduce cells lifetime variation.

In this line, figure 1 illustrates memory capacity degradation due to failures that cannot be tolerated as memory is written over its lifetime for error recovery mechanisms such as DRM, ECP, FREE-p [20], and SAFER. The percentage

978-1-4799-8720-7/15 $31.00 © 2015 IEEE

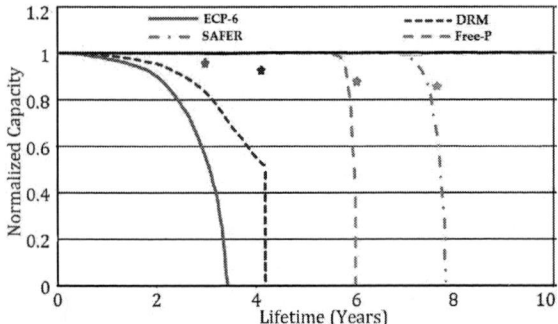

Figure 1. Memory capacity degradation as a function of lifetime under DRM, ECP, FREE-p, and SAFER schemes. Percentage of healthy cells in faulty lines are shown with "X" for each scheme when 50% of memory lines are corrupted.

of healthy cells in the lines marked as faulty under different mechanisms (shown by X for each technique) is also shown when memory size is reduced to half. It means when a line is marked as faulty, it still has many healthy cells; so, exploiting these healthy cells can result in longer memory lifetime and lower capacity degradation. In this paper, we propose a new scheme called *Intra-line Level Pairing* (ILP) to recover from stuck-at faults; it relies on pairing faulty parts of a line with other parts of that line which are healthy. This way, the line can continue its normal operation before it reaches to the state that faults cannot be tolerated by pairing different parts of the line. Note that ILP is orthogonal to all previous schemes working at line-level or page-level. In ILP, the target part of the line is programmed in Multi-Level Cell (MLC) mode to keep data of both faulty and target healthy parts. This way, when a part is paired to a healthy part, the faulty part can use the lifetime of healthy parts and results in improving overall memory lifetime while maintaining performance and energy at SLC level.

II. RELATED WORK

Line-level Error Protection Schemes. In ECP [17], errors are corrected by logging the bit-errors in a line. When bit failure occurs, position of the bit-error is determined by a pointer and an additional bit is used to hold the correct value. Moreover, ECP can handle up to 6 bit-errors which results in about 6.5 years lifetime and 12% storage overhead.

In order to extend the lifetime and reduce the overhead of all previous approaches, some proposals such as Free-p [20], SAFER [18], RePRAM [4], Zombie [2], Aegis [6] and LLS [8] are proposed. However, these studies also suffer from increasing bandwidth demand as a result of increasing the number of bit failures. They also impose higher latency and memory complexity due to sequential reads and necessity of good structures for managing extra reads and writes.

Page-level Error Protection Schemes. In DRM [7], when the first uncorrectable error occurs, OS-support technique for

further improving PCM lifetime is used. After each failure, the indication bit of the corresponding byte is set. Then, the corresponding page is added to a waiting list of faulty pages by OS. For page allocation, pair of faulty pages is selected in such a way that one of these two pages is maintained as the primary copy, and the other one is used as backup. It is more important that these selected pages should not have been the same faulty bit position.

Another scheme is PAYG error correction [13]. In PAYG, each line corresponds to one dedicated ECP entry and the remaining entries are kept in a common pool to be used on demand by all lines. It is notable that only few PCM lines in PAYG experience more than one permanent fault. Therefore, PAYG can handle more errors compare to ECP.

III. INTRA-LINE LEVEL PAIRING MECHANISM

This section describes our error detection and correction mechanisms which is used at each line. We then give a discussion about design trade-offs presented by our scheme.

A. Line Level Error Detection and Correction

In our proposed pairing scheme like other techniques, for error detection, we perform read-after-write to determine whether the write is done properly. Read-after-write has a negligible impact on system performance because read is much faster than write and PCM writes are much more expensive from latency and energy point of view compared to its reads.

For error correction, at each memory line, we use ECP-6. It means that ECP can correct up to 6 faulty bits per line. After issuing of read-after-write and detection of improper write operation. The ECP pointer can record the position of the bit error and the correct bit value is stored in its corresponding bit storage. Whenever all associated pointers of a line are allocated, any further bit failure is detected as an uncorrectable error and cause to detect that line as a faulty line. In this situation, our recovery mechanism is invoked by memory controller.

B. Proposed Method

A PCM line is assumed to tolerate up to 6 errors using ECP-6 for each line. It is seen that, even after applying ECP-6 and detection of the faulty status of a line, there exists some parts of line that are far from their endurance limits and are healthy. Therefore, we propose intra-line level mechanism as an alternative mechanism where a whole line is not marked as faulty at the time ECP cannot tolerate more faults. So, first we partition the line into some parts and then pair the faulty parts of it with some healthy parts of the line. When we cannot find any healthy part to host a faulty part, the line is marked as faulty; we can further pair such a faulty line with another healthy line in the same page.

Figure 2 illustrates an example for intra-line-level pairing where a line reaches its endurance limit and ECP-6

978-1-4799-8720-7/15 $31.00 © 2015 IEEE

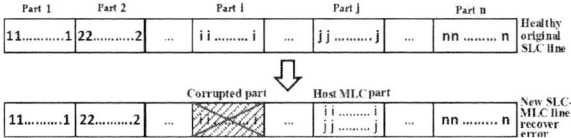

Figure 2. Pairing a faulty part i with a healthy part j; after pairing part i is marked as faulty and both i and j parts are stored in 2-bit MLC mode.

cannot help anymore. At this point, the controller invokes our pairing mechanism. The line is partitioned to n parts. Assume part i contains a number of faulty bits that cannot be tolerated by ECP. A healthy part should be found to pair with part i. Assume part j is found to be a healthy part, and is selected as the pairing target; then both parts (i and j) are stored in 2-bit MLC mode in the corresponding cells of part j. Note that, if there exists more than one pairing targets which are healthy, we have assumed to pair the faulty part with a target part which is farthest to the faulty part in the line. Indeed, spatial locality of write requests can make this decision. When a given part is written, write requests to the neighboring parts in upcoming cycles are more probable.

Our proposed method consists of determining the address of target healthy part in *healthy part selection*, and pairing process of faulty and healthy parts in *pairing mechanism*.

1) Healthy Part Selection.: It determines that a given faulty part with which healthy part of the line can be paired. From all healthy parts, we give the maximum freedom for choosing the target healthy part. However, If target part is near to its endurance limit, programming it in MLC mode lead to accelerate its failure, it means that target part could continue working well if it was not selected as a target part for pairing. Therefore, pairing the faulty part with a target healthy part that has a short time to wear out can be counterproductive. It is clear that we should select the target part which has not been under write stress recently and it should be the healthiest part. In this line, all parts include *valid/invalid* bit which indicates that this partition is *healthy/faulty* and it is practical to check all parts to find the healthiest one for pairing since the number of healthy parts of a line in a page is limited. More precisely, all parts of line can tolerate up to a limited number of cell failures, so one of the main important points is that the selection of pairing target should be based on cell failures in that part. Due to large storage overhead and metadata information, we cannot use dynamic partitioning. Based on our sensitivity analysis (Section IV-D), we fix the number of parts per line to 16 for 128B line size. We expect our method to achieve a better time-to-failure and performance compare to the others that uses some tables in order to find the most compatible pair and ECP-6 that imposes large storage overhead.

2) Meta-data Information.: Each part of a line uses two bits, named SLC/MLC and *valid/invalid*. More precisely, *valid/invalid* bit takes invalid state, when the ECP pointers of

the line cannot cover the bit failures in that part. Thereafter, the faulty part should be paired with a healthy part and stored in MLC mode. Therefore, each part needs to use a *SLC/MLC* bit to indicate if it is programmed in MLC mode. The address of the healthy part which is indicated as a pairing target should be specified; so the invalidated cells are used to keep the address of the healthy part hosting the data of the faulty part (if they cannot keep the address, we mark the whole line as faulty). As we have already discussed, healthy part selection indicates a healthy target part for pairing. However, it may fail to find a candidate when all parts are invalidated and are MLCed in the moment a pairing is requested. In that case, no intra-line-level pairing takes place and the line may be redirected to other healthy lines (using line-level mechanism). We assume having 16 parts each of 8 bytes for a 128B line; each partition has 2 bits (valid/invalid bit and *SLC/MLC* bit), so ILP imposes 3.12% storage overhead. It is intuitive that more partitions increase the chance of finding a healthy part, but it also increases the size of meta data information.

3) Pairing Algorithm.: After finding the healthy target part by selection algorithm, intra-line level pairing request is invoked. To store data bits of both faulty and target parts, healthy target part is programmed in 2-bit MLC mode. Our memory controller allows reallocating of both of parts to the other healthy parts when failure of a MLC part takes place.

Note that MLC, read/write requires repetitive sensing/programming steps which proportionally cost more access latency/energy. Furthermore, repetitive writes negatively impact lifetime and lead to rapid cell wear-out (in 2-bit-MLC, each cell failure causes two bit errors). When someone selects MLC PCM, his/her only concern is capacity; other parameters (e.g. lifetime, energy and performance) become worse. SLC is technology of choice when above-mentioned important parameters are considered. So, ILP only uses MLC scheme to make use of lifetime of healthy parts to improve overall memory lifetime while maintaining performance and energy at SLC level (end even enjoys better lifetime of individual SLC cells).

IV. EVALUATION

In this section, first we describe our simulation methodology and parameters. We then address the evaluation results of our proposed memory system compared to the other hard error recovery schemes for PCMs.

A. Evaluation Setting

Infrastructure. We simulated our proposed method by using microarchitectural level simulation of an out-of-order processor with UltraSPARCIII ISA using GEMS simulator [12] based on Simics toolset [11]. The entire memory hierarchy and memory controller is modeled in our simulator. Reads have higher priority compare to writes and only when there is no reads, write is scheduled. The integration

978-1-4799-8720-7/15 $31.00 © 2015 IEEE
529

Table I
MAIN CHARACTERISTICS OF SIMULATED CMPs.

Processor and Caches	
Cores	4-cores, SPARC-III, out-of-order, 4 GHz, Solaris 10 OS
L1 I/D	32kB private, 2-way, 64B, LRU, write through, 2-cycle hit
L2	2MB shared, 4-way, 64B, LRU, write back, 2-cycle tag, 5-cycle data hit, 16-cycle core to, L2MOESI directory
DRAM	128MB shared, off-chip, 8-way, 128B, LRU, write back,
L3	200-cycle hit, 64-cycle core to L3
Main Memory	
Controller	On-chip, 24-entry R/W queues, 64-cycle mc to bank, read-prioritized and write burst (if W queue full) scheduling
PCM Main Memory	4GB, 1KB page, 64B line, single-rank, 8 banks, SLC read 150ns (600 cycles), RESET: 125ns (500 cycles), 300uA, SET: 250ns (1000 cycles), T_{off}: 5ns, MLC read 250ns (1000 cycles), MLC write [9], [14]: "01": i/F1/F2 = 2/0.375/0.625, 8 iterations; "10": i/F1/F2 = 2/0.425/0.675, 6 iterations; "00": 1 iteration; "11": 1 iteration

Table II
CHARACTERISTICS OF THE EVALUATED WORKLOADS.

PARSEC-2 (2009), Multi-threaded workloads					
Workload	**WPKI***	**RPKI***	**Workload**	**WPKI***	**RPKI***
Blackscholes	0.003	0.03	Bodytrack	0.003	0.03
Caneal	1.12	1.31	Dedup	0.41	0.43
Facesim	0.65	0.24	Ferret	0.65	0.67
Fluidanimate	0.76	0.47	Freqmine	0.04	0.07
Raytrace	0.008	0.02	Streamcluster	0.01	0.03
Swaptions	0.002	0.02	Vips	0.05	0.07
X264	0.04	0.04			

SPECCPU (2006), 4-application multi-program workloads		
Workload	**WPKI***	**RPKI***
Mix1: xalancbmk,omnetpp,bzip2,mcf	2.4	2.23
Mix2: milc,leslie3d,GemsFDTD,lbm	6.64	10.11
Mix3: mcf,xalancbmk,GemsFDTD,lbm	3.69	3.08
Mix4: mcf,GemsFDTD,povray,perlbench	1.99	0.64
Mix5: mcf,xalancbmk,perlbench,gcc	2.2	0.8
Mix6: GemsFDTD,lbm,povray,namd	1.91	2.48
Mix7: gromacs,namd,dealII,povray	0.27	0.25
Mix8: perlbench,gcc,dealII,povray	0.53	1.08
Mix9: namd,povray,perlbench,gcc	0.53	0.66

*WPKI/RPKI represents memory write/read per kilo instructions.

Table III
TIME-TO-FAILURE OF DIFFERENT MECHANISMS.

CoV.	ECP-6	Free-P	SAFER	ILP
0.25	4.32	5.19	6.25	7.76
0.35	1.18	2.66	3.87	4.54

of our evaluated schemes with Start-and-Gap wear-leveling algorithm [15] and differential write scheme [5], [21] are considered. By using 3 levels of caches and PCM main memory, a 4-core 4GHz CMP system is modeled. Our simulation parameters are characterized in Table I. In order to alleviate the pressure on PCM main memory bandwidth, we use baseline with 128MB shared write back DRAM cache. Besides, single 4GB SLC PCM DIMM main memory is considered and divided into 8 banks which share 8 PCM chips.

We use both synthetic and real multi-threaded and multi-program (mix) workloads. Synthetic workloads are generated with different temporal and spatial write traffics. Also, complete set of multi-threaded programs in PARSEC-2 suite [3] and nine 4-application multi-programmed workloads from SPECCPU2006 benchmarks [19] are used as real workloads to evaluate different methods. These workloads are collected for different read/write accesses per kilo instructions (details are shown in Table II).

Evaluated Architectures. We implemented our proposed method against intra-line level error correction schemes presented in [17], [18], [20]. Two other error correction schemes that work at page level are also included [7], [13].

Metrics. We measured the elapsed time between the system startup time to the time that PCM capacity is reduced to less than 50% of its maximum capacity (i.e., 2GB in our 4GB system). We call this endurance metric as *time-to-failure* (or lifetime). We also consider Execution Time of the evaluated systems for performance comparison.

B. Synthetic Traffic

As large endurance limit of 10^8 per cell requires time-consuming simulation, past researches ignore accurate full-system simulation and adopt a practical alternative in order to evaluate the PCM capacity in a large memory. We also assume a PCM memory with 4K pages each with 1KB data

cells, and corresponding metadata information. Besides, we consider a model where the endurance of each individual cell is set by a random variable with a normal distribution of 10^8 writes (in average) and a standard deviation of 0.25×10^8 in order to model process variation. Lifetime versus memory capacity provided by the evaluated systems is shown in Table III. Results reveal that in contrast to ECP, Free-P and SAFER, our method provides a larger memory capacity for cell endurance. Table III (second row) shows the capacity versus lifetime results for CoV of 0.35. Evaluating the impact of high process variation also shows that our method offers better capacity and lifetime gain compared to other evaluated intra-line level error correction schemes.

By increasing the capability of pairing to inter-line level, we expect much better PCM lifetime and capacity. This is not a concern of this work as we are focusing on intra-line level error correction as an orthogonal way to other inter-line level or page-level correction schemes.

C. Real Workloads

In this part, we evaluate performance and endurance metrics under real workloads and conduct sensitivity analysis on partitioning size.

1) Performance: The effectiveness of our proposed method against Free-P, SAFER, and ECP schemes is depicted in Figure 3.a. We assumed a process variation with CoV of 0.25 for cell endurance. Figure 3 shows that our mechanism is lifetime and performance efficient compared to other evaluated systems. Mix8 and Mix9 are examples

978-1-4799-8720-7/15 $31.00 © 2015 IEEE

of heavy write traffics, in this situation, finding a target healthy part for pairing in ILP is more difficult due to the fact that all cells are under more write pressure. Compared to these write intensive traffics, blackscholes and raytrace workloads have few writes, so the capacity reduction due to write operations have little performance impact. Mix1 and Mix4 are examples of workloads which has many more reads than writes, so pairing mechanism has a small impact as a result of performance bottleneck shifts from writes to reads.

2) Endurance: We defined lifetime limit as a time that half of memory capacity cannot be used under different protection mechanisms. So, each application should be simulated up to the defined time and elapsed time is measured for ILP and depicted against ECP-6, Free-P and SAFER schemes in Figure 3.b. The gain achieved in lifetime and performance by ILP is due to the extra cell failures that ILP can revive. Figure 3.c shows the percentage of recovered cells per memory line for each system. ILP shows higher recovery rate with respect to other schemes.

D. Sensitivity to Partition Size

To evaluate the effect of partitioning on the efficiency of our proposed pairing scheme and its overhead, Figure 4 reports time-to-failure and storage overhead for different partitioning sizes: 4, 8, 16, 32 parts per 128B line. In less partitions, there are more parts of line which are faulty and we cannot find target healthy parts for pairing, so we'll have less recovery and time-to-failure but lower storage overhead. On the other hand, when more partitions are used, more opportunity exists in finding the target healthy part. This supports memory capacity, reliability issues and causes more performance improvement in cost of increasing storage overhead and complexity of finding suitable target for pairing within each line. Since partition size of 16 provides acceptable failures coverage with an acceptable storage overhead and searching complexity for finding healthy target parts, so we assume partitioning size of 16 for each line.

E. Comparison to Page-level Schemes

In this part, we compare ILP with other schemes such as DRM [7], and PAYG [13]. These methods tolerate faults at page level and consider a SLC PCM main memory as a baseline. The geometric average values for all workloads shows that ILP improves the execution time by 32% and 14%, and lifetime by 44% and 27% compare to DRM and PAYG, respectively.

V. CONCLUSION

We proposed an intra-line level pairing (ILP) scheme for PCMs that applies pairing on-demand when a part of PCM line reaches its first uncorrectable bit failure. ILP finds a healthy part of PCM line to be paired with a faulty part allowing both parts to be stored in the healthy cells in MLC mode. Experiments showed that our proposed

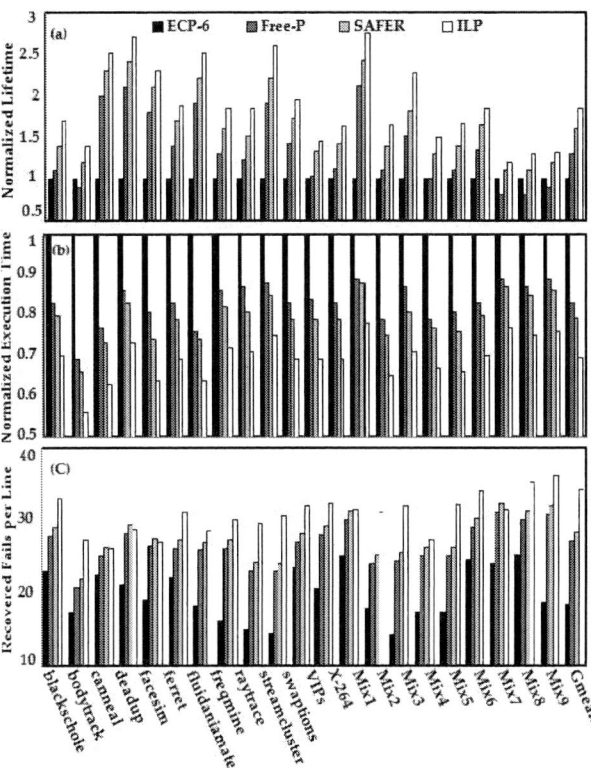

Figure 3. Evaluation results of ECP-6, SAFER, Free-P and ILP mechanisms: (a) normalized lifetime; (b) normalized execution time; and (c) average number of recovered errors per line.

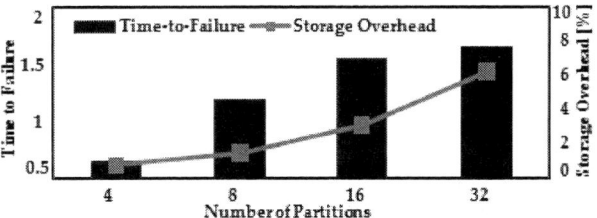

Figure 4. Partitioning effect on lifetime and storage overhead.

method improved system performance and resulted in lifetime improvement over the state-of-the-art error correction schemes.

REFERENCES

[1] "Emerging Research Devices, ITRS," Nov. 2011. [Online]. Available: http://www.itrs.net/

[2] R. Azevedo, J. D. Davis, K. Strauss, P. Gopalan, M. Manasse, and S. Yekhanin, "Zombie memory: Extending memory lifetime by reviving dead blocks," in *ACM SIGARCH Computer Architecture News*, vol. 41, no. 3. ACM, 2013, pp. 452–463.

[3] C. Bienia and K. Li, "PARSEC 2.0: A new benchmark suite for chip-multiprocessors," in *MoBS*, June 2009.

[4] J. Chen, G. Venkataramani, and H. Huang, "RePRAM: recycling PRAM faulty blocks for extended lifetime," in *DSN*, June 2012, pp. 1–12.

[5] S. Cho and H. Lee, "Flip-N-Write: a simple deterministic technique to improve PRAM write performance, energy and endurance," in *MICRO*, Dec. 2009, pp. 347–357.

[6] J. Fan, S. Jiang, J. Shu, Y. Zhang, and W. Zhen, "Aegis: Partitioning data block for efficient recovery of stuck-at-faults in phase change memory," in *Proceedings of the 46th Annual IEEE/ACM International Symposium on Microarchitecture*. ACM, 2013, pp. 433–444.

[7] E. Ipek, J. Condit, E. B. Nightingale, D. Burger, and T. Moscibroda, "Dynamically replicated memory: building reliable systems from nanoscale resistive memories," in *ASPLOS*, Mar. 2010, pp. 3–14.

[8] L. Jiang, Y. Du, Y. Zhang, B. Childers, and J. Yang, "LLS: cooperative integration of wear-leveling and salvaging for PCM main memory," in *DSN*, June 2011, pp. 221–232.

[9] L. Jiang, B. Zhao, Y. Zhang, J. Yang, and B. Childers, "Improving write operations in MLC phase change memory," in *HPCA*, Feb. 2012, pp. 1–10.

[10] B. C. Lee, E. Ipek, O. Mutlu, and D. Burger, "Architecting phase change memory as a scalable DRAM alternative," in *ISCA*, June 2009, pp. 2–13.

[11] P. Magnusson, M. Christensson, J. Eskilson, D. Forsgren, G. Hallberg, J. Hogberg, F. Larsson, A. Moestedt, and B. Werner, "Simics: A full system simulation platform," *Computer*, vol. 35, no. 2, pp. 50–58, Feb. 2002.

[12] M. M. K. Martin, D. J. Sorin, B. M. Beckmann, M. R. Marty, M. Xu, A. R. Alameldeen, K. E. Moore, M. D. Hill, and D. A. Wood, "Multifacet's general execution-driven multiprocessor simulator (GEMS) toolset," *CAN*, vol. 33, no. 4, pp. 92–99, Nov. 2005.

[13] M. K. Qureshi, "Pay-As-You-Go: low-overhead hard-error correction for phase change memories," in *MICRO*, Dec. 2011, pp. 318–328.

[14] M. K. Qureshi, M. Franceschini, and L. Lastras-Montano, "Improving read performance of phase change memories via write cancellation and write pausing," in *HPCA*, Jan. 2010, pp. 1–11.

[15] M. K. Qureshi, J. Karidis, M. Franceschini, V. Srinivasan, L. Lastras, and B. Abali, "Enhancing lifetime and security of PCM-based main memory with start-gap wear leveling," in *MICRO*, Dec. 2009, pp. 14–23.

[16] M. K. Qureshi, V. Srinivasan, and J. A. Rivers, "Scalable high performance main memory system using phase-change memory technology," in *ISCA*, June 2009, pp. 24–33.

[17] S. Schechter, G. H. Loh, K. Straus, and D. Burger, "Use ECP, not ECC, for hard failures in resistive memories," in *ISCA*, June 2010, pp. 141–152.

[18] N. H. Seong, D. H. Woo, V. Srinivasan, J. A. Rivers, and H.-H. S. Lee, "SAFER: Stuck-at-fault error recovery for memories," in *MICRO*, Dec. 2010, pp. 115–124.

[19] C. D. Spradling, "SPEC CPU2006 benchmark tools," *CAN*, vol. 35, no. 1, pp. 130–134, Mar. 2007.

[20] D. H. Yoon, N. Muralimanohar, J. Chang, P. Ranganathan, N. Jouppi, and M. Erez, "FREE-p: Protecting non-volatile memory against both hard and soft errors," in *HPCA*, Feb. 2011, pp. 466–477.

[21] P. Zhou, B. Zhao, J. Yang, and Y. Zhang, "A durable and energy efficient main memory using phase change memory technology," in *ISCA*, June 2009, pp. 14–23.

2015 IEEE Computer Society Annual Symposium on VLSI

Using Configurable Bit-Width Voters to Mask Multiple Errors in Integrated Circuits

Thiago Berticelli Ló

Departamento de Ensino, Pesquisa e Extensão
Instituto Federal de Educação, Ciência e Tecnologia Sul-rio-grandense, IFSUL
Charqueadas/RS, Brazil
thiagolo@charqueadas.ifsul.edu.br

Fernanda Lima Kastensmidt, Antonio Carlos Schneider Beck

Instituto de Informática – PGMICRO/PPGC
Universidade Federal do Rio Grande do Sul, UFRGS
Porto Alegre/RS, Brazil
{fglima,caco}@inf.ufrgs.br

Abstract— **N-Modular Redundancy (NMR) with majority voters has been widely used to increase reliability. While bit-voters perform bit by bit comparisons, which are the most basic and fast voting scheme, word-voters consider all bits in parallel to determine the final output, increasing data integrity but likewise the area. This paper proposes to merge the advantages of both aforementioned voters and fill the gap in between the existent designs, by grouping voters into a structure of configurable sets of bits (i.e.: the proposed system is composed of a set of X voters of Y bits, where X and Y are configurable). We explore the design space by considering scenarios with multiple errors and project restrictions such as area, delay, error rate, number of modules, and probability of corrupt outputs. We will show that there are many cases where the proposed voter fits the aforementioned restrictions better than the others.**

Index Terms—**majority voter; hardware modular redundancy; reliability; multiple error masking**

I. INTRODUCTION

Neutron-induced soft errors have been a problem in modern designs operating in dependable applications [1]. Due to the reduction in transistor sizing, voltage supply and increase of operating frequency, nanometer-scale technologies are more likely to malfunctions due to soft errors. Therefore, reliability becomes a major concern and fault tolerant techniques must be applied to ensure the correct operation of the circuit.

In order to achieve reliability, it is mandatory to mask soft errors by using hardware redundancy with voters. Hardware (or spatial) redundancy is the replication of the original circuit, resulting in a system composed of N identical redundant modules connected through majority voters. Usually N is an odd number to ensure majority, and the most common configuration of N-modular redundancy (NMR) is the well-known Triple Modular Redundancy (TMR or 3-MR).

Therefore, considering that majority voters can perform bit by bit or word by word comparisons, we propose a configurable bit-width voter to cope with multiple errors while ensuring failure detection in cases of majority values cannot be found. The proposed voter is composed of a set of X voters of Y bits, where X and Y are configurable. Requirements of area, delay, error masking capability and correctness are evaluated and the proposed configurable voter is verified under multiple error case scenarios. Its results are compared to the bit and word implementations and show that the use of NMR systems with the proposed configurable voter present higher masking coverage in different scenarios.

When a neutron particle strikes the integrated circuit, it can generate secondary charge particles that can ionize the silicon, depositing some charge. When a drain of a transistor in off-mode collects this charge, a transient voltage pulse can occur in that stroke node. This transient pulse can be seen as Single Event Transient (SET) in a logic gate (Fig. 1(a)), or it may generate the bit-flip of a memory or register cell, known as Single Event Upset (SEU), Fig. 1(b).

Even though SET or SEU are single event faults, they can generate multiple errors: Fig. 1 (a) illustrates a case when a single SET may provoke two errors (E) in the next register due to the fanout in the logic; and Fig. 1(b) illustrates a case when a single SEU in a register may also provoke two errors (E) in the next register, but now because the Boolean function characteristics of the circuit.

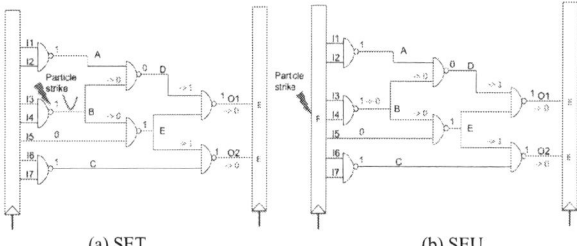

(a) SET   (b) SEU

Fig. 1. Examples of multiple errors due to single faults (a) SET (b) SEU.

In addition, when using hardware redundancy, multiple errors can also appear in multiple copies of the redundant circuits due to charge sharing or accumulation of faults over time. Single-event-induced charge sharing is characterized when one primary node receives the primary impact, and the neighboring secondary nodes collect part of charge deposited by the particle that struck the primary node [2]. These secondary nodes can be in the same logic gate or in adjacent logic gates in the floorplanning layout, but belonging to distinct circuit redundant copies, as shown in Fig. 2 (a). In this example, a TMR system is considered. Accumulation of faults is also common, mainly under high flux of particles and when system reset does not occur very often. It may lead to multiple errors in multiple circuit modules as shown in fig. 2(b). Consequently, in scenario with multiple errors, TMR may not always present an acceptable level of error masking [3]. Therefore, N-MR must be applied. In this paper, we consider NMR systems with a range of output bits with error (E) that may vary from 0% to 30%.

978-1-4799-8720-7/15 $31.00 © 2015 IEEE   533

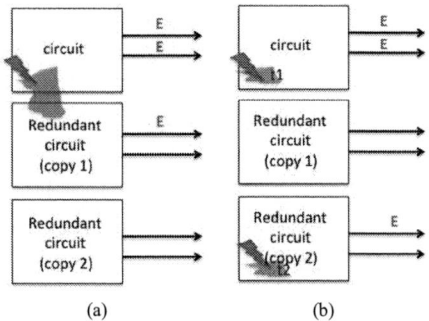

(a)   (b)

Fig. 2. Examples of multiple errors in TMR due to (a) charge sharing (b) accumulation of faults at time t1 and t2.

In this scenario, it is important to investigate different configurations of majority voters able to mask multiple errors that may accumulate in nano meter integrated circuits due to multiple faults and share sharing. There is a vast area to explore in terms of sizes of voters, bits distribution and number of redundancies as it will be presented and discussed in this paper.

II. MASKING ERRORS WITH MAJORITY VOTERS

A. Standard Implementations

There are different ways to implement majority voters. The most common models are the bit- and word-voters. Bit-voter is a majority voter that takes the majority of the inputs to be the output value, usually performed on a bit-by-bit basis [4][5]. For a system with *m* outputs, conventional NMR systems use N single-bit voters (Fig. 3). Usually odd numbers of voters (N) are used. Therefore, there will always be a result, even if the final value is incorrect. This kind is the most employed voter as its hardware implementation is quite simple. A large part of literature is devoted to reliability modeling of TMR systems [6][7].

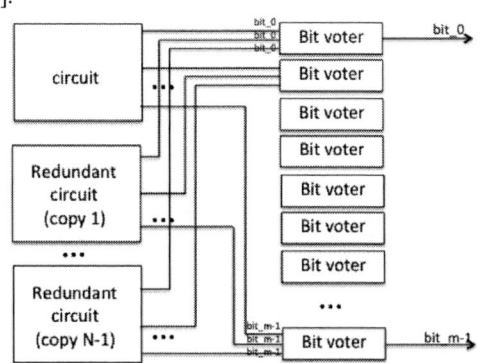

Fig. 3. Bit-voter of a NMR system with m bits outputs

The word-voter always considers the whole word output at once. Therefore, the chosen output is the word, which appears the most as output of the modules. In this model, there is an additional control signal (failure signal) that invalidates the output in cases there is no majority value. In response to the failure signal, appropriate action (depending on the application)

can be initiated (e.g.: re-execution starting from a checkpoint) [8]. In this way, a system with the word-voter is capable of maintaining data integrity (reduced probability of corrupt outputs). The authors in [8] have shown that the data integrity of a TMR system composed of a word voter improves at least by an order of magnitude over conventional systems with bit-by-bit voting. Thus, the word-voter has enhanced capabilities that can be used to design TMR systems that are protected against common-mode (CMFs) [9] and multiple module failures. Fig. 4 shows a word-voter of (with a word of m bits) in a NMR system.

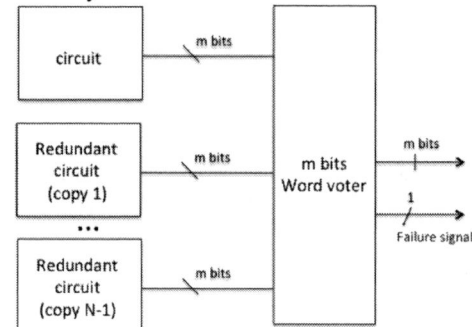

Fig. 4. Word-voter of a NMR system with m bits outputs

B. Tradeoff of using Bit- and Word- Majority Voters

To evaluate the tradeoff between bit and word-voters, three scenarios of error distribution at the outputs of a NMR system under multiple errors are investigated. A redundant system composed of five modules (5-MR) is considered, with outputs of 6-bits each. Each scenario has 7 output bits in error, out of a total of 30 (23%), but with different distributions. The gold output is "000000". Therefore, any ′1′ bit at the output means that faults have generated an error at the correspondent output bit.

Table I shows the results for the first scenario, where the errors are distributed in 3 of the 5 modules and the majority of the errors occur at the same bit: #4. One can observe that the word-voter model is able to mask the faults as expected, since the output of two modules (#3 and #4) are correct ("000000"), while the bit-voter incorrectly masked the errors at the bit #4.

TABLE I.   CASE-STUDY I: 5-MR SYSTEM WITH A 6-BIT OUTPUT VOTED BY BIT- AND WORD-VOTERS WITH 23% OF ERRONEOUS OUTPUT BITS. THE WORD-VOTER CAN SUCCESSFULLY MASK THE ERRORS

Module Number	Word (6 bits)						Value (dec.)
	Bit 5	Bit 4	Bit 3	Bit 2	Bit 1	Bit 0	
#1	0	1	0	0	1	0	18
#2	0	1	0	1	0	0	20
#3	0	0	0	0	0	0	0
#4	0	0	0	0	0	0	0
#5	1	1	0	0	0	1	49
Bit-Voter	0	1	0	0	0	0	**16**
Word-Voter	0	0	0	0	0	0	**0**

Table II shows the results for the second scenario, in which every module has at least one error bit at the output, but not necessarily at the same bit output. One can observe that the bit-voter is able to mask all the errors and to provide the correct output. However, the opposite happened with the word-voter, since the output of two modules (#1 and #2) presented the value of "001000" (8, in decimal). As they are the majority, the wrong value is selected. In this case, the bit-voter is able to mask the errors; while the word-voter is not. In addition, the word-voter informs the wrong output, which is a failure.

TABLE II. CASE-STUDY II: 5-MR SYSTEM WITH A 6-BIT OUTPUT VOTED BY BIT- AND WORD-VOTERS WITH 23% OF ERRONEOUS OUTPUT BITS. THE BIT-VOTER CAN SUCCESSFULLY MASK THE ERRORS

| Module Number | Word (6 bits) | | | | | | Value (dec.) |
	Bit 5	Bit 4	Bit 3	Bit 2	Bit 1	Bit 0	
#1	0	0	1	0	0	0	8
#2	0	0	1	0	0	0	8
#3	0	0	0	1	0	0	4
#4	1	0	0	0	1	0	34
#5	1	0	0	0	0	1	33
Bit-Voter	0	0	0	0	0	0	0
Word-Voter	0	0	1	0	0	0	8

It is important to note that the word-model voter has a limitation when an error occurs at least once in each of the modules. In such cases, it will not be able to mask the errors and obtain the correct value. However, the absence of a correct result will be signaled (when there is no majority), avoiding a failure situation. Chances of this happening increases as fewer modules are used; the larger is the word; or the higher is the probability of errors.

TABLE III. CASE-STUDY III: 5-MR SYSTEM WITH A 6-BIT OUTPUT VOTED BY BIT- AND WORD-VOTERS WITH 23% OF ERRONEOUS OUTPUT BITS. VOTERS CAN NOT MASK THE ERRORS

| Module Number | Word (6 bits) | | | | | | Value (dec.) |
	Bit 5	Bit 4	Bit 3	Bit 2	Bit 1	Bit 0	
#1	0	1	1	0	0	0	24
#2	0	1	0	0	0	0	16
#3	0	0	0	0	1	0	2
#4	0	0	0	0	1	0	2
#5	1	1	0	0	0	0	48
Bit-Voter	0	1	0	0	0	0	16
Word-Voter	0	0	0	0	1	0	2

Table III shows the results for the third scenario. One can observe that neither the word-voter model nor the bit-voter are able to mask the errors.

C. Experimental Analysis

A simulator was developed to evaluate the behavior and quantify the probability of correct and incorrect outputs considering multiple errors in the inputs of the bit- and word-voter models. In order to evaluate which voter model has the highest probability of masking errors, 3-MR, 5-MR and 7-MR systems with 32-bit word were considered.

A varying number of multiple errors were applied randomly at any of the redundant modules outputs. The number of errors varies between a single bit error to up to 30% of bit errors. For example, for the 5-MR redundant system with a 32-bit output each, there are 160 output bits. Therefore, 30% of this total would result in 48 erroneous output bits. For each one of the evaluated NMR system, we have performed 1,000,000 simulation runs.

The results are shown in Fig. 5(a) and 6(a), respectively. As expected, the system with the largest number of redundant modules is more likely to mask multiple errors both in the bit as the word-voter models. In Fig. 5 (a), we note that the 7-MR system achieves 80% of correction even in the presence of 30 errors (13.39% of all bits), while the 3-MR reaches this value with 5 (5.21%) errors at most. However, when one considers the word-voter model in Fig. 6 (a), the system with 7 modules achieves 80% of correction with at most only 7 faults (3.12%), while the 3-MR system has a correction of just 32% with 2 faults (2.08%). In the latter case (3-MR), when two faults occur in different modules and different positions of the output, each module will present a different value, so there will be no majority.

At first glance, the bit-voter model performs better, because it is able to mask more errors than the word-voter. However, the correctness of the output generated by the voters must also be considered. In many cases, despite the fact that the voter generates a result, this may not be the correct one. As an

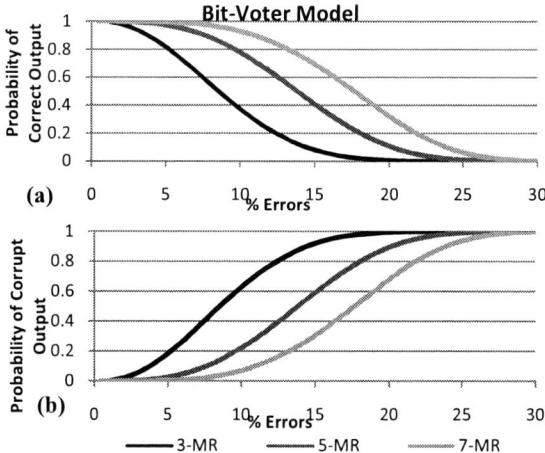

Fig. 5. (a)Probability of Correct and (b) Corrupt Outputs of a NMRs system with bit-voter.

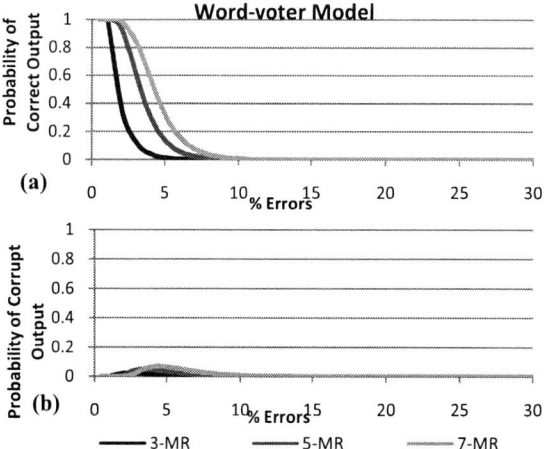

Fig. 6. (a)Probability of Correct and (b) Corrupt Outputs of a NMR system with the word-voter.

978-1-4799-8720-7/15 $31.00 © 2015 IEEE

example, let us consider a 3-MR system. If 2 faults occur in the first bit of two modules, both the bit- or word- voter models would generate an incorrect output. On the other hand, if the bits were in different positions, the bit-voter would still produce an incorrect result, while the word-voter would generate a signal to inform that the result is not correct.

Therefore, let us now analyze, in Fig. 5(b) and 6(b), the probability of the result being a corrupted (incorrect) output in the presence of multiple errors. For the bit-voter model, these values are complementary in respect to the probability of the generation of the correct output. However, when one compares Fig. 5 (b) and 6 (b), one can note that the word-voter has less chances of generating an incorrect output, which would lead to a system failure. This is the main advantage of this model: as already mentioned, it has an extra control signal (failure signal) to invalidate the output. Because of that, while the probability of the bit-voter to result in an incorrect value (and continue without being observed) is inversely proportional to the generation of the correct value, for the word-voter this value does not exceed 7% in the worst case, even with a large number of errors.

III. PROPOSED CONFIGURABLE BIT-WIDTH N-MR VOTER

The simulations presented in the previous section show that each voter model has its advantages and drawbacks. While the word-voter is better when it comes to data integrity and avoids system failure, the bit-voter masks more errors. The choice of which voter to use depends on the needs of the project. In this work a technique that combines the advantages of both voters is presented, enlarging the design space. The configurable bit-width NMR voter, which combines the two aforementioned voters, is described as follows. The set of output bits of each module is divided into smaller subsets. Each subset is evaluated using the same approach as the word-voter. The first subset of each module is compared with each other, as for the second subset and so on. The final result is obtained by recombining the result of each subset. In the case a subset does not present majority, a failure signal is indicated. The technique explores different amounts of subsets with different bit-widths, as shown in Fig. 7. If one considers one configurable NMR voter that contains one subset only with all the output bits, it would be the same as the word voter. On the

other extreme, a configurable NMR with the number of subsets that equals to the number of total output bits would have the same behavior as the bit-voter.

For example, let us consider again the case-study III from section III.b (repeated in Table IV): a 5-MR system with 6-bit outputs. In this case, both bit- and word-voters failed to mask the errors or even detect them. The configurable bit-width voter could be composed of 2 sets of 3-bit word-voters or 3 sets of 2-bit word-voters, as shown in the last two rows of Table IV. The former would be capable of masking the errors and provide the correct outputs, while the latter would be able to signal the failure (no majority). Therefore, in some error cases scenarios, it is necessary to have an intermediate approach in between the bit- and a word-voters.

TABLE IV. USING CONFIGURABLE 3-BIT AND 2-BIT VOTERS IN A 5-MR SYSTEM WITH 6-BITS OUTPUTS WITH 23% OF ERRONEOUS OUTPUT BITS

Module Number	Word (6 bits)						Value (dec.)
	Bit 5	Bit 4	Bit 3	Bit 2	Bit 1	Bit 0	
#1	0	1	1	0	0	0	24
#2	0	1	0	0	0	0	16
#3	0	0	0	0	1	0	2
#4	0	0	0	0	1	0	2
#5	1	1	0	0	0	0	48
Bit-Voter	0	1	0	0	0	0	16
Word-Voter	0	0	0	0	1	0	2
Configurable 3-bit voter	0	0	0	0	0	0	0 (0|0)
Configurable 2-bit voter	-	-	0	0	0	0	Failure (no majority)

A. Data Integrity (S)

By data integrity of a NMR system, we mean the probability of the system to not produce or propagate corrupted (incorrect) outputs. Therefore, to evaluate the voter models one should not only examine the probability of correction, but also the probability of maintaining data integrity, so the safety of the system is enhanced. Let us consider the possible voter outputs: *a) Correct* (C): When the output voter succeeds in obtaining the expected value. *b) Detected Failure* (F): When the voter is unable to correct the output, but detect errors (indicated by the failure signal). *c) Incorrect* (I): The voter produces a corrupt (incorrect) output value. Based on these possible outputs, we can define data integrity (S) as the sum of the correct results (C) and those results that were not corrected, but the failure was detected (F), as shown in (1) or (2).

$$S = C + F \qquad (1)$$
$$S = 1 - I \qquad (2)$$

B. Evaluation of the Voter Models

To estimate the probability of correct outputs (C) and data integrity (S) for the proposed technique, we have simulated the 3-MR, 5-MR and 7-MR system using the configurable model, with the number of errors ranging from 1% to 30% of the total number of bits. Six different forms the configurable voters were evaluated. They are represented in the figures as follows: (#subsets) x (#bits). In Figure 8, the x-axis refers to the number of errors (given as the percentage of total number of output bits

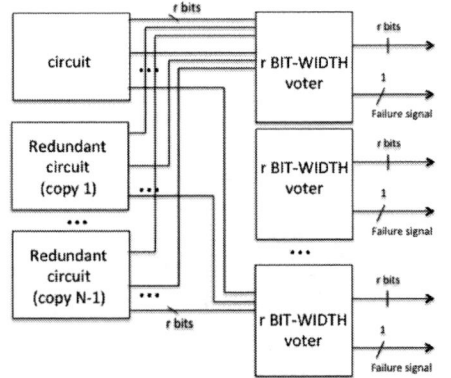

Fig. 7. Configurable bit-width voter in NMR system with m bits outputs

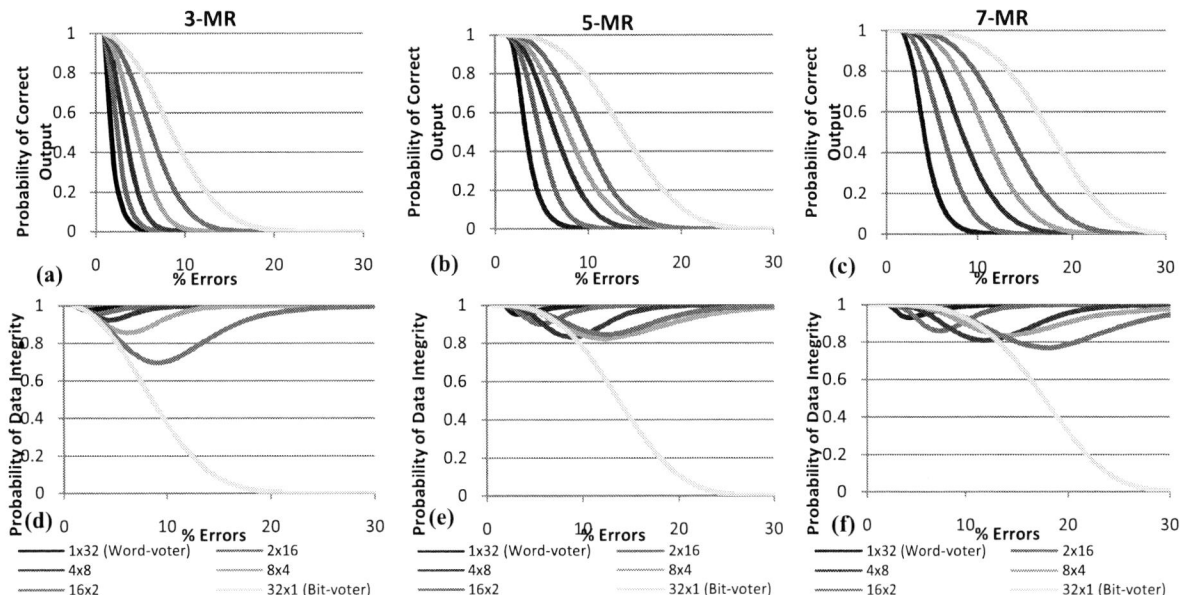

Fig. 8. (a-c) Probability of Correct output and (d-f) Data Integrity outputs of a 3-, 5- and 7-MR system with the proposed voter.

of all modules). The y-axis of the Fig. 8 a, b and c represents the probability of correcting the output (C), while the y-axis of the Fig. 8 (d, e, f) depicts the probability of maintaining the data integrity in the presence of multiple errors. At one extreme one can observe the behavior of the word-voter (1x32), which has a low correction probability (C), but high data integrity (S). At the other extreme, there is the behavior of a bit-voter (32X1), with high correction probability and low data integrity (S), which increases the number of errors when compared to the word-voter. This behavior can also be observed for the 5-MR e 7-MR systems. One can also observe that the curve for the probability to correct the output considering 32x1 model (bit-voter) is the same as for the probability of maintaining data integrity, for any of the NMR systems (Fig. 8). In this case, the detected failure rate always equals to zero (F = 0), so data integrity (S) always equals to the rate of correct outputs only (S=C).

With the proposed technique one can obtain intermediate values of (F) and (C), so it is possible to explore the trade-offs between both and data integrity (S). Therefore, at design time, one can estimate what is the percentage of correction needed and the acceptable error rate. With these requirements in hand, one can analyze which model of voter meets the project requirements.

IV. DESIGN SPACE EXPLORATION

In order to automatically explore the design space, the developed tool can find a set of solutions, using the configurable voter, that satisfy the requirements and constraints of each particular project. Its project parameters are the word size (output modules); area; delay; minimum percentage of correction; maximum percentage of corrupted outputs; number of redundant modules and expected range of errors.

A. Area and Delay

To gather data on area and delay, we implemented in HDL the 3-MR, 5-MR and 7-MR voters, considering 32-bit word sizes, and synthesized to the 65nm Virtex-5 [10] FPGA (6-input LUT architecture), using the ISE Design Suite [11]. Figure 9 shows the area utilization (y-axis) for the voters (x-axis). While 32 LUTs were required to implement the 3-MR e 5-MR bit-voters, 64 LUTs were needed to implement the 7-MR. In the first two cases, as every output bit does not depend on any other bit in the word, only a 6-input LUT is enough for each output bit, which is not true for the 7-MR. In this implementation, as the size of the subset increases, the number of partial-voters needed decreases. For example, in the case of the 16x2 voter, two bits are needed as inputs for each one of the 16 voters. In the case of the 3-MR, each partial-voter that

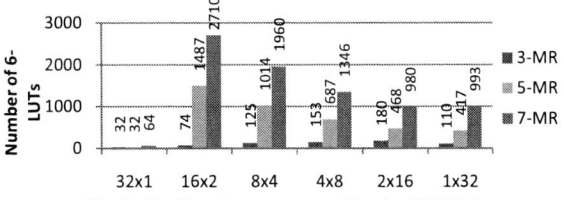

Fig. 9. Configurable-voter area utilization (6-LUTs).

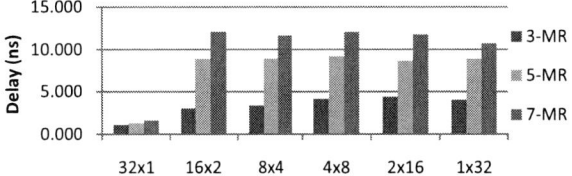

Fig. 10. Configurable-voter delay for different forms of organizing

978-1-4799-8720-7/15 $31.00 © 2015 IEEE

composes a subset has six entries, so only one six-entry LUT is necessary to implement each, which justifies the large difference in the use of LUTs when one compares 3-MR with the 5-MR or 7-MR implementations. The delay of the each voter is shown in Figure 10. The bit-voter (32x1) is faster, as expected. The average delay and standard deviation for the other forms of grouping were 3.8ns and 0.46ns (3-MR); 8.9ns and 0.11ns (5-MR); 11.6ns and 0.38ns, respectively.

B. Experiments

Four possible scenarios with particular requirements and restrictions were evaluated (Table V). The word size is always 32 bits, and the number of errors is expressed as a percentage of the total number of bits.

TABLE V. SPACE EXPLORATION SCENARIOS FOR A SET OF REQUIREMENTS UNDER MULTIPLE ERRORS

Case Scenarios	Requirements					
	Correct Output	Data Integrity	Modules Amount	%Errors	Area (#LUTs)	Delay (ns)
# I	> 75%	> 85%	<= 7	< 4%	-	-
# II	> 85%	> 95%	<= 5	< 6%	< 1500	< 9
# III	> 40%	> 80%	<= 7	< 14%	< 3000	< 15
# IV	> 50%	> 80%	<= 7	< 14%	-	-

TABLE VI. POSSIBLE SOLUTIONS FOR THE FIRST CASE-STUDY

Voter model	3-MR		5-MR		7-MR	
	Correct Output	Data Integrity	Correct Output	Data Integrity	Correct Output	Data Integrity
2x16	-	-	-	-	85.3%	97.3%
4x8	-	-	79.1%	96%	95.7%	99.4%
8x4	-	-	87.9%	98.1%	98.6%	99.8%
16x2	76.5%	88.4%	94.2%	98.5%	99.4%	99.9%
32x1	87.6%	87.6%	98.4%	98.4%	99.9%	99.9%

For the first exploration scenario, eleven possible solutions have been found (Table VI): 3, 5 or 7 modules are required and one of the five available adaptable voters: 2x16, 4x8, 8x4, 16x2 or 32x1. Although the word-voter (1x32) achieves the desired data integrity, it failed with the desired correctness rate, even with 7 redundant modules (bit-voters with an even number of modules have not been considered). In this case, the designer can opt for any one of the solutions: one can prioritize the ones with fewer modules, less LUTs (Fig. 9), shorter delay (Fig. 10) or those with higher masking rates. For the second exploration scenario, there is only one solution that satisfies the requirements: the 5-MR system with the 16x2 configurable voter (C = 85.3%, I = 4.5% and S = 95.5%). The 32X1 version (bit-voter) would meet the desired correctness rate, area and delay, but would fail with the desired data integrity. There is also one solution only that satisfies the requirements of the third case study: the 7-MR system, with the 16x2 configurable voter (C = 41.1%, I = 18. 5% S = 81.5%). In this case, if the bit-voter (32x1) model was used, the percentage of corrupt output would be higher (worse) than necessary (C = 74%, I = 26%, S = 74%). On the other hand, the word-voter model would meet the requirements of data integrity, but would fail with the correction rate.

In this last case exploration scenario, it has not been possible to find a configuration that would satisfy the project

requirements. So, given this result, the designer would have to try different alternatives, such as increasing the number of modules or even use other techniques of fault tolerance.

V. CONCLUSION AND FUTURE WORK

In this paper we presented a configurable bit-width NMR voter for multiple error masking. We demonstrated that, with this voter, it is possible to explore the trade-offs between correctness and data integrity, enlarging the design space so one can make the best decision when choosing the technique of fault tolerance to be applied. A tool that finds a set of solutions using the proposed technique that satisfies the requirements and constraints of each particular project was developed. As future work we will evaluate behavior assuming voter failures and using redundant voters.

REFERENCES

[1] "IBM journal research and development," Soft Errors in Circuits and Syst., vol. 52, no. 3, 2008.

[2] E. Ibe, H. Taniguchi, Y. Yahagi, K. Shimbo, and T. Toba, "Impact of Scaling on Neutron-Induced Soft Error in SRAMs From a 250 nm to a 22 nm Design Rule", Electron Devices, IEEE Transactions on (Volume:57 , Issue: 7), 2010 Page(s):1527 – 1538.

[3] A.D. Tipton, J.A. Pellish, J.M. Hutson, R. Baumann, X. Deng, A. Marshall, M.A. Xapsos, H.S. Kim, M.R. Friendlich, M.J. Campola, C. M. Seidleck, K.A. LaBel, M.H. Mendenhall, R.A. Reed, R.D. Schrimpf, R.A. Weller, and J.D. Black, "Device-Orientation Effects on Multiple-Bit Upset in 65 nm SRAMs", Nuclear Science, IEEE Transactions on (Volume:55 , Issue: 6), 2008, Page(s): 2880 – 2885.

[4] K Nikolic, A. Sadek, and M. Forshaw, "Fault-tolerant techniques for nanocomputers", Nanotechnology ,Volume 13 Number 3, 2002.

[5] D. Burlyaev, P. Fradet, and A. Girault, "Verification-guided voter minimization in triple-modular redundant circuits", in Design, Automation and Test in Europe Conference and Exhibition (DATE), 2014.

[6] P.R. Lorczak, Charles River Analytics Inc., Cambridge, MA, USA, A.K. Caglayan, and D.E. Eckhardt, "A theoretical investigation of generalized voters for redundant systems", Fault-Tolerant Computing, 1989. FTCS-19. Digest of Papers., Nineteenth International Symposium on, 1989. Page(s): 444 - 451

[7] L. Sterpone, M. Violante, A. Panariti, A. Bocquillon, F. Miller, N. Buard, A. Manuzzato, S. Gerardin, and A. Paccagnella, "Layout-aware multi-cell upsets effects analysis on TMR circuits implemented on SRAM-based FPGAs", IEEE Trans. Nucl. Sci., vol. 58, no. 5, pp. 2325–2332, Oct. 2011.

[8] S. Mitra and E. McCluskey, "Word-voter: A new voter design for triple modular redundant systems", in Proc. IEEE VLSI Test Symp. , May 2000,pp. 465– 470.3.

[9] S. Mitra, N. R. Saxena, and E. J. McCluskey,"Common-mode failures in redundant VLSI systems: A survey", IEEE Trans. Reliability, vol. 49, no. 3, pp. 285-295, Sep. 2000.

[10] Xilinx Inc. (2009, February). Virtex-5 Family Overview [Online]. Available: http://www.xilinx.com/.

[11] Xilinx Inc. (2014, September). ISE Design Suite [Online]. Available: http://www.xilinx.com/.

2015 IEEE Computer Society Annual Symposium on VLSI

Communication-aware Parallelization Strategies for High Performance Applications

(Invited Paper)

Imran Ashraf and Koen Bertels
Computer Engineering Lab,
Delft University of Technology, The Netherlands

Nader Khammassi and Jean-Christophe Le Lann
Lab-STICC UMR CNRS 6285,
ENSTA Bretagne, France

Abstract—With the advent of multicore processor architectures and the existence of a huge legacy code base, the need for efficient and scalable parallelizing compilers is growing. Where multi-core processors were seen as the way forward to address the known challenges such as the memory, power and ILP wall, efficient parallelization to make use of the multiple cores, is still an open issue. In this paper, we present two complementary tools, MCROF and XPU which provide an alternative development path to parallelise applications and that address the challenges of identifying potential parallelism and exploiting it in a different way. The MCROF tool provides a detailed profile of the data flowing inside an application and the XPU programming paradigm provides an intuitive and simple interface to express parallelism as well as the necessary runtime support. We demonstrate through two different use cases that better performance up to $4\times$ can be achieved than available commercial compilers.

Keywords—*Data-communication profiling, program parallelization, Multicore, Parallel Programming*

I. INTRODUCTION

The number of transistors per chip is growing due to technology scaling and increasing the clock rate of processors is becoming technologically less viable [1]. The current trend is therefore to integrate a growing number of processing cores on chip, forcing parallelizing compilers to mature rapidly and to provide efficient code for the multi-core processors. Most parallelizing compilers focus on loop parallelization as most of the execution time is spent in loops. However, scalable parallelism is in many cases not realizable because memory accesses and interprocessor communication are the bottlenecks. Recent research makes it clear that memory accesses and data transfers account for the majority of the power consumption [2] [3] and thus need to be addressed and handled more explicitly in order to achieve (power) efficient performance. This paper presents a tool chain that parallelizes an application based the data flowing inside an application and how this helps in mapping (manually) the algorithm on the architecture using intuitive parallel constructs. We present in detail a use case, Canny Edge Detection, as well as the performance numbers for a second application, fluid animate.

The rest of the paper is organized as follows. Section II discusses some of the available parallelizing compiler frameworks. Section III presents the results of the Canny Edge Detection algorithm when using two state-of-the-art commercial parallelizing compilers. Section IV introduces the tools used in

this paper and section V presents the proposed parallelization methodology and achieved results.

II. RELATED WORK

Though static-analysis tools [4] can also track datacommunication, a large number of tools [5], [6] utilize dynamic-analysis to collect producer-consumer relationship at runtime. These tools have high run-time overhead, which limits their use for realistic workloads affecting the quality of the generated information. Furthermore, the provided information lacks necessary dynamic details and is not linked to the source code, making it hard to utilize this information. A number of automatic parallelization compilers exist such as Par4All [7], Cetus [8], Parallware [9], Polaris [10] or PolyCC/PLUTO [11] which are source to source compilers that can produce parallel code after analyzing the sequential code using different parallelization techniques. The Intel ICC [12] is a popular compiler which provides an automatic parallelization feature which allows both instruction-level and thread-level parallelization of sequential regions of the input code.

Automatic parallelization of sequential code has had limited success [13]. Great advances have been made in automatic parallelism extraction at the instruction level, however, in order to exploit efficiently modern multicore platforms, compilers need to capture parallelism also at thread-level which is a challenging task. Pareon by Vector Fabrics [14] is an example of a tool, which assists the programmer and guides the parallelization process instead of performing automatic code parallelization.

III. PARALLELIZATION USING EXISTING COMMERCIAL COMPILERS

In this section, we present the parallelization of the Canny application by using two commercial compilers; we refer to them as CC1 and CC2 in the rest of the discussion. CC1 can be used both in automatic and semi-automatic way, whereas, CC2 uses only a semi-automatic approach. We attempted to use PolyCC/PluTo compiler [11] which uses polyhedral analysis to tile and parallelize loops in sequential programs. However, the compiler suggested significant manual modification of the code in order to make it processable by the compiler. For instance, the compiler requires removing all affine expressions from the inner loop to be able to process it. So the parallelization process is no longer transparent.

978-1-4799-8720-7/15 $31.00 © 2015 IEEE

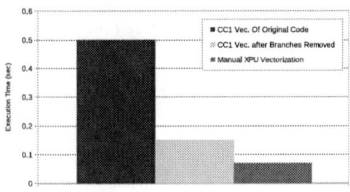

Fig. 1. Performance comparison of automatically vectorized code using CC1 and manually vectorized code.

a) CC1: CC1 allows automatic parallelization of sequential program at thread-level using OpenMP and at instruction-level through vectorization using SSE/AVX intrinsics. In order to parallelize a sequential program, the compiler searches for loops which do not expose cross-iteration dependence and are good candidate for parallel execution. A data flow analysis is performed to ensure correct and safe parallel execution. The compiler then uses OpenMP to specify the parallelism.

Default Automatic Mode: is a one shot fully automatic parallelization mode in which the program is parallelized using compile-time static analysis.

Run-time Controlled Mode: collects the run-time information (profiling data) and uses it to guide the program parallelization.

Guided Parallelization Mode: in which compiler can be used to perform run-time analysis to generate an advisory reports, suggesting ways (often code modifications) to the programmer to parallelize loops.

b) Parallelization of Canny Application using CC1:
For the canny application, the default automatic mode and run-time controlled mode did not show any significant speedup over the original sequential application, hence, we used the guided parallelization mode. By using the advisory reports, the loops in the `gaussian_smooth` were successfully parallelized, however, many other loops could not be parallelized due to false data-flow dependences. Listing 1 illustrates an example where a small manual modification made the loop parallelizable. We made several manual modifications such as nested loop fusion and iterator duplication to help the compiler. The resulting parallel program achieved better speedups (up to 4× for 16 threads on a platform with two Intel Xeon E5-2670 processors at 2.6 GHz).

CC1 is capable of performing instruction-level parallelism by vectorization. In this regard, we performed three experiments to evaluate the vectorization quality in each case:

1) In order to evaluate the vectorization quality, we measured the achieved performance of one of the two loops of the `gaussian_smooth` function. This loop was reported as auto-vectorized by the compiler.
2) In the original code, this loop contains control branches (`if`) to handle loop bounds, which limit vectorization efficiency. We removed these branches.
3) Finally, we vectorized the code manually using SIMD intrinsics (SSE4.2).

The results of these three versions are shown in Figure 1. It can be seen that the manual vectorization achieves signif-

icantly better performances than the automatic vectorization performed by the compiler even after vectorization-friendly code transformations.

c) CC2: CC2 aims at the parallelization of sequential C/C++ code for ARM Cortex A9 and x86 as target systems. To use CC2, an application is compiled with a C99 compliant compiler to perform instrumentation of the application. The instrumented application is then executed on the model of the target architecture to generate the profile of the application. The generated profile can be visualized in a GUI to select the loops which have high execution contribution as the candidates for parallelization. Furthermore, the dependencies are also reported in a GUI.

After selecting the number of cores in the architecture, CC2 predicts the achievable speedup at the loop-level as well as the application-level. Based on the selected parallelization, refactoring steps are presented by the tool to be applied on the sequential code by the programmer to parallelize it. CC2 suggests parallelizations in terms of a low level threading API and OpenMP pragmas.

```
1
2 // Loop carried dependency on pos in orignal sequential loop
3 for (int pos=0, r=0; r<rows; r++, pos++)
4     for (int c=0; c<cols; c++, pos++)
5         image[pos] = x;
6
7 // equivalent parallelizable loop
8 for (int pos=0; pos<rows*cols; pos++)
      image[pos] = x;
```

Listing 1. CC1 fails to resolve an easily removable data dependencies which prevents parallelization.

d) Parallelization of Canny Application by CC2: For the loops in `gaussian_smooth`, CC2 detected that there are no loop dependencies and suggested an OpenMP pragma. For a 4-core system, loop speedup of 4× was predicted, whereas the achieved speedup is only 2.58×. Similarly, a 2.2× application speedup was predicted, however, the actual application speedup is 1.64×.

For the loop in `magnitude_xy`, CC2 detected r as an induction variable, however, for `pos` variable, synchronization was suggested. This synchronization resulted in slowdown, instead of speedup. By changing the code, as already discussed in Listing 1, programmer can avoid this synchronization. Hence, a loop speedup of 2.84× and an application speedup of 1.7× was achieved. Finally, for the third loop in `non_max_supp`, CC2 predicted no speedup, hence, it was not parallelized.

e) Lessons Learned: It can be summarized from the discussion in this section that Parallel compilers, such as CC1, may significantly improve programmer's productivity by automatic code parallelization, however, such compilers suffers from inherent difficulties. Hence, manual code analysis and modifications are required.

CC2 performs run-time analysis to detect dependencies and suggest parallelization. However, there are cases in which even simple loop carried dependencies are not resolved automatically. Therefore, these dependencies are either resolved manually or synchronization is suggested. This synchronization can result in loop (and hence application) slow-down, instead of speedup, requiring manual inspection and modification of application by the programmer. Furthermore, the information is provided among loop iterations, not across loop

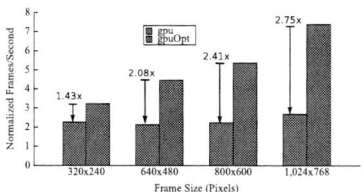

Fig. 2. Normalized Frames per seconds achieved by the GPU and data-communication optimized GPU implementation.

nests or functions in the application to exploit coarse-grained application parallelization and data-flow optimizations.

This outlines the need for tools which analyze the programs dynamically to provide data-flow information to highlight real data-dependences. This information should be provided at various granularity levels to extract and express various forms of parallelization available in the application.

IV. TOOLS DESCRIPTION

This section briefly introduces the tools involved in this discussion, namely MCROF and XPU.

MCPROF [15] is a runtime memory and communication profiler which generates detailed application profile in terms of memory access patterns and data-communication at function and loop-level granularity. It is based on Intel Pin Dynamic Binary Instrumentation (DBI) framework [16]. MCROF performs instruction-level instrumentation to track memory reads and writes by each instruction. Furthermore, routine-level instrumentation is utilized to keep track of the currently executing function. These are tracked by maintaining a call-stack of the functions executing in the application. To track the dynamic allocations in the application, image-level instrumentation is utilized to selectively instrument library images for memory (re)allocation/free routines.

The tracked memory reads/writes are associated with parts of the source code depending upon the selected granularity i.e. functions, loops or other marked regions in the source code. In this way, a producer-consumer relationship is established between functions/loop/objects in the source code and reported in the form of a communication graph. This production-consumption relation is expressed by edges in the graph where the color of the graph shows the intensity of the communication. Furthermore, the generated graph also shows percentage of the dynamically executed instructions and the number of calls to each function. In this way, communication as well as the computation intensive parts of the application are shown in the generated graph.

The information generated by MCROF has been utilized earlier [15] to perform data-communication aware partitioning of sequential applications on a heterogeneous architecture. Kanade-Lucas-Tomasi Feature Tracker (KLT) [17] application was mapped to a GPU-based platform. After applying the optimizations based on MCROF to the initial GPU implementation (gpu), a data-communication optimized version of the GPU implementation (gpu_{opt}) was obtained. Figure 2 shows the Frames Per Seconds (fps) achieved by both the implementations normalized to the CPU implementation. Increasing the frame size, results in an increase in the amount of computation

performed on the GPU. Increased computation results in better utilization of the available resources of the GPU, resulting in higher speedup as can be observed from this figure. On the other hand, increasing the frame size also increases the amount of frame data transferred to the GPU for processing and getting results back. This data-communication has been optimized in the case of gpu_{opt} based on the information provided by MCROF. Hence, gpu_{opt} implementation achieves up-to $2.75\times$ higher speedup as compared to gpu implementation where this communication is not optimized.

XPU [18] [19] is a structured parallel programming framework which aims to easily express and exploit parallelism. XPU allows the expression of different types of parallelism at different levels of granularity. Supported parallelism types includes data parallelism [20], task parallelism [19] and pipeline parallelism [21]. These different parallelism types can be composed hierarchically in the same application [22].

XPU utilizes C++ meta-programming techniques [23] [24] exploiting the potential of the standard C++ language and thus does not require any particular tool except standard C++ compiler. XPU provides a friendly and light weight programming interface which enables the programmer to design parallel applications or parallelize sequential applications with minimal code changes without any significant alteration.

XPU is mainly composed of a set of recurrent parallel execution patterns which specifies execution configuration of a group of tasks. In order to promote reuse of sequential legacy code, these tasks are designed to encapsulate different pieces of code including functions, class methods or lambda expressions. XPU's execution patterns handle transparently many issues such as communication, synchronization and task scheduling. An intelligent run-time system exploits the information which is extracted transparently from both the available hardware resources and perform the resource allocation through cache-aware and load-balanced task scheduling [20], [21].

V. MCROF-XPU APPROACH

The XPU parallel programming model is designed for easing explicit parallelism expression and therefore requires locating the hot-spots in the program and extracting parallelism by analyzing task-dependencies (producer-consumer relationships). Usually this analysis is performed manually by reading and analyzing the code and profiling the application, which is a time-consuming and error-prone task. MCROF can automate this analysis phase and provides a clear picture of the program with all the required information including task-dependencies, compute-intensive and communication-intensive hot-spots in the application. Hence, MCROF and XPU can form a parallelization tool-chain which offers a much smoother transition from the sequential-code to the parallel-code. The MCROF-XPU parallelization methodology follows the traditional parallelization steps which can be summarized as follows:

1) **Profiling sequential code:** to locate hot-spots.
2) **Extracting parallelism and granularity adjustment:** analyzing data and task dependencies and decomposing tasks to extract more parallelism, if available, at finer grain.
3) **Expressing parallelism**: using the parallel programming API.

978-1-4799-8720-7/15 $31.00 © 2015 IEEE

4) **Executing the parallel code efficiently:** parallel programming libraries are often based on a run-time which is
responsible of efficient scheduling to optimize execution.

A. Profiling The Sequential Application using MCROF

Figure 3 shows the MCROF output graph of Canny, where
nodes represent functions. Each node contains the function
name, the percentage of dynamically executed instructions by
this function with respect to the whole application, as well
as the total number of calls to this function. For instance,
`gaussian_smooth` is executed once and it is a computationally intensive function as it covers 80% of the instructions
executed by the whole application.

In addition, the inter-function data-flow is represented by
edges. Furthermore, the statically and dynamically allocated
objects involved in each flow are also detected by MCROF
and represented by rectangles containing object names and
allocation sizes. The communication intensity is quantitatively
shown by the number of bytes on each edge and also illustrated
by the color the edges from red (highest) to green (lowest). In
this way, programmer is able to visualize the computation- and
communication-intensive parts of an application, in a single
graph, without manual source code inspection.

To extract parallelism, coarse-grained functions should be
decomposed to extract potential finer-grain parallelism, we
refer to this decomposition step as granularity adjustment.

B. Granularity Adjustment

As seen in Figure 3, `gaussian_smooth` has 80% execution contribution in the application. In order to extract parallelism inside this function, we use the MCROF feature of splitting functions at loop level granularity. Figure 4 shows the simplified output graph generated by MCROF where the loops in
`gaussian_smooth` are split as separate nodes represented
by `gaussian_smooth1` and `gaussian_smooth2`. A
similar split of `derivative_x_y` is also shown as
`derivative_x_y1` and `derivative_x_y2`.

As depicted in the Figure 4, the `gaussian_smooth`
function exposes dependencies between its two loop nests,
thus they cannot be executed concurrently. However, each of
these loops can be parallelized individually since they operate
independently on columns and rows of the image. On the other
hand, the two parts of the `derivative_x_y` do not expose
any dependency and therefore can be executed in parallel,
since these loops are using a common read-only input while
producing separate outputs. A similar analysis is performed on
other functions to extract the available parallelism.

C. Parallelization Using XPU

After analyzing the data-flow dependencies and extracting
parallelism, we use the XPU framework to express the parallelism. In this regard, XPU skeletons are utilized to exploit
various forms of parallelism available in the application.

1) Thread-level parallelism: Thread-level data parallelism
can be achieved by parallelizing sequential loops. In XPU,
this is expressed by using `parallel_for` pattern. For
instance, Listing 2 shows how the `gaussian_smooth` is
parallelized using the XPU parallel loop skeleton. The task

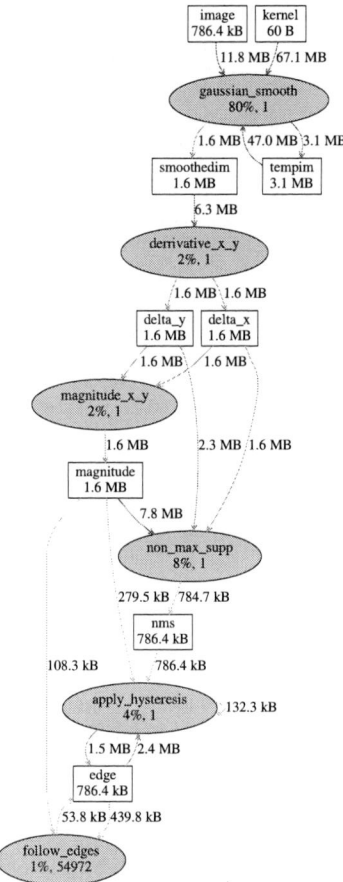

Fig. 3. Task dependency graph of Canny application generated by MCROF.

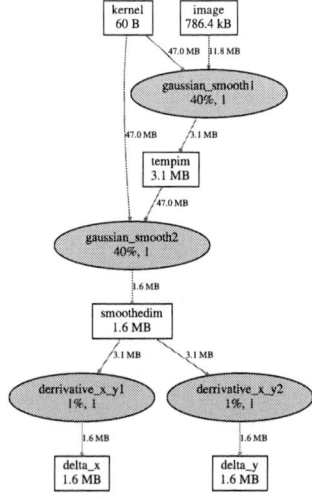

Fig. 4. MCROF profile at loop-level granularity showing independence of the
two loop nests inside the *derivative_x_y* function.

```
 1 int gaussian_smooth1(char *inImg, float *kernel, char *
       outImg, /* more args */)
 2 {    /* process an image row */ }
 3 int main(){
 4    // task definition
 5    xpu::task gauss_task(gaussian_smooth1, image, kernel,
          tempim, /* more args */);
 6    // parallel loop construction
 7    xpu::parallel_for parallel_gauss(0,size,1,&gauss_task);
 8    // parallel loop execution
 9    parallel_gaussian.run();
10 }
```

Listing 2.   XPU parallel for loop construction and execution.

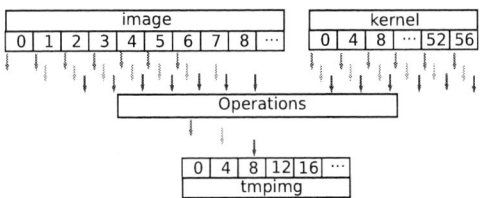

Fig. 5.   Fine-grain access-pattern of *gaussian_smooth1* loop reading *image*, *kernel* and writing *tmpimg* objects. Accesses in same iteration are represented by the same color.

which processes the data elements is constructed and named `parallel_gaussian`. We note that data partitioning and tasks scheduling are handled transparently by the XPU run-time to ensure dynamic forward-scalability and execution-efficiency across different platforms.

2) Instruction-level parallelism: Beside loop parallelization, vectorization can act as a great performance multiplier by allowing SIMD operations on the pixels of the image. XPU provides transparent vectorization through built-in vectorized types such as `vec4f` or `vec8s`, which are implemented using x86 SSE intrinsics. For instance, using `vec4f` allows vectorization of standard operations on single precision `float` and other composite operations which are not available natively in SSE instruction-sets, such as trigonometric functions. Similarly, by using the `vec8s` type, the programmer can operate implicitly on 4 floats simultaneously.

Understanding the data-access pattern is one of the major challenging task in the vectorization process especially in the case of irregular or non-contiguous memory-accesses (load and stores). Tracking data-accesses across loops iterations can be a time-consuming and error-prone task. To address this issue, MCROF can generate a graphical view of the data-access pattern of both input and output data in a user-delimited region of the code, particularly loops. Figure 5 shows the data-access patterns in the first three iterations of the gaussian loop. Each

```
 1 xpu::vec4f v1,v2,v3,acc,k1,k2,k3,k4;
 2 for (int i=begin; i<end; i++) {
 3    v1 = &in[i];    v2 = &in[i+4]; //load inputs
 4    v3 = &in[i+8]; v4 = &in[i+12];
 5    // vectorized operations :
 6    acc = k1*v1 + k2*v2 + k3*v3 + k4*v4;
 7    sum = (k1+k2+k3+k4).sum();
 8    acc /= sum;
 9    out[i-(kernel_size/2)] = acc.sum(); //store output
10 }
```

Listing 3.   Code Vectorization using XPU.

```
 1 // tasks definition
 2 xpu::task dx_t(derivative_x, smoothed_img, dx);
 3 xpu::task dy_t(derivative_y, smoothed_img, dx);
 4 // parallel execution
 5 xpu::parallel(&dx_t, &dy_t)->run();
```

Listing 4.   Task-level parallelism in *derivative_xy* expressed in XPU.

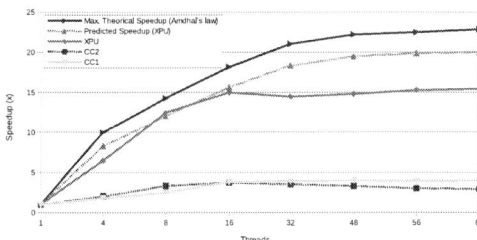

Fig. 6.   Achieved speedups over the sequential execution for the Canny application when parallelized by various approaches.

color corresponds to an iteration. In each iteration, we used the XPU vectorized type `vec4f` to load the required inputs from both `image` and `kernel` at the indicated position. Four multiplications and the sum is then performed simultaneously. This allowed us to achieve a significant speedup over both the sequential code and the automatically vectorized code generated by the compiler. Listing 3 shows an example of vectorized code using XPU.

3) Task Parallelism: As depicted in the MCROF output in Figure 4, `derivative_x` and `derivative_y` do not expose any producer-consumer dependencies and thus can be executed as parallel tasks. This can be easily specified using XPU as shown in Listing 4.

Figure 6 shows the speedup achieved for the Canny application parallelized by MCROF-XPU methodology, CC1 and CC2. The theoretical speedup based on Amdahl's law and estimated speedup while considering SIMD support is also shown. It can be seen that a speedup of $15\times$ is achieved for 16 cores on an `Intel Xeon E5-2670`, which is about $4\times$ higher than what achieved by CC1 and Pareon. Similar performances are achieved on 64 cores platform with four `AMD Opteron 6274` processors.

Apart from the Canny application, we have also parallelized *fluidanimate* application which is a part of the PARSEC Benchmark [25]. This benchmark includes three versions of the application: a serial version, a parallel version which uses POSIX Thread API and another parallel version which uses Intel Thread Building Blocks. Another parallel version using XPU has been developed in [18] [20]. We developed a new version using the MCROF-XPU methodology. In the original XPU-based *fluidanimate* version, all the five processing stages were parallelized. However, the MCROF analysis report have shown that some of these stages are not hot-spots which are not worth parallelization. Parallelizing these stages introduces a communication overhead which affects the overall performance. The granularity adjustment using MCROF showed that splitting the `computeForces` processing stage in three sub-stages clearly isolate the computationally intensive regions which should be parallelized.

978-1-4799-8720-7/15 $31.00 © 2015 IEEE

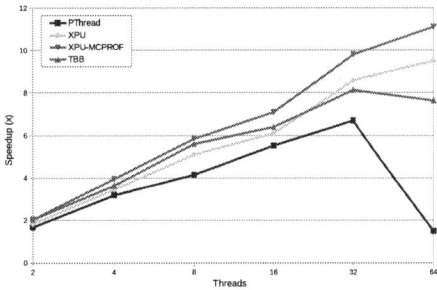

Fig. 7. Performance comparison of the PThread, TBB, XPU only and XPU-MCROF versions of the PARSEC Fluidanimate application.

The XPU parallel_for skeleton has been used to express the thread-level data parallelism exposed by different processing stages. The later skeleton provides a scalable data partitioning and uses a cache-aware scheduling policy which promote data reuse and improve spatial and temporal data locality. This scheduling techniques appeared to be particularly beneficial in this study case since each fluid cell is processed on the processor core. Furthermore, we replaced the fluid cells arrays by XPU *vec3f* arrays to take advantage of SSE vectorization. In the reference sequential code, these fluid cells are expressed using regular float arrays.

Figure 7 shows the achieved performance by the four versions. We observe that using MCROF profiling information improved the performances of the original XPU-only version. We note that the performances of the PThread version suffers from significant degradation when more than two processors (32 threads) on the AMD platform (64 cores/4 processors) are used, our investigation have shown that this degradation is caused by the use of barriers which results to an expensive many-to-many communication which affect the performances especially when the four processors are used. The XPU version uses a synchronization mechanism that follows a less-expensive one-to-many communication pattern.

VI. CONCLUSION

With the emergence of multicore processor architectures, we can no longer avoid parallelizing applications and making parallelizing compilers more efficient is an important objective. In this paper, we have presented the integrated use of two tools which are very complementary in their functionality. As data is in many cases the bottleneck blocking scalable use of multiple computing cores, the need for a detailed analysis of the data flowing through the application (and thus the hardware resources). MCROF provides such a detailed profile which can then be used by XPU, a parallel middle layer and programming approach which provides minimally invasive code changes to express and exploit in a natural way the available parallelism in an application. Not only does the combined approach provide better performance it also reduces substantially the overall time needed to parallelize sequential applications. Future research will allow to fully integrate the approach such that automatic code changes can be introduced.

ACKNOWLEDGMENT

This research is supported by Artemis EMC2 Project (grant 621429), Artemis Almarvi Project (grant 621439) and Artemis CRAFTERS Project (grant 295371). The authors would like to thank Valery Kritchallo for useful discussions.

REFERENCES

[1] M. Horowitz and W. Dally, "How scaling will change processor architecture," in *ISSCC*, 2004, pp. 132–133 Vol.1.

[2] S. Borkar, "Exascale computing - a fact or a fiction?" in *IPDPS*, 2013.

[3] R. Nair, "Active memory cube: A processing-in-memory approach to power efficiency in exascale systems," in *WoNDP*, 2014.

[4] M. D. Ernst, "Static and dynamic analysis: Synergy and duality," in *WODA*, Portland, Oregon, 2003, pp. 24–27.

[5] W. Heirman et al., "A communication profiler to optimize embedded resource usage," *ProRISC*, 2009.

[6] S. Ostadzadeh, "Quantitative application data flow charac. for heterogeneous multicore architectures," Ph.D. dissertation, TU Delft, Dec 2012.

[7] M. Amini et al., "Par4All: From Convex Array Regions to Heterogeneous Computing," Jan 2012.

[8] C. Dave et al., "Cetus: A Source-to-Source Compiler Infrastructure for Multicores," *Computer*, Dec. 2009.

[9] Appentra, "Parallware," http://www.appentra.com/products/parallware/.

[10] W. Blume et al., "Polaris: Improving the Effectiveness of Parallelizing Compilers," ser. LCPC, London, UK, UK, 1995.

[11] U. Bondhugula et al., "PLuTo: A Practical and Fully Automatic Polyhedral Parallelizer and Locality Optimizer," The Ohio State University, Tech. Rep., Oct 2007.

[12] Intel, "Automatic Parallelization with Intel Compilers," https://software.intel.com/en-us/articles/automatic-parallelization-with-intel-compilers.

[13] J. Shen and M. Lipasti, *Modern Processor Design: Fundamentals of Superscalar Processors*. McGraw-Hill Higher Education, 2002.

[14] Vector Fabrics, "Pareon profile." URL: {http://www.vectorfabrics.com}

[15] I. Ashraf et al., "MCProf: Memory and Communication Profiler," Delft University of Technology, Tech. Rep., November 2014.

[16] C. Luk and et al., "Pin: Building Customized Program Analysis Tools with Dynamic Instrumentation," in *PLDI*, 2005, pp. 190–200.

[17] B. D. Lucas and T. Kanade, "An Iterative Image Registration Technique with an Application to Stereo Vision," 1981, pp. 674–679.

[18] N. Khammassi, "High-Level Structured Programming Models For Explicit and Automatic Parallelization on Multicore Architectures." Ph.D. dissertation, ENSTA Bretagne, Lab-STICC, Brest, France, 2014.

[19] N. Khammassi et al., "MHPM: Multi-Scale Hybrid Programming Model: A Flexible Parallelization Methodology," ser. HPCC, Washington, DC, USA, 2012, pp. 71–80.

[20] N. Khammassi et al., "Design and implementation of a cache hierarchy-aware task scheduling for parallel loops on multicore architectures," in *PDCTA*, Sydney, Australia, 2014.

[21] N. Khammassi and J.-C. Le Lann, "A high-level programming model to ease pipeline parallelism expression on shared memory multicore architectures," ser. HPC, San Diego, CA, USA, 2014.

[22] Khammassi, N. and Le Lann, J.C., "Tackling Real-Time Signal Processing Applications on Shared Memory Multicore Architectures Using XPU," ser. ERTS, Toulouse, France, Feb 2014.

[23] J. Koskinen, "Metaprogramming in C++." URL: www.cs.tut.fi/~kk/webstuff/MetaprogrammingCpp.pdf

[24] H. Singh, "Introspective C++," Ph.D. dissertation, Virginia Polytechnic Institute, Virginia, VA, USA, 2004.

[25] C. Bienia et al., "The PARSEC Benchmark Suite: Characterization and Architectural Implications," in *PACT*, 2008.

978-1-4799-8720-7/15 $31.00 © 2015 IEEE

2015 IEEE Computer Society Annual Symposium on VLSI

Design of Fault-Tolerant and Reliable Networks-on-Chip

Junshi Wang
UESTC,
Chengdu, China

Masoumeh Ebrahimi
KTH, Sweden, and
University of Turku, Finland

Letian Huang
UESTC,
Chengdu, China

Axel Jantsch
TU Wien,
Vienna, Austria

Guangjun Li
UESTC,
Chengdu, China

Abstract—Networks-on-Chips (NoCs) are at the core of high performance multi-processor systems-on-chips. As the number of cores and sub-systems on chip grows, the size and complexity of NoCs increase as well. Due to the process variation, aging effects and soft-errors in current and expected future process generations, the probability of failure in the NoCs rises and has to be fought at all levels: circuit, architecture, and communication protocols.

This paper discusses appropriate fault models for NoCs and their effects on the architecture and network levels. A method to design fault-tolerant NoCs comprising of techniques at the link level, the routing level, and the end-to-end level of the communication is presented. In addition, the proposed method offers an isolation technique where the computing cores are decoupled from the faults in the network. This technique avoids or at least attenuates the severe impacts of faults on the network performance and functionality. These point techniques are combined together to design fault-tolerant and reliable NoCs.

Keywords—*Fault-tolerant design flow, Networks-on-Chip, Fault modelling*

I. INTRODUCTION

With technology scaling, process variation, wear-out effects, IR drop, thermal hotspots and other phenomena are increasingly challenging. Rather than striving for a perfect chip, it may be more effective to allow failures sometimes to happen and tolerate them. Three steps are required to provide fault tolerance: detection, diagnosis, correction/reconfiguration. In Network-on-Chip (NoC), faults can be detected and handled at several layers in the communication stack and at different places. Link-level error coding, fault-tolerant routing and end-to-end retransmission are three types of point techniques that are located in different layers and components. Although there is a plethora of publications describing these point techniques [1], attempts to combine several of them into a complete solution are rare.

In this work, we cover detection, diagnosis, and correction and formulate a comprehensive design flow for providing a complete solution for a fault-tolerant NoC (FT NoC). In particular we make the following contributions:

1) A layered fault model is described and a fault notation is proposed to describe the faults at the circuit layer.
2) We propose a comprehensive design flow to cover all types of faults, all communication layers and all components.

3) We demonstrate the flow by realizing a partial FT NoC that covers only the datapath links between routers.
4) We compare fault detection and fault-tolerant methods with respect to their capacity for dealing with transient, intermittent and permanent faults.

II. RELATED WORK

The common goal of all fault-tolerant techniques is to increase the reliability but at the cost of larger area, power, and delay. There are many proposed techniques in literature targeting isolated problems while there are few attempts proposing a comprehensive solution for the reliability of the NoC platform as a whole [1]. This section covers a small portion of different attempts.

Faults may change the content of a packet, for example a bit flip caused by the low noise margin [2], the electromagnetic coupling effects [3], or the crosstalk [4]. The faults on the data path of the routers can be detected and possibly corrected using Error Detecting Codes (EDC) and Error Correcting Codes (ECC) such as hamming codes [5][6]. Complex coding techniques provide a higher correctness and are even capable of correcting burst errors, but the increasing on execution time limits their protection capability [7]. Data redundancy mechanisms such as retransmission at the Link-to-Link or End-to-End levels are also well-known solutions to tackle faulty packets [8][6]. The End-to-End retransmission, however, significantly increases packet latency besides the need of a supporting infrastructure such as an Ack/Nack protocol.

Unlike the faults affecting the content of packets, faults in the control path of the routers are hard to detect and can barely be tolerated. A fault (e.g. stuck-at-1 or stuck-at-0) in the control path may lead to the miscalculation of the routing algorithm or a wrong connection between the input and output ports. When these faults occur, packets may be forwarded to a wrong direction and eventually lead to the deadlock or livelock. A common solution to overcome these types of faults is to disable the faulty components and employ a proper routing algorithm to bypass the disabled faulty links [9][10][11], routers[12], or both[13][14][15]. In almost all of these methods, to deal with the deadlock, the network operation has to be stopped and the system has to be reset. The new topology is then discovered during the setup period and then the network can start working again under a new configuration. Some proposals avoid the costly reset [12][15]. However, some packets might be lost in the transition phase but

978-1-4799-8720-7/15 $31.00 © 2015 IEEE   545

the network is protected from entering any deadlock situation for a small number of faults.

Redundancy techniques such as the triple modular redundancy or multiple data sampling are also well-known and can be adapted to NoCs [16]. A rigorous analysis on the use of spare wires and cores to replace defective ones is performed in [8], suggesting that the effective yield of the chip and its overall cost can be significantly improved by using redundant modules. The redundancy techniques are popularly used in safety-critical systems and typically avoided in non-critical systems due to their large overhead.

In [16], several approaches such as architectural solutions and retransmission techniques have been proposed and combined together to improve the reliability of the network against transient faults. To avoid costly end-to-end retransmission, the hop-to-hop retransmission scheme is used. In this approach, a flit transmission over a link takes 3 cycles including 1 cycle for detecting an error, 1 cycle for sending an acknowledgement, and 1 cycle for the transmission itself. In addition, to overcome the faults in the router logics, a mechanism to detect the deadlock and recovery from this situation is proposed, which has a considerable latency overhead. In general, although different approaches are carefully designed and combined to satisfy the low area and power overhead, they are not suitable to tackle permanent faults due to large latency penalties.

In Bulletproof [17], different techniques such as triple modular redundancy, end-to-end error detection and resource sparing are combined. Combinations of techniques at different levels such as the system level, component level, and gate level are examined and the tradeoff between area, power, latency, and reliability are reported accordingly. Bulletproof offers a design space exploration but it lacks a comprehensive method satisfying different tradeoffs.

Before dealing with a fault, it must be detected. Many of the fault detection methods rely on off-line testing mechanisms [18][19][20]. However, faults caused by temporary conditions such as overheating, IR-drop, and wear-out can be hardly detected by off-line testing. This is mainly because of the difficulty in creating similar working conditions in a dynamic environment and designing a proper testing mechanism for them. Built-in self-test (BIST) is a common mechanism in digital circuits to detect faults at run-time [9][19][21][22].

NoCAlert [23] provides a comprehensive on-line and real-time fault detection mechanism specific to the NoC platform. It employs a group of lightweight micro-checker modules that collectively implement real-time hardware assertions. The checkers operate concurrently with the normal NoC operation and are capable of detecting a wide range of faults instantaneously.

In this paper, we suggest a combination of fault-tolerant techniques at different layers which are complementary to each other in order to achieve reliability against transient, intermittent and permanent faults at run time. The area, power, and delay overheads are taken into consideration when choosing the techniques.

III. Fault Model

A. Layer Structure of Fault Model

Fault models are categorized according to three different layers (as shown in Table I): physical layer, circuit layer, and network layer.

TABLE I: Layer structure of fault model

Network Layer	misrouting, packet drop, deadlock, etc.	
Circuit Layer	Behavior	stuck-at-0, stuck-at-1, reversal
	Value	transient, intermittent, permanent
	Location	data path, control path
Physical Layer	Thermal, Process variation, IR-drop, Crosstalk, etc	

Physical level. Continuous models describe the physical phenomena leading to misbehavior. They are rooted in the physical laws governing the operation of the devices.

Circuit level. The errors of circuit output value are defined as faults in circuit layer. The faults in circuit layer have three aspects (as shown in Table I). The first aspect, called "Value", focuses on the changes of the value, including stuck-at-0, stuck-at-1 and reversal. The second aspect, called "Occurrence", describes the time aspect of faults (see examples in Fig. 1). "Location" distinguishes between data path and control path.

SEU (Single Event Upset) and thermal breakdown, lead to functional faults of transistors and result in wrong output values. High temperature, IR-drop, and process variation cause delay fluctuations, which may also result in the errors of output values. Aging phenomena typically start with increasing delays but eventually lead to functional faults.

The relationship between faults in the physical layer and circuit layer are a many-to-many mapping. For example, high temperature can lead to intermittent faults and can also result in permanent damage to a circuit. IR-drop can lead to transient faults when the computation load changes rapidly, or can lead to intermittent faults when the power grid has fluctuations.

Network layer. Network behavior disorder is the abstraction of circuit level faults considering the meaning behind the binary values. Many kinds of network faults have been defined, like misrouting, packet damage, data error, deadlock, and so on [23].

As network layer faults are highly diverse and depend on the particular behavior and component. Of concern, we have not identified a unified, elegant fault model. Hence, we find network layer fault models are unsuitable to guide the selection and design of NoC fault-tolerant methods. Circuit layer faults are abstractions of physical effects, but there is no simple and direct relation. Since circuit level faults have the virtue to be simple and can be uniformly applied to all components and functions in the network, we use a circuit level fault model as the basis for design and evaluation of a fault-tolerant NoC.

B. Fault Models in Circuit Layer

An error is described as $FM = \{P_O, P_L, P_R, e\}$, in which e is the type of fault, P_O, P_L, and P_R describe the temporal and spatial distribution of faults by defining the occupation probability, impact probability and recovery probability, respectively. These three parameters can be a constant value or a function. Faults can recover naturally when the causing condition (like high temperature or low voltage) ends. P_R describes duration before the recovery, by defining the expectation of the duration as $\frac{1}{P_R}$. Transient faults usually recover quickly, intermittent faults recovery may take longer, while permanent faults never recover. P_L is only relevant to intermittent faults and describes the probability that the fault actually changes the value of a signal. When an intermittent fault occurs and if P_L is 0.5, there is a 50% chance that a value is changed in a particular cycle. Three examples (Fig. 1) of fault notation are given as follow:

Example 1 (Transient fault) $FM_1 = \left\{1 \times 10^{-4}, 1, 0.9, Inverted\right\}$. The duration time of this transient fault is $1/0.9 \approx 1.1$ cycles on average.

Example 2 (Intermittent fault) $FM_2 = \left\{1 \times 10^{-7}, 0.5, 0.0625, Stack - at - 0\right\}$. The duration of this intermittent fault is $1/0.0625 = 16$ cycles on average. During this time, the value is impacted by the fault in half of the cycles on average.

Example 3 (Permanent faults) $FM_3 = \left\{1 \times 10^{-10}, 1, 0, Stack - at - 1\right\}$. There is a new stack-at-1 fault every 10^{10} cycles on average and the fault never recovers ($P_R = 0$). These faults are enable in every cycle after occurrence ($P_L = 1$).

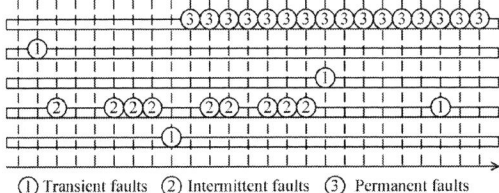

① Transient faults ② Intermittent faults ③ Permanent faults

Fig. 1: Fault model examples.

A given signal can suffer from more than one kind of faults. A *fault scenario* $S = \{FM_1, FM_2, FM_3, \ldots\}$ combines the fault notations of different kinds of faults.

IV. COMPREHENSIVE FAULT-TOLERANT DESIGN

In this section, we compare several fault detection and fault-tolerant methods and we propose a comprehensive fault tolerant design flow.

A. Choosing Fault Detection Methods

Fault detection methods are necessary to detect and diagnose faults in the NoC. First, they detect and report faults. Then, they locate the faulty components. For the data path, detection methods should point out the error bits. For the control path, they should be able to locate the faulty component (RC-route compute, VA-virtual buffer allocation, SA-switch allocation, and buffers) in a router. Usually, the capacity of location is lower than the capacity of detection.

Three typical fault detection methods are compared in Table II. Error Detection Coding is often used on the data path while assertions can detect errors on the control path by checking certain rules. Built-in Self-Tests (BIST) can provide detailed detection and diagnosis on every component.

TABLE II: Fault detection and location methods
(-: no coverage; P: partial coverage; A: Complete coverage)
(Ref.=Reference, Tran.=Transient, Inter.=Intermittent, Perm.=Permanent)

Method	Ref.	Data Path			Control Path		
		Tran.	Inter.	Perm.	Tran.	Inter.	Per.
Error Detection Code	[5][9]	P	P	P	-	-	-
Assertion	[23]	-	-	-	A	A	A
Built-In Self Test	[9]	P	P	A	P	P	A
	[21]	P	P	A	P	P	A
	[22]	P	P	A	P	P	A
	[19]	P	P	A	-	-	-

When choosing a fault detection method, the impact of the testing procedure is also an important metric. ECC and assertions introduce additional delays, and BISTs have to isolate the components under tests with wrappers. The method proposed by [9][21] isolates one router with its ports while [22] and [19] totally stop the network before the testing. When only testing the link, [9] blocks the packets on the link.

B. Choosing Fault-tolerant Methods

Fault-tolerant methods can be classified into real-time tolerance, reconfiguration and retry.

Real-time tolerance. Methods correct faults at each cycle, e.g. error correction coding (ECC) and multi-sampling (MS). These methods have the capacity for fault detection and fault tolerance at the same time and they can cover transient, intermittent and permanent faults.

Reconfiguration. Faults are tolerated using the redundant resources in the network by reconfiguring routers or paths. Typical methods of this kind include triple modular redundancy (TMR), spare-wire (SW), split-transmission (ST), and fault-tolerant routing algorithms (FTR). To use these methods, faults must already have been diagnosed, hence they cannot work for transient faults.

Retry tolerance. Hop-to-Hop (H2H) and End-to-End (E2E) retransmission cannot tolerate permanent faults when combined with static routing. The reason is that the same faulty path is chosen over and over. Retransmissions are more effective when transient or intermittent faults are introduced and the chance of the second successful delivery is high. Similarly, stochastic routing tries different paths at the same time and expects one of them can arrive correctly.

Typical methods are compared in Table III.

TABLE III: Fault-tolerant methods
(-: no coverage; P: partial coverage; A: complete coverage)
(Ref.=Reference, Tran.=Transient, Inter.=Intermittent, Perm.=Permanent)

Method	Ref.	Data Path			Control Path		
		Tran.	Inter.	Perm.	Tran.	Inter.	Perm.
Error Correcting Code	[5][9]	P	P	P	-	-	-
Fault-tolerant Routing	[24]	-	P	A	-	P	A
	[11][12][25]	-	P	P	-	P	P
Triple Modular Redundancy	[26]	P	P	P	P	P	P
Multi Samples	[26]	P	P	P	P	P	P
Stochastic Routing	[27]	A	A	A	A	A	A
Split-Link transaction	[6]	-	P	P	-	-	-
Spare-Wire transaction	[28]	-	P	P	-	-	-
Hop-to-Hop Retransmission	[16]	A	A	-	A	A	-
End-to-End Retransmission	[16]	A	A	-	A	A	-

C. Comprehensive Design

As shown in table II and table III, using only one method cannot address all challenges on reliability. It is necessary to

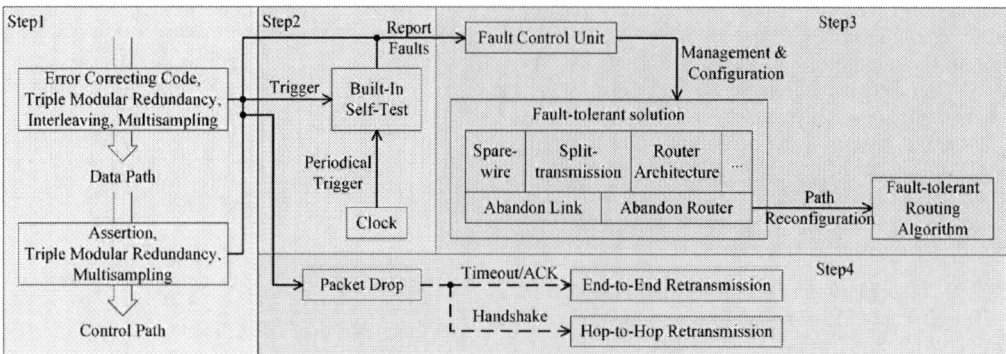

Fig. 2: General flow of fault tolerance

combine different fault detection and fault-tolerant methods following the general flow shown in Fig. 2.

The fault-tolerant flow contains four steps. At first, the faults in flits and control signals are detected and, if possible, corrected by real-time tolerance. Error correcting codes (on the data path), assertions (on the control path), triple modular redundancy and multi-sampling can be used in this step.

When faults are detected, a fault diagnosis process (step 2 in Fig. 2) is triggered. It can also be triggered periodically. BIST is the most common method to run exhaustive tests.

Faults detected by the first two steps will be reported to the fault control unit. The fault control unit manages and configures different fault-tolerant methods. The fault control unit can be deployed centrally or distributively and located in each router and even each port. Reconfiguration methods will be called in this step. With increasing number of faults in the network, the strategy moves from router reconfiguration to path reconfiguration, because router configuration can maintain the topology but its fault-tolerance capacity is limited. If the faults cannot be recovered by router reconfiguration, fault-tolerant routing algorithms are called to reconfigure paths by abandoning links or routers.

The fourth step deals with faulty flits. Damaged flits have to be dropped and retransmitted. The fault-tolerant flow only drops packets. Retransmission is activated by a time-out of not acknowledged packets or handshake signals.

To design reliable FT NoCs, all four steps are necessary but in each step only one or two candidate methods are chosen. For example, designers can choose one method for the data path and one method for the control path protection in the first step. Either hop-to-hop retransmission or end-to-end retransmission is enough to make sure that no packets are lost.

V. EVALUATION

We simulated a fault-tolerant design for an 8×8 mesh network where we implemented part of the comprehensive design flow, as shown in Fig. 3. The FT NoC tolerates faults on the links between routers. It does not provide a complete solution for a FT NoC but it serves to demonstrate the comprehensive design flow and studies the effectiveness of several methods with respect to transient, intermittent and permanent faults.

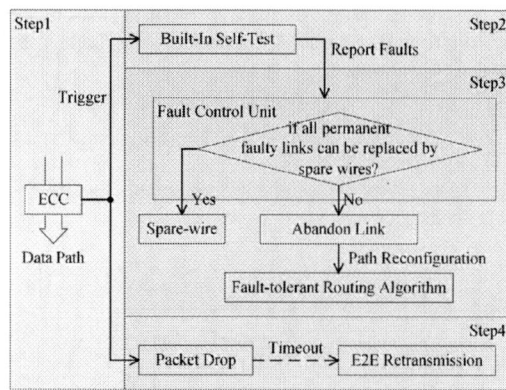

Fig. 3: Fault tolerance flow of simulation

A. Reliable Design

The fault-tolerant flow of our design is shown in Fig. 3, which is a simplified version of general flow in Fig. 2.

ECC. We implement a group Hamming(12,8) at each port, which can tolerate 1 faulty bit and detect 2 faulty bits in 12 bits [6]. This code word will be checked at each router. If there is only one faulty bit in a group of 12 bits, the codec will correct it. If there are two faulty bits, the codec will report faults to trigger a BIST and drop the packet.

BIST. Test vectors are injected from the output port and checked at the input port. Every port reuses the same testing process.

Fault-tolerant Solution. We select spare-wires and a fault-tolerant routing algorithm. The spare-wire architecture can replace 2 permanently faulty wires in 16 bits wires. A fault-tolerant routing algorithm, called HiPFaR [12], is designed to tolerate faulty links.

Fault Control Unit. The fault control unit is located at each port and makes distributed decisions. If the BIST diagnoses faults, the faulty wires are replaced by spare wires. If spare wires cannot tolerate all permanent faulty wires in a link, this link is marked as faulty and bypassed by the fault-tolerant routing algorithm.

Retransmission. End-to-End retransmission is triggered by a time-out. To reduce the number of retransmission packets

978-1-4799-8720-7/15 $31.00 © 2015 IEEE   548

and avoid congestion, only the first packets are retransmitted with timer overflow. In the simulation the time-out is set to 1000 cycles.

The original width of a flit is 128 bits and extended to 192 bits due to the Hamming(12,8) code protection. Including spare wires, the number of wires between two routers is 216 bits. The redundancy rates with and without spare-wire are 33.33% and 40.74% respectively.

B. Simulation Setup

The 8×8 network is simulated using a modified POPNET simulator. Each router contains five physical ports (Local, North, South, East, West). Ports on the Y-axis have two virtual channels and ports on X-axis have one virtual channel. Each virtual channel provides buffers for 12 flits at the input port. The network is simulated with a low injection of 0.01 packets/cycle/router in a random, uniform traffic pattern. A packet contains five flits.

Three fault models are exercised, each with only one kind of faults, transient, intermittent or permanent: $FM_t = \{P_O, 1, 0.9, Inverted\}$, $FM_i = \{P_O, 0.01, 0.001, Inverted\}$ and $FM_p = \{P_O, 1, 0, Inverted\}$, respectively. For the permanent case, we simulate the steady state after the faults have accumulated to compress the simulation time. Faults are injected according to P'_O, which is defined as the proportion of accumulated faults.

The following methods are simulated for comparison: 1) ECC; 2) ECC+FTR (Fault-tolerant Routing); 3) ECC+SW (Spare-wire); 4) ECC+SW+FTR, 5) ECC+SW+RT (Retransmission); 6) ECC+SW+RE+FTR. If SW is not used, any line with a permanent fault will be abandoned. XY-routing is implied in those methods without fault-tolerant routing algorithm.

The evaluation metrics are end-to-end latency, delivery rate, and retransmission time. End-to-end latency is the time from when a packet is injected into the network first time until this packet is successfully received, including all retransmissions. To measure the number of successful delivered packets, 30000 packets are injected into the network.

C. Simulation under Transient and Intermittent Fault

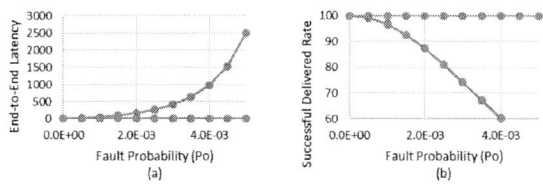

Fig. 4: Simulation with transient fault $FM = \{P_O, 1, 0.9, Inverted\}$.

Fig. 4 and Fig. 5 illustrate the performance under transient and intermittent faults, respectively. Because spare-wire and fault-tolerant routing are used only when there are permanent faults, the differences between the lines in each figure are due to retransmission. Thus, in the plots only two distinct lines are visible.

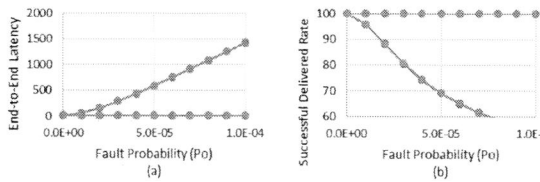

Fig. 5: Simulation with intermittent fault $FM = \{P_O, 0.01, 0.001, Inverted\}$.

Retransmission can provide 100% delivery rate when facing transient faults $P_O < 0.005$ and intermittent faults $P_O < 0.0001$. However, for transient faults with $P_O > 0.004$ and for intermittent faults with $P_O > 7.5 \times 10^{-5}$, the latency becomes excessive.

D. Simulation under Permanent Fault

In contrast to transient faults and intermittent faults, permanent faults accumulate over time. Faults are injected only at the beginning of the simulation according to the given P'_O.

The simulation results are drawn in Fig. 6. At first, SW significantly increases the fault-tolerance capacity in the network. The delivery rate of ECC and ECC+FTR drops lower than 60% for $P'_O > 0.008$ and $P'_O > 0.013$, respectively. The methods with SW achieve delivery rates higher than 90% up to $P'_O = 0.028$.

Comparing methods with and without FTR, the number of delivered packets can be increased by reconfiguring paths. In the best observed case, when $P'_O = 0.04$, 13% more packets are delivered due to FTR. Moreover, FTR can reduce the number of retransmissions and the end-to-end latency by choosing another link at the second time.

At the beginning of the simulation, the network knows nothing about faults and discovers them using ECC and BIST. So some packets are rescued by retransmission. On the other hand, the improvement is limited. The original paths and retransmission paths of most packets are the same, because the faults do not change during simulation. With $P'_O = 0.04$, only 0.6% packets are delivered due to retransmission.

Finally, comparing ECC+SW+RE+FTR to ECC+SW+RE with $P'_O > 0.028$, no improvement is observed, because of deadlocks. Thus, to combine FTR with other methods, deadlocks and livelocks must be avoided or handled properly.

VI. CONCLUSION

Fault-tolerant design takes an important role in NoCs to overcome failures introduced by aggressive technology scaling. In this paper, we have introduced a design flow for FT NoCs and we have reported a first study in combining several methods. From the simulation results, we draw the following conclusions: 1) Retransmission can help to increase the number of delivered packets significantly for transient and intermittent faults. 2) Spare-wire clearly increases the fault-tolerance capability but has significant overhead. 3) Fault-tolerant routing helps to tolerate more faults and reduces the overhead of retransmission and end-to-end latency, if deadlocks and livelocks are avoided.

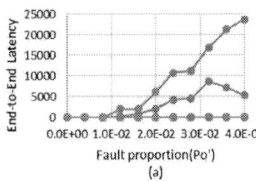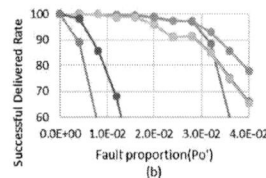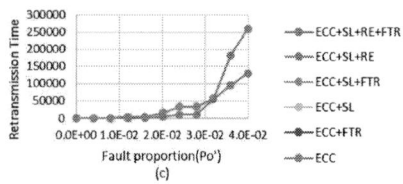

Fig. 6: Simulation with permanent fault

ACKNOWLEDGMENT

The research is support by National Natural Science Foundation of China No. 61006027 and No. 61176025, New Century Excellent Talents Program No. NCET-10-0297, the Fundamental Research Funds for the Central Universities No. ZYGX2012J003. This work is also supported by VINNOVA-MarieCurie within the CUBRIC and ERoT projects and Academy of Finland.

REFERENCES

[1] M. Radetzki, C. Feng, X. Zhao, and A. Jantsch, "Methods for fault tolerance in networks-on-chip," *ACM Computing Surveys (CSUR)*, vol. 46, no. 1, p. 8, 2013.

[2] S. Yang and M. Greenstreet, "Noise margin analysis for dynamic logic circuits," in *IEEE/ACM International Conference on Computer-Aided Design (ICCAD)*, 2005, pp. 406–412.

[3] I. Erdin, M. Nakhla, and R. Achar, "Circuit analysis of electromagnetic radiation and field coupling effects for networks with embedded full-wave modules," *IEEE Transactions on Electromagnetic Compatibility*, vol. 42, no. 4, pp. 449–460, 2000.

[4] Aniket and R. Arunachalam, "A novel algorithm for testing crosstalk induced delay faults in vlsi circuits," in *18th International Conference on VLSI Design*, 2005, pp. 479–484.

[5] H. Zimmer and A. Jantsch, "A fault model notation and error-control scheme for switch-to-switch buses in a network-on-chip," in *1st IEEE/ACM/IFIP International Conference on Hardware/Software Codesign and System Synthesis*, 2003, pp. 188–193.

[6] Q. Yu and P. Ampadu, "A dual-layer method for transient and permanent error co-management in noc links," *IEEE Transactions on Circuits and Systems II: Express Briefs*, vol. 58, no. 1, pp. 36–40, 2011.

[7] J. Collet, "A brief overview of the challenges of the multicore roadmap," in *Proceedings of the 21st International Conference on Mixed Design of Integrated Circuits Systems (MIXDES)*, 2014, pp. 22–29.

[8] S. Shamshiri, A.-A. Ghofrani, and K.-T. Cheng, "End-to-end error correction and online diagnosis for on-chip networks," in *IEEE International Test Conference*, 2011, pp. 1–10.

[9] A. DeOrio, D. Fick, V. Bertacco, D. Sylvester, D. Blaauw, J. Hu, and G. Chen, "A reliable routing architecture and algorithm for nocs," *IEEE Transactions on Computer-Aided Design of Integrated Circuits and Systems*, vol. 31, no. 5, pp. 726–739, 2012.

[10] K. Aisopos, A. DeOrio, L.-S. Peh, and V. Bertacco, "Ariadne: Agnostic reconfiguration in a disconnected network environment," in *International Conference on Parallel Architectures and Compilation Techniques (PACT)*, 2011, pp. 298–309.

[11] M. Ebrahimi, M. Daneshtalab, J. Plosila, and F. Mehdipour, "Md: minimal path-based fault-tolerant routing in on-chip networks," in *18th Asia and South Pacific Design Automation Conference (ASP-DAC)*, 2013, pp. 35–40.

[12] M. Ebrahimi, M. Daneshtalab, and J. Plosila, "High performance fault-tolerant routing algorithm for noc-based many-core systems," in *21st Euromicro International Conference on Parallel, Distributed and Network-Based Processing (PDP)*, 2013, pp. 462–469.

[13] M. Koibuchi, H. Matsutani, H. Amano, and T. M. Pinkston, "A lightweight fault-tolerant mechanism for network-on-chip," in *Proceedings of the 2nd ACM/IEEE international symposium on networks-on-chip*, 2008, pp. 13–22.

[14] C. Feng, Z. Lu, A. Jantsch, J. Li, and M. Zhang, "A reconfigurable fault-tolerant deflection routing algorithm based on reinforcement learning for network-on-chip," in *Proceedings of the 3rd International Workshop on Network on Chip Architectures*, 2010, pp. 11–16.

[15] M. Ebrahimi and M. Daneshtalab, "A light-weight fault-tolerant routing algorithm tolerating faulty links and routers," *Computing*, pp. 1–18, 2013.

[16] S. Murali, T. Theocharides, N. Vijaykrishnan, M. J. Irwin, L. Benini, and G. D. Micheli, "Analysis of error recovery schemes for networks on chips," *IEEE Design & Test of Computers*, vol. 22, no. 5, pp. 434–442, 2005.

[17] K. Constantinides, S. Plaza, J. Blome, B. Zhang, S. Bertacco, S. Mahlke, T. Austin, and M. Orshansky, "Bulletproof: a defect-tolerant cmp switch architecture," in *the 12th International Symposium on High-Performance Computer Architecture*, 2006, pp. 5–16.

[18] J. Raik, V. Govind, and R. Ubar, "An external test approach for network-on-a-chip switches," in *15th Asian Test Symposium*, 2006, pp. 437–442.

[19] E. Cota, F. Kastensmidt, M. Cassel, M. Herve, P. Almeida, P. Meirelles, A. Amory, and M. Lubaszewski, "A high-fault-coverage approach for the test of data, control and handshake interconnects in mesh networks-on-chip," *IEEE Transactions on Computers*, vol. 57, no. 9, pp. 1202–1215, 2008.

[20] A. M. Amory, E. Brião, É. Cota, M. Lubaszewski, and F. G. Moraes, "A scalable test strategy for network-on-chip routers," in *IEEE International Test Conference (ITC)*, 2005, pp. 1–9.

[21] M. R. Kakoee, V. Bertacco, and L. Benini, "At-speed distributed functional testing to detect logic and delay faults in nocs," *IEEE Transactions on Computers*, vol. 63, no. 3, pp. 703–717, 2014.

[22] X.-T. Tran, Y. Thonnart, J. Durupt, V. Beroulle, and C. Robach, "Design-for-test approach of an asynchronous network-on-chip architecture and its associated test pattern generation and application," *IET Computers & Digital Techniques*, vol. 3, no. 5, pp. 487–500, 2009.

[23] A. Prodromou, A. Panteli, C. Nicopoulos, and Y. Sazeides, "Nocalert: An on-line and real-time fault detection mechanism for network-on-chip architectures," in *45th Annual IEEE/ACM International Symposium on Microarchitecture (MICRO)*, 2012, pp. 60–71.

[24] E. Wachter, A. Erichsen, A. Amory, and F. Moraes, "Topology-agnostic fault-tolerant noc routing method," in *Proceedings of the Conference on Design, Automation and Test in Europe*, 2013, pp. 1595–1600.

[25] H. Hsin, E. Chang, C. Lin, and A.-Y. Wu, "Ant colony optimization-based fault-aware routing in mesh-based network-on-chip systems," *IEEE Transactions on Computer-Aided Design of Integrated Circuits and Systems*, vol. 33, no. 11, pp. 1693–1705, 2014.

[26] P. A. Frantz, M. Cassel, F. L. Kastensmidt, É. Cota, and L. Carro, "Crosstalk-and seu-aware networks on chips," *IEEE Design & Test*, vol. 24, no. 4, pp. 340–350, 2007.

[27] P. Bogdan, T. Dumitraş, and R. Marculescu, "Stochastic communication: A new paradigm for fault-tolerant networks-on-chip," *VLSI design*, vol. 2007, 2007.

[28] T. Lehtonen, D. Wolpert, P. Liljeberg, J. Plosila, and P. Ampadu, "Self-adaptive system for addressing permanent errors in on-chip interconnects," *IEEE Transactions on Very Large Scale Integration (VLSI) Systems*, vol. 18, no. 4, pp. 527–540, 2010.

978-1-4799-8720-7/15 $31.00 © 2015 IEEE

2015 IEEE Computer Society Annual Symposium on VLSI

Design Exploration For Next Generation High-Performance Manycore On-chip Systems: Application To big.LITTLE Architectures

Anastasiia Butko, Abdoulaye Gamatié, Gilles Sassatelli, Lionel Torres and Michel Robert
LIRMM (CNRS and University of Montpellier), Montpellier, France
Email: {firstname.lastname}@lirmm.fr

Abstract—Next generation embedded systems will massively adopt on-chip manycore architectures to provide both performance and energy-efficiency. This trend will definitely establish the convergence of embedded computing and high-performance computing. In such a context, one major design challenge will concern the choice of adequate architecture parameters given system requirements. Moreover, it will affect the way applications can suitably exploit architecture resources for an efficient execution. This paper deals with manycore on-chip system design exploration by using via simulation. It presents an approach enabling one to study central design parameters in an accurate and cost-effective manner. This approach is illustrated through the design exploration for ARM big.LITTLE heterogeneous multicore technology in the gem5 framework.

Keywords—*High-performance computing, energy-efficiency, manycore, heterogeneous, big.LITTLE, modeling, gem5, trace-driven.*

I. INTRODUCTION

The number of cores in future on-chip architectures will keep in increasing as already observed in state-of-the-art systems such as the Many Integrated Core Architecture (MIC) of Intel [1], TILE-Gx of Tilera [2], Multi-Purpose Processor Array (MPPA) of Kalray [3] and ARM big.LITTLE architecture [4]. MIC is the architecture adopted by Intel Xeon Phi coprocessors used as compute accelerators in the world's fastest supercomputer (Tianhe-2 [5]). It is composed of 61 cores interconnected by a bi-directional ring network. The TILE-Gx architecture is composed of 72 cores interconnected by a 2D mesh NoC using wormhole routing packets. MPPA is a manycore architecture integrates 256 cores, where cores are distributed across 16 compute clusters. The big.LITTLE technology promotes heterogeneous and adaptive architectures as illustrated in Samsung Exynos Octa 5410 and 5422 chips. It enables to dynamically migrate applications between two different clusters of ARM cores: a low-energy consumption cluster ("LITTLE") composed of four Cortex-A7 cores versus a high-performance cluster ("big") composed of four Cortex-A15 cores.

To illustrate the potential of on-chip big.LITTLE architecture, Table I reports energy-efficiency numbers measured on the Odroid XU3 board (see Figure 1) integrating the Exynos Octa 5422 chip. These numbers have been obtained by executing the high-performance Linpack benchmark to determine the number of floating-point operations per second (flops) and the corresponding power consumption, for different frequencies of the two clusters. Here, the configuration composed of four

Cortex-A15 cores running at 800MHz appears as the most energy-efficient.

Thanks to their energy-efficiency, the above on-chip systems are significantly contributing to draw the current convergence of embedded computing and high-performance computing. Considering this trend, an important challenge concerns the design of adequate energy-efficient manycore systems that will successfully fill the requirements of both computing domains.

To address this challenge, relevant design exploration frameworks are required for cost-effectiveness reason. Candidate frameworks must be accurate enough to enable designers to address architecture features with respect to the way applications can suitably exploit resources for an efficient execution.

In this paper, we consider the gem5 architecture exploration framework for demonstrating an approach for modeling state-of-the-art multicore on-chip systems and exploring their scalability. We consider the big.LITTLE architecture as a base template by devising a model of the Exynos Octa 5422 clusters in gem5 full-system mode. We evaluate the accuracy of the resulting model via a subset of the Rodinia compute-intensive benchmark suite [6]. We argue that the accuracy of our models, which is around a precision error of 20% is relevant enough to show that the models capture in a consistent way system execution performed on real computer boards including Exynos Octa 5422 chip. In order to accelerate the model-based exploration for large-scale designs, we apply a trace-driven abstraction in gem5 [7] to the big.LITTLE architecture.

The rest of this paper is organized as follows: Section II discusses a few related gem5-oriented ARM core modeling studies. Then, Section III and IV respectively present our big.LITTLE architecture modeling in gem5 and an assessment of the modeling accuracy. Section V explores the design of large-scale system scenarios including more than hundred cores. Finally, concluding remarks are given in Section VI.

II. RELATED WORK

In order to carry out our design exploration for manycore embedded systems, we consider the gem5 simulator system [8]. It is a quasi-cycle accurate simulation platform for computer system architecture research. The gem5 framework provides multiple architecture exploration features: system emulation and complete full-system simulation modes, different CPU models which represent in-order and out-of-order

978-1-4799-8720-7/15 $31.00 © 2015 IEEE   551

TABLE I.　EVALUATION OF HIGH-PERFORMANCE LINPACK ON THE ODROID XU3 BOARD.

A7@200MHz	A7@800MHz	A7@1.4GHz	A15@200MHz	A15@800MHz	A15@1.4GHz	A15@2GHz
Total board power (W)						
2.2	2.6	3.5	2.9	5.1	7.5	12.5
HPL score (Gflops)						
0.26	1.04	1.68	0.96	3.42	4.96	4.7
Energy-efficiency (Mflops/W)						
118	404	484	347	746	662	376

executions, a large set of ISAs (ALPHA, ARM, x86, SPARC, PowerPC, MIPS) and variety of memory configurations, e.g. cache coherence protocols, interconnects, memory controllers, etc. In the next paragraphs, we mainly discuss a few relevant studies on the use of gem5 for design analysis.

In an early study [9], we evaluated the accuracy of modeling real systems using gem5 simulator. By considering a range of benchmarks from scientific computing and media applications domains, we compared simulation results against a real hardware. The observed accuracy was promising enough to consider gem5 as an interesting architecture design exploration framework. Authors in In [10] design a gem5 model of CoreTile Express system-on-chip (SoC) and estimate the accuracy of Cortex-A15 core, memory system and interconnect. They deeply explore the micro-architectural simulation for the homogeneous dual-core system. The work presented in [11] deals with the modeling and simulation of Cortex-A8 and Cortex-A9 cores in gem5. A comparison in terms of execution time is achieved against a real hardware execution based on ten benchmarks. Authors claim that their core models are more accurate than similar micro-architectural simulators. A similar study has been acieved for Cortex-A7 and Cortex-A15 cores in [12] by focusing on the micro-architectural simulation of these cores. The gem5 and McPAT frameworks have been combined to validate area and energy/performance trade-offs against the published datasheet information. However, this work does not aim to multicore evaluation. It only demonstrates the difference between Cortex-A7 and Cortex-A15 cores running single-threaded applications. The current work rather focus on the multi- and manycore design exploration based on the same cores models.

III. MODELING OF A BIG.LITTLE ARCHITECTURE

We introduce the main features of the computer board integrating the considered big.LITTLE technology. Then, we describe the modifications applied to gem5 so as to infer a corresponding model that will serve later for design exploration.

A. Reference platform main features

To evaluate the accuracy of our big.LITTLE model we used the Odroid XU3 board with the embedded Exynos 5422 chip which is illustrated in Figure 1.

The Exynos 5422 processor includes two clusters known as "LITTLE" cluster with four Cortex-A7 in-order cores and "big" cluster with four Cortex-A15 out-of-order cores. In contrast to the previous chip versions, Exynos 5422 features the heterogeneous multiprocessing (HMP) solution also known as global task scheduling (GTS) thus all eight cores can run simultaneously. The LITTLE cluster supports 200MHz - 1.4GHz frequency range and the big cluster supports 200MHz

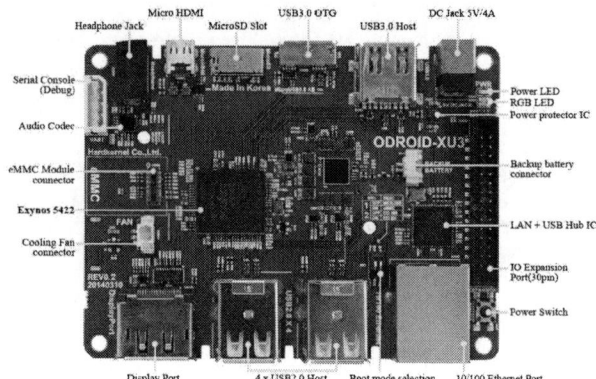

Fig. 1.　Odroid XU3 board [13].

- 2GHz. Each core has its 32Kb private L1 data and instruction caches with 2-way associativity and 4ns latency. LITTLE and big clusters contain two 512kB and 2MB L2 caches which are shared among four cluster cores respectively. The cache coherency between them as well as with the memory is maintained by the Cache Coherence Interconnect 400 (CCI) [14]. The main memory is 2GB LPDDR3 RAM running at 933MHz and is integrated by package on package (PoP) method. It has two 32-bit channels and achieves 14.9GB/s memory bandwidth.

B. ARM big.LITTLE model

In order to simulate heterogeneous big.LITTLE architecture in symmetric multiprocessing (SMP), e.g. only big or LITTLE clusters, as well as in HMP modes we calibrated our core model according to the Odroid XU3 reference platform:

- The key feature of the reference big.LITTLE processor is the ability to run eight cores simultaneously. The actual gem5 version does not support ARM full-system simulation with more than four cores out of snoop control unit (SCU) implementation. We by-passed this restriction by modifying the SCU component through not masking the real core number. Another issue relates to the in-order CPU model which is not devised yet to the ARM ISA. This problem is often discussed in the research community and according to gem5 developers there are three solutions: (i) *TimingSimpleCPU* model, (ii) *MinorCPU* model and (iii) *DerivO3CPU* model which can be modified to produce quasi-in-order execution [11] [12]. In our experiments we evaluated all three scenarios. The last modifications related to the heterogeneous nature of the considered architecture are performed in the gem5

978-1-4799-8720-7/15 $31.00 © 2015 IEEE

TABLE II.   RODINIA BENCHMARK DESCRIPTION.

Application/Kernel	Abbreviation	Domain	Problem size
Back Propagation	backprop	Pattern Recognition	65536
Breadth-First Search	bfs	Graph Algorithms	4096
Heart Wall	heartwall	Medical Imaging	test.avi, 1 frame
HotSpot	hotspot	Physics Simulation	64 x 64
K-means openmp/serial	kmeans	Data Mining	100
Lower Upper Decomposition	lud	Linear Algebra	256
k-Nearest Neighbors	nn	Data Mining	42760
Needleman-Wunsch	nw	Bioinformatics	1024
Speckle Reducing Anisotropic Diffusion	srad v1	Image Processing	1 x 502 x 458
	srad v2		512 x 512

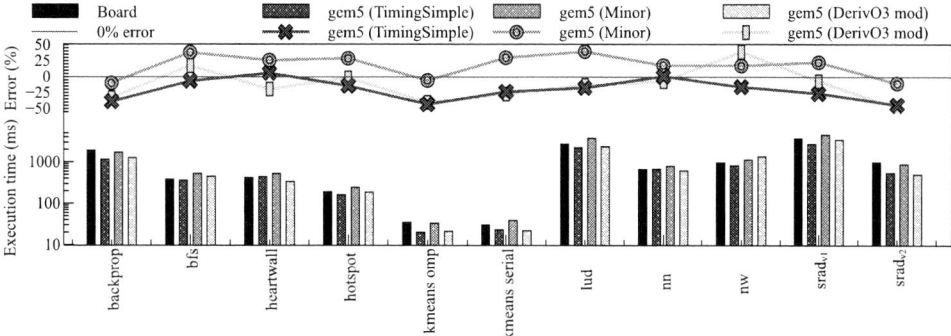

Fig. 2.   Execution time comparison for LITTLE Cortex-A7 cluster running at 200MHz.

full-system creation script so as to support multiple CPU models throughout a simulation.

- To support multiple clocks for LITTLE and big clusters we changed the *fs.py* script.

- Since each cluster has its individually shared among the cores L2 cache we added new option that identifies the number of L2 caches as well as their parameters. As gem5 does not contain the ARM CCI-400 interconnect component we ensured the cache coherency by connecting L2 caches and memory via the coherent crossbar (*CoherentXBar* [15]).

- The recent gem5 version provides multiple DRAM controller models. Thus we chose LPDDR3 with two 32-bit channels. Note that the LPDDR3 timing corresponds to 800MHz frequency and not 933MHz.

IV.   ACCURACY EVALUATION

Now we assess the accuracy of the previous big.LITTLE architecture model against the reference Odroid XU3 platform.

A. Rodinia benchmark suite

The reference Odroid XU3 board run Linux Kernel 3.10 which allows GTS. We modified the Linux Kernel 3.10 in order to run it on gem5 simulator.

The Rodinia benchmark suite for heterogeneous computing [6] is used to validate our big.LITTLE model. It contains twenty applications and kernels from different scientific domains which is parallelized with OpenMP for multicore CPUs and with CUDA API for GPUs. We used OpenMP

implementation with four threads per each cluster. Also, the `GOMP_CPU_AFFINITY` variable is used to ensure identical thread scheduling on the board and on the gem5 system. The following eleven applications and kernels are chosen: *backprop, bfs, heartwall, hotspot, kmeans openmp/serial, lud, nn, nw, srad v1/v2*. The complete set of application description and problem size are presented in Table II.

B. Analysis

We explored three available options to model ARM in-order processor and to identify the accuracy of each one:

1) *TimingSimpleCPU* is the simplest purely functional in-order model which uses timing memory accesses.
2) *MinorCPU* is an in-order processor model with a fixed pipeline but configurable data structures and execute behavior. It supports the Fetch (1,2), Decode and Execute pipeline stages.
3) *DerivO3CPU* (modified) is the most complex out-of-order model which has Fetch, Decode, Rename, Issue/Execute/Writeback and Commit pipeline stages.

The comparative results are presented in Figure 2. Note that the scale for the execution time is logarithmic. The figure shows the execution time for eleven Rodinia applications and kernels executed on the Cortex-A7 cluster running at 200MHz on: (i) reference board, (ii) gem5 *TimingSimpleCPU* model, (iii) gem5 *MinorCPU* and (iv) modified gem5 *DerivO3CPU*. As we can see, the absolute error percentage varies between 1% and 50%. The minimum and maximum errors as well as the absolute average error for each scenario are listed in Table III. The results show that the execution time absolute error for

978-1-4799-8720-7/15 $31.00 © 2015 IEEE

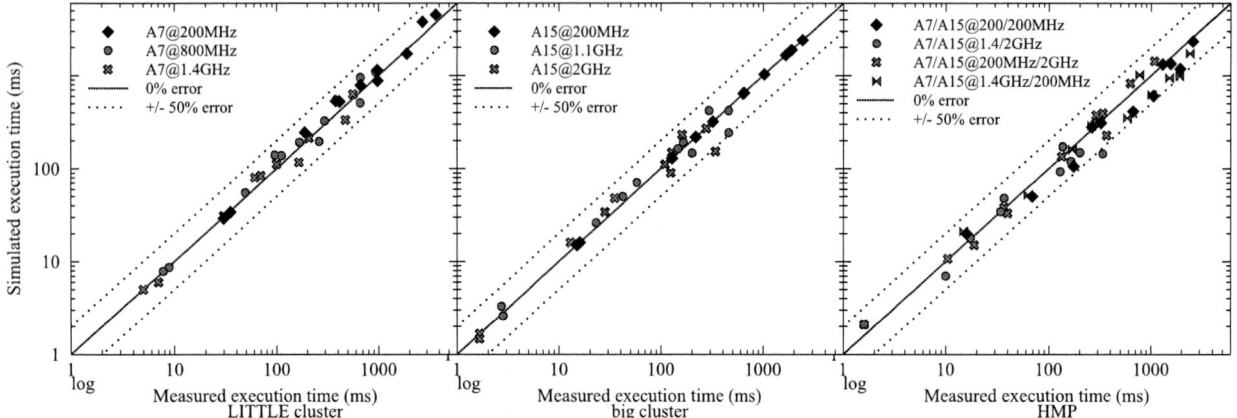

Fig. 3. Execution time comparison between gem5 model and Exynos Octa 5422.

all three models is around 22%. Thus, we conclude that for performance evaluation it is enough to use the *TimingSimpleCPU* model. However, for more detailed studies, such as microarchitectural or power consumption exploration, it is necessary to switch to a more detailed one.

TABLE III. CORTEX-A7 IN-ORDER MODEL EXECUTION TIME ERROR SUMMARY.

CPU model	Minimum error	Maximum error	Absolute average error
TimingSimpleCPU	0.6%	43.9%	21.4%
MinorCPU	5.7%	39.7%	22.5%
DerivO3CPU (mod)	2.2%	48.7%	21.6%

For further accuracy evaluation to simulate the in-order Cortex-A7 cores we selected the modified gem5 *DerivO3CPU* model. The following three scenarios are considered:

- Accuracy evaluation of the LITTLE Cortex-A7 cluster in SMP mode running at 200MHz, 800MHz and 1.4 GHz (LITTLE 1, 2, and 3 respectively).

- Accuracy evaluation of the big Cortex-A15 cluster in SMP mode running at 200MHz, 1.1GHz and 2GHz (big 1, 2, and 3 respectively).

- Accuracy evaluation of the big.LITTLE in HMP mode with Cortex-A7/A15 running at 200MHz/200MHz, 1.4GHz/2GHz, 200MHz/2GHz and 1.4GHz/200MHz (big.LITTLE 1, 2, 3 and 4 respectively).

The correlation results are shown in Figure 3. Each scenario has eleven points which correspond to the chosen Rodinia kernels and applications. Their execution time varies between milliseconds and seconds thus the scale is logarithmic. Two large-dotted lines show the -50% and 50% error edges. The comparative results, such as minimum/maximum and average absolute errors are presented in Table IV.

To summarize the average absolute errors are 19.3%, 20.1% and 22.9% for LITTLE cluster, big cluster and big.LITTLE in HMP mode respectively.

For a more detailed analysis we considered application output information over the time spent in different stages,

TABLE IV. EXECUTION TIME ERROR SUMMARY.

Scenario	Minimum error	Maximum error	Absolute average error
LITTLE 1	2.2%	48.7%	21.6%
LITTLE 2	2%	45.6%	19.7%
LITTLE 3	1%	38.7%	17%
big 1	2.7%	39.6%	17.9%
big 2	5.8%	46.7%	21.1%
big 3	0.9%	54.8%	21.4%
big.LITTLE 1	1.4%	43.2%	22.7%
big.LITTLE 2	0.8%	56.9%	23.3%
big.LITTLE 3	0.9%	37.8%	19.5%
big.LITTLE 4	2.4%	48.3%	26%

e.g. initial setup, I/O, kernel execution, etc. We chose three applications: *bfs*, *lud* and *srad v1*. The comparative results between reference measured and modeled Cortex-A15 cluster running at 1.1GHz are shown in Table V. We noticed that execution time precision error varies dramatically between 5% and 90% according to execution stages. The total execution time is compensated. We observed that throughout presented three examples the computation kernel stage error is low and amounts around 20%. At the same time, stages which are related to memory operations, e.g. *Store results*, *Read image from file*, *Save image into file*, etc., produce high error percentage. Thus, we conclude that the main source of error in our model is memory system. One of the reason is LPDDR3 model difference mentioned in Section III-B. Another possible cause is non-realistic cache coherence protocol used in classical memory model [15] as well as the lack of CCI-400 model. This observation can also explain the slight error raise by switching to the HMP mode as its memory communication become more complex and inaccurate cache coherence system provides a noticeable discrepancy.

V. ARCHITECTURE EXPLORATION

The presented big.LITTLE gem5 model allows us to explore important parameters as cache size, interconnect width, memory technology. However, due to current limitation of gem5 to simulate easily more than eight ARM cores, exploring large-scale ARM-based system models is not feasible. Thus to evaluate the scalability of the Rodinia benchmark running on

TABLE V.   APPLICATION DIFFERENT STAGE COMPARISON.

Stage	Execution time (ms)		Error
	Board	gem5	
bfs			
Read graph	0.58	0.04	-93.7%
Allocate memory	3.7	4.5	21.5%
Kernel	33.7	41.5	23%
Store results	5.1	3.3	-35.1%
Total	44.0	49.6	12.7%
lud			
Kernel	79.476	72.384	-8.9%
Verify	216.514	347.586	60.5%
Total	295.99	419.976	41.9%
srad v1			
Initial setup	0.31	0.06	-79.4%
Read image from file	140.5	259.4	84.6%
Resize image	3.5	3.3	-5.4%
Allocate memory	0.12	0.09	-29.5%
Extract image	90.3	8.8	-90.3%
Compute	14.4	11.1	-22.9%
Compress image	38.8	23.3	-40%
Save image into file	170.8	111.1	-35%
Free memory	2.4	0.9	-62.3%
Total	461.1	418.5	-9.2%

the big.LITTLE heterogeneous manycore we used the trace-driven approach [7]. To demonstrate the exploration flow we chose *hotspot* application with 1024 problem size.

A. Trace-driven simulation

The trace-driven simulation consists on three phases: (i) collection, (ii) reduction and (iii) simulation. The last phase implies the replacement of gem5 full-system cores by trace injectors (TIs) which main goal is to replay the traces obtained in the collection phase. The key advantages of such trace-driven approach is the significant reduction of the simulation time and also the ability to replicate the traces in order to evaluate system scalability [7].

To collect the Cortex-A7 traces we used the *TimingSimpleCPU*. As we showed in Section IV-B this model is suitable for performance evaluation and in addition it provides well-organized traces where each request is always followed by a response. The Cortex-A15 trace-driven simulation is a tedious task. Its out-of-order nature at times complicates trace injections and requires extra micro-dependency analysis. To solve this issue we decided to emulate the Cortex-A15 behavior by using collected Cortex-A7 traces.

In Figure 4 we illustrated the *hotspot* kernel runtime behavior captured on the Odroid XU3 board with Scalasca/Score-p instrumentation [16] and analyzed with Vampir tool [17]. The figure represents execution of four threads under two Cortex-A7 and two Cortex-A15 cores running at the same frequency. Expectedly the Cortex-A15 duration is less than the Cortex-A7 corresponding to 0.16s and 0.23s respectively. Based on these values we calculated an acceleration factor as 1.45x and applied it for big cluster trace-driven simulation. Consequently, the acceleration factor varies from one application to another.

Trace replication technique [7] relies on overlapping trace patterns with the increasing number of TIs. The *hotspot* kernel consists on two stages:

1) Read input data stage is performed by master thread successively and takes 80% of total execution time.
2) Parallel region stage is executed on all available cores evenly and takes the remaining 20% of total execution time.

The percentage values shown above are taken from the Scalasca/Vampir profile and correlate with the published analysis [6]. To obtain the replication pattern we captured the parallel region traces presented in Figure 5. We illustrated the trace pattern collected at the core#0 (Figure 5 a)) and at the core#1 (Figure 5 b)) on the system with four cores and 4 threads. Each kernel iteration is composed on two `pragma omp parallel for`: (i) compute temperature and (ii) store results. We observed that the results storage region has a significant raise of cache miss number. The further exploration is focus on the parallel region evaluation only.

B. Results

Based on previous trace-driven design, we present the ARM big.LITTLE heterogeneous manycore scalability analysis. We evaluated three scenarios:

- LITTLE cluster with 4, 8, 16, 32, 64 and 128 cores (injectors),
- big cluster with 4, 8, 16, 32, 64 and 128 cores (injectors),
- big.LITTLE in HMP mode with 4/4, 8/8, 16/16, 32/32 and 64/64 cores (injectors).

The execution time and speedup for each scenario are presented in Figure 6. As we can see, the best execution time, as well as the speedup obviously shows big cluster. The LITTLE cluster provides the worst execution time. The big.LITTLE speedup is normalized by the faster big cluster. We observed that the execution time in HMP mode is worst than in the big cluster and slightly better than in the LITTLE cluster. It explained by the OpenMP programming nature that we observed in Figure 4 where faster Cortex-A15 cores wait when the slower Cortex-A7 cores terminate. For all three scenarios the speedup reaches the plateau around 64 cores (injectors). It explained by the memory/interconnect saturation.

To address this common issue we propose to explore the big.LITTLE architecture with alternative network-based *Ruby* memory subsystem [15]. System includes two-level cache hierarchy. The consistency of the memory is maintained by the MESI coherence protocol. This protocol models inclusion between the L1 and L2 caches and has four stable states, M, E, S and I, hence the name. The interconnection network has the following features: Mesh topology, XY routing algorithm

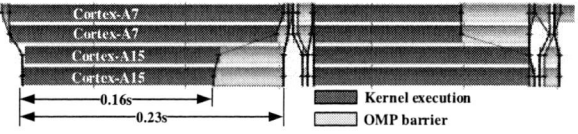

Fig. 4.   Hotspot parallel region runtime behavior running on the Odroid XU3 board.

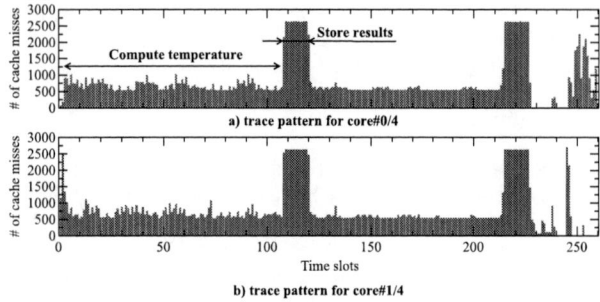

a) trace pattern for core#0/4

b) trace pattern for core#1/4

Fig. 5.   Hotspot parallel region trace pattern.

and detailed GARNET network micro-architecture model (16-byte links, 10 virtual networks, 4 virtual channels per virtual network, 4 buffers per virtual channel, 1 cycle on-chip link latency). Figure 6 b) shows the achieved speedup for the LITTLE cluster (Ruby) up to 128 cores. Application shows a plateau which originates from saturation of the external memory bandwidth that according to the gem5 statistic file is about 200 million DDR accesses per second. *Hotspot* parallel region investigation shows that system scalability can be improved by efficient network interconnect on around 30% of execution time speedup.

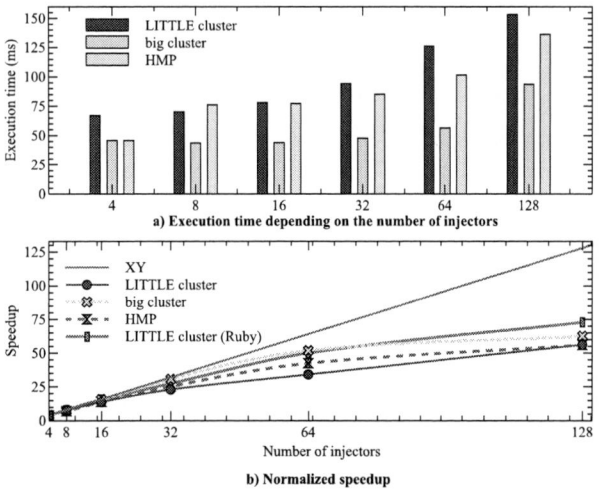

a) Execution time depending on the number of injectors

b) Normalized speedup

Fig. 6.   Execution time and speedup evaluation.

VI.   Conclusion

In this paper, we proposed ARM big.LITTLE cores models in the gem5 architecture exploration framework, which are accurate enough to enable a relevant design exploration of heterogeneous manycore systems. More precisely, models of Cortex-A7 and Cortex-A15 cores have been combined in full-system mode and evaluated against the Exynos Octa 5422 system-on-chip in order to assess the models accuracy. A reasonable precision error of about 20% has been obtained. Due to the limitation of gem5 full-system mode to simulate more than eight ARM cores, we applied our trace-driven approach [7] to explore the heterogeneous nature of big.LITTLE systems including more than one hundred cores. The scalability of such

systems has been addressed and compared with homogeneous system configuration. All the study has been achieved by using a subset of the Rodinia compute-intensive benchmark suite [6].

Future work includes a more detailed analysis of out-of-order Cortex-A15 core in gem5. Then, traces could be generated using this model for replication and further architecture parameter exploration, e.g., cache and memory configuration.

Acknowledgment

The research leading to these results has received funding from the European Community Seventh Framework Programme (FP7/2013-2016) under the Mont-Blanc 2 Project: http://www.montblanc-project.eu, grant agreement n° 610402.

References

[1] Intel, "Many Integrated Core Architecture," 2015. [Online]. Available: http://www.intel.com

[2] EZchip, "TILE-Gx Multicore ," 2015. [Online]. Available: http://www.tilera.com

[3] Kalray, "Multi-Purpose Processor Array," 2015. [Online]. Available: http://www.kalrayinc.com

[4] ARM, "big.LITTLE Technology," 2015. [Online]. Available: http://www.arm.com

[5] TOP500, "TOP500 Supercomputer Sites," 2015. [Online]. Available: http://www.top500.org/

[6] S. Che, M. Boyer, J. Meng, D. Tarjan, J. Sheaffer, S.-H. Lee, and K. Skadron, "Rodinia: A benchmark suite for heterogeneous computing," in *Workload Characterization, 2009. IISWC 2009. IEEE International Symposium on*, Oct 2009, pp. 44–54.

[7] A. Butko, R. Garibotti, L. Ost, V. Lapotre, A. Gamatié, G. Sassatelli, and C. Adeniyi-Jones, "A trace-driven approach for fast and accurate simulation of manycore architectures," in *Design Automation Conference (ASP-DAC), 2015 20th Asia and South Pacific*, Jan 2015, pp. 707–712.

[8] N. Binkert, B. Beckmann, G. Black, S. K. Reinhardt, A. Saidi, A. Basu, J. Hestness, D. R. Hower, T. Krishna, S. Sardashti, R. Sen, K. Sewell, M. Shoaib, N. Vaish, M. D. Hill, and D. A. Wood, "The gem5 simulator," *SIGARCH Comput. Archit. News*, vol. 39, no. 2, pp. 1–7, Aug. 2011. [Online]. Available: http://doi.acm.org/10.1145/2024716.2024718

[9] A. Butko, R. Garibotti, L. Ost, and G. Sassatelli, "Accuracy evaluation of gem5 simulator system," in *Reconfigurable Communication-centric Systems-on-Chip (ReCoSoC), 2012 7th International Workshop on*, July 2012, pp. 1–7.

[10] A. Gutierrez, J. Pusdesris, R. Dreslinski, T. Mudge, C. Sudanthi, C. Emmons, M. Hayenga, and N. Paver, "Sources of error in full-system simulation," in *Performance Analysis of Systems and Software (ISPASS), 2014 IEEE International Symposium on*, March 2014, pp. 13–22.

[11] F. Endo, D. Courousse, and H.-P. Charles, "Micro-architectural simulation of in-order and out-of-order arm microprocessors with gem5," in *Embedded Computer Systems: Architectures, Modeling, and Simulation (SAMOS XIV), 2014 International Conference on*, July 2014, pp. 266–273.

[12] F. A. Endo, D. Courousse, and H.-P. Charles, "Micro-architectural simulation of embedded core heterogeneity with gem5 and mcpat," in *Proceedings of the 2015 Workshop on Rapid Simulation and Performance Evaluation: Methods and Tools*, ser. RAPIDO '15. New York, NY, USA: ACM, 2015, pp. 7:1–7:6. [Online]. Available: http://doi.acm.org/10.1145/2693433.2693440

[13] "Odroid-xu3," 2015. [Online]. Available: http://www.hardkernel.com

[14] *CoreLink CCI-400 Cache Coherent Interconnect Technical Reference Manual*, ARM, November 16 2012, revision r1p1.

[15] gem5, "Classic Memory System," 2015. [Online]. Available: http://www.m5sim.org

[16] "Scalasca," 2015. [Online]. Available: http://www.scalasca.org/

[17] "Vampir - performance optimization," 2015. [Online]. Available: https://www.vampir.eu/

978-1-4799-8720-7/15 $31.00 © 2015 IEEE

2015 IEEE Computer Society Annual Symposium on VLSI

On Analysis of On-Chip DC-DC Converters for Power Delivery Networks

Ghizlane Mouslih, Aida Todri-Sanial, Pascal Nouet

CNRS-LIRMM/University of Montpellier, France

ghizlanemouslih@gmail.com, todri@lirmm.fr, nouet@lirmm.fr

Abstract— **This paper explores the benefits and costs of integrating an on-chip DC-DC converter as voltage regulator module for reliable power delivery networks (PDNs). We perform detailed time-domain and frequency analyses on PDNs to assess the impact of on-chip DC-DC converter on PDN impedance and overall supply noise behavior. As results show, the reliability of PDNs depends on the DC-DC converter design and its efficiency. Various switching scenarios are analyzed for multiple voltage distributions on circuits of different sizes to explore the benefits of on-chip DC-DC conversion.**

Keywords— DC-DC converter, power delivery network, voltage drop

I. INTRODUCTION

Many systems (cellular phones, laptops, MP3 players which are using batteries as their power supply) require that the primary source of DC power be converted to other voltages. Such systems usually include several sub-circuits that need to be supplied with different supply voltage levels (*i.e.* different voltage potentials might be needed for memory, disc drives, display and operating logic), which may not be the same as battery's voltage level. Employing DC-DC converters is a method to generate multiple voltage levels from a single DC supply voltage to be fed to different sub-circuits on a chip. This method of generating multiple voltage levels from a single source can reduce the chip area substantially [1]. However, DC voltage provided by battery or rectifier contains high voltage ripples and it is not constant enough, thus it is not applicable for most devices. This is where, DC-DC regulators are employed to attenuate the ripples regardless of change in the load current or input voltage. Thus, not only they allow providing different levels of voltage but they also ensure more stability on the supply voltage.

There exist several types of DC-DC converters such as linear regulators, buck converters and switched capacitor converters which are some of the most well-known and used converter circuits. A buck converter requires large passive elements for inductance and capacitance (*LC*) as an output filter, however it also has higher power efficiency than a linear regulator. A switched-capacitor converter also needs a large capacitors and one more drawback is that the output voltage levels are limited by the ratios of the capacitors, which makes them not very suitable for low-power systems using dynamic voltage scaling method.

Based on existing literature, inductive converters tend to have better performance and efficiencies [2], [3] and fast response control over wide load range [4]. DC-DC converters with faster switching frequencies are becoming popular due to their ability to decrease the size of the output capacitor and inductor to save chip space. Meanwhile, the demands from the power supply increase as processor core voltage drops below 1V, making lower voltage level difficult to achieve at faster frequencies due to the lower duty cycle. In this paper, a DC-DC buck converter for low power applications is designed and implemented in 45nm CMOS technology based on PTM models [5], with input and output voltage of 1.8V and 1.1V, respectively.

One of the objectives of this paper is to first study the switching-mode power supplies (SMPS) and different controlling methods, which are commonly used in SMPS. Second, we implement a buck converter and analyze its power efficiency on power delivery networks. Overall, we develop a framework for analyzing the benefits and cost of on-chip DC-DC converters, their efficiency and impact on power delivery network's reliability. Figure 1 illustrates the concept of such analysis framework. Such framework allows us to investigate the downsizing of passive components like inductor's size without sacrificing the system's overall performance. As a case study circuit, we utilize a 35-stage ring oscillator that is designed in 45nm CMOS technology using Cadence Virtuoso tool.

The rest of the paper is organized as follows. Section II describes briefly the model of on-chip power delivery networks that we use in this work. Section III describes the DC-DC converter design and parameter selection that is used on our analysis framework. In Section IV, we present the simulation results based on time-domain and frequency analyses on PDN to assess the changes in PDN impedance and overall supply noise behavior. Section V, concludes the paper.

I. PDN MODELING

In order to develop an analysis framework for on-chip DC-DC converters, we start by establishing a power delivery network model that can serve as reference for studying the efficiency of the converter. To do so, a distributed power model is developed for a chip size of 1μm-by-1μm. As a switching circuit, we apply a 35-stage ring oscillator designed in 45nm technology node that draws current from the power delivery network.

978-1-4799-8720-7/15 $31.00 © 2015 IEEE   557

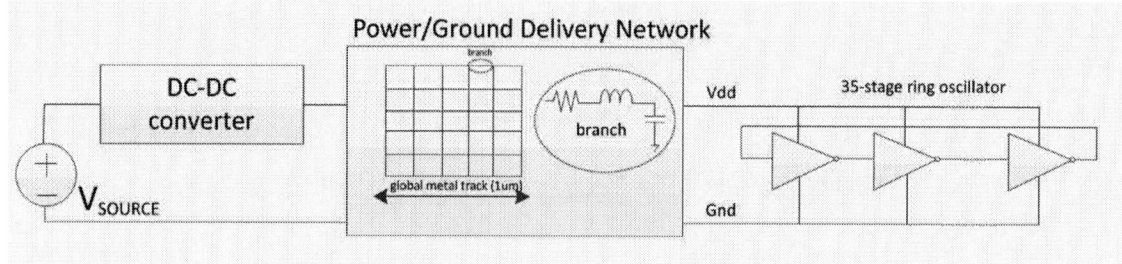

Figure 1. Overall description of analysis framework for investigating on-chip DC-DC conversion.

Power delivery networks are usually modeled as a mesh structure to represent their actual layout design with global metal tracks (i.e. top two metal layers) running in parallel and perpendicular where vias are inserted in their cross-sections. The mesh structure is represented as a distributed *RLC* network. Table I provides the electrical parameters used for the distributed power delivery model based on the PTM [5] interconnect model.

Table I: The electrical parameters used for the distributed power delivery mesh structure [5].

Typical wire dimensions	*RLC* values	*RLC* (per mm)
W=0.45µm s=0.45µm l=2500µm t=1.2µm h=0.2µm K=2.2	R=101.851 Ohms L=4.258nH M12=3.811nH M13=3.465nH M14=3.262nH (k12=0.895 k13=0.813 k14=0.766) C_{ground}=205.0775fF C_{couple}=183.055fF C_{total}=571.1875fF	R=40.7404 Ohm/mm L=1.7032 nH/mm Cg=82.031 fF/mm Cc=73.222 fF/mm Ctotal=228.48 fF/mm

The physical dimensions of the metal tracks for a chip size of 1µm-by-1µm are listed in Table II. The mesh size is represented as a 6-by-6 grid where each branch has the following *RLC* parasitics.

Table II: Power grid branch *RLC* parasitics used in this work.

PDN Branch Length	Resistance	Inductance	Capacitance
180 nm	7.2 m Ohm	306nH	0.4fF

II. DC-DC BUCK CONVERTER

In this section, we describe the design and chosen parameters for the DC-DC converter that we utilize for our analysis. A DC-DC contains two main blocks, a power processor and a feedback control part as shown in Figure 2. The feedback control loop is needed to assure the stability of the system and regulate the output voltage level by adjusting the switching duty cycle. Additionally, the feedback control consists of two main parts, a voltage error-amplifier (with compensation circuit) and voltage comparator to keep close the output voltage to a reference voltage. A circuit level representation of Figure 2 is shown in Figure 3.

The error-amplifier compares the feedback voltage (applied to inverting input) to a reference voltage (applied to non-inverting input) then their difference, which is called voltage error signal, is applied to non-inverting input of voltage comparator. The comparator circuit, which is simply $1/V_{osc}$ where V_{osc} is the peak-to-peak amplitude of the oscillator voltage, compares this error voltage to saw tooth ramp that is generated by ramp generator. If the error voltage is large then the output voltage of the comparator goes high but when the error voltage is small (i.e. negligible) then the output of the comparator goes low. According to this voltage, the driver determines the corresponding modulated signal duty cycle for control of the power switches.

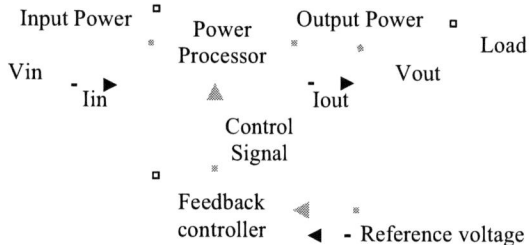

Figure 2. Block diagram of DC-DC regulator.

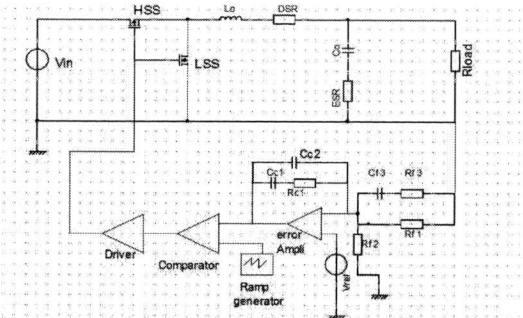

Figure 3. DC-DC converter circuit schematic.

In order to assure the stability of the system and derive the parameters of the converter, we modeled the circuit shown in Figure 3 with three blocks as presented in Figure 4. The power stage $G_p(s)$ includes the switches, the drivers and the output inductor and capacitor. The compensator block $H(s)$ represents the error-amplifier with the compensation network.

978-1-4799-8720-7/15 $31.00 © 2015 IEEE   558

For simulation purposes, we implement a DC-DC converter design procedure for typical synchronous buck converter with voltage-mode control and voltage-mode error-amplifier based on 45nm CMOS technology. Such converters tend to improve noise immunity and efficiency, but require a certain amount of loop compensation to achieve an acceptable performance. By selecting such design, the DC-DC converter can provide a loop gain function with a high bandwidth (high zero-crossover frequency) and adequate phase margin. As a result, fast load response and good steady state output can be achieved which is an important criterion for on-chip DC-DC conversion. We apply an appropriate compensation network based on [6] in order to assure this stability.

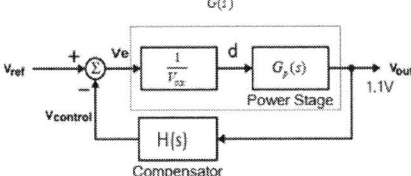

Figure 4. Block diagram model of the DC-DC converter.

Initially, the power and control stages are reviewed with respect to the small-signal model and relevant transfer functions. In the next subsections, we describe how the crossover frequency of the loop gain is significant in the context of the frequency range where feedback is effective in attenuating the closed-loop characteristics, followed by a detailed description of the compensator design procedure.

A. Power stage

The transfer function of the power stage can be represented as a second order system with a double pole at the resonance frequency F_{LC} of the LC filter and a zero produced by the ESR of the output capacitor as follows:

$$G_p(s) = \frac{V_{out}}{d}(s) = \frac{R_{Load}(C_0.ESR.s+1)}{L_0 C_0.s^2(R_{Load}+ESR) + s.(L_0+R_{Load}.C_0.ESR)+R_{Load}} V_{in} \quad (1)$$

$$G(s) = G_p(s).\frac{1}{V_{osc}} \quad (2)$$

where L_o, C_o and R_{load} are the inductance, capacitance and load resistance values, respectively. ESR is the equivalent series resistance of the output capacitor. The parameter values are derived and chosen to optimize the buck converter design following the design theory described in [7].

B. Loop gain

The loop gain of system is defined as the product of transfer functions along the closed control loop. Based on the model shown in Figure 4, we derive:

$$M(s) = \frac{1}{k} \times H(s) \times G(s) \quad (3)$$

where $1/k$ represents the gain of the resistor divider which is used in the feedback loop when $V_{out} > V_{ref}$. The cross-over frequency, also called bandwidth of the loop is when the loop gain equals unity. It determines F_0, the speed of the system response to load transients. Typically F_0 can be set between 1/10 to 1/5 of the switching frequency where the higher the

crossover frequency, the faster the load transient response would be. However, the crossover frequency should also be low enough to allow attenuation of the switching noise. The slope of the loop gain at 0Hz should be about –20dB in order to ensure a stable system. The phase margin should be greater than 45º for overall stability. As the double pole of the power stage causes the gain to fall within a slope of -40dB/dec up to the zero frequency (or F_{ESR}), which compensates one of the poles, a compensation network is needed to compensate the slope of -40dB to -20dB, so we can ensure a stable system.

C. Compensator

To have a stable closed loop buck converter with appropriate performance, a properly designed compensator is required. We utilize a Type III compensator as it provides enough phase margin by adding an extra zero to the system's closed loop transfer function to keep it stable, also shown in Figure 5.

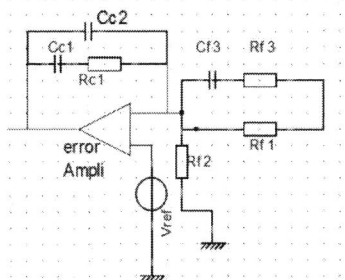

Figure 5. Type III compensation network.

The transfer function of type III compensator is given by:

$$H(s) = \frac{V_e}{V_{out}} = \frac{Z_c}{Z_f} \quad (4)$$

The pole generated by C_{c1} and R_{c1} is usually set at a much higher frequency as compared with the frequency of the zero generated by C_{c2} and R_{c1}. This means $C_{c1} < C_{c2}$. Therefore:

$$H(s) = -\frac{(1+R_{c1}.C_{c1}.s).[1+s.C_{f3}.(R_{f1}+R_{f3})]}{R_{f1}.s.C_{c1}.(R_{c1}.C_{c2}.s+1).(1+s.R_{f3}.C_{f3})} \quad (5)$$

The conventional compensation strategy [8], [9] employed with voltage mode control is to use two compensator zeros to counteract the LC double pole, one compensator pole to nullify the output capacitor ESR zero and one compensator pole located at one half switching frequency to attenuate high frequency noise. By performing the computations, $F_{z2}=F_{LC}=4.6MHz$, $F_{z1}=0.75F_{LC}=3.45MHz$, $F_{p2}=F_{ESR}=0.4GHz$, and $F_{p1}=F_s/2=1GHz$ are obtained.

III. SIMULATIONS AND RESULTS

Here, we perform two types of analyses: (1) with DC-DC converter, and (2) without DC-DC converter where an ideal supply source is used to supply voltage to the 35-stage ring oscillator. The ring oscillator has a 2GHz operating frequency based on 45nm CMOS technology.

A. Jitter and phase noise in ring oscillators

In general, CMOS circuits are sensitive to power supply fluctuations, as well as to noise generated in nearby switching circuits (such as noise transferred through power grid and substrate). Hence, the propagation delay t_d, is a function of supply voltage. Delay variations can be created due to variations on the rising and falling edges, which are also referred as jitter [10]. Figure 6 depicts a comparison between the transient analysis of the ring oscillator when powered by an ideal supply voltage and the DC-DC converter. It shows an operating frequency of 2GHz for the ring oscillator and no additional delay or phase noise is observed due to converter.

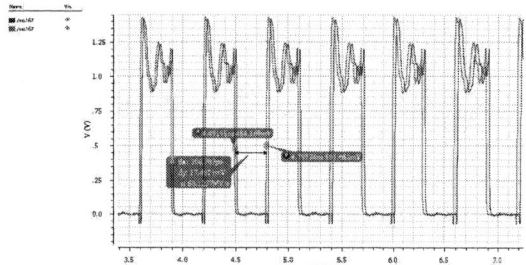

Figure 6. Output transient analysis of ring oscillator with and without DC-DC converter.

B. PDN impedance

Figure 7 shows the impedance plot of the PDN using an ideal voltage source versus using the DC-DC converter. The most noticeable is a resonant peak in the crossover frequency; this is due to the bandwidth of the converter. The converter has to have low bandwidth since parasitics appear at low frequencies.

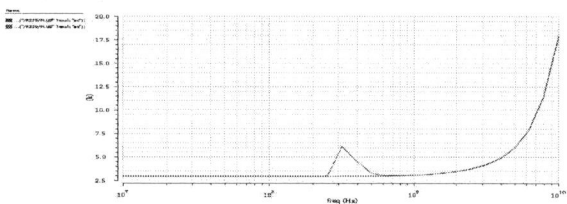

Figure 7. Comparison of the PDN impedance with and without DC-DC converter.

C. Voltage drop

Voltage drop is an important PDN metric because it is directly related to the delay increase and frequency degradation. Figure 8 shows the DC-DC converter output voltage supplied to the ring oscillator on various PDN nodes. The worst-case voltage drop is determined from the supply voltage transients. Figure 9 shows the voltage drop map on each PDN node when using the DC-DC converter versus the ideal voltage supply. The comparison between the two distributions shows a higher voltage drop when DC-DC converter is used.

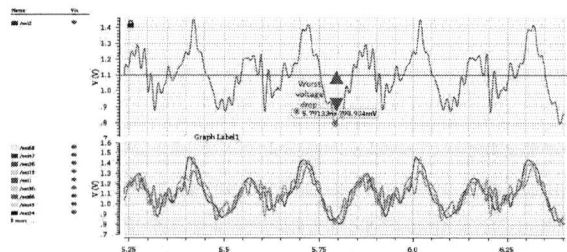

Figure 8. Worst-case voltage drop with DC-DC converter.

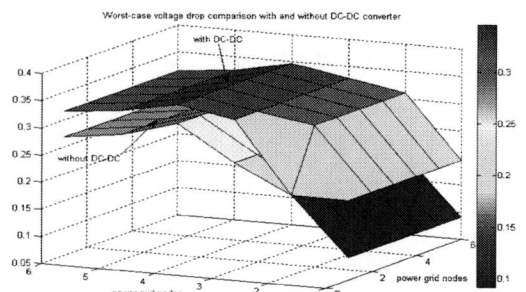

Figure 9. Voltage drop map on each PDN node with DC-DC converter versus ideal supply voltage.

IV. CONCLUSION

In this paper, an analysis framework was described for studying on-chip DC-DC converters for reliable power delivery. Simulation results show that while implementing on-chip DC-DC converter still requires research on downsizing of discrete inductive and capacitive components to be able to implemented on-chip, nevertheless it presents a viable solution for future low-power high-performance embedded systems. Integrating DC-DC converter shows no additional delay or phase noise on the system. As future work, we will investigate larger and more complex circuits with various on-chip DC-DC converters for multiple voltage domains and their impact on systems performance and PDN reliability.

REFERENCES

[1] N. Safari, "Design of a DC-DC buck converter for ultra-low power applications in 65nm CMOS Process." Master Thesis, 2012.

[2] X. Zhang et. al, "Monolithic/Modularized Voltage Regulator Channel," Power Electronics, IEEE Transactions on , vol.22, no.4, pp.1162-1176, 2007.

[3] Q. Zhao, et al., "Characterization of Cdv/dt Induced Power Loss in Synchronous Buck DC–DC Converters," IEEE Power Elec. Trans., vol.22, no.4, pp.1508-1513, 2007.

[4] P. Li et al., "A 90–240 MHz Hysteretic Controlled DC-DC Buck Converter With Digital Phase Locked Loop Synchronization," IEEE Journal of Solid States, vol.46, no.9, pp.2108-2119, 2011.

[5] http://ptm.asu.edu/

[6] G. Schrom, et al., "Optimal Design of Monolithic Integrated DC-DC Converters," IEEE ICICDT pp.1-3, 2006.

[7] http://www.eetimes.com/document.asp?doc_id=1225686

[8] National Semiconductor. "Compensating Voltage Mode Buck Regulators".

[9] National Semiconductor, "LM2853—3A 550 kHz Synchronous SIMPLE SWITCHER Buck Regulator from the PowerWise Family".

[10] G. Jovanovic, et al, "A CMOS Voltage Controlled Ring Oscillator with Improved Frequency Stability", IEEE Appl. Math. Infom and Mech, vol.2. pp. 1-9, 2010.

Multilevel Modeling Methodology For Reconfigurable Computing Systems Based On Silicon Photonics

Zhen Li, Sébastien Le Beux, Christelle Monat, Xavier Letartre, Ian O'Connor*

Lyon Institute of Nanotechnology (INL) - Ecole Centrale de Lyon, Ecully, F-69134, France

*{name.surname}@ec-lyon.fr

Abstract— **Reconfigurable computing systems, e.g. FPGA, represent an increasingly attractive architectural solution for high-end supercomputing due to high aggregated computational resources, high energy-efficiency and flexibility. However, to further increase the computing bandwidth of such systems while decreasing the energy consumption, emerging technologies such as silicon photonics, are urgently needed. In this context, we have proposed the OLUT architecture, which could serve as the core building block of prospective silicon photonic reconfigurable computing systems. To implement this concept, a modeling methodology is required to evaluate the impact of technological constraints on system metrics. In this paper, we propose a multi-level design methodology to optimize the performance of reconfigurable computing systems based on silicon photonics by exploring the design space from multiple perspectives of system dimensions, device parameters and technologies. This method allows us to efficiently implement a functional and energy-efficient reconfigurable computing architecture that could reach ~50fJ/bit logic operation at a BER of $<10^{-18}$, as a potential answer to the future demands of reconfigurable computing.**

Keywords— Multilevel modeling methodology, silicon photonics technology, reconfigurable computing architecture, OLUT

I. INTRODUCTION AND RELATED WORK

The current trend in high performance computing is to increase parallelism and flexibility through the use of reconfigurable computing architectures, e.g. Field programmable gate arrays (FPGAs)[1]. FPGAs are prefabricated semiconductor chips that typically consist of a large number of configurable logic blocks which are interconnected via a dynamic routing network, and configurable I/O (Input/Output) interfaces. An FPGA allows all its operational hardware resources to be configured and customized to suit a specific computing task through post-fabrication and user-defined programming.

Future mobile applications in the "Big Data" era require FPGA systems to have high computing bandwidth with low energy dissipation. While CMOS technology is approaching its fundamental scaling limits, traditional approaches in increasing FPGA computational power will ultimately lead to higher heat generation and more stringent energy/operation requirements. Concurrently, emerging technologies promise to provide significant improvements in computing performance, energy consumption, area and cost over conventional standard FPGA technology, including 3D integration, non-volatile memories and silicon photonics technology. In particular, integrated silicon photonics is very attractive to partially replace traditional, slow and power consuming, electrical computing circuits. It can be used to design new I/O interfaces, dynamic interconnect fabrics and configurable logic cells. A Reconfigurable Directed Logic (RDL) architecture [4][5] has been proposed to improve the computational latency of electrical computing architectures through interconnected optical switches. An implementation of a 1-bit half adder circuit using the RDL architecture with a computation performance of 100Mbit/s has also been demonstrated in [6]. However, the area cost of RDL architecture is extremely high due to its two-plane logic implementations.

In our previous work [2], the Optical lookup table (OLUT) has been proposed for performing parallel computations by taking advantage of silicon photonics, e.g. wavelength division multiplexing (WDM), thereby realizing a high-bandwidth and energy-efficient reconfigurable computing architecture. Furthermore, the method for cascading multiple OLUTs and an advanced OLUT scheme for increasing the level of parallelism by exploiting complementary logic are proposed in [7] and [10]. However, there is currently no modeling approach or tools able to guarantee that a functional reconfigurable computing system built from silicon photonic devices can reach a well-identified set of targeted system performance metrics and operational reliability. In the Electronic Design Automation (EDA) domain however, design space exploration techniques are widely used to facilitate the design of complex computing systems consisting of a large volume of building blocks, e.g. an IC composed billions of individual transistors. Our postulate in this paper is that a similar approach can be developed and integrated into the design cycle of photonic computing architecture such as OLUTs.

In order to best fit the silicon photonics technology to the implementation of the OLUT concept, we propose a multi-level modeling methodology. The objective is to show how to properly implement and optimize the reconfigurable photonic computing architecture based on OLUTs for reaching system-

level performance requirements by exploring the design space from the multiple perspectives of system dimensions, photonic device parameters and technology variations.

The paper is organized as follows. Section II of this paper presents a full description of the reconfigurable photonic computing architecture by distinguishing between three interacting levels, and the basic principle of OLUT architectures. Section III describes the multi-level modeling methodology for design space exploration and the associated models of the system and devices. Finally, the evaluation results obtained by using the modeling approach are illustrated through various computing examples implemented by OLUTs.

II. RECONFIGURABLE COMPUTING ARCHITECTURE BASED ON SILICON PHOTONICS

A. Structural overview of the OFPGA

The general architecture of the proposed silicon photonics based reconfigurable computing system is presented in Fig. 1and is a 3-tier architecture: i) the top-most level of the photonic reconfigurable computing system is called the "OFPGA tier" and is composed of a large number of OLUT blocks connected through a given interconnect topology. Its main function is to separate a complex logic function into smaller ones that can be implemented by OLUT blocks with a small I/O dimension. ii) In the middle layer ("OLUT tier"), a single OLUT block subordinating to the OFPGA tier is responsible for performing a basic logic operation, and its performance is also analyzed by taking into account the parameters of various silicon photonics devices at the lower level. iii) The lowest level ("OD tier") contains OLUT building blocks made from silicon photonics technology, which are needed to provide the required functionality for light, e.g. routing or switching, detection, etc. Each component at the lower level is subordinate to the next higher level. The main characteristics of components at the lower level are communicated to a higher level when performing design space exploration to reach the given requirements, while the resulting consequences can be used iteratively to refine the characteristics of low-level elements. This categorization of the reconfigurable computing architecture makes each functional element of the OLUT and Device tiers more reusable and it becomes more convenient to change them for system-level optimization.

B. Operating principles of the OLUT architecture

As previously stated, a conventional FPGA typically consists of a large number of configurable logic blocks which are interconnected via a dynamic routing network. Similarly, an OFPGA is composed of a large number of optical look-up tables (OLUTs), which is a critical element in the overall architecture.

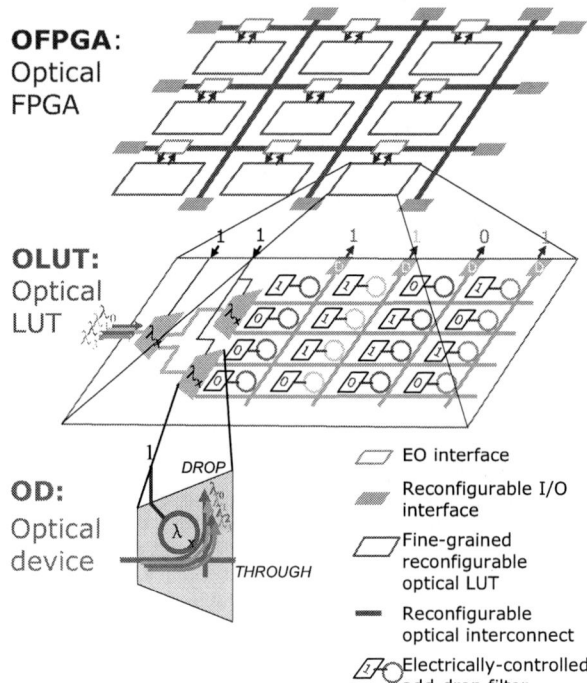

Fig. 1 A multi-level (3-tier) reconfigurable computing architecture based on silicon photonics

The n-m-OLUT interfaces n electrical data inputs to m data outputs (electrical or optical), using m optical signals at distinct wavelengths (λ_0,..., λ_{m-1}). Similarly to the electrical LUT, the OLUT is composed of two parts: *routing* and *memorization*. In the routing part, the m optical signals λ_i (i = 0...$m-1$) share the same optical path as specified by the electrical input data set. In the memorization part, they are driven into m memorization stages (represented by m distinct columns), each of which is composed of 2^n identical add-drop filters and interconnected by 2^n horizontal waveguides. Each stage of the memorization part performs a basic Boolean function through a specific wavelength, all the stages operating in parallel thanks to WDM. As in electrical LUTs, the executed Boolean function depends on the configuration stored in the SRAM memories.

In the OLUT architecture, WDM is implemented by using two different wavelength filtering schemes: (i) in the routing part, where all the optical signals, independently of their wavelength, are propagated along the same path; and (ii) in the memorization part, where each spectrally distinct optical signal is individually routed according to the configuration. For the 2-4-OLUT example:

Routing part: The behavior of the switch in the routing part is illustrated in Fig. 2(a) according to its *DROP* state (solid line) or *THROUGH* state (dashed line). The arrows represent the four incident optical signals for which the wavelength values λ_0, λ_1, λ_2 and λ_3 are either ideally aligned with the switch resonant wavelengths (represented by peaks in the transmission spectrum) in the *DROP* state, or detuned by a wavelength difference $\Delta\lambda$ in the *THROUGH* state. The wavelengths of the injected optical signal are regularly spaced consistently with the *Free Spectral Range* (FSR$_x$) of the switch. Hence, in the case where the add-drop filter is in the *DROP* state, all the signals are redirected to a given waveguide, while in the *THROUGH* state, all the signals propagate along the other waveguide.

Memorization part: Fig. 2(b) and (c) illustrate the operation of the add-drop filters in the memorization part as well as their transmission spectrum. Compared to those of the routing part, their FSR is slightly larger (see *FSR$_{m0}$* and *FSR$_{m1}$* in Fig. 2 (b) and (c)) so that at most one resonant wavelength is aligned with one wavelength of the injected optical signal: λ_0 in (b) and λ_1 in (c), respectively. In addition, their relative FSR should be slightly different to avoid the potential scenario where the add-drop resonances become aligned with the wavelength of the other optical signals after the tuning/detuning process. Having different FSR and distinct resonant wavelengths in the memorization part can be achieved by changing the device geometry (e.g. the microring radius) or through a thermal control. As a result, only one signal wavelength is redirected to the vertical waveguide when the add-drop is switched to the *DROP* state (solid line), the other ones propagating through the same waveguide. Similarly to the add-drop used in the routing part, all the signals propagate along the horizontal waveguide if the add-drop filter is in the *THROUGH* state (dashed line).

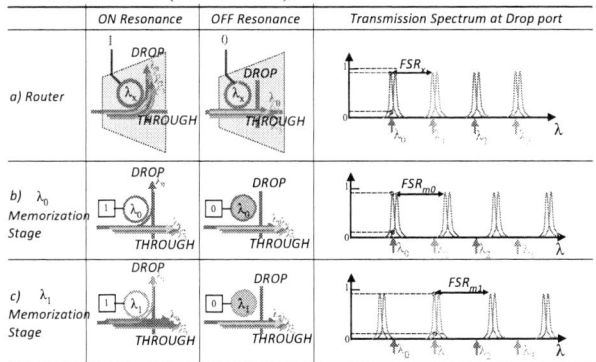

Fig. 2 Illustration of wavelength filtering schemes in the OLUT. In a) router: when the add-drop filter is in the DROP state, all signals are redirected to the DROP port and when in the THROUGH state, all optical signals propagate along the other waveguide. The transmission at the DROP port is represented in the right column. In b) λ_0 and c) λ_1 memorization stage, only one signal wavelength (i.e. λ_0 and λ_1) is redirected to the vertical waveguide when the add-drop is switched to the DROP state, the other ones propagating through the same waveguide. All the signals will be forwarded to the THROUGH port if the add drop is in the THROUGH state.

III. MULTI-LEVEL MODELING METHODOLOGY

This section describes the multi-level modeling methodology that efficiently generates the design space for the

OLUT architecture and allows the implementation of a functional and energy-efficient reconfigurable computing architecture based on OLUTs. However, complete OFPGA evaluation requires industry automated tools for mapping standard benchmark into OFPGA and is out of the scope of this paper.

A. System metrics and technological parameters

Here we first introduce the key metrics of the reconfigurable computing architecture:

- Bit error rate (BER): represents the operation reliability. A BER of 10^{-18} is chosen here, i.e. considering a single error occurs per 10^{18} data bits. BER can also be considered as a system constraint.
- Energy efficiency (E, in fJ/bit): the overall energy consumption of the OLUT to generate 1 bit of output is the sum of all the contributions from optical devices.
- Data rate (B, in Gbit/s): B=1Gbit/s is fixed here for an electro-optic OLUT implementation, which is mainly limited by the physical property (i.e. free carrier lifetime) of silicon electro-optic add-drop filters

B. Methodology Overview

Fig. 3 illustrates the modeling methodology for the 3-tier architecture. Basically, the design space exploration in the *OFPGA tier* (the top layer in Fig. 1) nests the energy efficiency analysis and design space exploration in the *OLUT tier* (the middle level in Fig. 1). Ideally, the modeling process starts with the top-most level for a given computing application (e.g. ALU or full adder) with predefined library information such as interconnect topology. It communicates to the *OLUT tier* with lower-level specifications including the number of OLUTs, the input and output dimensions of OLUTs and the OLUT complementary scheme option [10]. It then focuses on the design of a functional and energy-efficient OLUT block and progressively goes down to the device level (*OD tier*). Inside the *OD tier*, the input specifications include the geometry (ring radius), Q_c and $\Delta\lambda$ of the add-drop filters. Similar to the exploration process at high level, it needs to take into account test bench information specified by the higher level (*OLUT tier*) such as input data and memory configurations, and to fetch the characteristics of other components from the library of silicon photonics technology (e.g. lasers and photodetectors as well as the waveguides). The energy-efficiency analysis is then performed under a given BER value and the result is an energy map giving the feasible/infeasible design space of an OLUT as well as the optimal energy consumption of the target OLUT block, as shown as the branch desicision in Fig. 3. The obtained OLUT metrics can be further used by moving back up to the top-most level to evaluate the performance and the energy efficiency of the computing architecture according to various system-level design options and characteristics as mentioned previously. Note that the design space exploration of the overall OFPGA system is currently a manual process. It requires going back and forth between the results and the system-level specifications to investigate alternative architectural options that would provide the optimal results (e.g. best energy efficiency) for the given application. Such a design space exploration requires automated tools for mapping application benchmarks (e.g. MCNC), taking into account the main advantage of OLUTs, i.e. the parallel computation on a same set of data. The implementation of such a tool is part of future work.

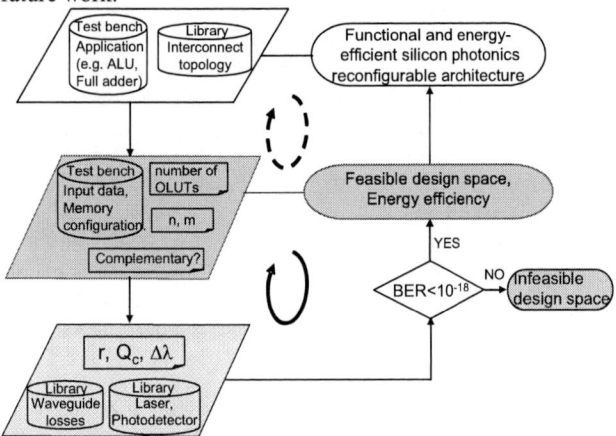

Fig. 3 Illustration of a multi-level modeling methodology for designing a functional and energy-efficient photonic reconfigurable computing architecture based on OLUTs

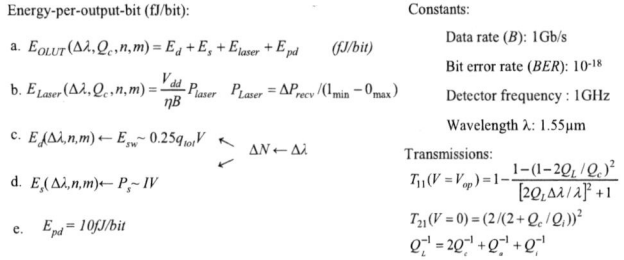

Fig. 4 Physical equations and constants used by the energy model

C. Energy model

This subsection provides the groundwork of the energy efficiency analysis for the proposed photonic reconfigurable computing architecture.

The equations and the constant values used by the model are listed in Fig. 4. We note here that the overall OLUT energy consumption E_{OLUT} (i.e. energy-per-output-bit, given by Equation (a) in Fig. 4) is the sum of the following contributions:

- E_d represents the dynamic energy dissipated by the add-drop filters for performing the state transition (Drop to Through) that needs carriers to be injected to tune the resonant wavelength (Equation (b) in Fig. 4).

- E_s is the static energy consumed by the add-drop filters and is determined by the electrical bias and current obtained

from the electrical simulation of PIN based silicon microring resonator (Equation (c) in Fig. 4).

- E_{laser} is the minimum input optical power delivered by the laser to distinguish the logic '1' from logic '0' (according to the laser energy dissipation within the OLUT) for reaching the target BER=10^{-18} (Equation (d) in Fig. 4).

- E_{pd} is an estimation of the energy dissipated by each photodetector and is set to be 10fJ/bit at a frequency of 1GHz.

D. Energy efficiency analysis based on device parameters

The energy model is built upon the physical parameters of critical devices, i.e. add-drop filters. According to [9], the parameters for fully describing the operation of the add-drop filters are: the coupling quality factor Q_c, the wavelength shift $\Delta\lambda$ between the resonance controled by the electrical bias and the incoming signal wavelength as well as the radius of the microring. Since the latter parameter defines the intrinsic quality factor Q_i of a microring for a given technology[8], the transmissions of the add-drop filter in the active regime are expressed according to the values of Q_c and $\Delta\lambda$, such that the feasible design space can be built based on the values of Q_c and $\Delta\lambda$ for an OLUT with a given number of input and output bits n and m.

The optimum value of E_{olut} can be achieved by making a trade-off between the above contributions, in particular the energy dissipation of lasers and within the add-drop filters, for different values of Q_c and $\Delta\lambda$. For a given number of input and output bits n and m:

- Higher E_d and E_s are required for operating the n-m-OLUT with an increasing wavelength shift $\Delta\lambda$, since more carriers are required to be injected.

- However, E_{laser} is inversely proportional to the minimum difference between the worst-case transmission values for logic '1' and logic '0' (1_{min} and 0_{max} in equation (b) in Fig. 4) and thus relies on the Q_c and the $\Delta\lambda$. On the one hand, for a larger wavelength shift $\Delta\lambda$, an increased transmission value is generally obtained at the *Through* port when the add-drop filter is in the *Through*-state, thereby facilitating the distinction of logic '1' and logic '0', which in turn should lead to a smaller source energy E_{laser} to be used to drive the OLUT. On the other hand, as shown by the transmission expressions in Fig, for an increased value of Q_c (or more precisely of the ratio Q_c/Q_i), it is more difficult to achieve the *Drop*-state when the add-drop filter is powered off (since T_{21} gets lower), but it does help the *Through*-state to be achieved (T_{11} increases) because an increased Q_c leads to a narrower spectral bandwidth of the resonance. This implies that the impact of Q_c on E_{laser} is not always the same and depends on which state (Drop or Through) of the add-drop filter mostly predominantly governs the worst-case transmission value. For instance, if the *Through*-state transmission is worse than the *Drop*-state one, a lower E_{laser} is required for increasing Q_c. Conversely, E_{laser} increases with Q_c

when the *Through*-state transmission is higher than the *Drop-state* one.

In the next section, the energy-efficiency analysis will be performed under given constants such as BER, and the resulting energy map will give the feasible/infeasible design space and the optimal energy consumption of the target OLUT block.

IV. DESIGN SPACE EXPLORATION

By carrying out variations on the critical parameters of the photonic device and applying the modeling methodology described in Section III according to various dimensions of the OLUT, the design space and the impact of device characteristics on the system performance are established.

A. 2-2-OLUT

1) Feasible design space

Fig. 5 represents the total energy dissipation E_{OLUT} in colour-scale for a 2-2-OLUT according to the wavelength shift $\Delta\lambda$ and the coupling quality factor Q_c of a 2µm radius ring resonator with intrinsic $Q_i=10^5$. We limit the exploration range for $\Delta\lambda$ up to 0.8nm as we aim to design low power devices, and for Q_c to a few 10^4. The feasible design space can be defined inside the following boundaries of the figure:

- For small values of Q_c, the resonance linewidth is too broad, such that the wavelength shift $\Delta\lambda$ must exceed the boundary value (e.g. $\Delta\lambda>0.8$nm for $Q_c=3000$) in order to switch the add-drop filters to pass from the *DROP* to the *THROUGH* state with sufficient extinction ratio.

- For large values of Q_c, the ratio of Q_c to the intrinsic Q_i becomes too large, inhibiting the resonant interference process in the resonators, and the *DROP* state cannot be achieved anymore (e.g. if $Q_c/Q_i>0.5$)

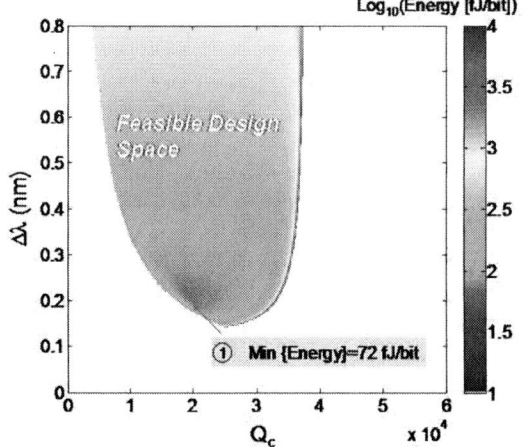

Fig. 5 Total energy dissipation (in fJ/bit) in 2-2-OLUTs according to $\Delta\lambda$ and Q_c for ring radius of 2µm ($Q_i\sim10^5$). The infeasible design space is represented by the white colour background. The color scale is in logarithmic.

Within the feasible design space in Fig. 5, i.e. between these boundaries, the minimum value of E_{olut} (72fJ/bit, see the marker #1 in the figure) is obtained for $Q_c\sim20,000$ and $\Delta\lambda\sim0.2$nm for a bias of $\{I,V\}=\{10\mu A, 0.87V\}$. This value results from the trade-off between the static energy dissipated by the add-drop filters E_s and that of the lasers E_{laser}. Because Q_c is fixed, the two contributions E_{laser} and E_s vary differently with $\Delta\lambda$, and each dominates alternately the OLUT energy consumption E_{olut}, as shown in Fig. 6. In addition, for a fixed value of $\Delta\lambda$, when E_s mostly determines the overall energy dissipation of the OLUT, E_{olut} becomes almost independent of Q_c. It happens that the minimum energy dissipation is obtained when the ratio of E_{pd} reaches its maximum (E_{pd} is fixed at 10fJ/bit, as presented previously).

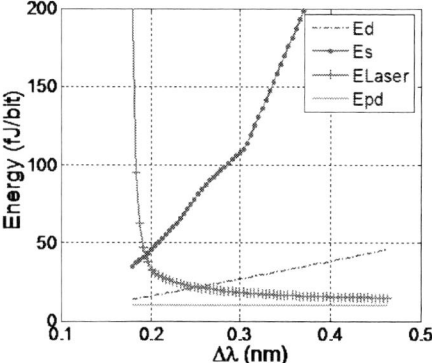

Fig. 6 Energy dissipation (fJ/bit) across various components for a 2-2-OLUT (ring radius =2µm, Q_c=20,000)

2) Energy optimization by exploration of ring radius

The minimum shift that can be used to drive the OLUT is determined by the ratio of Q_c to Q_i, i.e. intrinsically by Q_i for a given range of Q_c. Because Q_i strongly depends on the radius of the ring, the design space is extended for larger ring sizes (e.g. to Q_c above $6x10^4$ for $Q_i\sim5x10^5$) and a lower minimum value of the wavelength shift of ~0.07nm can be used to drive the OLUT. The minimum wavelength shift therefore depends on Q_i and consequently on the device geometry, i.e. the ring radius. Intuitively, since a higher Q_c value results in a spectrally narrower (more selective) add-drop transfer function, a smaller wavelength shift becomes sufficient for reaching a targeted signal-to-noise ratio (or bit error rate). As result, a lower minimum of the energy-per-output-bit figure of 48fJ/bit is obtained for $\Delta\lambda=0.1$nm when Q_c is $\sim43,000$. This is achieved with a reduced ratio of Q_c to Q_i as compared to the small ring case (in Table 1), which ensures a larger *DROP* state transmission of the add-drop filters, thereby leading to a reduced E_{laser} as compared with the 2-2-OLUT ($r=2\mu m$). However, although the minimum injected carrier concentration decreases for the smaller optimal wavelength shift that can be afforded for the larger ring case, it does not help reducing the total energy dissipation.

This indicates that an optimum value for the radius of the ring can be generated efficiently by using the proposed modeling methodology to minimize the total OLUT energy consumption. The minimum values of the total energy dissipation and the associated parameters of the add-drop filters in 2-2-OLUT for varying ring radius are summarized in the Table 1 and a minimum value for E_{olut} of 43fJ/bit is obtained for r=5µm (Q_i~4x10^5).

Tab.1 Minimum energy dissipation E_{olut} associated with specific values of the Q factors and the wavelength shift for 2-2-OLUTs with r=2µm, 4µm 5µm and 7µm microrings.

2-2-OLUT	min E_{olut}	$\Delta\lambda$	Q_c/Q_i
r=2µm	74 fJ/bit	0.2nm	0.2
r=4µm	51fJ/bit	0.12nm	0.15
r=5µm	43 fJ/bit	0.1nm	0.11
r=7µm	48fJ/bit	0.1nm	0.086

B. Scalability study: from 2 to m output bits

The proposed methodology can be applied to evaluate the scalability of OLUT architecture. By setting the ring radius value to 7µm (Q_i~5x10^5), which is the smallest value that can be used for up to a 2-8-OLUT (to have a FSR small enough to support 8 wavelength channels within a 100nm bandwidth), all OLUTs considered in the analysis can be driven properly. Fig. 7 shows the minimum values of energy-per-(output)-bit E_{olut} for the OLUTs with 2 input data channels that produce between 2 and 8 output bits. The lowest energy consumption of ~47 fJ/bit is obtained for 2-4-OLUTs driven by a shift $\Delta\lambda$~0.11nm. Moreover, it shows that while the parallelism increases with m for OLUTs, the minimum energy consumption is relatively constant, which demonstrates the scalability of the OLUT architecture.

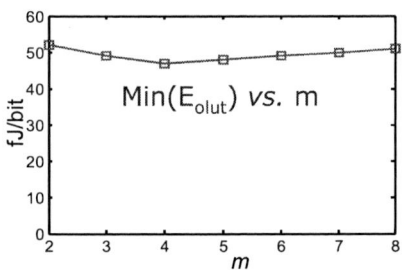

Fig. 7 Minimum energy dissipation (fJ/bit) for the OLUTs with two input data channels that produce between 2 and 8 output bits

C. Discussion

Similarly to the conventional EDA approach, the methodology described in this work can efficiently generate the validity range of the device parameters that allow a complex computing system built from silicon photonics devices to reach a well identified set of target energy efficiency or system performance metrics. This allows designers to evaluate the impact of technology options at various levels such as the device level, the architecture and application level. Indeed, this can help designers to accelerate the technology to architecture loop to

generate energy-efficient computing hardware platforms suited to emerging applications.

However, it should be pointed out that the proposed model does not include some energy dissipation sources that should be taken into account in a real environment, e.g. the static energy consumption for lasers, the thermal energy required by the add-drop filters for pre-calibration and real-time thermal tuning. In particular, this implementation of the OLUT (microring-based silicon photonics) is sensitive to temperature, and thermal tuning could potentially be a showstopper here. Moreover, the current model does not take into account the technology variation and their effects, such as fabrication error induced resonance shift, which is part of future work.

V. CONCLUSIONS

This paper has described a methodology for the design space exploration of the OLUT architecture. The methodology allows the identification of the best design tradeoffs to perform photonic reconfigurable computing with OLUTs, mainly according to energy efficiency and computing performance. Various design options such as the number and the size of OLUTs, the device physical characteristics are taken into account. The analytical results showed the potential of OLUT architectures to reach ~50fJ/bit logic operation, which is indeed comparable to the total energy dissipation per logic operation for current silicon CMOS devices (i.e. at the femtojoule level, according to ITRS). This work demonstrates the potential of silicon photonics computing architectures for future on-chip computation at high data rate and low energy consumption.

VI. REFERENCES

[1] J. Rose, et al., "Architecture of Programmable Gate Arrays: The Effect of Logic Block Functionality on Area Efficiency," IEEE J. Solid State Circuits, pp. 1217-1225, 1990.

[2] Z. Li, S. Le Beux, C. Monat, X. Letartre, I.O'Connor, "Optical Look Up Table", in Proc. of DATE, pp.483-486, 2013.

[3] I. O'Connor et al., "Systematic Simulation-Based Predictive Synthesis of Integrated Optical Interconnect," IEEE Trans. VLSIS, vol. 15, no. 8, pp. 927-940, Aug. 2007.

[4] J. Hardy and J. Shamir, "Optics Inspired Logic Architecture," Opt. Express 15(1), 150–165, 2007.

[5] Q. Xu and R. Soref, "Reconfigurable optical directed-logic circuits using microresonator-based optical switches," Opt. Express 19(6), 5244–5259, 2011.

[6] Y.Tian, L.Zhang, J.Ding, L. Yang, "Demonstration of electro-optic half adder using silicon photonic integrated circuits", Opt. Express 22(6), 6958-6965, 2014.

[7] S. Le Beux et al., "Reconfigurable photonic switching: Towards all-optical FPGAs", in Proc. VLSI-SoC, 180-185, 2013.

[8] M. Borselli, T. J. Johnson, and O. Painter, "Beyond the Rayleigh scattering limit in high-Q silicon microdisks: theory and experiment," Opt. Express 13(5), 1515-1530 2005.

[9] M. Lipson, "Compact electro-optic modulators on a silicon chip", IEEE J. Sel. Top. Quant. Elec.12 (6), 1520–1526, 2006

[10] Z. Li et al., "Complementary logic interface for high performance optical computing with OLUT", VLSI-SoC, 201

2015 IEEE Computer Society Annual Symposium on VLSI

Multi-Swarm Optimization of a Graphene FET Based Voltage Controlled Oscillator Circuit

Elias Kougianos
Electrical Engineering Technology
University of North Texas. USA.
Email: eliask@unt.edu

Shital Joshi
Computer Science and Engineering
University of North Texas, USA.
Email: ShitalJoshi@my.unt.edu

Saraju P. Mohanty
Computer Science and Engineering
University of North Texas, USA.
Email: saraju.mohanty@unt.edu

Abstract—**The lack of well defined abstraction levels and immature design tools have made the custom design and optimization of analog circuits slow, complex and laborious. Furthermore, CMOS technology beyond 10 nm faces fundamental limits which may restrict its applicability for future devices. In this paper, a Graphene Field Effect Transistor (GFET) based cross-coupled LC circuit is used as case study to design a mixed-signal integrated circuit. A Multi-Swarm Optimization (MSO) algorithm is used for fast exploration of the design space for the GFET based LC-VCO to maximize its operating frequency given phase noise and power dissipation as optimization constraints. The length (L) and width (W) of the graphene channel are selected as the design variables. The phase noise is kept below -80 dBc/Hz and power dissipation is less than 16 mW. The design optimization flow results in a maximum frequency of 2.58 GHz for $L = 3.35\mu$m and $W = 1.82\mu$m and the phase noise and power dissipation constraints are well below the guard limits.**

Keywords-**Graphene FET, LC-VCO, Multi swarm optimization (MSO)**

I. INTRODUCTION

Until recently, silicon based field effect transistors (FETs) were the only choice for the semiconductor industry. As complementary metal oxide semiconductor (CMOS) has hit the power wall i.e. non scaling of KT/q and hence of V_{th} and V_{DD}, it may not be the obvious choice for future smarter devices. With the recent discovery of graphene, the industry and academia have started focusing their attention to graphene based electronic systems. For technologies beyond the 10 nm node or below, graphene is considered to be a viable solution to the problem due to its high field-effect mobility (as high as 15000 cm$^2/(Vs)$ [1]–[4] and a high Fermi velocity ($\sim 10^8$ cm/s) even at room temperature.

In this paper a design flow for graphene FET based analog circuits is presented. An LC oscillator is considered as a case study to present the fast design optimization flow for a mixed-signal integrated circuit design. Various optimization techniques have been used in the literature in optimizing analog circuits. Swarm optimization, simulated annealing, and evolutionary algorithms are common [5]–[7]. Bee colony optimization and particle swarm optimization are examples of swarm optimization techniques. However in this paper, a multi-swarm optimization algorithm is used to explore the design space of the GFET based LC-VCO, where the objective

is to maximize the frequency with phase noise and power dissipation considered as constraints. The design variables for the optimization are the length (L) and width (W) of the graphene channel. Multiple swarms of particles are generated to search the design space. The oscillator is implemented in Verilog-A based on [8].

The rest of this paper is organized as follows: In section II new contributions are highlighted. In section III related research in the area of GFET based circuit design as well as swarm optimization techniques in optimizing analog circuits are discussed. In section V the proposed design and optimization flow of the LC oscillator circuit are presented. The baseline design of the LC-VCO is described in Section IV. The optimization results and final sized circuit are described in Section VI. In conclusion, some possible future directions are examined in Section VII.

II. NOVEL CONTRIBUTIONS OF THIS PAPER

To the best of the authors' knowledge, this is the first attempt to propose a design flow for GFET based cross-coupled version of an LC oscillator. A new optimization algorithm called multi-swarm optimization (MSO) is used in the design flow, which is also a new attempt. The output of this optimization flow gives the proper sizing of the GFET device so as to achieve the maximum frequency within given constraints. Thus, this paper is a forward step in the design and optimization of GFET based RF circuits where an LC oscillator is used.

III. RELATED PRIOR RESEARCH

After the successful preparation of a two dimensional structure of carbon atoms, graphene has drawn a lot of attention from the electronic device community [9], [10]. In the previous works, graphene based devices like low noise amplifier [11], mixer [12], high frequency graphene amplifier [13], and a frequency doubler [14] have been presented. In [15], the design and simulation of a GFET based LC-VCO for WLAN applications along with its process variation analysis is presented. Optimization is necessary at the final stage of the circuit design so as to meet the design goal. Swarm optimization has been widely used in analog and RF circuits. In this paper, the MSO technique is used, which is

978-1-4799-8720-7/15 $31.00 © 2015 IEEE

a variation of particle swarm optimization (PSO). Instead of a single swarm, multiple swarms are used to find the best global solution for the optimization problem. In the literature, various versions of MSO have been used for tracking the best solution for global optima from the generated multiple swarm, [16]. In [17], speciation based approaches were used where the number and size of the swarms is dynamically adjusted so as to construct an ordered list of particles. In [18], MSO has been presented for a dynamic environment which keeps track of the changing optimum value over time.

IV. Design and Characterization of a Graphene LC-VCO

Unlike semiconductors, graphene lacks an intrinsic bandgap and hence it cannot be turned off completely. The typical transfer and output characteristics of the GFET show a "kink" at the saturation region, which is due to the transition of carrier type (either positive holes or negative electrons depending upon the gate voltage). Due to the zero bandgap, graphene shows ambipolar conduction depending on the top gate voltage. Exploiting the ambipolar characteristics of GFET, analog and RF circuits can be built such as frequency doublers, RF mixers, digital modulators and phase detectors. However as a case study, a cross coupled LC oscillator using GFET [1], [15] is designed as shown in Fig. 1, where the ambipolar property of graphene is exploited. Due to the high electron mobility of the GFETs, the phase noise is reduced and the linear response is improved.

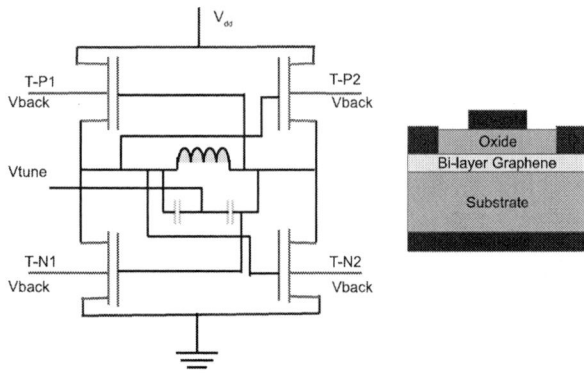

Fig. 1. GFET based LC-VCO with GFET-2D cross-section (left).

TABLE I
GFET DEVICE PARAMETERS [1], [15].

Device Parameters	Default Values
μ (Mobility)	0.3 m^2/(Vs)
Length	6.0 μm
Width	1.6 μm
R_s (p-channel)	600 Ω
R_s (n-channel)	4000 Ω
V_{supply}	9 V
I_{bias}	0.7 mA

In this paper, the GFET is modeled in Verilog-A. Fig. 2 shows the source drain $I - V$ characteristics for the NFET

and PFET around the operating region. Table I summarizes the GFET parameters used for the baseline design in this paper where the values for channel length (L) and width (W) are selected heuristically. Table II summarizes the characteristics of the baseline oscillator. The phase noise performance and the tuning characteristics at the operating condition are shown in Fig. 3(a) and Fig. 3(b).

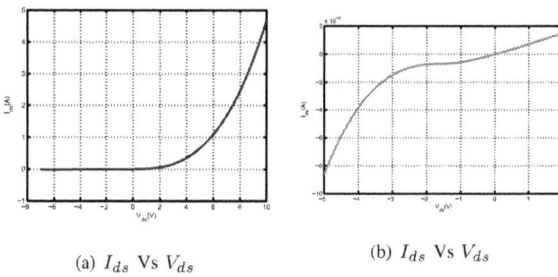

(a) I_{ds} Vs V_{ds} (b) I_{ds} Vs V_{ds}

Fig. 2. $I - V$ curve of N-type and P-type GFET around operating region.

TABLE II
CHARACTERIZATION OF BASELINE GFET BASED LC-VCO

LC-VCO Characteristics	Estimated Values
f_{center}	2.56 GHz
$V_{tank,p-p}$	0.8 V
I_{bias}	0.77 mA
Tuning Range	4.88%
Phase Noise (1MHz offset)	-88.25 dBc/Hz

V. Proposed Design Optimization Flow for GFET based LC-VCO

A. Overall Design Optimization Flow

Fig. 4 shows the proposed design flow for the LC-VCO circuit. The design process starts with determining the approximate design points from the behavioral analysis. The baseline design of the circuit is performed from which a netlist is generated. Various figures of merits (FoMs) such as frequency, phase noise and power dissipation are chosen to characterize the oscillator. After choosing the FoMs for the circuit, design variables (length (L) and width (W)) of the GFET are selected. The netlist obtained from the baseline design is then parameterized with respect to these design variables. In order to obtain the distribution of the variation of the characteristics due to the variation of design parameters, parametric analysis is performed on the FoMs. This gives the variable range and the constraints are then chosen based on the design requirements. In the final step of the design, optimization is performed to obtain the maximum frequency for the given constraints. After the successful completion of optimization, a netlist is obtained from this final sized circuit.

The Multi-swarm optimization (MSO) algorithm is used in this paper for optimization. In MSO, several swarms of simulated particle are used to estimate the solution of the optimization problem. Any optimization problem starts with problem formulation which includes problem constraints and

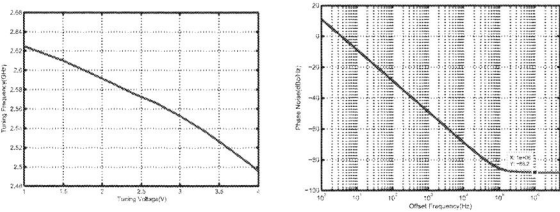

(a) Fequency Vs Tuning Voltage   (b) Phase Noise Vs Offset frequency

Fig. 3. Tuning range and Phase noise characteristics of the baseline GFET based LC-VCO.

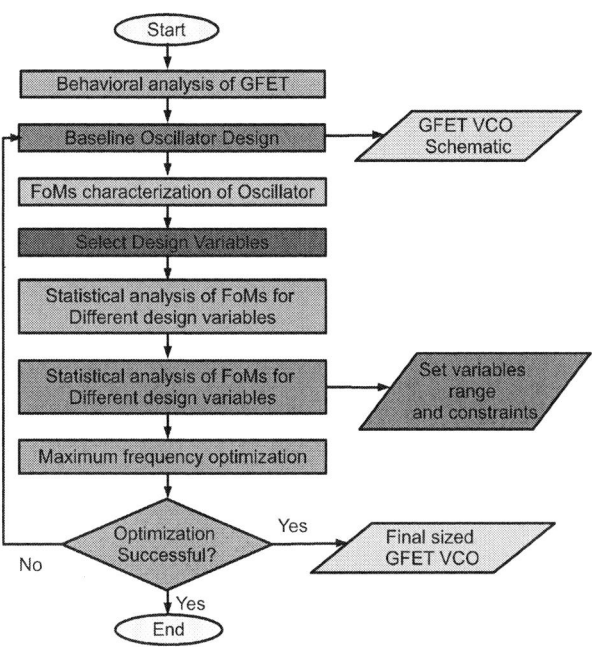

Fig. 4. Design Flow of GFET based LC-VCO Circuit Optimization.

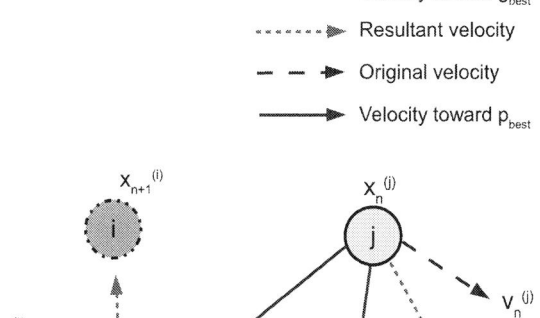

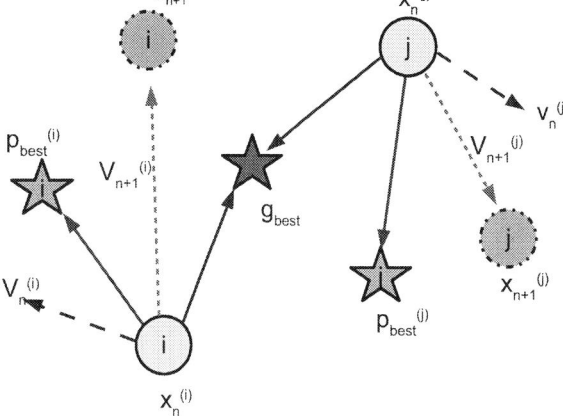

Fig. 5. Particle velocity updating where particles i and j are acceleretaed towards their best location, g_{best} and p_{best}

B. Multi-Swarm Optimization (MSO) Technique

PSO is an efficient and powerful heuristic technique to solve the global optimization. It has drawn a lot of attention in recent years due to its ease of implementation, robustness and efficiency. The PSO system consists of a single swarm of potential solutions called particles, each of them being associated with a fitness value. These particles move towards the optimal solution with a specified velocity, keeping track of the best position obtained so far. The particles' personal best position is referred as p_{best} while the swarm's best position is referred as g_{best} (or global best). Thus in each iteration, the particle move towards its g_{best} as well as to its p_{best}.

In the second iteration, the best particle from the previous iteration is used to calculate the particle velocity in order to move closer to the optimum solution in each successive iteration. The particle velocity is updated using the following expression:

$$
\begin{aligned}
v_{t+1} =\ & wv_t + c_1 r_1 (x_i - p_{best}) \\
& + c_2 r_2 (x_i - s_{best}) + c_3 r_3 (x_i - g_{best}).
\end{aligned} \quad (1)
$$

The particle position is updated using:

$$
x_i = x_i + v_{t+1}. \quad (2)
$$

In Eqn. 1 and Eqn. 2, w is the inertia weight that tends to retain the previous velocity [19] and c_1, c_2 and c_3 are the constants that tend to retain the particle behavior to its own previous position, to its swarm behavior and to the global behavior of multiple swarms respectively. r_1, r_2 and r_3 are

the design variable ranges as the main step. In this circuit optimization, length (L) and width (W) of the graphene channel are chosen as the design variables which are represented by the particle positions. The cost function for each particle is then evaluated . The current position of each particle is compared with the previous best position in order to make a decision on whether to retain the current best position or update the value with the previous best position. Since the objective is to maximize the the frequency of oscillation, the current frequency is observed at each value set of the design variables. Accordingly the value set or the coordinate of the best design variable is decided, based on the present best value and previous best value. From each swarm, a best swarm particle is selected and the best global particle is chosen from the total number of swarms. The maximum frequency is now obtained in the first iteration.

978-1-4799-8720-7/15 $31.00 © 2015 IEEE   569

three random numbers generated each time which reflect the natural phenomenon of swarm intelligence. In each iteration, the particle moves towards the global best particle, the best particle in its swarm as well as to its previous best position. In constrained optimization, the design variables must be within a specified range. If some particles go beyond the range then those particles need to be reflected back from the boundary so as to maintain their position within the swarm. After a specific number of iterations, based on the convergence criteria or designer's intuition, convergence to a global optimum can be achieved and the simulation is stopped. The optimization process can be considered to be completed if a relatively constant best particle position is obtained or if the frequency of the oscillator does not improve or it swings back and forth around some average value.

VI. EXPERIMENTAL RESULTS

The results obtained from the parametric analysis are shown in Fig. 6. Datas are extracted from this analysis and are then fed to the optimization algorithm to optimize the circuit. The values for phase noise and power dissipation are chosen to be less than -80 dBc/Hz and 16 mW respectively thus optimizing the circuit for maximum possible frequency.

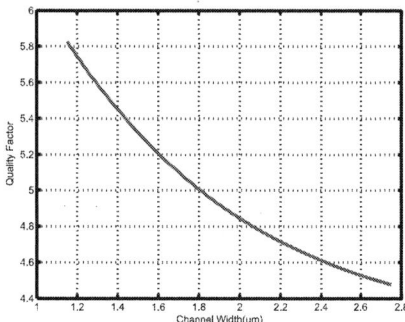

Fig. 7.   Quality Factor of the GFET based LC-VCO.

Table III shows the design variable range used. As can be seen in Fig. 6(a) and Fig. 6(b), there is a dip in the curves corresponding to a resonating condition. This dip is created in the tank circuit due to the GFET capacitances. The design variable range is selected in such a way that these variables avoid resonance. In Fig. 6(f), two peaks are observed.

In order to understand this behavior, it is necessary to analyze Fig. 6(d) and Fig. 7. In Fig. 7, the quality factor Q is measured as a function of varying channel width and it is observed that the Q value decreases with the channel width. However in Fig. 6(d), the reverse trend is seen with the power dissipation versus width. For an LC oscillator, the phase noise is dependent on both Q and signal power. Based on Leeson's equation, phase noise is proportional to the inverse of signal power and is inversely proportional to the square of Q [20]. So from Fig. 6(d) and Fig. 7, it becomes clear that phase noise has two peaks with channel width variation. Again the

design variable width (W) is selected to avoid these phase noise peaks as well. Thus it becomes clear from Fig. 6 that the MSO approach does not allow the optimization problem to oscillate in local optima, unlike PSO. In MSO, due to multiple swarms, which explore for the global optimum independently, the chance of getting stuck in a local optimum is always lower than that in PSO.

To facilitate the optimization, inertia constants need to be chosen based on the problem. In the absence of inertia constants, the particle may not converge [21]. When a small inertia constant is chosen then the algorithm becomes more like a local search algorithm. Similarly, a large inertia constant can make it a global search algorithm which tries to explore new areas. Both high and low values of inertia constant are not preferable. Thus there is need to have some balance between the local and the global search. For any optimization algorithm, it is generally preferred to have more exploitation ability at the beginning stages to find good seeds and then limit the search to the local area around these seeds. It is beneficial to make the inertia as constantly decreasing functions of time. In this paper, the value of inertia constant is gradually decreased form an initial value of 0.7 to a final value of 0.33. These values have been chosen after exhaustive trial and error until satisfactory results were obtained. Based on particle movement, the maximum iterations is set to be $N = 20$. Table IV summarizes the parameters in defining particle velocity.

TABLE IV
MULTI-SWARM OPTIMIZATION ALGORITHM PARAMETERS.

Algorithm Parameters	Values
w	0.7 - 0.33
$c1$	0.2
$c2$	0.2
$c3$	0.2

Similarly, critical parameters are the acceleration constants. Small values for these constants make the particles roam far away from the target global region and high values result in abrupt particle movement. The acceleration constants are chosen small compared to the values selected in [22] for a single swarm PSO. Table IV shows the values chosen for acceleration constant after several trial and error.

The optimization steps and results are shown in Fig. 8. The initial and final particle positions after 20 iterations are shown in Fig. 8(a) and Fig. 8(b), respectively. As can be seen in these figures, the initial particles are spread all over the design space but after 20 iterations, the particles of all swarms are converged around the global optimum particle position. The interpolated contours in Fig. 8(b) helps to visualize the convergence step where each line illustrates connected isometric values. The global best particle position is obtained by averaging across all steps and Table V illustrates the optimal oscillator values obtained after optimization. Fig. 9(a) and Fig. 9(b) show the tuning range and phase noise of the optimized LC-VCO.

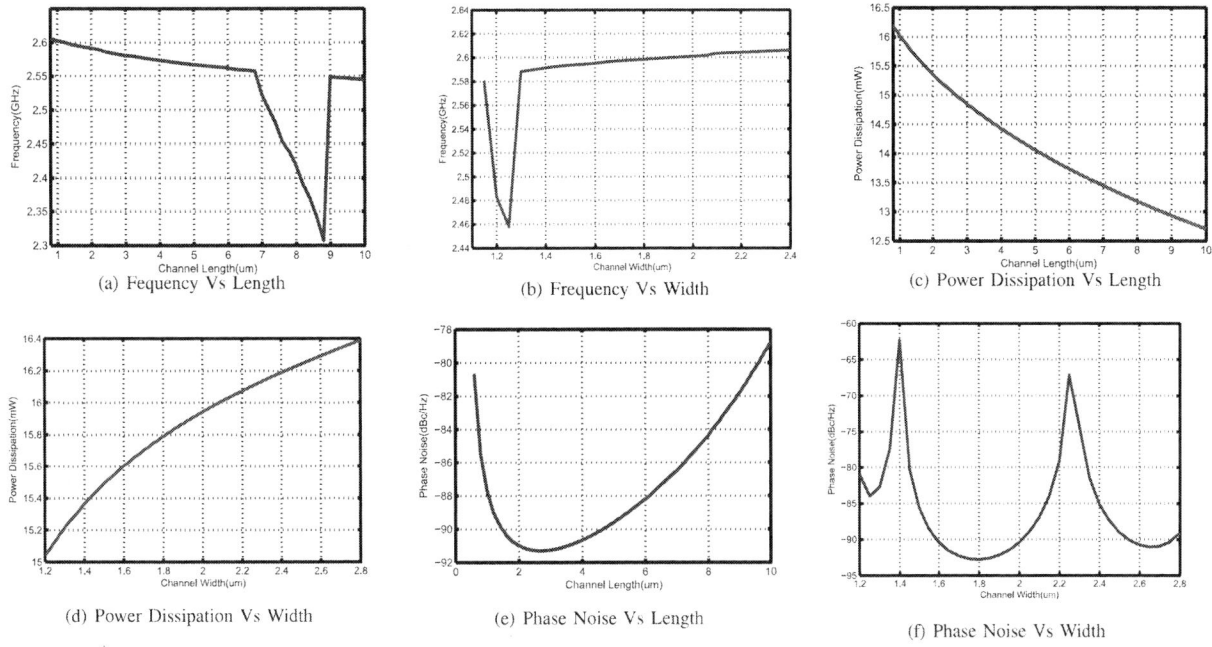

(a) Fequency Vs Length

(b) Frequency Vs Width

(c) Power Dissipation Vs Length

(d) Power Dissipation Vs Width

(e) Phase Noise Vs Length

(f) Phase Noise Vs Width

Fig. 6. Characteristics of the GFET based LC-VCO.

TABLE III
GFET-BASED LC-VCO DESIGN VARIABLES AND CONSTRAINTS.

LC-VCO Parameter	Parameter Type	Minimum Value	Maximum Value
L	Design Variable	$3\mu m$	$7\mu m$
W	Design Variable	$1.4\mu m$	$2.2\mu m$
Power Dissipation	Design Constraint	Minimize	$16mW$
Phase Noise	Design Constraint	Minimize	$-80dBc/Hz$

TABLE V
CHARACTERISTICS OF THE OPTIMAL LC-VCO.

LC-VCO Parameters	Estimated Values
Channel Length	3.35 μm
Channel Width	1.82 μm
Frequency	2.58 GHz
Power Dissipation	11.74 mW
Phase Noise (at 1MHz offset)	-92.92 dBc/Hz
Tuning range	4.26%
$V_{tank(p-p)}$	0.75 V
I_{bias}	0.83 mA

VII. CONCLUSIONS AND FUTURE RESEARCH

The required LC-VCO oscillator is successfully designed following the design flow and optimization algorithm presented in this paper, where the design constraints of phase noise and power dissipation are well met. The netlist obtained from the baseline oscillator is used to apply MSO for the given objective function, design variables and constraints. The baseline LC-VCO with center frequency of 2.56 GHz is initially designed and the final LC-VCO with center frequency of 2.58 GHz is obtained, where the power dissipation and phase noise are 26.625% and 16.15% below their maximum values.

As an alternate design approach, in future work, a surrogate model of the circuit will be created which will then be used to perform optimization instead of using a netlist, in order to reduce the optimization time. In order to perform parasitic aware design, parasitic extraction will be done and multi-objective optimization will be performed to obtain the final layout.

ACKNOWLEDGMENTS

The authors would like to thank UNT graduates and students Dr. Geng Zheng, Mr. Mahesh Gautam, and Mr. A. Khan for their help and inputs on the initial version of this paper.

REFERENCES

[1] S. P. Mohanty, *Nanoelectronic Mixed-Signal System Design.* McGraw-Hill Education, 2015.
[2] Y. M. Banadaki, K. M. Mohsin, and A. Srivastava, "A Graphene Field Effect Transistor For High Temperature Sensing Applications," in *Proceedings of SPIE, Nanosensors, Biosensors, and Info-Tech Sensors and Systems*, vol. 9060, 2014, pp. 90 600F–90 600F–7.
[3] S. V. Morozov, K. S. Novoselov, M. I. Katsnelson, F. Schedin, D. C. Elias, J. A. Jaszczak, and A. K. Geim, "Giant Intrinsic Carrier mobilities in Graphene and Its Bilayer," *Physical Review Letter*, vol. 100, 2008.
[4] A. K. Geim and K. S. Novoselov, "The Rise of Graphene," *Nature materials*, vol. 6, no. 3, pp. 183–191, 2007.

978-1-4799-8720-7/15 $31.00 © 2015 IEEE

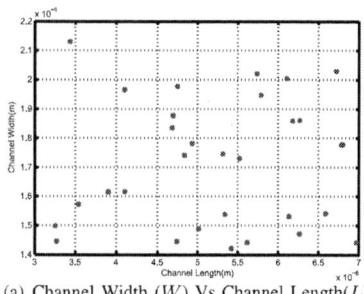

(a) Channel Width (W) Vs Channel Length(L)

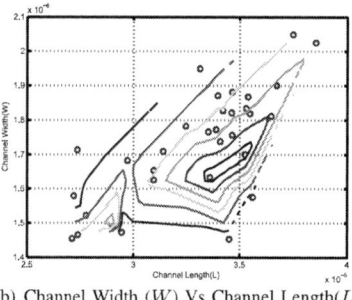

(b) Channel Width (W) Vs Channel Length(L)

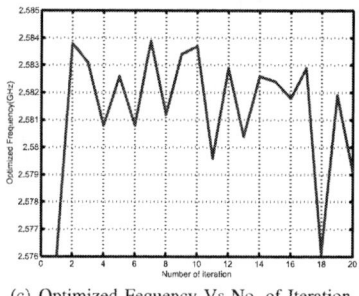

(c) Optimized Frequency Vs No. of Iteration

Fig. 8. GFET based LC-VCO Optimization: (a) Initial particles (x), (b) particles after 20th iteration (o), (c) Optimized frequency.

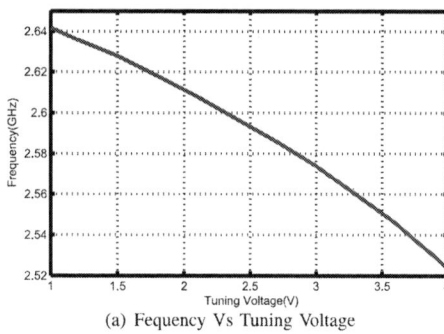

(a) Fequency Vs Tuning Voltage

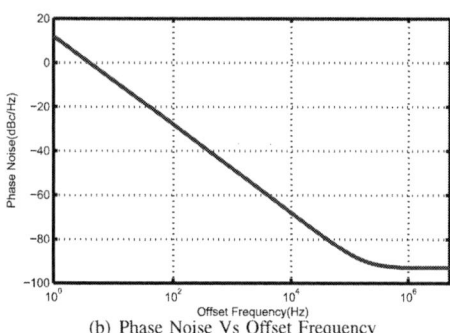

(b) Phase Noise Vs Offset Frequency

Fig. 9. Tuning range and Phase noise of Optimized GFET based LC-VCO.

[5] M. Fakhfakh, Y. Cooren, A. Sallem, M. Loulou, and P. Siarry, "Analog Circuit Design Optimization Through The Particle Swarm Optimization Technique," *Analog Integrated Circuits and Signal Processing*, vol. 63, no. 1, pp. 71–82, April 2010.

[6] O. Garitselov, S. P. Mohanty, and E. Kougianos, "Accurate polynomial metamodeling-based ultra-fast bee colony optimization of a nano-cmos phase-locked loop," *Journal of Low Power Electronics*, vol. 8, pp. 317–328, 2012.

[7] G. G. E. Gielen, H. C. C. Walsch, and W. M. C. Sansen, "Analog Circuit Design Optimization Based on Symbolic Simulation and Simulated Annealing," *IEEE Journal of Solid-State Circuits*, vol. 25, pp. 707–713, 1990.

[8] I. J. Umoh and T. J. Kazmierski, "VHDL-AMS model of a dual gate graphene FET," in *Forum on Specification and Design Languages (FDL)*, 2011, pp. 1–5.

[9] F. Schwierz, "Graphene transistors," *Nature Nanotechnology*, vol. 5, pp. 487–496, May 2010.

[10] K. S. Novoselov, A. K. Geim, S. V. Morozov, D. Jiang, Y. Zhang, S. V.

Dubonos, I. V. Grigorieva, and A. A. Firsov, "Electric Field Effect in Atomically Thin Carbon Films," *Science*, vol. 306, no. 5696, pp. 666–669, 2004.

[11] S. Das and J. Appenzeller, "An All-Graphene Radio Frequency Low Noise Amplifier," in *Proceedings of IEEE Radio Frequency Integrated Circuits Symposium (RFIC)*, 2011, pp. 1–4.

[12] H. Wang, A. Hsu, J. Wu, J. Kong, and T. Palacios, "Graphene-Based Ambipolar RF Mixers," *IEEE Electron Device Letters*, vol. 31, no. 9, pp. 906–908, 2010.

[13] S. J. Han, K. A. Jenkins, A. Valdes Garcia, A. D. Franklin, A. A. Bol, and W. Haensch, "High-Frequency Graphene Voltage Amplifier," *Nano Letters*, vol. 11, no. 9, pp. 3690–3693, 2011.

[14] M. E. Ramon, K. N. Parrish, S. F. Chowdhury, C. W. Magnuson, H. C. P. Movva, R. S. Ruoff, S. K. Banerjee, and D. Akinwande, "Three-Gigahertz Graphene Frequency Doubler on Quartz Operating Beyond the Transit Frequency," *IEEE Transactions on Nanotechnology*, vol. 11, no. 5, pp. 877–883, sept. 2012.

[15] M. A. Khan, S. P. Mohanty, and E. Kougianos, "Statistical Process Variation Analysis of a Graphene FET based LC-VCO for WLAN Applications," in *Proceedings of the 15th International Symposium on Quality Electronic Design*, 2014, pp. 569–574.

[16] C. Li and S. Yang, "Fast multi-swarm optimization for dynamic optimization problems," in *Proceedings of the Fourth International Conference on Natural Computation*, 2008, pp. 624–628.

[17] D. Parrott and X. Li, "A particle swarm model for tracking multiple peaks in a dynamic environment using speciation," in *Evolutionary Computation, 2004. . Congress on*, vol. 1, 2004, pp. 98–103.

[18] T. M. Blackwell and J. Branke, "Multi-Swarm Optimization in Dynamic Environments," *Application of Evolutionary Computing*, vol. 3005, pp. 489–500, 2004.

[19] D. Bratton and J. Kennedy, "Defining a standard for Particle Swarm Optimization," in *Proceeding of IEEE Swarm Intelligence Symposium*, 2007, pp. 120–127.

[20] T. H. Lee and A. Hajimiri, "Oscillator Phase Noise: A Tutorial," *IEEE Journal of Solid-State Circuit*, vol. 5, pp. 326–326, March 2000.

[21] R. C. Eberhart and Y. Shi, "A modified particle swarm optimizer," in *Proceedings of IEEE World Congress on Computational Intelligence*, 1998, pp. 69–73.

[22] ——, "Particle Swarm Optimization: Developments, Applications and Resources," in *Proceedings of World Congress on Evolutionary Computation*, 2001, pp. 81–86.

978-1-4799-8720-7/15 $31.00 © 2015 IEEE 572

2015 IEEE Computer Society Annual Symposium on VLSI

Fast Stimuli Generation for Design Validation of RTL Circuits Using Binary Particle Swarm Optimization

Prateek Puri and Michael S. Hsiao
Bradley Department of Electrical and Computer Engineering
Virginia Tech, Blacksburg, VA 24061, USA
Email: {prateekp, hsiao}@vt.edu

Abstract—Generating high quality test sequences for complex digital circuits is known to be extremely challenging. In this paper, we introduce a test generation algorithm using Binary Particle Swarm Optimization (BPSO) to generate high-quality test sequences that achieve high branch coverage in short execution times for synthesizable RTL designs. Initially, a global search is conducted using Binary Particle Swarm Optimization which is later supported by a controlled graphical search method to reach target corner cases. The controlled search uses the control-flow graph to provide hints at critical points in the state space to reach hard corner cases. The fast convergence of BPSO allows the proposed method to deliver high coverage while generating short final test sequences. Substantial speedups over the state of the art methods have also been achieved.

Keywords-test generation; swarm; hybrid algorithm

I. INTRODUCTION

Over the last two decades, chip design has been conducted at the register transfer (RT) Level using Hardware Descriptive Languages (HDL), such as VHDL and Verilog. The modeling at the behavioral level not only allows for better representation and understanding of the design, but also allows for encapsulation of the sub-modules as well, thus increasing productivity. Despite these benefits, validating a RTL design is not necessarily easier. Design validation has been recognized as a bottleneck in the design cycle before the final release can be made. Thus, design validation not only exacerbates the time to market factor but also consumes many resources. Previously, several coverage metrics such as state coverage, condition coverage, branch coverage, path coverage, etc. have been proposed and used to evaluate the quality of the test stimuli. In the ideal scenario, a high quality test sequence should be of minimal length and offer maximum coverage as per any adopted metric. Random stimuli generally fail to achieve a high coverage as some branches may require a specific test sequence in order to be covered. Even though several advancements have been made in coverage-directed test generation, derivation of high quality test sequences remains an arduous task.

In recent years, several proposed techniques such as HYBRO [1] and BEACON [2] have shown great potential in the field of test generation at the RTL. HYBRO unrolls the circuit and then applies a Satisfiability Modulo Theory (SMT) solver to determine an input assignment corresponding to a target

path. The computational cost of SMT solvers may limit both the length of paths and the number of branches that can be explored; thus, it may fail to cover those branches that require long sequences of vectors. Existing simulation-based methods, on the other hand, can scale to larger designs. For example, BEACON targets hard-to-reach branches without resorting to deterministic engines such as symbolic execution in HYBRO. However, simulation-based methods require effective guidance, without which the resulting sequence can be long and may still require much execution time.

Nature-inspired population based stochastic optimization techniques such as Genetic Algorithm (GA), Ant Colony Optimization (ACO), Particle Swarm Optimization (PSO), etc. are robust and computationally efficient optimization methods for solving large scale problems that contain several local optima. However, substantial computational effort is required in fine tuning the search. To improve convergence rate and solution quality, such global search methods are typically hybridized with local search techniques [3]. Local search starts by adopting a population member as its base solution. It then explores the local neighborhood around the base solution and feedbacks the additional information gained to the global search. The hybrid global-local search algorithms are comparatively faster and generally produce better results. With the aim of developing simple, fast and effective stimuli generation techniques for a variety of RTL designs, we propose an alternative approach: Particle Swarm Optimization based Functional Test generator (PSOFT). PSOFT hybridizes Particle Swarm Optimization technique with a controlled graphical search method in which each particle represents a candidate test sequence. Compared to GA and ACO, the strengths of PSO are in its inherent simplicity and typically faster convergence rate through its one way information sharing mechanism.

In PSOFT, the search starts from targeting easy branches to covering hard to reach branches in later stages by altering the objective function and gradually growing the size of both the swarm and the particles. When compared to conventional hybrid evolutionary techniques, the controlled search feedbacks information to the swarm to improve results, but starts from the reset state. Thus, the controlled search does not need to rely on an initial solution vector which is

978-1-4799-8720-7/15 $31.00 © 2015 IEEE   573

required by conventional local search techniques. In addition, in order to reduce computational effort, the controlled search is selectively activated. In comparison to previous methods, PSOFT either matches or surpasses branch coverage for all ITC'99 benchmark circuits with significant improvements in execution time and test set sizes.

The contributions in present work are:

- We introduce PSOFT, a hybrid Particle Swarm Optimization technique for fast and effective automatic test generation for RTL designs.
- PSOFT generates short input sequences quickly that achieve high coverage.
- Measures are taken to combat premature saturation of swarms fitness value. By modifying the objective function in later stages, the search is directed from easy branches to difficult targets.
- To save computational effort further, the swarm starts with a smaller population which is supplemented by controlled population explosions when needed. The size of particles is also dynamically adjusted to improve penetration while generating shorter test sequences.
- Global search is aided by a controlled graphical search which feedbacks critical structural information of DUT to swarm for better test generation.

The outline for the rest of the paper is as follows: In the next section, we briefly present prior relevant work in RTL design validation and explain a binary variant of Particle Swarm Optimization. Section 3 describes the proposed technique in detail. Simulation results for several benchmarks are discussed in Section 4 and Section 5 concludes the paper.

II. BACKGROUND WORK

In [4] Corno et al. introduced a suite of synthesizable RT level benchmarks and proposed ARTIST, a Genetic Algorithm (GA) based automatic test generator. ARTIST uses GA to analyze simulation traces in the instrumented code and then provides improved test sequences to a VHDL simulator. ARTIST demonstrated the feasibility of RT-Level test generator, thus improving the overall validation flow. In [5] Li et al. proposed ATCLUB for test generation. In ATCLUB, clusters of circuit states are formed which simplifies the finite state machine and makes test generation easier. ATCLUB reduces computation time and vector count by slightly compromising on coverage. Both ARTIST and ATCLUB failed to reach several hard targets. In recent years, HYBRO and BEACON have been proposed as a branch coverage driven technique for RT level test generation. HYBRO analyzes the Control Flow Graph (CFG) of the RTL statically, extracts symbolic expressions and passes them to a SMT solver. HYBRO performs circuit unrolling, but for branches that require long specific sequences, unrolling over large number of cycles becomes impractical. Also, the usage of SMT solvers usually adds high computational complexity thus increasing the computation time of overall

technique. BEACON combines the ant colony optimization (ACO) method with an evolutionary technique to improve its exploration capability. It covers hard to reach targets by removing more visited paths from the search space thus improving upon final coverage. However, BEACON uses a constructive meta-heuristic algorithm (ACO) where at each step ants have to look at an entire set of paths before selecting one. Thus, a significant portion of total computational effort is required to determine the utility of all available paths which inherently slows down its convergence rate and may result in long test sequences. Whereas HYBRO and BEACON use higher level of abstraction for design validation, population based stochastic techniques have been implemented to achieve gate level stuck-at coverage [6], [7] However, during test generation at the gate level, these metaheuristics miss the chance of using information available at higher level descriptions such as branching conditions or control paths. Recently, Gent and Hsiao in [8] proposed a mixed level test generation technique which uses a feedback mechanism between RT level and gate level. The technique used dominators in circuit graph to identify critical nodes which can lead away from target node. These critical nodes were used as guiding points for test generation.

A. Binary Particle Swarm Optimization

Since its inception in 1995 by Kennedy and Eberhart [9], PSO has attracted many researchers as a global optimization technique in the continuous domain because of its ease of implementation. Similar to other nature inspired algorithms such as GA, ACO, etc., PSO contains a population of candidate solutions which are individually known as particles and collectively called as a swarm. PSO tries to imitate the social interactions among the members of a swarm and uses algebraic operators to improve the particles. In PSO, every particle is associated with a d-dimensional velocity vector & a position vector & memory to store their best experiences. The global best experience of swarm is also memorized & updated during evolution. The position vector represents a feasible solution whereas the velocity vector controls its motion in the d-dimensional search space. Higher velocities promote exploration whereas lower velocities result in convergence. In a given iteration, every particle in the swarm tries to improve its position. The movement of particle in the search space is governed by its inertia, its own previous best learning and collective experience of the swarm.

From theoretical analysis of particles trajectory in both continuous & discrete domains, Clerc & Kennedy [10] found that each particle converges quickly to a weighted mean of its own best position and the global best position. The reason for fast convergence was attributed to the one way information sharing mechanism in PSO as the global best particle is the only agent that supplies information to all other particles. When compared to PSO, GA and ACO are significantly different in the way of sharing information.

While PSO possesses a direct global control, ACO is based on stigmergy. In ACO, each ant examines goodness of all available paths from a node in the graph before choosing one which in turn slows down ACO. GA when compared to PSO exhibits a mutual information sharing process. In GA, the entire population moves relatively uniformly towards optimal region whereas PSO is heavily influenced by the best solution. Although problem specific, PSO typically outperforms both GA and ACO especially in terms of convergence speed [11]. Hassan et al. in [12] validate the computational efficiency of PSO over GA using formal hypothesis testing approach on standard benchmark functions.

The performance of PSO for continuous domain problems motivated researchers to develop modified versions to handle discrete optimization problems. In this regard, Kennedy [13] proposed the discrete binary version of PSO (BPSO) algorithm. Later, different variants of discrete PSO were applied to solve combinatorial problems like Travelling Salesman, planar graph coloring, resource constrained job scheduling, financial ratio selection, etc. However, in the original version of BPSO, there were certain difficulties in the interpretation of continuous domain PSO to either the discrete or binary domain. Khanesar et al. in [14] address such issues and present a better interpretation of discrete/binary PSO. Another important component in the discrete PSO is the transfer function which is used for mapping the continuous domain velocity component into a discrete domain component known as bit changing or flipping probability. Traditional BPSO algorithms implemented an S-shaped transfer functions; however, in [15] it was found that using V-shaped transfer functions on standard benchmark functions improved performance of BPSO. The proposed test generation method in this paper uses a variant of binary PSO [14] with a V shaped transfer function. Here, each particle is associated with two velocities or bit flipping probabilities. During generation (t), the following equations are used to govern the movement of particle (i) in dimension (j) of search space.

$$V_{ij}^1(t+1) = w * V_{ij}^1(t) + d_{ij,1}^1 + d_{ij,2}^1 \qquad (1)$$

$$V_{ij}^0(t+1) = w * V_{ij}^0(t) + d_{ij,1}^0 + d_{ij,2}^0 \qquad (2)$$

$$V_{ij}^c = \begin{cases} V_{ij}^1(t+1), & \text{if } x_{ij}(t) = 0 \\ V_{ij}^0(t+1), & \text{if } x_{ij}(t) = 1 \end{cases} \qquad (3)$$

$$TF(x) = \left| \frac{2}{\pi} \tan^{-1}(\frac{x\pi}{2}) \right| \qquad (4)$$

$$X_{ij} = \begin{cases} (X_{ij}(t))', & \text{if } rand() < V_{ij}^c \\ (X_{ij}(t)), & \text{if } rand() \geq V_{ij}^c \end{cases} \qquad (5)$$

If $P_{ibest}^j = 1$ then $d_{ij,1}^1 = c_1 r_1$ & $d_{ij,1}^0 = -c_1 r_1$
If $P_{gbest}^j = 1$ then $d_{ij,2}^1 = c_2 r_2$ & $d_{ij,2}^0 = -c_2 r_2$
If $P_{ibest}^j = 0$ then $d_{ij,1}^1 = -c_1 r_1$ & $d_{ij,1}^0 = c_1 r_1$

If $P_{gbest}^j = 0$ then $d_{ij,2}^1 = -c_2 r_2$ & $d_{ij,2}^0 = c_2 r_2$

In the above equations, P_{ibest}^j, P_{gbest}^j represent the jth dimensional bit value of personal best, global best particle in the swarm, TF(x) is a V-shaped transfer function; w represents inertia weight, r_1, r_2, r_3, $rand()$ represents random number between $(0, 1)$ whereas c_1, c_2, c_3 are fixed constants.

III. METHODOLOGY

This section presents the proposed test generation framework and heuristics used. Figure 1 illustrates a high level flow of the test generation process. The HDL is first cross-compiled to C++ using Verilator [16] for faster simulation when compared to other public domain Verilog simulators. The compiled code is also instrumented which provides a one-to-one mapping between HDL source and C++ base; a database of counters related to branch activations is built. These database counters are used in determination of fitness values of particles at run time. Furthermore, Verilator provides an interface to access the internal variables and registers of DUT. The control flow graph (CFG) of the code extracted by Verilator is later used in the second stage for a controlled graphical search technique embedded in BPSO.

Since every primary input is mapped to a separate dimension in the search space, the value of individual element in an input test vector becomes binary in nature, hence a binary variant of PSO is implemented. The position of every particle represents a sequence of vectors with each vector mapped to a separate primary input and is initialized with random binary values for n clock cycles forming a d-dimensional candidate solution. To promote exploration during the initial stages, the d-dimensional velocity vector is randomly initialized to high values as a logarithmic function of particles dimension. During the first initial runs of PSOFT, the goal is to cover as many branches as possible. Thus, the fitness of an individual particle is directly related to the branch coverage achieved by it. Later, the focus shifts to covering the remaining hard-to-reach branches by adjusting the objective function and allowing for controlled explosions in the swarm population which enhances its exploration capability. Controlled explosion of the swarm in later stages is used to target hard branches. While adding more particles enhances the exploration capability, some branches may require longer test sequences to be reached. Thus, the size of particles (corresponding to the sequence length) can be dynamically increased to allow the swarm to penetrate deeper in the circuit covering those branches. Unlike deterministic methods that can only handle on the order of hundreds of cycles, we can allow the particles to represent tens of thousands of cycles. Moreover, the graphical controlled search is activated whenever the swarm fails to achieve the desired coverage despite increases in exploration and penetration capability. The controlled search specifically targets a given branch and feedbacks the structural information gained by it to the swarm. A set of valid and critical nodes is constructed to guide the search. It

978-1-4799-8720-7/15 $31.00 © 2015 IEEE

has been empirically observed that the uncovered branches in many cases are related. Thus, targeting a branch from the uncovered set of branches often leads to discovery of new states which results in covering several other previously uncovered branches. The presence of a central control allows

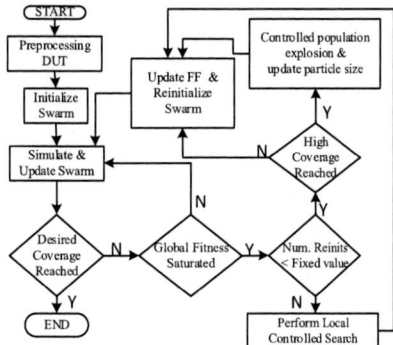

Figure 1.   PSOFT Test Generation Framework

the swarm to converge quickly but may sometimes lead to a premature convergence. In such cases, it is beneficial to reinitialize the swarm while removing the already covered branches from the target list. The updated target list is used for fitness determination in the reinitialized swarm, thus allowing the swarm to escape previously encountered local optima. A controlled graphical search is activated if the algorithm runs out of fixed number of re-initializations without reaching desired branch coverage. The information gained by this search is fed back to PSO for improved test generation. The details of the controlled graphical search algorithm is presented in Section 3B.

A. Overcoming Fitness Oscillation

The position vector of the global best particle is added to the final test set while dropping the branches covered by it from the target list. Each dimension of the particle's current position is updated using its personal best experience & the global best experience of the swarm (Eq. $1-5$). In a scenario where the personal best experience of the particle contradicts the global best, the net force exerted on particle can make its velocity oscillate (Figure 2). To handle such cases, the effect of the second best particle in the swarm is consulted while updating the velocity of the particle. This is done by adding parameters $d^1_{ij,3}$ and $d^0_{ij,3}$ to Eq.1 and Eq.2 respectively. In equations given below P^j_{sbest} represent the jth dimensional bit value of second best particle in the swarm.

$$\text{If } P^j_{gbest} = P^j_{ibest} \text{ then } d^1_{ij,3} = d^0_{ij,3} = 0$$

$$\text{else} \begin{cases} \text{If } P^j_{sbest} = 1 \text{ then } d^1_{ij,3} = c_3 r_3, \ d^0_{ij,3} = -c_3 r_3 \\ \text{If } P^j_{sbest} = 0 \text{ then } d^1_{ij,3} = -c_3 r_3, \ d^0_{ij,3} = c_3 r_3 \end{cases}$$

Figure 2 represent one dimensional movement of particle due to attractive/repulsive forces applied on it. Figure 3

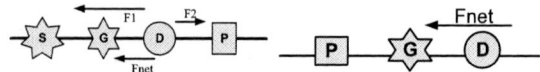

Figure 2.   Net force on D^{th} dimension due to P_{ibest}, P_{gbest} & P_{sbest}

shows a d-dimensional position update for a particle in an iteration. It is highly probable that a few branches in DUT will require longer test sequences. To handle such cases, the particles size is dynamically increased. The size of the swarm is also increased by adding more particles. The larger swarm in later stages saves the computational effort when the target branch is reached in few initial runs. In addition, as the number of uncovered branches decrease, the fitness function is modified (Eq. 6) such that the particles covering the hard to reach branches are favored. This drives the swarm towards the unexplored regions of the search space.

$$FF(i) = \sum_{k; SW[k] \neq 0} \frac{P[k]}{SW[k]} \qquad (6)$$

FF (i) is used to calculate the fitness of ith particle, where P[k] and SW[k] represent number of times k^{th} branch is covered by individual particle and swarm, respectively.

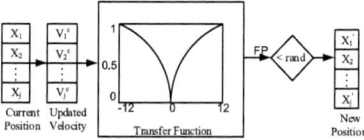

Figure 3.   Position update by mapping velocity in bit flipping probability

B. Controlled Graphical Search Method

The graphical search is selectively activated as some branches may require very long and/or specific sequence of input vectors. Many of such uncovered branches are inter-related, and reaching any one of them can often lead to the discovery of several previously uncovered branches. The detail of the search is illustrated in Figure 4. The controlled search starts with a target node, a set of valid nodes and a fitness function for progress evaluation. The valid node set (VNS) represents those nodes which can lead to target node obtained from the CFG. Beginning from the start node, all the outgoing links are analyzed by expanding the child nodes in a depth-first fashion until either the target node or a leaf node is reached. Nodes leading to the target node are added to VNS. All parent nodes with at least one child present in VNS are immediately added to VNS. The branching condition leading to the target node serves as the fitness function for the search. The search terminates if no improvement is seen in the fitness value for a period of time. The graphical search is effective especially in presence of complex loops in DUT.

The aim of the graphical search is to identify the critical nodes among the valid nodes. A critical node is the one

978-1-4799-8720-7/15 $31.00 © 2015 IEEE

which can lead the execution to a non-valid node during the current clock cycle. While the current implementation will only mark a node as critical if they lead to a non-valid node, the dominators based search in [8] will mark a valid node as a critical node if it can lead to either a non-valid node or a valid ancestral node, in case of loops. Hence, our approach does not require the critical node to be a dominator node. To reduce memory requirements, the control state of the circuit at a critical node is explored by eliminating control registers that do not affect the branching condition. It is feasible that multiple input vectors may lead to the same valid node from a critical node. In such a case, bit relaxation at input side is done. For example, vectors 111 and 110 can be merged to form 11x. The information obtained at critical nodes is stored in a database and is retrieved if the same circuit state is encountered in the future.

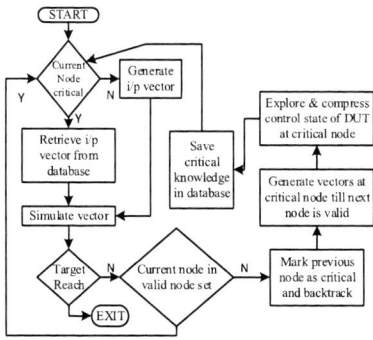

Figure 4. Controlled Graphical Search

To illustrate the process of the controlled graphical search, consider the b12 circuit which is a control intensive benchmark. B12 represents a Simon-says game in which the player has to guess a randomly generated sequence of numbers. A right guess will lead the player to node 63 (Figure 5) whereas a wrong guess will result in loss condition and termination of the game. To reach node 68 which represents the WIN state, the player has to correctly guess the number more than 500 consecutive times. Such stringent constraints demand a very long and specific sequence of vectors, making it very difficult for stochastic methods to reach the WIN state. A subsection of CFG for b12 is shown in Figure 5 where each node represents a basic code block and edges represent the branching condition. To save space, we omit instrumented C++ code for b12 & only show a subsection of its CFG.

In this example, we target node 68 by guiding the generation of vectors using valid and critical nodes and use the corresponding branch condition in computing the fitness. As evident from the CFG, the presence of several looping structures involving critical nodes makes it very difficult to reach the WIN state. Nodes L1, L2, L3, L4, L5 represent losing states and therefore help in determination of critical nodes: 50, 58, 59, 60 & 62. The test generation process

is guided by these determined valid and critical node sets. Proceeding from start node inputs vectors are generated and simulated while keeping track of the current and the last visited node. The node is marked critical if it can lead to a non-valid node (47, 49, 52, 54 & 56) during the search. The search backtracks and attempts to generate a useful input vector whenever the search does not advance towards a valid node from a critical node. The circuit state and useful input vector at critical nodes are saved in the database and retrieved during future encounters which saves test generation effort, especially in case of looping structures. It is seen that covering node 68 require a sequence of having more than 32, 000 vectors & helped discover several previously uncovered nodes: 97, 98, 99, 100, 101 & 102.

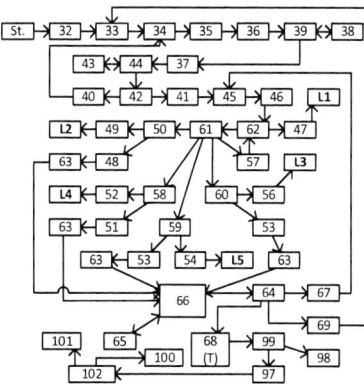

Figure 5. Subsection of CFG of b12 benchmark

IV. EXPERIMENTAL RESULTS

PSOFT is developed using C++ and evaluated on a 64 bit, 3.4 Ghz Intel Core $i7 - 2600$ CPU with 8GB memory running Ubuntu 12.04. Experiments were conducted on a set of ITC'99 [17] benchmark circuits using only one CPU core. The benchmark circuits used are control intensive & contain several hard to reach states which makes them good candidates for functional test generation. The results of the proposed algorithm is compared with HYBRO & BEACON.

A. Algorithmic Settings

PSOFT based approach uses the following parameters: Initial swarm size is set to 20 with an increment of 5 particles during controlled population explosion. The number of generations for a particular initialization of swarm is typically set to 20 whereas the swarm reinitializes if no fitness improvement is seen for 6 consecutive generations. The size of particles is determined by circuits size and is set to a maximum of 3000 cycles whereas the dynamic expansion rate is set to 30% of current size. The maximum number of swarm re-initializations is set to 10. The size of the swarm and individual particles are expanded if swarm ends with zero fitness for 2 consecutive re-initializations.

978-1-4799-8720-7/15 $31.00 © 2015 IEEE

Table I
BRANCH COVERAGE & COMPARISON WITH PRIOR WORK

Bench	Characteristics			PSOFT				BEACON			HYBRO		
	#Branch	#PI	#FF	BC(%)	T(s)	Verilator(s)	Size	BC(%)	T(s)	Size	BC(%)	T(s)	Size
b01	26	2	5	**100**	0.015	0.06	**20**	**100**	**0.0024**	113	94.44	0.07	200
b06	24	2	9	**95.83**	0.055	0.07	**22**	**95.83**	**0.0054**	1731	94.12	0.1	NA
b07	19	1	49	**90**	**0.093**	0.07	**53**	**90**	0.37	759	NA	NA	NA
b10	32	11	17	**96.87**	**0.384**	0.08	**58**	93.75	11.40	3547	96.77	52.14	6450
b11	32	7	31	**96.87**	3.574	0.09	2566	**96.87**	11.95	**1235**	91.30	326.85	4530
b12	105	5	121	**98.09**	3.782	0.12	35128	**98.09**	111.42	37006	NA	NA	NA
b13	63	10	53	**95.23**	4.463	0.11	**5765**	NA	NA	NA	NA	NA	NA
b14	211	32	245	**91.94**	**3.674**	0.12	**606**	**91.94**	204.65	4381	83.50	301.69	NA

B. Results

Table 1 compares the proposed approach with BEACON and HYBRO. For each circuit, the branch coverage, execution time, & test length are reported. For PSOFT, we also report the time used by Verilator. All PSOFT, BEACON and HYBRO were executed on comparable computing platforms so the difference in execution times can be attributed to the fundamental nature of underlying techniques. From the results, PSOFT achieves competitive branch coverages whereas dynamic adjustments in the particles size leads to shorter sequences. Fast convergence achieved by PSO results in significant speedup over BEACON. For example, in circuit b10, PSOFT achieves 96.77% branch coverage with just 58 vectors in 0.384 seconds, whereas BEACON achieved branch coverage of 93.75% with 3547 vectors in 11.40 seconds, & HYBRO achieves 96.77% branch coverage with 6450 vectors in 52.14 seconds. For b12, PSOFT achieves 98.09% branch coverage in only 3.782 seconds whereas BEACON takes 111.42 seconds to achieve same coverage. HYBRO did not report on this circuit. For b12, approximately 90% of the test sequence is generated using graphical search due to the difficulty of this circuit. The storage and retrieval of inputs from the feedback database saved significant test generation effort. In most circuits, our algorithm matches or surpasses branch coverage achieved by existing methods with orders of magnitude improvement in run time and test set size.

V. CONCLUSION

This paper presents a Particle Swarm Optimization based test generation technique for RTL design validation. The proposed method builds upon the fast converging nature of PSO to quickly generate useful test sequences. To aid the test generation process, a graphical search technique is embedded in PSO to provide important structural information of DUT. We have shown that our method is highly effective in exploring the control state space by achieving high branch coverage & offers significant advantage in terms of computational efficiency and test set size over existing methods.

REFERENCES

[1] L. Liu and S. Vasudevan, "Efficient validation input generation in rtl by hybridized source code analysis," in *Design, Automation & Test in Europe Conference & Exhibition (DATE), 2011.* IEEE, 2011, pp. 1–6.

[2] M. Li, K. Gent, and M. S. Hsiao, "Design validation of rtl circuits using evolutionary swarm intelligence," in *Test Conference (ITC), 2012 IEEE International.* IEEE, 2012, pp. 1–8.

[3] V. Kelner, F. Capitanescu, O. Léonard, and L. Wehenkel, "A hybrid optimization technique coupling an evolutionary and a local search algorithm," *Journal of Computational and Applied Mathematics*, vol. 215, no. 2, pp. 448–456, May 2008.

[4] F. Corno, M. S. Reorda, and G. Squillero, "Rt-level itc'99 benchmarks and first atpg results," *IEEE Design & Test of computers*, vol. 17, no. 3, pp. 44–53, Sept 2000.

[5] H. Li, Y. Min, and Z. Li, "An rt-level atpg based on clustering of circuit states," in *Test Symposium, 2001. Proceedings. 10th Asian.* IEEE, 2001, pp. 213–218.

[6] M. S. Hsiao, E. M. Rudnick, and J. H. Patel, "Sequential circuit test generation using dynamic state traversal," in *Proceedings of the 1997 European conference on Design and Test.* IEEE Computer Society, Mar 1997, p. 22.

[7] X. Chuanpei, L. Zhi, and M. Wei, "Study of differential evolution on atpg," in *Communications, Circuits and Systems Proceedings, 2006 International Conference on*, vol. 3. IEEE, 2006, pp. 2084–2087.

[8] K. Gent and M. S. Hsiao, "Dual-purpose mixed-level test generation using swarm intelligence," in *Test Symposium (ATS), 2014 IEEE 23rd Asian.* IEEE, 2014, pp. 230–235.

[9] R. C. Eberhart and J. Kennedy, "A new optimizer using particle swarm theory," in *Proceedings of the sixth international symposium on micro machine and human science*, vol. 1. New York, NY, 1995, pp. 39–43.

[10] M. Clerc and J. Kennedy, "The particle swarm-explosion, stability, and convergence in a multidimensional complex space," *Evolutionary Computation, IEEE Transactions on*, vol. 6, no. 1, pp. 58–73, Feb 2002.

[11] E. Elbeltagi, T. Hegazy, and D. Grierson, "Comparison among five evolutionary-based optimization algorithms," *Advanced engineering informatics*, vol. 19, no. 1, pp. 43–53, July 2005.

[12] R. Hassan, B. Cohanim, O. De Weck, and G. Venter, "A comparison of particle swarm optimization and the genetic algorithm," in *Proceedings of the 1st AIAA multidisciplinary design optimization specialist conference*, 2005, pp. 18–21.

[13] J. Kennedy and R. C. Eberhart, "A discrete binary version of the particle swarm algorithm," in *Systems, Man, and Cybernetics, 1997. Computational Cybernetics and Simulation., 1997 IEEE International Conference on*, vol. 5. IEEE, 1997, pp. 4104–4108.

[14] M. A. Khanesar, M. Teshnehlab, and M. A. Shoorehdeli, "A novel binary particle swarm optimization," in *Control & Automation, 2007. MED'07. Mediterranean Conference on.* IEEE, 2007, pp. 1–6.

[15] S. Mirjalili and A. Lewis, "S-shaped versus v-shaped transfer functions for binary particle swarm optimization," *Swarm and Evolutionary Computation*, vol. 9, pp. 1–14, April 2013.

[16] W. Snyder, P. Wasson, and D. Galbi, "Verilator," 2007.

[17] S. Davidson, "Itc'99 benchmark circuits-preliminary results," in *Test Conference, 1999. Proceedings. International.* IEEE, 1999, pp. 1125–1125.

2015 IEEE Computer Society Annual Symposium on VLSI

On the Performance Exploration of 3D NoCs with Resistive-Open TSVs

Charles Effiong, Vianney Lapotre, Abdoulaye Gamatié, Gilles Sassatelli, Aida Todri, Khalid Latif
LIRMM (CNRS and University of Montpellier), France
Email: {firstname.lastname@lirmm.fr}

Abstract—Three-dimensional Networks-on-Chip (3D NoCs) are based on Through-Silicon-Vias (TSV), which offer several advantages such as stacking, high throughput and energy efficiency. However, TSVs suffer from design process variations. On the other hand, designing purely asynchronous serializers enables reliable inter-tier communication with moderate performance overhead. A side benefit lies in the intrinsic delay insensitivity of asynchronous logic which exploits serialized TSV links to their full timing potential, thereby mitigating process variability impact. This paper explores similar impact on 3D NoCs. It considers randomly generated process variation maps for which the impact on performance is analyzed according to various design parameters, e.g. TSV probabilistic delay distributions, TSV size and serialization rate.

Keywords-Performance, 3D NoCs, resistive-open TSVs

I. INTRODUCTION

Process variation will continue to be a concern for scaled technology nodes and even more so for 3D integrated circuits due to additional processing and stacking steps. 3D integration can be implemented using Through-Silicon-Vias (TSVs) with different approaches such as via-first, -middle or -last. Regardless of the processing steps, process variations occur and designers are concerned to determine system worst-case performance and energy efficiency.

TSVs delays can vary significantly due to different defects and/or impurities that are introduced during manufacturing process [1]. From a performance evaluation point of view, it is imperative to develop techniques investigating process variation impact on communication architectures, e.g., Networks-on-Chip (NoCs). One possible approach to detecting TSV defects is post-silicon validation test. This involves prototyping on a real silicon but before the actual product release, which incurs prohibitive cost and design time overhead. Existing works [2] suggested the use of spare TSVs for recovering failed TSVs. These additional TSVs induce more processing, higher cost and less area for logic.

Another promising alternative approach lies in the use of delay-insensitive logic for inter-tier links [3]. Asynchronous logic makes it possible to exploit most TSVs regardless of their delay, including those with large resistive-open defects. This solution however results in asymmetric, process variation-dependent NoC communication performance. Therefore, it is necessary to study the impact of a wide range of possible process variation patterns to understand the 3D NoC performance degradations and evaluate the impact on computer architecture design. This paper aims to investigate

to what extend process technology has an impact on the communication network performance. More specifically, we aim to establish a duality principle between process technology and architecture design to evaluate communication performance. Such duality not only will serve for architecture performance analysis but also identify the architecture parameters (TSV bundle design, number of stacking tiers, number of cores/routers, communication protocol, etc.) that can be used as design knobs for alleviating the detrimental effect of defective TSVs.

To reach our goal, we consider: probability distribution functions (PDFs) for TSV delays to model TSVs with small or large resistive-open defects; TSV process technology, e.g., diameter or depth; process maps representing process variations stacked tiers; and serialization rate that represents the ratio of N-bit communication data to the number of TSV links between two communicating routers in a 3D NoC. Our main contributions are as follows: i) we present an algorithm for generating typical process variation maps in order to evaluate architecture design performance; ii) from those process variation maps, we generate 3D NoC RTL models in VHDL for performance analysis; iii) we extensively analyze the impact of process variation on communication performance depending on architecture parameters.

II. RELATED WORK

A variety of techniques has already been proposed to explore process variation impact in terms of 3D NoC performance. In [3] authors introduce a fully asynchronous 3D NoC design that employs serialization for inter-tier communication. They conclude that the communication performance is silicon-dependent. We adhere to this observation. In [4], authors have designed a 3D NoC exploiting a similar asynchronous serialization scheme, and compared the latency of the NoC with non-defective TSV delays, a partially connected network with TSV delays having resistive and open defects. The cases with serialization rates of 2 and 4 have been compared. The simulation results show that the average latency of the 3D NoC is significantly reduced for serialization rate 2 when compared to that of 4 for low injection load of about 10%.

In [5], the impact of open defects on TSVs is analyzed. To recover from open defects, an optimal method based on integer linear programming (ILP) to allocate a spare TSV to a functional one has been proposed. This method

978-1-4799-8720-7/15 $31.00 © 2015 IEEE

ensures that a spare TSV is not very close to a functional TSV (to prevent a case of both becoming faulty), and that a spare TSV is not too far from a functional TSV (to minimize additional interconnect length due to re-routing). In this paper, we investigate the impact of TSV process variation on 3D NoC communication performance. The work presented in [2] proposes a redundant TSV architecture with reasonable cost for ASICs by addressing the recovery rate and timing problem issues. Authors claimed that their design can successfully recover most of the failed chips and increase the yield of TSV bonding to 99.99%. This paper presents an analysis of TSV performance in context of process variation, which can guide the designer regarding placement of redundant TSVs.

We address the problem by considering a number of design parameters such as TSV delay distribution, process technology, and associated serialization rates. To analyze the impact of spatial distributions on 3D NoC performance, gate-level fully asynchronous (delay-insensitive) logic for inter-tier communication has been used.

III. PROCESS VARIATION MAPS

TSVs could suffer from various manufacturing defects [1] such as misalignment, impurities, voids, pinholes, cracks, and unfilled or delaminated TSVs exhibiting themselves as resistive opens or shorts. Here, we consider TSVs that are subject to resistive open defects. Open defects lead to discontinuity on a signal line between two nodes, while resistive defects still maintain electrical connection, but may present large variations in timing as compared to the defect-free TSVs [6]. On the other hand, TSV geometry, e.g. its diameter size, has also an impact on TSV delay.

A. 3D NoC process map pattern and instance

TSV delay ranges can be used to model process variations. To capture such a variation, we introduce the notion of *process map pattern*, which represents a generic shape of homogeneous TSV delay areas as shown in Figure 1.

In Figure 1(a), each small colored square in a map pattern symbolically denotes a given TSV delay range: *white* means non-defective TSV with low latency range, *grey* means open resistive defect leading to moderate latency range, and *black* means large resistive defect corresponding to TSVs with high latency range. Each pattern reflects a possible TSV delay distribution scenario, corresponding to a process variation for a 3D NoC. While some of these patterns are shown just for illustrating our approach, only pattern (i) in Figure 1(a) can be related (to the best of our knowledge) to a possible TSV defect scenario already identified in literature [7]. Hence, from now on, we will mainly consider this map pattern (unless another pattern is explicitly mentioned).

We infer so-called *process map instances* from above pattern by specifying the percentages of TSVs for each delay range. For instance, Figure 2(i) illustrates a map instance

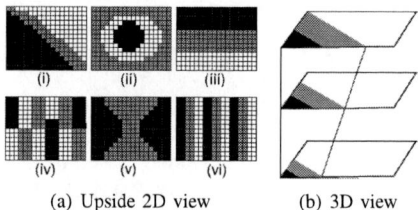

(a) Upside 2D view   (b) 3D view

Figure 1. Process variation map patterns: in Subfigure 1(a), each small colored square denotes a given TSV delay range; Subfigure 1(b) is a 3D perspective of the map pattern (i) in Subfigure 1(a).

relying on pattern (i) of Figure 1(a) composed of a homogeneous area of 35% non-defective TSVs, a homogeneous area of 20% TSVs with small open resistive defect and a homogeneous area 45% TSVs with large resistive defect. Figure 2(ii) shows a map instance composed of an area of 80% non-defective TSVs, an area of 15% small resistive open TSVs and an area of 5% large resistive open TSVs. Figure 2(iii) shows a map instance composed of an area of 15% non-defective TSVs, an area of 50% small resistive open TSVs and an area of 35% large resistive open TSVs. Finally, to provide map instances capturing more realistic effects of process variation, the previous map instances are refined by combining both defective and non-defective TSVs into heterogeneous areas. This is shown by the map instances in Figures 2 (iv) – (vi).

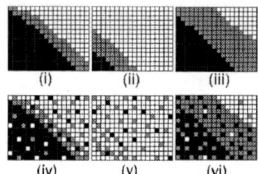

Figure 2. Instances of map pattern in Fig. 1(a)(i).

B. Process map instance generation

The aim of process map generation is to produce various process variation distributions on TSVs and tiers. The resulting process maps are supposed to reflect typical maps of 3D NoC fabrication processes. For this reason, they are defined according to a random delay distribution among TSVs given an initial map pattern as of Figure 1(a)(i). This is described in four steps via Algorithm 1.

In Step 1, input data are declared. They comprise 3D NoC topological properties such as its dimension, process map pattern and a probabilistic distribution function (PDF) for TSV delay. In Step 2, local variables comprise TSV counters, temporary list variables corresponding to each map pattern area, i.e., here white, grey and black. Here, a homogeneous (resp. heterogeneous) list contains TSVs of the same color (resp. different colors). These lists are used in Step 3 to swap different colored TSVs. Further local

978-1-4799-8720-7/15 $31.00 © 2015 IEEE   580

ALGORITHM 1: Process Map Generation Algorithm

```
// Step1: input: 3D NoC topology info; process map
   pattern; PDF info including percent. Pw, Pg &
   Pb of white, grey & black TSVs in 3D NoC and
   their delay ranges, given a size; percent. of
   white TSVs in grey & black areas: Pw_g, Pw_b;
   percent. of grey TSVs in black area: Pg_b;
// Step2: initialization of local variables:
   counters of white, grey & black TSVs:
   Wcnt,Gcnt,Bcnt; homogeneous lists of TSVs:
   Wtmp,Gtmp,Btmp; heterogeneous lists of TSVs:
   Wmix, Gmix, Bmix; total num. of white (resp.
   grey & black) TSVs in 3D NoC: Nw (resp. Ng &
   Nb); num. of white TSVs in grey & black map
   areas: Nw_g,Nw_b; num. of grey TSVs in white &
   black map areas: Ng_w,Ng_b; num. of black TSVs
   in white & grey areas: Nb_w,Nb_g.
// Step3: TSV swapping phase
```
1   $(Nw, Ng, Nb, Nw\_g, Nw\_b, Ng\_w, Ng\_b, Nb\_w, Nb\_g) =$ function (3D NoC topological properties, $Pw, Pg, Pb, Pw\_g, Pw\_b, Pg\_b)$;

2   **if** $(Nw\_g + Nw\_b < Nw)$ *and* $(Ng\_w + Ng\_b < Ng)$ *and* $(Nb\_g + Nb\_w < Nb)$ **then**

3     $Wcnt = Gcnt = Bcnt = 0$;

4   Store respectively all white, grey and black TSVs in $Wtmp$, $Gtmp$ and $Btmp$; Create empty lists $Wmix, Gmix, Bmix$;

5   **while** $(Nw\_g + Nw\_b > 0)$ **do**

6   **if** $(Nw\_g > 0)$ **then**

7   randomly pick a TSV from $Wtmp$ and randomly insert in $Gmix$; $Wcnt$- -; randomly pick a TSV from $Gtmp$ and randomly insert in $Wmix$;

8         $Gcnt$- -; Nw_g- -;

9   **end**

10   **if** $(Nw\_b > 0)$ **then**

11   randomly pick a TSV from $Wtmp$ and randomly insert in $Bmix$; $Wcnt$- -; randomly pick a TSV from $Btmp$ and randomly insert in $Wmix$;

12         $Bcnt$- -; Nw_b- -;

13   **end**

14   **end**

15   **while** $(Ng\_b > 0)$ **do**

16   randomly pick a TSV delay from $Gtmp$ and randomly insert in $Bmix$; $Gcnt$- -; randomly pick a TSV delay from $Btmp$ and randomly insert in $Gmix$;

17       $Bcnt$- -, Ng_b- -;

18   **end**

19   **end**

```
// Step4: production of map instance.
```
20   Write $Wmix$, $Gmix$ and $Bmix$ to PM_file according to input process map pattern; **return** PM_file as map instance;

---

variables are considered in order to check the validity of TSV distribution percentages expected from input data.

Step 3 represents the heart of our process map generation algorithm. From algorithm inputs, we compute on line 1: the total numbers of white (resp. grey and black) TSVs in 3D NoC, the expected number of white TSVs in grey and black map areas, the number of grey TSVs in white and black map areas, and the number of black TSVs in white and grey map areas. These numbers are used in the conditional statement on line 2 to check the validity of TSV swapping between input map pattern areas. Basically, the number of white TSVs to be swapped in grey and black areas must not exceed the total number of white TSVs available initially in the 3D NoC itself. The same constraint holds for grey and black TSVs w.r.t. their respective complementary colors here. If TSV swapping condition is satisfied, then local variables are initialized. Afterwards, the **while** statement on line 5 deals with white TSVs random swapping in grey and black areas respectively. Then on line 15 grey TSVs are randomly swapped in the black area.

Finally, in Step 4 the heterogeneous map areas resulting from previous random TSV swappings are merged into a single output map instance as those shown in Figure 2. This process map therefore reflects a typical process variation associated with a manufactured 3D NoC and can be used in our experiment to explore the network performances. The above algorithm is implemented in C++ and the average time for generating a map instance is 5 seconds.

IV. CONSIDERED 3D NoC ARCHITECTURE

Different 3D NoC architectures have been proposed, e.g. Hybrid 3D NoC-Bus solution or true 3D NoC [8]. The 3D Mesh NoC considered in the current paper, relies on an extension of 2D mesh NoC, where in addition to existing links, each node has two additional links (up and down) for inter-layer communications. Our NoC architecture template is Globally Asynchronous Locally Synchronous (GALS) as routers are synchronous but communicate asynchronously by means of: i) bisynchronous FIFOs for 2D intra-tier communication channels, ii) fully asynchronous logic serializers for 3D inter-tier communication channels. Figure 3 presents the inter-layer communication scheme between two routers, residing one hop away from each other.

Vertical channels are serialized and fully implemented using Quasi Delay Insensitive (QDI) asynchronous logic, as in [3]. The communication process is encoded using a QDI 4-phase 2-rail asynchronous protocol. The two data rails encode a bit value, with one acknowledge signal. Thus, three TSVs are necessary to transfer a single bit. This asynchronous implementation guarantees that the NoC channel functional correctness is insensitive to delay variation that can be caused by voltage, temperature, or manufacturing process variations. Up and down vertical channels between two routers are shown in Figure 3, where each grey arrow is a TSV. The serial channel is composed of 2:1 self-controlled multiplexers (SCM) and demultiplexers (SCD) building serialization and deserialization trees respectively. Let us consider a 3D NoC with N-bits communication data. The numbers of TSVs between two routers varies according to the serialization rate. Each group of 3 TSVs (i.e. bits b_i_w0, b_i_w1 and ack_i) characterizes 1 parallel bit communication in the channel. For sending N bits in parallel (i.e. no serialization), N groups are required, which means 3N TSVs. When serialization is used, bits are multiplexed

978-1-4799-8720-7/15 $31.00 © 2015 IEEE

in time according to the serialization rate. Thus, the number of groups of 3 TSVs is 3N/S where S is the serialization rate. In Figure 3, index p denotes the ratio N/S.

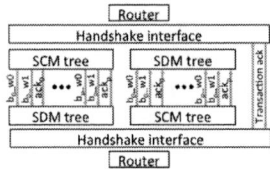

Figure 3.   Inter-layer communication.

The handshake interface provides a solution for connecting the asynchronous channel with the router and informs the router when a new transaction can be initiated. For this purpose, an additional TSV for each channel is instantiated, named transaction ack in Figure 3. This signal is valid if the receiver router is able to receive a data and if the data corresponding to the previous transaction has been sampled by the receiver router. Finally, the number of TSVs in a single channel is (3N/S) + 1. Since the architecture is purely asynchronous logic design, the serialization subsystem has a performance that solely depends on propagation delays of both gates and TSVs. Communication with the synchronous router islands is performed through asynchronous FIFOs not described here. The latency of each TSV is therefore accounted for in the gate-level description of each asynchronous serializer of the simulation model. This approach makes it possible to accurately analyze various metrics such as bandwidth and communication latencies in the NoC.

The process map instances generated by Algorithm 1 are considered together with our NoC template to generate corresponding 3D NoC models simulated for performance exploration. This generation exploits TSV delay ranges described through input PDF information to customize TSV links in NoC template according to their symbolic color.

V. SIMULATION RESULTS

In our experiments, the 3D NoC average packet latency is evaluated under the maximum injection rate of network load, decided by the network saturation point.

To perform the simulations, we generate a 3D NoC with 3×3×3 mesh topology that operates at 1GHz clock frequency. The packets have a fixed length of seven flits, the buffer size is eight flits, and the data width is set to 32 bits. The following protocols are used: a static ZXY wormhole routing algorithm, a credit based flow control for intra-layer communications, and asynchronous logic for inter-layer communications through the hand-shake flow control. In addition, a uniform traffic pattern is considered for simulation, according to which a node sends a packet to any other node with an equal probability [3]. Latency for each packet is logged and dynamically averaged. Simulation finishes when the processed average is stable.

For the sake of generality, we consider here three generic levels of TSV diameter sizes, referred to as *small-size* (SS), *medium size* (MS) and *large size* (LS). We also assume that large-size TSVs have smaller delays due to their lower resistance. Hence, for simulation we consider for SS (resp. MS and LS) a delay range going from 12 ps (resp. 8ps and 6ps) for defect-free TSVs to 400ps (resp. 190ps and 48ps) for large resistive open TSVs. Note however that possible effects of capacitance and inductance can make small-size TSVs faster than large-size TSV despite their low resistance [6]. This is simply captured in our approach by modifying the previously assumed delay range assignment accordingly. For all simulations, we use ModelSim running on a 2GHz Intel Xeon eight-core processor.

A. Average packet latency and saturation point

We consider a 3D NoC serialization rate of 2. All TSVs in this 3D NoC have medium diameter size, i.e., MS, with a delay range decomposed as follows: 8ps – 44ps for non defective TSVs, 56ps – 92ps for small resistive open TSVs, and 106ps – 190ps for large resistive open TSVs. The process map pattern used to generate map instances is provided in Figure 1(a)(i).

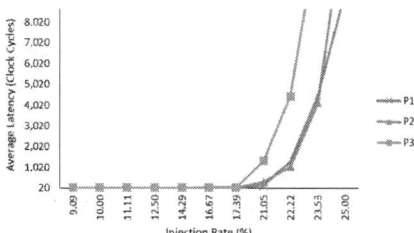

Figure 4.   Average packet latency vs. injection rates.

Then, we evaluate the impact of the TSV partitioning described as: (percentage of non defective TSVs, percentage of small resistive open TSVs, percentage of large resistive open TSVs). The partitionings studied in the next are: P1 = (80%, 15%, 5%), P2 = (5%, 80%, 15%) and P3 = (15%, 5%, 80%). To generate a process map instance, we used 3% as the percentage of white TSVs in grey (Pw_g) and in black (Pw_b) areas, and of grey TSVs in black (Pg_b) area.

The average packet latency curves for uniform pattern with variable injection rates for the previous partitionings are shown in Figure 4. They have been computed within 2 hours. It can be observed that the average packet latency is improved by 4.6% for setup P1 as compared to the setup P2. In a similar manner, this average latency is improved by 9.5% for setup P2 as compared to the setup P3. This is because setups P1 and P2 have more routers located next to non-defective TSV compared to setup P3. The latter has the worst performance both in terms of saturation point and average latency. This is due to the larger fraction of

978-1-4799-8720-7/15 $31.00 © 2015 IEEE   582

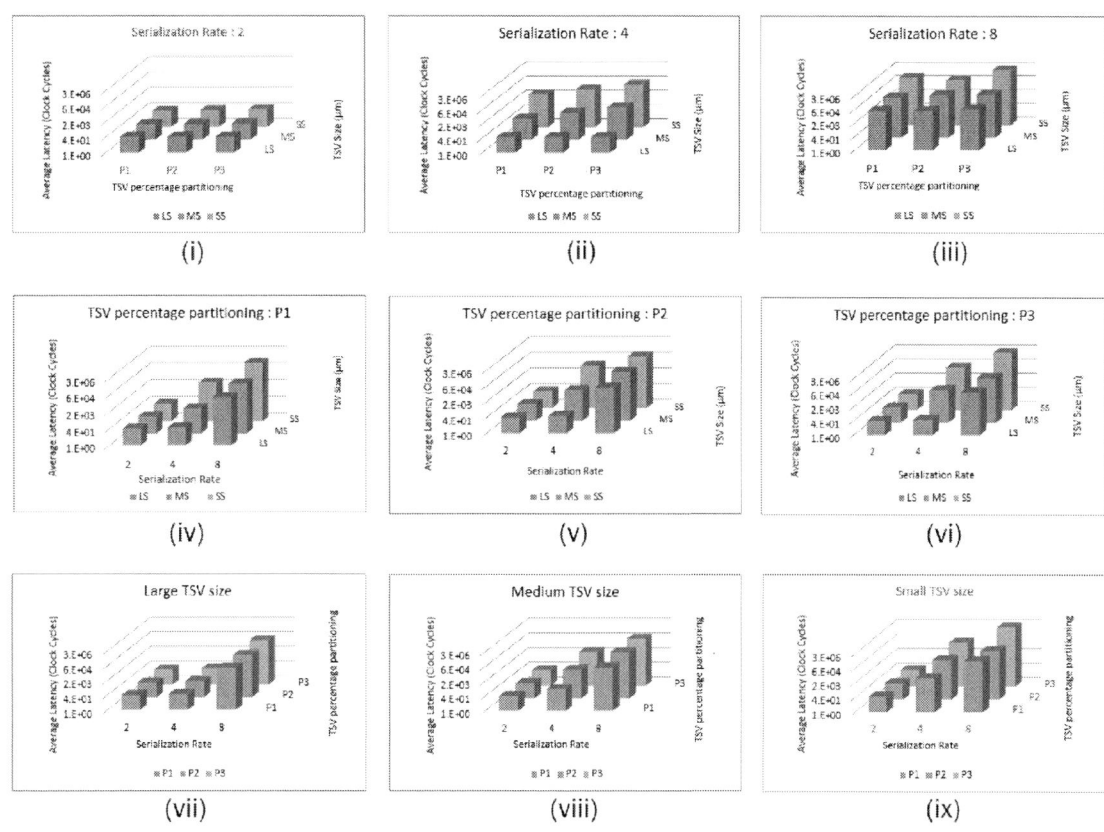

Figure 5. Exploration of 3D NoC design parameters impact on packet latency for 17.39% injection rate.

large open resistive defects as compared to P1 and P2. This suggests that large open resistive defects have significant impact on communication performance creating communication bottlenecks in the network.

From the above experiment, we observe that the minimum saturation point for any of configurations P1, P2 and P3 is reached at the fixed injection rate of 17.39% (which is the saturation point of P3) as shown in Figure 4. In the next section, we select this injection rate to conduct a thorough analysis on the considered 3D mesh NoC.

B. Further exploration results

We analyze the impact of various design parameters (serialization rate, PDF, and TSV size) in 3D mesh NoC at the fixed injection rate of 17.39%. The results discussed below were obtained within 6 hours simulation.

Impact of Serialization Rate. We analyzed three different serialization rates, 2, 4 and 8, for inter-layer communications in the 3D NoC to identify the best configuration according to TSV size and percentage partitioning. Figures 5(i), (ii) and (iii) respectively show the simulation results. For a serialization rate of 2, we observe that neither the percentage partitioning nor TSV size significantly affect the

performance of the 3D NoC. For other serialization rates, an increase in average latency is observed when the percentage of large resistive open TSVs grows, especially for medium and small size TSVs. These observations suggest that with a small serialization rate, i.e., a higher number of TSVs, the asynchronous logic used for inter-layer communication makes the NoC quasi-insensitive to performance variation.

Impact of TSV percentage partitioning. We consider the P1, P2 and P3 setups (see Section V-A) to investigate the outcome in function of TSV size and serialization rate. The results are described in Figures 5(iv), (v) and (vi) respectively. Globally, similar tendencies are seen for each setup, confirming the previous observations: small serialization rate offers best average latencies of 3D NoC while medium and small size TSVs induce worst performances.

Impact of TSV Size. In the last analysis, we explore the impact of the three TSV sizes, i.e., SS, MS and LS on the 3D NoC, in function of TSV percentage partitioning and serialization rate. The corresponding results are shown in Figures 5(vii), (viii) and (ix) respectively. The main observation concerns the large-size TSV case. Here, the results for serialization rates 2 and 4 are not significantly different, while for medium and small size TSVs the difference is more

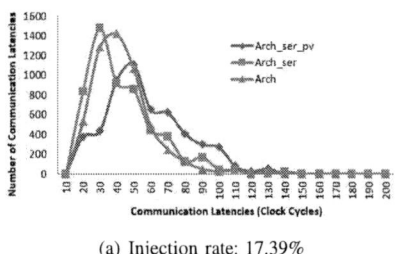 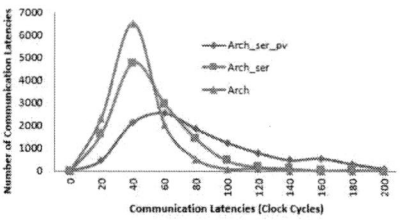

(a) Injection rate: 17.39%   (b) Injection rate: 19.05%

Figure 6.   Distributions of communication latencies in 3D NoC for two injection rates.

visible. For an efficient 3D NoC design, the serialization rate of 4 would be typically prefered to 2 in order to reduce the number of TSVs, therefore to save area.

General insights. From previous observations, we can notice that among considered parameter the serialization rate has the highest impact on the 3D NoC performance analysis. Smaller rates offer the best average latencies. This is particularly true for large-size TSVs (assuming that capacitance and inductance effects are negligible).

C. Distribution of communication latencies

The actual distribution of packet latencies also plays an important role in NoC efficiency. Though this has been studied before for homogeneous NoC, the specific asymmetric nature of our 3D NoC template may impact the standard deviation. In order to adequately analyze this, we devised three versions of the NoC as follows: i) a fully homogeneous unserialized 3D NoC, referred to as *Arch*; ii) a serialized 3D NoC with non defective TSVs only, referred to as *Arch_ser*; iii) a serialized asymmetric 3D NoC with both defective and non defective TSVs related to process variation, referred to as *Arch_ser_pv* (i.e., the template considered in this paper).

Figure 6 shows the histogram distribution of communication latencies observed for all three versions, with injection rates of 17.39% and 19.05%. The serialization rate for the *Arch_ser* and *Arch_ser_pv* is 2. A shift is observed towards higher latencies for these versions, which may create an issue from application / quality of service point of view.

Indeed, this can result in issues related to localized performance unbalance where cores located in affected NoC are exhibit higher communication latencies, as well as global performance overhead with larger standard deviation of communication latencies. Both issues are further prominent for specific process map patterns in which a higher fraction of communications are affected.

VI. CONCLUSIONS AND PERSPECTIVES

We presented an algorithm for deriving process variation maps that reflect TSVs defect in 3D stacking. These maps are used to systematically generate RTL models of 3D NoC for which we extensively analyzed the communication performances, by taking into account both architecture and tech-

nology parameters. Our investigations showed that though the use of asynchronous logic in NoC helps in mitigating variability, the resulting asymmetry of the communication architecture may pose performance problems. This work needs to be confronted with manufactured hardware variability information which is not readily available from industry.

Future work will aim at studying adaptive runtime techniques for smart data routing, taking into account the monitored link performance for mitigating the identified issues.

REFERENCES

[1] A. Topol, D. L. Tulipe, L. Shi, D. Frank, K. Bernstein, S. Steen, A. Kumar, G. Singco, A. Young, K. Guarini, and M. Ieong, "Three-dimensional integrated circuits," *IBM J. Res. Development*, vol. 50, no. 4–5, pp. 491 – 506, 2006.

[2] A.-C. Hsieh and T. Hwang, "TSV Redundancy: Architecture and Design Issues in 3-D IC," *IEEE Trans. on Very Large Scale Integration Systems*, vol. 20, no. 4, pp. 711–722, 2012.

[3] F. Darve, A. Sheibanyrad, P. Vivet, and F. Petrot, "Physical Implementation of an Asynchronous 3D-NoC Router Using Serial Vertical Links," in *ISVLSI'11*, 2011, pp. 25–30.

[4] A. Kologeski, F. Lima Kastensmidt, V. Lapotre, A. Gamatie, G. Sassatelli, and A. Todri-Sanial, "Performance exploration of partially connected 3D NoCs under manufacturing variability," in *Int'l New Circuits and Systems Conf.*, 2014, pp. 61–64.

[5] F. Ye and K. Chakrabarty, "TSV open defects in 3D integrated circuits: characterization, test and optimal spare allocation," in *Design Automation Conference*, June 2012, pp. 1024–1030.

[6] C. Bermond, L. Cadix, A. Farcy, T. Lacrevaz, P. Leduc, and B. Flechet, "High frequency characterization and modeling of high density TSV in 3D integrated circuits," in *Works. on Signal Propagation on Interconnects*, 2009, pp. 1–4.

[7] C. Metzler, A. Todri-Sanial, A. Bosio, L. Dilillo, P. Girard, A. Virazel, P. Vivet, and M. Belleville, "Computing detection probability of delay defects in signal line TSVs," in *18th IEEE European Test Symp.*, May 2013, pp. 1–6.

[8] J. Kim, C. Nicopoulos, D. Park, R. Das, Y. Xie, V. Narayanan, M. S. Yousif, and C. R. Das, "A Novel Dimensionally-decomposed Router for On-chip Communication in 3D Architectures," in *ISCA'07*, 2007, pp. 138–149.

978-1-4799-8720-7/15 $31.00 © 2015 IEEE

2015 IEEE Computer Society Annual Symposium on VLSI

SymmTop: A Symmetric Circuit Topology for Ultra Low Power Wide Temperature-Range Applications

Elena K. Weinberg and Mircea R. Stan

Charles L. Brown Department of Electrical and Computer Engineering, University of Virginia, Charlottesville, VA
{ekw7ej,mircea}@virginia.edu

Abstract—A major roadblock for ultra-low power (ULP) circuits lies in increased sensitivity to process variations at near- and sub-VT. Additionally, temperature variations can be detrimental to circuit functionality for energy harvesting applications operating in outdoor environments. This vulnerability to variations at the device level is exacerbated by imbalances between pull-up (PUN) and pull-down (PDN) networks at the circuit level. In this paper, we propose SymmTop, a symmetric circuit topology in conjunction with a biasing scheme that maintains symmetrically robust operation of ULP energy harvesting systems in a variety of climates, including extreme temperatures. Through noise-analysis and Monte Carlo (MC) simulations performed in a commercial 28nm Fully Depleted Silicon On Insulator (FDSOI) technology, we demonstrate the robustness to variations and extreme temperature ranges of the proposed symmetric 2-to-1 multiplexer design (mirror mux) with bias compared to conventional 2-input NAND and NOR-based circuits. The linearity between temperature and bias makes this scheme easily applicable for programmable sensors in a variety of applications requiring environmental adaptability.

Keywords—sub-threshold, wide temperature-range, nanoscale CMOS, ultra-low power, energy harvesting, 28nm FDSOI

I. INTRODUCTION

As ICs continue to scale, power consumption becomes an increasingly important concern, making sub-VT an attractive route for many ULP applications. Many such applications utilize energy harvesting, a mechanism by which energy already present in the environment is collected and stored by a device and used as its power supply [1]. Such energy may occur naturally (e.g. solar energy, body heat) or be available as a side-effect of some other technological activity (e.g. electromagnetic RF energy, mechanical vibrational energy, etc.). Because the energy is already present, it makes it essentially free, thus energy harvesting is a very promising avenue for designers with ULP applications. A major roadblock to usability of such applications lies in increased sensitivity to variations both at near- and sub-VT, especially at the low voltages typical for energy harvesting, and the wide temperature ranges encountered by outdoor systems.

We propose SymmTop, a symmetric circuit topology in conjunction with a back-gate biasing scheme for ULP energy harvesting applications operating in variable and extreme temperatures. We show that the relationship between temperature and required bias is linear, making this scheme easily applicable for programmable sensors in a variety of applications requiring environmental adaptability.

The contributions of this paper are as follows:

- Symmetric topology for robustness in sub-VT.

- Back-gate biasing scheme for balancing circuits that require adaptability to a variety of temperatures.

- Linear relationship between required bias and temperature, for ease of implementation in energy harvesting applications.

- Analysis of the importance of symmetry in sub-VT.

- Flow for implementing mux-based designs optimal for variation-tolerance in sub-VT.

II. CIRCUIT ASYMMETRY AND ITS IMPACTS IN SUB-VT

Circuit imbalances are amplified in sub-VT. This is because the degraded I_{on}/I_{off} ratio makes sub-VT circuits more prone to effects from threshold voltage shifts [2]. Transistor drive current in weak inversion is given by [3]:

$$I_{DS} = I_0 \exp((V_{GS} - V_T)/nV_{th}) \tag{1}$$

$$I_0 = m_0 \, C_{ox} \, (W/L)(n - 1)V_{th}^2 \tag{2}$$

where I_0 is the drain current when $V_{GS}=V_T$, V_{GS} is the voltage between gate and source, V_T is the transistor threshold voltage, V_{th} is the thermal voltage (given by kT/q), n is the sub-VT slope factor, and C_{ox} is the capacitance of the oxide. Note in (1), threshold voltage is an exponential factor, which illustrates the high sensitivity to threshold voltage shifts experienced by circuits in sub-VT. Moreover, thermal voltage V_{th} is a quadratic term in (2) and an exponential term in (1), illustrating the strong relationship between temperature and drive strength.

A. Asymmetric Circuit Topologies

The impact of parallel and stacked transistors in conventional CMOS gates is magnified in sub-VT and can influence functionality: the inherent asymmetry between pull-up (PUN) and pull-down (PDN) networks in most CMOS implementations is detrimental to circuit robustness. 50-point Monte Carlo (MC) simulations in Fig. 1 show the greater impact of variations on the asymmetric NAND and NOR topologies, in (b) and (c), respectively, as compared to a stacked inverter gate, in (a). We use a *stacked* inverter gate (an inverter with two pMOS devices in series as the PUN and 2 nMOS devices in series as the PDN) instead of a regular inverter so that our comparison between gates is consistent in terms of sizing, area and device parameter variations. It is clear from Fig. 1 that having a *balanced* circuit topology as

978-1-4799-8720-7/15 $31.00 © 2015 IEEE

585

well as balanced PDN-PUN drive strength, like for the inverter, is important in terms of robustness and reliability. Fig. 2 (a) and (b) show the asymmetric circuit topologies of conventional NAND and NOR CMOS gates

B. Functional Asymmetry and Temperature Fluctuations

Many energy-harvesting applications are designed to operate in outdoor environments. They have to be able to withstand a wide range of extreme temperatures. Moreover, with continued scaling, drastic gains in power density have made operation in extreme heat an important characteristic of current ICs. [4] discusses the strong impact that temperature variations have on sub-VT leakage. Variations are harmful to SRAM cells, which frequently have to tolerate high heat during operation. This makes adaptability to a variety of temperatures an important characteristic of many such designs. In addition to quadratic V_{th}, decreased I_{on}/I_{off} in sub-VT makes having a balanced design extremely important since the voltage swing consists of a much smaller margin. [5] discusses the impact of temperature on logic output swing in greater detail. Fig. 3 illustrates the impact of temperature variations on static noise margins (SNMs) for an inverter between VDD=50mV and VDD=400mV.

C. Related Work

Significant work has been done in the area of sub-VT optimization. [6] conducts an analysis of logic families optimal for sub-VT. Some work has been done in the exploration of symmetric topology and design methodology, though none like the one proposed here. [4] presents an analysis of the strong impact that temperature variations have on sub-VT leakage and describes how harmful this is to SRAM cells. [7] discusses at length the effectiveness of using body bias to mitigate the harmful effects from process variations. [8] proposes back-gate biasing for high performance applications with ultra-low power consumption. One of the more closely related works is [9], in which the authors design a symmetric SRAM cell to save area.

III. DEEP NANOSCALE FDSOI FOR ULP OPERATION

Due to the high amount of power consumed by bulk CMOS circuit implementations, alternative transistor designs have been introduced in recent years. A promising technology is Silicon on Insulator (SOI). SOI devices have a layer of silicon dioxide (SiO2) right beneath the surface of the device. This is referred to as the buried oxide (BOX). SOI technologies can either be Partially Depleted (PD) or Fully Depleted (FD). The difference is in the depth of the BOX layer [10]. The BOX layer for an FDSOI device is typically less than 50nm and scales with the SOI device. Fig. 4 (a) shows the structure of a typical bulk CMOS transistor and Fig. 4 (b) shows the structure of an FDSOI transistor. Due to its modified structure, biasing FDSOI devices is called "back-gate biasing" instead of "body biasing".

The most common way to balance circuits conventionally is by increasing the width of p-type devices (e.g. the 2-to-1 or 3-to-1 "rule of thumb" pMOS to nMOS sizing for CMOS inverters). However, especially in sub-VT, as devices become smaller the imbalance becomes much more significant, so that

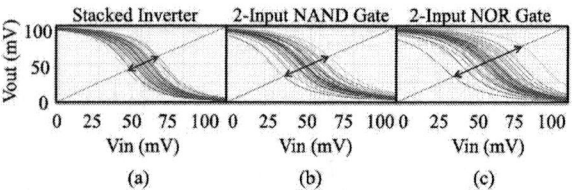

Fig. 1. The impact that asymmetric topology has on circuit robustness. 50-point MC simulations at VDD=100mV for (a) a stacked inverter gate, (b) a 2-input NAND gate, and (c) a 2-input NOR gate. Notice the large spread in values for the NAND and NOR gates shown in (b) and (c), respectively, as opposed to the stacked inverter gate in (a).

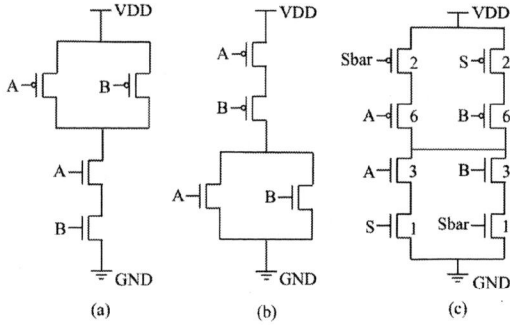

Fig. 2. (a) 2-input NAND gate and (b) 2-input NOR gate. The asymmetry inherent to these logic gates is harmful to robustness in sub-VT. (c) Our proposed 2-to-1 mirror mux topology annotated with sizing.

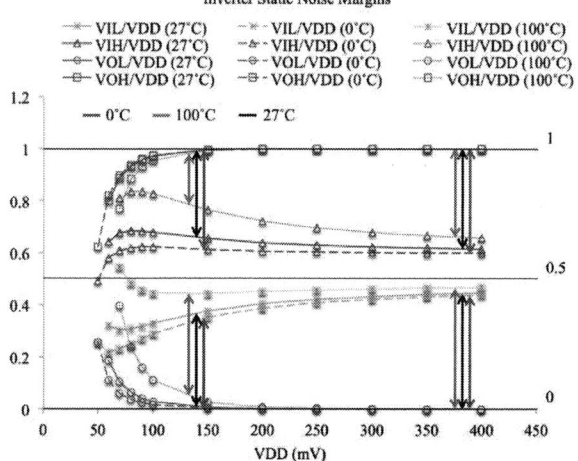

Fig. 3. The impact that temperature fluctuations have on device SNMs between VDD=50mV and VDD=400mV. Normalized VIL, VOL, VIH, and VOH are plotted against supply voltage here. Normalization is with respect to VDD as shown in the legend. Note the significance of the variation in SNMs between operation at 0˚C, 27˚C, and 100˚C, particularly at lower voltages.

the widths of PUN devices must be upsized much more than for super-VT. Fig. 5 shows the sizing required for balancing the inverter with finger width of 320nm and one finger for the nMOS is an eight-fingered pMOS. Thus, this method of circuit balancing results in area overhead and can also increase leakage power. Back-gate biasing controls the threshold

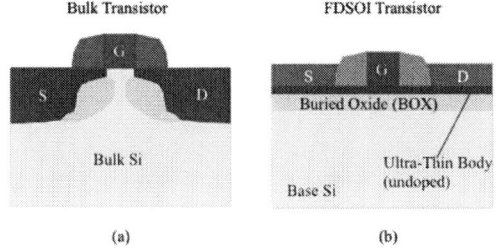

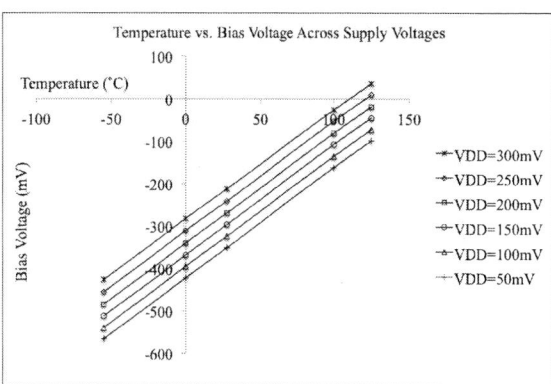

Fig. 4. (a) Bulk CMOS transistor and (b) FDSOI transistor (based on [15])

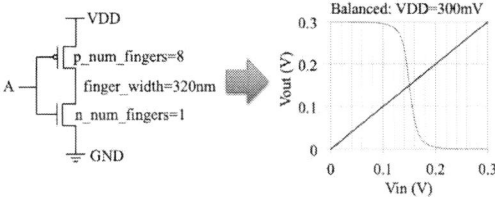

Fig. 6. The linear relationship between temperature and bias voltage for a given supply voltage can be expressed as V_{bias} (mV) = $2.6T + 0.56V_{DD} - 451$. The bias voltage here is the voltage at which the metastable voltage, $V_M=V_{DD}/2$. This is the point at which the PUN and PDN are balanced.

Fig. 5. The required sizing scheme for balancing an inverter at the sub-VT voltage, VDD=300mV. For the inverter with devices sized with a finger width of 320nm, where the nMOS has one finger, the pMOS must have 8 fingers to produce a balanced VTC. In super-VT, the inverter is balanced for pMOS devices sized only twice that of nMOS devices, not eight times. Thus, this method of circuit balancing has significant area overhead in sub-VT.

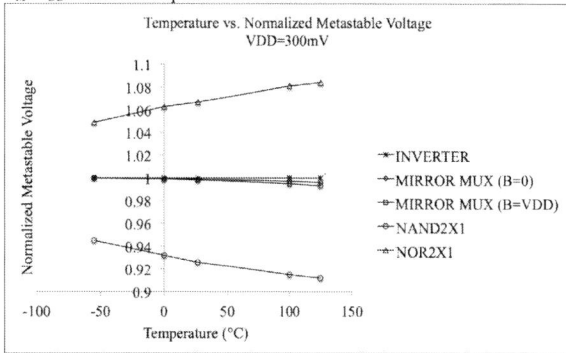

Fig. 7. The normalized metastable voltage (V_M) plotted against temperature for 2-input NAND and NOR gates and the mirror mux for both B=0 and B=VDD. The gates are normalized with respect to the i t ' V hi h f VDD 300 V i 150 V

voltage in FDSOI devices, and serves as an alternate knob for balancing the circuit. [11] discusses the benefits of FDSOI technology and its optimality for sub-VT in greater detail.

A. FDSOI as Optimal for Sub-VT

FDSOI technology has a number of advantages over standard bulk CMOS, including reduced stack effect and negligible drain to substrate capacitance [10]. [11] describes the advantages of using FDSOI for ULP applications in greater detail.

B. Back-Gate Bias for Balanced Sub-VT Circuits

Balance becomes all the more important in FDSOI technology where there is an increase in threshold voltage variation due to the dependence of threshold voltage on the BOX thickness, and the inevitable variation in wafer thickness [10]. Unlike bulk MOSFETs, FDSOI MOSFETs allow a wide range of back-gate bias voltages to be applied without concern for energy or capacitance parasitics [12].

IV.   MUX-BASED LOGIC FOR PUN AND PDN SYMMETRY

The first way we explore balance is in circuit topology. We discuss our exploration of a symmetric multiplexer gate design, a "mirror mux", as optimal for variation-tolerance in sub-VT. The second way in which we explore balance is in functionality and adaptability. From this we determine there is a linear relationship between the bias required to balance a design and the operating temperature.

A. Mux-Based Logic Design and Decision Diagrams

The Shannon Expansion Theorem states that any Boolean function can be expressed as follows [16]:

$$F(x_1, x_2, ..., x_i, ..., x_n) = x_i F(x_i = 1) + x_i' F(x_i = 0) \qquad (3)$$

where F is the Boolean function, n is the total number of inputs to F, and x_i is some input to function F. This relationship provides a way to express logic functions as a mux-based implementation using (3). Shannon's Expansion also laid the groundwork for binary decision diagrams (BDDs) and Boolean satisfiability solvers.

BDDs were first proposed in [17]. They are a means of defining the functionality of a given circuit such that the design can be verified with computational tools. The BDD is a tree-like structure that can be generated and used as a functional description. [18] takes this further to say that each node of the BDD can then in turn be mapped to a 2-to-1 mux. Moreover, logic sub-blocks can easily be combined with a BDD-based (i.e. mux-based) implementation. BDDs are used for a number of applications, including CAD and verification tools, as well as heuristics for Boolean satisfiabilty.

B. Sizing and Topology

In this section we present the topology for the mirror mux. We explored a number of options for sizing and we will discuss our findings and explain why we ultimately chose the sizing illustrated in Fig. 2 (c).

1) Topology

The mirror mux design is illustrated in Fig. 2 (c). Note the symmetry between the PUN and PDN. This design promotes variation-tolerance because it has equivalent stacked and parallel transistors. Since the effects from stacked and parallel transistors are amplified in sub-VT, this is an important characteristic for ULP circuit design.

2) Sizing

All devices utilize a finger width of 320nm. In the case where no transistors are stacked, nMOS devices consist of one finger and pMOS devices consist of two fingers. Sizing for

stacked devices is also accounted for. For example, for two transistors in series, the number of fingers for those devices is doubled. We balanced our gates using back bias instead of with sizing ratio. We explored six sizing options for the mux within the standard 2:1 ratio, all having the same overall area.

The first option we tested was all PDN devices sized the same, with two fingers for the stack and PUN devices all sized the same, having twice the number of fingers as PDN devices, accounting for the stack. The next option and the option we ultimately found offered the most robustness to variations is

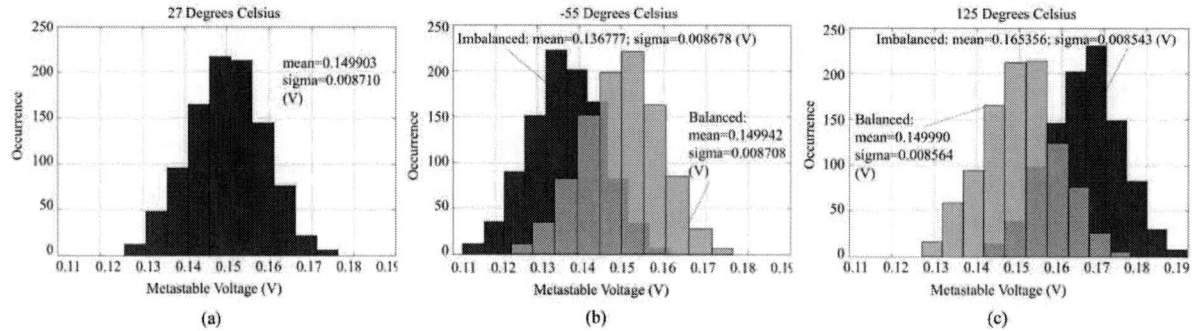

(a)   (b)   (c)

Fig. 8. Plots showing the distribution of the metastable voltage for an inverter at an operating voltage of 300mV from 1000 point MC simulations. (a) shows the distribution at nominal temperature (27˚C), (b) shows the distributions for balanced and imbalanced cases at -55˚C, and (c) shows the distributions for the balanced and imbalanced cases at 125˚C. For the imbalanced cases, the bias voltage is set to -212mV, the voltage at which the inverter VTC is balanced at room temperature (27˚C), and the temperature is raised up to 125 degrees Celsius and down to -55 degrees Celsius. Note the imbalanced cases share a similar standard deviation to the balanced cases but their distributions are not centered over $VDD/2$ as the balanced results.

TABLE I. DISTRIBUTION OF FREQUENCIES AT VDD=300MV

Standard deviation is of interest here. Low standard deviation indicates signal stability. The standard deviation in frequency is smallest for the mirror mux. This indicates that the signal produced by the mirror mux-based ring oscillator is the most stable of gates surveyed here, making it ideal for ULP applications.

Frequency—Balanced						Imbalanced				
Temperature	-55˚C		27˚C		125˚C		-55˚C		125˚C	
Gate	StdDev (Hz)	Mean (Hz)	StdDev (Hz)	Mean (Hz)	StdDev (Hz)	Mean (Hz)	StdDev (Hz)	Mean (Hz)	StdDev (Hz)	Mean (Hz)
INVERTER	2.67E+04	9.67E+04	6.29E+05	3.68E+06	4.79E+06	4.68E+07	2.35E+04	8.49E+04	4.83E+06	4.58E+07
MIRROR MUX B=0	5.185E+03	1.838E+04	1.30E+05	7.52E+05	1.01E+06	1.01E+07	4.562E+03	1.545E+04	9.795E+05	9.787E+06
MIRROR MUX B=VDD	5.476E+03	1.907E+04	1.38E+05	7.95E+05	1.068E+06	1.073E+07	4.721E+03	1.586E+04	1.071E+06	1.051E+07
NAND2X1	1.970E+04	6.847E+04	4.494E+05	2.621E+06	3.309E+06	3.363E+07	1.799E+04	6.248E+04	3.249E+06	3.215E+07
NOR2X1	1.451E+04	4.975E+04	3.215E+05	1.856E+06	2.341E+06	2.349E+07	1.302E+04	4.317E+04	2.349E+06	2.278E+07

TABLE II. DISTRIBUTION OF TOTAL POWER AT VDD=300MV

Lowest mean power is of interest here. For 27˚C and -55˚C the mirror mux has the lowest total power consumption. At high temperature (125˚C) the mirror mux has the highest power consumption, with the inverter having the lowest power consumption. This is likely due to simplicity of design.

Total Power—Balanced						Imbalanced				
Temperature	-55˚C		27˚C		125˚C		-55˚C		125˚C	
Gate	StdDev (W)	Mean (W)	StdDev (W)	Mean (W)	StdDev (W)	Mean (W)	StdDev (W)	Mean (W)	StdDev (W)	Mean (W)
INVERTER	1.75E-10	6.35E-10	4.38E-09	2.49E-08	5.05E-08	4.13E-07	1.54E-10	5.56E-10	5.24E-08	4.19E-07
MIRROR MUX B=0	1.39E-10	4.89E-10	4.57E-09	2.41E-08	9.64E-08	6.63E-07	1.200E-10	4.110E-10	1.05E-07	7.02E-07
MIRROR MUX B=VDD	1.366E-10	4.805E-10	4.52E-09	2.38E-08	9.507E-08	6.583E-07	1.179E-10	4.021E-10	1.048E-07	7.022E-07
NAND2X1	1.886E-10	6.601E-10	4.870E-09	2.698E-08	6.968E-08	5.313E-07	1.757E-10	5.948E-10	7.351E-08	5.416E-07
NOR2X1	1.974E-10	6.769E-10	5.117E-09	2.830E-08	6.829E-08	5.350E-07	1.759E-10	5.832E-10	7.099E-08	5.411E-07

TABLE III. DISTRIBUTION OF LEAKAGE POWER AT VDD=300MV

The mux consumes the most leakage power of all the gates on average. This can likely be attributed to its many parallel paths.

Leakage Power—Balanced						Imbalanced				
Temperature	-55˚C		27˚C		125˚C		-55˚C		125˚C	
Gate	StdDev (W)	Mean (W)	StdDev (W)	Mean (W)	StdDev (W)	Mean (W)	StdDev (W)	Mean (W)	StdDev (W)	Mean (W)
INVERTER	9.91E-13	8.27E-12	4.020E-10	1.496E-09	2.108E-08	1.170E-07	9.910E-13	8.272E-12	2.715E-08	1.498E-07
MIRROR MUX B=0	3.002E-12	1.581E-11	1.170E-09	4.312E-09	7.003E-08	3.923E-07	2.916E-12	1.578E-11	7.938E-08	4.375E-07
MIRROR MUX B=VDD	2.859E-12	1.536E-11	1.135E-09	4.241E-09	6.859E-08	3.883E-07	3.036E-12	1.606E-11	7.491E-08	4.179E-07
NAND2X1	1.972E-12	1.134E-11	7.056E-10	2.464E-09	4.008E-08	2.187E-07	1.970E-12	1.144E-11	4.504E-08	2.419E-07
NOR2X1	1.712E-12	1.070E-11	6.330E-10	2.320E-09	3.695E-08	2.090E-07	1.867E-12	1.111E-11	4.031E-08	2.238E-07

the one in Fig. 2 (c) where the data inputs are sized larger than the select inputs. The third option we tried was the dual of the sizing shown in Fig. 2 (c), where the select inputs are sized larger than the data inputs. We also tried each of these combinations with the select inputs on the inner transistors instead of the outer. We found that having the select inputs on the outside resulted in reduced active power, which is why we ultimately opted to use the design shown in Fig. 2 (c). In terms of performance, the sizing combination we selected is suboptimal, however, the aim of this work is improved stability, not performance.

C. Back-Gate Biasing Scheme

We utilize FDSOI back-gate biasing instead of sizing to tune the metastable voltage of an inverter to balance the drive currents of the pMOS and nMOS devices. This is done by selecting the bias that gives the inverter a metastable voltage, V_M, of $VDD/2$. We bias the pMOS and nMOS with equal voltages to allow for greater simplicity in the design space.

V. SYMMETRY RESULTS

Simulations are run for temperatures between -55°C and 125°C, the military standard for temperature.

A. Static Results and the Linear Correlation

To evaluate the steady-state behavior we conduct 1000-point MC simulations for symmetrically functioning designs at -55C, 27C, and 125C, and then for imbalanced designs that do not utilize our biasing scheme. We balance the voltage transfer characteristic (VTC) curve of a basic inverter gate at a number of supply voltages and temperatures. We define a VTC as balanced for metastable voltages of half the supply voltage, $V_M=VDD/2$. We measure bias voltages for -55°C, 0°C, 27°C, and 125°C. The results from this analysis for a supply voltage of 300mV are presented in Fig. 7 which shows that the mirror mux is indeed more balanced than the NAND and NOR gates for the same bias voltages that perfectly balance an inverter. The linear relationship between bias voltage and temperature illustrated in Fig. 6 is the basis for our proposed biasing scheme. This relationship holds for temperatures as low as

TABLE IV. DISTRIBUTION OF ACTIVE POWER AT VDD=300MV
The mux consumes the least amount of active power in almost every case. Thus, the majority of power consumption comes from the leakage.

Active Power—Balanced				Imbalanced	
Temperature	-55°C	27°C	125°C	-55°C	125°C
Gate	Mean (W)	Mean (W)	Mean (W)	Mean (W)	Mean (W)
INVERTER	6.20E-10	2.34E-08	2.96E-07	5.48E-10	2.69E-07
MIRROR MUX B=0	4.73E-10	1.98E-08	2.71E-07	3.95E-10	2.65E-07
MIRROR MUX B=VDD	4.65E-10	1.96E-08	2.70E-07	3.86E-10	2.84E-07
NAND2X1	6.49E-10	2.45E-08	3.13E-07	5.83E-10	3.00E-07
NOR2X1	6.66E-10	2.60E-08	3.26E-07	5.72E-10	3.17E-07

-100°C and for supply voltages as low as 50mV, well beyond existing inverter SNMs according to nominal simulations. From this we were able to derive the relationship between temperature, supply voltage, and bias voltage:

$$V_{bias} (mV) = 2.6T + 0.56V_{DD} - 451 \qquad (4)$$

where T is the temperature in °C and V_{DD} is the supply voltage in millivolts at which your design operates.

The results from our MC analysis are illustrated in Fig. 8. Note how the balanced bias voltage shifts the distribution of metastable voltages so that it is centered over $VDD/2$. While the standard deviation of the distribution does not seem to be heavily impacted by the threshold voltage shift, from the mean it is clear that the distributions employing our proposed biasing scheme are much more balanced than their counterparts employing use of a universal bias voltage.

B. Transient Results

Transient analyses conducted include frequency, active power, and leakage power on 49-stage ring oscillators (ROs) consisting of inverters, 2-input NAND and NOR gates, and our inverting mirror mux topology. 100-point MC simulations are run in order to gain some insight into the impacts of variation on this methodology. We present in this section statistical data with respect to frequency, active power, and leakage power from HSPICE simulations of MC with both random process and mismatch variations.

MC results show the mirror mux RO has the most consistently low variation in frequency of gates tested in both imbalanced and balanced cases. At low temperature the mux is best overall in terms of frequency stability across temperatures. The mux also has the lowest total power consumption of all gates. We also found that in 3 out of 20

Fig. 9. Proposed tool flow for mux-based design.

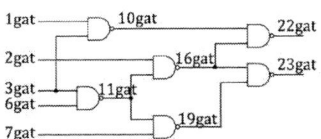

Fig. 10. ISCAS '89 c17 benchmark circuit (based on [22])

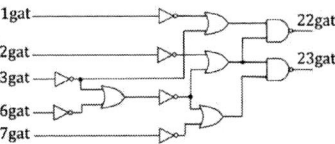

Fig. 11. Top-level synthesis of c17 benchmark circuit without mux-based design constraint.

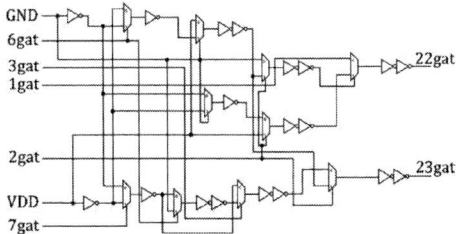

Fig. 12. Top-level synthesis of c17 benchmark with mux-based design constraint. The area of the design shown here is about 3x the area of the non-mux implementation shown in Fig. 11. However, this implementation has a number of unnecessary buffers inserted between gates that, if removed, would reduce the area.

MC runs at *VDD*=150mV, ring oscillators that are not properly back-gate biased fail at 125˚C. In general, topologies that are not balanced have much higher minimum supply voltages than those that are balanced.

Table I shows the standard deviation and mean of the RO frequency at 300mV for the gates tested. It can be seen that the mirror mux has the most stable frequency of all the gates. Key values are highlighted in light gray. Tables II, III, and IV show the distributions of total, active, and leakage power, respectively. We can see that the mux unfortunately has the highest leakage of the topologies, which can be attributed to the many parallel paths in the design.

VI. IMPLEMENTATION DETAILS AND APPLICATIONS

A. Bias Voltage Generation

A number of methods can be used to generate the bias voltages in our proposed scheme. Since the current consumption is low, there is very little load on the bias voltage node, thus, the voltages can be generated with a charge pump or with VDD and low dropout (LDO) regulator. [19] proposes circuitry for adaptive body biasing for NBTI resilience.

B. The Benefits of Symmetry in Layout

Symmetry in circuit design is not only important for circuit robustness, but also for physical implementation. With the small scale on which manufacturing now occurs, the more regularity the better when it comes to designs.

C. Mux-Based Design and Synthesis with BDDs

In order to utilize mux-based design we propose a tool flow that uses binary decision diagrams (BDDs) to synthesize arbitrary combinational logic into a mux-based hardware implementation. We implement our flow using the CUDD package [13]. Fig. 9 illustrates the flow. We begin with a Verilog script. We utilize ODIN II, a component of VPR [14], an FPGA synthesis tool, to convert the Verilog to Berkeley Logic Interchange Format (BLIF), the input file type for CUDD. CUDD is then used to convert arbitrary logic into a mux-based implementation. The output file is BLIF format and so needs to be converted to Verilog using ABC [20]. Physical design is attainable using commercial EDA tools.

As a test case, the ISCAS '89 c17 benchmark circuit is taken through the flow. The typical implementation of this circuit is illustrated in Fig. 10. Fig. 11 shows top-level synthesis of the Verilog code associated with c17. Fig. 12 shows the mux-based implementation generated by top-level synthesis, having gone through the entire tool flow. Implementing the mux-based design for this circuit requires two additional inputs that are held constant—one is tied to ground and the other is tied to VDD. The top-level synthesis tool inserts a number of unnecessary buffers into the design, thus the area of the mux-based design is about three times that of the non-mux design, and the delay is just under three times that of the non-mux design. If the unnecessary buffers are removed, the area and delay of the mux-based design will decrease.

VII. CONCLUSIONS AND FUTURE WORK

There are several directions still left to explore with respect to sub-VT and symmetry. This biasing scheme could be applied to transistors in an SRAM cell to bolster robustness and reliability to mitigate the problems presented in [4]. Though we did not test outside the military standards for device operating temperatures that is a potential area of work for the future as well. Lastly, the design flow in Fig. 9 can be made more robust and used on a real application.

REFERENCES

[1] Mitcheson, P., et al. "Energy harvesting from human and machine motion for wireless electronic devices." *Proceedings of the IEEE* 96.9 (2008): 1457-1486.

[2] Pu, Y., et al. "Vt balancing and device sizing towards high yield of subthreshold static logic gates." *Proceedings of the 2007 ISLPED*. ACM, 2007.

[3] Wang, A., et al. *Sub-threshold Design for Ultra Low-Power Systems*, Springer, 2006.

[4] Singh, R., et al. "Analysis of the Effect of Temperature Variations on Sub-threshold Leakage Current in P 3 and P 4 SRAM Cells at Deep Sub-Micron CMOS Technology." *International Journal of Computer Applications* 35.5 (2011).

[5] Zhao, W., et al. "Robustness-driven energy-efficient ultra-low voltage standard cell design with intra-cell mixed-V t methodology." *Proceedings of ISLPED*. IEEE Press, 2013.

[6] Zhang, Y.(2013). Synthesis Based Design Techniques for Robust, Energy Efficient Subthreshold Circuits. Retrieved from http://libra.virginia.edu/catalog/libra-oa:3899

[7] Narendra, S., et al. "Impact of using adaptive body bias to compensate die-to-die Vt variation on within-die Vt variation." *Proceedings of the 1999 ISLPED*. ACM, 1999.

[8] Choi, H., et al. "Improved current drivability with back-gate bias for elevated source and drain structured FD-SOI SiGe MOSFET." *Microelectronic Engineering* 86.11 (2009): 2165-2169.

[9] Saeidi, R., et al. "A subthreshold symmetric SRAM cell with high read stability." *IEEE Transactions on Circuits and Systems II*, 61.1 (2014): 26-30.

[10] Sakurai, T., et al. Fully-Depleted SOI CMOS Circuits and Technolgy for Ultralow-Power Applications. The Netherlands: Springer, 2006.

[11] Vitale, S., et al. "FDSOI process technology for subthreshold-operation ultralow-power electronics." *Proceedings of the IEEE* 98.2 (2010): 333-342.M. Young, The Technical Writer's Handbook. Mill Valley, CA: University Science, 1989.

[12] "Learn More About FD-SOI." *ST Microelectronics*. 1 Jan. 2015. Web. 20 Feb. 2015. <http://www.st.com/web/en/about_st/learn_fd-soi.html>.

[13] F. Somenzi, "CUDD: CU decision diagram package release 2.5.0," University of Colorado at Boulder, 1998.

[14] Luu, J., et al. "VTR 7.0: Next Generation Architecture and CAD System for FPGAs," ACM TRETS, Vol. 7, No. 2, June 2014, pp. 6:1 - 6:30.

[15] Cauchy, X. "Questions and Answers on Fully Depleted SOI Technology for next Generation CMOS Nodes." *SOI Industry Consortium*. 1 Apr. 2010. Web. 20 Feb. 2015.

[16] Rutenbar, R.A. (2013) *VLSI CAD: Logic to Layout* [1-63].

[17] Akers, S. "Binary decision diagrams." *Computers, IEEE Transactions on* 100.6 (1978): 509-516.

[18] Lee, D, et al. "Combinational logic design using six-terminal NEM relays." *Computer-Aided Design of Integrated Circuits and Systems, IEEE Transactions on* 32.5 (2013): 653-666.

[19] Qi, Z., et al. "NBTI resilient circuits using adaptive body biasing." *Proceedings of the 18th GLSVLSI*. ACM, 2008.

[20] Berkeley Logic Synthesis and Verification Group, ABC: A System for Sequential Synthesis and Verification, Release 5a3f4545e2fd. http://www.eecs.berkeley.edu/~alanmi/abc/

[21] "Circuit Modeling." Web. 29 Mar. 2015. <http://ssdi.di.fct.unl.pt/~fa/phd/Chapter2_Circuit_Modelling.pdf>.

2015 IEEE Computer Society Annual Symposium on VLSI

Energy-Aware Computing via Adaptive Precision under Performance Constraints in OFDM Wireless Receivers

Fernando Cladera
University of Rennes 1, IRISA, INRIA
fcladera@fcladera.com

Matthieu Gautier
University of Rennes 1, IRISA, INRIA
matthieu.gautier@inria.fr

Olivier Sentieys
INRIA, IRISA, University of Rennes 1
olivier.sentieys@irisa.fr

Abstract—To cope with rapid variations of channel parameters, wireless receivers are designed with a significant performance margin to reach a given Bit Error Rate (BER), even for the worst-case channel conditions. Indeed, one of the steps during the design phase is the choice of the architecture bit-width, and the smallest wordlength that ensures the correct behaviour of the receiver is usually chosen. In this paper, an adaptive precision OFDM receiver is proposed. Significant energy savings come from varying at run time processing bit-width, based on estimation of channel conditions, without compromising the BER constraints. To validate the energy savings, the energy consumption of basic operators has been obtained from real measurements for different bit-widths on a FPGA and a processor using soft SIMD. Results show that up to 63% of the dynamic energy consumption can be saved using this adaptive technique.

I. INTRODUCTION

Mobile wireless channels are characterized by time-varying multipath propagation, noise, and interference effects. To cope with these rapid variations of channel parameters, wireless receivers are designed with a significant performance margin to be able to reach a given quality of the link, even for the worst-case channel conditions. In particular, in the baseband processing, data and operator bit-widths are oversized to deal with these unfavorable conditions. Nevertheless, today's mobile wireless devices are often battery powered, hence the necessity of low energy consumption is a strong design constraint [1]. One of the keys for lowering the energy consumption is adaptive bit-width wireless receivers. Dynamic Precision Scaling (DPS) has been proposed in the literature [2][3] as an adaptive precision system for wireless receivers. In DPS systems, the influence of the bit-width under multiple channel conditions is analyzed in the design phase. During the execution, the wordlength is changed dynamically to fit the performance requirements. The use of variable precision reduces the energy consumption, compared to a receiver where the bit-width is constant. Indeed, in [2] the energy consumption of the digital baseband processing is decreased by 25% to 40% in a WCDMA receiver.

Orthogonal Frequency-Division Multiplexing (OFDM) is a widely used standard for modern wireless receivers (e.g. LTE, DVB, DAB, 802.11a/g/n). Prior literature [3][4][5][6] focused on the use of adaptive bit-width in OFDM receivers. In [4], intensive simulations have been used to optimize the Fast Fourier Transform (FFT), the most consuming block in an OFDM system. The potential of DPS applied to OFDM

receivers is shown, because up to 50% of the energy can be saved. However, in this work it is not specified how to switch between the fixed-point implementations during the execution. Indeed, a fixed-point implementation selector will decrease the announced energy savings. In [5], an adaptive wordlength OFDM receiver is developed without the need of an off-line simulation step. This method has the drawback of modifying the OFDM frame specified in the standards, because a search symbol is inserted in order to estimate the quality of the received signal. Some improvements are made in [6] with the addition of a Viterbi decoder. Yet, a high number of iterative operations is needed to find the correct wordlength, decreasing the amount of energy saved by the adaptive method. Another DPS approach has been presented in [3]. In this paper, analytical models have been used to determine the optimum size for each component of the FFT, reducing the time needed for intensive simulations. Nevertheless, only an Additive White Gaussian Noise (AWGN) channel is analyzed. As OFDM was designed to deal with multipath channels [7], the influence of non Gaussian channels must be considered. Our work in this paper is similar in concept with the recent work in inexact circuit design [8][9] or the philosophy of designing adequately-engineered systems [10]. However, in the case of adaptive precision it is possible to guarantee that the reduced accuracy respects a given performance constraint, based on estimation of external conditions.

In this paper, a new channel-aware variable wordlength method is presented for wireless OFDM systems. Energy consumption is reduced by modifying at run time the receiver bit-width based on the estimation of the channel conditions, without compromising the Bit Error Rate (BER) constraints. First, the BER is determined for various Signal-to-Noise Ratio (SNR) levels, channel types, and processing wordlengths. A simulation approach is used: for each channel type and SNR, a receiver with a specific fixed-point architecture is simulated and the BER is measured. Then, the best fixed-point implementation for each channel condition is determined through an off-line optimization process. At run time, the DPS algorithm is used (Fig. 1): the channel condition is determined by combining low-complex yet efficient estimation algorithms generating the metric p, which allows the choice of the fixed-point data format $f_{fp}(p)$ that minimizes the energy consumption subject to a given BER performance constraint. To validate the energy savings, the energy consumption of basic operators processing random data has been obtained from

978-1-4799-8720-7/15 $31.00 © 2015 IEEE

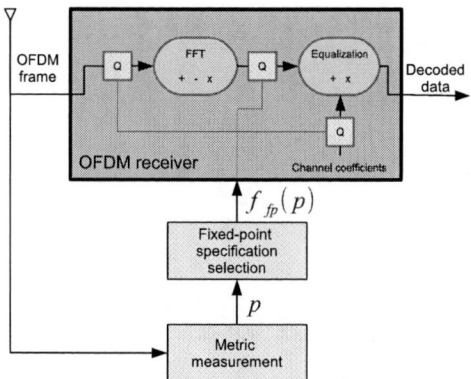

Fig. 1: DPS algorithm applied to an OFDM receiver.

real measurements. The energy spent by the operators depends on their bit-width. Two architectures have been considered: a Virtex-5 Field-Programmable Gate Array (FPGA) and an ARM processor using soft Single Instruction, Multiple Data (SIMD). Finally, the architecture consumption is estimated as a function of the number of operators and the wordlength.

To summarize, this paper studies an adaptive precision OFDM receiver, where a low complex selector is used to choose the processing wordlength at run time, estimating the energy savings with real measurements issued of experiments and taking into account not only AWGN, but also multipath channels.

This paper is organized as follows: Section II describes the OFDM system model used, as well as the energy reduction strategy. In Section III, the energy consumption estimation method is presented for both FPGAs and ARM processors. Section IV presents the fixed-point DPS receiver and the performance results are discussed in Section V. Finally, Section VI draws some conclusions.

II. OFDM SYSTEM MODEL

A. Floating-Point Receiver Model

In this study, an OFDM frame-based model has been defined using the following parameters: 512-point FFT (N_{FFT}) with 300 used subcarriers, 16-QAM modulation and 128-point Cyclic Prefix (N_{cyp})[1]. Each frame is composed of at least ten OFDM symbols. The channel coefficients for the equalizer are known and the synchronization is presumed perfect. Introduced in Fig. 1, the system model is reduced to three important blocks: the FFT, the simple one-tap equalizer and the hard decision block[2]. The blocks which execute intensive calculations during the operation, the FFT and the equalizer, will be optimized.

Two channel models are used: a simple AWGN channel and a Frequency Selective Fading (FSF) channel with AWGN. For both channels, the noise level varies from 0dB to 48dB, in steps

[1]These parameters have been extracted from [11] to emulate a long Cyclic prefix LTE receiver working at 5 MHz.

[2]In Fig. 1 some quantizers have been added to simulate the fixed-point behaviour. Those quantizers are obviously not present in the floating-point model.

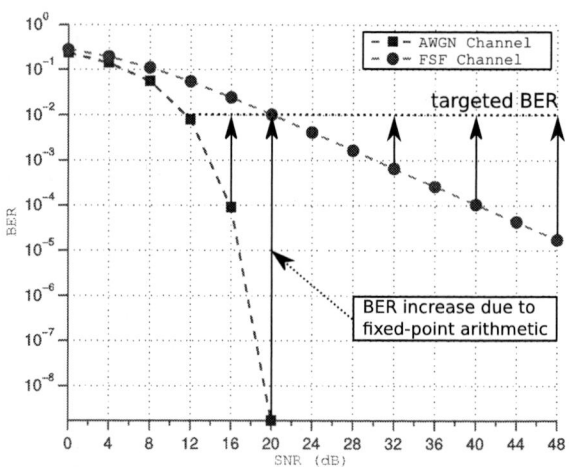

Fig. 2: Floating-point OFDM receiver performance showing the superlative quality for high SNR values and the energy saving strategy.

of 4 dB. The FSF channel has an exponential delay profile, with a length of $\approx 0.3\,N_{cyp}$, 9 paths and 16 dB between the first and the last path. This channel is similar to the Extended Vehicular A Model from [12]. In order to reduce the number of simulations, only one FSF channel has been considered in this study.

B. Energy Saving Strategy

Wireless channels are usually designed to target a specific application (voice, video, data). Each application imposes specific parameters such as a required bandwidth, a maximum latency or a level of reliability. These parameters are related with the link quality, which is evaluated using metrics like the BER, the Packet Error Rate (PER), and the Error Vector Magnitude (EVM). Indeed, these metrics are linked [5]. In our work, the BER is used for measuring the link quality. A BER of 10^{-2} is targeted as the desired quality, independently of the channel condition. This BER is a standard value for voice applications [13]. However, the same procedure of optimization can be applied for different BER values (may be lower), to target another applications.

Fig. 2 shows the simulated BER performance of the receiver for the channels specified in Sec. II-A. Low values of SNR will not be optimized, because the BER constraint cannot be achieved. High values of SNR imply a superlative quality, for both channels. The use of reduced precision will degrade the quality of the link, up to the objective value. However, it will decrease the energy consumption.

III. ENERGY CONSUMPTION ESTIMATION

To evaluate the amount of energy spent by the receiver, a high level estimation step is carried out. The energy consumption of the receiver is calculated based on the number of operations. To this purpose, the energy consumption cost function $C(wl)$ of simple arithmetic operators (multipliers, adders) is obtained from experiments. $C(wl)$ links the amount of energy spent to do one operation depending on the bit-width wl of this operator. In addition, a maximum activity

Operation		wl											
		4	5	6	7	8	9	10	11	12	13	14	15
ADD	UInt	1.55	1.57	1.58	1.60	1.61	1.63	1.64	1.66	1.67	1.69	1.70	1.72
	SIMD2	0.80	0.81	0.83	0.85	0.86	0.88	0.89	0.91	0.93	0.94	0.96	0.97
	SIMD3	0.55	0.56	0.58	0.60	0.61	0.63	-	-	-	-	-	-
MULT	UInt	1.97	1.99	2.01	2.03	2.04	2.06	2.07	2.08	2.10	2.11	2.12	2.13
	SIMD2	0.98	0.99	1.01	1.04	1.07	-	-	-	-	-	-	-

TABLE I: Energy consumed for simple arithmetic operations in an ARM7TDMI (in $[nJ]$).

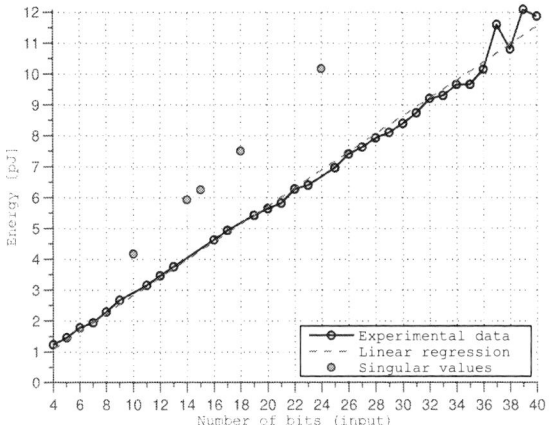

Fig. 3: Energy consumption of Virtex-5 adders with variable wordlengths.

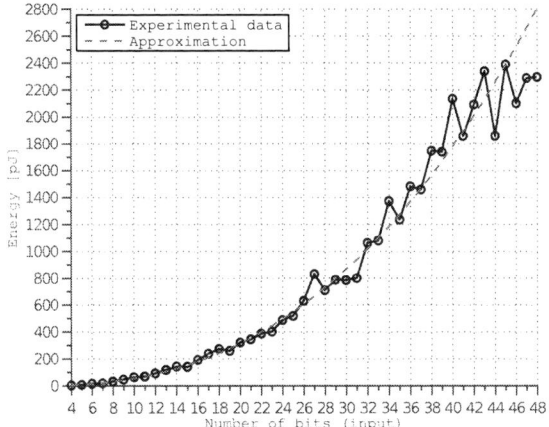

Fig. 4: Energy consumption of Virtex-5 LUT multipliers with variable wordlengths.

is considered by feeding the operators with random data. Even when this energy consumption computation is not highly accurate, these values can be used as a rough estimation of the gains achievable by our variable wordlength algorithm.

A. FPGAs

One of the targets chosen to estimate the $C(wl)$ was a Virtex-5 XC5VLX50T FPGA. Special architectures with a high number of operators were synthesized. Then, arithmetic operations were carried out using random data from a Pseudo-Random Binary Sequence (PRBS). The energy consumed by the operators can be measured using the technique described in [14].

Only the dynamic energy consumption (related to the activity in the circuit) is considered. Also, only Look-Up Table (LUT) operators have been used. Fig. 3 and Fig. 4 show $C(wl)$ obtained for adders and LUT multipliers respectively.

B. ARM

An instruction-level energy estimation has been carried out for an ARM7TDMI processor (SAM7X256). Indeed, in order to estimate the dynamic energy consumed to do an operation (addition, multiplication), the energy to load a pair of values from the memory, calculate and store the result in memory has been measured. In addition, the energy when the NOP (No OPeration) instruction is executed has also been measured. The NOP instruction does not imply an access to the data memory nor an arithmetic calculation. Thus, it can be used to estimate the static power consumption of the processor. Finally, both values are subtracted to obtain the dynamic energy consumption of the processor.

Single Instruction, Multiple Data (SIMD) architectures allow multiple parallel calculations per instruction. However, this processor is not capable of executing SIMD instructions natively. Thus, the soft-SIMD technique described in [15] was used to improve the parallelism in the processor. Table I summarizes the energy values obtained. SIMD2 implies that two operations are made simultaneously, whereas SIMD3 implies three simultaneous operations in a 32-bit word.

IV. FIXED-POINT DPS RECEIVER

A. Fixed-Point Receiver

After validating the floating-point model in Section II-A, quantizers have been introduced to simulate a fixed-point architecture, as shown by Fig. 1. The quantizers limit the dynamic and reduce the precision in the FFT input, the FFT output and the channel coefficients. Finding the appropriate fixed-point wordlength for each quantizer is a task composed of two steps: first, the dynamic range is found (the integer part in a fixed-point data). Secondly, the precision (the fractional part in a fixed-point data) has to be chosen.

The dynamic range can be easily obtained using a simulation approach: the floating point model is simulated for every channel type and SNR. The data in each block is stored and analyzed afterwards. The dynamic range that suits at least 99% of the values determines the number of bits in the integer part, for each channel condition.

Obtaining the fractional part implies a much more difficult work: for every channel type and SNR, multiple simulations have been executed using different fractional parts for each block. Each precision implies a BER, and the precision chosen is a tradeoff between quality of the link and the energy spent

978-1-4799-8720-7/15 $31.00 © 2015 IEEE

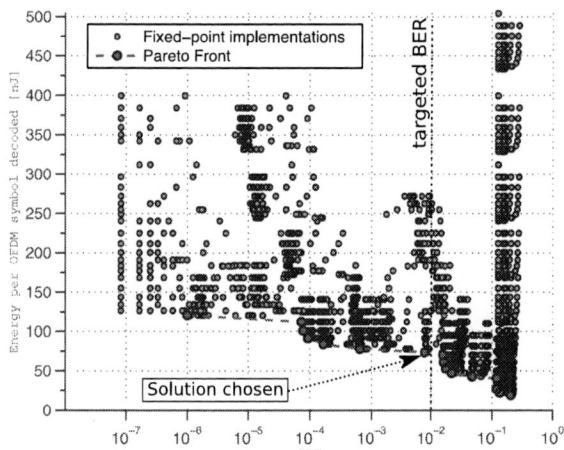

Fig. 5: Fixed-point implementation selection for an AGWN channel and a SNR of 24 dB, using $C(wl)$ of a Virtex-5 processor.

during the reception. Fig. 5 represents this method using the cost function of the Virtex-5 FPGA: the points represent different fixed-point implementations of the receiver, with an associated BER and energy consumption. The fixed-point implementation on the Pareto front closer to the objective BER is chosen. The cost of the solution is estimated using $C(wl)$ and knowing the number of operations needed to demodulate an OFDM symbol: 13×10^3 real additions and 9×10^3 real multiplications in the FFT, 600 real additions and 1200 real multiplications in the equalizer.

After choosing the fixed-point implementations for all the possible channel conditions, Fig. 6a is built using the cost function of the Virtex-5 FPGA, showing the energy consumption per OFDM symbol for each channel condition. As expected, there is an important difference depending on the channel type and SNR. The "worst-case" FPGA receiver will always use the largest wordlength, spending approximately $278\,nJ$ per OFDM symbol, independently of the working conditions. Oppositely, in a DPS receiver the energy consumption can be reduced down to $71\,nJ$ if better channel conditions are available. Therefore, up to 74% of the energy consumed can be saved, without taking into account the consumption of the implementation selector.

This analysis can be extended using the cost function of the ARM processor. The following hypotheses are made: the processor will run the most optimized SIMD mode available for a given bit-width, and the energy to bind multiple sub-word operands into a 32-bit word is not considered. Fig. 6b shows the energy consumption for each channel condition. The "worst-case" receiver consumes $30.2\,\mu J$, whereas the dynamic receiver can reduce its energy consumption down to $18.7\,\mu J$, saving 38% of the energy.

B. Fixed-Point Implementation Selector

The implementation selector is composed of two low complexity estimators with the aim to determine the SNR and the channel type. A 12-bit fixed-point architecture has

been implemented for both estimators[3]. In our receiver, the channel conditions are estimated for each OFDM symbol, and the system has no memory about previous channel conditions. Both estimators were optimized to target a detection error of 10^{-3}.

1) SNR Estimation: The Cyclic Prefix (CP) introduces some redundancy in the OFDM symbol, to prevent multipath channel effects. When the channel conditions are not too severe, this redundancy can be used to estimate the SNR [16]. This estimator is chosen due to its simplicity, good performance, and low energy consumption. The estimator is set to use the 64 last points of the 128 available in the CP. Thus, 513 real additions, 256 real multiplications and one division are needed. The energy consumption due to SNR estimation is $25.15\,nJ$ per OFDM symbol in the FPGA and $1.03\,\mu J$ in the ARM processor.

2) Channel Type Estimation: In an OFDM receiver, the equalizer coefficients are estimated dynamically during the reception using training symbols and reference signals. If the channel has only a single path, the equalizer coefficients are flat. But, if the channel presents multiple paths, the coefficients show peaks and depressions. Hence, the variance of the equalizer coefficients can be used as an indicator of the channel type. Only 32 points from the 300 available in the equalizer are used. With these parameters, 129 real additions, 64 real multiplications and 1 real division are needed. The energy consumption due to channel type detection is $6.84\,nJ$ per OFDM symbol for the FPGA and $0.27\,\mu J$ for the ARM. More points would allow a better channel detection, which is useful for distinguishing between multiple channel models.

In our receiver, the equalizer coefficients are presumed known. However, to ensure a correct behavior of the channel estimator, the coefficients have been tuned with a Normalized Linear Mean Squares (NLMS) equalizer in a floating point receiver.

V. PERFORMANCE RESULTS

A. Implementation Selector Performance

As it was pointed out in Section IV-B, the number of points for both selectors was chosen targeting a detection error of 10^{-3} at 16 dB. During the simulations, the average detection error was approximately $10^{-2.5}$ for the AWGN channel and $10^{-1.9}$ for the FSF channel proposed in Section II-A. This may be seen as a drawback because more than 1% of the frames are decoded with a wrong fixed-point implementation in the FSF channel. A deeper analysis shows that most of the errors for the AWGN channel are due to erroneous SNR detection. In contrast, for the FSF channel, the errors are due to both the channel ($10^{-2.1}$) and the SNR ($10^{-2.2}$) detection. Nonetheless, an erroneous detection does not imply necessarily a reduction in the BER: if a fixed-point implementation with a larger wordlength is chosen instead of the right one, the energy consumption will be increased but the quality constraint will be reached.

B. Global System Performance

The DPS system is built, linking the fixed-point OFDM receiver with the implementation selector. Simulations have

---

[3]A wordlength analysis has been carried out for the implementation selector, obtaining the best tradeoff with 12 bits.

978-1-4799-8720-7/15 $31.00 © 2015 IEEE

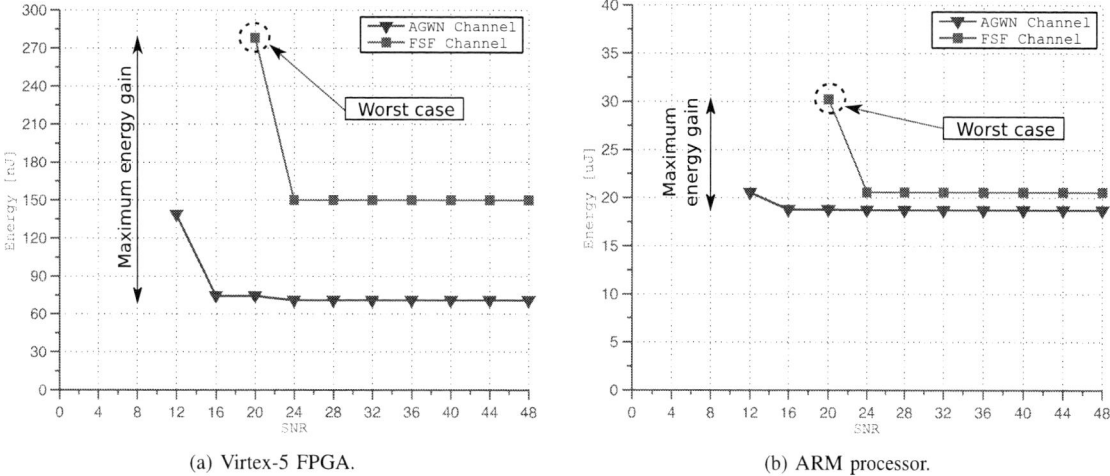

(a) Virtex-5 FPGA.　　　　　　　(b) ARM processor.

Fig. 6: Energy consumption per OFDM symbol for AGWN and FSF channels, depending on SNR values.

been run in time-varying channels to evaluate the global system performance. Fig. 7 shows the BER-SNR curve for the dynamic receiver (full line) compared with the "worst-case" receiver (dotted line). The system operates correctly: when the BER increases beyond 10^{-2}, the precision is reduced and the system follows the targeted constraint. In order to evaluate the energy savings, the global energy consumption is calculated by

$$E_{total} = E_{receiver} + E_{selector}, \qquad (1)$$

where $E_{receiver}$ are the values found after optimization in Section IV-A and $E_{selector}$ corresponds to the constant energy consumption of the selector, $32\,nJ$ for the FPGA and $1.3\,\mu J$ for the ARM. Hence, the performance of our system can be calculated easily: in the worst condition, the FPGA receiver uses the largest wordlength and, due to the energy needed by the selector, the consumption is 111% compared to the "worst-case" receiver. Moreover, in the best case (high SNR and AWGN channel), the consumption of the receiver is only 37% of the total consumption (63% of the energy saved). The ARM receiver uses 104% of the energy in the worst condition and 66% in the best one (34% of the energy saved).

Fig. 8 shows the real energy consumption of the receiver for the FPGA, compared with the "worst-case" solution. It also shows the accuracy of the implementation selector. For each channel condition 1000 simulations have been carried out. A receiver with a perfect selector would follow the dotted line, whereas the results of our receiver are shown using dots (correct channel condition estimation) and triangles (bad channel condition estimation). It can be seen in Fig.8a that for the AWGN channel, those errors are generally due to a wrong SNR estimation. In Fig. 8b, for the FSF channel, wrong estimations are due to erroneous channel and SNR estimations. However, the system performance is guaranteed as shown by Fig. 7.

C. Discussion and Further Work

Our system proves a correct behavior, reducing the energy consumed during the execution. A low-power yet efficient

selector is presented and tested in a fixed-point receiver. Compared with [4], our work uses a similar technique to find the correct fixed-point implementations, reaching similar energy reductions (50%). However, a selector to choose the right implementation is described in our work. In [5], modifications are introduced in the OFDM frame to save up to 30.2% of the energy. In [6], multiple calculations are needed to estimate the channel conditions, reducing the amount of energy saved (23.9%). Our savings are higher without modifying the OFDM frame. [3] proposes using the DPS offline optimization method too. Still, the channel type is not considered into the optimization, obtaining an average saving of 17%. The importance of analyzing multiple channel conditions, not only the SNR but also the channel type, is demonstrated in this work. Thus, our energy reductions are higher compared to state-of-the-art proposals.

In our system, the number of quantizers is reduced because of the simulation time needed. To solve this problem, analytic methods such as those proposed in [3] may be used

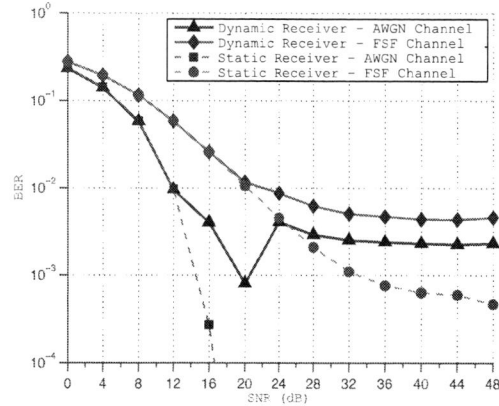

Fig. 7: Correct performance of the dynamic system, following the BER target.

978-1-4799-8720-7/15 $31.00 © 2015 IEEE

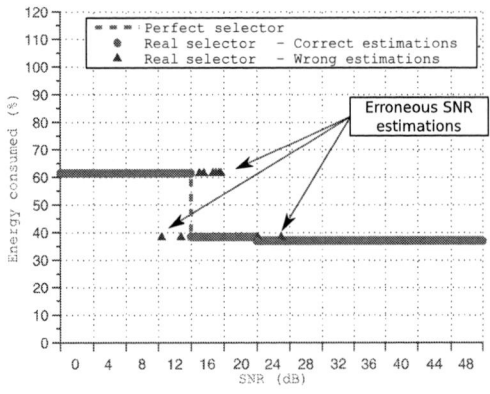

(a) AGWN channel (FPGA cost function).

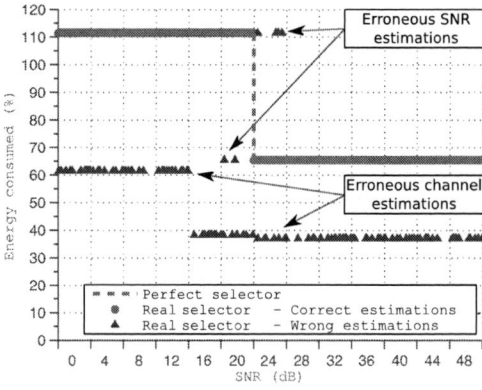

(b) FSF channel (FPGA cost function).

Fig. 8: Energy consumption in a DPS receiver compared to the "worst-case" static fixed-point implementation.

to determine the correct fixed-point implementation for each channel condition. The influence of omitted blocks, such as the channel coefficients estimation and the synchronization should be considered too.

Finally, our energy estimation only considers arithmetic operators. Other components, such as memories, should be considered, but their energy consumption will also be improved by our adaptive precision approach. Other parameters, such as the energy needed for the reconfiguration of the architecture, have to be measured in a real DPS OFDM receiver in order to ensure that the energy savings are still high to justify this approach.

VI. CONCLUSIONS

This work demonstrates a DPS OFDM receiver that can switch its fixed-point specification depending not only on the noise level but also on the channel type. After presenting an OFDM receiver with its corresponding parameters, a floating to fixed-point conversion process has been carried out. The influence of the channel type and SNR in the energy consumption was shown. A low-power implementation selector was built using two estimators of the channel type and noise level. The whole system was assembled and successfully tested. Compared to related work the energy savings were higher due to the use of a channel-aware DPS receiver. In addition, no modifications to existing standards nor intensive calculations are needed to select the fixed-point implementation. Estimations using FPGA show that up to 63% of the energy can be saved using this technique. Using an ARM7 processor, an energy saving of 34% is achieved. An ASIC implementation would probably take even more advantage of the energy savings offered by the proposed adaptive precision technique.

REFERENCES

[1] D. Banerjee, S. Devarakond, S. Sen, and A. Chatterjee, "Real-Time Use-Aware Adaptive MIMO RF Receiver Systems for Energy Efficiency under BER Constraints," in *Proc. Design Automation Conference (DAC'13)*. ACM, 2013, pp. 1–7.

[2] H. N. Nguyen, D. Menard, and O. Sentieys, "Dynamic precision scaling for low power WCDMA receiver," in *Proc. Int. Symp. Circuits Syst. (ISCAS'09)*, Taipei, Taïwan, China, May 2009, pp. 205–208.

[3] S. Lee and A. Gerstlauer, "Fine Grain Word Length Optimization for Dynamic Precision Scaling in DSP Systems," in *Proc. Int. Conf. Very Large Scale Integr. (VLSI-SoC'13)*, Istanbul, Turkey, Oct. 2013, pp. 266–271.

[4] D. N. Bruña, "Exploiting adaptive precision in software defined radios," Ph.D. dissertation, Katholieke Universiteit Leuven, Dec. 2007.

[5] S. Yoshizawa and Y. Miyanaga, "Use of a Variable Wordlength Technique in an OFDM Receiver to Reduce Energy Dissipation," *IEEE Trans. Circuits Syst. I*, vol. 55, no. 9, pp. 2848–2859, Oct 2008.

[6] J. Kim, S. Yoshizawa, and Y. Miyanaga, "Dynamic Wordlength Calibration to Reduce Power Dissipation in Wireless OFDM Systems," in *Proc. Asia Pacific Conf. Circuits Syst. (APCCAS'10)*, Kuala Lumpur, Malaysia, 2010, pp. 628–631.

[7] T. Chiueh and P. Tsai, *OFDM Baseband Receiver Design for Wireless Communications*. John Wiley & Sons, 2007.

[8] J. George, B. Marr, B. Akgul, and K. Palem, "Probabilistic Arithmetic and Energy Efficient Embedded Signal Processing," in *Proc. Int. Conf. Compilers, Archit. and Synth. for Embedded Syst. (CASES'06)*. ACM, 2006, pp. 158–168.

[9] A. Lingamneni, C. Enz, K. Palem, and C. Piguet, "Synthesizing Parsimonious Inexact Circuits through Probabilistic Design Techniques," *ACM Trans. on Embedded Comput. Syst. (TECS)*, vol. 12, no. 2s, p. 93, 2013.

[10] K. V. Palem, "Energy Aware Algorithm Design via Probabilistic Computing: from Algorithms and Models to Moore's Law and Novel (Semiconductor) Devices," in *Proc. Int. Conf. Compilers, Archit. and Synth. for Embedded Syst. (CASES'03)*. ACM, 2003, pp. 113–116.

[11] J. Zyren, "Overview of the 3GPP Long Term Evolution Physical Layer," Freescale, White Paper 3GPP EVOLUTION WP, 2007.

[12] ETSI, "LTE - Evolved Universal Terrestrial Radio Access (E-UTRA) - User Equipment (UE) radio transmission and reception," ETSI, Technical Specification 3GPP TS 36.101 v. 11.6.0-11, 2013.

[13] W. Fernando, R. Rajatheva, and K. Ahmed, "Performance of Coded OFDM with Higher Modulation Schemes," in *Proc. IEEE Int. Conf. Comm. Tech. (ICCT'98)*, vol. 2, Beijing, China, Oct. 1998.

[14] R. Bonamy, D. Chillet, S. Bilavarn, and O. Sentieys, "Power Consumption Model for Partial and Dynamic Reconfiguration," in *Proc. Int. Conf. Reconf. Comput. and FPGAs (ReConFig'12)*, Cancun, Mexico, Dec. 2012, pp. 1–8.

[15] S. Kraemer, R. Leupers, G. Ascheid, and H. Meyr, "SoftSIMD-Exploiting Subword Parallelism Using Source Code Transformations," in *Proc. Conf. Design, Automation & Test in Europe (DATE'07)*, Nice, France, 2007, pp. 1–6.

[16] L. Wilhelmsson, I. Diaz, T. Olsson, and V. Öwall, "Analysis of a Novel Low Complex SNR Estimation Technique for OFDM Systems," in *Proc. Int. Conf. Wireless Comm. Netw. (WCNC'11)*, Cancun, Mexico, Mar. 2011, pp. 1646–1651.

978-1-4799-8720-7/15 $31.00 © 2015 IEEE

2015 IEEE Computer Society Annual Symposium on VLSI

The Solar Cells and the Battery Charger System Using the Fast and Precise Analog Maximum Power Point Tracking Circuits

Yasuhiro Sugimoto

Dept. of Electrical, Electronic, and Communication Engineering
Chuo University, Tokyo 112-8551, Japan
sugimoto@sugi.elect.chuo-u.ac.jp

Abstract—We propose a simple, fast and precise power calculation circuit for photovoltaic (PV) power systems. The power is calculated according to the principle that the integration of the PV output current over the time interval proportional to the PV output voltage becomes the power. This power calculation circuit is realized using simple analog circuits; neither a microprocessor nor a DSP is necessary to perform the function of Maximum Power Point Tracker (MPPT). As the calculated result at the output of the power calculation circuit becomes the analog voltage, a power comparison between the previous result and the current result is easily done using analog comparators where the previous result is sampled-and-held. Using the comparison result, the duty ratio of the boost converter is controlled to find the maximum power point in the PV output voltage range by adopting the hill-climbing perturb and observe (P&O) method. The prototype of the PV power system was circuit-designed and laid out using a 0.18-μm CMOS IC process. The SPICE circuit simulation results indicated that the PV power calculation achieved within $\pm 2\%$ error for the PV output power from 50 mW to 300 mW, and $\pm 5\%$ for the PV output power from 10 mW to 350 mW. The operation of the whole power system was verified in a transient simulation by applying the current step from the current source of the equivalent PV-cell circuit. The result shows convergence to the maximum power point when the PV output currents are 100 mA, 300 mA and 500 mA with a 50-Ω load resistor at the output of the boost DC-DC converter.

Keywords-solar cells; battery charger; analog; hill-climbing; MPPT;

I. INTRODUCTION

Recently, photovoltaic power systems for the production of electricity from the sun's energy have gained importance. However, the ability of PV cells to generate power depends on irradiation of sun-light, temperature and loading conditions[1]. In order to extract the maximum power from the PV cell, it is necessary for its settings to be optimal. As a PV cell is considered to be a constant current source in combination with a pn junction diode when light irradiation is constant, the "optimal" setting described above becomes equal to finding the optimal PV output voltage. This is done by adjusting the load impedance of PV cells using a switching regulator such as a DC-DC converter[2]. Usually, the duty ratio of the power transistor of a DC-DC converter is controlled for this purpose. The whole system operation to find the maximum power point (MPP) is called MPPT and

the most popular control method is the hill-climbing P&O. This method of finding the maximum power point works by changing the PV output voltage slightly and comparing the PV output powers of the previous and current steps[3].

The calculation of power involves the multiplication of the PV output voltage and current. If this multiplication is done in a digital domain, the power converter needs to install microprocessors or Digital Signal Processors (DSPs) together with current sensors, anti-alias filters and ADCs to achieve sophisticated MPPT control. However, the use of a large digital function is not favorable in terms of power consumption, cost, amount of hardware required, area required and cumbersome software development[4]. Historically, multiplication in analog form has been simply done by using a Gilbert cell with only seven transistors. Therefore, it is better to realize the multiplication in analog form as much as possible. Many previous works have adopted the analog approach to realize the MPPT function[5],[6].

In this paper, an MPPT system using the proposed analog power calculation circuit that integrates the current proportional to the PV output current over the time interval proportional to the PV output voltage is introduced. The PV current detection is done by copying the current flowing in the power transistor of the DC-DC converter assuming that the duty ratio does not change drastically. As the output of the power calculation circuit becomes the analog voltage, a comparison between the PV output power in the previous and current steps is easily performed using an analog comparator if the calculated result in the previous step in an analog form has been sampled-and-held. Using the current comparison result and the memorized previous comparison result, now in logic form, the hill-climbing P&O method is easily realized. Therefore, there is no need to use a microprocessor, a DSP, ADCs or anti-alias filters, leading to a very simple implementation of the MPPT system.

II. CONVENTIONAL SYSTEM OF SOLAR CELLS AND THE BATTERY CHARGER

Figure 1 shows the output voltage and current characteristics of the theoretical photovoltaic (PV) array. The PV array is considered to be a constant current source when the incident light power is constant, and its output voltage is

978-1-4799-8720-7/15 $31.00 © 2015 IEEE   597

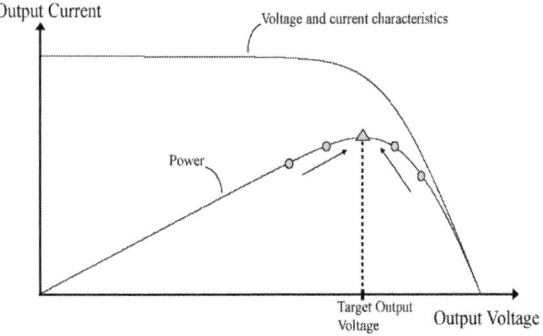

Figure 1. Voltage and current characteristics of the PV array and the hill-climbing P&O method.

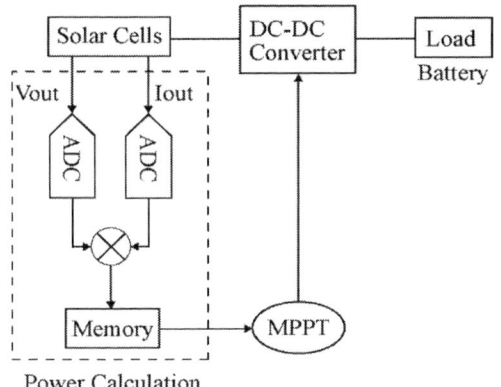

Figure 2. Block diagram of the conventional battery charger system.

determined by the load impedance at the output of the PV array. However, the output current decreases rapidly when the output voltage reaches the forward bias voltage of the pn junction diode due to the nature of the PV diode. Therefore, the maximum power point for the output voltage and current exists as shown by the triangle in Figure 1. Usually, the target output voltage is researched and decided by changing the load impedance at the output of the PV array. The method is called the hill-climbing perturb and observe (P&O) method. The power is calculated by multiplying the output voltage and current, and currently calculated power is compared with the previously calculated power. The impedance is changed toward the direction indicated by arrows as shown in Figure 1 after the results of the power comparison.

Figure 2 shows the conventional block realization of the battery charger system using solar cells[4]. The PV output voltage and current are digitized by ADCs and multiplied to obtain the power. The MPPT method uses the information of the calculated power in the memory, and the duty of the DC-DC converter is adjusted to obtain the target voltage for the maximum power output. However, the system uses ADCs and even a DSP to digitize, multiply and conduct the

MPPT method. Therefore, the overhead for performing the MPPT method is heavy.

III. STRATEGIES TO MAKE THE SYSTEM STRUCTURE SIMPLE

In this paper, the approach that is proposed uses analog and/or mixed-signal functions instead of using the conventional digital functions such as ADCs and a DSP to achieve power calculation, and to employ the MPPT method. The hardware count is projected to be greatly reduced by applying analog functions. Figure 3 shows a block diagram of whole the proposed power system including an equivalent circuit of PV array cells, a boost converter, the load and MPPT control circuits. V_{out} and I_{out} are detected at the output of PV array cells and the power is calculated at the power calculation block. The output of the power calculation block, which is denoted by P_c in Figure 3, is the voltage that has the current power value extracted from PV array cells. This voltage is stored in the sample-and-hold (S/H) block; therefore, its output P_p becomes the power information of the previous power calculation in analog voltage form. The result of comparison d using a comparator indicates whether the power in the current step has increased ($d = 1$) or decreased ($d = 0$) from the power in the previous step, and this d also becomes a part of decision that performs the hill-climbing P&O method. To complete the hill-climbing P&O method, information about the decision in the previous step is needed. This is dd, and this decision from the previous step was stored in the T flip-flop. Using d and dd of the current and previous power calculations, the logic to control the duty cycle of the driving pulse for the power transistor M_B is decided according to the algorithm of the hill-climbing P&O method. Table 1 shows this.

Table I
THE LOGIC FOR THE HILL-CLIMBING P&O METHOD IN MPPT.

Previous step decision dd	Current step d	Decision T
0	0	1
0	1	0
1	0	0
1	1	1

Decision T from the output of the logic in Figure 3 corresponds to whether the duty cycle of the switching DC-DC converter is modified or not. The "1" in decision T indicates that the control should be performed to increase the PV output voltage, while "0" in decision T prescribes a decrease of the PV output voltage. In the case of dd=0, d=1, for example, the power in the current step increased although the PV output voltage was decreased in the previous step. This means that the PV output voltage of the maximum power point is smaller than the current PV output voltage. So, decision T is that the PV output voltage should decrease; that is, T=0. The duty control signal of this decision T is

978-1-4799-8720-7/15 $31.00 © 2015 IEEE   598

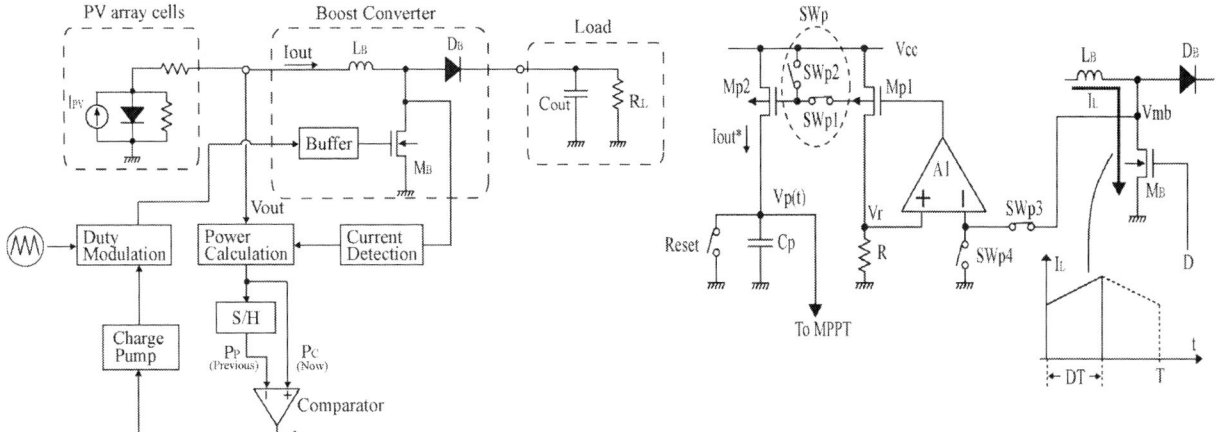

Figure 3. The whole circuit of the proposed power system.

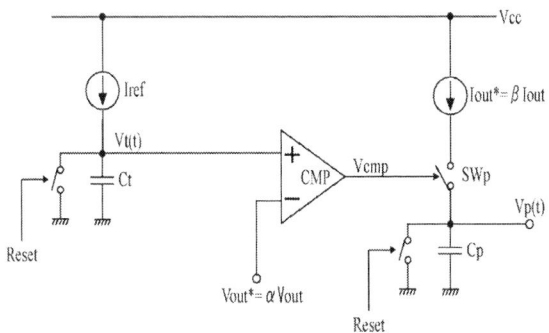

Figure 4. Detail of the proposed power calculation circuit.

realized by using a charge pump and a duty modulation function and producing the drive pulse of M_B as shown in Figure 3. Note that only a small amount of circuit is necessary to prepare.

IV. CIRCUIT DESIGN OF THE POWER CALCULATION AND CURRENT DETECTION BLOCKS

Figure 4 explains the circuit of the power calculation block that is implemented using the mixed analog and digital circuit. However, it is basically analog. In Figure 4, C_t integrates the constant current I_{ref} until the voltage $V_t(t)$ across the capacitor C_t reaches $V_{out}^* = \alpha V_{out}$ where α is a constant. Once the $V_t(t)$ surpasses V_{out}^*, the value of V_{cmp}, which is the voltage at the output of the comparator CMP, becomes high. If t_v is the time when $V_t(t)$ reaches V_{out}^*, then

$$V_t(t) = \frac{1}{C_t} \int_0^{t_v} I_{ref} dt = \frac{I_{ref}}{C_t} t_v \qquad (1)$$

Figure 5. Current detection circuit.

Equating $V_t(t)$ to V_{out}^*, we obtain

$$t_v = \frac{\alpha C_t}{I_{ref}} V_{out} \qquad (2)$$

As α, C_t and I_{ref} are all constant values, equation 2 indicates that t_v is proportional to the PV output voltage V_{out}. On the other hand, C_p stores charge which is proportional to I_{out} until V_{cmp} becomes high. When V_{cmp} becomes high, SW_p turns off and the current flow into C_p from I_{out}^* stops, such that the output voltage $V_p(t)$ is held constant. Therefore, $V_p(t)$ becomes

$$V_p(t) = \frac{1}{C_p} \int_0^{t_v} I_{out}^* dt = \frac{\beta I_{out}}{C_p} t_v \qquad (3)$$

Substituting equation 2 into equation 3, we obtain

$$V_p(t) = \frac{\alpha \beta C_t}{I_{ref} C_p} I_{out} V_{out} \qquad (4)$$

Equation 4 indicates that the voltage $V_p(t)$ across the capacitor C_p in the hold state is proportional to the product of the PV output voltage V_{out} and the current I_{out}, that is, the PV output power. Therefore, the power calculation is simply implemented using a comparator, analog switches, capacitors and current sources instead of using ADCs and a multiplier. This reduces the hardware count to a great extent. The subject here is how constants α and β are realized. It is clear that the voltage V_{out}^*, and therefore the constant value α, is obtained by the division of V_{out} using resistors.

I_{out} is the current that flows into the inductor L_B. The current in the inductor changes up and down during the control of one clock period. In the case of the boost converter, the power transistor M_B turns on to charge the energy in the inductor and turns off to release the energy to the load through a diode D_B in Figure 3. The current flowing in the load becomes the average of the current between the energy-charging phase and the energy-releasing phase. The circuit in this paper to detect the PV output current is shown

978-1-4799-8720-7/15 $31.00 © 2015 IEEE   599

in Figure 5 and gives the current which is approximately proportional to I_{out}[7].

In this circuit, the inductor current when the power transistor turns on is only monitored and read in. When M_B turns on, all the inductor current flows into the power transistor M_B. Due to the on resistance of M_B, the voltage V_{mb}, whose value becomes proportional to the inductor current value, appears at the drain terminal of M_B. As this voltage is copied to V_r in Figure 5 by the aid of the feedback configuration of an operational amplifier and a transistor M_{p1}, the current flowing in M_{p1} becomes

$$I_{Mp1} = V_{mb}/R \propto I_L/R \qquad (5)$$

I_{Mp1} is mirrored to I_{Mp2} and becomes the charging current I_{out}^* for charging the capacitor C_p in Figures 4 and 5. Here, the duty ratio control is slow and it is assumed that no sudden power change occurs during the calculation. Therefore, charges in C_p in consecutive control steps of MPPT are considered approximately proportional to I_{out} changes. SW_{p1} and SW_{p2} in Figure 5 consist of SW_p in Figure 4. When the comparator output becomes high, SW_{p1} turns off while SW_{p2} turns on. SW_{p3} and SW_{p4} are prepared to lead the V_{mb} voltage to the input terminal of the op-amp when the current I_L flows into M_B. I_{out}^* flows until SW_p turns off. The duration of the power calculation is around 200 μs and the operating frequency of the DC-DC converter is about 1 MHz.

V. EXPERIMENTAL RESULTS

In order to verify the proposed MPPT operation, we conducted the circuit design and simulation of the circuit shown in Figure 3. The circuit design is based on device parameters from a 0.18-um CMOS process. Figure 6(a) shows the terminal voltage $V_p(t)$ when the PV output current changes at three different output voltage settings. Linearity for the current change is needed. In this case, however, PV cells are replaced by a dc voltage source in order to examine linearity of the power calculation circuit by making voltage ranges wider than those limited by the forward-biased junction voltage. We call the dc voltage source the PV-equivalent source. In Figure 6(a), $V_p(t)$ change is displayed for the current change from 50 mA to 500 mA with PV-equivalent output voltages of 0.3 V, 0.5 V and 0.7 V. The figure shows that the PV output power is correctly calculated and output to $V_p(t)$. Figure 6(b) shows the error rate where the point at $V_{out} = 0.5V$ and $I_{out} = 200mA$ is the reference, and the error is measured as the percentage deviation from the ideal ratio of voltages that are derived from the calculation using the value of the reference point. The error rates become larger in the low current and high current regions; however, they are within 4% and are allowable to maintain linearity.

Figure 7(a) shows the terminal voltage $V_p(t)$ when the assumed PV-equivalent output voltage changes in the case with three different currents. Linearity for the voltage change

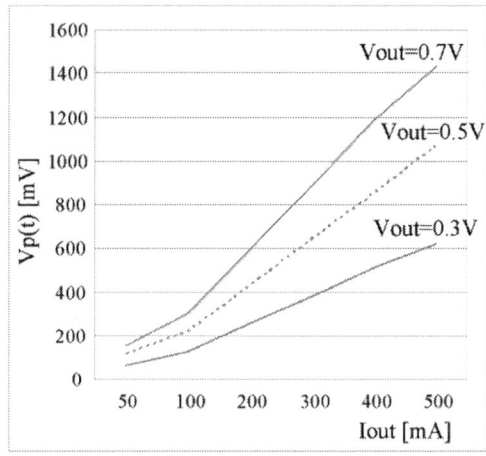

(a) The output voltage change of the proposed power calculation circuit when PV-equivalent output current changes.

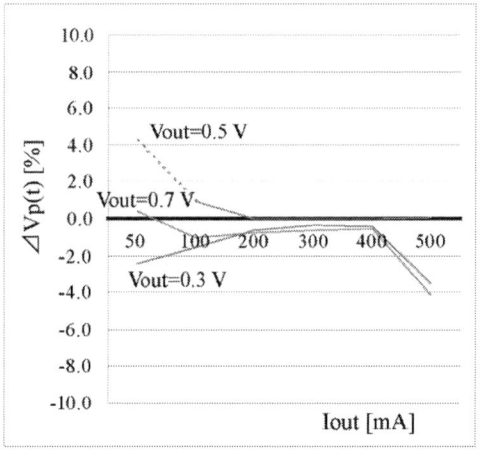

(b) Relative error of $V_p(t)$ for current changes.

Figure 6. Current characteristics of the proposed power calculation circuit.

is also needed. Again, PV cells are replaced by a dc voltage source. $V_p(t)$ change is displayed for the voltage change from 0.3 V to 1.1 V with assumed PV output currents of 100 mA, 200 mA and 300 mA. The figure shows that the PV-equivalent output power is correctly calculated and output to $V_p(t)$. As the supply voltage of the power calculation circuit is 1.5 V, the $V_p(t)$ becomes saturated at this voltage. Figure 7(b) shows the error rate where it is measured as the percentage deviation from the ideal ratio of currents. The error rate is within $\pm$2% and reveals good linearity.

Figure 8 shows the overall characteristics of the power

978-1-4799-8720-7/15 $31.00 © 2015 IEEE

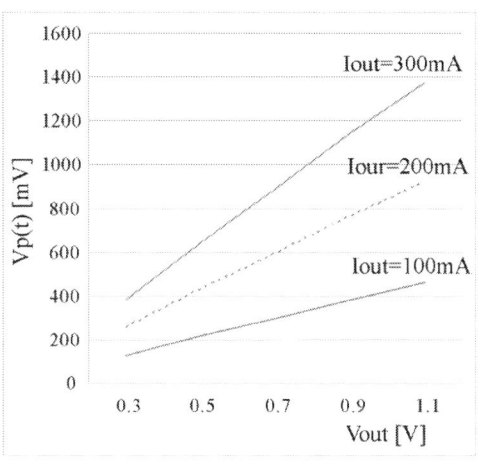

(a) The output voltage change of the proposed power calculation circuit when the PV-equivalent output voltage changes.

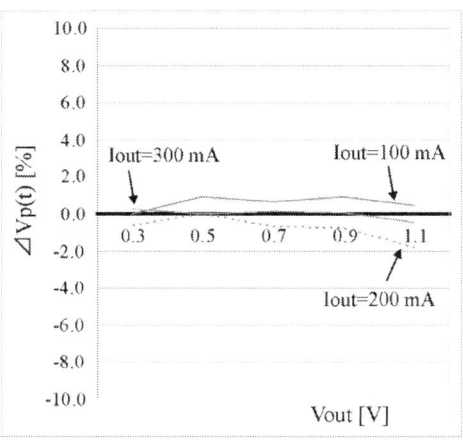

(b) Relative error of $V_p(t)$ for voltage changes.

Figure 7.   Voltage characteristics of the proposed power calculation circuit.

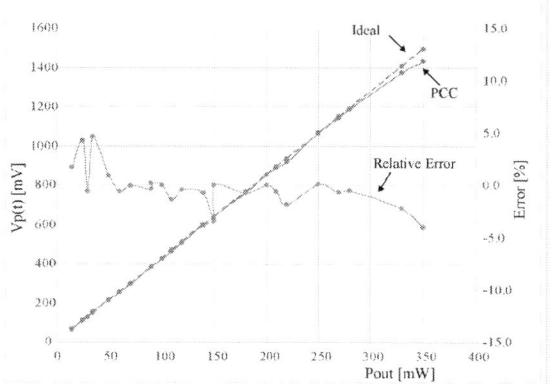

Figure 8.   The input power and the output voltage characteristics of the power calculation circuit.

calculation circuit which plots the relationship between the input power to the circuit and the output voltage $V_p(t)$ shown in Figures 4 and 5. The PV output power is the product of the PV-equivalent output voltage and current. The power calculation circuit (PCC) needs to generate an output voltage that is proportional to its input power, that is, the PV-equivalent output power which is shown as P_{out} in Figure 8. The curve labeled "PCC" is the simulated characteristics. Although the curve begins to depart from the ideal straight line at more than 300 mW of the PV output power, this is due to the low supply voltage of 1.5 V. At a power of 300 mW, $V_p(t)$ reaches around 1.3 V and there is no margin for

the circuit. The figure also plots the error between the ideal and PCC curves. For the range of assumed PV output power from 50 mW to 300 mW, the error is within ±2%, which means that excellent linearity has been obtained. The error in the power range below 50 mW becomes large because the 100 % power level itself becomes small. The increasing error in the power range greater than 300 mW is due to the low supply voltage.

Figure 9 shows the transient simulation result of the whole circuit shown in Figure 3. It demonstrates how the output power calculated by multiplying V_{out} and I_{out} of the actually designed circuit converges to the final power value in time when step current I_{PV} values of 100 mA, 300 mA and 500 mA are applied to the equivalent current source of PV cells. The load resistor R_L at the output of the DC-DC converter is chosen to be 50 Ω. The estimated final power values are 34 mW, 96 mW and 142 mW, while the observed power values in simulation are 34 mW, 93 mW and 141 mW, respectively. The output power converges in about 7 ms after the step change of currents, and deviates upside and down with the $\Delta P_{in}/2$ values of 9 mW, 11 mW and 9 mW, respectively, due to the nature of using the hill-climbing and P&O method.

Figure 10 further verifies whether the system can find the maximum power point in different power levels that are denoted by the PV currents I_{PV}s of 100 mA, 300 mA and 500 mA. The estimated maximum power levels are 34 mW, 96 mW and 142 mW, respectively, and they are denoted by triangles in Figure 10. The circles converge approximately to the maximum power points of the circuit with R_{out} values of 10 Ω, 50 Ω and 100 Ω, respectively. The simulated maximum power points are close to the estimated maximum power points, indicating the correct operation of the proposed MPPT system.

Figure 11 is the layout of the proposed whole MPPT system except for the PV and external inductor and the

978-1-4799-8720-7/15 $31.00 © 2015 IEEE   601

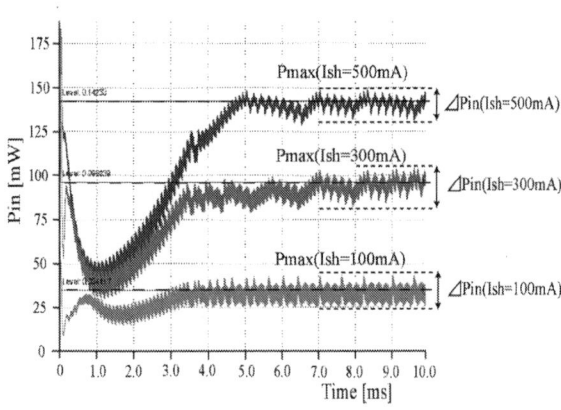

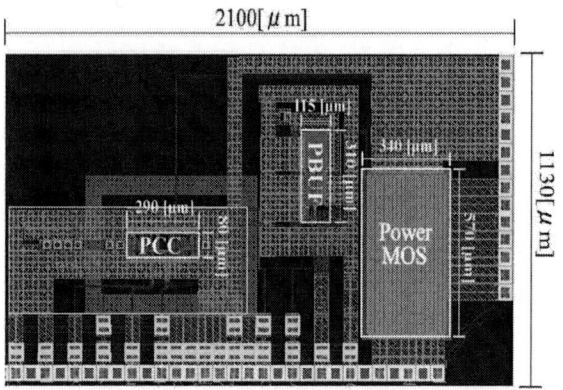

Figure 11. The layout pattern of the MPPT system.

Figure 9. The transient simulation result of the whole MPPT system shown in Figure 3.

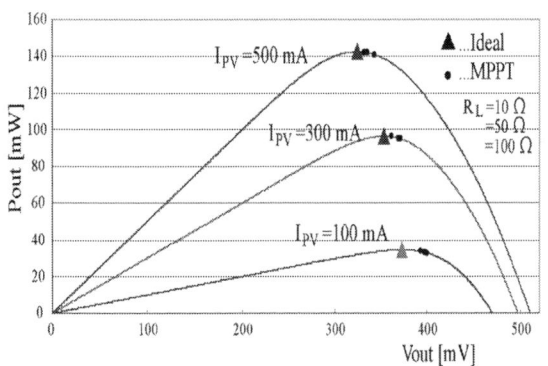

Figure 10. The plot of estimated and simulated maximum power points.

Schottky diode in Figure 3 utilizing a 0.18-μm CMOS process. A boost converter including a power MOS transistor and a driving buffer circuit PBUF for the power MOS are also installed. The active area of the PCC, which is the proposed MPPT circuit, is only 290 μm $\times$ 80 μm. From this layout example, the analog approach turns out to be cost effective.

VI. CONCLUSION

We developed a simple, fast and precise analog MPPT circuit for the PV power system without using ADCs, microprocessors or DSPs. The basic concept to measure the current PV power is that PV output current is integrated over the time interval which is proportional to the PV output voltage. The hill-climbing P&O method is adopted to find the maximum power point. We obtained less than ±5% deviation from linearity for the power calculation,

guaranteed convergence in transient response for the step current inputs, and a small layout area of 290 μm $\times$ 80 μm using a 0.18-μm CMOS.

ACKNOWLEDGMENT

We thank the continuous financial support by Chuo University, Tokyo for the research grant given for this work.

REFERENCES

[1] D.P. Hohm and M.E. Ropp, "Comparative Study of Maximum Power Point Tracking Algorithms", Prog. Photovolt: Res. Appl. 2003; II:47-62, Jhon Wiley & Sons, Ltd.

[2] Y. Qiu, C.V. Liempd, B.O. Veld, P.G. Blanken and C.V. Hoof, "5uW-to-10mW Input Power Range Inductive Boost Converter for Indoor Photovoltaic Energy Harvesting with Integrated Maximum Power Point Tracking Algorithm", ISSCC 2011, paper no.6.8, pp.118-119, Februay 2011.

[3] T. Esram and P.L. Chapman, "Comparison of Photovoltaic Array Maximum Power Point Tracking Techniques", IEEE Trans., Energy Conversion, vol.22, no.2, pp.439-449, June 2007.

[4] STMicroelectronics, "SPV1020 Technical Data", http://www.st.com/web/en/resource/technical/document/datasheet/

[5] R. Enne, M. Nikolic and H. Zimmermann, "Dynamic Integrated MPP Tracker in 0.35 μm CMOSA", IEEE Trans., Power Electronics, vol.28, no.6, pp.2886-2894, June 2013.

[6] T.H. Tsai and K. Chen, "A 3.4mW Photovoltaic Energy-Harvesting Charger with Integrated Maximum Power Point Tracking and Battery Management", ISSCC 2013, paper no.4.5, pp.72-73, Februay 2013.

[7] K. Umimura, H. Sakurai, and Y. Sugimoto, "A CMOS Current-mode DC-DC Converter with Input and Output Voltage-Independent Stability and Frequency Characteristics Utilizing a Quadratic Slope Compensation Scheme", ESSCIRC 2007, pp.178-181, September 2007.

978-1-4799-8720-7/15 $31.00 © 2015 IEEE

2015 IEEE Computer Society Annual Symposium on VLSI

3D DFT Challenges And Solutions

Yassine Fkih[1,2], Pascal Vivet[1]

Marie-Lise Flottes[2], Bruno Rouzeyre[2], Giorgio Di Natale[2]

Juergen Schloeffel[3]

(1) CEA-Leti, MINATEC Campus, F38054, Grenoble France

(2) LIRMM, Univ Montpellier II/CNRS, Montpellier France

(3) Mentor Graphics, Hamburg, Germany

Abstract—**Design-For-Test (DFT) of 3D stacked integrated circuits based on Through Silicon Vias (TSVs) is one of the hot topics in the field of test of integrated circuits. This is due to the test access complexity of dies' components that must be controlled/observed before and after bonding (especially for upper dies), and the high complexity of 3D systems where each die can embed hundreds of IPs. DFT of 3D circuits concerns all the components of the 3D system, including the dies and the inter-die interconnections. We address the problem of test architecture definition for both TSVs testing before bonding and cores testing before and after bonding. We present test solutions allowing to access the components under test while physical interconnects for test data propagation differ according to the stacking step. The paper also discusses core test scheduling issues.**

Key words: 3D-IC, Design-For-Testability, IEEE 1687, test scheduling

I. INTRODUCTION

The stacking process of integrated circuits using TSVs (Through Silicon Vias) is a promising technology that keeps the development of the integration more than Moore's law, where TSVs enable the tight integration of various dies in a 3D fashion. 3D stacking will allow a wide range of new applications thanks to high performance, smaller form factors, heterogeneous stacking (digital, memory, RF, MEMS), and interposers for multi-chip connection, which become similar to silicon boards. The first upcoming 3D applications are mainly the WideIO DRAM 3D memory interface for high throughput and low power memory-on-logic stacking [1].

3D Integrated Circuits (3D-ICs) present however new test challenges related to the new fabrication process. Indeed the test may have to be performed at pre-bond phase, mid-bond phase, and post-bond phase to guarantee the production quality. Pre-bond test targets the individual dies at wafer level, by testing not only classical logic (digital logic, IOs, RAM, etc.) but also non bonded TSVs. Mid-bond test targets the test of partially assembled 3D stacks, whereas post-bond test targets the final circuit. It is generally admitted that a 3D test flow [2] should involve test procedures at all stacking levels of the 3D components.

A first issue is regarding the test of new 3D elements themselves, such as TSV and μ-bumps, with associated 3D

test defaults. A second test issue concerns the overall 3D Design-For-Test (DFT) architecture. It must be compatible with existing test standards, allow test in pre-, mid- and post-bond phases, and offer enhanced test scheduling within the 3D stack. In addition, the possibility of easy test pattern retargeting from 2D (die-level) to 3D (stack-level) is definitely a plus. While test scheduling aims to optimize the test time on every stack, test pattern retargeting is of primary concern for industry to reduce test development efforts to translate test patterns from dies to stacks.

Many DFT architectures for testing 3D integrated circuits have been proposed in the past. The first papers treated pre-bond test of 3D processors using scan islands and the so-called Layer Test Controller (LTC) [3], scan chain optimization approaches [4], and other test issues like test cost optimization [5]. More recent works propose die level wrappers based either on IEEE 1500 [6] or IEEE 1149.1 [7] test standards. These test architectures have mainly three features: the use of dedicated probe pads for pre-bond testing on every stack tier, the usage of "TestElevators" [6, 7] to drive test signals up and down during post-bond test of non-bottom dies through the bottom die, and the use of a hierarchical WIR (Wrapper Instruction Register) chain to configure test paths. These features satisfy 3D circuits testing requirements in case of homogenous 3D-ICs where all dies have similar interfaces (e.g. IEEE 1149.1). A new standard IJTAG, or IEEE 1687, is currently methodology for accessing instrumentation embedded within a semiconductor device, without defining the instruments or their features themselves, it incorporates IEEE 1149.1-x and the design-for-test standard IEEE 1500. We note that proposals should be in line with the on-going 3D test working group IEEE P1838 [8].

Assuming dies from different sources, test interfaces must be define to provide compatible test access mechanisms between the different dies, facilitate exchange of test information between Design-for-Test engineers, Test CAD tools, and Test engineers, and allow implementation of various test schedules according to the test phase.

This paper discusses DFT issues for 3D-SICs from different perspectives including the pre-bond test of TSVs and the definition of the DFT architecture at system level taking into account test pattern retargeting and test scheduling flexibility challenges.

This work was funded thanks to the French national program 'programme d'Investissements d'Avenir, IRT Nanoelec' ANR-10-AIRT-05
And was conducted in the framework of the Catrene CT312
MASTER3D project and received financial support from the "Ministère du Redressement Productif"

978-1-4799-8720-7/15 $31.00 © 2015 IEEE

The paper also presents experiments performed from envisaged solutions. It is organized as follows: section II deals with the pre-bond test of TSVs, it highlights the challenges, presents the BIST solution we have developed for testing these vertical interconnects while they are not yet bonded, and shows its ability to distinguish faulty TSVs from fault-free ones on real experimental data. Section III presents the proposed 3D DFT architecture based on IEEE 1687 and details the main features that enable automatic test path configuration according to the test phase, its flexibility w.r.t. the stack under test, and its ability to provide test pattern retargeting. In Section IV, we make the link between the flexibility of our 3D DFT architecture and 3D test scheduling. Finally, section V concludes the paper.

II. PRE-BOND TEST OF TSVs

Testing TSVs at pre-bond level contributes towards production of Known Good Dies before stacking.

The testability of TSVs before bonding has been the subject of several research works which can be divided in two categories: the first category relies on fine probe heads to establish direct contacts with TSVs [9,10,11], and the second is based on indirect measurements using dedicated test pads and embedded test infrastructures [12,13,14,15,16]. Our proposed test architecture detailed in [17] belongs to this second approach.

The principle is to use ring oscillators to evaluate the electrical characteristics of TSVs (capacitance). A Built-In-Self-Test (BIST) infrastructure elaborates the measurement of the frequencies of the ring oscillators and generates signatures. Communication with the BIST logic is ensured by the mean of a JTAG interface.

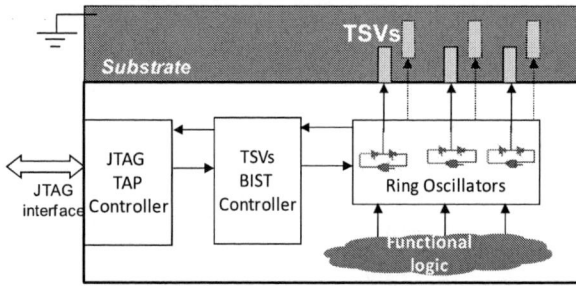

Figure 1. Pre-bond test of TSVs using ring oscillators

The ring oscillators are directly connected to the accessible end of TSVs (see Fig.1). Their frequencies depend directly on the capacitance of these vertical interconnects, and thus, a frequency variation is an image of a capacitance variation (defect) in the TSV under test. The detection of a frequency variation is done by comparing all ring oscillators in the circuit.

A. Design of the ring oscillators

The design of a ring oscillator for TSV testing depends on several parameters including the number of TSVs per ring oscillator (either a large ring oscillator charging many TSVs or a small ring oscillator charging only one TSV), and the electrical characteristics of the inverters (capacitance) used in

the ring to provide oscillations. Sensitivity to PVT (Process, Voltage, and Temperature) variations is also an issue.

As detailed in [17], our test method is based on a relative frequency variation measurement, which has been shown robust against global PVT variations. The best fault detection accuracy has been obtained for only one TSV connected per ring oscillator and inverters providing the smallest input capacitance.

B. Design of the BIST circuitry

The BIST architecture ensures the launch of the BIST procedure, the counting of the number of oscillations for each ring oscillator during a given time, the comparison between oscillator frequencies, and the generation of the final test result. In order to limit the area overhead, we chose to implement only one counter, the BIST structure thus also selects the ring oscillators to be measured one by one.

The proposed BIST supports 3 possibilities of feedback: pass/fail, diagnostic and debug. The pass/fail feature allows a quick test of the TSVs by comparing the frequencies of the ring oscillators. If the difference between the maximum and the minimum measured frequencies exceeds a user-programmable threshold, the BIST returns 0 (Fail), 1 otherwise (Fault-free). The 'diagnostic' option returns the total number of the failing TSVs, and the 'debug' option returns their positions in the chip.

The user-programmable threshold TTT (TSV Test Threshold) [17] can be set to values ranging from 5% to 50% (step=5%). Tuning the TTT parameter can be used for TSVs characterization: TTT is first set to a large value (50% of deviation) then gradually decreased until the test fail. The last TTT value for which the test passes corresponds to the dispersion of the TSV characteristics in the circuit.

C. Test chip implementation

We implemented the circuit shown in Figure 1, as a test chip prototype for one channel matrix of a WideIO memory including 276 TSVs. Synthesis results using the STMicroelectronics 65nm library shows that the area of our proposed BIST architecture, including the ring oscillators, represents only 2% of the TSVs area.

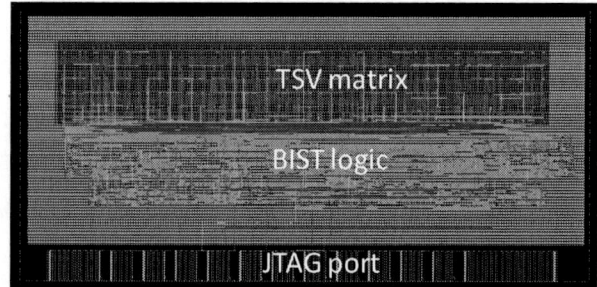

Figure 2. Layout of the BIST of pre-bond test of TSVs

The layout of the test chip is shown in Figure 2. The ring oscillators were placed as close as possible to the TSVs, within the TSVs matrix, in order to limit the area overhead and the impact of local (intra-chip) PVT variations on the measurement.

A simulation of the BIST procedure for 1000 TSVs shows a total test time of 2ms with a JTAG clock of 50 MHz and a BIST clock of 100 MHz.

D. Application on real technological data

TSV reliability measurements have been performed on TSV matrices from CEA-LETI with following characteristics: diameter=10μm, length=80μm (AR=1/8). The expected TSV resistance and capacitance are around 25 Ohm and 250 fF respectively.

The summary of the capacitance measurements are presented below. The plots present a histogram of the measured capacitances, and a color map at waver level shows the localization of potential TSV defects (values out of range). Measurements were done on 'correct' wafers and 'incorrect' wafers (with injected faults).

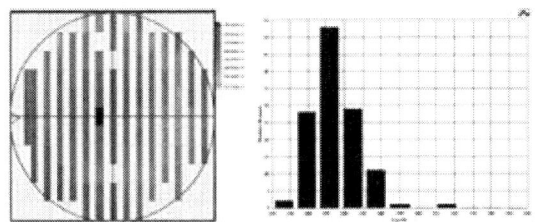

Figure 3.a Capacitance measurements of TSVs on correct wafers

Figure 3.a shows measurement results from a correct wafers, the measured mean capacitance of TSVs is around 250 fF, with a standard deviation value of 12fF, which give an overall yield of 95% in case of correct wafers.

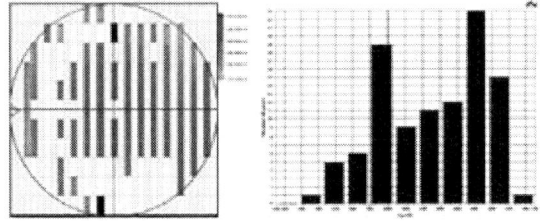

Figure 3.b Capacitance measurements of TSVs on incorrect wafers

Figure 3.b shows measurement results corresponding to incorrect wafers, the measured mean capacitance equals to 435 fF with a standard deviation (sigma) of 58 fF and an overall yield of 78%.

As a summary, the dispersion is clearly very large, and can be observed by the mean of our proposed BIST approach based on ring oscillators. In such context the detection of faulty TSVs is done using a comparison approach called "average approach", where the average oscillation frequency among all the TSVs is calculated by the BIST logic and the left and right limits are defined by the mean of the TTT user-programmable threshold (see fig. 4).

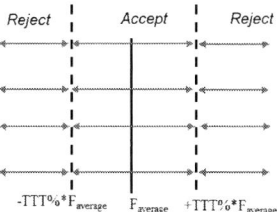

Figure 4. Binning strategy of the measured frequencies.

Applied in a case such the one shown in Figure 3.b where TSVs are considered as faulty if their capacitance exceeds 3 sigma. It is enough to define the TTT to be equal to 3 sigma to be able to detect the faulty TSVs. Applied to that case TTT= 174fF, i.e. frequencies higher than 250+174=424fF or less than 250-174=76fF are rejected and their associated TSVs considered as faulty.

III. 3D DFT ARCHITECTURES BASED ON IEEE 1687

The 3D DFT architecture heavily depends on the specifications of the 3D-IC including the number of stacked dies, the nature of the interposer if any (passive or active), and the test infrastructure of each die. In the state of the art, 3D ICs are assumed to be uniform in the sense that all dies embeds either IEEE 1500 DFT and/or IEEE 1149.1 wrapper and communication means. We propose 3D DFT architectures for both uniform and heterogeneous 3D ICs, where dies relies on heterogeneous test interfaces.

A. Uniform 3D DFT architecture

The first type of 3D DFT deals with regular structures as in [6, 7]. The test architecture uses the new IEEE 1687 instead of IEEE 1149.1 as test standard in each die (see fig.5). Multiplexing logic between test pads and test TSVs is used for switching test paths sinks and sources after stacking. For instance, the TDI pad on the top die is used to input test data to this tier before bonding, while test data are transported to this die from the die below after bonding.

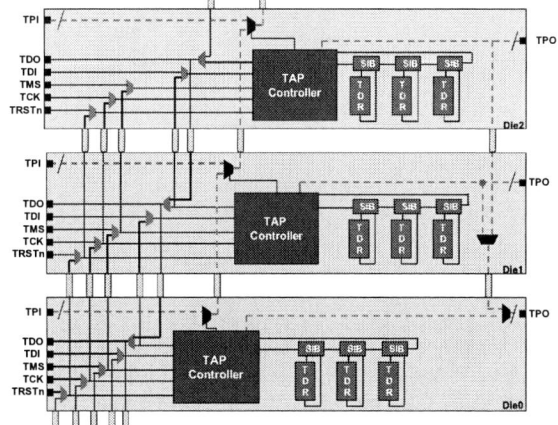

Figure 5. IEEE 1687 based 3D DFT architecture

Each individual die can embed a variety of DFT logic controlled from its TAP controller and associated Test Data Registers (TDRs). The proposed DFT architecture requires that all stacked dies are equipped (1) with a JTAG interface

(TDI, TDO, TMS, TCK, optional TRST) as a test access mechanism in order to build the 3D DFT chain, and (2) a TAP controller to build around it IEEE 1687 circuitry: Segment Insertion Bit (SIBs) and associated TDRs..

In order to provide many-bit test data to the tiers after stacking, and thus shorten the test time thanks to concurrent testing of several IPs, each die must be equipped with parallel test inputs. A boundary scan solution with a parallel test access mechanism can be found in [18].

The detailed control of the JTAG Multiplexers is out of the scope of this paper. It could be done with additional configurations registers as currently proposed in IEEE 1838 [8] or it could be optimized using an automatic die-detection mechanism as proposed in [19, 20] but with the limitation of having a static concatenated TAP serial chain.

B. Heterogeneous 3D DFT architecture

Stacked 3D-ICs may have an irregular test structure; i.e. dies do not embed the same test infrastructure. Figure 6 gives an example of a 2.5D circuit (dies stacked on an interposer) with 3 stacked dies: die_0 (left) is JTAG compliant, die_1 (middle) has a 3-bit test interface (test start, test enable, test result), and die_2 (right) has a IEEE 1500 wrapper.

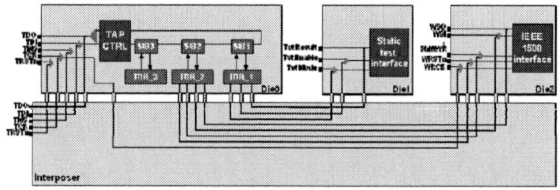

Figure 6. 3D DFT Architecture of a 3D circuit on passive interposer

Die_0 has been modified in order to manage the test of all the dies in the 2.5D system. A gateway register allows a dynamic configuration of the test infrastructure. Die_0, die_1 and die_2 can be concurrently or serially tested thanks to the SIBs. Die_0 embeds the IEEE 1687 infrastructure: the TAP controller, an IR and a decoder. Die_1 and Die_2 are considered as instruments. We note that the signals WSI and WSO of the IEEE 1500 interface of die_2 are connected directly to the active scan-path and not through latches of the TDR_2 [21].

C. Comparison between both architectures

As a summary, the first test architecture shown in Fig.5 manages the test of uniform 3D circuits where all stacked dies have the same test interface (IEEE 1149.1 in our case). Test signals are transmitted from the bottom die to the top die using TSVs as elevators. Each die embed IEEE 1687 infrastructure which enables test pattern retargeting and enhance test time optimization which will be explained in more details in the next section.

The second test architecture shown in Fig.6 manages the test of heterogeneous 3D circuits such as 2.5D circuits where dies are stacked on passive interposer and have different test interfaces. In this later case, one die should embed a IEEE 1149.1 test interface and a IEEE 1687 specific circuitry to be considered as the master die. The other dies will be considered

as slaves and are accessed through the master die as instruments.

The 2 proposed test architectures can be mixed for complex 3D circuits with multi-tower stacked dies: each tier should have an IEEE 1149.1 test interface and establish the serial connection to its adjacent tiers, and within the tier one die can be used as a master die to manage the test of its adjacent dies.

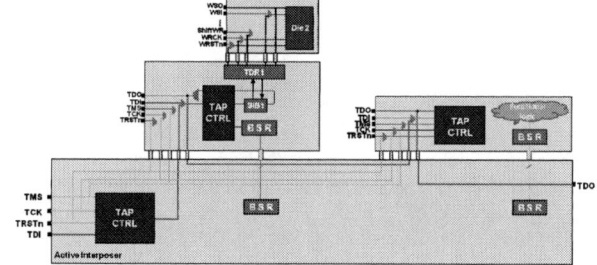

Figure 7. 3D DFT Architecture for a multi-tower 3D IC

An example of a multi-tower 3D SIC is shown in Fig.7, where two dies are stacked within an active interposer and one die stacked on the top of another die.

The interposer with the 2 dies above it forms a serial chain of JTAG TAP controllers. The top-left die embeds IEEE 1687 logic that manages the test of the die above it which has an IEEE 1500 wrapper and considered as a slave of that die.

The example shows that the uniform test architecture was used for the interposer with 2 dies and the heterogeneous test architecture was used to handle the test of a die stacked above another die. This proves the complementarities of the two proposed 3D DFT architectures.

D. Features of our proposed 3D DFT architecture based on IEEE 1687 and automatic die-detection mechanism

In addition to enabling the test of all the components of the 3D SIC at all binding levels, our proposed 3D DFT architecture allows new features. Thanks to the use of the IEEE 1687 logic and associated high level languages ICL and PDL, the 3D DFT architecture allows

➤ High flexibility of the test of instruments thanks to the dynamic control of SIBs and TDRs, which is advantageous in 3D context especially for test scheduling.

➤ Test pattern retargeting by the mean of the ICL and PDL languages at 2 different levels. The first level is intra-die level where patterns can be retargeted from core level to die level[21] and the second level is inter-die level where patterns are retargeted from die level (2D level) to 3D SIC level (3D level) which is also called 2D to 3D test pattern retargeting [22].

➤ Automatic reconfiguration of the test paths of the 3D system thanks to the use of an automatic die detection which allows a dynamic reconfiguration of test paths during any testing level: pre-bond, mid-bond or post-bond without requiring any configuration in the 3D SIC [20].

E. Implementation of our proposed 3D DFT architetcure on a test chip prototype

We implemented our proposed 3D DFT proposal in a real 3D active interposer called INTACT, designed in the CEA/Leti. The INTACT prototype is composed of an active interposer, on which 6 identical dies are stacked.

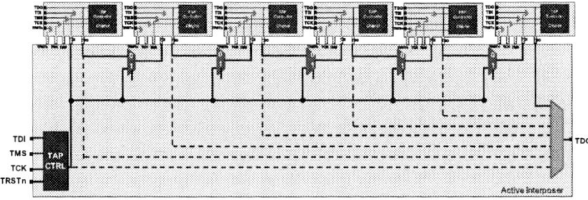

Figure 8. 3D DFT architecture for the INTACT prototype

The proposed 3D DFT architecture enables the test of all components at all bonding levels, for both the interposer and the stacked dies.

At interposer level, the pre-bond test is done by configuring the multiplexer of the output TDO to select out the interposer's TDO, and for post-bond (mid-bond) test the multiplexer is configured to drive out the last bonded die's TDO.

At die level, the pre-bond test is done through pads, multiplexers (in pink color) are configured to select inputs from pads. For post-bond (and mid-bond) multiplexers (in pink color) are configured to select signals from TSVs and then build the serial chain between TAP controllers. The signal TDO of all dies is always driven to the pad and to the TSV.

The JTAG signals: TCK, TRST, and TMS are common to all dies and the interposer. The TDI-TDO chain is formed serially from the interposer to the first dies, to the next, till the last bonded one.

IV. 3D DFT AND NECESSITY OF TEST SCHEDULING

In addition to the challenges of the Design For Test, there are other challenges for 3D SICs. Among them, the thermal issue is probably the most critical one. The vertical heat dissipation paths in 3D SICs are long and cause the increase of temperature in the 3D SIC. High temperature has adverse impacts on the performances of integrated circuits. The interconnect delay becomes slower while the driving strength of a transistor decreases with increasing temperature. Leakage power has an exponential dependence on the temperature and increasing on-chip temperature can even result in thermal runaways.

The aim of the 3D test scheduling task is to optimize (lower) the system test time with respect to power consumption and thermal dissipation constraints. For instance, if we consider a tier containing N instruments and M scan chains as shown in Figure. At pre-bond level, a test scheduling can be done by launching the test of all the scan chains and instruments in parallel, with respect to the thermal heating issue at tier-level.

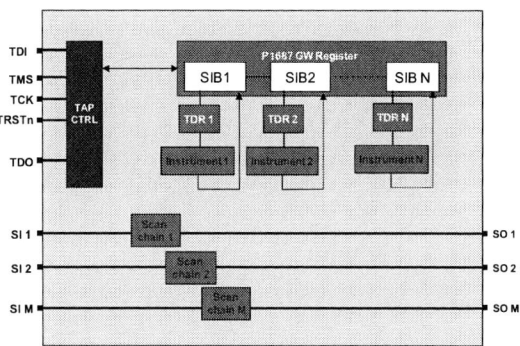

Figure 8. Pre-bond scheduling, testing all the components

At post-bond level, thermal issues are stronger and the test scheduling by launching all the tests in parallel can be too stressful and damage the circuit.

For instance, if the same tier including N instruments and M scan chains is the bottom die within a 3D stack. The test scheduling performed at pre-bond level can be not efficient at post-bond level because of the power and thermal issues that are accentuated in post-bond level.

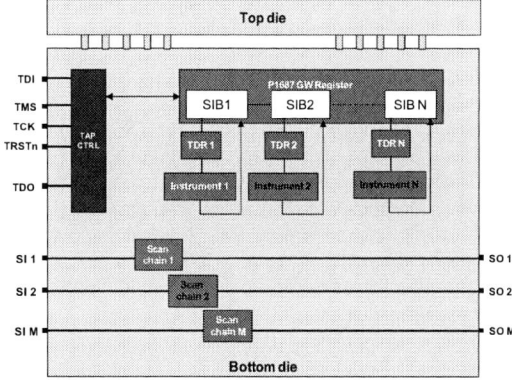

Figure 9. Post-bond scheduling, testing some instruments and some scan chains

In such a case, running the test of all the instruments and all the scan chains in parallel can be prohibited. The post-bond test scheduling needs to be updated by running the test of only some instruments and some scan chains within the bottom die.

3D scheduling must take into account 1/ the topology of the test bus for feeding test data to scan chains in the cores as discussed above, and 2/ the control architecture for selecting instruments (e.g. BIST engines) through SIBs .As explained in the previous chapter, our proposed 3D DFT architecture is flexible in terms of selecting instruments thanks to the use of the IEEE P1678 and thus does not constrain test concurrency.

Figure 9 presents the proposed design for controlling shift-in and out operations in scan chains with a scan enable signal (SE) if and only if the instrument embedding these scan chains is selected through its SIB.

978-1-4799-8720-7/15 $31.00 © 2015 IEEE

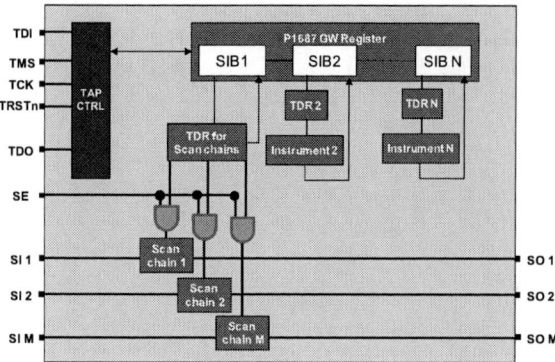

Figure 9. control of scan chains using IEEE 1687 TDR

The length of the associated TDR in this case is equal to the number of scan chains in the target instrument. As a result, each scan chain has a dedicated "SE". The advantage of the local generation of "SE" signals is to offer more flexibility in terms of test parallelism since scan chains can be used simultaneously or serially.

With this architecture in mind, a test scheduling procedure can be set up since it is now possible to evaluate the area overhead cost of any proposed schedule.

V. CONCLUSION AND FUTURE WORK

We covered different aspects of the DFT challenges to address for 3D ICs. The first aspect concerns the pre-bond test of TSVs where we presented a BIST solution based on ring oscillators. The proposed solution was implemented on a prototype test chip, showing its low implementation cost, and real characteristic measurements on TSV matrices showed that the proposed approach allows to differentiate faulty from fault-free TSVs.

The second aspect of the 3D DFT challenges concerns the test architecture. We proposed an architecture based on IEEE 1687 which allows the test of all the components at all the 3D bonding levels and fulfils other 3D test requirements including 3D test pattern retargeting a dynamic selection of the instruments under test. Moreover the proposed 3D DFT architecture is scalable and can be adapted to specific 3D ICs technologies especially with interposers (passive or active) and/or multi-tower 3D technologies. The last discussed aspect of 3D test challenges concerns the relation between DFT architecture and test scheduling. We have shown that the proposed 3D DFT architecture does not block any test scheduling at any test level and presented and illustrative example of 3D test scheduling that can be done at the different stages of the 3D stacking process. This work opens perspectives towards 3D test scheduling of various instruments "IPs" with physical constraints such as power and thermal issues.

REFERENCES

[1] WideIO JEDEC standard, see http://www.jedec.org/

[2] E.J. Marinissen and Y. Zorian, "Testing 3D chips containing through-silicon vias", in Proc. ITC, 2009, pp.1-11.

[3] Dean L. Lewis, HsienHsin S. Lee "A Scan Island Based Design Enabling Pre-bond Testability in Die Stacked Microprocessors", ITC 2007

[4] Xiaoxia Wu, Paul Falkenstern, Yuan Xie "Scan Chain Design for Three-dimensional Integrated Circuits (3D ICs)", ICCD 2007

[5] Li Jiang, Lin Huang and Qiang Xu, "Test Architecture Design and Optimization for Three-Dimensional SoCs", DATE'09

[6] E.J.Marinissen, J.Verbree, M.Konijnenburg, "A structured and scalable test access architecture for TSV-based 3D stacked ICs", in proceedings of VTS 2010, pp. 269 – 274.

[7] E.J.Marinissen, C.-C. Chi, J.Verbree, M.Konijnenburg, "A Standardizable 3D DFT Architecture", 3D-TEST'10

[8] http://grouper.ieee.org/groups/3Dtest/

[9] Brandon Noia and Krishnendu Chakrabarty, "Pre-Bond Probing of TSVs in 3D Stacked ICs", in proc International Test Conference 2011, pp 1-10

[10] Brandon Noia and Krishnendu Chakrabarty "Identification of Defective TSVs in Pre-Bond Testing of 3D ICs", in proc Asian Test Symposium ATS'11, pp 187-194

[11] Li Jiang, Qiang Xu, Bill Eklow and Krishnendu Chakrabarty "3D IC/TSV Probe Test using Adhesive Silicon Interposer" workshop 3D DATE'2012.

[12] Y. Lou, Z. Yan, F. Zhang, and P.D. Franzon, "Comparing Through-Silicon-Via (TSV) Void/Pinhole Defect Self-Test Methods", J. Electronic Testing, 2012, pp.27-38.

[13] Yuan Chen, Cheng-Wen Wu, Ding-Ming Kwai "On-Chip TSV Testing for 3D IC before Bonding Using Sense Amplification", in proceedings of ATS' 2009, pp 450-455

[14] Tsai M, Klooz A, Leonard A, Appel J, Franzon P, "Through silicon via (TSV) defect/pinhole self test circuit for 3D-IC", in proceedings of 3D'IC, 2009, pp 1-8

[15] Sergej Deutsch and Krishnendu Chakrabarty, "Non-Invasive Pre-Bond TSV Test Using Ring Oscillators and Multiple Voltage Levels" in proc DATE'2013, pp 1065-1070

[16] G.Di Natale, M.L.Flottes, B.Rouzeyre, H.Zimouche, "Built-In-Self-Test for Manufacturing TSV Defects before bonding" in proc VTS'2014, pp 1-6

[17] Y.FKIH, P.VIVET, B.ROUZEYRE, ML.FLOTTES, G.DINATALE "A 3D IC BIST for pre-bond test of TSVs using ring oscillators" IEEE International New Circuits and Systems Conference (NEWCAS), 2013

[18] Han Ke, Deng Zhongliang, Huang Jianming "Boundary Scan with Parallel Test Access Mechanism", ICEMI '09, pp 4-70 - 4-73

[19] Y.Fkih, P.Vivet, B.Rouzeyre, M-L.Flottes, G.Di Natale, J. Schloeffel, "3D Design for Test Architectures Based on IEEE P1687", workshop 3D Test'13

[20] Y.Fkih, P.Vivet, B.Rouzeyre, M-L.Flottes, G.Dinatale, "A JTAG based 3D DFT architecture using automatic die detection", IEEE PRIME'13, pp 341 - 344

[21] F.G.Zadegan, U.Inglesson, G.Carlsson, E.Larsson "Reusing and Retargeting On-Chip Instrument Access Procedures in IEEE P1687", IEEE CEDA 2012

[22] Y.FKIH, P.VIVET, B.ROUZEYRE, ML.FLOTTES, G.DINATALE, J.SCHLOEFFEL "2D to 3D Test Pattern Retargeting using IEEE P1687 based 3D DFT Architectures", in proc ISVLSI 2014, pp 386-391

2015 IEEE Computer Society Annual Symposium on VLSI

Thermal Aspects and High-Level Explorations of 3D stacked DRAMs

Christian Weis, Matthias Jung,
Omar Naji and Norbert Wehn
University of Kaiserslautern
Kaiserslautern, Germany

Cristiano Santos and Pascal Vivet
CEA Leti
Grenoble, France

Andreas Hansson
ARM Ltd
Cambridge, UK

Abstract—**DRAMs are very sensitive to temperature changes as they use capacitors as volatile and leaky bit storage elements. 3D stacking of heterogenous dies provokes more and more challenges, such as high power densities and thermal dissipation, and has a much stronger impact on the retention time of 3D stacked WIDE I/O DRAMs that are placed on top of an MPSoC. Consequently, it is very important to study the temperature behavior of WIDE I/O DRAMs and explore on high-level with advanced modeling how the thermal issues can be mitigated. In this paper, we demonstrate thermal modeling of 3D integrated ICs and we further provide detailed measurements on temperature-dependent bit error rates of WIDE I/O DRAMs based on real silicon (WIOMING chip) and high-level explorations with advanced modeling and tools, such as gem5 (full system-simulator), DRAMSys and DOCEAs AceThermalModeler^{TM}(ATM), to obtain mitigation paths for thermal problems in 3D stacked DRAMs.**

Keywords—*3D integration, thermal modeling, TSV, measurements, DRAM, wide I/O, retention time, error models, power down*

I. INTRODUCTION

In current or future tightly coupled processing and memory systems the energy and thermal dissipation are restricting today's application performance on smartphones as well as high-end servers. Advanced fabrication processes that use 3D packaging based on TSV technology enable tighter integration of systems. These start to break down the memory and bandwidth walls. However, this increases largely the power density and reduces the heat dissipation properties of the aggressively thinned dies to enable reliable TSV production. In addition, poorly conductive adhesive materials used to bond dies together considerably contribute to increase the vertical thermal resistance. These thermal issues of 3D ICs cannot be solved by tweaking the technology and circuits alone. In fact, a 3D stacked SoC aggravates the thermal crisis, which can provoke errors in circuits and especially in DRAMs as they are highly sensitive to temperature changes and have to be refreshed regularly due to their charge-based bit storage property (capacitor). Thus, advanced thermal models (detailed and compact models) and high-level simulators, such as gem5 [1] and DRAMSys [2], are required to investigate current or future processing and main memory systems (based on DRAMs). In this paper, we highlight three major aspects when tackling the issues of closely coupled 3D integrated circuits and systems:

- First, we present the WIOMING 3D test chip and the derived thermal compact model. We created the models with an advanced thermal modeling tool - the DOCEA AceThermalModeler^{TM}(ATM) - and verified these against real silicon.

- Second, we detail our investigations on a retention-aware DRAM bit error model that considers variable retention times (VRTs) and data pattern dependence.
- Finally, we show how the power consumption of the DRAM subsystem can be reduced with the help of a sophisticated power-down policy, which is beneficial to mitigate thermal issues in a 3D DRAM cube.

II. THERMAL MODELING

In this Chapter, we first introduce generally the test chip vehicle WIOMING 3D IC in Section II-A and we detail the advanced thermal model derived from several experiments. Section II-D highlights the truth about thermal TSV usage.

A. Memory-on-Logic 3D IC Test Circuit

The 3D test chip used for thermal measurements is the WIOMING SoC circuit [3]. This packaged memory-on-logic 3D circuit is fabricated in a 65nm technology where dies are stacked in a face-to-back configuration with an underfill material covering the gap between dies. The logic die (bottom) is thinned down to 80μm to expose the integrated TSVs (TSV middle). Four TSV arrays are placed at the center of the die, where each TSV array contains 254 TSVs connected through aligned μ-bumps to a WIDE I/O compatible DRAM memory. The WIDE I/O DRAM with 4 channels ($4\times128 = 512$ I/Os) on top of the WIOMING SoC die is based on the device from Samsung [4]. Figure 1 brings additional circuit and technology information (further technology details can be found in [5]).

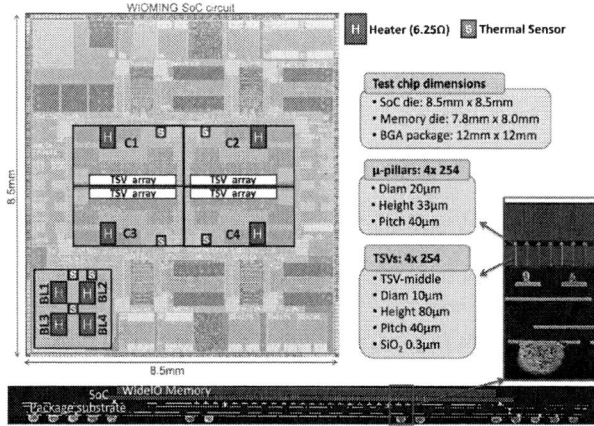

Fig. 1: Circuit and Technology details: Floorplan of the WIOMING 3D IC including 4 heaters and 4 thermal sensors in the center area (C1-C4) and 4 heaters and 3 thermal sensors in the bottom-left corner (BL1-BL4) & Cross-Sections of the 3D Package

978-1-4799-8720-7/15 $31.00 © 2015 IEEE

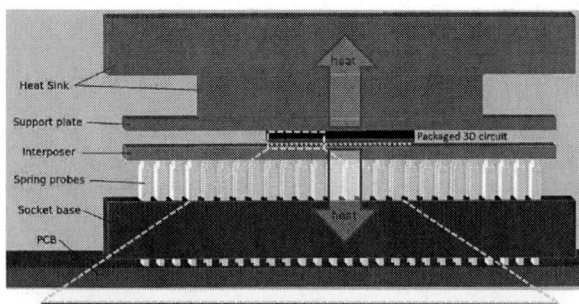

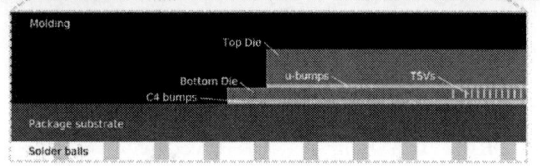

Fig. 2: Physical Representation of the Chip-Package-Board System in ATM

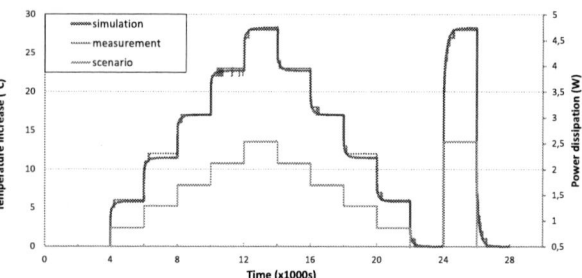

Fig. 3: Transient Temperature Response for a Staircase Power Dissipation Scenario with four Active Heaters (C1-C4)

This circuit is instrumented with eight resistive heaters (poly resistance) to emulate hotspot power dissipation. Each heater is individually controlled by the embedded software and can dissipate up to $0.659\,W\,(0.37\,kW/cm^2)$. The integrated thermal sensors are monitored in real time with $1\,°C$ of temperature resolution. Sensors present an accuracy of $\pm 1\,°C$ for the calibration temperature ($25\,°C$), degrading to $\pm 4\,°C$ at $100\,°C$. Heaters and sensors placed at the center area (C1-C4) are indeed to evaluate the impact of hotspots on the DRAM performance, while heaters placed at the bottom-left corner (BL1-BL4) are used in this work to emulate a hotspot behavior produced by a quad-core processor (such as an ARM Cortex A9, for instance).

The experimental setup used for thermal characterization includes the packaged thermal test circuits, a small PCB interposer and a BGA socket mounted on a PCB including additional peripheral circuitry. This experimental setup, which is fully controlled via embedded software, is placed inside a climatic chamber to keep the environment temperature stable during the thermal measurements.

B. Thermal Model Implementation

The DOCEA AceThermalModeler™ (ATM) software [6] is used to generate a dynamic compact thermal model for the chip-package-board system using a numerical finite difference method (FDM) approach. The generated thermal model is based on a complete geometric representation and thermophysical properties of the experimental setup including the packaged test circuits, socket and board elements. The heat transfer coefficients are applied on the system boundary interfaces to mimic the boundary conditions in the climatic chamber. The heat transfer is modeled via full 3D heat diffusion in solid materials, with support for anisotropic thermal conductivity properties and with no restrictions to the heat flow paths. ATM offers a Python API in order to automate and control the thermal model generation flow. Figure 2 shows the geometry representation of the complete modeled system with details of the packaged 3D circuit and the main heat flow paths.

Comparisons of the measurement data and simulation results for multiple power dissipation scenarios presents an average difference of 3.96% and worst case lower than 14% in case of steady-state response. Figure 3 shows that the simulated transient response is in good agreement with measured data. The generated CTM is able to properly model complete systems with multi-length scales, effectively accounting for the hotspot effects on the spatial temperature distribution. More details of the thermal modeling methodology and the silicon correlation results can be found in [7].

C. Temperature Map of the WIDE I/O DRAM Die

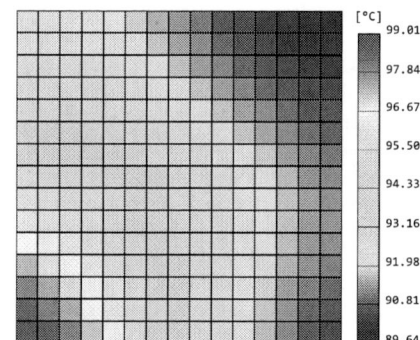

Fig. 4: Temperature Map of the WIDE I/O DRAM die

Due to in-appropriate temperature sensors within the DRAM (not correctly enabled from SoC and with too large inaccuracy), we cannot obtain an accurate thermal map by the embedded software. For that reason we used Aceplorer in conjunction with AceThermalModeler (ATM) from DOCEA Power [6] to create an accurate temperature distribution map of the WIDE I/O DRAM die. Such thermal model [8] has been proven to be fast enough for rapid system level exploration with a correct correlation between simulation and silicon measurements. Thus, the simulated thermal map is sufficiently accurate, to feed the next model, which is our proposed SystemC-TLM2.0 DRAM bit error model. To obtain the temperature map of the DRAM die (top die) we modified the ATM-model of the WIOMING 3D IC [8]. We set the power of the heaters to their highest values in order to achieve the highest temperature for Chip 1 (104°C). Figure 4 shows the temperature map of the DRAM die. We see an expected overall temperature decrease of top die (DRAM) compared to the sensors on the bottom die and as expected Channel 3 is by far the hottest area of the DRAM die.

978-1-4799-8720-7/15 $31.00 © 2015 IEEE

D. Thermal TSVs: Facts and Fiction

The use of TSV insertion techniques has been reported in previous studies for thermal mitigation in 3D stacked circuits [9], [10]. However, the thermal blockage effect [11] caused by the thin oxide layer isolating TSVs from the silicon substrate causes the TSVs to be not effective for thermal mitigation. In case of hotspot dissipation, the poor thermal conductivity properties of the SiO_2 layer used for TSV isolation dominate the thermal impact of the TSV arrays. Figure 5 brings simulation results to show the thermal impact of the TSV arrangement used in the WIOMING circuit for compatibility with the WIDE I/O protocol. A hotspot with power density of $3\,kW/cm^2$ is considered for simulations at two distinct positions: a) at the center of the silicon die and b) in between two TSV arrays. In order to distinguish individual contributions, we simulate cases with only TSVs and no μ-bumps, with only μ-bumps, and with both TSVs and μ-bumps. The reference case used for comparisons does not have TSVs or μ-bumps. Simulation results show a temperature increase of 1.24% for a) and 3.04% for b) in case of having only TSVs. Notice that the hotspot is more fenced by the TSV arrays in the second case, thus being more affected by the resulting lateral thermal blockage effect. In the case of the 3D stack with μ-bumps but without TSVs, the presence of μ-bumps in the underfill layer improves the thermal coupling between dies and brings a temperature reduction of 13% for hotspot b). These results confirm that temperature reduction comes, in fact, from the better thermal coupling between dies caused by the μ-bumps in cases where thermal TSVs are considered as vertical connections spanning across multiple stacked dies. Additionally, TSV arrays may even provoke exacerbated hotspots in some cases. Numerical simulations and experimental data from a dedicated thermal test chip are presented in [12] for an extensive study of the effectiveness of TSV arrays for heat dissipation in 3D integrated circuits.

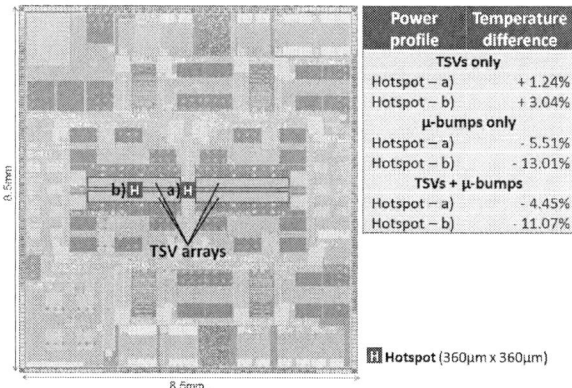

Power profile	Temperature difference
TSVs only	
Hotspot – a)	+ 1.24%
Hotspot – b)	+ 3.04%
μ-bumps only	
Hotspot – a)	- 5.51%
Hotspot – b)	- 13.01%
TSVs + μ-bumps	
Hotspot – a)	- 4.45%
Hotspot – b)	- 11.07%

Hotspot (360μm x 360μm)

Fig. 5: Simulation Results of the Thermal Impact of TSV arrays on the WIOMING Circuit for Hotspot Power Dissipation Profiles

III. MEASUREMENTS AND BIT ERROR MODELING

The 3D MPSoC with WIDE I/O DRAM presented in [3] (WIOMING chip) is used in this Chapter for our investigations. As described in Section II-A this 3D IC features thermal sensors and heaters, which are used for online monitoring of temperature and the tuning of thermal models. We employ

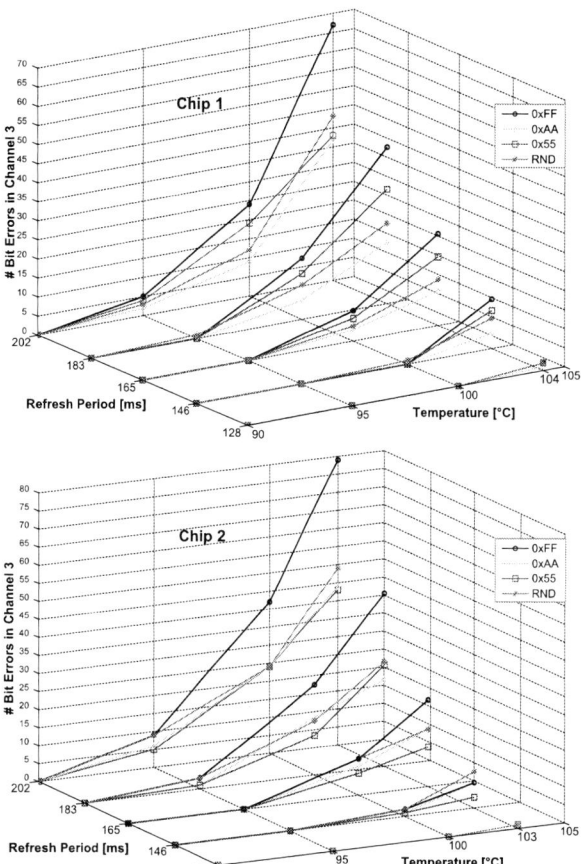

Fig. 6: Measurement Results with Fixed Refresh Periods using test b) for Chip 1 and 2

the MAGALI (SoC) evaluation board and two WIOMING devices (Chip 1 and 2) to investigate the temperature dependent retention errors (bit flips). The WIDE I/O DRAM of the WIOMING 3D IC is based on the chip presented by Samsung in [4]. We run two different tests: a) with a refresh pause of one second and b) with a fixed refresh period to obtain our measurement results. During the important test time it is ensured that the temperature is at least stable for 20 seconds. Additionally, we investigate different data pattern reaching from simple 0xFF, 0x55, 0xAA to random pattern (RND).

A. WIDE I/O DRAM Measurement Results

Figure 6 shows the measurement results of the two different chips. We see clearly in the two plots the data dependence of the error rates. Although the two chips behave differently for the exact error numbers, the overall trend can be seen

TABLE I: Coverage at a Refresh Period of 202ms and T=103/104°C

Data pattern	0xFF	0xAA	0x55	RND
Chip 1 (%)	82	45	46	52
Add. cov. errors (%)	0	9.8	0	11.0
Chip 2 (%)	84	47	46	52
Add. cov. errors (%)	0	6.4	4.3	7.4

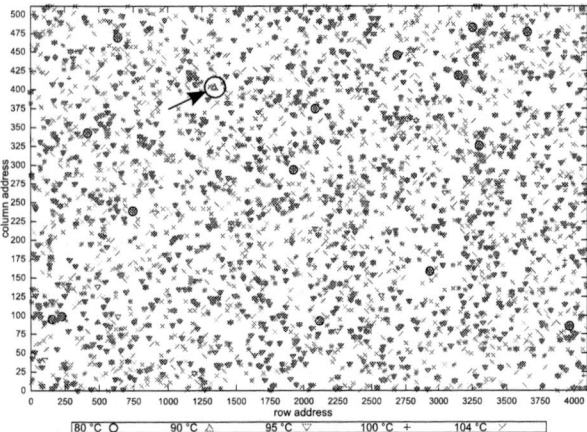

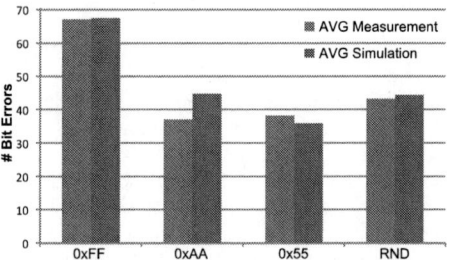

Fig. 8: Comparison of Simulation and Measurements for a Refresh Period of 202ms and a Temperature of 90 °C

Fig. 7: Scatter-plot of accumulated Bit Errors at different Temperatures for Channel 3, Bank 1 with data pattern `0xFF` and test a)

in Table I. The `0xFF` pattern detects ≈83% of the retention errors, the others are able to find only 46-52%. The additional error coverage of the test pattern for each chip compared to the `0xFF` test is shown in the rows "Add. cov. errors" in percentage. The `RND` and the `0xAA` test contribute most to increase the coverage.

In Figure 7 we see the effect of VRTs (variable retention times) for DRAM cells. Not all cells are permanent failing beginning from a specific temperature. For instance, the red triangle (90°C) fail shown in the circle is a typical VRT fail, as it disappears at higher temperature.

B. Retention Time Bit Error Modeling

DRAM cells use capacitors as volatile and leaky bit storage elements. The retention time is defined as interval between two refreshes in which the cell can safely maintain its data. It is well known that the retention time depends inverse exponentially on the temperature. In 3D stacking, the challenges of high power densities and thermal dissipation are exacerbated and have a much stronger impact on the retention time of 3D stacked WIDE I/O DRAMs that are placed on top of an MPSoC.

Consequently, a retention error aware DRAM model is key to analyze, for instance, the impact of lower refresh rates or disabled refresh on the executed application. Especially for error resilient applications this can be exploited, to save energy [13]. We measured the retention times of the WIDE I/O DRAM devices and observed data pattern dependencies (compare Figure 6) and variable retention times. These data are used to create a DRAM retention time error model [14]. Figure 8 shows the comparison of the averaged results of 30x repeated model simulations and real measurements of the WIDE I/O DRAM. We see that our error model implements the correct trend for the data pattern dependency and has bit error rates near to the measured values. The overhead of the retention-aware DRAM bit error model with respect to the simulation execution time of DRAMSys is in average only 30%. Thus, our proposed model can be used for Monte-Carlo-Simulations and is suitable for the early investigations on the

temperature vs. retention time trade-off in future 3D stacked MPSoCs with 3D DRAMs.

IV. HIGH-LEVEL INVESTIGATIONS AND SIMULATIONS

In modern DRAMs, intelligent power-down techniques are used to reduce the energy consumption of several applications. Unfortunately, in many cases, the saved energy comes at the cost of system performance. In this Chapter, we investigate the impact of a staggered power-down approach presented in [15] on the performance and energy consumption of server and mobile systems. Further, we extend this power-down approach to a bankwise-staggered policy for 3D stacked DRAMs, such as WIDE I/O, in Section IV-B. This policy can be beneficial to mitigate thermal issues in a 3D DRAM cube attached to an underlying multi-core processing system.

A. Staggered Power-Down for Standard DDR3 DRAMs

A realistic application in the server world is a server/client system where the client sends requests to the server and the server responds. As an example for such applications, this experiment considers memcached-Twitter with a server configuration in gem5. The system configured in gem5 is made up of two systems. One system represents the client system sending requests to the server system which in term processes the requests and responds back to the client. Since the experiment performed in this work explores the DRAM energy for the server system, Table II shows the configuration parameters for this system. The server system configured in gem5 is a made up of a single core 2GHz out-of-order (OoO) CPU with three levels of caches. The first level of cache consists of a 32KB data cache and a 32KB instruction cache. The second level and the third level are a 256KB and 1MB cache respectively. The DRAM is configured as a single channel single rank DDR3 with four devices per rank. This provides 2GB of main memory.

The power-down technique considered in this paper works as following: After a read or write, the DRAM which is in the active state switches directly to the active low-power (active power-down = `ACT_PDN`) state if no read or write request is buffered. When the next scheduled refresh arrives, the DRAM switches to precharge power-down (`PRE_PDN`) if the conditions for switching are still valid since before performing a refresh, all banks should be precharged. If there is still no read or write request, the DRAM issues a self-refresh (`SREF`) entry by the arrival of the next periodic refresh. The DRAM remains in the self-refresh state until woke up by an incoming read or write request.

978-1-4799-8720-7/15 $31.00 © 2015 IEEE

TABLE II: Gem5 Server and Mobile Systems Configuration Parameters

Parameter	Server	Mobile
CPU	1 (OoO) Core	1 (OoO) Core
CPU Frequency	2 GHz	1.7 GHz
OS	Ubuntu 12.04	Andriod Jellybeans
DL1 (Data Cache Level 1) Size	32 KB	32 KB
IL1 (Instruction Cache Level 1) Size	32 KB	32 KB
L2 (Cache Level 2) Size	256 KB	1 MB
L3 (Cache Level 3) Size	1 MB	no level 3 cache
Number of DRAM Channels	1	1
Number of DRAM Ranks	1	1
Number of DRAM Devices per Rank	4	1
DRAM Type	DDR3	LPDDR3
DRAM Size	2 GB	512 MB

Because of the relatively large wake-up latencies, the system performance is sensitive to the length of the DRAM idle periods. In other words, if the executed application (also depending on the CPU and its cache system) provokes many short idle periods, the DRAM keeps on sleeping and waking-up, thus a negative impact on performance can be observed. However, if the system offers few long idle periods, the DRAM has the opportunity to sleep for a long period and therefore the remarkable energy savings due to power-down will not be at the cost of system performance.

Figure 9 shows the distribution of the time spent in the power-down states while using the staggered power-down approach for the memcached-Twitter server system configured in gem5. As seen in Figure 9(a), for an open page policy, the DRAM spends 97% of its power-down time in the active power-down state. This indicates that the idle periods offered by the system are composed of many short periods since the idle periods where not long enough for the DRAM to switch to self-refresh. The same conclusion could be made while analyzing the results of a closed page policy presented in Figure 9(b). Due to the short idle periods, the DRAM spends most of its idle time in the precharge power-down state.

In the considered server system, switching to low-power leads to a remarkable energy saving as shown in Figure 10(b). A 25% and 19% saving is observed for an open and a closed page policy respectively. As expected, due to the relatively short idle periods, switching to power-down leaves a negative impact on performance in the analyzed server system. Figure 10(a) shows a 10% and 5% degradation in the number of analyzed requests while running the simulation for one second configured to an open page and closed page policy respectively.

We analyze further the impact of the mentioned power-down approach on the performance and DRAM energy of a mobile system configured in gem5 while running BBench.

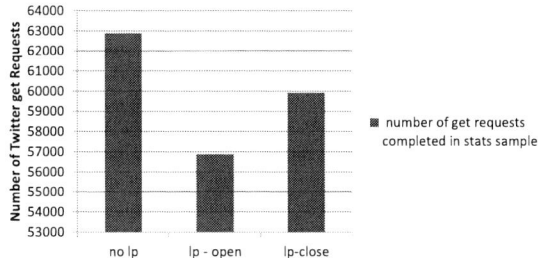

(a) Impact of Staggered Power-down on Performance

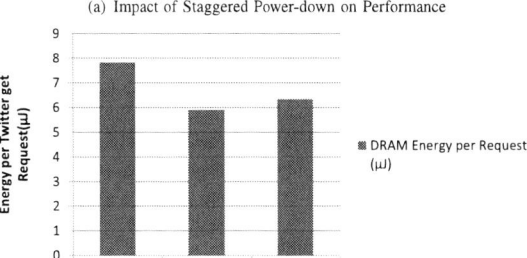

(b) Impact of Staggered Power-down on DRAM Energy

Fig. 10: Impact of Power-down on Server System Performance and DRAM Energy

The system is composed of a single out-of-order (OoO) core running at 1.7GHz with 2 level of caches; a 32KB level 1 data cache, a 32KB level 1 instruction cache and a 1MB level 2 cache. The operating system is Android Jellybeans and the considered DRAM is a LPDDR3 device (see Table II). Unlike the results of the analyzed server system, the DRAM spends most of the power-down time in the self-refresh state (76%) as depicted in Figure 11(a). This indicates that the idle periods are composed of long idle periods. Therefore, as expected, the large DRAM energy saving (40%) has no negative impact on the system performance since the BBench run-time remains constant as observed in Figure 11(b).

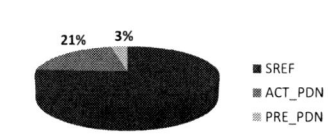

(a) Distribution of Power-down Time for the Mobile System

(b) Impact of Staggered Power-down on DRAM Energy and Performance

Fig. 11: Impact of Power-down on the Mobile System in gem5

(a) Staggered Power-down for Open Page Policy   (b) Staggered Power-down for Closed Page Policy

Fig. 9: Distribution of Power-down States for memcached-Twitter Server System

978-1-4799-8720-7/15 $31.00 © 2015 IEEE   613

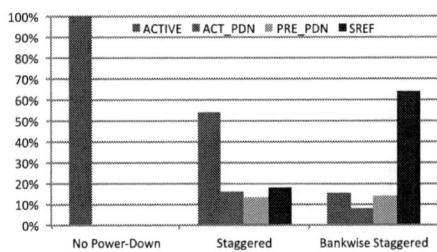

Fig. 12: Power-down Usage for Different Power-down Strategies

TABLE III: Staggered Bankwise Power-down

	No Power-Down	Staggered	Bankwise Staggered
Avg. Power	69.76 mW	65.00 mW	60.53 mW
Avg. Latency	29.80 ns	50.90 ns	46.30 ns

B. Bankwise Staggered Power-down for WIDE I/O DRAMs

The before presented staggered power-down policy seems to be contradicting to the bankwise refresh strategy presented in [16]. The bankwise refresh strategy tries to reduce the number of refreshes per bank, but the staggered power-down needs a non-bankwise refresh on all banks as trigger for switching the power-down states. However, both techniques can be combined when the DRAM is able to power down the banks independently.

We run a representative trace (chstone-mips) in three different modes (no power-down, staggered, bankwise staggered) to quantify the impacts for the staggered bankwise power-down approach. Figure 12 shows the power-down usage of the different strategies. We see that the active periods over all banks are largely reduced (down to 14%) while using bankwise staggered power-down. Contrary, the time the DRAM banks are in SREF increases to 63%. While a DRAM bank is in SREF another bank can operate on the interface (ACTIVE). Due to this behavior, the expected power savings are limited, since the I/O part of the DRAM device contributes significantly to the overall power consumption. In Table III the average power and request latency of the three modes are shown. We see for the bankwise staggered power-down an improvement in average power of 13.4% and 7.9% compared to no power-down and staggered, respectively. Additionally, the average request latency is recovered by 9.1% compared to the staggered policy.

V. CONCLUSION AND OUTLOOK

We presented the WIOMING 3D test chip and the derived thermal models. The created models are verified against real silicon and enable fast and accurate thermal coupled simulations. Further, we measured the temperature and refresh period depending DRAM bit error rates of two WIDE I/O DRAM chips. These investigations led to a retention-aware DRAM bit error model that considers variable retention times (VRTs) and data pattern dependence. This model permits early investigations and explorations on the temperature vs. retention time trade-off in future 3D stacked MPSoCs with 3D DRAMs, such as WIDE I/O. Finally, we demonstrated how the power consumption of the DRAM subsystem can be reduced with the help of a sophisticated power-down policy, which is beneficial to mitigate thermal issues in a 3D DRAM cube attached to a heterogenous MPSoC.

ACKNOWLEDGMENTS

The authors thank the companies DOCEA Power and SYNOPSYS for their great support. This work was partially funded by the German Research Foundation (DFG) as part of the priority program Dependable Embedded Systems SPP 1500 (http://spp1500.itec.kit.edu), the DFG grant no. WE2442/10-1.

REFERENCES

[1] Nathan Binkert, et al. *The gem5 simulator*. SIGARCH Comput. Archit. News, 39(2):1–7, August 2011.

[2] Matthias Jung, et al. *TLM modelling of 3D stacked wide I/O DRAM subsystems: a virtual platform for memory controller design space exploration*. In Proceedings of the 2013 Workshop on Rapid Simulation and Performance Evaluation: Methods and Tools, RAPIDO '13, pages 5:1–5:6, New York, NY, USA, 2013. ACM.

[3] Denis Dutoit, et al. *A 0.9 pJ/bit, 12.8 GByte/s WideIO memory interface in a 3D-IC NoC-based MPSoC*. In VLSI Technology (VLSIT), 2013 Symposium on, pages C22–C23. IEEE, 2013.

[4] Jung-Sik Kim, et al. *A 1.2V 12.8GB/s 2Gb mobile Wide-I/O DRAM with 4x128 I/Os using TSV-based stacking*. In Solid-State Circuits Conference Digest of Technical Papers (ISSCC), 2011 IEEE International, pages 496–498, Feb 2011.

[5] P. Coudrain, et al. *Towards efficient and reliable 300mm 3D technology for wide I/O interconnects*. In Electronics Packaging Technology Conference (EPTC), 2012 IEEE 14th, pages 330–335, Dec 2012.

[6] DOCEA Power. *AceThermalModeler and Aceplorer*. http://www.doceapower.com, 2014, last access: 20.01.2015.

[7] C. Santos, et al. *Thermal modeling methodology for efficient system-level thermal analysis*. In Custom Integrated Circuits Conference (CICC), 2014 IEEE Proceedings of the, pages 1–4, Sept 2014.

[8] C. Santos, et al. *System-level thermal modeling for 3D circuits: Characterization with a 65nm memory-on-logic circuit*. In 3D Systems Integration Conference (3DIC), 2013 IEEE International, Oct 2013.

[9] J. Cong et al. *Thermal via planning for 3-D ICs*. In Computer-Aided Design, 2005. ICCAD-2005. IEEE/ACM International Conference on, pages 745–752, Nov 2005.

[10] B. Goplen et al. *Placement of thermal vias in 3-D ICs using various thermal objectives*. Computer-Aided Design of Integrated Circuits and Systems, IEEE Transactions on, 25(4):692–709, April 2006.

[11] Yibo Chen, et al. *Analysis and mitigation of lateral thermal blockage effect of through-silicon-via in 3D IC designs*. In Low Power Electronics and Design (ISLPED) 2011 International Symposium on, pages 397–402, Aug 2011.

[12] C. Santos, et al. *Using TSVs for Thermal Mitigation in 3D Circuits: Wish and Truth*. In IEEE 3DIC, Cork, Ireland, 2014.

[13] Christina Gimmler-Dumont et al. *A Cross-Layer Reliability Design Methodology for Efficient, Dependable Wireless Receivers*. submitted to ACM Transactions on Embedded Computing Systems, 2013.

[14] Christian Weis, et al. *Retention Time Measurements and Modelling of Bit Error Rates of WIDE I/O DRAM in MPSoCs*. In Proceedings of the conference on Design, Automation & Test in Europe. European Design and Automation Association, 2015.

[15] M. Jung, et al. *Optimized active and power-down mode refresh control in 3D-DRAMs*. In Very Large Scale Integration (VLSI-SoC), 2014 22nd International Conference on, pages 1–6, Oct 2014.

[16] M. Sadri, et al. *Energy Optimization in 3D MPSoCs with Wide-I/O DRAM Using Temperature Variation Aware Bank-Wise Refresh*. In Design, Automation and Test in Europe Conference and Exhibition (DATE), 2014, pages 1–4, March 2014.

978-1-4799-8720-7/15 $31.00 © 2015 IEEE

Interconnect Challenges for 3D Multi-cores: from 3D Network-on-Chip to Cache Interconnects

P. Vivet, C. Bernard, E. Guthmuller, I. Miro-Panades, Y. Thonnart, F. Clermidy
CEA-LETI, MINATEC Campus, 17 rue des Martyrs, 38000 Grenoble, France

Abstract—With the era of massive multi-core architecture targeting cloud computing for high end performances or advanced consumer electronics with tighter power consumption constraints, 3D integration technology will allow to design large scale multi-core. Thanks to advanced available 3D technology, it will be possible to maintain overall power consumption budget, increase chip to chip bandwidth, and preserve overall system cost by smart system partitioning. One of the main challenge of such multi-cores is clearly the interconnect infrastructure. For designing such 3D multi-cores, it is required to address two primary concerns: the 3D physical link by itself, and advanced interconnects scaled to 3D. The paper present an overview of 3D interconnects with 3D asynchronous Network-on-Chip architectures, with focus on 3D asynchronous links, and advanced interconnect structures for memory caches in 3D.

Key words: 3D architecture, TSV, Network-on-Chip, Asynchronous communication, Serial Link, Cache protocol

I. INTRODUCTION

After the profound evolution of deep submicron technologies, we are facing today the wall of transistor shrinking. More Moore trend is facing severe difficulties: cost, variability, power density, timing uncertainties [1]. In order to pursue technology and design integration, More Than Moore technologies, such as 3D integration, are becoming more attractive. With 3D technologies, and so-called Through Silicon Vias (TSV) [2,22] it is possible to stack various dies together. The 3D technologies are opening a full scope of new application possibilities, by integrating more devices from potentially different technologies (CMOS, MEMS, DRAMs, etc.). Full 3D can be envisaged by stacking vertically dies on top of each other's or Interposers – also called 2.5D – by stacking horizontally dies onto silicon substrate.

The advantages of 3D are numerous: i) shorter communication distance between dies, thus reducing communication load and then reducing communication power consumption, this allows to pursue denser integration [3,12]; ii) possibility of stacking dies from various heterogeneous technologies, like stacking memory on top of logic, MEMs, image sensors, or even photonic, in order to benefit of the best technology where it best fits ; iii) improve system yield and cost by partitioning the system in a divide & conquer approach: multiple similar dies are fabricated, tested & sorted before the final 3D assembly, instead of fabricating ultra large dies with very reduced yield – this is the case for instance of the first commercial FPGA interposer-based from Xilinx [4].

All the forecasted 3D advantages will facilitate designing future many core architectures: faster and shorter communication, allowing denser computation & reduced cache sizes ; use the appropriate technology to efficiently partition the many-core architecture : stack DRAM on top of core arrays, like with WideIO [5,6,7], or use 3D DRAM, with Hybrid Memory Cube [8] ; implement power management and system IOs within an active interposer ; partition the many-core in regular 3D stackable tiles in order to improve system cost, with mask reuse, and overall yield reduction.

Nevertheless, 3D integration presents various design issues, such as testability, power delivery network integrity, power density, thermal dissipation [27], and timing closure problems such as 3D clock distribution [9]. For 3D multi-core, the main challenge is to design efficient 3D system level interconnects, offering large bandwidth, for reduced power consumption [10]. Network-on-Chip has already proved their efficiency to design 2D multi-core, and NoCs can be naturally extended to 3D [11]. Due to timing closure issues, and the lack of maturity of 3D technologies and 3D CAD tools, it will be impossible to implement global clocking in 3D NoC. The use of Globally Asynchronous Locally Synchronous (GALS) scheme will restrict synchronization problems within synchronous dies, while asynchronous NoC links will provide robust 3D interfaces [12].

3D memory caches are a natural way to partition a many-core architecture over multiple 3D tiers. Indeed, it does not increase too largely the power density and in the same time it solves the problem of the external memory bandwidth requirements of high performance many-core architectures.

In this paper, we present in section II an overview of 3D NoC topology for many core architecture, with two proposals: an homogeneous 3D NoC architecture and a hierarchical 3D NoC for interposer based many-core system. While extensive studies has been carried out on 3D NoC topologies implementation [11], a primary concern is also the design of the 3D physical link itself. We present in section III, the constraints of designing such a 3D physical link and two possibilities of asynchronous 3D link, firstly a serial link for trading-off TSV area versus link throughput, and secondly a protocol converter for 3D asynchronous 2 phase link communication. Finally, as an example of 3D many core architecture, we propose in section IV an adaptive 3D cache architecture and study impact of 3D NoC on performances.

This work was funded thanks to the French national program 'programme d'Investissements d'Avenir, IRT Nanoelec' ANR-10-AIRT-05

978-1-4799-8720-7/15 $31.00 © 2015 IEEE

II. 3D Multi-Core Interconnects

A. 3D NoC Principles : topology and clocking constraints

For power efficient embedded applications, Network-on-Chip (NoC) has already proven to be an advanced communication infrastructure for the current "2D" many-core architectures [13]. Based on structured packet based interconnect, NoC offers full modularity, scalability, high performance, and quality of service, compared to existing bus-based interconnect. For 3D-stacked circuits, 3D NoC, as a natural extension of current 2D NoC, is then a promising solution for increased modularity and scalability [11]. A 3D NoC will allow to seamlessly interconnect either i) different dies, using heterogeneous technologies, different IPs for distinct sub-applications, or ii) similar dies, using a pure tile based approach. In both cases, from the architecture and programmer point of view, the 3D NoC offers a unified protocol throughout the 3D stack.

Even if 3D technology introduces new application possibilities, it also incurs VLSI design problems. For example, delivering a clock to each die and dealing with clock synchronization and clock tree is a critical problem in the context of 3D circuits [14]. To avoid global clocking, Globally Asynchronous and Locally Synchronous (GALS) scheme must be adopted to implement efficiently a 3D NoC : processor tiles and individual dies must have separate clock domains, and efficient synchronizations must be performed at the 3D NoC interfaces. The 3D GALS NoC implementation can be achieved in various ways: either using a full robust asynchronous logic [15], or using a multi-synchronous clocking schemes (like mesochronous clocking, source-synchronous clocking, etc…) [16].

In the following sections, we will present two types of 3D NoC for multi-core: one homogenous 3D NoC based multi-core and one hierarchical 3D NoC for interposer based multi-core.

B. Homogeneous 3D-NoC for 3D Multi-core

In the first 3D architecture, a homogeneous 3D NoC is proposed to build a 3D Multi-Core, using similar dies and tiles. The objective is to reduce mask costs, to improve yield by partitioning the multi-core system in smaller dies, while reducing the overall system consumption and latency, due to shorter NoC paths.

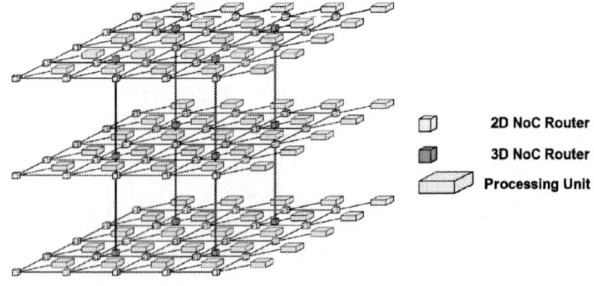

Figure 1. Homegenous 3D NoC multi-core architecture

Due to design limitation, the asynchronous 3D NoC router cannot be efficiently implemented using 7x7 ports (N, E, S, W, Up, Down, Local Resource). The 3D router can be implemented hierarchically [17], by splitting the initial 7x7 router in the respective 2D plan and 3D plan. As proposed figure 1, the 2D router 5x5 will handle the intra die 2D communications, while the 3D router 4x4 will handle the inter die 3D communications and with the local resource. As a result, the communication latency is increased, with two-hop for inter-die communication, but intra-die communication and router throughput is preserved with smaller routers.

Such approach is well adapted for regular multi-core architecture, but architecture partitioning in similar dies can also be done for less regular application. For example, for 4G wireless telecommunication application, such as 3GPP-LTE, it is required to handle multiple antennas in a Multiple-In-Multiple-Out (MIMO) approach (figure 2).

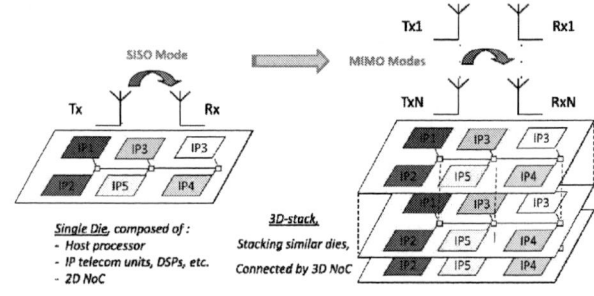

Figure 2. MIMO scheme, using several dies, implementing SISO

As proposed in [18], this can be achieved by designing a single die to control one antenna (SISO, equivalent to MIMO 1:1), while the stacking of similar dies will be able to control more antennas (MIMO 2:2, MIMO 4:2, etc…).

The proposed circuit is composed of several telecom processing units interconnected by a 3D NoC and controlled by a host processor. Compared to a 2D reference platform, the proposed 3D circuit keeps the same performance and power consumption, while reducing total mask cost.

C. Hierarchical 3D-NoC for interposer-based Multi-core

Since 3D technologies are not mature enough, stacking easily a large number (with a number of die above 2, #die>2) of homogenous processing dies is still not feasible.

Another limitation of homogeneous multi-core would come also from power densities that cannot exceed a certain limit to avoid thermal runaway. Another possibility to scale many-core architecture is to use Active Interposers.

Figure 3. Hierarchical 3D NoC for interposed-based multi-core

By splitting the multi-core system in multiple chiplets stacked onto an active interposer (figure 3), one can achieve:
- Yield optimization: chiplets are composed of clusters of cores, fabricated in advanced technology, tested and assembled on a large size interposer, fabricated in mature technology.
- Power & Thermal dissipation mitigation: compared to homogeneous 3D, power density is reduced, power delivery networks are simplified, and thermal dissipation is comparable to standard 2D dies.
- Smart Interposer features: by using active logic within the interposer, it is possible to provide advanced NOC interconnects and embedded power management, (compared to passive interposers limited to wire only [4]).

In such multi-core 3D architecture, the system level interconnect is hierarchical, and composed of multiple level of interconnects : 2D NoC within the chiplets in order to connect the clusters of cores, 2D NoC within the interposer in order to connect the chiplets, and 3D vertical links between the chiplets and the interposer.

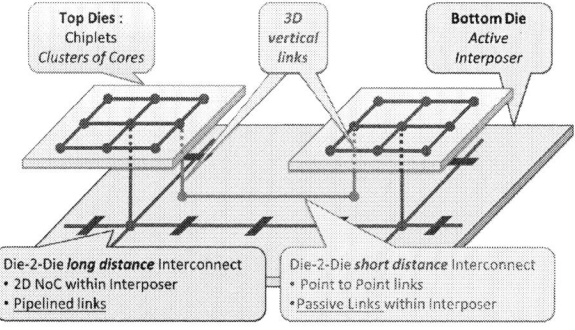

Figure 4. Pipelined and Passive NoC links within the interposer

Due to different traffic types and various communication distances between cores within any chiplets (figure 4), it is then possible to have two kind of communication links : i) direct chip-2-chip communication using pure passive links within the interposer (as a short cut within the 3DNoC topology), and ii) regular 2D NoC long distance link within the interposer, using pipelined NoC links, for long range core-to-core communication.

In all above NoC architectures, the homogeneous or hierarchical, the main challenge is to design efficient 3D NoC communication links for chip-2-chip 3D communication.

III. 3D ASYNCHRONOUS COMMUNICATION LINKS

In the 3D NoC architecture, it is required to design efficient 3D communication links. As presented before, since 3D technologies and associated CAD tools are not mature enough, it is impossible to design 3D clocked interfaces, using global clocking strategies. The GALS approach, using asynchronous 3D links will get rid of 3D interface clocking issues. By using robust QDI asynchronous logic [12], it allows mitigating any timing variations due to 3D stacking: thermal impact, TSV & Micro-bumps delays, variable or unknown delays due to heterogeneous dies stacking, dies from distinct corners, etc.. 3D clocking is a first challenge that can be addressed using asynchronous logic, but the 3D physical link still present physical implementation challenges.

A. 3D NoC Physical Link

The 3D NoC physical link is usually composed of the following elements : the NoC router, the NoC logical link, some additional DFT logic for testability, and the 3D physical interface itself, including TSV and/or μ-bumps, and lastly μ-buffer cells.

Some testability logic is added along the NoC link, using Boundary Scan cells, in order to test the 3D interface after 3D assembly fabrication [19]. On the NoC communication path, the test muxes will then present some additional delay (the Boundary Scan FF's being outside of the 3D critical path).

The 3D interface is also composed of specific 3D IO cells, commonly called *μ-buffer* cells. The μ-buffer cell is functionally equivalent to a regular 2D IO pad cell; it can be an input/output/bidir cell, and usually include the following features: a main buffer, ESD protection, level shifter, pull-down. Contrarily to 2D IO pad cell, this 3D μ-buffer cell is smaller (the buffer is sized according to TSV/μ-bump capacitance load) and the ESD protection is smaller, but the μ-buffer can be easily as large as 5 FlipFlop's.

Finally, the 3D interface is composed of a matrix of TSVs and/or μ-bumps, depending if it is a bottom or a top die, and if it is a Face2Face or Face2Back configuration. Due to Keep-Out-Zone rules, with today's 3D technology a TSV/μ-bump pitch of 40 μm down-to 20 μm pitch can be achieved, only the μ-buffer cells can be safely and easily implemented within the TSV/μ-bump matrix. The overall 3D link logic delay is then equivalent about 8 gates, evaluated as follow : 2*(DFT mux + μ-buffer + routing within the matrix + μ-bump/TSV load).

As a conclusion, in the overall NoC topology, the 2 routers seems close to each other's, with one hop NoC communication delay, compared to long 2D on-chip communication. But still a large impact occurs in the 3D NoC link : i) a large area footprint, due to TSV/μ-bump/μ-buffer matrix size, and ii) a large communication delay that can reach about *1ns* in a CMOS65nm technology, requiring careful physical design.

In order to optimize the 3D link, two options can be proposed, either optimize for area by using a link serialization, or optimize for delay by using more advanced asynchronous protocols. Two proposals of NoC logical links are respectively presented in the following sections.

978-1-4799-8720-7/15 $31.00 © 2015 IEEE

B. 3D Asynchronous Serial Link

To optimize the 3D link connection and associated TSV area cost, an asynchronous serial link proposes a tradeoff between number of TSV connection per link and the 3D link performance. As proposed in [20], an efficient asynchronous serial link is implemented using robust Quasi Delay Insensitive logic. The NoC logical link is composed of serialization logic, which transfer NoC data flits to a serial bit of a given size. The serial link, figure 5, is composed of asynchronous self-controlled muxes/de-muxes that will serialize/de-serialize the incoming/outcoming bits. With an input NoC data path of n bits, and an output 3D link of p bits, a serialization ratio of n/p is achieved, providing a trade-off between number of TSV/µ-bumps and 3D link throughput.

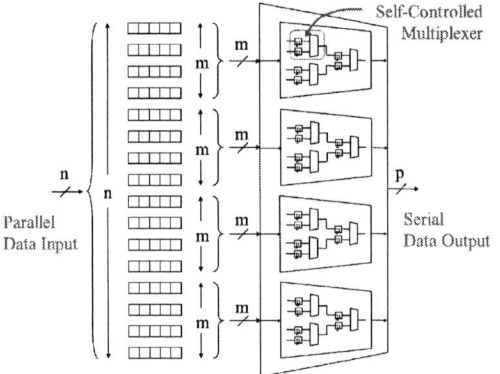

Figure 5.   NoC serial link architecture

For a NoC data path of 32-bits, by using QDI asynchronous logic and encoding, a NoC bi-directional link accounts for 180 signals [15]. A serialization ratio of 2 or 4, will allow saving TSV & µ-buffer matrix area by 2 or 4, but at the cost of ½ or ¼ throughput degradation, and with slight latency increase. Using asynchronous 4-phase protocol and a serialization ratio of 2, a link cycle time of $2 * 4 * 1$ ns = 8 ns can be achieved (~125 MHz), with 50% TSV number.

Depending on the 3D technology and the core technology, the serial link trade-off will clearly show different impact on the 3D link area, if taking into account both TSV and serial link area [20]. In case of standard TSV (TSV diam. 10µm,), serialization will always reduce area impact, while for more aggressive TSV technology (TSV diam. 1µm), serialization will have little impact even for advanced CMOS technology.

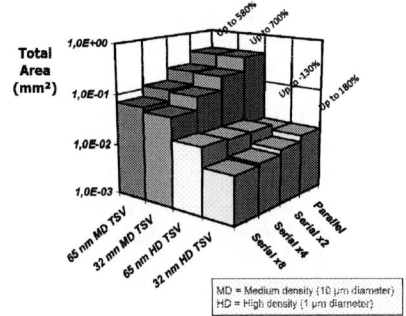

Figure 6.   3D serialization trade-offs for different nodes and ratio.

As a conclusion, serialization may reduce area impact of the link, but will strongly affect the link throughput. Another solution then consists in using asynchronous 2-phase protocol, instead of the commonly used 4-phase protocol.

C. 3D Asynchronous NoC Protocol Converter

In order to optimize the 3D NoC link, and get rid of the 3D link delay, we propose to use 2-phase asynchronous communication on the 3D link.

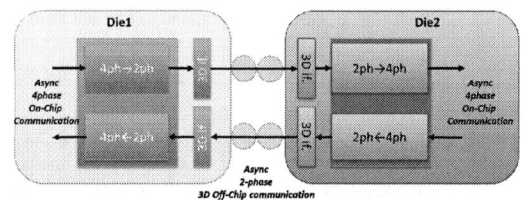

Figure 7.   Asynchronous Link 4-phase⇔2-phase Protocol Converter

On the 3D link proposed figure 7, the principle consists in using the regular 4-phase asynchronous protocol well-adapted for on-chip communication, and use 2-phase asynchronous protocol for the 3D off-chip communication, to reduce the impact of the 3D physical link delay. Efficient 4-phase ⇔ 2-phase protocol converters are proposed, using new 1T-of-n transition encoding, where the details can be found in [21].

Regarding communication performances (figure 8), 4-2 protocol conversion costs some little performance impact when the communication link is very small, where 4-phase is then well adapted, while offer a clear throughput increase for larger communication link, as can be found for the 3D physical link. For a link delay of 1ns, a throughput of about 450 MHz may be achieved.

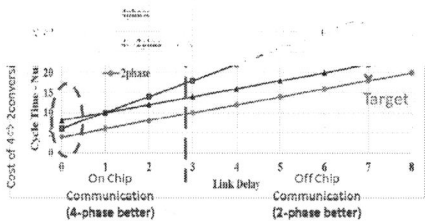

Figure 8.   4-2 Protocol Converter Throughput wrt. NoC link

Using 2-phase protocol, the number of signal transition is also reduced by 2, thus halving the dynamic power consumption of the NoC physical link. For a 4x4 2D NOC in a CMOS32nm tech [21], even if 2-phase link is faster, the dynamic power consumption is reduced by 20% (Table I).

TABLE I.   POWER CONSUMPTION REDUCTION OF NOC LINK
INCLUDING 4 –PHASE⇔2-PHASE CONVERTION (CMOS 32NM)

Implementation	Leakage Power	Dynamic Power	Total Power
Async NoC (standard 4-phase)	1.36 mW	7.415 mW	8.77 mW
Async NoC (4⇔2 Phase Converters)	1.39 mW	5.92 mW	7.32 mW
Overhead	+2.2%	-20.2%	-16.5%

IV. 3D STACKED CACHE INTERCONNECT

Following the presentation of 3D NoC physical link design and performances, we present a 3D many core cache coherent architecture, where the 3D NoC is used to transfer data cache information between core tiles and cache tiles.

A. 3D adaptive cache architecture

With the availability of high density TSVs in recent 3D technologies [2,22], we propose a distributed 3D cache architecture in [23]. This 3D NUCA cache with tiles interconnected by 3D NoC is depicted in Figure 9. Processing units (processors or dedicated hardware accelerators) send requests to cache controllers mapped in the global memory space. The cache controllers then dispatch these requests to the stacked cache tiles through a 3D version of the DSPIN NoC [24]. By using 3D NoC, the architecture is fully stackable and the number of cache tiers can be chosen at assembly time. So the total cache size can be modified according to application's needs late in the design process.

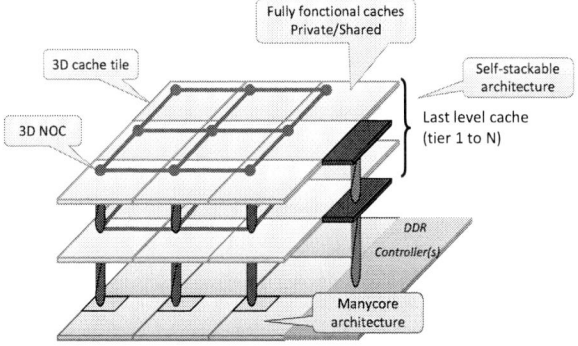

Figure 9. 3D adaptive cache

The cache is organized as a regular 3D mesh built around multiple 3D NoCs. The manycore architecture used here is the TSAR processing architecture [26]. The TSAR platform is a 2D mesh of processing tiles. Each tile contains four MIPS32 cores with their L1 caches and shared L2 caches, interconnected by a local crossbar interconnect. The processing tiles are connected to a NoC and cache coherency between L1 caches is ensured by L2 caches.

The 3D cache is used as a L3 last level cache in the TSAR platform. Each 3D cache tile implements a fully autonomous cache able to process TSAR requests and issue read or write requests to the external memory controller, in case of cache miss for example. Thus, the control is fully distributed in cache tiles ensuring the scalability when the size of the architecture grows. At the output of TSAR L2 caches, cache access controllers place the data in the 3D NUCA cache and can dynamically reallocate the cache in function of application's needs [23].

B. Performances of the 3D cache

Thanks to the dynamic reallocation of cache tiles to memory segment, the execution time and external memory traffic can be both reduced by up to 50%. Moreover, in a four tiers 3D cache with eight by eight tiles per tier (256 total tiles), this 3D cache can provide a bandwidth up to 745 GB/s [25].

We have implemented a 1 MB tile of this 3D cache architecture in order to evaluate its feasibility and its performances. We choose to study the most complete configuration where the cache tile includes vertical links for both 3D NoCs (cache access controllers to cache tiles and cache tiles to external memory controller).

This 3D physical implementation using the STMicroelectronics CMOS 28nm low power bulk technology is shown in Figure 10. The tile includes 284 signal TSVs and 280 power TSVs. The number of power TSVs has been computed from WIDEIO power distribution infrastructure. We use 5μm-wide TSVs with a 10μm pitch and the data of the cache were stored in SRAM memories. The copper pillars have a 20μm pitch which is bigger than the pitch of TSVs. So, to build compact TSVs arrays, back side redistribution layer (RDL) routing is needed between TSVs and back bumps.

The cache tile is a square of 1.5mm². The achieved useful memory density is as high as 77% of the tile area which validates the fine-grained 3D approach.

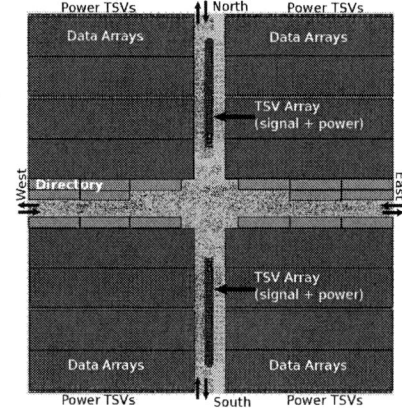

Figure 10. Physical view of the 3D cache tile

C. Impact of 3D NoC properties on performances

We first want to measure the impact of the number of vertical links on the execution time of applications in the underlying manycore architecture. We executed some of the Splash2 benchmarks as described in [24] on the TSAR architecture with 64 cores (16 clusters). There is 16 tiles of 1 MB L3 cache in each tier (16 MB of L3 cache for 1 tier and 64 MB for 4 tiers).

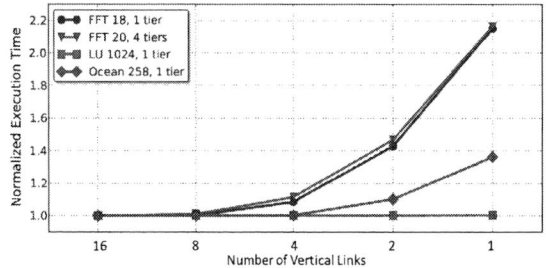

Figure 11. Execution time function of number of 3D vertical links

978-1-4799-8720-7/15 $31.00 © 2015 IEEE

Figure 11 shows that using one vertical link for two tiles does not lead to much degradation of execution time over the base architecture with one link per tile. But having less vertical links leads to significant increase of the execution time, in particular for memory intensive FFT applications. High density vertical interconnects are then mandatory to reach optimum performances.

We perform another experimentation to show that exploiting horizontal links of the 3D mesh is also important to reach the best performances. To that effect, we use two different routing strategies: a Z-first (ZXY) routing policy and a Z-last (XYZ) routing policy (equivalent to having horizontal routing in the manycore tier only). We also study 3D cache configurations with less but bigger cache tiles (constant total cache size) to mutualize the cache controller cost. The Figure 12 shows the execution time of Splash FFT for different configuration. It shows that when the number of tiles is reduced, the network is congested and thus the Z-first routing policy gives better results. It can be explained by the scalability of the aggregated horizontal bandwidth in this case as opposed to the Z-last routing policy.

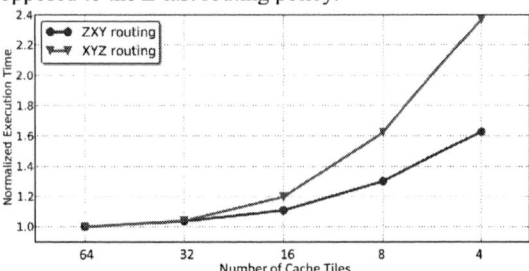

Figure 12. FFT Execution time, function of number of cache tiles for 2 topologies (full 3D mesh vs. vertical-only) with a 4 tiers 3D cache

V. CONCLUSION

As a More-Than-Moore scheme, 3D technology provides heterogeneous integration capabilities that will permit to design the future large scale multi-core architectures. 3D integration offers reduced communication distance, reduced power consumption, optimized yield by using homogenous tiles of processing cores, stacked vertically or horizontally. One of the main challenges is to design 3D optimized link, which still suffer from rather large TSV/μ-bumps area impact and 3D interface logic delay due to testability & IO buffering. We proposed two kind of 3D-Network-on-Chip for 3D multi-core, either a regular homogenous 3D NoC, or a hierarchical 3D NoC adapted for interposer based system. For robust 3D link & avoiding synchronization issues, we proposed two kind of asynchronous vertical links, either using serial link for TSV area reduction, and 4-2 asynchronous protocol converters to optimize the 3D link throughput, achieving 500 MHz in a 65nm technology. Finally, a 3D manycore architecture based on 3D tiles of L3 caches is proposed, allowing to fully exploit the opportunities offered by 3D technologies while leveraging their shortcomings. Experiments have shown how important is the 3D NoC topology. As further work, we plan to implement the proposed 3D tiles of caches by using an active interposer, and using the proposed optimized 3D NoC links using both asynchronous links and source-synchronous links.

REFERENCES

[1] International Technology Roadmap for Semiconductors, http://www.itrs.net/, 2014.

[2] J.U. Knickerbocker et all, "2.5D and 3D technology challenges and test vehicle demonstrations," *Electronic Components and Technology Conference (ECTC), 2012 IEEE 62nd* , pp.1068-1076, June 2012.

[3] T. Thorolfsson, S. Melamed, W. Rhett Davis, Paul D. Franzon, "Low-Power Hypercube Divided Memory FFT Engine Using 3D Integration", TODAES, Vol 16, Issue 1, Article 5, Nov. 2010 .

[4] http://www.xilinx.com/support/documentation/white_papers/wp380_Stacked_Silicon_Interconnect_Technology.pdf

[5] JEDEC web site: http://www.jedec.org

[6] J.-S. Kim et al., "A 1.2v 12.8gb/s 2gb mobile wide-i/o dram with 4x128 i/os using tsv-based stacking,", *in ISSCC 201*, feb. 2011, pp. 496 –498.

[7] D. Dutoit, & all, "A 0.9 pJ/bit, 12.8 GByte/s wideIO memory interface in a 3D-IC NoC-based MPSoC", *VLSI Symposium , June 2013.*

[8] HMC website : www.hybridmemorycube.org/

[9] Lim, S. K.; "TSV-Aware 3D Physical Design Tool Needs for Faster Mainstream Acceptance of 3D ICs." In: *DAC* 2010.

[10] F. Clermidy, D. Dutoit, E. Guthmuller, I.M. Panades, P. Vivet, "3D stacking for multi-core architectures: From WIDEIO to distributed caches". *ISCAS 2013*, Beijing, China, May 2013, pp. 537-540.

[11] A. Sheibanyrad, F. Petrot, A. Jantsch, "3D Integration for NoC-Based SoC Architectures", *Springer 2011*, ISBN 978-1-4419-7617-8.

[12] P. Vivet, D. Dutoit, Y. Thonnart and F. Clermidy, "3D NoC Using Through Silicon Via: an Asynchronous Implementation", *VLSI-SOC'2011*, Hong Kong, Oct'2011.

[13] D. Melpignano, & all, "Platform 2012, a many-core computing accelerator for embedded SoCs: Performance evaluation of visual analytics applications", *Design Automation Conference*, June 2012.

[14] V. F. Pavlidis, I. Savidis, and E. G. Friedman "Clock Distribution Networks for 3-D Integrated Circuits", *Proc. of IEEE Custom Integrated Circuits Conference*, CICC'08, 2008.

[15] Y. Thonnart, P. Vivet and F. Clermidy, "A Fully-Asynchronous Low-Power Framework for GALS NoC Integration", *Proc. of Design And Test in Europe*, DATE'10, Dresden, Germany, March 2010.

[16] I. Miro Panades, A. Greiner, "Bi-Synchronous FIFO for Synchronous Circuit Communication Well Suited for Network-on-Chip in GALS Architectures", *NOCS'2007, Princeton, Ne Jersey, USA*, 2007.

[17] W. Lafi, D. Lattard, A. Jerraya, "An asynchronous hierarchical router for networks-on-chip-based three-dimensional multi-processor system-on-chip". *Softw., Pract. Exper.*, Vol. 42, No. 7 Pg. 877-890, 2012.

[18] W. Lafi, D. Lattard and A. Jerraya "A Stackable LTE Chip for Cost-Effective 3D Systems", *in IPSJ Transactions on System LSI Design Methodology*, Vol 5, pg. 2-13, 21 fev 2012.

[19] Y. Fkih, P. Vivet, M.L. Flottes, B. Rouzeyre, G. Di Natale, J. Schloeffel, "3D DFT Challenges And Solutions", ISVLSI'2015, Montpellier, France, July 2015.

[20] F. Darve, A. Sheibanyrad, P. Vivet, F. Petrot, "Physical Implementation of an Asynchronous 3D-NoC Router using Serial Vertical Links", *IEEE ISVLSI'2011*, Chennai, India, July 2011.

[21] J. Hilgemberg Pontes, P. Vivet, Y. Thonnart, "Two-phase Protocol Converters for 3D Asynchronous 1-of-n Data Links", ASP-DAC'15, Tokyo, Japan, January 2015.

[22] J. Charbonnier et al., "Development and Characterisation of a 3D Technology Including TSV and Cu Pillars for High Frequency Applications", ECTC'2010, Las Vegas, USA, June 2010.

[23] E. Guthmuller, I. Miro-Panades, and A. Greiner, "Adaptive stackable 3d cache architecture for manycores," in IEEE Computer Society Annual Symposium on VLSI, ISVLSI 2012, aug. 2012, pp. 39 –44.

[24] I. Miro-Panades, A. Greiner, and A. Sheibanyrad, "A Low Cost Network-on-Chip with Guaranted Service Well Suited to the GALS Approach," *Nano-Networks and Workshops, 2006.*

[25] E. Guthmuller, I. Miro-Panades, and A. Greiner, "Architectural exploration of a fine-grained 3D cache for high performance in a manycore context,' in *VLSI-SoC 2013, Istanbul, Turkey, Oct 2013*.

[26] TSAR website: https://www-soc.lip6.fr/trac/tsar/

[27] C. Santos, P. Vivet, JP Colonna, P. Coudrain, R. Reis, "Thermal Performance of 3D ICs: Analysis and Alternatives", *3DIC 2014.*

A Framework for Efficient Implementation of Analog/RF Alternate Test with Model Redundancy

S. Larguech, F. Azais, S. Bernard, M. Comte, V. Kerzerho and M. Renovell

LIRMM - CNRS/Univ. Montpellier
161 rue Ada, 34392 Montpellier, France

Abstract— A promising solution to reduce the testing costs of analog/RF circuits is the alternate test strategy, which permits to replace costly specification measurements by simple low-cost indirect measurements. Despite the substantial test cost reduction offered by this strategy, its deployment in industry is today limited mainly because of a lack of confidence in alternate test predictions. A potential solution to improve test confidence is to exploit model redundancy. The idea is to build different regression models for each specification during the training phase, and then to verify prediction consistency between the different models during the production testing phase. In case of divergent predictions, the devices are removed from the alternate test tier and directed to a second tier where further testing may be applied. In this paper, we present a framework for efficient implementation of alternate test with model redundancy. Results are illustrated on a power amplifier case study for which we have experimental test data over 11,200 devices.

Keywords- alternate testing, indirect testing, test efficiency, test confidence, model redundancy, low-cost measurements, specifications, analog/RF ICs

I. INTRODUCTION

The conventional practice for testing analog/RF circuits relies on specification-based testing, i.e. the performances of each fabricated device are explicitly measured and compared to specification limits. This approach offers good test quality but at the price of high testing costs because it necessitates the use of dedicated equipment equipped with expensive analog/RF test resources.

In order to lessen the burden of specification-based testing, a promising solution, the concept of "alternate test" (also called "indirect test") has emerged as an attractive solution [1]. The objective is to replace the conventional analog/RF performance measurements by some simple and low-cost measurements. The fundamental principle is actually to learn during a training phase the correlation between conventional analog/RF performances and low-cost Indirect Measurements (*IMs*). This correlation is then exploited during the mass production testing phase in order to deduce the circuit performances using only those low-cost indirect measurements. This approach has been applied to various types of analog and RF circuits [2-5].

In this paper, we focus on test efficiency evaluation and we develop a generic framework that permits to define effective implementation of alternate test with redundancy. Results will be illustrated on a Power Amplifier (PA)

fabricated by NXP Semiconductors for which we have experimental test data from about 11,200 devices. These data include 37 indirect measurements and one conventional RF performance measurement. The typical measurement repeatability error ε_{meas} for this performance is around 3%.

The paper is organized as follows. Section II introduces the general background of alternate testing and briefly reviews previous works. Section III details the proposed framework for implementation of alternate test with model redundancy. Finally section IV presents results on the PA case study.

II. BACKGROUND ON INDIRECT TESTING

A. General principle

The underlying idea of alternate testing is that process variations that affect the conventional performance parameters of the device also affect non-conventional low-cost indirect parameters. If the correlation between the indirect parameter space and the performance parameter space can be established, then specifications may be verified using only the low-cost indirect signatures. However the relation between these two sets of parameters is complex and usually cannot be simply identified with an analytic function. Instead, machine-learning algorithms are used during an initial training phase in order to build regression models that map the circuit performances to the low-cost indirect measurements. Different regression tools can be employed, including polynomial regression, Multivariate Adaptive Regression Splines (*MARS*), feed-forward Artificial Neural Networks (*ANNs*), support vector machines (*SVMs*), etc. In this work, we use Multivariate Adaptive Regression Splines to build regression models.

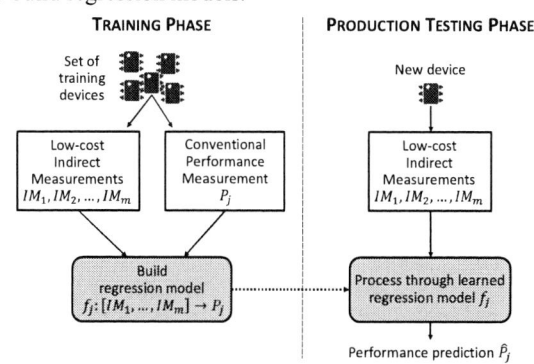

Figure 1. Alternate test synopsis

Figure 1 summarizes the indirect test synopsis, which involves two distinct phases: a training phase, and a production testing phase. For the sake of simplicity, we consider a single performance to be evaluated (denoted P_j), and a pattern of low-cost indirect measurements (denoted $IM = \{IM_1, \dots, IM_m\}$). In practice several performances have to be evaluated; a regression model is actually built for every performance, which may use different indirect measurements for each performance. During the training phase, both the conventional performance measurement and the low-cost indirect measurements are performed on a limited set of training devices; a machine-learning algorithm is then trained to build a regression model $f_j: \{IM_1, \dots, IM_m\} \to P_j$. During the production-testing phase, only the low-cost indirect measurements are made for every new device; the conventional performance is then predicted using the regression model learned in the initial training phase.

B. Previous works

One of the crucial issues associated with the indirect test approach consists in developing regression models that satisfy the following challenges:

- High model accuracy: the regression model learned during the training phase has to accurately represent the relationship between the indirect parameters and the specification to be predicted.

- High prediction reliability: the specification has to be correctly predicted for all devices evaluated during the production-testing phase, although the model is built on a limited training set.

Obviously, the quality of a regression model is strongly related to the choice of the indirect measurements used to build the model, which have to be information-rich and correlate well with the device performances. The choice of adequate indirect measurements actually remains an open problem. Very often, indirect measurements are defined ad-hoc, based on the expertise of the designer and the precise knowledge of the device under test. Another approach is to identify as many indirect measurements as possible and to search for pertinent subsets within the large set of IM candidates. A number of works can be found in the literature targeting this topic [6-12]. Note that a prerequisite for building efficient regression models is that data are described by a fixed probability density function. Outliers should therefore be excluded from the training phase since they are not consistent with the statistical characteristics of the remainder of the data; a defect filter is presented in [13] that permits to screen out outliers.

Another essential point for the deployment of alternate testing deals with confidence in test predictions. An interesting proposal is to implement a two-tier test procedure in which confidence estimation is established during the production testing phase: only devices for which confidence is sufficient are predicted using the learned regression models, other devices are directed to another tier where additional testing may be applied to characterize them. In

this way, it is expected that most of the devices are evaluated through the low-cost alternate test tier and only a small fraction of devices are evaluated through a more expensive test procedure. As a result, the overall test cost is reduced compared to standard specification testing while accuracy is preserved. This approach is explored in [14,15], based on guard-band allocation in the indirect measurement space in order to identify devices for which the alternate test decision is prone to error. Note that the guard-band allocation technique necessitates information on passing and failing devices.

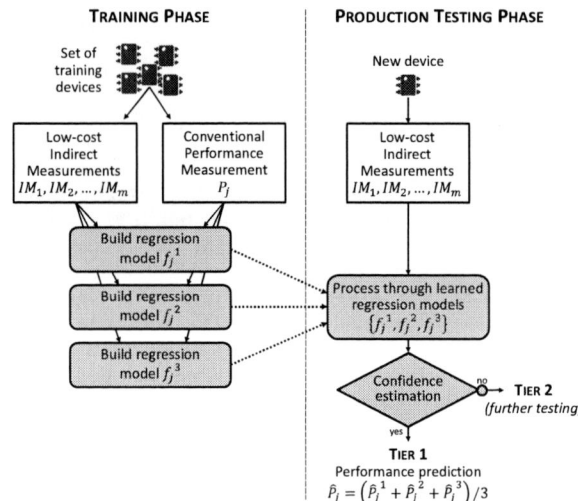

Figure 2. Two-tier alternate test synopsis with model redundancy [16]

In case pass/fail information is not available, another solution is proposed in [16]. The general idea is to exploit model redundancy in order to distinguish reliable predictions from suspect predictions. More precisely for a given performance to be evaluated, 3 regression models that involve different combinations of indirect measurements are built during the training phase for each specification. Then during the testing phase, confidence estimation is established by checking the consistency between the values predicted by the 3 different models. For this, the difference between the predicted values is computed for each pair of models and checked against a predefined threshold value. If one (or more) of these differences is superior to the divergence threshold ε_{div}, the prediction is considered suspect and the device is directed to the second tier where further testing may be applied. On the contrary, if all these differences are inferior to the divergence threshold ε_{div}, it means that there is no discrepancy between the values predicted by the different models and the prediction is considered reliable. In this case, the alternate test tier can be used to evaluate the device performance, which is computed as the mean of the values predicted by the different models. This strategy is illustrated in figure 2.

Finally another issue related to the alternate test strategy is test efficiency evaluation. Classical metric used in the literature to evaluate prediction accuracy is the average or rms error; maximal error is also sometimes reported.

However these metrics do not give any information on prediction reliability, i.e. how many of the predicted circuits are evaluated with a satisfying accuracy. In this work, we will use the Failing Prediction Rate (*FPR*) metric introduced in [11], which permits to quantify prediction reliability.

III. GENERIC FRAMEWORK

In this paper, our objective is to provide a generic framework for efficient implementation of the alternate test strategy with model redundancy proposed in [16]. More specifically, this framework involves (i) *IM* space exploration for multiple model generation, (ii) selection and construction of redundant models, and (iii) exploration of tradeoff between test accuracy and test cost. These different aspects are detailed in the following subsections.

A. IM space exploration for multiple model generation

Implementation of alternate test with model redundancy assumes that it is possible to build several accurate models involving different combinations of indirect measurements. This is not really a restrictive assumption as far as a sufficient number of indirect measurements are available, which is a realistic postulate. Indeed, standard DC tests typically performed on any device can be used as indirect measurements, the device may be equipped with internal observation points or include built-in sensors which offer additional indirect measurements, and finally indirect measurements can be performed under different test conditions (e.g. multi-Vdd settings) which allows to multiply the number of *IMs* by the number of different settings.

For the case study considered in this paper, we have actually 37 indirect measurements based on standard DC tests performed under different power supply conditions. The problem is therefore how to select, from this large set of available measurements, subsets of pertinent *IMs* that lead to accurate prediction models. This problem is a recurrent problem in the field of machine-learning, known as feature selection, and it has been addressed by many researchers [17]. Selection of a small number of features permits to lessen 'curse of dimensionality', reduce feature measurement cost and reduce computational burden. In the context of alternate testing, this problem is strongly related to the dimensionality of the indirect measurement space. Indeed the simplest approach that consists in performing an exhaustive search yields prohibitive computational cost. For our case study, if we want to select subsets composed of up to *p*=10 *IMs* among the set of *m*=37 available candidates, it involves $\sum_{k=1}^{p} C_m^k$ combinations corresponding to the construction of more than 500.10^6 models, which is not feasible in a reasonable processing time.

Heuristics have been employed to avoid exhaustive search, in particular an iterative procedure, also known as Sequential Forward Selection (*SFS*), which has been used in a number of works [9-12]. This procedure starts by building regression models for each individual *IM* and selecting the *IM* that generates the model with the minimum prediction error. The search is then extended to all possible pairs of *IMs* that include the previously selected one. This process is repeated until a stopping criterion is reached, for instance the number of selected *IMs* reaches a maximum target limit or no further improvement is obtained by adding new *IM*. This procedure is very efficient in terms of processing time. Indeed the number of models to be built with the *SFS* strategy in order to select a subset of up to *p IMs* among a set of *m* candidates is given by:

$$M_{SFS} = \sum_{k=1}^{p} C_{m-k+1}^1 = p \cdot (m+1) - \frac{p \cdot (p+1)}{2} \quad (1)$$

which corresponds to the construction of only 325 models for our case study if we want to select a subset of up to 10 *IMs* among the 37 candidates. However, the objective of *SFS* strategy is the selection of a single *IM* subset; therefore, it cannot be directly used in the context of alternate test with model redundancy.

Our proposal is to use a variation of the *SFS* strategy in order to perform multiple subset selection, as suggested in [18]. The procedure starts as the classical *SFS* by building regression models for each individual *IM*. However instead of selecting a single *IM* that generates the best model in terms of prediction error, we select *l IMs* that correspond to the *l*-best models in terms of prediction error. At the second iteration, we explore all possible children obtained by adding a new *IM* to each one of the *l* selected *IMs*. At this point, we have two possibilities: we can merge all children in a single list or we can keep individual lists related to the parents selected in the previous iteration. In the first case, a global ranking of all children is performed according to the prediction error of their associated model; the *l* subsets retained at the second iteration correspond to the *l*-best children overall. In the second case, ranking is performed in each individual list; only the first-best child of each list is retained. This process is then repeated for the following iterations until the stopping criterion is reached, i.e. no further improvement is obtained by adding new *IM* or maximum target size *p* of *IM* subsets is reached. Note that the prediction error is evaluated here in terms of rms error computed on devices of the training set:

$$\varepsilon_{rms}(f_j) = \frac{1}{P_j\big|_{nom}} \cdot \sqrt{\frac{1}{N}\sum_{i=1}^{N}\left(P_j(i) - \hat{P}_j(i)\right)^2} \quad (2)$$

where $P_j(i)$ the actual performance value of the ith instance, $\hat{P}_j(i)$ the predicted performance value of the ith instance using model f_j, and N the number of instances in the set. Note that this metric is expressed in percentage through normalization with the nominal performance value $P_j\big|_{nom}$.

The two versions of the extended *SFS* strategy for multiple subset selection are illustrated in figure 3. We call these two possibilities 'non-parental' and 'parental' methods, since in one case the link between a parent and his children is lost between two succeeding iterations, while it is preserved in the second case. The interest of the parental method is that it pushes towards a more diverse exploration of the space, which might be beneficial for the implementation of model redundancy. Indeed in this strategy, efficiency of confidence estimation relies on the assumption that it is very unlikely that regression models built using different combinations of indirect measurements will erroneously predict the device

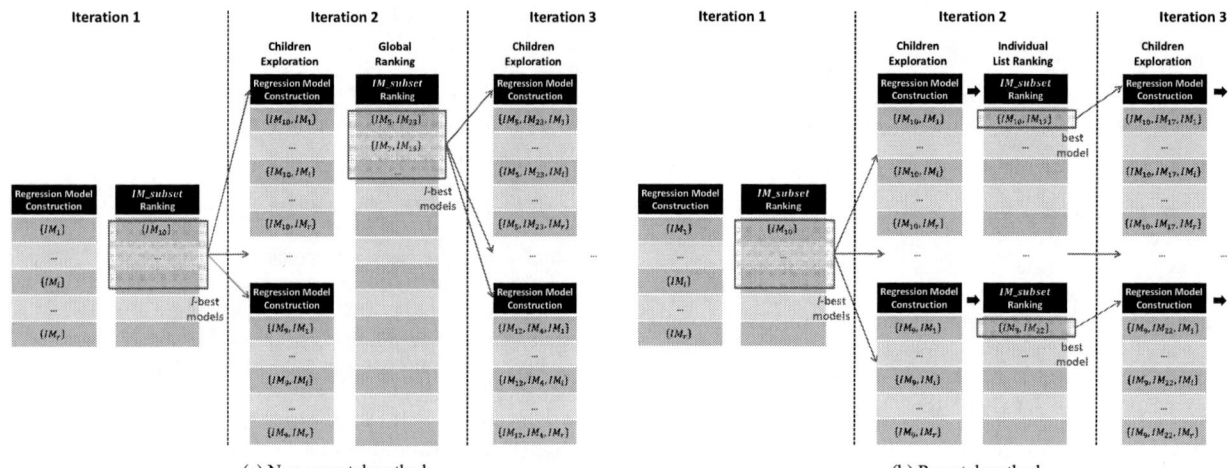

(a) Non-parental method   (b) Parental method

Figure 3. Extented *SFS* strategy for multiple subset selection

performance with the same error. The more diverse the models are, the more likely this assumption is.

The computational cost of the extended *SFS* strategy depends on the size m of the search space, on the number l of retained subsets at each iteration, and on the maximum size p of the retained subsets. The number of models to be built with this strategy is given by:

$$M_{extended-SFS} = l \left(\sum_{k=1}^{p} C_{m-k+1}^1 - m \right) + m$$
$$= l \cdot \left(p \cdot (m+1) - \frac{p \cdot (p+1)}{2} - m \right) + m \qquad (3)$$

For our case study with $m=37$ candidates, if we want to select subsets composed of up to $p=10$ *IMs* and to retain $l=10$ subsets at each iteration, it corresponds to the construction of 2,917 models, which is easily tractable in a reasonable processing time.

B. Selection and construction of redundant models

At the end of the previous phase, we have at our disposal several *IM* subsets of different sizes that can be used to generate prediction models for a given device performance. These subsets have been selected according to the ability of the generated models to accurately represent the relationship between the indirect measurements and the targeted performance for devices of the training set. The idea is now to evaluate the ability of these models to correctly predict the performance for devices of the validation set, in particular regarding prediction reliability. More precisely, the learned models are used to perform performance prediction for all devices of the validation set and we compute for each model the Failing Prediction Rate defined by:

$$FPR(f_j) = \frac{1}{N} \sum_{i=1}^{N} \left(|P_j(i) - \hat{P}_j(i)| > \varepsilon_{meas} \right) \qquad (4)$$

with $\left(|P_j(i) - \hat{P}_j(i)| > \varepsilon_{meas} \right) = 1$ if true

$\left(|P_j(i) - \hat{P}_j(i)| > \varepsilon_{meas} \right) = 0$ otherwise

where ε_{meas} corresponds to the typical repeatability error of the conventional performance measurement and N is the number of instances in the set. This metric permits to quantify prediction reliability since it expresses the percentage of circuits with a prediction error that exceeds the conventional measurement uncertainty.

Our proposal for the implementation of model redundancy is to select, for a given size of *IM* subsets, the 3 models with the best *FPR* value. These models will be used as redundant models during the production-testing phase in order to establish prediction confidence and estimate the device performance in case of satisfying confidence.

Furthermore in order to reinforce confidence, we investigate an original option, which consists in building the meta-models using ensemble learning [19]. Ensemble learning refers to a collection of methods that learn a target function by training a number of individual learners and combining their predictions. The resulting model is called ensemble model or meta-model. Numerous empirical and theoretical studies in various domains have demonstrated that ensemble models very often attain higher accuracy than single models. Our objective here is to investigate the use of meta-models in the context of the alternate test strategy.

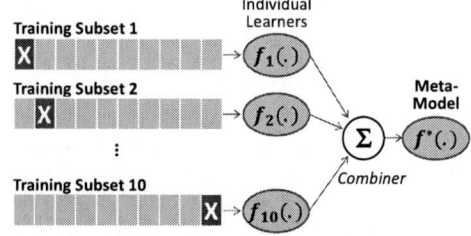

Figure 4. Meta-model construction with cross-validated committees [19]

Various methods exist in the literature for constructing ensembles, e.g. subsampling the training examples, manipulating the input/output features, injecting randomness… In this work we investigate a method based on subsampling the training examples, and more specifically a method based on "cross-validated committees". The principle of this method is illustrated in figure 4. The original training set is randomly divided into k partitions and k overlapping training subsets are constructed by dropping out one of these k partitions (typical value for k is 10). Individual models are then learned on each one of these training subsets and the meta-model is obtained by combining predictions of the individual learners with a uniform weighted average.

C. Tradeoff exploration: reliability vs. cost

The last phase of the test preparation is related to the choice of the divergence threshold ε_{div} used for confidence estimation and its impact on the tradeoff between test reliability and test cost. Indeed confidence estimation is established by comparing the difference between the predicted values for each pair of redundant values to a pre-defined threshold. If all these differences are inferior to the threshold, the device is predicted using the alternate test tier, otherwise it is directed towards the second tier for further testing.

Evidently, the choice of this threshold is crucial since it affects both the number of retested devices and the reliability of test predictions. A strict threshold may exclude many devices from the alternate test tier, resulting in higher test cost. On the contrary a relaxed threshold will direct only few devices to the second test tier, but may allow unreliable predictions for a number of devices. In other words, the choice of this divergence threshold enables the exploration of the tradeoff between test cost and test reliability.

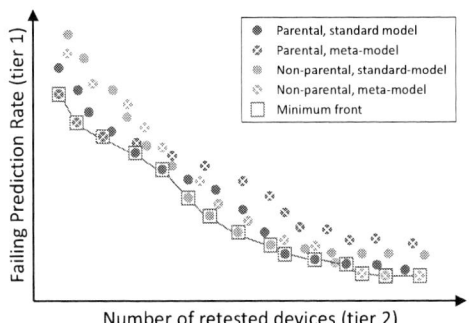

Figure 5. Exploration of cost-reliability tradeoff, for a given IM subset size

Practically our proposal is to vary the value of the divergence threshold and to evaluate the *FPR* corresponding to validation devices predicted using the alternate test tier and the percentage of validation devices directed towards the second tier, for each size of selected *IM* subsets. As illustrated on the conceptual example of figure 5, we can then plot the *FPR* against the number of retested devices for the different scenarios of redundant model generation and determine the minimum front, where each point of the optimum curve corresponds to a given scenario and a given divergence threshold. This evaluation is very important because it gives access to quantitative information on the expected performance of alternate test implementation. Based on this information, the test engineer can decide whether alternate test offers sufficient performance for his specific application context and choose the more appropriate implementation.

IV. RESULTS

The proposed generic framework has been used on our case study, with the objective to implement redundancy using models composed of up to p=10 *IMs*. The full dataset has been separated in a training set of 2,264 devices and a validation set of 8,943 devices, based on latin-hypercube sampling in order to ensure similar statistical characteristics between both sets.

Results of the *IM* space exploration for multiple model generation with *l*=10 are illustrated in figure 6, in terms of intrinsic model accuracy (evaluated on training set) and prediction accuracy and reliability (evaluated on validation set). Accuracy and reliability results of models generated using MARS built-in selection feature are also given as reference.

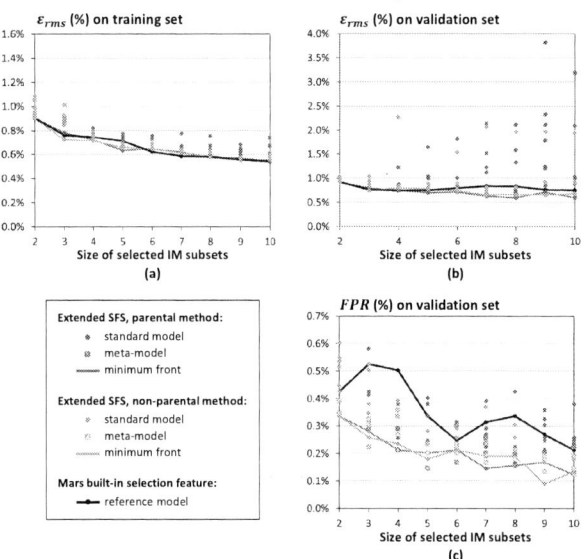

Figure 6. Accuracy and reliability metrics vs size of selected *IM* subsets

Several comments arise from these graphs. First the efficiency of the extended *SFS* strategy for multiple model generation is clearly demonstrated since for a given *IM* subset size, it permits to obtain several models with accuracy similar to the reference model. Moreover a number of these models exhibit lower *FPR* than the reference model, revealing better prediction reliability. Note that regarding the comparison between parental and non-parental methods, there is no evident trend of one method being superior to the other. The second comment is that ensemble learning is an interesting option but that does not necessarily yield to better prediction reliability since in some cases, meta-models exhibit a higher *FPR* than the corresponding standard models. Finally the last comment is more general and concerns the impact of the number of *IMs* used to build the models. The rms error evaluated on the training set slightly decreases when increasing the number of *IMs* while the rms error evaluated on the validation set is almost constant for models built with more than 2 *IMs*. Hence looking only at

978-1-4799-8720-7/15 $31.00 © 2015 IEEE

the classical accuracy metric, there is no interest in selecting a high number of indirect measurements. However the use of a higher number of *IMs* permits to slightly improve prediction reliability, as indicated by the *FPR* metric.

Results of cost-reliability tradeoff exploration are given in figure 7 which reports the minimum front of *FPR* obtained from the different selection scenarios with respect to the percentage of retested devices, for different sizes of selected *IM* subsets. This graph contains all the relevant information to help the test engineer choosing the more appropriate tradeoff for his specific application context. This graph also clearly illustrates the influence of the number of *IMs* used to build the redundant models. The general trend is that the higher this number, the lower the *FPR* that can be reached for a given percentage of devices directed to the second tier. Note that for this case study, interesting performances can be achieved. In particular it is possible to obtain an *FPR* below 0.05%, which means that more than 99.95% of the devices evaluated by the alternate test tier can be predicted with an error lower than the classical measurement repeatability error. This corresponds to a significant improvement of test reliability compared to the conventional implementation of alternate test with a single prediction model, which permits to reach only an *FPR* of about 0.2%. Moreover such a low *FPR* can be obtained with only a small fraction of devices directed to the second tier, incurring very limited test overhead (e.g. less than 0.15% of devices using redundant models built with selected subsets of 10 *IMs*).

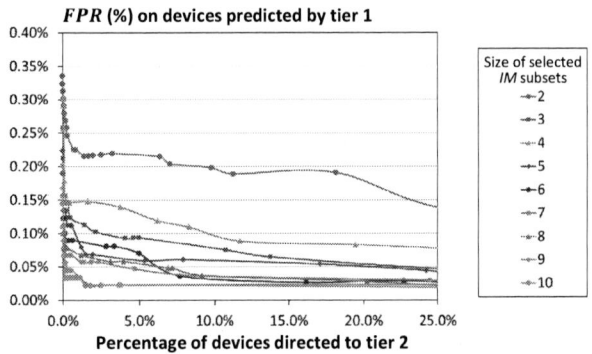

Figure 7. Tradeoff between test cost and test reliability for different sizes of selected *IM* subsets

V. Conclusion

In this paper, we have presented a generic framework for efficient implementation of alternate test with model redundancy, which involves (i) *IM* space exploration for multiple model generation, (ii) selection and construction of redundant models and (iii) exploration of cost-reliability tradeoff depending on the choice of the divergence threshold for confidence estimation. Different options have been investigated for *IM* space exploration and construction of redundant models. The proposed framework permits to select the more appropriate scenario and to have a quantitative evaluation of the alternate test performance in terms of prediction reliability.

The use of the proposed framework has been illustrated on a practical case study for which we have experimental test data. Results clearly demonstrate the benefit of implementing model redundancy, which permits to achieve very low failing prediction rate with only few devices directed to a second tier for further testing. More generally, this framework constitutes an essential element to guide the test engineer regarding practical aspects of alternate test implementation.

Acknowledgements

This work has been carried out within the framework of ENIAC JU project "ELESIS: European Library-based flow of Embedded Silicon test Instruments".

References

[1] P.N. Variyam, A. Chatterjee, "Enhancing test effectiveness for analog circuits using synthesized measurements, " Proc. VLSI Test Symp., pp.132-137, 1998

[2] R. Voorakaranam, et al., "Production deployment of a fast transient testing methodology for analog circuits: Case study and results," Proc. IEEE Int'l Test Conference, pp. 1174–1181, 2003

[3] S.S. Akbay, A. Halder, A. Chatterjee, D. Keezer, "Low-cost test of embedded RF/Analog/Mixed-signal circuits in SOPs," IEEE Trans on Advanced Packaging, vol. 27, no. 2, pp. 352–363, 2004

[4] A. Halder, A. Chatterjee, "Low-cost alternate EVM test for wireless receiver systems," Proc. VLSI Test Symp., pp. 255–260, 2005

[5] S. Goyal, A. Chatterjee, M. Purtell, "A low-cost test methodology for dynamic specification testing of high-speed data converters," Journal of Electronic Testing (JETTA), vol. 23, no. 1, pp. 95–106, 2006

[6] S. Ellouz et al., "Combining internal probing with artficial neural networks for optimal RFIC testing," Proc. Int'l Test Conf, p.9, 2006

[7] L. Abdallah et al., "Sensors for built-in alternate RF test," Proc. European Test Symp., pp.49-54, 2010

[8] H. Ayari et al., "Smart selection of indirect parameters for DC-based alternate RF IC testing," Proc. VLSI Test Symp., pp.19-24, 2012

[9] J. Liaperdos et al., "Adjustable RF mixers' alternate test efficiency optimization by the reduction of test observables," IEEE Trans. on CAD, vol. 32, no. 9, pp. 1383–1394, 2013

[10] M.J. Barragan, G. Leger, "Efficient selection of signatures for analog/RF alternate test," Proc. European Test Symp., p.6, 2013

[11] S. Larguech, et al., "Evaluation of indirect measurement selection strategies in the context of analog/RF alternate testing," Proc. Latin-American Test Workshop, p.6, 2014

[12] M. Barragan, G. Leger, "Feature selection for alternate test using wrappers: application to a LNA case study," Proc. Design Automation and Test in Europe Conf., p.6, 2015

[13] H.G. Stratigopoulos et al., "Defect Filter for Alternate RF Test," Proc. European Test Symp, pp.101-106, 2009

[14] H.G. Stratigopoulos, Y. Makris, "Error Moderation in Low-Cost Machine-Learning-Based Analog/RF Testing," IEEE Trans. on CAD, vol.27, no.2, pp.339-351, 2008

[15] N. Kupp, P. Drineas, M. Slamani, Y. Makris, "Confidence Estimation in Non-RF to RF Correlation-Based Specification Test Compaction," Proc. European Test Symp., pp.35-40, 2008

[16] H. Ayari, et al., "Making predictive analog/RF alternate test strategy independent of training set size," Proc. Int'l Test Conf., p.9, 2012

[17] H. Liu, H. Motoda, "Feature Extraction, Construction and Selection: A Data Mining Perspective," Kluwer Academic, MA, USA, 1998

[18] S. Larguech et al., "A Generic Methodology for Building Efficient Prediction Models in the Context of Alternate Testing," accepted at Int'l Mixed-Signal Test Workshop, p.6, 2015

[19] Z. Zhou "Ensemble Methods:Foundations and Algorithms," Boca Raton, FL: Chapman & Hall/CRC, 2012

978-1-4799-8720-7/15 $31.00 © 2015 IEEE

2015 IEEE Computer Society Annual Symposium on VLSI

Test and Calibration of RF Circuits Using Built-In Non-Intrusive Sensors

Athanasios Dimakos, Martin Andraud, Louay Abdallah,
Haralampos-G. Stratigopoulos, Emanuel Simeu, and Salvador Mir
Université Grenoble Alpes, TIMA, F-38000 Grenoble, France
CNRS, TIMA, F-38000 Grenoble, France

The test of the RF functions of mixed-signal Systems-on-Chip (SoCs) represents a very large fraction of the overall production cost [1]. In addition, RF circuits, when designed in the advanced technology nodes (e.g. 65nm and beyond), typically require some calibration mechanisms so as to guarantee a high yield [2]–[4].

Built-in test is a promising approach to facilitate test and calibration of RF circuits. It can alleviate the dependence on sophisticated and elaborate Automatic Test Equipment (ATE), it can drastically reduce the test time, and it can be inserted in an on-chip feedback loop that assigns an optimal tuning knob setting so as to achieve desired performance trade-offs.

However, built-in test structures that tap into sensitive RF signal paths may seriously degrade the performances of the RF circuit. To avoid this, the RF circuit and the associated built-in test structures need to be carefully co-designed. However, designers are rather reluctant to do so since it becomes very challenging, if possible at all, to meet the aggressive design objectives that exploit the full capabilities of the advanced technology node.

To this end, our work focuses on developing a non-intrusive built-in test approach based on sensors that are totally transparent to the RF circuit. We rely on two classes of sensors, namely variation-aware [5]–[7] and temperature sensors [8].

Variation-aware sensors consist of dummy layout components (i.e. transistors, resistors, capacitors, etc.) and dummy analog stages (i.e. bias stage, gain stage, current mirror, etc.). These dummy structures are copied from the topology of the RF circuit and are placed on the die in close physical proximity to the corresponding structures in the RF circuit such that they are well matched. The underlying idea is that these structures "witness" the same die-to-die and correlated intra-die process variations that the RF circuit exhibits. Therefore, it is expected that their outputs will track variations in the performances of the RF circuit. Essentially, they capitalize on the undesired phenomenon of process variations to infer implicitly variations in the performances. The cost reduction stems from the fact that the inputs and outputs of these dummy structures are DC or low-frequency and can be multiplexed to employ only two additional pins. Moreover, the RF circuit does not need to be powered on during testing.

Variation-aware sensors can also be utilized in the context of calibration since they offer an "image" of process variations [9]. A regression model can be trained to predict the performances of the RF circuit given a tuning knob setting and the outputs of variation-aware sensors. Since the variation-aware sensors are not electrically connected to the RF circuit, their outputs are invariant to the tuning knobs. In this way, their outputs need to be measured only once and then off-line we can search for the tuning knob setting that optimizes the performance predictions of the pre-trained regression model. Essentially, the variation-aware sensors offer the possibility for a quick one-shot calibration without requiring an iterative test/calibration loop.

However, variation-aware sensors cannot detect the presence of a random defect in the RF circuit since they are not electrically connected to it. For this purpose, we can employ temperature sensors. The underlying idea is that a defect will change the current that flows through a circuit branch and, thereby, the power dissipated by components in this branch. In turn, this will change the temperature profile on the die near a dissipating component. A temperature sensor can capture the temperature change indicating the presence of the defect.

Acknowledgement

This research has been carried out in the framework of the projects ENIAC ELESIS and ANR SACSO.

References

[1] F. Poehl, F. Demmerle, J. Alt, and J. Obermeir, "Production test challenges for highly integrated mobile phone SoCs," in *Proc. IEEE European Test Symposium*, 2010, pp. 17–22.

[2] N. Kupp, H. Huang, Y. Makris, and P. Drineas, "Improving analog and RF device yield through performance calibration," *IEEE Design & Test of Computers*, vol. 28, no. 3, pp. 64–75, 2011.

[3] A. Goyal, M. Swaminathan, A. Chatterjee, D. Howard, and J. Cressler, "A new self-healing methodology for RF amplifier circuits based on oscillation principles," *IEEE Transactions on Very Large Scale Integration (VLSI) Systems*, vol. 20, no. 10, pp. 1835–1848, 2012.

[4] C. Maxey, G. Creech, S. Raman, J. Rockway, K. Groves, T. Quach, L. Orlando, and A. Mattamana, "Mixed-signal SoCs with in situ self-healing circuitry," *IEEE Design & Test of Computers*, vol. 29, no. 6, pp. 27–39, 2012.

[5] L. Abdallah, H.-G. Stratigopoulos, S. Mir, and C. Kelma, "RF front-end test using built-in sensors," *IEEE Design & Test of Computers*, vol. 28, no. 6, pp. 76–84, 2011.

[6] ——, "Experiences with non-intrusive sensors for RF built-in test," in *Proc. IEEE International Test Conference*, 2012, paper 17.1.

[7] A. Dimakos, H.-G. Stratigopoulos, A. Siligaris, and s. Mir E. De Foucauld, "Non-intrusive built-in test for 65nm RF LNA," in *Proc. IEEE International Mixed-Signal, Sensors, and Systems Test Workshop*, 2014.

[8] L. Abdallah, H.-G. Stratigopoulos, S. Mir, and J. Altet, "Defect-oriented non-intrusive RF test using on-chip temperature sensors," in *Proc. IEEE VLSI Test Symposium*, 2013.

[9] M. Andraud, H.-G. Stratigopoulos, and E. Simeu, "One-shot calibration of RF circuits based on non-intrusive sensors," in *Proc. Design Automation Conference*, 2014.

This is an extended abstract for an invited talk in a special session organized in the IEEE Computer Society Annual Symposium on VLSI (ISVLSI), Montpelier, France, July 8-10, 2015.

978-1-4799-8720-7/15 $31.00 © 2015 IEEE

2015 IEEE Computer Society Annual Symposium on VLSI

Silicon Demonstration of
Statistical Post-Production Tuning

Yichuan Lu, Kiruba Subramani
The University of Texas at Dallas
Richardson, TX 75080

He Huang, Nathan Kupp
Yale University
New Haven, CT 06520

Yiorgos Makris
The University of Texas at Dallas
Richardson, TX 75080

Abstract—Towards meeting the continued demand for higher performance in analog/RF ICs, the semiconductor industry has resorted to aggressive design styles. In conjunction with the increased variation of modern processes, aggressive design jeopardizes yield and, by extension, profitability of the design. In an effort to facilitate aggressive high-performance design without compromising production yield, modern designs incorporate tuning knobs which are used after manufacturing to individually calibrate the performances of each chip, as well as some type of non-volatile memory to store the selected knob settings. In this work, we discuss a statistical post-production tuning method, which rapidly selects the appropriate knob positions using low-cost measurements, and we demonstrate it on a tunable Low Noise Amplifier which we designed and fabricated in IBMs 130nm RF CMOS process.

Keywords—knob tuning, low noise amplifier (LNA), post-production calibration, regression, testing, yield

I. INTRODUCTION

In analog/RF design, circuits are typically designed conservatively in order to ensure high yield. Indeed, an aggressive design approach would increase the risk of device performances falling below the circuit specifications due to the impact of process variation. As fabrication technologies continue to scale down, the impact of process variation on the performances of analog/RF circuits has become more prominent. Accordingly, circuit designers are compelled to sacrifice more performance in order to ensure a reasonable yield. On the other hand, the semiconductor industry continuously demands higher-performing ICs without compromising yield in order to remain competitive. As a solution towards facilitating aggressive high-performance design as well as high yield, modern designs incorporate tuning knobs which are used after manufacturing to individually calibrate the performances of each device.

The idea behind post-production tuning is that an aggressive design would have built-in knobs which can calibrate the performances of the circuit. Through the change of knob value during a post-production testing phase, we can effectively compensate the effects of process variation on device performances individually on every fabricated IC. By selecting an appropriate knob setting, the device can be made to operate within its specification range, even if initially it fails to abide by them. Various methods exist for selecting an appropriate knob setting. Once a knob setting that brings device performances within the design specification is selected, the corresponding value is stored in non-volatile memory before the IC is deployed [1]. Thereby, both aggressive design and high yield are achieved, while the customer is oblivious that each IC may have been tuned to a different knob setting.

Tuning knobs are designed to affect device performances, but the interplay between the impact of process variation and the effect of knob tuning complicates the task of measuring and calibrating device performances. Since tuning is performed during test time, while a device is interfaced with Automatic Test Equipment (ATE), each additional test performed increases the cost of the device. In addition to the number of tests performed, the cost of a single test also affects the overall test cost (e.g. the cost of taking a high-frequency measurement is much higher than taking DC measurements). While this introduces challenges for selecting the optimal knob value, it also provides fertile ground for leveraging low-cost test techniques such as alternate test.

Several previous publications attack the problem of performance calibration in analog/RF devices. These approaches fall mainly into two categories, iterative [2], [3], [4], which select a knob setting through a test-&-tune sequence of iterations and one-shot [5], [6], [7], which only test and tune once. One-shot methods have an advantage over the iterative ones due to lower tuning time and cost, though iterative methods can potentially be slightly more accurate. Among the existing one-shot methods, [7] assumed that the knob effects can be approximated as first order models. These models require a strict knob design, so that the knobs have no effect on each other, however this comes with increased design difficulty. This assumption is revoked in [6], in a method called mid-point alternative test-based calibration, where the new assumption is that the impact of process variations and knob positions is orthogonal. A similar method is proposed in [5], however it employs non-intrusive sensors which have the advantage of not affecting the signal path.

In this paper, we demonstrate the application of mid-point alternate test-based calibration using measured data from a tunable Low Noise Amplifier (LNA) which we designed and fabricated in IBM's 130nm RF CMOS process. The remainder of this paper is structured as follows. Existing knob selection and performance calibration methods are discussed in section II. Section III introduces the fabricated LNA design that is used to demonstrate the concept of tuning knobs. The proposed method is, then, experimentally evaluated in section IV with the help of results from the fabricated chips.

II. KNOB SELECTION AND PERFORMANCE CALIBRATION

A. Specification test

As the industrial standard testing method, specification test is the most accurate and highest costing approach to determining pass/fail of a device. Each performance of the device under test (DUT) is directly measured through the ATE. By comparing measurements to specifications, a pass/fail decision is made. Even though this method is accurate, employing it for the purpose of knob tuning, wherein a device may have a

978-1-4799-8720-7/15 $31.00 © 2015 IEEE

Configuration	Total Testing Cost	
	Test Set Term	Training Set Term
Exhaustive Specification Test	$C= N_{test} *N_k *P$	
Exhaustive alternate Test	$C= N_{test} *N_k *A$	$+ N_{train} *N_k *(A+P)$
Iterative approaches	$C= N_{test} *N_i *A$	$+ N_{train} *N_k *(A+P)$
One-shot approaches	$C= N_{test} *A$	$+ N_{train} *N_k *(A+P)$

Variable	Definition
N_{test}	Number of devices in test set.
N_{train}	Number of devices in training set.
N_k	Number of knob settings
N_i	Number of Iteration needed
P	Relative cost for measuring all performances
A	Relative cost for measuring all alternate tests

Fig. 1: Cost model for test and performance calibration methods

large number of settings for each knob, is impractical. Indeed, selection of the optimal knob setting requires specification testing for every performance and for every knob setting, the cost of which is prohibitive.

B. Alternate test

Alternate test is naturally suitable to address the problem of selecting an appropriate position for the tuning knobs while maintaining a low cost. In alternate test, instead of measuring the circuit performances which incurs high cost, a set of low-cost measurements are taken either externally or through on-chip sensors. These measurements are designed to correlate well with the circuit performances, while requiring far fewer test resources in order to be collected. Alternate test is a two-phase process. First, for an initial batch of devices, both the high-cost specification tests and the low-cost measurements are collected. From this data, regression models are then trained to correlate the low-cost measurements with the circuit performances. In the post-production phase, only the low-cost measurements are taken; performances are then predicted through the trained regression models and compared to the specifications in order to determine a pass/fail decision for the device. For a knob-tunable device, each knob setting would have its corresponding high-cost and low-cost measurements. A regression could then be trained to predict device performances for each knob setting based on the low-cost measurements for this or for all knob settings. This method can substantially reduce the overall test cost by replacing high-cost with low-cost measurements in the production phase.

C. Iterative Approaches

Following the general concept of alternate test, iterative methods incorporate a directed search approach with the help of low-cost measurements, in order to determine the optimal knob setting. Once regression models are trained in a post-production phase, these methods select a knob setting and perform an alternate test to determine the performances. The knob setting selection is performed iteratively, while the search function is optimized through each iteration. In short, a cycle of test-&-tune operations is repeated until a satisfying knob position is selected. If a device does not contain a passing knob setting, the algorithm stays in this test-&-tune loop until it determines that there is no passing knob setting or until a pre-specified iteration limit is reached. These methods reduce the cost of calibration from a factor of all possible knob settings to the number of iterations needed to converge or the limit.

D. One-shot approaches

The aim of these methods is to determine the optimal knob setting without repeating a test-&-tune cycle. These methods rely on a set of low-cost measurements for only a single knob setting, based on which they make a decision on the optimal

knob though statistical learning methods. By collecting low-cost measurements for a single knob setting, they minimize the time an integrated circuits spends on the ATE for the purpose of knob tunning. Herein, we implement one such one-shot calibration approach, namely mid-point alternate test, and we demonstrate its effectiveness using silicon data.

E. Cost Model

As we have discussed, one of the major obstacles of post-production calibration is the cost of testing/tunning. If the benefit of yield recovery does not cover the cost of testing/tunning, the calibration method should not be implemented. In Figure 1, we compare the cost model for different calibration methods we previously mentioned. Since all approaches except exhaustive specification utilize a regression model, the cost is divided into a test set term and a training set term. These models require a small training set where both alternate and specification tests are performed. The training set data is used to train regressions to correlate low-cost measurements with performances. Thereby, we eliminate the high-cost performance measurement P in the test set term. In high volume environments, $N_{test} \gg N_{train}$, therefore adaptation of alternate test substantially lowers the total test cost. For the test set term, one-shot approaches have the lowest testing cost due to the lack of need for a test-&-tune loop, and the ability to predict with one set of alternate tests.

F. Mid-point alternate test

Mid-point alternate test-based calibration [6] is the focus of this work, towards demonstrating statistical knob tunning. It is one of the prominent approaches in the one-shot calibration category due to its low test cost and high prediction accuracy. This method makes the assumption that the impact of process variations and tuning knobs on device performances are orthogonal. The overall performance model is:

$$\hat{P} = f(A, K) = \hat{\beta}_0 + A^T \hat{\beta}_a + K^T \hat{\beta}_k + \varepsilon \quad (1)$$

where $\hat{\beta}_0$ denotes the performance of a variation-free device, K^T is the vector of knob settings and all pairwise interaction terms, and $\hat{\beta}_k$ is the knob effect parameter estimated by:

$$K = (\underbrace{K_1, K_2, ..., K_p}_{\text{main effects}}, \underbrace{K_1 K_2, K_1 K_3, ...K_{p-1}K_p}_{\text{interaction terms}})^T \quad (2)$$

$$\hat{\beta}_k = (\underbrace{\beta_1, \beta_2, ..., \beta_p}_{\text{main effects}}, \underbrace{\beta_{1:2}, \beta_{1:3}, ...\beta_{(p-1):p}}_{\text{interaction terms}})^T \quad (3)$$

The term $A^T \hat{\beta}_a$ represents the correlation between low-cost measurement and process variations. Since the test is based on alternate test, we observe the process variation in terms of low-cost measurement. Each alternate test gives a direct measurement of the magnitude of process variation effects, and all of the alternate tests improve the estimation of the

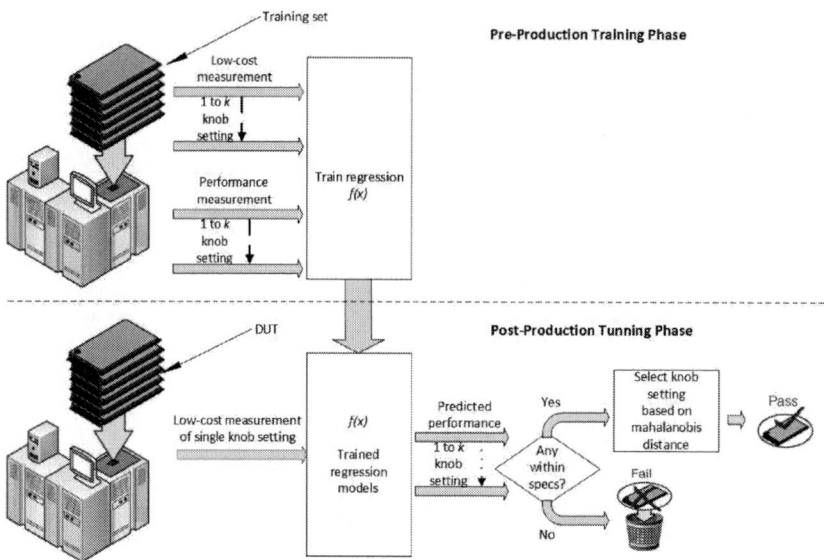

Fig. 2: Mid-point alternate test calibration flow

performances. As we can see in Equation 1, the term for reflecting the impact of process variations and tuning knobs are separate and assumed orthogonal. Following this model, we can generate regressions in order to predict performances.

With a basic understanding of how we model device performances, we can now proceed to the mid-point alternate test flow. Figure 2 demonstrates the complete test and calibration procedure. In the pre-production phase, the goal is to train a regression for each performance vector. Both low-cost and performance measurements for all knob settings are taken. The low-cost measurement are then used along with these functions in order to predict the performances of the device. The effect of changing the knobs on performance is modeled as an offset to the variation-free device performance. The trained regression takes low-cost measurements and knob settings as predictors, and gives a performance prediction for each knob setting as a response. In the post-production phase, low-cost measurements for the mid-point knob setting are taken from the DUT. The regressions then predict the performances for all knob settings of the device. If the predicted performances for all knob settings fail the design specifications, the device is classified as a failing device. If there exist more than one knob settings that would bring the device within its design specifications, the optimal knob setting is selected among them by using as a metric the normalized Mahalanobis distance, which maximizes distance from the specification planes.

III. TUNABLE DEVICE DESIGN

To demonstrate post-production tuning in silicon, we designed and fabricated a 2mm*2mm test chip in IBM's 130nm RF CMOS technology through the fabrication service provided by MOSIS. The chip contains RF tunable circuits, low-cost measurement sensors, a stimulus generator, and various peripheral and protection circuits. The chip is housed in a custom-designed evaluation board, which provides connection to external measurement instruments. Figure 3 shows a microphotograph of the fabricated die and Figure 4 shows the custom board housing the designed chip.

A. RF circuit

The RF circuit in this chip design is a cascode common source Low Noise Amplifier (LNA), which operates at a frequency of 1.575GHz. The LNA plays a critical role as it is the first component in the receiver chain so its noise figure is directly added to the receiver's overall noise, and the contribution of other components to receiver noise is scaled down by the gain of the LNA. As a result, a high performance LNA is a crucial part of high-end receiver design. This attribute makes the LNA an ideal candidate for demonstrating the tuning process in analog/RF circuits. We chose the cascode topology as our initial architecture, as it is one of the most universal RF LNA designs. The LNA schematic, as shown in Figure 5 (a), has three active devices: one amplifying common source transistor, one bias transistor, and one cascode transistor. The common source transistor provides the signal amplifying ability of the LNA. Its optimized width generates the minimal noise figure achievable for this fabrication technology for the given power consumption constraint. A cascode transistor of the same width is connected to the common source transistor to isolate the input from the load, therefore simplifying output matching. The bias voltage needed by the common source transistor is provided by the bias transistor, while the DC noise is minimized at the conjunction by a large resistor. A spiral inductor is placed at the drain of the cascode transistor along with other passives, for output stabilization and matching to a 50 ohm load. Input matching of 50 ohm is achieved through source degeneration provided by taking advantage of the parasitic bond-wire inductance. The specifications of the circuit are shown in Figure 5 (a).

B. Knob design

In the above circuit schematic, we were able to identify three key bias voltages as knobs that can be used for varying the performances of the LNA. Each of these three knobs has a different effect on the LNA performances and they are not orthogonal to each other. In other words, performance $\hat{y}$ increases proportionally with the voltage on $knob_1$ and

978-1-4799-8720-7/15 $31.00 © 2015 IEEE

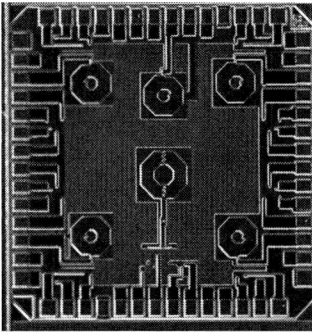

Fig. 3: Micro-photograph of fabricated die

Fig. 4: PCB housing chip and interface connections

$knob_3$ individually. Yet the simultaneous increase of voltage on both $knob_1$ and $knob_3$ will decrease the performance $\hat{y}$. This property provides more freedom for knob selection and simplifies the knob design process, as the knobs are no longer constrained by the need for orthogonality. However, it also calls for more sophisticated tuning algorithms. The IBM 130nm fabrication technology has a default supply voltage of 1.2V and a breakdown voltage of 1.6V. To avoid high voltage that may cause damage to the devices, we set the upper limit of the knob voltage as 1.4V. The lower limit of knob voltage is bound by the device's ability to meet the specifications. Evidently, we selected the range to be from .8V to 1.4V, with a step size of .1V to provide sufficient granularity and range in the choice of knob positions. With 3 knobs and 7 knob positions, we have $7\times3=343$ total sets of knob combinations.

C. Low cost measurement sensor

The low-cost measurement sensor is implemented in the form of a peak detector. The peak voltage measurement has been shown to be highly correlated with LNA performances. Its input is connected to the output of the LNA in order to measure the positive peak voltage of the signal. The sensor captures the amplified signal at the output of the LNA. A capacitor acts as a high-pass filter, blocking the DC value and allowing the AC value to go through. The positive swing of the filtered signal charges the output capacitor to the peak value of the AC signal swing; as a result, the peak voltage is stored. A mirrored sensor circuit provides a measurement of the output capacitor voltage when there is no signal input at the sensing point. The final peak voltage is measured as the difference between the voltage of the two output capacitors.

D. Stimulus generator

A simple sine wave is the stimulus that we use for test. We generate this signal with an on-die voltage-controlled oscillator circuit. A cross-coupled PMOS-NMOS topology helps in achieving symmetry of the rising and failing times of the oscillation. The initial oscillation is created by a parallel resonant LC-tank. It contains a spiral inductor and two MOS varactors, which provide a resonant frequency range between .9GHz and 1.6GHz through the tunning of the MOS varactors. To sustain the generated oscillation, the active devices behave as "negative resistance" and counterbalance the LC tank's internal resistance. An input voltage is applied to the varactors to tune the output frequency of the sine wave. The generated differential signal is, then, connected to a differential-output to single-ended output conversion stage, which can be connected to the RF input of an LNA as stimulus in the desired frequency.

E. Chip and board overview

Each chip contains 4 LNA circuits, 4 peak detector circuits and one VCO circuit. Each LNA has a dedicated peak detector. The LNAs are replicas of each other, with rotation in the x and/or y axis. Each LNA and peak detector combination occupies a corner of the chip to ensure that the routing and the distances to chip pins are the same. With this setup, we are able to minimize the difference between each LNA on the same die, avoid the introduction of design variation, and generate more samples for testing purposes. The chips are fabricated in IBMs 130nm RF CMOS process through MOSIS. A total of 40 chips are fabricated with the above setup, which gives us 160 LNAs. Each chip is then individually mounted on a custom printed circuit board with chip-on-board technology. With the chip-on-board structure, we are able to remove the introduction of parasitics and variation by board sockets and IC packages. The boards are designed to ensure minimal variation. They provide a simple interface between the die and the test equipment, as well as a matching network. After extensive testing, we concluded that out of 40 boards, 4 were defective; therefore, we have 36 boards, or 144 LNA samples for test purposes.

F. Data collection

Each of the 36 boards is tested through two different setups. In setup 1, we collect the device's actual performances at every knob setting. Three DC power supplies each connect to a knob of the LNA, providing the tunning voltage. The input and output ports of the LNA are connected to a network analyzer to collect the S parameters (S11, S12, S21, and S22). As the power supplies sweep through each of the 343 knob settings, the S parameters of the LNA are measured and stored into a computer. Setup 2 collects the device's low-cost measurement data at every knob setting. The on-die VCO functions as a stimulus generator and provides a sine wave to the LNA input, while the dedicated peak detector of each LNA functions as a peak measuring sensor. The DC power supplies function in the same way as in setup 1. Another DC supply changes the VCOs control voltage, thereby varying the stimulus frequency. The output voltage of the peak detector is recorded for each frequency and for every knob position. Using the above two testing setups, we populated the data sheet shown in Figure 6.

IV. TESTING AND TUNING

Using all the collected data, we conducted various experiments to evaluate the effectiveness of alternate testing and statistical post-production tuning. In all cases, we split the data

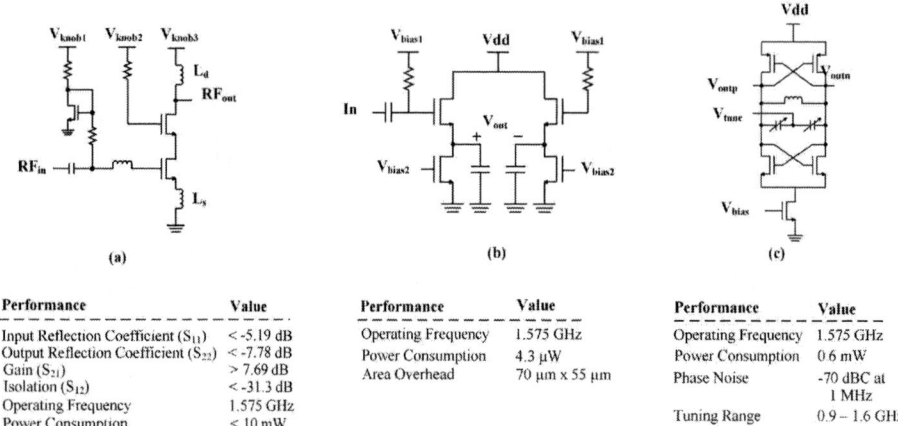

Fig. 5: Circuit design (a) LNA schematic (b) Peak detector schematic (c) VCO schematic

Performance	Value
Input Reflection Coefficient (S₁₁)	< -5.19 dB
Output Reflection Coefficient (S₂₂)	< -7.78 dB
Gain (S₂₁)	> 7.69 dB
Isolation (S₁₂)	< -31.3 dB
Operating Frequency	1.575 GHz
Power Consumption	< 10 mW

Performance	Value
Operating Frequency	1.575 GHz
Power Consumption	4.3 µW
Area Overhead	70 µm x 55 µm

Performance	Value
Operating Frequency	1.575 GHz
Power Consumption	0.6 mW
Phase Noise	-70 dBC at 1 MHz
Tuning Range	0.9 – 1.6 GHz

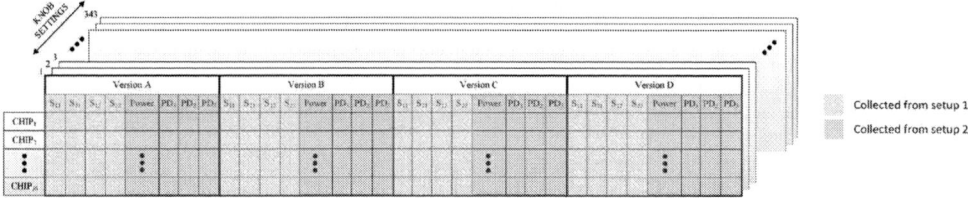

Fig. 6: Graphical depiction of dataset

into training and testing sets. Of the total of 144 LNAs, 80 were used for training and 64 were used for testing. To avoid any variation caused by the LNA position/orientation on the die, the training and the testing sets had a balanced number of LNAs from each of the four positions. For example, among the 80 LNAs in the training set, 20 were located in each of the top right, top left, bottom right, and bottom left corners of the die. To ensure statistical significance of the reported results, we performed 10 cross-validations for each experiment and averaged the results.

A. Specification test

The purpose of specification testing in our set of experiments is to emulate a knob free device, and define a baseline in order to compare the results of our statistical post-production calibration method. Specification testing reports the yield achieved by the process for this device, assuming that the design is knob-free. We emulate the knob-free device by using the center knob setting ($knob_1$ = 1.2V, $knob_2$ = 1.2V, $knob_3$ = 1.2V). The performances of each device are then compared to the design specifications. If the device does not meet its specifications, it is defined as a failing device. The pass/fail ratio of our 144 LNAs in the center knob setting is as follows:

Fail	Pass
31.1%	68.9%

As indicated by this result, without calibration capabilities the LNA production would experience a yield loss of 31.1%. We expect the result to improve as we incorporate knobs in the design to recover this yield loss.

B. Exhaustive specification test

Exhaustive specification test is performed to collect the performances of a device in every knob setting and compare to the design specifications. If a device has at least one knob setting that would make it operate within the design specifications, then the device is healable. If all knobs of a device fail, then the device is unhealable. This result, which is shown below, provides the ceiling for any post production tuning method, as it is the best achievable result by any possible test.

Unhealable	Healable
3.8%	96.2%

As may be seen in the above results, the minimal yield loss that can be achieved for our LNA circuit through exhaustive specification testing is 3.8%. Below, we also calculate the mean and standard deviation of the number of knob settings per device that make a device pass its specifications.

Mean of passing knob settings	Standard deviation
245	72

Out of a total of 343 choices, each device has on average 245 passing knob settings. On the other hand, the large standard deviation reflects a high amount of variation among different devices.

C. Exhaustive alternate test

This test evaluates the capability of alternate testing by utilizing all low-cost measurements from every possible knob setting. To predict the performances of the chip through low-cost measurements, we train a regression function for each dimension of the specification vector (S11,S21,S12,S22). In our case, we trained four MARS regressions, each with the low-cost measurements and the perspective knob settings as predictors. Then, the low-cost measurements are used to predict the performances for each knob setting for each LNA in the test set. Subsequently, among all possible knob settings for which all predicted specifications of the device are satisfied, we use the Mahalanobis distance in order to pick the optimal setting. This optimizes prediction robustness as it seeks to maximize the distance of the predicted device performances from their specifications and, thereby, tolerate prediction error. Finally, we compare the actual performances of every device for the chosen knob setting to their specifications, and we obtain the following confusion matrix.

		Actual	
		Fail	Pass
Predicted	Fail	0%	5.2%
	Pass	3.8%	90%

The results indicate that 90% of the devices are healed in this way, while 5.2% are healable but the method does not select a correct knob and 3.8% are unhealable yet the method still selects a setting which it predicts that it heals the device.

D. Mid-point alternate test

Evolved from the exhaustive test method, in mid-point alternate test we still train MARS regressions with low-cost measurements and perspective knob settings to predict each performance vector. However, in the testing phase, only the low-cost measurement for the center knob setting ($knob_1 = 1.2V$, $knob_2 = 1.2V$, $knob_3 = 1.2V$) is used to predict the performances for all knob settings of the device. This saves significant time in tuning, since only the mid-point alternate tests are collected, rather than alternate tests for every one of the 343 knob settings. The rationale of this method is that alternate tests for a single knob position are sufficient to reflect the impact of process variations on performances, while the impact of knob variation on performances can be statically learned and is orthogonal to the impact of process variations. The results of mid-point alternate test are reported in the following confusion matrix.

		Actual	
		Fail	Pass
Predicted	Fail	0%	5.5%
	Pass	3.8%	89.7%

As can be observed, mid-point alternate test is almost equally effective with respect to yield loss after statistical post-production tuning as the exhaustive alternate test approach, while being two orders of magnitude faster (i.e. only one set of alternate measurements instead of 343 need to be obtained and evaluated through regression models).

Figure 7 contrasts the effectiveness of the aforementioned methods for one device. The Y-axis shows the Mahalanobis distance between the performances of the device for a given knob setting and the optimal choice, while the X-axis lists the 343 knob settings, rank-ordered based on the corresponding Mahalanobis distance. As may be observed, exhaustive specification testing picks the knob setting that maximizes

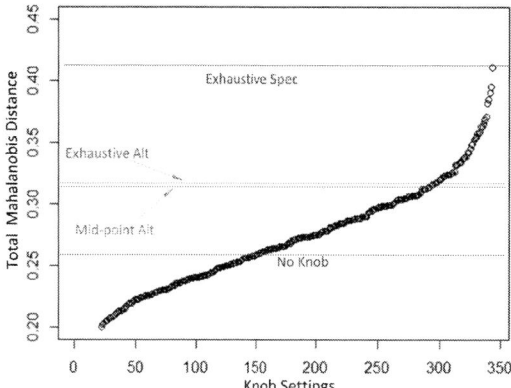

Fig. 7: Comparison of the four methods

that distance, thereby optimizing robustness of the device. Exhaustive alternate test and mid-point alternate test both select a knob setting that is within the top-25% of the available choices. The Mahalanobis distance for these two methods is very close. In contrast, the neutral knob setting represented by the "No Knob" line results in a far worse choice.

V. CONCLUSION

In this paper, we demonstrated a statistical post-production tuning method using actual silicon data. For the purpose of this research, a 1.575 GHz LNA with tuning knobs was fabricated in IBM's 130nm RF/CMOS process. Through extensive characterization, the performances of the device and the sensors were recorded for all possible 343 knob settings. Our experimental results demonstrated that the mid-point alternate test is a very cost-effective way of calibrating an analog/RF device after its production. Specifically, this method operates in one shot, avoiding costly test-&-tune iterations, thereby reducing cost, while achieving similar yield recovery results as the more expensive exhaustive alternate test approach.

REFERENCES

[1] T. Das and P. R. Mukund, "Self-calibration of gain and output match in LNAs," in *IEEE International Symposium on Circuits and Systems*, 2006, pp. 4–9.

[2] V. Natarajan, A. Banerjee, S. Sen, S. Devarakond, and A. Chatterjee, "Yield recovery of RF transceiver systems using iterative tuning-driven power conscious performance optimization," *IEEE Design and Test of Computers*, vol. 32, no. 1, pp. 61–69, 2015.

[3] S. Sun, F. Wang, S. Yaldiz, X. Li, L. Pileggi, A. Natarajan, M. Ferriss, J.-O. Plouchart, B. Sadhu, B. Parker *et al.*, "Indirect performance sensing for on-chip self-healing of analog and RF circuits," *IEEE Transactions on Circuits and Systems*, vol. 61, no. 8, pp. 2243–2252, 2014.

[4] S. K. Devarakond, V. Natarajan, S. Sen, and A. Chatterjee, "BIST-assisted power aware self healing RF circuits," in *IEEE Mixed-Signals, Sensors, and Systems Test Workshop*, 2009, pp. 1–4.

[5] M. Andraud, H.-G. Stratigopoulos, and E. Simeu, "One-shot calibration of RF circuits based on non-intrusive sensors," in *ACM/IEEE Design Automation Conference, 2014*, 2014, pp. 1–6.

[6] N. Kupp, H. Huang, P. Drineas, and Y. Makris, "Post-production performance calibration in analog/RF devices," in *IEEE International Test Conference*, 2010, pp. 1–10.

[7] D. Han, B. Kim, and A. Chatterjee, "DSP-driven self-tuning of RF circuits for process-induced performance variability," *IEEE Transactions on Very Large Scale Integration Systems*, vol. 18, no. 2, pp. 305–314, 2010.

2015 IEEE Computer Society Annual Symposium on VLSI

Toward adaptation of ADCs to operating conditions through on-chip correction

V. Kerzérho, L. Guillaume-Sage, F. Azaïs, M. Comte, M. Renovell and S. Bernard

LIRMM – University of Montpellier 161, rue ADA, 34 392 Montpellier, France

name@lirmm.fr

Abstract—this paper discusses the need for on-chip correction of mixed-signal integrated circuits, and particularly Analog-to-Digital Converters (ADCs), for in-situ adaptation to the operating conditions. By operating conditions, we mean the environmental conditions and the electrical settings applied to the ADC by the complex system in which it is implemented. This discussion is supported by experimental measurements of the impact of temperature and sampling frequency on the non-linearity of the ADC and its correction using a so-called LUT-based (Look-Up Table) correction technique. These experimental results enlighten the need for in-situ correction and an architectural solution is presented. The feasibility of this solution is discussed in terms of silicon area overhead and LUT-filling time.

Index Terms— mixed-signal circuits, ADC, LUT-based correction, in-situ correction

I. INTRODUCTION

Nowadays, it becomes difficult for designers to meet the required performances for the devices in demand and the time-to-market. This issue is particularly critical for mixed-signal circuits. A potential solution consists in using, after-manufacturing, a correction scheme. However corrections schemes after manufacturing are usually set for specific and limited operating conditions. Moreover correction techniques are generally architecture-dependent and cannot cover all possible user cases in the application or drift due to aging effects.

In this paper, we address the particular case of Analog-to-Digital Converters (ADCs), an essential component between the analog world and the digital processing circuitry.

The paper is organized as follows. Section II gives an overview of market trends and induced constraints on design and manufacturing of integrated systems, and briefly reviews existing correction techniques for ADCs. Section III provides a state-of-the-art of LUT-based correction techniques. Some experimental validations of LUT-based correction are presented in section IV, which highlight the need to go from a conventional correction scheme to an adaptive solution to the operating conditions. Finally section V presents an architectural solution for the implementation of in-situ LUT-based correction and discusses its feasibility in terms of area overhead and processing time.

II. MARKET TRENDS AND INDUCED CONSTRAINTS

In the Integrated Circuits (ICs) domain, Moore's law used to be the leitmotiv. Nowadays more-than-Moore is already reached

thanks to 3D stacking technologies. At the system level, a common trend is to relax constraints induced by manufacturing of analog and mixed-signal circuits. For instance, it is now common to use the same manufacturing process for both digital and analog parts of a system. From the architecture point-of-view, another trend is to limit the number of analog components as much as possible. We can expect that Software Designed Radio (SDR) will be the standard architecture in the near future and that most of the functions of a system will be operated by digital circuits.

If we go up of one level, it is now commonly accepted that we tend to have less electronic devices but performing more diverse applications. The obvious example is the smart phone that is not only used to communicate but also to take pictures or as a geographical positioning appliance. From IC point-of-view, it means that the same components will be integrated in a number of different systems with various operating conditions, systems in turn integrated in a number of different electronic devices. If we take the example of ADC, the same design will be integrated in different System-on-Chips (SoC) or electronic systems. These systems require the same function but in different operating conditions (power supply, sampling frequency, bandwidth, temperature, humidity…). ADCs are already designed for a range of these settings but there is a strong interest to develop ADCs able to function under a number of different operating environments.

Besides these market trends, technology scaling has also to be taken into account since it exacerbates process variations [1] and reliability failure mechanisms [1,2]. These aspects are traditionally addressed during the development of manufacturing process, but they can also be addressed during circuit design. For instance in [3], reliability concerns are addressed by linking the degradation of basic elements to the performances of the circuit. Other works focus on identification of dominant failure mechanisms [4] or location of weak spots [5] so that appropriated countermeasures can be implemented at the early design stage.

Regarding ADCs which are the focus of this paper, it is clear that design requirements become more and more stringent as new generations provide increasing speed and resolution. As a result, it becomes harder to ensure high production yield. At the same time, the high demand for integrated systems suitable for different standards or application fields has forced designers to develop wideband ADCs. Moreover, the sensitivity of new technologies to operating conditions requires careful attention. At the end, the designer must take all possible user scenarios into

978-1-4799-8720-7/15 $31.00 © 2015 IEEE

account, which means increasing the design constraints and leads designers to search for new strategies.

One attractive solution to make both the design and manufacturing of high performance devices easier is to correct the ICs. Such a correction can be undertaken at different times: (I) after manufacturing, (II) before use, (III) periodically during the circuit life, or (IV) on-line. From (I) to (IV), we have an increase in the quality of the IC adaptation to the current usage, which is counterbalanced by an increasing impact on the required circuitry.

In this paper we focus on case (III), as a good compromise between performance enhancement and additional cost in terms of hardware resources. Indeed, in case (II), the correction only takes into account the starting point conditions. Therefore, such a correction is insensitive to variation of these conditions (e.g. temperature, power supply, etc.). In case (IV), online correction of the ADC requires considerable additional circuitry as illustrated in existing solutions dedicated to specific ADC architectures [6-8].

Reference [9] gives a global overview of existing correction techniques. The ADC error correction methods are separated into four categories, i.e. methods based on: (a) Architecture, (b) Model Inversion, (c) Dithering and (d) Look-Up Tables (LUT). Category (a) includes: Successive Approximation Register Converters, Sigma-Delta Converters, Pipelined Converters, etc. For all of these, the correction is applied inside the architecture. Thus, the adaptation is effective only within the specific architecture. In (b), the ADC's non-linearity is mathematically modeled (e.g. Volterra) and an inverse model then compensates the imperfections. The limiting factor for such a method is the design of the inverse model in a reduced silicon area. In (c), correction relies on dithering, which is a method for randomizing the quantification error by adding an input-uncorrelated stimulus to the input signal. None of solutions developed in these three categories is sufficient for handling dynamic variations of integral non-linearity (INL). Finally in (d), correction relies on the use of a Look-Up Table that stores the correlation between the imperfect ADC's output values and the correct digitalized code. For our study, this last approach is chosen because of its flexibility and reduced silicon area overhead. However it is important to mention that none of the LUT-based techniques listed in [9] have been developed for in-situ self-correction.

An in-situ self-correction has already been published in [10] It uses an efficient INL measurement technique known as Stimulus Error Identification and Removal (SEIR), which aims to compensate the lack of linearity of the test stimulus by digital post-processing. Nevertheless, the amount of dedicated resources required for full-range correction of the ADC makes impossible the complete integration of the SEIR method. The technique might be adapted for some particular ADC architectures which necessitates the correction of only a reduced number of codes. However this technique cannot be adapted to changes in the conditions of use of the ADC, which is one of the best advantages that can offer an in-situ method as it will be demonstrated in the section IV.

III. LUT-BASED CORRECTION FOR ADCS

A. Basic principle

As mentioned in the introduction, several techniques have been developed for ADC correction. The LUT-based approach [11-15] is a very efficient solution because it works only in the digital domain, i.e. after conversion of the signal. The fundamental principle is illustrated in Figure 1, which simply consists in substituting the ADC output code by a pre-computed corrected code stored in a LUT.

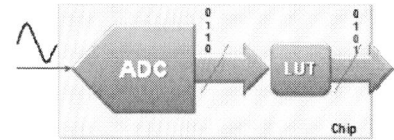

Figure 1: basic principle of LUT-based correction of ADC

The LUT is usually filled during the production phase of the ADC, after estimation of its non-linearity. The first step of the process is therefore INL estimation of the ADC. INL is defined for each code i and actually consists of several independent components that can be modelled using a multiple component equation [16,17]:

$$INL(i,d,k) = LCF_s(i,k) + LCF_d(i,d,k) + HCF(i,k) \quad (1)$$

where LCF_s is the static part of the low-code frequency component, LCF_d is the dynamic part of the low-code frequency component, HCF is the high-code frequency component, $d = \delta Vin/\delta t$ is the slope of the input signal at the amplitude corresponding to code i, and k represents external parameters such as temperature, drift due to aging or new application setups. Note that in the context of LUT-based correction, the objective is not to perform a perfect measurement but to estimate the INL with the required accuracy for the correction process.

B. Static versus dynamic correction

In classical LUT-based correction techniques, there are two different approaches, i.e. static and dynamic corrections. Static correction of an n-bit ADC necessitates a LUT of 2^n words. Indeed in the nominal operation of the ADC, the LUT is simply addressed by the current output code [8-10], which implies a one-to-one correspondence between the ADC and the LUT. Some solutions exist to reduce the size of the LUT by correcting only a selected subset of the ADC codes. Note that static correction is efficient only if the input frequencies are close to the frequency used to fill the LUT [13] or if the ADC INL is relatively insensitive to the input frequency.

Dynamic correction was developed to overcome the limitations of static one. In order to account for dynamic non-linearity, it is necessary to consider both current and previous output codes. Consequently, additional computed data must be used to fill and address the LUT. Two different dynamic correction approaches have been proposed in the literature. The first technique called phase-plan technique [13] uses the input signal slope as an additional addressing data; the resulting table has therefore two dimensions. The main challenge is to obtain a good estimation of the input signal slope. The second dynamic correction method is called state-space correction [14,15]. The basic principle is to consider not only the current ADC output code, but also a number of preceding codes, which may be representative of the input slope. The size of the correction table is directly related to the number considered output codes, i.e. a table of dimension K+1 in cased of K previous samples. The drawbacks of this method are the huge size of the table and the long time required to address it. Again, solutions have been proposed to reduce this size: the samples may be under-sampled and truncated [6,7] or certain bits can be masked after a learning phase. These different addressing techniques have been compared in [7,8] and it appears that the effectiveness of the correction decreases as the size of the LUT decreases to the point where "dynamic" correction performs no better than static correction.

C. Summary

Static LUT-based correction is very efficient because it is able to achieve good correction with quite a small LUT. However, static correction cannot take into account the dynamic part of the INL. On the other hand, although dynamic correction is a promising technique, the overall benefits are debatable because of its complexity and required silicon area. The critical aspect of additional silicon area is addressed in section V, which compares different LUT-based correction solutions for a given technology.

IV. FROM CORRECTION TO ADAPTATION OF ADCs TO THE OPERATING CONDITIONS

A. Experimental set-up

In order to investigate the performance of LUT-based correction approaches through actual measurements, experiments have been performed on a commercial converter. This converter is a 12-bit pipelined ADC [18] that embeds a digital correction block which ensures no missing codes and low Differential Non-Linearity (DNL), as illustrated in the block diagram of Figure 2. According to its datasheet, it presents an INL in the range of ±4LSBs. The purpose of the experiments is not to provide a discussion on the performances of this particular converter but to use it as a support for evaluation of LUT-based correction solutions.

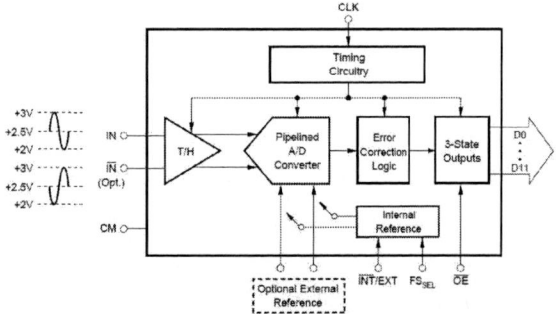

Figure 2: ADC block diagram (from product datasheet)

The experimental set-up developed around this converter is given by Figure 3. The purpose of this experimental set-up is to perform INL measurement using the ramp-based histogram method. Three external instruments are used, i.e. (i) a multi-function data acquisition device (NI6353) with 16-bit resolution resources for analog signal generation and capture, (ii) a logic analyzer (Agilent 16821A) for ADC output code capture, and (iii) an arbitrary waveform generator (TTI TG5011) for clock signal generation.

This setup is controlled by a PC using Labview environment. The filter set on the application board is specified in the datasheet and is mainly used to adapt the single-ended output of the signal generator to the differential input the ADC. Note that the analog signal delivered at the ADC input is also captured in order to monitor the quality of the stimuli. Indeed histogram-based method requires a stimulus with a resolution significantly higher than the one from the Device-Under-Test (DUT). This setup can be used to perform INL measurement at different sampling frequencies. The slope of the input ramp stimulus is adapted according to the sampling frequency in order to have the required number of samples by code for the histogram, i.e. 10 samples by code in our experiments.

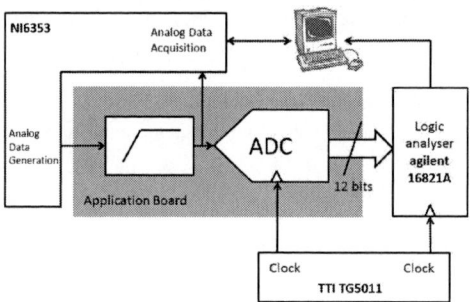

Figure 3: experimental set-up

B. Impact of operating conditions on INL

Generally an ADC is supposed to work in wide-range operating conditions. In order to examine the impact of operating conditions on the converter INL, we have performed experiments considering different sampling frequencies and different temperatures. More precisely, the INL of the ADC has been measured for three different sampling frequencies of 5MHz, 10MHz and 15 MHz at eight different temperatures from 20°C to 80°C. Results are presented in Figure 4, 5 and 6 for the three values of the sampling frequency. Note that in order to present relevant comparison between the eight measured INL curves at different temperatures, a polynomial interpolation has been done on each INL curve. This interpolation prevents from computing the conventional INL parameter which is defined as the maximum INL value observed over all the ADC codes, but it gives the general trend of the INL curve which corresponds to the LCF_s member of equation 1.

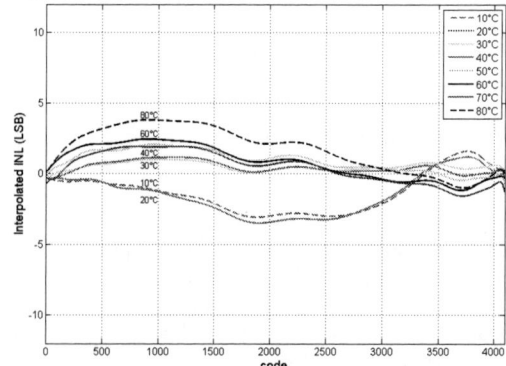

Figure 4: INL f_s=5MHz, t°: 20→80°C

978-1-4799-8720-7/15 $31.00 © 2015 IEEE   636

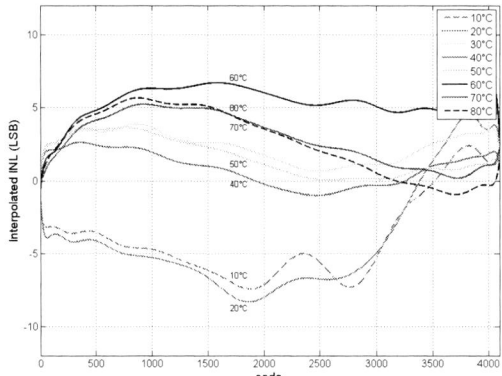

Figure 5: INL f_s=10MHz, t°: 20→80°C

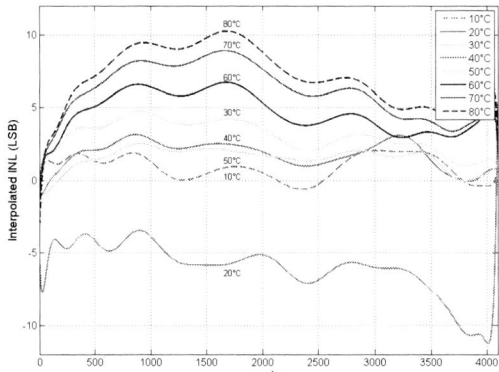

Figure 6: INL f_s=15MHz, t°: 20→80°C

Analyzing these results, we can observe that both the sampling frequency and the temperature have a significant impact on the general INL trend. However it appears that the influence of temperature is higher than the influence of sampling frequency and that the influence of temperature also depends on the value of the sampling frequency. To illustrate these points, we have computed for each code i, the maximum variation observed over the different temperatures $\Delta I\widetilde{N}L_{t°}(i)$ and the maximum variation observed over the different sampling frequencies $\Delta I\widetilde{N}L_{f_s}(i)$. Results are summarized in Figure 7.

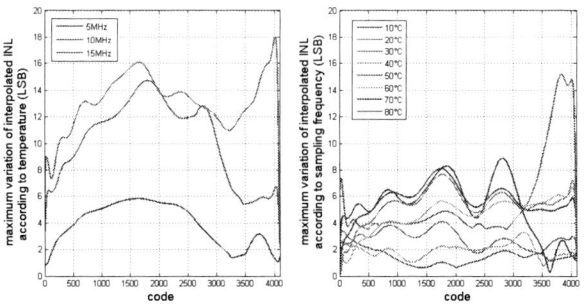

Figure 7: maximum variation of interpolated INL for each code according to (a) temperature and (b) sampling frequency

According to Figure 7, the variation of INL curve trend over the temperature range of 10°C to 80°C is very important, with a maximum value around 6 LSB in case of 5MHz sampling

frequency, 15 LSB in case of 10MHz sampling frequency, and 18 LSB in case of 15MHz sampling frequency. In the same way, a significant variation of INL curve trend is observed over the different sampling frequencies, with a maximum value that reaches 15 LSB in case of 20°C temperature.

More generally these experiments clearly illustrate that both electrical and environmental conditions (sampling frequency and temperature in this particular case) can have a significant impact of the performance of the converter during its operation. Moreover, these parameters cannot be considered as independent factors since the influence of the sampling frequency depends on the value of the temperature and vice-versa.

C. Correction of ADC

As explained in section III, INL measurement (or estimation) is the first step for the implementation of LUT-based correction approaches. In case of static correction, the LUT has only one-dimension and corrected codes are easily computed based on a rounding of the estimated INL. During the normal operation of the converter, the current output code simply addresses the LUT, which substitutes this code with the one in the table. This solution has been implemented for the particular case of the ADC considered in the previous experiment working at a sampling frequency of 5MHz for an external temperature of 20°C. Results of INL measurement before and after correction are presented in Figure 8. The efficiency of the correction is clearly illustrated since the maximum INL reduces from 4.5 down to 0.5LSB.

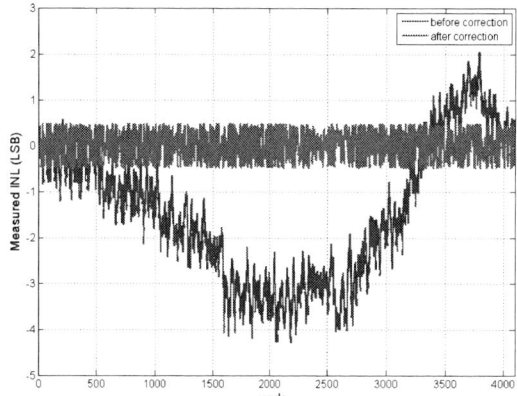

Figure 8: INL f_s=5MHz, 20°C corrected at 5MHz, 20°C

In order to estimate the dynamic performances, ADC models defined according to the measured INL before and after correction have been simulated. The Total Harmonic Distortion (THD) and Spurious Free Dynamic Range (SFDR) before and after the correction are given in Table 1.

Table 1: dynamic parameters of the simulated ADC before and after correction

	THD (dB)	SFDR (dB)
Before correction	-60.2	61.4
After correction	-82.8	86.3

D. The need for in-situ correction

The previous experiment has clearly demonstrated the benefit of LUT-based static correction when the operating conditions are similar to the ones used for the computation of the correction table. Let us now assume that the converter is integrated in a system with operating conditions different form the ones used by the IC provider for correction implementation.

First we consider the case of a correction table computed for a sampling frequency of 5MHz and a temperature of 20°C, and we assume than the converter is integrated in a complex system warming the environment to 80°C. Figure 9 shows the INL measured with the experimental setup before and after correction; the maximum INL actually increases from 4LSB up to 7LSB. Similarly, the dynamic performances are lowered as reported in Table 2. These results clearly show that the correction process is not effective in this case and even induces a degradation of the ADC performances.

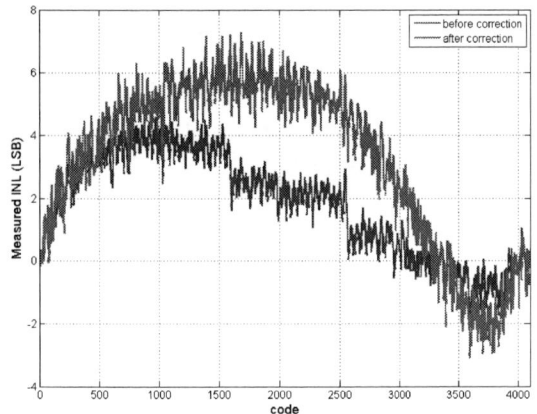

Figure 9: INL f_s=5MHz, 80°C corrected at 5MHz, 20°C

Table 2: ADC dynamic performances at f_s=5MHz, 80°C before and after correction, with correction at 20°C

	THD (dB)	SFDR (dB)
Before correction	-62.3	65.4
After correction	-56.4	57.5

Let us now consider a different scenario, i.e. the correction table has been computed for a sampling frequency of 15MHz and a temperature of 20°C. The ADC is then integrated in a complex system requiring a sampling frequency of 5MHz. Figure 10 gives the measured INL before and after correction and Table 3 reports the dynamic parameters of the corresponding simulated ADCs. Here again, the correction decreases the quality of the ADC.

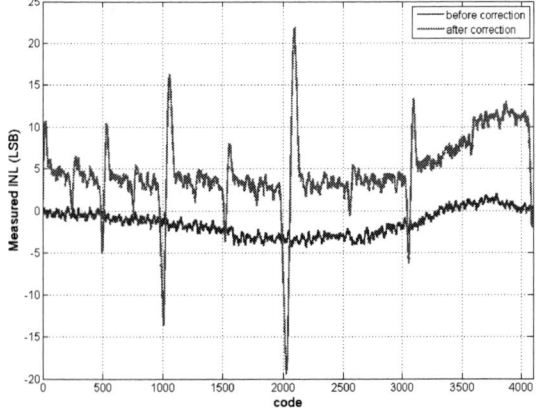

Figure 10: INL f_s=5MHz, 20°C corrected at 15MHz, 20°C

Table 3: ADC dynamic performances at f_s=5MHz, 20°C before and after correction, with correction at 15MHz

	THD (dB)	SFDR (dB)
Before correction	-60.6	61.7
After correction	-55.6	60.6

These experiments demonstrate that because ADC performances are sensitive to the operating conditions, LUT-based correction is effective only if the correction is computed under similar operating conditions than the ones used in the application.

V. PROPOSED SOLUTION

A. Principle

A potential approach to alleviate the limitation of conventional LUT-based correction is to perform in-situ correction. Indeed by completing the LUT *in-situ*, i.e. directly in the application, the corrected codes are obviously computed according to both electrical and environmental conditions. An interesting solution has been introduced in [17,19], in which static LUT-based correction is associated with on-chip capability for computing and filling the LUT. The general architecture of this adaptive solution is depicted in Figure 11. Besides the LUT that stores corrected codes, it also includes additional circuitry for on-chip signal generation and on-chip INL estimation.

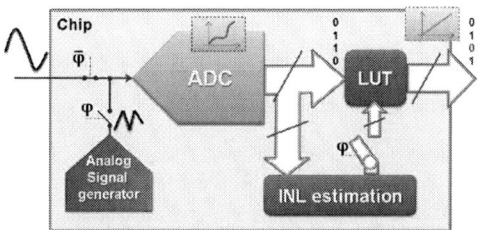

Figure 11: architecture for in-situ adaptive LUT-based correction

During the correction process, a control signal disconnects the primary input of the ADC and connects the integrated signal generator. The output INL is estimated and the computed corrected codes are stored in the LUT. After this correction phase, the ADC is disconnected from the embedded generator and the ADC runs with its normal primary input; output codes are then corrected according to the values stored in the LUT. In this solution, since the correction circuitry is integrated in the chip, the above correction process can be repeated at any time or location according to the operating conditions.

To be able to cover diverse applications, the generator must be flexible and able to generate an input signal in the corresponding frequency range of the targeted application. [17,19] demonstrate the feasibility of such adaptive correction. In this paper, we provide additional data to discuss the implementation regarding the area overhead and processing time.

B. Area estimation

The area overhead is one of the main issues concerning ADC correction. Table 4 gives the area overhead estimation in AMS 0.35 μm technology for different ADC resolutions and different correction solutions: static, dynamic (with 2D and 3D LUTs) and adaptive correction. For the four cases, the LUT and circuitry required for implementing the correction have been described in VHDL and synthetized using the previously quoted technology.

In case of the adaptive solution, the area of the embedded signal generator has been also included to compute the area overhead.

Table 4: Area estimation of the different LUT-based techniques for several ADC resolutions

	Area (mm²)			
n	Static	Dynamic 2D	Dynamic 3D	Adaptive
6	0.161	0.335	0.496	0.182
8	0.232	0.477	0.709	0.255
10	0.820	1.653	2.473	0.848
12	2.924	5.861	8.785	2.951

These results show that compared to the conventional static correction, the area overhead is roughly 2 times higher for 2D dynamic correction and 3 times higher for 3D dynamic correction. However the benefit in terms of adaptation to the operating conditions is limited since only 2 or 3 conditions can be taken into account. In contrast, the adaptive solution permits to handle any operating conditions and induces only a small increase of the area overhead. Note that the increase in the area overhead depends on the converter resolution and diminishes for higher resolution: in case of a 12-bit converter, the increase is less than 1%.

C. LUT-filling time estimation

The adaptive LUT-based technique allows filling the LUT in the application. To be practical, it is clear that the time required for completing the correction process should be as short as possible since the circuit is not operational during this procedure.

The correction technique is based on an estimation of the INL using the LH-BIST technique [19]. Most of the processing time is actually dedicated to the acquisition of the samples required for INL estimation, the time for computing corrected codes and loading the LUT being negligible. An estimation of this time is given by the following equation:

$$Time \cong H_{ideal}/f_s + 4/f_s \qquad (2)$$

where H_{ideal} is the ideal number of occurrences for each code in the histogram, f_s is the sampling frequency and n the converter resolution. Table 5 summarizes this processing time for different converter resolutions and different sampling frequencies, in case of an ideal number of occurrences for each code equal to 32.

Table 5: Estimation of processing time for several ADC resolutions

	Processing Time (ms)		
n	fs=5MHz	fs=50MHz	fs=100MHz
6	0.41	0.04	0.02
8	1.64	0.16	0.08
10	6.55	0.66	0.33
12	26.22	2.62	1.31

These results show that the processing time increases for higher resolution and lower sampling frequency. However, even for a relatively low-frequency 12-bit converter, the processing time does not exceed few tens of ms, which seems acceptable in the context of a periodical in-situ correction.

VI. CONCLUSION

In this paper, we have discussed the need for on-chip correction of mixed-signal circuits in order to perform in-situ adaptation to the operating conditions. Based on experiments on a commercial converter, we have demonstrated that LUT-based correction is efficient only if computed according to the operating conditions, in our case sampling frequency and temperature. As a consequence, it becomes mandatory to integrate the resources in the ADC to make it able to run itself the correction after integration in the system. An architectural solution has been evaluated in terms of area overhead and processing time. Results have shown the feasibility of this approach. This adaptive correction solution has to be more deeply investigated regarding other operating conditions. It would be also interesting to evaluate the impact of wearing-out mechanisms on the effectiveness of the correction.

REFERENCES

[1] B. Vaidyanathan, A.S. Oates, "Technology Scaling Effect on the Relative Impact of NBTI and Process Variation on the Reliability of Digital Circuits," IEEE Trans on Device & Materials Reliability , vol.12, no.2, pp.428,436, June 2012.

[2] J. Srinivasan, S.V Adve, P. Bose, J.A. Rivers, "The impact of technology scaling on lifetime reliability," Int'l Conference on Dependable Systems and Networks, 2004, pp.177,186, 2004.

[3] S.A. Oates, "Reliability challenges for the continued scaling of IC technologies," Custom Integrated Circuits Conference (CICC), pp.1,4, 2012

[4] Baoguang Yan, et al., "Reliability Simulation and Circuit-Failure Analysis in Analog and Mixed-Signal Applications," IEEE Trans on Device & Materials Reliability, vol.9, no.3, pp.339,347, Sept. 2009.

[5] V.Kerzérho, H G. Kerkhoff, G.-J. Bollen, and Y. Xing, "The search for resilience weak spots in automotive mixed-signal circuits", Int'l Mixed-Signals, Sensors & Systems Test Workshop (IMS3TW), pp.137-142, 2011

[6] J. Tsimbinos and K. V. Lever, "Improved error-table compensation of A/D converters," Inst. Elect. Eng. Circuits, Devices Systems, Vol. 144, n°6, pp. 343-349, Dec. 1997.

[7] H. Lundin, M. Skoglund and P. Handel, "A criterion for optimizing bit-reduced postcorrection of AD converters", IEEE Trans on Instrumentation & Measurement, Vol. 53 (4), pp.1159-1166, 2004.

[8] A. J. Davis, et al., "Digital correction of circuit imperfections in cascaded Σ-Δ modulators composed of 1st-order sections", Int'l Symposium on Circuits & Systems (ISCAS), Vol.5, pp.689-692, 2000.

[9] E. Balestrieri, P. Daponte and S. Rapuano, "A State of the Art on ADC Error Compensation Methods," IEEE Trans on Instrumentation & Measurement, Vol. 54, pp. 1388-1394, Aug. 2005.

[10] L. Jin, D. Chen, R.L. Geiger, "A digital self-calibration algorithm for ADCs based on histogram test using low-linearity input signals", Int'l Symposium on Circuits & Systems (ISCAS), pp. 1378-1381, 2005.

[11] H. Lundin, "Post-correction of analog-to-digital converters", Royal Institute of Technology (KTH), Stockholm, Sweden, Licentiate thesis TRITA–S3–SB–0324, May 2003.

[12] F.H. Irons, D.M. Hummels and S.P. Kennedy, "Improved compensation for analog-to-digital converters", IEEE Trans. on Circuits & Systems, Vol.38 (8), pp. 958-961, 1991.

[13] H. Lundin, T. Andersson, M. Skoglund and P. Handel, "Analog-to-digital converter error correction using frequency selective tables", Radio Vetenskap och Kommunikation, pp. 487-490, 2002.

[14] S. Acunto, P. Arpaia, D.M. Hummels and F.H. Irons, "A new bidimensional histogram for the dynamic characterization of ADCs", IEEE Trans on Instrumentation & Measurement, Vol. 52, pp. 38-45, Feb. 2003.

[15] H. Lundin, M. Skoglund and P. Handel, "Optimal index-bit allocation for dynamic postcorrection of analog-to-digital converters", IEEE Trans. on Signal Processing, Vol. 53 (2), pp. 660-671, 2005.

[16] S. Medawar, P. Händel, N. Björsell and M. Jansson, "Input-dependent integral nonlinearity modeling for pipelined analog-digital converters", IEEE Trans. on Instrumentation & Measurement, Vol. 59, N° 10, pp. 2609-2620, October 2010.

[17] S. Bernard, F. Azaïs, M. Comte, O. Potin, V. Kerzérho, and M. Renovell, "Adaptive LUT-based system for insitu ADC auto-correction," Int'l Mixed-Signals, Sensors & Systems Test Workshop (IMS3TW), pp. 1-6, 2010.

[18] ADS807 datasheet, http://www.ti.com/lit/ds/symlink/ads807.pdf

[19] V. Kerzérho et al., "A novel implementation of the histogram-based technique for measurement of INL of LUT-based correction of ADC", Microelectronics Journal, Vol.44, No.9, pp.840-843, Sept. 2013.

2015 IEEE Computer Society Annual Symposium on VLSI

A Full-swing CMOS Current Steering DAC with an Adaptive Cell and a Quaternary Driver

Yanghyeok Choi, Seonghyun Park, Jieun Yoo, Seol Namgung, and Minkyu Song

Dept. of Semiconductor Science, Dongguk Univ-Seoul,100-715, Korea
E-mail: mksong@dongguk.edu

Abstract— This paper describes a CMOS current steering digital-to-analog converter (DAC) with a full-swing output signal. Generally, a normal current steering DAC cannot have a full swing output signal, because conventional DACs have an inevitable voltage drop at the output current cell. In order to improve the drawbacks, we propose a new scheme of quaternary driver and an adaptive output current cell composed of both nMOS and pMOS. As well as the nMOS operates from the power supply to the half of the power supply, the pMOS operates independently from the half of the power supply to the ground voltage. Then, the final output voltage is obtained through a multiplexer that is driven by a quaternary driver which selects the optimized current cell. A 6-bit 1GS/s current steering DAC has been fabricated with Dongbu 0.11um 1-poly 6-metal CMOS technology to verify the performance of the proposed full-swing DAC. The effective chip area is 0.46mm² and power consumption is about 19.1mW. The measured results reveal that the DAC has a full swing output signal.

Keywords— *full-swing; current steering DAC; adaptive cell; quaternary driver*

I. INTRODUCTION

Recently, the current trends of system-on-chip (SOC) are towards for integrating digital circuits and analog circuits in a chip. As a result, a data converter, which is part of a vital interface within those systems, is becoming an increasingly more important block. A digital-to-analog converter (DAC) is a representative circuit that a digital code is converted into an analog signal [1]-[14]. Normally, the kinds of DAC are mainly divided into two categories: one is voltage-steering type, and the other is current-steering type [1]. Since the settling time of the output voltage depends on the slew rate of the op-amp, the voltage-steering type based on the op-amp

for the DAC output is not suitable for high-speed applications. In case of the current-steering type, by the way, the current generally flows directly through off-chip resistors or termination resistors inside the chip to obtain a fast operating speed. However, the output voltage at the termination resistor cannot have a full-swing, since the inevitable voltage drop is generated between the drain and the source of the output current cell. Therefore, it is almost impossible to have a full-swing output voltage at the current steering type DACs, even though a rail-to-rail output voltage is needed for the applications such as display drivers, medical equipment, and so on.

In this paper, a current-steering DAC with a full-swing output voltage from the ground voltage to the power supply is proposed. The DAC has an architecture that follows the thermometer code method which has excellent monotonicity and low glitch energy. In addition, the conventional logic and latch circuits have been simplified in order to reduce the power consumption and to correct mismatches. In order to implement the full-swing output voltage, a new scheme of quaternary driver and an output current cell composed of both nMOS and pMOS are discussed. Firstly, the nMOS current cell operates from the power supply to the half of the power supply. Secondly, the pMOS current cell operates independently from the half of the power supply to the ground voltage. Then, the final output voltage is obtained through a multiplexer that is driven by a quaternary driver which selects the optimized current cell. The contents of the paper are as follows. In section II, the circuit design technique of the full-swing DAC is described. Measurement results and conclusions are summarized in Section III and IV, respectively.

978-1-4799-8720-7/15 $31.00 © 2015 IEEE   640

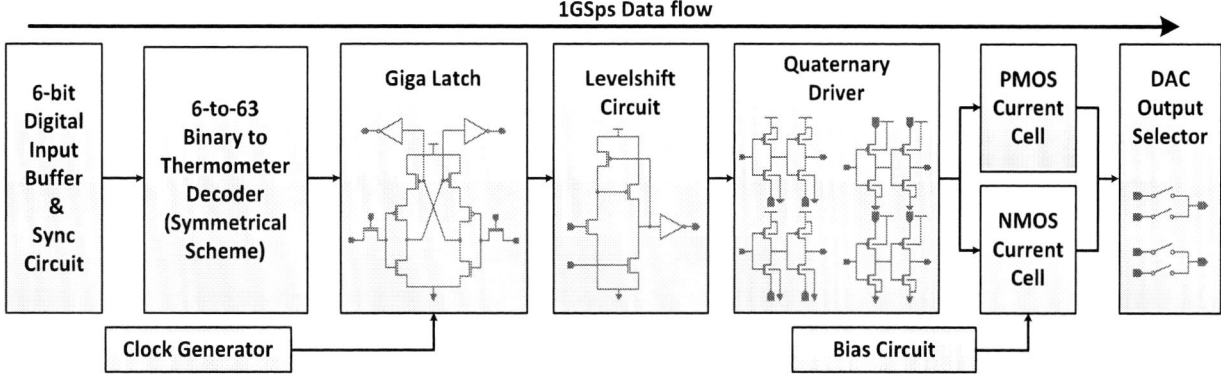

Fig. 1. Block diagram of the full-swing DAC

II. DESIGN OF A FULL-SWING CURRENT-STEERING DAC

A. Architecture

Fig. 1 shows the block diagram of the full-swing DAC. The 6-bit digital input codes are arranged at the input data sync block to obtain the same delay time. At the block of 6-to-63 binary to thermometer decoder, the binary digital codes are converted into thermometer codes to improve the linearity and decrease the glitch noise. Then, the thermometer codes are transferred to the proposed switching logic, giga latch, and level shifter. Finally, the output currents are converted into analog voltages by the termination resistors. Generally, the 6-bit current steering DAC is designed with a full matrix structure for high-speed operation and accuracy. Further, there are many advantages such as a simple design, low INL and DNL error, accurate monotonicity, low noise analog output, and so on. Thus this structure is now widely used. However, since the signals to drive the output current cell are entirely thermometer code, the complexity of binary to thermometer decoder increases exponentially with DAC resolution.

B. Circuit Design

The output current cell is the most important circuit that dominantly determines the performance of the current steering DAC. Thus it needs a careful design to consider many factors. In order to improve the drawbacks of the conventional full-swing current cell, a novel current cell with a digital driver is proposed. Fig. 2 shows the circuit diagram. The digital driver is composed of four types which drive the output current cell, respectively. Thus it is called quaternary driver. Dependent on the digital code, the quaternary driver drives the appropriate digital value to the output current cell. The quaternary driver generates weak high positive, weak high negative, weak low positive, and weak low negative, respectively. This is because the switching MOS should be

operated in the saturation mode, not in the linear mode. If the switching MOS is operated in the linear mode, there exists a voltage drop at the switching MOS. In order to minimize the voltage drop at the switching MOS, therefore, the proposed quaternary driver is designed. At the right side of Fig.2, there is the circuit diagram of the current cell. The operation region of nMOS current source is from the power supply to the half of the power supply. On the contrary, the operation region of pMOS current source is from the half of the power supply to the ground voltage. For example, when the power supply voltage is 3.3V and the binary input code is 000000, the voltage of P_OUTP node is 0V, P_OUTN node is 1.65V, N_OUTP node is 3.3V, and N_OUTN node is 1.65V. When the binary input code is 011111, the voltage of P_OUTP node is 1.65V, P_OUTN node is 0V, N_OUTP node is 1.65V, and N_OUTN node is 3.3V. When the binary input code is 111111, the output voltages are the same as those of the binary input code 000000, because we have the symmetrical thermometer decoder.

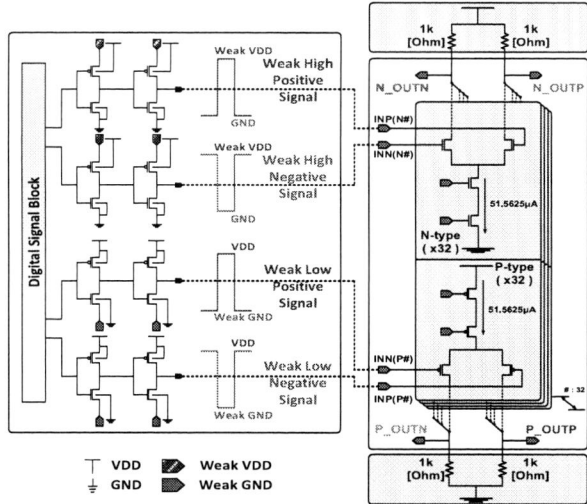

Fig. 2. Circuit diagram of quaternary driver and current cell

978-1-4799-8720-7/15 $31.00 © 2015 IEEE

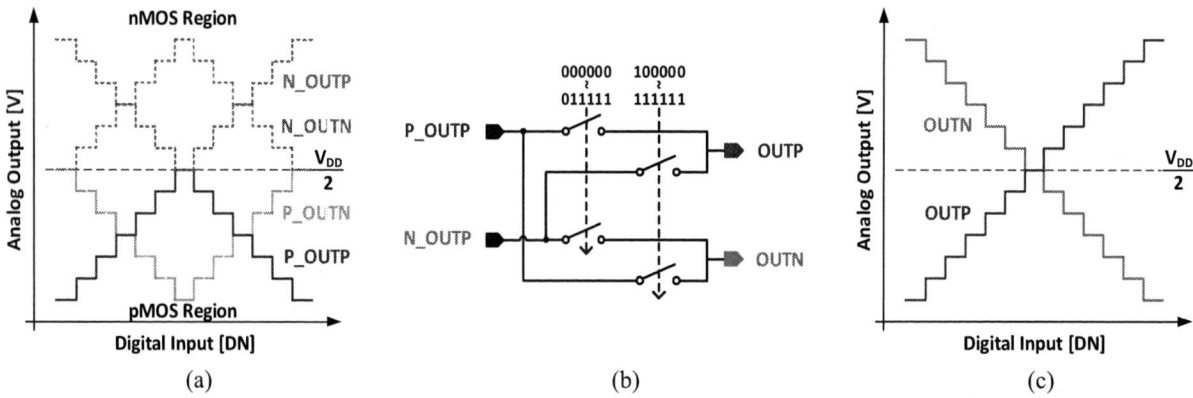

(a)   (b)   (c)

Fig. 3 (a) output waveforms of the current cell   (b) circuit diagram of output selector
(c) final full-swing output waveform

The output waveforms of current cells are shown in Fig. 3(a). Fig. 3(b) shows the circuit diagram of output selector. According to the input signal, the optimized output voltage is chosen. Using both P_OUTP and N_OUTP, we can obtain the final fully differential output voltage, OUTP and OUTN. Therefore, the final full-swing analog output voltage is obtained as shown in Fig. 3(c), even in a current steering type DAC. When the current cell shown in Fig.2 is designed, the consistency between the pMOS current cell and the nMOS current cell must be carefully considered. Specially, at the transition code from 011111 to 100000, even though the selected output current cell is changed, each current cell must keep the desired value without any errors. In order to implement the required condition, we set the termination resistors are chosen to be higher, and the gate voltage of cascade current mirror is controlled by a very high impedance technique with an op-amp. Further, to keep the linearity and consistency of each cell, external impedance control pads are also included. However, as the operating frequency is increased, the inconsistency between the pMOS current cell and the nMOS current cell becomes serious. It will be discussed in the experimental results.

III. MEASURED RESULTS

The proposed DAC has been fabricated with Dongbu 0.11 μm 1P6M CMOS technology. Fig. 4 shows the chip layout and photograph of the fabricated DAC. For the accurate data synchronization, the clock signal uses a layout method of tree structure which is a helpful routing technique to equalize the parasitic resistance and the delay time. Excluding the input and output pads of the DAC, the core area is 0.46 mm^2, and the power consumption is about 19.1mW with 1.2 V for the digital and 3.3 V for the analog. Fig. 5 shows a photograph of the printed circuit board (PCB) used for measurement. With

the fabricated PCB, many experimental results are taken to verify the performance. Fig.5 also shows the experimental environments of the proposed DAC. The performance of the proposed DAC has been tested with the Compuscope 3200 (Gage Applied Technologies). The Compuscope 3200 has a built-in hardware that can monitor the digital code through a digital capture board, including INL, DNL and SNR.

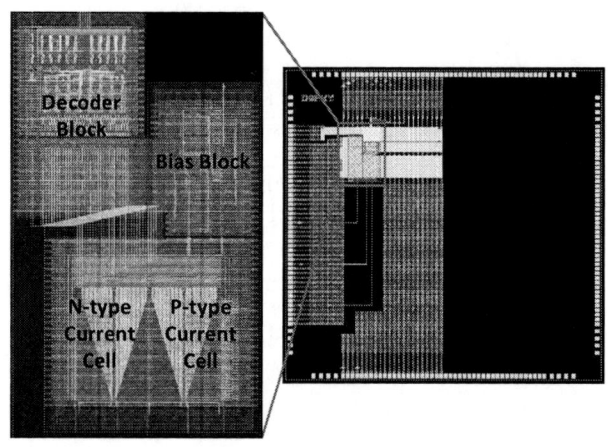

(a)

(b)

Fig.4 (a) Layout implementation and (b) chip photograph

978-1-4799-8720-7/15 $31.00 © 2015 IEEE   642

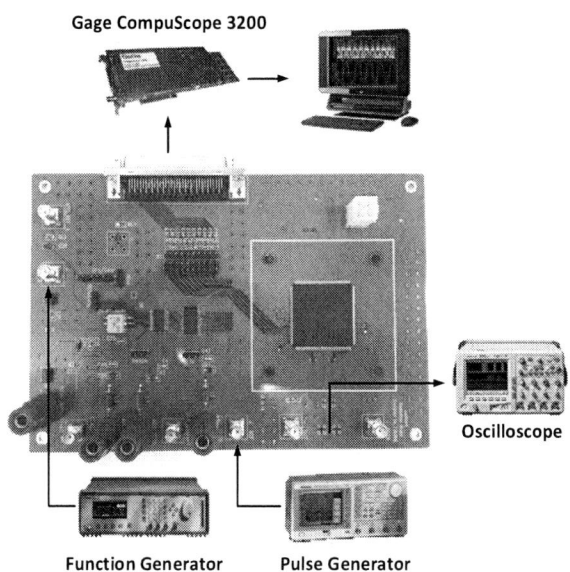

Fig.5. Photo for testing PCB and experimental environments

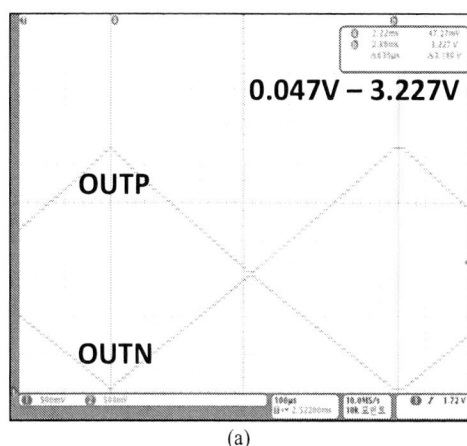

(a)

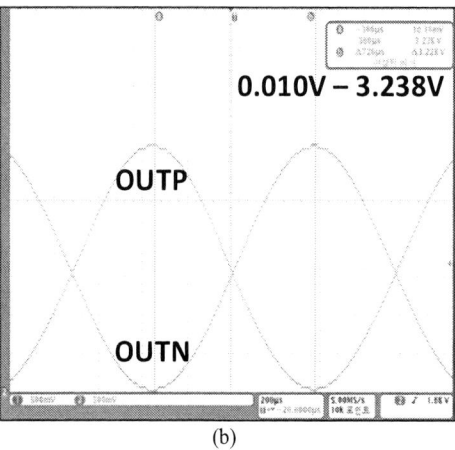

(b)

Fig.6. Measured resuls for full-swing output voltage
(a)ramp signal (b)sine waveform

Fig.6 shows the measured results of the full-swing output voltage at a low frequency. The measured full-swing ramp signal and sine waveform are shown in Fig. 6(a) and (b), respectively. Fig.6(a) is a full-swing ramp signal with the input frequecny of 800Hz at the sampling frequency of 10MS/s. Fig.6(b) is a sine waveform with the input frequecny of 700Hz at the sampling frequency of 5MS/s. Thus it is verified that the proposed DAC has a full-swing output voltage. Fig.7 shows the measured results for INL and DNL. They are within ±0.7LSB and ±0.7LSB, respectively. The measured results of INL and DNL are satisfied with the desired system specifications. Fig.8 shows the measured results of frequency response, when the input frequency is 100MHz and the sampling frequency is 1GHz. Fig.9 shows the measured SNDR and SFDR, when the input frequency is increased at the sampling frequency of 1GHz. The measured results are getting worse, when the input frequency is increased. This is because the inconsistency between the pMOS current cell and the nMOS current cell is becoming serious, and the output resistance and parasitic capacitances are larger than those of the conventional ones. However, the measured results are also satisfied with the overall specifications of a normal DAC. Tab.1 shows the performance of the proposed DAC and the comparison among the conventional ones. Even though the other performances are superior to the proposed one, the proposed one has a full-swing advantage.

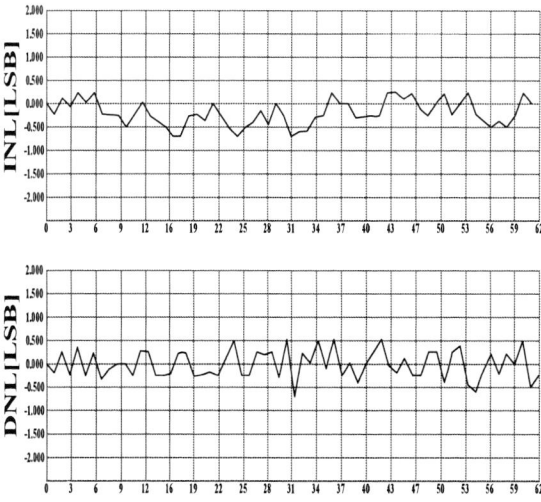

Fig. 7. Measured INL(top) and DNL(bottom)

978-1-4799-8720-7/15 $31.00 © 2015 IEEE

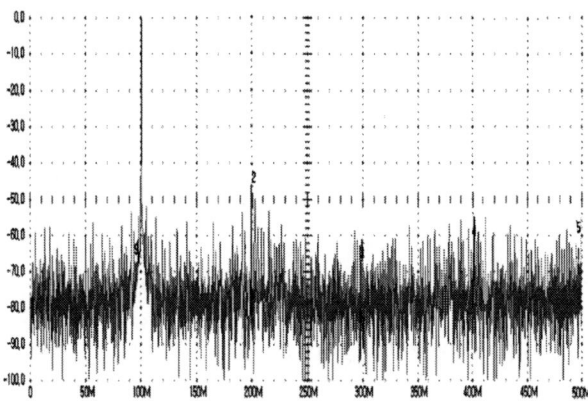

Fig. 8. Measured FFT result (fin=100MHz, fs=1GHz)

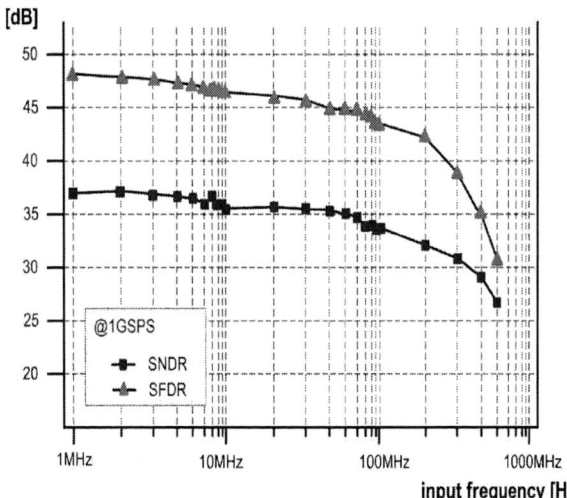

Fig. 9. Measured SNDR & SFDR vs input frequency (@1GS/s)

Tab.1 Performance summary and comparison

	[12]	[13]	[14]	This Work
Full-swing	YES (buffer)	NO	NO	YES
Resolution	10-b	6-b	6-b	6-b
Technology	45nm CMOS	90nm CMOS	0.13μm CMOS	0.11μm CMOS
Sampling Frequency	300MS/s	3GS/s	3GS/s	1GS/s
SFDR	73dB	36.24dB	36.2dB	43dB (100MHz)
INL DNL	±0.6LSB ±0.4LSB	±0.1LSB ±0.1LSB	±0.02LSB -	±0.7LSB ±0.7LSB
Power consumption	476mW	8.32mW	29mW	19.1mW
Core area	2.25 mm²	0.045mm²	0.2mm²	0.46mm²
FOM	1550f [J]	43.3f [J]	151.04f [J]	298.4f [J]

IV. CONCLUSION

In this paper, a current steering DAC with the full-swing output voltage was designed. In order to support a rail-to-rail output voltage, a new current cell and quaternary driver have been described. The proposed current cell was composed of both pMOS and nMOS. Further, the output voltage was obtained through a multiplexer that was driven by a quaternary driver which selects the optimized current cell. To verify the performance of the proposed techniques, a 6-bit 1GS/s CMOS current steering DAC with the full-swing output voltage has been designed and fabricated with Dongbu 0.11um technology. The area of the designed chip was 0.46 mm², and the static and the dynamic performance were measured on the PCB. The measured INL and DNL were within ±0.7 LSB and ±0.7 LSB, the measured SFDR was about 43dB, and the power consumption was 19.1mW, when the clock frequency was 1GS/s at the input frequency of 100MHz. Tab.1 shows the measured performance summary of the fabricated chip, and the comparison with the conventional ones.

ACKNOWLEDEGMENT

This work was supported by BK21 Plus, by the Center for Integrated Smart Sensors funded by the Ministry of Science, ICT & Future Planning as Global Frontier Project (CISS-2013M3A6A6073718), and by the MOTIE(Ministry of Trade, Industry, and Energy) supervised by the KEIT(Korea Evaluation Institute of Industrial Technology) and Exicon Co. Ltd.

REFERENCES

[1] Georgi I. Radulov et al., "An on-chip self-calibration method for current mismatch in D/A Converters," ESSCIRC., pp. 169-172, Sept. 2005.

[2] Q. Huang et al., "A 200MS/s 14b 97mW DAC in 0.18um CMOS." ISSCC Dig. Tech. Papers, pp. 364-365, Feb., 2004.

[3] Jussi Prikkalaniemi, et al., "A14b 40 MSPS DAC with current mode deglitcher," Proc. IEEE ISCAC 2002, vol. 1, 2002. pp. 121-124.

[4] Park, K., et al., "A digital-to-analog converter based on differential-quad switching", Solid-State Circuits, IEEE Journal of, Volume 37, Issue : 10, pp 1335-1338, Oct. 2002.

[5] K. O'Sullivan et al., "A 12-bit 320Msample/s Current Steering CMOS D/A Converter in 0.44mm2," IEEE, J. Solid-State Circuits, vol. 38, no. 7, pp. 1064-1072, Jul., 2004

[6] Van Den Bosch et al., "A 12 b 500MSample/s current-steering CMOS D/A converter," ISSCC. pp, 366-367. Feb. 2001.

[7] G. Van der Plas, et al., "A 14-bit Intrinsic Accuracy Q2 Random Walk CMOS DAC," IEEE J. Solid-State Circuits, vol. 32, pp. 1708-1718, Dec. 1999.

[8] Yonghua Cong, et al., "A 1.5V 14b 100MS/s Self Calibrated DAC." Solid-State Circuits, IEEE Journal of, Volume 38, Issue : 12, pp. 2051-2060, Dec. 2003.

[9] A. Van den Bosch, et al., "SFDR Bandwidth Limitations for High Speed High Resolution Current Steering CMOS D/A Converters," Proc. ICECS,

978-1-4799-8720-7/15 $31.00 © 2015 IEEE

pp. 1193-1196, 1999.

[10] Luschas S., et al., "Output impedance requirements for DACs," Proceedings of the 2003 ISCAS, Volume:1, pp. I-512-515, May, 2003.

[11] M. Clara, W. Klatzer, D. Gruber, A. Marak, B. Seger, W. Pribyl, "A 1.5V 13bit 130-300MS/s self-calibrated DAC with active output stage and 50MHz signal bandwidth in 0.13µm CMOS," 34th European Solid-State Circuits Conference 2008 (ESSCIRC 2008), pp.262-265, Sept. 2008.

[12] Mohammad S. Mehrjoo, and James F. Buckwalter, "A 10 bit, 300 MS/s Nyquist Current-Steering Power DAC With 6 V Output Swing, IEEE " IEEE J. Solid-State Circuits, vol. 49, no. 6, pp.1408-1418, June 2014.

[13] Ren-Li Chen, Soon-Jyh Chang, "A 6-bit Current-Steering DAC With Compound Current Cells for Both Communication and Rail-to-Rail Voltage-Source Applications," IEEE Transactions on Circuits and Systems II: Express Briefs, , vol.59, no.11, pp.746,750, Nov. 2012.

[14] Xu Wu, P. Palmers, M.S.J. Steyaert, "A 130 nm CMOS 6-bit Full Nyquist 3 GS/s DAC," IEEE Journal of Solid-State Circuits, vol.43, no.11, pp.2396-2403, Nov. 2008.

2015 IEEE Computer Society Annual Symposium on VLSI

Flexible Ultra-Low-Voltage CMOS Circuit Design applicable for digital and analog circuits operating below $300mV$

Yngvar Berg
Department of Micro- and Nanosystem Technology
Buskerud and Vesfold University College
Borre, Norway
Email: Yngvar.Berg@hbv.no

Omid Mirmotahari
Department of Informatics
University of Oslo
Oslo, Norway
Email:omidmi@ifi.uio.no

Abstract—A generic ultra low-voltage (ULV) CMOS design approach is presented. By applying a floating capacitor to the gate terminal of the enhanced driving transistors, obtained by using a charge injection technique, we may change the ON and OFF currents. The delay in circuits where the enhanced transistors are utilized can be reduced significantly compared to complementary CMOS. The current level of the transistors may be increased for high speed and decreased for low power applications. The design approach may be used to implement ultra low-voltage and high-speed digital logic and Flip-Flops. In addition, the generic technique can be used to implement multiple-valued and analog ultra low-voltage CMOS circuits. For ultra low-voltage dogital applications the delay may be reduced to less than 10% compared to static CMOS. The high-speed Flip-FLOP presented shows a similar increase in speed compared to conventional Flip-Flops for low supply voltages. For the analog circuit presented the increased current level is used to obtain rail-to-rail operation at higher frequencies than conventional analog circuits.

Keywords-CMOS; Low-Voltage; High-Speed; Floating-Gate; Domino logic; Flip-Flop; Analog; Multiple-Valued logic.

I. INTRODUCTION

The International Technology Roadmap for Semiconductors (ITRS) asserts that power consumption is now the major technical problem facing the semiconductor industry. The exploiting market of portable electronic appliances fuels the demand for complex integrated system that can be powered by lightweight batteries with larger recharge time. In recent years, the power problem has emerged as one of the fundamental limits facing the future of CMOS integrated circuit design.

The use of voltage scaling along with pipelining makes Flip-Flops particularly important at ultra low-voltages. Especially in high-speed and low-voltage digital systems the timing details of the Flip-Flops represents a significant challenge. In low-voltage applications, i.e low drive currents, the relative delay variations may be quite large and will impose an additional problem. In order to reduce the problems associated with low-voltage operation we increase the current level and thereby reduce the delay and relative delay variation.

Depending upon the application, there are numerous methods that can be used to reduce the power consumption

of VLSI circuits, these can range from low-level measures based upon fundamental physics, such as using a lower power supply voltage or using high threshold voltage transistors; to high-level measures such as clock-gating or power-down modes. The power consumption in digital circuits, which mostly use complementary metal-oxide semiconductor (CMOS) devices, is proportional to the square of the power supply voltage; therefore, voltage scaling is one of the important methods used to reduce power consumption. To achieve a high transistor drive current and thereby improve the circuit performance, the transistor threshold voltage must be scaled down in proportion to the supply voltage. However, reducing the transistor threshold voltage V_t results in a significant increase in the subthreshold leakage current. For correct operation of a CMOS gate the theoretical minimum supply voltage is close to 2 times the thermal voltage $V_T = kT/q$ where k is Boltzmanns's constant, T is temperature in Kelvin and q is electronic charge [1]. In most digital applications the supply voltage will be determined based on a trade off between performance, i.e speed and power, and robustness.

Energy-efficiency is one of the most required features for modern electronic systems designed for high-performance and/or portable applications. In one hand, the ever increasing market segment of portable electronic devices demands the availability of low-power building blocks that enable the implementation of long-lasting battery-operated systems. On the other hand, the general trend of increasing operating frequencies and circuit complexity, in order to cope with the throughput needed in modern high-performance processing applications, requires the design of very high-speed circuits.

The power-delay product (PDP) metric relates the amount of energy spent during the realization of a determined task, and stands as the more fair performance metric when comparing optimizations of a module designed and tested using different technologies, operating frequencies and applications [2]. Minimum energy for a CMOS gate, i.e optimal supply voltage, is close to the threshold voltage of the nMOS transistor (V_{tn}) for a specific process [3]–[6]. For high performance digital design the energy-delay product (EDP) is a widely used metric for comparison of different

978-1-4799-8720-7/15 $31.00 © 2015 IEEE   646

logic styles. Several approaches to high speed and low voltage digital CMOS circuits have been presented [7], [8]. The use of multiple threshold voltages devices and supply voltages have been examined for low-voltage design [8], [9]. In addition, bootstrapping [10] and floating-gate (FG) [11] techniques, have been exploited for dynamic threshold voltage and supply voltage scaling to improve performance in low-voltage digital logic. Floating-gate logic implemented in a modern CMOS process requires frequent initialization to avoid significant leakage. By using floating capacitances to the transistor gate terminals, the semi-floating-gate (SFG) nodes can have different DC levels than provided by the supply voltage headroom and is often referred to as a threshold voltage shift. There are several approaches to FG CMOS logic [12], [13]. In section II the flexible ULV enhanced transistor is presented. The ULV transistor is characterized by a precharge and an evaluate phase similar CMOS precharge logic. In section III the flexible ULV transistor is used in several different application, including ULV domino logic [14] which can be either high-speed or low-power, high-speed Flip-Flop, Multiple-Valued logic (MVL) and analog circuits. Finally, a conclusion is included in section IV.

II. FLEXIBLE ULTRA-LOW-VOLTAGE ENHANCED TRANSISTOR

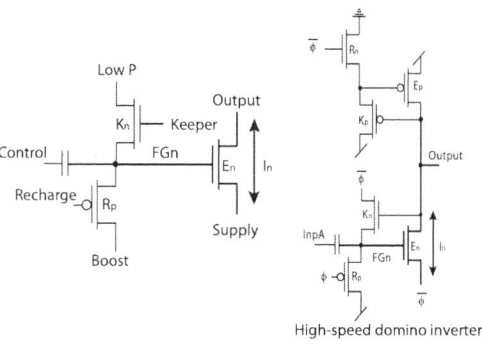

Figure 1: Ultra low-voltage high-speed and low-power transistor applied in high-speed domino logic.

The enhanced nMOS evaluation transisor E_n in Figure 1 is used to obtain a higher current level than provided by an ultra low supply voltage directly applied to the transistor gate terminal. By seperating the gate, i.e semi floating gate, by use of a floating capacitor we can apply an offset to the gate terminal and obtain different effective voltages. This is equivalent to a threshold voltage shift of the evaluation transistor E_n seen from the control input. If the evaluation transistor is used in analog- or multiple-valued logic design an additional feedback floating capacitor can be used between the *output* and *FGn*.

The different transistors in Figure 1 serve different purpouses:

1) **Evaluate transistor** E_n. The evaluation transistor are provinding the current driving an output.
2) **Recharge transistor** R_p. The recharge transistor are used to set the initial voltage on the semi floating-gate FGn or to provide the input signal to a Flip-Flop.
3) **Keeper transistor** K_n. The keeper transistor is used to provide a static output or to drive the transistor into deep sleep mode.

The control and inut signals can be applied to the different transistor terminals to implement different functions:

1) **Control terminal**. The Control terminal may be used to provide the input to precharge logic or to provide a clock edge for latching an input signal in a Flip-Flop.
2) **Recharge terminal**. The Recharge terminal is used to enable the recharge of the FGn node in recharge/precharge mode.
3) **Boost terminal**. The Boost terminal is used to set the initial state of the semi floating-gate FGn to a specific voltage. If the Boost voltage is equal to V_{DD} the evaluate transistor will be enhanced and will provide a large current if a positive input transition is applied at the Control input in evaluation mode. If the Boost voltage is equal to 0V the evaluate transistor will be driven into sleep mode. The Boost input may also be used to provide the D input to a Flip-Flop.
4) **Keeper terminal**. The Keeper terminal is typically used to drain the semi floating-gate FGn to reduce static current if the output is high. The Keeper signal is normally connected to the output.
5) **Low P terminal**. The Low P terminal signal is used to provide the semi floating-gate FGn voltage for low static current in low power mode.
6) **Supply terminal**. The Supply terminal is typically driven by a clock driver inverter and used both to precharge and evaluate.

The maximum effective gate voltage of the nMOS evaluate transistor, i.e *enhanced* transistor in Figure 1 is equal to $V_{FGn} = V_{Boost} + k \times \Delta V_{Control}$ trough the floating capacitor, where k is the capacitive division factor given by the floating capacitance relative to the total capacitance seen by the gates of the evaluate transistors. An optimal value for k is $\frac{1}{2}$, which will force the gate of the evaluate transistors to $V_{Boost} + \frac{1}{2}V_{DD}$. For the nMOS transistor E_n this voltage increase will increase the current running through the transistor and pull the output to 0.

The recharge and evaluation mode of the ULV logic are characterized by:

- **Precharge and recharge.** The precharge and recharge phase starts when ϕ switches from 0 to 1. The recharge transistors, labeled R, are turned ON and will recharge the gate of the evaluating transistors labeled E. More specifically, the gate of the nMOS evaluating transistors will be forced to V_{NBoost} and the gate of

978-1-4799-8720-7/15 $31.00 © 2015 IEEE   647

the pMOS evaluating transistors will be recharged to $V_{PBoost} (= GND)$. Typical offset voltages applied are $V_{NBoost} = V_{DD}$.

- **Evaluate.** In the evaluation phase, determined by $\phi = 0$, the recharge transistors are turned OFF and the gate of the evaluating transistors are temporarily floating allowing an input transition to affect the current running through the transistors.

III. Typical applications

The flexible ULV transistor can be utilized for both high-speed or low-power ultra low-voltage digital applications and analog low-voltage design:

1) **Domino logic.** Either GND or a clock signal is applied to the *Supply* terminal. The *Supply* terminal is used to provide a virtual ground in evaluation mode and a precharge value in the precharge mode.
2) **Pass transistor logic.** The pass input, assumed to be precharged to logical 1, is applied to the *Supply* terminal. The *Output* is precharged to 1 and only a negativ input transition, hence *Supply* terminal, in the evaluation mode will affect the *Output*, and only when the evaluate transistor is enhanced through a positive input transistion at the *Control* terminal.
3) **Flip-Flop.** The input D can be applied directly to the *Supply* and be latched into a Flip-Flop at the positive clock edge applied to the *Control* terminal. $\overline{D}$ can also be latched through the semi floating gate $V_{FG} = V_{\overline{D}} + V_{DD}/2$ if the *Supply* is connected to GND.
4) **Multiple valued logic and Analog.** The *Supply* terminal is connected to a clock signal. In evaluation mode virtual ground is applied to the *Supply* and in (p)recharge mode virtual V_{DD} is applied to the *Supply* providing an Aouto-Zero (AZ) operation.
5) **Analog.** The *Supply* terminal can be used as for multiple valued logic. In addition it can be used to provide a Common Source (CS) voltage for a differential pair configuration.

More than one floating capacitor may be used to mimic a transistor stacking effect. The recharge transistor R_p in Figure 1 is used to provide an offset, i.e threshold voltage shift, to the the evaluation transistor. When the recharge transistor is ON the Boost voltage is recharged onto the gate, $V_{FGn} = V_{Boost}$, of the evaluation transistor. This is defined by precharge or recharge mode. The recharge signal is typically a clock signal and has to an inverted signal compared to a clock signal applied to *Supply*. All circuits operating in the same clock phase are (p)recharged simultaneously.

If $V_{Boost} > V_{DD}/2$ in the (p)recharge mode the evaluate transistor will be ready to enter an enhanced mode when a positive transition is applied to the *Control* terminal.

Otherwise the evaluate transistor can be used as in a low-power and low-speed or sleep mode. Typically, the *Recharge* input is a clock signal and the *Boost* signal is V_{DD}. When the logical depth is large the *Control* signal can be applied to the the *Recharge* terminal to prevent a severe leakage of the floating gate FG_n. The leakage can be reduced further by connect the Output to the *Boost* terminal. If we apply $\overline{D} = 1$ to the *Boost* terminal in a Flip-Flop and a clock signal to the *Recharge* and *Control* then $\overline{D} = 1$ will be latched into the Flip-Flop at the positive clock edge, i.e Output = D.

The keeper transistor K_n in Figure 1 is primarily used to provide a low static power option for domino logic and pass transistor logic. By using the Output and/or differential output the static power consumption can be reduced significantly.

A. ULV Domino Logic

The precharge to 1 ULV domino inverter is shown in Figure 1 right side.

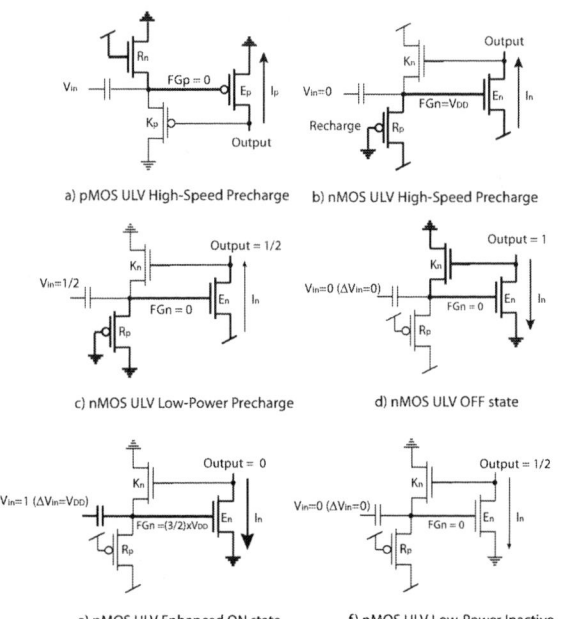

Figure 2: Ultra low-voltage high-speed and low-power transistor applied in high-speed domino logic.

Different configurations of the ULV transistor exploited for digital domino circuits are shown in Figure 2. The high-speed pMOS ULV transistors in precharge mode is shown in a), the high speed nMOS ULV in b) and the low-power nMOS ULV in c). The recharge transistors are used to set the initial voltage on the semi-floating-gates. The output of the pMOS ULV transistor will be precharged to gnd and the output of the nMOS ULV transistor will be precharged to V_{DD}. Note that the supply terminal of the evaluate transistor are reverse biased in precharge mode. The keeper transistors

978-1-4799-8720-7/15 $31.00 © 2015 IEEE

$K_{n/p}$ will contribute to the recharge of the semi-floating-gates in the precharge mode and will remain active if the output is not changing.

Table I: Mode of operation. The increase, i.e boost, or decrease in maximum current for an nMOS ULV transistor for a supply voltage equal to $250mV$ is shown.

V_{Boost} nMOS	V_{bulk} nMOS	Current Boost	Mode
V_{DD}	V_{DD}	x42	Most active (MA), Fig. 2 b) and e)
V_{DD}	$float$	x26	Medium active (mA), Fig. 2 b) and e)
V_{DD}	gnd	x10	Moderate active (A), Fig. 2 b) and e)
gnd	$float$	x0.034	Sleep (S), Fig. 2 c) and f)
gnd	gnd	x0.018	Deep sleep (DS), Fig. 2 c) and f)

For the ULV transistors in Figure 2 d), e), and f) the precharged value is logic 1. The nMOS ULV transistor in enhanced OFF mode is shown in Figure 2 d). The OFF state is defined by no input transition and the output is supposed to stay high. The keeper transistor will start to reset the semi-floating-gate to 0 and thereby turn off the E_n transistor in order to reduce the leakage current. In the enhanced ON state shown in e) the input switches from 0 to 1 and the output toggles. When the output is pulled to 0 a pMOS ULV evaluating transistor providing the initial precharge value 1 will be turned off by a pMOS keeper transistor activated by the low output. The low-power or sleep mode ULV transistor shown in f) will not be active due to a significantly lower current level than active ULV transistors. The operating frequency of the ULV logic gates operating at a supply voltage equal to $250mV$ is between 10 and 42 times the maximum frequency of complementary CMOS, whereas the maximum frequency of ULV low-power gates is 1.8% of complementary CMOS. By using the the sleep mode configuration combined with high-speed ULV circuits we can effectively achieve dynamic sleep mode operation without adding circuitry. The current level of the sleeping ULV transistors will be less than 0.2% of the enhanced ULV transistors and the activity will be blocked. The different mode of operation of the ULV transistors are given in Table I.

The current boost can be utilized to implement ultra low-voltage and high speed domino logic or exploited in low power implementations. In the latter, the increased current of the ULV logic has to be matched by increased supply voltage of complementary CMOS or NP domino logic. If a specific clock frequency is required the appropriate supply voltage is dependent on the logic style applied. In this paper we examine the potential energy savings of ULV domino logic compared to complementary CMOS and NP domino logic.

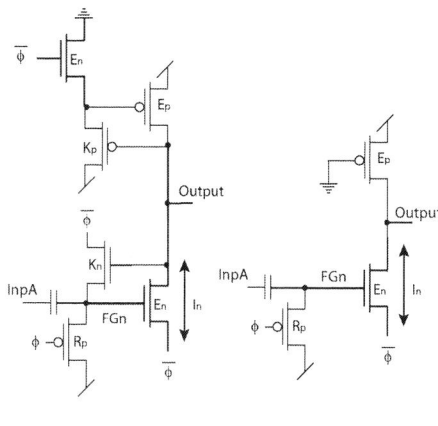

Figure 3: a) Low-power precharge to 1 (N domino) inverter and b) High-speed precharge to 1 domino inverter.

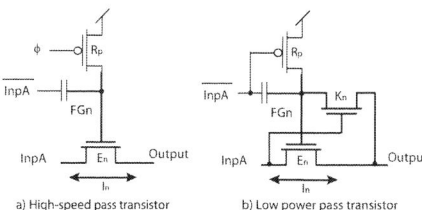

Figure 4: a) Low-power pass transistor and b) High-speed pass transistor.

B. High Speed ULV Pass Transistor Circuits

The high-speed pass transistors in Figure 4 a) can not be turned properly OFF due to the capacitive input. The voltage level of the gate (FG_n) of the nMOS pass transistor (E_n) is limited to V_{DD} and $3V_{DD}/2$. These gate voltages determine the OFF and ON current provided by the nMOS pass transistor. By adding an nMOS feedback transistor K_n from the output to the semi floating floating gate FG_n we obtain the nMOS low-power pass transistor as shown in Figure 4 b). The feedback transistor K_n is controlled by the input and will only be turned on if the input is 1 and the output is not dependent on the pass transistor.

C. High Speed ULV Flip-Flop

A fully differential ULV pass transistor D-Flip-Flop with reset/keeper transistor is shown in Figure 5 b) . The ULV Flip-Flop consists of 16 transistors and 4 floating capacitors. The floating capacitors may be poly-poly, metal-metal or even MOS parasitic capacitance. The capacitance is typically less than 1 femto farads, i.e approximately the same level as parasitic capacitances of a MOS transistor.

The increased current level will yield an improved robustness due to reduced effect of mismatches and process variations. By forcing the effective voltage of the pass transistors above the low supply voltage the enhanced transistors

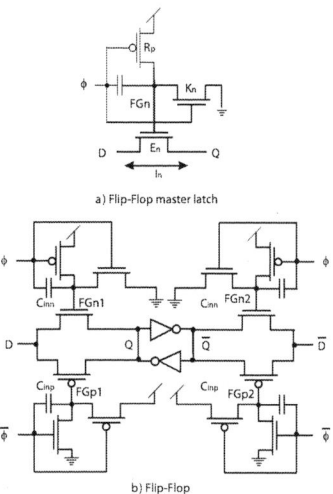

a) Flip-Flop master latch

b) Flip-Flop

Figure 5: a) Flip-Flop master latch and b) Flip-Flop.

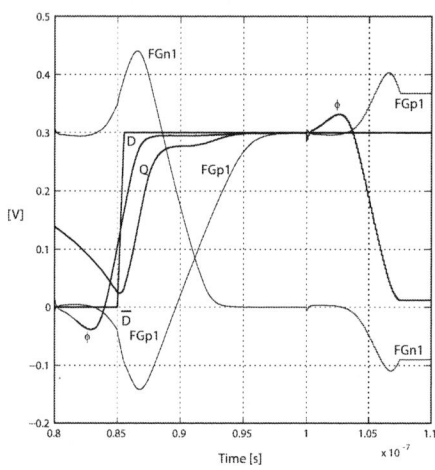

Figure 6: Simulated response of the ULV differential D Flip-Flop operating at $300mV$.

will be temporarily in strong inversion whereas the standard transistors operating at a supply voltage equal to $300mV$ will be in weak or moderate inversion assuming at $90nm$ CMOS process. Transistors operating in weak or moderate inversion will not only provide a low driving current, but will be strongly dependent on mismatches due to an exponential relationship between the effective voltage and the driving current. We can observe the enhanced mode at the positive clock edge, i.e the gate of the nMOS pass transistors $FGn1$ and $FGn2$ are increased above $400mV$ and the gate of the pMOS pass transistors $FGp1$ and $FGp2$ are pulled down to $-110mV$. This enhanced mode is the only time that the pass transistors can set the output of the Flip-Flop. Hence, the Flip-Flop is edge triggered and the delay is only dependent on the enhanced pass transistors. The input switches at the positive clock edge and the set-up time is approximately $100ps$.

Table II: *Timing details for the differential ULV Flip-Flop operating at $300mV$ compared to the Nikolic Flip-Flop [15].*

NIK.	This		Relative
26 transists	16 transistors		
t_{dq}	t_{setup}	t_{cq}	
8.55ns	100ps	600ps	12.2%

Simulated response of the ULV differential D Flip-Flop operating at $300mV$ is shown in Figure 6. Note the increased effective voltage of the Evaluate transistors. $FGn1$ and $FGn2$ is increased from $300mV$ to more than $400mV$ and the $FGp1$ and $FGp2$ nodes are pulled below $-100mV$. The Nikolic Flip-Flop [15] operating at $300mV$ will have a data D to Q delay of $8.55ns$ which is more than 10 times the set-up and clock to Q delay of the ULV Flip-Flop. Timing details for the differential ULV Flip-Flop operating

at $300mV$ compared to the Nikolic Flip-Flop [15] are given in Table II.

D. ULV Analog and Multiple Valued (MV)Circuits

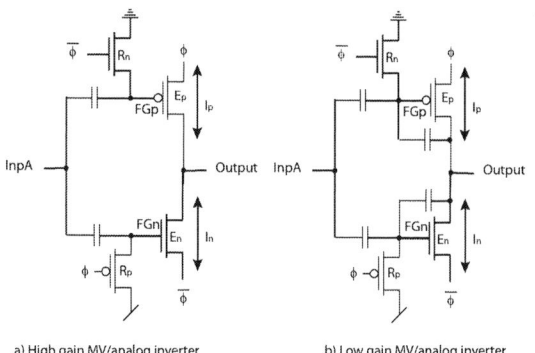

a) High gain MV/analog inverter   b) Low gain MV/analog inverter

Figure 7: a)High-gain MV/analog inverter and b) low-gain MV/analog inverter.

The analog inverter and a double input binary inverter can be used to design a simple ULV transconductance amplifier. The pass-band ULV transconductance amplifier is shown in Fig. 8. The recharge transistors are connected to offset voltages which can be used to tune the frequency response of the amplifier. If we increase the offsets, i.e. $V_{offset+} > V_{DD}$ and $V_{offset-} < 0V$, the amplifier can process higher frequency input signals. The most practical offset voltages are V_{DD} and $gnd(0V)$. Other offset voltages can be be applied if external voltages or internally generated voltages are provided.

The frequency response of the ULV transconductance amplifier with offset voltages equal to 250mV and 0V is shown in Fig. 9. The passband is 1.8MHz to $60MHz$ and the gain is close to 30dB.

978-1-4799-8720-7/15 $31.00 © 2015 IEEE   650

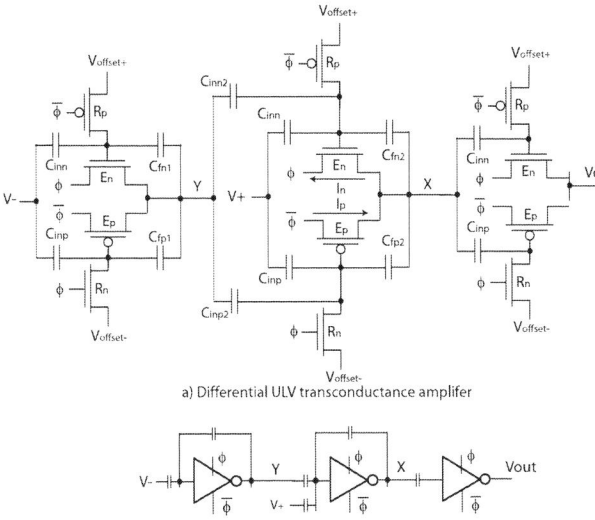

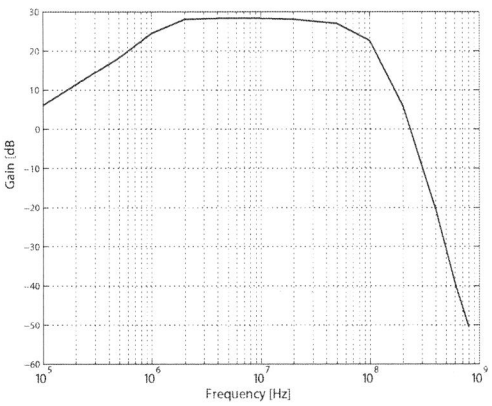

Figure 8: The ULV transconductance amplifier. a) transistor schematics, b) simplified representation.

Figure 9: The frequency response of the ULV transconductance amplifier with offset voltages equal to 250mV and 0V.

IV. CONCLUSION

The flexibility of a ULV transistor and different ULV CMOS circuits are exploited. The ULV circuits discussed in this paper includes low-voltage high-speed domino, pass transistor logic and Flip-Flop. The speed improvement of the ULV circuits are significant and in most applications more than 10 times compared to complementary CMOS. The design technique can also be used to implement low-power digital gates and low-voltage analog amplifiers. Simulated data for a 90nm AMS CMOS process are included.

REFERENCES

[1] Zhai, B.; Blaauw, B.; Sylvester, D.; Flautner, K.; Theoretical and Practical Limits of Dynamic Voltage Scaling. *Proc. 41st*

Design Automation Conference. **2004**, 873.

[2] Chandrakasan A.P.; Sheng S.; Brodersen R.W.; Low-power CMOS digital design. *IEEE Journal of Solid-State Circuits*, **1992**, *volume 27, issue 4*, 473 - 484.

[3] Gonzalez, B.; Gordon,M.; Horowitz, M. A.; Supply and Threshold Voltage Scaling for Low Power CMOS. *IEEE Journal of Solid-State Circuits.* **1997**, *volume 32, no 8*, 1210 -1216.

[4] Hanson, S.; Zhai,B.; Bernstein, K.; Blaauw, D.; Bryant, A.; Chang, L.; Das, K.K.; Haensch, W.; Nowak, E.J.; Sylvester, D.M.; Ultralow-voltage, minimum-energy CMOS. *IBM J. RES. & DEV.* **2006**, *vol . 50 no. 4/5*, 469 - 490.

[5] Wang, A.; Chandrakasan A.P.; Kosonocky, S.V.; Optimal Supply and Threshold Scaling for Subthreshold CMOS Circuits. *Proceedings of the IEEE Computer Society Annual Symposium on VLSI (ISVLSI).* **2002**.

[6] Meindl, J.D.; Davis, J.A.; The fundamental limit on binary switching energy for terascale integration (TSI). *IEEE Journal of Solid-State Circuits.* **2000**, *vol 35, no. 10*, 1515 - 1516.

[7] Verma, N.; Kwong, J.; Chandrakasan A.P.; Nanometer MOSFET Variation in Minimum Energy Subthreshold Circuits. *IEEE Transactions on Electron Devices*, **2008**, *volume 55, no. 1*, 163 - 174.

[8] Kim, K.; Agrawal, D.; Ultra Low Energy CMOS Logic Using Below-Threshold Dual-Voltage Supply. *Journal of Low Power Electronics.* **2011** *vol. 7, no. 4*, 1 - 11.

[9] Nguyen, D.; Davare, A.; Orshansky, M.; Chinnery, D.; Thompson, B.; Keutzer, K.; Minimization of dynamic and static power through joint assignment of threshold voltages and sizing optimization, *Proc. International Symposium on Low Power Electronics and Design.* **2003**, 158 163.

[10] Chen, P. C.; Kuo, J.B.; Sub-I V CMOS large capacitive-load driver circuit using direct bootstrap technique for low-voltage CMOS VLSI. *Electronic Letters.* **2002**, *vol. 39, no. 6*, 265 - 266.

[11] Mirmotahari, O.; Berg, Y.; Robust Low-Power CMOS Precharge Logic. *In proceedings of the IEEE Faible Tension Faible Consommation (FTFC).* **2013**.

[12] Kotani, K.; Shibata, T.; Imai, M.; Ohmi, T; Clocked-Neuron-MOS Logic Circuits Employing Auto-Threshold-Adjustment. *In Proceedings of the IEEE International Solid-State Circuits Conference (ISSCC).* , **1995**, 320-321.

[13] Shibata, T.; Ohmi, T.; A Functional MOS Transistor Featuring Gate-Level Weighted Sum and Threshold Operations. *IEEE Transactions on Electron Devices*, **1992**, *volume 39*.

[14] Berg, Y.; Novel ultra low-voltage and high speed domino cmos logic. *In proceedings of the IEEE International Conference on VLSI and System-on-chip.* **2010**, 225-228.

[15] Nikolic, B.; Oklobdzija, V. G.; Stojanovic, V.; ia, W.; J. Chiu, K.S.; and Leung, M.T.T; Improved Sense-Amplifier-Based Flip-Flop: Design and Measurements. *IEEE J. Solid-State Circuits.* **2006**. *vol. 35*, 867-877.

978-1-4799-8720-7/15 $31.00 © 2015 IEEE

A Linear Comparator-based Fully Digital Delay Element

Afshin Seraj and Mohammad Maymandi-Nejad
Electrical Engineering Department of Ferdowsi University of Mashhad
Mashhad, Iran
maymandi@um.ac.ir

Parvin Bahmanyar
Electrical Engineering Department of Ferdowsi University of Mashhad
Chair for Circuit Design and Computer Engineering of Technische Universität Braunschweig
Braunschweig, Germany

Manoj Sachdev
Electrical and Computer Engineering Department of University of Waterloo
Waterloo, Canada

Abstract— A linear delay element is proposed in 0.18 μm CMOS technology with a power supply of 1.8V. The proposed delay element maintains linearity over a relatively large input voltage range of 1.2V and its delay range (sensitivity) can be tuned through a bias voltage. Its power dissipation is 50μW at a clock frequency of 1GHz and its robustness in different process corners has been shown through simulations. Additionally, a 6-bit 107MS/s Fully Digital ADC with 1.2 V input range has been implemented using the proposed delay element. The simplicity of design and functioning of the proposed delay element contributes to its improved power and energy consumption.

Keywords— *delay line; linear delay; delay-locked-loop; fully digital ADC; time-based ADC; pulse width modulator;*

I. INTRODUCTION

Delay Element (DE) or voltage-to-time converter (VTC) is an integral part of many digital circuits such as digital VCOs, DLLs, PLLs and Fully Digital ADCs [1]. It can also be applied in Pulse Width Modulators and Pulse Position Modulators [2]. In a DE, the input-output delay for a given transition is usually controlled by an analog input voltage, V_i. Often a periodic input such as a clock (Clk_{in}) and its delayed output clock (Clk_{out}) are controlled by the controlling voltage. A linear VTC (having a relationship between the delay and V_i) is required in all above-mentioned applications. In Fully Digital ADCs (FD-ADC), which can be formed by connecting a DE to a Time-to-Digital converter, the Linearity range of DE is a pivotal concern and improves the ADCs input range, speed, SNDR and THD [2-5].

Also in DLLs, DE's lack of linearity causes false locking and also increases the jitter, while reducing the lock-in speed [6-9]. The DEs presented so far provide linearity over a relatively small range of V_i [3]. The proposed DE provides linearity within a range of 1.2V and can accept two independent inputs

for controlling the rising and falling edge delays separately. This can provide more flexibility in some applications. The simplicity of the design and transistors sizing is another advantage of the proposed DE. Moreover, It is possible to alter the sensitivity of the delay to the controlling voltage (V_i), even after the DE in implanted on chip.

This article is organized as follows. In II, the proposed DE is discussed in detail and its robustness to different process conditions is shown through simulations. In III, the double-edged version of the proposed DE is presented and compared to other types of DE, in regard of linearity. Ultimately, In IV the functioning of a FD-ADC is briefly explained and it is shown that the use of the proposed DE leads to an increase in the input range and speed of FD-ADCs.

II. PROPOSED DELAY ELEMENT

The functioning of the proposed DE is based on a current source charging or discharging a capacitor. The basic part of the proposed DE is shown in Fig. 1. The proposed DE operates as follows. Transistor M1 behaves as a capacitor (C). The size of this capacitor depends on the value of delay range the designer is seeking. It can be even eliminated if very short delays are desirable. Transistor M3 operates as a constant current source and transistor M2 serves as a comparator. The W/L ratio of M2 is chosen to be much larger than that of M3 and M4 so that the discharging current of C is determined mainly by M3. Transistor M5 charges the capacitor to V_{dd} when the clock (Clk_{in}) is low.

When the Clk_{in} goes high, capacitor C starts discharging through M2, M3 and M4 and the voltage at V_S starts falling at a constant rate determined by C and the current of M3. Note that during this stage of operation, the voltage of the drain of M4 is kept close to 0V due to M4's large W/L ratio.

978-1-4799-8720-7/15 $31.00 © 2015 IEEE

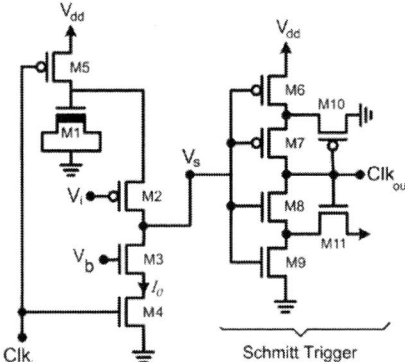

Fig. 1. Proposed delay element

When the source voltage of M2 reaches the value of (V_i+V_{thp}) transistor M2 turns off and the voltage V_S drops to GND. In this way, the capacitor's voltage is compared with V_i+V_{thp}. As V_s reaches the triggering voltage of the Schmitt trigger (0.5V in this work), it makes a sharp positive transition. Fig. 2a depicts the transient response of V_s (solid lines) and the capacitor's voltage (dotted lines) for 13 different values of V_i ranging from 0V to 1.2V with a step size of 0.1 V. Similarly, Fig. 2b shows the transient response of Clk_{out} for 13 different values of V_i. As can be seen in this figure, the delays are linearly changing with the input voltage. The Schmitt trigger circuit in the proposed DE plays an important role. First of all it sharpens the edges of the output voltage. As can be seen in Fig. 2a the voltage V_s has a relatively large fall time and it can cause problem in the succeeding stages. Using the Schmitt trigger eliminates this problem. Second, it provides two well separated switching thresholds (V_{trp}^{+} and V_{trp}^{-}). In our design V_{trp}^{+}=1.3V and V_{trp}^{-}=0.5V.

According to the above explanation, the delay can be found from the following equation.

$$I_0 \times t_d = C \times \left(V_{dd} - V_{thp} - V_i\right) \qquad (1)$$

Where I_0 is the current of M3 discharging capacitor C, t_d is the delay that the voltage of C takes to reach to $(V_{dd}-V_{thp}-V_i)$ from V_{dd}.

Fig. 3a shows the simulation results of the characteristic curve of the proposed DE, displaying its delay versus V_i. The slope of the characteristic curve is -4.2 ns/V within a linear region of 1.2V. This slope can be altered by manipulating the V_b and the sizes of the current source and the capacitor. The characteristic curve is drawn for five different process corners (TT, FF, FS, SF, SS) in Fig. 3b. According to this figure, the DE manages to maintain linearity in different process corners. The current of M3, and consequently the slope of the characteristic curve, can be easily changed by bias voltage V_b. This is illustrated in

Fig. 4. This mechanism provides a very useful means to the designer to program the delay element. As a result, the proposed DE can be modified to incorporate programmability through a digital vector. In order to achieve this, the bias voltage V_b should be adjusted by the digital vector.

Besides the clock frequency and the Schmitt Trigger circuit, the power consumption of the proposed DE is dependent on the size of M1 as the capacitor and the W/L ratio of M3 as the current source. Reduction in both these items results in lower power dissipation. In this work, simulations show a power consumption of 50µW for a clock of 1GHz.

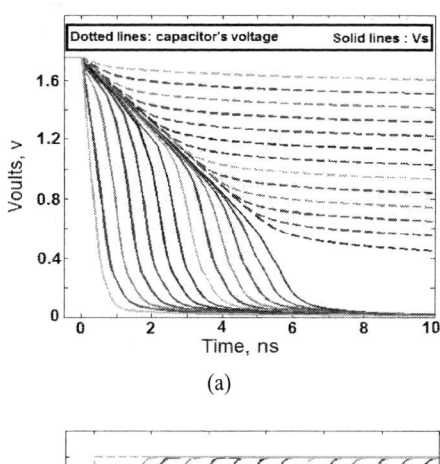

(a)

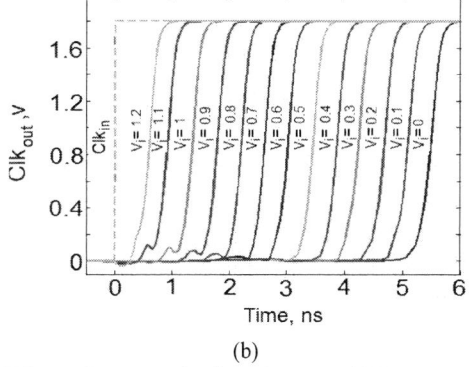

(b)

Fig. 2. *Voltages of some crucial nodes in the proposed DE's circuit*
a Dotted lines show the voltages of V_s and solid lines show the voltages of the capacitor for thirteen different V_i from 0V to 1.2V.
b Voltage of Clk_{out} for thirteen different V_i from 0V to 1.2V.

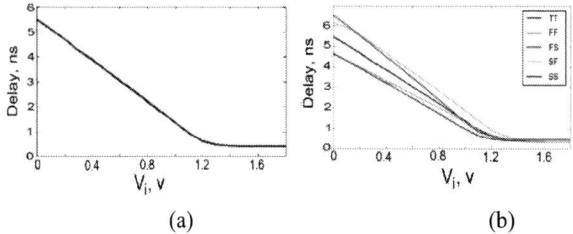

(a)   (b)

Fig. 3. Proposed DE's delay versus its input voltage (V_i) (V_b=1.5V)
a Characteristic curve of the proposed DE
b Characteristic curve of the proposed DE in different process corners

978-1-4799-8720-7/15 $31.00 © 2015 IEEE   653

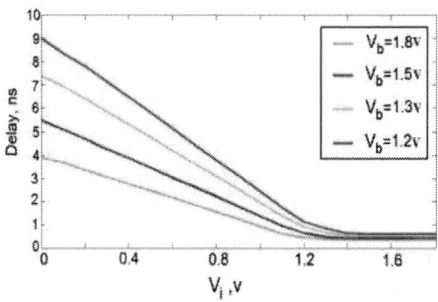

Fig. 4. Effect of V_b on the characteristic curve of the proposed DE

It is worth mentioning that although the capacitance of a MOSCAP changes, depending on whether its voltage is below or above the threshold voltage, this does not significantly interfere with the correct functioning of the proposed DE. This is because the voltage across the MOSCAP in the proposed DE is always more than V_{thn}, even if V_i=0V.

III. DOUBLE-EDGED VERSION OF THE PROPOSED DE

The DE in Fig. 1 operates linearly with the rising edge of the Clk_{in}. The falling-edge version of this DE can be designed in the same manner. To have a linear DE working with both edges of the clock, the circuits in Fig .1 and its falling-edge counterpart can be combined as depicted in Fig. 5. $V_{i, rise}$ and $V_{i, fall}$ are two separate independent voltages that determine the rising-edge delay and the falling-edge delay of the DE, respectively. Therefore, The rising-edge delay and falling-edge delay can be set independently. Transistors Ms1 and Ms2 are switches connecting either the rising-edge DE or the falling-edge DE to the Schmitt trigger, depending on the state of Clk_{in}. The characteristic curves of the DE in Fig. 5 is shown in Fig .6. Note that the circuit of Fig. 5 has two separate voltages to control the slope of the delay curves. This provides a desirable flexibility in the design of fully digital circuits.

Table. I compares the linearity of the proposed DE with other DEs'. All the DE's in this table are simulated within an input range of 1.2V and their characteristic curves (controlling voltage versus delay) are obtained. The curves are then estimated with linear curves using curve fitting and the maximum deviation of the characteristic curve from the linear curve is considered as the INL. The ratio of INL on 1.2V (the controlling voltage range) is taken for comparing the linearity. Clearly, the proposed DE has a superior performance in terms of linearity as is evident by lower value of this ratio for the proposed DE.

IV. FULLY DIGITAL ADC

A simple structure of FD-ADC is shown in Fig. 7, which consists of a linear Voltage-to-Time converter and a Time-to-Digital converter. The delay line consists of several DEs (64

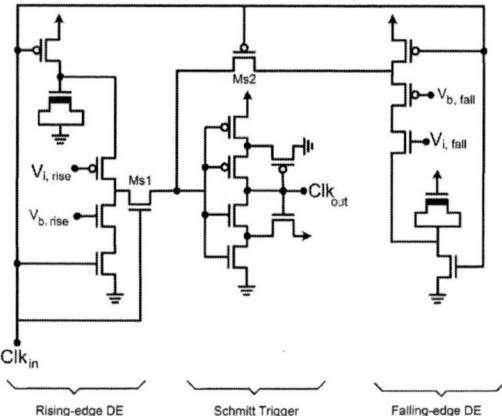

Fig. 5. Proposed DE working with both edges of the clock.

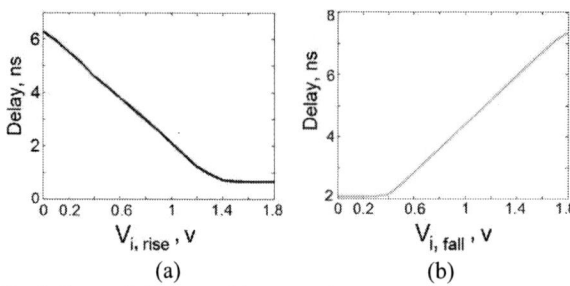

Fig. 6. Characteristic curves of the proposed DE in Fig. 5
a Delay between the rising edges of Clk_{in} and Clk_{out}
b Delay between the falling edges of Clk_{in} and Clk_{out}

TABLE I.   COMPARISION BETWEEN DIFFERENT TYPES OF DE

	Controlling Voltage Range (V)	INL/1.2V(%)
Shunt Capacitor DE [1]	[0.6 1.8]	35
Current Starved DE [1]	[0.6 1.8]	40
Power-based DE in [4]	[0.6 1.8]	42.5
Proposed DE in [4]	[0.6 1.8]	12.5
[8]	[0 1.2]	14.1
[2]	[0.3 1.5]	5.8
[6]	[-0.4 0.8]	9.1
[12]	[0 1.2]	25
Proposed DE	[0 1.2]	1.3

for a 6-bit ADC) connected in series. Each of them, which is usually constructed of two series digital inverters, is connected to a positive edge-triggered register. Initially, each DE holds the voltage of 0V at both its input and output. As the Clk_{in} signal goes high and passes through the calibration block, which generates a constant delay, it propagates along the delay line.

978-1-4799-8720-7/15 $31.00 © 2015 IEEE   654

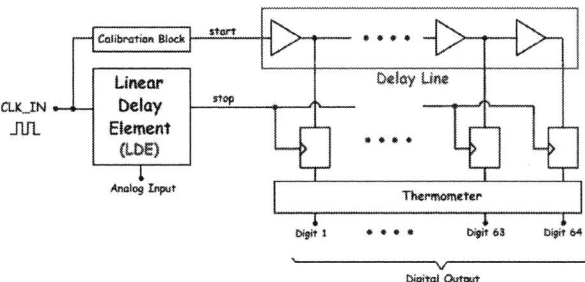

Fig. 7. A simple structure of FD-ADCs

As the output of the LDE(stop) rises in reaction to the rise of Clk_{in}, the registers sample the outputs of all the DEs in the delay line. These outputs can either be 0 or 1, depending on whether the clock signal has reached them till that moment or not. Ultimately, the recorded data in the registers is inputted into a thermometer to be transformed into a digital code. As a result, the digital output is proportional to the delay of the LDE which is proportional to the Analog Input.

In order to provide a linear delay element to use as the LDE, Many designs have used the existing nonlinear DEs within a relatively narrow range of analog input in order to emulate the linear functioning of an LDE. This approach leads to a small input range (small LSB) for the FD-ADC [2,5] and can also limit the speed of the FD-ADC [3].

To examine the performance of the proposed DE in this article, A 6-bit FD-ADC with the structure depicted in Fig. 7 has been implemented with the proposed DE in Fig. 1 as the LDE. In the simulation, a clock with the frequency of 107MHz samples a 1MHz sinusoidal analog input voltage and the digital output is used to calculate the SNDR. For measuring the linearity, a ramp analog input voltage spanning the entire input range (from 0.6V to 1.8V) is used to obtain the INL for this ADC. The FD-ADC's characteristics are shown in table II. The simulated FD-ADC only operates during the positive cycle of the clock, but with some alterations in the structure, it is possible to exploit both of the cycles of the clock and double the speed of this FD-ADC.

TABLE II. CHARACTERISTICS OF THE IMPLEMENTED FD-ADC

CMOS Technology	0.18μm
Power Supply	1.8V
Sampling Rate	107MS/s
Number of Bits	6
Input Range	1.2V
SNDR	31.1dB (for a 1MHz Input)
INL	0.9LSB
Power Dissipation	600μW

CONCLUSION

A tunable delay element that is linear over an input voltage range of 1.2V is proposed and simulations in different process corners assert its robustness. Low area and power dissipation, proper linearity and simplicity in designing make the proposed delay element a good candidate to be used in applications like DLLs, PLLs, VCOs and Fully Digital ADCs where linearity of delay element improves the overall performance.

REFERENCES

[1] M. Maymandi-Nejad, M. Sachdev, "A Monotonic Digitally Controlled Delay Element," IEEE journal of Solid State Circuits, vol. 40, Issue. 11,pp. 2212–2219, 2005, doi:10.1109/JSSC.2005.857370

[2] M. Ismail, H. Mostafa, "A New Design Methodology for Voltage-to-Time Converters (VTCs) Circuits Suitable for Time-based Analog-to-Digital Converters (T-ADC)," 27th IEEE International System-on-Chip Conference (SOCC), USA, pp. 103–108, 2014, doi: 10.1109/SOCC.2014.6948908

[3] Li. Guansheng, Y.M. Tousi, A. Hassibi, E. Afshari, "Delay-Line-Based Analaog-to-Digital Converters," IEEE Transaction on Circuits and Systems, vol. 56, Issue. 6, pp. 464–468, 2009, doi: 10.1109/TCSII.2009.2020947

[4] H. Farkhani, M. Maymandi-Nejad, M. Sachdev, "A Fully Digital ADC Using A New Delay Element With Enhanced Linearity," IEEE International Symposium on Circuits and Systems, pp. 2406–2409, 2008, doi: 10.1109/ISCAS.2008.4541940.

[5] H. Pekau, A. Yousif, J.W. Haslett, "A CMOS Integrated Linear Voltage-to-Pulse-Delay-Time Converter for Time Based Analog-to-Digital Converters," IEEE International Symposium on Circuits and Systems(ISCAS), 2006, doi: 10.1109/ISCAS.2006.1693099

[6] G. Jovanovic, M. Stojcev, D. Krstic, "Delay Locked Loop with Linear Delay Element," 7th International Conference on Telecommunication in Modern Satellite, Cable and Broadcasting Services, vol. 40, Issue. 11, pp. 397-400, 2005, doi:10.1109/TELSKS.2005.1572136.

[7] M. Gharib, A. Abrishamfar, "A Novel low-power and high-performance dual-loop DLL with linear delay delay element," 51th IEEE Midwest Symposium on Circuits and Systems, pp. 763–766, 2008, doi: 10.1109/MWSCAS.2008.4616911

[8] M. Moazedi, A. Abrishamfar, A.M. Sodagar, "A Highly Linear Modified Pseudo-differential Current Starved Delay Element with Wide Tuning Range," 19th Iranian Conference on Electrical Engineering, 2011

[9] M. Moazedi, A. Abrishamfar, A.M. Sodagar, "A Delay-Locked-Loop with Quasi-linear Modified Differential Delay Element," International Journal of Scientific & Engineering Research, 2013

[10] L. Chang-Ming, C. Yi-Chung, H. Po-Chiun, "Time-domain analog-to-digital converters with domino delay lines, " International Symposium on VLSI Design, Automation, and Test (VLSI-DAT), April 2013, pp.1,4,22-24, doi: 10.1109/VLDI-DAT.2013.6533841

[11] H. Mostafa, Y.I, Ismail, "Highly-linear voltage-to-time converter (VTC) circuit for time-based analog-to-digital converters (T-ADCs)," 20th IEEE international conference on Electronics, Circuits and Systems, December 2013, pp.149-152, doi: 10.1109/ICECS.2013.6815376

[12] H. Pekau, A. Yousif, J.W. Haslett, "A CMOS integrated linear voltage-to-pulse-delay-time converter for time based analog-to-digital converters," IEEE International Symposium on Circuits and Systems (ISCAS), May 2006, pp.,2376,21-24, doi: 10.1109/ISCAS.2006.1693099

[13] S. Schidl, K. Schweiger, W. Gaberl, H. Zimmermann, "Analogously tunable delay line for on-chip measurements with sub-picosecond resolution in 90 nm CMOS," Electronics Letters, July 2012, vol. 48, no. 15, pp.910-911

2015 IEEE Computer Society Annual Symposium on VLSI

Built-In Self Optimization for Variation Resilience of Analog Filters

Jiafan Wang, Congyin Shi, Edgar Sanchez-Sinencio and Jiang Hu
Department of Electrical and Computer Engineering, Texas A&M University
jiafan0420@tamu.edu, scytulip@tamu.edu, sanchez@ece.tamu.edu, jianghu@ece.tamu.edu

Abstract—At deep sub-micron integrated circuit technologies and beyond, characteristics of analog circuits are increasingly affected by process variations and device aging. Although design-time approaches are able to take these effects into account, it is very difficult for a single static design to efficiently handle all kinds of variation scenarios. In this work, a built-in self optimization technique is developed such that each analog IC chip can autonomously improve its performance in the presence of significant variations. As a demonstration, an on-chip analog self-test platform is implemented to test an active-RC band-pass filter, whose components can be reconfigured to obtain different frequency responses. The desired response is found by a digital circuit implementation of multi-start meta-heuristic optimization. The proposed technique does not rely on external test equipment or any general purpose digital signal processor. Its effectiveness is confirmed by simulations as well as measurement results from a test-chip fabricated in 180 nm CMOS technology.

I. INTRODUCTION

Nowadays, analog integrated circuit (IC) design continues to be a great challenge. Different from the streamlined digital design using standard cells and automatic tools, analog IC design still relies on hand calculation, SPICE simulation, and designers' personal experiences. In order to improve the efficiency of analog design, automatic optimization methods have been explored. For example, [1] studied analog circuit sizing, which can changes the dimension of transistors, capacitors or resistors. However, as the IC process technology scales down into deep submicron regime, manufacturing process variations become pronounced and often result in remarkable performance deviation from specifications. Moreover, device aging [2], such as bias temperature instability (BTI) and hot carrier injection (HCI), causes additional characteristic changes over time. Circuit optimization techniques have attempted to address these issues [3], [4]. However, design-time optimization implies a uniform solution, which is difficult to achieve for all different variation cases.

To address the individual process variations, [5] studied post-silicon tuning techniques where circuits are designed with a certain configuration, which is performed at manufacturing testing. Each chip instance is tested and tuned by test equipment to compensate for its own variations. As testing is performed only once before chips are integrated into their systems, such tuning is not adequate for tackling aging effects that change over time. An alternative approach is self-tuning, which can be performed at any time in product life. One such approach [6] is self-tuning in communication circuits where the configuration search is undertaken by the baseband digital signal processor (DSP). Evidently, the dependence on a DSP restricts applications of this technique. A built-in self tuning technique for A/D converter design is described in [7]. The tuning objective function is a parasitic mismatch, which is relatively simple compared to the overall analog system performance, such as gain, linearity, phase margin, etc. A recent work [8] attempts to achieve a built-in self tuning of general analog circuit performance. The tuning is controlled by a neural network circuit, whose training needs external assistance.

In this work, we explore a general framework of build-in self optimization for variation resilience of analog ICs. By "optimization" instead of "tuning," our framework contains a digital circuit that implements an optimization algorithm rather than simple if-then rules. This is a powerful approach that can handle cases where both performance function and variation-performance relation are complicated. The study platform here is band-pass filter, which is a common analog module existing in many analog IC designs. The filter is designed with configurable parasitics, such as capacitors and resistors. We define a cost function that can capture the mismatch of its frequency response from the specification. The configuration search is realized by a simulated annealing (SA) based approach. Both the cost function and the SA-based configuration search are implemented by digital circuits. As transistor feature size keeps shrinking, more silicon estates become available to support these digital self optimization circuits. Even so, we still strive hard to minimize circuit overhead. The effectiveness of conventional SA is hampered by the limited area budget. To this end, we devise algorithmic techniques that significantly improve search quality. Since the digital optimizer circuit is invoked occasionally, it can be power gated to minimize its power overhead. The main advantages of our approach includes the following:

- Our approach does not require test equipment, host processor or training, and is therefore a truly autonomous approach.

- It captures compounding variation effects without relying on any models.

- It is flexible when used in different kinds of analog circuits with only change of the cost function.

The effectiveness of our approach is confirmed by simulations and measurement results from a test chip of 180 nm CMOS technology.

The rest of the paper is organized as follows. Section II overviews the architecture of proposed built-in self optimization system. The post-silicon circuit tuning method and the frequency response of an ideal 2-order Butterworth band-pass filter is introduced in Section III. The simulated annealing-based built-in self optimization technique is discussed in Section IV including the implementation of cost function and multi-start meta-heuristic. Section V covers the functionality verification and performance comparison by chip measurement and simulation. Finally, conclusions are given in Section VI.

II. SYSTEM ARCHITECTURE

An overview of the proposed built-in self optimization architecture is depicted in Fig. 1. It consists of an on-chip analog test bench and a digital optimization circuit. In the analog part, a stimulus generator produces a clock signal and a sinusoidal waveform, both of which have the same frequency. The sinusoidal waveform is fed into the circuit-under-test (CUT) and stimulates the CUT to generate an output. A multiplier, which adopts the self-mixing technique in

978-1-4799-8720-7/15 $31.00 © 2015 IEEE

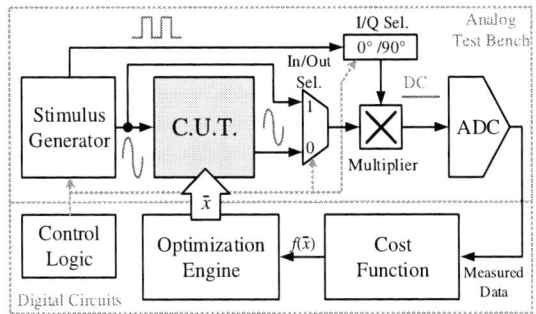

Fig. 1. Overview of the proposed built-in self optimization system architecture.

[9], and an analog-to-digital converter (ADC) are employed for measuring the output response. In addition, the measurement path can be selected by an input/output selector. Moreover, coherent detection is used, and an I/Q selector can shift the multiplier's clock by 0 or 90 degrees. I/Q data, which is related to the signal modulation in communication systems, can be retrieved by the ADC, indicating the phase and amplitude changes of the input and the output waveform. Later, the amplitude of the waveform is calculated by the digital part. Compared to the pure analog approach in [9], the digital approach reduces the complexity of the analog parts and is more scalable for advanced IC technologies.

The analog and digital circuits cooperate in a closed loop to perform the self optimization procedure. Based on the CUT's output responses measured by the ADC, a main part of the digital circuit computes the cost function $f(\vec{x})$, which is to be minimized by the optimization. The value of the cost function depends on a vector of CUT configuration variables, $\vec{x}$, whose values are to be decided by the optimization engine. We developed a digital circuit optimizer based on a hybrid multi-start meta-heuristic. The optimization is an iterative procedure where various configurations of $\vec{x}$ are applied to CUT till the one that minimizes $f(\vec{x})$ is found or the limit to iterations is reached. Apart from the cost function and the core optimization engine, a control logic is needed to coordinate the timing of the entire system. In the proposed approach, the optimization engine is generic for almost any type of analog circuits, while the cost function circuit needs be customized for different analog circuits.

III. PROGRAMMABLE BANDPASS FILTER AND ITS NON-IDEAL EFFECT

The second order band-pass filter (BPF) is chosen as the analog circuit platform for our study, because it is a very common module in many analog circuit systems, and its performance description as well as its relation with variations are not straightforward. The schematic of a fully programmable Tow-thomas active-RC band-pass filter (BPF) biquad is illustrated in Fig. 2. All the resistors and capacitors are digitally controlled arrays. In this design, a 5-bit resistor bank and a 3-bit ($k = 3$) capacitor bank are implemented. From Fig.2, we can derive the transfer function of the whole BPF, assuming the amplifiers are ideal.

$$
\begin{aligned}
H(s) &= \frac{V_{out}}{V_{in}} = -\frac{G_{BPF}\frac{\omega_0}{Q}s}{s^2 + \frac{\omega_0}{Q}s + \omega_0^2} \\
&= -\frac{\frac{1}{R_K C_1}s}{s^2 + \frac{1}{R_Q C_1}s + \frac{1}{R_1 R_2 C_1 C_2}}
\end{aligned} \tag{1}
$$

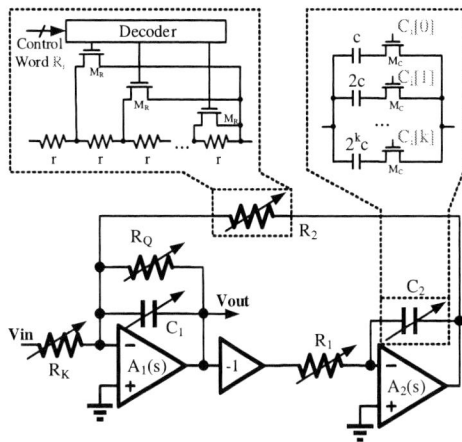

Fig. 2. Second-order band-pass filter and the configurable resistance/capacitor array structure.

and

$$
G_{BPF} = -\frac{R_Q}{R_K}, \quad \omega_0 = \frac{1}{\sqrt{R_1 R_2 C_1 C_2}}, \quad Q = \sqrt{\frac{R_Q^2 C_1}{R_1 R_2 C_2}}
$$

where G_{BPF} is the gain, ω_0 is the central angular frequency, and Q is the quality factor of the BPF. Particularly, if we set $R_Q = R_K = R_{x_1}$, $R_1 = R_2 = R_{x_2}$, and $C_1 = C_2 = C$; then, we theoretically have $G_{BPF} = -1$, $\omega_0 = 1/\sqrt{R_{x_1}C}$, and $Q = R_{x_1}/R_{x_2}$. Here, R_{x_1} and R_{x_2} are chosen as the "tuning knobs" of the BPF, which can control the shape of the frequency response. In our design, they're digitally controlled resistor arrays with 5-bit control words; thus, later integer numbers x_1 and x_2 will be used to present the control words instead of the resistance values. Additionally, a fixed C is selected to reduce the number of dimensions for the optimization problem, because x_1 is sufficient to control the BPF central frequency.

Ideally, the frequency response of a BPF is well defined by (1), and deterministically depends on the resistor and capacitor values, which can be properly chosen at design time. But unfortunately, non-idealities in the IC chips will introduce more complexity. Firstly, the amplifiers are not ideal. They have limited gains, limited output impedance and their own frequency responses, which are usually low-pass. The frequency response of the amplifiers are annotated as $A_1(s)$ and $A_2(s)$ in Fig. 2, which should be inserted into (1). But even worse is that $A_1(s)$ and $A_2(s)$ will change due to the aging effects of the MOSFETs in the amplifiers [2]. Moreover, as illustrated in Fig. 2, MOS switches M_R in the resistor bank and M_C in the capacitor bank introduce ON resistance in series with the resistor and capacitor separately. The parasitic capacitance in these switches also need to be considered. Nevertheless, the most important thing is the process variation. On the one hand, the transistors are affected by the variations. Simulation reveals 6.4% deviation on the bandwidth of the amplifiers, and 2.7% deviation on the amplifiers' DC gain among chips. However, feedback loop technique is used in the BPF design and, thus, the circuit is robust against variations of MOSFETs. But on the other hand, the standard deviation of the absolute resistance value is about 8% of the model parameter, and it is 3.5% for the capacitance value. These are the main error sources for an on-chip BPF. In other words, deterministic resistor/capacitor values, i.e., fixed x_1 and x_2 values, provide little assurance for the desired frequency response. Therefore, a self optimization

978-1-4799-8720-7/15 $31.00 © 2015 IEEE

mechanism, which finds a proper set of component values and can operate for individual chip instances, is necessary.

The effect of variations, compounded with non-ideal conditions, can be quite complex. It may distort the frequency responses in various ways such as deviating the gain at the central frequency, shifting the central frequency, changing the bandwidth. Moreover, the Q-factor may be affected as well. It is very difficult, if not impossible, to find analytical functions to describe how the performance deviation depends on the variations. This is why we take a model-free multi-start meta-heuristic optimization approach, in this framework.

IV. OPTIMIZATION AND ITS DIGITAL CIRCUIT IMPLEMENTATION

A. Optimization Cost Function and Its Implementation

In general, *optimization* is a procedure to find values for a set of decision variables $\vec{x}$ such that certain objective function $\Phi(\vec{x})$, which is also called a cost function, is minimized subject to certain constraints. The definition of the cost function plays a fundamental role in guiding the optimization solution search. For an analog circuit system, we wish the actual circuit characteristics to match well with the specifications. Hence, a general template of cost function is the mismatch between the actual characteristics and corresponding specifications, and can be described by

$$
\begin{aligned}
\Phi(\vec{x}) &= \|\beta_1\phi_1(\vec{x}), \beta_2\phi_2(\vec{x}), \ldots, \beta_n\phi_n(\vec{x})\|_2 \quad (2)\\
\phi_i(\vec{x}) &= g_i(\vec{x}) - \alpha_i G_i
\end{aligned}
$$

where $\phi_i(\vec{x})$ indicates the mismatch between the actual characteristic $g_i(\vec{x})$ and corresponding specification G_i, and α_i and β_i are constant weighting factors.

For the BPF design described in Section III, the decision variables $\vec{x}$ determine the resistor configuration. Frequency response is a main performance metric for the BPF. According to the transfer function (1). The highest gain g_{s_2} should be obtained at the central frequency s_2. $3dB$ gain degradation is expected at the two bandwidth frequencies, s_1 and s_3, and we have $s_1 < s_2 < s_3$. Thus, the cost function for the BPF is defined to be the mismatch of gains g_{s_1}, g_{s_2} and g_{s_3}.

Instead of directly implementing the cost function as in (2), we make a couple of changes in order to limit the circuit implementation overhead. First, the L2-norm $\|\cdot\|_2$ is replaced by $|\cdot|$ so as to reduce the circuit complexity. Second, our cost function focuses on the normalized frequency response curve, where only the central frequency location and the gain drops at s_1 and s_3 matter. In fact, we just enforce based on the gain at s_2 being $3dB$ greater than that at s_1 and s_3. The vertical offset of the response curve can be easily handled, and we merely include a penalty term in the cost function to make the measured gain at s_2 to be greater than 1. Therefore, the cost function for the BPF is defined as

$$
\begin{aligned}
\Phi_{BPF}(\vec{x}) &= \beta_1\left|\frac{A(s_2)_{out}}{A(s_2)_{in}} - \alpha_1\frac{A(s_1)_{out}}{A(s_1)_{in}}\right|\\
&+\beta_2\left|\frac{A(s_2)_{out}}{A(s_2)_{in}} - \alpha_2\frac{A(s_3)_{out}}{A(s_3)_{in}}\right|\\
&+\beta_3\left|\frac{A(s_1)_{out}}{A(s_1)_{in}} - \alpha_3\frac{A(s_3)_{out}}{A(s_3)_{in}}\right| + P \quad (3)
\end{aligned}
$$

where $A(s_i)_{out}$ and $A(s_i)_{in}$ are the signal amplitudes at the ith frequency point for the output and the input, respectively. The ratio $A(s_i)_{out}/A(s_i)_{in}$ is the measured gain $g_{s_i}(\vec{x})$ at frequency s_i. The last term, P, in the right-hand side of (3) is

to penalize any violation to the constraint that the gain at s_2 must be no less than 1. It is described by

$$
P = \begin{cases} 0 & \text{if } \frac{A(s_2)_{out}}{A(s_2)_{in}} \geqslant 1,\\ 1 - \dfrac{A(s_2)_{out}}{A(s_2)_{in}} & \text{otherwise.} \end{cases} \quad (4)
$$

In the optimization theory, penalty function is a technique for transforming a constrained optimization problem into an unconstrained one.

The constant coefficients in (3) are decided as follows. We set $\alpha_1 = \alpha_2 = 1/\sqrt{2}$ and $\alpha_3 = 1$ such that the gains at s_1 and s_3 are the same and both have 3dB drop compared to the gain at s_2. We set $\{\beta_1, \beta_2, \beta_3\} = \{1, 1, 1\}$ for our case of BPF. By changing the β vector values, the cost function can be configured for other types of filters. For example, $\{0, 1, 0\}$ is for a low-pass filter, and $\{1, 0, 0\}$ is for a high-pass filter.

The value of $A(s_i)$ is obtained by computing $\sqrt{I^2 + Q^2}$, where the I/Q data is retrieved from the measurement, which has been discussed in Section III. A direct computing of $\sqrt{I^2 + Q^2}$ with square and square-root units entails a large circuit area. Instead, we adopt the CORDIC (COordinate, Rotation DIgital Computer) algorithm [10], which is an iterative procedure of addition/subtraction and shifting. The pseudo code for computing $\sqrt{I^2 + Q^2}$ is given by Algorithm 1 where N is the number of iterations, and i indicates current step.

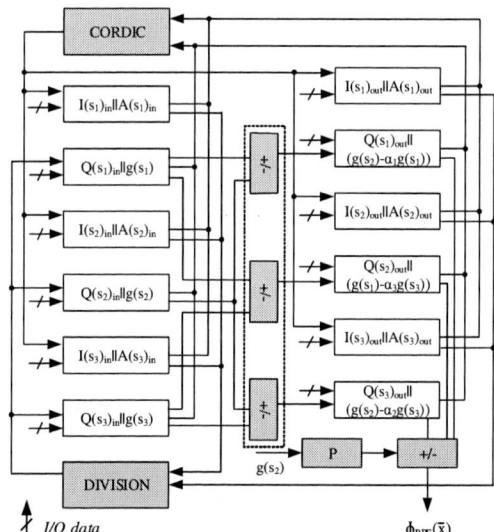

Fig. 3. Cost function block diagram. Each grey rectangle is an arithmetic block, and each white one is a buffer. $Q(s_1)\|g(s_1)$ means $Q(s_1)$ is first stored in the buffer and then $g(s_1)$.

The main idea of Algorithm 1 is to change the point (I_0, Q_0) by a series of rotations along the circumference of a circle, which has the radius of $\sqrt{I_0^2 + Q_0^2}$ and is centered at $(0,0)$, until its y-coordinate reaches 0. Then the absolute value of its x coordinate is equal to the desired result. The entire rotation process takes the total N iterations. In each iteration, the point (I_i, Q_i) is rotated by an angle of $arctan(2^{-i})$, and the direction of the rotation is determined by the sign of I_iQ_i. The rotated angle keeps decreasing and, finally, I_i converges to $K\sqrt{I_0^2 + Q_0^2}$, where K is a constant and can be omitted since we care only the amplitude ratio in (3). It should be

Input : I/Q data stream I_0 and Q_0
Output: Amplitude of the I/Q modulation

1 **if** $|Q_0| > |I_0|$ **then**
2 $\quad$ $temp \leftarrow |I_0|; I_0 \leftarrow |Q_0|; Q_0 \leftarrow temp;$
3 **end**
4 **for** $i \leftarrow 0$ to $N-1$ **do**
5 $\quad$ $d_i = -sign(I_iQ_i);$
6 $\quad$ $I_{i+1} = I_i - 2^{-i}d_iQ_i;$
7 $\quad$ $Q_{i+1} = Q_i + 2^{-i}d_iI_i;$
8 **end**
9 return I_N;

Algorithm 1: CORDIC-based Square Root of Power Sum.

noted that the swap of operands in Steps 1 and 2 is critical for an implementation with limited precision. If $I_0 < Q_0$ and the swap are not performed, after several right shifts in Step 7, I_i would reduce to to 0, and Q_i would not converge. Empirically, we set $N = 8$ and find it suffices to make the algorithm converge.

To further reduce the circuit size, the divisions in (3) are implemented by serial operations of shifting and subtractions. Additions and subtractions in the cost function are realized by carry-ripple adders. Moreover, the first three terms on the right-hand side of (3) share the same operation. Hence, the circuit for one term is implemented and it is reused for all three terms by properly selecting operand inputs using multiplexers. The detailed block diagram of the cost function is shown in Fig. 3.

B. The Optimization Engine

The proposed optimization engine is a multi-start meta-heuristic composed by simulated annealing (SA) and sensitivity-based search. SA [11] is a famous stochastic search algorithm that can reduce the chance of being trapped at a local optimal solution. The stochastic nature of SA requires many iterations to obtain good coverage of the solution space. To overcome this drawback, we repeat multiple SA procedures with different initial solutions that are evenly distributed in the solution space. This is why our optimization is a multi-start approach. In general, it often takes many iterations for SA to reach near global optimal solution. In order to accelerate the convergence, we run a limited number of SA iterations and then take the best solution to start a sensitivity-based search. A sensitivity-based search is good at carefully searching a local solution space. Although it can be easily trapped at local optima in general, this weakness is largely avoided when combined with SA. Neither SA nor sensitivity-based searches depend on system models. Therefore, they can be directly applied with measurement based cost function.

The pseudo code for the multi-start meta-heuristic is provided in Algorithm 2. Vector $\vec{x}$ denotes decision variables $\{x_1, \ldots, x_l, \ldots\}$, where x_l is a multi-bit control signal for the lth tuning knob. The two tuning knobs for the BPF are R_{x_1} and R_{x_2} as described in Section III. $\Phi(\vec{x})$ is the cost function to be minimized and is defined in (3). T is the temperature in simulated annealing, and t is the temperature decrease at each SA iteration. If the cost function value is smaller than threshold θ, the SA search is terminated.

Algorithm 2 first divides the entire solution space into K subspaces and then repeats K SA procedures with initial solutions at the centers of these subspaces. The core part of SA is in Steps 8-18. In Step 9, a neighbor of $\vec{x}_j$, which is T distance away from $\vec{x}_j$, is randomly selected for the subsequent evaluation. Steps 14 and 15 are the key components of SA, and they accept $\vec{x}_{\_new}$ with probability $P(\Phi(\vec{x}_j), \Phi_{\_new}, T)$, which is defined as

$$P(\Phi(\vec{x}_j), \Phi_{\_new}, T) = \min(1, \exp(-\Delta\Phi/T)) \quad (5)$$

Input : Initial temperature T_0, max number of
$\qquad$ iteration limit J and cost function threshold θ
Output: The best solution $\vec{x}_{\_best}$ and corresponding
$\qquad$ cost function value $\Phi_{\_best}$

1 Evenly divide solution space into K subspaces
2 **for** $i = 1; i \leq K; i{+}{+}$ **do**
3 $\quad$ Initialize $T \leftarrow T_0; j \leftarrow 0;$ // Start one SA
4 $\quad$ $\vec{x}_0 \leftarrow$ central point in ith subspace
5 $\quad$ **if** $\Phi(\vec{x}_0) \leqslant \Phi_{\_best}$ **then**
6 $\quad\quad$ $(\vec{x}_{\_best}, \Phi_{\_best}) \leftarrow (\vec{x}_0, \Phi(\vec{x}_0));$
7 $\quad$ **end**
8 $\quad$ **while** $j < J$ *and* $\Phi(\vec{x}_j) > \theta$ **do**
9 $\quad\quad$ $\vec{x}_{\_new} \leftarrow Neighbor(\vec{x}_j, T);$
10 $\quad\quad$ $\Phi_{\_new} \leftarrow \Phi(\vec{x}_{\_new});$
11 $\quad\quad$ **if** $\Phi_{\_new} \leqslant \Phi_{\_best}$ **then**
12 $\quad\quad\quad$ $(\vec{x}_{\_best}, \Phi_{\_best}) \leftarrow (\vec{x}_{\_new}, \Phi_{\_new});$
13 $\quad\quad$ **end**
14 $\quad\quad$ **if** $P(\Phi(\vec{x}_j), \Phi_{\_new}, T) \geqslant RANDOM$
$\quad\quad\quad$ **then**
15 $\quad\quad\quad$ $(\vec{x}_{j+1}, \Phi_{j+1}) \leftarrow (\vec{x}_{\_new}, \Phi_{\_new});$
16 $\quad\quad$ **end**
17 $\quad\quad$ $T \leftarrow T - t; j{+}{+};$
18 $\quad$ **end**
19 **end**
20 **while** *There is decrease on* $\Phi_{\_best}$ **do**
21 $\quad$ **for** *each neighbor* $\vec{x}_j$ *of* $\vec{x}_{\_best}$ **do**
22 $\quad\quad$ **if** $\Phi(\vec{x}_j) < \Phi_{\_best}$ **then**
23 $\quad\quad\quad$ $(\vec{x}_{\_best}, \Phi_{\_best}) \leftarrow (\vec{x}_j, \Phi(\vec{x}_j));$
24 $\quad\quad$ **end**
25 $\quad$ **end**
26 **end**
27 return $(\vec{x}_{\_best}, \Phi_{\_best});$

Algorithm 2: Multi-start meta-heuristic.

where $\Delta\Phi = \Phi_{\_new} - \Phi(\vec{x}_j)$. To avoid large circuit area, we implement the exponential function $\exp(-\Delta\Phi/T)$ by Taylor expansion $\sum_{i=0}^{m} (-\Delta\Phi/T)^m/m!$. By keeping only the 0th and the first order term, i.e., $m = 1$, it is approximated by $1 - \Delta\Phi/T$. As the probability precision is not critical to the solution search, such approximation is reasonable. The RANDOM in Step 14 is a random number, whose generator is realized by an 8-bit linear feedback shift register in our design.

Steps 20-26 in Algorithm 2 constitute the sensitivity-based search. Its main difference from SA is at Step 21, where *every immediate* neighbor of current solution is examined. By "immediate," we mean the Hamming distance between $\vec{x}_j$ and $\vec{x}_{\_best}$ is 1. Another difference from SA is that only solutions bringing cost function reduction are accepted.

The optimization engine is designed as a high-level state machine depicted in Fig. 4. CF_ready and $\Phi_{\_new}$ are the input from cost function, $output x$ is the solution delivered to BPF and others are all registers. This process starts from "Idle" when Reset signal is low and then keeps on waiting for the $\Phi_{\_new}$ in "Tuning_SA" after $\vec{x}_{\_new}$ is sent. Once $\Phi_{\_new}$ is available for reading by a notice of CF_ready signal, it's used to update $\vec{x}_{\_best}$ in "Update_SA." The acceptance probability is computed in "Compare" and shows whether to update $\vec{x}$ by $\vec{x}_{\_new}$ in "Accept_SA." "Judge_SA" works as a controller to select "Neighbor" to continue in the current SA or "Multi_Start" to open a new SA or "Sensitivity" to begin the local search. While in the "Sensitivity" state, $\Phi_{\_best}$ will be backup in Φ before enumerating each 1-distance neighbor in "Immediate." The $\Phi_{\_new}$ of each immediate neighbor is recorded during "Tuning" state and helps to improve $\vec{x}_{\_best}$ in

978-1-4799-8720-7/15 $31.00 © 2015 IEEE

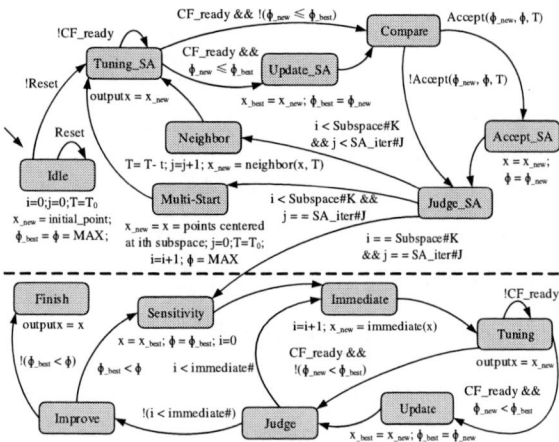

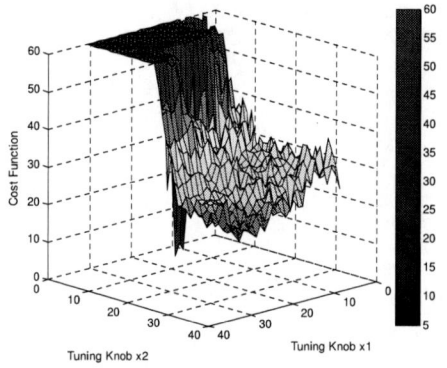

(a) Cost function $\Phi(\vec{x})$ for decision variables $\vec{x} = (x_1, x_2)$ corresponding to R_{x_1} and R_{x_2} in the BPF.

Fig. 4. The high-level state machine of multi-start meta-heuristic, in which the upper multi-start SA and the lower sensitivity-based searches are divided by the horizontal dashed line.

"Update." "Judge" guarantees that all the neighbors are covered and then moves ahead to "Improve". "Improve" compares the $\Phi_{\_best}$ with its old backup Φ and then decides to go on to "Sensitivity" or stop in "Finish" in which $\vec{x}$ is reported as the best found solution.

V. EXPERIMENT RESULT

A. Test Chip Measurement Results

The proposed built-in self optimization system with BPF as CUT was fabricated using 180 nm IBM process technology. Measurement was performed on-chip to confirm that the system works as expected. Set the central frequency $f_c = 31MHz$ and the bandwidth $BW = 8MHz$. We first enumerated all combinations of the tuning knobs x_1 and x_2 and measured the BPF frequency response for each configuration. Based on the measurement results, we plotted the cost function $\Phi_{BPF}(\vec{x})$ in Fig. 5 (a). We found that the global minimum of the cost function is at $\vec{x} = (22, 13)$. The chip testing results showed that our optimization engine was able to find this global minimum solution. We set the BPF according to the global minimum cost function where $\vec{x} = (22, 13)$, and measured the frequency response from $20MHz$ to $42MHz$. The results are plotted as the red curve with small circles in Fig. 5 (b). Its central frequency is near $31MHz$ and the $3dB$ drop frequencies are at $27.5MHz$ and $36MHz$. Thus, the bandwidth is $8.5MHz$, which is very close to the specification. We also measured and plotted frequency responses for two other solutions at $\vec{x} = (20, 13)$ and $\vec{x} = (12, 26)$. They are shown as the blue curve with small triangles and the black curve with small squares, respectively, in Fig.5 (b). The solution at $\vec{x} = (20, 13)$ has a bandwidth of $7MHz$, which implies a small deviation compared to the optimal solution, while the frequency response for $\vec{x} = (12, 26)$ is not only far from the optima, it is also not a BPF response. These results confirmed that our cost function definition leads to desired BPF performance.

B. Simulation Results

To validate the effectiveness of our approach on handling variations, applying statistical results to different instances of the CUT are necessary. We performed the statistical analysis through Monte Carlo simulations. Based on the variation data collected from the circuit simulation of the SPICE model, a

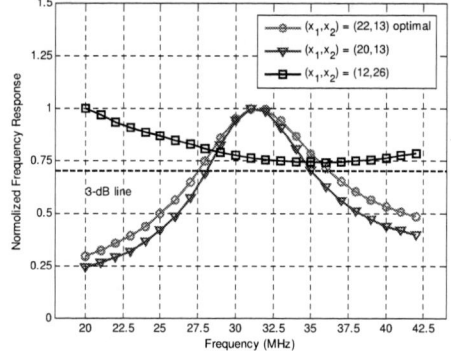

(b) Normalized frequency response from $20MHz$ to $42MHz$.

Fig. 5. Cost function and normalized frequency responses of the BPF with $f_c = 31MHz$ and $BW = 8MHz$, based on measurement of the test chip.

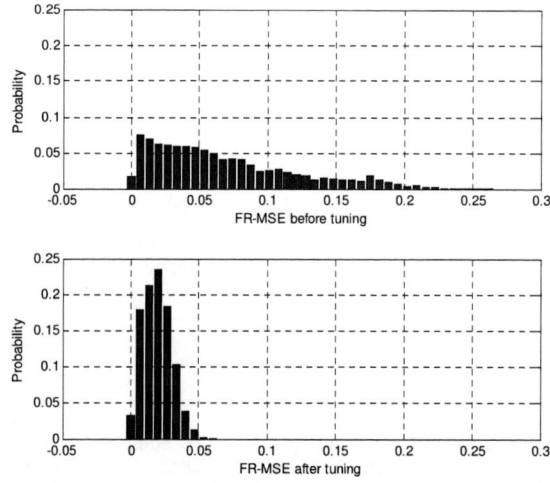

Fig. 6. Probability density functions of frequency response mean squared error (MSE) before and after the self optimization from the $5,000$-run Monte Carlo simulation on BPF with $f_c = 25MHz$ and $BW = 15MHz$.

5,000-run Monte Carlo simulation for the BPF design was performed. We evaluated the mean squared error (MSE) of frequency response, which is similar to results of (2). The probability density functions of the MSE before and after the self optimization are plotted in Fig. 6. On average, the self optimization can reduce the mean of MSE by around 71.3%.

We further compared our multi-start meta-heuristic with two other optimization approaches - standalone multi-start sensitivity-based search (Mul-Sen) and standalone SA. Fig. 7 shows the cost function value changes over iterations for these methods. It can be found that the proposed hybrid approach converges to a better solution in less evaluation iterations.

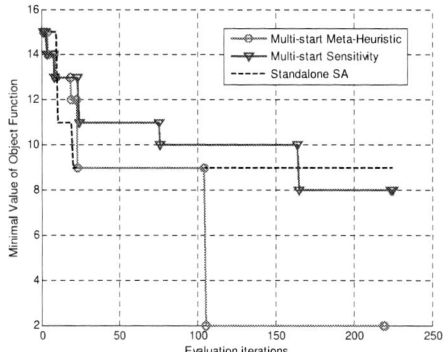

Fig. 7. Cost function changes over iterations for our multi-start meta-heuristic, multi-start sensitivity and standalone SA on BPF with $f_c = 23MHz$ and $BW = 6MHz$.

As a heuristic algorithm, the proposed method has some probability of failing to match the desired frequency response. The failure rate and the error of the outcome were evaluated by performing on 4,000 BPFs, whose central frequencies increase from $12MHz$ to $31MHz$ and Q-factors change from 1 to 4. The cumulative distribution functions from all three methods are plotted in Fig. 8, where the horizontal axis indicates the percentage error from the optimal solution. The results showed that the proposed approach produced more accurate solutions with lower error. Particularly, 77.6% of the solutions given by the hybrid algorithm have an error rate of less than 1%. In contrast, the standalone SA and the standalone sensitivity-based search had only 52.7% and 49.6% solutions with an error rate of less than 1%, respectively.

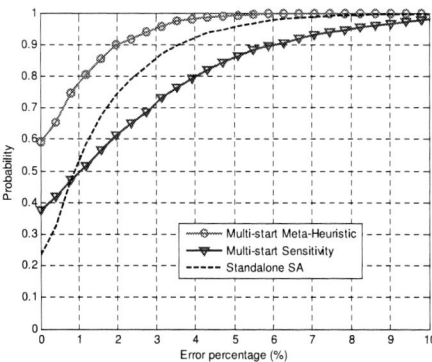

Fig. 8. Cumulative distribution functions of solution errors vs. the optimal solution from the simulation performed on 4,000 BPFs.

Other comparisons are summarized in Table I, where each gate count includes 3509 logic gates in the cost function circuit. A $10MHz$ clock is used to drive the simulation, and only the processing time for optimization engine is recorded. Considering that a medium performance ASIC chip may have millions of gates, thousand gates is a very small chip area overhead. Besides, since the self optimization is conducted very occasionally for a chip, the power is also acceptable. As for the processing time, although the proposed method is almost three times slower than Mul-Sen, it is still feasible because the total time for analog circuits being stable after each tuning is the dominant part in the whole process.

TABLE I.   GATE COUNT, POWER AND PROCESSING TIME COMPARISON.

	Gate Count	Power	Processing Time
Multi-start Meta-Heuristic	6744	$1.15mW$	$352.8\mu s$
Multi-start Sensitivity	4817	$0.56mW$	$95.6\mu s$
Standalone SA	5439	$0.93mW$	$335.8\mu s$

VI. CONCLUSION

In this work, we proposed a truly autonomous approach for each analog circuit chip to optimize itself for compensating process variations and aging effects. This approach is flexible to different kinds of analog circuits including but not limited to BPF with only change of the cost functions. It can also handle complex performance metrics, and complicated performance-variation dependence with meta-heuristic optimization instead of simple if-then rules. The effectiveness of this approach is supported by simulation and test chip results.

ACKNOWLEDGMENT

This work is partially supported by NSF (CCF-1255193) and SRC (2013-TJ-2421).

REFERENCES

[1] B. Liu, G. Gielen, and F. V. Fernández. *Automated Design of Analog and High-frequency Circuits.* Springer, 2013.

[2] P. D. Wit and G. Gielen. Degradation-resilient design of a self-healing xDSL line driver in 90 nm CMOS. *IEEE Journal of Solid-State Circuits,* 47(7):1757–1767, July 2012.

[3] Y. Xu, K. Hsiung, and X. Li. Opera: Optimization with ellipsoidal uncertainty for robust analog IC design. In *Proceedings of the ACM/IEEE Design Automation Conference,* pages 632–637, 2005.

[4] T. McConaghy and G. G. E. Gielen. Globally reliable variation-aware sizing of analog integrated circuits via response surfaces and structural homotopy. *IEEE Transactions on Computer-Aided Design,* 28(11):1627–1640, November 2009.

[5] X. Li, B. Taylor, Y. Chien, and L.T. Pileggi. Adaptive post-silicon tuning for analog circuits: Concept, analysis and optimization. In *Proceedings of the IEEE/ACM International Conference on Computer-Aided Design,* pages 450–457, 2007.

[6] D. Han, B. S. Kim, and A. Chatterjee. DSP-driven self-tuning of RF circuits for process-induced performance variability. *IEEE Transactions on VLSI Systems,* 18(2):305–314, February 2010.

[7] S. Ray and B. Song. A 13-b linear, 40-MS/s pipelined ADC with self-configured capacitor matching. *IEEE Journal of Solid-State Circuits,* 42(3):463–474, March 2007.

[8] D. Maliuk and Y. Makris. On-chip intelligence: A pathway to self-testable, tunable, and trusted analog/RF ICs. In *Proceedings of the Midwest Symposium on Circuits and Systems,* pages 1077–1080, 2014.

[9] A. Valdes-Garcia, F.A.-L. Hussien, J. Silva-Martinez, and E. Sanchez-Sinencio. An integrated frequency response characterization system with a digital interface for analog testing. *IEEE Journal of Solid-State Circuits,* 41(10):2301–2313, October 2006.

[10] L. Vachhani, K. Sridharan, and P.K. Meher. Efficient cordic algorithms and architectures for low area and high throughput implementation. *Circuits and Systems II: Express Briefs, IEEE Transactions on,* 56(1):61–65, Jan 2009.

[11] J. Haddock and J. Mittenthal. Simulation optimization using simulated annealing. *Computers and Industrial Engineering,* 22(4):387–395, April 1992.

Author Index

Abdallah, Louay	627	Bezerra, Eduardo Augusto	161
Ahangari, Hamzeh	149	Billoint, Olivier	350
Ahmadi, Mehrnaz	428	Bornat, Y.	227
Ajay, Arathi	68	Bosio, Alberto	515
Alanwar, Amr	131	Bossert, Martin	143
Alefragis, P.	119	Bossuet, Lilian	210
Alizadeh, Bijan	428, 243	Brown, Walter	113
Alkabani, Yousra	131	Brugger, Christian	262
Amarù, Luca	101	Burleson, Wayne P.	204
Andraud, Martin	627	Butko, Anastasiia	551
Ansaloni, Giovanni	268	Cairo, F.	286
Asadinia, Marjan	527	Calazans, Ney L. V.	27
Ashraf, Imran	539	Camus, Vincent	476
Asokan, Anu	515	Cane-Wissing, William	333, 303
Atienza, David	268	Carro, Luigi	485
Axelos, Nicholas	509	Casseau, Emmanuel	440
Azaïs, F.	621, 634	Cathébras, Guy	228
Aziz, Ahmedullah	333	Causapruno, G.	286
Babu, Hafiz Md. Hasan	187	Cavalheiro, David	309
Bahmanyar, Parvin	652	Cera, Márcia C.	410
Bandyopadhyay, Soumyadip	195	Chakrabarti, Amlan	80
Banerjee, Kunal	183	Chandramoorthy, Nandhini	339
Banerjee, Swapna	172	Chattopadhyay, Santanu	392
Barbareschi, Mario	200	Chattopadhyay, Subhasish	80
Basu, Soumya	268	Chen, Hu	398
Batrakov, K. G.	447	Chen, Shaoming	191
Beck, Antonio C. S.	485	Chen, Shenggang	398
Beck, Antonio Carlos Schneider	533, 410	Chen, Shuming	398
Becker, Georg T.	204	Chen, Xiaowen	398
Becker, Laurent	315	Cheng, Yuanqing	461
Bedour, Hassan	131	Cheng, Yun	368
Beerel, Peter A.	27	Cheng, Zenghua	255
Behjat, Laleh	38	Chew, Daniel	221
Beigne, Edith	321	Chiarulli, Donald M.	231, 125
Bekal, Anush	178	Chible, Hussein	56
Benini, Luca	280	Choi, Seonho	216
Benoit, Pascal	200	Choi, Yanghyeok	640
Berg, Yngvar	646, 86	Ciesielski, Maciej	113, 101
Bernard, C.	615	Cladera, Fernando	591
Bernard, S.	621, 634	Clarke, Chris	221
Bernard, Serge	228	Clement, Jeremie	356
Bernard-Granger, Fabrice	315	Clermidy, F.	615
Bertels, Koen	539	Clermidy, Fabien	350

978-1-4799-8720-7/15 $31.00 © 2015 IEEE

Author Index

Cofano, M.	286
Colombier, Brice	210
Comte, M.	621, 634
Cristal, Adrian	19, 149
Dadashi, Ali	86
Darav, Nima Karimpour	38
Datta, Suman	333, 303
de Castro, S.	362
De Micheli, G.	491
De Micheli, Giovanni	101
De Schryver, Christian	262
del Valle, P. Garcia	268
Dévigne, Clément	434
Di Natale, G.	362
Di Natale, Giorgio	603, 200, 467
Di Pendina, Gregory	321
Diény, Bernard	315
Dilillo, Luigi	515
Dimakos, Athanasios	627
Donaldson, Nick	221
Dong, Qing	167
Drechsler, Rolf	1
Du, Gaoming	404
Duric, Milovan	19
Dutertre, J.-M.	362
Ebrahimi, Masoumeh	545
Effiong, Charles	579
Efstathiou, Constantinos	91
Eftaxiopoulos, Nikolaos	509
El-Kharashi, M. Watheq	131
Emeretlis, A.	119
Enoiu, Eduard Paul	380
Enz, Christian	476
Fang, Yan	231, 125
Farulla, Giuseppe Airò	386
Fey, Görschwin	50
Fkih, Yassine	603
Flottes, M.-L.	362
Flottes, Marie-Lise	603
Forouzandeh, Behjat	428
Gaillardon, P. E.	491
Gaillardon, Pierre-Emmanuel	101
Gallifuoco, Vincenzo	386
Gamatie, Abdoulaye	579, 460
Gamatié, Abdoulaye	551
Gammaitoni, Luca	274
Garcia, Lucas	434
Garcia-Ortiz, Alberto	50, 416
Gautier, Matthieu	591
Ghandali, Samaneh	113
Ghasemzadeh, H.	491
Ghosal, Prasun	345
Girard, Patrick	515
Goswami, Manish	178
Graziano, M.	286
Gregorek, Daniel	416
Greiner, Alain	434
Grigorovici, Valentin	262
Gros, Stéphane	315
Große, Daniel	1
Grossi, Alessandro	327
Guillaume-Sage, L.	634
Guiraud, David	228
Guo, Yang	398
Gupta, Sumeet K.	333, 303
Gupta, Sumeet Kumar	339
Guthmuller, E.	615
Hansson, Andreas	609
Hartmann, R. R.	456
Heffernan, Donal	374
Hély, David	210
Higami, Yoshinobu	503
Hiller, Matthias	143
Hsiao, Michael S.	573
Hu, Jiang	656
Huang, He	628
Huang, Letian	545
Ibrahim, Ali	56
Indaco, Marco	386
Jabeur, Kotb	315
Jaiswal, Manish Kumar	249
Jamal, Lafifa	187
Jantsch, Axel	398, 545
Javadi, M. H. Seyed	33
Javerliac, Virgile	315
Jennings, Brandon	125

Author Index

Jiang, Yande	44	Li, Zhen	561
Joshi, Rohit	178	Liu, Chuangwen	44
Joshi, Shital	567, 292	Liu, Duo	113
Jung, Edward	216	Liu, Hao	434
Jung, Matthias	609, 262	Liu, Zonglin	398
Kalayappan, Rajshekar	137	Ló, Thiago Berticelli	533
Kang, Wang	461	Lopes, Jeremy	321
Kar, Bapi	107	Lorenzon, Arthur F.	485, 410
Karakonstantis, Georgios	268	Lourde, R. Mary	68
Kastensmidt, Fernanda	485	Lu, Jianzhuang	398
Kastensmidt, Fernanda Lima	533, 521	Lu, Yichuan	628
Kennings, Andrew	38	Lu, Zhonghai	398, 422, 404
Kerber, Andreas	199	Ma, Kaisheng	339
Kerzérho, V.	621, 634	MacNamee, Ciaran	374
Khammassi, Nader	539	Madsen, Jens Kargaard	74
Khan, Gul N.	62	Maffucci, Antonio	450
Khan, Mozammel H. A.	297	Mahdiani, H. R.	33
Kibis, O. V.	456	Mahzoon, Alireza	243
Kim, Moon S.	333	Makris, Yiorgos	628
Kim, Moon Seok	303	Maksimenko, S. A.	447
Klein, Jacques-Olivier	461	Malík, Peter	97
Kobayashi, Shin-Ya	503	Malik, Shweta	204
Kougianos, Elias	567, 292	Mandal, Chittaranjan	183, 195
Kroening, Daniel	7, 13	Mandal, Chittarnjan	107
Kuang, Jian	44	Mandal, Swagata	80
Kühne, Ulrich	1	Manna, Kanchan	392
Kupp, Nathan	628	Marinescu, Raluca	380
Kürzinger, Ludwig	143	Marongiu, Andrea	280
Kuzhir, P. P.	447	Martins, Victor M. Gonçalves	161
Lapotre, Vianney	579	Maymandi-Nejad, Mohammad	652
Larguech, S.	621	Melham, Tom	7, 13
Latif, Khalid	579	Métairie, Jérémy	440
Layer, Christophe	315	Metcalfe, Benjamin	221
Le Beux, Sébastien	561	Meunier, Quentin	434
Le Lann, Jean-Christophe	539	Mir, Salvador	627
Letartre, Xavier	561	Mirmotahari, Omid	646, 86
Levitan, Steven P.	231, 125	Miro-Panades, I.	615
Lewis, N.	227	Mishchenko, Alan	101
Li, Guangjun	545	Mohanty, Saraju P.	567, 292, 191
Li, Huawei	368	Moll, Francesc	309
Li, Xiaowei	368	Monat, Christelle	561
Li, Xueqing	339	Moradi, Farshad	74
Li, Yang	398	Moreira, Matheus T.	27

Author Index

Mouslih, Ghizlane	557	Ratković, Ivan	19
Müelich, Sven	143	Reis, Ricardo	521
Mukherjee, Rajdeep	7, 13	Renaud, S.	227
Murooka, Daijiro	167	Renovell, M.	621, 634
Mussard, Bruno	356, 460	Reorda, M. Sonza	491
Naccache, David	356	Riel, Heike	446
Naji, Omar	609	Riente, F.	286
Nakatake, Shigetoshi	167	Robert, Michel	551
Namgung, Seol	640	Roch, M. Ruo	286
Narayanan, Vijaykrishnan	333, 303, 339	Rossel, Olivier	228
Neto, Horácio C. C.	161	Rossi, Davide	280
Nielsen, Thomas	221	Rouzeyre, B.	362
Nigam, Tanya	199	Rouzeyre, Bruno	603
Noli, Luca	56	Roy, Kaushik	473
Nouet, Pascal	557	Russo, Ludovico O.	386
O'Connor, Ian	561	Sachdev, Manoj	652
Olivo, Piero	327	Saggio, Alberto	404
Orfei, Francesco	274	Saha, Mousumi	497
Oveis-Gharan, Masoud	62	Saini, Jogendra	80
Ozturk, Ozcan	149	Saluja, Kewal K.	503
Paar, Christof	204	Sampson, Jack	303
Pal, Sushanta Kumar	80	Sampson, John	339
Pala, D.	286	Sanchez, E.	491
Palomar, Oscar	19	Sanchez-Sinencio, Edgar	656
Paoli, Pierre	315	Santoro, G.	286
Park, Seonghyun	640	Santos, Cristiano	609
Pekmestzi, Kiamal	91, 509	Sarangi, Smruti R.	137
Peng, Huisheng	255	Sarbazi-Azad, Hamid	527
Perillat, Renaud	200	Sarhan, Hossam	350
Pešić, Djordje	19	Sarkar, Dipankar	183, 195
Piguet, Christian	481	Sarkar, Mayukh	345
Pomeranz, Irith	155	Sarkar, Santanu	172
Pons, Marc	481	Saroka, V. A.	456
Portnoi, M. E.	456	Sartor, Anderson L.	485, 410
Pravossoudovitch, Serge	515	Sassatelli, Gilles	551, 579, 460
Prinetto, Paolo	467	Sau, Suman	80
Puchinger, Sven	143	Sayed-Ahmed, Amr	1
Puri, Prateek	573	Schlachter, Jeremy	476
Quotb, A.	227	Schloeffel, Juergen	603
Raghavan, Praveen	74	Seceleanu, Cristina	380
Raghunathan, Anand	473	Seel, Andrew	125
Rakai, Logan	38	Seghaier, Ibtissem	237
Rampon, Jerome	200	Sengupta, Indranil	392

Author Index

Senni, Sophiane	460	Tsoumanis, Kostas	91
Sentieys, Olivier	591	Tu, Peishan	44
Seraj, Afshin	652	Turvani, G.	286
Séverac, Daniel	481	Unsal, Osman	19, 149
Shi, Congyin	656	Vacca, M.	286
Shuba, M. V.	447	Valero, Mateo	19
Sigl, Georg	143	Valle, Maurizio	56
Sikdar, Biplab K.	497	Valtchev, Stanimir	309
Simeu, Emmanuel	627	Varma, B. Sharat Chandra	249
Singh, Babu R.	178	Vatajelu, Elena Ioana	467
Singhvi, Ajay	27	Villa, Paulo R. C.	161
Slepyan, G. Y.	447	Virazel, Arnaud	515
So, Hayden K. H.	249	Vivet, P.	615
Song, Minkyu	640	Vivet, Pascal	603, 609
Soulier, Fabien	228	Voros, N.	119
Srivas, Mandayam	13	Voyiatzis, Ioannis	91
Srivastava, Ashok	191	Wajsbürt, Franck	434
Srivatsava, Ashok	178	Walczyk, Christian	327
Stan, Mircea R.	585	Wan, Jianghua	398
Stanić, Milan	19	Wang, Jiafan	656
Stratigopoulos, Haralampos-S.	627	Wang, Junshi	545
Subramani, Kiruba	628	Wang, Senling	503
Sugimoto, Yasuhiro	597	Wang, Ying	368
Sun, Hongbin	255	Weinberg, Elena K.	585
Sun, Shuwei	398	Weis, Christian	609, 262
Sur-Kolay, Susmita	107	Wenger, Christian	327
Swartz, William	38	When, Norbert	609, 262
Tabrizi, Aysa Fakheri	38	Wong, Stephan	485
Tadros, Ramy N.	27	Wu, Pangbo	44
Tagliavini, Giuseppe	280	Xie, Yuan	339
Tahar, Sofiène	237	Yalcin, Gulay	149
Takahashi, Hiroshi	503	Yashin, Victor V.	231
Tang, Haomo	44	Yoo, Jieun	640
Taylor, John	221	Young, Evangeline F. Y.	44
Teja, Vadapalli Shanmukha Sri	392	Yu, Cunxi	113
Thapliyal, Himanshu	297	Zaki, Mohamed H.	237
Theodoridis, G.	119	Zambelli, Cristian	327
Thole, Niels	50	Zamboni, M.	286
Thonnart, Y.	615	Zeinali, Behzad	74
Thuries, Sebastien	350	Zeng, Lang	461
Tisserand, Arnaud	440	Zhai, Baolu	255
Todri-Sanial, Aida	579, 557	Zhang, J.	491
Torres, Lionel	551, 356, 321, 460, 200, 467	Zhang, Liuyang	461

Author Index

Zhang, Xuchong.. 255

Zhang, Youguang...................................... 461

Zhang, Yu... 167

Zhao, Weisheng.. 461

Zhao, Xueqian.............................. 398, 422, 404

Zhao, Zhou... 191

Zheng, Nanning.. 255

Ziad, M. Tarek Ibn...................................... 131

Zianbetov, Eldar.. 321

Zimpeck, Alexandra L. 521

Zweig, Katharina Anna............................... 262

IEEE Computer Society
Technical & Conference
Activities Board

T&C Board Vice President
Cecilia Metra
Università di Bologna, Italy

IEEE Computer Society Staff
Evan Butterfield, *Director of Products and Services*
Lynne Harris, *CMP, Senior Manager, Conference Support Services*
Patrick Kellenberger, *Supervisor, Conference Publishing Services*

IEEE Computer Society Publications
The world-renowned IEEE Computer Society publishes, promotes, and distributes a wide variety of authoritative computer science and engineering texts. These books are available from most retail outlets. Visit the CS Store at *http://www.computer.org/portal/site/store/index.jsp* for a list of products.

IEEE Computer Society *Conference Publishing Services* (CPS)
The IEEE Computer Society produces conference publications for more than 300 acclaimed international conferences each year in a variety of formats, including books, CD-ROMs, USB Drives, and on-line publications. For information about the IEEE Computer Society's *Conference Publishing Services* (CPS), please e-mail: cps@computer.org or telephone +1-714-821-8380. Fax +1-714-761-1784. Additional information about *Conference Publishing Services* (CPS) can be accessed from our web site at: *http://www.computer.org/cps*

Revised: 18 January 2012

CPS Online is our innovative online collaborative conference publishing system designed to speed the delivery of price quotations and provide conferences with real-time access to all of a project's publication materials during production, including the final papers. The ***CPS Online*** workspace gives a conference the opportunity to upload files through any Web browser, check status and scheduling on their project, make changes to the Table of Contents and Front Matter, approve editorial changes and proofs, and communicate with their CPS editor through discussion forums, chat tools, commenting tools and e-mail.

The following is the URL link to the ***CPS Online*** Publishing Inquiry Form:
http://www.computer.org/portal/web/cscps/quote